Advanced Organic Chemistry

David E. Lewis

University of Wisconsin–Eau Claire

New York Oxford

OXFORD

UNIVERSITY PRESS

Oxford University Press is a department of the University of Oxford.
It furthers the University's objective of excellence in research,
scholarship, and education by publishing worldwide.

Oxford New York
Auckland Cape Town Dar es Salaam Hong Kong Karachi
Kuala Lumpur Madrid Melbourne Mexico City Nairobi
New Delhi Shanghai Taipei Toronto

With offices in
Argentina Austria Brazil Chile Czech Republic France Greece
Guatemala Hungary Italy Japan Poland Portugal Singapore
South Korea Switzerland Thailand Turkey Ukraine Vietnam

For titles covered by Section 112 of the US Higher Education Opportunity Act, please visit www.oup.com/us/he for the latest information about pricing and alternate formats.

Published by Oxford University Press
198 Madison Avenue, New York, New York 10016
http://www.oup.com

Cataloging-in-Publication data is on file with the Library of Congress

ISBN 978-0-19-975897-5

Printing number: 9 8 7 6 5 4 3 2 1

Printed in the United States of America
on acid-free paper

CONTENTS

CHAPTER FIVE

Frontier Orbitals and Chemical Reactions

CHAPTER SIX

Organic Reactions I: Pericyclic Reactions and the Conservation of Orbital Symmetry

CHAPTER SEVEN

Aromaticity: The 150-Year Riddle

CHAPTER EIGHT

Physical Organic Chemistry and Reaction Mechanisms

CHAPTER SEVENTEEN

Organic Reactions VI: Metal-Catalyzed Reactions for C—C Bond Formation

CHAPTER EIGHTEEN

Redox Reactions I: Oxidation

CHAPTER NINETEEN

Redox Reactions II: Reduction with Molecular Hydrogen or Its Equivalent

CHAPTER TWENTY

Redox III: Reduction with Complex Metal Hydrides and Active Metals

PREFACE

Over the past three decades of my career, I have seen introductory organic chemistry change from a foundation course for chemists and chemical professionals, to a service course for students planning on careers in the medical field. This change in emphasis has been accompanied, in my opinion, by an erosion of the level of the course material, so that students who do go on to graduate school in chemistry are no longer well prepared for the introductory courses in organic chemistry at the graduate level. At the same time, the amount and level of material that confronts new graduate students in organic chemistry has become more demanding, and this has further widened the gap between the introductory courses at the undergraduate and graduate levels.

This book specifically targets advanced undergraduate students and entering graduate students, and is an attempt to bridge that gap—to provide upper-level undergraduate students with an advanced textbook that addresses a number of important topics in more depth than now occurs in most introductory textbooks, while providing both a useful review, and then a much more substantial discussion of these topics for beginning graduate students.

Organization

This book has been written explicitly as a textbook, and its organization reflects this. The book is divided loosely into four sections: 1) introductory materials and general concepts of bonding and stereochemistry; 2) physical organic chemistry, reactive intermediates, and their reactions; 3) redox reactions; and 4) modern asymmetric reactions and synthesis. This last section also contains a chapter on the chemistry of the elements in the lower periods of the main groups, as well as a chapter on modern catalytic reactions for the formation of carbon–carbon bonds.

Each chapter concludes with a list of key terms introduced in the chapter. In chapters where new organic reactions are introduced, there are Reaction Synopses that contain the general reaction and typical reagents and reaction conditions to effect the reaction. I also feel that the humanity of our science is often overlooked in an effort to impart the most factual information possible. To counter this, I have included footnotes where biographical information about the chemists who developed the science can be found. For the most part, these footnotes contain references to longer biographical works or obituaries, but on a few occasions, these biographies are somewhat more expansive because additional information is so scarce.

Among the introductory chapters are discussions of stereochemistry and molecular orbital theory, including frontier orbital theory and the question of aromaticity. This section also contains a chapter (Chapter 3) containing the most frequently used acronyms in organic chemistry. In many books, the acronyms list is relegated to an appendix, but their use has become so widespread in modern synthetic organic chemistry that I have chosen to include them explicitly in the body of the book.

Traditionally, there have been three core courses in organic chemistry: 1) physical organic chemistry, which encompasses the study of the structure and properties of reactive intermediates; 2) organic reactions, in which the individual reactions and reaction types are taught; and 3) organic synthesis, where the design of syntheses—the art of combining organic reactions in sequence—is the focal point. In most cases, these three courses are sufficiently different that they can be taught separately, but this does lose the benefits that accrue to the student when new concepts are applied immediately after they are learned. In this book, I have deliberately chosen to follow each chapter introducing a new reactive intermediate with a chapter discussing reactions that involve those intermediates, so the student can see immediately just how these reactions can be used in the construction of complex molecules. For example, the chapter on modern molecular orbital theory and frontier orbital theory is followed immediately by a chapter on pericyclic reactions, where the application of these concepts in synthesis is illustrated.

Chapter 1 is essentially a review of concepts of structure and bonding from the introductory course in organic chemistry, with an emphasis on arrow-pushing and resonance theory. It is my belief that this will constitute a useful review for upper-level undergraduate students, especially, but in my experience it may not be completely unnecessary for entering graduate-level students. In my own upper-level undergraduate organic chemistry course, Modern Organic Chemistry, I have covered this material explicitly in class, and I have also set this as advanced reading, rather than explicitly including it in the lecture material.

Chapter 2 focuses on stereochemistry, and introduces students to both axial and planar chirality, and the Cahn-Ingold-Prelog method for assigning configurations to chiral molecules and to geometric isomers is introduced. The conformations of open-chain and cyclic molecules is discussed. The measurement of optical purity and the determination of enantiomeric excess (e.e.) or enantiomer ratio (e.r.) are discussed in the light of methods to achieve asymmetric synthesis. The Cram and Felkin-Anh models for predicting the stereochemistry of additions to chiral aldehydes and ketones are introduced.

Chapter 3 contains a list (already far from exhaustive) of acronyms used in modern synthetic organic chemistry.

Chapter 4 introduces the modern orbital model of atoms and molecules, including the LCAO method for obtaining molecular orbitals. Hybridization theory is extended to include variable hybridization, and the modeling of π-bonding systems by Hückel molecular orbital theory is introduced. Much of this forms the basis for the subsequent discussion of frontier orbital theory in **Chapter 5**. In Chapter 5, frontier orbital theory is introduced as an organizing principle for organic reactions, and as a basis for interpreting regiochemistry and stereochemistry in organic reactions, as well as stereoelectronic effects such as the anomeric effect in organic molecules. **Chapter 6** introduces pericyclic reactions. These reactions are now often a "special topic" in introductory texts, so this chapter is focused on how these reactions are used in modern synthesis to build complex molecules and transfer chirality from one part of a molecule to another. The chapter concludes with a discussion of "click" chemistry, and tandem and domino reactions.

Chapter 7 explores questions of aromaticity and antiaromaticity, and discusses the various criteria that have been developed to measure aromatic character, including resonance stabilization energy, and the related REPE, REPB and DRE parameters, HOMA, and NICS. The complementary Hückel and Möbius criteria for aromaticity are introduced, as is the Clar model of aromatic compounds, and its use in predicting reactivity in polycyclic aromatic systems.

Chapter 8 introduces physical organic chemistry. Reaction kinetics and the kinetic order of reactions are discussed, and the use of the rate law in making deductions about reaction mechanism is explored. The role of trapping and crossover experiments in deducing a mechanism is discussed. The Arrhenius equation and the activation parameters of a reaction are discussed, as is the Hammett equation and the various parameters arising from the original σ and ρ constants. The Grunwald-Winstein relationship for solvent effects on reactions, the Brønsted catalysis law, the salt effect, and element and isotope effects are introduced. The distinction made between primary and secondary isotope effects is pointed out, and the discussion is extended to include the concepts of the variable transition state and a mechanistic spectrum. The chapter concludes with a short discussion of the reactivity-selectivity principle.

Chapters 9–14 are all involved in the properties and chemistry of reactive intermediates. **Chapter 9** introduces the discussion of carbocations and their stabilization, including the topics of hyperconjugation, the Baker-Nathan effect, resonance and σ-bridging to give non-classical carbocations. Anchimeric assistance in reactions involving carbocations, and Wagner-Meerwein rearrangements of carbocations are also discussed. **Chapter 10** continues the discussion of carbocations, but now the focus shifts to synthetic reactions in which carbocations are key intermediates. These include Friedel-Crafts reactions and related formylations. Reactions involving carbocations stabilized by an adjacent lone pair-bearing heteroatom such as the Prins, aza-Prins, and Mannich reactions, as well as the Mukaiyama alkylation and aldol reactions, are introduced in this chapter, also.

Chapter 11 introduces the structure and reactivity of carbanions, and the related metal alkyls. The stabilization of carbanions by adjacent π bonded functional groups (e.g. carbonyl and cyano) is discussed, as is the generation of enolate anions by deprotonation. This discussion addresses questions of enolate regiochemistry and stereochemistry, and kinetic versus thermodynamic enolate generation. The stabilization of carbanions by third- and higher-row main group elements is introduced, as is the concept of *umpolung*. Anionic rearrangements (Wittig, Stevens, and Sommelet-Hauser) are discussed. The discussion of carbanion chemistry continues in **Chapter 12**, where the focus is now the use of carbanion nucleophiles in synthesis. This discussion includes some of the most important modern synthetic reactions, including the aldol addition and its congeners, and the Michael addition. The stereochemistry of these reactions is discussed at some length. The alkylation of organometallic nucleophiles, especially organocuprates, with halides and epoxides is discussed, and enamines and sulfinylimine anions are introduced as enolate anion surrogates.

Chapter 13 introduces free radicals, and important electrophilic reactive intermediates: carbenes, arynes and nitrenes. The structures of these species are discussed, and the reactions of carbenes, arynes and nitrenes are discussed in some detail. The involvement of metallocarbenoids in synthetic reactions is introduced. The synthetically useful reactions of free radicals are discussed in **Chapter 14**. This discussion includes site-specific methods for the generation of free radicals, and important radical reactions, such as radical cyclization reactions.

Students are introduced to the terminology and practice of modern synthesis in **Chapter 15**. This introduction includes discussions of selectivity in organic synthesis, retrosynthetic analysis, and the Evans-Lapworth consonant and dissonant difunctional relationship analysis. Also in this chapter, the use of protecting groups in synthesis is introduced, and the concept of orthogonal protection is discussed. The most widely used protecting groups for a number of the most common functional groups are introduced.

Important condensation and cascade reactions are the topic of **Chapter 16**. In this chapter, the Claisen condensation and its congeners (including the Robinson annelation), the Wittig reaction, and its analogues (including reactions involving sulfur ylides), the Ramberg-Bäcklund olefination and its analogues, and the Seyferth-Gilbert acetylene synthesis are all discussed. Thus, this chapter includes some of the most important methods for the synthesis of products containing C=C π bonds.

Chapter 17 concentrates on an area of synthetic organic chemistry that has undergone explosive growth in the last three decades: the use of transition metal-catalyzed reactions in synthesis. After a brief introduction to some basic organometallic chemistry, the chapter moves to discuss such important reactions as transition metal-catalyzed cross-coupling reactions such as the Stille, Heck, Negishi, and Suzuki couplings, olefin metathesis using transition metal carbene complexes, and the Tsuji-Trost and similar reactions of allylic substrates. The chapter also discusses catalyzed hydrometallation reactions, as well as carbometallation reactions.

The next four chapters are substantial, and are focused on redox reactions. **Chapter 18** is the largest single chapter in this book, and it covers oxidation reactions. **Chapter 19** is focused on reductions with molecular hydrogen or its surrogates (e.g. formic acid), and **Chapter 20** continues with a discussion of reduction with complex metal hydrides and active metals. The final chapter of this quartet, **Chapter 21**, is devoted to modern asymmetric redox reactions, with major emphasis on asymmetric oxidation reactions, includiong the Sharpless asymmetric epoxidation and dihydroxylation reactions, the Katsuki and Shi oxidations, and asymmetric hydroxylation of ketone enolates with chiral oxaziridines. The formation of chiral products by the desymmetrization of *meso* compounds is also addressed, as is the asymmetric hydrogenation of alkenes, and the reduction of carbonyl compounds and imines with chiral hydride transfer reagents.

Chapter 22 is devoted to a discussion of the organic chemistry of compounds of silicon, phosphorus, sulfur, selenium, and tin. Key discussions of the organic chemistry of compounds of these elements focus on the stabilization of carbocationic and carbanionic intermediates by the heteroatom, and on the control of regiochemistry exerted by the heteroatomin on many reactions. Reactions such as the Hosomi-Sakurai reaction, the Mislow-Evans, Pummerer, and Arbuzov rearrangements are also discussed.

Chapter 23 is, in many ways, the capstone chapter in organic synthesis because it focuses on asymmetric organic synthesis. Among the topics discussed are chiral auxiliary, chiral reagent, and chiral catalyst approaches in asymmetric synthesis, and the concepts of simple and double diastereoselection, along with the concepts of matched and mismatched reacting pairs of stereoisomers.

The chapters discussing the reactive intermediates are autonomous enough that they can be used as support for a course in physical organic chemistry (Chapters 1, 2, 4, 5, 7–9, 11 and 13 are primarily concerned with topics in physical organic chemistry, and I have used them to cover physical organic chemistry in our Modern Organic Chemistry course). Likewise, the chapters that describe modern organic reactions are also autonomous enough that they, too, can be used to support a stand-alone Organic Reactions course (Chapters 6, 10, 12, 14, and 16–22 are primarily concerned with organic reactions, and I have used them to support Modern Organic Chemistry when I have taught it with an organic reactions focus during the development of this book). For much the same reason, I have incorporated the references as footnotes on the page where they occur. While this has meant a small amount of duplication of literature citations, I feel that it helps the student to have the reference immediately available, rather than having to flip back through the previous pages to find it.

In many ways, modern organic synthesis has evolved much faster than other sub-disciplines of the science, and the tools available to the modern synthetic chemist are now truly staggering. From a beginning where problems of regioselectivity and chemoselectivity limited what chemists could accomplish, organic synthesis has become a truly asymmetric science. The chapters on synthesis (Chapters 15 and 23) have been written with this modern focus on the control over the absolute configuration of chiral products in mind. The chapter on modern catalysis (Chapter 17) and the final chapter on redox reactions (Chapter 21) have a strong focus on modern asymmetric synthesis.

There are three areas where I have made the decision, in the interests of keeping the size of this volume within reasonable bounds, to leave the discussion to more specialized texts: polymer chemistry, heterocyclic chemistry, and bioorganic chemistry. In all three areas, there are a number of excellent specialist books that will serve as resources. The first two have been the most difficult, especially given the importance of heterocycles in modern medicinal chemistry, and the ubiquitous nature of polymers in modern life, and it may be that they will be incorporated into a later edition of this book, depending on demand. I have also taken the difficult choice not to include spectroscopy in this book because, once again, there are excellent specialist books available for students at this level.

Every chapter of the book (except Chapter 3, for obvious reasons) has a number of worked problems. There are also copious end-of-chapter problems that should test the student, and for many it will be necessary to consult the original literature to obtain the complete picture—it was certainly necessary for the author!

Acknowledgments

This has been a huge undertaking, and I could not have accomplished it without the support of my family. My wife, Debbie, has had to cope with a preoccupied husband for years, and my children (Graeme and Veronica) have also had to cope with less than complete attention from their father. My students at UW-Eau Claire have been willing and effective guinea pigs during the development of the manuscript over the past several years, and many of their insights have found their way into the final book. Another individual whom I must thank is my friend and Mentor, Angas Professor Emeritus John H. Bowie of the University of Adelaide, Australia, who has mentored me throughout my career. The editorial staff at Oxford University Press assembled an impressive group of reviewers whose contribution to the final product cannot be overemphasized. Their reviews have been cogent and extremely useful, and I would be remiss without thanking them for their intellectual contribution to this book. Finally, I must thank the editorial staff at Oxford University Press who have been involved with this project over its lifetime: Jason Noe (Senior Editor), Andrew Heaton (Assistant Editor, Life Sciences and Chemistry), and Melissa Rubes (Editorial Assistant, Life Sciences and Chemistry).

Ancillary materials available include an extensive Solutions Manual, where the worked answers to every problem in the book are contained. An Instructor's Resource CD contains all the numbered Figures from the text in electronic format, as well as most other graphics from the chapters. It also contains the solutions in editable Word format.

CHAPTER ONE

The Fundamentals Revisited

Structure, Bonding, and Reactivity

1.1 Organic Structure: A Brief History

The origins of organic chemistry as a separate science are popularly traced to the synthesis of urea by Friedrich Wöhler[1] in 1828,[2] but this is actually an oversimplification of the progress of the science. Several authors have held the view that Wöhler's synthesis of urea was the seminal point in the overthrow of vital force theory.[3] Others believe that this was not the case. However, it was not until later (usually attributed to Kolbe's unambiguous synthesis of acetic acid from its elements[4]) that vital force theory was finally abandoned.[5]

The rapid development of organic chemistry during the second half of the 19th century can be traced to the development of the concept of ***chemical structure***.[6] Prior to the development of the structural theory of organic chemistry, chemists struggled to find common threads between different compounds and different reaction types. Then, in 1858, two papers appeared within 2 months of each other, one by a German chemist, Friedrich August Kekulé,[7] and the other by a young Scotsman, Archibald Scott Couper,[8] who was working in Paris in the laboratory of Charles Adolphe Wurtz.[9] These pioneering papers laid the groundwork for the spectacular advances to be made later. Of course, any new

---

1. **Friedrich Wöhler (1800-1892)** received his MD from Heidelberg, and after positions at the polytechnic schools at Berlin and Kassel, he became Professor at Göttingen, where he spent the remainder of his career. For a relatively modern, more complete biography, see: Kauffman, G.B.; Chooljian, S.H. (2001). *The Chemical Educator* 2001 6, 121.

2. (a) Wöhler, F. *Ann. Physik* [2] **1828**, *12*, 253. (b) Wöhler, F. *Ann. Chim.* **1828**, *37*, 330.

3. (a) Hofmann, A.W. *Ber. Deut. chem. Ges.* **1882**, *15*, 3127, 3152. (b) Armstrong, H.E. *Introduction to the Study of Organic Chemistry* (Longmans: London, 1884), p. 1. (c) Warren, W.H. *J. Chem. Educ.* **1928**, *5*, 1539. (d) Campaigne, E. *J. Chem. Educ.* **1955**, *32*, 403.

4. Kolbe, H. *Ann. Chem. Pharm.* **1845** *54*, 145.

5. (a) McKie, D. *Nature* **1944**, *153*, 608. (b) Hartmann, L. *J. Chem. Educ.* **1957**, *34*, 141.

This view, with supporting references, is given in several advanced organic chemistry textbooks, for example: (b) Wheland, G.W. *Advanced Organic Chemistry*, 3rd, Ed. (John Wiley & Sons: New York, 1960), p. 3-5, and references therein.

6. Lewis, D.E. In Giunta, C.J., Ed. *Atoms in Chemistry: From Dalton's Predecessors to Complex Atoms and Beyond. ACS Symp. Ser.* **2010**, *1044*, 35.

7. **Friedrich August Kekulé von Stradonitz (1829-1896)** was educated at Giessen and Paris, and held faculty positions at Heidelberg, Ghent and Bonn. At Ghent, he proposed the tetravalent carbon atom and the structure of benzene. "Kekulé's dream", a source of controversy since the 19th century, has even been the subject of a lawsuit! For a more complete biography, see: Alyea, H. In Halsey, W.D.; Friedman, E. Eds., *Collier's Encyclopedia*, (Macmillan: New York, NY, 1985), Vol. 14, p. 15.

8. **Archibald Scott Couper (1831-1892)** was born and educated in Scotland before moving to Paris to study with Wurtz. Independently of Kekulé, he proposed a theory of organic structure, but the delay in publication of his paper meant loss of priority for this theory. He suffered a nervous breakdown and returned to Scotland a broken man. For more complete biographies, see: (a) Dobbin, L. *J. Chem. Educ.* **1934**, *11*, 331. (b) Anschütz, R. *Proc. Roy. Soc. Edinburgh* **1909, 29,** 193.

9. **Charles Adolphe Wurtz (1817-1884)** was born in Strasbourg and educated, in part, in Justus Liebig's Giessen laboratory. He became a major figure in the development of French organic chemistry, and his Paris laboratory became an important center for research in Europe. See: Rocke, A.J. *Nationalizing Science. Adolphe Wurtz and the Battle for French Chemistry* (MIT Press: Cambride, MA, 2001).

theory must have its advocates, and the most ardent advocate for structural theory was neither Couper nor Kekulé but a young Russian chemist by the name of Aleksandr Mikhailovich Butlerov,[10] who saw and articulated more clearly than Kekulé, at least, the true potential of this new theory.[11] Butlerov had met Kekulé and Wurtz and had been struck by the ability of the new theory to reconcile many of the apparent contradictions still prevalent in organic chemistry. His textbook, the first based entirely on structural theory, was published first in Russian and then in German.[12] It set the trend for all subsequent textbooks in organic chemistry.

Lewis Structures

The modern classification of compounds as ionic or covalent is usually attributed to American chemist G. N. Lewis,[13] who in 1916 proposed a simple theory to account for the bonding in practically every type of compound known. At the same time, a similar theory was developed independently by German physicist Walther Kossel.[14] Both theories derive ultimately from the 1904 principle proposed by German chemist Richard Abegg[15] that each element possesses a positive and a negative valency whose sum is eight. Lewis expanded this principle and noted, in particular, that the noble gases were all characterized by having eight electrons in their valence shell.[16] In addition, he noted that the stable ions formed by atoms such as sodium and chlorine were characterized by having either lost enough electrons to lay bare an inner shell containing eight electrons (thus giving the ion a valence shell containing a complete octet) or by having gained enough electrons to complete the valence shell octet. From these observations, Lewis deduced that *a valence shell containing eight electrons is associated with extraordinary chemical stability.* Stated in its simplest terms, the Lewis theory is as follows:

> Whenever atoms react to form molecules or ions, they do so in such a way as to ensure that every non-hydrogen atom possesses a complete octet of electrons in the valence shell.

---

10. **Aleksandr Mikhailovich Butlerov (1828-1886)** was a Russian organic chemist who was educated at Kazan, where he became a faculty member after 3 years in western Europe, including a year with Liebig. In 1869 he became Professor at the Medical-Surgical Academy of St. Petersburg. For more details, see Lewis, D.E. *Early Russian Organic Chemists and Their Legacy* (Springer: Heidelberg, 2012), pp. 32, 47.

11. Some seminal early papers by the major players on the structural theory of organic chemistry: (a) Kekulé, F. *Liebigs Ann. Chem. Pharm.* **1858**, *106*, 129. (b) Couper, A.S. *Compt. rend.* **1858**, *46*, 1157. (c) Butlerow, A.M. *Ann. Chem. Pharm.* **1859**, *111*, 5146; *Z. Chem.* **1861**, *4*, 549. (d) Boutlerow, A. *Bull. Soc. Chim. Paris, Nouv. Sér.* **1864**, *1*, 100.

12. Butlerov, A.M. *Introduction to the Study of Organic Chemistry* (Kazan, 1864) [in Russian]; *Lehrbuch der organischen Chemie zur Einführung in das specielle Studien derselben* (Leipzig, 1867).

13. **Gilbert Newton Lewis (1875-1946)** was educated at Harvard, Leipzig, and Göttingen, then after a brief period at the Massachusetts Institute of Technology, he moved to Berkeley. Here he became one of the giants of physical chemistry of the 20th century, and yet he never won the Nobel Prize. For more details, see: Coffey, P. *Cathedrals of Science* (Oxford: New York, 2008).

14. **Walther Ludwig Julius Paschen Heinrich Kossel (1888-1956)**, the son of Albrecht Kossel, was educated at Heidelberg. His first positions were in Munich and Kiel, and he then moved to Danzig (Gdansk); he fled to Tübingen ahead of the Russians in 1945. For a more complete biography, see: Andrade, E.N.da C. *Nature,* **1956**, *178*, 4533.

15. **Richard Wilhelm Heinrich Abegg (1869-1910)** was educated at Berlin and served as assistant to both Nernst and Arrhenius after graduation. He became Professor at the Wrocław University of Technology in 1900, but he was killed in a ballooning accident a decade later. For more complete biographies, see: (a) Hills, W. *J. Chem. Soc., Trans.* **1911**, *99*, 599. (b) Nernst, W. *Ber. dtsch. chem. Ges.* **1913**, *46*, 619.

16. Seminal early papers on the development of the covalent and ionic bonding models:

(a) Lewis, G.N. *J. Am. Chem. Soc.* **1916**, *38*, 762. (b) Lewis, G.N. *Valence and the Structure of Atoms and Molecules* (The Chemical Catalog Co.: New York, 1923); reprinted by Dover Press in 1966 with a foreword by Kenneth Pitzer. (c) Kossel, W. *Ann. Physik* **1916**, *49*, 229. (d) Abegg, R. *Z. Anorg. Chem.* **1904**, *39*, 330-380.

These structural theories were consolidated, refined and extended by Irving Langmuir:

Langmuir, I. *J. Am. Chem. Soc.* **1919**, *41*, 868; *J. Am. Chem. Soc.* **1920**, *42*, 274.

This principle is called the ***Lewis octet rule***. It is one of the most useful concepts available to the organic chemist for rationalizing the structures and reactivity of organic compounds and the course of organic reactions.

The Lewis electron-pair covalent bond was the first truly effective generalization about chemical bonding used by organic chemists. Indeed, Lewis theory has formed the basis for numerous other theories used in organic chemistry, so that it occupies a central position in modern organic chemistry. It is applied throughout modern organic chemistry, and many, if not all, reactions are rationalized on the basis of the Lewis structures of the molecules involved. In addition, it is a testimonial to its durability that it remains one of the cornerstone theories of modern organic chemistry, despite the advances in atomic and molecular theory that have been made since it was first proposed in 1916. Because organic chemistry is intimately concerned with the fates of chemical bonds in molecules, it is critical that one have mastered an understanding of the Lewis structures of organic molecules.

The simple Lewis covalent bond model involves (in the formal sense, at least) the contribution of one electron to the bonding electron pair from each partner. Although this simple model does go a long way in explaining the bonding in the vast majority of covalent compounds, there are compounds known whose bonding is not completely explained by the simple Lewis covalent model. This is particularly the case for metal complexes, but this problem also occasionally arises in organic chemistry. In such cases, Lewis proposed that one of the two partners in the bond formally contributes *both* electrons to the bond. This bond, which was called a ***dative bond***, a ***semipolar bond***, or a ***coordinate covalent bond***, is frequently designated in older (especially British) books by an arrow from the donor atom to the acceptor atom, but this practice is now largely obsolete. Instead, modern applications of Lewis theory use the concept of ***formal charge*** to indicate that both electrons in a covalent bond formally originate on only one of the two atoms. The concept of formal charge is often more useful than the dative bond model—many coordinate covalent bonds exhibit properties normally associated with ionic bonds, so the formal charge model allows the chemist to make predictions of properties that are closer to the experimental values.

Determining the formal charge on any atom in a Lewis structure is actually fairly simple. Electrons in a molecule fall into two major types: (1) nonbonding lone pairs, which are under the formal control of a single nucleus, and (2) electrons in a covalent bond, which are under the formally equal control of two nuclei—one electron from each covalent bond is formally under the control of each nucleus. The total number of electrons in a molecule formally controlled by each nucleus, therefore, is equal to the sum of the number of electrons in lone pairs and half the number of electrons in covalent bonds.

Formal charge = (number of valence electrons in the neutral atom)
- (number of covalent bonds)
- (number of electrons in lone pairs)

Drawing the Lewis structures of polyatomic molecules and ions is relatively simple provided that one remembers two simple rules:

1. *Electrons in a molecule have no memory.* One cannot assign an electron in a molecule as originating on any particular component atom, even if it may be convenient to adopt such a view for understanding the course of a reaction.
2. *The valence shell of the elements of the second period of the periodic table has a maximum capacity of eight electrons.* This restriction does not apply to elements in the third or lower periods of the periodic table (the capacity of the third electron shell is 18 electrons). These elements can, and do, expand their outer-shell octets when necessary.

The process of drawing the Lewis structure of a molecule involves five fairly straightforward steps:

1. Determine the total number of valence electrons.
2. Establish the connectivity of the atoms.
3. Connect the atoms using single covalent bonds so that every atom is connected to at least one other by a covalent bond.
4. Distribute the remaining electrons (if any).
5. Assign formal charges to atoms as required.

Worked Problem

1-1 Complete the Lewis structures of each of the following compounds by distributing the electrons appropriately among the atoms, which are arranged as shown. Note that in some cases, there may be more than one valid Lewis structure that can be written. In those cases, draw all the valid Lewis structures.

(a) 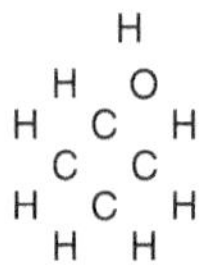 (b) 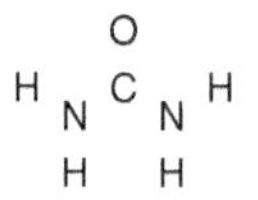 (c) H N C O

§Answer below.

Representing Organic Compounds: Structural Formulas

Because of the complexity of the structure of many organic compounds, as well as the importance of structure in organic chemistry, the development of shorthand methods for representing the structures of organic compounds has been almost mandatory to allow the drawing of structures in a reasonable amount of space and time. By way of reviewing

§ Answers for Worked Problem:

(a) 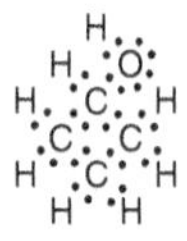 (b) 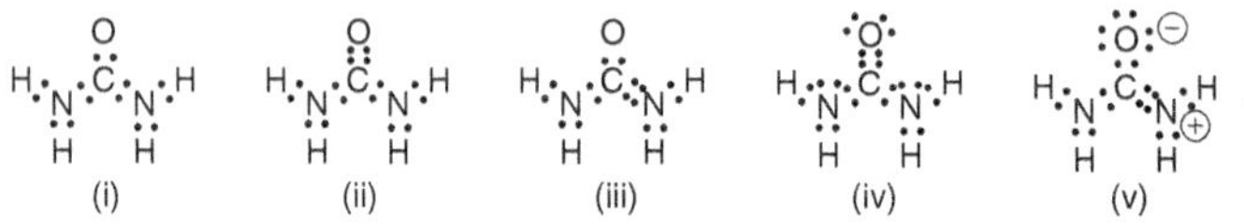

(c) 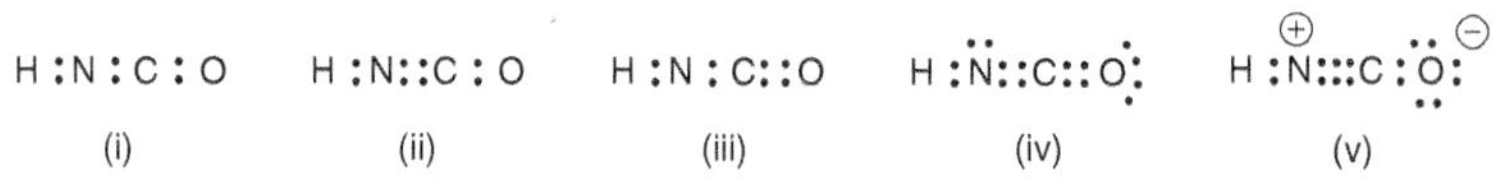

(a) In this example, there are 4(4) + 1(6) + 8(1) = 30 valence electrons to distribute. When we place a pair of electrons between each adjacent pair of atoms, all the valence shells have complete octets, and there are no electrons left over.

(b) In this example, there are 1(4) + 2(5) + 4(1) + 1(6) = 24 valence electrons to distribute. After placing one pair of electrons between each adjacent pair of atoms (i), there are 14 valence electrons used and 10 left. To simple fill the octets of the remaining atoms with lone pairs will take 12 valence electrons, so we must first place a double bond between two of the electron-deficient atoms. This gives us either a C=O bond (ii) or a C=N bond (iii). Now we have exactly enough electrons left to satisfy all the octets, and we can distribute them. Structure (iv) has no charged atoms, but when we test the nitrogen and oxygen atoms of structure (v), we find that the doubly bonded nitrogen atom has a formal positive charge, and the oxygen atom has a formal negative charge.

(c) This problem is identical in all respects to that posed by (b), except that we now need to add two double bonds (or one triple bond) to the structure to complete the octets of all. Again, the structures must be tested for the presence of charged atoms.

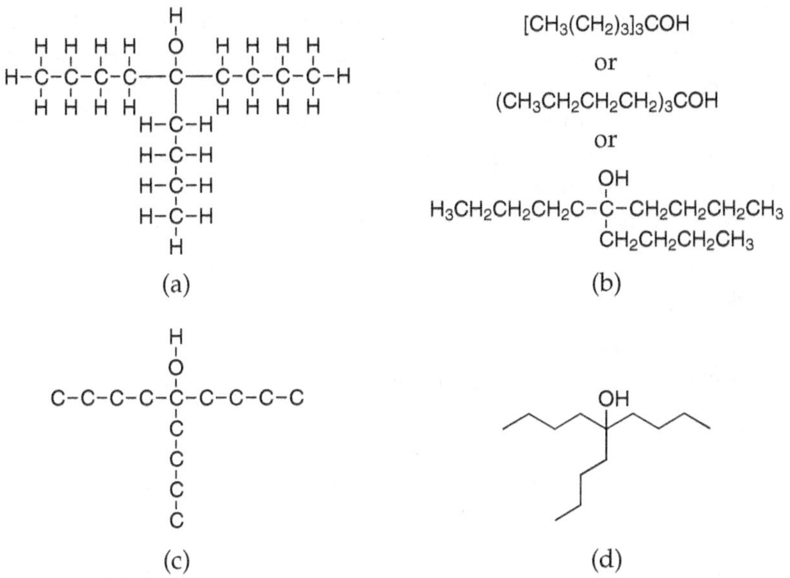

Figure 1.1 The different representations of tributylcarbinol found in organic chemistry are the complete structural formula (a), the condensed structural formula (b), the connectivity formula (c), and the line formula (d). Of these, the line formula has found the most widespread use in modern organic chemistry.

the conventions for drawing organic structures, let us use tributylcarbinol (5-butyl-5-nonanol, Figure 1.1) as an example.

The molecular formula of this compound is $C_{13}H_{28}O$, and its full structure is shown in Figure 1.1 as (a). Even this structure is a simplification over what we saw in Worked Problem 1-1; now the bonding electron pairs are represented by lines instead of pairs of dots, and the lone pairs are omitted. This particular representation, in which every atom of the molecule is explicitly drawn, is often referred to as a ***complete structural formula***. It is essentially the same as the formula first proposed by Crum Brown[17] in 1861 and published in 1864.[18]

Even in a molecule as small as tributylcarbinol, however, it takes considerable space to draw this structure. The first simplification, therefore, is to condense the various groups of atoms about each carbon to a simple molecular formula for each carbon atom. This gives rise to formula (b) in Figure 1.1, which is usually referred to as a ***condensed structural formula***, or simply a ***condensed formula***. Such formulas are most frequently encountered in printed works because they are very amenable to typesetting, although they convey little or no information about molecular shapes. An alternative simplification of formula (a) can be made by omitting the hydrogen atoms. Although such formulas (c) are very good for showing connectivity, they are very poor at showing molecular shape. By omitting the symbols for the carbon atoms, one can simplify the formula (c) even further to give (d), which is known as a ***line formula***, or, if a ring of atoms is involved, a ***polygon formula***.

In a line formula, the end of each line, as well as each vertex, is occupied by a carbon atom bearing sufficient hydrogen atoms to complete its octet (unless it is charged). Where

---

17. **Alexander Crum Brown (1838-1922)** was educated at Edinburgh (MA, 1858; MD,1861) and London (DSc, 1862). His MD thesis contained the first truly modern line formulas. After postdoctoral study with Bunsen and Kolbe, he returned to Edinburgh, where he remained, a much-loved teacher, until his retirement in 1908. For a more complete biography, see: *J. Chem. Soc., Trans.* **1923**, 3422.

18. Crum Brown, A. MD Thesis, University of Edinburgh, 1861; *Trans. Roy. Soc. Edin.* **1864**, *23*, 707-719; *Proc. Roy. Soc. (Edinburgh)* **1866-1867**, *6* (#73), 89.

fewer than four lines meet at a vertex, the remainder of the four valencies of the carbon atom are assumed to be occupied by hydrogen atoms. Atoms that are neither carbon nor hydrogen are called ***heteroatoms***, and all heteroatoms are explicitly shown in line formulas. In general, only non-hydrogen and noncarbon atoms are explicitly indicated in a line formula, although hydrogen atoms bonded to heteroatoms are frequently shown, as shown in Figure 1.1 (d). In modern usage, lone pairs on heteroatoms are often omitted unless specifying their presence makes an important point for the chemist.

Worked Problem

1-2 Draw the structures of the compounds in Worked Problem 1-1 using lines to represent covalent bonds and condensed formulas for parts of the molecule as appropriate.

§Answer below.

Problems

1-1 One compound, with the molecular formula C_5H_8, is the hydrocarbon isoprene, which is the basis of the "isoprene rule" for terpenoid natural products. Draw the complete Lewis dot structures for all the compounds with the molecular formula C_5H_8 and the complete Lewis structures with lines for each covalent bond, as well as the line or polygon formula for each.

1-2 What is the molecular formula of each of the two compounds below?

(a)

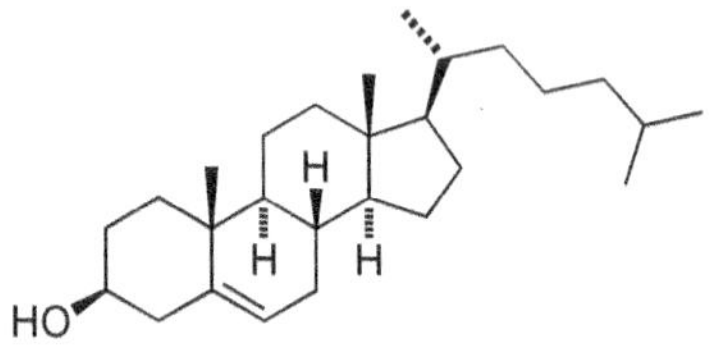

(b)

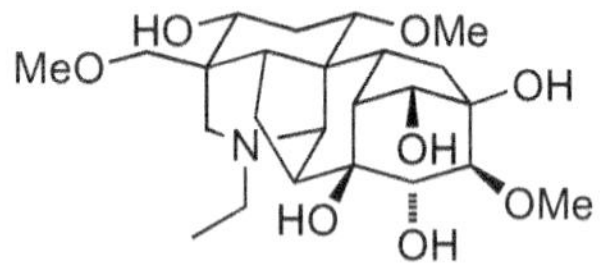

§ Answers for Worked Problems:

(a)
(b)
(c)

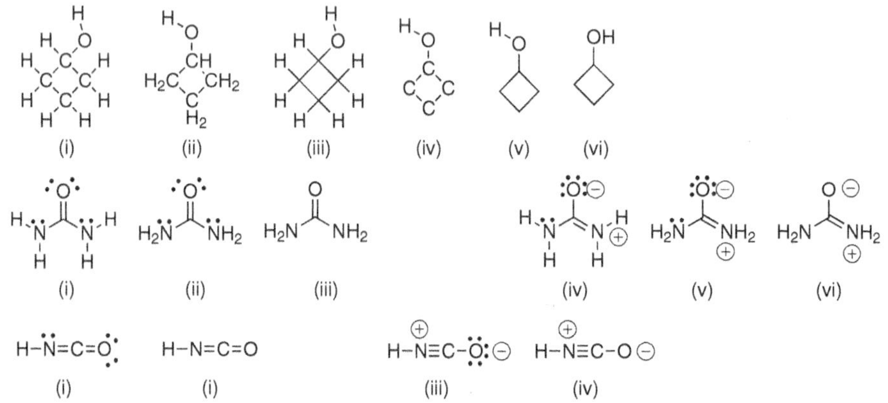

(a) The structures, from left to right, are the complete structural formula (i), a condensed structural formula (ii), a simplified complete structural formula (iii), the connectivity formula (iv), the polygon formula with the OH group explicitly drawn (v), and the polygon formula as it is usually used (vi). Note how, in these representations, the lone pairs have been omitted, as is often done for clarity.

(b) The formulas, from left to right, are the complete structural formula (i, note how the double bond is so much more easily spotted by even this simplification relative to the Lewis dot structure), the line formula with condensed groups (ii), and the line formula with the lone pairs omitted (the usual structure). The next three structures represent the same representations for the form of the molecule with atoms carrying formal charges.

(c) The four structures in this case are the same as the corresponding structures in (b).

1.2 Proton Transfers Revisited: Acids and Bases

The concepts of acid and base are central to chemistry as a whole. A vast majority of organic reactions are either promoted by acids or bases or involve the participation of acids or bases at some stage of the reaction pathway. Moreover, the application of acid-base chemistry in organic chemistry is wide reaching; unlike the other subdisciplines of chemistry, an overwhelming majority of organic reactions are carried out in media other than water. This means that the concepts of acids and bases must be applied to a much broader range of conditions than those that applied when most of us learned about acids and bases for the first time. However, let us begin by reviewing the definitions of acids and bases in aqueous solution.

Arrhenius Acids and Bases

The earliest operational definitions of acids and bases were related to the neutralization of acids by bases given by the relatively simple equation that follows:

$$\text{Acid} + \text{Base} \rightarrow \text{Salt} + \text{Water} \qquad \text{(Eq. 1.1)}$$

This simple equation defines neutralization: the destruction of an acid by an equivalent amount of base to give a salt. The formation of the water itself comes from the neutralization process, a situation first recognized by the great Swedish chemist, Svante Arrhenius. As part of his dissociation theory of electrolytes (that when ionic compounds dissolve in water the ions dissociate and become free to move within the solution), Arrhenius looked at electrolytes that were originally nonconductors and proposed the first definition of an acid.

> An acid is a substance that functions as a source of hydrogen ions in aqueous solution.

Once acids had been defined in this manner, the extension of this definition to bases was relatively straightforward. The Arrhenius definition of a base is stated below.

> A base is a substance that functions as a source of hydroxide ions in aqueous solution.

The strength of an acid is related to the ease with which it functions as a hydrogen ion (or, more accurately, hydronium ion) source in aqueous solution. The functional definitions of strong and weak acids are still based largely on the Arrhenius definition. Strong acids ionize (dissociate) completely in aqueous solution. Many are familiar to most students in chemistry; sulfuric acid (H_2SO_4), nitric acid (HNO_3), hydrochloric acid (HCl), hydrobromic acid (HBr), hydriodic acid (HI), and perchloric acid ($HClO_4$) are among the more familiar strong acids. All these acids are covalent when pure but dissolve in water to give a strongly conducting solution containing hydronium (H_3O^+) ions. The pH of 1 *M* solutions of these acids is typically quite low—close to zero. Weak acids, on the other hand, do not dissolve to give complete dissociation. Instead, much of a weak acid remains unionized in aqueous solution. Typical weak acids are compounds such as acetic acid (CH_3COOH), which ionizes only to the extent of a few percent in a 1 *M* solution. The pH of a 1 *M* solution of a weak acid is much more likely to be close to 2 or 3, indicating a hydronium ion concentration only a few percent of that of a strong acid.

In a similar vein, the strength of a base is related to the ease with which it functions as a source of hydroxide ions in aqueous solution; strong bases are good sources of hydroxide ions in aqueous solution and weak bases are not. Most of the strong bases commonly used in aqueous solution are ionic hydroxides that already have the hydroxide ion present, and so need only dissolve to dissociate and liberate their hydroxide ion. The ionic hydroxides of the Group IA and heavier Group IIA metals, especially those of the Group IA metals, dissolve freely in water to give solutions containing high concentrations of hydroxide ions.

The substances LiOH, NaOH, and KOH are all widely used as strong bases. Ammonia, on the other hand, dissolves extremely well in water, but it does so without extensive ionization. Ammonia solutions, often labeled as "ammonium hydroxide," are actually best described as ammonia molecules dissolved in water, with only a few percent being converted to ammonium and hydroxide ions. Like the aqueous solutions of weak acids, aqueous solutions of weak bases such as ammonia are relatively poor electrolytes.

There are two structural features that one can use to predict the probable strength of an acid: the atom to which the acidic hydrogen is bonded and the number of multiple bonds between the central atom and oxygen in oxyacids. In general, as the electronegativity of the element bonded to the acidic hydrogen atom increases, acid strength increases. For example, O–H hydrogens are more acidic than N–H hydrogens, which are in turn more acidic than C–H hydrogens. Acid strength also increases dramatically as the element to which the hydrogen is bonded becomes larger. Although hydrogen fluoride is a relatively weak acid ($pK_a \approx 2$), hydrogen iodide is the strongest binary (two-element) acid known. In addition, the effects of multiple bonds between the central atom and oxygen atoms appear to be cumulative. For example, hypochlorous acid (HClO), which has no chlorine-oxygen double bonds, is a very weak acid, and perchloric acid ($HClO_4$), which has three oxygen atoms doubly bonded to the central chlorine atom, is one of the strongest acids known.

Lowry-Brønsted Acids and Bases

Not all chemistry is carried out in aqueous solution, and not all acid-base reactions result in the formation of water. There are, in fact, many cases where an acid-base reaction has obviously occurred but where no water is formed. For example, hydrogen chloride gas reacts with ammonia gas to form ammonium chloride, a salt. Because hydroxide ion is never involved in this reaction, it does not fit well with the Arrhenius definition of an acid-base reaction. And yet this reaction clearly leads to the neutralization of the hydrogen chloride, the acid, and the ammonia, the base, leading to the formation of the salt. A more widely applicable definition of acids and bases is that due to Lowry[19] and Brønsted,[20] who extended the Arrhenius definition of acids and bases to cover just such situations.

> In the Lowry-Brønsted formalism, an acid is a hydrogen ion donor; a base is a hydrogen ion acceptor.

A simple example of an acid-base reaction by the Lowry-Brønsted definition is the protonation of the methanol molecule by hydrogen chloride to give the methyloxonium ion (Figure 1.2). The oxygen atom of the methanol molecule is functioning as the proton

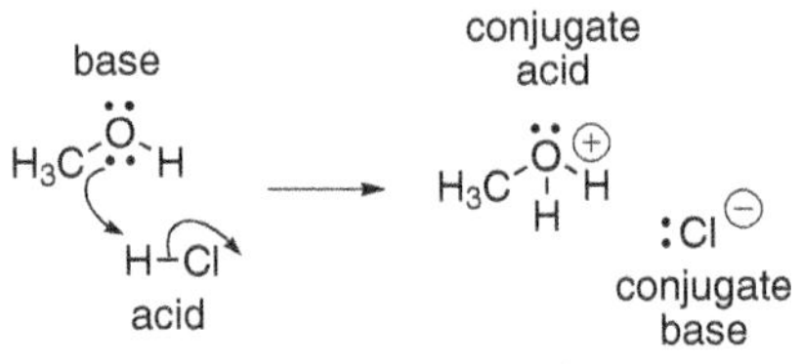

Figure 1.2 The transfer of a proton to a proton acceptor is an acid-base reaction under the Lowry-Brønsted definition. Here, the proton is transferred from a molecule of hydrogen chloride to a molecule of methanol to generate its conjugate acid, methyloxonium ion.

---

19. **Thomas Martin Lowry (1874-1936)** took his DSc under H. E. Armstrong in 1899, and he remained with him until 1913. His career took him through positions at Westminster Training College and Guy's Hospital Medical School, to the inaugural Chair in Physical Chemistry at Cambridge. Lowry, who wrote several books, was elected a Fellow of the Royal Society in 1914. For more biographical information, see: Pope, W.J. *Obit. Notices Fellows Roy. Soc.* **1938**, *2*, 287.

20. **Johannes Nicholaus Brønsted (1879-1947)** was educated at Copenhagen (PhD, 1908), where he remained as Professor of Physical and Inorganic Chemistry. His principal research was in thermodynamics, especially the strengths and catalytic properties of acids and bases. Throughout World War II, he firmly opposed the Nazis, a stand that won him election to the Danish Parliament in 1947. Unfortunately, by the time of his election, he was too ill to take his seat.

acceptor, and is therefore the base. The hydrogen chloride molecule is the proton donor, and is therefore the acid. When a proton is removed from an acid, HA, the anion formed, A^-, is termed the ***conjugate base*** of the acid. When a proton is added to a base, the product of the reaction is termed the ***conjugate acid*** of the base. Thus, during the course of the reaction in Figure 1.2, the acid (hydrogen chloride) is converted to its conjugate base (chloride ion), while the base (methanol) is converted to its conjugate acid (methyloxonium ion).

Because of the frequency and importance of proton-transfer reactions in organic chemistry, some familiarity with the acidity constant, K_a, is required. The acidity constant of a Lowry-Brønsted acid was originally defined in terms of its ability to form hydronium ion, H_3O^+, from water and is given in Equation 1.3:

$$HA + H_2O \rightarrow H_3O^+ + A^- \qquad \text{(Eq. 1.2)}$$

$$K_a = \frac{[H_3O^+][A^-]}{[HA]} \qquad \text{(Eq. 1.3)}$$

However, with the extension of acid-base chemistry into nonaqueous reaction solvents, this definition is rather too narrow, and we are now in a position where Equation 1.1 is too limiting. Nevertheless, we have retained the concept of the acidity constant, K_a, and its negative $\log_{10}$, the pK_a, as useful measures of the relative acid strength of a species in solution. Many of these values for organic compounds, however, have been measured indirectly—by competition reactions in a nonaqueous solvent, for example.

As the acid (or base) becomes stronger, its conjugate base (or acid) becomes weaker. Thus, for a strong acid such as hydrogen chloride (or hydrochloric acid), the conjugate base (chloride anion) is weak, whereas for a weak acid, such as water, the conjugate base (hydroxide anion) is strong. All acid-base reactions proceed to give the weaker of the two conjugate acids and the weaker of the two conjugate bases as the products of the reaction.

Let us examine the case of a simple proton-transfer reaction between an acid, HA, and a base, B^-:

$$HA + B^- \rightarrow HB + A^- \qquad \text{(Eq. 1.4)}$$

The equilibrium constant for this reaction is given by Equation 1.5.

$$K_{eq} = \frac{[HB][A^-]}{[HA][B^-]} \qquad \text{(Eq. 1.5)}$$

By multiplying both the numerator and the denominator of this expression by $[H_3O^+]$, Equation 1.6 is obtained.

$$K_{eq} = \frac{[HB][A^-][H_3O^+]}{[HA][B^-][H_3O^+]} \qquad \text{(Eq. 1.6)}$$

Reorganization of Equation 1.6 gives the particularly useful Equation 1.7.

$$K_{eq} = \frac{[H_3O^+][A^-]}{[HA]} \bullet \frac{[HB]}{[H_3O^+][B^-]} = \frac{K_a^{HA}}{K_a^{HB}} \qquad \text{(Eq. 1.7)}$$

What Equation 1.7 tells us is (1) that if HA is a stronger acid than HB (i.e., if the K_a of HA is greater than the K_a of HB), then the equilibrium for the reaction of HA with B^- will lie to the right, and (2) that the equilibrium constant for the reaction will be given by the ratios of the K_a values for the two acids. The other measure of acidity, the pK_a, which was defined earlier as $-\log_{10}(K_a)$, is a much more commonly used measure of acid strength. Stronger acids have an algebraically smaller value of the pK_a than weaker acids. Using pK_a values for the two acids, Equation 1.7 can be rewritten as the useful Equations 1.8 and 1.9:

$$\log(K_{eq}) = pK_a^{HB} - pK_a^{HA} \quad \text{(Eq. 1.8)}$$

$$K_{eq} = 10^{(pK_a^{HB} - pK_a^{HA})} \quad \text{(Eq. 1.9)}$$

As pointed out earlier, acid the strength of a covalent compound varies with the element to which the acidic hydrogen is bonded. As the covalent radius of that element increases, the acid strength increases, and as the electronegativity of the element bonded to the acidic hydrogen increases, acid strength increases. This allows us to observe that Se—H bonds are more acidic than S—H bonds; that S—H bonds are, in turn, more acidic than O—H bonds; and that O—H bonds are more acidic than N—H bonds, which are, in turn, more acidic than C—H bonds. One consequence of this is that we can expand our view of acids and, more particularly, bases. Bases such as amide anion, NH_2^-, the conjugate base of ammonia (NH_3, $pK_a \approx 35$) and methyllithium, CH_3Li, the conjugate base of methane (CH_4, $pK_a \approx 45–60$), are much stronger bases than hydroxide ion, the conjugate base of water. It is worth repeating here that although the pK_a scale was originally defined for Lowry-Brønsted acids in aqueous solution, we will use it for a wide variety of acids and bases which are much too strong to exist in aqueous solution without reacting with the water. Some typical pK_a ranges are given in Table 1.1.

Let us now examine some examples of acid-base reactions in organic chemistry and see just how to approach the problems associated with proton transfer chemistry.

Example 1. $CH_3Li + H_2O \rightarrow$

In this reaction, as in all potential acid-base reactions, we must first identify the acid and the base. While doing so, we must keep in mind that there is one acid and one base on each side of the arrow. In this example, the two species reacting are methyllithium, CH_3Li,

Table 1.1 Typical Ranges of pK_a Values for Representative Acids

Functional Group	Acidic Bond	Acid	Examples	Conjugate Base	Examples	pK_a Range
Alcohol	O–**H**	RO–**H**	$(CH_3)_3C$–O**H**, **H**$_2$O	RO^-	$(CH_3)_3C–O^-$, OH^-	15–19
Carboxylic acid	COO–**H**	RCOO–**H**	HCO–O**H**, CH_3CO–O**H**	RCO–O^-	HCO–O^-, $CH_3CO–O^-$	4–6
Amine	N–**H**	R_2N–**H**	$(CH_3)_2$CH–N**H**$_2$, N**H**$_3$	R_2N^-	NH_2^-, $(CH_3)_2CH–NH^-$	33–38
Amide	CON–**H**	RCONR–**H**	CH_3CO–N**H**$_2$	$RCONR^-$	$CH_3CO–NH^-$	15–17
Alkane	C–**H**	R–**H**	C**H**$_4$, (C**H**$_2$)$_6$	R^-	CH_3^-, $[(CH_2)_5CH]^-$	45–60
Alkyne	C≡C–**H**	RC≡C–**H**	CH_3–C≡C–**H**	RC≡C^-	CH_3–C≡C	25–30
Aldehyde, ketone	COC–**H**	$RCOCR_2$–**H**	CH_3COC**H**$_3$	$RCOCR_2^-$	$CH_3COCH_2^-$	25–30

and water, H_2O. The methyllithium is a compound that contains a Group IA metal, lithium, and we are usually safe in considering that such compounds will react as ionic compounds containing Li^+ cations. This means that methyllithium also contains the CH_3^- anion. This ion is capable of accepting a proton to become the neutral molecule, methane, CH_3—H (or CH_4); it is the conjugate base of methane. More importantly, it is highly unlikely that CH_3^- will readily lose a proton to become CH_2^{2-}. Therefore, the water is almost certain to be the source of protons in this reaction. At this point, we can fill out the table of reacting species in Table 1.2.

Table 1.2 Reacting Species

	Reacting Species	Conjugate
Acid	H_2O	OH^-
Base	CH_3^-	CH_4

This now allows us to write the balanced equation for this reaction:

$$CH_3^- + Li + H_2O \rightarrow CH_4 + Li^+ \, OH^-$$

From the data in Table 1.1, we find that the pK_a of H_2O is in the of 15 to 18 and that the pK_a of CH_4 is in the range of 45 to 60. On substituting these values in Equation 1.9, we obtain:

$$K_{eq} = 10^{(45-18)} = 10^{27} \text{ and } K_{eq} = 10^{(60-15)} = 10^{45}$$

as the minimum and maximum values for the equilibrium constant for this reaction. These are both huge numbers, so we expect that methyllithium will react completely with water to give methane gas and lithium hydroxide. It does—methyllithium is rapidly (explosively!) and quantitatively destroyed by water.

Example 2. $LiN(CH_3)_2 + (CH_2)_5 \rightarrow$

In this reaction, the first step is also to identify the acid and the base. In this example, the two species reacting are lithium dimethylamide, $LiN(CH_3)_2$, and cyclopentane, $(CH_2)_5$. Again, we may consider that the lithium dimethylamide will react as an ionic lithium compound, which means that it contains the $(CH_3)_2N^-$ anion. This ion is capable of accepting a proton to become the neutral molecule, dimethylamine, $(CH_3)_2N$—H; it is the conjugate base of dimethylamine. So the cyclopentane, which cannot accept a proton, is forced to be the source of protons in this reaction. At this point, we can fill out the table of reacting species (Table 1.3).

This now allows us to write the balanced equation for this reaction:

From the data in Table 1.1, we find that the pK_a of $(CH_2)_5$ is in the range of 45 to 60 and the pK_a of $(CH_3)_2NH$ is in the range of 33 to 38. On substituting these values in Equation 1.9, we obtain:

$$K_{eq} = 10^{(45-38)} = 10^{-5} \text{ and } K_{eq} = 10^{(60-33)} = 10^{-27}$$

as the maximum and minimum values for the equilibrium constant for this reaction. These are both very small numbers, so we expect that lithium dimethylamide will not react with cyclopentane. In fact, these numbers suggest that the reverse reaction (between cyclopentyllithium and dimethylamine to give cyclopentane and lithium dimethylamide) should be the one that actually occurs, and this is observed.

Table 1.3 Reacting Species

	Reacting Species	Conjugate
Acid	$(CH_2)_5$	$[(CH_2)_4CH]^-$
Base	$(CH_3)_2N^-$	$(CH_3)_2NH$

Worked Problem

1-3 Complete the following equations using the approximate pK_a values given in Table 1.1. If no reaction should occur, write "N/R" and give the reason.

(a) $CH_3CH_2CH_2CH_2Li + CH_3OH \rightarrow$

(b) $NaNH_2 + HC°CCH_2CH_3 \rightarrow$

(c) $LiN(CH_3)_2 + (CH_3)_4C \rightarrow$

§Answer below.

Leveling Effect

There is one very important consequence of Equations 1.6 and 1.7, and this is known as the ***leveling effect***. Simply stated, the leveling effect states that one cannot use an acid or base in a solvent that has a stronger conjugate base or acid than the acid one is trying to use. In other words, the solvent limits both the strongest acid you can use (the conjugate acid of the solvent) and the strongest base (the conjugate base of the solvent). In water, for example, this means that any attempt to use an acid stronger than hydronium ion, H_3O^+, is doomed to failure, as is any attempt to use a base stronger than hydroxide ion, OH^-, because the stronger acid will simply react with the water to give hydronium ion, and the stronger base will, likewise, react with the water to give hydroxide ion. As an illustration, because methane (CH_4) has a pK_a of 45 to 60, one cannot use its conjugate base, methyllithium (CH_3Li) in any solvent with a smaller pK_a. Using the equation of Worked Problem 1-3 (a) above, we see that if one tries to use methyllithium in water, the solvent reacts with the base to give lithium hydroxide (the conjugate base of water) and methane. For similar reasons, one cannot use concentrated sulfuric acid ($pK_a \approx -7$) in aqueous solution, because it reacts with the water to give the conjugate acid of water, hydronium ion, H_3O^+. *Beginning students attempting to use H_3O^+ and OH^- in the same solution frequently commit this error (usually when answering a mechanism question).* Of course, these two react to give water. Similar errors are frequently made using other incompatible acid-base combinations.

Problem

1-3 The reactant, product, and intermediate in the two-step reaction below are all correct, but the mechanism as written is wrong. What is the error in the mechanism as written? What changes are required to correct it (assume that the reaction occurs in D_2SO_4/D_2O)?

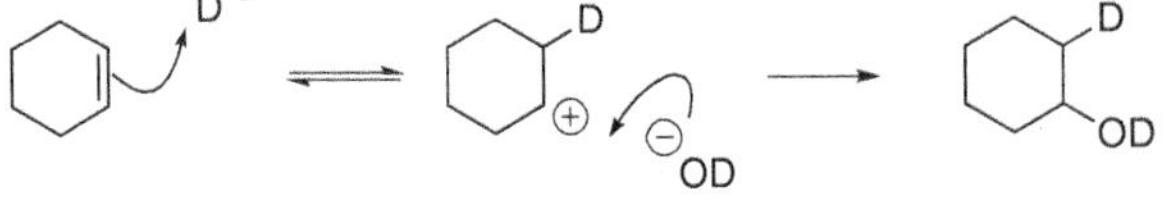

§ Answers for Worked Problems:

(a) $CH_3CH_2CH_2CH_2Li + CH_3OH \rightarrow CH_3CH_2CH_2CH_3 + Li^+OCH_3^-$

(b) $NaNH_2 + HC{\equiv}CCH_2CH_3 \rightarrow Na^+\,^-C{\equiv}CCH_2CH_3 + NH_3$

Note that in this reaction, only the hydrogen directly attached to the triple bond is attacked. Its pK_a from Table 1.1 is approximately 25 to 30, making it a stronger acid than ammonia; the approximate pK_a values of the other hydrogens are predicted to be in the 45 to 60 range.

(c) $LiN(CH_2CH_3)_2 + (CH_3)_4C \rightarrow$ N/R

The conjugate acid of $LiN(CH_2CH_3)_2$ is $H{-}N(CH_2CH_3)_2$, which is a stronger acid than $(CH_3)_4C$. Using Equation 1.9 and the pK_a values in Table 1.1, we can calculate that the equilibrium constant in favor of products will be less than 10^{-5}, corresponding to no reaction; the equilibrium strongly favors the weaker acid and the weaker base.

1.3 Reactions and Reaction Mechanisms: "Electron Pushing"

Every organic reaction occurs by an ordered series of steps that convert the reactants into products. The detailed step-by-step description of this process is known as the ***reaction mechanism***, and the purpose of a mechanism—as written by organic chemists—is to show in a clear manner how the electron reorganization between reactants and products may be rationalized. In the process, the mechanism provides a framework for discussing the reaction and for proposing modifications to a reaction.

Writing Mechanisms

All reactions involve, at their core, the breaking and making of chemical (usually covalent) bonds. When we are dealing with a covalent bond, there are only two ways in which this may be accomplished. In the first, the bond may break in such a fashion that both atoms retain one unpaired electron, a process known as ***homolysis***. Alternatively, the bond may break in such a fashion that one of the two atoms retains both electrons of the bonding pair, a process known as ***heterolysis***.

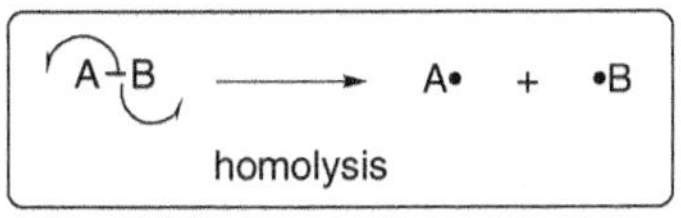

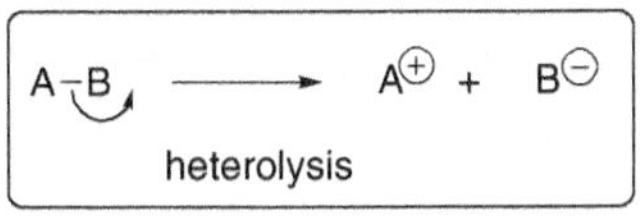

In writing modern mechanisms for organic reactions, the focus is on the movement of electrons, although this emphasis arose only slowly during the middle decades of the 20th century, due largely to the work of Sir Robert Robinson[21] and Sir Christopher Ingold.[22] Today, arrows are used to specify the relocation of electrons during a reaction step.[23] When the movement of an unpaired electron is shown, the arrowhead has a single "barb" (this type of arrow is often called a "fish-hook"); when the movement of a pair of electrons is shown, the arrowhead has both "barbs." This modern differentiation between the two types of arrows did not become commonplace until the 1960s. Prior to that time, the double-headed arrow was used to designate the movement of electrons without regard to the number of electrons.[24] In either case, the arrow shows the movement of the electrons from their initial location to their final location during that particular reaction step.

---

21. **Sir Robert Robinson (1886-1975)** was educated at Manchester (BSc, 1906; DSc, 1910) and in 1912 became the first Professor of Pure and Applied Organic Chemistry at Sydney. After time in industry following his return to Britain in 1915, eventually becoming Waynflete Professor of Chemistry at Oxford University. He received the Nobel Prize in 1947; for a more complete biography see: Todd, Lord; Cornforth, J.W. *Biogr. Mem. Fellows Roy. Soc.* **1976**, *22*, 414.

22. **Sir Christopher Kelk Ingold (1893-1970)** was educated at Southampton and London, where he also spent the major part ofhis career. Ingold was a pioneer of physical organic chemistry with his student and collaborator, E. D. Hughes. For a more complete biography, see: Shoppee, C.W. *Biogr. Mem. Fellows Roy. Soc.* **1972**, *18*, 348.

23. Early papers on use of "curly arrows" to rationalize reactivity used the arrows to designate movement of one or two electrons at a time: (a) Kermack, W.O.; Robinson, R. *J. Chem. Soc.* **1922**, *121*, 427. (b) Robinson, R. *J. Soc. Chem. Ind.* **1924**, *43*, 1297. (c) Allan, J.; Oxford, A.E.; Robinson, R.; Smith, J.C. *J. Chem. Soc.* **1926**, 401.

The use of the two-barbed arrow to indicate movement of an electron pair was proposed by Lowry and expanded on by Ingold: (d) Lowry, T.M. *J. Chem. Soc.* **1923**, *123*, 822, 1886. (e) Lowry, T.M. *Nature* **1925**, *114*, 376. (f) Ingold, C.K.; Ingold, E.H. *J. Chem. Soc.* **1926**, 1310.

24. Ingold and Robinson entered into a lifelong dispute over priority claims for this useful concept; although history shows that Robinson first used the "curly arrows," it was Lowry and Ingold who gave the concept its modern usage. The differentiation between arrow types did not appear in advanced textbooks in organic chemistry until after 1962; until then, the movement of one or two electrons was commonly shown by the same type of arrow.

Homolysis

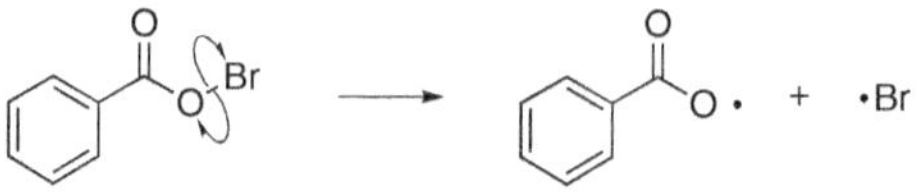
(1.1)

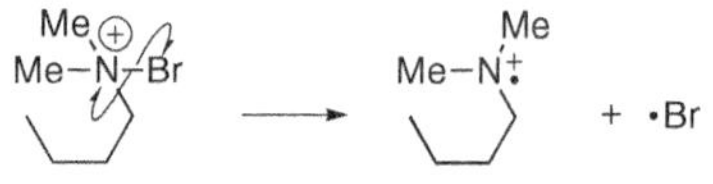
(1.2)

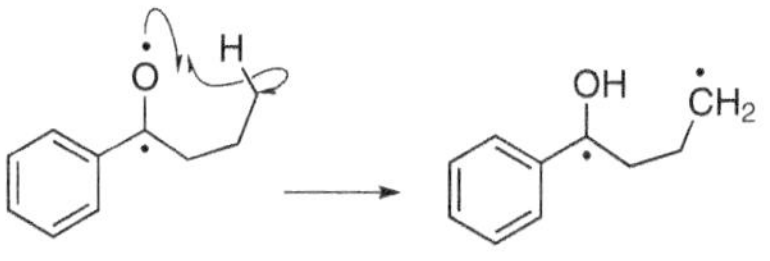
(1.3)

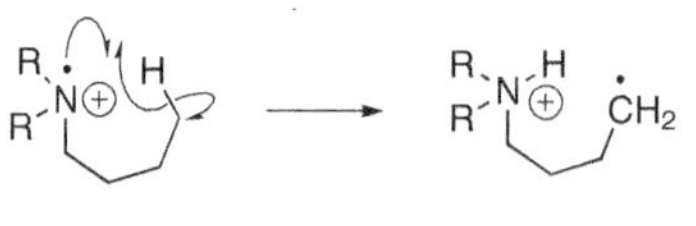
(1.4)

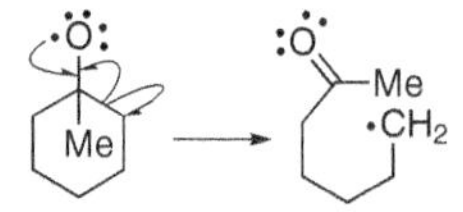
(1.5)

During a homolysis, the two electrons of the rupturing (or forming) bond are distributed equally—basically the same situation as in the intact bond. Therefore, *the formal charges on both atoms of the bond being cleaved remain the same*, because the two atoms formally control one electron of the pair when bonded, and each has one electron when the bond is broken.

The simplest case of a homolysis reaction is the homolytic cleavage of one bond, as shown in Examples **1.1** and **1.2**. In Example **1.1**, the atoms at both ends of the reacting bond are uncharged in the starting compound, so they remain uncharged in the final free radicals. In Example **1.2**, the nitrogen carries a formal positive charge in the starting ion, and it retains that formal charge in the product.

The same situation—that the charge on the atoms involved does not change during the course of the reaction step—holds true when an atom is transferred from one site in a radical to another, or when a radical fragments. In Example **1.3**, which is taken from photochemistry, the hydrogen atom is transferred from carbon to oxygen; none of the atoms involved is charged in the starting species, and no atom is charged in the final product. In the hydrogen atom transfer reaction shown as Example **1.4**, the nitrogen atom carries a formal positive charge in the starting radical, and it retains that formal positive charge in the product. In the radical fragmentation reaction shown as Example **1.5**, the same principles obtain.

Problem

1-4 Write mechanisms to account for each of the following homolysis reactions.

(a)

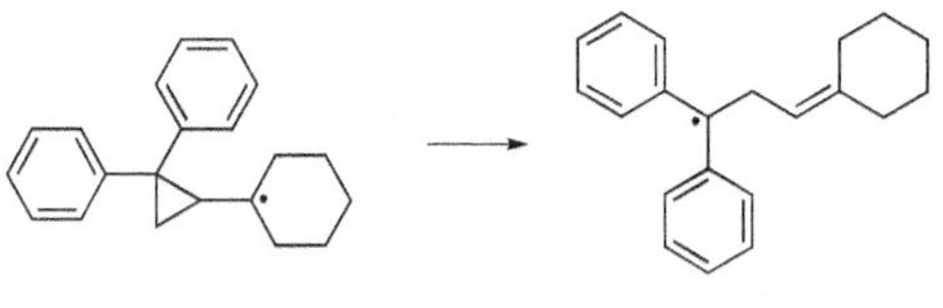

(b)

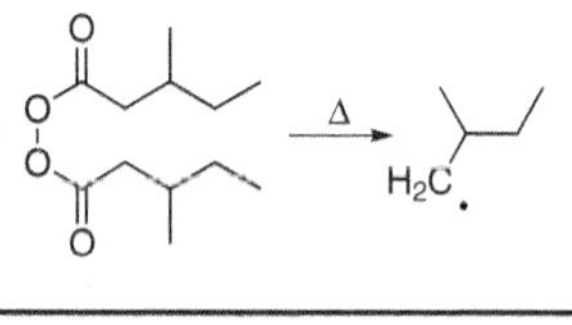

(c)

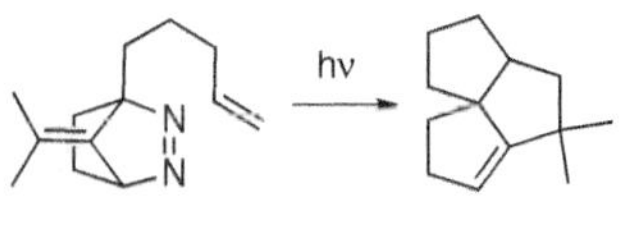

Heterolysis

In contrast to the homolysis case, heterolysis of a covalent bond leads to a change in the formal charge of *both participating atoms, with the atom retaining the two electrons from the bond gaining a unit of formal negative charge and the atom losing the two electrons of the bond gaining a unit of formal positive charge.*

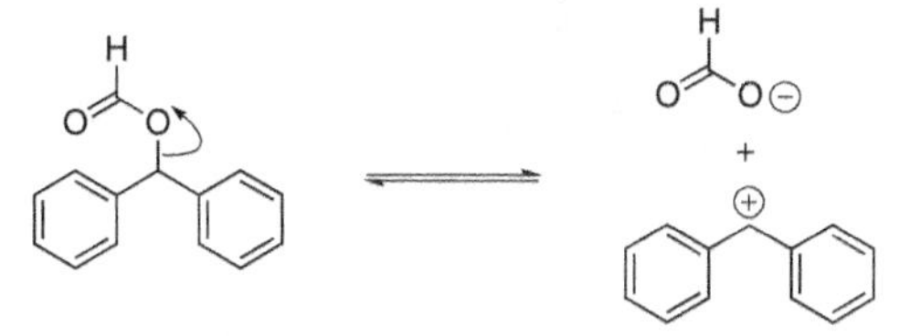

(1.6)

(1.7)

In the first simple heterolysis shown as Example **1.6**, the carbon-oxygen bond is broken. The carbon atom, which is at the origin of the arrow, gains one unit of formal positive charge to become a carbocation. The oxygen atom, which is at the terminus of the arrow, gains a full unit of formal negative charge to become an anion.

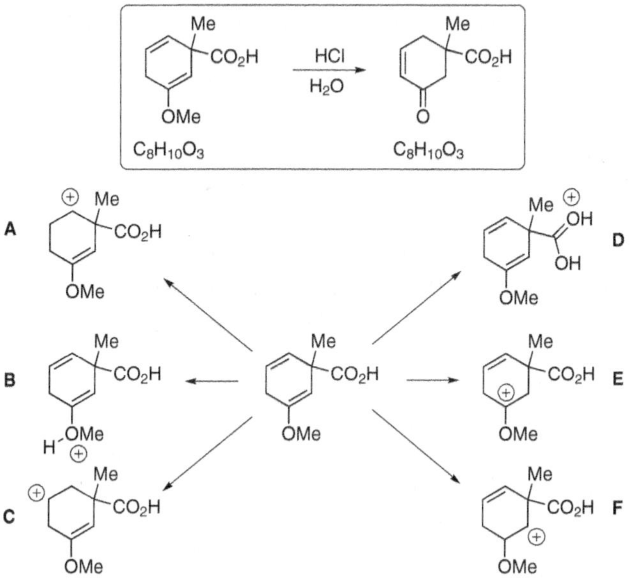

Figure 1.3 Potential initial protonation sites of the starting compound.

In Example **1.7**, two bonds—one being formed, and one being broken—are involved at the same time. The negatively charged sulfur, which is at the origin of one arrow, loses one unit of formal negative charge, and the bromine, which is at the terminus of gains one unit of formal negative charge, becoming a bromide ion. The formal charge on the carbon atom, which is at the origin of one arrow and at the terminus of another, remains unchanged.

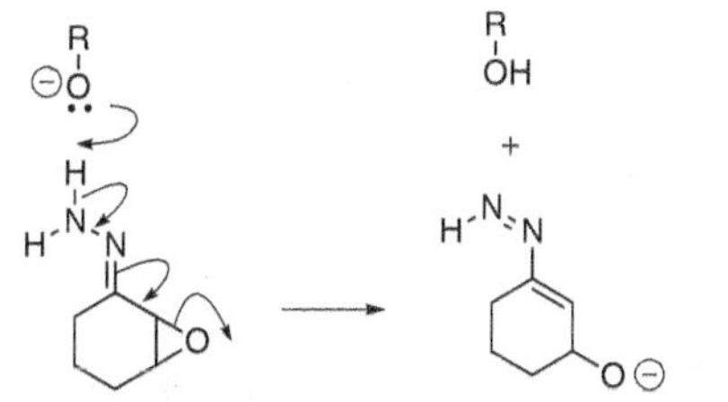

(1.8)

The same situation applies in more complex mechanistic steps where electrons around carbon reorganize (e.g., in the fragmentation reaction shown in Example **1.8**, the negative

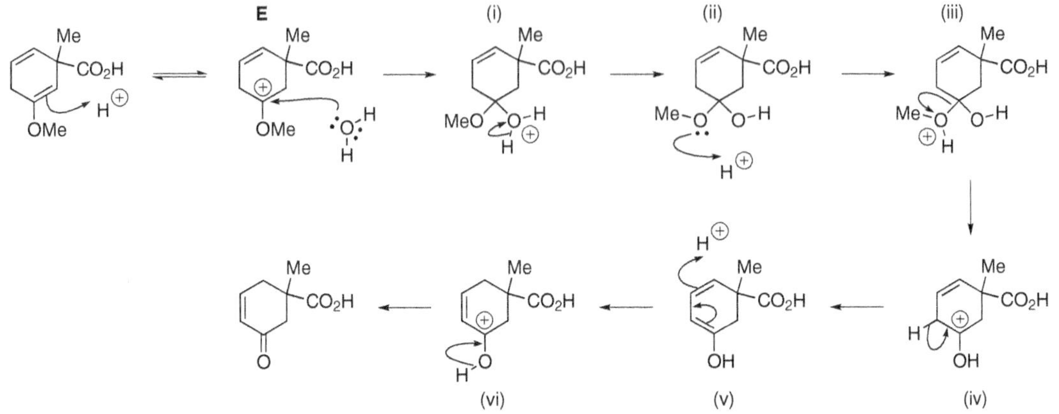

Figure 1.4 The complete mechanism.

charge on the alkoxide oxygen eventually ends up on the other oxygen atom; because the nitrogen and carbon atoms are at the origin of one arrow and the terminus of another, none of these atoms undergoes a change of formal charge).

Let us now look at how one approaches writing mechanisms for reactions, by using the reaction highlighted in Figure 1.3. A quick examination of the molecular formulas of the reactant and the product shows us that CH_2 is lost—the methyl group of the ether is now gone. Checking the molecular formulas this way is often very helpful in determining what might have happened during a reaction (the author of this book has used the technique more than once the past year in reading papers in catalytic synthesis). The remaining parts of the molecule show that the substituents have not moved (with the exception of the double bond in the ring), so we do not have skeletal rearrangements to contend with. There are six sites where the proton from the HCl may attack the ring, and the results of these attacks are shown in Figure 1.3, but only one of them has the effect of changing the structure adjacent to the OCH_3 group that is going to be lost and that is intermediate **E**, which places a new positive charge adjacent to the group that is going to be lost. If that cation reacts with water (not *hydroxide anion, which is totally incompatible with the hydrochloric acid*), we get the reaction in Figure 1.4 to give intermediate cation (i). This cation now has relatively few options for reaction, and the simplest is to lose a proton to give the hemiacetal (ii). Reprotonation of this hemiacetal can occur on either the OH or the OMe oxygens; protonation of the OH is nonproductive, so we protonate on the other one to give the new oxonium ion (iii). This now allows us to lose the methyl group as methanol by heterolysis of the C–O bond, giving us the new cation (iv). If this cation loses a proton from carbon, as in the second step of the E1 mechanism), we get a diene (v). Protonation of the diene at the terminal carbon gives an allyl cation (vi) that can now lose a proton to give the final product. Note how in every heterolysis of a bond during this mechanism, one of the atoms gains a unit of formal positive charge relative to its charge in the reactant state, and the other gains a unit of formal negative charge relative to its charge in the reactant state.

Worked Problem

1-4 Write a mechanism that accounts for the formation of the product below. (Hint: CF_3CO_2H behaves like a strong acid with a non-nucleophilic anion.)

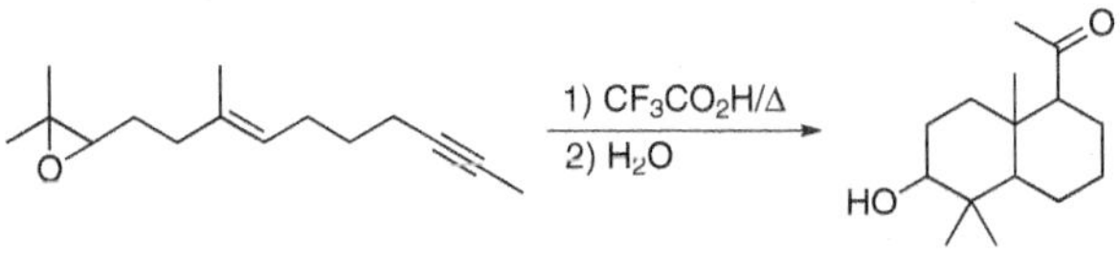

§Answer on next page.

Problems

1-5 Write mechanisms for each of the transformations shown below:

(a) 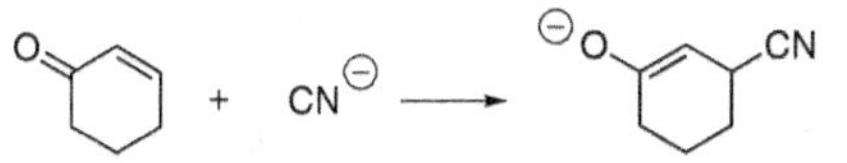

(b) 

(c) 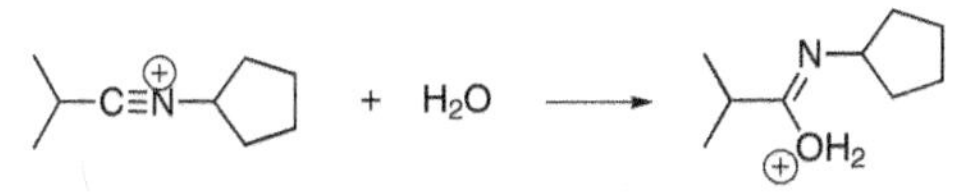

(d) 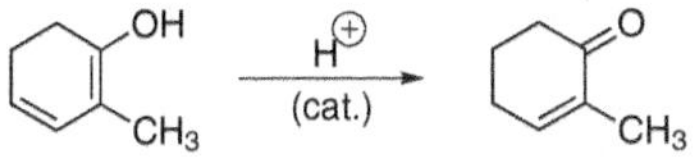

(e) 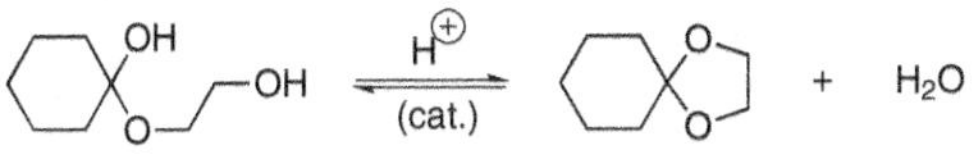

(f) 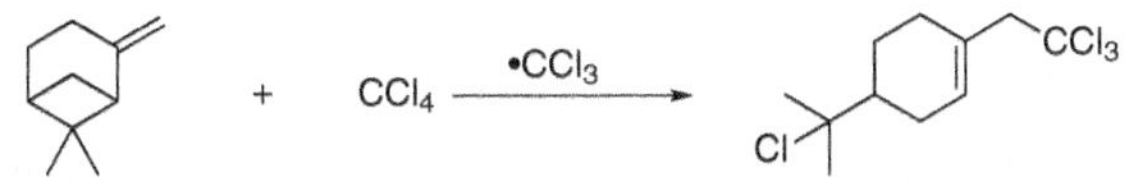

(g) 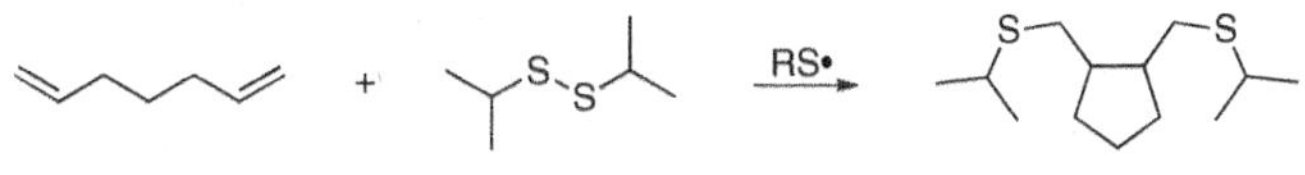

(h) 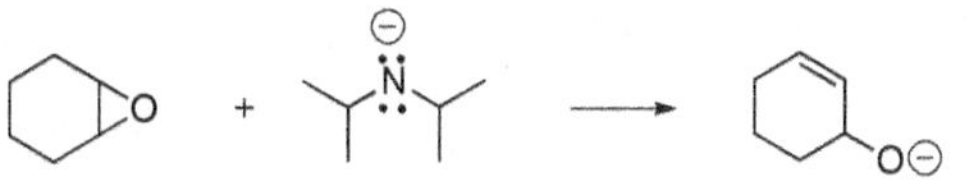

(i) 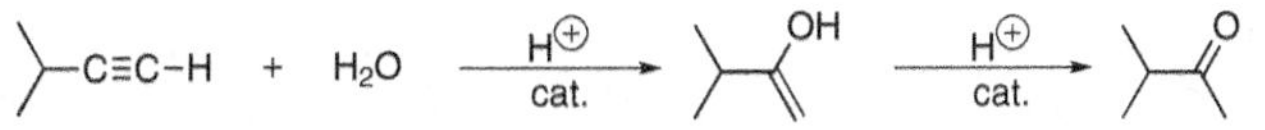

(j) 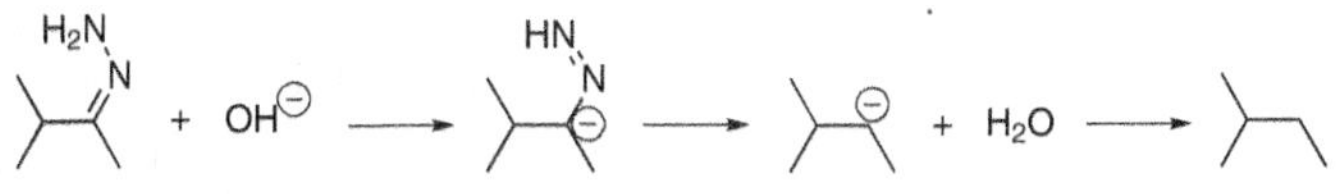

(continues)

§ Answers for Worked Problem:

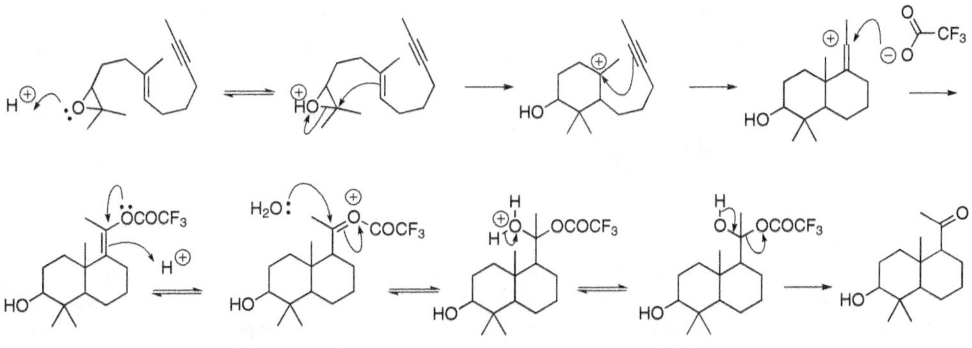

Although the trifluoroacetate anion is generally considered a non-nucleophilic anion, it traps the vinyl cation as a vinyl trifluoroacetate because the vinyl cation is such a high-energy cation.

Problems (continued)

1-6 Write a mechanism that accounts for the formation of each of the products in the reactions below:

(a)
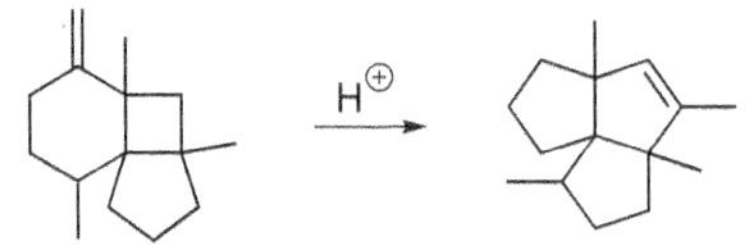

(b)
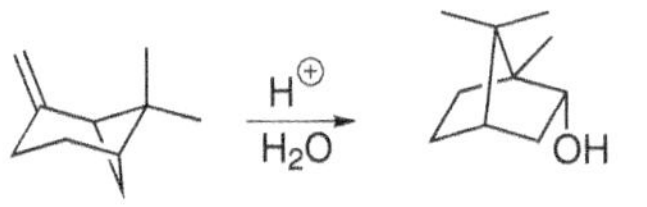

(c)
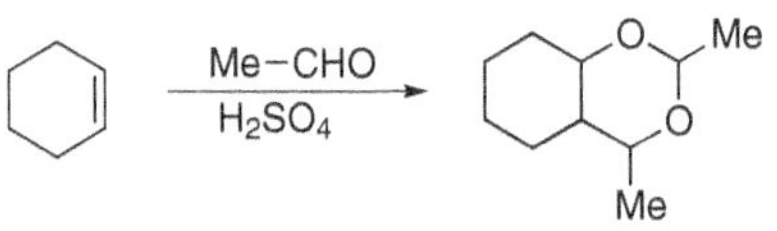

(d)
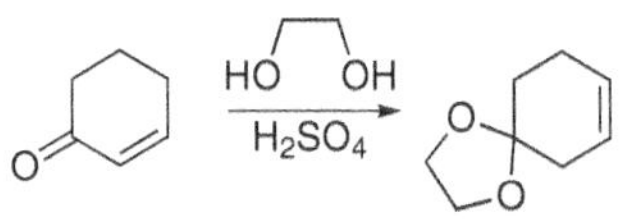

(e)
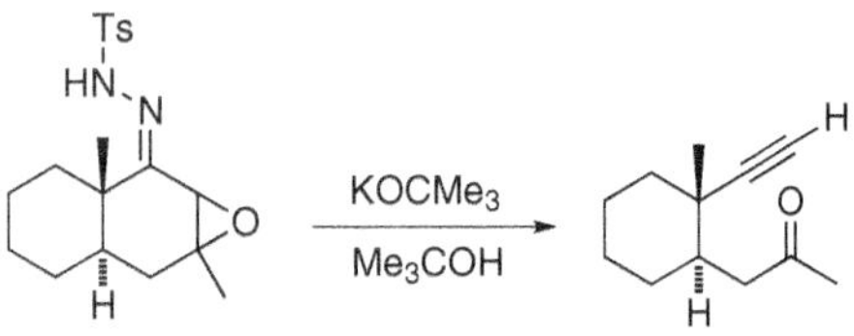

(f)
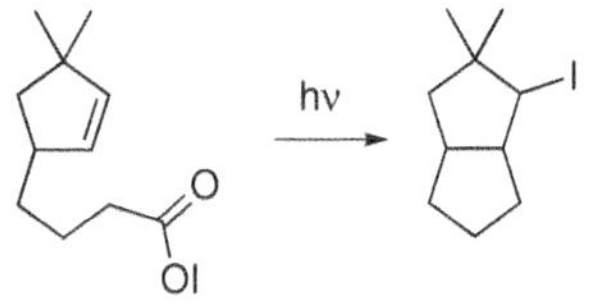

(g)
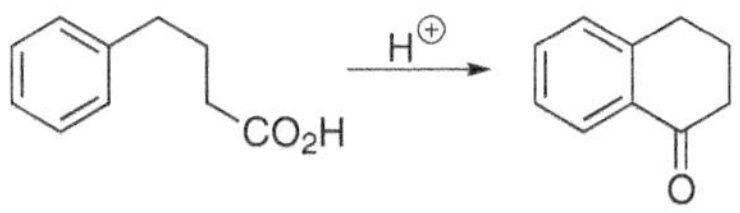

Kinetic Order, Molecularity, and Mechanistic Type

Much of the experimental evidence for the mechanism of an organic reaction is deduced from kinetics. Kinetics is the study of the relationship between the rate of a chemical reaction and two experimental parameters that the chemist can vary: the concentrations of the reactants and the reaction temperature. Kinetic measurements give the chemist a ***rate law***, which relates the rate of the reaction to reactant concentrations. Rate laws are all of the form

$$\text{Rate} = k[A]^m[B]^n\ldots \qquad \text{(Eq. 1.10)}$$

where [A], [B], and so on, are the concentrations of the reactants and m, n, and so on, are mathematical exponents. Only those reactants that participate in the reaction before or during the rate-determining step appear in the rate law.

In 1889, the great Swedish chemist, Svante Arrhenius,[25] studied the mutarotation of glucose in aqueous solution, and from his studies, he derived what is now known as the ***Arrhenius rate equation***:

$$k = A \bullet \exp(-E_a/RT) \qquad \text{(Eq. 1.11)}$$

where k is the rate constant, A is the "pre-exponential" factor and is related to entropy, E_a is the activation energy of the reaction, and T is the Kelvin temperature. Inherent in this equation is the observation that all reactions have to surmount an activation energy in order to occur.[26]

One way to visualize the processes occurring in a reaction is to use a ***reaction coordinate diagram*** or ***reaction energy profile***.

25. **Svante August Arrhenius (1859-1927)** was educated at Uppsala (PhD, 1884, but with the lowest possible passing grade). After being appointed to the physics faculty at Uppsala, he studied with Ostwald, van't Hoff, and Boltzmann. In 1903, he received the Nobel Prize in Chemistry. A more complete biography is available at the Nobel website, and at: Snelders, H.A.M. *Dictionary of Scientific Biography* (Charles Scribner's and Sons: New York, 1970); vol. 1, p. 296.

26. Arrhenius, S. *Z. physik. Chem.* **1889**, 4, 226.

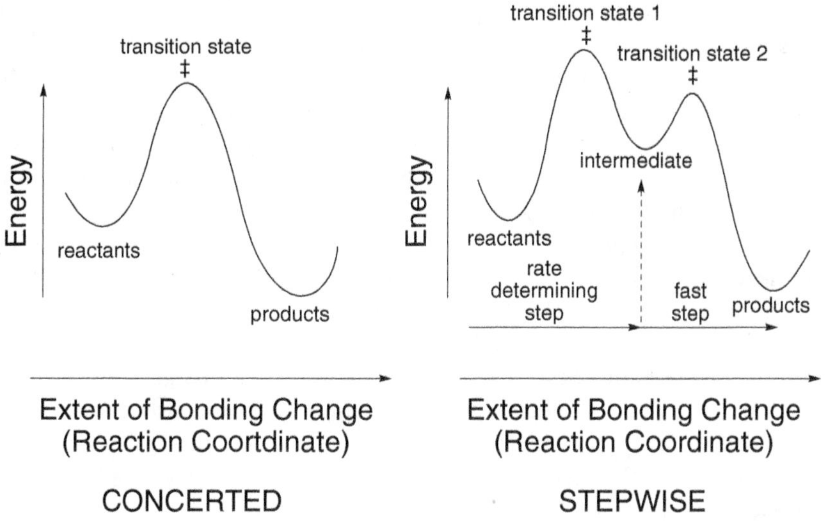

Figure 1.5 Reaction coordinate diagrams for concerted and stepwise reactions. The concerted reaction occurs in a single step through a single activated complex. In the stepwise reaction, more than one activated complex is involved, and at least one intermediate is formed. In this example, the first step is rate determining because the energy of transition state 1 is higher than the energy of transition state 2.

This diagram has its ultimate origin in the Arrhenius equation and followed from the construction of reaction potential energy surfaces[27] by American chemist, Henry Eyring,[28] in the early 1930s. In these diagrams, the pathway between reactants and products lay along a trench or "*valley*" in the surface, with the high-energy point being a saddle point. The simplification that is made to give the modern reaction coordinate diagram was to use the path along the bottom of the valley as the x-axis and energy as the y-axis of the diagram. In this way, the "reaction coordinate" now corresponds to the progress of each molecule individually through the bonding changes needed to be converted to products.[29]

The high-energy point on the diagram is unusual, in that it represents a structure that is inherently unstable, as well as the highest energy point of the reaction pathway. Eyring called this the ***activated complex***,[30] a name that emphasizes its structure, whereas Evans and Polanyi called it the ***transition state***,[31] which emphasizes its ephemeral existence.

Reactions can be divided into two broad categories based on the timing of the bonding changes involved. Reactions that occur in a single step—in which all the bonding changes occur simultaneously—are called ***concerted*** reactions. They proceed through a single activated complex, and there is only one transition state in the reaction coordinate diagram. In contrast to a concerted reaction, a ***stepwise*** reaction proceeds through two activated complexes, and an ***intermediate*** is formed during the reaction. In stepwise reactions, only those species involved prior to the highest energy step, known as the ***rate-determining step***, appear in the rate law of the reaction. The reaction coordinate diagrams for a typical concerted and stepwise reaction are compared in Figure 1.5.

---

27. (a) Eyring, H.; Polanyi, M. *Z. physik. Chem.* **1931**, *B12*, 279. (b) Eyring, H. *J. Am. Chem. Soc.* **1931**, *53*, 2537; **1932**, *54*, 3191. (c) Eyring, H. *Chem. Rev.* **1932**, *10*, 103.

28. **Henry Eyring (1901-1981)** was born in Juarez, Mexico, and educated at Arizona and Berkeley (PhD, 1927). After graduating, he worked at Wisconsin, Berlin, and the University of California at Berkeley before moving to Princeton in 1931. In 1946 he moved to Utah, where he spent the rest of his career. For a more complete biography, see: Kauzmann, W. *Biogr. Mem. Nat. Acad. Sci.* (National Academies Press: Washington, D.C., 1996), 47.

29. One of the first appearances of the reaction coordinate diagram in the modern form in a textbook is in Glasstone, S.; Laidler, K.J.; Eyring, H. *The Theory of Rate Processes* (McGraw-Hill: New York, 1941), pp. 97-100.

30. The activated complex was first proposed by Eyring, who applied statistical methods to the calculation of reaction rates: (a) Eyring, H. *J. Chem. Phys.* **1935**, *3*, 107. (b) Eyring, H. *Chem. Rev.* **1935**, *17*, 65. (c) Eyring, H. *Trans. Faraday Soc.* **1938**, *34*, 41.

31. (a) Evans, M.G.; Polanyi, M. *Trans. Faraday Soc.* **1935**, *31*, 875; **1937**, *33*, 448. (b) Polanyi, M. *J. Chem. Soc.* **1937**, 629.

Two particularly important parameters used to describe a reaction are its ***kinetic order*** and its ***molecularity***. It is important to distinguish between these two. The kinetic order of a reaction is the sum of the exponents in the rate law. It is an *experimental* quantity that is directly related to the algebraic form of the rate law, it can be determined by appropriate kinetic measurements, and it relates the rate of the reaction to the concentrations of the reactants. The molecularity, on the other hand, is a *theoretical* concept that is used to describe the relationship of the activated complex to the reactants. Reactions where the activated complex is derived from a single distinct chemical species are ***unimolecular***, those where two distinct chemical species combine to form the activated complex are ***bimolecular***, those where three distinct chemical species form the activated complex are termed ***termolecular***, and so on. Unlike kinetic order, molecularity depends on the assumed mechanism of the reaction. In some ways, this gives molecularity an advantage compared to the kinetic order of a reaction, and it is constant for all reactions of a given mechanism type. However, one must remember the caveat that, unlike the kinetic order of the reaction, molecularity may be subject to reinterpretation as the proposed mechanism changes.[32]

Most of the time, the kinetic order and the molecularity of a reaction correspond, but this is not universally the case, as the acid-catalyzed reaction between *tert*-butyl alcohol and bromide ion shows. As shown in Figure 1.6, the reaction mechanism involves the formation of two reactive intermediates—the oxonium ion, which carries the actual leaving group that is replaced, and the carbocation, which reacts with the nucleophile. Of these two, the carbocation is the higher-energy intermediate, so its formation is more endothermic.

In this reaction, the slow step is the heterolysis of the carbon-oxygen bond of the oxonium ion, and only the oxonium ion is involved in the slow step. This makes this reaction unimolecular because only one particle is involved in the rate-determining step. However,

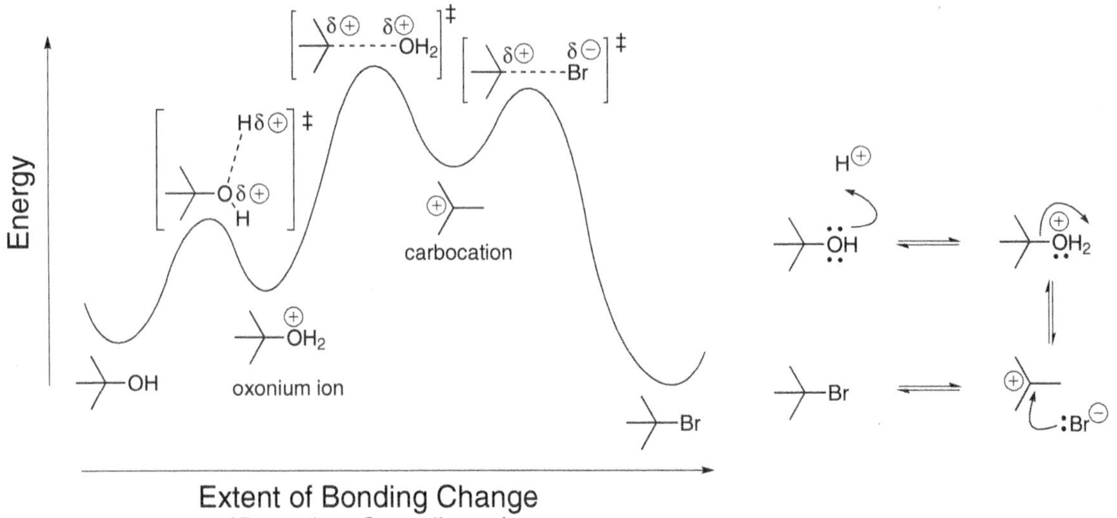

Figure 1.6 The reaction between *tert*-butyl alcohol and bromide ion in the presence of a strong acid catalyst. Note how the oxonium ion is formed in a rapid pre-equilibrium step.

32. Molecularity: Hughes, E.D.; Ingold, C.K.; Patel, C.S. *J. Chem. Soc.* **1933**, 526. For discussions of molecularity and kinetic order, see Ingold, C.K. *Structure and Mechanism in Organic Chemistry* (Cornell University Press: Ithaca, N.Y., 1953), pp. 308-316; Gould, E.S. *Mechanism and Structure in Organic Chemistry* (Holt, Rinehart and Winston: New York, 1959), p. 164.

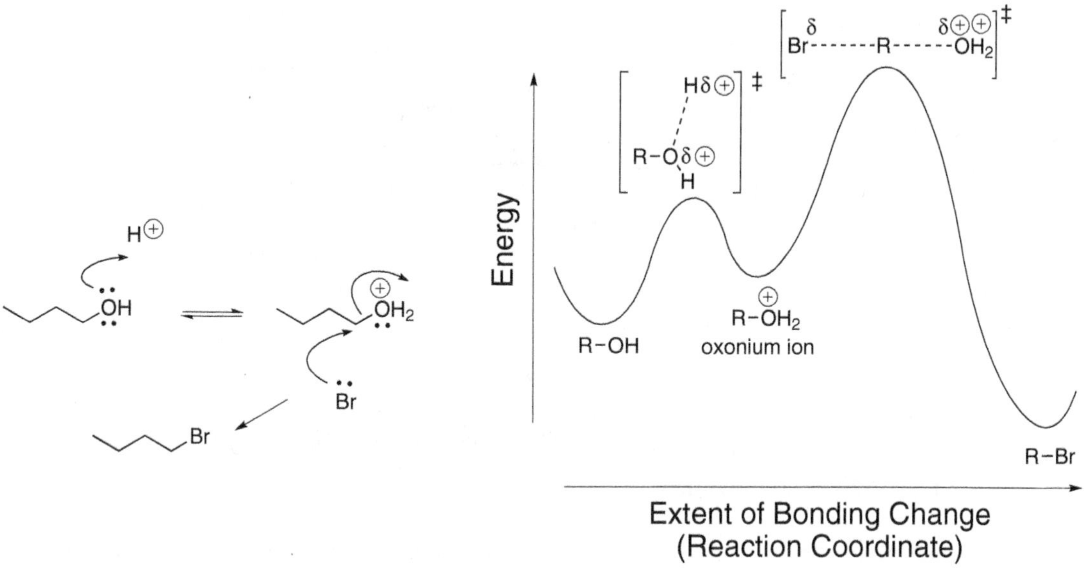

Figure 1.7 The acid-catalyzed reaction between 1-butanol and bromide ion.

because two reactants are required to make the species that participates in the rate determining step, the rate law for the reaction is:

$$\text{Rate} = k[\text{ROH}][\text{H}^+] \qquad (\text{Eq. } 1.12)$$

This makes the reaction ***second-order***. Here we see the effects of a ***pre-equilibrium***, where the initial starting compound is first converted in a rapid equilibrium step to the actual substrate for the reaction, which then participates in the slow step of the reaction.

By comparison, the acid-catalyzed reaction between 1-butanol and bromide anion (Figure 1.7) proceeds by the S_N2 mechanism. It is a ***bimolecular*** reaction between the oxonium ion and bromide ion, with ***third-order*** kinetics, and a rate law:

$$\text{Rate} = \text{k}[\text{ROH}][\text{H}^+][\text{Br}^-] \qquad (\text{Eq. } 1.13)$$

The Arrhenius equation permits the effects of temperature on a reaction to be predicted, but it is often more useful to be able to predict the effects of small changes in the reacting species on relative reaction rates. One important tool used by organic chemists to predict the relative rates of closely related reactions was proposed in 1955[33] by American chemist George S. Hammond.[34] His proposal, which has become known as the ***Hammond postulate***, states that:

> If two states, as for example a transition state and an unstable intermediate, occur consecutively during a reaction process and have nearly the same energy content, their interconversion will involve only a small reorganization of molecular structure.

Thus, we expect that in reactions where the activation energy is low (usually exothermic reactions), the activated complex and the reactant are separated by only a small energy

33. Hammond, G.S. *J. Am. Chem. Soc.*, **1955**, *77*, 334.

34. **George Simms Hammond (1921-2005)** was educated at Bates College and Harvard University (PhD, 1947). He held faculty positions at Iowa State University, the California Institute of Technology, and the University of California at Santa Cruz, and he served as Foreign Secretary of the National Academy of Sciences. For a more detailed biography, see: Wamser, C.C. *J. Phys. Chem. A* **2003**, *107*, 3149.

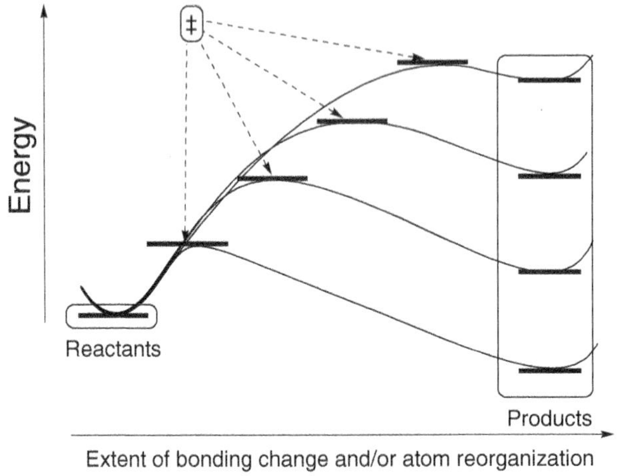

Figure 1.8 The change in transition state structure with the enthalpy of the reaction.

gap, and it will therefore require only a small reorganization of the atomic positions to convert the reactants to the activated complex. In a highly endothermic reaction, with a large activation energy, the energy gap between reactants and activated complex is large, but that between the activated complex and the product is small; in these cases, we expect that the conversion of product to activated complex, and vice versa, will involve only a small reorganization of the atomic positions. This can be stated in a generalized form (shown schematically in Figure 1.8).

> The more endothermic a reaction, the higher its activation energy, and the more closely the structure of the activated complex resembles the structure of the product (or intermediate). The more exothermic a reaction, the lower its activation energy, and the more closely the structure of its activated complex resembles the structure of the reactants.

A more recent theoretical basis for the Hammond postulate arises as an extension of the Marcus theory[35] for outer-sphere electron transfer reactions.[36] To apply this extension of Marcus theory, the potential energy surface of the reactant and product are represented by parabolas that are separated along the reaction coordinate. The positions are labeled in Figure 1.9 as being at the reactant geometry and the product geometry. The fact that these outer-sphere electron transfer reactions are not accompanied by structural changes greatly simplifies the geometric picture and allows the effects of energy to be isolated—only the parameter along the z axis changes.

In an endothermic reaction, the product parabola is located at a higher Gibbs free energy than that of the reactant (i.e., higher up the vertical axis). In a thermoneutral reaction, the two parabolas are at the same energy (i.e., the same height up the axis. In an exothermic reaction, the product parabola occurs at a lower Gibbs free energy than the reactant. The points where the reactant and product parabolas cross (the black dots in Figure 1.9) correspond to the transition state energy and geometry. As is evident from Figure 1.9, the predictions of the Marcus theory and the Hammond postulate correlate well when the reaction is not highly exothermic.

---

35. (a) Marcus, R.A. *J. Chem. Phys.* **1956**, *24*, 966, 979; **1965**, *41*, 2654; *43*, 679. (b) Siders, P; Marcus, R.A. *J. Am. Chem. Soc.* **1981**, *103*, 741, 748.

36. **Rudolph Arthur Marcus (1923-)** was educated at McGill University in Montreal, and after postdoctoral work in North Carolina, he joined the Polytechnic Institute of Brooklyn in 1951. In 1964 he moved to the University of Illinois at Urbana-Champaign, and in 1978 he joined the California Institute of Technology. He received the Nobel Prize in 1992 for his theoretical work in electron transfer reactions and unimolecular reactions. For more biographical information, see the Nobel Foundation web site.

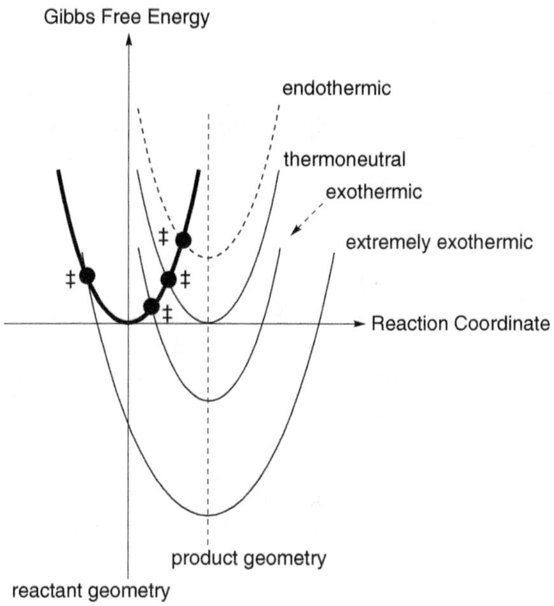

Figure 1.9 Marcus theory approach to the prediction of transition state energy and structure.

Worked Problem

1-5 The electrophilic substitution of benzene by bromine in the presence of ferric bromide and the sulfonation of benzene by sulfur trioxide in concentrated sulfuric acid follow a common mechanism. The first step is addition of the electrophile to the benzene π-bond system, and the second step is loss of a proton from the benzenonium ion intermediate. However, bromination is irreversible, whereas sulfonation is reversible (i.e., it comes to an equilibrium). Draw appropriate reaction coordinate diagrams to explain this observation. The mechanism is drawn below:

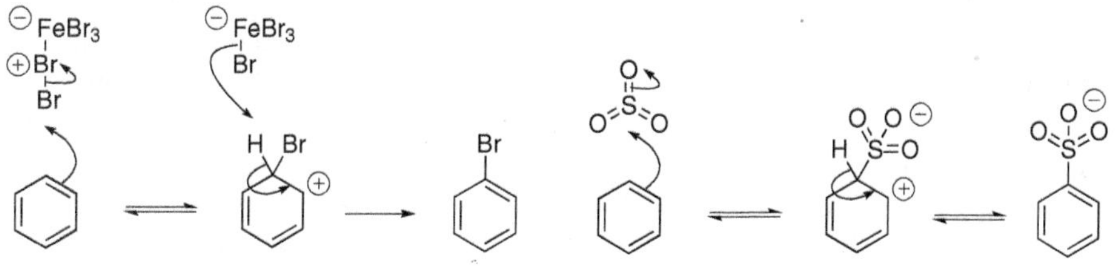

§Answer below.

§ Answers for Worked Problem:

The first step of the reaction is the same in both, so any difference in the outcome most likely comes in the second step of the reaction. In bromination, the activation energy leading from the intermediate cation to the product must be lower than the activation energy leading back to reactant, whereas in sulfonation, the activation energy leading from the intermediate cation to the product must be comparable to the activation energy for the reaction leading back to the reactant. In sulfonation, also, the relative energies of the reactant and product are probably comparable, thus permitting the reaction rates converting reactant to product and product back to reactant to be similar.

Problems

1-7 The ring opening of three-membered rings by strong, non-nucleophilic bases is a method for the formation of allylic compounds. The reaction occurs in a single step.

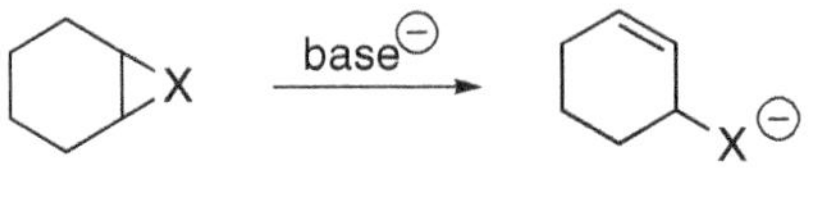

Arrange the compounds with X = N, X = O, and X = S in the expected order of the rate of this reaction under comparable conditions. Draw appropriate reaction coordinate diagrams to support your answer. You may find the following pK_a values useful: H_2O, 15.74; NH_3, 33; H_2S, 7.00.

1-8 Arrange the compounds with values of n from 1 to 3 in order of the ease of their ring opening reactions by hydrogen sulfide in the presence of strong acid. Draw appropriate reaction coordinate diagrams for the reaction.

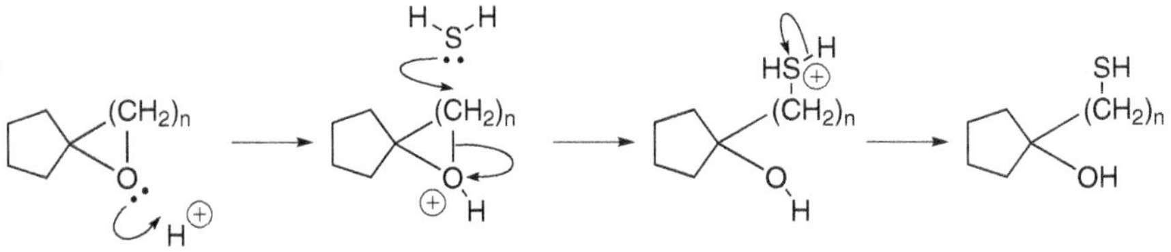

1.4 Resonance

Occasionally, the known facts about the structure of a molecule or ion cannot be accommodated by a single Lewis structure. These compounds are best described in terms of ***resonance theory***,[37] first developed by the American chemist Linus Pauling.[38] In a resonance-stabilized system, there are at least two valid Lewis structures that can be drawn for the molecule or ion, and these are related by having a different arrangement of the electrons between the nuclei. The compliant structures that can be drawn using the Lewis rules are known as ***resonance contributors***, ***limiting structures***, or ***canonical forms***, and the correct structure is a ***resonance hybrid*** of these three. In order not to confuse resonance with equilibrium, canonical forms (which have no independent existence) are linked by a double-headed arrow[39] rather than by the double arrows used to designate equilibria (which link structures that ***do*** exist).

Writing structures of canonical forms and predicting the structure and reactivity of resonance hybrids follows the set of general precepts below. Of course, these generalizations are like most generalizations—they work in most cases, but it is always possible to find exceptions to the rule.

---

37. Early papers describing resonance: (a) Pauling, L.; Wheland, G.W. *J. Chem. Phys.* **1933**, *1*, 362-374. (b) Pauling, L.; Sherman, J. *J. Chem. Phys.* **1933**, *1*, 606-617.

The subject is dealt with at length in Wheland, G.W. *Resonance in Organic Chemistry* (John Wiley & Sons: New York, 1955).

38. **Linus Carl Pauling (1901-1994)** was educated at the California Institute of Technology (PhD, 1925) and then studied in Europe with Sommerfeld, Bohr, Schrödinger, and Bragg before returning to "Caltech" in 1927 as a member of the faculty. Pauling made numerous contributions to physical and theoretical chemistry. He received the Nobel Prize twice (Chemistry, 1954; Peace, 1962). For more complete biographies, see the Nobel Foundation web site, and the Linus Pauling Institute web site (http://lpi.oregonstate.edu/lpbio/lpbio2.html, accessed November 21, 2012).

39. (a) Bury, C.R. *J. Am. Chem. Soc.* **1935**, *57*, 2115. (b) Eistert, B. *Angew. Chem.* **1936**, *49*, 33.

1. Resonance stabilization always requires delocalized electrons in a molecule. It must be possible to write at least two different, valid Lewis structure for the molecule.
2. The number of atoms carrying formal positive or negative charges is minimized (i.e., the number of contributors in which there is charge separation is minimized).
3. Nuclei may not move during the interconversion of canonical forms—only electrons may move.
4. Maximum resonance stabilization occurs when the number of equivalent canonical forms is maximized.
5. The structure of the resonance hybrid will most strongly resemble the average of the structures predicted for the major canonical forms.
6. The reactivity of the resonance hybrid will most often resemble the reactivity predicted for the minor canonical form.

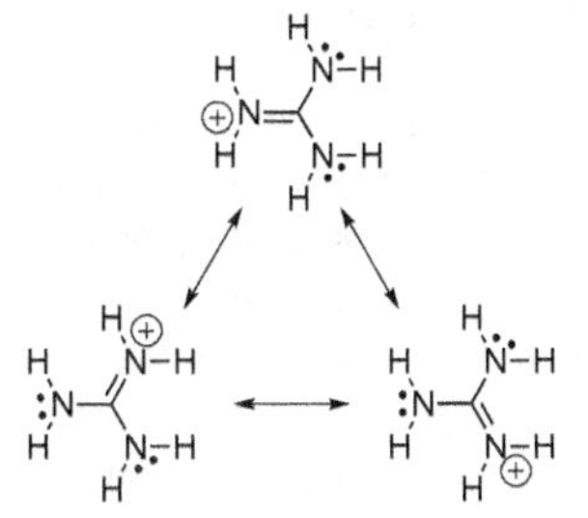

(1.9)

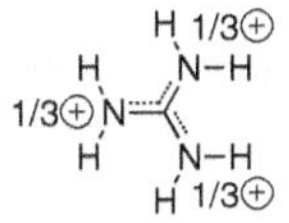

(1.10)

We can illustrate the principles of resonance using the guanidinium cation, shown in Example **1.9**. This ion is known to have three equivalent nitrogen atoms. Using the rules for writing correct Lewis structures, one cannot, however, write a compliant structure in which more than two nitrogen atoms are equivalent. The third nitrogen is always different; one is double-bonded and carries a formal positive charge, whereas the two are single-bonded and neutral. It is possible to write two other structures for this ion by simply rearranging the way that the electrons in lone pairs and π bonds are distributed, so one obtains three different structures containing the same set of nitrogen atoms. However, because each of these compliant structures has one unique nitrogen atom, *none* can be the correct structure of the guanidinium cation. The true structure must lie somewhere between these three structures—it must be a hybrid structure of the three with elements of all. A more accurate representation of the structure of the guanidinium cation is shown as Example **1.10**.

It is important to remember that *canonical forms are simply useful representations of limiting Lewis structures for a compound and do not exist in the real world.*[40] Nevertheless, resonance theory has the important ability to be used to predict both structure and reactivity of reactive intermediates in organic chemistry. The structure of a resonance-stabilized species may be described in terms of equivalent or nonequivalent canonical forms. The guanidinium cation above is an example of a species whose canonical forms are equivalent, as are the carboxylate anion, and the nitro group, each of which has two equivalent canonical forms (Figure 1.10).

Many resonance-stabilized systems have canonical forms that are not equivalent (i.e., do not contribute equally to the resonance hybrid); the structure of such systems looks more like a weighted average of the contributors. To predict the structure and reactivity of

40. The analogy that is widely used for describing resonance—that a rhinoceros (a real animal) is the resonance hybrid of a unicorn (a fictional creature) and a dragon (another fictional creature)—is due to J. D. Roberts [cited in Wheland, G.W. *Advanced Organic Chemistry*, 3rd. Ed. (John Wiley & Sons: New York, 1960), p. 91].

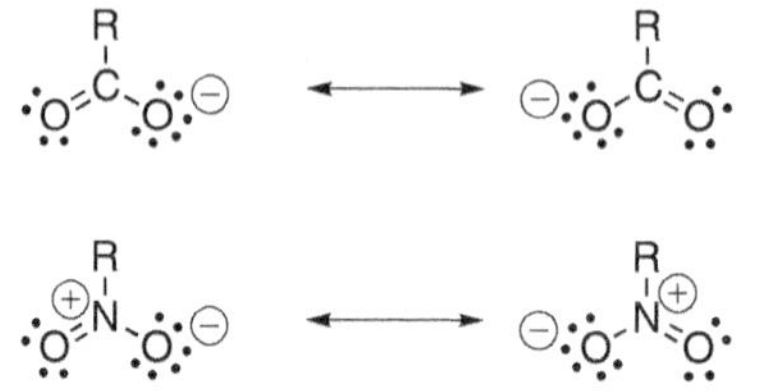

Figure 1.10 The two equivalent canonical forms of the structure of the carboxylate anion (upper) and the nitro group (lower) have the same set of bonds and electron pairs. No nuclei move during the conversion of one canonical form to the other.

such species, one needs to be able to predict the relative importance of the canonical forms. This can be done on the basis of a few simple rules:

1. Canonical forms where all non-hydrogen atoms have a complete octet are more important than canonical forms where one atom lacks a complete octet.
2. Canonical forms with the maximum number of covalent bonds tend to be most important.
3. When two or more resonance structures are possible, the major contributors to the overall resonance hybrid will be those in which all non-hydrogen atoms possess a complete outer-shell octet of electrons.
4. In anions, the canonical forms with the negative charge on an electronegative atom are preferred.
5. In cations where the charged atom must have only six electrons, canonical forms with a positive charge on an atom of low electronegativity are preferred.
6. In cations for which one can draw canonical forms where every non-hydrogen atom has a complete octet, these canonical forms are more important even if this requires placement of the positive charge on an electronegative atom.

Unequally contributing canonical forms can be used to predict the chemical reactivity, structure, and stability of the resonance-stabilized species.

In general, the chemical reactivity of a resonance-stabilized molecule or ion most often resembles the reactivity that one would predict for the canonical form making the minor contribution to the resonance hybrid (minor resonance contributor). In contrast, the molecular structure of a resonance-stabilized molecule most often resembles the structure that one would predict for the canonical form making the major contribution to the resonance hybrid (major resonance contributor). It is also observed that the chemical stability of a resonance-stabilized molecule or ion is best predicted by the structure of the major resonance contributor.

Two good examples of resonance-stabilized charged systems are the enolate anions, $R_2C{=}CRO^-$, and acylium cations, RCO^+. There are two reasonable canonical forms of the structure of the enolate anion (Figure 1.11), and the acylium ion (Figure 1.12).

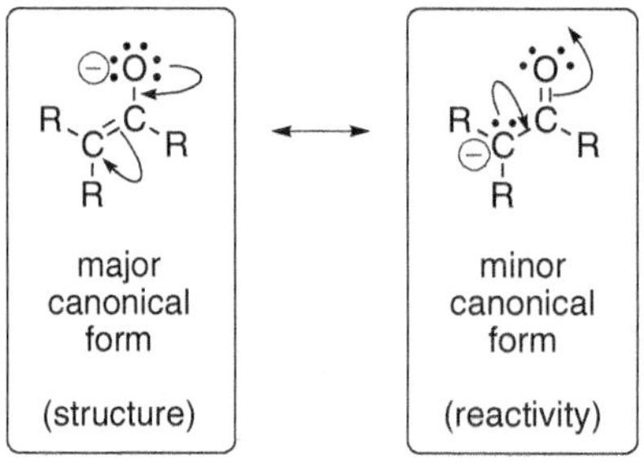

Figure 1.11 The enolate anion is an example of an anion in which the two resonance contributors are no longer equivalent. The major contributor to the structure of this ion is the one on the left, where the negative charge is on the more electronegative oxygen atom; the minor contributor (which frequently dictates the normal chemical reactivity) has the negative charge on the carbon atom.

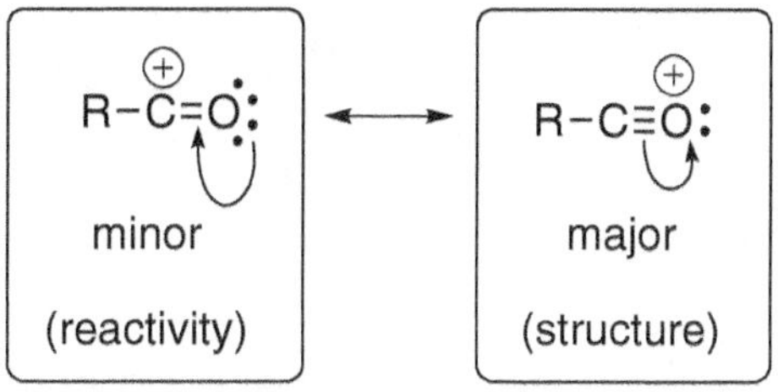

Figure 1.12 In the acylium cation, only the contributor on the right has a complete outer-shell octet on every non-hydrogen atom. Therefore, even though it requires that the more electronegative oxygen atom carry a formal positive charge, it is still the major contributor to the resonance hybrid.

In the left-hand canonical form in Figure 1.11, there is a carbon-carbon double bond, and the negative charge is on the oxygen atom. In the right-hand canonical form, there is a carbon-oxygen double bond, and the negative charge is on the carbon atom. Based on the arguments above, one would predict that the contributor on the left would be the more thermodynamically stable because the negative charge is on the most electronegative atom. Accordingly, we would expect that this should contribute to the overall structure of the anion to a greater extent than the contributor with the negative charge on carbon. Indeed, the structures of several metal enolates chemically similar to this have been determined, and in every case the carbon-carbon bond is a double bond, and the length of the carbon-oxygen bond is exactly that expected for a carbon-oxygen single bond.

The two most important canonical forms of the structure of an acylium ion are shown in Figure 1.12. The major difference between the two canonical forms is that the positively charged carbon atom in the left-hand contributor has only six valence electrons, whereas *all* the non-hydrogen atoms in the right-hand contributor have complete valence shell octets. The major contributor is the right-hand structure, in spite of the fact that it requires that a positive charge be placed on the most electronegative atom (the oxygen atom).

The canonical forms of the acylium ion emphasize another important precept of resonance theory—when two or more canonical forms can be drawn for a structure, the major contributor(s) to the resonance hybrid will be the one(s) in which every non-hydrogen atom possesses a complete valence shell octet of electrons.

All arguments about structure based on resonance arguments must be prefaced with one important restriction. These arguments apply only if one is dealing with *reasonable* resonance contributors. In other words, all resonance contributors must obey the rules for writing Lewis structures (the number of atoms with complete outer-shell octets must be maximized, the number of atoms with incomplete outer-shell octets must be minimized). It is always possible to write resonance contributors for the structure of any molecule or ion, but most of these will be patently unreasonable. They will contain several (i.e., more than two) atoms without complete outer-shell octets, they will contain several atoms carrying opposite charges that could be neutralized by simply moving electrons, and they will contain several atoms with unpaired electrons.

Although resonance contributors in which there is charge separation are occasionally used in discussions of the reactivity of organic compounds, such contributors are invariably minor contributors to the overall resonance hybrid. By far the most common versions of this situation involve polar covalent bonds, which may be viewed as the resonance hybrid of a nonpolar contributor and an ionic contributor. The formaldehyde and formamide molecules provide good examples of this situation (Figure 1.13). In both of these molecules, one can write at least one resonance contributor to the structure in which two atoms bear opposite formal charges. In the formaldehyde molecule, this is done by choosing to leave the carbon atom with six electrons, and placing three lone pairs around the

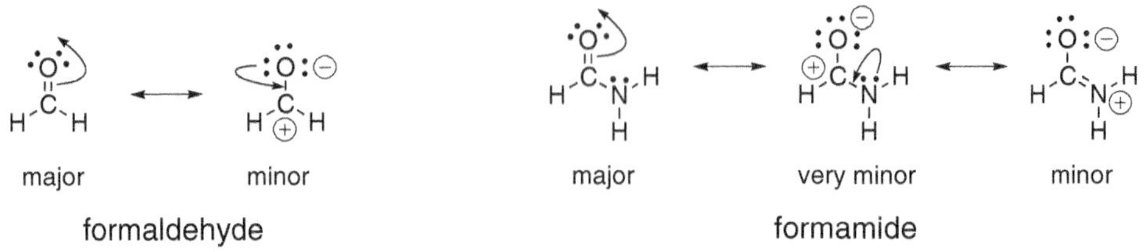

Figure 1.13 Resonance in formaldehyde and formamide molecules illustrates charge separation in generating the minor contributing structures. In both cases, the structures carrying formal charges are minor contributors, but the reactivity of the molecules may be best rationalized in terms of the minor contributors.

oxygen atom; in formamide, one can write a similar canonical form, and one where no non-hydrogen atom lacks a complete outer-shell octet, but the oxygen is again negatively charged, and the nitrogen atom carries a formal positive charge. This process, where one generates a minor canonical form carrying two atoms with opposite charges, is known as ***charge separation***. Canonical forms where there is charge separation are always minor contributors to the resonance hybrid.

Problems

1-9 Which of the following molecules and ions is/are resonance-stabilized? Draw the resonance contributors for those that are. Of those that are resonance stabilized, for which ions are there two or more equivalent resonance contributors? For those resonance-stabilized species where the contributors are not equivalent, indicate the major and minor contributors.

(a) 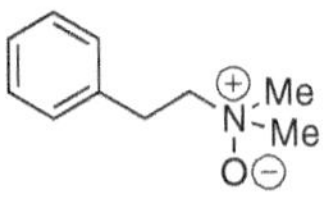

(b) 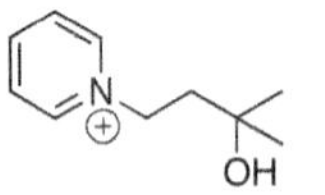

(c) 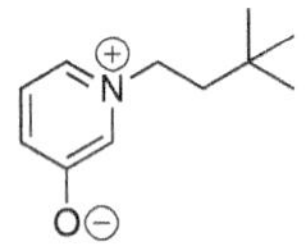

(d) 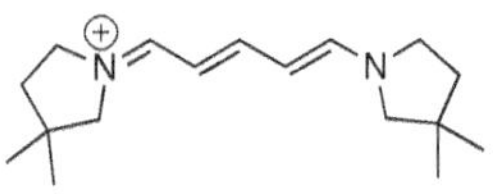

(e) 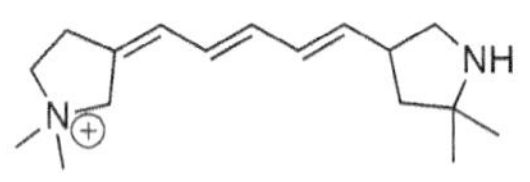

(f) 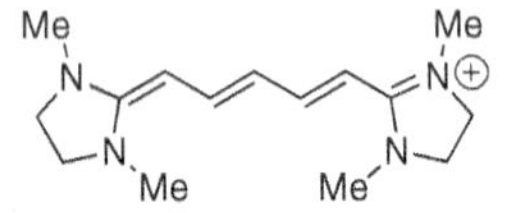

(g) 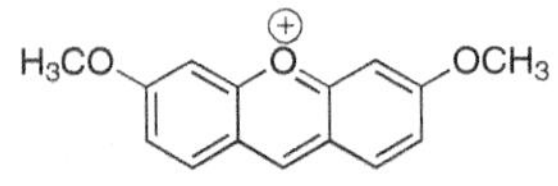

(h) 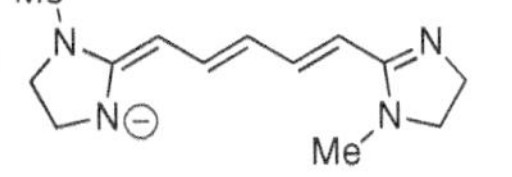

(i) 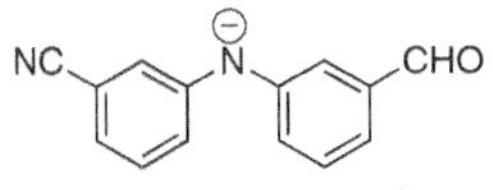

(j) 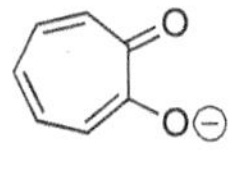

(k) 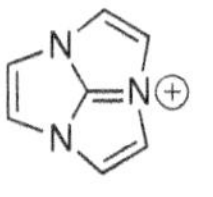

(l) 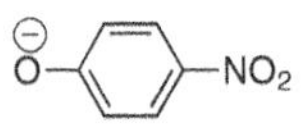

(m) 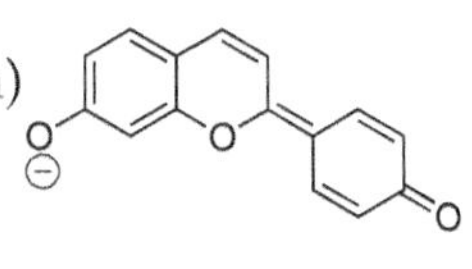

1-10 The diazomethane molecule, H_2CN_2, and the azide ion, N_3^-, are both linear, and both contain the N–N group. Draw the reasonable canonical forms of these two species and indicate which make the major contributions to the resonance hybrid of each. The diazomethane molecule reacts with acid to give a resonance-stabilized ion with the molecular formula $CH_3N_2^+$. Draw the reasonable canonical forms of this ion, and designate the major and minor contributors to the resonance hybrid.

1-11 Benzoic acid, $C_6H_5CO_2H$, dissolves in concentrated sulfuric acid to give a cation with the formula $C_6H_5CO_2H_2^+$, where an oxygen atom has been protonated. There are two possible oxygen atoms for protonation in this molecule, but the experimental evidence suggests that only one protonated species is formed in measurable amounts. What is the structure of the ion formed, and why is it the only major product of the reaction?

1-12 Dicyclohexylcarbodiimide, whose structure is given below, is not strongly resonance-stabilized despite having two nitrogen atoms and two π bonds. Why not?

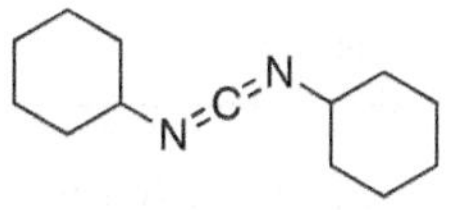

1-13 The compounds below are weak bases that react with strong acid to give cations in which the nitrogen atoms are not protonated. Suggest a structure for the protonated species and suggest why the protonation does not occur at nitrogen.

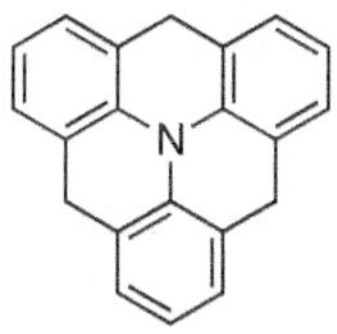

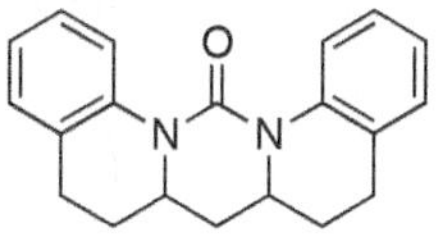

1-14 The sydnones are an interesting class of compounds known as "mesoionic" compounds. A typical sydnone is shown below. Draw the structures of all reasonable canonical forms contributing to the resonance hybrid of this compound, and show the electron movement that interconverts them. Can you draw a structure without any formally charged atoms?

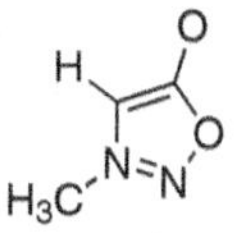

1-15 The cations below are strongly resonance stabilized. Draw all the possible reasonable resonance contributing forms for each ion and show the electron movement that interconverts them.

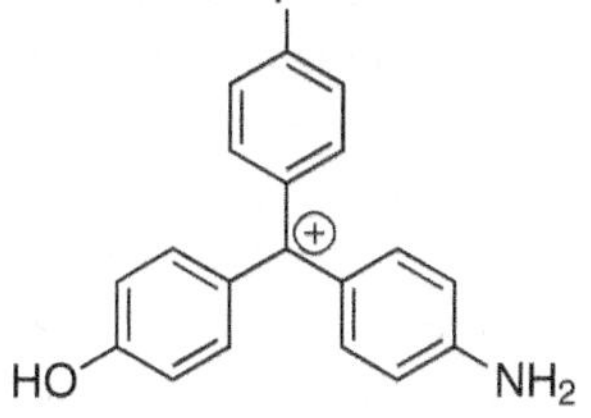

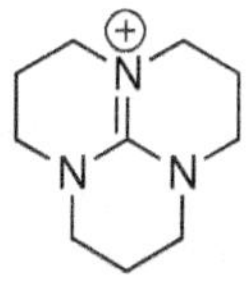

1-16 The classes of compounds below have the approximate pK_a values shown. How can these values be rationalized? (Remember the Hammond postulate; you might also want to consider resonance.)

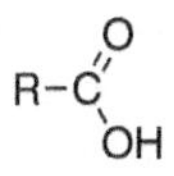

pK_a 4-6

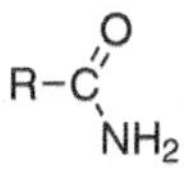

pK_a 15-16

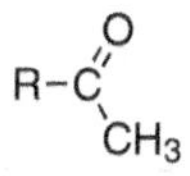

pK_a 25-27

1-17 Suggest a rationalization for the relative rates of nitration of the series of substituted benzene derivatives, C_6H_5R, below, based on an analysis of the possible benzenonium ion intermediates involved.

Isomer R:	H	CH_3	CO_2Et	Cl	Br	CH_2CO_2Et	CH_2Cl
ortho	1	42	0.0026	0.030	0.037	4.62	0.200
meta	1	2.5	0.0079	0.00	0.00	1.16	0.140
para	1	58	0.0009	0.139	0.106	10.4	0.951

1-18 The protonation of an epoxide leads to the resonance-stabilized carbocation shown below.

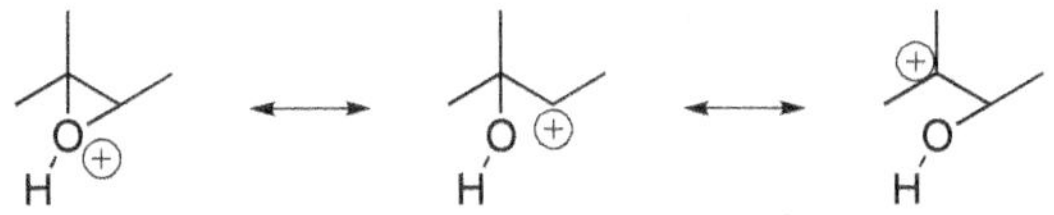

The equilibria below would also delocalize the positive charge to the same carbon atoms.

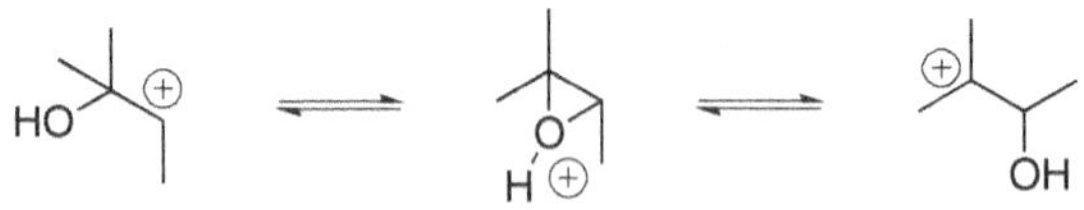

How does the equilibrium model differ from the resonance model?

1.5 Spectroscopy and Chromatography

Nothing changed organic chemistry during the 20th century so much as the development of spectroscopy and chromatography. With the availability of techniques such as hplc (high performance liquid chromatography), we are able to separate mixtures that were once considered impossible to separate. With the advent of modern techniques of NMR (nuclear magnetic resonance) spectroscopy and mass spectrometry, we are now able to deduce structural details in molecules that just a decade ago would have been viewed as hopelessly complex.

This book will not explicitly treat modern spectroscopy except superficially. To obtain a better grasp of the subject, you should consult introductory books in organic chemistry and specialized books in spectroscopy. That being said, some form of review is still appropriate here.

Infrared (IR) spectroscopy (or, more properly, spectrophotometry) records the absorption of IR radiation by the sample. The absorption of radiation occurs by exciting molecular vibrations of the molecule, and it is useful because many functional groups in organic compounds absorb IR radiation within a narrow, characteristic range of frequencies. These characteristic functional group frequencies occur in the range from 4000 to 500 cm^{-1}, with the most useful vibrations occurring above 1500 cm^{-1}. A correlation chart for IR frequencies is given as Figure 1.14.

The 20th century was certainly the decade of NMR spectroscopy, and it saw the rise of ^1H NMR and ^{13}C NMR spectroscopy as routine methods for examining the structures of organic compounds. The proton spectrum covers a fairly narrow range, with most resonances appearing between –2 and 12 ppm. As with the IR spectrum, the ^1H NMR spectrum can be subdivided into ranges where one can make structural inferences about the resonances. The

Frequency (cm^{-1})

3000	2500	2000	1500	1000	500
X—H (str.) X=O, N	X—H (str.) X=S, etc.	X≡Y (str.) X=Y=Z (str.)	X=Y (str.)	fingerprint	out-of-plane
OH alcohols 3400-3200 acids 3200-2500 NH amines 3300-3100 amides 3300-3100 CH alkynes 3300 alkenes 3100-3000 arenes 3150-3050	OH acids 3200-2500 O=C—H aldehydes 2720 2820 SH 2550	C≡N 2260-2150 C≡C 2250-2100	C=O aldehydes 1720-1740 ketones 1705-1725 carboxylic acid 1700-1725 esters 1730-1750 amides 1660-1690 anhydrides 1810, 1760 acid chlorides 1800 arenes 3150-3050 C=C alkenes 1600-1660 arenes 1600, 1500	skeletal vibrations complex	NMR spectroscopy does this better

Figure 1.14 Correlation chart for infrared spectra. Note how the fingerprint region is left blank (this region contains skeletal vibrations that complicate interpretation), and the out-of-plane region is also omitted (this is are better served by NMR spectroscopy).

chemical shift range of ^{13}C resonances is wider, and again, one can subdivide the spectrum into ranges that contain the resonances of particular kinds of carbon atoms.

NMR spectroscopy has the added bonus that the nuclear spins of the NMR active nuclei allow them to couple, and this spin-spin coupling allows one to derive both stereochemical and regiochemical information by the application of an increasingly useful array of techniques. Today, one can literally trace atomic connectivity through the backbone of a molecule and so derive the complete structure by NMR. A chemical shift correlation chart for ^{1}H NMR spectra is given in Figure 1.15.

A chemical shift correlation chart for ^{13}C NMR spectra is given in Figure 1.16.

Chapter Summary

This chapter has discussed chemical bonding—from (1) ionic bonding, where there is complete transfer of an electron from one reactant to the other; to (2) nonpolar covalent bonding, where the electron pair is shared essentially equally; to (3) polar covalent bonding, which lies between the two extremes. Lewis structures of organic compounds have been considered, with the concepts of formal charge and the dative (coordinate covalent) bond. The Arrhenius and Lowry-Brønsted theories of acids and bases have been introduced as a prelude to later discussions of proton transfer reactions. The general classes of organic reactions (homolysis and heterolysis) have also been discussed, along with writing mechanisms for them. Kinetics of reactions, including rate laws and the Arrhenius equation, have been introduced, and the relationship between mechanism and kinetics has been briefly discussed. The concept of resonance and the significance of major and minor canonical forms of resonance-stabilized species, have also been introduced.

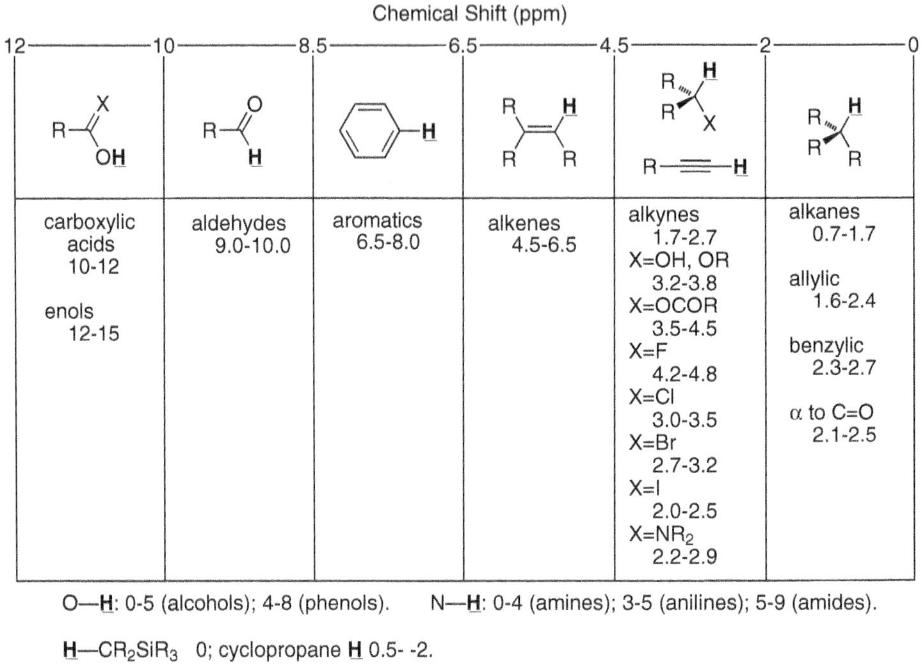

Figure 1.15 Correlation chart for ^{1}H NMR spectroscopy. The chemical shift ranges are such that methyl protons tend to occur at the high-field end of the range and methine protons at the low end of the range, with methylene protons in the middle. CH resonances typically occur ≈ 0.4 ppm downfield from the CH_2 resonances, which in turn resonate ≈ 0.4 ppm downfield from the CH_3 resonances.

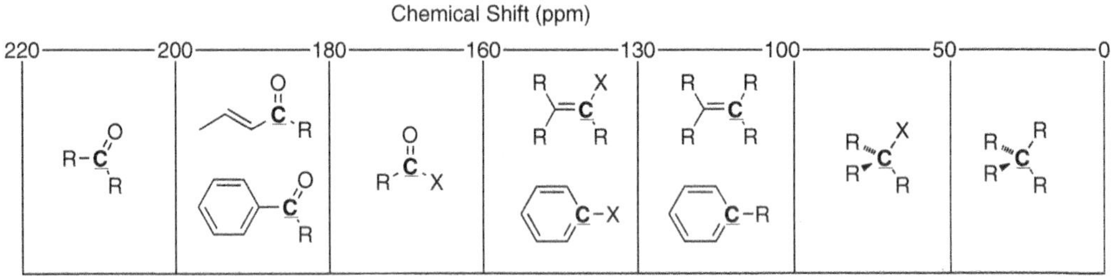

Figure 1.16 Correlation chart for ^{13}C nuclear magnetic resonance chemical shifts. The ranges are approximate.

Key Terms

acid
Arrhenius rate equation
base
canonical forms
complete structural formula
condensed structural formula
connectivity formula
coordinate covalent bond
covalent bond
dative bond
formal charge
Hammond postulate
heterolysis
hemolysis
ionic bond
kinetic order
kinetics
line formula
major contributor
mechanism
minor contributor
molecular formula
molecularity
polygon formula
rate law
reaction coordinate diagram
reaction energy profile
resonance
resonance contributors
resonance hybrid

Additional Problems

1-19 The addition of Br_2 to 1,3-butadiene in carbon tetrachloride solution gives 3,4-dibromo-1-butene as the major product of the reaction at 0°C, and 1,4-dibromo-2-butene when the reaction is carried out at 77°C. Provide an explanation for this observation.

1-20 Write an acceptable mechanism for the reaction below.

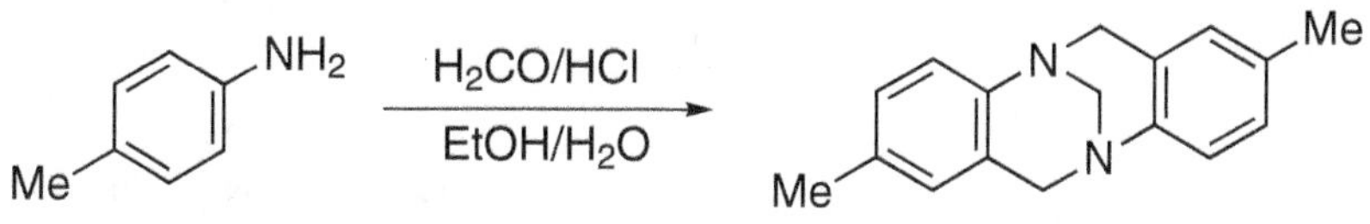

$$2\ C_7H_9N + 3\ CH_2O \longrightarrow C_{17}H_{18}N_2 + 3\ H_2O$$

1-21 Why does the anion below decompose rapidly to give the products shown, even at low temperature?

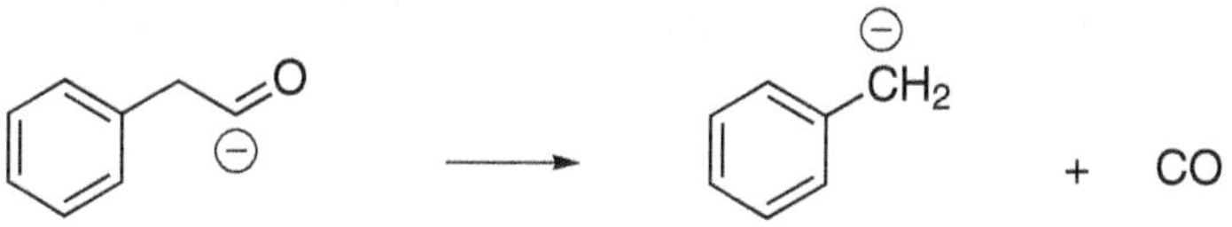

1-22 The Sharpless asymmetric epoxidation of geraniol at 25°C gives the two enantiomeric products in a ratio of 39:1. What is the difference in the activation energies of the two competing reactions (in kcal mol^{-1}) if the Arrhenius pre-exponential (A) factor is assumed to be the same for both reactions?

[R = 1.98717 cal mol^{-1} K^{-1}]

1-23 (a) The amine in the figure is a considerably stronger acid (pKa ≈ 15) than one would expect. Using resonance arguments to support your answer, provide a reasonable rationalization of this observation. You should consider both the amine and its conjugate base in your answer.

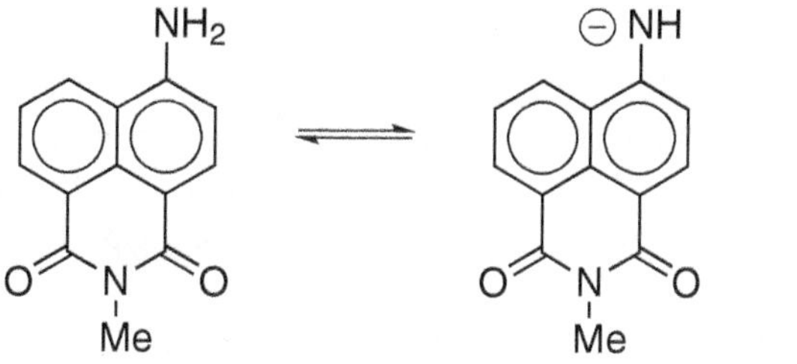

(b) Draw a reaction coordinate diagram that rationalizes the effects of resonance on this reaction.

1-24 Suggest reasonable mechanisms for the transformations below:

(a)

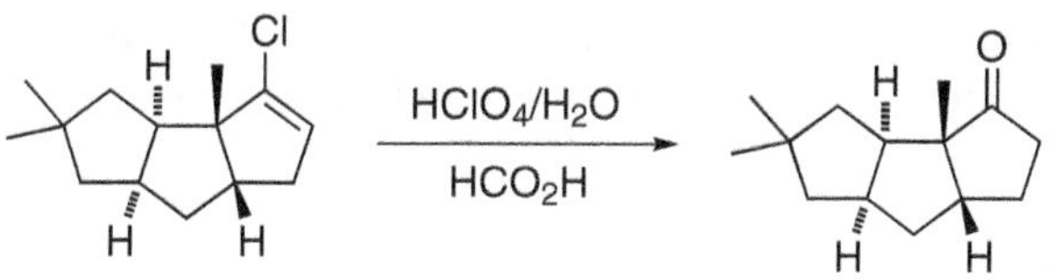

(b)

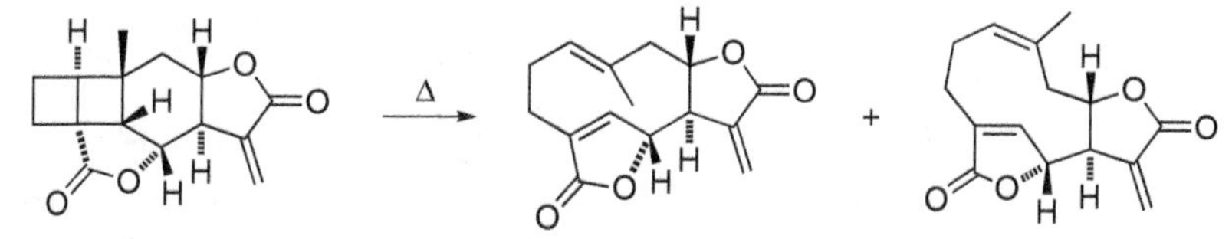

(c) 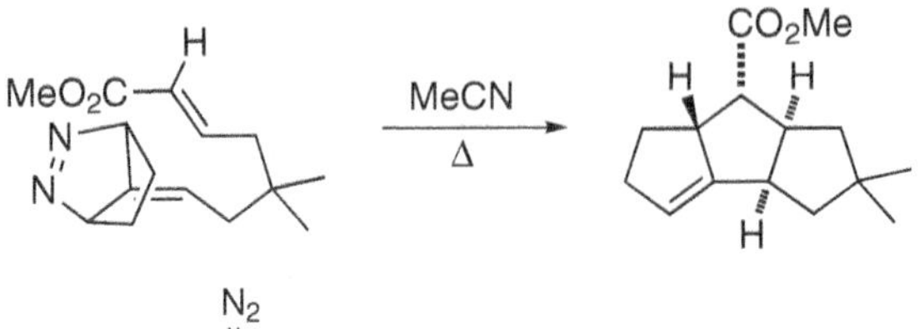

(d) 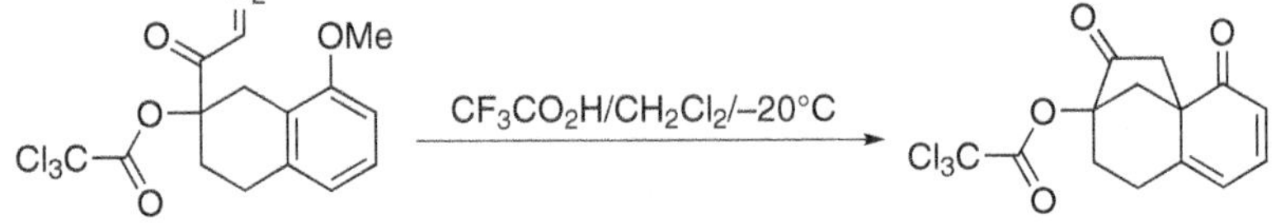

(e) 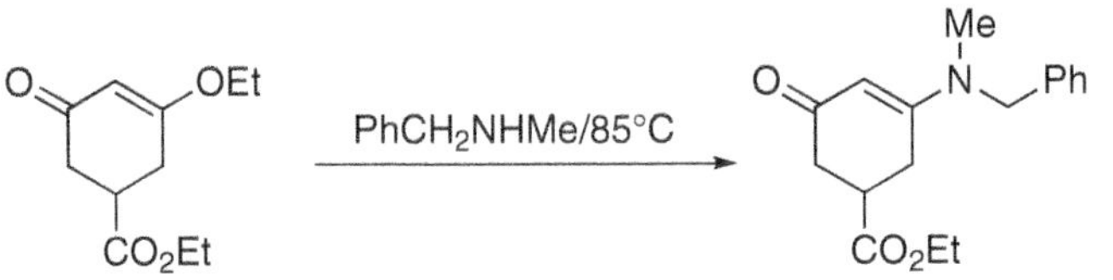

(f) 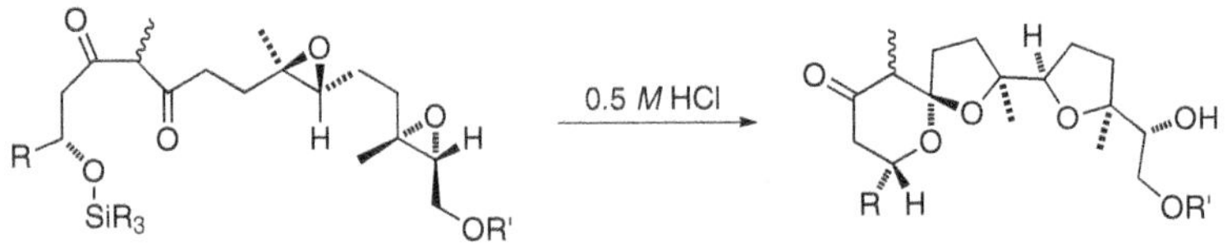

(g) 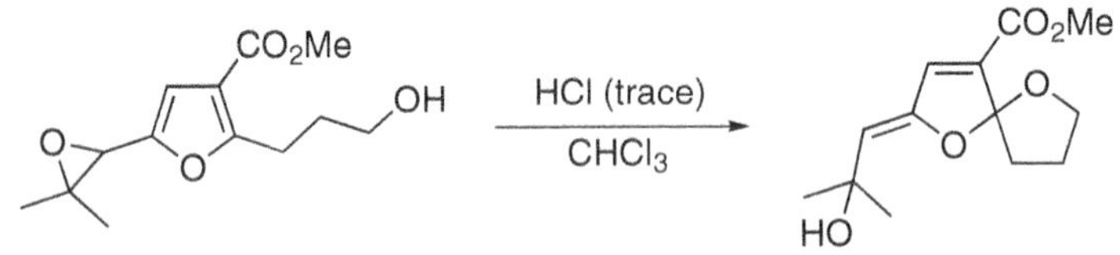

1-25 The mechanism of the reaction in the figure was used as a worked example earlier in this chapter (Worked Problem 1-4). Based on the mechanism given in that Worked Problem 1-4, draw a labeled reaction coordinate diagram for this reaction.

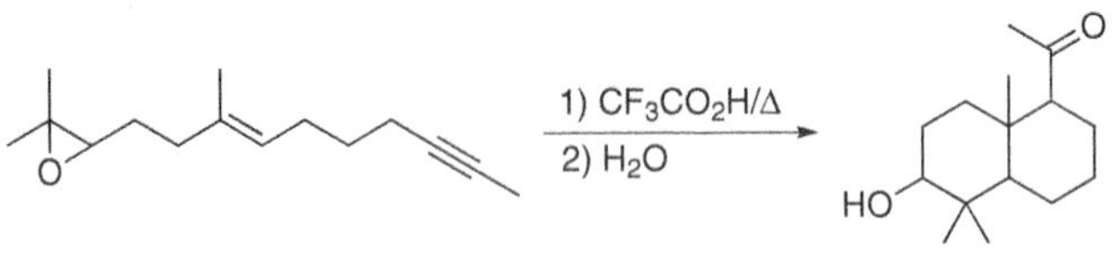

1-26 The addition of bromine to an alkene in the presence of other nucleophiles leads to products that have bromine and another group. Explain how this reaction might be used to develop a scale of nucleophilic power.

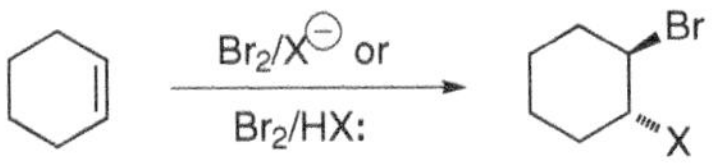

1-27 Mononitration (HNO_3–AcOH or HNO_3–CF_3CO_2H) of the citrus flavonoid pigment diosmetin (shown in the figure) occurs only to give the two mononitro products shown, where nitration has occurred in the phenyl ring, and none of the product of nitration in the other aromatic ring despite it being highly activated. Suggest a reason why this should be so. [*Tetrahedron Lett.* 2010, *51*, 1145].

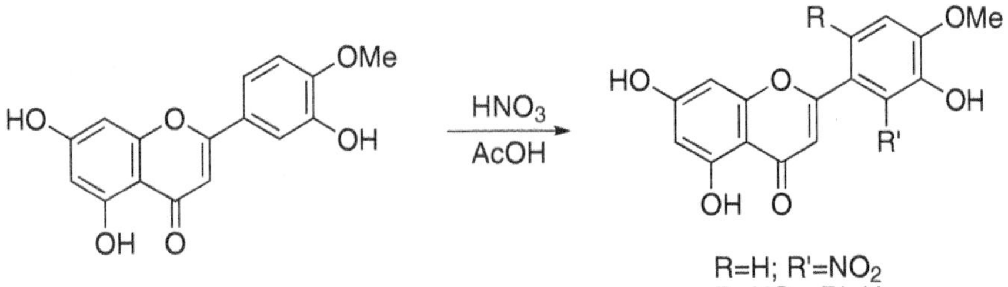

R=H; R'=NO_2
R=NO_2; R'=H

1-28 Each of the species below is a 1,3-dipole that will react with an alkene to give a five-membered heterocyclic product. Each of the 1,3-dipoles is stabilized by resonance. Draw the contributors to the resonance hybrid of each 1,3-dipole and write a reasonable one-step mechanism for each reaction given below, using the resonance contributor shown.

(a) 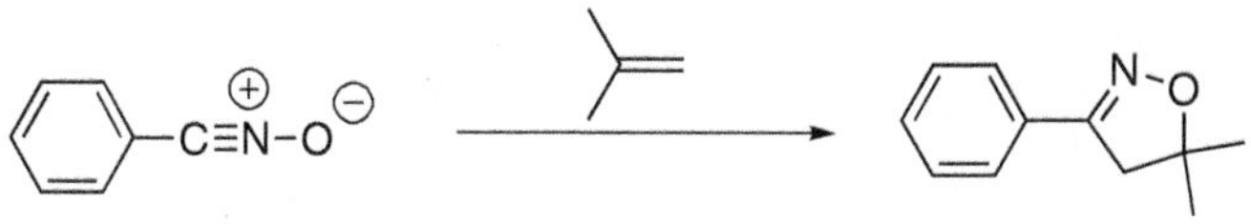

(b) 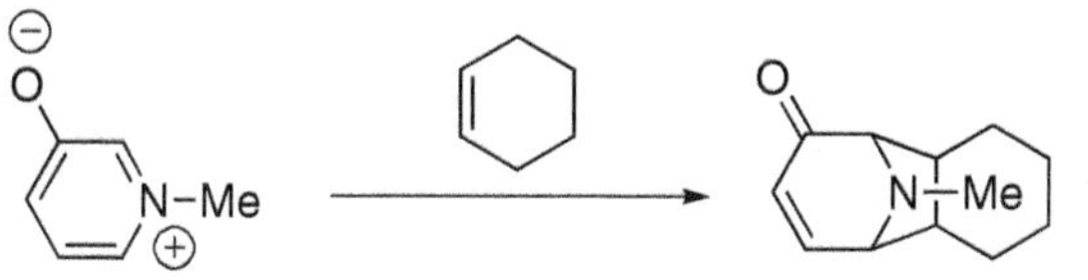

(c) 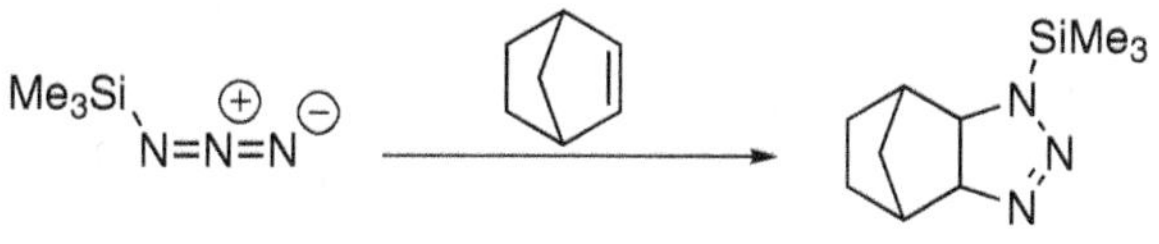

(d) 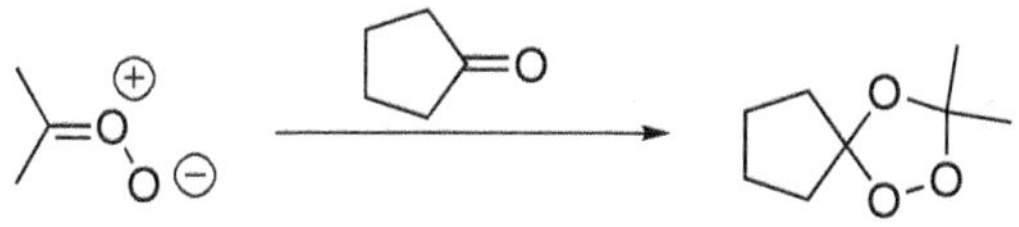

(e) 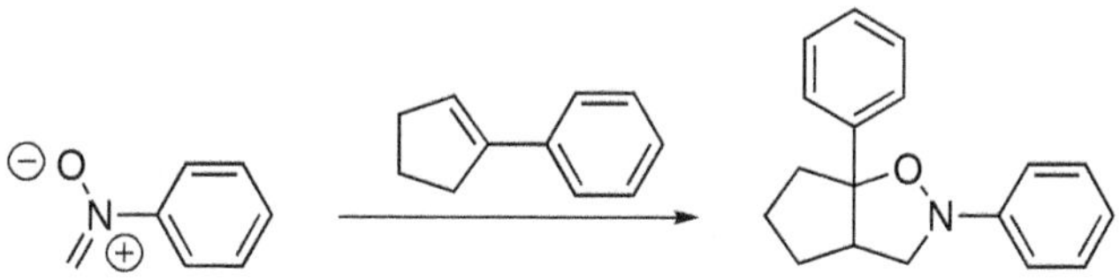

CHAPTER TWO

Stereochemistry

2.1 Stereoisomers and Chirality[1]

The three-dimensional shape of a molecule can be a critical factor in its reactivity. The stereochemistry of compounds is usually considered under two major categories: ***conformation*** and ***configuration***. The conformations of a molecule are dynamic—they are constantly interconverted by rotation about σ bonds. Thus, the term *preferred conformation* of a molecule refers to the *weighted average conformation of a population*, not a single conformation of a single molecule. Configurations, on the other hand, cannot be interconverted without cleaving and reforming covalent bonds. Stereoisomers with differing configurations are further defined as ***enantiomers***, which are ***chiral*** molecules that are nonsuperimposable mirror images, and ***diastereoisomers***, which are not mirror images. The term ***stereoisomers*** is usually reserved for compounds differing in configuration—conformers are not commonly designated as stereoisomers.[2]

Representing Stereochemistry

A molecule is a three-dimensional object, but we almost always have to represent it in two dimensions. As with any three-dimensional object, how it will appear depends on the perspective from which we view it. There are several different conventions that organic chemists use to represent molecules in two dimensions, and all of them are derived at some point from the same kinds of approaches that an artist might adopt to paint a three-dimensional object. Those most commonly used by modern organic chemists are shown in Figure 2.1.

Reduced to their simplest principles, each of the conventions shown in Figure 2.1 requires that we first take up a position in space at some point from which we can see the whole molecule and that we then draw the molecule in perspective from that point. From this kind of approach, we obtain representations known as ***projection formulas***, such as the line formula (A); the ***bow-tie*** projection (B); and its analog, the ***Fischer*** projection,[3] named for Emil Fischer[4] (C); and the ***Newman*** projection[5] (D), named for

---

1. The definitive book on this subject is Eliel, E.L.; Wilen, S.H.; Mander, L.N. *Stereochemistry of Organic Compounds* (Wiley-Interscience: New York, 1994). This book is encyclopedic in coverage and should be the first reference work consulted for more details than can be provided in this chapter.

2. (a) McCasland, G.E. *A New General System for the Naming of Stereoisomers.* (Chemical Abstracts Service: Columbus, OH, 1950), 2. (b) Mislow, K.; Siegel, J. *J. Am. Chem. Soc.* **1984**, *106*, 3319. (c) Eliel, E.L. *Chirality*, **1997**, *9*, 48.

3. Fischer, E. *Ber. Deut. chem. Ges.* **1891**, *24*, 2683.

4. **Emil Fischer (1852-1919)** studied at Bonn, then Strasbourg (PhD, 1874). After 8 years as assistant to Baeyer, he became professor of chemistry at Berlin. His pioneering work in synthetic chemistry and stereochemistry won him the Nobel Prize in 1902. A more complete biography may be found in: *Nobel Lectures, Chemistry, 1901-1921* (Elsevier Publishing Company, Amsterdam, 1966)

5. Newman, M.S. *J. Chem. Educ.* **1955**, *32*, 344.

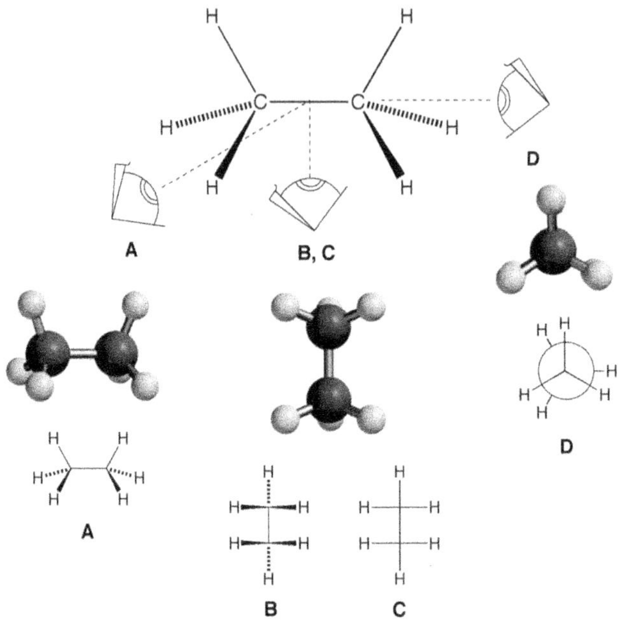

Figure 2.1 Conventions for drawing representations of the stereochemistry of organic molecules and the perspectives that give rise to each representation

Melvin S. Newman.[6] These projection formulas are, of course, stylized representations of what the viewer actually sees. The convention of omitting the symbol for the carbon atoms is common (just as in drawing line or polygon formulas) and can often lead to a dramatic simplification of the diagram—one that actually draws the reader's attention to the shape of the molecule.

The convention in the line formula (A) and the bow-tie projection (B) is that all bonds drawn as simple straight lines are assumed to lie in the plane of the page, whereas atoms at the thick end of the wedges are projecting out of the plane of the page toward the reader and those at the end of the hatched lines project away from the viewer behind the plane of the page. Thus, in the line formula B for ethane, there are four atoms in the plane of the page (the two carbon atoms and two of the six hydrogens).

In the Fischer projection (C), there are no wedges or hatched lines to indicate the location of the four groups about each carbon atom, but a Fischer projection corresponds to a strictly defined orientation of the molecule where the horizontal bonds project out toward the viewer, and the vertical bonds project away from the viewer, as shown in the bow-tie projection (B). Although this is a useful convention in the right circumstances, it does have some rather unfortunate consequences because simple rotations of the Fischer projection do not correspond to simple rotations of the molecule *unless the rotation is through a multiple of 180°*. Rotations of a Fischer projection through odd multiples of 90° invert the absolute configuration of the molecule being represented.

The Newman projection (D) is possibly the best representation for illustrating the orientation of groups about σ bonds, but it suffers from the fact that one is restricted to looking at pairs of atoms instead of the molecule as a whole. Like the other projections, the Newman projection is stylized—the central carbon atoms are omitted and the electron density of the σ bond is represented by a circle. Because the electron density of the σ bond

6. **Melvin S. Newman (1908-1993)** was educated at Yale (PhD, 1932). He joined the Ohio State University, where he spent the remainder of his professional career, in 1935. He is best remembered for his development of the Newman projection formulas for illustrating conformations of organic compounds. For more detail, see: Paquette, L.A.; Orchin, M. *Biogr. Mem. Nat. Acad. Sci.* **1998**, *73*, 334.

should obscure part of each bond to the back carbon atom, only the bonds to the front carbon are visible for their entire length.

Problem

2-1 Draw the structure of each of the following molecules using projection formulas (A) to (D) from Figure 2.1. Is there more than one way to represent each molecule?

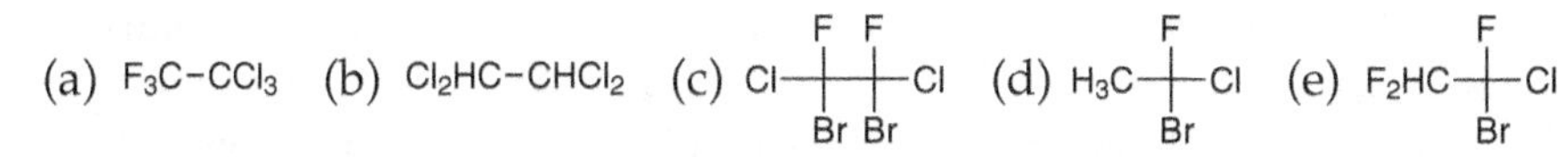

Symmetry Elements in Organic Compounds

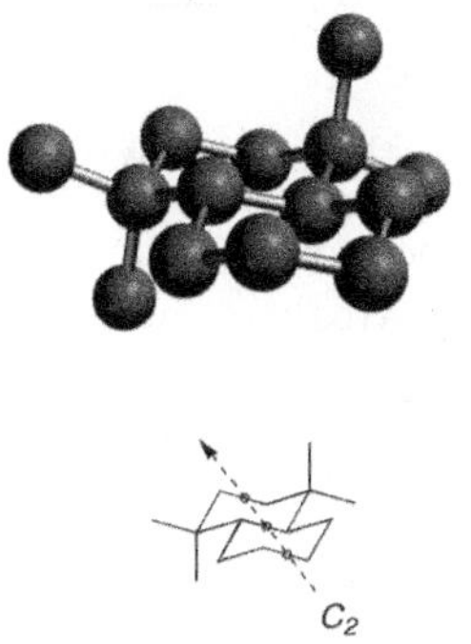

(2.1a)

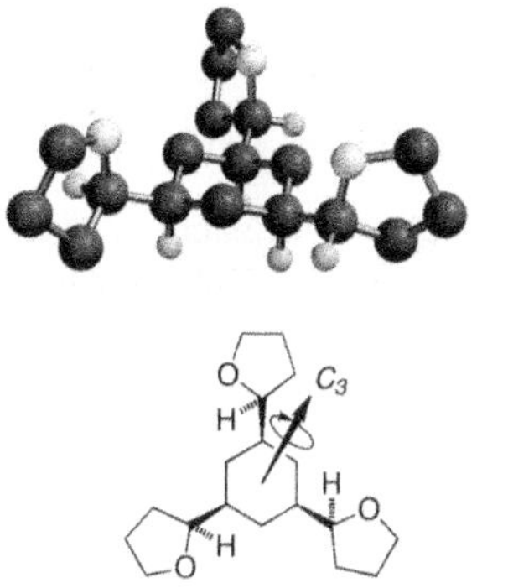

(2.1b)

The symmetry of an organic molecule determines, among other things, whether the molecule will be chiral. For this reason, we need to be well acquainted with the most common symmetry elements found in organic molecules and with their effects on the stereochemical properties of the molecule.

The simplest symmetry element in an organic compound is a proper axis of symmetry, designated C_n, where n is the order of the symmetry axis. The effect of this symmetry operation is to rotate the molecule through $360/n$ degrees. In molecules that possess this symmetry element, the rotation converts the molecule into a totally equivalent representation. Example **2.1a** has a C_2 axis of symmetry through the midpoints of the three bonds indicated by heavy dots on the axis, and Example **2.1b** has a C_3 axis of symmetry perpendicular to the page in the polygon representation.

Problem

2-2 Locate all axes of symmetry in the following molecules: (a) terephthalonitrile; (b) 1,1-dichlorocyclobutane; (c) 1,3,5-tribromo-2,4,6-trichlorobenzene; and (d) mellitic anhydride.

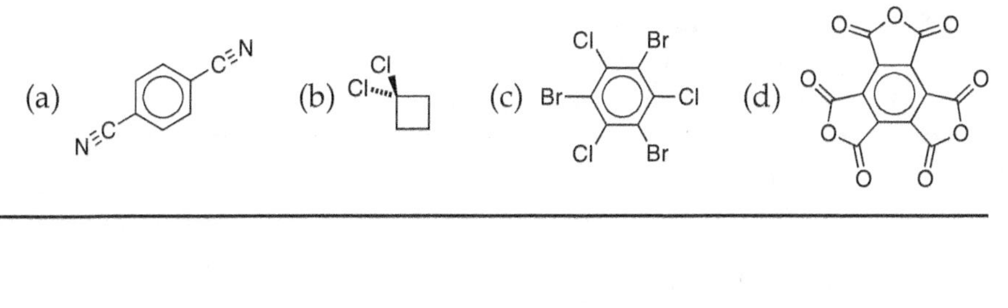

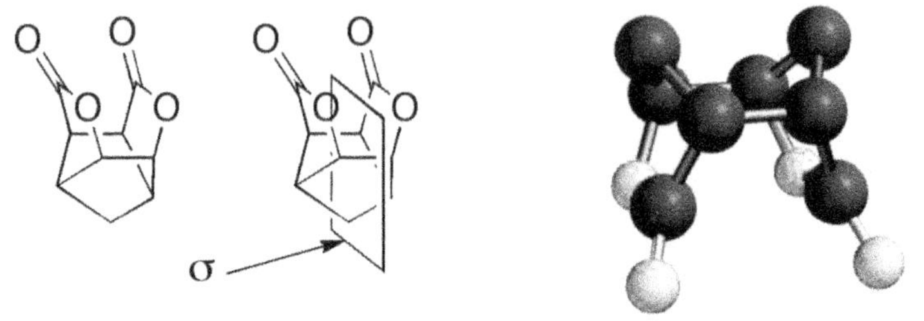

(2.2)

Another common symmetry element in an organic molecule is the mirror plane of symmetry, σ. The effect of this symmetry operation is to reflect one half of the molecule through the mirror. In molecules that possess this symmetry element, one half of the molecule reflects exactly onto the other half. In the molecule in Example **2.2**, for example, every atom in the left half of the molecule is reflected through the mirror plane onto a corresponding atom in the right half of the molecule.

Problems

2-3 Locate all the mirror planes of symmetry in the molecules in Problem 2-2.

2-4 Draw the Newman projections of the compounds in Problem 2-1 to show how they may have a mirror plane of symmetry. Are there any compounds for which no mirror plane of symmetry can be drawn?

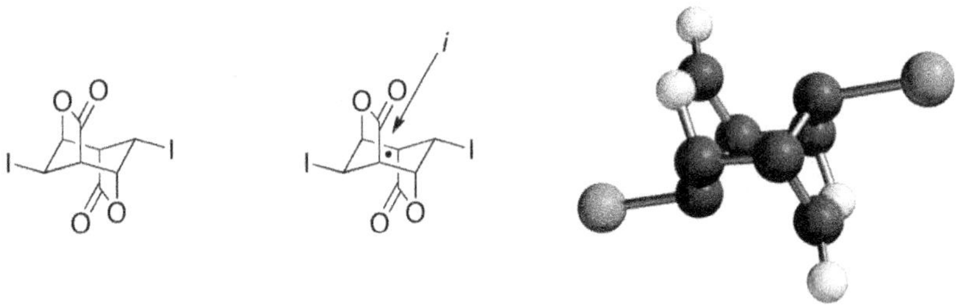

(2.3)

The third symmetry element that can be present in an organic molecule is an inversion center of symmetry, *i*. The effect of this symmetry operation is to reflect each atom of a molecule through a single point (the center of symmetry). This symmetry element is common in inorganic compounds but much less common in organic compounds than the proper axes and mirror planes of symmetry. But it is still common enough to remember it. In the molecule in Example **2.3**, for instance, there is a center of inversion designated by the dot right in the middle of the six-membered ring. Every atom in this molecule has a corresponding atom of the same type at the same distance from the inversion center on the opposite side of the center.

Problem

2-5 Draw the Newman projections of the compounds in Problem 2-1 to show how they may have an inversion center of symmetry. Are there any compounds for which no inversion center of symmetry can be drawn?

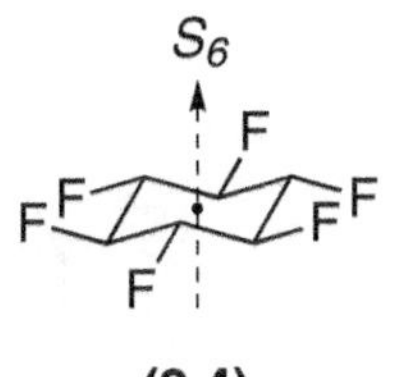

(2.4)

A fourth symmetry element that is *occasionally* found in organic molecules is an improper axis of rotation, S_n, where n is an even number. The operation of this symmetry element is more complex than the simple symmetry elements described above because it involves the sequential application of an n-fold axis of rotation and a mirror plane of symmetry perpendicular to it. This symmetry operation is always accompanied by a proper axis of rotation of half the order of the improper axis. In the molecule in Example **2.4**, for instance, there is a threefold axis through the center of the ring. Rotation of the molecule by 60° (360 ÷ 6) around this axis and reflection in the plane perpendicular to the axis restores the original appearance. This molecule has a sixfold improper axis of rotation (S_6).

Worked Problem

2-1 Identify the symmetry elements present in the molecule below. Is the compound chiral? Give reasons for your answer.

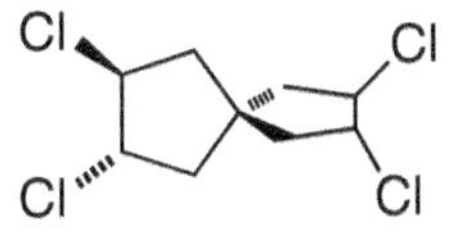

§Answer below

Problems

2-6 Identify the symmetry elements present in each of the following molecules (you will find molecular models helpful).

(a) 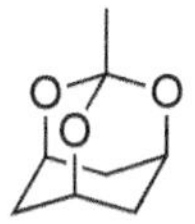

(b) 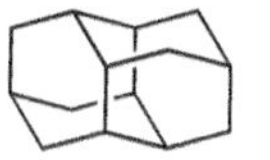
(c) 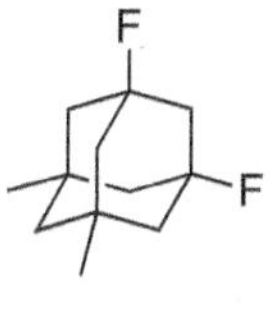

(d) 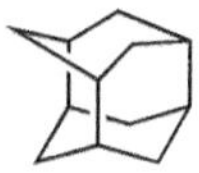

(continued)

§ Answer for Worked Problem:

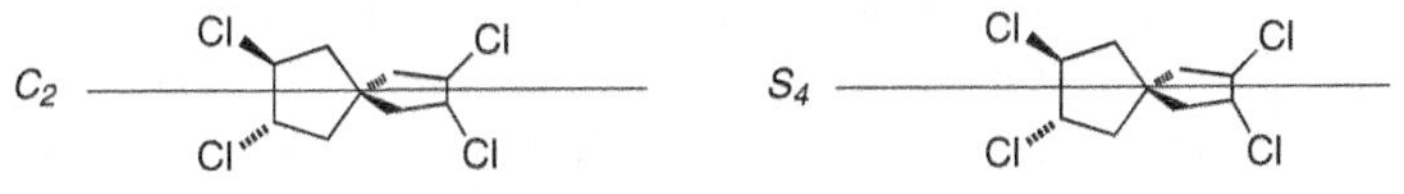

This molecule has a twofold axis of symmetry (C_2) and a fourfold improper axis of symmetry (S_4). The S_4 operation is detailed below:

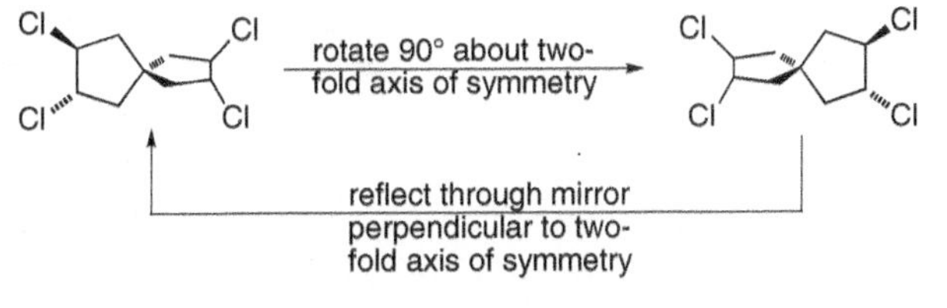

Although the molecule possesses neither a mirror plane of symmetry nor an inversion center of symmetry, it *does* possess an improper axis of rotation (S_4), so it cannot be chiral.

Problems (continued)

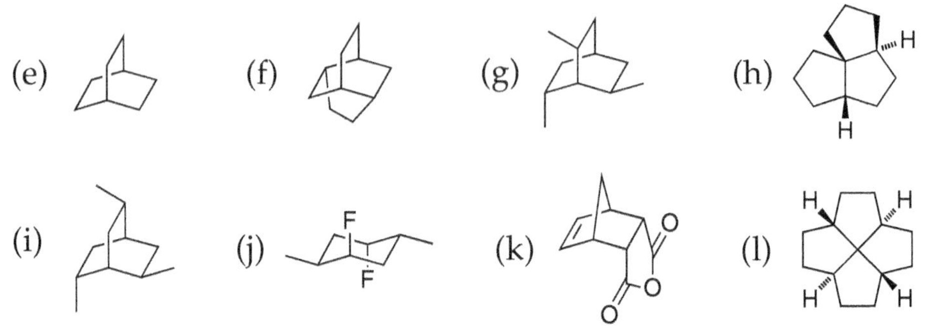

2-7 Are the compounds below chiral? Give reasons for your answer.

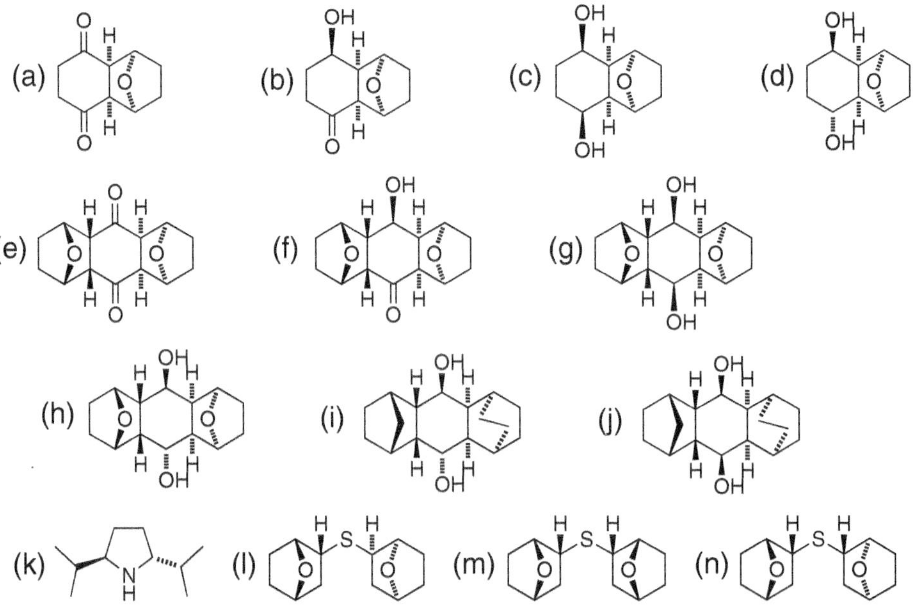

Protocenters of Stereochemistry

For a molecule to exist as stereoisomers, it must have at least four groups that can be arranged in two or more configurationally stable, nonequivalent ways. Stereochemistry is most simply discussed in terms of ***protocenters of stereochemistry***, which we will define as a *single atom or group within a molecule, bonded to two different substituent groups in a nonlinear fashion.*[7] The bisector of the bond angle at the protocenter defines the axis of the protocenter, and the three atoms define the plane of the protocenter. Representative examples of protocenters are shown in Figure 2.2.

Figure 2.2 Representative types of protocenters

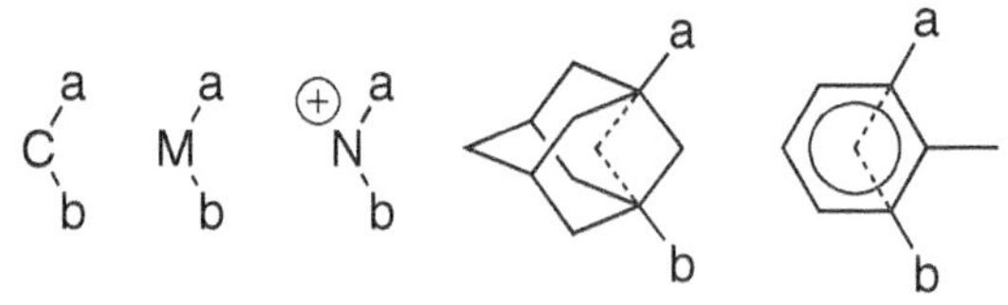

7. Lewis, D.E. *J. Chem. Educ.* **2010**, *87*, 604.

It requires two protocenters in a molecule to give rise to stereoisomers. Protocenters may be centered on the same axis (coaxial) or on different axes. By far the most common situation in organic compounds involves coaxial protocenters, and the discussion that follows will focus on that. There are two parameters that define the relationship between two coaxial protocenters: the distance between them, *r*, and the angle between the two planes, θ. Protocenters may be coincident ($r = 0$), in which case they share a common atom, or noncoincident ($r > 0$). They may also be coplanar ($\theta = 0$) or noncoplanar ($\theta \neq 0$). When the two protocenters are coplanar, the molecule cannot be chiral, but *E/Z* isomerism may result. When the two protocenters are not coplanar, the molecule may or may not be chiral, depending on the separation between them.

Table 2.1 Combinations of Two Coaxial Protocenters That Can Give Rise to Stereoisomers in a Molecule

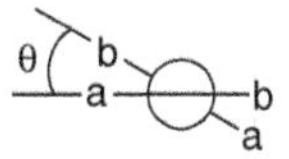

θ	Protocenters	r = 0	r > 0
0	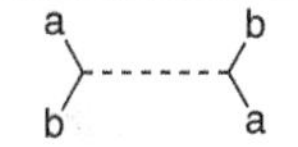	Achiral (*E/Z*)	Achiral (*E/Z*)
≠ 0		Achiral	Chiral
≠ 0		Achiral	Chiral
≠ 0		Chiral	Chiral

The possible stereochemical outcomes arising from a pair of coaxial protocenters are summarized in Table 2.1. When the two protocenters are not coincident ($r > 0$) and not coplanar ($\theta \neq 0$), the molecule will be chiral without restriction on the identities of the groups bonded to the protocenters, meaning that as few as two different groups are sufficient to give rise to chirality in the molecule. The situation becomes more restrictive when the two protocenters are coincident. In this case, both protocenters must carry different substituents in order for the molecule to be chiral. Under these circumstances, the result is a center of chirality. When the two protocenters are coplanar, the molecule is not chiral but exists as geometric (*E/Z* or *cis/trans*) isomers (e.g., coincident, coplanar protocenters correspond to the situation in a square planar compound.

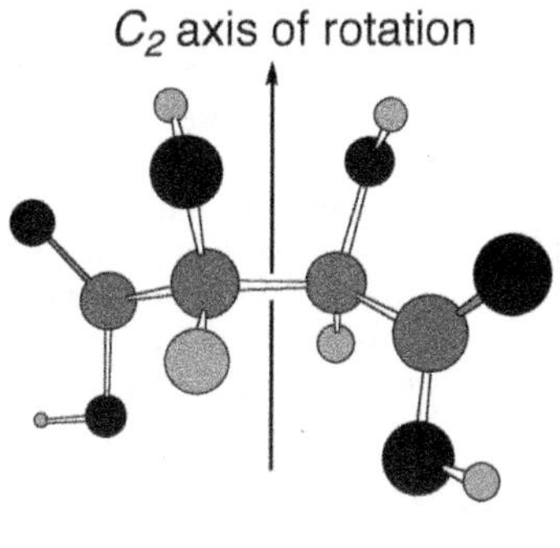

The minimum requirement for a molecule to be able to exist as enantiomers is that it lack reflection or inversion symmetry. A mirror plane of symmetry or an inversion center is not permitted in a chiral molecule. All the symmetry requirements for chirality can be met by a single criterion—*for a molecule to be chiral, it may not have an improper axis of rotation.* This does not mean that a chiral molecule must lack any symmetry; proper axes of rotation are not only permitted, but quite common in chiral molecules. A good example of this is tartaric acid, which has a twofold axis of symmetry but which is still chiral. In fact, the existence of a twofold axis of symmetry in the molecule is often a key feature of chiral molecules used in the synthesis of chiral catalysts (e.g., BINAP,[8] the binaphthyl shown next page).

8. Reviews: (a) Noyori, R. *Science*, **1990**, *248*, 1194. (b) Noyori, R.; Takaya, H. *Acc. Chem. Res.* **1990**, *23* 345.

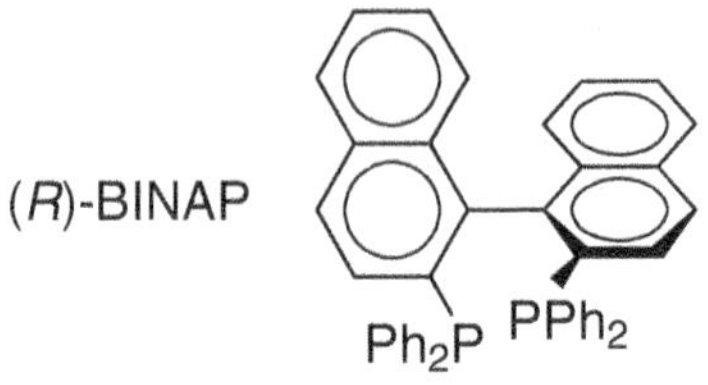

In 1815, the French chemist Jean-Baptiste Biot[9] observed that solutions of certain organic compounds rotate the plane of plane-polarized light.[10] Forty-five years later,[11] Louis Pasteur[12] suggested that molecules of the same substance with opposite rotations might be related to each other as object to mirror image, much like a right- and left-handed helix. The importance of configuration was reinforced by the work of Johannes Wislicenus,[13] who suggested in 1873 that the differing properties of stereoisomers could be traced to differing three-dimensional arrangements of their atoms.[14] This was a radical proposal at the time, because few chemists believed that it was possible to deduce the physical locations of atoms within a molecule. This proposal was taken further the next year, when Jacobus Henricus van't Hoff[15] and Joseph-Achille Le Bel[16] independently proposed the concept of the tetrahedral carbon.[17,18,19] Interestingly, although these two chemists had worked together in Wurtz's laboratory in Paris during the year before their publications,

---

9. **Jean Baptiste Biot (1774-1862)** studied at the École Polytechnique under Lagrange and Berthollet and taught at the Collège de France. His study of the properties of light included plane-polarized light. He was made chevalier (1814), and later commander (1849), of the Legion of Honour. For further biographical details, see Crosland, M.P. *Jean Baptiste Biot.* In *Dictionary of Scientific Biography* (Charles Scribner's Sons: New York, 1970), vol. 2, p. 133.

10. Biot, J.B. *Bull. soc. philomath. Paris* **1815**, 190; **1816**, 125; *Mém. acad. roy. sci. inst. France* **1817**, *2*, 41, 114.

11. See: (a) Pasteur, L. *Researches on the Molecular Asymmetry of Natural Organic Products* Alembic Club Reprints (University of Chicago Press, Chicago, 1902). (b) *Oevres de Pasteur* (Masson et Cie.: Paris, 1922), vol. 1, p. 315, 329.

12. **Louis Pasteur (1822-1895)** studied chemistry under Dumas and Balard, and his resolution of tartaric acid still stands as one of the seminal events in the history of organic chemistry. His later work in microbiology made him one of the greatest scientists of his generation. Biographies of Pasteur abound, such as Debré, P. *Louis Pasteur*; E. Forster, transl. (Johns Hopkins University Press: Baltimore, 1998).

13. **Johannes Wislicenus (1835-1902)** was educated at Halle but was forced to flee Germany to avoid arrest for his religious activism. After 2 years in the United States, he moved to Zürich; in 1857 he returned to Halle, as a private assistant. During his career, he worked at the ETH, Würzburg, and Leipzig. For a more complete biography, see: Beckmann, E. *Ber. dtsch. Chem. Ges.* **1905**, *37*, 4861.

14. Wislicenus, J. *Liebigs Ann. Chem.* **1873**, *167*, 302.

15. **Jacobus Henricus van't Hoff (1852-1911)** was educated at Delft, Leiden, Bonn, Paris, and finally at Utrecht (PhD, 1874). His contributions to thermodynamics helped to found the discipline of physical chemistry and won him the first Nobel Prize in Chemistry. After initial positions at Utrecht and Amsterdam, he became Professor at Berlin. More biographical details are available at the Nobel Prize web site.

16. **Joseph-Achille Le Bel (1847-1930)** was born in Péchelbronn, France, and he was educated at the École Polytechnique in Paris. He inherited the family's petroleum works, making him independently wealthy. He proposed the tetrahedral carbon and predicted, correctly, that optical activity could arise in compounds of elements other than carbon. For more biographical details, see: Leicester, H.M. (1973). *Le Bel, Joseph Achille.* In Gillispie, C.M., Ed. *Dictionary of Scientific Biography,* Vol. 8, (New York: Scribners: New York, 1973)

17. (a) van't Hoff, J.A. *Voorstel tot Uitbreiding der tegenwoordig in de scheikunde gebruitke Structuur-Formules in die ruimte; benevens een daarmeê samenhangende opmerkung omtrent het verband tusschen optisch actief Vermogen en Chemische Constitutie van Organsiche Verbindingen* (Utrecht, 1874); *Arch. Néerland. Sci. Exactes Natur.* **1874**, 9, 445; *Bull. soc. Chim. Paris* **1875**, [2] *23*, 295; *La chimie dans l'espace* (Bazendijk: Rotterdam, 1875); *Die Lagerung der Atome im Raume*, Hermann, F., trans. (Vieweg: Braunschweig, 1877).

18. Le Bel, J.A. *Bull. soc. chim. Paris* **1874**, [2] *22*, 337.

19. For an interesting history of the development of stereochemistry, see Ramberg, P.J. *Chemical Structure, Spatial Arrangement. The Early History of Stereochemistry 1874-1914.* (Ashgate Publishing Company: Aldershot, U.K., 2003).

it appears that they developed their ideas independently and did not even discuss them with each other.[20]

The initial reception of the concept—or the idea that it was even possible to establish the positions of atoms in space—was lukewarm, and Hermann Kolbe, editor and founder of the *Journal für praktische Chemie*, and a highly respected (and conservative) chemist, sought to prevent its circulation by an infamous diatribe[21] against the idea.

Chirality in simple molecules can be in the form of a center of chirality, an axis of chirality, or a plane of chirality. A ***center of chirality*** is a position in a molecule with more than three different substituents arranged in three dimensions. The concept is most commonly seen in tetrahedral species, but the transition state for an S_N2 reaction of a chiral secondary halide, which is a trigonal bipyramid, also possesses a chiral center. Some examples of chiral centers in a molecule are gathered in Figure 2.3.

---

20. Walker, J. *J. Chem. Soc.* **1913**, *103*, 1127.

21. Kolbe, H. *J. Prakt. Chem. [2]* **1877**, *15*, 473.

In a recently published paper with the same title [*Signs of the Times*], I pointed out that one of the causes of the present-day retrogression of chemical research in Germany is the lack of general and, at the same time, fundamental chemical knowledge; under this lack no small number of our professors of chemistry are laboring, with great harm to the science. A consequence of this is the spread of the weed of the apparently scholarly and clever, but actually trivial and stupid natural philosophy, which was displaced fifty years ago by exact natural science, but which is now brought forth again, out of the storeroom harboring the errors of the human mind, by pseudoscientists who try to smuggle it, like a fashionably dressed and freshly rouged prostitute, into good society, where it does not belong.

Anyone to whom this concern seems exaggerated may read, if he is able to, the book by Messrs. van't Hoff and Herrmann on *The Arrangement of Atoms in Space*, which has recently appeared and which overflows with fantastic foolishness. I would ignore this book, as [I have] many others, if a reputable chemist had not taken it under his wing and warmly recommended it as an excellent accomplishment.

A Dr. J.H. van't Hoff, of the Veterinary School at Utrecht, has no liking, it seems, for exact chemical investigation. He has considered it more convenient to mount Pegasus (apparently borrowed from the Veterinary School) and to proclaim in his *La chimie dans l'espace* how the atoms appear to him to be arranged in space, when he is on the chemical Mount Parnassus which he has reached by bold flight.

The prosaic chemical world had little liking for these hallucinations. Therefore, Dr. F. Herrmann, Assistant at the Agricultural Institute at Heidelberg, undertook to give them wider vogue by means of a German edition. This carries the title, "*The Arrangement of Atoms in Space*, by Dr. J.H. van't Hoff; translated into German from the author's monograph *La chimie dans l'espace* by Dr. F. Herrmann, Assistant at the Agricultural Institute at Heidelberg; with a foreword by Dr. Johannes Wislicenus, Professor of Chemistry at the University of Würzburg . . ."

It is not possible to criticize this work even half-way thoroughly because the play of fantasy therein dispenses completely and entirely with factual basis and is absolutely unintelligible to the sober scientist.

It is indicative of the present day, in which critics are few and hated, that two practically unknown chemists, one from a veterinary school and the other from an agricultural institute, judge with such assurance the most important problems of chemistry, which may well never be solved—in particular, the question of the spatial arrangement of atoms—and undertake their answer with such courage as to astonish real scientists.

As I have said, I would have taken no notice of that work if Wislicenus had not inconceivably written a foreword for it and, not jokingly but in complete seriousness, warmly recommended it as a worthwhile contribution—whereby many young inexperienced chemists might be misled into assigning some value to these shallow speculations . . .

It is a sign of the times that the modern chemists feel themselves qualified and able to give an explanation for everything, and, when the results of experience are not sufficient, they seize upon supernatural explanations. Such treatment of scientific questions, which is not far removed from the belief in witches and from ghost-rapping, even Wislicenus considers to be admissible . . .

Wislicenus thereby makes it clear that he has left the ranks of exact scientists and has gone over to the camp of the natural philosophers of ominous memory, who are only by a narrow "medium" separated from the spiritualists.

[Translation taken from Wheland, G.W. *Advanced Organic Chemistry*, 3rd, ed. (John Wiley & Sons: New York, 1960), pp. 197-198]

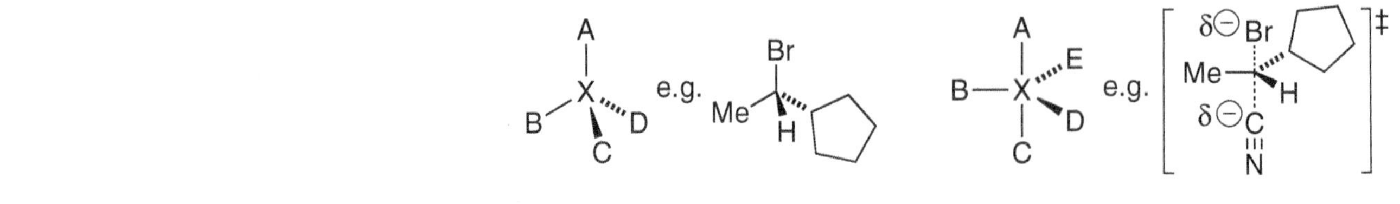

Figure 2.3 Typical centers of chirality. The atom X is usually carbon in organic chemistry, but it may be any element (including nonmetals such as sulfur and phosphorus, or metals).

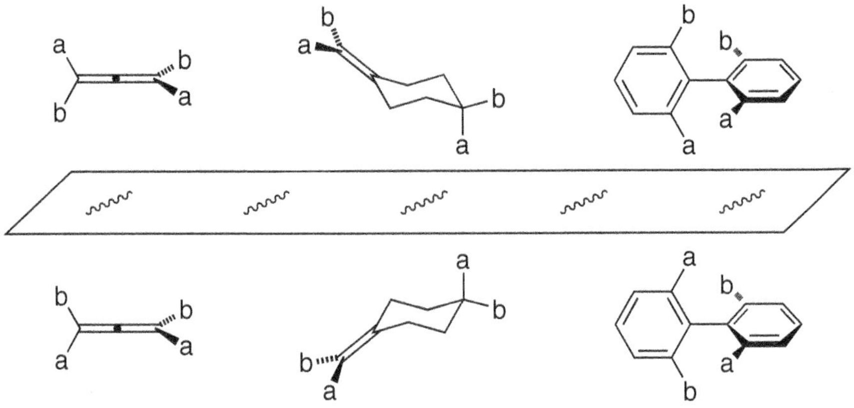

Figure 2.4 Typical compounds that have an axis of chirality but no chiral centers. Note how each molecule and its mirror image are not superimposable.

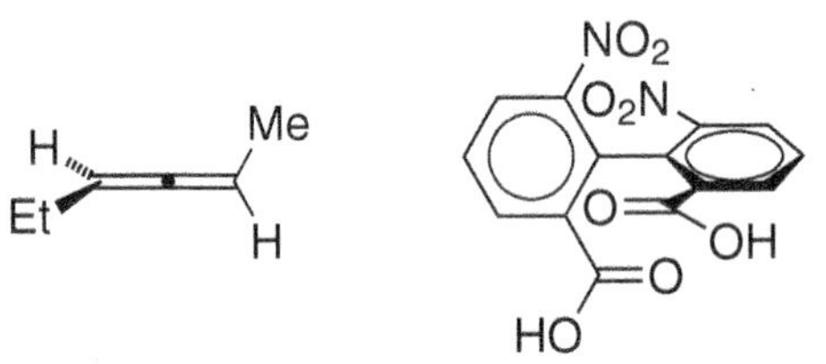

(2.5)

An ***axis of chirality*** occurs in a molecule with two noncoplanar protocenters located at opposite ends of the axis; it occurs in alkylidenecycloalkanes or substituted allenes where two mutually perpendicular protocenters are separated by two double bonds or a ring (Figure 2.4). In a molecule with an axis of chirality, it is not necessary to have a chiral center for the molecule to be able to exist as enantiomers. A third kind of molecule with an axis of chirality where the two protocenters are not necessarily orthogonal is exemplified by biphenyls carrying bulky ortho substituents that prevent free rotation about the σ bond between the two aromatic rings. These isomers are also called ***atropisomers***. The two structures in Example **2.5** are examples of an allene[22] and a biphenyl[23] that have been prepared in optically active form.

2.2 Absolute Configuration

The ***absolute configuration*** of a chiral molecule is a description of the three-dimensional arrangement of its atoms such that the two enantiomers are distinguished. The configuration of a chiral molecule is designated using one of three conventions. In the

22. (a) Walbrich, J.M.; Wilson, J.W.; Jones, W.M. *J. Am. Chem. Soc.* **1968**, *90*, 2895. (b) Borden, W.T.; Corey, E.J. *Tetrahedron Lett.* **1969**, 6741.

23. Christie, G.H.; Kenner, J. *J. Chem. Soc.* **1922**, *121*, 614. (b) Newman, P.; Rutkin, P.; Mislow, K. *J. Am. Chem. Soc.* **1958**, *80*, 465.

oldest of these conventions, the configuration of the molecule is related to its optical rotation. If the specific rotation is right-handed, or clockwise, it is the *d* (Latin *dextro*, to the right) or (+) isomer, and if the specific rotation is left-handed, or counterclockwise, it is the *l* (Latin *laevo*, to the left) or (–) isomer. Racemates are designated as *dl* or (±) under this system, and this part of the system has persisted; racemic compounds are still referred to as *dl* or (±) forms or ***modifications*** of the compound. It is important to recognize that the *dl* system is empirical and that there is no simple relationship between the rotation of a molecule and its absolute configuration (indeed, lactic acid and ethyl lactate with the same absolute configuration actually have specific rotations of opposite sign).

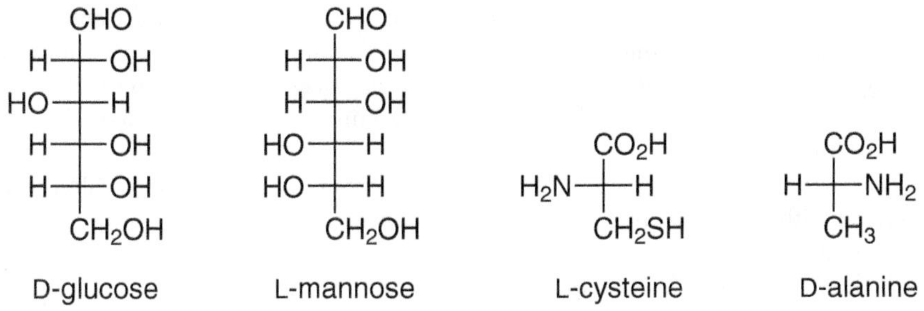

In biochemistry, as well as in the related fields of carbohydrate chemistry and peptide chemistry, the absolute configurations of molecules are assigned based on the arrangement of groups about a single (defining) chiral center of the molecule in one specific Fischer projection (the standard orientation). For carbohydrates, the molecule is written vertically with the carbonyl closer to the top of the representation. In this orientation, the configuration is D when the hydroxyl group on the lowest chiral center is to the right and L when it is to the left. Note that this configuration is specified using small capital letters to distinguish it from the rotation, which uses lower case letters. Amino acids are written with the carboxyl group at the top and the alkyl group at the bottom; the amino acid is L when the amino group is to the left and D when it is to the right.

The most widely used method for specifying the absolute configuration of a chiral molecule today is the *R/S* convention developed by Cahn,[24] Ingold,[25] and Prelog[26] based on the application of sequence rules.[27] These sequence rules were developed to

---

24. **Robert Sidney Cahn (1899-1981)** was Editor of the *Journal of the Chemical Society*, but biographical details about him are hard to obtain, so this brief biography is longer than most in this book. Cahn began his graduate education at Manchester, where he studied the chemistry of the morphine alkaloids under Robert Robinson. Following his MA, conferred by Cambridge, he moved to Cairo, where he spent 4 years on the faculty of the Egyptian University studying the constituents of cannabis (he had developed an allergy to morphine derivatives that stopped his research into these compounds). In 1928, he received the PhD from Frankfurt. In 1930, he moved to University College of North Wales at Bangor. He remained there until 1936, when he joined Cooper, McDougall & Robertson as a member of the Cooper Technical Bureau, working with insecticides of natural origin. In 1949, he joined the Chemical Society (now the Royal Society of Chemistry) as Editor of the *Journal of the Chemical Society*. In this capacity he was a prime moving force in the development of organic nomenclature. He remained in this position until 1963, when he retired, although he continued working as a consultant for the Society.

25. **Sir Christopher Kelk Ingold (1893-1970)**: see Chapter 1.

26. **Vladimir Prelog (1906-1997)** was educated at the Czech Institute of Technology in Prague (doctorate, 1929). He taught at Zagreb until 1942, when he moved to the ETH, in Zürich, where he became Professor in 1950 and Head of the Laboratory of Organic Chemistry in 1957. For a more complete biography see the Nobel Prize web site.

27. (a) Cahn, R.S.; Ingold, C.K. *J. Chem. Soc.* **1951**, 612. (b) Cahn, R.S.; Ingold, C.K.; Prelog, V. *Experientia* **1956**, *12*, 81; *Angew. Chem. Int. Ed. Engl.* **1966**, *5*, 385. (d) Cahn, R.S. *J. Chem. Educ.* **1964**, *41*, 116. (e) Fernelius, W.C.; Loening, K.; Adams, R.M. *J. Chem. Educ.* **1974**, *51*, 735. (f) Prelog, V.; Helmchen, G. *Angew. Chem. Int. Ed. Engl.* **1982**, *21*, 567.

allow the specification of absolute configuration unambiguously by text alone, so they do not necessarily correlate with other nomenclature rules (e.g., International Union of Pure and Applied Chemistry [IUPAC] rules). The sequence rules for chiral centers follow:

1. Atoms directly attached to the chiral center are ranked in increasing order of their atomic numbers (e.g., H < C < N < O ...).
2. If two or more atoms directly attached to the chiral center have the same atomic number, these atoms are ranked in increasing order of their mass numbers (e.g., ^{1}H < ^{2}H < ^{3}H).
3. If the application of these two sequence rules does not result in a differentiation between all four groups attached to the chirality center, rules 1 and 2 are applied in sequence to all atoms one bond further removed from the chirality center in those groups that have not yet been differentiated. This process is repeated until either all four groups are differentiated or until it is determined that the center in question is not chiral.
4. Once the four groups attached to the chiral center are ranked, the molecule is oriented with the lowest ranked group oriented away from the viewer (the standard orientation).
5. If the rankings of the three higher ranked groups in this orientation decrease in a clockwise direction, the configuration is *R*. if the rankings of the three higher ranked groups decrease in a counter-clockwise direction, the configuration is *S*.

When one or more of the groups attached to the chiral center has multiple bonds, the rules are modified slightly. The two bond types in the multiple bond—σ and π—are counted individually, but only the σ bonds are used if the rules must be applied to atoms further from the chiral center; the atoms at the ends of the π bonds are terminal "phantom atoms."

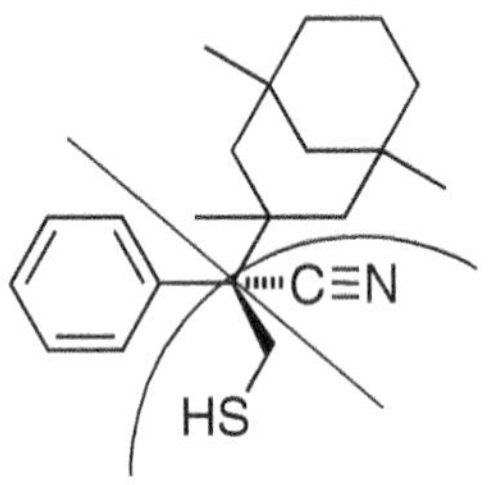

(2.6)

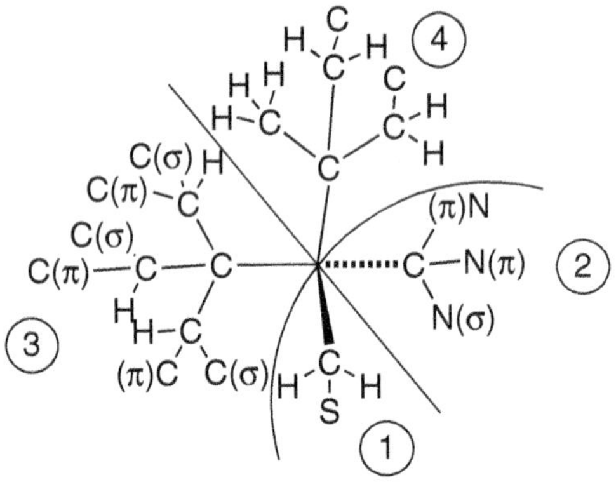

Figure 2.5 Applying the sequence rules to determine the absolute configuration of a chiral molecule. Note how all four atoms directly attached to the chiral center are carbon, so we must move one bond further out to begin to apply the rules. The single sulfur atom at the next level has the highest atomic number, followed by nitrogen, so these elements are ranked 1 and 2, respectively. To distinguish between the remainder, we need to move another two levels out before a difference is found.

There are numerous papers describing mnemonics to aid in determining the absolute configurations of complex molecules: (f) Mandal, D.K. *J. Chem. Educ.* **2007**, *82*, 274; *J. Chem. Educ.* **2000**, *77*, 866. (g) Wade, L.G., Jr. *J. Chem. Educ.* **2006**, *83*, 1793. (ch Gawley, R.E. *J. Chem. Educ.* **2005**, *82*, 1009. (i) Lloyd-Williams, P.; Giralt, E. *J. Chem. Educ.* **2005**, *82*, 1031; *J. Chem. Educ.* **2003**, *80*, 1178. (j) Hart, H. *J. Chem. Educ.* **2001**, *78*, 1632. (k) Luján-Upton, H. *J. Chem. Educ.* **2001**, *78*, 475. (l) Siloac, E. *J. Chem. Educ.* **1999**, *76*, 798. (m) Neeland, E.G. *J. Chem. Educ.* **1998**, *75*, 1573. (n) Baker, R.W.; George, A.V.; Harding, M.M. *J. Chem. Educ.* **1998**, *75*, 853.

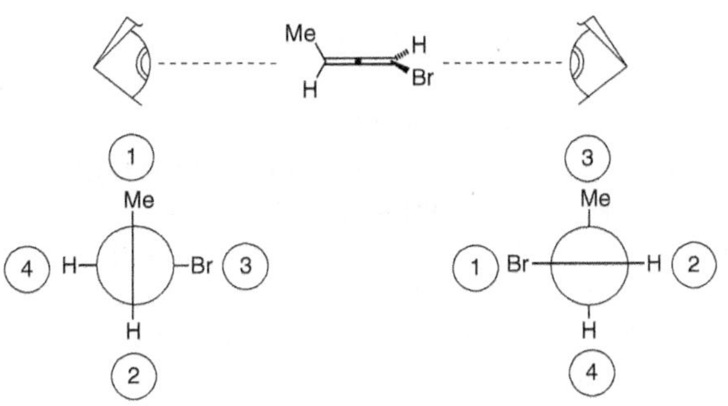

Figure 2.6 Determining the absolute configuration of a chiral allene

Using Example **2.6**, we can illustrate the application of these rules to a simple chiral compound. All four groups are bonded to the chiral center through carbon, so we need to move outward one bond to attempt to rank them. The results of this are summarized in Figure 2.5. In the right diagram, we see that the two smallest groups, physically, are actually the two highest ranked due to the application of rule number 1. Note also that *the rules apply to atoms individually*: One simply looks for the single atom with the highest atomic number; the aggregate total atomic number of atoms does not affect the outcome, so three nitrogen atoms do not outrank one sulfur atom. Based on this analysis, the configuration of the compound is *R*.

To apply these rules to compounds with an axis of chirality, two modifications are made to the rules: (1) in the standard orientation, these compounds are viewed in terms of the Newman projection along the chiral axis; and (2) groups on the protocenter at the near end of the axis are ranked higher than groups on the protocenter at the far end of the axis.

The direction in which the rankings of the top three ranked groups decrease in the standard orientation still decides the configuration: clockwise is *R*, and counterclockwise is *S*. Applying these rules to the simple allene shown in Figure 2.6 gives the name, *S*-1-bromo-1,2-butadiene.

Planar Chirality[28]

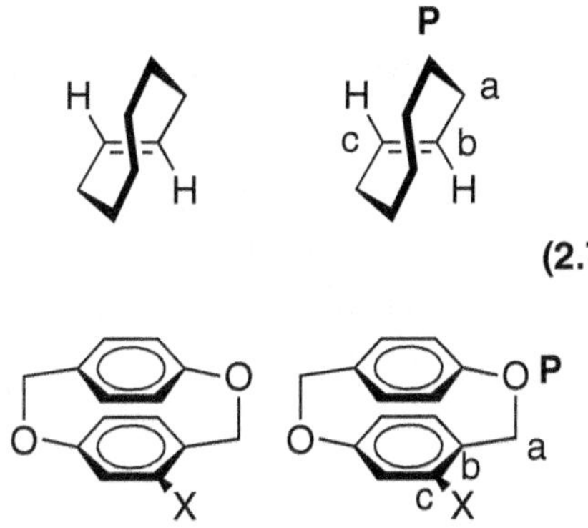

(2.7)

Molecules such as *trans*-cyclooctene or the *p*-cyclophane shown in Example **2.7** are chiral because they do not superimpose on their mirror images. However, these molecules possess neither a chiral center nor a chiral axis. In these cases, the molecules are dissymmetric with respect to a plane, and they are said to exhibit ***planar chirality***. Assigning the configuration to a molecule of this type requires that we first define the chiral plane. In the case of *trans*-cyclooctene, this will be the plane of the double bond,

28. For a discussion of assigning the configuration to molecules with planar chirality, see Eliel, E.L.; Wilen, S.H.; Mander, L.N. *Stereochemistry of Organic Compounds* (Wiley-Interscience: New York, 1994), ch. 14-1, p.1119.

and in the case of the cyclophane, this will be the plane of the substituted (higher priority) aromatic ring. In each molecule, an atom known as the ***pilot atom*** is now chosen. This atom is not in the chiral plane but it lies closest to it; if more than one atom is in the same location relative to the plane, the one with the higher priority according to the sequence rules is chosen (in the cyclophane, for example, the oxygen is chosen over the carbon). The pilot atom is labeled **P** in the structures. To assign the configuration of the molecule, one starts at the pilot atom and moves one bond at a time into the chiral plane through the first three atoms (labeled **a**, **b**, and **c**). If there is a choice for any of these atoms, the one with the higher priority according to the sequence rules is chosen (i.e., the carbon over the hydrogen in cyclooctene and the C-X carbon over the C—H carbon in the cyclophane). When one now views the atoms **a**, **b**, and **c** from the position of the pilot atom (**P**), the isomer is designated as *P* (for *Plus*; originally, this was *R*, but it has been replaced) if **a-b-c** describes a clockwise direction; if **a-b-c** describes a counter-clockwise direction, the enantiomer is *M* (for *Minus*; originally *S*). Thus, in the examples we have been using, the *trans*-cyclooctene has the *P* configuration and the cyclophane has the *M* configuration.

Diastereoisomers

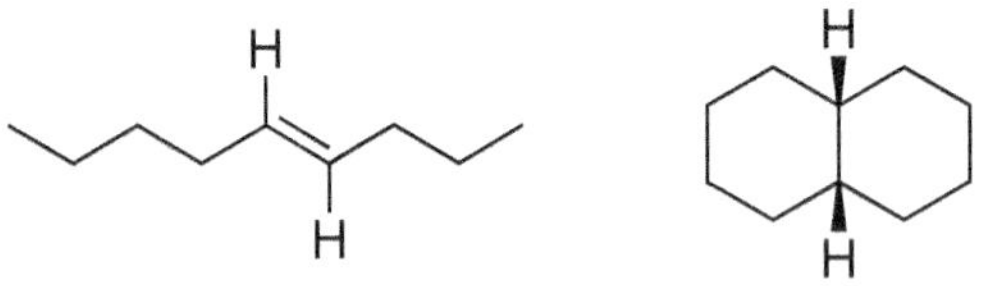

trans-4-nonene *cis*-decalin **(2.8)**

Stereoisomers not related to each other as enantiomers are diastereoisomers. There are two categories of diastereoisomers with which you should already be acquainted. The first of these has the special name ***geometric isomers***. These diastereoisomers differ from each other in the arrangement of protocenters about a rigid framework such as a ring or a double bond (e.g., Example **2.8**). These stereoisomers are named according to the Cahn-Ingold-Prelog *E*/*Z* system or according to the older *cis*/*trans* system. The older system is now generally restricted to alkenes where both protocenters carry a hydrogen atom (e.g., *trans*-4-nonene) or cyclic hydrocarbons (e.g., *cis*-decalin, for which the *E*/*Z* system has not been approved by IUPAC).

In the Cahn-Ingold-Prelog *E*/*Z* system, the protocenters are treated one at a time. The groups attached to each protocenter are ranked using the sequence rules, and the location of the two higher-ranked groups is compared. If the two higher ranked groups are on the same side of the double bond, the configuration is *Z*; if they are on opposite sides, the isomer is *E*.

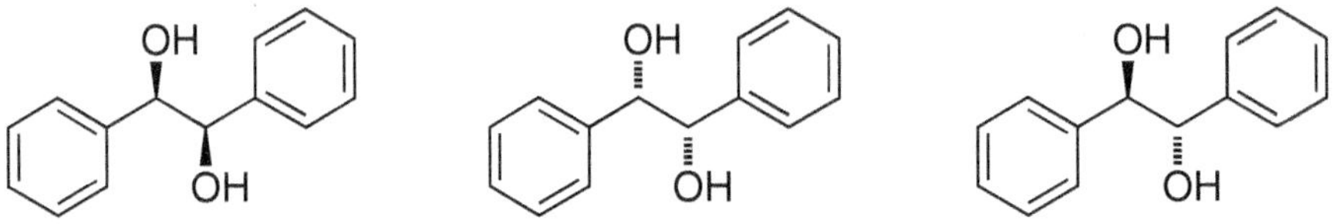

(1*R*, 2*R*) (1*S*, 2*S*) (1*S*, 2*R*), or (1*R*, 2*S*), or *meso*

(2.9)

When a molecule contains an even number of chiral centers, with pairs of chiral centers carrying the same four groups, whether or not the molecule as a whole is chiral depends on the absolute configuration of the individual chiral centers. This is illustrated by the three isomeric 1,2-diphenylethane-1,2-diols shown in Example **2.9**. In the two structures on the left, the two chiral centers of the molecule have the same absolute configuration, so the molecules are chiral, and enantiomers of each other. In the right structure, however, the two chiral centers have opposite absolute configurations, which means (1) that the left half of the molecule is the mirror image of the right half, and (2) that the molecule may be numbered from either end, giving two different correct names once the stereochemistry is included. The molecule is no longer chiral, despite the fact that it has two chiral centers. This is a ***meso* compound.**

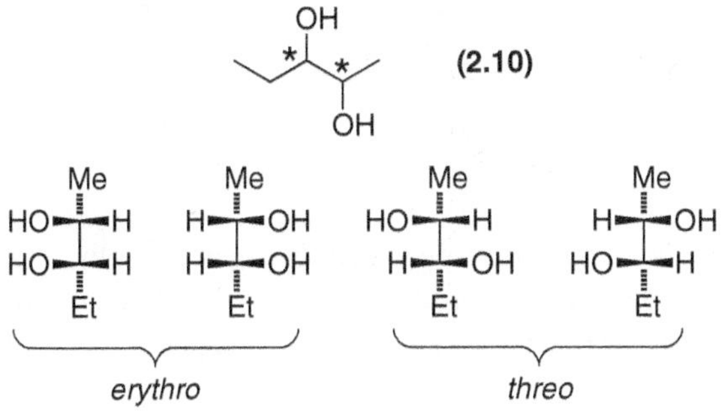

(2.10)

Another type of chiral compound is represented by 2,3-pentanediol (Example **2.10**). The chiral centers at C-2 and C-3 in this molecule have the same pair of substituents (H, OH), with the third substituent at each carbon (Me, Et) being similar, but different. In compounds of this type it is least ambiguous to specify the absolute configuration of this compound by specifying the absolute configuration of each chiral center individually. However, it is sometimes useful to specify the relative configurations of the chiral centers of the molecule (e.g., as in a racemate). The configuration of a racemate can be specified using the R^*/S^* terminology, where the asterisk indicates that we are dealing with *relative* configurations rather than absolute configurations (i.e., we do not specify which enantiomer of the chiral compound we are dealing with). In the case of 2,3-pentanediol, however, there is also a system of nomenclature where one can specify the relative configuration based on the presence or absence of a plane approximating a plane of symmetry. When the molecule is drawn in a conformation that shows eclipsing of the maximum number of identical groups attached to the chiral centers, the isomer with the greater number of eclipsing interactions of identical groups (the isomer more closely mimicking a *meso* compound) is the *erythro* isomer, and the other isomer is the *threo* isomer. There are two enantiomers of each diastereoisomer.

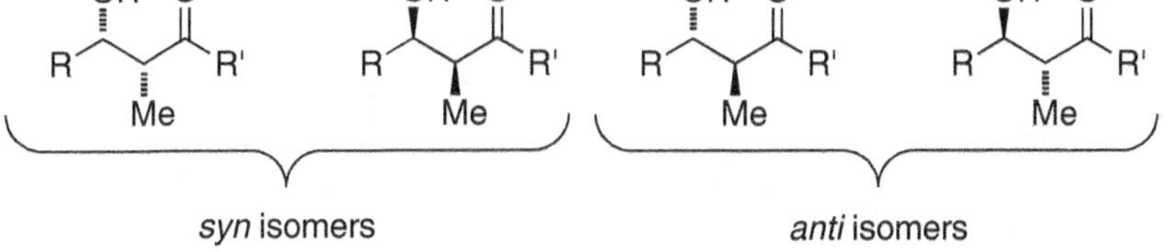

(2.11)

Work directed toward the synthesis of the macrolide antibiotics led to the need for a simple, general method for designating the relative stereochemistry of straight-chain polyhydroxycarboxylic acid derivatives in a way that allows the direct comparison of similar compounds. In such systems (e.g., Example **2.11**), the stereochemistry of the compounds (and, by extension, the reactions producing them) is designated using the *syn*/*anti* nomenclature system.[29] With the compound drawn with the straight-chain backbone in the extended ("zig-zag") conformation, the isomer is *syn* if the substituent groups are pointed in the same direction relative to the chain and *anti* if they are pointed in opposite directions.

α,β Nomenclature

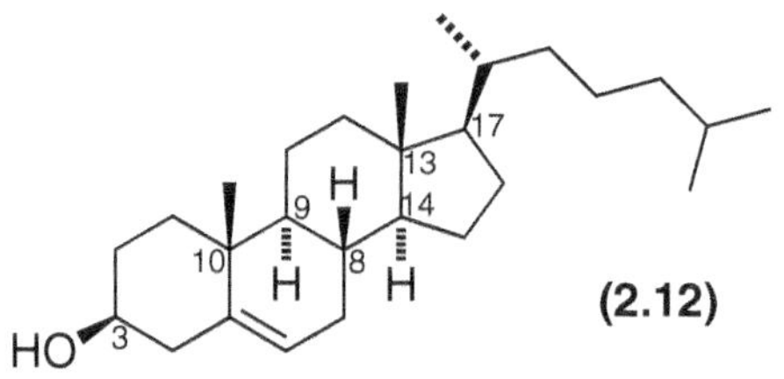

(2.12)

Another convention that is still in widespread use occurs in the semisystematic or trivial names of steroids and related natural products. Under this convention, the steroid (or other) ring system is written as the plane projection, which places each of the substituents either above (β) or below (α) the plane of the paper. In this system, the hydroxyl group at position 3 in the cholesterol molecule (Example **2.12**) is β, as are the methyl groups at positions 10 and 13, and the hydrogens at positions 9 and 14 are in the α position.

Epimers

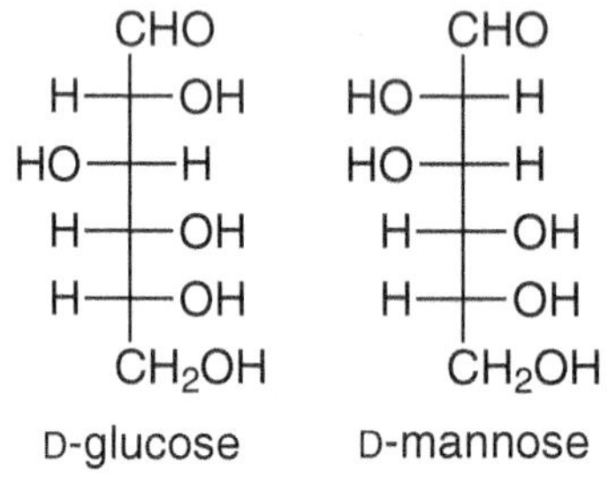

(2.13)

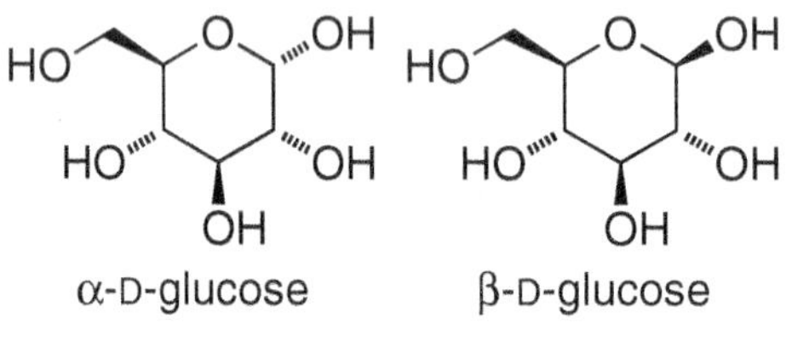

(2.14)

A special situation arises when two compounds with multiple chiral centers have the same absolute configuration of all but one of them. Examples of this are especially common in carbohydrates—the Fischer projections of the open-chain forms of D-glucose and D-mannose, for example (shown as Example **2.13**), have three of the four chiral centers in common. Diastereoisomers that differ only by a single chiral center are referred to as ***epimers***. Epimers can also arise when a single chiral compound reacts to generate a new chiral center, as occurs in the cyclization of D-glucose to give the two isomeric hemiacetals (Example **2.14**). In these molecules, the newly generated hydroxyl group may be above the plane of the ring, or below it. This is another place where the α,β system for naming stereoisomers is widely used. Epimers of

29. Masamune, S.; Kaiho, T.; Garvey, D.S. *J. Am. Chem. Soc.* **1982**, *104*, 5521.

this type, based on acetals or ketals or similar compounds are also known as ***anomers***, so one speaks of the α-anomer of D-glucose, for example. The equilibrium between anomers is subject to a stereoelectronic effect referred to as the ***anomeric effect***, which we will discuss further in Chapter 4.

Pseudochiral Centers

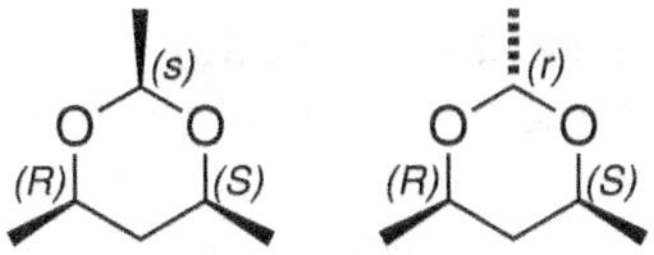

(2.15)

A special case arises when a *meso* compound has an odd number of protocenters, as illustrated by the two acetals in Example **2.15**. Both these compounds have a mirror plane of symmetry, with the acetal carbon being a stereochemical center but not a chiral center. This type of stereochemical center is referred to as a ***center of pseudochirality***. There is a subrule of the Cahn-Ingold-Prelog sequence rules that allows one to assign a configuration to this type of pseudochiral center. In a molecule of type $CabR_RR_S$, where R_R and R_S are two alkyl substituents with the same groups, differing only in absolute configuration, R_R (with the *R* configuration) takes precedence over R_S, and the configuration of the pseudochiral center is designated *r* or *s*, applying the rules—with this subrule—as usual.[30] The two compounds are, therefore, (2*s*,4*S*,6*R*)-2,4,6-trimethyl-1,3-dioxane (left) and (2*r*,4*S*,6*S*)-2,4,6-trimethyl-1,3-dioxane (right).

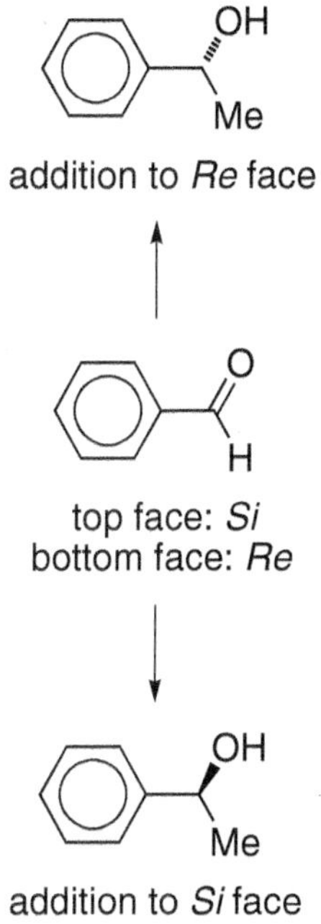

(2.16)

Prochiral Centers

The carbonyl group of an aldehyde is not chiral (it is planar and lacks the three-dimensional character necessary for chirality), but the addition of a Grignard reagent to the aldehyde can convert it into a new chiral center (as in Example **2.16**). This makes the aldehyde group a ***prochiral center***—one that has the potential to be converted into a chiral center by an addition reaction. The absolute configuration of the product will depend on which of the two faces of the carbonyl group participates in the reaction. We designate the two faces of a prochiral center using the same Cahn-Ingold-Prelog sequence rules as for chiral centers. In this case, there will be only three groups attached to the prochiral center, so we do not need to worry about a standard orientation. If the rankings of the groups decrease in the clockwise direction as one looks at the prochiral center, it is the *Re* face, if counterclockwise, it is the *Si* face.

Stereotopism

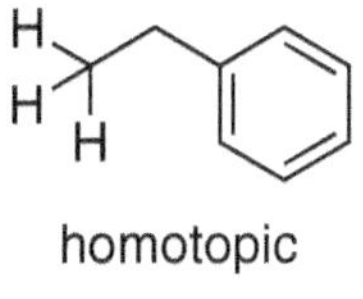

homotopic

(2.17)

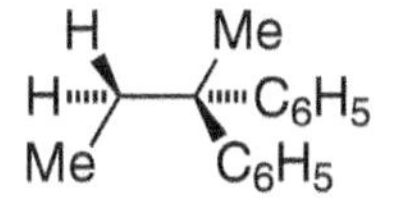

enantiotopic

(2.18)

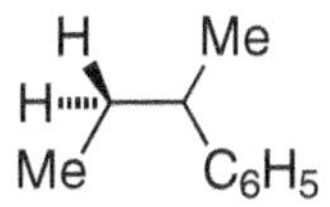

diastereotopic

(2.19)

Identical groups attached to a common atom may still have different stereochemical properties (often evident when, for example, their nuclear magnetic resonance (NMR)

30. (a) Cahn, R.S.; Ingold, C.K.; Prelog, V. *Angew. Chem. Int. Ed. Engl.* **1966**, *5*, 385. (b) Goldberg, S.I.; Bailey, W.D. *J. Am. Chem. Soc.* **1971**, *93*, 1046. (c) Nourse, J.G. *J. Am. Chem. Soc.* **1975**, *93*, 4594. (d) IUPAC 1974 Recommendations for Section E, "Fundametal Stereochemistry" *Pure Appl. Chem.* **1976**, *45*, 13.

spectra are recorded). This type of stereochemistry is known as ***stereotopism***. Groups are ***homotopic*** when the replacement of any leads to exactly the same product; the simplest examples are the three hydrogens of a methyl group (e.g., Example **2.17**), or the methylene hydrogens of a simple cycloalkane. Homotopic groups are equivalent under all conditions. ***Enantiotopic*** groups are groups attached to the same atom where replacement of one gives the enantiomer of the compound obtained when the other is replaced; an example of enantiotopic groups are the phenyl groups and the methylene hydrogens of 2,2-diphenylbutane (Example **2.18**). Enantiotopic groups are different only in a chiral environment (e.g., enantiotopic hydrogens have the same chemical shifts in the NMR spectrum and the same reaction rate in proton abstraction reactions by achiral bases). ***Diastereotopic*** groups are groups attached to the same atom where the replacement of one gives rise to a diastereoisomer of the compound obtained by replacement of the other. For groups to be diastereotopic, the molecule must already have a chiral center; an example is the methylene hydrogens of 2-phenylbutane (Example **2.19**). Diastereotopic groups are not chemically equivalent; they have different chemical shifts in the NMR spectrum, and they react at different rates with the same reagent.

Geometric Isomers

Compounds need not have a chiral center to exist as stereoisomers, a fact that was known by the time that van't Hoff and Le Bel published their pioneering work. Much of the work with these compounds was carried out by German chemist, Johannes Wislicenus. He found that oximes existed in two forms, and first used the term, ***geometric isomers***, to describe them. Geometric isomers occur in alkenes, oximes, and cycloalkanes. Today, geometric isomers in alkenes and oximes and similar compounds are designated using the Cahn-Ingold-Prelog *E/Z* system.

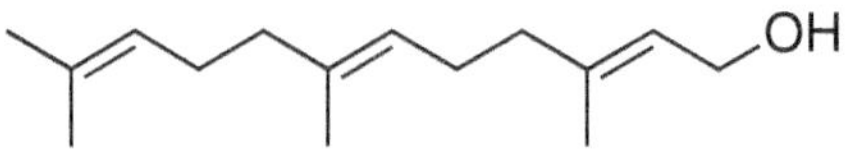

(2*E*,6*E*)-3,7,11-dodeca-2,6,10-trien-1-ol
(farnesol)

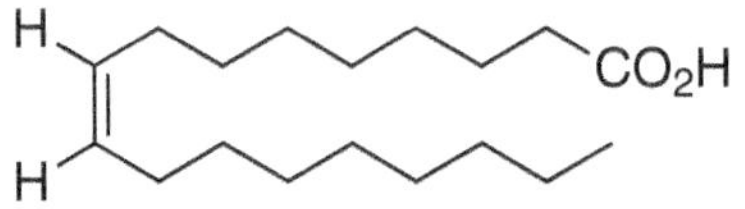

Z-octadec-9-enoic acid
(oleic acid)

(2.20)

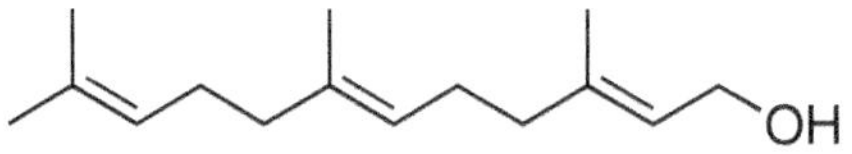

trans, *trans*-farnesol

(2.21)

The *E/Z* system gives an unambiguous designation of stereochemistry for alkenes (and it would for cycloalkanes, except that IUPAC does not permit the use of this system for cycloalkanes at this time). In systems where both protocenters carry a hydrogen atom, it is still common to find the older *cis/trans* nomenclature used (e.g., oleic acid is also *cis*-9-octadecenoic acid, Example **2.20**), but this is gradually being replaced even in these cases by the newer system. There are also compounds (e.g., *trans,trans*-farnesol, Example **2.21**) where the stereochemical designation a trisubstituted double bond by the *cis/trans* system is fixed by long-standing convention.

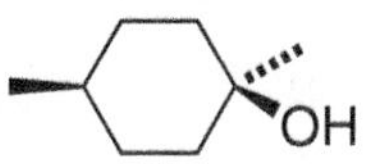

(2.22)

1-*cis*-4-dimethylcyclohexan-*r*-1-ol

With substituted cycloalkanes, the use of the *R/S* system gives the most unambiguous nomenclature, but this cannot be applied to cyclic systems that do not contain chiral centers. Although the application of the *E/Z* nomenclature system to such cyclic compounds would provide the simplest solution to this problem in nomenclature (e.g., compound **2.22** would be *Z*-1,4-dimethylcyclohexanol), this is not approved by IUPAC. At this time, one selects one substituent on the ring (usually the highest ranked substituent) as the reference substituent and designates this by the descriptor, *r*. One then specifies the locations of the other substituents relative to this. Under this system, compound **2.22** becomes 1,*cis*-4-dimethyl-cyclohexan-1-*r*-ol.

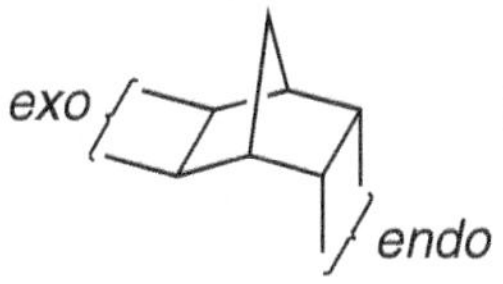

(2.23)

In bridged-ring systems, another system of nomenclature is widely established by common usage. This is the *exo/endo* system of stereochemical designation, which is especially important in bicyclic compounds. In these compounds, substituents that are *cis* to the shortest bridge of the bicyclic system are designates as *exo* substituents, whereas substituents that are *trans* to the shortest bridge are *endo* substituents, as shown in Example **2.23**.

Problems

2-8 How many ways can one select the two protocenters that combine to generate a chiral center? Illustrate your answer using bromochlorofluoromethane as an example.

2-9 Draw the 24 possible Fischer projections of CHClBrF and assign each an absolute configuration in the R/S system. How does this number correlate with the answer to Problem 2-8?

2-10 Does either of the following two molecules possess an improper axis of rotation? Give your reasoning.

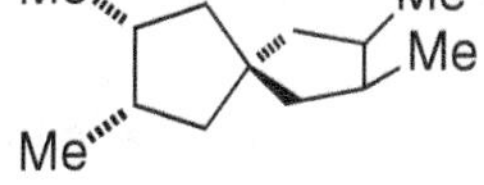

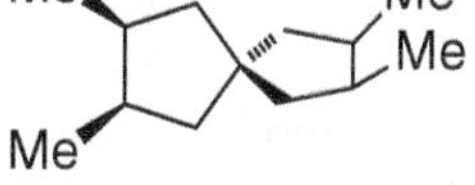

2-11 Identify all the sets of homotopic, enantiotopic, and diastereotopic atoms or groups in *tert*-butylcyclohexane.

(continues)

(**Problems** continued)

2-12 What is the type of chirality (center, axis, plane) and the absolute configuration of each chiral element in each of the following compounds? If the molecule as a whole is not chiral, why not?

(a) 

(b) 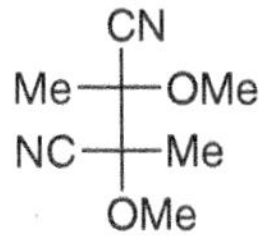

(c) 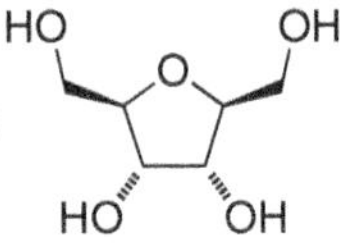

(d) 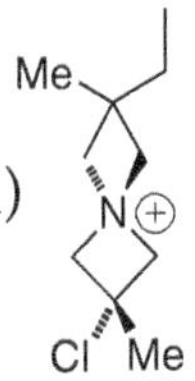

(e) 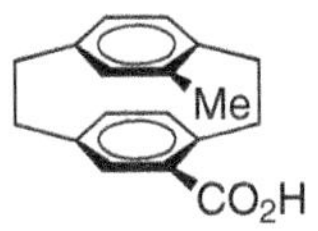

(f) 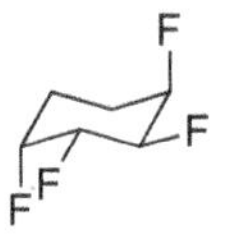

(g) 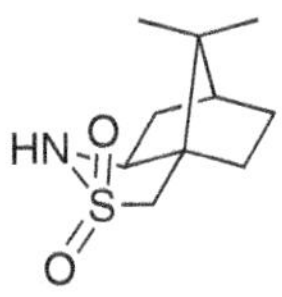

(h) 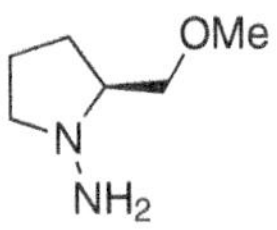

(i) 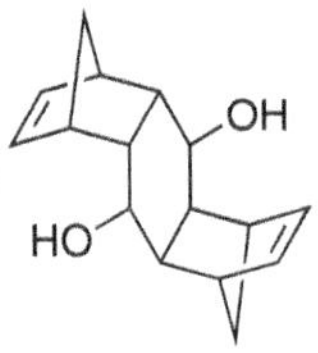

(j) 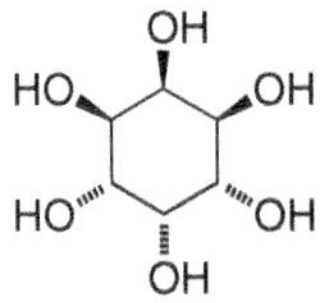

(k) 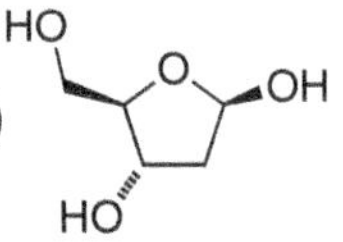

(l) 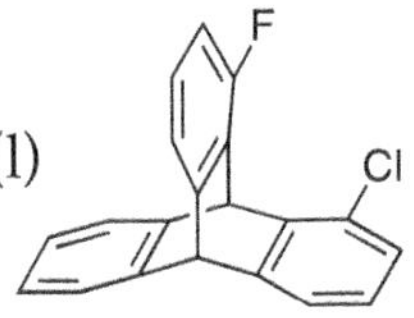

Stereochemistry in Reactions: Defining Terms

Many organic reactions occur with stereochemical consequences, and this means that we need to define some terms to allow the discussion of reaction stereochemistry that comes later in this book. Reactions may be stereospecific, stereoselective, or stereorandom. A ***stereospecific*** reaction is one in which a single stereoisomer of the reactant is constrained by the mechanism of the reaction to give a single stereoisomer in the product, and in which *changing the stereochemistry of the reactant changes the stereochemistry of the product.* In other words, one can reliably predict the stereochemistry of the reaction product, knowing the stereochemistry of the reactant and vice versa. A ***stereoselective*** reaction is one where more than one stereoisomer of the product may be formed. By convention, the use of this term is reserved for reactions where one stereoisomer of the product predominates by a large amount over the other possible stereoiosmers. Unlike a stereospecific reaction, the stereochemistry of the major product is not necessarily determined by the stereochemistry of the starting material. A ***stereorandom*** process is one in which both stereoisomers are produced in approximately equal amounts.

Stereospecific reactions abound in organic chemistry: the S_N2 reaction occurs with inversion of configuration at the carbon bearing the leaving group, and Baeyer-Villiger rearrangements (e.g., Example **2.24**[31]) occur with retention of configuration at the migrating carbon. Where there are two reacting centers in a molecule, the reaction may also occur with defined stereochemistry: in catalytic hydrogenation, the two hydrogen atoms are delivered to the same face of the reacting π bond, a situation similar to the hydroboration and hydroxylation reactions (e.g., Example **2.25**[32]); the addition of halogens and halogen-like electrophiles to alkenes proceeds to give the product where one part of the reagent adds to one face of the π bond, and the remainder of the reagent

31. Turner, R.B. *J. Am. Chem. Soc.* **1950**, *72*, 878.
32. Suzuki, T.; Ghozati, K.; Katoh, T.; Sasai, H. *Org. Lett.* **2009**, *11*, 4286.

adds to the other face. In the E2 elimination, the hydrogen and the leaving group must adopt an *anti* conformation in order for the reaction to proceed. Addition reactions in which both parts of the reagent add to the same face of the π bond are known as ***suprafacial*** or ***syn*** additions, and reactions such as catalytic hydrogenation or osmium tetroxide hydroxylations belong to this class. Addition reactions where the two parts of the reagent add to opposite faces of the π bond are known as ***antarafacial*** or ***anti*** additions, and reactions such as the addition of bromine to alkenes belong to this class.

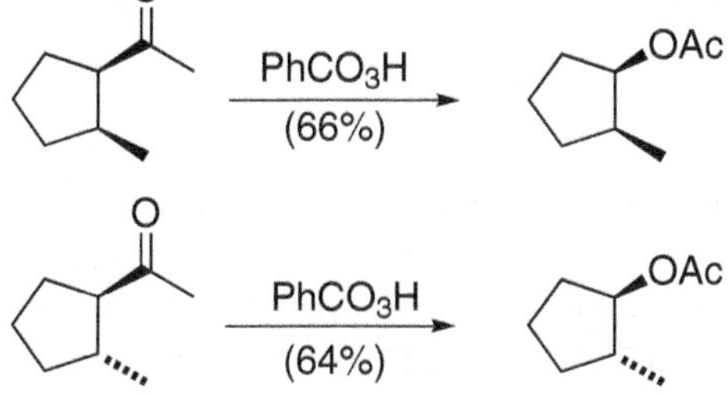

(2.24)

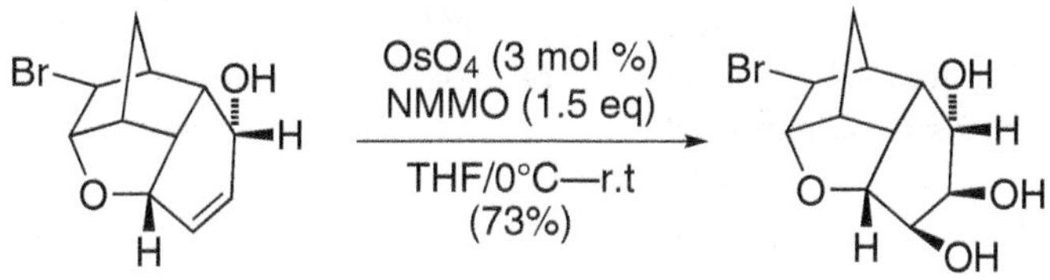

(2.25)

2.3 **Optical Purity and Configuration**

A sample of a compound in which all the molecules are of the same enantiomer is now "optically pure," although the terms "enantiopure" and "homochiral" are still encountered.[33] A sample of a compound containing more of one enantiomer than the other is now "enantiomerically enriched." It is still most usual to describe such compounds in terms of the ***enantiomeric excess, ee***, although a case has now been made for "***enantiomer ratio***," ***er***, in its place.[34] This is gaining in popularity. The enantiomeric excess of a compound is defined as the ratio of the specific rotation of the sample to the maximum specific rotation of the same compound, where the specific rotation is

$$[\alpha]_\lambda^T = \frac{\alpha}{cl} \text{ for solids, } [\alpha]_\lambda^T = \frac{\alpha}{l\rho} \text{ and for liquids,}$$

where α is the observed rotation, is the path length in dm, *c* is the concentration of the solute in g mL^{-1}, and ρ is the density of a pure liquid in g mL^{-1}. Because the behavior or the specific rotation is not necessarily linear with concentration, it is important to note the concentration of the sample. This corresponds to the difference between the fractions of the two enantiomers in the sample or to the amount of one enantiomer present in excess of the racemate:

$$\text{Optical purity} = \{[\alpha]_\lambda^T\,(\text{obs})/\ [\alpha]_\lambda^T\,(\text{max})\} \times 100\% = \{(\%\text{ major enantiomer}) - (\%\text{ enantiomer})\} = \text{major enantiomer in total} - \%\text{ racemate in total}$$

33. (a) Review: Raban, M.; Mislow, K. *Top. Stereochem.* **1967**, *2*, 199. (b) For a discussion of the use and abuse of stereochemical terms, see Eliel, E.E. *Chirality* **1997**, *9*, 428.
34. Gawley, R.E. *J. Org. Chem.* **2006** *71*, 2411.

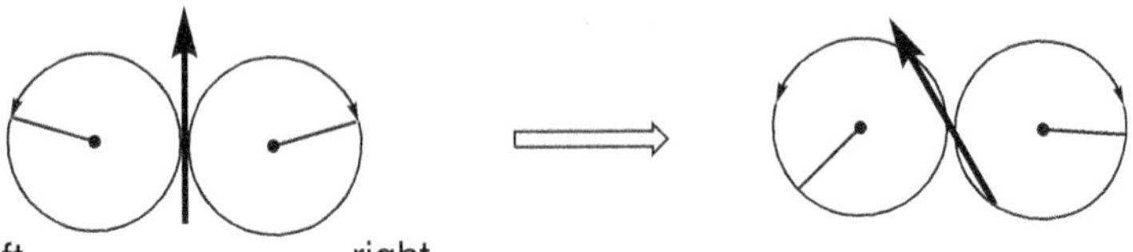

Figure 2.7 Differential refraction of waves of circularly polarized light leads to observed rotation of the plane of plane-polarized light.

For example, optically pure (+) glyceraldehyde has a specific rotation of +14°. Thus, a sample of glyceraldehyde with a specific rotation of +7.0° is (7.0/14) 100 = 50% optically pure. It may be viewed as consisting of a mixture of 50% (+) glyceraldehyde and 50% (±) glyceraldehyde, or 75% (+) glyceraldehyde and 25% (–) glyceraldehyde.[35] The optical purity of a compound can be measured by a variety of methods, of which NMR spectroscopy[36] and chromatography on a chiral stationary phase[37] have become the most widely used. The use of these techniques, which provide the ratio of the enantiomers (or diastereoisomers) directly, prompted a reevaluation of the use of the terms enantiomeric excess (e.e.) and diastereoisomeric excess (d.e.)[38] as well as the suggestion that the e.e. be replaced by the enantiomer ratio (e.r.) as a more useful measure of the composition of a mixture of enantiomers.

Before we proceed further, we need to discuss the nature of plane-polarized light, which is responsive to chirality, in more detail. In plane-polarized light, the electric vector of the radiation oscillates in a single plane, but its chiral nature may be viewed as arising from the superimposition of two waves of circularly polarized light, one left-handed, and one right-handed (Figure 2.7). In circularly polarized light, the electric vector describes a helix along the direction of propagation of the wave. If the two circularly polarized waves are passed through an achiral medium, they emerge unchanged at the end, and the plane of polarization remains unchanged. If the same plane-polarized wave is passed through a chiral medium, however, the two circularly polarized waves will interact differently with the medium. If the medium has different refractive indices for right and left circularly polarized light, for example, the wave emerging from the medium will again be plane polarized, but the plane of polarization will have been rotated by an amount proportional to the difference between the two refractive indices. The angle of rotation per unit path length is $\alpha = \frac{(n_L - n_R)\pi}{\lambda}$, where n_L and n_R are the refractive indices for the left and right circularly polarized light, respectively, and λ is the wavelength of the light.[39] Just as the refractive indices of right circularly and left circularly polarized light

35. This method does require that be independent of concentration: Horeau, A. *Tetrahedron Lett.* **1969**, 3121.

36. (a) Raban, M.; Mislow, K. *Tetrahedron Lett.* **1965**, 4249; **1966**, 3961. (b) Jacobus, J.; Raban, M. *J. Chem. Educ.* **1969**, *46*, 351. (c) Tokles, M.; Snyder, J.K. *Tetrahedron Lett.* **1988**, *29*, 6063. (z) Review: Parker, D. *Chem. Rev.* **1991**, *91*, 1441.

37. Gas chromatography: (a) Halpern, B.; Westley, J.W. *Chem. Commun.* **1965**, 246. (b) Vitt, S.V.; Saporovskaya, M.B.; Gudkova, I.P.; Belikov, V.M. *Tetrahedron Lett.* **1965**, 2575. (c) Guett, J.; Horeau, A. *Tetrahedron Lett.* **1965**, 3049. (d) Westley, J.W.; Halpern, B. *J. Org. Chem.* **1968**, *33*, 3978.

High-performance liquid chromatography: Pirkle, W.H.; Finn, J. In Morrison, J.D., Ed. *Asymmetric Synthesis*, Vol. 1 (Academic Press: New York, 1983), p. 87.

38. (a) Selke, R.; Facklam, C.; Foken, H.; Heller, D. *Tetrahedron: Asymmetry* **1993**, *4*, 369. (b) Seebach, D.; Beck, A.K.; Schmidt, B.; Wang, Y.M. *Tetrahedron* **1994**, *50*, 4363. (c) Gallagher, D.; Du, H.; Long, S.A.; Beak, P. *J. Am. Chem. Soc.* **1996**, *118*, 11391.

39. Fresnel, A. *Ann. chim. phys. [2]* **1825**, *28*, 147.

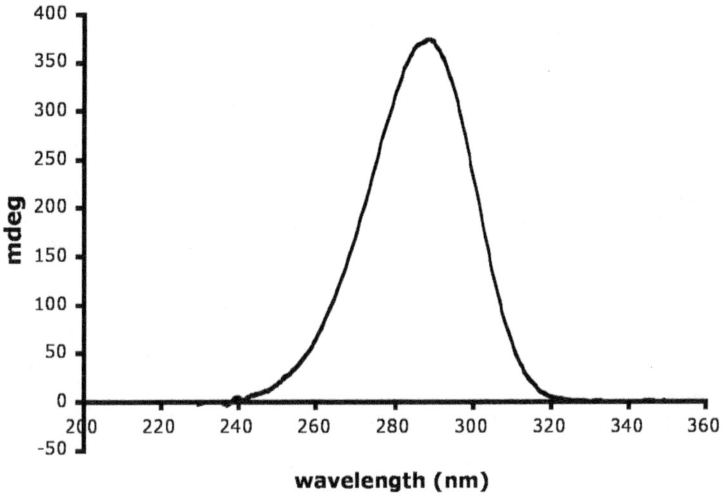

Figure 2.8 Circular dichroism spectrum of camphorsulfonic acid (1 mg mL^{-1}). The vertical scale is specified as millidegrees, which is a reasonable approximation for small values of θ because tan θ ≈ θ when θ is small.

are different in a chiral medium, so, too, are their extinction coefficients for absorption. This results in plane-polarized light becoming elliptically polarized on passing through a chiral medium. The measurement of differential absorption of right and left circularly polarized light is the basis of the technique of ***circular dichroism*** (CD). CD is expressed in terms of the molar ellipticity, θ, which is related to the difference in absorbance (ΔA) for right and left circularly polarized light by $\theta = 3238(\Delta A)$. The tangent of the ellipticity, tan θ, is related directly to ΔA. CD is especially important in biochemistry, where it is used to probe the conformations of biological molecules such as nucleic acids and proteins. The CD spectrum of (1*R*)-camphor-10-sulfonic acid (1 mg/mL in water) is shown in Figure 2.8.

The specific rotation of a substance depends both on the temperature and wavelength at which the measurement is made. Most measurements are made using the sodium D line (589.0 nm). This sharp, bright line in the sodium atomic spectrum has long been easy to obtain. The measurement of the specific rotation over the range of wavelengths from the visible to the ultraviolet (UV) gives a spectrum known as the ***optical rotatory dispersion*** (ORD) curve. ORD spectra of a group of four steroids[40] are shown in Figure 2.9.

In a **plain ORD curve**, the specific rotation changes monotonically with wavelength. In ORD curves with an **anomalous Cotton effect**, the magnitude of the rotation initially increases as the wavelength becomes shorter, and it then changes sign, passing through another maximum (or minimum).[41] Typical examples of the anomalous Cotton effect are provided by the ORD spectra shown in Figure 2.9. Enantiomers exhibit mirror image ORD curves (i.e., if the ORD curve of one enantiomer exhibits a positive Cotton effect, the ORD curve of other will exhibit a negative Cotton effect).

The use of ORD curves in studying the chiroptical properties of chiral ketones led to the development of the ***octant rule***.[42] This empirical rule, which was originally developed to allow the prediction of the sign of the Cotton effect from a known absolute configuration, can also be applied in the corollary sense, to determine absolute

40. Djerassi, C.; Closson, W.; Lippman, A.E. *J. Am. Chem. Soc.* **1956**, *78*, 3163.

41. (a) Cotton, A. *Ann chim. phys. [7]* **1896**, *8*. 347. (b) Mitchell, S. *The Cotton Effect* (G. Bell & Sons)

42. (a) Moffitt, W.; Woodward, R.B.; Moscowitz, A.; Klyne, W.; Djerassi, C. *J. Am. Chem. Soc.* **1961**, *83*, 4013. (b) Mislow, K.; Glass, M.A.W.; Moscowitz, A.; Djerassi, C. *J. Am. Chem. Soc.* **1961**, *83*, 2771.

Figure 2.9 Optical rotatory dispersion spectra of a series of steroidal ketones

(a)

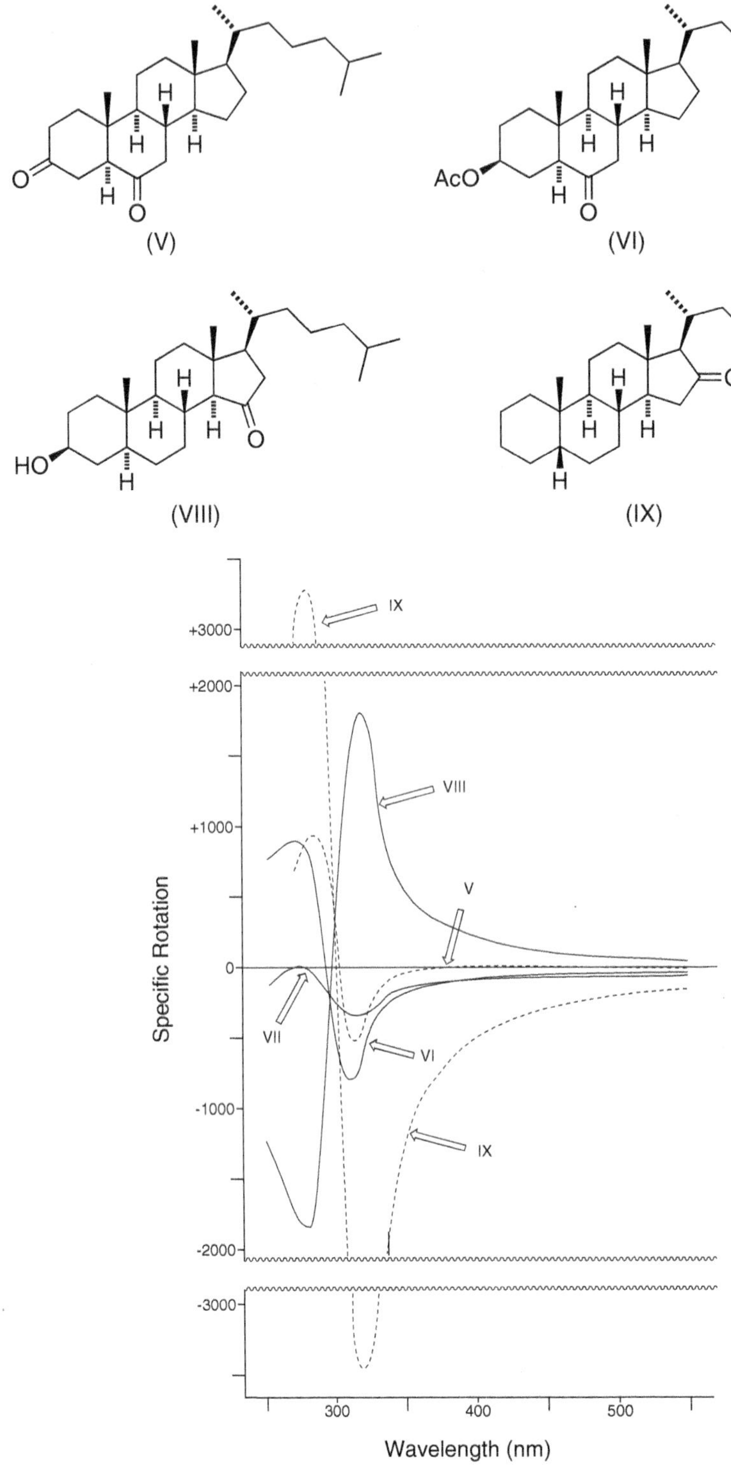

configuration of a known structure from the sign of the Cotton effect. To apply the rule, the carbonyl group of the ketone is oriented as if to draw the Newman projection with the carbonyl oxygen atom toward the observer (Figure 2.10). The middle of the C = O bond is the origin, with the plane of the sp^2 orbitals and the plane perpendicular to it and containing the C = O bond (the two lines in Figure 2.10) as the first two planes. The plane perpendicular to the C = O bond is the third plane. These three planes divide the

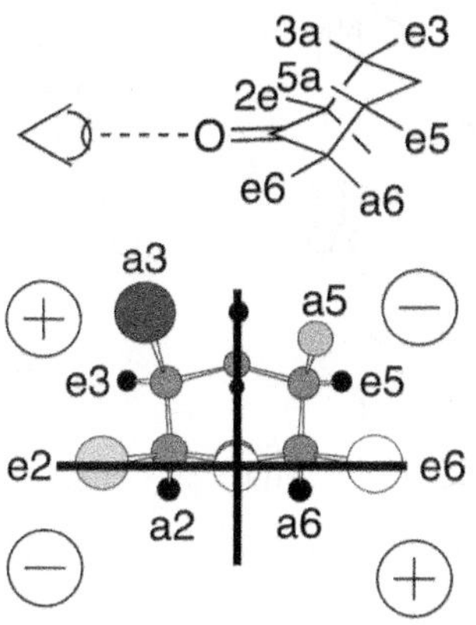

Figure 2.10 Octant rule for cyclohexanone derivatives

region about the C = O group into eight octants: four rear octants and four front octants. The four rear octants make positive (lower right and upper left) and negative (lower left and upper right) contributions to the observed Cotton effect, with the corresponding front octants making the opposite contributions. However, because groups in the front octants must necessarily project forward beyond the carbonyl oxygen, they are seldom involved.[43]

When one examines the cyclohexanone ring using these rules, one can make some fairly straightforward predictions. Thus, by numbering the carbonyl carbon 1 and then proceeding in clockwise order around the cyclohexane ring (so that the α carbon to the left becomes position 2, and the α carbon to the right becomes position 6), one predicts that:

1. Substituents at position 3 make a positive contribution to the Cotton effect and that substituents at position 5 make a negative contribution
2. Substituents at position 4 make a minimal contribution to the Cotton effect
3. Equatorial substituents at positions 2 and 6 make a minimal contribution to the Cotton effect
4. Axial substituents at position 2 make a negative contribution to the Cotton effect, and axial substituents at position 6 make a positive contribution

Exciton Chirality and Absolute Configuration

In the early 1980s, a method known as ***exciton chirality method*** was developed for determining the absolute configuration of cyclic 1,2-diols where the two OH groups are gauche to each other. It involves measuring the CD spectra of derivatives that have a groups that can absorb UV-visible light (a ***chromophore***).[44] The derivatives are defined to have positive or negative chirality based on the dihedral angle between the two chromophores, as illustrated by structures **2.26** and **2.27** in Figure 2.11. When the dihedral angle from front to back is clockwise, the chirality is defined as positive. The technique can be used on very small scale (micro to nanomolar). The diol is esterified with an achiral aromatic acid (*p*-aminobenzoate esters have been especially popular), and the CD spectrum of the resulting diester (which need not even be purified) is then measured. The spectrum shows two Cotton effects near the maximum absorbance of the ester, one positive and one

43. Review of chiroptical spectroscopy: Kirk, D.N. *Tetrahedron,* **1986**, *42*, 777.
44 Monograph: Harada, N.; Nakanishi, K. *Circular Dichroic Spectroscopy. Exciton Coupling in Organic Stereochemistry* (University Science Books: Mill Valley, CA, 1983).

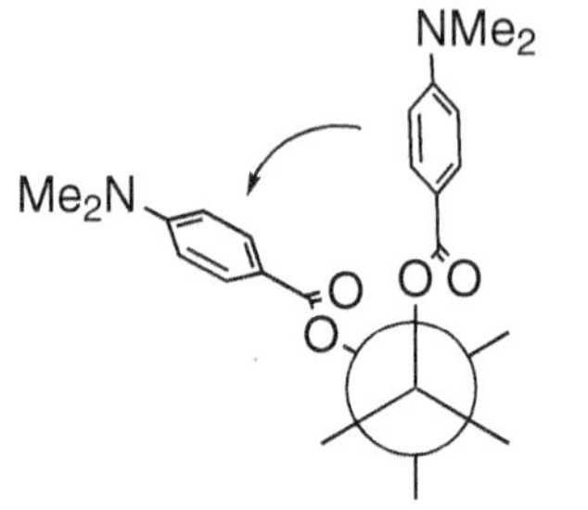

counterclockwise: negative

(2.26)

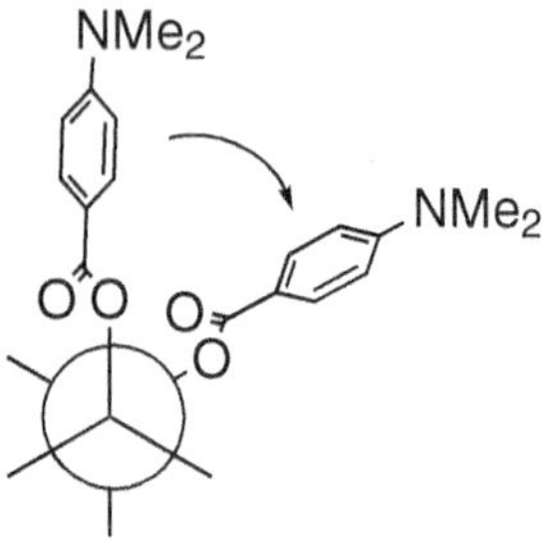

clockwise: positive

(2.27)

Figure 2.11 Defining positive and negative chirality

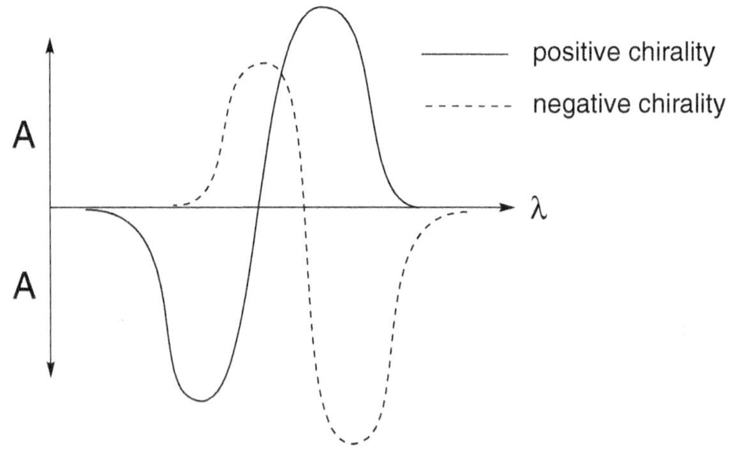

Figure 2.12 Predicted Cotton effects in chirality exciton method

negative. For molecules with "positive chirality," the longer wavelength Cotton effect is positive, and the shorter wavelength Cotton effect is negative. The reverse is true for molecules with negative chirality (Figure 2.12).

Problems

2-13 What is the relationship (enantiomers, diastereoisomers, constitutional isomers) between the two molecules in each of the following pairs of compounds?

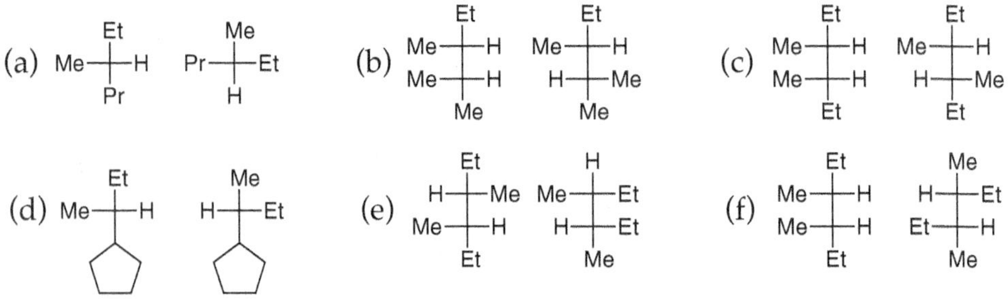

2-14 What is the relationship (enantiomers, diastereoisomers, epimers, anomers, constitutional isomers) between the two molecules in the following pairs of compounds?

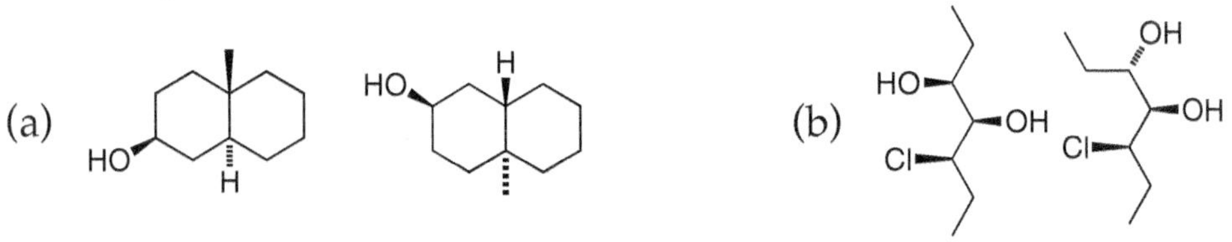

(c) 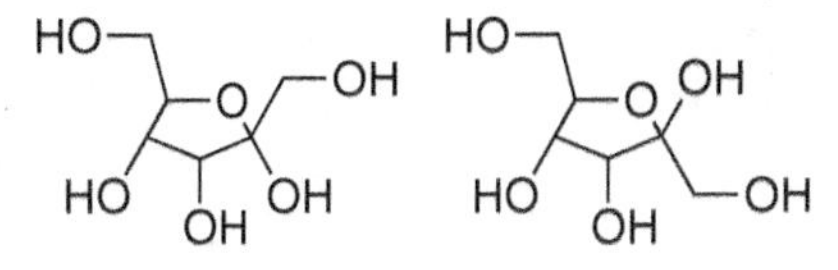

(d) 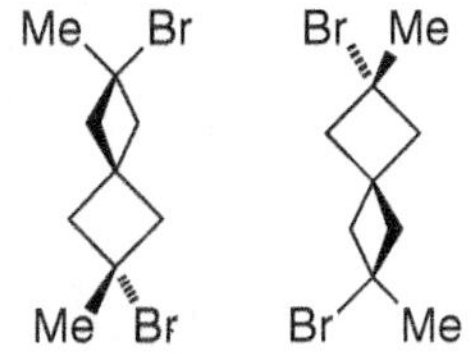

(e) 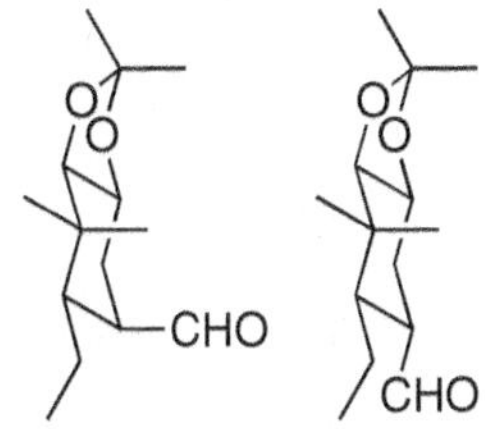

(f) 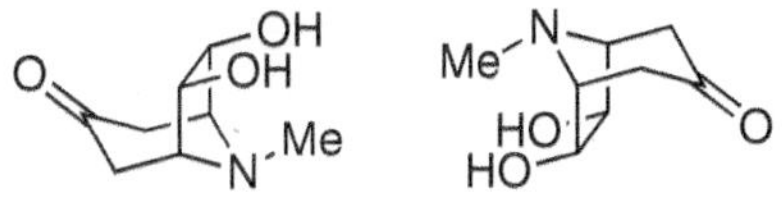

2-15 Define the sign of the chirality of each of the following 1,2-diols as needed to predict the Cotton effects in the CD spectra of their *bis-p*-dimethylaminobenzoate derivatives. Note that a single OH group distant from the diol has little effect on the Cotton effects.

(a) 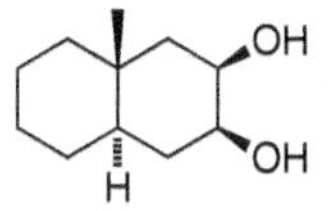

(b) 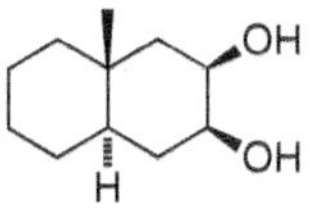

(c) 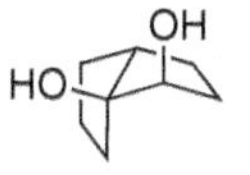

(d) 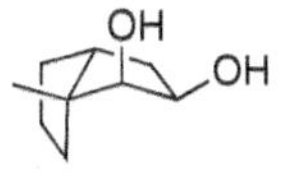

(e) 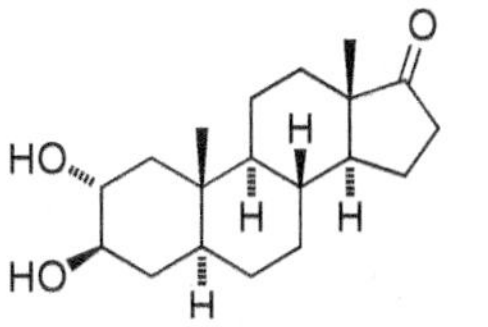

(f) 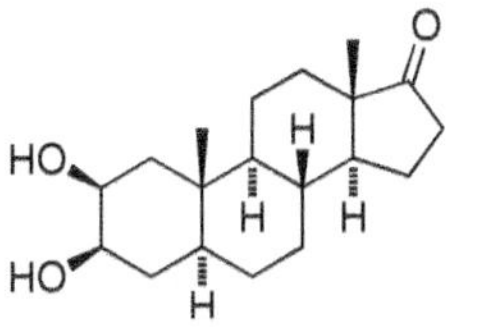

(g) 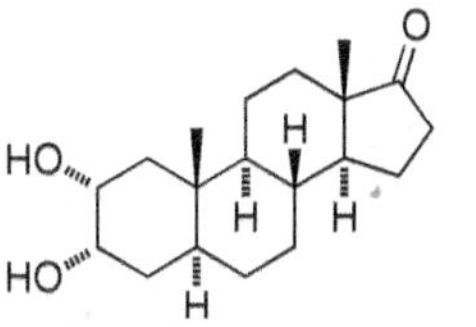

(h) 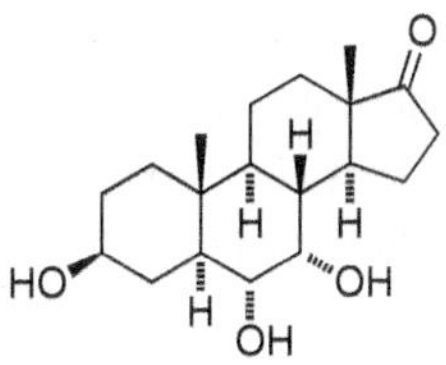

(i) 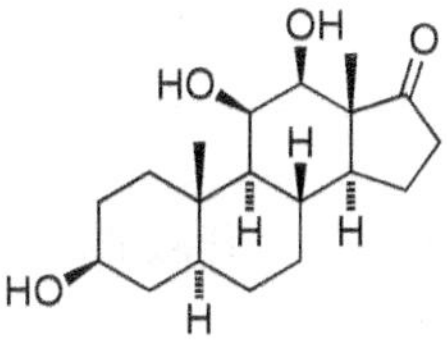

(j) 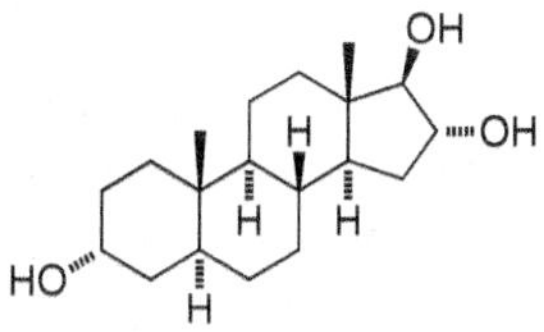

2.4 **Conformation**

The field of conformational analysis was founded by British chemist D. H. R. (later Sir Derek) Barton[45] and Norwegian chemist Odd Hassell[46] who shared the Nobel Prize in 1969 for the development. Conformations of open-chain saturated compounds are interconverted by rotation about the σ bonds. The rotation is not free, but every σ bond has a barrier to rotation that increases with both the number and size of groups at the two ends of the bond, except in extreme circumstances. All σ bonds pass through two extreme conformations during rotation: the eclipsed conformation and the staggered conformation. For σ

45. **Sir Derek Harold Richard Barton (1918-1997)** was educated at Imperial College (PhD, 1942). His career spanned more than four decades on two continents, beginning at Imperial College in London as an assistant lecturer in 1945. He shared the Nobel Prize in 1969, and was knighted in 1972. In 1978, Barton moved to the Institut pour la Récherche des Substances Naturelles in Gif-sur-Yvette, outside Paris. He moved to Texas A&M University in 1986. For more information, see: Ley, S.V.; Myers, R.M. *Biogr. Mem. Fellows Roy. Soc.* **2002**, *48*, 1, and the Nobel Foundation web site

46. **Odd Hassell (1897-1981)** was educated in Oslo, Munich and Berlin. He returned to the University of Oslo in 1925, where he spent the remainder of his professional career. From 1943 to 1945, he was imprisoned by the Nazis. He shared the 1969 Nobel Prize in Chemistry. For more biographical details, see the Nobel Foundation web site.

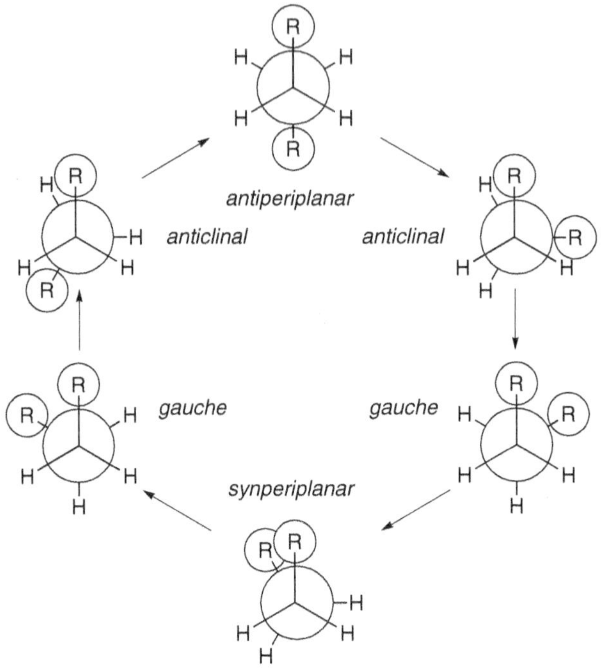

Figure 2.13 Conformations about nonterminal C–C bonds in straight-chain alkanes

bonds in an alkane, every staggered conformation is lower in energy than any eclipsed conformation.

Bonds in the middle of a chain (e.g., the 2,3 bond of butane) have conformations available to them beyond the simple eclipsed and staggered conformations.

When we examine the internal C—C bonds of *n*-alkanes, we can see that there are six different extreme conformations, three eclipsed, and three staggered, that can be described in terms of the dihedral angles (θ) between the methylene groups. In the eclipsed conformations, the dihedral angles in degrees fit the equation $\theta = 120n$, where n is an integer, and in the staggered conformations, the dihedral angles fit the equation $\theta = 120n + 60$. The conformation where $\theta = 0°$ is known as the ***synperiplanar*** (***sp***, also called the ***cis*** or ***syn***) conformation, and those where $\theta = 120°$ or $240°$ are known as ***anticlinal*** conformations. The conformation where $\theta = 180°$ is known as the ***anti*** (***a***) conformation, and those where $\theta = 60°$ or $300°$ are known as ***gauche*** conformations. In 1974, IUPAC recommended the names ***synclinal*** for the *gauche* conformation, and ***antiperiplanar*** for the *anti* conformation; neither has yet come into widespread use. These conformations are shown in Figure 2.13.

The conformational energy of a series of saturated open-chain compounds is summarized in Figure 2.14.[47] Calculations show that the barrier to rotation in the series methanol (one eclipsing H-H interaction in the highest energy conformation), methylamine (two H-H interactions), and ethane (three H-H interactions) increases by approximately 0.9 kcal mol^{-1} per H-H eclipsing interaction. In alkanes, the barrier to rotation has a minimum value in ethane and increases as the number of alkyl groups eclipsed increases to a maximum in neopentane. The barrier to rotation increases by approximately 0.6 kcal mol^{-1} as one moves from ethane to propane to isobutane to neopentane, which corresponds to a CH_3-H eclipsing energy of approximately 0.6 kcal mol^{-1} greater than the H-H eclipsing energy.

The *anti* conformation is the lowest energy conformation of the butane molecule. The *synperiplanar* conformation is the highest energy conformation—some 20 kJ

47. All calculations in this section were performed by the author at the HF/6-31G* level of theory, using Spartan 08 for the Mac.

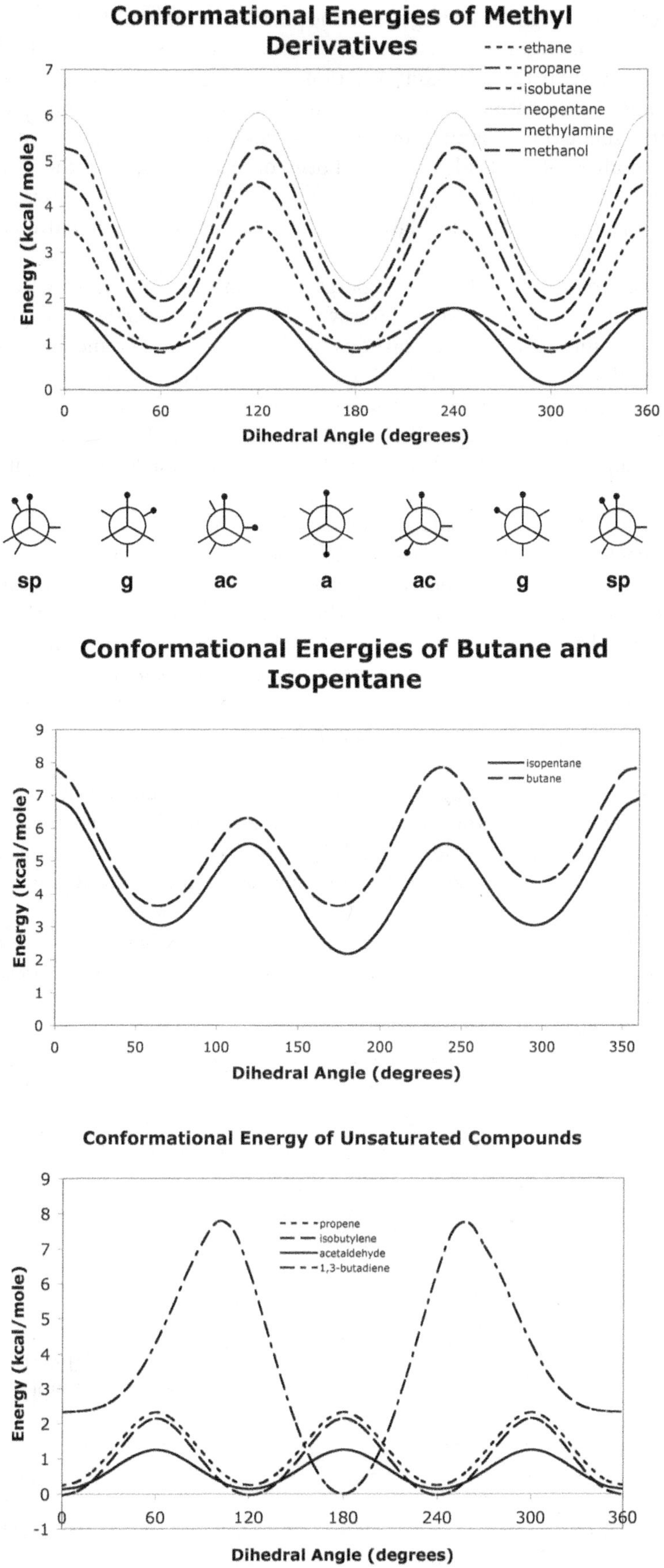

Figure 2.14 Conformational energy of open-chain compounds

mol^{-1} (4.8 kcal mol^{-1}) higher in energy than the *anti* conformation. The other eclipsed conformation, the *anticlinal* conformation, is approximately 14 kJ mol^{-1} (3.0 kcal mol^{-1}) higher in energy than the *anti* conformation. The energy difference between the *gauche* and *anti* conformations of butane is only 3.6 kJ mol^{-1} (0.87 kcal mol^{-1}), and this is sufficiently small that a sample of butane at room temperature should contain significant numbers of molecules close to both conformations, with those near the *anti* conformation predominating by a factor of slightly more than five. Neither eclipsed conformation contributes significantly to the conformer population at room temperature.

In saturated compounds, the lowest energy conformations are all staggered, as the top two diagrams in Figure 2.14 show. However, the angular dependence of conformational energy in unsaturated hydrocarbons (and carbonyl compounds) is markedly different from that of these saturated systems. In the case of all nonconjugated alkenes, as well as the structurally similar carbonyl compounds, the lowest energy conformation is, in fact, one in which the allylic or α-hydrogen eclipses the double bond. In conjugated compounds, on the other hand, the *anti* conformation once again has the lowest energy.

Although they have fewer conformations available than their open-chain counterparts, cyclic compounds still have some flexibility, although the levels of flexibility vary. For example, the ***chair*** conformation of the six-membered ring, is much less conformationally flexible that similar five-membered rings. In the chair conformation, each ring atom has one bond parallel to the ring's threefold axis of symmetry (the ***axial*** bonds) and one bond disposed generally in the plane of the ring (the ***equatorial*** bonds). The alternative relatively low-energy conformation of the six-membered ring is the skew-boat, or skew conformation. This conformation, which is flexible, is some 4 to 5 kcal mol^{-1} higher energy than the chair.

The cyclohexane chair can undergo conformational inversion, known as "ring flipping," through a high-energy transition state known as a half-chair (Figure 2.15). In the process of this inversion, all equatorial bonds become axial, and all axial bonds become equatorial. Thus, in simple monosubstituted cyclohexanes, the substituent may occupy both an axial and an equatorial position, with the two in equilibrium. Equatorial substituents are lower energy than axial substituents, so a substituent will occupy the equatorial position preferentially; the energy difference between the two chair conformers is known

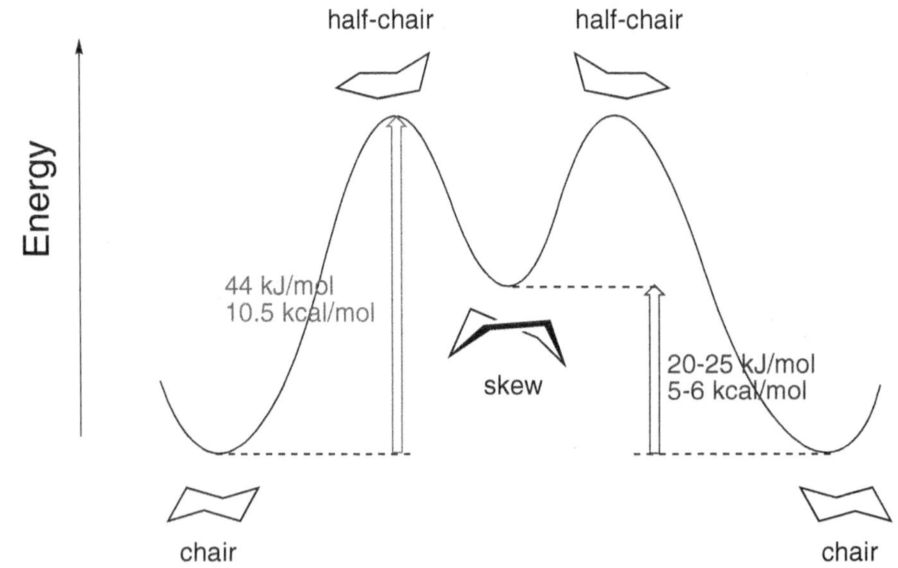

Figure 2.15 Conformations of the cyclohexane chair

as the ***conformational preference*** of the substituent. Values for a number of common substituents are gathered in Table 2.2.

In 1980, Corey and Feiner[48] developed a program for analyzing the conformation of polysubstituted cyclohexanes. They defined a destabilization energy, E_D, which is composed of three components: an A_R value for R–H 1,3-diaxial interactions, a U_R value for R–R' 1,3-diaxial interactions, and a G_R value for R–R' 1,2-diequatorial gauche interactions. Values for a series of common substituent fragments are collected in Table 2.3.

Table 2.2 Conformational Preference of Cyclohexane Substituents

X	Conformational Preference (kcal mol^{-1})	X	Conformational Preference (kcal mol^{-1})
CH_3	1.7	OH	0.7
C_2H_5	1.8	OCH_3	0.7
$CHMe_2$	2.1	Cl	0.4
CMe_3	5	Br	0.5
CH_2CMe_3	2.0	I	0.4
C_6H_5	3.1	NH_2	1.8
Cyclohexyl	2.15	NMe_2	2.1
COOH	1.2	CN	0.2

Problem

2-16 Calculate the destabilization energy of the two conformations of 1,1,2-trimethylcyclohexane given below using the Corey-Feiner constants. The measured conformational preference in this compound differs from that of methylcyclohexane by approximately 0.2 kcal mol^{-1}. How does this compare with the predicted value?

2.5 An Introduction to Asymmetric Synthesis: Asymmetric Induction

Table 2.3 A_R, U_R and G_R Values for Cyclohexane Derivatives (kcal mol^{-1})

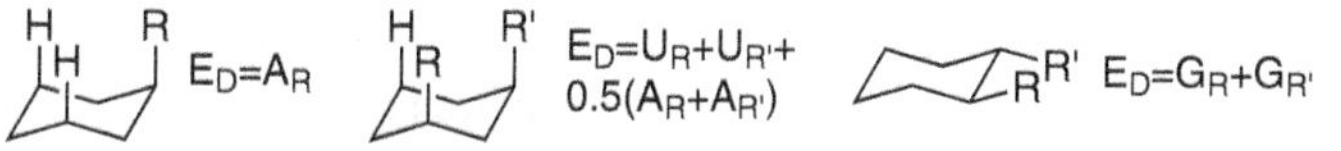

R or R'	A_R	U_R	G_R	R or R'	A_R	U_R	G_R
F	0.2	0	0	Aryl	3.0	1.1	1.2
Cl	0.4	0.4	0.5	C≡	0.2	1.2	0
Br	0.4	0.4	0.8	CO_2H, CO_2R	1.2	1.2	0.5
I	0.4	0.4	1.0	CHO	0.8	0.8	0.3
OH, OR	0.9	0.9	0.2	CH_3, CH_2R	1.8	1.8	0.4
NHR	1.3	1.3	0.3	CHR_2	2.1	2.1	0.8
NR_2	2.1	2.1	0.5	CR_3	6.0	6.0	2.5
C=	1.3	0.9	0.2				

Creating New Chiral Centers in a Reaction

An achiral reagent may add to either face of a prochiral center to generate a product with a new chiral center. An example of such a reaction is the reaction between the prochiral 2-butyl cation, and chloride ion (Figure 2.16). The two possible activated complexes for this reaction are enantiomers; outside a chiral environment, the energy of enantiomers does not depend on their absolute configurations, so they must be of equal energy. Therefore, the reaction leading to the *R* enantiomer of the product must proceed at exactly the same rate as the reaction leading to the *S* enantiomer, and an equal number of molecules of the *R* and *S* enantiomers will be formed. This is an important observation.

If the reaction between a prochiral compound and an achiral reagent generates a new chiral center, the product will be obtained as the racemate.

48. Corey, E.J.; Feiner, N.F. *J. Org. Chem.* **1980**, *45*, 765.

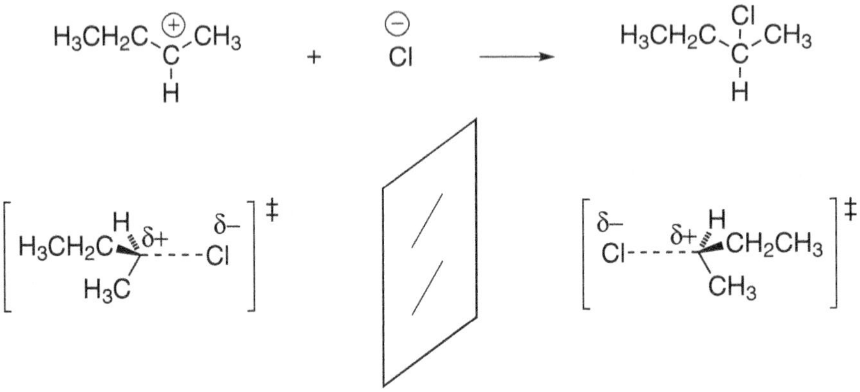

Figure 2.16 Generating a new chiral center from an achiral precursor

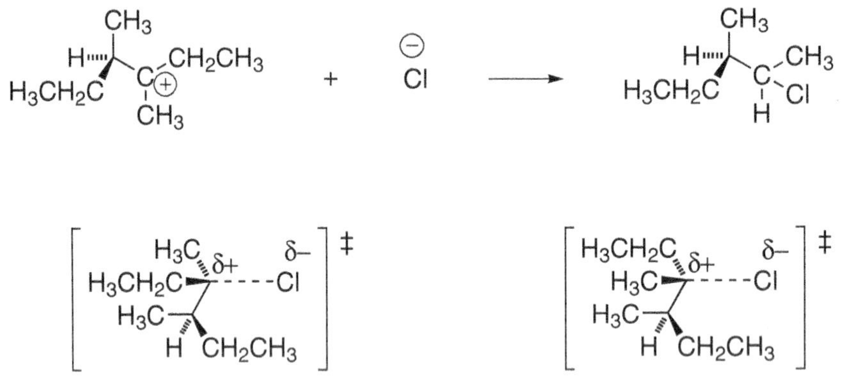

Figure 2.17 Generating a new chiral center in the presence of another

In other words, *one cannot form an optically active product from the reaction of two achiral starting compounds alone*, but one will always obtain the racemic mixture.

If there is a chiral center already present in the starting molecule, however, the situation is different because one is generating the new chiral center in the chiral environment generated by the presence of the first chiral center. For example, if one carries out the same kind of reaction as in the example above but now uses a chiral cation as the electrophilic reagent, the two products formed are no longer enantiomers but diastereoisomers. In these circumstances, the activated complexes for the competing reactions are diastereoisomers (Figure 2.17); they are not mirror images, they have different enthalpies, and the activation energies of the competing reactions are not equal. Consequently, the two products will be formed at different rates, and they will be formed in unequal amounts. In practically all reactions of this type, the two products are formed in different amounts; one isomer almost always predominates—sometimes to the almost complete exclusion of the other.

If a reaction of a chiral compound leads to the generation of a new chiral center, the product will almost always be an unequal mixture of the possible diastereoisomeric products.

The application of these two principles to the direct synthesis of enantiomerically enriched chiral compounds has led to the field of ***asymmetric synthesis***.[49] From its

49. The topic of asymmetric synthesis is the subject of several multivolume monographs, books, and reviews: (a) Morrison, J.D., Ed. *Asymmetric Synthesis*, 5 vols. (Academic Press: New York, 1983-1985). (b) Morrison, J.D.; Mosher, H.S. *Asymmetric Organic Reactions* (Prentice-Hall: New York, 1971). (c) Eliel, E.L.; Otsuka, S. *Asymmetric Reactions and Processes in Chemistry, ACS Symposium Series 185* (American Chemical Society: Washington, D.C., 1982). (d) Aitken, R.A.; Kilényi, S.N., Eds. *Asymmetric Synthesis* (Blackie Academic & Professional: Glasgow, 1992). (e) Gawley, R.E.; Aubé, J. *Principles of Asymmetric Synthesis* (Pergamon:: Oxford, 1996). (f) Procter, G. *Asymmetric Synthesis* (Oxford University Press: Oxford, 1996). (g) Ward, R.S. *Chem. Soc. Rev.* **1990**, *19*, 1. (h) Whitesell, J.K. *Chem. Rev.* **1989**, *89*, 1581. (i) Oppolzer, W. *Tetrahedron* **1987**, *43*, 1969. (j) Quinkert, G.;

beginnings in the last three decades of the 20th century, asymmetric synthesis has now grown to be a major theme of organic synthesis, and a huge amount of effort has gone into developing methods for making enantiomerically enriched chiral compounds. This, in turn, briefly led to a whole new vocabulary of stereochemistry, although many of the terms are now no longer in use, with a gradual reversion to the earlier terminology.

Asymmetric synthesis is the synthesis of a single enantiomer of a chiral product from achiral precursors. A reaction where an achiral substrate is converted into an unequal mixture or enantiomeric products is an ***enantioselective*** reaction, and the process involved is known as ***asymmetric induction***. The synthesis of a new chiral center with high levels of asymmetric induction requires that it be formed in a chiral environment so that the two transition states are not enantiomeric but diastereomeric. Two main strategies have been adopted by organic chemists for accomplishing this: covalent modification of one of the achiral substrates (the reactant or the reagent) before the formation of the new chiral center and carrying out the reaction in a chiral environment without covalently modifying either substrate (using a chiral catalyst).

What factors affect the addition of nucleophiles to a chiral aldehyde or ketone may be among the most intensely studied questions in organic chemistry, with two major models for predicting the stereochemical outcome emerging during the last half of the 20th century. These major models take into account the influences of steric and electronic factors on the stereochemical course of the addition reaction, with the balance between the two determining, to a degree, the prediction.

The first empirical rule for predicting the stereochemistry of addition reactions to carbonyl groups was proposed by American chemist, Donald J. Cram in 1952.[50] ***Cram's rule*** (Figure 2.18) assigns the control of asymmetric induction in the addition of nucleophiles to chiral aldehydes and ketones to steric factors. The model is based on a reactive conformation in which the carbonyl oxygen is oriented *anti* to the largest group, which eclipses the hydrogen in aldehydes, and the alkyl group in ketones. The complexation of the metal of an organometallic reagent to the carbonyl oxygen was expected to

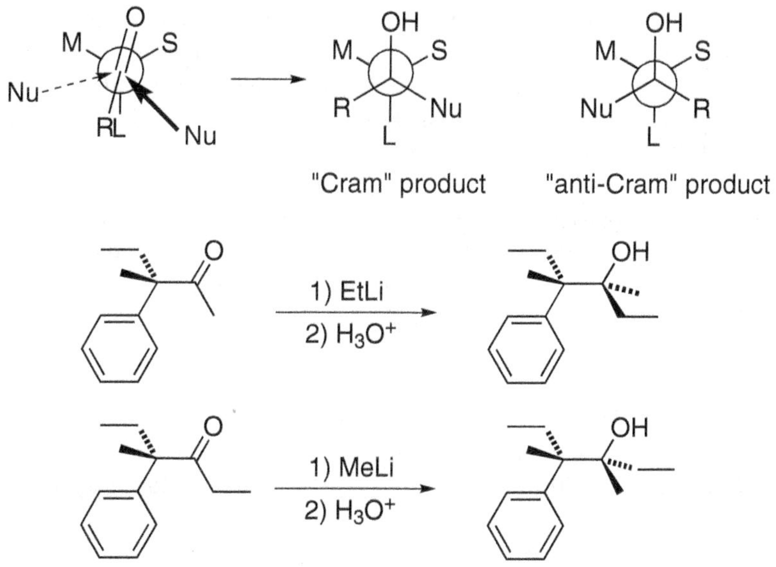

Figure 2.18 The Cram rule for addition to chiral ketones

Stark, H. *Angew. Chem.Int. Ed. Engl.* **1983**, *22*, 637. (k) Mosher, H.S.; Morrison, J.D. *Science* **1983**, *221*, 1013. (l) Tramontini, M. *Synthesis* **1982**, 605. (m) Kagan, H.; Fiaud, J.C. *Top. Stereochem.* **1978**, *10*, 175.

50. (a) Cram, D.J.; Abd Elhafez, F.A. *J. Am. Chem. Soc.* **1952**, *74*, 5828. (b) Cram, D.J.; Knight, J.D. *J. Am. Chem. Soc.* **1952**, *74*, 5835. (c) Curtin, D.Y.; Harris, E.E.; Meislich, E.K. *J. Am. Chem. Soc.* **1952**, *74*, 2901. (d) Leitereg, T.J.; Cram, D.J. *J. Am. Chem. Soc.* **1968**, *90*, 4019.

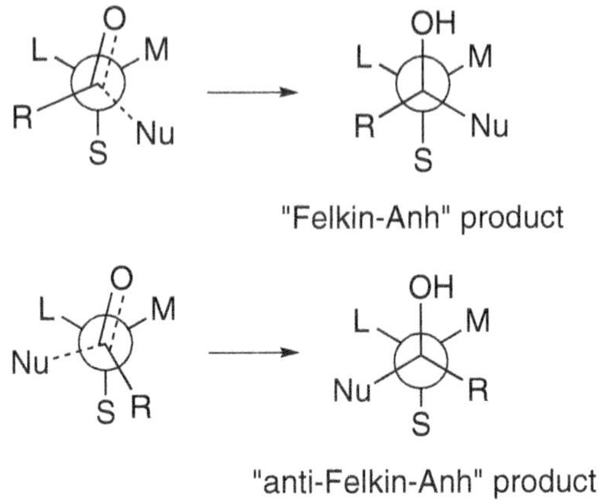

Figure 2.19 The Felkin-Anh rule for addition to chiral aldehydes

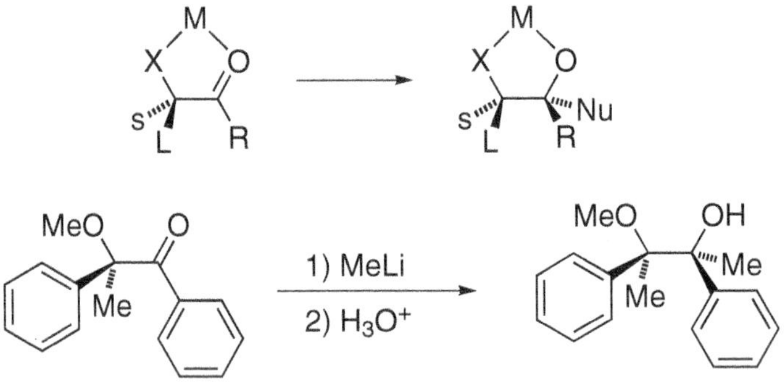

Figure 2.20 Chelation control in addition to chiral aldehydes

increase the effective size of the carbonyl oxygen and make this conformation most favorable. The major diastereoisomer of the product is then predicted to be that where the nucleophile adds to the carbonyl group from the same side as the small group at the α carbon.

A similar rule, known as the ***Felkin-Anh rule***, was proposed in 1968. In the Felkin-Anh model (Figure 2.19), the addition is presumed to occur through a reactant-like transition state where the conformation has the carbonyl oxygen between the medium and large groups, and, in carbonyl compounds with electronegative α substituents, as far from the substituent as possible. This model is actually a combination of two different models, one of which (the Felkin model[51]) has steric interactions predominating, and the other (the Anh model[52]) has stereoelectronic effects dominating.

When the α carbon carries a group capable of coordinating the metal of an organometallic reagent (Figure 2.20), the addition is proposed to occur by ***chelation control***,[53] where the metal is chelated by the carbonyl oxygen and the substituent, and the nucleophile again attacks from the same side as the small α substituent.

Chiral Auxiliaries

The first approach for obtaining asymmetric induction in the formation of a new chiral center that we will discuss involves modification of the reactant by a chiral molecule called a ***chiral auxiliary***.[54] The chiral auxiliary converts the previously achiral substrate

---

51. Chérest, M.; Felkin, H.; Prudent, N. *Tetrahedron Lett.* **1968**, 2199.
52. Anh, N.T.; Eisenstein, O. *Nouv. J. Chem.* **1977**, *1*, 61.
53. (a) Cram, D.J.; Kopecky, K.R. *J. Am. Chem. Soc.* **1959**, *81*, 2748. (b) Frye, S.V.; Eliel, E.L. *J. Am. Chem. Soc.* **1988**, *110*, 484. (c) Chen, X.; Hortelano, E.R.; Eliel, E.L.; Frye, S.V. *J. Am. Chem. Soc.* **1992**, *114*, 1778. (d) Reetz, M.T. *Acc. Chem. Res.* **1993**, *26*, 462.
54. Seyden-Penne, J. *Chiral Auxiliaries andLigands in Asymmetric Synthesis* (John Wiley & Sons: NewYork, 1995).

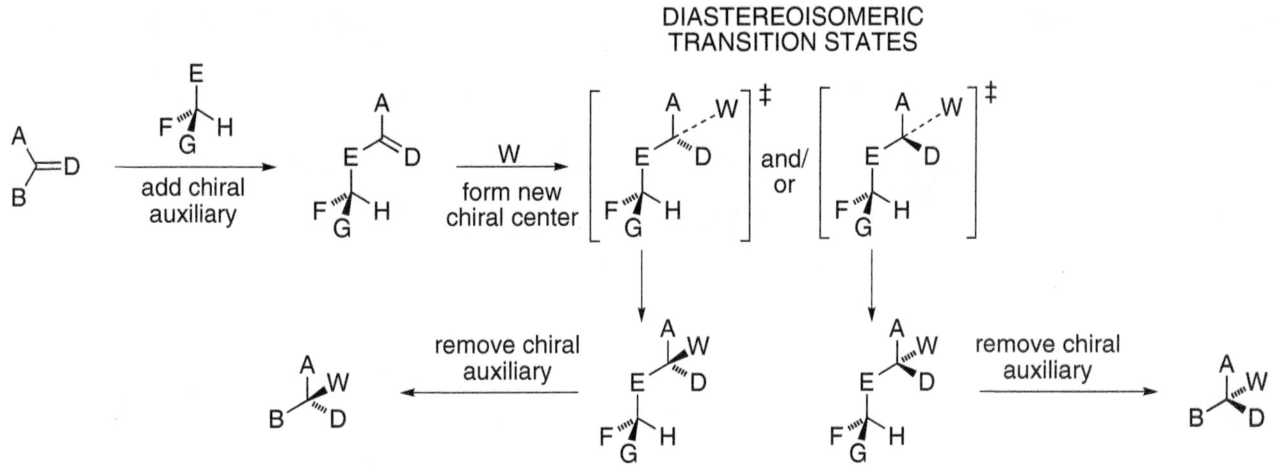

Figure 2.21 The chiral auxiliary approach to asymmetric induction

into a chiral compound, so that generating a new chiral center now involves the formation of diastereoisomers rather than enantiomers. Using the chiral auxiliary approach, organic chemists have been able to carry out important reactions with d.e.'s above 99.9% (d.r.'s > 1999:1) in favorable cases. After the new chiral center has been generated, the chiral auxiliary is then removed to leave the product, enriched in one enantiomer (Figure 2.21).

One of the oldest and most striking examples of asymmetric induction in a reaction involving the use of a chiral auxiliary is the Evans asymmetric aldol addition[55] between an aldehyde and an enol borinate, where the two new chiral centers generated during the course of the reaction are both generated with e.e.'s at or above 99.8% (e.r.'s ≥ 999:1), as illustrated below. This reaction, and its stereochemical nuances, will be discussed in much more detail later in this book, when we discuss the aldol addition reaction as a method for stereoselective synthesis.

Chiral Reagents

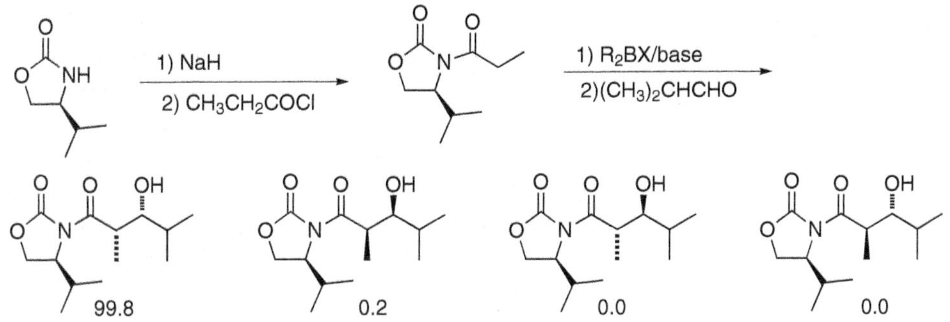

(2.29)

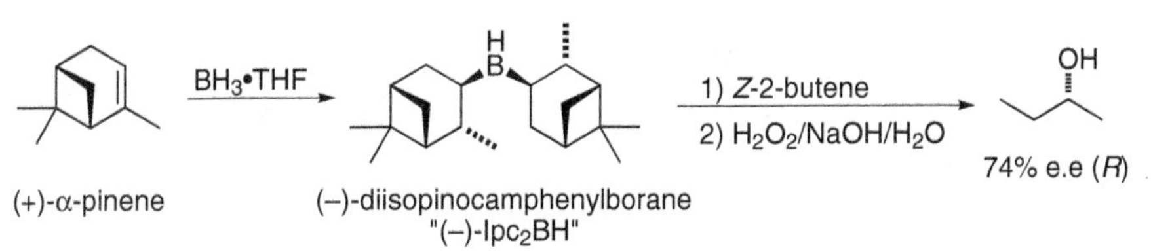

(2.29)

55. Evans, D.A.; Bartroli, J.; Shih, T.L. *J. Am. Chem. Soc.*, **1981**, *103*, 2127.

Figure 2.22 Chiral reagent approach to asymmetric induction

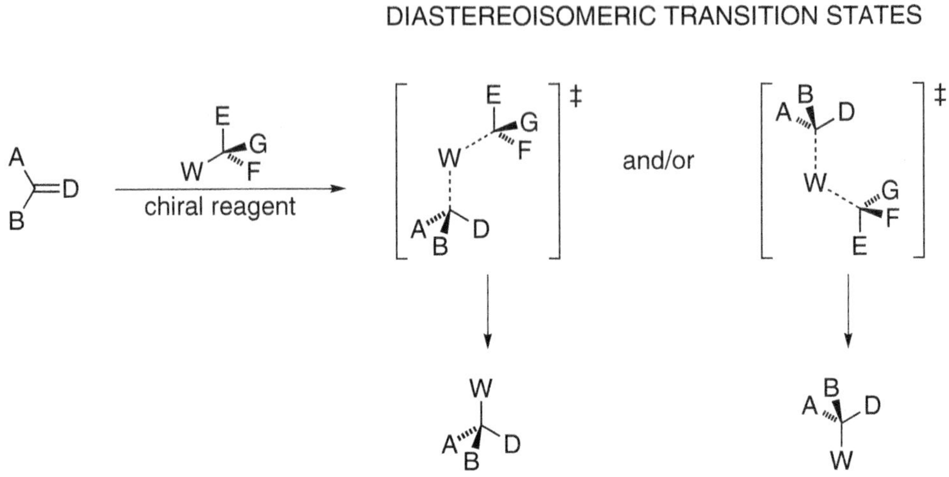

The same success in terms of asymmetric induction might have been anticipated by moving the chiral auxiliary to the reagent species rather than the reactant species (Figure 2.22). Here, however, the levels of asymmetric induction have been less spectacular than with the chirally modified substrates. This relatively lower level of success may be traced to the fact that when the chiral group is part of the reagent instead of the substrate, it tends to be further from the developing chiral center. This means that it exerts less influence over the geometry and transition state energy (and, therefore, the activation energy of the reaction).

To achieve high levels of asymmetric induction with a chiral reagent, it is important to use a reagent that will form the shortest possible bonds between the reacting species during the reaction. For this reason, many of the successful methods of using chiral reagents have been based to date on boron reagents. B–C, B–N, and B–O bonds are quite short, and this brings the two reacting species much closer together than with reagents based on other elements (e.g., lithium). For example, chiral borane reagents (diisopinocampheyl-borane, in particular) have been used successfully for the hydroboration of *cis* alkenes, as shown in Example **2.29**.[56]

Chiral Catalysts[57]

The second major strategy for constructing a new chiral center enriched in one enantiomer is to form it in a chiral environment that strongly favors one of the diastereoisomeric activated complexes. The simplest way to accomplish this without using at least one chiral reagent is by means of a chiral catalyst (Figure 2.23). This approach has been successfully applied by living systems to the construction of chiral molecules for countless millennia; it is used by every living cell to form chiral molecules from achiral precursors. In a living cell, the key to this is the enzyme—a catalyst that is formed from chiral α-amino acids. In almost every enzyme-catalyzed reaction, at least one of the reacting species is bound to the enzyme so that the new chiral center cannot be formed in one of the two enantiomeric forms.

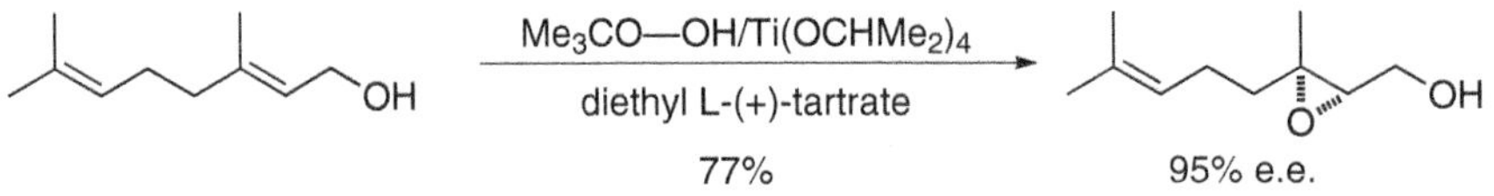

(2.30)

56. (a) Brown, H.C.; Zweifel, G.; *J. Am. Chem. Soc.* **1961**, *83*, 486. (b) Brown, H.C.; Yoon, N.M. *Isr. J. Chem.* **1977**, *15*, 12. (c) Brown, H.C.; Desai, M.C.; Jadhav, P.K. *J. Org. Chem.* **1982**, *47*, 5065. (d) Srebnik, M.; Ramachandran, P.V. *Aldrichim. Acta* **1987**, *20*, 9.

57. (a) Ojima, I., Ed. *Catalytic Asymmetric Synthesis*, 3nd. ed. (Wiley-VCH: New York, 2010) and earlier editions. (b) Noyori, R. *Asymmetric Catalysis in Organic Synthesis* (Wiley-Interscience: New York, 1994).

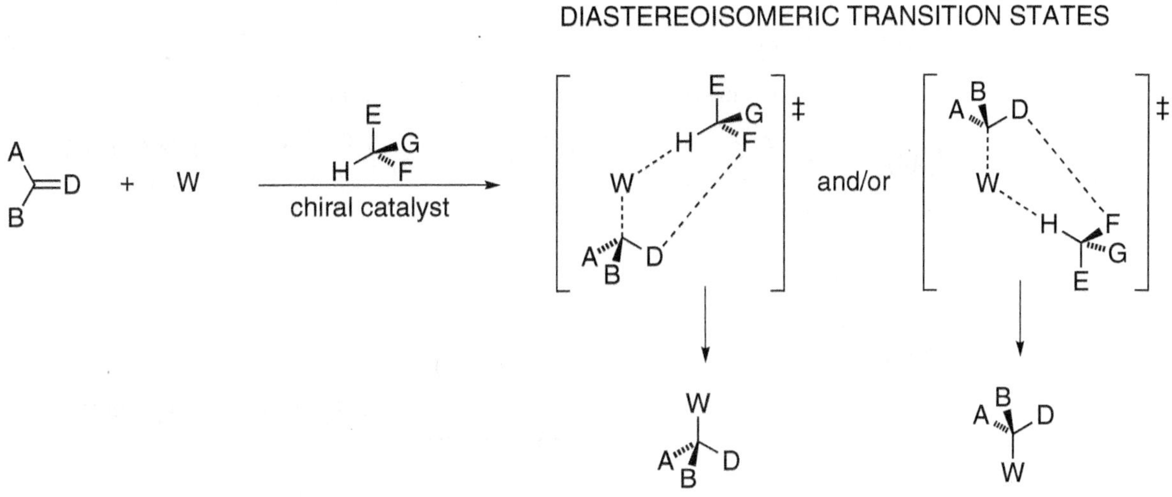

Figure 2.23 Chiral catalyst approach to asymmetric induction

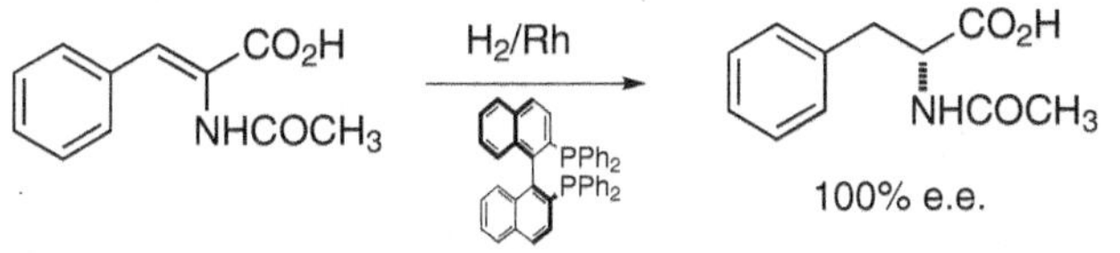

(2.31)

Until the 1980s, organic chemists had been less successful in adapting organic reactions to asymmetric synthesis by the use of chiral catalysts, but there were two spectacular successes: the ***Sharpless asymmetric epoxidation reaction***, in which an achiral allylic alcohol is oxidized to an epoxide,[58,59] and the emergence of ***asymmetric homogeneous hydrogenation*** catalysts[60] developed by Noyori and Knowles.[61] In many cases, the Sharpless asymmetric epoxidation gives e.e's of 95% or higher (e.r's ≥ 39:1), and in especially favorable cases its performance is even better. In similar vein, the asymmetric hydrogenation of α-aminocinammates is now the method of choice for the preparation of chiral phenylalanine derivatives. The development of chiral catalysts is now a major emphasis of synthetic organic chemistry.

---

58. Reviews: (a) Rossiter, B.E. In Morrison, J.D., Ed. *Asymmetric Synthesis, Vol. 5* (Academic Press: Orlando, 1985), 194. (b) Finn, M.G.; Sharpless, K.B. In Morrison, J.D., Ed. *Asymmetric Synthesis, Vol. 5* (Academic Press: c, 1982), 247. (c) Johnson, R.A.; Sharpless, K.B. In Ojima, I., Ed. *Catalytic Asymmetric Synthesis*, 2nd. ed. (Wiley-VCH: New York, 2000), p. 231.

59. **K. Barry Sharpless (1940-)** was educated at Dartmouth (BS, 1963) and Stanford (PhD, 1968). He joined the Massachusetts Institute of Technology (MIT) in 1970, moved to Stanford in 1977, and returned to MIT in 1980. In 1986 he moved to the Scripps Institute in La Jolla. Sharpless shared the 2001 Nobel Prize for his work. For a more detailed biography, see the Nobel Foundation web site

60. Reviews: (a) Okhuma, T.; Kitamara, M.; Noyori, R. In Ojima, I., Ed. *Catalytic Asymmetric Synthesis*, 2nd. ed. (Wiley-VCH: In Morrison, J.D., Ed. *Asymmetric Synthesis, Vol. 5* (Academic Press: New York, 1982), 247, 2000), p.1. (b) Halpern, J. In Morrison, J.D., Ed. *Asymmetric Synthesis, Vol. 5* (Academic Press: In Morrison, J.D., Ed. *Asymmetric Synthesis, Vol. 5* (Academic Press: New York, 1982), 247, 1982), 41. (c) Koenig, K.E. In Morrison, J.D., Ed. *Asymmetric Synthesis, Vol. 5* (Academic Press: New York, 1982), 71.

61. **Ryoji Noyori (1938-)** was educated at Kyoto University (D Eng, 1967). From 1963 to 1968, he was an instructor at Kyoto University. In 1968, he took the Chair in Organic Chemistry at Nagoya. He served as Dean of the Graduate School of Science at Nagoya from 1997 to 1999 and is currently Director of the Research center for Materials Science. Noyori shared of the 2001 Nobel Prize in Chemistry. For a more detailed biography, see the Nobel Foundation web site.

William Standish Knowles (1917-) was educated at Harvard (B.A. 1939) and Columbia (PhD, 1942). Knowles joined the Thomas and Hochwalt Laboratories, shortly after their acquisition by Monsanto. He spent his entire career at Monsanto, except for nine months with R. B. Woodward in 1959. He shared the Nobel Prize in Chemistry for 2001. For a more detailed biography, see the Nobel Foundation web site.

Worked Problem

2-2 Predict the major organic product of each of the following addition reactions.

(a) 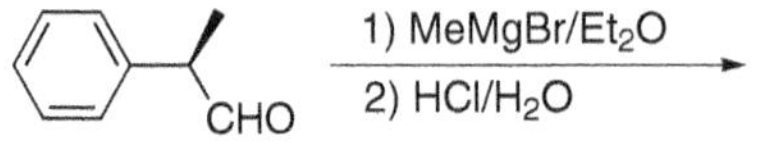

(b) 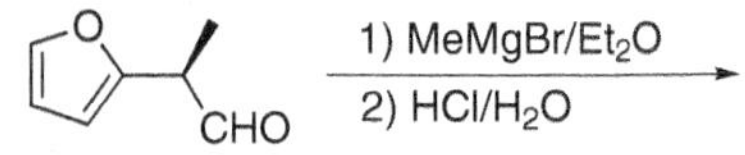

§Answer on below

Problem

2-17 What will be the major organic product of each of the following reactions?

(a) 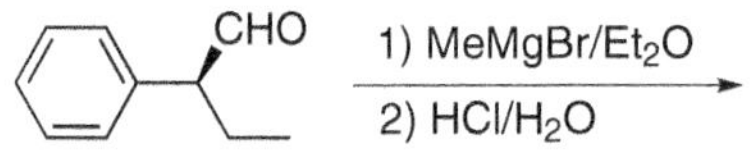

(b) 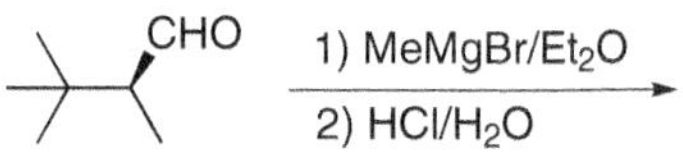

(c) 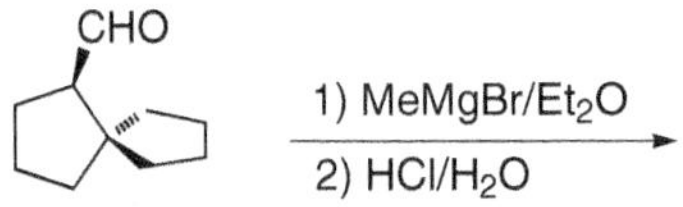

(d) 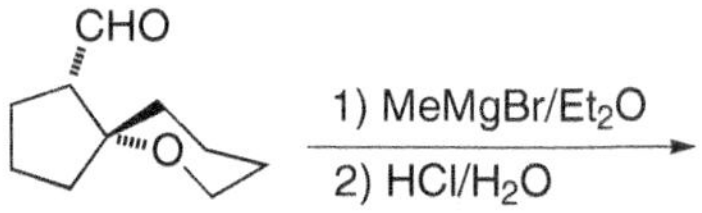

(e) 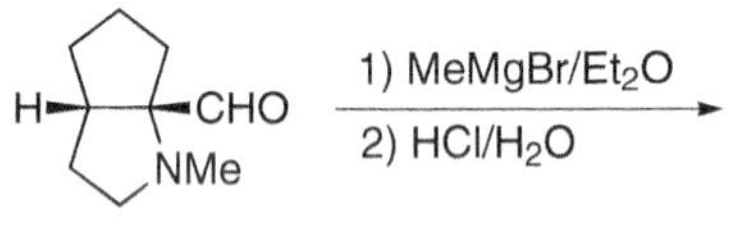

(f) 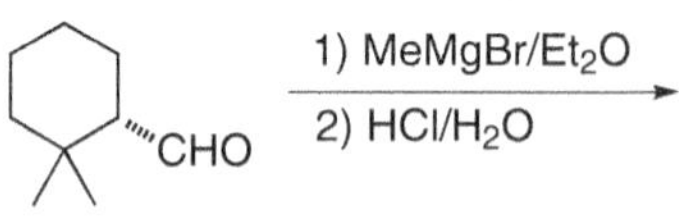

§ Answer for Worked Problem:

(a) This compound has a chiral center adjacent to the carbonyl group. The groups attached to that chiral center, in terms of increasing size, are H, Me, and C_6H_5 (Ph). The Cram and Felkin-Anh models for predicting the outcome the reaction are:

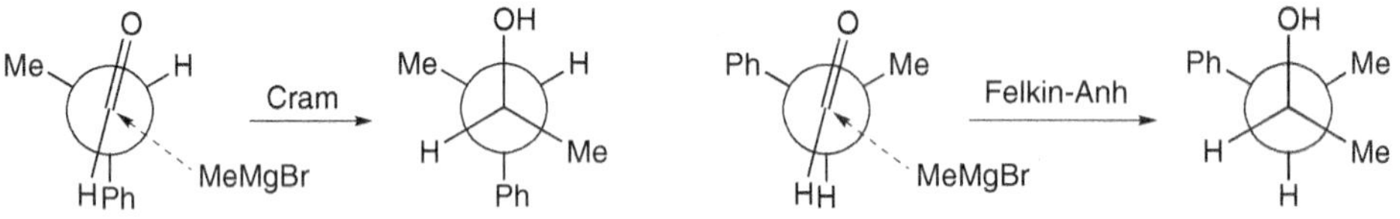

Both models predict that the major product of the reaction should be:

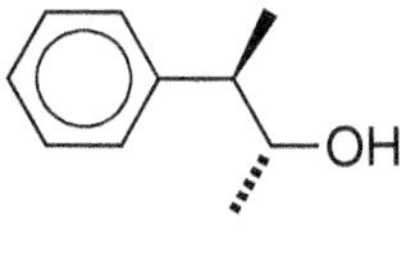

(b) This compound has a chiral center adjacent to the carbonyl group. The groups attached to that chiral center, in terms of increasing size, are H, Me, and furyl(C_4H_3O). The oxygen atom in the furan ring means that this reaction should be under chelation control:

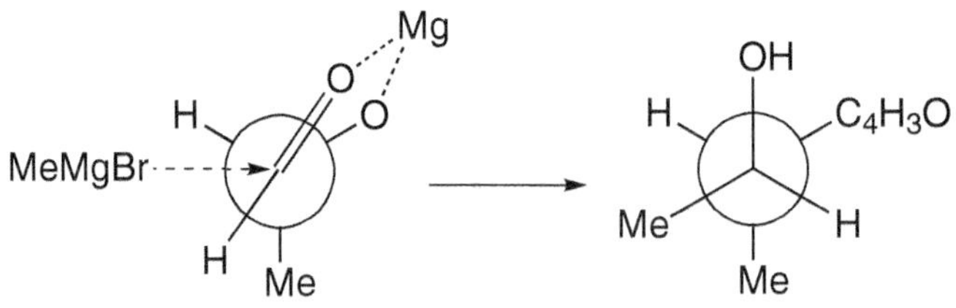

This makes the predicted major product of the reaction:

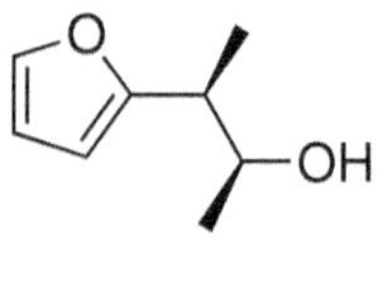

Using Conformational Analysis in Synthesis

Conformational analysis of reactive intermediates and activated complexes has become an integral part of analyzing an organic reaction, particularly when one is looking to control stereochemistry. This has been especially the case when the transition state for the reaction involves a six-membered cyclic structure, where one can use the conformational preferences of the substituent groups to help make predictions about preferred conformations and stereochemistry.

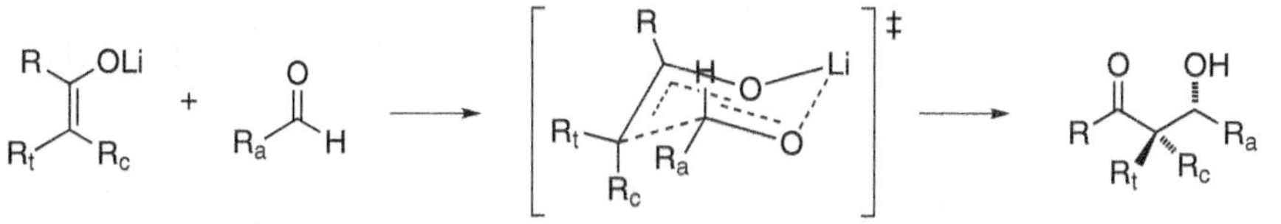

(2.32)

A good example of this type of analysis is provided by the aldol addition of preformed lithium enolates to aldehydes. The stereochemistry of this reaction is commonly analyzed in terms of the preferred conformation of the transition state. The first such model that has been widely used for this reaction is known as the ***Zimmerman-Traxler transition state*** (Example **2.32**),[62] which is a chairlike six-membered transition state. In the Zimmerman-Traxler transition state for the aldol addition of enolate anions to aldehydes, the largest groups are in the equatorial position.

In the Evans asymmetric aldol addition,[63] this type of analysis allows one to predict the absolute configuration of the major product. The enol borinate intermediate is formed as the *Z* isomer under the reaction conditions. What Evans and his coworkers found is that the absolute configuration of the chiral center adjacent to the nitrogen in the heterocyclic ring totally controls the face-selectivity of the reaction, rendering moot the differences leading to Cram selectivity. They rationalized this finding in terms of the Zimmerman-Traxler transition state model shown in Figure 2.24, where the conformation of the ring of the chiral auxiliary places the isopropyl group over the (smaller) oxygen atom.

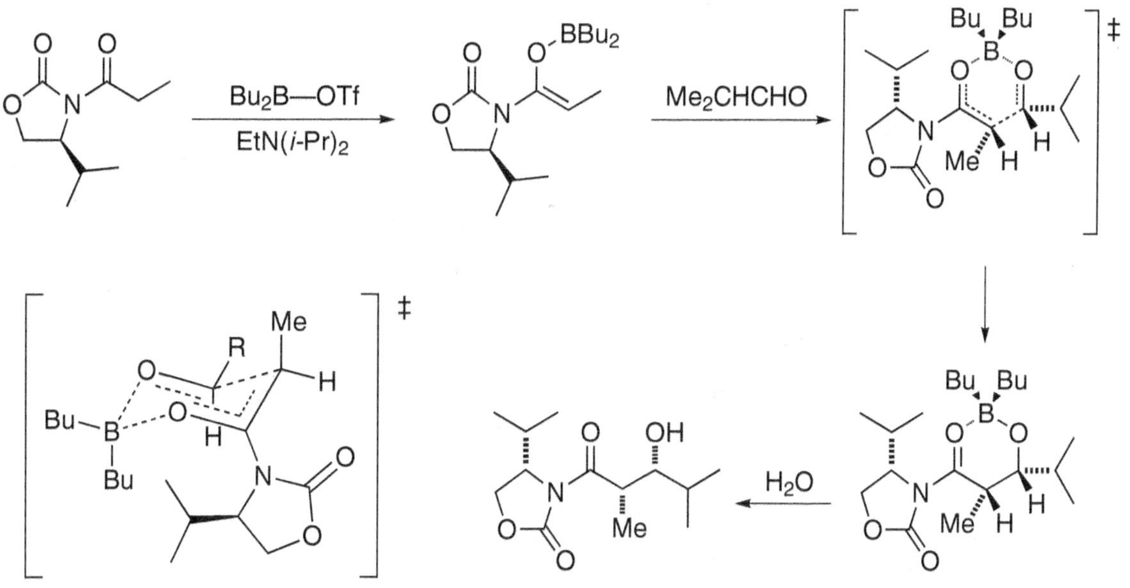

Figure 2.24 Predicting the stereochemistry of the Evans asymmetric aldol addition reaction using a Zimmerman-Traxler transition state model

62. Zimmerman, H.; Traxler, M. *J. Am. Chem. Soc.* **1957**, *79*, 1920.
63. Review: Evans, D.A. *Aldrichimica Acta* **1982**, *15*, 23.

One of the especially useful consequences of cyclic transition states having a preferred conformation that can be predicted on the basis of conformational analysis of similar stable species occurs in reactions leading to **chirality transfer**. In these reactions, a chiral center at one site in the reacting molecule is lost, while its stereochemistry is transferred to a prochiral center elsewhere in the molecule.[64]

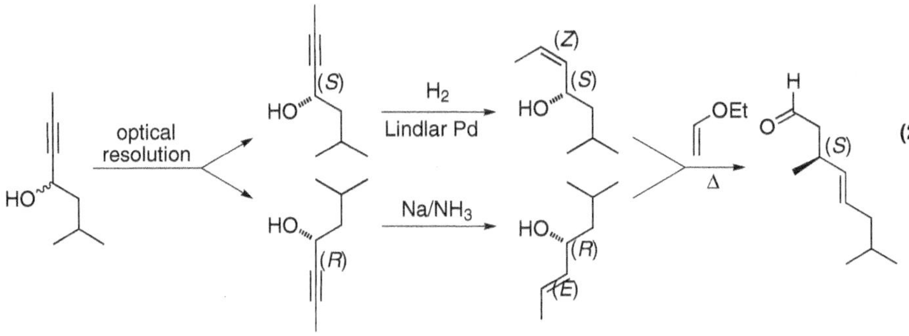

(2.33)

We will discuss this type of reaction in more detail when we discuss pericyclic reactions (e.g., the Cope and Claisen rearrangements), but a simple example of the strategy will serve here. In the synthesis of the side chain of α-tocopherol, Saucy and coworkers used an especially elegant method to obtain a single isomer of the product from the racemic starting alcohol by using the stereochemistry of the intermediate compounds.[65] In their procedure (Example **2.33**), the acetylenic alcohol was resolved, and the different enantiomers were reduced to give different geometric isomers of the double bond.

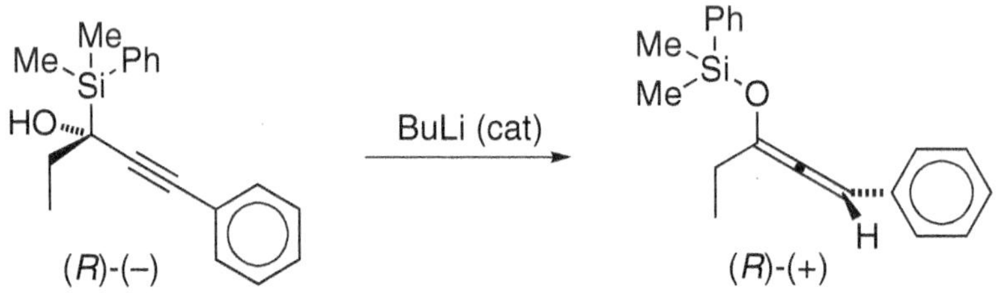

(2.34)

A cyclic transition state is not always necessary for good levels of chirality transfer, as is illustrated by the reaction shown as Example **2.34**.[66] In this reaction, the original chiral center is the carbinol carbon of the silyl alcohol. When this alcohol is treated with a catalytic (5 mol %) amount of strong base, the oxygen is deprotonated, and there is a migration of the silicon from carbon to oxygen. This is accompanied by transfer of a proton from another molecule of the alcohol, to give the chiral allene. This is, therefore, an example of a center of chirality in the reactant being converted into an axis of chirality in the product.

Worked Problem

2-3 The Claisen rearrangement shown in Example **2.33** predicts the stereochemistry of the reaction based on a chairlike transition state. How would he prediction change if the reaction were to proceed through a boatlike transition state?

§Answer on next page

64. Review: Hill, R.K. In Morrison, J.D., Ed. *Asymmetric Synthesis* (Academic: New York, 1984), vol. 3, ch. 8, p. 503.
65. Chan, K.K.; Cohen, N.; DeNoble, J.P.; Specian, A.C., Jr.; Saucy, G. *J. Org. Chem.* **1976**, *41*, 3497.
66. Reynolds, T.E.; Bharadwaj, A.R.; Scheidt, K.A. *J. Am. Chem. Soc.* **2006**, *128*, 15382.

Problems

2-18 Account for the appearance of the conformational energy profile of isopentane ($CH_2CH_2CHMe_2$), which is given in Figure 2.18.

2-19 *Cis*-1,4-di-*tert*-butylcyclohexane should have one axial and one equatorial tert-butyl group. Conformational analysis of the compound suggests that in the preferred conformation, it does not. How does this occur, and why?

2-20 The Cope rearrangement is similar to the Claisen rearrangement, with the exception that there is now one more alkene π bond, with its accompanying stereochemistry. Use the same analysis of stereochemistry and conformation used in Sample Problem 2-3 to predict the stereochemistry of the product in the Cope rearrangement, and the effects of changing the conformation of the transition from chairlike to boatlike.

2-21 Predict the major organic product expected in each of the following organic reactions:

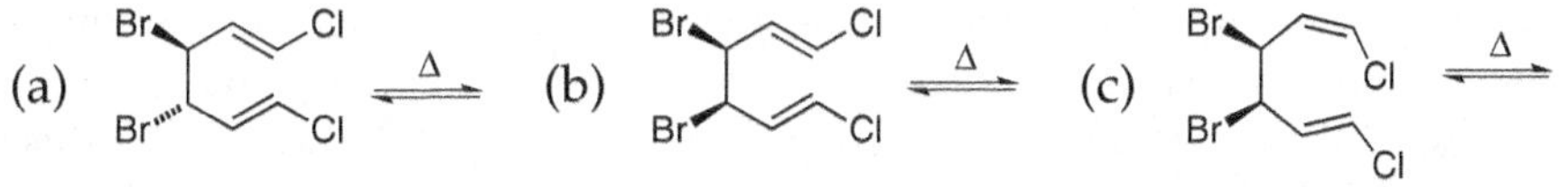

2.6 How Well Did It Go? Measuring Enantiomer Ratios

The final measure of the success of an asymmetric synthesis is the enantiomer ratio (or enantiomeric excess) of the products of the reaction. When the e.r. is small, one can use the optical rotation to see how well the reaction has worked. However, as one attains the kind of e.r. now expected in an asymmetric synthesis, this method is much less reliable

§ Answer for Worked Problem:
The reaction above proceeds through a chairlike transition state.

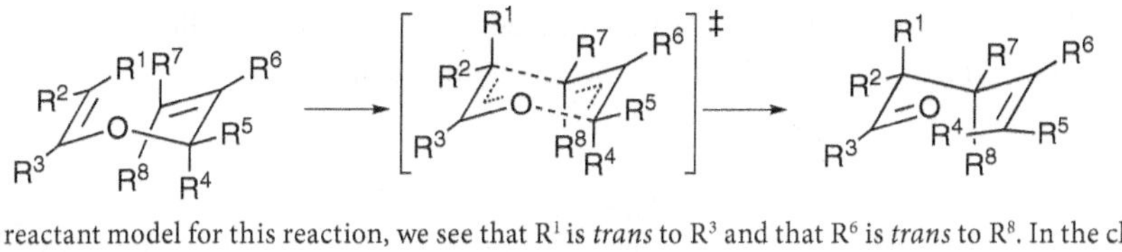

In the reactant model for this reaction, we see that R[1] is *trans* to R[3] and that R[6] is *trans* to R[8]. In the chairlike transition state model, we see that R[1] and R[8] adopt pseudoaxial positions in the chair, with the result that they become *anti* to each other in the product as drawn. In the model as drawn, R[4] is axial and R[5] is equatorial, which leads to the new double bond geometry with R[4] *trans* to R[6].

In the boatlike transition state, the two alkene π bonds have the same geometry as above. In the transition state model, the atoms whose hybridization remains unchanged at sp^2 are the most likely atoms to occupy the positions shown below, because this removes any flagpole-flagpole interactions.

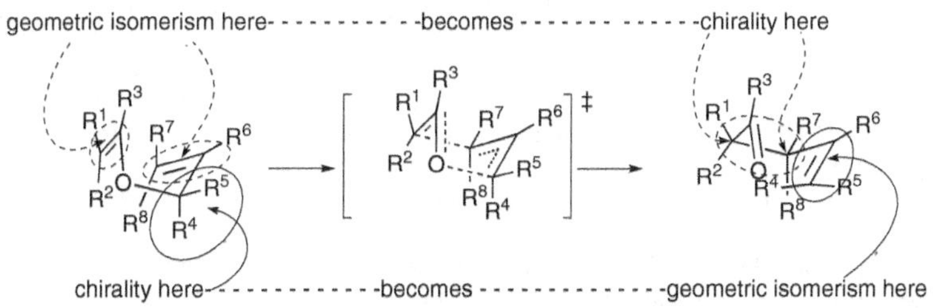

In this model, the conformation of the final product places R[2] *anti* to R[7], with R[4] *trans* to R[6]. This is the same stereochemistry as predicted by the chairlike transition state model.

unless the final product has a very large specific rotation. Under these conditions, other techniques must be used to give a measure of the e.r. of the product.

This has been accomplished by two major approaches: NMR spectroscopy, and chiral high-performance liquid chromatography (HPLC). Beginning in the late 1970s with the work of Pirkle,[67] chiral HPLC has become the method of choice for separating mixtures of enantiomers as both an analytical and a preparative technique. The basis for the separation is, as one might expect, the formation of diastereoisomeric species as the enantiomers interact with the chiral column material.

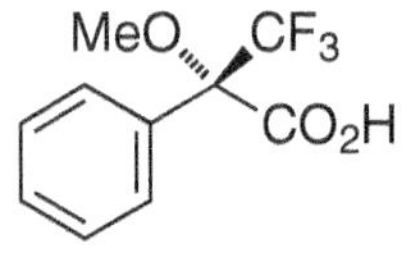

Mosher's acid

The other way to analyze mixtures of enantiomers involves the formation of a derivative of the mixture with a single enantiomer of a chiral derivatizing reagent and then observing the formation of diastereoisomeric products by looking at the NMR spectrum. One especially useful chiral derivatizing agent is known as Mosher's acid.[68] This reagent has the added advantage that it contains a trifluoromethyl group, which allows ^{19}F NMR spectroscopy to be used to measure the enantiomer ratio (a considerable simplification because the protons of the molecules do not interfere with the analysis).

2.7 **Chirality at Atoms Other Than Carbon: Inversion of Pyramidal Centers**

The heteroatoms of silanes, quaternary ammonium and phosphonium salts, and amine-*N*-oxides and phosphine oxides are tetrahedral; therefore, they are structurally analogous to the carbon-based chiral centers of the type that we have been discussing in this chapter. As such, these compounds should be resolvable into configurationally stable enantiomeric forms, and they are.[69] One particularly simple method for resolving quaternary ammonium salts is by means of their (*S*)-BINOL complexes (Example **2.35**) if these can be crystallized.[70] This involves simply partitioning the precipitated complex between dichloromethane and water separates the quaternary salt from the resolving agent.

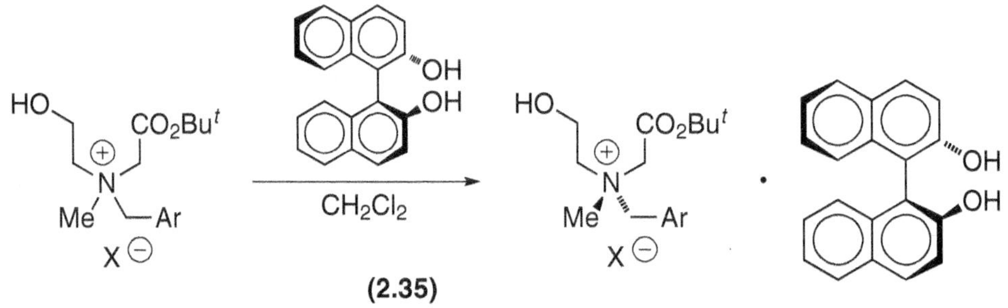

(2.35)

67. (a) Pirkle, W.H.; House, D.W. *J. Org. Chem.* **1979**, *44*, 1957. (b) Pirkle, W.H.; House, D.W. Finn, J.M. *J. Chromatogr.* **1980**, *192*, 143. (c) Pirkle, W.H.; Finn, J.M. *J. Org. Chem.* **1981**, *46*, 2935. (d) Pirkle, W.H.; Schreiner, J.L. *J. Org. Chem.* **1981**, *46*, 4988.

68. (a) Dale, J.A.; Dull, D.L.; Mosher, H.S. *J. Org. Chem.* **1969**, *34*, 2543. (b) Dale, J.A.; Mosher, H.S. *J. Am. Chem. Soc.* **1973**, *95*, 512.

69. (a) Barbachyn, M.R.; Johnson, C.R. In Morrison, J.D.; Scott, J.W., Eds. *Asymmetric Synthesis* (Academic: New York, 1984), Ch. 2, p. 227. (b) Valentine, D., Jr.; In Morrison, J.D.; Scott, J.W., Eds. *Asymmetric Synthesis* (Academic: New York, 1984), Ch. 3, p. 263. (c) Davis, F.A.; Jenkins, R.H., Jr. In Morrison, J.D.; Scott, J.W., Eds. *Asymmetric Synthesis* (Academic: New York, 1984), Ch. 4, p. 313. (d) Maryanoff, C.A.; Maryanoff, B.E.; In Morrison, J.D.; Scott, J.W., Eds. *Asymmetric Synthesis* (Academic: New York, 1984), Ch. 2, p. 355.

70. Tayama, E.; Otoyama, S.; Tanaka, H. *Tetrahedron: Asymmetry* **2009**, *20*, 2600.

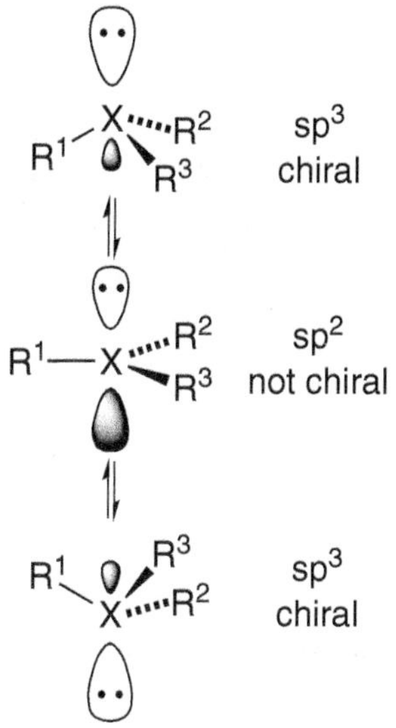

(2.36)

Like the heteroatoms in chiral ammonium and phosphonium salts, the heteroatoms in amines, phosphines, sulfoxides, and sulfonium salts may also be chiral. In these cases, the fourth group about the chiral center is the lone pair on the heteroatom. Potentially, therefore, such compounds ought to be resolvable into the two enantiomers. There is a mechanism for inversion of configuration available to these compounds that is not available to chiral centers lacking a lone pair. In the absence of lone pairs, the inversion of configuration of the chiral center requires the cleavage and reformation of a covalent bond. In the case of a heteroatom carrying a lone pair, however, the inversion may occur by means of a rehybridization of the heteroatom from sp^3 (chiral) to sp^2 (not chiral) and back again (Example **2.36**). This is the same hybridization change that occurs at carbon during the S_N2 reaction, but the process generally has a lower activation energy than the S_N2 process.

The barrier to inversion of open-chain amines is small[71] (as measured by dynamic NMR,[72] it is typically 5–7 kcal mol^{-1}). This makes this inversion an especially facile process, because a barrier to inversion of at least 23 kcal mol^{-1} is needed for enantiomers to be configurationally stable at room temperature.[73] Acyclic amines rapidly racemize at room temperature. The barrier to inversion of amine nitrogen atoms is raised substantially by (1) incorporating the nitrogen atom into a small ring (e.g., an aziridine, where attaining the sp^2 hybridization results in increasing the already large angle strain in the three-membered ring) and by (2) attaching an electronegative substituent to the nitrogen. Thus, although simple *N*-alkylaziridines themselves are not configurationally stable at room temperature, *N*-chloro-2,2-diphenylaziridine (Example **2.37**) has been prepared in optically active form.[74] Its barrier to inversion of 24.4 kcal mol^{-1} is modest, so it still racemizes completely in 4 days at 0°C. The effects of small ring size and electron-withdrawing substituents are even more pronounced in rings such as the oxaziridine ring system, which can fairly readily give stereochemically stable enantiomeric compounds.

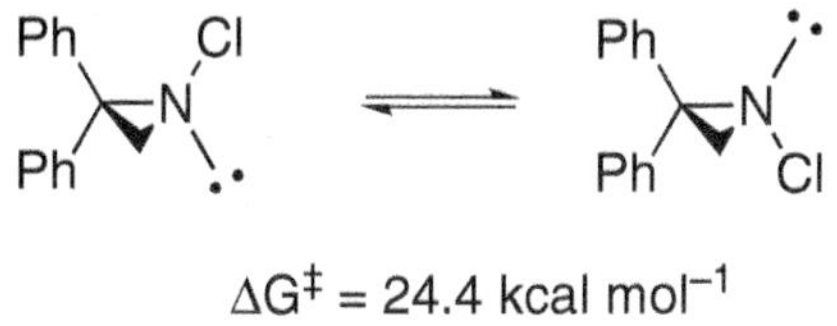

(2.37)

Unlike tricoordinate nitrogen, the barrier to inversion of tricoordinate sulfur and phosphorus is large enough that chiral sulfoxides and phosphines are configurationally stable at room temperature and may be resolved and used. The barriers to inversion in simple diaryl sulfoxides and aryl methyl sulfoxides are typically in the range of 35 to 42 kcal mol^{-1}, which means that the racemization of these compounds typically has a half-life of 6 hours at 200°C.[75] The activation parameters of this inversion are consistent with its being the "umbrella" inversion at sulfur, rather than some other mechanism. The corresponding barrier to inversion of phosphines is dependent on the substituents on phosphorus. For simple trialkylphosphines not bearing electronegative substituents or having the phosphorus in a small ring, the barrier to inversion is typically in the

71. Reviews: (a) Rauk, A.; Allen, L.C.; Mislow, K. *Angew. Chem. Int. Ed. Engl.* **1970**, *9*, 400. (b) Lambert, J.B. *Top. Stereochem.* **1971**, *6*, 19. (c) Jennings, W.B.; Boyd, D.R. In lambert, J.B.; Takechi, Y., Eds. *Cyclic Organonitrogen Stereodynamics* (VCH: Cambridge, 1992), 105.

72. (a) Stevenson, P.E.; Burkey, D.L. *J. Am. Chem. Soc.* **1974**, *96*, 3061. (b) Bushweller, C.H.; Anderson, W.G.; Stevenson, P.E.; Burkey, D.L.; O'Neil, J.W. *J. Am. Chem. Soc.* **1974**, *96*, 3892. (c) Bushweller, C.H.; Lourandos, M.Z.; Brunelle, J.A. *J. Am. Chem. Soc.* **1974**, *96*, 1591.

73. Kessler, H. *Angew. Chem. Int. Ed. Engl.* **1970**, *9*, 219.

74. Annunziata, R.; Fornasier, R.; Montanari, F. *J. Chem. Soc., Chem. Commun.* **1972**, 1133.

75. Rayner, D.R.; Gordon, A.J.; Mislow, K. *J. Am. Chem. Soc.* **1968**, *90*, 4854.

range of 29 to 36 kcal mol^{-1}.[76] The incorporation of an electronegative substituent on phosphorus raises the barriers to inversion significantly (as happens with the corresponding amines).[77]

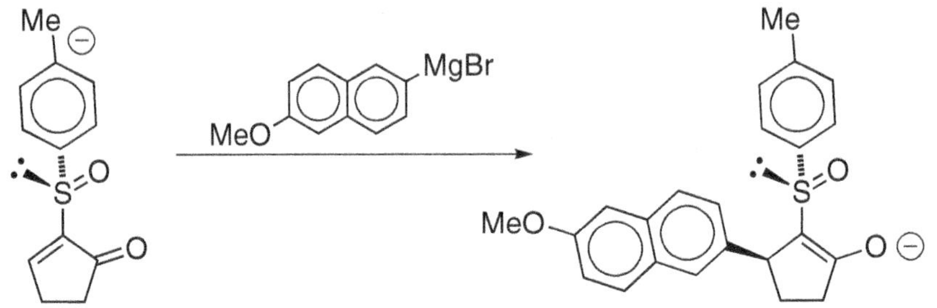

(2.38)

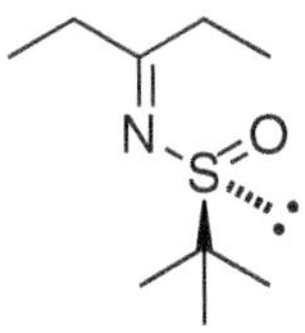

(2.39)

Chiral sulfoxides are readily available by the reaction between optically active menthyl *p*-toluenesulfinate with alkyllithium nucleophiles; the reaction occurs with inversion at sulfur. Indeed, chiral sulfoxides have been used as chiral directing groups in a variety of reactions, including conjugate addition (e.g., Example **2.38**)[78] and reduction.[79] In both these reactions, chelation by a metal ion involving the sulfoxide oxygen restricts the conformation of the reactant and thus facilitates chirality transfer from sulfur to carbon. Nitrogen bound to sulfur makes these sulfur derivatives especially configurationally stable. For this reason, chiral sulfinylamines such as the *tert*-butylsulfinylimine (Example **2.39**), which carries a chiral sulfur atom, have become widely used chiral auxiliaries in asymmetric synthesis.

Chapter Summary

This chapter has considered organic stereochemistry—the shape of molecules in three dimensions. The discussion has ranged from enantiomers (molecules that are nonsuperimposable mirror images) to diastereoisomers (molecules that are stereoisomers but not enantiomers), including considerations of geometric isomers and *meso* compounds. The symmetry requirements compatible with chirality (axes of symmetry) and those incompatible with chirality (mirror plane, inversion center, improper axis of rotation) have been discussed. The protocenter model for the basis of stereoisomerism has been introduced. The difference between configuration and conformation has been discussed, and the Cahn-Ingold-Prelog method for assigning configurations to chiral molecules and to geometric isomers has been introduced. The conformations of open-chain and cyclic molecules have been discussed. The measurement of optical purity and the determination of enantiomeric excess (e.e.) or enantiomer ratio (e.r.) have been discussed in the light of methods to achieve asymmetric synthesis. The three conceptually different methods for obtaining asymmetric induction in a reaction (chiral auxiliaries, chiral reagents, and chiral catalysts) have been introduced. The Cram and Felkin-Anh models for predicting the stereochemistry of additions to chiral aldehydes

76. Baechler, R.D.; Mislow, K. *J. Am. Chem. Soc.* **1970**, *92*, 3090.

77. Baechler, R.D.; Andose, J.D.; Stackhouse, J.; Mislow, K. *J. Am. Chem. Soc.* **1972**, *94*, 8060.

78. Posner, G.H. *Acc. Chem. Res.* **1987**, *20*, 72.

79. Bode, M.L.; Gates, P.J.; Gebretnsae, S.Y.; Vleggaar, R. *Tetrahedron* **2010**, *66*, 2026.

and ketones have been introduced. Chirality at atoms other than carbon has been briefly discussed, and the configurational stability of chiral compounds carrying a lone pair on the chiral atom has been briefly addressed. The concepts of topism and prochirality have been discussed.

Key Terms

absolute configuration
anomers
anti, gauche
asymm etric induction
atropisomers
axial, equatorial
axis of chirality
boat, chair
Cahn-Ingold-Prelog systems
center of chirality
center of inversion
center of pseudochirality
chiral auxiliary
chirality
circular dichroism
conformation
Cram's rule
diastereoisomers
eclipsed, staggered
conformational preference
enantiomeric excess
enantiomers
epimers
Felkin-Anh rule
geometric isomers
improper axis of rotation
meso isomers
mirror plane of symmetry
Newman projection
optical activity
optical rotatory dispersion
plane of chirality
prochiral center
proper axis of rotation
protocenters of stereochemistry
racemate
topism

Additional Problems

2-22 Draw the Newman projection of the most stable conformation about the indicated bond of each of the following compounds.

(a) *meso*-3,4-dimethylhexane (the 3–4 bond)
(b) (3*R*,4*R*)-3-chloro-4-methylheptane (the 3–4 bond)
(c) *S*-1,2-dicyclopentylpropane (the 1–2 bond)
(d) *meso*-2,3-dicyclohexylbutane (the 2-3 bond)
(e) (1*S*,2*S*)-1,2-dichlorocyclohexane (the 1-2 bond)
(f) *meso*-1,2-divinylcyclopropane (the 1–2 bond)

2-23 Each of the following reactions gives a product in which at least one new chiral center has been formed. The stereochemistry of the final product has not been given explicitly in any of these reactions, but it can be deduced from the reactants and the reagents used. Draw all the products in each of the reactions (including pairs of enantiomers, if appropriate). Are the products formed in equal amounts? Draw the lowest energy and highest energy conformations about the bond to the new chiral center in each product (use the Newman projection for acyclic compounds).

(a) 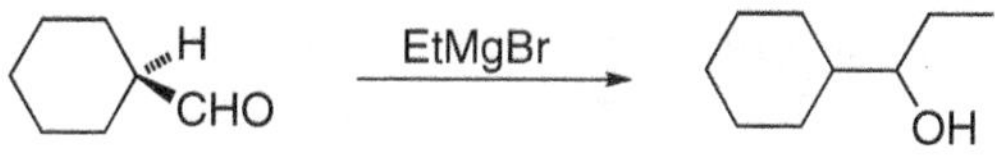

(b) 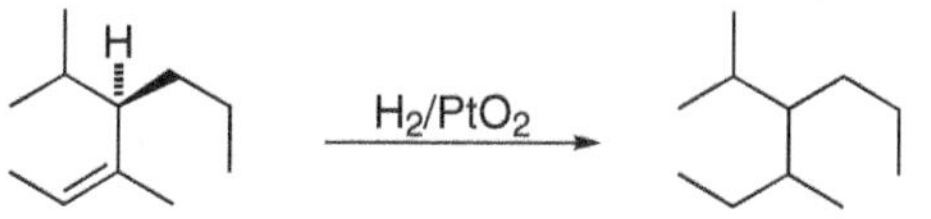

(c) 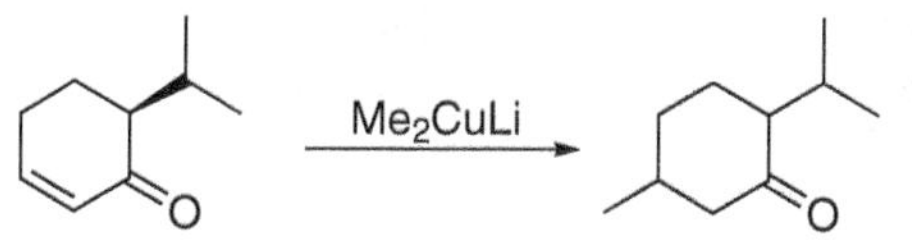

(d) 

(e) 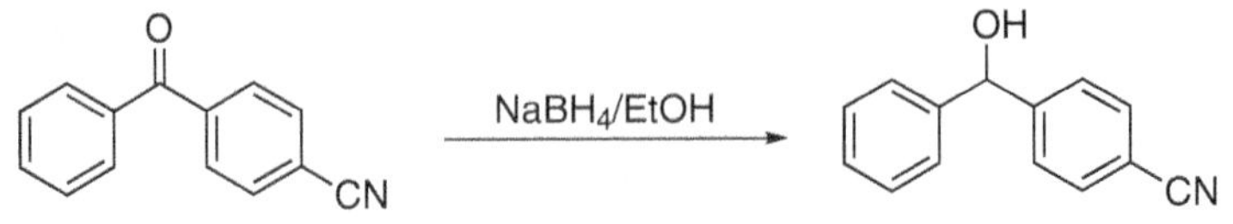

(f) 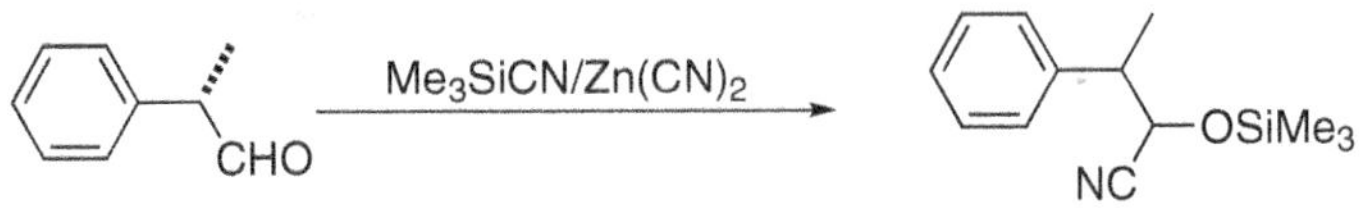

2-24 What is the major product of each of the following reactions?

(a) 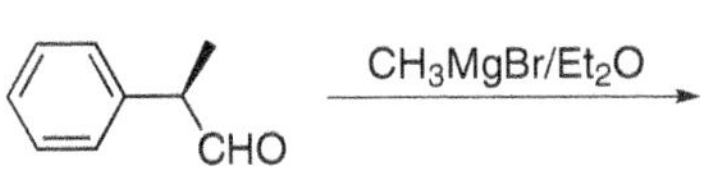

(b) 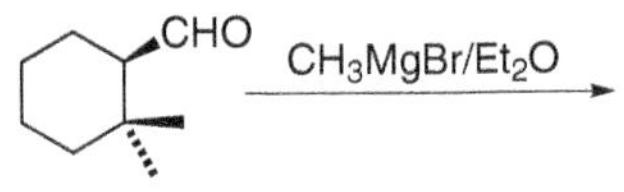

(c) 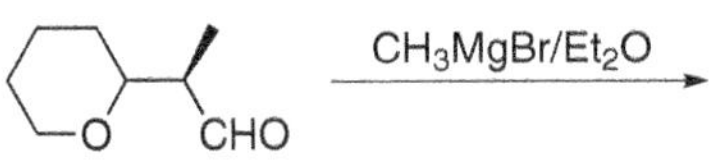

(d) 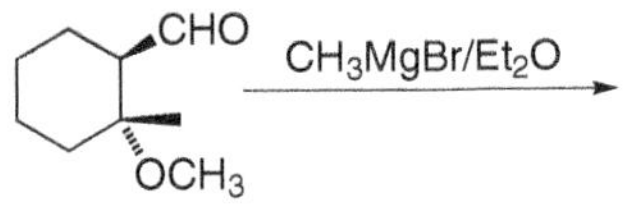

(e) 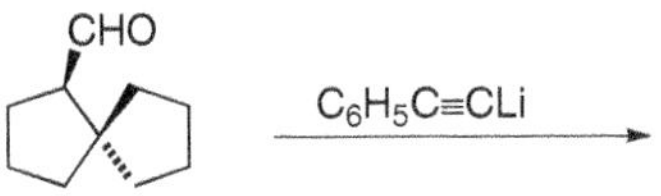

(f) 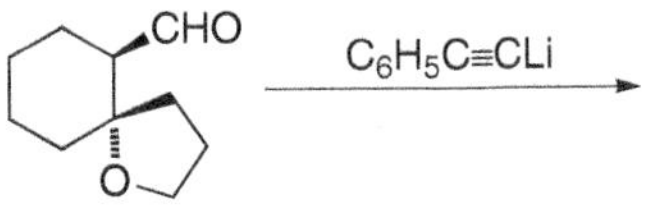

(g) 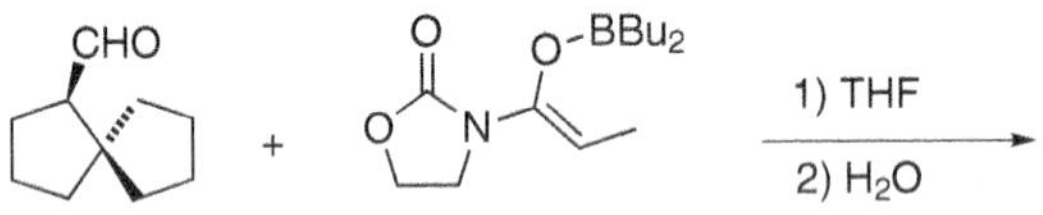

(h) 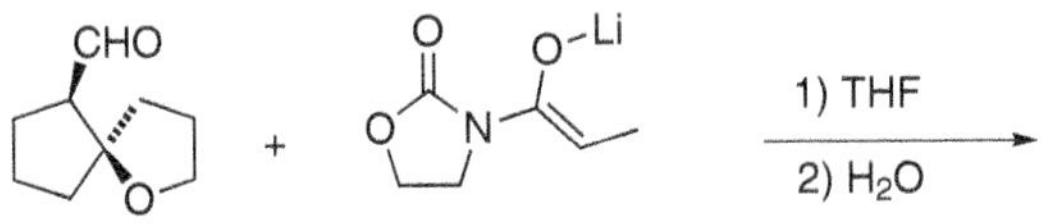

(i) 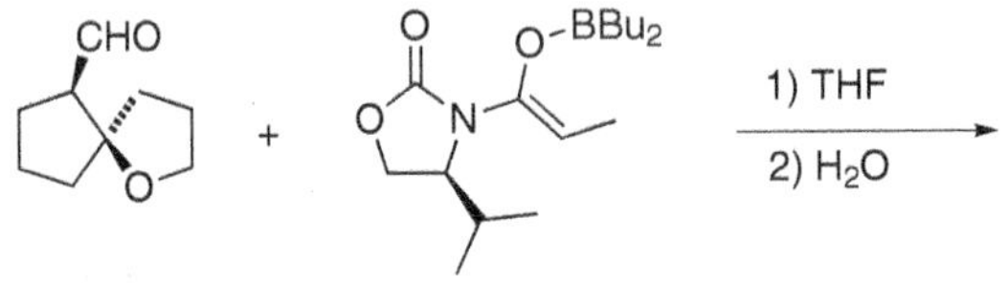

(j) 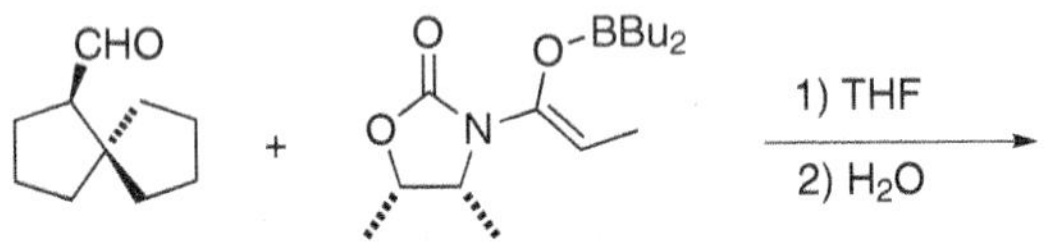

2-25 Using dynamic NMR spectroscopy to study the inversion of aziridines, it has been shown that the inversion of the aziridine nitrogen in the upper compound at left (a 1,4-benzoquinone) is more than 50 times as fast as the inversion of its reduction product (a 1,4-hydroquinone derivative, shown as the lower example). Provide a rationalization of these observations. Based on your answer, how could you slow down the inversion of the quinone?

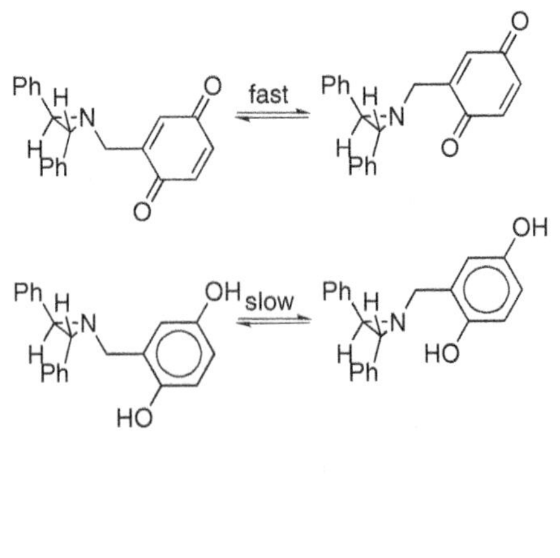

2-26 In carbon tetrachloride solution, *cis*-cyclohexane-1,3-diol exists mainly in the conformer with two axial OH groups, but in methanol solution, the major conformer has both OH groups equatorial. Why?

2-27 Calculate the populations of the two chair conformers of each of the following cyclohexane derivatives using the Corey-Feiner parameters.

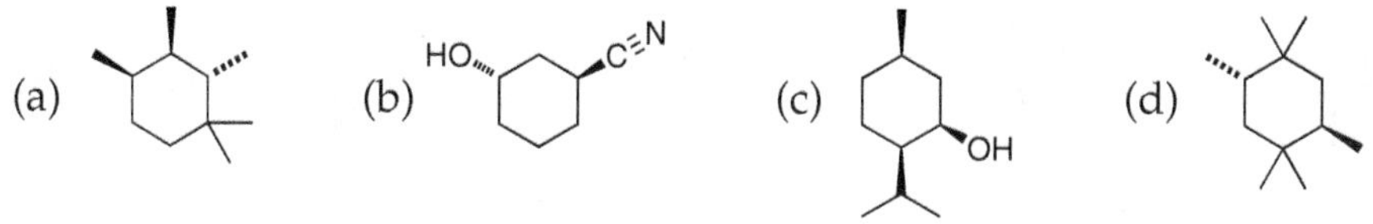

2-28 The diketone in the figure is a *meso* compound, with a plane of symmetry through the molecule, and therefore it cannot be resolved into two enantiomers. Reduction of the diketone with a metal hydride reducing agent leads to the formation of two diastereoisomeric diol products. The diol in the figure has a mirror plane of symmetry like the starting diketone, and is therefore a *meso* compound incapable of optical resolution. The diastereoisomeric diol shown does not have a mirror plane of symmetry (nor, in fact, does it have a simple axis of rotation). Can it be resolved into enantiomers? Give your reasons.

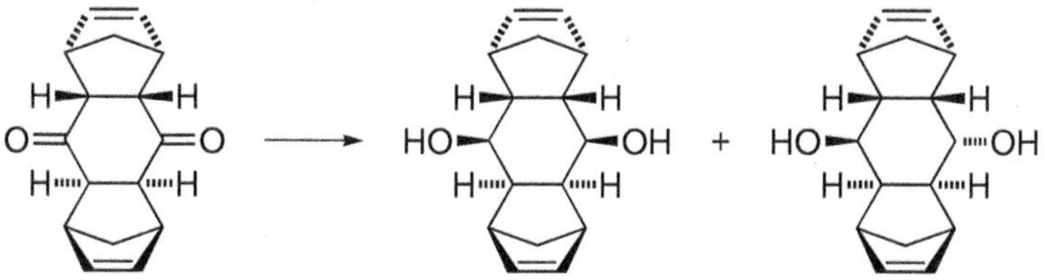

2-29 The following synthesis of acetate anion with a chiral methyl group was completed by the research group of Sir John Cornforth in 1969. Analyze the stereochemistry of each step of the sequence. Then deduce both the regiochemistry and the stereochemistry of each reaction used in the sequence.

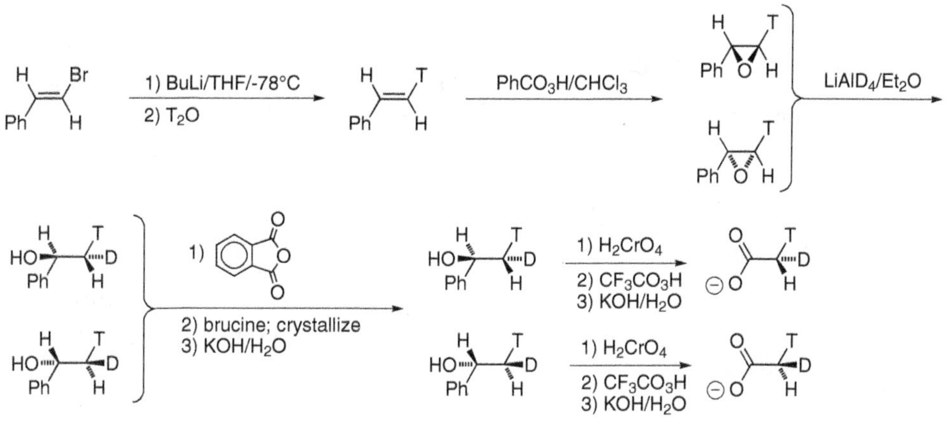

2-30 The barrier to inversion at sulfur in benzylic sulfoxides is much less than for similar simple dialkyl or diaryl sulfoxides. In addition, sulfoxides such as the one at right, which is chiral at both sulfur and carbon, racemize at both carbon at sulfur. Suggest a mechanism for the racemization that is consistent with these observations.

2-31 Tröger's base is a fascinating molecule that has a rigid skeleton containing two chiral nitrogen atoms. In dilute acid, Tröger's base racemizes. Two mechanisms, shown below as mechanisms A and B, have been proposed for this reaction.

Mechanism A

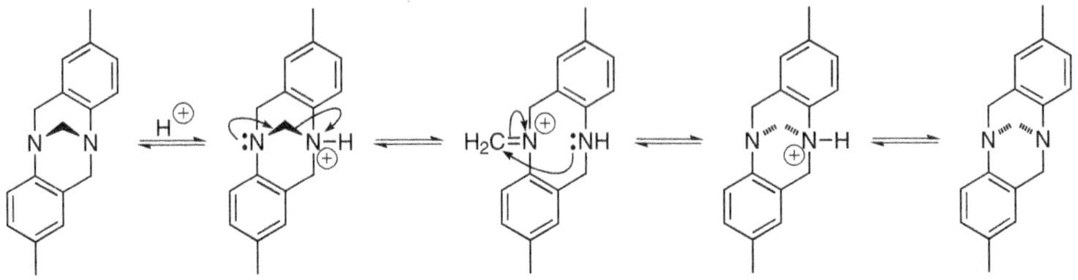

Mechanism B

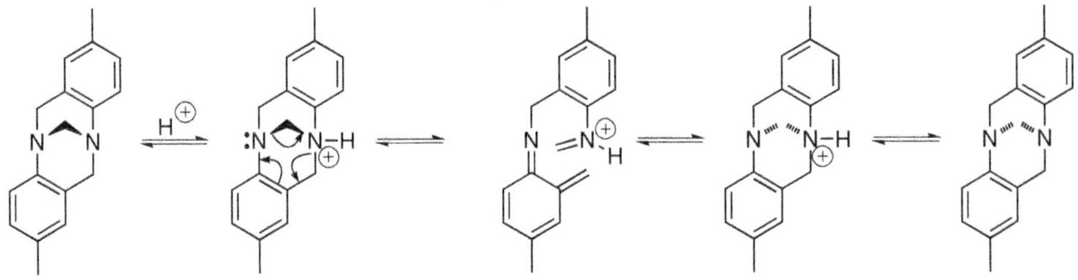

Where bonds are broken, the racemization involves inversion of configuration as the bonds are reformed. Consider the racemization of the Tröger's base derivative at left and suggest the outcome of its racemization by both mechanisms. (As part of your answer, it may be useful to analyze the topism of the various groups on the Tröger's base molecule.)

2-32 The polynactins are a class of ionophore antibiotics based on the same ring system. They differ in the substitution pattern on the ring. Analyze each of the five polynactins below for their symmetry elements. Which is (are) the most symmetrical (i.e., has [have] the highest order symmetry element)?

	R_1	R_2	R_3	R_4
Nonactin	Me	Me	Me	Me
Monactin	Et	Me	Me	Me
Dinactin	Et	Me	Et	Me
Trinactin	Et	Et	Et	Me
Tetranactin	Et	Et	Et	Et

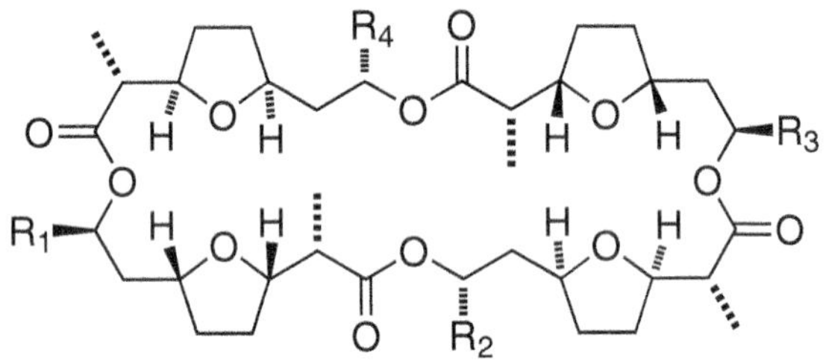

Does this give any hints about how the synthesis of any of these compounds might be approached?

2-33 All the L-amino acids have the same general structure and the same Fischer projection. What is the absolute configuration of each of the amino acids shown in the *R/S* system?

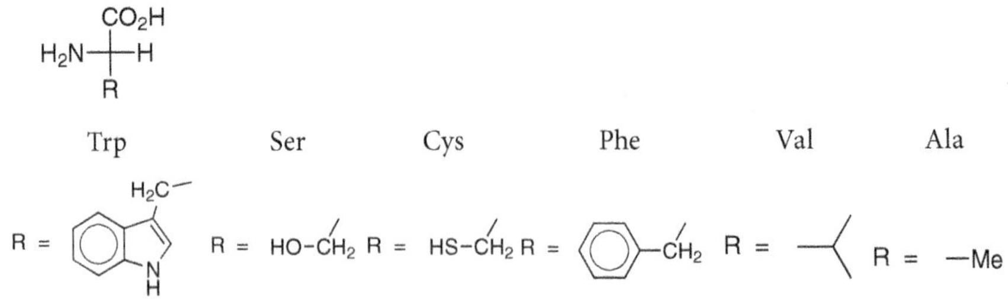

Does this suggest any caveats when using the *R/S* protocol in biochemical systems?

2-34 The synthesis of 8-phenylmenthol (or phenmenthol) from pulegone is carried out using the sequence of reactions shown.

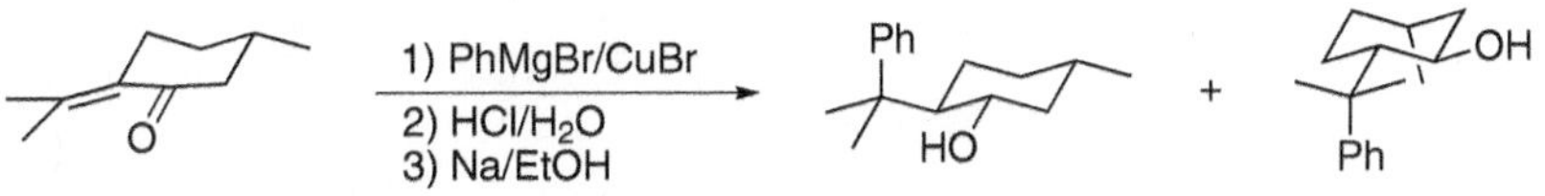

Analyze the stereochemistry of the reactions that give these two products. What fixes the stereochemistry of the carbinol carbon? The minor product of the reaction is *epi-ent*-phenmenthol. Which of the two structures above is most likely to be this minor product? Why?

2-35 Deduce the stereochemical course of each of the following reactions. Where more than one organic reactant is involved, deduce the reaction stereochemistry with respect to each reactant.

(a) 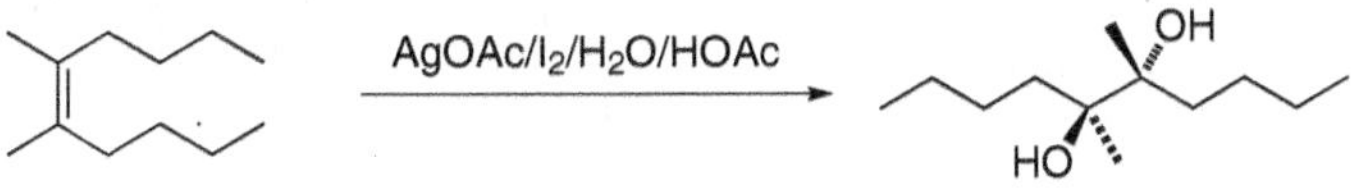

(b) 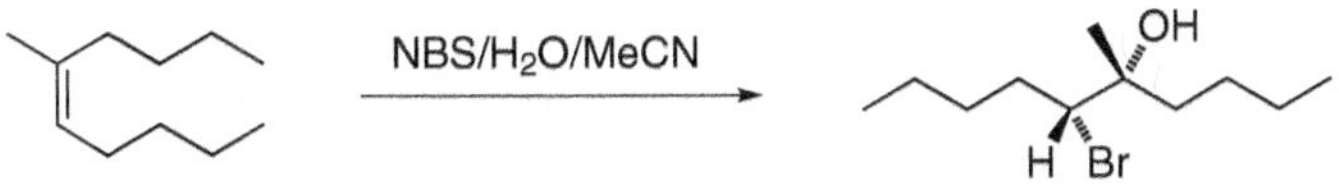

(c) 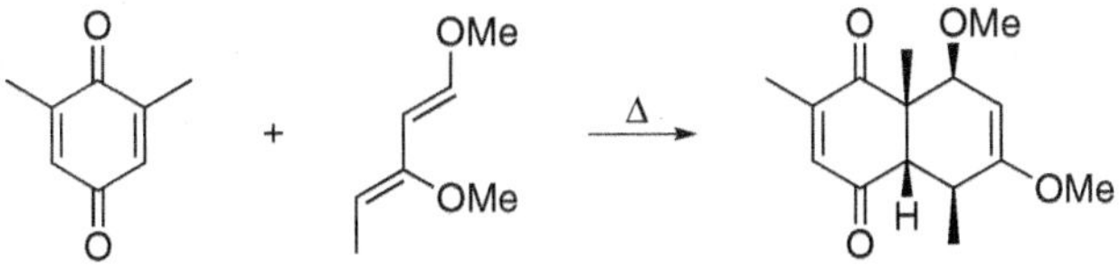

(d) 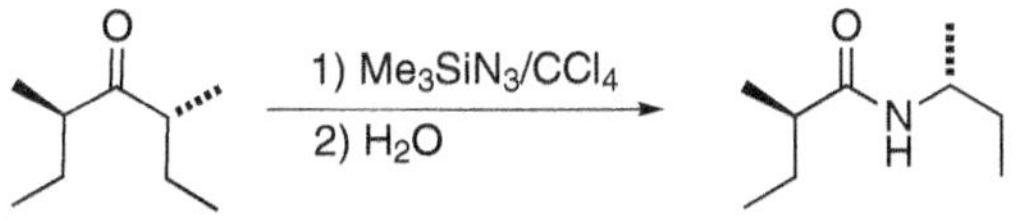

(e) 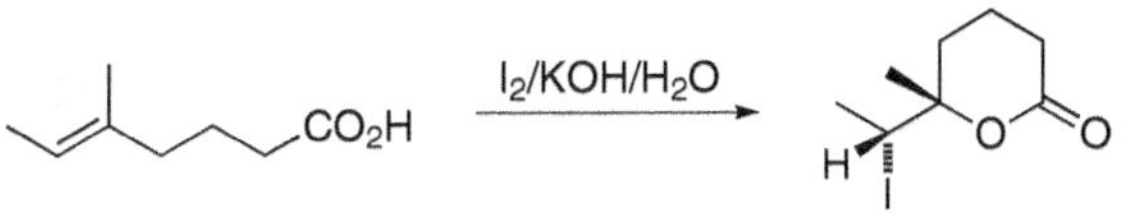

2-36 Why is the 400 MHz ^{1}H NMR signal from the methylene group indicated in the compound shown much more complex than the simple quartet that would be expected by splitting by the three protons of the adjacent methyl group?

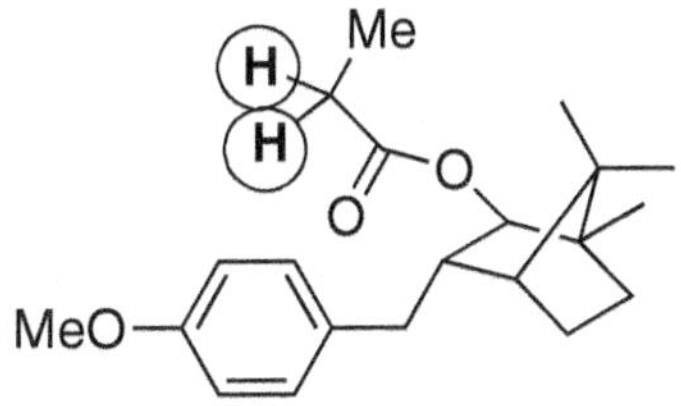

CHAPTER THREE

Organic Shorthand: Acronyms and Name Reactions

3.1 Organic Synthesis: A Brief Introduction

In this author's opinion, organic synthesis is at the heart of organic chemistry, because everything that we do as organic chemists ultimately requires the synthesis of organic compounds. Compounds are synthesized for a variety of reasons: to confirm structures assigned on the basis of spectroscopic studies or chemical degradation, to elucidate the stereochemistry of a compound whose stereochemistry cannot be established by other means, to modify an existing natural compound to make it more active, or to develop methods for the preparation of important medicinal compounds that may be in short supply from natural sources. Synthesis has been used to explore the limits of structure and bonding. For example, the synthesis of cyclobutadiene[1] was an important target for testing theories of aromaticity. Also, the synthesis of prismane[2] provided insights into the synthesis of highly strained molecules and also showed that this fully saturated isomer of benzene could, in fact, exist.

In this book, we will be looking at a wide range of organic reactions—how they work and how they can be assembled into sequences to achieve the synthesis of complex organic compounds. As organic chemistry has grown, so have the number of reactions and reagents available to the synthetic chemist to accomplish the tasks at hand. The increasing complexity of modern synthetic reactions has resulted in a need for a shorthand method for discussing or representing reagents and/or reactions.

3.2 Name Reactions: A Historical Overview

The first type of such shorthand to become widely used by organic chemists is the ***name reaction***. At first, name reactions were used sparingly and almost always in reference to reactions that were already widely used and well known. In this way, a name reaction was more widely viewed as a tribute to the discoverer of a highly useful reaction. For example, within the dozen years after its first appearance in the literature, Victor Grignard's synthesis of alcohols by the reaction between an alkyl- or arylmagnesium halide and a carbonyl compound had become so widely used that almost all chemists knew what was meant by the term, "Grignard's synthesis." Within the same time period, the Grignard reaction had so completely displaced the previous methods—also name reactions using organozinc reagents—that it was the reaction of first choice for the synthesis of alcohols.

This general principle—that a reaction on which the discoverer's name was conferred should be both widely known and widely used—stood for much of the 20th century, and

1. Review: Pettit, R. *Pure Appl. Chem. 1968*, 17, 253.
2 Katz T.J.; Acton N. *J. Am. Chem.* Soc. *1973*, 95, 2738.

conferring a name on a reaction was thus sparingly done. Toward the end of that century, however, the proliferation of variations on a theme and the need for using a specific variation of a particular reaction meant that the use of name reactions and named reagents began to be used more as a shorthand to describe the explicit version of the base reaction being used rather than as a tribute to the discoverer of the generic reaction. This trend still continues.

Unfortunately, this very proliferation of named reactions and reagents has made a chapter like this increasingly necessary in modern organic synthesis—few people can keep the full set of named reactions and reagents clear in their minds (and that includes this author!).[3] Over the course of this book, we will discuss many name reactions. Some of the oldest named reactions—all from before World War I—are gathered (in alphabetical order) in Table 3.1. The list is not exhaustive, and it has expanded by a large amount in the century that followed.

When you examine the list, you will find many reactions that you have already learned in your introductory course in organic chemistry; it is worthwhile taking some time to review them. Unless you have been asked for a superhuman effort, you will not have studied all the reactions in Table 3.1, so do not be discouraged if you find reactions that you are not familiar with. When you examine Table 3.1, you will find two rather obvious omissions from the list—Markovnikov's rule[4] for addition and Zaitsev's rule[5] for elimination. Likewise, substituent effects on regiochemistry in aromatic substitution are not in the table.

Table 3.1 Long-standing Name Reactions

Reaction Name	Reaction	Year	Ref.
Baeyer-Villiger oxidation	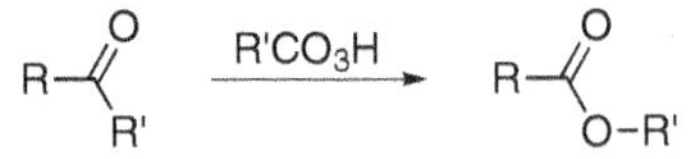	1899	6
Beckmann rearrangement	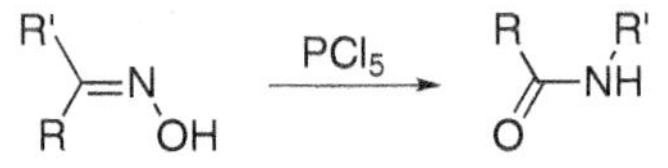	1886	7
Blaise reaction	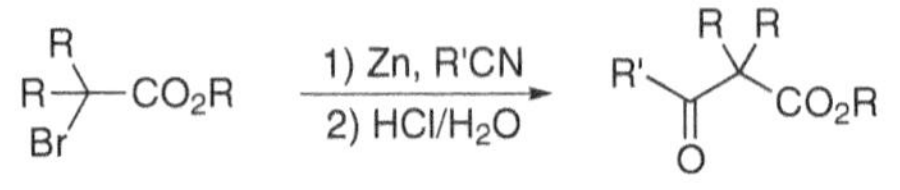	1901	8
Borodin-Hunsdiecker reaction	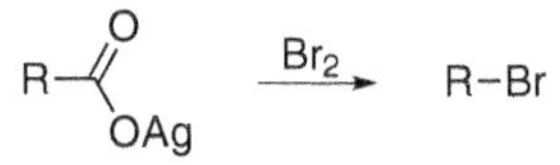	1861	9

3. There are several books and monographs devoted to name reactions that have appeared in recent years: (a) Kürti, L.; Czakó, B. *Strategic Applications of Named Reactions in Organic Synthesis* (Academic Press: New York, 2005). (b) Li, J.J. *Name Reactions. A Collection of Detailed Mechanisms and Synthetic Applications*, 4th ed. (Springer-Verlag: Berlin, 2009). (c) Mundy, B.P.; Ellerd, M.G.; Favaloro, F.G. Name Reactions and Reagents in Organic Synthesis, 2nd ed. (Wiley-Interscience: New York, 2005).

4. (a) Markownikoff, W. *Ann. Chem. Pharm.* **1870**, 133, 228. (b) Markovnikov, *V. Compt. Rend.* **1875**, 82, 668, 728, 776.

5. Saytzeff, A. *Justus Liebigs Ann.* Chem. **1875**, 179, 296.

6. Baeyer, A.; Villiger, V. *Ber. dtsch. chem. Ges.* **1899**, *32*, 3625; **1900**, *33*, 858.

7. Beckmann, E. *Ber. dtsch. chem. Ges.* **1886**, *19*, 988.

8. Blaise, E.E. *Compt. rend.* **1901**, *132*, 478.

9. Borodine, A. *Ann. Chem. Pharm.* **1861**, *119*, 121; Z. *Chem.* **1861**, *4*, 5; **1869**, *12*, 342; *Ann. Chem. Pharm.* **1869**, *121*, 119.

Reaction Name	Reaction	Year	Ref.
Bouveault-Blanc reduction	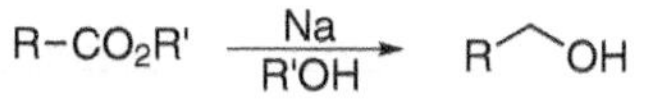	1903	10
Claisen condensation	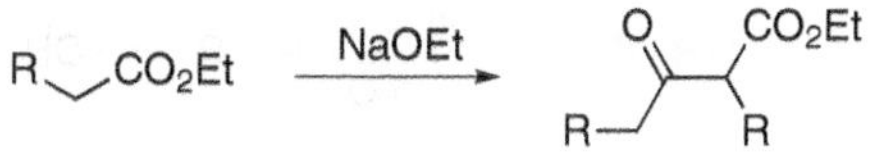	1881	11
Claisen rearrangement		1912	12
Clemmensen reduction	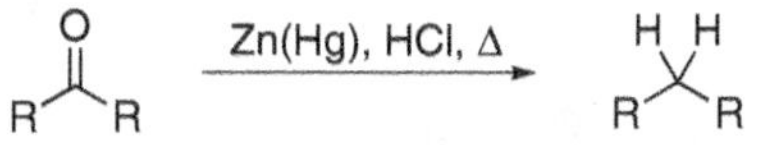	1913	13
Darzens condensation	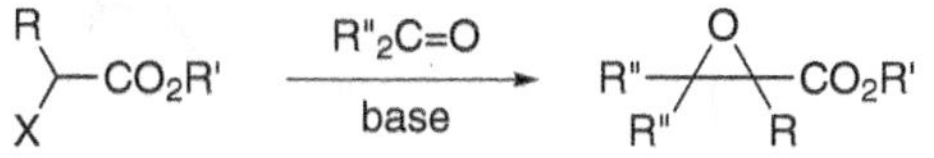	1904	14
Demyanov rearrangement	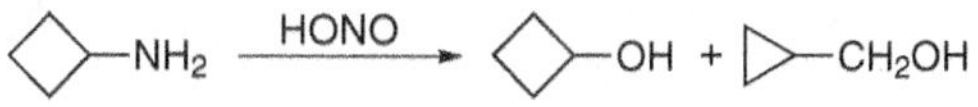	1903	15
Dieckmann condensation	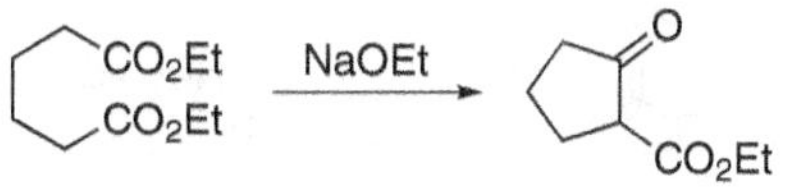	1894	16
Diels-Alder reaction		1929	17
Étard oxidation	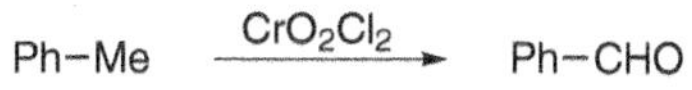	1880	18
Friedel-Crafts acylation	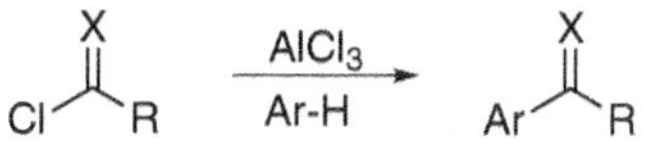	1877	19
Fries rearrangement	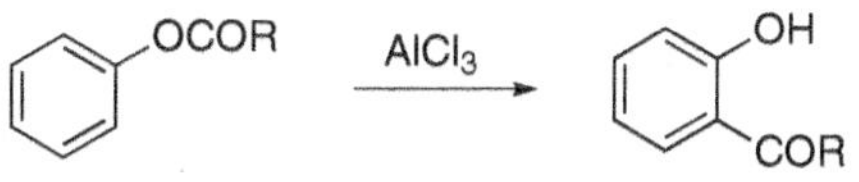	1908	20
Gattermann-Koch reaction	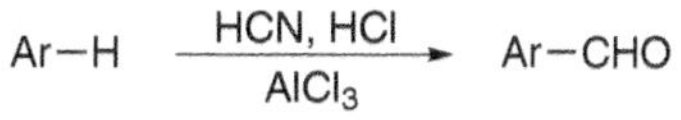	1897	21
Glaser coupling	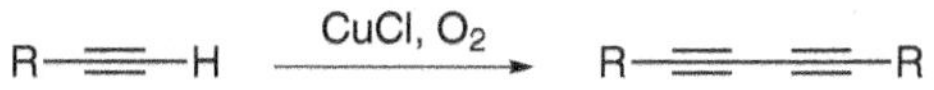	1869	22

(continues)

10. Bouvealt, L.; Blanc, G. *Compt. Rend.* **1903**, *136*, 1676; *Bull. Soc. Chim. France* [3], **1904**, *31*, 666.
11. Claisen, L.; Claparède, A. *Ber. dtsch. chem. Ges.* **1881**, *14*, 2460.
12. Claisen, L. *Ber. dtsch. chem. Ges.* **1912**, *45*, 3157.
13. Clemmensen, E. *Ber. dtsch. chjem Ges.* **1913**, *46*, 1837; **1914**, *47*, 51, 681.
14. Darzens, G. *Compt. rend.* **1904**, *139*, 1214; **1905**, *141*, 766; **1906**, *142*, 214.
15. (a) Demyanov, N.Ya. *Zh. Russ. Fiz.-Khim. O-va.* **1903**, *35*, 26; **1904**, *36*, 186. (b) Demjanov, N. *Chem. Zentr.* **1903**, *I*, 828; **1904**, *I*, 1214; *Ber. dtsch. chem. Ges.* **1907**, *40*, 4393, 4961; **1908**, *41*, 43.
16. Dieckmann, W. *Ber. dtsch. chem. Ges.* **1894**, *27*, 102, 965; **1900**, *33*, 595, 2670; *Ann. Chem. Pharm.* **1901**, *317*, 51, 93.
17. Diels, O.; Alder K. *Justus Liegigs Ann. Chem.* **1929**, *460*, 98.
18. Étard, A.L. *Compt. rend.* **1880**, *90*, 534; *Ann. Chim. Phys.* [5] **1881**, *22*, 218.
19. Friedel, C.; Crafts, J.M. *Compt. rend.* **1877**, *84*, 1392, 1450.
20. (a) Fries, , K.; Fink, G. *Ber. dtsch. chem. Ges.* **1908**, *41*, 4271. (b) Fries, K.; Pfaffendorf, W. *Ber. dtsch. chem. Ges.* **1910**, *43*, 212.
21. Gattermann, L. *Ber. dtsch. chem. Ges.* **1898**, *31*, 1149; *Ann. Chem. Pharm.* **1907**, *357*, 313.
22. Glaser, C. *Ber. dtsch. chem. Ges.* **1869**, *2*, 422; *Ann. Chem. Pharm.* **1870**, *154*, 159.

(**Table 3.1** continued)

Reaction Name	Reaction	Year	Ref.
Grignard reaction	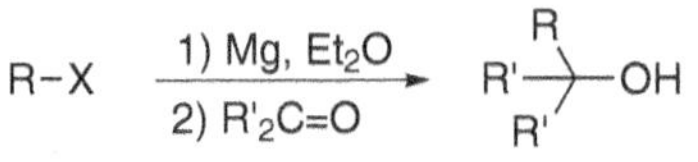	1900	23
Hantzsch pyridine synthesis	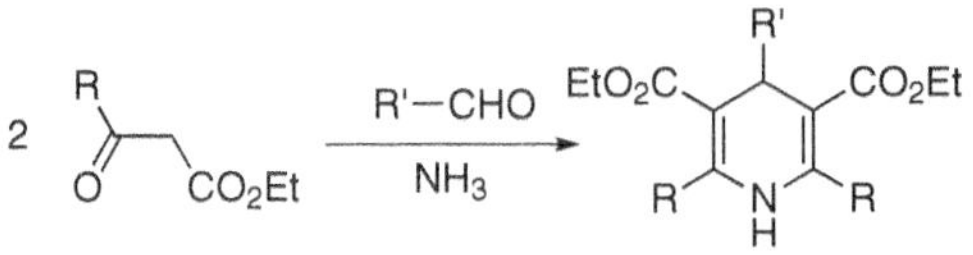	1882	24
Hell-Volhard-Zelinskii reaction	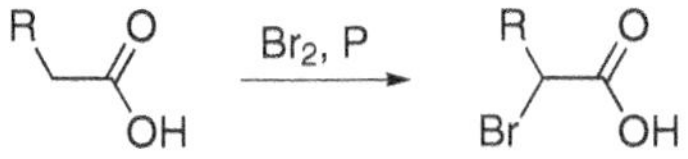	1881	25
Hofmann rearrangement	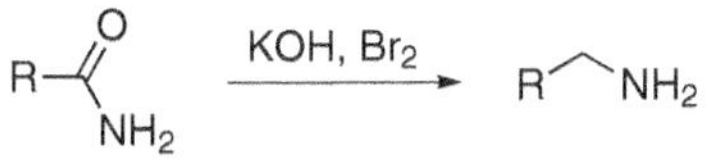	1881	26
Hofmann, Curtius, Lossen rearrangements		1874–1894	27
Kishner cyclopropane synthesis	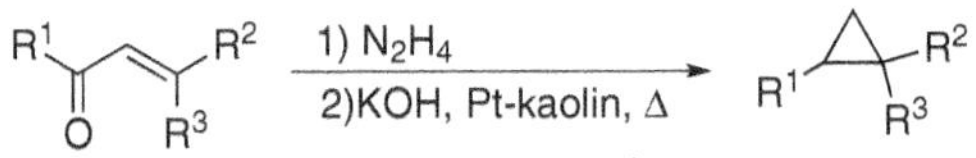	1911	28
Kolbe carbonation	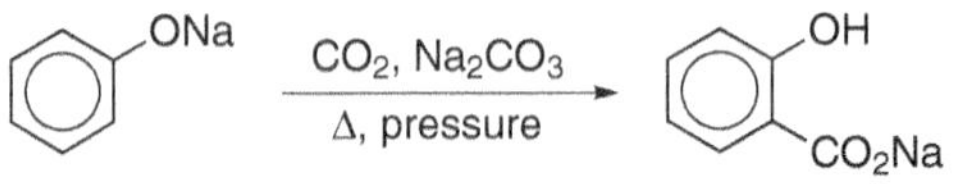	1860	29
Michael addition	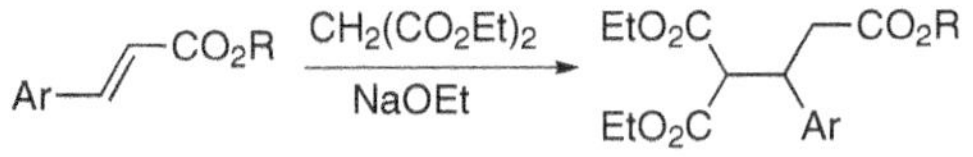	1887	30
Perkin condensation	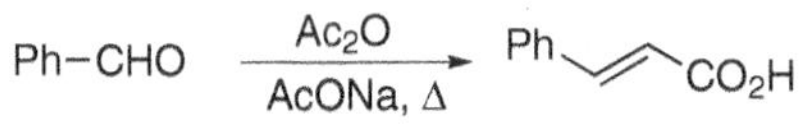	1868	31
Prilezhaev reaction	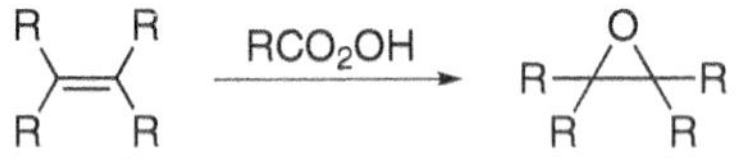	1909	32

23. (a) Barbier, P. *Compt. rend.* **1899**, *128*, 110. (b) Grignard, V. *Compt. rend.* **1900**, *130*, 1322; **1901**, *133*, 336, 558; **1902**, *134*, 849; *Chem. Zentr.* **1901**, II, 622; *Ann. Chim. Phys.* **1901**, *24 [vii]*, 433. (c) Grignard, V.; Tissier, L. *Compt. rend.* **1901**, *133*, 835; **1902**, *134*, 107.

24. (a) Hantzsch, A. *Ann. Chem. Pharm.* **1882**, *215*, 1, 72; *Ber. dtsch. chem. Ges.* **1885**, *18*, 1744; **1886**, *19*, 289.

25. (a) Hell, C. *Ber. dtsch. chem. Ges.* **1881**, *14*, 891. (b) Volhard, J. *Ann. Chem. Pharm.* **1887**, *242*, 141. (c) Zelinskii, N. *Ber. dtsch. Chem. Ges.* **1887**, *20*, 2026.

26. Hofmann, A.W. *Ber. dtsch. chem. Ges.* **1881**, *14*, 2725.

27. (a) Curtius, T. *J. prakt. Chem.* [2] **1894**, *50*, 275. (b) Lossen, W. *Ann. Chem. Pharm.* **1874**, *161*, 347; **1875**, *175*, 271, 313.

28. (a) Kizhner, N.; Zavadovskii, A. *Zh. Russ. Fiz.-Khim. O-va.* **1911**, *43*, 1132. (b) Kizhner, N. *Zh. Russ. Fiz.-Khim. O-va.* **1912**, *434*, 1132165, 849; **1913**, *45*, 949, 957, 987; **1915**, *47*, 1102.

29. (a) Kolbe, H. *Ann. Chem. Pharm.* **1860**, *113*, 125. (b) Schmitt, R. *J. prakt. Chem.* [2] **1885**, *31*, 397.

30. Michael, A. *J. prakt. Chem.* [2] **1887**, *35*, 349.

31. Perkin, W.H. *J. Chem. Soc.* **1868**, *21*, 53, 181; *J. Chem. Soc.* **1877**, *31*, 388.

32. Prileschajew, N. *Ber. dtsch. chem. Ges.* **1909**, *42*, 4811.

Reaction Name	Reaction	Year	Ref.
Reformatskii reaction	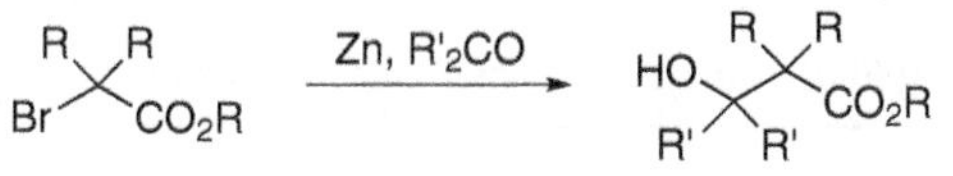	1887	33
Reimer-Tiemann reaction	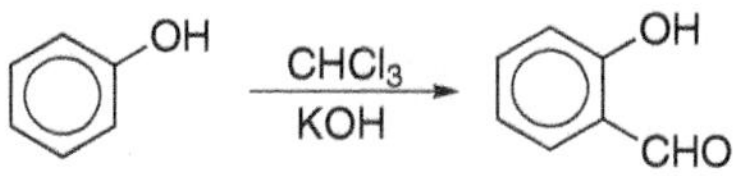	1876	34
Stobbe condensation	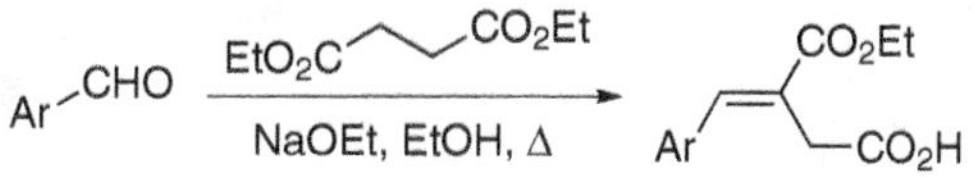	1893	35
Strecker amino acid synthesis	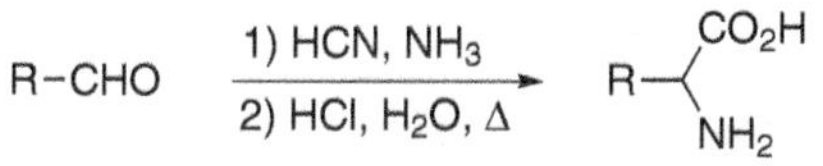	1850	36
Tollens oxidation	R–CHO $\xrightarrow{Ag(NH_3)_2OH}$ R–CO_2H	1882	37
Ullmann Coupling	Ar–Br $\xrightarrow[Na]{R-I}$ Ar–R	1881	38
Wagner-Meerwein rearrangement	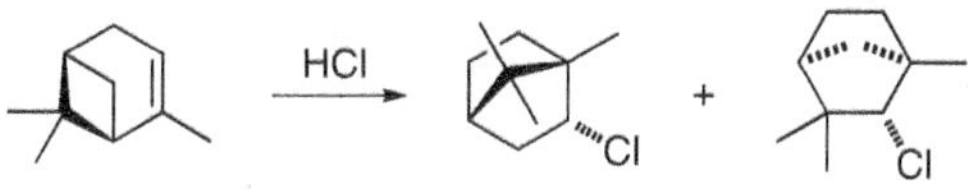	1899	39
Wolff-Kishner reduction	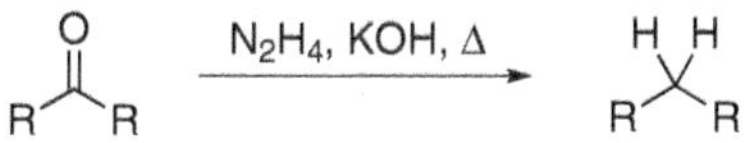	1911	40
Wurtz coupling	R–X $\xrightarrow{Na}$ R–R	1855	41

3.3 Acronyms and Abbreviations

The dramatic growth in the number and complexity of reagents in modern organic chemistry has led to an equally dramatic increase in the use of acronyms for reagents, so that since the last third of the 20th century or so, the use of acronyms in organic chemistry has blossomed. There are now acronyms for parts of a molecule, such as alkyl and acyl groups, functional groups (especially protecting groups), solvents, and an increasing number of common reagents. These are summarized in Tables 3.2 through 3.6. The entries in the tables are organized in alphabetical order.

33. (a) Reformatsky, S. *Ber. dtsch. chem. Ges.* **1887**, *20*, 2110; **1895**, *28*, 2838, 2842, 3262. (b) Reformatsky, S.; Plesconossoff, B. *Ber. dtsch. chem. Ges.*, **1895**, *28*, 2838.
34. Reimer, K.; Tiemann, F. *Ber. dtsch. chem. Ges.* **1876**, *9*, 824, 1268, 1285.
35. Stobbe, H. *Ber. dtsch. chem. Ges.* **1893**, *26*, 2312; *Ann. Chem. Pharm.* **1894**, *282*, 280.
36. Strecker, A. *Ann. Chem. Pharm.* **1850**, *75*, 27; *Ann. Chem. Pharm.* **1854**, *91*, 349.
37. Tollens, B. *Ber. dtsch. chem. Ges.* **1882**, *15*, 1635.
38. (a) Tollens, B.; Fittig, R. *Ann. Chem. Pharm.* **1864**, *131*, 301. (b) Fittig, R.; König, J. *Ann. Chem. Pharm.* **1867**, *144*, 277.
39. (a) Vagner, E.E. *Zh. Russ. Fiz.-Khim. O-va.* **1899**, *31*, 690. (b) Wagner, G. *Ber. dtsch. chem. Ges.* **1899**, *32*, 2302; **1900**, *33*, 2121. (c) Meerwein, H. *Liebigs Ann. Chem.* **1914**, *405*, 129.
40. (a) Kizhner, N.M. *Zh. Russ. Fiz.-Khim. O-va.* **1911**, *43*, 582. (b) Wolff, L. *Ann. Chem. Pharm.* **1912**, *394*, **86.**
41. Wurtz, C.A. *Ann. Chim. Phys. [3]* **1855**, *44*, 275; *Ann. Chem. Pharm.* **1855**, *96*, 364.

Table 3.2 Acronyms for Alkyl, Acyl, and Substituent Groups, Including Protecting Groups for Alcohols and Amines

Acronym	Name/Structure	Acronym	Name/Structure
Ac	acetyl, CH_3CO–	Ar	aryl
Ad	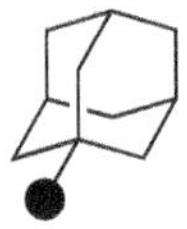 1-adamantyl	Adoc	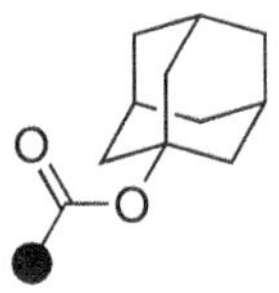 1-adamantyloxycarbonyl
Alloc	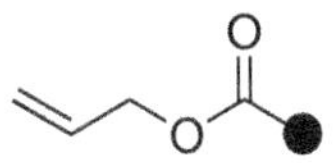 allyloxycarbonyl	Boc	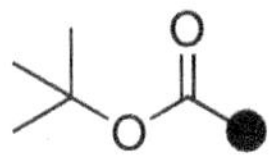 *tert*-butoxycarbonyl
Bom	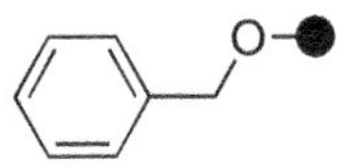 benzyloxymethyl	Bs	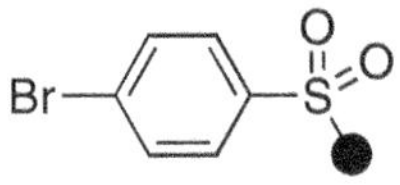 *p*-bromobenzenesulfonyl, brosyl
Bt	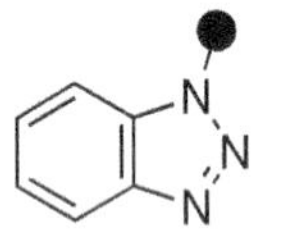 benzotriazol-1-yl	Btc	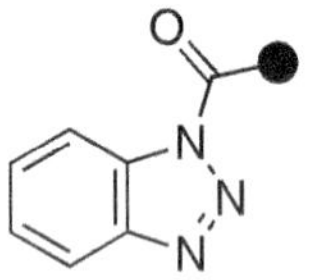 benzotriazolylcarbonyl
Bu or *n*-Bu	*n*-butyl, –$(CH_2)_3CH_3$	*i*-Bu, or Bu^i	isobutyl, –$CH_2CH(CH_3)_2$
s-Bu, or Bu^s	*sec*-butyl, $CH(CH_3)CH_2CH_3$	*t*-Bu or Bu^t	*tert*-butyl, –$C(CH_3)_3$
Bz or PhCO	benzoyl, C_6H_5CO–	Bzl	benzyl, $C_6H_5CH_2$–
Cp	 η^5-cyclopentadienyl	Cp*	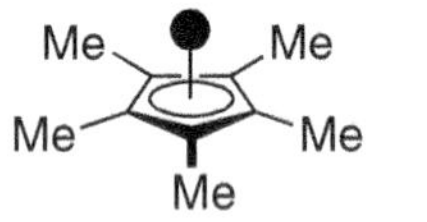 η^5-pentamethylcyclopentadienyl
Cy or Chx	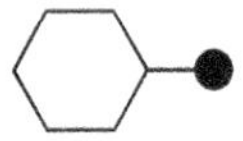 cyclohexyl	dimsyl	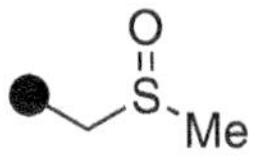 methylsulfinylmethyl
DNP	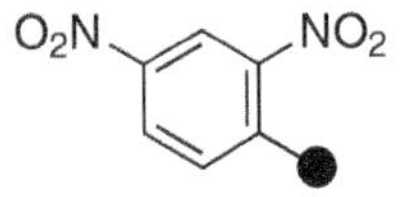 2,4-dinitrophenyl	EE	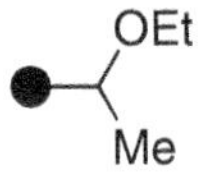 1-ethoxyethyl
Et	–C_2H_5 ethyl	Me	–CH_3 methyl
Fm	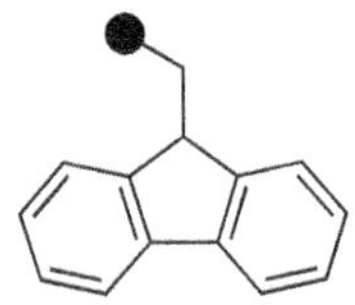 9-fluorenylmethyl	Fmoc	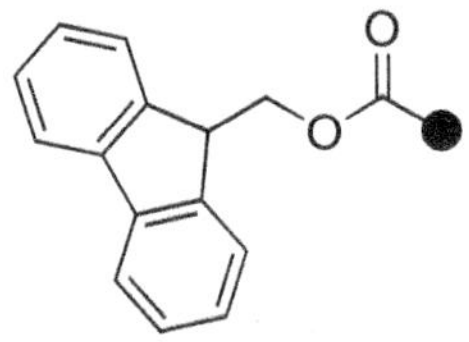 9-fluorenylmethyloxycarbonyl

Acronym	Name/Structure
MEM	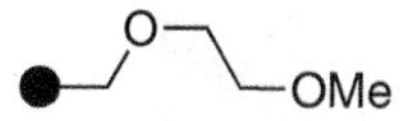 (2-methoxyethoxy)-methyl
MOM	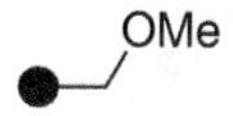 methoxymethyl
Nf	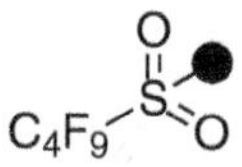 nonaflyl, or perfluorobutanesulfonyl
PEG	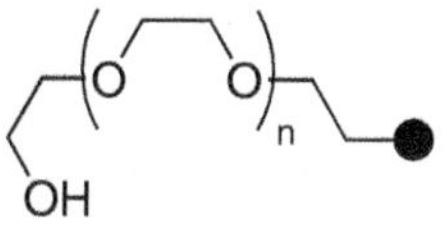 polyethylene glycol or polyethylene glycol substituent
Ph	$-C_6H_5$ phenyl
PMB	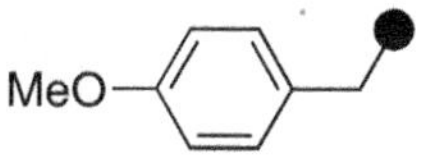 *p*-methoxybenzyl
Pnb	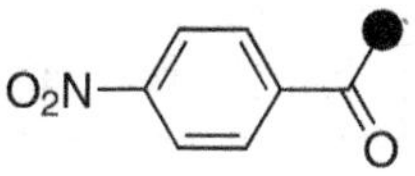 *p*-nitrobenzoyl
i-Pr or Pr^i	$-CH(CH_3)_2$ isopropyl
sisyl	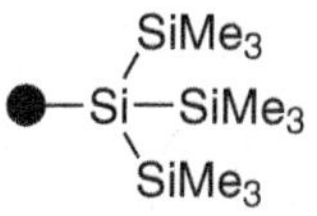 tris(trimethylsilyl)silyl
TBS or TBDMS	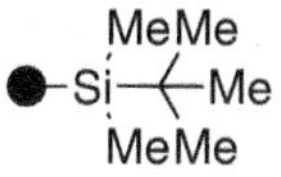 *tert*-butyldimethylsilyl

Acronym	Name/Structure
Mes	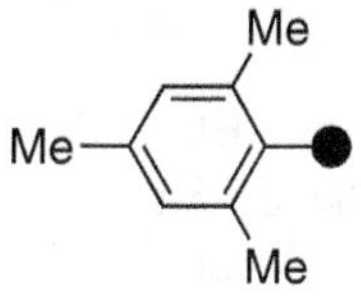 mesityl
Ms	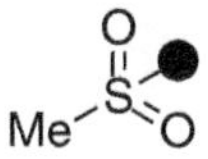 methanesulfonyl, mesyl
Ns	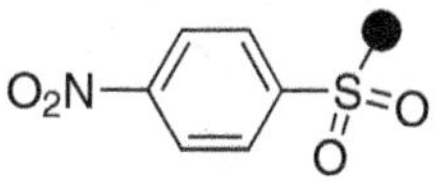 *p*-nitrobenzenesulfonyl, nosyl
Pfp	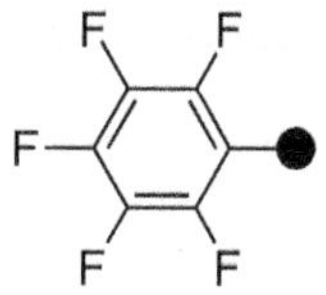 pentafluorophenyl
Pht, PhthN	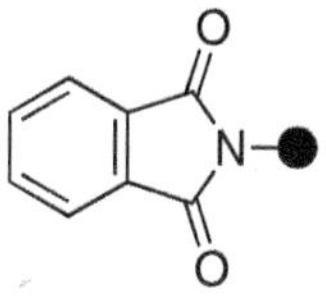 phthalimidoyl
PMP	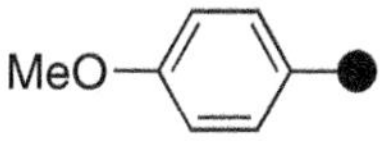 *p*-methoxyphenyl
Pr	$-(CH_2)_2CH_3$ *n*-propyl
SEM	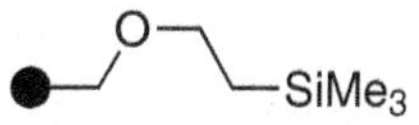 (2-trimethylsilyl-ethoxy)methyl
TBDPS	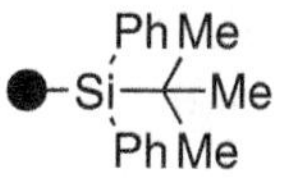 *tert*-butyldiphenylsilyl
Teoc	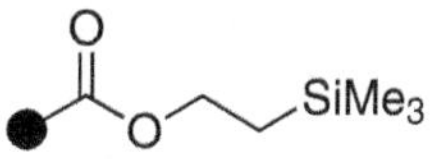 2-(triemthylsilyl) ethoxycarbonyl

(continues)

(**Table 3.2** continued)

Acronym	Name/Structure	Acronym	Name/Structure
TES	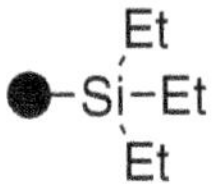 triethylsilyl	Tf	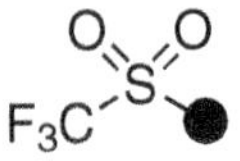 trifluoromethanesulfonyl, or triflate
TFA	CF_3CO- trifluoroacetyl	Th	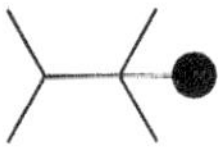 *tert*-hexyl, or thexyl
THP	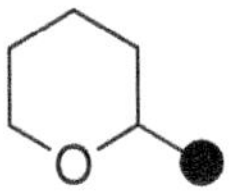 tetrahydropyranyl	Tibs , Tris, trisyl	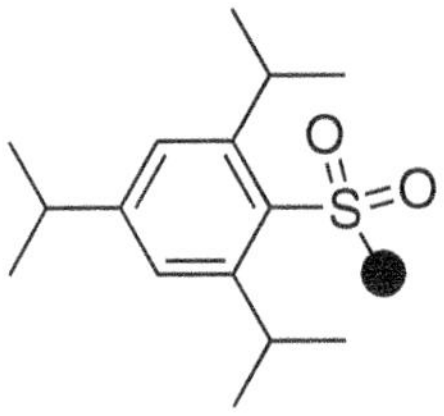 2,4,6-triisopropyl-benzenesulfonyl
TMS	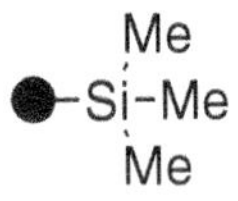 trimethylsilyl	Tr	$-CPh_3$ trityl, or triphenylmethyl
Troc, Tce, Tcec	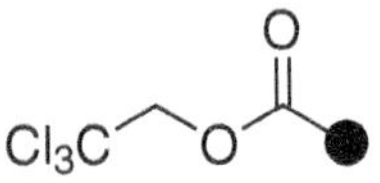 2,2,2-trichloroethoxycarbonyl		
Ts, Tos	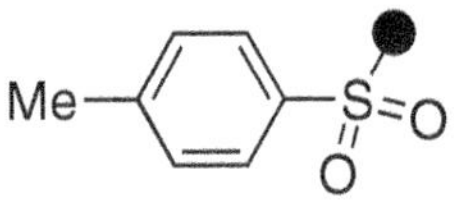 *p*-toluenesulfonyl	Z or Cbz	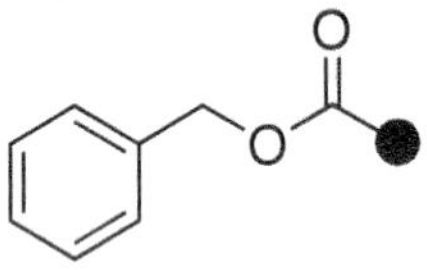 benzyloxycarbonyl

Table 3.3 Acronyms for Solvents

Acronym	Name/Structure	Acronym	Name/Structure
An, ACN	MeCN acetonitrile	bmimCl	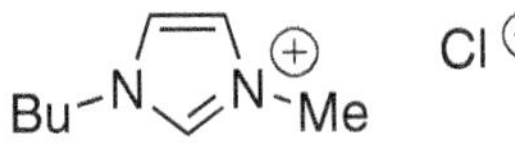 *N*-butyl-*N'*-methyl-imidazolium chloride
DCE	$ClCH_2CH_2Cl$ ethylene dichloride, or 1,2-dichloroethane	DCM	CH_2Cl_2 dichloromethane
DEG	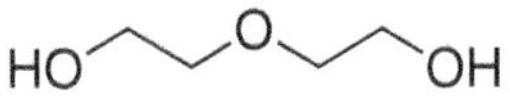 diethylene glycol	DMA	$MeCONMe_2$ dimethylacetamide

DME	$MeOCH_2CH_2OMe$ 1,2-dimethoxyethane	DMEU	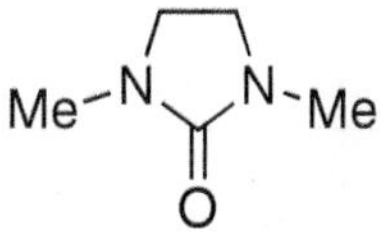 *N,N'*-dimethylethyleneurea *N,N'*-dimethylimidazolidin-2-one
DMF	$HCONMe_2$ dimethylformamide	DMI	see DMEU
DMPU	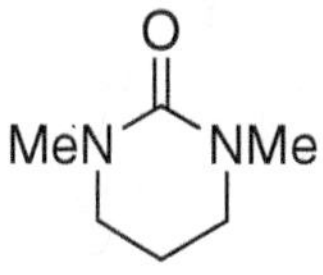 *N,N'*-dimethylpropylene urea	DMSO	Me_2SO dimethyl sulfoxide
EDC	See DCE	emimCl	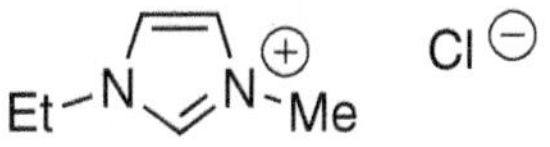 *N*-ethyl-*N'*-methylimidazolium chloride
EtOAc	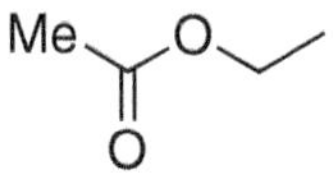 ethyl acetate	hmim BF_4	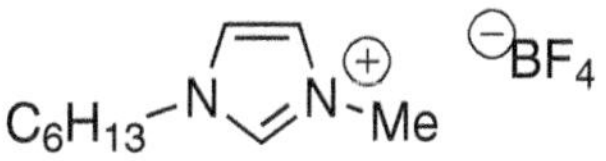 *N*-hexyl-*N'*-methylimidazolium tetrafluoroborate
HMPA	$(Me_2N)P = O$ hexamethylphosphoric triamide	MBE, MTBE	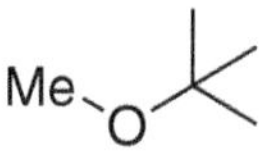 methyl tert-butyl ether
MEK	MeCOEt methyl ethyl ketone, or 2-butanone	MIBK	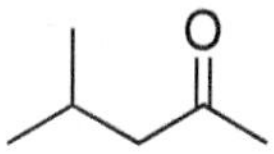 methyl isobutyl ketone, or 4-methyl-2-pentanone
MTBE	See MBE	NMP	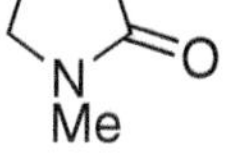 *N*-methylpyrrolidone
py	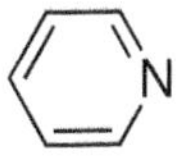 pyridine	THF	 tetrahydrofuran

Table 3.4 Acronyms for Oxidizing Reagents

Acronym	Name/Structure	Acronym	Name/Structure
BAIB, DAIB, PIDA	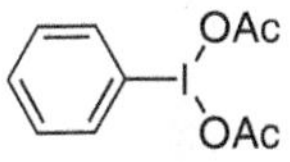 diacetoxyiodobenzene, or iodosobenzene diacetate, or phenyliodonium diacetate	BIDC	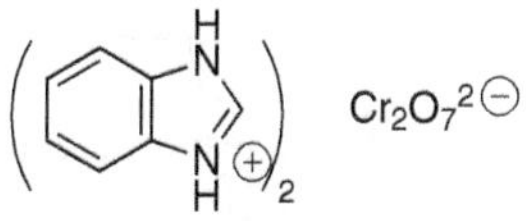 benzimidazolium dichromate

(continues)

(**Table 3.4** continued)

Acronym	Name/Structure	Acronym	Name/Structure
BPCC	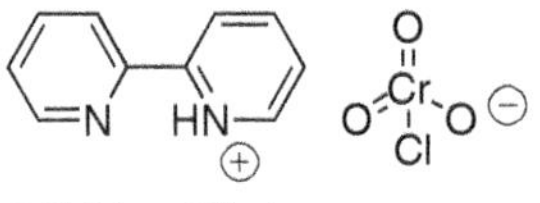 2,2'-bipyridinium chlorochromate	BTI, BTIB, PIFA	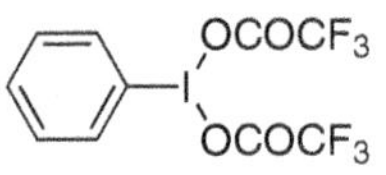 *bis*-(trifluoroacetoxy)-iodobenzene, or phenyliodonium *bis*-(trifluoroacetate)
CAN	$(NH_4)_2Ce(NO_3)_6$ ceric ammonium nitrate	DDQ	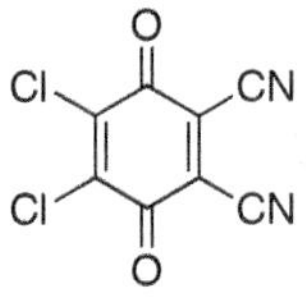 2,3-dichloro-5,6-dicyano-1,4-benzoquinone
DMD, DMDO	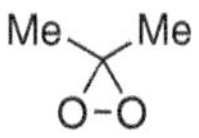 dimethyldioxirane	DMP	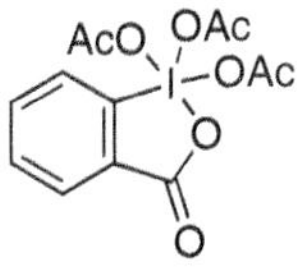 Dess-Martin periodinane
HTIB	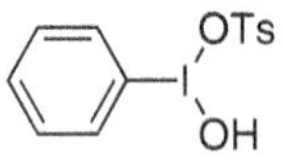 Koser's reagent, hydroxy(tosyloxy)iodobenzene	IBX	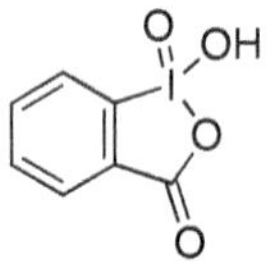 iodoxybenzoic acid
ICC	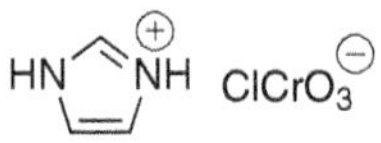 imidazolium chlorochromate	IFC	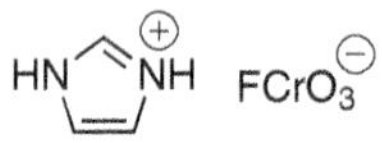 imidazolium fluorochromate
LTA	lead tetraacetate $Pb(OAc)_4$	MCPBA *m*-CPBA	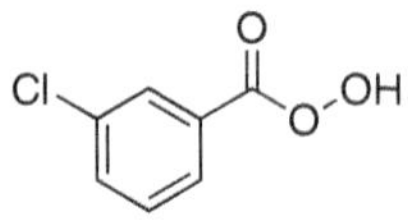 m-chloroperbenzoic acid
MMPP	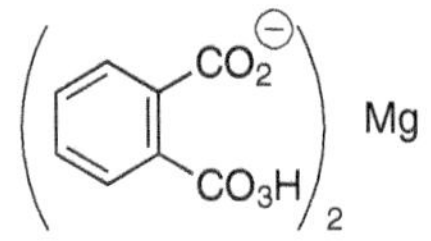 magnesium monoperphthalate	MPCC	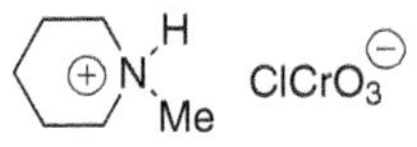 *N*-methylpiperidinium chlorochromate
NBS	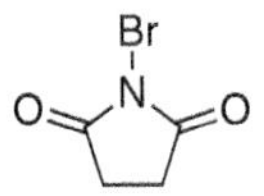 *N*-bromosuccinimide	NCS	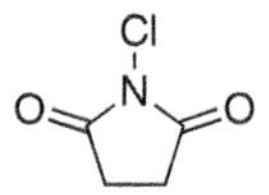 *N*-chlorosuccinimide
NIS	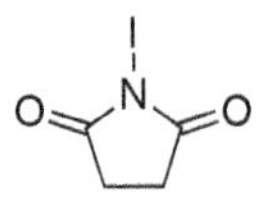 *N*-iodosuccinimide	NMO, NMMO	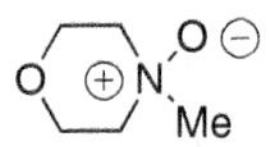 *N*-methylmorpholine N-oxide
PCC	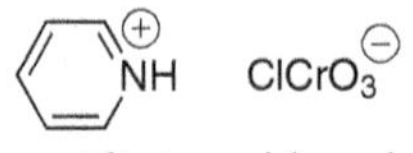 pyridinium chlorochromate	PDC	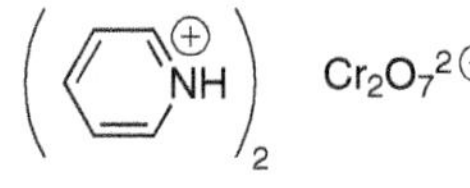 pyridinium dichromate

Acronym	Name/Structure	Acronym	Name/Structure
PFC	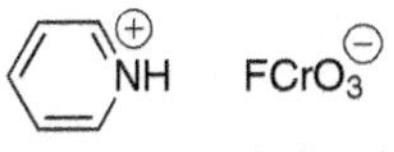 pyridinium fluorochromate	PIDA	see DAIB
PIFA	see BTI	QDC	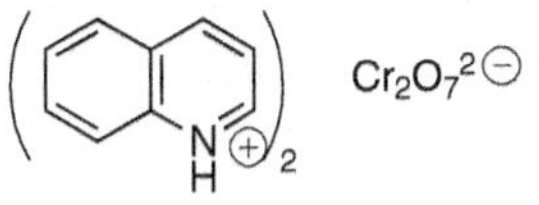 quinolinium dichromate
TBACC	$Bu_4N\ ClCrO_3$ tetrabutylammonium chlorochromate	TMACC	$Me_4N\ ClCrO_3$ tetramethylammonium chlorochromate
TMNO, TMANO	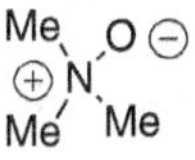 trimethylamino *N*-oxide		

Table 3.5 Acronyms for Reducing Reagents

Acronym	Name/Structure	Acronym	Name/Structure
Alpineborane®	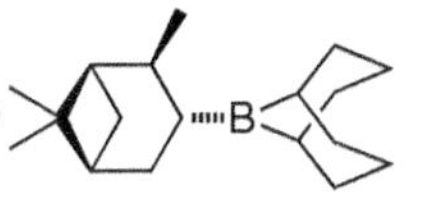 *B*-isopinocampheyl 9-BBN	9-BBN	 9-borabicyclo[3.3.1]nonane
BMS	$Me_2S{\bullet}BH_3$ borane-dimethyl sulfide	DIBAL-H	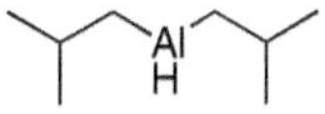 diisobutylaluminum hydride
Ipc-BH_2	 isopinocampheylborane	Ipc_2BH	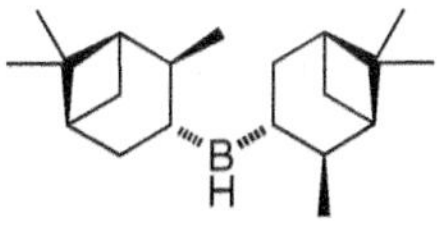 diisopinocampheylborane
K-Selectride	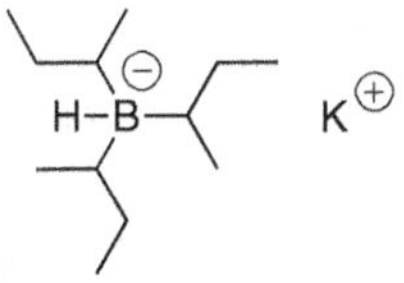 potassium tri-*sec*-butylborohydride	L-Selectride	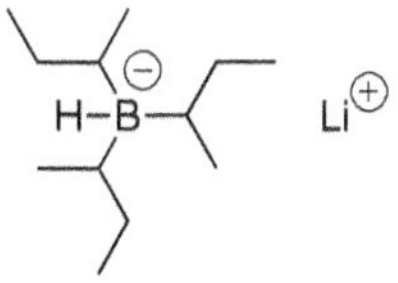 lithium tri-*sec*-butylborohydride
PBH, pinBH	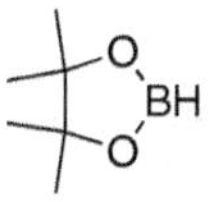 pinacolylborane	Red-Al®	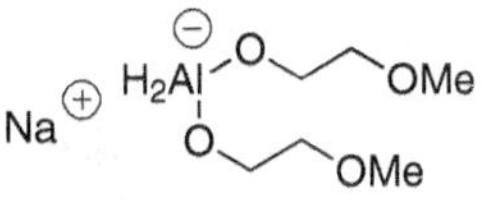 sodium *bis*-(2-methoxy-ethoxy)aluminum hydride
Sia_2BH	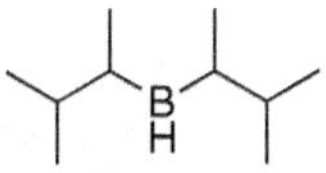 disiamylborane	$ThBH_2$	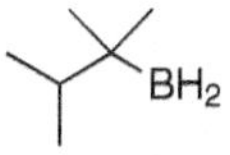 thexylborane

Table 3.6 Acronyms for Ligands, Acids, Bases, Catalysts, and Other Substances

Acronym	Name/Structure	Acronym	Name/Structure
ABCN, ACN	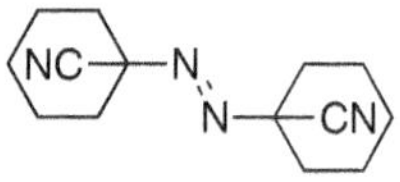 azo-*bis*(cyclohexane-carbonitrile)	acac	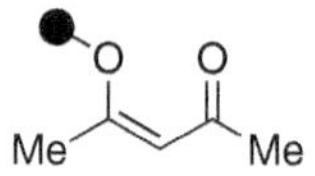 Acetylacetonato
AIBN	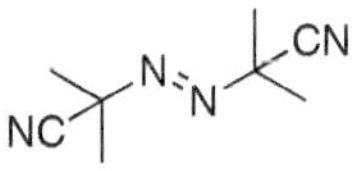 azo-*bis*(isobutyronitrile)	BHT	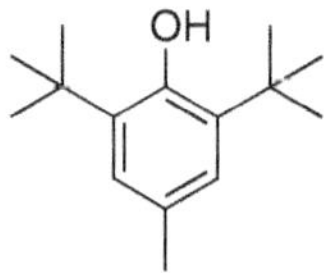 butylated hydroxytoluene
BINAP	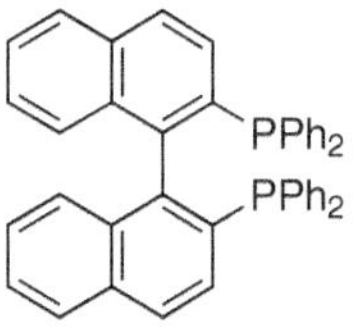 2,2'-bis(diphenylphosphino)-1,1'-binaphthyl	BINOL	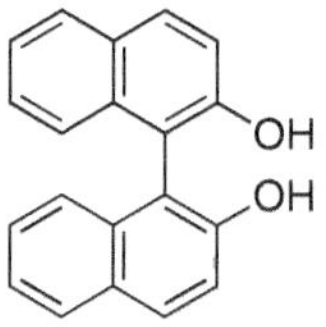 1,1'-binaphthol
Bipy, bpy	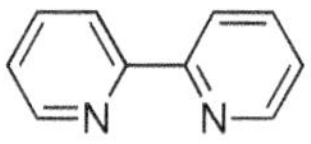 2,2'-bipyridyl	bpe	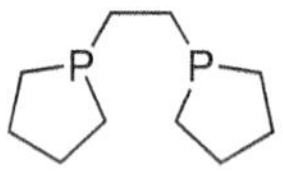 1,2-bis(phospholano)ethane
BPO	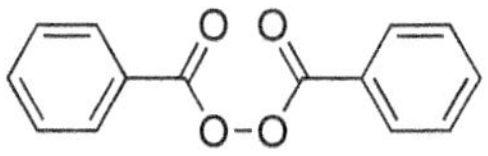 benzoyl peroxide	(-)-BPPFA	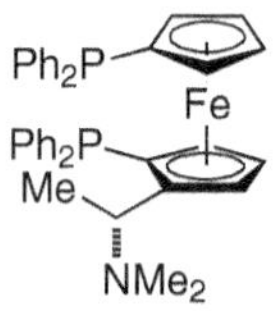
BPPM	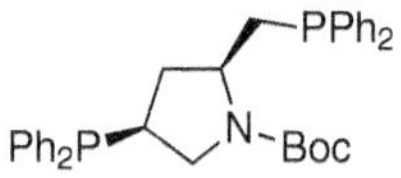	BTC	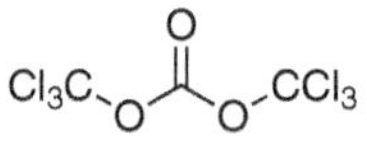 *bis*(trichloromethyl) carbonate
BTEA(B,C,F)	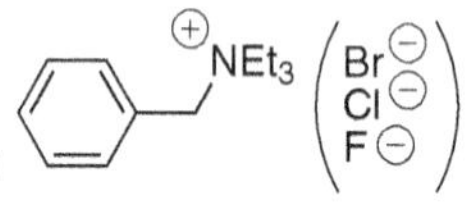 benzyltriethylammonium (bromide, chloride, fluoride)	BTF	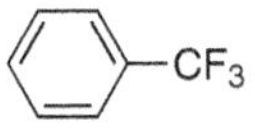 (trifluoromethyl)benzene, or benzotrifluoride
12-C-4	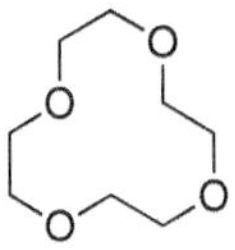 12-crown-4, or 1,4,7,10-tetraoxacyclodecane	18-C-6	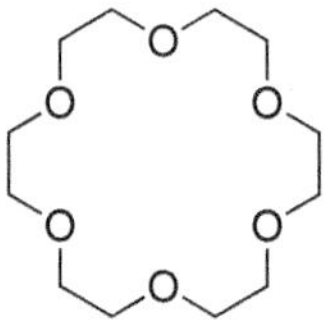 18-crown-6, or 1,4,7,10,13,16-hexaoxa-cyclooctadecane
CBS catalyst	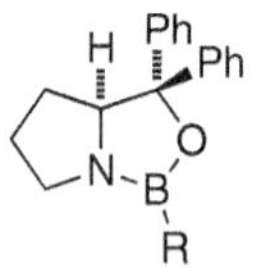 (*S*)-alkyl-Corey-Bakshi-Shibata catalyst	CDI	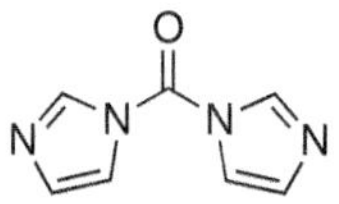 carbonyl *bis*(imidazole)

Acronym	Name/Structure	Acronym	Name/Structure
Chiraphos	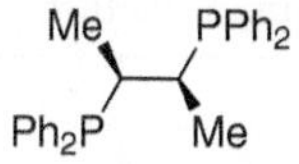 (*S*,*S*)-2,3-*bis*(diphenylphosphino)-butane	cod	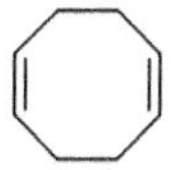 1,5-cyclooctadiene
cot	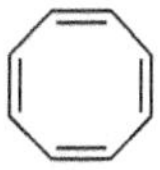 cyclooctatetraene	CSA	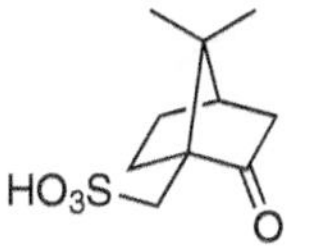 camphorsulfonic acid
CSI	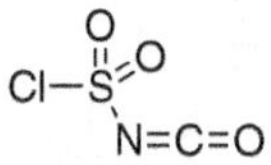 chlorosulfonyl isocyanate	CTA(B,C)	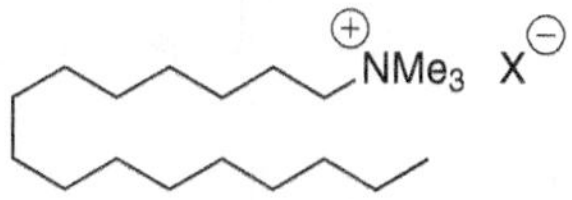 cetyltrimethylammonium (bromide, chloride)
DABCO	 1,4-diazabicyclo[2.2.2]-octane	DavePhos	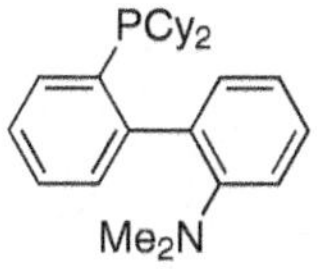 2-dimethylamino-2'-diphenylphosphinobiphenyl
dba	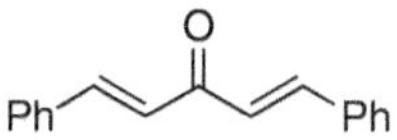 dibenzylideneacetone	DBN	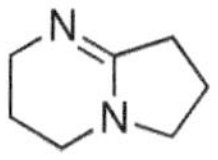 1,5-diaza-bicyclo[4.3.0]-non-5-ene
DBU	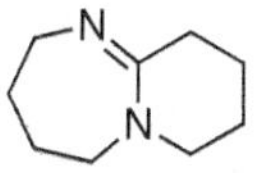 1,6-diaza-bicyclo[5.4.0]-undec-6-ene	DCC	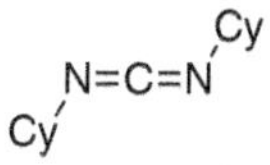 dicyclohexylcarbodiimide
DCE, EDC	$ClCH_2CH_2Cl$ 1,2-dichloroethane	DEAC	Et_2AlCl diethylaluminum chloride
DEAD	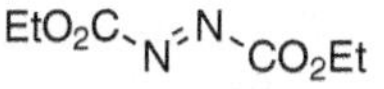 diethyl azodicarboxylate	DETA, dien	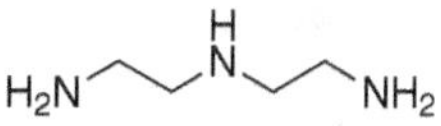 diethylenetriamine
DHP	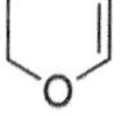 dihydropyran	DIAD	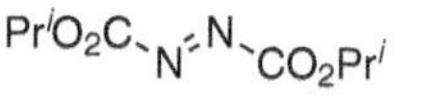 diisopropyl azodicarboxylate
DIC, DICI, DIPCDI	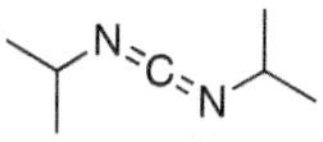 diisopropylcarbodiimide	DIEA, DIPEA, EDA, Hünig's base	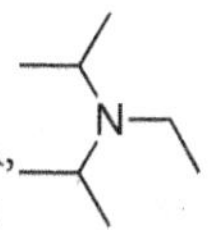 diisopropylethylamine

(continues)

(**Table 3.6** continued)

Acronym	Name/Structure	Acronym	Name/Structure
DIOP	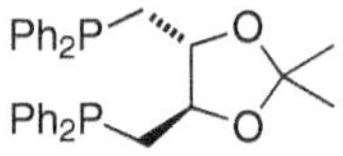 (−) 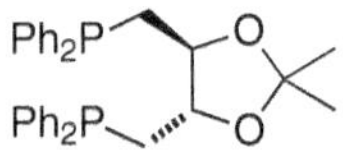 (+)	DIPA	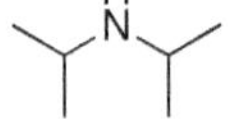 diisopropylamine
(*R*,*R*)-DIPAMP	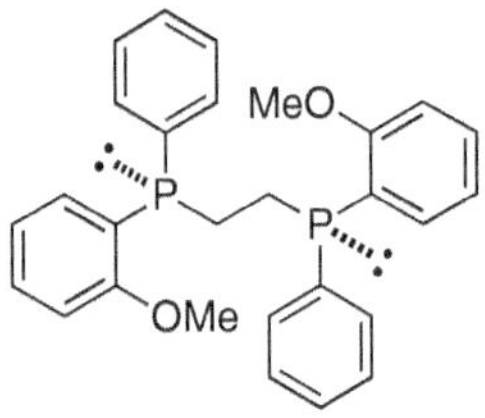	(−)-DIPT	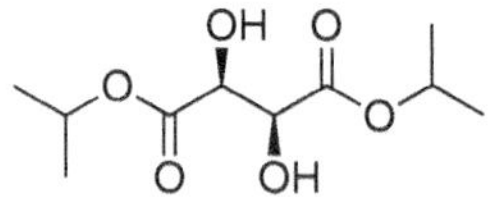 *S*,*S*-(−)-diisopropyl tartrate
(+)-DIPT	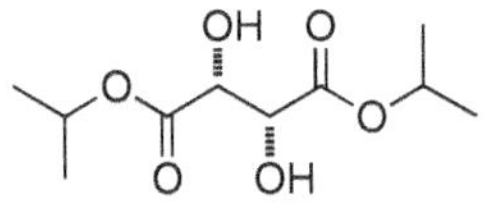 *R*,*R*-(+)-diisopropyl tartrate	DMAP	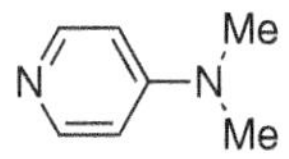 4-dimethylaminopyridine
DMS	Me_2S dimethyl sulfide	DNPH	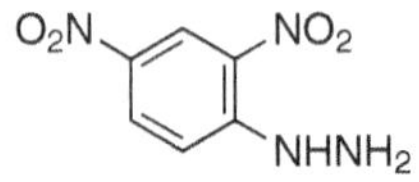 2,4-dinitrophenyl-hydrazine
dpbp	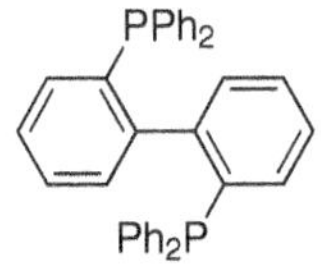 2,2'-*bis*(diphenyl-phosphino) biphenyl	dppe, or diphos	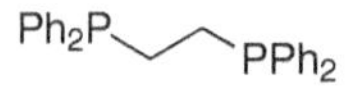 1,2-*bis*(diphenyl-phosphino) ethane
dppf	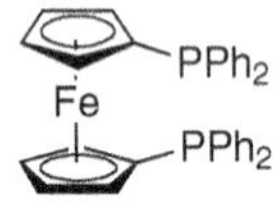 1,1'-*bis*-(diphenylphosphino)-ferrocene	EDCI, EDC	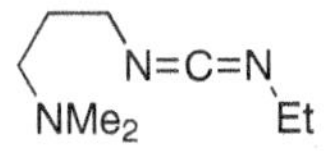 *N'*-(3-dimethylaminopropyl)-*N*-ethylcarbodiimide
EDTA	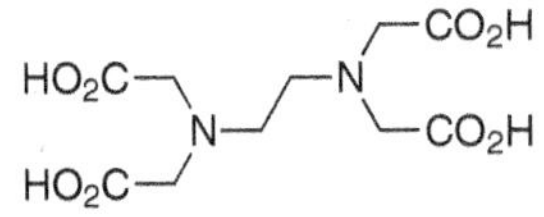 ethylenediaminetetraacetic acid	Grubbs-I	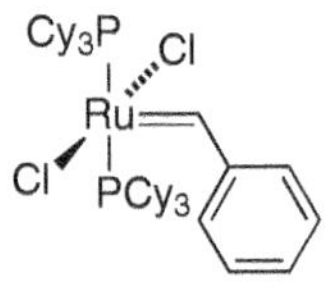
Grubbs-II	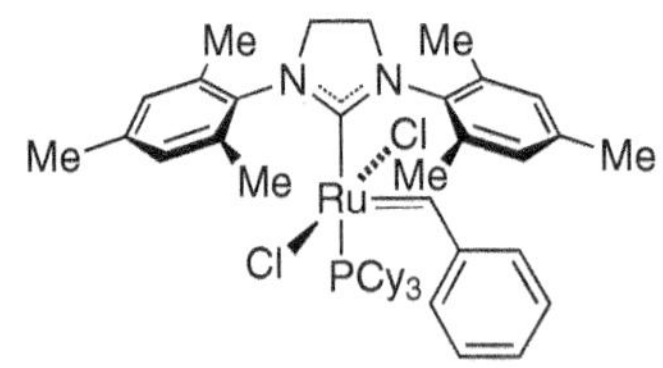	Grubbs-Hoveyda	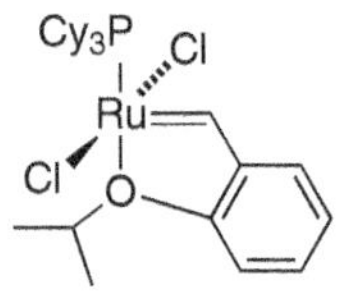

Acronym	Name/Structure	Acronym	Name/Structure
H_2TPP TPP	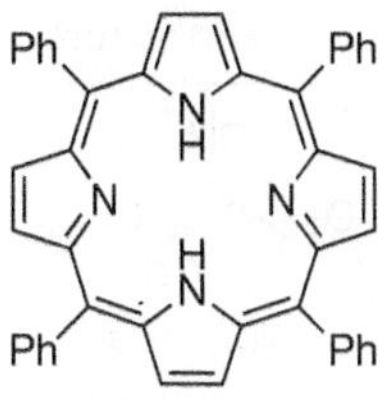 tetraphenylporphyrin	HBT, HOBt	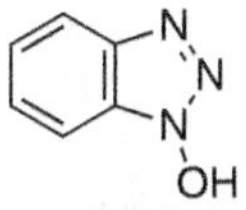 1-hydroxybenzotriazole
HFIP	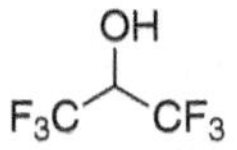 hexafluoroisopropyl alcohol	HMDS	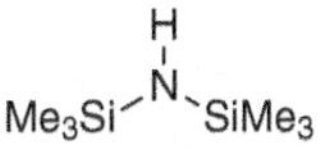 hexamethyldisilazane
Hünigs base	see DIEA	ICy	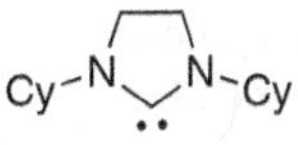 1,3-dicyclohexylimidazol-2-ylidene
IMes	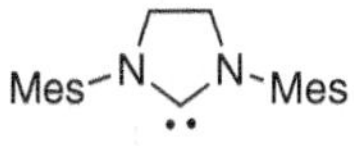 1,3-dimesitylimidazol-2-ylidene	KAPA	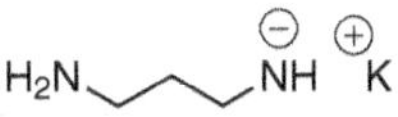 potassium 3-aminopropylamide
KHMDS, KHDS	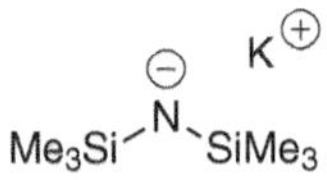 potassium hexamethyldisilazide	LDA	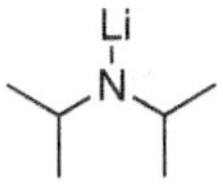 lithium diisopropylamide
LHMDS, LHDS	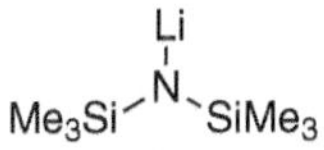 lithium hexamethyldisilazide	LICA	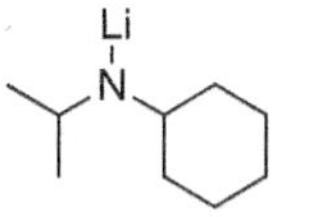 lithium isopropylcyclohexylamide
LTMP, LiTMP	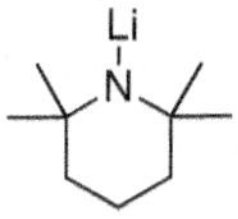 lithium 2,2,6-6-tetramethylpiperidide	Me-BPE, (*R,R*-Me-BPE, *S,S*-MeBPE)	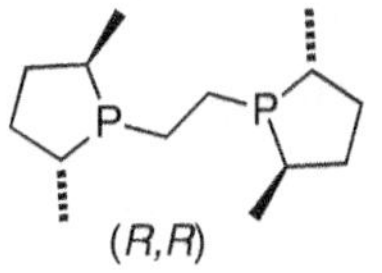 (*R,R*) 1,2-bis(2,5-dimethyl-phospholano)ethane
(+)-MIB	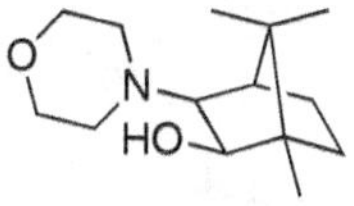 (1*S*,3*S*)-(+)-3-(morpholino) isoborneol	(-)-MIB	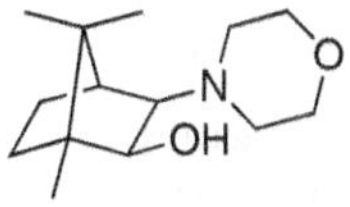 (1*R*,3*R*)-(–)3-(morpholino) isoborneol
MoOPH, Vedejs' reagent	MoO5•py•HMPA	MTPA, Mosher's acid	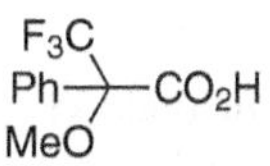 α-methoxy-α-(trifluoromethyl)-phenylacetic acid

(continues)

(**Table 3.6** continued)

Acronym	Name/Structure	Acronym	Name/Structure
MVK	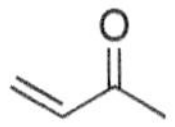 methyl vinyl ketone	NHS, HOSu	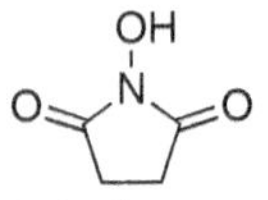 *N*-hydroxysuccinimide
NMM	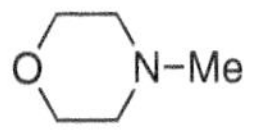 *N*-methylmorpholine	PABA, PAB	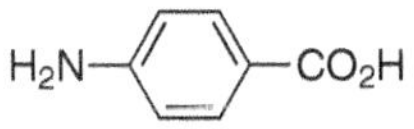 *p*-aminobenzoic acid
PAH	polycyclic aromatic hydrocarbon	PAMP	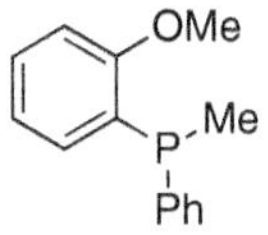 phenyl(*o*-anisyl)-methylphosphine
PET	photoelectron transfer	phen	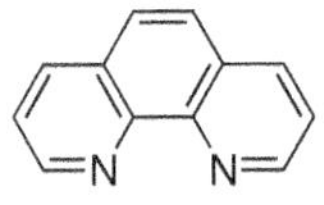 1,8-phenanthroline
PMHS	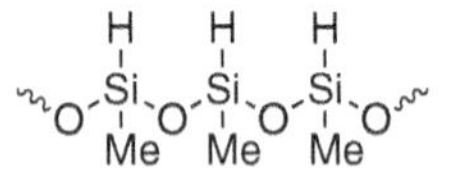 poly(methylsiloxane)	PPA	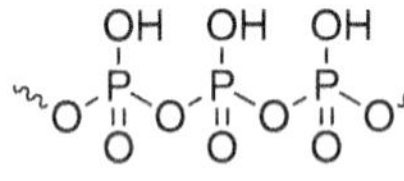 polyphosphoric acid
PPTS	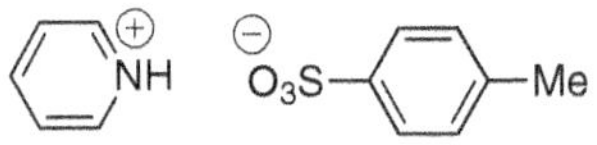 pyridinium *p*-toluenesulfonate	PTC	phase transfer catalyst
pybox	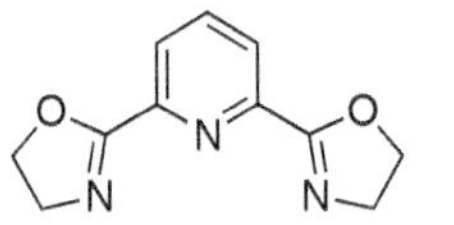 2,6-*bis*(oxazolin-2-yl)pyridine	RAMP	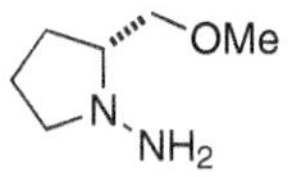 *R*-2-methoxymethyl-1-aminopyrrolidine
ROM	ring opening methathesis	ROMP	ring opening metathesis polymerization
SAMP	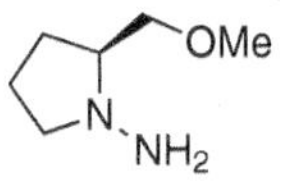 *S*-2-methoxymethyl-1-aminopyrrolidine	Schrock	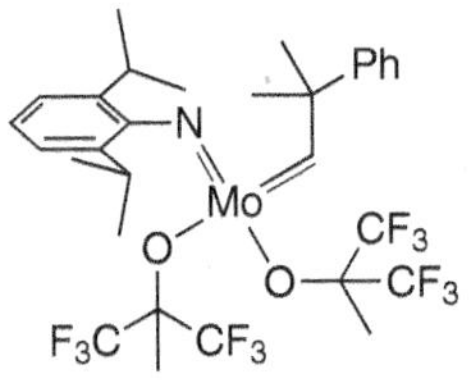
SDS	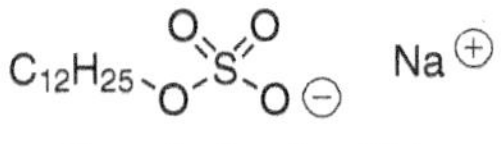 sodium dodecyl sulfate (sodium lauryl sulfate)	SET	single electron transfer
TASF	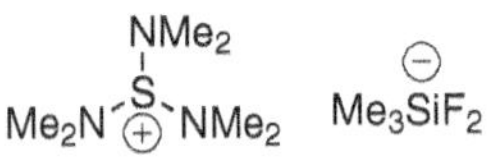 *tris*(dimethylamino)sulfonium difluorotrimethylsilicate	TBAB (C, F)	Bu_4N Br (Cl, F) tetra-*n*-butylammonium bromide (chloride, fluoride)

Acronym	Name/Structure	Acronym	Name/Structure
TFA	CF_3CO_2H trifluoroacetic acid	TFAA	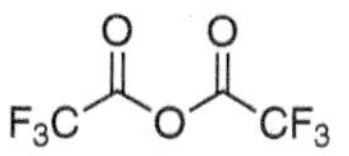 trifluoroacetic anhydride
TIP	Ti(O-i-Pr)$_4$ titanium tetra-isopropoxide	TMAB(C, F)	tetramethylammonium bromide (chloride, fluoride) Me_4N Br (Cl, F)
TMAH	Me_4N OH tetramethylammonium hydroxide	TMEDA	$Me_2NCH_2CH_2NMe_2$ tetramethylethylenediamine
TMG	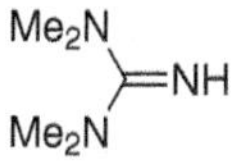 *N,N,N',N'*-tetramethylguanidine	TosMIC	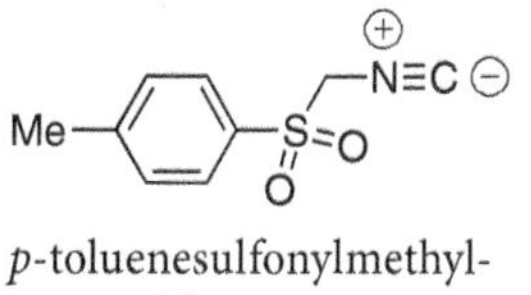 *p*-toluenesulfonylmethyl-isocyanide
TPP	Ph_3P triphenylphosphine	TPP	tetraphenylporphyrin, see H_2TPP
TPPO	$Ph_3P = O$ triphenylphosphine oxide		

The lists in the tables are not exhaustive by any means, and the number of acronyms in general use will almost certainly have grown by the time you are reading this. Still, the use of this type of shorthand to refer to reagents and reactions is now the norm in modern synthesis, so one needs to know what the latest acronyms are. There are four reasonably good web sites to find the acronym or name reaction you are seeking:

- www.cheminform.com/acronyms/
- www.chem.wisc.edu/areas/reich
- orgchem.chem.uconn.edu/namereact/named.html
- www.organic-chemistry.org/reactions.htm

The Cheminform web site is maintained by Wiley-VCH and Fiz Chemie Berlin (Fachinformationszentrum Berlin GmbH). Limited to reagent acronyms, it has been quite reliable since being posted to the web, but one should be careful to visit it regularly—the URL changed during 2014.The Reich web site at the University of Wisconsin is also a reliable resource, especially for name reactions, but also for acronyms of important reagents. The University of Connecticut web site, maintained by Michael B. Smith, is focused on name reactions, as is the fourth web site, which is part of the Organic Chemistry Portal maintained, in part, by Douglass Taber, of the University of Delaware.

Remember, however, the lifetimes of web sites are limited by a number of factors beyond the user's control, and they are not always maintained as well later on as they are when they are new.

CHAPTER FOUR

Orbitals and Reactivity

4.1 Introduction

Until the mid-1920s, the Lewis theory of bonding[1] was the only truly general rationalization of how bond formation occurs in chemical compounds. Now, a century later, it still provides a good qualitative description of bonding in organic compounds, and for that reason it is still in use. However, Lewis bonding theory does have its limitations, not the least of which is its failure to accurately predict the structures, bonding, and chemistry of certain types of compounds (e.g., nitromethane and the allyl cation). Many of these limitations were alleviated by expanding Lewis theory to include the theory of resonance,[2] but what was really required was a new theory that would account for the observed chemical and physical properties of compounds without requiring such empirical approaches as resonance. This theory, developed over the decades of the 1920s and 1930s, was quantum mechanics, which was applied to problems in chemistry in the form of ***valence bond theory;***[3] ***molecular orbital theory;***[4] and, in the last two decades of the 20th century, ***density functional theory.***[5]

---

1. Lewis, G.N. *J. Am. Chem. Soc.* **1916**, *38I*, 762.

2. (a) Pauling, L.; Wheland, G.W. *J. Chem. Phys.* **1933**, *1*, 362. (b) Pauling, L.; Sherman, J. *J. Chem. Phys.* **1933**, *1*, 606-617. (c) Wheland, G.W. *Resonance in Organic Chemistry* (John Wiley & Sons: New York, 1955).

3. (a) Heitler, W.; London, F. *Z. Phys.* **1927**, *44*, 455. (b) Segiura, Y. *Z. Phys.* **1927**, *45*, 484. (c) Wang, S.C. *Phys. Rev.* **1928**, *31*, 579. (d) Rosen, N. *Phys. Rev.* **1931**, *38*, 2099. (e) Weinbaum, S. *J. Chem. Phys.* **1933**, *1*, 593. (f) James, H.M.; Coolidge, A.S. *J. Chem. Phys.* **1933**, *1*, 825.

For an extended discussion: Coulson, C.A. *Valence*, 2nd ed. (Oxford University Press: Oxford, 1961), ch. 5, ch. 6.

4. For leading references, see: (a) Hund, F. *Z. Phys.* **1931**, *73*, 1, 565. (b) Mulliken, R.S. *J. Chem. Phys.* **1933**, *1*, 492. (c) Lennard-Jones, J.E. *Trans. Faraday Soc.* **1929**, *25*, 6668. (d) Hückel, E. *Z. Phys.* **1931**, *72*, 310; **1932**, *76*, 628; **1933**, *83*, 632. (e) Hückel, E. *Trans. Faraday Soc.* **1934**, *30*, 40.

Textbooks: (f) Pauling, L.; Wilson, E.B., Jr. *Introduction to Quantum Mechanics with Applications to Chemistry* (McGraw-Hill: London, 1935); reprinted by Dover Books, New York, 1985. (g) Coulson, C.A. *Valence*, (Oxford/Clarendon Press: London, 1952), ch. 4, ch. 6, ch. 7. (h) Liberles, A. *Introduction to Molecular Orbital Calculations* (Holt, Rinehart and Winston: New York, 1966). (i) Roberts, J.D. *Noptes of Molecular Orbital Theory* (W.A. Benjamin: New York, 1962). (j) Streitwieser, A., Jr. *Molecular Orbital Theory for Organic Chemists* (Wiley-Interscience: New York, 1961). (k) Dewar, M.J.S. *The Molecular Orbital Theory of Organic Chemistry* (McGraw-Hill: New York, 1969). (l) Dewar, M.J.S.; Dougherty, R.C. *The PMO Theory of Organic Chemistry* (Plenum: New York, 1975). (m) Zimmerman, H.E. *Quantum mechanics for Organic chemists* (Academic Press: New York, 1975). (n) Richards, W.G.; Horsley, J.A. *Ab Initio Molecular Orbital Calculations for Chemists* (Oxford University Press (Clarendon Press): New York, 1970). (o) Pople, J.A.; Beveridge, D.L. *Approximate Molecular Orbital Theory* (McGraw-Hill: New York, 1970). (p) Jorgensen, W.L.; Salem, L. *The Organic Chemist's Book of Orbitals* (Academic Press: New York, 1973).

5. (a) Parr, R.G.; Yang, W. *Density Functional theory of Atoms and molecules* (Oxford University Press: Oxford, 1989). (b) Hohenberg, P.; Kohn, W. *Phys. Rev. A* **1964**, *136*, 864. (c) Becke, A.D. *Phys. Rev. A* **1988**, *38*, 3098. (d) Becke, A.D. *J. Chem. Phys.* **1992**, *96*, 2155; **1992**, *97*, 9173; **1993**, *98*, 5648.

4.2 Atomic Orbitals: The History of the Modern Atomic Model

The equation for determining the energy of a photon is due to Max Planck,[6] who first introduced the idea that light was transmitted in the form of small energy packets, which he called ***quanta***. The energy could be straightforwardly related to the frequency of the radiation.[7] Planck's equation, $E = h\nu$, has become one of the cornerstones of modern atomic physics and modern chemistry.

The idea that energy in an atom is quantized was a revolutionary one, reinforced when the Danish physicist Niels Bohr[8] proposed a model of the nuclear atom in which the quantization of energy levels was an integral part of the model.[9] From Bohr's work first emerged the concept of ***quantum numbers***, which have become so important in more recent theories of atomic structure and bonding.

The electron had been discovered in the late 19th century, and its behavior had marked it as a particle (e.g., it possesses both mass and momentum). However, electron beams can be diffracted—behavior restricted to waves—so that electrons also exhibit some characteristics of a wave. This initial radical departure from the principles of classical mechanics—that particles could exhibit wave behavior and that waves could exhibit particle-type behavior—was first suggested in 1924 by Louis de Broglie,[10] who proposed the term ***wave-particle duality***[11] to describe the equivalence of matter and energy at the atomic and subatomic level.

The dual nature of the electron solved some problems in atomic physics, but more problems were added by the work of Werner Heisenberg,[12] who pointed out that many of the ideas of classical physics could not be applied at the atomic level. In his ***uncertainty principle***,[13] he stated that the position and momentum of a particle cannot be known simultaneously to such precision that the product of the uncertainty in each quantity is less than $h/2\pi$. Equation 4.1, due to Kennard[14] and Weyl,[15] uses the standard deviations of position (σ_x) and momentum (σ_p):

$$\sigma_x \bullet \sigma_p \geq h/2\pi \qquad (4.1)$$

---

6. **Max Karl Ernst Ludwig Planck (1858-1947)** was educated at Munich and Berlin. In 1889, he joined the faculty of Physics at Berlin; he received the 1918 Nobel Prize in Physics. After the war, Planck resumed the presidency of the Kaiser Wilhelm Society, which was renamed the Max Planck Society in his honor. For more biographical details, see the web site of the Nobel Foundation.

7. Planck, M. *Ann. Phys.* **1901**, *4*, 553.

8. **Niels Hendrik David Bohr (1885-1962)** was educated in Copenhagen and Cambridge. He returned to Copenhagen in 1916 as professor of physics. Bohr escaped the Nazi occupation of Denmark and eventually became an important contributor to the Manhattan Project. Bohr returned to Copenhagen in 1945 and devoted the rest of his life to finding peaceful uses for atomic energy. There are numerous biographies of Bohr available, including Blaedel, N. *Harmony and Unity: The Life of Niels Bohr* (Science Tech.: Madison, Wisconsin, 1988).

9. Bohr, N. *Phil. Mag*, **1913**, *26*, 1; **1914**, *27*, 506.

10. **Louis Victor Pierre Raymond, 7th duc de Broglie (1892-1987)** took his PhD in theoretical physics at the Sorbonne in 1924. De Broglie was professor of physics at the Université de Paris from 1928 until his retirement in 1962. He received the Nobel Prize in Physics in 1929. In 1960, he succeeded his brother as duc de Broglie. For more biographical information, see: Abraham, A. *Biogr. Mem. Fellows Roy. Soc.* **1988**, *32*, 22.

11. De Broglie, L. Thesis, Université de Paris (Sorbonne), 1924; *Ann. phys.* (10) **1925**, *3*, 22.

12. **Werner Karl Heisenberg (1901-1975)** was educated at Munich. He became professor of theoretical physics at Leipzig in 1928 and was Professor in Berlin and Director of the Kaiser Wilhelm Institute during World War II. Heisenberg was awarded the 1932 Nobel Prize in Physics. His work on the Nazi atomic bomb is still a source of controversy. For more biographical details, see: Rechenberg, H.; Wiemers, G. *Werner Heisenberg (1901–1976), Schritte in die neue Physik* (Sax-Verlag Beucha, 2001).

13. Heisenberg, W. *Z. Phys.* **1927**, *43*, 172.

14. Kennard, E.H. *Z. Phys.* **1927**, *44*, 326.

15. Weyl, H. *Gruppentheorie und Quantenmechanik* (Hirzel: Leipzig, 1928).

The insight of genius required to consolidate the theories of atomic structure and bonding was supplied by the Austrian physicist Erwin Schrödinger,[16] who combined a number of principles and theories extant in the mid-1920s and unified them into what we now term ***wave mechanics***, or ***quantum mechanics***.[17] The English physicist P. A. M. Dirac[18] combined the work of de Broglie, Heisenberg, and Schrödinger into one unified theory.[19] In simple terms, Schrödinger took the approach that electrons about an atomic nucleus will behave exactly like a standing wave in three dimensions. However, he also incorporated into his theory the Heisenberg uncertainty principle, so that instead of addressing the positions and energies of electrons in atoms in classical terms, he addressed the problem in terms of probability. Because the spectra of atoms can be measured in very high precision, the energy levels of the electrons in atoms can be determined with a very small uncertainty. Accordingly, the position of the electron must be highly uncertain. Although it had been shown that quantization of the energy levels of the atom led to a model that could be used to predict atomic properties, one of the major questions which Bohr's theory of the atom had left unaddressed was *why* the energy levels of atoms should be quantized. Wave mechanics answered this question.

Standing Waves and Atomic Orbitals

At the heart of quantum mechanics is the premise that one can reasonably treat the electrons of an atom or molecule as a standing wave in three dimensions.

The simplest standing wave is the standing wave in one dimension, exemplified by a vibrating string. The displacement, r, of any part of a vibrating string at time t can be calculated by using the standard equation for a simple harmonic oscillator if one knows two fundamental properties of the vibration: the amplitude of the vibration, A, which is defined as the maximum displacement of the string from its equilibrium position, and its frequency, ω

$$r = A \sin(2n\pi\omega t) \qquad (4.2)$$

Equation 4.2 is the ***wave function*** for a standing wave in one dimension and is simply an algebraic expression that one can use to describe a vibrating string. The variable n in this equation is an integer that tells if one is dealing with the fundamental vibration ($n = 1$), or an overtone or harmonic ($n = 2, 3, 4, \ldots$). This number is strictly analogous to the quantum numbers of atomic physics. The wave functions for the fundamental and first three harmonic vibrations of a standing wave in one dimension are plotted in Figure 4.1.

In Figure 4.1, there are several important points marked as ***nodes***. A node is defined as *a point within the standing wave where the displacement from the equilibrium position is always zero*. A standing wave in one dimension must have a minimum of two nodes, one at each

---

16. **Erwin Schrödinger (1887-1961)** was educated in Vienna and remained there until 1920. In 1921, he went to Zürich, and in 1927, he joined the faculty at Berlin but left Germany in 1933 over the Nazi treatment of the Jews. From 1940 to 1956, he taught at Dublin before returning to Vienna. He shared the 1933 Nobel Prize in Physics. For more detail, see: Moore, W.J. *A Life of Erwin Schrödinger* (Canto ed.) (Cambridge University Press: Cambridge, 2003).

17. Schrödinger, E. *Ann. Phys.* **1926**, *79*, 361, 489, 734; **1926**, *80*, 437; **1926**, *81*, 109.

18. **Paul Adrien Maurice Dirac (1902-1984)** was educated at Bristol and Cambridge, where he was appointed Lucasian Professor of Mathematics at the age of 30; he was Lucasian Professor until 1969. Dirac shared the 1933 Nobel Prize in Physics with Heisenberg. Following his retirement from Cambridge, Dirac became Professor of Physics at Florida State University, a post he held until his death. For more biographical detail, see: Dalitz, R.H.; Peierls, R. *Biogr. Mem. Fellows Roy. Soc.* **1986**, *32*, 138.

19. Dirac, P.A.M. *Proc. Roy. Soc. A* **1927**, *113*, 621; **1927**, *114*, 243; **1928**, *117*, 610.

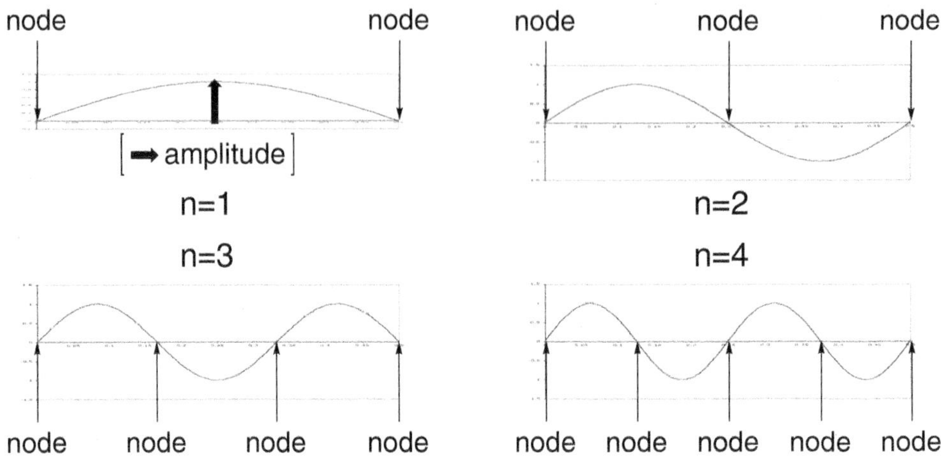

Figure 4.1 A vibrating string provides an example of a standing wave in one dimension

end of the wave. For convenience, we shall call these nodes terminal nodes. For the fundamental vibration, these are the only nodes; for the harmonics, however ($n = 2, 3, 4, \ldots$ in Equation 1.6), there are $(n - 1)$ internal nodes in addition to the two terminal nodes. In a one-dimensional standing wave, all nodes are ***point nodes***, or ***nodal points***.

The wave equation of a standing wave in two dimensions is more complex than the wave equation for a vibrating string because the vibration is occurring in more dimensions. When reduced to its simplest form, the wave equation for a standing wave in two dimensions requires two, rather than one, "quantum number"—one for each dimension—to describe the vibration completely.

The best analogy for a standing wave in two dimensions is probably a vibrating drumhead. Some typical vibrations of a drumhead are illustrated in Figure 4.2. In two dimensions, the nodes are no longer nodal points, but are, instead, nodal lines; there are two major types of nodes in two dimensions—circular nodes and linear nodes—which divide the vibrations into two classes (symmetrical vibrations and antisymmetrical vibrations). The symmetrical vibrations of a drumhead have only circular nodes; they are shown as the top line of drumhead vibrations. The antisymmetrical vibrations of a drumhead have at least one linear node, so that half of the drumhead is above the average plane, while half of the drumhead is below it. The antisymmetrical vibrations are the lower ones in Figure 4.2. Exactly what type of vibration one is examining will be determined by the values of the two "quantum numbers" in the wave equation.

Because of the additional dimension, the picture in three dimensions becomes even more complex, although the same rules apply. In three dimensions, the nodes now become surfaces, rather than lines or points, and three "quantum numbers," rather than two, are required to describe the vibration. Just as there were two types of nodal lines in standing waves in two dimensions (a vibrating drumhead), there are two types of nodal surfaces in three dimensions: spherical nodes, which are analogous to the circular nodes in two dimensions, and planar nodes, which are analogous to the linear nodes in two dimensions.

Schrödinger's equation for the energy of an electron in an atom is a second-order differential equation for a standing wave in three dimensions. As with the string and the drumhead, the fact that this equation describes a *standing* wave means that only certain solutions are allowed (there may not be a nonintegral number of waves within the atom). The energy of the atom can have only certain values—it must be quantized. Here, for the first time, the origin of the quantization of atomic energy levels was explained.

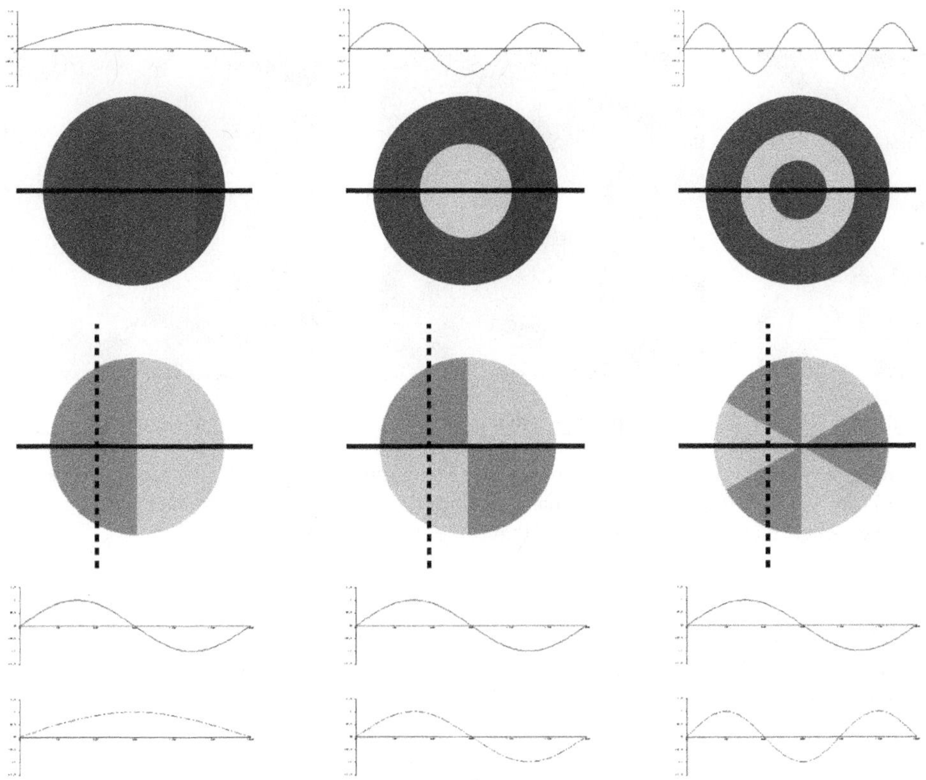

Figure 4.2 Standing waves in two dimensions. The dark areas and light areas are moving in opposite directions with respect to the viewer. Standing waves in two dimensions have nodal lines (circles or straight lines) rather than nodal points. The displacement of the drumhead with respect to the dashed lines drawn on each diagram is shown above the diagram for the top set of vibrations and below the diagram for the lower set of vibrations.

Each of the solutions to the Schrödinger equation is in the form of an algebraic function—the ***wave function*** in three dimensions. Each of these wave functions (usually designated by the Greek letter ψ for atomic orbitals, and by the Greek letter ϕ for molecular orbitals) is properly called the ***probability amplitude*** of an electron of a given energy in any region of space around the nucleus of the atom. The square of the wave function (ϕ^2 or ψ^2) gives the probability of finding an electron within a given volume around the nucleus of the atom. One of the important things to note about wave functions is that the sum of the squares of all the orbital wave functions in a given subshell is the equation of a sphere.

Because this is the simplest set of solutions, we usually describe the atom in terms of orbitals that are mathematically orthogonal (i.e., the product of the wave functions of any pair of orbitals is zero). The individual orbitals of an atom are described in terms of three integer quantum numbers. These quantum numbers are the ***principal quantum number***, ***n***, which defines the electron shell in which the orbital occurs and specifies the total number of nodes in that orbital; the ***azimuthal quantum number***, ***l***, which defines the subshell to which the orbital belongs and specifies the type of orbital by specifying the number of planar nodes (there are $l - 1$); and the ***magnetic quantum number***, m_l, which defines the orientation of that particular orbital in space. There are a total of n^2 orbitals in an energy level for which the principal quantum number has the value n.

We can describe the shapes of orbitals by specifying their symmetry. This is a useful way of describing orbitals because it allows us to identify a directional component of the atomic orbital that is important in the formation of molecular orbitals. The electron probabilities for a 2p and a 4s orbital are given in Figure 4.3.

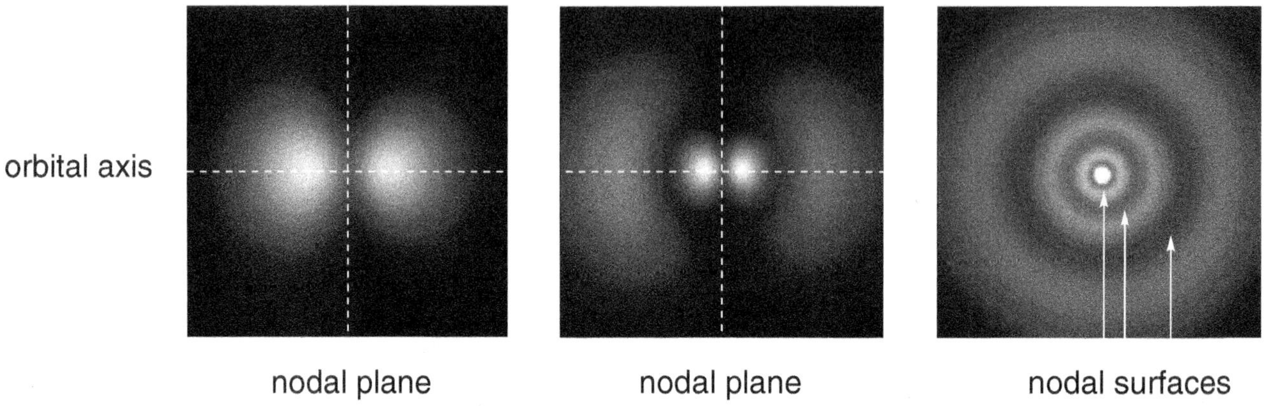

Figure 4.3 Projections of the electron probability distributions for 2p (left), 3p (center), and 4s (right) orbitals. Note that the 4s orbital has only spherical nodal surfaces, that the 2p and 3p orbitals have a single planar nodal surface, and that there are additional nodes in the 3p.

All s orbitals are spherically symmetrical; they have no directional characteristics. In contrast, p orbitals have a unique axis perpendicular to a single planar node passing through the nucleus, so unlike s orbitals, p orbitals have a directional character. When a p orbital is viewed from a perspective perpendicular to the nodal plane (i.e., along the axis), it always appears circular in cross-section—it is cylindrically symmetrical.[¶]

The fourth quantum number for electrons is the ***spin quantum number***, m_s, which takes one of only two permitted values, ± 1/2. Because of this restriction on the possible values for the spin quantum number of an electron, the capacity of an orbital is restricted to two electrons. The origins of this restriction on the electron capacity of an orbital were first set forth in 1925 by Austrian-born physicist Wolfgang Pauli,[20] in the form of the principle that bears his name, the ***Pauli Exclusion principle***. This principle states that *no two electrons in any atom may simultaneously have the same values for all four quantum numbers.*[21] In other words, because the electrons in any single orbital must have the same values of n, *l*, and m_l, they must differ in the value of m_s. Because there are only two possible values of m_2, *the orbital can hold only two electrons.* The *Pauli Exclusion principle applies not only to electrons in the orbitals of isolated atoms but also to electrons in the orbitals of molecules.* Two electrons occupying the same orbital are said to be *paired.*

¶. ***Spatial Properties of Wave Functions***
One of the important things to note about wave functions is that the sum of the squares of the orbital wave functions in a given subshell is the equation of a sphere. Let us take the wave functions for the orbitals of the 2p subshell, where ρ is the ***radial distribution function***, which is independent of orientation:

Let us take the wave functions for the orbitals of the 2p subshell, where ρ is the ***radial distribution function***, which is independent of orientation:

$$2p_x: \rho(3/4\pi)^{1/2}\sin\theta\cos\phi = \rho(3/4)^{1/2}x$$
$$2p_y: \rho(3/4\pi)^{1/2}\sin\theta\sin\phi = \rho(3/4)^{1/2}y$$
$$2p_z: \rho(3/4\pi)^{1/2}\cos\theta = \rho(3/4)^{1/2}z$$

When squared and summed, this becomes $\psi_x^2 + \psi_y^2 + \psi_z^2 = (3/4\pi)\rho^2(x^2 + y^2 + z^2)$, which is the equation for a sphere. A similar situation holds true for both subsets of the five orbitals of a d subshell (d_{xy}, d_{yz}, d_{xz} and $d_{x^2-y^2}$, d_{z^2}).

20. **Wolfgang Pauli (1900-1958)** was educated at Munich. He taught physics at Göttingen, Copenhagen, and Hamburg before joining the ETH in Zürich. He received the 1945 Nobel Prize in Physics for his work in atomic physics. In 1940, Pauli joined the Institute for Advanced Study at Princeton University, and in 1946 was naturalized a US citizen. After World War II, he returned to Zürich, where he spent the remainder of his life. Biographies of Pauli abound; see, for example, Enz, C.P. *No Time to be Brief, A scientific biography of Wolfgang Pauli.* (Oxford University Press: Oxford, 2002).

21. Pauli, W. *Z. Phys.* **1925**, *31*, 765.

The filling of the orbitals around the nucleus is carried out according to the ***Aufbau principle*** (from the German *Aufbauprinzip*, "principle of building up"). This principle, named by Bohr in 1922,[22] derives its name from the German word to build up. It states that *when filling orbitals around a nucleus of an atom, the lowest energy orbitals are filled first. Like the Pauli Exclusion principle, the Aufbau principle also holds true for orbitals in molecules.*

When the application of the Pauli Exclusion and Aufbau principles leads to more than one possible electron configuration for an atom or molecule (i.e., when the highest energy orbital containing electrons is part of a degenerate set of orbitals), the distinction between the possible configurations is made possible by ***Hund's rule***,[23] first promulgated by Friedrich Hund.[24] This rule states that *the lowest energy electron configuration of an atom or molecule where the highest energy electrons partially fill a set of degenerate orbitals is the one in which there is the maximum possible number of unpaired electrons.*

Worked Problem

4-1 The molecular orbitals of cyclobutadiene (C_4H_4) fall into the pattern below, where there are three sets of degenerate bonding orbitals (C–H σ, C–C σ, and C–C π), with their antibonding counterparts, as well as a degenerate set of nonbonding (n) orbitals. There are 20 valence electrons in the cyclobutadiene molecule. Complete the diagram by distributing the electrons to the appropriate orbitals.

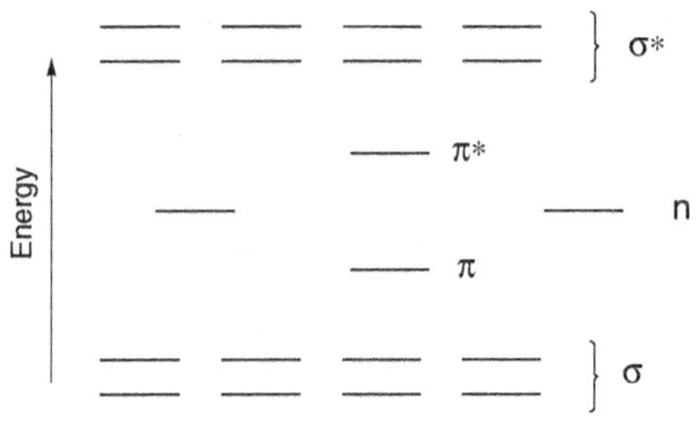

§Answer on next page.

4.3 Covalent Bonding and Molecular Orbitals

In the Lewis theory of bonding, a covalent bond is a pair of electrons shared between two nuclei. In the more modern view, this has changed relatively little, except that the electron pair is now placed into a molecular orbital, which encompasses both nuclei. Molecular orbitals differ from atomic orbitals in one very important respect. *Atomic orbitals are associated with one and only one nucleus; molecular orbitals are associated with at least two nuclei.* Atomic orbitals are always labeled with the Roman alphabet (e.g., s, p, d, f, g), whereas molecular orbitals are always labeled with Greek letters (e.g., σ, π, Δ).

22. Bohr, N. *Z. Phys.* **1922**, *9*, 1.

23. Hund, F. *Z. Pys.* **1925**, *33*, 345; *Linienspektren und periodisches Systeme der Elements* (Springer: Berlin, 1927), p. 124.

24. **Friedrich Hund (1896-1997)** was educated at Göttingen and Marburg (PhD, Göttingen, 1922). His academic career began at Göttingen, and he held positions at Rostock, Leipzig, Jena, and Frankfurt before returning to Göttingen to finish his career. Leading sources for more biographical information on this centenarian scientist can be found at the Leipzig University web site: http://www.uni-leipzig.de/unigeschichte/professoren katalog/leipzig/Hund_67/

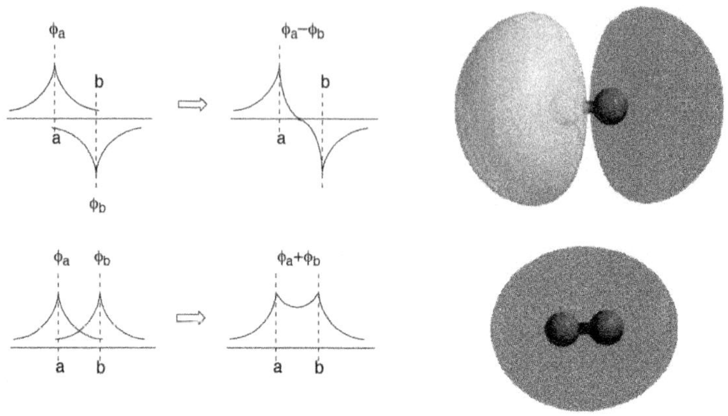

Figure 4.4 The overlap of hydrogen atomic orbitals. Subtracting the wave functions, or *out-of-phase overlap*, gives an antibonding molecular orbital. Adding the wave functions, or *in-phase* overlap, gives a bonding orbital.

The formation of molecular orbitals from atomic orbitals is usually represented as involving overlap of the two atomic orbitals to form the new molecular orbitals. Mathematically, this involves the linear combination (addition or subtraction) of the wave functions for the atomic orbitals according to a strict set of rules: the ***linear combination of atomic orbitals*** (LCAO) method. One of the most stringent rules for combining wave functions (orbitals) can be set forth as a "rule of conservation of orbitals."

> During the formation of a set of molecular orbitals from a set of atomic orbitals on two or more atoms, the mathematics mandates that the number of molecular orbitals produced must be exactly equal to the number of atomic orbitals used. Orbitals can be altered but neither created nor destroyed.

Molecular orbitals can be generated by coaxial overlap of orbitals to give cylindrically symmetrical σ and σ^* orbitals or by coplanar overlap of orbitals to give π and π^* orbitals with mirror plane symmetry. In forming two molecular orbitals by overlap of two atomic orbitals (Figures 4.4 and 4.5), the only permitted combinations of atomic wave functions (designated ϕ_a and ϕ_b for the two atoms, a and b) are (1) the sum, $(1/\sqrt{2})(\phi_a + \phi_b)$, termed

§ Answer for Worked Problem:
The 20 valence electrons of the molecule are distributed into this set of orbitals according to the Aufbau principle. This gives the following electron configuration for the molecule, with two unpaired electrons, each in one of the two degenerate n orbitals.

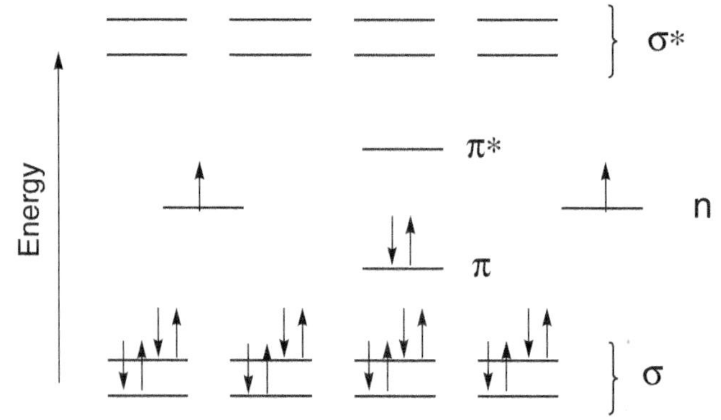

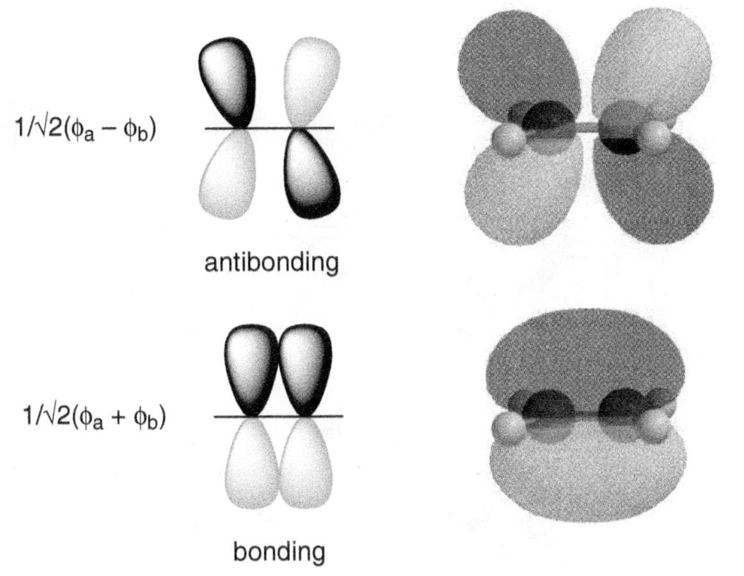

Figure 4.5 Out-of-phase coplanar overlap of p orbitals gives an antibonding π^* orbital. In-phase coplanar overlap of p orbitals gives a bonding π orbital.

in-phase overlap) and (2) the difference, $(1/\sqrt{2})(\phi_a - \phi_b)$, termed ***out-of-phase overlap***). The factor $1/\sqrt{2}$ is called a ***normalization factor*** and simply scales the wave function so that its square is equal to 1.

These two combinations arise from the mathematical constraints on orbital overlap. The molecular orbitals formed must be orthogonal (i.e., the product of their wave functions must be zero), and the sum of the squares of the molecular orbitals must be equal to the sum of the squares of the atomic orbitals from which they are formed. In-phase overlap gives a molecular orbital lower in energy than either of the atomic orbitals used in its formation; this is the ***bonding molecular orbital***.

Out-of-phase overlap gives a molecular orbital higher in energy than either of the atomic orbitals used in its formation; this is the ***antibonding molecular orbital***. *The Aufbau principle, the Pauli Exclusion principle, and Hund's rule apply to molecules in exactly the same way they apply to atoms.* Thus, electrons occupy the bonding orbitals first, and antibonding orbitals remain empty.

Up to this point, we have not addressed the effects of orbital energies on the efficiency of orbital overlap. Fortunately, the relationship is both simple and logical—the efficiency of orbital overlap increases as the energies of the two orbitals involved become more closely matched—or more nearly equal. In particular, orbital overlap is most efficient when orbitals in the same electron shell are involved. A carbon 2p orbital, for example, will form a very strong π bond when it overlaps with the 2p orbital of an oxygen atom, and it will form only a relatively weak π bond when it overlaps with the 3p orbital of a phosphorus atom.

Hybrid Atomic Orbitals: Valence Bond Theory

The methane molecule, CH_4, is known to be tetrahedral, and any bonding model that we use to describe the structure of methane should produce this geometric result. However, the atomic orbitals of the valence shell of carbon are not oriented at 109.471 . . .° to each other but rather at 90° to each other. This problem is solved quite elegantly one way by noting that one can inscribe a tetrahedron in a cube, which permits the formation of a tetrahedral molecule from a set of orbitals that are perpendicular to each other. In this way, one can obtain four bonding and four antibonding orbitals, as shown in Figure 4.6.

The result of this approach in methane is one σ-symmetric bonding orbital and its corresponding σ^* antibonding orbital as well as a set of three degenerate π-symmetric

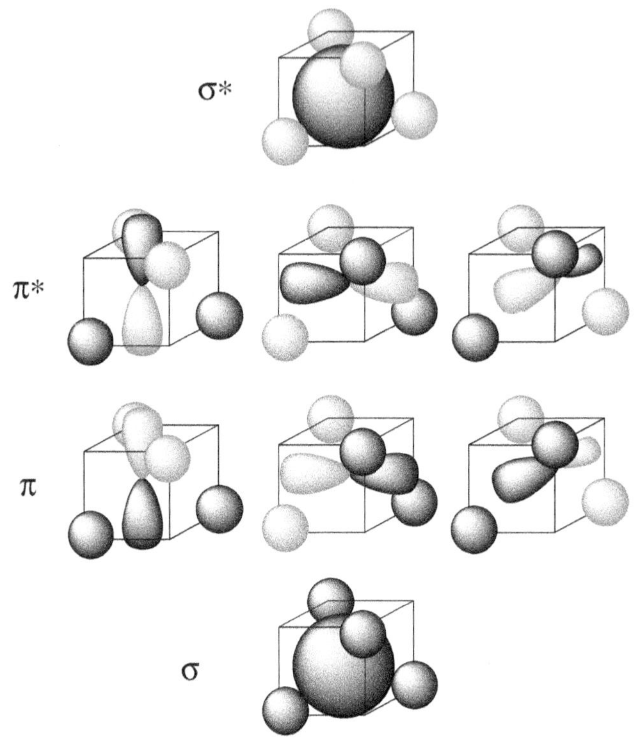

Figure 4.6 Linear combinations of atomic orbitals to generate one set of molecular orbitals for methane.

orbitals and their corresponding π^* antibonding orbitals.[¶] Photoelectron spectroscopy of the methane molecule (which gives information about the electron energies of the molecule) shows two peaks in the ratio 3:1, which lends support to this model of the methane molecule.[25]

A similar set of wave functions can be derived from the 12 atomic orbitals of carbon and hydrogen for the 12 LCAO molecular orbitals of the ethylene molecule; these orbitals are shown in Figure 4.7. There is again a pair of unique orbitals with σ and σ^* symmetry derived from the carbon 2s orbitals and the hydrogen 1s orbitals, with two sets of two π-symmetry orbitals and their corresponding π^* antibonding orbitals. These antibonding orbitals are formed from the carbon $2p_x$ orbitals and the hydrogen 1s orbitals, and a corresponding set is formed from the carbon $2p_y$ orbitals and the hydrogen 1s orbitals. The ethylene molecule also has a pair of unique π and π^* orbitals formed from the carbon $2p_z$ orbitals (compare with Figure 4.5).

This LCAO method for describing the bonding in molecules is well adapted to computers, but the molecular orbitals rapidly become too complex for most individuals to

¶. ***Wavefunctions of Simple Molecules from Atomic Basis Orbitals***
The wave functions of the molecular methane shown above are:

$$\phi_{\sigma} = (1/2\sqrt{2})(2\phi_s + \phi_a + \phi_b + \phi_c + \phi_d)$$
$$\phi_{\pi(x)} = (1/2\sqrt{2})(2\phi_x + \phi_a + \phi_b - \phi_c - \phi_d)$$
$$\phi_{\pi(y)} = (1/2\sqrt{2})(2\phi_y + \phi_a - \phi_b + \phi_c - \phi_d)$$
$$\phi_{\pi(z)} = (1/2\sqrt{2})(2\phi_z + \phi_a - \phi_b - \phi_c + \phi_d)$$
$$\phi_{\pi^*(x)} = (1/2\sqrt{2})(2\phi_x - \phi_a - \phi_b + \phi_c + \phi_d)$$
$$\phi_{\pi^*(y)} = (1/2\sqrt{2})(2\phi_y - \phi_a + \phi_b - \phi_c + \phi_d)$$
$$\phi_{\pi^*(z)} = (1/2\sqrt{2})(2\phi_z - \phi_a + \phi_b + \phi_c - \phi_d)$$
$$\phi_{\sigma^*} = (1/2\sqrt{2})(2\phi_s - \phi_a - \phi_b - \phi_c - \phi_d)$$

25. Hamrin, K.; Johansson, G.; Gelius, U.; Fahlman, A.; Nordling, C.; Siebahn, K. *Chem. Phys. Lett.* **1968**, *1*, 613.

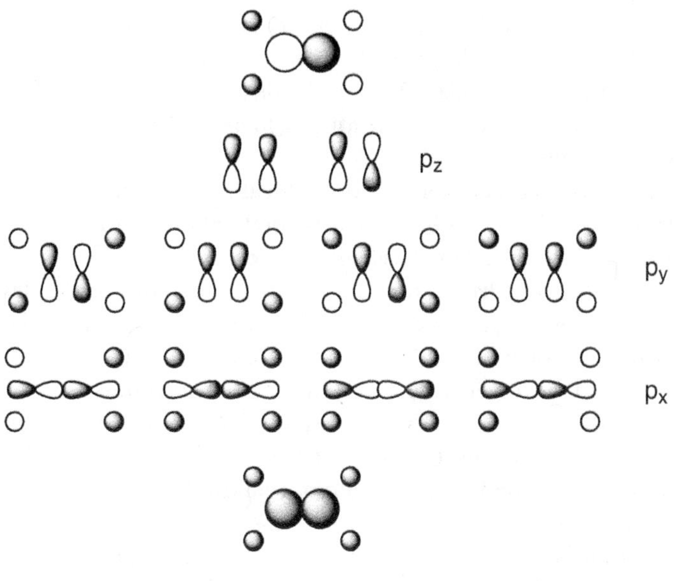

Figure 4.7 Linear combinations of atomic orbitals to generate one set of molecular orbitals for ethylene.

visualize without one. The most widely used model for visualizing bonding in molecules describes it in terms of two types of bond: σ bonds, formed by coaxial overlap of atomic orbitals, and π bonds, formed by coplanar overlap of p orbitals. The most stable arrangement of electrons in a molecule corresponds to the one in which all atoms are linked by σ bonds, with the minimum number of σ bonds being used. In this model, multiple bonds consist of one—and only one—σ bond, and one or more π bonds. π Orbitals differ from σ orbitals in one very important way; because the p atomic orbitals used in their formation have a nodal plane through the nucleus of the atom, π molecular orbitals also have a nodal plane through the two nuclei—*π electrons do not occupy the space directly between the two nuclei.* This means that π electrons are always further from the nuclei defining the bond than are σ electrons, so that they are not so tightly bound, and they are more polarizable and more responsive to external electronic influences. π Bonds are more reactive than σ bonds.

The side-by-side overlap of the p orbitals required to form the π and π* orbitals also imposes another very stringent requirement on molecules containing π bonds. Because π overlap is most effective when the two p orbitals are parallel and coplanar, any rotation of the atoms at the two ends of the π bond relative to each other will weaken, and eventually break, the π bond. Therefore, unless a compound has two π bonds between the same two atoms (i.e., a triple bond), such rotation about the π bond is forbidden.

Problems

4-1 Draw a set of LCAO molecular orbitals for acetylene, HC ≡ CH, similar to those drawn in Figure 4.7 for ethylene.

4-2 Draw a set of molecular orbitals for ethane similar to those in Figure 4.7 for ethylene. (Hint: put the two carbon atoms along the x axis, with two hydrogen atoms in the xz plane.)

The molecular orbitals generated by the LCAO approach above are all ***delocalized***—every molecular orbital encompasses every nucleus of the molecule. Although this approach is ideally suited for the generation of molecular orbitals by computer programs, it is rather less intuitive for most organic chemists, who find it much easier to visualize the localized bonds of the Lewis model. The same holds true when it comes to discussing the chemistry of compounds; it is often less confusing to use the simple Lewis model of bonding with its localized covalent bonds.

An alternative to this process of forming delocalized molecular orbitals is provided by ***valence bond theory***, which is basically a theoretical form of Lewis bonding theory. At its core, it involves the overlap of ***hybrid atomic orbitals*** to form localized molecular orbitals. The process of ***hybridization***,[26] which is one of the simplest models for obtaining directed atomic orbitals capable of forming localized molecular orbitals, involves combining the wave functions of orbitals of a single atom according to simple mathematical principles to generate a set of new atomic orbitals. The mathematics of hybridization results in a set of wave functions that are also valid solutions to the Schrödinger equation and a set of orbitals whose overall symmetry, in terms of the electron probability distribution of the atom, remains unchanged. Hybridization of carbon and other second-row elements always involves a 2s orbital, with one or more 2p orbitals. The unused p orbitals remain unchanged, orthogonal to the hybrid orbitals. In the discussion that follows, we will refer to the p orbitals by their cartesian designations: x, y, and z.

As we discuss hybridization, it will be easy to get the impression that atoms undergo this real, physical process *before* they form molecules.[27] This is not now generally accepted to be the case, although the case to the contrary has been made quite eloquently by Alabugin.[28] As with all versions of molecular orbital theory, *hybridization theory is a useful model that allows us to rationalize certain properties of the chemical bonds in molecules*—in this case, it is the strength, length, and orientation of covalent bonds. It is not a description of a physically real process.

There are three common hybrids formed by second-row elements: sp, sp^2, and sp^3 hybrid orbitals. Hybrid orbitals are characterized by the ***hybridization index***, which is n for an sp^n hybrid orbital.[29] Although the wave function for an s orbital is the equation for a sphere with no directional component, the wave functions for p orbitals all contain an important directional component (e.g., the directional component of the p_x orbital causes the wave function to have a value of zero at all points where $x = 0$). Consequently, all hybrid orbitals have a directional component.

The simplest hybrid orbitals are sp hybrids, formed from the s orbital and a single p orbital. The hybridization of the s orbital with the p_x orbital gives a pair of wave functions that are oriented along the x axis; sp hybrid orbitals are colinear, with the front lobes oriented at 180° to each other (this geometry is described as linear or, occasionally, as digonal). Note that the two unused p orbitals (the p_y and p_z orbitals in this case) remain unchanged and are oriented along the y and z axes, orthogonal to the hybrid orbitals.

The set of sp^2 hybrid orbitals is formed from the s orbital and two of the degenerate p orbitals. Using the p_x and p_y orbitals in the hybridization process gives three degenerate sp^2

---

26. (a) Pauling, L. *Proc. Natl. Acad. Sci.* **1928**, *14*, 359; *J. Am. Chem. Soc.* **1931**, *53*, 1367; *The Nature of the Chemical Bond, 3rd. ed.* (Cornell University Press: Ithaca, NY: 1960), p.111*ff.* (b) Slater, J.C. *Phys. Rev.* **1931**, *37*, 481. (c) van Vleck, J.H. *J. Chem. Phys.* **1933**,*1*, 1778. (d) Hultgren, R. *Phys. Rev.* **1932**, *40*, 891.

27. A state known as the "valence state," an excited form of the atom has been proposed, although it is pointed out that the carbon atom, for example, never attains this state: (a) van Vleck, J.H. *J. Chem. Phys.* **1933**, *1*, 177, 219; **1934**, *2*, 20. (b) Mulliken, R.S. *J. Chem. Phys.* **1934**, *2*, 782; *J. Phys. Chem.* **1952**, *56*, 295. (c) Moffitt, W.E. *Proc. Roy. Soc. A* **1950**, *202*, 534, 548.

28. (a) Alabugin, I.V.; Manoharan, M. *J. Comp. Chem.* **2007**, *28*, 373. (b) Alabugin, I.V.; Manoharan, M.; Peabody, S.; Weinhold, F. *J. Am. Chem. Soc.* **2003**, *125*, 5973.

29. (a) Bingel, W.A.; Lüttke, W. *Angew. Chem. Int. Ed. Engl.* **1981**, *20*, 899. (b) Lowry, T.H.; Richardson, K.S. *Mechanism and Theory in Organic Chemistry* (Harper & Row: New York, 1976), pp. 43-49.

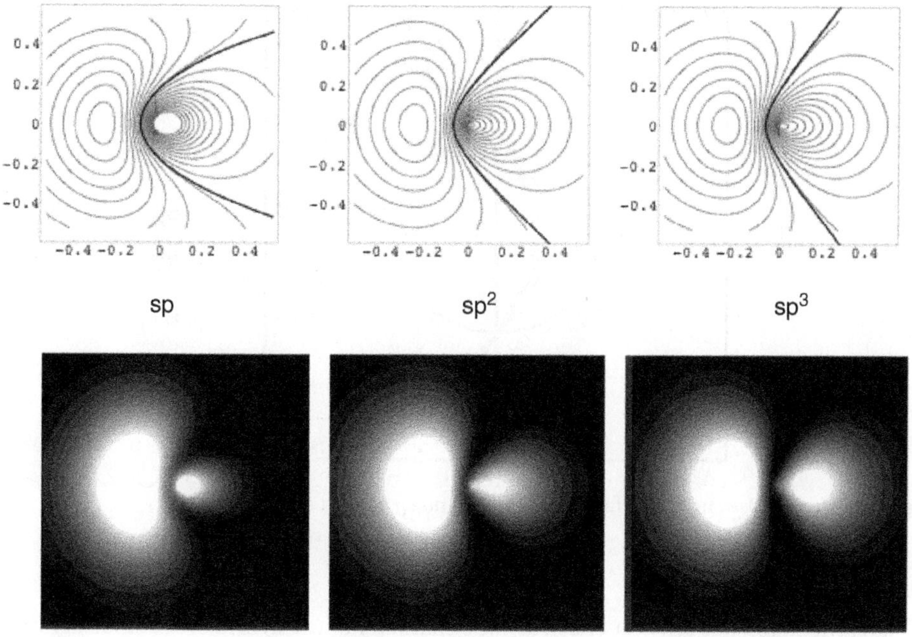

Figure 4.8 Wave functions for hybrid orbitals of carbon projected on a plane containing principal axis of the orbital. Orbital wave function calculations courtesy of Dr. Frederick W. King, University of Wisconsin-Eau Claire. Projection diagrams were prepared using *Atom in a Box* (Dauger Research, Inc.).

hybrids in the *xy* plane, with one oriented along the *y* axis, and the other two at 120° to it; sp^2 orbitals lie in a common plane, with the front lobes oriented toward the corners of an equilateral triangle (this geometry is most often referred to as trigonal planar). The unused p orbital (the p_z orbital in this case) remains unchanged and lies along the axis orthogonal to the plane of the sp^2 orbitals (in this case, the *z* axis).

The four degenerate sp^3 hybrid orbitals derived from the s orbital and the complete set of degenerate p orbitals (the p_x, p_y, and p_z orbitals) are oriented toward the corners of a tetrahedron that are defined by the Cartesian coordinates (1, 1, 1), (–1, 1, 1), (1, –1, –1), and (–1, 1, –1). This geometry is almost always referred to as tetrahedral.

The projections of the wave functions of the three types of hybrid atomic orbitals for a carbon atom projected on the plane containing the orbital axis are plotted on the same scale in Figure 4.8, which also shows a superimposed plot of projection of the nodal surfaces on the plane containing the principal axis of the orbital.

Hybrid atomic orbitals still retain some of the character of the pure atomic orbitals from which they are derived. Thus, one can define a property of the orbital, its ***s character***, which for an sp^n orbital is $1/(1 + n)$. The amount of s character relates the properties of the hybrid orbital to the extent to which the s orbital wave function contributes to the wave function of the hybrid. All hybrid orbitals have a larger front lobe and a smaller back lobe, with the relative size (i.e., volume) of the front lobe increasing as the s character of the orbital increases. The sp orbital has the largest front lobe and smallest back lobe, whereas the sp^3 orbital has the two lobes closer to equal in size.

Figure 4.8 highlights another difference between hybrid orbitals and the atomic orbitals from which they are derived. The nucleus is located on the nodal plane of a p orbital and at the center of the spherical nodes of the s orbitals, but the nodal surface of a hybrid orbital neither passes through the atomic nucleus nor is located symmetrically with respect to it—the nucleus is located within the back lobe of the hybrid orbital. The diagram in Figure 4.9 shows that the nucleus most closely approaches the nodal surface in the sp^3 hybrid and is further from the nodal surface as the s character of the hybrid orbital increases.

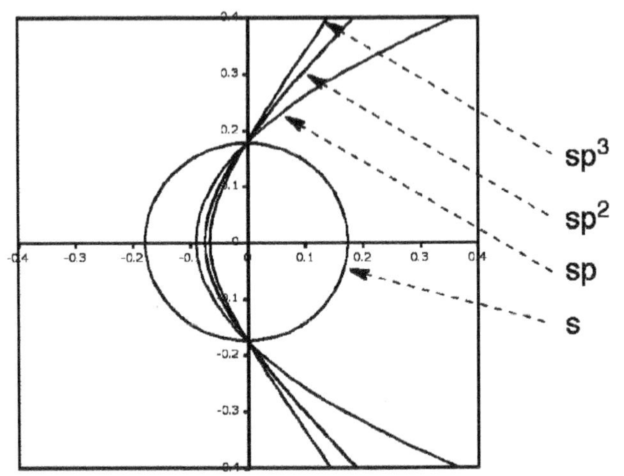

Figure 4.9 Location of nodal surfaces of s, sp, sp^2, and sp^3 hybrid orbitals

Hybridization has another effect: as the s character of the hybrid orbital increases, the front lobe becomes shorter. That is, the distance from the nucleus to where there is the maximum probability of finding an electron becomes shorter. This, of course, puts the electrons closer to the nucleus, so that they are held more strongly by the nucleus. It is worth remembering at this stage that one defines an increased hold by the nucleus on the electrons as increased electronegativity. In other words, *the higher the s character of the hybrid orbitals, the higher the electronegativity of the atom.*[30] The electronegativity of carbon decreases with hybridization in the order sp > sp^2 > sp^3. It also means that the length of the bond decreases (and the covalent radius of the atom decreases) as the hybridization changes from sp^3 to sp^2 to sp.[31]

The most efficient overlap of the hybrid orbital always occurs through the front lobe, which makes σ bonds formed from hybrid atomic orbitals highly directional. In addition, the hybridization model results in the σ bonds that a second-row atom forms by using hybrid orbitals being stronger than those formed by the same atom from unhybridized p orbitals.[32] This means that this model predicts that an atom will form its strongest and most directional σ bonds when it uses hybrid atomic orbitals to form them.¶ It is also

---

30. (a) Sanderson, R.T. *J. Am. Chem. Soc.* **1983**, *105*, 2259; *J. Chem. Educ.* **1988**, *65*, 112, 223. (b) Walsh, A.D. *Disc. Faraday Soc.* **1947**, *2*, 18.

31. (a) Allen, F.H.; Kennard, O.; Watson, D.G.; Brammer, L.; Orpen, A.G.; Taylor, R. *J. Chem. Soc, Perkin Trans. 1* **1987**, *2*, S1-S19. (b) Smith, M.B.; March, J. *March's Advanced Organic Chemistry. Reactions, Mechanisms, and Structure*, 5th ed. (Wiley-Interscience, New York, 2001), p. 20. (c) Pauling, L. *The Nature of the Chemical Bond*, 3rd ed. (Cornell University Press: Ithaca, 1960), p. 224.

32. Pauling, L.; Sherman, J. *J. Am. Chem. Soc.* **1937**, *59*, 1450.

¶. ***Hybridization and Molecular Orbital Formation***

Hybrid orbitals are formed by the combination of atomic orbitals of the same atom. Mathematically, the process involves the linear combinations of atomic wave functions below. The process leads to degenerate orbitals whose orientation is determined by the coefficients of the p wave functions used in the hybridization process. X, y, and z represent the p_x, p_y, and p_z wave functions.

sp	sp^2	sp^3
$\phi_a=(1/\sqrt{2})[s + x]$	$\phi_a=(1/\sqrt{3})[s + (\sqrt{2})y]$	$\phi_a=(1/2)[s + x + y + z]$
$\phi_a=(1/\sqrt{2})[s + x]$	$\phi_b=(1/\sqrt{3})[s + (1/\sqrt{2})x - (\sqrt{6}/2)y]$	$\phi_b=(1/2)[s - x - y + z]$
unchanged: y, z	$\phi_c=(1/\sqrt{3})[s - (1/\sqrt{2})x - (\sqrt{6}/2)y]$	$\phi_c=(1/2)[s + x - y - z]$
	unchanged: z	ϕ_d (1/2)[s - x + y - z]
		unchanged: none

generally observed that *lone pairs on second-row elements tend to occupy hybrid atomic orbitals in preference to unhybridized atomic orbitals unless the atom adjacent to the atom carrying the lone pair is* sp^2 *or sp hybridized.*

When one compares the strength of σ bonds formed from hybrid orbitals, the strongest bonds are formed from sp hybrid orbitals and the weakest from sp^3 hybrid orbitals. The relative strengths of σ bonds can be related to the s character of the hybrid orbitals used to form them. *As the s character of the hybrid atomic orbitals used to form them increases, σ bonds become stronger.*[33] In contrast to the σ bond case, the π-type overlap of hybrid atomic orbitals to form π and π* molecular orbitals is seldom observed. Thus, all π bonds in a molecule are formed from unhybridized p orbitals.

Some important generalizations about hybrid atomic orbitals and molecular orbitals (covalent bonds) hold true for most organic molecules.

1. All σ bonds to carbon in organic compounds are formed from hybrid atomic orbitals, π bonds are formed from unhybridized p orbitals, and lone pairs usually occupy hybrid orbitals.
2. Every atom of a molecule is involved in at least one σ bond.
3. Atoms in a molecule will form the fewest σ bonds consistent with rule 2.
4. The strongest σ bonds are formed from hybrid atomic orbitals with the highest possible s character.
5. All multiple bonds are formed from a single σ bond, with one or more π bonds.
6. The shape of a molecule is determined only by the σ bonding framework only; π bonds have no effect on molecular shape.

The τ Bond Model of Multiple Bonds: Variable Hybridization

Multiple bonds in organic compounds can be described in two ways, the σ/π description, which is the one that will be used in this book, and the τ bond system,[34] which is actually older and in which the bonds between the carbon atoms are "bent" bonds. This description of bonding has been used very effectively to rationalize the bonding in cyclopropanes,[35] where the electron density of the σ bonds is found outside the internuclear axis of the ring, and it has been shown to work just as well as the σ/π model.[36]

The wave functions of the molecular methane generated from hybrid wave functions are of the simplified form:

$$\psi_{\sigma} = (1/\sqrt{2})(\phi_H + \phi_C) \qquad \psi_{\sigma^*} = (1/\sqrt{2})(\phi_H - \phi_C)$$

where ϕ_C is the wave function of a carbon hybrid atomic orbital, and ϕ_H is the wave function of a hydrogen 1s orbital.

This can be expanded to (e.g., ϕ_x is the carbon $2p_x$ wave function):

$$\psi_{\sigma a} = 1/(2\sqrt{2})[2\phi_H + \phi_s + \phi_x + \phi_y + \phi_z] \qquad \psi_{\sigma^* a} = 1/(2\sqrt{2})[2\phi_H - \phi_s - \phi_x - \phi_y - \phi_z]$$

$$\psi_{\sigma b} = 1/(2\sqrt{2})[2\phi_H + \phi_s - \phi_x - \phi_y + \phi_z] \qquad \psi_{\sigma^* b} = 1/(2\sqrt{2})[2\phi_H - \phi_s + \phi_x + \phi_y - \phi_z]$$

$$\psi_{\sigma c} = 1/(2\sqrt{2})[2\phi_H + \phi_s + \phi_x - \phi_y - \phi_z] \qquad \psi_{\sigma^* c} = 1/(2\sqrt{2})[2\phi_H - \phi_s - \phi_x + \phi_y + \phi_z]$$

$$\psi_{\sigma d} = 1/(2\sqrt{2})[2\phi_H + \phi_s - \phi_x + \phi_y - \phi_z] \qquad \psi_{\sigma^* d} = 1/(2\sqrt{2})[2\phi_H - \phi_s + \phi_x - \phi_y + \phi_z]$$

33. (a) Pauling, L. *Proc. Natl. Acad. Sci.* **1949**, *35*, 229. (b) Walsh, A.D. *Trans. Faraday Soc.* **1947**, *43*, 60. (c) Mulliken, R.S. *J. Am. Chem. Soc.* **1955**, *72*, 4493; *J. Chem. Phys.* **1951**, *19*, 900. (d) Coulson, C.A. *Valence* (Oxford Press/Clarendon Press: London, 1952), p. 198.

34. (a) Wintner, C.E. *J. Chem. Educ.* **1987**, *64*, 587, and references therein. (b) Palke, W.E. *J. Am. Chem. Soc.* **1986**, *108*, 6543. (c) Carroll, F.A. *Perspectives in Structure and Mechanism in Organic Chemistry* (Brooks/Cole: Pacific Grove, 1998), p. 47.

35. (a) Coulson, C.A.; Moffitt, W.E. *J. Chem. Phys.* **1947**, *15*, 151; *Phil. Mag.* **1949**, *40*, 1. (b) Walsh, A.D. *Trans. Faraday Soc.*, **1949**, *45*, 179. (c) Hamilton, J.G.; Palke, W.E. *J. Am. Chem. Soc.* **1993**, *115*, 4159. (d) de Meijer, A. *Angew. Chem. Int. Ed. Engl.* **1979**, *18*, 809. (e) Wiberg, K.B. In Rappoport, Z., Ed. *The Chemistry of the Cyclopropyl Group* (John Wiley: New York, 1987), ch. 1. (f) Rozsondai, B. In Rappoport, Z., Ed. *The Chemistry of the Cyclopropyl Group* (John Wiley: New York, 1995), vol. 2, ch. 3.

36. Schultz, P.A.; Messmer, R.P. *J. Am. Chem. Soc.* **1988**, *110*, 8258; **1993**, *115*, 10925, 10943.

In the original form of this system, the hybridization of the two carbon atoms at each end of a multiple bond remains at sp^3, with the multiple bonds formed by noncolinear overlap of two pairs of sp^3 hybrid orbitals. The molecular geometry predicted by this model may be improved by introducing the concept of ***variable hybridization***,[37] which allows hybrid orbitals to have varying degrees of s and p character (i.e., a variable hybridization index).¶ For example, increasing the p character of the hybrid orbitals used to form the C—C bonds in cyclopropane allows the smaller (60°) bond angles to be accommodated.

Problems

4-3 The angle between two hybrid orbitals on the same atom cannot be less than 90°. Why?

4-4 What type of orbital carries the lone pairs in each of the following anions? Give your reasoning.

(a) 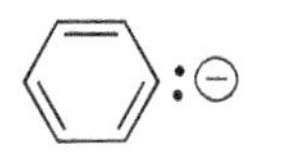
(b) 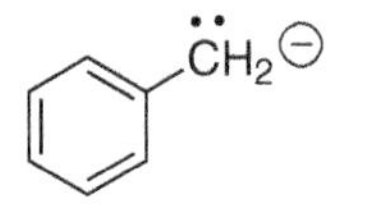
(c) 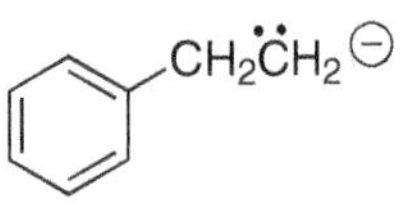

4-5 Describe the bonding in the following ions in terms of the hybridization of the atoms involved.

(a) 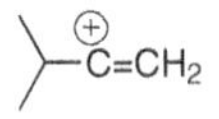
(b) 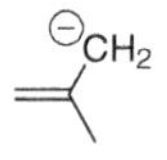
(c) 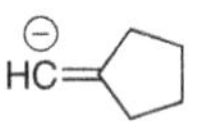
(d) 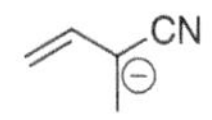

Extended π Systems: Delocalized π Orbitals of Aliphatic Compounds

LCAO molecular orbitals that are confined to two adjacent nuclei are ***localized*** molecular orbitals. However, when there are more than two sp^2 or sp hybridized atoms adjacent to each other, the potential exists for the formation of ***delocalized*** π orbitals,

37. For a useful discussion of variable hybridization, see: Carroll, F.A. *Perspectives in Structure and Mechanism in Organic Chemistry* (Brooks/Cole: Pacific Grove, 1998), pp. 38-45.

¶. ***Variable Hybridization***

The concept of variable hybridization can also be discussed in terms of a coefficient of mixing, λ, where $\lambda = \sqrt{m}$ for an sp^m orbital. The angle between sp^m and sp^n hybrid orbitals, α, is related to the values of λ by the following expression,[a] where $\mu = \sqrt{n}$:

$$1 + \lambda\mu \cos\alpha = 0, \text{ or } \cos\alpha = -1/\lambda\mu$$

When the two hybrid orbitals are equivalent, this expression simplifies to

$$1 + \lambda^2 \cos\alpha = 0, \text{ or } \cos\alpha = -1/\lambda^2, \text{ or } \cos\alpha = -1/m$$

Based on the H—C—H bond angle in cyclopropane as 115.1°,[b] the carbon atom uses $sp^{2.36}$ hybrids for the C—H bonds. If one s orbital and n p orbitals are used in to obtain $n + 1$ hybrid orbitals about an atom, the hybridization indices of hybrid orbitals around that atom must obey the two relationships below:

$$\sum_i \frac{1}{1+\lambda_i^2} = 1 \text{ and } \sum_i \frac{\lambda_i^2}{1+\lambda_i^2} = n$$

This now allows us to calculate the hybrid orbitals used for forming the C—C bonds of the cyclopropane ring; the result is that $sp^{4.92}$ hybrids are used, which predicts an angle between the hybrid orbitals of 101.7°. A similar variable hybridization model of the carbon atoms in ethylene[c] leads to the τ orbitals being formed from sp^5 hybrid orbitals oriented at 101.5° to each other.

[a] For the derivation of this equation, see: Coulson, C.A. *Valence* (Oxford/Clarendon Press; London, 1952), p. 193.

[b] Bastiansen, O.; Fritsch, F.N.; Hedberg, K. *Acta Crystallogr.* **1964**, *17*, 538.

[c] Liberles, A. *Introduction to Theoretical Organic Chemistry* (Macmillan: New York, 1968), p. 181.

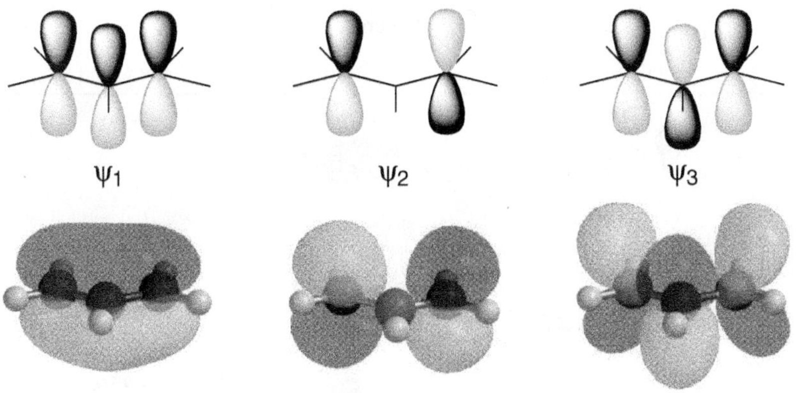

Figure 4.10 Linear combinations of p orbitals to generate the π molecular orbitals of the isolated allyl system. The π molecular orbitals of the isolated allyl system. Orbital energies increase with the number of internuclear nodes (from left to right).

where each orbital of the π system encompasses more than two nuclei. Excellent examples of this are provided by an isolated allyl system, which we used as an example for our review of resonance theory, and by 1,3-butadiene, the prototypical conjugated diene. In an isolated allyl group, all three carbon atoms are sp^2 hybridized, with a single unhybridized p orbital at each position. Side-by-side overlap of these three orbitals will give three new π orbitals (ψ_1, ψ_2, and ψ_3), each of which encompasses all three nuclei (Figure 4.10).

1,3-Butadiene is a typical conjugated diene. The four carbon atoms are all sp^2 hybridized, so that each has one unhybridized p orbital to use in the formation of the π orbital system. Overlap of these four atomic orbitals ("basis orbitals") must give four molecular orbitals, and we designate these orbitals as ψ_1, ψ_2, ψ_3, and ψ_4, in order of increasing energy (Table 4.1).

The simplest rules for combining p basis orbitals into delocalized π molecular orbitals were formulated by Hückel in 1931.[38] In this method, the energies of the orbitals are expressed in terms of two variables, α and β, where α represents the Coulomb integral, a measure of how tightly the electron is bound to the nuclei of the atoms (and which has the same value for all atoms of a particular element), and β represents the resonance integral, a measure of the electron energy in the molecular orbital relative to the electron in the isolated atoms. Both α and β are negative because they identify a stabilizing energy. The rules themselves are quite simple for open-chain systems.

1. The number of delocalized π orbitals is equal to the number of p atomic orbitals (contributing p basis orbitals) used to make them.
2. Unless the π bonding system is cyclic and delocalized, there are no degenerate orbitals.
3. The lowest energy orbital has no internuclear nodes (i.e., all π-type overlap of the contributing p basis orbitals is bonding in character). The highest energy orbital has an internuclear node between every adjacent pair of atoms (i.e., all, π-type overlap of the contributing p basis orbitals is antibonding in character).
4. Each orbital has one more internuclear node than the orbital immediately below it in terms of energy.
5. All nodes are distributed as symmetrically as possible between the nuclei of the π system.
6. Where there is more than one way to distribute nodes in accordance with rule 4, bonding overlap is concentrated in the middle of the system.

38. (a) Hückel, E. *Z. Phys.* **1931**, *70*, 204; **1931**, *72*, 310; **1932**, *76*, 628; **1933**, *83*, 632; *Trans. Faraday Soc.* **1934**, *30*, 40. (b) Isaacs, N.S. *Physical Organic Chemistry* (Longman Scientific & Technical: London, 1987), §1.2, pp. 3-24. (c) Carey, F.A.; Sundberg, R.D. *Advanced Organic Chemistry. Part A: Structure and Mechanisms*, 4th ed. (Kluwer Academic/Plenum Publishers: New York, 2000), §1.4, pp. 31-36.

Table 4.1 Overlap of p Basis Orbitals in 1,3-Butadiene

Orbital	p Orbital Overlap	π Molecular Orbital	Orbital Energy (eV)
ψ_4	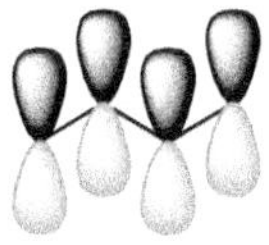	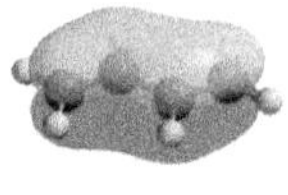	+2.05
ψ_3	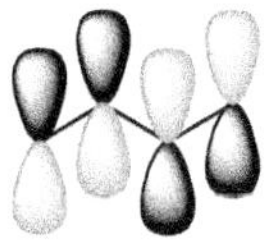	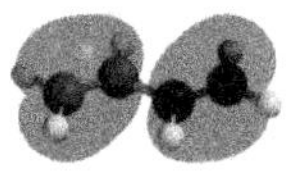	+0.22
ψ_2	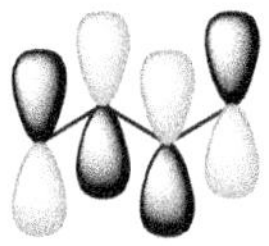	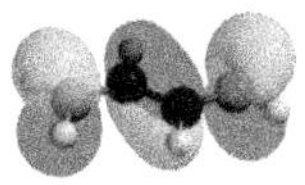	-9.56
ψ_1	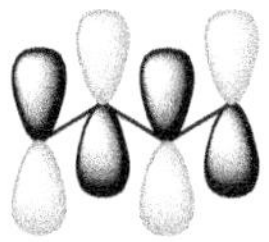		-12.29

Implicit in the forgoing discussion about the formation of molecular orbitals from atomic orbitals, is an observation seldom stated explicitly.

For every bonding orbital, there is a corresponding antibonding orbital.

This also leads to another definition of a covalent bond.

A covalent bond encompasses two or more nuclei, and it consists of an occupied bonding orbital and an unoccupied antibonding orbital.

The application of these rules to conjugated systems larger than 1,3-butadiene can be seen in Table 4.2, which contains the five π molecular orbitals for the 2,4-pentadienyl cation and the six π molecular orbitals of 1,3,5-hexatriene. Careful examination of the hexatriene orbitals shown in Table 4.2 shows that the six 2p basis orbitals do not contribute equally to each of the six π orbitals; instead, the orbital coefficients at the individual atoms change from one π orbital to the next. The orbital coefficients of the π orbitals of 1,3,5-hexatriene, as calculated using simplified Hückel molecular orbital theory, are given in Table 4.3 (and used in Table 4.2). These coefficients give a reasonable approximation to the appearance of the orbital, although not anywhere close to the high level of precision available from modern computer programs. Still, as qualitative estimates, they are more than adequate for most purposes.

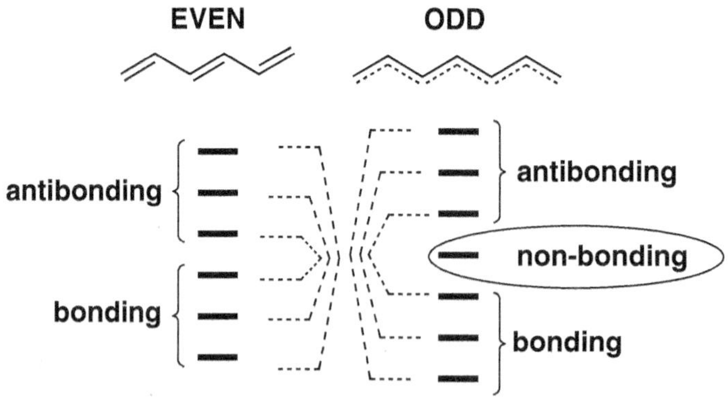

Table 4.2 π Molecular Orbitals of 2,4-Pentadienyl Cation and 1,3,5-Hexatriene

	2,4-Pentadienyl Cation		1,3,5-Hexatriene	
Orbital	p Basis Orbital Overlap*	π Molecular Orbital	p Basis Orbital Overlap*	π Molecular Orbital
ψ_6	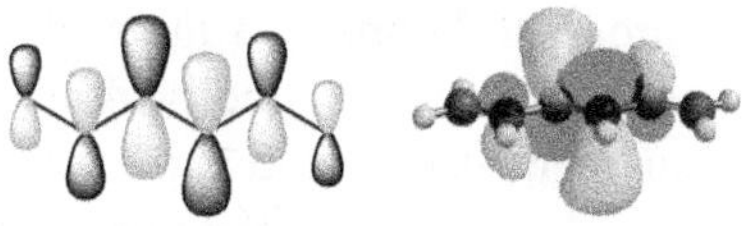			
ψ_5	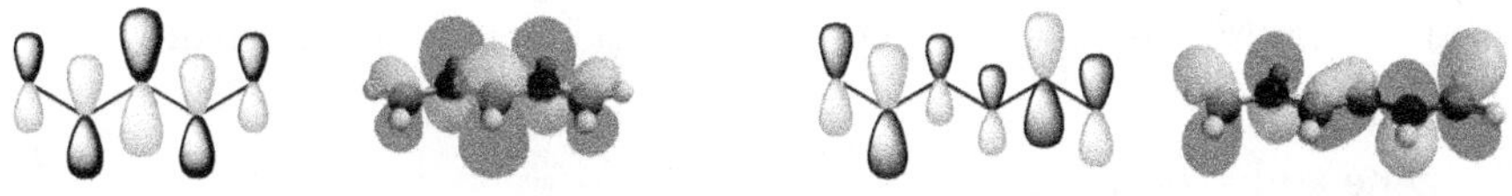			
ψ_4				
ψ_3	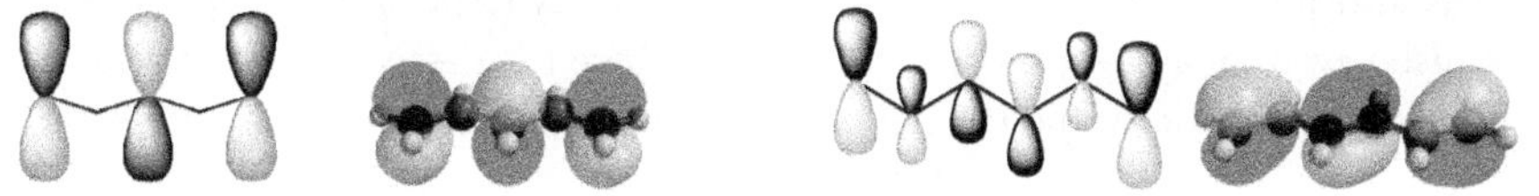			
ψ_2				
ψ_1				

*The basis p orbitals have been drawn to reflect their relative coefficients in the molecular orbital. They are not drawn to scale.

Table 4.3 Hückel Orbital Coefficients of the π Orbitals of 1,3,5-Hexatriene

	ϕ_1	ϕ_2	ϕ_3	ϕ_4	ϕ_5	ϕ_6
ψ_1	0.232	0.418	0.521	0.521	0.418	0.232
ψ_2	0.418	0.521	0.232	−0.232	−0.521	−0.418
ψ_3	0.521	0.232	−0.418	−0. 418	0. 232	0. 521
ψ_4	0.521	−0.232	−0. 418	0. 418	0. 232	−0. 521
ψ_5	0.418	−0. 521	0. 232	0. 232	−0. 521	0. 418
ψ_6	0.232	−0. 418	0. 521	−0. 521	0. 418	−0. 232

In extended open-chain (but not conjugated cyclic) π orbital systems, when there are an even number of orbitals, there will be equal numbers of bonding and antibonding orbitals, and when the π system contains an odd number of orbitals, the extra orbital occupies the middle position in energetic terms (e.g., ψ_2 of the allyl system) and is a nonbonding orbital. This is illustrated here for two typical conjugated systems: the 1,3,5-hexatriene and

2,4,6-heptatrienyl systems. The mathematics of calculating Hückel molecular orbital coefficients is shown below.[¶]

Bond Orders and Occupancy of Molecular Orbitals

The definition of a covalent bond given above also provides a convenient way to assign bond orders to bonds in a molecule. This is accomplished quite simply by adding the number of electrons in bonding orbitals between two atoms in a molecule and subtracting the number of electrons in antibonding orbitals between the same atoms. The bond order of the bond is one half of the total. Because the strength of a bond is directly related to its bond order, this is an important parameter of bonding to be able to predict.

Benzene and Aromaticity

The obvious next problem to address using molecular orbital theory is that of the structure of benzene and the electronic nature of aromaticity. The pursuit of the answer to this question has been long and arduous. The fact that it is not yet over is attested to by the fact that is still holds a fascination for organic chemists (especially computational organic chemists) even today[39]—almost two centuries since the discovery of benzene itself.[40] We will discuss aromaticity and the electronic structure of aromatic compounds in Chapter 7.

¶. ***Hückel Molecular Orbital Theory***

In this theory, the energies of the π molecular orbitals are related to α, the Coulomb integral (a measure of how tightly the electron is bound to the nuclei of the atoms—and which has the same value for all atoms of a particular element), and β, the resonance integral (a measure of the electron energy in the molecular orbital relative to the electron in the isolated atoms).

Hückel's major simplification of molecular orbital theory was to refer the energies of orbitals to α and to make the rather important assumption that β was zero for all nonadjacent atoms of a molecule. This is a difficult assumption to justify in these days of fast computers, but it was an important simplification when all the mathematics had to be done by hand. This leads to the following secular determinant, which can be further simplified as shown by dividing the entire matrix by β to give the alternative form, where $x = (\alpha - E)/\beta$ (the example is for a 2,4-pentadien-1-yl system):

$$\begin{vmatrix} \alpha-E & \beta & 0 & 0 & 0 \\ \beta & \alpha-E & \beta & 0 & 0 \\ 0 & \beta & \alpha-E & \beta & 0 \\ 0 & 0 & \beta & \alpha-E & \beta \\ 0 & 0 & 0 & \beta & \alpha-E \end{vmatrix} = 0 \Rightarrow \begin{vmatrix} x & 1 & 0 & 0 & 0 \\ 1 & x & 1 & 0 & 0 \\ 0 & 1 & x & 1 & 0 \\ 0 & 0 & 1 & x & 1 \\ 0 & 0 & 0 & 1 & x \end{vmatrix} = 0$$

The solution of this secular determinant gives the coefficients of the π orbitals. For acyclic, conjugated systems, the solutions are all of the form:

$$\psi_j = \sum C_{r,j}\phi_r \text{ and } C_{r,j} = \left(\sqrt{\frac{2}{N+1}}\right)\sin[rj\pi(N+1)]$$

where N is the number of atoms in the conjugated system (e.g., 4 for butadiene, 5 for pentadienyl, 6 for hexatriene), r designates the location of the contributing atomic 2p basis orbital within the conjugated system (i.e., which atom along the conjugated chain: 1, 2, 3, or 4 for butadiene; 1, 2, 3, 4, or 5 for pentadienyl), and j designates which π molecular orbital of the set of delocalized π orbitals is under study (e.g., ψ_1, ψ_2, ψ_3). Using this theory, one may also calculate the approximate energy of the orbital:

$$E = \alpha + 2\beta\cos\frac{j\pi}{N+1}$$

39. There is a huge volume of literature on the subject of aromaticity: see, for example, the recent special issues of *Chem. Rev.*: *Chem. Rev.* **2001**, *101* (5) and *Chem. Rev.* **2005**, *105* (7).

40. Faraday, M. *Phil. Trans. Roy. Soc. London* **1825**, *115*, 440.

Bonds Involving Third-Row and Higher Row Elements

The s atomic orbitals of atoms in the third and higher rows of the periodic table have (n−1) spherical nodes at finite distances from the nucleus, and the p atomic orbitals of the third and higher rows of the periodic table have (n−2) spherical nodes at finite distances from the nucleus in addition to the planar node through the nucleus. This has the effect of reducing the efficiency of (especially π) overlap between these elements and elements in the second row. In fact, the most efficient overlap of atomic orbitals to form molecular orbitals usually occurs when both elements are in the same row of the periodic table (hydrogen being the one exception to this rule). We can see this nicely if we examine the π overlap between the 2p orbital on carbon and the 2 p orbital on oxygen or the 3p orbital on sulfur.

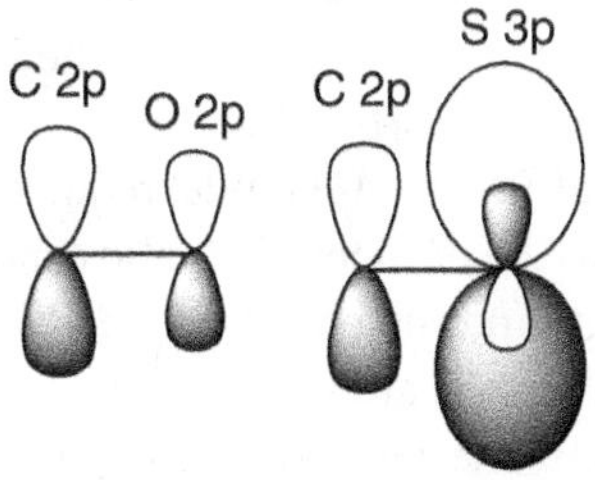

As is evident from the diagram at right, the overlap between the 2p orbitals of carbon and oxygen is in phase at all points, and this should result in a low-energy bonding molecular orbital. In contrast, there are areas where the overlap between the 2p and 3p orbitals are in phase, and other areas where the overlap is out of phase (i.e., antibonding in character). Thus, we expect that this π bond will be weaker, and it is (the C=O bonds in CO_2 have dissociation energies of 192 kcal mol^{-1}; the C=S bonds of CS_2 have dissociation energies of 138 kcal mol^{-1}).

Problems

4-6 Sketch the π molecular orbitals of 1,3,5,7-octatetraene, paying particular attention to the phases of the overlap of the p basis orbitals contributing to each π molecular orbital and the locations of nodal planes.

4-7 It is frequently important to characterize the π orbitals in terms of their symmetry with respect to simple symmetry elements. Symmetric orbitals are those whose lobes have phases that are symmetric with respect to the symmetry element (e.g., the π molecular orbital of ethylene is symmetric with respect to the mirror plane of symmetry perpendicular to the σ plane). Antisymmetric orbitals undergo a phase change on application of the symmetry element (e.g., the π* orbital of ethylene is antisymmetric with respect to the same mirror plane). Characterize each of the π orbitals of (a) 1,3,5-hexatriene and (b) the 2,4-pentadienyl radical as either symmetric or antisymmetric with respect to (1) a C2 axis of rotation, and (2) a mirror plane of symmetry perpendicular to the σ plane through the center of the molecule. Does a pattern emerge?

4-8 Consider the nonbonding π molecular orbitals of the allyl 2,4-pentadienyl, 2,4,6-heptatrienyl and 2,4,6,8-nonatetraenyl systems. What general feature is common to all these orbitals?

Chapter Summary

The modern orbital model of the atom describes the electrons around the nucleus of an atom as three-dimensional standing waves. This wave is characterized by three quantum numbers—n, the principal quantum number (electron shell); l, the azimuthal quantum number (electron subshell); and m_l, the magnetic quantum number (orbital within a subshell)—defining each orbital. Orbitals may be described using only their quantum numbers, but chemists generally defer to using the terms s ($l = 0$), p ($l = 1$), d ($l = 2$), and f ($l = 3$) to describe the subshells of the electron shell. s Orbitals are spherically symmetrical; p orbitals have one planar node, with all other nodal surfaces being spherical; and four of the five d orbitals have two planar nodes, with the rest being spherical. The Pauli Exclusion principle limits the occupancy of any orbital to two electrons, the Aufbau principle dictates that the lowest-energy orbitals are filled first, and Hund's rule states that when electrons are distributed into degenerate orbitals, the number of unpaired electrons should be maximized.

Molecular orbitals are generated by overlap of atomic orbitals in a coaxial (σ) or coplanar (π) fashion; mathematically, this process involves the LCAO. Every σ bonding orbital has a corresponding σ^* antibonding orbital, and every π bonding orbital has a corresponding π^* antibonding orbital. The LCAO model requires that the number of molecular orbitals formed must be equal to the number of atomic orbitals used to generate them. The geometry of most organic compounds can be addressed by using the hybridization model of the atom; hybrids orbitals of carbon are sp (linear), sp^2 (trigonal planar), and sp^3 (tetrahedral). Bond angles other than the idealized values for these hybrid orbitals may be accommodated by means of variable hybridization. Extended π-bonding systems may be modeled using Hückel molecular orbital theory, which allows a qualitative (or simplified, semiquantitative) description of the molecular orbitals to be obtained.

Key Terms

atomic orbital
- s orbitals
- p orbitals
- d orbitals

Aufbau principle
Hund's rule
Hückel theory
hybrid atomic orbitals
- sp hybrid orbitals
- sp^2 hybrid orbitals
- sp^3 hybrid orbitals

LCAO
molecular orbital
- σ orbitals
- π orbitals

node
Pauli Exclusion principle
quantum numbers
- principal quantum number, n
- azimuthal quantum number, l

magnetic quantum number, m_l
spin quantum number, m_s

Additional Problems

4-9 The favored conformation of the ester functional group is as shown below. Suggest a reason why this conformation should be the favored one in the ester, based on considerations of the molecular orbitals involved. (Hint: Is it easier to delocalize electron density from the lone pairs on oxygen into the π^* orbital of the carbonyl group in one conformation than in the other? What effect would this electron delocalization have?)

4-10 Use similar arguments to those used in Problem 4-9 to suggest why the eclipsed conformation of ethane should be higher in energy than the staggered conformation.

4-11 (a) The simplified Hückel molecular orbital analysis of the benzene molecule returns the result that the six π orbitals of benzene are all completely delocalized over all six atoms of the molecule. In the lowest energy π orbital, there are no nodes perpendicular to the plane of the ring. The next two π orbitals are degenerate (equal energy), and each has a single nodal plane perpendicular to the ring. The next two orbitals are also degenerate, and each has two nodal planes perpendicular to the ring. The highest energy π orbital has three nodal planes perpendicular to the ring. Draw the hexagonal ring of the benzene molecule (e.g., the highest energy π orbital is shown at right) and show the phases of the p basis orbitals in each of the six π orbitals with the location of each of their nodal planes indicated by straight lines.

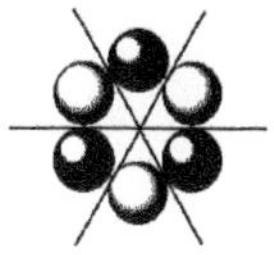

(b) Designate each of these π orbitals as symmetric (*S*) or antisymmetric (*A*) with respect to (i) a mirror plane of symmetry perpendicular to the plane of the ring and (ii) a C_2 axis of symmetry through the plane of the ring of the benzene molecule.

4-12 The same pattern that occurs in the nonbonding orbital of odd-number open-chain compounds also occurs in certain cyclic compounds with conjugated π-bonding systems ("alternant" systems) as well. Referring to the answer to Problem 4-8 above, sketch the nonbonding orbital (ψ_4) of the benzyl cation (at left). How does this molecular orbital correlate with the reactivity of the cation?

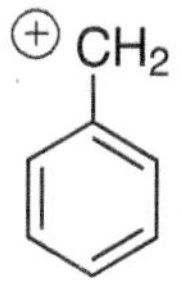

4-13 The preferred conformation of 1,2-difluoropropane is the one in which the two fluorine atoms are *gauche* to each other and in which the methyl group is *anti* to one of the fluorine atoms. This staggered conformation places the two fluorine atoms as close together as possible while retaining the staggered arrangement of σ bonds. Provide a reasonable explanation for why this should be the preferred conformation.

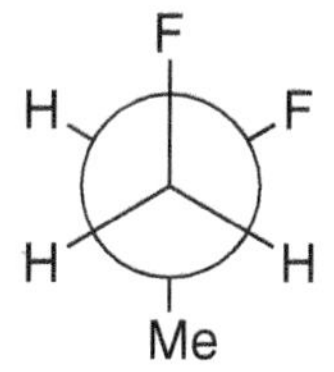

4-14 The complete set of molecular orbitals formed from the atomic orbitals of the valence shells of two oxygen atoms to give an oxygen molecule is shown below. What is the bond order in the oxygen molecule? What other noteworthy feature of the oxygen molecule is revealed by this orbital diagram?

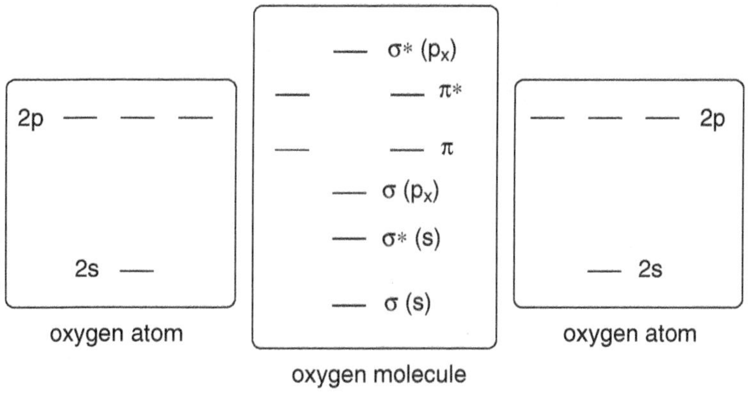

The nitric oxide (NO) and carbon monoxide (CO) molecules have the same set of molecular orbitals with the same sequence of energies. What is the bond order in these molecules, and are there any noteworthy features of these molecules revealed by the orbital diagrams?

4-15 Oxygen-centered free radicals fragment according to the scheme below. What does this reaction tell about the stabilities of the radicals and the bond energies of the products?

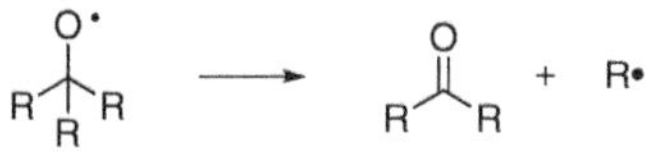

4-16 Terminal alkynes are much stronger acids than any other type of hydrocarbon. Allylic hydrogens and benzylic hydrogens, also, tend to be much more easily removed by strong bases than similar hydrogens in saturated hydrocarbons. Suggest an explanation for these observations.

4-17 The H—C—H angle in cyclopropane is 114.97°, the C—H bond length is 1.0786Å, and the C—C bond distance is 1.5030Å. The corresponding values for ethane are 1.094Å for the C—H bond and 1.537Å for the C—C bond. Do these values tend to support or provide support against the theory of variable hybridization? How?

4-18 Theoretical calculations show that the ions $^{-}CH_2OH$ and $^{-}CH_2NH_2$ cannot be isolated even in the gas phase but that they are unstable with respect to electron loss (i.e., the corresponding radicals have a positive electron affinity). The anion $^{-}CH_2SH$ is detectable in the gas phase, which means that the corresponding radical has a negative electron affinity. Explain these observations using molecular orbital and/or valence bond arguments.

4-19 The α effect is the name given to the observation that nucleophiles where the nucleophilic atom is directly bonded to another atom carrying a lone pair (e.g., hydrazine, H_2NNH_2, and hydroxylamine, $HONH_2$) are much stronger nucleophiles than their structural analogs that do not have this structural feature (e.g., methylamine, H_3CNH_2), even though they tend to be weaker bases. Give a reasoned explanation for why this might be so.

4-20 The antineoplastic compounds calicheamicin and esperamicin both undergo the Bergman cyclization to give a diradical after they are activated by nucleophilic attack. Why is the nucleophilic attack necessary for the Bergman cyclization to occur in these molecules?

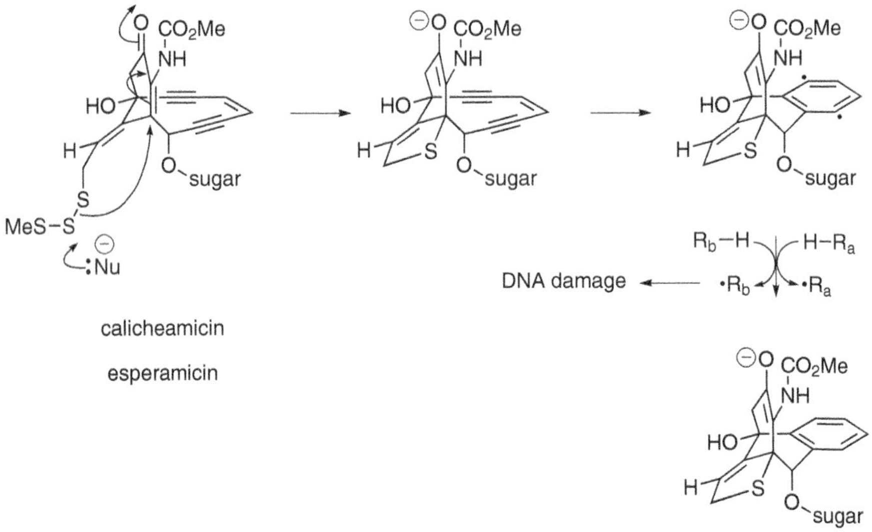

calicheamicin

esperamicin

CHAPTER FIVE

Frontier Orbitals and Chemical Reactions

5.1 Chemical Reactions: Frontier Orbitals

Now that we have discussed the bonding and molecular orbitals in individual molecules and other entities, it is time to examine how the molecular orbitals in the reacting species participate in a reaction to become those of the species produced.

Ultimately, all chemistry can be reduced to a question of the formation and cleavage of chemical bonds during chemical reactions. One of the major successes of the Lewis theory of bonding was the way in which it provided a clear qualitative picture of the bonding in chemical compounds, so that the bonding changes occurring during chemical reactions could be rationalized. In order for molecular orbital theory to be equally useful, it must be capable of doing at least what Lewis theory was able to accomplish, and we are probably justified in expecting more. Fortunately, molecular orbital theory does allow us to rationalize reactions that cannot be explained at all using Lewis theory alone.

The heart and soul of organic chemistry is how molecular orbitals in different molecules (or different parts of the same molecule) interact and how electrons reorganize themselves during chemical reactions. This is the reason why we have spent as much time as we have reviewing molecular orbital theory. Let us begin by restating an extremely important fundamental principle.

> The Pauli Exclusion principle applies to molecules as well as to atoms and absolutely forbids placing an electron into any orbital already containing two electrons.

This means that any time electrons move from one molecule to another, they must move from an occupied orbital on the first molecule into an *empty* orbital of the second. If this electron reorganization is to result in the formation of a new covalent bond between atoms in two different molecules, electrons from one molecule must move into a position where they can be shared by the two nuclei between which the new bond is going to be formed. Because all the electrons of a molecule are in orbitals, one must overlap orbitals on both molecules for the reaction to occur. Thus, the Pauli Exclusion principle becomes the first principle of chemical reactions.

> Chemical reactions can only be initiated by overlapping a filled or half-filled (occupied) orbital on one reactant with an *empty* (unoccupied) orbital on the other.

Having decided that the initiation of a chemical reaction involves the overlap of an empty orbital with a filled orbital, we must now answer the key question—*which* unoccupied orbital and *which* occupied orbital. Clearly, the reaction will proceed most readily if the two orbitals are closely matched in energy. The filled orbitals in most molecules are either bonding orbitals containing bonding electron pairs or nonbonding orbitals containing lone pairs. The unoccupied orbitals are either antibonding orbitals or empty atomic orbitals. The Aufbau principle demands that the unoccupied orbitals of a molecule or ion are always at higher energy than its occupied orbitals. Therefore, the two orbitals closest to each other in energy will be the ***highest energy occupied molecular orbital*** (HOMO), on one reactant, and the ***lowest energy unoccupied molecular orbital*** (LUMO), on the other. These orbitals are called the ***frontier orbitals*** of the reacting molecules, and it is the frontier orbitals that actually govern much of the reactivity of organic compounds.

Since their emergence in the mid-1960s,[1] the concepts of frontier orbitals and the conservation of orbital symmetry[2] have revolutionized the way in which organic chemists look at reactions, and we will cast many of our discussions of chemical reactivity in terms of the frontier orbital concepts. As with any theory, the conservation of orbital symmetry has limitations, especially when used to predict the outcomes of reactions of excited state reactions. Nevertheless, the value of the theory has been shown by half a century of successful application.

By knowing the shapes and relative energies of the orbitals involved in most organic reactions, one can actually predict much of the reactivity of organic molecules. As implied in our earlier discussions, π bonds are usually weaker than σ bonds (i.e., the π orbital is usually higher in energy than the σ orbital, and the π^* orbital is usually lower in energy than the σ^* orbital). Therefore, the energies of frontier molecular orbitals (FMOs), which have important consequences for reactivity of organic compounds, generally increase in the order (worth the effort of committing to memory):

$$\sigma < \pi < \mathbf{n} < \pi^* < \sigma^*$$

With the exception of the hydrogen cation, H^+, which has no electrons at all and cannot therefore have a HOMO, all molecules and ions have both a HOMO and a LUMO. In principle, therefore, there are always two possible HOMO-LUMO combinations for reactions in which H^+ is not a participant. Fortunately, one of these combinations is usually obviously right, and the other is usually obviously wrong. We can best illustrate the concept of HOMO and LUMO by looking at some examples. Let us begin with the three different reactive intermediate species with the molecular formula CH_3: the methyl cation, CH_3^+, the methyl radical, $CH_3\bullet$, and the methyl anion, CH_3^-.

Before we begin, a caveat must be sounded; much of the discussion below is at a rather simplistic level, but this allows us to use it in the context of much more complicated examples without the need for recourse to a computer. Of course, if higher levels of rigor are needed, then one must go to the computer. However, provided that one understands the limitations of the theory one is using, one can usually obtain reliable results without doing so.

We will begin with the methyl cation. The Lewis structure of this cation, as well as its orbital energy level diagram and electron configuration, and its frontier orbitals are given in Figure 5.1. There are six valence electrons in this cation, which means that the valence electrons are all used in the formation of the three C–H σ bonds and that the central carbon atom carries a formal positive charge.

There are only three σ bonds about the central carbon atom, the central carbon atom is sp^2 hybridized: the cation is trigonal planar in shape. The remaining orbital on the central carbon atom is the 2p orbital not used in the formation of the sp^2 hybrid orbitals, and it is empty. We can think of the HOMO simplistically as a carbon-hydrogen σ orbital, but the HOMO is actually a linear combination of all three C–H σ orbitals; the LUMO is fairly well approximated by

---

1. (a) Woodward, R.B.; Hoffmann, R. *J. Am. Chem. Soc.* **1965**, *87*, 395; *The Conservation of Orbital Symmetry* (Verlag Chemie/Academic Press: Weinheim, 1971)—this book is a stand-alone reprint of *Angew. Chem.Int. Ed. Engl.* **1969**, *8*, 781. (b) Hoffmann, R.; Woodward, R.B. *Acc. Chem. Res.* **1968**, *1*, 17. (c) Fukui, K.; Fujimoto, H. *Bull. Chem. Soc. Jpn.* **1967**, *40*, 2018; **1969**, *42*, 3399. (d) Fukui, K. *Fortschr. Chem Forsch.* **1970**, *15*, 1; *Acc. Chem. Res.* **1971**, *4*, 57; *Angew. Chem. Int. Ed. Engl.* **1981**, *21*, 801. (e) Longuet-Higgins, H.C.; Abrahamson, E.W. *J. Am. Chem. Soc.* **1965**, *87*, 2045.

2. (a) Fleming, I. *Frontier Orbitals and Organic Chemical Reactions* (Wiley-Interscience: New York, 1976). (b) Gilchrist, T.L.; Storr, R.C. *Organic Reactions and Orbital Symmetry*, 2nd ed. (Cambridge University Press: Cambridge, 1979). (c) Lehr, R.E.; Marchand, A.P. *Orbital Symmetry* (Academic Press: New York, 1970). (e) Pearson, R.G. *J. Chem. Educ.* **1981**, *58*, 753.

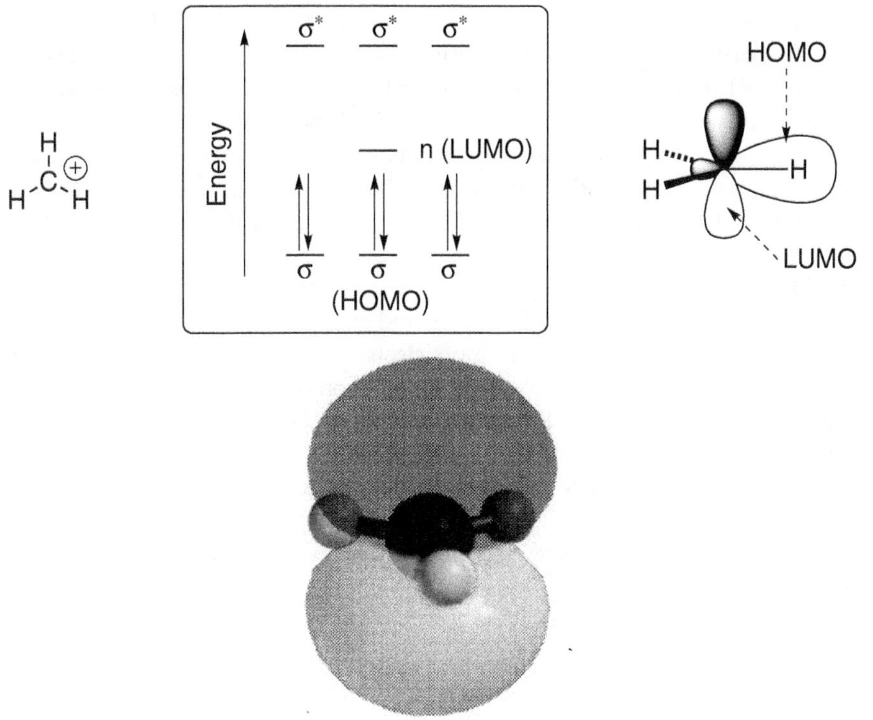

Figure 5.1 The orbital energy level diagram for the methyl cation. The highest energy occupied molecular orbital (HOMO) is one of the carbon-hydrogen σ orbitals, and the lowest energy unoccupied molecular orbital (LUMO) is the empty p orbital not used in forming the sp^2 hybrid orbitals of the central carbon atom.

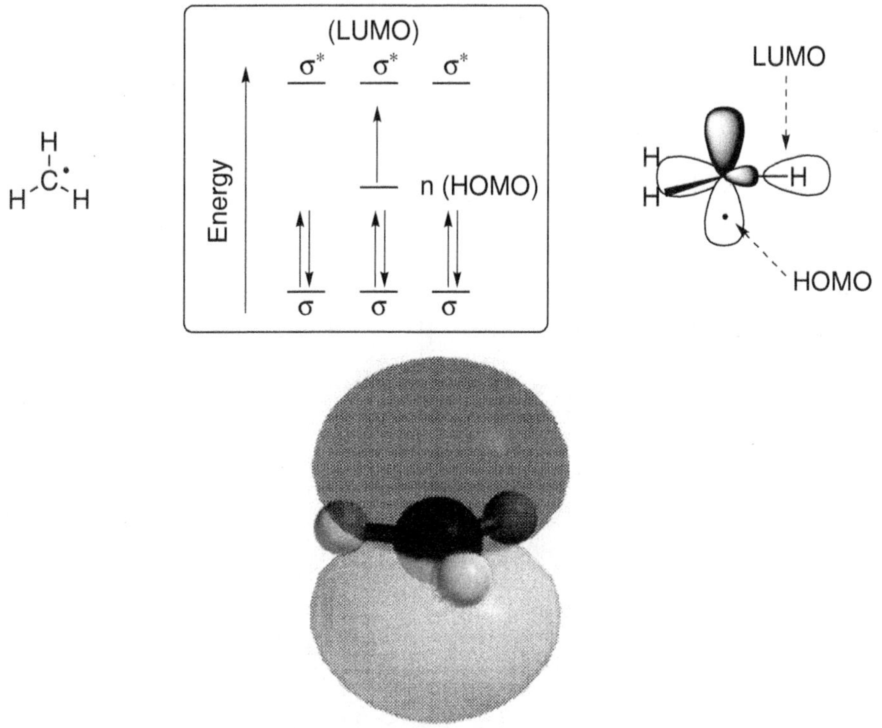

Figure 5.2 The orbital energy level diagram and the frontier molecular orbitals of the methyl free radical. HOMO, highest energy occupied molecular orbital; LUMO, lowest energy unoccupied molecular orbital.

the 2p orbital on the carbon. This means that that the LUMO of the cation is an *empty* p orbital. Although it is unoccupied, because this orbital is neither bonding nor antibonding, it must be a nonbonding orbital: *not all nonbonding orbitals are occupied.* Often in this book, we will use the symbol **a** to denote an unoccupied nonbonding molecular orbital.

The methyl free radical has the formula CH_3•, and therefore it has just one electron more than the methyl cation for a total of seven valence electrons. Its Lewis structure, orbital energy level diagram and electron configuration, and its FMOs are given in Figure 5.2. These may be compared with those of the cation.

In the free radical, the unpaired (nonbonding) electron is in the fourth valence orbital of the central carbon atom. However, whether this odd electron is in an unhybridized p orbital or an sp^3 hybrid orbital depends on several factors, and examples of free radicals of both structural types are known. The methyl radical itself is planar. The HOMO of the methyl radical is the p orbital containing the unpaired electron, whereas the LUMO of the methyl radical is a C–H σ* antibonding orbital. Because the HOMO of every free radical is occupied by only one electron instead of two, it is called the ***singly occupied molecular orbital*** (SOMO).

The third case that we will examine here is the case of the methyl anion, CH_3^-, whose Lewis structure, orbital energy level diagram and electron configuration, and FMOs are given in Figure 5.3. This anion has one more electron than the free radical, and so it has eight valence electrons and a formal negative charge on the central carbon atom. The valence electrons are distributed as three C–H σ bonds and one lone pair. To accommodate this arrangement of valence electrons, the central carbon atom requires four hybrid orbitals (one for each σ bond and one for the lone pair), and must therefore be sp^3 hybridized. The anion therefore, has a trigonal pyramidal shape, just like the ammonia molecule. The HOMO of the methyl anion is the (nonbonding) sp^3 hybrid orbital containing the lone pair of electrons, whereas the LUMO is a C–H σ* antibonding orbital (or, more accurately, a linear combination of all three C–H σ* orbitals).

The presence of heteroatoms can complicate the picture somewhat because of the different energies of the orbitals involved. There is, however, a general rule of thumb that one can use to decide what the relative energies of the orbitals will be. In general, *the more polar the bond, the lower the energy of all the orbitals—bonding* and *antibonding—associated with it.*

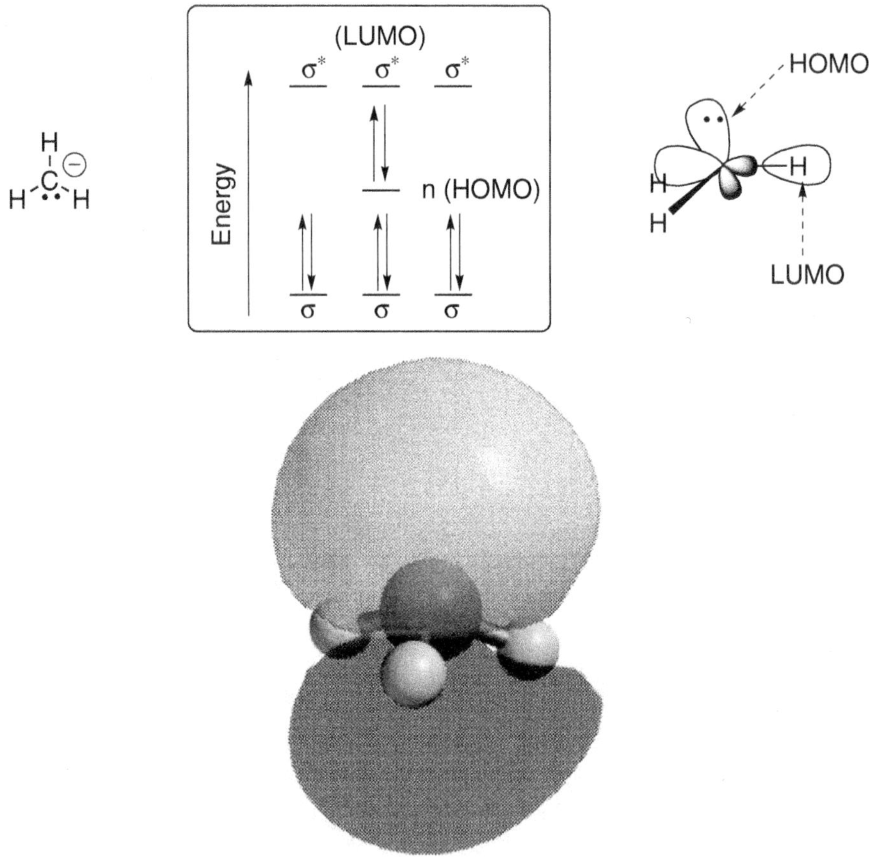

Figure 5.3 The orbital energy level diagram and the frontier molecular orbitals of the methyl anion. HOMO, highest energy occupied molecular orbital; LUMO, lowest energy unoccupied molecular orbital.

Of course, the picture is not really this simple.

From the perspective of the discussion we have just had, the fortunate thing about methyl cations, anions, and free radicals is that they have only one carbon atom and three σ bonds. This makes the methyl system a very simple one to analyze, and what we have done up to now is usually more than sufficient. However, as the systems become more complex, the level of sophistication that we need to discuss them also rises (although not to prohibitive heights). Let us look at the corresponding ethyl species. In Table 5.1, the FMOs of the ethyl cation, the ethyl radical, and the ethyl anion are shown as obtained from computer calculations at the B3LYP/6-31+G* level. This level of sophistication in calculations is actually quite low, but it still provides a wealth of qualitative information that we can use in discussing chemical reactivity.

When we consider the chemistry of the ethyl cation, we know that the electron-deficient carbon atom dominates its reactivity: it usually behaves as the electrophile in reactions, which means that the LUMO is by far the more important of its two FMOs. Likewise, the reactions of the ethyl free radical all involve the unpaired electron, which makes the SOMO (HOMO) the most important FMO in its reactions. The same analysis of the predicted reactivity of the ethyl anion, which we know is both a powerful nucleophile and a powerful base, leads to the conclusion that the orbital carrying the lone pair is going to dominate the reactivity of this species: its reactions involve the HOMO as the most important frontier orbital.

If we now examine these most important FMOs carefully, a remarkable consistency is revealed. In every one of these orbitals, not only is the orbital on carbon an important component of the orbital but also the carbon-hydrogen σ bonds on the adjacent carbon as well. In other words, *hyperconjugation plays an extremely important role in each of these FMOs*. We can generalize this observation to the molecular orbitals of most species with sp^2-hybridized atoms. In almost every one of these species, the C–H σ orbitals on the carbon atom adjacent to the sp^2-hybridized atom are involved in the HOMO or LUMO of the species. This accounts, for example, for why carbonyl compounds with α hydrogens

Table 5.1 Frontier Orbitals of Ethyl Species, $CH_3CH_2^*$

Frontier Orbital	$(CH_3CH_2^+)$†	$(CH_3CH_2^\bullet)$	$(CH_3CH_2^-)$
LUMO	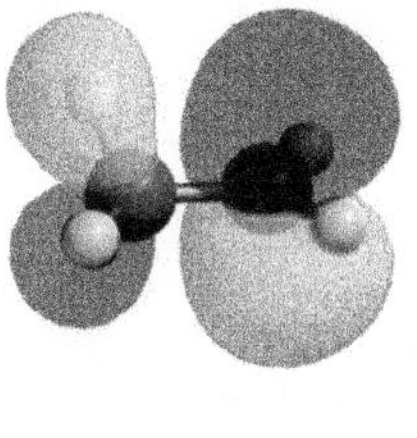	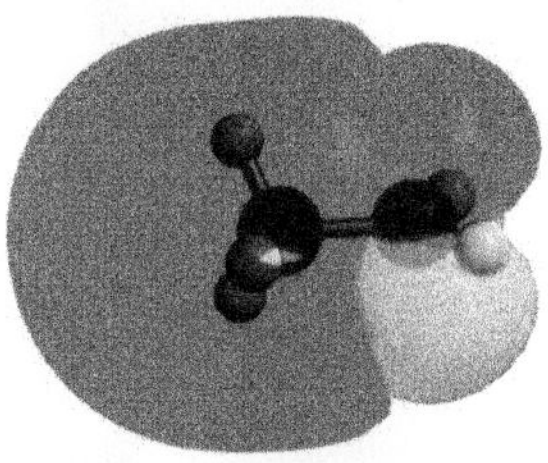	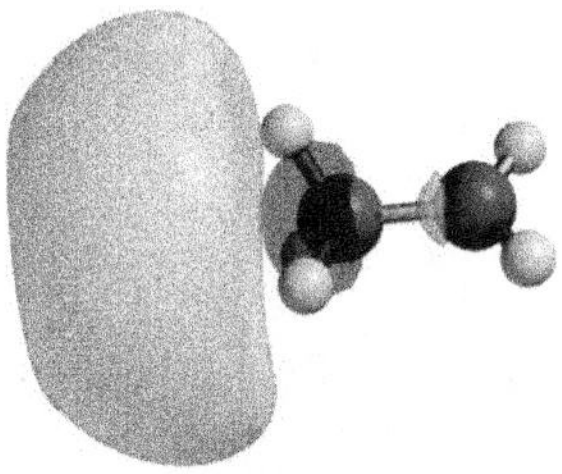
HOMO (SOMO)			

†The actual structure of this ion is bridged, but computations at this low level return orbitals that can be compared directly with those of the radical and the anion. The highest energy occupied molecular orbital (HOMO) of this cation does not involve the p orbital on the cation carbon, so the orientation has been changed to show the σ orbitals that participate in the HOMO.

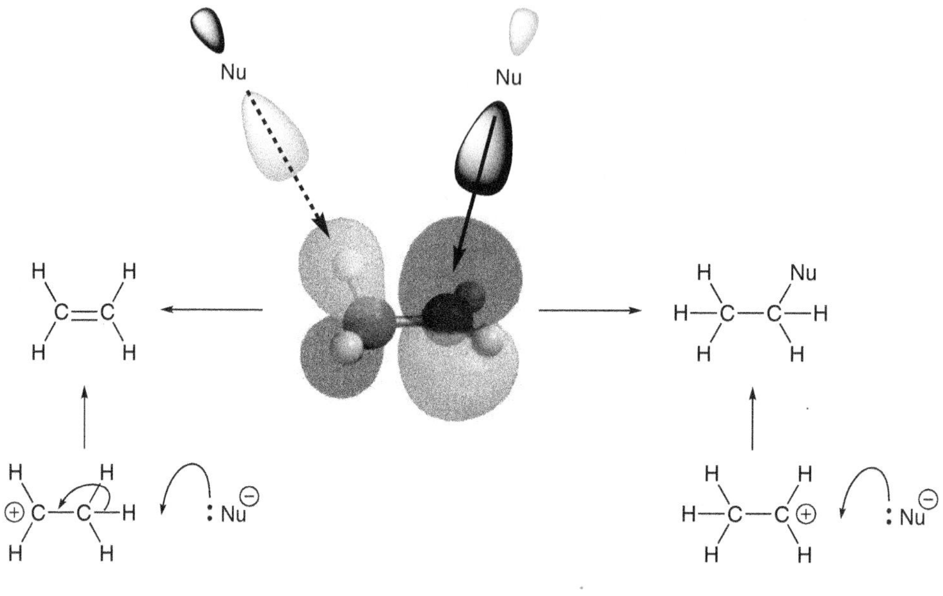

Figure 5.4 The reactions of the ethyl cation depend on the point at which the nucleophile attacks the cation lowest energy unoccupied molecular orbital.

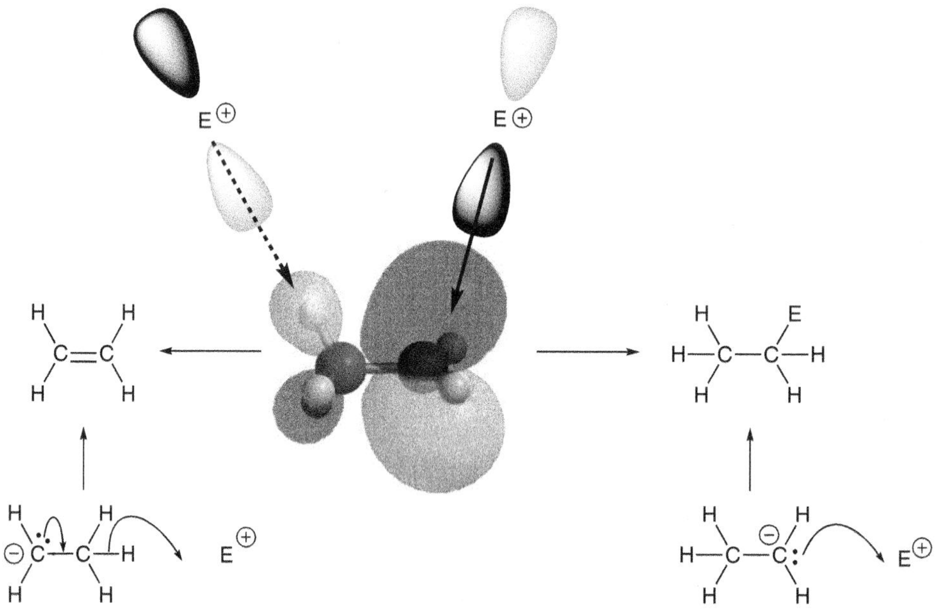

Figure 5.5 The reactions of the ethyl anion depend on the point at which the electrophile attacks the anion highest energy occupied molecular orbital.

can react with nucleophilic species to give either addition products (nucleophilic addition) or enolate anions (acid-base reactions), and why carbocations give products of both S_N1 substitution and E1 elimination.

Let us now look at the LUMO of the ethyl cation again. Here we see that there are actually two places for the HOMO, a nucleophile, to overlap effectively with the cation—the cation carbon and two of the β hydrogens. It is more than a coincidence that these two positions correspond to the two possible reactions of the ethyl cation with which we are already familiar (Figure 5.4). Overlap of the cation LUMO with the nucleophile HOMO at the carbon atom gives us the familiar second step of the S_N1 substitution reaction, with the formation of a new carbon-nucleophile bond. If that overlap is moved to one of the β-hydrogen atoms, however, we get the second step of the E1 elimination reaction. You

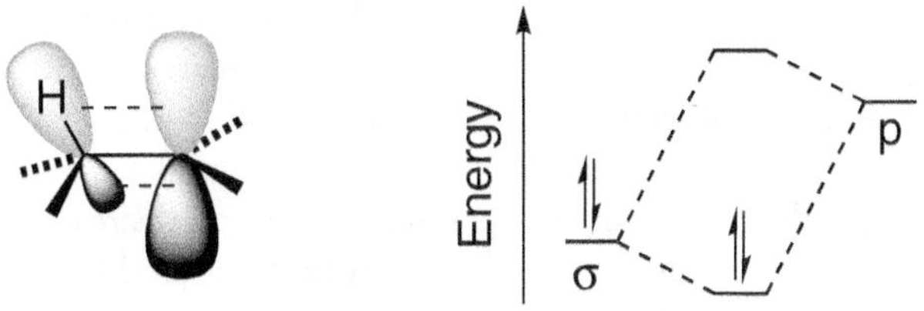

Figure 5.6 p-Overlap between the filled C—H σ orbital and the empty 2p orbital on the carbocation carbon leads to stabilization of the cation.

may recall that the S_N1 and E1 reactions always compete—now you have a reason why (it simply depends on exactly where the HOMO-LUMO overlap occurs).

Likewise, when we examine the HOMO of the ethyl anion, we see that the electrophile may attack at the carbanion carbon or at the β hydrogen (Figure 5.5). In this case, the hydrogen atom is transferred as a hydride ion, rather than as a proton. However, in all other ways this reaction is strictly analogous to the cation case. The transfer of a hydride anion from a carbanion nucleophile has become a very important synthetic method for the reduction of carbonyl compounds.[3]

Conformational and Stereoelectronic Effects

The interaction of filled orbitals with empty orbitals is not restricted to reactions between molecules but is also manifested in the overlap between a filled and an empty orbital on adjacent atoms in a molecule. Just as when atomic orbitals on adjacent atoms overlap, the product of the overlap of molecular orbitals on adjacent atoms leads to a change in the respective orbital energies, with the lower energy orbital becoming even lower in energy and the higher energy orbital becoming higher in energy. The extent to which the energy of the orbitals involved changes depends on the efficiency of the overlap.

One major form of this type of overlap is the overlap in carbocations that leads to stabilization of the cation by hyperconjugation. This type of overlap, and the effect on the energy of the system, is shown in Figure 5.6. The π-type overlap between the empty π orbital on carbon and the filled C—H σ orbital on the adjacent carbon has the effect of lowering the energy of the filled orbital and raising the energy of the empty orbital. However, because the energy of the carbocation depends only on the total energy of the electrons in the filled orbitals, this leads to a lowering of the energy of the cation.

Problem

5-1 A pair of ethyl radicals may react with each other in two different ways. Using the SOMO above as a working model, predict the products of these two possible reactions.

Another example of this type of overlap, between a filled orbital on one atom and an empty orbital on the adjacent atom, is provided by the ethane molecule. When one carries out high-level calculations on the conformers of ethane, one finds that not only does the

3. For reviews on the reduction of ketones by hydride transfer from trialkylboranes, for example: (a) Srebnik, M.; Ramachandran, P.V. *Aldrichimica Acta* **1987**, *20*, 9. (b) Dhar, R.K. *Aldrichimica Acta* **1994**, *27*, 43. (c) Cho, B.T. *Aldrichimica Acta* **2002**, *35*, 3. (d) Midland, M.M. *Chem. Rev.* **1989**, 89, 1553. (e) Midland, M.M.; Greer, S.; Tramontano, A.; Zderic, S.A. *J. Am. Chem. Soc.* **1979**, *101*, 2352. (f) Midland, M.M.; McDowell, D.C.; Hatch, R.L.; Tramontano, A. *J. Am. Chem. Soc.* **1980**, *102*, 867. (g) Midland, M.M.; Graham, R.S. *Org. Syn.* **1985**, *63*, 57.

dihedral angle change during the rotation about the carbon-carbon σ bond but that the bond distances also change. Thus, one finds that the carbon-carbon σ bond in the staggered conformation is slightly shorter than in the eclipsed conformation, whereas the carbon-hydrogen bonds are slightly longer. This may be viewed as reflecting an increased carbon-carbon bond order in the staggered conformation and a decreased carbon-hydrogen bond order relative to the values in the eclipsed conformation. This may be visualized in terms of a π-type delocalization of electron density from the filled σ orbital on one atom into the empty σ^* orbital on the adjacent atom, as shown schematically in Figure 5.7. The computed molecular orbitals show that even when a more sophisticated treatment us used, the overlap is more efficient in staggered ethane.

The antiperiplanar arrangement of the two C—H bonds results in the most efficient geometry for delocalizing electrons from the σ orbital to the adjacent σ^* orbital, so we expect that this arrangement of bonds will be the lowest energy, leading to the staggered conformation being the most stable. By comparison, the overlap between the same orbitals in the synperiplanar geometry is much less efficient, with a partial antibonding component.

A similar situation holds when we look at the conformations of alkenes and carbonyl compounds. In this case, it is overlap between the σ orbitals of the C—H bonds and the π^* orbital that influences conformational energy. This arrangement, where both C—H σ orbitals can overlap with the C—C π^* orbital, is predicted to be the lowest energy, which leads to the conclusion that the eclipsed conformation is lowest energy (Figure 5.8).

Lone pairs can replace the σ orbitals in the forgoing discussion. This leads to two effects that have been observed empirically. The first of these is the ***gauche effect***,[4] where the gauche conformer of a disubstituted ethane becomes more favored as the electronegativity of the two substituents increases.[5] The second is the ***anomeric effect***,[6] which is the observation that the conformational preference of substituents at the ring carbon adjacent to the heteroatom in a saturated six-membered heterocycle is generally lower than the conformational preference of the same substituent in cyclohexane, leading to a higher percentage of the axial conformer in the equilibrium mixture (Figure 5.9). The theoretical basis for the anomeric effect has been discussed at length, with most authors proposing the model just discussed, where the electron density from a lone pair is delocalized into the σ^* orbital of the adjacent carbon-heteroatom bond.[7] This rationalization is not, however, universally accepted,[8] and other rationalizations based on dipole-dipole interactions have also been proposed.[9] The anomeric effect has been found to be particularly applicable in carbohydrates, where the β (equatorial) isomer of the six-membered hemiacetal is frequently found to be much less favored than the α (axial) isomer. These isomers are known as ***anomers***.

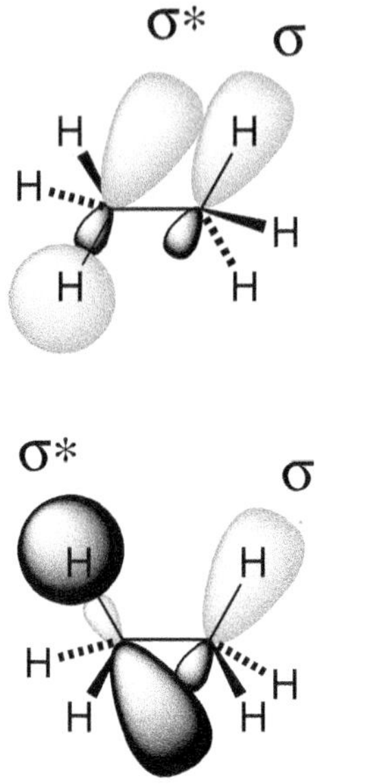

Figure 5.7 π-Overlap between adjacent filled and empty orbitals in ethane is most efficient when the bonds are antiperiplanar. Synperiplanar overlap is much less efficient.

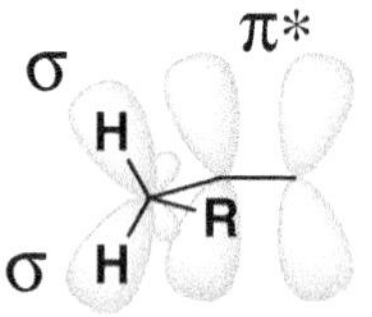

Figure 5.8 π-Overlap between adjacent filled σ and empty π^* orbitals in propylene is most efficient when the alkene conformation is eclipsed.

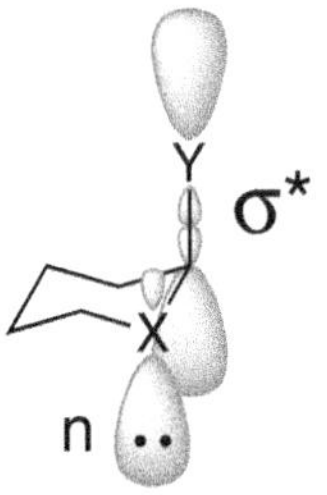

Figure 5.9 π-Overlap between a filled nonbonding orbital on the heteroatom and the empty σ^* orbital on the adjacent atom may account for the extra stability of axial substituents in these six-membered rings.

---

4. (a) Huang, J.; Hedberg, K. *J. Am. Chem. Soc.* **1990**, *112*, 2070. (b) Dixon, D.A.; Matsuzawa, N.; Walker, S.C. *J. Phys. Chem.* **1992**, *96*, 10740. (c) Wolfe, S. *Acc. Chem. Res.* **1972**, *5*, 102.

5. Phillips, L.; Wray, V. *J. Chem. Soc. Chem. Commun.* **1973**, 90.

6. Monographs: (a) Kirby, A.J. *The Anomeric Effect and Related Stereoelectronic Effects at Oxygen* (Springer-Verlag: Berlin, 1983). (b) Szarek, W.A.; Horton, D. *Anomeric Effect* (American Chemical Society: Washington, D.C., 1979). (c) Deslongchamps, P. *Stereoelectronic Effects in Organic Chemistry* (Pergamon: Oxford, 1983).

Reviews: (d) Zefirov, N.S. *Tetrahedron* **1977**, *33*, 3193. (e) Lemieux, R.U. *Pure Appl. Chem.* **1971**, *27*, 527. (f) Angyal, S.J. *Angew. Chem. Int. Ed. Engl.* **1969**, *8*, 157.

7. (a) Juarista, E.; Cuevas, G, *Tetrahedron* **1992**, *48*, 5019. (b) Salzner, U.; Schleyer, P.v.R. *J. Am. Chem. Soc.* **1993**, *115*, 10231. (c) Romers, C.; Altona, C.; Buys, H.R,; Havinga, E,. *Top. Stereochem.* **1969**, *4*, 39. (d) Wolfe, S.; Whangbo, M.; Mitchell, D.J. *Carbohydrate Res.* **1979**, *69*, 1. (e) Fuchs, B.; Ellencweig, A.; Tartakovsky, E.; Aped, P. *Angew. Chem. Int. Ed. Engl.* **1986**, *25*, 287. (f) Praly, J.; Lemieux, R.U. *Can. J. Chem.* **1987**, *65*, 213. (g) Booth, H.; Khedair, K.A.; Readsha, S.A. *Tetrahedron* **1987**, *43*, 4699.

8. Box, V.G.S. *Heterocycles* **1990**, *31*, 1157.

9. (a) Pearson, R.G. *J. Am. Chem. Soc.* **1988**, *110*, 7684. (b) Hati, S.; Dhatta, D. *J. Org. Chem.* **1992**, *57*, 6056.

5.2 Using Frontier Molecular Orbitals to Categorize Reactions and Reagents

Organizing chemical reactions and compounds into classes is a long-standing hallmark of organic chemistry. Recall how compounds are grouped into classes based on their functional groups and how reactions are classified according to their outcomes (e.g., addition, elimination, substitution, oxidation, reduction) and mechanisms. In this chapter, we will examine the common classes of organic reactions and categorize them in terms of the frontier orbitals involved. Later in this book, we will expand our discussion of synthetic reactions to retrosynthetic analysis, and the design of syntheses. For now, however, let us focus on the frontier orbitals of the reacting species and on how the combination of frontier orbitals involved directs the outcome of the reaction.

Bond-forming reactions can be broadly characterized as being unimolecular or bimolecular. Because of the Pauli Exclusion principle, bimolecular reactions must always involve the overlap of an *empty* orbital on one reactant with a *filled* orbital on the other: *the HOMO in one reactant (which may contain either one or two electrons) must overlap with the LUMO of the other* unless two free radicals are involved. Unimolecular reactions, on the other hand, involve the HOMO only. Bimolecular reactions leading to bond formation are much more common than unimolecular bond-forming reactions and include some of the most widely used synthetic reactions known.

In simple organic compounds, there are only three types of orbitals that can function as a HOMO—an occupied σ orbital, an occupied π orbital, or an occupied nonbonding (n or n_π) orbital (carrying either a lone pair or single electron). Similarly, there are only three possible types of orbitals that can function as a LUMO—an empty nonbonding molecular or atomic (a or a_π) orbital (e.g., the 2p orbital on CH_3^+), an empty π^* orbital, or an empty σ^* orbital. Some typical examples of these orbitals are given in Table 5.2.

The n and n_π orbitals in Table 5.1 are actually subsets of the same type of frontier orbital—as are the a and a_π orbitals. Those orbitals designated with the subscript π are nonbonding orbitals in conjugated systems. There are nine and only nine possible combinations of the three types of HOMO and the three types of LUMO; they are tabulated in Table 5.3.[10]

Table 5.2 Representative Frontier Orbitals

Frontier Orbital	Examples
a	Empty p orbital on B, C Al, and so on; empty d orbitals on Si, Sb, and so on ***Electrophilic***
a_π	ψ_2 of allyl cation, and so on; ψ_4 of benzyl cation, and so on ***Electrophilic***
n	Lone pairs in hybrid atomic orbitals ***Nucleophilic***
n_π	ψ_2 of allyl anion, and so on; ψ_4 of benzyl anion, anisole, aniline, and so on ***Nucleophilic***
π	π orbital of alkene, carbonyl group, cyano group, and so on; ψ_2 of diene, conjugated carbonyl compound or nitrile, and so on; ψ_3 of benzene, and so on ***Nucleophilic***
π^*	π^* orbital of alkene, carbonyl group, cyano group, and so on; ψ_3 of diene, conjugated carbonyl compound or nitrile, and so on; ψ_4 of benzene, and so on ***Electrophilic***
σ	σ orbital (usually C—H σ orbital) ***Nucleophilic***
σ^*	σ^* orbital (usually C—X σ^* orbital of bond to a leaving group) ***Electrophilic***

10. (a) Lewis, D.E. *J. Chem. Educ.* **1999**, *76*, 1718. (b) Jensen, W.B. *J. Chem. Educ.* **2001**, *78*, 727.

Table 5.3 Possible Frontier Molecular Orbital Combinations in Organic Chemistry

HOMO	LUMO	Examples of Typical Outcome
n	a	Bond formation in step 2 of S_N1 substitution
n	π^*	Bond formation + bond rupture in nucleophilic or free radical addition
n	σ^*	Bond formation + bond rupture in S_N2 substitution; anomeric effects in cyclic systems
π	a	Bond formation + bond rupture in electrophilic addition
π	π^*	Bond formation + bond rupture in cycladdition
π	σ^*	Bond formation + bond rupture in electrophilic addition
σ	a	Bond formation + bond rupture in cation rearrrangements
σ	π^*	Stabilization of conformations of unsaturated compounds
σ	σ^*	Stabilization of conformations of saturated compounds

*HOMO, highest energy occupied molecular orbital; LUMO, lowest energy unoccupied molecular orbital.

Every organic reaction known in which a new bond is formed by transfer of an electron pair from one reactant to another fits one of these HOMO-LUMO (FMO) combinations. In general, we look at organic reactions in terms of the bonds being formed. However, on occasion we will need to look at reactions in which a bond is broken and those in which no bond is formed. Although such reactions can (with some difficulty) be fitted into the classifications above, the relationships are less clear-cut than in reactions in which a bond is formed; we will address such bond rupture reactions at the time we encounter them.

The transfer of a pair of electrons from the HOMO of one molecule to the LUMO of the other corresponds to the reaction between a Lewis acid and a Lewis base.[11] The compound that reacts via a filled HOMO is functioning as an electron pair donor (i.e., a Lewis base or a nucleophile). *A nucleophile, or Lewis base, participates in reactions via its HOMO.* In such reactions, the pair of electrons from the HOMO is placed into the LUMO, so that the compound that reacts via its LUMO is functioning as an electron pair acceptor (i.e., a Lewis acid or an electrophile). *An electrophile, or Lewis acid, participates in reactions via the LUMO.* What this means, of course, is that much of organic chemistry can actually be reduced to the reaction of a Lewis acid with a Lewis base—or the reaction of an electrophile with a nucleophile.

The process of examining a reaction from the perspective of what orbitals are overlapping and what new orbitals are formed can be thought of as a type of "orbital inventory." In generating an orbital inventory for a reaction, one need only adhere to the rules for orbital combination that we have already discussed for atomic orbitals:

1. Orbitals may be combined in a variety of different ways but neither created nor destroyed.
2. For every bonding molecular orbital there is a corresponding antibonding orbital.

One may take this orbital inventory to its logical conclusion and map the correspondence between the orbitals of the reactants and the orbitals of the product. Such a mapping is called an ***orbital correlation diagram***. As the occasion arises, we will make use of both

11. (a) Jensen, W.B. *Lewis Acid-Base Concepts: An Overview* (Krieger Publishing Co.: Melbourne, FL, 1979). (b) Jensen, W.B. *Chem. Rev.* **1978**, *78*, 1. (c) Jensen, W.B. *Chemtech* **1982**, *12*, 755.

the pictorial orbital diagrams and the simplified energy level diagrams when discussing organic reactions as well as using orbital inventories and correlation diagrams. First, you should become familiar with the basic concepts so that you can use them later.

Perturbation Approximations for Orbital Coefficients

The use of FMOs for studying organic reactions can be enhanced, especially in reactions involving π-bonded systems, by approximating second-order effects on the coefficients of the contributing basis orbitals in those π bonding systems. This method, which is really a combination of simple Hückel theory and resonance concepts, gives the chemist a way to approximate the FMOs in a more "exact" way, and this allows them to be used with greater confidence in predicting reaction regiochemistry. The basic concepts of this approach and its applications have been discussed in much greater detail than is possible in this single chapter in the book by Fleming.[12]

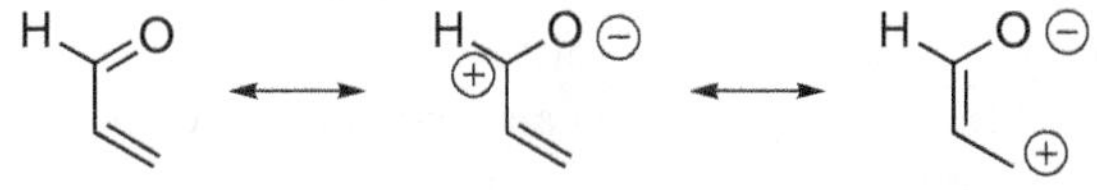

major (5.1) minor (5.2) minor (5.3)

At its heart, the perturbation method for approximating the FMOs of a reacting species leads to a prediction of the relative sizes (larger or smaller) of the lobes of the orbital at a particular position in the orbital. Let us take acrolein (Example **5.1**) as an example. The aldehyde molecule can be represented using traditional resonance models as shown at left. In this case, there is one major (Example **5.1**; neutral), and two minor (Examples **5.2** and **5.3**; dipolar) canonical forms.

When discussing the chemistry of compounds containing π bonds between unlike atoms (e.g., a carbonyl group), one must be careful when talking about the HOMO, because the HOMO of a carbonyl group, for example, is actually a lone pair (or linear combination of lone pairs), and not the C=O π orbital. For this reason, when talking about reactions of π-bonded systems, we will use the term HOMO (π) when referring to the highest energy occupied π molecular orbital.

When we look at the π orbital system of acrolein, we note that it is composed of four atoms (the three sp^2-hybridized carbon atoms and the sp^2-hybridized oxygen atom) and four electrons. The system will have two occupied and two unoccupied molecular orbitals. To predict its reactivity, we now need to approximate its FMOs. This can be done by separating the orbitals into (1) a ***base system***, which is the hydrocarbon that most closely approximates the neutral canonical form; and (2) the ***perturbing influence***, which is the hydrocarbon ion that most closely resembles the bipolar canonical form.

If we apply this approach to approximating the π molecular orbitals of the acrolein molecule, we infer that the base π molecular orbitals of the acrolein molecule should be modeled on those of 1,3-butadiene (Example **5.4**) and that the perturbing influence should be the π orbitals of the allyl cation (Example **5.5**). The HOMO (π) should be ψ_2 of 1,3-butadiene modified by ψ_1 of the allyl cation, and the LUMO should be ψ_3 of 1,3-butadiene modified by ψ_2 of the allyl cation. In similar fashion, one can approximate the π molecular orbitals of 1-methoxy-1,3-butadiene (Example **5.6**) by perturbing the molecular orbitals of 1,3-butadiene (Example **5.4**) with those of the 2,4-pentadienyl anion (Example **5.7**). In this case, the HOMO (π) should be ψ_2 of 1,3-butadiene modified by ψ_3 of the pentadienyl anion, and the LUMO should be ψ_3 of 1,3-butadiene modified by ψ_4 of the pentadienyl anion (Figure 5.10).

12. Fleming, I. *Frontier Orbitals and Organic Chemical Reactions* (Wiley-Interscience: New York, 1976).

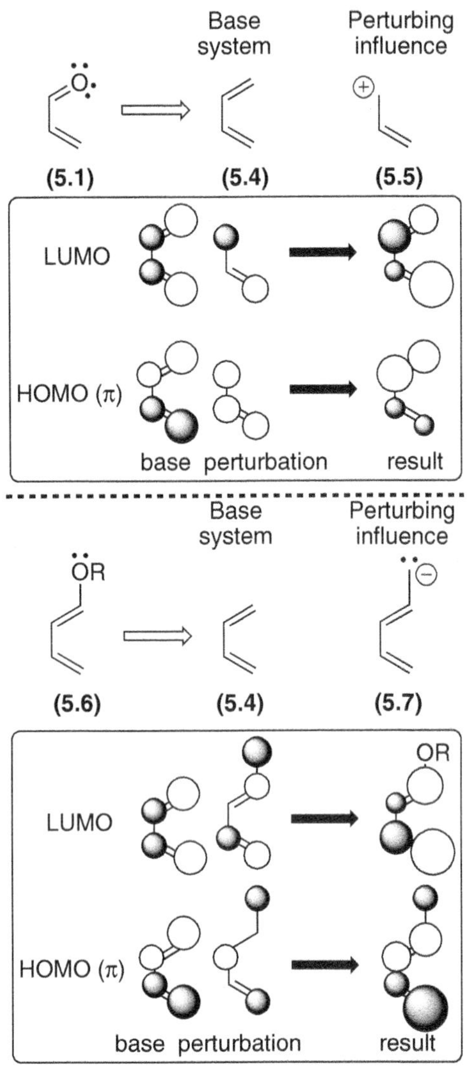

Figure 5.10 Construction of the approximate frontier molecular orbitals of acrolein (left) and 1-alkoxy-1,3-butadiene (right). HOMO, highest energy occupied molecular orbital; LUMO, lowest energy unoccupied molecular orbital.

This removes the symmetry of the orbital coefficients and leads to the situation where there are now preferred sites of attack. It is clear that the superimposition of ψ_1 of the allyl cation on ψ_2 of 1,3-butadiene gives an orbital where the largest orbital coefficient is at the β carbon (Figure 5.11). The characteristic reaction of acrolein and other conjugated carbonyl compounds is nucleophilic addition. The LUMO that we have just approximated strongly suggests that the preferred regiochemistry of this reaction will be either 1,2- or 1,4-. We expect that nucleophiles attacking the alkene π bond will do so preferentially at the β position. Again, because FMO interactions become more important as the reacting species become softer, we expect that soft nucleophiles will tend to give more attack at the β carbon of acrolein.

Perturbation methods can also permit us to estimate the relative energies of the FMOs of a molecule. Based on the same model that we have just used, we can calculate that the energy of the HOMO (π) of acrolein will lie between those of the allyl cation [$\alpha + 2\beta\cos(\pi/4)$] and 1,3-butadiene [$\alpha + 2\beta\cos(2\pi/5)$]. Likewise, we can predict that the LUMO energy of acrolein will lie between those of the allyl cation [$\alpha + 2\beta\cos(\pi/2)$] and 1,3-butadiene [$\alpha + 2\beta\cos(3\pi/5)$]. This allows us to obtain the following energies for the two frontier orbitals of acrolein:

$$E_{HOMO} = \alpha + 2\beta\cos(2\pi/5) + c[\alpha + 2\beta\cos(\pi/4)] \quad (5.1)$$
$$E_{LUMO} = \alpha + 2\beta\cos(3\pi/5) + c[\alpha + 2\beta\cos(\pi/2)] \quad (5.2)$$

where c is a coefficient that determines the contribution of the allyl cation wave function to the complete wave function (in other terms, here we see the importance of the minor contributor to the resonance hybrid). This analysis, simplified as it is, allows us to predict that the conjugation of an alkene π bond with a carbonyl group will lead to a lowering of both the HOMO (π) and LUMO energies. We can carry out a similar analysis with ethyl vinyl ether, which can now be approximated by an ethylene molecule perturbed by an allyl anion. This leads to the conclusion that both π molecular orbitals have their energy raised by an electron-donating group, and that the β carbon has the larger orbital coefficient in the HOMO (π).

When a π bond is formed between a pair of like atoms, the magnitudes of the orbital coeficients are equal at the same location relative to the center of the orbital. This is not the case, however, when the two atoms participating in the π bond are different. One example of this is provided

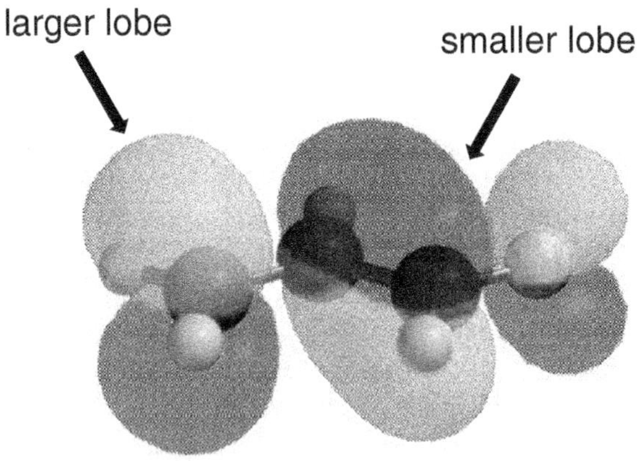

Figure 5.11 The asymmetry in the basis orbital contributions to the lowest energy unoccupied molecular orbital of acrolein leads to different outcomes with different nucleophilic additions

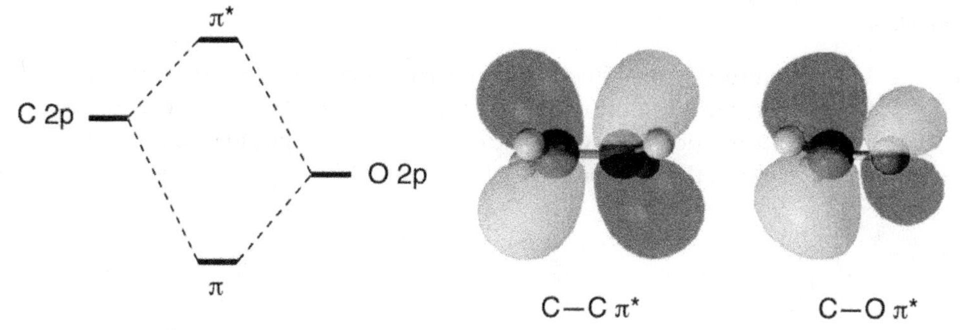

Figure 5.12 Orbital overlap to generate the π orbitals of the C=O group, and a comparison of the lowest energy unoccupied molecular orbitals of C=C and C=O bonds.

by the carbonyl group (Figure 5.12). Here, the overlap is between the carbon 2p orbital and the oxygen 2p orbital. Because the carbon 2p orbital is higher in energy, we find that the π orbital is closer in energy to the oxygen atomic orbital and the π* orbital closer in energy to the carbon atomic orbital. This extends to the orbital coefficients as well. The π orbital has a larger coefficient on the oxygen, and the π* orbital has a larger coefficient on carbon. This is illustrated for the LUMOs of C—C and C—O π bonds in Figure 5.12. As we see, the LUMO of an alkene π bond is symmetric, as expected. The LUMO of the carbonyl group, on the other hand, is quite asymmetric, with large lobes on carbon and small lobes on oxygen, consistent with attack of a nucleophile on a carbonyl group occurring at carbon.

Worked Problem

5-1 The Diels-Alder reaction is a cycloaddition reaction where the participating orbitals are ψ_2 of the diene and π* of the dienophile. Using perturbation molecular orbital theory, deduce the qualitative coefficients of the FMOs in Diels-Alder reaction below. What is the regiochemistry of the reaction?

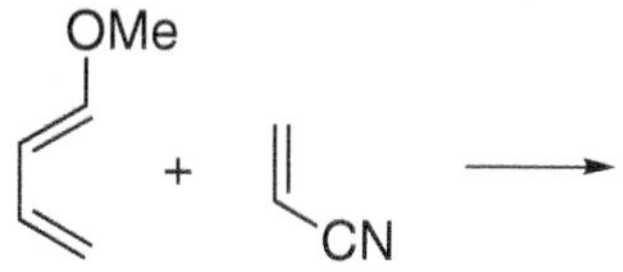

§Answer below

§ Answer to Worked Problem:
The diene orbital may be viewed as ψ_2 of 1,3-butadiene perturbed by ψ_2 of the allyl cation. This means that the largest lobe of the diene is at the end of the conjugated system opposite to the methoxy group.

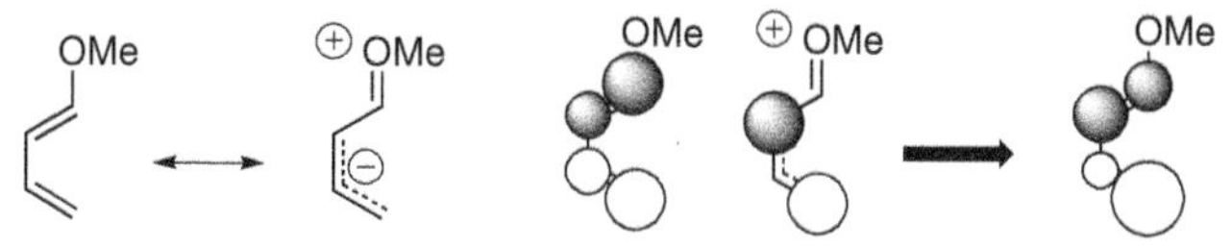

The dienophile orbital may be viewed as the π* orbital of ethylene perturbed by ψ_2 of the allyl cation, so a similar analysis gives the result that the larger lobe of the dienophile is at the end of the conjugated system opposite to the cyano group.

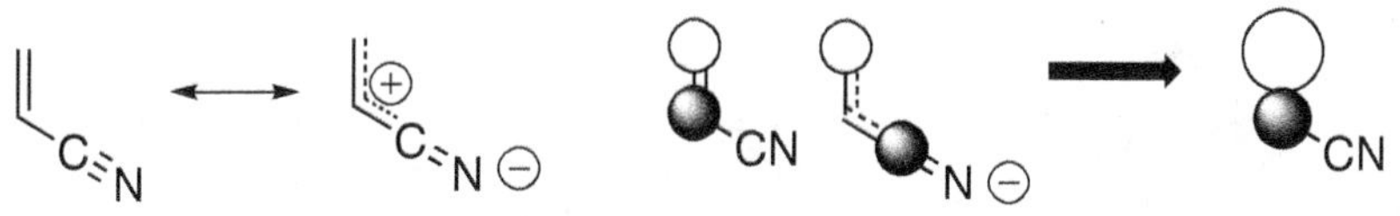

Problem

5-2 Using perturbation molecular orbital theory, deduce the qualitative orbital coefficients in the diene ψ_2 orbital and the dienophile π^* orbital in each of the Diels-Alder reactions below. Use the results of this analysis to predict the regiochemistry of the reactions:

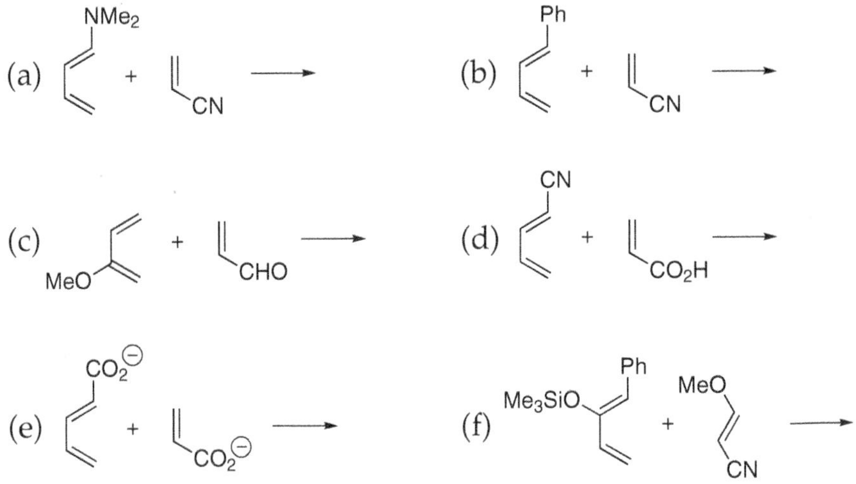

Let us now look at the nine possible FMO combinations of Table 5.2, which are all represented by real reactions in organic chemistry, although the first six entries represent much more common reaction types.

5.3 Categorizing Reactions Using Frontier Molecular Orbital Pairings

Frontier Molecular Orbital Overlap Between *n* and *a* Orbitals

This pairing of FMOs leads to the formation of a new σ bond by permitting transfer of electrons from a filled nonbonding orbital to an empty nonbonding orbital. It is the simplest of all the nine possible HOMO-LUMO combinations, because it does not involve any antibonding orbitals. In this case, the overlap of the two nonbonding orbitals leads to the formation of two new orbitals (one bonding, one antibonding), as shown in Figure 5.13. The reaction between trimethylborane and ammonia is a typical example. In the most common case, two electrons are transferred and the net result is the formation of a new bond without the rupture of a bond in either reactant. Overall, the reaction converts what are in essence two atomic orbitals to a σ orbital and a σ^* orbital.

When we overlap these two orbitals, the most efficient overlap is *large lobe to large lobe*, so we expect that the regiochemistry of the product will be most favorable when the two substituents on the six-membered ring are on adjacent carbons. The stereochemistry of the reaction follows the Alder *endo* rule, which places both substituents on the same face of the six-membered ring:

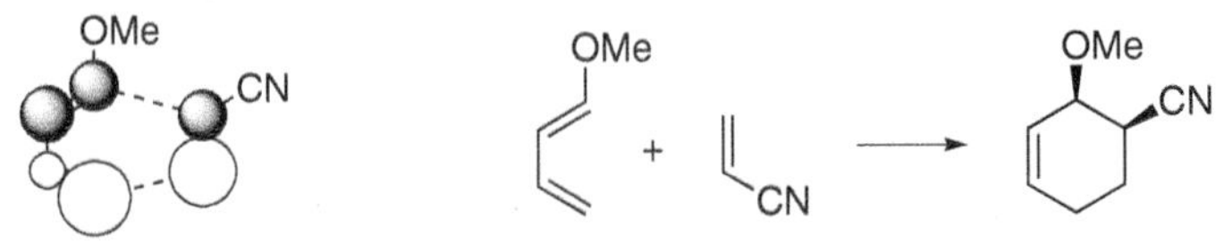

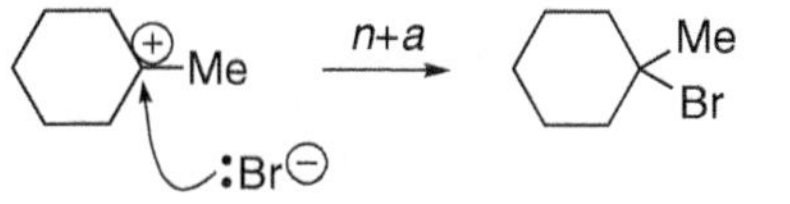

(5.8)

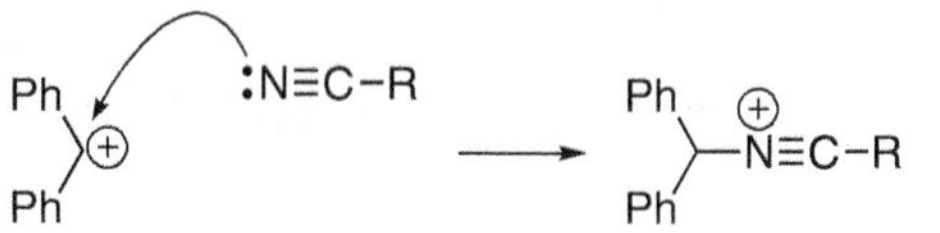

(5.9)

The most common occurrence of this pairing of orbitals in organic chemistry is during the fast step of the S_N1 reaction, where the carbocation (the Lewis acid) reacts with the nucleophile (the Lewis base) (Example **5.8**). It is noteworthy that the slow step of the S_N1 reaction corresponds to the reverse of this process. This is also the key bond-forming step in the Ritter reaction of nitriles (Example **5.9**).[13]

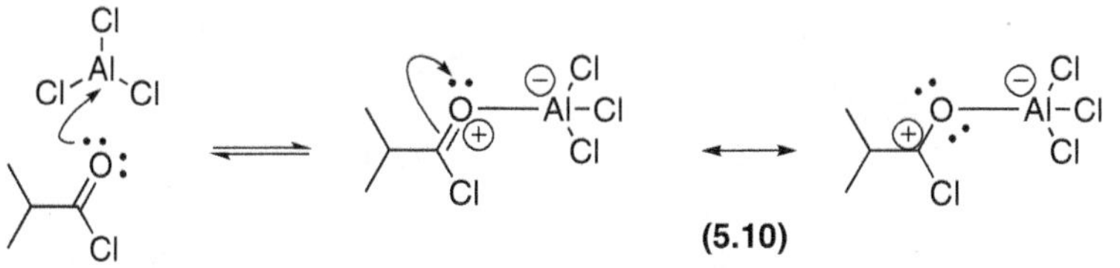

(5.10)

The free (unbound) proton does not occur in condensed media, although the reaction between a proton and a Lewis base belongs to this class of reactions. More importantly, perhaps, so do the reactions of Lewis bases with Lewis acids such as aluminum chloride (e.g., Example **5.10**), titanium tetrachloride (e.g., Example **5.11**), and a wide range of boron reagents (e.g., Example **5.12**). These $n + a$ reactions are often important steps in the mechanisms of key synthetic reactions. For example, the complexation of the carbonyl oxygen atom by a Lewis acid is a key step in many reactions that we most commonly categorized as nucleophilic additions: magnesium in the Grignard addition, titanium in the Mukaiyama aldol addition reaction,[14] and boron in the Evans asymmetric aldol addition,[15] as well as the addition of allylboranes to carbonyl compounds. Complexation of the carbonyl oxygen of an acid chloride by aluminum chloride (Example **5.10**) is a key stage in the Friedel-Crafts acylation of arenes. Similar reactions of alkyl halides with Lewis acids (e.g., Example **5.13**) are key steps in the Friedel-Crafts alkylation of arenes and alkenes (especially enol trimethylsilyl ethers).[16]

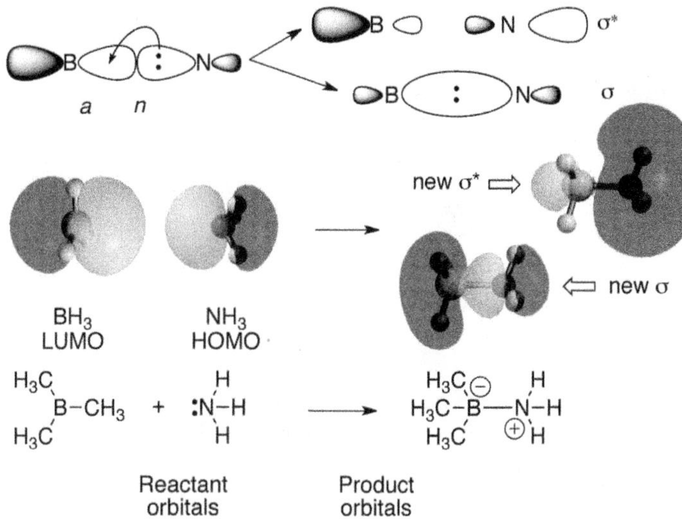

Figure 5.13 Frontier molecular orbital overlap between *n* and *a* orbitals

13. Krimer, L.I.; Cota, D. *Org. React.* **1969**, *17*, 213.
14. Mukaiyama, T.; Banno, K.; Narasaka, K. *J. Am. Chem. Soc.* **1974**, *96*, 7503.
15. Evans, D.A.; Bartroli, J.; Shih, T.L. *J. Am. Chem. Soc.*, **1981**, *103*, 2127.
16. (a) Reetz, M.T.; Chatziiosifidis, I.; Löwe, U.; Maier, W.F. *Tetrahedron Lett.* **1979**, 1427. (b) Reetz, M.T.; Chatziiosifidis, I.; Hübner, F.; Heimbach, H. *Org. Syn.* **1984**, *62*, 95.

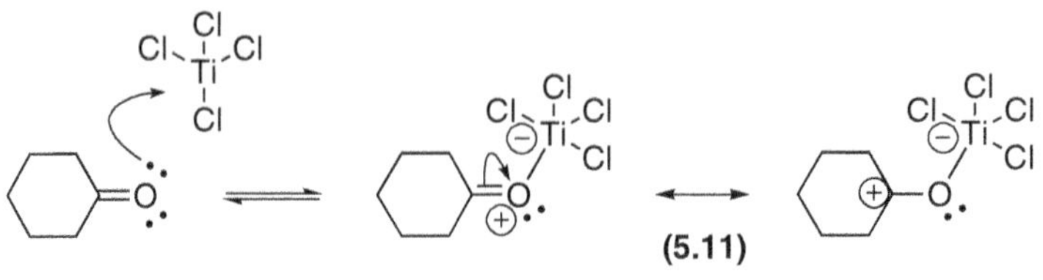

(5.11)

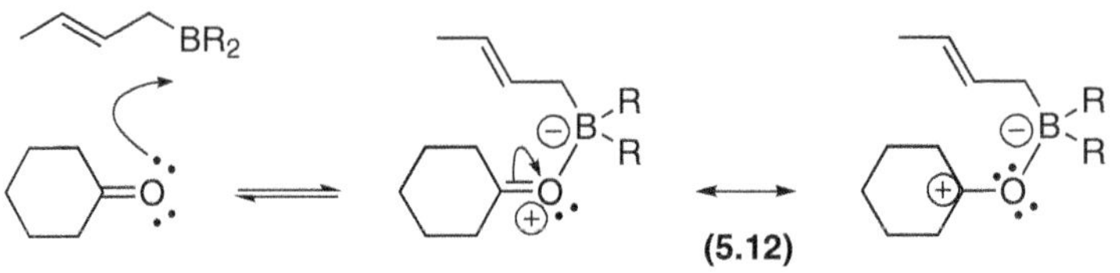

(5.12)

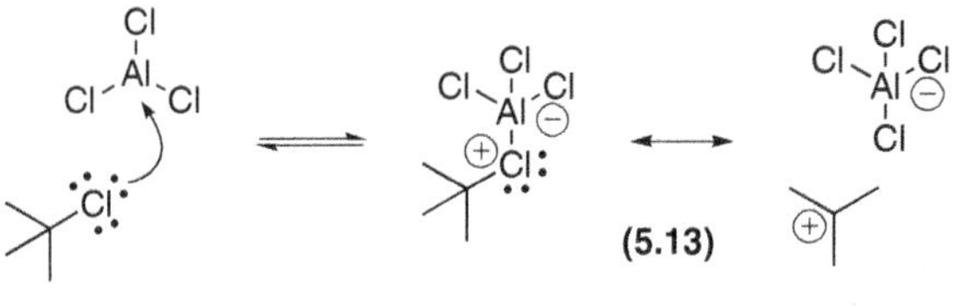

(5.13)

Because this overlap of FMOs leads to the formation of a bond without requiring concomitant bond rupture, the relative energies of the *n* and *a* orbitals are frequently of relatively little consequence in comparison to reactions where bond rupture is required. Nevertheless, in general, the lower the energy of the LUMO, the more reactive the electrophile or Lewis acid is, and the higher the energy of the HOMO, the more reactive the nucleophile or base; lowering the energy of its LUMO increases the strength of a Lewis acid, and raising the energy of its HOMO increases the strength of a Lewis base. In the same reaction, however, these two effects are often incompatible with each other. Consequently, we seldom find a strong Lewis base involved in reactions where a strong Lewis acid is used—the Lewis acid is capable of accepting a pair of electrons from even weak electron-pair donors, and a strong Lewis base tends to be incompatible with reaction conditions where a strong Lewis acid is generated. Conversely, a strong Lewis acid is seldom involved in reactions where a strong Lewis base is used—the Lewis base is capable of donating a pair of electrons to even weak electron-pair acceptors, and a strong Lewis acid tends to be incompatible with reaction conditions that lead to the formation of a strong Lewis base.

Conjugated Systems: a_π and n_π Orbitals

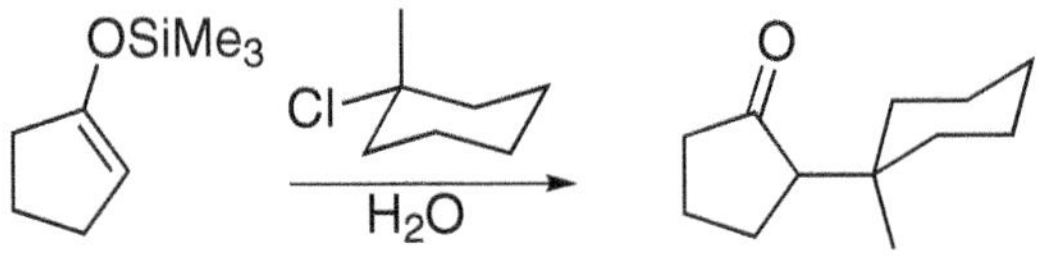

(5.14)

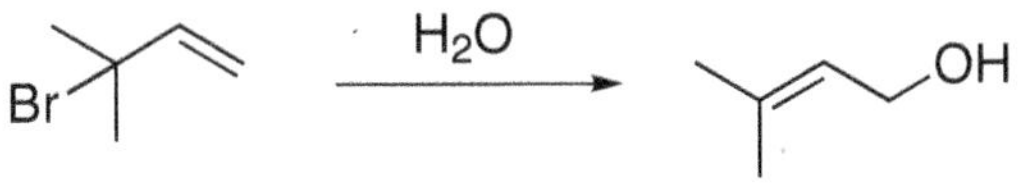

(5.15)

Recall that conjugated π bonding systems with an odd number of atoms have a nonbonding orbital between the bonding and antibonding orbitals. To distinguish this type of orbital form those that are localized on a single atom, we will designate them as a_π and n_π orbitals. The major difference between these orbitals and the localized *a* and *n* orbitals that we have been discussing is that these orbitals are delocalized, with lobes (i.e., sites for reaction) at more than one location, thus permitting regioisomers to be formed in their reactions. In a species such as the allyl cation, this nonbonding orbital is the LUMO, whereas in systems such as an enolate anion, this orbital is the HOMO. Thus, the second

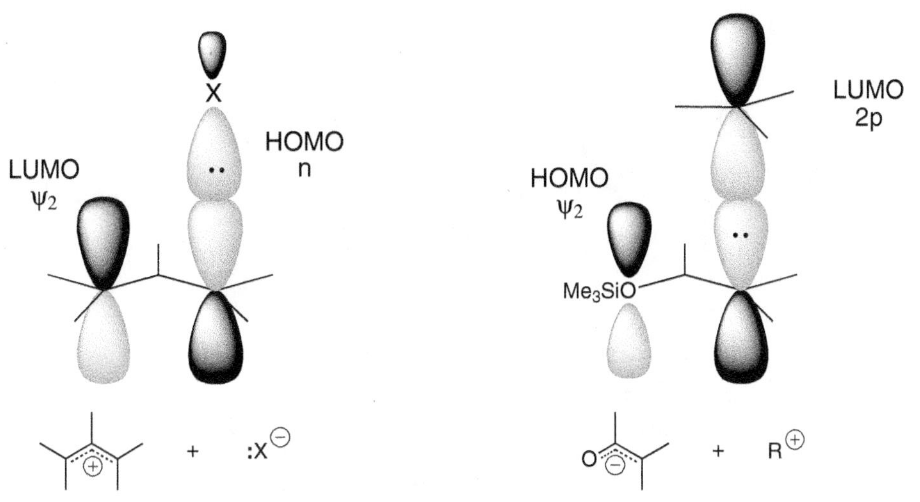

Figure 5.14 Reactions involving conjugated cations and/or anions with frontier molecular orbital overlap between *n* and *a* orbitals. HOMO, highest energy occupied molecular orbital; LUMO, lowest energy unoccupied molecular orbital.

step of the S_N1 reactions of allyl halides (the left-hand example in Figure 5.14) also belongs to the $n + a_\pi$ subclass of reactions, with ψ_2 of the allyl cation as the a_π orbital. The Friedel-Crafts alkylation of an enol trimethylsilyl ether[16,17] (the right-hand example in Figure 5.14) belongs to the $n_\pi + a$ subclass or reactions, where the carbocation (a) interacts with ψ_2 of the silyl ether as the n_π orbital. The interaction between an allyl cation and an enol trimethylsilyl ether or an enamine would represent the $n_\pi + a_\pi$ subclass of reactions. As we shall see later in this chapter, systems such as the ones in Example **5.14** may also be treated as $a + \pi$ overlaps.

Problems

5-3 The Friedel-Crafts acylation proceeds through an electrophile generated by the reaction between aluminum chloride and an acyl chloride. Example **5.10** above discusses one type of electrophile that can be formed by this combination of reactants. What is the other, and what FMOs are involved in its formation?

5-4 The reaction below can give more than one chloride as the product. What chloride should be the major product formed, and overlap of what FMOs initiates its formation?

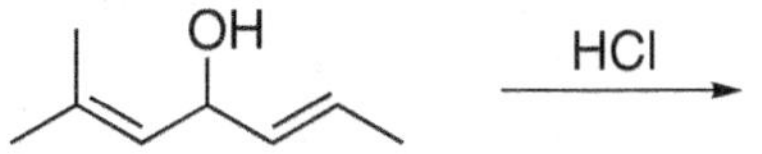

5-5 The reaction between a silyl ether and fluoride anion occurs in two stages. Write a mechanism consistent with this and specify the orbital overlap that initiates the reaction (i.e., which orbitals are involved?).

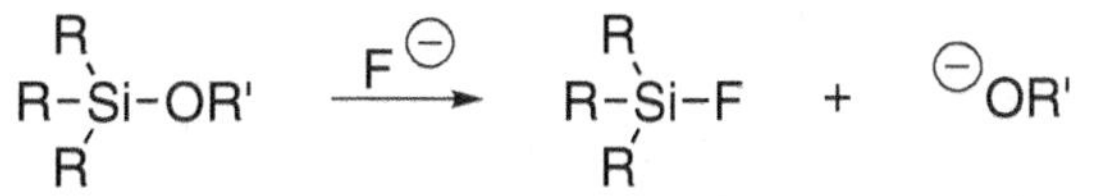

17. Chan, T.H.; Paterson, I.; Pinsonnault, J. *Tetrahedron Lett.* **1977**, 4183.

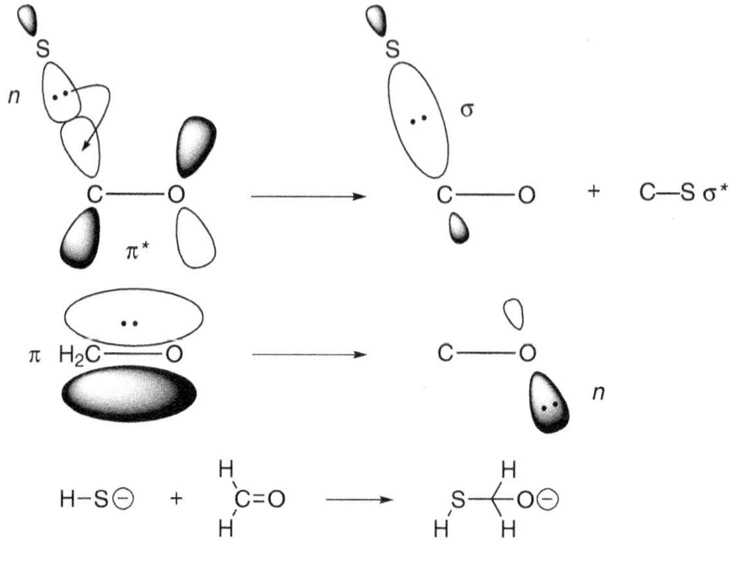

Figure 5.15 Frontier molecular orbital overlap between *n* and *p** orbitals

Frontier Molecular Orbital Overlap Between *n* and π^* Orbitals

This combination of FMOs initiates nucleophilic addition to a polar π bond, such as a carbonyl group, which means that it is the FMO overlap responsible for initiating some of the most important synthetic organic reactions available to the organic chemist. The simplest illustration of a reaction initiated by this pairing of FMOs is given by the addition of a heteroatom nucleophile to a carbonyl group (Figure 5.15). In this reaction, a new σ bond is formed at the expense of a π bond by overlap of a filled n orbital of the nucleophile (sulfur in the example above) with the unoccupied π^* orbital of the carbonyl group. In nucleophilic addition, the HOMO-LUMO overlap that initiates the reaction puts electrons into an antibonding orbital, which results in rupture of the π bond. As their name implies, antibonding orbitals are the absolute opposite of bonding orbitals, and they destabilize the association of the two nuclei.

> Whenever electrons are placed into an antibonding orbital, the bond corresponding to that orbital is broken.

Thus, the HOMO-LUMO overlap now affects *three* orbitals—the n orbital of the nucleophile and the π^* and π orbitals of the electrophile—and the fate of all three must be taken into account when deciding what the overall bonding change is. In this reaction, the electrons of the nucleophile lone pair become the electrons of the new σ bond, and the electrons in the π bonding orbital become the new nonbonding electron pair. For clarity, the new σ^* orbital has been omitted from Figure 5.6.

Reactions initiated by $n + \pi^*$ overlap are among the most important bond-forming reactions in organic chemistry, and they include a wide diversity of reactions. The π^* orbital, for example, can be the C—O π^* orbital of an isolated carbonyl group; the C—N π^* orbital of an isolated nitrile group or imine; or ψ_3 of an α,β-unsaturated carbonyl system, imine, or nitrile. It can also, however, be the lowest unoccupied π orbital of an electron-deficient aromatic ring system such as nitrobenzene, where it is the functional equivalent of ψ_3 of an α,β-unsaturated carbonyl system, or be pyridine, where it is the functional equivalent of the C—N π^* orbital of an isolated imine group.

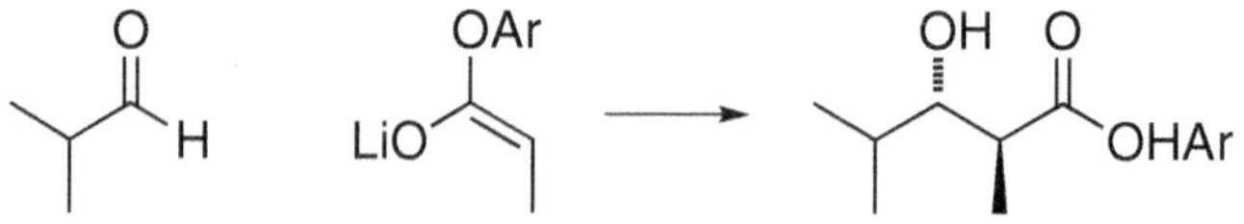

(5.16)

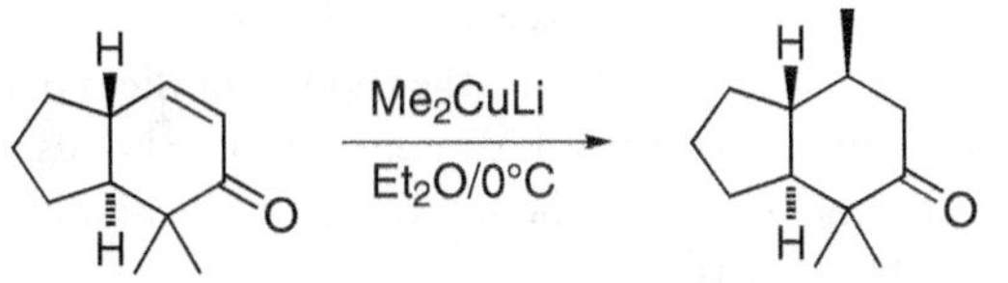

(5.17)

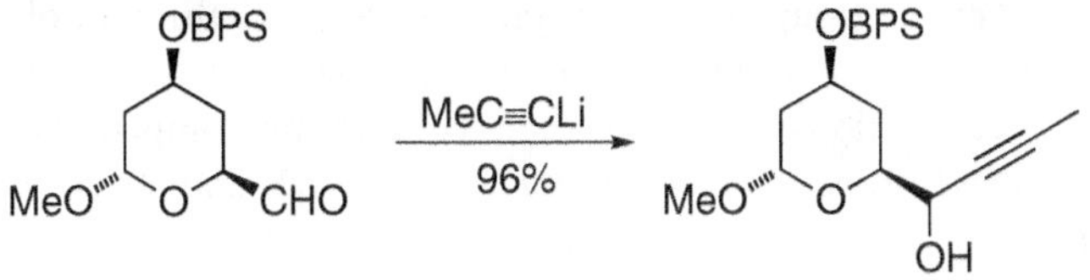

(5.18)

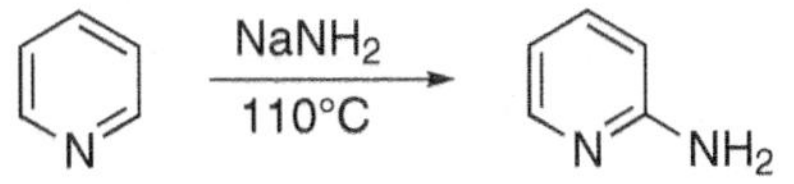

(5.19)

When the π^* orbital is the antibonding orbital of an aldehyde or ketone, the reactions include the addition of metal enolates (e.g., Example **5.16**), metal alkyls (e.g., Example **5.17**) and alkali metal alkynides (e.g., Example **5.18**) to the carbonyl group to give alcohols, and the addition of heteroatom nucleophiles and cyanide or azide anion to the carbonyl group to give addition products. In conjugated systems, we see both 1,2- and 1,4-addition reactions arising from this combination of FMOs, with the regiochemistry of the reaction often being controlled by the hardness and softness of the two reacting species (i.e., the HOMO-LUMO gap), as we saw in Section 5.2, above.

This combination of FMOs also corresponds to the first step in nucleophilic substitution at sp^2-hybridized centers by the addition-elimination mechanism, including nucleophilic acyl substitution and nucleophilic aromatic substitution as well as the Chichibabin reaction of pyridines (e.g., Example **5.19**).

Problems

5-6 Using the curved arrow formalism, write a mechanism to account for the formation of the product in each of the four examples above.

5-7 Identify the FMOs used by each of the reactants in the examples above.

Frontier Molecular Orbital Overlap Between *n* and σ^* Orbitals

The σ-type overlap of these two FMOs is the pairing in the S_N2 reaction (Figure 5.16) and in the transfer of a proton from a Lowry-Brønsted acid to a base. In essence, this reaction is

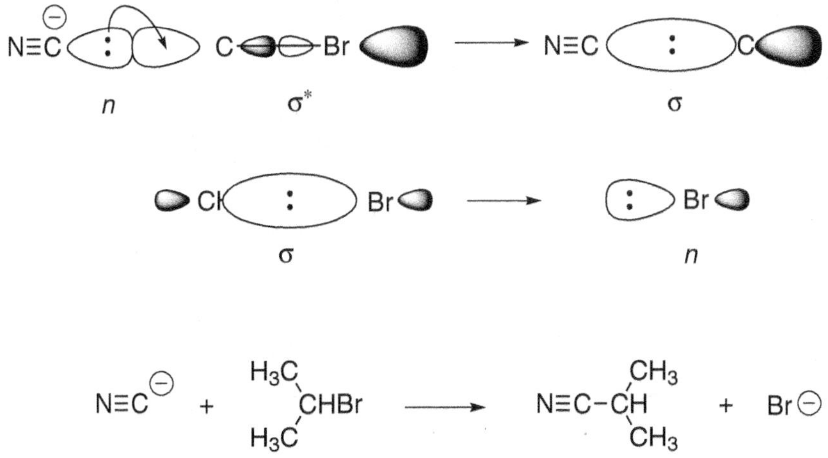

Figure 5.16 Frontier molecular orbital overlap between *n* and *s** orbitals

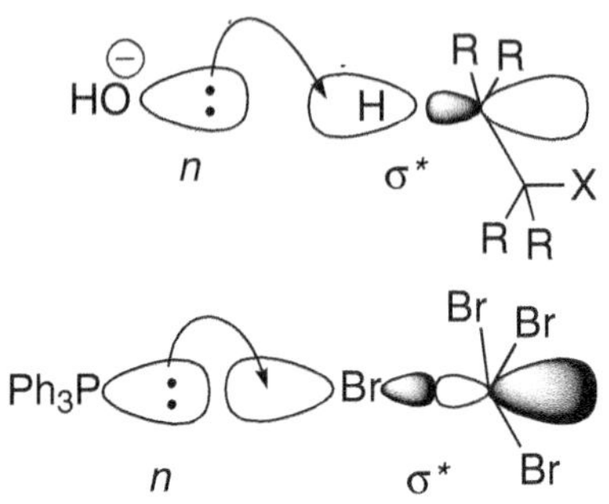

Figure 5.17 Initial frontier molecular orbital overlap involving an *n* orbital and an *s** orbital

basically the same as the reactions in the previous section, except that the LUMO is now a σ^* orbital rather than a π^* orbital—the rupture of the σ bond produces two separate species rather than simply breaking one of the two bonds of a double bond. All the remaining arguments remain the same. Again, the new σ^* antibonding orbital has been omitted for clarity.

The second major group of reactions that is initiated by this type of FMO overlap is the deprotonation of organic compounds by strong bases. There are many examples of such reactions: the deprotonation of carbonyl compounds to give enolate anions, base-promoted elimination by the E2 and E1cb mechanisms (Figure 5.17), and the fast deprotonation step in the E1 mechanism. The useful reaction between triphenylphosphine and carbon tetrabromide, also shown in Figure 5.8, proceeds by overlap of the nonbonding orbital on phosphorus with a C—Br σ^* orbital at the bromine end.

Problem

5-8 What will be the products formed in the two reactions in Figure 5.17? Assume that there is no further movement of electrons.

Frontier Molecular Orbital Overlap Between π and *a* Orbitals

The addition of a localized electrophile to a carbon-carbon π bond is an important reaction that is involved not only in the electrophilic addition to alkenes but also in electrophilic aromatic substitution reactions. It is initiated by the overlap of the empty *a* orbital of the electrophile with the filled π orbital of the alkene, as illustrated in Figure 5.18.

Figure 5.18 Frontier molecular orbital overlap between *a* and *p* orbitals

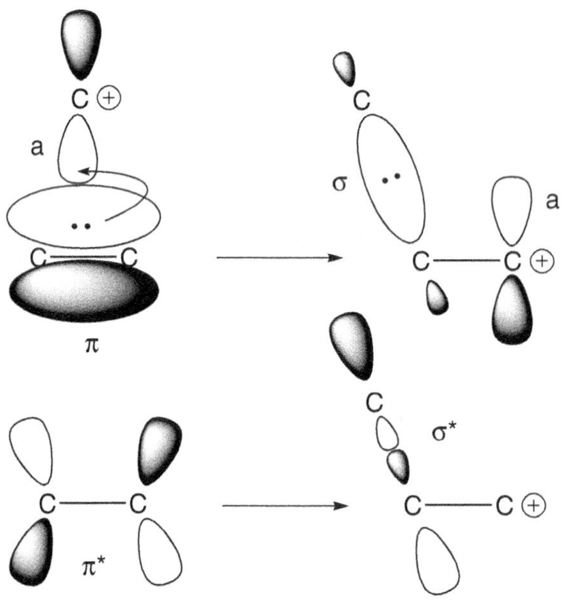

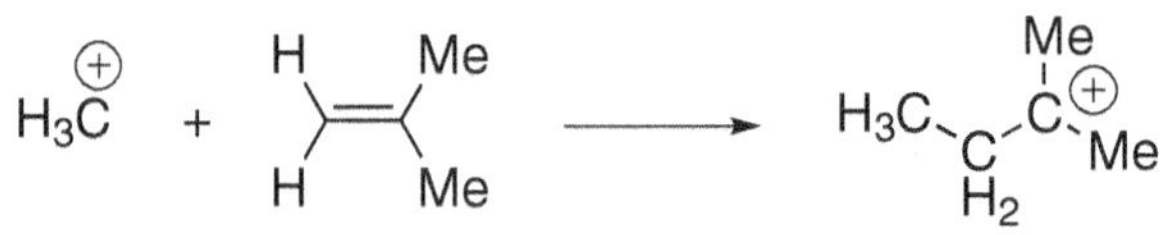

In this reaction, three orbitals are affected: the two FMOs as well as the antibonding orbital corresponding to the HOMO. In this case, the electrons originally in the π orbital are used to form the new σ bond, so that the second terminus of the original π bond becomes an atom bearing an empty nonbonding orbital.

This combination of FMOs is the pairing in the addition of carbocations to alkenes, as in the Johnson polyene cyclization,[18,19] the alkylation of enol silyl ethers by alkyl halides and titanium tetrachloride,[20] and, formally, the hydroboration of alkenes.[21] It is also the first step of the electrophilic substitution reactions of arenes and alkenes. Although reactions initiated by protonation may be viewed as belonging to this class, one should also bear in mind that the free proton does not actually exist in condensed media and that protonation reactions really belong to the class initiated by $\sigma^* + \pi$ overlap.

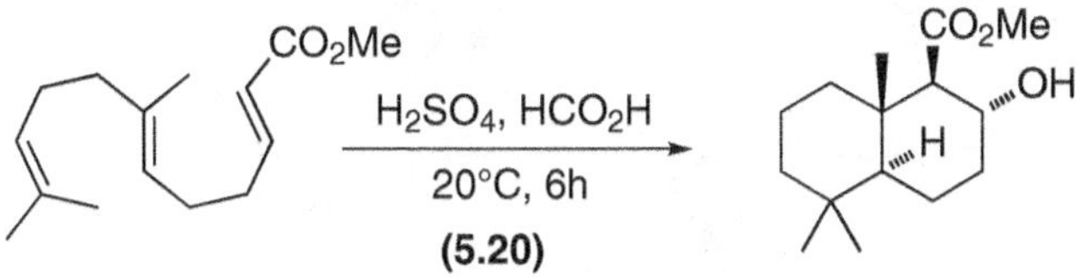

(5.20)

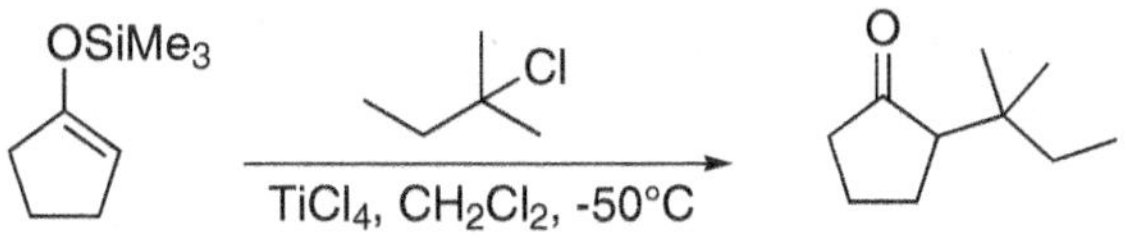

(5.21)

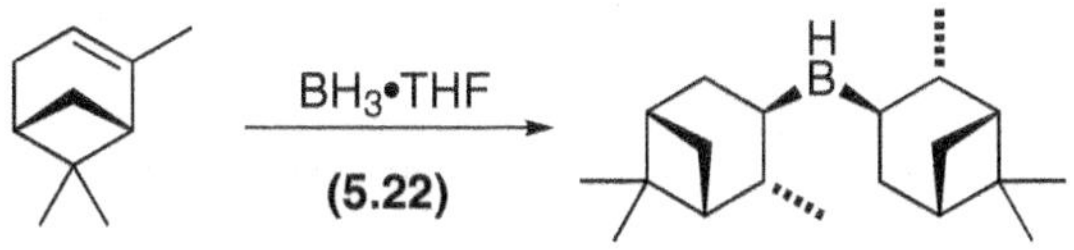

(5.22)

Frontier Molecular Orbital Overlap Between π and π^* Orbitals

This pairing of FMOs occurs in two major reaction classes: electrophilic addition to simple alkenes and arenes (as in the first step of electrophilic aromatic substitution) as well as in cycloaddition reactions.

Electrophilic Addition

The critical part of the HOMO-LUMO overlap is obviously the formation of the new σ orbital (and its associated σ^* orbital, which has been omitted from Figure 5.19 for clarity). The electrons in the complementary π orbital of the electrophile become a nonbonding lone pair in the product, and the complementary π^* orbital of the nucleophile becomes an empty nonbonding orbital in the product. From the perspective of both participating bonds, these reactions are addition reactions because both π bonds are broken and replaced by σ bonds. From the perspective of the LUMO, this is a nucleophilic addition reaction, and from the perspective of the HOMO, it is an electrophilic addition reaction.

18. Johnson, W.S. *Acc. Chem. Res.* **1968** *1*, 1.
19. Stadler, P.A.; Nechvatal, A.; Frey, A.J.; Eschenmoser, A. *Helv. Chim. Acta* **1957**, *40*, 1373.
20. (a) Reetz, M.T.; Chatziiosifidis, I.; Löwe, U.; Maier, W.F. *Tetrahedron Lett.* **1979**, 1427. (b) Reetz, M.T.; Chatziiosifidis, I.; Hübner, F.; Heimbach, H. *Org. Syn.* **1984**, *62*, 95.
21. (a) Brown, H.C. *Hydroboration* (W.A. Benjamin: New York, 1962). (b) Brown, H.C. *Boranes in Organic Chemistry* (Cornell University Press: Ithaca, NY, 1972). (c) Brown, H.C. *Organic Synthesis via Boranes* (Wiley-Interscience: New York, 1975).

Figure 5.19 Frontier molecular orbital overlap between *p* and *p** orbitals in electrophilic addition

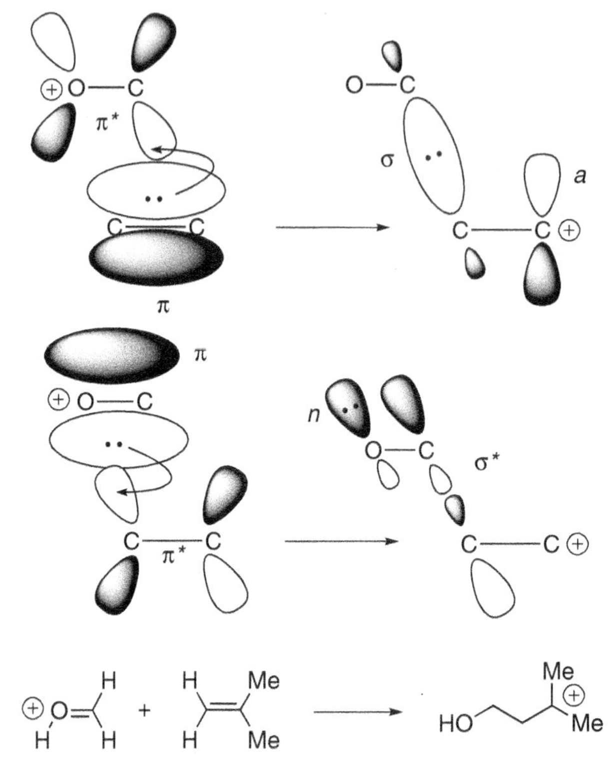

This pairing of FMOs is the overlap that initiates electrophilic addition of carbonyl and iminium groups to electron-rich alkenes, as in the Mannich reaction,[22] the Mukaiyama aldol reaction of enol trimethylsilyl ethers,[23] and the Prins reaction[24] of aldehydes with alkenes in the presence of acid. It is also the first step of the acylation of vinylsilanes[25] and stannanes, and the first step of electrophilic aromatic substitution, as illustrated by the examples below.

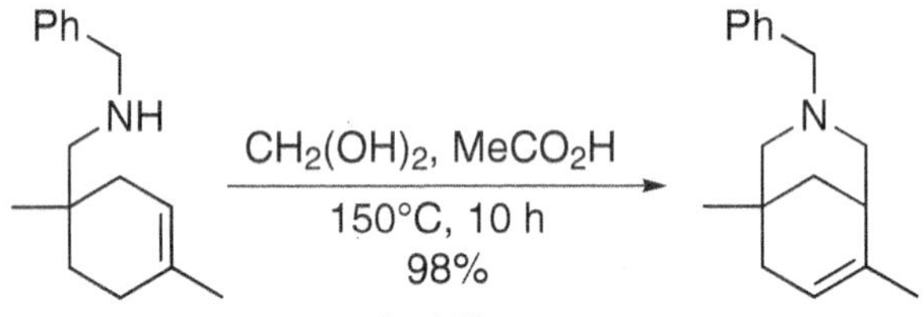

(5.23)

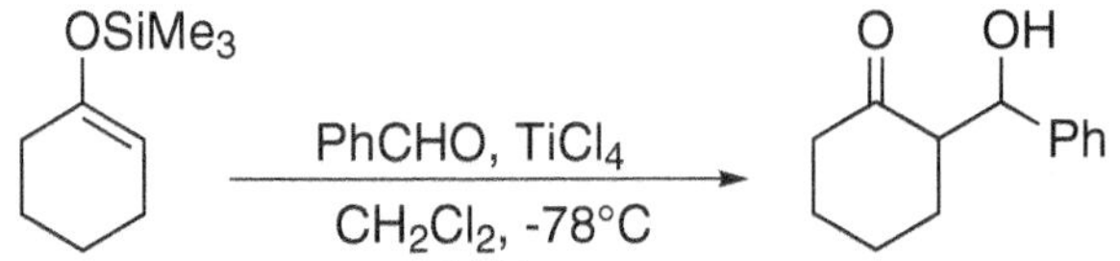

(5.24)

---

22. Reviews: (a) Tramontini, M. *Synthesis*, **1973**, 703. (b) Thompson, B.B. *J. Pharm. Sci.* **1968**, *57*, 715. (c) Arend, M.; Westermann, B.; Risch, N. *Angew. Chem. Int. Ed. Engl.* **1998**, *37*, 1044.
23. Mukaiyama, T.; Banno, K.; Narasaka, K. *J. Am. Chem. Soc.* **1974**, *96*, 7503.
24. Review: Adams, D.R.; Bhatnagar, S.P. *Synthesis* **1977**, 661.
25. Fleming, I.; Pearce, A. *J. Chem. Soc., Chem. Commun.* **1975**, 633.

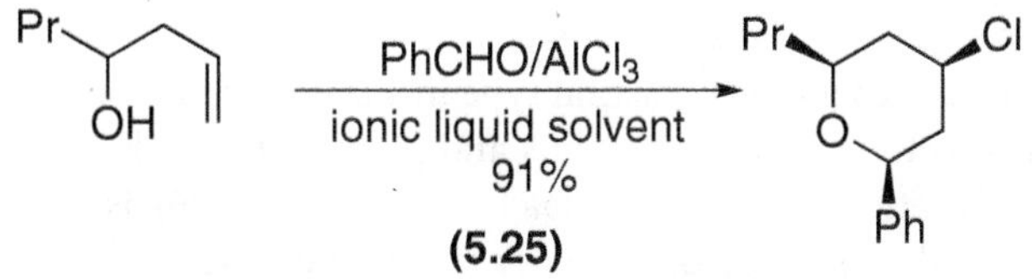

(5.25)

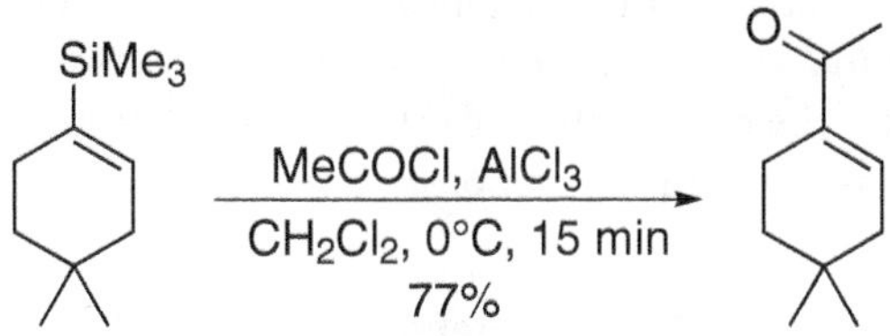

(5.26)

Problem

5-9 Using the curved arrow formalism, write reasonable mechanisms for reactions **5.23**, **5.24**, **5.25**, and **5.26**.

Worked Problem

5-2 Using the curved arrow formalism, write a reasonable mechanism for reaction **5.20** and provide reasons for each step where appropriate:

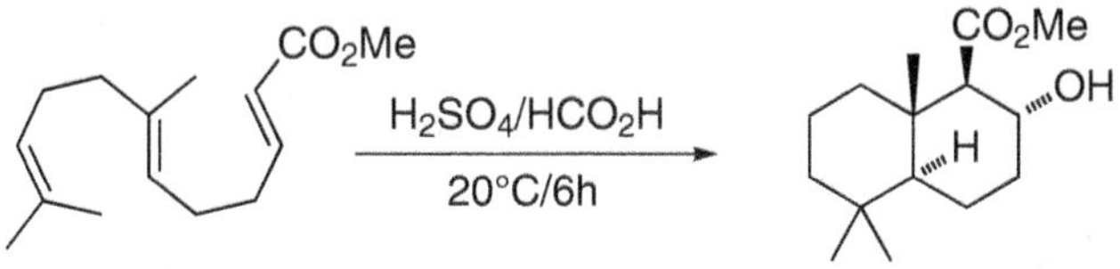

§Answer below

§ Answer to Worked Problem:

The reaction starts with the protonation of the terminal alkene π bond to give the tertiary carbocation, which then adds to the alkene π bond in the middle of the chain to give a new tertiary carbocation. This cation adds to the final alkene π bond in the direction to place the carbocation at the α, rather than the β position of the ester (why is this regiochemistry preferred? Think about the electron distribution in the carbonyl group). The cation is then trapped by water to give the final alcohol.

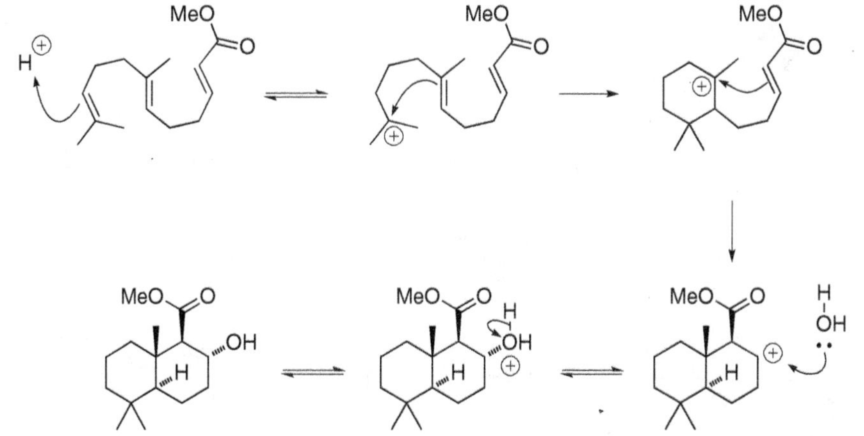

Cycloadditions

Cycloaddition reactions are one reaction type in a class of reactions, known as ***pericyclic reactions***, that involve π-bonded systems and proceed through cyclic transition states. The Diels-Alder reaction, or [4 + 2] cycloaddition (Figure 5.20), is a typical cycloaddition reaction. The orbital correlation diagram for this reaction shows that there are six orbitals modified during the reaction. It is developed according to principles, now known as the ***conservation of orbital symmetry***, first set forth by Woodward and Hoffmann, and by Fukui. We will look at the conservation of orbital symmetry and at pericyclic reactions in general, in much more detail in Chapter 6.

Frontier Molecular Orbital Overlap Between π and σ* Orbitals

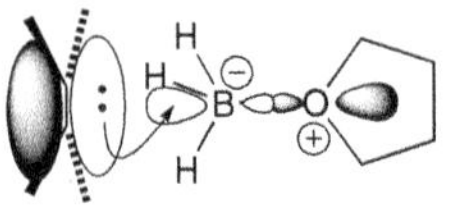

(5.27)

This situation has characteristics of both the $a + \pi$ and the $\pi + \pi^*$ cases. The similarity to the $a + \pi$ case becomes evident when one considers that the typical reagent in this type of reaction (the oxonium ion has been used in Figure 5.21) is normally the immediate precursor to the carbocation electrophile in reactions like the Johnson polyene cyclization. Similarly, because the proton does not occur free in condensed medium, all protonations of alkenes actually involve the σ* orbital of the bond to hydrogen. In fact, in a formal sense, the same set of examples used in the section on may also be used here, including the hydroboration reaction (Example **5.22**), where one may, in fact, view the initial electrophilic

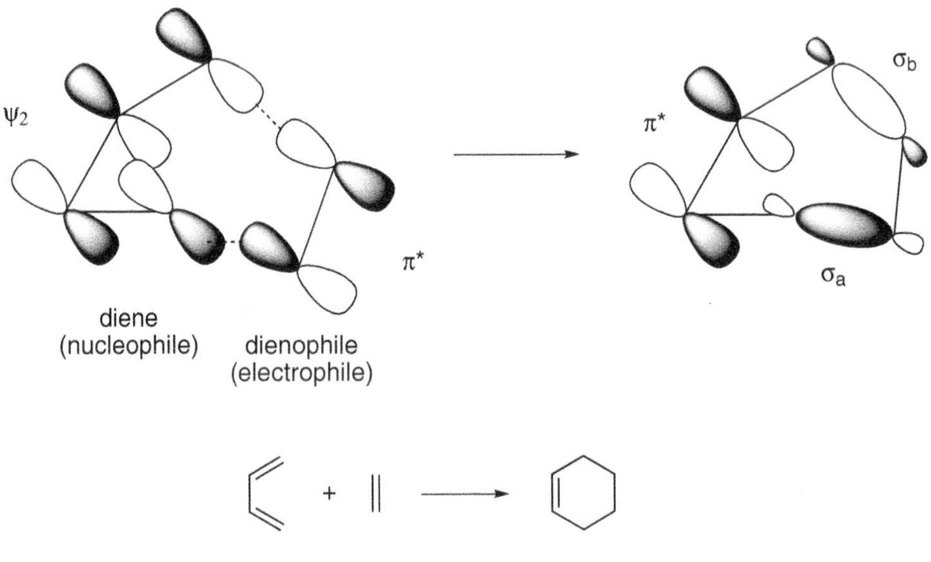

Figure 5.20 Frontier molecular orbital overlap between π and π* orbitals in cycloaddition

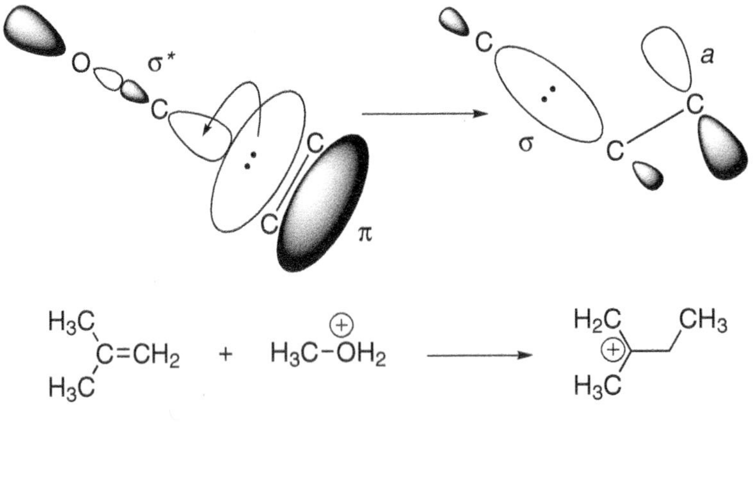

Figure 5.21 Frontier molecular orbital overlap between π and σ* orbitals

attack of the alkene as involving not free BH_3 but rather the Lewis acid-Lewis base complex with THF (Example **5.27**; the added steric bulk due to the THF molecule may also rationalize the high sensitivity of this reaction to steric hindrance).

The reaction with the electrophile results in the rupture of the π bond, as in the $a + \pi$ case, and in the rupture of the existing σ bond, similar to the $\pi + \pi^*$ case. All other arguments remain the same as those made for the $\pi + \pi^*$ case. Like the $\pi + \pi^*$ case, the characterization of this reaction depends on the perspective that one adopts when carrying out the characterization. From the perspective of the LUMO, this reaction is a nucleophilic substitution reaction because one σ bond is broken as the new one is formed. From the perspective of the HOMO, this reaction is electrophilic addition because the π bond is broken and replaced by σ bonds.

Frontier Molecular Orbital Overlap Between *a* and σ Orbitals

This orbital pairing usually occurs with π-type symmetry, in which case it corresponds to the orbital overlap that gives rise to both hyperconjugation in carbocations and Wagner-Meerwein rearrangements of carbocations. In the case of hyperconjugation, the transfer of electrons from the HOMO to the LUMO is incomplete, resulting in the energy of the σ orbital being lowered and the energy of the *a* orbital being raised. Wagner-Meerwein rearrangements involve the same orbital overlap, but now the electron transfer is complete, with the formation of three new orbitals. One possible energy correlation diagram for the Wagner-Meerwein rearrangement is shown in Figure 5.22.

This orbital overlap is also responsible for the initiation of reactions other than rearrangement when the electron density is freely available to the external electrophile. There are two situations in which this is the case: when the σ bond is a C—H or other σ bond to hydrogen, in which case a very strong electrophile is usually required, or when the bond is in a cyclopropane or cyclobutane ring, in which case a weaker electrophile can be used. The electron pair in σ orbitals involving hydrogen is much more accessible in the extranuclear space than is the σ bonding pair in other covalent bonds. Consequently, this type of orbital overlap is most easily achieved with C—H σ bonds. The electron density in

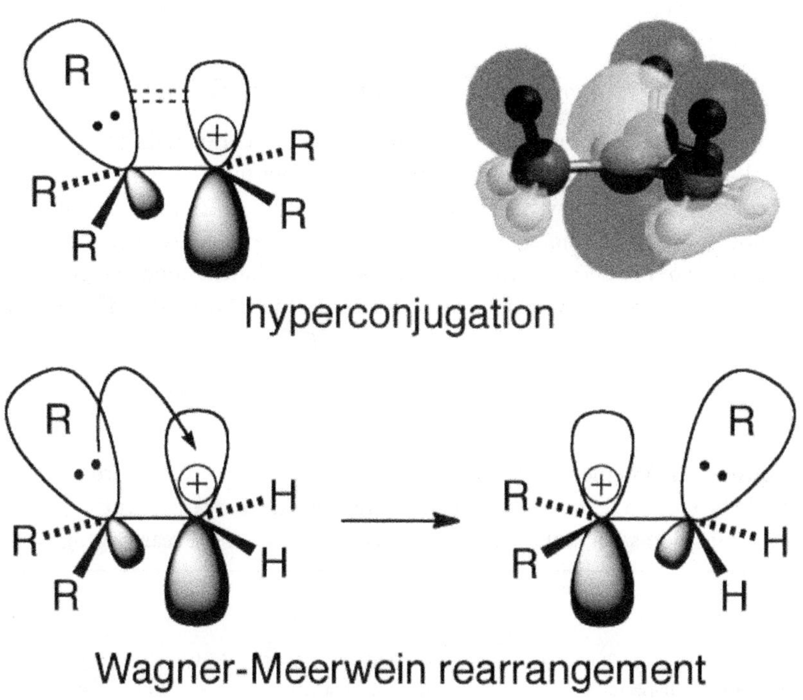

Figure 5.22 Frontier molecular orbital overlap between *a* and σ orbitals in hyperconjugation and Wagner-Meerwein rearrangements

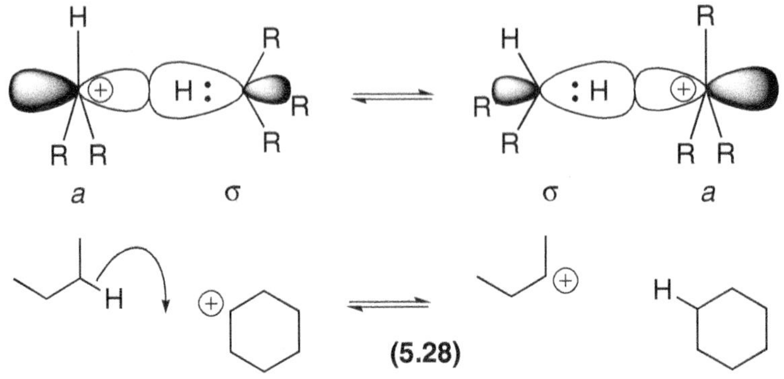

(5.28)

Figure 5.23 Frontier molecular orbital overlap in the hydride anion transfer between carbocations

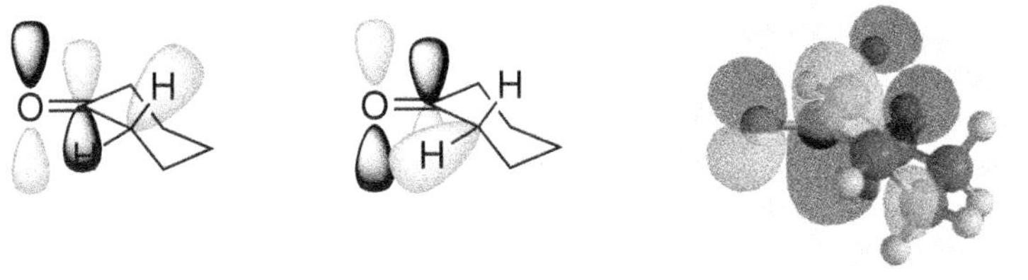

Figure 5.24 Frontier molecular orbital overlap in cyclohexanone is apparent in the lowest energy unoccupied molecular orbital of cyclohexanone.

cyclopropane C—C bonds is also much more accessible to electrophiles because it does not lie directly between the two carbon atoms.

The overlap involving a C—H σ orbital is illustrated by the transfer of a hydride anion from an alkane to a strong Lewis acid, such as a carbocation, which is a key step in alkane rearrangements (Example **5.28**, Figure 5.23).[26] Similar orbital overlap, between the C—H σ orbital and the a_π orbital of the triphenylmethyl cation is responsible for initiating oxidation of alcohols by this reagent. The reactions of cyclopropanes with simple proton acids to give ring-opened products have been known for well over a century.

Frontier Molecular Orbital Overlap Between σ and π* Orbitals

The π-type overlap of adjacent σ and π* orbitals is a major contributing factor to the conformational preferences of alkenes and carbonyl compounds. It also contributes to the unusual acidity of the α hydrogens of carbonyl compounds (and the axial α hydrogens of cyclohexanones) by providing a mechanism for delocalization from the C—H σ bond onto the carbonyl oxygen. These orbitals are shown for cyclohexanone in Figure 5.24. Note how the equatorial hydrogens do not overlap well with the π* orbital and how the lobes at the equatorial α hydrogens in the cyclohexanone LUMO are extremely small.

Problem

5-10 Consider the Wagner-Meerwein rearrangement. (a) What pair of frontier orbitals could be used to describe the transition state for the rearrangement? (b) What does this transition say about the favored stereochemistry of the rearranging group?

---

26. (a) Pines, H.; Wackler, R.C. *J. Am. Chem. Soc.* **1946**, *68*, 525, 2518. (b) Wackler, R.C.; Pines, H. *J. Am. Chem. Soc.* **1946**, *68*, 1642. (c) Pines, H.; Abraham, B.M.; Ipatieff, V.N. *J. Am. Chem. Soc.* **1948**, *70*, 1742. (c) Pines, H.; Aritstoff, E.; Ipatieff, V.N. *J. Am. Chem. Soc.* **1949**, *71*, 749.

Frontier Molecular Orbital Overlap Between σ and σ^* Orbitals

The π-type overlap of adjacent σ and σ^* orbitals has been implicated in stabilization of staggered conformations of open-chain compounds, as we discussed in Chapter 2 (Example 5.29).[27] In saturated alkane chains, it has been proposed that the stabilization of the staggered conformation by delocalization by π-type interaction between the bonding orbital with the antibonding orbital on the adjacent atom is more important than the destabilization of the eclipsed conformation by H-H repulsion.

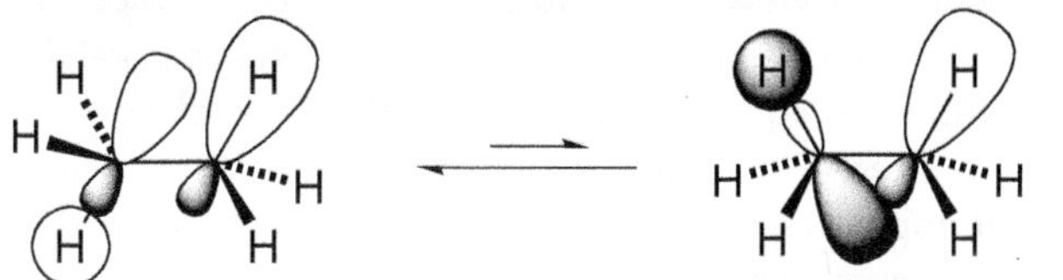

(5.29)

Likewise, the bromination of cyclopropanes (Figure 5.25) involves this FMO overlap between the σ^* orbital of the bromine molecule and the σ orbital of the C—C bond (which you recall, does not lie symmetrically around the C—C) axis. In addition, the molecular orbitals in Figure 5.25 suggest that the attack of bromine, the electrophile, on the cyclopropane ring is not at the midpoint of the C—C ring bond but displaced toward one vertex of the ring.

Problems

5-11 Which FMOs are involved in each of the following reactions? Categorize each reaction under a reaction type (e.g., nucleophilic addition).

(a) 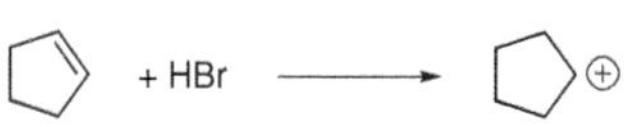

(b) 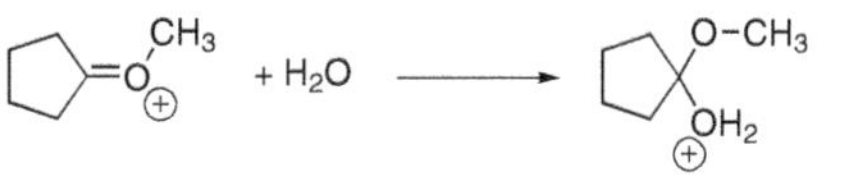

(c) 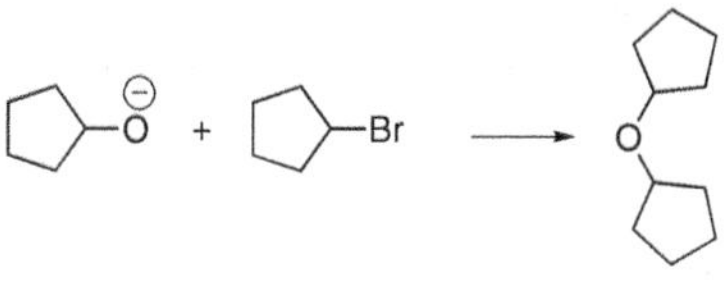

(d) 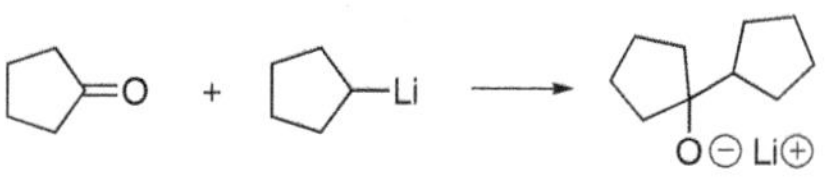

(e) 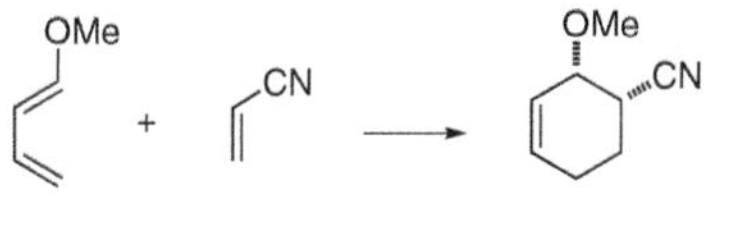

(f) 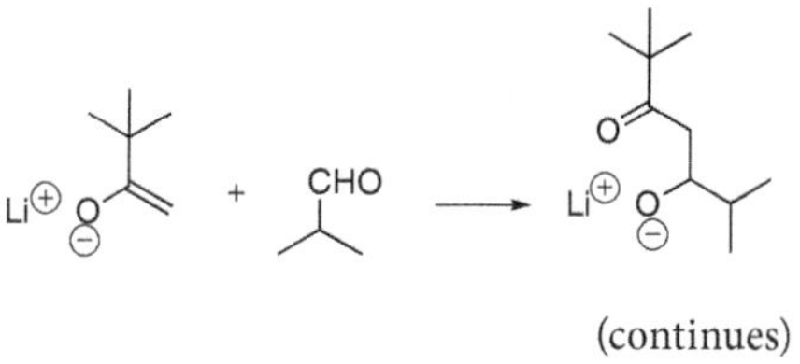

(continues)

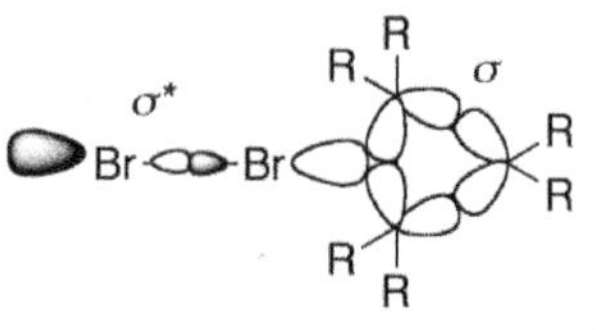

(5.30)

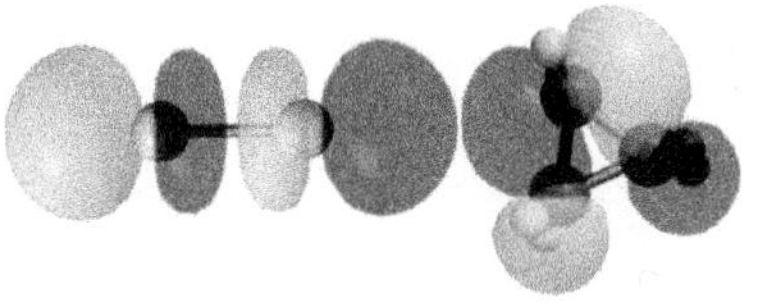

Figure 5.25 Frontier molecular orbital overlap in the addition of bromine to cyclopropane

---

27. (a) Weinhold, F. *Nature* **2001**, *411*, 539. (b) Weinhold, F. *J. Chem. Educ.* **1999**, *76*, 1141. (c) Weinhold, F. *Angew. Chem. Int. Ed. Engl.* **2003**, *42*, 4188. For a dissenting view, see: Bickelhaupt, F.M.; Baerends, E.J. *Angew. Chem. Int. Ed. Engl.* **2003**, *42*, 4183.

Problems (continued)

5-12 Using the curved arrow formalism, write mechanisms for each of the reactions in Problem 5-11.

5.4 Stereoelectronic Effects

In the preceding section, FMOs were used to categorize organic reactions according to the FMO overlap responsible for the reaction. However, this overlap also has both regiochemical and stereochemical consequences because the orbitals involved actually have a shape that determines the three-dimensional relationship between the reacting species. This spatial requirement imposed by the shape and orientation of the participating orbitals is often referred to as a ***stereoelectronic effect***. A stereoelectronic effect may be defined as a stereochemical and/or regiochemical limitation on an organic reaction that arises from the requirements of the FMO overlap in that reaction.

For example, the cylindrical symmetry of the σ^* orbital means that reactions involving this orbital as one of the participating FMOs will lead to the requirement that the other reagent must approach along the axis of the σ bond. One obvious consequence of this coaxial, σ-type overlap of the FMOs (Figure 5.26, upper example) is that S_N2 reactions proceed with inversion of configuration—because the nucleophile is constrained to approach the large (back) lobe of the σ^* orbital along the bond axis, inversion of configuration must result. In similar vein, the overlap of adjacent σ and σ^* orbitals is what allows the E2 elimination reaction (Figure 5.26, lower example) to be concerted—for all bonding changes to occur in a single step.

The Anomeric Effect

The π-type overlap of a filled n orbital with a σ^* orbital is implicated in the anomeric effect[28] in tetrahydropyran derivatives (Figure 5.27). The anomeric effect refers to the increased preference for axial orientation of an electronegative substituent at position 2 of a

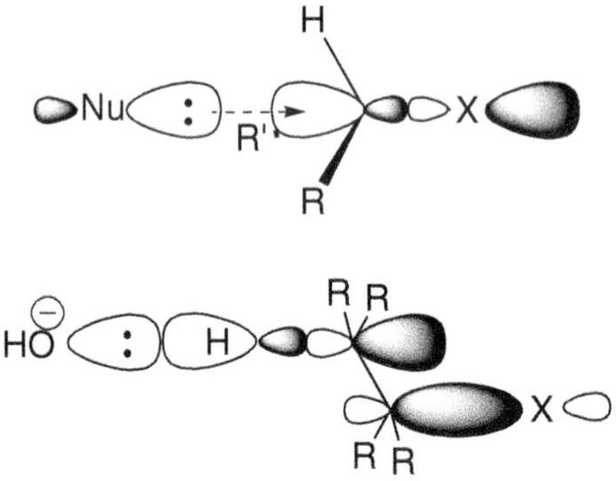

Figure 5.26 Coaxial frontier molecular orbital overlap in the SN2 reaction leads to inversion of configuration at carbon, and coplanar overlap of the adjacent s and s* orbitals allows the E2 reaction to be concerted.

Figure 5.27 Frontier molecular orbital overlap in the anomeric effect

28. (a) Deslongcamps, P. *Stereoelectronic Effects in Organic Chemistry* (Pergamon Press: Oxford, 1983), p. 7. (b) Carey, F.A.; Sundberg, R.J. *Advanced Organic Chemistry. Part A: Structure and Mechanisms*, 4th. ed. (Kluwer Academic/Plenum Publishers: New York, 2000), p. 151. (c) Lemieux, R.U. *Pure Appl. Chem.* **1971**, *25*, 527, (d) David, S.; Eisenstein, O.; Hehre, W.J.; Salem, L.; Hoffmann, R. *J. Am. Chem. Soc.* **1973**, *95*, 3806.

tetrahydropyran ring. This has been proposed to result from π-type overlap between the lone pair with the σ* orbital of the bond to the electronegative substituent.

Problem

5-13 The four structures A–D below are chair-chair conformations of the molecule at the left. Arrange them in approximate order of stability and give reasons for your choice.

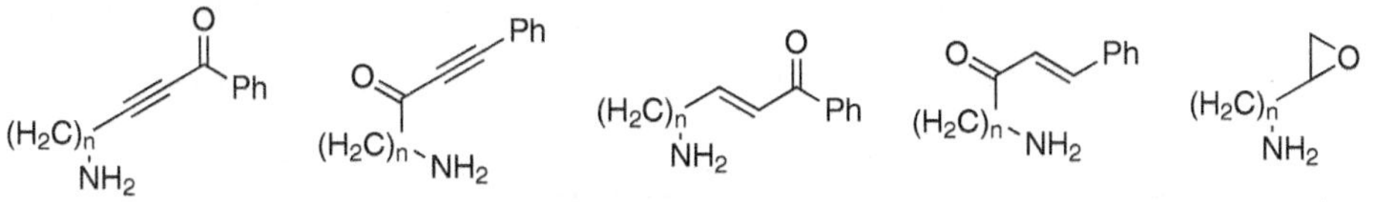

One important consequence of the shapes of the orbitals involved comes from looking at cyclization reactions involving π* orbitals. In 1976, J. E. Baldwin and his students proposed a series of empirical rules for predicting the ease of regioisomeric nucleophilic cyclization reactions. These rules are now known as Baldwin's rules.[29]

Nucleophilic Cyclizations—Baldwin's Rules

The closure of a ring by the intramolecular reaction between a nucleophile and an electrophile may or may not proceed as designed, depending on whether or not the FMOs involved can adopt a favorable orientation with respect to each other. The ramifications of this were first recognized by J. E. Baldwin, who, beginning in 1976, outlined a set of empirical rules, summarized in Table 5.4, that permit one to predict the probable success of intramolecular nucleophilic cyclization reactions.

The principle underlying Baldwin's rules may be traced to the work of Bürgi and Dunitz, who studied nucleophilic addition to carbonyl compounds and observed that there was a favored trajectory of approach of the nucleophile to the carbonyl group.[30] (Figure 5.28 illustrates this approach for the reaction between cyanide ion and acetone.) In this preferred approach trajectory, the nucleophile makes an O—C—Nu angle of approximately 110° with the C=O axis, perpendicular to the plane of the carbonyl group. This has been attributed to the type of overlap between the nucleophile HOMO and the carbonyl LUMO. As we can see from Figure 5.28, the lobes of the LUMO at the carbon atom are canted back away from the oxygen atom, which makes orthogonal approach to the C=O axis less effective than this nonlinear approach.

In the Baldwin system, the attack is defined as *endo* if the breaking bond is endocyclic in the transition state of the cyclization and *exo* if the breaking bond is exocyclic in the transition state of the cyclization. The electrophilic atom is designated as *tet* if it is sp^3 hybridized, *trig* if it is sp^2 hybridized, and *dig* if it is sp hybridized. Based on this system, Baldwin designated certain cyclizations as being favored and others as being disfavored. An extensive review of Baldwin's rules with their limitations and an updating of the rules with respect to alkynes has been published by Alabugin.[31] The most important conclusion

29. (a) Baldwin, J.E. *J. Chem. Soc., Chem.Commun.* **1976**, 734. (b) Baldwin, J.E.; Cutting, J.; Dupont, W.; Kruse, L.; Silberman, L.; Thomas, R.C. *J. Chem. Soc., Chem.Commun.* **1976**, 736. (c) Baldwin, J.E.; Reiss, J.A. *J. Chem. Soc., Chem.Commun.* **1977**, 77. (d) Baldwin, J.E.; Kruse, L.I. *J. Chem. Soc., Chem.Commun.* **1977**, 233. (e) Baldwin, J.E.; Lusch, M.J. *Tetrahedron* **1982**, *38*, 2939. (f) Deslongchamps, P. *Stereoelectronic Effects in Organic Chemistry* Pergamon Press: Oxford, 1983), pp. 234-240.

30. (a) Bürgi, H.B.; Dunitz, J.D.; Schefter, E. *J. Am. Chem. Soc.* **1973**, *95*, 5065. (b) Bürgi, H.B.; Dunitz, J.D.; Lehn, J.M.; Wipff, G. *Tetrahedron* **1974**, *30*, 1563. (c) Bürgi, H.B.; Dunitz, J.D. *Acc. Chem. Res.* **1983**, *16*, 153.

31. Gilmore, K.; Alabugin, V.F. *Chem. Rev.* **2011**, *111*, 6513.

Table 5.4 Baldwin's Rules for Nucleophilic Ring Closure

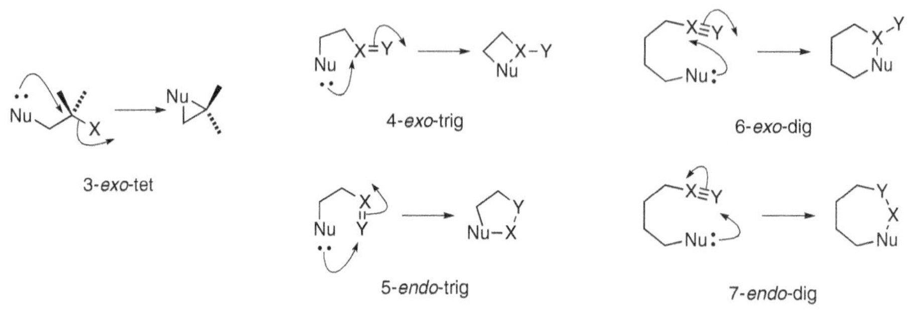

Ring Size	Tet	trig	dig
3	*exo* favored	*exo* favored	*exo* disfavored
		endo disfavored	*endo* favored
4	*exo* favored	*exo* favored	*exo* disfavored
		endo disfavored	*endo* favored
5	*exo* favored	*exo* favored	*exo* favored
	endo disfavored	*endo* disfavored	*endo* favored
6	*exo* favored	*exo* favored	*exo* favored
	endo disfavored	*endo* favored	*endo* favored
7	*exo* favored	*exo* favored	*exo* favored
		endo favored	*endo* favored

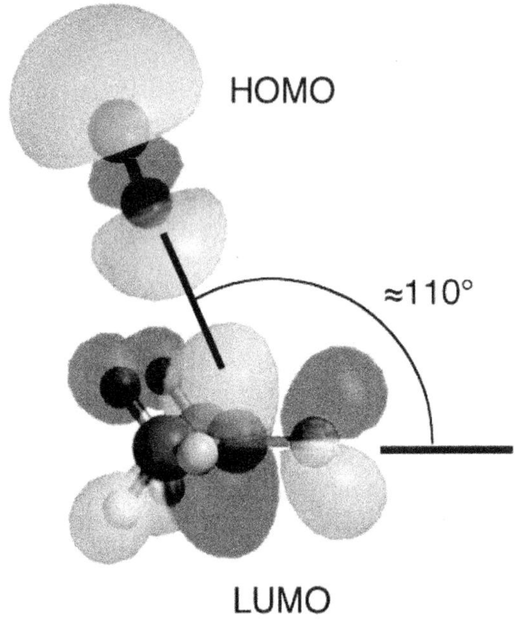

Figure 5.28 Bürgi-Dunitz trajectory for the addition of cyanide ion to acetone

of that study has been to revise the predictions with respect to the additions to dig systems to replace the "acute angle" attack postulated by Baldwin with a trajectory more in accord with the Bürgi-Dunitz angle.

Note that Baldwin's rules extend only to the formation of relatively small rings—up to seven atoms. Medium-sized rings present a special problem due to intramolecular nonbonded interactions, and when one is forming large rings by intramolecular cyclizations, the chain is usually flexible enough to allow the correct orientation of the participating orbitals to be attained for either cyclization outcome.

5.5 Radical Reactions

The analysis above can also be applied to reactions where free radicals are involved, but because radical reactions occur by transfer of single electrons, rather than electron pairs, the picture is often not quite so clear-cut. The FMO of a free radical is always the SOMO, which requires only one electron to become completely filled.

These limitations notwithstanding, the analysis above can be applied to free radical reactions. However, in this case, because one of the participating orbitals must be the SOMO, there should be only three types of pairings available if a new covalent bond is to be formed; the SOMO is not an empty orbital.

Free radical reactions fall into just two common classes, and these are initiated by either SOMO-π^* or SOMO-σ^* overlap. The former of these two SOMO-LUMO pairings leads to addition of a free radical to an alkene π bond, as shown in Figure 5.29. The latter pairing of FMOs characterizes the intramolecular rearrangements of hydrogen atoms to oxygen- or nitrogen-centered free radicals, as occurs in the Barton[32] and Hofmann-Löffler-Freytag[33] reactions (shown in Figure 5.30).

The difficulties in applying the simple form of FMO theory to free radicals also highlights the fact that its performance when used to analyze reactions involving excited states of the reactants is often not as good as the same analysis for reactions between ground-state molecules. As with any theory, one should interpret the results of any prediction with care, bearing in mind the limitations imposed by the basis of the theory itself.

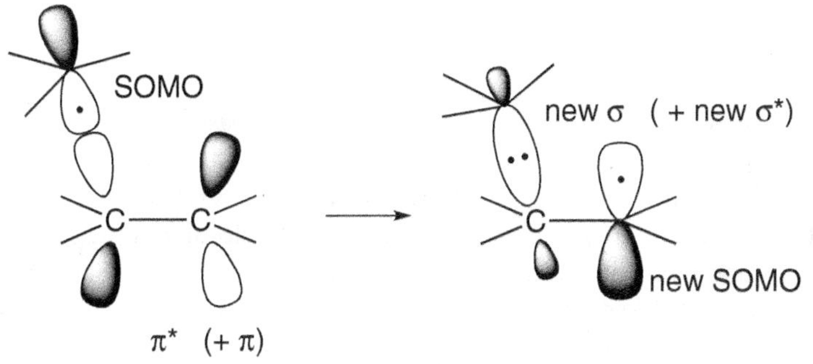

Figure 5.29 Singly occupied molecular orbital (SOMO)–lowest energy unoccupied molecular orbital overlap in free radical addition

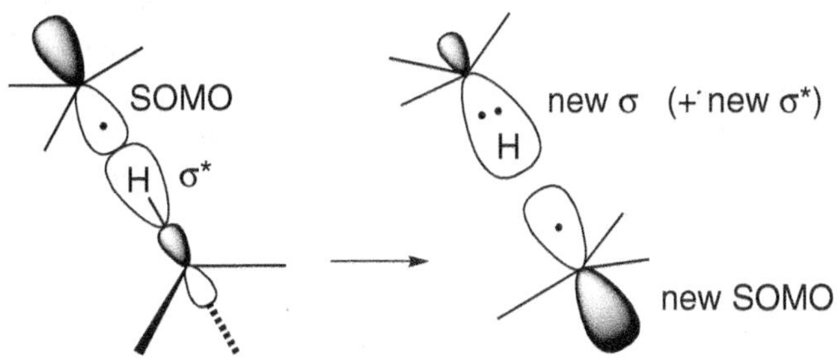

Figure 5.30 Singly occupied molecular orbital (SOMO)–lowest energy unoccupied molecular orbital overlap in hydrogen atom transfer

32. Barton, D.H.R.; Beaton, J.M.; Geller, L.E.; Pechet, M.M. (1961). *J. Am. Chem. Soc.* **1961**, *83*, 4076.

33. (a) Hofmann, A.W. *Ber. dtsch. chem. Ges.* **1879**, *12*, 984. (b) Löffler, K.; Freytag, C. *Ber. dtsch. chem. Ges.* **1909**, 42, *3431*.

Problems

5-14 Draw the FMO overlap for each of the following cyclization modes: 3-*endo*-dig, 3-*endo*-trig, 3-*endo*-tet, 4-*exo*-dig, 4-*exo*-trig, 4-*exo*-tet, 5-*endo*-dig, 5-*exo*-dig, 5-*endo*-trig, 5-*exo*-trig, 5-*endo*-tet, 5-*exo*-tet, 6-*endo*-dig, 6-*exo*-dig, 6-*endo*-trig, 6-*exo*-trig, 6-*endo*-tet, and 6-*exo*-tet. Where appropriate, use one of the compounds below to illustrate your answer (it may be necessary to use another compound in some cases):

5-15 The cyclization of the 5-hexenyl cation and the 5-hexenyl radical lead to very different products, as shown below. Rationalize the differences between these two reactions.

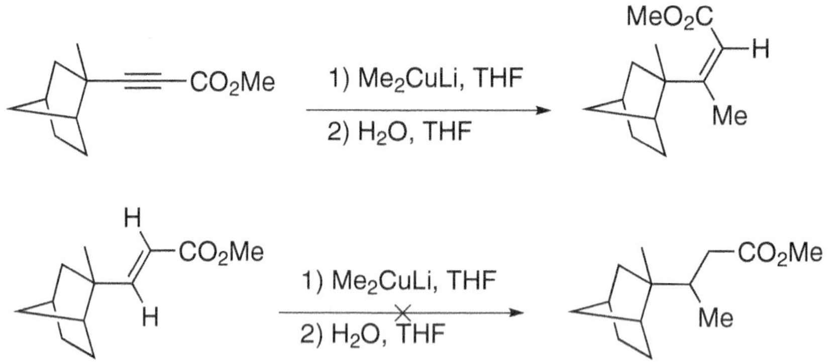

Chapter Summary

In this chapter, a method for categorizing organic reactions based on the FMOs involved has been introduced. The Pauli Exclusion principle, the Aufbau principle, and Hund's rule apply to molecular orbitals and molecules, as well as atoms, which leads to the occurrence of FMOs. These orbitals are the HOMO; (highest energy occupied molecular orbital), the LUMO; (the lowest energy unoccupied molecular orbital), and, in free radicals, the SOMO (singly occupied molecular orbital). These FMOs determine much of the chemistry of the molecule.

Using a very simple model for reactions not involving free radicals, we find that there are three types of filled orbitals, or HOMOs: a filled σ orbital, a filled π orbital, and a filled n orbital. Likewise, there are only three types of empty orbitals that can be involved: an empty nonbonding (atomic) a orbital, a π^* orbital, and a σ^* orbital. This gives only nine combinations of HOMO and LUMO for the initiation of the reaction. The application of these principles to a range of reactions is illustrated.

The effects of the FMOs involved leads to predictions about the regiochemistry and stereochemistry of a number of reactions, phenomena such as the anomeric effect, and Baldwin's rules for nucleophilic ring closure.

Key Terms

anomeric effect	FMO	perturbing influence
Baldwin's rules	HOMO	SOMO
base system	LUMO	stereoelectronic effect
Bürgi-Dunitz trajectory	perturbation theory	

Additional Problems

5-16 Provide a reasonable explanation for the observations shown below. Why does the alkynoic ester react, whereas the alkenoic ester does not?

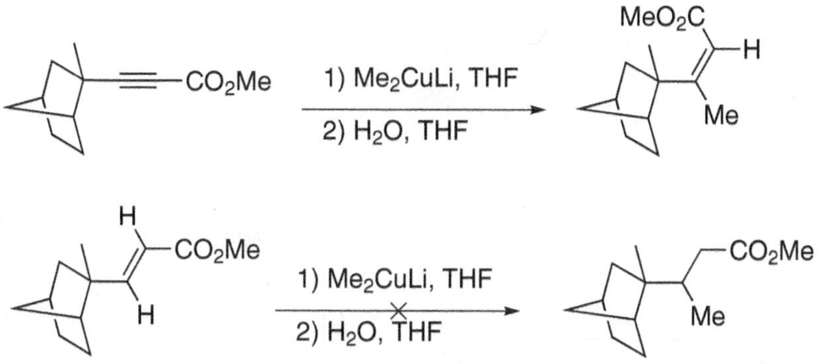

5-17 The reaction below is known to give an alcohol by intramolecular ring opening of the epoxide ring. What is the structure of the major product and why is this the only product formed?

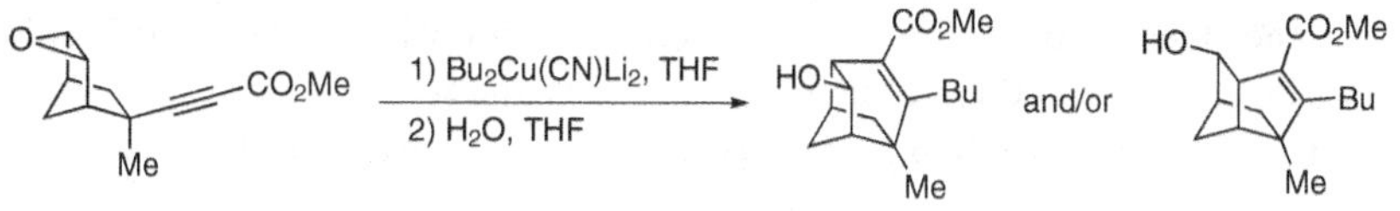

(A) (B)

(Hint: Consider the probable mechanism of this reaction and the frontier orbitals of the cyclization step/epoxide ring opening.)

5-18 Based on your knowledge of FMO theory and Baldwin's rules, predict the major organic product of each of the following reactions. Where appropriate, also specify the stereochemistry of the final product.

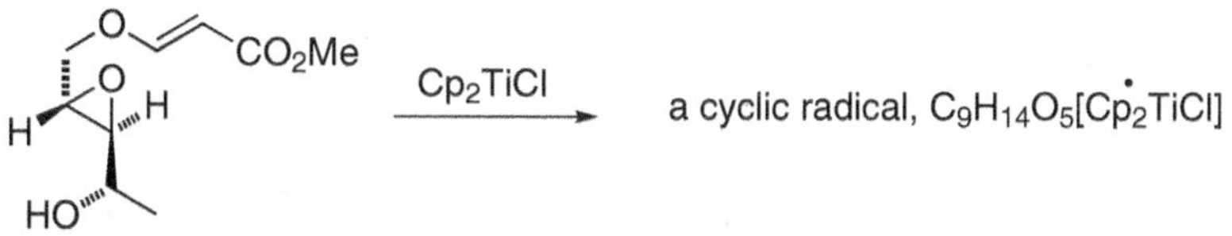

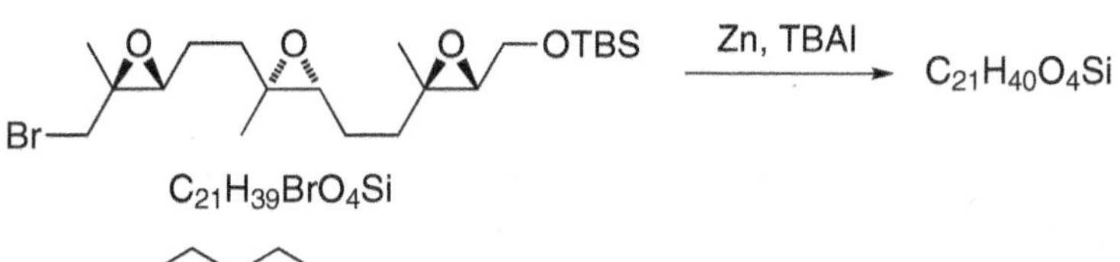

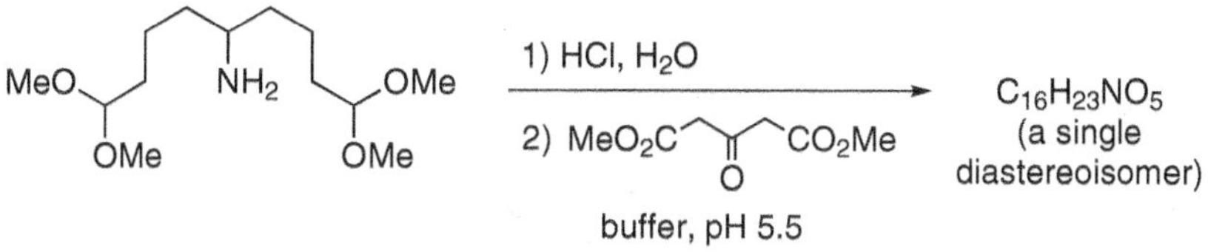

5-19 The reaction shown here proceeds readily when any attempt is made to add a Grignard reagent or to reduce the sulfonate with lithium aluminum hydride. Write a reasonable mechanism for the transformation and give reasons why it occurs so readily. [*Can. J. Chem.* **1979**, *57*, 3348]

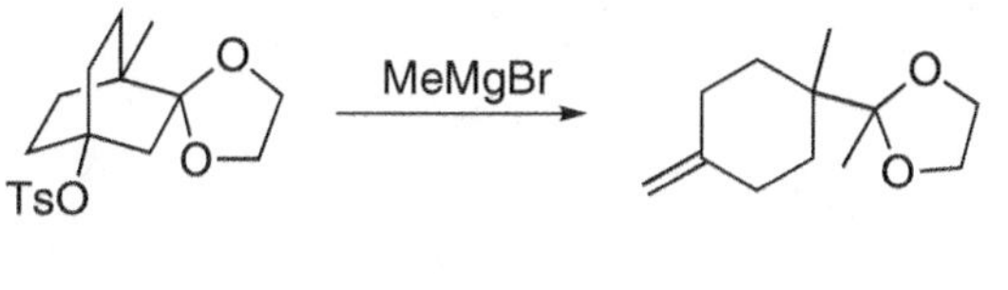

5-20 The Birch reduction of benzene proceeds through the conjugated anion shown to give 1,4-cyclohexadiene when the reaction is carried out in the presence of a proton donor such as ethanol or *tert*-butyl alcohol, but not the more stable, conjugated isomer. When the proton source is omitted, the major product is cyclohexene. Why?

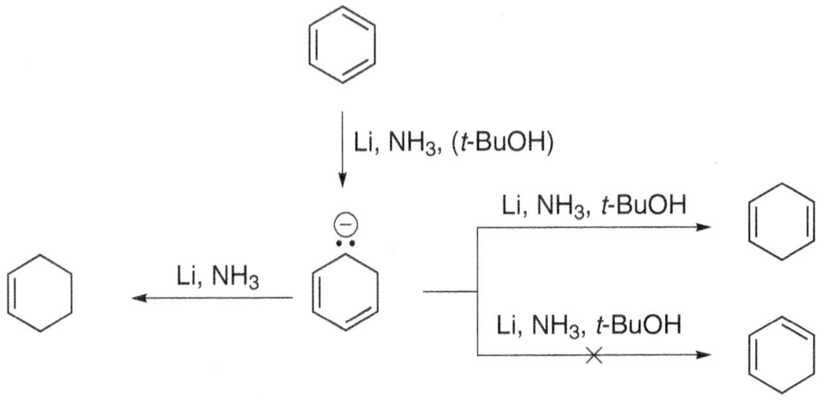

5-21 The answer to Sample Problem 5-2 involved protonation of the starting material at the alkene π bond. This is not the only position where protonation could occur. Where else could the protonation take place, and why? Should this (these) be a major or minor reaction pathway(s)? What would be the products generated by protonation at this (these) position(s) and why do we not observe them?

CHAPTER SIX

Organic Reactions I

Pericyclic Reactions and the Conservation of Orbital Symmetry

6.1 **Introduction**

In Chapter 5, we began our exploration of organic reactions by using the pairings of frontier molecular orbitals (FMOs) as a tool for categorizing them. We now continue the discussion by looking at some important synthetic reactions that lead to the formation of new carbon-carbon bonds. In this chapter, we will focus on pericyclic reactions, which include such important carbon-carbon bond-forming reactions as the Diels-Alder cycloaddition and the Cope and Claisen rearrangements. Historically, these reactions drove the development of frontier orbital theory and the theory of conservation of orbital symmetry.

Definitions

Concerted reactions that proceed through a fully conjugated, cyclic array of continuously bonded atoms, with reorganization of electron density at every atom of the activated complex, are known as *pericyclic reactions*. The transition state for a pericyclic reaction will contain either $4n$ electrons, in which case it is antiaromatic, or $4n + 2$ electrons, in which case it is aromatic. We will discuss the question of aromaticity in much more detail in the next chapter of this book. The aromatic character of the transition state determines the stereochemistry of the reaction. One of the most important characteristics of pericyclic reactions is that their stereochemistry can be predicted by analyzing the symmetry of the frontier orbitals involved. Pericyclic reactions can be either unimolecular or bimolecular. Some representative examples are shown in Table 6.1.

The most common unimolecular pericyclic reactions are ***sigmatropic rearrangements***, where migration of an atom to a remote site in the molecule occurs during the reaction with concomitant migration of π bonds, and ***electrocyclizations***, where an open-chain

Table 6.1 Representative Pericyclic Reactions

Unimolecular

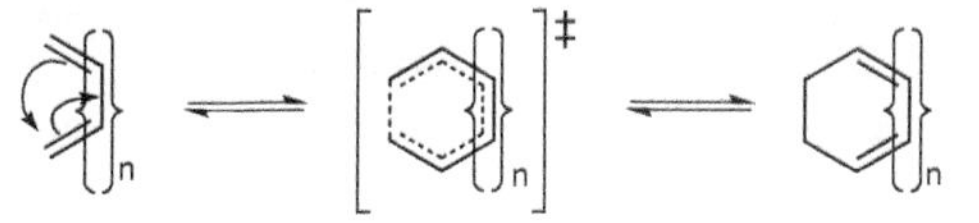

(6.1)

Electrocyclization

Bimolecular

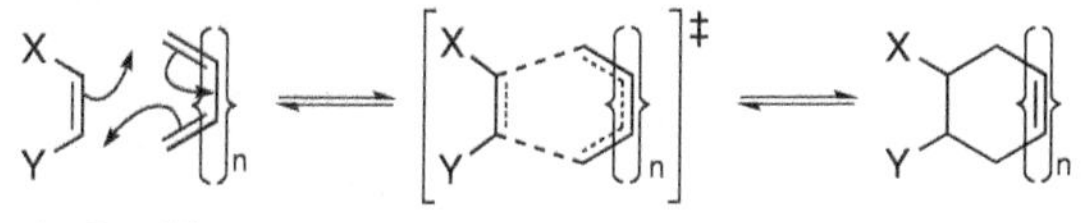

(6.2)

Cycloadditions

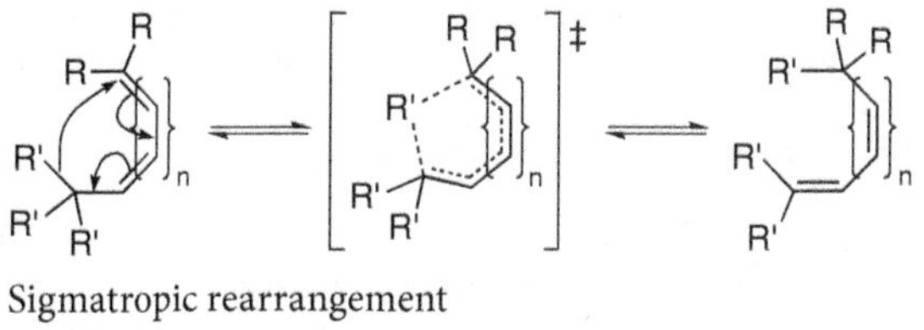

(6.3)

Sigmatropic rearrangement

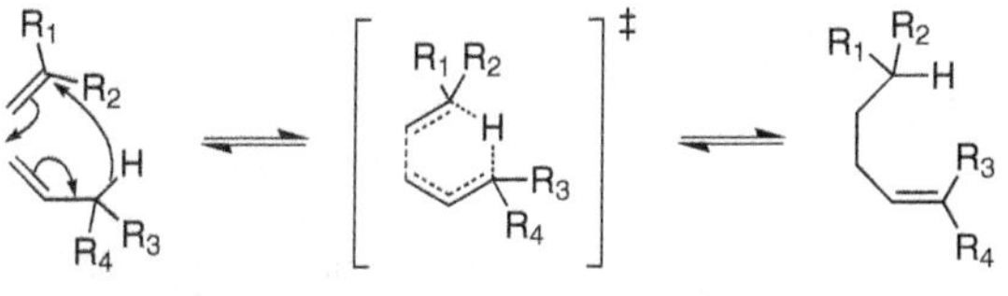

(6.4)

Ene reaction

conjugated system is converted into a cyclic system with one less π bond. Both these reactions involve the highest energy occupied molecular orbital (HOMO) of the starting reactant. The most common bimolecular pericyclic reactions are ***cycloadditions***, where two species react to give a new cyclic product, and ***ene reactions***, where a new σ bond linking the two reactants is formed with concomitant rearrangement of a hydrogen atom.

The most common cycloaddition reactions are the Diels-Alder and carbene cycloaddition reactions. The limitations imposed by the Pauli exclusion principle mandate that, in cycloaddition reactions the HOMO of one reactant ***must*** be paired with the lowest energy unoccupied molecular orbital (LUMO) of the other.

One of the most important properties of pericyclic reactions to emerge during the last half of the 20th century was the concept that, in certain cases, these reactions are subject to an inherent symmetry-imposed barrier. Reactions lacking this barrier are designated ***symmetry-allowed***, whereas those subject to this barrier are designated ***symmetry-forbidden***.

In 1965, Woodward[1] and Hoffmann[2] set out the principles of conservation of orbital symmetry—the major product of a pericyclic reaction that may give more than one product will be that one in which the symmetry of the FMOs is preserved throughout the reaction—to rationalize the stereochemistry of electrocyclic reactions.[3] Similar conclusions[4] were drawn by Japanese theoretical chemist Kenichi Fukui,[5] who shared the Nobel Prize with Hoffmann in 1981.

This principle, which permits the construction of orbital correlation diagrams for some pericyclic reactions, can be used in simplified form to determine whether a particular reaction is symmetry-allowed or symmetry-forbidden under given stereochemical constraints. In the simplified model, the overlap leading to the new bond formation is the key factor. If it is bonding in nature at all locations where new σ bonds are being formed, the reaction is symmetry-allowed, and if there is any antibonding overlap involved, the reaction is symmetry-forbidden.

Problem

6-1 Classify each of the following pericyclic reactions as a member of one of the four classes defined in Table 6.1.

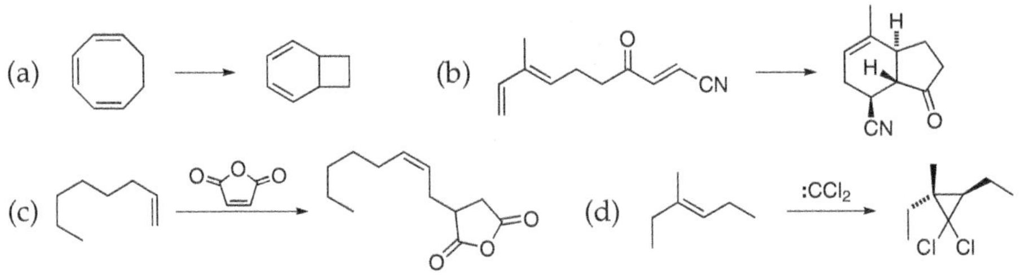

---

1. **Robert Burns Woodward** (1917–1979) was educated at the Massachusetts Institute of Technology (BS, 1937; PhD, 1938) and spent almost all his career, except for a short postdoctoral at Illinois, at Harvard University. A synthetic organic chemist of the first water, Woodward received the Nobel Prize in 1965. More biographical details can be obtained from the following: (a) Blout, E. *Biogr. Mem. Nat. Acad. Sci.* **2001**, *80*, 1. (b) Todd, A.; Cornforth, J.W. *Biogr. Mem. Fell. Roy. Soc.* **1981**, *27*, 628.

2. **Roald Hoffmann (1937–)** was educated at Columbia (BA, 1958) and Harvard (MA, 1960; PhD, 1962). His family escaped from a Nazi labor camp and settled in the United States. After three years at Harvard as a Junior Fellow, Hoffmann joined the faculty of Cornell University, where he remains today. In addition to his Nobel Prize-winning chemistry, Hoffmann is also a published poet and playwright. For more biographical information, see the Nobel Foundation web site.

3. (a) Woodward, R.B.; Hoffmann, R. *J. Am. Chem. Soc.* **1965**, *87*, 395. (b) Longuet-Higgins, H.C.; Abrahamson, E.W. *J. Am. Chem. Soc.* **1965**, *87*, 2045.

4. Fukui, K. *Bull. Chem. Soc. Japan* **1966**, *39*, 498.

5. **Kenichi Fukui (1918–1998)** was educated at Kyoto Imperial University and returned as a faculty member after spending World War II as an army fuel chemist. He was Professor of Physical chemistry until 1982, President of Kyoto Institute of Technology from 1982–1988, and Director of the Institute for Fundamental Chemistry until his death. For more biographical detail, see: Buckingham, A.D.; Nakatsuji, S. *Biogr. Mem. Fellows Roy. Soc.* **2001**, *47*, 223.

(e) 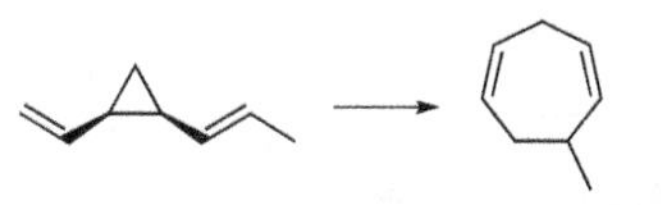

(f) 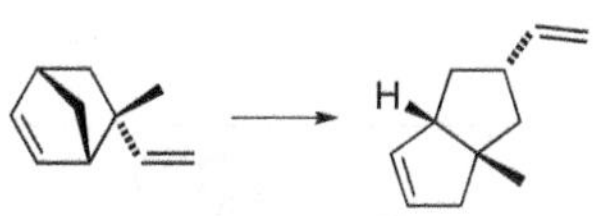

6.2 π Molecular Orbitals of Conjugated Systems and Orbital Symmetry

In order to discuss pericyclic reactions, we must first revisit the LCAO π molecular orbitals of conjugated systems, and examine the symmetry properties of those orbitals. We discussed the molecular orbitals of conjugated systems at some length in Chapter 4, where we focused on the Hückel molecular orbital theory[6] for predicting the orbital coefficients of polyenes. One of the results of Hückel theory is that the 2p orbitals do not contribute equally to each of the π orbitals of the conjugated system; orbital coefficients at the individual atoms change from one π orbital to the next.

Orbital Symmetry

One feature of Hückel π molecular orbitals is the alternation of symmetry properties as the energy of the orbital increases. For example, using the π molecular orbitals of 1,3,5-hexatriene (Figure 6.1), we see that ψ_1, the lowest energy orbital, is symmetric with respect to a mirror plane of symmetry (the phase sign of *no* lobe changes when the symmetry operation is applied), as are ψ_3 and ψ_5. On the other hand, ψ_2, ψ_4 and ψ_6 are antisymmetric (the phase sign of *every* lobe changes when the symmetry operation is applied) with respect to the same symmetry operation. The symmetry properties of the orbitals are exactly reversed when one uses the C_2 axis of symmetry as the symmetry operation.

The lowest energy Hückel π orbital is symmetric with respect to a mirror plane and antisymmetric with respect to a C_2 axis of symmetry as a result of the mathematics of

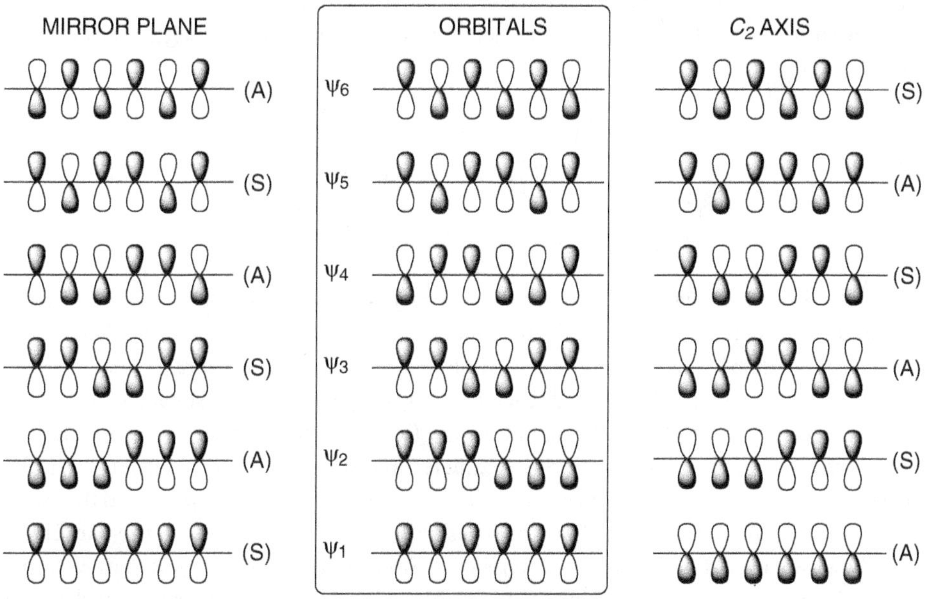

Figure 6.1 Symmetry properties of Hückel π molecular orbitals of 1,3,5-hexatriene

6. (a) Hückel, E. *Z. Phys.* **1931**, *70*, 204; **1931**, *72*, 310; **1932**, *76*, 628; **1933**, *83*, 632. (b) Hückel, E. *Trans. Faraday Soc.* **1934**, *30*, 40. (c) Isaacs, N.S. *Physical Organic Chemistry* (Longman Scientific & Technical: London, 1987), §1.2, pp. 3–24. (d) Carey, F.A.; Sundberg, R.D. *Advanced Organic Chemistry. Part A: Structure and Mechanisms*, 4th ed. (Kluwer Academic/Plenum Publishers: New York, 2000), §1.4, pp. 31–36.

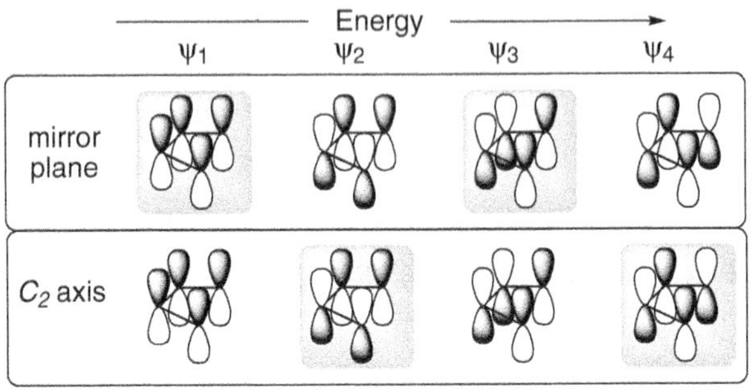

Figure 6.2 Symmetry of the π orbitals of 1,3-butadiene (orbitals that are symmetric with respect to the symmetry operation are in highlighted boxes)

Hückel theory. The same symmetry is exhibited by the terminal lobes of these orbitals. In a symmetric orbital, the terminal lobes have the same phase after application of the symmetry operation. In an antisymmetric orbital, the terminal lobes have the opposite phase sign after application of the symmetry operation.

For a reaction to occur by a concerted mechanism, the overlap of orbital lobes at the location where the new σ bond is to be formed must be in phase (bonding). When the overlap is out of phase (antibonding), the reaction almost always proceeds by a stepwise and not a concerted mechanism. Reactions leading to in-phase overlap are symmetry-allowed, and reactions leading to out-of-phase overlap are symmetry-forbidden.

The electrocyclization of (2*E*,4*E*)-2,4-hexadiene (Example **6.1**) will serve as an example for discussing this reaction. The π molecular orbitals in 1,3-butadiene are designated ψ_1, ψ_2, ψ_3, and ψ_4, in increasing order of energy (Figure 6.2). In the diene, ψ_2 is the HOMO and ψ_3 is the LUMO. Figure 6.2 shows how the phase signs of the lobes of the four π orbitals change or do not change when a particular symmetry element is applied to the *s-cis* conformation of the diene (the required conformation to give the cyclic product). The highlighted structures in Figure 6.2 are symmetric with respect to the symmetry operation—there is no change of sign—in the other structures, which are antisymmetric with respect to the symmetry element, every phase sign changes on applying the symmetry operation. As we can see, all the phase signs of ψ_2 change during the mirror plane symmetry operation, and they remain the same after the C_2 axis is applied: ψ_2 is symmetric (S) with respect to the C_2 axis of symmetry, and antisymmetric (A) with respect to the mirror plane.

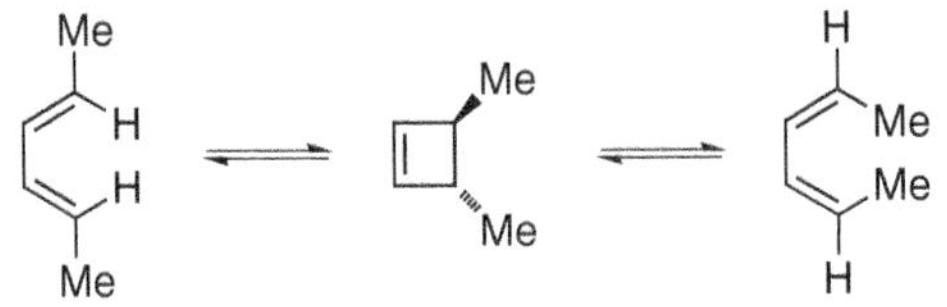

(6.1)

Figure 6.2 also shows that the symmetry properties of each orbital as a whole are identical to the symmetry properties of the terminal lobes alone, so *only the terminal lobes of the participating orbitals need explicitly be used to predict the outcome of a reaction.* One does not have to know the phases of every lobe in the FMO to predict the stereochemical outcome of an electrocyclization—it is not necessary (in most cases) to use the entire orbital to decide whether a particular reaction is symmetry allowed or symmetry forbidden. Thus, the stereochemical consequences of pericyclic reactions are adequately predicted by a simplified process where the symmetry properties of just the terminal lobes of the participating orbitals are used. For example, the stereochemistry of the electrocyclization in Example **6.1** may be predicted using this method, as shown in Figure 6.3. In principle, the cyclization may occur by pathways that retain either the mirror plane of symmetry, or the C_2 axis of symmetry. Because this reaction is unimolecular, the reaction proceeds *via* the HOMO (ψ_2), which is symmetric with respect to the C_2 axis of symmetry, and antisymmetric with respect to the mirror plane. Consequently, the reaction will proceed by the pathway that

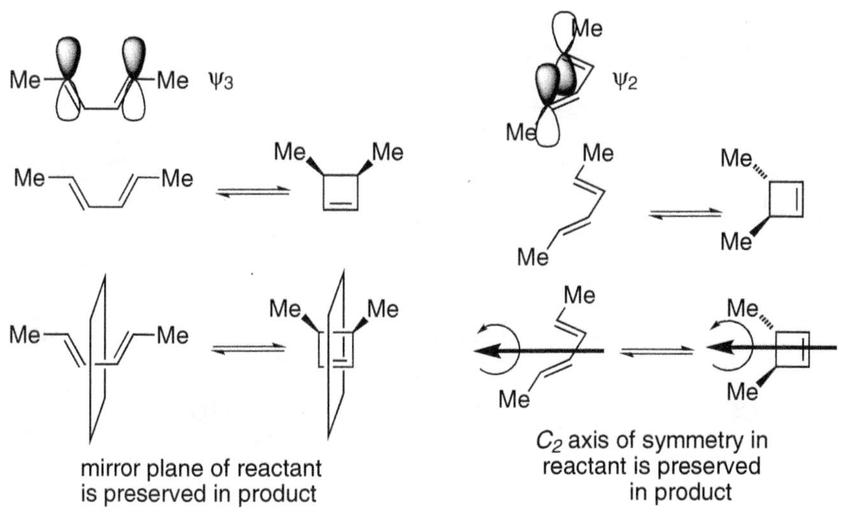

Figure 6.3 Possible electrocyclic ring closures of (2*E*,4*E*)-2,4-hexadiene

preserves the C_2 axis of symmetry of the HOMO (Figure 6.2) over the entire course of the reaction, and we predict that the *trans*-disubstituted product will be formed.

6.3 **Frontier Orbital Analysis: Stereochemical Consequences**

Electrocyclizations

Now let us answer the question, "How is the new bond formed?" In the case of an electrocyclic reaction, this involves the rotation of the two ends of the conjugated system to permit σ-type overlap of the 2p orbitals at the ends of the chain, as shown in Figure 6.4. This rotation can be in the same direction (clockwise for both or counterclockwise for both), when it is known as a ***conrotatory*** ring closure, or in opposite directions, when it is known as a ***disrotatory*** ring closure. If the FMO is symmetric with respect to a twofold axis of symmetry, the C_2 axis of symmetry is preserved throughout the reaction by a conrotatory process and is lost during a disrotatory process, and if the FMO is symmetric with respect to a mirror plane, the reverse is true.

When the reaction symmetry matches the symmetry of the FMO involved, the reaction is favorable because it leads to bonding overlap of the terminal lobes—it is ***symmetry-allowed*** in the ground state. When the reaction symmetry does not match the symmetry of the FMO, however, the reaction becomes very unfavorable because it leads to antibonding overlap of the terminal lobes—it is ***symmetry-forbidden*** in the ground state. This is illustrated in Figure 6.4, which shows that, if the lobes at the end of the conjugated system are out of phase, a conrotatory ring closure is required to give bonding overlap, and that if the lobes are in phase, a disrotatory ring closure is required to give bonding overlap. Because ψ_2 for the 1,3-diene system is symmetric with respect to a C_2 axis, we expect that (2*E*,4*E*)-2,4-hexadiene should give *trans*-3,4-dimethylcyclobutene as the product of the reaction and that heating *trans*-3,4-dimethylcyclobutene should give (2*E*,4*E*)- and/or (2*Z*,4*Z*)-2,4-hexadiene.

Cycloaddition reactions are the prototypical bimolecular pericyclic reactions. By far the most widely used cycloaddition reaction is the Diels-Alder reaction between a diene and a dienophile to give a cyclohexene-based product. To discuss the Diels-Alder reaction, we first need to define some stereochemical terms. When an addition occurs to a π-bonded compound, if the two new σ bonds are formed on the same face of the molecule, the addition is said to be ***suprafacial*** with respect to that reactant. This term is gradually replacing the older one, *syn* addition, in discussions of the stereochemistry of additions to unsaturated systems.

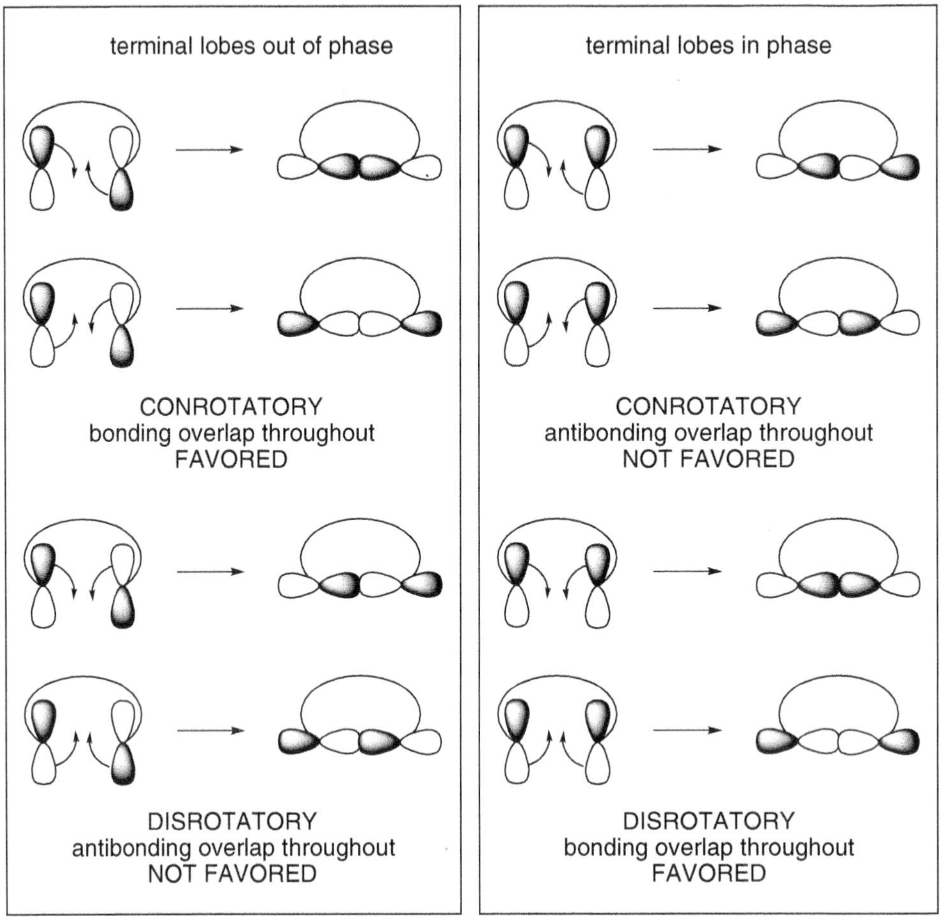

Figure 6.4 Orbital symmetry and electrocyclization

When the addition occurs such that the two new σ bonds are formed on opposite faces of the reactant, the addition is ***antarafacial***; this term is gradually replacing the older one, *anti* addition. Other addition reactions that we have studied already fall into these two classes: hydroboration is a suprafacial addition to an alkene, whereas the addition of bromine to an alkene is antarafacial.

In a cycloaddition reaction, two new σ bonds are formed at the expense of two π bonds. This means that the reaction may be suprafacial with respect to both reactants, suprafacial with respect to one and antarafacial with respect to the other, or antarafacial with respect to both (this combination is quite rare and can be ignored in most cases). The stereochemical consequences of each of these reaction pathways are shown in Figure 6.5. Note how suprafacial addition leads to retention of relative configuration at the reacting centers, whereas antarafacial addition leads to inversion of relative configuration.

Cycloadditions

Cycloaddition reactions are occasionally described using a system that shows the number of π electrons in each reacting species as well as the stereochemistry of the addition with respect to that reacting species. For example, the Diels-Alder reaction is the reaction between a diene (with four π electrons) and a dienophile (with two π electrons). Using this system, it is designated as a $_4\pi + {}_2\pi$ cycloaddition. In addition, the reaction is suprafacial with respect to the diene ($_4\pi_s$) and the dienophile ($_2\pi_s$), so the complete representation of the reaction in this system is $[_4\pi_s + {}_2\pi_s]$. A simpler system denotes the cycloaddition by the number of atoms in the two reacting species. The Diels-Alder reaction is a [4 + 2] cycloaddition in this system, also.

Cycloaddition reactions are very amenable to the same simplified FMO analysis that we used for electrocyclizations. In other words, only the terminal lobes of the reacting species

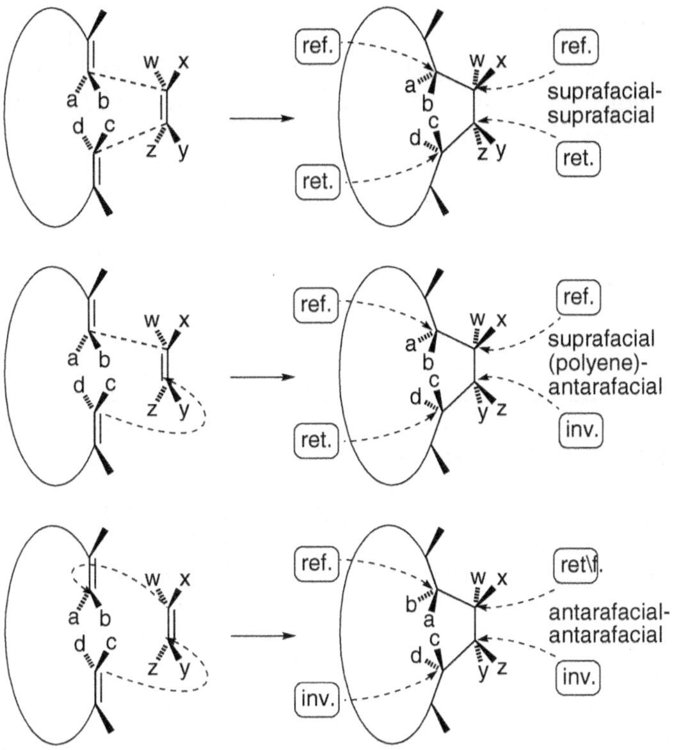

Figure 6.5 Stereochemical outcomes of cycloadditions: "ret." indicates retention of configuration and "inv." indicates relative inversion of configuration relative to the reference centers, labeled "ref."

need be considered explicitly. The two possible HOMO-LUMO combinations for the [4 + 2] and [6 + 2] cycloadditions are shown in Figure 6.6, which illustrates this simplified approach. Based on this simple analysis, we anticipate that the [4 + 2] cycloaddition will be symmetry-allowed when suprafacial with respect to both participating reactants, whereas the related [6 + 2] cycloaddition, which forms a cyclooctadiene, will be symmetry-forbidden unless the reaction is suprafacial with respect to one participant and antarafacial with respect to the other. In both reactions, the two species react through complementary FMOs; one uses the HOMO, and the other uses the LUMO. This observation may be extended to state that *in all multiorbital pericyclic reactions, one unoccupied orbital is used in the overlap that initiates the reaction.* Note how the suprafacial overlap at both ends of each reacting species in the [6 + 2] cycloaddition involves one antibonding interaction; this indicates that the reaction with this stereochemistry has a symmetry-imposed barrier—it is symmetry-forbidden. Extending this analysis to cycloadditions in general reveals that $[m + n]$ cycloadditions are symmetry-allowed in the ground state with suprafacial-suprafacial (or antarafacial-antarafacial) stereochemistry when $m + n = 4n + 2$ (i.e., 2, 6, 10, 14) and symmetry-forbidden when $m + n = 4n$ (i.e., 4, 8, 12).

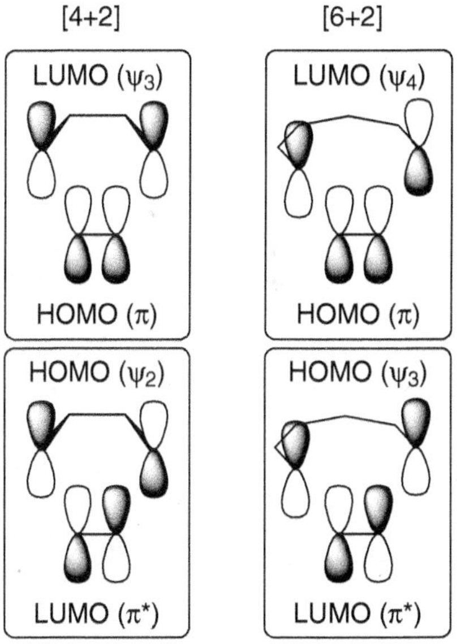

Figure 6.6 Simplified frontier orbital analysis for cycloaddition reactions

Problems

6-2 The cycloaddition reaction below is symmetry-allowed either *suprafacial-suprafacial* or *antarafacial-antarafacial*. What would be the difference observed in the products of the two reactions?

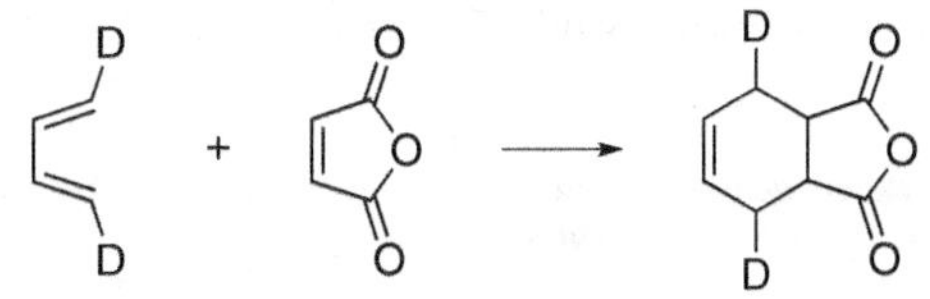

(continues)

(**Problems** continued)

6-3 The *antarafacial-antarafacial* overlap in Figure 6.5 is shown as given below. What are the other possible stereochemical outcomes of an *antarafacial-antarafacial* cycloaddition in this system?

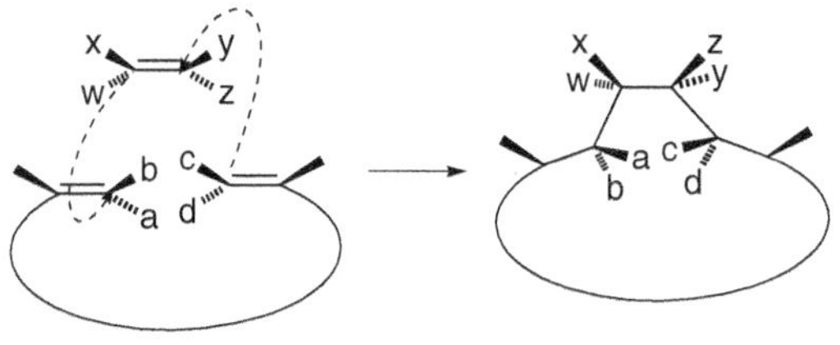

Thermal and Photochemical Reactions

In the discussion that follows, the ideas will be developed in terms of thermal reactions: reactions that occur in the ground state of the molecule. In addition, most of the discussion will be primarily concerned with the lobes at the end of the conjugated π systems (the terminal lobes).

However, the activation energy of organic reactions may be supplied by visible or ultraviolet light as well as by heat. When carried out photochemically, these reactions may be treated as occurring through the excited state of one of the reacting species.[7] The absorption of light by a conjugated system is accompanied by promotion of an electron from a lower energy ground-state orbital to the ground-state LUMO, making the ground-state LUMO the HOMO of the excited state. This results in a reversal of the symmetry types of the HOMO and LUMO in the excited state compared to the ground state. As we shall see, this has important effects on the stereochemistry of a pericyclic reaction.

6.4 **Sigmatropic Rearrangements: More Details**

Sigmatropic rearrangements (Example **6.2**) are concerted reactions that occur through a cyclic activated complex. They most often involve a hydrogen atom or an alkyl group as the migrating group (R'). Regardless of the migrating group, sigmatropic rearrangements always involve the relocation of π bonds as well as σ bonds. It is important to remember that there are two σ bonds that participate in the sigmatropic rearrangement—one being formed and one being broken. In the activated complex, however, the migrating group is bonded to both ends of the conjugated π system simultaneously, so in-phase (bonding) overlap to the migrating group is required at both ends of the conjugated π system for the reaction to be favored.

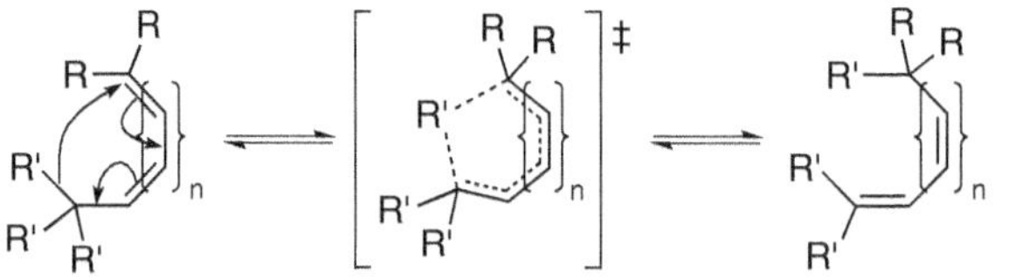

(6.2)

Like cycloadditions and electrocyclizations, sigmatropic rearrangements can be analyzed in terms of simplified frontier orbital analysis rather than a full orbital correlation

7. For a review containing experimental data for photochemical pericyclic ring openings, see: Lawless, M.K.; Wickham, S.D.; Mathies, R.A. *Acc. Chem. Res.* **1995**, *28*, 493.

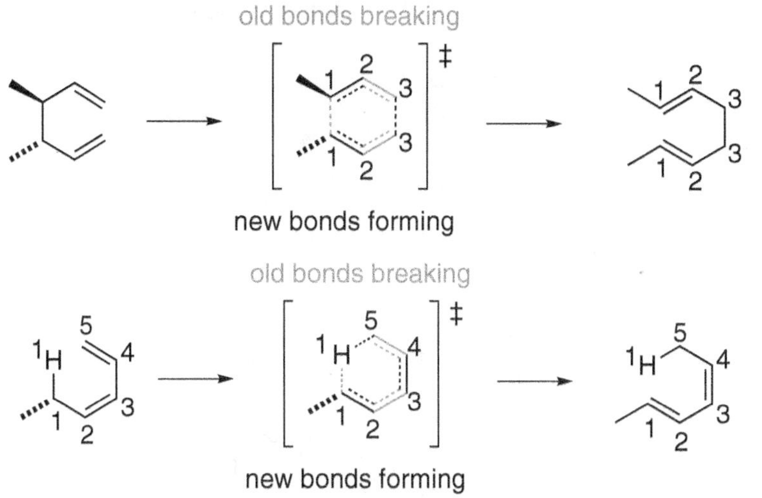

Figure 6.7 Nomenclature of sigmatropic rearrangements

analysis. In the activated complex for a sigmatropic rearrangement, the two halves of the activated complex are linked by two partial σ bonds—the remnant of the σ bond that is broken during the reaction and the incipient stage of the new σ bond that is formed during the reaction. To designate the rearrangement type, the two parts of the reactant molecule are numbered from the σ bond that is broken toward the σ bond that is formed. The rearrangement is then named for the two loci that specify the place where the new σ bond is formed, with the two numbers in square brackets: an [*m*,*n*]-sigmatropic rearrangement. Thus, in the upper example in Figure 6.7, the new bond is being formed between position 3 of the top half of the activated complex and position 3 of the bottom half, making this is a [3,3] sigmatropic rearrangement. In the lower example, the hydrogen atom has number 1, and it migrates to position 5 of the other half of the activated complex, so this is a [1,5]-sigmatropic rearrangement.

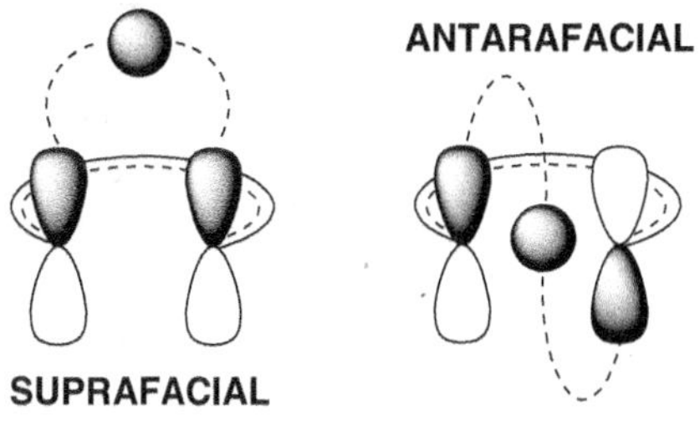

Figure 6.8 Stereochemistry of rearrangement

By analogy with cycloadditions, the stereochemistry of sigmatropic rearrangements is defined as either suprafacial, where the breaking bond and the forming bond are on the same face of the conjugated system, or antarafacial, where the these bonds are on opposite faces of the conjugated system (Figure 6.8). The stereochemistry of any particular reaction can be predicted from an analysis of the FMOs involved.

[1,n]-Sigmatropic Rearrangements of Hydrogen

Let us begin by discussing the simplest types of this reaction, [1,3]- and [1,5]-sigmatropic rearrangements of hydrogen (Examples **6.3** and **6.4**, respectively). The simplest approach to use to predict the stereochemistry of sigmatropic rearrangements is to divide the activated complex into two complementary systems: a cation and an anion. Thus, the stereochemistry of sigmatropic rearrangement of hydrogen can be predicted using the orbitals of a proton (LUMO, 1s) and the conjugated anion (HOMO, ψ_2), or the orbitals of a hydride anion (HOMO, 1s) and the conjugated cation (LUMO, ψ_2). One may also view the reaction as occurring by overlap of the ***singly occupied molecular orbitals*** (***SOMOs***) of two free radicals; the prediction remains the same because, again, the FMOs of the two participating species remain the same (1s, ψ_2).

(6.3)

(6.4)

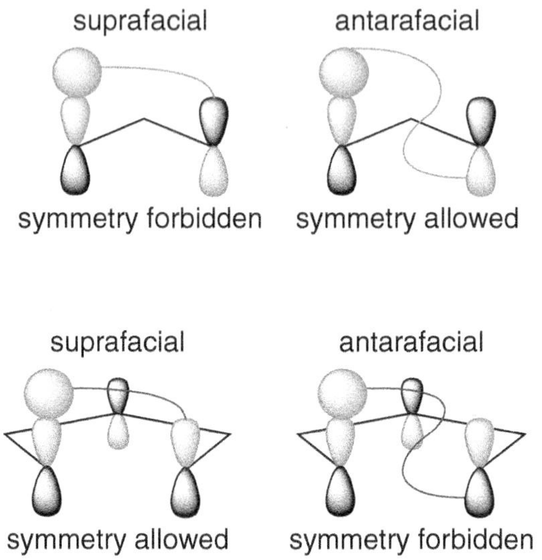

Figure 6.9 Symmetry properties of sigmatropic rearrangements of hydrogen

For the [1,3]-sigmatropic rearrangement of hydrogen, the pair of orbitals involved is the same regardless of the choice made: the 1s orbital of hydrogen and ψ_2 of the allyl system. Of the two possible combinations, the [1,3]-suprafacial rearrangement is symmetry-forbidden, and the [1,3]-antarafacial rearrangement is symmetry-allowed (Figure 6.9). However, geometrical constraints prevent the concerted antarafacial migration, so we expect that this reaction will not proceed by a concerted mechanism.

For the [1,5]-sigmatropic rearrangement, the FMOs are the hydrogen 1s orbital and ψ_3 of the pentadienyl system. This reaction is now symmetry-allowed suprafacially and symmetry-forbidden antarafacially. Because the suprafacial reaction is not prevented by geometric constraints, concerted [1,5]-sigmatropic rearrangements of hydrogen often occur readily.

Extending this analysis, we see that suprafacial [1, *m*] sigmatropic rearrangements of hydrogen are symmetry-allowed when $m + 1 = 4n + 2$ and symmetry-forbidden when $m + 1 = 4n$, with the corresponding antarafacial sigmatropic rearrangements obeying the opposite symmetry rules.

[1,n]-Sigmatropic Rearrangements of Alkyl

Things change when the migrating group becomes alkyl, rather than hydrogen (Figure 6.10) because, unlike hydrogen, carbon atoms may react using either an sp^3 hybrid orbital (suprafacial reactions, like hydrogen) or a 2p orbital (antarafacial reactions). Here, if the migrating carbon atom retains its sp^3 hybridization (i.e., its stereochemistry), the analysis of the reaction is strictly analogous to that of sigmatropic rearrangements of hydrogen. The suprafacial migration is symmetry-forbidden in the [1,3]-sigmatropic rearrangement, and symmetry-allowed in the [1,5]-sigmatropic rearrangement. *[1,5]-Sigmatropic rearrangements occur with retention of configuration at the migrating carbon atom.*

Recall that the suprafacial [1,3]-sigmatropic rearrangement of hydrogen is symmetry-forbidden under all conditions. This is not the case for carbon, however, which can and does participate in [1,3]-sigmatropic rearrangements. This is because the FMO of the hydrogen atom is the spherical 1s orbital, with a single lobe available, and an sp^3 carbon atom can rehybridize to sp^2, to give a 2p orbital, which has two lobes of equal size and shape, *but opposite phase*. Under these circumstances, a suprafacial [1,3]-sigmatropic rearrangement can occur, but it must do so with inversion of configuration at the migrating carbon to be symmetry-allowed. If we think about the transition state for this reaction, it looks something like the transition state for the S_N2 reaction: the migrating carbon atom is sp^2

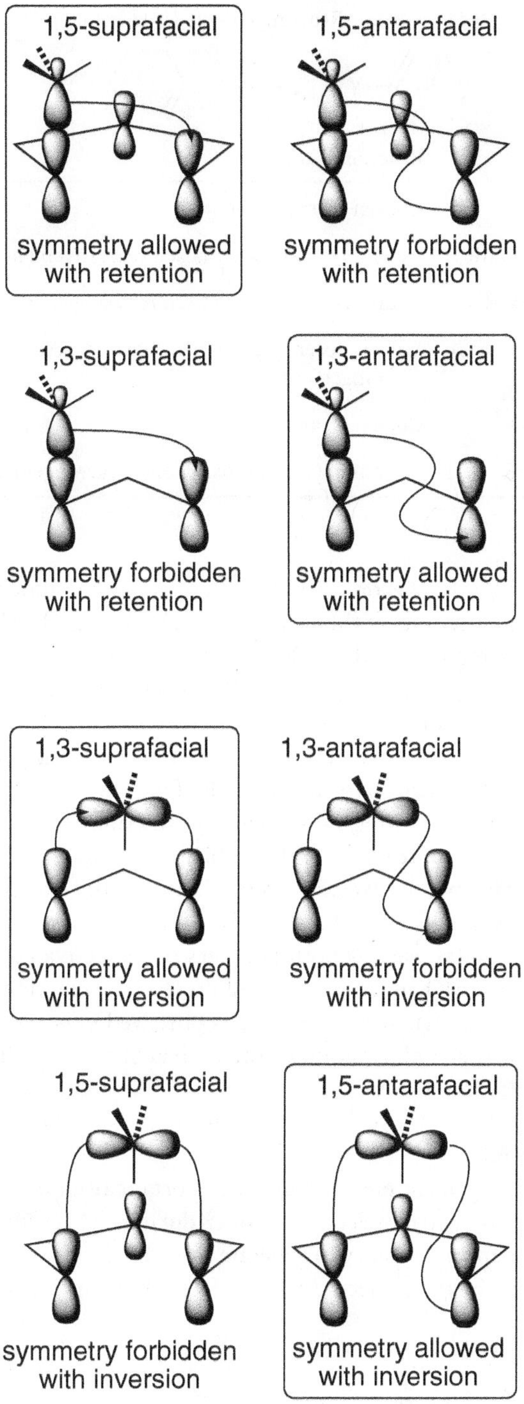

Figure 6.10 Stereochemistry and symmetry properties of [1,3]- and [1,5]-sigmatropic rearrangements of alkyl groups. Symmetry-allowed reactions are highlighted in boxes.

hybridized with the "nucleophile" partially bonded to one lobe of the 2p orbital and the "leaving group" partially bonded to the other. Thus, [1,3]-sigmatropic rearrangements of carbon groups can and do occur in a suprafacial manner; *[1,3]-sigmatropic rearrangements occur with inversion of configuration at the migrating carbon atom.*

[3,3]-Sigmatropic Rearrangements

The [3,3]-sigmatropic rearrangement is a reaction that has been very effectively used to transfer chirality from one part of a molecule to another, as we will discuss in more

Table 6.2 [3,3]-Sigmatropic Rearrangements Used in Organic Synthesis

X	Y	Reaction Name
O	R	Claisen rearrangement
O	$OSiR_3$	Ireland ester enolate Claisen rearrangement
O	OR	Johnson orthoester Claisen rearrangement
O	NMe_2	Eschenmoser ketene *N,O*-acetal Claisen rearrangement
CR_2	R	Cope rearrangement
CR-O^-	R	Oxyanion Cope (or oxy-Cope) rearrangement

detail below. Therefore, it has become a widely used reaction in modern organic synthesis, and there are many versions of this reaction that have been developed since the first report of the rearrangement by Ludwig Claisen in 1912.[8] The more common named versions of the reaction are gathered in Table 6.2. Many of the extensions of the Claisen rearrangement have originated from the study of the effects of substituents on the vinyl ether moiety. Generally, it has been found that electron-releasing substituents on the vinyl ether accelerate the reaction, so much of the focus of research has been involved with the study of electron-donating substituents at this site. This work has given rise to several variants of the parent reaction (Table 6.3).

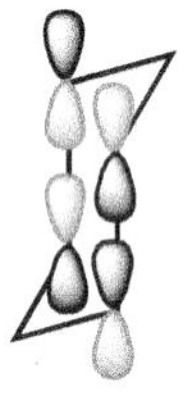

This rearrangement may be treated—very simplistically—as proceeding through complementary ions (an allyl cation and an allyl anion) or as a pair of free radicals. In either case, ψ_2 is the orbital used by both partners in the activated complex, as shown at left. Based on simple FMO analysis, one predicts that the reaction is suprafacial with respect to both allylic substructures, and this is observed experimentally. However, it is important to know the conformation of the activated complex in this reaction, because reactions proceeding through a chairlike transition state will give stereochemical results different from those occurring through boatlike transition states, as illustrated in Figure 6.11.

The Claisen rearrangement emerged as an important reaction for the formation of carbon-carbon bonds in a stereoselective manner during the last third of the twentieth century. This resulted from its ability to generate new chiral centers with predictable stereochemistry at a site in the molecule remote from existing chiral centers ("chirality transfer").[9]

8. Claisen, L. *Ber. Deut. chem. Ges.* **1912**, *45*, 3157.

Ludwig Claisen (1851–1930) studied at Bonn (PhD, 1874), with an interruption for the Franco-Prussian War. He taught at Bonn (until 1882), at Owens College in Manchester (1882–1885), Munich (1886–1890), Aachen (1890–1897), Kiel (1897–1904), and Berlin (1904–1907) before retiring to Godesberg and his private laboratory. For more biographical details, see: (a) Pötsch., W. *Lexikon bedeutender Chemiker* (VEB Bibliographisches Institut: Leipzig, 1989). (b) http://www.uni-kiel.de/ps/cgi-bin/fo-bio.php?nid=claisen&lang=e

9. Reviews: (a) Rhoads, S.J.; Raulins, N.R. *Org. React.* **1975**, *22*, 1. (b) Bennett, G.B. *Synthesis* **1977**, 589. (c) Ziegler, F.E. *Chem. Rev.* **1988**, *88*, 1423. (d) Blechert, S. *Synthesis* **1989**, 71. (e) Wipf, P. In Fleming, I.; Trost, B.M., Eds. *Comprehensive Organic Synthesis*, (Pergamon: Oxford, 1991), vol. 5, p. 827. (f) Nowicki, J. *Molecules* **2000**, *5*, 1033.

Monographs and textbooks: (f) Smith, M.B. *Organic Synthesis*, 2nd. ed. (McGraw-Hill: Boston, 2002), p. 1021. (g) Carruthers, W.; Coldham, I. *Modern Methods of Organic Synthesis*, 4th ed. (Cambridge University Press: Cambridge, 2004), p. 244. (h) Zwiefel, G.S.; Nantz, M.H. *Modern Organic Synthesis: An Introduction* (W.H. Freeman & Co.: New York, 2007), p. 390. (i) Hiersmann, M.; Nubbemeyer, U., Eds. *The Claisen Rearrangement. Methods and Applications* (Wiley-VCH: Weinheim, 2006).

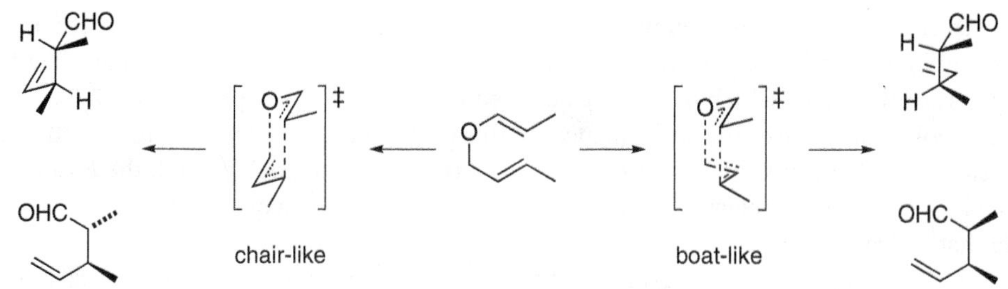

Figure 6.11 Stereochemical outcomes of suprafacial [3,3]-sigmatropic rearrangement of the *E,E* enol ether through chair and boat transition states

Table 6.3 Representative Claisen-type Rearrangements

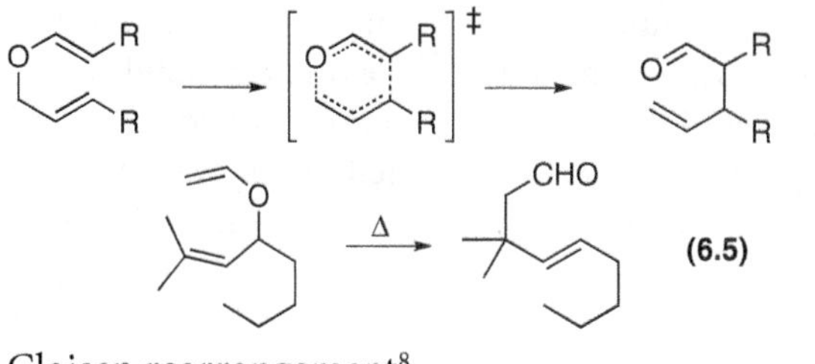

Claisen rearrangement[8]

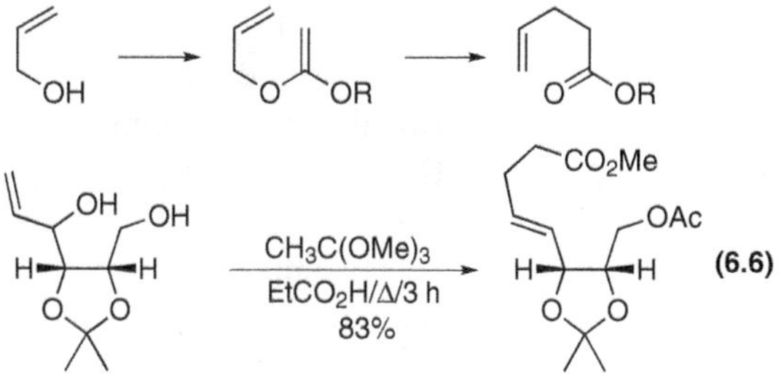

Johnson orthoester variant[10]

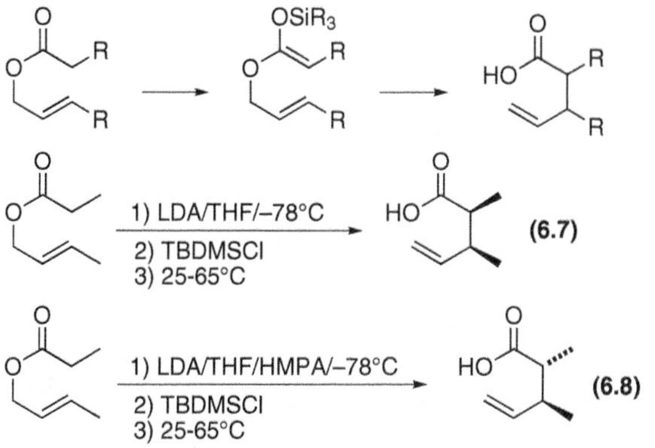

Ireland variant[11]

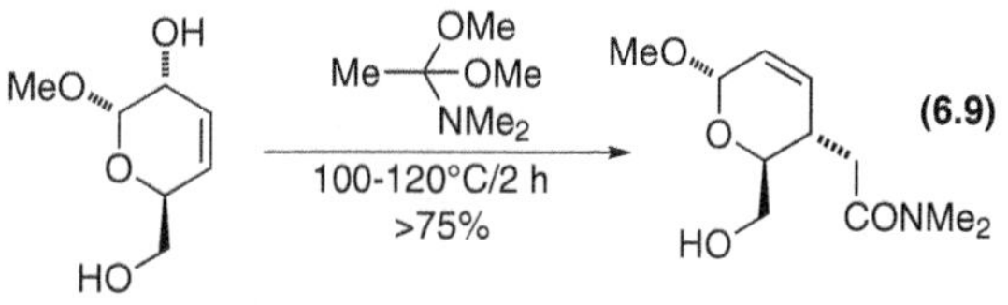

Eschenmoser variant[12]

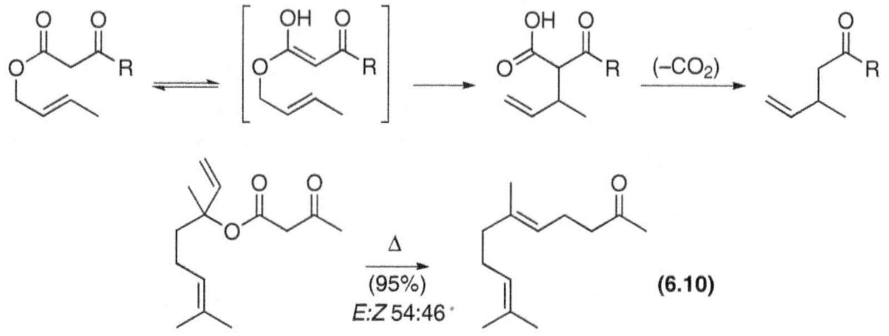

Carroll variant[13]

10. (a) Johnson, W.S.; Werthemann, L.; Barrett, W.R.; Brockom, T.J.; Li, T.; Faulkner, D.J.; Petersen, M.R. *J. Am. Chem. Soc.* **1970**, *92*, 741. (b) Trust, R.I.; Ireland, *R.E. Org. Syn. Coll. Vol. VI* **1988**, 1781.

11. (a) Ireland, R.E.; Mueller, R.H. *J. Am. Chem. Soc.* **1972**, *94*, 5897. (b) Ireland, R.E.; Mueller, R.H.; Willard, A.K. *J. Am. Chem. Soc.* **1976**, *98*, 2868. (c) Ireland, R.E.; Wilcox, C.S. *Tetrahedron Lett.* **1977**, 2839. (d) Pereira, S.; Srebnik, M. *Aldrichimica Acta* **1993**, *26*, 17.

12. Wick, A.E.; Felix, D.; Steen, K.; Eschenmoser, A. *Helv. Chim. Acta* **1964**, *47*, 2425.

13. (a) Carroll, K.F. *J. Chem. Soc.* **1940**, 704. (b) Kimel, W.; Cope, A.C. *J. Am. Chem. Soc.* **1943**, *65*, 1992. (c) Samochvalov, G.I.; Preobazhenskii, N.A. *Zh. Obshch. Khim.* **1957**, *27*, 2501. (d) Nazarov, I.N. *Zh, Obshch. Khim.* **1958**, *28*, 1444.

In this reaction, an allyl vinyl ether (or a substituted variant of an allyl vinyl ether) is heated to form a carbonyl compound. In general, the reaction proceeds through a chair-like transition state, which has been calculated to be favored by approximately 2-5 to 3.0 kcal mol^{-1}.[14] It is one of the consequences of this preferred stereochemistry that in simple systems capable of producing both the *E* and *Z* isomer of the double bond, the *E* isomer tends to be strongly preferred, because the substituent occupies a quasiequatorial position in that activated complex.[15]

The [3,3]-sigmatropic rearrangement of 1,5-dienes[16] is known as the Cope[17] rearrangement. Unlike the Claisen rearrangement, where the equilibrium almost always favors the carbonyl product, the Cope rearrangement and certain of its congeners (e.g., the aza-Cope rearrangement, where the carbon atom at the 2- position of the 1,5-diene is replaced by the nitrogen atom of an iminium ion) tend to be more easily reversed. When a 1,5-diene is heated, a new 1,5-diene is formed where the σ and π bonds of the two allyl groups comprising the two halves of the diene system have reversed their positions. The stereospecificity of these rearrangements is illustrated by the two reactions shown here: *meso*-3,4-dimethyl-1,5-hexadiene gives (2*E*,6*Z*)-2,6-octadiene as the product (Example **6.11**), whereas (±)-3,4-dimethyl-1,5-hexadiene gives (2*E*,6*E*)-2,6-octadiene (Example **6.12**).

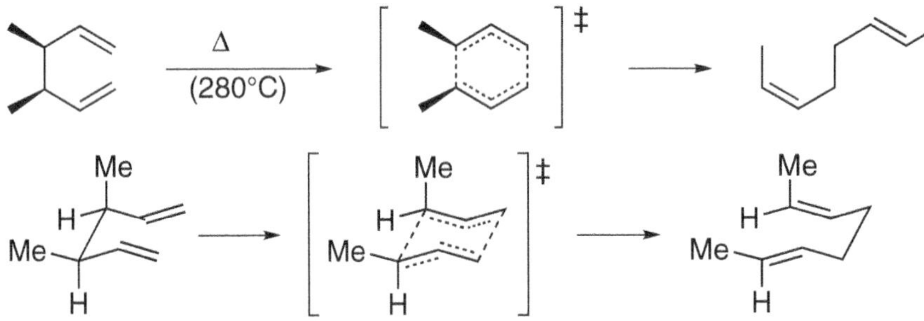

(6.11)

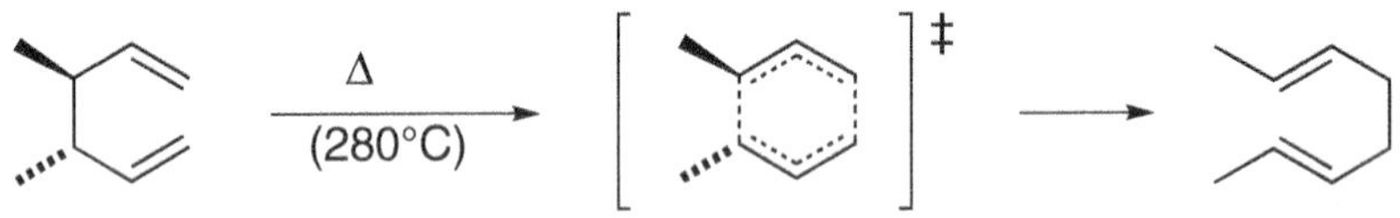

(6.12)

Like the Claisen rearrangement, the Cope rearrangement proceeds through a chair-like transition state,[18] so one need only draw the transition state in the most stable chair conformation to assign the stereochemistry of the two π bonds in the product (e.g., *meso*-3,4-dimethyl-1,5-hexadiene gives the *E*,*Z* isomer of the product).

The Cope rearrangement is generally more facile when the reaction leads to the reduction of strain in the molecule, as illustrated in Figure 6.12. Thus, the energy barrier for

14. (a) Vitorelli, P.; Winkler, T.; Hansen, J.-H.; Schmid, H. *Helv. Chim. Acta* **1968**, *51*, 1457. (b) Hansen, H.J.; Schmid, H. *Tetrahedron* **1974**, *30*, 1959. (c) Vance, R.L.; Rondan, N.G.; Houk, K.N.; Jensen, F.; Borden, W.T.; Komornicki, A.; Wimmer, E. *J. Am. Chem. Soc.* **1988**, *110*, 2314.

15. (a) Faulkner, D.J.; Petersen, M.R. *J. Am. Chem. Soc.* **1969**, *95*, 553. (b) Faulkner, D.J.; Petersen, M.R. *Tetrahedron Lett.* **1969**, 3243.

16. (a) Cope, A.C.; Hardy, E.M. *J. Am. Chem. Soc.* **1940**, *62*, 441. (b) Rhoads, S.J.; Raulins, N.R. *Org. React.* **1975**, *22*, 723. (c) Schröder, G.; Oth, J.F.M.; Meréný, R. *Angew. Chem. Int. Ed. Engl.* **1965**, *4*, 752.

17. **Arthur Clay Cope (1909–1966)** was educated at Butler University (BS, 1929) and the University of Wisconsin (PhD, 1932). After postdoctoral study at Harvard University, he was on the faculty of Bryn Mawr College (1934–1941), Columbia (1942–1945), and MIT (1945–1965). Cope was President of the American Chemical Society in 1961. For more details, see: Roberts, J.D.; Sheehan, J.C. *Biogr. Mem. Nat. Acad. Sci.* **1981**, *60*, 17.

18. (a) Doering, W.v.E.; Roth, W.R. *Tetrahedron* **1962**, *18*, 67. (b) Doering, W.v.E.; Roth, W.R. *Angew. Chem.* **1963**, *75*, 27.

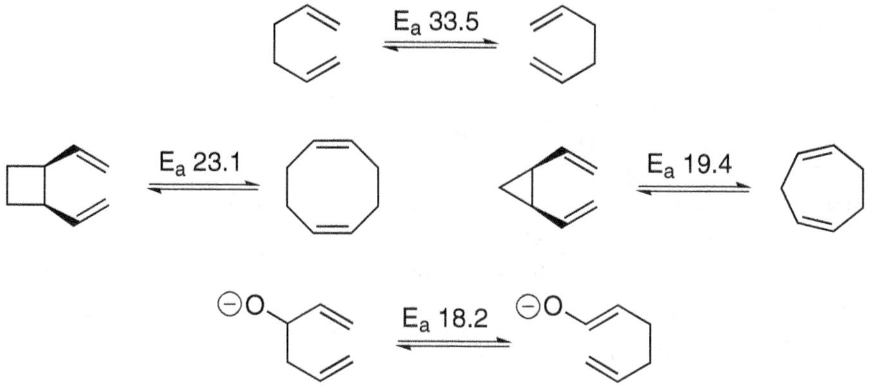

Figure 6.12 Variation of activation energy of the Cope rearrangement

the Cope rearrangement of 1,5-hexadiene is 33.5 kcal mol^{-1},[19] whereas the corresponding values for *cis*-1,2-divinylcyclobutane[20] and *cis*-1,2-divinylcyclopropane[21] are 23.1 kcal mol^{-1} and 19.4 kcal mol^{-1}, respectively. Substituents capable of conjugating with the newly formed double bonds in the product are also effective at lowering the activation energy for the reaction. For example, the Cope rearrangement of 1,5-hexadien-3-ol initially leads to 1,5-hexadien-1-ol, which immediately tautomerizes to the aldehyde, thus also making the reaction irreversible. This reaction, known as the oxy-Cope rearrangement,[22] becomes even faster (by a factor in excess of 10^{10}) when the conjugate base of the alcohol is the reactant.[23] The activation energy for the rearrangement of the conjugate base of 1,5-hexadien-3-ol is 18.2 kcal mol^{-1},[24] even lower than the activation energy for the rearrangement of *cis*-1,2-divinylcyclopropane.

Reaction Synopses

Sigmatropic Rearrangements

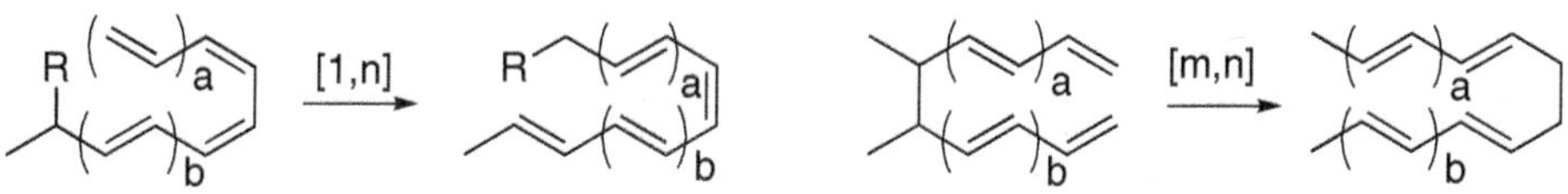

Rearrangement designated as [m.n]-sigmatropic rearrangement, where m and n indicate loci of new σ bond formation based on locus of broken σ bond in starting material

Stereochemistry: suprafacial if [m + n]=4x + 2; antarafacial if [m+n] = 4x

(continues)

19. Doering, W.v.E.; Toscano, V.G.; Beasley, G.H. *Tetrahedron* **1971**, *27*, 5299.
20. Hammond, G.S.; Deboer, C.D. *J. Am. Chem. Soc.* **1964**, *86*, 899.
21. Brown, M.J.; Golding, B.T.; Stofko, J.J., Jr. *J. Chem. Soc., Chem. Commun.* **1973**, 319.
22. (a) Berson, J.A.; Walsh, E.J., Jr. *J. Am. Chem. Soc.* **1968**, *90*, 4729. (b) Viola, A.; Padilla, A.J.; Lennox, D.M.; Hecht, A.; Proverb, R.J. *J. Chem. Soc., Chem. Commun.* **1974**, 491.
Reviews: (c) Marvell, E.N.; Whalley, W. In Patai, S., Ed. *The Chemistry of the Hydroxyl Group*, Part 2 (Wiley: New York, 1971), p. 738. (d) Paquette, L.A. *Angew. Chem. Int. Ed. Engl.* **1990**, *29*, 609. (e) Paquette, L.A. *Synlett* **1990**, 67.
23. (a) Evans, D.A.; Nelson, J.V. *J. Am. Chem. Soc.* **1980**, *102*, 774. (b) Miyashi, H.; Tazato, A.; Mukai, T. *J. Am. Chem. Soc.* **1978**, *100*, 1008. (c) Paquette, L.A.; Pegg, N.A.; Toops, D.; Maynard, G.D.; Rogers, R.D. *J. Am. Chem. Soc.* **1990**, *112*, 277. (d) Gajewski, J.J.; Gee, K.R. *J. Am. Chem. Soc.* **1991**, *113*, 967. (e) Wender, P.A.; Ternansky, R.J.; Sieburth, S.M. *Tetrahedron Lett.* **1985**, *26*, 4319.
24. Evans, D.A.; Golob, A.M. *J. Am. Chem. Soc.* **1975**, *97*, 4765.

Cope Rearrangement

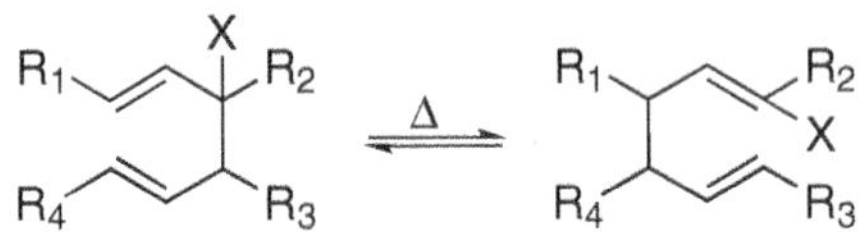

Stereochemistry: suprafacial with respect to both halves of the activated complex
Reaction occurs through a chairlike transition state.
Accelerated by electron-rich substituents (e.g., OH, O^-, NR_2) at position X (e.g., oxy-Cope rearrangement)

Claisen Rearrangement

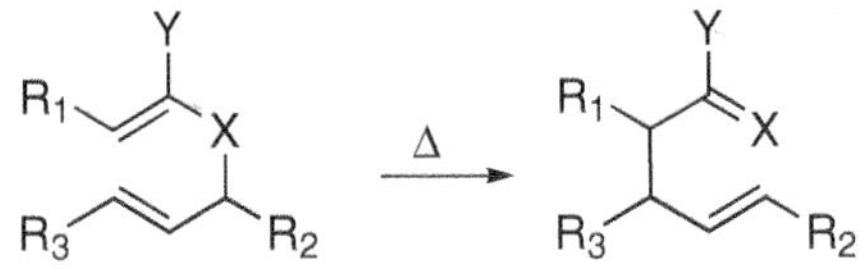

Stereochemistry: suprafacial with respect to both halves of the activated complex
Reaction occurs through a chairlike transition state.
Variants: Ireland (X=O, Y=$OSiMe_3$); Johnson (X=O, Y=OR); Eschenmoser (X=O, Y=NR_2); Carroll (X=O, Y=OH, R_1=RCO)

Worked Problem

6-1 Under what conditions should the following reaction occur?

§Answer below

Problems

6-4 Electrocyclization of dienes is a poor method for forming stereochemically pure cyclobutenes. Why?

6-5 When the conjugated tetraene below is heated, the bicyclic product shown is the only product of the reaction. Write a mechanism that accounts for its formation and rationalize the stereochemistry.

§ Answer to Worked Problem:
This reaction is a [1,5]-sigmatropic shift of hydrogen.

The reaction is symmetry-allowed in the ground state with suprafacial stereochemistry, so the reaction will proceed under thermal conditions. One only needs to heat the diene to cause the isomerization. Note how the product is the *Z* diene.

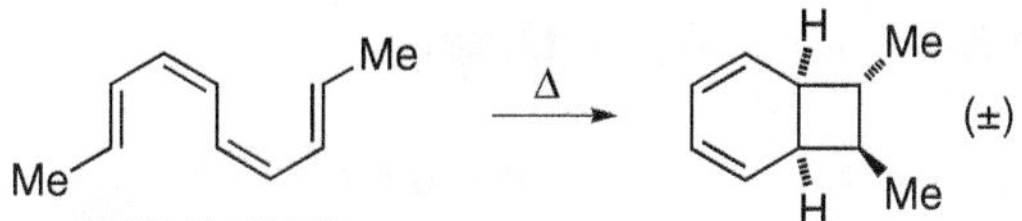

(2E,4Z,6Z,8E)

6-6 What is the major organic product of each of the following reactions? Designate the stereochemistry of the major product.

(a) 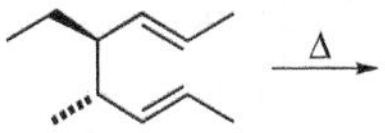 (b) 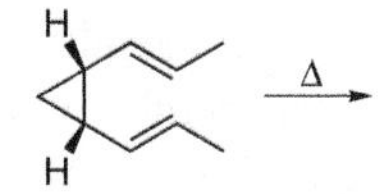 (c) 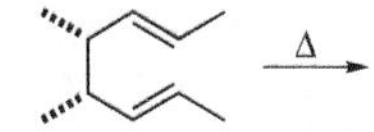

(d) 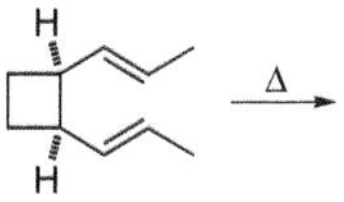 (e) 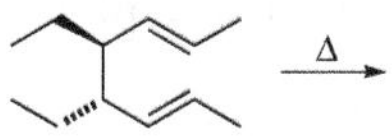 (f) 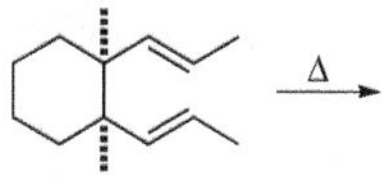

(g) 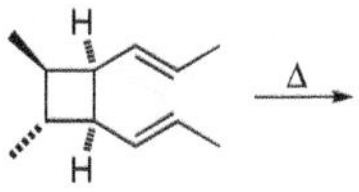 (h) 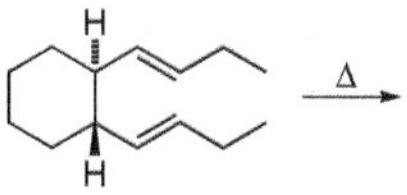 (i) 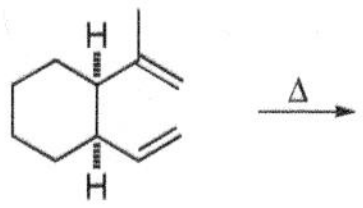

6-7 5-Methyl-1,3-cyclopentadiene is formed from the reaction between the sodium salt of cyclopentadiene and methyl iodide at −78°C. If this diene is stored at room temperature, it becomes a mixture of 1-methyl-1,3-cyclopentadiene, 2-methyl-1,3-cyclopentadiene, and 5-methyl-1,3-cyclopentadiene in just a few hours. Why? Which isomer should be formed in greatest amount and which in least?

6-8 The reaction below is concerted. Show the stereochemistry of the final product and rationalize its formation.

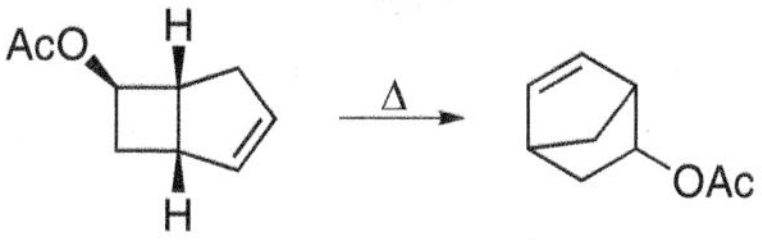

6-9 Predict the products of each of the following reactions.

(a) 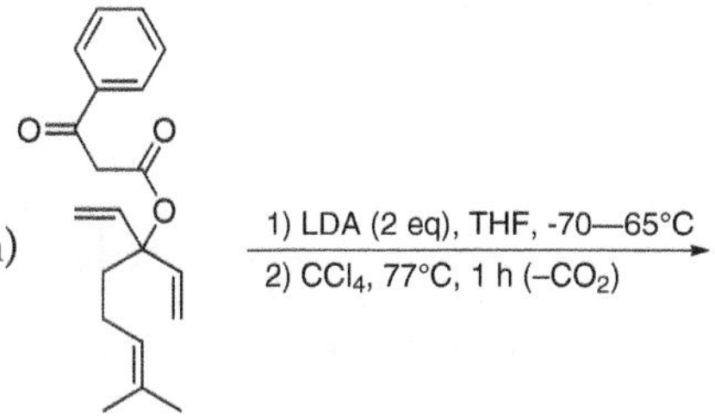

(b) 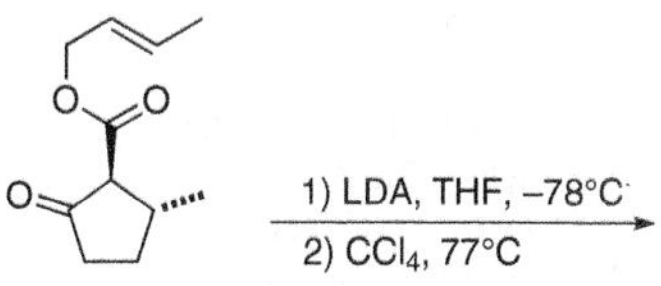

(c) 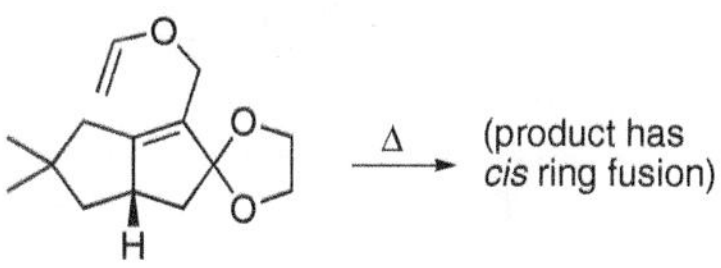

(d) 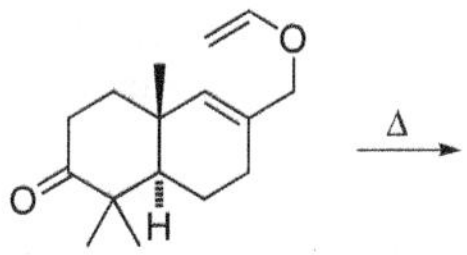

(e) 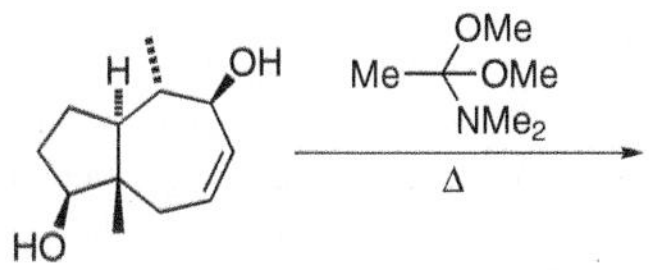

(f) 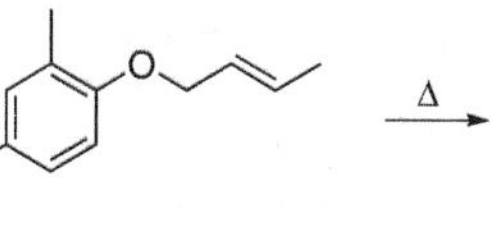

(g) 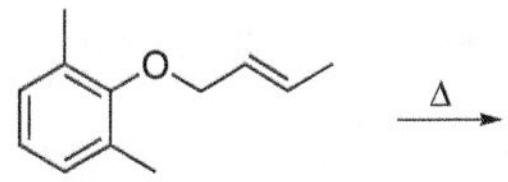

6.5 Cycloaddition Reactions: More Details

In 1906, Wieland discovered that 1,3-dienes would react with alkenes to form a cyclohexenes,[25] and Albrecht reported that 1,3-cyclopentadiene reacted with 1,4-benzoquinone to give both 1:1 and 2:1 adducts.[26] However, the structures of many of the products remained unknown until the 1920s, when Otto Diels and Kurt Alder[27] finally determined them.[28] Their systematic exploration of the scope of the "diene synthesis," as it was then called established the reaction—now almost universally known as the ***Diels-Alder reaction***—as a major method for the synthesis of six-membered rings.[29] Part of their legacy is the terminology that we now use to describe the two participating reactants: the diene and the dienophile (from "diene" and the Greek, φιλος, philos, loving).

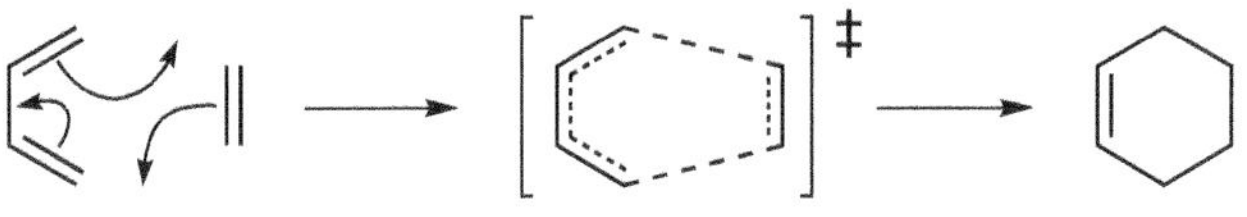

Categorization of Cycloadditions

Cycloaddition reactions are commonly categorized according to one of two conventions that specify either the number of electrons used by each of the two participating species or the number of atoms used by each of the two participating species. Cycloaddition reactions share many of the traits of sigmatropic rearrangements. They are concerted reactions (although alternative nonconcerted mechanisms have been proposed) that are both highly regioselective and stereoselective (or, more usually, stereospecific).

The Diels-Alder reaction, which is the prototypical [4 + 2] cycloaddition reaction, has been the subject of considerable research since the pioneering studies of Diels and Alder in the 1920s, and its predictable stereochemical and regiochemical outcomes have made the it one of the cornerstone reactions of modern organic synthesis. The [2 + 2] cycloaddition corresponding to the Diels-Alder reaction is most commonly used in the form of photochemical addition of an alkene to an enone or in the form of the addition of a ketene to an alkene.

To achieve effective HOMO-LUMO overlap, the lobes at the ends of the diene must overlap with the lobes at the ends of the dienophile. At the same time, the geometry of the activated complex geometry must allow the simultaneous formation of the two new σ

---

25. Wieland, H. *Ber. Deut. chem. Ges.* **1906**, *39*, 1492.

26. Albrecht, W. *Justus Liebvigs Ann. Chem.* **1906**, *348*, 31.

27. **Otto Paul Hermann Diels** (1876–1954) received his PhD from Berlin, where he eventually became professor. In 1916 he moved to Kiel, where he spent the rest of his professional life, retiring in 1945. Diels shared the 1950 Nobel Prize in Chemistry with his student, Kurt Alder. For more biographical detail, see: Schmauderer, E. "Otto Diels." In: Gillispie, C.C., Ed., *Dictionary of Scientific Biography*, (Charles Scribner's Sons: New York, 1972), vol. 4, p. 90.

Kurt Alder (1902–1958) received his PhD under Otto Diels at Kiel in 1926. He remained at Kiel until 1936, then became director of research for the German chemical giant I.G. Farben. In 1940, he moved to Cologne as professor of chemistry, and he remained there for the rest of his life. Alder was quite young when he died in 1958—less than three weeks before his 56th birthday. For more biographical details, see: Ihde, A.J. "Kurt Alder" *Dictionary of Scientific Biography* (Charles Scribner's Sons: New York, 1970), vol. 1 p. 105.

28. Diels, O.; Alder, K. *JustusLiebigs Ann. Chem.* **1928**, *460*, 98.

29. Reviews: (a) Kloetzel, M.C. *Org. React.* **1948**, *4*, 1. (b) Holmes, H.L. *Org. React.* **1948**, *4*, 61. (c) Butz, L.W.; Rytina, A.W. *Org. React.* **1949**, *5*, 136. (d) Walling, C. In Brooks, B.T., Ed. *The Chemistry of Petroleum Hydrocarbons* (Reinhold: New York, 1955), vol. 3, ch. 47. (e) Huisgen, R.; Grashey, R.; Sauer, J. In Patai, S., Ed. *The Chemistry of Alkenes* (Wiley-Interscience: New York, 1964), p. 878. (f) Wassermann, A. *Diels-Alder Reactions: Organic Background and Physicochemical Aspects* (Elsevier: Amsterdam, 1965). (g) Sauer, J. *Angew. Chem. Int. Ed. Engl.* **1966**, *5*, 211. (h) Wollweber, H. *Diels-Alder Reaktion* (Verlag, G.T.: Stuttgart, 1972). (i) Danishefsky, S. *Acc. Chem. Res.* **1981**, *14*, 400.

bonds and the new π bond in a six-membered ring. For these reasons, the Diels-Alder reaction must occur through the *s-cis* conformer of the diene. Reacting through the *s-trans* conformation requires HOMO-LUMO overlap at unreasonably large distances and results in the reaction proceeding through an activated complex resembling a *trans*-cyclohexene, which is prohibitively strained. Cyclic dienes that are locked by the ring in the *s-cis* conformation (e.g., 1,3-cyclopentadiene and 1,3-cyclohexadiene) tend to be much more reactive than acyclic dienes. Dienes that cannot achieve the *s-cis* geometry (e.g., bicyclo[4.4.0] deca-1,6-diene) do not react in the Diels-Alder reaction at all.

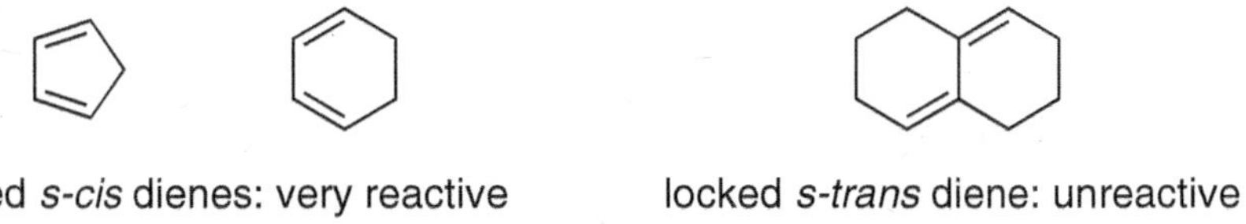

Dienes such as 4-methyl-1,3-pentadiene, which can attain the *s-cis* conformation only with difficulty (the *s-cis* conformation is some 6 kcal mol^{-1} higher energy than the *s-trans* conformation), tend to react only extremely slowly—if at all. This particular diene, for example, reacts with tetracyanoethylene to give a 11% yield of the [4 + 2] adduct, which is accompanied by the [2 + 2] adduct as the major (60%) product (Example **6.13**).[30] The reaction with maleic anhydride (Example **6.14**) gives erratic results, giving two monomeric products in varying ratios, depending on the purity of the reactants.[31]

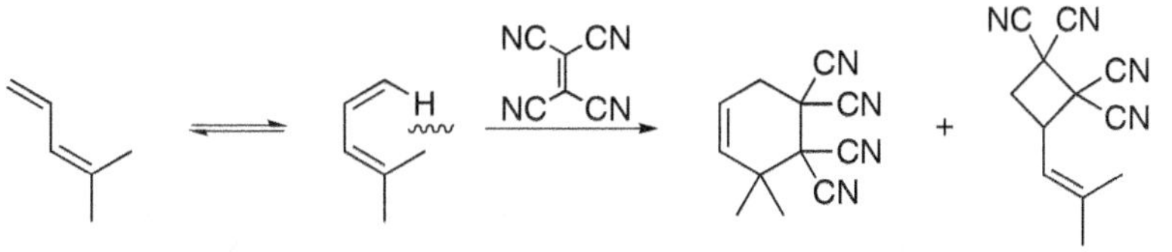

(6.13)

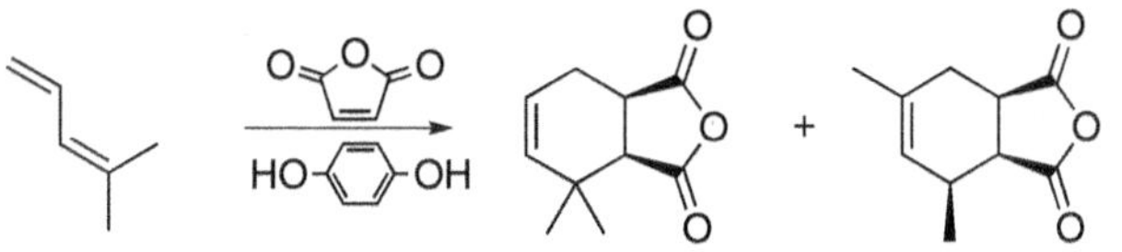

(6.14)

For a concerted cycloaddition reaction to proceed, the overlap at both ends of both reacting orbitals must be between lobes of like phase. In the [4 + 2] cycloaddition, the FMOs are of like symmetry type—both symmetric or both antisymmetric. To meet the in-phase overlap requirement, therefore, the addition must be suprafacial with respect to both reactants; the stereochemistry of the product of the Diels-Alder reaction will reflect its formation by suprafacial addition with respect to both reactants. There are two different HOMO-LUMO combinations (Figure 6.13) that can be used to initiate the reaction: the diene HOMO and the dienophile LUMO or the diene LUMO and the dienophile HOMO.

Of the two possible combinations of FMOs in the Diels-Alder reaction, the dominant FMO overlap is the combination of the diene HOMO and the dienophile LUMO. Therefore, any structural change that lowers the energy of the dienophile LUMO will accelerate the reaction and any structural change that raises the energy of the dienophile LUMO will retard it. By similar reasoning, any structural change that raises the energy of the diene HOMO will accelerate the reaction and any change that lowers the diene HOMO energy will retard it. Electron-withdrawing groups tend to make the π bond more polar and lower the energy of its bonding and antibonding orbitals; electron-releasing groups have the

30. Stewart, C.A., Jr. *J. Am. Chem. Soc.* **1962**, *84*, 117.

31. (a) Goldman, N.L. *Chem. Ind. (London)* **1963**, 1036. (b) Ichizaki, I.; Avai, A. *Bull. Chem. Soc. Japan* **1964**, *37*, 432.

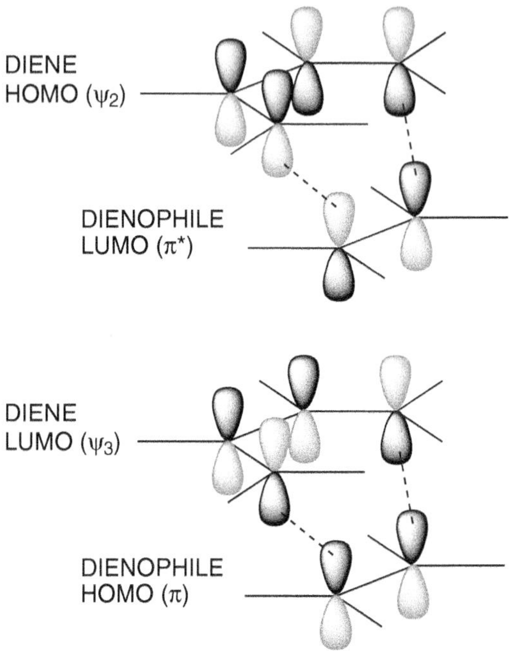

Figure 6.13 Frontier molecular orbital overlaps possible in [4 + 2] cycloadditions. HOMO, highest energy occupied molecular orbital; LUMO, lowest energy unoccupied molecular orbital.

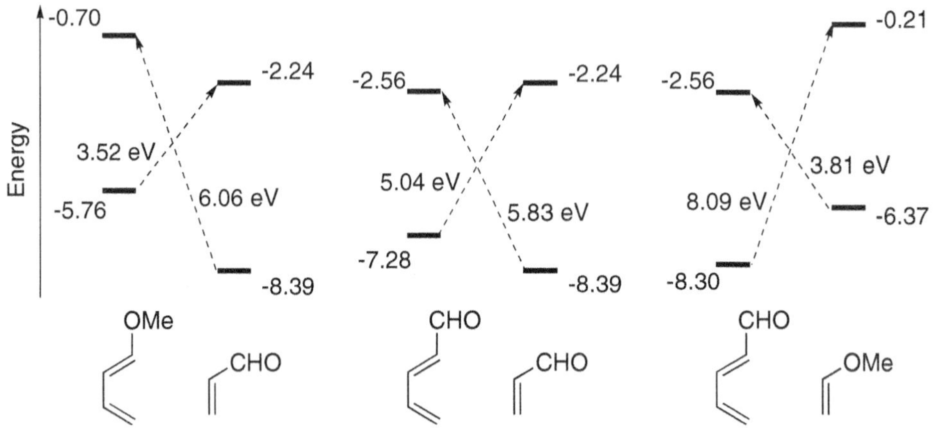

Figure 6.14 Frontier orbital energy gaps for some representative Diels-Alder reactions

opposite effect. Therefore, electron-withdrawing substituents accelerate the Diels-Alder reaction if they are on the dienophile and retard it if they are on the diene; electron-releasing substituents accelerate the reaction if they are on the diene, and retard it if they are on the dienophile. Conjugating groups (e.g., vinyl groups or aromatic rings) raise the energy of bonding orbitals and lower the energy of antibonding orbitals, so they accelerate the reaction if they are on either reactant.

The HOMO-LUMO energy gap is inversely related to the rate of the Diels-Alder reaction,[32] so the larger the gap, the slower the reaction. The energy gap between FMOs for the diene HOMO—dienophile LUMO and diene LUMO—dienophile HOMO pairs, obtained by simple calculations at the B3LYP/6-311++G** level in the gas phase, are shown in Figure 6.14 for three pairs of reactants.

As is evident, when the diene carries an electron-releasing group and the dienophile carries an electron-withdrawing group, the difference in the energy gap is more favorable for the diene HOMO—dienophile LUMO combination of FMOs (the "normal electron

32. (a) Fleming, I. *Frontier Orbitals and Organic Chemical* Reactions (Wiley-Interscience: London, 1976), p. 27. (b) Fleming, I. *Molecular Orbitals and Organic Chemical Reactions*; Reference ed. (John Wiley & Sons: London, 2010), p. 297.

demand" Diels-Alder reaction) by approximately 2.5 eV. When both reacting species carry the same type of substituent, the normal electron demand Diels-Alder reaction is still favored, but now only by approximately 0.8 eV. When the complementary pair is a diene carrying an electron-withdrawing substituent, and a dienophile carrying an electron-releasing substituent, reaction by diene LUMO–dienophile HOMO overlap ("inverse electron demand") now becomes favored, by approximately 4.3 eV, although the HOMO-LUMO gap (3.81 eV) is still larger than that for the normal electron demand Diels-Alder reaction between reactants carrying complementary substituents (3.52 eV).

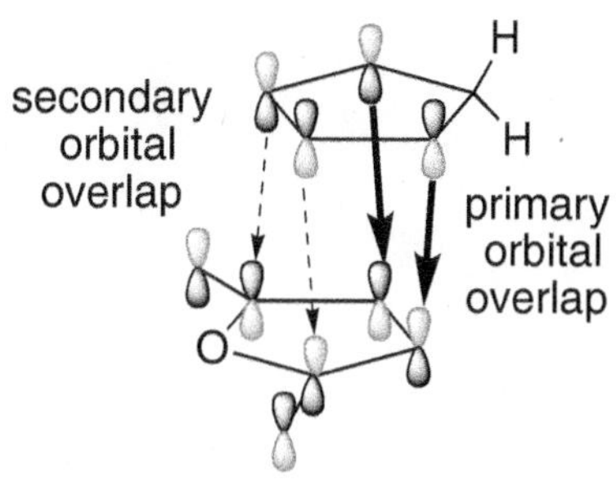

Figure 6.15 Secondary orbital overlap in the Diels-Alder reaction

The substituent on the dienophile, in particular, has a defining effect on the stereochemistry of the final cycloadduct, as was first reported by Alder and Stein.[33] Stated most simply, it is found that conjugating substituents on the dienophile tend to become *endo* substituents in the cycloadduct. In a bridged ring system with both a convex and a concave surface, the convex surface is called the *exo* surface, and the concave surface, where all substituents project into the cavity, is called the *endo* surface. The origin of the *endo* preference is generally taken to be due to secondary orbital interactions (i.e., bonding-type interactions between the lobes of the FMOs that are not directly involved in the formation of new bonds in the product),[34] although this view has been challenged in favor of a variety of alternative effects.[35] Figure 6.15 shows the primary (heavy arrows) and secondary (dashed arrows) orbital overlaps between the HOMO of cyclopentadiene (top) and the LUMO of maleic anhydride (bottom) in the *endo* orientation. Note how the secondary orbital overlap in this reaction is bonding in nature and therefore expected to be stabilizing.

Qualitative Perturbation Theory: Approximating Frontier Orbitals of Substituted Alkenes and Dienes

The incorporation of heteroatoms or other species into a conjugated system affects both the energies of the molecular orbitals of the system, and the coefficients of the π molecular orbitals of the system. Although it is useful to be able to calculate the orbital coefficients in order to predict the outcome of reactions, it is actually not absolutely necessary to do so. Recall that the formation of covalent bonds by overlap of hybrid atomic orbitals is most efficient when overlap occurs through the large front lobes of the orbitals. This allows us to predict reaction regiochemistry with only a qualitative or semiquantitative idea of the coefficients of the FMOs, so long as we can predict the largest lobes of the FMOs of the reactants.

So, we have three questions.

1. How does the presence of a substituent on a π bond affect the energies of the associated FMOs?
2. How does the presence of a substituent on a π bond affect the orbital coefficients of the frontier orbitals?
3. Does the charge on an atom affect the energy and coefficients of the molecular orbitals? A series of simple calculations at the B3LYP/6-31++G** level provide us with data to answer this question (Tables 6.4 and, Table 6.5; Figure 6.16).

Table 6.4 Energies of π Molecular Orbitals of $[H_2C{=}NH_x]^*$

Species	π	π^*	Energy Difference (eV)
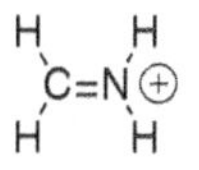	−17.90	−8.88	9.02
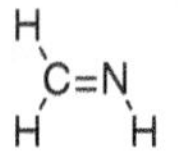	−9.36	−0.93	8.43
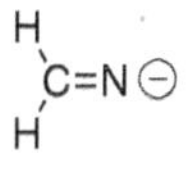	−2.37	+4.62	6.99

33. Alder, K.; Stein, G. *Angew. Chem.* **1937**, *50*, 510.

34. (a) Arrieta, A.; Cossío, F.P.; Lecea, B. *J. Org. Chem.* **2001**, *66*, 6178, and references cited therein. (b) Apeloig, Y.; Matzner, E. *J. Am. Chem. Soc.* **1995**, *117*, 5375.

35. (a) Ruiz-López, M.F.; Assfeld, X.; García, J.I.; Mayoral, J.A.; Salvatella, L. *J. Am. Chem. Soc.* **1993**, *115*, 8780. (b) García, J.I.; Mayoral, J.A.; Salvatella, L. *Acc. Chem. Res.* **2000**, *33*, 658. (c) Gajewski, J.J. *J. Org. Chem.* **1992**, *57*, 5500. (c) Xidos, J.D.; Gosse, T.L.; Burke, E.D.; Poirier, R.A.; Burnell, D.J. *J. Am. Chem. Soc.* **2001**, *123*, 5482.

Table 6.5 Frontier Molecular Orbitals of Substituted Alkenes in the Gas Phase

Molecule	HOMO Energy	LUMO Energy	$E_{LUMO}-E_{HOMO}$	Molecule	HOMO Energy	LUMO Energy	$E_{LUMO}-E_{HOMO}$
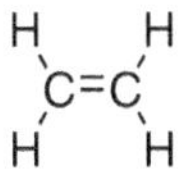	−7.65	−0.30	7.35	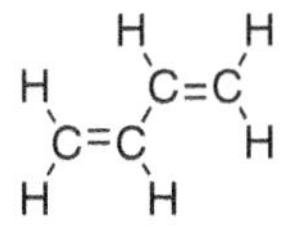	−6.61 (ψ_2)	−1.16 (ψ_3)	5.45
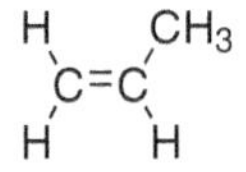	−7.15	+0.03	7.18	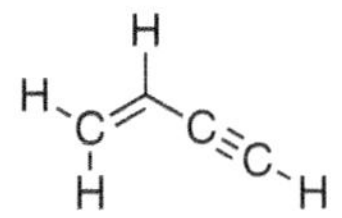	−7.00 (ψ_2)	−1.29 (ψ_3)	5.71
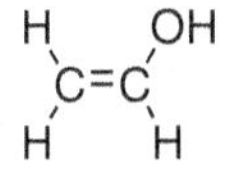	−10.83	−0.31	10.52	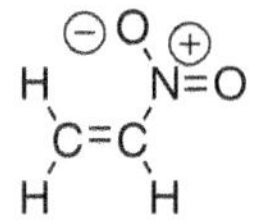	−9.01 (ψ_2)	−3.12 (ψ_3)	5.89
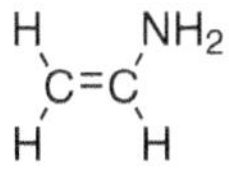	−9.47	−0.02	9.45	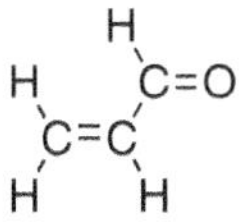	−8.39 (ψ_2)	−2.24 (ψ_3)	6.15
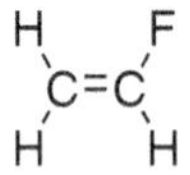	−12.76	−0.27	12.49	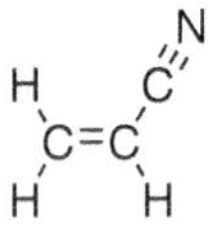	−8.25 (ψ_2)	−2.01 (ψ_3)	6.24

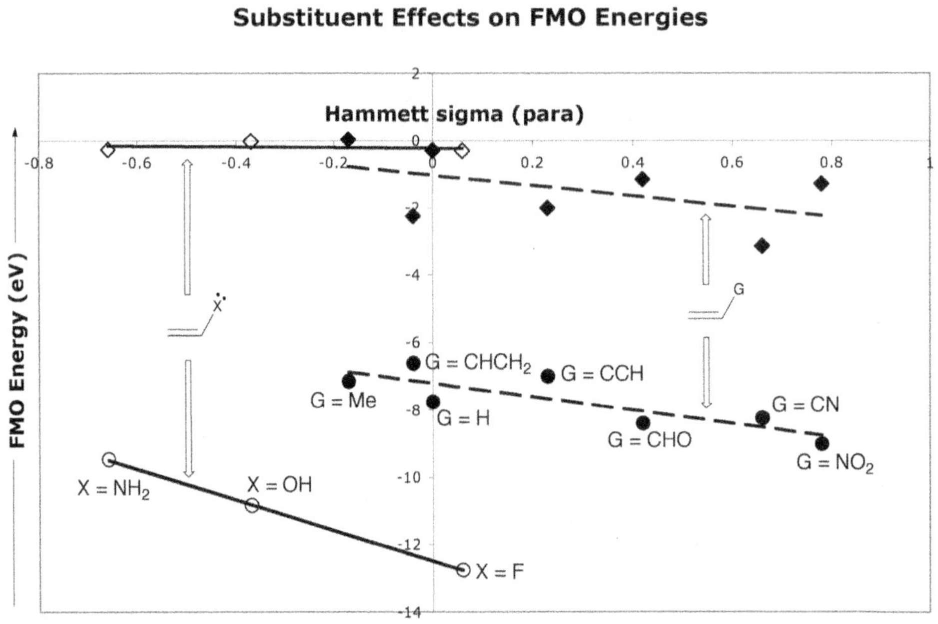

Figure 6.16 Hammett plot of the variation of the energy of the highest energy π orbital (lower two traces) and lowest energy unoccupied molecular orbital (upper two traces) energies [B3LYP/6-311+G**] for substituted ethylenes. Open symbols and dashed lines correspond to substituents that carry a lone pair on the atom directly bonded to the alkene π bond. Filled symbols and lines correspond to electron-withdrawing substituents that conjugate with the alkene π bond by pπ-pπ conjugation. FMO, frontier molecular orbital; MO, molecular orbital.

The data in Table 6.4 show that a positive charge on the nitrogen atom lowers the energy of both the π and π* orbitals and the LUMO of the ion, and a negative charge raises the energy of both orbitals. The energy gap between π and π* gradually increases as the charge on nitrogen becomes more positive. The net effect of the charge changes is to bring the HOMO of the anion much closer in energy to the LUMO of the cation (ΔE = 6.51 eV), which is in line with what we know about their chemistry; the anion is a base and nucleophile, and the iminium ion is an electrophile and acid.

The molecules in Table 6.5 have been chosen to illustrate the effects of substituent electronegativity and substituent electron withdrawing or releasing propensity. By careful analysis of the data, we can draw three general conclusions about substitution of hydrogen of the ethylene molecule.

1. The substitution of carbon by an electron-withdrawing conjugating group lowers the energy of both the π and π* orbitals. The effect on the π* orbital is more pronounced.
2. The substitution of hydrogen by a heteroatom carrying a lone pair lowers the energy of the π orbital substantially as the electronegativity of the substituent increases but has little effect on the energy of the π* orbital (it very slightly lowers it).
3. The substitution of hydrogen by a conjugating hydrocarbon group reduces the separation between HOMO and LUMO. The effects of triple-bonded groups on the LUMO energy are slightly less than those of doubly bonded groups.

Predicting Regiochemistry

Often, a Diels-Alder reaction can give more than one product, especially when it is possible for both reactants to react in more than one way. Let us take the Diels-Alder dimerization of (3*E*)-1,3,5-hexatriene as a typical example. As shown here, there are three possible products for this reaction, and the question "Can we predict the major monocyclic regioisomer produced in this reaction?" should be asked. Based on our earlier arguments above about the efficiency of orbital overlap, the answer is "yes."

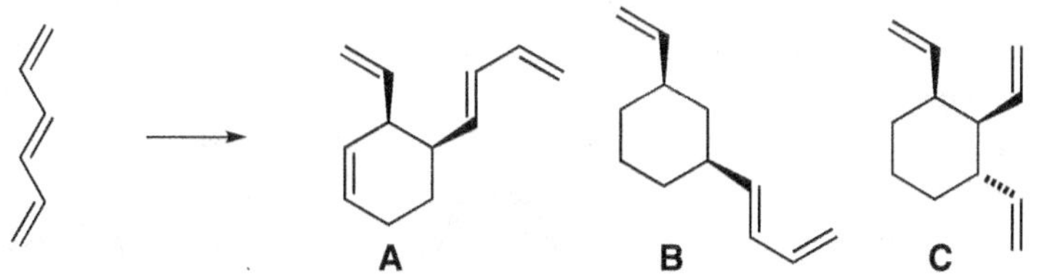

The HOMO-LUMO overlap leading to these products involves either ψ_3 of the diene (positions 2 and 4) and ψ_4 of the dienophile (positions 1 and 2 or 3 and 4)—or ψ_4 of the diene (positions 1 and 4) and ψ_3 of the dienophile (positions 1 and 2 or 3 and 4). Thus, there are six potential modes of frontier orbital overlap that we must consider. Using the Hückel coefficients in Table 2.3, we obtain the results collected in Table 6.6.

The results are definitive: the most efficient orbital overlap occurs when the terminal atoms of the two reacting species form a new bond—when the overlap occurs with the two lobes with the largest orbital coefficients superimposed. We thus predict that the major product of this reaction would be compound A. It is reassuring to find that this is, indeed, the observed major bimolecular cycloaddition product of this reaction.[36]

Table 6.6 Predicting Regiochemistry of a Diels-Alder Reaction

Diene	Dienophile	A	B	C
ψ_3	ψ_4	(0.52)(0.52) + (−0.42)(−0.23) = 0.367	(0.52)(0.42) + (0.52)(0.23) = 0.338	(0.52)(0.23) + (0.42)(0.23) = 0.216
ψ_4	ψ_3	(0.52)(0.52) + (0.42)(0.23) = 0.367	(0.52)(0.23) + (0.42)(0.52) = 0.338	(0.52)(0.42) + (0.23)(0.42) = 0.315

36. Kharasch, M.S.; Sternfeld, E. *J. Am. Chem. Soc.* **1939**, *61*, 2318.

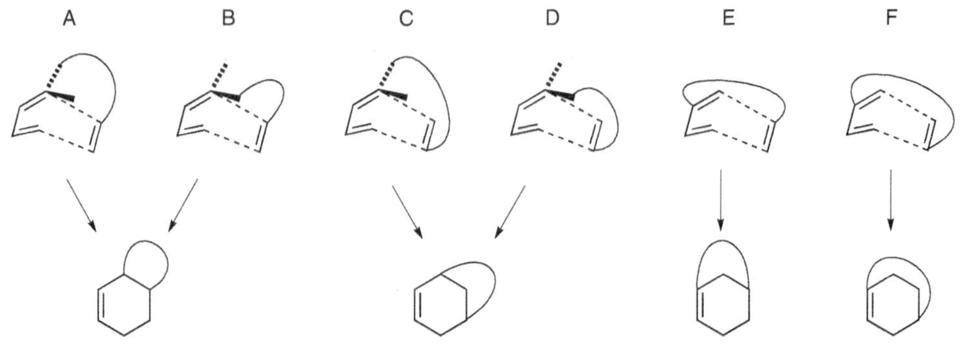

Figure 6.17 Possible cyclization modes in intramolecular Diels-Alder reactions

Intramolecular Diels-Alder reactions

The idea of tethering the diene to the dienophile was first proposed by Alder in 1953,[37] but the idea languished until the development of reliable methods for the synthesis of the necessary trienes some two decades later. When the two reacting species in the Diels-Alder reaction are tethered to one another, additional constraints are imposed on the course of the Diels-Alder reaction. This has resulted in a large volume of literature that we will simply summarize here.[38]

The reaction can lead to the formation of either fused-ring or bridged-ring products, depending on the length and location of the tether. It is generally accepted that the six examples in Figure 6.17 represent the possible cyclization modes. The first two of these (types A and B) lead to the formation of fused-ring products; the next two (types C and D) lead to bridged-ring systems, containing a 3,5-bridged cyclohexene ring; and the last two represent bridgehead alkenes, where the cyclohexene is bridged 1,3- (type E) or 1,4- (type F).

The outcome of an intramolecular Diels-Alder reaction is frequently determined by the length and type of the linker between the diene and the dienophile. When this tether is short, the dominant reaction pathway is either type A or type B, to give fused-ring products. In addition to structural features that assign an intramolecular Diels-Alder reaction to one of the six classes above, the outcome of the reaction is affected by the nature and location substituents on the system, and the length of the bridge between the diene and the dienophile.

Fused-ring products are generally favored as products, with the proportion of fused-ring product increasing as the bridge length is decreased. Generally, the intramolecular and intermolecular variants of the reaction are very similar. Thus, it is observed that the intramolecular reaction is suprafacial with respect to the diene and the dienophile moieties, and that *E* dienes tend to react faster than *Z* dienes. When the stereochemistry is not suprafacial, this generally indicates that the diene or dienophile isomerizes before the cycloaddition itself. The diene stereochemistry also affects the stereochemistry of the product: whereas an *E* diene can attain the reactive conformation leading to both the *cis* and *trans* isomers of the product, the *Z* isomer is usually restricted to the conformation leading to the *cis* isomer.[39] This is illustrated in Figure 6.18 for the intramolecular Diels-Alder reaction of 1,3,8-nonatriene.

37. Alder, K.; Schumacher, M. *Fortschr. Chem. Org. Naturst.* **1953**, *10*, 66.

38. (a) Taber, D.F. *Intramolecular Diels-Alder and Alder Ene Reactions* (Springer Verlag: Berlin, 1984). (b) Ciganek, E. *Org. React.* **1984**, *32*, 1. (c) Oppolzer, W. *Angew. Chem. Int. Ed. Engl.* **1977**, *16*, 10. (d) Brieger, G.; Bennett, J.N. *Chem. Rev.* **1980**, *80*, 63. (e) Fallis, A.G. *Can. J. Chem.* **1984**, *62*, 183. (f) Craig, D. *Chem. Soc. Rev.* **1987**, *16*, 187. (g) Roush, W.R. In Trost, B.M.; Fleming, I., Eds. *Comprehensive Organic Synthesis* (Pergamon Press: Oxford, 1991); vol. 5, ch. 4.4.

39. (a) Pyne, S.G.; Hensel, M.J.; Byrn, S.R.; McKenzie, A.T.; Fuchs, P.L. *J. Am. Chem. Soc.* **1980**, *102*, 5960. (b) Yoshioka, M.; Nakai, H.; Ohno, M. *J. Am. Chem. Soc.* **1984**, *106*, 1133.

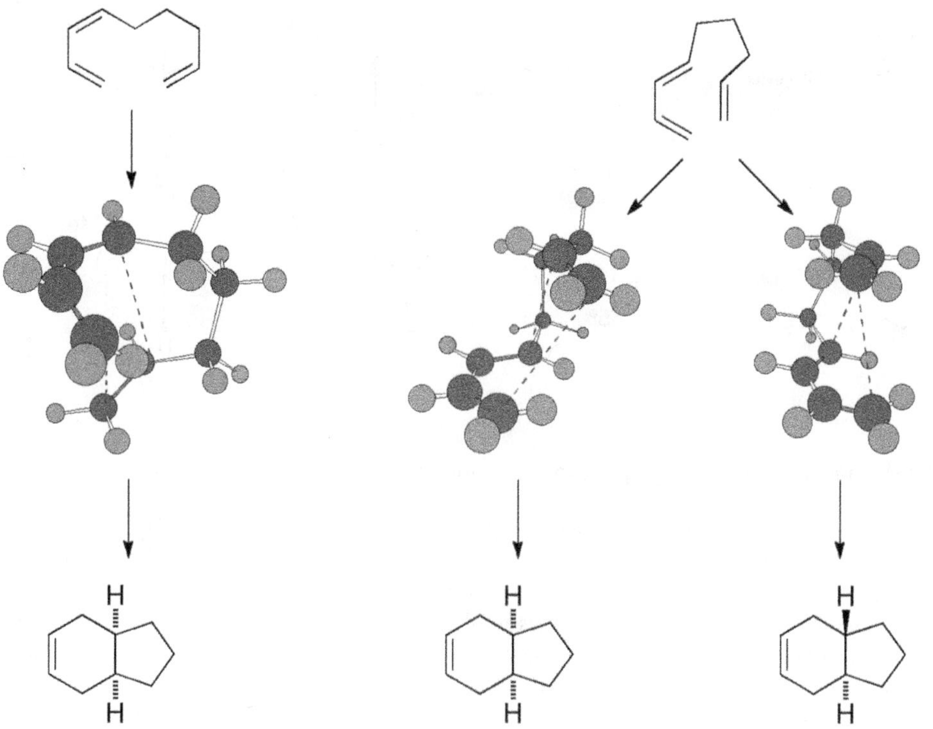

Figure 6.18 Intramolecular Diels-Alder reactions of *Z* and *E* 1,3,8-nonatrienes

The question of *exo* versus *endo* orientation is more complicated in the intramolecular reaction because the bridge often confers a strong preference for *exo* addition. A flexible bridge can allow the formation of both *cis*- and *trans*-fused products, but predicting which will predominate is not necessarily a simple task, as the two examples on the right in Figure 6.18 show. As is observed with the intermolecular reaction, the addition of a Lewis acid increases the rate of the reaction, and, at the same time, improves the *endo* selectivity of the reaction.[40]

In similar fashion, changing the length of the bridge usually shifts the *cis/trans* ratio, but the direction of the change is not always predictable at this time, although shorter bridges tend to favor the *trans* ring fusion in the product. Examples **6.15** and **6.16**[41] show the effects of changing the length of the bridge in two closely related systems. Note how, in both these systems, the *trans* product is favored, although there is more *cis* product formed when the bridge contains four carbons rather than three. Note also how, in the system above, which has a five-carbon bridge, the *cis* isomer is now the favored product.

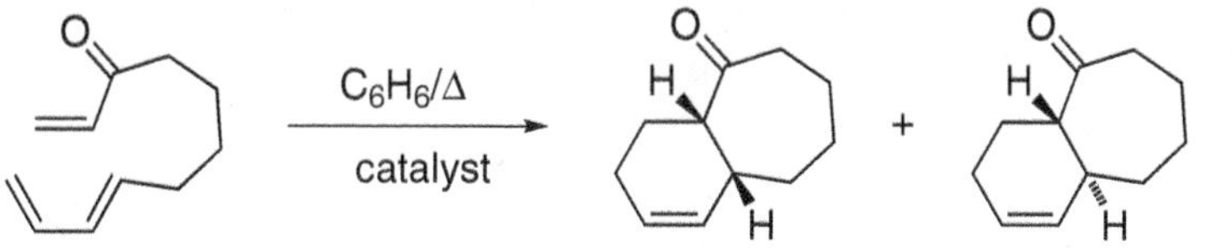

(6.15)

Conditions	Yield	*cis*	:	*trans*
no catalyst, 155°C, 5h	90%	62		38
Et_2AlCl (4 eq), 25°C, 3h	88%	100		0
$LiBF_4$ (1.1 eq), 25°C, 3h	100%	100		0

40. Smith, D.A.; Houk, K.N. *Tetrahedron Lett.* **1991**, *32*, 1549.
41. Wulff, W.D.; Powers, T.S. *J. Org. Chem.* **1993**, *58*, 2381.

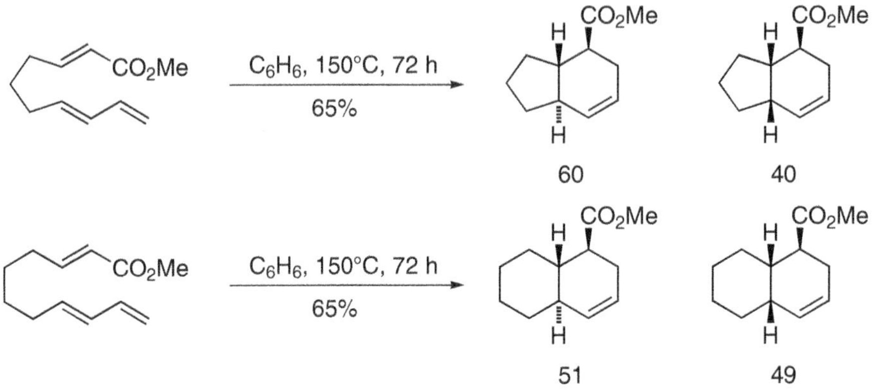

(6.16)

Carbene Additions: [2 + 1] Cycloadditions[42]

Few reactive intermediates have generated the interest that methylene, CH_2, which is the prototypical example of a carbene, has engendered among organic chemists, as we will see in much more detail in Chapter 13. The most useful addition reactions of carbenes are the concerted additions of singlet carbenes to alkenes. In these reactions, two carbon atoms from the alkene are involved in the reaction, and one carbon from the carbene; the reaction is usually called a [2 + 1] cycloaddition.

The HOMO of the singlet carbene is the sp^2 hybrid orbital carrying the lone pair, and the LUMO is the empty 2p orbital (Figure 6.19); the energy difference between the HOMO and LUMO of this singlet carbene is one of the smallest of any organic species. Because of its empty p orbital, the singlet carbene shows reactivity similar to a carbocation. It is an electrophile, and it will add to nucleophilic sites in a molecule. The lone pair of electrons in the sp^2 orbital confers the nucleophilic reactivity of a carbanion on the carbene, also, and carbenes can react with appropriate electrophiles in a molecule (although this reaction is less frequently observed). The addition itself is a concerted reaction, and it can be viewed as proceeding through the HOMO-LUMO overlap shown in Figure 6.19. In accord with the fact that alkenes do not usually react with nucleophiles, the initiating FMO overlap involves the carbene LUMO.

The [2 + 1] cycloaddition of singlet carbenes to alkenes is a suprafacial addition, so that the *E* alkene gives the *E* cyclopropane, and the *Z* alkene gives the *Z* cyclopropane. You may notice the similarity of the FMO overlap between the alkene and the singlet carbene to overlap between an alkene and a halogen or a peracid to generate a three-membered cyclic heteronium ion or an epoxide. These reactions, also, are suprafacial [2 + 1] cycloadditions where the LUMO is an antibonding orbital of a halogen-halogen or oxygen-oxygen bond instead of an empty atomic orbital.

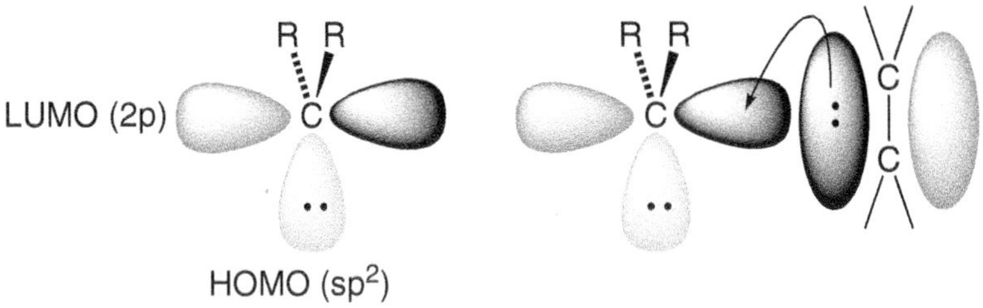

Figure 6.19 The frontier molecular orbitals (FMOs) of a singlet carbene (left) and initiating FMO overlap in the addition of a singlet carbene to an alkene. HOMO, highest energy occupied molecular orbital; LUMO, lowest energy unoccupied molecular orbital.

42. For a review of stereoselective cyclopropanation reactions, see Lebel, H.; Marcoux, J.-F.; Molinaro, C.; Charette, A.B. *Chem. Rev.* **2003**, *103*, 977.

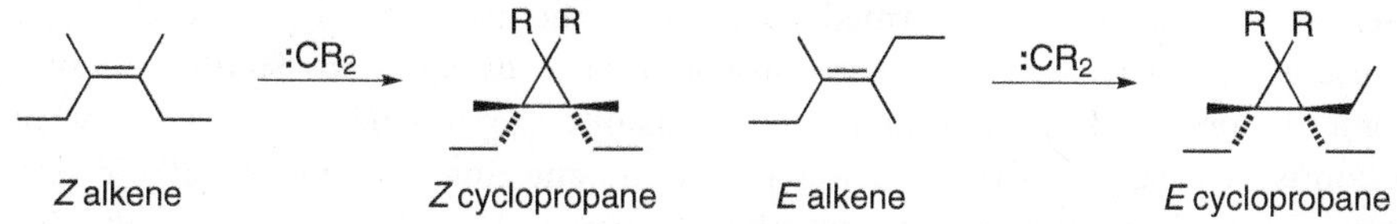

The most common method for the generation of singlet carbenes is α-elimination, where both groups are lost from the same carbon atom instead of adjacent carbons, as in the eliminations we have studied so far. The simplest example of an α elimination reaction that leads to a carbene is the reaction between chloroform and a strong base (Example **6.17**). When chloroform reacts with a strong base such as potassium *tert*-butoxide in the presence of an alkene such as cyclohexene, a *gem*-dichlorocyclopropane is formed. The reaction proceeds through :CCl_2, dichlorocarbene, which is generated by the α-elimination of chloride from the CCl_3^- ion, itself formed by deprotonation of chloroform.[43]

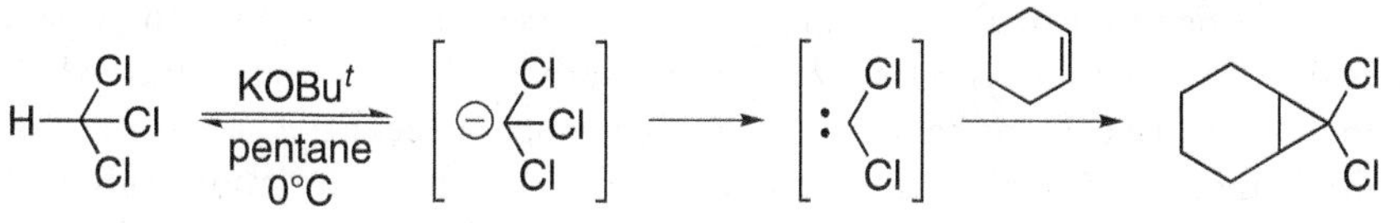

(6.17)

A second method for the production of carbenes is α-elimination from *gem*-dihalides (Example **6.18**). When a *gem*-dihalide reacts with butyllithium, a metal-halogen exchange reaction occurs to give an intermediate α-haloorganolithium reagent. At quite low temperatures (above −80°C), this intermediate loses the lithium halide to give the carbene:[44]

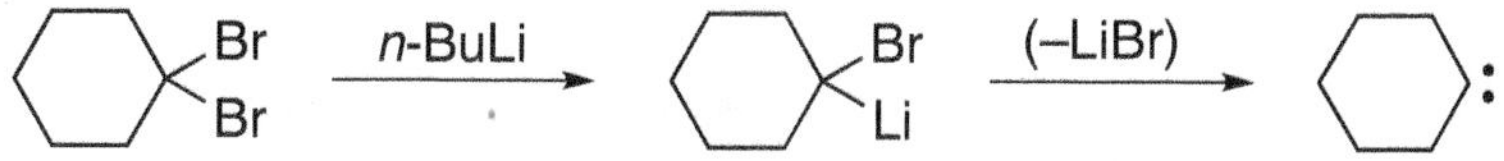

(6.18)

A closely related method for the production of carbenoids is the Simmons-Smith reaction,[45] named for the two chemists who first developed it.[46] The Simmons-Smith cyclopropanation reaction (Example **6.19**) involves an α-halomethylzinc halide,[47] which is generated by the reaction of the dihalide with activated zinc metal (in the form of a zinc-copper couple) or a dialkylzinc.[48] The most common form in which this reaction is used is in the addition of methylene itself, CH_2, to an alkene. Thus, when methylene iodide is treated with the zinc-copper couple in a suitable solvent (diethyl ether is commonly used), the α-iodoorganozinc iodide analogous to a Grignard reagent is formed. In the

---

43. (a) Hine, J. *J. Am. Chem. Soc.* **1950**, *72*, 2438. (b) Hine, J.; Dowell, A.M., Jr. *J. Am. Chem. Soc.* **1954**, *76*, 2688. (c) Hine, J.; Dowell, A.M., Jr.; Singley, J.E., Jr. *J. Am. Chem. Soc.* **1956**, *78*, 479.

44. (a) Hoeg, D.F.; Lusk, D.I.; Crumbliss, A.L. *J. Am. Chem. Soc.* **1965**, *87*, 4147. (b) Taylor, K.G.; Chaney, J. *J. Am. Chem. Soc.* **1972**, *94*, 8924. (c) Goldstein, M.J.; Dolbier, W.R., Jr., *J. Am. Chem. Soc.* **1965**, *87*, 2293.

45. Simmons, H.E.; Smith, R.D. *J. Am. Chem. Soc.* **1959**, *81*, 4256.

46. **Howard Ensign Simmons, Jr. (1929–1998)** established the existence of benzyne under J.D. Roberts at the Massachusetts Institute of Technology (PhD, 1954). He joined E.I. DuPont de Nemours and Company and rose to become Vice President and Senior Science Advisor of the company. He received the American Chemical Society's 1994 Priestley Medal. For more biographical information, see: Roberts, J.D.; Collette, J.W. *Biogr. Mem. Nat. Acad. Sci.* **1999**, *76* (e-book; http://www.nap.edu/readingroom.php?book=biomems&page=hsimmons.html; retrieved July 3, 2013).

Ronald Dean Smith (1930–) was educated at the Massachusetts Institute of Technology (PhD, 1955). After graduating, he followed his fellow student, Howard Simmons, to DuPont. He ended his career at the DuPont Merck Pharmaceutical Company, a joint venture of Merck and Du Pont, and served on the board of editors of *Drug Development Research* from 1994–2004.

47. (a) Emschwiller, G. *Compt. rend.* **1929**, *188*, 1555. (b) Emschwiller, G. *Compt. rend.* **1926**, *183*, 665.

48. (a) Furukawa, J.; Kawabata, N.; Nishimura, J. *Tetrahedron Lett.* **1966**, 3353. (b) Furukawa, J.; Kawabata, N.; Nishimura, J. *Tetrahedron* **1968**, *24*, 53. (c) Ito, Y.; Fujii, S.; Masashi, N.; Kawamoto, F.; Saegusa, T. *Org. Syn. Coll. Vol. 6* **1988**, 327. (d) Charette, A.B.; Lebel, H. *Org. Syn. Coll.Vol. 8* **2004**, 613.

presence of an alkene, this intermediate transfers methylene to give the cyclopropane. Because of its particularly mild reaction conditions, the Simmons-Smith reaction has become the preferred method for the synthesis of cyclopropanes. The Simmons-Smith reaction is especially useful for fusing a cyclopropane ring onto a cyclic allylic alcohol because the product is the isomer in which the carbene adds to the same face of the ring as the hydroxyl group (which coordinates to the zinc atom).

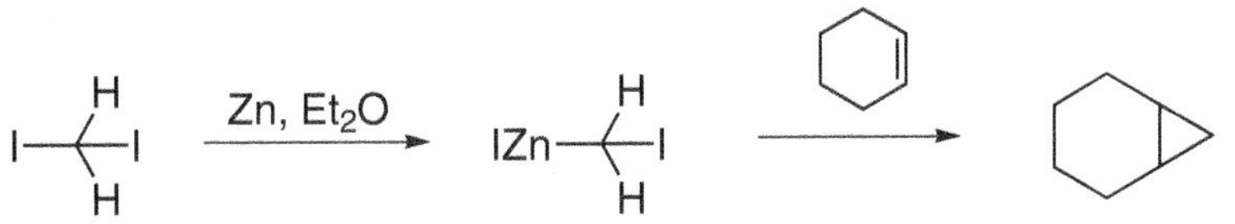

(6.19)

The final method for the formation of singlet carbenes or carbenoids that we will discuss here is the decomposition of diazoalkanes, $R_2C{=}N_2$, by ultraviolet light. Under such conditions, a very reactive singlet carbene is produced that readily adds to any alkene to give the cyclopropane product, although the reaction course may be complicated by participation of the triplet carbene. A rather less reactive carbenoid species is obtained when one decomposes the diazoalkane catalytically using a copper (free metal, as in Example **6.20**, or salt) or rhodium (usually the diacetate) catalyst. Recently, chromium carbonyl complexes (see Example **6.21**) have found use as catalysts for decomposing diazoalkanes, giving carbenoid species with approximately the same reactivity as the copper carbene complexes.

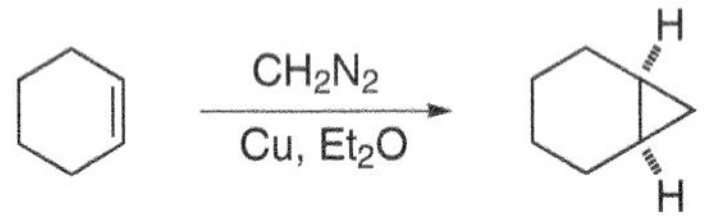

(6.20)

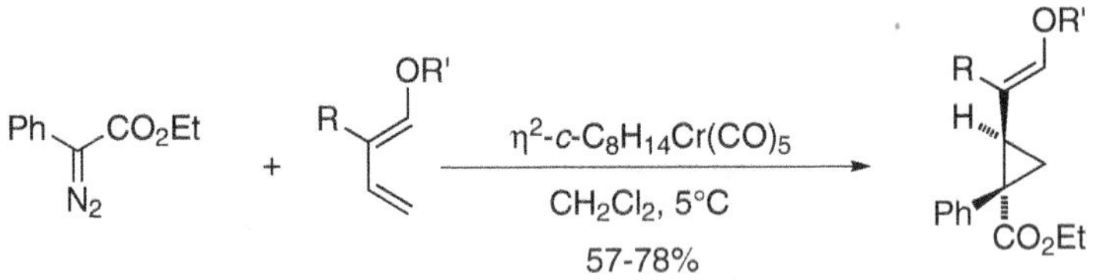

(6.21)

Intramolecular addition of a carbene to a diene (Example **6.22**), proceeds to give the products of [2 + 1] cycloaddition—not the products of [4 + 1] cycloaddition,[49] a result that is also observed in the intermolecular reaction.[50]

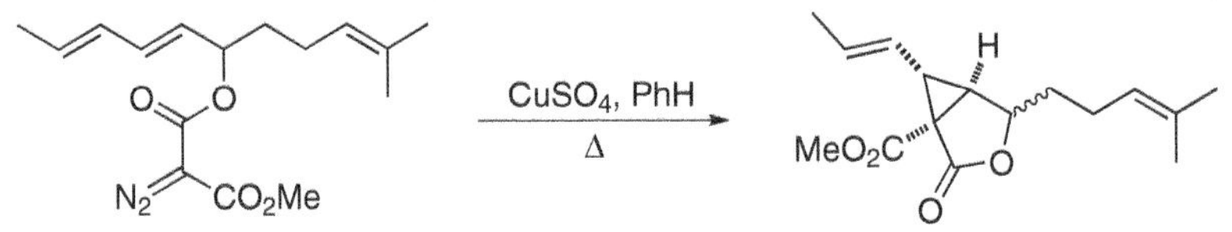

(6.22)

Ketene Additions: [2 + 2] Cycloadditions[51]

Under most circumstances, [2 + 2] cycloadditions to form four-membered rings do not occur readily at room temperature. There is one exception to this, however, and that occurs

49. Hudlicky, T.; Govindan, S.V.; Frazier, J.O. *J. Org. Chem.* **1985**, *50*, 4166.

50. (a) Hahn, N.D.; Nieger, M.; Dötz, K.H. *Eur. J. Org. Chem.* **2004**, 1049. (b)Aggarwal, V.K.; de Vicente, J.; Bonnert, R.V. *Org. Lett.* **2001**, *3*, 2785. (c) Doyle, M.P.; Dorow, R.L.; Buhro, W.E.; Griffin, J.H.; Tamblyn, W.H.; Trudell, M.L. *Organometallics* **1984**, *3*, 44. (d) Anciauy, A.J.; Demonceau, A.; Noels, A.F.; Warin, R.; Hubert, A.J.; Teyssié, P. *Tetrahedron* **1983**, *39*, 2169. (e) Doyle, M.P.; Dorow, R.L.; Tamblyn, W.H.; Buhro, W.E. *Tetrahedron Lett.* **1982**, *23*, 2261. (f) Wenkert, E.; Goodwin, T.E.; Rau, B.C. *J. Org. Chem.* **1977**, *42*, 2137.

51. Review: Hyatt, J.; Raynolds, P.W. *Org. React.* **1994**, *45*, 159.

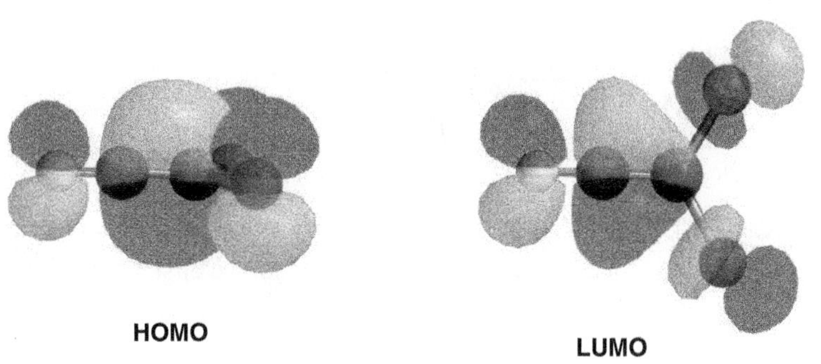

Figure 6.20 Frontier orbitals of dichloroketene. The molecule is oriented perpendicular to the page in the HOMO, and parallel to the page in the LUMO.

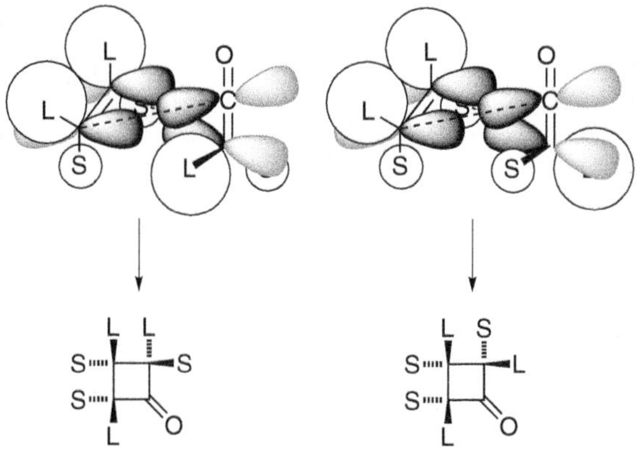

Figure 6.21 Stereochemistry of the cycloadducts of ketenes and alkenes is determined by the orthogonal approach of the two reactants. Only one lobe of the highest energy occupied molecular orbital and one of the lowest energy unoccupied molecular orbital are shown for clarity.

when the more electrophilic component of the reaction has a pair of cumulated double bonds. There are two carbonyl compounds that fulfill this structural requirement: the ketenes and isocyanates. In this chapter, we will concentrate our discussion on the ketenes, although most of what is said applies also to the additions of isocyanates to alkenes.

The reactivity of ketenes comes from the presence of the sp-hybridized carbon atom in the center of the functional group. This carbon, which is formally part of a carbonyl group, is particularly reactive towards nucleophiles and electron-rich π bonds.

The FMOs of dichloroketene, a widely used ketene for cyclobutane synthesis, are shown in Figure 6.20. It is worthwhile noting that the HOMO and LUMO of this species are orthogonal to each other, and this is critical to the reactivity of these species, since the addition is actually not a [2 + 2] cycloaddition, but rather, a $[_{\pi}2_s + _{\pi}2_s + _{\pi}2_a]$ cycloaddition that involves the HOMO of the alkene and both FMOs of the ketene.

The stereochemistry of the reaction is one of the most interesting aspects of the cycloadditions of ketenes to alkenes. Because both FMOs of the ketene are involved in the reaction, the cycloadditions of ketenes to alkenes require an orthogonal relationship between the two reacting species at the beginning of the reaction. This leads to the two possible orientations of approach shown in Figure 6.21, where "L" represents the larger of the two groups attached to a carbon atom, and "S" represents the smaller of the two groups. What becomes clear from examining Figure 6.21 is that this approach leads to the more hindered product because this initial approach places the smallest groups in closest proximity during the approach of the reactants.

Ketenes are almost always prepared in situ from acid chlorides by elimination of hydrogen chloride using triethylamine as the base, or from α-haloacid chlorides by reductive elimination using zinc metal as the reducing agent (Example **6.23**). The reactivity of the ketene in the cycloaddition reaction increases as the number of electron-withdrawing groups on the ketene double bond increases. For this reason ketenes such as dichloroketene are often used as the electrophilic reagent in [2 + 2] cycloadditions (Example **6.24**).

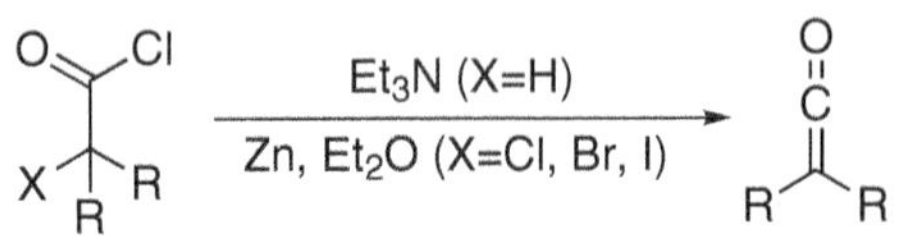

(6.23)

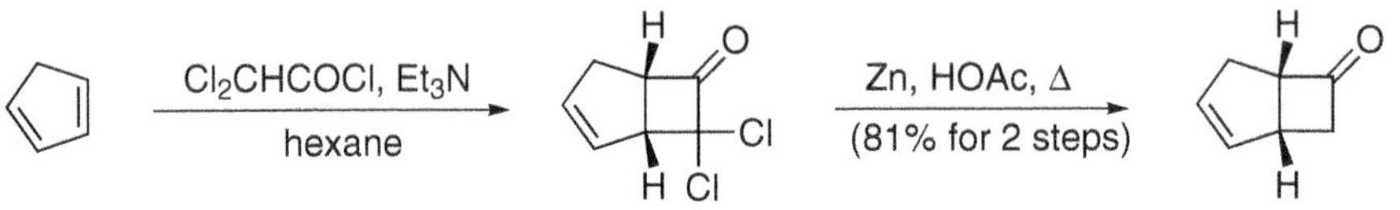

(6.24)

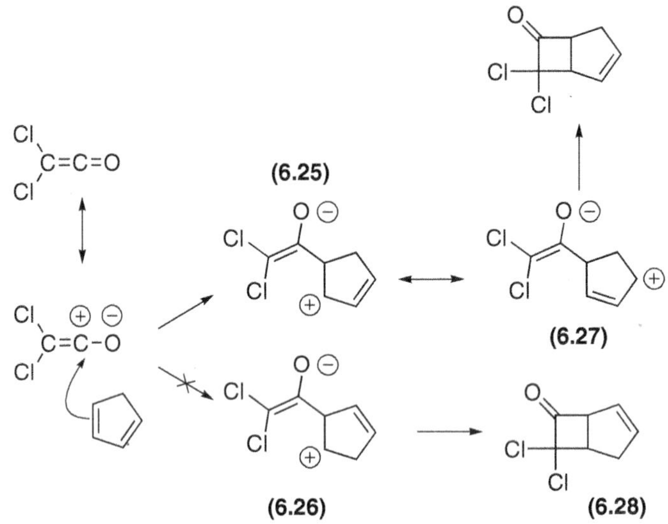

Figure 6.22 Regiochemistry of ketene cycloadditions to cyclopentadiene

Like the Diels-Alder and carbene cycloaddition reactions we have already discussed, ketene additions give the products of suprafacial addition to the alkene: *Z*-alkenes give *cis*-cyclobutanone products, and *E* alkenes give *trans*-cyclobutanone products. Like the Diels-Alder reaction, the ketene cycloaddition can also give rise to regioisomeric products, and, like the Diels-Alder reaction, one can usually predict the major product of the cycloaddition by using resonance arguments similar to those which we used above.

One can predict the regiochemistry of the ketene [2 + 2] cycloaddition of a ketene to an alkene by invoking of the addition of the carbocation carbon of the minor canonical form to the alkene in a stepwise reaction: The addition of the carbonyl carbon should obey Markovnikov's rule, so the carbonyl carbon should add to the alkene to give the more stable of the two possible carbocation intermediates (Figure 6.22). The addition of dichlorocarbene to cyclopentadiene is illustrative. The carbonyl carbon adds to the diene to give the resonance-stabilized allylic cation rather than the simple secondary cation, so the top regioisomer is formed almost exclusively.

Ketenes are highly reactive species, and in the absence of an electron-rich alkene or other nucleophilic species, they dimerize. Although ketene itself dimerizes to the β-lactone derivative known as diketene (Example **6.29**), alkylated ketenes dimerize to cyclobutanedione derivatives (Example **6.30**). It is for this reason that ketenes are usually prepared in the presence of the ketenophile.

(6.29)

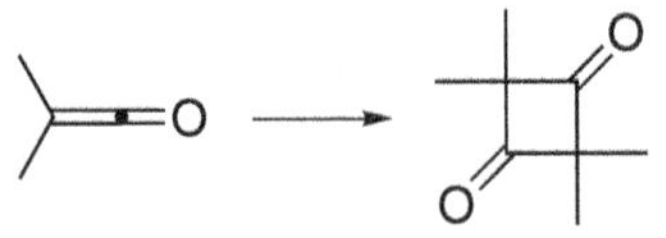
(6.30)

Worked Problem

6-2 What combination of diene and dienophile is needed to make each of the following compounds? Where appropriate, be sure to specify stereochemistry.

(a) 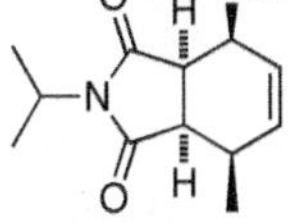

(b) 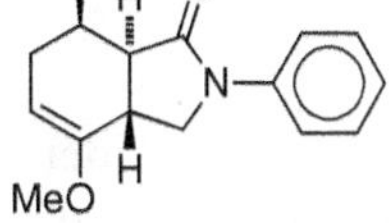

§Answers below

Problems

6-10 Predict the major product of each of the following Diels-Alder reactions.

(a) 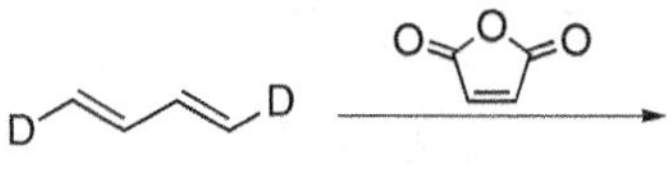

(b) 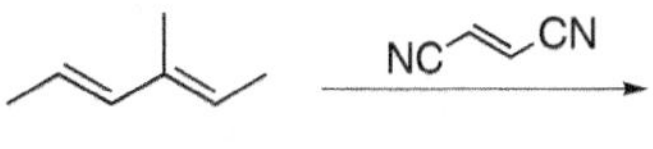

(c) 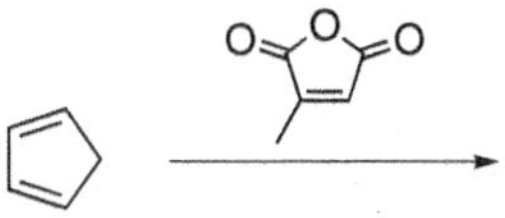

(d) 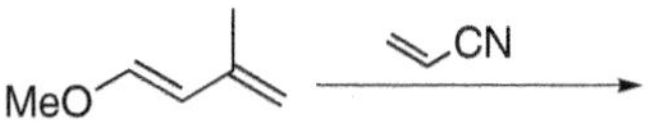

(e) 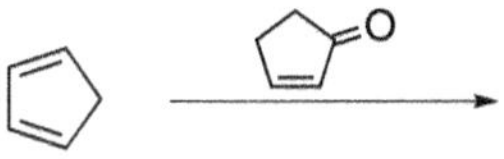

(f) 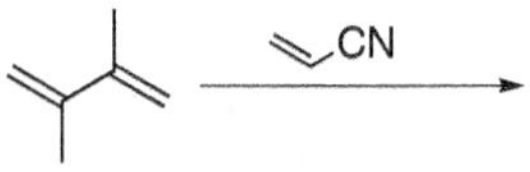

(g) 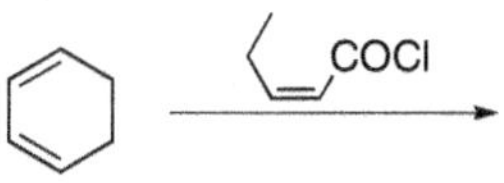

(h) 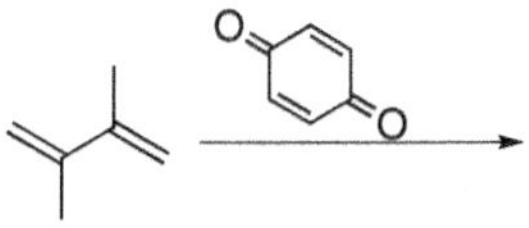

1 mole 1 mole

(i) 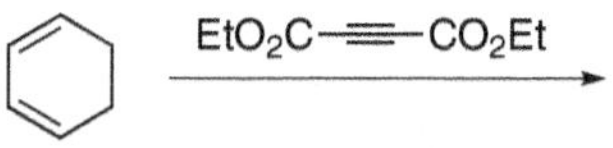

1 mole 1 mole

(j) 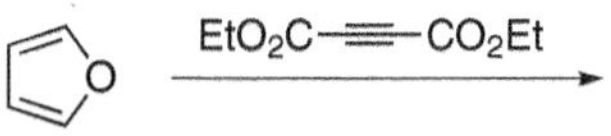

1 mole 1 mole

6-11 What combination of diene and dienophile is needed to make each of the following compounds?

(a) 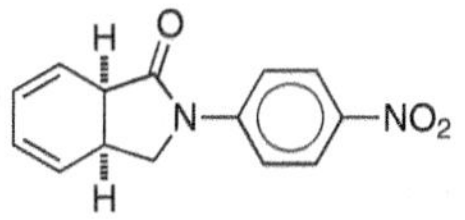

(b) 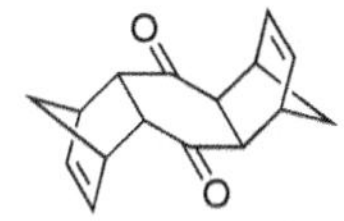

(c) 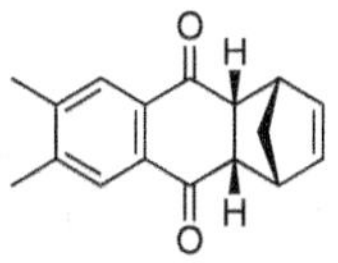

(continues)

§ Answer to Worked Problem:

(a) 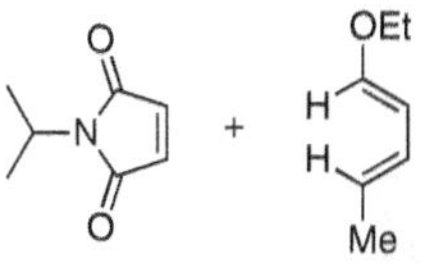

(b) 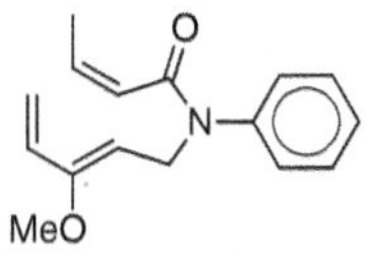

(a) This product is formed by the intermolecular Diels-Alder reaction between the imide, which must have the *Z* stereochemistry, and the diene. From the stereochemistry of the product, we see that if the reaction proceeds with suprafacial stereochemistry, the diene must have the *E*,*E* stereochemistry. The product shown is the expected *endo* product.

(b) This product is formed by the intramolecular Diels-Alder reaction shown. Note that the stereochemistry of the ring junction is *trans*; this means that the ring junction cannot come from just one part of the reactant. The one stereochemical deduction we can make unequivocally is that the dienophile part of the molecule has *Z* stereochemistry.

(**Problems** continued)

(d) 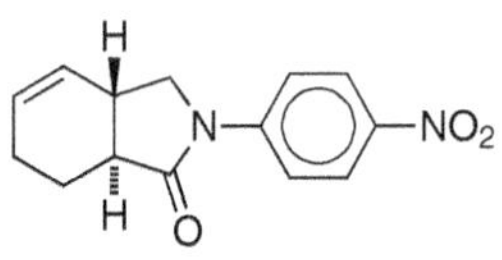

(e) 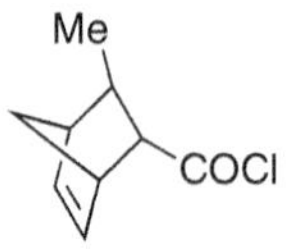

(f) 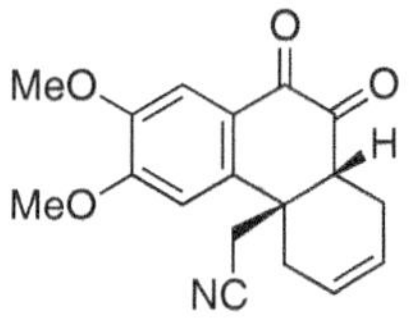

6-12 The attempted Diels-Alder reactions below do not give the products one might expect but the products shown, instead. Write a mechanism to rationalize the formation of this product and indicate what reactions are involved at each step.

(a) 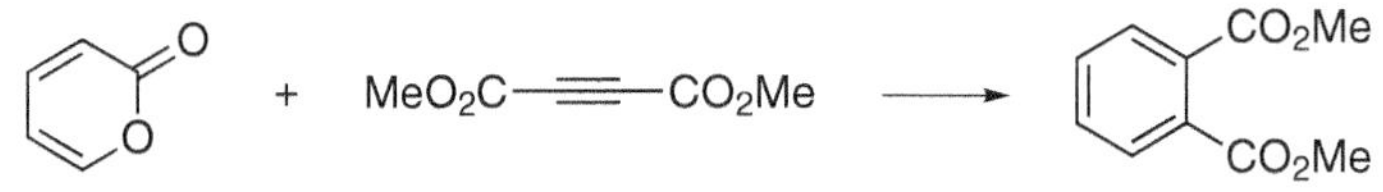

(b) 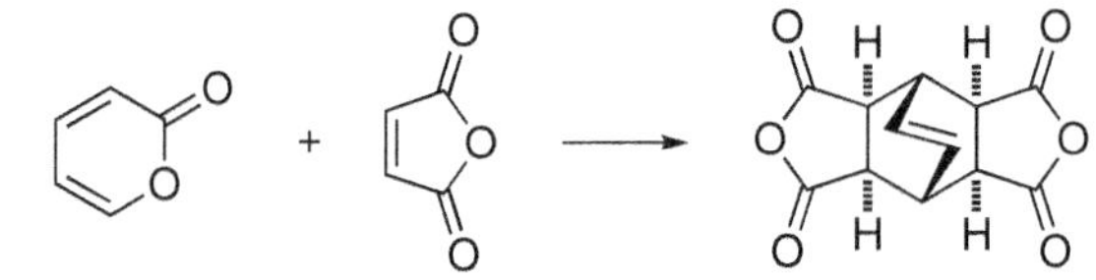

(c) 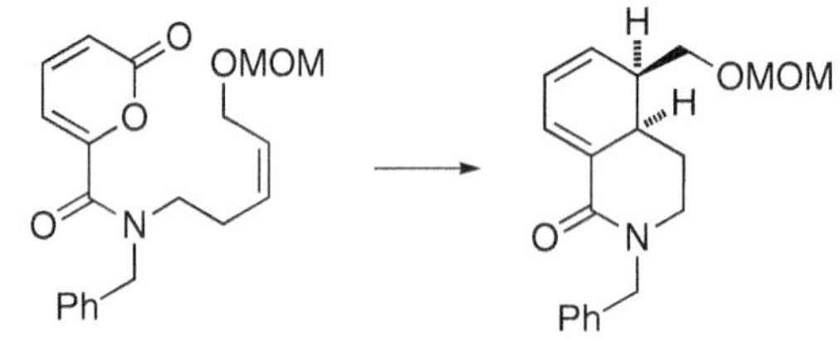

6-13 What is the major organic product of each of the following reactions? Give all isomers where more than one may form.

(a) 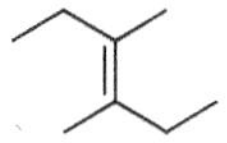 $CHCl_3$, KOBu^t / pentane →

(b) 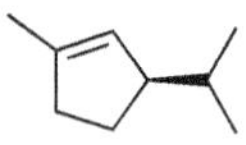 $CHCl_3$, KOBu^t / pentane →

(c) 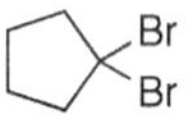 BuLi, THF, −78°C / cyclohexene →

(d)  CH_2N_2 / Cu, Et_2O →

(e) 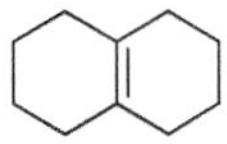 $CHCl_3$, KOBu^t / cyclohexane →

(f) 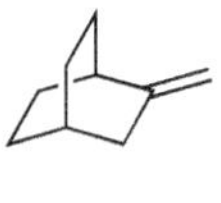 CH_2I_2, Zn(Cu) / Et_2O →

(g) 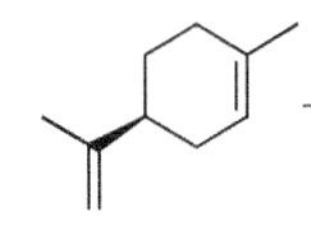 $CHCl_3$, KOBu^t / cyclohexane →

(h)  CH_2I_2, Zn(Cu) / Et_2O →

(i) 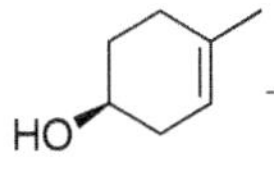 CH_2I_2, Et_2Zn / Et_2O →

(j) 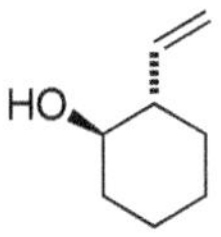 CH_2N_2 / $Rh_2(OAc)_4$ →

(k) 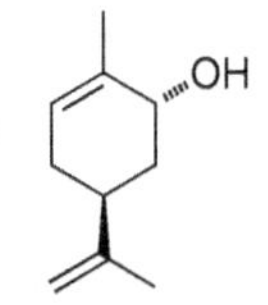 CH_2I_2, Et_2Zn / Et_2O (1 equiv.) →

(l) 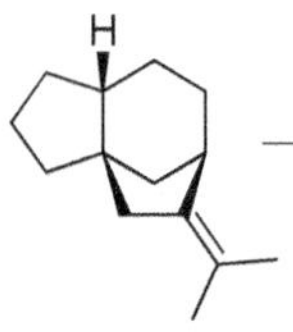 CH_2N_2 / $Rh_2(OAc)_4$ →

6-14 What is the major organic product of each of the following reactions? Where more than one regio- or stereoisomer is possible, give all isomers (draw only one enantiomer of racemates).

(a) 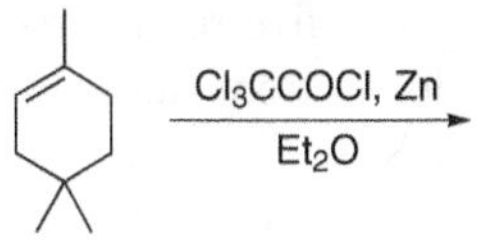

(b) 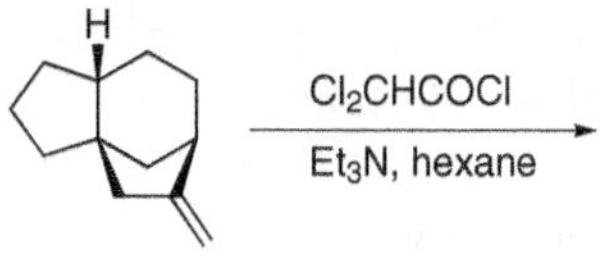

(c) 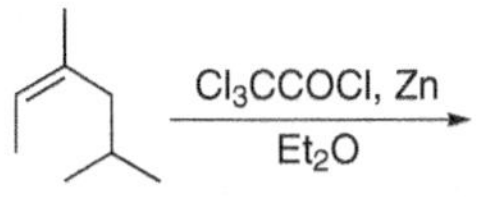

(d) 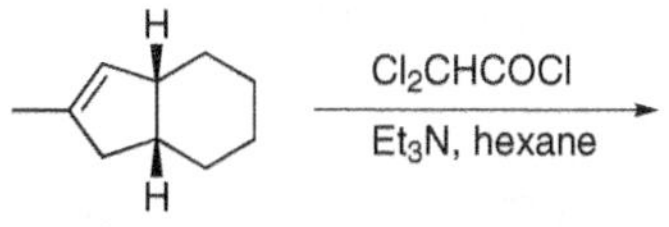

(e) 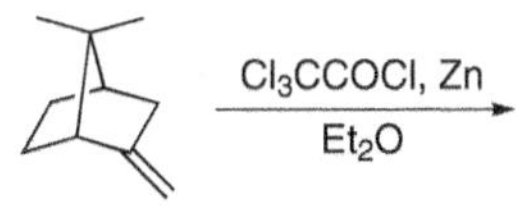

(f) 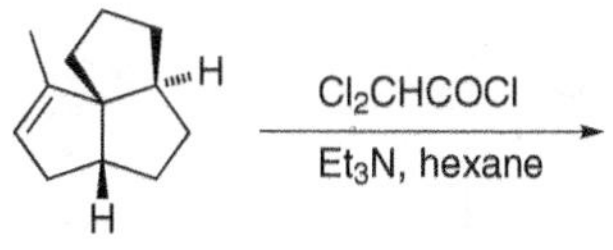

Reaction Synopses

Diels-Alder Reaction ([4 + 2] Cycloaddition)

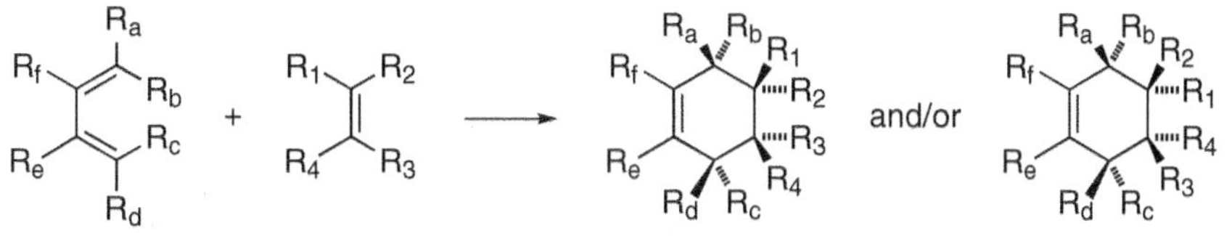

Reagents: 1,3-dienes; usually with electron-donating groups (e.g., R, RO, R_2N); alkenes (dienophile); usually with electron-withdrawing groups (e.g., C=O, C≡N, NO_2, SO_2)

Stereochemistry: suprafacial with respect to both participants; *endo* orientation preferred for conjugated dienophiles

Regiochemistry: unsymmetrical dienophiles react with unsymmetrical dienes to give mainly 1,4- or 3,4-disubstituted cyclohexenes

Carbene Cycloaddition ([2 + 1] Cycloaddition)

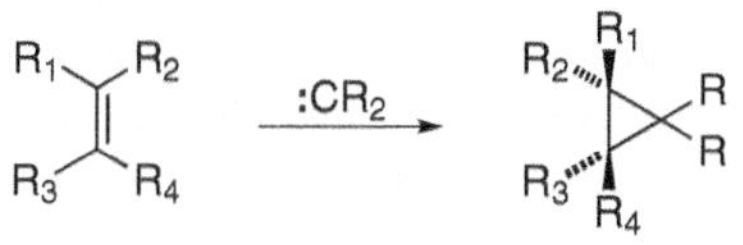

Reagents: $CHCl_3$, $Me_3CO^-K^+$; R_2CBr_2, BuLi, Et_2O; and so on (α elimination); or CH_2N_2, Cu; CH_2N_2, $[Rh(OCOCH_3)_2]_2$ and so on; or CH_2I_2, Zn(Cu), Et_2O; CH_2I_2, Et_2Zn; etc. (Simmons-Smith Reaction)

Stereochemistry: suprafacial with respect to alkene

[2 + 2] Cycloaddition of Ketenes

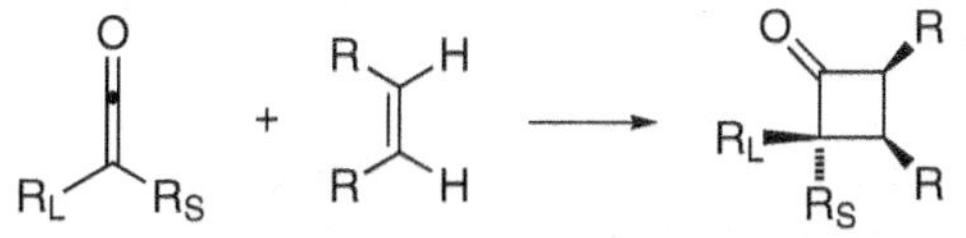

Reagents: $R_2CHCOCl$, Et_3N, Et_2O; $R_2CClCOCl$, Zn, Et_2O
Stereochemistry: suprafacial with respect to alkene
Regiochemistry: Markovnikov addition based on ketene carbonyl group as the electrophile

1,3-Dipolar Additions[52]

An important type of cycloaddition reaction that we have not yet discussed involves the concerted addition of a resonance-stabilized zwitterion to an alkene (e.g., Examples **6.31** and **6.32**) or alkyne (e.g., Examples **6.33** and **6.34**) to give a five-membered heterocyclic product. These additions are also known as [3 + 2] cycloadditions, although, in terms of the electrons involved, they are actually $[_{\pi}4_s + _{\pi}2_s]$ additions, with the zwitterionic reactant acting at the four-electron component. The zwitterionic reactant is known as a ***1,3-dipole***, and a list of common 1,3-dipoles is gathered in Table 6.7.

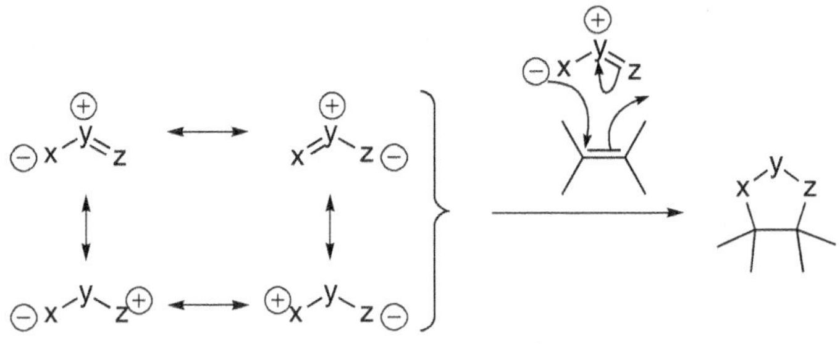

(6.31)

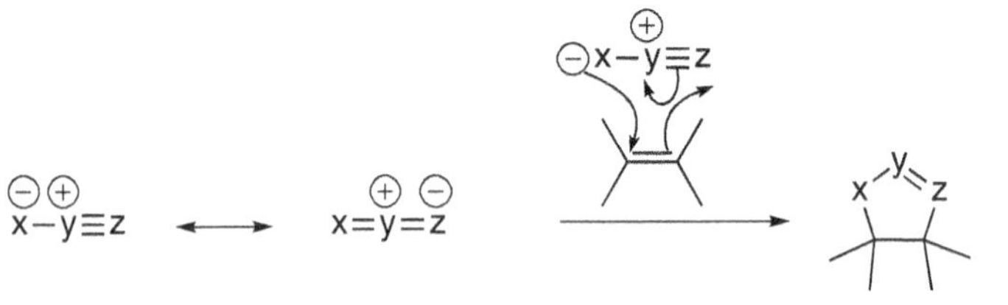

(6.32)

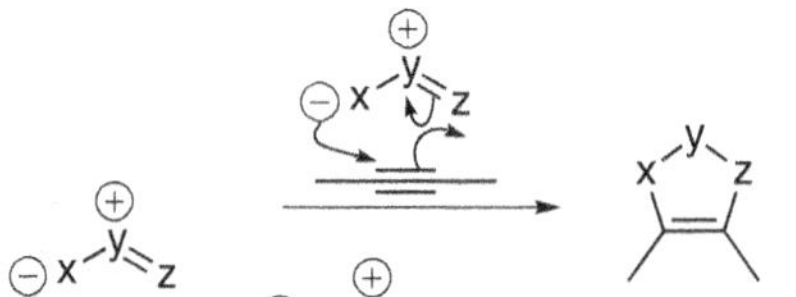

(6.33)

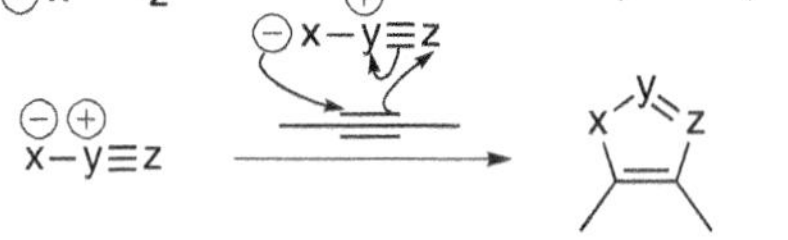

(6.34)

1,3-Dipolar cycloadditions are suprafacial with respect to the alkene component (the "dipolarophile"), as illustrated by the addition of *p*-anisyl azide to dimethyl fumarate (Example **6.35**)[53] and the addition of *p*-nitrophenyl azide to the two geometric isomers of 1-propoxy-1-propene (Example **6.36**).[54] Like the Diels-Alder reaction, these cycloadditions tend to be favored by electron-withdrawing substituents on the alkene (the dipolarophile). The facile inversion of configuration of nitrogen means that we cannot establish the theoretically predicted suprafacial nature of the reaction with respect to the dipole.[55]

---

52. (a) Huisgen, R. *Angew. Chem. Int. Ed. Engl.* **1963**, *2*, 565. (b) Huisgen, R. *Angew. Chem. Int. Ed. Engl.* **1963**, *2*, 633. (c) Huisgen, R.; Grashey, R.; Sauer, J. In Patai, S., Ed. *The Chemistry of Alkenes* (Wiley-Interscience: NewYork, 1964), ch. 11. (d) Huisgen, R. *Bull. Soc. chim. France* **1965**, 3431. (e) Huisgen, R. *J. Org. Chem.* **1976**, *41*, 403. (e) Firestone, R.A. *Tetrahedron* **1977**, *33*, 3009. (f) Bianchi, G.; de Micheli, C.; Gandolfi, R. In Patai, S., Ed. *Chemistry of Double-Bonded Functional Groups Part 1* (Wiley-Interscience: New York, 1977), p. 369. (g) Black, D. St.C.; *Synthesis* **1975**, 205. (h) L'Abbé, G. *Chem. Rev.* **1969**, *69*, 345. (i) Kauffmann, T. *Angew. Chem. Int. Ed. Engl.* **1965**, *4*, 553. (i) Gothelf, K.V.; Jørgensen, K.A. *Chem. Rev.* **1998**, *98*, 863.

53. Huisgen, R.; Szeimies, G.; Möbius, L. *Chem. Ber.* **1966**, *99*, 475.

54. Huisgen, R.; Szeimies, G. *Chem. Ber.* **1965**, *98*, 1153.

55. For a cogent discussion of the stereochemistry of this addition and the controversy leading up to general acceptance of the concerted mechanism, see: Houk, K.N.; Gonzalez, J.; Li, Y. *Acc. Chem. Res.* **1995**, *28*,1 81.

Table 6.7 Representative 1,3-Dipoles

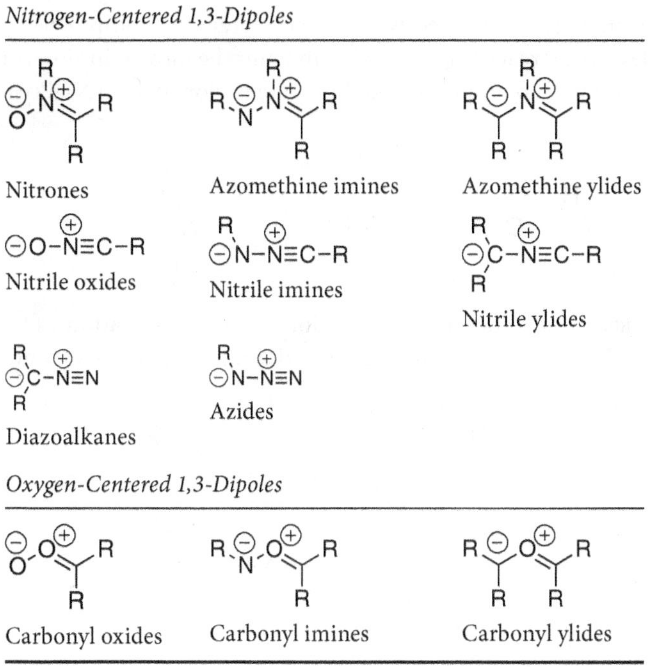

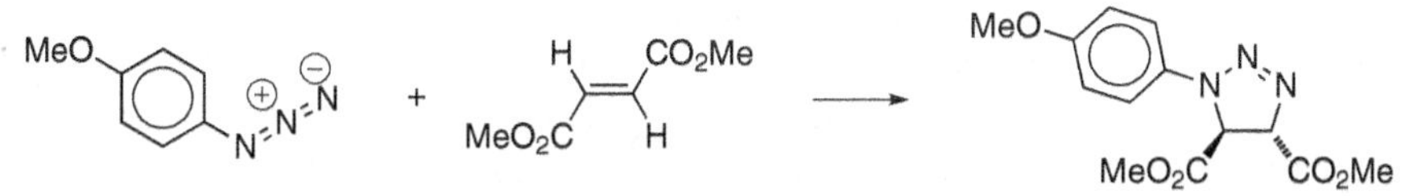

(6.35)

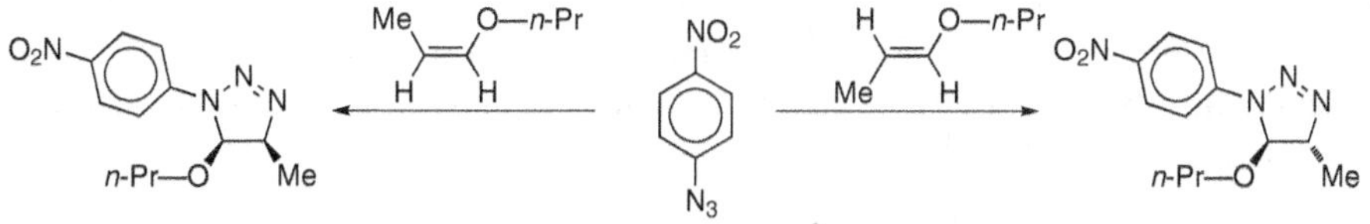

(6.36)

Intramolecular 1,3-cycloaddition reactions are useful reactions for the construction of heterocyclic systems, as illustrated by the two examples below, which are from Kozikowski's synthesis of prostaglandins (Example **6.37**),[56] and from Tufiarello's synthesis of cocaine.[57]

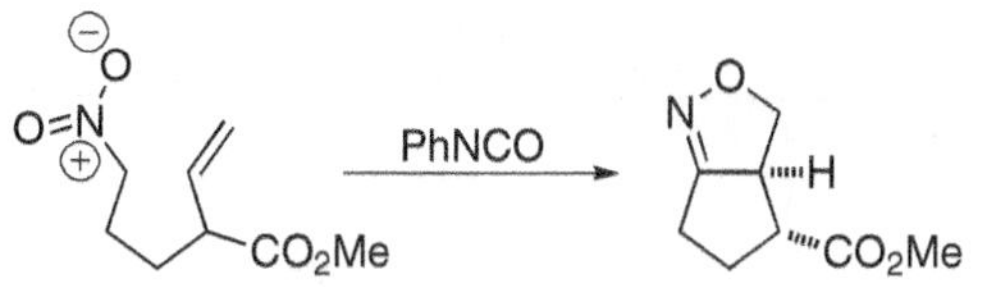

(6.37)

Kozikowski

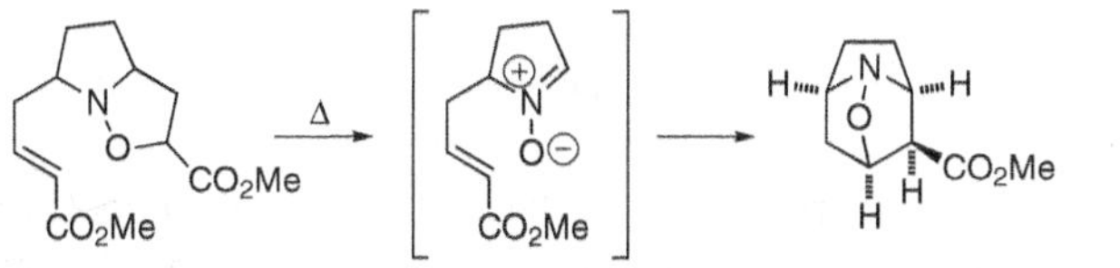

(6.38)

Tufiarello

56. Kozikowski, A.P.; Stein, P.D. *J. Am. Chem. Soc.* **1982**, *104*, 4023
57. Tufiarello, J.J.; Mullen, G.B. *J. Am. Chem. Soc.* **1978**, *100*, 3638.

Generation of 1,3-Dipoles

Azides and nitrones are stable species that can be prepared and purified before use, but other 1,3-dipoles are too reactive to be stored and must be formed in situ. Nitrones are normally prepared from carbonyl compounds by condensation with an *N*-alkylhydroxylamine (Example **6.39**).

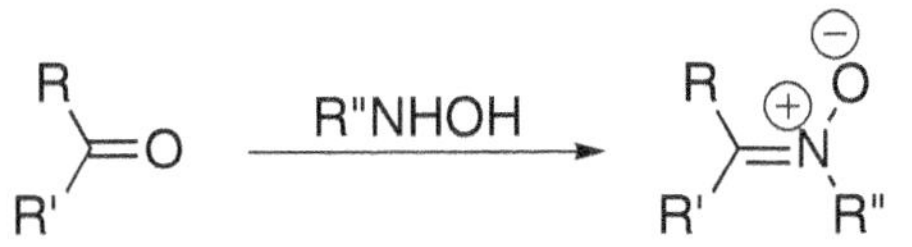

(6.39)

Nitrile oxides are generated by two major methods: oxidation of an aldoxime by hypochlorite or dehydration of a nitroalkane with phenyl isocyanate and triethylamine (Example **6.40**).

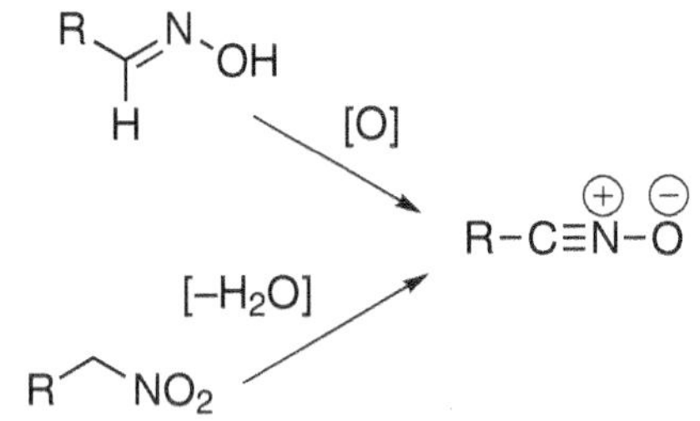

(6.40)

Azomethine ylides (Example **6.41**) and carbonyl ylides (Example **6.42**) are formed by the same reaction: the electrocyclic ring opening of a three-membered heterocycle. When heated above 100°C, aziridines carrying electron-withdrawing substituents on the ring carbons give azomethine ylides, and similar epoxides give carbonyl ylides by a conrotatory ring opening. These ring systems undergo disrotatory ring opening photochemically. The 1,3-dipoles can be trapped in situ with appropriate dipolarophiles. 1-Alkyl-3-hydroxypyridinium ions (Example **6.43**) are representatives of a particularly accessible class of azomethine ylides, and these compounds have been used in the synthesis of tropane derivatives.

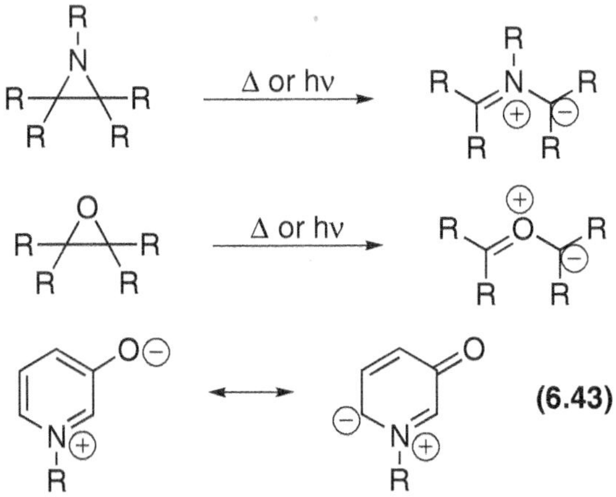

(6.41)

(6.42)

(6.43)

Diazoalkanes[58] are generally prepared by the base-promoted decomposition of a nitrosoamine amide. For example, the treatment of *N*-methyl-*N*-nitroso-*p*-toluenesulfonamide (Example **6.44**, Z = Ts, R = H) with potassium hydroxide yields diazomethane.

---

58. (a) Regitz, M.; Maas, G. *Diazo Compounds* (Academic Press: New York, 1986). (b) Black, T.H. *Aldrichimica Acta* **1983**, *16*, 3. (c) Cowell, G.W.; Ledwith, A. *Quart. Rev. Chem. Soc.* **1970**, *24*, 119. (d) Smith, P.A.S. *Open-Chain Nitrogen Compounds* (W.A. Benjamin: New York, 1966).

An alternative approach to diazoalkanes is the oxidation of hydrazones with HgO, Ag_2O, MnO_2 or $Pb(OAc)_4$ (Example **6.45**).

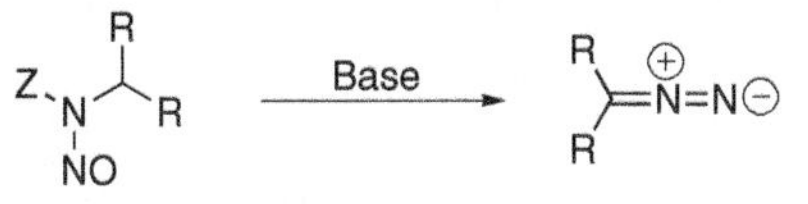

(6.44)

Z = RCO, ROCO, H_2NCO, $ArSO_2$, etc.

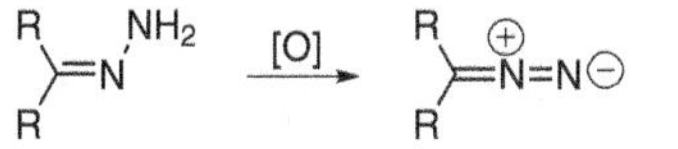

(6.45)

1,3-Dipolar Cycloadditions and "Click Chemistry"

In 2001, Sharpless and his coworkers introduced the term "click chemistry" to describe synthetic reactions that conformed to the set of criteria in Table 6.8.[59] These reactions, which were included because of their usefulness in both large- and small-scale reactions and their adaptability to modular assembly of complex molecules have become indispensable methods for the formation of heterocyclic pharmaceuticals.

Table 6.8 Criteria for "Click" Reactions

Reaction Characteristic
1. Modular
2. Wide scope
3. High yield
4. Inoffensive byproducts or no byproducts
5. Stereospecific (enantiospecific if possible)
6. Simple reaction conditions (insensitive to water and oxygen)
7. Readily available starting materials and reagents
8. No solvent or benign solvent (e.g., water) that is easily removed
9. Simple product isolation (no chromatography)
10. High thermodynamic driving force ($\Delta H° \leq -20$ kcal mol^{-1})

Among the most widely used "click" reactions are ring openings of strained compounds such as epoxides and aziridines, and—important in this present discussion—1,3-dipolar cycloadditions. Any online search under "click chemistry" now reveals hundreds of recent examples of these reactions, many of them syntheses of pharmaceutical intermediates. It is interesting to note that in a very short period of time, this new terminology has replaced Huisgen's original nomenclature.

The reaction between azides and terminal alkynes or cyano groups activated by an electron-withdrawing group[60] is a particularly useful reaction for the synthesis of triazoles or tetrazoles, which are important ring systems in pharmaceuticals. The control of the regiochemistry of the reaction between terminal alkynes and azides had been problematic until the discovery that this reaction is catalyzed by copper (I) salts.[61] The reaction has become even more useful with the discovery that the copper (I) catalyst can be generated in situ from copper (II) sulfate and sodium ascorbate, avoiding the formation of diacetylene coupling products and other byproducts. The effects of the copper (I) catalyst on the course of the reaction are illustrated by Reactions **6.46** and **6.47**. In the absence of the catalyst, a mixture of regioisomers is obtained, while in the presence of the catalyst, only a single regioisomer is formed.

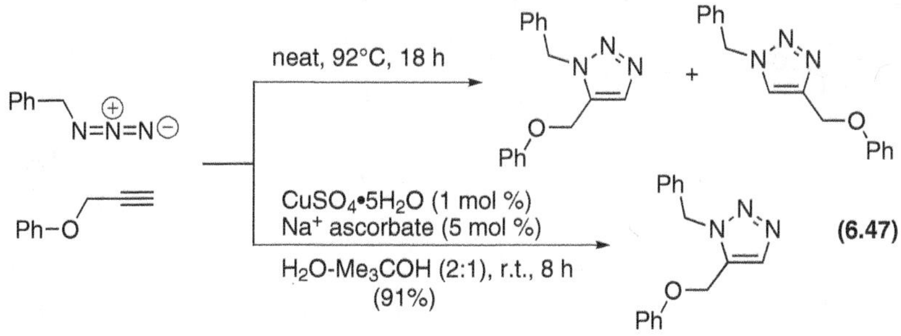

(6.46)

(6.47)

59. Kolb, H.C.; Finn, M.G.; Sharpless, K.B. *Angew. Chem. Int. Ed. Engl.* **2001**, *40*, 2004.

60. (a) Demko, Z.; Sharpless, K.B. *Angew. Chem. Int. Ed. Engl.* **2002**, *41*, 2110. (b) Demko, Z.; Sharpless, K.B. *Angew. Chem. Int. Ed. Engl.* **2002**, *41*, 2113.

61. (a) Rostovtsev, V.; Green, L.G.; Fokin, V.V.; Sharpless, K.B. *Angew. Chem. Int. Ed. Engl.* **2002**, *41*, 2596. (b) Tornøe, C.W.; Christensen, C.; Meldal, M. *J. Org. Chem.* **2002**, *67*, 3057.

Reaction Synopses

1,3-Dipolar Cycloaddition ("Click")

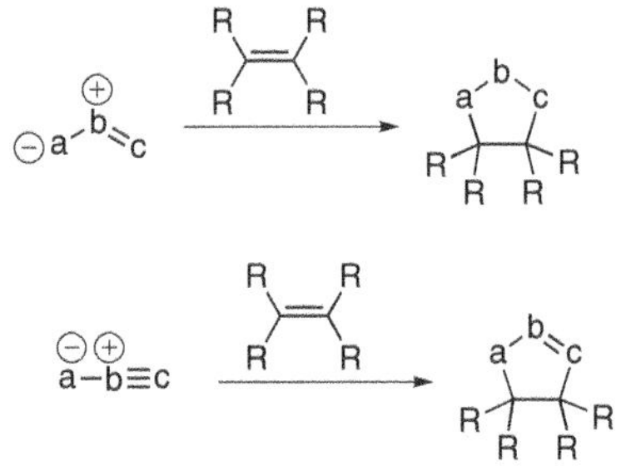

Reagents:	RN_3; RN_3, Cu^+, and so on
	R_2CHNO_2, PhNCO; $R_2C{=}NOH$, OCl^- (nitrile oxide)
	$R_2C{=}O$, RNHOH (nitrone)
	$R_2C{=}N_2$
Stereochemistry:	suprafacial

Problems

6-15 Predict the products of each of the following reactions.

(a) 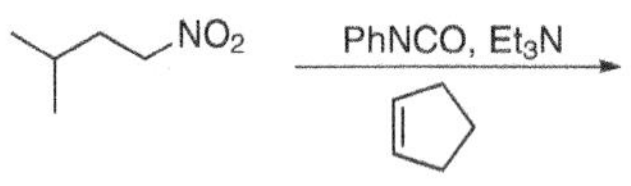

(b) 

(c) 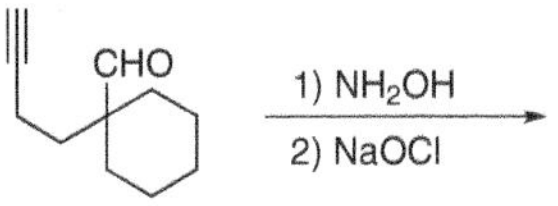

(d) 

(e) 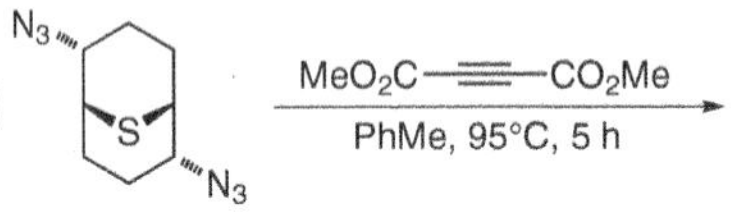

6-16 The following reaction sequence involves a 1,3-dipole intermediate. Suggest the structure of the intermediate, and the structure of the final product of the reaction sequence.

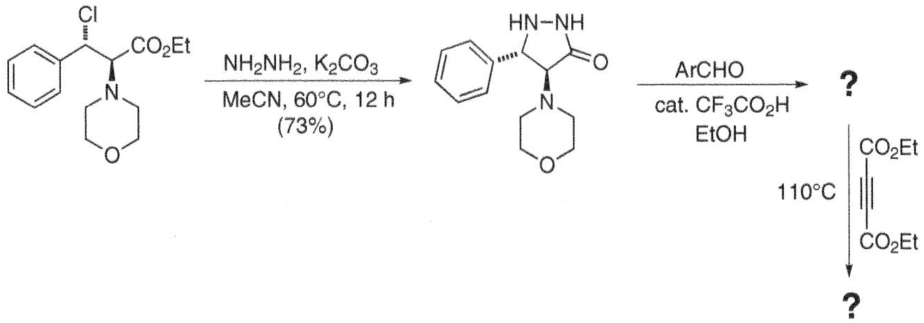

6.6 Ene, Retro-Ene, and Similar Reactions

The ene reaction (Example **6.48**), also known as the Alder ene reaction,[62] may be viewed as a hybrid of the Diels-Alder [4 + 2] cycloaddition reaction and a [1,5]-sigmatropic rearrangement, in that it is a pericyclic reaction leading to formation of a new C–C σ bond through a six-membered cyclic transition state with concomitant group transfer of the group at the allylic position. The compound carrying the allylic group is known as the ***ene***, and the compound with the π bond to which the allylic group is transferred is known as the ***enophile***. The group that is transferred in the ene reaction is usually electropositive, with hydrogen and metals (especially magnesium) being most widely used. The permissible structural variation in the enophiles is quite large, with alkenes, alkynes, and carbonyl compounds having been particularly widely used as enophiles in this reaction.

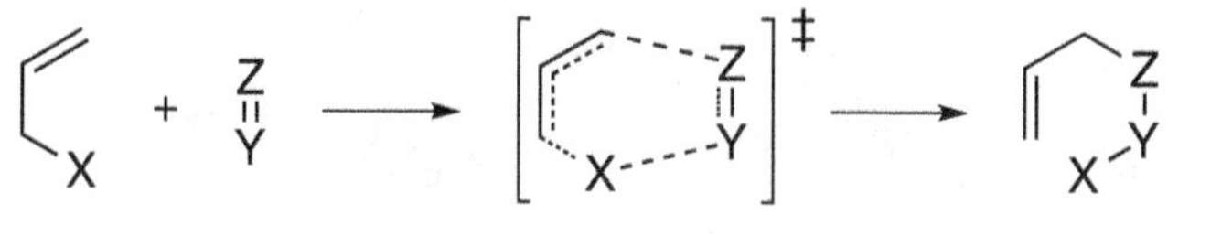

(6.48)

X = H, metal
XY= C=C, C≡C, C=O, C=S, C=N, N=N, N=O, Se=O, etc.

In common with the Diels-Alder reaction, where the reaction is favored by electron-deficient dienophiles and electron-rich dienes, the ene reaction is favored by electron-deficient enophiles and electron-rich enes. For this reason, ene reactions are often carried out using electron-deficient carbonyl groups (as in formaldehyde or α-dicarbonyl compounds) or sulfonylimines as enophiles, and are promoted by Lewis acid catalysts. Alkynes make better enophiles than alkenes, whereas enols, being very electron-rich, are also highly reactive in the ene reaction, as evidenced by the synthesis of the interesting natural product, modhephene.[63] Some typical examples of the ene reaction are provided by Examples **6.49**, **6.50**, and **6.51**. Example **6.51** is interesting because it involves the tautomerization of the ketone to the enol before the ene reaction.

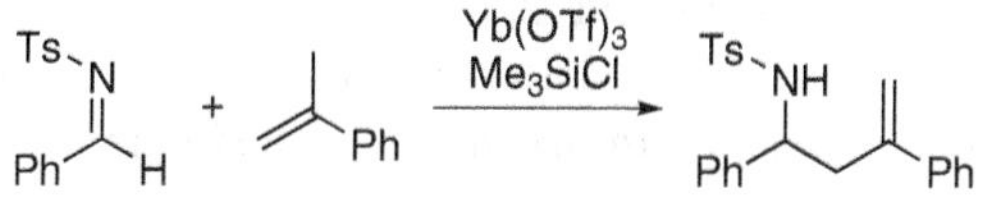

(6.49)[63]

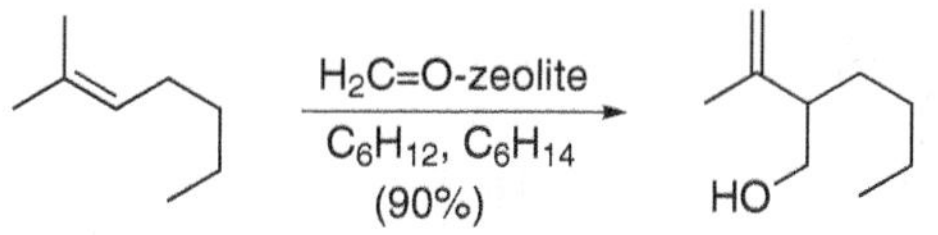

(6.50)[64]

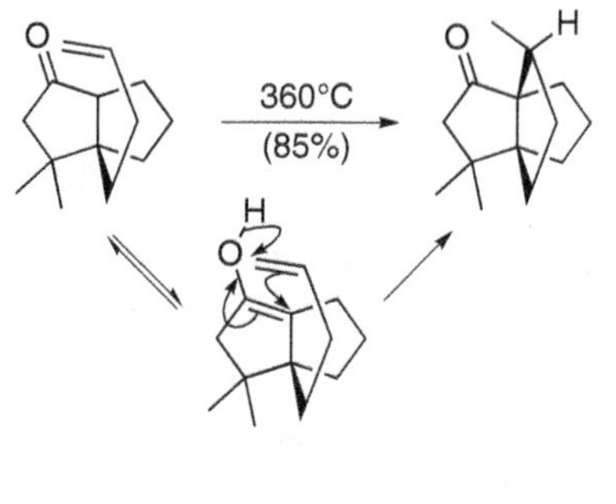

(6.51)[65]

62. (a) Alder, K.; Pascher, F.; Schmitz, A. *Ber. dtsch. chem. Ges.* **1943**, *76*, 27.
Reviews: (b) Oppolzer, W. *Angew. Chem. Int. Ed. Engl.* **1989**, *28*, 38. (c) Oppolzer, W. In Trost, B.M.; Fleming, I., Eds. *Comprehensive Organic Synthesis* (Pergamon Press: Oxford, 1991), *5*, 29.
63. Yamanaka, M.; Nishida, A.; Nakagawa, M. *Org. Lett.* **2000**, *2*, 159.
64. Okachi, T.; Onaka, M. *J. Am. Chem. Soc.* **2004**, *126*, 2306.
65. Schostarez, H.; Paquette, L.A. *Tetrahedron* **1981**, *37*, 4431.

One especially important class of ene reactions involves the reaction of an alkene with singlet oxygen to give an allyl hydroperoxide (Example **6.52**).[66] The reaction does proceed with suprafacial stereochemistry, as required by the concerted ene mechanism, but there is strong evidence that the reaction actually occurs by initial formation of a perepoxide (or an epoxide *O*-oxide, analogous to an amine *N*-oxide) intermediate.[67] By either mechanism, the net result of the reaction is the same: transfer of hydrogen from the allyl position to oxygen and allylic rearrangement of the alkene π bond. It is also generally observed that singlet oxygen tends to abstract hydrogen from more congested side of the alkene, as shown in the example at left, where the hydrogen is removed from the more hindered benzyl group, rather than the more sterically accessible methyl group. This phenomenon is known as the *cis* effect.[68] The suprafacial stereochemistry of the reaction has been shown by using a chiral deuterated substrate.[69]

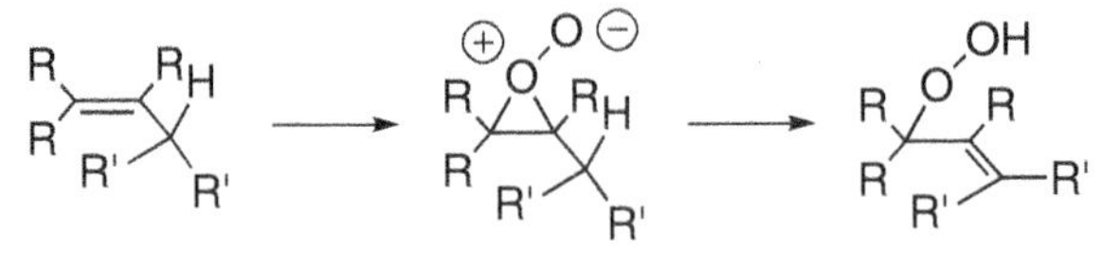

(6.52)

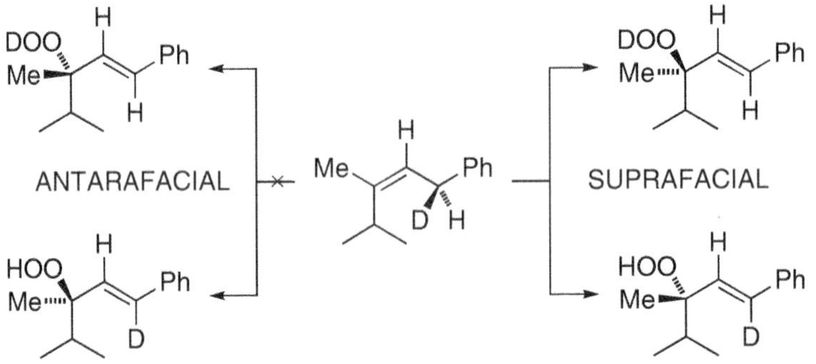

Another ene reaction of importance in organic synthesis is the ene reaction between selenium dioxide and alkenes (Example **6.53**). In this reaction, the product isolated is the allylic alcohol, formed by sequential ene and [2,3]-sigmatropic rearrangement reactions. The fact that two sequential pericyclic reactions are involved in the overall reaction means that the position of the double bond remains unaltered in the final product.

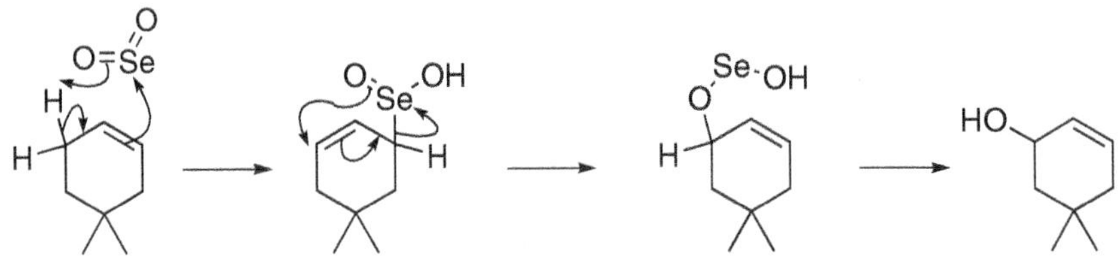

(6.53)

Cyclic Eliminations Through Six-Membered Transition States

Although we have focused our attention to this point on pericyclic reactions that form new C–C σ bonds, these are not the only useful reactions of this type. Pyrolysis

66. Reviews: (a) Wasserman, H.H.; Ives, J.L. *Tetrahedron* **1981**, *37*, 1825. (b) Clennan, E.L. *Tetrahedron* **2000**, *56*, 9151. (c) Prein, M.; Adam, W. *Angew. Chem. Int. Ed. Engl.* **1996**, *35*, 477.

67. (a) Hurst, J.R.; Wilson, S.L.; Schuster, G.B. *Tetrahedron* **1985**, *41*, 2191. (b) Gorman, A.A.; Hamblett, I.; Lambert, C.; Spencer, B.; Standen, M.C. *J. Am. Chem. Soc.* **1988**, *110*, 8053. (c) Song, Z.; Chrisope, D.R.; Beak, P. *J. Org. Chem.* **1987**, *52*, 3938. (d) Song, Z.; Beak, P. *J. Am. Chem. Soc.* **1990**, *112*, 8126.

68. (a) Orfanopoulos, M.; Bellarmine Grdina, Sr. M.; Stephenson, L.M. *J. Am. Chem. Soc.* **1979**, *101*, 275. (b) Schulte-Elte, K.H.; Rautenmstrauch, V. *J. Am. Chem. Soc.* 1980, *102*, 1738.

69. Orfanopoulos, M.; Stephenson, L.M. *J. Am. Chem. Soc.* **1980**, *102*, 1417.

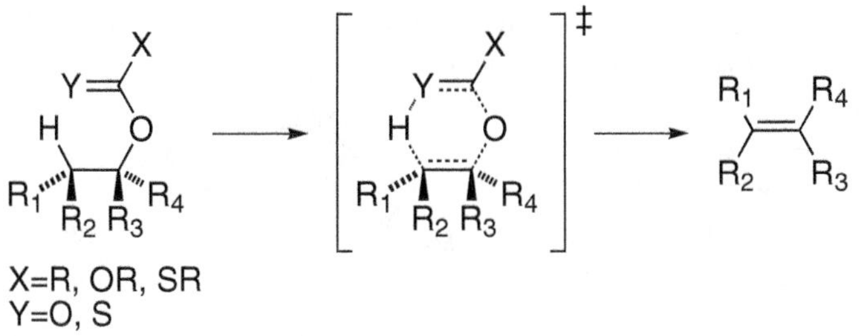

Figure 6.23 Pyrolysis reactions to give alkenes that may proceed by a retro-ene reaction involving [1,5]-sigmatropic rearrangement of hydrogen

reactions to give alkenes have been known for more than a century and a half.[70] Pyrolytic elimination reactions of esters and similar species are generally believed to proceed by a concerted, cyclic mechanism[71] and are therefore best described as retro-ene reactions. The concerted mechanism is not, however, unchallenged; Wertz and Allinger proposed a stepwise mechanism through a carbocation stabilized by the surface on which the reaction occurs.[72]

The retro-ene reaction of esters and similar species follows the general course shown in Figure 6.23. The classes of compounds that participate in this reaction are carboxylate,[73] xanthate esters,[74] and carbonate esters.[75]

Pyrolysis of esters proceeds with suprafacial stereochemistry to give the product of *syn* elimination. Thus, the two labeled esters in Examples **6.54** and **6.55** give different labeled *trans*-stilbenes as the sole products of the elimination.[76] The stereospecificity of this reaction makes it clear that the removal of the proton involves a *syn* elimination, and the simplest rationalization for this is provided by the retro-ene mechanism.

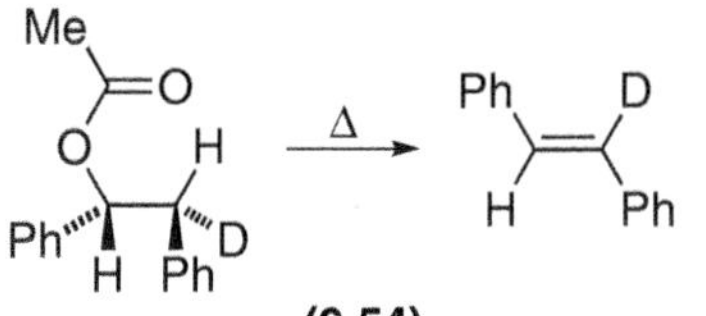

(6.54)

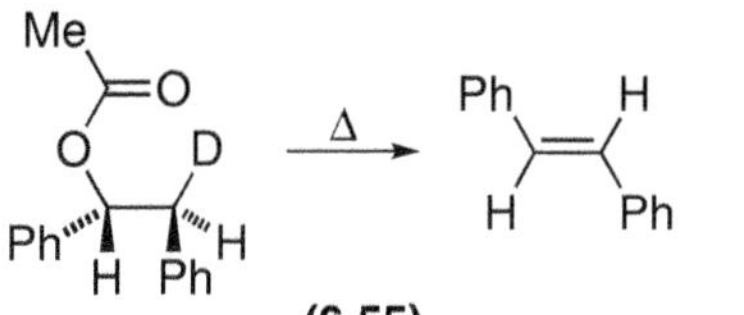

(6.55)

The sulfur analog of the ester pyrolysis reaction was reported by Russian chemist Lev Aleksandrovich Chugaev[77] in 1900.[74] In the Chugaev elimination, the simple ester is replaced by the xanthate ester, which is prepared from the alcohol by treatment of the alkoxide anion sequentially with carbon disulfide and methyl iodide. The advantage of the Chugaev elimination is that it occurs at substantially lower temperatures that the pyrolytic elimination of the corresponding acetate esters. The pyrolysis of the two xanthates in Example **6.56** by

70. Heintz, W. *Pogg. Ann. Phys. Cghem.* **1853**, *93*, 519.

71. Reviews: (a) DePuy, C.H.; King, R.W. *Chem. Rev.* **1960**, *60*, 431. (b) Smith, G.G.; Kelly, F.W. *Prog. Phys. Org. Chem.* **1971**, *8*, 75.

72. Wertz, D.H.; Allinger, N.L. *J. Org. Chem.* **1977**, *42*, 698.

73. (a) Ratchford, W.P.; Rehberg, C.E. *Org. Synth.* **1949**, *29*, 2; *Org. Synth.* **1955**, *Coll. Vol. 3* 30. (b) Benson, R.E.; McKusick, B.C.; Grummitt, O.; Budewitz, E.P.; Chudd, C.C. *Org. Synth.* **1958**, *38*, 78; *Org. Synth.* **1963**, *Coll. Vol. 4*, 746. (c) Putnam, R.E; Anderson, B.C.; Sharkey, W.H. *Org. Synth.* **1963**, *43*, 17; *Org. Synth.* **1973**, *Coll. Vol. 5*, 235.

74. Tschugaeff, L. *Ber. dtsch. chem. Ges.* **1900**, *33*, 3118.

75. O'Connor, G.L.; Nace, H.R. *J. Am. Chem. Soc.* **1953**, *75*, 2118.

76. Curtin, D.Y.; Kellom, D.B. *J. Am. Chem. Soc.* **1953**, *75*, 6011.

77. **Lev Aleksandrovich Chugaev (1873–1922)** was educated at Moscow University, (M Chem, 1903, Dr Chem, 1906) where he was a student of Markovnikov. After Markovnikov's death in 1904, he was appointed Adjunct at the Bacteriological Institute of Moscow University, and then, in 1908, Professor at St. Petersburg University. For more biographical details, see: Lewis, D.E. *Early Russian Organic Chemists and Their Legacy* (Springer: Heidelberg, 2012), p. 111.

Cram[78] provided clear evidence that this reaction also proceeds with *syn* stereochemistry and is therefore very likely a retro-ene reaction.

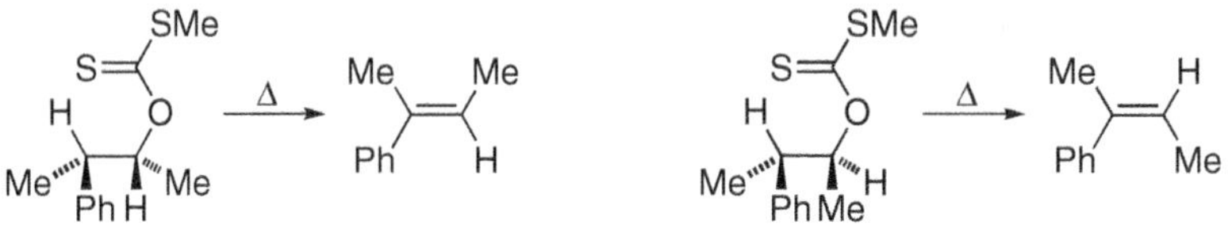

(6.56)

The ease with which pyrolytic eliminations of xanthates and related compounds occur varies with the identity of the leaving group, as shown in Example **6.57**. As the data show, the incorporation of a heteroatom on both sides of the carbonyl or sulfonyl group lowers the temperature for the reaction: simple esters eliminate at 400°C, whereas xanthates, carbonates, and sulfites all eliminate below 300°C.

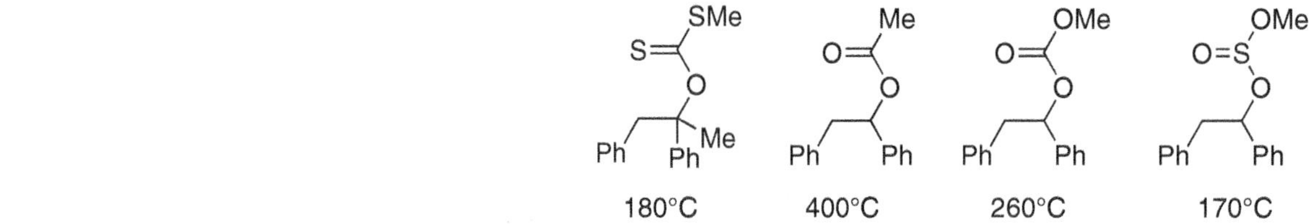

(6.57)

Problems

6-17 The decarboxylation of β-ketoacids may occur through a retro-ene mechanism. Write a reasonable mechanism for this reaction as a retro-ene reaction and indicate what is (are) the initial product(s) of the reaction.

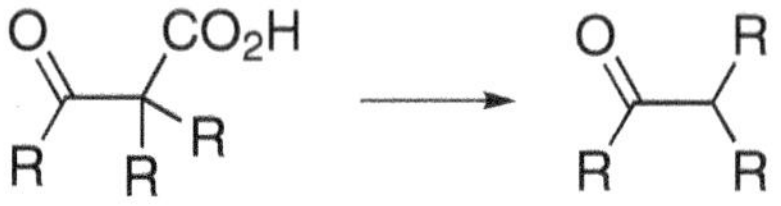

6-18 Predict the structure of the major organic product of the following reaction [*Tetrahedron Lett.* **1975**, 4053]:

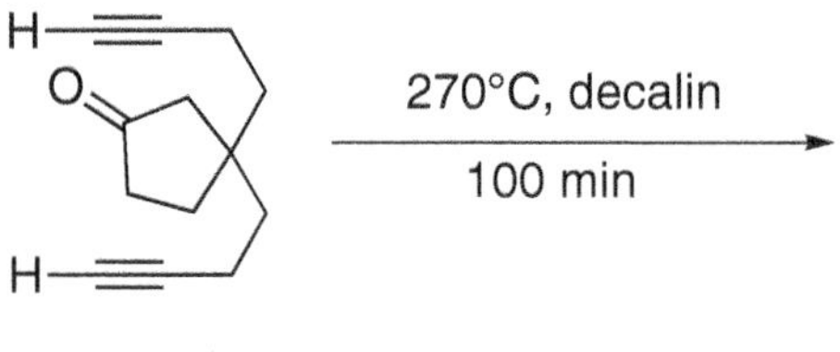

Cyclic Eliminations Through Five-Membered Transition States

There are three important elimination reactions that formally occur through a five-membered cyclic transition state. They share the common feature of rearrangement of hydrogen from carbon to a negatively charged atom, with concomitant neutralization of a positive charge by bond cleavage and formation of a double bond. The three species most widely used in this type

78. Cram, D.J. *J. Am. Chem. Soc.* **1949**, *71*, 3883.

of elimination are amine *N*-oxides (the Cope elimination [X=NR_2],[79] sulfoxides [X=S],[80] and selenoxides [X=Se]).[81]

Neither the Cope elimination nor the sulfoxide elimination is stereospecific, and both operate at some level by two competing mechanisms.[80a] At low temperatures, the cyclic mechanism (Example **6.58**) operates, and the reaction is stereospecific and *syn*. At higher temperatures, however, there is evidence that a free radical mechanism (Example **6.59**) also operates and that this mechanism is not stereospecific. The end result is that the stereospecificity of the reaction is lowered as the reaction temperature rises. The elimination of selenoxides occurs at or below room temperature, so these reactions are stereospecific and *syn*, as required by the cyclic mechanism. Then radical mechanism is especially favored by conjugating groups β to sulfur, as demonstrated by crossover experiments (which we will discuss at length in Chapter 8).

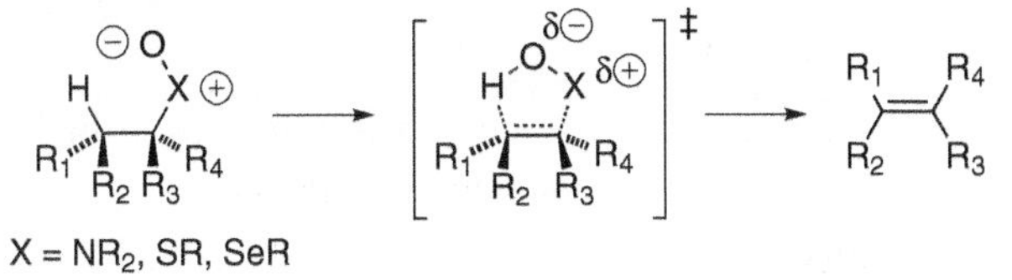

(6.58)

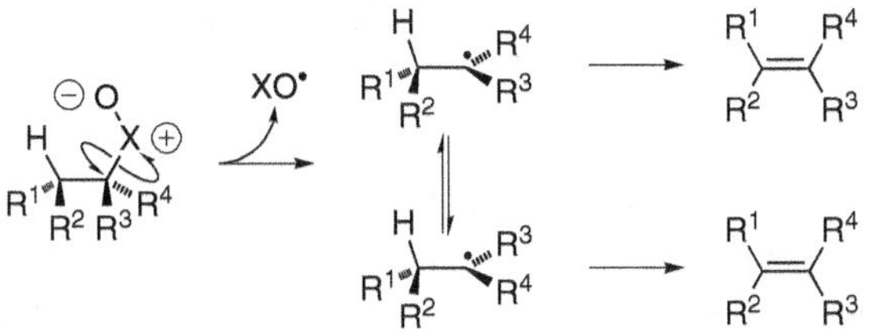

(6.59)

Reaction Synopses

Ene Reaction

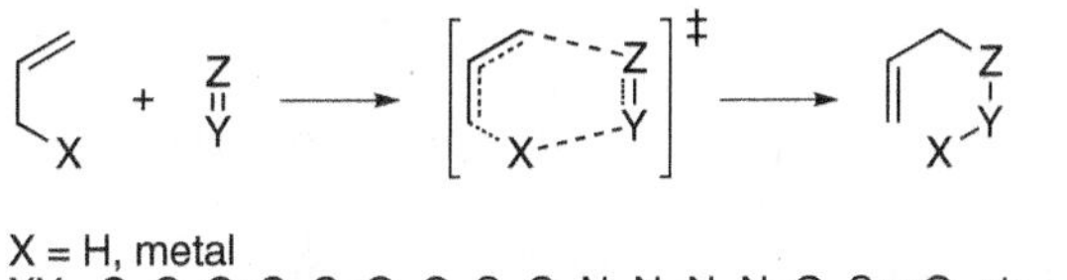

Reagents:

enophiles: $R_2C{=}CR_2$; RC≡CH; RCHO, L.A.; R_2CO, L.A; RNO; ArNO; SeO_2; and so on
O_2, rose Bengal, hν; and so on

enes: $R_2C{=}CH_2$; $R_2C{=}CR(OH)$; and so on
$R_2C{=}CR\text{-}CR_2M$ (e.g., M=Zn, Mg, Li)

Stereochemistry: suprafacial (*syn*)

Regiochemistry: hydrogen migration from most congested alylic position usually preferred in reactions with singlet oxygen

(continues)

79. Cope, A.C.; Foster, T.T.; Towle, P.H. *J. Am. Chem. Soc.* **1949**, *71*, 3929; **1953**, *75*, 3212.

80. (a) Kingsbury, C.A.; Cram, D.J. *J. Am. Chem. Soc.* **1960**, *82*, 1810. (b) Emerson, D.; Craig, A.; Potts, I. *J. Org. Chem.* **1967**, *32*, 102.

81. Sharpless, K. B.; Young, M. W.; Lauer, R. F. *Tetrahedron Lett.* **1973**, 1979.

(**Reaction Synopses** continued)

Pyrolytic Eliminations

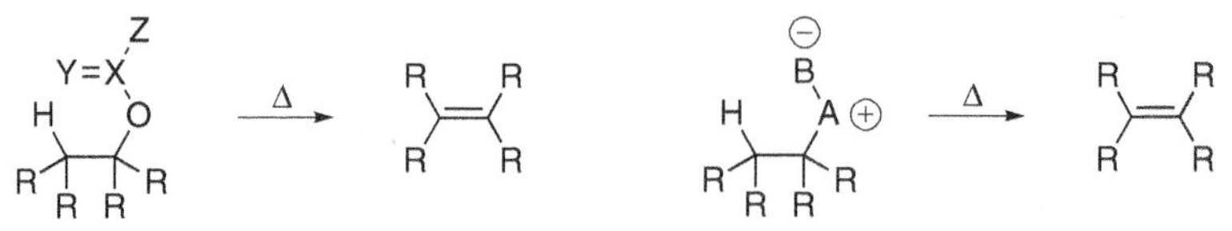

Ester pyrolysis: X=C, Y=O, Z=R
Carbonate pyrolysis: X=C, Y=O, Z=OR
Chugaev elimination (xanthate pyrolysis): X=C, Y=S, Z=SR
Sulfite pyrolysis: X=S, Y=O, Z=OR
Cope elimination: A=NR_2, B=O
Sulfoxide elimination: A=SR, Y=O
Selenoxide elimination: A=SeR, Y=O

Mechanism: concerted through a cyclic transition state; at high temperatures a stepwise mechanism may operate

Stereochemistry: suprafacial (*syn*)

6.7 Orbital Correlation Diagrams

In the area of pericyclic reactions more than any other, the use of ***orbital correlation diagrams*** can provide useful insights into the stereochemical and regiochemical course of the reaction. The basic premise behind these diagrams is simple: every orbital in the reacting species should correlate with a corresponding orbital in the product(s) of the reaction. Let us take the electrocyclization reaction as a first example. The diagrams for this reaction are the most easily constructed, and most widely used for discussing pericyclic reactions.

Electrocyclization

As for many of these reactions, it is usually simplest to generate the orbital correlation diagram from the perspective of the forward (bond-forming) reaction rather than the reverse reaction. So, in the electrocyclization reaction, we view the reaction as a cyclization, even if the reaction we are actually looking at may be the ring-opening. In this reaction, the net bonding change is to reduce the number of π bonds by one and to form a new σ bond, so the set of orbitals, ψ_1, ψ_2, ψ_3, and ψ_4, is lost and is converted into the new set of orbitals, σ, σ^*, π, and π^*. We can draw these orbitals in order of increasing energy, and we can also label them in terms of their symmetry with respect to the symmetry operation being conserved in the reaction. In a disrotatory reaction, the mirror plane of symmetry is preserved throughout the reaction, so the symmetry properties of the orbitals of the diene alternate, beginning with the symmetric orbital ψ_1. In the product, the two bonding orbitals are both symmetric with respect to the mirror plane of symmetry, and the two antibonding orbitals are antisymmetric. The labeling of the π orbitals is exactly the opposite for the conrotatory process, where the C_2 axis of symmetry is preserved, whereas the labels of the σ orbitals remain the same.

To complete the diagram, one simply correlates the orbitals of the reactant with the lowest energy available orbital of the same symmetry in the product, beginning with the lowest energy orbital of the reactant. This leads to the diagram in Figure 6.24, which shows that the HOMO of the reactant correlates with an antibonding orbital in the product in the disrotatory reaction, regardless of whether one looks at the reaction in the direction of

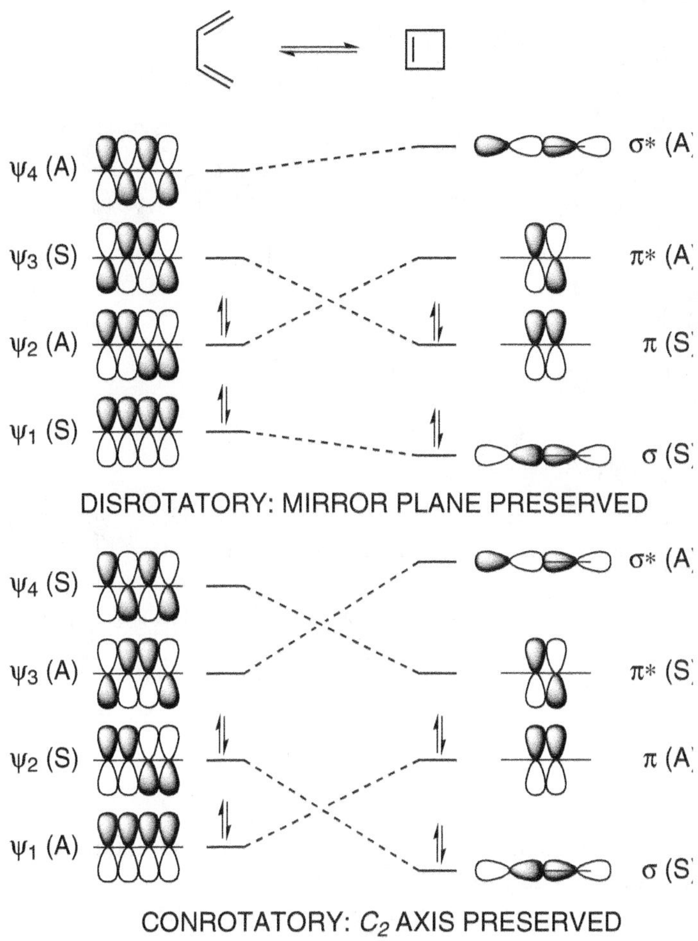

Figure 6.24 Orbital correlation diagrams for disrotatory (top) and conrotatory (bottom) electrocyclizations of 1,3-butadiene to cyclobutene

ring closure or the ring opening—an inherently endothermic process. This contrasts with what is found for the conrotatory process, where all the bonding orbitals in the reactant correlate with bonding orbitals in the product. Thus, if we were to predict the outcome of this reaction, we would predict that the disrotatory process would have a high activation energy—it would be symmetry-forbidden—and the conrotatory process would have a lower activation energy—it would be symmetry-allowed. This is what is found experimentally: conrotatory processes involving $4n$ electrons occur under thermal (ground-state) conditions, and disrotarory processes occur photochemically.

Problem

6-19 Construct orbital correlation diagrams analogous to Figure 6.23 for the electrocyclization of 1,3,5-hexatriene to 1,3-cyclohexadiene and the cyclization of 1,3,5,7-octatetraene to 1,3,5-cyclooctatriene.

Cycloadditions

Orbital correlation diagrams for cycloadditions can be constructed by a similar process. We will examine both the [2 + 2] cycloaddition and the [2 + 4] cycloaddition as the prototypes of cycloadditions involving $4n$ and $(4n + 2)$ π electrons. The cycloaddition is bimolecular, which means that there are now stereochemical consequences for both participating molecules, with three combinations of stereochemical outcomes being

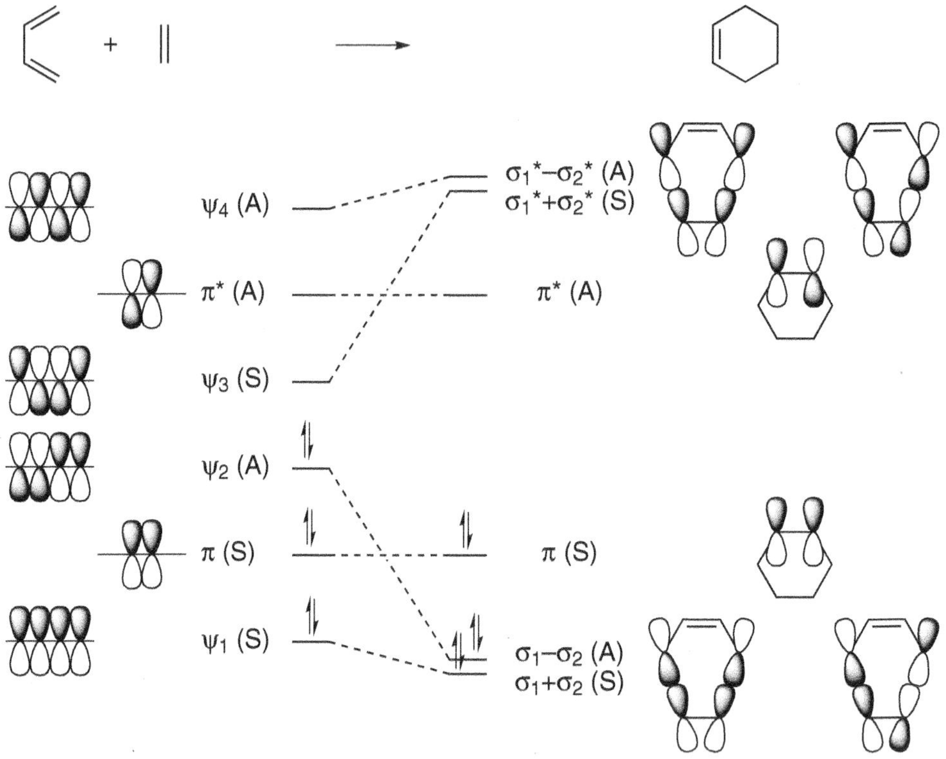

Figure 6.25 The orbital correlation diagram for the Diels-Alder ([4 + 2] cycloaddition) reaction

possible: suprafacial-suprafacial, suprafacial-antarafacial, and antarafacial-antarafacial. Of these three, the suprafacial-suprafacial process is easiest to visualize, and because it then allows the other stereochemical outcomes to be deduced, we need only try it.

[4 + 2] Cycloadditions

This cycloaddition reaction, which is typified by the Diels-Alder cycloaddition, gives rise to the orbital correlation diagram shown in Figure 6.25. Again, we begin the process by ordering the reacting π-type orbitals in increasing order of energy. However, we are now forming two new σ bonds on adjacent atoms, and we need to be able to distinguish them. We do this by using linear combinations of the orbitals ($\sigma_1^{(*)} + \sigma_2^{(*)}$) and ($\sigma_1^{(*)} - \sigma_2^{(*)}$), which are symmetric and antisymmetric with respect to the preserved mirror plane of symmetry for this reaction. In the diagram, we see the effects of conjugation on π orbital energies. The π orbital of the dienophile lies between ψ_1 and ψ_2 of the diene, and the π^* orbital of the dienophile lies between ψ_3 and ψ_4 of the diene. As we can see, in this reaction, the bonding orbitals of the reactants all correlate with bonding orbitals in the product, so we expect the suprafacial-suprafacial reaction to be symmetry-allowed in the ground state, as is observed experimentally.

[2 + 2] Cycloadditions

The [2 + 2] cycloaddition differs from the [4 + 2] cycloaddition (and, in fact, all other cycloadditions) by producing a final product that has lost all the π bonds of the reactant. The other way in which [n + n] cycloadditions are unique lies in the observation that the orbitals involved can be described in terms of not one but *two* mirror planes of symmetry. Thus, a pair of orbitals may be symmetrical with respect to one of the mirror planes and antisymmetric with respect to the other. By convention, we describe the symmetry within the individual molecules first and the symmetry between molecules second, as illustrated in Figure 6.26. In this pairing of orbitals, however, we see that one of the bonding molecular orbital combinations in the reactants correlates to an antibonding orbital

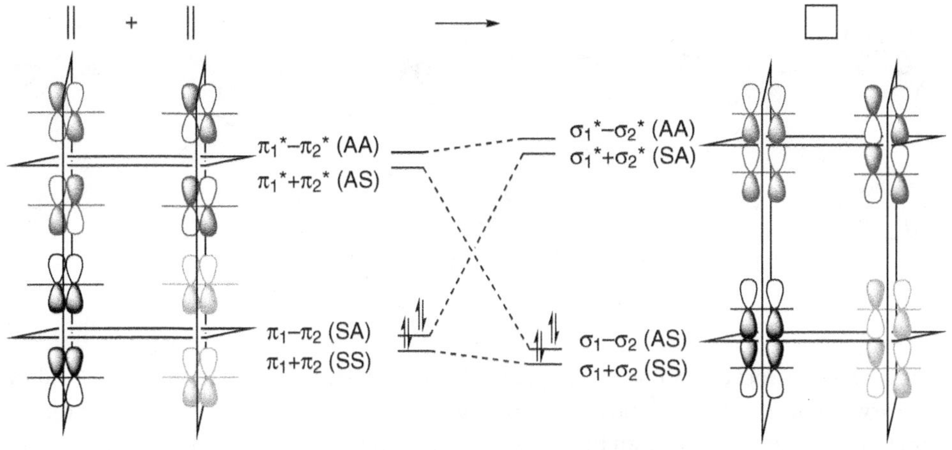

Figure 6.26 The orbital correlation diagram for the suprafacial-suprafacial [2 + 2] cycloaddition reaction

combination in the product. This makes the concerted reaction symmetry-forbidden, and we therefore expect that it will be allowed photochemically.

Problem

6-20 Construct orbital correlation diagrams analogous to Figures 6.24 and 6.25 for the [4 + 4] cycloaddition and the [4 + 6] cycloaddition reactions.

Sigmatropic Rearrangements and Ene Reactions

As noted at the beginning of this section, orbital correlation diagrams for these reactions are rather less straightforward to construct than the correlation diagrams for electrocyclizations and cycloadditions. For this reason, they are rather less widely used, in spite of the symmetry being retained in some reactions—the Cope rearrangement, for example, retains a C_2 axis of symmetry throughout the reaction if it proceeds through a chairlike transition state. It does. In general, the simpler interpretations of the Woodward-Hoffmann rules described in Section 6.2 suffices for these types of reactions.

6.8 Combining Pericyclic Reactions in Sequence

Another Look at Electrocyclizations[82]

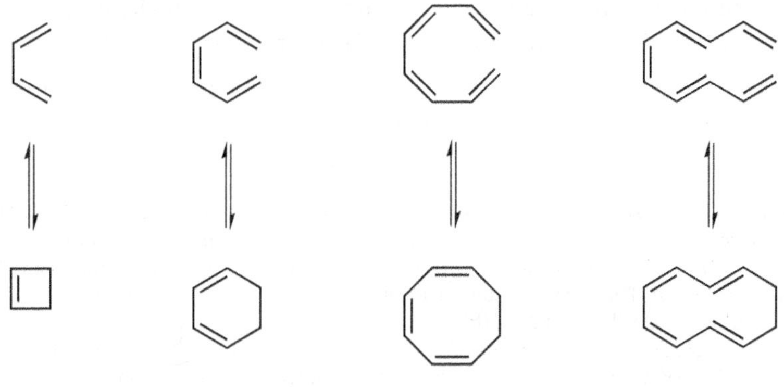

82. Monograph: Marvell, E.N. *Thermal Electrocyclic Reactions* (Academic Press: London, 1980).

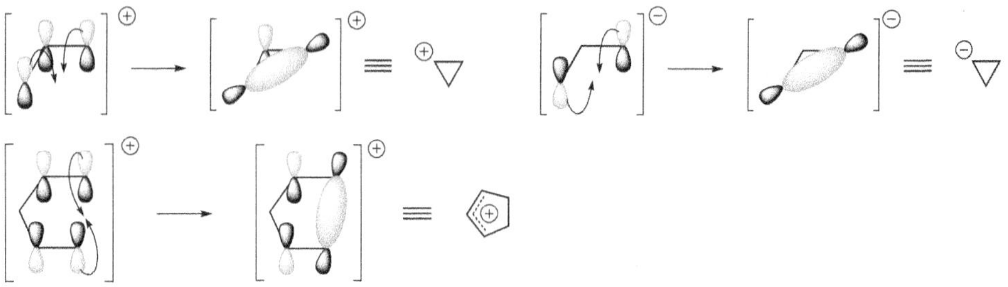

Figure 6.27 Electrocyclic ring closures of ions

Almost all electrocyclizations are, in principle, reversible, so the reaction can be used in the forward direction to prepare cyclic products, or it can be used in the reverse direction to prepare polyenes. Given that they are the simplest of the unimolecular pericyclic reactions, we used them in Section 6.2 to introduce many of the common features of this type of reaction. As we discussed in this section, the thermally allowed electrocyclizations of conjugated dienes (and, therefore, the electrocyclic ring opening of cyclobutenes) are conrotatory processes in the ground state. Photochemically, the reaction occurs with disrotatory stereochemistry. When one considers 1,3,5-hexadiene derivatives, however, the stereochemical consequences are reversed: the electrocyclic ring closure is disrotatory in the ground state (the thermal reaction) and conrotatory in the excited state (the photochemical reaction). In general, systems with $4n$ participating electrons are conrotatory in the ground state and disrotatory in the excited state, and systems with $(4n + 2)$ participating electrons are disrotatory in the ground state and conrotatory in the excited state.

The analysis we used earlier in this chapter to analyze the electrocyclization of polyenes also applies to the cyclization of ionic species with odd numbers of atoms (Figure 6.27). The electrocyclic ring closure of the allyl cation (three atoms with two π electrons) to the cyclopropyl cation, for example, is a disrotatory process, whereas the ring closure of the allyl anion (three atoms with four π electrons) to the cyclopropyl anion is a conrotatory process. The Nazarov[83] cyclization[84] (cyclization of cross-conjugated dienones under acid catalysis) formally involves the conrotatory cyclization of the 3-hydroxy-2,4-pentadienyl cation. In general, it is easier to predict the stereochemistry of electrocyclizations of ions in terms of the ring closure reaction, and here, just as with the polyenes, we find that $[4n]$ electron systems undergo conrotatory electrocyclization and that $[4n + 2]$ electron systems undergo disrotatory electrocyclizations.

In conrotatory electrocyclizations, the substituent at one end of the conjugated system rotates inward and the other rotates outward. In disrotatory reactions, both substituents rotate inward or outward. This affects the stereochemistry of the product; in the electrocyclic ring opening of 3-monosubstituted cyclobutenes, for example, inward rotation of the larger substituent gives the more hindered *Z* isomer of the diene.

---

83. **Ivan Nikolaevich Nazarov (1905–1957)** was educated at the K. A. Timiryazev Moscow Agricultural Academy (Dr Chem, 1934). He was appointed to the Institute of Organic Chemistry of the USSR Academy of Sciences and in 1947 became Professor at the Lomonosov Institute of Fine Chemicals in Moscow. In 1953, he became a full Academician. For more biographical details, see: Lewis, D.E. *Early Russian Organic Chemists and Their Legacy* (Springer: Heidelberg, 2012), p. 118.

84. (a) Nazarov, I.N.; Zaretskaya, I.I. *Izv. Akad. Nauk SSSR Otd. Khim. Nauk* **1941**, 211 [*Chem. Abstr.* **1943**, *37*, 6243[7]]; **1942**, 200 [*Chem. Abstr.* **1945**, *39*, 1619[4]]. Reviews; (b) Santelli-Rouvier, C.; Santelli, M. *Synthesis* **1983**, 429. (c) Habermas, K.L.; Denmark, S.E.; Jones, T.K. *Org. React.* **1994**, *45*, 1.

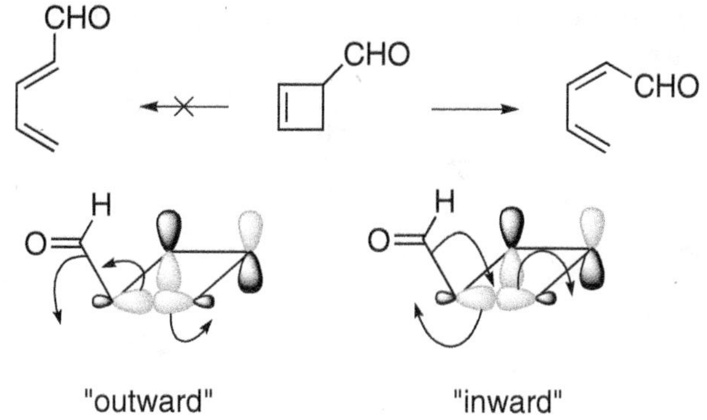

Figure 6.28 Possible modes of conrotatory ring opening of 2-cyclobutene-1-carboxaldehyde.

It was predicted[85] in 1985 that more electron-withdrawing substituents (e.g., the aldehyde group) should prefer to rotate inward during the reaction. This was confirmed experimentally: the conrotatory ring opening of 3-formyl-cyclobutene at 20–70°C gave *Z*-2,4-pentadienal as the sole product (Figure 6.28).[86] This has been exploited in synthesis, as shown by Example **6.60**, where an inward conrotatory ring opening is followed by a disrotatory ring closure to give the cyclic ether. Outward rotation during the electroreversion would not give the correct stereochemistry for the second ring closure to occur. An expansion of the early predictions has led to the results in Table 6.9, which contains calculated activation energies for electrocyclic ring opening of some substituted cyclobutenes. Some typical examples of synthetically useful electrocyclizations are shown by Examples **6.61** to **6.63**, and many more are in the *Organic Reactions* review.[87]

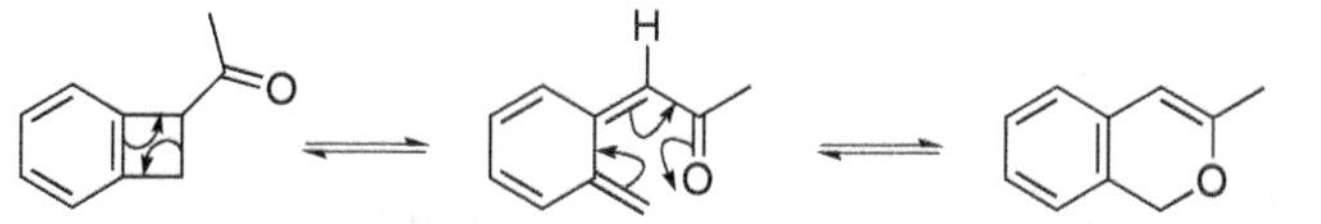

(6.60)

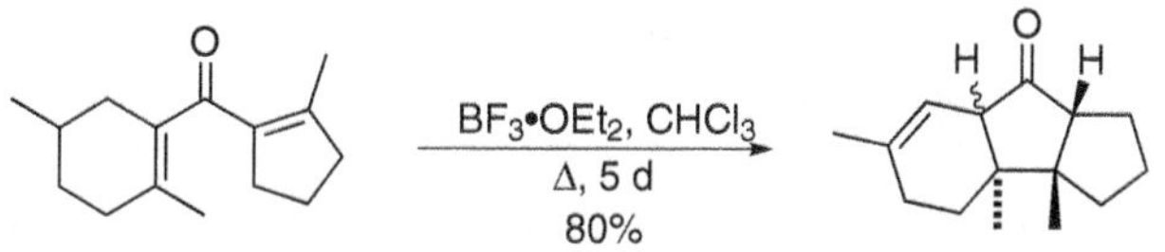

(6.61)[88]

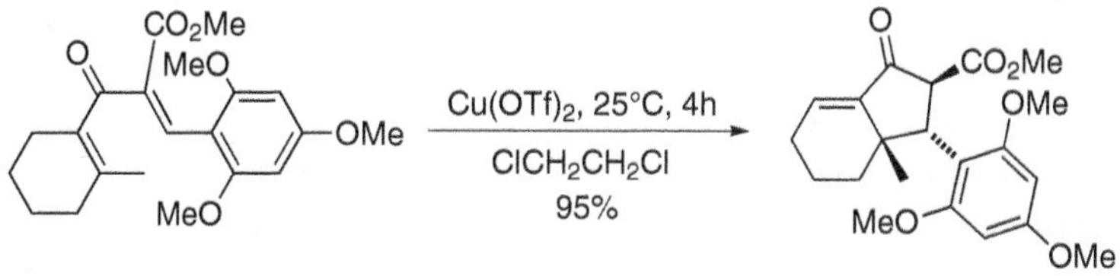

(6.62)[89]

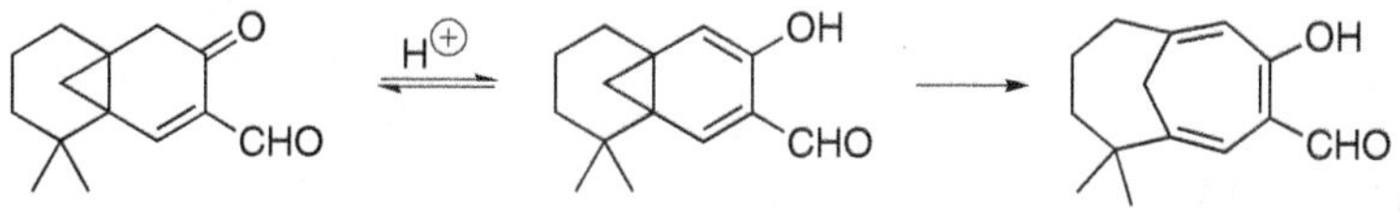

(6.63)[90]

---

85. (a) Kirmse, W.; Rondan, N.G.; Houk, K.N. *J. Am. Chem. Soc.* **1984**, *106*, 7989. (b) Rondan, N.G.; Houk, K.N. *J. Am. Chem. Soc.* **1985**, *107*, 2099.
86. Rudolf, K.; Spellmeyer, D.C.; Houk, K.N. *J. Org. Chem.* **1987**, *52*, 3708.
87. Habermas, K.L.; Denmark, S.E.; Jones, T.K. *Org. React.* **1994**, *45*, 1.
88. Harding, K.E.; Clement, K.S. *J. Org. Chem.* **1984**, *49*, 3870.
89. He, W.; Sun, X.; Frontier, A.J. *J. Am. Chem. Soc.* **2003**, *125*, 14278.
90. Marshall, J.A.; Conrow, R.E. *J. Am. Chem. Soc.* **1980**, *102*, 4274.

Table 6.9 Calculated Activation Energies for Electrocyclic Ring Opening of Substituted Cyclobutenes (in kcal mol^{-1})[91]

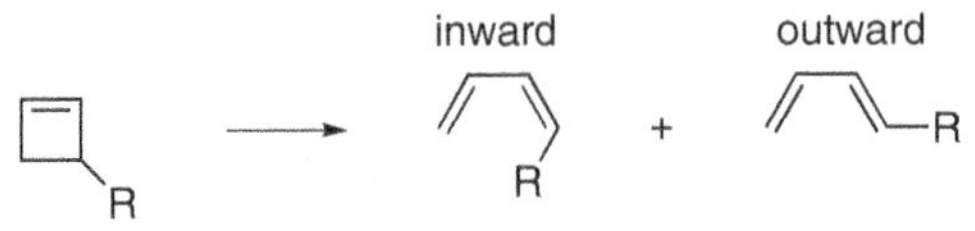

R	E_{act}^{in} (A)	E_{act}^{out} (A)	$E_{act}^{in} - E_{act}^{out}$	E_{act}^{in} (B)	E_{act}^{out} (B)	$E_{act}^{in} - E_{act}^{out}$
F	44.3	29.2	+15.1	58.9	42.0	+17.9
CH_3	38.8	33.5	+5.3	51.3	44.5	+6.8
CN	35.2	31.0	+4.2	47.2	42.9	+4.3
CHO	26.5	31.2	−4.7	38.0	42.6	−4.6
NO	25.1	29.8	−4.7	39.1	41.7	−2.6
CF_3				50.9	48.3	+2.6
NH_2				52.2	34.7	+17.5
OH				54.6	37.4	+17.2
C≡CH				48.2	40.6	+7.6
NO_2				50.1	42.8	+7.3
$COCH_3$				42.6	41.4	+1.2
CO_2H				44.2	41.9	+2.3

A: MP2/6/6-31G//3-21G+ZPE. B: 6-31G*//3-21G

Coupling Pericyclic Reactions: Tandem and Domino Reactions[92]

Pericyclic reactions provide an extremely powerful arsenal for the synthetic organic chemist, and, with recent advances in methods for the synthesis of dienes and alkenes, it is now possible to use these reactions in concert by appropriately designing the starting material. An example of the application of the tandem Cope-Claisen rearrangement is provided by a 1984 synthesis[93] of a key ketoacid intermediate (Example **6.68**) for steroid synthesis. The sequence begins with the homoallyl vinyl ether **6.65**, which contains a 1,5-diene moiety. When this is heated, it first undergoes a Cope rearrangement, to give the allyl vinyl ether **6.66**, which then undergoes a subsequent Claisen rearrangement to give the aldehyde **6.67**. Note how the relative stereochemistry of two of the three new chiral centers is fixed by the stereochemistry of the original tertiary ether.

The tandem Diels-Alder reaction is nicely demonstrated by the synthesis of the complex polycyclic product **6.70** from an acyclic precursor by a tandem Diels-Alder sequence.[94] In the first Diels-Alder reaction, the alkyne dienophile reacts to give the 1,4-cyclohexadiene derivative **6.69**, which then undergoes the second Diels-Alder reaction to give the final product.

91. Niwayama, S.; Kallel, E.A.; Spellmeyer, D.C.; Sheu, C.; Houk, K.N. *J. Org. Chem.* **1996**, *61*, 2813.
92. The first issue of *Chemical Reviews* in 1996 was devoted to tandem, domino and cascade reactions. See especially: (a) Parsons, P.J.; Penkett, C.S.; Shell, A.J. *Chem. Rev.* **1996**, *96*, 1. (b) Tietze, L.F. *Chem. Rev.* **1996**, *96*, 115. (c) Denmark, S.E.; Thorarensen, A. *Chem. Rev.* **1996**, *96*, 137. (d) Winkler, J.D. *Chem. Rev.* **1996**, *96*, 167.
93. Ziegler, F.E.; Lim, H. *J. Org. Chem.* **1984**, *49*, 3278.
94. Goldberg, D.; Hansen, J.; Giguere, R. *Tetrahedron Lett.* **1993**, *34*, 8003.

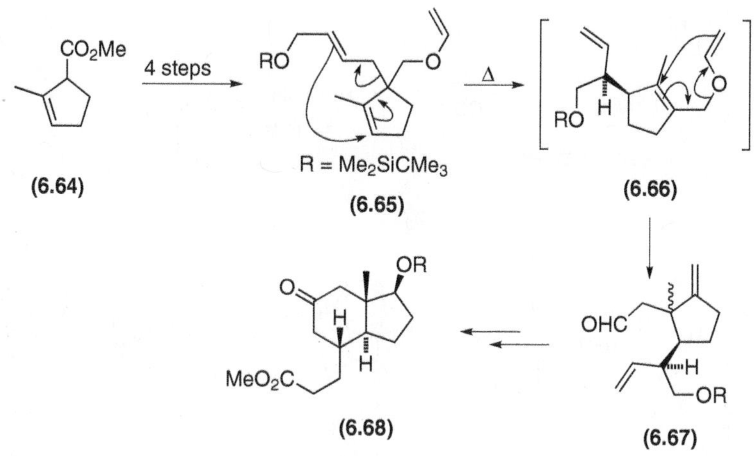

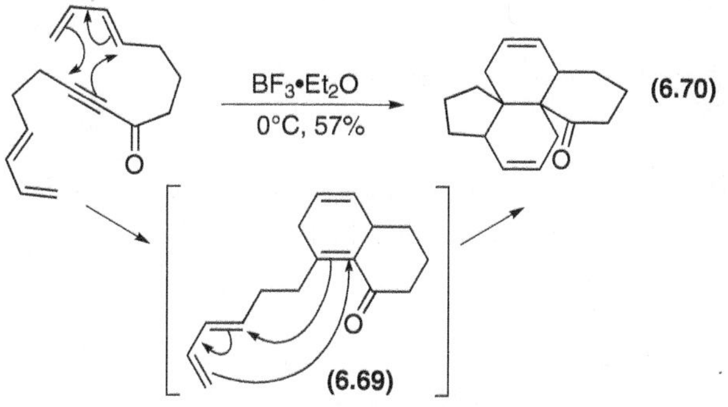

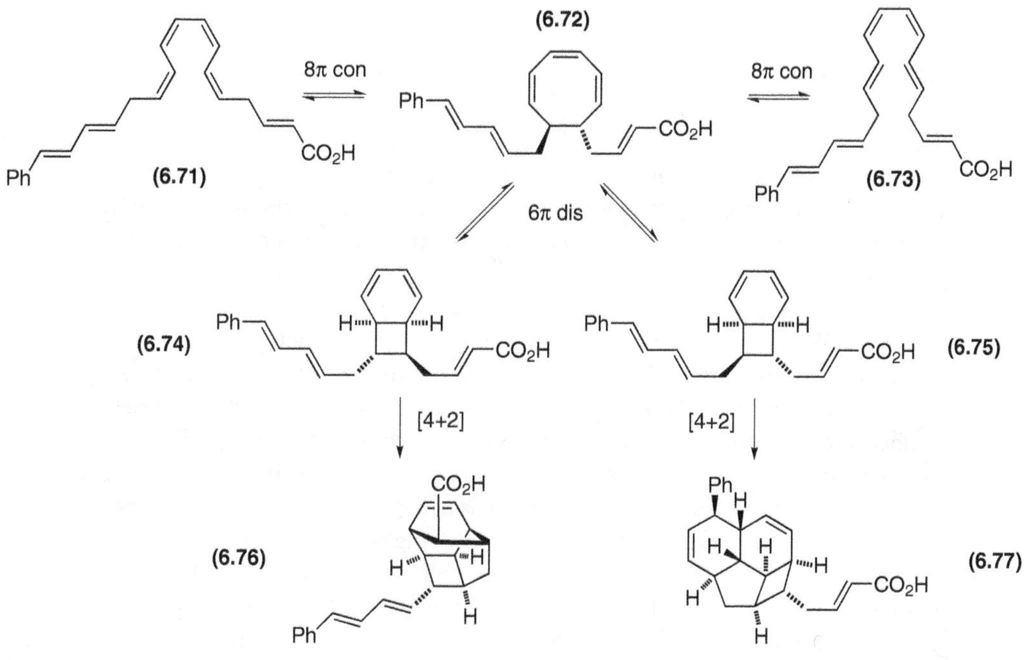

Figure 6.29 Black's proposal for the biosynthesis of the endiandric acids

Perhaps the most spectacular application of tandem electrocyclic reactions in synthesis is exemplified by the confirmation of the possibility of Black's proposal[95] (Figure 6.29)

95. Bandaranayake, W.M.; Banfield, J.E.; Black, D. St.C. *J. Chem. Soc., Chem. Commun.* **1980**, 902.

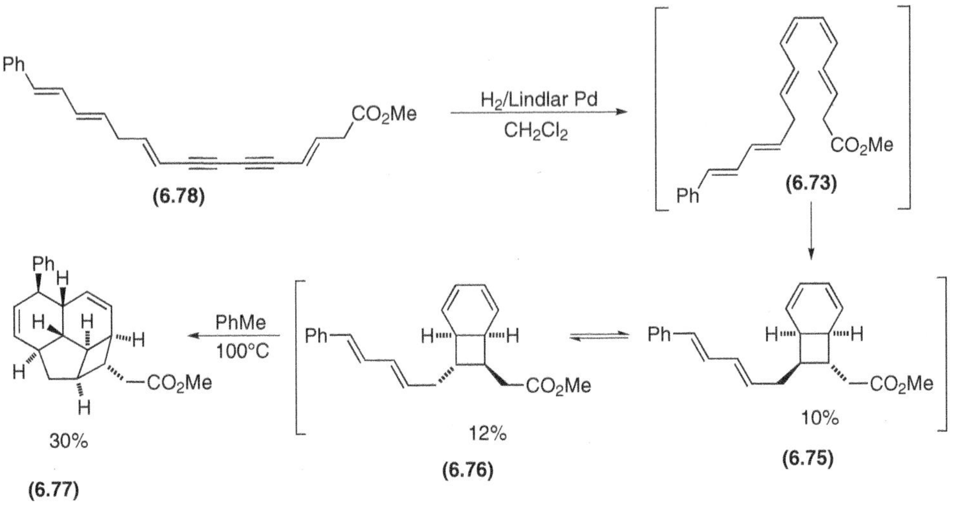

Figure 6.30 Nicolaou's biomimetic synthesis of the endiandric acids

for the biosynthesis of the endiandric acids (natural products isolated as racemic compounds isolated from the Australian tree, the Dorrigo plum) by Nicolaou's "biomimetic" synthesis.[96]

Nicolaou's synthesis (Figure 6.30) involves the synthesis of a diyne precursor, which is hydrogenated over Lindlar palladium to give the *Z,Z*-diene, which then undergoes the electrocyclic cascade to give the methyl esters of endiandric acids D and E at room temperature, and the methyl ester of endiandric acid A on thermolysis.

Chapter Summary

This chapter has been devoted to reactions that demonstrate the power of the concept of the conservation of orbital symmetry: the pericyclic reactions, which comprise electrocyclizations, cycloadditions, sigmatropic rearrangements, and ene reactions (the chelotropic reactions have not been discussed in this chapter). The symmetry of the π molecular orbitals has been described in terms of the two major symmetry operations (C_2 axis of rotation and σ), and described appropriately as symmetric or antisymmetric with respect to the symmetry operation. Pericyclic reactions can be considered as symmetry-allowed or symmetry-forbidden. Symmetry-forbidden reactions have inordinately high activation energies and usually occur by a stepwise mechanism. The stereochemistry of pericyclic reactions is described as conrotatory or disrotatory for electrocyclizations and as suprafacial or antarafacial for cycloadditions and sigmatropic rearrangements. The approximation of the FMO coefficients by perturbation theory has been described. The following major synthetic reaction types have been discussed: (1) electrocyclic ring opening and closure; (2) sigmatropic rearrangements of hydrogen; (3) [1,*n*]-sigmatropic rearrangements of alkyl, with both retention of configuration ($4n + 2$ electrons) and inversion of configuration ($4n$ electrons) at the migrating carbon; (4) the Cope and Claisen ([3,3]-sigmatropic) rearrangements; (5) the Diels-Alder ([4 + 2] cycloaddition) reaction; (6) cycloadditions of carbenes (the Simmons-Smith reaction) and ketenes; (7) cycloadditions of 1,3-dipoles and "click" chemistry; (8) ene and retro-ene reactions, including pyrolytic elimination reactions; and (9) tandem and domino pericyclic reactions. Finally, this chapter introduced the concept of orbital correlation diagrams for analyzing electrocyclization and cycloaddition reactions.

96. (a) Nicolaou, K.C.; Zipkin, R.E.; Petasis, N.A. *J. Am. Chem. Soc.* **1982**, *104*, 5558. (b) Nicolaou, K.C.; Petasis, N.A.; Zipkin, R.E. *J. Am. Chem. Soc.* **1982**, *104*, 5560.

Key Terms

1,3-dipole
antarafacial
carbine
Carroll variant
Claisen rearrangement
click chemistry
conrotatory
Cope rearrangement
cycloaddition
Diels-Alder
disrotatory
electrocyclization
ene reaction
Eschenmoser variant
HOMO
Ireland ester enolate variant
Johnson orthoester variant
ketene
ketene addition
LUMO
orbital correlation diagram
oxyanion Cope rearrangement
pericyclic reactions
pyrolytic elimination
retro-ene reaction
sigmatropic rearrangement
Simmons-Smith reaction
SOMO
suprafacial
symmetry-allowed
symmetry-forbidden
tandem and domino reactions

Additional Problems

6-21 Rationalize the stereochemistry of the reaction below. [*J. Am. Chem. Soc.* **1971**, *93*, 1777.] Why does the reaction fail when the ester groups are replaced by methyl?

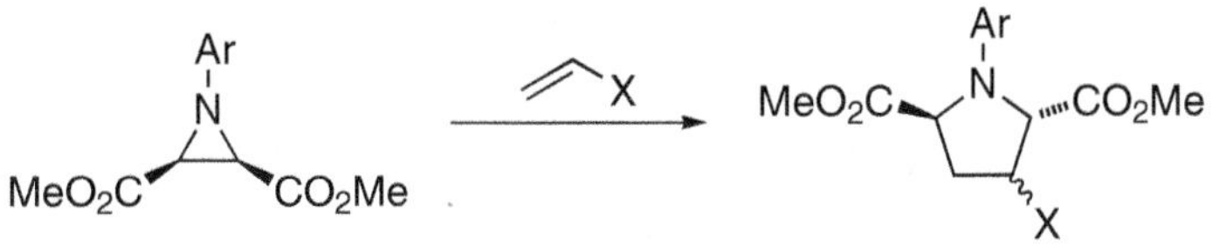

6-22 The reaction below was used as a key step in Tufiarello's synthesis of cocaine [*J. Am. Chem. Soc.* **1978**, *100*, 3638]. Provide a mechanism that rationalizes this transformation.

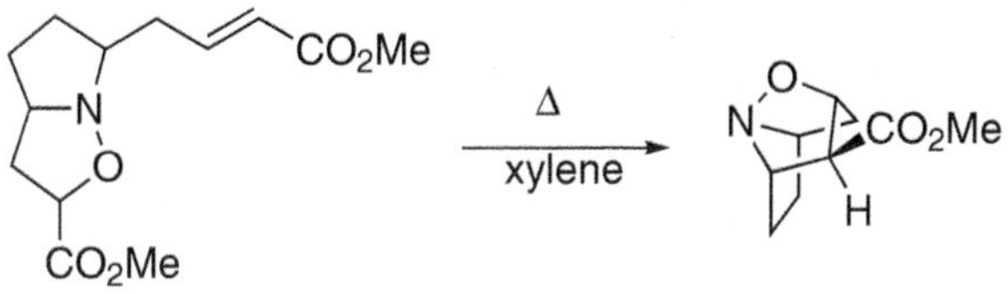

6-23 The reactions below have been used as key steps in the synthesis of a series of natural products. Predict the major product of each reaction. Where appropriate, give your reasoning (e.g., is it necessary to explain why one regioisomer or stereoisomer of the product is expected to dominate the product mixture?).

(a) 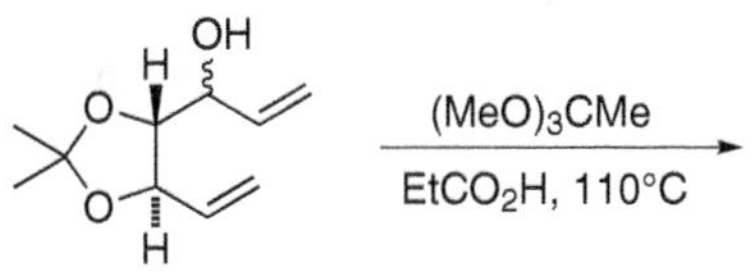

[*J. Am. Chem. Soc.* 2006, *128*, 15960]

(b) 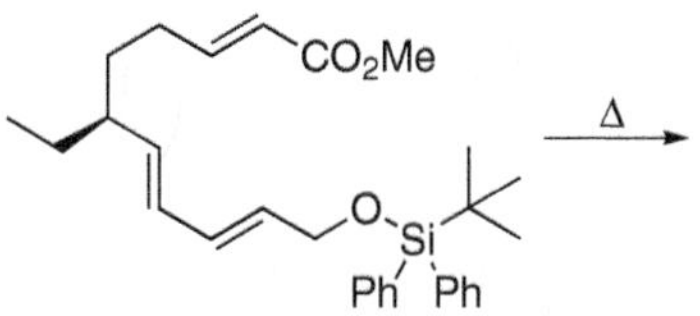

[*J. Am. Chem. Soc.* **1981**, *103*, 6967]

(c) 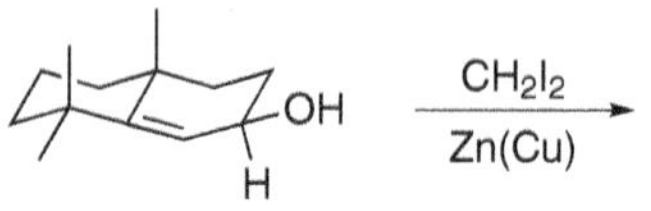

[*J. Am. Chem. Soc.* **1963**, *85*, 3673]

(d) 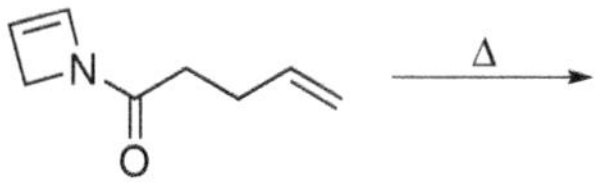

[*J. Org. Chem.* **1991**, *56*, 6729]

(e) 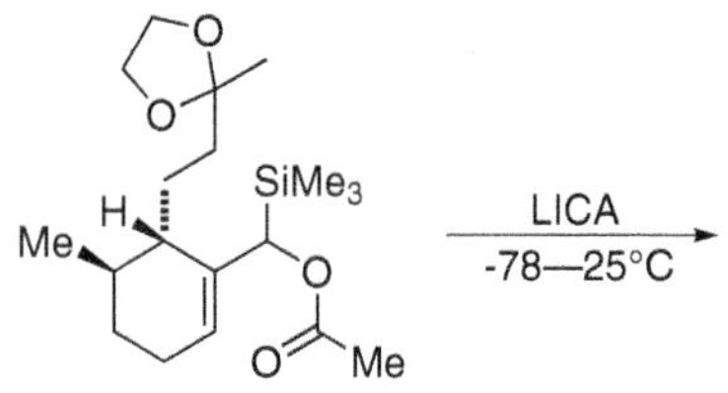

[*Tetrahedron Lett.* **1987**, *28*, 4629]

(f) 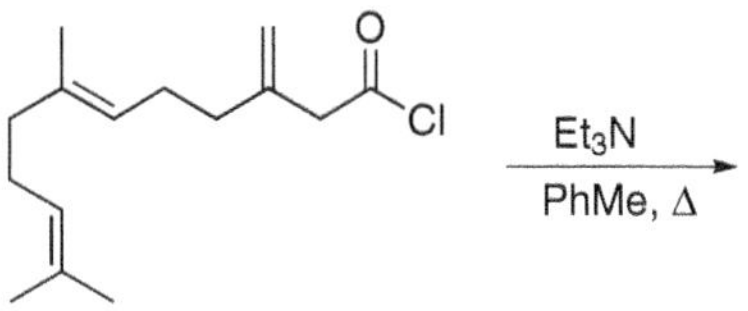

[*Tet. Lett.* **1985**, *26*, 3535]

(g) 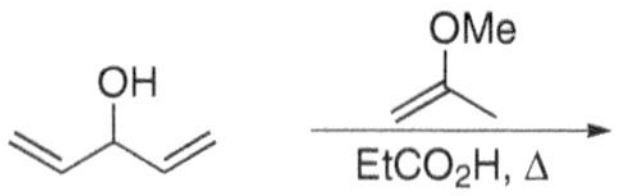

[*Org. Lett.* **2004**, *6*, 3345]

(h) 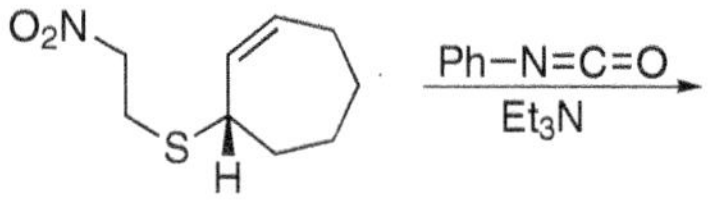

[*J. Am. Chem. Soc.*, **1978**, *100*, 6291]

(i) 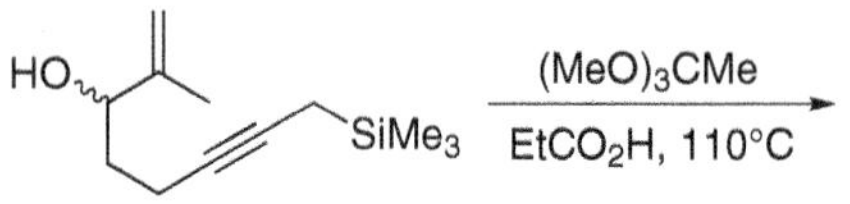

[*J. Am. Chem. Soc.* **1980**, *102*, 5122]

(j) 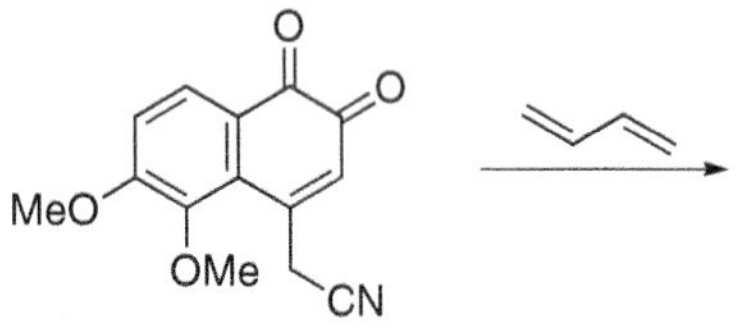

[*J. Am. Chem. Soc.*, **1956**, *78*, 1380]

(k) 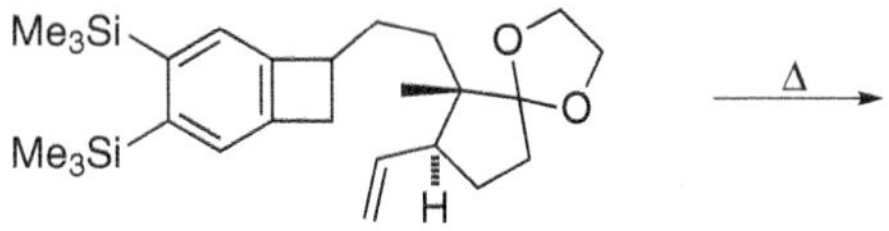

[*J. Am. Chem. Soc.* **1979**, *101*, 215]
(tandem reaction)

(l) 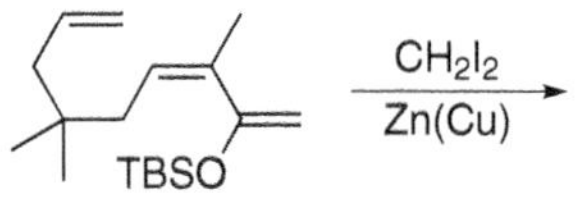

[*J. Am. Chem. Soc.* **2008**, *130*, 4421]

(m) 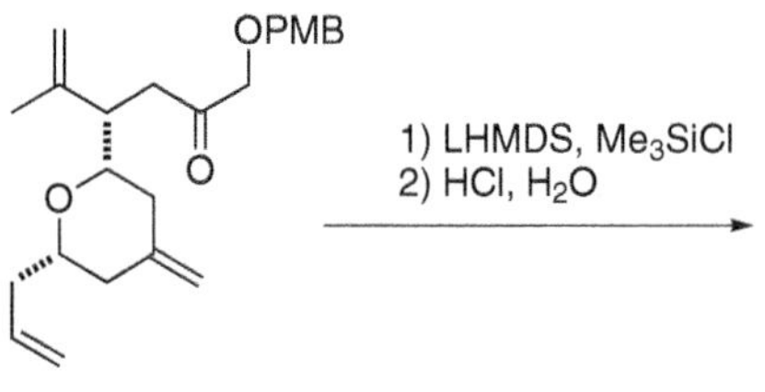

[*Org. Lett.* **2006**, *8*, 1117]

(n) 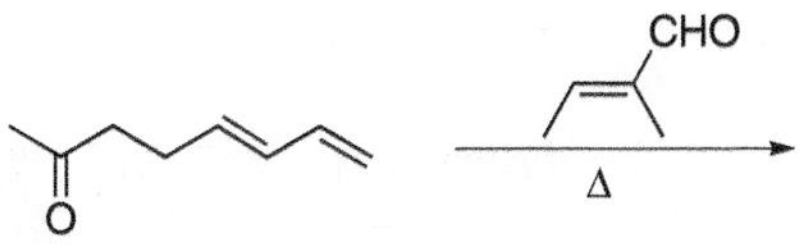

[*Org. Lett.* **2004**, *6*, 3345]

(o) 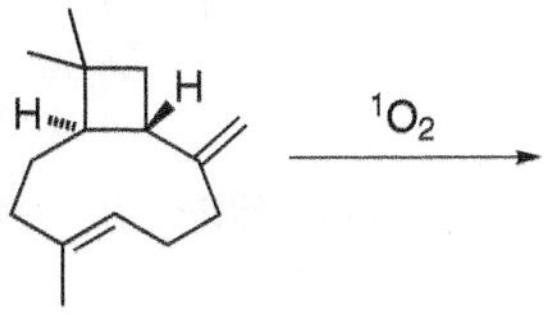

[*J. Am. Chem. Soc.* **1980**, *102*, 1738]
Tetrahedron Lett. **1968**, 689]

(p) 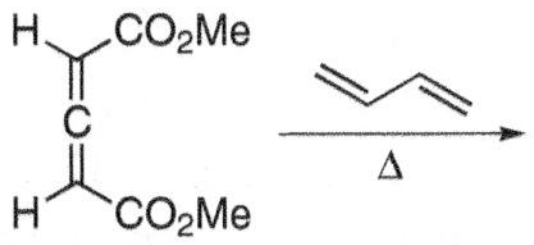

[*J. Am. Chem. Soc.* **1968**, *90*, 5336]

(q) 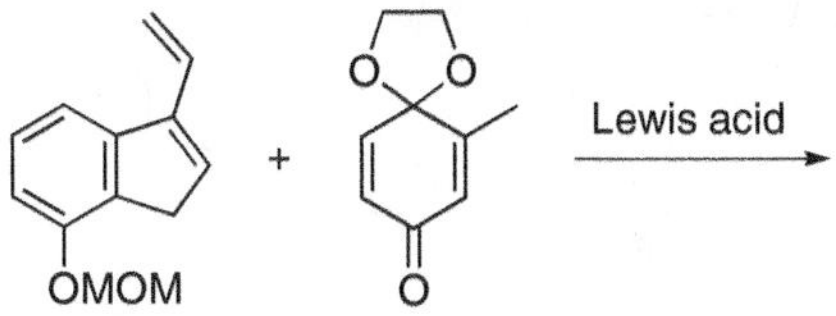

[*J. Am. Chem. Soc.* **2008**, *130*, 2783]

(r) 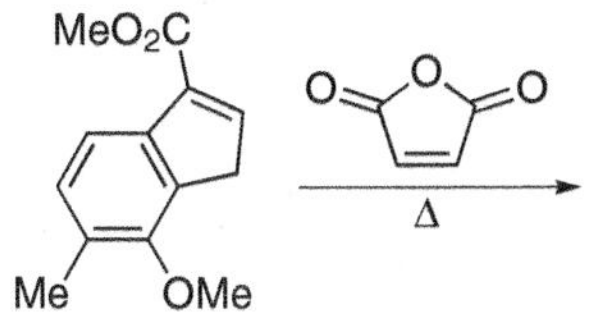

[*Pure Appl. Chem.* **1979**, *51*, 689]

(s) 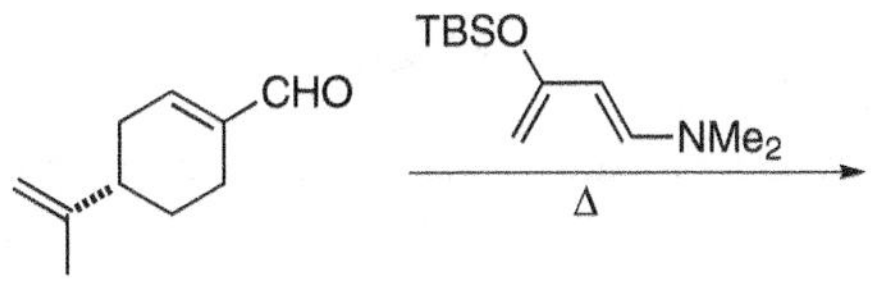

[*Angew. Chem. Int. Ed.* **2008**, *47*, 6199]

(t) 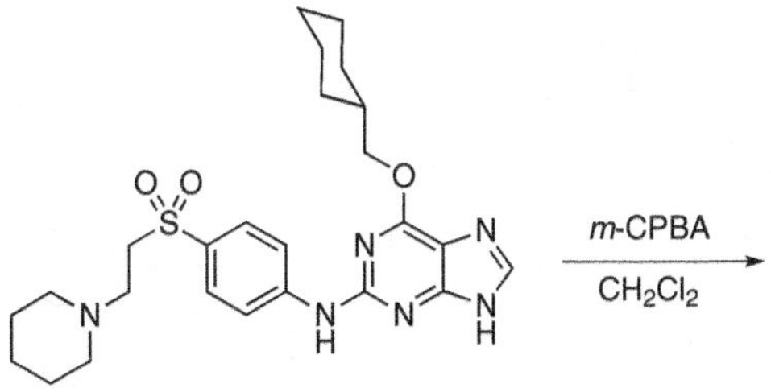

[*J. Am. Chem. Soc.* **2006**, *128*, 6012]

(u) 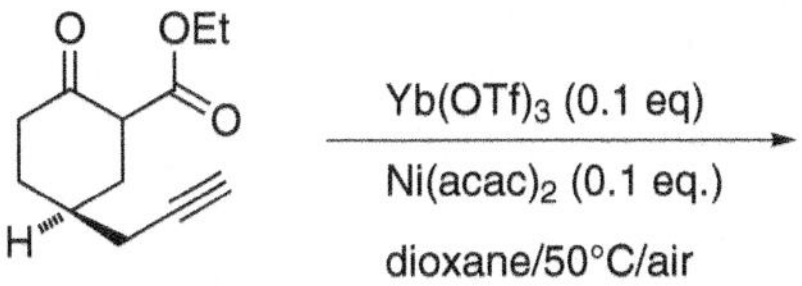

[*Org. Lett.* **2005**, *7*, 2185.

(v) 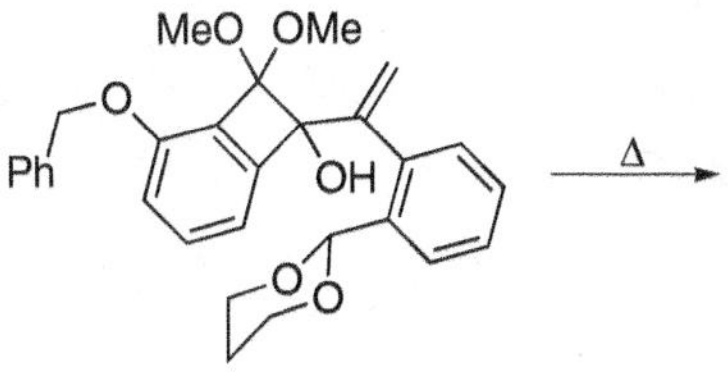

[*Org. Lett.* **2004**, *6*, 2503] (tandem reaction)

(w) 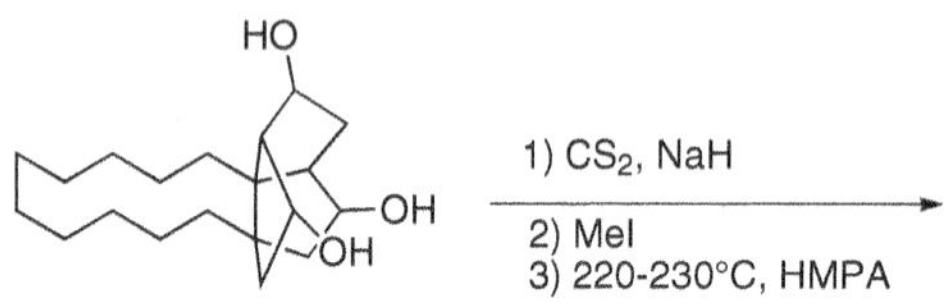

[*Tetrahedron Lett.* **1990**, *31*, 3409]

(x) 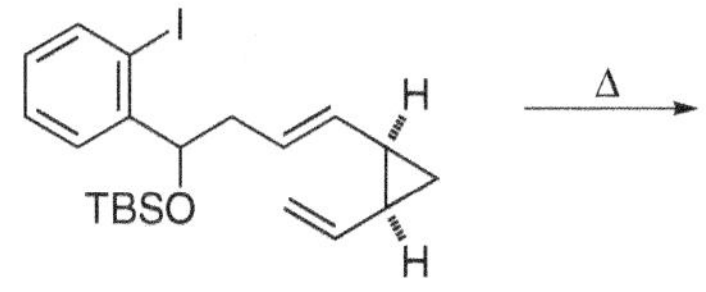

[*J. Am. Chem. Soc.*, **1993**, *115*, 2042]

6-24 The reaction below proceeds by a combination of pericyclic reactions. Give the course of this reaction and rationalize the stereochemistry of each step in the pathway. [*J. Am. Chem. Soc.* **1980**, *102*, 6353]

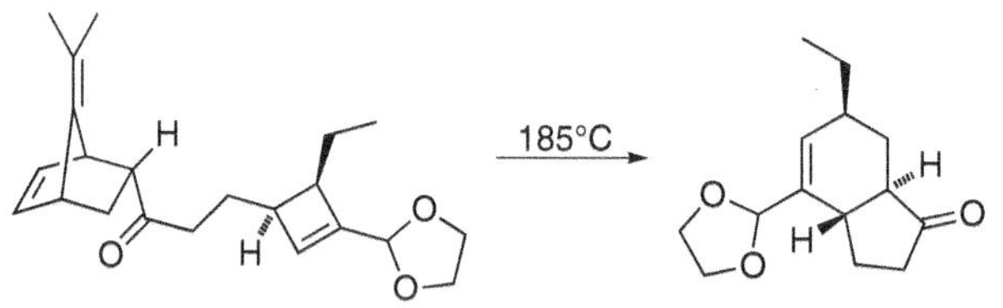

6-25 Suggest a mechanism for the reaction below.

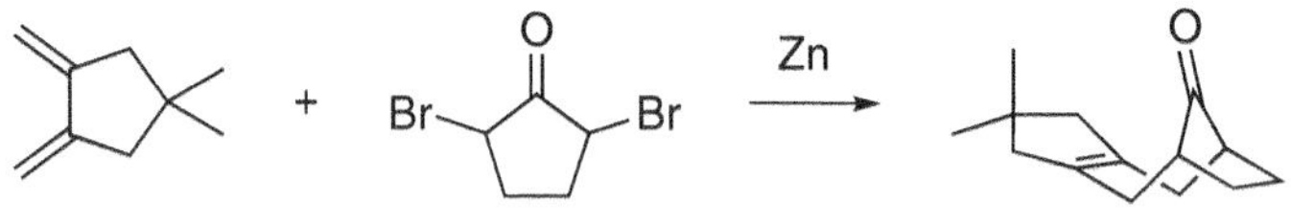

6-26 Each of the following compounds can be formed by means of a pericyclic reaction. Give the reactants required for its formation.

(a) 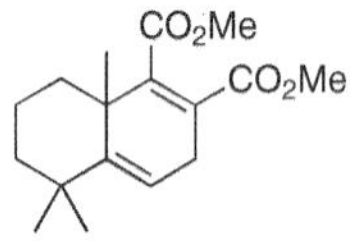

(b) 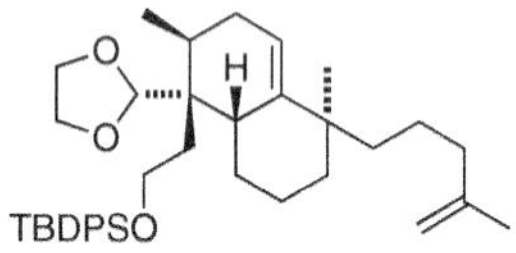

(c) 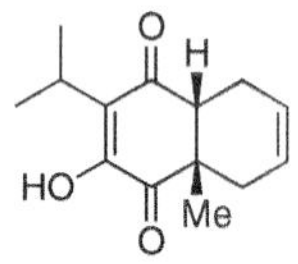

(d) 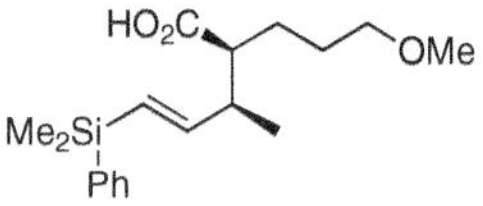

(e) 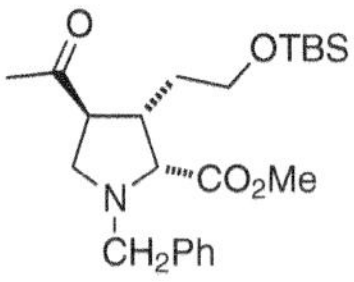

(f) 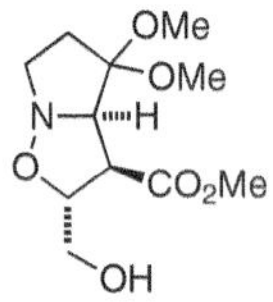

(g) 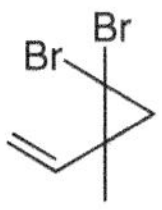

(h) 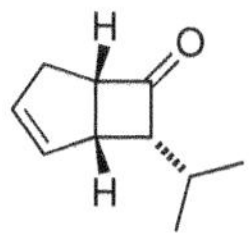

(i) 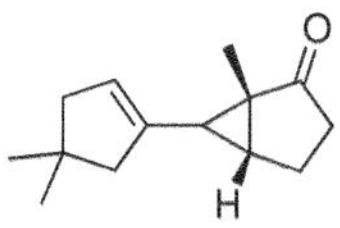

(j) 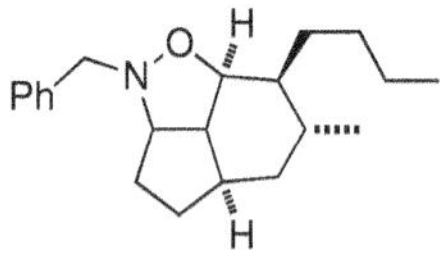

(k) 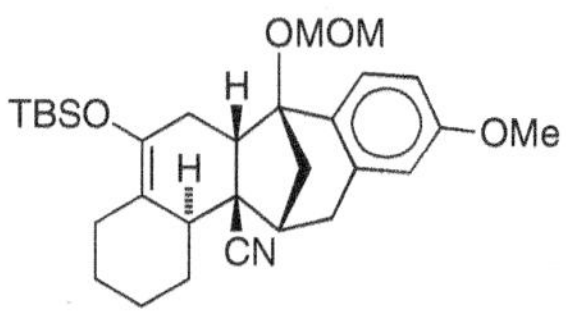

(l) 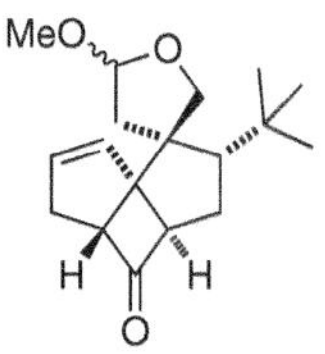

6-27 The reaction below has been proposed to occur as a concerted [14 + 2] cycloaddition. What is the final product, including stereochemistry? [*Tetrahedron Lett.* **1981**, *22*, 215]

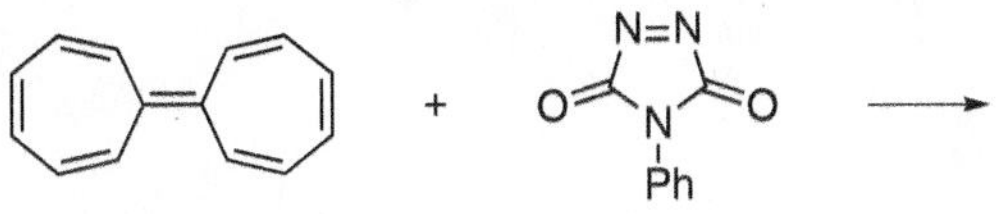

CHAPTER SEVEN

Aromaticity: The 150-Year Riddle

7.1 Benzene: The Beginning

Within 10 years of its isolation by Faraday[1] in 1825,[2] benzene (Example **7.1**) had been synthesized from other compounds, and Mitscherlich[3] had established its molecular formula as C_6H_6 by means of vapor density measurements. In 1834, he synthesized benzene from benzoic acid and showed that it was identical to Faraday's hydrocarbon.[4] The empirical formula of benzene (CH) shows quite clearly that it is a highly unsaturated compound—even more unsaturated than ethylene. And herein lay the problem for organic chemistry—it did not fit the normal reactivity patterns that had begun to appear as a means of making organic chemistry less of a jungle and more a rational science. Benzene has held organic chemists in thrall ever since.[5]

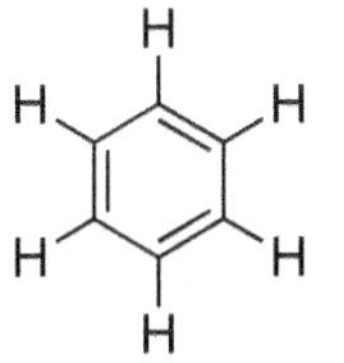

(7.1)

Like many concepts that modern organic chemists use (such as electronegativity), aromaticity is not particularly amenable to a single, simple definition. Most chemists have adopted the functional definition of aromaticity being the consequence of cyclic π-electron delocalization, as proposed Jackman and Elvidge[6] a half century ago. This simple statement recognized three major manifestations of this delocalization: (1) bond lengths are averaged, (2) the system is stabilized by resonance, and (3) magnetic properties due to the delocalization of the π electrons in a cyclic system. Sondheimer stated it even more simply, defining aromaticity as "the set of properties associated with cyclic conjugation."[7]

1. **Michael Faraday (1791–1867)** was largely self-taught. In 1813 he became the protégé of Sir Humphrey Davy at the Royal Institution. He spent the rest of his life there, becoming Fullerian Professor in 1833. Biographies of Faraday abound; for a recent biography, see: Hamilton, J. *A Life of Discovery: Michael Faraday, Giant of the Scientific Revolution.* (Random House: New York, 2004).

2. Faraday, M. *Phil. Trans. Roy. Soc. London* **1825**, *115*, 440.

3. **Eilhardt Mitscherlich (1794–1863)** was educated at Heidelberg and Göttingen, where he studied medicine. In 1821, he succeeded Klaproth at Berlin, after spending 2 years in Berzelius' laboratory. For more complete biographies, see: *Obituary notices of Fellows deceased. J. Chem. Soc.* **1864**, *13*, ix-xvi (near the end of the volume) and Schutt, H.-W. *Eilhard Mitscherlich: Prince of Prussian Chemistry* (American Chemical Society and Chemical Heritage Foundation, 2004).

4. Mitscherlich, E. *Ann. Chem. Pharm.* **1834**, *9*, 39.

5. There is a huge volume of literature on the subject of aromaticity: see, for example, the special issues of *Chem. Rev.*: *Chem. Rev.* **2001**, *101* (5) and *Chem. Rev.* **2005**, *105* (7).

6. Jackman, L.M.; Elvidge, J.A. *J. Chem. Soc.* **1961**, 859.

7. Sondheimer, F. *Pure Appl. Chem.* **1964**, *7*, 363.

Modern spectroscopic techniques, especially nuclear magnetic resonance (NMR) spectroscopy, have also played an important part in defining what is and is not aromatic, as we shall see later in this chapter. The effects of the aromaticity of the ring can be seen in the ^{1}H NMR spectrum of the benzene molecule: the protons of the benzene molecule resonate at 7.27 ppm, 1.4 ppm downfield from those of the vinyl protons of 1,3-cyclohexadiene (5.86 ppm). The "outer" protons of the aromatic macrocycle, [18]annulene[8] (Example **7.2**) resonate at 10.75 ppm, with the inner protons resonating at –4.22 ppm, indicative of a strong ***paratropic ring current***.

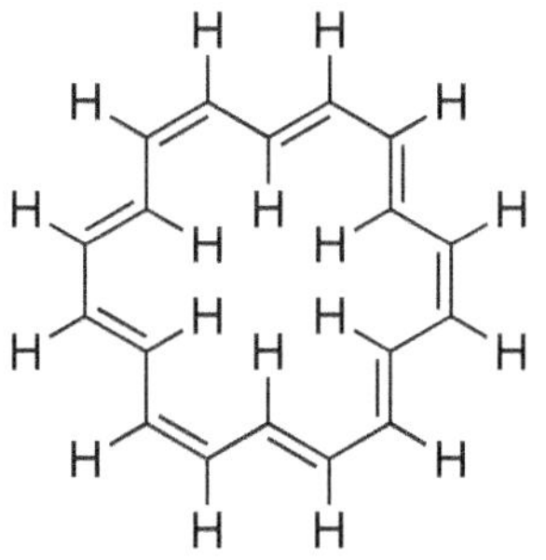

(7.2)

Despite the advances of the last century and a half, the chemist's view of aromaticity as a concept may still best be described using the language of Justice Potter Stewart's statement (about hard-core pornography): "I shall not today attempt further to define the kinds of material I understand to be embraced within that shorthand description; and perhaps I could never succeed in intelligibly doing so. But I know it when I see it,..."[9] Most organic chemists have an intrinsic understanding of what is meant by the term, "aromatic," including its effects on reactivity (e.g., substitution by electrophiles, rather than addition), but remain rather hard-pressed to provide an objective, universal definition—practically every definition proposed to date will raise objections from some members of the organic chemistry community.

7.2 Aromaticity and Antiaromaticity: The Hückel Molecular Orbital Model of Cyclic Polyenes

The structure of benzene was a riddle that occupied some of the finest minds of 19th-century organic chemistry. A somewhat more complete description of the history of the planar hexagonal structure for benzene can be found at the end of this chapter, in Section 7.6. The upshot of six decades of work was that, by the end of the 19th century, the planar hexagon formula for benzene had become firmly established, and the term ***aromatic*** had come to represent a compound derived from benzene or a compound whose reactivity was similar to that of benzene.

The first hesitant steps toward a theoretical basis for predicting when a molecule would be aromatic began with the Armstrong-Baeyer[10] (Example **7.3**) and Thiele[11] (Example **7.4**) formulas for benzene. In the centric formula for benzene, each ring has a sextet of "free"

8. (a) Sondheimer, F.; Wolovsky, R. *Tetrahedron Lett.* **1959**, 3. (b) Sondheimer, F.; Wolovsky, R.; Amiel, Y. *J. Am. Chem. Soc.* **1962**, *84*, 274.

9. *Nico Jacobellis v. Ohio, 378* **U.S.** 184 (1964).

10. (a) Armstrong, H.E. *J. Chem. Soc.* **1887**, *51*, 264. (b) Baeyer, A. *Ann. Chem. Pharm.* **1888**, *245*, 103.

11. Theile, J. *Ann. Chem.* **1899**, *306*, 87; *Ann. Chem.* **1899**, *308*, 333.

Friedrich Karl Johannes Thiele (1865–1918) was educated at Halle (PhD, 1890), and remained there until 1893; he taught at Munich (1893–1902), and Strasbourg (1902–1910); in 1910 he became Rector of the university. For a more complete biography, see: Straus, F. *Ber. dtsch. chem. Ges.* **1927**, *60*, A75.

valencies, indicated by the short bonds directed toward the center of the ring, which Armstrong[12] proposed was associated with aromatic character (actually, the structure had been proposed 6 years earlier[13] by Lothar Meyer). These centric bonds, he reasoned, meant that there were no double bonds in the molecule, which explained the troubling lack of reactivity of the benzene ring toward reagents that attack alkenes. Although the Armstrong theory was attractive in its relative simplicity, its empiricism did not satisfy most organic chemists, who wanted to know just what these "centric bonds" really were. The Thiele model was slightly more satisfying but still had its limitations. Interestingly, the "centric bonds" of the Armstrong-Baeyer model and the partial bonds of the Thiele model both presaged the predictions of modern resonance theory. Without knowing what these bonds really are, the critical question—"What is needed for a molecule to be aromatic, and why does it make a molecule aromatic?"—could not be answered. The search for an answer to this question is as intriguing a story as any in organic chemistry.

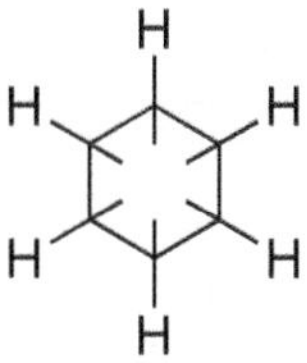

(7.3)

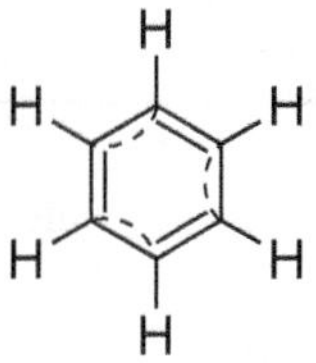

(7.4)

Worked Problem

7-1 Resonance can be used to predict the bond lengths and bond orders in aromatic compounds. What bond orders are predicted for the C—C bonds of naphthalene, based on equal contributions to the resonance hybrid from each canonical form?

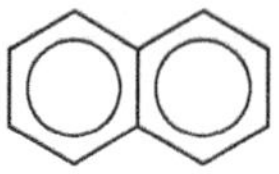

§Answer below.

---

12. **Henry Edward Armstrong (1848–1937)** studied in London, then Leipzig (PhD, 1869). During his subsequent academic career he became a Fellow of the Royal Society and a major force in the Chemical Society of London. For more biographical details, see: *J. Chem. Soc.* **1940,** 1418.

13. Meyer, J.L. *Die Modernen Theorien der Chemie* (Maruschke and Berendt: Breslau, 1872).

§ Answer for Worked Problem:

There are three reasonable resonance contributors that we can write for the naphthalene molecule. These are shown below:

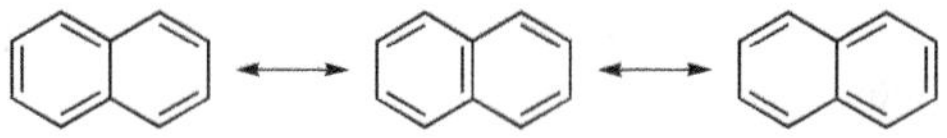

Assuming equal contributions from each structure, we can now look at the bonds of the naphthalene molecule in turn:

1 – 2 = 3 – 4 = 5 – 6 = 7 – 8 = (1/3)(2 + 2 + 1) = 1.67; 2 – 3 = 6 – 7 = (1/3)(2 + 1 + 1) = 1.33

1 – 8a = 4 – 4a = 4a – 5 = 8 – 8a = (1/3)(2 + 1 + 1) = 1.33 4a – 8a = (1/3)(2 + 1 + 1) = 1.33

This gives a molecule with the bond orders shown:

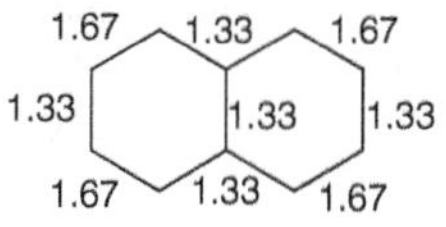

Problem

7-1 Write all the major resonance contributors for each of the following polycyclic aromatic hydrocarbons, and predict the bond orders of their C—C bonds.

(a) phenanthrene 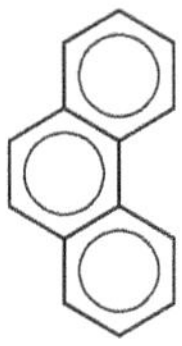

(b) benzo[*a*]pyrene 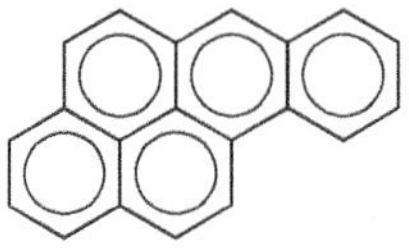

(c) tetracene 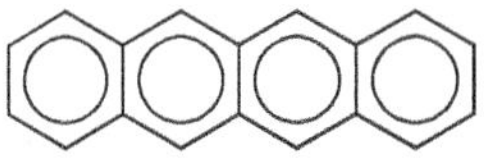

(d) pyrene 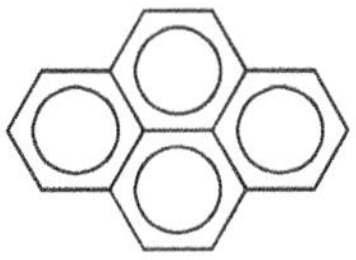

It took nearly another half century before the German physicist Erich Hückel[14] provided a theoretical basis for aromaticity[15] that satisfied most chemists. This work was an outcome of his work on the π molecular orbitals of conjugated polyenes. Hückel's molecular orbital calculations, described in Chapter 4, can be applied to cyclic, planar, conjugated polyenes as well as open-chain polyenes. These calculations revealed an unexpected dichotomy. Planar molecules containing a cyclic conjugated π orbital system with $(4n + 2)$ π electrons, where n is an integer (0, 1, 2, 3,...), were predicted to be unusually stable; these compounds are aromatic. In contrast, planar molecules containing a cyclic conjugated π orbital system with $4n$ π electrons were predicted to be unusually unstable; these compounds are ***antiaromatic***. The Hückel models led chemists to test the limits of aromaticity as predicted by this theory. The results of these studies modified how aromaticity was defined, and, as our instrumental and computational capabilities have grown, so the defining characteristics of an aromatic system have continued to change.

Hückel's calculations revealed that delocalized, cyclic π molecular orbitals have a regular pattern in their energies. Unlike the orbitals of acyclic conjugated π systems, where each orbital is at a different energy level, cyclic, delocalized π orbitals occur as a single lowest energy π orbital, with all higher energy π orbitals occurring as degenerate pairs. In molecules with an odd number of atoms in the ring, there is one unique π orbital at the lowest energy, and in molecules with an even number of atoms in the ring, the same pattern holds, with the exception that there is also one unique π orbital at the highest energy. From Hückel's work, a set of four basic structural and electronic requirements for aromaticity emerged:

1. All aromatic compounds are planar.
2. All aromatic compounds are cyclic.
3. All aromatic compounds have a conjugated, delocalized cyclic π bond system.
4. All aromatic compounds have an even number of π electrons not divisible by 4 (i.e., $4n + 2$) in the delocalized π bonding system. (Compounds with $4n$ π electrons are antiaromatic.)

Because almost all known aromatic compounds meet all four of these structural requirements, determining whether or not a compound is aromatic simply becomes a

14. **Erich Armand Arthur Joseph Hückel (1896–1980** was educated at Göttingen (PhD, 1921, under Debye). After postdoctoral study with mathematician David Hilbert and with Max Born, his independent career took him from Zürich (1922), to the Technische Hochschule Stuttgart (1930) and Marburg (1937). For more information, see: Karachalios, A. *Erich Hückel (1896–1980): From Physics to Quantum Chemistry* (Springer: Heidelberg,) 2010.

15. Hückel, E. *Z. Phys.* **1931**, *70*, 204.

problem of determining whether a compound possesses all four of these characteristics. Let us now use these four characteristics structural characteristics of aromaticity to determine whether or not the four hydrocarbons below are aromatic, antiaromatic, or non-aromatic.

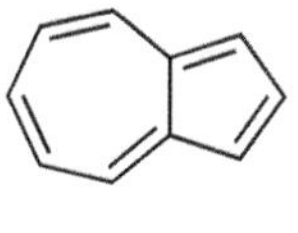

Azulene

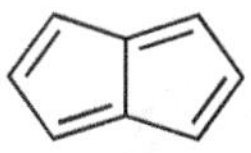

Pentalene

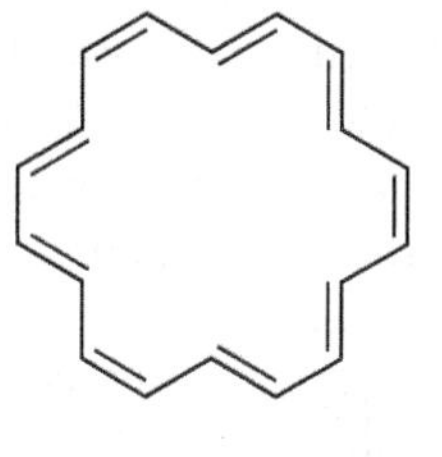

18-Annulene (cyclooctadeca-1,3,5,7,9,11,13,15,17-nonaene)

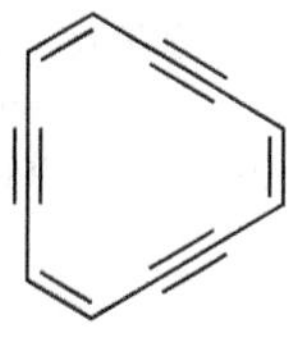

Cyclododeca-1,5,9-trien-3,7,11-triyne

Each of these hydrocarbons contains a conjugated π-bonding system but not all are aromatic. In azulene ($C_{10}H_8$), all ten carbon atoms are sp^2 hybridized; the molecule is planar. In a fused-ring system such as azulene, one writes the structure so that the conjugated π system—if one is present—is located around the periphery of the molecule; σ bonds between rings are ignored. If one can write the full conjugated system around the perimeter, the system is cyclic; if not, it is not. When the azulene molecule is written in this way, one can write the conjugated π electron system so that all five of the π bonds—all ten of the π electrons – are involved in a linearly conjugated system. There are $4n + 2$ π electrons ($n = 2$), so the azulene molecule satisfies all the requirements for aromaticity; azulene is aromatic. When the same analysis is applied to pentalene (C_8H_6), we find that there are eight π electrons ($4n$, $n = 2$), which means that this hydrocarbon is antiaromatic. Although derivatives of pentalene had been prepared as early as 1962,[16] the parent hydrocarbon was prepared for the first time in an argon matrix only in 1997.[17]

The next two molecules are typical representatives of a class of compounds in which there is a single large ring with alternating π and σ bonds. These compounds, which are known as **annulenes** (Latin *annulus*, a ring) have names that indicate the ring size; usually they have only single and double bonds in the ring, as in [18]annulene, but they may also have triple bonds in the ring, in which case they are usually referred to as **dehydroannulenes**. Earlier in this chapter, we encountered the [18]annulene molecule, which satisfies all the structural requirements for aromaticity: it is planar and conjugated, the conjugated π electron system is cyclic, and it has $4n + 2$ ($n = 4$) electrons. It should be aromatic. In its chemistry, [18]annulene is not as unreactive as benzene, but it does still react as an aromatic hydrocarbon. The final compound is tridehydro[12]annulene.[18] Here, we have triple bonds in the ring, and one must remember that only one of the two orthogonal π bonds of the triple bond can be used in determining aromaticity; under these rules, this hydrocarbon is antiaromatic because it has $4n$ ($n = 3$) π electrons.

Another class of aromatic species consists of aromatic ions, all of which are characterized by the same planar, cyclic delocalized π electron system containing $4n + 2$ π electrons. Thus, one finds that the cyclopropenium ion (Example **7.5**), which has two π electrons delocalized over its three-membered ring, should be aromatic, and derivatives of this ion do, indeed, exhibit unusual stability. The cyclopentadienide anion (Example **7.6**), contains

---

16. Le Goff, E. *J. Am. Chem. Soc.* **1962**, *84*, 3935.
17. Bally, T.; Chai, S.; Neuenschwander, M.; Zhu, Z. *J. Am. Chem. Soc.* **1997**, *119*, 1869.
18. (a) Untch, K.G.; Wysocki, D.C. *J. Am. Chem. Soc.* **1966**, *88*, 2608. (b) Sondheimer, F.; Wolovsky, R.; Garratt, P.J.; Calder, I.C. *J. Am. Chem. Soc.* **1966**, *88*, 2610.

six π electrons delocalized over its planar five-membered ring, is also especially stable as evidenced by the unusually high pK_a (≈ 15) of the parent diene. The cycloheptatrienylium ion (Example 7.7, also known as the **tropylium** ion) contains six π electrons delocalized over its planar seven-membered ring; it is also unusually stable. We find that cycloheptatriene derivatives carrying a leaving group all react rapidly under S_N1 conditions. Cyclooctatetraene, which is tub-shaped and non-aromatic, reacts with potassium as shown in Example **7.8** to give the radical anion. This anion then disproportionates to give cyclooctatetraene and the planar cyclooctatetraene dianion,[19] which has ten π electrons delocalized over its eight carbon atoms and is quite stable.

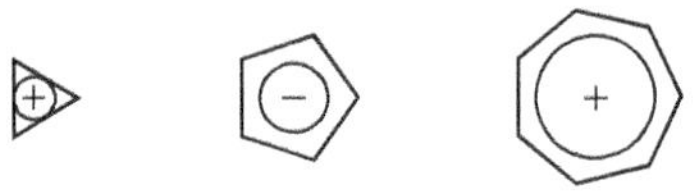

(7.5) **(7.6)** **(7.7)**

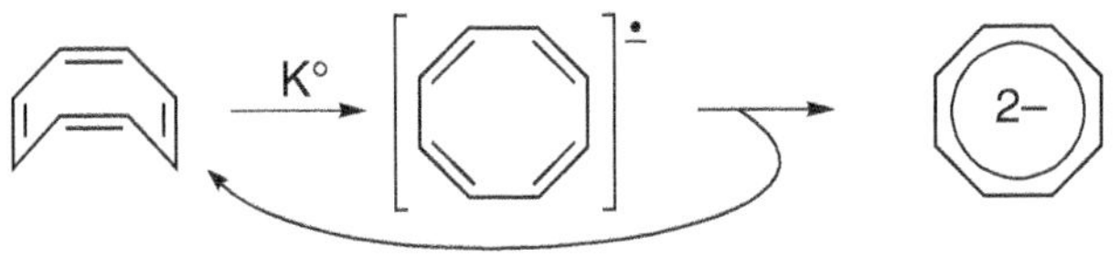

(7.8)

Whether or not a compound is aromatic depends on the total number of electrons in the π orbitals. We can illustrate this by looking at the cyclopentadienyl system. This system is common to the cyclopentadienyl cation, the cyclopentadienyl free radical, and the cyclopentadienide anion. All three of these species can be represented by five contributing structures. The electron configurations of these three species are shown in Figure 7.1. As we can see, the cyclopentadienyl cation, which has four π electrons, has two unpaired electrons in a pair of degenerate π orbitals. In the fully delocalized form, it is antiaromatic. The cyclopentadienide anion, on the other hand, has six π electrons, which are all paired; it is aromatic. The free radical, by having the unpaired electron, is strongly delocalized, but it cannot be aromatic or antiaromatic. The free radical is non-aromatic. The cyclopentadienyl cation is difficult to prepare from simple cyclopentadienyl halides by solvolysis: 5-iodo-1,3-cyclopentadiene does not react with silver perchlorate in propionic acid,[20] for example, and the reactions of derivatives that do solvolyze occur extremely slowly.[21] The cation has been prepared by the reaction between 5-bromocyclopentadiene and antimony pentafluoride at 78K, and shown to be a triplet in the ground state by electron paramagnetic spin resonance spectroscopy.[22] In contrast to this, the unusually high acidity of cyclopentadiene and the stability of its conjugate base were recognized as early as 1900.[23] Its gas-phase acidity relative to ethylene has been determined experimentally[24] as –55 ± 3 kcal mol^{-1}, which is an indication of how strongly stabilized the anion is.

The difference between the effects of electron delocalization in aromatic and antiaromatic compounds raises an especially important point about molecular energies. It is axiomatic that the ground state of a molecule will be at its lowest possible energy. For an

19. Katz, T.J. *J. Am. Chem. Soc.* **1960**, *82*, 3784.
20. (a) Breslow, R.; Hoffman, R.J. *J. Am. Chem. Soc.* **1972**, *94*, 2110. (b) Breslow, R. *Acc. Chem. Res.* **1973**, *6*, 393.
21. Allen, A.D.; Sumonja, M.; Tidwell, T.T. *J. Am. Chem. Soc.* **1999**, *119*, 2371.
22. (a) Saunders: M.; Berger, R.; Jaffe, A.; McBride, J.M.; O'Neill, J.; Breslow, R.; Hoffman, J.M., Jr.; Perchonock, C.; Wasserman, E.; Hutton, R.S.; Kuck, V.J. *J. Am. Chem. Soc.* **1973**, *95*, 3017. (b) Wasserman, I.; Hutton, R.S. *Acc. Chem. Res.* **1977**, *10*, 27.
23. Thiele, J. *Ber Deut. chem. Ges.* **1900**, *33*, 666. (b) Thiele, J. *Chem. Ber.* **1901**, *34*, 68.
24. (a) Rogers, D.W.; McLafferty, F.W.; Podosenin, A.V. *J. Phys. Chem.* **1996**, *100*, 17148. (b) Rogers, D.W.; McLafferty, F.W.; Podosenin, A.V. *J. Phys. Chem.* **1997**, *101*, 4776.

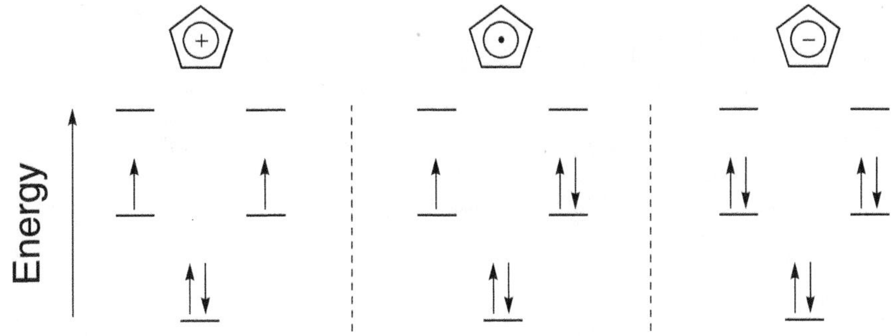

Figure 7.1 Electron configurations of cyclopentadienyl species

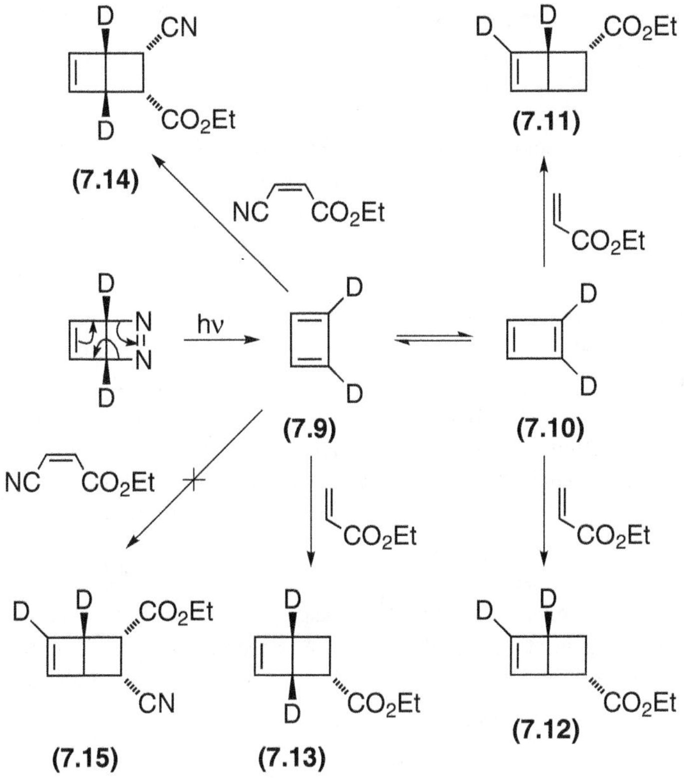

Figure 7.2 Cycloaddition reactions providing evidence of the rectangular shape of cyclobutadiene in the ground state

aromatic molecule, this means that the fully delocalized, aromatic structure will be the ground-state structure of the molecule because it is especially stable. However, what are the options for an antiaromatic compound?

In antiaromatic compounds, the effects of electron delocalization are to destabilize the molecule (i.e., to raise its energy), which means that the delocalized (antiaromatic) form of the molecule is actually higher in energy than the localized (non-aromatic) form. This means that the ground-state energy of these molecules will be lower if they are not delocalized. Evidence that this is actually the case was provided by spectroscopy in an argon matrix[25] and confirmed by the cycloaddition reactions of doubly deuterated cyclobutadiene (Figure 7.2), which led to the conclusion that it is a rectangular (localized, non-aromatic)

25. Masamune, S.; Souto-Bachilier, F.A.; Machiguchi, T.; Bertie, J.E. *J. Am. Chem. Soc.* **1978**, *100*, 4889.

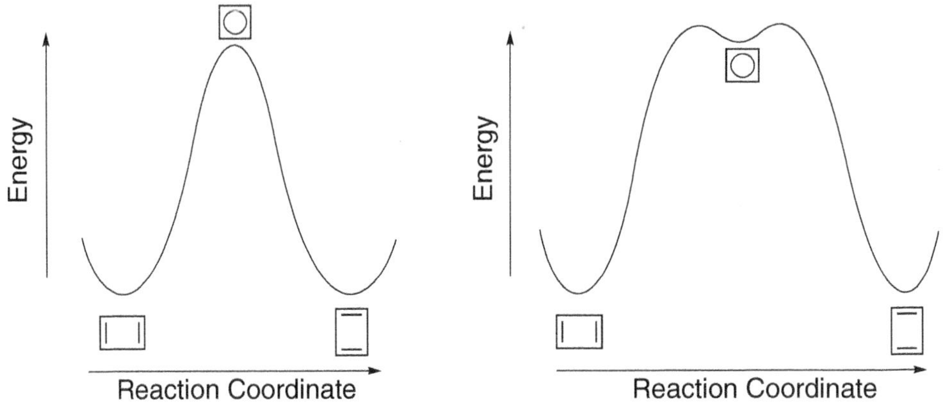

Figure 7.3 Possible reaction pathways interconverting rectangular ground states of the cyclobutadiene molecule

rather than a square (delocalized, antiaromatic) molecule.[26] Trapping experiments with cyclobutadiene-1,2-d_2 (**7.9**, **7.10**) using ethyl acrylate gave all three possible products (**7.11**, **7.12**, **7.13**), which did not rule out the possibility that the cyclobutadiene molecule could be square. However, the same trapping reaction, using the much more reactive dienophile, ethyl *Z*-3-cyanoacrylate, led to the formation of a single product (**7.14**), which is only consistent with the molecule being rectangular in the ground state. The results show that rate of equilibration of the two rectangular forms is fast compared to the rate of the Diels-Alder reaction with ethyl acrylate but slow compared to the rate of the same reaction with the more reactive dienophile. The X-ray crystal structure of 1,3-dimethylcyclobutadiene with a closely associated carbon dioxide trapped in a crystal matrix has revealed a square ring,[27] but only three carbons are planar, and the CO_2 is bent, which raises the question of whether the molecule is actually the cyclobutadiene.

The results we have just discussed suggest that the antiaromatic form of the molecule may not be the ground state, but may, in fact, correspond to a transition state between two or more isomeric, non-aromatic ground states, or to a reactive intermediate in a shallow well between the non-aromatic ground states. These possibilities are shown schematically for rectangular cyclobutadiene in Figure 7.3.

Heterocyclic Compounds: Aromatic or Not? Where is the Lone Pair?

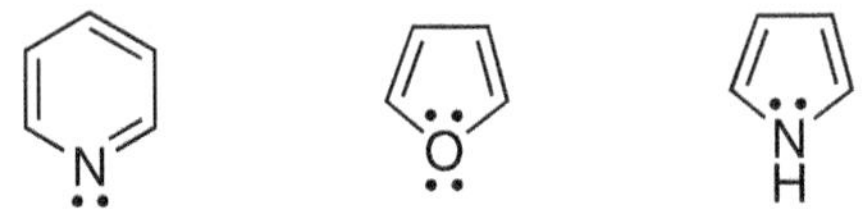

Heterocyclic compounds can also be subjected to the Hückel analysis for aromaticity, bearing in mind that rehybridization of an atom to achieve aromaticity is an energetically favorable process. The presence of lone pairs on the heteroatom thus provides a potential extra source for the needed π electrons, but only one lone pair from any one heteroatom can contribute to the aromatic π electron system. Lone pairs involved in the aromatic π

26. Whitman, D.W.; Carpenter, B.K. *J. Am. Chem. Soc.* **1980**, *102*, 4272.
27. Legrand, Y.-M.; van der Lee, A.; Barboiu, M. *Science* **2010**, *399*, 299.

electron system are usually not basic enough to be protonated by acids. Thus, in the pyridine molecule, the three π bonds in the ring are sufficient to make the molecule aromatic, so the lone pair is localized on nitrogen, in an sp^2 hybrid orbital, and it can be protonated. In the pyrrole molecule, on the other hand, the lone pair on nitrogen is required to make up the aromatic π electron system. Thus, the electrons will formally be in the p orbital that becomes part of the conjugated π system, with the N—H bond being formed by overlap of the nitrogen sp^2 hybrid orbital with the hydrogen 1s orbital. Because the lone pair is not localized on the nitrogen atom, pyrrole is not protonated on nitrogen; pyrrole reacts with acids by protonation on carbon. In the furan molecule, there are two lone pairs on oxygen. One of the two lone pairs on oxygen is required to make up the aromatic π electron system, but the other lone pair is not. This second lone pair, which is in an sp^2 hybridized orbital, can, in principle, be protonated by strong acids (although it usually is not).

Problems

7-2 Which of the following hydrocarbons will be aromatic, and which will be antiaromatic if completely delocalized and planar?

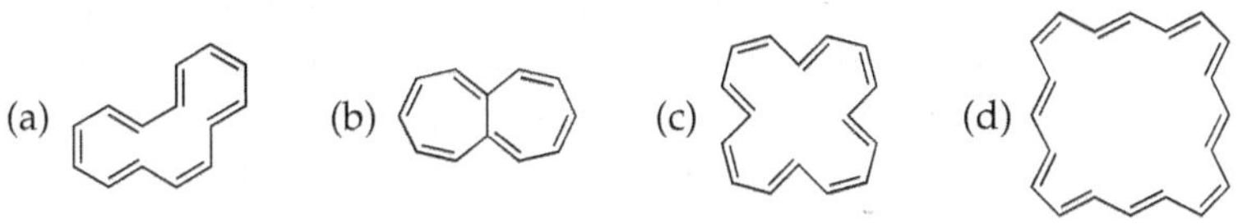

7-3 Which of the following compounds are aromatic? In the aromatic compounds, are the electrons of the lone pair required to give the aromatic π electron system? What will be the hybridization of the heteroatoms in the molecules that are not aromatic? Why?

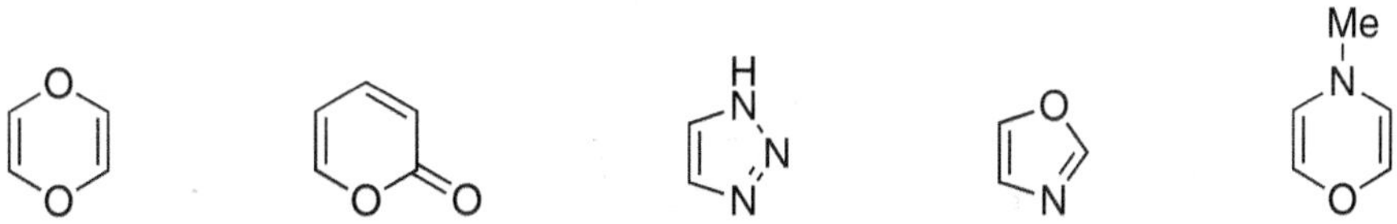

7-4 Anthracene reacts with bromine in carbon disulfide to give the dibromide shown. This compound slowly loses hydrogen bromide on standing to give 9-bromoanthracene.

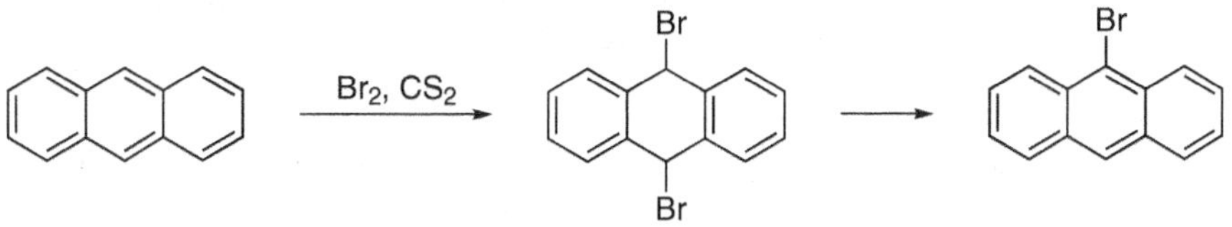

The alternative reaction shown below does not occur.

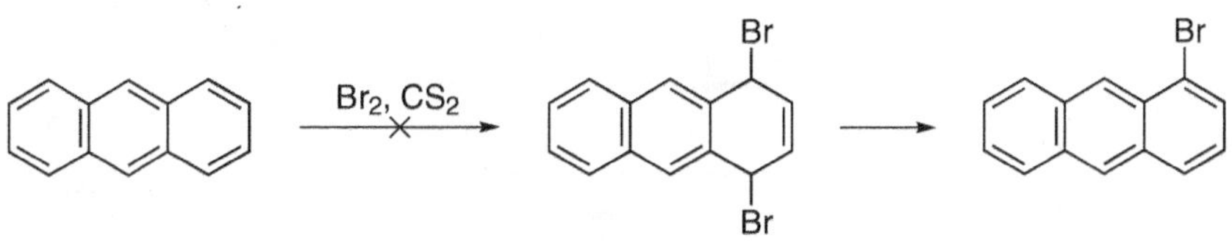

Rationalize why the addition reaction occurs the way it does and why the slow elimination then follows spontaneously.

(continues)

Problems (continued)

7-5 Provide a reasonable explanation for the dipole moments of the series of fulvenes below.

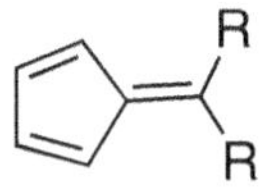

R=H 1.20 D
R=Ph 1.34 D
R=p-ClC_6H_4 0.64 D

Orbital Energies of Hückel π Orbitals of Cyclic Systems

The orbital energies of the π orbitals of an aromatic ring can be calculated directly from Hückel theory, but one particularly useful simplification of the process is to use the circle method of Frost and Musulin.[28] In this method, the ring of the molecule is drawn as a regular polygon inscribed point-down in a circle centered at α, and of radius 2β, where α and β are as defined in Chapter 3 (α represents the Coulomb integral, a measure of how tightly the electron is bound to the nuclei of the atoms, and is constant for all atoms of a particular element; β represents the resonance integral, a measure of the electron energy in the molecular orbital relative to the electron in the isolated atoms). Recall that both α and β are negative because they identify a stabilizing energy. The Hückel π orbital energies for several cyclic C_nH_n systems are gathered in Table 7.1. The population of these orbitals may be determined by filling the orbitals in accord with the Aufbau principle and Hund's rule.

One of the important outcomes of the Hückel analysis of cyclic π systems is the deduction that when there are $4n + 2$ π electrons in the system, the occupied orbitals are all filled, and the molecule is diamagnetic. On the other hand, when the system has $4n$ π

Table 7.1 Energies of Hückel π Orbitals of Cyclic C_nH_n Systems

$$E = \alpha + \beta\cos\left(\frac{2\pi i}{n}\right)$$

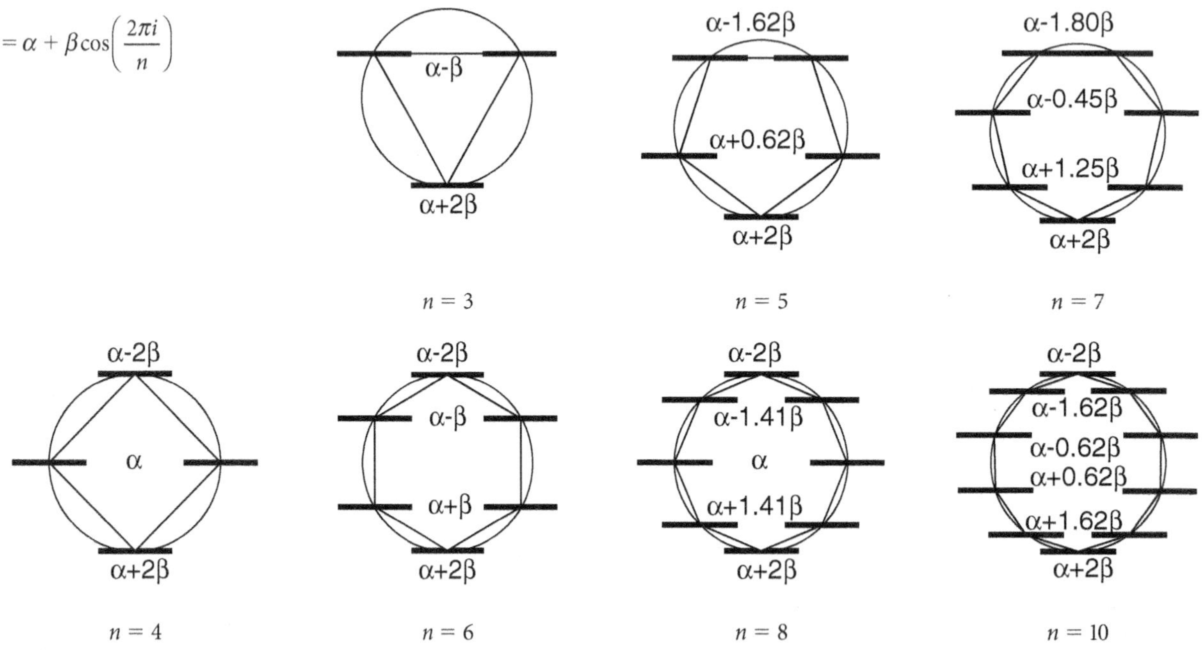

28. Frost, A.A.; Musulin, B. *J. Chem. Phys.* **1953**, *21*, 572.

electrons, the system will have two singly occupied π orbitals, with the result that the compound is predicted to be a diradical and highly reactive. These compounds are designated as ***antiaromatic.*** This analysis also points out another important difference between aromatic and antiaromatic compounds—aromatic compounds are diamagnetic and are therefore repelled by an external magnetic field, whereas antiaromatic compounds are paramagnetic and are therefore attracted by an external magnetic field.[¶]

Hückel Molecular Orbitals in Benzene and Aromatic Compounds

Every π orbital of an aromatic compound is completely delocalized over the entire ring system. The overlap of the six 2p basis orbitals to give the six π orbitals of benzene is shown schematically in Figure 7.4, and the overlap of the five 2p basis orbitals to give the five π orbitals of the cyclopentadienyl system is shown in Figure 7.5.

These orbitals illustrate one characteristic of the Hückel molecular orbitals of cyclic, delocalized π electron systems that contrasts with their open-chain analogs. Recall that none of the Hückel molecular orbitals of open-chain polyenes is degenerate. The number of nodal planes between the 2p basis orbitals, and hence the number of phase dislocations in the π molecular orbital, increases by one as one moves from one π orbital to the next in the set, in ascending order of energy. Thus, in 1,3,5-hexatriene, the lowest energy π molecular orbital, ψ_1, has no phase dislocations between adjacent 2p basis orbitals; ψ_2, the next

---

[¶]. ***Calculating Hückel Orbital Coefficients and Energies of Aromatic Compounds***

The major difference between the Hückel π molecular orbitals of aromatic systems and those of open-chain π-bonded systems is the fact that the π system of aromatic compounds is cyclic and conjugated. This leads to the following secular determinant, which is usually shown in the form where $x = (\alpha - E)/\beta$:

$$\begin{vmatrix} \alpha-E & \beta & 0 & \cdots & 0 & \beta \\ \beta & \alpha-E & \beta & 0\vdots & \vdots & 0 \\ 0 & \beta & \alpha-E & \beta & 0 & \vdots \\ \vdots & 0 & \beta & \ddots & \beta & 0 \\ 0 & \vdots & 0 & \beta & \alpha-E & \beta \\ \beta & 0 & \cdots & 0 & \beta & \alpha-E \end{vmatrix} = 0 \Rightarrow \begin{vmatrix} x & 1 & 0 & \cdots & 0 & 1 \\ 1 & x & 1 & 0 & & 0 \\ 0 & 1 & x & 1 & 0 & \vdots \\ \vdots & 0 & 1 & \ddots & 1 & 0 \\ 0 & & 0 & 1 & x & 1 \\ 1 & 0 & \cdots & 0 & 1 & x \end{vmatrix} = 0$$

This secular equation always has one unique solution at $x = -2$, corresponding to an orbital energy, $E = \alpha + 2\beta$. When the system has an odd number of atoms, the remaining solutions occur as degenerate pairs of orbitals whose energies are given by:

$E_j = \alpha + 2\beta\cos\left[\frac{2(j-1)\pi}{N}\right]$, where N is the number of atoms in the cyclic π molecular orbital system and j is the specific orbital in that system. When the system has an even number of atoms, the same set of degenerate orbitals is obtained, but there is also there is a second unique solution at $x=2$, corresponding to an orbital energy, $E = \alpha - 2\beta$.

The coefficients of the orbitals are not as simply obtained as in the acyclic system, but one simple generalization, at least, can be made: in one of the orbitals of each degenerate set, at least one planar node will pass through one atomic nucleus.

The wave functions of the Hückel orbitals for benzene are:

$\psi_1 = (1/\sqrt{6})(p_1 + p_2 + p_3 + p_4 + p_5 + p_6)$ $\quad E_1 = \alpha + 2\beta$

$\psi_2 = (1/2)(p_2 + p_3 - p_5 - p_6)$ $\quad E_1 = \alpha + \beta$

$\psi_3 = (1/\sqrt{3})(p_1 + \frac{1}{2} p_2 - \frac{1}{2}p_3 - p_4 - \frac{1}{2}p_5 + \frac{1}{2}p_6)$ $\quad E_1 = \alpha + \beta$

$\psi_4 = (1/2)(p_2 - p_3 + p_4 + p_5 - p_6)$ $\quad E_1 = \alpha - \beta$

$\psi_3 = (1/\sqrt{3})(p_1 - \frac{1}{2} p_2 - \frac{1}{2}p_3 + p_4 - \frac{1}{2}p_5 - \frac{1}{2}p_6)$ $\quad E_1 = \alpha - \beta$

$\psi_6 = (1/\sqrt{6})(p_1 - p_2 + p_3 - p_4 + p_5 - p_6)$ $\quad E_1 = \alpha - 2\beta$

The wave functions of the Hückel orbitals for cyclopentadienyl are:

$\psi_1 = (1/\sqrt{5})(p_1 + p_2 + p_3 + p_4 + p_5)$ $\quad E_1 = \alpha + 2\beta$

$\psi_2 = (1/2)(p_2 + p_3 - p_5 - p_6)$ $\quad E_2 = \alpha + 0.618\beta$

$\psi_3 = [\sqrt{(2/5)}](p_1 + 0.309p_2 - 0.809p_3 - 0.809p_4 + 0.309p_5)$ $\quad E_3 = \alpha + 0.618\beta$

$\psi_4 = (1/2)(p_2 - p_3 + p_4 - p_5)$ $\quad E_4 = \alpha - 1.618\beta$

$\psi_3 = [\sqrt{(2/5)}]((p_1 - 0.809p_2 + 0.309p_3 + 0.309p_4 - 0.809p_5)$ $\quad E_5 = \alpha - 1.618\beta$

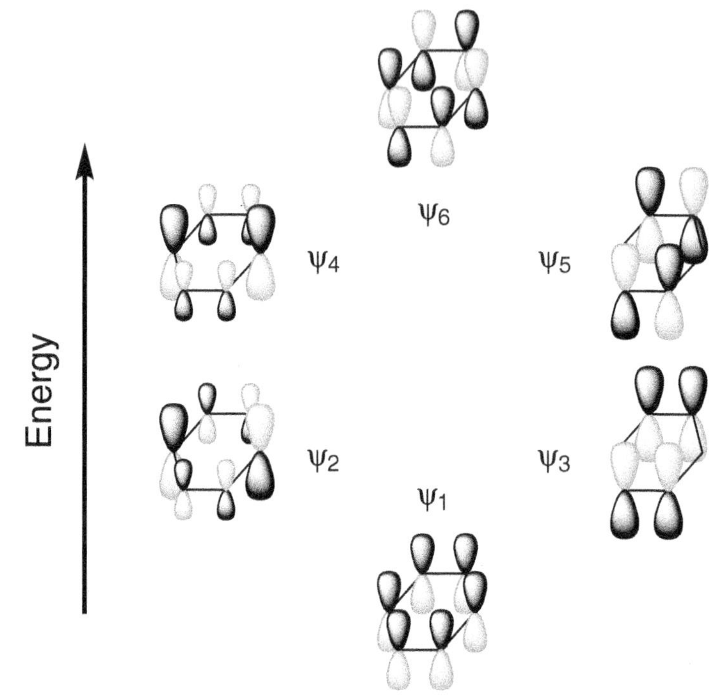

Figure 7.4 The Hückel π molecular orbitals of the benzene molecule. Four of the orbitals occur as two degenerate pairs of orbitals, which differ only in the orientation of the nodal planes perpendicular to the plane of the ring. The basis p orbitals are drawn to convey the relative contribution of each to the molecular orbital, but are not drawn accurately to scale.

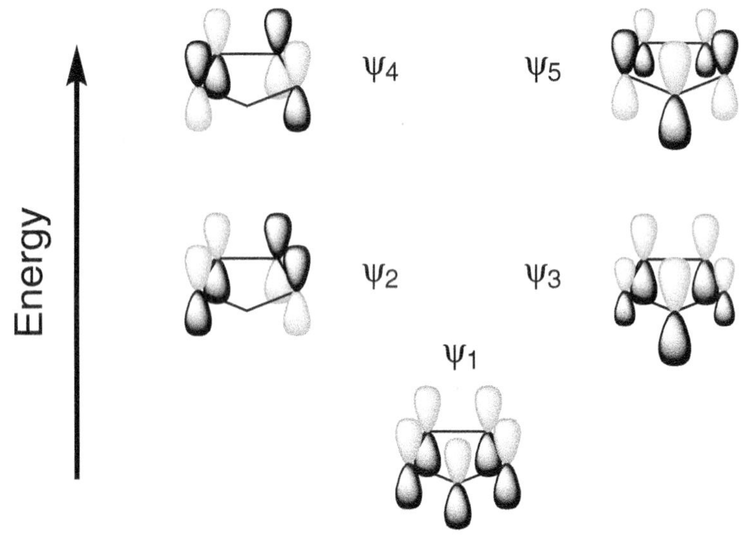

Figure 7.5 The Hückel π molecular orbitals of the cyclopentadienyl system. Four of the orbitals occur as two degenerate pairs of orbitals, which differ only in the orientation of the nodal planes perpendicular to the plane of the ring. The basis p orbitals are drawn to convey the relative contribution of each to the molecular orbital, but are not drawn accurately to scale.

orbital in energy, has one phase dislocation, located such that the C_2 symmetry of the orbital relative to this node is maintained. The highest energy orbital, ψ_6, has five phase dislocations (one between every pair of adjacent 2p basis orbitals).

In the cyclic, delocalized π orbitals in Figures 7.4 and 7.5, we see the differences from the open-chain systems emerge once we get to the degenerate pairs of orbitals. In both ψ_2 and ψ_3, we see that there is a single nodal plane that crosses the ring. But this leads to a pair of phase dislocations in the orbital, rather than one. In ψ_4 and ψ_5, there are two nodal planes that cross the ring and therefore four phase dislocations. The two orbitals in each degenerate pair differ only in the way that the nodal plane is oriented in each. This leads to another characteristic of the π molecular orbitals of cyclic systems—one of each degenerate pair of orbitals has a nodal plane that passes through the nucleus of at least one of the ring atoms.

7.3 How Does One Measure Aromaticity? Criteria for Aromaticity and Its Quantification

There are several different criteria that may be used to determine whether a compound is or is not aromatic. Each has its limitations and drawbacks, but most are of sufficiently general applicability that they can be used with relative confidence. The oldest criterion used to quantify the aromatic character of a molecule is thermodynamic. It rests on the observation that reactions of benzene and other aromatic compounds leading to nonaromatic products are generally less exothermic than one would predict on the basis of the simple Lewis structures of the molecules. Aromatic compounds are relatively easy to form, unusually stable, and less reactive than one would predict a priori. The second criterion is structural; one fundamental property of aromatic compounds is the total delocalization of the π electrons in the cyclic, conjugated π system, with the result that the alternation of the σ bond lengths apparent in open-chain conjugated polyenes is either absent completely or significantly reduced. The third criterion comes from the magnetic properties of aromatic rings. In a magnetic field, the cyclic π electrons of the ring enter into motion that generates a sustained ring current. This leads to observable effects on the ^{1}H NMR spectrum, in particular, where the proton resonances of the aromatic ring are deshielded significantly compared to the protons of a structurally similar alkene.

Based on each of these criteria, antiaromatic compounds are expected to be perfect corollaries of their aromatic counterparts. One, they are expected to be difficult to form, relatively unstable, and highly reactive. Two, they are expected to exist in a localized form rather than a delocalized form to minimize the destabilizing effects of bond delocalization. Three, they are expected to exhibit shielding of nuclei at the periphery of the ring and strong deshielding of nuclei in the center of the ring.

Thermodynamics and Aromatic Resonance Stabilization Energy

All aromatic ring systems are stabilized by resonance. The degree of stabilization can be measured experimentally a number of ways. For example, the resonance stabilization energy of the benzene molecule can be determined by measuring the heat of hydrogenation of benzene and comparing it with the heats of hydrogenation of cyclohexene and 1,3-cyclohexadiene. The specific value obtained depends on the assumptions made during the calculation, but the most widely used methods yield values near 36 kcal mol^{-1} for benzene.

The calculation for benzene relies on the fact that cyclohexene and benzene both produce cyclohexane on exhaustive hydrogenation. The enthalpies of hydrogenation for cyclohexene, 1,3-cyclohexadiene, and benzene have been measured to a high degree of precision.[29] The enthalpy for hydrogenation of cyclohexene is −28.59 kcal mol^{-1}, which one may extrapolate to the enthalpy of hydrogenation of the hypothetical, localized 1,3,5-cyclohexatriene by tripling this value to −85.77 kcal mol^{-1}. When benzene is hydrogenated, however, the enthalpy of the reaction is found to be −49.73 kcal mol^{-1}. Thus, the hydrogenation of benzene is 36.1 kcal mol^{-1} less exothermic than the hydrogenation of the hypothetical 1,3,5-cyclohexatriene.

---

29. (a) Kistiakowski, G.B.; Ruhoff, J.R.; Smith, H.A.; Vaughan, W.E. *J. Am. Chem. Soc.* **1936**, *58*, 137. (b) Kistiakowski, G.B.; Ruhoff, J.R.; Smith, H.A.; Vaughan, W.E. *J. Am. Chem. Soc.* **1936**, *58*, 146.

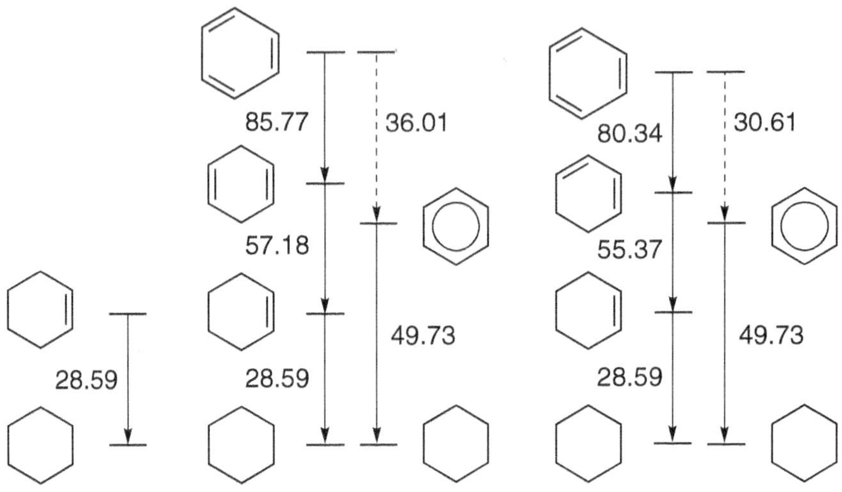

Figure 7.6 Estimating the aromatic resonance stabilization energy of benzene by hydrogenation

This model does not take into account the effects of conjugation, which are apparent when one compares the enthalpy of hydrogenation of cyclohexene (–28.59 kcal mol^{-1}) with that of 1,3-cyclohexadiene (–55.37 kcal mol^{-1}). This value is 1.81 kcal mol^{-1} less exothermic than twice the value for cyclohexene. We may assume that this represents the extra stabilization from (non-aromatic) conjugation of two adjacent π bonds. If we assume that such conjugation must be taken into account for the three pairs of conjugated alkene π bonds in the hypothetical 1,3,5-cyclohexatriene ring, the value of the aromatic resonance stabilization energy of benzene must be reduced by three times this value, which leads to a value of 30.6 kcal mol^{-1} for the aromatic resonance stabilization energy of the benzene molecule. It is interesting to note that the enthalpy of hydrogenation of the first π bond in benzene is *endothermic*. These calculations are summarized in Figure 7.6.

An estimate of the resonance stabilization energy of benzene can also be obtained from bond energies by measuring the heats of combustion.[30] The average bond energies of a wide range of hydrocarbon molecules have been determined with sufficient accuracy that they can be used with considerable confidence in determining ΔH_f of hydrocarbons from the enthalpy of combustion of those hydrocarbons. When the bond energies are used to calculate the ΔH_f of benzene, the calculated value is found to be –4180 kJ mol^{-1}. When the value is calculated from the measured ΔH for combustion, the value is found to be –4343 kJ mol^{-1}. Thus, the aromatic resonance stabilization energy of benzene found by this method is 163 kJ mol^{-1} (or 39.0 kcal mol^{-1}). This discrepancy between the values obtained from catalytic hydrogenation and combustion is reproduced for a range of aromatic compounds. The values obtained from combustion are typically higher by approximately 10% of the value, and they are usually considered to overestimate the actual value by this quantity.

Regardless of which calculations one uses, they come with a caveat. Although the heat of hydrogenation, or the heat of combustion of benzene is a measurable quantity, *the aromatic resonance stabilization energy is not;* it must be deduced from the experimental data. However, the changes necessary to convert 1,3,5-cyclohexatriene to benzene require not only the complete delocalization of the π electrons but also the lengthening of the double bonds and the shortening of the σ bonds. This means that inferring the resonance

30. Pauling, L. In Gilman, H., Ed. *Organic Chemistry. An Advanced Treatise*, 2nd Ed., (John Wiley & Sons: New York, 1943), ch. 26, p1943.

Table 7.2 Aromatic Resonance Stabilization Energies

Compound	Stabilization Energy (kcal mol^{-1})	Compound	Stabilization Energy (kcal mol^{-1})
	36*; 41†	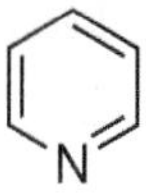	27*; 40†
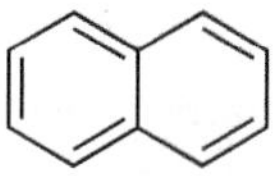	61*; 77†	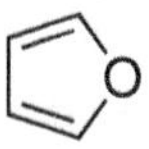	16*; 24†
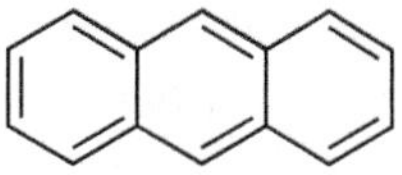	83.5[a]; 116†	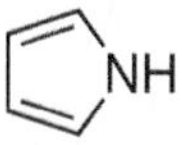	21*; 24†
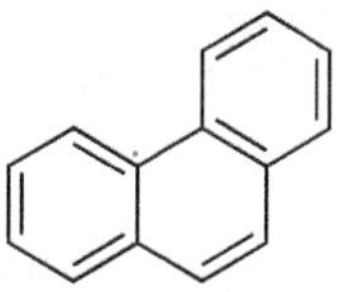	91.3*; 130†	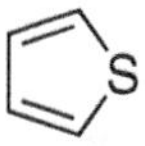	28*; 29†
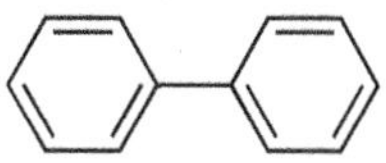	70*		

*Values taken from Reference 30.
†Values taken from Reference 31.

stabilization energy of benzene is subject to the details of the model of the hypothetical 1,3,5-cyclohexatriene that is used. We will discuss the limitations of that model later in this chapter, when we examine computational approaches to answering the question of just what constitutes aromaticity. Aromatic resonance stabilization energies have been calculated for a number of organic compounds, and a selection of these are gathered in Table 7.2.[31,32]

Recognizing that the values obtained experimentally all depend on the model chosen for the non-aromatic species, a more modern approach has been to calculate the enthalpies for ***homoisodesmic reactions***: reactions in which the numbers and types of bonds in the "reaction" are conserved. These reactions do not need to be real—capable of actually being carried out in the laboratory—but they do give reasonable estimates of the stabilization energy. One of the most popular types of homoisodesmic reactions involves hydrogen transfer from one reactant to another. For example, one such reaction involves the conversion of benzene and butane to 1,3-cyclohexadiene and *cis*-2-butene,[33] which gives an estimate of the resonance stabilization energy for benzene of 34.0 ± 0.4 kcal mol^{-1}, well in line with the values obtained by other methods.

31. Wheland, G.W. *Resonance in Organic Chemistry* (John Wiley & Sons: New York, 1955), p. 98 *ff.*
32. Ingold, C.K. *Structure and Mechanism in Organic Chemistry* (Cornell University Press: Ithaca, N.Y., 1953), p. 183.
33. Enthalpies used were from Pedley, J.B. *Thermochemical Data and Structures of Organic Compounds*; Thermodynamics Research Center: College Station, TX, 1994; vol. 1

Structural Evidence of Aromaticity

The bond distances and angles of the benzene molecule are known with a very high degree of precision from a variety of physical methods including neutron and electron[34] diffraction, X-ray crystallography,[35] and Raman spectroscopy.[36] The ring is a planar hexagon, with all carbon-carbon bond lengths equal at 1.397 Å and all carbon-hydrogen bond lengths equal at 1.084 Å. The C—C bond lengths can be compared to 1.532 Å for ethane and 1.32 Å for ethylene. The single bond length in 1,3-butadiene is 1.48 Å. Thus, we find that there is a substantial bond alternation in the non-aromatic 1,3-butadiene (the bond distances alternate between 1.32 and 1.48 Å). This may be contrasted with benzene, where the bond distances are all the same, and intermediate between these two values. This loss of C—C bond alternation is generally viewed as a diagnostic feature of aromatic systems, although only benzene and similar totally symmetrical, monocyclic aromatic species can achieve perfect uniformity of C—C bond distances. In naphthalene and other polycyclic aromatic hydrocarbons, the C—C bond distances vary somewhat, but the range is still quite restricted (typically the values lie between 1.36 and 1.45 Å). This also holds true in the aromatic annulenes. In [18]annulene, for example, the C—C bond lengths are not uniform but are 1.38 and 1.42 Å.[37] Computations have shown that the return to localized rather than delocalized structures for $(4n + 2)$ systems occurs at [30]annulene, although the molecule still retains significant aromatic stabilization.[38]

In 1952, Clar suggested[39] that the six-membered ring with three double bonds in it (i.e., the true "benzene" ring) is especially stable, and that it therefore ought to contribute more importantly to the overall resonance hybrid in polycyclic aromatic hydrocarbons. According to the Clar model of polycyclic hydrocarbons, the contributing structures with the largest number of disjoint aromatic sextets will be the most important. If we apply the Clar rule to two of the five Kekulé canonical forms of the structure of phenanthrene, for example (Figure 7.7), we find that in one of the canonical forms (the lower one), only the middle ring is a benzene-like aromatic sextet, and that in the other canonical form (the upper one), there are two benzene-like aromatic sextets. This means that canonical forms of the upper type are expected to contribute more to the overall resonance hybrid than are structures like the lower canonical form. This, in turn, leads to the conclusion that the 9,10 bond in phenanthrene ought to exhibit a greater level of reactivity than any of the other π bonds. It does.

The greatest strength of the Clar model is its simplicity. Electrons in Clar structures may only be attributed to one ring, so the Clar model requires that no two aromatic sextets in a simple fused-ring aromatic compound may share a common side—Clar sextets must always be in disjoint rings. In a Kekulé structure where two rings sharing a common side simultaneously have a sextet, the sextet must be assigned to one of the rings; once electrons have been assigned to one ring of a structure, they cannot simultaneously be assigned to another—thus, only one of the rings is a true sextet (and therefore is written with the circle in it), and the other ring is not (and is therefore written with two double bonds, instead). Clar analyses of the canonical forms of a large number

34. (a) Jones, P.R.L. *Trans. Faraday Soc.* **1935**, *31*, 1036. (b) Schomaker, V.; Pauling, L. *J. Am. Chem. Soc.* **1939**, *61*, 1769. (c) Bacon, G.E.; Curry, N.A.; Wilson, S.A. *Proc. Roy. Soc.* **1964**, *A279*, 98.

35. (a) Cox, E.D.; Cruickshank, D.W.J.; Smith, J.A.S. *Proc. Roy. Soc.* **1958**, *A247*, 1. (b) Piermarini, G.J.; Mighell, A.D.; Weir, C.E.; Block, S. *Science* **1969**, *165*, 1250. (c) Fourme, R.; André, D.; Renaud, M. *Acta Crystallogr.* **1971**, *B27*, 1275.

36. Langseth, A.; Stoicheff, B.P. *Can. J. Phys.* **1956**, *34*, 350.

37. (a) Bregman, J.; Hirshfeld, F.L.; Rabinovich, G.; Schmidt, G.M.J. *Acta Crtstallogr.* **1965**, *19*, 227. (b) Hirshfeld, F.L.; Rabinovich, G. *Acta Crystallogr.* **1965**, *19*, 235.

38. Choi, C.H.; Kertesz, M. *J. Chem. Phys.* **1998**, *108*, 6681.

39. (a) Clar, E. *Aromatische Kohlenwasserstoffe*, 2nd ed. (Springer: Berlin, 1952). (b) Clar, E. *Polycyclic Hydrocarbons* (Academic: London, 1964). (c) Clar, E. *The Aromatic Sextet* (John Wiley: London, 1972.).

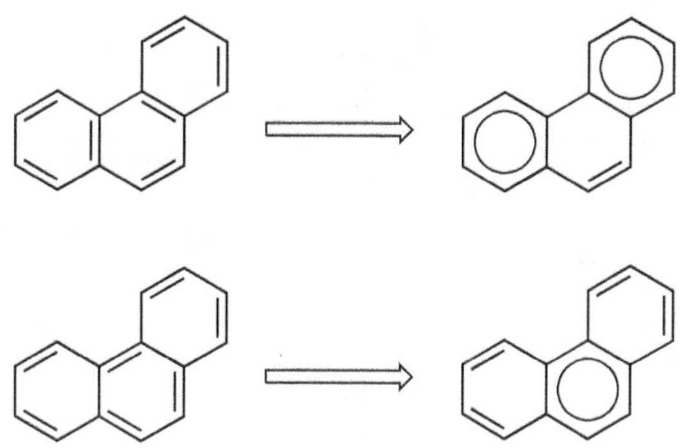

Figure 7.7 Clar resonance structures for two canonical forms of phenanthrene compared to classical Kekulé canonical forms. The structure with two full six-electron rings is the more important contributor to the resonance hybrid.

of polycyclic aromatic hydrocarbons quickly illustrates the usefulness of this simple model in predicting reactivity in these compounds.

Magnetic Evidence of Aromaticity

One of the first experimental measurements that could be associated with aromaticity in a quantitative sense is derived from the values of the magnetic susceptibility. This criterion for aromaticity is known as the ***magnetic exaltation***, or, earlier, the ***incrément de délocalisation***.[40] In homogeneous medium (e.g., in solution or the gas phase), one may define the mean molar magnetic susceptibility, $\overline{\chi}^{\mathrm{m}}$, which is defined as one-third of the trace of the magnetic anisotropy tensor:

$$\overline{\chi}^{\mathrm{m}} = \frac{1}{3}\{\chi_{xx}^{\mathrm{m}} + \chi_{yy}^{\mathrm{m}} + \chi_{zz}^{\mathrm{m}}\} = \frac{1}{3}\mathrm{tr}(\chi^{\mathrm{m}}) \qquad (7.1)$$

There are a large number of mean molar magnetic susceptibilities that have been measured, and by 1910, it had been found that they generally follow a simple set of additivity rules.[41] The magnetic exaltation, $\overline{\chi}^{\mathrm{m}}$, is the difference between the measured value of and the value calculated on the basis of Pascal's additivity principle ($\overline{\chi}^{\mathrm{m}}_{\mathrm{calc}}$):

$$\Lambda^{\mathrm{m}} = x^{\mathrm{m}} - x^{\mathrm{m}}_{calc} \qquad (7.2)$$

where the value of is $\overline{\chi}^{\mathrm{m}}_{\mathrm{calc}}$ calculated on the assumption that there is no delocalization of the π electrons in the molecule. The values of the molar magnetic susceptibility are all negative for diamagnetic compounds, so positive value of the magnetic exaltation indicates that the measured magnitude of the molar magnetic susceptibility of the molecule is greater than the calculated value; this has been proposed as being an indicator of aromaticity.[42]

The origins of the magnetic exaltation in aromatic compounds can be traced to the ***ring current***. The cyclic, delocalized nature of the π electron system of an aromatic compound leads to a highly characteristic interaction with an external magnetic field. When an aromatic compound is placed in a strong external magnetic field, the electrons in the π

40. (a) Pacault, A. *Ann. Chim. (12e série)* **1946**, *1*, 527. (b) Hoarau, A. *Ann. Chim. (13e série)* **1956**, *1*, 544. (c) Bergmann, E.D.; Hoarau, J.; Pacault, A.; Pullman, A.; Pullman,B. *J. Chim. Phys. Phys.-Chim. Biol.* **1952**, *49*, 474. (d) Pacault, A.; Hoarau, J.; Marchand, A. In *Advances in Chemical Physics* (Wiley: New York, 1961); Vol. 3, pp. 171–238.

41. (a) Pascal, P. *Ann. Chim. Phys.* **1910**, *19*, 5. (b) Pascal, P. *Ann. Chim. Phys.* **1912**, *25*, 289. (c) Pascal, P. *Ann. Chim. Phys.* **1913**, *29*, 218. The values for several substituents and compounds are gathered in: (d) Ingold, C.K. *Structure and Mechanism in Organic Chemistry* (Cornell University Press: Ithaca, NY, 1953), pp. 185–196.

42. (a) Benson, R.C.; Norris, C.L.; Flygare, W.H.; Beak, P. *J. Am. Chem. Soc.* **1971**, *93*, 5591. (b) Dauben, H.J.; Wilson, J.D.; Laity, J.L. *J. Am. Chem. Soc.* **1968**, *90*, 811.

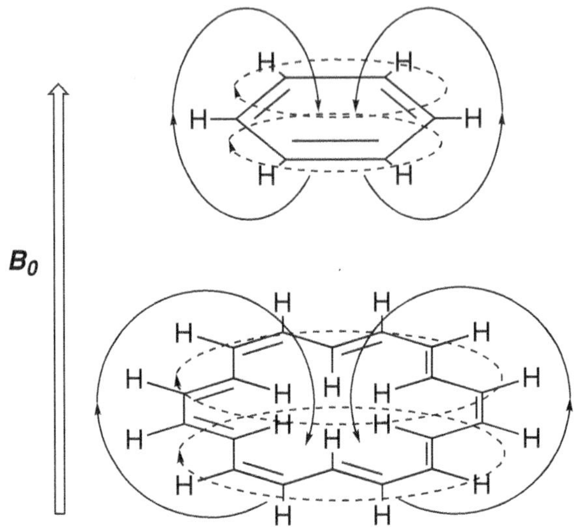

Figure 7.8 Ring currents in aromatic molecules. The direction of the ring current is indicated by the dashed lines, and the induced field by solid lines.

orbitals begin to cycle, generating what is known as a ***ring current*** (Figure 7.8). In an aromatic compound, this ring current opposes the external magnetic field in the center of the aromatic ring, with the result that it reinforces the external field at the periphery. The effect of the ring current is to deshield the protons attached to an aromatic ring by some 2 ppm relative to simple alkenes of similar structure. Aromatic compounds are said to exhibit a ***diatropic ring current***, where the ring current generates an induced magnetic field that opposes the external field at the center of the ring, while antiaromatic compounds exhibit a ***paratropic ring current***, where the ring current generates an induced magnetic field that reinforces the external field at the center of the ring.

The real test of using ring currents to probe aromatic character came with the ^1H NMR examination of [18]annulene, which is known to be aromatic. The spectrum is temperature-sensitive, and at −70°C, it shows two resonances in the ratio of 2:1. The more intense resonance is at 9.28 ppm and the less intense resonance is at −2.99 ppm. The low-field resonance is attributed to the twelve protons on the outside of the ring, where the ring current reinforces the external field, and the high-field resonance is attributed to the six protons on the inside of the ring, where the ring current opposes the external magnetic field.[43]

The first simple model for estimating the magnitude of the ring current (and, therefore, the resonance energy) of a compound based on the chemical shifts of the protons in the molecule was developed by Hess, Schaad, and Kakagawa.[44] In their model, the ring current is proportional to the difference between the chemical shifts of the "inside" protons and the "outside" protons in the series of compounds shown in Figure 7.9, which include compounds with cyclic π electron systems containing from 14 to 30 π electrons. The ^1H NMR spectra of these compounds show a remarkable regularity in the chemical shifts of the inside and outside protons. In compounds with $(4n + 2)$ π electrons, the outside protons are considerably downfield from the inside protons. In compounds with $(4n)$ π electrons, the reverse is true. In addition, as the number of π electrons increases, the difference in the chemical shifts of the inside and outside protons decreases, which is consistent with the effects of aromatic resonance stabilization decreasing as the number of π electrons increases.

43. Haddon, R.C.; Haddon, V.R.; Jackman, L.M. *Curr. Top. Chem. Res.* **1971**, *16*, 105.
44. Hess, B.A., Jr.; Schaad, L.J.; Nakagawa, M. *J. Org. Chem.* **1977**, *42*, 1669.

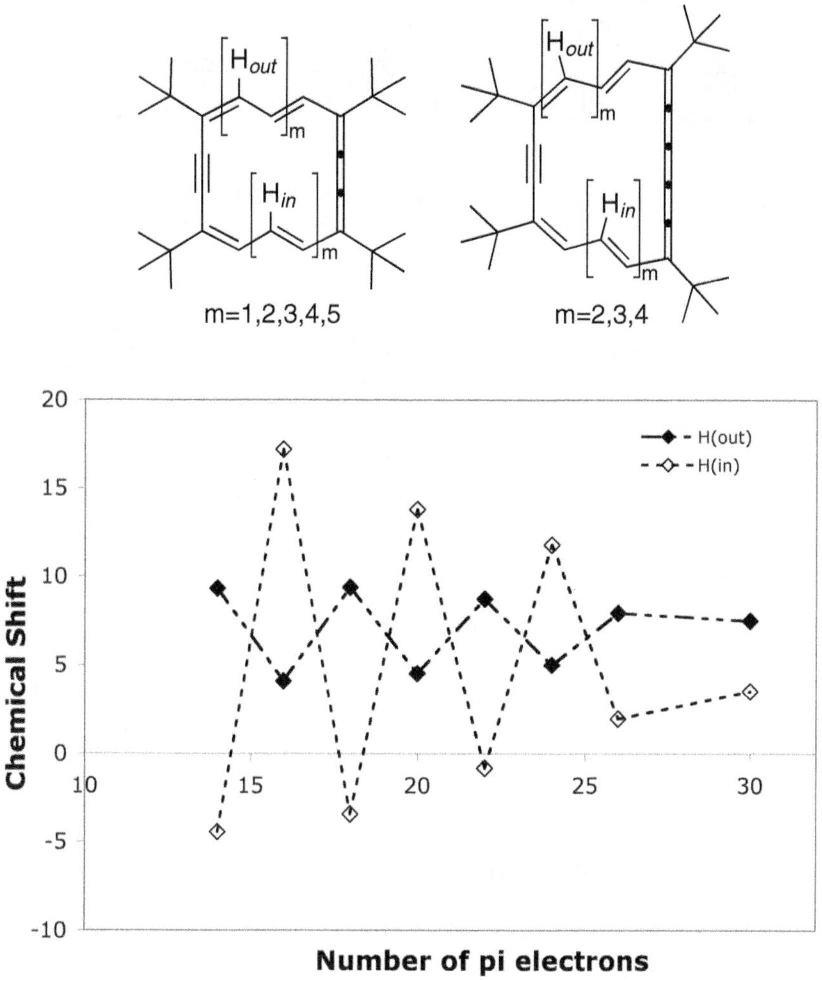

Figure 7.9 Variations in the chemical shifts of the inside and outside protons of dehydro-[n]annulenes

Computational Evaluation of Aromaticity

The advent of high-speed computing and the high degree of confidence that can be placed in the results of high-level computational results have revolutionized the ways that questions of aromaticity are now decided.

Within the past two decades, the difficulties of working with the ring current, which is not directly measurable, have been overcome by using the concept of the ***nucleus-independent chemical shift***, or ***NICS***, which is defined as the magnetic shielding at the geometric center of the aromatic ring (or at a location above the center of the ring).[45] This value, which can be readily calculated to high levels of precision and is only modestly dependent on the basis set used in the computation, was one of the first computational variables that found widespread applicability in defining aromaticity. In an aromatic system, the NICS value is negative (typically the more negative the value, the more strongly aromatic the ring), and in antiaromatic systems, the NICS value is positive. A number of NICS values for representative aromatic and antiaromatic compounds, as well as for compounds that are not aromatic, are presented in Figure 7.10.

Homoisodesmic reactions, in particular, lend themselves well to the computational determination of aromatic resonance stabilization energies because the deficiencies in the

45. (a) Schleyer, P.v.R.; Maerker, C.; Dransfeld, A.; Jiao, H.; van Eikema Hommes, N.J.R. *J. Am. Chem. Soc.* **1996**, *118*, 6317. (b) Jiao, H.; Schleyer, P.v.R.; Mo, Y.; McAllister, M.A.; Tidwell, T.T. *J. Am. Chem. Soc.* **1997**, *119*, 7075.

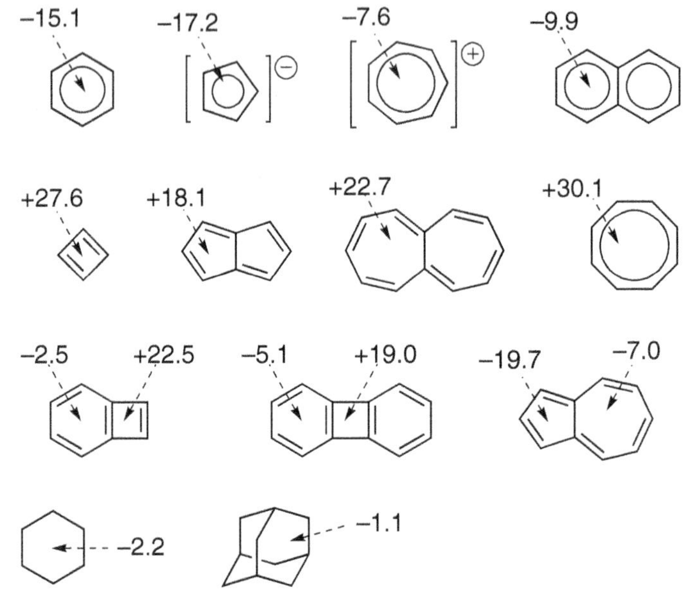

Figure 7.10 Nucleus-independent chemical shift values for a series of aromatic, antiaromatic, and non-aromatic rings. The values are taken from Reference 45 (a).

individual calculations tend to cancel out.[46] With the level of confidence in modern basis sets for computation, one can obtain quite good estimates of the resonance stabilization energy. For example, the reaction between isopropyl cation and a series of conjugated and saturated hydrocarbons has been used to obtain computed energies for the hydride transfer reaction that are in excellent agreement with the observed values.[47]

Another computational measure of aromaticity is the harmonic oscillator model of aromaticity (HOMA) index.[48] This is a geometry-based index, given by the relationship:

$$\text{HOMA} = 1 - \frac{\alpha}{n}\sum(R_{opt} - R_i)^2 \qquad (7.3)$$

where n is the number of bonds in the ring, R_{opt} is the optimal bond distance in an aromatic system (equal to 1.388Å), and α is a normalization factor (for CC bonds, $\alpha = 257.7$) that gives HOMA = 0 for a non-aromatic system (the localized Kekulé structure of benzene) and HOMA = 1 for a system in which all bond distances are equal to the optimal value. For benzene, the HOMA is calculated to be 0.9965. The HOMA can be dissected into two independent contributions, which describe whether the decrease in π-electron delocalization is due to bond alternation or bond elongation. The effect of bond alternation is designated as GEO and is defined as:

$$\alpha(R_{opt} - R_{av})^2 \qquad (7.4)$$

The effect of bond elongation is designated as EN and is defined as:

$$\frac{\alpha}{n}\sum(R_{av} - R_i)^2 \qquad (7.5)$$

Thus, the overall decrease in aromaticity is given by HOMA = 1 − EN − GEO. The HOMA index has generally been a useful index of aromaticity, and it correlates well with NICS values.

46. George, P.; Trachtman, M.; Boch, C.W.; Brett, A.M. *Theor. Chim. Acta* **1975**, *38*, 121.

47. Wiberg, K.B. *Chem. Rev.* **2001**, *101*, 1317.

48. (a) Kruszewski, J.; Krygowski, T.M. *Tetrahedron Lett.* **1972**, 3839. (b) Krygowski, T.M. *J. Chem. Inf. Comput. Sci.* **1993**, *33*, 70. (c) Bodwell, G.J.; Bridson, J.N.; Cyrański, M.K.; Kennedy, J.W.J.; Krygowski, T.M.; Mannion, M.R.; Miller, D.O. *J. Org. Chem.* **2003**, *68*, 2089.

Worked Problem

7-2 Using the Clar structures, predict the structure of the most likely carbocation to be produced by the protonation of anthracene with deuterium.

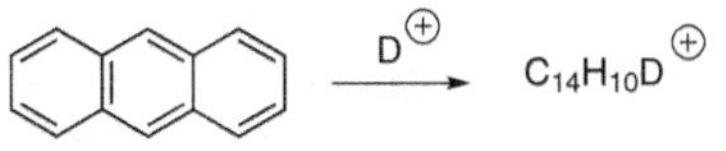

§Answer below

Problems

7-6 Draw the Clar structures for each of the polycyclic aromatic hydrocarbons shown. Benzenoid aromatic rings are highly resistant to reduction by dissolving metals, so the major Clar contributing structures tend to show where the most chemical resistance in the molecule resides. Based on the Clar structures, deduce what will be the most likely dihydro compounds formed from each of these polycyclic aromatic hydrocarbons by reduction with sodium in ethanol.

(a) 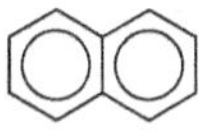 (b) 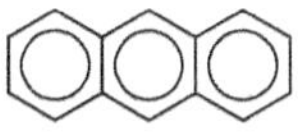 (c) 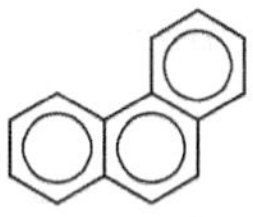 (d) 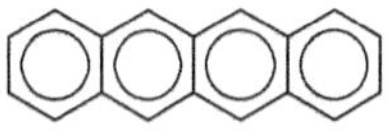

(e) 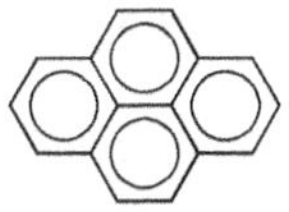 (f) 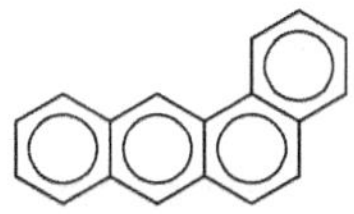

(continues)

§ Answer for Worked Problem:
The solution to this problem is best approached from the perspective of the cation since the Clar structures for the parent hydrocarbon do not give a clear difference between choices.

(a) Protonation at position 1:

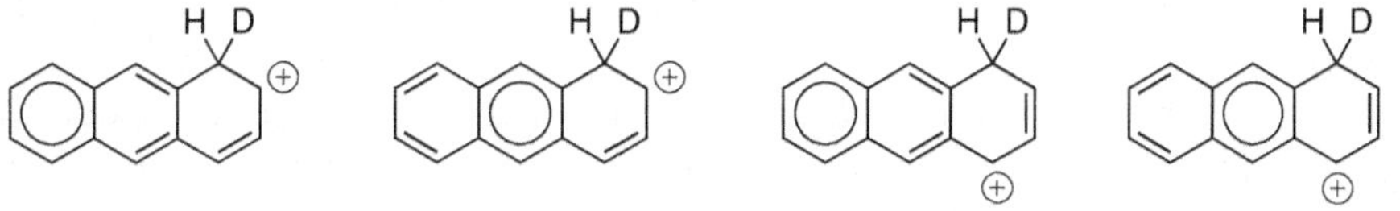

There are four structures, each with one Clar ring, but there are two different major contributing structures with each cation where neither of these sextets is interrupted.

(b) Protonation at position 2:

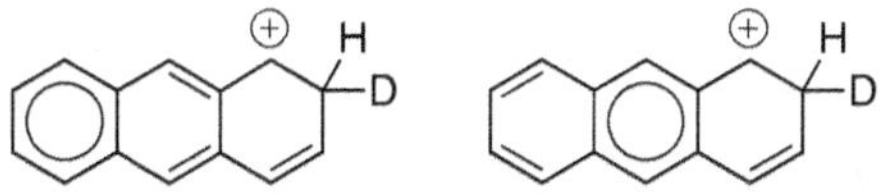

There are two structures, each with one Clar ring, but there is only one major contributing structures for each cation where neither of these sextets is interrupted.

(c) Protonation at position 9:

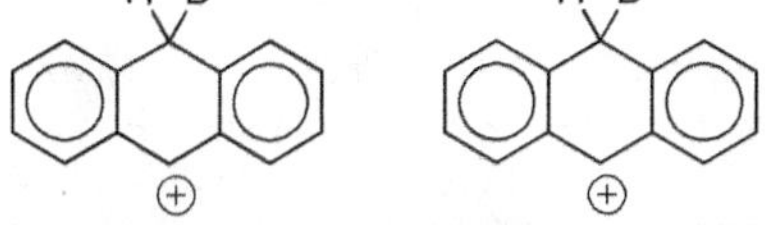

This ion is the only one for which there is a structure with *two* Clar rings. This is expected to be the major product of protonation.

Problems (continued)

7-7 What can be deduced about the structures of the following molecules from the chemical shifts of their protons?

(a) 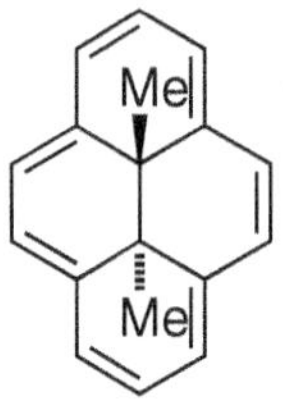

CH_3: δ -4.25
ring H: δ 8.0-8.7

(b) 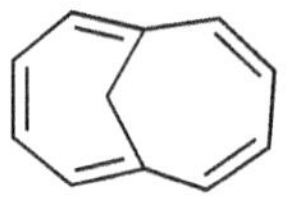

CH_2: δ -0.5
vinyl H: δ 7.8

(c) 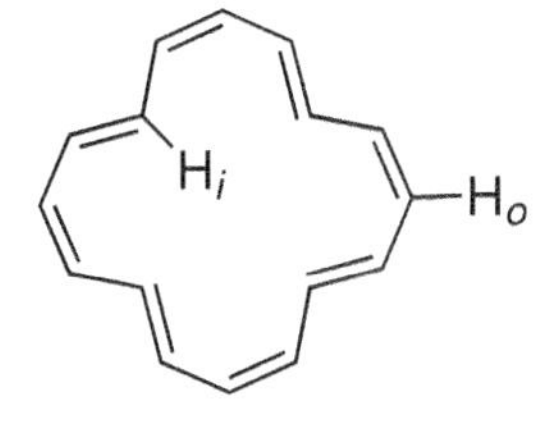

H_i: δ 10.4; H_o: δ 5.4

Another computational measure of aromaticity is the ***Dewar resonance energy*** (DRE).[49] This quantity is simply the difference between the measured enthalpy of formation and the same enthalpy calculated for the completely localized system. The computational work of Dewar has shown that the calculated values for the enthalpy of formation of the localized systems are actually quite dependable, so these resonance stabilization energies are actually fairly reliable. For totally unsaturated systems containing only carbon, hydrogen, and oxygen, $C_nH_mO_p\ddot{O}_q$, where O is an oxygen atom linked by two σ bonds and Ö is a carbonyl oxygen, Baird[50] has shown that the DRE can be calculated using Equation 7.6:

$$\text{DRE} = 7.435n - 0.605m - 32.175p - 29.38q - \Delta H_f^0\ (\text{experimental}) \qquad (7.6)$$

Using this approach, one finds that a compound is aromatic if the DRE is positive, antiaromatic if the DRE is negative, and non-aromatic if the DRE is zero. Compounds with DRE values close to zero are considered to be neither aromatic nor antiaromatic. Baird has calculated values of the DRE for a number of compounds, and they illustrate this point quite well. For benzene, the DRE is calculated to be +21 kcal mol^{-1}; for cyclodecapentaene ([10]annulene), it is +6 kcal mol^{-1}; for cyclobutadiene, it is –17 kcal mol^{-1}; and for planar cyclooctatetraene, it is –10 kcal mol^{-1}.

The DRE can be placed on a more useful comparative basis by dividing the total DRE by the number of π electrons in the π system. This gives the ***resonance energy per π electron* (REPE)**.[51] Using this parameter, one finds that a series of simple polycyclic aromatic hydrocarbons have comparable REPE values, despite their total DREs being quite different. Shortly after this, a modified parameter, the ***resonance energy per π bond* (REPB)** was proposed as being more widely applicable, especially in charged systems.[52]

Another value that can be used as a criterion for aromaticity is the ***absolute hardness***, η, of the compound in question. This is defined as one half of the energy gap between the HOMO and LUMO of the molecule in question. Zhou, Parr, and Garst have shown that this correlates with a number of other indexes of aromaticity,[53] including the REPE.

---

49. Dewar, M.J.S.; de Llano, C. *J. Am. Chem. Soc.* **1969**, *91*, 789.

Michael James Steuart Dewar (1918–1997) was educated at Oxford University (DPhil, 1942). His career in theoretical organic chemistry spanned over half a century on two continents, at Queen Mary College, Chicago, Texas at Austin, and Florida at Gainesville. For more biographical detail, see: (a) Murrell, J. N. *Biogr. Mem. Fell. Roy. Soc.* **1998**, *44*, 129; (b) Dewar, M.J.S. *A semiempirical life* (American Chemical Society: Columbus, 1992).

50. Baird, N.C. *J. Chem. Educ.* **1971**, *48*, 509.

51. (a) Hess, B.A., Jr.; Schaad, L.J. *J. Am. Chem. Soc.* **1971**, *93*, 305. (b) Hess, B.A., Jr.; Schaad, L.J. *J. Am. Chem. Soc.* **1971**, *93*, 2413. (c) Hess, B.A., Jr.; Schaad, L.J. *J. Am. Chem. Soc.* **1973**, *95*, 3907. (d) Hess, B.A., Jr.; Schaad, L.J. *J. Org. Chem.* **1971**, *36*, 3418. (e) Hess, B.A., Jr.; Schaad, L.J.; Holyoke, C.W., Jr. *Tetrahedron* **1972**, *28*, 3657. (f) Hess, B.A., Jr.; Schaad, L.J.; Holyoke, C.W., Jr. *Tetrahedron* **1972**, *28*, 5299. (g)) Hess, B.A., Jr.; Schaad, L.J.; Holyoke, C.W., Jr. *Tetrahedron* **1975**, *31*, 295. (h) Schaad, L.J.; Hess, B.A., Jr. *Chem. Rev.* **2001**, *101*, 1465.

52. Aihara, J.-I. *Bull. Chem. Soc. Japan* **1977**, *50*, 3057.

53. (a) Zhou, Z.; Parr, R.G.; Garst, J.F. *Tetrahedron Lett.* **1988**, 4843. (b) Zhou, Z.; Parr, R.G. *J. Am. Chem. Soc.* **1989**, *111*, 7371.

Bird,[54] however, found that this relationship does not always work well with heterocyclic compounds. In these cases, he found that the absolute hardness scale used by Zhou, Parr, and Garst needed to be modified and a value obtained on the basis of the molar refractivity used. This restored a good correlation with the REPE values for as wide range of heterocyclic aromatic compounds.

7.4 Möbius Aromaticity[55]

In 1964, Heilbronner[56] published a seminal paper in which he suggested that it might be possible for a sufficiently large cyclic π system to twist without losing conjugation, such that there would be a phase dislocation in the system (Figure 7.11) and that such a system could exhibit aromatic character. Möbius aromaticity has since been defined by the International Union for Pure and Applied Chemistry (IUPAC)[57] as a monocyclic array of orbitals in which there is a single out-of-phase overlap (or, more generally, an odd number of out-of-phase overlaps) that reveals the opposite pattern of aromatic character to Hückel systems. Topologically, this is the same as a Möbius strip, which is a three-dimensional object with only a single surface. In a Möbius annulene, there is a gradual twist of the π orbital system so that the lobes of the p orbitals at the ends of the cycle become exactly 180° out of phase with each other, giving a phase dislocation. One of the fundamental stereochemical aspects of the Möbius system is that the ring possesses a C_2 axis of symmetry. This means that in the conjugated system, the p basis orbitals on one side of the ring are orthogonal to those on the opposite side of the ring, and this orthogonality of orientation is now the simplest way to recognize a Möbius system.

One of the consequences of this type of topology is the extra stability of cyclic, conjugated π systems with $(4n + 2)$ π electrons with Hückel (untwisted) topology might be enjoyed by cyclic, conjugated π systems with $(4n)$ π electrons with Möbius topology. In other words, in Möbius annulenes, molecules with $4n$ π electrons should be aromatic, and molecules with $4n + 2$ π electrons should be antiaromatic. This intriguing possibility has led to much research in an effort to obtain just such a system. In 1966, Zimmerman[58] proposed a mnemonic for Möbius annulenes similar to the Frost-Musulin diagram[59] for Hückel annulenes. In the Zimmerman mnemonic, the polygon is drawn with one side down in a circle of radius 2β, in contrast to the Frost model, where the polygon is drawn with its vertex down.

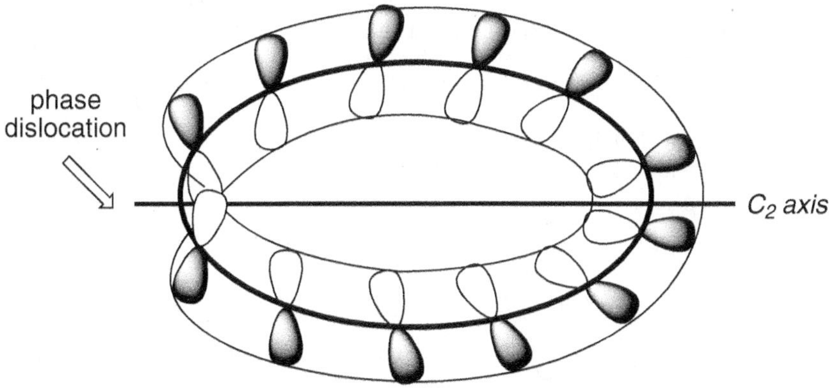

Figure 7.11 A Möbius annulene, showing the C_2 symmetry and the phase dislocation in the π bonding system. Note how the p basis orbitals on the right are orthogonal to those on the left.

54. Bird, C.W. *Tetrahedron* **1997**, *53*, 3319.
55. Reviews: (a) Rzepa, H.S. *Chem. Rev.* **2005**, *105*, 3697. (b) Herges, R. *Chem. Rev.* **2006**, *106*, 4820.
56. Heilbronner, E. *Tetrahedron Lett.* **1964**, 1923.
57. Muller, P. *Pure Appl. Chem.* **1994**, *66*, 1077.
58. Zimmerman, H.E. *J. Am. Chem. Soc.* **1966**, *88*, 1564.
59. Frost, A.A.; Musulin, B. *J. Chem. Phys.* **1953**, *21*, 572.

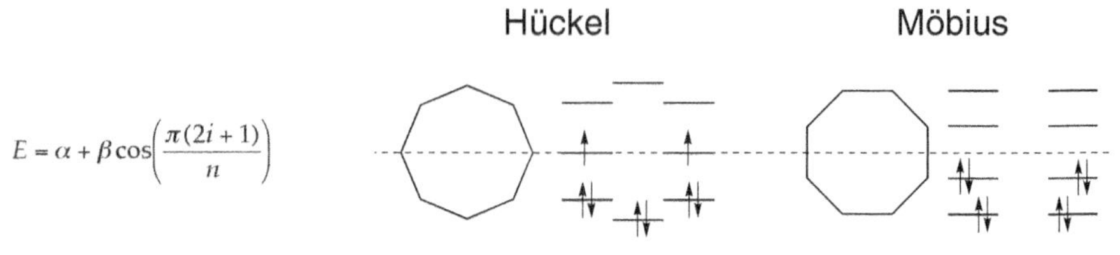

Figure 7.12 Comparison of Frost and Zimmerman models for predicting orbital energies in Hückel and Möbius planar cyclooctatetraene

The comparison of the Hückel and Möbius planar cyclooctatetraenes is shown in Figure 7.12, where α, β, *I*, and *n* are as defined earlier for aromatic systems, as well as in Chapter 4.

Zimmerman's paper introduced Möbius-type aromatic systems as a theoretical possibility, and efforts to identify Mobius-type aromatic species began soon thereafter. One of the early observations to support the involvement of a Möbius-type aromatic species in a reaction was the solvolysis of precursors of the cyclononatetraenyl cation, shown as Examples **7.16** and **7.17**. In both cases, the products of the reaction had the deuterium label uniformly distributed throughout the product, a situation that can occur only if a symmetrical intermediate is involved, thus providing indirect evidence that the cyclononatetraenyl cation is fully delocalized. Computations by Schleyer revealed that for this cation, the molecule adopts a twisted, C_2-symmetric structure (Example **7.18**), consistent with Möbius-type aromaticity, rather than a planar, C_{2v}-symmetric structure.[60,¶]

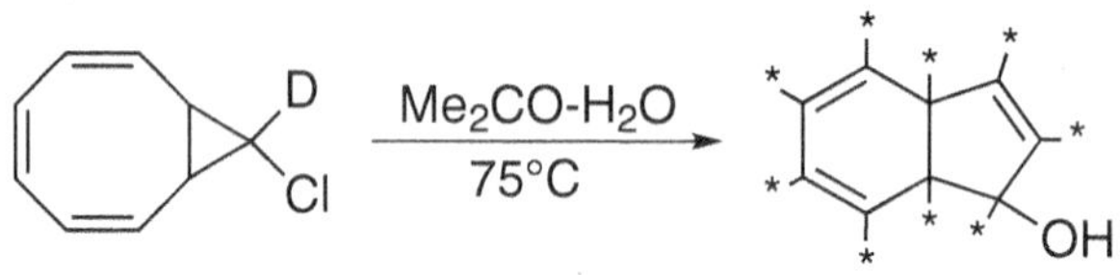

(7.16)[61]

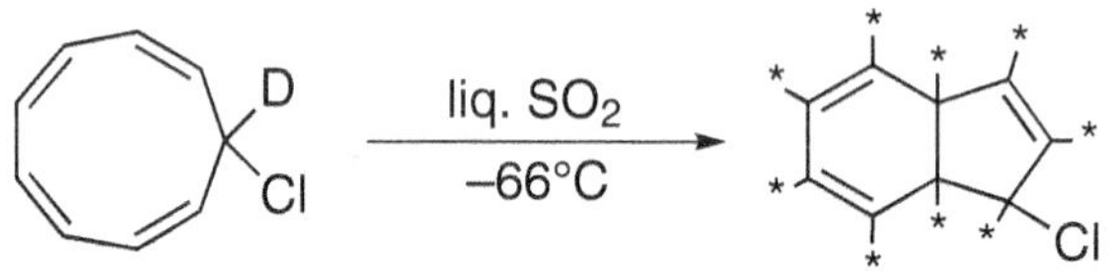

(7.17)[62]

60. Mauksch, M.; Gogonea, V.; Jiao, H.; Schleyer, P.v.R. *Angew. Chem. Int. Ed.* **1998**, *37*, 2395.

¶. ***Calculating Hückel Orbital Coefficients and Energies of Möbius Annulenes***

The π molecular orbitals of Hückel systems have a secular determinant that has only values of β because there is no phase dislocation in the π system. In contrast to this, the secular determinant of Möbius systems has a single −β element (actually two, because the matrix is symmetrical), due to the single phase dislocation. This leads to the following secular determinant, which is usually shown in the form where $x = (\alpha - E)/\beta$:

$$\begin{vmatrix} \alpha - E & \beta & 0 & \cdots & 0 & -\beta \\ \beta & \alpha - E & \beta & & & 0 \\ 0 & \beta & \alpha - E & & & \vdots \\ \vdots & & & \ddots & \beta & 0 \\ 0 & & & \beta & \alpha - E & \beta \\ -\beta & 0 & \cdots & 0 & \beta & \alpha - E \end{vmatrix} = 0 \Rightarrow \begin{vmatrix} x & 1 & 0 & \cdots & 0 & -1 \\ 1 & x & 1 & & & 0 \\ 0 & 1 & x & & & \vdots \\ \vdots & & & \ddots & 1 & 0 \\ 0 & & & 1 & x & 1 \\ -1 & 0 & \cdots & 0 & 1 & x \end{vmatrix} = 0$$

When the system has an even number of atoms, the solutions occur as degenerate pairs of orbitals whose energies are given by:

$E_j = \alpha + 2\beta\cos\left[\frac{(2j-1)\pi}{N}\right]$, where *N* is the number of atoms in the cyclic π molecular orbital system and *j* is the specific orbital in that system. The highest-energy orbitals in this system are at $E = \alpha - 2\beta$. When the system has an odd number of atoms, the same set of degenerate orbitals is obtained, but the highest-energy orbital, at $E = \alpha - 2\beta$, is now unique.

61. Schleyer, P.v.R.; Baborak, J.C.; Su, T.M.; Boche, G.; Schneider, G. J. *Am. Chem. Soc.* **1971,** *93,* 279.

62. Anastassiou, A,G.; Yakali, E. *J. Chem. Soc., Chem. Commun.* **1972,** 92.

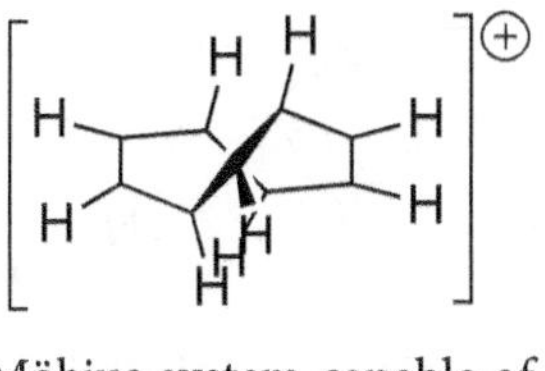

(7.18)

The search for a stable Möbius system capable of being studied was the object of intense research for decades. It finally came to fruition in 2003, when Herges and his research group reported[63] the synthesis of a Möbius (4*n*) aromatic hydrocarbon (**7.23**), although the conclusions of the Herges paper were challenged by Schleyer in 2005,[64] when a computational analysis of the Möbius C_{16} macrocycle and several less benzannulated analogs were analyzed. The successful synthesis, which provided experimental support for Heilbronner's predictions, is shown in Figure 7.13.

The sequence involved an initial [2 + 2] photocycloaddition of *syn*-tricyclooctadiene to the 9,9'-bianthracene (**7.19**), which possesses two strained, pyramidalized alkene π bonds. The structure of the ladder-like photoproduct (**7.20**) was established by X-ray structure analysis. The photochemical ring expansion reactions led through an unstable intermediate (**7.21**) to a cyclohexadiene (**7.22**), which was obtained as two isomers, a C_2-symmetric isomer with a *trans* ring junction and a C_s-symmetric isomer with a *cis* ring junction. The irradiation of the mixture at shorter wavelength over a longer time period led to a mixture of isomers from which the C_2-symmetric isomer with Möbius topology was isolated and characterized by single crystal X-ray structure analysis. The X-ray analysis revealed that the C—C bond distances in the Möbius hydrocarbon (**7.23**) varied from 1.361-1.456Å, corresponding to a HOMA value of 0.50 for the nine bonds of the polyene bridge of the molecule and an overall HOMAl of 0.35. By comparison, the C_s-symmetric (Hückel) isomer showed much greater variability in the bond distances, with a HOMA index of 0.05 for the same nine bonds in the polyene bridge. This indicates that the Möbius isomer has the hallmarks of aromaticity; in contrast, the Hückel isomer does not. The twist in the polyene bridge also shows a remarkable difference: the amount of twist in the Möbius (aromatic) ring system is not more than 35°, while the Hückel system escapes antiaromaticity by twisting the central π bond of the polyene bridge by 103° relative to the neighboring π bonds of the bridge.

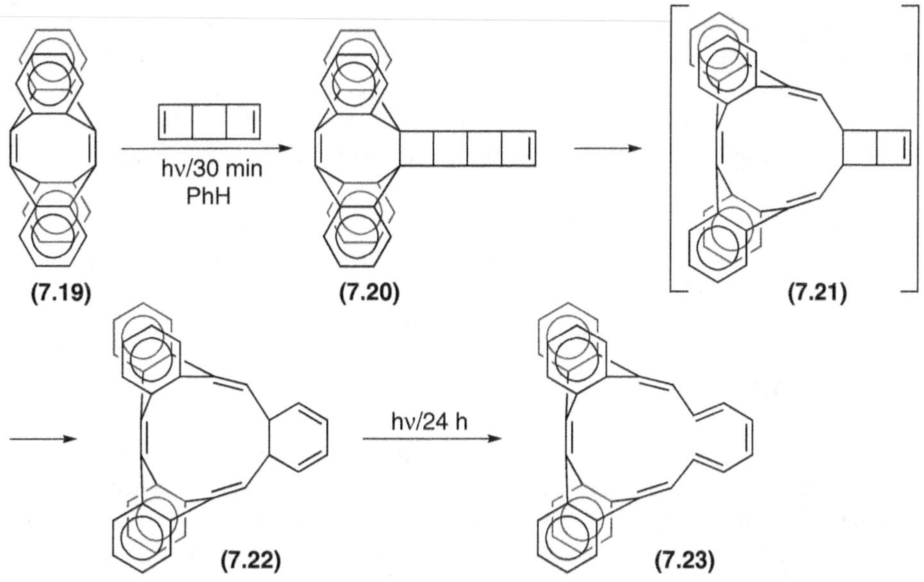

Figure 7.13 The first synthesis of a Möbius aromatic hydrocarbon

63. Ajami, D.; Oeckler, O.; Simon, A.; Herges, R. *Nature* **2003**, *426*, 819.

64. Castro, C.; Chen, Z.; Wannere, C.S.; Jiao, H.; Karney, W.L.; Mauksch, M.; Puchta, R.; van Eikema Hommes, N.J.R.; Schleyer, P.v.R. *J. Am. Chem. Soc.* **2005**, *127*, 2425.

Figure 7.14 An expanded porphyrin showing Möbius-Hückel switching

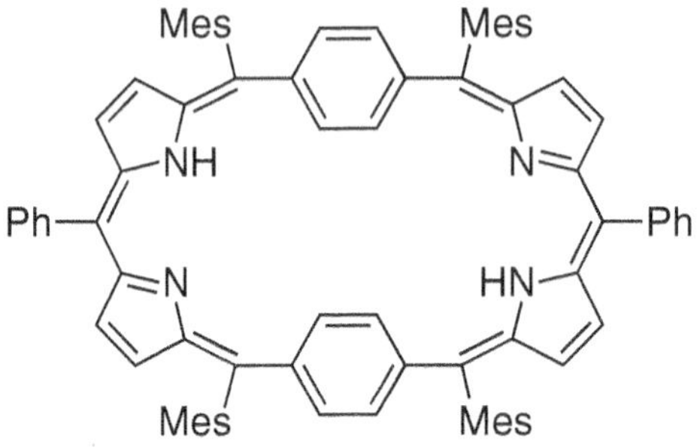

(7.24)

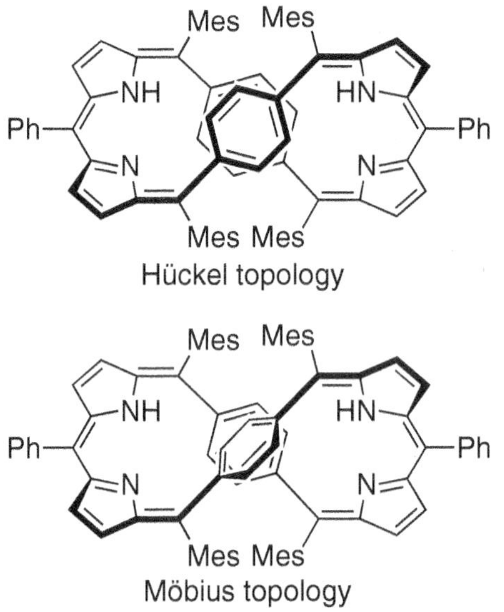

A synthesis of a compound capable of unambiguous existence as a Möbius aromatic system was provided by the expanded porphyrin **7.24**,[65] whose properties are both solvent- and temperature-dependent. One of the two conformers of the macrocycle has the phenylene rings stacked face-to-face—Hückel topology; in contrast, the other conformer has them stacked face-to-edge—Möbius topology. Figure 7.14 shows the conformations of the two phenylene rings and an approximate macrocycle geometry. In both topological forms, the ring is twisted; this gives the Möbius system the required C_2 symmetry and the Hückel conformer the required C_{2v} symmetry. The structural and spectroscopic evidence all indicate that the Möbius conformer of this 28 π e^- macrocycle is preferred at low temperature, in polar solvents, and that it is, in fact, aromatic at 203 K.

7.5 **Homoaromaticity**

In systems where two π bonds are sufficiently close spatially but not necessarily adjacent to each other in the molecule, it is still possible for a significant interaction between them to occur, provided that both the orientation and the separation of the p orbitals allows significant overlap. This process, which is analogous to the conjugation of adjacent π bonds in a molecule, is termed ***homoconjugation***.

65. Stępień, M.; Latos-Grażyński, L.; Sprutta, N.; Chwalisz, P.; Szterenberg, L. *Angew. Chem. Int. Ed.* **2007**, *46*, 7869.

(7.25)

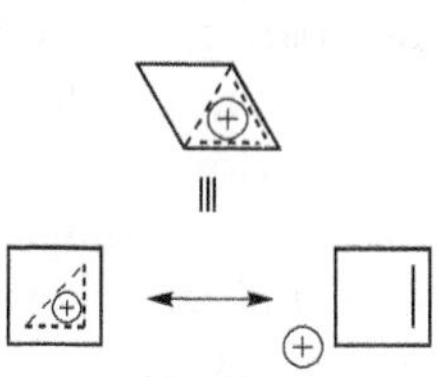

(7.26)

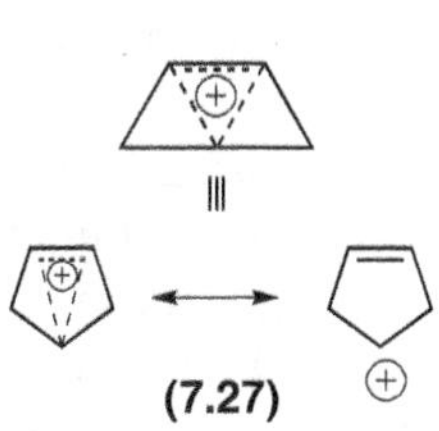

(7.27)

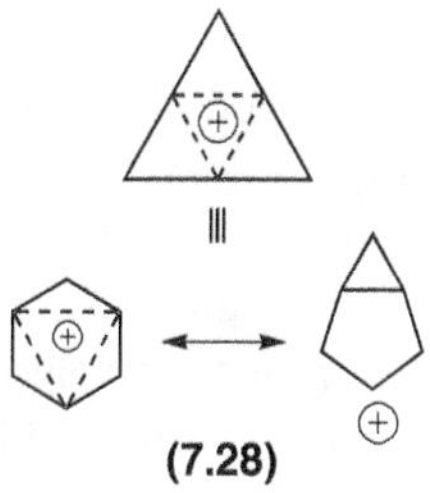

(7.28)

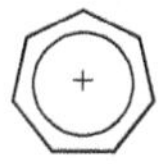

(7.29)

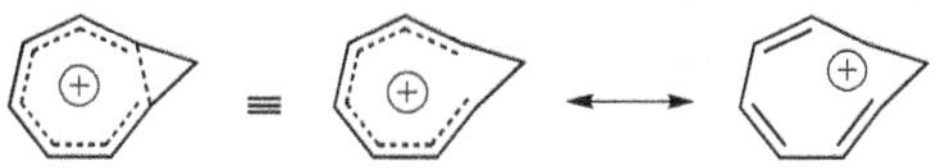

(7.30)

Figure 7.15 Aromatic and homoaromatic carbocations: aromatic cyclopropenium cation (**7.25**), monohomocyclopropenium cation (**7.26**), bishomocyclopropenium cation (**7.27**), trishomocyclopropenium cation (**7.28**), aromatic tropylium cation (**7.29**), and monohomotropylium cation (**7.30**)

In 1956, Doering suggested that the terminal sp^2 carbons in cycloheptatriene should be close enough for a homoconjugative interaction to occur between them, and he argued that this interaction would constitute the formation of an aromatic π electron system.[66] This type of interaction, where the formal cyclic conjugation is interrupted by an intervening sp^3 hybridized atom, is termed ***homoaromaticity***.[67] The subject of homoaromaticity has been reviewed many times since its proposal in 1956.[68]

Clear criteria for homoaromaticity were proposed by Childs and Cremer and their coworkers in the mid-1990s.[69] Homoaromaticity requires both the presence of homoconjugation and manifestations of "aromaticity." This includes electron delocalization in a closed cyclic system, bond orders that approach those of a true aromatic system, charge delocalization in charged systems, a stabilizing resonance energy greater than 2 kcal mol^{-1}, and magnetic effects that tend toward equalization of the chemical shifts of carbon atoms in the homoaromatic ring with effects of a diatropic ring current.

Where it can occur, homoaromaticity is especially evident in carbocations, where it leads to (often dramatic) stabilization of the cation, or enhancement in the rate of its formation. A wide range of homoaromatic cations known, and some of these types are shown in Figure 7.15. The carbocations (**7.26**), (**7.27**), and (**7.28**) in Figure 7.14 are all based on the aromatic two-electron cyclopropenium cation (**7.25**). They simply involve inserting a methylene group into one of the ring bonds of the cation, thus expanding the ring by one carbon and formally interrupting the conjugation by one carbon. One can continue this process until all three of the ring bonds are replaced to give a six-membered ring. The cation (**7.30**) in the lower row is derived from the aromatic six-electron tropylium ion (**7.29**) by the same process.

---

66. Doering, W.v.E.; Laber, G.; Vonderwahl, R.; Chamberlain, N.F.; Williams, R.B. *J. Am. Chem. Soc.* **1956,** *78*, 5448.

67. (a) Winstein, S. *J. Am. Chem. Soc.* **1959**, *81*, 6524. (b) Winstein, S.; Sonnenberg, J.; deVries, L. *J. Am. Chem. Soc.* **1959**, *81*, 6523.

68. (a) Winstein, S. *Chem. Soc. Spec. Publ.* **1967**, *21*, 5. (b) Winstein, S. *Q. Rev. Chem. Soc.* **1969**, *23*, 1411. Reprinted in *Carbonium Ions*; Olah, G.A.; Schleyer, P.v.R., Eds. (Wiley-Interscience: New York, 1972); Vol. III, p 965. (c) Story, P. R.; Clark, B. C., Jr. In *Carbonium Ions*; Olah, G.A.; Schleyer, P.v.R., Eds.; (Wiley-Interscience: New York, 197)2; Vol. III, p 1007. (d) Warner, P. In *Topics in Nonbenzenoid Aromatic Chemistry*; Nozoe, T., Breslow, R., Hafner, K., Ito, S., Murata, I., Eds. (Hirokawa Publishing Co.: Tokyo, 1977); Vol. II, p 283. (e) Paquette, L.A. *Angew. Chem., Int. Ed. Engl.* **1978**, *17*, 106. (f) Childs, R.F. *Acc. Chem. Res.* **1984**, *17*, 347. (g) Williams, R.V.; Kurtz, H.A. *Adv. Phys. Org. Chem.* **1994**, *29*, 273. (h) Cremer, D.; Childs, R.F.; Kraka, E. In *The Chemistry of the Cyclopropyl Group*; Rappoport, Z., Ed. (J. Wiley & Sons: Chichester, 1995); Vol. 2, p 339. (i) Childs, R.F.; Cremer, D.; Elia, G. In *The Chemistry of the Cyclopropyl Group*; Rappoport, Z., Ed. (J. Wiley & Sons: Chichester, 1995); Vol. 2, p 411. (j) Williams, R.V. *Chem. Rev.* **2000**, *100*, 1185.

69. (a) Cremer, D.; Childs, R.F.; Kraka, E. In *The Chemistry of the Cyclopropyl Group*; Rappoport, Z., Ed. (J. Wiley & Sons: Chichester, 1995); Vol. 2, p 339. (b) Childs, R.F.; Cremer, D.; Elia, G. In *The Chemistry of the Cyclopropyl Group*; Rappoport, Z., Ed. (J. Wiley & Sons: Chichester, 1995); Vol. 2, p 411.

The trishomocyclopropenium cation was the first homoaromatic species to be prepared and characterized. Winstein and Sonnenberg found that solvolysis of the *trans* tolylate, Example **7.31**, was faster than the solvolysis of the *cis* isomer, Example **7.32**, and that the solvolysis of the *trans* isomer (but not the *cis* isomer) led to complete scrambling of the deuterium among the three possible positions.[70] Similar observations were made in the case of solvolysis of 7-norbornenyl systems: the solvolysis of 7-*anti*-norbornenyl tosylate (Example **7.34**) is much faster than that of the 7-*syn* isomer (Example **7.33**).[71] Again, this rate increase was attributed to the homoaromatic character of the carbocation (in this case, a bishomocyclopropenium cation).

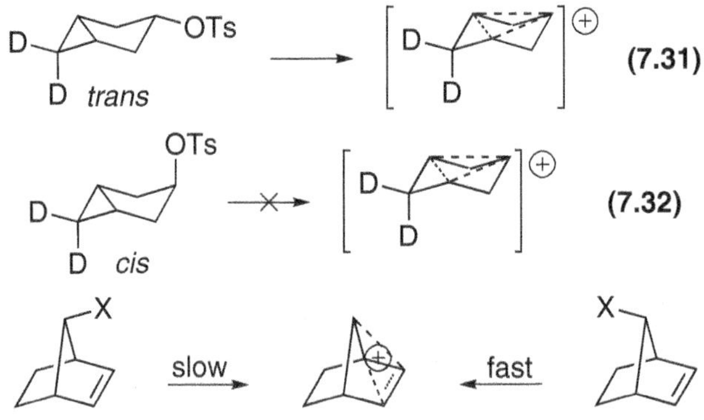

(7.31)

(7.32)

(7.33) **(7.35)** **(7.34)**

Like aromatic compounds, homoaromatic species are expected to be unusually stable, and the preponderance of evidence suggests that this is the case in homoaromatic cations. One may expect that homoaromaticity should be accompanied by bond distance changes in keeping with the bond distance equalization characteristic of aromatic species, and that there may also be distortion of bond angles, also, if this improves the homoconjugation. This type of geometric effect has been observed in a series of bishomocyclopropenium ions.[72]

The one anion that has been repeatedly cited as homoaromatic is the bicyclo[3.2.1]octa-3,5-dien-1-yl anion (Example **7.36**). In the parent hydrocarbon, the rate of hydrogen-deuterium exchange is 10^5 times faster than the corresponding allyl system, and homoaromaticity in the anion has been inferred to rationalize this observation.[73] The large upfield shift of the methylene protons in this anion suggests that there is a diatropic ring current, which would support claims of homoaromaticity for this anion.[74] These conclusions, however, are not without dispute, and Werstiuk and Ma[75] have carried out computations of the same anion. They found no evidence for a ring current, which would, of course, argue against homoaromaticity in this anion.

---

70. (a) Winstein, S.; Sonnenberg, J. *J. Am. Chem. Soc.* **1959**, *81*, 6523. (b) Winstein, S.; Sonnenberg, J. *J. Am. Chem. Soc.* **1961**, *83*, 3235. (c) Winstein, S.; Sonnenberg, J. *J. Am. Chem. Soc.* **1961**, *83*, 3244.

71. Woods, W.G.; Carboni, R.A.; Roberts, J.D. *J. Am. Chem. Soc.* **1956**, *78*, 5653.

72. (a) Laube, T. *Acc. Chem. Res.* **1995**, *28*, 399. (b) Laube, T. *J. Am. Chem. Soc.* **1989**, *111*, 9224. (c) Laube, T.; Lohse, C. *J. Am. Chem. Soc.* **1994**, *116*, 9001. (d) Evans, W.J.; Forrestal, K. J.; Ziller, J.W. *J. Am. Chem. Soc.* **1995**, *117*, 12635.

73. (a) Brown, J.M.; Occlowitz, J.L. *Chem. Commun.* **1965**, 376. (b) Jiao, H.; Schleyer, P.v.R. In *AIP Conference Proceedings 330: E.C.C.C. 1, Computational Chemistry*; Bernardi, F., Rivail, J.-L., Eds. (American Institute of Physics: Woodbury, NY, 1995); p. 107. (c) Freeman, P.K.; Pugh, J.K. *J. Org. Chem.* **2000**, *65*, 6107.

74. (a) Brown, J.M. *Chem. Commun.* **1967**, 638. (b) Winstein, S.; Ogliaruso, M.; Sakai, M.; Nicholson, J.M. *J. Am. Chem. Soc.* **1967**, *89*, 3656.

75. (a) Werstiuk, N.H.; Ma, J. *Can. J. Chem.* **1999**, *77*, 752. (b) Werstiuk, N.H.; Ma, J. *Can. J. Chem.* **1996**, *74*, 875.

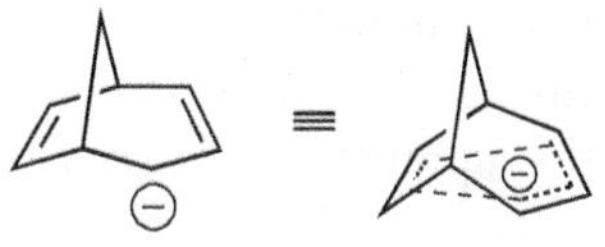

(7.36)

As is illustrated by the case of the anion above, the study of homoaromaticity is facilitated by modern computational methods for analysis of data, even more than aromaticity. In part, this is because the replacement of conjugation by homoconjugation means that the properties one associates with a fully aromatic system are, of necessity, attenuated, which means, in turn, that the experimental data can sometimes be ambiguous. In addition, one is often dealing with species that are reactive, transient, or not capable of being isolated pure or in crystalline form, and whose spectra can be difficult to interpret. Computational results in such cases can be a powerful adjunct to the experimental data that helps to make the case for a homoaromatic system.

Aromatic Transition States

In pericyclic reactions, which we discussed in detail in the previous chapter of this book, the concerted movement of electrons during the reaction involves, at some point in the reaction pathway, the delocalization of the electrons in a closed, cyclic system. This situation represents a logical extension of the concept of homoaromaticity, and it leads, in turn, to the concept of the ***aromatic transition state***, which is defined as a transition state in which there is a closed, delocalized system of $4n + 2$ electrons.

Based on the work of Dewar[76] and Zimmerman,[77] one can deduce that ground-state pericyclic reactions will proceed by Hückel transition states if $4n + 2$ electrons are involved, and by Möbius transition states if $4n$ electrons are involved. Common reactions proceeding via Hückel aromatic transition states are the Claisen and Cope rearrangements (Example **7.37**), the Diels-Alder addition (Example **7.38**), and suprafacial [1,5]sigmatropic rearrangements of hydrogen (Example **7.39**). The Cope rearrangement, especially, has been the object of intense study.[78]

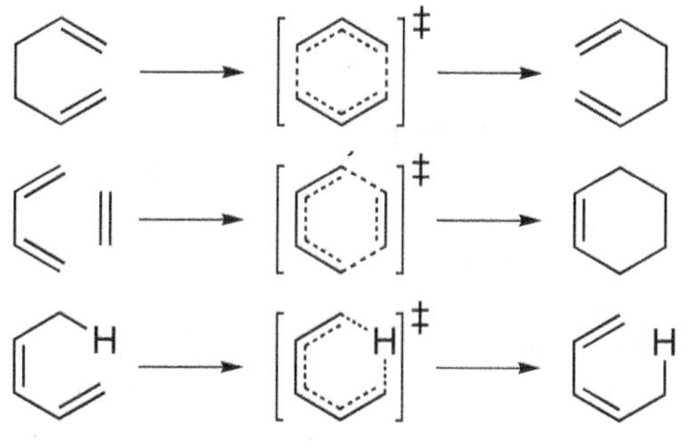

(7.37)

(7.38)

(7.39)

76. Dewar, M.J.S. *Tetrahedron* **1966**, *Suppl. 8*, 75.
77. Zimmerman, H.E. *Acc. Chem. Res.* **1971**, *4*, 272.
78. (a) Black, K.A.; Wilsey, S.; Houk, K.N. *J. Am. Chem. Soc.* **1998**, *120*, 5622. (b) Roth, W.R.; Gleiter, R.; Paschmann, V.; Hackler, U.E.; Fritzsche, G.; Lange, H. *Eur. J. Org. Chem.* **1998**, 961. (c) Hrovat, D.A.; Duncan, J.A.; Borden, W.T. *J. Am. Chem. Soc.* **1999**, *121*, 169. (d) Hrovat, D.A.; Beno, B.R.; Lange, H.; Yoo, N.-Y.; Houk, K.N.; Borden, W.T. *J. Am. Chem. Soc.* **1999**, *121*, 10529. (e) Doering, W.v.E.; Wang, Y. *J. Am. Chem. Soc.* **1999**, *121*, 10967. (f) Oliva, J.M. *Theor. Chem. Acc.* **1999**, *103*, 1. (g) Staroverov, V.N.; Davidson, E.R. *J. Am. Chem. Soc.* **2000**, *122*, 186.

Concerted reactions proceeding through Möbius transition states are frequently carried out photochemically. They include [2 + 2] photocycloadditions of alkenes to give cyclobutanes (Example **7.40**) and the suprafacial [1,7]sigmatropic rearrangement of hydrogen (Example **7.41**).

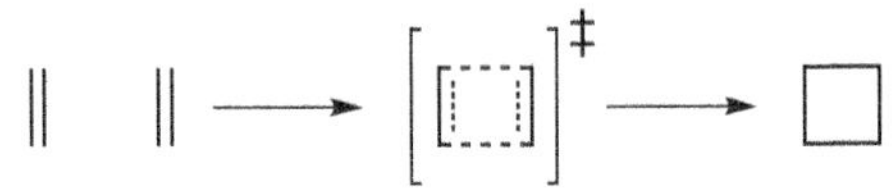

(7.40)

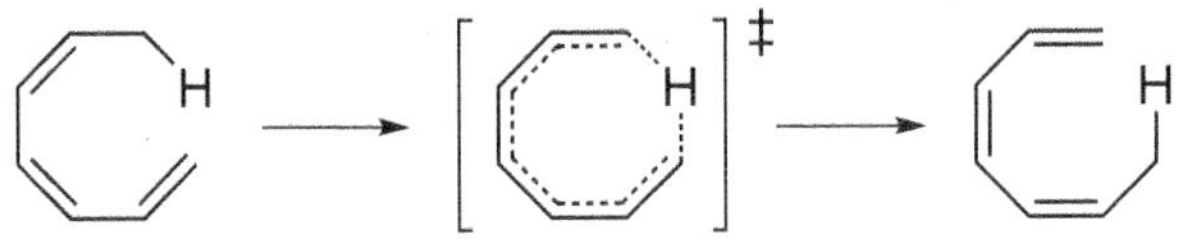

(7.41)

7.6 The Rest of the Story: The History and Mythology of Benzene

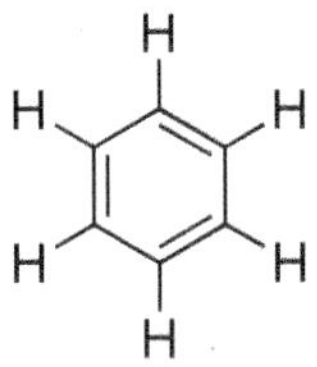

Kekulé 1865

As indicated at the beginning of this chapter, solving the structure of benzene occupied the finest minds of 19th-century organic chemistry. The first to propose the cyclic structure for benzene was August Kekulé, in 1865.[79,80] Kekulé's claim is rendered the more romantic by the story, which he himself promulgated about his theory, as told by his eulogist, Francis Robert Japp,[81] during the *Kekulé Memorial Lecture* of the Chemical Society of London in 1898:

> According to his own version, the idea came to him during a dream. "I was sitting, writing at my text-book; but the work did not progress, my thoughts were elsewhere. I turned my chair to the fire and dozed. Again the atoms were gambolling before my eyes. This time the smaller groups kept modestly in the background. My mental eye, rendered more acute by repeated visions of the kind, could now distinguish larger structures, of manifold conformation: long rows, sometimes more closely fitted together; all twining and twisting in snake-like motion. But look! What was that? One of the snakes had seized hold of its own tail, and the form whirled mockingly before my eyes. As if by a flash of lightning I awoke; and this time also I spent the rest of the night in working out the consequences of this hypothesis."
>
> "Let us learn to dream, gentlemen," adds Kekulé, "then perhaps we shall find the truth... but let us beware of publishing our dreams before they have been put to the proof by the waking understanding."

Whether or not the Kekulé dream story is fact or apocryphal romanticism is still the topic of hot—even vitriolic—debate (some versions of the story have him

79. Kekulé, A. *Bull. Soc. Chim.* **1865**, 3, 98.

80. Kekulé's original formula was of the "sausage formula" variety, which made the molecule difficult to visualize as a cyclic compound. The formula that we now routinely use to represent the benzene molecule—and call the Kekulé structure—was actually introduced by Albert Ladenburg.

81. Japp, R. *J. Chem. Soc.* **1898**, *73*, 100.

daydreaming while riding a London bus), and the truth will probably never be known. In the 1990s—nearly a century after Kekulé's death—it even resulted in a lawsuit against the Division of the History of Chemistry of the American Chemical Society! Still, Kekulé's dream is now firmly entrenched as part of the folklore of organic chemistry. More importantly, Kekulé's 1865 paper supplied the first widely disseminated structure for benzene—dreamt up or not—that provided a rational basis for discussion of its chemistry.

Immediately, chemists recognized that there were problems fitting the known chemistry of benzene to the Kekulé structure because of its double bonds. Based on the Kekulé structure, benzene should react as a highly unsaturated compound. It does not. Furthermore, the Kekulé structure predicts that there should be two *ortho* isomers of dibromobenzene—one where both bromine atoms are on the same double bond and one where they are on different double bonds—as Albert Ladenburg[82] (then a junior faculty member at Heidelberg) pointed out. But only one *ortho* isomer exists.

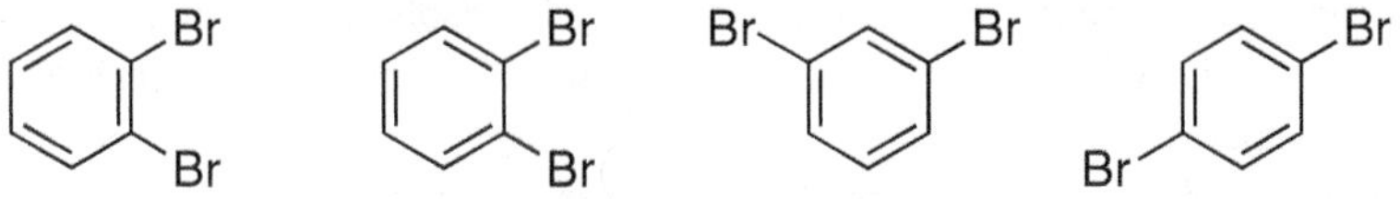

ortho *ortho* *meta* *para*

The attempts to resolve these inconsistencies between the structure and the chemistry of the benzene molecule led to many benzene structures being proposed within 5 years of Kekulé's paper. New structures continued to be proposed until the end of the 19th century. Some of the structures are given below.

At this point, it is worthwhile dispelling some quasi-historical inaccuracies that have crept into the benzene story over the years. One of these concerns an earlier structure proposed by Austrian physicist Johann Loschmidt that has been interpreted as being the first cyclic structure proposed for this molecule. It was not, as Loschmidt himself admitted in his monograph, Loschmidt's intention to represent a cyclic structure for the benzene molecule because he did not know the arrangement of the six carbon atoms of the benzene molecule. He simply used a larger version of the Dalton symbol for carbon to represent them and he arranged the hydrogens symmetrically about it (Loschmidt's representation contains a mirror plane and two C_2 axes and a pair of C_2 axes, but no C_6 axis).[83]

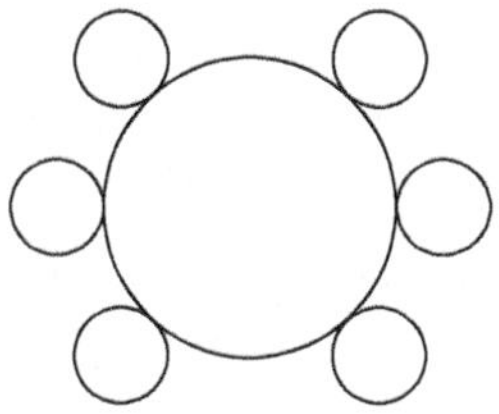

In similar fashion, the structure designated as Dewar benzene was not seriously proposed as a structure for benzene by Sir James Dewar at all, but it was just one of seven possible structures for benzene he pointed out could be built with a set of models that he had developed![84] It was proposed as a reasonable structure for benzene by (1) Georg

82. **Albert Ladenburg (1842–1911)** was educated at Heidelberg (PhD, 1863). His career began in Heidelberg in 1868; he moved to Kiel in 1872 and Breslau in 1889. He received many honors during his lifetime, including the Davy medal of the Royal Society. For his autobiography, published posthumously, see: Ladenburg, A. *Lebenserinnerungen*, (Trewendt & Granier, Breslau, 1912).

83. For a lucid account, see: Roth, H.D. *Bull. Hist. Chem.* **2013**, *38*, 29.

84. Dewar, J. *Proc. Roy. Soc. Edinburgh* **1867**, 6, 82.

Städeler[85] of the University of Zürich, in 1868,[86] and (2) Hermann Wichelhaus[87] of the University of Berlin, in 1869.[88] Likewise, the structure now known as the Armstrong-Baeyer centric formula[89] was actually first proposed by Lothar Meyer in 1872.[90]

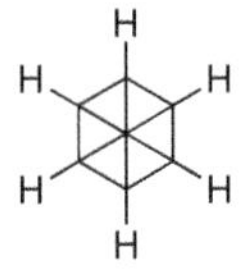

Claus 1867

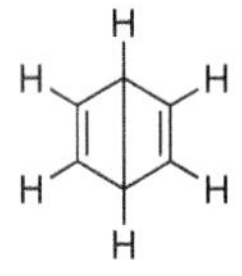

[Dewar 1867[
Städeler 1868
Wichelhaus 1869

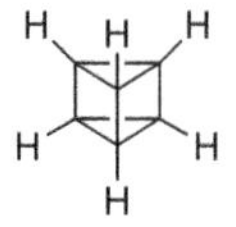

Ladenburg 1869

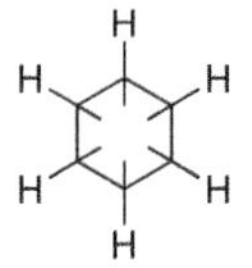

(7.3)

L. Meyer 1872
Armstrong-Baeyer
1877-1878

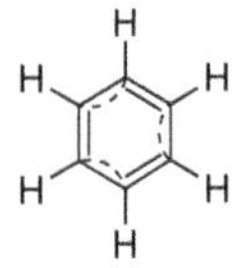

(7.4)

Thiele 1899

Except for the Ladenburg formula,[91] all the formulas proposed retain the original structural feature of the Kekulé formula: the planar six-membered ring of carbon atoms with a hydrogen atom at each vertex. The problem to be resolved, of course, was to explain the lack of reactivity of the benzene molecule and the number of isomeric substituted benzenes that exist.

The Claus,[92] Ladenburg, and Armstrong-Baeyer formulas for benzene sought to address the problem of the lack of typical alkene reactivity by assigning structures devoid of any double bonds. In the Claus formula,[93] all the "diagonal" bonds are long bonds; in the Ladenburg formula, the molecule is not planar but prismatic; and in the Armstrong-Baeyer formula, the fourth valence of each carbon atom is directed toward the center of the ring in a rather undefined way. Thiele modified the Kekulé structure by adding the "partial valencies" between the double bonds. His own research with conjugated dienes had shown him that these compounds often react by 1,4- rather than 1,2-addition, and he proposed the partial valence concept to rationalize this reactivity.[94] By being cyclic, benzene should be unusually unreactive.

Of course, someone with Kekulé's ego could not remain passive in the face of the objections to his structure. In order to answer them, Kekulé proposed that the double bonds were in a continual state of flux, an idea that approximates, but does not attain, the modern concept of resonance. Kekulé believed that what we now would call canonical forms actually represented real compounds that were in rapid equilibrium with each other. Städeler's modification of Kekulé's structure is really a compromise between the original Kekulé

---

85. **Georg Karl Andreas Städeler (1821–1871)** was educated at Göttingen (PhD, 1845). Städeler built his independent career at Göttingen and then at Zurich, working in both organic chemistry and inorganic analysis. For a more complete biography, see: Kraut, G.S. *Ber. dtsch. chem. Ges.* **1871**, *4*, 425.

86. Staedeler, G. *J. prakt. Chem.* **1868**, *103*, 105.

87. **Carl Hermann Wichelhaus (1842–1921)** studied at Bonn, Göttingen, Heidelberg (PhD, 1863), Ghent, and London. He joined the faculty at Berlin in 1867and became Professor of Chemical Technology in 1871. He retired to Heidelberg in 1916; from 1877–1880 he was also Director of the Patent Office. For more biograhical details, see: Deutsches Biographisches Lexikon. *Biographisches Handbuch deutscher Männer und Frauen der Gegenwart*; (Verlagsbuchhandlung Schulze & Co.: Leipzig, 1905); p. 1564.

88. Wichelhaus, H. *Ber. Deut. chem. Ges.* **1869**, *2*, 197.

89. (a) Armstrong, H.E. *J. Chem. Soc.* **1887**, *51*, 264. (b) Baeyer, A. *Ann. Chem. Pharm.* **1888**, *245*, 103.

90. Meyer, J.L. *Die Modernen Theorien der Chemie* (Maruschke and Berendt: Breslau, 1872).

91. Ladenburg, A. *Ber. Deut. chem. Ges.* **1869**, *2*, 272. Ladenburg actually proposed a planar structure with the same linkages, but he stated that the prismatic version was preferred.

92. **Karl Ludwig Claus (1838–1900)** was educated at Marburg and Göttingen (PhD, 1862). He immediately joined the faculty of the University of Freibourg, where he spent the rest of his life. Claus was one of the first to recognize the importance of structure in organic chemistry; his planar structure for benzene was unique in not requiring any double bonds. For a more complete biography, see: Vis, G.N. *J. Prakt. Chem.* **1900**, *62*, 127.

93. Claus, A. *Theoretische Betrachtungen un den Anwendung zur Systematik der Organischen Chemie* (Freiborg, 1867).

94. Thiele, J. *Ann. Chem.* **1899**, *306*, 87; *Ann. Chem.* **1899**, *308*, 333.

structure and the Claus formula, but he also espoused Kekulé's idea of constant flux to avoid the objections raised by the presence of the two double bonds in his structure. Thanks to the resonance theory developed by Linus Pauling, today we represent the benzene molecule as the resonance hybrid of the two canonical forms shown; frequently, the π bonds of the ring are represented by a circle.

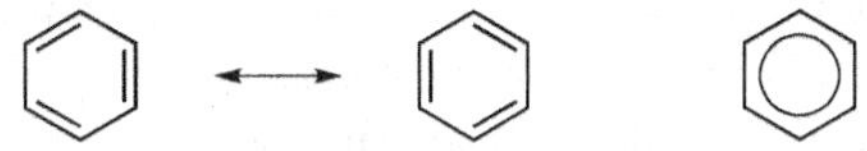

There are 12 possible isomeric substitution patterns of benzene by one or more identical substituents. As the number of chemically different substituents on a ring increases, the number of possible isomers also increases. As shown, the bromination of *o*-dichlorobenzene gives only two compounds, while the bromination of *o*-chlorotoluene, on the other hand, gives four compounds. It was by studies of this type that Wilhelm (Gugliemo) Körner[95] was able to establish that all six carbons of the benzene ring are equivalent, as well as being able to identify the substitution patterns in benzene derivatives.[96]

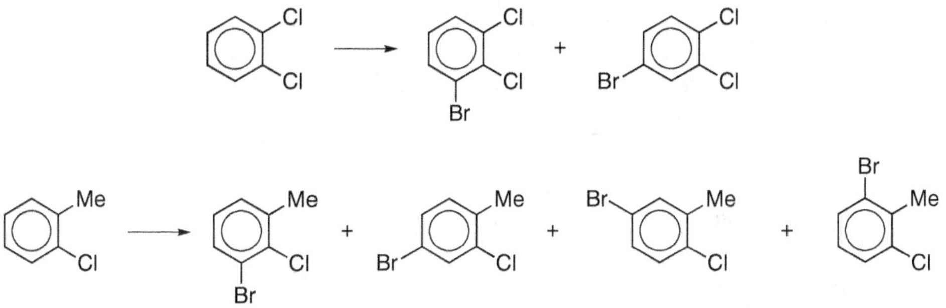

Problems

7-8 How many monobromobenzenes would be expected for (1) Kekulé's formula, (2) Claus' formula, (3) von Baeyer's formula, (4) Dewar's (Städeler's) formula, and (5) Ladenburg's formula for benzene if the bonds do not move? How many dibromobenzenes? How many tribromobenzenes?

7-9 How many isomers of dibromochlorobenzene could possibly be formed from each of the isomeric dibromobenzenes in Problem 7-8?

7-10 How many isomers of chlorobromofluorobenzene are possible? How many of these could be formed from each of the three isomers of bromofluorobenzene?

7-11 An isomer of xylene is treated with nitric acid under conditions that give a mononitroxylene. Separation of the reaction mixture gives three different products. What was the structure of the original xylene?

95. **Gugliemo (Wilhelm) Körner (1839–1925)** was educated at Giessen (PhD, 1864) and he then studied with Kekulé until 1867. When Kekulé left Ghent for Bonn, Körner moved to Cannizzaro's laboratories in Palermo. In 1870 the "Scuola Superiore di Agricoltura" was founded at Milan, and Körner became the first Professor of Organic Chemistry. For a more complete biography, see: *J. Chem. Soc. Trans.* **1925**, *127*, 2973.

96. Körner, W. *Gazz. Chim. Ital.* **1874**, *4*, 303.

Chapter Summary

The topic of this chapter, aromaticity, is one of the oldest problems in organic chemistry, and it is still not fully solved. The first general theoretical model for aromaticity was developed by Erick Hückel. Hückel aromatic molecules are planar, cyclic, delocalized and have $(4n + 2)$ π electrons. Molecules with the same structural characteristics but with $4n$ π electrons are Hückel antiaromatic. In aromatic compounds, C—C bond lengths tend toward an average value, with bond-length alternation becoming more pronounced as the size of the ring increases. This bond alternation may be quantified by means of the HOMA parameter, which ranges from 1.0 for aromatic compounds to 0.0 for non-aromatic compounds. The stability of an aromatic compound is measured by aromatic resonance stabilization energy, which must be determined indirectly from experimental data or computationally (e.g., by homoisodesmic reactions). Magnetic methods to quantify aromatic character include (1) a diatropic (aromatic) or paratropic (antiaromatic) ring current in the ^{1}H NMR spectrum and (2) magnetic exaltation, which is the increase in the molar magnetic susceptibility. Computational methods for quantifying aromaticity include (1) the NICS at the center of the ring or above the center of the ring can define aromaticity or antiaromaticity of the ring (negative NICS values correspond to aromaticity in the ring); (2) the DRE, which is positive for an aromatic compound and negative for an antiaromatic compound; (3) the REPE, which is similar to the DRE, but is approximately constant for any system; (4) the REPB, which is similar to the REPE, but is more reliable in charged systems; and (5) the absolute hardness, which is defined as half the HOMO-LUMO gap. The energies of the Hückel π molecular orbitals of aromatic annulenes can be predicted using the Frost-Musulin circle, and the stability of aromatic compounds can be predicted using the Clar model. Möbius aromaticity was predicted in 1964, and the relative energies of orbitals can be predicted using the Zimmerman circle. Möbius complements Hückel aromaticity; $4n$ systems are aromatic, and $(4n + 2)$ systems are antiaromatic.

Key Terms

absolute hardness
antiaromatic
aromatic
aromatic resonance stabilization energy
Clar sextet
Dewar benzene
Dewar resonance energy (DRE)
diatropic ring current
Frost-Musulin circle
Harmonic oscillator model of aromaticity (HOMA)
homoaromatic
homoisodesmic reaction
Hückel antiaromaticity
Hückel aromaticity
Kekulé structure
Körner proof
Ladenburg benzene
magnetic exaltation
Möbius antiaromaticity
Möbius aromaticity
nucleus-independent chemical shift (NICS)
paratropic ring current
resonance energy per π bond (REPB)
resonance energy per π electron (REPE)
Zimmerman circle

Additional Problems

7-12 The left-hand compound shown below is predicted to be aromatic according to Hückel's rule. There are two extreme structures that one can draw for the molecule; the canonical form on the left consists of three furan rings linked by three vinyl bridges, and the canonical form on the right consists of an [18]annulene bridged by three oxygen atoms. The compound shows proton signals at δ 8.66 and 8.88. The "normal" chemical shift for furan β protons is 7.0 ppm. Which of these canonical forms do you expect to make a greater contribution to the resonance hybrid, and why? The sulfur analog of the left-hand compound shows proton resonances at δ 6.73 and 6.77. What does this observation suggest? How does your answer compare if the left-hand compound is changed to the molecule on the right?

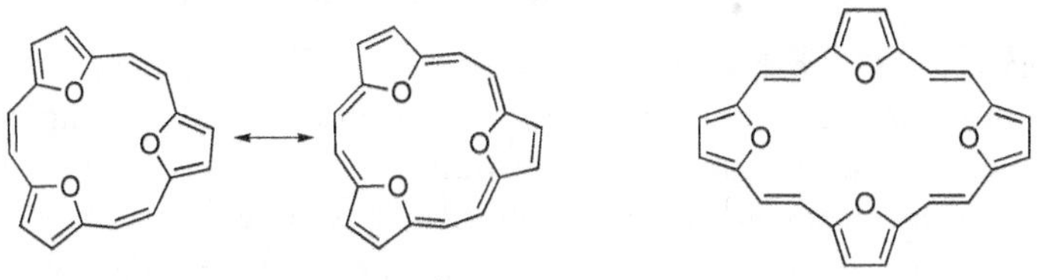

7-13 Would the compounds below be expected to exhibit Hückel or Möbius aromaticity? Give your reasons.

(a)  (b) 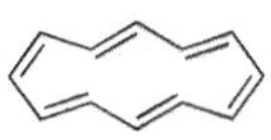 (c) 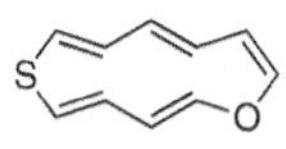 (d) 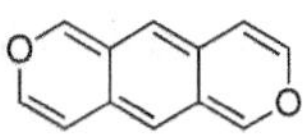

7-14 The compounds below can all be described as bridged annulenes, but there is evidence in some cases that this may not be the most accurate description of the molecules. What alternative description is consonant with the structures drawn?

(a) 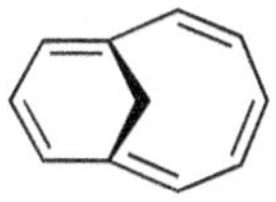 (b) 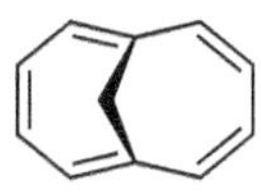 (c) 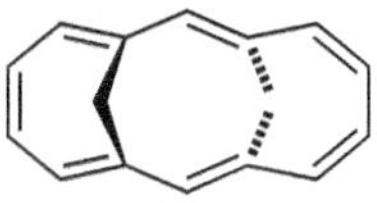

7-15 The cyclic allene below is aromatic. Does it exhibit resonance of the type shown by simple benzene derivatives? If it does, draw at least one other resonance contributor.

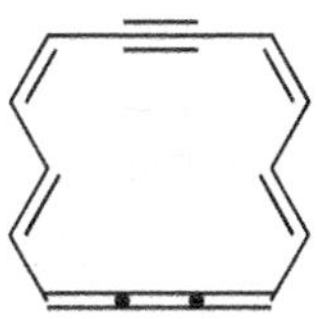

7-16 The Hückel analysis of [14]annulene indicates that it should be aromatic, since it has 14 π electrons ($4n + 2$, $n = 3$). However, the spectra and reactivity of this molecule at room temperature suggest that it is essentially a simple polyene, with little or no aromatic character. At −60°C, however, the ^{1}H NMR spectrum splits into two resonances at δ 7.6 and 0.0, in the ratio 5:2. The single crystal X-ray structure of [14]annulene shows that the molecule is significantly distorted from planarity. Rationalize these observations and answer the question, "is [14]annulene aromatic or not?"

7-17 The ester below exhibits only one resonance in the ^{1}H NMR spectrum other than the methyl resonances. This resonance occurs at δ 8.45. Provide a reasonable explanation for this observation. What other properties would you predict for this compound on the basis of your analysis?

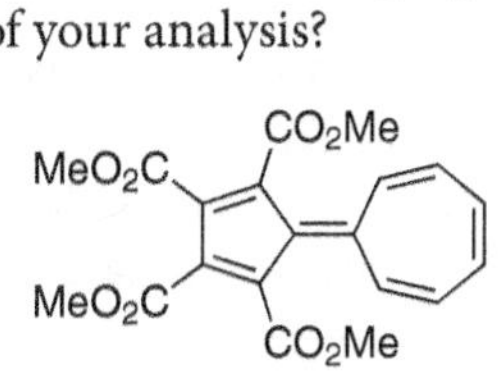

7-18 Diphenylcyclopropenone has a dipole moment of 5 D, and the stretching frequency of its carbonyl group is 1650 cm^{-1}. The carbonyl stretching frequency of cyclopropanone is greater than 2000 cm^{-1}. Provide a reasonable explanation for these observations. What does your answer suggest about the infrared stretching frequency of the α-diketone shown below the diphenylcyclopropenone? (The infrared stretching frequency of α-diketones is typically near 1750 cm^{-1}.)

7-19 In Example **7.15**, we discussed the formation of a Möbius aromatic cation from deuterium-labeled 9-chlorocyclonona-1,3,5,7-tetraene by dissolving in liquid sulfur dioxide (a strongly ionizing medium). The reaction leads to products in which the deuterium is statistically scrambled. Is there a reasonable rationalization of the observed result that does not involve a Möbius aromatic cation?

7-20 When 5-bromo-1,3-cyclopentadiene reacts with antimony pentafluoride at 78K, a carbocation is formed that exhibits an electron paramagnetic spin resonance (EPR) spectrum consistent with a triplet diradical. What inferences do these results allow to be made about the cation formed in this reaction? Are these deductions in accord with predictions? Why or why not?

7-21 The cyclobutadiene dianion, $C_4H_4^{2-}$, is predicted by simple Hückel analysis to be aromatic, and the dilithium salt of some derivatives of this ring system is, indeed, planar [*Organometallics* **2002**, *21*(6), 1072]. However, the structure of the free dianion (in the absence of the counterions) is calculated to be bent, not planar, with one of the negative charges localized. What would cause this difference?

7-22 The pK_a values for the three hydrocarbons below in DMSO solution are as shown below each structure. Rationalize these values.

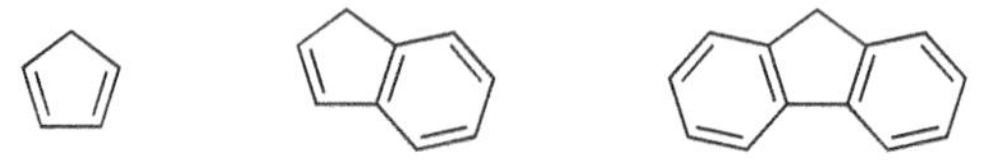

pK_a	18.0	20.1	22.6

7-23 The compounds below are all 14 π e^- species and so all should, in principle, be aromatic, with a diatropic ring current. The ^1H NMR spectra of the compounds are summarized in the table.

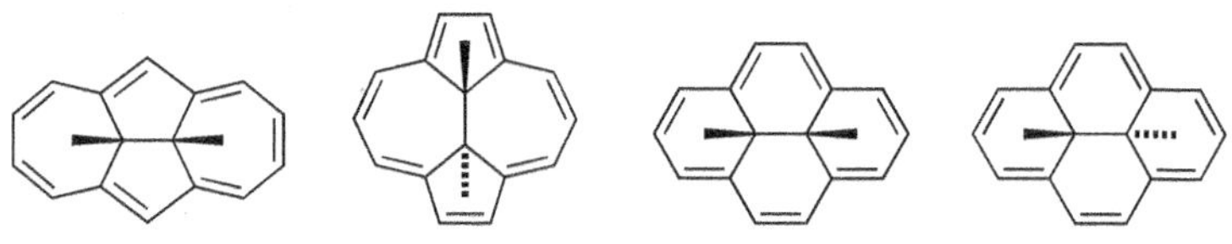

δ_{Me} (ppm)	-1.82	-4.53	-2.06	-4.25
δ_H (ppm)	8.74	8.17	8.77	8.64
δ_H (ppm)	8.24	8.00	8.74	8.60
δ_H (ppm)	7.50	7.82	8.04	8.11

The resonances of the ring protons are all consistent with the ring current being similar in all for compounds, but there is a clear difference when it comes to the methyl resonances. Using models to analyze the shapes of the molecules, suggest why the chemical shifts of the methyl resonances vary so widely.

7-24 Based on the observation that the square planar ring in the sequestered 1,3-dimethylcyclobutadiene structure is not completely planar and the observation that the carbon dioxide molecule is bent, the authors of the paper [*Science* **2010**, *329*, 299] propose that the interaction between the two molecules trapped in the host lattice is a significant van der Waals interaction rather than a covalent bond. After reading the paper, is this position reasonable or not? You may find the following papers useful in formulating your answer to this question: *Science* **2010**, *330*, 1047c-e, available online, and *Chem. Commun.* **2011**, *47*, 1851. The following blog also contains a very instructive dialog among the major players on this question: http://www.ch.ic.ac.uk/rzepa/blog/?p=2828 (retrieved July 11, 2013).

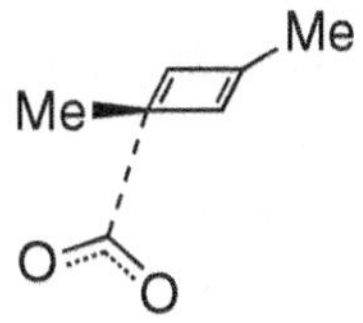

CHAPTER EIGHT

Physical Organic Chemistry and Reaction Mechanisms

8.1 Introduction

In Chapter 1, we briefly reviewed the process of writing reaction mechanisms and the elementary processes in the steps of most common organic reactions (heterolysis and homolysis of bonds). In this chapter, we will continue expand on these discussions, with more focus on how evidence for the existence of reactive intermediates is garnered and on how this evidence is integrated into a model for the mechanism of the reaction.[1]

In modern usage, a reaction mechanism is defined as the set of elementary reactions required to convert reactants to products. In other words, it is a step-by-step description of the transformation of reactants to products. As part of this description, a reaction may be designated as ***concerted***, in which case the conversion of reactants to products takes place in a single step, or it may be ***stepwise***, in which case at least one reactive intermediate is involved. The individual steps of stepwise reactions seldom proceed at the same rate, which means that one of the steps becomes the ***rate-limiting step*** or ***rate-determining step***. In order for a proposed mechanism to merit serious consideration, it must fit the observed stoichiometry, it must explain the observed kinetics, and it must make predictions that pass the test of experiment.

All organic reactions can be discussed in terms of the ***reaction coordinate diagram***, or ***reaction energy profile***, and deducing this diagram for a reaction is also an important part of elucidating the overall reaction mechanism. The reaction coordinate diagram simply traces the change in potential energy of a single reacting molecule (or set of reacting molecules) through the bonding changes from reactant to product. Stable species—species formally capable of being isolated and studied, including reactive intermediates—appear as valleys along the reaction coordinate; the depth of the valley determines if this is what is termed a ***metastable intermediate***, which is usually difficult or impossible to isolate, or a stable intermediate, which can sometimes be isolated and studied. If the valley between peaks in the reaction coordinate diagram is very shallow, this is usually interpreted as indicating that the intermediate is metastable. Unstable species (species that have lifetimes less than one molecular vibration) appear as peaks in the reaction coordinate diagram. These species are ***transition states*** and designated by the cross of Lorraine (‡). They cannot be isolated under any circumstances.

---

1. General textbooks discussing mechanism and theory in chemistry: (a) Ingold, C.K. *Structure and Mechanism in Organic Chemistry* (Cornell University Press: Ithaca, 1953). (b) Lowry, K.S.; Richardson, T.M. *Mechanism and Theory in Organic Chemistry*, 3rd. ed. (Harper & Row: New York, 1987). (c) Sykes, P. *A Guidebook to Mechanism in Organic Chemistry* , 2nd ed. (John Wiley & Sons: New York, 1965). (d) Gould, E.S. *Mechanism and Structure in Organic Chemistry* (Holt,, Rinehart and Winston: New York, 1959). (e) Gardiner, W.C., Jr. *Rates and Mechanisms of Chemical Reactions* (Benjamin: New York, 1969). (f) Carroll, F.A. *Perspectives on Structure and Mechanism in Organic Chemistry* (Brooks/Cole: Pacific Grove, 1998). (g) Carey, F.A.; Sundberg, R.J. *Advanced Organic Chemistry. Part A: Structure and Mechanism*, 4th ed. (Kluwer Academic/Plenum: New York, 2000).

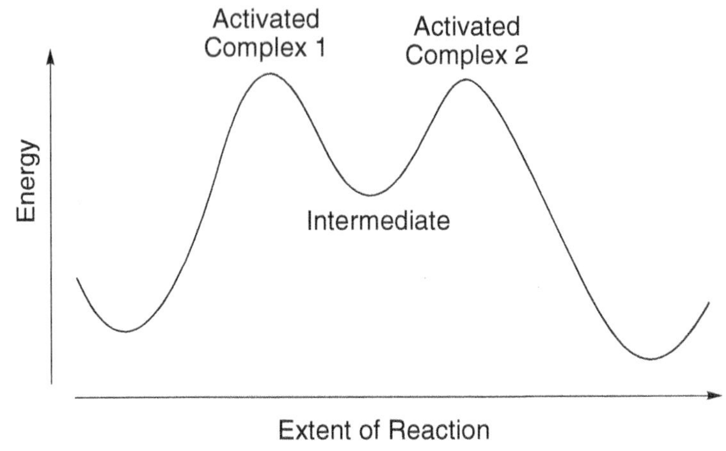

Figure 8.1 Reaction coordinate diagram showing intermediate

Reactive intermediates are, as their name implies, chemically reactive species that are formed during the course of a chemical reaction but which are neither the reactant nor the product. They represent minima on the potential surface of a reaction (valleys in the typical reaction coordinate diagram). The difference between an intermediate and an activated complex is that the potential minimum for the intermediate is deep enough to contain at least one stable vibrational energy level; an activated complex occurs at a saddle point on the surface and does not contain any stable vibrational level (Figure 8.1).

There are five reactive intermediates where the reactivity is centered on carbon, that commonly occur in organic reactions: carbocations, carbanions, free radicals, carbenes, and arynes; and one where the reactivity is centered on nitrogen: nitrenes, which are nitrogen analogs of carbenes. Of these, the carbanions are electron-rich electron-pair donors; the free radicals have an unpaired electron; and the others are electron-deficient, electron-pair acceptors.

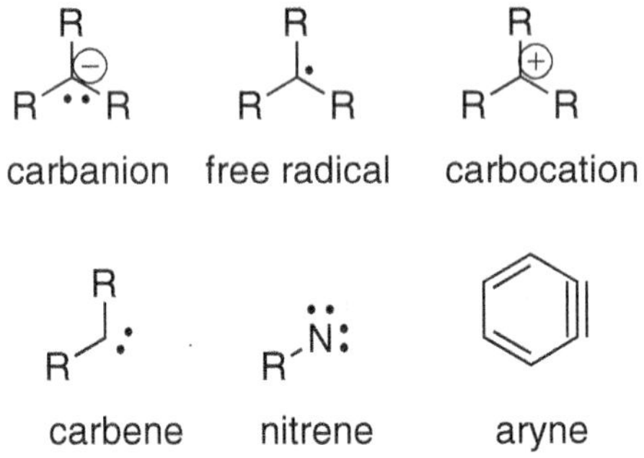

Very often, the formation of these intermediates is endothermic, which usually results in the formation of the reactive intermediate being the rate-determining step of the reaction in which they occur. This makes their formation amenable to study by kinetic measurements. More recently, it has become possible to study these intermediates directly by spectroscopy, and rapid advances in computer speed and the sophistication and reliability of the predictions of computer programs has made their study by molecular orbital calculations reliable, also.

All reactions may be described in terms of a ***potential energy surface*** that allows one to trace the path from reactants to products. A simple potential surface for a concerted, two-bond reaction is shown as Figure 8.2 (a). The path between reactants and products lies in a valley on the potential surface. When the complete potential surface is examined, stable species appear as ***potential wells***, and transition states as ***saddle points***. In Figure 8.2(a), there are two potential wells, corresponding to the reactant (labeled R) and product (labeled P), and a single saddle point (labeled with the cross of Lorraine). Although organic chemists do not often use the potential energy surface as a whole, a

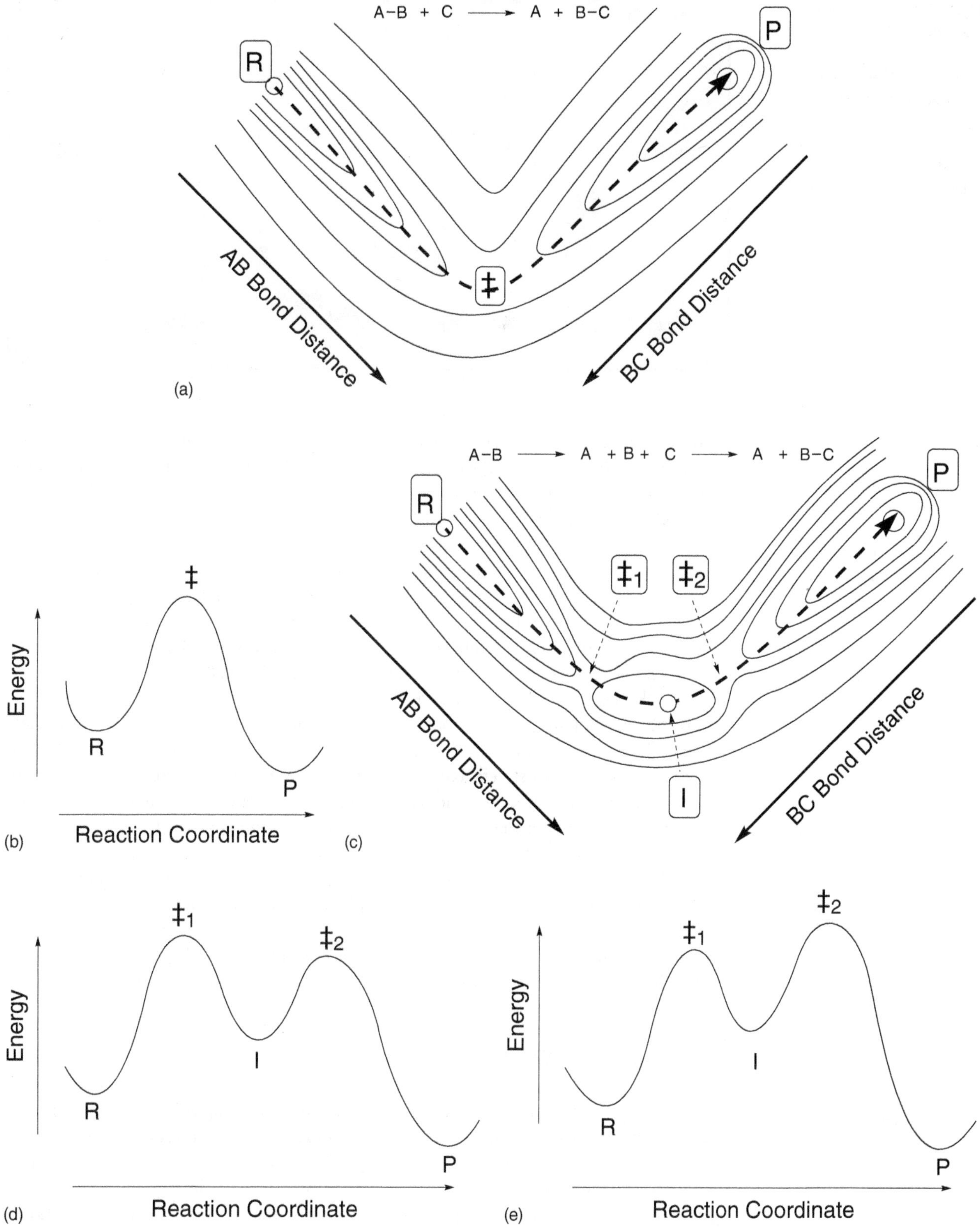

Figure 8.2 (a) Potential energy surface for a concerted reaction, (b) reaction coordinate diagram for a concerted reaction, (c) potential energy surface for a reaction with (d) reaction coordinate diagram for a two-step reaction with the first step rate determining, and (e) reaction coordinate diagram for a two-step reaction with the second step rate determining

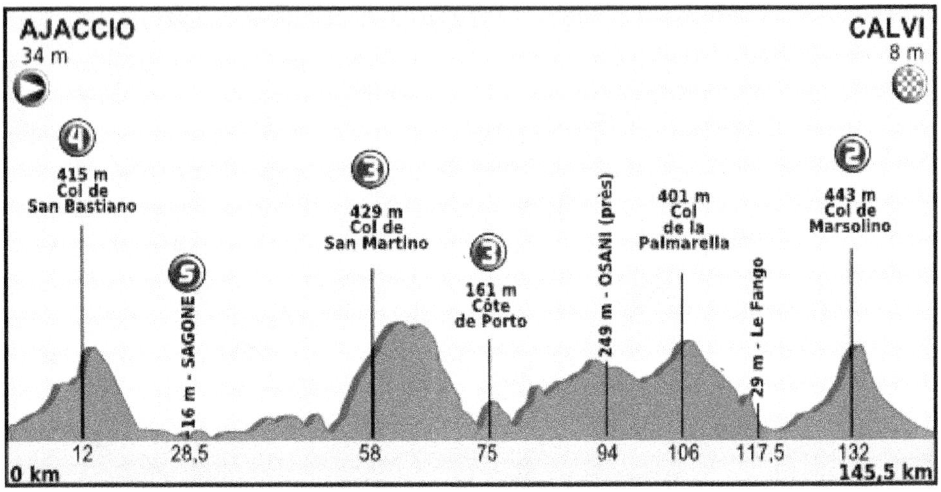

Figure 8.3 The Corsican stage of the 2013 Tour de France. Image from Wikimedia Commons, licensed under the Creative Commons Attribution-Share Alike 3.0 unported. http://en.wikipedia.org/wiki/File:Tour_de_France_2013_stage_03.png

modified (simplified) form has been used in discussing the timing of bonding changes and in making predictions about the effects of changes in the reactants on the structure of the activated complex, or transition state.

Reactions can also be described by means of a ***reaction coordinate diagram***, or a ***reaction energy profile***. This diagram shows the variation in the potential energy of the system along the floor of the energy valley between reactants and products, much like the altitude maps that are used on television to describe stages of the Tour de France (e.g., Figure 8.3, from the 2013 Tour). The potential surface in Figure 8.2 (a) has only one saddle point, so there is only one activated complex for this reaction, and the reaction coordinate diagram has only a single maximum and two minima—Figure 8.2 (b).

Not all reactions are concerted, as the lower three examples in Figure 8.2 show. In Figure 8.2 (c), a potential surface for a stepwise reaction passing through a reactive intermediate (labeled I) is shown. In this case, there are two saddle points (i.e., two activated complexes or transition states), labeled $‡_1$ and $‡_2$. These two transition states may be of equal energy, or, much more commonly, one will be of higher energy than the other. When the energy of $‡_1$ is higher than the energy of $‡_2$, the reaction coordinate looks like Figure 8.2 (d); if the opposite is true, the reaction coordinate looks like Figure 8.2 (e).

At the end of this introduction, it is worth repeating that, of course, one must always bear in mind that one cannot *prove* a mechanism, but only *disprove* it.[2] All mechanisms, as written, are simply the best model available to explain the experimental observations at the time they are proposed and are always subject to modification if better data are obtained that contradict one or more details of the mechanistic model.

8.2 Solvents and Solubility

It is a rare organic reaction, indeed, that is not carried out in a solvent. When one considers the fundamental processes involved, the process of forming a solution is, in many ways, a form of chemical reaction between the solvent and the solute: the solvent forms new solvent-solute bonds at the expense of solvent-solvent and solute-solute bonds. The function of a solvent in a reaction is twofold. It provides a common medium where the two reactants can undergo the collision required to initiate the reaction, and it exerts a moderating on the rate

2. This assertion is not universally accepted: (a) Buskirk, A.; Baradaran, H. *J. Chem. Educ.* **2009**, *86*, 551. (b) Brown, T.L. *J. Chem. Educ.* **2009**, *86*, 552. (c) Lewis, D.E. *J. Chem. Educ.* **2009**, *86*, 554. (d) Yoon, T. *J. Chem. Educ.* **2009**, *86*, 556. (e) Wade, P.A. *J. Chem. Educ.* **2009**, *86*, 558.

of the reaction, allowing one to modulate the rates of extremely exothermic reactions by changing reactant concentrations. There is a large body of literature that deals with the effects of solvent on reaction rates, beginning with the work of Nikolai Aleksandrovich Menshutkin[3] on the formation of tetraethylammonium iodide,[4] and encompassing a wide range of nucleophilic displacements.[5]

There are four major types of interactions between solvent and solute: London dispersion forces, dipole-dipole interactions, hydrogen bonding, and ion-dipole interactions. Which of these forces operates depends on the nature of both the solvent and the solute, but the solvent polarity may be the most important solvent property because many reactions involve both covalent and ionic reagents.[6]

The weakest intermolecular forces are *London dispersion forces*,[7] which are short-range, uniformly attractive forces between molecules due to temporary dipoles arising from asymmetric instantaneous electron distribution, and first proposed by German-born physicist, Fritz London.[8] London dispersion force occurs between all molecules, polar and non-polar. Polar molecules can also participate in dipole-dipole interactions, which are slightly stronger attractive forces between permanent dipoles.

London dispersion forces and dipole-dipole interactions are commonly referred to under the collective term, *van der Waals forces*, after Johannes Diderik van der Waals,[9] the Dutch physicist who first explained deviations from ideal behavior by gases by proposing that attractive forces exist between molecules of gases. The radius at which these forces of attraction are strongest is called the van der Waals radius of an atom or molecule; molecules compressed together within their van der Waals radii strongly repel each other.

Quantifying the effects of solvent on the rate of a reaction is not a simple task, which means that a variety of scales for solvent polarity have been developed over the years. The simplest measures of solvent polarity are its dipole moment (μ) and dielectric constant (ε). However, neither of these measures is, by itself, sufficient to describe the effects of solvent on many reactions. Several scales for solvent polarity, in particular, have been developed, and we will examine these in detail in the next chapter.

Many times, one of the reagents in a reaction is ionic. The solubility of ionic compounds in a solvent frequently correlates with the *dielectric constant* of the solvent, which is a measure of how effectively the solvent shields ions from the effect of adjacent charges, and how

---

3. **Nikolai Aleksandrovich Menshutkin (1842–1907)** was educated at St Petersburg University (Dr Chem, 1869), and remained there as Professor after graduation. He rose to become Dean of the faculty, and in 1902 he was appointed to the Petersburg Technological Institute. For more biographical details, see: Menschutkin, B. *Ber. dtsch. chem. Ges.* **1907**, *40*, 5087.

4. Menshutkin, N. *Z. Phys. Chem.* **1890**, *5*, 589; **1891**, *6*, 41.

5. The effects of solvent on the rates of S_N1 and S_N2 reactions have been the subject of especially intense investigation. For general discussions, see: (a) Ingold, C.K. *Structure and Mechanism in Organic Chemistry* (Cornell University Press: Ithaca, NY, 1953), pp. 345–355. (b) Gould, E.S. *Mechanism and Structure in Organic Chemistry* (Holt, Rinehart and Winston: New York, 1959), pp. 184–185. (c) Carroll, F.A. *Perspectives on Structure and Mechanism in Organic Chemistry* (Brooks-Cole: Pacific Grove, 1998): pp. 329–332. (d) Carey, F.A.; Sundberg, R.L. *Advanced Organic Chemistry, Part A: Structure and Mechanisms*, 4th ed. (Kluwer Academic/Plenum Publishers; New York, 2000), pp. 290–295.

6. An excellent discussion of the various solvent parameters, and the process of solvation is given in Isaacs, N.S. *Physical Organic Chemistry* (Longman Group UK: Harlow, Essex, 1987), Ch. 5, pp. 171–209.

7. (a) London, F. *Z. Physik* **1930**, *63*, 245. (b) London, F. *Trans. Faraday Soc.* **1937**, *33*, 8.

8. **Fritz Wolfgang London (1900–1954)** was educated at Bonn, Frankfurt and Munich (PhD, 1921 in philosophy). He and his brother fled the Nazi persecution, and after the war, London became Professor at Duke University. For more details, see: Gavroglu, K. *Fritz London: A Scientific Biography* (Cambridge University Press: Cambridge, 2005).

9. **Johannes Diderik van der Waals (1837–1923)** was educated at Leiden (PhD, 1873). In 1877, he was appointed as Professor of Physics at the University of Amsterdam, where he developed his Nobel Prize-winning equation for gas behavior in 1881. For more biographical details, see the Nobel Foundation web site.

Table 8.1 Dielectric Constants (D) and Dipole Moments (μ) of Common Solvents

Solvent	D	μ
Acetic acid	6.17	1.68
Acetone	20.6	2.69
Acetonitrile	35.9	3.53
Benzene	2.27	0
Chloroform	4.80	1.15
Dichloromethane	8.9	1.14
Diethyl ether	4.2	1.15
Dimethylformamide (DMF)	36.7	3.24
Dimethyl sulfoxide (Me_2SO)	46.5	4.06
Ethanol	24.5	1.66
Ethyl acetate	6.02	1.82
Hexane	1.89	0.085
Hexamethylphosphoric triamide (HMPA)	29.3	4.31
Methanol	32.7	2.87
Pyridine	12.9	2.37
Tetrahydrofuran	7.58	1.75
Water	78.4	1.8

Table 8.2 Boiling Points of the Covalent Hydrides

IIIA	IVA	VA	VIA	VIIA
B -92.59	C -161.6	N -33.35	O 100.0	F 19.74
Al	Si -112	P -87.7	S -60.7	Cl -85.09
Ga	Ge -88	As -62.5	Se -41.2	Br -66.78
In	Sn -52.5	Sb -17	Te -2	I -35.41
Tl	Pb	Bi ≈22	Po	At

Table 8.3 Boiling Points of Methyl Compounds

VA	VIA	VIIA
$Me{-}NH_2$ -6.3°C	Me-OH 65°C	Me-F -78.4
$Me{-}PH_2$ -14°C	Me-SH 6.2°C	Me-Cl 24.2

strongly the solvent solvates an ion. The force of attraction between ions separated by a distance r is given by

$$F = -(1/D)(q_1q_2/r^2) \qquad \text{(Eq. 8.1)}$$

where D is the dielectric constant of the medium (the dielectric constant of a vacuum is 1) and q_1 and q_2 are the charges on the ions. In hexane, the value of D is low, which means that the force of attraction between the ions is not reduced very much by intervening solvent molecules. In water, on the other hand, the value of D is 80—the force between ions in water is only 1.25% of that between the same ions separated by the same distance in a vacuum. Generally, as the dielectric constant decreases, the reactivity of ionic species becomes more extreme. Anions become stronger bases and stronger nucleophiles, and cations become stronger acids and stronger electrophiles. The solvent dielectric constant can have a dramatic effect on reactivity, but one must be careful to remember that it is not the only solvent property that is important, as has been shown by carrying out the hydrolysis of acid chlorides in several different mixtures of solvents with the same overall dielectric constant.[10] Dipole moments and dielectric constants of a range of common solvents are gathered in Table 8.1.

The effects of hydrogen bonding and dipole-dipole interactions on molecular properties are illustrated by the data in Table 8.2, where the boiling points of the covalent hydrides of the main-group elements are collected. Among these hydrides, ammonia (–33°C), water (100°C), and hydrogen fluoride (22°C) have remarkably high boiling points, especially given their low molecular weights. This effect is also shown (albeit to a lesser extent) by the boiling points of the six methyl compounds in Table 8.3; methanol (CH_3OH) and methylamine (CH_3NH_2) have two of the highest boiling points, despite having two of the lowest molecular weights.

When a solute dissolves in a solvent to form a solution, two sets of forces must be broken and replaced by attractive forces between solute and solvent particles (*solvation*): the forces between particles (molecules or ions) in the solute and the intermolecular forces between molecules of the pure solvent. Thus, forming a solution is ultimately a compromise between energy loss because of the rupture of solute-solute bonds and solvent-solvent bonds and energy gain due to the formation of solute-solvent bonds. The balance between these forces determines the solubility of a compound in a solvent.

For example, the water solubility of an ionic compound is determined by the balance between the loss of energy due to the rupture of ionic bonds in the solute lattice and hydrogen bonds in the solvent, and the energy gain due to the formation of solvating

10. (a) Brown, D.A.; Hudson, R.F. *J. Chem. Soc.* **1953**, 3352. (b) Hudson, R.F.; Saville, B. *J. Chem. Soc.* **1955**, 4114.

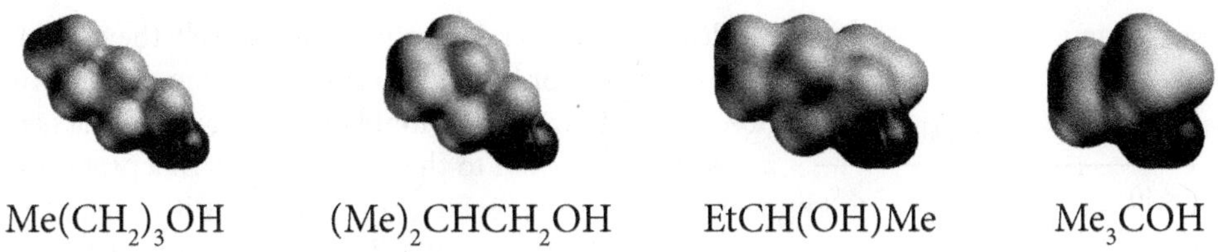

Figure 8.4 Models of the isomeric butyl alcohols showing the partial charges. Non-polar regions are light grey negative regions are black and positive regions are dark grey

ion-dipole bonds. Because a non-polar (low dielectric) solvent cannot form such ion-dipole bonds to stabilize the individual ions by solvation, most ionic compounds are insoluble in non-polar solvents. For similar reasons, most non-polar molecules are not very soluble in water because the energy loss due to rupture of the hydrogen bonds between water molecules (which must be broken to allow room for the solute molecules) is not compensated for by any sufficiently strong solvating dipole-dipole bonds.

The solvent solvates all of the reactants in the reaction, which means that reactions are not as we simplistically write them, but are between solvated species instead. In turn, this means that in order for the reaction to proceed, we must remove solvent molecules from around each reactant to permit contact between the reacting species—we must break solvent-solute bonds. The stronger the force of attraction between the solvent molecule and the solute, the more difficult this is to accomplish, and therefore, the slower the reaction will be.

These effects can be visualized by considering the water solubility of the isomeric butyl alcohols (structures in Figure 8.4). The polar part of the molecule is the O—H group, which can form hydrogen bonds with water molecules to compensate for the water-water hydrogen bonds that are lost when the alcohol dissolves. The remaining part of all four molecules is a non-polar hydrocarbon group, and this will participate only in much weaker van der Waals interactions with the water molecules, so we expect that the alcohol with the most compact alkyl group will probably have the highest water solubility. In 1-butanol, the alkyl group is elongated, which requires that many more water-water hydrogen bonds be broken to accommodate the alkyl group than in *tert*-butyl alcohol, where the alkyl group is very compact. The other two alcohols both have more compact alkyl groups attached to the carbon bearing the OH group, and so their water solubility (*sec*-butyl alcohol ≈ 12.5 g/100 mL H_2O; isobutyl alcohol ≈ 10 g/100 mL H_2O) is intermediate between that of 1-butanol (≈ 9 g/100 mL H_2O) and *tert*-butyl alcohol (miscible with water in all proportions).

Polar solvents used in organic reactions are usually characterized as being ***protic*** or ***aprotic dipolar***. Protic solvents are solvents containing an OH group capable of forming strong hydrogen bonds: water, acetic acid, formic acid, or aqueous alcohol mixtures are most widely used. They tend to solvate all ionic species—both cations and anions—very strongly, and this strong solvation stabilizes both types of ionic solutes. Conversely, aprotic dipolar solvents, such as acetone, dimethyl sulfoxide, and hexamethylphosphoric triamide, typically have a strong dipole whose negative end is easily accessible for solvating cations but whose positive end is sterically congested, preventing it from being available for strong solvation of anions.[11] They lack an O—H bond capable of solvating anions.

11. (a) Sheehan, J.C.; Bolhofer, W.A. *J. Am. Chem. Soc.* **1950**, *72*, 2786. (b) Zaugg, H.E.; Horrom, B.W.; Borgwardt, S. *J. Am. Chem. Soc.* **1960**, *82*, 2895. (c) Zaugg, H.E.; Dunnigan, D.A.; Michaels, R.J.; Swett, L.R.; Wang, T.S.; Sommers, A.H.; DeNet, R.W. *J. Org. Chem.* **1961**, *26*, 644. (d) Barlow, G.H.; Zaugg, H.E. *J. Org. Chem.* **1972**, *37*, 2246. (e) Zaugg, H.E.; Rataczyk, J.F.; Leonard, J.E.; Schaefe, A.D. *J. Org. Chem.* **1972**, *37*, 2249. (f) Zaugg, H.E.; Leonard, J.E. *J. Org. Chem.* **1972**, *37*, 2253. (g) Smiley, R.A.; Arnold, C. *J. Org. Chem.* **1960**, *25*, 257. (h) Friedman, L.; Shechter, H. *J. Org. Chem.* **1960**, *25*, 877. (i) Paquette, L.A.; Philips, J.C. *Tetrahedron Lett.* **1967**, 4645. (j) Magnera, T.F.; Caldwell, G.; Sunner, J.; Ikuta, S.; Kebarle, P. *J. Am. Chem. Soc.* **1984**, *106*, 6140. (k) Mitsuhashi, T.; Yamamoto, G.; Hirota, H. *Bull. Chem. Soc. Jpn.* **1994**, *67*, 831. (l) Okamoto, K. *Adv. Carbocation Chem.* **1989**, *1*, 181.

Table 8.4 Relative Activity of Chloride Ion in 1.0*M* Tetraethylammonium Chloride in Common Solvents

Solvent	Activity	ΔE_a (kcal mol^{-1})
Acetonitrile	2.00×10^6	12.0
Dimethylformamide	3.16×10^6	12.3
Dimethyl sulfoxide	3.16×10^5	10.9
Methanol	1.00	3.40
Water	0.0032	0.00

So although they tend to solvate cations well, they tend to solvate anions only poorly. This results in a less pronounced stabilization of ionic reactants and products in a reaction compared to the same reaction in a protic solvent. Non-polar species, or species with dispersed charges, are better solvated by aprotic dipolar solvents. Stronger solvation of the reactants of a reaction than the activated complex will tend to raise the activation energy of a reaction and retard its rate, whereas stronger solvation of an activated complex than the reactants will lower the activation energy and increase the reaction rate. In 1961, Australian chemist A. J. Parker[12] summarized the results of several workers by suggesting that larger, more polarizable anions were better solvated by aprotic dipolar solvents than smaller, less polarizable anions, leading to the observed reversal of reactivity of anions in aprotic dipolar solvents compared to protic solvents like methanol.[13] Parker's proposals are corroborated by the activity coefficients of anions in aprotic dipolar solvents.[14] Representative values for chloride anion are gathered in Table 8.4.

The aprotic dipolar solvents were one solution to the problem of having to dissolve ionic and non-polar compounds in the same solution to effect reactions, but they are by no means the only way to accomplish this. In the late 1960s and early 1970s,[15] the technique of ***phase transfer catalysis*** was introduced, especially by the work of Makosza,[16] Brändström,[17] and Starks.[18] This technique relies on the fact that it is possible to prepare ionic compounds that are more soluble in hydrocarbon (e.g., benzene) and relatively non-polar (e.g., chloroform, ethyl acetate) than in water. These ionic compounds are characterized by having a central atom with a formal charge surrounded by large alkyl groups: tetraalkylammonium salts where the total number of carbon atoms in the cation is greater than about 12 to 16 are quite soluble in relatively non-polar solvents: tetrabutylammonium chloride is soluble in benzene, for example. The surface potentials of two cations (tetrabutylammonium and octadecyltrimethylammonium) used as phase transfer catalysts are shown. In both cases, the positively charged nitrogen atom is relatively inaccessible and cannot be easily solvated directly, with the result that the solubility of compounds with this ion is dominated by the non-polar alkyl groups on the nitrogen.

---

12. **Alan James Parker (1933–1982)** was educated at the University of Western Australia (UWA) (PhD, 1957). After five years abroad, he returned to Australia to positions at UWA, the Australian National University, and Murdoch University, in turn. He made major contributions in metallurgy. For more details, see: Cole, A.R.H., Watts, D.W., *Hist. Records Aust. Sci.*, **1986**, 6, 399.

13. (a) Miller, J.; Parker, A.J. *J. Am. Chem. Soc.* **1961**, *83*, 117. (b) Parker, A.J. *Quart. Rev. Chem. Soc.* **1962**, *16*, 163. (c) Parker, A.J. *Adv. Org. Chem.* (Wiley: New York, 1965), Vol. 5, p. 1.

14. (a) Parker, A.J. *Chem. Rev.*, **1969**, *69*, 1. (b) Parker, A.J. *Pure Appl. Chem.* **1981**, *53*, 1437.

15. Reviews: (a) Dockx, J. *Synthesis* **1973**, 441. (b) Dehmlow, E.V. *Angew Chem. Int. Ed. Engl.* **1974**, *13*, 170; **1977**, *16*, 493. (c) Starks, C.M.; Liotta, C.L.; Halpern, M. *Phase Transfer Catalysis: Fundamentals, Applications and Industrial Perspectives* (Chapman and Hall: New York, 1994). (d) Makosza, M.; Fedorynski, M. *Adv. Catal.* **1987**, *35*, 375. (e) Dehmlow, E.V.; Dehmlow, S.S. *Phase Transfer Catalysts*, 3rd ed. (VCH: New York, 1993). (f) Halpern, M.E., Ed. *Phase-Transfer Catalysis: Mechanisms and Syntheses* (ACS: Washington, D.C., 1997). (g) Sasson, Y.; Neumann, R., Eds. *Handbook of Phase Transfer Catalysis* (Blackie: London, 1997). (h) Nelson, A. *Angew. Chem. Int. ed. Engl.* **1999**, *38*, 1583.

16. (a) Makosza, M.; Serafinowa, B. *Rocz. Chem.* **1965**, *39*, 1223. (b) Makosza, M.; Wawrzyniewicz, W. *Tetrahedron Lett.* **1969**, 4569.

17. Brändström, A.; Gustavii, K. *Acta Chem. Scand.* **1969**, *23*, 1215.

18. Starks, C.M. *J. Am. Chem. Soc.* **1971**, *93*, 195.

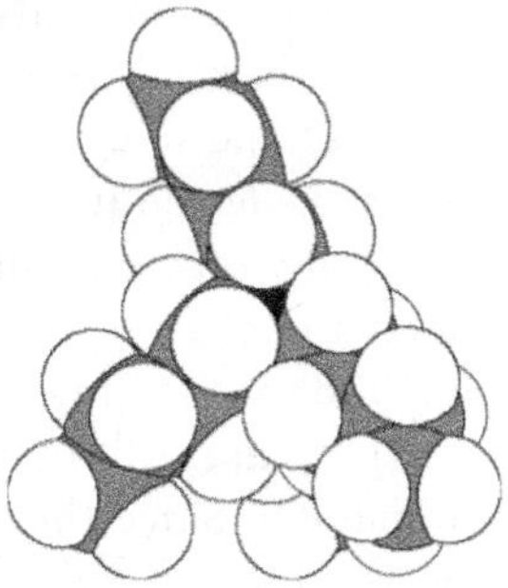

Bu_4N^+

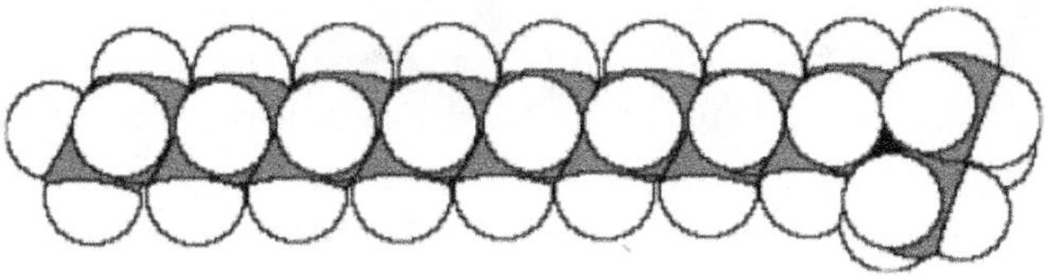

$C_{18}H_{35}NMe_3^+$

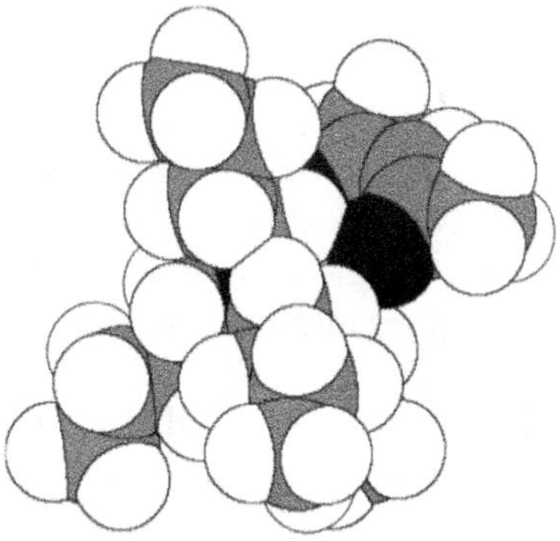

Bu_4N^+ $(MeCO)_2CH^-$

When these compounds dissolve in non-polar solvents, they do so as solvated *ion pairs*, where the cation and the anion are in contact with each other, and so the bonds between the ions are not completely broken (the ion pair tetrabutylammonium acetylacetonate is shown). However, in the ion pair, the anion is much more accessible to another molecule than it would be in water, making it a much more reactive species than in aqueous solution. We also see the effect of solvent dielectric on reactivity: as the dielectric constant decreases, the reactivity of ionic species becomes more pronounced: anions become stronger bases and nucleophiles, cations become stronger acids and electrophiles.

By vigorously mixing a solution of a sodium salt (e.g., sodium cyanide) in water with a solution of a tetraalkylammonium chloride or hydrogen sulfate in an organic solvent that is immiscible in water, one can establish the equilibrium shown in Example **8.1**, which allows one to dissolve the nucleophilic anion in the non-polar organic solvent, where it is a much more powerful nucleophile.

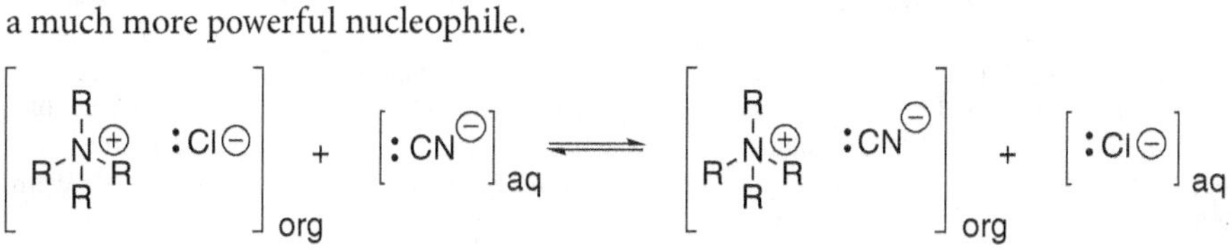

(8.1)

An alternative method for increasing the solubility of ionic reagents in relatively non-polar solvents is the use of crown ethers. These compounds, whose ability to enhance the solubility of metal ions in organic solvents by complex formation was first reported by Nobel Laureate, Charles J. Pedersen in 1967,[19] are cyclic ethers that selectively bind alkali metal cations, especially. They are named according to a simple mnemonic in which the number of atoms in the ring is separated from the number of oxygen atoms by the word, "crown." The space-filling model of the potassium-selective ether, 18-crown-6, is shown as Example **8.2**; the lithium complex of the lithium-selective 12-crown-4 is shown as Example **8.3**. Since their discovery, crown ethers have found an important place in organic synthesis.[20]

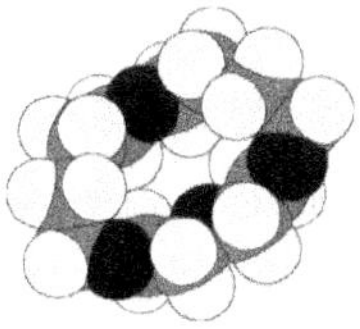

(8.2)

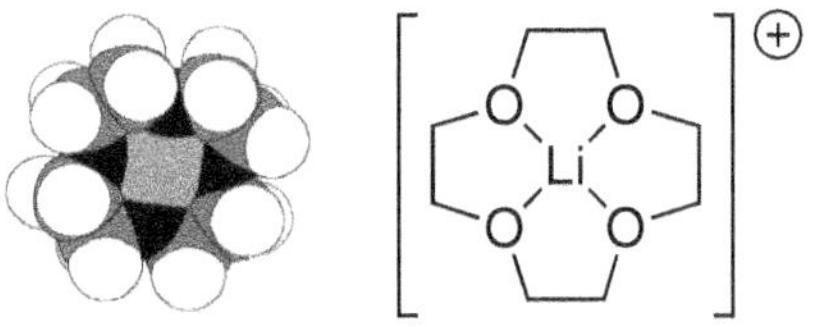

(8.3)

Ionic Liquids

The push toward "green" chemistry in the last decade of the 20th century saw the rise of a new class of solvents for use in organic reactions. These solvents consist of low-melting ionic solids and are generally referred to as *ionic liquids*. The advantages of ionic liquids as solvents in "green" chemical reactions quickly became apparent. Because the liquid is composed solely of ions, it is nonvolatile, which eliminates the use of (often toxic and usually regulated) volatile organic solvents.[21] Although predicting the melting points of ionic organic salts is rather hit-and-miss, there is now a sufficiently wide range of low-melting salts available for use as solvents in organic reactions that the use of these ionic liquids is now relatively straightforward.

Ionic liquids have a number of especially useful properties that make them attractive as solvents for a wide range of reactions:

1. They are excellent solvents for a wide range of solutes, including both ionic and molecular species, which allows unusual pairings of solutes to be brought into solution together.
2. They are highly polar.
3. They are composed of noncoordinating ions, which means that they can be used with solutes that would suffer from coordination by the solvent.

---

19. Pedersen, C.J. *J. Am. Chem. Soc.* **1967**, *89*, 2495.

20. For representative reviews, see: (a) Pedersen, C.J.; Frensdorff, H.K. *Angew Chem. Int. Ed. Engl.* **1972**, *11*, 16. (b) Christensen, J.J.; Eatough, D.J.; Izatt, R.M. *Chem. Rev.* **1974**, *74*, 351. (c) Gokel, G.W.; Durst, H.D. *Synthesis* **1976**, 168. (d) Izatt, R.M.; Christensen, J.J. *Synthetic Multidentate Macrocyclic Compounds* (Academic Press: New York, 1978). (e) Hiroaka, M. *Crown Compounds: Their Characteristics and Applications* (Kondasha: Tokyo, 1982; Elsevier: Amsterdam, 1982).

21. Reviews and Monographs: (a) Welton, T. *Chem. Rev.* **1999**, *99*, 2071. (b) Wasserscheid, P.; Keim, W. *Angew. Chem., Int. Ed.* **2000**, *39*. 3772. (c) Sheldon, R.A. *Chem. Commun.* **2001**, 2399. (d) Dupont, J.; de Souza, R.F.; Suarez, P.A.Z. *Chem. Rev.* **2002**, *102*, 3667. (e) Rogers, R.D.; Seddon, K.R. *Ionic Liquids; ACS Symp. Ser*, **2002**, *818*. (f) Rogers, R.D.; Seddon, K.R. *Ionic Liquids as Green Solvents; ACS Symp. Ser*, **2003**, *856*. (g) Rogers, R.D.; Seddon, K.R. *Ionic Liquids IIIB: Fundamentals, Progress, Challenges and Opportunities; ACS Symp. Ser*, **2005**, *902*. (h) Plechkova, N.V.; Rogers, R.D.; Seddon, K.R. *Ionic Liquids: From Knowledge to Application; ACS Symp. Ser*, **2009**, *1030*. (i) Wasserscheid, P., Welton, T., Eds. *Ionic Liquids in Synthesis*, 2nd ed.; (Wiley-VCH: Weinheim, 2007).

4. They are often immiscible with organic solvents and with water, which makes them a useful highly polar alternative to both.
5. They do not evaporate under conditions of high vacuum, which makes them unique among solvents.

The most common room-temperature ionic liquids in use today are *N,N′*-dialkylimidazolium salts, where the anion is typically a fluorinated anion such as BF_4^-, PF_6^-, Tf_2N^- (or similar anions). The alkyl groups of the imidazolium ion are frequently methyl and a higher alkyl group. One of the first room-temperature ionic liquids was 1-ethyl-3-methylimidazolium tetrafluoroborate ([emim][BF_4]; Example **8.4**), prepared by heating the imidazolium chloride with silver tetrafluoroborate in methanol.[22] The corresponding *n*-butyl compound, [bmim][BF_4] (Example **8.5**), has become popular because its preparation is especially easy.[23]

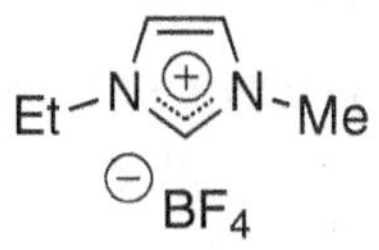

[emim][BF_4] **(8.4)**

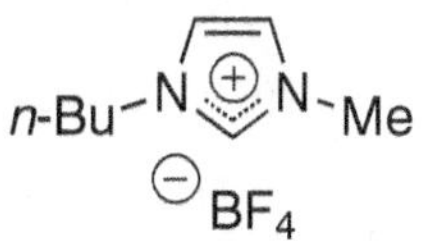

[bmim][BF_4] **(8.5)**

These solvents, from their composition, might reasonably be expected to be quite polar: one expects a high dielectric constant, for example, from a liquid composed solely of ions. However, polarity is a notoriously difficult concept to quantify, so just how polar these liquids are considered to be depends on the measure being used to define the solvent polarity. If one uses the simplest such measure, the ability to dissolve and stabilize ionic salts, ionic liquids are obviously highly polar, but if one uses a more quantitative scale, one often finds that these solvents (especially the solvents with larger organic groups) resemble highly polar organic liquids such as dimethyl sulfoxide. This is nicely illustrated by the nucleophilicity of anions toward *n*-alkyl sulfonates in ionic liquids: the second-order rate constants of the S_N2 reactions of azide, halide, and isothiocyanate anions with 1-hexyl or 1-octyl methanesulfonate in the ionic liquid all lie between the rate constants for the corresponding reactions in methanol and dimethyl sulfoxide.[24]

Ionic solvents lend themselves to a wide range of applications. They have been used as solvents in widely disparate reactions, including Friedel-Crafts reactions,[25] catalytic hydrogenation,[26] hydroformylation,[27] Suzuki coupling,[28] and oxidation of benzylic alcohols by sodium hypochlorite.[29] One fascinating application of ionic liquids in synthesis involves the osmium tetroxide hydroxylation of alkenes. Osmium tetroxide is expensive, toxic, and highly volatile, but it has been found that—remarkably—solutions of osmium tetroxide in the ionic liquid [emim][BF_4] do not lose osmium tetroxide even over the course of several months, and that the osmium tetroxide solution can be used and reused with an organic co-oxidant several times over without losing activity (Example **8.6**).[30]

---

22. Wilkes, J.S.; Zaworotko, M.J. *J. Chem. Soc., Chem. Commun.* **1990**, 965.
23. Dyson, P.J.; Grossel, M.C.; Srinivasan, N.; Vine, T.; Welton, T.; Williams, D.J.; White, A.J.P.; Zigras, T. *J. Chem. Soc., Dalton Trans.* **1997**, 3465.
24. Landini, D.; Maia, A. *Tetrahedron Lett.* **2005**, *46*, 3961.
25. Boon, J.A.; Levisky, J.A.; Pflug, J.L.; Wilkes, J.S. *J. Org. Chem.* **1986**, *51*, 480.
26. Brown, R.A.; Pollet, P.; McKoon, E.; Eckert, C.A.; Liotta, C.L.; Jessop, P.G. *J. Am. Chem. Soc.* **2001**, *123*, 1254.
27. (a) Chauvin, Y.; Mussmann, L.; Olivier, H. *Angew. Chem., Int. Ed. Engl.* **1995**, *34*, 2698. (b) Favre, F.; Olivier-Bourbigou, H.; Commereuc, D.; Saussine, L. *Chem. Commun.* **2001**, 1360.
28. Matthews, C.J.; Smith, P.J.; Welton, T. *Chem. Commun.* **2000**, 1249.
29. Xie, H.; Zhang, S.; Duan, H. *Tetrahedron Lett.* **2004**, *45*, 2013.
30. Yanada, R.; Takemoto, Y. *Tetrahedron Lett.* **2002**, *43*, 6849.

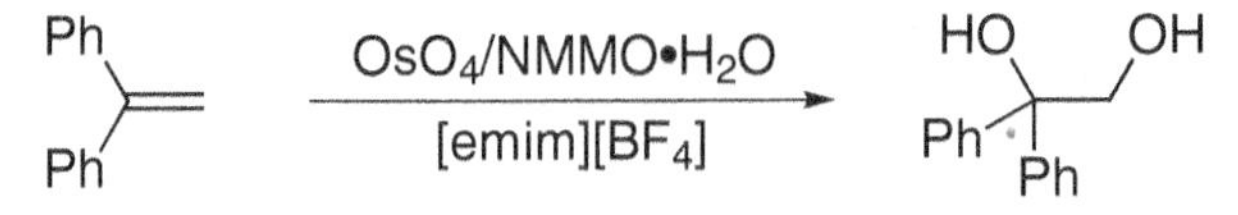

(8.6)

Run:	1	2	3	4	5
Yield:	95	93	96	95	93

Problems

8-1 Arrange the following molecules in increasing order of their expected boiling points. Give reasons for your choice.

(a) $CH_3(CH_2)_3CH_3$; $(CH_3)_4C$; $(CH_3)_2CHCH_2CH_3$

(b) $CH_3CH_2CH_2CH_2N(CH_3)_2$; $N(CH_2CH_3)_3$; $(CH_3)_2NC(CH_3)_3$

(c) $CH_3(CH_2)_3N(CH_3)_2$; $(CH_3)_2CH(CH_2)_2NHCH_3$; $(CH_3)_3CH(CH_2)_2NH_2$

8-2 Arrange the following compounds in increasing order of their solubility in water. Give reasons for your rankings.

(a) $CH_3(CH_2)_3OH$; $(CH_3)_3COH$; $CH_3CH_2CH(OH)CH_3$; $(CH_3)_2CHCH_2OH$

(b) $CH_3(CH_2)_4OH$; $CH_3(CH_2)_2CH(OH)CH_3$; $(CH_3CH_2)_2CHOH$; $CH_3CH_2C(CH_3)_2OH$

8-3 *n*-Butyl alcohol ($CH_3CH_2CH_2CH_2OH$) and diethyl ether ($CH_3CH_2OCH_2CH_3$) have the same solubility in water, but the boiling point of diethyl ether is 35°C while the boiling point of *n*-butyl alcohol is 118°C. Explain these observations.

8-4 Sodium cyanide is a strong nucleophile in ethanol solution, and an even more powerful nucleophile in dimethylformamide (DMF) solution. However, the reaction between 1-bromobutane and sodium cyanide in hexane proceeds extremely slowly, despite the fact that the solvent is aprotic. Why?

8.3 Reaction Kinetics[31]

The first step in the process of elucidating a reaction mechanism usually involves measuring the kinetics for the reaction under study. Kinetic measurements at a single temperature provide the rate law of the reaction. This gives the kinetic order of the reaction, from which one may deduce the molecularity of the reaction; the rate constant, *k*, for the reaction; and the identity of those species participating in the reaction up to and including the rate determining step.

The determination of kinetics for a reaction involves the measurement of the rate of change of the concentration of the reactants or products with respect to time. One can write the differential rate law for a typical reaction in the form:

$$-\frac{d[R]}{dt} = k[A]^a[B]^b... \quad \text{or} \quad \frac{d[P]}{dt} = k[A]^a[B]^b... \qquad \text{(Eq. 8.2)}$$

where [R] is the concentration of the reactant being monitored; [P] is the concentration of the product; *k* is the rate constant for the reaction; [A], [B], and so on, are the

31. For general textbooks on kinetics, see: (a) House, J.E. *Principles of Chemical Kinetics* (Wm. C. Brown: Dubuque, 1997). (b) Frost, A.A.; Pearson, R.G. *Kinetics and Mechanism* (John Wiley & Sons: New York, 1953).

concentrations of the individual reactants; and a, b, and so on, are the exponents for the individual reactants.

First-Order Reactions

The simplest case of a rate law is for a reaction where the rate varies directly with the concentration of one of the reactants (a ***first-order reaction***):

$$-\frac{d[A]}{dt} = k[A] \text{ or } \frac{d[P]}{dt} = k[A] \qquad \text{(Eq. 8.3)}$$

Rearranging this equation gives an alternative form that can be integrated to give a more useful form of the rate equation, ***integrated rate law*** for the reaction. For a first-order reaction, the integrated rate law is of the form:

$$\ln[A]_t = \ln[A]_0 - kt \qquad \text{(Eq. 8.4)}$$

In other words, a plot of ln[A] versus time for a first-order reaction will be linear, with a slope of $-k$ and a y intercept of ln $[A]_0$.

Second-Order Reactions

Reactions exhibiting second-order kinetics fall into two classes: those where only one reactant appears in the rate law and those where two reactants appear in the rate law. The former reactions obey a rate law of the form:

$$-\frac{d[A]}{dt} = k[A]^2, \text{ which integrates to } \frac{1}{[A]_t} = kt + \frac{1}{[A]_0} \qquad \text{(Eq. 8.5)}$$

In this case, a plot of 1/[A] versus time will be linear, with a slope of k and a y intercept of $1/[A]_0$.

The second type of second-order reaction has a rate law that is first order in both reactants. These reactions obey a rate law of the form:

$$-\frac{d[A]}{dt} = -\frac{d[B]}{dt} = k[A][B] \qquad \text{(Eq. 8.6)}$$

In this case, the kinetics are less simple, but one can obtain the integrated rate law:

$$\ln\frac{[A]_t}{[B]_t} = \{k([B]_0 - [A]_0)\}t + \ln\frac{[A]_0}{[B]_0} \qquad \text{(Eq. 8.7)}$$

This gives a linear graph when $\ln([A]_t/[B]_t)$ is plotted against time, with a slope of $k([B]_0\text{-}[A]_0)$ and a y intercept of $\ln([A]_0/[B]_0)$.

More frequently, reactions such as this are run under conditions where one reactant is in great excess over the other. Under these conditions, the kinetics simplify to approximate first-order kinetics, because the change in concentration of the excess reagent is small enough to be neglected (i.e., $[B]_t \approx [B]_0$). Reaction kinetics measured under these conditions are said to be measured under ***pseudo-first-order conditions***. The pseudo-first-order rate constant when B is in excess, for example, is related to the true second-order rate constant by the simple relationship:

$$k = k_{obs}[B] \approx k_{obs}[B]_0 \qquad \text{(Eq. 8.8)}$$

This approach may also be applied to higher order reactions in two reagents by using a large excess of one reagent over the other. In this case, the order of the reaction with

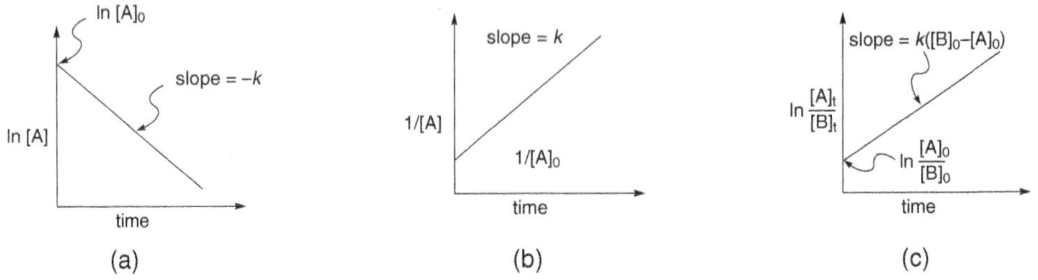

Figure 8.5 Plots of the integrated rate laws of (a) a first-order reaction, (b) a second-order reaction with only one reactant appearing in the rate law, and (c) a second-order reaction with two reactants appearing in the rate law

respect to each reactant must be determined separately, of course, by using each of the reactants as the excess reagent.

The most common forms of the integrated rate laws for first- and second-order reactions are shown in Figure 8.5.[ʕ]

Making Deductions from the Rate Law

Simple nucleophilic substitution reactions at saturated carbon provide useful models for how one approaches elucidating a mechanism from a rate law. The first mechanistic proposals about these reactions were set forth in the first decades of the 20th century by a variety of investigators.[32] The process of finding out whether or not a reactive intermediate is involved in a reaction has changed little since the pioneering work of Hughes and Ingold between 1927 and 1933.

If any of the reactants present in the balanced, stoichiometric equation for a reaction is not represented in the rate law, at least two steps must be required to complete bond rupture and bond formation. The case of the S_N1 substitution reaction is illustrative. For this reaction, only one of the stoichiometric reactants appears in the rate law. If we use the S_N1 reaction between *tert*-butyl bromide and water as the example,

---

ʕ ***The Mathematics***:

First-order reactions:

$$-\frac{d[A]}{dt} = k[A] \Rightarrow -\frac{d[A]}{[A]} + k\,dt \Rightarrow -\int_0^t \frac{d[A]}{[A]} = \int_0^t k\,dt \Rightarrow \ln\frac{[A]_0}{[A]_t} = kt \Rightarrow \ln[A]_t = \ln[A]_0 - kt$$

Second-order reactions

(i) *Rate laws involving a single reagent:*

$$-\frac{d[A]}{dt} = k[A]^2 \Rightarrow -\int_0^t \frac{d[A]}{[A]^2} = \int_0^t k\,dt \Rightarrow \frac{1}{[A]_t} - \frac{1}{[A]_0} = kt \Rightarrow \frac{1}{[A]_t} = kt + \frac{1}{[A]_0}$$

(ii) *Rate laws involving two different reagents:*

$$-\frac{d[A]}{dt} = -\frac{d[B]}{dt} = k[A][B]$$

In this case, the rate equation is most simply solved by denoting the amount of excess reagent ($[B]_0 - [A]_0$) by the term x. This now changes the rate law to:

$$-\frac{d[A]}{dt} = k[A]([A]+x) \Rightarrow -\int_0^t \frac{d[A]}{[A]([A]+x)} = \int_0^t k\,dt \Rightarrow kt = -\left(\frac{1}{x}\right)\ln\frac{[A]_t + x}{[A]_t} + \left(\frac{1}{x}\right)\ln\frac{[A]_0 + x}{[A]_0} \Rightarrow$$

$$(kx)t = \ln\left(\frac{[A]_t}{[A]_t + x}\right)\left(\frac{[A]_0 + x}{[A]_0}\right) \Rightarrow (kx)t = \ln\left\{\left(\frac{[A]_t}{[B]_t}\right)\left(\frac{[B]_0}{[A]_0}\right)\right\} \Rightarrow \ln\frac{[A]_t}{[B]_t} = \{k([B]_0 - [A]_0)\}t + \ln\frac{[A]_0}{[B]_0}$$

32. (a) Fischer, E. *Justus Liebigs Ann. Chem.* **1911**, *381*, 123. (b) Werner, A. *Ber. Deut. chem. Ges.* **1911**, *44*, 873. (c) Werner, A. *Justus Liebigs Ann. Chem.* **1912**, *386*, 1. (d) Pfeiffer, P. *Justus Liebigs Ann. Chem.* **1912**, *386*, 92. (e) Gadamer, J. *J. prakt. Chem.* **1913**, *78*, 312. (f) Meisenheimer, J. *Justus Liebigs Ann. Chem.* **1912**, *386*, 126. (g) Le Bel, J.-A. *J. chim. phys.* **1911**, *9*, 323. (h) Lewis, G.N. *Valence and Structure of Atoms and Molecules* (Chemical Calatog Co.: New York, 1923), p. 113. (h) London, F. *Z. Elektrochem.* **1929**, *35*, 552. (i) Lowry, T.M. *Inst. inter. chim. Solvay, Conseil chim.* (Brussels), **1925**, 130.

it is clear that both reactants are required to form the C—O bond, but the rate law (Equation 8.9) is first-order:

$$Me_3C\text{-}Br + H_2O \rightarrow Me_3C\text{-}OH + HBr$$

$$\text{Rate} = k[Me_3CBr] \qquad \text{(Eq. 8.9)}$$

Taken together, these two facts require that this reaction must occur in a minimum of two steps: (1) the cleavage of the carbon-halogen bond, which occurs before or during the slow step of the reaction, and (2) the formation of the carbon-oxygen bond, which must occur in a subsequent kinetically fast step. The reaction coordinate diagram for this reaction is shown as Figure 8.6; the reaction is a multistep reaction composed of the three elementary reactions shown on the left-hand side of the diagram. This diagram reinforces an important point. Kinetics can tell us that the reaction proceeds in a stepwise manner involving at least one reactive intermediate and allows us to pinpoint the slow bonding change, but *kinetics alone cannot tell us how many steps are required to convert reactants to products* (i.e., about the complete mechanism). To complete the picture, other information is required.

The foregoing discussion shows us that the rate law may allow us to deduce that a reaction proceeds in a stepwise manner. Unfortunately, a similar simple deduction cannot be made in the case of reactions where every reactant appears in the rate law. In these cases, one cannot deduce from the rate law alone whether or not a reaction is concerted, but additional evidence from other studies (e.g., stereochemical studies) is needed before one can make that determination. This is illustrated nicely by base-promoted elimination reactions.

$$R_2CH\text{-}CR_2X + RO^- \rightarrow R_2C{=}CR_2 + ROH + X^-$$

$$\text{Rate} = k[RX][RO^-] \qquad \text{(Eq. 8.10)}$$

In the case of both the E2 and E1cb mechanisms, the rate law is of the form shown in Equation 8.10, where both the alkyl halide and the base appear in the rate law. The

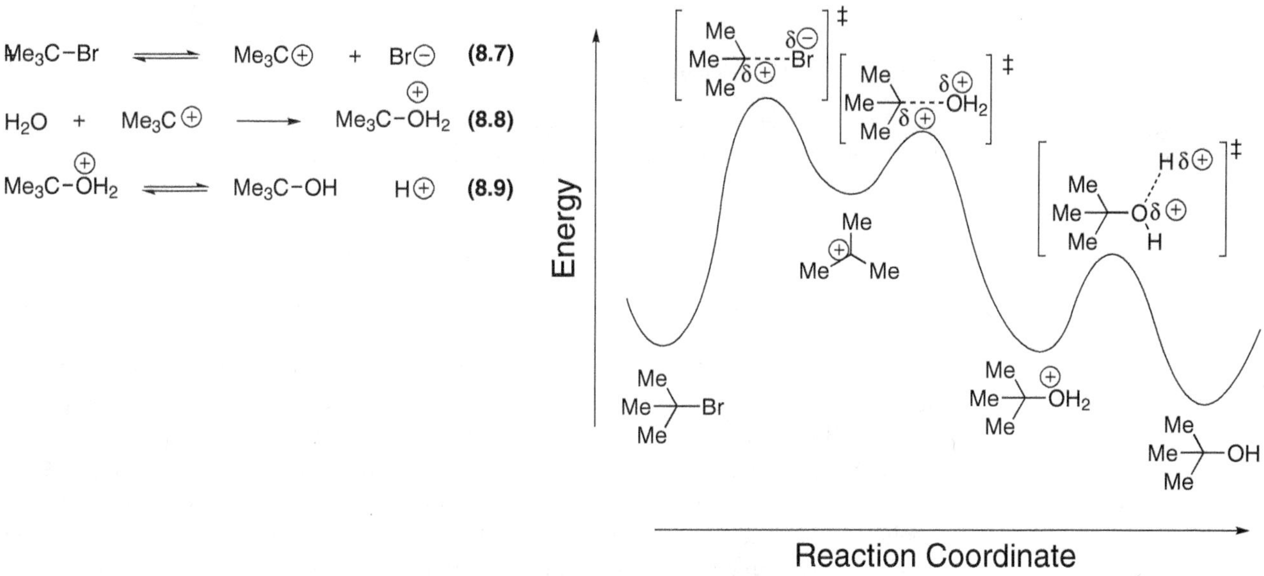

Figure 8.6 Reaction coordinate diagram for the S_N1 substitution of *tert-butyl bromide*

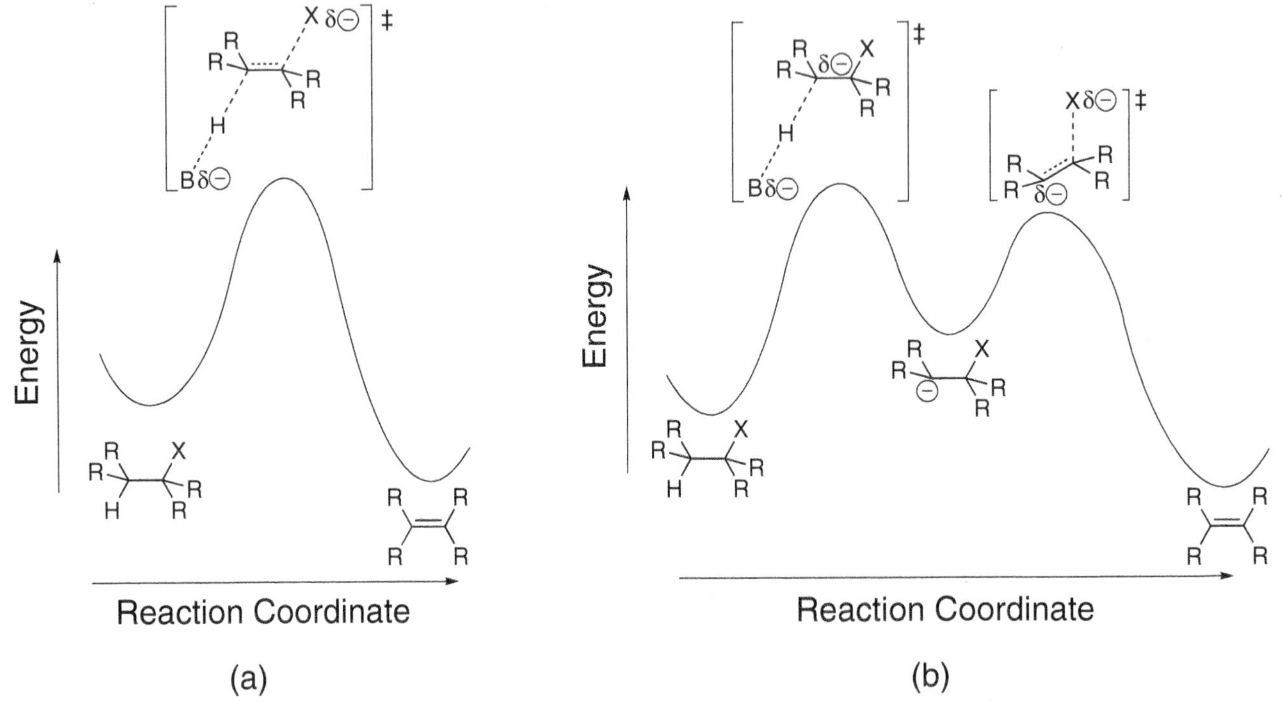

Figure 8.7 Reaction coordinate diagrams for base-promoted (a) E2 and (b) E1cb elimination from an alkyl halide. Both reactions have the same rate law: rate = k[RX][base].

difference between the two mechanisms is that the E2 reaction is concerted, whereas the E1cb reaction is stepwise, with proton transfer to the base to give a carbanion intermediate as the rate-determining step. These two mechanisms can, however, be distinguished by stereochemical and isotope exchange studies—the concerted E2 reaction lacks a pathway for exchange of hydrogen and has a strict stereochemical constraint (the elimination is antarafacial), whereas the E1cb reaction does not have the a strict stereochemical constraint and does have a pathway for hydrogen exchange between the carbanion intermediate and the solvent. The reaction coordinate diagrams for these two related reactions are shown in Figure 8.7.

Preequilibrium and Catalysis

The E1 elimination reaction between *tert*-butyl alcohol and phosphoric acid at elevated temperatures gives isobutylene and water, and it obeys a second-order rate law, shown as Equation 8.11.

$$Me_3COH \rightarrow Me_2C{=}CH_2 + H_2O$$

$$\text{Rate} = k[Me_3COH][H^+] \qquad \text{(Eq. 8.11)}$$

In the reaction, the concentration of acid appears in the rate law, but the acid does not appear in the stoichiometric equation for the reaction. This is the classical definition of a catalyst. The rate law tells us that the reaction between *tert*-butyl alcohol and the strong acid either precedes the rate-determining step or is part of the rate-determining step. A variety of other experiments can be carried out to demonstrate that the former situation applies to this reaction and that the alcohol reacts with the protons in a rapid, reversible reaction to give an intermediate that then participates in the rate-determining step of the reaction. This situation is also often referred to as a ***pre-equilibrium***. The reaction coordinate diagram for this reaction is shown in Figure 8.8. This reaction also shows one

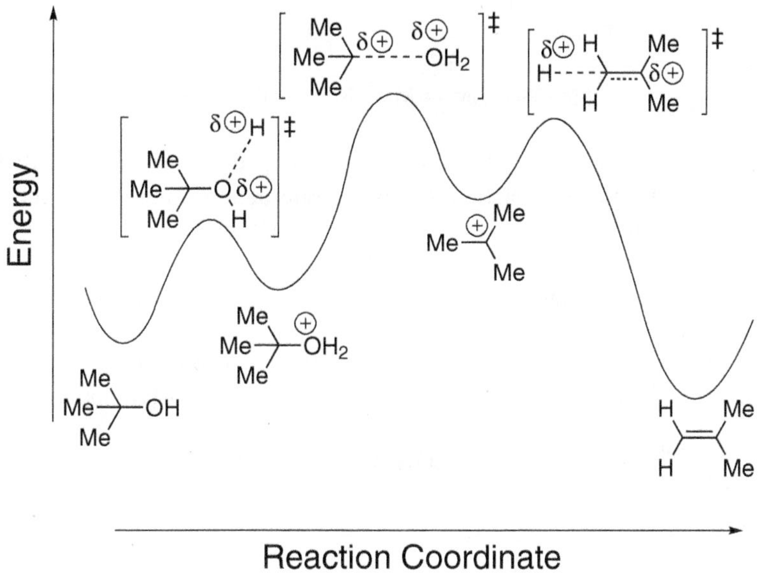

Figure 8.8 Reaction coordinate diagram for the acid-catalyzed dehydration of *tert-butyl alcohol*

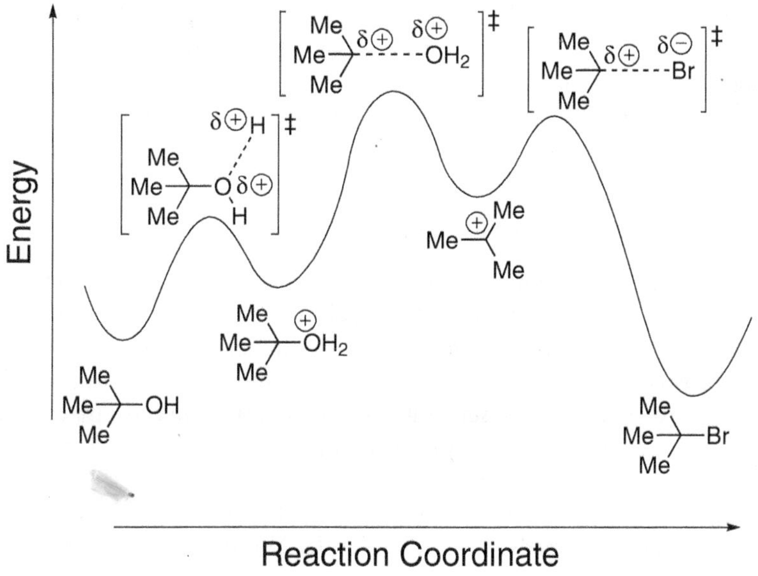

Figure 8.9 Reaction coordinate diagram for the reaction between tert-butyl alcohol and bromide ion in the presence of strong acid

important characteristic of catalysts: they always appear in the rate equation of a reaction, even if they do not appear in the overall stoichiometric equation.

Deducing a Rate Law from a Reaction Coordinate Diagram

The reaction coordinate diagram for the reaction between *tert*-butyl alcohol and bromide ion in the presence of a strong acid is shown in Figure 8.9. The reaction coordinate is very similar to that of the E1 elimination that we have just discussed, and we will now use this diagram to make some deductions about the form of the rate law for the reaction. There are three steps required in the reaction to convert the *tert*-butyl alcohol to *tert*-butyl bromide, and these can be represented by the elementary reaction in reactions 8.12 to 8.14.

$$Me_3COH + H^+ \rightarrow Me_3COH_2^+ \qquad \text{(Eq. 8.12)}$$

$$Me_3COH_2^+ \rightarrow Me_3C^+ + H_2O \qquad \text{(Eq. 8.13)}$$

$$Me_3C^+ + Br^- \rightarrow Me_3CBr \qquad \text{(Eq. 8.14)}$$

The second transition state is the highest point in the diagram, so we can deduce that the unimolecular heterolysis reaction 8.13 is the rate-determining step of the reaction. Based on this, we can write a rate law for the reaction of the form:

$$\text{Rate} = k[Me_3COH_2^+] \qquad \text{(Eq. 8.15)}$$

We cannot measure the concentration of the *tert*-butyl alcohol conjugate acid directly, but we can deduce it from the expression for the equilibrium constant for reaction 8.12:

$$K_{eq} = \frac{[Me_3COH_2^+]}{[Me_3COH][H^+]} \Rightarrow [Me_3COH_2^+] = K_{eq}[Me_3COH][H^+] \qquad \text{(Eq. 8.16)}$$

which leads to:

$$\text{Rate} = kK_{eq}[Me_3COH][H^+], \text{ or Rate} = k_{obs}[Me_3COH][H^+] \qquad \text{(Eq. 8.17)}$$

In this reaction, the observed rate constant k_{obs} is a composite constant that is the product of the first-order rate constant for reaction 8.13 and the equilibrium constant for reaction 8.12 (the pre-equilibrium reaction). Note that bromide anion does not participate in the reaction until after the rate-determining step, so its concentration does not appear in the rate law.

Predicting a Rate Law from a Mechanism

The base-promoted elimination reactions of certain halides and similar compounds can be viewed under some circumstances as taking place in a stepwise manner, where the reactive intermediate is the carbanion formed by deprotonation of the halide, as shown in Figure 8.10. This sequence of reactions is characterized by three rate constants: k_1, the rate constant for deprotonation of the substrate by base; k_{-1}, the rate constant for deprotonation of the conjugate acid of the base by the carbanion; and k_2, the rate constant for the loss of the leaving group from the carbanion.

Based on this mechanistic scenario, we can write the rate expression for the formation of the product, which is formed by unimolecular decomposition of the carbanion.

$$\text{Rate} = k_2[R^-] \qquad \text{(Eq. 8.18)}$$

Clearly, we may or may not be able to determine the carbanion concentration, but it would be better to write a rate law in terms of species that we can easily measure (RX and B^-). This means that we will need to find expressions for the concentration of the anion in terms of the concentrations of the other species. One way to accomplish this is to apply the

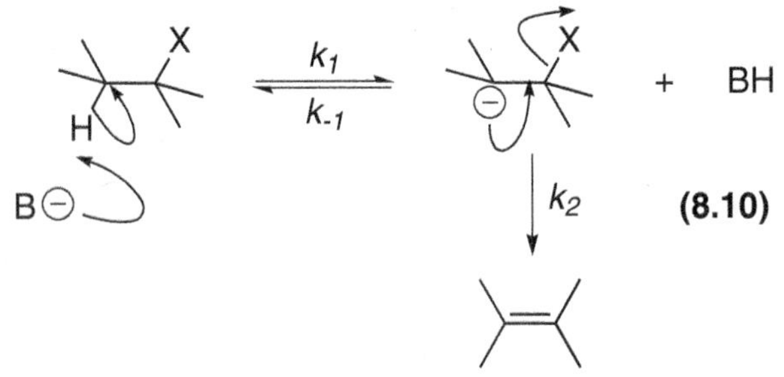

(8.10)

Figure 8.10 The E1cb mechanism for elimination

steady-state approximation to the reaction analysis. This approximation relies on the fact that because the carbanion is likely to be a high-energy species, it is most likely to react as soon as it is formed, so that only a small, constant (or very nearly so) concentration of the intermediate is ever present in the reaction mixture after a short initiation period.

The carbanion intermediate is formed in a bimolecular reaction with a rate equal to $k_1[RX][B^-]$. It can be lost in two ways: it can continue on to product, with a rate equal to $k_2[R^-]$, or it can react with the conjugate acid of the base to revert to starting material, with a rate equal to $k_{-1}[R^-][BH]$. At the steady state, the change in concentration of the carbanion is zero, which leads to the expression below:

$$k_1[RX][B^-] = k_2[R^-] + k_{-1}[R^-][BH] \qquad \text{(Eq. 8.19)}$$

This can be rearranged to

$$\left[R^-\right](k_2 + k_{-1}[BH]) = k_1[RX]\left[B^-\right], \text{ or } \left[R^-\right] = \frac{k_1[RX][B^-]}{(k_2 + k_{-1}[BH])} \qquad \text{(Eq. 8.20)}$$

which gives the final form of the rate law:

$$\text{Rate} = \frac{k_2 k_1[RX][B^-]}{(k_2 + k_{-1}[BH])} \qquad \text{(Eq. 8.21)}$$

Depending on the relative values of the rate constants, this expression can be simplified further. For example, if k_2 is much greater than either of the other two rate constants, this equation simplifies to:

$$\text{Rate} = k_1[RX][B^-] \qquad \text{(Eq. 8.22)}$$

In another possible scenario, k_1 is much greater than the other two rate constants. This would lead to complete deprotonation of the substrate before the rate-determining step, making the concentrations of B^- and HB constant and simplifying the kinetics to:

$$\text{Rate} = k_{exp}[RX], \text{ where } k_{exp} = \frac{k_2[B]}{k_2 + k_{-1}[HB]} \qquad \text{(Eq. 8.23)}$$

Worked Problem

8-1 The reaction of 1,1-dimethoxycyclohexane when heated in the presence of strong acid is:

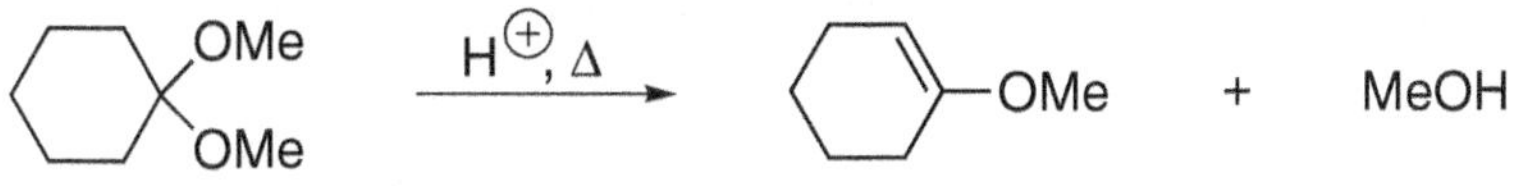

The rate law for this reaction is: Rate $= k[C_8H_{16}O_2][H^+]$

Write a mechanism, and draw a reaction coordinate diagram consistent with this rate law.

§Answer on next page.

Problems

8-5 There are six major mechanisms for base-promoted β-elimination to give alkenes: E1, E2, E1cb$_I$, E1cb$_R$, E1cb$_{ip}$, and E1$_{anion}$. The rate laws for these mechanisms are as follows:

(continues)

Problems (continued)

E1: Rate = k[RX]

E2: Rate = $k[RX][B^-]$

E1cb$_I$: Rate = $k[RX][B^-]$

E1cb$_R$: Rate = $k[RX][B^-]$ if the solvent is B—H

E1cb$_{ip}$: Rate = $k[RX]B^-]$

E1$_{anion}$: Rate = k_{exp}[RX] provided that $[B^-]>[RX]$

Draw reaction coordinate diagrams similar to Figures 8.6 to 8.8 for the E1cb elimination of Br^- from Ph_2CHCH_2Br by each of the E1cb mechanisms above.

8-6 You are asked to carry out an elimination reaction that proceeds by one of the six mechanisms (you do not know which) in Problem 8-5. Under the conditions you are using, the reaction is complete in one 3-hour laboratory period. You now find out that that you can actually use one half the amount of solvent, which makes the reaction much less expensive to carry out. How long will it now take for the reaction to be complete, based on each of the mechanisms in Problem 8-5? How would you tell if the reaction is E1, E2, or E1cb?

8-7 Draw appropriate reaction coordinate diagrams for the experimental conditions specified below and use them to predict the effects on the rate of those experimental conditions (you may want to use the Hammond postulate here):

(a) Increased solvation of the reactants.

(b) Increased solvation of the activated complex.

(c) Increased solvation of the products.

(d) Preferential solvation of the reactants over the products.

(e) Preferential solvation of the products over the reactants.

§ Answer for Worked Problem:
Although H^+ does not appear in the overall balanced equation for this reaction, it does appear in the rate law. This very strongly suggests that the proton is involved in the reaction as part of a pre-equilibrium. This gives the mechanism below, with the corresponding reaction coordinate diagram. [*Note how the high-energy carbocation intermediate is resonance-stabilized.*]

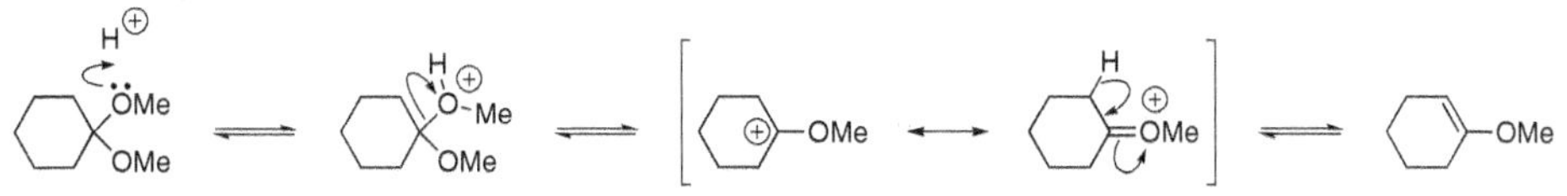

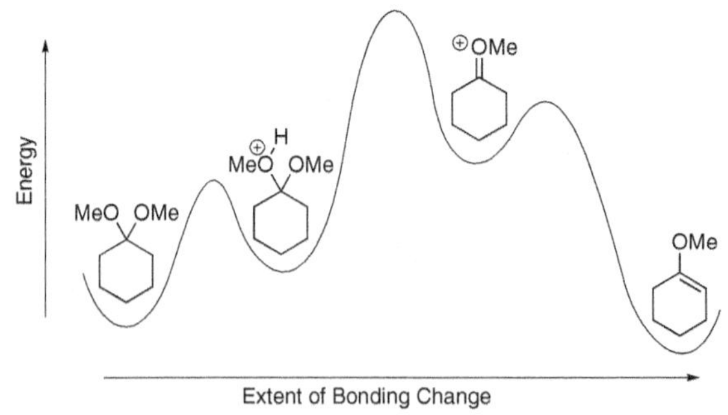

8-8 The two reactions below occur by the same concerted mechanism, but the effects of solvent on the rate constant for the reaction are opposite in the two reactions. Provide a reasonable rationalization for these observations.

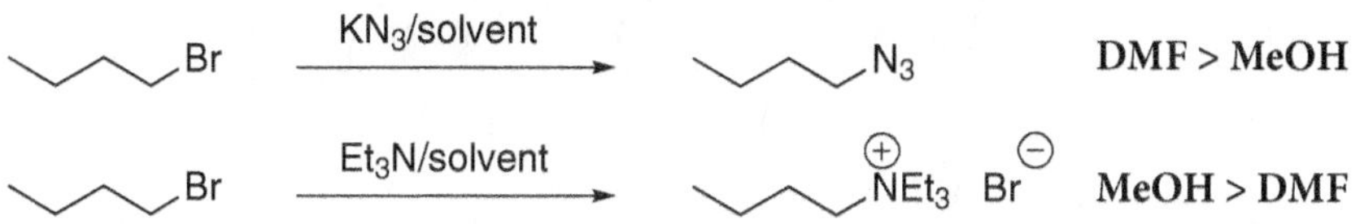

8-9 The formation of alkali metal complexes with crown ethers proceeds best in non-polar solvents (benzene or dichloromethane), but the complexes are unstable when heated in alcohols. Why?

8-10 Like carbocations, free radicals can be stabilized by adjacent heteroatoms bearing lone pairs of electrons. Provide a reasonable explanation of this by drawing the contributors to the resonance hybrid of the acetone radical anion (MeCOMe)$\bullet^-$.

8.4 Activation Parameters from Kinetic Studies

The rate constant for a reaction is temperature-dependent. The relationship between the value of the rate constant and temperature is the Arrhenius equation,[33] whose most usual form is

$$k = Ae^{-E_a/RT} \qquad \text{(Eq. 8.24)}$$

where k is the rate constant for the reaction; E_a is the activation energy for the reaction; A is the pre-exponential factor, or frequency factor, for the reaction (related to the probability that a given collision will actually result in a reaction); R is the ideal gas constant; and T is the kelvin temperature. By plotting the logarithm of the rate constant k against (1/T), one obtains a straight line with slope ($-E_a$/R) and y intercept log A. This gives one the activation energy E_a for the reaction and the Arrhenius pre-exponential factor A, which is related to the entropy of activation.

On the basis of transition state theory,[34] two simple relationships emerge: (1) between the activation energy E_a and the enthalpy of activation $\Delta H^\ddagger$ and (2) between the Arrhenius pre-exponential factor A and the entropy of activation $\Delta S^\ddagger$:

$$E_a = \Delta H^\ddagger + RT \qquad A = \left(\frac{RT}{N_0 h}\right) e^{\frac{\Delta S^\ddagger}{R}} \qquad \text{(Eq. 8.25)}$$

where N_0 is Avogadro's number and h is Planck's constant. The entropy of activation, in particular, gives useful information about the formation of the activated complex for a reaction: reactions for which $\Delta S^\ddagger$ is negative always involves bringing greater order to the system as the activated complex is formed, and this is usually interpreted in terms of a bimolecular activated complex or loss of rotational degrees of freedom. However, this still needs to be interpreted carefully, because the organization of solvent in the activated complex may reverse the sign of the activation entropy in reactions where the disorder of the

33. Arrhenius, S. *Z. physik. Chem.* **1889**, *4*, 226.

34. (a) Eyring, H. *J. Chem. Phys.* **1935**, *3*, 107. (b) Evans, M.G.; Polanyi, M. *Trans. Faraday Soc.* **1935**, *31*, 875. (c) Glasstone, S.; Laidler, K.;. Eyring, H. *The Theory of Rate Processes* (McGraw-Hill: New York, 1941). (d) Laidler, K.J.; King, M.C. *J. Phys. Chem.* **1983**, *87*, 2657. (e) Truhlar, D.G.; Hase, W.L.; Hynes, J.T. *J. Phys. Chem.* **1983**, *87*, 2664. (f) Pechukas, P. In Miller, W.H., Ed. *Dynamics of Molecular Collisions. Part B* (Plenum Press: New York, 1976), p. 239. (g) Pechukas, P. *Ann. Rev. Phys. Chem.* **1981**, *32*, 159.

Table 8.5 Standard Entropies of Activation for Some Organic Reactions

Reaction		$\Delta S^{\ddagger}$ (e.u.)	Ref.
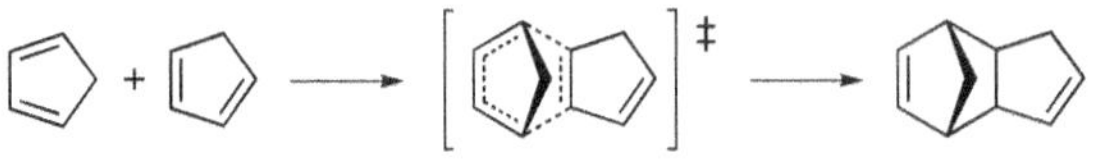	(8.11)	−34 −26	35 36
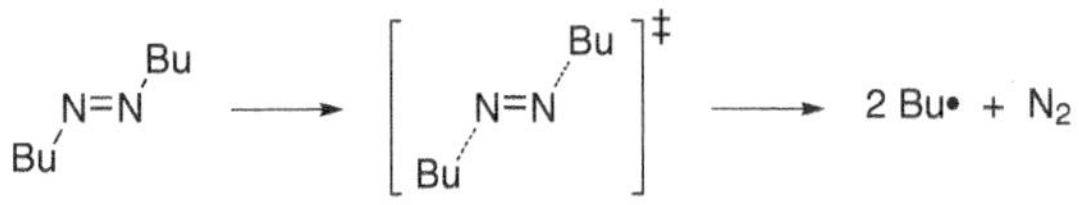	(8.12)	+19	37
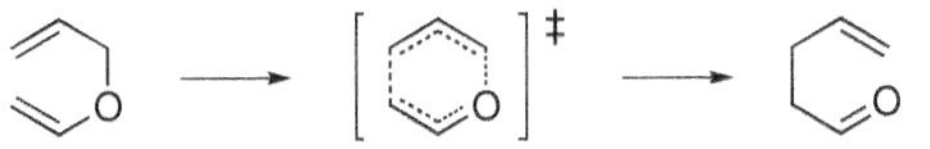	(8.13)	−8	38
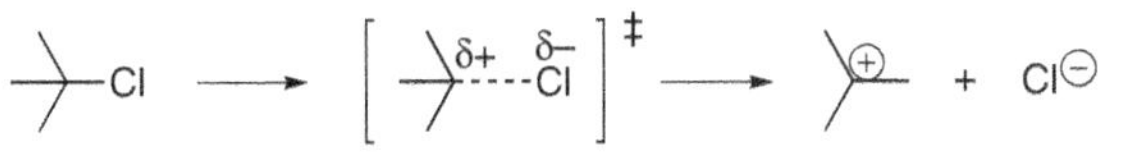	(8.14)	−6.6	39
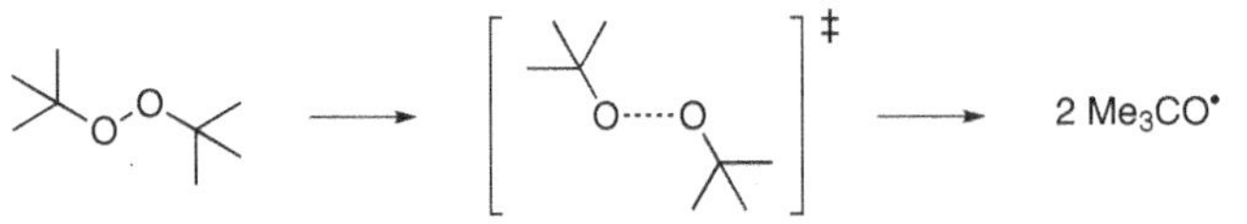	(8.15)	+13.8	40
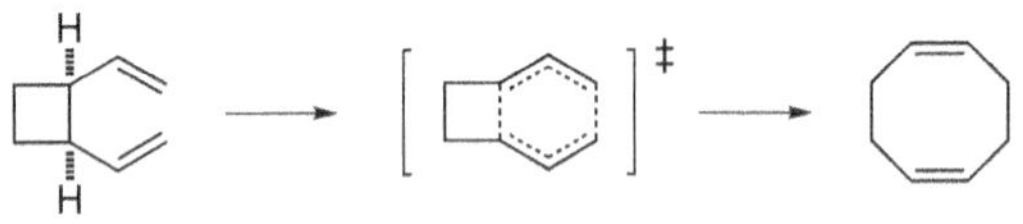	(8.16)	−11.7	41

*The negative value reflects the greater order imposed on the solvent by solvation of the incipient positive and negative charges.

reacting species actually increases. Some standard entropies of activation for several organic reactions are given in Table 8.5.

The volume of activation for a reaction may be calculated from the dependence of the rate constant on pressure. But, a plot of ln k versus P has a slope of $-\Delta V^{\ddagger}$, which is very small (the effects of pressure are orders of magnitude smaller than the corresponding effects of temperature). This necessitates making measurements over a range of tens or hundreds of atmospheres. In addition, interpreting volumes of activation is complicated by solvation and desolvation during the reaction, just like interpreting entropies of activation in solution. However, in non-polar, poorly solvating solvents, one may conclude that reactions with negative volumes of activation generally proceed through a highly ordered, compact transition state (e.g., the Diels-Alder reaction), while reactions with a positive volume of activation often proceed through a looser, less ordered transition state.

8.5 Correlating Reaction Rates: Linear Free Energy Relationships

The idea that one might be able to predict the rate of an organic reaction based on the rate of a known reaction is an attractive one to most organic chemists. Similarly, being able to

35. Wassermann, A. *Monatsh. Chem.* **1952**, *83*, 543.
36. (a) Kistiakowsky, G.B.; Lacher, J.R. *J. Am. Chem. Soc.* **1936**, **58**, 123. (b) See, also, Table 1 in Frost, A.A.; Pearson, R.G. *Kinetics and Mechanism* (John Wiley & Sons: New York, 1953), p. 101.
37. Blackham, A.U.; Eatough, N.L. *J. Am. Chem. Soc.* **1962**, *84*, 2922.
38. Schuler, F.W.; Murphy, G.W. *J. Am. Chem. Soc.* **1950**, *72*, 3155.
39. Grunwald, E.; Winstein, S. *J. Am. Chem. Soc.* **1948**, *70*, 846.
40. Bartlett, P.D.; Hiatt, R.R. *J. Am. Chem. Soc.* **1958**, *80*, 1398.
41. Hammond, G.S.; DeBoer, C.D. *J. Am. Chem. Soc.* **1964**, *86*, 899.

predict the effects of a small structural change on the equilibrium constant for a reversible reaction is also attractive. However, it is also obvious that any differences between the system used as the basis for the prediction and the system under scrutiny must be small—the reactions must be closely related—for the predicted result to engender confidence.

The idea that changing the substituents directly attached to a reacting center will change the rate constant for a reaction or the equilibrium constant for an equilibrium reaction is reasonably intuitive. We expect that if we take the S_N2 reaction of butyl halides (Example **8.17**), for instance, that 1-bromobutane, a typical primary halide, will react faster than 2-bromobutane, its secondary isomer, while replacing the 1-bromobutane with 1-chlorobutane will slow the reaction down even further due to the stronger bond to the leaving group. It is comforting to find that these predictions are borne out by experiment.[42]

$$\text{R-X} \xrightarrow[\text{40°C}]{\text{NaI, Me}_2\text{CO}} \text{R-I} \qquad \textbf{(8.17)}$$

$$R = n\text{-Bu, X = Cl: } k = 1.55 \times 10^{-5}\ \text{L mol}^{-1}\text{s}^{-1}$$
$$R = n\text{-Bu, X = Br: } k = 5.09 \times 10^{-3}\ \text{L mol}^{-1}\text{s}^{-1}$$
$$R = s\text{-Bu, X = Br: } k = 8.79 \times 10^{-5}\ \text{L mol}^{-1}\text{s}^{-1}$$

What may be less intuitive is the thought that changing substituents at a location remote from the reacting center—especially if those substituents are not capable of conjugating with the reacting center—should have an effect of reaction rates or equilibria. We can see evidence that they do, by looking at the ionization constants of acids: halogenated carboxylic acids have pK_as that rise as the halogen is moved further from the carbonyl group along the alkyl chain (Example **8.18**), and substituted benzoic acids also have pK_a values that vary with the substituent.

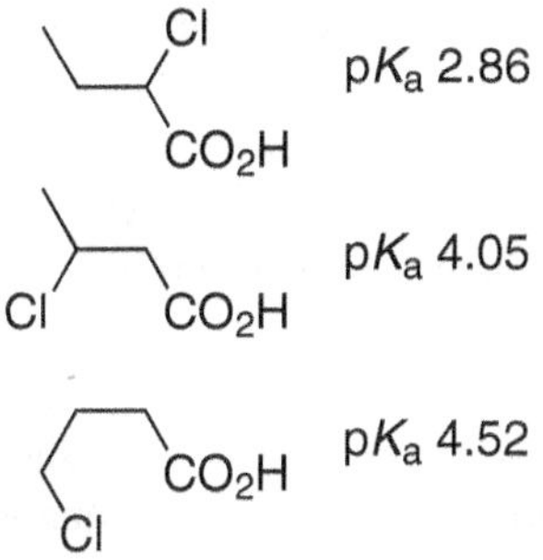

(8.18)

In 1935,[43] Louis P. Hammett[44] addressed the possibility of a relationship between reaction rates and equilibrium constants in the following way.[45] "*The idea that there is some sort of relationship between the rate of a reaction and the equilibrium constant is one of the most persistently held and at the same time most emphatically denied concepts in chemical theory. Many organic chemists accept the idea without question and use it, frequently with considerable success, but practically every treatise on physical chemistry points out that such a relationship has no theoretical basis and that it is in fact contradicted in many familiar cases.*" He then proceeded to show that for certain reactions, at least, there actually *is* a relationship between the rate constants and the equilibrium constants.

---

42. Pace, R.D.; Regmi, Y. *J. Chem. Educ.* **2006**, *83*, 1344.
43. Hammett, L.P. *J. Am. Chem. Soc.* **1935**, *57*, 125.
44. **Louis Plack Hammett (1894–1987)** was educated at Harvard, Zürich, and Columbia (PhD, 1922). After graduation, he remained at Columbia, where he spent his entire career. Hammett was a pioneer of physical organic chemistry, which term he coined, and was responsible for both the Hammett equation and the Hammett acidity constant. For more biographical details, see Westheimer, F.H. *Biogr. Mem. Nat. Acad. Sci.* **1997**, *72*, 136.
45. Hammett, L.P. *J. Am. Chem. Soc.* **1937**, *59*, 96.

In 1937, he went further and analyzed the rate and equilibrium data for 37 organic reactions of compounds with an aromatic ring, which led to him taking the first steps to place the qualitative observations ("conventional wisdom") on a quantitative footing. Since these first steps, the area of physical organic chemistry devoted to correlating the rates of organic reactions has blossomed. Today, a variety of linear free energy relationships is available to the organic chemist to quantify the effects of ring substituents and solvent on reaction rates.

It should be borne in mind during the following discussions that writing a linear free energy relationship between two reactions has implications that should be remembered when interpreting the results of such a relationship. First and foremost, there is an implicit assumption that the two reactions are reasonable models for each other—the choice of model can be critical in obtaining a successful correlation (e.g., the choice of σ, σ^+, or σ^- can have a dramatic effect on the fit of the data in certain reactions). This also means that the ability of the linear free energy relationship to correlate experimental data depends on the appropriateness of the model reaction chosen.

Substituent Effects: the Hammett Equation

The effects of remote substituents on the free energy of reaction ΔG are exercised as both enthalpy and entropy contributions. Hammett suggested that in reactions where the substituent is remote from the reaction center, the effects of substituents would be derived almost exclusively from the enthalpy contribution ΔH. He based this conclusion on the postulate that because the substituted aromatic ring is basically unchanged during the reaction, one would expect changes at the molecular level in the region of the substituent (e.g., solvation) to be very small. This means that the entropy contributions from the substituent to the free energy change would be close to zero. This is illustrated schematically in Figure 8.11, where the shaded region shows the region where the greatest changes during the reaction are expected to occur. It is worth noting that *ortho* substituents are located close enough to this region that we may expect that they will not obey the same simple relationships as the *meta* and *para* substituents because of steric effects. Of course, these arguments do not hold if there is strong resonance between the substituent and the reacting center, because that will lead to major changes in the electron density (and hence, the solvation) of the ring during the reaction.

Hammett's original proposal was that, in a series of substituted aromatic compounds undergoing the same reversible reaction in the side chain, it should be possible to relate the equilibrium constants for the reactions by the following equation:

$$\log_{10} (K/K_0) = \sigma\rho \qquad \text{(Eq. 8.26)}^{\S}$$

In this equation, which now bears his name, there are two constants characteristic of the reacting system. There is a substituent constant, σ, which measures the effect of the

Figure 8.11 Schematic of the region affected during a side-chain reaction of substituted aromatic compounds

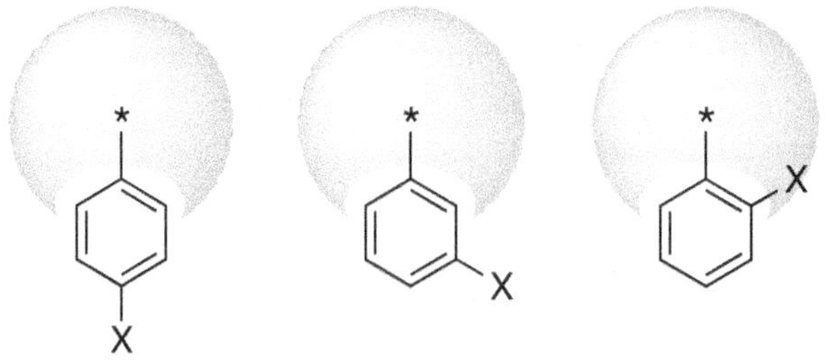

§. The logarithms in almost all of the equations in this chapter are in base 10. The term "log" will be used to designate $\log_{10}$ throughout the chapter; for natural, or Napierian logarithms, the term "ln" will be used.

substituent on the position of equilibrium, and a reaction constant, ρ, which measures the effect of the reaction type on the position of the equilibrium.

Showing that this equation represents a linear free energy relationship is fairly straightforward. We begin by expanding the log term to get Equation 8.27:

$$\log (K/K_0) = \log K - \log K_0 = \sigma\rho \qquad \text{(Eq. 8.27)}$$

Using the relationship between log K and ΔG, we can re-write equation 8.27 in the form:

$$-\Delta G/2.30\ 3RT - (-\Delta G_0/2.303RT) = \sigma\rho \qquad \text{(Eq. 8.28)}$$

which can be rearranged to:

$$\Delta G = \Delta G_0 - (2.303RT)\sigma\rho \qquad \text{(Eq. 8.29)}$$

Equation 8.29 relates the free energy of the reaction where the reactant carries the substituent of interest to the free energy of the same reaction where the reactant carries a standard substituent.

By replacing the free energy of reaction with the free energy of activation, it is possible to obtain a second form of the Hammett equation that relates reaction rates:

$$\log (k/k_0) = \sigma\rho, \text{ or } \Delta G^{\ddagger} = \Delta G_0^{\ddagger} - (2.303RT)\sigma\rho \qquad \text{(Eq. 8.30)}$$

The Hammett equation has four unknowns, so a standard system is required to calibrate the equation and make it useful. The standard reaction chosen by Hammett was the ionization of substituted benzoic acids (**8.19**, reaction constant $\rho = 1$) with hydrogen as the standard substituent ($\sigma = 0$). The ρ value for a reaction gives an estimate of the type and amount of charge development at the reaction center. Reactions with $\rho > 1$ tend to be characterized by more negative charge at the reacting center; reactions for which $\rho < 1$ tend to be characterized by increased positive charge at the reacting center. Some typical reaction ρ values are collected in Table 8.6.

There is a vast literature[46] discussing the effects of substituents on the rates of chemical reactions, and the construction of ***Hammett plots*** often provides an early indication of the nature of the reaction under study. In its original form, the Hammett equation works well for reactions in the side chain when no new interaction by resonance between the substituent and the side chain develops during the reaction. Thus, *meta* substituents tend to give the closest agreement with the Hammett equation.

As the importance of conjugation between the reaction center and the substituent increases, deviations from the Hammett equation become more pronounced. As the result of work by a number of research groups, two new substituent constants, σ^+ and σ^-, were proposed. The former is useful for reactions in which positive charge is developed in a

46. Reviews: (a) Hammett, L.P. *Physical Organic Chemistry* (McGraw-Hill: New York, 1940), ch. 6. (b) Jaffé, H.H. *Chem. Rev.* **1953**, *53*, 191. (c) van Bekkum, H.; Verkade, P.E.; Wepster, B.M. *Rec. trav. chim* **1959**, *78*, 815. (d) Topsom, R.D. *Prog. Phys. Org. Chem.* **1976**, *12*, 1. (e) Unger, S.H.; Hansch, C. *Prog. Phys. Org. Chem.* **19776**, *12*, 91. (f) Levitt, L.S.; Widing, H.F. *Prog. Phys. Org. Chem.* **1976**, *12*, 119. (g) Ritchie, C.D.; Sager, W.F. *Prog. Phys. Org. Chem.* **1964**, *2*, 323. (h) Ehrenson, S. *Prog. Phys. Org. Chem.* **1962**, *2*, 195. (i) Charton, M. *Prog. Phys. Org. Chem.* **1981**, *13*, 119. (j) Pross, A.; Radom, L. *Prog Phys. Org. Chem.* **1981**, *13*, 1. (k) Taft, R.W.; Topsom, R.D. *Prog. Phys. Org. Chem.* **1987**, *16*, 1. (l) Topsom, R.D. *Prog. Phys. Org. Chem.* **1987**, *16*, 85. (m) Topsom, R.D. *Prog. Phys. Org. Chem.* **1987**, *16*, 125. (n) Charton, M. *Prog. Phys. Org. Chem.* **1987**, *16*, 287. (o) Ehrenson, S.; Brownlee, R.T.C.; Taft, R.W. *Prog. Phys. Org. Chem.* **1973**, *10*, 1. (p) Stock, L.M.; Brown, H.C. *Adv. Phys. Org. Chem.* **1963**, *1*, 35.

Books and monographs: (q) Hine, J. *Structural Effects on Equilibria in Organic Chemistry* (Wiley: New York, 1975). (r) Johnson, K.F. *The Hammett Equation* (Cambridge University Press: New York, 1973). (s) Wells, P.R. *Linear Free Energy Relationships* (Academic Press: New York, 1968).

Table 8.6 Hammett ρ Values for Some Representative Organic Reactions*

Reaction		ρ Value
	**(8.19)**	1 exactly
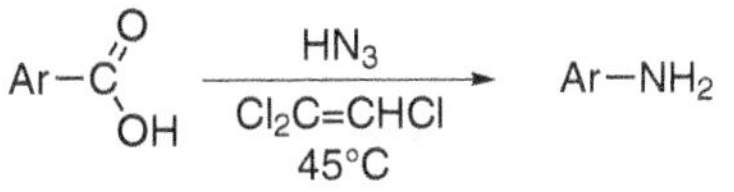	**(8.20)**	−1.415
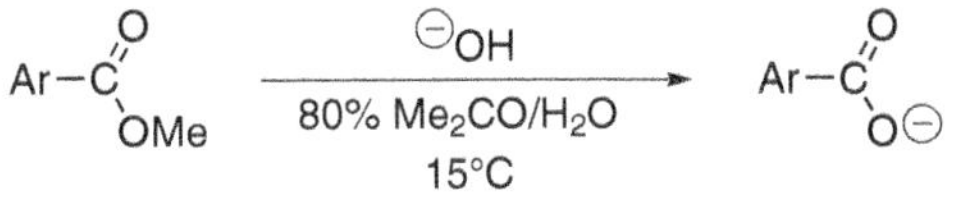	**(8.21)**	+2.299
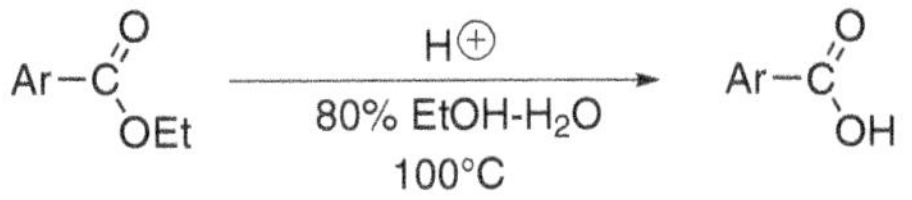	**(8.22)**	+0.144
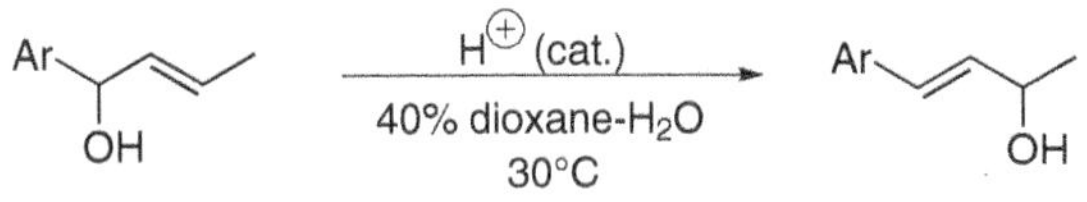	**(8.23)**	−4.672
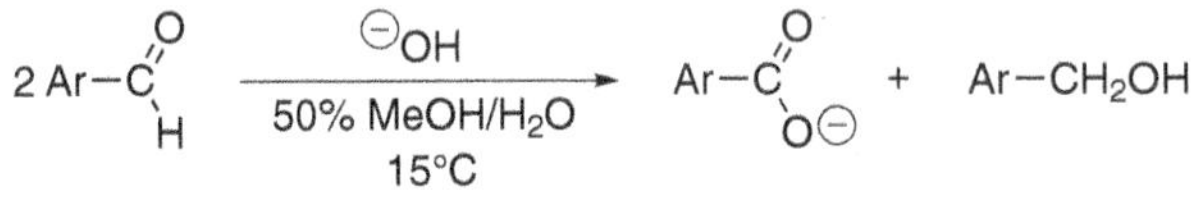	**(8.24)**	+3.63

*Values taken from Jaffé, H.H. *Chem. Rev.* **1953**, *53*, 191.

location capable of conjugating directly with the substituent; it is based on the solvolysis of substituted cumyl chlorides in 90% acetone-water.[47] The latter is useful for reactions in which negative charge is developed in a location capable of conjugating directly with the substituent; it is based on the ionization of substituted phenols.[48] Because the substituent at the *meta* position seldom interacts strongly by resonance with the side chain, σ^+ and σ^- constants are usually only quoted for *para* substituents.

The use of the σ^+ constant is nicely illustrated by the plots for the ethanolysis of triarylmethyl chlorides in 40% ethanol–60% ether solution.[49] The data for the reactions are plotted versus σ and σ^+ in Figure 8.12. When the data are plotted against the original Hammett σ values, the plot is curved or has a clear inflection point when the sign of the σ constant changes. In contrast to this, the plot versus σ^+ is linear over the whole range of reactions studied.

As the study of substituent effects on chemical reactions matured, the number and type of substituent constants also increased from the original set of three (σ, σ^+, and σ^-). There was a realization that substituents interact with the aromatic ring by both inductive and resonance mechanisms, which, of course, meant that σ was almost certainly a composite variable containing both inductive, σ_I, and resonance, σ_R, contributions, a suggestion first made by Taft.[50] The relationship between these substituent constants is:

$$\sigma = \sigma_I + \sigma_R \qquad \text{(Eq. 8.31)}$$

47. Brown, H.C.; Okamoto, Y. *J. Am. Chem. Soc.* **1958**, *80*, 4979.
48. (a) Miller, J. *Aust. J. Chem.* **1956**, *9*, 61. (b) Fujita, T.; Iwasa, J.; Hansch, C. *J. Am. Chem. Soc.* **1964**, *86*, 5175. (c) Cohen, L.A.; Jones, W.M. *J. Am. Chem. Soc.* **1963**, *85*, 3397.
49. Nixon, A.C.; Branch, G.E.K. *J. Am. Chem. Soc.* **1936**, *58*, 492.
50. (a) Taft, R.W. In Newman, M.s., Ed. *Steric Effects in Organic Chemistry* (John Wiley & Sons: New York, 1956), ch. 3. (b) Taft, R.W., Jr. *J. Am. Chem. Soc.* **1957**, *79*, 1045.

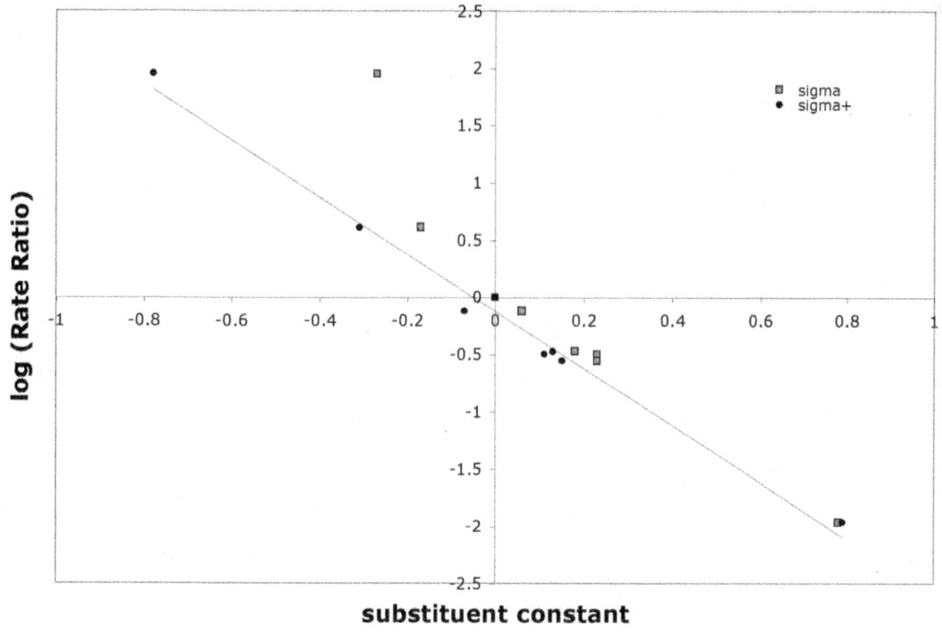

Figure 8.12 Hammett plots for the ethanolysis of triarylmethyl chlorides

The inductive effect of the substituent is a through-bond effect due largely to electronegativity of the substituent. It is almost always accompanied by a field effect, which is a through-space effect, and which may be, in fact, the dominant effect.[51] This is especially the case when the substituent is charged (e.g., NR_3^+ or CO_2^-). The measurement of the inductive component, σ_I, required a system incapable of interacting by resonance but still geometrically rigid. To isolate the inductive term (although still mixed with the field effect of the substituents), the pK_a values of 4-substituted bicyclo[2.2.2]-octane-1-carboxylic acids (Example **8.25**) were used as the standard.

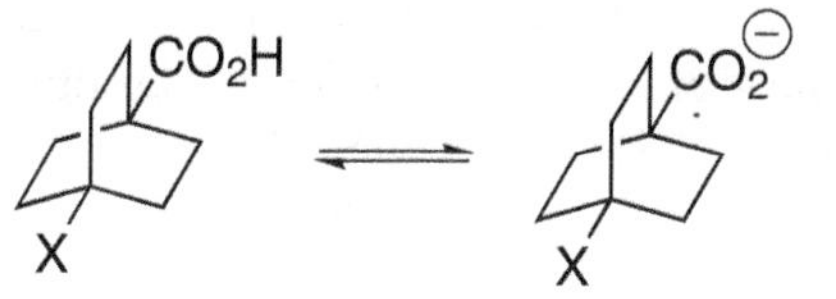

(8.25)

An alternative interpretation of the dual character of the Hammett σ constant is that of Swain and Lupton,[52] who proposed that the σ constant was related to a field constant (F) and a resonance constant (R) by the relationship:

$$\log (k/k_0) = fF + r\mathrm{R} \qquad \text{(Eq. 8.32)}$$

where f is the contribution to σ from field effects and r is the contribution from resonance. An earlier approach to separating the inductive/field effects and resonance effects is that proposed by Yukawa and Tsuno,[53] whose equation:

$$\log(k_x/k_0) = \rho[\sigma + r(\sigma^+ - \sigma)] \qquad \text{(Eq. 8.33)}$$

51. E.g., the field effect appears to dominate the substituent effects on the pK_a's of 4-substituted bicyclo[2.2.2] octane-1-carboxylic acids and cubane-1-carboxylic acids: (a) Cole, T.W., Jr.; Mayers, C.J.; Stock, L.M. *J. Am. Chem. Soc.* **1974**, *96*, 4555. (b) Wilcox, C.F.; Leung, C. *J. Am. Chem. Soc.* **1968**, *90*, 336.

52. Swain, C.G.; Lupton, E.C., Jr. *J. Am. Chem. Soc.* **1968**, *90*, 4328.

53. (a) Yukawa, Y.; Tsuno, Y. *Bull. Chem. Soc. Japan* **1959**, *32*, 965. (b) Yukawa, Y.; Tsuno, Y. *Bull. Chem. Soc. Japan* **1959**, *32*, 971.

Table 8.7 Substituent Constants for Linear Free Energy Relationships

Substituent	σ_m	σ_p	σ^{+*}	σ^{-*}	σ_I	σ_R	F	R
NH_2	−0.16	−0.66	−1.3		0.12	−0.82	0.38	−2.52
Me	−0.07	−0.17	−0.31		−0.04	−0.11	0.01	−0.41
Ph	0.06	−0.01	−0.17		0.10	−0.11	0.25	−0.37
OH	0.12	−0.37	−0.92		0.25		0.46	−1.89
OMe	0.12	−0.27	−0.78	−0.2	0.27	−0.61	0.54	−1.68
F	0.34	0.06	−0.07	0.02	0.50	−0.45	0.74	−0.60
I	0.35	0.18	0.13		0.39	−0.16	0.65	−0.12
CHO	0.35	0.42		1.04	0.25			
CO_2H	0.37	0.45	0.42				0.44	0.66
Cl	0.37	0.23	0.11		0.46	−0.23	0.72	−0.24
COMe	0.38	0.50		0.87	0.28	0.16	0.50	0.90
Br	0.39	0.23	0.15		0.44	−0.19	0.72	
CO_2R	0.37	0.45	0.48	0.68	0.30	0.14	0.47	0.67
CF_3	0.43	0.54			0.45	0.08	0.64	0.76
CN	0.56	0.66	0.66	0.90	0.56	0.13	0.90	0.71
NO_2	0.71	0.78	0.79	1.24	0.65	0.15	1.00	1.00

*These values are for *para* substituents only.

contains the original Hammett equation with a correction for the resonance effect of the substituent for strongly interacting substituents. The parameter r is the contribution from resonance to stabilization of the transition state. In practice, one determines the value of ρ from the Hammett plot using the substituents for which $\sigma^+ = \sigma$, and then one obtains the value of r by plotting $[(\log k/k_0)/\rho - \sigma]$ versus $(\sigma^+ - \sigma)$ for the substituents for which $\sigma \neq \sigma^+$. The slope of this plot is r. The various σ values of a number of common substituents are gathered in Table 8.7. (The values in Table 8.3 are taken from compilations in Reference 54.)

Worked Problem

8-2 In a competition experiment where exchange does not occur, one can construct kinetic data for use in a Hammett plot. An example of this is provided by the selective activation of the benzyl position of substituted toluenes under rhodium catalysis (*Organometallics* **2010**, *29*, 624). Construct a Hammett plot from these data using the σ_p values from Table 8.3; what can you deduce about the nature of the transition state for the reaction from this plot?

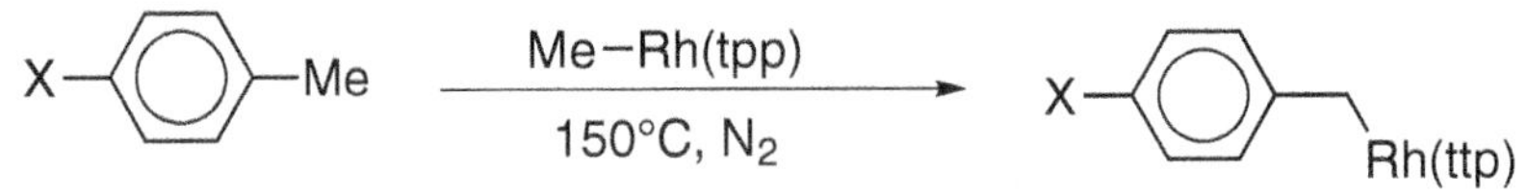

54. (a) Ritchie, C.D.; Sager, W.F. *Prog. Phys. Org. Chem.* **1964**, *2*, 323. (b) Hansch, C.; Leo, A.; Unger, S.; Kim, K.H.; Nikaitani, D.; Liem, E.J. *J. Med. Chem.* **1973**, *16*, 1207. (c) Swain, C.G.; Unger, S.H.; Rosenquist, N.R.; Swain, M.S. *J. Am. Chem. Soc.* **1983**, *105*, 492. (d) Hine, J. *Structural Effects on Equilibria in Organic Chemistry* (Wiley-Interscience: New York, 1975). (e) Ehrenson, S.; Brownlee, R.T.C.; Taft, R.W. *Prog. Phys. Org. Chem.* **1973**, *10*, 1.

The ratios obtained are as follows:

Substituent	Me	*t*-Bu	F	Cl	CF_3
σ_p	−0.14	−0.15	0.06	0.23	0.54
$\log_{10}(k_X/k_H)$	0.45	0.45	0.16	−0.19	−0.28

§ Answer below.

Solvent Effects

Reactions where charge is developed in the activated complex (e.g., the rate determining step of the S_N1 and E1 reactions) are especially sensitive to the effects of solvent due to the importance of solvation in stabilizing the incipient ions. The effects of solvent were the subject of a systematic study by Grunwald and Winstein,[55] who proposed a linear free energy relationship similar to the Hammett relationship for the effects of solvent:

$$\log(k/k_0) = m\mathbf{Y} \qquad \text{(Eq. 8.34)}$$

In the Grunwald-Winstein equation, which is restricted to protic (hydroxylic) solvents, k is the rate constant for the reaction in the solvent of interest, k_0 is the rate constant for the same reaction in the standard solvent, **Y** is the solvent ionizing power, and m is the sensitivity of the substrate to solvent ionizing power. As with the Hammett equation, the Grunwald-Winstein equation is a multiparameter reaction that must be based on a standard substrate and solvent. For this purpose, the standard reaction, for which m was assigned the value of 1, was the solvolysis of *tert*-butyl chloride, and the standard solvent, assigned a **Y** value of 0, was 80% v/v ethanol-water. To obtain the **Y** values of other solvents, the same solvolysis reaction was carried out in those other solvents, with the **Y** values then being assigned on the basis of the simple equation:

$$\mathbf{Y} = \log \frac{k_{t\text{-BuCl, solvent}}}{k_{t\text{-BuCl, 80\% EtOH}}} \qquad \text{(Eq. 8.35)}$$

The original Grunwald-Winstein relationship is based on the assumption that the solvolysis of *tert*-butyl chloride proceeds by a limiting S_N1 mechanism. This has several important implications for the mechanism. One, there is no nucleophilic assistance to the displacement by solvent (i.e., there is effectively no covalent interaction between the solvent and the developing carbocation center in the activated complex). Two, the rate-determining step in the reaction is the formation of a solvated carbocation, not an oxonium ion. Three, the nucleophile attacks a fully developed carbocation to form the substitution product (i.e., Path **8.26** rather than Path **8.27**).

55. (a) Grunwald, E.; Winstein, S. *J. Am. Chem. Soc.* **1948**, *70*, 846. (b) Winstein, S.; Grunwald, E.; Jones, H.W. *J. Am. Chem. Soc.* **1951**, *73*, 2700. (c) Fainberg, A.H.; Winstein, S. *J. Am. Chem. Soc.* **1956**, *78*, 2770.

§ Answer for Worked Problem:

The Hammett plot has a slope of −1.14, but the plot omitting the CF_3 substituent (which does not carry a lone pair capable of delocalizing a positive charge and which does not hyperconjugate well with a cation center,) has a slope of −1.57 and is a better straight line. These observations are both consistent with the development of a substantial positive charge on the benzylic carbon that is resonance-stabilized by the lone pair on the substituent. The Hammett plot works equally well with Taft σ^+ parameters,; in this case, the slope is −1.50, which leads to the same conclusions as with the simple Hammett σ_p parameters.

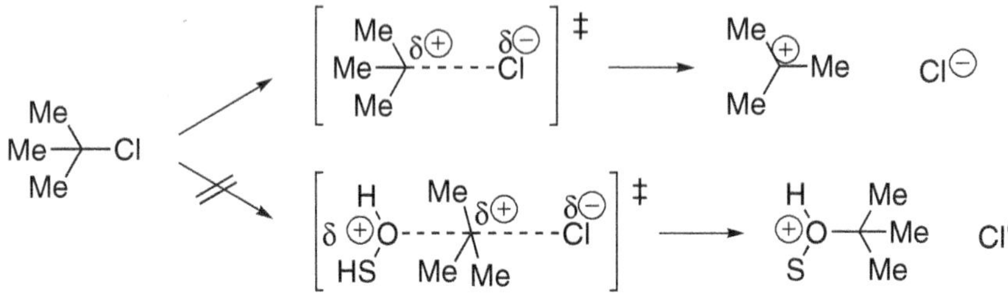

(8.26)

(8.27)

Like the Hammett equation before it, the Grunwald-Winstein equation has also been the subject of considerable investigation, especially with respect to the covalent participation or nonparticipation of solvent. The reason for this is that many, if not most, S_N1 substitution reactions do not follow the limiting mechanism but have some degree of nucleophilic assistance from solvent. Not surprisingly, the relationship has undergone refinement as our understanding of the role of solvent in organic reactions has grown.

Studies of rearrangements during solvolysis reactions of norbornyl halides led Schleyer to the conclusion that many solvolysis reactions of tertiary halides involve, to some degree, nucleophilic assistance from the solvent. As a result, Schleyer and his coworkers introduced the $\mathbf{Y}_{OTs}$ scale, based on the solvolysis of 1-adamantyl tosylate (Example **8.28**).[56] The choice of this substrate was predicated on two important considerations: (1) nucleophilic assistance by solvent from the back side is prevented by the adamantane ring system[57] and (2) any alkene formed would be a bridgehead alkene in violation of Bredt's rule,[58] which would suppress elimination during solvolysis. These workers then proposed a modified equation, where the solvent ionizing power, $\mathbf{Y}_{OTs}$, and the solvent nucleophilicity, **N**, are treated separately[59]:

$$\log(k/k_0) = l\mathbf{N}_{OTs} + m\mathbf{Y}_{OTs} \qquad \text{(Eq. 8.36)}$$

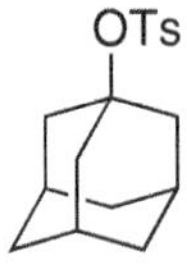

(8.28)

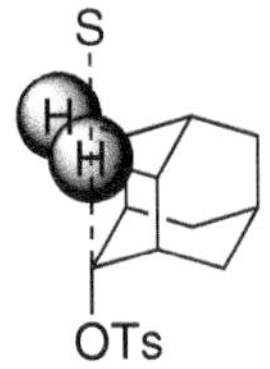

(8.29)

In this equation, m is a measure of the susceptibility of the substrate to the ionizing power of the solvent ($\mathbf{Y}_{OTs}$) and l is a measure of the sensitivity of the substrate to solvent nucleophilicity.

As pointed out above, the initial substrate for establishing the $\mathbf{Y}_{OTs}$ scale was 1-adamantyl tosylate. Subsequently, it was found that, within experimental error, $\mathbf{Y}_{OTs}$ values obtained from the solvolysis of 2-adamantyl tosylates (Example **8.29**) were equal to the values obtained from the solvolysis of 1-adamantyl tosylates. This allowed the scale to be extended to solvents for which $\mathbf{Y}_{OTs}$ was negative or zero. It had been argued that the axial hydrogens in the 1,3-positions relative to the leaving group would prevent the solvent from providing nucleophilic assistance from the back side; the experimental results support this argument—although the absolute rates for the solvolysis reactions were different, the $\mathbf{Y}_{OTs}$ values were the same, within experimental error.

The value of the nucleophilicity parameter, **N**, was calculated by comparing the solvolysis of 1-adamantyl tosylate, which is assumed to be immune to nucleophilic participation by the solvent, with the solvolysis of methyl tosylate, which should be especially susceptible to nucleophilic participation by solvent, in the same series of solvents. By setting the value of l (the susceptibility of the substrate to solvent nucleophilicity) to unity for

56. (a) Fry, J.L.; Lancelot, C.J.; Lam, L.K.M.; Harris, J.M.; Bingham, R.C.; Raber, D.J.; Hall, R.E.; Schleyer, P.v.R. *J. Am. Chem. Soc.* **1970**, *92*, 2538. (b) Raber, D.J.; Bingham, R.C.; Harris, J.M.; Fry, J.L.; Schleyer, P.v.R. *J. Am. Chem. Soc.* **1970**, *92*, 5977.

57. Fort, R.C., Jr.; Schleyer, P.v.R. *Adv. Alicyc. Chem.* **1966**, *1*, 283.

58. (a) Bredt, J. *Liebigs Ann. Chem.* **1924**, *437*, 1. (b) Buchanan, G.L. *Chem. Soc. Rev.* **1974**, *3*, 41. (c) Szeimies, G. In Abramovich, R.A., Ed. *Reactive Intermediates* (Plenum Press: New York, 1983), vol. 3, p. 299.

See also: (d) Maier, W.; Schleyer, P.v.R. *J. Am. Chem. Soc.* **1981**, *103*, 1891. (e) Harnisch, J.; Baumgärtel, O.; Szeimies, G.; Van Meerssche, M.; Germain, G.; Declercq, J. *J. Am. Chem. Soc.* **1979**, *101*, 3370. (f) Engel, P.S.; Keys, D.E.; Kitamura, A. *J. Am. Chem. Soc.* **1985**, *107*, 4964.

59. Schadt, F.L.; Bentley, T.W.; Schleyer, P.v.R. *J. Am. Chem. Soc.* **1976**, *98*, 7667.

the solvolysis of methyl tosylate, one can determine the solvent nucleophilicity parameter **N**.[60]

$$\mathbf{N} = \log\left(\frac{k}{k_0}\right)_{\text{MeOTs}} - 0.3\mathbf{Y} \qquad \text{(Eq. 8.37)}$$

It would be convenient if the range of solvent ionizing power and solvent nucleophilicity could be expressed using a single scale, but the effects of solvent are very susceptible to the identity of *both* the substrate *and* the leaving group. Thus, there are a variety of **Y** scales that have been set up for use with a variety of substrates and leaving groups. The $\mathbf{Y}_{Cl}$ and $\mathbf{Y}_{Br}$ scales are based on solvolysis of 1-adamantyl chloride and 1-adamantyl bromide,[61] and the $\mathbf{Y}_{I}$ scale is based on 1-adamantyl iodide.[62] For a more complete discussion of a number of scales of solvent ionizing power, one should consult the review by Bentley.[63] Values of **Y**, $\mathbf{Y}_{OTs}$ and $\mathbf{N}_{OTs}$ are collected in Table 8.8.

Table 8.8 Solvent Ionizing Power and Nucleophilicity Parameters

Solvent	Y	Y_{OTs}	N_{OTs}
F_3CCO_2H	1.84	4.57	−5.56
H_2O	3.49	4.1	−0.44
$(F_3C)_2CHOH$-H_2O (97:3)	2.46	3.61	−4.27
F_3CCH_2OH	1.04	1.80	−3.00
F_3CCH_2OH-H_2O (85:15)		1.92	−2.01
F_3CCH_2OH-H_2O (50:50)		2.14	−0.93
HCO_2H	2.05	3.04	−2.35
$MeCO_2H$	−1.64	−0.61	−2.35
EtOH	−2.03	−1.75	0.00
EtOH-H_2O (80:20)	0.00	0.00	0.00
EtOH-H_2O (50:50)	1.66	1.29	−0.09
EtOH-H_2O (20:80)	3.05	3.32	−0.34
MeOH		−0.92	−0.04
MeOH-H_2O (80:20)		0.47	−0.05
MeOH-H_2O (50:50)		2.00	−0.19
MeOH-H_2O (20:80)		3.39	−0.35
Me_2CO-H_2O (80:20)		−0.94	−0.42
Me_2CO-H_2O (50:50)	1.40	1.26	−0.39
Me_2CO-H_2O (20:80)	2.91	3.05	−0.38
Dioxane-H_2O (90:10)	−2.03	−2.41	−0.51
Dioxane-H_2O (80:20)		−1.30	−0.29
Dioxane-H_2O (20:80)	2.88		

The Salt Effect[64]

When ionic salts are added to a reaction solvent, the ionic strength and dielectric constant of the solution are raised, so that the solvent effectively becomes more polar as the salt concentration increases. Provided that neither of the ions present in the added salt duplicates an ion produced in the reaction ("common ions"), the addition of a salt to a reacting solution has the effect of accelerating reactions where polar solvents are used. This is known as the ***salt effect***, and it can provide useful information about the possible mechanisms of a reaction. Depending on the reaction and the solvent, the salt effect can vary directly with the added salt concentration,[65] the logarithm of the added salt concentration,[66] or the square root of the added salt concentration.[67]

Salt effects in non-polar solvents can be especially dramatic. Thus, Winstein[68] found that the solvolysis of 4-methoxyneophyl *p*-toluenesulfonate in diethyl ether containing dissolved lithium perchlorate was increased by a factor of 10^5 relative to the same reaction

60. (a) Schadt, F.L.; Bentley, T.W.; Schleyer, P.v.R. *J. Am. Chem. Soc.* **1976**, *98*, 7667. (b) Bentley, T.W.; Carter, G.E. *J. Org. Chem.* **1983**, *48*, 579. (c) Bentley, T.W. *Adv. Chem. Ser.* (American Chemical Society: Washington, D.C., 1987), No. 215, ch. 18.

The value, m=0.3, in this equation is contested: (d) Kevill, D.N.; Lin, G.M.L. *J. Am. Chem. Soc.* **1979**, *101*, 3916. (e) Kevill, D.N. *Adv. Chem. Ser.* (American Chemical Society: Washington, D.C., 1987), No. 215, ch. 19. (f) Kevill, D.N.; Anderson, S.W. *J. Org. Chem.* **1986**, *51*, 5029. (g) Kevill, D.N.; Rissman, T.J. *J. Org. Chem.* **1985**, *50*, 3062.

61. (a) Bentley, T.W.; Carter, G.E. *J. Am. Chem. Soc.* **1982**, *104*, 5741. (b) Bentley, T.W.; Bowen, C.T.; Parker, W.; Watt, C.I.F. *J. Am. Chem. Soc.* **1979**, *101*, 2486.

62. Bentley, T.W.; Carter, G.E.; Roberts, K. *J. Org. Chem.* **1984**, *49*, 5183.

63. Bentley, T.W.; Llewellyn, G. *Prog. Phys. Org. Chem.* **1990**, *17*, 121.

64. Monograph: Loupy, A.; Tchoubar, B. *Salt Effects in Organic and Organometallic Chemistry* (VCH: Weinheim, 1992).

65. (a) Fainberg, A.H.; Winstein, S. *J. Am. Chem. Soc.* **1956**, *78*, 2763. (b) Fainberg, A.H.; Winstein, S. *J. Am. Chem. Soc.* **1956**, *78*, 2780. (c) Perrin, C.L.; Pressing, J. *J. Am. Chem. Soc.* **1971**, *93*, 5705.

66. (a) Bateman, L.C.; Church, M.G.; Hughes, E.D.; Ingold, C.K.; Taher, N.A. *J. Chem. Soc.* **1940**, 979. (b) Winstein, S.; Friedrich, E.C.; Smith, S. *J. Am. Chem. Soc.* **1964**, *86*, 305.

67. Winstein, S.; Klinedinst, P.E., Jr.; Robinson, G.C. *J. Am. Chem. Soc.* **1961**, *83*, 885.

68. Winstein, S.; Smith, S.; Darwish, D. *J. Am. Chem. Soc.* **1959**, *81*, 5511.

in the absence of the added salt. In fact, the acceleration of the reaction in ether becomes so great that above 0.036 *M* lithium perchlorate, ether becomes a better ionizing solvent than acetic acid. In this system, the effects of added salt could be fit to a simple first-order linear relationship:

$$k = k_0 (1 + b[\text{LiClO}_4]) \qquad \text{(Eq. 8.38)}$$

where k is the first-order rate constant in the presence of added lithium perchlorate, and k_0 is the same rate constant in the absence of lithium perchlorate. Pocker[69] made similar observations on the ionization of trityl chloride in lithium perchlorate/ether solution. In fact, this solution has become a popular high ionic strength, nonaqueous solution for carrying out a wide range of organic reactions.[70] It is particularly interesting to note that this solution actually promotes the ionization of allyl vinyl ethers.[71]

The Brønsted Catalysis Law

When the transfer of a proton from one reactant to another is an important part of the reaction mechanism, the rate of the reaction will generally reflect the participation of the acid. The timing of the proton transfer (before the rate-determining step or during the rate-determining step) affects the terminology used to discuss the participation of acid. As we discussed earlier, if the proton transfer does not occur during the rate-determining step but during an earlier step, the reaction is said to be subject to specific acid catalysis. Specific acid catalysis involves only the proton H^+ as the acid, so one finds that reactions carried out in different buffers at the same pH (and, therefore, the same $[H^+]$) proceed at the same rate, regardless of the identity of the acid in the buffer.

Reactions that are catalyzed by strong acid (solvated protons) with proton transfer occurring before the rate-determining step are said to undergo ***specific acid catalysis***. When the reaction involves the transfer of a proton from one molecule to another (as against a simple proton) during the rate-determining step, the reaction is subject to ***general acid catalysis***. In such reactions, the rate of the proton transfer will depend on the identity of the acid, and will reflect the rates of proton transfer from an acid to the substrate. In these cases, each acid will appear as an individual term in the rate law:

$$k_{obs} = k_0 + k_{H^+}[\text{H}] + k_1[\text{HA}_1] + k_2[\text{HA}_2] + ... \qquad \text{(Eq. 8.39)}$$

Reactions subject to general acid catalysis often obey what is known as the ***Brønsted catalysis law***:[72]

$$k = CK_a^{\alpha} \text{ or } \log k = \alpha \log K_a + \log C = \log C - \text{p}K_a \qquad \text{(Eq. 8.40)}$$

The value of α, which according to the original theory, has a maximum value of 1 and a minimum value of zero, is often interpreted as correlating with the sensitivity of the reaction to the change in the acid catalyst, just as the Hammett ρ value correlates with the sensitivity of the reaction to changes in the substituent.[73] When the value of α is close to 1, this may be taken as evidence of specific acid catalysis of the reaction by H^+. When the value of α is close to zero, this is frequently taken to mean that catalysis of the reaction is by solvent alone.

69. Pocker, Y.; Buchholz, R.F. *J. Am. Chem. Soc.* **1970**, *92*, 2075.

70. Reviews: (a) Grieco, P.A. *Aldrichimica Acta* **1991**, *24*, 59. (b) Balasubramanian, K.K. *Acros Organics Acta* **1999**, *6*, 12. (c) Gunanathan, C. *Synlett* **2002**, 649.

71. (a) Grieco, P.A.; Clark, J.D.; Jagoe,, C.T. *J. Am. Chem. Soc.* **1991**, *113*, 5488. (b) Palani, N.; Chadha, A.; Subramanian, K.K. *J. Org. Chem.* **1998**, *63*, 5318.

72. (a) Brönsted, J.N.; Pedersen, K.J. *Z. physik. Chem.* **1924**, *108*, 185. (b) Brönsted, J.N. *Chem. Rev.* **1928**, *5*, 231.

73. (a) Bender, M.L. *Chem. Rev.* **1960**, *60*, 53. (b) Lewis, E.S. *J. Phys. Org. Chem.* **1990**, *3*, 1.

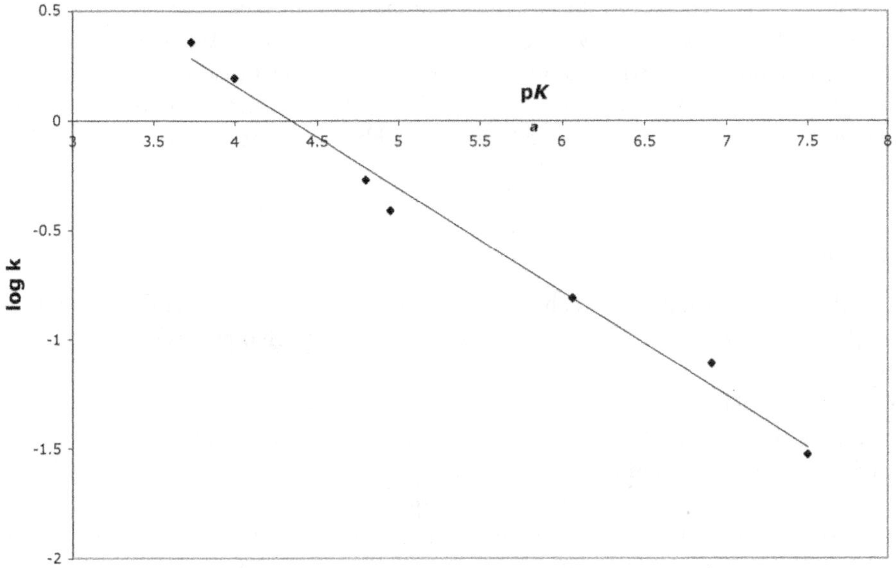

Figure 8.13 Brønsted plot for the acid-catalyzed hydrolysis of phenyl diethyl orthoformate at constant ionic strength

Values of α intermediate between the two extremes are usually taken as evidence of general acid catalysis of the reaction. The value of α has also come to be widely interpreted as conveying the extent of proton transfer in the transition state,[74] although this interpretation is not universally accepted, especially following the discovery of reactions for which $\alpha > 1$.[75]

An excellent example of a reaction subject to general acid catalysis is the acid-catalyzed hydrolysis of orthoesters (Example **8.30**).[76] In the hydrolysis of phenyl diethyl orthoformate in 1:1 THF-water, the rate constants for hydrolysis in the presence of a wide range of weak acids ranging from cyanoacetic acid (pK_a 3.74) to cacodylic acid (dimethylarsonic acid, pK_a 7.50) obey the Brønsted rate law, with an α value of 0.47, as shown in Figure 8.13. The α value of 0.47 is consistent with general acid catalysis of this reaction.

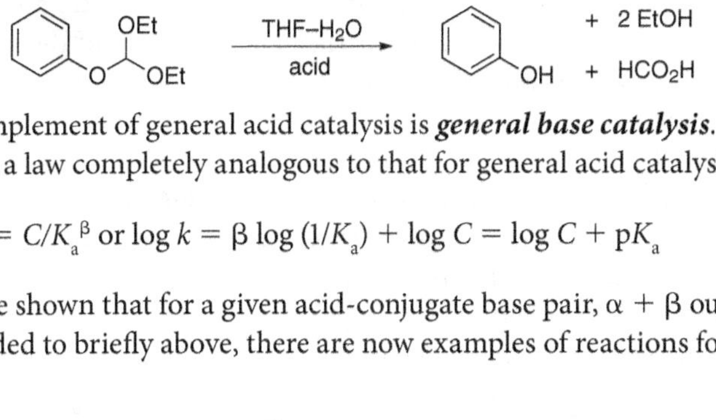

(8.30)

The complement of general acid catalysis is ***general base catalysis.*** General base catalysis follows a law completely analogous to that for general acid catalysis:

$$k = C/K_a^{\beta} \text{ or } \log k = \beta \log (1/K_a) + \log C = \log C + pK_a \qquad \text{(Eq. 8.41)}$$

It can be shown that for a given acid-conjugate base pair, $\alpha + \beta$ ought to be equal to 1, but, as alluded to briefly above, there are now examples of reactions for which $\alpha > 1$.

8.6 **Isotope and Element Effects**

If a bond that undergoes a change during the rate-determining step of a reaction is replaced by either a bond to a heavier isotope of the same element, or a bond to a different element, the rate of the reaction will almost always change, or the position of an equilibrium will

74. Leffler, J.E. *Science* **1953**, *117*, 340.

75. (a) Streitwieser, A., Jr.; Kaufman, M.J.; Bors, D.A.; Murdoch, J.R.; MacArthur, C.A.; Murphy, J.T.; Shen, C.C. *J. Am. Chem. Soc.* **1985**, *107*, 6983. (b) Wiseman, F.; Keestner, N.R. *J. Phys. Chem.* **1984**, *88*, 4354. (c) Bordwell, F.G.; Boyle, W.J., Jr.; Yee, K.C. *J. Am. Chem. Soc.* **1970**, *92*, 5926. (d) Pross, A. *J. Org. Chem.* **1984**, *49*, 1811. (e) Jencks, W.P.; Brant, S.R.; Gandler, J.R.; Fendrisch, G.; Nakamura, C. *J. Am. Chem. Soc.* **1982**, *104*, 7045. (f) Hupe, D.J.; Wu, D. *J. Am. Chem. Soc.* **1977**, *99*, 7653. (g) Hupe, D.J.; Pohl, E.R. *J. Am. Chem. Soc.* **1984**, *106*, 5634. (h) Agmon, N. *J. Am. Chem. Soc.* **1980**, *102*, 2164. (i) Murdoch, J.R. *J. Am. Chem. Soc.* **1972**, *94*, 4410.

76. Anderson, E.; Fife, T.H. *J. Org. Chem.* **1972**, *37*, 1993.

Table 8.9 Element Effect in Substitution and Elimination of Alkyl Halides

Halide	Relative Rate
$Me_3C–F$	1
$Me_3C–Cl$	$\approx 10^5\ k_{RF}$
$Me_3C–Br$	$\approx 25–60\ k_{RCl}$
$Me_3C–I$	$\approx 1.5–4.5\ k_{RBr}$

almost always shift. This technique provides a corollary to the measurement of substituent and solvent effects by directly probing the bonding changes in the transition state. When the reacting bond is replaced by a bond to a heavier isotope, the ratio of rate or equilibrium constants is known as an ***isotope effect.*** When the reacting bond is replaced by a bond to a different element (i.e., the leaving group is replaced by a different atom), the ratio of rate constants is known as the ***element effect***.

The Element Effect

When the leaving group atom is changed, the strength of the bond being broken is also changed. This means that the ΔG^0 for the reaction will also change, making the reaction either more less endothermic than the original reaction. Using arguments based on the Hammond postulate, we would predict that, the weaker the bond to the leaving group, the less endothermic the reaction, the lower its activation energy, and the faster its rate. Although this argument is somewhat simplistic, it is still useful in allowing us to make some statements about transition state structure: a large element effect implies that there is substantial rupture of the bond to the leaving group in the transition state. Conversely, a ***null element effect*** (where the element substitution has no effect on the rate of the reaction) strongly implies that the bond is not broken in the rate-determining step. The element effect is well illustrated by the unimolecular nucleophilic substitution and elimination reactions of alkyl halides; as the leaving group is changed from fluoride to chloride to bromide to iodide, the rate constant for the reaction increases (Table 8.9).[77]

Isotope Effects

The situation with isotope effects is rather more subtle, but the same effect is observed overall. In introductory courses in chemistry, the statement is often made that all isotopes of a particular element are chemically identical and that they participate in the same reactions to give the same products. As far as it goes, this statement is true, but it is a major oversimplification. In fact, isotopic substitution does affect the chemistry and physics of a compound in one very important way: it affects the energy content of the molecule at a given temperature, and so affects both the position of equilibrium in reversible reactions (an ***equilibrium isotope effect***) and the rate constant of the reaction (a ***kinetic isotope effect***) in all reactions. Kinetic isotope effects, for example, are determined by measuring the rates of the same reaction for two ***isotopic isomers***, or ***isotopomers*** (molecules differing only in the isotopic constitution, but not in structure or stereochemistry).

The origins of the isotope effect lie in the vibrational energies of the molecules and transition states involved. Applying the Schrödinger equation to a simple harmonic oscillator (the simplest model of a vibrating bond, like a vibrating spring that obeys a Hooke's Law relationship), we find that the energy of the system is described by:

$$E_v = h\nu_0(v + ½) \qquad \text{(Eq. 8.42)}$$

where v is the quantum number specifying the vibrational energy level of the molecule. This shows that even at the lowest vibrational energy level, the vibrational energy of the molecule is ½ $h\nu_0$, which is known as the ***zero point energy*** of the molecule, and is vibrational energy that the molecule possesses even at 0 K.

77. (a) Hughes, E.D.; Shapiro, U.G. *J. Chem. Soc.* **1937**, 1177. (b) Cooper, K.A.; Hughes, E.D. *J. Chem. Soc.* **1937**, 1183. (c) Cooper, K.A.; Hughes, E.D.; Ingold, C.K. *J. Chem. Soc.* **1937**, 1280. (d) Hughes, E.D.; MacNulty, B.J. *J. Chem. Soc.* **1937**, 1283. (e) Church, M.G.; Hughes, E.D.; Ingold, C.K. *J. Chem. Soc.* **1940**, 966. (f) Shorter, J.; Hinshelwood, C.N. *J. Chem. Soc.* **1949**, 2412. (g) Brown, H.C.; Fletcher, R.S. *J. Am. Chem. Soc.* **1949**, *71*, 1849. (h) Brown, H.C.; Stern, A. *J. Am. Chem. Soc.* **1950**, *72*, 5068.

One can calculate the vibrational frequency of the zero level vibration from the relationships:

$$\nu_0 = \frac{1}{2\pi}\sqrt{\frac{k}{\mu}} \quad \text{and} \quad \frac{1}{\mu} = \frac{1}{m_1} + \frac{1}{m_2} \qquad \text{(Eq. 8.43)}$$

where k is the vibrational force constant for the bond, μ is the ***reduced mass*** of the molecule, and m_1 and m_2 are the masses of the atoms at the ends of the bond. This shows that the frequency of vibration, and therefore the zero-point energy of the molecule, is sensitive to isotopic substitution.

The same situation applies to vibrations of the activated complex, with the exception that the separation between vibrational energy levels is usually much smaller. This means that the activation energy for the reactions of the two isotopomers will be different if the bond to the labeled atom is broken (endothermic—high activation energy) or formed (exothermic—low activation energy) in the rate-limiting step (Figure 8.14).

The situation in polyatomic molecules is basically the same as in a simple diatomic molecule, with these exceptions. The vibrations of bonds more than about three bonds removed from the labeled atom are not affected and those not directly involving the labeled atom are often only marginally affected. Thus, we may focus only on the labeled bond without introducing major errors into the discussion.[78] For those who are interested in a more mathematical discussion of the theoretical basis of the kinetic isotope effect, several reviews and monographs are available.[79]

Kinetic isotope effects are categorized as ***primary isotope effects*** when the bond to the labeled atom is broken or formed and ***secondary isotope effects*** when the bond to the labeled atom remains intact. Typically, primary isotope effects are predominantly due to the loss of the zero point energy of the stretching vibration of the bond being broken, and secondary isotope effects arise from the changes in the frequency of bending vibrations

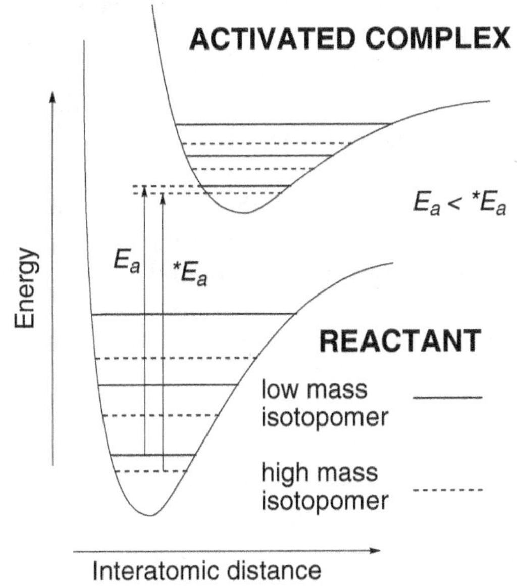

Figure 8.14 The potential energy diagram for two isotopomers of a diatomic molecule in the ground state and the transition state of a reaction involving rupture of the labeled bond

78. Sims, L.B.; Lewis, D.E. In Buncel, E.; Lee, C.C., Eds. *Isotopes in Organic Chemistry* (Elsevier: Amsterdam, 1984); Vol. 6; p. 161.

79. (a) Kresge, A.J. *J. Am. Chem. Soc.* **1980**, *102*, 7797. (b) Wiberg, K.B. *Chem. Rev.* **1955**, *55*, 713. (c) Westheimer, F.H. *Chem. Rev.* **1961**, *61*, 265. (d) Collins, C.J.; Bowman, N.S., Eds. *Isotope Effects in Chemical Reactions* (Van Nostrand Reinhold: New York, 1970). (e) Melander, L. *Isotope Effects on Reaction Rates* (Ronald Press: New York, 1960). (f) Melander, L.; Saunders, W.H., Jr. *Reaction Rates of Isotopic Molecules* (John Wiley: New York, 1980). (g) Willi, A.V. *Isotopeneffekte bei chemischen Reaktionen* (Thieme Verlag: Stuttgart, 1983).

involving the labeled atom. Unless they are specified otherwise, most isotope effects are assumed to be primary isotope effects.

In a reaction where the bond to the labeled atom is completely broken, the difference in activation energies may be expected to be at a maximum. We can calculate the maximum expected value of the isotope effect in such a reaction by calculating the difference in zero point energy for the two isotopomers from the Planck relationship ($E = h\nu_0/2$) and the Arrhenius relationship for the rate constant:

$$k = Ae^{-E_a/RT}\text{, which gives } k / k^* = e^{\Delta E_a/RT} = e^{hc(\Delta\nu)/2\mathbf{k}T} \qquad \text{(Eq. 8.44)}$$

where c is the velocity of light that converts the frequency in cm^{-1} to frequency in s^{-1} and $\mathbf{k}$ is the Boltzmann constant (which is the value of the gas constant in Joules per molecule).

Using these relationships, we can obtain approximations to the maximum value of the isotope effect for a series of isotopic substitutions. As can be seen from the data in Table 8.10, the most dramatic changes in rate are greatest for substitutions with hydrogen isotopes, but measurable differences can be observed for many other atoms ("heavy-atom isotope effects").[80]

When the reaction proceeds by simple cleavage of the bond to the labeled atom, the value of the isotope effect is expected to be near the maximum value. However, atom *transfer* reactions (reactions where the rupture of one bond is compensated by formation of a new bond to the transferring atom) are much more common in organic chemistry. Examples of such reactions are S_N2 substitution (the transferring atom is carbon), E2 and E1cb elimination (the transferring atom is the β hydrogen), deprotonation α to a carbonyl group (the transferring atom is the α hydrogen), and nucleophilic addition to carbonyl groups (the transferring atom is the carbonyl carbon).

In 1961, Westheimer predicted that the isotope effect of a reaction where the labeled atom is transferred from one atom to another should pass through a maximum value as the structure of the transition state changes from reactant-like to product-like (now known as the ***Westheimer maximum***).[82] The maximum value occurs when the force constants of the two partial bonds to the atom being transferred are equal, because the frequency of the symmetrical stretching vibration involving the labeled atom becomes independent of the mass of the labeled atom ($\nu = \nu^*$) at this point.

We can see this from a less mathematical viewpoint in Figure 8.15. The three reaction coordinate diagrams represent an endothermic reaction, where the transition state occurs late along the reaction coordinate and strongly resembles the product; a thermoneutral reaction, where the transition state is "central" on the reaction coordinate and resembles neither the reactant nor the product strongly; and an exothermic reaction, where the transition state occurs early along the reaction coordinate and strongly resembles the reactant.

---

80. The value of a kinetic isotope effect can be calculated from vibration frequencies of the reactant and transition state alone:

$$\frac{k}{k^*} = \left(\frac{\nu_L}{\nu_L^*}\right)^{\ddagger} \prod_i^{3N-7}\left(\frac{\nu}{\nu^*}\exp\left[-\frac{1}{2}(u_i - u_i^*)^{\ddagger}\right]\frac{1-\exp(-u_i^*)^{\ddagger}}{1-\exp(-u_i)^{\ddagger}}\right)\prod_i^{3N-6}\left(\frac{\nu}{\nu^*}exp\left[\frac{1}{2}(u_i - u_i^*)\right]\frac{1-\exp(-u_i)}{1-\exp(-u_i^*)}\right)$$

where $u = h\nu/\mathbf{k}T$ and the asterisk refers to the heavy isotope. Vibration frequencies labeled with the cross of Lorraine (‡) refer to the transition state, and the others refer to the reactants. The quantity $(\nu_L)^{\ddagger}$ is the reaction coordinate frequency, and corresponds to the mixed translation-vibration through the transition state. See (a) Bigeleisen, J. *J. Chem. Phys.* **1949**, *17*, 675. (b) Bigeleisen, J.; Goeppert-Mayer, M. *J. Chem. Phys.* **1947**, *15*, 261. (c) Wolfsberg, M.; Stern, M.J. *Pure Appl. Chem.* **1964**, *8* 325. (d) Bell, R.P. *The Proton in Chemistry*, 2nd ed.; (Cornell University Press: Ithaca, 1979), ch. 11. (e) Sims, L.B.; Lewis, D.E. In Buncel, E.; Lee, C.C., Eds. *Isotopes in Organic Chemistry* (Elsevier:Amsterdam, 1984), vol. 6, ch. 4.

81. Streitwieser, A., Jr.; Jagow, R.H.; Fahey, R.C.; Suzuki, S. *J. Am. Chem. Soc.* **1958**, *80*, 2326.

82. Westheimer, F.H. *Chem. Rev.* **1961**, *61*, 265.

Table 8.10 Maximum Theoretical Values of Primary Kinetic Isotope Effects

Isotope change	$\sqrt{\mu_{light}/\mu_{heavy}}$	Limiting ratio	ν_{light} (cm^{-1})	ν_{heavy} (cm^{-1})	Maximum k/k^*
H→D	$\sqrt{\frac{(m_2+2)}{2(m_2+1)}}$	$\sqrt{\frac{1}{2}}(0.7071)$	≈ 3000	≈ 2120	8.3‡
$^{12}C \rightarrow ^{14}C$	$\sqrt{\frac{12(m_2+14)}{14(m_2+12)}}$	$\sqrt{\frac{12}{14}}(0.9258)$	≈ 1400	≈ 1300	1.3
$^{14}N \rightarrow ^{15}N$	$\sqrt{\frac{14(m_2+15)}{15(m_2+14)}}$	$\sqrt{\frac{14}{15}}(0.9661)$	≈ 1350	≈ 1300	1.12
$^{16}O \rightarrow ^{18}O$	$\sqrt{\frac{16(m_2+18)}{18(m_2+16)}}$	$\sqrt{\frac{16}{18}}(0.9428)$	≈ 1250	≈ 1180	1.19
$^{32}S \rightarrow ^{34}S$	$\sqrt{\frac{32(m_2+34)}{34(m_2+32)}}$	$\sqrt{\frac{32}{34}}(0.9701)$	≈ 1000	≈ 970	1.08
$^{35}Cl \rightarrow ^{37}Cl$	$\sqrt{\frac{35(m_2+37)}{37(m_2+35)}}$	$\sqrt{\frac{35}{37}}(0.9726)$	≈ 700	≈ 680	1.05

‡The maximum limiting ratio observed experimentally for H/D isotopomers actually close to 1.35, which leads to a maximum value of k/k^* nearer to 5.5 (see Reference 81).

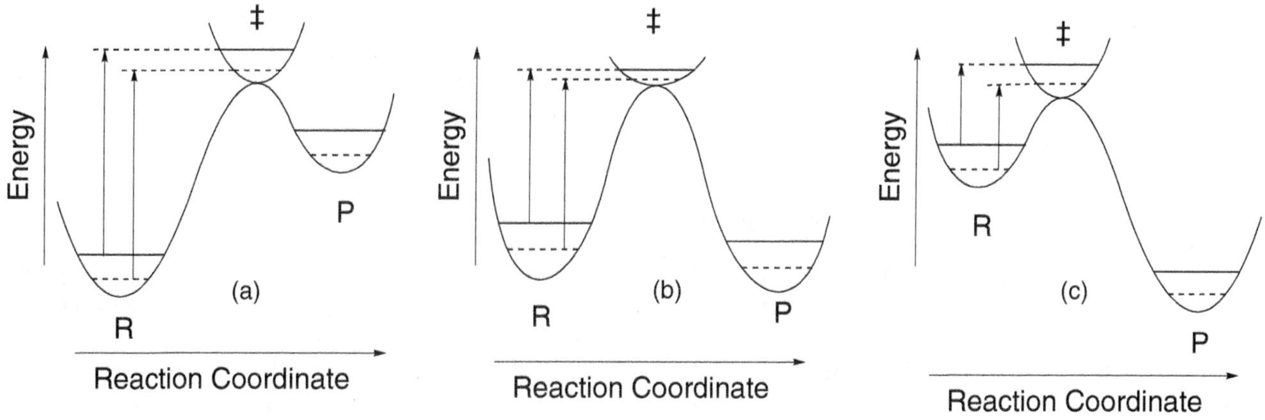

Figure 8.15 Reaction coordinate diagrams for (a) endothermic reactions, (b) thermoneutral reactions, and (c) exothermic reactions

In the endothermic reaction, the transition state occurs late and strongly resembles the product. The zero point energy of the bond to the transferring atom in this transition state will be close to the zero point energy of the completely formed bond in the product. In the exothermic reaction, the transition state is early and strongly resembles the reactant. The zero point energy of the bond to the transferring atom in this transition state will be close to the zero point energy of the unbroken bond in the reactant. In the thermoneutral reaction, there is substantial loss of bonding compared to both the reactant state and the product state, so the loss of zero point energy is at a maximum, leading to a maximum kinetic isotope effect.

Secondary Isotope Effects

Secondary isotope effects are isotope effects that are measured at a position where the bond to the isotopically labeled atom is neither broken nor formed. Except for isotopes of hydrogen, secondary isotope effects are usually negligible in magnitude. However, substitution by deuterium at a position where there is bonding change at carbon can give useful information about the hybridization of the carbon atom in the transition state. Secondary isotope effects reflect changes in the frequencies of *bending* vibrations, rather than stretching vibrations, as happens with primary isotope effects. Any change that tightens the bonding around the atom bonded to the labeled hydrogen (e.g., going from an sp^2 to an sp^3 hybridized state at carbon) usually leads to increases in the frequency of bending vibrations involving the labeled atom and hence to an inverse secondary isotope effect ($^{h}k/^{D}k < 1$). These kinds of isotope effects are seen, for example, in the isomerization of maleic acid to fumaric acid catalyzed by thiocyanate anion (Example **8.31**), for which $^{H}k/^{D}k = 0.86$ at 25°C.[83] Similar effects are seen in S_N2 reaction, where the trigonal bipyramidal transition state is much more crowded than the sp^3-hybridized starting compound; the S_N2 reaction of methyl bromobenzenesulfonate with methoxide ion to give dimethyl ether has a secondary deuterium isotope effect of 0.96.[84] The reverse situation also holds true: when the bonding change loosens the bonding to the labeled hydrogen, the frequencies of the bending vibrations decrease, and the isotope effect is normal. The heterolysis of the carbon-oxygen bond of the 4,4'-dimethoxybenzhydryloxonium ion (Example **8.32**) has a secondary isotope effect of 1.15.[85]

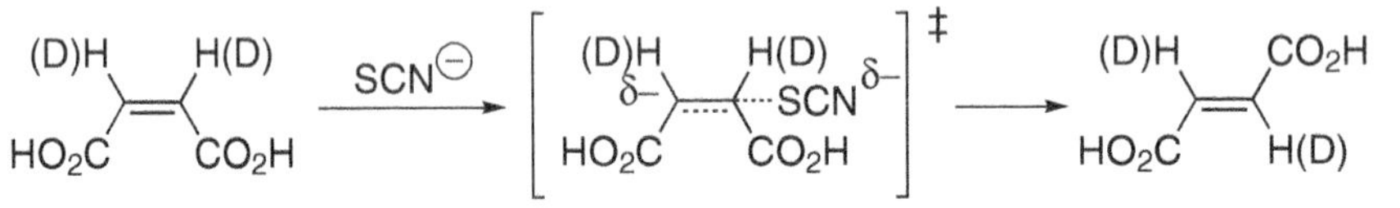

(8.31)

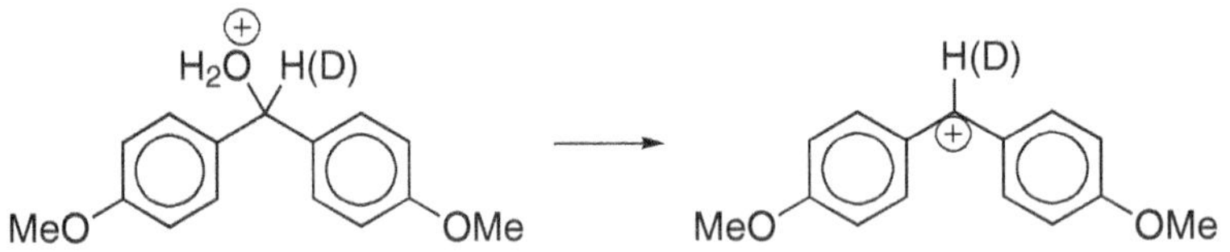

(8.32)

Using Element and Isotope Effects to Postulate Mechanism and Transition State Structures

Element and isotope effects provide clear evidence of specific bond cleavages in the rate determining steps of a reaction. A good example of the application of both element effects and kinetic isotope effects to a mechanistic problem is provided by the base-promoted dehydro-halogenation of 2,2-diaryl-1,1,1-trihaloethanes.[86] For this reaction, there are data available (Table 5.7) for both the element effect (k_{Br}/k_{Cl}) and the deuterium isotope effect ($^{H}k/^{D}k$). In addition, the reaction is subject to general base catalysis, with Brønsted β values for the chloride of 0.61 for alkoxide ion bases, and 0.54 when the phenoxide bases are included, and 0.57 and 0.50 for the corresponding bromide.

In this reaction, the values of the deuterium isotope effect and the element effect are both substantial with all bases, indicating that the cleavage of both the C—H and the C—X bonds is important in the transition state. Several mechanisms might be consistent with these observations: the E2 mechanism, where both bonds are broken in a single step,

---

83. Seltzer, S. *J. Am. Chem. Soc.* **1961**, *83*, 1861.
84. Johnson, R.R.; Lewis, E.S. *Proc. Chem. Soc.* **1958**, 52.
85. Stewart, R.; Gatzke, A.L.; Mocek, M.; Yates, K. *Chem. Ind. (London)* **1959**, 331.
86. Fontana, G.; Frenna, V.; Gruttadauria, M.; Natoli, M.C.; Noto, R. *J. Phys. Org. Chem.* **1998**, *11*, 54.

Table 8.11 Base-Promoted Elimination of Diaryltrihaloethanes

Base	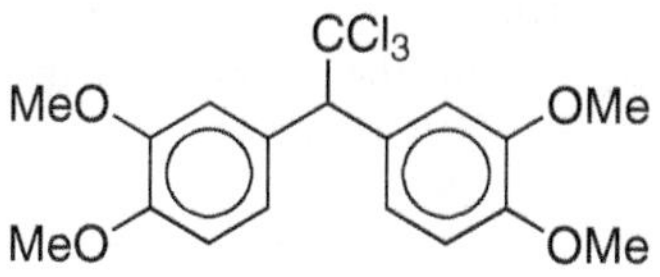	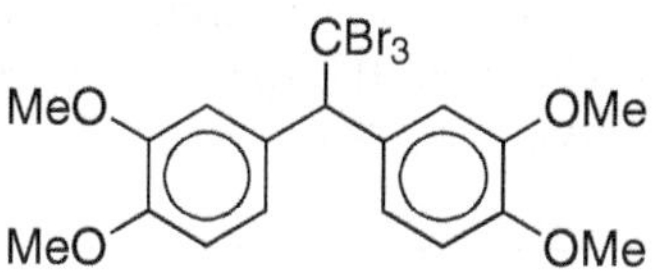
$MeO^-/MeOH$	$^Hk/^Dk$ 4.91	k_{Br}/k_{Cl} 26.1
$EtO^-/EtOH$	$^Hk/^Dk$ 5.66	k_{Br}/k_{Cl} 27.5
Me_3CO^-/Me_3COH	$^Hk/^Dk$ 3.53	k_{Br}/k_{Cl} 22.0
$PhO^-/EtOH$	$^Hk/^Dk$ 4.55	k_{Br}/k_{Cl} 40.4
$4\text{-}O_2NC_6H_4O^-/EtOH$	$^Hk/^Dk$ 3.5	k_{Br}/k_{Cl} 52.2

and the Elcb mechanisms, where the two bonds are broken sequentially. Based on the data in Table 8.11, the most likely candidate for this reaction is the E2 mechanism, because the deuterium isotope effect and the concomitant element effect give effective proof that there is simultaneous C—H and C—X bond cleavage in the transition state for the reaction.

This process has been applied especially widely to enzymatic reactions, as evidenced by recent reviews of aspects of the subject.[87] In a typical procedure, the transition state structure is first deduced by bond order–vibrational analysis methods[88] and then followed by completing the model with more modern computational methods. The procedure advocated by Schramm[89] is a modern extension of the successive labeling-theoretical approaches advocated earlier.[90] Schramm's synopsis of the method is instructive.

Element and isotope effects provide clear evidence of specific bond cleavages in the rate determining steps of a reaction. A good example of the application of both element effects and kinetic isotope effects to a mechanistic problem is provided by the base-promoted dehydrohalogenation of 2,2-diaryl-1,1,1-trihaloethanes.[91] For this reaction, there are data available (Table 8.11) for both the element effect (k_{Br}/k_{Cl}) and the deuterium isotope effect ($^Hk/^Dk$). In addition, the reaction is subject to general base catalysis. Brønsted β values for the chloride are 0.61 for alkoxide ion bases and 0.54 when the phenoxide bases are included and 0.57 and 0.50 for the corresponding bromide.

1. Synthesize compounds labeled at locations where the possibility of bonding change is to be probed (***successive labeling***[92]).
2. Measure the kinetic isotope effects at these positions to a reasonably high degree (≈ 0.5%) of precision.
3. Compute a truncated ("cutoff") transition state model, complete with bond lengths and bond angles, using bond order–vibrational analysis methods.

---

87. (a) Schramm, V.L. *Acc. Chem. Res.* **2006**, *36*, 588. (b) Berti, P.J.; McCann, J.A.B. *Chem. Rev.* **2006**, *106*, 506.

88. (a) Sims, L.B.; Lewis, D.E. In Buncel, E.; Lee, C.C., Eds. *Isotopes in Organic Chemistry* (Elsevier: New York, 1984), vol. 6, p. 161. (b) Berti, P.J.; Schramm, V.L. In Truhlar, D.G.; Morokuma, K., Eds. *Transition state modeling for catalysis* (Amercan Chemical Society: Washington, D.C., 1999), p. 473. (c) Sims, L.B.; Burton, G.W.; Lewis, D.E. *BEBOVIB-IV, QCPE No. 337* (Quantum Chemistry Program Exchange, Department of Chemistry, Indiana University, Bloomington, IN, 1977). (d) Huskey, W.P. *J. Am. Chem. Soc.* **1996**, *118*, 1663. (e) Casamassina, T.E.; Huskey, W.P. *J. Am. Chem. Soc.* **1993**, *115*, 14.

89. Schramm, V.L. *Ann. Rev. Biochem.* **1998**, *67*, 693.

90. Sims, L.B.; Lewis, D.E.; Netherton, L.T. In Duncan, W.P.; Susan, A.B., Eds. *Synthesis and Applications of Isotopically Labeled Compounds* (Elsevier: Amsterdam, 1982), p. 261.

91. Fontana, G.; Frenna, V.; Gruttadauria, M.; Natoli, M.C.; Noto, R. *J. Phys. Org. Chem.* **1998**, *11*, 54.

92. Fry, A. *Pure Appl. Chem.* **1964**, *8*, 409.

4. Transfer these parameters as fixed values to a semiempirical calculation for optimizing the complete transition state model.
5. Compute the wave function to determine the electron distribution at the van der Waals surface.

Problems

8-11 The Hammett ρ constants for elimination reactions from a series of substituted 2-phenylethyl derivatives is given below.

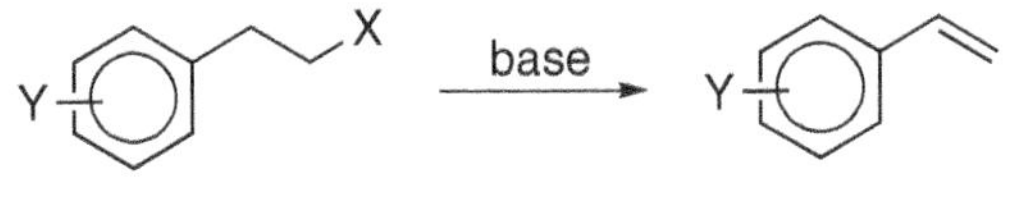

X = I, ρ = 2.07; X = Br, ρ = 2.14; X = OTs, ρ = 2.27; X = Cl, ρ = 2.61;

X = $^{+}SMe_2$, ρ = 2.75; X = F, ρ = 3.12; X = $^{+}NMe_3$, ρ = 3.77

What conclusions do you draw about the mechanisms of these reactions from these data? What experiments would you carry out to test these conclusions?

8-12 The Hammett ρ value for the acid catalyzed hydrolysis of substituted benzamides in water at 100°C is +0.118, and the ρ constant for the corresponding alkaline hydrolysis in water at 100°C is +1.055. Rationalize the difference in these two values in terms of what the ρ value tells us about the electron density at the carbonyl carbon during the rate-limiting step of the reaction.

8-13 The Grunwald-Winstein *m* value for the solvolysis of 1-adamantyl bromide in 40% aqueous ethanol is 1.20. The activation parameters for the same reaction are ΔH‡ = +21.0 kcal mol^{-1}, and ΔS‡ = -6.2 eu. Which mechanism (S_N1 or S_N2) do these data best support for this solvolysis? Based on this conclusion, explain the activation parameters for the reaction.

8-14 Consider the E1cb mechanisms in Problem 8-5. Which of these reactions will exhibit general base catalysis, and which will exhibit specific base catalysis? Why?

8-15 Predict the probable magnitudes of the primary deuterium isotope effects (large, medium, small, or null) and element effects (large, small, or null) for the six mechanisms of elimination in Problem 8-1. Assume that the substrate is an ethyl halide.

8-16 The rate constants for the unimolecular Cope elimination of a series of substituted 2-phenylethyldimethylamine *N*-oxides in 90% dimethyl sulfoxide-10% water at 65°C are given as follows:

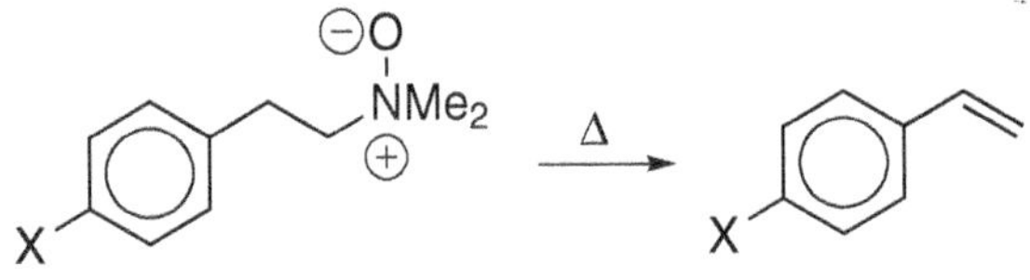

X = OMe, $k = 0.67 \pm 0.05 \times 10^{-5}$; X = Me, $k = 1.4 \pm 0.2 \times 10^{-5}$;

X = H, $k = 3.1 \pm 0.7 \times 10^{-5}$; X = Cl, $k = 8.2 \pm 1.2 \times 10^{-5}$;

X = NO_2, *k* = "too fast for convenient measurement"

Construct a Hammett plot from these data and deduce the likely arrangement of partial charges in the transition state.

8.7 Trapping and Crossover Experiments

Detecting and characterizing reactive intermediates is an important part of elucidating a reaction mechanism. Trapping experiments, or competition experiments, often provide the most direct information about a reactive intermediate involved in a reaction; they involve carrying out the reaction in the presence of a second molecule capable of reacting with the intermediate and diverting it from the normal pathway.[93] The structure of the reactive intermediate is then deduced from the structure of the product.

One of the oldest examples of this type of trapping reaction is provided by the addition of bromine to alkenes.[94] In this reaction, which is now agreed to proceed through a three-membered bromonium ion,[95] the existence of the ion was deduced from two important facts: the addition to simple alkenes was *anti*, and when the reaction was carried out in the presence of added nucleophiles, mixed products were obtained. Interpretation of the ratio of products from added nucleophiles has led to the conclusion that the bromonium ion from 2-butene has a lifetime on the order of 10^{-10} seconds in methanol.[96]

This same type of intermediate also demonstrates the utility of steric hindrance in stabilizing reactive species by preventing subsequent reactions well enough to allow its isolation and characterization. An excellent example of this is provided by the case of the bromonium ion. Although most are very short-lived, that from adamantylidene-adamantane has been crystallized as the tribromide salt (Example **8.33**) and had its structure determined by single crystal X-ray structure analysis.[97]

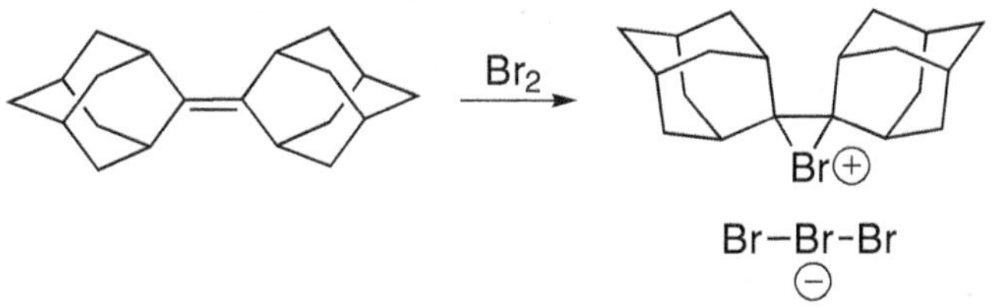

(8.33)

In similar fashion, strong evidence for the existence of *o*-benzyne was obtained by carrying out the reaction believed to form benzyne in the presence of reactive dienophiles such as furan (Example **8.34**),[98] tetraphenylcyclopentadienone (Example **8.35**),[99] or anthracene (Example **8.36**).[100] All of these reactions provided crystalline products whose structures were consistent with their formation by a Diels-Alder reaction with benzyne. Benzyne is, in fact reactive enough to give similar products with benzene[101] and naphthalene.[102]

---

93. For a discussion of competition methods, see: Huisgen, R. *Angew. Chem. Int. Ed. Engl.* **1970**, *9*, 751.

94. (a) Francis, A.W. *J. Am. Chem. Soc.* **1925**, *47*, 2340. (b) Biilmann, E. *Rec. trav. chim.* **1917**, *36*, 313. (c) Gomberg, M. *J. Am. Chem. Soc.* **1919**, *41*, 1414. (d) Read, J.; Williams, M.M. *J. Chem. Soc.* **1920**, *117*, 359. (e) Read, J.; Hook, R.G. *J. Chem. Soc.* **1920**, *117*, 1214. (f) Read, J.; Andrews, A.C.P. *J. Chem. Soc.* **1921**, 1774. (g) Read, J.; Hurst, E. *J. Chem. Soc.* **1922**, 989. (h) Read, J.; Hurst, E. *J. Chem. Soc.* **1922**, 2550. (i) Read, J.; Reid, W.G. *J. Chem. Soc.* **1928**, 745. (j) Conant, J.B.; Jackson, E.L. *J. Am. Chem. Soc.* **1924**, *46*, 1727. (k) Jackson, E.L. *J. Am. Chem. Soc.* **1926**, *48*, 3166. (l) Meinel, K. *Ann. Chem.* **1934**, *510*, 129. (m) Bartlett, P.D.; Tarbell, D.S. *J. Am. Chem. Soc.* **1936**, *58*, 466. (m) Backer, H.J.; Strating, J. *Rec. trav. chim.* **1934**, *53*, 525. (n) Bockemüller, W.; Hoffmann, F.W. *Ann. Chem.* **1935**, *519*, 165.

95. Roberts, I.; Kimball, G.E. *J. Am. Chem. Soc.* **1937**, *59*, 947.

96. Nagorski, R.W.; Brown, R.S. *J. Am. Chem. Soc.* **1992**, *114*, 7773.

97. Slebocka-Tilk, H.; Ball, R.G.; Brown, R.S. *J. Am. Chem. Soc.* **1985**, *107*, 4504.

98. Wittig, G.; Pohmer, L. *Angew. Chem.* **1955**, *67*, 348.

99. Fieser, L.F.; Haddadin, M. *J. Org. Syn.* **1966**, *46*, 107.

100. Friedman, L.F.; Logullo, F.M. *J. Org. Chem.* **1969**, *34*, 3089.

101. Brewer, J.P.N.; Heaney, H. *Tetrahedron Lett.* **1965**, 4709.

102. Miller, R.G.; Stiles, M. *J. Am. Chem. Soc.* **1963**, *85*, 1798.

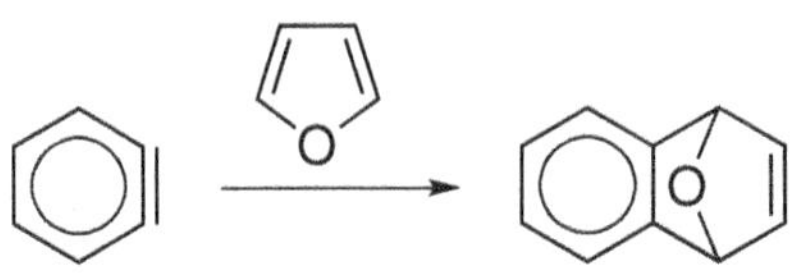

(8.34)

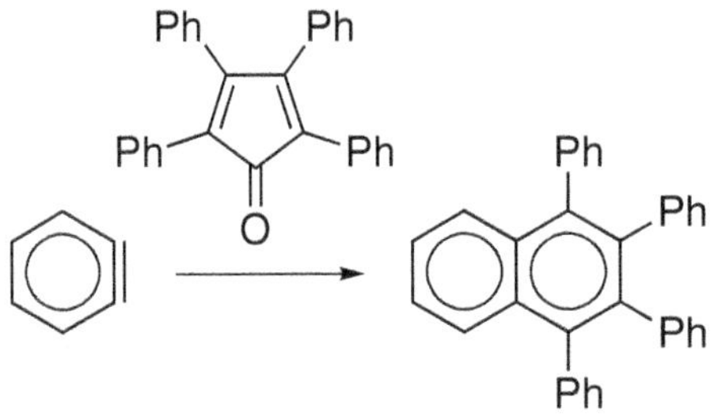

(8.35)

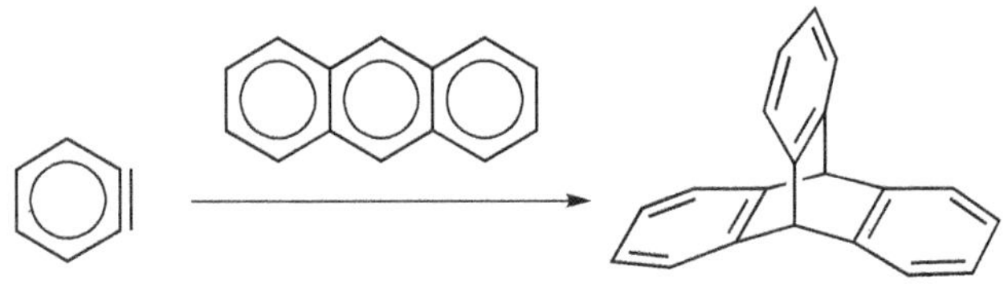

(8.36)

Free radicals have been trapped by the use of nitroso compounds (e.g., Example **8.37**) or nitrones (e.g., Example **8.38**) to give very stable nitroxide free radicals that can be detected by electron paramagnetic resonance spectroscopy. This process, known as ***spin trapping***, gives a product free radical from whose spectrum the structure of the original free radical can often be deduced.

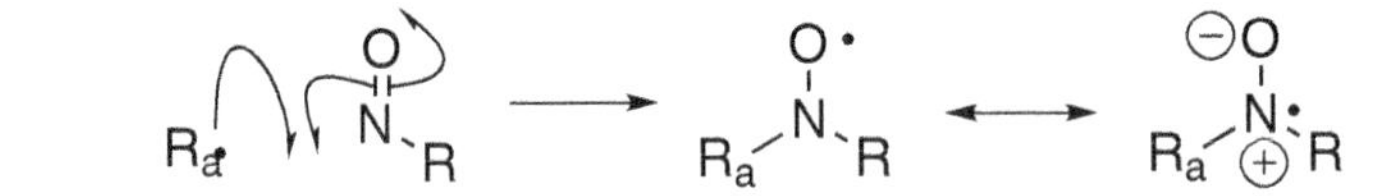

(8.37)

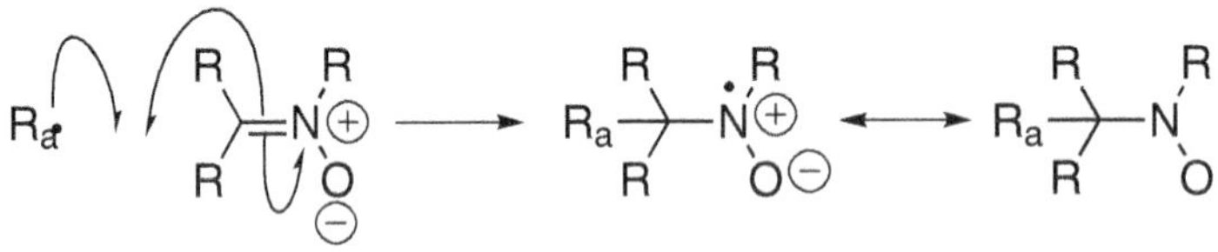

(8.38)

Crossover experiments provide a method to distinguish between intramolecular and intermolecular mechanisms for a reaction. This type of experiment is performed by carrying out the reaction with a mixture of two slightly different reactants and then analyzing the product mixture. If the reaction proceeds by an intramolecular (nondissociative) mechanism, no possibility exists for the exchange of groups between reactants. Therefore, only two products will be formed, and each will be clearly derived from only one of the two reactants. If the reaction involves the formation of reactive intermediates, such as cations, anions, or radicals (i.e., it proceeds through a dissociative mechanism), the possibility exists for the intermediate from one of the starting compounds to react with the complementary intermediate from the other. Thus, the reaction allows for the possibility of four different products, two of which are *crossover products*, where there has been an exchange of groups between reactants, to be formed. Crossover experiments have been especially useful in the study of rearrangement reactions. A typical example of a reaction studied by using crossover experiments is the Claisen rearrangement of allyl phenyl ethers (Examples **8.39** and **8.40**). These studies show that the cross-over products, Examples **8.43** and **8.44**, are not produced, suggesting that the mechanism for this reaction is not dissociative.[103]

103. Tarbell, D.S. *Org. React.* **1944**, *2*, 1.

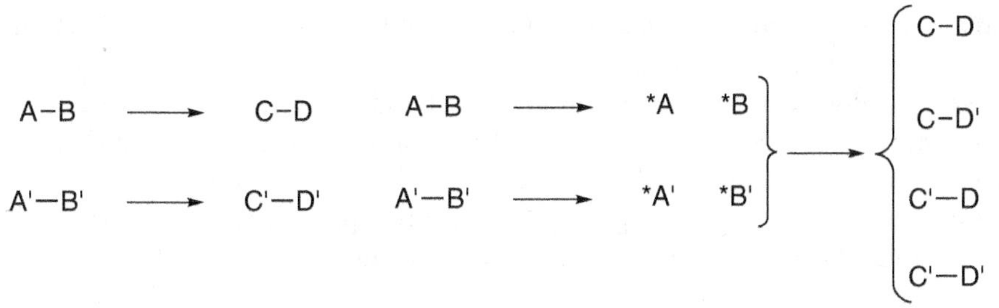

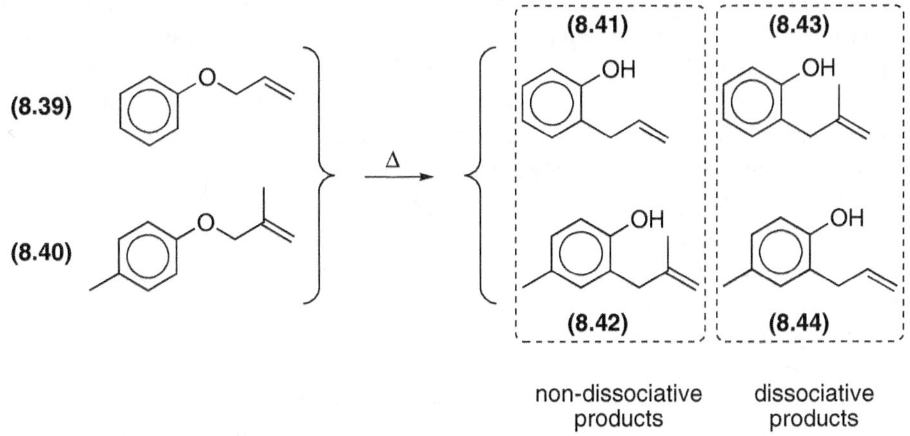

One must bear in mind a caveat when interpreting the results of crossover experiments. If crossover products are observed, it is unambiguous that they must be formed in some type of dissociative process. However, this does not mean that all the products are formed in a dissociative reaction. The non-crossover products may be formed in either a dissociative reaction or in a competing nondissociative reaction. The problem is more subtle in cases where crossover products are not observed, because this does not necessarily rule out the dissociative pathway. It is possible that the products of the initial dissociation are reactive enough that they recombine to give the product before they can diffuse away from each other. In this case, what are called ***cage effects*** operate: the two reactive intermediates are held in the same solvent cage long enough to react completely with each other.

Problems

8-17 Draw the structures of the expected products when benzyne is trapped by (a) benzene; (b) naphthalene; (c) 1,3-cyclohexadiene.

8-18 In reactions that proceed by multistep mechanisms, it is sometimes possible to prepare one of the key reactive intermediates by an alternative route. This was done by Porter and Zuraw (*J. Chem. Soc., Chem. Commun.* **1985**, 1472) in their study of the rearrangement of allylic hydroperoxides. Using this paper as a guide, discuss what information this kind of experiment might provide. Are there any limitations?

8.8 The Variable Transition State and the Concept of the Mechanistic Spectrum

The idea of the variable transition state is not new. It forms the basis, for example, for the Hammond postulate,[104] where the effects of structural changes in the reactants and

104. Hammond, G.S. *J. Am. Chem. Soc.*, **1955**, *77*, 334.

products are related to the structure and energy of the transition state. In such discussions, we speak of "early" and "late" transition states.

We have already defined the terms concerted and stepwise, but on occasion we will need to make even finer distinctions about the relative timing of bonding changes in a reaction—the extent of bond making and bond breaking in the transition state. Of course, by definition all the bond-making and bond-breaking processes in a concerted reaction must all occur during the same *step*, but this does not necessarily imply that they occur at the same *rate*.[105] The structure of the activated complex may reveal that bond breaking is well in advance of or lags behind bond making, in which case the reaction may be termed ***non-synchronous***, or that the two processes do, in fact, occur at the same rate, in which case the reaction may be termed ***synchronous***.[106] Differences in the relative extent of bond making and bond breaking in the transition state leads to the concept of a ***mechanistic spectrum*** for a reaction—illustrating the effect of substituents on transition state structure. Isotope effects represent a particularly powerful method for the elucidation of transition state structure when they are measured for a series of different isotopomers of the reactant, the process known as successive labeling.[107]

Quantitative or semiquantitative discussions about transition state structure are facilitated through the use of transition state maps, also known as More O'Ferrall-Jencks diagrams.[108] These are simplified representations of the potential energy surfaces, with the bonding changes defined in terms of the bond orders of the participating bonds. These diagrams allow one to make predictions about the effects of substitutions on the transition state for the reaction and to show the dynamics of the activated complex as it passes through the transition state. A typical example of the More O'Ferrall-Jencks diagram is shown as Figure 8.16.

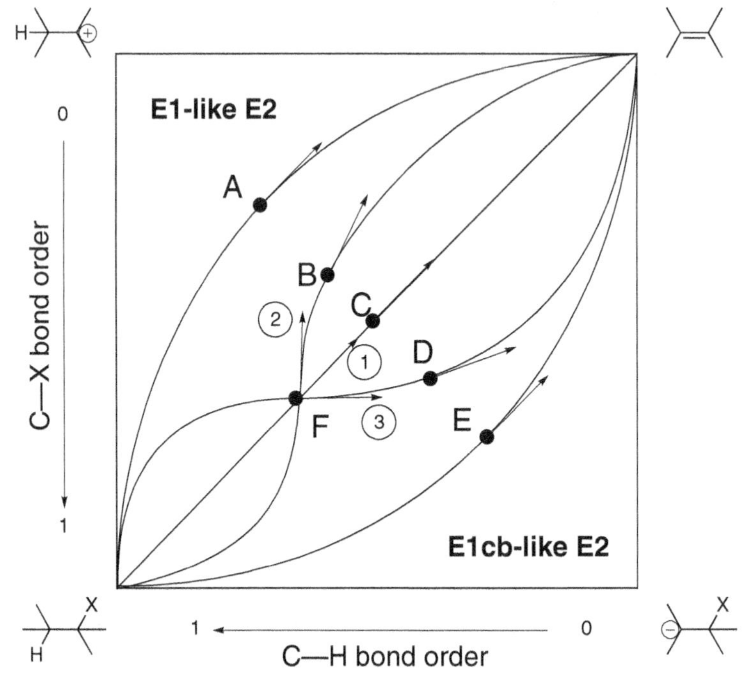

Figure 8.16 More O'Ferrall-Jencks diagram for β-elimination reactions

105. This question and the question of synchronicity in bond formation and rpture has been reviewed: *Adv. Phys. Org. Chem.* **1992**, *29*.

106. Reviews: (a) Bernasconi, C.F. *Adv. Phys. Org. Chem.* **1992**, *27*, 119. (b) Bernasconi, C.F. *Adv. Phys. Org. Chem.* **2010**, *44*, 223.

107. Fry, A. *Pure Appl. Chem.* **1964**, *8*, 409.

108. (a) More O'Ferrall, R.A. *J. Chem. Soc. B* **1970**, 274. (b) Jencks, D.A.; Jencks, W.P. *J. Am. Chem. Soc.* **1977**, *99*, 7848.

In this figure, the mechanistic spectrum for β-elimination reactions is represented. The axes of the diagram represent the bond orders of the two bonds being broken. Vertical movement from the starting compound along the axis corresponds to C—X bond rupture to give the carbocation intermediate characteristic of the E1 mechanism of elimination. Horizontal movement eventually leads to C—H bond rupture to give the carbanion intermediate characteristic of the E1cb reaction. These stepwise mechanisms therefore follow a path around the edges of the diagram. All other activated points within the diagram have some degree of both C—H and C—X bond rupture, and correspond to activated complexes for the concerted reactions. It is evident that there are many possible options for concerted transition states, as illustrated by points A–E on the diagram. Positions A and B locate transition states where C—X bond rupture is in advance of C—H bond rupture: they are transition states for E2 reactions with different degrees of E1 character. Conversely, positions D and E locate transition states where C—H bond rupture is in advance of C—X bond rupture: they are transition states for E2 reactions with different degrees of E1cb character. The points C and F locate transition states where C—H and C—X bond rupture have occurred to the same extent ("central" transition states).

The lines in the diagram show possible reaction pathways for a variety of concerted reactions. These allow us to discuss the other important feature of the activated complex: its dynamics as it passes through the transition state. This is best illustrated by the activated complex F, where there are three different possible reaction coordinate pathways shown. In pathway 1, all points between the reactant and activated complex are characterized by having the same amount of C—H and C—X bond rupture, with no charge development at either carbon, and the reaction coordinate motion is a combination of equal contributions from C—H and C—X bond rupture. In pathway 2, the reaction coordinate motion is dominated by C—X bond rupture, with little contribution from C—H bond rupture. In pathway 3, on the other hand, the reaction coordinate motion is dominated by C—H bond rupture, with little contribution from C—X bond rupture. The details of the reaction coordinate motion are best answered by a combined successive labeling-theoretical approach.[109] One example of a reaction for which such an approach has been very successful is the E2 elimination of trimethylamine from 2-phenylethyltrimethtlammonium cation. The available isotope effect data for this reaction are gathered in Table 8.12.[110]

The data reveal some important features of the transition state for this reaction immediately. One, there are substantial deuterium and ^{14}C isotope effect at the β carbon, which indicates that there is partial cleavage of this bond in the transition state for the reaction. Two, there are small ^{15}N and α-^{14}C isotope effects, indicating cleavage of the carbon-nitrogen bond in the transition state. Three, there is an almost null secondary deuterium isotope effect and an almost constant ^{14}C isotope effect at the α carbon, which is consistent with a situation where σ bonding loss to nitrogen is largely compensated for by π-bond formation to the β-carbon (i.e., there is little charge development at this site).

When we take all these observations together, we find that the best transition state model for the data is an E2 transition state but that carbon-hydrogen bond rupture is in advance of carbon-nitrogen bond rupture: the reaction is an E1cb-like E2 reaction. Theoretical calculations reveal an E1cb-like E2 transition state with a partial charge of approximately 0.2 e^- developed at the β carbon.[111]

109. Sims, L.B.; Lewis, D.E.; Netherton, L.T. In Duncan, W.P.; Susan, A.B., Eds. *Synthesis and Applications of Isotopically Labeled Compounds* (Elsevier: Amsterdam, 1982), p. 261.

110. Data in the table are taken from: (a) Smith, P.J.; Bourns, A.N. *Can. J. Chem.* **1974**, *52*, 749. (b) Bourns, A.N.; Smith, P.J. *Proc. Chem. Soc.* **1964**, 366. (c) Eubanks, J.R.I.; Sims, L.B.; Fry, A. *J. Am. Chem. Soc.* **1991**, *113*, 8821.

111. Lewis, D.E.; Sims, L.B.; Yamataka, H.; McKenna, J. *J. Am. Chem. Soc.* **1980**, *102*, 7411.

Table 8.12 Isotope Effects for the Hofmann Elimination of 2-Phenylethyl-Trimethylammonium Salts ($Ar—CH_2—CH_2—NMe_3^+$)

Ar	RO^-	Isotopomer	k/k*
C_6H_5	$EtO^-/EtOH/40°C$	β-D_2	3.23 ± 0.04
C_6H_5	$HO^-/H_2O/90°C$	α-D_2	1.02 ± 0.01
C_6H_5	$EtO^-/EtOH/40°C$	^{15}N	1.0142 ± 0.0004
C_6H_5	$EtO^-/EtOH/40°C$	α-^{14}C	1.044 ± 0.002
C_6H_5	$EtO^-/EtOH/40°C$	β-^{14}C	1.041 ± 0.001
4-$MeOC_6H_4$	$EtO^-/EtOH/40°C$	β-D_2	2.64 ± 0.04
4-$MeOC_6H_4$	$EtO^-/EtOH/40°C$	^{15}N	1.0137 ± 0.0009
4-$MeOC_6H_4$	$EtO^-/EtOH/40°C$	α-^{14}C	1.050 ± 0.002
4-$MeOC_6H_4$	$EtO^-/EtOH/40°C$	β-^{14}C	1.040 ± 0.002
4-ClC_6H_4	$EtO^-/EtOH/40°C$	β-D_2	3.48 ± 0.06
4-ClC_6H_4	$EtO^-/EtOH/40°C$	^{15}N	1.0114 ± 0.0009
4-ClC_6H_4	$EtO^-/EtOH/40°C$	α-^{14}C	1.041 ± 0.002
4-ClC_6H_4	$EtO^-/EtOH/40°C$	β-^{14}C	1.044 ± 0.002
4-$CF_3C_6H_4$	$EtO^-/EtOH/40°C$	β-D_2	4.15 ± 0.05
4-$CF_3C_6H_4$	$EtO^-/EtOH/40°C$	^{15}N	1.0088 ± 0.0006
4-$CF_3C_6H_4$	$EtO^-/EtOH/40°C$	α-^{14}C	1.032 ± 0.005
4-$CF_3C_6H_4$	$EtO^-/EtOH/40°C$	β-^{14}C	1.044 ± 0.004
4-$O_2NC_6H_4$	$EtO^-/EtOH/40°C$	α-^{14}C	1.019 ± 0.005
4-$O_2NC_6H_4$	$EtO^-/EtOH/40°C$	β-^{14}C	1.042 ± 0.003

Between the 1960s and 1970s, Swain and Thornton set forth a number of rules[112] that can be used in conjunction with the More O'Ferrall-Jencks diagram to make predictions of substituent effects on transition state structure. The upshot of their work was that we can divide the effects of substituents into two components: *parallel effects, which are the effects of the structural change on the mixed translation*–vibration motion along the reaction coordinate (these are the effects one would predict on the basis of the Hammond postulate, for example), and *perpendicular effects, which are the effects of the structural change on the true vibrations of the activated complex.*

These effects are indexed to the reaction coordinate and so are relatively easy to discuss in terms of the More O'Ferrall-Jencks diagram for the reaction. Parallel effects result in a shift of transition state along the direction of the reaction coordinate either toward reactant or product. Perpendicular effects result in a shift of the transition state structure perpendicular to the reaction coordinate. If we use the β-elimination reaction, as in Figure 8.16, as

112. (a) Swain, C.G.; Thornton, E.R. *J. Am. Chem. Soc.* **1962**, *84*, 817. (b) Swain, C.G.; Thornton, E.R. *J. Am. Chem. Soc.* **1962**, *84*, 922. (c) Thornton,, E.R. *J. Am. Chem. Soc.* **1967**, *89*, 2915. (d) Steffa, L.J.; Thornton, E.R. *J. Am. Chem. Soc.* **1967**, *89*, 6149. (e) Winey, D.A.; Thornton, E.R. *J. Am.Chem. Soc.* **1975**, *97*, 3102.

our example, we see that the four corners of the diagram correspond to the four stable species: the reactant, the product alkene, the carbanion formed by abstraction of the β-hydrogen, and the carbocation formed by loss of the leaving group. These are four bonds that are involved in the reaction (i.e., whose stretching motions become part of the reaction coordinate): the σ bond between the base and the β-hydrogen, the σ bond between the β-hydrogen and the β-carbon, the σ bond between the α-carbon and the leaving group, and the carbon-carbon π bond. In the course of the reaction, the B—H bond is formed, the C—H bond breaks, the C—C π bond forms, and the C—X bond breaks, which makes the overall reaction coordinate motion (Example **8.45**) a coupled asymmetric stretching motion of these four bonds.

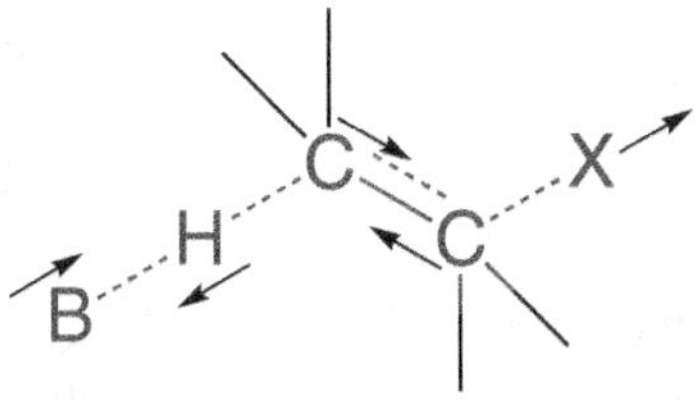

(8.45)

Let us begin by looking at a classical E2 reaction with a central transition state, indicated by a heavy black dot in Figure 8.17. If the bond to the leaving group becomes stronger, this will lower the energy of both the reactant and the carbanion, because both these species have the C—X bond, but will not affect the energy of the carbocation or the product, which do not. This will reduce the energy of the entire bottom edge of the diagram and will tend to move the structure of the transition state away from the diagonal toward the bottom edge of the diagram, as predicted by the Hammond postulate. We expect to see the reaction become more Elcb-like as the bond to the leaving group becomes stronger. At the same time, because we have lowered the energy of the reactant but not the energy of the product alkene, we expect to see the transition state move toward the product, as predicted by the Hammond Postulate. We predict that the perpendicular component of the shift will therefore be toward the bottom edge of the diagram and that the parallel component will be toward the product. The resultant is that we predict that the E2 reaction will, indeed, become more Elcb-like and that the structural change will affect mainly the C—H bond (toward the open circle in Figure 8.17).

The situation changes somewhat when we examine an Elcb-like E2 reaction because the reaction coordinate motion changes (Figure 8.18), becoming dominated by proton transfer until after the transition state is passed.[113] Stretching of the C—X bond is relatively unimportant at the transition state, so we expect the parallel effect of increasing the

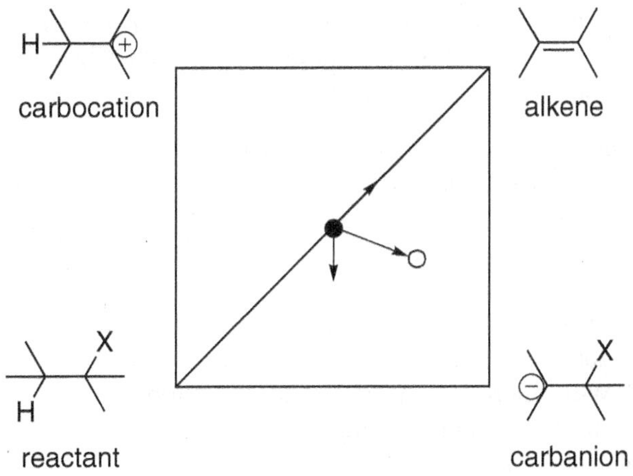

Figure 8.17 Increasing bond strength to the leaving group in an E2 reaction

113. Smith, P.J.; Bourns, A.N. *Can. J. Chem.* **1974**, *52*, 749.

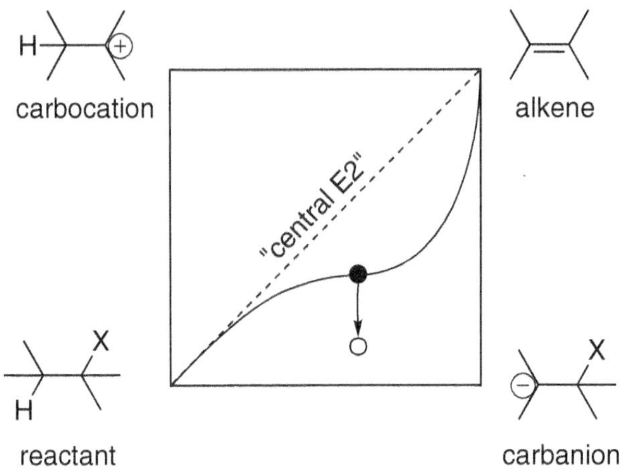

Figure 8.18 Increasing bond strength to the leaving group in an Elcb-like E2 reaction

strength of the bond to the leaving group to be small. The perpendicular effect of the change should be to move the transition state toward the bottom edge of the diagram. Again, the position of the predicted new transition state is indicated by an open circle. We predict that the transition state will become even more strongly anionic, with less C—X bond rupture in the transition state.

Problems

8-19 Modify the More O'Ferrall-Jencks diagram in Figure 8.17 to represent each of the following reactions:

(a) nucleophilic substitution of an alkyl halide

(b) oxidation of a secondary alcohol to a ketone

(c) addition of singlet carbene to an aldehyde to give an epoxide

(d) nucleophilic acyl substitution of an acid chloride by methoxide anion

(e) addition of dichloroketene ($Cl_2C{=}C{=}O$) to an alkene to give a cyclobutanone. (Hint: ketenes often react as though they were zwitterions $R_2C{=}C^+{-}O^-$.)

8-20 In the discussion of the element effect in Section 8.6, we came to the conclusion that the reaction probably proceeds by the E2 mechanism. The authors of the paper (*J. Phys. Org. Chem.* **1998**, *11*, 54) reach the conclusion that the reaction proceeds by the $Elcb_I$ mechanism. Evaluate both positions ($Elcb_I$ and E2), and suggest experiments that could be used to distinguish between the two mechanisms.

8-21 The element effect for the reaction between a halobenzene and amide ion in liquid ammonia reveals the order of reactivity as:

$$PhBr > PhI > PhCl >> PhF$$

When deuterated halobenzenes react with amide ion in liquid ammonia (NH_3), fluorobenzene-2,6-d_2 undergoes 100% loss of deuterium before any aniline is formed. With chlorobenzene-2-*d*, aniline formation occurs at approximately twice the rate of deuterium exchange, and bromobenzene-2-*d* gives aniline without loss of deuterium.

Rationalize these results and sketch the reaction coordinates for the reaction of each of the halobenzenes with amide anion in liquid ammonia.

8.9 Using Regiochemistry and Stereochemistry

The stereochemistry and regiochemistry of organic reactions has been a subject of study for well over a century, beginning with the two empirical rules for predicting the regiochemistry of reactions that are still taught in introductory organic chemistry courses: Markovnikov's rule for addition[114] and Zaitsev's rule for elimination.[115]

The regiochemistry and stereochemistry of reactions can provide strong evidence about reaction mechanism in appropriate circumstances. For example, concerted reactions generally occur stereospecifically: one stereoisomer of the reactant gives one stereoisomer of the product, and a different stereoisomer of the reactant gives a different stereoisomer of the product. The reason why concerted reactions tend to be stereospecific is ***stereoelectronic*** in origin. In order for a concerted reaction to proceed, all the molecular orbitals involved must be aligned correctly to permit the reorganization of electrons during the single step of the reaction. Thus, the observation that a reaction is stereospecific can often be used to suggest that the mechanism of the reaction might be concerted. We see this in the E2 and S_N2 reactions: the E2 reaction occurs with antarafacial stereochemistry, whereas the S_{N2} reaction occurs with inversion of configuration.

The corollary of this is not, however, true: the observation of stereospecificity alone is not always an indicator of concertedness. One must be careful about inferring that a reaction is concerted because it occurs stereospecifically. A good example of such a reaction is the addition of a halogen to an alkene π bond, for which one can propose both stepwise and concerted mechanistic pathways. Due to geometric constraints, the concerted reaction, which would proceed through a four-membered transition state, is forced to proceed with suprafacial (*syn*) stereochemistry. Because this stereospecific reaction actually occurs with antarafacial (*anti*) stereochemistry, the reaction stereochemistry cannot be reconciled with a concerted mechanism. One must conclude that this reaction occurs by a stepwise mechanism involving at least one reactive intermediate.

The regiochemistry of organic reactions is often less useful than stereochemistry, but there are occasions where the observation of differences in regiochemistry can be used fruitfully in making deductions and predictions about reactions. For example, Markovnikov's rule was originally defined in structural terms of the alkene, but it was later defined in terms of the relative energies of the two possible regioisomeric carbocation intermediates. Knowing that the addition of electrophiles to alkene π bonds must obey Markovnikov's rule places certain limitations on the structure and type of intermediate involved in the reaction.

Probably the most famous regiochemical problem that led to mechanistic information was the addition of hydrogen bromide to alkenes, eventually solved by Kharasch.[116] It had been observed that although both HCl and HI added to alkenes according to Markovnikov's rule, the addition of HBr was rather random. Kharasch and Mayo noted that the amount of anti-Markovnikov addition correlated with the ability of the allyl bromide in their study to oxidize ferrous ion; they termed the observation "the peroxide effect." The reversal of regiochemistry, they reasoned, meant that in the reactions carried out in the presence of peroxides, the bromine must add first to the alkene π bond to give the reactive intermediate, whereas in the absence of peroxides, the proton adds first to give the more stable of the two possible carbocations. This ultimately led to the deduction that the function of the peroxides was to generate bromine atoms from the hydrogen bromide and that these bromine atoms added to the alkene π bond to give the more stable free radical.

---

114. (a) Markownikoff, V. *Ann. Chem. Pharm.* **1870**, *133*, 228. (b) Markownikoff, V. *Compt. rend.* **1875**, *82*, 668. (c) Markownikoff, V. *Compt. rend.* **1875**, *82*, 728. (d) Markownikoff, V. *Compt. rend.* **1875**, *82*, 776.

115. Saytzeff, A. *Ann. Chem. Pharm.* **1875**, *179*, 296.

116. Kharasch. M.S.; Mayo, F.R. *J. Am. Chem. Soc.* **1933**, *55*, 2468.

Problem

8-22 Draw a More O'Ferrall-Jencks diagram for an S_N2 reaction and use it to deduce the probable effects of each of the following changes in bond strengths on the structure of a transition state located in the exact center of the diagram. Assume that the reaction coordinate motion of the reference reaction lies along the diagonal of the diagram.

(a) Increasing the C—nucleophile (C—Nu) bond strength.

(b) Increasing the C—leaving group (C—X) bond strength.

(c) Decreasing the strength of both the C—Nu and C—X bonds.

(d) Increasing the strength of both the C—Nu and C—X bonds.

(e) Decreasing the strength of the C—Nu bond and increasing the strength of the C—X bond.

(f) Increasing the strength of the C—Nu bond and decreasing the strength of the C—X bond

8.10 The Reactivity-Selectivity Principle[117]

Most chemists have an intuitive feel for the reactivity-selectivity principle: that ***the more reactive a species is, the less selective it is.*** We see this in the reactions of free radicals with alkanes, where the more reactive radicals are less selective in abstracting hydrogen atoms than the less reactive radicals. This can be seen from the data in Table 8.13, where the reactivity of radicals toward C—H bonds is compared to abstraction of a methyl hydrogen.

However, the reactivity-selectivity principle is far from universal. Although it works with simple one-bond and two-bond reactions such as hydrogen atom transfer reactions, it frequently fails in more complex reactions such as E2 eliminations, where there are four bonds changing in the activated complex or reactions where changes in solvation make a major contribution to the reaction coordinate.

Table 8.13 Selectivity of Radicals Toward C—H Bonds

Radical (X•)	CH_3—H	RCH_2—H	R_2CH—H	R_3C—H	D (X—H)
F•	0.5	1	1.2	1.4	135
Cl•	0.004	1	4.3	7.0	103
Br•	0.002	1	780	1700	87
I•		1	18		71
HO•	0.04	1	7	45	118
H•		1	4.8	40	105
CF_3•	0.06	1	8.3	90	105
CH_3•	0.09	1	10	80	104

Data from Kerr, J.A. In Kochi, J., Ed. *Free Radicals* (Wiley-Interscience, New York, 1973), vol. 1, ch. 1.

117. Review: Pross, A. *Adv. Phys. Org. Chem.* **1977**, *14*, 69.

Chapter Summary

This chapter has introduced the field of physical organic chemistry, which may be broadly defined as physical chemistry applied to the properties and reactions of organic compounds. In this chapter, mechanistic types have been more rigorously defined as concerted or stepwise, and the concept of the rate-limiting step has been introduced. The relationship between the potential energy surface for the reaction and the reaction coordinate diagram (reaction energy profile) has been defined, as have the meanings of activated complexes and transition states. In practice, these terms are often used as full synonyms, but in formal terms, the term activated complex, refers to a static structure, and the term, transition state, refers to a dynamic species. Reaction kinetics and the kinetic order of reactions have been discussed, and the use of the rate law in making deductions about reaction mechanism has been explored. The role of trapping and crossover experiments in deducing a mechanism has been discussed. The Arrhenius equation and the activation parameters of a reaction have been considered, as have the Hammett equation and the various parameters arising from the original σ and ρ constants. The Grunwald-Winstein relationship for solvent effects on reactions has been discussed, as has the Brønsted catalysis law and the salt effect. Element and isotope effects have been introduced, and the distinction made between primary and secondary isotope effects has been pointed out; this discussion was extended to the concepts of the variable transition state and a mechanistic spectrum. The chapter concluded with a short discussion of the reactivity-selectivity principle.

Key Terms

activated complex
activation energy
aprotic dipolar solvent
Arrhenius equation
Brønsted catalysis law
catalysis
concerted
crossover experiment
dielectric constant
dipole moment
element effect
equilibrium isotope effect
first-order kinetics
general acid catalysis
general base catalysis
Grunwald-Winstein relationship
Hammett constants
Hammett equation
ionic solvent
isotopomers
kinetic isotope effect
linear free energy relationships
London dispersion force
mechanistic spectrum
molecularity
More O'Ferrall-Jencks diagram
potential energy surface
pre-equilibrium
primary kinetic isotope effect
pseudo–first-order kinetics
rate constant
rate law
rate-limiting step (rate-determining step)
reaction coordinate diagram
reaction energy profile
reaction kinetics
reactive intermediate
reactivity-selectivity principle
solvent parameters N, Y

Additional Problems

8-23 The Morita-Baylis-Hillman reaction is a reaction between an aldehyde and a Michael acceptor that is catalyzed by a tertiary amine or phosphine base to give a densely functionalized product. Early work on the mechanism of this reaction showed second-order kinetics and led to the mechanism shown below being proposed for the reaction.

More recent kinetic work has shown that the reaction proceeds according to the rate law:

$$\text{Rate} = [\text{RCHO}]^2[\text{R}_3\text{N}][\text{H}_2\text{C}{=}\text{CH-EWG}]$$

(a) What changes does this rate law require in the activated complex for the reaction? Suggest a structure for the activated complex for the reaction.
(b) Based on your proposed activated complex, what would the rate-determining step of the reaction be? Would this reaction be subject to a kinetic isotope effect and which atom should be labeled to see that isotope effect if it is expected?
(c) When the Morita-Baylis-Hillman reaction is carried out in the presence of an alcohol, the rate law changes to:

$$\text{Rate} = [\text{RCHO}][\text{R}_3\text{N}][\text{H}_2\text{C}{=}\text{CH-EWG}][\text{ROH}]$$

How is this change in the rate law reflected in the structure of the activated complex?

8-24 Consider the effects of changing the substituent on the aromatic ring of the benzyl halide undergoing S_N2 substitution in the reaction corresponding to the More-O'Ferrall-Jencks diagram.

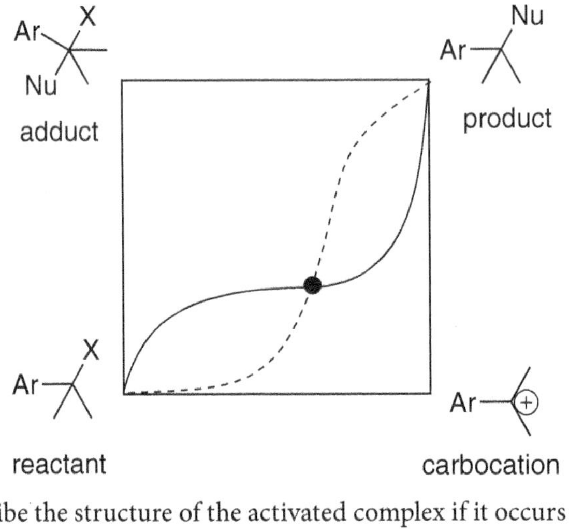

(a) Describe the structure of the activated complex if it occurs at the point indicated by the heavy dot.
(b) If the reaction coordinate motion corresponds to the full line in the diagram, what will be the effects on the activated complex structure be of changing the substituent at the *para* position on the ring to a more electron-withdrawing group?
(c) How does the effect of changing the substituent at the *para* position to a more electron-withdrawing group change the structure of the activated complex if the reaction coordinate motion corresponds to the dashed line in the diagram?

8-25 The [2 + 2] cycloaddition of diphenylketene to styrene has been studied extensively by the successive labeling technique. The primary and secondary isotope effects given in the table below have been measured for the reaction, which has been shown to be stereospecific and *sym* (suprafacial) with respect to the styrene by the use of specifically monodeuterated styrenes.

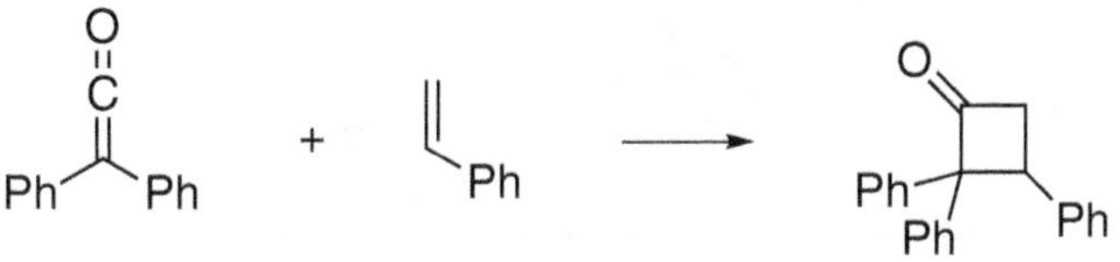

(a) Based on the available data, suggest a structure for the activated complex of this reaction. Give reasons for your answer.

(b) What would you expect the activation parameters for the reaction ($\Delta H^{\ddagger}$ and $\Delta S^{\ddagger}$) to be in qualitative terms (large, small, positive, negative)?

Isotopomer	kie	Isotopomer	kie	Isotopomer	kie
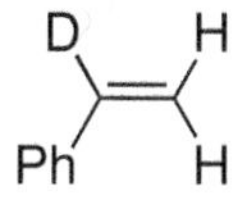	1.23 ± 0.03	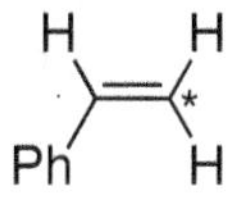	1.0055 ± 0.0005	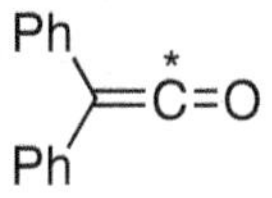	1.08 ± 0.005
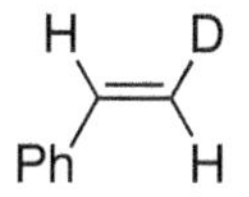	0.879 ± 0.004	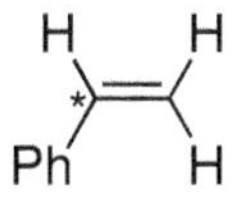	1.007 ± 0.003	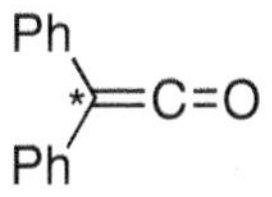	1.01 ± 0.003
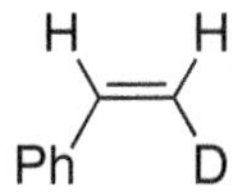	0.889 ± 0.004				

8-26 Using the Hammond postulate and the expression for Gibbs free energy ($\Delta G = \Delta H - T\Delta S$), provide a reasonable explanation for why E2 elimination is favored over S_N2 substitution and why E1 elimination is favored over S_N1 substitution, at high temperatures. Are the effects of temperature the same in both unimolecular and bimolecular reaction types?

8-27 The rate of the E2 elimination reaction between alcoholic potassium hydroxide and 1-bromobutane or 2-bromobutane under reflux can be measured using a gas buret to measure the volume of butenes formed in the reaction. If the same quantities of alkyl halide and base are used in the same volume of ethanol, one finds that 1-bromobutane gives only about 30% as much butene as 2-bromobutane. Why?

8-28 The reaction shown may proceed by a free radical chain mechanism or an ionic mechanism involving a tertiary carbocation.

(a) Draw both mechanisms.

(b) One way to probe this mechanistic question is to study the effects of changing solvent on the rate of the reaction. If the solvent is changed from benzene to acetic acid, what will be the effect on the rate of the reaction if it proceeds through each of these mechanisms?

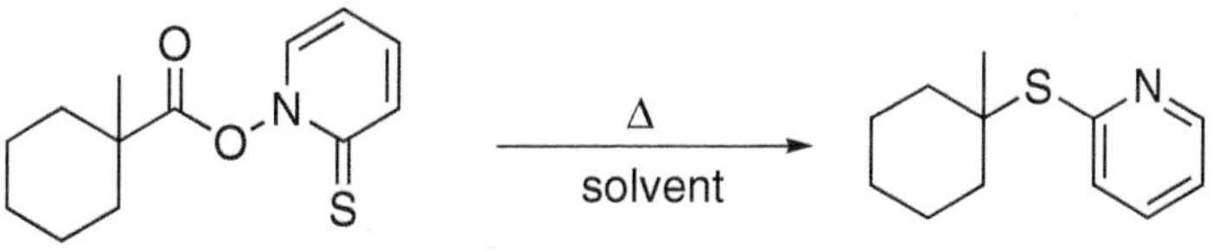

8-29 The hydrolysis of strained β-lactones can be accomplished under conditions of both acid and base catalysis. When the reaction is carried out using an optically active lactone and ^{18}O labeled water, the data in the table are obtained.

Reactant	pH −2 to 0	pH 3 to 8	pH 10 to 12
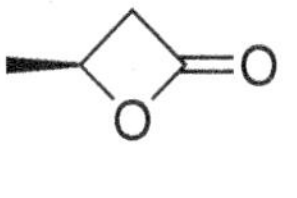	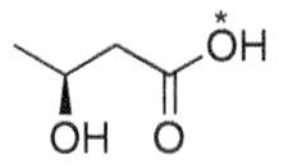	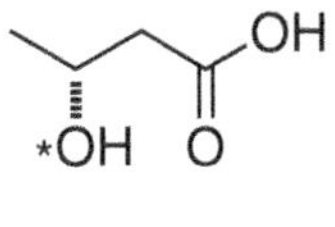	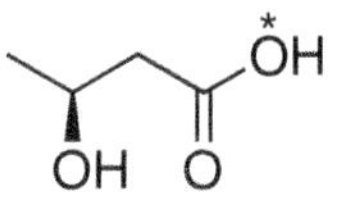

Is the mechanism constant over the entire range of pH? If it is not, what mechanisms operate under what conditions, and how does the mechanism change?

8-30 In aqueous solution, diacetone alcohol $Me_2C(OH)CH_2COMe$ decomposes into acetone in a process catalyzed by sodium hydroxide. The observed first-order rate constant is dependent on the concentration of the catalyst, as shown in the table. Rationalize these observations. [Data from *J. Am. Chem. Soc.* **1929**, *51*, 3215.]

[NaOH] (mM)	k_{obs}
5.0	2.32×10^{-3}
10.0	4.67×10^{-3}
20.0	9.40×10^{-3}
40.0	1.92×10^{-2}
100.	4.79×10^{-2}

8-31 When the initial concentrations of two reactants (A and E) in a bimolecular reaction are not equal, the integrated rate law takes the form:

$$kt = \frac{1}{[E]_0 - [A]_0}\left[\ln\left(\frac{[E]_0}{[A]_0}\right) - \ln\left(\frac{[E]_t}{[A]_t}\right)\right],$$ where $[E]_0$ and $[A]_0$ are the initial

concentrations of the two reacting species, and $[E]_t$ and $[A]_t$ are their concentrations at time *t*. The reaction is carried out using ^1H NMR as a tool for monitoring concentrations. Based on the data below, where the internal standard peak refers to a resonance from a compound that is not consumed in the reaction, calculate the rate constant for the reaction.

Initial concentrations: $[A]_0 = 0.558\ M$; $[E]_0 = 0.730\ M$.

Time	Area of peak A (H_1)	Area of Peak E (H_3)	Area of Internal Standard Peak
21	2.69	10.34	43.03
42	1.93	7.61	37.80
60	1.73	7.04	39.22
79	1.48	6.25	39.53
97	1.31	5.56	39.55
100	1.28	5.56	39.61
119	1.15	5.08	39.42

8-32 The quantity h_0 is related to the Hammett acidity parameter H_0 as $h_0 = 10^{-H_0}$, which makes it a good approximation to the activity (or concentration) of H^+ (aq) in aqueous solutions of strong acids up to 5 *M*. The observed first-order rate constant for the acid-catalyzed hydrolysis of the strained compound, β-propiolactone, changes with the concentration of the acid. Provide an interpretation of the effects of acid. Include in your answer a mechanism for the reaction, a reaction coordinate diagram, and a rate law for the reaction. [Data from *J. Am. Chem. Soc.* **1950**, *72*, 3267.]

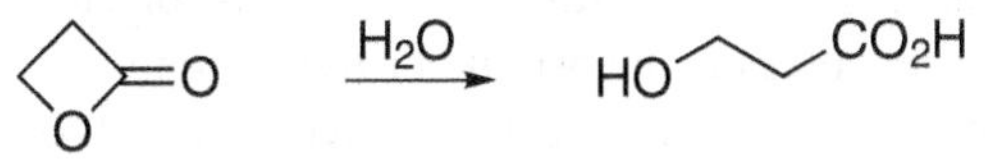

H_0	k_{obs}	H_0	k_{obs}
−0.58	3.65 × 10⁻³	−2.28	69.9 × 10⁻³
−0.94	5.38 × 10⁻³	−1.72	17.6 × 10⁻³
−1.01	5.21 × 10⁻³	−0.73	5.39 × 10⁻³
−1.24	8.89 × 10⁻³	−1.28	9.52 × 10⁻³
−1.55	14.41 × 10⁻³	−1.57	13.1 × 10⁻³
−1.84	32.2 × 10⁻³	−1.64	18.9 × 10⁻³
−1.85	32.6 × 10⁻³	−2.22	57.3 × 10⁻³

8-33 When optically active *threo*-3-*p*-anisylbutyl tosylate reacts with water in 50% aqueous ethanol, the product of the reaction is (±)-*threo*-3-*p*-anisyl-2-butanol. Explain these observations.

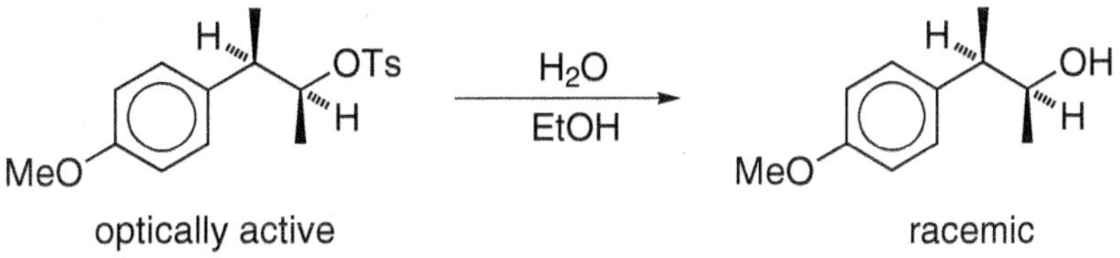

8-34 When 2,4,6-trinitroanisole is heated with sodium ethoxide, the product obtained is a mixture of 1-ethoxy-2,4,6-trinitrobenzene and starting material. When the reaction mixture is not heated, a red solid that gives the following microanalysis data is obtained: C, 34.74; H, 3.24; N, 13.50; Na, 7.39%. What is the solid and how is it formed?

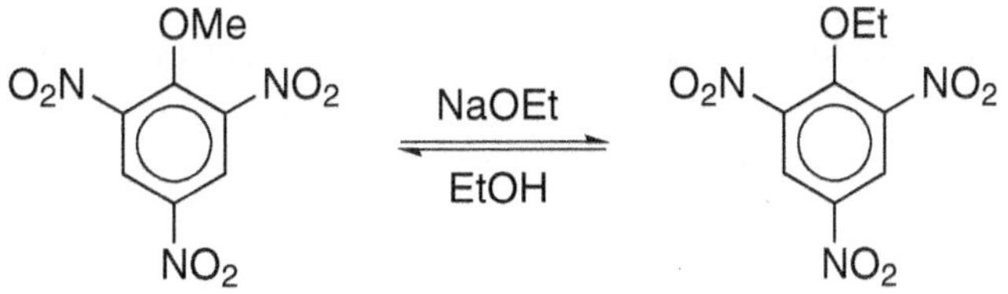

8-35 When cyclohexenone is converted to its cyclic ethylene ketal under standard conditions, the product obtained is the rearranged ketal shown. Suggest a mechanism for this transformation and provide a reasonable rationalization for the observed results.

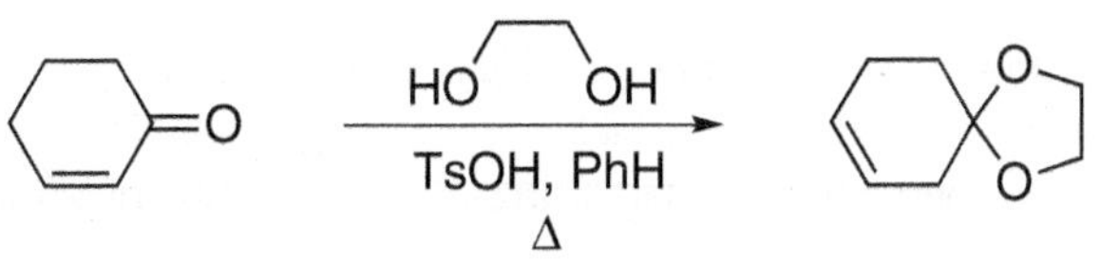

8-36 In a 1957 review (*Chem. Rev.* **1957**, *57*, 935], Long and Paul refer to a conclusion reached by Zucker and Hammett that specifies that if a reaction has a rate law with a linear dependence on H_0, the activated complex involves only the substrate and a proton, whereas if the reaction has a rate law that depends on $[H^+]$, the activated complex is composed of the substrate, a proton, and a molecule of water. What are the major assumptions leading to this hypothesis and under what conditions might it not apply?

8-37 The $\mathbf{Y}_{OTs}$ values for solvents do not give a good correlation for reaction rates of alkyl chlorides. What is the origin of this apparent limitation of this solvent parameter? (Hint: Consider the solvation requirements of the products of the solvolysis reaction.)

8-38 The regiochemistry of protonation of the guanidine derivative shown is controlled by the substituent X. When a complete series of protonations is carried out using both electron-releasing and electron-withdrawing substituents, the Hammett plot of the equilibrium constants for the protonation is linear with a relatively large negative ρ value. What does this tell about the regiochemistry of the protonation? What would be the results of the Hammett plot if protonation occurred at the alternative site?

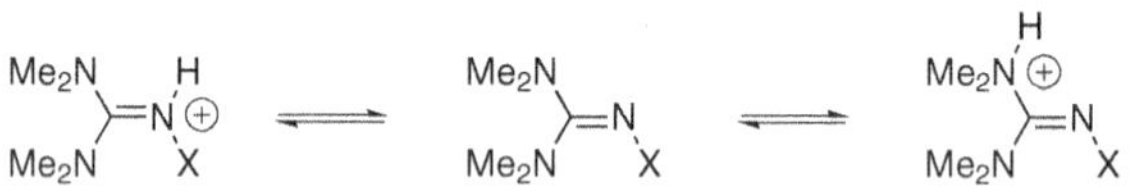

8-39 The hydrolysis of 2-deoxy-adenosine monophosphate, *d*AMP, is catalyzed by acid. Kinetic isotope effects for the reaction have been measured for labeling at a number of locations on the *d*AMP molecule, as shown at right.

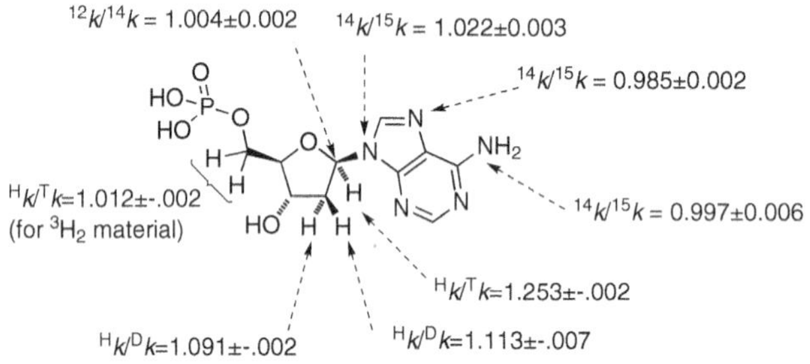

Based on these isotope effects, what qualitative deductions can be made about the transition state for this reaction?

8-40 The rate law for the substitution of 2-benzyloxytetrahydropyran (BzTHP, shown) by methanol in toluene containing a catalytic amount of *p*-toluenesulfonic acid (TsOH) is of the form:

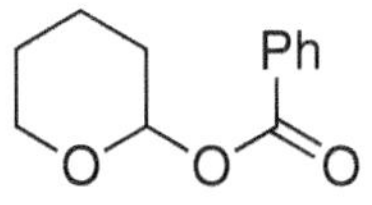

$$\text{Rate} = k[\text{BzTHP}][\text{TsOH}]$$

(a) Is this reaction concerted or stepwise? Why or why not?
(b) Draw a reasonable reaction coordinate diagram for this reaction.

8-41 The reaction between the phthalimide *N*-oxyl radical and a variety of substituted toluenes gives the data in the table below (*J. Org. Chem.* **2003**, *68*, 9364). Construct Hammett plots from these data using all three substituent parameters: σ, σ^+, and σ^-.

Which of the plots gives the best fit to a straight line? What does this tell you about the nature of the transition state for the hydrogen atom transfer?

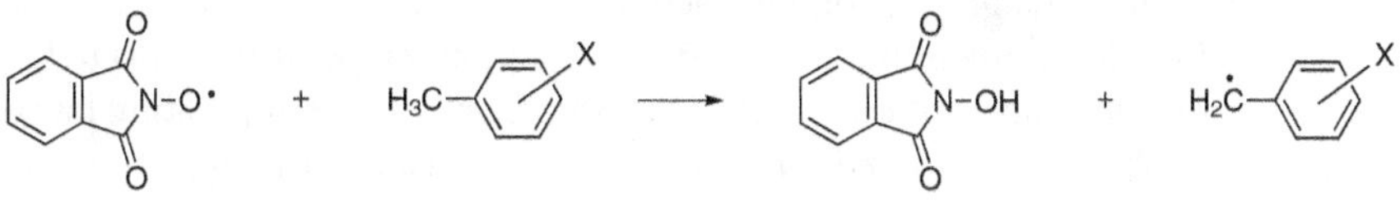

Substituent	k_2 (L mol^{-1} s^{-1})	Substituent	k_2 (L mol^{-1} s^{-1})
p-OMe	10.3	*p*-Br	0.595
p-Me	5.95	*p*-CO_2H	0.272
m-Me	3.06	*m*-CO_2H	0.200
H	0.620	*p*-NO_2	0.119

A temperature study of the same reaction with toluene in acetic acid gives the following results:

Temperature	17°C	25°C	40°C	55°C	70°C
k_2 (L mol^{-1} s^{-1})	0.81	1.2	2.4	4.4	7.8

What is the activation energy of this reaction? What are the values of $\Delta H^{\ddagger}$ and $\Delta S^{\ddagger}$ for this reaction?

8-42 The heterocyclic ring of *N*-aryl-4-chloro-1,8-naphthalimides had been assumed to be inert to nucleophilic attack until 2013, when the these substances were reported to react with butylamine to give *N*-butyl-4-chloro-1,8-naphthalimide [*Org. Biomol. Chem.* **2013**, *11*, 4390]. The kinetic data for the reactions of a series of substituted *N*-aryl-4-chloro-1,8-naphthalimides with butylamine are given in the left-hand table below. What do these data allow one to deduce about the transition state structure of the reaction?

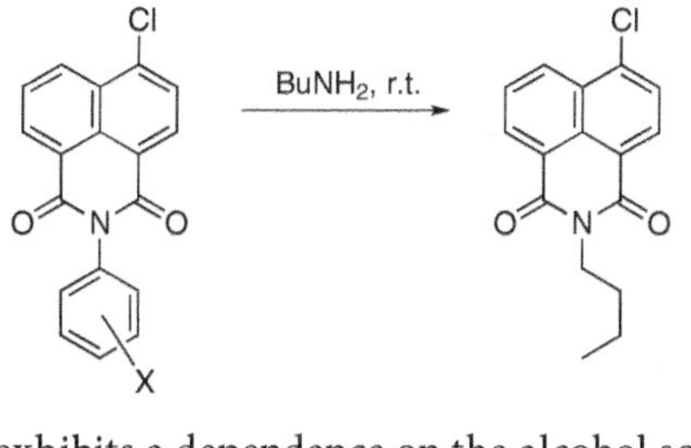

The reaction also exhibits a dependence on the alcohol solvent, as shown in the right-hand table. Construct a Brønsted plot based on the data given. Is this reaction subject to any type of acid catalysis?

Substituent	k_{obs} (min^{-1})	Alcohol	k_{obs} (min^{-1})
m-Cl	$2.02 \pm 0.04 \times 10^{-3}$	MeOH	$5.02 \pm 0.06 \times 10^{-3}$
p-Br	$7.9 \pm 1.2 \times 10^{-4}$	*n*-PrOH	$3.06 \pm 0.04 \times 10^{-3}$
m-OMe	$3.21 \pm 0.6 \times 10^{-4}$	*i*-PrOH	$2.07 \pm 0.13 \times 10^{-3}$
H	$1.64 \pm 0.07 \times 10^{-4}$	*t*-BuOH	$1.14 \pm 0.10 \times 10^{-3}$
		$MeOCH_2CH_2OH$	10.4×10^{-3}

8-43 The S_N1 and E1 reactions share a common rate law and reaction mechanism through the slow step of the reaction. These two mechanisms frequently operate together, so the product of the reaction is a mixture of alkene and the substitution product. It is known that increasing the reaction temperature tends to favor elimination over substitution and that keeping the reaction temperature low tends to favor substitution over elimination. Provide a reasonable explanation for these observations using both kinetic and thermodynamic arguments as well as appropriate reaction coordinate diagrams.

CHAPTER NINE

Reactive Intermediates I

Carbocations

9.1 **Carbocations**[1]

The ionization of organic compounds was established in 1902, when Walden showed that triphenylmethyl chloride would dissolve in sulfur dioxide to give an electrically conducting solution (Example **9.1**).[2] Carbocations are electrophiles characterized by an sp^2-hybridized, positively charged central carbon atom with three σ bonds (Example **9.2**). They are typically formed by one of two major processes: heterolysis of a bond to a leaving group and protonation of a π bond. Although electrochemical oxidation of a neutral precursor such as a free radical may be used to generate a carbocation,[3] it is less common.

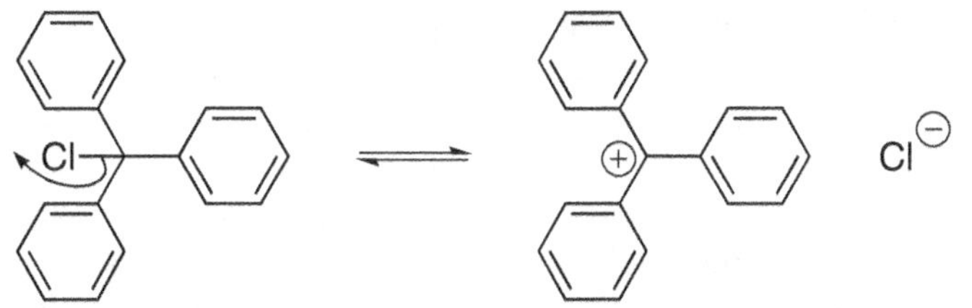

(9.1)

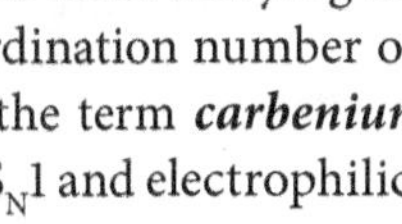

(9.2)

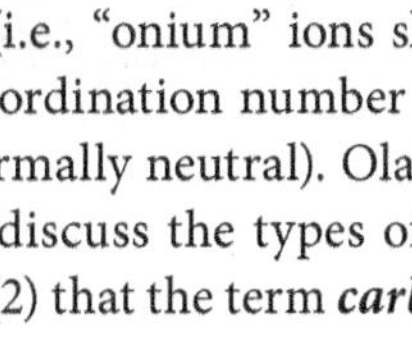

(9.3)

Carbocations were called ***carbonium ions*** until the 1970s, but in 1971 American chemist George Olah, winner of the 1994 Nobel Prize in Chemistry, suggested that this term should be reserved for cations such as CH_5^+ (Example **9.3**). Here, there are more than four groups simultaneously bonded to the carbon atom (i.e., "onium" ions should be cations where the atom carrying the formal charge has a coordination number one higher than the coordination number of the same atom when formally neutral). Olah also suggested (1) that the term ***carbenium ion*** should be used to discuss the types of cations formed during S_N1 and electrophilic addition reactions[4] and (2) that the term ***carbocation*** be used

1. Monographs: (a) Olah, G.A.; Schleyer, P.v.R. *Carbonium Ions*, 5 vols. (Wiley: New York, 1968–1976). (b) Bethell, D.; Golg, V. *Carbonium Ions* (Academic Press: London, 1963).

Reviews: (c) Deno, N.C. *Prog. Phys. Org. Chem.* **1964**, *2*, 129. (d) Stang, P.J. *Prog. Phys. Org. Chem.* **1973**, *10*, 205. (e) Schleyer, P.v.R.; Allen, L.C. *Top. Stereochem.* **1973**, *7*, 253. (f) Olah, G.A. *Top. Curr. Chem.* **1979**, *80*, 19. (g) Kebarle, P. *Ann. Rev. Phys. Chem.* **1977**, *28*, 445.

2. Walden, P. *Ber. dtsch. chem. Ges.* **1902**, 35, 2018.

3. (a) Corey, E.J.; Bauld, N.l.; LaLonde, R.T.; Casanova, J., Jr.; Kaiser, E.T. *J. Am. Chem. Soc.* **1960**, *82*, 2645. These authors suggest that this occurs when the Kolbe electrolysis fails to give the expected radical dimer: (b) Walker, J.; Carmack, M. *J. Chem. Soc.* **1900**, *77*, 374. (c) Linstead, R.P.; Shepard, B.R.; Weedon, B.C.L. *J. Chem. Soc.* **1951**, 2854. (d) Finkelstein, M.; Peterson, R.C. *J. Org. Chem.* **1960**, *25*, 136.

4. Olah, G.A. *CHEMTECH* **1971**, *1*, 566. In 1902, Moses Gomberg suggested that "carbonium ion" may be inappropriate: Gomberg, M. *Behrett.* **1902**, *35*, 2397. For a history of the name, "carbonium ion" see Traynham, J.G. *J. Chem. Educ.* **1986**, *63*, 930.

Table 9.1 Calculated Enthalpies (kcal mol^{-1}) of Hydride Transfer from Fluorinated Hydrocarbons to CH_3^+ in the Gas Phase and Lowest Energy Unoccupied Molecular Orbital (LUMO) Energies of Fluorinated Cations

R_1	R_2	R_3	ΔE (B3LYP/6-31+G*)	E_{LUMO} (eV)
CH_3	CH_3	CH_3	−239.3	−8.8
CH_3	CH_3	CH_2F	−229.2	−9.2
CH_3	CH_3	CHF_2	−212.9	−9.8
CH_3	CH_3	CF_3	−204.2	−10.1
CH_3	CH_2F	CF_3	−192.0	−10.6
CH_3	CHF_2	CF_3	−176.5	−11.1
CH_3	CF_3	CF_3	−168.7	−11.3
CH_2F	CF_3	CF_3	−156.0	−9.2
CF_3	CF_3	CF_3	−126.2	−12.9
$CH_3CH_2CH_2$	CH_3	CH_3	−243.3	−8.5
CH_3CH_2CHF	CH_3	CH_3	−235.1	−8.8
CH_3CHFCH_2	CH_3	CH_3	−238.7*	−8.5
$FCH_2CH_2CH_2$	CH_3	CH_3	−237.1†	−8.6
$FCH_2CH_2CH_2$	CH_3	CH_3	−247.1‡	−8.0

*The conformation of this ion places the fluorine atom physically closer to the cation carbon than the other fluorinated 1,1-dimethylbutyl cations.
†Extended-chain conformation.
‡Cyclic conformation.

as a generic term to indicate a carbon-centered cation.[5] Although the International Union of Pure and Applied Chemistry (IUPAC) has accepted these definitions,[6] the organic chemistry community has been slow to embrace the term "carbenium ion," choosing instead to adopt Olah's second suggestion and to call intermediates with a positively charged carbon atom "carbocations."

Carbocations are electrophiles. Their reactivity is due to the presence of a carbon atom lacking a complete octet and is modulated by the groups bonded to the electron-deficient carbon atom. As would be predicted, electron-releasing groups stabilize the cation and reduce its reactivity, and electron-withdrawing groups destabilize the cation and increase its reactivity. The modulation of carbocation reactivity can be discussed in terms of two competing effects: the ***inductive effect***[7] and the ***resonance effect*** or ***mesomeric effect***.[8] The inductive effect transmits the effects of electronegativity through the σ bonds of the cation, and it is attenuated as the number of intervening σ bonds increases.

The operation of the inductive effect (actually comprised of two components: the inductive effect itself, which passes through bonds, and the ***field effect***,[9] which passes through space) is illustrated by the calculated enthalpies of the gas phase hydride transfer reactions from fluorinated isoalkane derivatives to methyl cation to give the fluorinated *tert*-alkyl carbocations summarized in Table 9.1. The energy of the lowest energy unoccupied molecular orbital (LUMO) of each cation mirrors the trend in reaction energy for its formation.[§] The reaction energies are also shown in Figure 9.1.

The first ten rows of Table 9.1 show the cumulative effects of increasing fluorine substitution on the energy of the *tert*-butyl cation; each fluorine atom makes the overall reaction less exothermic (i.e. reduces the stability of the carbocation). Density functional theory (DFT) calculations at the B3LYP/6-31+G* level show an linear decrease in the exothermicity of the reaction of 12.3 kcal mol^{-1} per fluorine atom as fluorine atoms are added to the cation. The last four rows of Table 9.1 show the attenuating effects of

5. Olah, G.A. *J. Am. Chem. Soc.* **1972**, *94*, 808.

6. Gold, V.; Loening, K.L.; McNaught, A.D.; Sehmi, P. *Compendium of Chemical Terminology: IUPAC Recommendations* (Blackwell Scientific Publications: Oxford, 1987).

7. (a) Lewis, G.N. *Valence and the Structure of Atoms and Molecules* (Chemical Catalog Company: New York, 1923), p. 139. (b) Ingold, C.K. *Structure and Mechanism in Organic Chemistry* (Cornell University Press: Ithaca, 1953), pp. 67–71. (c) Gould, E.S. *Mechanism and Structure in Organic Chemistry* (Holt, Rinehart and Winston: New York, 1950), pp. 200–209. (d) Friedl, Z.; Hapala, H.; Exner, O. *Coll. Czech. Chem. Comm.* **1979**, 2928.

8. (a) Ingold, C.K. *Structure and Mechanism in Organic Chemistry* (Cornell University Press: Ithaca, 1953), pp. 73–90. (b) Gould, E.S. *Mechanism and Structure in Organic Chemistry* (Holt, Rinehart and Winston: New York, 1950), pp. 212–220, esp. footnote 23 on p. 217. (c) Wheland, G.W. *Advanced Organic Chemistry*, 3rd, ed. (John Wiley & Sons: New York, 1960), p. 86. (d) Ingold, C.K.; Ingold, E.H. *J. Chem. Soc.* **1926**, 1310. (e) Robinson, R. *Outline of an Electrochemical Theory of the Course of Organic Reactions* (The Institute of Chemistry of Great Britain and Ireland: London, 1932). (f) Robinson, R. *I. Soc. Dyers Cluorists, Jubilee Issue* **1934**, 65. (g) Ingold, C.K. *J. Chem. Soc.* **1933**, 1120. (h) Ingold, C.K. *Chem. Rev.* **1934**, *15*, 225.

9. Roberts, J.D.; Moreland, W.T., Jr. *J. Am. Chem. Soc.* **1953**, *75*, 2167.

§. Calculations in this chapter were performed with Spartan 08 version 1.2.1 for Mac.

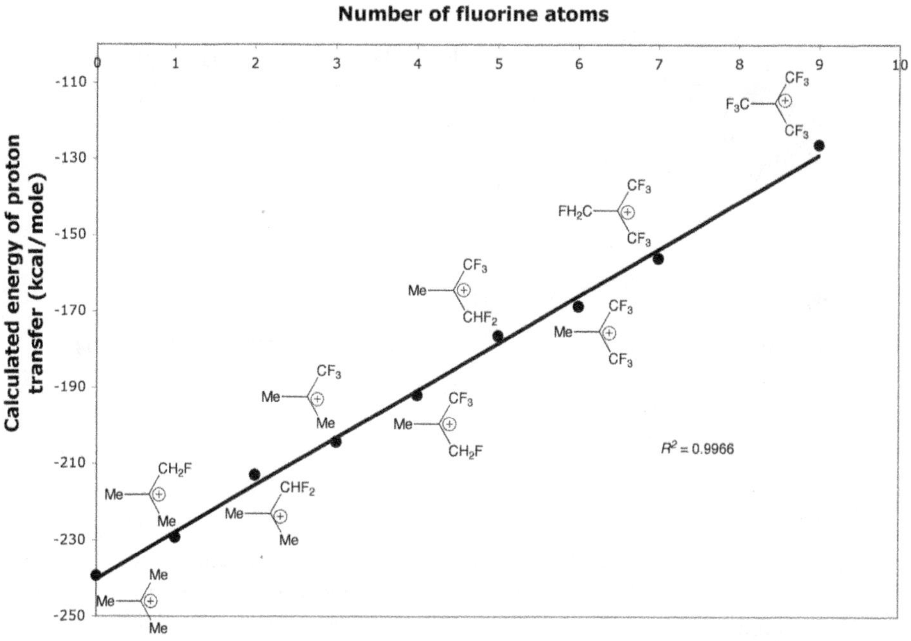

Figure 9.1 Calculated enthalpies of hydride transfer reactions between methyl cation and fluorinated isobutanes to give methane and fluorinated tert-butyl carbocations in the gas phase

intervening σ bonds, and we see that the enthalpy of the gas-phase reaction decreases as the fluorine is moved further from the position of the cation carbon (i.e., the carbocation becomes more stable, or less destabilized as the fluorine is moved further from the cation carbon). The data also reveal that placing the fluorine in a location physically close to the cation carbon lowers the energy of the cation. When the fluorine is in a location where it can form a cyclic heteronium ion, the reaction is actually more favorable than the deprotonation of the non-fluorinated compound.

Simple alkyl groups are electron-releasing by the inductive and field effects in the order:

$$CH_3— < CH_3CH_2— < (CH_3)_2CH— < (CH_3)_3C—$$

When the hybridization changes, the inductive effect changes direction, and unsaturated alkyl groups are generally weakly electron-withdrawing by the inductive effect (as expected from the change in electronegativity with hybridization), in the following order:

$$R_2C{=}CR— < RC{\equiv}C—$$

All major mechanisms for stabilizing carbocations delocalize the positive charge by π-type overlap of the empty 2p orbital of the charged carbon atom of the carbocation with an occupied orbital. This charge delocalization is usually described in terms of three common mechanisms: (1) ***conjugation***, where the occupied orbital is an adjacent π bond or lone pair; (2) ***hyperconjugation***, where the occupied orbital is an adjacent σ bond; or (3) ***bridging***, where the occupied orbital is a non-adjacent σ or π bond. Charge delocalization by conjugation or hyperconjugation generates a ***classical cation***, whereas delocalization by the third mechanism generates a ***non-classical cation***. In principle, each of these mechanisms of charge delocalization may operate, although one usually dominates.

Problem

9-1 Suggest reasons for the relative orders of energy of the fluorinated 1,1-dimethylbutyl cations shown in Table 9.1.

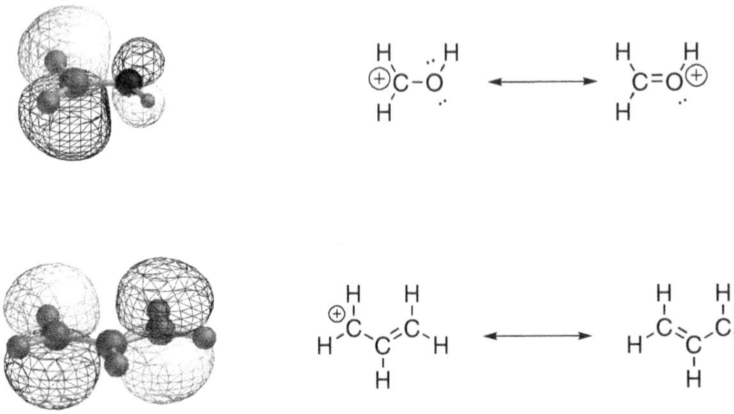

Figure 9.2 The hydroxymethyl cation (top) and the allyl cation (bottom). The orbitals are the lowest energy unoccupied molecular orbitals of the two cations.

Conjugation

Conjugation with a lone pair on an atom directly bonded to the cation center is the most effective process for stabilizing carbocations. The effectiveness of this conjugation is due to the fact that conjugation results in a π-bonded heteronium ion, rather than the simple substituted carbocation. This is illustrated in Figure 9.2 by the hydroxymethyl cation, H_2COH^+, which is shown computationally to be the conjugate acid of formaldehyde. In contrast to this, conjugation with a π bond allows the delocalization of the positive charge but always results in a structure that has one formally electron deficient atom. This is illustrated in Figure 9.2 for the allyl cation, $H_2CCHCH_2^+$.

One can use resonance arguments to predict this same relative order of stability—all non-hydrogen atoms in the major canonical form of the heteroatom-stabilized cation have complete outer-shell octets, and at least one non-hydrogen atom in the π bond–stabilized cation lacks a complete outer-shell octet.

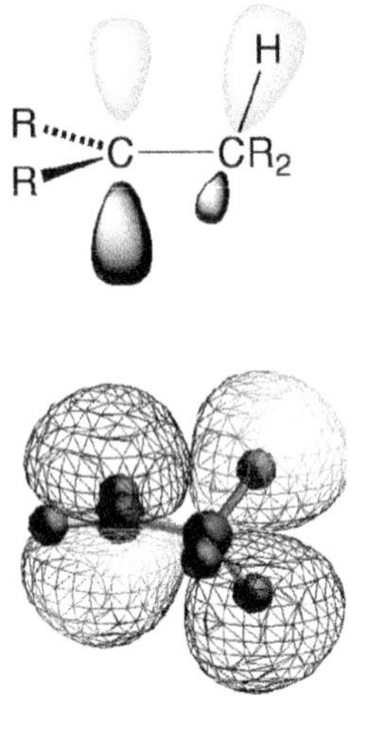

Figure 9.3 Orbital overlap for hyperconjugation in cations (top) and the lowest energy unoccupied molecular orbital of the ethyl cation (bottom)

Hyperconjugation

Where there is no lone pair or π bond available for stabilizing a carbocation by conjugation, stabilization can occur through hyperconjugation,[10] a π-type delocalization of the electrons from the σ orbitals of the bonds at the β carbon into the empty p orbital on the charged (α) carbon. This is shown in Figure 9.3, where the orbital overlap for hyperconjugation is shown. Below the orbital overlap diagram is the LUMO of the ethyl cation calculated at the AM1 level.[11] In this orbital, which the simple valence-bond approach predicts to be the empty 2p orbital on the cation carbon, we see what resembles the involvement of two of the C—H σ orbitals of the methyl group simultaneously in the delocalization of the charge by π-type overlap with the 2p orbital.

The view of hyperconjugation as π-type delocalization of σ electrons can be used not only as a mechanism for stabilizing electron-deficient species such as carbocations but also as a rationale for electron delocalization (from the σ orbital into an adjacent σ* orbital) to stabilize the staggered conformation of alkanes. The effectiveness of hyperconjugation has

10. (a) Wheland, G.W. *J. Chem. Phys.* **1934**, *2*, 474. (b) Mulliken, R.S.; Riecke, C.A.; Brown, W.G. *J. Am. Chem. Soc.* **1941**, *63*, 41. (c) Lofthus, A. *J. Am. Chem. Soc.* **1957**, *79*, 24. (d) Deasy, C.L. *Chem. Rev.* **1945**, *36*, 145. (e) Baker, J.W. *Hyperconjugation* (Oxford University Press: London, 1952). (f) Becker, F. *Angew. Chem.* **1953**, *65*, 97. (g) Wheland, G.W. *Resonance in Organic Chemistry* (John Wiley & Sons: New York, 1955), §3.10, 9.27. (h) Dewar, M.J.S. *Hyperconjugation* (Ronald Press:New York, 1962). (i) Shiner, V.J.; Campaigne, E., Eds. *Conference on Hyperconjugation* (Pergamon: New York, 1959); *Tetrahedron* **1958**, *5*, 105. (j) White, J.C.; Cave, R.J.; Davidson, E.R. *J. Am. Chem. Soc.* **1988**, *110*, 6308.

11. This very unsophisticated level of theory permits us to look at the primary carbocation structure. At higher levels of sophistication, the bridged-ion structure is obtained, as we will discuss in greater detail in Section 9.2.

been correlated with the reactivity of substituted aromatic hydrocarbons in electrophilic aromatic substitution. Based on the rates of electrophilic aromatic substitution, one can deduce that C—H σ bonds are much more effective at stabilizing a positive charge than C—C σ bonds, which is consistent with the observed order of reactivity of benzenes substituted by simple alkyl groups toward electrophiles.[12] This phenomenon is also known as the ***Baker-Nathan effect***,[13] after the authors[14] of the first paper to document that alkyl groups attached to an aromatic ring facilitated displacements of benzylic halides in an order opposite to that of the inductive effects of the alkyl groups.

$$CH_3\text{—} > CH_3CH_2\text{—} > (CH_3)_2CH\text{—} > (CH_3)_3C\text{—}$$

Comparison of Conjugation and Hyperconjugation

Let us now compare the effectiveness of the two mechanisms, conjugation and hyperconjugation, in stabilizing cations. The first fact to bear in mind is that the effectiveness of charge delocalization in carbocations is decided by the *occupied* molecular orbital involved in the charge delocalization; it is generally observed that a π orbital or a nonbonding electron pair is more effective than a σ orbital. This means that, as a general rule, conjugation is more effective at stabilizing a carbocation than hyperconjugation. When hyperconjugation is the only mechanism available for delocalizing the positive charge, it most effective when the occupied molecular orbital is a σ orbital involving an electropositive element (e.g., H or Si). Taking both of these factors into account, one may predict that, when the number of alkyl groups attached to the charged carbon is the same, conjugated cations are more stable than saturated cations and that saturated alkyl cations become more stable as the number of carbon-hydrogen σ bonds at the β carbons increases. The stability of simple hydrocarbon carbocations increases in the order:

$$CH_3^+ < RCH_2^+\ (1°) < R_2CH^+\ (2°) < R_3C^+\ (3°) \approx H_2C{=}CHCH_2{}^+\ (1°\ \text{allyl})$$

The order of cation stability given above can be seen nicely in the series of isomeric carbocations in Table 9.2, where the enthalpies calculated for the gas-phase deprotonation of *trans*-4-ethyl-1-vinylcyclohexane and 1,4-diethylcyclohexene by methyl cation using are gathered. These values were obtained using DFT methods at the B3LYP/6-31+G* level.

Based on the data in Table 9.2, we can see that saturated secondary cations are around 10 kcal mol^{-1} less stable than saturated tertiary cations in the gas phase and that the incorporation of a double bond at the allylic position lowers the cation energy by another 10 to 16 kcal mol^{-1} in this system. Although the size of the energy differences is reduced by solvation, the general order of stability remains the same in solution as in the gas phase. Computations of the structure and energy of the primary carbocation in Table 9.2 return a bridged structure instead of the simple primary carbocation, and this structure has an energy some 14 kcal mol^{-1} above that of the corresponding secondary carbocation. Other

12. (a) de la Mare, P.B.; Robertson, P.W. *J. Chem. Soc.* **1943**, 279. (b) Berliner, E.; Berliner, F. *J. Am. Chem. Soc.* **1949**, *71*, 1195.

13. (a) Baker, J.W.; Nathan, W.S. *J. Chem. Soc.* **1935**, 1844. (b) Baker, J.W. *J. Chem. Soc.* **1939**, 1150. (c) Hughes, E.D.; Ingold, C.K.; McNulty, B.J. *J. Chem. Soc.* **1940**, 909.

14. **John William Baker (1898–1967)** was educated at Imperial College (PhD, 1925, DSc, 1928), where he came under the influence of Ingold. He spent his entire independent career at Leeds, but, surprisingly, never rose to the rank of Professor. He published over 100 papers and monographs in physical organic chemistry.

Wilfred Samuel Nathan (1910–1961) was educated at University College, Bangor (PhD, 1935), and after a year at Leeds with Baker, he moved to University College, London, to work with Ingold. Within a year, he had joined B.P., and he remained with this company for the rest of his life.

For more biographical details on Baker and Nathan, and a detailed description of the discovery of the Baker-Nathan effect, see: Saltzman, M.D. *Bull. Hist. Chem.* **2012**, *37*, 82.

Table 9.2 Calculated ΔH°_{react} for Formation of Isomeric Carbocations $C_{10}H_{17}^{+}$ by Hydride Transfer to Methyl Cation in the Gas Phase

Cation	Type	ΔH°_{react} (B3LYP/6-31+G*)
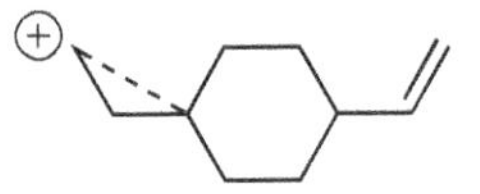	1°	−221.8*
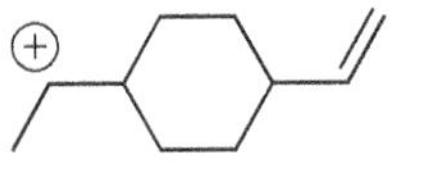	2°	−235.2
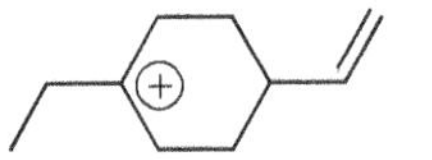	3°	−246.3
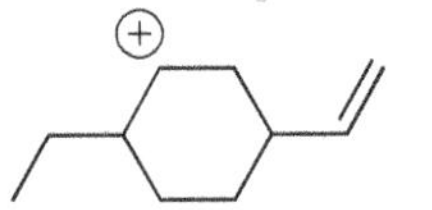	2°	−238.6
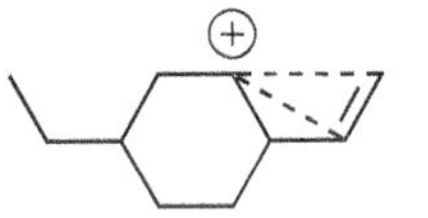	Homoallylic	−240.7*
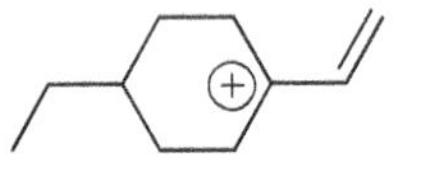	Allylic (1° ↔ 3°)	−255.1
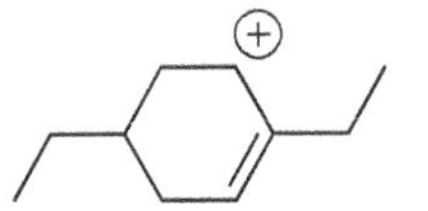	Allylic (2° ↔ 2°)	−254.3
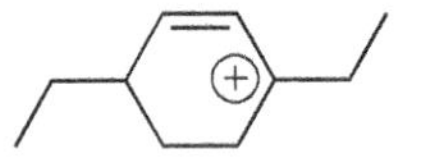	Allylic (2° ↔ 3°)	−264.4
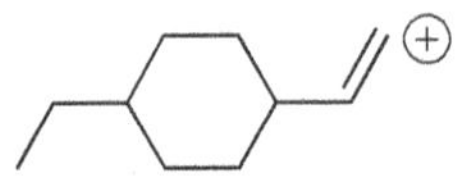	Vinyl (1°)	−203.6
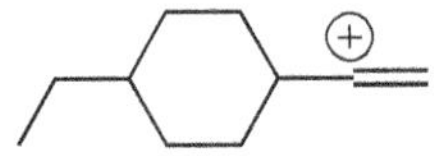	Vinyl (2°)	−224.7

*This energy is for the bridged, non-classical (see §9./4) cation shown.

calculations at this level do not return the primary carbocation or this bridged ion but return the secondary cation, which would indicate that the barrier to rearrangement in this system may be especially low. This is consistent with two facts. One, to date, no bona fide observation of a primary carbocation has been made. Two, the calculated minimum energy of even the ethyl cation (which must be—formally, at least—a primary carbocation) occurs for a structure in which the two carbon atoms are bridged by a hydrogen.[15] As expected, the change in hybridization of the carbocation carbon from sp^2 to sp raises the energy of the vinyl carbocations significantly; vinyl carbocations are more difficult to make, and much more reactive than their saturated counterparts.

Similar calculations for the reaction between methyl cation and a series of 2-substituted propanes, summarized in Table 9.3 and shown graphically in Figure 9.4, give an interesting picture of the relative efficiency of stabilization of a carbocation by an adjacent double bond or heteroatom. What emerges confirms the general statement above. Hyperconjugation is less efficient as a method for stabilizing a cation than conjugation with an alkene or an aromatic ring (in this case, the difference is 6.5–18 kcal mol^{-1}) but more stable than conjugation with an alkyne triple bond (the effective electronegativity of the sp-hybridized carbon is quite high, so this is an electron-withdrawing substituent).

Stabilization by a second-row element bearing a lone pair is considerably more efficient, with the least electronegative element (N) being most efficient at stabilizing the cationic charge. The halogen series, which combines electron withdrawal by the inductive effect with electron release by conjugation is also informative: fluorine, despite being the most electronegative, and therefore the most strongly electron-withdrawing, halogen, is actually not the least efficient at stabilizing the adjacent positive charge. Recall that the efficiency of π overlap decreases as the p orbitals come from different electron shells, and here it would appear that the effectiveness of delocalization of the cation changes with the efficiency of the π overlap: 2p-2p (C−F) > 2p-3p (C−Cl) > 2p−4p (C−Br).

Aromatic and Homoaromatic Carbocations

In Chapter 7, we discussed aromatic carbocations at some length. In these ions, the stabilization of the cation by conjugation is supplemented by the additional stabilization conferred by the aromaticity of the cation, and this results in an equalization (or close to it) of the bond lengths in the carbocation. This is revealed by the X-ray crystal structures of the

15. (a) Hoffmann, R.; Radom, L.; Pople, J.A.; Schleyer, P.v.R.; Hehre, W.J.; Salem, L. *J. Am. Chem. Soc.* **1972**, *94*, 6221. (b) Raghavachari, K.; Whiteside, R.A.; Pople, J.A.; Schleyer, P.v.R. *J. Am. Chem. Soc.* **1981**, *103*, 5649. (c) Ausloos, P.; Rebbert, R.E.; Sieck, L.W.; Tiernan, T.O. *J. Am. Chem. Soc.* **1972**, *94*, 8939. (d) Vorachek, J.H.; Meisels, G.G.; Geanangel, R.A.; Emmel, R.H. *J. Am. Chem. Soc.* **1973**, *95*, 4078. (e) Klopper, W.; Kutzelnigg, W. *J. Phys. Chem.* **1990**, *94*, 5625.

Table 9.3 Calculated Enthalpies of Hydride Transfer to Methyl Cation in the Gas Phase to Form Substituted Isopropyl Cations

Product Cation	ΔH (B3LYP/6-31+G*)	Product Cation	ΔH (B3LYP/6-31+G*)
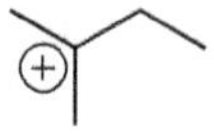	−242.4	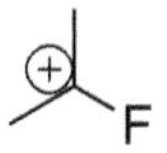	−227.0
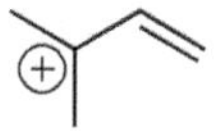	−249.9	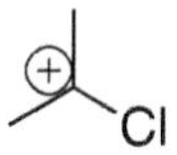	−226.2
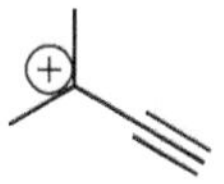	−238.6	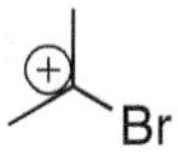	−226.7
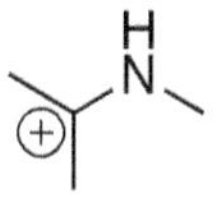	−286.6	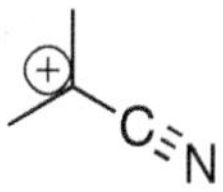	−209.6
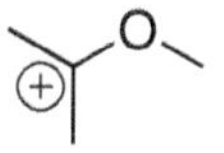	−262.8	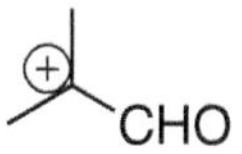	−214.8
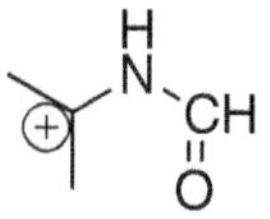	−257.3	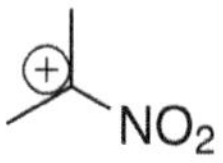	−199.7
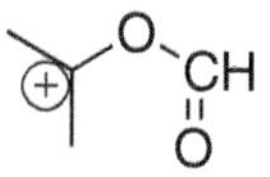	−236.2	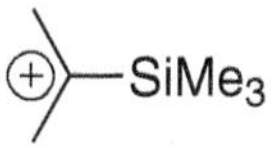	−241.9
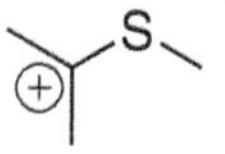	−258.7	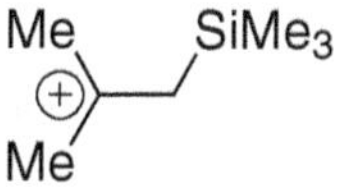	−267.8

tropylium (Example **9.4**) and triphenylcyclopropenium (Example **9.5**) cations. The formation and manipulation of aromatic cations is rendered especially easy by their high stability. Thus, the tropylium ion is sufficiently unreactive that the ultraviolet (UV) spectrum of its tetrafluoroborate salt can be measured in aqueous solution.

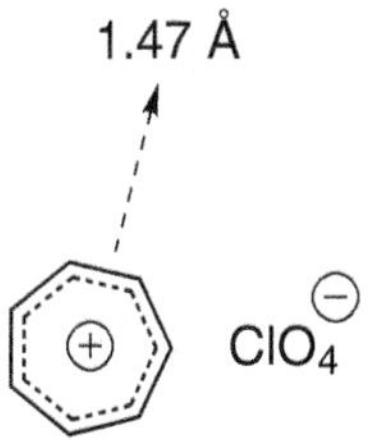

(9.4)[16]

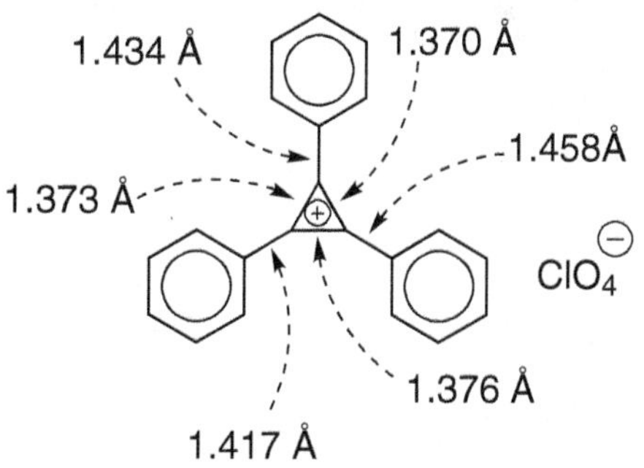

(9.5)[17]

16. Kitaigorodskii, A.I.; Strichkov, Y.T.; Khotsyanova, T.L.; Vol'pin, M.E.; Kursanov, D.N. *Izv, Akad. Nauk SSSR* **1960**, 39.
17. Sundaralingam, M.; Jensen, L.H. *J. Am. Chem. Soc.* **1966**, *88*, 198.

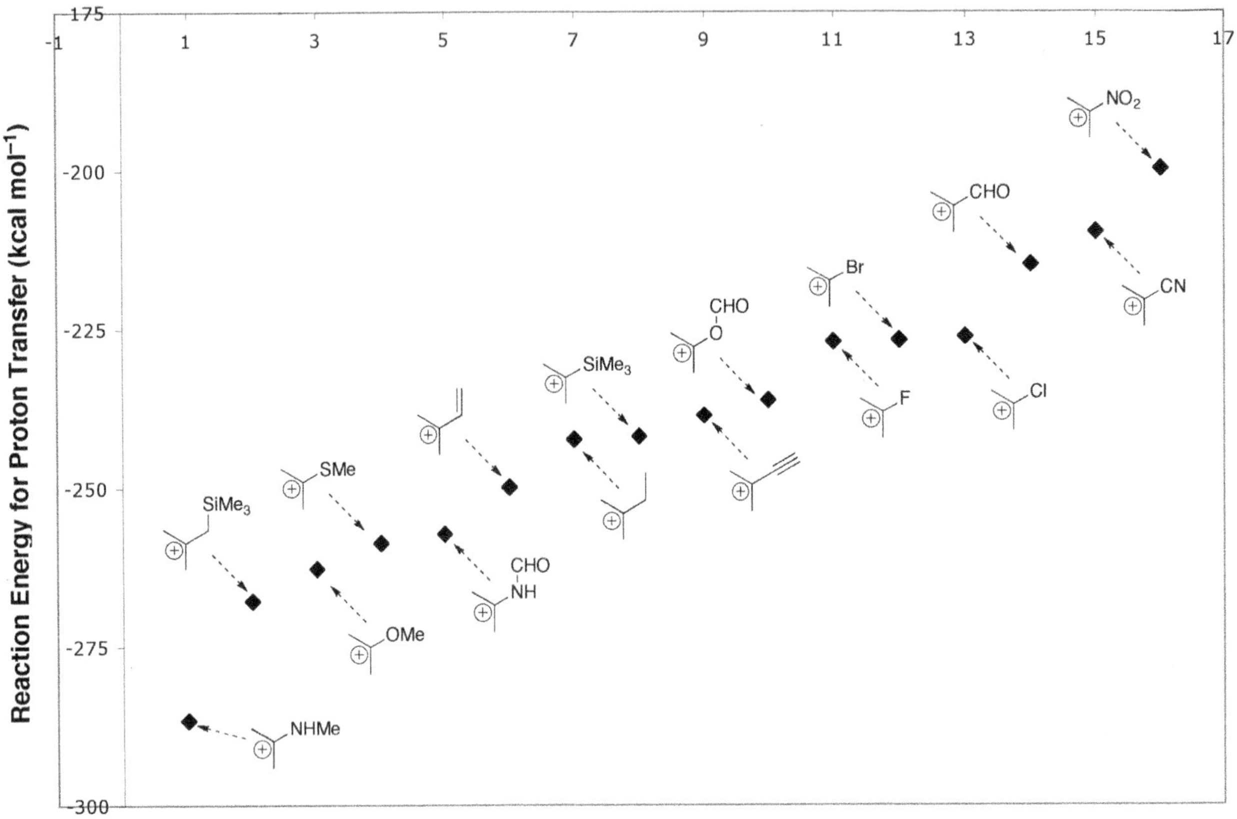

Figure 9.4 Calculated enthalpies of the reaction for the formation of substituted isopropyl carbocations by hydride transfer reactions to methyl cation

Structure of Carbocations

The structures of a number of carbocations have been established by X-ray crystallography, and the structures all clearly show that the carbon carrying the formal charge is planar—sp^2 hybridized—in alkyl cations. Some typical carbocations are shown as Examples **9.6** to **9.8**. The structures of carbocations are also in accord with the discussion we have had about the mechanisms for their stabilization. The structures of the two acylium ions (Examples **9.7** and **9.8**), in particular, show clearly that the formally charged carbon is sp-hybridized and that the carbon-oxygen bond is a triple bond rather than a double bond, in agreement with the simple resonance picture of these cations.

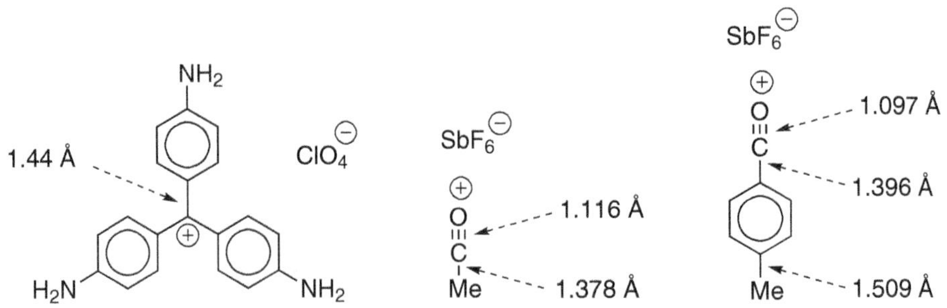

(9.6)[18] **(9.7)**[19] **(9.8)**[20]

18. Gomes de Mesquita, A.H.; McGillavry, C.H.; Eriks, K. *Acta Crystallogr.* **1965**, *18*, 437.
19. Boer, F.P. *J. Am. Chem. Soc.* **1966**, *88*, 1572.
20. Chevrier, B.; Le Carpentier, J.M.; Weiss, R. *J. Am. Chem. Soc.* **1972**, *94*, 5718.

Spectroscopy of Carbocations

Carbocations, being among the oldest known reactive intermediates, have been the subject of study by almost every method used by organic chemists. The fact that the earliest carbocations prepared were colored (the triphenylmethyl cation is yellow) made them candidates for study by UV-visible spectrophotometry. In general, an extended π bonding system is required for a molecule to absorb in the easily accessible regions of the UV-visible spectrum (λ_{max}> 200 nm), and carbocations (even the *tert*-butyl cation) show sufficient electron delocalization to meet this criterion. The absorption maxima and intensity of absorption of a number of common cations are given in Table 9.4. Carbocations formed by protonation of alkenes have been used as indicators for determining H_0 values for acids.[21]

Carbon-13 nuclear magnetic resonance (NMR) spectroscopy is easily the most powerful spectroscopic technique for the study of carbocations. Stable solutions of many types of carbocations can be prepared in superacid media (e.g., FSO_3H containing the extremely powerful Lewis acid SbF_5—an acid strong enough to protonate an alkane[22]). The ^{13}C NMR spectra of carbocations not stabilized by conjugation with either a π bond or a lone pair show a resonance for the positively charged carbon near 330 ppm, consistent with the high degree of electron deficiency at this position. The α carbons of tertiary cations appear at chemical shifts close to those of ketones with similar structures (typically 20 to 35 ppm); the α carbons of secondary carbocations appear some 20 to 40 ppm downfield from those of tertiary carbocations (close to 60 ppm).[23]

Table 9.4 Ultraviolet-Visible Spectra of Selected Carbocations

Cation	Counterion	Solvent	λ_{max} (nm)	$\log_{10} \varepsilon$	Ref.
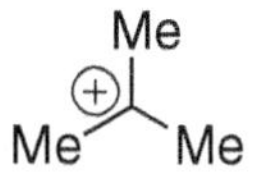	SbF_6^- (FSO_3^-)	FSO_3H/SbF_5	< 210	—	24
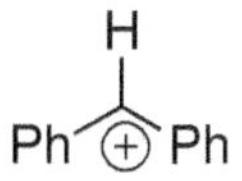	SbF_6^- (FSO_3^-)	FSO_3H/SbF_5	292 440	3.46 4.58	20
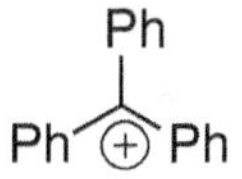	SbF_6^- (FSO_3^-)	FSO_3H/SbF_5	403 429	4.59 4.59	20
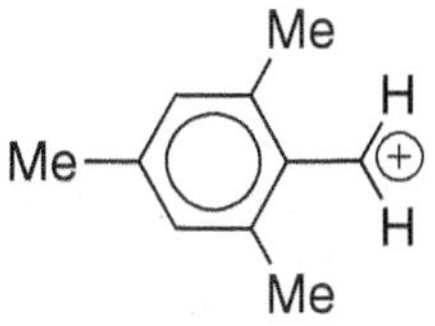	SbF_6^- (FSO_3^-)	FSO_3H/SbF_5	318 370 480	4.04 3.64 3.08	
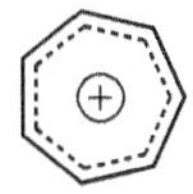	Br^-	H_2O/HBr	247 275	3.06 3.64	25
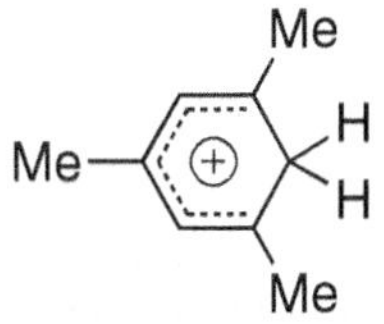	BF_4^-	HF/BF_3	256 355	3.94 4.04	

21. Deno, N.C. *Surv. Progr. Chem.* **1964,** *2,* 155.
22. Olah, G.A.; Schlosberg, R.H. *J. Am. Chem. Soc.* **1968,** *90,* 2726.
23. (a) Olah, G.A.; White, A.M. *J. Am. Chem. Soc.* **1969,** *91,* 5801. (b) Olah, G.A. *Angew. Chem. Int. Ed. Engl.* **1973,** *12,* 173.
24. Olah, G.A.; Pittman, C.U., Jr.; Waack, R.; Doran, M. *J. Am. Chem. Soc.* **1966,** 88, 1488.
25. Doering, W.v.E.; Knox, L.H. *J. Am. Chem. Soc.* **1954,** 76, 3203.

The spectra of non-classical carbocations (whose structures we will discuss in much more detail in Section 9.5) show the resonance of the positively charged carbon close to 100 ppm, some 200 ppm upfield from the carbon resonance of simple carbocation. For example, the cyclopentyl carbocation has a single resonance at 97.1 ppm in SbF_5-$ClSO_2F$, and the positively-charged carbon in the norbornyl carbocation resonates at 90.7 ppm[26] in SbF_5-SO_2. In keeping with the delocalization of the positive charge and the reduction in the overall electron-deficient character of the cation carbon, substituted benzyl cations resonate near 230 ppm, with strongly electron-releasing groups in a position to conjugate with the charge moving the resonance up to 70 ppm upfield from the simple benzyl cation position.[27]

Worked Problem

9-1 The structure calculated for the ion $[Me_2C{-}CHO]^+$ is not planar but has the carbonyl group twisted out of the plane of the carbocation carbon. Suggest a reasonable rationalization of this result.

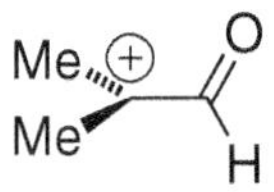

§Answer below.

Problems

9-2 Provide a reasonable explanation for why the calculated ΔH_{react} for the formation of the cation $[Me_2C{-}CHO]^+$ by hydride transfer to CH_3^+ is more exothermic than that for the formation of the cation $[Me_2C{-}CN]^+$.

9-3 Provide a reasonable explanation for the order of energies for the ions Me_2CF^+, Me_2CCl^+ and Me_2CBr^+ in Table 9.3 and Figure 9.4.

26. Olah, G.A.; White, A.M. *J. Am. Chem. Soc.* **1969**, *91*, 3954.
27. (a) Olah, G.A.; Mo, Y.K.; Halpern, Y. *J. Am. Chem. Soc.* **1972**, *94*, 3551. (b) Olah, G.A.; Halpern, Y.; Mo, Y.K.; Liang, G. *J. Am. Chem. Soc.* **1972**, *94*, 3554. (c) Olah, G.A.; Schlosberg, R.H.; Porter, R.D.; Mo, Y.K.; Kelly, D.P.; Mateescu, G.D. *J. Am. Chem. Soc.* **1972**, *94*, 2034.

§ Answer to Worked Problem:
In order to answer this question, we need to examine both possible conformers: the planar conformer and the non-planar, and we need to determine what factors in these conformers stabilize or destabilize these conformations. In the planar conformer, the empty 2p orbital on the positively charged carbon atom is in perfect alignment with the π bonding system, which would result in the generation of a delocalized π orbital system. If we look at this π orbital system in terms of resonance, we see that this places a positive charge on an electron-deficient oxygen in the minor contributor. Such an interaction is inherently destabilizing.

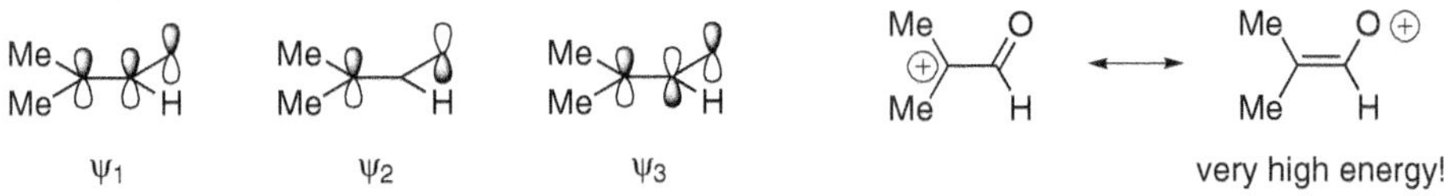

In the orthogonal conformer, the π-type interaction is no longer a factor. In this conformer, instead, the possibility exists that the lone pair on oxygen may be able to interact (at least to some extent) with the empty 2p orbital and so lead to a small level of stabilization similar to that observed in three-membered onium ions. This effect, even small, is stabilizing.

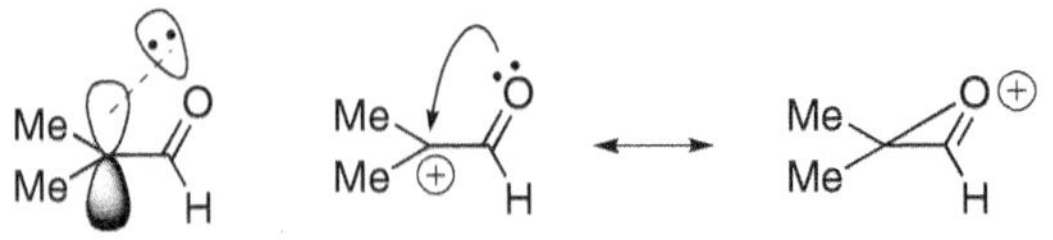

9.2 Rearrangements

Up to now, we have discussed structure of carbocations without considering how their stability might affect the outcome of a reaction in which they may be involved, except for the rate of the reaction. However, the relative stability of the carbocation intermediate also affects such reactions in another very important way: *reactions involving carbocations often give products with rearranged carbon skeletons*. This occurs because the rearrangement of a less stable carbocation to a more stable carbocation is extremely favorable. The S_N1 reaction of neopentyl bromide in aqueous ethanol (Example **9.9**)[28] as well as the E1 dehydration of pinacolyl alcohol with anhydrous oxalic acid (Example **9.10**)[29] both offer examples of reactions that give products with rearranged carbon skeletons.

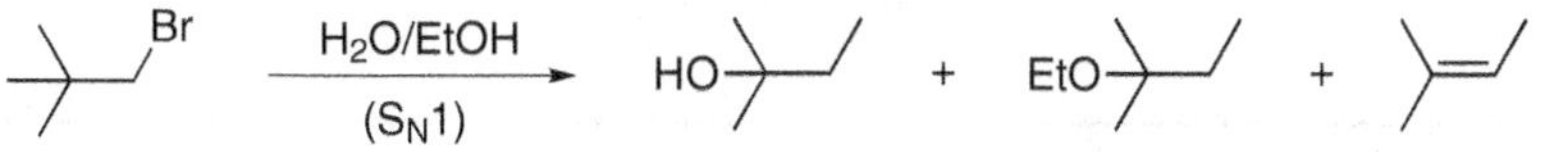

(9.9)

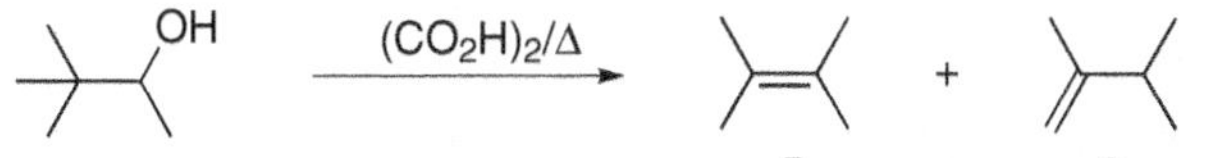

5 : 2

(9.10)

The data in Table 9.2 suggest that the enthalpy difference between "primary" and secondary carbocations is more than 14 kcal mol^{-1}, the enthalpy difference between secondary and tertiary carbocations is typically another 10 kcal mol^{-1}, and the enthalpy difference between saturated and allylic carbocations is a typically a further 10 kcal mol^{-1}. These large differences in enthalpy mean that carbocation rearrangements are very exothermic.

Table 9.2 **(redux)** Calculated $\Delta H°_{react}$ for Formation of Isomeric Carbocations $C_{10}H_{17}^+$ by Hydride Transfer to Methyl Cation in the Gas Phase

Cation	Type	$\Delta H°_{react}$ (B3LYP/6-31+G*)
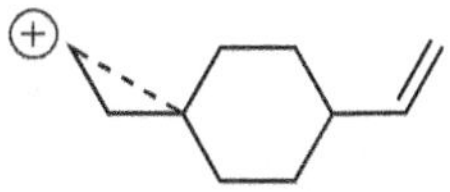	1°	−221.8*
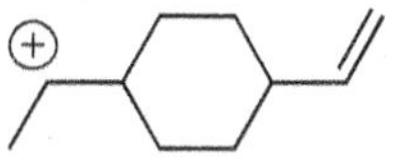	2°	−235.2
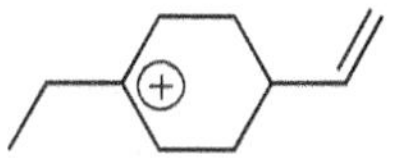	3°	−246.3
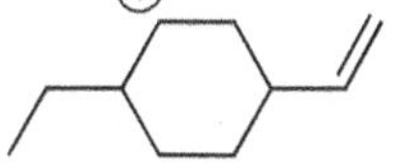	2°	−238.6
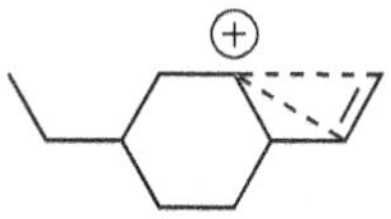	Homoallylic	−240.7*
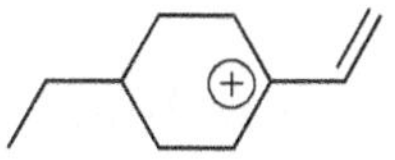	Allylic (1° ↔ 3°)	−255.1

(continues)

28. Dostrovsky, I.; Hughes, E.D. *J. Chem. Soc.* **1946**, 166.
29. (a) Whitmore, F.C.; Rothrock, H.S. *J. Am. Chem. Soc.* **1933**, *55*, 1100. (b) Whitmore, F.C.; Rothrock, H.S. *J. Am. Chem. Soc.* **1932**, *54*, 3431.

Table 9.2 (redux) Calculated ΔH°_{react} for Formation of Isomeric Carbocations $C_{10}H_{17}^{+}$ by Hydride Transfer to Methyl Cation in the Gas Phase

Cation	Type	ΔH°_{react} (B3LYP/6-31+G*)
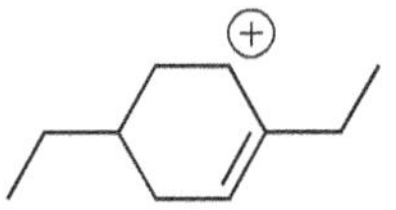	Allylic (2° ↔ 2°)	−254.3
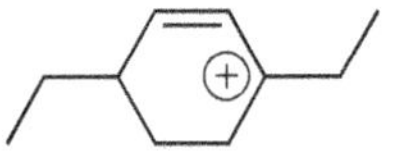	Allylic (2° ↔ 3°)	−264.4
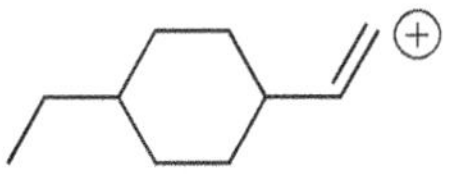	Vinyl (1°)	−203.6

*This energy is for the bridged, non-classical (see §9./4) cation shown.

Thus, the Hammond postulate[30] predicts that the activation energies for the exothermic rearrangements should be extremely low, so that the reaction should be rapid. The computational results for the primary carbocation in Table 9.2 certainly suggest that this is the case, because, depending on the exact starting geometry, computations at the B3LYP/6-31+G* level can return either the secondary carbocation as the minimum-energy structure or the bridged ion.

There are, of course, limitations on carbocation rearrangements: (1) groups usually rearrange only to the adjacent carbon atom and (2) rearrangements occur rapidly only if the rearrangement leads to a more stable cation—if there is no net gain in stability from the rearrangement, it is usually slow. This is nicely illustrated by the dehydration of 4-methyl-2-pentanol at 140°C, a reaction that gives a mixture of alkenes (Figure 9.5).[31]

In this reaction, we see that the secondary cation formed initially undergoes direct elimination to the alkene just over 50% of the time, and that slightly less than 50% rearranges by hydride shift to an isomeric secondary cation. Approximately 70% of this secondary cation, in turn, undergoes a hydride shift to the more stable tertiary cation, while only 10% of the same cation undergoes a methide shift to give an isomeric secondary carbocation: the more exothermic rearrangement is strongly favored.

Rearrangements of an alkyl group or hydrogen atom to an adjacent atom in a carbocation are called 1,2-rearrangements or Wagner-Meerwein rearrangements (after Egor Egorevich Vagner,[32] or Georg Wagner, who first reported them in 1899,[33] and Hans Leb(e)recht Meerwein,[34]

30. Hammond, G.S. *J. Am. Chem. Soc.*, **1955**, *77*, 334.

31. This reaction forms the basis of an undergraduate experiment in gas chromatography: Nienhouse, E.J. *J. Chem. Educ.* **1969**, *46*, 765. The experimental percentages reported here are the averages of those reported by students at the University of Wisconsin–Eau Claire in 2005. Compare with the literature values: 4-methyl-1-pentene, 5.2%; 4-methyl-2-pentene (*E* and *Z*) 69.4%; 2-methyl-1-pentene, 3.8%; 2-methyl-2-pentene, 19.9%; 3-methyl-2-pentene (*E* and *Z*), 1.9%

32. **Egor Egorevich Vagner (Georg Wagner, 1849–1903)** was educated under Zaitsev at Kazan and under Butlerov at St. Petersburg, where he also became assistant to Menshutkin. His career was in Poland, first at the Novo-Aleksandriya Institute of Agriculture and Forestry, then at Warsaw University. Vagner was a brilliant chemist and a bon vivant of the first water. For more biographical details, see: Lewis, D.E. In Patterson, G.D.; Rasmussen, S.C., Eds. *Characters in Chemistry; A Celebration of the Humanity of Chemistry. ACS Symp. Ser.* **2013**, *1136*, ch. 9.

33. (a) Wagner, E.E. *Zh. Russ. Phys. Khim. Obshch.* **1899**, *31*, 680. (b) Wagner, G.; Brickner, W. *Ber. dtsch. chem. Ges.* **1899**, *32*, 2302. (c) Wagner, G. *Ber. dtsch. chem. Ges.* **1900**, *33*, 2121.

34. **Hans Leb(e)recht Meerwein (1879–1965)** was educated at Bonn (PhD, 1903). His career took him from Charlottenburg to Bonn to Königsberg to Marburg, where he became the first Director of the Chemical Institute. Meerwein's name is celebrated in two reactions, and the Meerwein salts. For more biographical details, see: Jaenicke, L. *Chem. uns. Zeit* **1992**, *20*, 187.

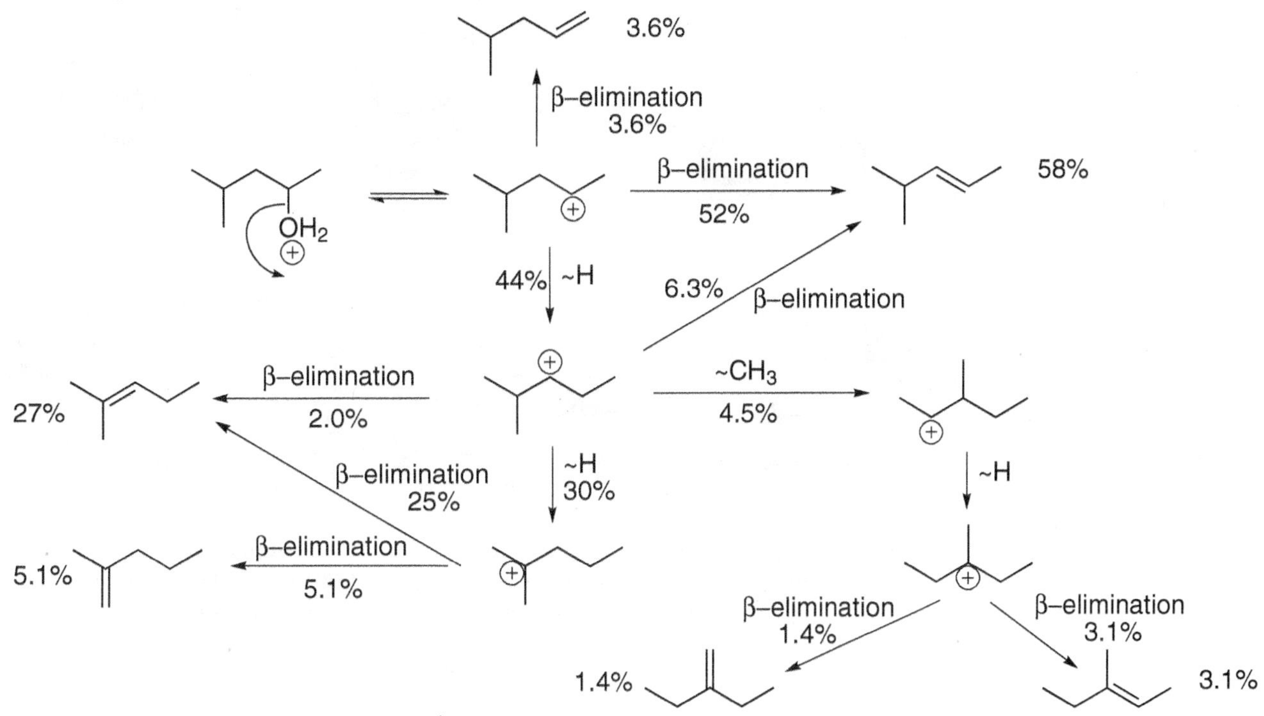

Figure 9.5 Acid-catalyzed dehydration of 4-methyl-2-pentanol at 140°C. The percentages indicate the average percent of the original alcohol that reacts by that particular pathway.

whose research[35] expanded Wagner's original observations). Establishing the involvement of carbocations in these rearrangement reactions is generally attributed to American chemist Frank Whitmore,[36] who proposed the carbocation-based mechanism for the Wagner-Meerwein rearrangement.[37]

Generally, the rearrangement of a hydrogen atom occurs more rapidly than the rearrangement of an alkyl group. During the course of the Wagner-Meerwein rearrangement, the migrating group remains bound to the rest of the cation; chiral alkyl groups migrate with retention of configuration. A careful examination of the activated complex in Figure 9.6 reveals that the activated complex for a 1,2-rearrangement reaction is very similar to the activated complex for electrophilic addition of an electrophile (proton or carbocation) to an alkene, and this means that Markovnikov addition and carbocation rearrangement have the same driving force governing them: the more stable cation is the preferred final product. At the same time, note how the activated complex for the rearrangement bears a striking resemblance to the bridged-ion structure for the primary carbocation in Table 9.2.

---

35. (a) Meerwein, H.; Unkel, W. *Liebigs Ann. Chem.* **1910**, *376*, 152. (b) Meerwein, H. *Liebigs Ann. Chem.* **1913**, *396*, 200; **1914**, *405*, 129; **1918**, *417*, 255. (c) Meerwein, H.; van Emster, K. *Ber. dtsch. chem. Ges.* **1920**, *53*, 1815; **1922**, *55*, 2500. (d) Meerwein, H.; Gérard, L. *Liebigs Ann. Chem.* **1923**, *435*, 174. (e) Meerwein, H.; Hammel, O.; Serini, A.; Vörster, J. *Liebigs Ann. Chem.* **1927**, *453*, 16.

36. **Frank Clifford Whitmore (1887–1947)** was educated at Harvard and worked at Minnesota, Northwestern, and Pennsylvania State University. His major contributions were to the theory of carbocation rearrangements. For more biographical details, see: Marvel, C.S. *Biogr. Mem. Nat. Acad. Sci.* (National Academy of Science: Washington, D.C., 1954), p. 287.

37. (a) Whitmore, F.C. *J. Am. Chem. Soc.* **1932**, *54*, 3274. (b) Whitmore, F.C.; Johnston, F. *J. Am. Chem. Soc.* **1933**, *55*, 5020. (c) Whitmore, F.C.; Rothrock, H.S. *J. Am. Chem. Soc.* **1932**, *54*, 3431. (d) Whitmore, F.C.; Fleming, G.H. *J. Am. Chem. Soc.* **1933**, *55*, 4161. (e) Whitmore, F.C.; Wittle, E.L.; Popkin, A.H. *J. Am. Chem. Soc.* **1939**, *61*, 1586.

It is interesting that Whitmore refused call his reactive intermediate a "carbonium ion," and that, in fact, he explicitly excluded the equivalency of his neopentyl intermediate and the carbonium ion proposed by Meerwein: Whitmore, F.C.; Fleming, G.H. *J. Chem. Soc.* **1934**, 1269.

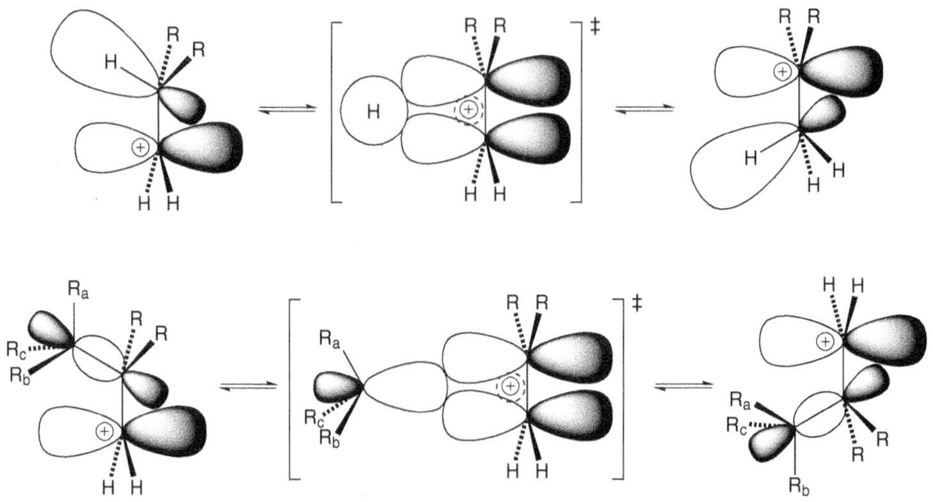

Figure 9.6 The Wagner-Meerwein shift of hydrogen (top) and alkyl (bottom). Both reactions pass through an activated complex whose orbital arrangement closely resembles the overlap of the π bond of an alkene with the empty orbital of an electrophile during an electrophilic addition reaction.

Problems

9-4 What is the product of each of the following reactions if each is carried out under ideal S_N1 conditions? Write the mechanism of the reaction with the structures of all intermediates involved in the formation of the final product.

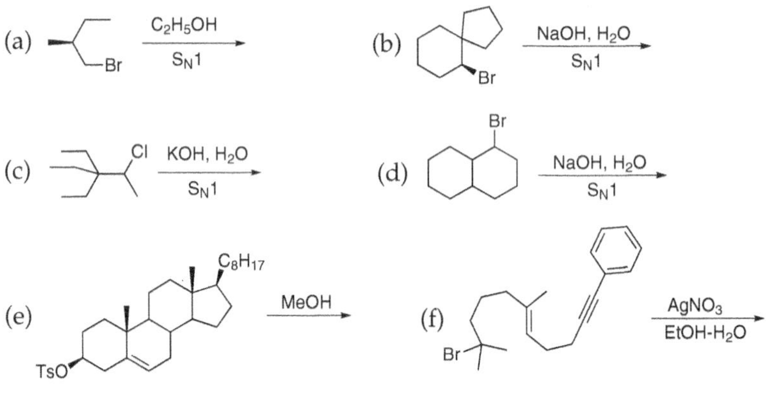

9-5 Write a reasonable mechanism for each of the following reactions.

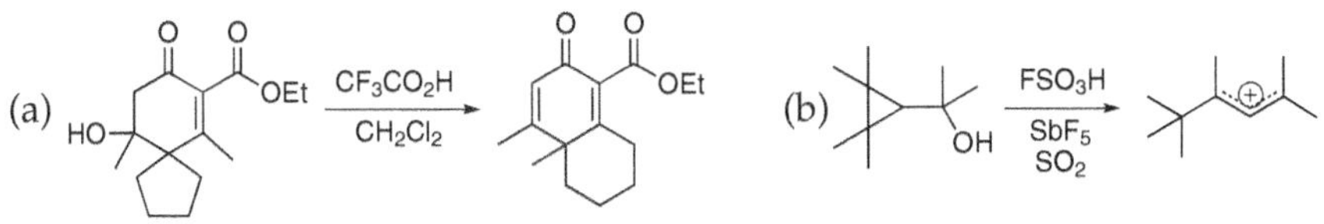

The foregoing discussion tacitly assumes that carbocation rearrangements lead to the more stable cation as the product, and this is generally the case. It is not, however, always the case, as has been illustrated by Nishizawa on the rearrangement of cations generated from certain tertiary alcohols (e.g., Example **9.11**).[38] In this work, Nishizawa has observed

38. Nishizawa, M.; Asai, Y.; Imagawa, H. *Org. Lett.* **2006**, 8, 5793.

that the counterion is critical to determining the product of the rearrangement, because the counterion affects the conformation of the rearranging carbocation. The rearranging group must fulfill a stereoelectronic requirement that permits the type of activated complex shown in Figure 9.6 to be formed (i.e., the σ bond to the migrating group must be able to achieve coplanarity with the empty 2p orbital on the cation carbon). Therefore, anything that restricts the conformation of the cation will affect the outcome of the reaction.

(9.11)

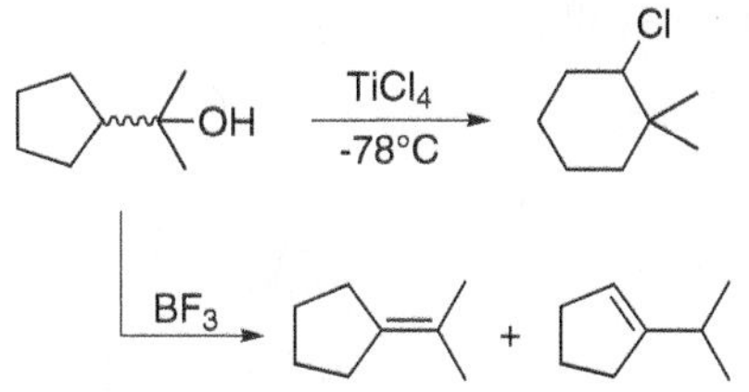

(9.12)

(9.13)

Transannular Rearrangements

Transannular rearrangements are rearrangements from one position of a reactive intermediate to a remote position of the same intermediate by transfer of a group across a medium-sized ring. Thus, in carbocations where the cation carbon is part of a medium-sized ring, we find that the cation may be isomerized by rearrangement of the hydrogen from one side of the ring to the other. This type of rearrangement is facilitated by the fact that there is severe steric crowding of one side of the medium-sized ring by the other, bringing the groups on opposite sides of the ring into van der Waals contact with each other in many cases. This same steric crowding is why (1) medium-sized rings are difficult to make by most cyclization methods and why (2) the ring strain in medium-sized rings rises to a maximum before falling as the ring size gets larger.

One example of this is provided by the reaction of the alcohol in Example **9.14**,[39] where the hexacyclic compound (Example **9.16**), that one would predict to be the major product of the reaction, formed by cyclization of the initial carbocation (Example **9.15**), was not. Instead, the carbocation underwent a 1,4- (transannular) rearrangement of hydrogen from the other benzylic position to give a tertiary carbocation (Example **9.17**) that lost a proton to give the observed product (Example **9.18**). Note the proximity of the migrating hydrogen to the largest lobe of the LUMO of the cation in Example **9.15**.

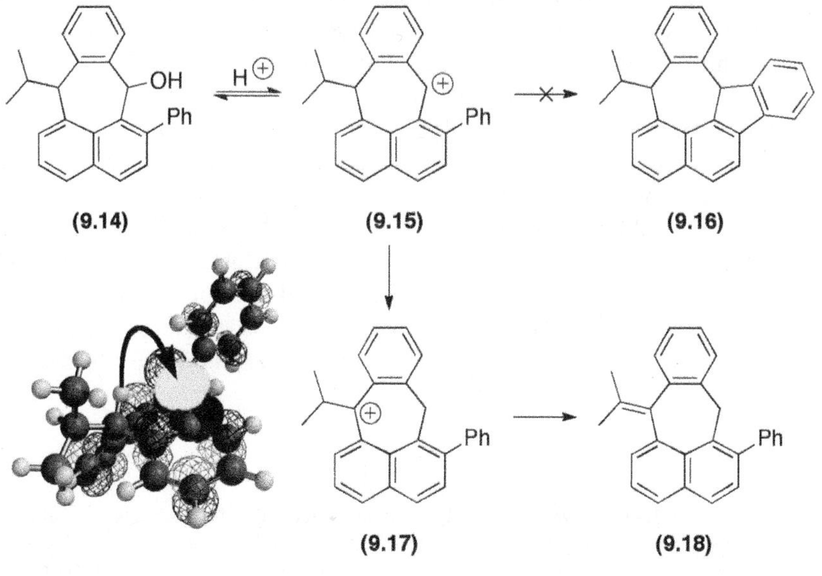

(9.14) (9.15) (9.16)

(9.17) (9.18)

39. Lansbury, P.T. *Acc. Chem. Res.* **1969**, *2*, 210.

Problem

9-6 The following reaction has been used as part of a synthesis of a precursor to 1-deoxypaclitaxel. Suggest a mechanism for the reaction. The workup of the reaction involved the sequential addition of methanol and then sodium potassium tartrate solution. Suggest a mechanism for the reaction and give reasons why the reaction should occur at all.

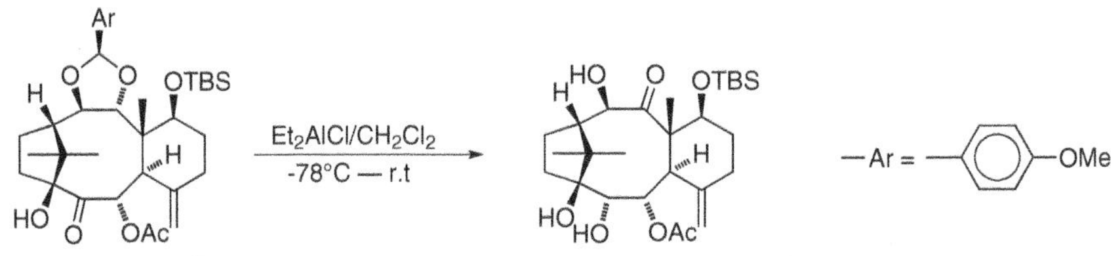

Reaction Synopses

Carbocation Rearrangements

X: OH, OR, NR_2, SR, etc.
Regiochemistry: more stable carbocation almost always predominates
Transannular rearrangements: can occur in medium-sized (8- to 12-membered) cyclic carbocations

9.3 Neighboring Group Participation

The carbocations that we have discussed so far are ***classical carbocations***, where the positive charge is formally localized on a single carbon or delocalized by conjugation (π-type overlap of the empty p orbital with an *adjacent* π bond or lone pair) or hyperconjugation (π-type overlap of the empty p orbital with a σ bond on the *adjacent* atom). However, in the late 1950s, American chemist Saul Winstein[40] noted that certain alkyl halides and similar compounds with good leaving groups underwent S_N1 reactions several hundred times more rapidly than would be predicted on the basis of the structures of the carbocations that should form as intermediates, whereas other structurally similar compounds reacted at the rates one would predict. The one common structural feature of these species was the presence of a nearby group that had an available electron pair—either a π bond or a lone pair—to assist the heterolysis of the carbon-leaving group bond. This group was not in the position to participate in hyperconjugation or conjugation with the cation center. Winstein proposed that the heterolytic step of the rapid S_N1 reactions was assisted by this nearby group in a process known as ***neighboring group participation*** and introduced the term ***anchimeric assistance*** to describe the

40. **Saul Winstein (1912–1969)** was born in Montreal, and educated at the University of California, Los Angeles (UCLA) (AB, MA) and the California Institute of Technology (PhD, 1938). After a postdoctoral fellowship at Harvard, Winstein began his career at the Illinois Institute of Technology in Chicago; he moved to UCLA in 1941. He was embroiled in an often vitriolic controversy with H.C. Brown over the non-classical carbocation that lasted the rest of his life. For more biographical details, see: Young, W.G.; Cram, D.J. *Int. J. Chem. Kinetics* **1970**, *2*, 167.

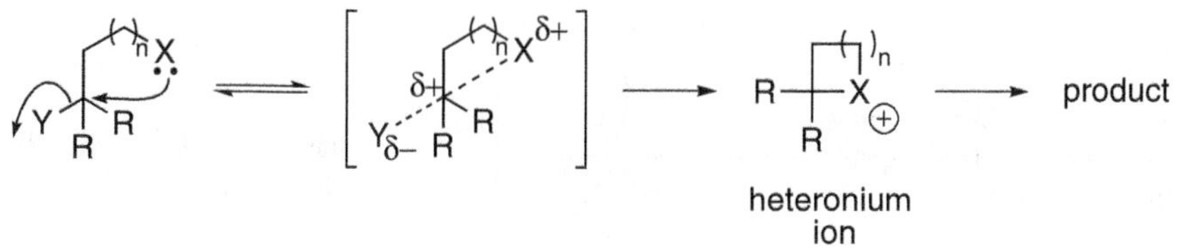

(9.19)

Figure 9.7 Neighboring group participation in the heterolysis of a carbon-leaving group bond by a nearby heteroatom

phenomenon.[41] It is generally observed that anchimeric assistance occurs only when absolutely needed (either the carbocation being formed is relatively high energy, or the leaving group is poor). It is seldom found to occur when the starting compound is tertiary, for example. This phenomenon has been described this phenomenon in terms of a *tool of increasing electron demand.*[42]

The net result of neighboring group participation by atoms with lone pairs is to form a transient intermediate ***heteronium ion***, where the positive charge is localized on the heteroatom (Figure 9.7). It is important to remember that the slow step of these reactions is still the unimolecular ionization of the alkyl halide or sulfonate, so the reaction is still an S_N1 reaction. However, reactions involving heteronium ions occur much more rapidly than similar reactions where anchimeric assistance does not occur.

The intermediacy of a heteronium ion in such reactions has been inferred from the stereochemistry of the S_N1 reactions of the two stereoisomers of 3-bromo-2-butanol,[43] which give different stereoisomers of the dibromide. In some cases,[44] the heteronium ion has been intercepted, and a salt of one has been isolated. When anchimeric assistance occurs during an S_N1 reaction, the stereochemical outcome also differs from what is observed in unassisted reactions: the product is no longer racemic but is formed with retention of configuration at the reacting carbon.

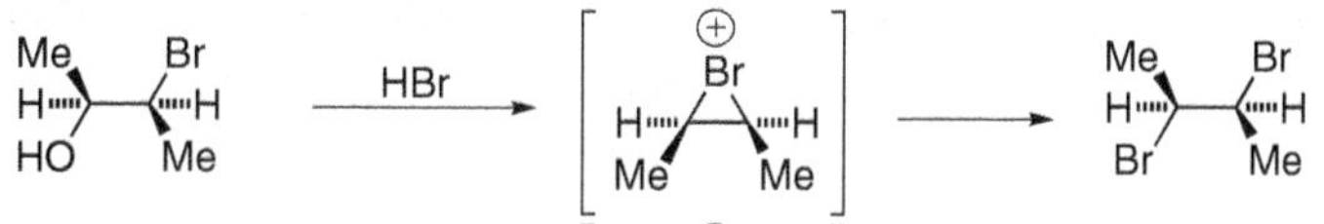

(9.20)

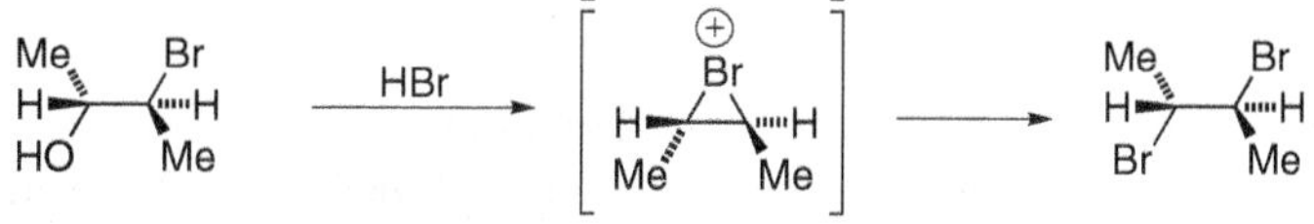

(±) (9.21)[45]

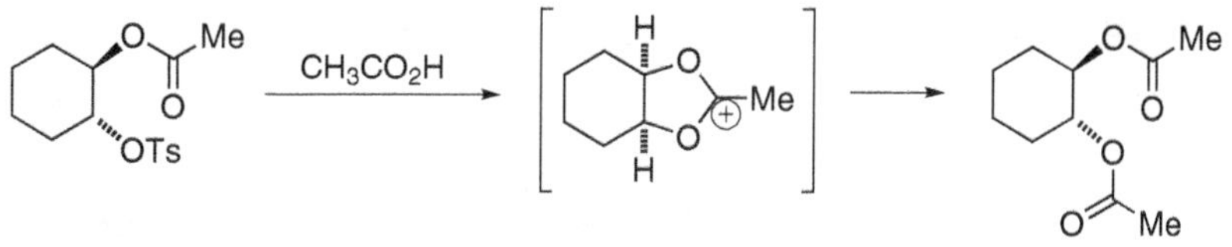

(±) (9.22)[46]

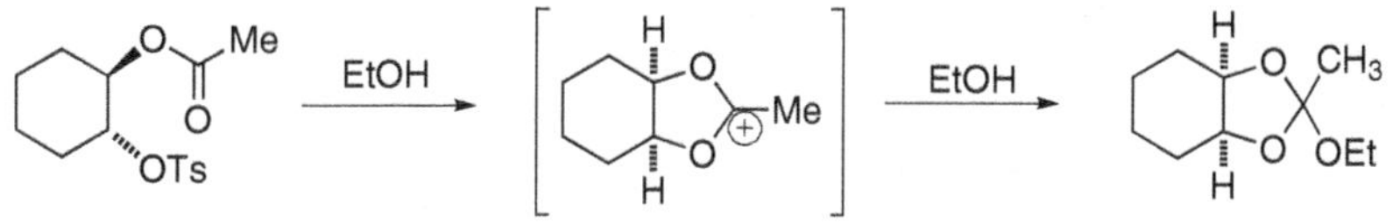

(9.23)[45, 46]

41. (a) Winstein, S. *Bull. Soc. Chim. France* **1951**, *18*, C55. (b) Streitwieser, A., Jr. *Chem. Rev.* **1956**, *56*, 675. (c) Capon, B. *Quart. Rev. Chem. Soc.* **1964**, *18*, 45. (d) Gould, E.S. *Mechanism and Structure in Organic Chemistry* (Holt, Rinehart, and Winston: New York, 1959), ch. 14. (e) Isaacs, N.S. *Physical Organic Chemistry* (Longman Scientific & Technical: Burnt Mill, Harlow, 1987), pp. 589–595. (f) Allred, E.L.; Winstein, S. *J. Am. Chem. Soc.* **1967**, *89*, 3991, 3998. (g) Winstein, S.; Allred, E.; Heck, R.; Glick, R. *Tetrahedron* **1968**, 3, 1. (h) Winstein, S.; Grunwald, E. *J. Am. Chem. Soc.* **1948**, *70*, 828. (i) Capon, B.; McManus, S.P. *Neighboring Group Participation* (Plenum: New York, 1976).

42. Lambert, J.B.; Mark, H.W.; Holcomb, A.G.; Magyar, E.S. *Acc. Chem. Res.* **1979**, *12*, 317.

43. Winstein, S.; Lucas, H.J. *J. Am. Chem. Soc.* **1939**, *61*, 1576.

44. (a) Winstein, S.; Grunwald, E.; Buckles, R.E.; Hanson, C. *J. Am. Chem. Soc.* **1948**, *70*, 816. (b) Winstein, S.; Hanson, C.; Grunwald, F. *J. Am. Chem. Soc.* **1948**, *70*, 812. (c) Winstein, S.; Hess, H.V.; Buckles, R.V. *J. Am. Chem. Soc.* **1942**, *64*, 2796.

45. Winstein, S.; Buckles, R.E. *J. Am. Chem. Soc.* **1942**, *64*, 2780, 2787; **1943**, 65, 613, 2196.

46. Winstein, S.; Anderson, C.B.; Friedrich, E.C. *Tetrahedron Lett.* **1963**, 2037.

Figure 9.8 S_N1 substitutions of stereoisomeric 3-bromo-2-butanols

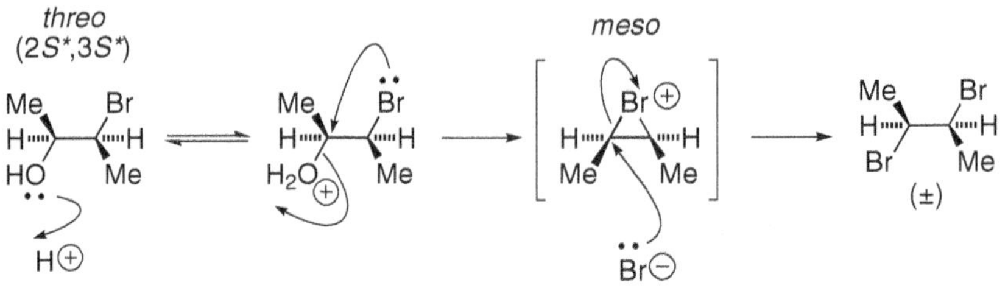

(9.24)

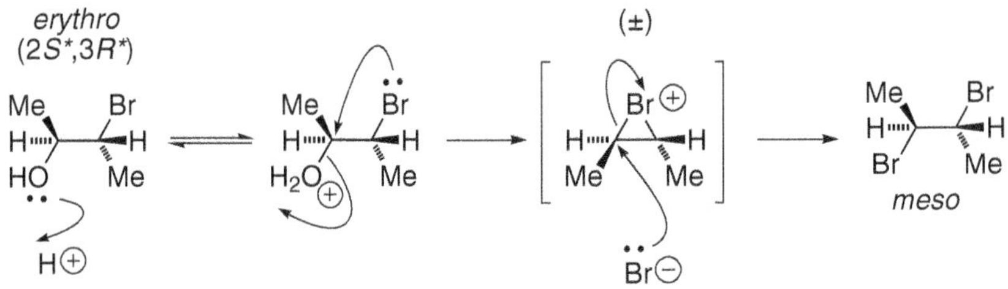

(9.25)

This stereochemical outcome is a result of two inversions (Figure 9.8). The first inversion occurs during the ionization of the alkyl halide: the lone pair on the assisting atom attacks the developing carbocation from the back side as the leaving group departs. In the second step, this is reversed: as the bond to the assisting atom is broken, the nucleophile attacks the incipient carbocation from the back side, leading to net retention of configuration at the reacting carbon atom—the second inversion reverses the first. Retention of configuration, especially when coupled with the evidence of the rate law and solvent effects, and so on, is usually sufficient to indicate that anchimeric assistance takes place during a reaction. It is also noteworthy that both enantiomers of *threo*-3-bromo-2-butanol (Example **9.24**) gave racemic product, strongly implying that *meso* stereochemistry of the intermediate bromonium ion is *meso*.

Worked Problem

9-2 The two optically active sulfonates below undergo solvolysis in acetic acid to give the outcomes shown. Provide a reasonable rationalization for this observation.

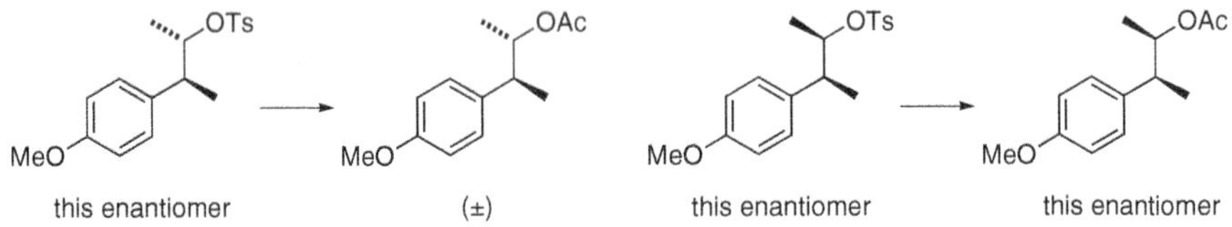

§Answer below.

§ Answer to Worked Problem:
In both reactions, the ionization of the sulfonate is assisted by the participation of the π electrons of the electron-rich aryl group to give a bridged benzenonium in the first step of the reaction. This participation in the solvolysis leads to inversion of configuration at the carbon carrying the leaving group, so the stereochemistry of the benzenonium ion will reflect the stereochemistry of the starting sulfonate:

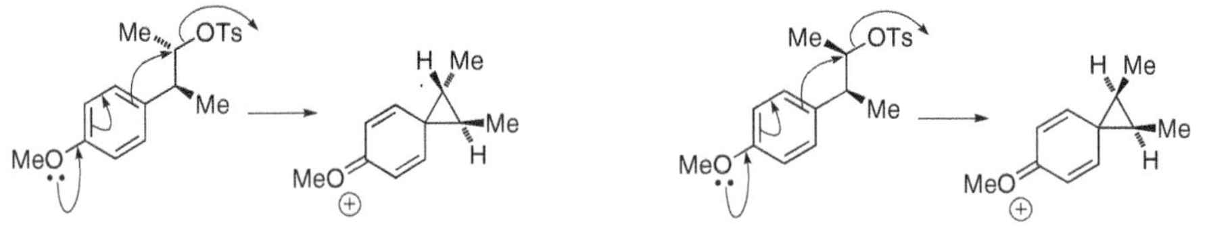

chiral: this enantiomer only

achiral: *meso*

In the second step, the attack of the acetate anion on the benzenonium ion leads to the racemic product when the benzenonium ion is *meso*, and it returns an optically active product with overall retention of configuration when the benzenonium ion is chiral.

Problems

9-7 What will be the major product(s) outcome of each of the following reactions? Where more than one product may form, indicate which is expected to be the major product. Assuming that each chiral compound is used as the enantiomer shown, what will be the stereochemistry of the final product(s)?

(a) 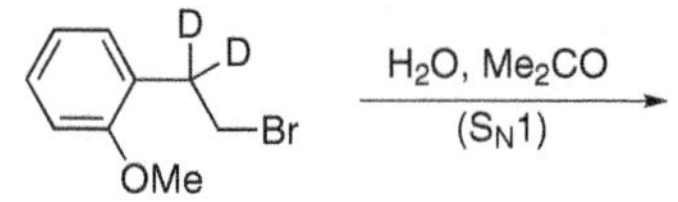

(b) 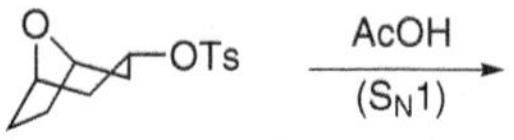

(c) 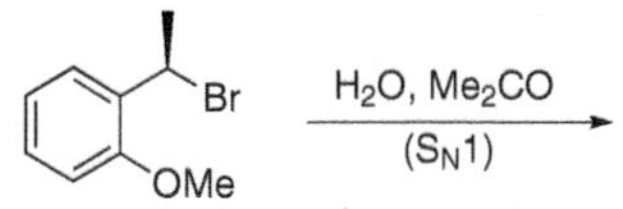

(d) 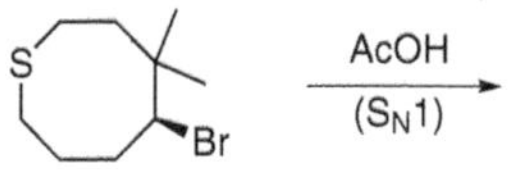

9-8 Is neighboring group assistance involved in the following reaction? Support your answer with a mechanism.

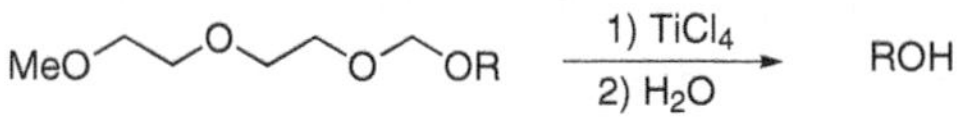

9-9 Write a reasonable mechanism for the transformations below:

(a) 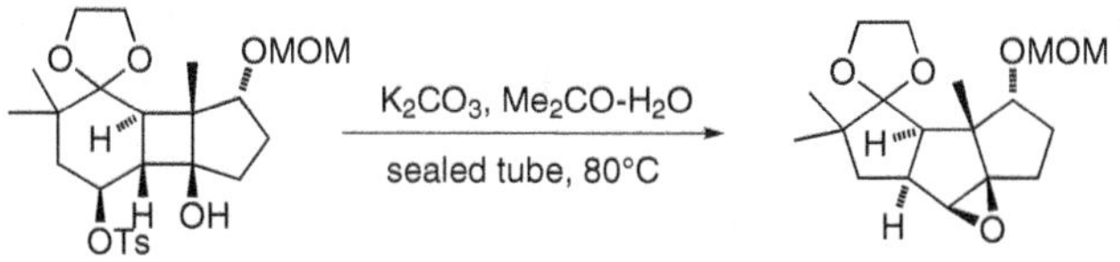

(b) 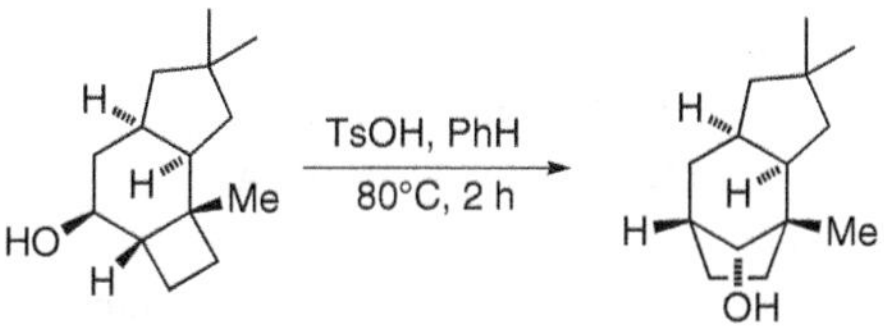

Reaction Synopses

Neighboring Group Participation (Anchimeric Assistance)

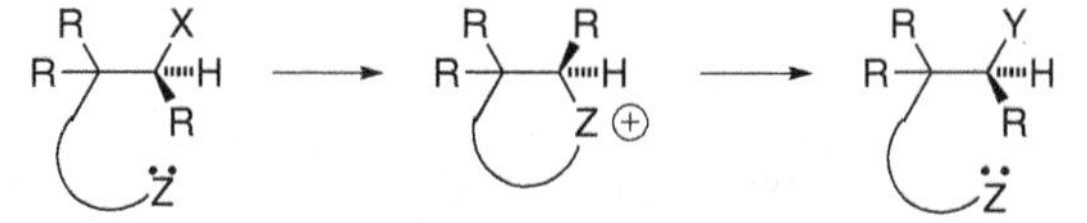

Z: OH, OR, OCOR, RCONR', SR, RCO, etc.
Stereochemistry: always results in retention of configuration at carbon with leaving group
Reaction speed: usually much faster than similar unassisted reactions

9.4 Nonclassical Carbocations

The question of modes of stabilization of carbocations arose in the late 1940s when Winstein and Trifan noted that the solvolysis of *exo*-2-norbornyl sulfonates (Example **9.26**) in acetic acid occurred some 350 times faster than the same solvolysis of the *endo* isomers (Example **9.27**) or of the corresponding cyclohexyl compounds.[47] In addition, they observed that the same product—exclusively the *exo* acetate—was obtained regardless of

47. Winstein, S.; Trifan, D.S. *J. Am. Chem. Soc.* **1949**, *79*, 2953.

which diastereoisomer of the starting sulfonate ester was used—and that the product from optically active starting sulfonates was completely racemized.

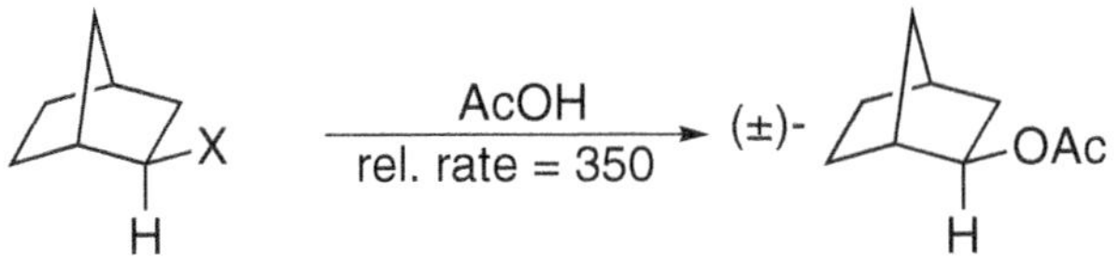

(9.26)

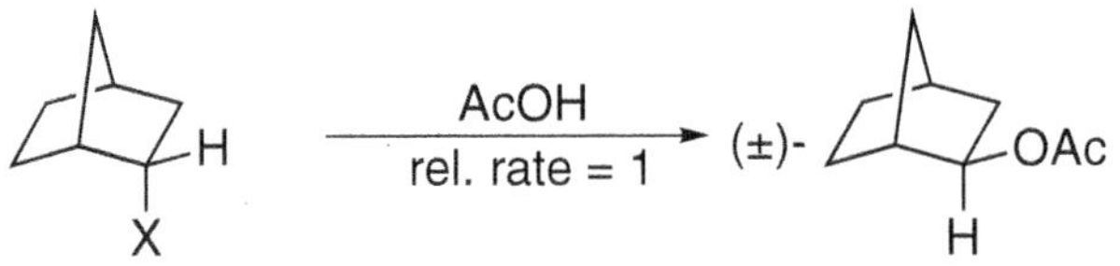

(9.27)

X = OTs, OBs

These observations were rationalized in terms of a mechanism involving participation by the electrons of the σ bond *anti* to the carbon-leaving group bond to give a bridged carbocation that was C_s-symmetric. In 1951, Roberts[48] first used the term ***non-classical*** to describe these σ-bridged carbocations. The non-classical carbocation concept ignited a firestorm of controversy in organic chemistry,[49] involving the research groups of Brown[50] and Winstein, and, after Winstein's death, of Olah.[51]

The question is still not completely resolved, but when the neighboring group that participates in the heterolysis step is a non-adjacent C—C σ or π bond, the cation formed is today generally (but not universally) accepted as being non-classical. In a non-classical cation, the charge is delocalized by σ-type overlap of the empty p orbital on the charged carbon atom with a σ bond or with π bond that is not adjacent to it. Clearly, this stabilization mechanism has rather stringent structural requirements, so it may not be unreasonable to view this type of stabilization as a last resort—usually a bridged cation will not form unless there is no alternative. This typically restricts non-classical carbocations to primary or secondary cations. Most tertiary carbocations are sufficiently stabilized by hyperconjugation that they do not require stabilization by bridging; tertiary cations are classical.

---

48. (a) Roberts, J.D.; Mazur, R.H. *J. Am. Chem. Soc.* **1951**, *73*, 3542. (b) Roberts, J.D.; Lee, C.C. *J. Am. Chem. Soc.* **1951**, *73*, 5009.

49. For leading papers and monographs, see: (a) Winstein, S.; Clippinger, E.; Howe, R.; Vogelfanger, E. *J. Am. Chem. Soc.* **1965**, *87*, 376. (b) Colter, A.; Friedrich, E.C.; Holness, N.J.; Winstein, S. *J. Am. Chem. Soc.* **1965**, *87*, 378. (c) Winstein, S. *J. Am. Chem. Soc.* **1965**, *87*, 379, 381. (d) Brown, H.C. *Pure Appl. Chem.* **1982**, *54*, 1783. (e) Sargent, G.D. In Olah, G.A.; Schleyer, P.v.R., Eds. *Carbonium Ions*, vol. 3 (John Wiley: New York, 1972), ch. 24. (f) Bartlett, P.D. *Non-Classical Ions: Reprints and Commentary* (Benjamin: New York, 1965). (g) Olah, G.A.; Jewett, C.L.; Kelley, D.P.; Porter, R.D. *J. Am. Chem. Soc.* **1972**, *94*, 147. (h) Berson, J.A. In DeMayo, P., Ed. *Molecular Rearrangements, Part One* (Wiley: New York, 1963). (i) Brown, H.C. *The Non-Classical Ion Problem* (Plenum: New York, 1977). (j) Sargent, G.D. *Quart. Rev. Chem. Soc.* **1966**, *20*, 301. (k) Saltzmann, M.D.; Wilson, C.L. *J. Chem. Educ.* **1980**, *57*, 289.

50. **Herbert Charles Brown (1912–2004)** was born in London, and educated at the University of Chicago (PhD, 1938), where he stayed as a postdoctoral researcher and instructor. He moved to Wayne State in 1943, and to Purdue in 1947. His development of hydroboration as a synthetic method revolutionized synthetic organic chemistry. Brown shared the 1979 Nobel Prize in Chemistry. For more biographical details, see the Nobel Foundation website.

51. **George Andrew Olah (1927–)** was educated at the Budapest Technical University (PhD, 1949), where he remained until 1954. In 1954 he became Associate Director of the Central Research Institute of the Hungarian Academy of Sciences and head of the organic chemistry laboratory. In 1956, he fled to Canada, and he moved to Case Western Reserve University in 1965 and to the University of Southern California in 1977. He received the 1994 Nobel Prize in Chemistry for his work in carbocation chemistry. For more biographical details, see Olah's autobiography: Olah, G.A. *A Life Of Magic Chemistry: Autobiographical Reflections of a Nobel Prize Winner.* (Wiley-Interscience: New York, 2000).

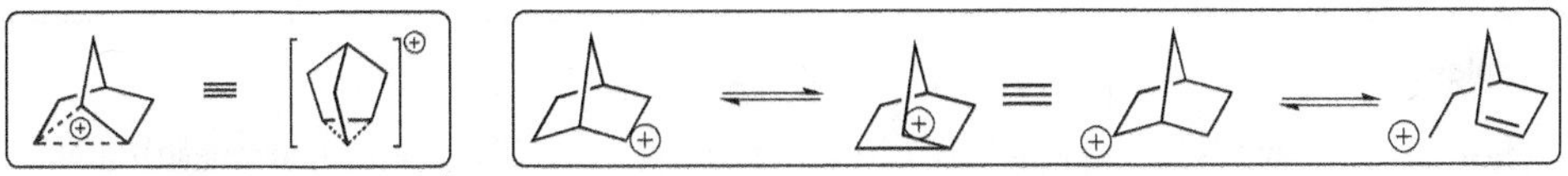

non-classical (**9.28**) classical (**9.29**)

Figure 9.9 The norbornyl cation is the prototypical non-classical carbocation.

non-classical (**9.30**)

classical (**9.31**)

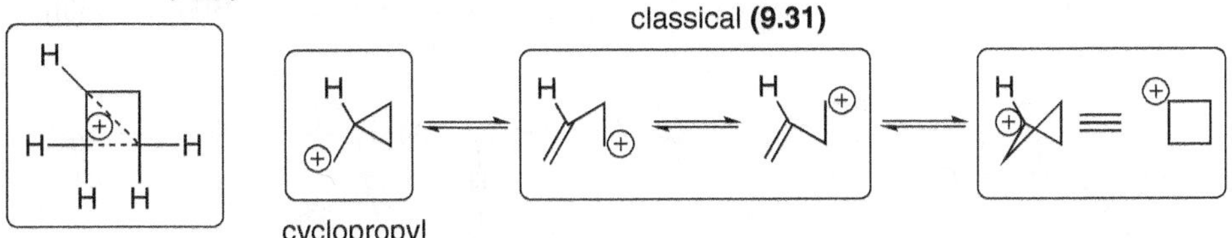

cyclopropyl-carbinyl (**9.32**) homoallyl (**9.33**) cyclobutyl (**9.34**)

Figure 9.10 The homoallyl carbocation

The most famous non-classical cation is the norbornyl cation, about which hundreds of research articles have been written.[52] The structure of the norbornyl cation can be written as a bridged non-classical cation, as a pair of rapidly equilibrating classical ions, or as a π complex of a primary cation and an alkene π bond (Figure 9.9). By 1990, sufficient experimental evidence[53] had accumulated to convince the vast majority of organic chemists that the non-classical cation is legitimate but not before more than three decades of (often rancorous[54]) debate. Even so, the question still draws attention today,[55] but the low-temperature X-ray crystal structure determination of the tetrabromoaluminate salt in 2013[56] clearly reveals that the non-classical structure is correct.

Another group of cations that are, in a formal sense, at least, non-classical are the homoallyl cation[57] (Figure 9.10) and the structurally similar homoaromatic carbocations. In the homoallyl cation, there is involvement of the π bond through space to allow delocalization of the positive charge over more atoms in the ion. This delocalization is different from classical conjugation because the delocalization is actually σ delocalization rather than π delocalization. In homoaromatic carbocations, the same situation applies: the formal cyclic conjugated π electron system is interrupted by at least one sp^3 hybridized carbon, and the electron delocalization has at least some σ character.

All the precursors that would lead to non-classical cations react unusually rapidly in S_N1 reactions, and unusually fast reactions are often interpreted as experimental evidence that a nonclassical carbocation may be involved as a reaction intermediate in S_N1 reactions. In addition, modern computations at a relatively high level of sophistication predict that primary carbocations also have a σ-bridged structure that leads to electron delocalization. These structures are, therefore, entirely analogous to the structure of the non-classical norbornyl cation shown in Figure 9.9, so primary carbocations, also, may reasonably be designated as nonclassical carbocations. A similar situation holds true for many secondary, cyclic carbocations (e.g., the cyclopentyl carbocation).

---

52. For historical perspectives on this most famous of questions, see: (a) Davenport, D.A. *Aldrichimica Acta* **1987**, *20*, 25. (b) Weininger, S.J. *Bull. Hist. Chem.* **2000**, *25*, 123.

53. The key evidence came from ^{13}C NMR spectroscopy, which showed that the two interconverting classical carbocations could not be separated by a barrier of more than 0.2 kcal mol^{-1}, based on classical rate theory: Yannoni, C.S.; Macho, V.; Myhre, P.C. *J. Am. Chem. Soc.* **1982**, *104*, 7380.

54. Commentary from an "innocent bystander": Walling, C. *Acc. Chem. Res.* **1983**, *16*, 448

55. (a) Werstiuk, N.H.; Muchall, H.M. *J. Mol. Struct. (Theochem.)* **1999**, *463*, 225. (b) Werstiuk. N.H.; Muchall, H.M. *J. Phys. Chem. A* **2000**, *104*, 2054. (c) Smith, K.; Pan, H.; Hondrogiannis, G.; Mamantov, A.; Pagni, R.M. *ARKIVOC* **2009**, 68.

56. Scholz, F.; Himmel, D.; Heinemann, F.W.; Schleyer, P.v.R.; Meyer, K.; Krossing, I. *Science* **2013**, *341*, 62.

57. Roberts, J.D.; Mazur, R.H. *J. Am. Chem. Soc.* **1951**, *73*, 3542.

Problems

9-10 The solvolysis in acetic acid of *exo*-2-norbornyl-2,3-$^{14}C_2$ *p*-bromo-benzenesulfonate (below) gives the labeling pattern shown (the heavy dots indicate the labeled positions, although the label does not occur at positions 1 and 4 simultaneously). What does this pattern of labeling in the product tell you about the structure and dynamics of the cation intermediate in the solvolysis? [*J. Am. Chem. Soc.* **1954**, *76*, 4501].

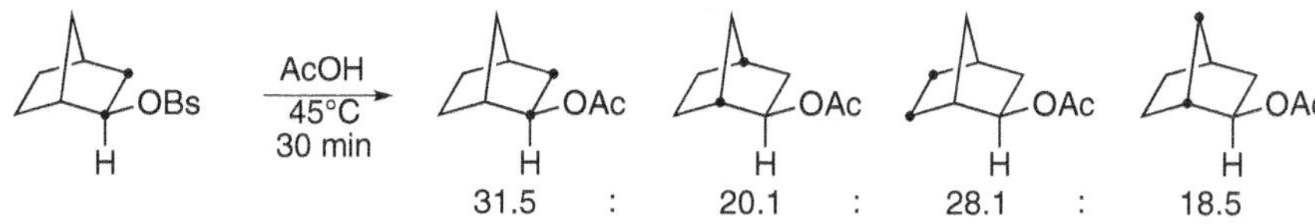

9-11 S_N1 nucleophilic substitution of the enantiomerically pure bromides below by water is expected to occur unusually rapidly due to probable anchimeric assistance. Write a mechanism that shows the structure of the intermediate heteronium ions that would be formed if the oxygen atom participates in the reaction. Are any of these heteronium ions further stabilized by resonance? What is the stereochemistry of the final product of each reaction?

(a) 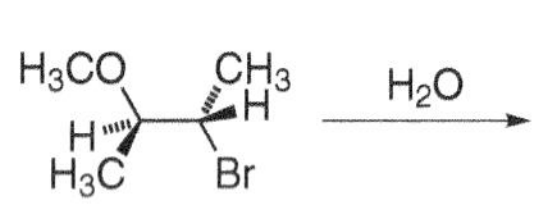

(b) 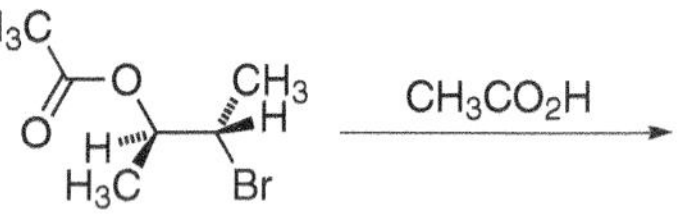

9-12 When the two regioisomeric alkyl tosylates below are solvolyzed in aqueous ethanol, the same mixture of alcohols is obtained. Draw the structures of the expected products and rationalize their formation.

(a) 

(b) 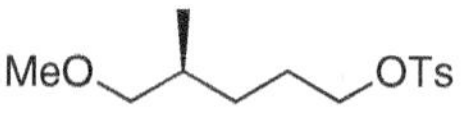

9-13 One of the two halides below is expected to react much faster than the other in an S_N1 solvolysis in 80% aqueous ethanol to give an alcohol product. Which isomer is expected to react faster—and why?

(a) 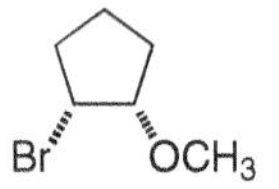

(b) 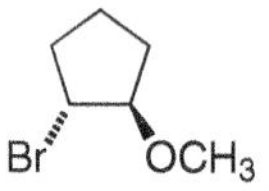

9-14 When cyclohexane is treated with HBr-$AlBr_3$, methylcyclopentane is formed; the reaction is accelerated by a catalytic amount of cyclohexene. Write a mechanism for this transformation and show how non-classical carbocations may be involved in the process.

9-15 The formation of 1,2-diols from alkenes is an important oxidation reaction. Osmium tetroxide gives the *cis*-1,2-diol where the reagent has attacked the alkene from the less-hindered side. R. B. Woodward developed a method for the synthesis of the same product where the two OH groups are added to the more hindered face of the double bond (e.g., giving a diol where there is a 1,3-diaxial interaction with the methyl group in the example shown). The reaction involves the addition of an acyl hypoiodite to an alkene in moist acetic acid. Suggest a reasonable mechanism for this reaction.

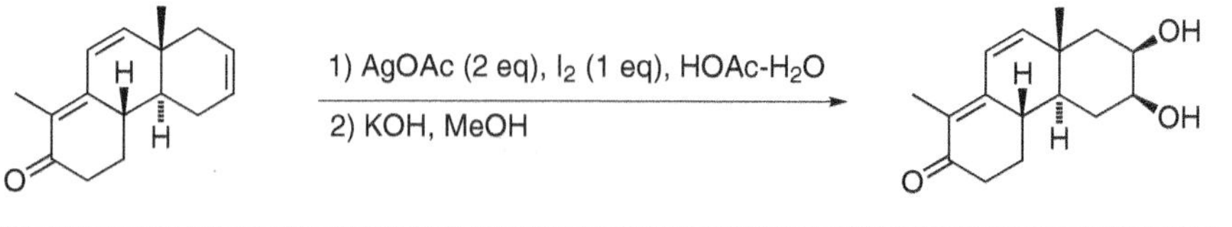

9.5 Methods for Generating Carbocations

Methods for the generation of carbocations from non-ionic precursors can be subdivided into four major classes. The first two of these are the most widely used, and they consist of (1) addition of an electrophile to a π bond (Example **9.35**) and (2) heterolysis of a bond to a leaving group (Example **9.36**). Strong Lewis acids such as aluminum chloride can—formally, at least—abstract a hydride anion from alkanes to generate carbocations (Example **9.37**). A final method, based on one-electron oxidation of a neutral precursor, is of less general utility (Example **9.38**).

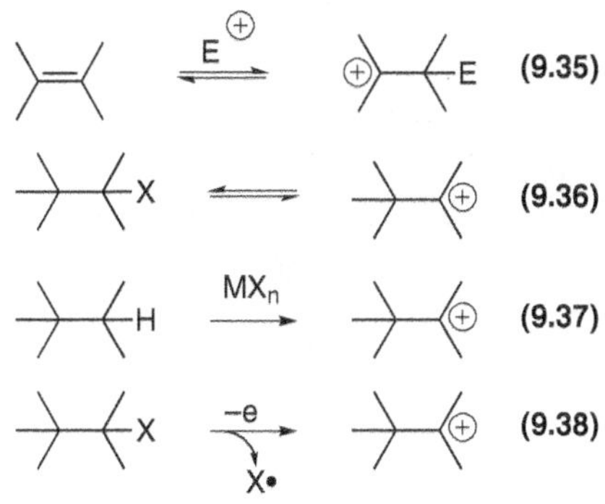

The loss of a leaving group from a cation precursor is by far the most common method for preparing carbocations. Leaving groups in general may be categorized as good or poor, with groups such as the diazonium ion or trifluoromethanesulfonate ester (triflate, OTf) being excellent leaving groups in most solvents. Vinyl triflates, for example, can serve as precursors to the otherwise difficult-to-form vinyl cations. At the other end of the leaving group spectrum are groups such as fluoride, which is a poor leaving group under normal conditions, and groups such as hydroxide, which acts as leaving group only under exceptional circumstances. In recent years, there has been an effort to quantify the ability of a leaving group to leave, and this has included the development of approximate scales of intrinsic nucleofugality in the gas phase.[58] An approximate ranking of leaving group ability derived from solution-phase reactions is given in Table 9.5.

The one feature of the data in Table 9.5 is that the best leaving groups are those that leave as the weakest bases, so that any method that will increase the strength of the conjugate acid of the leaving group should also lead to an improvement in the ability of the group to function as a leaving group. This is observed. We can see this from the experimental rate data in Table 9.5. These data show that trifluoroacetate esters solvolyze more rapidly than do acetates by several orders of magnitude and that *p*-nitrobenzoate esters solvolyze some four times faster than the acetate. These values are in accord with the pK_a values of the conjugate acids of the leaving groups (−0.25 to 0.3, depending on the source of the data; 3.44; 4.76, respectively).

The logical conclusion that one can draw from the data in Table 9.5 is that decreasing the basicity of the leaving group increases the effectiveness of the leaving group. The simplest method for doing this is to conjugate the leaving group with a strong Lewis acid, and H^+, BF_3, $ZnCl_2$, and $TiCl_4$ have all been used for this purpose. This concept has been taken to its limit by Olah, who showed that one can convert a modest leaving group into a powerful leaving group by complexation with SbF_5 (one of the few Lewis acids strong enough to have a bridging fluorine atom in the solid phase). Using SbF_5, Olah was able to generate carbocations from alkyl chlorides and fluorides under conditions where they could be studied spectroscopically. The SbF_5X^- ion is non-nucleophilic, and so the carbocations were not intercepted, as is usually the case in solvolysis reactions.

58. Jaramillo, P.; Domingo, L.R.; Pérez, P. *Chem. Phys. Lett.* **2006**, *420*, 92–99.

Table 9.5 Leaving Groups in Approximate Decreasing Order of Ability to Leave at sp^3 Carbon

Substrate	Conjugate Base of Leaving Group	Relative Rate Constant for Solvolysis	Ref.
$R{-}N_2^+$	N_2		
R—OTf	$CF_3SO_2^-$	1.4×10^8	59
$R{-}OSO_2F$	FSO_3^-		
$R{-}OSO_2R'$	$R'SO_3^-$	3.0×10^4 (R′ = Me); 3.7×10^4 (R′ = p-C_6H_4Me)	43
R—I	I^-	91	43
R—Br	Br^-	14	43
$R{-}OH_2^+$	H_2O		
$R{-}OCOCF_3$	$CF_3CO_2^-$	2.1	43
R—Cl	Cl^-	1.0	43
$R{-}O(R')H^+$	R′OH		
$R{-}OPO_3H$	H_3PO_4		
$R{-}SR_2'^+$	$R_2'S$		
$R{-}NR_3'^+$	$R_3'N$		
R—F	F^-	9×10^{-6}	43
R—OCOR′	$R'CO_2^-$	1.4×10^{-6} (R′ = Me); 5.5×10^{-6} (R′=p-$C_6H_4NO_2$)	43

Another method for increasing the effectiveness of a leaving group is to incorporate it into a small ring (Example **9.39**). Thus, although ethers are protonated only under strongly acidic conditions, epoxides are protonated by dilute aqueous acids, and the three-membered oxonium ions react as carbocations in solution. A similar situation occurs with the aziridines (where X=NR).

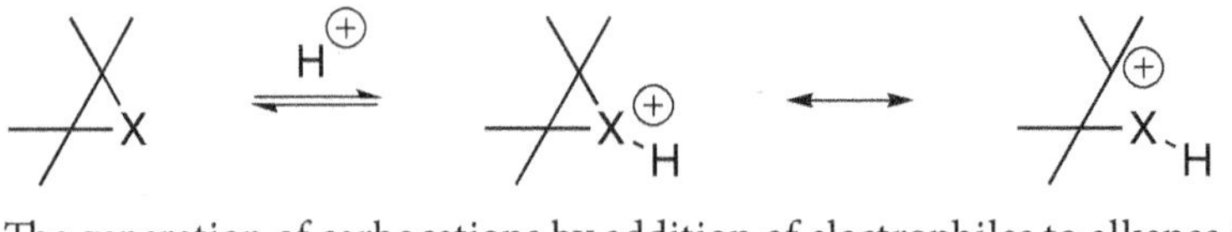

(9.39)

The generation of carbocations by addition of electrophiles to alkenes is an important step in the overall addition of electrophilic reagents to alkenes, and this underlies the basis for the fact that Markovnikov regiochemistry predominates in such reactions. The stronger the acid, the more favorable the protonation, and this has led to the development and use of superacids (acids whose H_0 values are more negative than concentrated H_2SO_4). The strongest acid known is hexafluoroantimonic acid, $HSbF_6$, which is formed by dissolving SbF_5 in HF. Olah developed "magic acid" (HSO_3F/SbF_5) for the study of carbocations. Both these acids are strong enough to protonate even saturated hydrocarbons to give onium ions that lose molecular hydrogen to give trivalent carbocations.[60] Some idea of the types of reactions that occur when alkanes are teated with "magic acid" can be gained from Figure 9.11. In addition to the alkane reactions, once alkenes are formed, the cations can add to the alkenes to generate larger, more complex carbocations. The reaction between

---

59. Solvolysis of PhCHXMe in 80% EtOH at 75°C: Noyce, D.S.; Virgilio, J.A. *J. Org. Chem.* **1972**, *37*, 2643.
60. Olah, G.A.; Schlosberg, R.H. *J. Am. Chem. Soc.* **1968**, *90*, 2726.

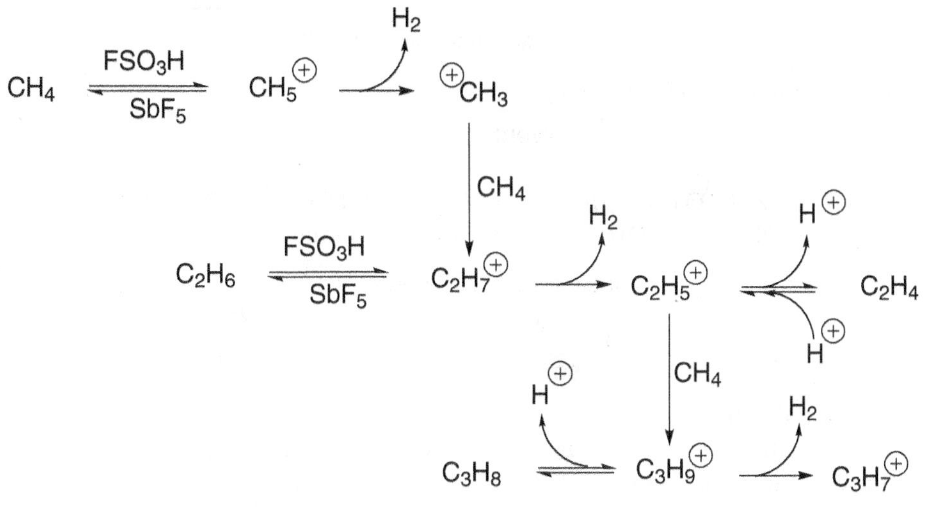

Figure 9.11 The reaction between methane and other saturated hydrocarbons and "magic acid." (Adapted from Olah, G.A. *A Life of Magic Chemistry: Autobiographical Reflections of a Nobel Prize Winner.* (Wiley-Interscience: New York, 2000.)

alkenes and strong acids in a strongly ionizing solvent such as liquid sulfur dioxide gives carbocations under conditions that allow them to be studied by NMR, for example. In a similar vein, the treatment of alkanes with strong Lewis acids such as aluminum bromide results in carbocation formation (ostensibly by hydride abstraction from the alkane), with concomitant formation of an anionic "ate" complex from the Lewis acid. The resultant cations may rearrange before reacting with the hydride "ate" complex to regenerate a rearranged alkane, as illustrated by the synthesis of adamantane from the tricyclic precursor, tricyclo[5.2.1.$0^{2,6}$]decane (Example **9.40**).[61]

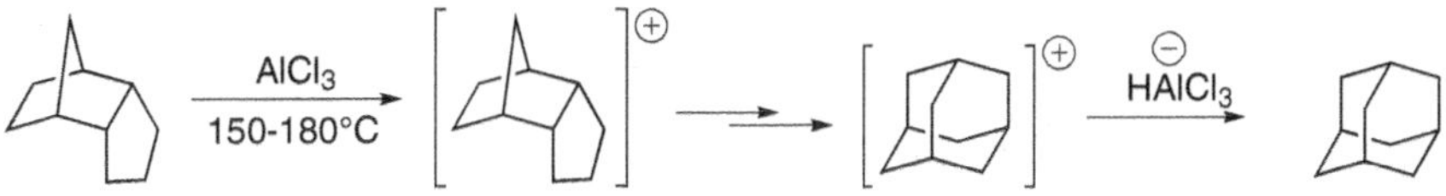

(9.40)

The remaining methods for the generation of carbocations are not suited for their large-scale generation. Carbocations can be formed by ionization of neutral molecules, either by electron impact or by radioactive decay. In the mass spectrometer, for example, the impact of a high-energy electron on a neutral molecule in the gas phase gives rise to a radical cation, which is an odd-electron species. The excess energy in the radical cation from the electron impact leads to fragmentation to give a more stable even-electron species (a simple cation). Based on the potential at which various ions are formed (the appearance potential of the ion), one can deduce a number of parameters about the molecule undergoing ionization, including, among other things, bond strengths in the molecule.

The final method for the generation of a carbocation comes from the use of tritiated compounds as precursors to the cation. In this case, the radioactive β decay of the tritium atom leads to a species bearing a positive charge, with what is formally, at least, a carbon-helium bond to the helium-3 atom generated by the nuclear reaction. This immediately loses a helium atom to give the free cation. Of course, because tritiated compounds are used in a tracer mode, this method is of very limited applicability.

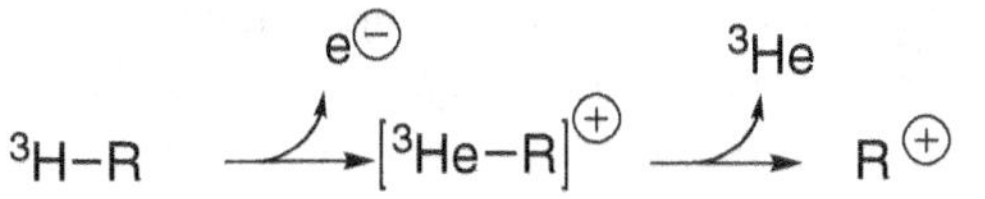

(9.41)

61. Schleyer, P.v.R.; Donaldson, M.M.; Nicholas, R.D.; Cupas, C. *Org. Syn.* **1973**, *Coll. Vol. 5*, 16.

Reaction Synopses

Forming Cations by Heterolysis or Solvolysis

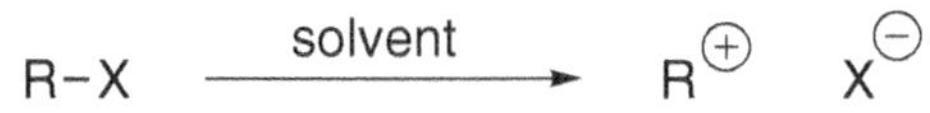

X: $N_2^+ > OTf > I > OTs \approx OMs \approx Br > OH_2^+ > Cl \approx OTFA > F \approx NR_3^+ > OR$
Solvent: EtOH-H_2O; AcOH; HCO_2H; CF_3CO_2H; H_2O-dioxane; etc.

Forming Cations by Heterolysis Assisted by Lewis Acids

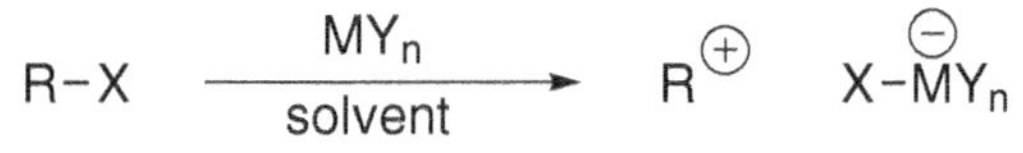

X: Cl, F, OH, OR
MY_n: BF_3; BBr_3; $AlCl_3$; $AlBr_3$; Me_3SiI; $SnCl_4$; $TiCl_4$; $ZnCl_2$; SbF_5; etc.
Solvent: usually an alkane or CH_2Cl_2

Forming Cations with Superacids

X: OH, OR
HY: H_2SO_4; FSO_3H; HF-SbF_5; etc.
Solvent: usually FSO_3H, SO_2 or similar strongly ionizing solvent

Forming Cations from Alkenes

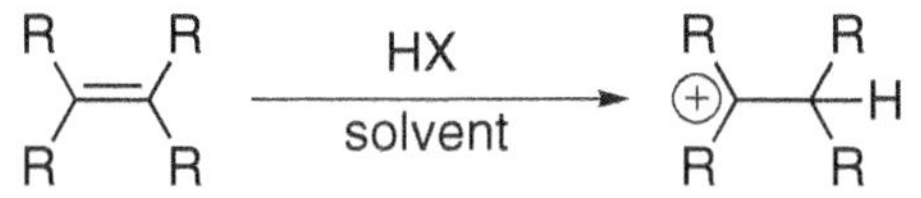

X: Cl, Br, I, OSO_3H, ClO_4; BF_4; etc.
Regiochemistry: Markovnikov addition usually occurs

Forming Cations by Hydride Abstraction

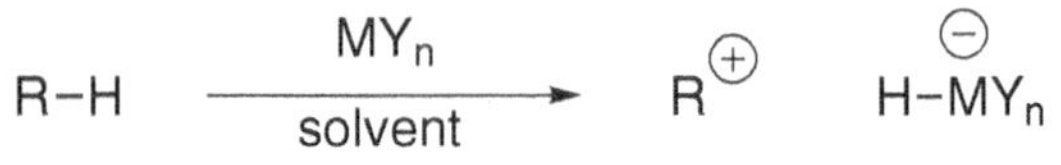

MY_n: $AlBr_3$; SbF_5; etc.

Chapter Summary

This chapter introduced the carbocations, one of the oldest known classes of reactive intermediates that participate in organic reactions. Carbocations possess a tricoordinate carbon atom carrying a formal positive charge. They can be generated by heterolysis, electrophilic addition, radiolysis, or electron impact, with the first two methods being the most common.

Carbocations are stabilized by any mechanism that leads to delocalization of the charge from carbon. There are three such mechanisms. In descending order of effectiveness, they are conjugation with the lone pair on a heteroatom, conjugation with a C—C π bond, and hyperconjugation. One manifestation of hyperconjugation is the Baker-Nathan effect. The ^{13}NMR spectroscopy of carbocations generated in superacid medium has been discussed.

Stabilization of a carbocation by a lone pair on a heteroatom located where it can form a cyclic heteronium ion is known as neighboring group participation or anchimeric assistance. Anchimeric assistance leads to retention of configuration in S_N1 reactions, for example. Stabilization of carbocations by non-adjacent σ or π bonds gives rise to non-classical carbocations. Primary and secondary carbocations are almost always non-classical, but tertiary carbocations are stabilized enough by hyperconjugation that bridging is unnecessary.

The characteristic reaction of carbocations is the Wagner (Vagner)-Meerwein rearrangement, which is a 1,2-rearrangement of hydrogen or alkyl. During the Wagner-Meerwein rearrangement, the migrating alkyl group retains its configuration.

Key Terms

anchimeric assistance
Baker-Nathan effect
carbenium ion
carbocation
carbonium ion
classical carbocation
non-classical cation
heteronium ion
neighboring group participation
solvolysis
superacid
Wagner-Meerwein rearrangement

Additional Problems

9-16 The homoallyl cation seldom rearranges to the allyl cation, even though the latter is stabilized by resonance. What inferences can you make from this observation—beyond the obvious one that the activation energy for this rearrangement must be too high to permit rapid rearrangement?

9-17 The closure of the tetracyclic carbon skeleton of steroids involves a cascade of carbocation reactions that is initiated by protonation of the epoxide oxygen of squalene epoxide. The starting epoxide is shown, along with the intermediate tetracyclic carbocation, and the final product, lanosterol.

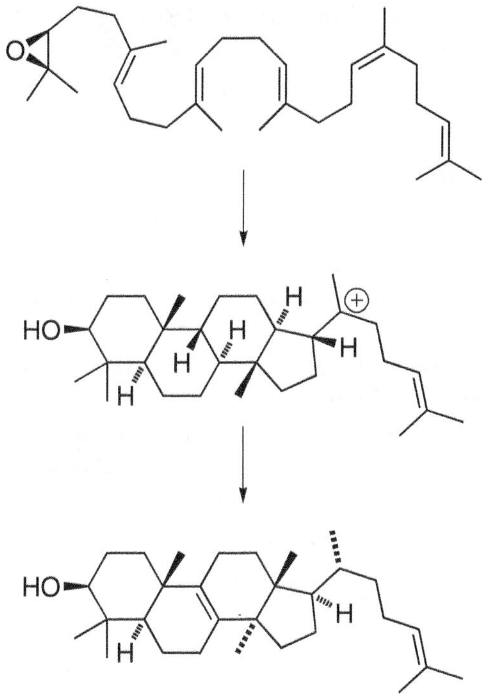

(a) Write a mechanism that accounts for the formation of the tetracyclic intermediate carbocation.

(b) Does any part of the mechanism of formation of the tetracyclic carbocation involve an unusual stereochemical or regiochemical outcome?

(c) The final tetracyclic product is formed by the rearrangement of several groups. Where it is possible to deduce the stereochemistry of the reaction and what is the stereochemistry of each rearrangement?

9-18 1-Adamantyl halides and sulfonates are bridgehead compounds that react under solvolytic conditions to give substitution products, and the bicyclo[2.2.1]hept-1-yl halides and sulfonates are resistant to both solvolysis and bimolecular nucleophilic substitution. Rationalize the difference in the reactivity of these two systems.

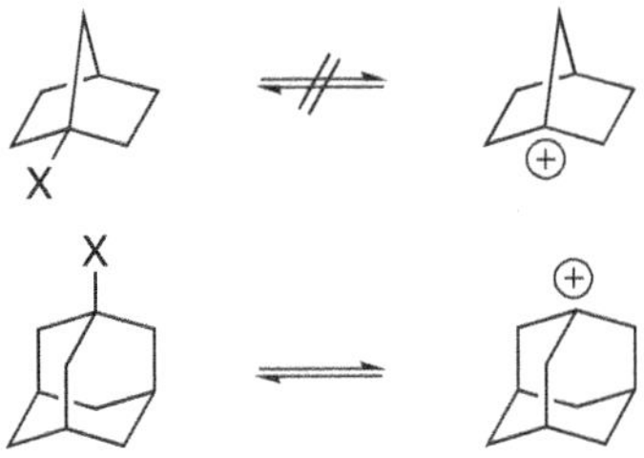

9-19 The reaction of the amine shown with nitrous acid gives a diazonium ion. This ion loses nitrogen to give a carbocation that gives compounds (e.g., the alcohol shown) based on the tricyclic skeleton shown as the major products of the reaction. Give the structure of this cation intermediate, and rationalize the formation of this product from it.

9-20 How many mechanistically different ways are there, potentially, to generate the 2-norbornyl cation? If all these methods give a product with the same ^{1}H and ^{13}C NMR spectra, regardless of the origin of the cation, what deductions can one make about the cation structure? The ^{1}H NMR spectrum of the 2-norbornyl cation in SbF_5-FSO_2Cl-SO_2F_2 solution at −100°C and −158°C is summarized in the table.

Temp	δ (area)	δ (area)	δ (area)	δ (area)	δ (area)
− 100°C	4.92 (4H)		2.82 (1H)		1.93 (6H)
−158°C	6.75 (2H)	3.17 (2H)	2.82 (1H)	2.13 (4H)	1.37 (2H)

9-21 When the alcohol shown is heated with triphenylphosphine in carbon tetrachloride, chloroiceane, a molecule with a carbon skeleton that mimics the structure of ice, is formed. The oxophosphonium ion is an intermediate in this reaction. Suggest a mechanism for the formation of the iceane skeleton.

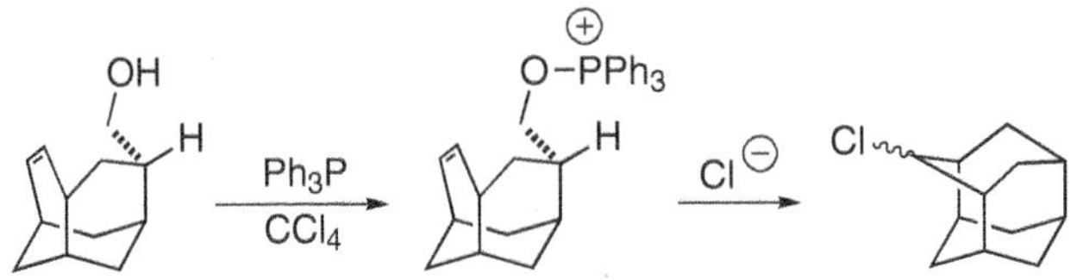

9-22 Predict the major carbocation formed when each of the following alkenes is protonated. If the major cation formed is not the first cation formed, provide a reason.

(a) 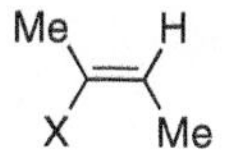 (i) X=Ph; (ii) X=Me; (iii) X=OMe; (iv) X=NMe_2; (v) X=$SiMe_3$; (vi) X=SPh

(b) 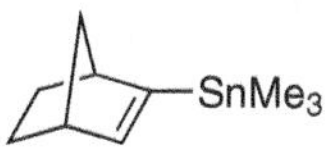

(c) 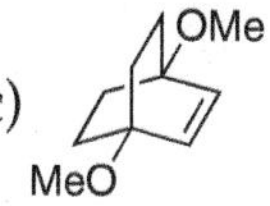

(d) 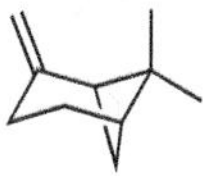

9-23 The alcohol shown undergoes facile S_N1 substitution, catalyzed by acid. The ^{13}C NMR spectrum of the cation intermediate formed in this reaction shows that it possesses an alkene double bond. What is the structure of the cation, and why does it form? What is the mechanism of the second step of the S_N1 reaction?

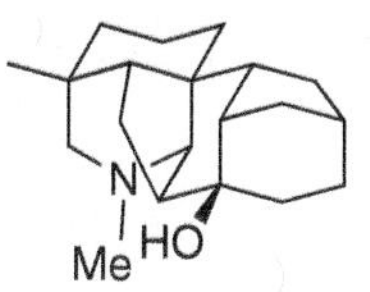

9-24 In the reaction shown, the Grignard reagent does not open the epoxide by the expected nucleophilic attack to give the tertiary alcohol. Instead, the product shown is formed. Write a reasonable mechanism for this reaction and provide a reasonable explanation for why the reaction proceeds as observed.

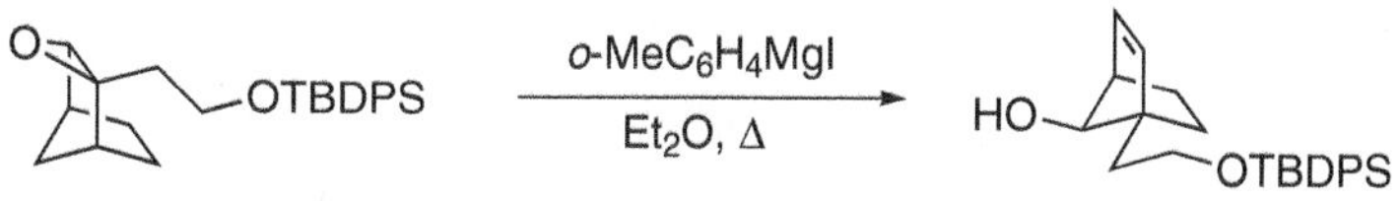

9-25 How does the reaction shown occur? How would you expect changing the solvent from toluene to DMF to affect the product ratio?

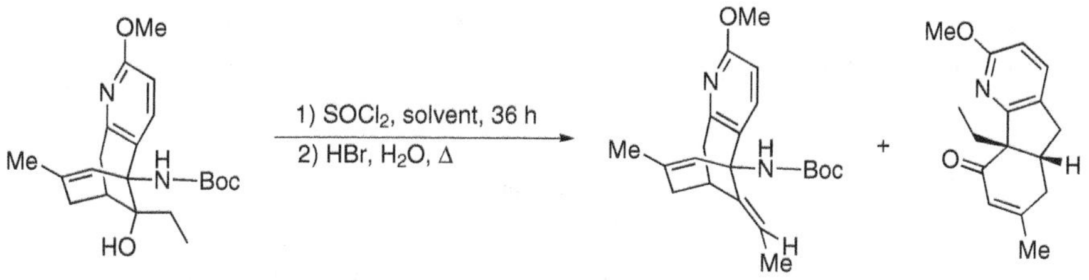

CHAPTER TEN

Organic Reactions II

Synthesis Using Carbocations to Form C—C Bonds

10.1 Introduction

In the last chapter, we focused our attention on the preparation, structure, and properties of carbocations. In this chapter, we will shift the focus to reactions where those carbocations are key participants. The most important types of these reactions are addition to alkene π bonds as well as electrophilic substitution in alkenes and arenes.

As we have seen, carbocations can be stabilized by three different mechanisms: (1) hyperconjugation, (2) conjugation with a carbon-carbon π bond or an aromatic ring (Example **10.1**), and (3) conjugation with an atom bearing a lone pair of electrons (Example **10.2**). All three types of carbocations are key intermediates in synthetically useful C—C bond-forming reactions. Simple carbocations stabilized by hyperconjugation or conjugation with a C—C π bond are important in the Friedel-Crafts alkylation and Mukaiyama alkylation reactions. Carbocations stabilized by an adjacent oxygen or nitrogen are important in the Friedel-Crafts acylation and its derivative reactions, as well as the Mukaiyama aldol addition, the Mannich reaction, and the Prins reaction.

It is worthwhile noting that both classes of electrophiles are represented in variants of the Friedel-Crafts and Mukaiyama reactions.

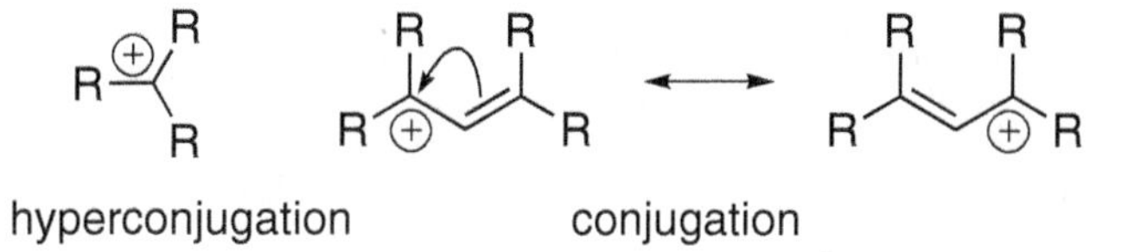

(10.1)

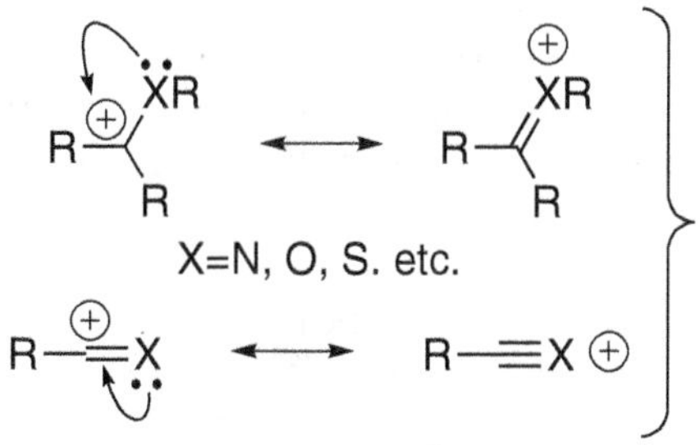

(10.2)

The relative strength of electrophilic reagents can be approximated by examining the rates of their reactions with aromatic compounds. The strongest electrophiles (e.g., nitronium ions) react with all aromatic compounds, whereas less reactive electrophiles (e.g., alkyl cations) react with aromatic compounds provided that they do not carry any strongly electron-withdrawing substituents. The weakest electrophiles (carbenes and arenediazonium ions) react only with the most electron-rich aromatic compounds.

10.2 Friedel-Crafts Reactions

While studying the effects of metallic aluminum on alkyl halides in benzene, French chemist Charles Friedel[1] and his American student James Mason Crafts[2] observed that (1) copious quantities of hydrogen chloride were evolved after a brief induction period and that (2) the benzene was converted to a mixture of alkylbenzenes.[3] Later they found that aluminum chloride was actually the catalyst of the reaction. Subsequently, they found that this reaction will also work with acid chlorides to give ketones, and that it is, in fact, quite a general reaction when a halide is treated with an aromatic compound in the presence of a Lewis acid. Since the pioneering efforts of Friedel and Crafts, a plethora of analogous reactions has been discovered, but all are variations on the same theme: the reaction between a carbocation (or stabilized carbocation) electrophile with an aromatic ring to give a substitution product.

This reaction is quite general and adaptable to most alkyl halides that will undergo S_N1 or S_N2 reactions. It is now known as the ***Friedel-Crafts alkylation***,[4] and it is a very important method for the formation of carbon-carbon bonds between an aromatic ring and an alkyl group.

The reaction between an aromatic hydrocarbon and an acid chloride or anhydride[5] (especially a mixed trifluoroacetic, phosphoric, or trifluoromethanesulfonic anhydride) in the presence of a Lewis acid or a strong Lowry-Brønsted acid produces an aromatic ketone. The mechanism of the reaction involves either a Lewis acid complex of the acid derivative, or an acylium ion. The reaction, which is known as ***Friedel-Crafts acylation***, is viewed as being more useful than the alkylation reaction for two reasons: (1) it gives the product of monosubstitution only and (2) it gives a product with a functional group in the side chain.

The first step in these electrophilic aromatic substitutions is always the same: the addition of the electrophile to the conjugated π bond system of the aromatic ring (Figure 10.1). Thus, there are two major factors that control the outcome of any substitution of this type: (1) the nature of the electrophile, which may be a simple alkyl cation or its equivalent (**10.3**), an oxonium or iminium ion (**10.4**), or an acylium or nitrilium ion (**10.5**) and (2) the nature of the arene. We will now examine these in turn.

The Electrophile: Generating Carbocation Electrophiles

Simple carbocation electrophiles can be formed by the heterolysis of carbon-halogen bonds (Examples **10.6** and **10.7**), promoted by strong Lewis acids such as $AlCl_3$, $AlBr_3$, $TiCl_4$, $FeBr_3$ or $SnCl_4$. Allylic carbocations can be formed by heterolysis of halides in the presence of silver ions. The conversion of alcohols to carbocations (Example **10.8**) can be accomplished in one step by means of strong acids—the poor nucleophilicity of their conjugate bases has made polyphosphoric acid and trifluoroacetic acid popular for this;

1. **Charles Friedel (1832–1899)** entered Wurtz' laboratory in 1854 (Dr ès-sc, 1859) and rose through the École des Mines from conservator in 1856 to Professor of Mineralogy in 1876. He became Director of the Paris Academy in 1878 and succeeded Wurtz as Professor of Organic Chemistry at the Sorbonne (1884). For more biographical details, see: Willemart, A. *J. Chem. Educ.* **1949**, *26*, 3.

2. **James Mason Crafts (1839–1917)** was educated at Harvard (SB, 1858) and then traveled to Europe where he studied under Bunsen and Wurtz. In Paris, he met Charles Friedel, with whom he collaborated for most of the next three decades (1860–1867; 1874–1891). He was Professor at Cornell (1867–1871) and MIT (1871–1874). He returned to the United States permanently in 1891—as Professor (1891–1898) and President (1898–1900) at the Massachusetts Institute of Technology. For more biographical details, see: Ashdown, A.A. *J. Chem. Educ.* **1928**, *5*, 911.

3. Friedel, C.; Crafts, J.M. *Comptes Rend.* **1877**, *84*, 1392, 1450.

4. (a) Olah, G.A.; Krishnamurti, R.; Prakash, G.K.S. In *Comprehensive Organic Synthesis*; Trost, B. M., Fleming, I., Eds. (Pergamon Press: Oxford, 1991) Vol. 3, Chapter 1.8, p 293. (b) Roberts, R.M.; Khalaf, A.A. *Friedel-Crafts Alkylation Chemistry* (Marcel Dekker: New York, 1984). (c) Olah, G. A., Ed. *Friedel-Crafts Chemistry* (Wiley: New York, 1973). (d) Olah, G. A., Ed. *Friedel-Crafts and Related Reactions* (Wiley-Interscience: New York, 1963–65); Vols. 1–4.

5. Review: Berliner, E. *Org. React.* **1949**, *5*, 229.

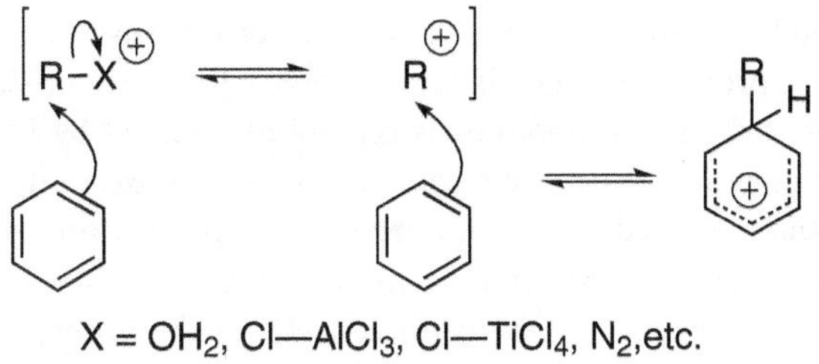

(10.3)

Figure 10.1 The first step in the Friedel-Crafts class of electrophilic aromatic substitutions

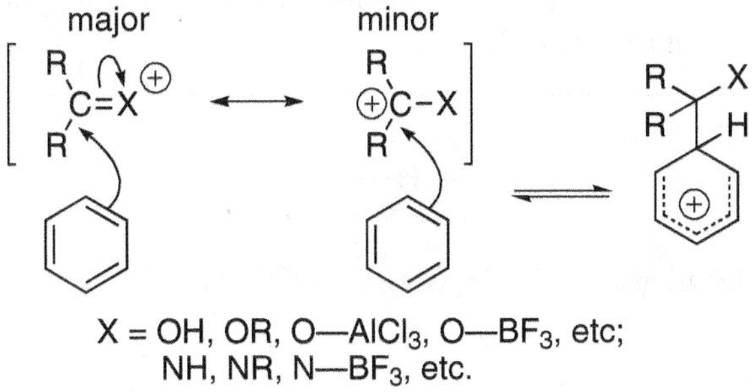

(10.4)

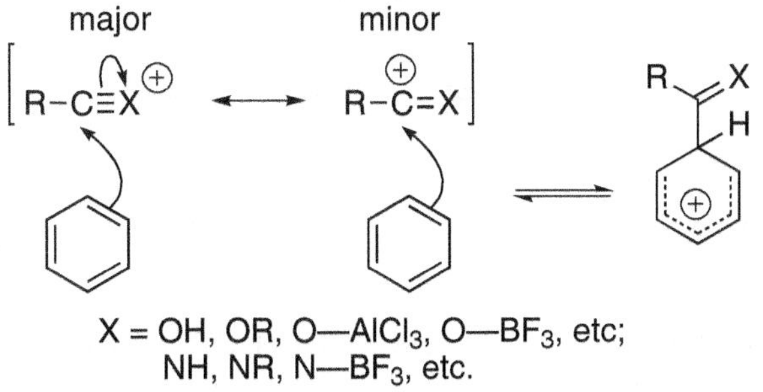

(10.5)

sulfuric acid tends to give higher amounts of tarry by-products and so is less frequently used. Primary alkylamines react with nitrous acid to give alkyldiazonium ions that rapidly lose molecular nitrogen to give a "hot" carbocation (Example **10.9**).

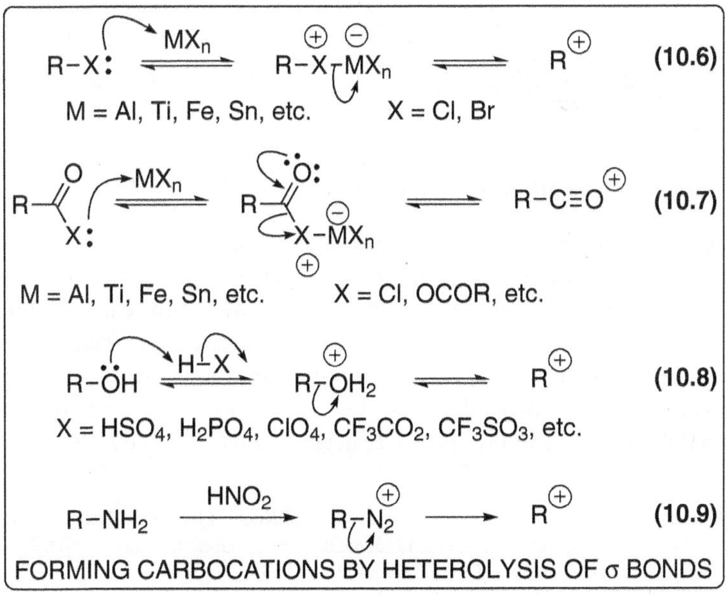

Protonation of π-bonded Lewis bases also provides a convenient route to carbocation electrophiles: alkenes add protons and Lewis acids to give carbocations (Example **10.10**), and the oxygen atom of carbonyl compounds is readily protonated by Lowry-Brønsted acids, or complexed by Lewis acids to give a resonance-stabilized oxonium ion (Examples **10.11** and **10.12**). This is the preferred process for generating carbocation equivalents from acids and their derivatives. A similar reaction occurs with the nitrogen atom of nitriles to give resonance-stabilized nitrilium cations (Example **10.13**) or their more complex equivalents (Example **10.14**). Less commonly used methods for generating carbocations include the oxidation of carboxylic acids by lead tetraacetate in the presence of pyridine[6] and the electrochemical oxidation of free radicals.

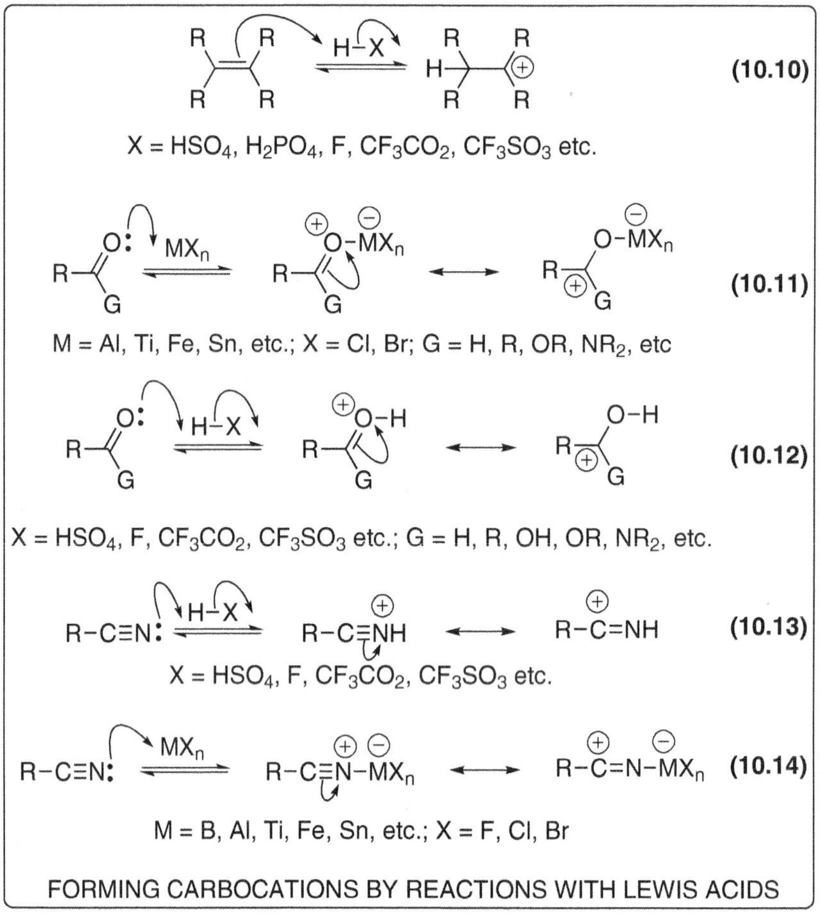

The Electrophile: Strength and Reactivity

The electrophile in the Friedel-Crafts alkylation using an alkyl chloride and aluminum chloride, for example, may be viewed as either the Lewis acid complex of the alkyl halide for halides that do not react well by the S_N1 mechanism[7] and as the carbocation for halides that do. The reacting species is therefore determined by the equilibrium constant of the dissociation reaction in Example **10.6**. In acylation reactions, using acid chlorides and aluminum chloride, for example, the electrophile may also be either a Lewis acid complex of the acyl halide (Example **10.15**) or the free acylium ion (Example **10.16**).

---

6. (a) Sheldon, R.A.; Kochi, J.K. *Org. React.* **1970**, *19*, 279. (b) Kochi, J.K. *J. Am. Chem. Soc.* **1965**, *87*, 3609. (c) Corey, E.J.; Casanova, J., Jr. *J. Am. Chem. Soc.* **1963**, *85*, 165, 169. (d) Gream, G.E.; Wege, D. *Tetrahedron Lett.* **1967**, 503. (e) Theine, A.; Traynham, J.G. *J. Org. Chem.* **1974**, *39*, 153.

7. Brown, H.C.; Grayson, M. *J. Am. Chem. Soc.* **1953**, *75*, 6285.

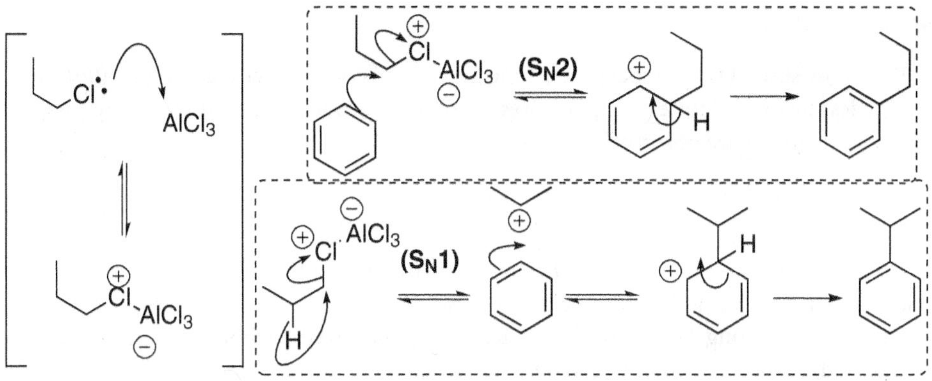

Figure 10.2 Competing S_N2 and S_N1 pathways for the Friedel-Crafts alkylation of benzene with 1-chlropropane

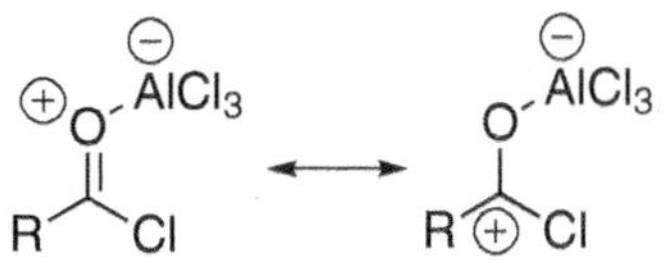

Lewis acid complex

(10.15)

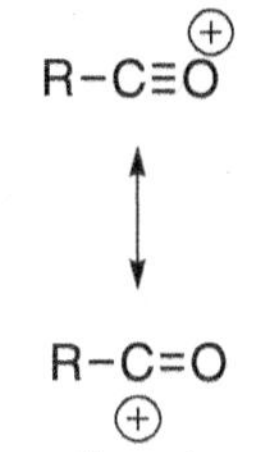

acylium ion

(10.16)

Depending on whether the active electrophile is the free alkyl carbocation or the Lewis acid–Lewis base complex of the alkyl halide, the Friedel-Crafts alkylation reaction can be viewed in two ways. It may be considered either an S_N2 reaction, with the arene acting as the nucleophile to displace the halide ion complex of the Lewis acid or an S_N1 reaction, with the leaving group being lost first, followed by trapping of the carbocation with the arene, which again functions as the nucleophile (Figure 10.2).

Benzene and alkylbenzenes react with *n*-propyl chloride in the presence of aluminum chloride to give mainly the *n*-propylarene at low temperature, where little of the free carbocation is formed (Example **10.17**) and mainly the isopropylarene at high temperatures, where the formation of the free carbocation from the complex is more favored (Example **10.18**).[8] The reaction of isobutyl chloride and benzene catalyzed by aluminum chloride (Example **10.19**) gives only *t*-butyl-benzene.[9] The amount of the rearranged alkylbenzene is higher when less reactive arenes are used, consistent with a competition between the dissociation of the Lewis acid complex of the halide (the S_N1 pathway) and the direct displacement by the arene (the S_N2 pathway).

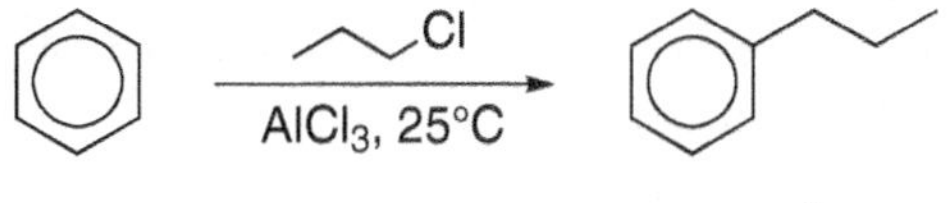

(10.17)

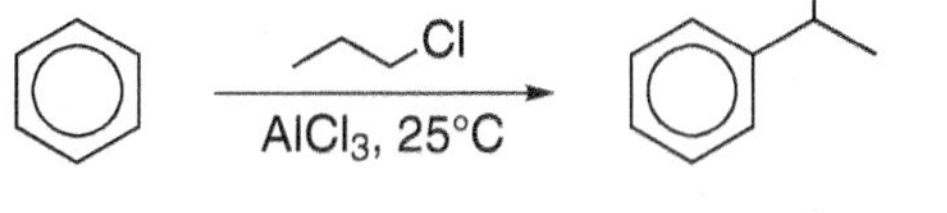

(10.18)

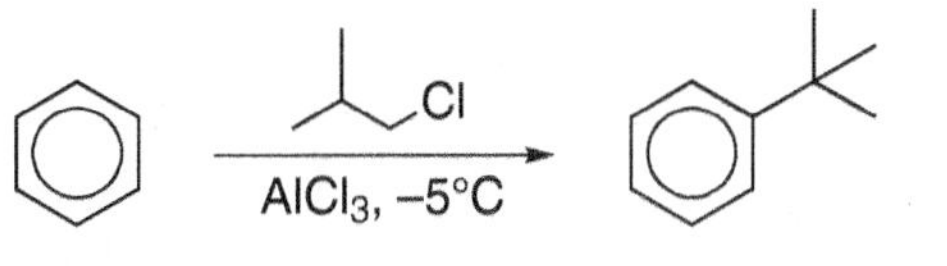

(10.19)

8. (a) Ipatieff, V.N.; Pines, H.; Schmerling, L. *J. Org. Chem.* **1940**, *5*, 253. (b) Roberts, R.M.; Shienthong, D. *J. Am. Chem. Soc.* **1960**, *82*, 2851. However, it has also been reported that 1-chloropropane gives predominantly isopropylbenzene, even at low temperatures: (c) Roberts, R.M.; Shienthong, D. *J. Am. Chem. Soc.* **1960**, *82*, 732. (d) Marsi, K.L.; Wilen, S.H. *J. Chem. Educ.* **1963**, *40*, 214.

9. Konovalov, M.I. *J. Russ. Phys.-Chem. Soc.* **1895,** *27*,156. (b) Konowaloff, M.I. *Bull. Soc. Chim. France* **1896**, *16*, 864.

Worked Problem

10-1 Why should the dissociation of the Lewis acid complex with an alkyl halide be less favored at lower temperatures, so that the products alkylation without rearrangement are observed?

§Answer below.

The active electrophile can also be generated by the reaction between alkenes or alcohols with strong Lowry-Brønsted acids. The reactions of carbonyl compounds with strong Lewis acids or strong Lowry-Brønsted acids give stabilized carbocations. Cations from conjugated carbonyl compounds react through the β carbon to give saturated ketones, as illustrated by the reaction (Example **10.20**) between mesityl oxide and benzene.[10] An example of the reaction of a saturated ketone with an arene is the synthesis of bisphenol A from phenol and acetone (Example **10.21**).

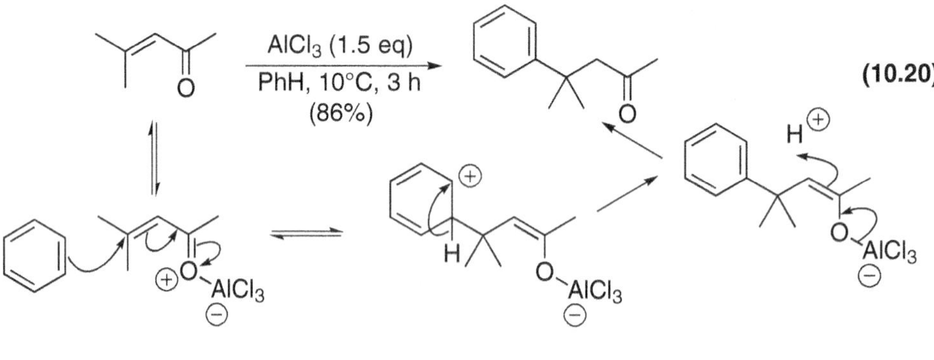

(10.20)

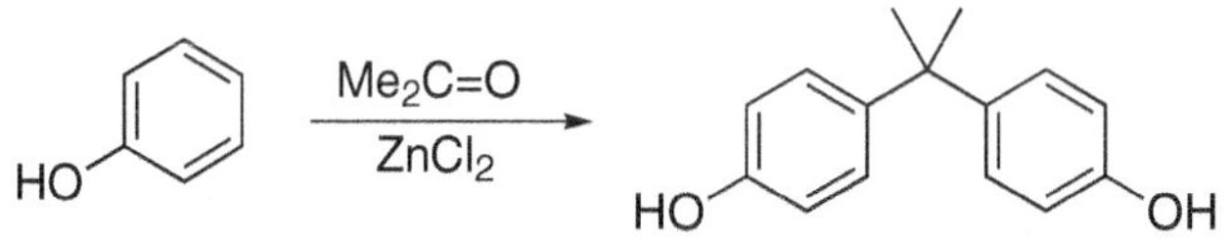

(10.21)

The Arene: Reactivity and Orientation

The Friedel-Crafts reaction requires an aromatic ring that is not excessively deactivated toward electrophilic substitution: arenes carrying *m*-directing groups do not react, so nitrobenzene and *o*-dichlorobenzene can be used as solvents for the reaction. Generally speaking, aromatic compounds react more rapidly as the electron density in the ring increases. Thus, five-membered heterocyclic compounds (furan, pyrrole, indole, and thiophene) all react more rapidly than benzene with electrophiles, because the average electron density in the ring is 1.2 π electrons per position.

As with simple benzenoid aromatics, the orientation of substitution in these heterocycles can be predicted on the basis of resonance theory. Thus, furan, pyrrole, and

10. Koelsch, C.F.; LeClaire, C.D. *J. Org. Chem.* **1941**, 6, 516.

§ Answer to Worked Problem:
In the complex, all atoms have complete valence shell octets, but in the simple alkyl carbocation, the charged carbon atom does not. Thus, the ΔH_{react} for dissociation is positive. The dissociation of the complex, on the other hand, converts a single particle into two, so the ΔS_{react} for dissociation is also positive. Thus, we find that at low temperature, where enthalpy dominates, the non-dissociated complex is favored, and at higher temperatures, where entropy becomes more important, the free carbocation becomes more favored.

thiophene all react with electrophiles to give substitution at the 2-position of the ring (Example **10.22**). The major canonical form of the intermediate onium ion in these reactions is an ion with linear conjugation in the conjugated system and three major canonical forms. Substitution at the 3-position gives a cross-conjugated intermediate cation (Example **10.23**) with only two major canonical forms. In indoles, on the other hand, the major product of electrophilic substitution is at the 3-position. In this case, only one (Example **10.24**) of the two major isomeric cations (Examples **10.24** and **10.25**) that have an iminium ion structure, only **10.24** retains the benzenoid sextet (the Clar sextet[11] that we discussed in Chapter 7), so this predicts that the major product will be the 3-substituted indole.

(10.22) 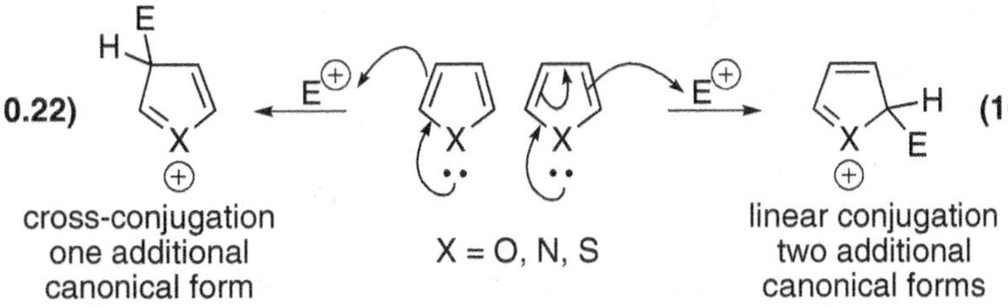 (10.23)

(10.24) 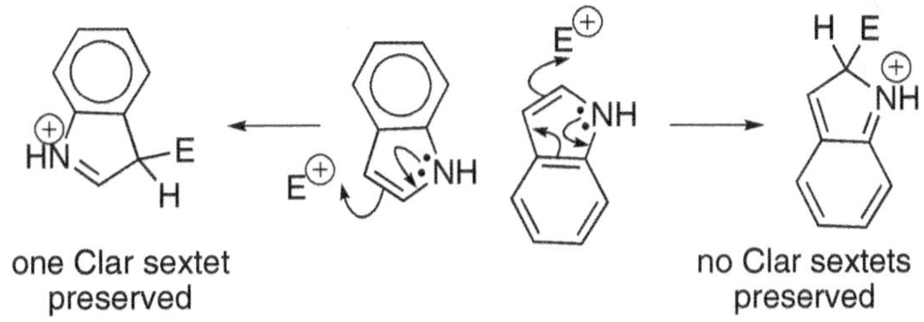 (10.25)

A similar situation occurs in fused aromatic hydrocarbons. Attack of the electrophile at position 1 of a naphthalene ring system gives a carbocation in which there are two contributing structures (Examples **10.26** and **10.27**) that retain the Clar sextet. Attack at position 2, on the other hand, gives a carbocation (Example **10.28**) with only one such structure. We predict that electrophilic aromatic substitution in naphthalene under kinetic conditions will give the 1-substituted naphthalene as the major product. It does; electrophilic aromatic substitution of naphthalene at low temperatures gives the 1-substituted naphthalene (Example **10.29**). The reaction is reversible, however, and the same reaction under thermodynamic conditions gives the much less sterically hindered 2-substituted naphthalene (Example **10.30**). Likewise, Friedel-Crafts acylation of 1-substituted naphthalenes leads to substitution at the 4-position.[12] 2-substituted derivatives give products where the substitution has occurred at the 8-position,[13] unless the electrophile is hindered, in which case the 2,6-disubstituted naphthalene may be obtained as the major product. The effects can be fairly subtle, as illustrated by Example **10.31**, where the product formed is the kinetic 1-alkyl product, and Example **10.32**,[14] where the product formed is the thermodynamic product. All five-membered heterocycles and all fused-ring aromatic hydrocarbons react more rapidly with electrophiles than benzene does—the approximate order of reactivity is pyrrole > furan > thiophene > anthracene > naphthalene > benzene—and all will react with weaker electrophiles than does benzene. By contrast, the six-membered

11. (a) Clar, E. *Aromatische Kohlenwasserstoffe*, 2nd ed. (Springer: Berlin, 1952). (b) Clar, E. *Polycyclic Hydrocarbons* (Academic: London, 1964). (c) Clar, E. *The Aromatic Sextet* (John Wiley: London, 1972.).

12. Jacobs, T.L.; Winstein, S.; Ralls, J.W.; Robson, J.H. *J. Org. Chem.* **1945**, *11*, 27.

13. (a) Leonard, N.J.; Hyson, A.M. *J. Org. Chem.* **1947**, *13*, 164. (b) Winstein, S.; Jacobs, T.L.; Day, B.F. *J. Org. Chem.* **1947**, *13*, 171.

14. For an example, see: Arai, K.; Ohara, Y.; Iizumi, T.; Takakuwa, Y. *Tetrahedron Lett.* **1983**, *24*, 1531.

nitrogen heterocycles, pyridine, quinoline, and isoquinoline are all electron-deficient, and they are resistant to electrophilic attack by carbon electrophiles.

(10.26) (10.27) (10.28)

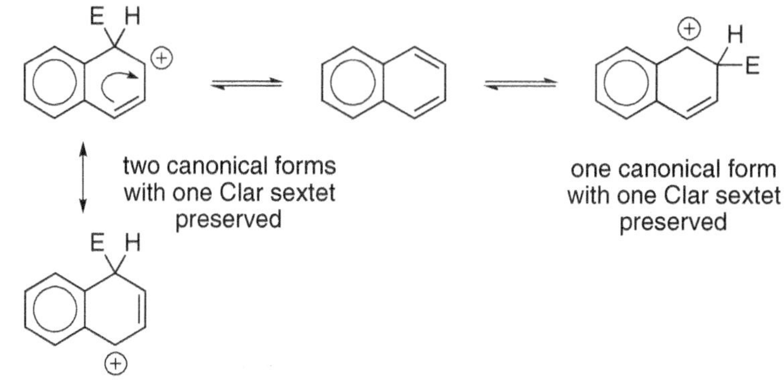

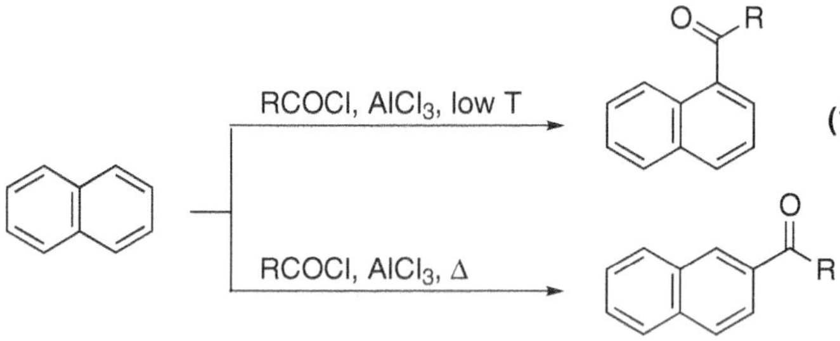

(10.29)

(10.30)

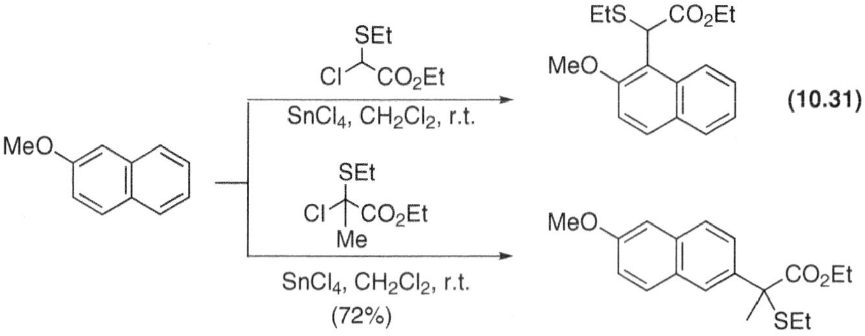

(10.31)

(10.32)

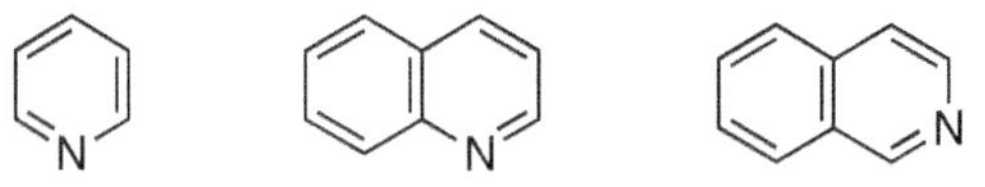

pyridine quinoline isoquinoline

Problems

10-1 Predict the major product of the acylation of anthracene and phenanthrene under kinetic conditions (CH_3COCl, $AlCl_3$, CS_2, 0°C).

anthracene 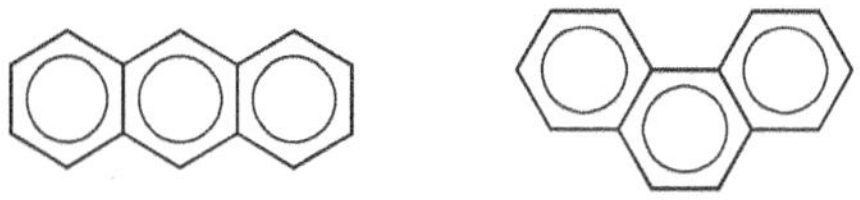 phenanthrene

10-2 Provide a rational explanation for the differences in the reactions leading to the products shown in **10.31** and **10.32**. Note that the reaction conditions are the same.

Complicating Factors in Alkylation: Dealkylation, Disproportionation, and Rearrangement of Products

The intermediate in the Friedel-Crafts alkylation is a benzenonium cation, which usually loses proton to reestablish aromaticity in the ring. Other than proton loss, there are two major reactions that these cations exhibit: (1) rearrangement (e.g., Example **10.33**), which changes the isomer distribution; and (2) dealkylation (e.g., Example **10.34**), where an alkyl cation is lost or transferred to another molecule of arene rather than a proton.[15] In an extensive study of the alkylation of anisole and toluene, Olah showed that anisole and toluene give predominantly *o,p* substitution under kinetic conditions, but rearrangement of the kinetic product occurs under thermodynamic conditions.[16] Although these reactions are often undesirable in laboratory syntheses, the dealkylation reaction is important in the disproportionation of toluene into benzene and xylene,[17] both of which have much more commercial value.

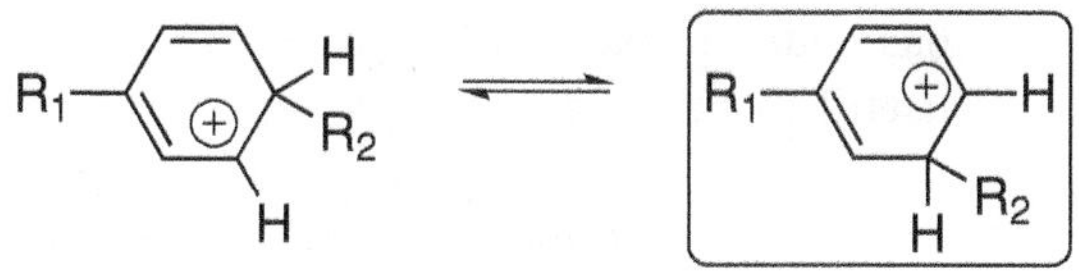

(10.33)

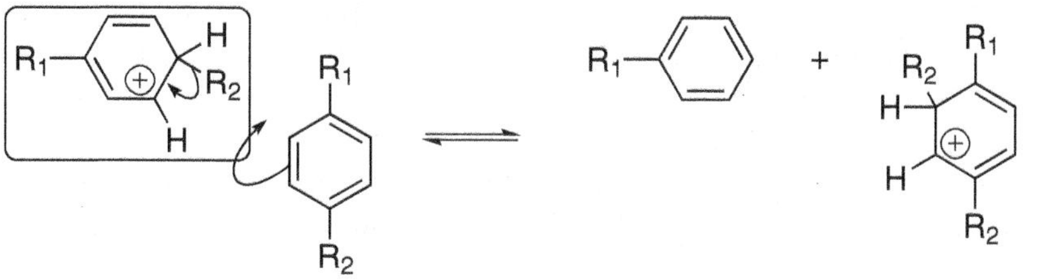

(10.34)

Acylation

In solution, alkylation by the Friedel-Crafts method often leads to polyalkylation due to a combination of factors, including the greater reactivity of the alkylbenzenes than benzene itself toward the electrophile. For this reason, Friedel-Crafts acylation has frequently been preferred over alkylation, because the product does not undergo further substitution.

"Greening" the Friedel-Crafts Acylation

The use of traditional metal halide catalysts for the Friedel-Crafts acylation (e.g., $AlCl_3$, $TiCl_4$, $SnCl_4$, $FeBr_3$) requires the use of more than one equivalent of the catalyst, because the formation of a strong complex between the Lewis acid and the ketone product removes one equivalent of the catalyst from the reaction mixture. This complex must then be hydrolyzed during isolation of the product; therefore, the catalyst cannot be reused and a large quantity of hazardous waste is generated.

Reduction of hazardous wastes is one of the core principles of "green" chemistry, and the Friedel-Crafts acylation, which generates large quantities of hazardous wastes, is a prime candidate for testing new, "greener" methods of synthesis. One of the earliest ways to reduce the amount of hazardous waste was developed for the Hoechst-Celanese synthesis of *p*-isobutylacetophenone by using anhydrous hydrogen fluoride as the acid

15. (a) Nightingale, D.F. *Chem. Rev.* **1393**, *25*, 329. (b) Nightingale, D.; Smith, L.I. *J. Am. Chem. Soc.* **1939**, *61*, 101. (c) Nightingale, D.; Carton, B., Jr. *J. Am. Chem. Soc.* **1940**, *62*, 280.

16. Olah, G.A.; Olah, J.A.; Ohyama, T. *J. Am. Chem. Soc.* **1984**, *106*, 5284.

17. E.g., see: Kaeding, W.W.; Chu, C.; Young, L.B.; Butter, S.A. *J. Catal.* **1981**, *69*, 392.

Table 10.1 Effects of Substrate and Catalyst on Friedel-Crafts Acylation

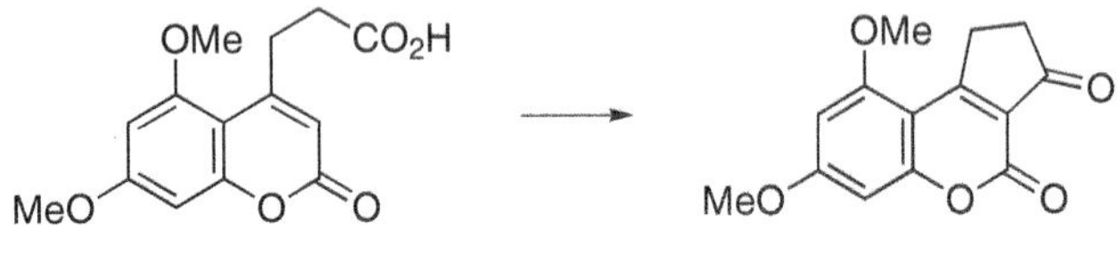

Reagent	Yield
(1) $(COCl)_2$; (2) $AlCl_3$ (3 eq.), CH_2Cl_2, r.t.	31
(1) $(COCl)_2$; (2) $Sc(OTf)_3$ (1 eq.), CH_2Cl_2, r.t.	46
(1) $(COCl)_2$; (2) $Sc(OTF)_3$ (1 eq.), TfOH (1 eq), CH_2Cl_2, r.t.	64
(1) $(CF_3CO)_2O$; (2) $Sc(OTf)_3$ (1 eq.), CH_2Cl_2, 40°C	0
(1) $(CF_3CO)_2O$; (2) $ZnCl_2$ (1 eq.), HOAC, 60°C	0
HF, –78°C to r.t.	36
$Sc(OTf)_3$ (1 eq.), $MeNO_2$, 60°C	0
$Sc(OTf)_3$ (1 eq.), $LiClO_4$ (5 eq.), $MeNO_2$, 60°C	79

catalyst (Example **10.35**).[18] Unlike metal salts, this catalyst can be recycled and reused, which makes HF—despite being highly toxic and corrosive—a "green" catalyst for the reaction.

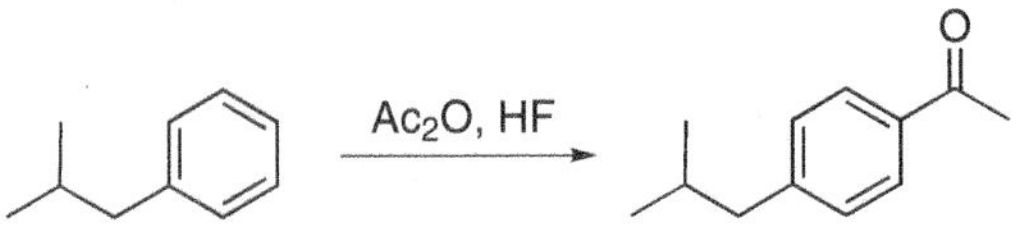

(10.35)

Other approaches to "greening" this reaction have focused on using scandium,[19] lanthanide,[20] gallium,[21] or bismuth[22] triflates as catalysts. A catalyst composed of hafnium (IV) triflate and lithium perchlorate in nitromethane has proved to be a highly efficient catalyst for alkylation and acylation reactions,[23] and a recent report has added samarium (II) iodide to the list of catalysts useful in the Friedel-Crafts acylation.[24] As part of the "green revolution" in synthetic organic chemistry, recent efforts have been focused on the development and use of ecocompatible catalysts—solid catalysts in particular[25]—in Friedel-Crafts reactions. The effects of a variety of catalyst and reactant combinations on a vinylogous intramolecular Friedel-Crafts acylation are shown in Table 10.1.[26] Note how in the reaction in Table 10.1, the electrophile attacks the alkene π bond that is conjugated with the methoxy substituent on the aromatic ring. The attack of an acylium ion on an alkene is not an uncommon reaction, especially when the carbocation formed is tertiary. An example is provided by the spontaneous cyclization (Example **10.36**) of the acid chloride from Corey's synthesis of a dolabellane diterpene.[27]

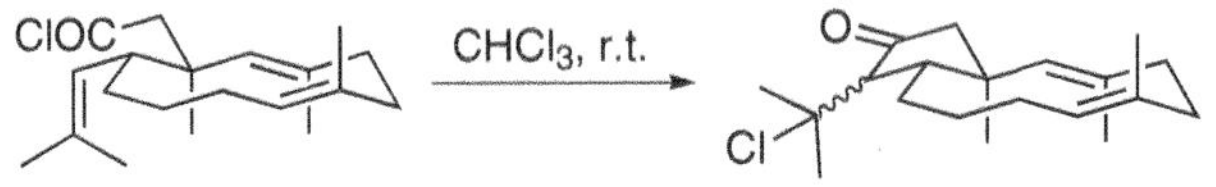

(10.36)

Worked Problem

10-2 What is the product of each of the reactions below?

(a) 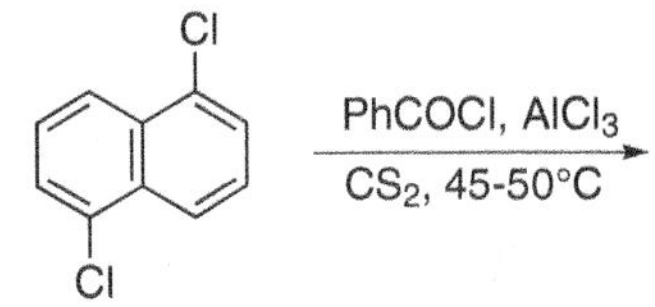

(b) 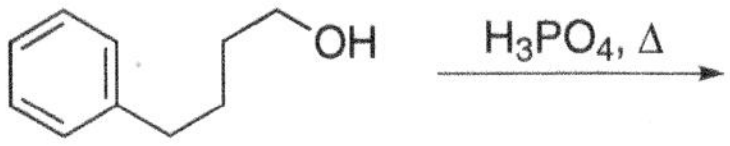

(continues)

18. Lindley, D.D.; Curtis, D.A.; Ryan, T.R.; de la Garza, E.M.; Hilton, C.B.; Kenesson, T.M. **U.S. Patent No. 5,068,448** [Nov. 26, 1991].
19. Kawada, A.; Mitamura, S.; Kobayashi, S. *Synlett* **1994**, 545.
20. Kobayashi, S.; Sugiura, M.; Kitagawa, H.; Lam, W.W.-L. *Chem. Rev.* **2002**, *102*, 2227.
21. Kobayashi, S.; Komoto, I.; Matsuo, J. *Adv. Synth. Catal.* **2001**, *343*, 71.
22. (a)Desmurs, J.-R.; Labrouillère, M.; Le Roux, C.; Gaspard, H.; Laporterie, A.; Dubac, J. *Tetrahedron Lett.* **1997**, *38*,8871. (b) Répichet, S.; Le Roux, C.; Dubac, J.; Desmurs, J.-R. *Eur. J. Org. Chem.* **1998**, 2743.
23. Hachiya, I.; Moriwaki, M.; Kobayashi, S. *Bull. Chem. Soc. Japan* **1995**, *68*, 2053.
24. Soueidan, M.; Collin, J.; Gil, R. *Tetrahedron Lett.* **2006**, *47*, 5467.
25. Sartori, G.; Maggi, R. *Chem. Rev.* **2006**, *106*, 1077.
26. Trost, B.M.; Toste, F.D. *J. Am. Chem. Soc.* **2003**, *125* 3090.
27. Corey, E.J.; Kania, R.S. *J. Am. Chem. Soc.* **1996**, *118*, 1229.

(c) 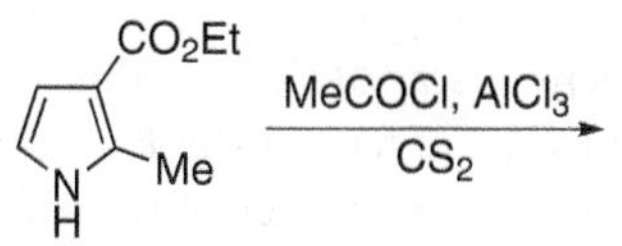 (d) 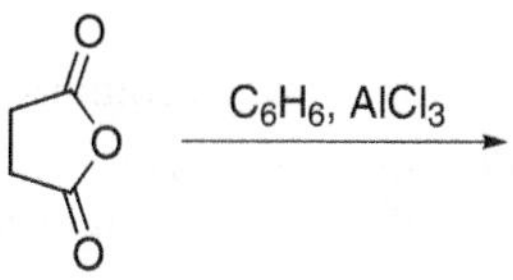

§Answers below.

Asymmetric Friedel-Crafts Alkylation

The rather severe conditions of the traditional Friedel-Crafts reaction tends to raise difficulties in carrying out such a reaction under asymmetric conditions. However, if the strong metallic Lewis acids are replaced by a proton from a modestly strong acid, one can successfully generate a chiral product in reasonable enantiomeric excess.

The two major successful approaches to inducing asymmetry in the Friedel-Crafts reaction have involved a C_2-symmetric catalyst to generate a chiral electrophile that then is then selectively attacked on one face of the alkene. In reactions involving proton transfer to generate the electrophile, polyphosphoric acid has been an especially popular reagent, because sulfuric acid often leads to undesired tar and resin formation and polyphosphoric acid does not present the hazards of anhydrous hydrogen fluoride. In an extension of this reaction, chiral phosphoric acid derivatives[28] such as Example **10.37** have been developed as catalysts for the (formal) asymmetric Friedel-Crafts alkylation of electron-rich arenes.[29] The common structural features of most of these chiral phosphoric acids are a 1,1'-binaphthyl core with sterically large hydrophobic groups (e.g., the 9-anthryl groups in the example shown) flanking the phosphate ester groups. This type of reaction is restricted to substitution in electron-rich arenes (e.g., pyrroles, as in Example **10.38**, and indoles, as in Example **10.39**). It has been proposed that the activated complex of the reaction is organized by means of specific hydrogen bonding (Figure 10.3). Although these reactions could equally well be described as conjugate addition of a nucleophilic arene to the Michael acceptor, it has become standard practice to refer to them

28. For reviews, see: (a) Schreiner, P.R. *Chem. Soc. Rev.* **2003**, *32*, 289. (b) Seayad, J.; List, B. *Org. Biomol. Chem.* **2005**, *3*, 719. (c) Pihko, P.M. *Angew. Chem. Int. Ed.* **2004**, *43*, 2062. (d) Taylor, M.S.; Jacobsen, E.N. *Angew. Chem. Int. Ed.* **2006**, *45*, 1520.

29. (a) Cai, Q.; Zhao, Z.-A.; You, S.-L. *Angew. Chem. Int. Ed.* **2009**, *48*, 7428. (b) Sheng, Y.-F.; Gu, Q.; Zhang, A.-J.; You, S.-L. *J. Org. Chem.* **2009**, *74*, 6899.

§ Answers to Worked Problem:

(a) 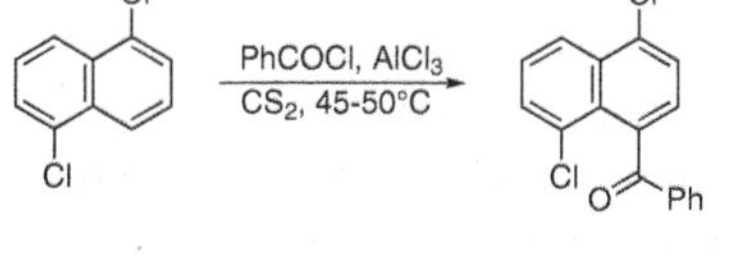

(b) 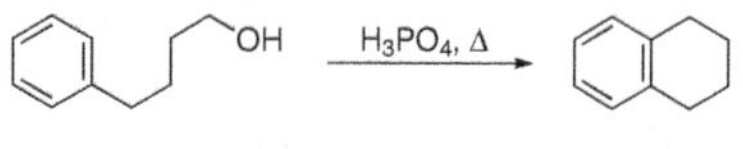

(c) 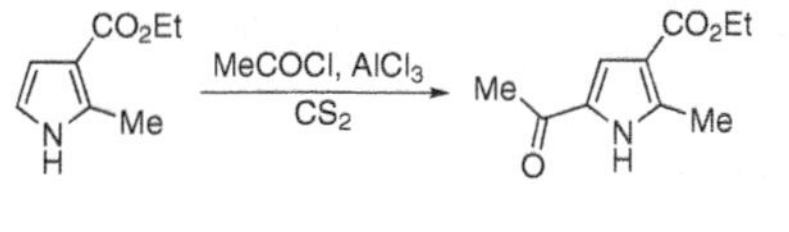

(d) 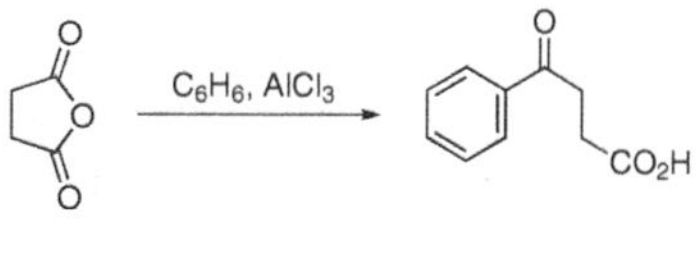

(a) In this reaction, the solvent is non-polar and the reaction temperature is relatively low; these are kinetic conditions. We expect that the major product will be the 4-benzoyl compound. (b) The phosphoric acid will protonate the alcohol to give the oxonium ion, which can then attack the aromatic ring to give a benzenonium ion with a six-membered ring fused to it, so the product is tetralin. Note how this intramolecular reaction allows the S_N2-like pathway to dominate the reaction. (c) This pyrrole has available 2- and 3-positions; electrophilic attack is favored at the 2-position, so the product will be the 5-acetylpyrrole. (d) Here the electrophile is generated from an anhydride; this Friedel-Crafts acylation will give the ketoacid shown.

as Friedel-Crafts alkylations because a new C—C bond is formed by substitution of a hydrogen atom on the aromatic ring. The stereochemistry of these additions is such that the *S* phosphoric acid catalyst leads to addition to the *re* face of the *E* alkene (where the group R_2 is assigned the highest priority), and the *R* enantiomer of the catalyst leads to addition to the *si* face.

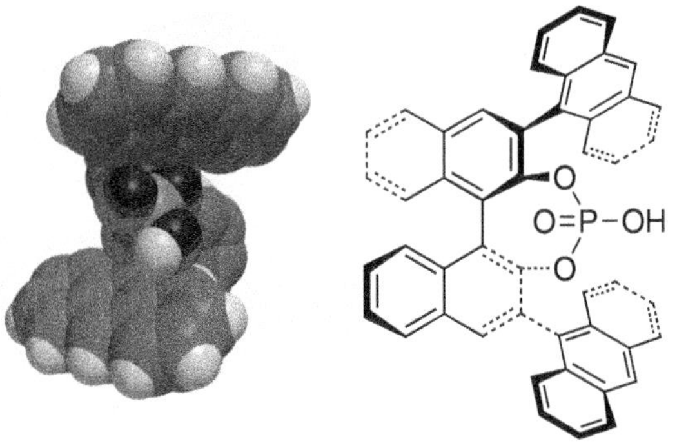

(10.37)

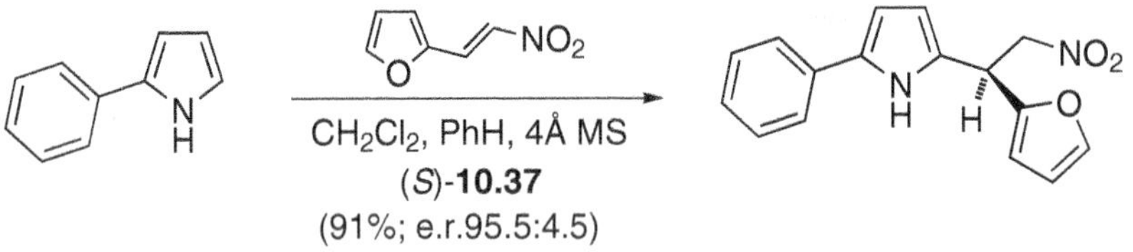

(10.38)[30]

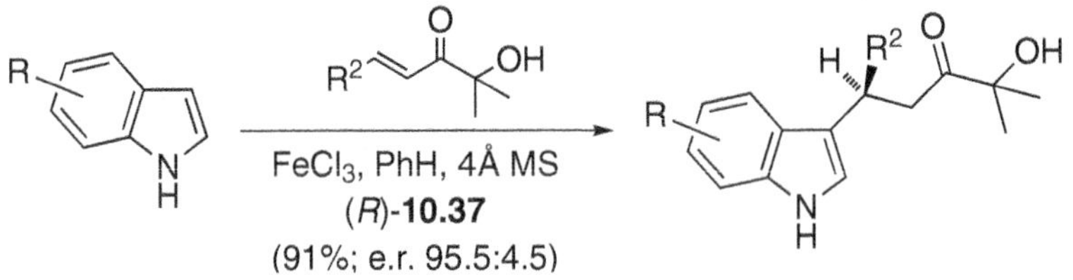

(10.39)[31]

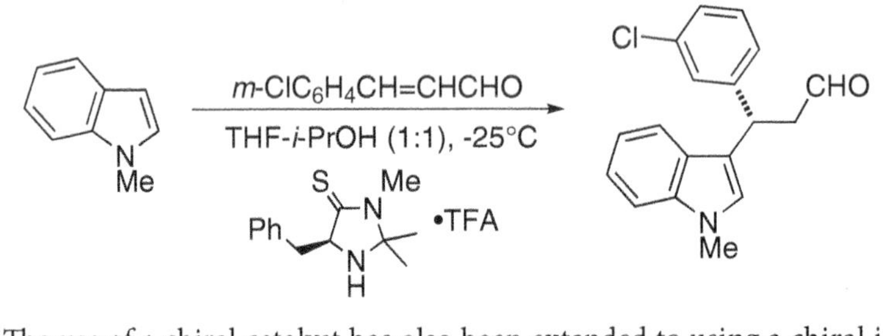

(10.40)[32]

The use of a chiral catalyst has also been extended to using a chiral imidazolidinone derivative to form chiral iminium ion electrophiles that then attack the electron-rich arene.[33] A recent report (Example **10.40**) shows the use of the conjugate acid of an imidazolidinethione as a catalyst for the chiral alkylation of indoles by cinnamaldehydes. In Example **10.40**, the *S* catalyst leads to addition to the *si* face of the alkene π bond of *E*-*m*-chlorocinnamaldehyde. It is worthwhile noting that the face selectivity of the two catalytic systems (the chiral phosphoric acids and the chiral imidazolidinethiones) is opposite for the same configuration in the catalyst.

---

30. For reviews, see: (a) Schreiner, P.R. *Chem. Soc. Rev.* **2003**, *32*, 289. (b) Seayad, J.; List, B. *Org. Biomol. Chem.* **2005**, 3, 719. (c) Pihko, P.M. *Angew. Chem. Int. Ed.* **2004**, 43, 2062. (d) Taylor, M.S.; Jacobsen, E.N. *Angew. Chem. Int. Ed.* **2006**, *45*, 1520.
31. Yang, L.; Zhu, Q.; Guo, S.; Qian, B.; Xia, C.; Huang, H. *Chem. Eur. J.* **2010**, *16*, 1638.
32. Liang, X.; Li, S.; Su, W. *Tetrahedron Lett.* **2012**, *53*, 289.
33. Paras, N.A.; MacMillan, D.W.C. *J. Am. Chem. Soc.* **2001**, *123*, 4370.

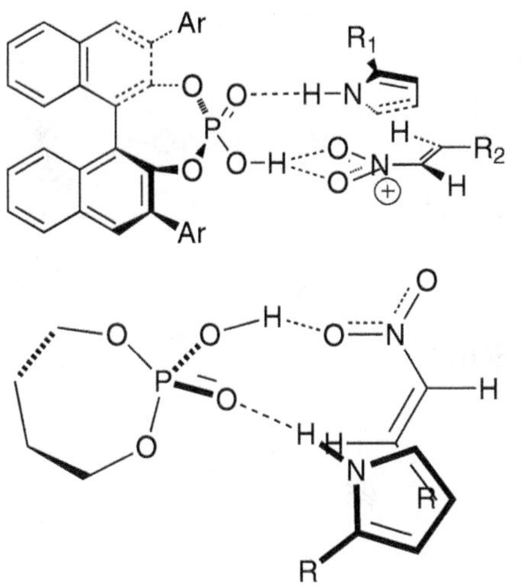

Figure 10.3 Hydrogen bonding organizes the activated complex of a Friedel-Crafts alkylation catalyzed by a chiral phosphoric acid derivative.

A variety of chiral C_2-symmetric *bis*-oxazoline ("box") ligands, which share the general structure **10.41**, have been used to chelate metal triflates as well as to provide effective chiral catalysts for the addition of electron-rich aromatic compounds to electron-deficient alkenes based on metals that include scandium,[34] copper (II),[35] and zinc.[36] Palomo and coworkers have used the copper (II) triflate complex (**10.42**) shown in Figure 10.4 for the alkylation of electron-rich aromatic compounds with α'-hydroxyenones; the reaction is proposed to proceed via chelated electrophile such as **10.43**.[37] In this square planar copper chelate, the *tert*-butyl group of one of the oxazoline rings blocks the *si* face of the β carbon of the enone, so the reaction occurs through the *re* face.

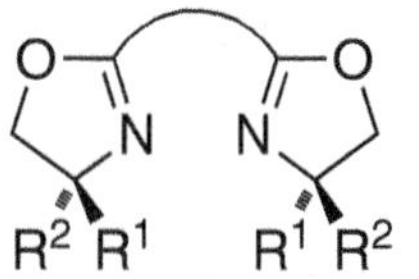

(10.41)

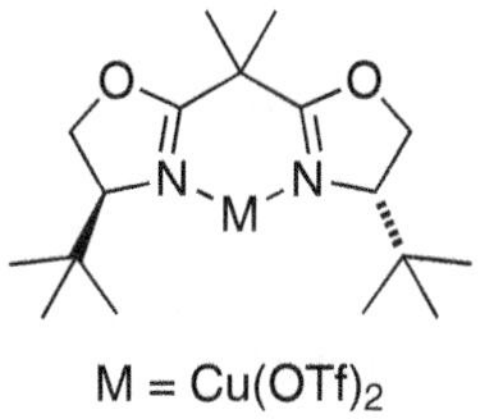

(10.42)

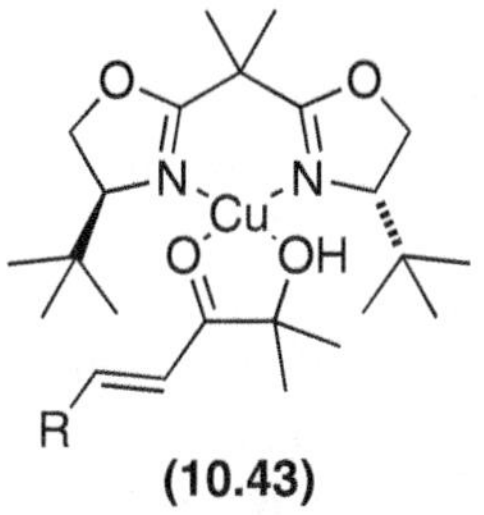

(10.43)

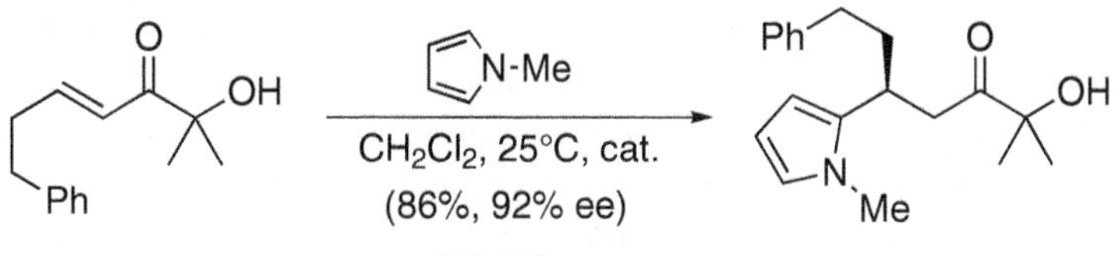

(10.44)

Figure 10.4 Use of a chiral copper (II) triflate "box" chelate to catalyze the asymmetric Friedel-Crafts alkylation

34. (a) Evans, D.A.; Fandrick, K.R.; Song, H.-J. *J. Am. Chem. Soc.* **2005**, *127*, 8942. (b) Evans, D.A.; Scheidt, K.A.; Fandrick, K.R.; Lam, H.W.; Wu, J. *J. Am. Chem. Soc.* **2003**, *125*, 10780.

35. (a) Palomo, C.; Oiarbide, M.; Kardak, B.G.; García, J.M.; Linden, A. *J. Am. Chem. Soc.* **2005**, *127*, 4154. (b) Zhou, J.; Ye, M.-C.; Huang, Z.-Z.; Tang, Y. *J. Org. Chem.* **2004**, *69*, 1309.

Review: (c) Poulsen, T.B.; Jørgensen, K.A. *Chem. Rev.* **2008**, *108*, 2903.

36. Lu, S.-F.; Du, D.-M.; Xu, *J. Org. Lett.* **2006**, *8*, 2115.

37. Palomo, C.; Oiarbide, M.; Kardak, B.G.; García, J.M.; Linden, A. *J. Am. Chem. Soc.* **2005**, *127*, 4154.

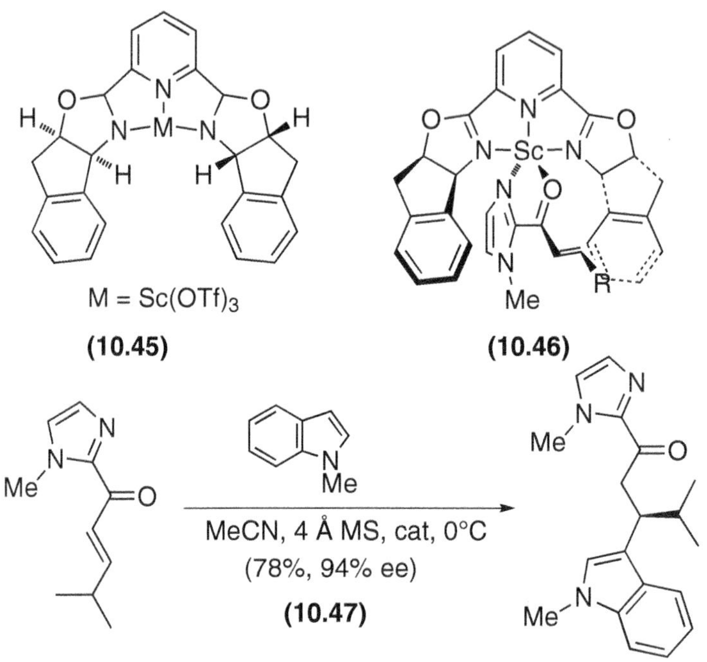

Figure 10.5 Use of a chiral scandium (III) triflate complex to catalyze the asymmetric Friedel-Crafts alkylation

Evans has used 2-acylimidazoles as substrates in similar alkylations of indoles catalyzed by a chiral (pybox)scandium (III) triflate complex (**10.45**, Figure 10.5)[38] and sees similar high levels of enantioselectivity in the reaction. In this reaction, the stereochemical outcome can be rationalized in terms of the *s-cis* conformation of the enone in the scandium chelate, **10.46**. The electrophiles in the reactions in Figures 10.4 and 10.5 share two structural features: (1) a group that restricts the conformation of the enone system to the *s-cis* conformation and (2) a group at the α' position that permits the formation of a chelate with the metal atom in the catalyst.

The Role of Solvent

It is reasonable that a reaction in which there are ionic intermediates should be facilitated by a polar solvent. For this reason, there has been a significant amount of activity investigating the effects of solvent on the Friedel-Crafts and similar electrophilic aromatic substitutions.[39] For example, using $Cu(OTf)_2$ as the catalyst in the ionic liquid [bmim][BF_4] permitted the benzoylation of a range of electron-rich aromatic hydrocarbons (e.g., Example **10.48**, 2-methoxynaphthalene).[40] Most interestingly, this reaction system permitted the synthesis of the kinetic product from 2-methoxynaphthalene rather than the thermodynamic product (6-methoxy-2-naphthyl phenyl ketone) obtained in conventional solvents. Similar acceleration of the reaction has been observed using indoles (e.g., Example **10.49**) and the ionic liquid [emim][Cl]–$AlCl_3$ to promote the reaction.[41] One useful advance in the use of ionic liquids has been the development of inter- (Example **10.50**) and intramolecular (Example **10.51**) alkenylation reactions of arenes by alkynes, using metal triflate catalysts.[42] In this reaction, the ionic liquid is proposed to stabilize the highly reactive vinyl cation, thus lowering the activation energy of the reaction. However, the

38. (a) Evans, D.A.; Fandrick, K.R.; Song, H.-J. *J. Am. Chem. Soc.* **2005**, *127*, 8942. (b) Evans, D.A.; Scheidt, K.A.; Fandrick, K.R.; Lam, H.W.; Wu, J. *J. Am. Chem. Soc.* **2003**, *125*, 10780.
39. Review: Earle, M.J.; Seddon, K.R. *Pure Appl. Chem.* **2000**, *72*, 1391.
40. Ross, J.; Xiao, J. *Green Chem.* **2002**, *4*, 129.
41. Yeung, K.-S.; Farkas, M.E.; Qiu, Z.; Yang, Z. *Tetrahedron Lett.* **2002**, *43*, 5793.
42. Song, C.E.; Jung, D.-a.; Choung, S.Y.; Roh, E.J.; Lee, S.-g. *Angew. Chem. Int. Ed.* **2004**, *43*, 6183.

intermolecular reaction is neither stereospecific nor regiospecific. Mixtures of (*o,p*-) regioisomers and (*E/Z*) stereoisomers are obtained in the alkenylation of anisole and chlorobenzene by 1-phenylpropyne, which limits its intermolecular applications.

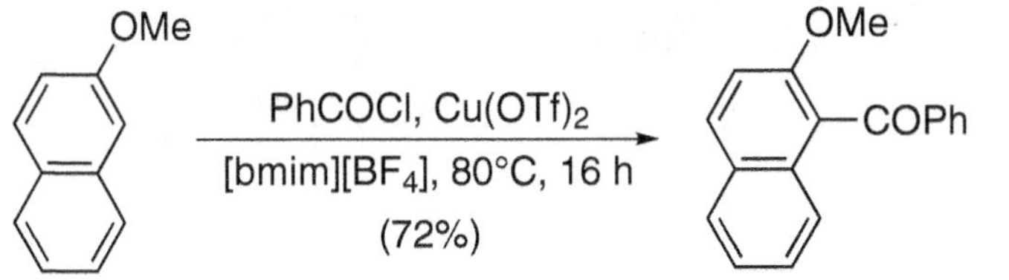

(10.48)

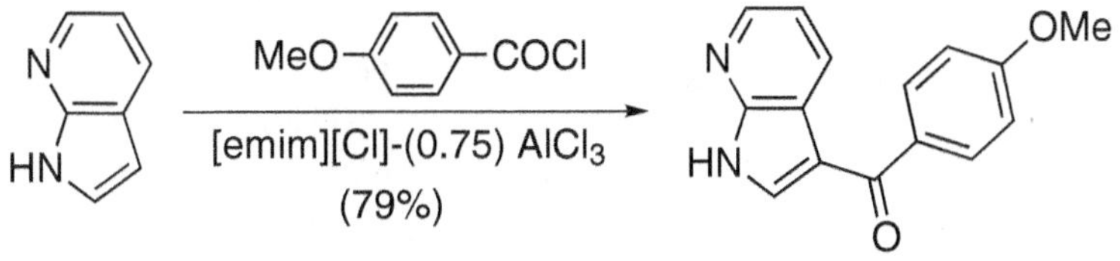

(10.49)

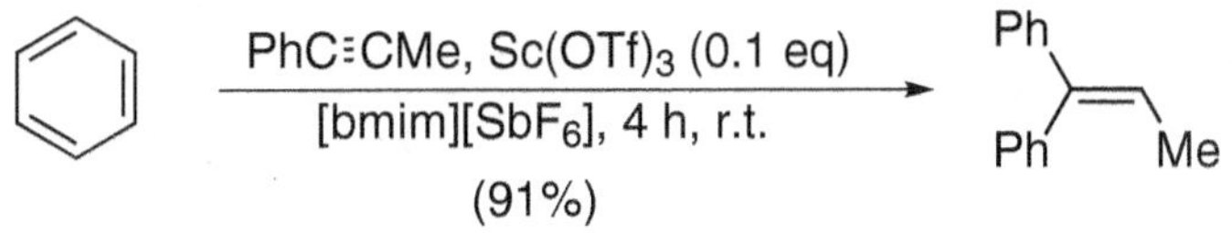

(10.50)

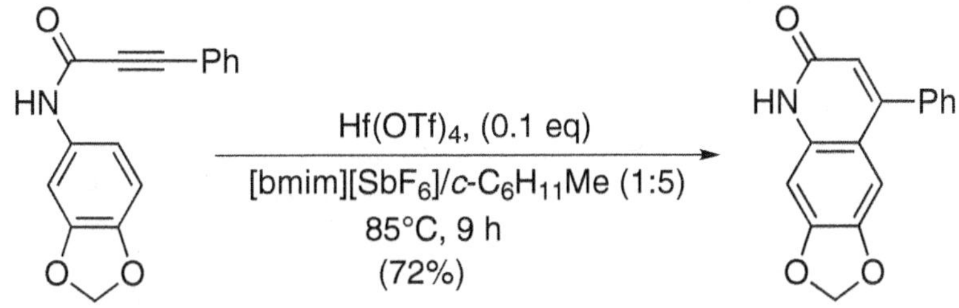

(10.51)

Houben-Hoesch Acylation

The reaction analogous to the Friedel-Crafts acylation using a nitrile instead of an acid chloride or anhydride is known as the ***Houben-Hoesch reaction***.[43] Although the electrophile may be viewed—formally, at least—as the resonance-stabilized nitrilium ion, research has shown that the mechanism is almost certainly more complex.[44] Evidence has been obtained that the actual electrophile may be the diprotonated form of the *sym*-triazene (**10.52**) formed by cyclotrimerization of the nitrile.[45] A mechanism involving this species in the acylation of benzene is shown in Figure 10.6; the final step of the reaction, which is not shown explicitly, is the acid-catalyzed hydrolysis of the imine. A modified Houben-Hoesch reaction (Example **10.57**), in which an *N*-alkylnitrilium ion is generated in situ, proved superior to alternative methods of acylation in the synthesis of *O*-methylsterigmatocystin. When the electrophile can be generated in an intramolecular manner (as in Example **10.58**), the Houben-Hoesch reaction provides a convenient method for the formation of nitrogen heterocycles.

---

43. (a) Hoesch, K. *Ber. dtsch. chem. Ges.* **1915**, *48*, 1122. (b) Houben, J. *Ber. dtsch. chem. Ges.* **1926**, 59, 2878. (c) Arkhipov, V.V.; Smirnov, M.N.; Khilya, V.P. *Chem. Heterocycl. Compd.* **1997**, 33, 515. (d) Kawecki, R.; Mazurek, A.P.; Kozerski, L.; Maurin, J.K. *Synthesis* **1999,** 751. Review.

44. Reviews: (a) Ruske, W. In Olah, G.A., Ed. *Friedel-Crafts and Related Reactions*; (Wiley-Interscience: New York, 1964), Vol. 3, Part I, Ch 32. (b) Spoerri, P.E.; DuBois, A.S. *Org. React.* **1949**, *5*, 387.

45. Amer, M.I.; Booth, B.L.; Noori, G.F.M.; Proença, M.F.R.P. *J. Chem. Soc. Perkin Trans. 1* **1983,** 1075.

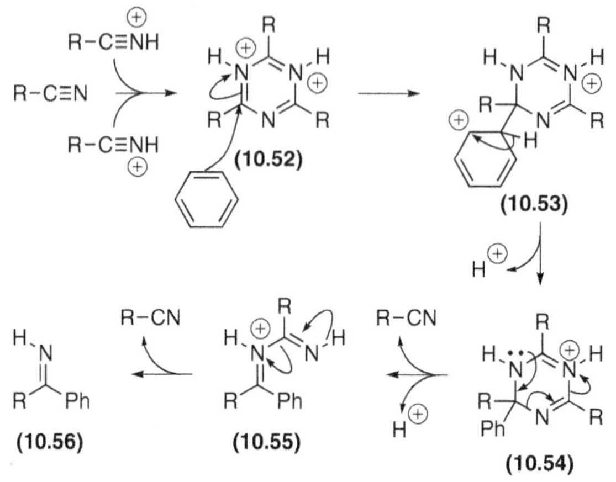

Figure 10.6 A mechanism for the Houben-Hoesch acylation reaction

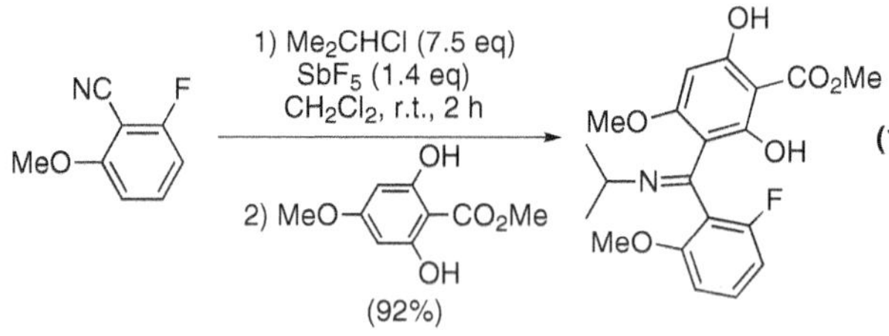

(10.57)[46]

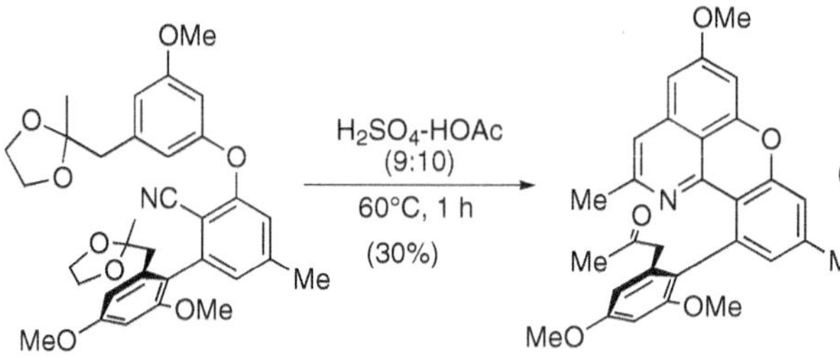

(10.58)[47]

Drill Problem

What will be the major organic product from the reaction of each of the compounds in the first list with each of the reagents in the second? Assume one equivalent of each reactant.

Compounds:

(a) 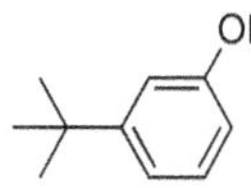 (b) 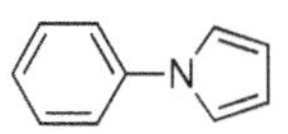 (c) 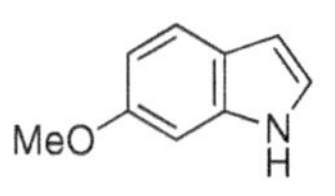 (d) 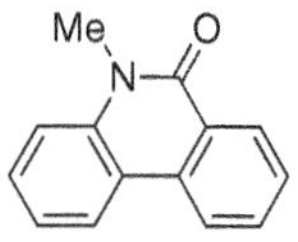

Reagents:

(A) MeCOCl, $AlCl_3$, CS_2. (B) Me_3CCl, $TiCl_4$, CH_2Cl_2, –78°C
(C) Me_2C=CH-$COCH_3$, $AlCl_3$. (D) *c*-$C_6H_{11}OH$, H_3PO_4, Δ

(continues)

46. Casillas, L.K.; Townsend, C.A. *J. Org. Chem.* **1999**, *64*, 4050.
47. Deguchi, J.; Hirahara, T.; Oshimi, S.; Hirasawa, Y.; Ekasari, W.; Shirota, O.; Honda, T.; Morita, H. *Org. Lett.* **2011**, *13*, 4344.

(E) PhCN, $AlCl_3$. (F) 1-methylcyclohexene, CF_3SO_3H
(F) *R*-**10.37**, *E*-PhCH=$CHNO_2$, PhH, 4Å MS

Reaction Synopses

Friedel-Crafts Alkylation

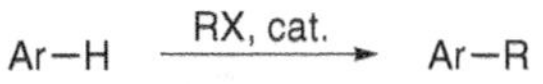

Reagents:

	RX:RCl, RBr, ROH, ROTf, etc.
or	R_2C=CHR, RCH=CHCOR, RCH=$CHNO_2$, etc.
Catalyst:	$AlCl_3$, $AlBr_3$, BF_3, $ZnCl_2$, $TiCl_4$, etc.
or	$Sc(OTf)_3$, $Cu(OTf)_2$, $Hf(OTf)_4$, etc.
or	H_3PO_4, CF_3CO_2H, CF_3SO_3H, H_2SO_4, HF, etc.
ArH:	must not be more deactivated toward electrophilic aromatic substitution than chlorobenzene
Solvent:	kinetic selectivity enhanced by nonpolar solvents; thermodynamic product favored by nitromethane or nitrobenzene. Ionic liquids enhance rate and kinetic selectivity.

Friedel-Crafts Acylation

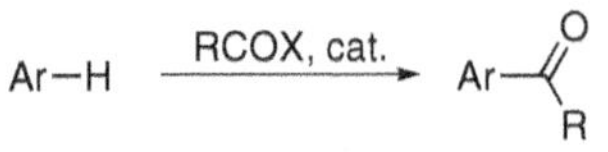

Reagents:

RCOX:	RCOCl, $(RCO)_2O$, RCO-O-SO_2R', RCO_2COCF_3, etc.
Catalyst:	$AlCl_3$, $AlBr_3$, HF, BF_3, $ZnCl_2$, $TiCl_4$, etc. or $Sc(OTf)_3$, $Cu(OTf)_2$, $Hf(OTf)_4$, etc. or H_3PO_4, CF_3CO_2H, CF_3SO_3H, H_2SO_4, HF, etc.
ArH:	must not be more deactivated towards electrophilic aromatic substitution than chlorobenzene
Solvent:	kinetic selectivity enhanced by nonpolar solvents; thermodynamic product favored by nitromethane or nitrobenzene. Ionic liquids as solvents enhance rate and kinetic selectivity.

Asymmetric Friedel-Crafts Alkylation

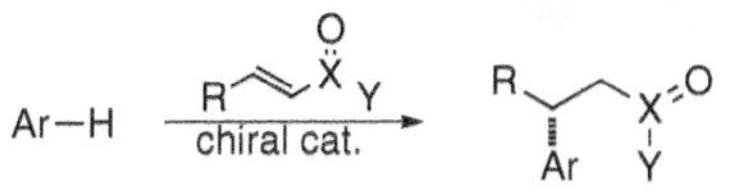

X = C, N, S; Y = O, CMe_2OH, 2-imidazolyl, etc.

Catalysts:

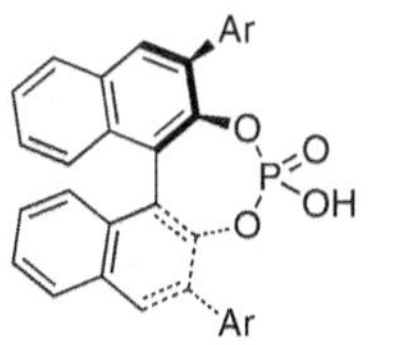

chiral Brønsted acid

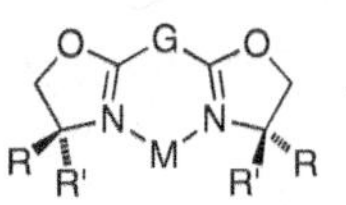

chiral box complex

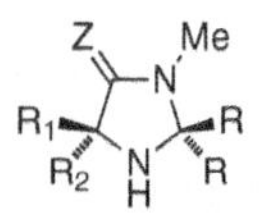

chiral imidazoline derivative

Ar = 9-anthryl, 9-phenanthryl, 3,5-dimesitylphenyl, etc.;
M = $Sc(OTf)_3$, $Cu(OTf)_2$, $Zn(OTf)_2$, $Hf(OTf)_4$, etc.

(continues)

(**Reaction Synopses** continued)

R, R' = H, *tert*-butyl, etc.
R_1, R_2 = H, alkyl, etc.; R_3 = Me, $(CH_2)_n$, etc.

ArH: pyrrole, indole, furan, thiophene, etc. The reaction generally fails with benzene and simple substituted benzenes.

Houben-Hoesch Acylation

Reagents: RCN, H_2SO_4; RCN, $AlCl_3$; RCN, $TiCl_4$; etc. or
(1) RCN, *i*-PrCl, SbF_5; (2) H_2SO_4, H_2O, Δ; etc.

Reaction: may be used to form imines or nitrogen heterocycles

---

Problem

10-3 What is the major organic product of each of the following reactions?

(a) 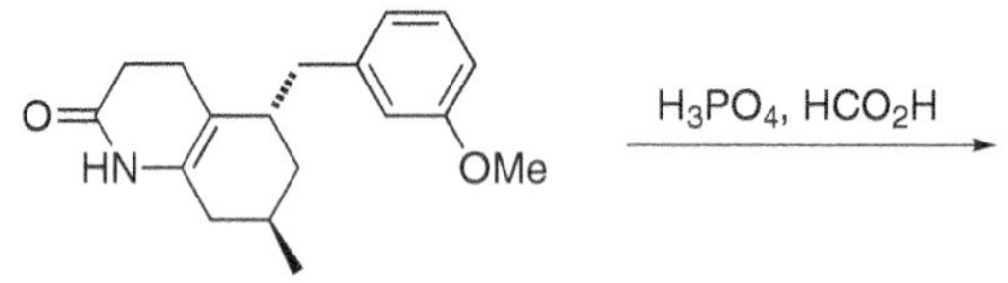

(b) 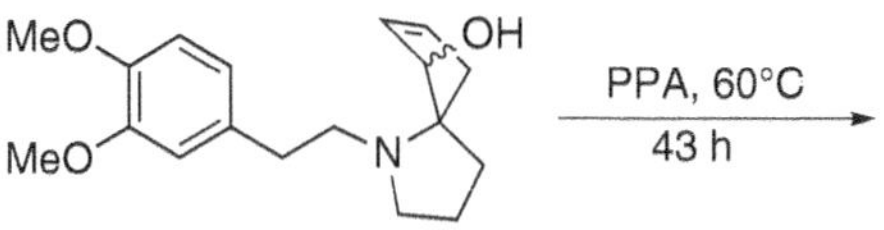

(c) 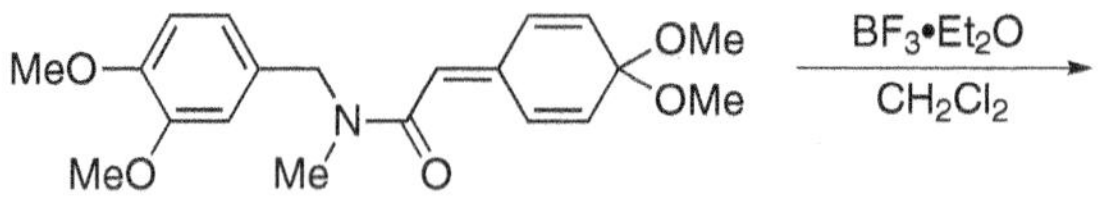

(d) 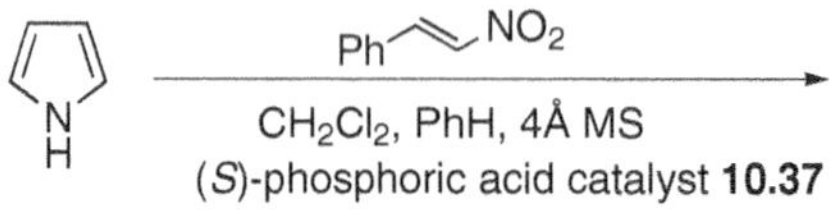

References: (a) *J. Am. Chem. Soc.* **1968**, *90*, 1647. (b) *J. Org. Chem.* **1996**, *60*, 115. (c) *J. Am. Chem. Soc.* **1978**, *100*, 1548. (d) *J. Org. Chem.* **2009**, *74*, 6899.

10.3 **Formylation**

Acylation is usually used to describe the formation of ketones, with the corresponding reactions to form aldehydes being known as ***formylation***. Although they are formally analogous to acylation reactions, formyl chloride (H—COCl) decomposes to HCl and CO above −60°C, so one cannot carry out the reaction by treating an aromatic compound with formyl chloride and a Lewis acid.[48] Consequently, surrogates for this unstable compound are needed, and the search for such surrogates has given us a number of useful formylation reactions. The simplest surrogate is (obviously?) a mixture of hydrogen chloride and carbon monoxide, which is used in the presence of a strong Lewis acid; this is the basis of the Gattermann-Koch formylation.[49] Replacement of the carbon monoxide in this

48. Ferguson, L.N. *Chem. Rev.* **1946**, *38*, 227.
49. For a review of formylations with carbon monoxide, see: Willemse, J.A. M.Sc. Dissertation, Rand Afrikaans University, Johannesburg, 2003.

mixture by a metal cyanide gives the Gattermann aldehyde synthesis. The use of a substituted iminium ion as the surrogate provides the basis of the Vilsmeier-Haack and Duff formylations. We will examine each of these reactions in turn.

Gattermann, Gattermann-Koch, and Gattermann-Olah Formylations

Formyl chloride is not stable, but subjecting the aromatic compound to a mixture of carbon monoxide and hydrogen chloride in the presence of aluminum chloride will give the aldehyde; this reaction is known as the ***Gattermann-Koch reaction***. Treating the aromatic compound with hydrogen cyanide and hydrogen chloride is known as the ***Gattermann aldehyde synthesis***.[50] The Gattermann-Koch reaction generally proceeds well only under high pressures of carbon monoxide, so it has largely been supplanted by the Gattermann synthesis, which does not require high temperatures. The Gattermann synthesis itself was improved by Adams,[51] who replaced the hydrogen cyanide with zinc cyanide; a small quantity of potassium chloride is usually needed in this version of the reaction, also.

In the Gattermann-Koch reaction, the reactive electrophile may be viewed, in a formal sense, as the formyl cation, $[H-C{\equiv}O]^+$, although the need for a Cu (I) cocatalyst (e.g., Example **10.59**[52]) suggests that the formation of the active electrophile may involve a transient copper carbonyl or formyl copper species. In the Gattermann reaction (Examples **10.60** and **10.61**), the active electrophile may be viewed as the conjugate acid of hydrogen cyanide, $[H-C{\equiv}N-H]^+$, although this, too, is almost certainly an oversimplification. A study of the reaction in acidic media using trimethylsilyl cyanide as the cyanide source[53] has found that the rate of product formation is directly related to the acidity of the reaction medium. The authors proposed that the diprotonated form of HCN, $[H-C{=}NH_2^{2+}]$, is the active electrophile. The final step of the Gattermann reaction is hydrolysis of an iminium ion to give the aldehyde.

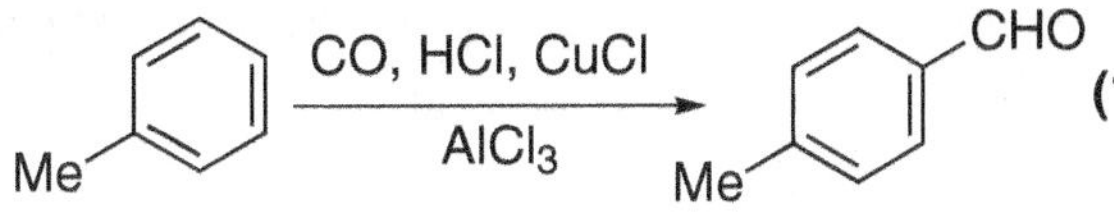

(10.59)

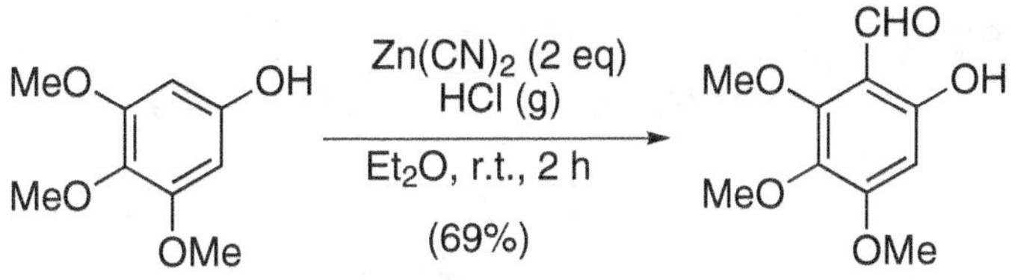

(10.60)[54]

50. (a) Gattermann, L.; Koch, J.A. *Ber. dtsch. chem. Ges.* **1897**, *30*, 1622.

Reviews: (b) Crouse, N.N. *Org. React.* **1949**, *5*, 290. (c) Olah, G. A. *Friedel-Crafts and Related Reactions* (Wiley- Interscience: New York, 1964) Vol.3, 1153. (d) Truce, W.E. *Org. React.* **1957**, *9*, 37. (e) Olah, G.A,; Ohannesian, L.; Arvanaghi, M. *Chem. Rev.* **1987**, *87*, 671.

Friedrich August Ludwig Gattermann (1860–1920) was educated at Göttingen (PhD 1885) under Victor Meyer, whom he followed to Heidelberg in 1889. He concluded his career at Freiburg 1899. For more biographical detail, see: Jacobson, P. *Ber. dtsch. chem,. Ges.* **1921**, *54*, A115.

Julius Arnold Koch (1864–1956) was an American chemist born in Germany. He graduated from the Pittsburgh College of Pharmacy in 1884 and spent seven years in the drug business. He became Chair of Pharmacy and was Dean at his alma mater in 1891. He remained there until 1932. In 1896 he traveled to Germany to study with Baeyer at Munich and with Gattermann and Victor Meyer in Heidelberg. Koch became Professor of Organic Chemistry at Pittsburgh in 1900. He was President of the American Pharmaceutical Association in 1923.

51. (a) Adams, R.; Levine, I. *J. Am. Chem. Soc.* **1923**, *45*, 2373. (b) Adams, R.; Montgomery, E. *J. Am. Chem. Soc.* **1924**, *46*, 1518.

52. Coleman, G.H.; Craig, D. *Org. Synth.* **1943**, *Coll. Vol. 2*, 583; *Org. Synth.* **1932**, *12*, 80.

53. Yato, M.; Ohwada, T.; Shudo, K. *J. Am. Chem. Soc.* **1991**, *113*, 691.

54. Burke, J.M.; Stevenson, R. *J. Nat. Prod.* **1986**, *49*, 522.

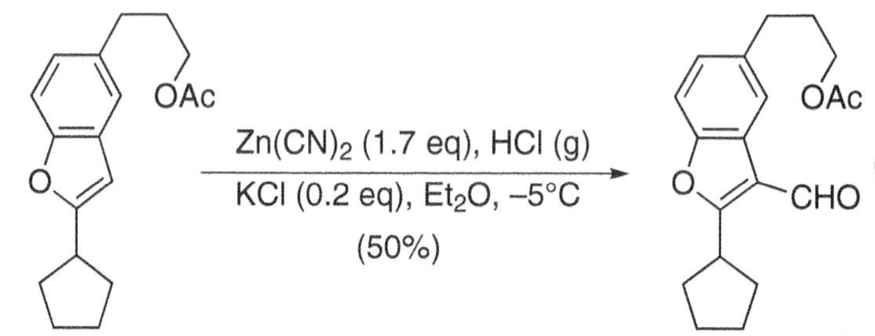

(10.61)[55]

In 1960,[56] Olah showed that formyl fluoride, the most stable of the formyl halides, could form a stable complex with boron trifluoride that would formylate benzene and more electron-rich aromatic hydrocarbons. The reaction is occasionally referred to as the ***Olah formylation***, or the ***Gattermann-Olah reaction***. The actual electrophile does not appear to the formyl cation but a boron trifluoride complex of the formyl fluoride (e.g., Example **10.62**).

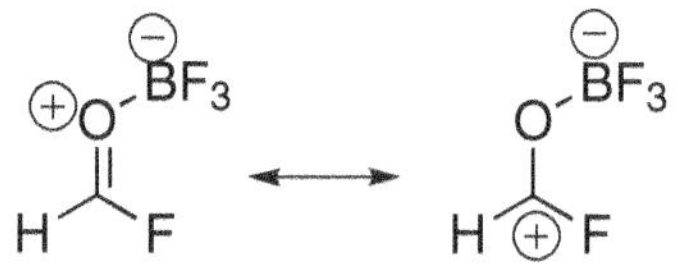

(10.62)

Vilsmeier-Haack Formylation

The reaction between dimethylformamide (DMF) and phosphoryl chloride gives an iminium ion (Example **10.63**) that is electrophilic enough to attack electron-rich aromatic species such as furans, pyrroles, indoles, and thiophenes (e.g., Examples **10.64** and **10.65**), as well as polycyclic hydrocarbons and simple benzenes with an electron-releasing substituent. This reaction is known as the ***Vilsmeier-Haack formylation*** after the two chemists who first reported it.[57] Hydrolysis of the initial product during the workup procedure then gives the aromatic aldehyde.

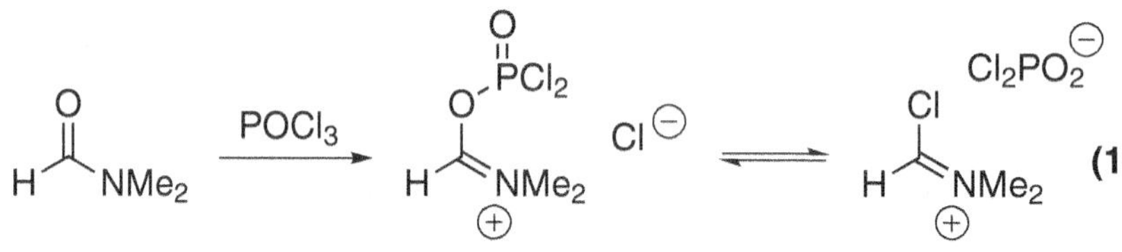

(10.63)

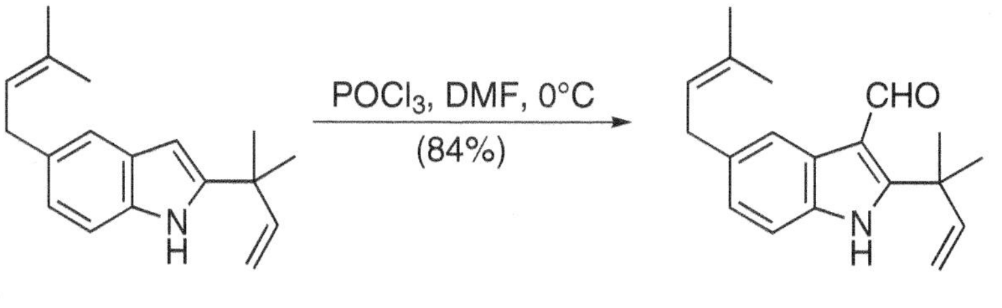

(10.64)[58]

55. Wong, H.N.C.; Niu, C.R.; Yang, Z.; Hon, P.M.; Chang, H.M.; Lee, C.M. *Tetrahedron* **1992**, *43*, 10339
56. Olah, G.A.; Kuhn, S.J. *J. Am. Chem. Soc.* **1960**, *82*, 2380.
57. Vilsmeier, A.; Haack, A. *Ber. dtsch. chem. Ges.* **1927**, *60*, 119. This paper refers to an earlier report of the reaction between *N*-methylacetanilide and phosphoryl chloride: Fischer, O.; Müller, A.; Vilsmeier, A. *J. Prakt. Chem.* **1925**, *109*, 69.

Anton Vilsmeier (1894–1962) graduated from Regensburg in 1914 and entered the German army. In 1920, he resumed his education and took his PhD at Erlangen under Otto Fischer. He remained at Erlangen as an Assistant until 1927, when he moved to I.G. Farbenindustrie (later BASF) as a dye chemist. He retired from BASF in 1959.

Albrecht Haack (1898–1976) received his PhD at Erlangen in 1926 under Fischer and Vilsmeier. From 1928–1941 he worked as an industrial chemist at "Milk Central" in Karlsruhe/Baden. After military service (1941–1945), he left chemistry and entered government service. He retired in 1969.

58. Kuttruff, C.A.; Zipse, H.; Trauner, D. *Angew. Chem. Int. Ed.* **2011**, *50*, 1402.

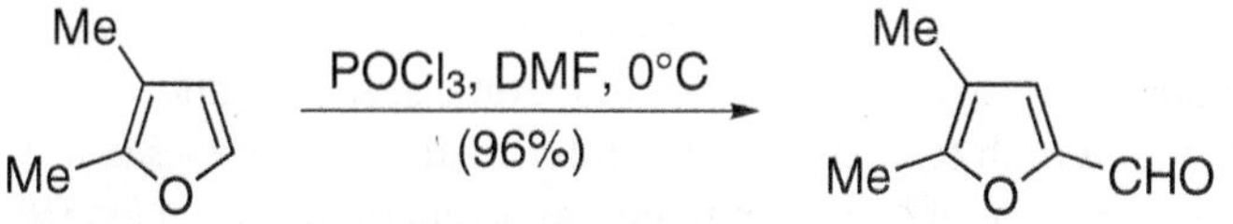

(10.65)[59]

The Vilsmeier-Haack reaction is a popular reaction because it avoids the use of strongly acidic reagents and is thus suitable for use with compounds carrying sensitive functionality. This reagent has been modified by replacing the phosphoryl chloride with oxalyl chloride, and this reagent has been reported[60] to give higher yields with sensitive compounds such as the electron-rich *N*-vinylpyrrole in Example **10.66**, which gave only 28% yield of the aldehyde when phosphoryl chloride was used.[61]

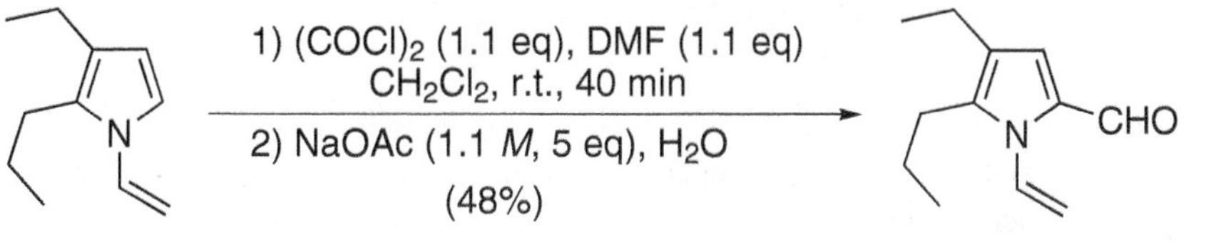

(10.66)

Duff Formylation Reaction

The ***Duff formylation***, first described in a series of four papers by Duff and Bills,[62] is a reaction between an electron-rich aromatic compound and hexamethylenetetramine in acetic acid at elevated temperature. The reaction proceeds by way of an intermediate

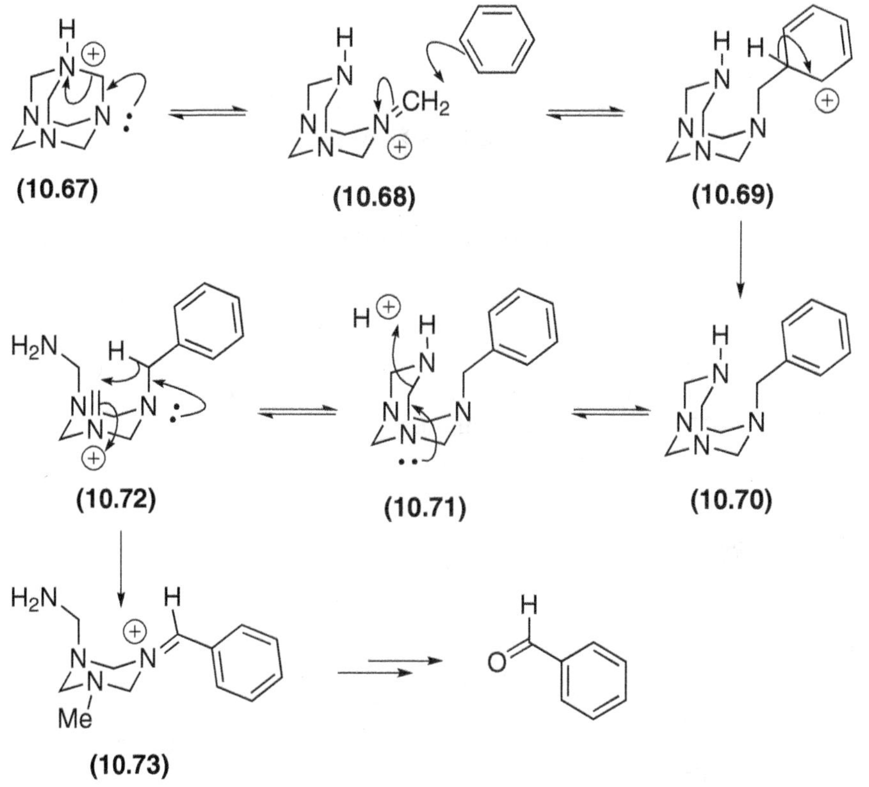

Figure 10.7 The mechanism of the Duff formylation reaction

59. Martin, S.F.; Gluchowski, C.; Campbell, C.L.; Chapman, R.C. *J. Org. Chem.* **1984**, *49*, 2512.
60. Mikhaleva, A.I.; Ivanov, A.V.; Skital'tseva, E.V.; Ushakov, I.A.; Vasil'tsov, A.M.; Trofimov, B.A. *Synthesis* **2009**, 587.
61. Mikhaleva, A.I.; Zaitsev, A.B.; Ivanov, A.V.; Schmidt, E. Yu.; Vasil'tsov, A.M.; Trofimov, B.A. *Tetrahedron Lett.* **2006**, *47*, 3693.
62. Duff, J.C.; Bills, E.J. *J. Chem. Soc.* **1932**, 1987; **1934**, 1305; **1941**, 547; **1945**, 276.

iminium ion (**10.68**) that attacks the aromatic ring to give a benzylamine derivative (**10.69**, Figure 10.7). If the reaction is carried out at low temperature, the product obtained is, in fact, a benzylamine, but at high temperature, the intermediate undergoes an assisted intramolecular 1,5-rearrangement of hydrogen to give an isomeric iminium ion (**10.73**) that can then be hydrolyzed to give the aromatic aldehyde. An alternative intermolecular hydrogen transfer has also been proposed as part of the mechanism.[63] It is worthwhile noting that the closely related ***Sommelet reaction***,[64] which is a formal oxidation of a benzyl halide to a benzaldehyde by hexamethylenetetramine and water, as in Example **10.74**,[65] almost certainly follows the same general mechanism as the Duff reaction.

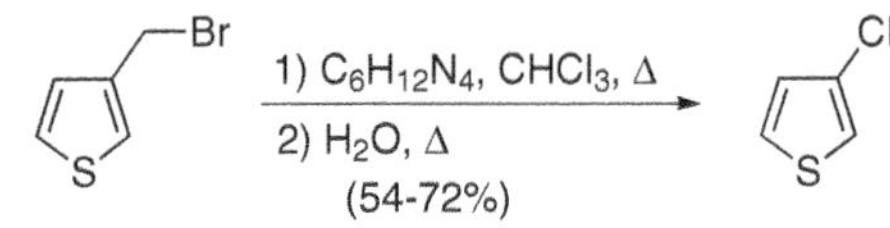

(10.74)

Like the Vilsmeier-Haack reaction, the Duff reaction is restricted to electron-rich aromatic compounds such as phenols. The reaction is not widely used in its original form because of the reaction conditions, but it has been used for the synthesis of simple phenolic aldehydes (e.g., Example **10.75**).[66] The replacement of acetic acid by trifluoroacetic acid allows the reaction to be carried out under much less forcing conditions and allows a larger range of aromatic hydrocarbons—including benzene itself—to be formylated by the reagent.[67]

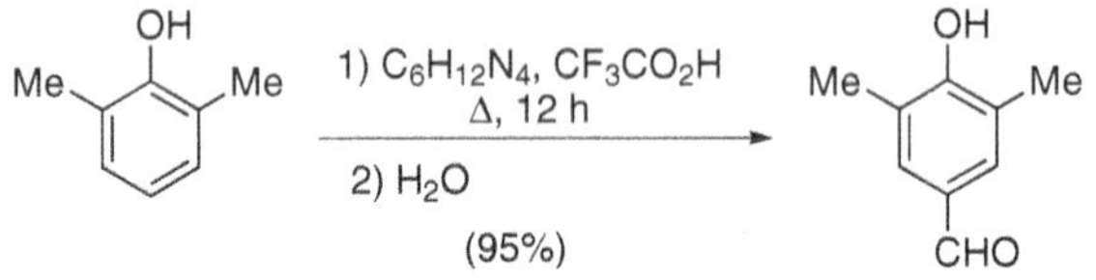

(10.75)

Reaction Synopses

Gattermann-Koch, Gattermann, and Gattermann-Olah Reactions

Ar–H ⟶ Ar–CHO

Reagents:
: CO, HCl, CuCl, $AlCl_3$ (Gattermann-Koch); or
: HCN, HCl, $AlCl_3$ (Gattermann); or
: $Zn(CN)_2$, HCl, KCl, Et_2O (Adams modification); or
: HCOF, BF_3 (Gattermann-Olah)

ArH: must be at least as reactive as benzene toward electrophilic aromatic substitution

Vilsmeier-Haack formylation

Ar–H ⟶ Ar–CHO

Reagents:
: DMF, $POCl_3$; PhN(Me)CHO, $POCl_3$; etc. or
: $(COCl)_2$, DMF; etc.

(continues)

---

63. Ogata, Y.; Sugiura, F. *Tetrahedron* **1968**, *24*, 5001.
64. Sommelet, M. *Compt. Rend.* **1913**, *157*, 852.
65. Campaigne, E.; Bourgeois, R.C.; McCarthy, W.C. *Org. Synth.* **1963**, *Coll. Vol. 4*, 918; **1953**, *33*, 93.
66. (a) Larrow, J.F.; Jacobsen, E.N. *Org. Synth.* **2004**, *Coll. Vol. 10*, 96; **1998**, *75*, 1. (b) Allen, C.H.F.; Leubner, G.W. *Org. Synth.* **1963**, *Coll. Vol. 4*, 866; **1951**, *31*, 92.
67. Smith, W.E. *J. Org. Chem.* **1972**, *37*, 3972.

ArH: must be more reactive than benzene toward electrophilic aromatic substitution

Duff Reaction

Ar–H ⟶ Ar–CHO

Reagents: $C_6H_{12}N_4$, HOAc, H_2O, Δ; or
(1) $C_6H_{12}N_4$, CF_3CO_2H, Δ; (2) H_2O Δ; etc.

ArH: must be more reactive than benzene toward electrophilic aromatic substitution; if CF_3CO_2H is used, benzene will react.

Worked Problem

10-3 What is the major monosubstitution product expected from each of the following reactions? Account for the regiochemistry of the reactions.

(a) 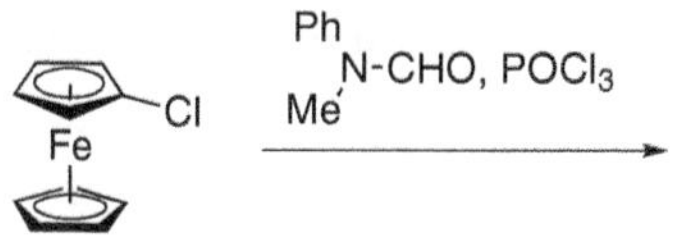

(b) 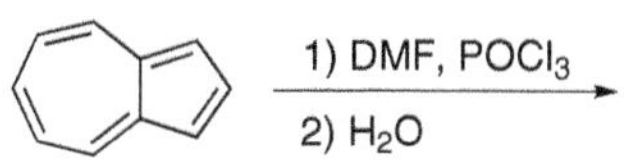

§Answers below.

---

§ Answers to Worked Problem:

(a) 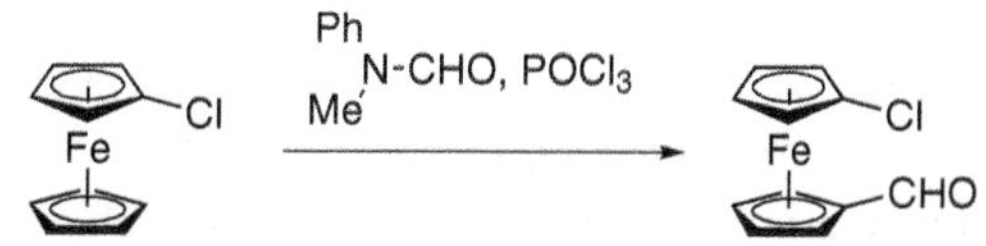

In this reaction, the ferrocene derivative has two aromatic rings, one substituted with chlorine (which is electron-withdrawing), and one unsubstituted. Because the unsubstituted ring is more electron-rich, it will be formylated first. Note how in this example, *N*-methylformanilide is used in place of DMF. This reagent occasionally provides advantages over DMF itself.

(b) 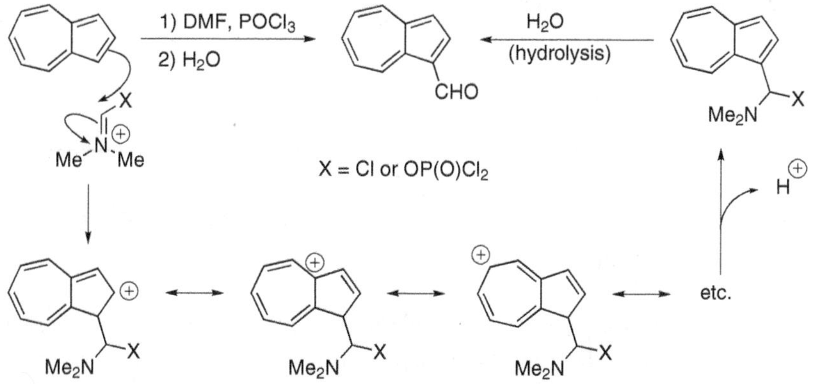

In this reaction, the electrophile attacks the five-membered ring because this leads to an aromatic tropylium ion contributor to the resonance hybrid of the intermediate carbocation. Attack on the seven-membered ring leads to complete loss of aromaticity of the azulene. Loss of the proton and hydrolysis of the initial product complete the reaction.

References: (a) *J. Am. Chem. Soc.* **1957**, *79*, 3416. (b) *Angew. Chem.* **1957**, *69*, 533.

Drill Problem

What are the products expected when each of the arenes below is treated with each of the reagents in the list below them?

(a) 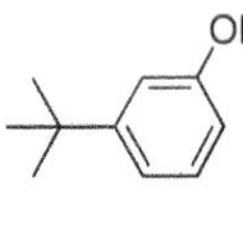

(b) 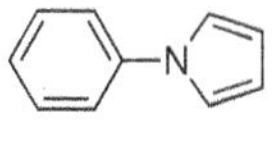

(c) 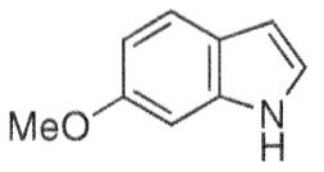

(d) 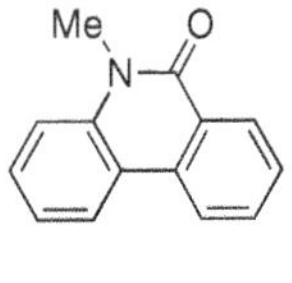

Reagents:

(A) $Zn(CN)_2$, HCl. (B) 1) PhN(Me)CHO, $POCl_3$; 2) H_2O

(C) 1) Cl_2CHOMe, $AlCl_3$; 2) H_2O. (D) $C_6H_{12}N_4$, CF_3CO_2H, Δ

Problems

10-4 Supply the structure of the major organic product of each of the following reactions.

(a) 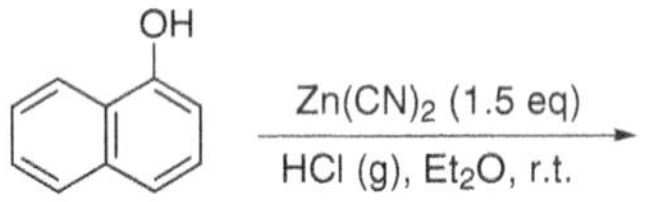

(b) 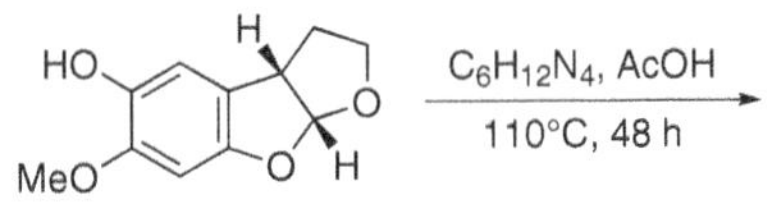

(c) 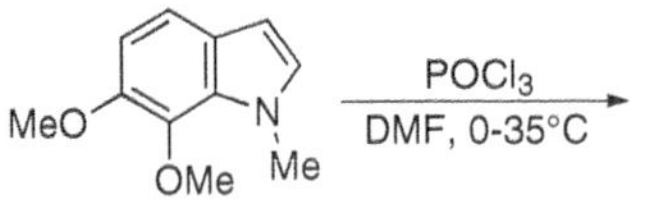

References: (a) *J. Org. Chem.* **1990**, *55*, 2913. (b) *J. Am. Chem. Soc.* **2005**, *127*, 11958. (c) *J. Am. Chem. Soc.* **1999**, *121*, 6771.

10-5 A reaction similar to the Vilsmeier-Haack formylation is the Reiche-Gross formylation [*Chem. Ber.* **1960**, *93*, 88; *Org. Synth.* **1973**, *Coll. Vol. 5*, 49], which involves treating an aromatic compound that is more electron-rich than benzene with methoxydichloromethane and a Lewis acid. Using the example below, give a reasonable mechanism for this reaction.

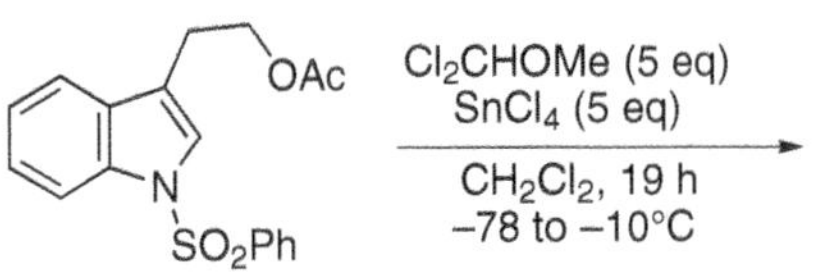

Reference: J. Am. Chem. Soc. **2011**, *133*, 2864.

10.4 Addition of Stabilized Carbocations: The Prins, Mannich, and Mukaiyama Reactions

Up to now, our discussion of the use of carbocation electrophiles in synthesis has been confined to their use in electrophilic aromatic substitution reactions. However, the overall usefulness of electrophilic addition reactions also extends to addition reactions, especially when a stabilized carbocation and an alkene carrying a substituent with a lone pair are involved. The addition of stabilized carbocations (oxonium ions) to alkenes is known as

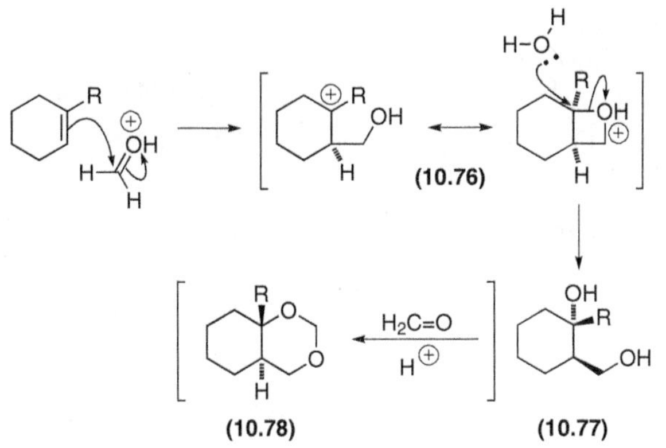

Figure 10.8 The mechanism of the acid-catalyzed Prins reaction

the ***Prins reaction***, after the Dutch chemist[68] who discovered it. The similar addition of iminium ions to enols is known as the ***Mannich reaction***, after its discoverer, German chemist Carl Mannich.[69] The use of alkenes that can form stabilized carbocations is exemplified by the ***Mukaiyama reactions***.[70]

The Prins Reaction[71]

In the Prins reaction, the conjugate acid of formaldehyde adds to an alkene to form a new carbon-carbon σ bond. This addition leads to an intermediate carbocation that is believed to be stabilized as a protonated oxetane by neighboring group participation of the lone pair of electrons on the oxygen (**10.76**). Trapping of the cation by a water molecule completes the addition reaction. The overall stereochemistry of the addition is usually *anti*, and this is what one would expect if the intermediate cation were stabilized by the neighboring group. The course of the Prins reaction is actually quite complex (there are numerous side reactions), but the formation of the major product of the reaction (**10.77**) may be rationalized as shown in Figure 10.8 for the sulfuric acid–catalyzed addition of formaldehyde to a 1-alkylcyclohexene. As expected on the basis of this mechanism, the overall addition occurs with Markovnikov regiochemistry. A major by-product of the reaction, especially in the presence of excess formaldehyde, is the 1,3-dioxane (**10.78**). The course of the reaction can be controlled to some degree by the solvent; when the solvent is bulky, or

---

68. **Hendrik Jacobus Prins (1889–1958)** was educated at the University of Delft (Dr Ir, 1912). He joined the N.V. Polak & Schwartz Essencefabrieken, in Hilversum immediately after his graduation, then, in 1925, he joined Nederlandse Thermochemische Fabrieken, being appointed Director in 1932. In 1951, he was made an Officer of the Order of Orange-Nassau. Prins continued his research into the reactions of cations and alkenes through World War II, although publication of his work, which was always in Dutch journals, was interrupted by the war.

69. **Carl Mannich (1887–1947)** was educated at Marburg and Berlin, where he graduated with a degree in pharmacy in 1900. He then studied at Berlin (PhD, 1903); his career took place in Berlin (1903), Göttingen (1911), and Frankfurt (1919). For more biographical details, see: Priesner, C. *Neue Deutsche Biographie* (Duncker & Humblot: Berlin, 1990), vol. 16, p. 71.

70. **Teruaki Mukaiyama (1927–)** graduated from Tokyo Institute of Technology in 1948. In 1952, he joined the faculty of Gakushuin University as Assistant Professor of Chemistry. He continued research toward his graduate degree, and in 1956, he was awarded the DSc degree from the University of Tokyo. In 1962, he moved to Tokyo Institute of Technology as Professor of Chemistry, and in 1974, he moved to the University of Tokyo in the same capacity. For more biographical detail, see: Kobayashi, S. *Chem. Asian J.* **2008**, *3*, 148.

71. (a) Prins, H.J. *Chem. Weekblad* **1917**, *14*, 627, 932; **1919**, *16*, 64, 1072, 1510. Reviews: (b) Arundale, E.; Mikeska, L.A. *Chem. Rev.* **1952**, *51*, 505.

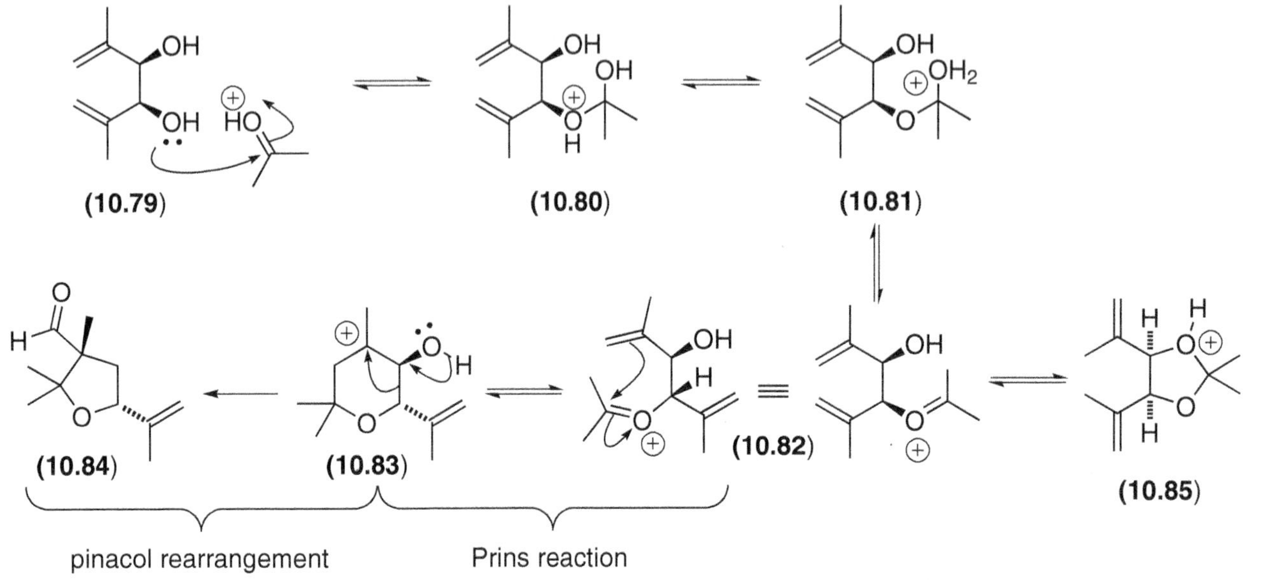

Figure 10.9 The pinacol-terminated Prins reaction

significantly less nucleophilic than water, the cation intermediate may undergo deprotonation to yield an unsaturated alcohol.

In addition to proton acids, Lewis acids such as $SnCl_4$, $AlCl_3$, Me_2AlCl, and $EtAlCl_2$ can also be used as catalysts in the Prins reaction. When diethylaluminum chloride is used as the catalyst, the Prins reaction can also be extended to other aldehydes, although the yields are generally not as good as with formaldehyde.

One of the more useful applications of the Prins reaction in synthesis has been the ***pinacol-terminated Prins cyclization*** reaction (Figure 10.9), discovered by Mousset,[72] and extended into a useful synthetic method for the synthesis of cyclic ethers by Overman.[73] The reaction consists of an initial addition to give the oxonium ion **10.80**, followed by dehydration to form the oxonium ion **10.82**—strictly analogous to forming a cyclic ketal. This is followed by an intramolecular Prins reaction to give a β-hydroxy-carbocation (**10.83**) that then undergoes a pinacol rearrangement to give the carbonyl product (in this case, the aldehyde **10.84**). The original Mousset example is shown in Figure 10.9. As the mechanism implies, at should be possible to apply this reaction to 1,3-dioxolanes (e.g., **10.85**). One can, and tetrahydrofurans are obtained as the major reaction products (e.g., Examples **10.86** and **10.87**).

---

72. (a) Martinet, P.; Mousset, G.; Michel, M. *C. R. Acad. Sci. Paris, Ser. C* **1969**, *268*, 1303. (b) Martinet, P.; Mousset, G. *Bull. Soc. Chim. Fr.* **1970**, 1071. (c) Martinet, P.; Mousset, G. *Bull. Soc. Chim. Fr.* **1971**, 4093. (d) Mousset, G. *Bull. Soc. Chim. Fr.* **1971**, 4097. (e) Chambenois, D.; Mousset, G. *C. R. Acad. Sci. Paris, Ser. C* **1972**, *274*, 715. (f) Chambenois, D.; Mousset, G. *C. R. Acad. Sci. Paris, Ser. C* **1972**, *274*, 2088. (g) Chambenois, D.; Mousset, G. *Bull. Soc. Chim. Fr.* **1974**, 2969. (h) Malardeau, C.; Mousset, G. *Bull. Soc. Chim. Fr.* **1977**, 988.

73. (a) Hopkins, M.H.; Overman, L.E. *J. Am. Chem. Soc.* **1987**, *109*, 4748. (b) Hopkins, M.H.; Overman, L.E.; Rishton, G.M. *J. Am. Chem. Soc.* **1991**, *113*, 5354; **1992**, *114*, 10093. (c) Gasparski, C.M.; Herrinton, P.M.; Overman, L.E.; Wolfe, J.P. *Tetrahedron Lett.* **2000**, *41*, 9431. (d) Cohen, F.; MacMillan, D.W.C.; Overman, L.E.; Romero, A. *Org. Lett.* **2001**, *3*, 1225. (e) Herrinton, P.M.; Hopkins, M.H.; Mishra, P.; Brown, M.J.; Overman, L.E. *J. Org. Chem.* **1987**, *52*, 3711. (f) Brown, M.J.; Harrison, T.; Herrinton, P.M.; Hopkins, M.H.; Hutchinson, K.D.; Mishra, P.; Overman, L.E. *J. Am. Chem. Soc.* **1991**, *113*, 5365. (g) Overman, L.E.; Rishton, G.M. *Org. Synth.* **1998**, *Coll. Vol. 9*, 4. (h) Cloninger, M.J.; Overman, L.E. *J. Am. Chem. Soc.* **1999**, *121*, 1092. (i) MacMillan, D.W.C.; Overman, L.E.; Pennington, L.D. *J. Am. Chem. Soc.* **2001**, *123*, 9033. (j) Overman, L.E.; Pennington, L.D. *J. Org. Chem.* **2003**, *68*, 7143.

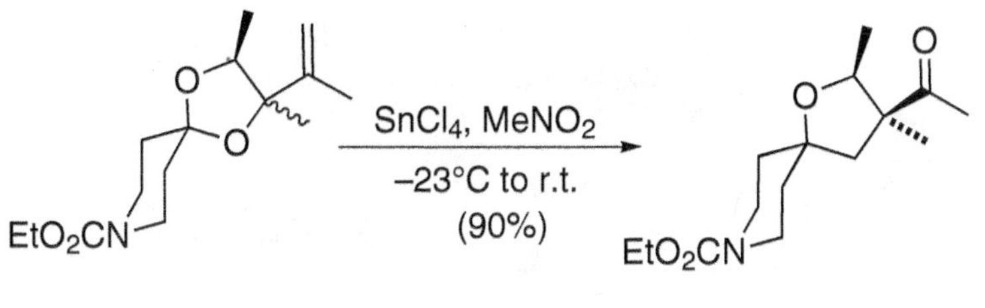

(10.86)[74]

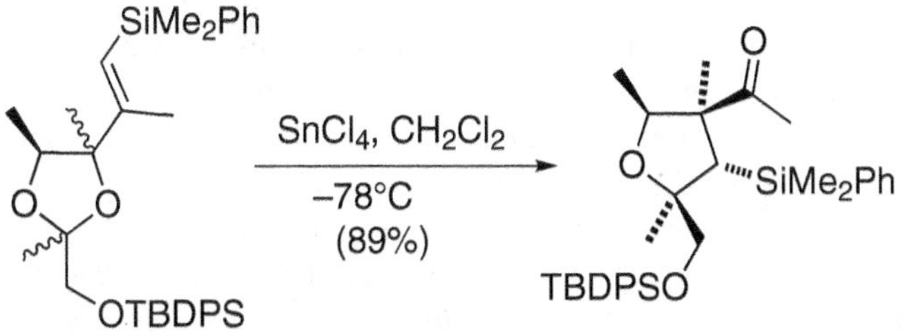

(10.87)[75]

Although the intermolecular Prins reaction does not proceed as well with aldehydes other than formaldehyde, the intramolecular cyclization of aldehydes and ketones and their conjugated counterparts onto alkenes is a facile reaction. In an extensive study,[76] Snider investigated the use of alkylaluminum chlorides to promote an intramolecular Prins-type addition to alkenes. The final product of the reaction was found to be determined by the amount of the Lewis acid catalyst used in the reaction, a result that was not expected. With one equivalent or less, the reaction followed the expected course and gave the alkene with the isopropenyl group (Example **10.88**). With two equivalents of the Lewis acid, however, the initial cyclization was followed by an unprecedented series of Wagner-Meerwein rearrangements to give the enone product (Example **10.89**).

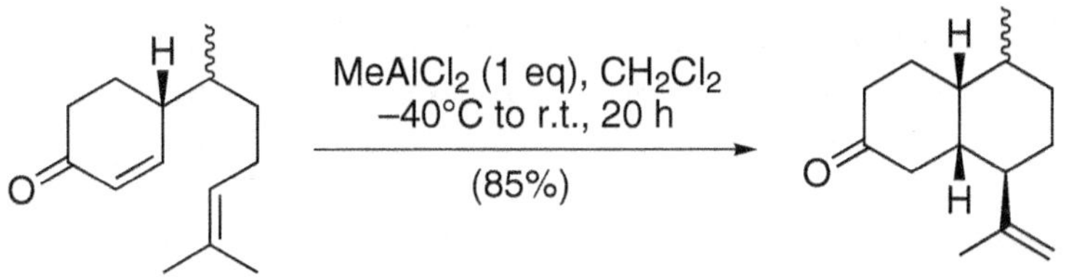

(10.88)

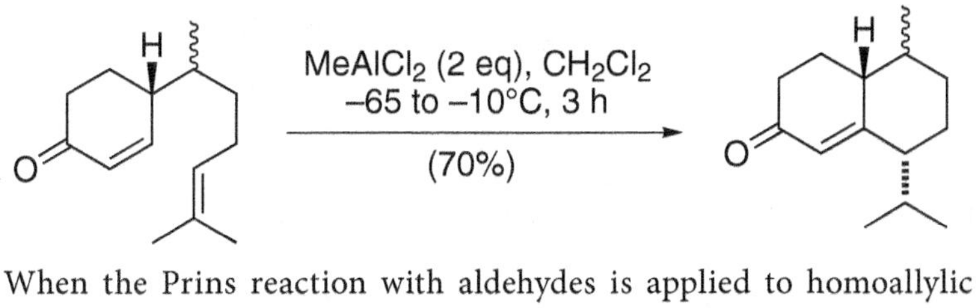

(10.89)

When the Prins reaction with aldehydes is applied to homoallylic alcohols, the product is a 4-hydroxytetrahydropyran with substituents at the 2- and 6-positions. In the product, all the substituents in the tetrahydropyran ring are equatorial. Example **10.90**[77] illustrates this highly diastereoselective Prins reaction under "green" conditions, using a heteropolymolybdic acid catalyst in aqueous solution. A similar reaction to give a 4-chlorotetrahydropyran (Example **10.91**) with all groups equatorial by means of an electrophilic ionic liquid as the solvent was earlier reported by the same research group.[78]

---

74. Overman, L.E.; Rishton, G.M. *Org. Synth.* **1998**, *Coll. Vol. 9*, 4.
75. Nemoto, H.; Ishibashi, H.; Nagamochi, M.; Fukumoto, K. *J. Org. Chem.* **1992**, *57*, 1707.
76. Snider, B.B.; Rodini, D.J.; van Straten J. *J. Am. Chem. Soc.* **1980**, *102*, 5872.
77. Yadav, J.S.; Reddy, B.V.S.; Kumar, G.G.K.S.N.; Aravind, S. *Synthesis*, **2008**, 395.
78. Yadav, J.S.; Reddy, B.V.S.; Reddy, M.S.; Niranjan, N.; Prasad, A.R. *Eur. J. Org. Chem.* **2003**, 1779.

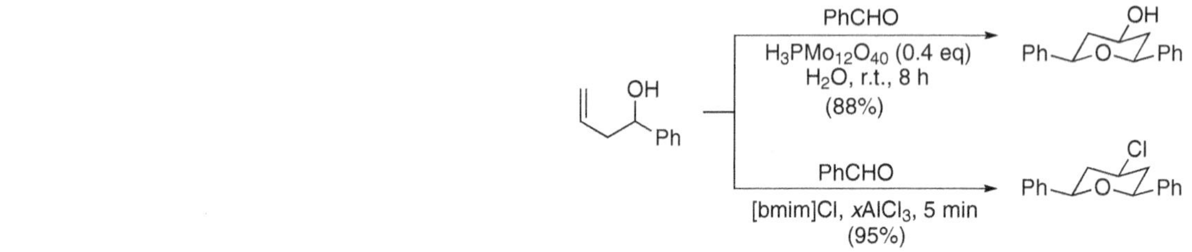

(10.90)

(10.91)

The addition of a carbocation to an alkene occurs fairly readily, but the process is often not especially regiospecific. This can be addressed by using an allylsilane in place of the simple alkene acceptor. The C—Si bond is much more efficient than a C—C or C—H bond at stabilizing the adjacent cation center by hyperconjugation, so additions become regiospecific, and, critically, the desilylation of the intermediate carbocation allows the regiospecific generation of an alkene double bond. It is an important feature of silane chemistry that *silanes can override the Zaitsev regiochemical preference for elimination from a carbocation.* A good example of the regiospecific addition of a carbocation to an allylsilane, and subsequent desilylation is provided by the key cyclization step from the synthesis of the hydroazulene species in Example **10.92**.[79] This reaction illustrates the directing effect of the silicon: in the absence of the silane directing group, the product would have been an endocyclic alkene.

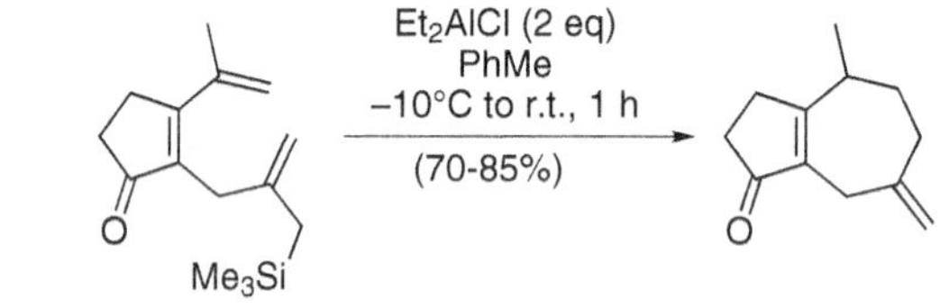

(10.92)

Aza-Prins Reactions

In an intramolecular reaction, the protonated carbonyl group of the Prins reaction can be replaced by an iminium ion (Example **10.93**). Markovnikov addition of the iminium ion to the alkene gives a cation (Example **10.94**) that then deprotonates to give an allylic amine (Example **10.95**) or reacts with a nucleophile to give the addition product (Example **10.96**). The electrophile is even more effective when the nitrogen atom is acylated. The resultant ***aza-Prins reaction*** is a useful carbon-carbon bond-forming reaction.

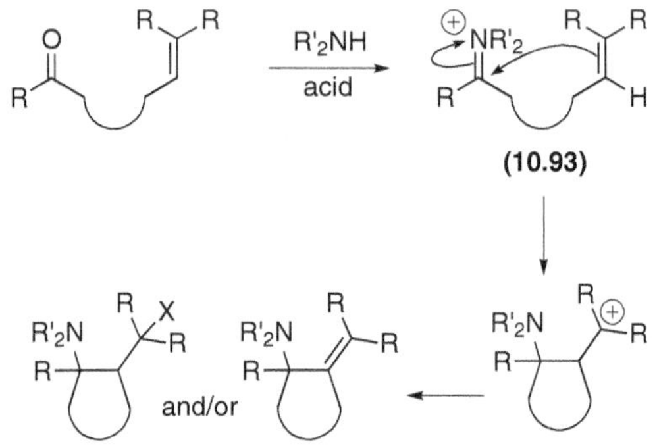

(10.93)

(10.96) (10.95) (10.94)

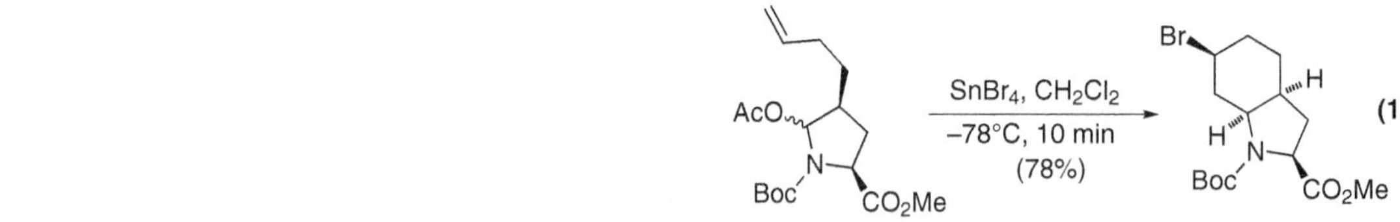

(10.97)[80]

79. Majetich, G.; Song, J.-S.; Leigh, A.J.; Condon, S.M. *J. Org. Chem.* **1993**, *58*, 1030.
80. Hanessian, S.; Tremblay, M.; Petersen, J.F.W. *J, Am. Chem. Soc.* **2004**, *126*, 6064.

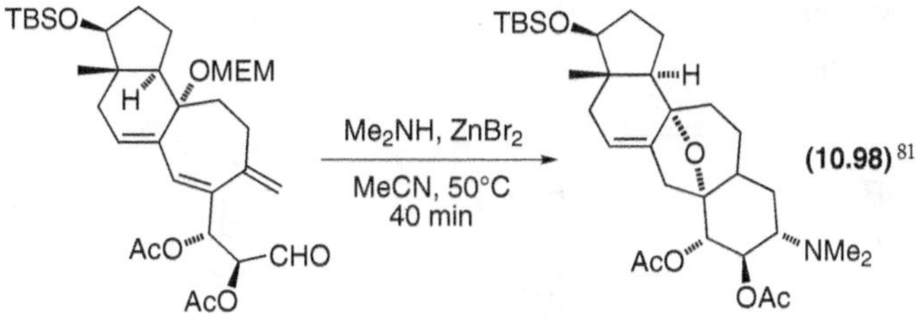

Drill Problem

(i) What will be the major organic product from the reaction of each of the compounds in the first list with each of the reagents in the second? You may assume one equivalent of the reagent (how would the product differ if an excess of the reagent were available?).

Compounds:

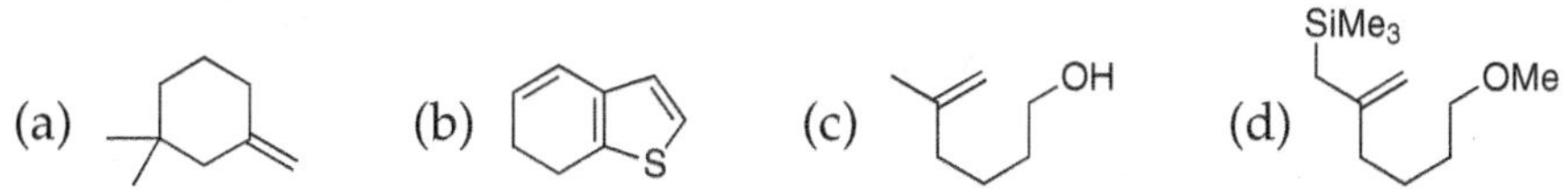

Reagents:

(A) $(H_2CO)_n$, H_2SO_4. (B) PhCHO, $EtAlCl_2$. (C) Me_2NH, $(CH_2O)_n$, CF_3CO_2H

(ii) What will be the major organic product from the reaction of each of the compounds below with a strong Lewis acid or a strong protic acid?

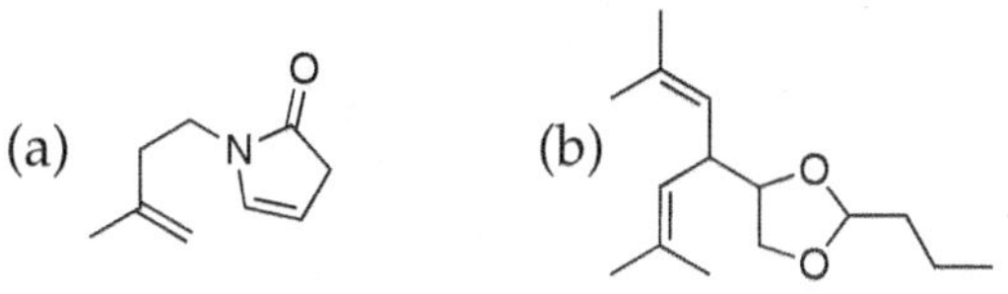

Reaction Synopses

Prins and Aza-Prins Reactions

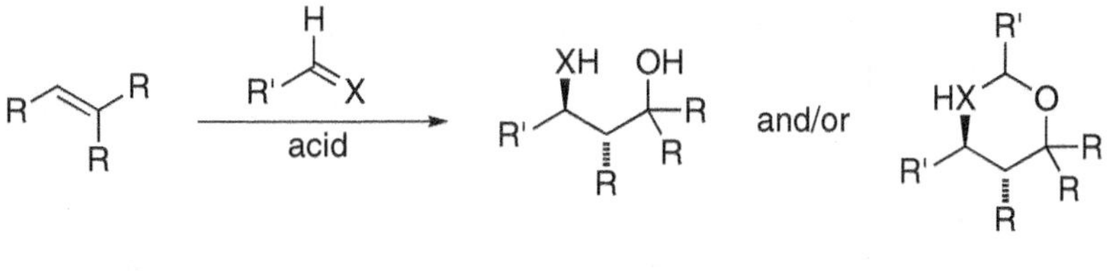

R: alkyl, aryl, H; R': H, alkyl, aryl; X=O, N;
Acid: H_2SO_4, Me_2AlCl, $EtAlCl_2$, BF_3, $ZnBr_2$, $SnBr_4$, $TiCl_4$, etc.

The reaction can be carried out under "green" conditions, using $H_3PMo_{12}O_{40}$ in water, or the electrophilic ionic solvent, [bmim]Cl•$x$$AlCl_3$.

The regiochemistry of the reaction can be controlled using an allylsilane as the alkene partner.

(continues)

81. Lee, H.M.; Nieto-Oberhuber, C.; Shair, M.D. *J. Am. Chem. Soc.* **2008**, *130*, 16864.

Figure 10.10 Mechanism of the Mannich reaction

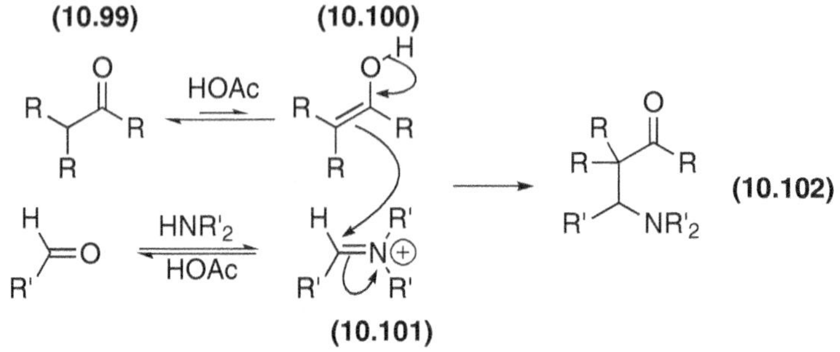

(**Reaction Synopses** continued)

Pinacol-Terminated Prins Reaction.

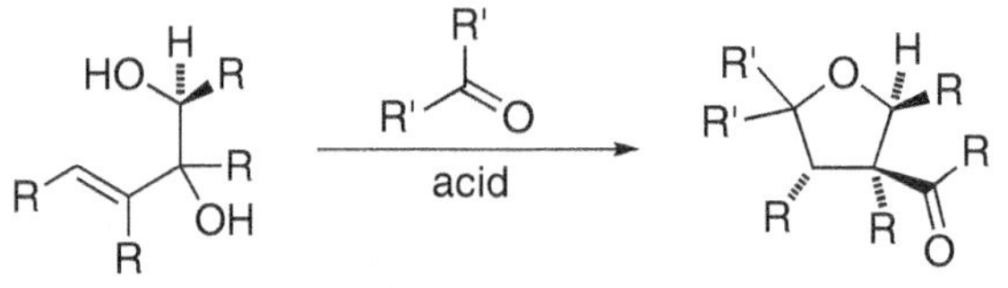

Reagents: $SnCl_4$, $MeNO_2$, –20°C; $SnCl_4$, CH_2Cl_2, –78°C; etc.

Relative stereochemistry of the product: can be predicted on the basis of a cyclohexane-like activated complex for the Prins step

Mannich Reaction: An Iminium Equivalent of the Aldol Addition

Few reactions have seen the rise to prominence that the Mannich reaction[82] has undergone since the last decade of the 20th century. In large part, this is due to the emergence of chiral secondary amines as catalysts in carbon-carbon bond-forming reactions involving aldehydes, in particular, through chiral iminium ions. In a formal sense, the Mannich reaction is an aza-aldol reaction in which the neutral carbonyl electrophile is replaced by a more electrophilic iminium ion (**10.101**). Because the reaction is carried out in the presence of a weak acid catalyst (typically acetic acid), the nucleophilic participant in the reaction is the thermodynamic enol of the second carbonyl component (**10.100**, Figure 10.10).

The iminium ion is most reactive when derived from formaldehyde, or an aldehyde rather than a ketone. In the first step, the amine condenses with the reactive carbonyl compound in the presence of the acid catalyst to form an iminium ion, and the acid also catalyzes the conversion of some of the ketone to its enol form. The enol then adds to the iminium ion to generate an intermediate oxonium that loses a proton to give the Mannich base. An intramolecular Mannich reaction (Example **10.103**) was the final step in Evans' synthesis of luciduline.[83]

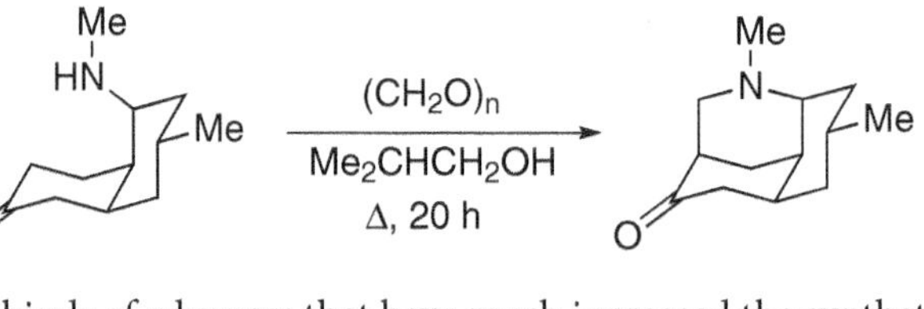

(10.103)

The same kinds of advances that have much increased the synthetic utility of the aldol addition reaction have also made the Mannich reaction even more useful in synthesis. It is now possible to preform the iminium ion (an example is ***Eschenmoser's salt***, $[H_2C{=}NMe_2]I$, formally derived from formaldehyde and dimethylamine (e.g., Example **10.104**).[84] Iminium

82. (a) Mannich, C.; Krösche, W. *Arch. Pharm.* **1912**, *250*, 647. Review: (b) Blicke, F.F. *Org. React.* **1942**, *1*, 303.
83. Scott, W.L.; Evans, D.A. *J. Am. Chem. Soc.* **1972**, *74*, 4779.
84. Schreiber, J.; Maag, H.; Hashimoto, N.; Eschenmoser, A. *Angew. Chem. Int. Ed. Engl.* **1971**, *10*, 330.

ions can also be formed by protonation of enamines at carbon (Example **10.105**). In addition, enol trimethylsilyl ethers have now made kinetic enolate surrogates readily available, so that one is no longer restricted to the thermodynamic enol as the reactant.

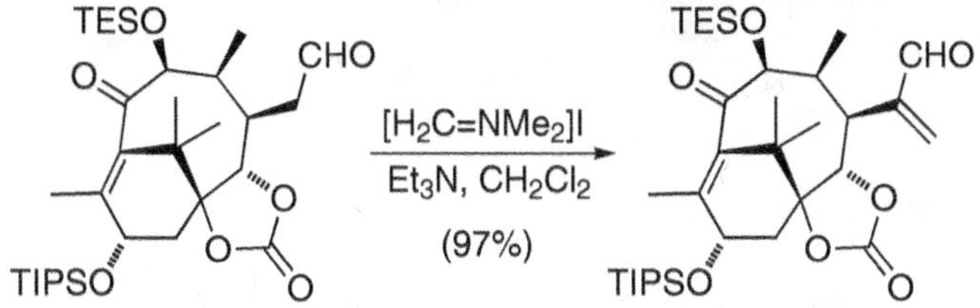

(10.104)[85]

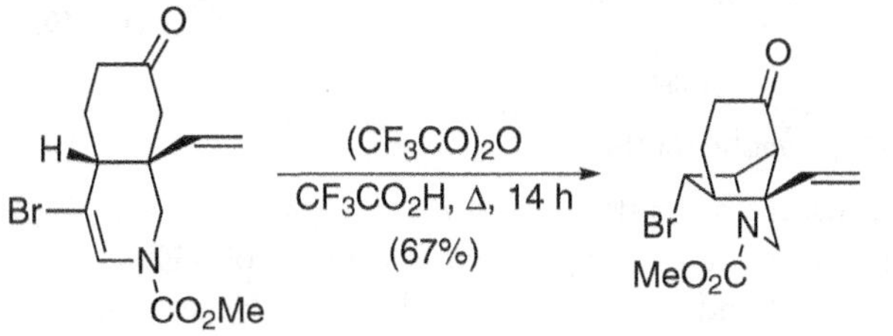

(10.105)[86]

A creative way to generate the enol is illustrated by an early step in the Woodward synthesis of vitamin B_{12}, where the acid-catalyzed hydration of a homopropargyl imine (Example **10.106**) provides an intermediate (Example **10.107**) containing both the enol and the iminium ion moieties.[87]

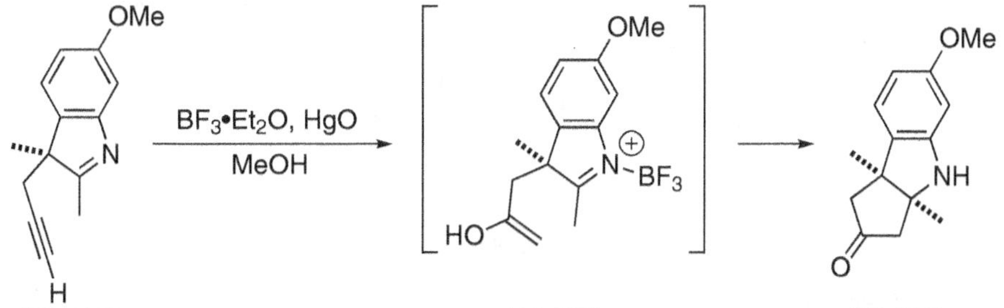

(10.106) **(10.107)** **(10.108)**

Enantioselective Mannich Reactions

As implied in the introductory paragraph of this section, the Mannich reaction can be adapted to asymmetric synthesis by appropriate choice of chiral catalysts. The reaction has generally been most effective when the imine is derived from a glyoxylate ester.[88] In a 2005 report,[89] (*S*)-2-(5-tetrazolyl)pyrrolidine (as in Example **10.109**) was shown to be an efficient chiral catalyst for the *syn*-selective, highly enantioselective Mannich reaction between aldehydes or ketones and the *p*-methoxyphenyl imine of ethyl glyoxylate. Branching at the β carbon reduces both the diastereoselectivity and the enantioselectivity of the reaction. Three years later,[90] a

85. Wender, P.A.; Badham, N.F.; Conway, S.P.; Floreancig, P.E.; Glass, T.E.; Houze, J.B.; Krauss, N.E.; Lee, D.; Marquess, D.G.; McGrane, P.L.; Meng, W.; Natchus, M.G.; Shuker, A.J.; Sutton, J.C.; Taylor, R.E. *J. Am. Chem. Soc.* **1997**, *119*, 2757.

86. Becker, M.H.; Chua, P.; Downham, R.; Douglas, C.J.; Garg, N.K.; Hiebert, S.; Jaroch, S.; Matsuoka, R.T.; Middleton, J.A.; Ng, F.W.; Overman, L.E. *J. Am. Chem. Soc.* **2007**, *129*, 11987.

87. Woodward, R.B. *Pure Appl. Chem.* **1968**, *17*, 519.

88. (a) Ferraris, D.; Young, B.; Dudding, T.; Lectka, T. *J. Am. Chem. Soc.* **1998**, *120*, 4548 (Ag-catalyzed). (b) Hagiwara, E.; Fujii, A.; Sodeoka, M. *J. Am. Chem. Soc.* **1998**, *120*, 2474. (c) List, B. *J. Am. Chem. Soc.* **2000**, *122*, 9336. (d) Juhl, K.; Gathergood, N.; Jorgensen, K.A. *Angew. Chem., Int. Ed.* **2001**, *40*, 2995. (e) Kobayashi, S.; Matsubara, R.; Kitagawa, H. *Org. Lett.* **2002**, *4*, 143. (f) Cordova, A.; Notz, W.; Zhong, G.; Betancort, J.M.; Barbas, C.F. *J. Am. Chem. Soc.* **2002**, *124*, 1842. (g) Nakamura, Y.; Matsubara, R.; Kiyohara, H.; Kobayashi, S. *Org. Lett.* **2003**, *5*, 2481–2484.

89. Cobb, A.J.A.; Shaw, D.M.; Longbottom, D.A.; Gold, J.B.; Ley, S.V. *Org. Biomol. Chem.* **2005**, *3*, 84.

90. Zhang, H.; Mitsumori, S.; Utsumi, N.; Imia, M.; Garcia-Delgado, N.; Mifsud, M.; Albertshofer, K.; Cheong, P.H.-Y.; Houk, K.N.; Tanaka, F.; Barbas, C.F., III *J. Am. Chem. Soc.* **2008**, *130*, 875.

similar, *anti*-selective reaction between an aldehyde and the same imine, catalyzed by chiral pyrrolidine-3-carboxylic acid derivatives (as in Example **10.110**), was reported.

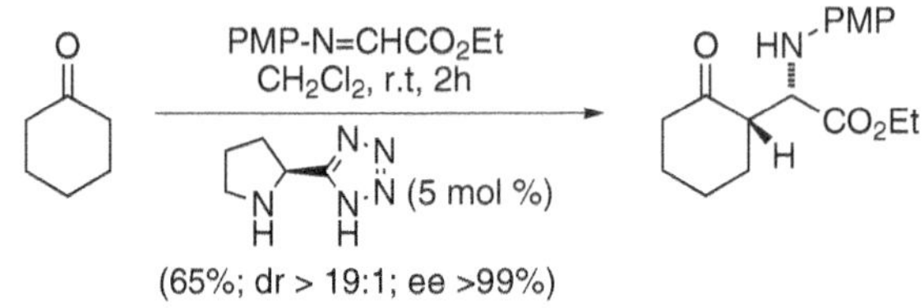

(10.109)

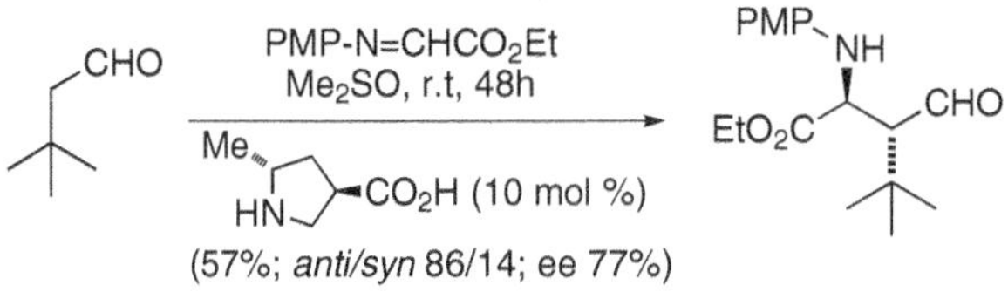

(10.110)

This reaction is also susceptible to steric effects in the form of β-branching in the aldehyde: straight-chain aldehydes routinely give *anti/syn* ratios of 93:7 or higher and ee's above 95%. Other hydrogen-bonding catalysts based on the thiourea skeleton have also been developed as efficient chiral catalysts for the *anti*-selective asymmetric Mannich reaction of nitroalkanes with imines, as in Example **10.111**,[91] although a model for predicting the stereochemistry in these reactions has not been proposed.

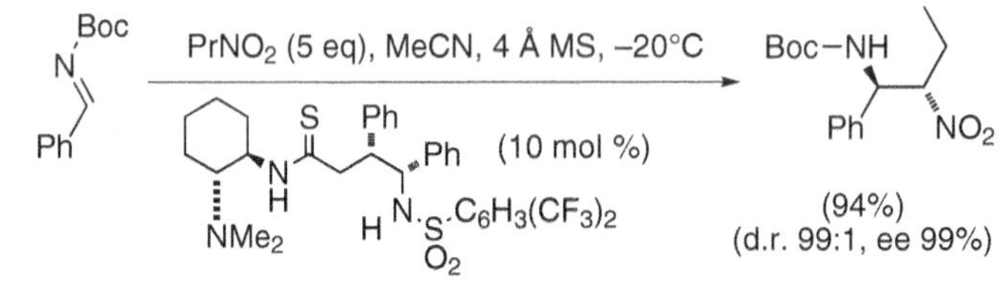

(10.111)

Reaction Synopses

Mannich Reaction

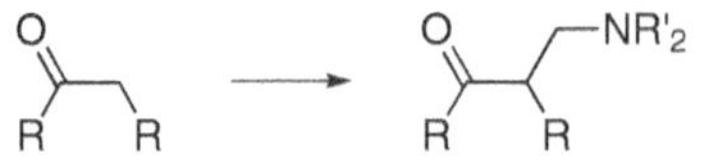

Reagents:

R'$_2$NH, CH$_2$O, AcOH, Δ; etc. or
(1) LDA, THF, –78°C; (2) [R'$_2$N=CH$_2$]I; etc.

Asymmetric Mannich Reaction

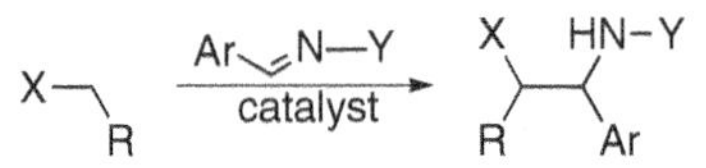

X: CHO, COR, NO$_2$Y = CO$_2$R, Boc, etc.

Catalysts:

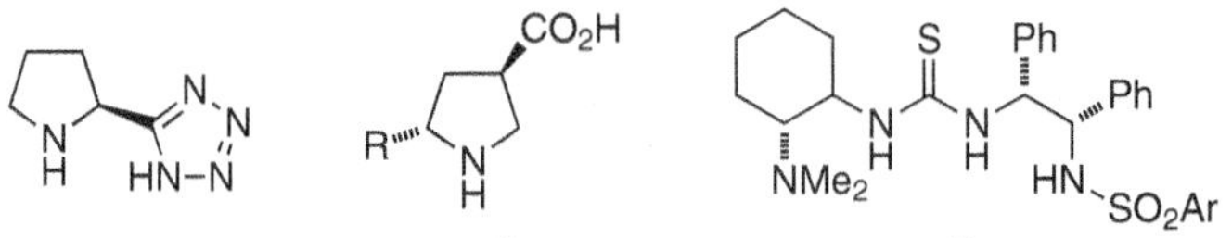

syn *anti* *anti*

Absolute configuration of product: can be predicted with pyrrolidine-3-carboxylic acid derivatives as catalysts.

---

91. Wang, C.-J.; Dong, X.-Q.; Zhang, Z.-H.; Xue, Z.-Y.; Teng, H.-L. *J. Am. Chem. Soc.* **2008**, *130*, 8606.

Worked Problem

10-4 Complete the following reactions.

(a) 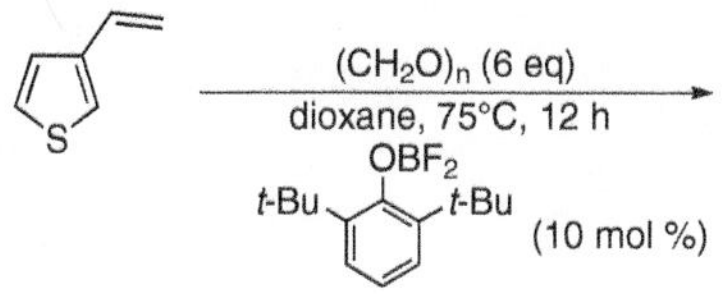

(b) ? 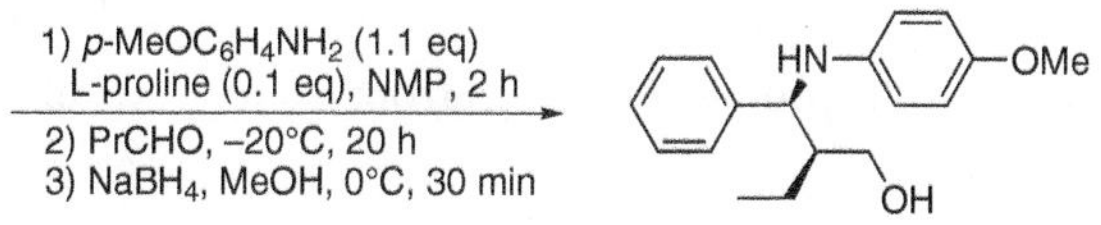

§Answers below.

Drill Problem

What is the major organic product from each of the following reactions?

Compounds:

(a) 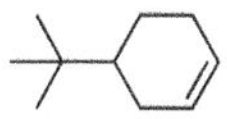 (b) 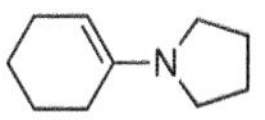 (c) 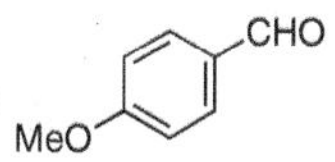

(continues)

§ Answers to Worked Problem:

(a) 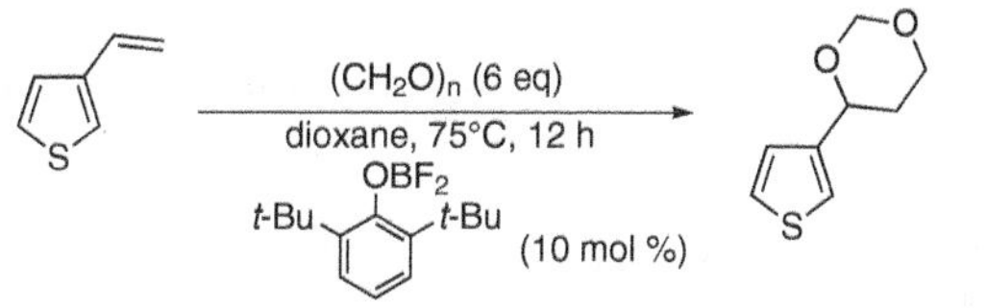

This is a Prins reaction catalyzed by a hindered Lewis acid. As is common, acids where the acidic site is sterically hindered by large, hydrophobic alkyl groups tend to be more effective acid catalysts. The large excess of formaldehyde used in this reaction strongly suggests that the final product will be the 1,3-dioxane rather than the 1,3-diol.

(b) 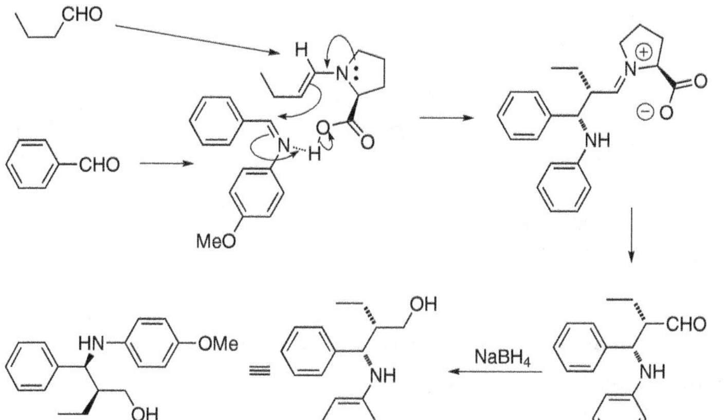

The missing reactant is benzaldehyde. The final product of this reaction sequence is a 3-aminoalcohol, and the most likely source of the primary alcohol functional group is an aldehyde. Thus, this reaction is a Mannich reaction, because the intermediate compound would be a β-aminoaldehyde. The two substituent groups other than the amine group at the β carbon are the two groups bonded to the carbonyl group of the missing reactant. The stereochemistry can be predicted on the basis of a model where a hydrogen bond formed between the aldimine and the proline enamine organizes the activated complex.

References: (a) *Synthesis* **2002**, 2521. (b) *Angew. Chem. Int. Ed.* **2003**, *42*, 3677.

(**Drill Problem** continued)

Reagents:

(A) $(CH_2O)_n$, H_2SO_4. (B) $EtO_2CCH{=}NPh$,

(C) PhCHO, Me_2AlCl.

Problems

10-6 Provide the major organic product expected from each of the following reactions.

(a)

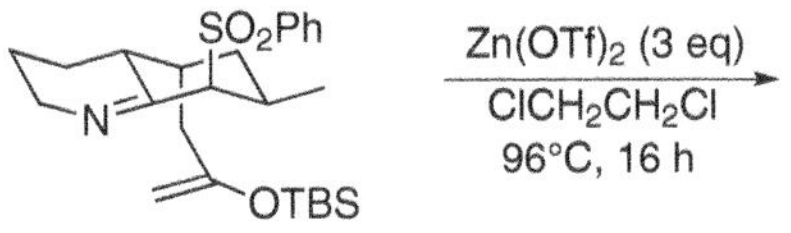

(b)

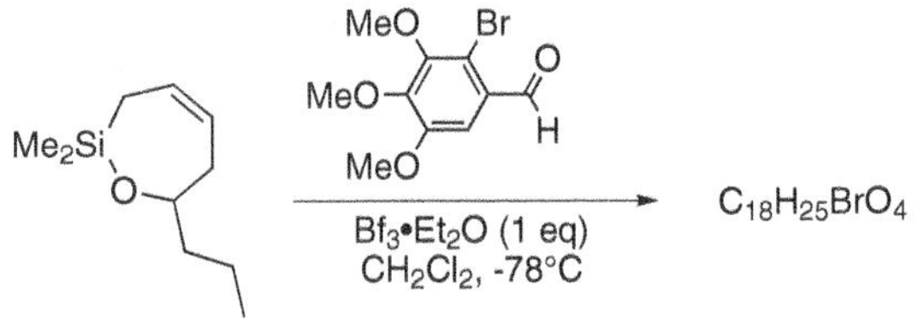

(c)

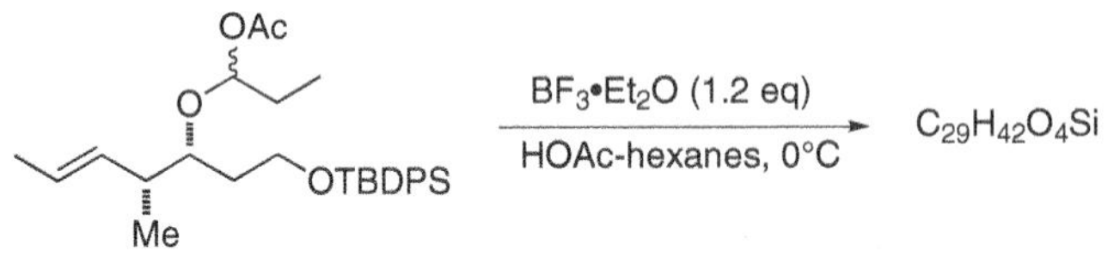

References: (a) *J. Am. Chem. Soc.* **2008**, *130*, 9238. (b) *Org. Biomol. Chem.* **2006**, *4*, 4118. (c) *Org. Lett.* **2000**, *2*, 1217.

10-7 Provide a reasonable mechanism that explains the regiochemistry of the reaction below.

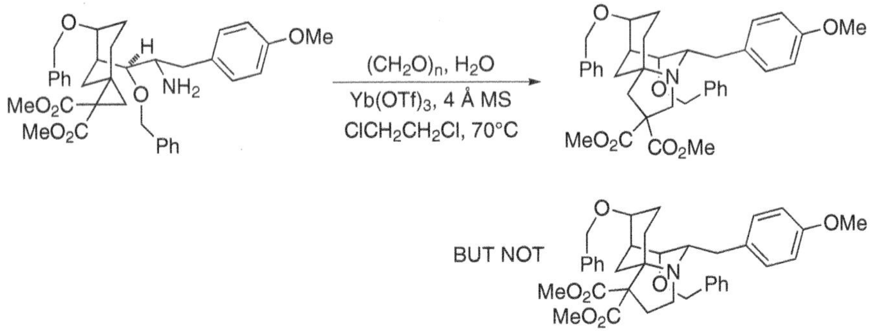

Reference: Org. Lett. **2009**, *11*, 777.

Electron-Rich Alkenes: Surrogates of Enolate Anions for Reactions with Carbocations

The nucleophilic partner of the electrophile in the Mannich and Mukaiyama reactions is an enol or the surrogate of an enolate anion. We will discuss the chemistry of enolate anions in much more detail in Chapter 12. At this point in time, one need only know that they are strong nucleophiles and strong bases (the pK_a values of the carbonyl compounds that one deprotonates to give the enolate anions are typically around 26), which makes them basically incompatible with carbocation electrophiles, or their analogous oxonium or iminium cations.

One defining structural feature of a useful enolate equivalent is a double bond–bearing substituent with a lone pair (Example **10.112**). One effect of such substituents is to make the alkene π bond more susceptible to the addition of electrophiles because the intermediate carbocation formed during these additions can be stabilized by conjugation of the positive charge with the lone pair. In resonance terms, these alkenes have a minor canonical form in which the β carbon carries a formal negative charge (Example **10.113**). Such alkenes react readily with carbocations to give products containing a new carbon-carbon σ bond. Depending on the substituent, the second step of the addition may resemble the second step of the S_N1 reaction, where the positive charge on carbon is neutralized by σ bond formation to the nucleophile, or it may resemble the second step of the E1 reaction, where the positive charge on carbon is neutralized by π bond formation.

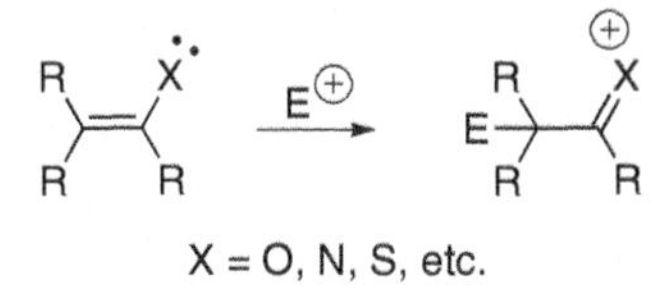

(10.112)

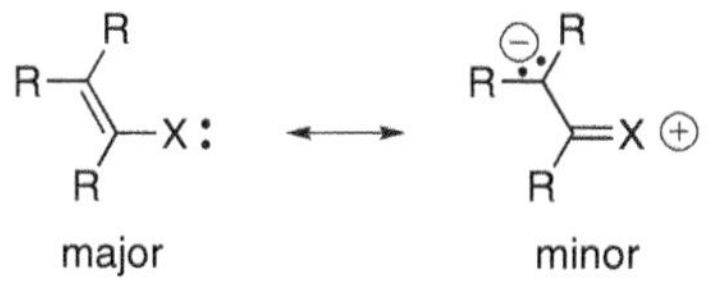

(10.113)

Enamines

One of the first enolate surrogates to be widely used was the ***enamine***, shown as Example **10.114**, which is formed by condensing a carbonyl compound with a secondary amine. The resultant compound has the nitrogen directly bonded to one of the sp^2-hybridized carbons of the alkene. Enamines are not widely used in reactions with carbocations, so we will reserve our discussion of these enolate surrogates to Chapter 12.

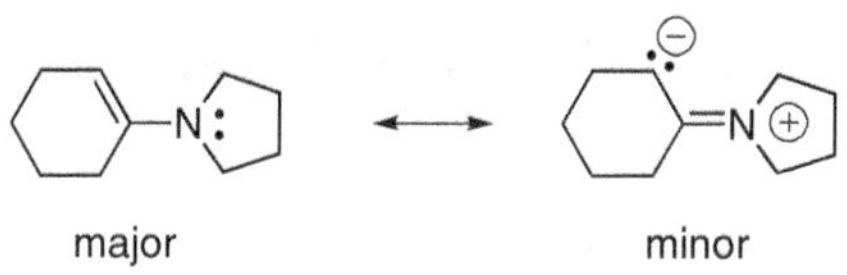

(10.114)

Enol Silyl Ethers

One of the most useful alkenes of this type is the enol trimethylsilyl ether (Example **10.115**).[92] An enol trimethylsilyl ether reacts with both unstabilized alkyl carbocations and stabilized carbocations such as those used in the Prins reaction to give an intermediate carbocation stabilized by the oxygen (Example **10.116**). Attack of the nucleophile in the second step of the addition usually occurs at the silicon atom to give a carbonyl compound (Example **10.117**) as the ultimate product. An enol trimethylsilyl ether reacts with an alkyl halide in the presence of a Lewis acid (a way of generating an alkyl cation equivalent) to give a carbonyl compound as the final product.

---

92. For reviews, see: (a) Mukaiyama, T. *Angew. Chem., Int. Ed. Engl.* **1977**, *16*, 817; *Org. React.* **1982**, *28*, 203. (b) Rasmussen, J.K. *Synthesis* **1977**, 91. (c) Brownbridge, P. *Synthesis* **1983**, 1, *85*. (d) For a general method of preparing enol trimethylsilyl ethers, see: Walshe, N.D.A.; Goodwin, G.B.T.; Smith, G.C.; Woodward, F.E. *Org. Synth.* **1986**, *62*, 1.

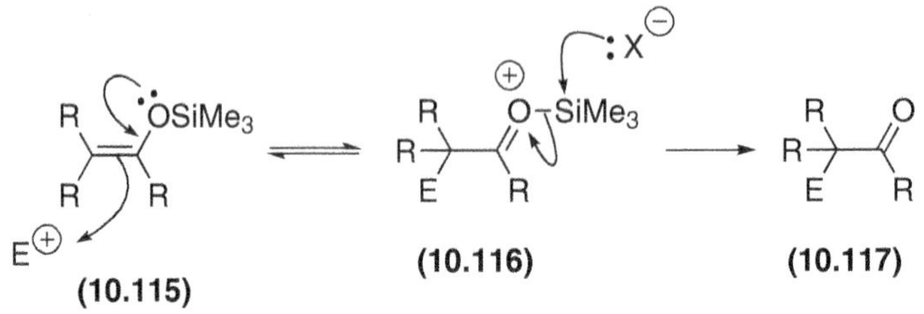

(10.115) (10.116) (10.117)

The first general solution to the long-standing problem of α-*tert*-alkylation of ketones was obtained by using the enol trimethylsilyl ether and a Lewis acid catalyst in place of the enolate anion.[93] Examples **10.118** and **10.119**, where the enol trimethylsilyl ether of a ketone is alkylated with a *tert*-alkyl chloride or an allylic chloride in the presence of titanium tetrachloride as a Lewis acid catalyst, are typical of this type of reaction. In both reactions, the titanium tetrachloride serves to generate the carbocation from the alkyl chloride.

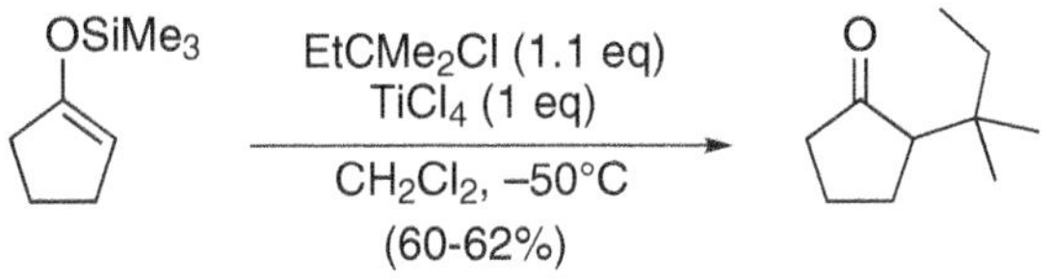

(10.118)[94]

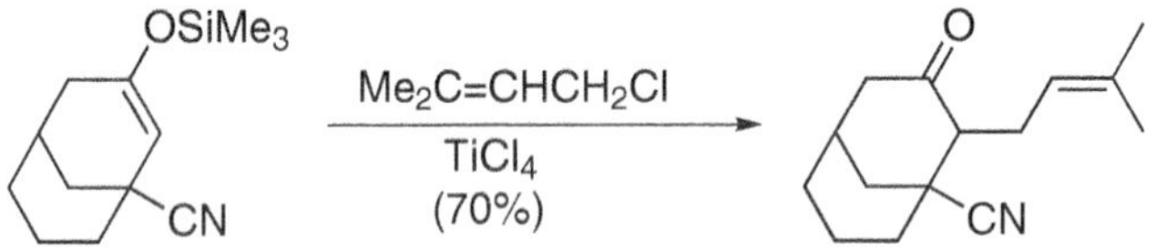

(10.119)[95]

The reaction of silyl enol ethers with allylic halides where the allyl cation is not symmetrical (e.g., the prenyl halides), give regioisomeric products, and the ratio of the products depends on a number of factors, including the structure of the halide, the identity of the Lewis acid, and the solvent (Table 10.2).

Enol trimethylsilyl ethers also react with aldehydes and ketones under acid catalysis in what is formally a Prins-type reaction. The reaction also corresponds to an acid-catalyzed

Table 10.2 Regiochemistry of Prenylation of 1-Trimethylsilyloxycyclopentene

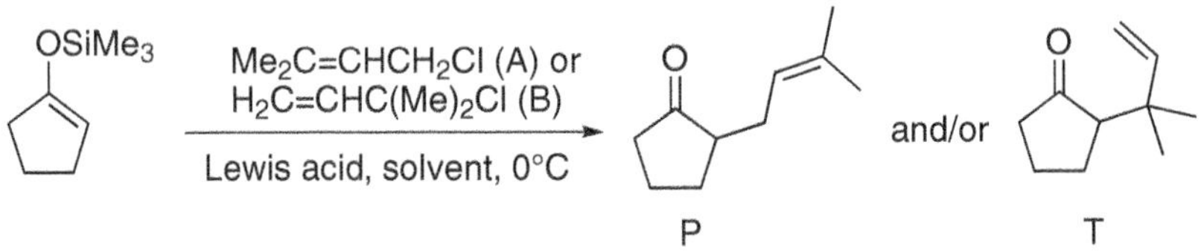

Halide	Lewis Acid	Solvent	P:T Ratio	Yield (crude)
B	$ZnCl_2$	$MeNO_2$	60:40	100%
A	$ZnCl_2$	$MeNO_2$	60:40	95%
A	CuI	CH_2Cl_2	88:12	85%
A	$ZnCl_2$	none	80:20	>85%

93. (a) Chan, T.H.; Paterson, I.; Pinnsonnault, J. *Tetrahedron Lett.* **1977**, 4183. (b) Reetz, M.T.; Maier, W.F. *Angew. Chem. Int. Ed. Engl.* **1978**, *17*, 48. (c) Reetz, M.T. *Angew. Chem. Int. Ed. Engl.* **1982**, *21*, 96. (d) Reetz, M.T.; Maier, W.F.; Heimbach, H.; Giannis, A.; Anastassiou, G. *Chem. Ber.* **1980**, *113*, 3734. (e) Reetz, M.T.; Hüttenhain, S.; Waltz, P.; Löwe, U. *Tetrahedron Lett.* **1979**, 4971. (f) Paterson, I. *Tetrahedron Lett.* **1979**, 1519.

94. Reetz, M.T.; Chatziiosifidis, I.; Hübner, F.; Heimbach, H. *Org. Synth.* **1984**, *6*, 95.

95. Kraus, G.A.; Kirihara, M. *Tetrahedron Lett.* **1992**, *33*, 7727.

version of the aldol condensation. When a protic acid is used to catalyze this reaction, the product is almost always the conjugated carbonyl compound derived by dehydration of the resultant β-hydroxycarbonyl compound. However, when a Lewis acid such as titanium tetrachloride is used, the β-hydroxycarbonyl compound can often be isolated. The reaction, which was developed into a useful synthetic reaction largely as a result of the work of Japanese chemist Teruaki Mukaiyama, is now known as the ***Mukaiyama aldol reaction***.[96] Unlike many aldol addition reactions, the Mukaiyama aldol addition is not always highly stereoselective, but the fact that it does not need to be carried out under basic conditions and its high level of chemoselectivity still render it a valuable reaction. For example, it can be used to prepare the aldol **10.120**,[97] which reverts to acetophenone and acetone under the "normal" aldol addition conditions, from the enol trimethylsilyl ether. The stereoselectivity of the reaction has been improved by the use of "super silyl" groups, $Si(SiMe_3)_3$, which are very large, and allow facile silyl transfer reactions; their reactivity has permitted the assembly of molecules with arrays of stereocenters, as in Example **10.121**.[98] In addition, the reaction can also be used with ketals and acetals. In these reactions, the Lewis acid converts the acetal or ketal into an oxonium ion that then adds to the silyl ether. This reaction is exemplified by the closure of the bridged 8-membered ring of the taxane skeleton (Example **10.122**).[99]

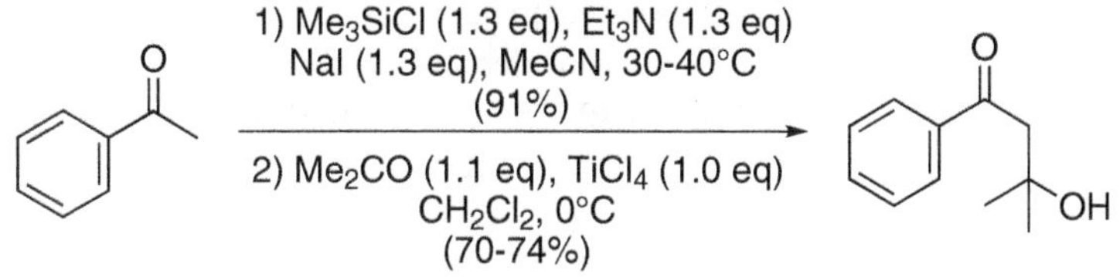

(10.120)

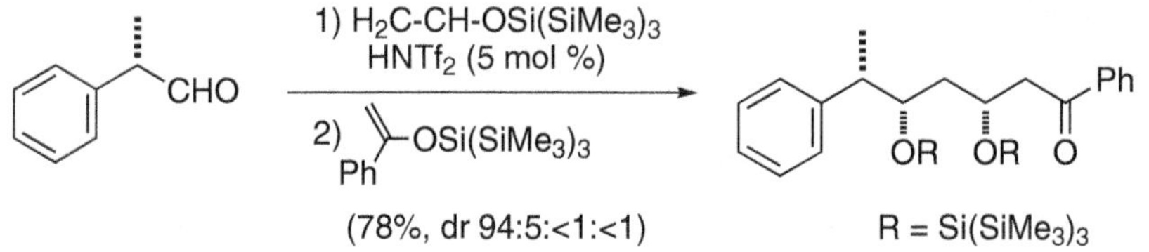

(10.121)

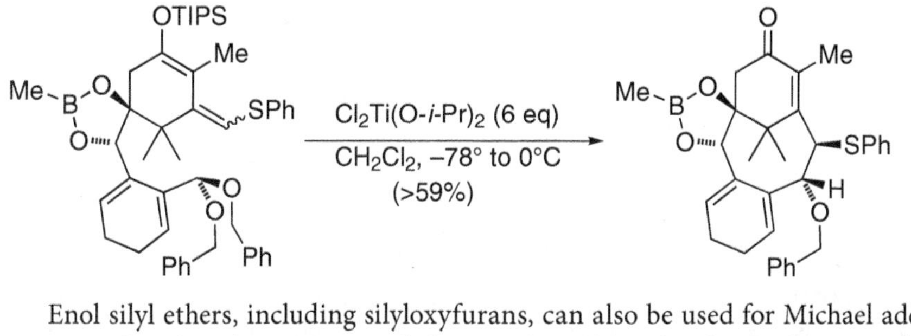

(10.122)

Enol silyl ethers, including silyloxyfurans, can also be used for Michael additions to conjugated carbonyl acceptors and the acetals and ketals of α,β-unsaturated carbonyl compounds.[100] α-Silyloxyfurans give a vinylogous Michael addition with enones, as shown in Example **10.123**.[101]

---

96. Mukaiyama, T.; Narasaka, K.; Banno, K. *Chem. Lett.* **1973**, 1011. (b) Mukaiyama,T.; Banno, K.; Narasaka, K. *J. Am. Chem. Soc.* **1974**, *96*, 7503.

97. Mukaiyama, T.; Narasaka, K. *Org. Synth.* **1986**, *62*, 6.

98. Boxer, M.B.; Yamamoto, H. *J. Am. Chem. Soc.* **2007**, *129*, 2762.

99. Kusama, H.; Hara, R.; Kawahara, S.; Nishimori, T.; Hajime Kashima, H.; Nakamura, N.; Morihira, K.; Kuwajima, I. *J. Am. Chem. Soc.* **2000**, *122*, 3811.

100. (a) Narasaka, K.; Soai, K.; Mukaiyama, T. *Chem. Lett.* **1974**, 1223. (b) Narasaka, K.; Soai, K.; Aikawa, Y.; Mukaiyama, T. *Bull. Chem. Soc. Japan* **1976**, *49*, 779.

101. Chabaud, L.; Jousseaume, T.; Retailleau, P.; Guillou, C. *Eur, J. Org. Chem.* **2010**, 5471.

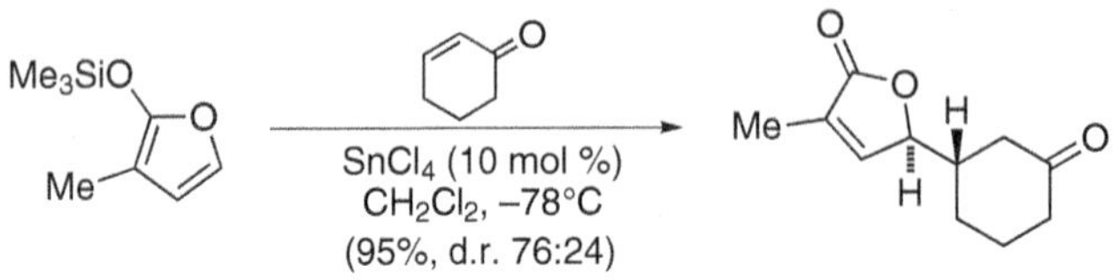

(10.123)

Allylboranes and Allylborinates

Boranes are Lewis acids that complex well to carbonyl groups to generate oxonium ions. What is especially interesting about the interaction of allylboranes with carbonyl groups is that the complexation of the carbonyl oxygen serves two purposes: (1) it makes the carbonyl carbon significantly more electron-deficient; and (2) it makes the allylic group attached to boron significantly more nucleophilic. This sets up a reaction that can be described as an analog of an intramolecular Prins reaction, proceeding through a cyclic transition state (e.g., Example **10.124**).

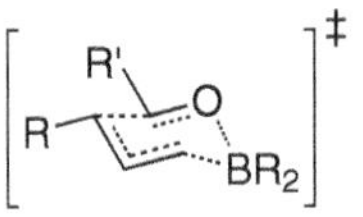

(10.124)

The addition reaction occurs with allylic rearrangement through a six-membered cyclic transition state, which allows two new chiral centers to be incorporated stereoselectively into the product. By the use of diisopinocampheylborane derivatives, these allylborane additions can also be made enantioselective (e.g., Example **10.125**). The required allylboranes are conveniently prepared from alkenes by deprotonation with "LICKOR" (effectively butylpotassium; we will discuss this reagent in much more detail in Chapter 11), followed by treatment with a methoxydialkylborane, as illustrated in Example **10.126.** This borane was subsequently used in an allylation reaction as part of the total synthesis of archazolid A and B.[102]

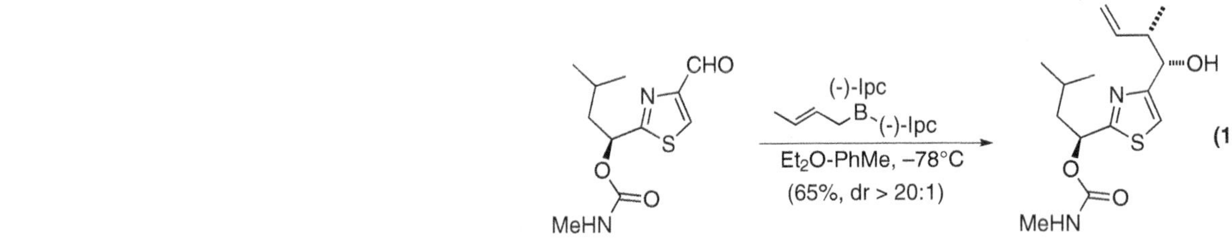

(10.125)

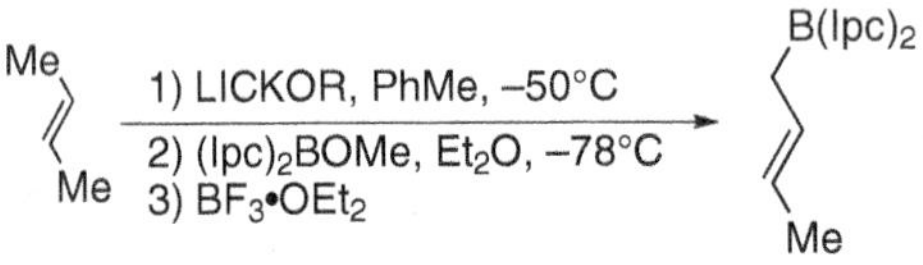

(10.126)

The addition of allylboranes to aldehydes is highly stereoselective, and the relative stereochemistry of the reaction can be predicted on the basis of a chairlike transition state. The observed stereochemistry is that the *E* borane gives the *anti* alcohol, whereas the *Z* borane gives the *syn* alcohol. The *Z* allylboron reagent is consistently more stereoselective than the *E* isomer. The absolute configuration of the major product can be predicted for simple allylboronates derived from tartrate esters using the model in Example **10.127**,[103] which predicts that the simple allylboronate derived from dialkyl (2*S*,3*S*)-tartrates will add to the *si* face of the aldehyde. Roush has shown that the chirality of the boron reagent dominates the enantioselectivity of the reaction.[104] In **10.128**, for example, the enantiomer shown of the *syn* isomer accounts for 95% of the product mixture.[105]

---

102. Mensche, D.; Hassfeld, J.; Li, J.; Mayer, K.; Rudolph, S. *J. Org. Chem.* **2009**, *74*, 7220.

103. Roush, W.R.; Hoong, L.K.; Palmer, M.A.J.; Park, J.C. *J. Org. Chem.* **1990**, *55*, 4109.

104. Roush, W.R.; Palkowitz, A.D.; Palmer, M.J. *J. Org. Chem.* **1987**, *52*, 316.

105. (a) Roush, W.R.; Adam, M.A.; Walts, A.E.; Harris, D.J. *J. Am. Chem. Soc.* **1986**, *108*, 3422. (b) Huckins, J.R.; de Vicente, J.; Rychnovsky, S.D. *Org. Lett.* **2007**, *9*, 4757.

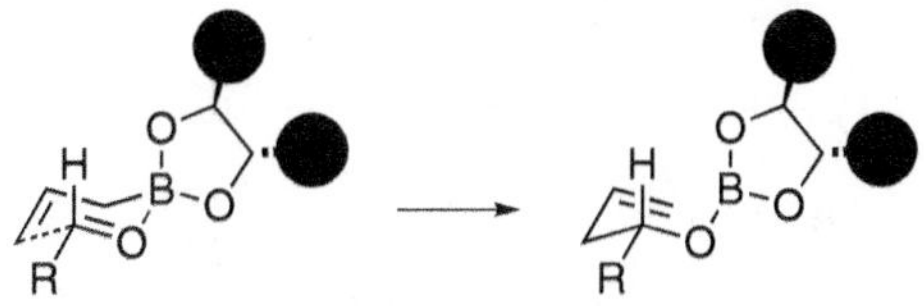

(10.127)

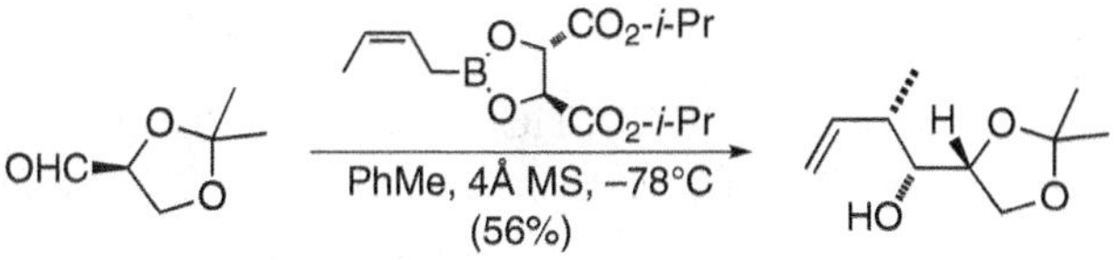

(10.128)

Worked Problem

10-5 What reagent or sequence of reagents should be used to effect the following transformations? Give your reasoning.

(a) 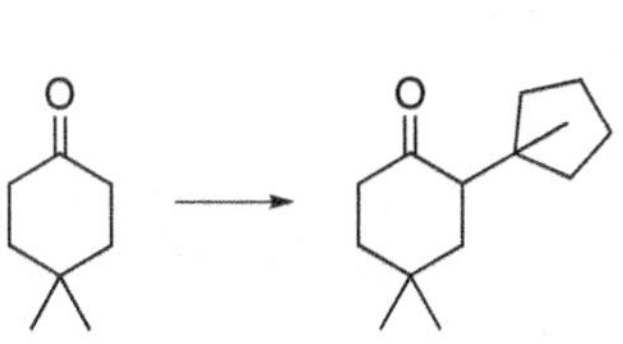

(b) 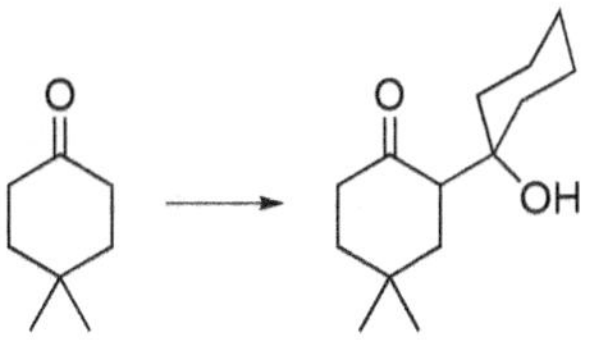

§Answers below.

Reaction Synopses

Alkylation of Enol Silyl Ethers

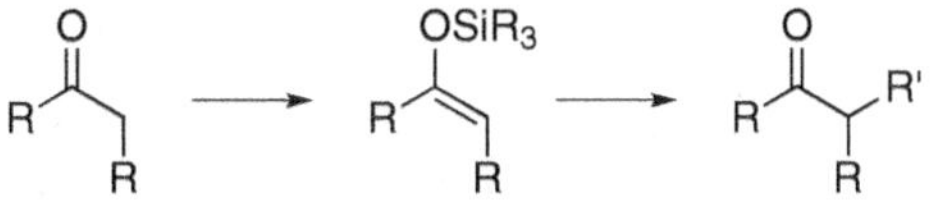

Reagents: (1) Me_3SiCl, Et_3N; (2) R'Cl, $TiCl_4$, CH_2Cl_2, -78°C; or
(1) LDA, THF, -78°C; (2) R'Cl, $TiCl_4$, CH_2Cl_2, -78°C; etc,

Best method for incorporating a tertiary alkyl group α- to a carbonyl group

(continues)

§ Answers to Worked Problem:

(a) This product may be prepared by alkylation of the trimethylsilyl enol ether, which should work well in this instance since the alkyl group is tertiary.

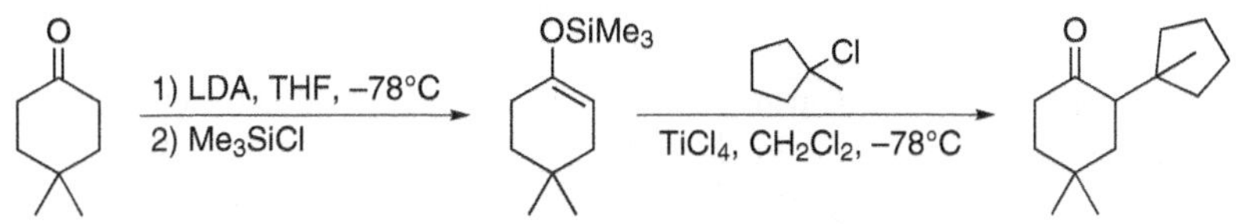

(b) Because the involvement of a carbocation is required, the Muikaiyama aldol addition should be used to prepare this product.

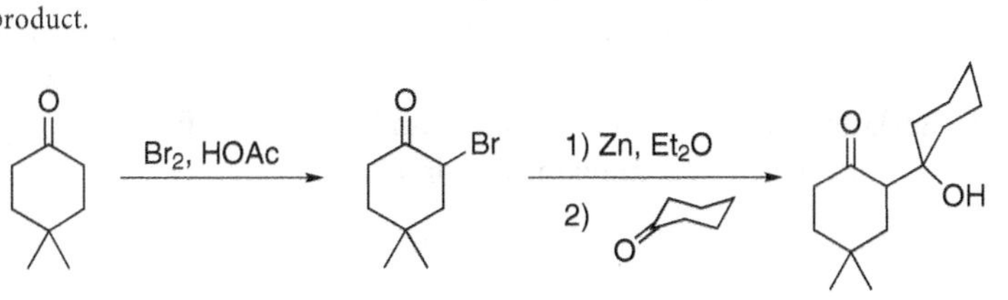

(**Reaction Synopses** continued)

Mukaiyama-Directed Aldol Addition

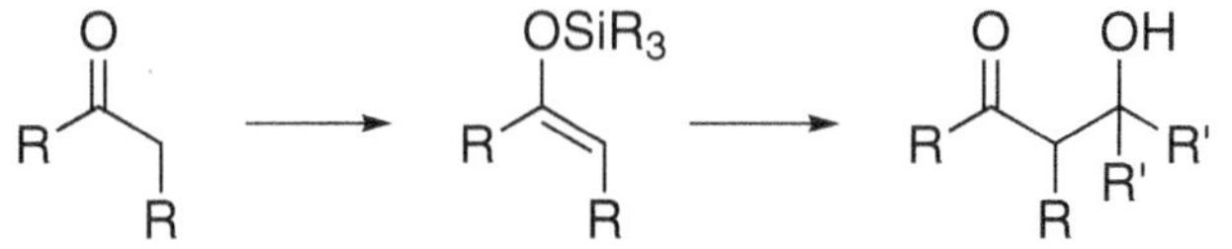

Reagents:	(1) Me_3SiCl, Et_3N; (2) R'_2CO, $TiCl_4$, CH_2Cl_2, –78°C; or (1) LDA, THF, –78°C; (2) Me_3SiCl; (3) R'_2CO, $TiCl_4$, CH_2Cl_2, –78°C; etc.
Conditions:	permit the isolation of labile aldol addition products

Addition of Allylboron Electrophiles: Allylation

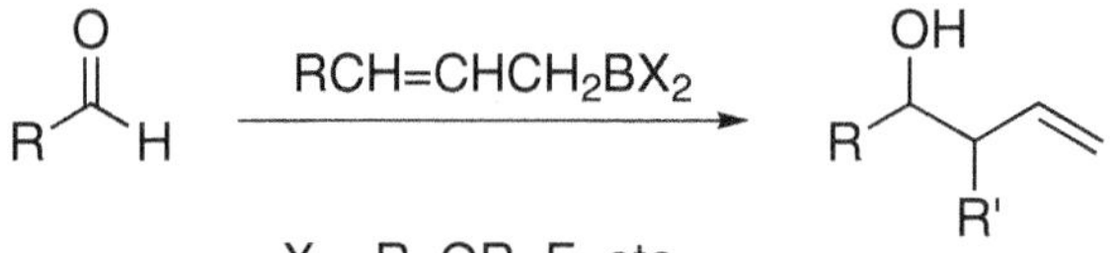

Reagents:	$RCH{=}CHCH_2BX_2$ CH_2Cl_2, –78°C; etc.
Diastereoselectivity:	determined by geometry of allylboron reagent: *Z* gives mainly *syn* product; *E* gives mainly *anti* product. Stereoselectivity of *Z* allylboron reagents are higher (often much higher).
Enantioselectivity:	fixed by chirality of the allylboronate in cyclic derivatives of dialkyl tartrates. Enantioselectivity is often high.

Chapter Summary

This chapter has introduced major carbon-carbon bond-forming reactions that involve carbocations as key intermediates. These reactions include the Friedel-Crafts alkylation and acylation reactions, as well as the related formylations that include the Gattermann, Gattermann-Koch, Houben-Hoesch, Vilsmeier-Haack, Duff, and Reiche-Gross reactions. The dealkylation and rearrangements occurring during Friedel-Crafts alkylation, and the asymmetric form of that reaction have been discussed. Reactions involving carbocations stabilized by an adjacent lone pair–bearing heteroatom include the Prins, aza-Prins, and Mannich reactions, as well as the Mukaiyama alkylation and aldol reactions. Enamines and enol ethers have been introduced as enolate surrogates; enol trimethylsilyl ethers have been introduced as nucleophilic partners in the Mukaiyama and related reactions, and the use of allylboranes for the formation of new carbon-carbon bonds by a form of intramolecular Prins reaction has been discussed.

Key Terms

allylboron allylation
aza-Prins reaction
dealkylation
Duff reaction
enamines
formylation
Friedel-Crafts acylation
Friedel-Crafts alkylation
Gattermann formylation
Gattermann-Koch formylation
Gattermann-Olah formylation
Houben-Hoesch acylation
Mannich reaction
Mukaiyama aldol addition
Mukaiyama alkylation
Pinacol-terminated Prins reaction
Prins reaction
Reich-Gross formylation
silyl enol ether
Vilsmeier-Haack formylation

Additional Problems

10-8 What is the structure of the major organic product of each of the following reactions?

(a) 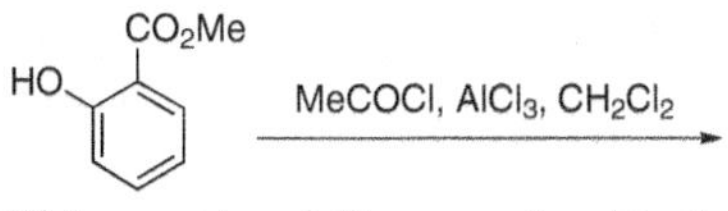

[*Org. Lett.* **2006**, *8*, 831]

This reaction fails to work with the methoxy compound. Why?

(b) 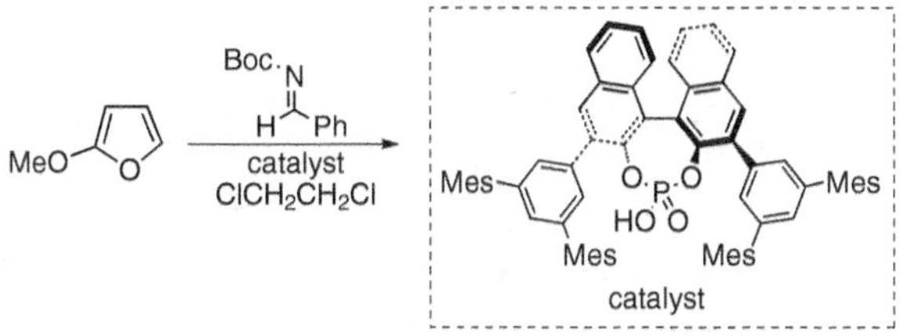

[*J. Am. Chem. Soc.* **2004**, *126*, 11804]

(c) 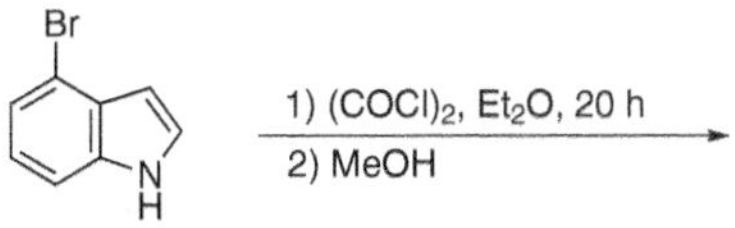

[*J. Am. Chem. Soc.* **2007**, *129*, 13794]

(d) 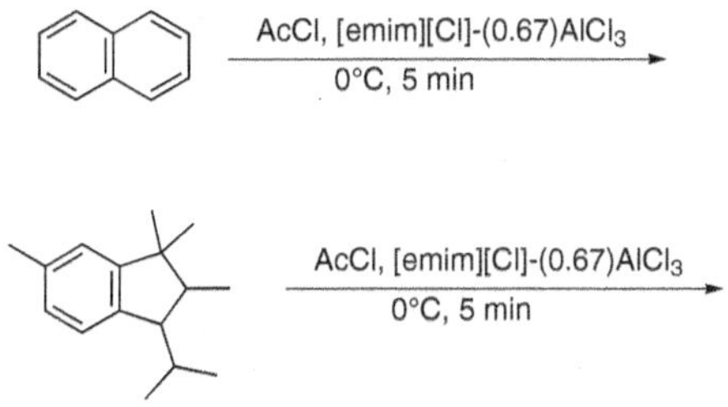

[*Chem. Commun.* **1998**, 2097]

(e) 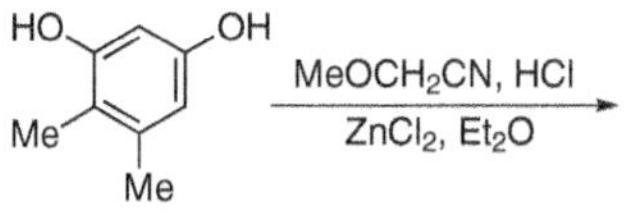

[*J. Med. Chem.* **1998**, *41*, 2333]

(f) 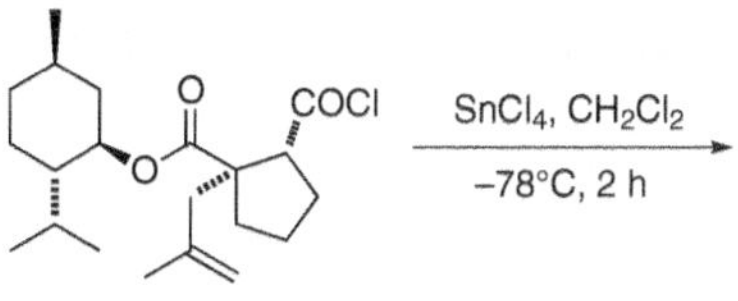

[*J. Am. Chem. Soc.* **1989**, *111*, 2302]

(g) 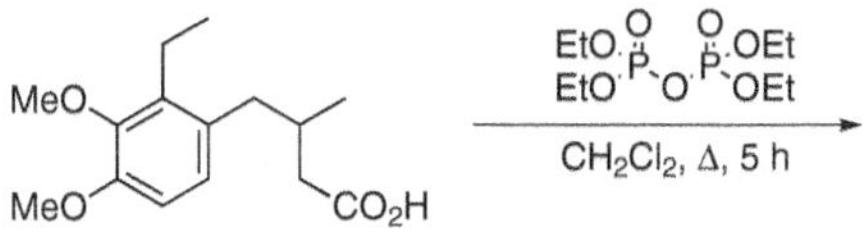

[*Tetrahedron Lett.* **2012**, *53*, 373]

(h) 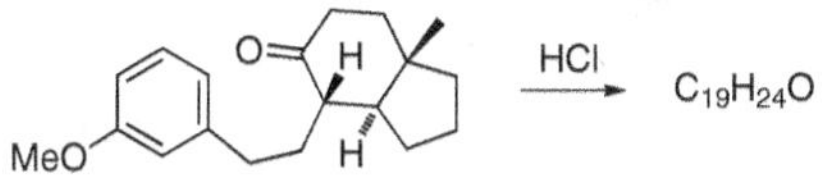

[*J. Org. Chem.* **1977**, *42*, 3214]

(i) 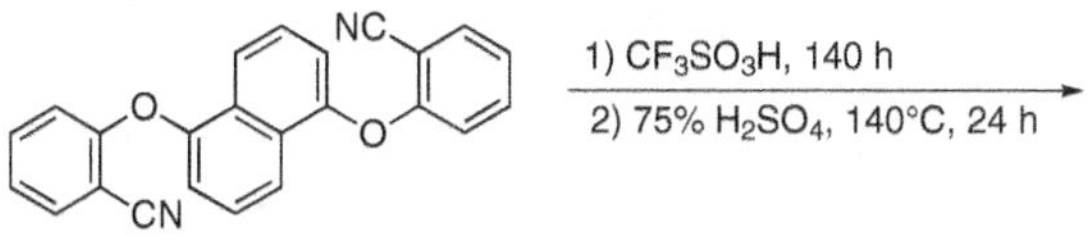

[*Org. Lett.* **2001**, *3*, 2337]

(j) 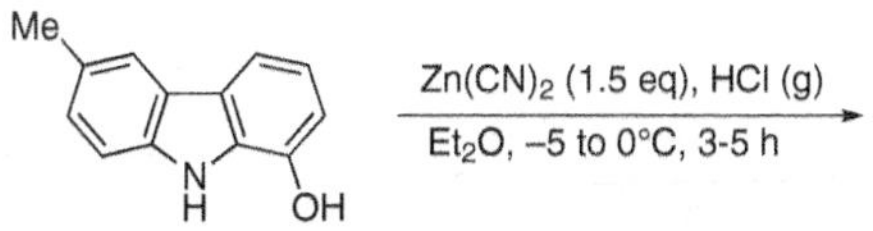

[*Heterocycles* **1999**, *51*, 2163]

(k) 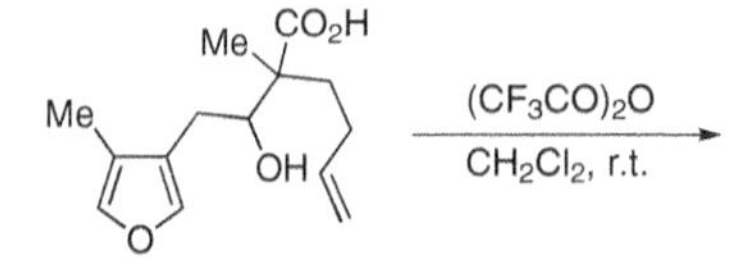

[*Tetrahedron* **2001**, *57*, 6935]

(l) 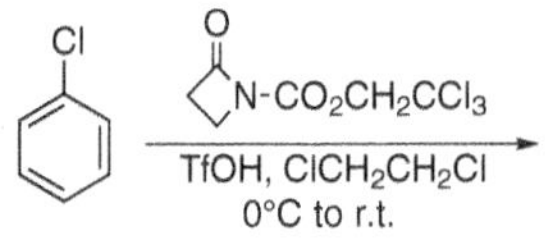

[*Tetrahedron* **2002**, *58*, 8475]

10-9 Give the structure of the major organic product expected from each of the following reactions.

(a) 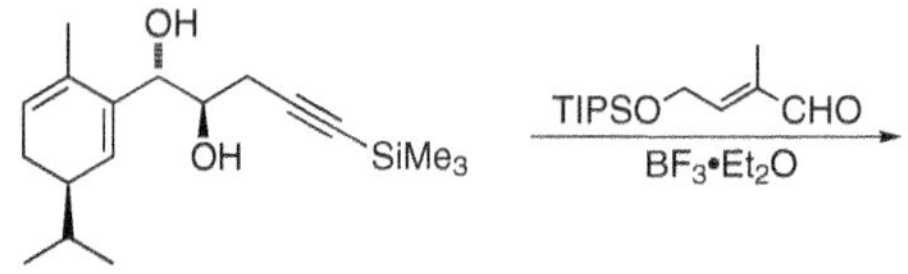

[*J. Am. Chem. Soc.* **1995**, *117*, 10391]

(b) 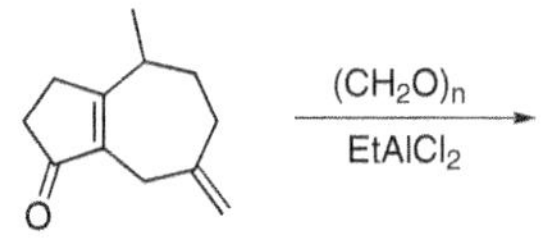

[*J. Org. Chem.* **1993**, *58*, 1030]

(c) 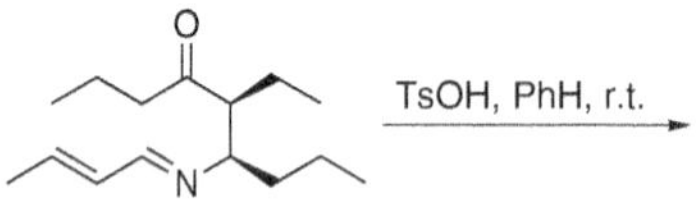

[*J. Am. Chem. Soc.*, **2005**, *127*, 8398]

(d) 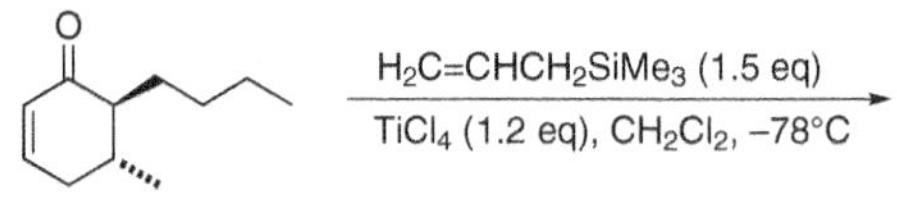

[*J. Am. Chem. Soc.* **1984**, *106*, 721]

(e) 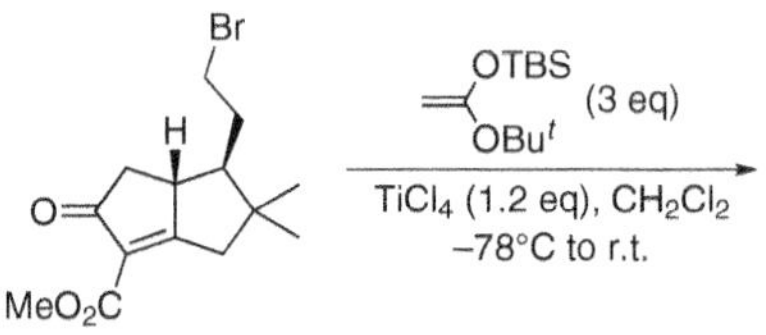

[*J. Am. Chem. Soc.* **1981**, *103*, 4136]

(f) 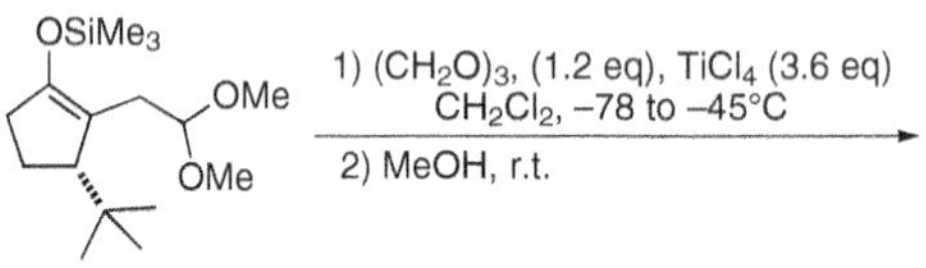

[*J. Am. Chem. Soc.* **1988**, *110*, 649]

(g) 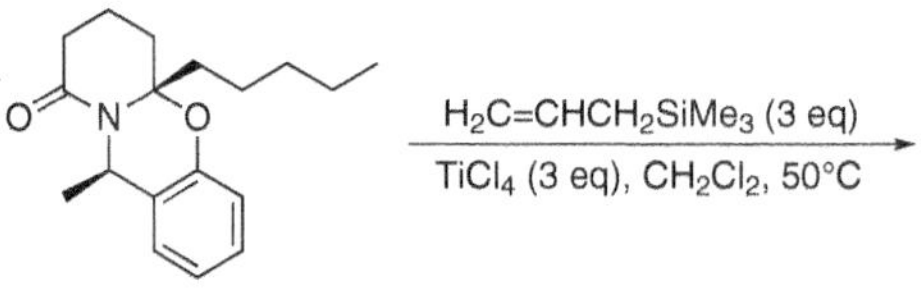

[*Tetrahedron Lett.* **1999**, *40*, 739]

(h) 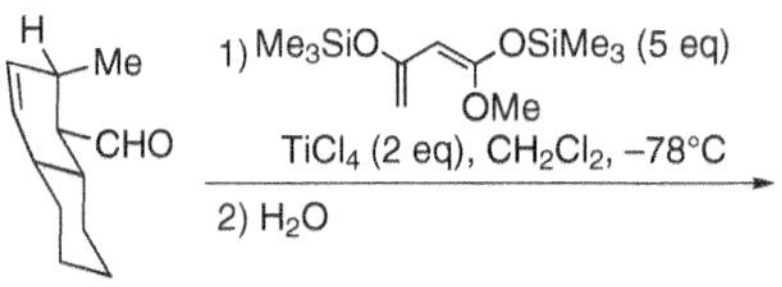

[*J. Org. Chem.* **2002**, *67*, 5969]

(i) 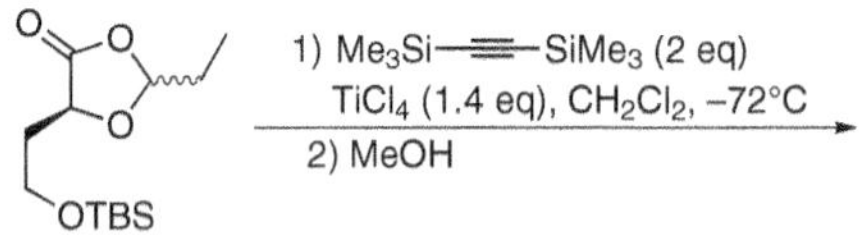

[*J. Am. Chem. Soc.* **2004**, *126*, 12432]

(j) 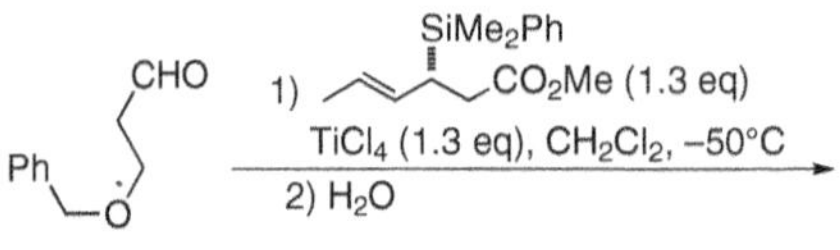

[*Org. Lett.* **2004**, *6*, 3533]

(k) 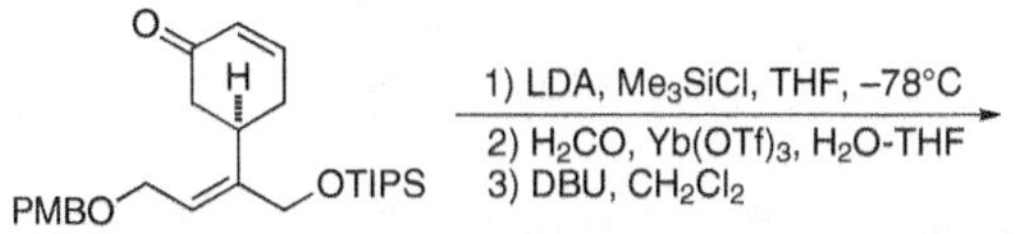

[*J. Am. Chem. Soc.* **2002**, *124*, 14546]

What is the reason for using the DBU?

(l) 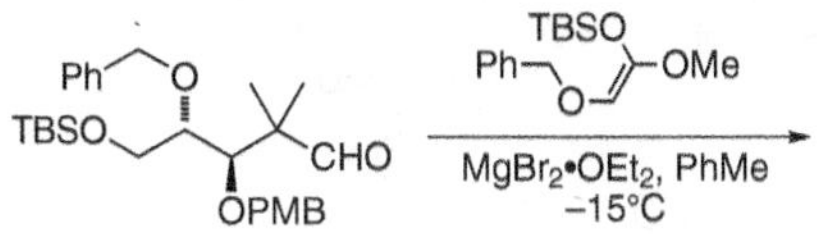

[*Chem. Eur. J.* **1999**, 5, 121]

(m) 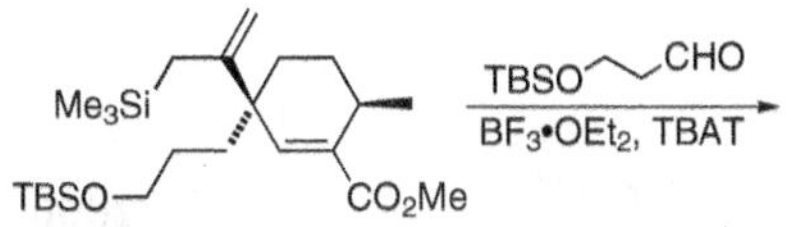

[*Angew. Chem. Int. Ed.* **2009**, *48*, 9315]

(n) 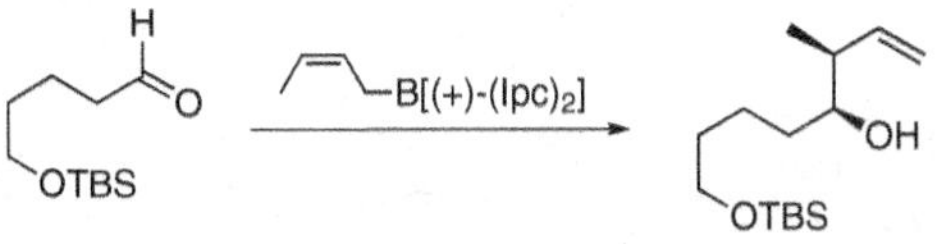

[*J. Am. Chem. Soc.* **2004**, *126*, 998]

(o) 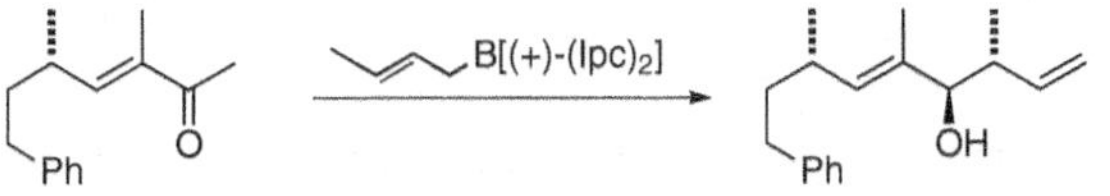

[*Org. Lett.* **2004**, 6, 3245]

(p) 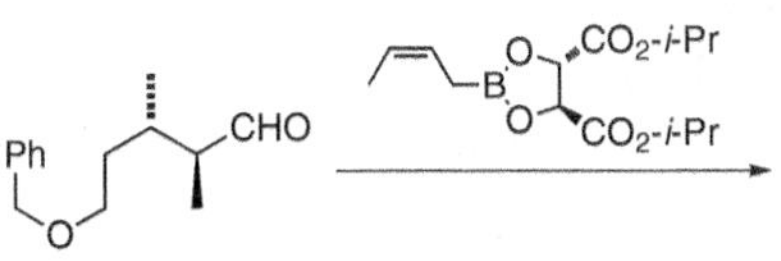

[*Org. Lett.* **2004**, 6, 3533]

(q) 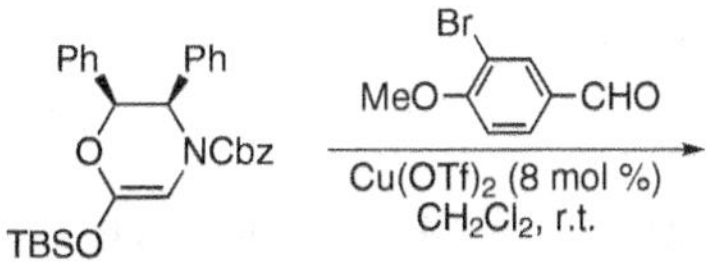

[*Tetrahedron Lett.* **2012**, *53*, 380]

10-10 Suggest a reasonable mechanism for each of the reactions below, all of which have been used a key steps in the total synthesis of natural products.

(a) 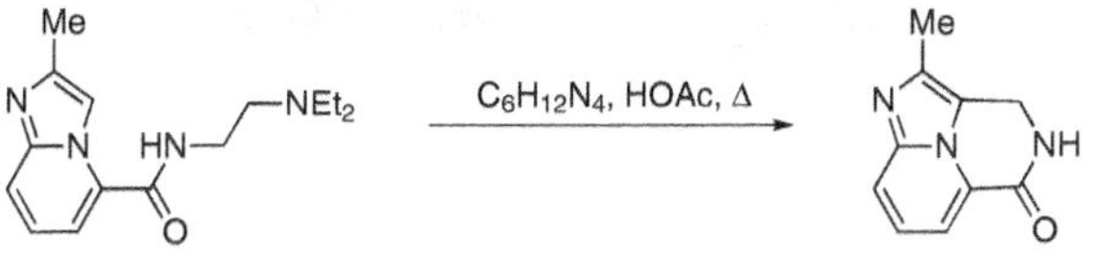

[*J. Org. Chem.* **2008**, *73*, 5989]

What light does this reaction shed on the mechanism of the Duff reaction?

(b) 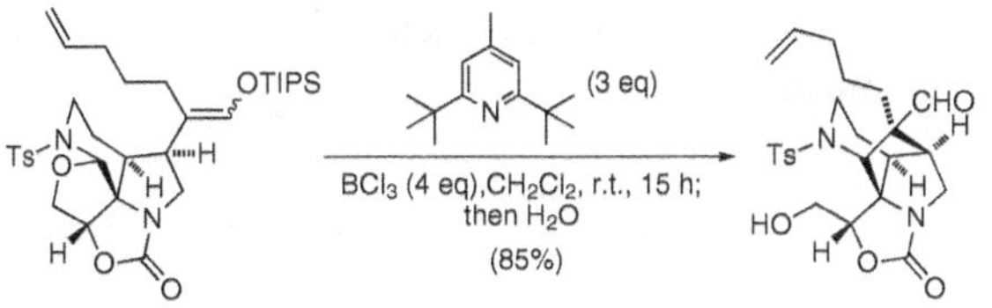

[*J. Am. Chem. Soc.* **2007**, *129*, 11987].

What is the function of the 2,6-di-*tert*-butyl-4-methylpyridine?

(c) 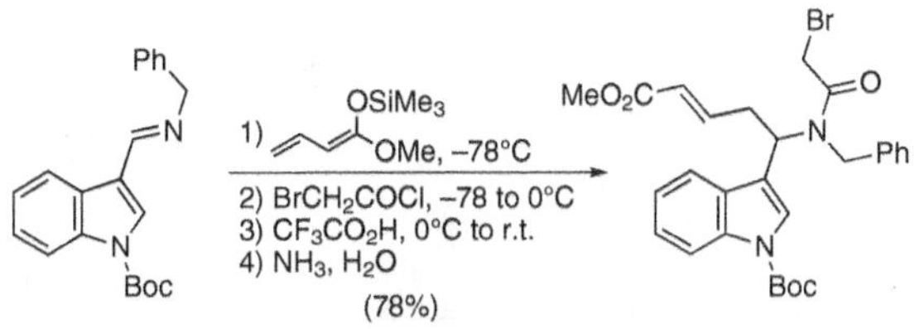

[*J. Org. Chem.* **2010**, *75*, 3529]

(d) 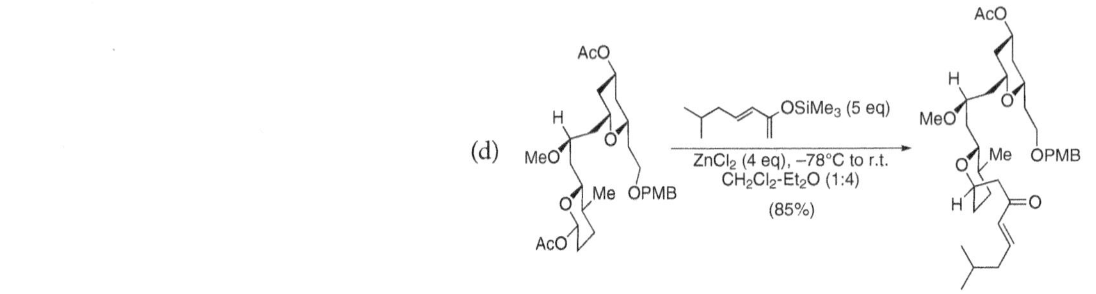

[*Org. Lett.* **2003**, *5*, 5035]

(e) 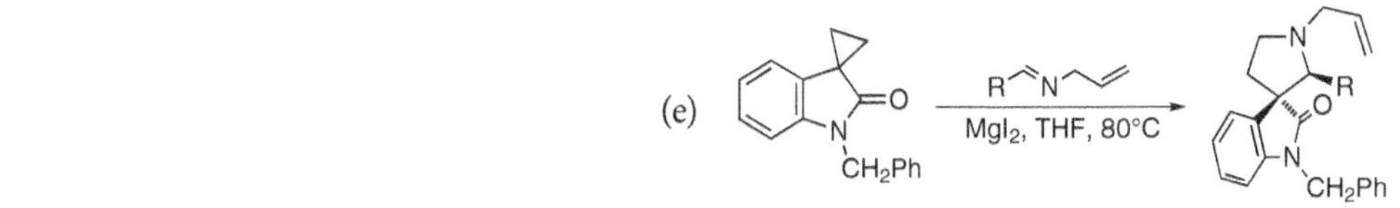

[*Chem. Eur. J.* **2006**, *12*, 8208]

(f) 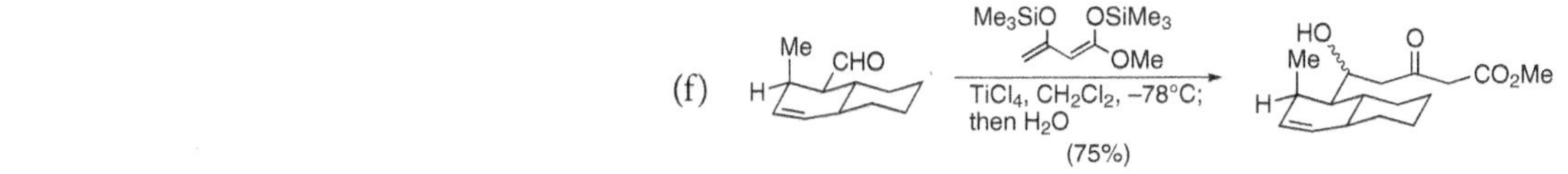

[*J. Am. Chem. Soc.* **2005**, *127*, 11616]

10-11 The bromide shown reacts with trimethylsilyl cyanide to give the nitrile shown. Because this halide is a tertiary bridgehead halide, it cannot react by the S_N2 mechanism, and it should not react by a simple S_N1 mechanism. Suggest a mechanism for the reaction (Hint: trimethylsilyl cyanide is an oxophilic Lewis acid.)

10-12 Account for the regiochemistry of the C—C bond formation in following reaction. [*J. Am. Chem. Soc.* **2003**, *125*, 2400; *Aust. J. Chem.* **2010**, *63*, 742].

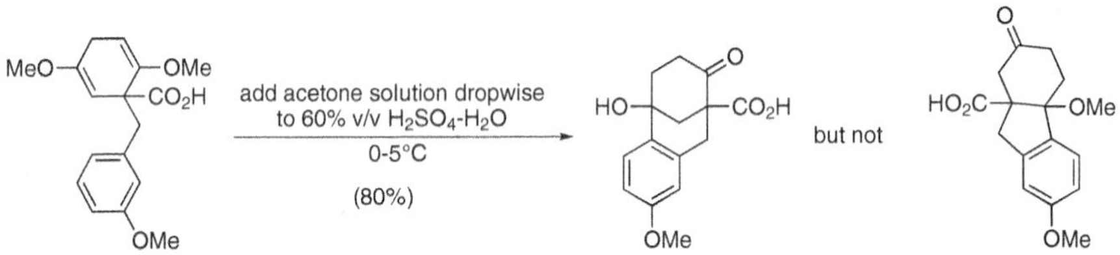

10-13 Write mechanisms that will account for the formation of products **10.88** and **10.89** in the reactions below. Your mechanism should address why the two reactions give different products, because both reactions involve the same starting compound and the same reagent.

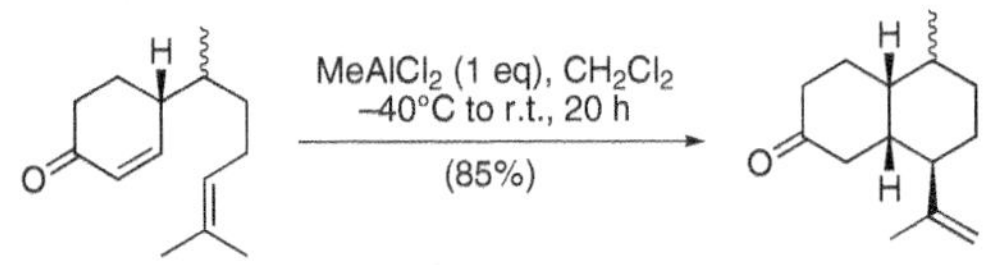

(10.88)

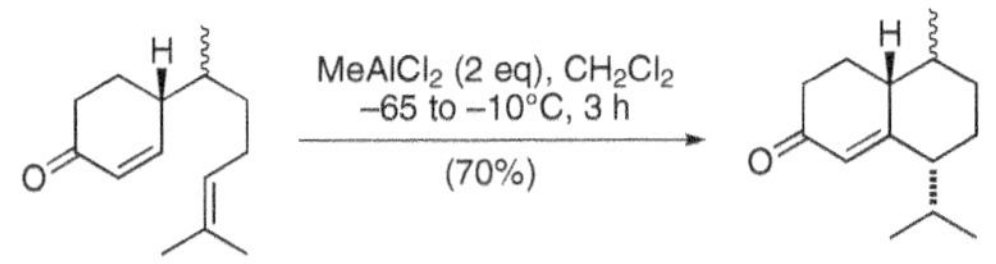

(10.89)

CHAPTER ELEVEN

Reactive Intermediates II

Carbanions and Their Reactivity

11.1 Carbanions: Introduction and Overview

Carbanions are species where a formally trivalent carbon carries a lone pair and a formal negative charge.[1] As a consequence of this, a carbanion is isoelectronic with the corresponding amine, and one can draw conclusions about carbanion structure by analogy with the amines. Thus, we expect that simple saturated alkyl anions (**11.1**) will be pyramidal, with the carbanion carbon sp^3 hybridized. Also, placing a double or triple bond or an aromatic ring in a position that allows conjugation should change the shape to trigonal planar and the hybridization of the anion carbon to sp^2 (**11.2**) by analogy with the enamines and anilines.

The simplest example of this type of nucleophilic carbon compound is the ***metal alkyl***, where the compound contains a metal-carbon bond and reacts as a Lewis base. Metal alkyls occupy a central position in the development of chemistry. It was the discovery of the zinc alkyls by English chemist Sir Edward Frankland[2] in 1849 that led to his publication in 1852[3] of the theory that all elements have a "saturation capacity" (i.e., they are limited in the number of bonds that they can form) and to the development of the concept of valence.

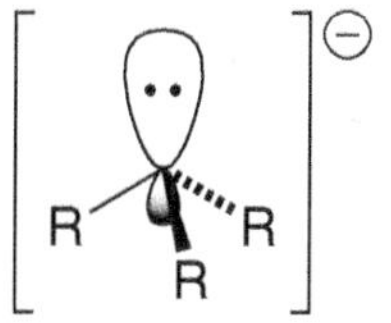

(11.1)

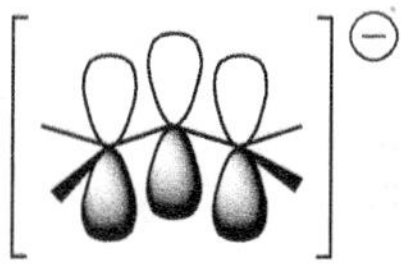

(11.2)

Free alkyl and alkenyl carbanions, where the carbon atom carries a full negative charge, are actually quite rare and occur only when the metal is very electropositive. Group IA salts of these anions are typical ionic compounds: they have ionic lattices and they react as strong nucleophiles with alkyl halides. For example, CH_3K has the same

1. Monographs: (a) Buncel, E.; Durst, T. *Comprehensive Carbanion Chemistry, Parts A, B, C* (Elsevier: New York, 1980–1987). (b) Stowell, J.C. *Carbanions in Organic Synthesis* (Wiley: New York, 1979). (c) Ogle, C.A. *Carbanion Chemistry* (Springer: New York, 1987). (d) Cram, D.J. *Fundamentals of Carbanion Chemistry* (Academic Press: New York, 1965). (e) Bates, R.B.; Ogle, C.A. *Carbanion Chemistry* (Springer-Verlag: Berlin, 1983). (f) Gronert, M. In Moss, R.A.; Platz, M.S.; Jones, M., Jr. *Reactive Intermediate Chemistry* (John Wiley & Sons: New York, 2004), Part 1, p. 69.

2. **Sir Edward Frankland (1825–1899)** was apprenticed to a pharmacist before extending his education in London and Marburg (PhD, 1849). He served as Professor of Chemistry at Owens College in Manchester, then at the Royal College of Chemistry, in London, where he provided regular reports on the pollution levels of the water (novel in the 1860s but very topical today). Frankland was knighted in 1897. For more biographical details, see: Tilden, W.A. *Famous Chemists. The Men and Their Work* (George Routledge & Sons: London; and E.P. Dutton & Co: New York, 1921;), p. 216.

3. (a) Frankland, E. *Ann. Chem. Pharm.* **1849**, *71*, 213. (b) Frankland, E. *J. Chem. Soc.* **1849**, *2*, 297. (c) Frankland, E. *Phil. Trans. Roy. Soc.* **1852**, *142*, 417. (d) Frankland, E. *Ann. Chem. Pharm.* **1853**, *85*, 329.

crystal lattice as NiAs, with alternating sheets of K^+ and CH_3^- ions in the lattice,[4] and methylsodium reacts with alkyl chlorides as an anionic reagent[5] (although its reactions with alkyl bromides and iodides are complicated by metal-halogen exchange reactions[6]). Butylpotassium can be prepared by a metathesis reaction between potassium *tert*-butoxide and butyllithium at low temperature, a mixture known as "LICKOR." This has found use as an especially powerful base, and we will discuss its use in more detail in Chapter 12. The only other "free" carbanions are alkynide anions, and here the stability is a result of the sp hybridization of the anion carbon.

The C—M bond in most metal alkyls is best described in terms of resonance (Example **11.3**), where the major contributor to the hybrid has a non-polar covalent bond, and the minor contributor has an ionic bond between carbon and the metal. As the polarity of the C—M bond decreases, the tendency to form aggregates also decreases. Alkyllithiums (Example **11.4**) occur as tetramers and hexamers, depending on the alkyl group and the solvent, and Grignard reagents Example (**11.5**)[7] occur as dimers; on the other hand, dialkylzinc reagents (e.g., Example **11.6**), where the C—M bond is even less polar, occur as discrete molecules (dimethylzinc is a linear molecule). When the metal in a metal alkyl lacks a complete octet, the structure is often an aggregate with bridging alkyl groups (e.g., Examples **11.7** and **11.8**). Trimethylaluminum, for example, occurs as a dimer with bridging methyl groups,[8] and dimethylberyllium as extended chains with all the methyl groups as bridging ligands.[9] Organometallic compounds based on practically every metal in the periodic table are now known. A completely accurate description of the bonding in many (most?) organometallic compounds requires delocalized, multicenter bonds (e.g., the three-center, two-electron bonds of the bridging alkyl groups in alkylmetal aggregates—Examples **11.4**, **11.5**, **11.7**, and **11.8**).

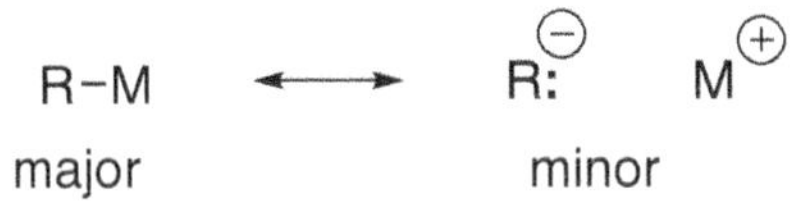

(11.3)

(11.4)

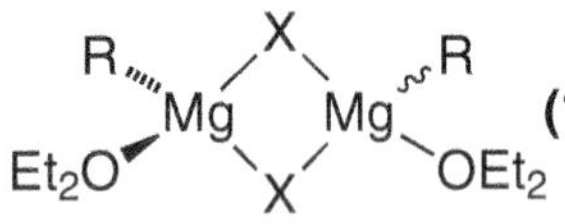

(11.5)

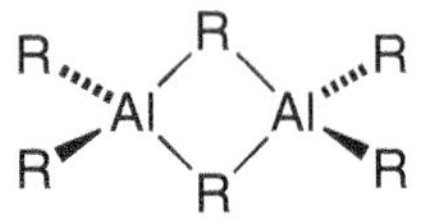

(11.7)

R-Zn-R **(11.6)**

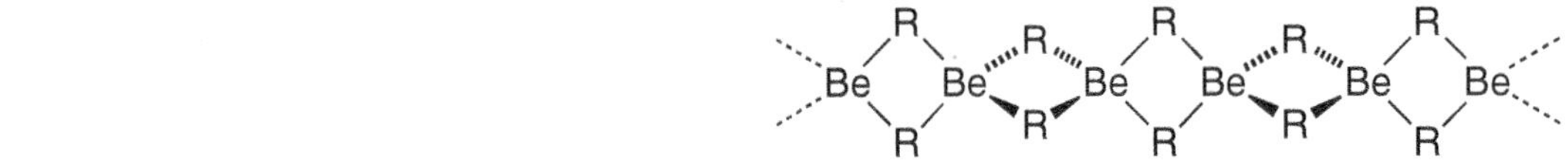

(11.8)

Fortunately, organometallic compounds of electropositive metals such as lithium and magnesium react as if the compound were ionic, so we generally introduce a negligible error into our discussion if we treat these compounds as the ionic (minor) contributor to

4. (a) Weiss, G.; Sauerman, G. *Angew. Chem. Int. Ed. Engl.* **1968**, *7*, 133. (b) Weiss, E.; Sauerman, G. *Chem. Ber.* **1970**, *103*, 265.
5. (a) Bergmann, E. *Helv. Chim.Acta* **1937**, *20*, 590. (b) Letsinger, R.T. *J. Am. Chem. Soc.*, **1948**, *70*, 406. (c) Letsinger, R.L.; Maury, L.G.; Burwell, R.G., Jr. *J. Am. Chem. Soc.*, **1951**, *73*, 2373.
6. Brink, N.G.; Lane, J.F.; Wallis, E.S. *J. Am. Chem. Soc.* **1943**, *65*, 943.
7. Structure: Spek, A.L.; Voorbergen, P.; Schat, G.; Blomberg, C.; Bickelhaupt, F. *J. Organomet. Chem.* **1974**, *77*, 147.
8. Vranka, R.G.; Amma, E.L. *J. Am. Chem. Soc.* **1967**, *89*, 3121.
9. Snow, A.I.; Rundle, R.E. *Acta Crystallogr.* **1951**, *4*, 348.

the resonance hybrid. Of course, in doing so, we must recognize that free carbanions actually do not exist in solution in the presence of metal ions. One should also bear in mind that this simplification may not be insubstantial. There is evidence from ^{7}Li and ^{13}C nuclear magnetic resonance (NMR) spectroscopy of alkyllithiums that the partial negative charge on the carbon may not be as large as would be predicted on the basis of this simple model.[10]

The variation in reactivity of the organometallic compounds is mainly due to the polarity of the carbon-metal bond. As the electronegativity of the metal increases, the percent ionic character of the bond decreases, and the polarity of the carbon-metal bond decreases. The amount of ionic character in a polar covalent bond is approximated by:

$$\text{Ionic character} = 1 = e^{-0.25|x_A - x_B|} \qquad \text{(Eq. 11.1)}$$

where x_A and x_B are the electronegativities of the two atoms at the ends of the bond.[11] The values given by this relationship are very approximate only, and they tend to overestimate the percent ionic character in molecules such as hydrogen fluoride. Nevertheless, the values obtained are still useful in a qualitative sense; the approximate percent ionic character of several carbon-metal bonds of importance in organic chemistry are given in Table 11.1.

Table 11.1 Percent Ionic Character in Metal-Carbon σ Bonds*

Metal	% Ionic Character	Metal	% Ionic Character
K†	35	Cu	15
Na	33	Ni	15
Li	32	Sn	14
Mg	26	Hg	13
Ti	21	Mo	9
Al	20	Pd	8
Zn	20	Ru	8
Cd	19	W	5

*Values are calculated using Equation 11.1 and a value of 2.55 for x_C.
†Shaded regions are ionic metal alkyls.

Table 11.2 Relative Importance of Ionic Forms of Metal Alkyls

	R–M	⟷ $R:^{-}$ M^{+}
M=Na, K	negligible	dominant
M=Mg, Zn, Al	major	minor, but substantial
M=Cu, Pd	major	minor
M=Hg	dominant	negligible

The more polar the metal-carbon bond, the more important the ionic contributor to the resonance hybrid becomes. When the ionic character rises above 32%—the shaded region of Table 11.1—the compound is best described as ionic, but in all other metal alkyls, the compound is best described as polar covalent, with the carbon atom is at the negative end of the dipole. The qualitative importance of the ionic contributor to the resonance hybrid is given in Table 11.2. As the importance of the ionic resonance contributor increases, the metal alkyl becomes both a stronger base and a more reactive nucleophile. Many metal alkyls function as good Lewis bases and also as good carbon nucleophiles. For this reason they have long held a place of importance in synthetic organic chemistry, especially because it is possible to modulate the reactivity of the metal alkyl by appropriately choosing the metal. Alkyllithiums react explosively with water and add to even the least reactive carbonyl group, dialkylmercury compounds dissolve in water and react with only the most reactive carbonyl compounds under forcing conditions, and many reactions based on palladium or ruthenium catalysts can be carried out in water.

The hybridization of the carbon atom can also play a role in determining the ionic character of the carbon-metal bond. The relative acidities of C—H bonds increase as the hybridization (and hence, the electronegativity) of carbon changes from sp^3 (p$K_a \approx$ 40–50) to sp^2 (p$K_a \approx$ 35–40) to sp (p$K_a \approx$ 25–30). As the s character of the orbital carrying the lone pair on the anion increases, the ionic character of the organometallic compound increases, also. Thus, the carbon-metal bond in metal alkynides is more ionic than corresponding

10. (a) McKeever, L.D.; Waack, R. *J. Chem. Soc. D, Chem. Commun.* **1969**, 750. (b) Review of NMR methods for strudying carbanions: O'Brien, D.H. in Buncel, E.; Durst, T., Eds. *Comprehensive Carbanion Chemistry, Part A. Structure and Reactivity* (Elsevier: Amsterdam, 1980), p. 271.

11. Pauling, L. *The Nature of the Chemical Bond* 3rd ed. (Cornell University Press: Ithaca, N.Y., 1960), section 3-9, pp. 97–102.

metal-carbon bonds in the corresponding metal aryls and metal alkyls—sodium alkynides are easily prepared by the reaction between sodium hydride and the alkyne.

The simple valence-bond view of the alkyl anion provides a good first approximation for discussing the bonding in metal alkyls. Thus, even in aggregates involving multicenter bonding, the carbon atom of the carbon-metal bond in saturated metal alkyls is sp^3 hybridized with the electron pair formally in an sp^3 hybrid orbital (e.g., **11.1**). However, if there is a π bond adjacent to this carbon atom (e.g., the allyl anion), the anion becomes delocalized and the carbon atom becomes sp^2 hybridized (e.g., **11.2**).

Stabilization of Carbanions: Comparison and Contrast with Cations

The Pauling electronegativity of carbon is 2.5, so carbon does not support a negative charge as well as later non-metals in the second row of the periodic table. Species where carbon carries a localized negative charge are high-energy, like those where carbon carries a localized positive charge, and mechanisms for stabilizing the ion are important in determining reactivity—even more important than for carbocations. Like cations, anions are stabilized by delocalization of the charge; the more effective the delocalization, the more stable the anion.

Stabilization of Carbanions

In contrast to carbocations, where the field effect lowers the energy of the cation and, along with hyperconjugation, is a stabilizing influence, the field effect of simple alkyl groups raises the energy of the anion, with the result that the stability of simple saturated carbanions (as measured by solution-phase reactions) follows the order shown in Figure 11.1. In the gas phase, competitive proton transfer reactions have been used to measure relative acidities of simple hydrocarbons.[12] These experiments give a slightly different order for carbanion energies:

$$C_2H_5^- > (CH_3)_2CH^- > CH_3^- > (CH_3)_3C^-$$

strongest base weakest base

Unlike carbocations, most simple, saturated carbanions do not rearrange—the driving force for rearrangements in carbocations comes from the need to disperse the region of electron deficiency; carbanions are not electron-deficient, so this does not apply to anions. This is not, of course, a totally exclusionary statement, and some anions do give rise to synthetically useful rearrangements.[13]

Stabilization by Third-Row and Higher Heteroatoms

Simple alkyl carbanions can be stabilized by third- and fourth-row heteroatoms[14] but not by electronegative second-row heteroatoms. The electron affinity of these anions is negative, which makes their generation in the gas phase problematic.[15] Carbanions stabilized

12. (a) DePuy, C.H.; Bierbaum, V.M.; Damrauer, R. *J. Am. Chem. Soc.* **1984**, *106*, 4051. (b) DePuy, C.H.; Gronert, S.; Barlow, S.E.; Bierbaum, V.M.; Damrauer, R. *J. Am. Chem. Soc.* **1989**, *111*, 1968.

13. For example, the [1,2]-Wittig rearrangement: (a) Marshall, J.A. In Trost, B.M.; Fleming, I, Eds. *Comprehensive Organic Synthesis* (Pergamon: London, 1991), vol. 3, p. 975. (b) Tomooka, K.; Yamamoto, H.; Nakai, T. *Liebigs Ann./Receuil* **1977**, 1275. Also, the [2,3]-Wittig rearrangement: (c) Nakai, T.; Mikami, K. *Chem. Rev.* **1986**, *86*, 885. (d) Nakai, T.; Mikami, K. *Org. React.* **1994**, *46*, 105.

14. Stabilization by silicon, phosphorus and sulfur: (a) Peterson, D.J. *J. Organomet. Chem.* **1967**, *9*, 373. (b) Magnus, P. *Aldrichimica Acta* **1980**, *13*, 43. (c) Ager, D.; Cookson, R.C. *Tetrahedron Lett.* **1980**, *21*, 1677. (d) Mathey, F.; Mercier, F. *J. Organomet. Chem.* **1979**, *17*, 255.

15. Downard, K.M.; Sheldon, J.C.; Bowie, J.H.; Lewis, D.E.; Hayes, R.N. *J. Am. Chem. Soc.* **1989**, *111*, 8112.

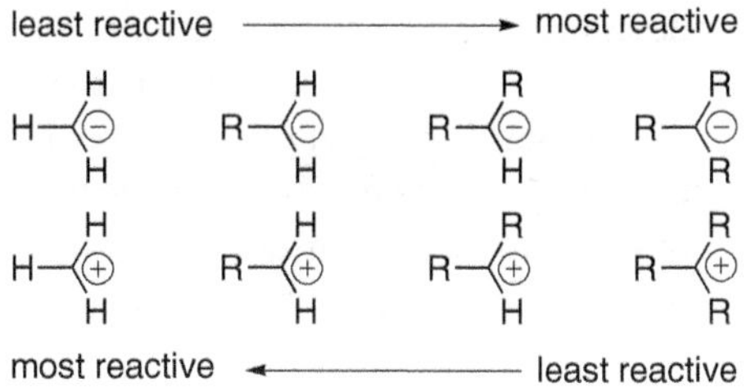

Figure 11.1 The effects of alkyl substitution on the reactivity of carbocations and carbanions are opposite.

by phosphorus are the key reagents in the Wittig,[16] Wadsworth-Emmons,[17] and Horner[18] reactions. Carbanions stabilized by adjacent sulfur (sulfoxide or sulfonyl groups, and dithianes) and silicon are also important synthetic intermediates.

The stabilization of carbanions by third-row elements, in particular, is often attributed to the presence of $d_\pi - p_\pi$ bonding (e.g., Example **11.9**) between the carbanion carbon and the heteroatom.[19] There is some support for this view when one observes that third-row elements (which would use 3d orbitals) stabilize carbanions better than the corresponding fourth-row elements (which would use 4d orbitals). The operation of $d_\pi - p_\pi$ bonding is not, however, universally accepted, and other mechanisms for stabilization of anions by heteroatoms are possible, including "inverse hyperconjugation" and stabilization arising from the polarizability of the third-row element.[20]

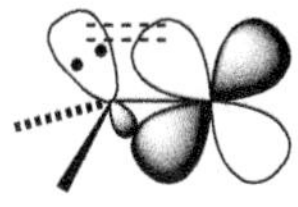

(11.9)

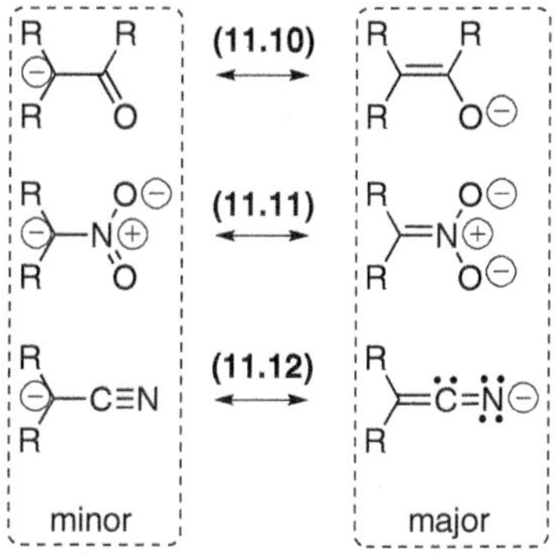

The most efficient stabilization of anions occurs by conjugation of the lone pair with a π bond to an electronegative element (e.g., Examples **11.10** to **11.12**), which allows the delocalization of the negative charge onto the most electronegative element. This leads to the formation of an extended π bonding system, giving resonance-stabilized anions that are much better

---

16. Reviews: (a) Maercker, A. *Org. React.* **1965**, *14*, 270. (b) Schlosser, M. *Top. Stereochem.* **1970**, *5*, 1. (c) Bestmann, H.J.; Vostrowsky, O. *Top. Curr. Chem.* **1983**, *109*, 85. (d) Murphy, P.J.; Brennan, J. *Chem. Soc. Rev.* **1988**, *17*, 1. (e) Maryanoff, B.E.; Reitz, A.B. *Chem. Rev.* **1989**, *89*, 863.

17. Reviews: (a) Wadsworth, W.S., Jr. *Org. React.* **1977**, *25*, 73. (b) Boutagy, J.; Thomas, R. *Chem. Rev.* **1974**, *74*, 87.

18. (a) Horner, L.; Hoffmann, H.; Wippel, H.G. *Chem. Ber.* **1958**, *91*, 61–63. (b) Horner, L.; Hoffmann, H.; Wippel, H.G.; Klahre, G. *Chem. Ber.* **1959**, *92*, 2499–2505.

19. (a) Kwart, H.; King, K. *d-Orbitals in the Chemistry of Silicon, Phosphorus, and Sulfur* (Springer: New York, 1977). (b) Wolfe, S.; LaJohn, L.A.; Bernardi, F.; Mangini, A.; Tonachini, D. *Tetrahedron Lett.* **1983**, *24*, 3789. (c) Wolfe, S.; Stolow, A.; LaJohn, L.A. *Tetrahedron Lett.* **1983**, *24*, 4071.

20. (a) Bernardi, F.; Csizmadia, I.G.; Mangini, A.,; Schlegel, H.B.; Whangbo, M.; Wolfe, S. *J. Am. Chem. Soc.* **1975**, *97*, 2209. (b) Lehn, J.M.; Wipff, G. *J. Am. Chem. Soc.* **1976**, *98*, 7498. (c) Borden, W.T.; Davidson, E.R.; Andersen, N.H.; Denniston, A.D.; Epiotis, N.D. *J. Am. Chem. Soc.* **1978**, *100*, 1604. (d) Bernardi, F.; Bottoni, A.; Venturini, A.; Mangini, A. *J. Am. Chem. Soc.* **1986**, *108*, 8171. (e) Bernasconi, C.F.; Kittredge, K.W. *J. Org. Chem.* **1998**, *63*, 1944.

Table 11.3 Effects of Solvent and Cation on Fluorenide Salts in Solution*

Cation	Solvent	Mol % $[R^-]_{solv}[M^+]_{solv}$
Li^+	Dimethoxyethane (DME)	100
Na^+	DME	95
K^+	DME	≈10
Cs^+	DME	0
Li^+	PhMe	0
Li^+	Dioxane	0
Li^+	Tetrahydrofuran	80
Li^+	Pyridine	100
Li^+	Me_2SO	100

*Data taken from Bernardi F., Csizmadia I.G., Mangin, A., Schlegel H.B., Whangbo M., Wolfe S. *J. Am. Chem. Soc.* **1975**, *97*, 2209.

described as enolate-type anions. As is common with resonance-stabilized species where the two canonical forms do not contribute equally to the resonance hybrid, the minor contributing form (the carbanion form) is the best predictor of reactivity. The stabilization of carbanions in solution by functional groups is in the following order:

$NO_2 > CHO > RCO > CO_2R > SO_2 > CN \approx CONR_2 > X > H > R$

Generally speaking, the negatively charged carbon of conjugated carbanions is sp^2 hybridized, and that of simple, non-conjugated carbanions, including those stabilized by sulfur or phosphorus, is sp^3 hybridized.

The reactivity of carbanions is dramatically affected by the identity of the counterion. This can be illustrated using the 9-fluorenyl anion, which is stable enough to exist in solution in ethers as the solvated ion $[R^-]_{solv}$. In ethereal solvents, the fluorenyl anion exists mainly as a solvent-separated ion pair, $[R^-]_{solv}[M^+]_{solv}$ (Example **11.13**), when the cation is lithium, and as a contact ion pair, $[R^-][M^+]$ (Example **11.14**), when the cation is cesium (Table 11.3).[21] This has been interpreted as reflecting the competition between solvent and the anion for the cation. Because the ion-dipole interaction energy is inversely proportional to the *square of the distance* between the ion and the dipole, and the ion-ion interaction energy is inversely proportional to the *distance* between the ions, the ion-ion energy becomes more important than the ion-dipole energy as the ion becomes larger. In other words, solvation of the cation becomes more important as the cation changes from cesium to lithium, so the solvent-separated ion pair becomes favored.

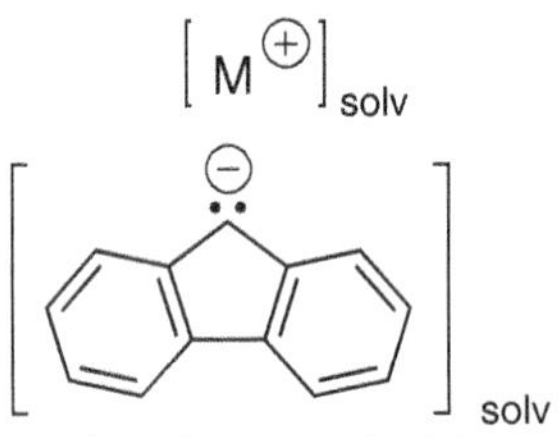

solvent-separated ion pair

(11.13)

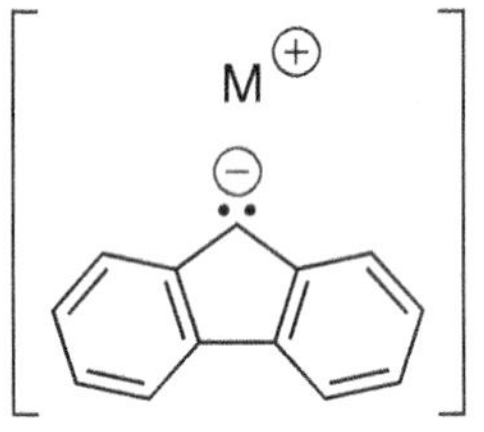

contact ion pair

(11.14)

Problem

11-1 Ylides are compounds having the general structure below. Ylides of third- and higher-row elements (X=P, S) are stable, useful nucleophiles. Ylides based on second-row elements (e.g., X=N) are not. Suggest a reason why this is so.

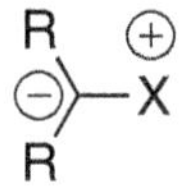

21. (a) Hogen-Esch, T.E.; Smid, J. *J. Am. Chem. Soc.* **1966**, *88*, 307. (b) Smid, J. In Szwarc, M., Ed. *Ions and Ion Pairs in Organic Reactions* (Wiley-Interscience: New York, 1971), vol. 1, ch. 3, p. 86. (c) Hogen-Esch, T.E.; Smid, J. *J. Am. Chem. Soc.* **1972**, *94*, 9240. (d) Smid, *J. J. Poly. Sci., Part A: Polymer Chem.* **2004**, *42*, 3655.
Review: (d) Hogen-Esch, T.E. *Adv. Phys. Org. Chem.* **1977**, *15*, 153.

11.2 Metal Alkyls[22]

There are three major methods for the formation of metal alkyls:

1. ***Direct synthesis or oxidative addition***—the reaction of the metal with an alkyl halide (Equation 11.2)
2. ***Transmetalation***—the metathesis reaction of an organometallic reagent with a metal halide or an alkyl halide. This can be either ***metal-halogen exchange*** (Equation 11.3) or ***metal-metal exchange*** (Equations 11.4 and 11.5).
3. Metalation of hydrocarbons (Equation 11.6)

$$\text{R–X} + 2\,\text{M} \rightarrow \text{R–M} + \text{M–X} \qquad \text{(Eq. 11.2)}$$

$$\text{R–X} + \text{R'–M} \rightarrow \text{R–M} + \text{R'–X} \qquad \text{(Eq. 11.3)}$$

$$\text{R–M} + \text{R'–M'} \rightarrow \text{R–M'} + \text{R'–M} \qquad \text{(Eq. 11.4)}$$

$$\text{R–M} + \text{M'} \rightarrow \text{R–M'} + \text{M} \qquad \text{(Eq. 11.5)}$$

$$\text{R–H} + \text{R–M} \rightarrow \text{R–M} + \text{R'–H} \qquad \text{(Eq. 11.6)}$$

Direct Synthesis

The formation of metal alkyls by direct combination of an alkyl halide with a metal involves the two-electron reduction of the carbon-halogen bond, or oxidative addition of the carbon-halogen bond to the metal. There is strong experimental evidence that this reaction occurs in two steps with an alkyl free radical formed as an intermediate,[23] although more recent work has evidence that no freely diffusing radicals are detectable during the formation of arylmagnesium halides.[24] Simple alkyl free radicals are sp^2-hybridized—they are not chiral. Therefore, *a chiral alkyl halide usually gives the organometallic compound in which the configuration of the chiral center is lost* (e.g., Example **11.15**). NMR studies of Grignard reagents have shown that the carbon-magnesium bond is sufficiently ionic in nature to render it configurationally labile.[25]

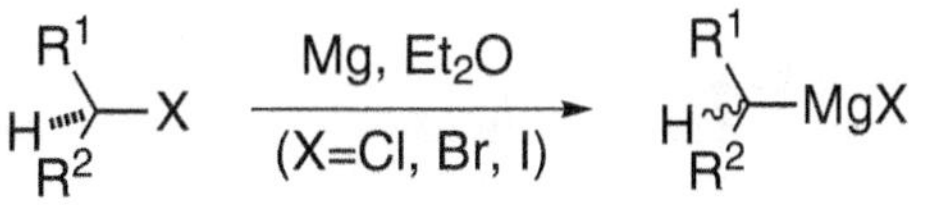

(11.15)

The direct synthesis of metal alkyls from the elemental metal is restricted to the more active metals such as the alkali metals, magnesium, aluminum, and zinc, although metal alkyls of less reactive metals can be formed readily from the alkyl halide with an alloy of the less electropositive metal and an alkali metal. Historically, the dialkylzinc reagents were prepared first, and despite the greater reactivity of similar organomagnesium reagents, organozinc reagents long overshadowed these compounds in synthesis. Useful syntheses based on the addition of alkylzinc reagents to carbonyl compounds were developed over three decades by the chemists of Kazan University in Russia: Butlerov, Zaitsev, Wagner, and Reformatskii.[26] It was problems with the preparation of certain organozinc

22. Monographs: (a) Stowell, J.C. *Carbanions in Organic Synthesis* (John Wiley & Sons: New York, 1979). (b) Bates, R.B.; Ogle, C.A. *Carbanion Chemistry* (Springer-Verlag: Brelion, 1983).

23. Ashby, E.C. *Quart. Rev. Chem. Soc.* **1967**, *21*, 259.

24. Walter, R.I. *J. Org. Chem.* **2000**, *65*, 5014.

25. (a) Whitesides, G.M.; Roberts, J.D. *J. Am. Chem. Soc.* **1965**, *87*, 4878. (b) Fraenkel, G.; Dix, D.T. *J. Am. Chem. Soc.* **1966**, *88*, 979. (c) Fraenkle, G.; Dix, D.T.; Adams, D.G. *Tetrahedron Lett.* **1964**, 3155.

26. For a synopsis of the work of these chemists, see Lewis, D.E. *Early Russian Organic Chemists and Their Legacy* (Springer: Heidelberg, 2012), pp. 32. 47, 75, 95 and 101, and references therein.

reagents that led French chemist Philippe Antoine Barbier[27] and his student Victor Grignard[28] to explore replacing the zinc with the more reactive magnesium. The Grignard reagent revolutionized organic synthesis by making organomagnesium nucleophiles easily and widely available for the first time. The importance of Grignard's contribution was quickly recognized; his reaction had been used in more than 800 published papers by the time he shared the Nobel Prize for Chemistry in 1912 with Paul Sabatier. Many organolithium compounds and Grignard reagents are now commercially available, making their use in the organic chemistry laboratory routine.

The use of transition metals other than zinc in organic chemistry is of more recent origin. Organocopper reagents, for example, were first prepared by Henry Gilman in 1936,[29] but their extensive exploitation as reagents in organic synthesis really dates from the 1960s and 1970s.[30]

Metathesis Between Alkyl Halides and Metal Alkyls

Metal-halogen exchange reactions are, in a formal sense, at least, displacement reactions of the alkyl anion from an alkyl halide by a nucleophile. The only nucleophiles powerful enough to do this are the alkyllithiums, which are very strong Lewis bases. The products of these reactions, which are reversible, are also alkyllithiums. The exchange reaction behaves, in some ways, as an acid-base reaction: the major products of the equilibrium reaction are usually the less basic metal alkyl and the alkyl halide derived from the more basic of the two alkyl anions.[31] The relative acidities of C—H bonds increase with increasing s character of the hybrid orbitals on carbon: sp^3 ($pK_a \approx 40-50$)[32] to sp^2 ($pK_a \approx 35-40$)[33] to sp ($pK_a \approx 25-30$).[34] Thus, alkyllithiums react with aromatic halides and vinyl halides to give aryllithiums and vinyllithiums.

---

27. **Philippe Antoine Barbier (1848–1922)** destroyed all records of his life's work shortly before he died, so this brief biography will be longer than most. He was born in Luzy, Nièvre, and in 1869 he moved to Paris, where he studied under Berthelot at the Collège de France. On graduating with his Dr-ès-sc in, he was appointed to teach mineralogy in the pharmacy college, and in 1878 he moved to Lyon as "Maître des Conférences" in the faculty of science. In 1879, he moved to Besançon as a junior faculty member in applied chemistry, moving back to Lyon in 1880 as Professor of Chemistry and Director of the Agronomy Station. At Lyon he discovered that the zinc used in the Zaitsev-Vagner synthesis of alcohols could be replaced by magnesium, thus introducing the use of magnesium intermediates into organic synthesis. The method proved to be unreliable, however, and he did not pursue it further personally. At Lyon, Barbier was responsible for the education of several influential chemists of the next generation, including Louis Bouveault and Victor Grignard (whose project with organomagnesium reagents he suggested and supervised). His student, Grignard, had planned to write the biography of his mentor, but never found the time to do the necessary research.

28. **François Auguste Victor Grignard (1871–1935)** was educated at Lyon (PhD, 1901). At the suggestion of his doctoral mentor, Barbier, he prepared the organomagnesium reagents that now bear his name. He shared the Nobel Prize in Chemistry for 1912 with Paul Sabatier (incidentally, Grignard often made the point that his mentor, Barbier, should have shared the Prize). Grignard served as Professor of Chemistry at Nancy, then at Lyon. For more biographical information, see the Nobel Foundation web site.

29. Gilman, H.; Straley, J.M. *Rec. Trav. Chim. Pays-Bas* **1936**, *55*, 821.

30. Reviews: (a) Posner, G.H. *Org. React.* **1972**, *19*, 1; **1975**, *22*, 253. (b) Normant, J.F. *Synthesis* **1972**, 63. (c) Normant, J.F. *Pure Appl. Chem.* **1978**, *50*, 709. (d) Jukes, A.E. *Adv. Organomet. Chem.* **1974**, *12*, 215. (e) Kaufmann, T. *Angew. Chem. Int. Ed. Engl.* **1974**, *13*, 291. (f) Posner, G.H. *An Introduction to Synthesis Using Organocopper Reagents* (Wiley: New York, 1980).

31. (a) Parham, W.E.; Bradshher, C.K. *Acc. Chem. Res.* **1982**, *15*, 300. (b) Bailey, W.F.; Punzalan, E.R. *J. Org. Chem.* **1990**, *55*, 5404. (c) Negishi, E.; Swanson, D.R.; Rousset, C.J. *J. Org. Chem.* **1990**, *55*, 5406. (d) Neumann, H.; Seebach, D. *Chem. Ber.* **1978**, *111*, 2785. (e) Milller, R.B.; McGarvey, G. *Synth. Commun.* **1979**, *9*, 831.

32. (a) Streitwieser, A., Jr.; Caldwell, R.A.; Young, W.R. *J. Am. Chem. Soc.* **1969**, *91*, 529. (b) Streitwieser, A., Jr.; Young, W.R.; Caldwell, R.A. *J. Am. Chem. Soc.* **1969**, *91*, 527. (c) Streitwieser, A., Jr.; Taylor, D.R. *J. Chem. Soc. D* **1970**, 1248. (d) Jaun, B.; Schwarz, J.; Breslow, R. *J. Am. Chem. Soc.* **1980**, *102*, 5741.

33. (a) Streitwieser, A., Jr.; Scannon, P.J.; Niemeyer, H.M. *J. Am. Chem. Soc.* **1972**, *94*, 7936. (b) Maskornick, M.J.; Streitwieser, A., Jr. *Tetrahedron Lett.* **1972**, 1625.

34. (a) Wooding, N.S.; Higginson, W.E.C. *J. Chem. Soc.* **1952**, 774. (b) Streitwieser, A., Jr.; Reuben, D.M.E. *J. Am. Chem. Soc.* **1971**, *93*, 1794. (c) Matthews, W.S.; Bares, J.E.; Bartmess, J.E.; Bordwell, F.G.; Cornforth, F.J.; Drucker, G.E.; Margolin, Z.; McCallum, R.J.; McCollum, G.J.; Vanier, N.R. *J. Am. Chem. Soc.* **1975**, *97*, 7006.

The available experimental data indicate that, despite its superficial resemblance to an S_N2 displacement of alkyl from the halogen, there is at least one single electron transfer step in the mechanism of the reaction.[35] Some typical examples of metal-halogen exchange reactions are given as Examples **11.16** to **11.18**.

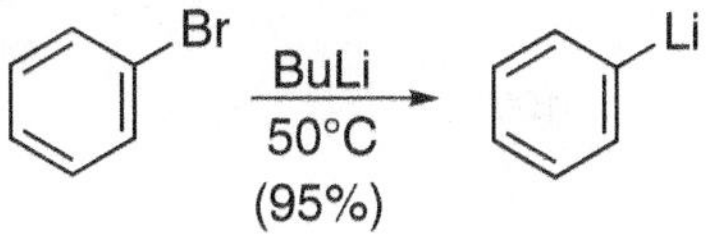

(11.16)[36]

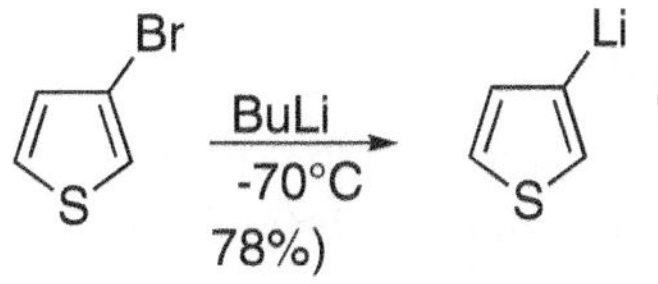

(11.17)[37]

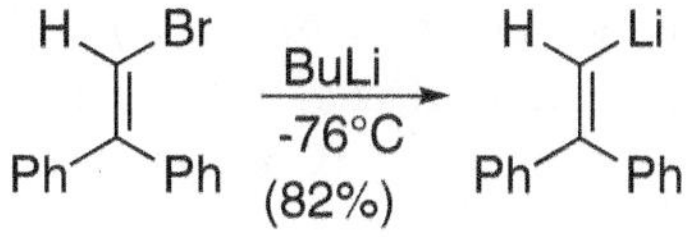

(11.18)[38]

Metathesis Between Metal Halides and Metal Alkyls

Exchange reactions where the metal alkyl of a more electropositive metal is allowed to react with the halide of a less electropositive metal constitute the most common method by which organometallic compounds can be prepared. In this way, one can prepare alkylcopper, alkylcadmium, and alkyltin (stannane) reagents, for example. In fact, the reaction of alkyllithium reagents or Grignard reagents with cuprous salts is the major method for the production of alkylcopper nucleophiles.

Whenever metal alkyls are formed from alkyllithium or Grignard reagents, the alkyl halide to be converted to the organometallic reagents is still subject to the same limitations as a halide that will be converted to a Grignard reagent—even though the product metal alkyls may react quite slowly with hydroxyl groups and other functional groups. In general, *the functional groups permitted on an organometallic reagent are restricted to those compatible with the most reactive organometallic reagent used in its preparation*, as illustrated in Examples **11.19** and **11.20**. However, at sufficiently low temperatures, the problem caused by the reactivity of the metal alkyl can be alleviated, as shown by Examples **11.21** and **11.22**; this still does not, of course, work for more reactive functional groups (e.g., aldehydes).

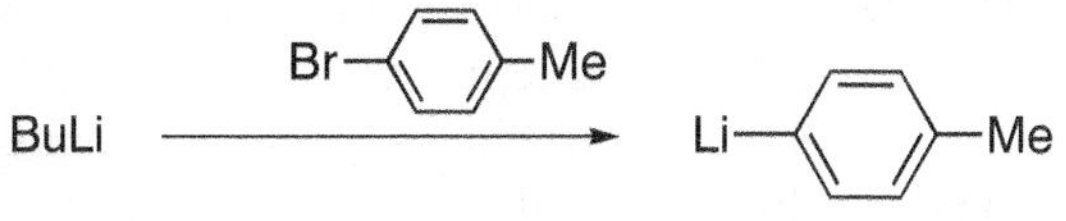

(11.19)

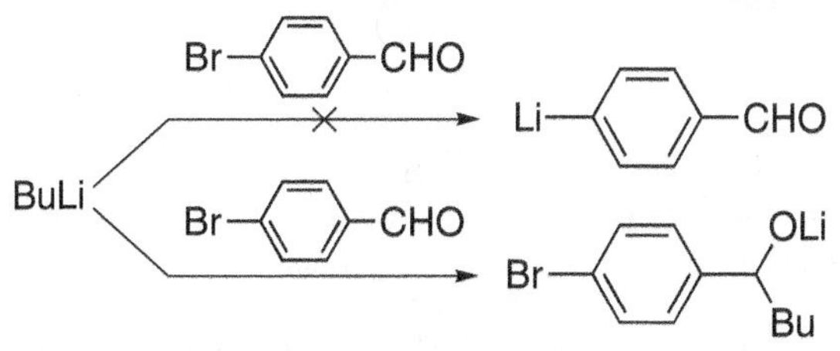

(11.20)

35. (a) Ward, H.R.; Lawler, R.G.; Loken, H.Y. *J. Am. Chem. Soc.* **1968**, *90*, 7359. (b) Ward, H.R.; Lawler, R.G.; Cooper, R.A. *J. Am. Chem. Soc.* **1969**, *91*, 746. (c) Lepley, A.R.; Landau, R.L. *J. Am. Chem. Soc.* **1969**, *91*, 748. (d) Ashby, E.C.; Pham, T.N. *J. Org. Chem.* **1987**, *52*, 1291. (e) Bailey, W.F.; Patricia, J.J. *J. Organomet. Chem.* **1988**, *352*, 1.

36. Trepka, W.J.; Sonnenfeld, R.J. *J. Organomet. Chem.* **1969**, *16*, 317.

37. Moses, P.; Gronowitz, S. *Arkiv. Kemi* **1962**, *18*, 119.

38. Köbrich, G.; Stöber, I. *Chem. Ber.* **1970**, *103*, 2744.

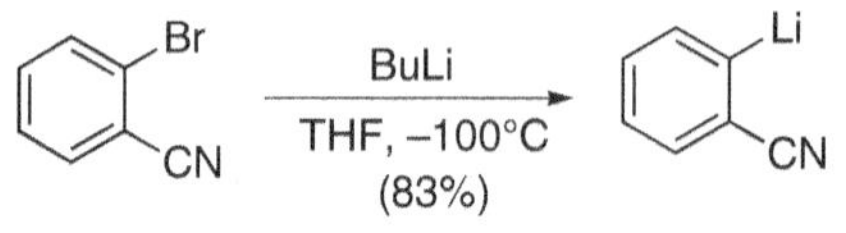

(11.21)[39]

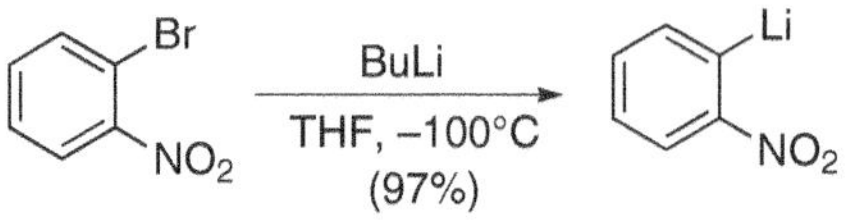

(11.22)[40]

Metathesis Between Metal Alkyls

One major problem in the synthesis of lithium alkyls by the direct synthesis method is the tendency of the reaction with reactive halides such as allyl and benzyl halides to give products of Wurtz coupling[41] and for the tendency of other halides (e.g., vinyl bromide) to undergo metal- halogen exchange too slowly for practical use. This can be avoided by generating the alkyllithium from a stannane reagent,[42] which is, available, in turn, by hydrostannylation of an alkyne (e.g., Example **11.23**), or from the alkyl halide by reaction with a stannane (e.g., Example **11.24**).[43]

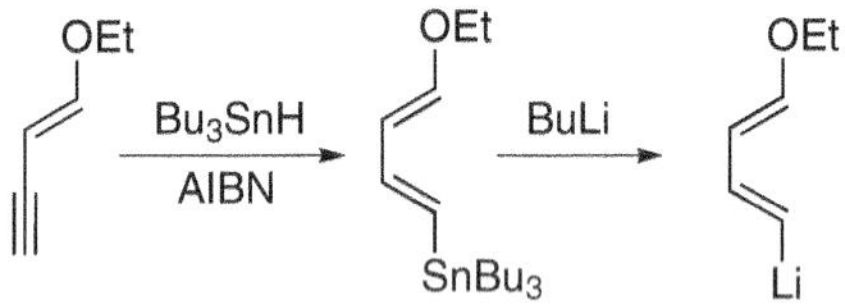

(11.23)[43b,c]

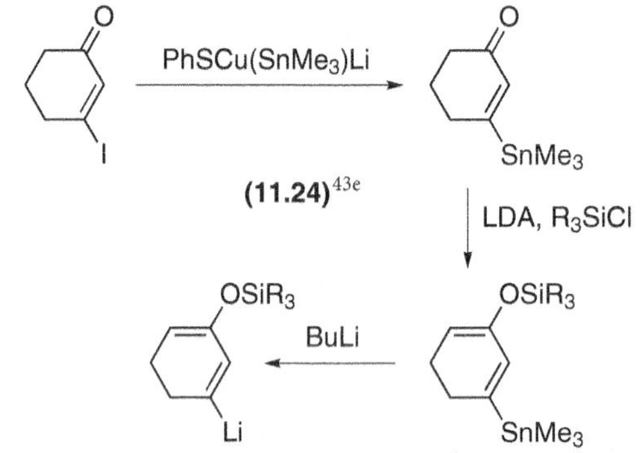

(11.24)[43e]

Exchange reactions where the metal alkyl of a more electronegative (i.e., less electropositive) metal is reduced by a less electronegative (i.e., more electropositive) metal are not widely used in the synthesis of metal alkyls except in reactions used to form the more nucleophilic metal alkyl under "halide-free" conditions. The most common type of this reaction involves the reduction of alkylmercury and alkyltin reagents with an alkali metal (e.g., Example **11.25**).[44] Because the reduction reactions are typically slow and reversible, it

39. Gilman, H.; Melstrom, D.S. *J. Am. Chem. Soc.* **1948**, *70*, 4177.

40. Köbrich, G.; Buck, P. *Chem.Ber.* **1970**, *103*, 1412.

41. (a) Wurtz, A. *Ann. Chim. Phys.* [3] **1855**, *44*, 275. (b) Wurtz, A. *Ann. Chem. Pharm.* **1855**, *96*, 364. (c) Kosolapoff, G.M. *Org. React.* **1951**, *6*, 326. (c) Lampman, G.M.; Aumiller, J.C. *Org. Syn., Coll. Vol. 6* **1988**, 133. (d) Pincock, R.E.; Torupka, E.J. *J. Am. Chem.Soc.* **1969**, *91*, 4593.

42. (a) Seyferth, D.; Weiner, M.A. *J. Am. Chem. Soc.* **1961**, *83*, 3583; *J. Org. Chem.* **1961**, *26*, 4797. (b) Seyferth, D.; Jula, T.F. *J. Orgamomet. Chem.* **1974**, *66*, 195. (c) Seyferth, D.; Mammarella, R.E. *J. Organomet. Chem.* **1979** , *177*, 53.

43. (a) Corey, E.J.; Williams, D.R. *Tetrahedron Lett.* **1977**, 3847. (b) Wollenberg, R.H.; Albizati, K.F.; Peries, R. *J. Am. Chem. Soc.* **1977**, *99*, 7365. (c) Wollenberg, R.H. *Tetrahedron Lett.* **1978**, 717. (d) Chen, S.L.; Schaub, R.E.; Grudzinkas, C.V. *J. Org. Chem.* **1978**, *43*, 3450. (e) Piers, E.; Morton, H.E. *J. Org. Chem.* **1979**, *44*, 3437.

44. (a) Fraenkel, G.; Dix, D.T.; Carlson, M. *Tetrahedron Lett.* **1968**, 579. (b) O'Brien, D.H.; Russell, C.R.; Hart, A.J. *J. Am. Chem. Soc.* **1979**, *101*, 633. (c) Barluenga, J.; Fañanás, F.; Yus, M. *J. Org. Chem.* **1979**, *44*, 4798. (d) Barluenga, J.; Fañanás, F.J.; Yus, M. *J. Org. Chem.* **1981**, *46*, 1281.

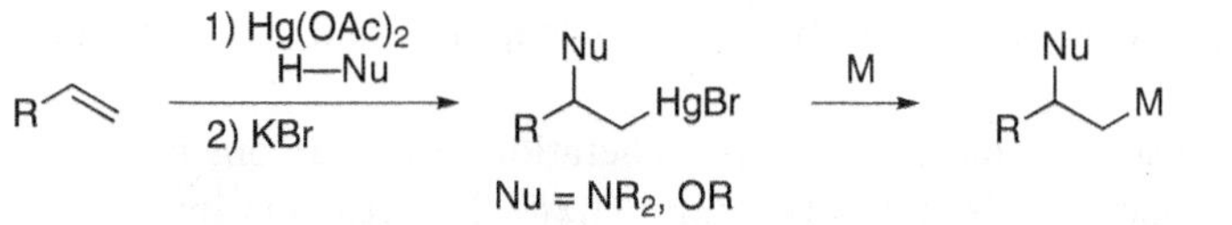

(11.25)[44c,d]

is usually important that the reducing metal be substantially more electropositive than the metal of the starting organometallic compound.[45]

Sample Problem

11-1 What is the major product of each of the following reactions?

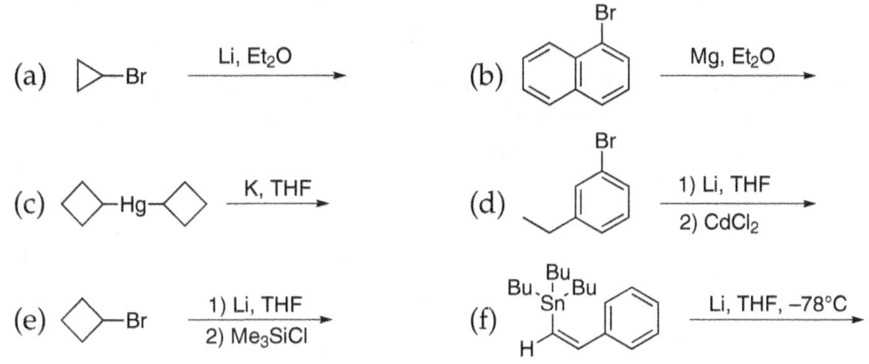

§Answers below.

Metalation of Hydrocarbons

The conjugate bases of simple hydrocarbons are usually too strongly basic to allow them to be generated by deprotonation. For this reason, direct metalation of hydrocarbons by either the reaction between the hydrocarbon and a reactive metal or by the deprotonation of a hydrocarbon by a metal alkyl (typically an alkyllithium or Grignard reagent) is generally restricted to the most acidic hydrocarbons. By far the most useful of these are the terminal alkynes, which are readily deprotonated by strong bases—sodium hydride, alkyllithiums or Grignard reagents—to give the corresponding metal

45. (a) Coates, G.E.; Wade, K. *Organometallic Compounds, 1, The Main Group Elements* (Methuen: London, 1967). (b) Makarova, L.G.; Nesmeyanov, A.N. *Methods of Elemento-Organic Chemistry* (North-Holland: Amsterdam, 1967), vol. 4.

§ Answers for Worked Problem:

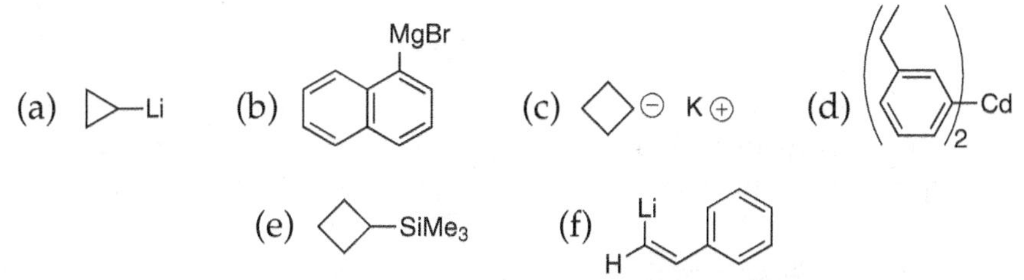

Reactions (a) and (b) are examples of direct synthesis of the metal alkyl from an alkyl halide; reactions (c) and (f) are examples of metal-metal exchange between an active metal and the metal alkyl of a less reactive metal; and reactions (d) and (e) are examples of the synthesis of the metal alkyl of a less reactive metal by metal-halogen exchange.

alkynides. These reagents are extremely useful nucleophiles that can be used in a variety of reactions.

An interesting reaction that involves the generation of carbanions is known as the "acetylene zipper" reaction (Examples **11.25** and **11.26**). This reaction involves the treatment of an unbranched internal alkyne with strong base (usually potassium 3-aminopropanamide, or "KAPA") and results in the rapid migration of the triple bond to the terminal position by means of a series of acid-base equilibria. The generation of the alkynide anion terminates the reaction series.[46] The driving force for this reaction appears to be the irreversible formation of the alkynide anion in the final step of the isomerization. With bases that are not strong enough to deprotonate the terminal alkyne (e.g., hydroxide anion or metal alkoxides), the equilibrium favors the internal alkynes and allenes rather than the terminal alkyne.

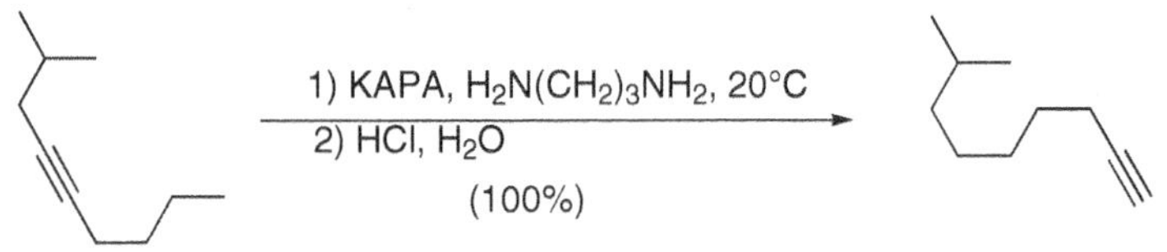

(11.26)

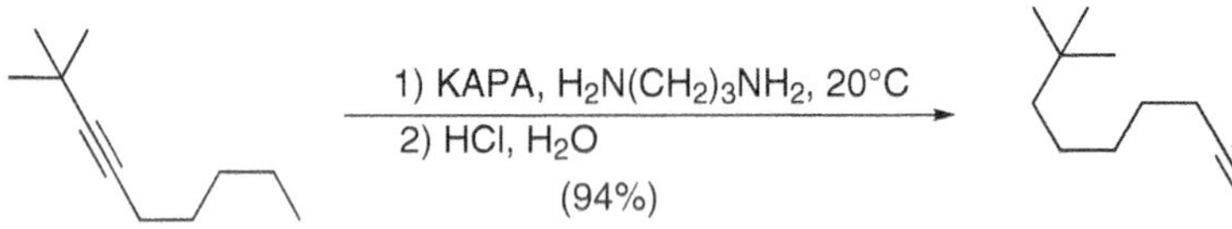

(11.27)

With bases that are not strong enough to completely deprotonate the terminal alkyne (e.g., hydroxide anion or metal alkoxides), the equilibrium favors the internal alkynes and allenes rather than the terminal alkyne. This reaction almost always results in a mixture of regioisomers, but anchoring the alkoxy anion at one location in the molecule tends to improve regioselectivity and permits the isomerization of acetylenic alcohols to allenic alcohols by bases such as butyllithium at low temperature, as illustrated in Examples **11.28** and **11.29**.[47] These reactions go through the lithium alkoxides, of course.

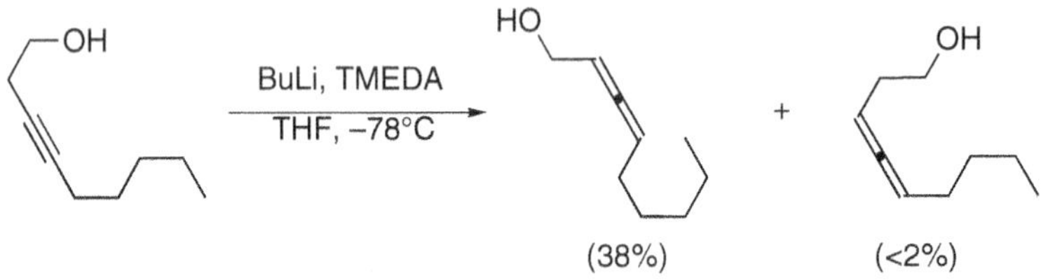

(11.28)

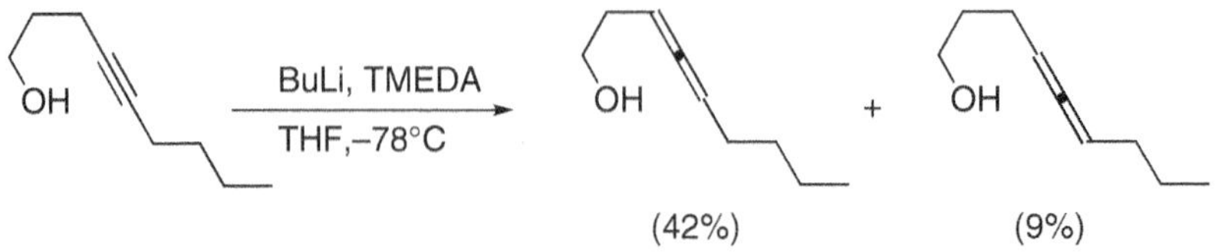

(11.29)

A second class of simple metal alkyls that can be easily prepared from the hydrocarbon is that in which the carbanion formed is strongly stabilized by resonance. The two representative types of these anions are the deep red triphenylmethide anion (Example **11.30**),

---

46. (a) Bushby, R.A. *Quart. Rev.* **1970**, *24*, 585. (b) Brown, C.A.; Yamashita, A. *J. Am. Chem. Soc.* **1975**, *97*, 891. (c) Brown, C.A.; Yamashita, A. *J. Chem. Soc. Chem. Commun.* **1976**, 959. (d) Macaulay, S.R. *J. Org. Chem.* **1980**, *45*, 734. (e) Abrams, S.R. *Can. J. Chem.* **1984**, *62*, 1333.

47. Enomoto, M.; Katsuki, T.; Yamaguchi, M. *Tetrahedron Lett.* **1986**, *27*, 4599.

which is readily available by deprotonation of triphenylmethane with strong base, and widely used as an indicator in titrations of alkyllithium reagents. The stability of the cyclopentadienide anion (Example **11.31**) is due to its aromaticity. Both these anions can be formed by treatment of the hydrocarbon with an alkyllithium or sodium hydride. Cyclopentadienide anion may also be formed by the reaction between cyclopentadiene and sodium metal.

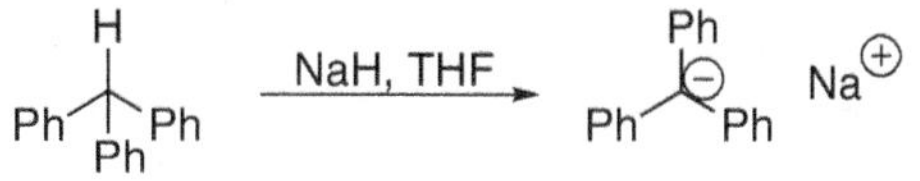

(11.30)

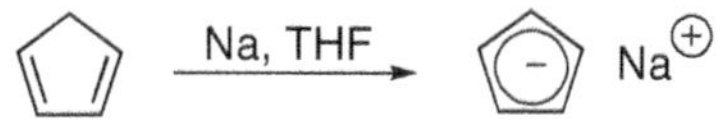

(11.31)

Reactivity of Metal Alkyls

One can modulate the reactivity of metal alkyls by choosing the metal appropriately. The reactivity of the metal alkyl as a base and a nucleophile can be predicted on the basis of the electronegativity of the metal—the less electronegative the metal, the more reactive the metal alkyl. Recall that the approximate percent ionic character of a number of carbon-metal σ bonds was given in Table 11.1.

Metal alkyls (other than sodium and potassium) are best described as polar covalent compounds with the carbon atom at the negative end of the polar carbon-metal bond. The variation in reactivity of the metal alkyls is mainly due to the variation in the polarity of the carbon-metal bond, and the more polar the carbon-metal bond, the more reactive the nucleophile. This can be illustrated using the examples in Table 11.4, where the addition of metal alkyls to carbonyl groups of gradually increasing reactivity is gathered.

Alkyllithiums are the only metal alkyls that will add to carboxylate anions to give the dialkoxide of the *gem*-diol. Grignard reagents add to carbon dioxide to give carboxylic acids, but dialkylzinc reagents do not (in fact, carbon dioxide was used as an inert atmosphere for the early work with these nucleophiles). Dialkylzinc reagents do react with aldehydes and ketones, but the less reactive dialkylcadmiums (just like the lithium dialkylcuprates) require an acid chloride as the carbonyl partner before reacting. Only alkyllithiums and Grignard reagents add to the cyano group of nitriles.

Effects of Overall Charge on Nucleophilicity

Unlike the metal alkyls of the alkali and alkaline earth metals, the transition metals may form both neutral and anionic metal-based species. Generically, it is found that the more negative the charge on a nucleophile, the stronger the nucleophile. This facet of nucleophile reactivity is nicely illustrated by the reactions of organozinc and organocopper nucleophiles with carbonyl electrophiles. Sodium trimethylzincate, $NaZnMe_3$, was first prepared by Wanklyn, who treated dimethylzinc with sodium metal[48]—based on its reactivity, he thought he had prepared sodium methyl. The resulting reagent reacted with carbon dioxide to give acetic acid: the first synthesis of a carboxylic acid by carbonation.[49] Because the Butlerov and Zaitsev syntheses of alcohols (dialkylzincs and alkylzinc iodides with acid chlorides) can be carried out under an atmosphere of carbon dioxide, the anionic ate complex is clearly a much stronger nucleophile.

48. Wanklyn, J.A. *Ann. Chem. Pharm.* **1858**, *108*, 67.
49. Wanklyn, J.A. *Ann. Chem. Pharm.* **1858**, *107*, 125; **1859**, *111*, 234.

Table 11.4 Reactivity of Metal Alkyls with Carbonyl Compounds

Substrate	R'Li	R MgX	R'_2Zn	R'_2Cd	R'Cu
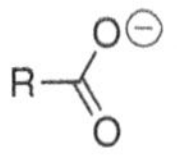	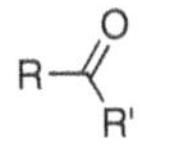	–	–	–	–
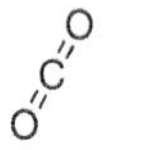	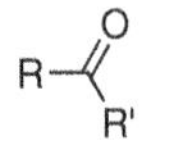	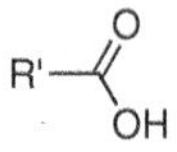	–	–	–
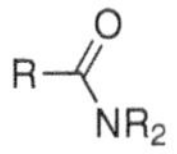	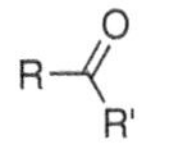	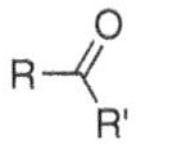	–	–	–
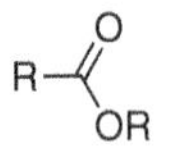	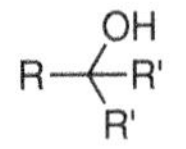	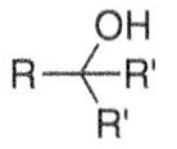	–	–	–
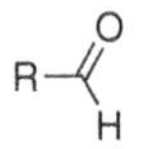	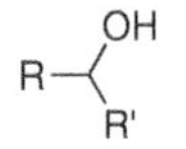	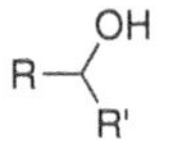	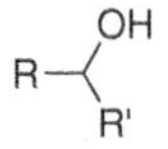	–	–
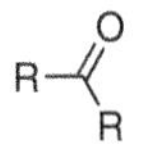	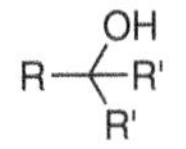	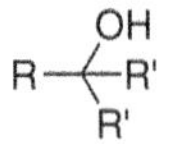	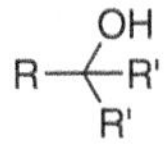	–	–
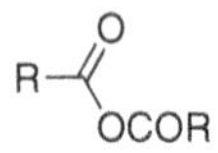	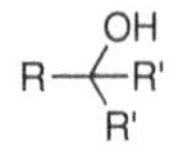	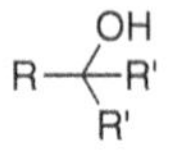	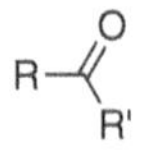	–	–
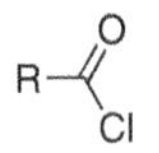	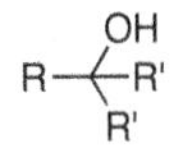	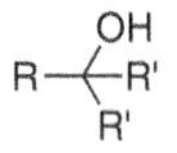	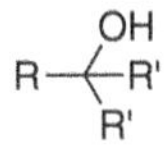	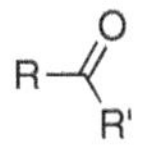	–
	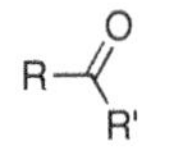	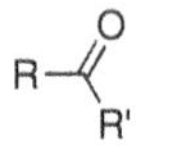	–	–	–
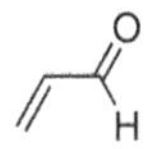	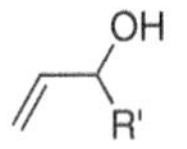	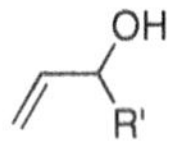	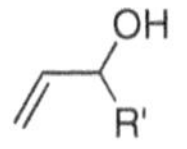	–	–
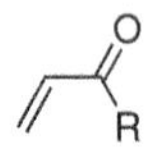	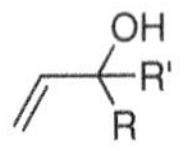	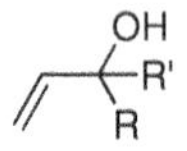	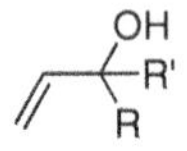	–	–

The behavior of organocopper nucleophiles illustrates this variation of nucleophilicity even better. The reaction of one equivalent of an alkyllithium with cuprous chloride yields an alkylcopper reagent (RCu)—a reagent that is relatively unreactive unless catalyzed by a Lewis acid[50] (RCu/BF_3 is a widely used reagent). The addition of a second equivalent of the alkyllithium converts this initial copper alkyl into a lithium dialkylcuprate, or ***Gilman reagent***, (R_2CuLi), which is named for American chemist Henry Gilman[51] and which is a

50. (a) Williams, D.R.; Nold, A.L.; Mullins, R.J. *J. Org. Chem.* **2004**, *69*, 5374. (b) Yamamoto, Y. *Angew. Chem. Int. Ed. Engl.* **1986**, *25*, 947.

51. **Henry Gilman (1893–1986)** was educated at Harvard University (PhD, 1918), and joined Iowa State University in 1919; he spent his entire career there. Gilman published more than 1,000 papers in organometallic chemistry—more than half after he had become legally blind. For more biographical details, see: (a) Earborn, C. *Biogr. Mem. Fellows Roy. Soc.* **1990**, *36*, 153. (b) Roberts, J.D. *Org. Synth.* **1987**, *66*, xiii.

much stronger nucleophile capable of adding to acid chlorides and conjugated carbonyl compounds. Gilman reagents are the preferred reagents for carrying out conjugate additions to α,β-unsaturated esters and ketones, but they are less effective when the conjugated system is β,β-disubstituted. Gilman reagents also tend to react poorly with secondary halides (see Example **11.32**)[52] or epoxides. When the alkyllithium is added to cuprous cyanide, an especially reactive cuprate reagent, the ***Lipshutz cuprate***[53] is formed. This nucleophile is capable of adding to a β,β-disubstituted-α,β-unsaturated carbonyl compound and reacting well with epoxides and secondary bromides.

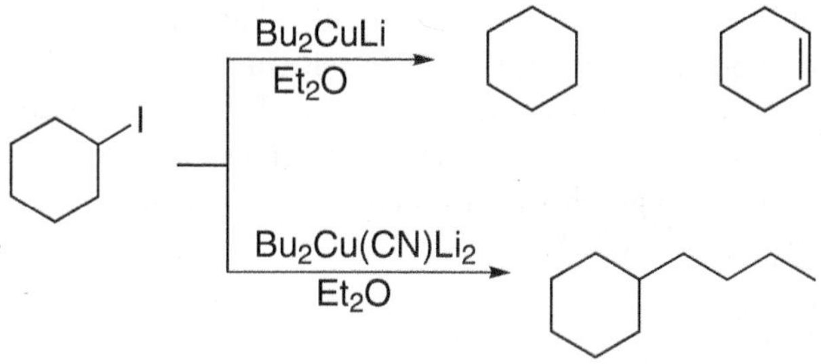

(11.32)

The structures of several copper alkyl reagents show these compounds to be oligomers based on a planar or bent square array of copper atoms bridged by alkyl ligands. The structure of a Gilman reagent (Example **11.34**) is analogous to the alkylcopper structure (Example **11.33**), except that two copper atoms on opposite sides of the array are replaced by lithium. The copper and lithium atoms in this structure are still bridged by alkyl groups.

(11.33) **(11.34)**

The structure of the Lipshutz "higher order" cuprates was the subject of intense debate during the last decades of the 20th century.[54] The structures of two "higher order" cyanocuprates were finally determined, and it was revealed that the cyanide anions bridge a pair of lithium, not copper, atoms. These structures may be seen as the consequence of cyanide ion disrupting the structure of the Gilman reagent by inserting between the lithium ions (**11.35**). In one case (**11.36**),[55] the structure consists of a zigzag arrangement of alternating -R–Cu–R- and -Li–C ≡ N–Li- chains. In the other case (**11.37**),[56] the structure consists of linear R_2Cu^- and $[LiCNLi]^+$ ions (the logical extreme of the zigzag structure as the lithium ions become solvated by donor solvents such as tetrahydrofuran (THF). The enhanced reactivity of these reagents compared to traditional Gilman reagents may be due to the closer resemblance of the Lipshutz cuprates to the free R_2Cu^- ion.

---

52. (a) Corey, E.J.; Posner, G.H. *J. Am. Chem. Soc.* **1968**, *90*, 5615. (b) Lipshutz, B.H.; Wilhelm, R.S.; Floyd, D.M. *J. Am. Chem. Soc.* **1981**, *103*, 7672.

53. Lipshutz, B.H.; Wilhelm, R.S.; Kozlowski, J.A.; Parker, D. *J. Org. Chem.* **1984**, *49*, 3928.

Bruce H. Lipshutz (1951–) was educated at Yale (PhD, 1977), and joined the University of California, Santa Barbara after postdoctoral study with Corey at Harvard. Lipshutz is known for his organocopper chemistry; he received a Presidential Green Chemistry Award in 2011. More biographical detail is available at Lipshutz' web site: http://web.chem.ucsb.edu/~lipshutzgroup/bio/index.shtml

54. Leading references to this debate: (a) Lipshutz, B.H.; James, B. *J. Org. Chem.* **1994**, *59*, 7585. (b) Bertz, S.H.; Miao, G.; Eriksson, M. *J. Chem. Soc., Chem. Commun.* **1996**, 815.

55. Kronenburg, C.M.P.; Jastrzebski, J.T.B.H.; Spek, A.L.; van Koten, G. *J. Am. Chem. Soc.* **1998**, *120*, 9688.

56. Boche, G.; Bosold, F.; Marsch, M.; Harms, K. *Angew. Chem. Int. Ed. Engl.* **1998**, *37*, 1684.

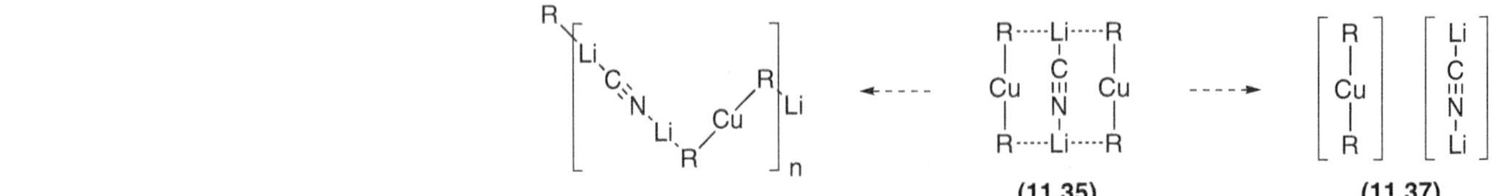

(11.36) (11.35) (11.37)

Copper-based organometallic reagents are important intermediates in a wide range of reactions leading to the formation of new carbon-carbon σ bonds.[57] Among these reactions is the reaction of lithium dialkylcuprates with alkyl halides, which proceeds well with primary alkyl bromides and iodides, but which seldom occurs in satisfactory yield with secondary halides; with chlorides, it only proceeds well with primary chlorides, and only in THF.[58] The reaction proceeds much better with the Lipshutz cuprates, which react well with secondary bromides to give hydrocarbon products with almost complete inversion of configuration (e.g., Example **11.38**).

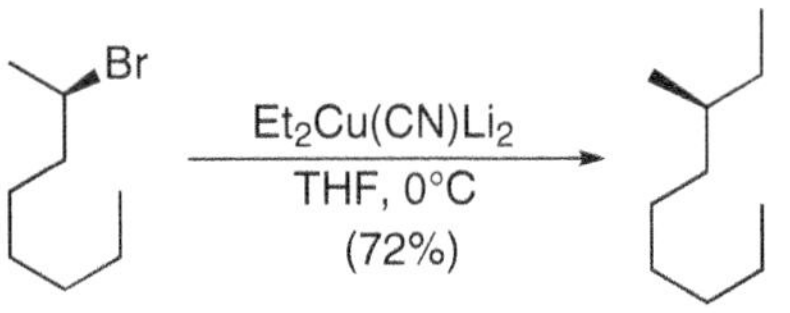

(11.38)[58]

Epoxides, also, react with Lipshutz cuprates to give alcohols (e.g., Example **11.39**) in higher yield than the Gilman reagents themselves.[59] It is important to note that although the reaction between a cuprate and an alkyl halide bears a superficial resemblance to the S_N2 reaction, it almost certainly involves single electron transfer steps and free radical intermediates.

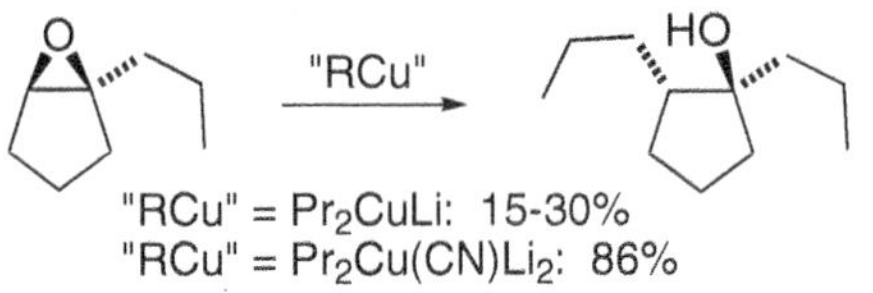

(11.39)[59]

Unlike the coupling of organocuprates with saturated alkyl halides and epoxides, the coupling of organocuprates with vinyl halides occurs with predominant retention of configuration (e.g., Examples **11.40** and **11.41**).[60] The formally similar coupling of aryl halides in the presence of a copper catalyst (the ***Ullmann reaction***) likely proceeds through an arylcopper intermediate.[61]

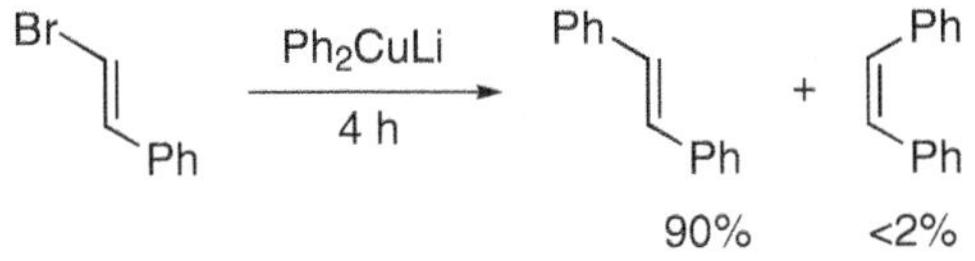

(11.40)

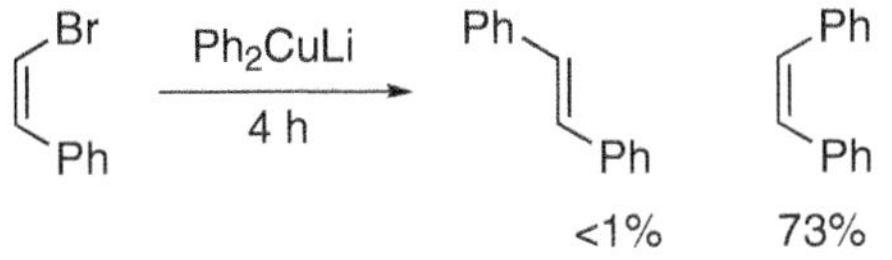

(11.41)

57. Monograph: (a) Posner, G.H. *An Introduction to Synthesis Using Organocopper Reagents* (Wiley-Interscience: New York, 1980). Reviews: (b) Posner, G.H. *Org. React.* **1975**, *22*, 253. (c) Normant, J.F. *Synthesis* **1972**, 63. (d) Jukes, A.E. *Adv. Organomet. Chem.* **1974**, *12*, 215.

58. Whitesides, G.M.; Fischer, W.F., Jr.; San Filipo, J., Jr.; Bashe, R.W.; House, H.O. *J. Am. Chem. Soc.* **1969**, *91*, 4871.

59. (a) Lipshutz, B.H.; Kozlowski, J.; Wilhelm, R.S. *J. Am. Chem. Soc.* **1982**, *104*, 2305. (b) Lipshutz, B.H.; Kotsuki, H.; Lew, W. *Tetrahedron Lett.* **1986**, *27*, 4825.

60. House, H.O. *Acc. Chem. Res.* **1976**, *9*, 59.

61. Reviews: (a) Bacon, R.G.R.; Hill, H.A.O. *Quart. Rev.* **1965**, *19*, 95. (b) Fanta, P.E. *Synthesis* **1974**, 9.

Sample Problem

11-2 What reagent or reagents are needed to complete the following transformations?

(a) 

(b) 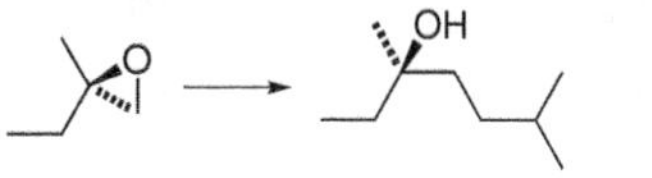

(c) 

(d) 

(e) 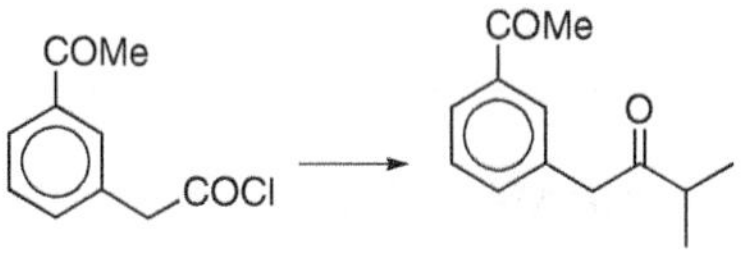

(f) 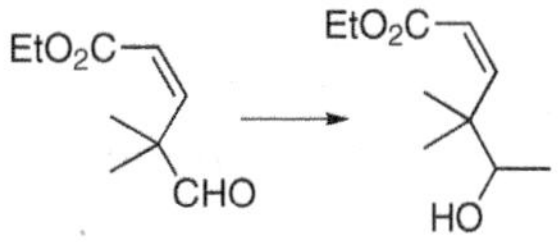

§Answers below.

Acid-Base Chemistry of Metal Alkyls—A Limitation on Their Use

Formally, metal alkyls are the conjugate bases of hydrocarbons, so they are strong bases. This leads to their strong tendency to react with protic compounds by simple proton transfer, a tendency that increases with the ionic character of the carbon-metal bond. Thus, alkali metal and alkaline earth metal organometallic compounds all react vigorously with water to generate the corresponding hydrocarbon, as do dialkylzinc reagents. Alkylcopper reagents, including lithium dialkylcuprates (but not the Lipshutz cuprates), react relatively slowly with water and alcohols to give hydrocarbons, whereas dialkylmercury reagents do not react appreciably with water at all (which is one reason for their biological toxicity).

Because proton transfers are among the most rapid reactions known, this places a limitation on the generation and use of metal alkyls in synthesis—one cannot use a metal alkyl in a reaction where there is an acidic proton capable of quenching it. Likewise, one must use a solvent that will not react with the more reactive organometallic compound when making a less reactive organometallic reagent by a metathesis reaction.

As a rule of thumb, alkali metal alkyls will react with ammonia and amines (pK_as ≈ 33–36) at temperatures near 0°C to give the corresponding amide salts (for example, lithium diisopropylamide [LDA], a very popular hindered base, is made by the reaction between butyllithium and diisopropylamine between -30°C and 0°C). Grignard reagents are slowly protonated by amines to give the magnesium amide salts—and alkylzinc reagents even more slowly. Organocopper reagents are not protonated by amines. Changing the proton source to water or an alcohol (pK_as ≈ 15–18) results in Grignard reagents and organozinc reagents being quenched rapidly. Gilman reagents are still not rapidly quenched by water or alcohols—although the more reactive Lipshutz cuprates are. The rapid quenching of less reactive metal alkyls usually requires ammonium ions or strong acids. The relative rates for protonolysis of some typical metal alkyls are shown in Table 11.5.

§ Answers for Worked Problem:

(a) This reaction involves the stereospecific replacement of the halogen of a vinyl halide by an alkyl group, so a cuprate should be used: Et_2CuLi, THF. (b) This is the ring opening of an epoxide, which can be accomplished with any organometallic reagent of high nucleophilicity: *i*-BuLi, THF or *i*-BuMgBr, Et_2O or *i*-Bu_2CuLi, or *i*-$Bu_2Cu(CN)Li_2$, THF. (c) This involves the attack on the carbonyl group of an α,β-unsaturated ketone and requires a more reactive nucleophile: MeLi, THF (preferred) or MeMgBr, Et_2O. (d) This involves the attack on the conjugated alkene π bond of an α,β-unsaturated ketone, so a cuprate must be used: Et_2CuLi, THF. (e) This requires a nucleophile that will react with an acid chloride but not with a ketone: *i*-Pr_2Cd or *i*-Pr_2CuLi. (f) This requires a nucleophile that will react with an aldehyde but not an ester: Me_2Zn.

Table 11.5 Effects of Bond Polarity on the Ease of Protonolysis of Metal Alkyls

	R_2NH	ROH	NH_4Cl (aq)	HCl (aq)
R–Li	Fast	Fast	Fast	Fast
R–MgX	Slow	Fast	Fast	Fast
R_2Zn	Slow	Moderate	Fast	Fast
$R_2Cu(CN)Li_2$	—	Moderate	Fast	Fast
R_2CuLi	—	Slow	Fast	Fast
RPdX	—	—	Slow	Slow

Some Final Words on the Reactivity of Simple Metal Alkyls

Generally speaking, the more ionic a metal alkyl, the more likely it is to react well with alkyl halides to give the products of carbon-carbon bond formation. Alkali metal alkynides, which are ionic metal alkyls, react well with alkyl halides by the S_N2 mechanism to give internal alkynes, but Grignard reagents, in general, do not (Example **11.42**). The treatment of an alkyl halide with sodium metal or a sodium-potassium alloy usually gives the product of Wurtz coupling. This reaction, which is generally agreed to proceed through sodium alkyls, does not involve a simple S_N2 displacement of the halide ion, but appears, instead, to involve free radicals at some point in the reaction.[62] This may be a general occurrence when metal alkyls react with alkyl halides.

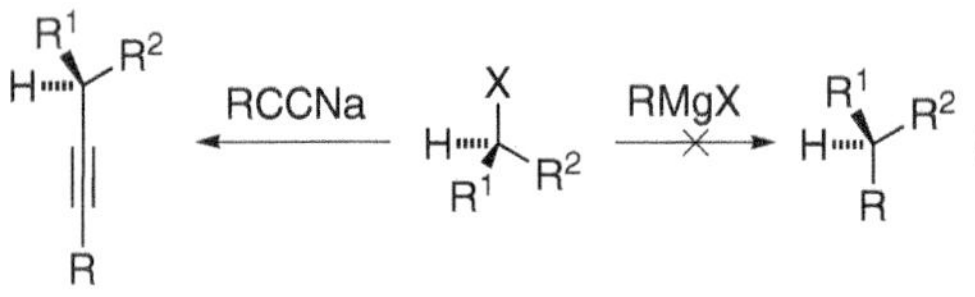

(11.42)

The ionic character of the carbon-metal bond is also revealed in characteristic reactivity patterns with carbonyl compounds (especially α,β-unsaturated carbonyl compounds; Example **11.43**). Ionic metal alkyls are strong bases that deprotonate the carbonyl compound to give an enolate anion. Ionic metal alkynides, which are much weaker bases, add to the carbonyl group to give allylic carbinols. The covalent metal alkyls react as nucleophiles rather than bases, so addition dominates their reactions. As the C—M bond becomes less polar, the nucleophile becomes softer and shows a greater tendency to add at the β carbon of α,β-unsaturated carbonyl compounds—alkyllithiums give the products of 1,2-addition only, Grignard reagents give mainly 1,2-addition but some 1,4-addition, and cuprates react to give products of 1,4-addition only.

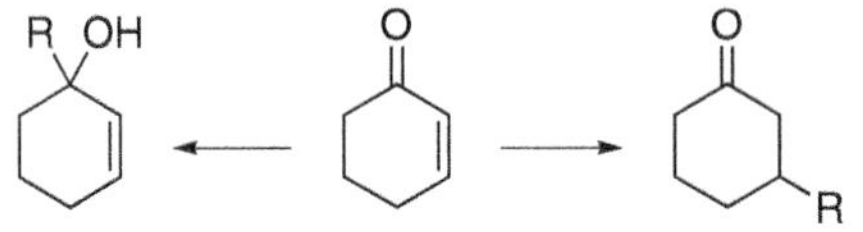

(11.43)

62. Garst, J.F.; Cox, R.H. *J. Am. Chem. Soc.* **1970**, *92*, 6389.

Afterword: Considering the reactivity of organolithiums and Grignard reagents, it might seem inconceivable that metal alkyls could exist within the cells of a living being. And yet vitamin B12, which was discovered in 1926, is an organocobalt compound that is essential for the normal maturation of erythrocytes; the cobalt-carbon bond is critical for the activity of the vitamin.

Not all metal alkyls are as benign as vitamin B12, however. For decades, tetraethyllead was used as an anti-knock additive to improve the octane rating of gasoline. Its use in gasoline was banned in the 1970s but not before it had been implicated as a major causative factor in raising the blood lead levels in children. Similarly, for many decades, certain organomercury compounds had been used as topical antiseptics, and it was believed that organomercury compounds were not a serious health hazard (in contrast to inorganic mercury compounds, whose toxicity had been well established). However, an outbreak of a mysterious, crippling disease in Minamata, Japan, during the early 1950s was subsequently traced to the large quantities of fish in the diet and to the high levels of mercury in the fish in Minamata Bay. Organomercury antiseptics are no longer freely available in the United States.

Reaction Synopses

Direct Synthesis of Organometallic Reagents

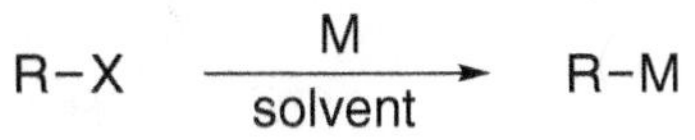

M = Li, Mg, Zn, Al, etc.;
X = Cl, Br, I

Reagents: Li, Et_2O, Li, THF; Mg, Et_2O; Zn, Et_2O, Δ; etc.

Stereochemistry: Chiral alkyl halides give partially or totally racemic organometallic reagents. The method works only for very electropositive metals.

Transmetalation

(a) Metal-halogen exchange

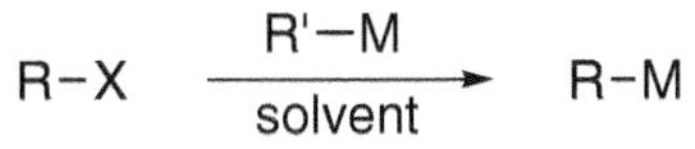

Reagents: BuLi, THF, –78°C; *t*-BuLi, THF, –100°C; etc.
Any halide may be used, including aryl and vinyl halides. The organometallic reagent (R'—M) may not be methyllithium.

(b) Metal-metal exchange

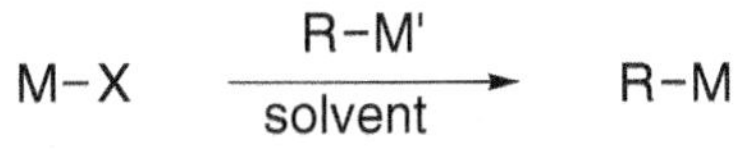

M': Li; Mg; etc. M: Cu; Cd; Hg; etc.

Reagents:

R–M': RLi, RMgX; R_2Zn, R_3Al; etc.

M–X: CuCl, CuBr, CuI, CuCN; $CdCl_2$, $CdBr_2$; $SiCl_4$, R_3SiCl; $SnCl_4$, R_3SnCl; $MnCl_2$, $ZnCl_2$, $HgCl_2$, etc.

M is always less electropositive than M'.

Metalation of Hydrocarbons (especially alkynes)

Reagents: BuLi, NaH, KH, LDA, RMgX, etc.

Hydrocarbon usually needs to have a pK_a value below 28.

R—C ≡ C—M can be formed with transition metals: M^{n+}, NH_3, H_2O (M=Cu, Ag, etc).

Acetylene Zipper

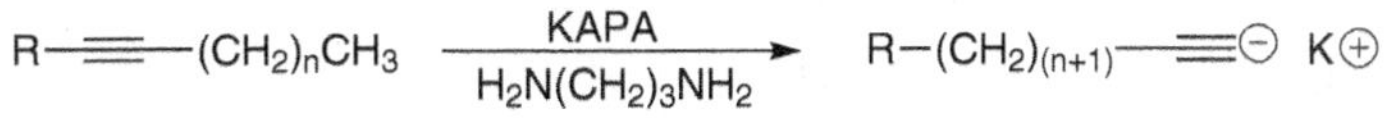

Reagents: (1) $KNH(CH_2)_3NH_2$ (KAPA); (2) H_3O^+

Note: the chain may not be branched; when the chain is branched, a mixture of acetylenes results.

(continues)

(**Reaction Synopses** continued)

Organocuprate Coupling

R–X ⟶ R–R'

RX: *Not* 3° alkyl; reaction is slow with 2° alkyl;

Epoxide: ring opening occurs at less substituted ring carbon.

Reagents: R_2CuLi; $R_2Cu(CN)Li_2$; etc.

Stereochemistry:

Inversion of configuration with alkyl bromides
Retention of configuration with vinyl halides
Inversion of configuration with epoxides

Drill Problems

(a) What is the major organic product formed when each of the compounds below is treated with the reagent specified?

(A) 3-pentanone (B) 2-methylpropanal

(C) cyclohexanone (D) acetophenone

(E) 2-cyclohexenone (F) cinnamonitrile

(F) methyl crotonate (G) (*S*)-2-ethyloxirane

(H) (*R*)-2-bromo-1-phenylethane

(I) (*R*)-2-octyl *p*-toluenesulfonate

(a) 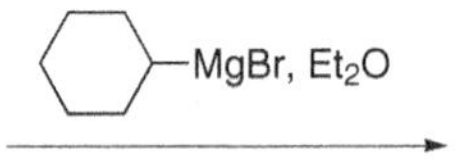

(b) 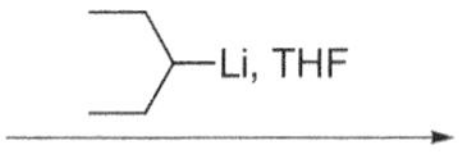

(c) 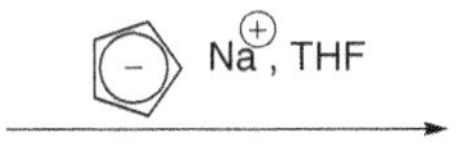

(d) 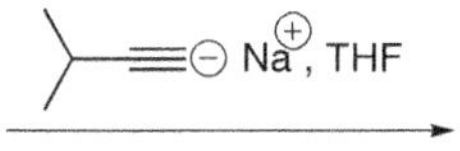

(e) $\xrightarrow{LiCuEt_2,\ THF}$

(b) Draw the structure of the major organic product that will be formed when the each of the alkynes below reacts with each of the following sequences of reagents.

Reagents:
(a) (1) $NaNH_2$, NH_3; (2) $BrCH_2CH{=}CH_2$, THF
(b) (1) LDA, THF, −78°C; (2) CO_2; 3) HCl, H_2O
(c) (1) EtMgBr, Et_2O, Δ; (2) cycloheptanone; (3) NH_4Cl, H_2O
(d) (1) BuLi, THF, −78°C; (2) $ClCO_2Et$, THF, −78°C; (3) NH_4Cl, H_2O
(e) (1) KNH_2, NH_3; (2) cyclopentene oxide; (3) HCl, H_2O
(f) (1) KNH_2, NH_3; (2) $Me(CH_2)_4Br$, THF; (3) KAPA, THF; (4) NH_4Cl, H_2O.

Compounds:
(A) 1-pentyne (B) 1-octyne
(C) phenylacetylene (D) 3-ethyl-1-pentyne
(E) 3,3-dimethyl-1-butyne (F) 1-ethynylcyclopentanol
(G) 3-methoxy-3-methyl-1-butyne

Problems

11-2 When one needs to prepare allylmagnesium bromide according to the reaction below, it is important to use highly activated magnesium to prevent the formation of a major by-product whose molecular formula is C_6H_{10}. Draw the structure of this compound and propose a reasonable mechanism for its formation.

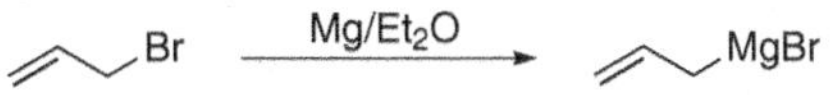

11-3 What is the final product of each of the following reactions or sequences of reactions?

(a)
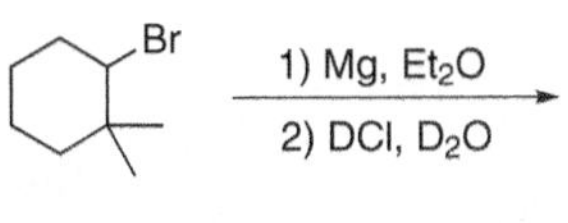

(b)
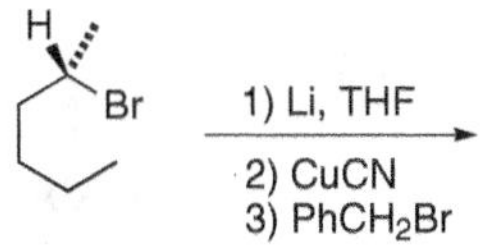

(c)
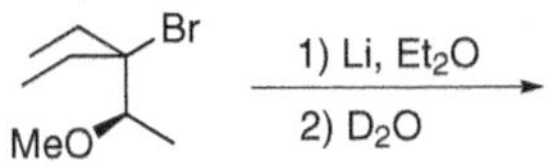

(d)
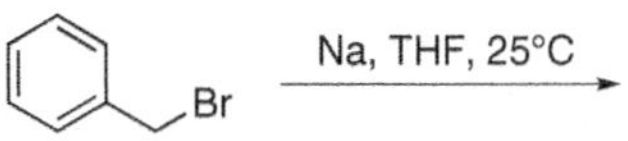

(e)
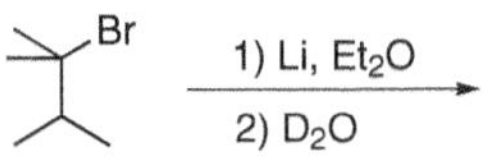

(f)
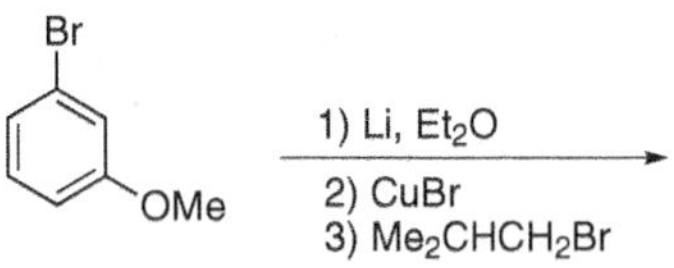

(g)
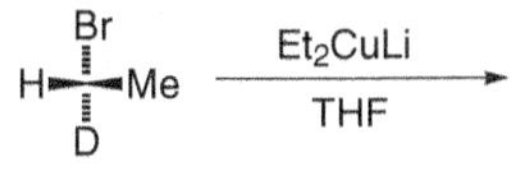

(h)
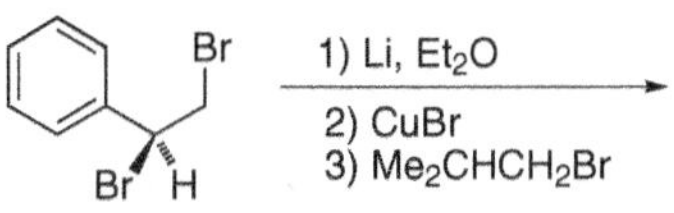

11.3 Carbanions Stabilized by Heteroatoms

A carbanion center may be stabilized by an adjacent heteroatom provided that the heteroatom occurs in the third row of the periodic table or lower. Second-row heteroatoms do not usually stabilize carbanions well (in fact such anions are not even detectable in the gas phase[63]), and when anions are formed adjacent to a second-row heteroatom, the resultant anion almost always rearranges by a homolytic mechanism, as occurs in the Wittig rearrangement of ethers.[64]

Stabilization by the adjacent heteroatom is, however, an important part of the stabilization of anions next to sulfur, phosphorus, selenium, and silicon. There are two mechanisms by which this stabilization can occur (Figure 11.2). In the first, the stabilization of the negative charge is attributed to delocalization of the negative charge into the empty d orbitals of the heteroatom by $d_\pi - p_\pi$ back-bonding, resulting in formation of a carbon-heteroatom π bond.[65] In the alternative view, which is particularly applicable to the carbon-phosphorus bond of phosphoranes, the extra electron density on carbon is delocalized into the

63. Downard, K.M.; Sheldon, J.C.; Bowie, J.H.; Lewis, D.E.; Hayes, R.N. *J. Am. Chem. Soc.* **1989**, *111*, 8112.

64. (a) Wittig, G.; Lohman, L.; Happe, W. *Ann.* **1942**, *550*, 260. (b) Wittig, G.; Lohman, L.; Happe, W. *Ann.* **1947**, *557*, 205. (c) Hauser, C.R.; Kantor, S.W. *J. Am. Chem. Soc.* **1951**, *73*, 1437. (d) Baldwin, J.E.; Patrick, J.E. *J. Am. Chem. Soc.* **1971**, *93*, 3556. (e) Felkin, H.; Frajerman, C. *Tetrahedron Lett.* **1977**, 3485. (f) Chérest, M.; Felkin, H.; Frajerman, C. *Tetrahedron Lett.* **1977**, 3489.

65. (a) Wolfe, S.; LaJohn, L.A.; Bernardi, F.; Mangini, A.; Tonachini, G. *Tetrahedron Lett.* **1983**, *24*, 3789. (b) Wolfe, S.; Stolow, A.; LaJohn, L.A. *Tetrahedron Lett.* **1983**, *24*, 4071.

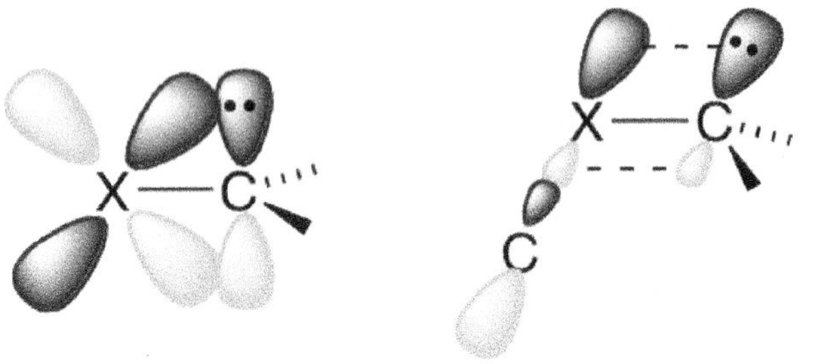

Figure 11.2 The models of the C—X bond in heteroatom-stabilized carbanions based on $d_\pi - p_\pi$ back-bonding (left), and delocalization of the excess electron density into low-energy antibonding orbitals (right).

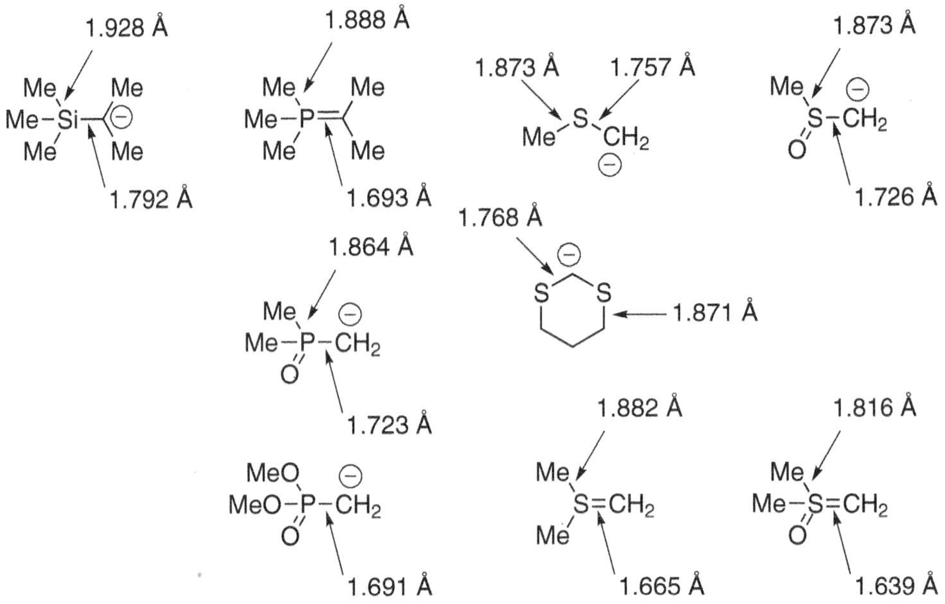

Figure 11.3 Bond lengths in stabilized anions. The bond between the carbanion carbon and the heteroatom is considerably shorter than the bond between a neutral carbon and the heteroatom in the same molecule. The P–C bond length in dimethyl methylphosphonate is 1.810 Å.

relatively low-energy antibonding orbitals of the adjacent carbon-heteroatom bond.[66] In either case, the electron density on carbon is delocalized in part onto the electronegative heteroatom, thus stabilizing the anion, and increasing the bond order (reducing the bond length) of the carbon-heteroatom bond. The values of the bond lengths computed for a number of carbanions stabilized by heteroatoms, calculated at the B3LYP/6-31+G* level, are gathered given in Figure 11.3. They reveal the same general trend: removing a proton from the carbon atom shortens the bond.

The enhanced stability of carbanions adjacent to heteroatoms is reflected in strength of the base required to generate the carbanion. This, in turn, depends on the stabilizing group: anions stabilized by just a single heteroatom usually require butyllithium or a similar base. As the stabilizing group becomes more electron withdrawing, or as the number of these groups attached to the carbon atom carrying the acidic proton increases, the stability of the anion increases, and a weaker base can be used for the deprotonation step. Some typical ranges of pK_as for a range of precursors to silicon-, phosphorus- and

66. (a) Bernardi, F.; Csizmadia, I.G.; Mangini, A.; Schlegel, H.B.; Whangbo, M.; Wolfe, S. *J. Am. Chem. Soc.* **1975**, *97*, 2209. (b) Lehn, J.M.; Wipff, G. *J. Am. Chem. Soc.* **1976**, *98*, 7498. (c) Borden, W.T.; Davidson, E.R.; Andersen, N.H.; Denniston, A.D.; Epiotis, N.D. *J. Am. Chem. Soc.* **1978**, *100*, 1604. (d) Bernardi, F.; Bottoni, A.; Venturini, A.; Mangini, A. *J. Am. Chem. Soc.* **1986**, *108*, 8171.

Table 11.6 Typical p*K*as for Silicon, Phosphorus, and Sulfur Compounds

Compound	pK_a	Compound	pK_a
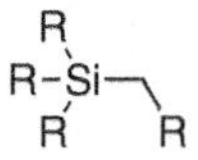	22–23 (R=fluorenyl)	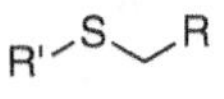	>30 (R=alkyl, aryl) 17-20 (R=acyl, CN)
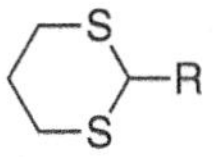	29–30 (R=alkyl) 19–21 (R=acyl, CN)	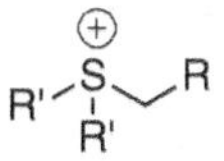	16–18
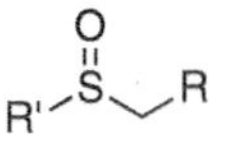	29–35	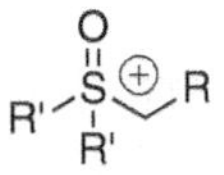	≈18
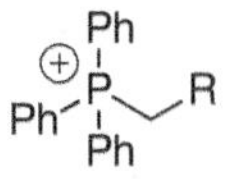	17–23 (R=alkyl, aryl) 6-9 (R=acyl, CN)	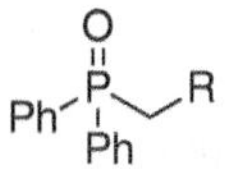	≈25 (R=alkyl, SR) ≈17 (R=CN)
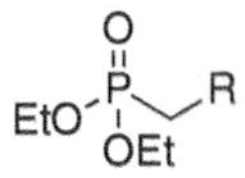	27–30 (R=alkyl) 16–18 (R=acyl, CN)		

sulfur-stabilized anions are collected in Table 11.6.[67] Compounds with pK_a values below 25 can typically be deprotonated with sodium hydride or LDA; compounds with pK_a values above 30 usually require butyllithium for effective deprotonation.

Ylides

The ylides differ from the simple unstabilized carbanions by having a group that could function as a leaving group bonded directly to the carbanion carbon. This positively charged group helps to stabilize the carbanion center by simple polar effects. When the heteroatom is in the third row or lower, stabilization also occurs by π-type delocalization of the electron density from carbon onto the heteroatom. Because of this π-type delocalization of the electron density of the carbanion, ylides can be represented by the resonance contributors shown at in Example **11.44**.

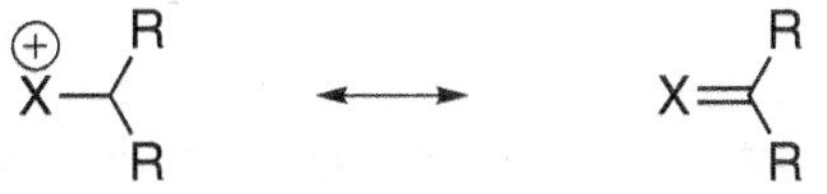

(11.44)

X = NR_3, SR_2, PR_3, AsR_3 X = SR_2, PR_3, AsR_3

The two types of ylides most frequently used in modern organic synthesis are the phosphorus ylides, or ***phosphoranes***, and the sulfur ylides, or ***sulfuranes***. These ylides are

67. These and many other acidity constants can be found in the following compilations: (a) Bordwell, F.G. *Acc. Chem. Res.* **1988**, *21*, 456. (b) Bordwell, F.G.; Matthews, W.S. *J. Am. Chem. Soc.* **1974**, *96*, 2116. (c) Matthews, W.S.; Bares, J.E.; Bartmess, J.E.; Bordwell, F.G.; Cornforth, F.J.; Drucker, G.E.; Margolin, Z.; McCallum, R.J.; McCollum, G.J.; Vanier, N.R. *J. Am. Chem. Soc.* **1975**, *97*, 7006. (d) Bordwelll, F.G.; Fried, H.E. *J. Org. Chem.* **1981**, *46*, 4327. (e) Smith, M.B.; March, J. *March's Advanced Organic Chemistry. Reactions, Mechansims, and Structure*, 5th ed. (Wiley: New York, 2001), ch. 8, pp. 329–331, and references therein. (f) Isaacs, N.S. *Physical Organic Chemistry* (Longman Scientific & Technical: London, 1987), ch. 6, contains several useful tables of acidity constants.

formed from the corresponding phosphonium or sulfonium ions by reaction with a strong base such as sodium hydride or butyllithium. The sulfonium and phosphonium ions are themselves formed by S_N2 displacement reactions of the neutral sulfides or phosphines, as shown in Examples **11.45** and **11.46**.

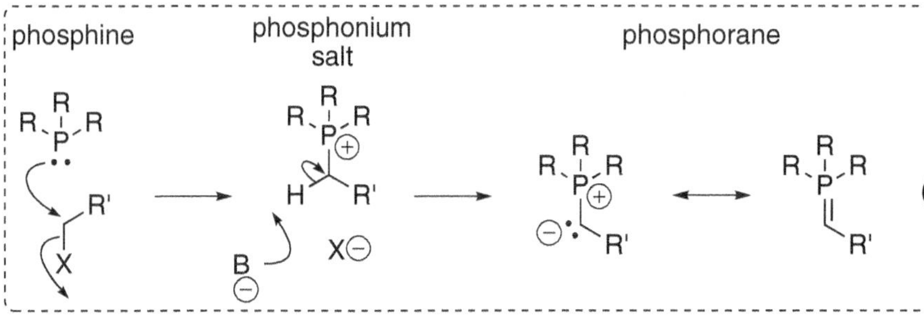

(11.45)

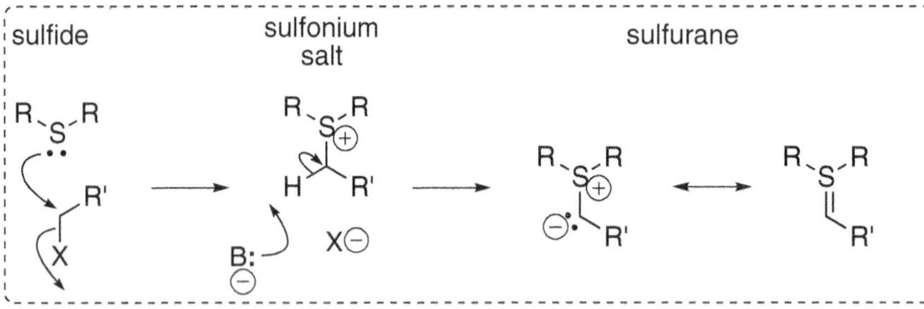

(11.46)

Phosphorus Ylides

Because the formation of a carbon-heteroatom π bond involves the overlap of a 2p orbital on carbon with a 3p or a 3d orbital on the heteroatom, the formation of the π bond is less efficient than the formation of a π bond between carbon and elements of the second row. In addition, the contribution from the dipolar canonical form in a carbon-phosphorus π bond of a phosphorane is higher than in a carbon-nitrogen π bond. The effect of having to overlap orbitals from different electron shells to form a bond is even more pronounced in π bond formation than in σ bond formation. Stable compounds with carbon-silicon π bonds are unknown, despite much effort to prepare one. Unlike the π bonds between carbon and the second-row elements, π bonds between carbon and third-row elements are much weaker than the corresponding σ bonds, so elimination from *gem*-disubstituted alkane derivatives is seldom favorable.

The X-ray crystal structures of those phosphorus ylides whose structures have been determined strongly suggest that the non-polar, double-bonded canonical form is the dominant contributor to the resonance hybrid. The P=C bond in methylenetriphenylphosphorane (Example **11.47**) has a length of 1.661 Å,[68] which corresponds to an approximate bond order of 2.0, similar to values from electron diffraction of trimethyl(methylene) phosphorane in the gas phase.[69] The bond lengths of the P=C bond in stabilized ylides vary between 1.7 and 1.74 Å, corresponding to bond orders between 1.5 and 1.7, with a concomitant reduction in the C—C bond length. The carbon atom in $Ph_3P=CH_2$ is very close to planar, consistent with sp^2 hybridization. Thus, there is solid evidence that there is substantial π bonding between phosphorus and carbon in phosphonium ylides. Non-stabilized ylides have a bond close to a full double bond, with minimal dipolar character, whereas stabilized ylides have a lower carbon-phosphorus bond order and a correspondingly higher contribution from the dipolar canonical form.

---

68. Bart, J.C.J. *J. Chem. Soc. (B)* **1969**, 350.
69. Ebsworth, E.A.V.; Fraser, T.E.; Rankin, D.W.H. *Chem. Ber.* **1977**, *110*, 3494.

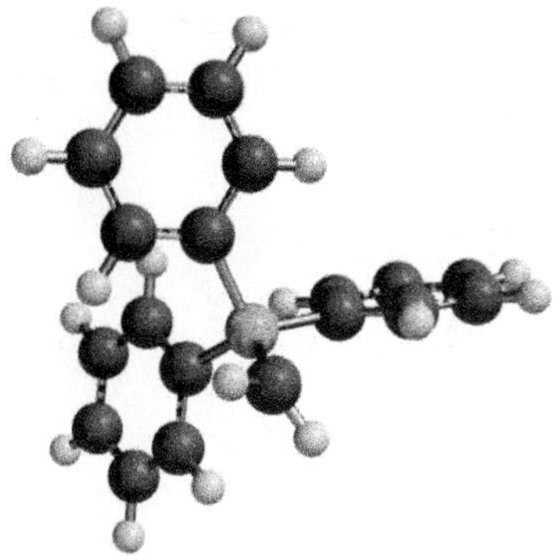

(11.47)

The most recognized reaction of phosphorus ylides is probably the ***Wittig reaction***[70] to give alkenes. Simple phosphorus ylides react with aldehydes to give *Z*-alkenes as the major product (Example **11.48**). As the anion is stabilized by groups in addition to the phosphorus, the proportion of the *E* alkane rises.

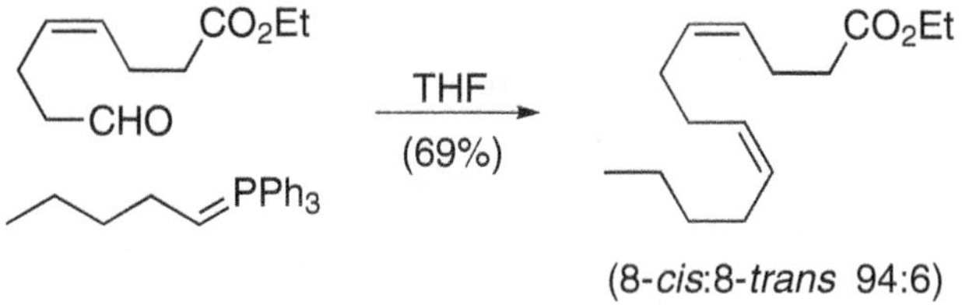

(11.48)[71]

Sulfur Ylides

Deprotonating trialkylsulfonium salts or trialkylsulfoxonium salts gives sulfur ylides. Computations at relatively high level of theory show that, as with the phosphorus ylides, the bond order of the carbon-sulfur bond in the sulfonium ylides (**11.49**) is close to 2.0 and that the ylide carbon is planar, indicating that the non-polar, double-bonded contributor dominates the resonance hybrid. In the sulfoxonium ylides (**11.50**), on the other hand, the C—S bond order is closer to 1.7, and the ylide carbon is now pyramidal, indicating that the dipolar form contributes significantly to the resonance hybrid (Figure 11.4). Sulfonium ylides are more strongly basic and more strongly nucleophilic than the sulfoxonium ylides. Sulfonium ylides seldom add reversibly to a carbonyl group, and their reactions are generally under kinetic control, whereas reactions of sulfoxonium ylides are often under thermodynamic control. This can have useful consequences, synthetically. Both types of ylides

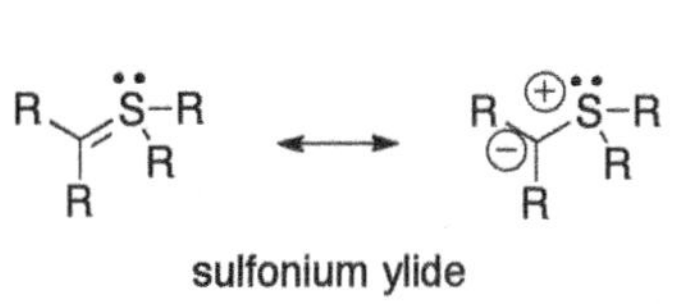

(11.49)

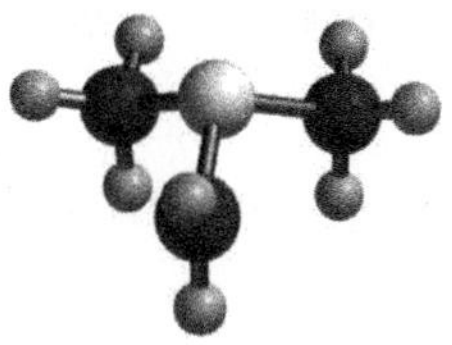

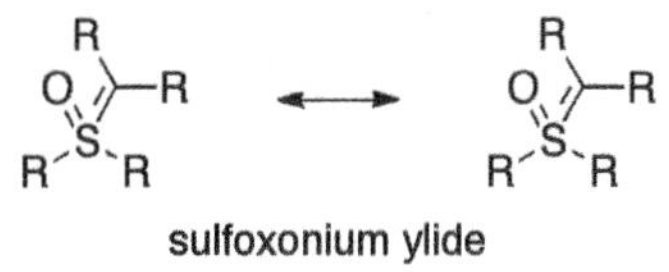

(11.50)

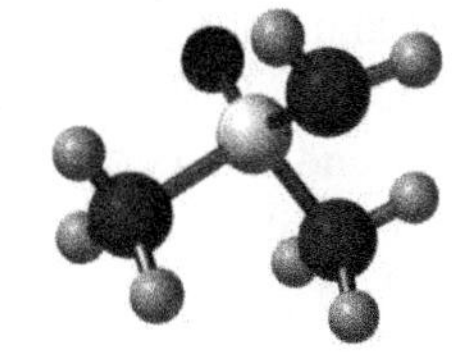

Figure 11.4 Formation and structures of sulfur ylides at the B3LYP/6-311**G(2df,2p) level of theory

70. Reviews (Wittig reaction): (a) Maercker, A. *Org. React.* **1965**, *14*, 270. (b) Schlosser, M. *Top. Stereochem.* **1970**, *5*, 1. (c) Bestmann, H.J.; Vostrowsky, O. *Top. Curr. Chem.* **1983**, *109*, 85. (d) Murphy, P.J.; Brennan, J. *Chem. Soc. Rev.* **1988**, *17*, 1. (e) Maryanoff, B.E.; Reitz, A.B. *Chem. Rev.* **1989**, *89*, 863.

71. Anderson, R.J.; Henrick, C.A. *J. Am. Chem. Soc.* **1975**, *97*, 4327.

react with saturated carbonyl compounds to give epoxides (Example **11.51**). In contrast, sulfoxonium ylides react with α,β-unsaturated carbonyl compounds to give cyclopropanes (Example **11.52**).[72]

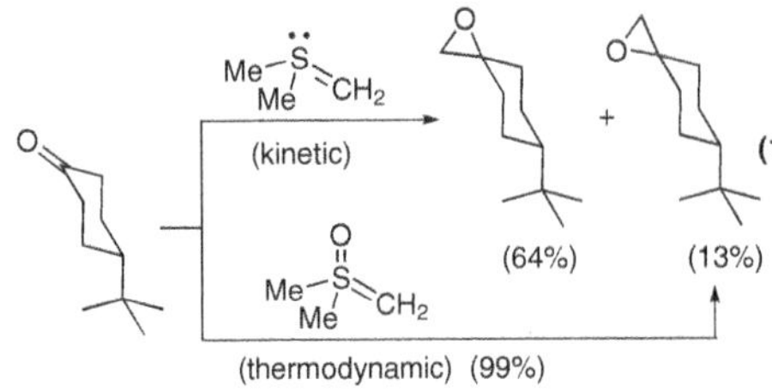

(11.51)

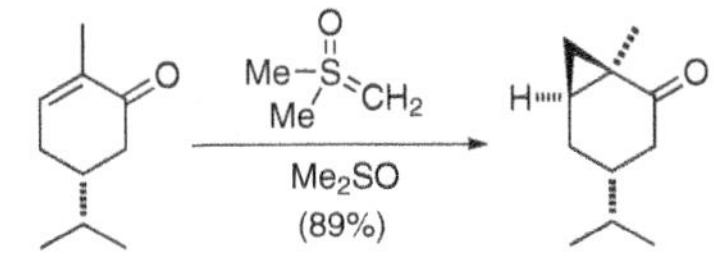

(11.52)

Dithiane Anions

With two sulfur atoms on the same carbon atom, the anions of thioacetals are readily prepared. The most widely used thioacetal carbanions are derived from 1,3-dithianes (Example **11.53**). The synthetic utility of dithiane anions arises from the fact that the nucleophilic carbon in the dithiane anion was originally a carbonyl carbon, so that the dithiane anion corresponds in a formal sense to an acyl anion. You will recall that the carbon atom of a carbonyl group is *electrophilic*, not nucleophilic, so one cannot prepare acyl anions directly from carbonyl compounds. Dithiane anions are important ***umpoled synthons***,[73] a term derived from the German *umpolung*, denoting reversal of the normal electrophilic or nucleophilic character of a carbon atom. Dithiane anions react rapidly with alkyl halides by the S_N2 mechanism, as shown, where the anion of the dithiane **11.54** reacts with the alkyl halide to give the dithiane **11.55**.[74] They also add rapidly to aldehydes and ketones at low temperatures to give alcohols. On hydrolysis of the dithiane, the carbonyl group is recovered, as illustrated by the mercury-assisted hydrolysis of Examples **11.55** to **11.56**.

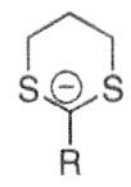

(11.53)

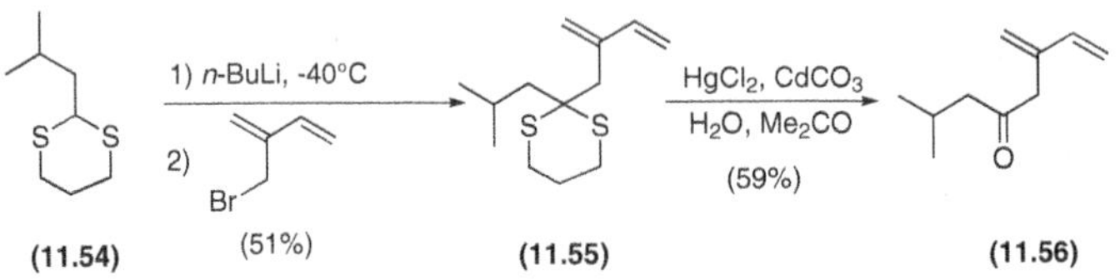

(11.54) **(11.55)** **(11.56)**

Alkylsulfinyl Anions, Phosphonate Anions, and Anions from Phosphine Oxides

The sulfoxides, sulfones, phosphine oxides, and phosphonate esters share a common structural feature: highly polar bonds to oxygen that are best represented by the hybrid of non-polar and polar canonical forms (Example **11.57**). In both types of compounds, the polar canonical form makes a significant contribution to the resonance hybrid, leading to X—C bond orders close to 1.5 and a pyramidal anion carbon atom. Not surprisingly, the conjugate bases of these compounds share some of the reactivity of the corresponding ylides.

---

72. Corey, E.J.; Chaykovsky, M. *J. Am. Chem. Soc.* **1965**, *87*, 1353.

73. (a) Seebach, D. *Angew. Chem. Int. Ed. Engl.* **1979**, *18*, 239. (b) Gröbel, B.T.; Seebach, D. *Synthesis* **1977**, 357. (c) Seebach, D.; Corey, E.J. *J. Org. Chem.* **1975**, *40*, 231.

74. Reece, C.A.; Rodin, J.O.; Brownlee, R.G.; Duncan, W.G.; Silverstein, R.M. *Tetrahedron* **1968**, *24*, 4249.

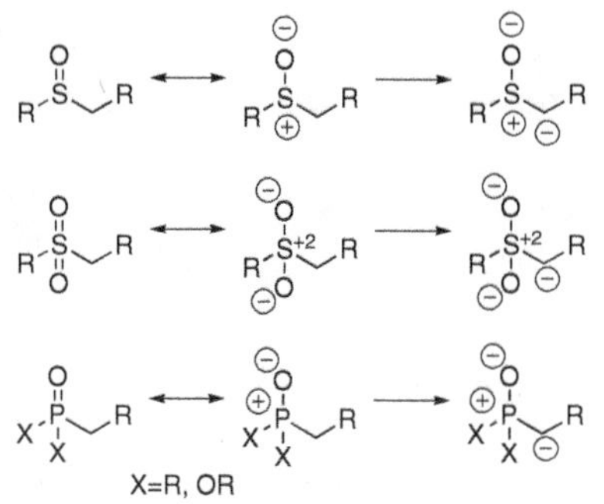

(11.57)

Sulfoxides are converted to their conjugate bases by treatment with strong bases. Dimethyl sulfoxide, for example, is converted into its conjugate base, the ***dimsyl anion*** ($CH_3SOCH_2^-$), by heating to 70°C with sodium hydride.[75] Sulfoxide anions are both good nucleophiles and strong bases. They react with esters to give β-ketosulfoxides. The β-ketosulfoxides may be prepared by the nucleophilic acyl substitution of esters by sulfoxide anion nucleophiles. Because of its ability to stabilize an adjacent carbanion, the sulfoxide group has achieved popularity as an activating group for the regioselective formation of enolate anions. Strong bases such as LDA or the dimsyl anion deprotonate β-ketosulfoxides to give the enolate anion adjacent to sulfur.

These enolate anions participate readily in S_N2 reactions to give alkylated sulfoxides that may then be reduced to the hydrocarbon by active metals. Like the sulfinyl group, the phosphoryl group also confers enhanced acidity on the hydrogens α to it. Phosphine oxides and phosphonate esters can both be converted to their conjugate bases with sodium hydride or butyllithium or lithium amide bases, and these conjugate bases are important nucleophiles that react with aldehydes and ketones to give alkenes. The reactivity of phosphoryl anions was explored by Horner[76] as well as by Wadsworth and Emmons.[77] The reaction between a phosphoryl anion and an aldehyde or ketone is generally known today as the ***Horner-Wadsworth-Emmons*** (**HWE**) reaction. The α-anions of phosphonate esters are the critical nucleophiles in the most widely used version of the reaction. In contrast to the Wittig reaction with unstabilized ylides, the HWE reaction favors the formation of the *E* isomer, as illustrated by Example **11.58**.

75. Corey, E.J.; Chaykovsky, M. *J. Am. Chem. Soc.* **1965**, *87*, 1345.

76. (a) Horner, L.; Hoffmann, H.; Wippel, H.G. *Chem. Ber.* **1958**, *91*, 61–63. (b) Horner, L.; Hoffmann, H.; Wippel, H.G.; Klahre, G. *Chem. Ber.* **1959**, *92*, 2499–2505.

Leopold Horner (1911–2005) was educated at Munich (PhD, 1937) and began his independent career in 1942 at the Plastics Research Institute in Frankfurt am Main. In 1952, he moved to Mainz as Professor. Horner's work in organophosphorus chemistry led to his enduring fame. For more biographical details, see: Kuntz, H. *Angew. Chem. Int. Ed.* **2005**, *44*, 7664.

77. Reviews: (a) Wadsworth, W.S., Jr. *Org. React.* **1977**, *25*, 73. (b) Boutagy, J.; Thomas, R. *Chem. Rev.* **1974**, *74*, 87.

William S. Wadsworth, Jr. (1927–). Because there is no published biography of Wadsworth, this biographical note will be somewhat longer than most in this book. Wadsworth was born in Farmington, Connecticut, to a long-established New England family, and was educated at Trinity College (site of the Wadsworth Library) and at Pennsylvania State University, where he received his PhD in 1956. He then joined the Rohm and Haas Company, and it was while in Emmons' group there that Wadsworth discovered the reaction that bears their joint names. His project had been to prepare new acrylate monomers containing a phosphonate group—such compounds had been proposed as potential flame retardants in synthetic fibers—and he set about preparing them by the condensation of α-phosphonoacetonitrile with formaldehyde. To his chagrin, the major product was an insoluble polymer that was totally devoid of phosphorus; by continued work, however, he was able to determine that the polymer was poly(acrylonitrile) formed by polymerization of the initial product. In 1963, Wadsworth joined the faculty of South Dakota State University, where he remained until his retirement in 1991.

William David Emmons (1924–2001) was educated at the Minnesota and Illinois (PhD, 1951). He joined the Rohm & Haas Company, where he became Director of Pioneering Research in 1973. From 1961–1969 he was an editor of *Organic Syntheses*. For more biographical details, see; Freeman, J.P. *Org. Synth.* **2003**, *80*, xxvi.

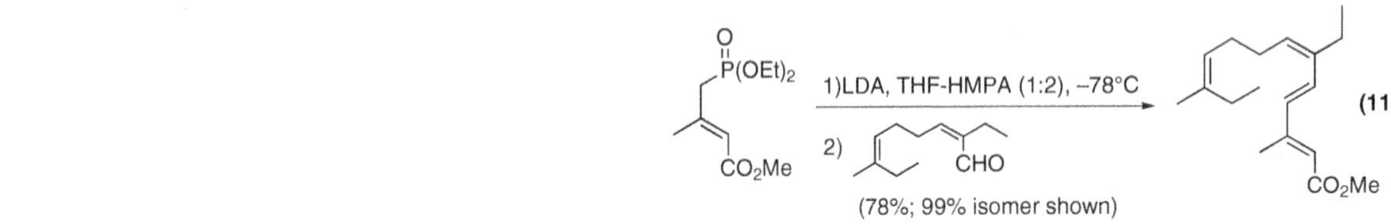

(11.58)[78]

Silane Anions: The Peterson Olefination

Carbanions stabilized by silicon are useful nucleophiles that participate in typical nucleophilic reactions: they react with alkyl halides to give products of S_N2 reactions, and they add to carbonyl compounds to give alcohols. When treated with acid or base, the adducts of silicon-stabilized anions and carbonyl compounds undergo a facile elimination to give alkenes (Example **11.59**). This two-step reaction converting a carbonyl compound to an alkene is known as the ***Peterson olefination***.[79] When the β-hydroxyalkylsilane is treated with strong base in the second step of the Peterson olefination, the elimination occurs with *syn* stereochemistry; when acid is used in the second step, the elimination occurs with *anti* stereochemistry.

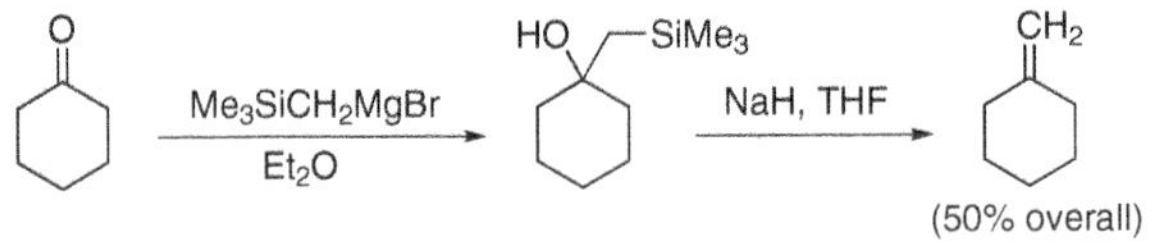

(11.59)

The Peterson olefination is less affected by steric hindrance than the Wittig reaction, so it offers a useful alternative to the Wittig reaction in reactions of hindered ketones. For example, the formation of the exocyclic methylene double bond in gorgonene was accomplished as shown in Example **11.60** using a Peterson reaction; the Wittig reaction led to epimerization at the bridgehead carbon in this particular case.[80]

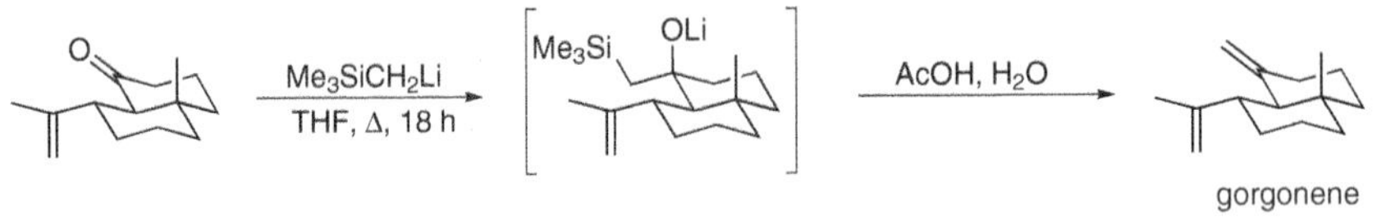

(11.60)

Reaction Synopses

Wittig Reaction

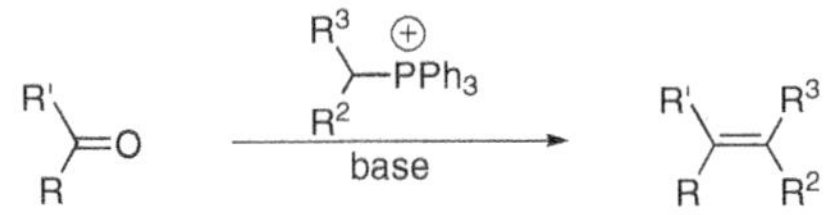

Base: NaH, Me_2SO; BuLi; NaH, THF; LDA, THF; etc.

Stereochemistry: R=alkyl, predominantly *Z*
R= COR, CO_2R, CHO, CN, Ph, etc, predominantly *E*

---

78. Wadsworth, W.S., Jr. *Org. React.* **1977**, *25*, 73.
79. (a) Peterson, D.J. *J. Org. Chem.* **1968**, *33*, 780. Reviews: (b) Chan, T.H. *Acc. Chem. Res.* **1977**, *10*, 442. (c) Birkofer, L.; Stiehl, O. *Top. Curr. Chem.* **1980**, *88*, 58. (d) Ager, D. J. *Synthesis* **1984**, 384; *Org. React.* **1990**, *38*, 1.
Donald John Peterson (1935–) received his graduate education at Iowa State University (PhD, 1962, under Gilman), and immediately joined the Procter and Gamble Company, where he rose to become Director of the Corporate Technology Division. Peterson's independent work with heteroatom-stabilized carbanions dates to his early years at P&G; his seminal publication appeared in 1968. In what may presage the future, information about Peterson is most conveniently found on the Facebook page devoted to his reaction:
D.J.Peterson,PhD/the Peterson Olefination Reaction (https://www.facebook.com/groups/209574029106964/)
80. Boeckmann, R.K., Jr.; Silver, S.M. *J. Org. Chem.* **1975**, *40*, 1755.

Horner-Wadsworth-Emmons Reaction

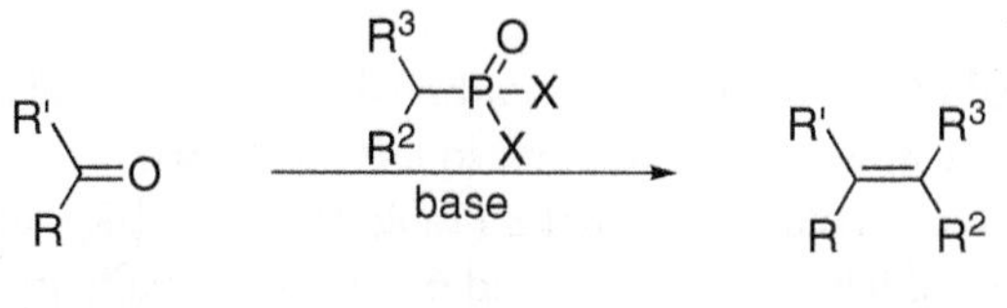

Base: NaH, Me_2SO; BuLi; NaH, THF; LDA, THF; etc.
X = alkyl, Ph (Horner reaction); X = alkoxy (Wadsworth-Emmons)

Stereochemistry: predominantly *E*

Peterson Olefination

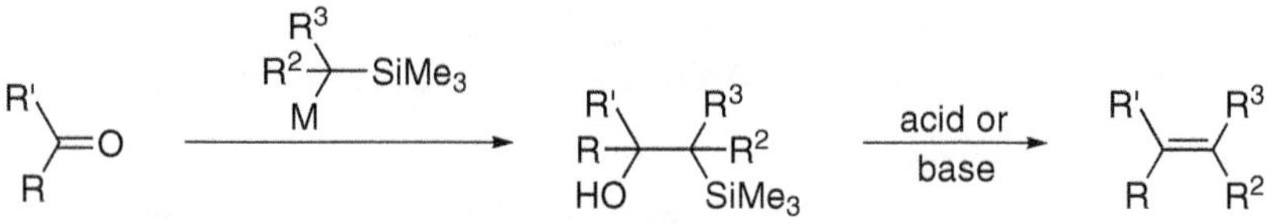

M = Li, Mg, etc.

Stereochemistry: elimination by base is *syn*; elimination by acid is *anti*.

Epoxide Synthesis Using Sulfonium Ylides

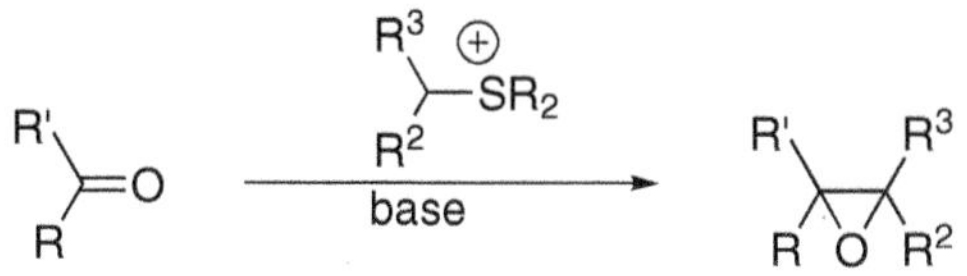

Base: NaH, Me_2SO; LDA, THF; etc.

Reagents: Me_3SI, NaH, Me_2SO; RCH_2SPh_2Br, NaH, THF; etc.

Reaction with sulfonium ylide is under kinetic control; reaction with sulfoxonium ylide is under thermodynamic control.

Cyclopropanation Using Sulfoxonium Ylides

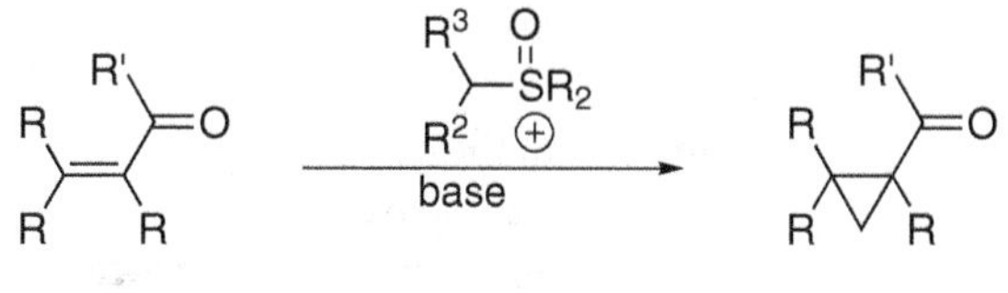

Base: NaH, Me_2SO; LDA, THF; etc.

Reagents: $Me_3S(O)I$, NaH, Me_2SO; $RCH_2S(O)Ph_2Br$, NaH, THF; etc.

Umpolung and Dithiane Anions

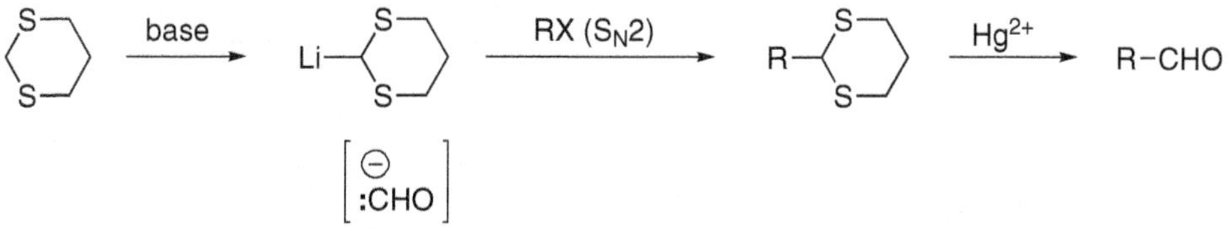

Base: BuLi, THF, –78°C; etc.

11.4 Formation of Enolate Anions

Enolate anions formed by deprotonation of carbonyl compounds have been extremely important nucleophiles for more than a century, so it is well worth knowing the approximate pK_as of some simple representative examples (Table 11.7). The pK_as of the α hydrogens of simple monofunctional aldehydes, ketones, and nitriles are usually in the range of 20 to 31, depending on the solvent. All are considerably weaker acids than water (pK_a 15.5), but stronger acids than ammonia and amines (pK_a 32–35).[81] As is apparent from the values in Table 11.7, the effects of multiple carbonyl or cyano groups are cumulative: β-dicarbonyl and α-cyanocarbonyl compounds as well as α-cyanonitriles are all stronger acids than water. Nitro groups have an even larger effect on the acidity of the α hydrogens, so nitromethane is approximately as strong an acid as pentane-2,4-dione.

Structure and Bonding in Enolate Anions and Nitrile α-Anions

The removal of an α-hydrogen from a carbonyl compound or a nitrile gives a resonance-stabilized carbanion with an extended π electron system. As shown in Example **11.61**, the major contributor to the resonance hybrid of these anions is the one with the negative charge on the heteroatom.

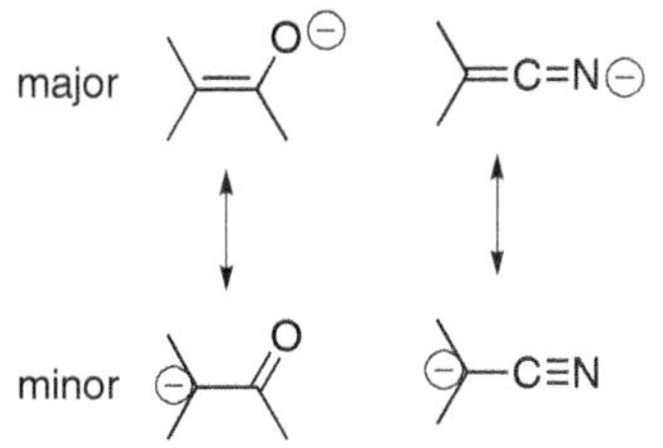

(11.61)

The effects of resonance stabilization of the anion can be viewed in terms of the reaction coordinate diagrams for the deprotonation reactions of an alkane and a carbonyl compound (Figure 11.5). Both reactions involve the cleavage of a carbon-hydrogen bond. Because neither the hydrocarbon nor its conjugate base is delocalized, the alkane provides a baseline energy profile that the deprotonation of the carbonyl compound can be compared against. In contrast to the hydrocarbon, the carbonyl compound is slightly stabilized by resonance, although much less so than the enolate anion due to the charge separation in the minor canonical form. Thus, compared to the ΔH for the deprotonation of the hydrocarbon, which gives a localized anion, the ΔH for deprotonation of the carbonyl compound is less positive: the carbonyl compound is more acidic.

Because the major contributor to the resonance hybrid for these anions is the one in which the heteroatom carries the negative charge, one would predict that structure of these anions should resemble an alkene,

Table 11.7 Typical pK_a Values of Carbonyl Compounds, Nitriles, and Common Bases

Compound	pK_a (H_2O)	pK_a (Me_2SO)
R-CH(COR')$_2$	≈ 9	≈ 13
R-CH(CN)$_2$	≈ 11–12	≈ 11–12
H_2O	15.74	31.4
ROH	15–19	29–32
R-CH$_2$-COR'	19–20	24–27
R-CH$_2$-CN	≈ 25	≈ 31
NH_3	32.5	≈ 35
RNH_2, R_2NH	≈ 33–43	≈ 35–40

81. For compilations of acidity constants, see (a) Bordwell, F.G. *Acc. Chem. Res.* **1988**, *21*, 456. (b) Bordwell, F.G.; Matthews, W.S. *J. Am. Chem. Soc.* **1974**, *96*, 2116. (c) Matthews, W.S.; Bares, J.E.; Bartmess, J.E.; Bordwell, F.G.; Cornforth, F.J.; Drucker, G.E.; Margolin, Z.; McCallum, R.J.; McCollum, G.J.; Vanier, N.R. *J. Am. Chem. Soc.* **1975**, *97*, 7006. (d) Bordwell, F.G.; Fried, H.E. *J. Org. Chem.* **1981**, *46*, 4327. (e) Smith, M.B.; March, J. *March's Advanced Organic Chemistry. Reactions, Mechansims, and Structure*, 5th ed. (Wiley: New York, 2001), ch. 8, pp. 329–331, and references therein. (f) Isaacs, N.S. *Physical Organic Chemistry* (Longman Scientific & Technical: London, 1987), ch. 6, contains several useful tables of acidity constants.

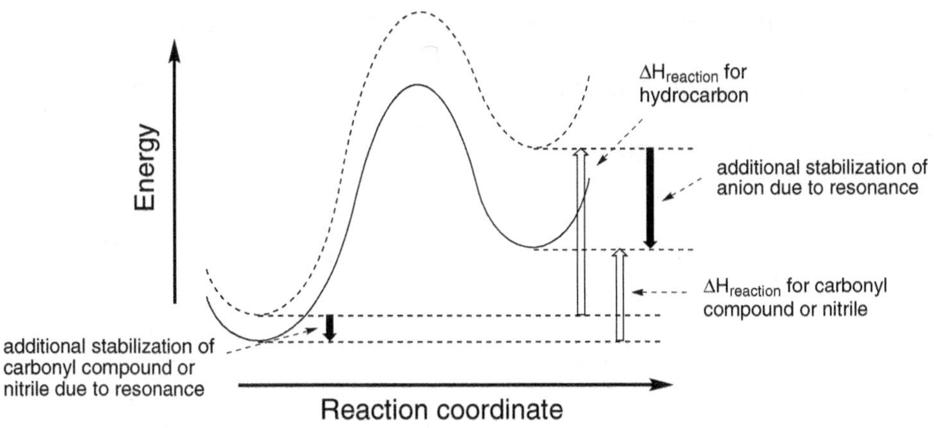

Figure 11.5 Reaction coordinate diagrams for deprotonation of a hydrocarbon (dashed line) and a carbonyl compound (solid line)

with a C=C double bond and a C—O single bond. The X-ray crystal structures of several metal enolates have shown that this is the case (Figure 11.5 gives the most widely cited example). The X-ray structures have also shown that the lithium enolates are aggregates of two or more enolate anions with their lithium counterions. The enolate oxygen atom in these aggregates is simultaneously coordinated to more than one lithium atom, and each lithium is coordinated by four oxygens. For example, in the lithium enolate in Figure 11.6, the enolate is tetrameric based on a Li_4O_4 scaffold, and each lithium atom is coordinated to three enolate oxygens and one oxygen atom from the solvent.[82]

Enolate anions can exhibit geometric (E/Z) isomerism like all other alkenes. For example, if we take the enolate anion derived from 3-pentanone, there are two isomeric enolates possible. Just like any other alkene, the stereochemistry of the enolate anions can be assigned as either *E* or *Z* by using the Cahn-Ingold-Prelog sequence rules. This can, however, pose problems when one is comparing the stereochemical outcome from reactions of geometrically identical enolates with differing metals, as shown by the two ester enolates in Example **11.62**. Although the spatial disposition of the four groups around the enolate double bond is the same in terms of the chemical nature of the groups, its *E/Z* configuration changes with the identity of the metal bonded to oxygen. To circumvent this problem, the *E/Z* system for enolates is modified to give the atom bound to the metal the higher priority, regardless of what the other substituent at this position is (Example **11.63**). This

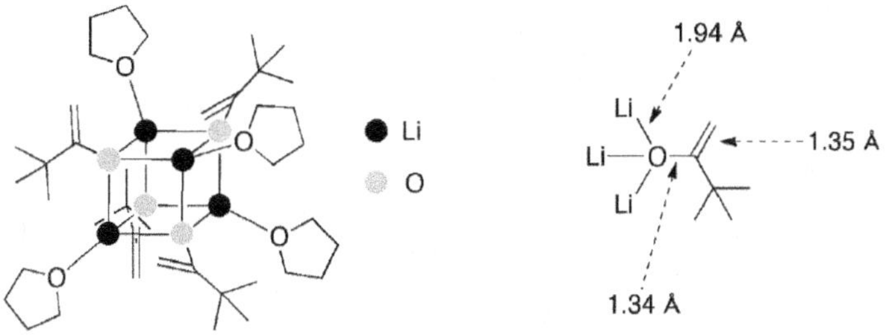

Figure 11.6 The structure of the lithium enolate of 3,3-dimethyl-2-butanone (pinacolone) shows clearly that (1) the carbon-carbon bond is a double bond (a carbon-carbon single bond would be 1.54 Å long) and that (2) the carbon-oxygen bond is a single bond (a double bond would be 1.21 Å long).

82. (a) Seebach, D. *Angew. Chem. Int. Ed. Engl.* **1988**, *27*, 1624. (b) Williard, P.G.; Liu, Q.-Y. *J. Am. Chem. Soc.* **1993**, *115*, 3380.

allows the reaction stereochemistry of enolates of ketones, esters, amides, and thioesters, for example, to be compared directly. Enolate configurations assigned in this way are designated *E(O)* and *Z(O)* configurations.[83]

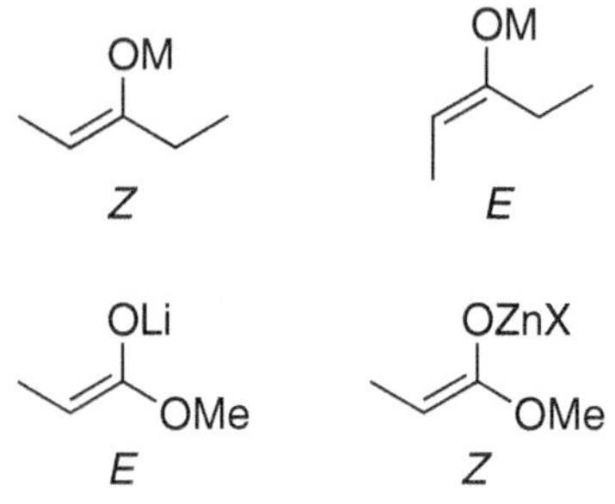

(11.62)

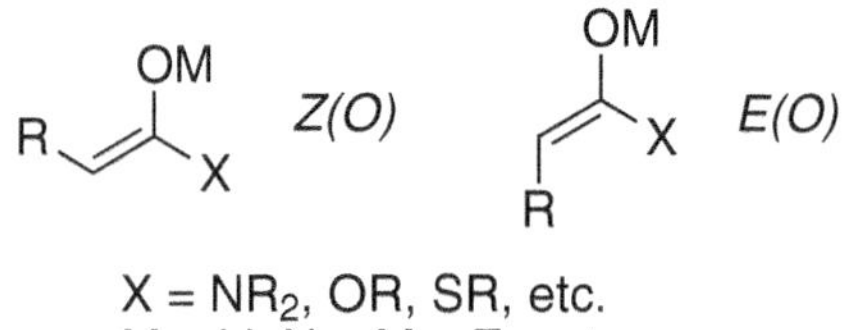

(11.63)

X = NR_2, OR, SR, etc.
M = Li, Na, Mg, Zn, etc.

The generation of a specific enolate anion involves two separate tasks: controlling the regiochemistry of enolate formation (in ketones) and controlling the stereochemistry of the enolate double bond. Both of these enolate properties—regiochemistry and stereochemistry—have important consequences in their reactions, so it is important to have methods that can reproducibly generate the specific enolate anion of choice.

Controlling Regiochemistry During Formation of Ketone Enolates

Aldehydes, nitriles, and derivatives of carboxylic acids (e.g., Example **11.64**) have only one α position and are therefore unable to form more than one regioisomeric enolate. However, ketones with hydrogen atoms at the α and α' positions are capable of forming two regioisomeric enolates. In the case of unsymmetrical ketones, which enolate is formed is decided by the reaction conditions—whether the ketone is deprotonated under ***kinetic*** or ***thermodynamic control*** (Example **11.65**).

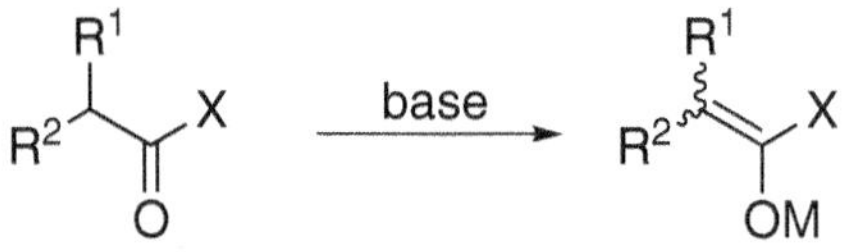

(11.64)

X=OR, SR, NR_2, Ar, etc.

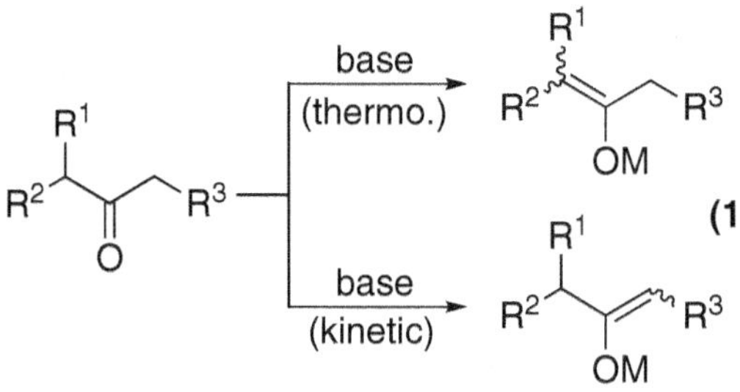

(11.65)

The stability of an alkene increases as the substitution of the carbon-carbon double bond increases. The same is true for enolates, and generating a metal enolate under

83. Masamune, S.; Choy, W. *Aldrichimica Acta* **1982**, *15*, 47.

equilibrium conditions will lead to the formation of the more substituted enolate as the major product. However, the kinetic acidity of most α hydrogens decreases in the order $CH_3 > CH_2 > CH$, so that methyl ketones are almost invariably deprotonated at the methyl group to give the less substituted enolate under conditions of kinetic control.[84] In similar fashion, the kinetic deprotonation of an α,β-unsaturated ketone (Example **11.66**) occurs preferentially at the sp^3-hybridized α' carbon[85] rather than at the γ carbon, despite the latter reaction giving the more stable enolate[86] (the thermodynamic enolate of an α,β-unsaturated carbonyl compound is, indeed, formed by deprotonation at the γ carbon).

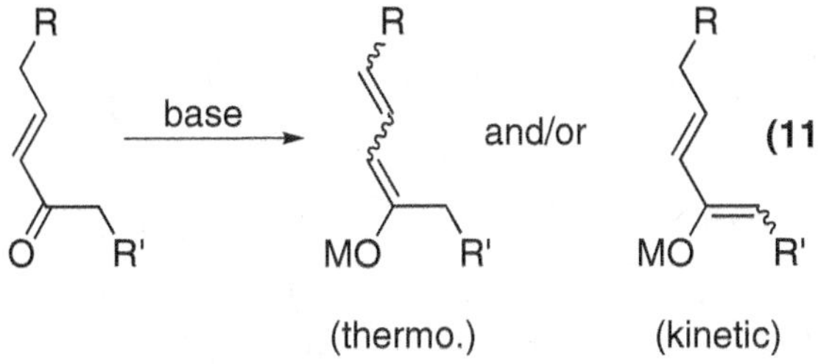

(11.66)

There are four experimental parameters that can affect the regiochemistry of enolate anion formation over which the chemist has control: the counterion, the solvent, the reaction temperature, and the reaction time. Because enolate anions undergo equilibration at room temperature and equilibration can occur over time even at low temperature, kinetic deprotonation is almost always carried out at low temperature and over short reaction times. The presence of a protic solvent provides a means for the enolate anion to equilibrate, so thermodynamic control of enolate regiochemistry is often provided by adding a proton source such as *tert*-butyl alcohol to the deprotonation reaction, as in Example **11.67**. Kinetic deprotonation is usually carried out by treating the carbonyl compound with a strong, hindered base containing a coordinating cation (e.g., Li^+), in an aprotic solvent and at low temperature. The bases that are most widely used are amide anion bases, shown as their lithium ("L") forms in Example **11.68**. The bases shown are lithium diisopropylamide (LDA),[87] lithium 2,2,6,6-tetramethylpiperidide (LTMP),[88] lithium isopropylcyclohexylamide (LICA), and lithium (or sodium or potassium) hexamethyldisilazides (LHDS or LHMDS; and thus NaHDS or NaHMDS, and KHDS or KHMDS).[89] These bases exist in solution as aggregates (often dimers),[90] so they are even more hindered than their monomeric structures would suggest. All these bases can be formed by the reaction between the amine and butyllithium or a metal hydride, and most are commercially available. All are soluble in organic solvents such as THF, which is a distinct advantage over strong bases such as potassium hydride and sodium amide, which are not, because it allows the

84. (a) House, H.O.; Trost, B.M. *J. Org. Chem.* **1965**, *30*, 1341. (b) House, H.O.; Czuba, L.J.; Gall, M.; Olmstead,H.D. *J. Org. Chem.* **1969**, *34*, 2324. (c) House, H.O.; Gall, M.; Olmstead, H.D. *J. Org. Chem.* **1971**, *36*, 2361. (d) Vedejs, E. *J. Am. Chem. Soc.* **1974**, *96*, 5944. (e) Reich, H.J.; Renga, J.M.; Reich, I.L. *J. Am. Chem.Soc.* **1975**, *97*, 5434. (f) Clark, R.D.; Heathcock, C.H. *J. Org. Chem.* **1976**, *41*, 1396. (g) Brown, C.A. *J. Org. Chem.* **1974**, *39*, 3913.

Reviews & monographs: (h) D'Angelo, J. *Tetrahedron* **1976**, *32*, 2979. (i) Stowell, J.C. *Carbanions in Organic Synthesis* (Wiley-Interscience: New York, 1979), pp. 127–216.

85. Lee, R.A.; McAndrews, C.; Patel, K.M.; Reusch, W. *Tetrahedron Lett.* **1973**, 965.

86. Büchi, G.; Wuest, H. *J. Am. Chem. Soc.* **1974**, *96*, 7573.

87. (a) Hammell, M.; Levine, R. *J. Org. Chem.* **1950**, *15*, 162. (b) Albarella, J.P. *J. Org. Chem.* **1977**, *42*, 2009. (c) Sasson, I.; Labovitz, J. *J. Org. Chem.* **1975**, *40*, 3670.

88. Olofson, R.A.; Dougherty, C.M. *J. Am. Chem. Soc.* **1973**, *95*, 581.

89. Amonoo-Neizer, E.H.; Shaw, R.A.; Skokvlin, D.O.; Smith, B.C. *J. Chem. Soc.* **1965**, 2997.

90. LHDS: Lucht, B.L.; Collum, D.B. *Acc. Chem. Res.* **1999**, *32*, 1035.

LDA: (a) Jackman, L.M.; DeBrosse, C.W. *J. Am. Chem. Soc.* **1983**, *105*, 4177. (b) Bauer, W.; Seebach, D. *Helv. Chim. Acta* **1984**, *67*, 1972. (c) Williard, P.G.; Salvino, J.M. *J. Org. Chem.* **1993**, *58*, 1. (d) Galiano-Roth, A.S.; Collum, D.B. *J. Am. Chem. Soc.* **1989**, *111*, 6772.

LICA: (a) DePue, J.S.; Collum, D.B. *J. Am. Chem. Soc.* **1988**, *110*, 5518. (b) DePue, J.S.; Collum, D.B. *J. Am. Chem. Soc.* **1988**, *110*, 5524.

deprotonation reaction to occur in solution instead of just at the surface of the undissolved solid. (Despite their base strength, the hydride bases give the thermodynamic enolate due to rapid proton transfer between the enolate and the excess carbonyl compound during the kinetically slow deprotonation step.) All of these amide bases will deprotonate a carbonyl compound or nitrile at temperatures well below 0°C.

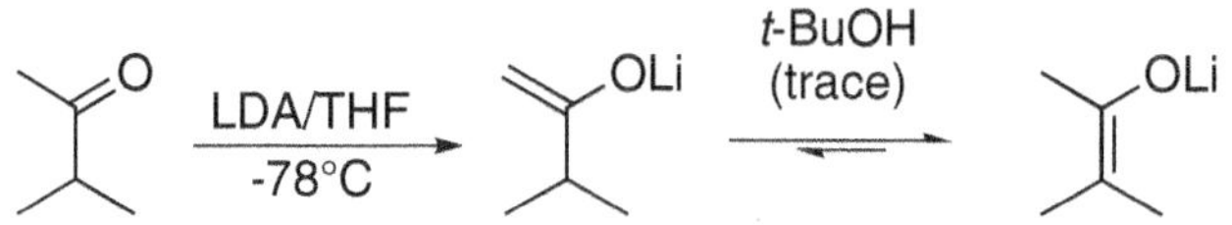

(11.67)

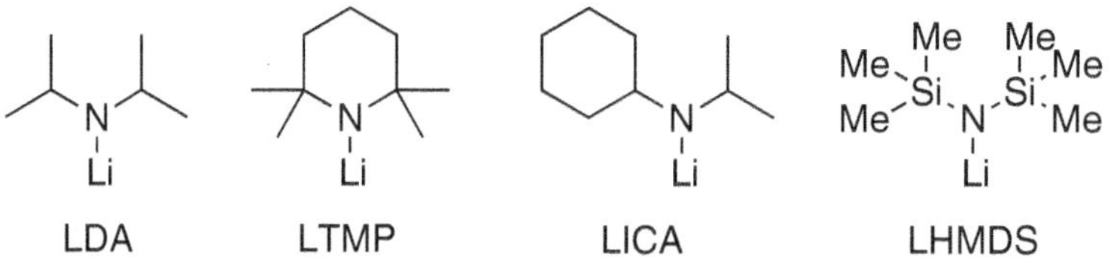

(11.68)

The regiochemistry of the enolate anion is also influenced by the conformation of the α hydrogen. Before the hydrogen atom can be abstracted by the base, the σ orbital of the C—H bond must be coplanar with one lobe of the π* orbital of the carbonyl group. In other words, the carbon-hydrogen bond and the carbonyl group must be in an orientation that allows the most effective hyperconjugation. In cyclohexanones, for example, the axial α hydrogens are removed much faster than the corresponding equatorial hydrogens (Figure 11.7).

Deprotonation of Conjugated Systems

The deprotonation of α,β-unsaturated ketones such as 2-cyclohexenone can occur at two sites: the α' carbon and the γ carbon (Figure 11.8). The deprotonation at the α' carbon leads to a cross conjugated enolate anion, which is a 2-substituted-1,3-butadiene. This is a process that is favored under kinetic control.[91] The deprotonation at the γ carbon is slower, but it gives an enolate anion that has linear conjugation, a 1-substituted-1,3-butadiene. This dienolate anion is more stable (lower energy) than the cross-conjugated isomer. Formation of this enolate anion is favored under thermodynamic control.[92] Interestingly, this preference for kinetic deprotonation at the α' carbon is still present, even when the α' carbon already carries an alkyl substituent and the γ carbon does not.

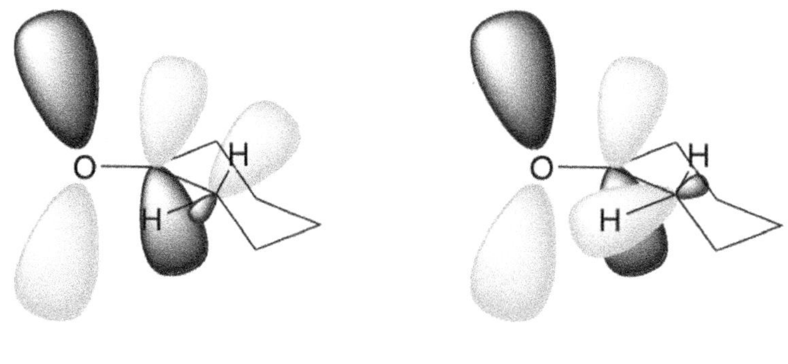

Figure 11.7 The axial hydrogens of cyclohexanones are more acidic than the equatorial hydrogens due to more efficient overlap of the C-H σ orbital with the π* orbital of the carbonyl group.

91. (a) Grieco, P.A.; Ferrino, S.; Oguri, T. *J. Org. Chem.* **1979**, *44*, 2593. (b) Stork, G.; Danheiser, R.L. *J. Org. Chem.* **1973**, *38*, 1775.

92. Lee, R.A.; McAndrews, C.; Patel, K.M.; Reusch, W. *Tetrahedron Lett.* **1973**, 965.

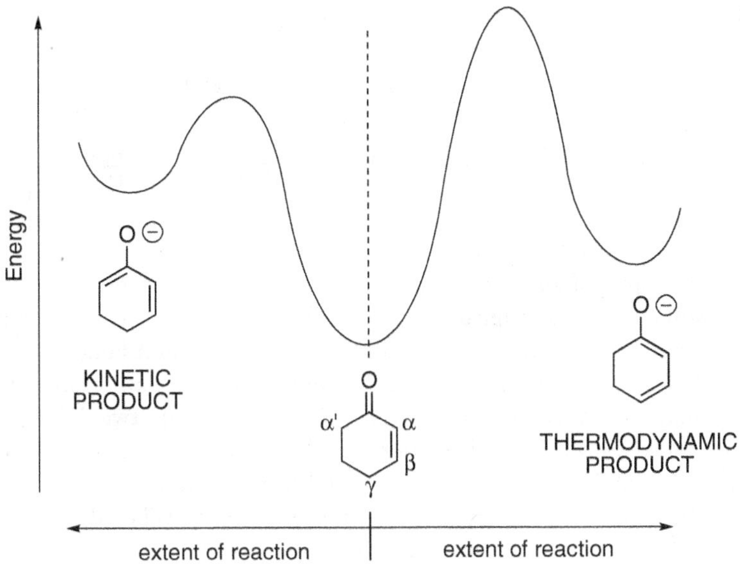

Figure 11.8 The deprotonation of 2-cyclohexenone under conditions of kinetic and thermodynamic control

Problems

11-4 Draw the geometric isomers of the major contributor to the resonance hybrid of the conjugate base of each of the following compounds. Label each isomer either *E* or *Z* according to the Cahn-Ingold-Prelog sequence rules. For ester and amide enolates, use the *Z(O)* and *E(O)* variant of this system.

(a) 3-pentanone

(b) propanal

(c) 2,2-dimethyl-3-hexanone

(d) 2-methylpentanal

(e) ethyl butanoate

(f) *N*,*N*-dimethylbutanamide

(g) *N*-propanoylpyrrolidine

(h) cyclohexyl propanoate

11-5 What is the structure of the enolate anion expected to be major product of each of the following reactions? Indicate whether each is formed under kinetic or thermodynamic control.

(a) 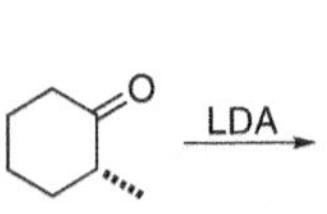

(b) 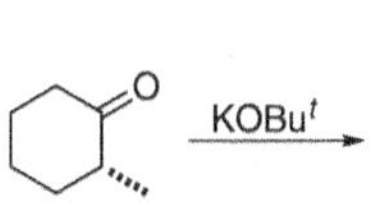

(c) 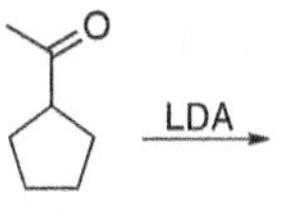

(d) 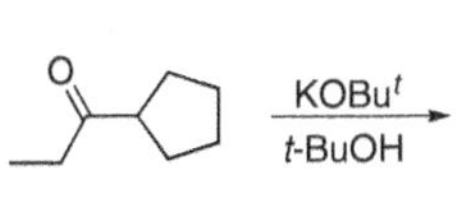

(e) 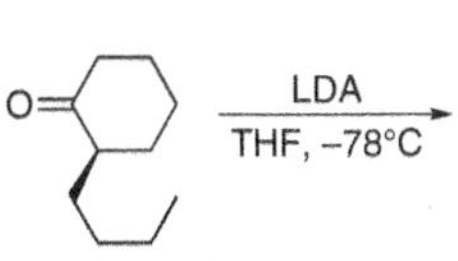

(f) 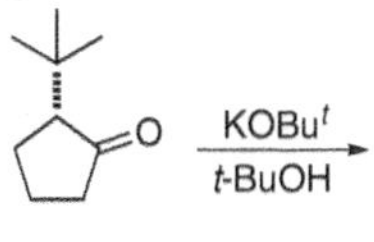

(g) 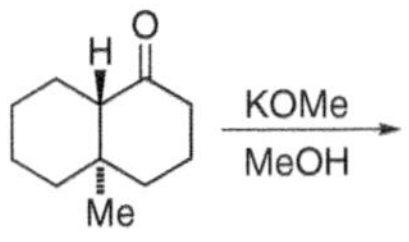

(h) 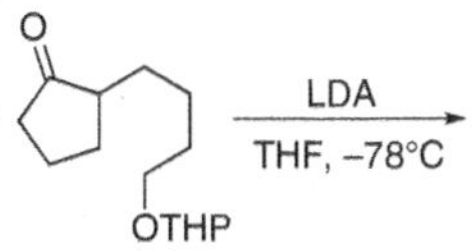

(i) 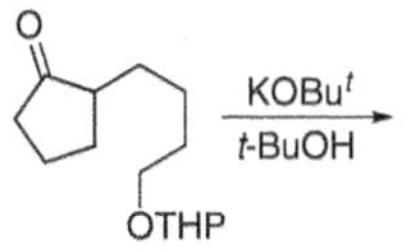

(j) 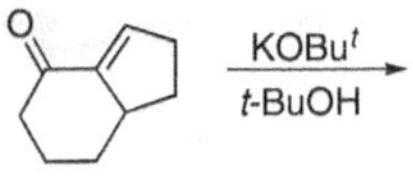

(k) 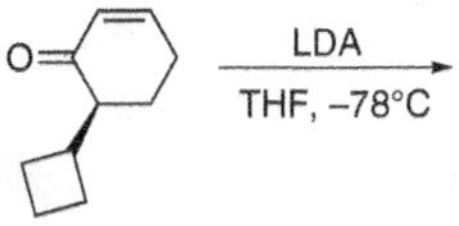

(l) 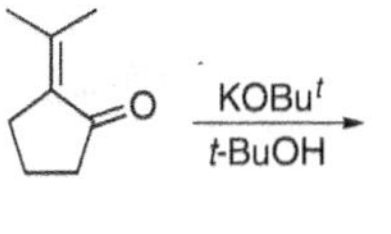

(continues)

(**Problems** continued)

(m) 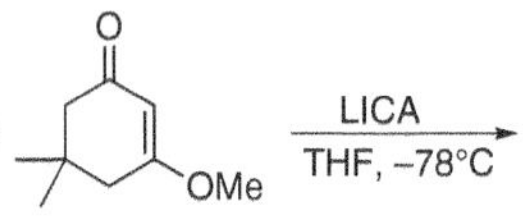

(n) 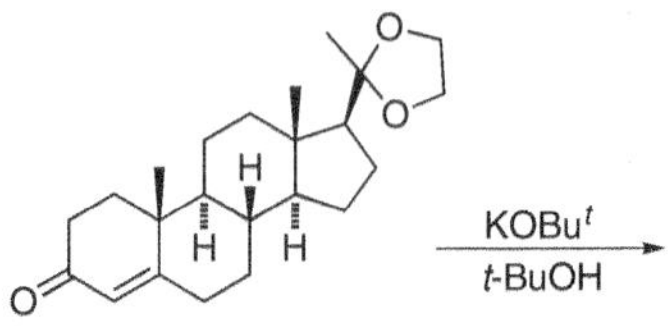

The π molecular orbitals of a ketone enolate anion are shown in Figure 11.9. In this anion, the lowest energy unoccupied molecular orbital is the antibonding π orbital ψ_3 and the highest energy occupied molecular orbital (HOMO) is the non-bonding π molecular orbital ψ_2. Because the terminal atoms of the HOMO of an enolate anion are not the same, electrophiles may react at carbon or at oxygen. Nucleophiles with two different nucleophilic sites are called ***ambident nucleophiles***, from the Latin words *ambo* (both) and *dens* (tooth)—the nucleophile can "bite" in two places. The reaction between an enolate anion and a proton illustrates the ambident nature of the nucleophile. When the proton reacts at the carbon atom, the product possesses a carbonyl group—we call this product the ***keto*** form of the compound. When the proton reacts at the oxygen atom, the product possesses an olefinic double bond bearing a hydroxyl group—we call this isomer the ***enol*** form of the compound (hence the term ***enolate anion***—the conjugate base of an enol). In general, protonation at oxygen is the kinetically favored process.

The keto and enol forms of carbonyl compounds are isomers that are in equilibrium (e.g., Example **11.69**) and are related to each other by interchanging the positions of a hydrogen atom and a π bond. Such isomers are known as ***tautomers***, and the equilibrium reaction interconverting them is known as ***tautomerization***. In the keto tautomer, the hydrogen is bonded to the α carbon, and the π bond is between carbon and oxygen. In the enol tautomer, the hydrogen is bonded to oxygen, and the π bond is between the two carbon atoms. The interconversion of tautomers is catalyzed by either acid or base.

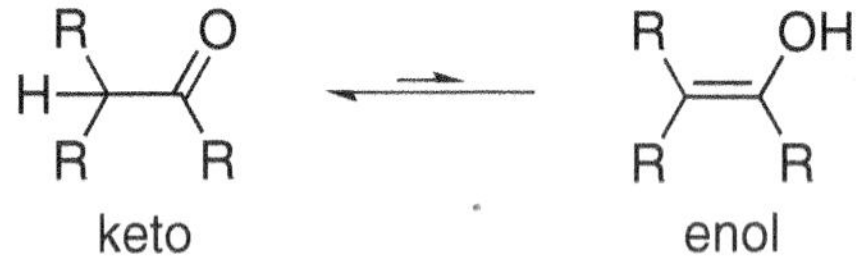

(11.69)

Deuterium Exchange and Epimerization

Enolate anions are powerful nucleophiles, and they readily participate in typical reactions of nucleophiles, especially S_N2 reactions and nucleophilic additions. Because of its ambident nature, an enolate anion can react with an electrophile at either oxygen or carbon; which site is more reactive depends on the electrophilic reagent—hard or soft. The oxyanion contributor

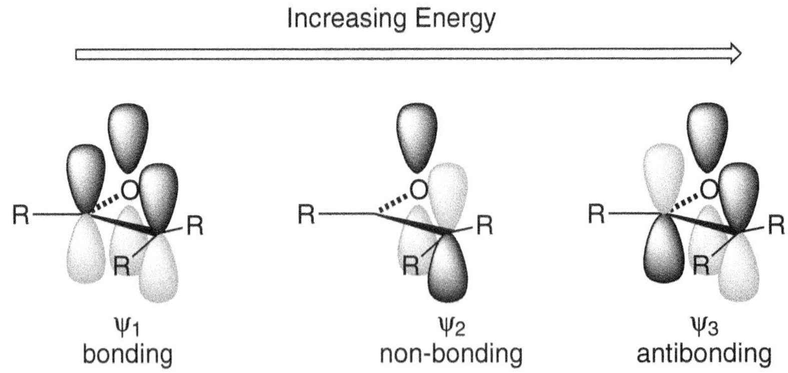

Figure 11.9 The π molecular orbitals of a ketone enolate anion. There are two different non-bonding orbitals that may function as the highest energy occupied molecular orbital: the lone pairs on oxygen and the non-bonding electron pair in the π molecular orbital ψ_2.

to the resonance hybrid is reasonably viewed as a hard base or hard nucleophile, and hard electrophiles tend to react with enolate anions at oxygen. Protonation of an enolate anion occurs most rapidly at oxygen to give the enol (Example **11.71**): the hard electrophile (the proton) reacts most rapidly at the hard location of the enolate (the oxygen atom). Despite protonation at oxygen being kinetically favored over protonation at carbon, the keto-enol equilibrium in simple aldehydes, ketones, nitriles, and esters strongly favors the keto tautomer (Example **11.70**), and the enol content of these compounds is quite small (Table 11.8). The keto tautomer does not dominate the equilibrium mixture for all carbonyl compounds, however. In compounds where there are two carbonyl groups bonded to the same carbon atom (β-dicarbonyl compounds), the amount of enol present at equilibrium can be quite substantial. Note that the enol content of aldehydes tends to be higher than the enol content of structurally similar ketones.

attack at carbon (slower)

keto **(11.70)**

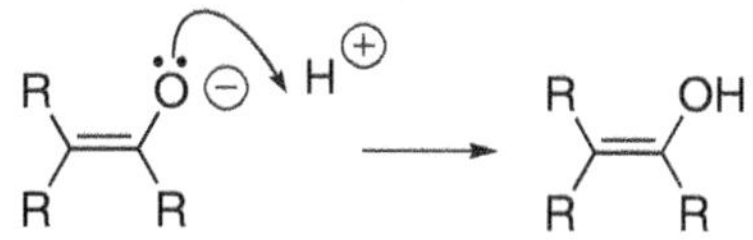
enol **(11.71)**

attack at oxygen (faster)

Table 11.8 Enol Content of Some Carbonyl Compounds in Protic Solvents*

Carbonyl Compound	% Enol at 25°C	Carbonyl Compound	% Enol at 25°C
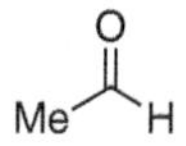	5×10^{-4}	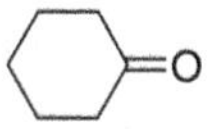	2×10^{-4} 3.9×10^{-5}
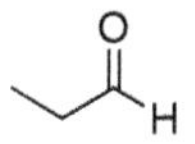	1.2×10^{-2}	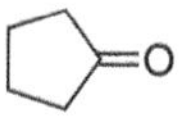	6.3×10^{-6} 1.7×10^{-6}
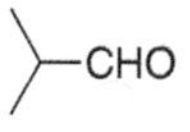	1.4×10^{-2}	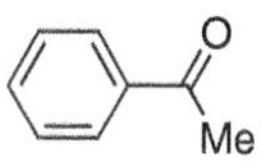	2×10^{-5} 1.2×10^{-6}
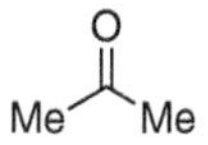	6.3×10^{-6} 6×10^{-7}	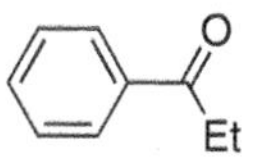	2×10^{-5}
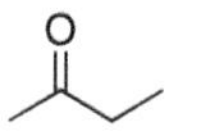	3.6×10^{-6} 5×10^{-7}	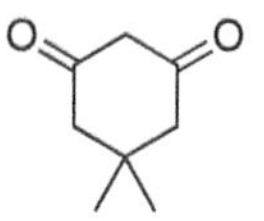	≈ 95
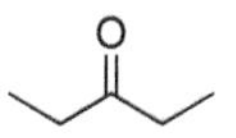	3.7×10^{-6} 1.6×10^{-6}	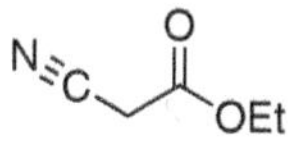	0.25
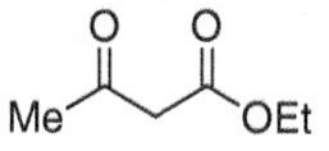	18.7	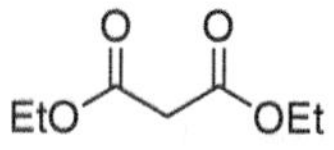	0.16

*The enol content of simple esters and nitriles is usually too low to measure (i.e., < 10^{-7}%).

Recall that diastereoisomers that are identical except for the configuration of one chiral center are called ***epimers***, and the interconversion of such stereoisomers by the inversion of a single chiral center is called ***epimerization***. The treatment of a carbonyl compound or a nitrile with a chiral α carbon leads to loss of stereochemical integrity at that position—epimerization occurs (e.g., Example **11.72**). When the carbonyl compound or nitrile is treated with deuterium oxide or a deuterated alcohol in the presence of a catalytic amount of base (e.g., Example **11.73**), the α hydrogens are gradually replaced by deuterium. When the α carbon is chiral, as in (*S*)-2-methyl-1-phenyl-1-butanone, the product is racemic. When there is more than one chiral center in the molecule undergoing the hydrogen exchange reaction, a mixture of stereoisomers is often obtained, with the most stable stereoisomer being the major one isolated.

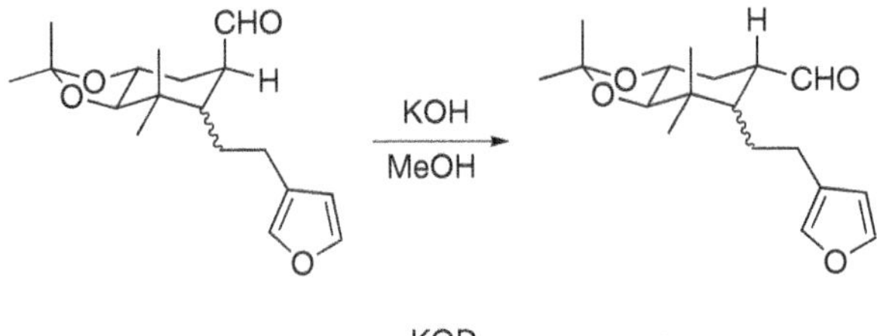

(11.72)

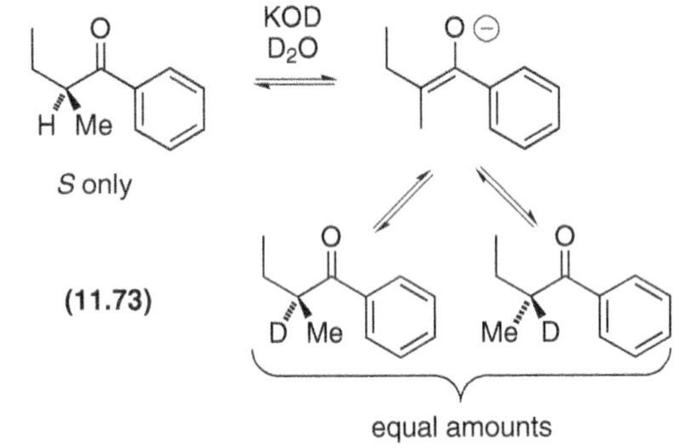

(11.73)

Exchange in conjugated ketones is rather more complex. When we examine the contributors to the resonance hybrid of the two regioisomeric anions from a conjugated enone, we see that exchange can, in principle, occur at the α position, the α' position, and the γ position. In practice, we see exchange at all the possible positions in a conjugated system. The contributors to the resonance hybrid of the dienolate anions from an α,β-unsaturated carbonyl system are shown in Figure 11.10. Protonation of a dienolate anion may occur at any atom that bears a negative charge in one of the contributing structures.

Figure 11.10 Enolate anions in conjugated carbonyl compounds

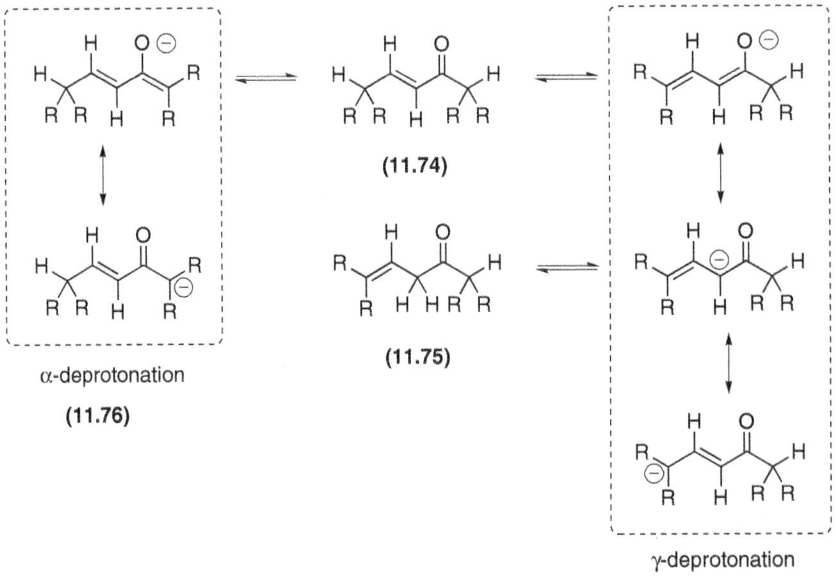

(11.74) **(11.75)** **(11.76)** **(11.77)**

An especially illustrative example of this is provided by the naturally occurring compound mitchelladione (Example **11.78**), in which fully ten hydrogens are replaced by deuterium after treatment with sodium ethoxide in deuterated ethanol (ethanol-*O*-*d*).[93]

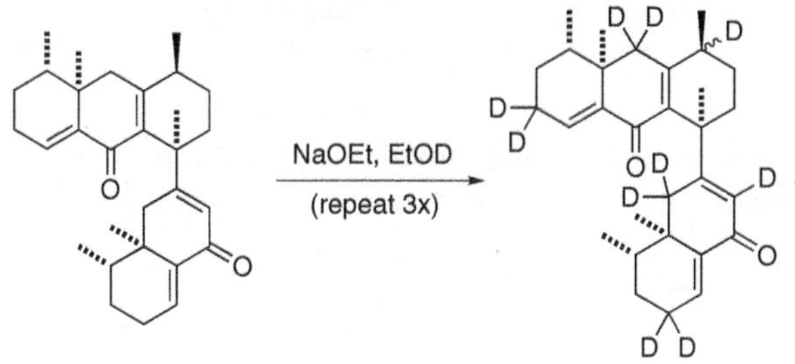

(11.78)

Problems

11-6 Draw the enolate anion that can be formed from each of the following compounds. Where more than one enolate may form, draw both and indicate which one is the more stable.

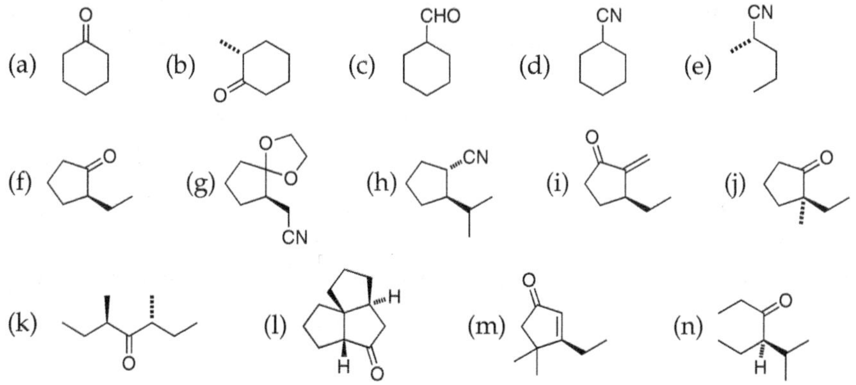

11-7 The enol content of the aldehydes MeCHO, EtCHO, and Me_2CHCHO increases from 5×10^{-4} to 0.16 at 25°C. Explain this trend.

Reaction Synopses

Formation of Enolate Anions

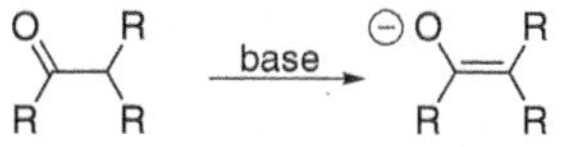

Base: KOBu^t, Bu^{t}OH; NaH, Me_2SO; NaH, THF; LDA, THF; LHMDS, THF; NaHMDS, THF; LiTMPP, THF; etc.

Kinetic deprotonation with strong amide bases at low temperature

Thermodynamic deprotonation with oxyanion bases and NaH, THF

Regiochemistry: kinetic deprotonation favors less substituted enolate; thermodynamic deprotonation favors more substituted enolate; kinetic deprotonation of α,β-unsaturated ketones occurs at α'-carbon; thermodynamic deprotonation of α,β-unsaturated ketones occurs at γ-carbon.

(continues)

93. Lewis, D.E.; Massy-Westropp, R.A.; Ingham, C.F.; Wells, R.J. *Aust. J. Chem.* **1982**, *35*, 809.

(**Reaction Synopses** continued)

Epimerization and α-Exchange

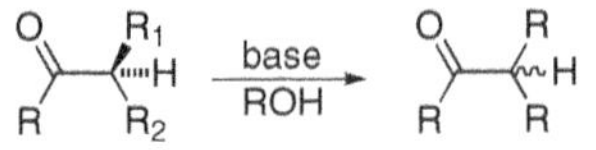

Base: $KOBu^t$, Bu^tOH; NaOEt, EtOH; K_2CO_3, MeOH; etc.
Stereochemistry: racemization of compounds with one chiral center; most stable diastereoisomer predominates in product mixture from compound with more than one chiral center.

11.5 Rearrangements of Carbanions

Rearrangements of carbanions are much less common than rearrangements of carbocations because carbanions lack an electron-deficient center to provide the driving force for rearrangements: every atom in a carbanion possesses a complete octet. Nevertheless, carbanions where the carbanion center is adjacent to a second-row heteroatom (especially oxygen and nitrogen) do undergo rearrangements. In these cases, the major driving force for the rearrangement appears to come from the transfer of the formal negative charge to the more electronegative element.

The two major classes of these rearrangements differ in mechanism and in the structural requirements of their substrates (see Example **11.79**). In [1,2]-rearrangements (which are structurally analogous to the rearrangement of carbocations), the only structural requirement is the heteroatom adjacent to the carbanion carbon. In [2,3]-rearrangements, however, one of the alkyl groups attached to the heteroatom is an allyl or benzyl group, and the rearrangement also involves the rearrangement of the π bond in the allyl group.

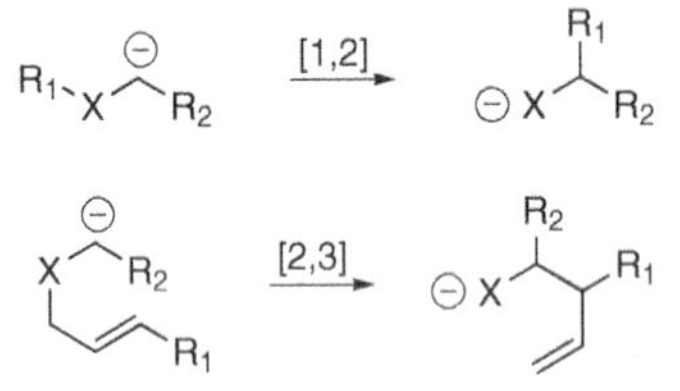

(11.79)

X=NR, O, S, NR_2^{+}, SR^{+}, etc.

By far the most widely studied of these reactions are the [1,2]- and [2,3]-Wittig rearrangements of ethers,[94] and the [2,3]-Wittig rearrangement of allyl ethers has become a useful reaction in stereoselective synthesis. Both reactions are initiated by the treatment of an ether with a strong base (typically an amide or alkyllithium base or a mixture of butyllithium and potassium *tert*-butoxide). At lower temperatures (below about –30°C), the [1,2]-rearrangement is slow, and one can use these strong bases in ether solvents. However, at higher temperatures (around 0°C), the [1,2]-Wittig rearrangement becomes competitive with the [2,3]-rearrangement.

The [1,2]-Wittig rearrangement (Example **11.80**) occurs through a radical pair mechanism,[95] constrained largely within the solvent cage. Support for this mechanism comes from a number of lines of evidence. One, the migratory aptitudes of substituents

94. Reviews, [1,2]-Wittig rearrangement: (a) Marshall, J.A. In Trost, B.M.; Fleming, I, Eds. *Compregensive Organic Synthesis* (Pergamon: London, 1991), vol. 3, p. 975. (b) Tomooka, K.; Yamamoto, H.; Nakai, T. *Liebigs Ann./Receuil* **1977**, 1275. Reviews, [2,3]-Wittig rearrangement: (c) Nakai, T.; Mikami, K. *Chem. Rev.* **1986**, *86*, 885. (d) Nakai, T.; Mikami, K. *Org. React.* **1994**, *46*, 105.

95. Review: Schöllkopf, U. *Angew. Chem. Int. Ed. Engl.* **1970**, *9*, 763.

(allyl ≈ benzyl > ethyl > methyl > phenyl)[96] reflect the stability of the free radicals, not of the carbanions.[97] Two, the migrating alkyl group largely retains its configuration.[98] Three, crossover products, which can only come from a dissociative mechanism, have been detected in the reaction.[99]

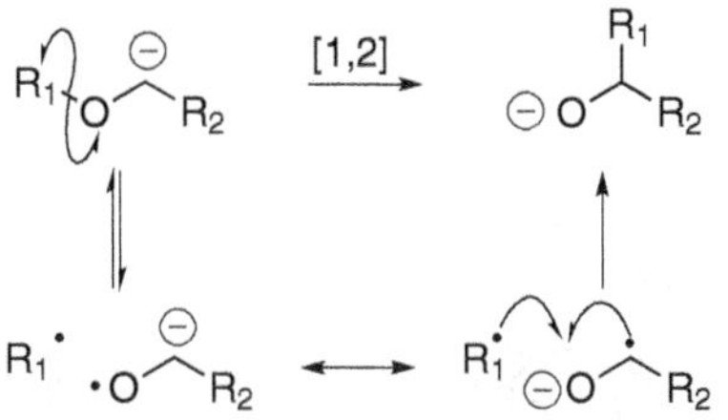

(11.80)

The analogous [1,2]-rearrangement of a nitrogen ylide (e.g., Example **11.81**) is known as the ***Stevens rearrangement.***[100] ^{1}H NMR evidence (chemically induced dynamic nuclear polarization [CIDNP], a technique used to determine whether radicals are involved in a reaction) indicates that it, too, proceeds through a radical pair mechanism.[101] There is also CIDNP evidence for the involvement of radicals in the Stevens rearrangement of sulfur ylides.[102]

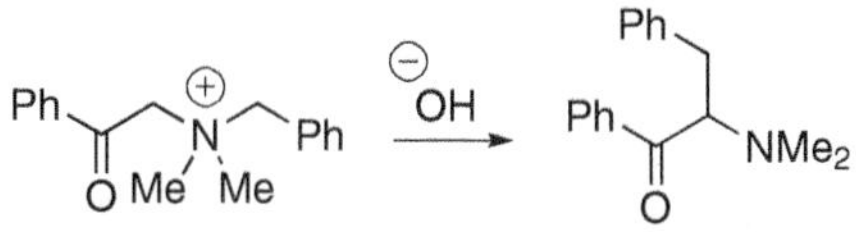

(11.81)

By contrast, the [2,3]-Wittig rearrangement of ethers (**11.82**) is a concerted rearrangement occurring through a cyclic transition state with concomitant rearrangement of a π bond—this is a [2,3]-sigmatropic rearrangement (Figure 11.11). The reaction is very stereoselective, and it is believed to occur through an envelope-type transition state with the large groups preferentially in the pseudoequatorial positions of the five-membered ring. The rearrangement is suprafacial with respect to the allyl group. Analogous sigmatropic rearrangements of allylic ammonium ylides (**11.83**) and allylic sulfur ylides (**11.84**) are also known. The [2,3]-sigmatropic rearrangements of benzyl ammonium ylides include the ***Sommelet*** (or ***Sommelet-Hauser***) rearrangement (**11.85**),[103] which provides a method for the synthesis of tertiary *o*-methylbenzylamines. Some examples of the use of these [2,3]-sigmatropic rearrangements in synthesis are gathered in Table 11.9.

---

96. (a) Wittig, G. *Angew. Chem.* **1954**, *66*, 10. (b) Slov'yanov, A.A.; Ahmed, E.A.A.; Beletskaya, I.P.; Reutov, O.A. *J. Chem. Soc., Chem. Commun.* **1987**, *23*, 1232.

97. Lansbury, P.T.; Pattison, V.A.; Sidler, J.D.; Bieber, J.B. *J. Amn. Chem. Soc.* **1966**, *88*, 78. (b) Schöfer, H.; Schöllkopf, U. *Tetrahedron Lett.* **1968**, 2809.

98. (a) Schöllkopf, U.; Schöfer, H. *Liebigs Ann. Chem.* **1963**, *663*, 62. (b) Felkin, H.; Frajerman, C. *Tetrahedron Lett.* **1977**, 3485. (c) Hebert, E.; Welvart, Z. *J. Chem. Soc., Chem. Commun.* **1980**, 1035. (d) Hebert, E.; Welvart, Z. *Nouv. J. Chim.* **1981**, *5*, 327.

99. (a) Lansbury, P.T.; Pattison, V.A. *J. Org. Chem.* **1962**, *27*, 1933. (b) Lansbury, P.T.; Pattison, V.A. *J. Am. Chem. Soc.* **1962**, *84*, 4295.

100. (a) Stevens, T.S.; Creighton, E.M.; Gordon, A.B.; MacNicol, M. *J. Chem. Soc.* **1928**, 3193. (b) Stevens, T.S. *J. Chem. Soc.* **1930**, 2107. (c) Stevens, T.S.; Snedden, W.W.; Stiller, E.T.; Thomson, T. *J. Chem. Soc.* **1930**, 2119. (d) Thomson, T.; Stevens, T.S. *J. Chem. Soc.* **1932**, 55. (e) Dunn, J.L.; Stevens, T.S. *J. Chem. Soc.* **1932**, 1926. (f) Thomson, T.; Stevens, T.S. *J. Chem. Soc.* **1932**, 1932. (f) Brewster, J.H.; Kline, M.W. *J. Am. Chem. Soc.* **1952**, *74*, 5179. (g) Fenney, E.F.; Druey, J. *Angew. Chem. Int. Ed. Engl.* **1962**, *1*, 155. (h) Review: Pine, S.H. *Org. React.* **1970**, *18*, 403.

101. (a) Schöllkopf, U.; Ludwig, U.; Ostermann, G.; Patsch, M. *Tetrahedron Lett.* **1969**, 3415. (b) Dolling, U.H.; Closs, G.L.; Cohen, A.H.; Ollis, W.D. *J. Chem. Soc., Chem. Commun.* **1975**, 545.

102. Iwamura, H.; Iwamura, M.; Nishida, T.; Yoshida, M.; Nakayama, J. *Tetrahedron Lett.* **1971**, 63.

103. (a) Sommelet, M. *Comptes Rend.* **1927**, *205*, 56. (b) Kantor, S.W.; Hauser, C.R. *J. Am. Chem. Soc.* **1951**, *73*, 4122. (c) Hauser, C.R.; van Eenam, D.N. *J. Am. Chem. Soc.* **1956**, *78*, 5698. (d) Bunnett, J.F. *Quart. Rev.* **1958**, *12*, 15. (e) Huisgen, R. *Angew. Chem.* **1960**, *72*, 3572.

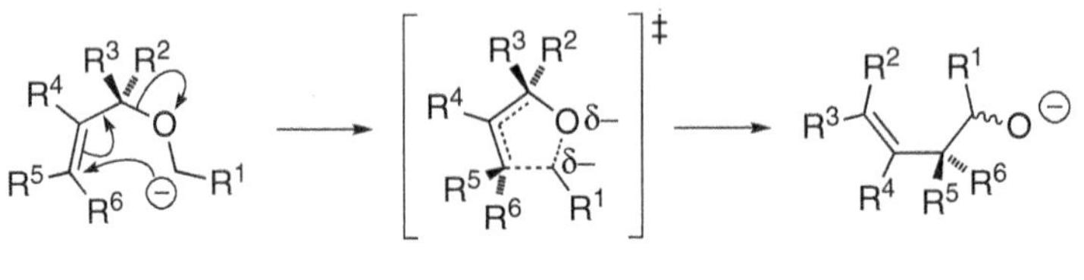

(11.82)

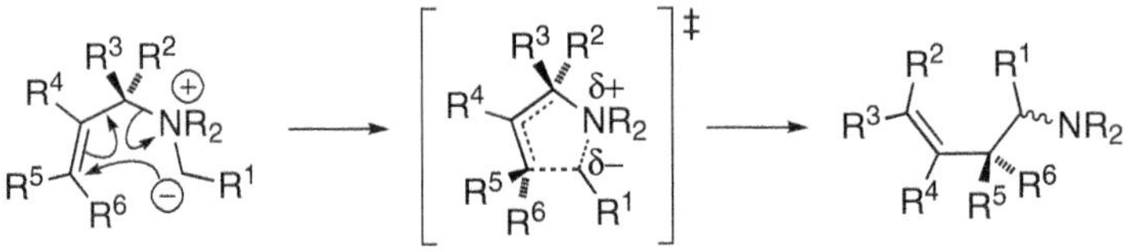

(11.83)

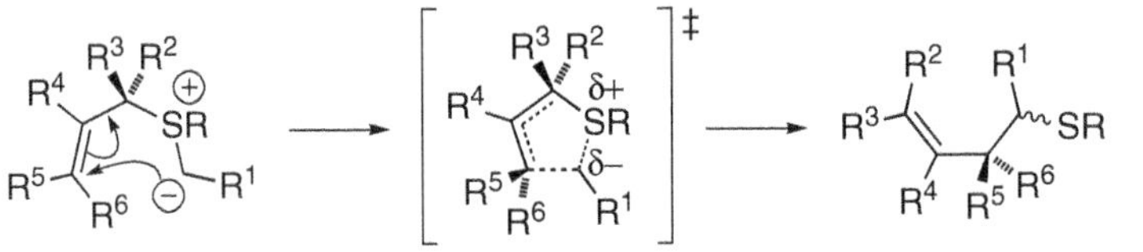

(11.84)

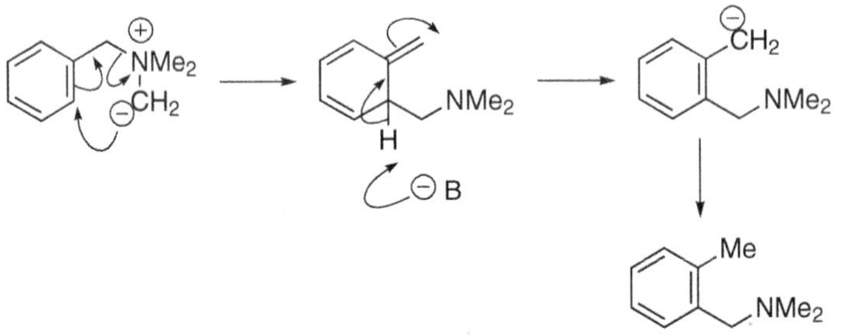

(11.85)

Figure 11.11 The variations on the theme of [2,3]-sigmatropic A of carbanions all occur through a five-membered transition state.

Table 11.9 Representative [2,3]-Sigmatropic Rearrangements in Synthesis

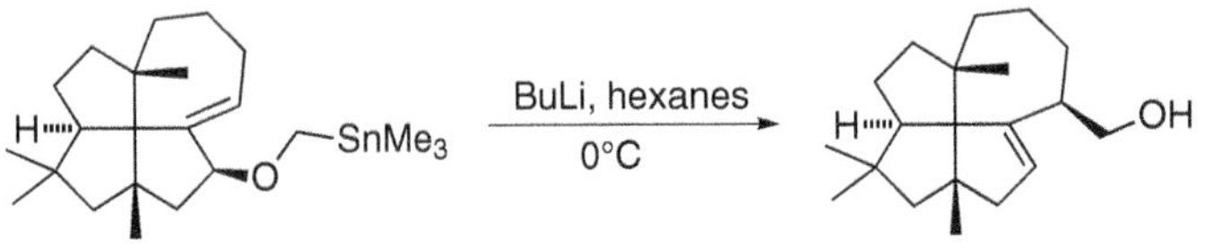

(11.86)[104]

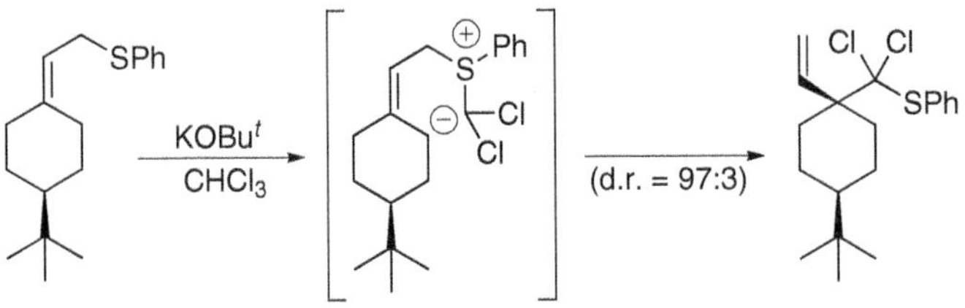

(11.87)[105]

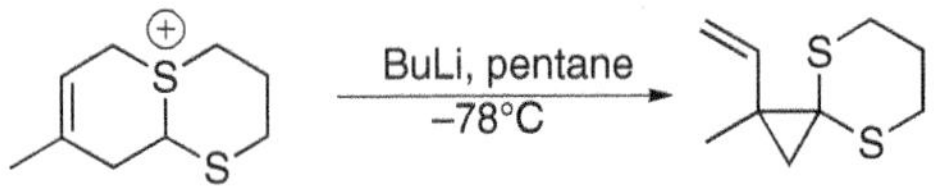

(11.88)[106]

104. Crimmins, M.T.; Gould, L.D. *J. Am. Chem. Soc.* **1987**, *109*, 6199.
105. Andrews, G.; Evans, D.A. *Tetrahedron Lett.* **1972**, 5121.
106. Corey, E.J.; Walinsky, S.W. *J. Am. Chem. Soc.* **1972**, *94*, 8932.

Reaction Synopses

[1,2]-Wittig Rearrangement

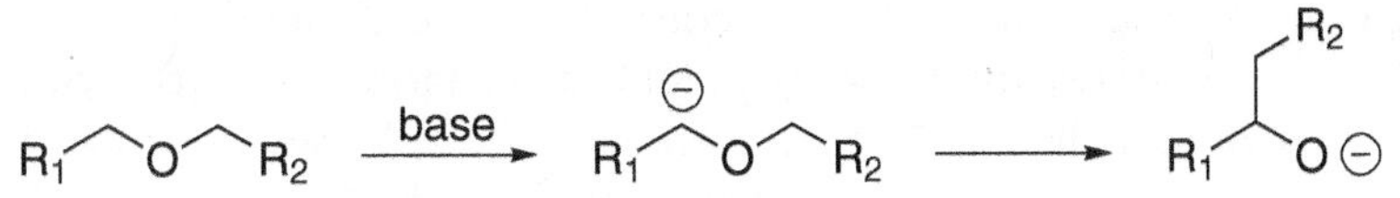

Base: BuLi; BuLi, KOBu^t; BuLi; TMEDA; etc.

Stereochemistry: mainly retention at migrating center; mainly inversion at terminal center

Stevens Rearrangement

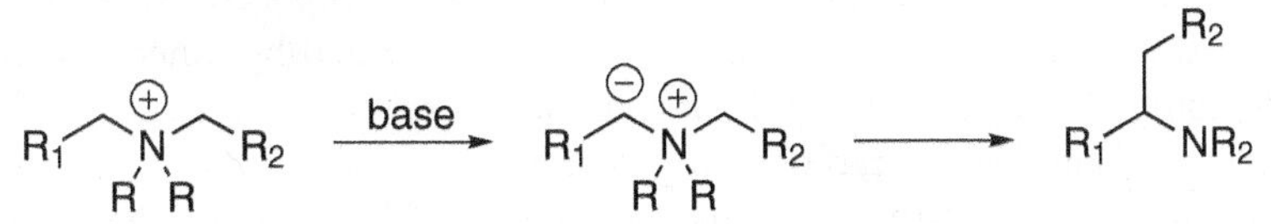

Base: Na_2CO_3; KOH; NaOEt; etc.

Reaction facilitated by R_1=acyl

[2,3]-Wittig Rearrangement

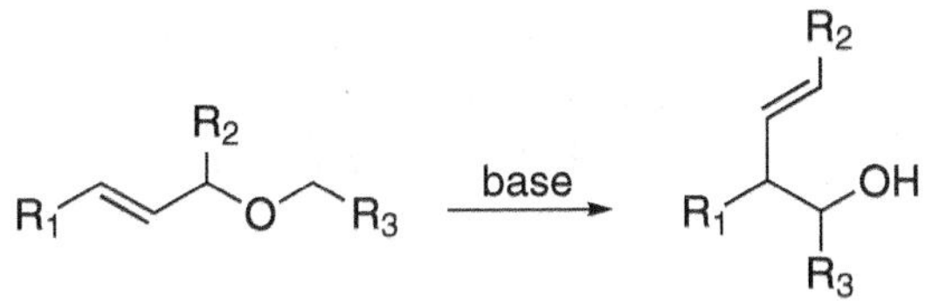

R_1=alkyl, phenyl; R_2=alkyl; R_3=alkyl, acyl, etc.

Base: KOEt, KOBu^t, etc. (R_3=alkyl); K_2CO_3, KOH, etc. (R_3=acyl)

Stereochemistry: product alkene is mainly *E*; reaction occurs with suprafacial stereochemistry.

Stevens Rearrangement

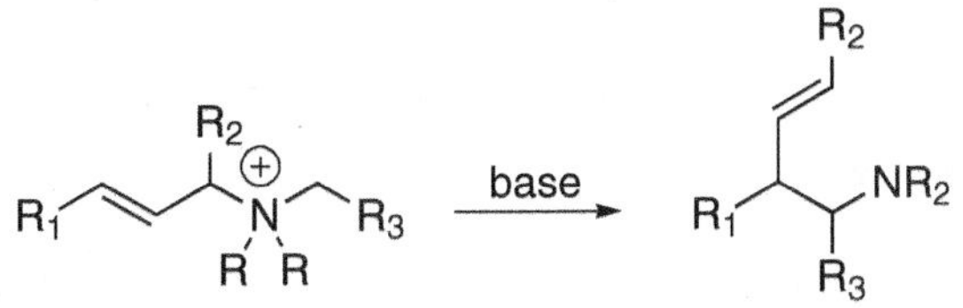

R_1=alkyl, phenyl; R_2=alkyl; R_3=alkyl, acyl, etc.

Base: KOEt, KOBu^t, etc. (R_3=alkyl); K_2CO_3, KOH, etc. (R_3=acyl)

Stereochemistry: product alkene is mainly *E*; reaction occurs with suprafacial stereochemistry.

Sommelet or Sommelet-Hauser Rearrangement

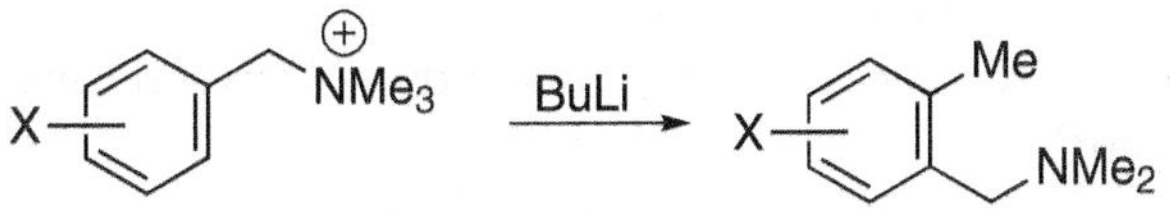

Chapter Summary

This chapter has focused on the carbanions. Free carbanions are actually rare, and the bonding in most metal alkyls is best described by means of multicenter, two-electron bonds. Metal alkyls of reactive metals (e.g., Mg) can be formed by direct synthesis from the metal and an alkyl halide; others are accessible by metathesis reactions between a metal alkyl of an electropositive element and either a metal halide or metal alkyl of the less electropositive metal. The nucleophilicity and base strength of the metal alkyl increases with increasing polarity of the C—M bond. The effect of substitution by alkyl groups of hydrogen on the carbanion carbon is approximately the reverse of that in carbocations. Delocalization of the charge by conjugation with an adjacent π bond is an important mechanism for carbanion stabilization, especially if the π bond is electron-deficient (e.g., a carbonyl, cyano or nitro group). These enolate anions are important synthetic intermediates and can exhibit stereoisomerism. They may be formed by kinetic or thermodynamic deprotonation of the carbonyl compound, and regioisomeric enolates can result from ketones. Epimerization can occur on deprotonation of a chiral α-carbon of a carbonyl compound. The carbanion can also be stabilized by an adjacent heteroatom from the lower rows of the main group (e.g., Si, P, S, Sn); carbanions stabilized by this mechanism are important nucleophiles. One subset of this type of carbanion comprises the phosphorus and sulfur ylides. Dithiane anions represent an *umpoled* carbonyl group and are equivalent to a carbonyl anion. Carbanions do not rearrange as readily as carbocations, but the conjugate bases of allylic ethers, amines, sulfonium ions and ammonium ions do (the Wittig, Stevens, or Sommelet-Hauser rearrangement). The Wittig rearrangement of ethers may be either [1,2] or [2,3]. The [1,2]-Wittig rearrangement proceeds through free radical intermediates, and the [2,3]-Wittig rearrangement is a pericyclic reaction.

Key Terms

alkynide anion
"ate" complexes
dimsyl anion
dithiane
Gilman reagent
Horner-Wadsworth-Emmons (HWE) reaction
Lipshutz cuprate
Peterson olefination
Sommelet-Hauser rearrangement
Stevens rearrangement
transmetalation
umpolung
Wittig reaction
Wittig rearrangement
ylide

Additional Problems

11-8 Consider the carbonyl compounds in Table 11.8. Rank these compounds in descending order of the expected rates of their exchange reactions with NaOD in D_2O. Is there any caveat when answering this question?

11-9 Alkyllithium solutions often contain lithium alkoxides and undissolved lithium hydride. The butyllithium content can be determined by titration using triphenylmethane as an indicator.

(a) What species is formed by the indicator? Why is it formed?
(b) What does this reaction tell you about the acidity of the hydrogen in the indicator?

11-10 The reaction of *cis*-2-bromo-1-methylcyclopropane with isopropyllithium in mixed ether-pentane at –70°C gives the corresponding 3-methylcyclopropyllithium. When this compound is trapped at –70°C with carbon dioxide, only *cis*-2-methylcyclopropanecarboxylic acid is obtained. When molecular bromine is used as the trapping agent, the stereochemistry of the product is scrambled.

(a) Are these results consistent with the cyclopropyllithium being configurationally stable at –70°C or not? Give your reasons.
(b) Provide a rationalization for the different stereochemical outcomes from using the two trapping reagents (CO_2 and Br_2).

11-11 In an experiment to generate and trap an optically active alkyllithium (–)-2-iodooctane was added over 20 minutes to a fourfold excess of 1.04 *M s*-butyllithium in petroleum ether (hexanes) and 25 mL of ether at –70°C, and the product formed was trapped with carbon dioxide. The product, 2-methyloctanoic acid, exhibited the following rotations:

Iodide Rotation	Temperature	Wait Before Quenching	Acid Rotation
–44.36°	–70°C	1 minute	–1.18°
–41.00°	–70°C	60 minutes	–0.96°
–39.86°	–70°C→E0°C	20 minutes	0°
–45.86°	–70°C	2 minutes	–0.78°*

*The iodide was added over 2 hours at –70°C.

If the absolute configurations of the acid and the halide are the same, what deductions can you make about the stereochemistry of the reactions involved?

11-12 Cyclopentadiene has a pK_a of 16, and triphenyl-methane has a pK_a of 31. The two compounds shown have structural features of both. Provide a rationalization for the pK_a values for cyclopentadiene, triphenylmethane, and the two compounds.

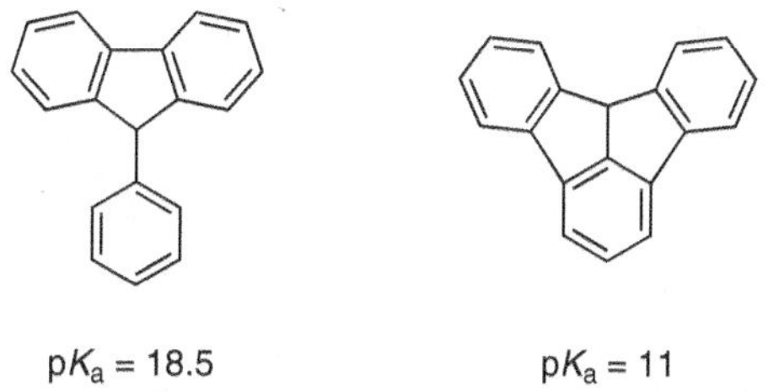

11-13 (a) Carbon-to-carbon rearrangements of carbanions are rare, but they have been observed, as exemplified by the reaction shown [*J. Am. Chem. Soc.* **1961**, *83*, 412, 2537]. When lithium is used as the metal, the rearrangement takes place only above 0°C. Provide a reasonable rationalization for the formation of the product in this reaction.

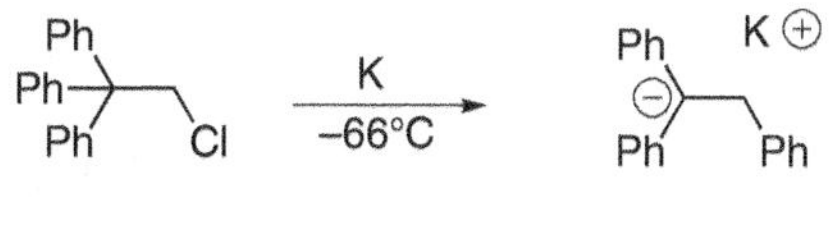

(b) When the corresponding anion is generated from the methylated analog shown, preferential migration of the phenyl group is observed. How does this modify your answer to part (a)?

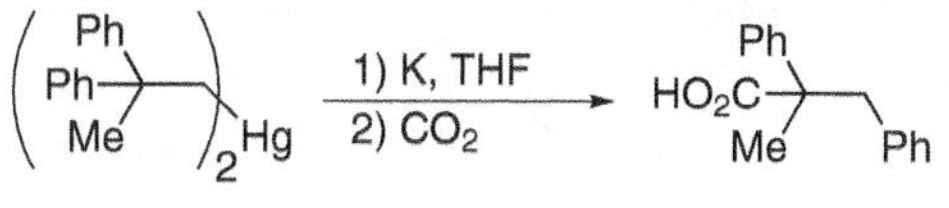

(c) The Grignard reagent, the alkyllithium, and the alkylpotassium with the same alkyl group as in part (b) show very different propensities to rearrange: the Grignard reagent does not rearrange, even in refluxing dioxane, the alkyllithium is stable at 0°C but rearranges at 40°C, and the alkylpotassium rearranges rapidly at room temperature. What does this tell you about the species actually rearranging? [*J. Am. Chem. Soc.* **1961**, *83*, 1196]

11-14 What is the final product of the sequence of reactions shown? Rationalize the regiochemistry and stereochemistry of the reaction. [*J. Org. Chem.*, **1973**, *38*, 2915].

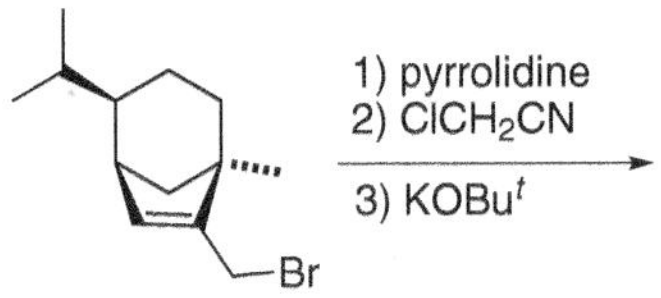

11-15 Sulfur ylides can be formed by the reaction between a sulfide and a diazoalkane, catalyzed by rhodium (II) acetate. Write a mechanism to account for the transformation pictured [*Org. Lett.* **2006**, *8*, 2511]. The product shown is the major product of the reaction.

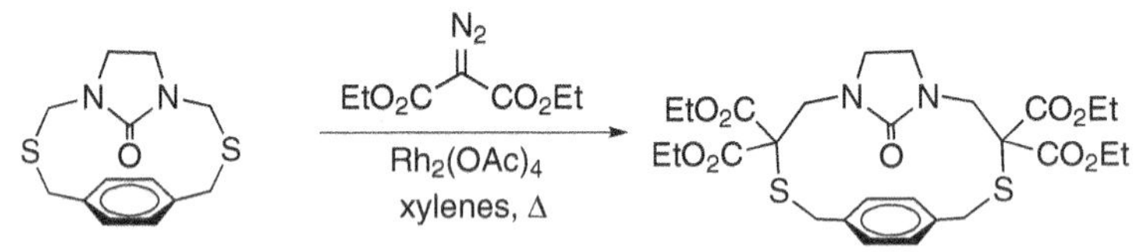

In the analogous reaction, shown below, the regiochemistry of the rearrangement changes, and the isomeric cyclophane is formed [*Org. Lett.* **2000**, *2*, 3785]. Provide a reasonable explanation for the change in regiochemistry caused by the structural changes in the starting compound.

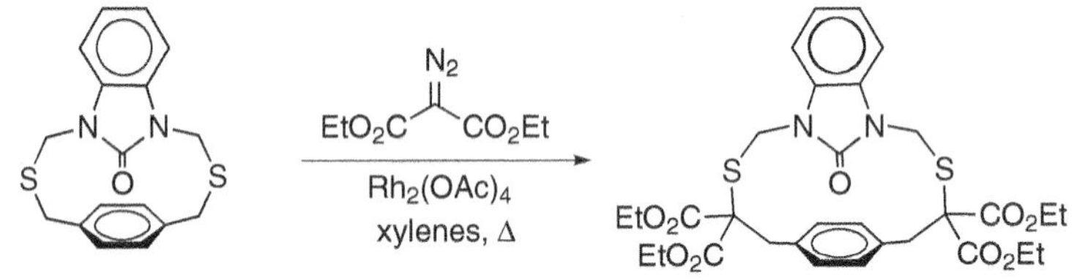

11-16 Identify which of the of the compounds in each pair shown will undergo H/D exchange more rapidly with a base such as KOBu^t, Bu^{t}OH? Give reasons for your answers.

(a) 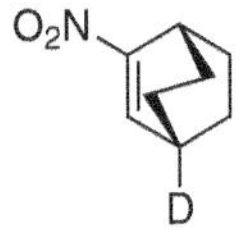 and 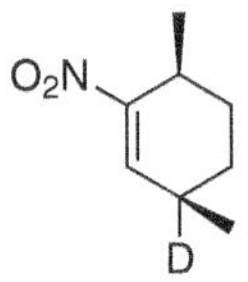

(b) 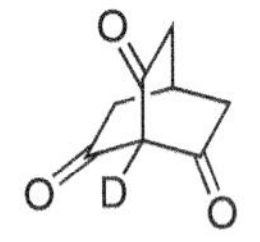 and 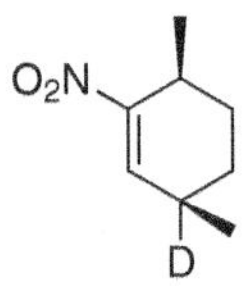

11-17 The bases typically favored for carrying out the Stevens rearrangement are strong amide bases such as $NaNH_2$, NH_3, LDA, THF rather than alkoxide bases such as potassium *tert*-butoxide or alkyllithium reagents. Why might this be the case? (Hint: think about the anion that will be formed.)

11-18 When an internal alkyne (e.g., 4-octyne) is treated with potassium *tert*-butoxide, it is largely isomerized to an allene. When the same internal alkyne is treated with KAPA, the product is the potassium salt of the terminal alkyne. Account for the difference in the outcomes of these two reactions.

11-19 The effective base strength of *n*-butyllithium is greatly enhanced if a mixture of potassium *tert*-butoxide and *n*-butyllithium in THF is used: the *n*-butyllithium behaves as a superbase. A similar enhancement of base strength in *n*-butyllithium is observed when *N, N'*-tetramethylethylenediamine (TMEDA) is added to the solution of the butyllithium. Suggest reasons why these effects are observed in these two different cases.

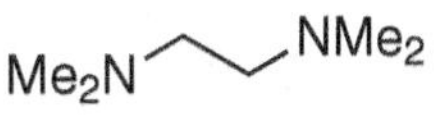

TMEDA

11-20 The control of regiochemistry of reactions of allylic anions is often problematic. The transformation shown is a tandem Michael addition-ring closure that involves the sulfone anion as the initiating nucleophile. What does the fact that the product shown is the only regioisomer of the product obtained, tell you about the initiating nucleophile? [*Pure Appl. Chem.* **2000**, *72*, 1671.]

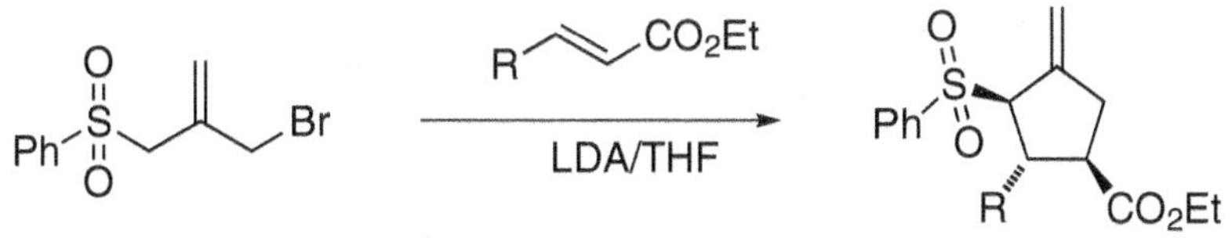

CHAPTER TWELVE

Organic Reactions III

Synthetic Reactions of Carbon Nucleophiles: Substitution and Addition

12.1 Carbon-Carbon Bond Formation: Carbon Nucleophiles and Electrophiles

Methods for the formation of carbon-carbon bonds are of pivotal importance in organic synthesis. In Chapter 6, we discussed the formation of carbon-carbon bonds using pericyclic reactions, and in Chapter 10, we discussed C—C bond-forming reactions involving carbocations as important intermediates. In this chapter, we will discuss the formation of C—C bonds by means of reactions involving carbon nucleophiles. Between them, reactions between a carbanion nucleophile and a carbon electrophile account for the vast majority of C—C bond-forming reactions in organic synthesis. These reactants can be classified as strong or weak, as summarized in Table 12.1.

Combinations of these reagents give rise to a multitude of synthetic organic reactions, as we have already seen for the reactions of carbocation electrophiles. It is important to remember that reactions between strong nucleophiles and strong electrophiles are usually precluded because the reaction conditions required to generate one of the reacting species are almost always incompatible with the other. It is also quite unusual—but not impossible—to find a productive reaction between a weak nucleophile and a weak electrophile. Such reactions often have an impossibly high activation barrier to overcome. The one common exception to this generalization involves cycloaddition reactions, especially of carbenes to alkenes. We will discuss this reaction further in Chapter 13. By far the most common bond-forming reactions of this type are between a strong nucleophile and a weak electrophile, or a weak nucleophile and a strong electrophile.

12.2 Substitution and Addition with Metal Alkyl Reagents Having C—M σ Bonds

Since the 1970s, no area of organic synthesis has expanded as rapidly as the preparation and use of organotransition metal compounds. These compounds now form the basis for a wide range of catalytic reactions, and they have revolutionized the formation of carbon-carbon bonds in compounds and the control of absolute configuration. Compounds of the Group VIIIB (Groups 8–10) metals (Fe, Co, Ni, Pd, Pt, Ru, Rh, and Ir, in particular) have become mainstays of modern organic synthesis. The preeminent position of these compounds in modern synthesis may be gauged by noting that two Nobel prizes in chemistry have been awarded to chemists who have developed methods for C—C bond formation using transition metal catalysts. This area is so important that we will devote Chapter 17 to a (too brief) introductory discussion of these reactions.

In Chapter 11, we discussed the metal alkyls, which occur in two different classes (Example **12.1**): simple metal alkyls, where the alkyl nucleophile is bonded to a metal bearing a formal positive charge, and "ate" complexes, where the alkyl group is attached to a

Table 12.1 Classifying Carbon Nucleophiles and Carbon Electrophiles

	Nucleophile	Electrophile
Strong	R−M R−≡−M metal alkyl, metal alkynide M = Li, Mg, Zn, Cu, etc. 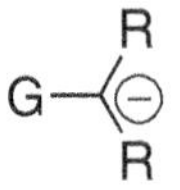 stabilized carbanion (including ylides) G = SiR_3, SR, $^{\oplus}SR_2$, $^{\oplus}PR_3$, etc. 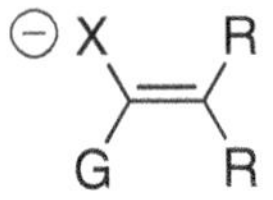 enolate anions and equivalents X = O, S, NR, =N G = R, OR, NR_2, etc.	$R-C\equiv X^{\oplus}$ acylium ions, nitrilium ions X = O, NR 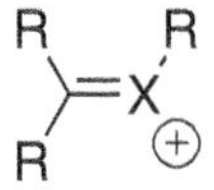 oxonium ions, iminium ions X = O, NR, S, etc. $R^{\oplus}$ $R-X^{\oplus}$ alkyl cation, Lewis acid complex X = $Cl\overset{\ominus}{Al}Cl_3$, $Cl\overset{\ominus}{Ti}Cl_4$, $Br\overset{\ominus}{Fe}Br_3$, etc.
Medium strength	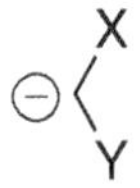 β-dicarbonyl anions, etc. X, Y = CHO, COR, CO_2R, CN, NO_2, etc.	$R-X^{\oplus}$ heteronium ions X = OR_2, NR_3, SR_2, etc. 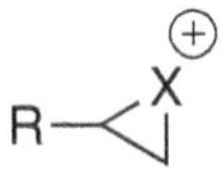 cyclic heteronium ions X = OH, NH_2, NHR, SH, SR, Cl, Br, I etc.
Weak	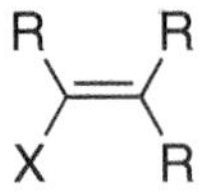 alkenes, including enol ethers, enamines, etc. X = R, Ar, $OSiR_3$, NR_2, etc 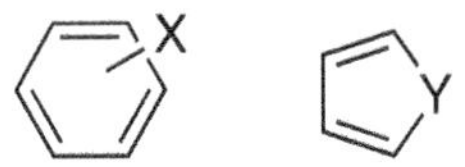 arenes X = OR, NR_2, etc. Y = O, NR, S, etc.	R−X alkyl halide, sulfonate, etc. X = halogen, OSO_2R, etc. 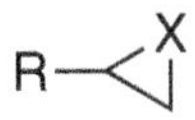 epoxides, aziridines, etc. X = O, NH, NR, S, etc. 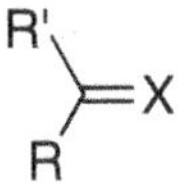 R'−C≡N carbonyl compounds imines, nitriles etc. R' = alkyl, aryl, C=C, C≡C, OR, NR_2, Cl, etc X = O, NR, S, etc. 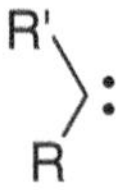 carbenes

metal bearing a formal negative charge. The "ate" complexes are usually prepared by the addition of a metal alkyl of a group IA metal or group IIA metal to a metal alkyl of another metal (e.g., aluminum, zinc, or copper).

$$\underset{\text{metal alkyl}}{R{-}M} \qquad \underset{\text{"ate" complex}}{\overset{\ominus}{R{-}M{-}R}} \qquad \textbf{(12.1)}$$

As strong nucleophiles, simple metal alkyls will react with weak carbon electrophiles to generate new carbon-carbon bonds. The types of reactions that these species enter into depend to some degree on the hardness or softness of the nucleophile. Generally speaking, more ionic (harder) nucleophiles are better for use in S_N2 reactions with alkyl halides. Alkali metal alkynides, cyanide anion, and lithium dialkylcuprates or trialkylzincates react with alkyl halides with inversion of configuration, as do anions stabilized by heteroatoms, such as dithiane, α-silylalkyl, and sulfoxide α-anions. Depending on the carbon nucleophile and its electrophilic partner, carbon-carbon σ bond formation in allylic systems can take place by simple S_N2 (Example **12.2**) or S_N2' (Example **12.3**) displacement of the leaving group, with softer nucleophiles tending to favor the S_N2' route. In general, nucleophiles with more ionic character are preferred for S_N2 substitutions, and nucleophiles with less ionic character are preferred for addition to carbonyl groups. Carbanion nucleophiles also add to electrophilic π bonds, as in carbonyl compounds and their conjugated analogues (Examples **12.4** to **12.8**). As we saw in the Chapter 11, the more ionic the metal-carbon bond, the greater the propensity of the metal alkyl to add in the 1,2- mode to conjugated carbonyl compounds.

(12.2)

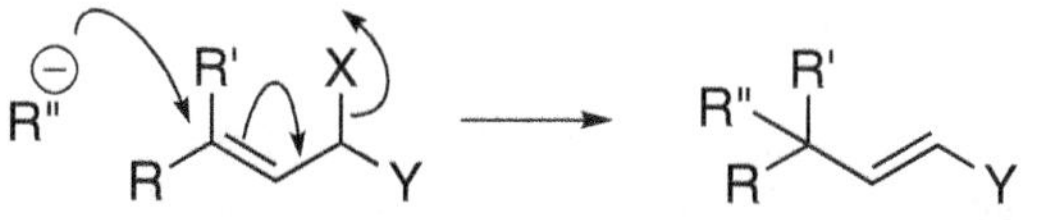
(12.3)

(12.4)

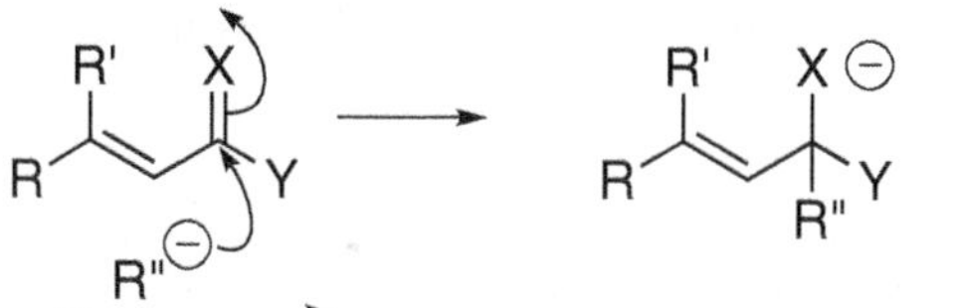
(12.5)

(12.6)

(12.7)

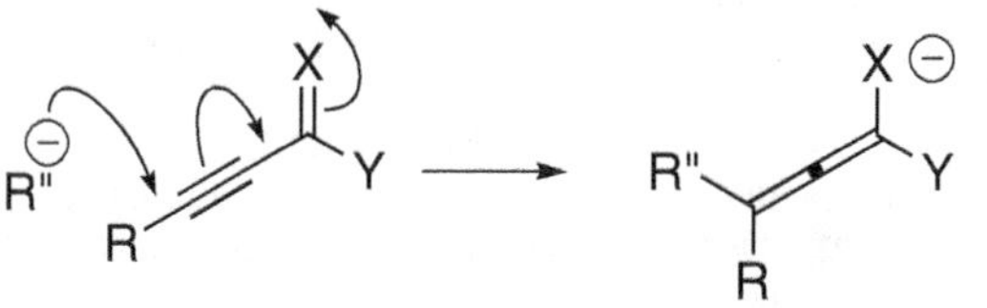
(12.8)

Nucleophilicity and basicity tend to be complementary characteristics of metal alkyls, with the result that it is often possible to modulate both forms of reactivity by judiciously choosing the structural features of the compound. More ionic nucleophiles are harder bases and tend to react more frequently as bases. The nucleophilicity of a metal alkyl is modulated by three factors: (1) the electronegativity of the metal, (2) the formal charge on the metal, and (3) the hybridization of the carbon atom. The more electronegative (less electropositive) the metal, the softer the nucleophile and the less basic the metal alkyl; the higher the formal negative charge on the metal, the stronger the nucleophile and the more strongly basic the compound; and the higher the s character of the hybrid orbitals on the carbon bonded to the metal, the more likely the compound is to be ionic and the harder and more powerful the nucleophile.

Using Metal Alkyls as Bases

Group IA and IIA metal alkyls react with organic compounds with pK_a values below about 30, especially alcohols, which limits the usefulness of these reagents in synthesis. However, these metal alkyls generally do not deprotonate simple hydrocarbons other than alkynes. As we discussed in Chapter 11, there is a clear-cut division of reactivity among alkali metal alkyls because alkyllithiums are aggregates of a polar covalent species, the metal alkyls of the higher alkali metals are effectively ionic salts with a free carbanion.[1] The reactivity of alkyllithiums can be modulated by additives and by changing the structure of the alkyl groups. If we assume *n*-butyllithium as the standard, we find that branching the alkyl group increases the tendency of the alkyllithium to act as a base. This is probably due, in part, to steric hindrance preventing the close approach required for it to act as a nucleophile and to the inherent increase in base strength with branching. Likewise, adding ligands that can disrupt the aggregation state of the alkyllithium also increases its tendency to act as a base. *N,N,N',N'*-tetramethylethylenediamine (TMEDA; Example **12.9**) is very widely used for this purpose.[2]

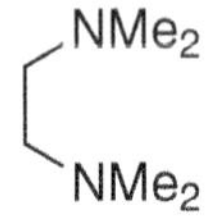

(12.9)

Alkylsodium or alkylpotassium reagents react with alkenes only by proton abstraction or hydrogen-metal exchange to give the allylmetal products.[3] Butylpotassium deprotonates alkenes at allyl and vinyl positions at the same time.[4] Branched alkyllithium reagents such as *sec*- and *tert*-butyl-lithium *add* to unactivated alkenes such as propene.[5] Deprotonation as the sole reaction also occurs with the 1:1 mixture of butyllithium and potassium *tert*-butoxide, a reagent known as ***Schlosser's reagent***[6] or "LICKOR," which metallates alkenes preferentially at the allylic position, before deprotonating at an available vinyl

1. Leading references: (a) Finnegan, R.A. *Tetrahedron Lett.* **1963**, 429. (b) Schade, C.; Bauer, W.; Schleyer, P.v.R. *J. Organomet. Chem.* **1985**, *295*, C25. (c) Pi, R.; Bauer, W.; Brix, B.; Schade, C.; Schleyer, P.v.R. *J. Organomet. Chem.* **1986**, *306*, C1.

2. (a) Eberhardt, G.G.; Butte, W.A. *J. Org. Chem.* **1964**, *29*, 2928. (b) Klein, J.; Medlik, A. *J. Chem. Soc., Chem. Commun.* **1973**, 275. (c) Akiyama, S.; Hooz, J. *Tetrahedron Lett.* **1973**, 4115. (d) Bates, R.B.; Beavers, W.A. *J. Am. Chem. Soc.* **1974**, *96*, 5001.

3. Reviews: (a) Schlosser, M. *Angew. Chem.* **1964**, *76*, 124, 258; *Angew. Chem. Int. Ed. Engl.* **1964**, *3*, 287, 362. (b) Ebel, H.F. in Müller, E., Ed. *Houben Weyl : Methoden der organischen Chemie* (G. Thieme Verlag: Stuttgart 1970); vol. 13/1, 261. (c) Schlosser, M.; *Struktur und Reaktivität polarer Organometalle* (Springer Verlag: Berlin, 1973). (d) Schlosser, M. *Mod. Synth. Methods* **1992**, *6*, 227. (e) Schlosser, M.; Faigl, F.; Franzini, L.; Geneste, H.; Katsoulos, G.; Zhong, G. *Pure Appl. Chem.* **1994**, *66*, 1439.

4. (a) Broaddus, C.D.; Logan, T.J.; Flautt, T.J. *J. Org. Chem.* **1963**, *28*, 1174. (b) Broaddus, C.D. *J. Org. Chem.* **1964**, *29*, 2689.

5. Bartlett, P.D.; Friedman, S.; Stiles, M. *J. Am. Chem. Soc.* **1953**, *75*, 1771.

6. (a) Schlosser, M. *J. Organomet. Chem.* **1967**, *8*, 9. (b) Schlosser, M.; Hartmann, J. *Angew. Chem.* **1973**, *85*, 544; *Angew. Chem. Int. Ed. Engl.* **1973**, *92*, 439; *Angew. Chem.* **1974**, *86*, 751; *Angew. Chem. Int. Ed. Engl.* **1974**, *13*, 701. (c) Schlosser, M. *Pure Appl. Chem.* **1988**, *60*, 1627.

position. Despite the presence of both metals (lithium and potassium), the metal alkyl product contains not lithium, but potassium.[7] This has been interpreted as support for the view of the reagent as the complex **12.10**.

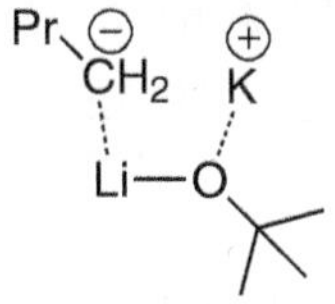

(12.10)

The allylmetal compounds obtained in this way are powerful nucleophiles in S_N2 reactions and can be alkylated to give new carbon-carbon bonds. Schlosser has reported a strong stereochemical bias in the allylpotassium products of "LICKOR" deprotonation of alkenes: the deprotonation to give the *endo* metal alkyl (Example **12.11**) is favored over the *exo* metal alkyl (Example **12.12**) by at least 15:1—and more often by a much higher ratio.

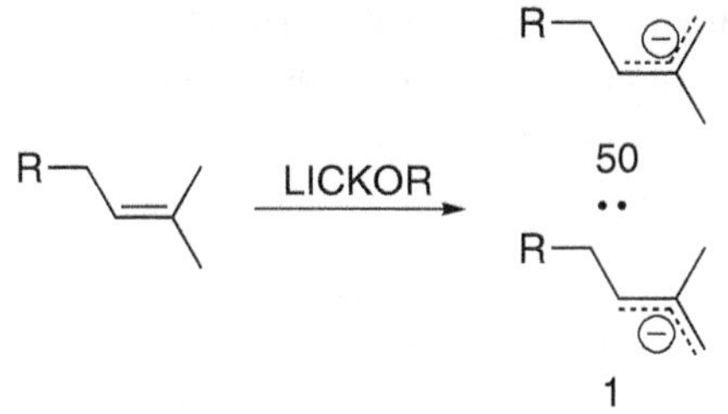

(12.11)

(12.12)

Alkylation with Metal Alkyls

Alkali Metal Alkyls and Alkynides

The Schlosser allylmetals and the metal alkynides are basically ionic nucleophiles that participate readily in nucleophilic displacement reactions. For example, trapping the Schlosser allyl-metal nucleophile from 2-methyl-2-butene with an alkyl halide gives a trisubstituted alkene with well-defined stereochemistry (Example **12.13**).[8] The usefulness of these anionic nucleophiles is illustrated by Examples **12.14** and **12.15**.

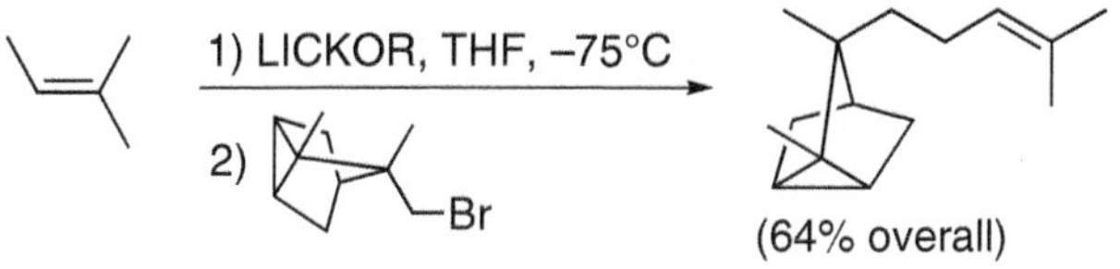

(12.13)[9]

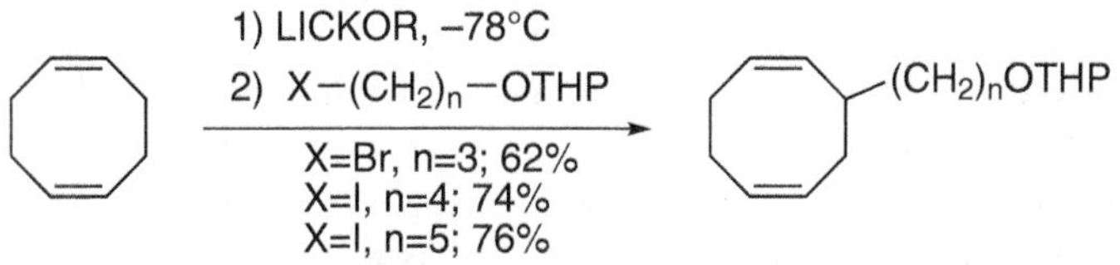

(12.14)[10]

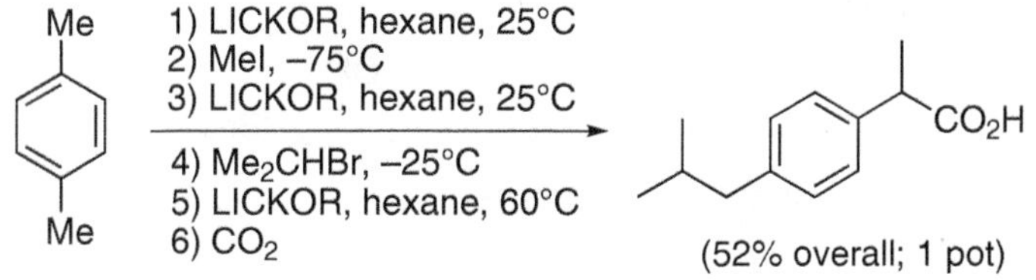

(12.15)[11]

7. Boche, G.; Etzrodt, H. *Tetrahedron Lett.* **1983**, *24*, 5477.
8. Review: Schlosser, M.; Desponds, O.; Lehman, R.; Moret, E.; Rauchschwalbe, G. *Tetrahedron* **1993**, *49*, 10175.
9. Schlosser, M.; Zhong, G.-f. *Tetrahedron Lett.* **1993**, *34*, 5441.
10. Revell, J.D.; Ganesan, A. *J. Org. Chem.* **2002**, *67*, 6250.
11. Faigl, F.; Schlosser, M. *Tetrahedron Lett.* **1991**, *32*, 3369.

In contrast to the higher alkali metal alkyls, lithium and magnesium alkyls do not react well with alkyl halides and sulfonates unless the halide is allylic or benzylic, although vinyllithium reagents (where the carbon is now sp^2 hybridized) *will* react with primary alkyl iodides, as shown in Example **12.16.**[12]

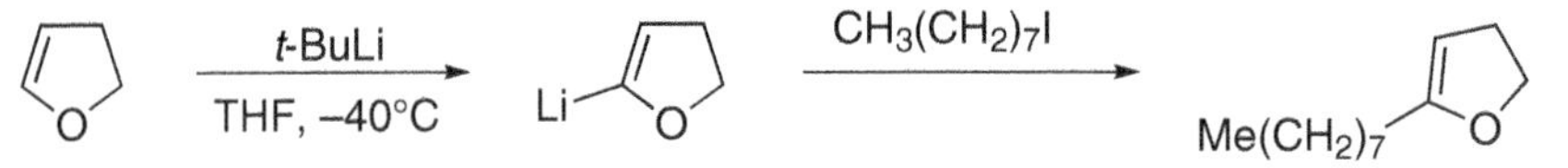

(12.16)

The alkyne carbon atom is sp-hybridized, and terminal alkynes are easily deprotonated by metal hydrides or alkyls to give the metal alkynide (Example **12.17**). Alkali metal alkynides react essentially as free carbanion nucleophiles with a wide range of electrophilic partners, including alkyl halides, alkyl sulfonates, epoxides, and carbonyl compounds. With alkyl halides, the reaction is restricted to primary halides (preferably bromides) that are not branched at the β carbon (i.e., RCH_2CH_2X).[13] This restriction may be traced to the fact that the alkynide anion is still a strong base and that there is competing E2 elimination. It is interesting that the use of allyl halides leads to the formation of compounds consistent with polyalkylation. In nucleophilic substitution reactions, alkynide anions react by the S_N2 mechanism.

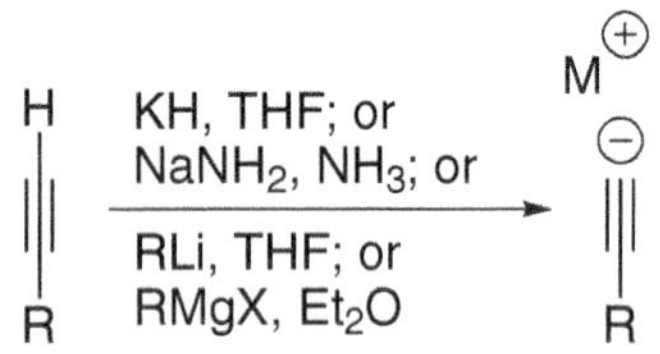

(12.17)

Electrophiles with functional groups containing oxygen atoms capable of functioning as ligands for a metal ion will react well with metal alkyls derived from lithium and magnesium. In these reactions, there is evidence that the oxygen atom of the functional group displaces a molecule of solvent to form a complex with the metal ion that serves both to increase the electrophilicity of the carbon and to anchor the nucleophile in a position conducive to group transfer (e.g., Example **12.18**). Thus, although alkyl halides and sulfonates do not react well with alkyllithiums and Grignard reagents, carbonyl compounds and epoxides do.

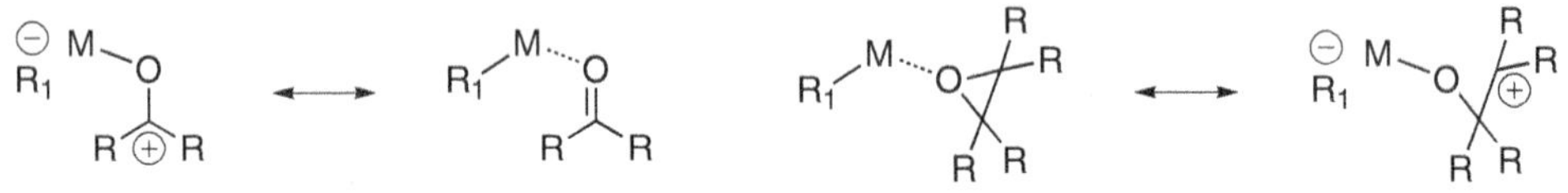

(12.18)

The reactions between an epoxide and alkylmetal nucleophiles give the products of *anti* ring opening with the metal alkyl attacking the less substituted carbon. The reaction may therefore be considered as an S_N2 epoxide ring opening. In certain cases (e.g., Example **12.19**) the reaction is facilitated by the addition of a Lewis acid other than the alkyllithium to complex the epoxide oxygen.

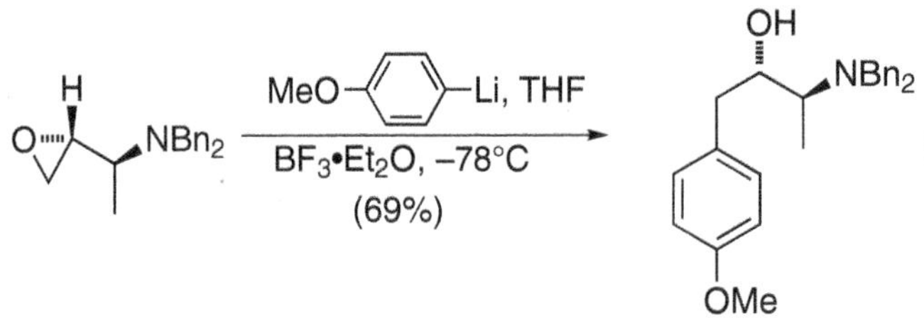

(12.19)[14]

12. Hedenström, E.; Högberg, H.-E. *Tetrahedron* **1994**, *50*, 5225.
13. Jacobs, T.L. *Org. React.* **1949**, *5*, 1.
14. Mena, M.; Valls, N.; Borregán, M.; Bonjoch, J. *Tetrahedron* **2006**, *62*, 9166.

Worked Problems

12-1 Trapping of an alkylpotassium reagent with a borate ester provides a useful entry into boronic acids that can be oxidized to alcohols. What is the major product of the reaction below? Provide a rationalization of the stereochemistry.

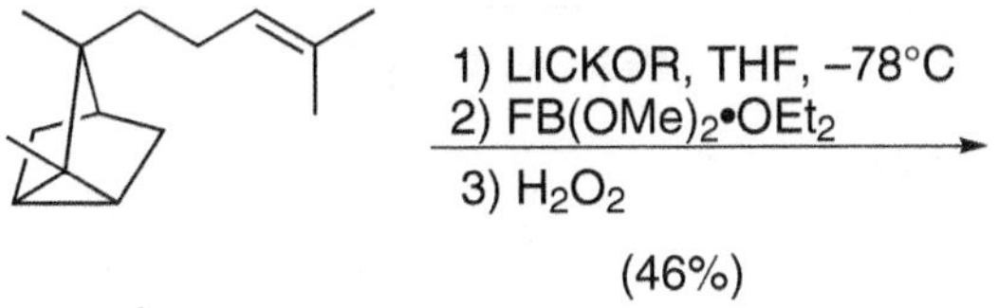

§Answer below.

12-2 What is (are) the major product(s) of the following reaction sequence?

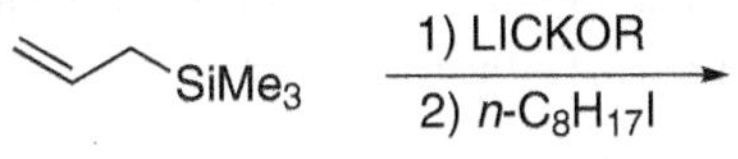

§§Answer below.

Drill Problems

What are the products of the treatment of the compounds in list A with each of the reagents in list B?

List A:

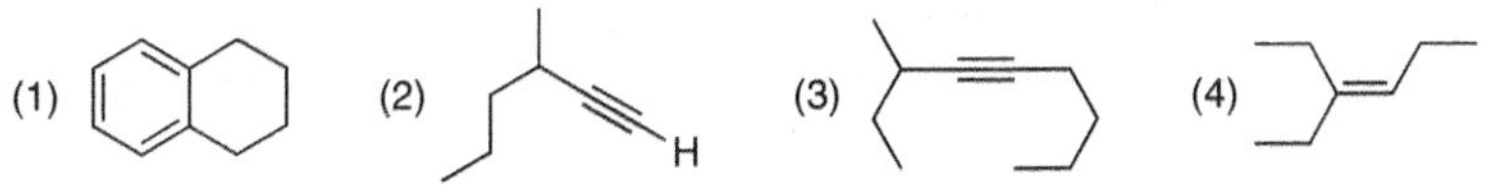

List B:

(i) *t*-BuLi, THF-HMPA, –78°C

(ii) KAPA, $H_2N(CH_2)_3NH_2$

§ Answer to Worked Problem:

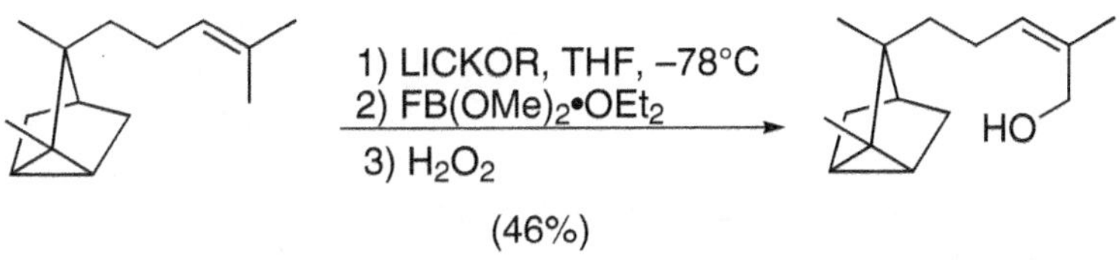

The deprotonation of this alkene occurs to give mainly the *endo* allylpotassium reagent. Trapping of this reagent with the boronic acid fluoride gives the *Z* borane that reacts with hydrogen peroxide to give the alcohol. [*Tetrahedron Lett* **1993**, *34*, 5441] The second step is identical to the oxidation step of the hydroboration-oxidation reaction and proceeds with retention of configuration at carbon.

§§ Answer to Worked Problem:

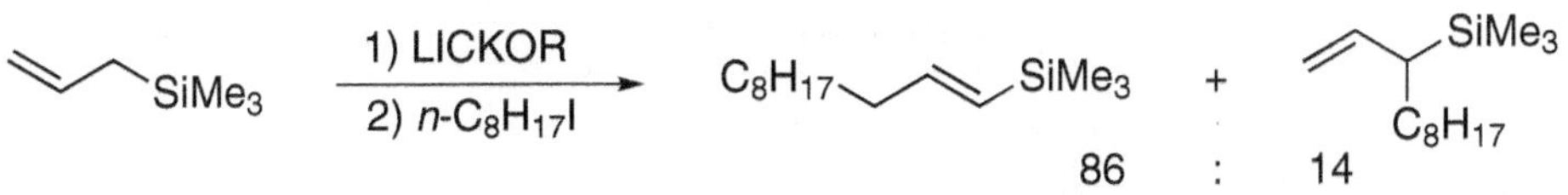

The deprotonation of the alkene gives an allylpotassium reagent that reacts preferentially through the less substituted end of the anion. [*Tetrahedron Lett.* **1984**, *25*, 717]

(iii) KOBu^t, THF

(iv) (1) LICKOR, hexane, –70°C; (2) EtI, THF, –70°C

(v) (1) LICKOR, hexane, –70°C; (2) *S*-2-iodobutane, THF, –70°C

Reaction Synopses

Deprotonation of Hydrocarbons

$$\text{R–H} \xrightarrow{\text{R'M}} \text{R'–H}$$

Reagents:

RH = alkyne: $NaNH_2$, NH_3; LDA, THF; RMgX, Et_2O; RLi, THF; etc.
RH = alkene: *t*-BuLi, TMEDA; *n*-BuLi, KOBu^t, –78°C; etc.

Alkenes deprotonated at allylic position if hydrocarbon only; deprotonated adjacent to heteroatom in vinyl ethers, sulfides

Alkylation with Metal Alkyls

$$\text{R–M} \xrightarrow{\text{R'X}} \text{R–R'}$$

Reagents: RM, Et_2O or THF, R'X (R' must be 1° or 2°)
Mechanism: S_N2 (M = Na, K), or S_N2' (M = Cu)

Stereochemistry: usually inversion at carbon; stereochemistry at carbanion carbon may not be retained
M = Na, K: X = OTs, Cl, Br, I; etc.
M = Li: X = OTs, epoxide; Br, I (R = vinyl, aryl); etc.
M = Mg: X = epoxide, etc.
M = Zn, Cu: X = OTs, Br, I, etc.

Problem

12-1 What is the major organic product in each of the following reactions or reaction sequences?

(a)

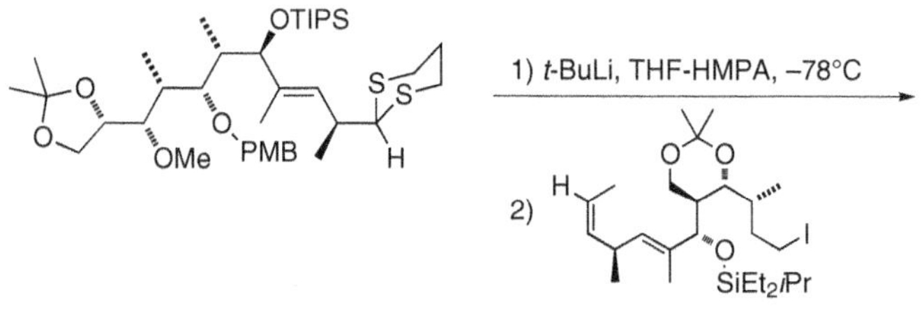

(b)

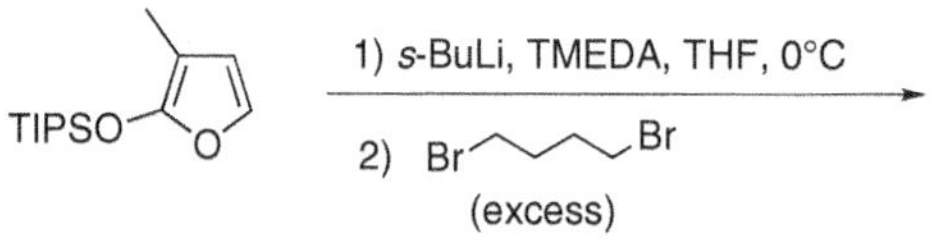

(Why is the order of addition of reagents—what is added to what and when—important in this reaction?)

(c)

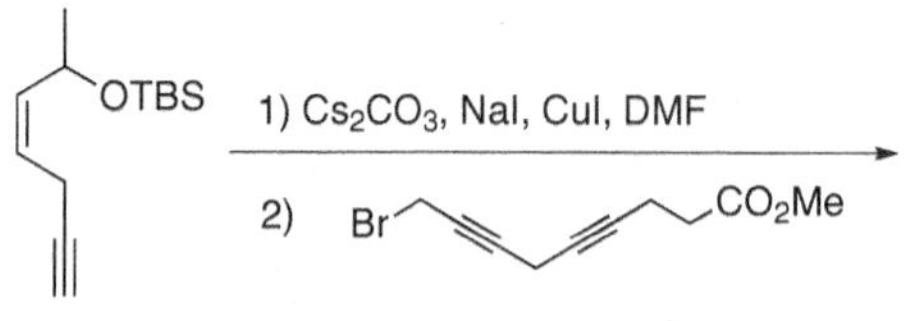

References: (a) *J. Am. Chem. Soc.* **2003**, *125*, 350. (b) *J. Am. Chem. Soc.* **1996**, *118*, 3299. (c) *Tetrahedron* **2003**, *59*, 7787.

Addition to Carbonyl Compounds

Carbonyl and similar electrophiles react with most simple metal alkyls by addition. Recall, however—alkylsodium and alkylpotassium reagents derived from saturated alkyl halides are such strong bases that competing deprotonation becomes the dominant reaction. For this reason, they are generally inferior to the corresponding lithium reagents. A similar situation often occurs in the reactions of ketones with strongly basic nucleophiles such as simple (unstabilized) phosphoranes,[15] where acid-base reactions often occur to give competing or concomitant epimerization at the α carbon, especially when the phosphorane is generated in dimethyl sulfoxide (DMSO).[16]

Table 12.2 Addition Products with Carbonyl Compounds

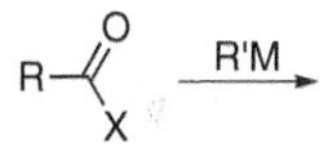

X	Product	X	Product
H	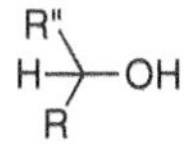	OR''	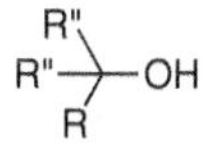
R''	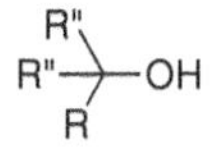	NR''_2	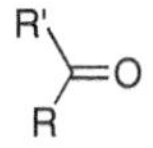
Cl	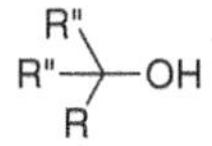	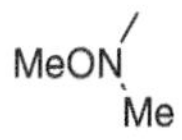	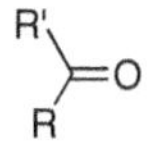

The course of the reaction between saturated carbonyl compounds and organolithium compounds or Grignard reagents (Table 12.2) is generally straightforward. Aldehydes and ketones give alcohols; esters and more reactive carboxylic acid derivatives give tertiary alcohols carrying two identical alkyl groups by an addition-elimination-addition sequence; and amides and nitriles give ketones (the adduct of the amide cannot eliminate a leaving group, so only one addition step occurs). Weinreb amides,[17] in particular, have been popular species for the formation of ketones from acyl derivatives (e.g., 2-acyloxy-4,6-dimethoxy-1,3,5-triazines, as in Example **12.20**).

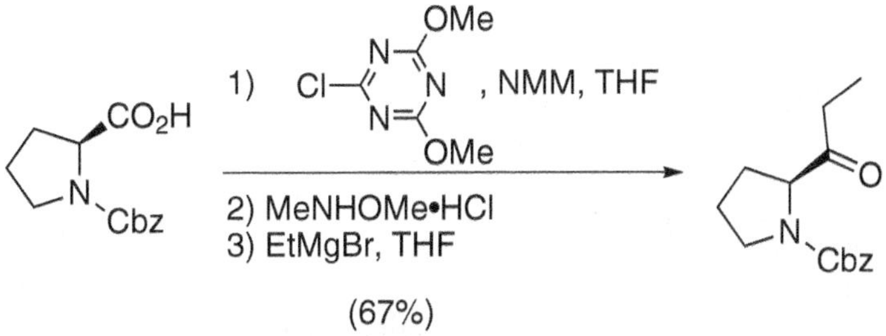

(12.20)[17]

Conjugation of the carbonyl group adds a regiochemical component to the nucleophilic addition reaction with metal alkyls, as summarized in Table 12.3, with the result that one can prepare compounds by 1,2-addition or 1,4-addition to the conjugated system by choosing the nucleophilic and electrophilic species appropriately. In general, conjugated aldehydes and acid chlorides react with nucleophiles by the 1,2-addition pathway, whereas conjugated esters and amides usually react preferentially by the 1,4-addition pathway (especially if the nucleophile can add reversibly). In contrast, conjugated ketones may react by either pathway or both. Conjugate addition of carbon nucleophiles to nitriles has long been known as a more difficult task than conjugate additions to other carboxylic acid derivatives, with the result that this methodology is rather less developed than additions to conjugated carbonyl compounds.[18] In additions to conjugated carbonyl systems, harder nucleophiles such as alkyllithium reagents or alkali metal alkynides tend to react with 1,2- regiochemistry. In general, additions to carbonyl groups are best carried out using nucleophiles of intermediate hardness, and conjugate additions are favored by soft nucleophiles. Alkyllithiums and alkali metal alkynides usually give 1,2-addition to conjugated carbonyl compounds, Grignard reagents tend to give a mixture of 1,2- and 1,4-addition

15. Adlerkreutz, P.; Magnusson, G. *Acta Chem. Scand. B* **1980**, *34*, 647.
16. Early examples: (a) Soffer, M.D.; Burk, L.A. *Tetrahedron Lett.* **1971**, 211. (b) Marshall, J.A.; Pike, M.T.; Carroll, R.D. *J. Org. Chem.* **1966**, *31*, 2933. (c) Heathcock, C.H.; Ratcliffe, R. *J. Am. Chem. Soc.* **1971**, *73*, 1746.
17. Nahm, S.; Weinreb, S.M. *Tetrahedron Lett.* **1981**, *22*, 3815.
18. Fleming, F.F.; Wang, Q. *Chem. Rev.* **2003**, *103*, 2035.

Table 12.3 Products of Nucleophilic Addition to Conjugated Carbonyl Compounds

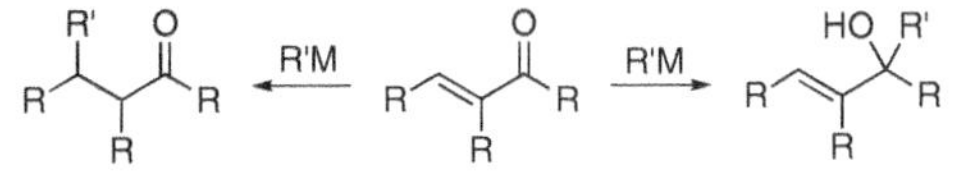

R'M (1:4)	R'M (1:2)
	RLi
	RC≡CM
RMgX	RMgX
R_2CuLi, etc.	
R_3ZnLi, etc.	R_2Zn

products, and organocuprates and organozincates give exclusive 1,4-addition. Thus, as the metal-carbon bond becomes more covalent (i.e., the nucleophile becomes softer), the proportion of conjugate addition rises. In a recent report, Knochel has shown that the lanthanide-lithium chloride complex $LnCl_3$•2LiCl can dramatically improve the yield of 1,2-additions of Grignard reagents to conjugated, hindered, and readily enolizable ketones, while at the same time suppressing reduction of the ketone by the Grignard reagent.[19]

Alkylcopper, Alkylzinc, and Alkylaluminum Reagents

Although copper and zinc are transition metals, their organometallic compounds tend to react more like those of the main group metals. The neutral organometallic derivatives of copper, zinc, and aluminum are usually non-nucleophilic or modest nucleophiles that can be induced to react with carbonyl compounds but that do not react with alkyl halides. Dialkylzinc reagents, for example, will add to acid chlorides, aldehydes, and ketones to give alcohols, as will alkylzinc halides.[20] However, these reagents will not attack carbon dioxide or conjugated ketones, and the reactions of dialkylzinc reagents where the alkyl group is larger than ethyl often lead to reduction rather than addition in their reactions with ketones (see Example **12.21**).

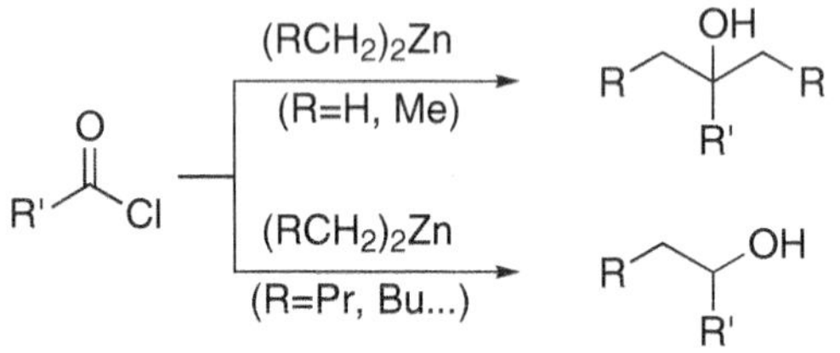

(12.21)

Zinc holds a central place in the history of organic synthesis, because dialkylzinc reagents were the first simple organometallic compounds prepared[21] (although Bunsen's "cacodyl," $Me_2As—AsMe_2$,[22] was the first organometallic compound prepared if one includes the metalloids). The first general syntheses of secondary and tertiary alcohols were accomplished by the addition of alkylzinc nucleophiles to acid chlorides and aldehydes, and the first synthesis of a carboxylic acid was accomplished by carbonation of sodium triethylzincate (mistaken at the time for ethylsodium) with carbon dioxide. This reaction also nicely illustrates the difference in reactivity between a metal alkyl and an ate complex: the additions of dialkylzinc or alkylzinc halides to aldehydes and acid chlorides can, in fact, be carried out under carbon dioxide as an inert atmosphere.

The pioneering work of Noyori in the 1990s showed that additions of dialkylzinc reagents to aldehydes are catalyzed by (especially chiral) β-aminoalcohol ligands;[23] this work resurrected the importance of organozinc nucleophiles. From this early work, it became apparent that the nucleophilicity of the alkylzinc reagent is highly dependent on the hybridization of zinc. Simple dialkylzinc reagents, where the zinc atom is sp hybridized, are relatively poor nucleophiles that require quite reactive electrophilic partners (e.g., acid

19. Krasovskiy, A.; Kopp, F.; Knochel, P. *Angew. Chem. Int. Ed.* **2006**, *45*, 497.

20. These reactions of organozinc nucleophiles are of long standing: (a) Butlerow, A. *Bull. Soc. Chim. Paris* **1863**, 5, 582; **1864**, 2, 106.; *Jahresb.* **1863**, 475; **1864**, 496; *Ann. Chem. Pharm.* **1867**, *144*, 132. (b) Saytzeff, A. *Ann. Chem. Pharm.* **1875**, *175*, 351, 374; **1877**, *185*, 148, 151, 175;. *J. prakt. Chem.* **1885**, *31*, 319. (c) Vagner, E.E. *Zh. Russ. Khim. Obshch. Fiz. Obshch.* **1876**, *8*, 290, *Zh. Russ. Fiz.-Khim. Obshch.* **1884**, *16*, 283. (d) Wagner, G. *Ann. Chem. Pharm.* **1875**, *175*, 351; *179*, 302, 313; *Ber. dtsch. chem. Ges.* **1894**, *27*, 2436.

21. Frankland, E. *Quart. J. Chem. Soc.* **1850**, *2*, 297.

22. Bunsen, R.W. *Ann. Pharm.* **1839**, *31*, 175;. **1841**, *37*, 1; **1842**, *42*, 14; **1843**, *46*, 1.

23. Noyori, R.; Suga, S.; Kawai, K.; Okada, S.; Kitamura, M.; Oguni, N.; Hayashi, M.; Kaneko, T.; Matsuda, Y. *J. Organomet. Chem.* **1990**, *382*, 19.

chlorides) for the reaction to occur. However, when the zinc atom is complexed by a hard donor ligand, the hybridization of zinc changes, and the p character of the orbitals increases; the resultant dialkylzinc complexes are much stronger nucleophiles. These observations have led to a resurgence of organozinc chemistry, as illustrated by the number of review articles in this area since 1990.[24]

One of the earliest generally useful reaction of organozinc nucleophiles is the Reformatsky (Reformatskii) reaction (Example **12.22**),[25] discovered in 1887.[26] In this reaction, an α-halocarbonyl compound (originally an α-haloester) reacts with zinc to give an organozinc halide, and this then adds to the carbonyl group of an aldehyde or ketone to give a β-hydroxycarbonyl product. What is important about the reaction is that it allows the construction of highly hindered β-hydroxycarbonyl compounds, which can be difficult to obtain by other methods. The crystal structures of two Reformatsky reagents (Figure 12.1) show that the zinc atom is approximately sp[3] hybridized, with the carbonyl oxygen of the ester and an oxygen atom from solvent coordinated to the metal.[27] In the two examples shown, the reagent is dimeric with an eight-membered ring that adopts either a chairlike (**12.23**) or tublike (**12.24**) conformation.

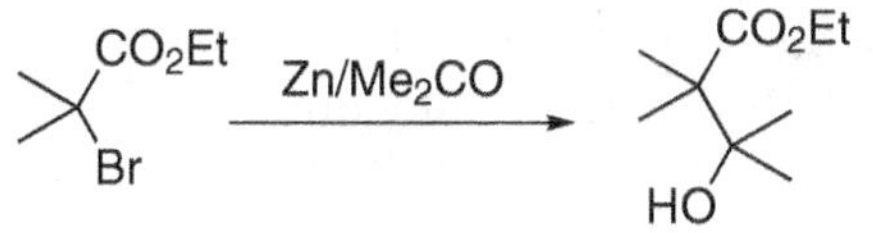

(12.22)

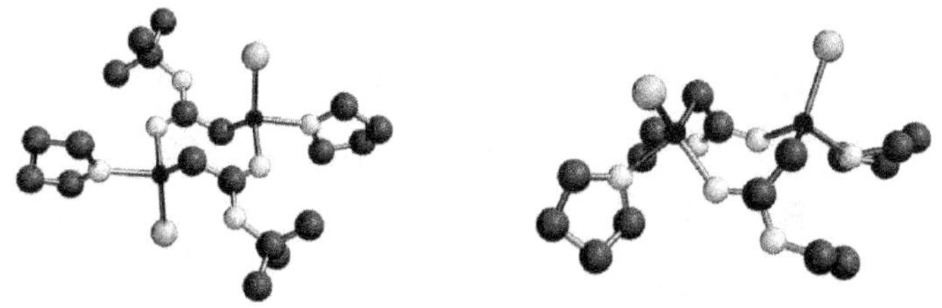

Figure 12.1 Crystal conformations (structures) of two Reformatskii reagents

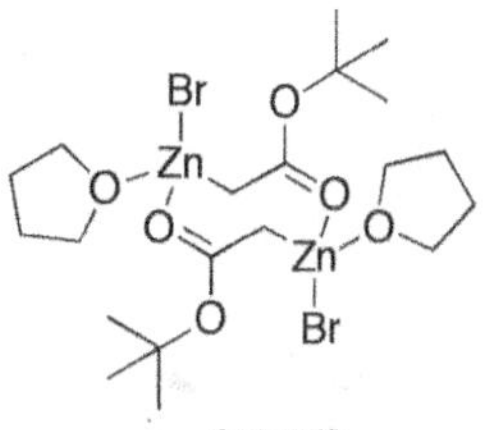

(12.23)

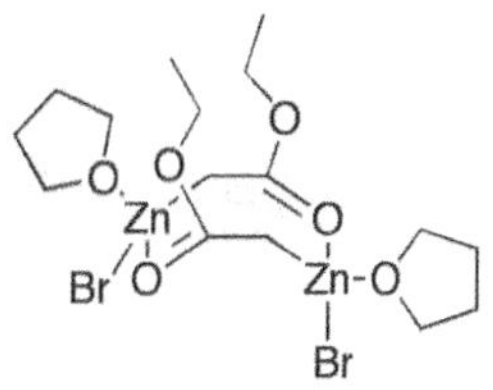

(12.24)

24. Selected reviews: (a) Noyori, R.; Kitamura, M. *Angew. Chem. Int. Ed.* **1991**, *30*, 49. (b) Soai, K.; Niwa, S. *Chem. Rev.* **1992**, *92*, 833. (c) Knochel, P.; Singer, R.D. *Chem. Rev.* **1993**, *93*, 2117. (d) Knochel, P.; Jones, P., Eds., *Organozinc Reagents: A Practical Approach*, (Oxford University Press: Oxford,1999). (e) Knochel, P.; Vettel, S.; Eisenberg, C. *Appl. Organomet. Chem.* **1995**, *9*, 175. (f) Knochel, P.; Almena, J.; Jones, P. *Tetrahedron* **1998**, *54*, 8275. (g) Erdik, E. *Organozinc Reagents in Organic Synthesis* (CRC Press: Boca Raton, FL, 1996). (h) Soai, K. *Enantiomer* **1999**, *4*, 591. (i) Frantz, D.E.; Faessler, R.; Tomooka, C.S.; Carreira, E.M. *Acc. Chem. Res.* **2000**, *33*, 373. (j) Boudier, A.; Bromm, L.O.; Lotz, M.; Knochel, P. *Angew. Chem. Int. Ed.* **2000**, *39*, 4414. (k) Pu, L.; Yu, H.-B. *Chem. Rev.* **2001**, *101*, 757. (l) Knochel, P.; Millot, N.; Rodriguez, A.L.; Tucker, C.A. *Org. React.* **2001**, *58*, 417. (m) Knochel, P. *Science of Synthesis* (Georg Thieme Verlag: Stuttgart, 2004), *3*, 5.

25. Reviews: (a) Shriner, R.L. *Org. React.* **1942**, *1*, 1. (b) Diaper, D.G.M.; Kukis, A. *Chem. Rev.* **1959**, *59*, 89. (c) Rathke, M.W. *Org. React.* **1975**, *22*, 423. (d) Ocampo, R.; Dolbier, W.R., Jr. *Tetrahedron* **2004**, *60*, 9325.

26. Reformatsky, S. *Ber. dtsch. chem. Ges.* **1887**, *20*, 1210.

Sergei Nikolaevich Reformatskii (1860–1934) graduated from the Kostroma Spiritual Seminary, then studied under Zaitsev at Kazan University (Dr Chem, 1890, Warsaw). After a year of study under Meyer and Ostwald, he returned to Russia in 1891 and was appointed Professor of Chemistry at the Kiev University, a post he held until his retirement. For more information, see: Lewis, D.E. *Early Russian Organic Chemists and Their Legacy* (Springer: Heidelberg, 2012), p. 101.

27. (a) Dekker, J.; Budzelaar, P.H.M.; Boersma, J.; Van der Kerk, G.; Spek, A. *J. Organometallics* **1984**, *9*, 1403. (b) Miki, S.; Nakamoto, K.; Kawakami, J.; Handa, S.; Nuwa, S. *Synthesis* **2008**, 409.

Alkylcopper reagents are generally unreactive toward electrophiles, but when coupled with boron trifluoride, they will add to enones to give products of conjugate addition, as shown by a key step (Example **12.25**) of an early total synthesis of the interesting sesquiterpene modhephene.[28] The active reagent is not known, but it may be formed by transfer of the alkyl group from copper to boron to give a copper (I) alkyltrifluoroborate.

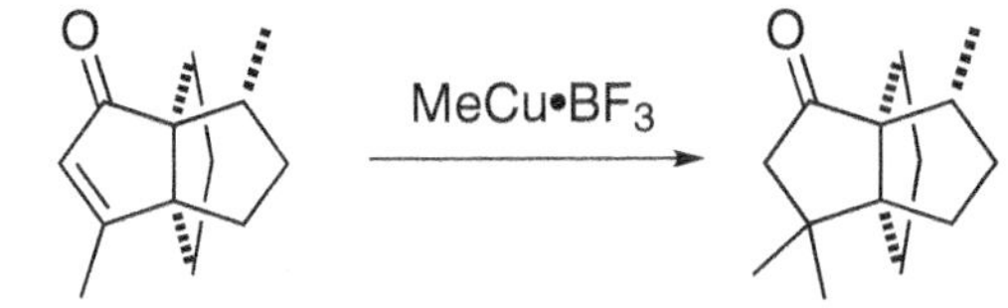

(12.25)

With "Ate" Complexes

The lack of success when most simple metal alkyls are treated with alkyl halides and sulfonates contrasts starkly with the success in the substitution reactions when "ate" complexes are used as the nucleophiles. Recall that an "ate" complex is an anionic complex of a transition metal alkyl in which the metal atom is coordinated to one or more alkyl groups more than the neutral metal alkyl.[29] This is most commonly accomplished by treating the metal alkyl with an alkyllithium or Grignard reagent (e.g., Example **12.26**). Converting the simple metal alkyls of copper, zinc and aluminum into anionic "ate" complexes transforms them into much stronger nucleophiles that can displace the halide anion from primary and secondary halides (e.g., Examples **12.27** and **12.28**).

$$RCu + RLi \longrightarrow R_2CuLi$$
$$R_2Zn + RLi \longrightarrow R_3ZnLi \qquad \textbf{(12.26)}$$
$$R_3Al + RLi \longrightarrow R_4AlLi$$

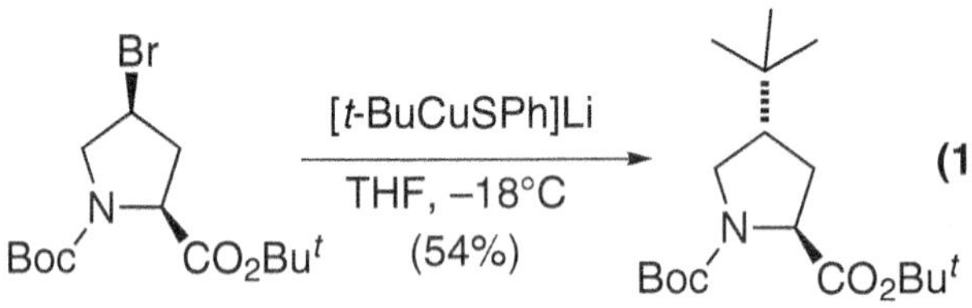

(12.27)[30]

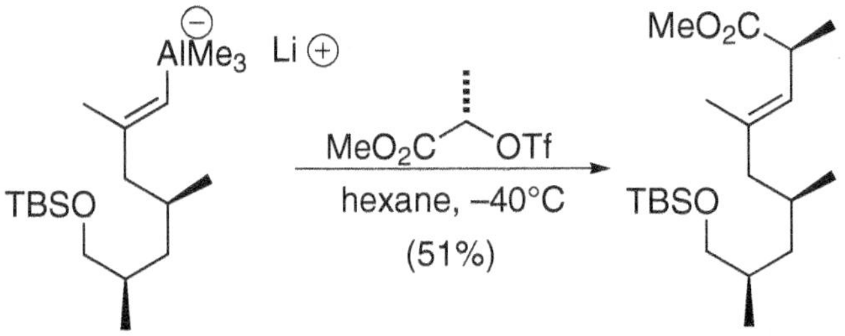

(12.28)[31]

Nucleophilic Substitution with Cuprates

The discovery of cuprate complexes (also known as ***Gilman reagents***) provided the organic chemist with a particularly useful and versatile nucleophile for use in organic

28. Karpf, M.; Dreiding, A.S. *Tetrahedron Lett.* **1980**, *21*, 4569.
29. (a) Wittig, G. *Quart. Rev.* **1966**, 191. (b) Tochtermann, W. *Angew. Chem., Int. Ed. Engl.* **1966**, *5*, 351.
30. Koskinen, A.M.P.; Helaja, J.; Kumpulainen, E.T.T.; Koivisto, J.; Mansikkamäki, H.; Rissanen, K. *J. Org. Chem.* **2005**, *70*, 6447.
31. Hoye, T.R.; Tennakoon, M.A. M.A. *Org. Lett.* **2002**, *2*, 1481.

synthesis.[32] Their impact on the development of modern synthesis may be gauged by the fact that among the four *Organic Reactions* reviews published between 1971 and 1992, more than 2200 references are cited.[33] A more recent review shows the continuing versatility of these reagents in controlling both regiochemistry and stereochemistry of the product.[34]

It was observed fairly early on that when ***homoleptic*** Gilman reagents (i.e., Gilman reagents where both the alkyl groups bound to copper are identical, such as Example **12.29**) react with electrophiles, only one of the two alkyl groups is transferred from copper to carbon. Thus, when working with cuprate reagents, especially if the ligand to be transferred is costly or complex, it is frequently desirable to use a ***heteroleptic*** reagent that contains a "non-transferring" ligand (designated by "L" in Example **12.30**). Typical non-transferring ligands are alkynide anions, cyanide anion, and thiolate anions, all of which form exceptionally strong bonds to copper.[35]

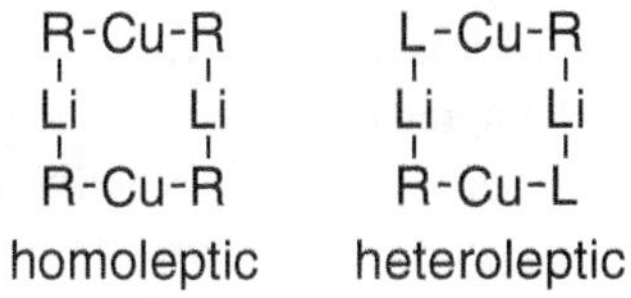

(12.29) **(12.30)**

Although the stereochemical result of these reactions is inversion of configuration, the mechanism is more complex than a simple S_N2 displacement.[36] After more than a decade of debate about a possible Cu (III) intermediate,[37] ^{1}H nuclear magnetic resonance evidence revealing the presence of a Cu(III) complex during an alkylation settled the question in 2007.[38] A very simplified mechanistic model is shown as Example **12.31**.

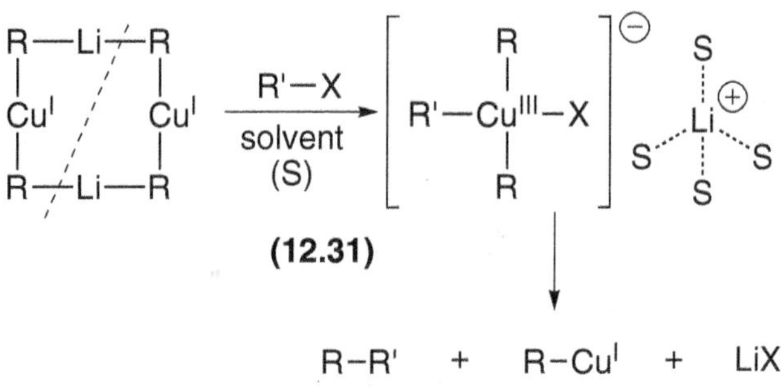

(12.31)

Just as the reactivity of alkylcopper reagents can be modified by the addition of Lewis acids to the reagent, so we find that cuprates become more useful nucleophiles

32. Monograph: Posner, G.H. *An Introduction to Synthesis Using Organocopper Reagents* (Wiley-Interscience: New York, 1980).

33. (a) Posner, G.H. *Org. React.* **1971**, *19*, 1; **1975**, *22*, 253. (b) Chapdelaine, M.J.; Hulce, M. *Org. React.* **1990**, *38*, 225. (c) Lipshutz, B.H.; Sengupta, S. *Org. React.* **1992**, *41*, 135.

34. Breit, B.; Schmidt, Y. *Chem. Rev.* **2008**, *108*, 2928.

35. (a) Corey, E.J.; Beames, D.J. *J. Am. Chem. Soc.* **1972**, *94*, 7210 and references therein. (b) Mandeville, W.H.; Whitesides, G.M. *J. Org. Chem.* **1974**, *39*, 400.

36. Review: Yoshikai, N.; Nakamura, E. *Chem. Rev.* **2012**, *112*, 2339.

37. (a) Dorigo, A.E.; Wanner, J.; Schleyer, P.v.R. *Angew. Chem. Int. Ed. Engl.* **1995**, *34*, 476. (b) Snyder, J.P. *J. Am. Chem. Soc.* **1995**, *117*, 11025. (c) Nakamura, E.; Mori, S.; Morokuma, K. *J. Am. Chem. Soc.* **1998**, *120*, 8273. (d) Nakamura, E.; Mori, S. *Angew. Chem. Int. Ed.* **2000**, *39*, 3751. (e) Mori, S.; Hirai, A.; Nakamura, M.; Nakamura, E. *Tetrahedron* **2000**, *56*, 2805. (f) Mori, S.; Nakamura, E. *Tetrahedron Lett.* **1999**, *40*, 5319. (g) Nakamura, E.; Yamanaka, M.; Yoshikai, N.; Mori, S. *Angew. Chem. Int. Ed.* **2001**, *40*, 1935. (h) James. P.F.; O'Hair, R.A. *J. Org. Lett.* **2004**, *6*, 2761.

38. (a) Gärtner, T.; Henze, W.; Gschwind, R.M. *J. Am. Chem. Soc.* **2007**, *129*, 11362. (b) Bertz, S.H.; Cope, S.; Murphy, M.; Taylor, B.J. *J. Am. Chem.Soc.* **2007**, *129*, 7208. (c) Hu, H.; Snyder, J.P. *J. Am. Chem. Soc.* **2007**, *129*, 7210.

when complexed with Lewis acids such as boron trifluoride. Thus, lithium or magnesium dialkylcuprates complexed with boron trifluoride are very useful nucleophiles for carrying out S_N2' displacements of allylic halides and acetates (e.g., Example **12.32**).[39]

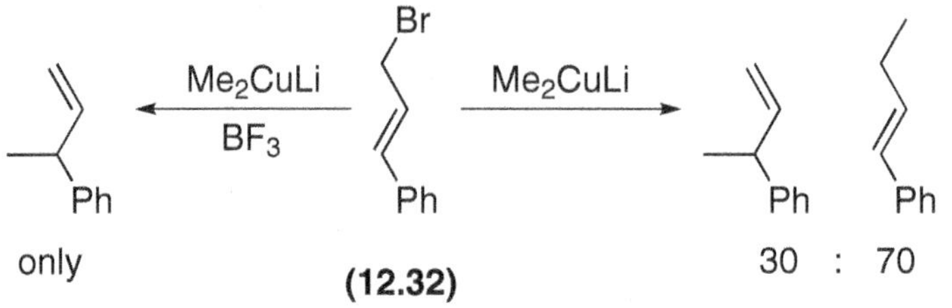

(12.32)

Since 1984, when Lipshutz reported the synthesis and applications of what he termed "higher order" cuprate reagents from copper (I) cyanide,[40] the use of Lipshutz cuprates has been widespread. These Lipshutz cuprates are characterized by greater reactivity than the simple Gilman reagents, and the origins of this reactivity sparked a long polemical debate in the literature, as discussed in Chapter 11. From the viewpoint of using them in organic synthesis, the fact that these reagents are more reactive nucleophiles than the simple Gilman cuprates, makes them useful in displacements of bromides or secondary alkyl iodides, as well as ring opening of epoxides and aziridines. The reactions of Lipshutz cuprates with bromides and epoxides proceed with inversion of configuration, whereas the reactions with alkyl iodides proceed with almost complete racemization. These results are consistent with free radicals being involved in the reactions with secondary alkyl iodides.[41] A study of reactions of Lipshutz cuprates with vinylaziridines, where both the S_N2 and S_N2' pathways are available (e.g., Example **12.33**), revealed a strong preference for the S_N2' pathway.[42]

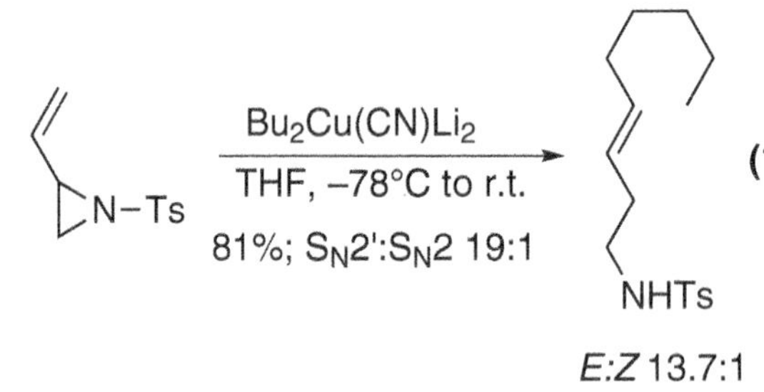

(12.33)

The reactions of organocuprate nucleophiles with vinyl halides constitute a useful method for the formation of carbon-carbon σ bonds. The reaction with vinyl iodides and bromides proceeds with retention of configuration, as illustrated by the formation of *Z*-1-phenylpropene from *Z*-2-bromo-1-phenylethene (Example **12.34**).[43]

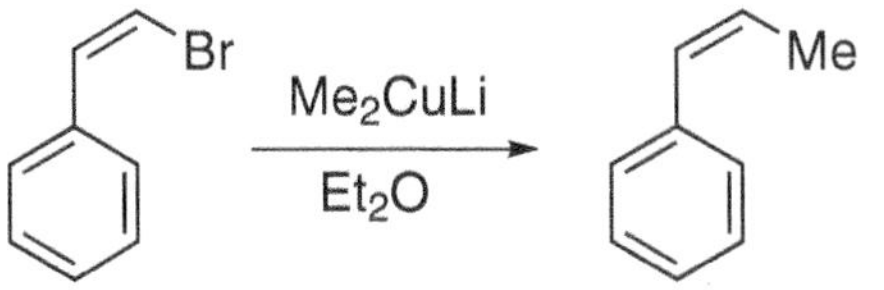

(12.34)

39. Maruyama, K.; Yamamoto, Y. *J. Am. Chem. Soc.* **1977**, *99*, 8068.

40. (a) Lipshutz, B.H.; Wilhelm, R.S.; Kozlowski, J.A.; Parker, D. *J. Org. Chem.* **1984**, *49*, 3928. (b) Lipshutz, B.H.; Wilhelm, R.S.; Kozlowski, J.A. *J. Org. Chem.* **1984**, *49*, 3938. (c) Lipshutz, B.H.; Kozlowski, J.A.; Wilhelm, R.S. *J. Org. Chem.* **1984**, *49*, 3943.

41. Ashby, E.C.; DePriest, R.N.; Tuncay, A.; Srivastava, S. *Tetrahedron Lett.* **1982**, *23*, 5251.

42. Cunha, R.L.O.R.; Diego, D.G.; Simonelli, F.; Comasseto, J.V. *Tetrahedron Lett.* **2005**, *46*, 2539.

43. Corey, E.J.; Posner, G.H. *J. Am. Chem. Soc.* **1967**, *89*, 3911.

Conjugate Addition with Cuprates

Gilman reagents do not add to ketone or ester carbonyl groups but do add to reactive carbonyl compounds such as acid chlorides[44] and thioesters,[45] with which they give the saturated ketone as the major product (e.g., Example **12.35**, from the synthesis of myodesmone[46]).

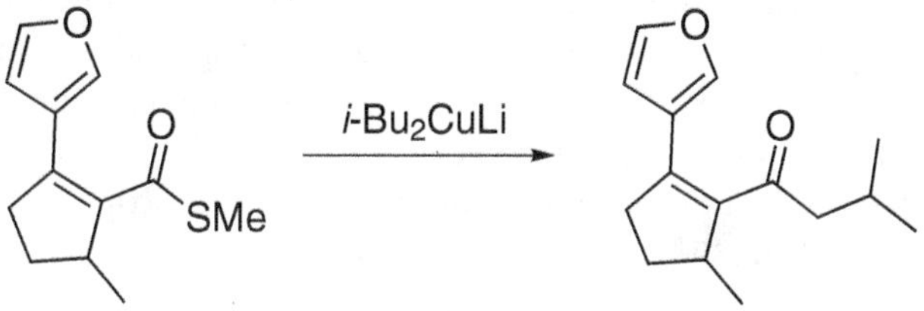

(12.35)

In contrast to their lack of reactivity toward most saturated carbonyl compounds by the 1,2-addition mode, Gilman reagents do react with conjugated carbonyl compounds to give 1,4-addition products, although the reaction is susceptible to steric effects and tends to be sluggish when the β carbon is disubstituted. In fact, copper is one of the few elements whose metal alkyls consistently add in the 1,4-mode to conjugated carbonyl compounds. However, its toxicity has proved to be a major stumbling block to its use in industrial applications. For this reason, considerable effort has been expended in developing reactions based on less toxic organometallic compounds, such as organozinc compounds, and using copper in catalytic amounts. The success of these reactions depends on copper's propensity to undergo metal-metal exchange (***transmetallation***) reactions with metal alkyls of less electronegative metals (e.g., zirconium and zinc, as in Example **12.36**.[47]).

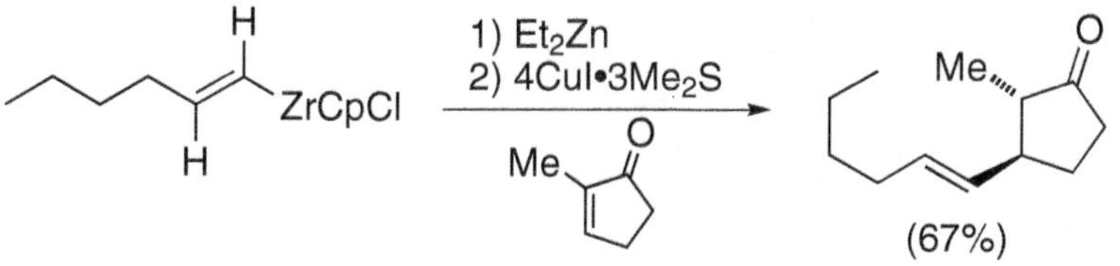

(12.36)

The competition between conjugate addition and substitution in thioesters is nicely illustrated by Example **12.37**. In the starting thioester, where the β position is monosubstituted, the propensity for alkylcuprate reagents is still for 1,4-addition to dominate, as shown by the copper-catalyzed addition of methylmagnesium bromide to the unsaturated thioester. This reaction gives the product of conjugate addition rather than the product of addition to the carbonyl group.[48] It is interesting to note that the reaction of simple Gilman reagents with saturated thioesters is economical—both alkyl groups bound to copper are actually transferred to the electrophile, in contrast to the general observation with these reagents.

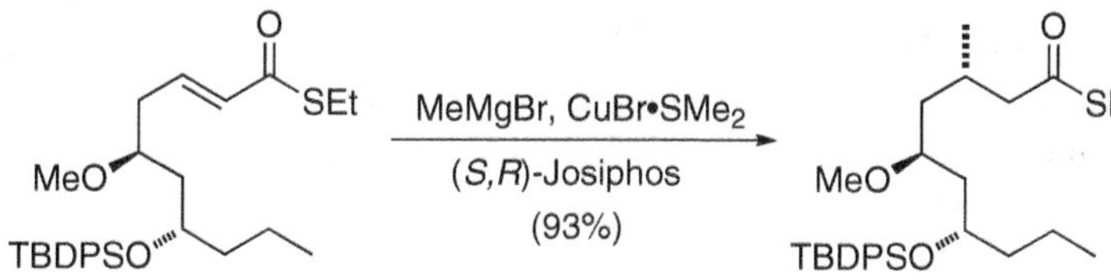

(12.37)

44. (a) Jallabert, C.; Luong-Thi, N.T.; Rivière, H. *Bull. Soc. Chim. Fr.* **1970**, 797. (b) Posner, G.H.; Whitten, C.E. *Tetrahedron Lett.* **1970**, 4647. (c) Luong-Thi, N.T.; Rivière, H.; Bégué, J.-P.; Forestier, C. *Tetrahedron Lett.* **1971**, 2113. (d) Posner, G.H.; Whitten, C.E.; McFarland, P.E. ***J.** Amer. Chem. Soc.* **1972** *94*, 5106.
45. Anderson, R.J.; Henrick, C.A.; Rosenblum, L.D. *J. Am. Chem. Soc.* **1974**, *96*, 36543.
46. Dieter, R.K.; Dieter, J.W. *Chem. Commun.* **1983**, 1378.
47. El-Batta, A.; Bergdahl, M. *Tetrahedron Lett.* **2007**, *48*, 1761.
48. Vintonyak, V.V.; Maier, M.E. *Org. Lett.* **2008**, *10*, 1239.

Organozinc Reagents with Catalytic Copper

Diethylzinc does not react with cyclohexenone in hydrocarbon solvents. However, Noyori and his coworkers discovered that (1) the combination of diethylzinc and a copper salt did accelerate the reaction, and that (2) the addition of *N*-benzylbenzenesulfonamide (using only 0.01 mol % Cu and 0.03 mol % sulfonamide; Example **12.38**) dramatically accelerated the conjugate addition. They also found that this provided a useful entry into zinc enolates that could be trapped further by electrophiles such as π-allylpalladium species and carbonyl compounds.[49] Recently, it has been shown that the zinc enolates formed by copper-catalyzed conjugate addition of dialkylzinc reagents to enones can be allylated to give the *trans* dialkylated product by means of a diiron nonacarbonyltriphenylphosphine catalyst, as shown in Example **12.39**.[50]

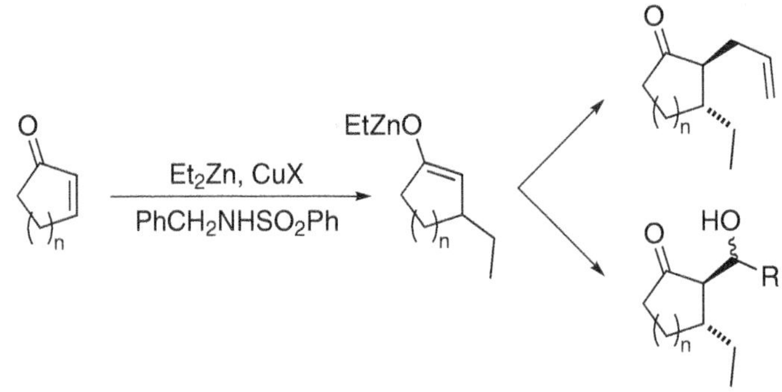

(12.38)

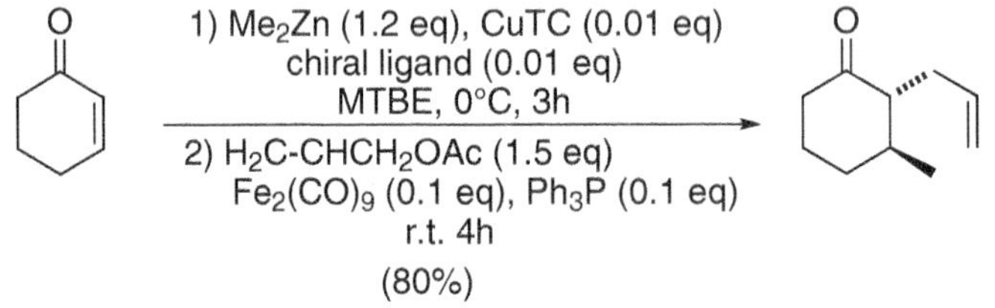

(12.39)

As with alkylation reactions, Lipshutz cuprates perform much better in conjugate additions to enones that do not react with simple cuprate reagents.[51] There is both experimental[52] and computational[53] evidence that the rate determining step of the conjugate addition is the last step of the reaction: C—C bond formation by reductive elimination from a copper (III) intermediate.

The effect of zinc on the regiochemistry of additions to enones is dramatic: while methyllithium adds exclusively (or nearly so) to enones to give products of 1,2-addition, lithium trimethylzincate gives almost exclusively 1,4-addition products. More importantly, perhaps, when a mixed zincate reagent is used, it is found that methyl transfer occurs more slowly than transfer of primary or secondary alkyl groups; the transfer of phenyl occurs much more slowly to give a mixture of products, including 1,2-addition products. This makes mixed zincates, formed by the reaction between zinc chloride TMEDA complex, methyllithium, and an alkylmagnesium halide (e.g., Example **12.40**), especially useful reagents for effecting conjugate additions to enones.[54]

---

49. Kitamura, M.; Miki, T.; Nakano, K.; Noyori, R. *Tetrahedron* **1996**, 5141.
50. Jarugumilli, G.K.; Cook, S.P. *Org. Lett.* **2011**, *13*, 1904.
51. (a) Yamamoto, Y.; Maruyama, K. *J. Am. Chem. Soc.* **1978**, *100*, 3240. (b) Brown, J.D.; Foley, M.A.; Comins, D.L. *J. Am. Chem. Soc.* **1988**, *110*, 7445.
52. Frantz, D.E.; Singleton, D.A.; Snyder, J.P. *J. Am. Chem. Soc.* **1997**, *119*, 3383.
53. (a) Nakamura, E.; Mori, S.; Morokuma, K. *J. Am. Chem. Soc.* **1997**, *119*, 4900. (b) Mori, S.; Nakamura, E. *Chem. Eur. J.* **1999**, *5*, 1534. (c) Nakamura, E.; Yamanaka, M. *J. Am. Chem. Soc.* **1999**, *121*, 8941. (d) Krauss, S. R.; Smith, S. G. *J. Am. Chem. Soc.* **1981**, *103*, 141. (e) Nakamura, E.; Yamanaka, M.; Mori, S. *J. Am. Chem. Soc.* **2000**, *122*, 1826.
54. Kjonaas, R.A.; Hoffer, R.K. *J. Org. Chem.* **1988**, *53*, 4133.

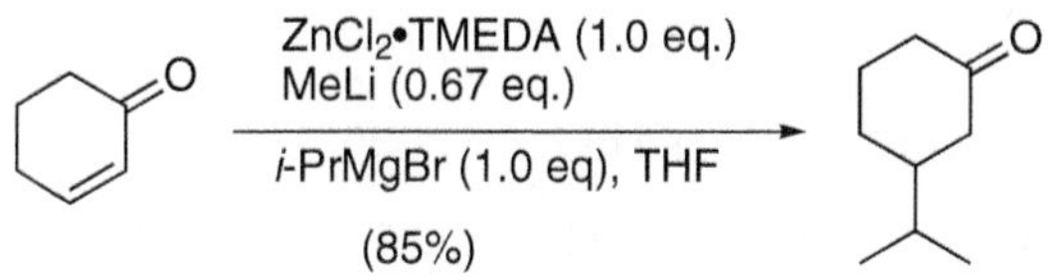

(12.40)

Worked Problem

12-3 What is the major organic product of each of the following reactions?

(a) 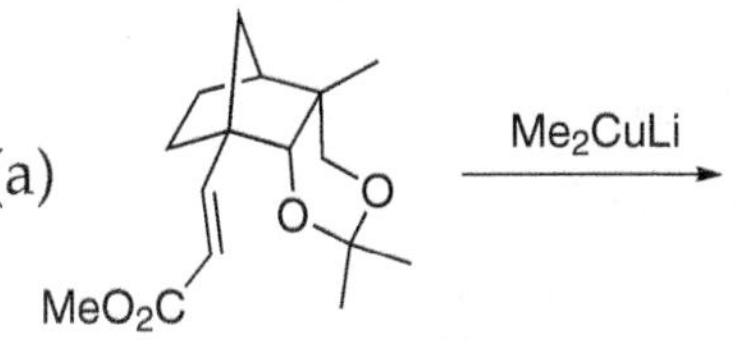

(b) 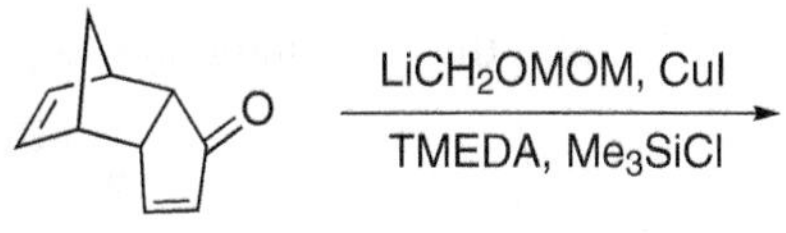

§Answers below.

Drill Problems

What are the products of the treatment of the compounds in list A with each of the reagents in list B? Note that not all combinations will give a useful reaction.

List A:

(i) 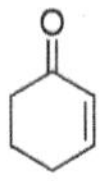
(2) 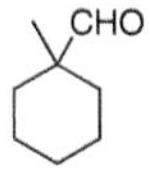

(3) 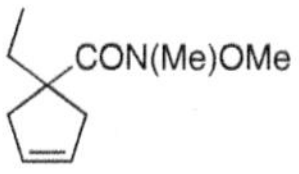

(4) 
(5) 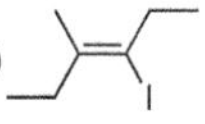

List B:

(i) (1) MeC ≡ CNa, THF, 0°C; (2) H_2O

(ii) (1) Bu_2CuLi, THF, 0°C; (2) NH_4Cl, H_2O

(iii) (1) EtMgBr, Et_2O, 0°C; (2) NH_4Cl, H_2O

(continued)

§ Answers to Worked Problem:

(a) 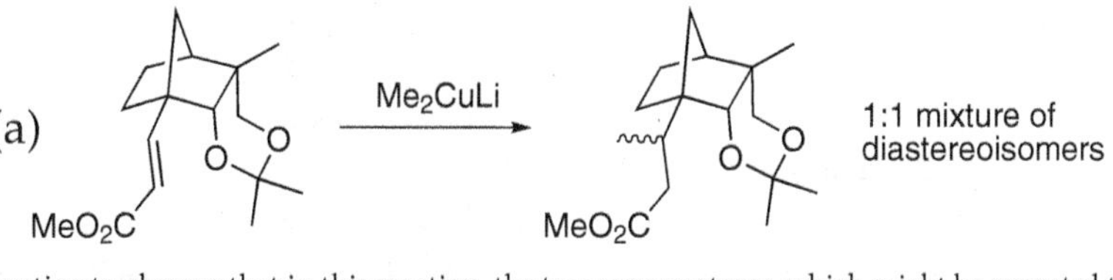

1:1 mixture of diastereoisomers

It is interesting to observe that in this reaction, the two oxygen atoms, which might be expected to coordinate the lithium, and thus direct the copper reagent to one face of the alkene selectively, do not bias the stereochemistry of the addition.

(b) 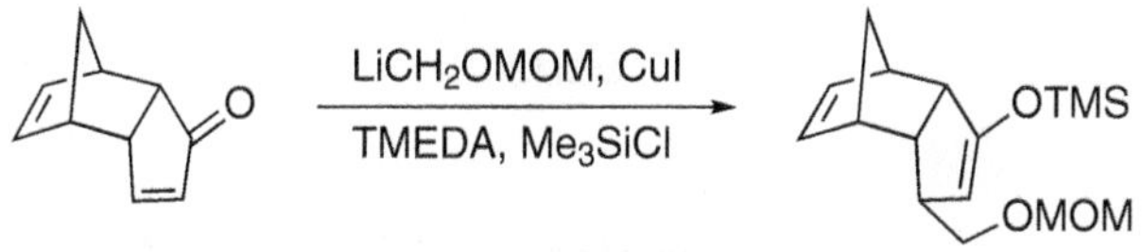

In this reaction, the enolate formed by addition is trapped with trimethylsilyl chloride to give the silyl enol ether. This is a widely used technique to prevent this enolate from entering into competing reactions.

References: (a) *J. Org. Chem.* **1978**, *43*, 1750. (b) *Org. Lett.* **2003**, *5*, 1321.

(Drill Problems continued)

(iv) (1) Me_2Zn, CH_2Cl_2; (2) NH_4Cl, H_2O

(v) (1) PhMgBr, CuCN, Et_2O, 0°C; (2) NH_4Cl, H_2O

(vi) (1) $BrZnCH_2CO_2Et$, Et_2O; (2) NH_4Cl, H_2O

Problems

12-2 What is the expected major organic product of each of the following reactions?

(a)

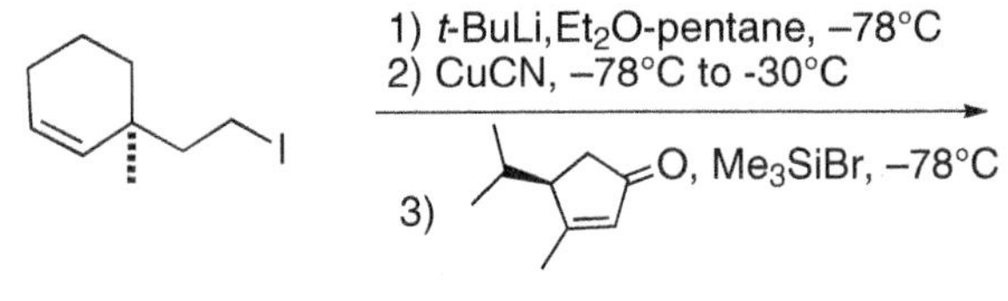

(b)

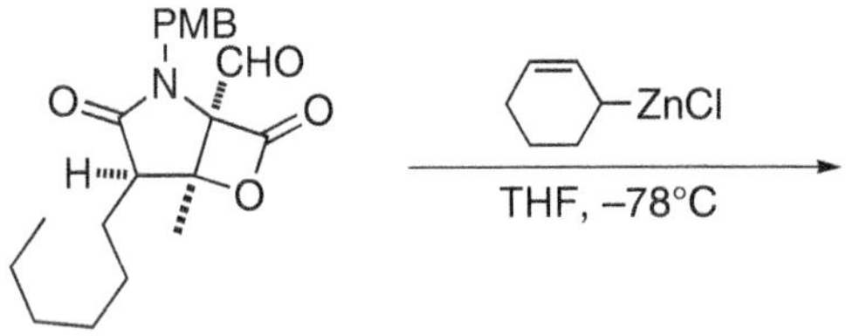

(c)

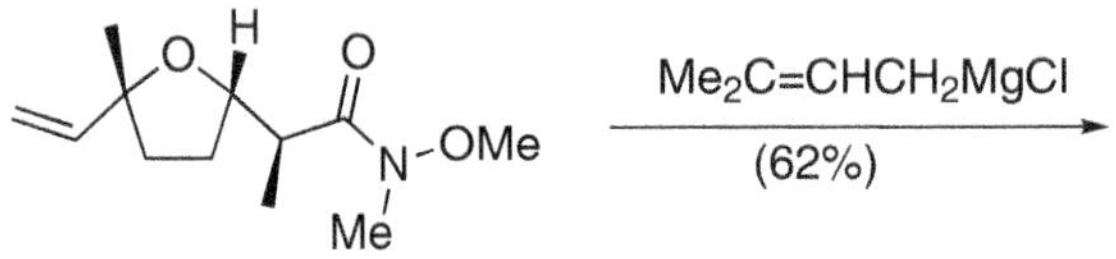

(d)

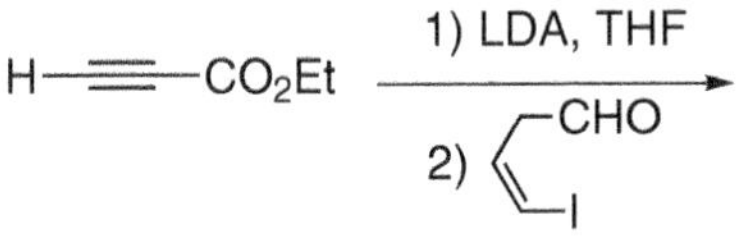

(e)

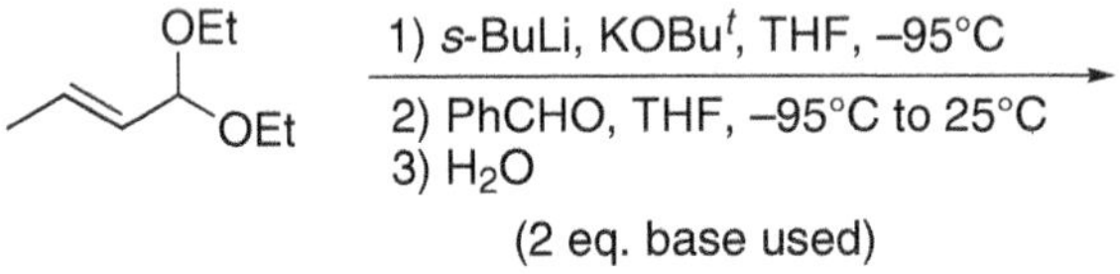

References: (a) *J. Am. Chem. Soc.* **2006**, *128*, 13095. (b) *Org. Lett.* **2007**, *9*, 2143. (c) *Org. Lett.* **2009**, *11*, 2217. (d) *Org. Lett.* **2006**, *8*, 345. (e) *Tetrahedron* **1994**, *50*, 12463.

12-3 The two reactions below would appear to contradict two generalizations made above about (1) the regiochemistry of deprotonations with "LICKOR," and (2) nucleophilic displacements of alkyl halides with alkyllithiums. If so, how can the generalizations be modified to make them useful again? What caveats does this raise about using this reagent in synthesis? [*J. Am. Chem. Soc.* 1987, *109*, 7553].

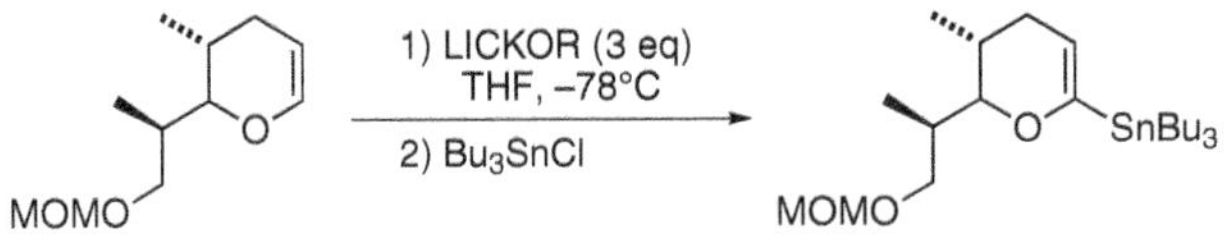

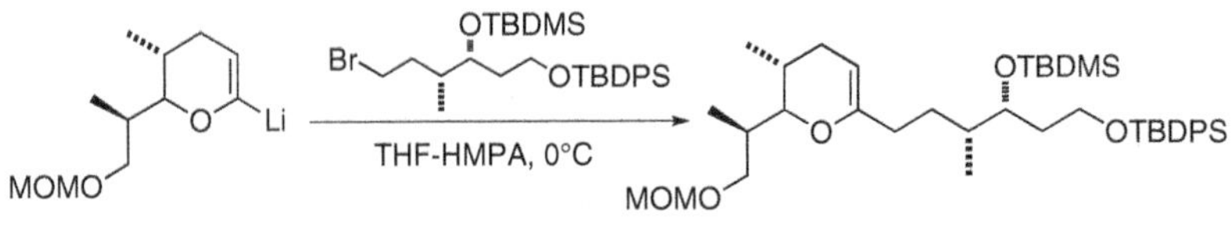

12-4 What are the major organic products of each of the following reaction sequences?

(a) 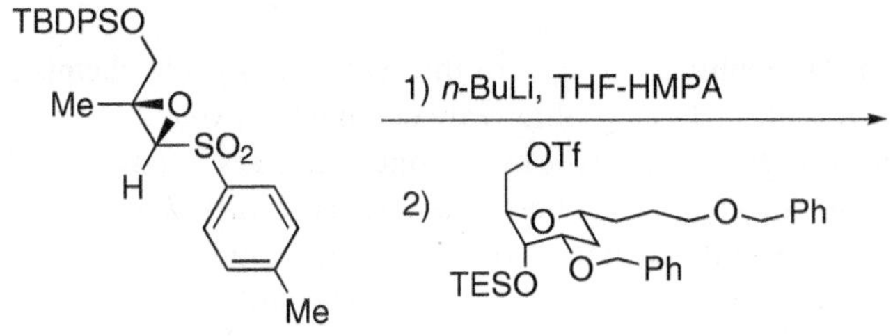

(b) 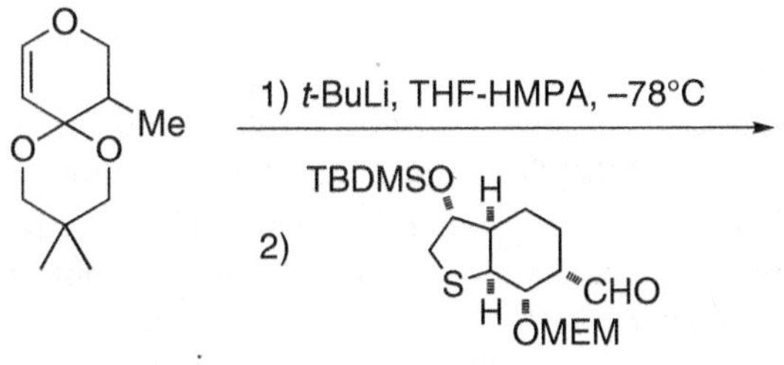

(c) 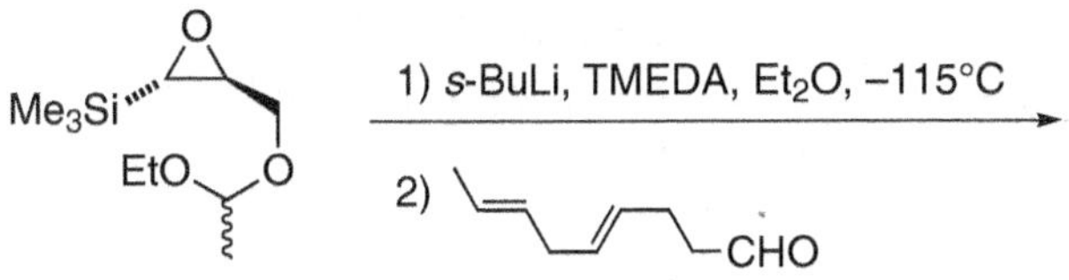

References: (a) *J. Am. Chem. Soc.* **1997**, *119*, 4557. (b) *J. Am. Chem. Soc.* **1991**, *113*, 4037. (c) *J. Org. Chem.* **1997**, *62*, 636.

Metal Alkyl Complexes of Other Transition Metals

As pointed out at the beginning of this section, this area of organic synthesis is now so important that it merits its own chapter. As noted earlier in this chapter, we will discuss the uses of transition metal catalysts in synthesis in Chapter 17.

Reaction Synopses

1,2-Addition of Metal Alkyls to Aldehydes, Ketones, and Imines

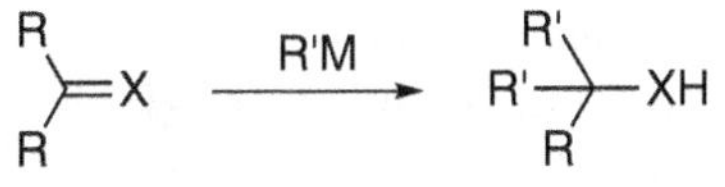

X = O, NR

Reagents:

X = O: R'MgX, Et_2O; R'Li, Et_2O; R_3ZnLi, Et_2O; etc.
X = NR: RNa, THF, –75°C; R_3ZnLi, THF; etc.

1,2-Addition of Metal Alkyls to Acid Derivatives

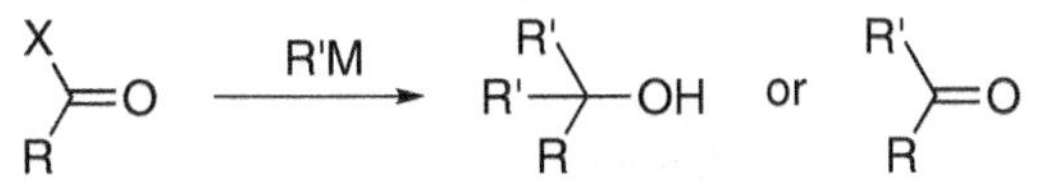

X = Cl, OR, NR_2

Reagents:

X = Cl: R'Li, THF; R'MgX, Et_2O

12.3 Metal Enolates: Versatile Carbon Nucleophiles

Of all the carbon nucleophiles available to the synthetic organic chemist, few rival the enolate anion for versatility. Enolate anions are obtained in two major ways: (1) deprotonation of a carbonyl compound or nitrile by a strong base and (2) reduction of a conjugated carbonyl system with an active metal under conditions of the Birch reduction. Of the two methods, deprotonation of the carbonyl precursor with a hindered amide base has become the routine method for generating these valuable nucleophiles.

Enolate Regiochemistry

In general, enolate regiochemistry is fairly easy to control.[55] The deprotonation reaction has a stereoelectronic requirement that favors the removal of the proton that best overlaps the π^* orbital of the carbonyl group; this proton is usually close to perpendicular to the plane of the carbonyl group. In open-chain compounds, which are conformationally flexible, this requirement has no effect on the regiochemistry of deprotonation because this conformation can, in principle, be attained by all α-hydrogens. The ensuing result is that *deprotonation under kinetic conditions leads to the less substituted enolate*—deprotonation at the less substituted α-carbon. Under thermodynamic (equilibrating) conditions, however, the more substituted enolate is formed as the major product (Example **12.41**).

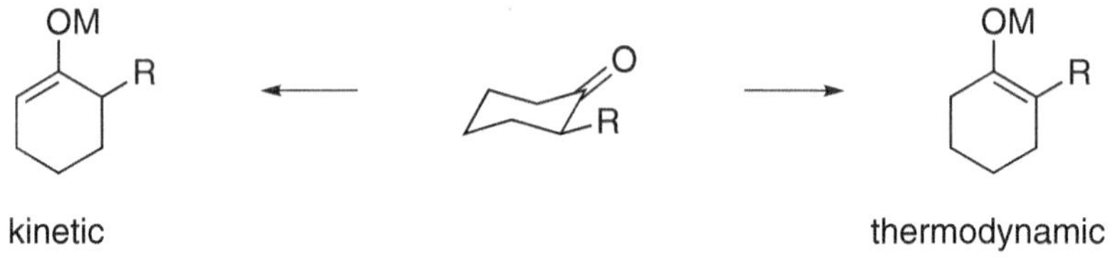

(12.41)

In cyclic compounds such as cyclohexanones, kinetic deprotonation occurs by attack at the axial, rather than at the equatorial α-hydrogen. In monosubstituted cyclohexanones, both α-carbons carry an axial hydrogen, so the "normal" regiochemical outcome under kinetic conditions is observed, and the deprotonation occurs at the unsubstituted α-carbon. On occasion, however, the stereoelectronic requirement will override the normal regiochemical outcome, and kinetic deprotonation occurs at the more substituted site. This is illustrated by Examples **12.42** and **12.43.** In the kinetic deprotonation (**12.42**), the bridgehead hydrogen overlaps best with the carbonyl π^* orbital, so despite forming a bridgehead alkene, this proton is the one removed during deprotonation by slow addition of ketone to excess base at –78°C. Under thermodynamic conditions (**12.43**), the less stable bridgehead enolate is able to equilibrate with the more stable non-bridgehead enolate, despite it being less substituted.[56]

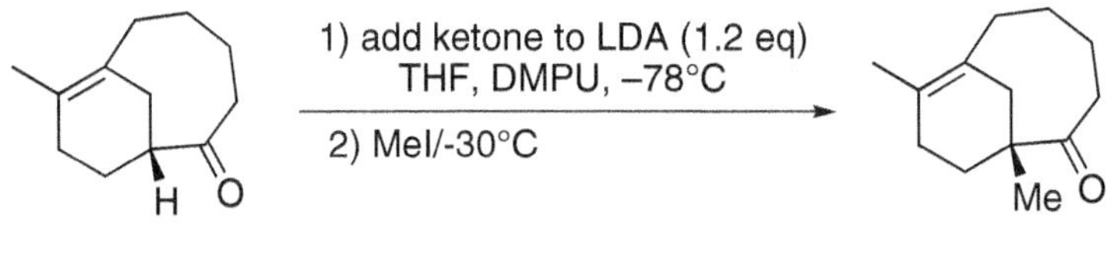

(12.42)

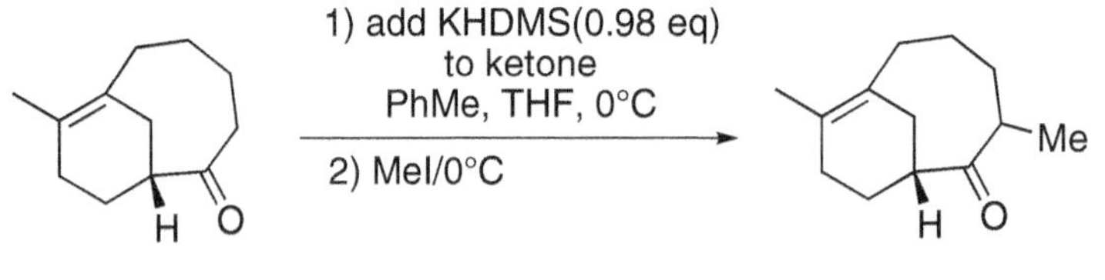

(12.43)

55. Review: D'Angelo, J. *Tetrahedron* **1976**, *32*, 2979.
56. Shea, K.J.; Sakata, S.K. *Tetrahedron Lett.* **1992**, *33*, 4261.

Enolates from Conjugated Ketones: Dienolate Anions

The conjugation of the carbonyl group of a ketone introduces a new regiochemical factor into the deprotonation reaction. In cyclohexenone, for example, the deprotonation may occur at either the α' position or the γ position (Example **12.44**). Deprotonation at the α' carbon gives a cross-conjugated dienolate and is the favored process under conditions of kinetic control (low temperatures, excess lithium amide bases). Deprotonation at the γ carbon, on the other hand, gives a dienolate anion that is conjugated in a linear manner. This dienolate is lower in energy than the cross-conjugated isomer and is therefore the dienolate favored under thermodynamic or equilibrating conditions (sodium or potassium alkoxide bases, higher temperatures, no excess base).

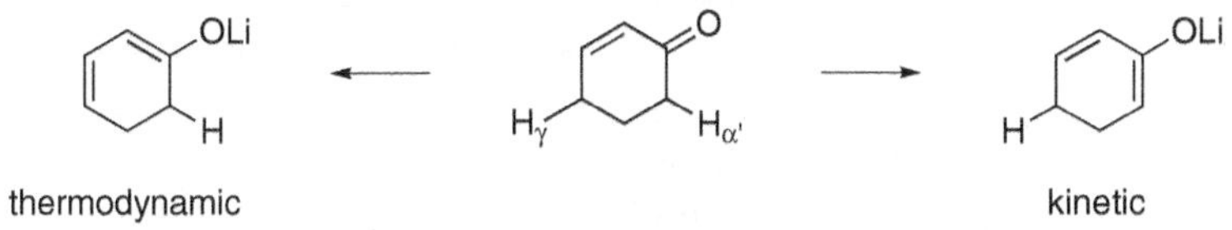

(12.44)

Cross-conjugated dienolate anions react like simple enolates, and there is no regiochemical ambiguity in their reactions: all reactions where the enolate acts as a carbon nucleophile occur through the α' carbon. Dienolates that are linearly conjugated, however, may react as carbon nucleophiles through either the α carbon or the γ carbon. Which position is attacked by the electrophile is influenced by both the exact nature of the substituents on the dienolate anion and by the associated metal. For example, although lithium dienolates tend to react with allyl halides at the α carbon by an S_N2 mechanism to give the β,γ-unsaturated carbonyl product, the more covalent (softer) copper (I) enolate tends to react through the γ carbon by the S_N2' mechanism to give the conjugated carbonyl product. This was especially the case when the dianions of carboxylic acids were used as the nucleophiles.[57]

Enolate Stereochemistry

The stereochemical outcome of many important reactions of enolates is determined by the stereochemistry of the enolate. For this reason, controlling enolate stereochemistry is often a key task in designing a synthesis. There are a number of factors that influence enolate stereochemistry, as revealed by the study of symmetrical open-chain ketones (which eliminate complications due to enolate regiochemistry).

Effect of Base and Carbonyl Substituent

The deprotonation of ethyl ketones and propionyl derivatives by strong bases such as lithium diisopropylamide (LDA) is fairly well accounted for by the model proposed by Ireland.[58] In this model, hydrogen transfer occurs through a chairlike transition state, as shown in Figure 12.2. The enolate geometry is then determined by the balance of the two possible sources of torsional strain: (1) a pseudo-1,3-diaxial interaction between the methyl group and R_1 or R_2

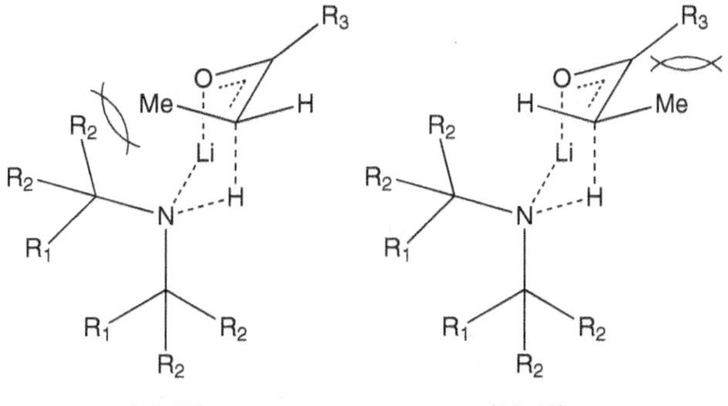

(12.45) **(12.46)**

Figure 12.2 Transition state models for the deprotonation of ethyl ketones

57. Katzenellenbogen, J.A.; Crumrine, A.L. *J. Am. Chem. Soc.* **1976**, *98*, 4925.
58. Ireland, R.E.; Muller, R.H.; Willard, A.K. *J. Am. Chem. Soc.* **1976**, *98*, 2868.

Table 12.4 Effects of Base and Substrate Structure on Enolate Geometry

R_1	R_2	R_3	*Z/E* ratio	Ref.
H	Me	OMe	5 : 95	59
H	Me	$OCMe_3$	5 : 95	59
H	Me	$SCMe_3$	10 : 90	60
$-(CH_2)_3-$	Me	Et	14 : 86	61
H	Me	Et	23 : 77 (30 : 70)	59, 62
H	Me	$CHMe_2$	60 : 40	59
$-(CH_2)_3-$	Me	$CHMe_2$	32 : 68	59
H	Me	CMe_3	>98 : 2	59
H	Me	Ph	>98 : 2	59
H	Me	NEt_2	≥97 : 3	63
H	Me	$-(CH_2)_4-$	≥97 : 3	63

in the activated complex leading to the *Z* enolate and (2) the buildup of allylic strain between the methyl group and R_3 as the reaction leading to the *E* enolate proceeds. The net result of these strain effects is that when the group R_3 is small (e.g., alkoxy), the allylic strain is less than the 1,3-diaxial interaction, and *E* enolate predominates (**12.45**). When R_3 is large (e.g., *tert*-butyl), the opposite is true, and the *Z* enolate predominates (**12.46**). Note how the planarity of the amide group leads to high levels of allylic strain and therefore to high proportions of the *Z* enolate. The results of a number of studies are summarized in Table 12.4.

As the base becomes more hindered, the proportion of the *Z* enolate in the product mixture increases, as shown in Figure 12.3 for a range of strong lithium amide bases. This result is not compatible with the chairlike model of deprotonation but is compatible with a boatlike conformation instead (Figure 12.4). Thus, it appears that as the base becomes more hindered, the chair conformation (which would require a large group in an axial position) becomes disfavored relative to the boat.

Figure 12.3 Effects of base size on enolate geometry

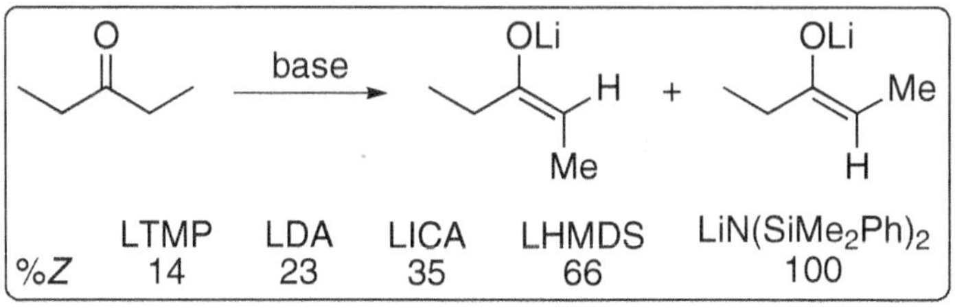

Figure 12.4 The boat transition state for deprotonation is favored when the base is very hindered.

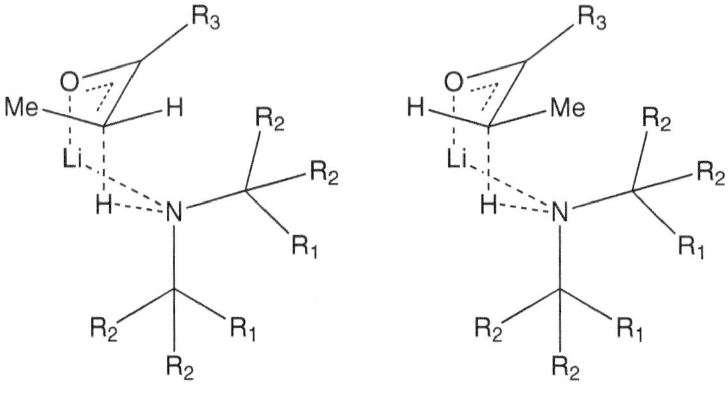

(12.47) (12.48)

59. Heathcock, C.H.; Buse, C.T.; Kleschick, W.A.; Pirrung, M.A.; Sohn, J.E.; Lampe, J. *J. Org. Chem.* **1980**, *45*, 1066.

60. Evans, D.A.; McGee, L.R. *Tetrahedron Lett.* **1980**, 3975.

61. (a) Fataftah, A.A.; Kopka, I.E.; Rathke, M.W. *J. Am. Chem. Soc.* 1980, *102*, 3959. (b) Nakamura, E.; Hashimoto, K.; Kuwajima, I. *Tetrahedron Lett.* **1978**, 2079.

62. Ireland, R.E.; Muller, R.H.; Willard, A.K. *J. Am. Chem. Soc.* **1976**, *98*, 2868.

63. Takacs, J.M. Ph.D. Thesis, California Institute of Technology, Pasadena, CA, 1981; cited in Evans, D.A. In Morrison, J.D., Ed. *Asymmetric Synthesis. Stereodeifferentiating Addition Reactions, Part B.* (Academic Press: New York, 1984), vol. 3, ch. 1, p. 1.

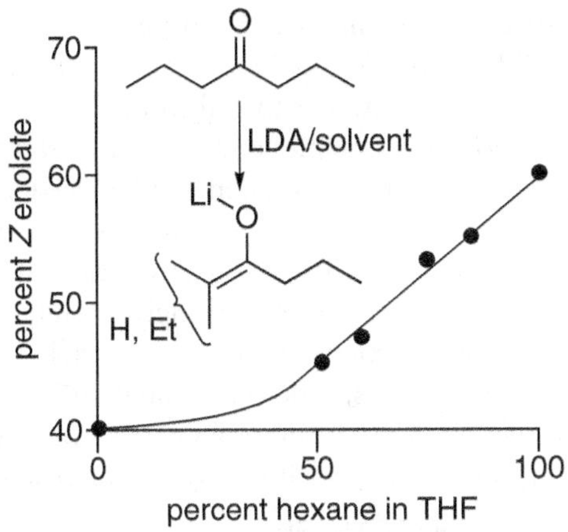

Figure 12.5 Effects of solvent composition on stereochemistry of enolate from 4-heptanone

Solvent

In 1976, Ireland and his coworkers discovered a remarkable effect of solvent on enolate geometry in the deprotonation of esters.62 In tetrahydrofuran (THF) as the solvent (or, more correctly, THF-hexane, because the butyllithium used to deprotonate the diisopropylamine was in hexane), the deprotonation of methyl propionate gave the *E(O)* enolate as the major component of a 95:5 mixture. When the solvent was changed to 23% hexamethylphosphoramide (HMPA)-THF (or the same solvent containing the hexane from the butyllithium), the stereochemistry was reversed, and the *Z(O)* enolate now became the major component of an 84:16 mixture. Four years later, the same effect was observed in the deprotonation of 3-pentanone, by Rathke and his coworkers,[64] who also noted that the presence of a small excess of 3-pentanone in the mixture was able to effect the same reversal of stereochemistry. The effect of the hexane on the geometry of the enolate obtained by LDA deprotonation of 4-heptanone was reported by Heathcock and Munchhoff.[65] They observed that, as the proportion of hexane in the solvent system increased, the proportion of the *Z* enolate increased (Figure 12.5), leading to a reversal of stereochemistry of the major enolate; from 50 to 100% hexane, the increase in the amount of the *Z* enolate was approximately linear with the concentration of hexane in the solvent, and below approximately 33% hexane, a leveling effect is observed.

The effects of solvent have long been proposed to be the result of changes in aggregation of the lithium amide base,[66] but a series of studies of the solution structure of lithium amide bases reveals that the addition of HMPA to a THF solution of LDA, for example, does not change the aggregation state of the base.[67] Instead, the structure of the activated complex appears to be very dependent on the nature of the substrate and the solvent together, and there is evidence that the most active form of the base may, in fact, be a dimer. The results of two decades of work in this area have been summarized by Collum in a useful review that provides guidelines for optimizing reactions of LDA with a range of substrates.[68]

Alkylation of Enolates

The highest energy occupied molecular orbital (HOMO) of the enolate anion is ψ_2 of the conjugated system, which makes the enlate anion an ambident nucleophile, exhibiting both hard (oxygen-centered) and soft (carbon-centered) reactivity. Enolate anions react

64. Fataftah, A.A.; Kopka, I.E., Rathke, M.W. *J. Am. Chem. Soc.* **1980**, *102*, 3959.
65. Munchhoff, M.J.; Heathcock, C.H. *Tetrahedron Lett.* **1982**, *33*, 8005.
66. (a) Wardell J.L. In Wilkinson, G.; Stone, F.G.A.; Abel, E.W., Eds. *Comprehensive Organometallic Chemistry*, Vol. 1 (Pergamon: New York, 1982), chap. 2 (b) Szwarc, M., Ed. *Ions and Ion Pairs in Organic Reactions*, (Wiley: New York, 1972), Vols. 1–2.
67. Collum, D.B. *Acc. Chem. Res.* **1993**, *26*, 227.
68. Collum, D.B.; McNeil, A.J.; Ramirez, A. *Angew. Chem. Int. Ed.* **2007**, *46*, 3002.

with hard electrophiles such as silyl halides and phosphoryl halides to give derivatives of the enol as the product. With softer nucleophiles, such as alkyl halides and carbonyl compounds, the major product of the reaction is the substituted carbonyl compound.

Enolate anions are powerful nucleophiles and generally react well with alkyl halides and epoxides. The reactions are typical S_N2 reactions, with inversion of configuration at the electrophilic carbon. The alkylation of an enolate anion is a strongly exothermic reaction, which means that the transition state for alkylation occurs early along the reaction coordinate, with the result that the activated complex strongly resembles the starting enolate.[69] This is manifest in the alkylation reactions of 4-*tert*-butylcyclohexanone (Example **12.49**). Unlike the reduction of cyclohexanones, which occur predominantly from the axial direction, the alkylation of the enolate of this ketone does not exhibit high levels of diastereoselectivity that one might expect by analogy with the reduction reaction.

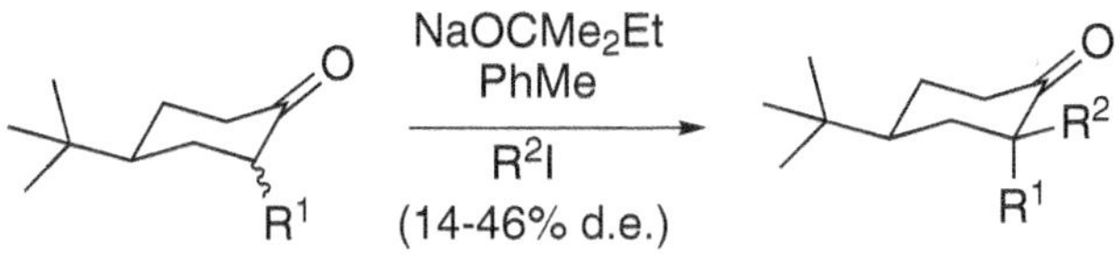

(12.49)

Conjugation of the Enolate: Dienolate Anions

Linear conjugation of the enolate anion with a further double bond gives rise to an additional regiochemical complication to the alkylation action, because the *C*-alkylation may now occur at either the α carbon or the γ carbon. Attack at both positions has been observed. However, the largest lobes in the HOMO of the dienolate are at the center of the system, which favors attack of the electrophilic reagent at the center of the conjugated system.

Which position is attacked by the alkyl halide is influenced by both the exact nature of the substituents on the dienolate anion and by the associated metal. Alkali metal dienolates—whether kinetic (Example **12.50**) or thermodynamic (Examples **12.51** and **12.52**)—tend to react with alkyl halides at the α carbon by an S_N2 mechanism to give the β,γ-unsaturated carbonyl product. By contrast, more covalent (softer) enolates (e.g., copper (I) enolates) tend to react through the γ carbon by the S_N2' mechanism to give the conjugated carbonyl product. This is especially the case when the dianions of carboxylic acids were used as the nucleophiles.[70]

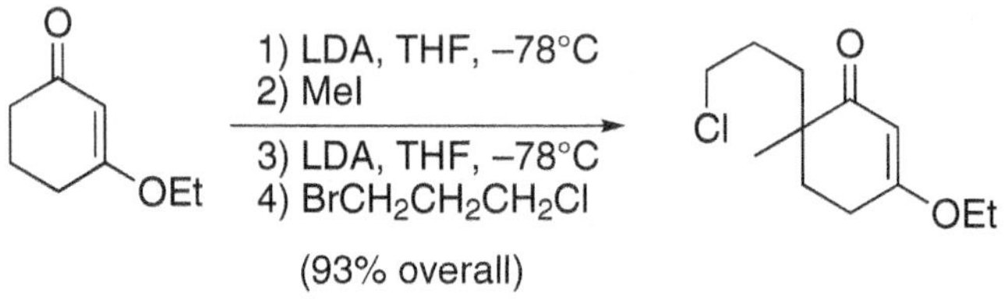

(12.50)[71]

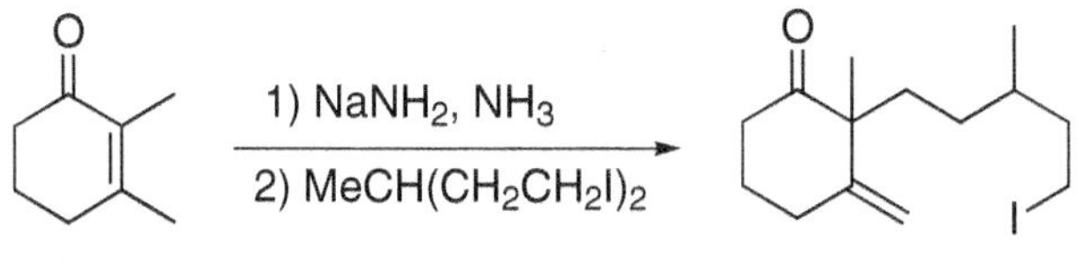

(12.51)[72]

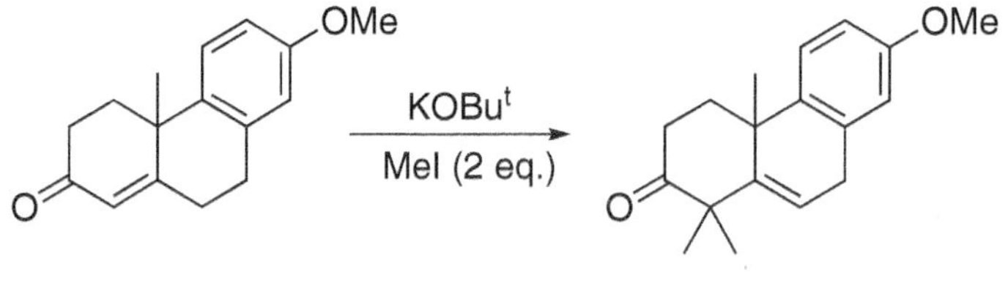

(12.52)[73]

69. House, H.O.; Tefertiller, B.A.; Imstead, H.D. *J. Org. Chem.* **1968**, *33*, 935.
70. Katzenellenbogen, J.A.; Crumrine, A.L. *J. Am. Chem. Soc.* **1976**, *98*, 4925.
71. Schultz, A.G.; Dittani, J.P. *J. Org. Chem.* **1983**, *48*, 2318.
72. Fukamiya, N.; Kato, M.; Yoshikoshi, A. *Chem. Commun.* **1971**, 1120.
73. Stork, G.; Meisels, A.; Davies, J.E. *J. Am. Chem. Soc.* **1963**, *85*, 3419.

O-Alkylation and Acylation of Enolates

As the electrophiles become harder, enolates react through oxygen rather than carbon. Thus, the reaction between an enolate and an acid chloride (which, although formally soft, is a relatively hard electrophile) generally gives the enol ester, rather than the α-acyl product, as shown by Example **12.53**, which gives a mixed enol carbonate ester.[74]

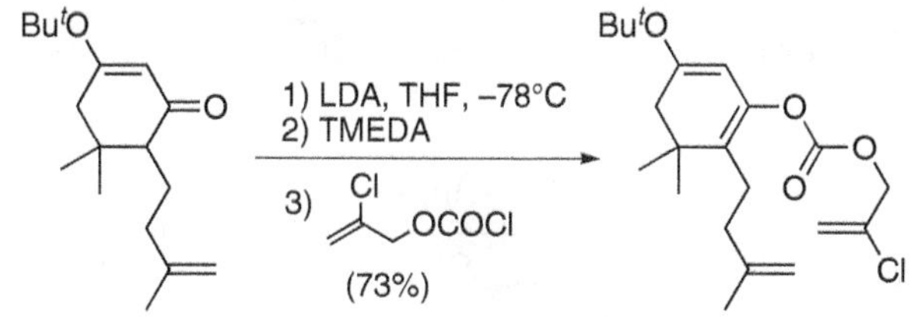

(12.53)

By far the most commonly used reagents to obtain enol derivatives are silyl halides, and a wide range of trialkylsilyl chlorides have been used to convert an enolate anion to the corresponding enol trialkylsilyl ether. The same reagents are widely used in the selective protection of alcohols.[75] We will discuss these reactions in much more detail when we discuss the use of protecting groups in organic synthesis.

The other class of acid chlorides that convert an enolate anion to the corresponding enol ester is the phosphoric acid chlorides, often diethyl phosphorochloridate.[76] This reagent was used in one key step of the synthesis of zoapatanol (Example **12.54**).[77] The enol phosphates derived in this manner are most frequently then subjected to Birch reduction to give the corresponding alkene.

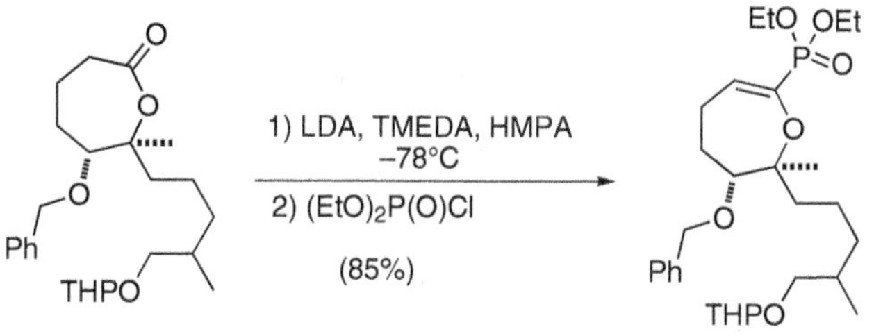

(12.54)

Drill Problems

What are the products of the treatment of the compounds in list A with each of the reagents in list B?

List A:

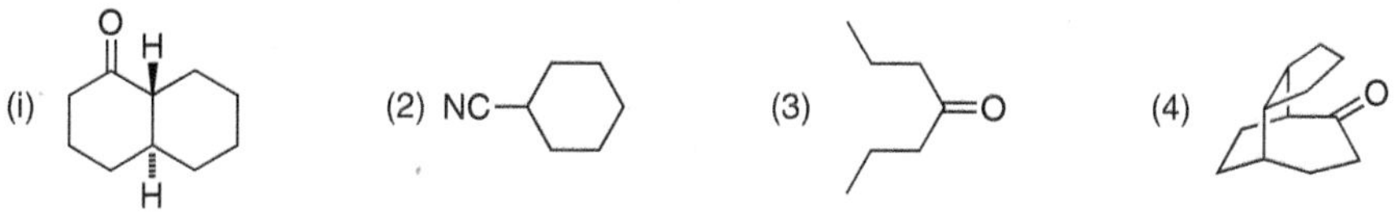

List B:

(i) (1) LDA, THF, –78°C; (2) CD_3I, –78° to r.t.

(ii) KOBu^t, THF, MeI (excess)

(iii) (1) LDA, THF, –78°C; (2) PhCHO, –78°C, 2 min; (3) NH_4Cl, H_2O

(iv) (1) LICA, THF, –78°C; (2) $ICH_2C(=CH_2)CH_2CH_3$

---

74. White, D.E.; Stewart, I.C.; Grubbs, R.H.; Stoltz, B.M. *J. Am. Chem. Soc.* **2008**, *130*, 810.
75. (a) Nelson, T.D.; Crouch, R.D. *Synthesis* **1996**, 1031. (b) Crouch, R.D. *Tetrahedron* **2004**, *60*, 5833. (c) Greene, T.W.; Wuts, P.G.M. *Protecting Groups in Organic Synthesis*, 3rd. ed. (Wiley: New York, 1991). (d) Higashibayashi, S.; Shinko, K.; Ishizu, T.; Hashimoto, K.; Shirahama, H.; Nakata, M. *Synlett* **2000**, 1306.
76. Ireland, R.E.; Pfister, G. *Tetrahedron Lett.* **1969**, 2145.
77. Kane, V.V.; Doyle, D.L. *Tetrahedron Lett.* **1981**, *22*, 3027.

Problems

12-5 What is the major organic product of each of the following reactions?

(a) 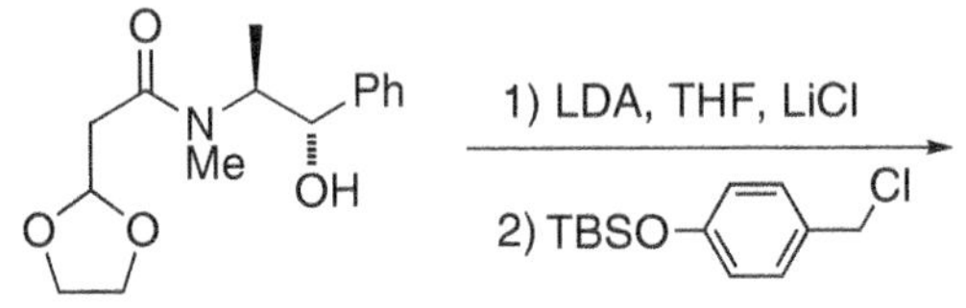

(b) 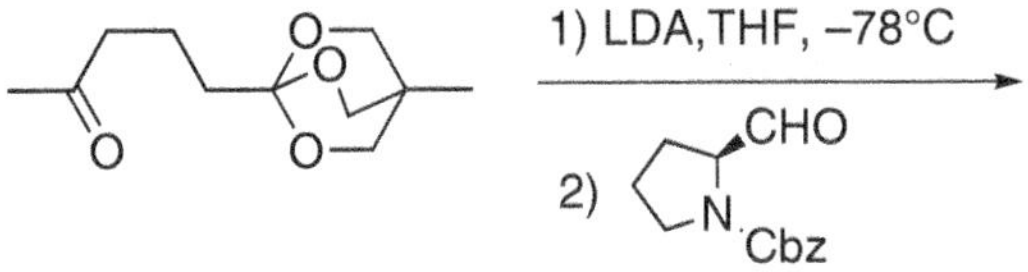

(c) 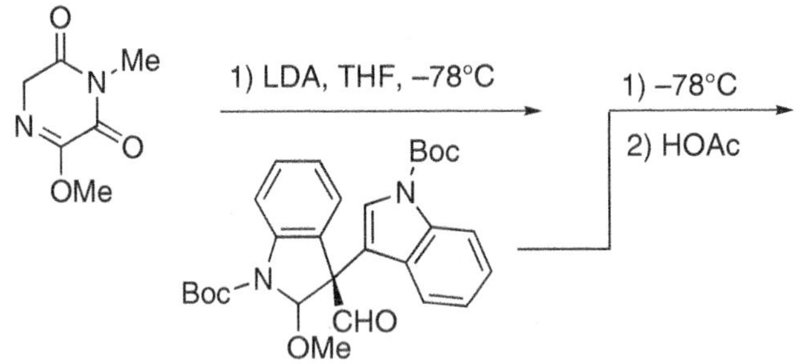

(d) 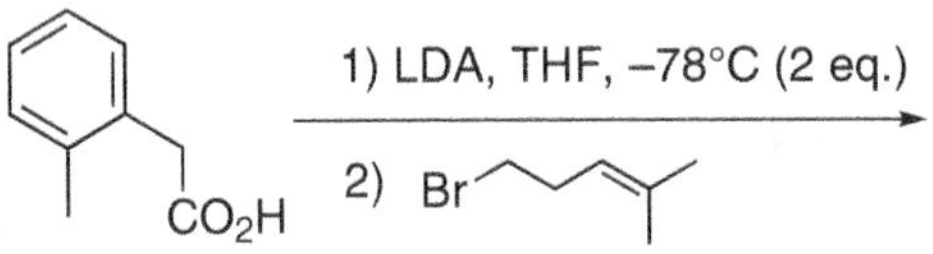

(e) 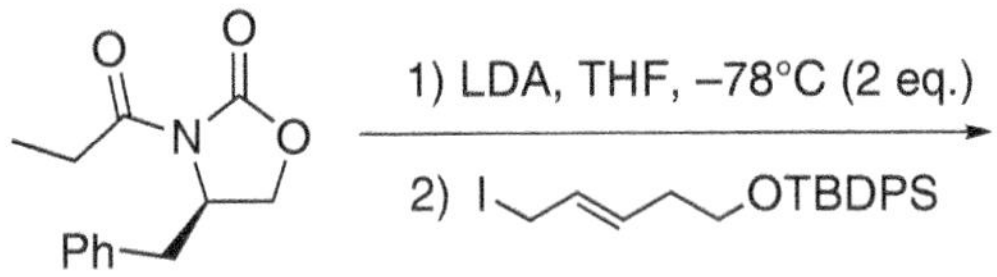

(f) 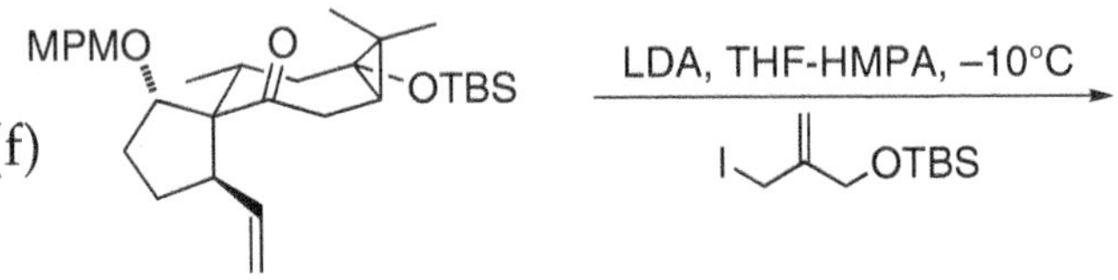

(g) 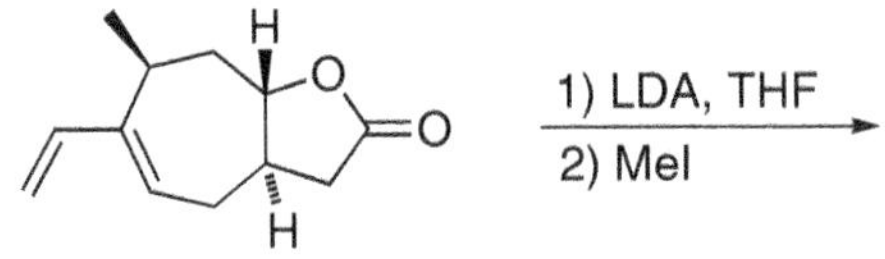

References: (a) *J. Am. Chem. Soc.* **2009**, *131*, 16905. (b) *Org. Lett.* **2005**, *7*, 4419. (c) *J. Am. Chem. Soc.* **2011**, *133*, 6549. (d) *J. Am. Chem. Soc.* **2009**, *131*, 452. (e) *Org. Lett.* **2008**, *10*, 1247. (f) *Org. Lett.* **2011**, *13*, 2160. (g) *Org. Lett.* **2005**, *7*, 3371.

12-6 What is the major organic product expected from each of the following reactions?

(a) 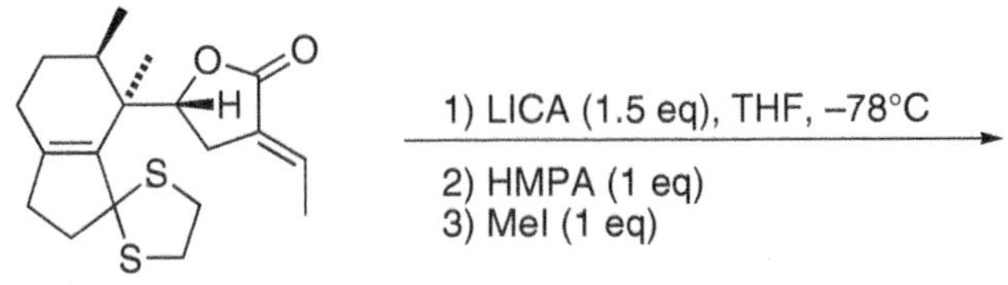

(b) 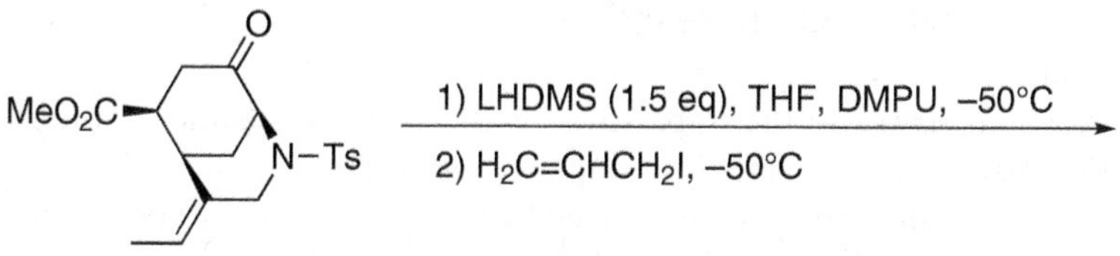

This reaction gives a single regioisomer of the product. Why?

(c) 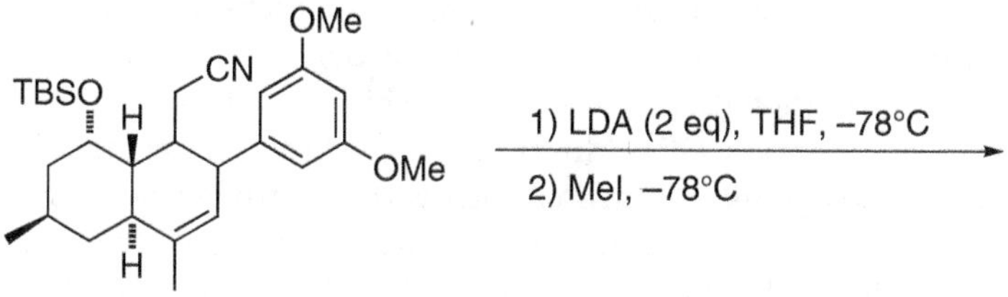

References: (a) *J. Org. Chem.* **1988**, *53*, 1441. (b) *J. Am. Chem. Soc.* **2011**, *133*, 8877. (c) *Org. Lett.* **2007**, *9*, 1461.

Reaction Synopses

Enolate Generation

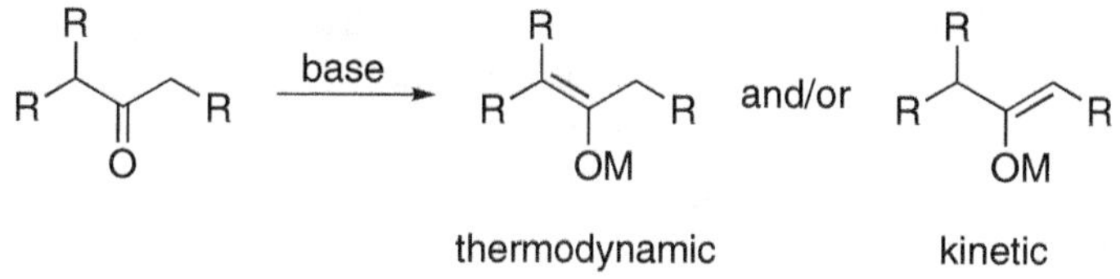

Kinetic enolate usually less substituted; favored by amide base

Thermodynamic enolate more substituted; favored by alkoxide base

Effect of base on regiochemistry:
Thermodynamic enolate: NaH, PhH; $NaNH_2$, THF; KOBu^t, *t*-BuOH; etc.
Kinetic enolate: LDA, THF, –78°C; *t*-BuLi, THF, –100°C; etc.

Sterically hindered bases favor *Z(O)* enolate.

Small groups (e.g., Me, OR) bonded to carbonyl carbon favor *E(O)* enolate; large groups (e.g., *t*-Bu, NR_2) favor *Z(O)* enolate.

Trapping of Enolates

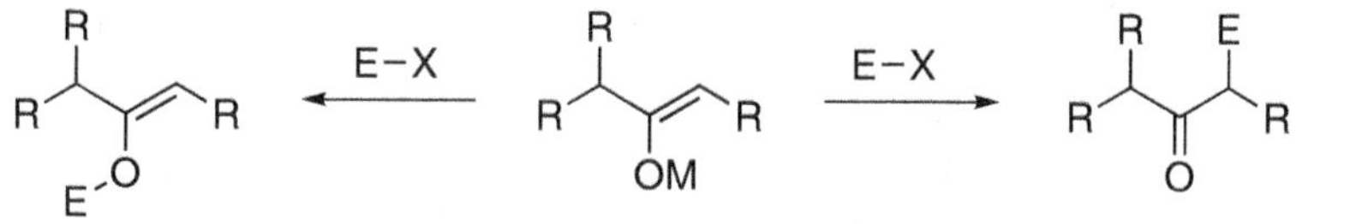

E = C=O, P=O, SiR_3, etc. E = R, RS, RSe, etc.

Reagents: Me_3SiCl, *t*-$BuMe_2SiCl$, *t*-$BuPh_2SiCl$, *i*-Pr_3SiCl, etc.;
or RCOCl, $(RO)_2P(O)Cl$, etc. (*O*-acylation);
or RI, RBr, ROTs, TOTf, etc. (*C*-alkylation);
or RSeSeR, RSeBr, RSBr, etc.

Hard electrophiles tend to favor reaction at oxygen; soft electrophiles favor reaction at carbon.

12.4 The Aldol Addition Reaction[78]

Few reactions have had the impact on the direction of organic synthesis that the aldol addition reaction had during the last quarter of the 20th century. The aldol addition is an old reaction, having first been published by the Irish chemist Robert Kane,[79] explored by Russian composer-chemist Aleksandr Borodin in 1864,[80] and extended by Wurtz[81] and Kekulé.[82] Although these early workers did isolate the β-hydroxyaldehydes from their reaction mixtures (i.e., they carried out aldol *addition* reactions; Example **12.55**) for more than a century after the discovery of the aldol reaction, the reaction was used in the form of the aldol *condensation* (Example **12.56**), in which an α,β-unsaturated carbonyl compound is the major product. The difficulties inherent in the quantitative conversion of simple carbonyl compounds to their enolates with the strong bases typically available to organic chemists before 1970 meant that little attention was paid to using the reaction for the stereocontrolled formation of new chiral enters. That changed in the late 1970s.

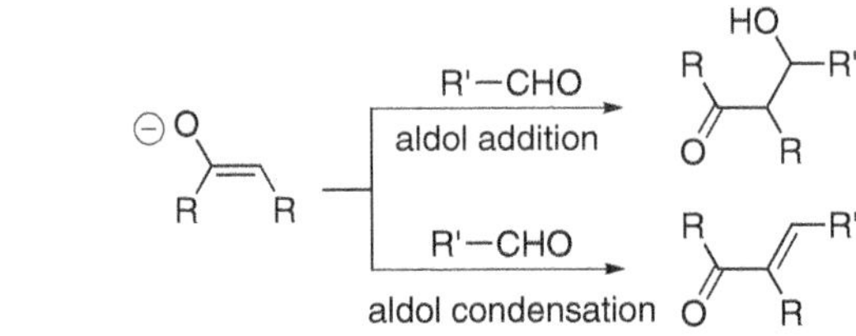

(12.55)

(12.56)

The development of the modern aldol addition has been driven by efforts directed toward the synthesis of polyketide natural products, especially macrolide antibiotics such as erythromycin (Example **12.57**). An examination of the backbone of macrocyclic ring reveals that practically every second carbon atom (marked by asterisks) carries an oxygen atom. Early on, this substitution pattern, and a similar substitution pattern in aromatic natural products, led J. N. Collie to propose that these natural products might be formed by a polymerization leading to products with repeating (-CH_2—CO-) (or [-CHR—CO-]) units.[83] This is exactly the pattern that one obtains when one adds an enolate anion to a carbonyl compound, which has made the aldol addition an important synthetic method.

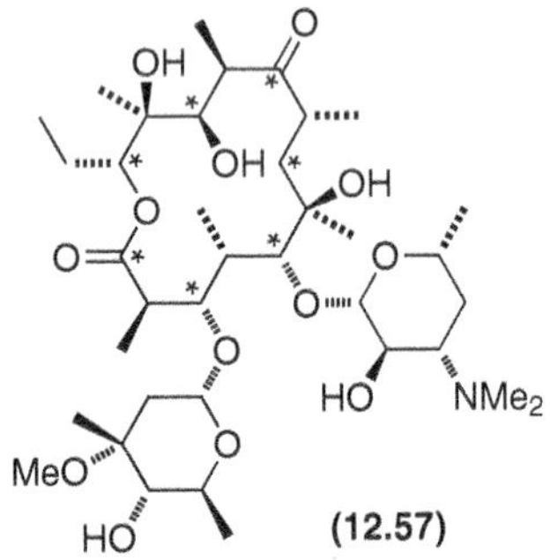

(12.57)

78. Reviews and monographs: (a) Mukaiyama, T. *Org. React.* **1982**, *28*, 203. (b) Heathcock, C.H. In Morrison, J.D., Ed. *Asymmetric Synthesis* (Academic Press: Orlando, FL, 1984); Vol. 3., Ch. 2, p. 111. (c) Franklin, A.S.; Paterson, I. *Contemp. Org. Synth.* **1994**, *1*, 317. (d) Cowden, C.J.; Paterson, I. *Org. React.* **1997**, *51*, 1. (e) Machajewski, T.D.; Wong, C.-H. *Angew. Chem. Int. Ed*. **2000**, *39*, 1352. (f) Carriera, E.M. In Otera, J., Ed. *Modern Carbonyl Chemistry* (Wiley-VCH: Weinheim, 2000); Chapter 8, p. 227. (g) Paterson, I.; Cowden, C.J.; Wallace, D.J. In Otera, J., Ed. *Modern Carbonyl Chemistry* (Wiley-VCH: Weinheim, 2000); Chapter 9, p. 249. (h) Mahrwald, R. Ed. *Modern Aldol Reactions* (Wiley-VCH: Weinheim, 2004). (i) Vicarion, J.L.; Badia, D.; Carillo, L.; Reyes, E.; Etxebarria, J. *Curr. Org. Chem.* **2005**, *9*, 219. (j) Mahrwald, R. *Aldol Reactions* (Springer: Dordrecht, 2009).

79. (a) Kane, R. *Ann. Phys. Chem.* [2] **1838**, *44*, 475; *J. Prakt. Chem.* **1838**, *15*, 129.

80. Borodin, A. *J. Prakt. Chem.* **1864**, *93*, 413; *Z. Chem. Pharm.* **1864**, *7*, 353; *Ber. dtsch. Chem. Ges.* **1873**, *6*, 973.

81. (a) Wurtz, A. *Bull. Soc. chim. Paris* **1872**, *17*, 436–442; *J. Prakt. Chem.* **1872**, *5*, 457; *Compt. Rend. Acad. Sci.* **1872**, *74*, 1361.

82. (a) Kekulé, A. *Ber. dtsch. Chem. Ges.* **1869**, *2*, 365–368; **1870**, *3*, 135–137.

83. (a) Collie, J.N. *J. Chem. Soc.* **1893**, *63*, 329. (b) Collie, J.N. *Proc. Chem. Soc.* **1907**, *23*, 230. (c) Collie, J.N. *J. Chem. Soc.* **1907**, *91*, 1806. (d) Collie, J.N.; Wilsmore, N.T.M. *J. Chem. Soc* **1896**, *69*, 293. (e) Collie, N.; Myers, W.S. *J. Chem. Soc.* **1893**, *63*, 122.

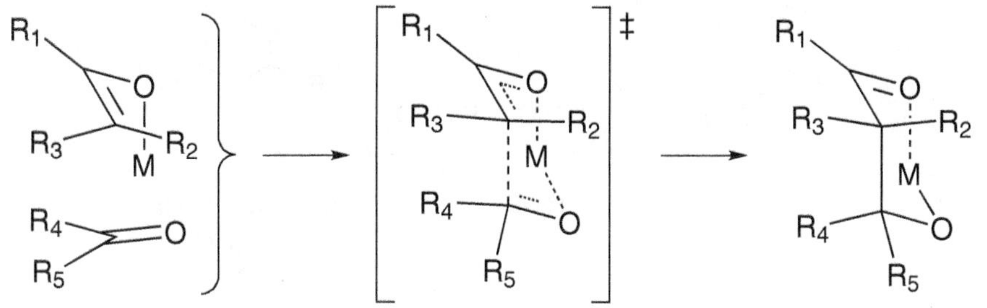

Figure 12.6 The use of the Zimmerman-Traxler transition state model to predict the stereochemistry of the major product formed in an aldol addition

Much of the modern fascination with this old reaction stems from the fact that one can incorporate two new chiral centers into a molecule with defined stereochemistry. The possibility of predicting the stereochemistry of the addition reaction was improved by the discovery of Zimmerman and Traxler, who noted a stereochemical bias in the Ivanov addition of the dianion of phenylacetic acid to aldehydes and rationalized it on the basis of a six-membered cyclic transition state.[84] Today, the Zimmerman-Traxler transition state model is a cornerstone of the discussions of stereochemistry in the additions of metal enolates to aldehydes (and, to a lesser extent, ketones) under conditions of kinetic control. This model (Figure 12.6) predicts that the stereochemistry of the enolate and the conformation of the chairlike transition state both affect the final stereochemistry. The conformation in the transition state is generally accepted to be that in which the larger group in each of the pairs (1) R_2 and R_3 and (2) R_4 and R_5 preferentially occupies an equatorial position in the transition state. Provided that the reaction is carried out under conditions of kinetic control (short reaction time, low temperature), this model generally does an adequate job of predicting the stereochemistry of the major product. However, one should always heed the caveat that:

When dealing with the aldol addition reaction, generalizations frequently fail.

Stereochemistry in the Aldol Addition

The modern aldol addition is very much concerned with the control of both relative and absolute stereochemistry in the final product. When two new chiral centers are formed in the reaction, four stereoisomers are possible (Figure 12.7), existing as two diastereomeric pairs of enantiomers (**12.58** to **12.61**). Thus, one may control the absolute stereochemistry of the product (enantioselection), the relative stereochemistry of the product (diastereoselection), or both. The clear target of the research since 1980 has been the control of both.

Diastereoselection: Factors Affecting the Diastereoisomer Ratio

Kinetic versus Thermodynamic Control

Much of the discussion that follows will focus on aldol addition reactions that are carried out under conditions of kinetic control, an area that was pioneered by American chemist Clayton Heathcock.[85] This occurs because under these conditions, diastereoselectivity is maximized. The aldol addition is, however, reversible, which allows the *syn* and *anti* isomers of the aldol to equilibrate, and this process is generally facile. A good example of just how facile this reaction can be is provided by the reaction in Example **12.62**, where the

---

84. Zimmerman, H.; Traxler, M. *J. Am. Chem. Soc.* **1957**, *79*, 1920.

85. **Clayton H. Heathcock (1936–)** received his BSc from Abilene Christian College, Texas in 1958 and then joined the Champion Paper and Fiber Company in Pasadena, Texas. In 1960 he entered the University of Colorado (PhD, 1963). After a year of postdoctoral work with Gilbert Stork at Columbia, he joined the University of California, Berkeley, where he spent his entire career. He was elected to the National Academy of Sciences in 1995, and received the Centenary Medal of the Royal Society of Chemistry in 1996. For more details, see: http://chemistry.berkeley.edu/faculty/chem/emeriti/heathcock.

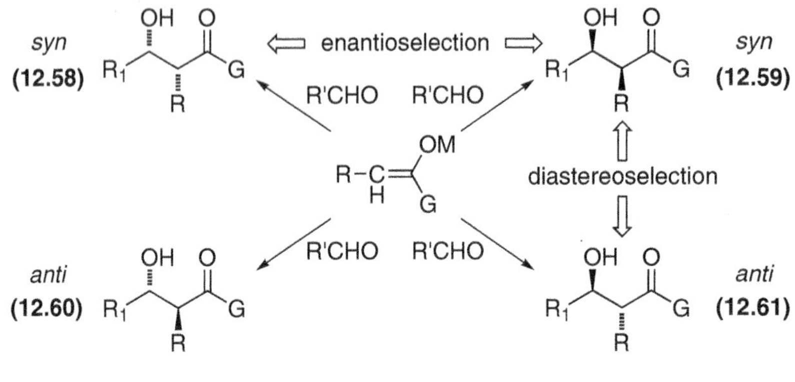

Figure 12.7 Stereochemical terminology in the aldol addition. Diastereoselection distinguishes isomers vertically, and enantioselection distinguishes isomers horizontally.

equilibration is rapid even at −60°C (the half-life of the reaction is approximately 1 minute under these conditions,[86] which corresponds to an activation energy for the equilibration of just over 1 kcal mol^{-1}). The factors affecting the *syn-anti* equilibrium are as one would expect: the size of the groups attached to the carbonyl carbon of the enolate precursor and the aldehyde, the identity if the counterion, and the enolate geometry.

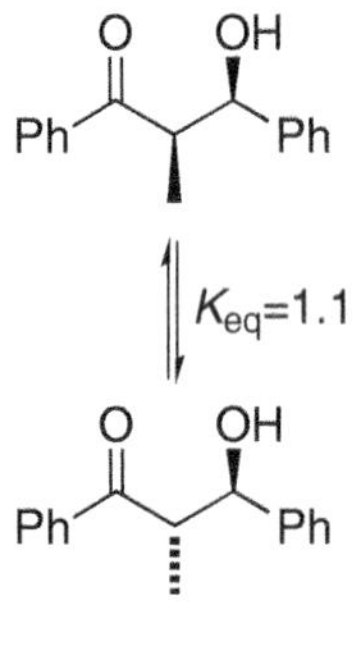

(12.62)

Effects of Enolate Geometry

In keeping with the predictions of the Zimmerman-Traxler model, it is found that *Z(O)* enolates often give higher levels of diastereoselectivity than *E(O)* enolates. If we examine the model, shown in Newman projection form in Figure 12.8, we can see one possible reason for this. In the reaction of the *E(O)* enolate, the group R is equatorial and gauche to R" in both transition states, leading to the diastereoisomeric aldol products, which means that the difference in conformational energy between the *syn* and *anti* aldol transition states is relatively low and that diastereoselectivity, also, should be low. On the other hand, in the addition involving the *Z(O)* enolate, the group R is forced to adopt an axial position.

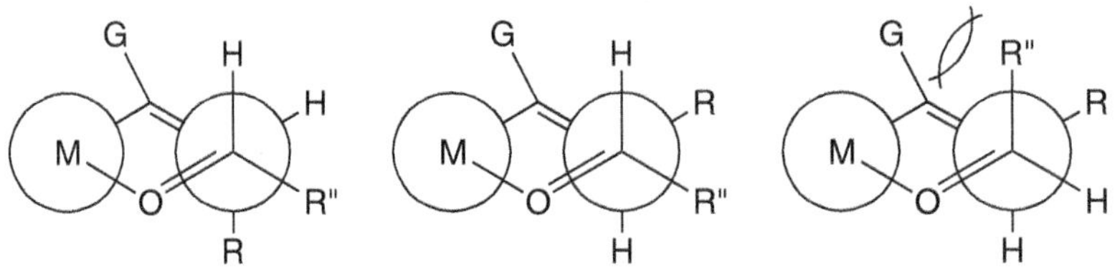

(12.63) **(12.64)** **(12.65)**

Figure 12.8 Newman projections for Zimmerman-Traxler transition states leading from Z(O) enolate to *syn* aldol product (**12.63**), E(O) enolate to *anti* aldol product (**12.64**), and from E(O) enolate to syn aldol product (**12.65**). Note the pseudo-1,3-diaxial interaction in (**12.65**).

86. Heathcock, C.H. In Nozaki, H, Ed. *Current Trends in Organic Synthesis. IUPAC Symp. Ser.* (Pergamon: Oxford, 1983), p. 27.

This means that if the *Z(O)* enolate is to give the *anti* aldol, the alkyl group of the aldehyde must also be in a pseudo-axial position. It is gratifying to find that the *Z(O)* enolate is much more diastereoselective than the *E(O)* enolate and that it gives predominantly the *syn* aldol adduct.

Table 12.5 summarizes the stereochemical outcome of aldol addition reactions of the *E(O)* and *Z(O)* enolates of ethyl ketones with benzaldehyde at –78°C. The same data are summarized graphically below the table. With the exception of the *Z(O)* enolate of propionaldehyde, the data fit a reasonable straight line, with an intercept at 45% *syn* and a slope of just over 0.5, suggesting a fairly strong intrinsic preference for the *syn* aldol and that *anti* aldols might be difficult to prepare with high levels of stereoselectivity.

Table 12.5 Stereochemical Outcome of Kinetic Aldol Additions of Lithium Enolates*

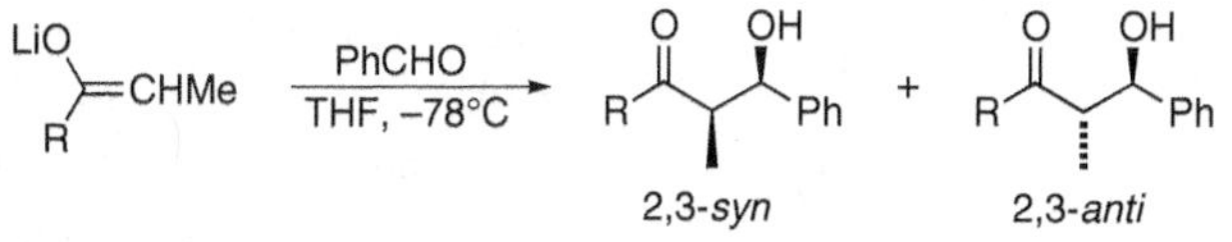

R	Enolate Geometry	2,3-*syn*	2,3-*anti*
H	*Z(O)*	50	50
H	*E(O)*	60	40
Et	*Z(O)*	90	10
Et	*E(O)*	40	60
i-Pr	*Z(O)*	90	10
i-Pr	*E(O)*	50	50
t-Bu	*Z(O)*	99	1

*Data taken from Ref. 87

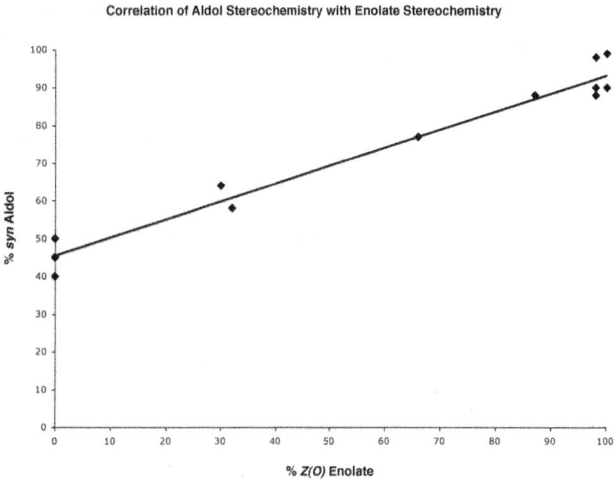

The Counterion

There is good evidence that the reversibility of the aldol addition may be modulated to some degree by the ability of the cation to chelate the two oxygens of the aldol adduct. Thus, no aldol adduct was isolated from the reaction of benzaldyde with the thermodynamic enolate of phenylacetone when the non-chelating *tris*-(diethylamino)sulfonium cation was the counterion. This presumably occurred because the equilibrium for formation of the aldol adduct was unfavorable. However, Noyori did observe that trimethylsilyl fluoride, present as an impurity in the base, trapped the minor component of the equilibrium mixture to give the silyloxyketone (Example **12.66**).[88] With lithium ion, which is a modest Lewis acid capable of chelating the two oxygen atoms, the equilibrium shifts to the product side. The reversibility of the reaction is now manifested in the stereochemical consequences of the reaction, rather than in the inability to isolate the product of the reaction.

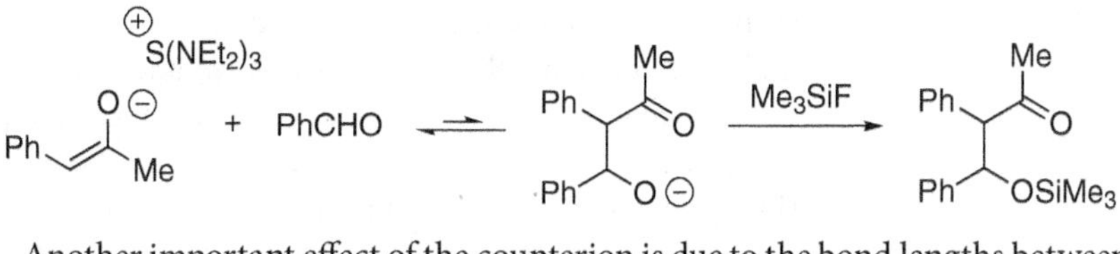

(12.66)

Another important effect of the counterion is due to the bond lengths between the metal and oxygen. Although the alkali metals are small, their metal-oxygen bonds are quite long (Li, 1.92–2.00Å; Na, 2.24–2.33Å; K, 2.57–2.67Å),[89] which makes the transition state fairly loose, and reduces the effects of enolate stereochemistry. Good Lewis acids that also have long bond distances to oxygen are zinc (1.975–1.982Å)[90] and magnesium (1.908–1.921Å).[91] These bond distances loosely correlate with a greater propensity to undergo equilibration at low temperatures. The aluminum aldolates (Al—O bond lengths 1.85–2.06)[92] appear to form

87. Heathcock, C.H. In Trost, B.M.; Fleming, I., Eds. *Comprehensive Organic Synthesis* (Pergamon: Oxford, 1991), vol. 2, p. 181.
88. Noyori, R.; Nishida, I.; Sakata, J.; Nishizawa, M. *J. Am. Chem. Soc.* **1980**, *102*, 1223.
89. Williard, G.P.; Carpenter, G.B. *J. Am. Chem. Soc.* **1986**, *108*, 462
90. Hlavinka, M.L.; Hagadorn, J.R. *Organometallics* **2006**, *25*, 3501.
91. Dove, A.P.; Gibson, V.C.; Marshall, E.L.; White, A.J.P.; Williams, D.J. *Chem. Commun.* **2002**, 1208.
92. van der Steen, F.H.; van Mier, G.P.M.; Spek, A.L.; Kroon, J.; van Koten, G. *J. Am. Chem. Soc.* **1991**, *113*, 5742.

strong chelates, which prevents the isomerization of the aldolate. In such cases, the stereochemistry of the aldol additions can be reliably predicted on the basis of the Zimmerman-Traxler model. The stereochemical predictions in this case are shown in Examples **12.67** to **12.70**.[93] In contrast to most metal enolates, the boron-oxygen bond in boron enolates is short, and boron aldolates can also form a stable chelate ring. Thus, the aldols formed from these nucleophiles do not equilibrate. The stereochemistry of the aldol addition of a variety of metal enolates with benzaldehyde are gathered in Table 12.6.

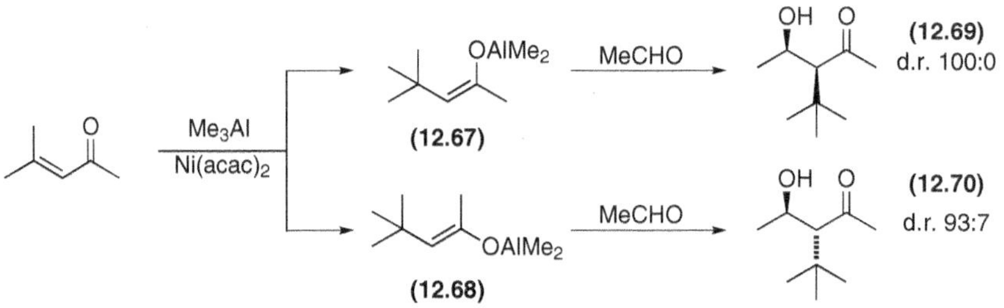

The information in Table 12.6 can also be obtained in visual form. Some data for aldol addition reactions of various metal enolates are gathered in the form of stereochemistry plots in Figure 12.9. In these plots, the vertical axis records the diastereoisomeric excess (d.e.) relative to the *syn* isomer. Thus, in this diagram, *anti*-selective reactions project in the negative direction, whereas *syn*-selective reactions project in the positive direction.

In Figure 12.9A, the nonparticipating groups, X, are arranged with ketones at the left, alkyl esters and thioesters next, and amides last. In the diagram on the right, the ketones are located on the left, and the carboxylic acid derivatives on the right. In the diagram on the left, the first thing that is obvious is that almost none of the reactions have a negative d.e. with respect to the *syn* stereoisomer—*anti*-selective reactions of metal enolates are quite rare. In addition, only six of the reactions have d.e.'s of 80% or greater, which is the benchmark that reactions stereoselective enough to be used routinely in synthesis now need to attain. In fact, most metal enolates do not participate in highly selective aldol addition reactions, and when they do, they tend to be *syn*-selective. Lithium enolates, especially, are not highly diastereoselective, with the exception of reactions where the group X is large (*tert*-butyl or 2,6-dialkylphenoxy); such a group may direct the protonation of an enolate intermediate as shown in Example **12.71.** Figure 12.9 also reveals that enolates of stronger Lewis acids (such as Zr) tend to give higher levels of *syn* selectivity with acid derivatives, but seem to make little difference with ketone enolates.

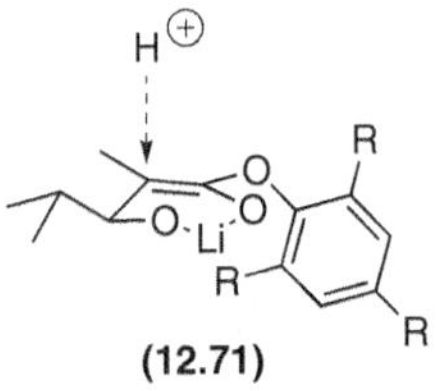

(12.71)

Table 12.6 Effect of Counterion on the Stereochemistry of Aldol Additions of Metal Enolates of Acid Derivatives

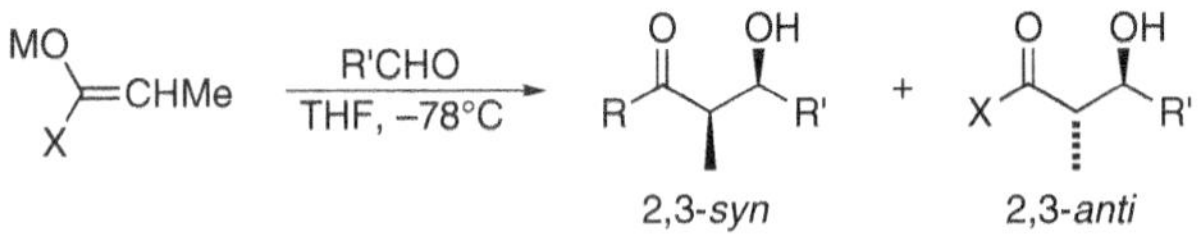

M	X	R'	*E(O):Z(O)*	*syn:anti*
Li	SBu^t	Ph	90:10	63:37
Li	$N(CH_2)_4$	Ph	< 5:95	60:40
Mg	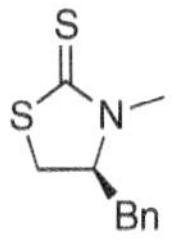		—	5:95
$AlEt_2$	Ph	Ph	—*	< 1:99
$AlEt_2$	Ph	Ph	—†	62:38
BBu_2	SBu^t	Ph	95:5	10:90
BBu_2	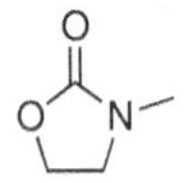	Ph	< 3:97	98:2
Zn	c-C_5H_7O‡	Ph	*E*	1.2:1
$TiCl_3$	SBu^t	Ph	—	86:14
Cp_2TiCl	$N(CH_2)_4$	Ph	—	2:98
Cp_2ZrCl	SBu^t	Ph	90:10	93:7
Cp_2ZrCl	$N(CH_2)_4$	Ph	< 5:95	95:5
Sn^{II}	NMe_2§	Ph	*Z*	92:8

*0°C, thermodynamic control.
†–78°C, kinetic control.
‡Cyclopentanone enolate.
§Thioamide rather than amide.

93. Jeffery, E.A.; Meisters, A.; Mole, T. *J. Organomet. Chem.* **1974**, *74*, 365, 373.

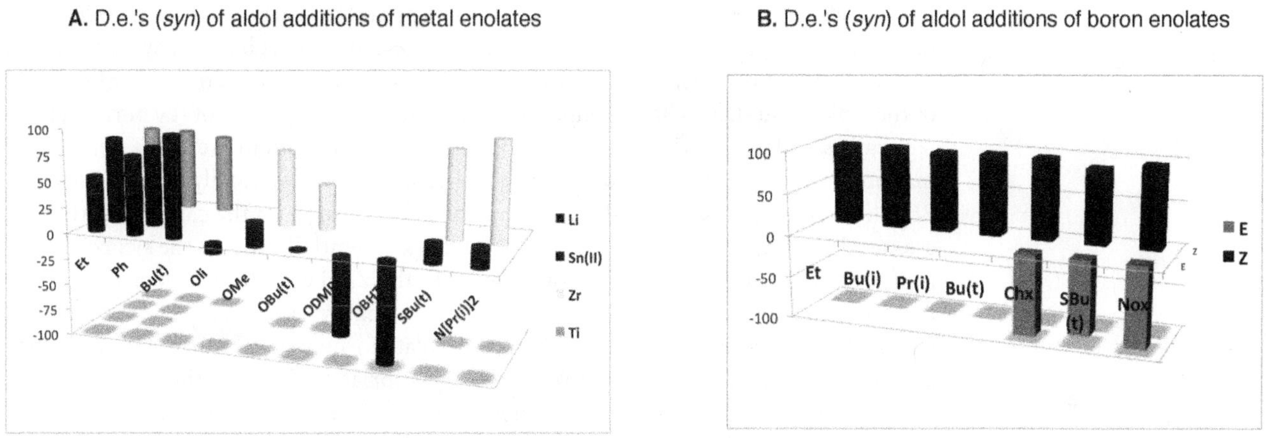

Figure 12.9 Stereochemistry of the aldol addition reactions of enolates of propionyl derivatives (**A**) and geometric isomers of boron enolates (**B**). On the left, ODMP represents a 2,6-dimethylphenyl ester, and OBHT represents a 2,6-di-*tert*-butyl-4-methylphenyl ester. On the right, Nox represents an *N*-oxazolidinyl group.

Boron Enolates

The case with the boron enolates (Figure 12.9B) presents a stark contrast to the metal enolates. The stereochemistry of the boron enolate exercises an *absolute* control over the aldol stereochemistry: *Z(O)* boron enolates give *syn* aldols, and *E(O)* boron enolates give *anti* aldols. Both geometric isomers of the enolate react with synthetically useful (d.r. ≥ 47) diastereoselectivity. More importantly, it appears that other factors (e.g., the size of the non-reacting group bonded to the carbonyl group) have only a small effect on the *syn/anti* ratio—in contrast to the metal enolates.

Boron enolates are formed by treating a carbonyl compound with a suitable boron species and a non-nucleophilic base. For example, the reaction between dicyclopentylboron triflate or 9-borabicyclo[3.3.1]non-9-yl triflate and a ketone in the presence diisopropylethylamine (Hünig's base, a hindered amine) gives the boron enolate of the ketone.[94]

The stereochemistry of the boron enolate formed in this reaction is determined in large part, by the size of the alkyl group bonded to the carbonyl carbon and by the steric requirements of the boron triflate used in the reaction. The formation of the boron enolate is believed to follow the same course (Figure 12.10), where there is a rapid preequilibrium

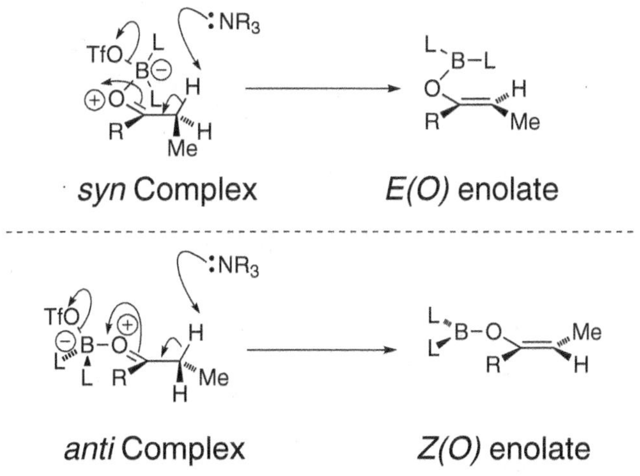

Figure 12.10 Models for predicting the stereochemistry of the boron enolate by a concerted pathway

94. (a) Evans, D.A.; Nelson, J.V.; Vogel, E.; Taber, T.R. *J. Am. Chem. Soc.* **1981**, *103*, 3099. (b) Hirama, M.; Masamune, S. *Tetrahedron Lett.* **1979**, *20*, 2225, 3937.

Lewis acid–Lewis base reaction between the borane and the carbonyl oxygen, followed by rate-limiting deprotonation. The loss of the leaving group from boron may be concerted with deprotonation, or it may occur prior to deprotonation. The preferred conformation of the Lewis acid–Lewis base complex depends, in part, on the size of the borane group. Sterically small groups (BL_2=BBu_2 or 9-BBN, for example) tend to prefer the *anti* conformation in the complex, but as the group becomes more sterically demanding (e.g., BL_2=(c-$C_5H_9)_2B$), eclipsing of the BL_2 group and the R group on the carbonyl carbon increases the energy to the point that the complex preferentially adopts the *syn* conformation. Deprotonation of the *anti* complex leads to the *Z(O)* enolate, and deprotonation of the *syn* complex leads to the *E(O)* enolate. An alternative explanation for the observed enolate stereochemistry posits that the *Z(O)* enolate formed by a hindered borane isomerizes fairly readily. Whatever the reason, by appropriately choosing the borane reagent, one can fix the enolate geometry, and hence the aldol stereochemistry, as shown by Examples **12.72** and **12.73**; both reactions give yields in excess of 80% with a range of aldehydes.[95]

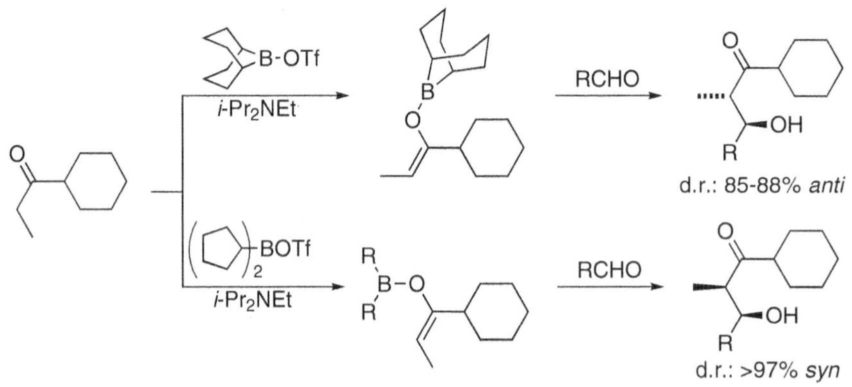

(12.72)

(12.73)

Addition of Enolates to Chiral Aldehydes

The strong tendency of the *Z(O)* enolates of ketones with large nonparticipating groups to undergo stereoselective addition to give the *syn* aldol adduct led to the development of a number of surrogates for propanal or propionate esters (where the group bonded to the carbonyl carbon is small). The common feature of these surrogates is their ability to be degraded to an aldehyde or carboxylic acid under mild oxidative or reductive conditions. Most of these surrogates are α,α-dialkyl-α-hydroxyketones or α,α-dialkyl-α-vinylketones. If the aldehyde component of the aldol addition has a chiral center at the α position, the aldol addition conforms to Cram's rule, as illustrated by the reactions leading to aldols in Examples **12.74** to **12.77**.

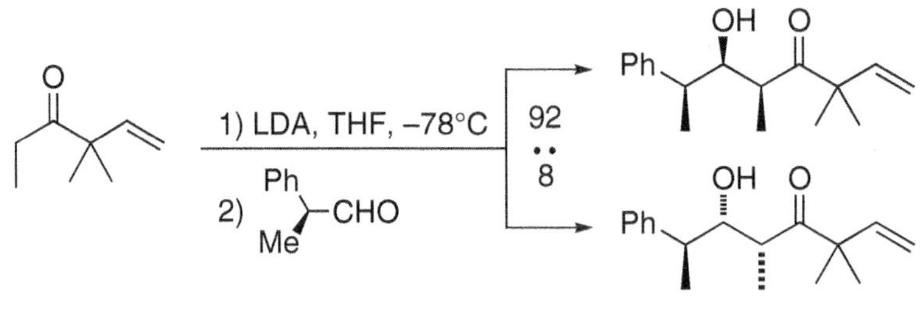

(12.74)

(12.75)[96]

95. Van Horn, D.E.; Masamune, S. *Tetrahedron Lett.* **1979**, 2229.
96. Mori, I.; Ishihara, K.; Heathcock, C.H. *J. Org. Chem.* **1990**, *55*, 1114.

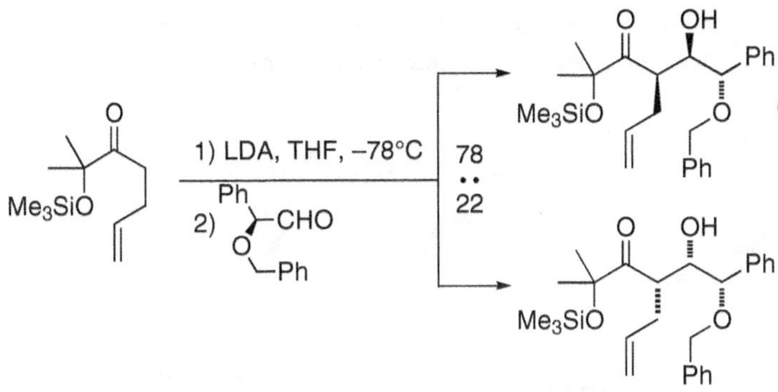

(12.76)

(12.77)[97]

Asymmetric Aldol Additions

In the early 1980s, Evans[98] extended the utility of boron enolates by developing a set of chiral of boron enolates capable of inducing high levels of enantioselection and diastereoselection.[99] This work revolutionized the application of the aldol addition reaction in organic synthesis. For such powerful chiral auxiliaries, the Evans chiral oxazolidinones are remarkably simple compounds, formed from an appropriate β-aminoalcohol (*S*-valinol, Example **12.78**, or 1*S*,2*R*-norephedrine, Example **12.79**) by treatment with a carbonate or chloroformate ester.

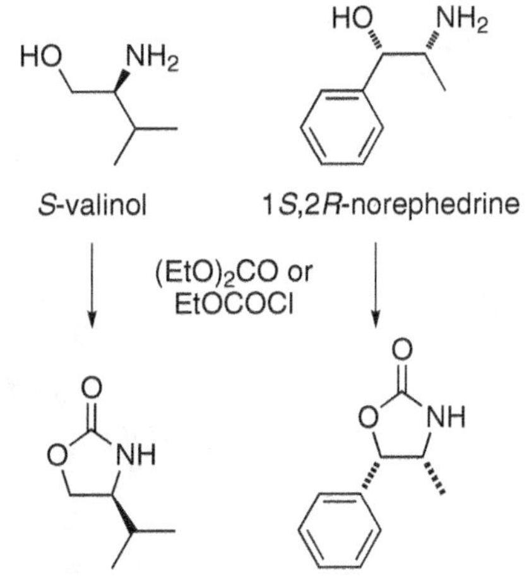

(12.78) **(12.79)**

The treatment of the chiral oxazolidinone with butyllithium and propionyl chloride gives the *N*-acyloxazolidinone, which reacts with dibutylboron triflate and triethylamine to give the *Z(O)* enolate with a d.r. greater than 100:1. The aldol addition of the boron enolate of the chiral *N*-propanolyoxazolidinone gives a mixture of the four possible diastereoisomeric aldol adducts, of which all but one are formed in negligible amounts. The *Z(O)* enolate of the *N*-propanoyl-oxazolodinone based on *S*-valinol gives the 2*S*,3*R* (*syn*) aldol,

97. Heathcock, C.H.; Montgomery, S.H. *Tetrahedron Lett.* **1983**, *24*, 4637.

98. **David A. Evans (1941–)** was educated at Oberlin College (AB, 1963) and the California Institute of Technology (Caltech; (PhD, 1967). He was immediately appointed to the faculty of UCLA and subsequently to Caltech (1974) and Harvard (1983). Evans has been a pioneer in developing a practical asymmetric aldol addition reaction. He was elected to the National Academy of Sciences in 1985 and won the Tetrahedron Prize in 1998. For an autobiographical sketch with some fascinating details, see: Evans, D.A. *Tetrahedron* **1999**, *55*, 8589.

99. Evans, D.A.; Bartroli, J.; Shih, T.L. *J. Am. Chem. Soc.* **1981**, *103*, 2127.

as in Example **12.80.** The *Z(O)* enolate of the *N*-propanoyloxazolidinone based on 2*S*,3*R*-norephedrine gives the enantiomeric aldol, as in Example **12.81.**[100]

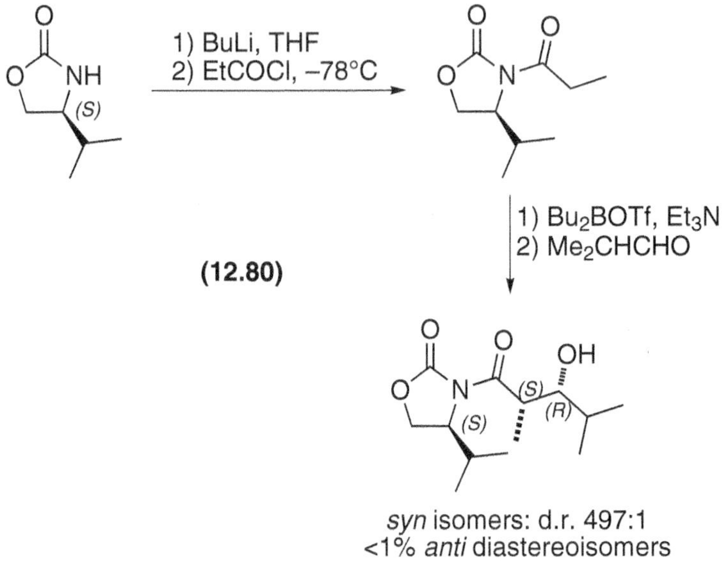

(12.80)

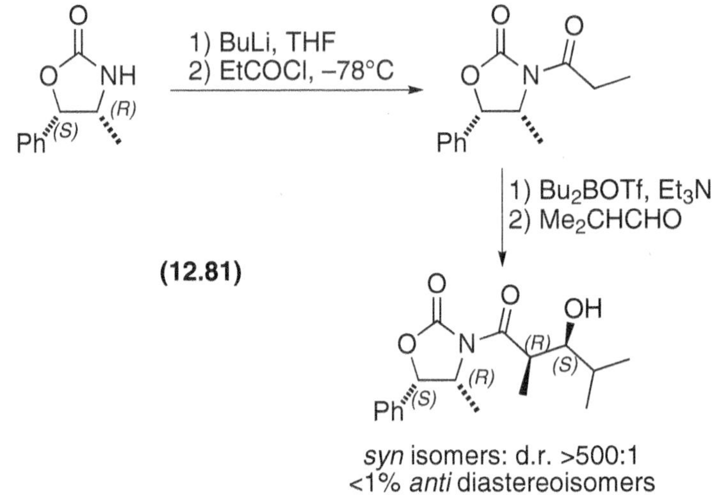

(12.81)

Studies[101] have shown that the conformation of the exocyclic C—N bond is the key stereochemistry-controlling feature in the enolate and that by controlling the conformation of this bond, one can control the enantioselectivity of the reaction. The system from the carbonyl oxygen to the end of the enolate is planar to maximize conjugation, so there are two extreme conformations possible for these metal enolates: a non-chelated conformation, where the carbonyl group and the enolate π bond are *cisoid* to each other (Example **12.82**), and a chelated conformation, where the same two π bonds are *transoid* to each other (Example **12.83**). In these two different conformations, the shielding group (shown as the large black circle) inhibits the approach to opposite faces of the enolate, thus leading to the opposite stereochemistry in the aldol product.

100. Evans, D.A.; Bartroli, J.; Shih, T.L. *J. Am.Chem. Soc.* **1981,** *103,* 2127.

101. Evans, D.A.; Takacs, J.M.; McGee, L.R.; Ennis, M.D.; Mathre, D.J.; Bartroli, J. *Pure Appl. Chem.* **1981,** *53,* 1109. For more recent computational work on this problem, see: (a) Li, Y.; Paddon-Row, M.N.; Houk, K.N. *J. Org. Chem.* **1990,** *55,* 481. (b) Bernardi, A.; Capelli, A.M.; Comotti, A.; Gennari, C.; Gardner, M.; Goodman, J.M.; Paterson, I. *Tetrahedron* **1991,** *47,* 3471.

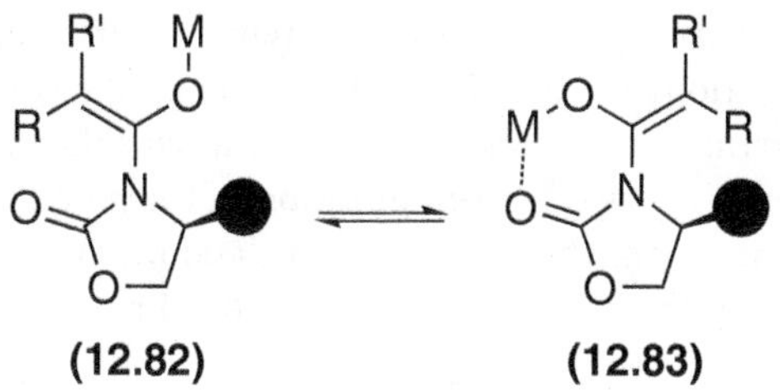

(12.82) non-chelated   **(12.83)** chelated

Being a second-row element, boron cannot expand its octet, so boron may only be bonded to two oxygens at once in the aldol reaction of enol borinates. Because one of these oxygens must be the enolate oxygen and the other must be the carbonyl oxygen of the aldehyde partner, the boron enolate must react through a non-chelated transition state. This allows the stereochemistry of the reaction of the boron enolate to be predicted on the basis of the model shown in Figure 12.11 for the *Z(O)* enolate. In the non-chelated conformation, the isopropyl group of the *S*-valinol auxiliary is placed so that it interferes with the axial alkyl group on boron in one of the two transition states but not in the other; the favored conformation is enclosed in the frame. Thus, this reaction strongly favors one aldol product over the other. In the chelated enolate, the opposite stereochemical finding is true; in this case, the favored conformation of the enolate has the *si* face of the enolate more accessible to the aldehyde. Thus, it has been found that the enantioselectivity of the titanium enolates of *N*-acyloxazolidinones is reversed compared to the boron enolates, although both *Z(O)* enolates are highly selective for the *syn* aldol.[102]

Chiral species have an intrinsic diastereoselectivity in reactions where new chiral centers are formed. This intrinsic diastereoselectivity is illustrated by Cram's rule selectivity in addition reactions, for example. When both reacting species are of this type, their intrinsic diastereoselectivities may be ***matched*** (e.g., Example **12.84**), in which case the reaction usually proceeds with very high levels of diastereoselectivity. Alternatively, they may be ***mismatched*** (e.g., Example **12.85**), in which case the reaction proceeds with lower (often quite low) levels of diastereoselectivity. The enolates derived

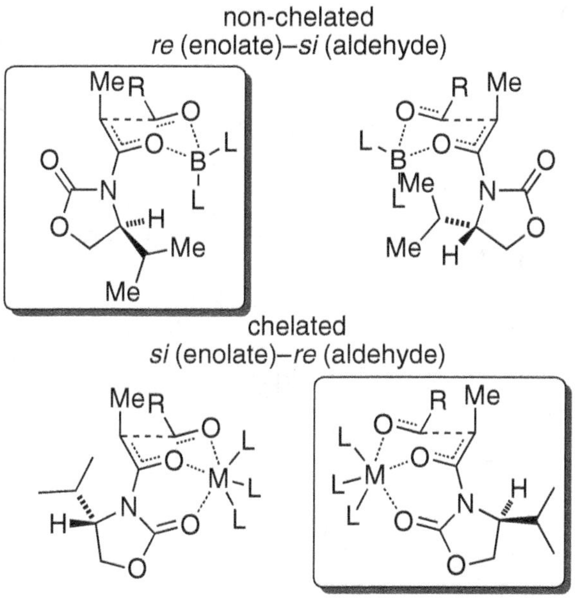

Figure 12.11 Predicting the stereochemistry of the aldol reaction of chiral Z(O) enolates for non-chelated (boron) enolates and chelated metal (e.g., Ti) enolates, using the same chiral oxazolidinone

102. Nerz-Stormes, M.; Thornton, E.R. *J. Org. Chem.* **1991**, *56*, 2489.

from the Evans chiral oxazolidinones exhibit extremely high levels of intrinsic diastereoselectivity. In the reactions of their *Z(O)* zirconium enolates with a chiral α-substituted aldehyde, the intrinsic diastereoselectivity of the oxazolidinones *almost completely overrode Cram's rule selectivity*—the configuration of the new chiral centers was determined *solely* by the chirality of the oxazolidinone, and not by the chirality of the aldehyde.[103] Note how even in the mismatched pair of chiral reactants, the diastereoisomer ratio was close to 20:1.

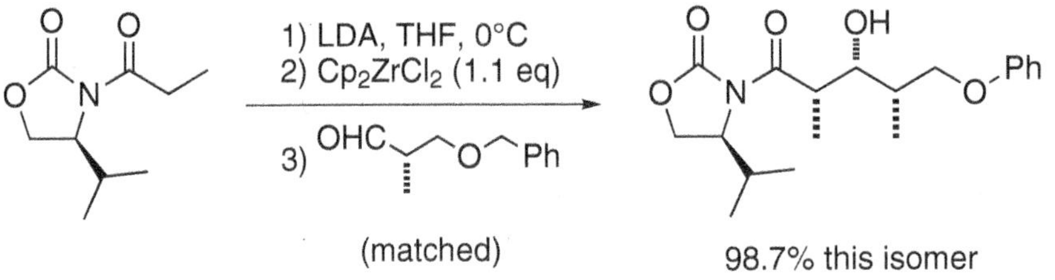

(12.84)

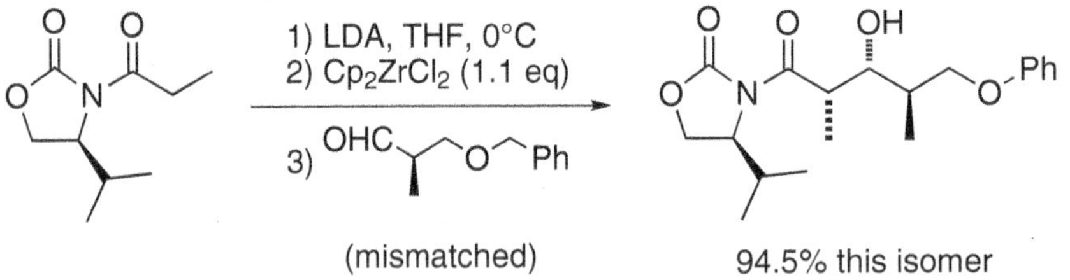

(12.85)

As pointed out in the previous section (revisit Figure 12.9 to recall this), obtaining *anti* aldols is a rather more difficult task than obtaining the *syn* aldols. Although this task can be accomplished using boron enolates, it is frequently experimentally simpler to make metal enolates, so considerable effort has gone into finding ways to obtain the *anti* aldol from a metal enolate. This remains true for the asymmetric reaction as well. There has been some success, however. An early report appeared in which the addition of a chiral titanium enolate to the titanium tetrachloride complex of an aldehyde gives the *anti* aldol with high levels of both enantio- and diastereoselectivity (Example **12.86**).[104]

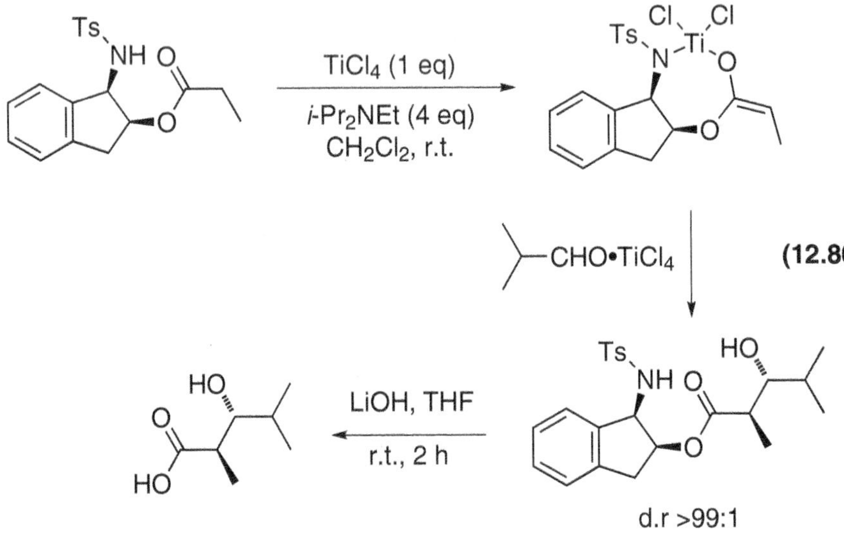

(12.86)

103. Evans, D.A.; McGee, L.R. *J. Am. Chem. Soc.* **1981**, *103*, 2876.
104. Ghosh, A.; Onishi, M. *J. Am. Chem. Soc.* **1996**, *118*, 2527.

Organocatalysis in Enolate Additions

The Wieland-Meischer ketone[105] has become an important starting material in a wide range of syntheses, including steroids and terpenoids. The synthesis of the racemic compound is carried out in two stages. The first stage involves an acid- or base-catalyzed Michael addition of 2-methyl-1,3-cyclohexanedione to methyl vinyl ketone. In the synthesis of the racemic ketone, the intramolecular aldol condensation of the diketone is catalyzed by pyrrolidine in refluxing benzene. In 1977, the synthesis of optically active ketone was reported, using L-proline in dry dimethyl sulfoxide (Example **12.87**) as an ***organocatalyst*** in the second step.[106] This reaction was an early example of what has now become a flood (the original Gutzwiller paper had been cited more than 100 times at the writing of this chapter, for example). The most successful catalysts share a common pair of structural features that are characteristic of proline: a five-membered cyclic secondary amine and a group at the α-carbon of the five-membered ring capable of functioning as a base (Example **12.88**). The group labeled Nu is frequently anionic in these organocatalysts, and large groups R tend to enhance the stereoselectivity in reactions where they are used. The group X may be another heteroatom (usually sulfur or oxygen).[107]

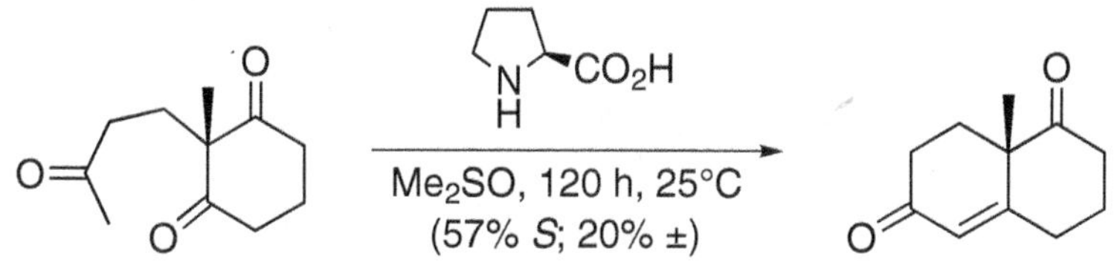

(12.87)

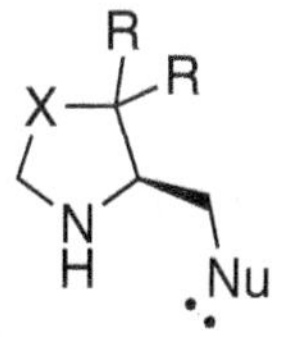

(12.88)

Stereochemistry of Organocatalytic Nucleophilic Additions

Unlike the additions of metal enolates to aldehydes, where the formation of the *anti* aldol is the greater challenge, in organocatalytic reactions, the reverse is often true. The dimerization of propionaldehyde catalyzed by L-proline in dimethylformamide, for instance (Example **12.89**), leads to a 3:1 mixture of *anti* and *syn* aldols. Each diastereoisomer is obtained in high enantiomeric excess.[108] Changing the solvent to a less polar solvent tends to raise the *anti:syn* ratio at the expense of making the reaction slower. Extending the reaction time leads to the formation of a trimeric product where, again, the major product (Example **12.90**) is that of *anti* addition at each stage, and the minor product (Example **12.91**) is that formed by *anti* addition to the *syn* aldol. However, one should be careful in making generalizations: in this reaction; for example, it has been reported that the product ratios show epimerization of the initial aldol during the second aldol addition.[109] Interestingly, the

105. (a) Wieland, P.; Miescher, K. *Helv. Chim. Acta* **1950**, *33*, 2215. (b) Ramachandran, S.; Newman, M.S. *Org. Synth., Coll. Vol. V* **1973**, 486.

106. (a) Gutzwiller, J.; Buchschacher, P.; Fürst, A. *Synthesis* **1977**, 167. (b) Buchsacher, P.; Fürst, P.A.; Gutzwiller, J. *Org. Synth., Coll. Vol. 7*, **1990**, 368.

107. For examples of successful organocatalysts for additions of enolates and enolate surrogates, see: (a) Notz, W.; Tanaka, F.; Barbas, C.F., III *Acc. Chem. Res.* **2004**, *37*, 580. (b) List, B. In Mahrwald, R., Ed. *Modern Aldol Reactions. Vol. 1: Enolates, Organocatalysis, Biocatalysis and Natural Product Synthesis*; (Wiley-VCH: Weinheim, 2004); Ch. 4, p. 161.

108. Northrup, A.B.; MacMillan, D.W.C. *J. Am. Chem. Soc.* **2002**, *124*, 6798.

109. Chowdari, N.S.; Ramachary, D.B.; Córdova, A.; Barbas, C.F., III *Tetrahedron Lett.* **2002**, *43*, 9591.

product of the first aldol addition is less reactive than propionaldehyde itself, which makes isolation of the dimer or trimer possible simply by changing the reaction conditions.

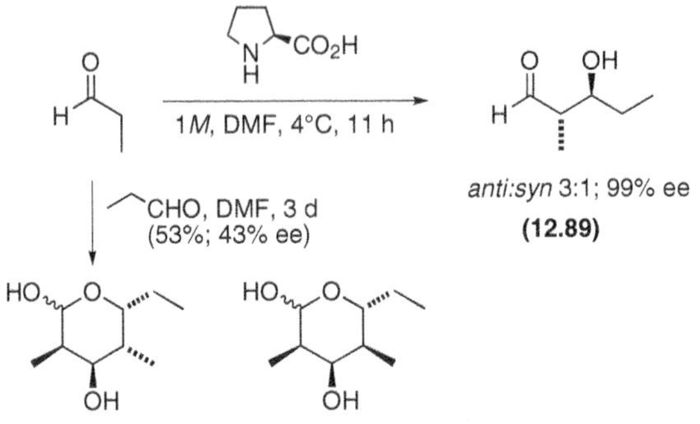

(12.89)

8 (12.90)

1 (12.91)

Reaction Synopses

Aldol Addition

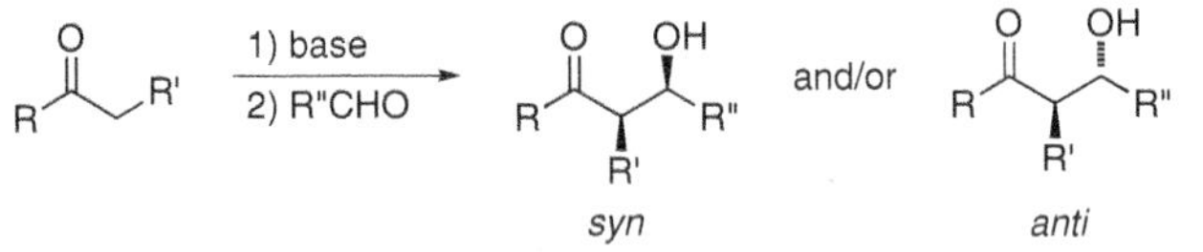

Base:

Metal enolate, thermodynamic: NaH, PhH; $NaNH_2$, THF; $KOBu^t$, *t*-BuOH; etc.
Lithium enolate, kinetic: LDA, THF, –78°C; LICA, THF, 8°C; *t*-BuLi, THF, –100°C; etc.
Metal enolate: (1) LDA, THD, –78°C; (2) MCl_n
Z(O) Boron enolate: Bu_2BOTf, *i*-Pr_2NEt; 9-BBN-OTf, *i*-Pr_2NEt; Bu_2BOTf, Et_3N; etc.
E(O) Boron enolate: $(c\text{-}C_5H_9)_2BOTf$, *i*-Pr_2NEt; etc.

Stereochemistry: may be predicted using Zimmerman-Traxler transition state model, but addition is reversible, limiting reliability of stereochemical predictions; addition of metal enolates to chiral aldehydes follows Cram's rule.

Addition of boron enolates can be predicted with confidence using the Zimmerman-Traxler model: *Z(O)* gives *syn* aldol, *E(O)* enolate gives *anti* aldol.

Evans Asymmetric Aldol Addition

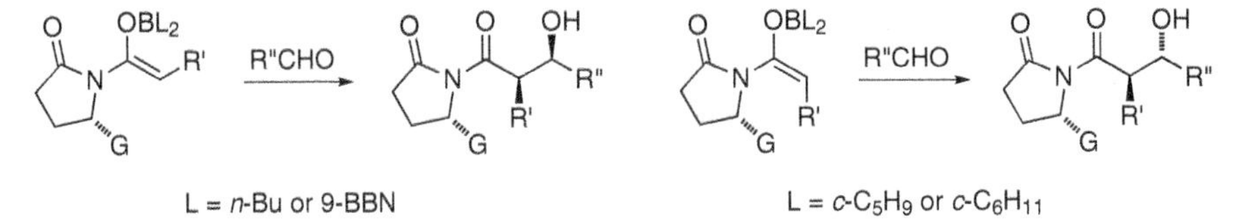

L = *n*-Bu or 9-BBN L = *c*-C_5H_9 or *c*-C_6H_{11}

Stereochemistry fixed by stereochemistry of boron enolate

Stereochemistry predicted with confidence by Zimmerman-Traxler model

Chirality of enolate fixes stereochemistry of product; Cram's rule is overridden in this reaction.

Chirality (but not *syn/anti* diastereoselectivity) of product aldol is reversed using coordinating metal enolate (e.g., Ti, Zr) instead of boron enolate.

Organocatalysis of Aldol Addition

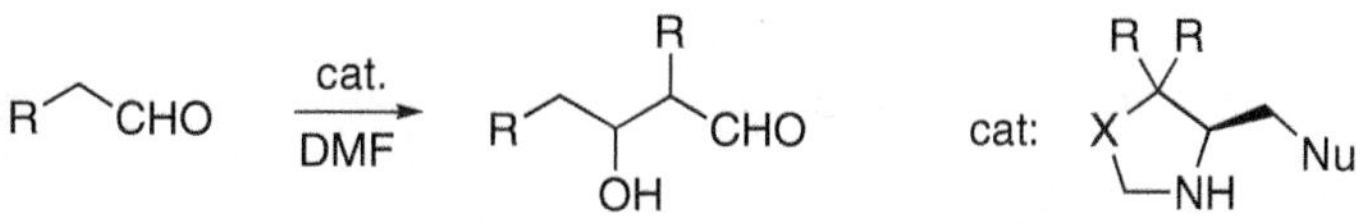

Catalyst, R = H, alkyl, aryl; X = CH_2, O, S, etc.; Nu = OR, NR_2, O^-, etc.

Stereochemistry: strong tendency to *anti* stereochemistry

Worked Problem

12-4 What is the major product expected in the following reaction sequence? The product is obtained as a mixture of diastereoisomers—which stereoisomer should predominate, and why?

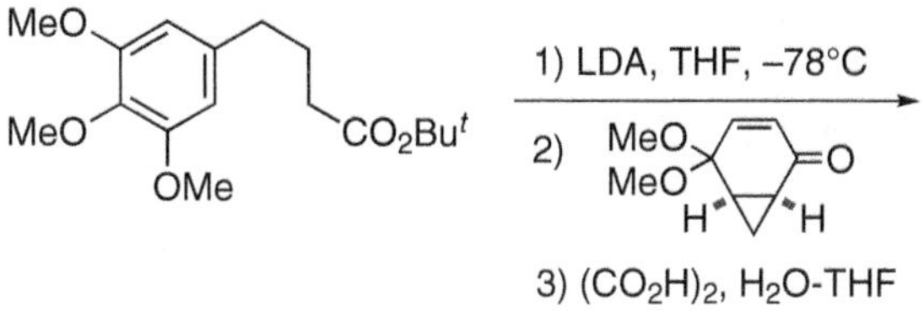

§Answer below.

Drill Problems

What is the major organic product of each of the following reaction sequences? Where appropriate, specify the absolute configuration of the major product.

(a) 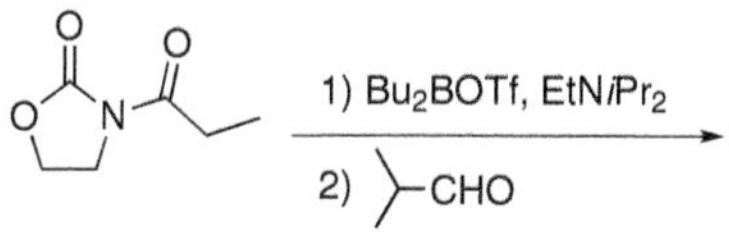

(b) 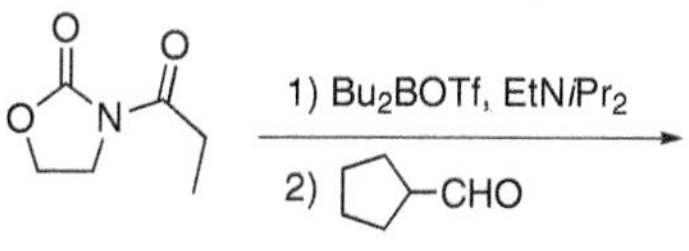

(c) 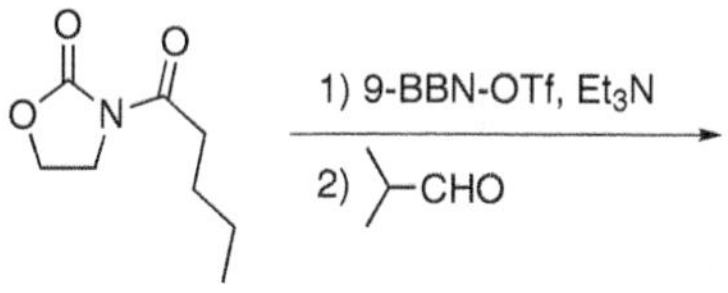

(d) 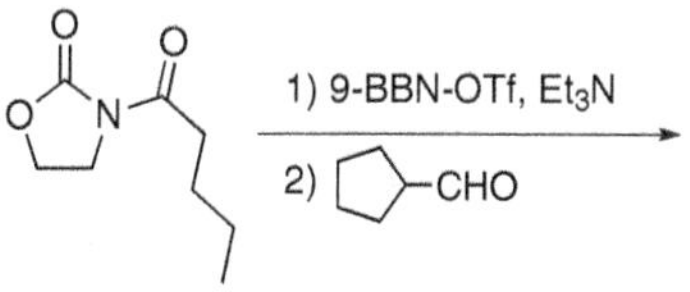

(continued)

§ Answer to Worked Problem:
The two stereoisomeric major products of this reaction are related to each other as *syn* and *anti* aldol adducts, but the difference (and, in fact, the designation of each) is complicated by the fact that the electrophilic component is a ketone instead of an aldehyde. However, one can say with some degree of certainty that the lithium enolate will add to the carbonyl group from the *exo* face of the bicyclic system, giving the *endo* alcohol as shown.

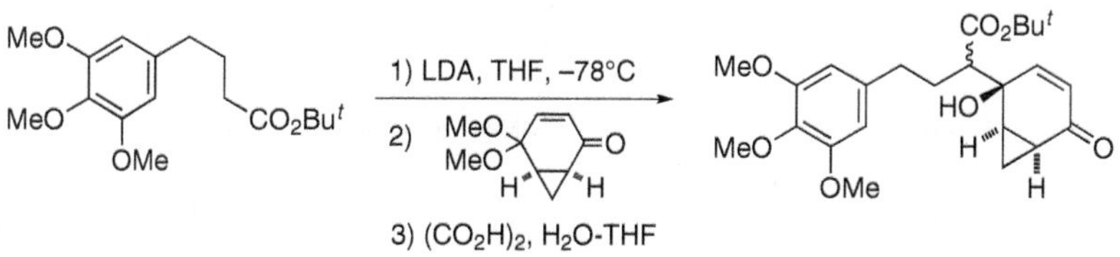

[*J. Am. Chem. Soc.* **1981**, *103*, 5813]

(Drill Problems continued)

(e) 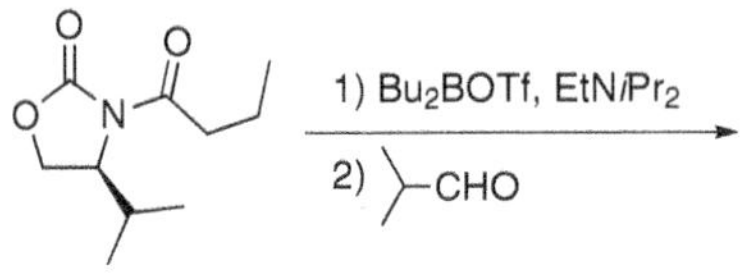

(f) 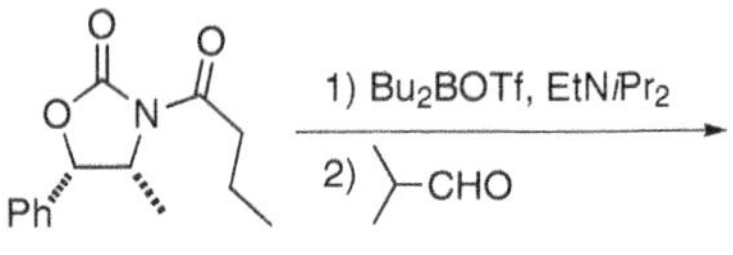

(g) 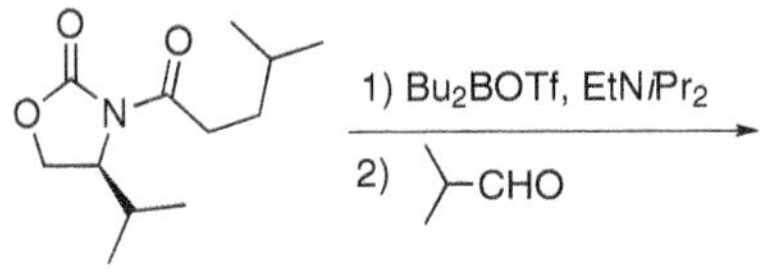

(h) 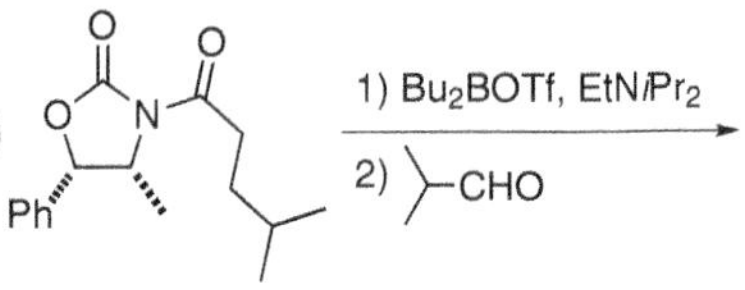

(i) 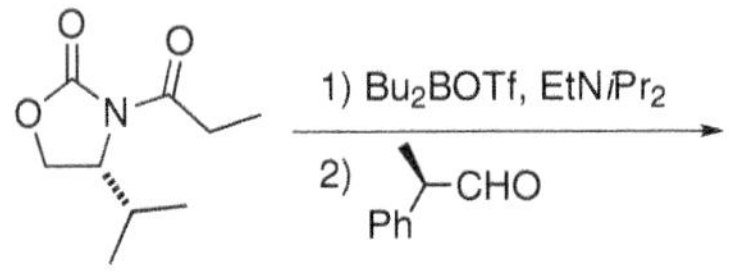

(j) 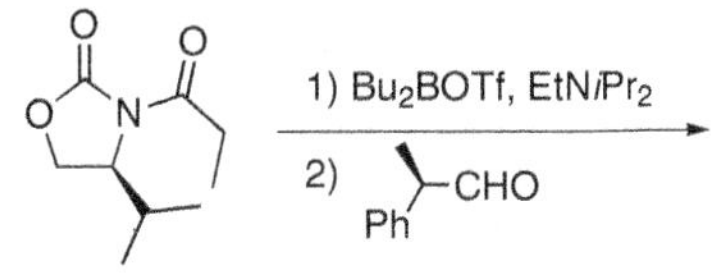

(k) 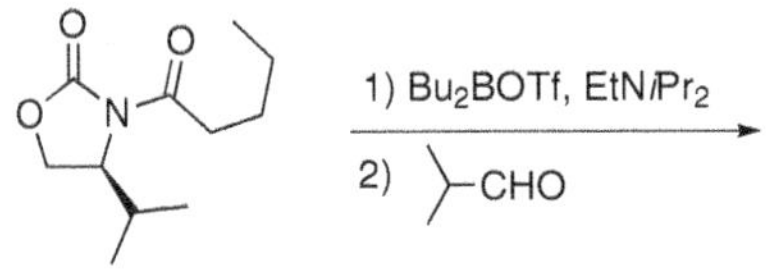

(l) 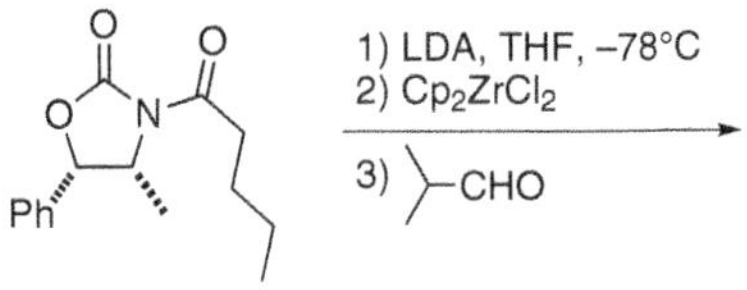

Problems

12-7 What reaction conditions or reagents should be used to complete the following transformations?

(a) 

(b) 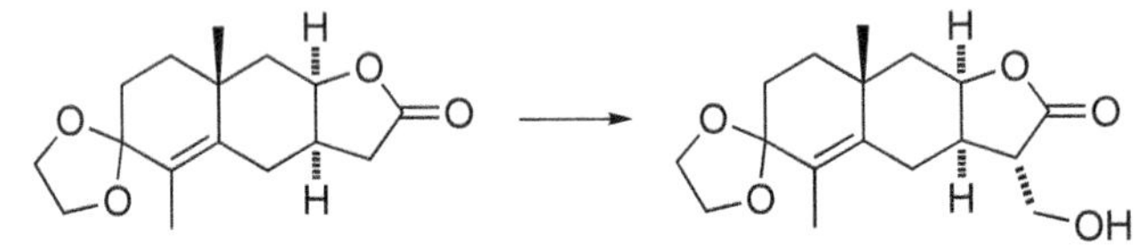

(c) 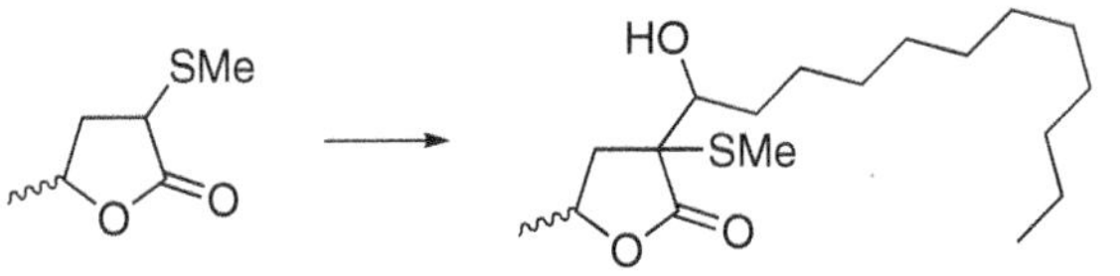

(d) 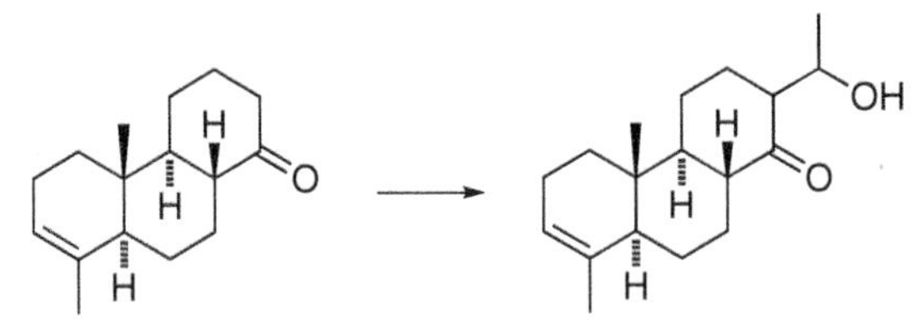

(e) 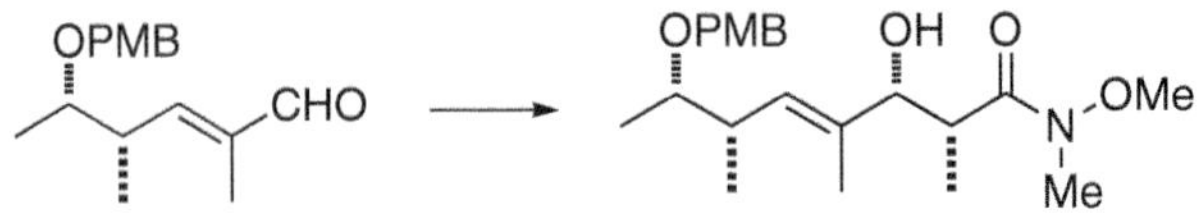

(f) 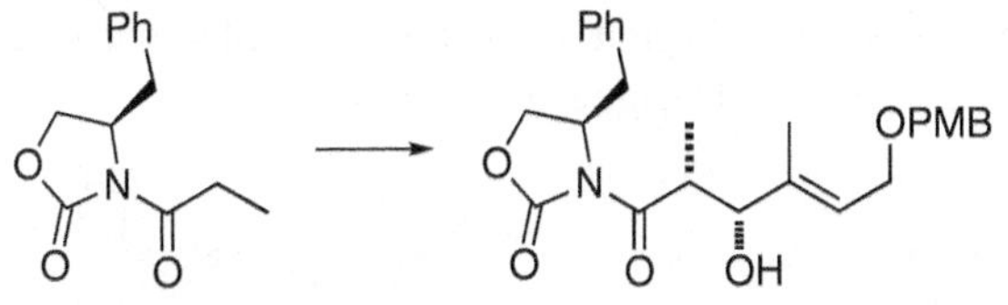

References: (a) *J. Org. Chem.* **1974**, *39*, 1322. (b) *Tetrahedron Lett.* **1975**, 743. (c) *Tetrahedron Lett.* **1980**, *21*, 3139. (d) *J. Am. Chem. Soc.* **1980**, *102*, 5085. (e) *Synthesis* **2005**, 1183. (f) *Org. Lett.* **2004**, *6*, 1445.

12-8 What is the major organic product of each of the following reactions, which are steps from some classic syntheses?

(a) 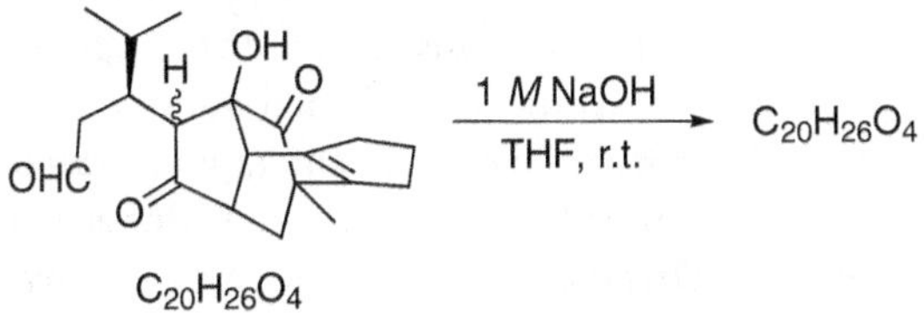

(b) 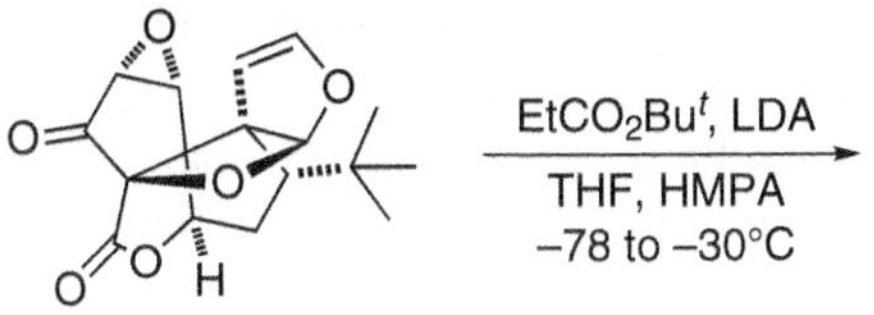

(c) 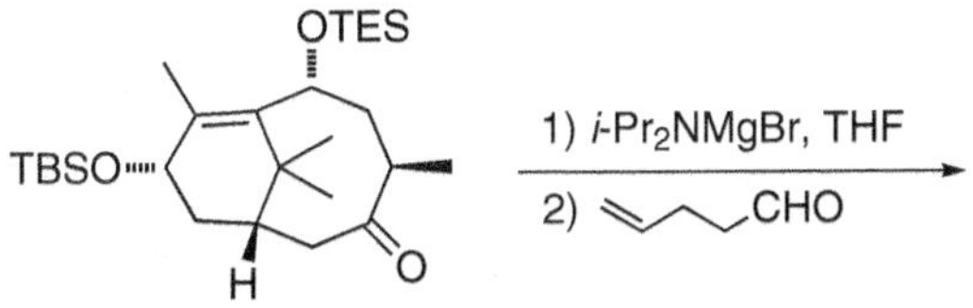

(d) 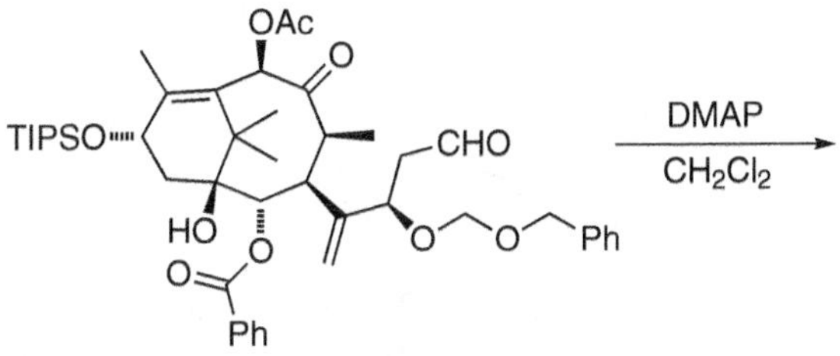

(e) 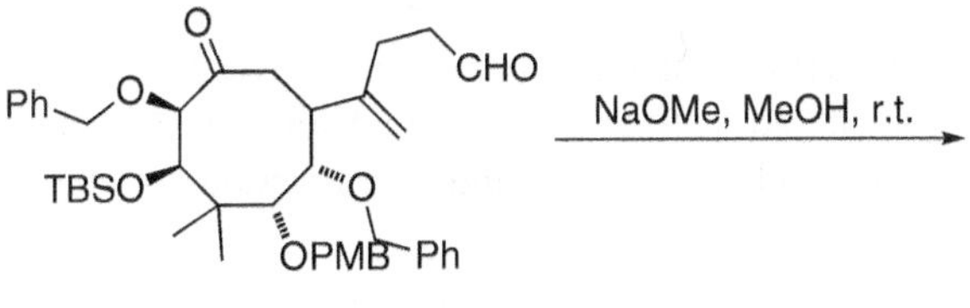

References: (a) *Can. J. Chem.* **1979**, *57*, 3348. (b) *J. Am. Chem. Soc.* **1988**, *110*, 649. (c) *J. Am. Chem. Soc.* **1994**, *116*, 1597. (d) *J. Am. Chem. Soc.* **1997**, *119*, 2757. (e) *Chem. Eur. J.* **1999**, *5*, 121.

12.5 **Conjugate Addition of Enolates and Similar Compounds**

Michael Addition of Enolate Anions

Conjugated carbonyl compounds react with nucleophiles to give either 1,2- or 1,4-adducts (Example **12.92**). Generally, the 1,4-adduct is favored thermodynamically, so where the addition to the carbonyl group is reversible, the 1,4-adduct reaction dominates the product mixture. This situation applies to the addition of most nucleophiles, with the exception of metal hydrides, alkali metal alkyls, and Grignard reagents.

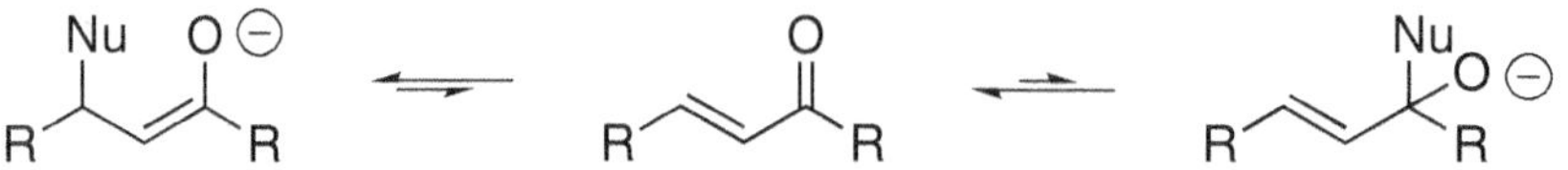

(12.92)

Enolate anions add to the β carbon of an α,β-unsaturated carbonyl compound or nitrile as soft carbon nucleophiles, rather than the harder oxyanion nucleophile, to give a new C—C bond; this reaction is known as the ***Michael addition***.[110] The product of the Michael addition is the enolate anion of a 1,5-dicarbonyl compound, so it is important to take precautions to reduce the reactivity of the product enolate anion to avoid polymerization of the conjugated carbonyl compound. For this reason, the reaction has often been carried out using the conjugate base of a β-dicarbonyl compound, an α-cyanocarbonyl compound, or a nitroalkane as the nucleophile because the rapid intramolecular proton transfer to the enolate anion prevents this side reaction. The usefulness of this form of the reaction is attested by the number of *Organic Syntheses* preparations using it.[111]

The same effect may also be accomplished by incorporating a third-row atom (usually silicon or sulfur) into the conjugated system at the α position to stabilize the enolate anion formed. A good example is provided by the reaction in Example **12.93**, from the total synthesis of vincamine,[112] where the intermediate lithium enolate is stabilized by the adjacent sulfur, making it a less reactive nucleophile than the initial amide (lactam) enolate anion.

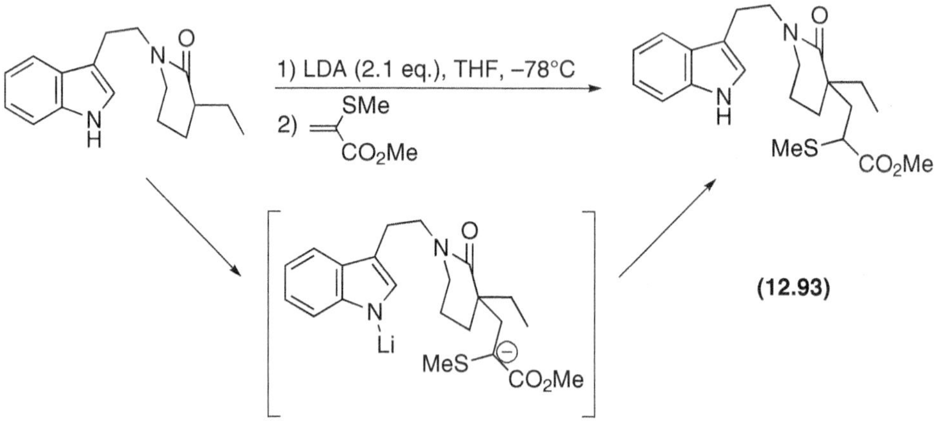

(12.93)

As with the aldol addition, the availability of preformed enolates by deprotonation of a suitable precursor expanded the potential scope of the Michael addition. As noted in the beginning of this section, the addition of a nucleophile to an enone can occur by the 1,2- or 1,4- manifold, so that there are three possible products in each reaction: the 1,2-adduct, the *syn*-1,4-aduct, and the *anti*-1,4-adduct (Examples **12.94** and **12.95**). The results of the Heathcock group's extensive study of Michael addition reactions of enones under both

110. (a) Michael, A. *J. Prakt. Chem.* **1887**, *35*, 349; **1894**, *49*, 20. (b) Bergman, E.D.; Ginsburg, D.; Pappo, R. *Org. React.* **1959**, *10*, 179.

Arthur Michael (1853–1942), an American chemist, never received a university degree but studied under Hofmann (1871 and 1873–1876), Bunsen (1872–1873), and Wurtz (1876–1877); during this year, he visited Mendeleev in St. Petersburg. He was twice Professor of Chemistry at Tufts College (1880–1891 and 1894–1907) and at Harvard (1912–1936). Independently wealthy, he built and worked in his own research laboratory in Britain. He was elected to the National Academy of Sciences in 1889. For more biographical information, see: Fieser, L.F. *Biogr. Mem. Natl. Acad. Sci.* **1975**, 331.

111. (a) Clarke, H.T.; Murray, T.F. *Org. Synth., Coll. Vol. 1* **1941**, 272. (b) Shriner, R.L.; Todd, H.R. *Org. Synth., Coll. Vol. 2* **1943**, 200. (c) Cason, J. *Org. Synth., Coll. Vol. 4* **1963**, 630. (d) Moffett, R.B. *Org. Synth., Coll. Vol. 4* **1963**, 652. (e) Horning, E.C.; Finelli, A.F. *Org. Synth., Coll. Vol. 4* **1963**, 776. (f) McMurry, J.E.; Melton, J. *Org. Synth., Coll. Vol. 6* **1988**, 648.

112. Herrmann, J.L.; Cregge, R.J.; Richman, J.E.; Semmelhack, C.L.; Schlessinger, R.H. *J. Am. Chem. Soc.* **1974**, *96*, 3702.

kinetic and thermodynamic conditions,[113] using amide enolates **12.96** (*Z(O)* enolate) and **12.97** (*E(O)* enolate) are summarized in Figures 12.12 and 12.13.

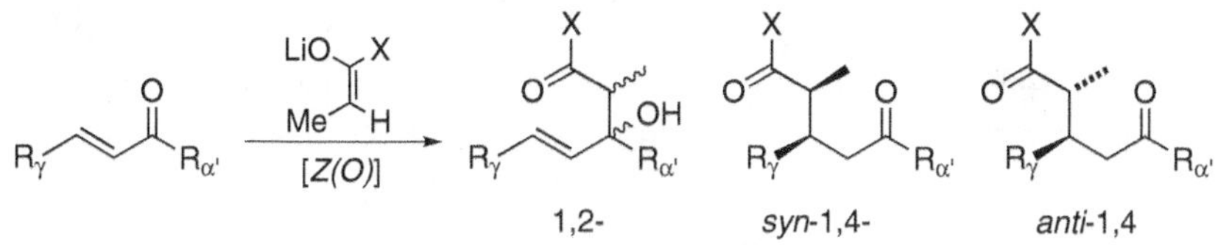

(12.94)

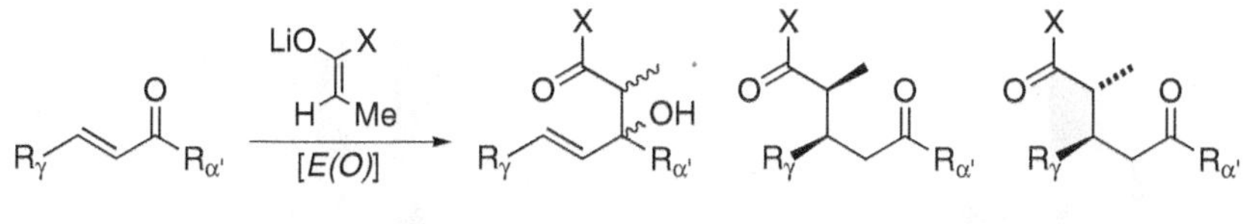

(12.95)

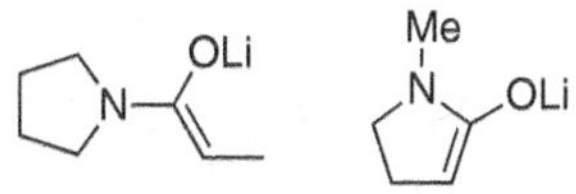

(12.96) **(12.97)**

In the reactions in **12.98** and **12.99**, contained in Figure 12.12, the enone moiety of the electrophile is kept constant and the α' group is varied through the series: Et, *i*-Pr, *t*-Bu, Chx, and Ph). The right-hand diagram shows that in most cases with these reactions, the reaction favors the formation of the 1,4-adduct, more especially (but not always) under thermodynamic control. The picture with respect to stereochemistry is much more straightforward. The *Z(O)* enolate is generally only modestly stereoselective, except for reactions involving aromatic ketones, where it is strongly *syn*-selective, and the *E(O)* enolate is uniformly strongly *anti* selective under both kinetic and thermodynamic conditions.

In the reactions in **12.100** and **12.101**, contained in Figure 12.13, the α' group is kept constant at *tert*-butyl, and the γ group of the Michael acceptor is varied through the series: Et, *i*-Pr, *t*-Bu, and Ph). The right diagram shows that in most cases with these reactions, the reaction favors the formation of the 1,4-adduct, more especially under thermodynamic control. The picture with respect to stereochemistry is much less straightforward than in the reactions in **12.98** and **12.99.** The *Z(O)* enolate is still usually only modestly stereoselective, except for reactions involving aromatic ketones, but now the *E(O)* enolate also exhibits lowered stereoselectivity (only four of the cases studied gave d.e.'s in excess of 90%).

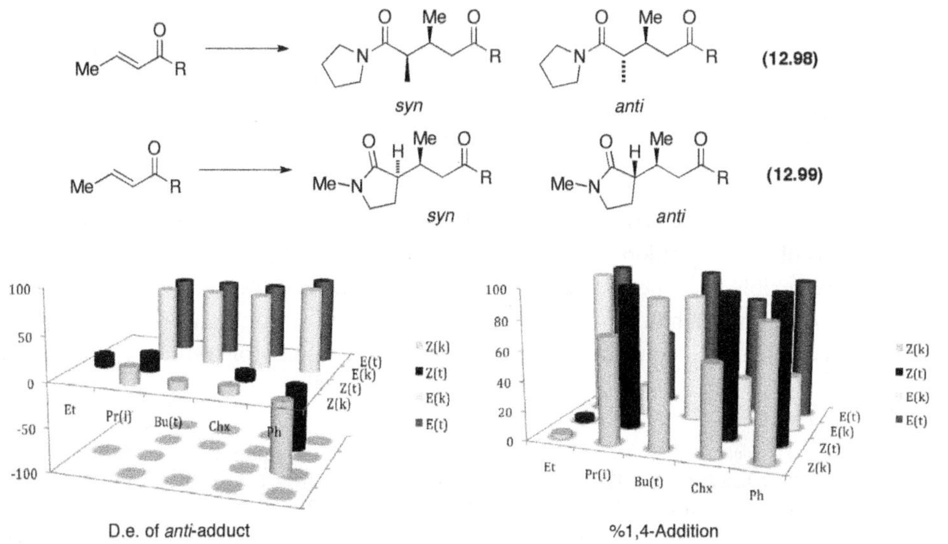

Figure 12.12 Effects of enone structure on the stereochemistry of the Michael addition of the amide lithium enolate to enones of the structure R_γ-CH = CH-CO-$R_{\alpha'}$ under kinetic and thermodynamic conditions. The reactions with the *Z(O)* enolate (**12.96**) are labeled Z(k) [kinetic] and Z(t) [thermodynamic], corresponding to reaction with the corresponding reactions of the *E(O)* enolate (**12.97**) being labeled E(k) and E(t).

113. Heathcock, C.H.; Henderson, M.A.; Oare, D.A.; Sanner, M.A. *J. Org. Chem.* **1985**, *50*, 3019.

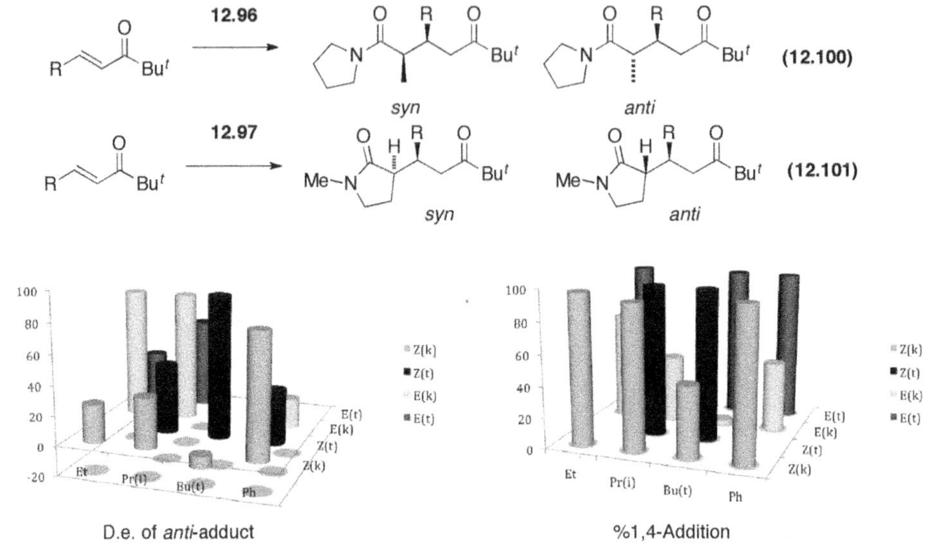

Figure 12.13 Effects of enone structure on the stereochemistry of the Michael addition of the amide lithium enolate to enones of the structure R_γ-CH = CH-CO-$R_{\alpha'}$ under kinetic and thermodynamic conditions. The reactions with the *Z(O)* enolate (**12.96**) are labeled Z(k) [kinetic] and Z(t) [thermodynamic], corresponding to reaction with the corresponding reactions of the *E(O)* enolate (**12.97**) being labeled E(k) and E(t).

Asymmetric Michael Additions

As with most important C—C bond-forming reactions, the Michael addition has also been the subject of intense activity aimed at controlling the absolute stereochemistry of the reaction. Numerous methods have been developed for a wide range of substrates and nucleophiles.[114] The reaction between enones and malonate esters, which is of both historical and economic importance, has also been the subject of intense activity.[115] Ironically, we have already discussed asymmetric Michael additions—during our discussion of asymmetric Friedel-Crafts alkylations! Here, in fact, we see the dichotomy of the language much of modern organic synthesis. Let us ask the simple question: is the reaction in Example **12.102** electrophilic aromatic substitution of the pyrrole by an iminium ion electrophile—or nucleophilic addition of the electron-rich π electron cloud of the pyrrole to an iminium ion as a Michael acceptor, followed by subsequent deprotonation? The best answer that this author can come up with is—BOTH!

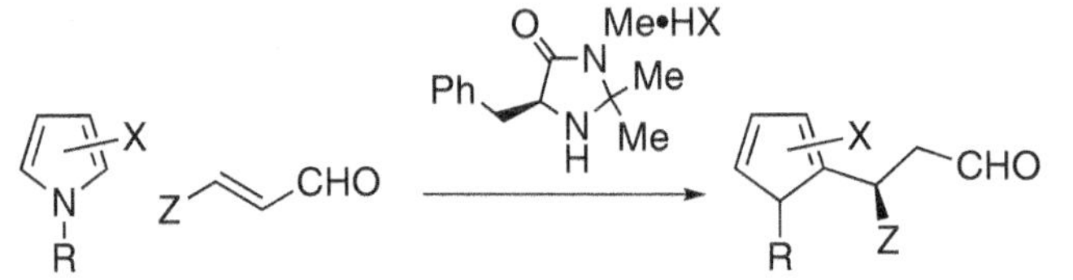

(12.102)

Conjugate Addition of Cyanide

Cyanide anion is often referred to as a pseudohalide because in many ways its chemistry mimics that of fluoride anion. It is, for example, the only simple carbanion nucleophile that adds reversibly to carbonyl groups, and, as such, it adds preferentially to conjugated carbonyl compounds to give γ-ketonitriles. However, simple cyanide salts often give poor yields with conjugated ketones because the equilibrium constant for addition is not always

---

114. Reviews: (a) Tsogoeva, S.B. *Eur. J. Org. Chem.* **2007**, 1701. (b) Ballini, R.; Bosica, G.; Fiorini, D.; Palmieri, A.; Petrini, M. *Chem. Rev.* **2005**, *105*, 933. (c) Berner, O.M.; Tedeschi, L.; Enders, D. *Eur. J. Org. Chem.* **2002**, 1877. (d) Jha, S.C.; Joshi, N.N. *Arkivoc* **2002**, *vii*, 167. (e) Sibi, M.P.; Manyem, S. *Tetrahedron* **2000**, *56*, 8033.

115. (a) Miyazaki, T.; Maekawa, H.; Yonemura, K.; Yamamoto, Y.; Yamanaka, Y.; Nishiguchi, I. *Tetrahedron* **2011**, *67*, 1598. (b) Somaiah, S.; Sashikanth, S.; Raju, V.; Reddy, K.V. *Tetrahedron: Asymmetry* **2011**, *22*, 1. (c) Fang, H.; Wu, X.; Nie, L.; Dai, X.; Chen, J.; Cao, W.; Zhao, G. *Org. Lett.* **2010**, *12*, 5366. (d) Sun, Y.; Fan, R. *Chem. Commun.* **2010**, *46*, 6834. (e) Figueiredo, R.M.; Christmann, M. *Eur. J. Org. Chem.* **2007**, 2575.

favorable. This can be remedied quite simply by changing the nucleophile to a dialkylaluminum cyanide (e.g., Example **12.103**).

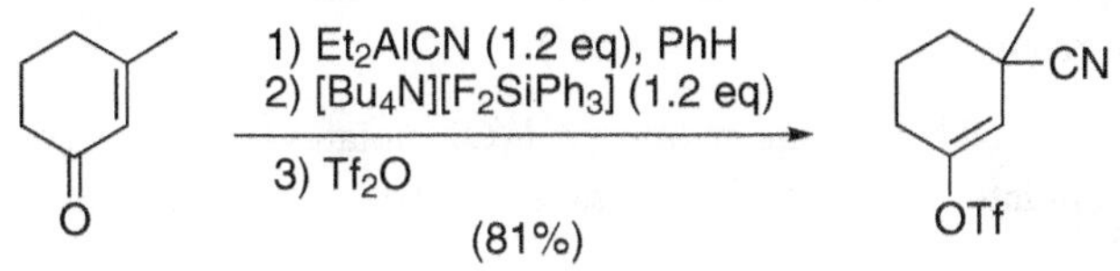

(12.103)[116]

Problem

12-9 What is the major organic product of each of the following reactions, which are steps from some classic syntheses?

(a)

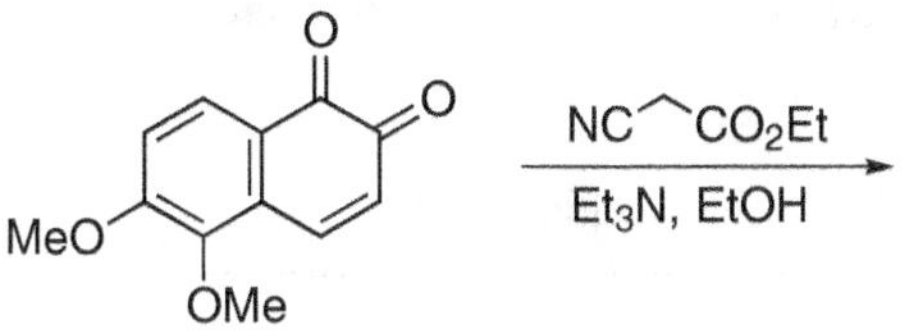

(b)

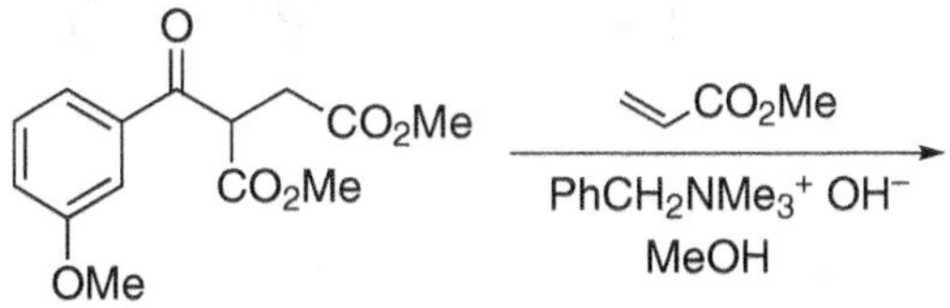

(c)

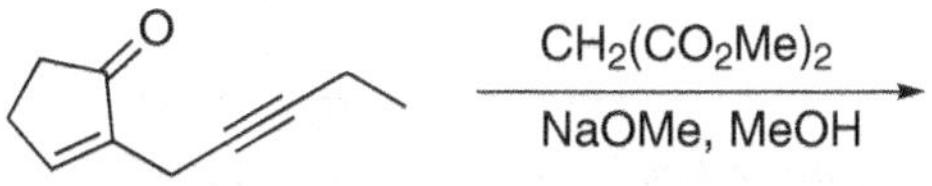

(d)

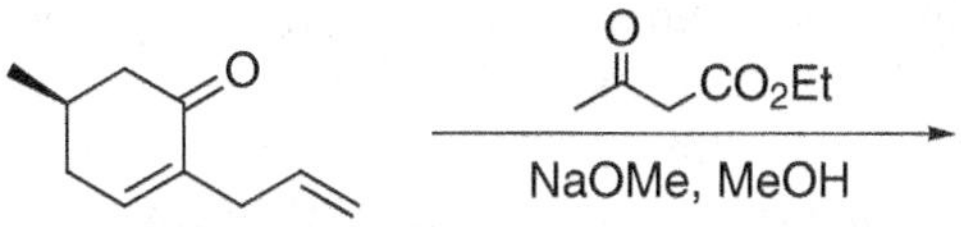

References: (a) *J. Am. Chem. Soc.* **1950**, *72* 228. (b) *J. Am. Chem. Soc.* **1962**, *84*, 3222. (c) *J. Org. Chem.* **1971**, *36*, 2021. (d) *J. Am. Chem. Soc.* **1985**, *107*, 4341.

Reaction Synopses

Michael Addition

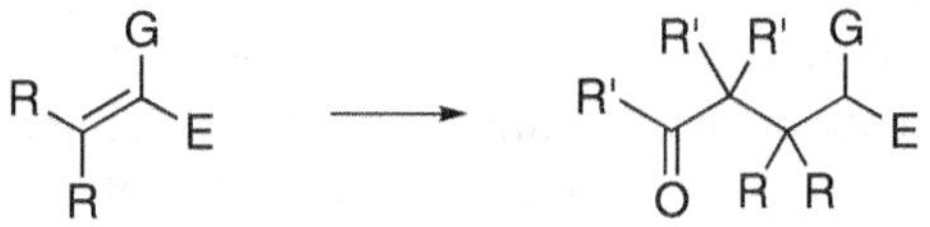

E: COR, CO_2R, $CONR_2$, CN, NO_2, SO_2R, etc.

G: H, R, SR, SeR, SiR_3, etc.

Reagents: (1) $RCOCH_2R$, LDA, THF, (2) $R'_2C = C(R)\text{-}COR'$; etc.

or (1) $R''_2N\text{-}C(R')\text{=}CR'_2$, (2) HOAc, H_2O, THF; etc.

(continued)

116. Peese, K.M.; Gin, D.Y. *J. Am. Chem. Soc.* **2006**, *128*, 8734.

(Reaction Synopses continued)

Polymerization suppressed in enolate additions if G is SR, SeR, SiR_3, or other third-row or higher group

Stereochemistry influenced by enolate geometry: *E(O)* enolates tend strongly to *anti* products; *Z(O)* enolates tend to be less stereoselective

Conjugate Addition of Cyanide

E:	COR, CO_2R, $CONR_2$, CN, NO_2, SO_2R, etc.
Reagents:	KCN, EtOH, Δ; Et_2AlCN, PhH; etc.

Aluminum reagent tends to give higher yields

12.6 **Azaenol Derivatives: Imines, Hydrazones, and Enamines**

Metalated Imines and Hydrazones

Enolate anions from carbonyl compounds are always subject to undesired self-condensation by the aldol or Claisen mechanism, and alkylation of ketone enolates is frequently difficult to control so that the monoalkylated product is obtained. Depending on the rate of proton transfer from the alkylated ketone to the unalkylated enolate, it may be impossible to prevent polyalkylation. For these reasons, a need was seen for a surrogate for the carbonyl group in these types of reactions. The obvious surrogates were compounds in which the carbonyl oxygen is replaced by nitrogen: imines and hydrazones.

It was quickly found, however, that although there is a structural homology between the C = O and C = N groups, homology does not extend to their reactivities. Enolization of cyclohexanone, for example, is too fast to monitor, even at −78°C, and the deprotonation of the corresponding imines by LDA occurs slowly at temperatures below 0°C. In addition, the much lower electrophilicity of the C = N group compared to the C = O group means that self-condensation is virtually nonexistent in reactions of imine anions.

Work by Collum[117] has shown that the deprotonation of imines is not always amenable to a simple explanation, particularly given the relatively easy *syn-anti* isomerization in these and similar compounds. Nevertheless, these systems are amenable to enantioselective alkylation reactions with high levels of asymmetric induction. A preference for deprotonation of ketimines at the less substituted α-carbon was observed by Meyers in his asymmetric alkylation work, but the observation is not universal. As shown in Example **12.104**,[118] Meyers found that it was necessary to heat the initial product under reflux to induce *syn/anti* isomerization of the imine and obtain good levels of deprotonation (only ≈ 33% of the imine deprotonates without heating).

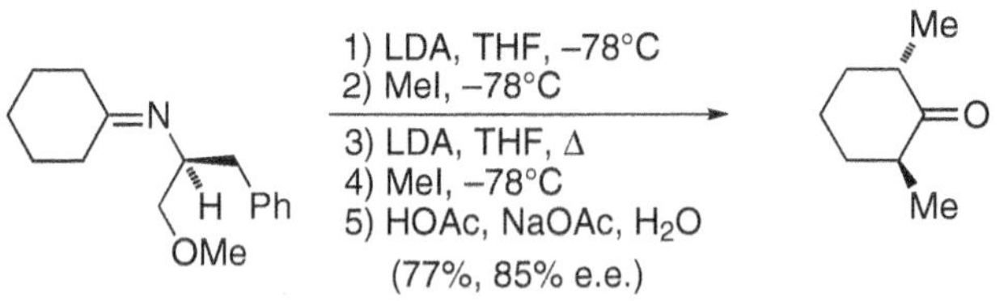

(12.104)

117. Liao, S.; Collum, D.B. *J. Am. Chem. Soc.* **2003**, *125*, 15114 and references therein.
118. Meyers, A.I.; Williams, D.R.; Erickson, G.W.; White, S.; Druelinger, M. *J. Am. Chem. Soc.* **1981**, *103*, 3081.

One class of imines that can be alkylated with a strong preference for axial alkylation are the *N,N*-dialkylhydrazones.[119] Although the alkylation of the *N,N*-dimethylhydrazone of 2-methylcyclohexanone (Example **12.105**) was reported to give the *trans* product as the major component of a 10:1 mixture,[120] later workers found that the *cis* isomer dominates the product mixture under kinetic conditions of short reaction times and low temperatures.[121]

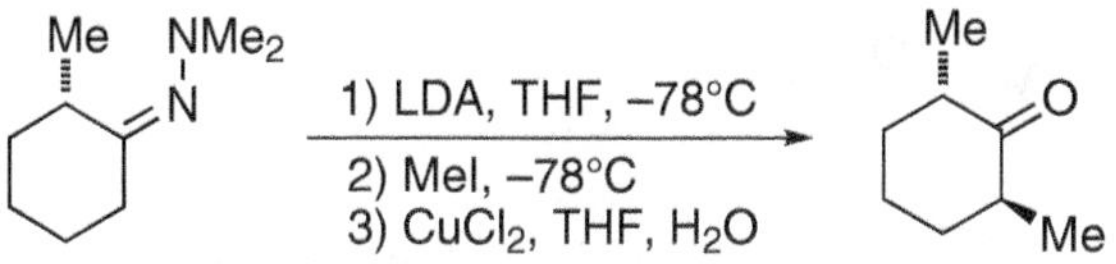

(12.105)

The metalated hydrazone method has been transformed into one of the most effective and general enantioselective alkylation procedures: the SAMP/RAMP hydrazone procedure of Enders.[122] These hydrazones can be deprotonated by LDA to give an anion that will react with a wide range of electrophiles with relatively high enantioselectivity levels (above 95% e.e. for alkylation reactions). The carbonyl compound can be recovered by exhaustive *N*-alkylation and hydrolysis; oxidative cleavage by ozonolysis (as shown in Example **12.106**) or neutral hydrogen peroxide; or by acid-catalyzed hydrolysis, depending on the functionality in the product.

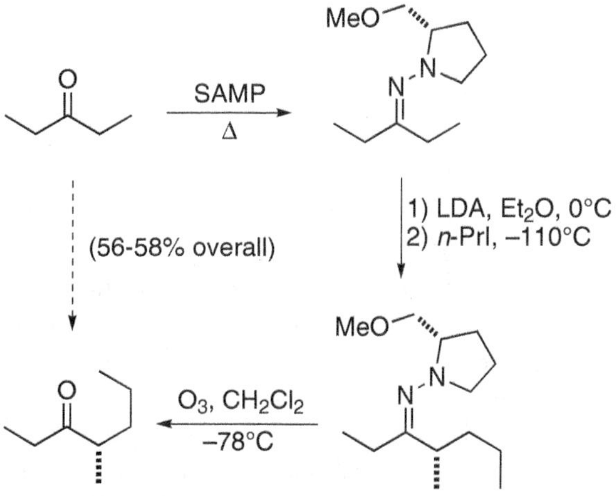

(12.106)

(97-98% e.e.)

The stereochemistry of the alkylation of a SAMP hydrazone may be rationalized in terms of the transition state model in Figure 12.14. In this model, the lithiated hydrazone adopts a conformation in which the pyrrolidine ring and the lithium chelate ring approximate a *trans*-hydrindane with the six-membered ring in a boatlike conformation, and the side chain in a pseudoequatorial position. In this conformation, alkylation of the *si* face of the alkene π bond (as designated in Figure 12.14) requires the alkyl halide to approach along a trajectory eclipsed by the adjacent methylene group of the pyrrolidine ring; alkylation of the *re* face requires the much less hindered approach eclipsed by the lone pair on nitrogen. Thus, one predicts that the product of this alkylation will be the ketone where the alkyl group is bonded to the *re* face of the enolate.

A more recent development is the use of sulfinylimines, sulfur-based chiral compounds that have been found to be good substitutes for aldehydes and ketones in both alkylation and aldol addition reactions. They also react with simple organometallic nucleophiles to give amine derivatives. In this case, the chirality transferred to the product is from a chiral

119. Review: Hickmott, P.W. *Tetrahedron* **1982**, *38*, 1975.

120. Corey, E.J.; Enders, D. *Chem.Ber.* **1978**, *111*, 1337.

121. Collum, D.B.; Kahne, D.; Gut, S.A.; DePue, J.S.; Mohamadi, F.; Wanat, R.A.; Clardy, J.; Van Duyne, G. *J. Am. Chem. Soc.* **1984**, *106*, 4865.

122. Enders, D.; Kipphart, H.; Fey, P. *Org. Synth., Coll. Vol. 8* **1993**, 403.

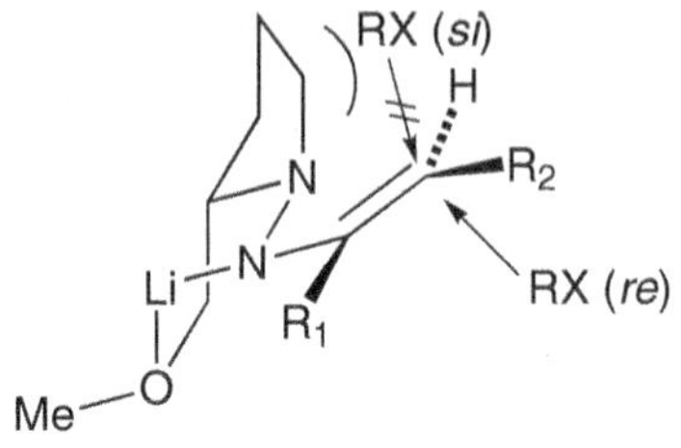

Figure 12.14 Transition state model for SAMP hydrazone alkylation.

sulfur, rather than chiral carbon. The chiral sulfinylimines derived from *tert*-butylsulfinylamine (Example **12.108**)[123] possess a chiral sulfur atom, making them some of the most (if not the most) widely used chiral auxiliaries today. They permit the formation of β-hydroxycarbonyl compounds from the methyl ketones or acetaldehyde with high levels of diastereoselectivity—something the Evans oxazolidinones cannot do. Importantly, the sulfur in this compound and its derivatives is quite configurationally stable, so it can be used as a chiral auxiliary in a wide range of applications. One of these is the asymmetric addition of the corresponding metal azaenolates to aldehydes to give chiral β-hydroxyaldimines that can be elaborated to a wide range of other products.

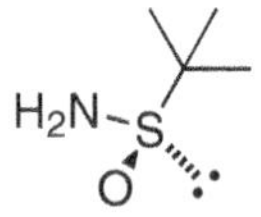

(12.108)

The aldimine or ketimine required (**12.109**) is formed from the chiral *tert*-butylsulfinylamine and the carbonyl compound under Lewis acid catalysis by anhydrous copper sulfate (aldehydes) or titanium tetraethoxide (ketones).[124] Note how both Lewis acid catalysts also serve as water scavengers. Deprotonation of the imine by LDA occurs on the less substituted side and is usually followed by exchange with magnesium bromide to give a magnesium azaenolate that then adds to the carbonyl group through a chairtype transition state (**12.110**) to give the β-hydroxyimine (**12.111**). Yields in this reaction are typically above 50% (usually above 75%), and d.r.'s range from 73:27 to 98:2 depending on the exact substitution pattern of the ketone and aldehyde partners. The β-hydroxyimine products can be converted to a wide range of products by relatively simple chemistry, as shown in Example **12.112**, where the β-hydroxycarbonyl product is shown.

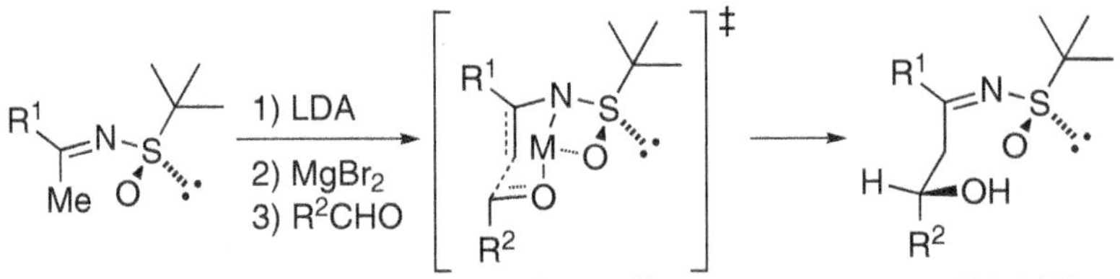

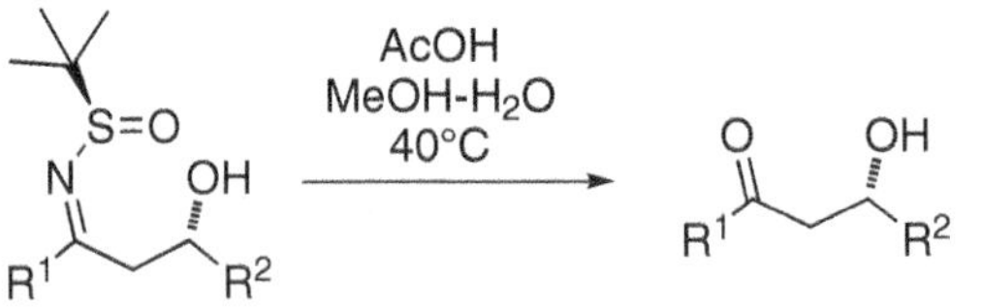

(12.112)

123. Reviews: (a) Ellman, J.A.; Owens, T.D.; Tang, T.P. *Acc. Chem. Res.* **2002**, *35*, 984. (b) Ellman, J.A. *Pure Appl. Chem.* **2003**, *75*, 39. (c) Ferreira, F.; Botuha, C.; Chemla, F.; Pérez-Luna, A. *Chem. Soc. Rev.* **2009**, *38*, 1162. (d) Roback, M.T.; Herbage, M.A.; Ellman, J.A. *Chem. Rev.* **2010**, *110*, 3600.

124. (a) Liu, G.; Cogan, D.A.; Ellman. J.A. *J. Am. Chem. Soc.* **1997**, *119*, 9913. (b) Cogan, D.A.; Ellman, J.A. *J. Am. Chem. Soc.* **1999**, *121*, 268.

One of the most important applications of chiral sulfinylimines is in the synthesis of chiral amines by addition of organometallic nucleophiles. Unlike most imines, sulfinylimines react with simple organometallic reagents to give sulfinylamine products that can be hydrolyzed to amines. The high levels of diastereoselectivity in the addition reactions come from the pronounced conformational preference of the sulfinylimine group itself. Computations[125] reveal that there is a strong conformational preference for the S = O bond and the N = C bond to be *syn*-coplanar (Example **12.113**). In this conformation, there is the possibility of significant π delocalization of the lone pair on nitrogen into the S–O σ* orbital. A second low-energy conformation (Example **12.114**) places the two lone pairs in an *antiperiplanar* arrangement; this is consistently some 6 kcal mol^{-1} higher in energy than the conformation in **12.113**.

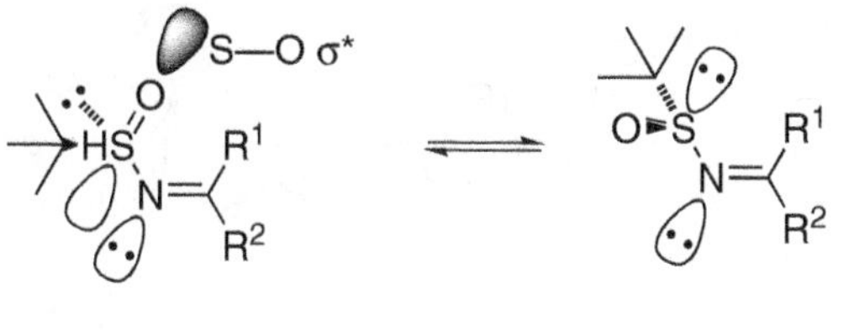

(12.113) **(12.114)**

The addition of the Grignard reagent occurs through a cyclic activated complex similar to that shown in Example **12.115**. The formation of this type of chelated activated complex suggests that the enantioselectivity of this reaction in solvents capable of coordinating the magnesium atom should be lower than in hydrocarbon or chlorocarbon solvents. It is reassuring to find that this is the case.[126] Interestingly, the stereochemistry of the addition is reversed when alkyllithium reagents are used as the nucleophiles in the presence of Lewis acid additives and in coordinating solvents. It is proposed that in this case, the reaction proceeds through an open transition state, making the addition intermolecular rather than intramolecular. The complexation of the sulfinyl oxygen by the Lewis acid should further stabilize the conformation shown in Example **12.117**, with the result that the metal alkyl now attacks from the face opposite the *tert*-butyl group, giving the stereochemistry shown in Example **12.118**. One example from drug synthesis is shown in Example **12.119**.[127] The free amine is liberated from the sulfinylamine as the hydrochloride salt by treatment with methanolic hydrogen chloride.

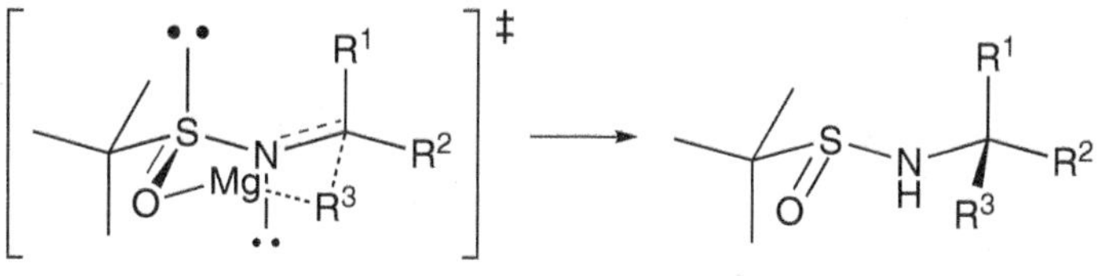

(12.115) **(12.116)**

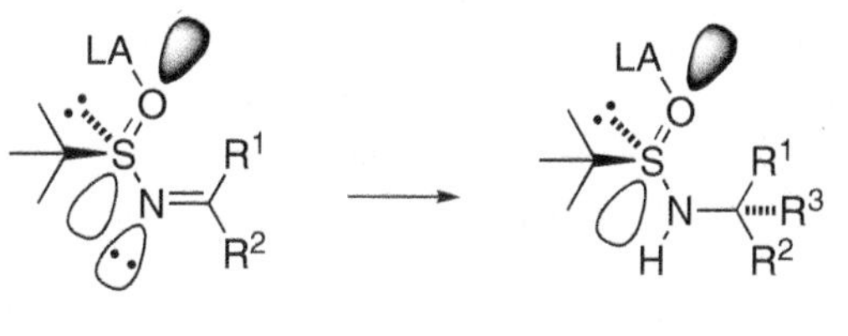

(12.117) **(12.118)**

125. (a) Bharatam, P.V.; Uppal, P.; Kaur, A.; Kaur, D. *J. Chem. Soc., Perkin Trans. 2*, **2000**, 43. (b) Bharatam, P.V.; Kaur, A.; Kaur, D. *J. Phys. Org. Chem.* **2002**, *15*, 197.

126. Cogan, D.A.; Liu, G.; Ellman, J. *Tetrahedron* **1999**, *55*, 8883.

127. Pflum, D.A.; Krishnamurthy, D.; Han, Z.; Wald, S.A.; Senanayake, C.H. *Tetrahedron Lett.* **2002**, *43*, 923.

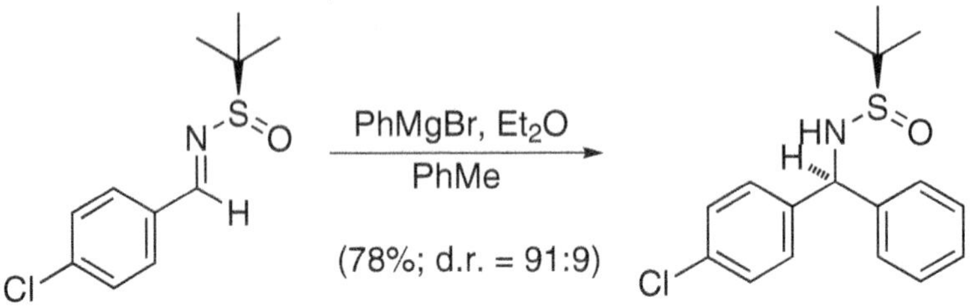

(12.119)

It is worthwhile noting that conjugated sulfinylimines do undergo 1,4-addition of alkyl groups in the presence of copper (I) cyanide and a Lewis acid catalyst, as shown in Example **12.120**.[128]

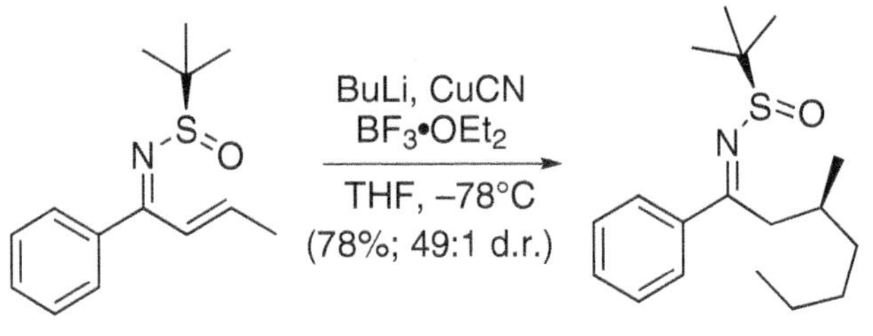

(12.120)

Enamines

Another alternative to the enolate anion is the enamine, formed by condensing a carbonyl compound with a secondary amine. The resultant compound, which has the nitrogen directly bonded to one of the sp^2-hybridized carbons of the alkene, is known as an ***enamine***.

The synthetic utility of enamines was recognized by Belgian-born American chemist Gilbert Stork,[129] whose work helped to develop enamines into powerful intermediates for organic synthesis. The enamine double bond is very electron-rich, so that it reacts as a nucleophile—effectively as a much less basic, neutral equivalent of an enolate anion. This reactivity has been rationalized on the basis of resonance (Example **12.121**). The nitrogen atom of the enamine is sp^2 hybridized, and its lone pair of electrons is delocalized into the carbon-carbon π bond. The minor contributor to the resonance hybrid has carbanionic character, and we therefore expect that the enamine will be a carbon nucleophile.

(12.121)

The alkylation of an enamine occurs predominantly at carbon to give an intermediate iminium ion that can be hydrolyzed under mild acidic conditions (e.g., dilute aqueous acetic acid) to give the carbonyl compound. An early example of the formation and use of enamines is provided by the steps from the synthesis of atrylacton, in Example **12.122**.[130] The enamine alkylation has two major advantages over enolate alkylation: (1) there is no propensity for self-condensation of the enamine as occurs with cyclopentanone enolates,

128. McMahon, J.P.; Ellman, J.A. *Org. Lett.* **2005**, *7*, 5393.

129. **Gilbert Stork (1921–)** was born in Brussels, and received his early education in Paris. His family moved to the United States early in World War 2. Stork graduated from The University of Florida (BS, 1942) and the University of Wisconsin (PhD, 1945). In 1953, after spending time in industry in Milwaukee and then at Harvard, Stork joined the faculty at Columbia University in New York. He spent his entire career there, and was instrumental in developing that department into one of the best in the world. For more biographical details, see: Hoffman, F. *Aldrichimica Acta* **1982**, *15*, 3.

130. Minato, H.; Nagasaki, T. *Chem. Comm.* **1965**, 377.

for example,[131] and (2) it avoids polyalkylation in especially enolizable ketones such as 2-tetralones.[132] However, because the enamine is also a less nucleophilic species, it often requires a more electrophilic partner to give reasonable reaction rates.

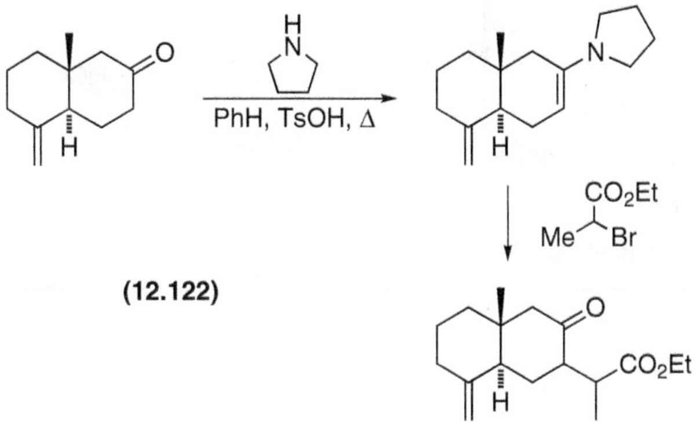

(12.122)

Reaction Synopses

Enamine Formation

Reagents: R'_2NH, TsOH, PhH, Δ; etc.

SAMP/RAMP Hydrazones

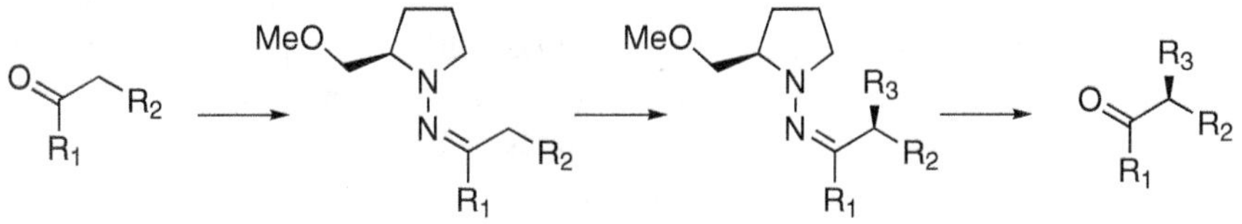

Reagents: (1) SAMP or RAMP, Δ, (2) LDA, THF, –78°C, (3) RX or other electrophile, (4) O_3, CH_2Cl_2;
or (1)–(3) the same, then (4) H_2O_2, MeOH;
or (1)–(3) the same, then (4) HCl, H_2O, THF;
or (1)–(3) the same, then (4) MeI (excess), (5) HCl, H_2O
Electrophiles: RX, RCHO, $R_2C = CR–E$, etc.

Sulfinylimine Anion Reactions

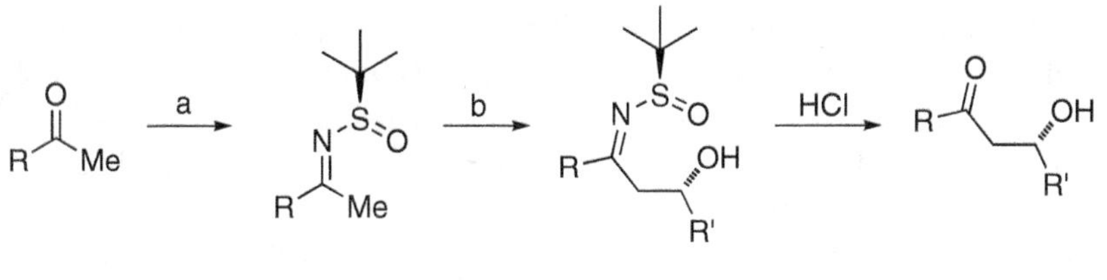

Reagents: (a) Bu^tSONH_2, $CuSO_4$; etc.
(b) (1) LDA, THF; then (2) $MgBr_2$; then (3) R'CHO

Most useful as a method for generating a functional chiral enolate of a methyl ketone

(continued)

131. There is a significant self-condensation of cyclopentanone under conditions of the Michael addition: (a) Prelog, V.; Metzler, O. *Helv. Chim. Acta* **1947**, *30*, 878. (b) Burkhalter, J.H.; Kurath, P. *J. Org. Chem.* **1959**, *24*, 990.

132. Stork, G.; Brizzolara, A.; Landesman, H.; Szmuszkowicz, J.; Terrell, R. *J. Am. Chem. Soc.* **1963**, *85*, 207.

(Reaction Synopses continued)

Additions to Sulfinylimines

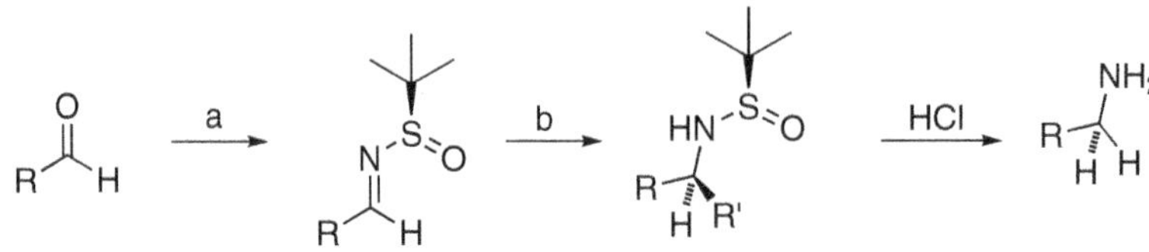

Reagents: (a) Bu^{t}SONH$_2$, CuSO$_4$; etc.
(b) R'MgX, Et$_2$O

Chapter Summary

This chapter has been focused on displacement and addition reactions of carbon nucleophiles. In nucleophilic substitution reactions of alkyl halides, the metal alkyls of the higher alkali metals, or "ate" complexes of other metals (especially Lipshutz cuprates) work best. Nucleophilic ring openings of epoxides, and nucleophilic addition to carbonyl compounds are facilitated by the presence of a cation capable of acting as a Lewis acid (e.g., Li, Mg, Zn). Used catalytically or stoichiometrically, copper (I) facilitates 1,4- addition to conjugated carbonyl compounds. The Reformatskii reagents, prepared from α-haloesters or α-haloketones and zinc metal, are excellent reagents for the preparation of β-hydroxycarbonyl compounds, especially in hindered systems.

Enolate anions are prepared by deprotonation of the α carbon of carbonyl compounds and similar species (nitro compounds, nitriles). Enolate anions can be formed under kinetic conditions using strong amide bases such as LDA, lithium isopropylcyclohexylamide (LICA), and lithium hexamethyldisilazide (LHMDS); kinetic deprotonation gives the less substituted enolate. Thermodynamic deprotonation is carried out with oxyanion bases (e.g., KOBu^t) at higher temperatures and in the presence of a catalytically small amount of excess ketone. Thermodynamic enolates have the more substituted double bond. Enolate anions are ambident nucleophiles that react with (1) harder electrophiles (acid halides, phosphoryl halides, silicon halides) through oxygen and (2) softer nucleophiles (alkyl halides, carbonyl compounds) through carbon.

Enolates are capable of exhibiting *E/Z* isomerism; the stereochemistry is best described using the *E(O)/Z(O)* system. The stereochemistry of the enolate can be affected by the size of the groups attached to the carbonyl group and by the size of the deprotonating base. Large bases and large groups attached to the carbonyl group favor the *Z(O)* enolate. The aldol addition reaction under kinetic conditions may be viewed in terms of the Zimmerman-Traxler model for the transition state, which predicts that *Z(O)* enolates will give higher levels of diastereoselectivity than the corresponding *E(O)* enolates. The highest levels of diastereoselectivity in aldol additions are obtained by using boron enolates, and Evans chiral boron enolates also override the Cram's rule selectivity of chiral aldehydes to allow the prediction of the absolute configuration of the product. 1,4-Regiochemistry in additions to conjugated carbonyl compounds is favored by softer nucleophiles and occurs best under thermodynamic conditions rather than kinetic conditions when the nucleophile is an enolate anion (when the reaction is known as a Michael addition). Michael additions under thermodynamic conditions tend to give *anti* products. *E(O)* amide enolates are more stereoselective than their *Z(O)* counterparts.

The oxygen in an enolate may be replaced by nitrogen to give an aza-enolate. Aza-enolate anion can be formed from imines or hydrazones, or from sulfinylimines; enamines are neutral aza-enolate nucleophiles. The use of aza-enolates in asymmetric synthesis is facilitated by means of Enders SAMP/RAMP hydrazones and chiral *tert*-butylsulfinylimines.

Key Terms

aldol addition
enamine
Evans asymmetric aldol addition
Michael addition
Reformatskii reaction
SAMP/RAMP hydrazones
sulfinylamines
sulfinylimines
Zimmerman-Traxler model

Additional Problems

12-10 What is the major organic product of each of the following reactions?

(a)

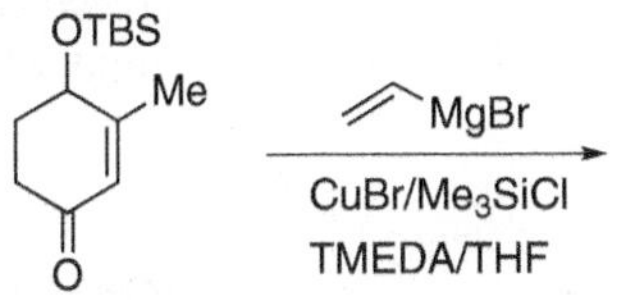

[*Tetrahedron Lett.* **2011**, *52*, 2037]

(b)

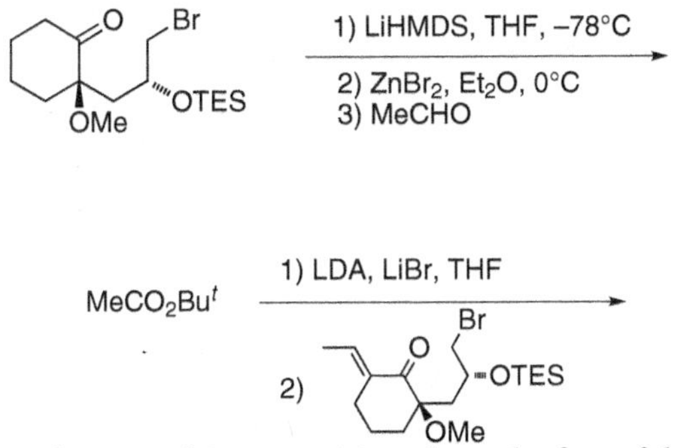

[*J. Am. Chem. Soc.* **2003**, *125*, 1498]

Is there anything surprising about the first of these two reactions?

(c)

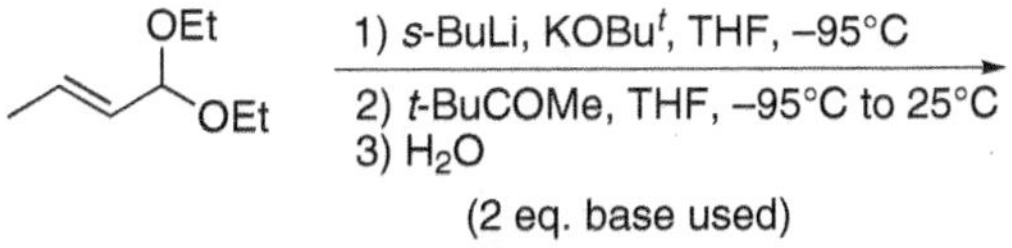

[*Tetrahedron* **1994**, *50*, 12463]

(d)

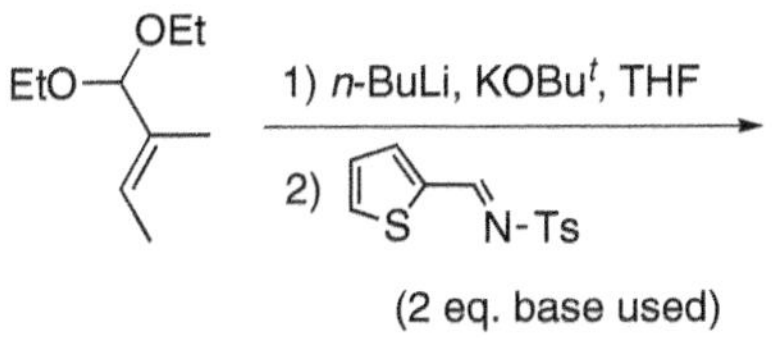

[*Org. Lett.* **2009**, *11*, 3914]

(e)

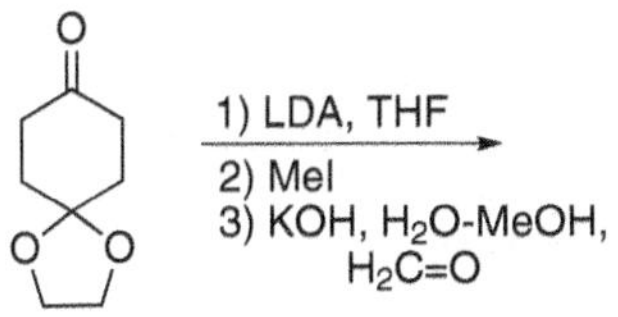

[*J. Am. Chem. Soc.* **2004**, *126*, 14358]

(f)

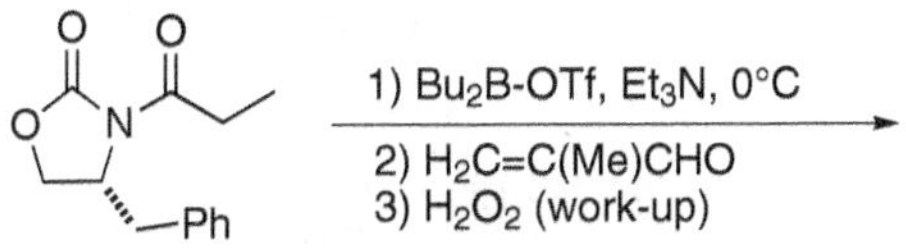

[*J. Org. Chem.* **1997**, *62*, 454]

(g)

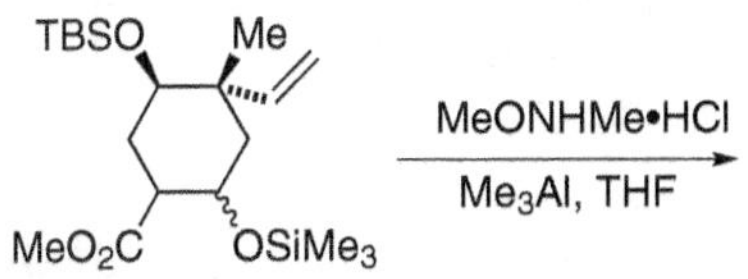

[*Tetrahedron Lett.* **2011**, *52*, 2037]

(h) 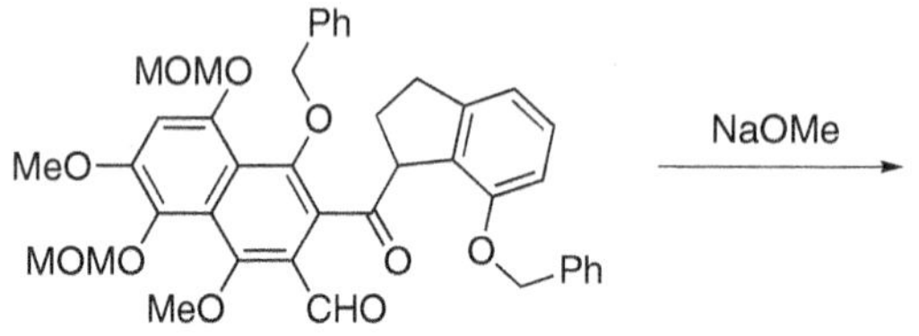

[*J. Org. Chem.* **1991**, *56*, 2115]

(i) 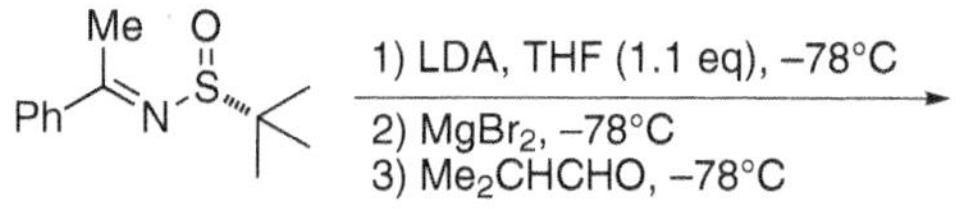

[*J. Am. Chem. Soc.* **2002**, *124*, 6518; **2003**, *125*, 11276]

(j) 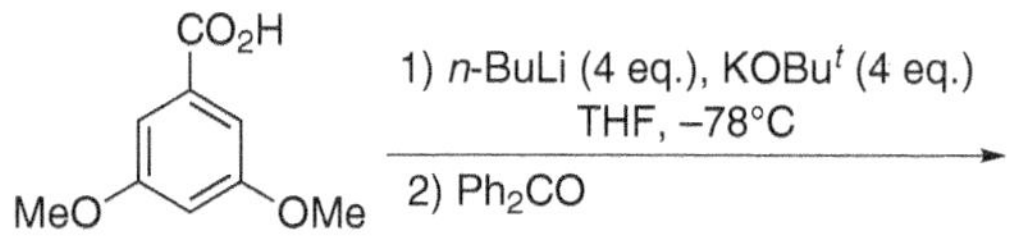

[*Tetrahedron Lett.* **2000**, *41*, 3157]

(k) 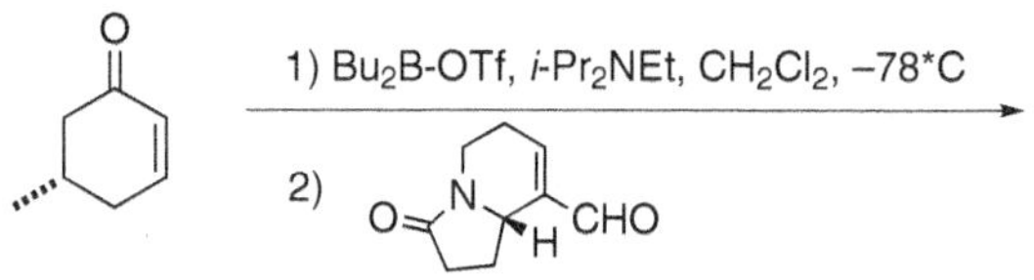

[*Org. Lett.* **2009**, *11*, 1179]

(l) 

[*Org. Lett.* **2011**, *13*, 6018]

(m) 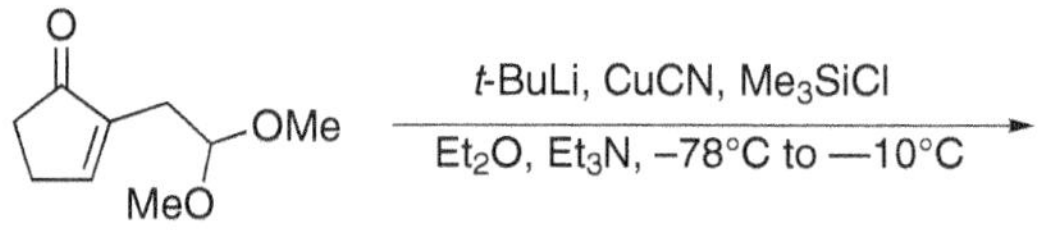

[*J. Am. Chem. Soc.* **1988**, *110*, 649]

(n) 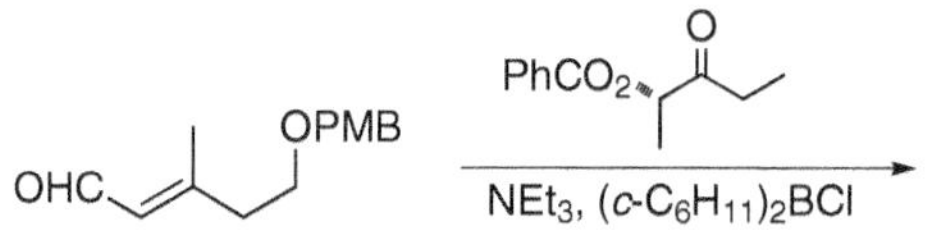

[*J. Org. Chem.* **2009**, *74*, 7220]

(o) 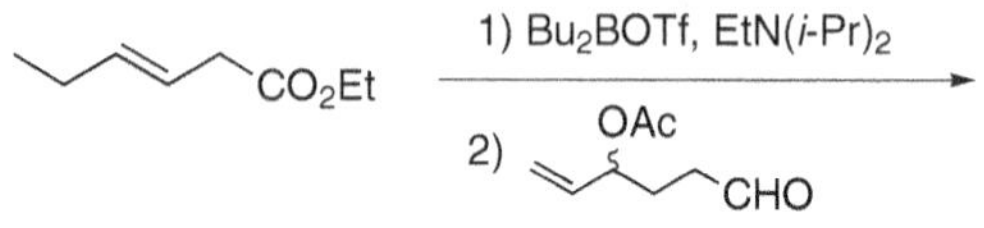

[*J. Org. Chem.* **2006**, *72*, 1235]

(p) 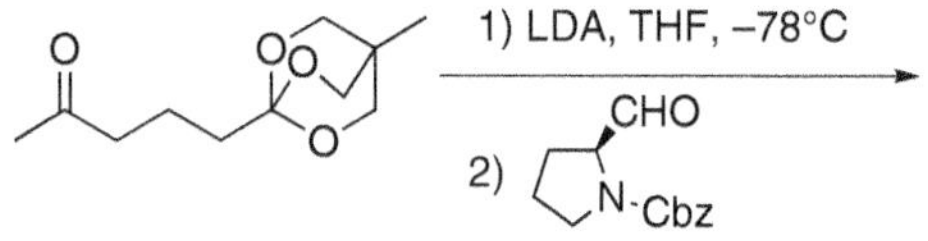

[*Org. Lett.* **2005**, *7*, 4419]

(q) 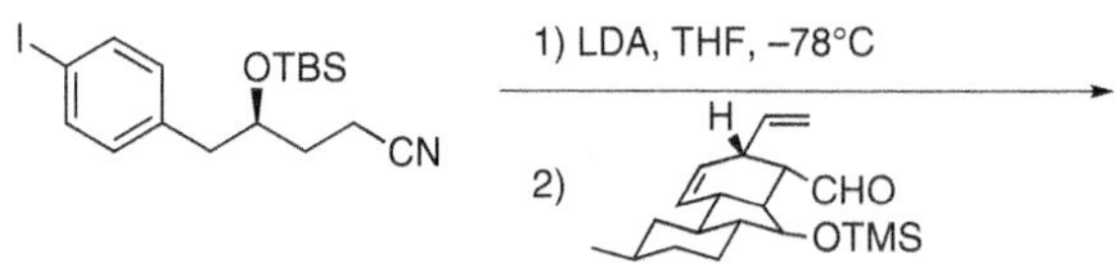

[*Org. Lett.* **2011**, *13*, 6268]

12-11 What reagent, or sequence of reagents, should be used to complete the following transformations?

(a) 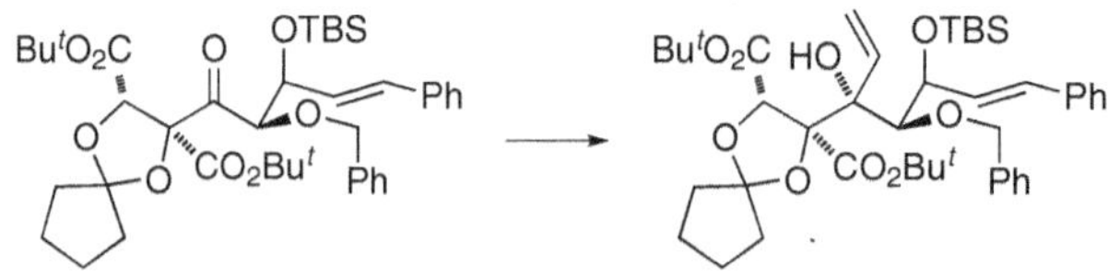

[*J. Am. Chem. Soc.* **1994**, *116*, 12111]

(b) 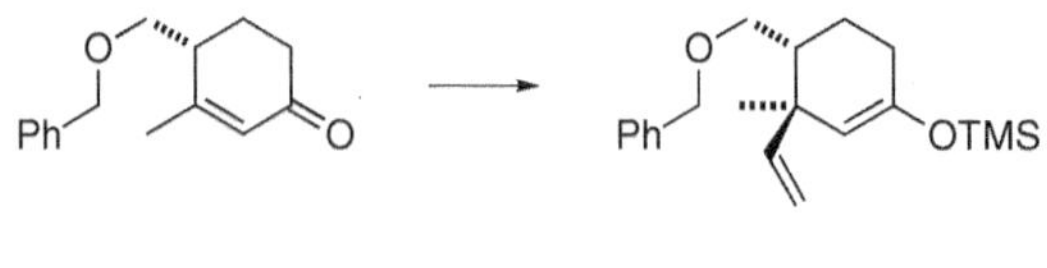

[*J. Org. Chem.* **1995**, *60*, 7837]

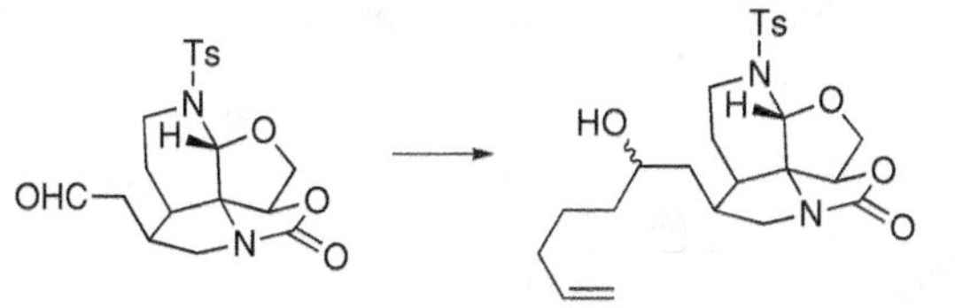

(c)

[*J. Am. Chem. Soc.* **2007**, *129*, 11987]

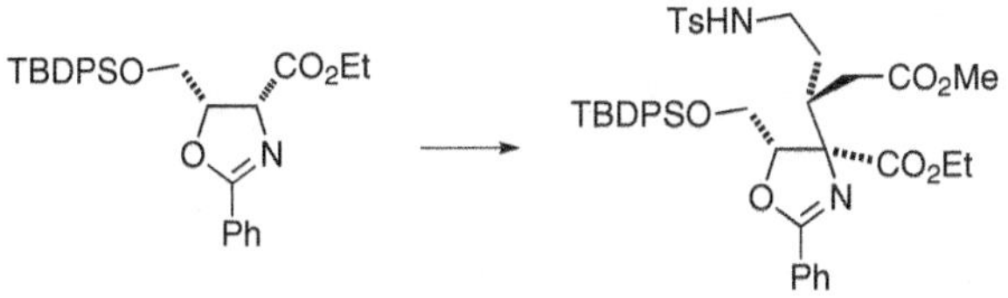

(d)

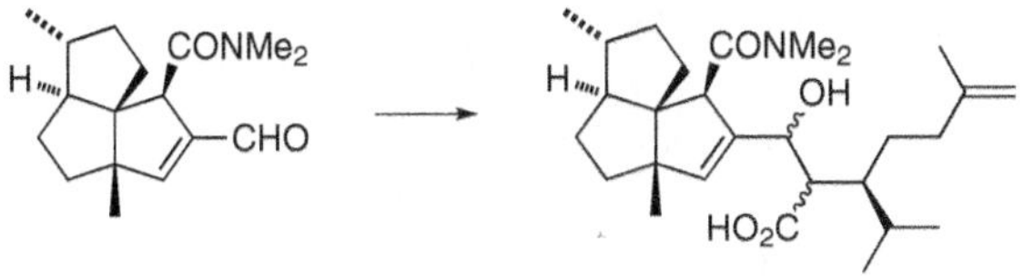

[*Tetrahedron Lett.* **1990**, *31*, 2517]

(e)

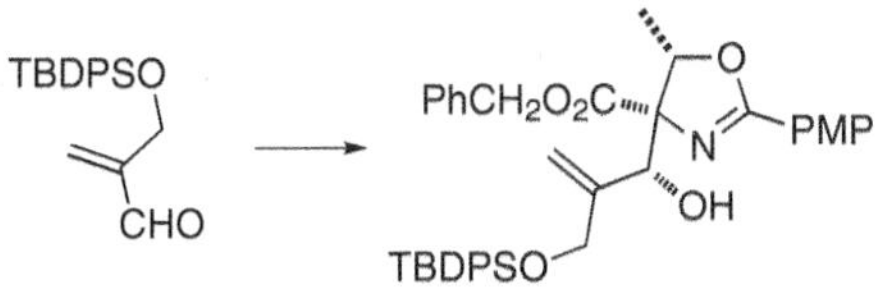

[*Angew. Chem. Int. Ed.* **2011**, *50*, 3497]

(f)

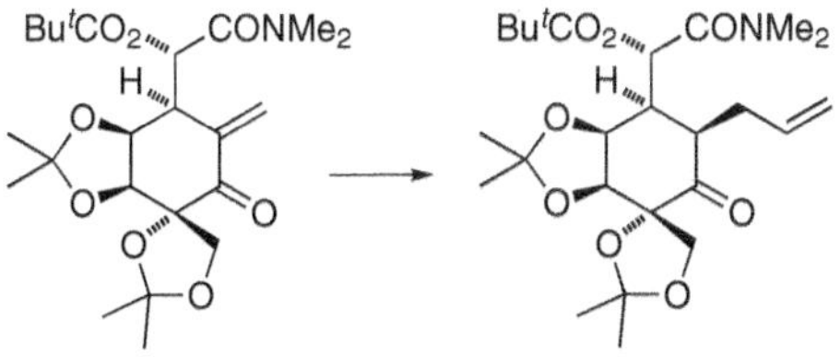

[*J. Am. Chem. Soc.* **2003**, *125*, 11510]

(g)

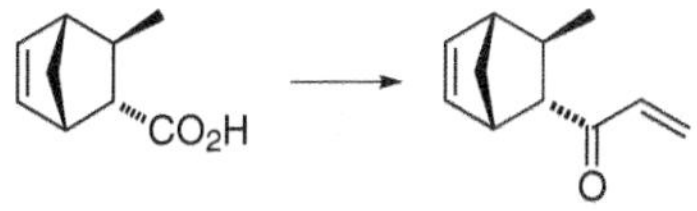

[*J. Am. Chem. Soc.* **2002**, *124*, 9974]

(h)

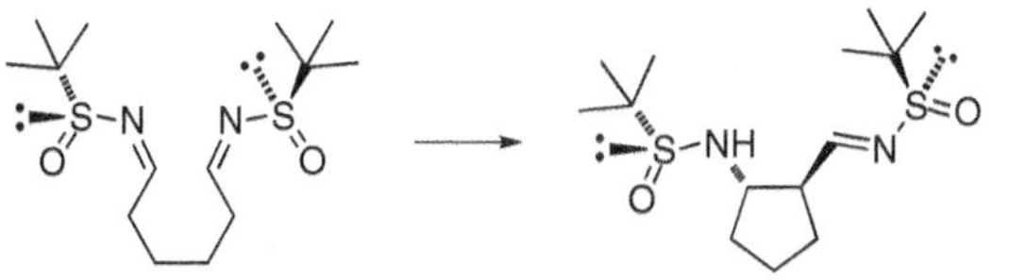

[*Org. Lett.* **2004**, *6*, 3621]

12-12 Answer the following questions about the reaction shown.

(a) What reagent is required to couple the two reactants?
(b) In what order should the reactants be added to the reaction flask?
(c) Provide a reasonable explanation for the regiochemistry of the reaction.

Reference: J. Am. Chem. Soc. **1984**, *106*, 4186.

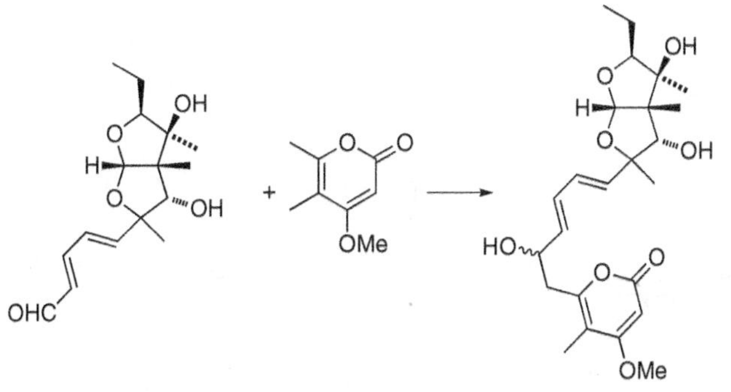

12-13 Write a reasonable mechanism for the transformation shown. [Reference: *J. Am. Chem. Soc.* **1997**, *119*, 2755.]

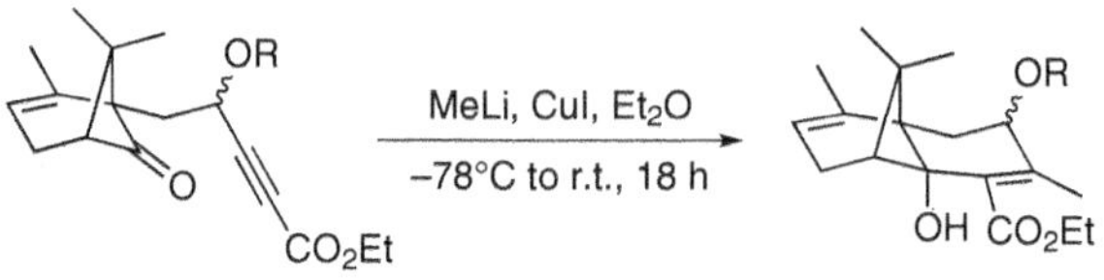

CHAPTER THIRTEEN

Reactive Intermediates III

Free Radicals, Carbenes, Arynes, and Nitrenes

13.1 Free Radicals[1]

Free radicals are species where at least one atom carries an unpaired electron; simple free radicals contain an odd total number of electrons. The chemistry of free radicals in solution has been studied since 1900, when Russian-born American chemist Moses Gomberg[2] attempted to prepare hexaphenylethane. What he obtained was a white solid that dissolved in solvents such as benzene to give a pale yellow solution that reacted with species such as oxygen to give products that could not be explained on the basis of its assumed identity as hexaphenylethane. Gomberg deduced that these products were consistent with a free trivalent carbon species and suggested that the colorless hexaphenylethane dissociated as it dissolved to give a yellow solution of free triphenylmethyl.[3] The structure of the dimeric compound became a source of much perplexity, dubbed "the hexaphenylethane riddle," over the almost seven decades before its structure was unambiguously deduced to be the dimeric compound shown,[4,5] first proposed by Paul Jacobson in 1905.[6] Gomberg's free radical hypothesis was confirmed experimentally when conclusive evidence for the existence of free radicals was provided by Paneth and Hofeditz in 1929 in studies of the pyrolytic decomposition of tetramethyllead (Figure 13.1).[7]

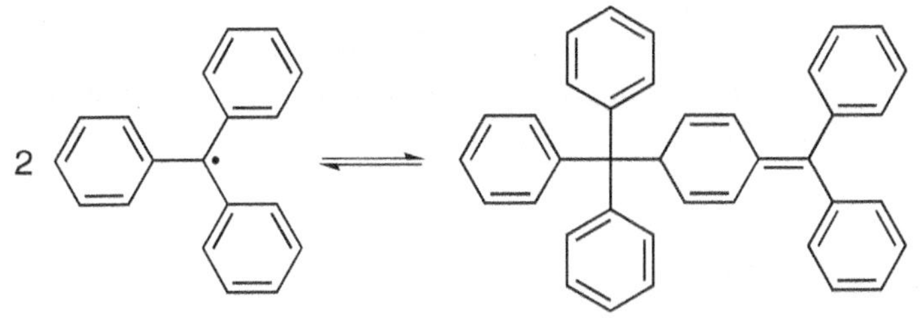

(13.1)

1. Monographs: (a) Pryor, W.A. *Free Radicals* (McGraw-Hill: New York, 1966). (b) Kochi, J.K. *Free Radicals*, 2 vols. (Wiley-Interscience: New York, 1973). (c) Nonhebel, D.C.; Walton, J.C. *Free-Radical Chemistry* (Cambridge University Press: Cambridge, 1974). (d) Nonhebel, D.C.; Tedder, J.M.; Walton, J.C. *Radicals* (Cambridge University Press: Cambridge, 1979).
Reviews: (e) Griller, D.; Ingold, K.U. *Acc. Chem. Res.* **1976**, *9*, 13. (f) Kaplan, L. *React. Intermed. (Wiley)* **1978**, *1*, 163; **1981**, *2*, 251; **1985**, *3*, 227.

2. **Moses Gomberg (1866–1947)** was born in Russia, which he left around 1884. He was educated at the University of Michigan (Michigan; PhD, 1894) and began a 43-year tenure on the faculty there in 1893, eventually becoming Professor and Chair. He spent 1896–1897 in Heidelberg and Munich. At Michigan, he rose to Professor and became Chairman of the Department of Chemistry. Gomberg's observations of triphenylmethyl initiated the field of free radical chemistry. For more biographical details, see: (a) Schoepple, C.S.; Bachmann, W.E. *J. Am.Chem. Soc.* **1947**, *69*, 2921; (b) Bailar, J.C., Jr. *Biogr. Mem. Nat. Acad. Sci.* **1970**, 141.

3. (a) Gomberg, M. *J. Am. Chem. Soc.* **1900**, *22*, 757. (b) Gomberg, M. *Ber. dtsch. chem. Ges.* **1900**, *33*, 3150.

4. (a) Lankamp, H.; Nauta, W.T.; MacLean, C. *Tetrahedron Lett.* **1968**, 249. (b) Smith, W.B. *J. Chem. Educ.* **1970**, *47*, 535.

5. For a history of the hexaphenylethane riddle, see: McBride, J.M. *Tetrahedron*, **1974**, *30*, 2009.

6. Jacobson, P. *Ber. dtsch. chem. Ges.* **1905**, *38*, 196.

7. Paneth, F.; Hofeditz, W. *Ber. dtsch. chem. Ges.* **1929**, *62*, 1335.

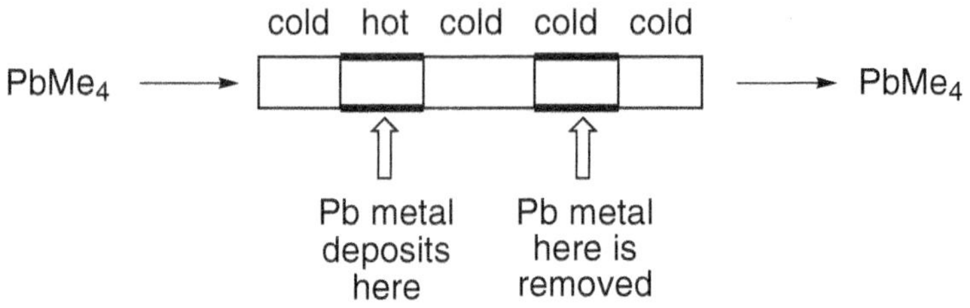

Figure 13.1 The Paneth and Hofeditz experiment to demonstrate the existence of free radicals. The tetramethyllead is decomposed thermally and deposits lead metal in the front part of the apparatus. Lead metal in the cold region of the tube downstream is gradually removed by the radicals in the effluent gas.

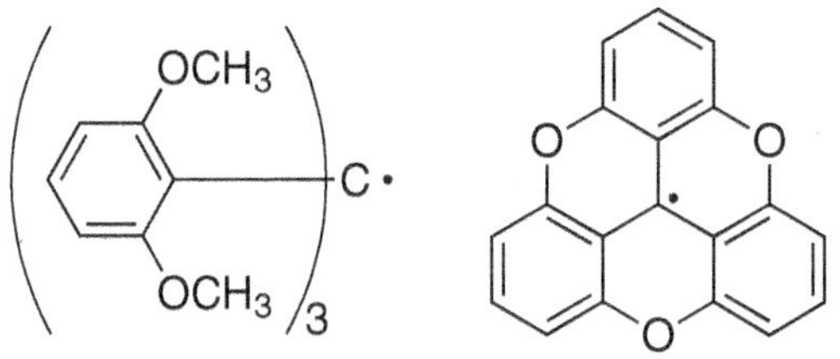

(13.2) **(13.3)**

Subsequently, much work was focused on the dimerization of triarylmethyl free radicals, and it has become apparent that the major causes of the stability of such radicals are steric rather than electronic. The *tris*-(2,6-dimethoxyphenyl)methyl radical (**13.2**) is completely unassociated even in the solid state, but the sesquixanthydryl radical (**13.3**) dimerizes to such an extent that there is no evidence of anything but dimer, even in solution.[8]

Worked Problem

13-1 The sesquixanthydryl radical dimerizes, and it has been suggested that it can do so because steric factors do not prevent the dimerization. How could the structure of this radical be altered to provide evidence for this assertion?

§Answer below.

8. Sabacky, M.J.; Johnson, C.S.; Smith, R.G.; Gutowsky, H.S.; Martin, J.C. *J. Am. Chem. Soc.* **1967**, *89*, 2054.

§ Answer to Worked Problem:

Oxygen is a very small atom, so gradually replacing the oxygen bridges by more sterically demanding groups (e.g., CMe_2 groups) might prevent dimerization or gradually change the molecule from one that dimerizes to one that does not. One example of such a series of molecules might be the three molecules below:

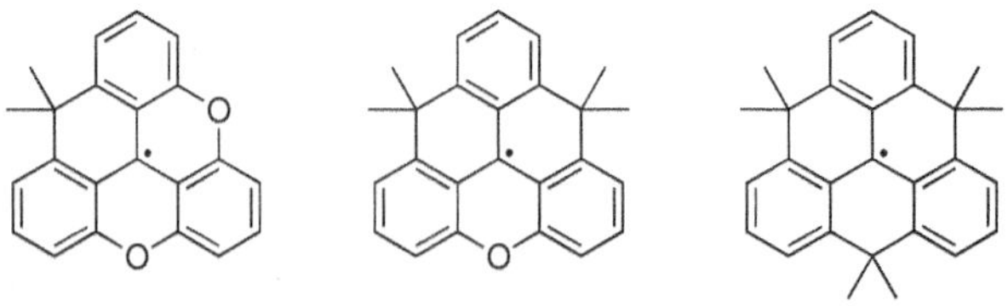

Structure and Bonding in Free Radicals

Organic free radicals are trivalent carbon species with an unpaired non-bonding electron on the carbon atom. Because the carbon atom carries one electron, the radical represents a species that is intermediate between the carbocation, where the central carbon carries a formal positive charge and is devoid of non-bonding electrons, and the carbanion, where the central carbon atom carries a formal negative charge and has a pair of non-bonding electrons. The electronic structure of free radicals represents a stage intermediate between the carbocation and the carbanion; therefore, free radicals might reasonably be expected to exhibit properties intermediate between the extremes defined by the cation and the anion. This appears to be the case.

The most direct method for obtaining data about the structure and bonding in free radicals is by ***electron paramagnetic resonance (EPR) spectroscopy***. This technique is analogous to nuclear magnetic resonance spectroscopy, with one exception. It is the orientation of the unpaired electron spin in an external magnetic field that is monitored rather than the orientation of nuclear spins. Based on EPR spectra, the evidence suggests that the preferred geometry of the radical center in simple hydrocarbon free radicals is planar with the unpaired electron occupying a 2p orbital.

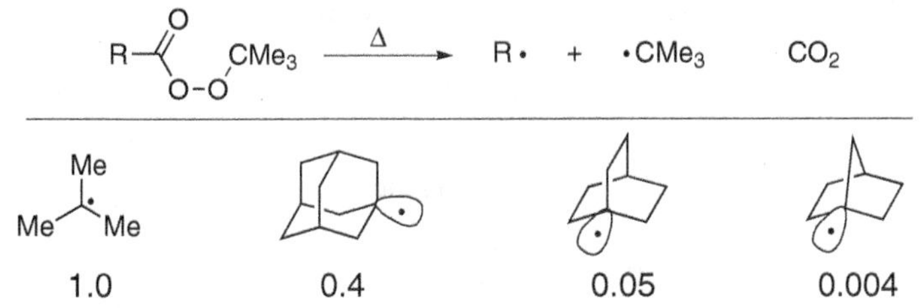

(13.5)

However, there is also abundant experimental evidence that free radicals are not *forced* to be planar. Many bridgehead radicals are known where the geometry of the carbon skeleton cannot accommodate an sp^2 hybridized carbon, and an excellent example of just how much a free radical may deviate from planarity is provided by the cubyl radical (**13.4**), a key intermediate in the first total synthesis of the highly strained hydrocarbon cubane.[9] However, the preference for planar geometry is shown by the relative rates of formation of the series of radicals shown (**13.5**) by pyrolysis of the *tert*-butyl peresters of the corresponding carboxylic acids.[10] Of the set, only the *tert*-butyl radical can easily attain planarity.

(13.4)

The relative stability of free radicals may be obtained by comparing bond dissociation energies of the corresponding hydrocarbons: the greater the bond dissociation energy, the less stable the radical. Typical values of the bond dissociation energies for a few representative hydrocarbons are presented in Table 13.1.

It is evident from the values in Table 13.1 that alkyl free radicals are easier to form (i.e., less reactive) than vinyl or aryl free radicals, and the stability of radicals increases as the number of substituents at the radical center increases. This order of stability is the same as that observed in carbocations. In the saturated carbocations, the order of stability was rationalized in terms of hyperconjugation, so it may be reasonable to suggest that free radicals are similarly stabilized by hyperconjugation, as shown in Example **13.6**. Conjugated free radicals (e.g., the allyl and benzyl radicals) are more stable than non-conjugated radicals, and the stability increases as the extent of conjugation increases.

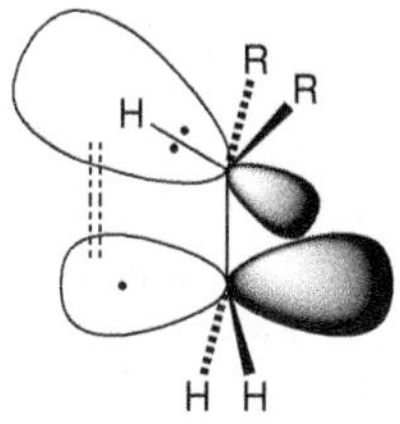

(13.6)

---

9. Eaton, P.E.; Cole, T.W., Jr. *J. Am. Chem. Soc.* **1964**, *86*, 3157.

10. (a) Lorand, J.P.; Chodroff, S.D.; Wallace, R.W. *J. Am. Chem. Soc.* **1968**, *90*, 5266. (b) Fort, R.C.; Franklin, R.E. *J. Am. Chem. Soc.* **1968**, *90*, 5267.

Table 13.1 Dissociation Energies* of C—H Bonds in Hydrocarbons

Bond	(kcal mol^{-1})	(kJ mol^{-1})
C_6H_5—H	110†	461†
H_2C=CH—H	≥ 108†	≥ 452†
H_3C—H	104	435
$(CH_3)_2$CH—H	101	421
RCH_2—H	98	410
R_2CH—H‡	95	398
R_3C—H	92	385
H_2C=CH-CH_2—H	89	372
H_2C=CH-CH(R)—H	83	347
$C_6H_5CH_2$—H	85	356

*Experimental error in values is ± 1 kcal mol^{-1} (± 4 kJ mol^{-1}).
†± 2 kcal mol^{-1} (± 8 kJ mol^{-1}).
‡Includes cyclopentyl and cyclohexyl groups.

When compared with carbon-centered free radicals, heteroatom-centered free radicals where the heteroatom is a second-row nonmetal are extremely energetic. At the same time, the presence of a heteroatom in a molecule tends to stabilize a carbon-centered free radical if the heteroatom is located on the α carbon of the radical (e.g., **13.7**). It is generally accepted that this stabilization occurs by a hyperconjugative mechanism similar to the stabilization of carbocations by an adjacent heteroatom (e.g., **13.8**; Figure 13.2).

The effects of a heteroatom are illustrated by the data in Table 13.2, where bond dissociation energies of representative C—H and O—H bonds of hydroxy compounds and ethers are gathered. The data in Table 13.2 show that incorporation of an sp^3-hybridized oxygen atom adjacent to the radical center reduces the dissociation energy of the C—H bond compared to the hydrocarbon, whereas the homolysis of the O—H bond consistently requires more energy than homolysis of the C—H bond in the hydrocarbon. In addition, it is clear that as the stabilization of the radical by hyperconjugation or conjugation increases, the amount of additional stabilization conferred by the heteroatom diminishes. There is one exception to the general rule that oxygen-centered radicals are extremely high energy—phenol, where the energy required for the homolysis of the O—H bond compares favorably with that required for the homolysis of the benzyl C—H bond of toluene. In the case of phenol, the homolysis of the O—H bond leads to a resonance-stabilized phenoxy radical, thus making it much more stable than the saturated oxygen-centered radicals.

The much higher energy of non-conjugated oxygen-centered free radicals is manifested in much higher reactivity. For example, hydroxy or alkoxy radicals will abstract a hydrogen atom from most hydrocarbons to give an alkyl free radical. This is one of the reasons that organic peroxides, which decompose to alkoxy radicals, are used to initiate many free radical reactions.

Reactivity of Free Radicals

As the only reactive intermediates with an unpaired electron, free radicals might be expected to exhibit some unique types of reactivity. This is most obvious in the mechanisms of free radical reactions: unlike ionic reactions, where it is possible for a reaction to proceed by a concerted mechanism, this is not possible in most free radical reactions.

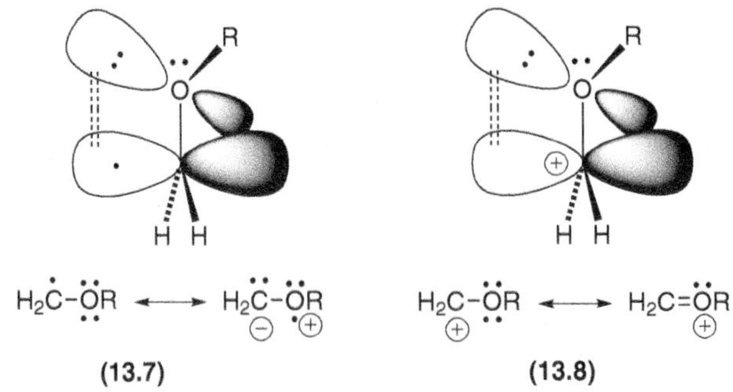

(13.7) (13.8)

Figure 13.2 Comparison between the stabilization of a free radical by an adjacent oxygen atom (left) and the stabilization of a carbocation by an adjacent oxygen atom (right). Both involve delocalization of electron density from oxygen to carbon.

Table 13.2 Bond Dissociation Energies in Hydrocarbon Derivatives

Bond	(kcal mol^{-1})	(kJ mol^{-1})	Bond	(kcal mol^{-1})	(kJ mol^{-1})
CH_3–H	104	435	HO–H	119	498
			HOO–H	90	377
RCH_2–H	98	410	RCH_2O–H	104	437
			$HOCH_2$–H	94	393
			$MeOCH_2$–H	93	389
R_3C–H	92	385	$HOCMe_2$–H	91	381
C_6H_5–H	110	461			
$C_6H_5CH_2$–H	85	356	C_6H_5O–H	88	368
H_2C=CH-CHR–H	83	347	H_2C=CH–CH(OH)–H	82	341
O=CR–H	87	364	RCOO–H	103–110	431–469

Free radical reactions differ from ionic reactions in one very important way: they tend to occur as chain reactions.[11] In a free radical chain reaction there are three distinct phases.

1. **Initiation**, the reaction or reactions where an initial source of free radicals to begin the reaction is generated
2. **Propagation**, the steps where the products of the reaction are formed
3. **Termination**, the reaction or reactions where the free radicals that carry the chain reaction are removed from the reaction mixture

Neither the initiation nor the termination steps contribute appreciably to the overall product of the radical reaction, and virtually all the products are generated during the propagation steps. In fact, *the balanced equation for the net chemical reaction corresponds to the sum of the balanced equations for the propagation steps only.* However, a free radical reaction will not occur until at least one free radical has been generated in an initiation step—*until an initiating radical is available, no formation of products can take place.*

Propagation

The differences between ionic and radical reactions are probably best illustrated by making a side-by-side comparison of a reaction that proceeds by an ionic mechanism (e.g., the addition of bromine to cyclohexene) and a reaction that proceeds by a free radical mechanism (the allylic bromination of cyclohexene).

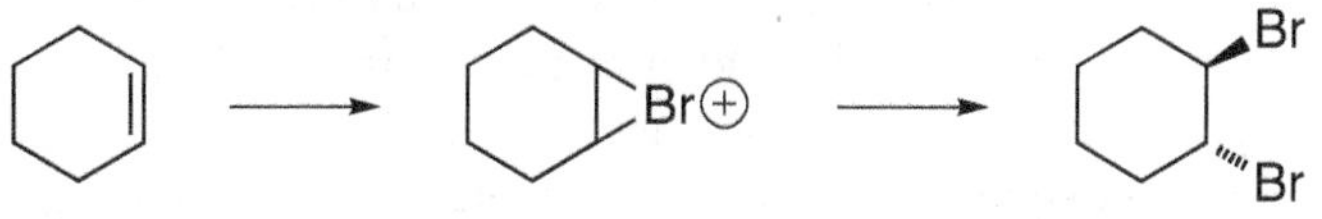

(13.9)

In the ionic reaction, bromine adds to the alkene to give the 1,2-dihalide. The reaction, which requires no discrete initiation step (beyond providing sufficient energy to overcome the activation energy), proceeds by two steps (Example **13.9**). In the first step, the alkene

11. Huyser, E.S. *Free Radical Chain Reactions* (Wiley-Interscience: New York, 1970).

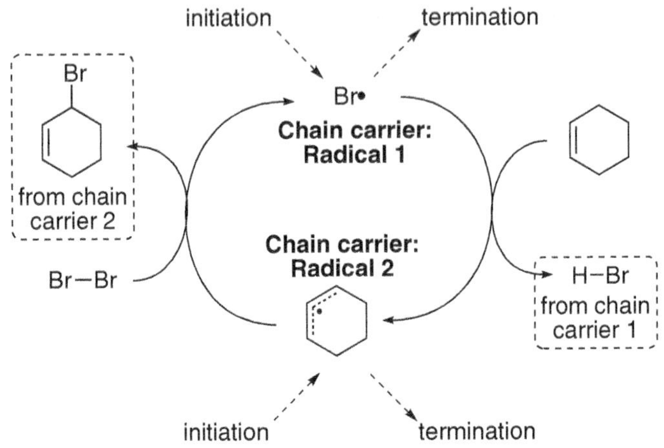

Figure 13.3 Cycle diagram for the free radical bromination of cyclohexene.

displaces bromide ion from the bromine molecule to give a cyclic bromonium ion that then reacts with a bromide ion to give the observed product. In the second step, the bromide ion may come from the same bromine molecule or a different one, so it is possible for both bromine atoms in the product to come from the same bromine molecule. More importantly, this means that *the two bromine atoms in the product do not* have to *come from different bromine molecules.*

In a free radical reaction, product formation requires at least two steps, and it is the hallmark of a free radical reaction that one of the chain carrying radicals is consumed and another chain-carrying radical is formed in each propagation step. In the bromination of cyclohexene, the first step involves the abstraction of the allylic hydrogen from cyclohexene by a bromine atom to give hydrogen bromide (one of the observed products of the reaction) and the 2-cyclohexenyl radical. The second step of the reaction involves the reaction between the 2-cyclohexenyl radical and molecular bromine to give 3-bromo-1-cyclohexene and a bromine atom. *Note that the two bromine atoms in the product must come from different bromine molecules.*

Free radical chain reactions can be represented as cycle ("carousel") diagrams, and this representation does a good job of showing the dynamics of free radical reactions, as illustrated by the free radical bromination of cyclohexene (Figure 13.3). In this diagram, the free radical chain carriers (the reactive intermediates) are located within the cycle, whereas the stable compounds (reactants, products, or initiators) are located outside the cycle. It is important for the success of a free radical reaction that the relative rates of the steps permit highly selective reactions. For this reason, much effort has been expended in measuring the rates of radical reactions.[12]

Initiation

The initiation of free radical chain reactions requires the production of an initiating free radical from an easily homolyzed precursor, known as an ***initiator***. Most initiators function by homolysis of a relatively weak bond or by the formation of an exceptionally stable molecule during the homolysis. The most widely used initiators are organic azo compounds (that homolyze to give N_2, an extremely stable molecule) and organic peroxides (where it is the weak O—O bond that homolyzes). Two of the most popular radical initiators are azo-*bis*-isobutyronitrile (AIBN **13.10**) and benzoyl peroxide (**13.11**), both of which undergo homolysis under thermal or photochemical reaction conditions.

12. For a leading review, see: Newcomb, M. *Tetrahedron* **1993**, *49*, 1151.

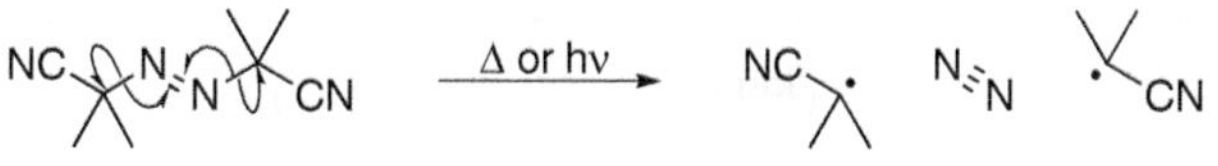

AIBN

(13.10)

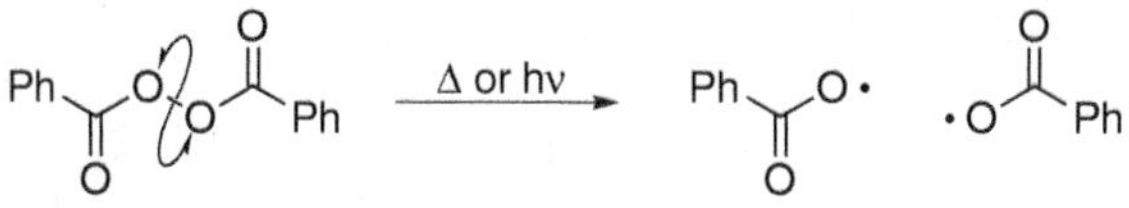

benzoyl peroxide

(13.11)

Oxygen-centered free radicals often undergo fragmentation to give simple alkyl free radicals and a carbonyl compound. The bond energy of a carbonyl group is typically between 170 and 180 kcal mol^{-1}, and the average bond dissociation of the C—O σ bond is approximately 86 kcal mol^{-1}. This means that the dissociation energy carbonyl π bond is quite high (≥ 85 kcal mol^{-1}), which makes the formation of the carbonyl compound and an alkyl free radical (e.g., **13.12**) thermodynamically favorable. In similar fashion, carboxy radicals lose carbon dioxide to give alkyl free radicals (**13.13**).

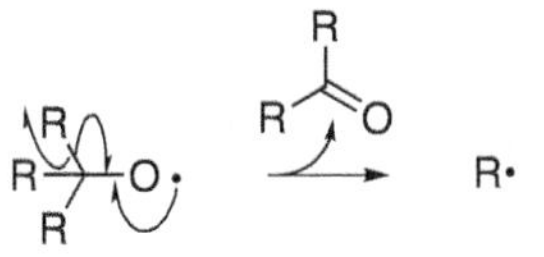

(13.12)

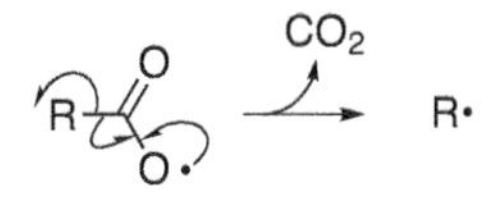

(13.13)

The initiation step in a free radical reaction is very important if the reaction is to be used in synthesis, so the site-selective formation of free radicals will be examined in much more detail in Chapter 14.

Termination

At some point during a free radical chain reaction, two radicals will be generated in close enough proximity to each other for as radical-radical reaction to occur. Such reactions, which remove a free radical from a chain reaction without replacing it, will terminate the chain reaction. The most common termination reactions are radical recombination reactions (**13.14**), where a new σ bond is formed, or atom-transfer reactions (**13.15, 13.16**) between radicals, which lead to the formation of a new π bond and a new σ bond. Most importantly, both reactions generate only non-radical products.

Atom transfer

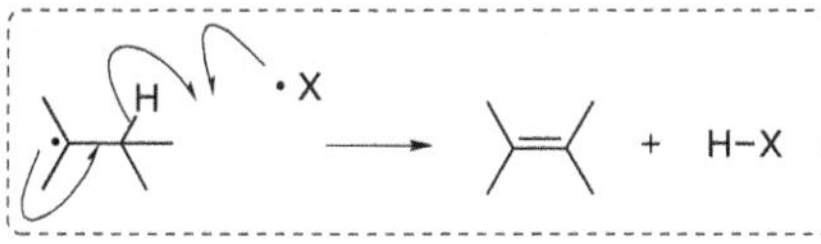

(13.15)

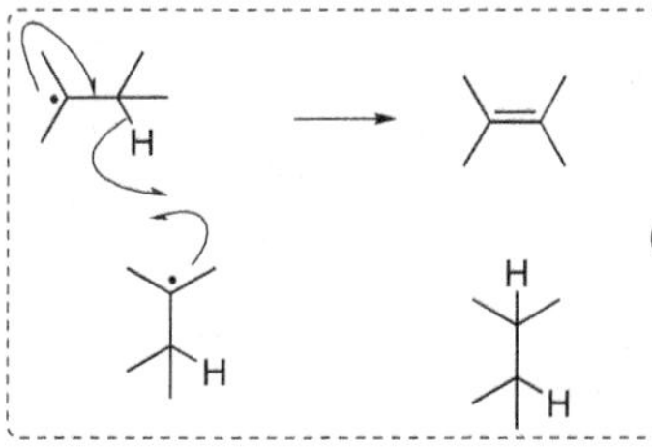

(13.16)

Radical recombination

R• + R• ⟶ R–R

R• + X• ⟶ R–X

X• + X• ⟶ X–X

(13.14)

Radicals in Synthesis

To reiterate: a pointed out just above, free radicals have become quite useful intermediates in organic synthesis, and we will discuss their use in synthesis in much more detail in Chapter 14.

Reaction Synopses

Methods for Generating Free Radicals

Peroxide initiators: $(PhCO_2)_2$; Me_3CO—OH; H_2O_2/Fe^{2+}; etc.

Azoalkane initiators: AIBN, etc.

Radical Fragmentation

Fragmentation of alkoxy radicals produces an aldehyde or ketone.

Fragmentation of acyloxy radicals produces carbon dioxide.

Worked Problem

13-2 The initiation of free radical reactions by hydroxy or alkoxy radicals usually occurs by hydrogen atom abstraction and by the 2-cyanopropyl radical usually occurs by addition to a carbon-carbon π bond. What is the reason for these differences?

hydroxy radical   2-cyanopropyl radical

§Answer below.

Problems

13-1 When he treated the dimer of triphenylmethyl radical with HCl, Russian chemist Aleksei Chichibabin obtained a new hydrocarbon with the formula $C_{38}H_{30}$. What is the most likely structure of this hydrocarbon and how is it formed?

§ Answer to Worked Problem:

The bond energy of the O—H bond is more than 100 kcal mol^{-1}, so abstraction of a hydrogen atom from sp^3-hybridized carbon (bond energy ≈ 85–95 kcal mol^{-1}) by the oxygen-centered hydroxy or alkoxy radicals is usually exothermic. This makes the hydrogen atom abstraction reaction a favorable and rapid reaction for these radicals. In contrast to this, the 2-cyanopropyl radical is carbon-centered. It will not usually participate in hydrogen atom abstraction reactions because they are not strongly exothermic. This leaves the predominant reaction of these radicals to be addition to a carbon-carbon π bond; because the C—C σ bond formed is stronger than the C—C π bond broken, this addition is almost always exothermic, making the addition favorable.

The addition of the hydroxy or alkoxy radical to the alkene is a possible reaction pathway, but with these highly reactive radicals, hydrogen atom abstraction, which leads to the formation of a new C—O σ bond, is considerably less exothermic and therefore less favorable.

13-2 The radical $(Me_3C)_3C\bullet$ is unusually unreactive for a non-conjugated free radical. For example, it exists in the solid state as the free radical, rather than the dimer, and it does not react with oxygen. Why should it be so unreactive?

13-3 The following free radical, which has only two aromatic rings in a position to conjugate with the free radical center, is a stable solid that does not react, even with oxygen. Why is this radical so stable?

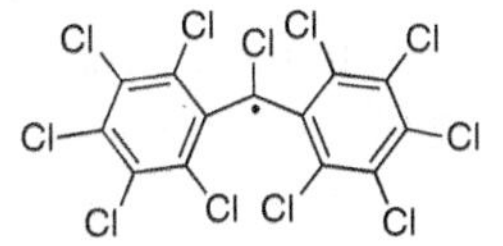

13-4 The final step in the Eaton synthesis of cubane [*J. Am. Chem. Soc.* 1964, *86*, 3157] involved heating a *tert*-butyl perester in diisopropylbenzene, as illustrated below. Suggest a mechanism for this transformation.

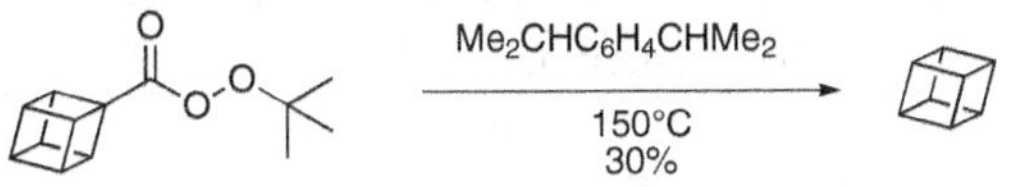

13-5 The Barton reaction is a reaction of a nitrite ester that led to the formation of an oxime of an aldehyde in Barton's synthesis of the corticosteroid, aldosterone. The oxime formed is the enol tautomer of a nitrosoalkane. Suggest a mechanism for the transformation.

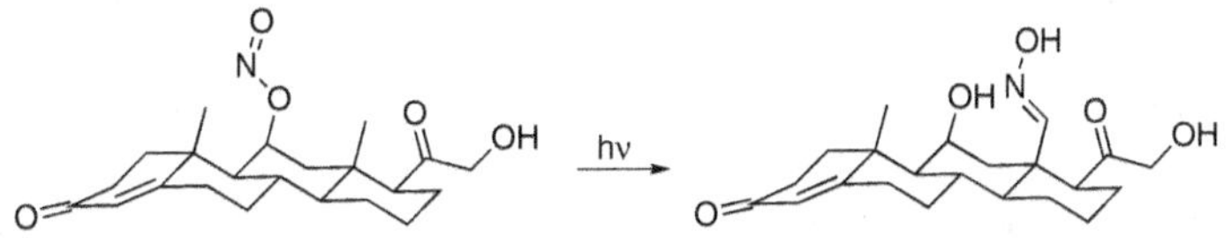

An isomeric product with a three-membered ring is formed in the same reaction. Based on your proposed mechanism, what is this product?

13-6 Oxygen reacts with ethers in the presence of light to generate explosive ether peroxides. What is the most probable structure of these ether peroxides, and by what mechanism are they formed? Draw a carousel diagram of your proposed mechanism.

13.2 Carbenes[13,14]

In 1950, Hine surprised the organic chemistry community with the first solid evidence that a neutral divalent species, a carbene, was involved in the reaction of aqueous hydroxide ion with chloroform.[15] A carbene is a neutral carbon electrophile with a formally divalent carbon bearing only six valence electrons, $:CR_2$. It can, therefore, act as an electron-pair

13. Monographs: (a) Hine, J. *Divalent Carbon* (Ronald Press: New York, 1964). (b) Jones, M., Jr.; Moss, R.A. *Carbenes*, 2 vols. (Wiley-Interscience, New York, 1973–1975). (c) Kirmse, W. *Carbene Chemistry*, 2nd ed. (Academic Press: New York, 1971). (d) Rees, C.W.; Gilchrist, T.L. *Carbenes, Nitrenes, and Arynes* (Nelson: London, 1969).
Reviews: (e) Moss, R.A.; Jones, M., Jr. *React. Intermed. (Wiley)* **1978**, *1*, 69; **1981**, *2*, 59; **1985**, *3*, 45. (f) Bethell, D. *Adv. Phys.Org. Chem.* **1969**, *7*, 153. (g) Closs, G.L. *Top. Stereochem.* **1968**, 3, 193. (h) Herold, B.J.; Gaspar, P.P. *Fortschr. Chem. Forsch.* **1966**, *5*, 89. (i) Kirmse, W. *Angew. Chem. Int. Ed. Engl.* **1965**, *1*, 4. (j) Schreck, J.O. *J. Chem. Educ.* **1965**, *42*, 260. (k) Köbrich, G. *Angew. Chem. Int. Ed. Engl.* **1967**, *5*, 41. (l) Buehle, C.A. *J. Chem. Educ.* **1972**, *49*, 239.

14. The term, "carbene" was, to quote footnote 9 of Doering, W.v.E.; Knoz, L.H. *J. Am. Chem. Soc.* **1956**, *78*, 4947, "collaborately conceived by W. von E. Doering, S. Winstein, and R.B. Woodward in a nocturnal Chicago taxi, and later delivered diurnally in Boston."

15. Hine, J. *J. Am. Chem. Soc.* **1950**, *72*, 2438.

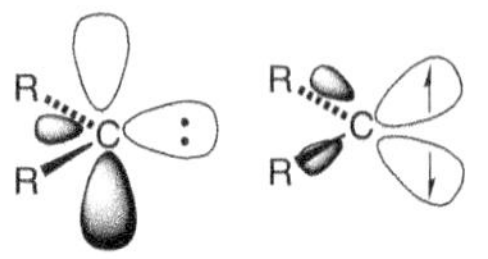

(13.17a) (13.17b)

singlet carbene

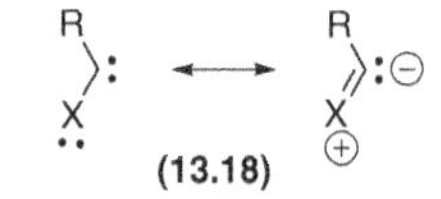

(13.18)

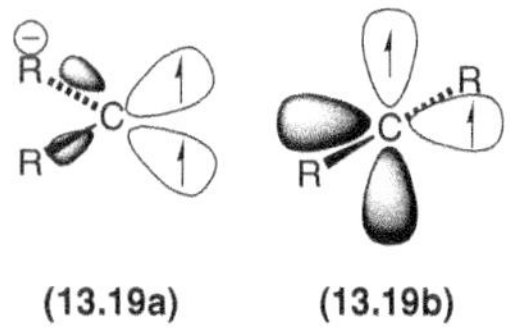

(13.19a) (13.19b)

triplet carbene

acceptor, and it is a Lewis acid. Carbenes exist in two forms: ***singlet carbenes***, in which all the electrons are paired, and ***triplet carbenes***, in which there are two unpaired electrons.

Singlet carbenes have two possible structures, with the center carbon being sp^2 hybridized and the lone pair of electrons in an sp^2 hybrid orbital (**13.17a**), or the center carbon is approximately sp^3 hybridized with each electron occupying a hybrid orbital (**13.17b**). Both models predict a bent structure. Herzberg[16] first isolated singlet methylene, and his subsequent work resulted in general acceptance of the sp^2 hybridized model for this species. This model, which allows the carbene to be represented formally as a zwitterion (**13.18**) allows a rationalization of the effect of heteroatoms on carbene reactivity: the delocalization of the lone pair from the heteroatom to the electron-deficient carbon reduces the electrophilicity of the carbene.

Likewise, there are two models for triplet carbenes. In one model (**13.19b**), the carbene is linear, with the center carbon being sp hybridized and the two unpaired electrons in the two unhybridized p orbitals. The other model (**13.19a**) resembles the sp^3 model of the singlet carbene with one spin reversed. The structure is predicted to be linear by one model and bent by the other. Both experimental[17] and theoretical[18] work suggests that triplet methylene is bent, with an H—C—H angle of 136°.

As their electronic structure implies, triplet carbenes are diradicals that react by a free radical mechanism; this means that reactions involving triplet carbenes cannot proceed by a concerted mechanism. Singlet carbenes, on the other hand, have an empty p orbital, like a carbocation, and an sp^2 orbital carrying a lone pair, like a carbanion. Thus, *a singlet carbene behaves in much of its chemistry as a zwitterion*: a hybrid of an electrophilic carbocation and a nucleophilic carbanion. In most of its reactions, it behaves predominantly as an electrophile.

Carbenes or their equivalents are most commonly generated from their precursors by the process of α-elimination (Table 13.3). During the reaction, a carbanion carrying a leaving group on the α carbon is formally generated, and the loss of the leaving group formally results in production of the carbene. The carbenes may be free, as occurs when they are generated from α-haloalkylmercury halides,[19] or they may be transferred directly from the organometallic precursor, without the intervention of the free carbene, as happens with the analogous α-haloalkylzinc halides.[20] Photolysis of a wide range of compounds, including diazoalkanes and diazirines, remains a standard method for the production of carbenes.

Table 13.3 Generation of Carbenes by α-Elimination

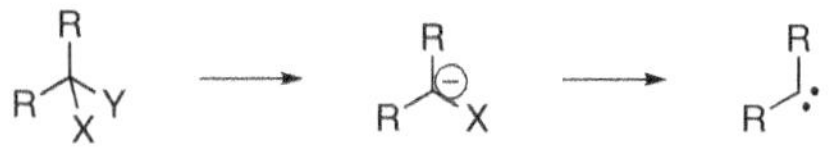

X	Y	Reaction Conditions
H	X (Cl, Br, I)	Base
X (Cl, Br, I)	X (Cl, Br, I)	RLi
RHg	X (Cl, Br, I)	Δ
$=N_2$ (diazo compound)	None	Δ or hν
$=N_2$ (diazirine)	None	Δ or hν

An alternative method for the generation of carbenes involves the removal of (1) a carbonyl oxygen atom from an ester or amide (X=O) or (2) a thiocarbonyl sulfur atom from a thionoester or thioamide (X=S) by heating the appropriate substrate with a trivalent phosphorus compound (usually a trialkyl phosphite). This reaction is critical in the Corey-Winter olefin synthesis (**13.20**) by *syn* elimination of the cyclic carbonate or thiocarbonate of a 1,2-diol.[21] Evidence that a carbene is involved as an intermediate in this reaction was provided by

16. (a) Herzberg, G.; Shoosmith, J. *Nature* **1959**, *183*, 1801. (b) Herzberg, G. *Proc. Roy. Soc.* **1961**, *A 262*, 291.
17. (a) Wasserman, E.; Yager, W.A.; Kuck, V.J. *Chem. Phys. Lett.* **1970**, *7*, 409. (b) Bunker, P.R.; Sears, T.J.; McKellar, A.R.W.; Evenson, K.M.; Lovas, F.J. *J. Chem.Phys.* **1983**, *79*, 1211. (c) McKellar, A.R.W.; Yamada, C.; Hirota, E. *J. Chem. Phys.* **1983**, *79*, 1220. (d) Bunker, P.R.; Jensen, P. *J. Chem. Phys.* **1983**, *79*, 1224.
18. (a) Bender, C.F.; Schaefer, H.F., III *J. Am. Chem. Soc.* **1970**, *92*, 4984. (b) Schaefer, H.F., III *Science*, **1986**, *231*, 1100; **1986**, *232*, 1319. (c) Wasserman, E.; Schaefer, H.F., III *Science* **1986**, *233*, 829.
19. (a) Seyferth, D.; Mui, Y.-P.; Burlitch, J.M. *J. Am. Chem. Soc.* **1967**, *89*, 4953. (b) Seyferth, D. *Acc. Chem. Res.* **1972**, *5*, 65.
20. Charette, A.B.; Marcoux, J.-F. *J. Am. Chem. Soc.* **1996**, *118*, 4539.
21. (a) Corey, E.J.; Winter, R.A.E. *J. Am. Chem. Soc.* **1963**, *85*, 2677. (b) Corey, E.J.; Hopkiss, P.B. *Tetrahedron Lett.* **1982**, *23*, 1797.

the work of Horton and Tyndall.[22] The driving force for the reaction almost certainly comes from the highly exothermic formation of the P=O or P=S groups in the phosphorus-containing product.

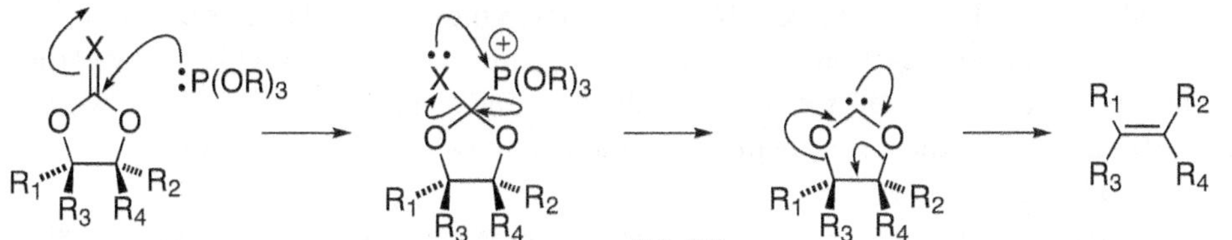

(13.20)

Carbenoids

Often, a carbene formally appears to be involved in a reaction, but there is evidence that no free carbene is actually produced in the reaction. The reagent in such reactions is called a ***carbenoid.*** Almost always it is an organometallic halide capable of eliminating a metal halide during the reaction, or it may be a species where there is a metal-carbene complex with a formal metal-carbon double bond. These species are most often formed by the reaction of a diazoalkane with rhodium (e.g., **13.21**) or copper (e.g., **13.22**).[23] The reactive intermediate in the Simmons-Smith cyclopropanation reaction is iodomethylzinc iodide, I—Zn—CHI_2.[24] Copper- and rhodium-catalyzed decomposition of sulfonium and iodonium ylides also appears to involve the formation of metal carbenoid intermediates.[25]

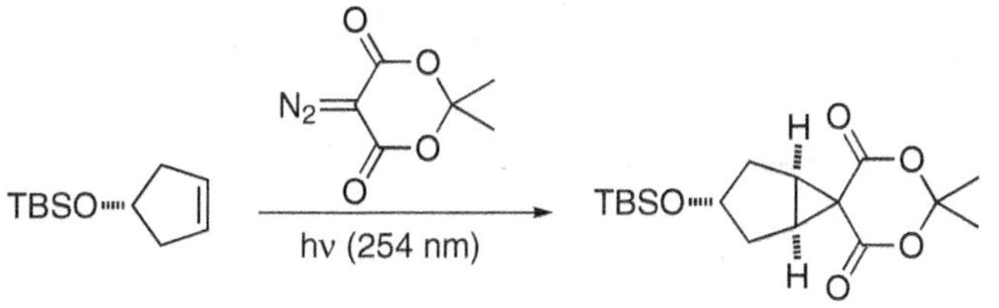

(13.21)[26]

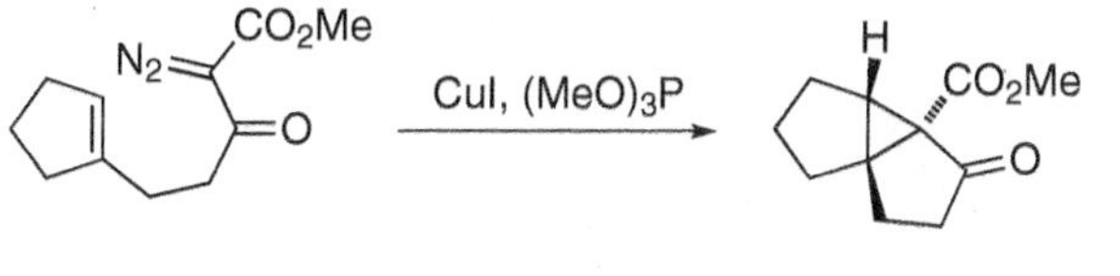

(13.22)[27]

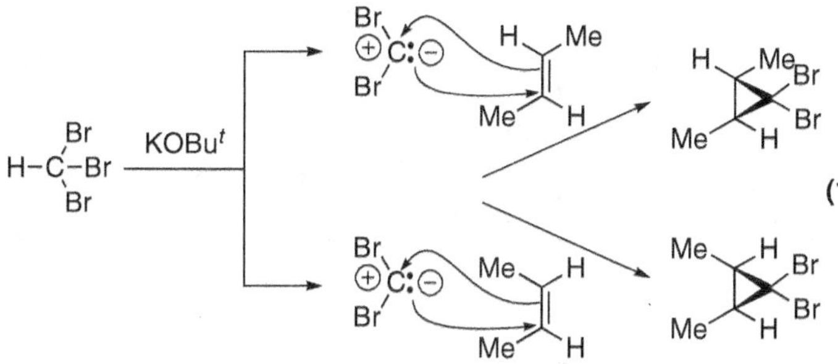

(13.23)[28]

22. Horton, D.; Tyndall, C.G., Jr. *J. Org. Chem.* **1970**, *35*, 3558.
23. (a) Moser, W.R. *J. Am. Chem. Soc.* **1969**, *91*, 1135. (b) Doyle, M.P. *Chem. Rev.* **1986**, *86*, 919. (c) Brookhart, M. *Chem. Rev.* **1987**, *87*, 411.
24. Charette, A.B.; Marcoux, J.-F. *J. Am. Chem. Soc.* **1996**, *118*, 4539.
25. Müller, P.; Fernandez, D.; Nury, P.; Rossier, J.C. *J. Phys. Org. Chem.* **1998**, *11*, 321.
26. Livinghouse, T.; Stevens, R.V. *Chem. Commun.* **1978**, 754.
27. Nakayama, M.; Ohira, S.; Okamura, Y.; Soga, S. *Chem. Lett.* **1981**, 731.
28. Skell, P.S.; Garner, A.Y. *J. Am. Chem. Soc.* **1956**, *78*, 3409

Carbene Addition

In their reactions, carbenes almost always form two bonds to the carbene carbon to complete its octet. Probably the most characteristic reactions of carbenes involve addition to a π bond to give a three-membered ring and insertion into a σ bond to give a pair of new σ bonds. The addition reactions of carbenes to π bonds can be distinguished by carbene type. Singlet carbenes add stereospecifically in a one-step, concerted reaction with *syn* stereochemistry, so the stereochemistry of the starting alkene is retained in the cyclopropane product (e.g., **13.23**).[29]

The reactivity of singlet carbenes toward alkenes reveals their electrophilic nature, paralleling the reactivity of known electrophilic reagents such as bromine[30] and peracids[31] toward alkenes. The results of a series of competitive cyclopropanations of alkenes and isobutylene by halogenated carbenes[32] are presented in Table 13.4.

The addition of triplet carbenes to alkenes requires addition to the π bond in a two-step reaction, which leads to loss of stereochemistry of the original alkene; this is shown schematically for addition of triplet CH_2**:** to a *cis* alkene in Figure 13.4. Thus, the closure of the three-membered ring after the initial addition of the carbene to the π bond requires spin inversion, which allows the rapid rotation about the carbon-carbon σ bond of the original alkene to occur. This results in the stereochemistry of the cyclopropane product becoming unrelated the stereochemistry of the starting alkene.

The other reaction characteristic of carbenes is the insertion into σ bonds (e.g., **13.24**).[33] The reactions of singlet methylene with saturated hydrocarbons is essentially random, while the reaction of the triplet shows selectivity for more substituted C—H bond insertion, as would be expected for a two-step reaction involving initial C—H homolysis. Thus, the reaction between propane and triplet carbene, generated by photolysis of ketene, gives a

Table 13.4 Relative Rates of Additions of Halocarbenes to Alkenes

Alkene	CFCl	CCl_2	CBr_2	Alkene	CFCl	CCl_2	CBr_2
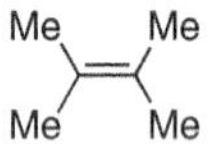	31	6.5	3.5	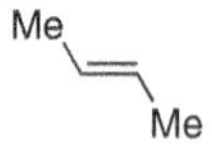	0.097	0.15	—
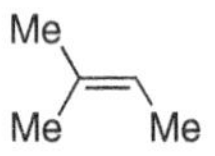	6.5	2.8	3.2	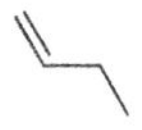	0.0087	0.011	—
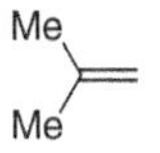	1	1	1	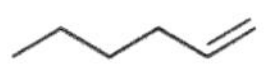	—	0.023	0.07
	0.14	0.23	—		—	0.12	0.4

29. (a) Doering, W.v.E.; LaFlamme, P. *J. Am. Chem. Soc.* **1956**, 78, 5447. (b) Skell, P.S.; Woodworth, R.C. *J. Am. Chem. Soc.* **1956**, *78*, 4496.

30. (a) Ingold, C.K.; Ingold, E.H. *J. Chem. Soc.* **1931**, 2354. (b) Anatakrishnan, S.V.; Ingold, C.K. *J. Chem. Soc.* **1935**, 984, 1396.

31. Swern, D. *J. Am. Chem. Soc.* **1947**, *69*, 1692.

32. (a) Moss, R.A.; Gerstl, R. *J. Org. Chem.* **1967**, *32*, 2268. (b) Moss, R.A.; Gerstl, R. *Tetrahedron*, **1967**, *23*, 2549. (c) Skell, P.S.; Garner, A.Y. *J. Am.Chem. Soc.* **1956**, *78*, 5430. (d) Doering, W.v.E.; Henderson, W.A., Jr. *J. Am. Chem. Soc.* **1958**, *80*, 5274.

33. (a) Kirmse, W. *Carbene Chemistry* (Academic Press: New York, 1964), ch. 2. (b) Hine, J. *Divalent Carbon* (Ronald Press: New York, 1964), ch. 2. (c) Ring, D.F.; Rabonovitch, B.S. *Can. J. Chem.* **1968**, *46*, 2435. (d) Seyferth, D.; Mai, V.A.; Gordon, M.E. *J. Org. Chem.* **1970**, *35*, 1993. (e) Seyferth, D.; Burlitch, J.M.; Yamamoto, K.; Washburn, S.S.; Attridge, C.J. *J. Org. Chem.* **1970**, *35*, 1989.

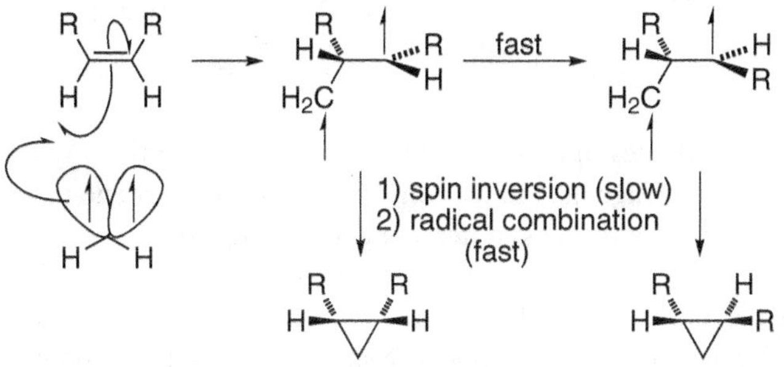

Figure 13.4 Addition of a triplet carbene to an alkene π bond results in stereochemical equilibration.

mixture of butane and isobutane in the ratio 63:37.[34] On the other hand, the reaction between pentane and singlet carbene, generated by photolysis of diazomethane, gives a mixture of hexane, 2-methylpentane, and 3-methylpentane in the ratio 49:34:17, which is close to the statistical ratio of 6:4:2,[35] and insertion into heptane is close to statistical.[36] The reaction between singlet methylene ($:CH_2$) and cyclohexene gives 40% addition to the π bond and 60% insertion products divided among 1-methylcyclohexene (10%), 3-methylcyclohexene (25%), and 4-methylcyclohexene (25%). This shows that for this very reactive carbene, there is an approximate 8:1 preference for addition over intermolecular insertion and that insertion into sp^3 C—H bonds is favored by a factor of approximately 1.25 over insertion into sp^2 C—H bonds. The insertion of singlet dichlorocarbene into (*S*)-(+)-2-phenylbutane (**13.25**) gives products with at least 97% retention of configuration at carbon.[37] Similar observations have been made with respect to carbene insertion into Si—H bonds.[38]

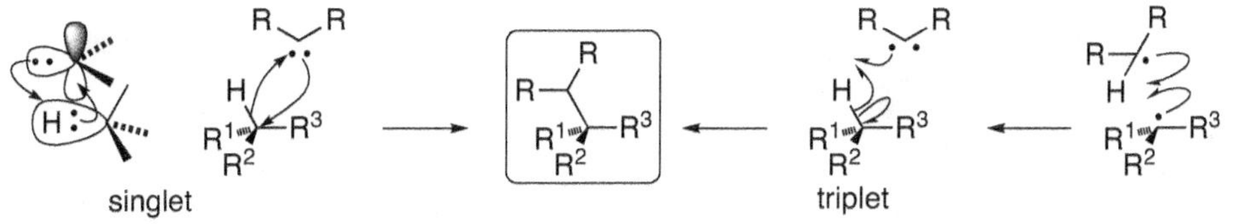

(13.24)

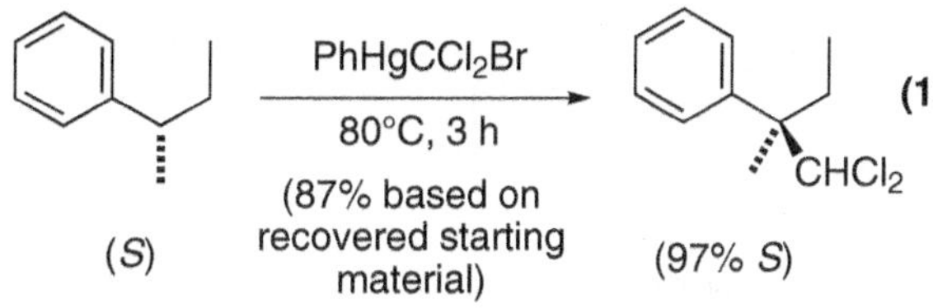

(13.25)

Carbenes and Carbenoids in Synthesis

The advent of the olefin metathesis reaction, which is catalyzed by a transition metal-carbene complex, has revolutionized the way that carbenes are used in organic synthesis. Copper and zinc carbenoids react predictably by addition to alkenes to give cyclopropanes or by insertion into bonds to give products of homologation.

---

34. Frey, H.M.; Kistiakowski, G.B. *J. Am. Chem. Soc.* **1957**, *79*, 6373.
35. (a) Doering, W.v.E.; Laughlin, R.G.; Buttery, R.G.; Chaudhuri, N. *J. Am. Chem. Soc.* **1956**, *78*, 3224. (b) Frey, H.M. *J. Am. Chem. Soc.* **1958**, *80*, 50005.
36. Richardson, D.B.; Simmons, M.C.; Dvoretzky, I. *J. Am. Chem. Soc.* **1961**, *83*, 1934.
37. (a) Seyferth, D.; Cheng, Y.M. *J. Am. Chem. Soc.* **1971**, *93*, 4072. (b) Seyferth, D.; Cheng, Y.M. *J. Am. Chem. Soc.* **1973**, *95*, 6763. These results contradict an earlier report that the insertion of carbenes into 2-phenylbutane occurs with racemization (Franzen, V.; Edens, R. *Justus Liebigs Ann. Chem.* **1969**, *729*, 33) by noting that the product formed has a very small specific rotation.
38. Sommer, L.H.; Ulland, L.A.; Parker, G.A. *J. Am. Chem. Soc.* **1972**, *94*, 3469.

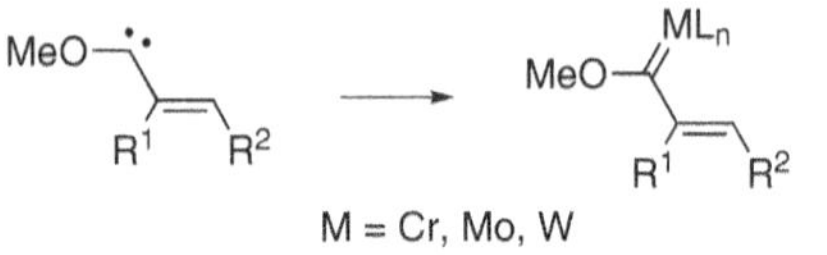

(13.26)

More recently, carbenes in the form of their transition metal complexes have become important synthetic reagents. Early on, the Fischer carbene complexes (**13.26**; ML_n=$Cr(CO)_5$) found use in the synthesis of a wide range of phenolic compounds from alkynes by addition of the carbene and carbon monoxide (itself, formally a carbene) to an alkyne.[39,40] The reaction proceeds by a multistep mechanism that involves a metallocycle intermediate—a common finding in many reactions involving organometallic complexes of the Group VIB (Group 6) metals. The currently accepted mechanism for the formation of phenolic products from the reaction of a Fischer carbene complex with an alkyne is shown in Figure 13.5.

The carbene complexes of molybdenum, developed especially by the research group of Richard Schrock,[41] and ruthenium, developed especially by the research group of Robert Grubbs,[42] react with alkenes to give metallacyclobutanes (**13.32**). These then fragment to give a new carbene complex and the metathesized alkene (**13.33**).[43]

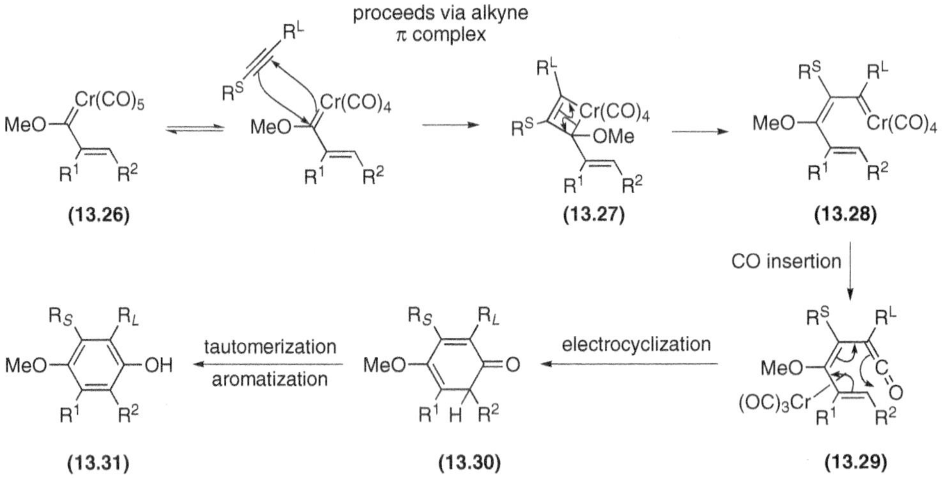

Figure 13.5 Formation of a phenol by addition of a Fischer carbene complex to an alkyne.

39. **Ernst Otto Fischer (1918–2007)** was educated at the Technische Universität München (PhD, 1952) after "work service" under the Nazis and military service during World War II. He spent his career at the University of Munich (1957–1964), the Technical University (1964–1984). He shared the 1973 Nobel prize in Chemistry. For more details, see the Nobel Foundation website.

40. Reviews and monographs on carbenes in synthesis: (a) Wulff, W. D. In Abel, E.W.; Stone, R.G.A.; Wilkinson, G., Eds. *Comprehensive Organometallic Chemistry II* (Pergamon Press: New York, 1995); Vol. 12, p 469. (b) Bernasconi, C.F. *Chem. Soc. Rev.* **1997**, *26*, 299. (c) Hegedus, L.S. *Tetrahedron* **1997**, *53*, 4105. (d) Wulff, W.D. *Organometallics* **1998**, *17*, 3116. (e) Dötz, K.H.; Tomuschatt, P. *Chem. Soc. Rev.* **1999**, *28*, 187. (f) Herndon, J.W. *Coord. Chem. Rev.* **1999**, *181*, 177. (g) Dörwald, F.Z. *Metal Carbenes in Organic Synthesis* (Wiley-VCH: New York, 1999).

41. (a) Schrock, R.R. *Acc. Chem. Res.* **1990**, *23*, 158. (b) Schrock, R.R.; Hoveyda, A.H. *Angew. Chem. Int. Ed.* **2003**, *42*, 4592. (c) Schrock, R.; Rocklage, S.; Wengrovius, J.; Rupprecht, G.; Felmann, J. *J. Mol. Catal.* **1980**, *8*, 73.

Richard Royce Schrock (1945–) obtained his PhD from Harvard (1971), and after postdoctoral study at Cambridge (1971–1972) and working at Du Pont (1972–1975), he joined the Massachusetts Institute of Technology, where he remains today. For his contributions to the development of olefin metathesis, Schrock shared the Nobel Prize in Chemistry for 2005. For more biographical details, see the Nobel Foundation web site.

42. (a) Grubbs, R.H. *Comp. Organometal. Chem.* **1982**, *8*, 499. (b) Grubbs, R.H.; Chang, S. *Tetrahedron* **1998**, *54*, 4413. (c) Chatterjee, A.K.; Morgan, J.P.; Scholl, M.; Grubbs, R.H. *J. Am. Chem. Soc.* **2000**, *122*, 3782. (d) Lee, C.W.; Grubbs, R.H. *Org. Lett.* **2000**, *2*, 2145. (e) Fürstner, A. *Angew. Chem. Int. Ed.* **2000**, *39*, 3102. (f) Trnka, T.M.; Grubbs, R.H. *Acc. Chem. Res.* **2001**, *34*, 18.

Robert Howard Grubbs (1942–) obtained his PhD from Columbia in 1968. After a year of postdoctoral study, he began his independent career at Michigan State University (1969–1978) before moving to the California Institute of Technology (1978–). Grubbs shared the Nobel Prize for Chemistry in 2005 for his work developing catalysts for olefin metathesis. For more biographical detail, see the Nobel Foundation web site.

43. Hérisson, J.-L.; Chauvin, Y. *Macromol. Chem.* **1971**, *141*, 161.

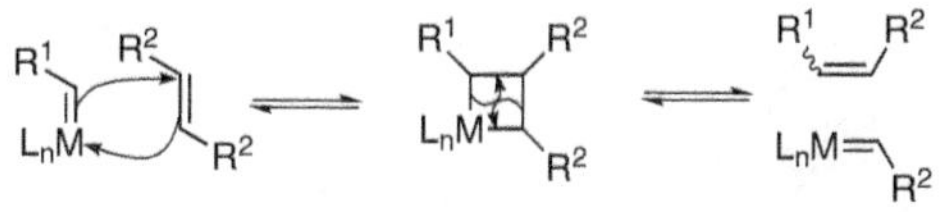

(13.26) (13.32) (13.33)

Rearrangements in Carbenes

Like carbocations, with which they share the structural feature of an empty p orbital on the carbene carbon, singlet alkylcarbenes often undergo 1,2-rearrangement of hydrogen to give alkenes. In most alkylcarbenes, the rearrangement is much faster than competing intramolecular insertion reactions to give cyclic products, as shown in the example below, where the carbene (**13.36**) is generated by thermolysis of the conjugate base of a sulfonylhydrazone (**13.35**).[44] Note how the hydrogen on the more substituted adjacent carbon rearranges preferentially to give the alkene **13.37** as the major product of the reaction, although not to the extent observed in carbocations.

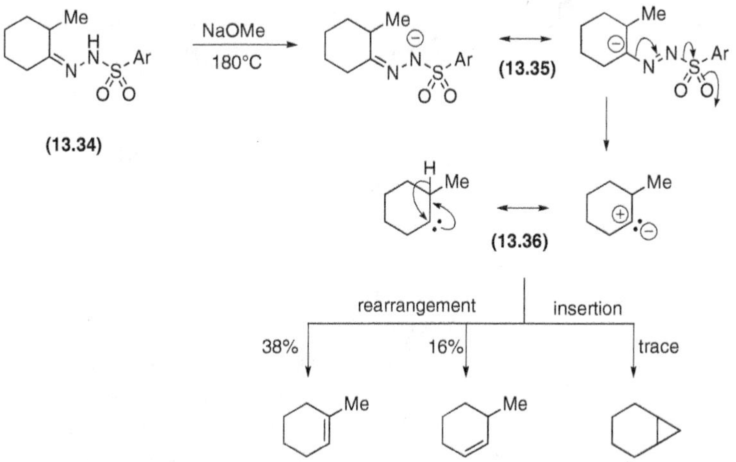

Although the rearrangement of hydrogen is usually preferred, where there is no hydrogen atom available, rearrangement of alkyl has been observed. In the rearrangement of the neophyl carbene (in Example **13.40**), for example, where either phenyl or methyl may migrate, the major product is formed by migration of phenyl although there is now substantial insertion into the adjacent methyl C—H bond to give cyclopropane products.[45] This relative order of migration correlates with the rearrangement of the corresponding cation.[46] Similar rearrangements of carbenes have been used, for example, in the synthesis of highly strained bridgehead alkenes.[47]

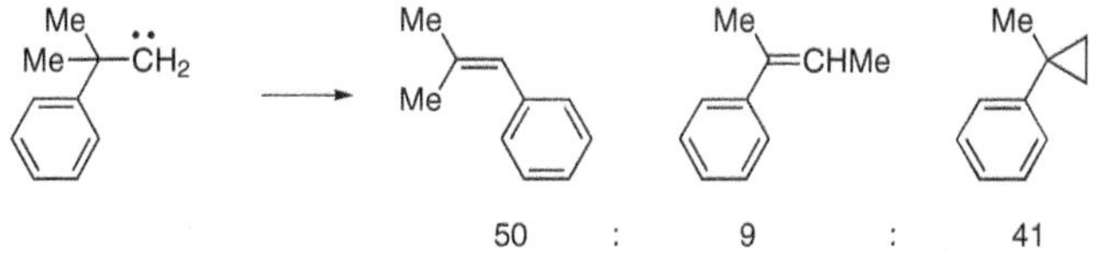

(13.40)

In contrast to saturated alkylcarbenes, allylcarbenes (e.g., **13.41**) usually give cyclopropenes (e.g., **13.42**[48]). This reaction may be viewed as either a carbene rearrangement, or as an intramolecular carbene addition to the π bond.

44. Wilt, J.W.; Wagner, W.J. *J. Org. Chem.* **1964**, *29*, 2788.
45. Philip, H.; Keating, J. *Tetrahedron Lett.* **1961**, 523.
46. See, for example, Ando, T.; Kim, S.-G.; Matsuda, K.; Yamataka, H.; Yukawa, Y.; Fry, A.; Lewis, D.E.; Sims, L.B.; Wilson, J.C. *J. Am. Chem. Soc.* **1981**, *103*, 3505.
47. Gudipati, M.S.; Radziszewski, J.G.; Kaszynski, P.; Michl, J. *J. Org. Chem.* **1993**, *58*, 3668.
48. York, E.J.; Dittmar, W.; Stevenson, J.R.; Bergman, R.G. *J. Am. Chem. Soc.* **1973**, *95*, 5680.

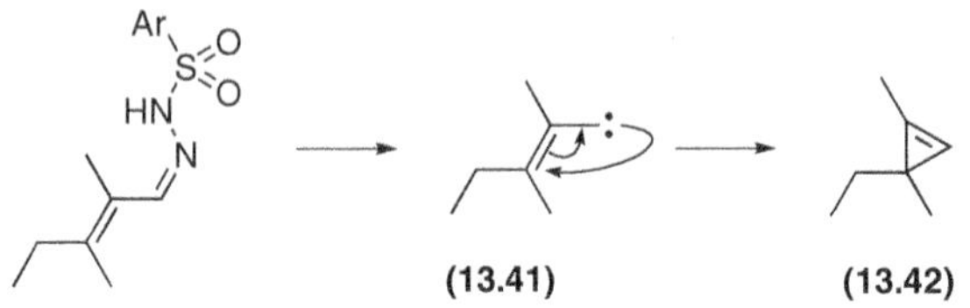

(13.41) (13.42)

Stable Carbenes[49]

During the last decades of the 20th century, much effort was expended in an effort to obtain a carbene stable enough to be isolated. This work was crowned with success in 1991, when the synthesis of two stable singlet carbenes, **13.43** and **13.34**, was reported.[50] Of the two groups of carbenes reported, the *N*-heterocyclic carbenes (NHCs; e.g., **13.43**) prepared by Arduengo's group have assumed a dominant position in the field.

(13.43) (13.44)

The idea that heterocyclic carbenes might function as nucleophilic catalysts had been suggested early on by Breslow. In 1958, he hypothesized that the conjugate base of a thiazolium ion (**13.45**, which is formally a heterocyclic carbene—see Figure 13.6) functioned as a nucleophile in promoting the benzoin condensation of aromatic aldehydes.[51] He proposed that the reaction occurs by initial addition of the carbene to the carbonyl group to give the adduct (**13.46**) that then isomerizes to the enamine **13.47**. This nucleophile then adds to a second molecule of the aldehyde to give a new adduct (**13.48**) that eventually eliminates the acyloin (**13.50**) to regenerate the starting carbene. Extending Breslow's observations that NHCs are nucleophilic in nature, the deduction that these compounds could be useful ligands for transition metal cations was a logical one to test. This has proved to be correct, and stable NHCs have become widely used ligands for transition metals in a wide variety of catalytic organic reactions.[52]

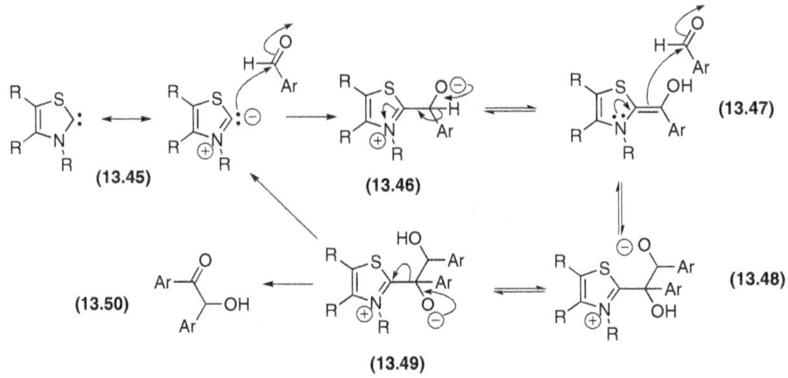

Figure 13.6 Reactivity of a thiazolium ion as a heterocyclic carbene.

49. Reviews: (a) Arduengo, A. *Acc. Chem. Res.* **1999**, *32*, 913. (b) Bourissou, D.; Gueret, O.; Gabbaï, F.P.; Bertrand, G. *Chem. Rev.* **2000**, *100*, 39.

50. (a) Arduengo, A.J., III; Harlow, R.L.; Kline, M. *J. Am. Chem. Soc.* **1991**, *113*, 361. (b) Regitz, M. *Angew. Chem. Int. Ed. Engl.* **1991**, *30*, 674.

51. Breslow, R. *J. Am. Chem. Soc.* **1958**, *80*, 3719.

52. (a) Herrmann, W.A. *Angew. Chem. Int. Ed. Engl.* **2002**, *41*, 1290. (b) Herrmann, W.A.; Köcher, C. *Angew. Chem. Int. Ed. Engl.* **1997**, *36*, 2162. (c) Weskamp, T.; Schattenmann, W.C.; Spiegler, M.; Herrmann, W.A. *Angew. Chem. Int. Ed. Engl.* **1998**, *37*, 2490. (d) Nolan, R.A.; Kelly, R.A., III; Navarrro, O. *J. Am. Chem. Soc.* **2003**, *125*, 16194. (e) Herrmann, W.A.; Elison, M.; Fischer, J.; Köcher, C.; Artus, G.R.J. *Angew. Chem. Int. Ed. Engl.* **1995**, *34*, 2371. (f) Scholl, M.; Trnka, T.M.; Morgan, J.P.; Grubbs, R.H. *Tetrahedron Lett.* **1999**, *40*, 2247. (g) Choi, T.-L.; Grubbs, R.H. *J. Chem. Soc., Chem. Commun.* **2001**, 2648. (h) Fürstner, A.; Ackermann, L.; Gabor, B.; Goddard, R.; Lehmann, C.W.; Mynott, F.; Stelzer, F.; Thiel, O.R. *Chem. Eur. J.* **2001**, *7*, 3236.

Worked Problem

13-3 The Corey-Winter reaction shown below is one step of a synthesis of a methylenecepham derivative by intramolecular cyclization [*Bioorg. Med. Chem.* **2003**, *11*, 1957].

(a) What carbene is formed in this reaction?

(b) What alternative carbenes could have been formed in this reaction?

(c) Provide a rationalization for why this particular carbene is the one that gives the observed product.

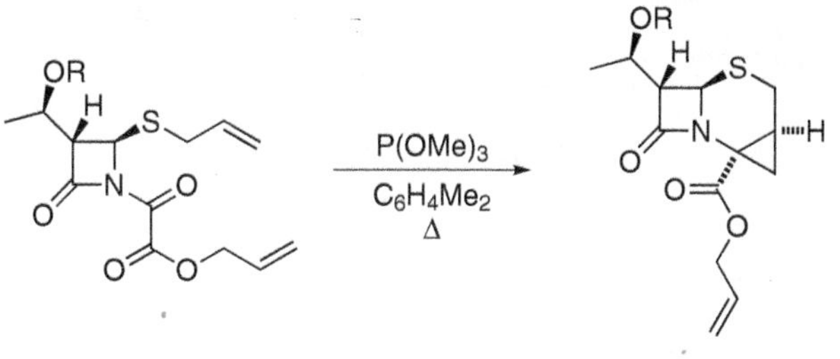

§Answers below.

§ Answers to Worked Problem:

(a) The location of the cyclopropane specifies the location of the carbene carbon in the deoxygenation product:

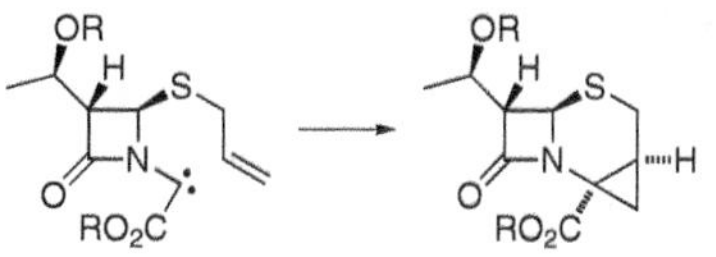

(b) There are three carbonyl groups in the starting compound, and each is part of a carboxylic acid derivative. Consequently, three carbenes could form:

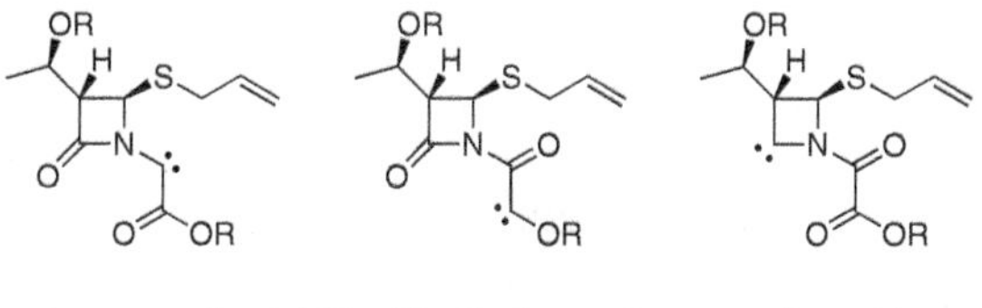

(c) All three of these carbenes can be stabilized by the lone pair on an adjacent heteroatom. When we examine the three carbenes and take into account resonance stabilization of the carbenoid center, we observe the following:

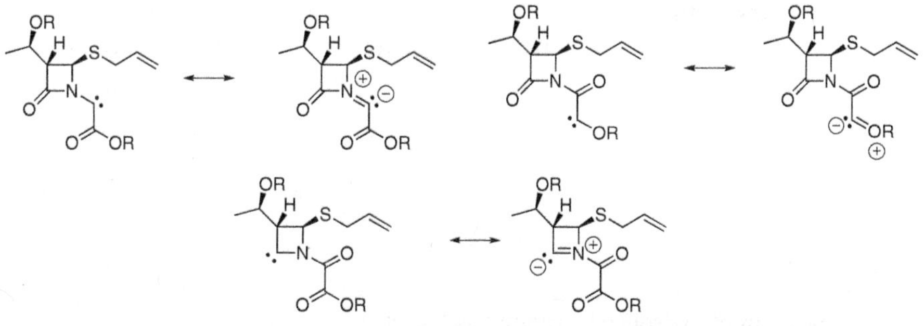

Of the three possible carbenes, we see that the first (top left above) gives rise to the zwitterionic contributing structure with the maximum conjugation and the minimum strain, so we expect that this carbene should be the one preferentially formed.

Reaction Synopses

Carbene Preparation

Carbenes from α-Elimination

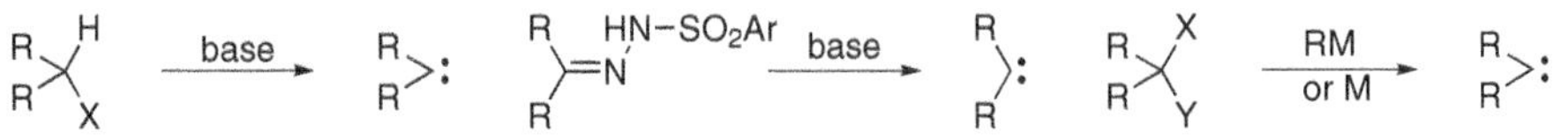

X = Cl, Br, I; Y=F, Cl, Br, I M=Li, Zn base=RO^-, NR_2^-

Reagents: BuLi, THF; Et_2Zn, THF; Zn(Cu), Et_2O; etc.
KOBu^t, pentane; $NaNH_2$, THF; LDA, THF; etc.

Carbenes by Deoxygenation

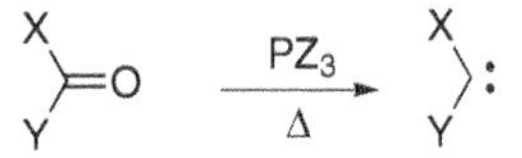

X, Y = OR, NR_2, etc. Z=alkyl, alkoxy

Carbenes by Photolysis or Thermolysis

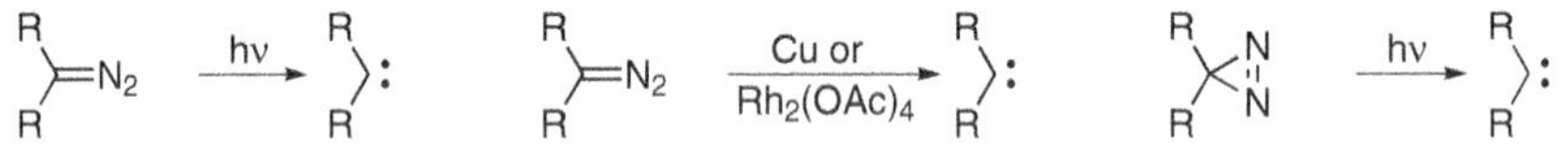

Carbene is often triplet when generated photochemically and singlet when generated via the metal carbenoid.

Carbene Addition

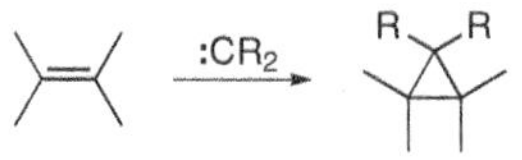

Reagents: $CHCl_3$, KOBu^t; CH_2I_2, Et_2Zn; etc.

Stereochemistry: addition is *syn* (suprafacial); stereochemistry of alkene is retained in cyclopropane product.

Carbene Insertion

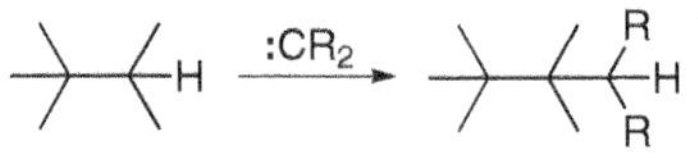

Stereochemistry: insertion occurs with retention of stereochemistry.

Regiochemistry: insertion into C—H bonds is usually preferred over insertion into C—C bonds; insertion into sp^3-C—H bonds is preferred over insertion into sp^2-C—H bonds.

Carbene Rearrangements

Rearrangement of hydrogen is usually preferred.

Problems

13-7 Carbenes can be formed by each of the three methods below. Suggest mechanisms for these reactions.

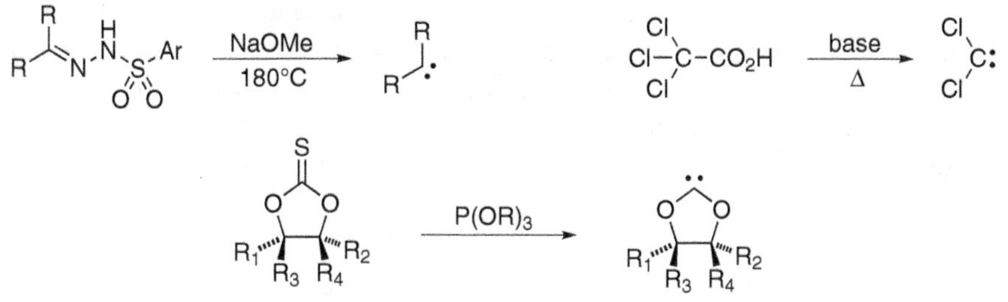

13-8 In the two reactions that follow, diphenylsulfonium methylide reacts with 2-octene in the presence of a copper (II) acetylacetonate catalyst to give cyclopropane products. The two stereoisomers of 2-octene give different products. Provide a mechanism that rationalizes these observations.

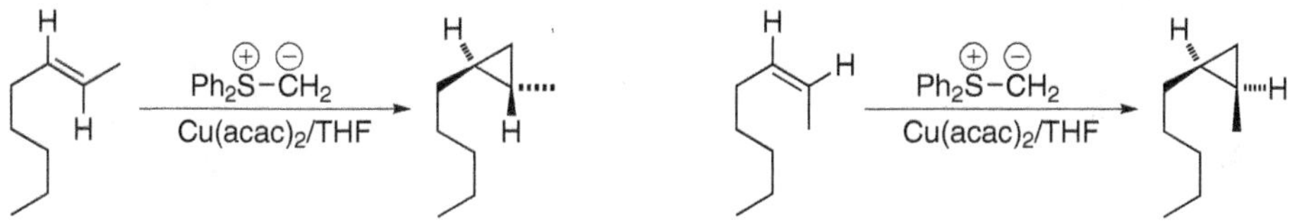

13-9 The Wolff rearrangement is a reaction of α-diazoketones that leads to the formation of a ketene.

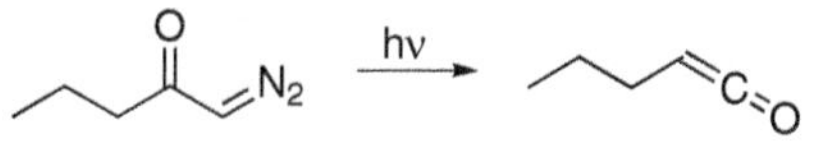

Write a mechanism for this reaction. Does your mechanism change if the reaction of the diazoketone labeled in the carbonyl carbon leads to a product where the label is distributed through two positions, as indicated below? If so, how?

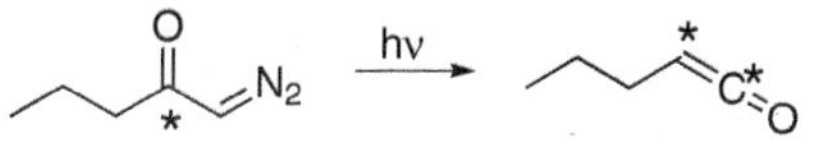

13-10 The reaction between 1,1-dibromoalkenes and butyllithium gives an alkyne as shown below. Write a mechanism that accounts for this transformation.

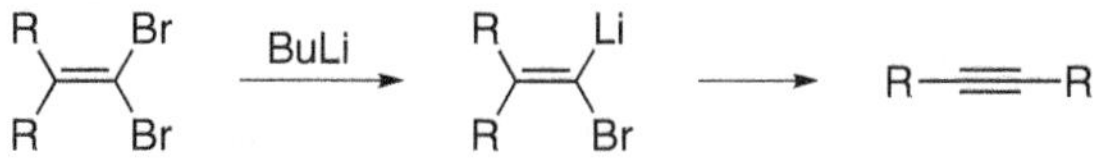

13-11 Diazirines are frequently used as substrate analogs for photochemical tagging of proteins. For example, the compound below was used as a means for probing the biochemistry of anesthesia. When this photolabel was activated, three amino acid residues were labeled: one glutamate residue, and two serine residues [*J. Biol. Chem.* **2004**. *179*, 17640]. Suggest probable structures for the photolabeled amino acids if the diazirine gives the triplet carbene on photolysis.

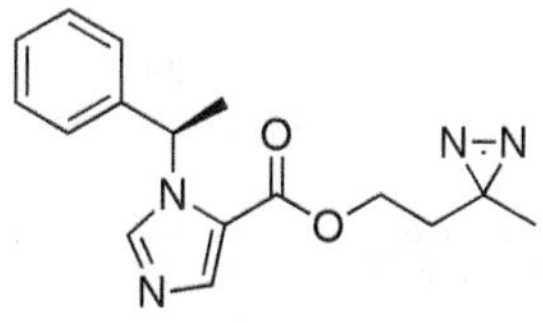

13.3 Nitrenes

The monovalent nitrogen species with an open sextet on nitrogen, structurally analogous to carbenes, are known as ***nitrenes***.[53] It is not surprising that, because they carry only one group attached to the nitrogen and are therefore not amenable to the exploitation of steric hindrance, no stable nitrenes have yet been isolated. In fact, nitrenes are, in general, more reactive than the corresponding carbenes. In addition, nitrenes are less well known. Like carbenes, nitrenes can exist as both singlet (**13.51**) and triplet (**13.52**) species. Reactivity attributable to both forms has been observed, although the singlet nitrene is the more usual reacting species.

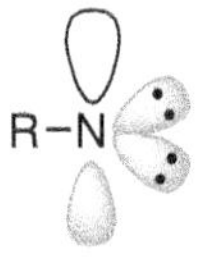

(13.51)

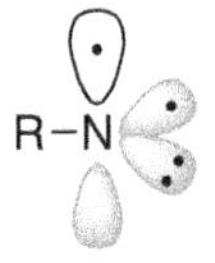

(13.52)

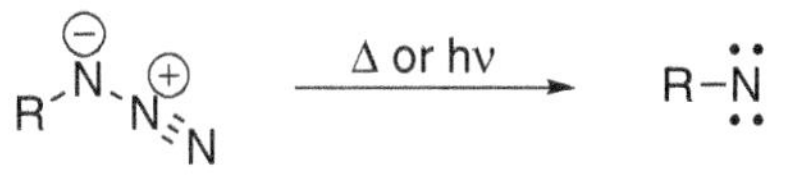

(13.53)

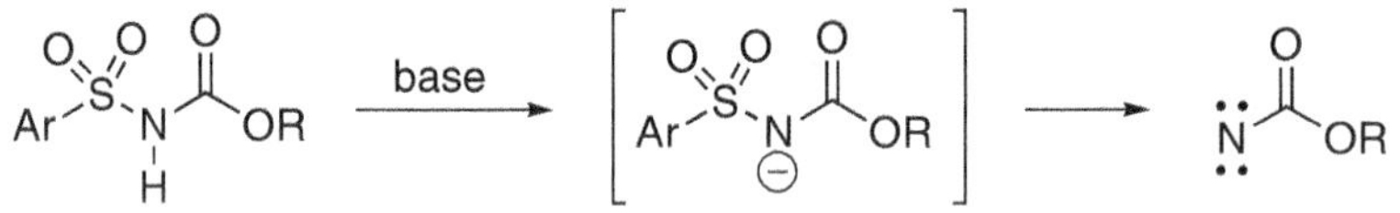

(13.54)

Nitrenes are formed by only two truly general methods. One is the decomposition of azides (**13.53**) by either photolysis or thermolysis, which is the more general method. The other is the base-promoted α-elimination from *N*-(*p*-nitrobenzenesulfonyl)urethanes (**13.54**), which provides a method for the preparation of alkoxycarbonylnitrenes.

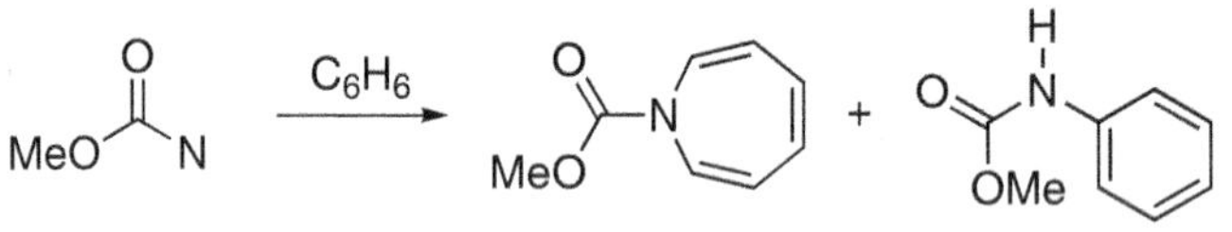

(13.55)

Nitrenes undergo both addition reactions to give aziridines and insertion reactions to give amines, and the two frequently compete. For example, methoxycarbonylnitrene reacts with benzene to give a mixture of *N*-carboxymethylazepine and methyl *N*-phenylurethane, as in Example **13.55**. Cyclohexene reacts with the same nitrene to give a mixture of the aziridine by addition and the isomeric urethanes formed by insertion.[54] The insertion of a nitrene into a C—H bond is a stereospecific process and occurs with retention of configuration, as illustrated by Examples **13.56** and **13.57** for intermolecular[55] and intramolecular[56] insertion reactions, respectively. The facility with which nitrenes insert into C—H bonds

---

53. Monographs: (a) Scriven, E.F.V. *Azides and Nitrenes* (Academic Press: New York, 1984). (b) Lwowski, W. *Nitrenes* (Wiley: New York, 1970).
Reviews: (c) Scriven, E.F.V. *React. Intermed.* (Plenum: New York, 1982), vol. 2,p. 1. (d) Lwowski, W. *React. Intermed.* (Wiley: New York, 1978), vol. 1,p. 197. (e) Lwowski, W. *React. Intermed.* (Wiley: New York, 81), vol. 2,p. 315. (f) Lwowski, W. *React. Intermed.* (Wiley: New York, 1985), vol. 3,p. 305. (g) Lwowski, W. *Angew. Chem. Int. Ed. Engl.* **1967**, *6*, 897. (h) Abramovitch, R.A. In McManus, S.P. *Reactive Intermediates* (Academic Press: New York, 1973), p. 127. (i) Hünig, S. *Helv. Chim. Acta* **1971**, *54*, 1721. (j) Belloli, R. *J. Chem. Educ.* **1971**, *48*, 422. (k) Meth-Cohn, O. *Acc. Chem. Res.* **1987**, *20*, 18. (l) Horner, L.; Christmann, A. *Angew. Chem. Int. Ed. Engl.* **1963**, *2*, 599. (m) Abramovitch, R.A.; Davis, B.A. *Chem. Rev.* **1964**, *64*, 149.
54. (a) Berry, R.S.; Cornell, D.; Lwowski, W. *J. Am. Chem. Soc.* **1963**, *85*, 1199. (b) Cornell, D.W.; Berry, R.S.; Lwowski, W. *J. Am. Chem. Soc.* **1965**, *87*, 3626.
55. Simson, J.M.; Lwowski, W. *J. Am. Chem. Soc.* **1969**, *91*, 5107.
56. Smolinsky, G.; Feuer, B.I. *J. Am. Chem. Soc.* **1964**, *86*, 3085.

has made azides popular reagents for the photochemical labeling of proteins. Addition reactions to alkenes are stereospecific suprafacial reactions, consistent with the singlet nitrene as the reacting species.

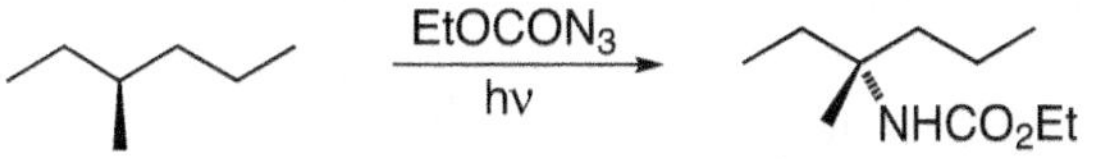

(13.56)

(13.57)

Like carbenes, whose formation and use can be catalyzed by transition metals through the medium of metallocarbenes, nitrenes can also be prepared catalytically using appropriate precursors and transition metal catalysts (e.g., **13.58**). In 1991, for example, Evans[57] reported the copper-catalyzed aziridination of alkenes using (*N*-(*p*-toluenesulfonyl) imino)phenyliodinane (PhINTs) as the source of the nitrene and a copper catalyst; both Cu (I) and Cu (II) were effective catalyzing the reaction. The reaction has spurred considerable research into the use of this reagent and metal catalysts.[58] Iron and manganese compounds can also serve as catalysts, but Evans found that the copper catalysts were the most efficient. A recent review shows how oxidation of a sulfonamide by phenyliodine diacetate in the presence of rhodium (II) carboxylates to give products from nitrenes.[59]

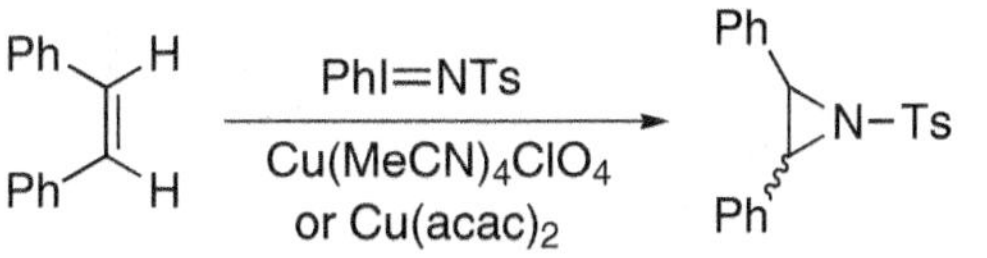

(13.58)

The utility of nitrene insertion into C—H σ bonds received a significant boost when Che[60] and Du Bois[61] found that sulfamate esters would react in the presence of a rhodium (II) catalyst and a terminal oxidant. The reaction produced the necessary iodinane in situ to give products of insertion with complete retention of configuration at carbon (e.g., **13.59**). This reaction has been elegantly applied by Du Bois in the synthesis of complex natural products such as tetrodotoxin[62] and saxitoxin.[63]

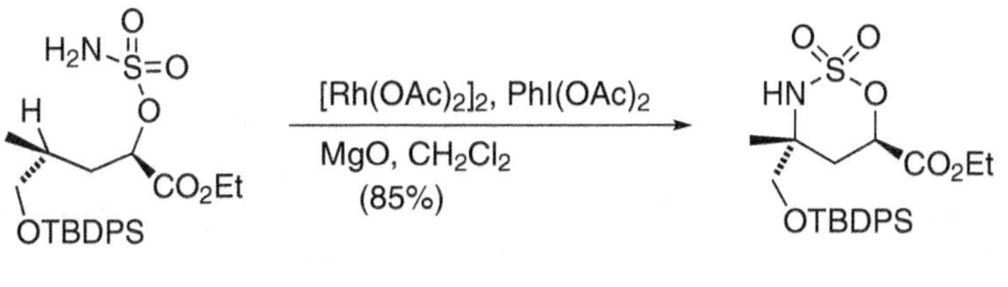

(13.59)

57. Evans, D.A.; Faul, M.M.; Bilodeau, M.T. *J. Org. Chem.* **1991**, *56*, 6744.
58. Review: Halfen, J.A. *Curr. Org. Chem.* **2005**, *9*, 657.
59. Dauben, P.; Dodd, R.H. *Synlett.* **2003**, 1571.
60. (a) Liang, J.-L.; Yuan, S.-X.; Chan, P.W.H.; Che, C.-M. *Org. Lett.* **2002**, *4*, 4507. (b) Yu, X.-Q.; Huang, J.-S.; Zhou, X.-g.; Che, C.-M. *Org. Lett.* **2000**, *2*, 2233. (c) Au, S.-M.; Zhang, S.-B.; Fung, W.-H.; Yu, X.-Q.; Che, C.-M.; Cheung, K.-K. *Chem. Commun.* **1998**, 2677. (d) Au, S.-M.; Huang, J.-S.; Che, C.-M.; Yu, W.-Y. *J. Org. Chem.* **2000**, *65*, 7858.
61. (a) Espino, C.G.; Wehn, P.M.; Chow, J.; Du Bois, J. *J. Am. Chem. Soc.* **2001**, *123*, 6935. (b) Espino, C.G.; Du Bois, J. *Angew. Chem., Int. Ed.* **2001**, *40*, 598. (c) Fiori, K.W.; Fleming, J.J.; Du Bois, J. *Angew. Chem., Int. Ed.* **2004**, *43*, 4349. (d) Espino, C.G.; Du Bois, J. *Angew. Chem., Int. Ed.* **2001**, *40*, 598. (e) Kim, M.; Mulcahy, J.V.; Espino, C.G.; Du Bois, J. *Org Lett.* **2006**, *8*, 1073. (f) Espino, C.G.; Fiori, K.W.; Kim, M.; Du Bois, J. *J. Am. Chem. Soc.* **2004**, *126*, 15378. (f) Wehn, P.M.; Lee, J.; Du Bois, J. *Org. Lett.* **2003**, *5*, 4823.
62. Hinmann, A.; Du Bois, J.H. *J. Am. Chem. Soc.* **2003**, *125*, 11510.
63. Fleming, J.J.; Du Bois, J.H. *J. Am. Chem. Soc.* **2006**, *128*, 3926.

Reaction Synopses

Methods for Generating Nitrenes
From Nitrogen Anion Precursors

X = OSO_2Ar, Cl, Br, N_2^+, etc.

Reagents: RN_3, hν; RNHOTs, $KOBu^t$, Δ; etc.

From Iodinanes

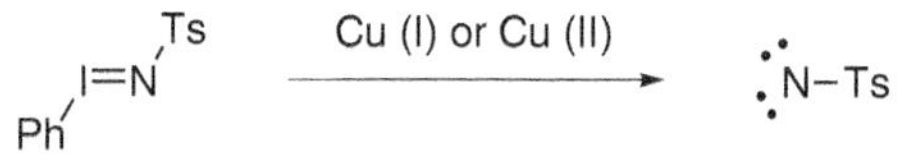

Reagents: PhINTs, $Cu(MeCN)_4ClO_4$; $Cu(acac)_2$; Mn(TPP)Cl; Fe(TPP)Cl; $Rh_2(OAc)_4$; $Co(acac)_2$; etc.

or $PhI(OAc)_2$, RSO_2NH_2, $Rh_2(OCOR)_4$

Nitrene Addition

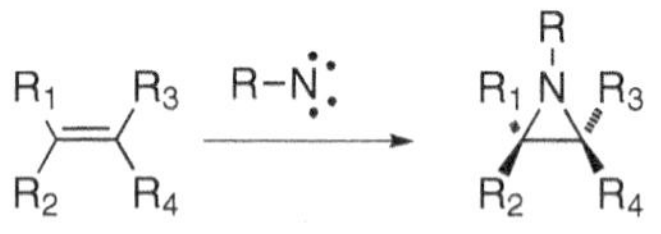

Stereochemistry: addition is *syn*; stereochemistry of alkene is retained.

Nitrene Insertion

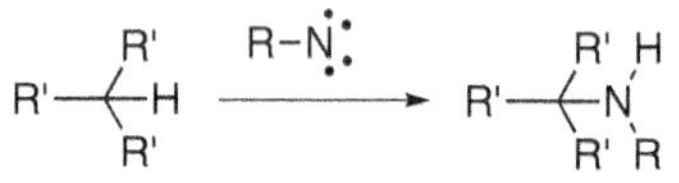

Stereochemistry: reaction occurs with retention of configuration.

13.4 Arynes[64]

In 1953, John D. Roberts[65] obtained unequivocal proof of the existence of a symmetrical intermediate formed by the elimination of HCl from chlorobenzene-1-^{14}C with sodium amide in liquid ammonia **(13.60)**.[66] The possibility that this same symmetrical neutral intermediate might be formed in the reaction between fluorobenzene and phenyllithium **(13.61)** had been considered by Wittig in 1942,[67] but his report that *o*-fluoroanisole and

64. Reviews: (a) Bunnett, J.F. *J. Chem. Educ.* **1961**, *38*, 278. (b) Wittig, G. *Angew. Chem. Int. Ed. Engl.* **1965**, *4*, 731. (c) Kauffmann, T. *Angew. Chem. Int. Ed. Engl.* **1965**, *4*, 543.

A large part of a special issue of the Australian Journal of Chemistry has been dedicated to reviews and papers on benzyne chemistry: (d) Wentrup, K. *Aust. J. Chem.* **2010**, *63*, 979. (e) Kitamura, T. *Aust. J. Chem.* **2010**, *63*, 987. (f) Brown, R.F.C. *Aust. J. Chem.* **2010**, *63*, 1007. (g) Winkler, M.; Sander, W. *Aust. J. Chem.* **2010**, *63*, 1013.

Monographs: (i) Hoffmann, R.W. *Dehydrobenzene and Cycloalkynes* (Academic Press: New York, 1967).

65. **John Dombrowski Roberts (1918–)** was educated at the University of California, Los Angeles (BA, 1941; PhD, 1944) and began his independent career at Harvard (1945–1947) and the Massachusetts Institute of Technology (1947–1950) before moving to the California Institute of Technology (1950–1988). Roberts made major contributions in physical organic chemistry and organic spectroscopy. His autobiography contains more detail: Roberts, J.D. *The Right Place at the Right Time* (American Chemical Society: Washington, D.C., 1990).

66. Roberts, J.D.; Simmons, H.E.; Carlsmith, L.A.; Vaughan, C.W. *J. Am. Chem. Soc.* **1953**, *75*, 3290.

67. Wittig, G. *Naturwiss.* **1942**, *30*, 696.

m-fluoroanisole gave different products appeared to eliminate the possibility. It was not until the work of Huisgen,[68] 12 years later, that this error was corrected. Huisgen found that not only do these two halides give the same product mixture, but that this is not a unique situation.

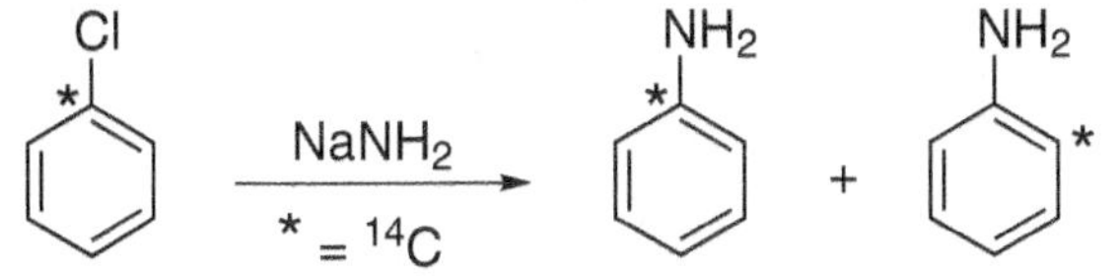

(13.60)

X = Cl	43%	53%
X = I	46%	52%

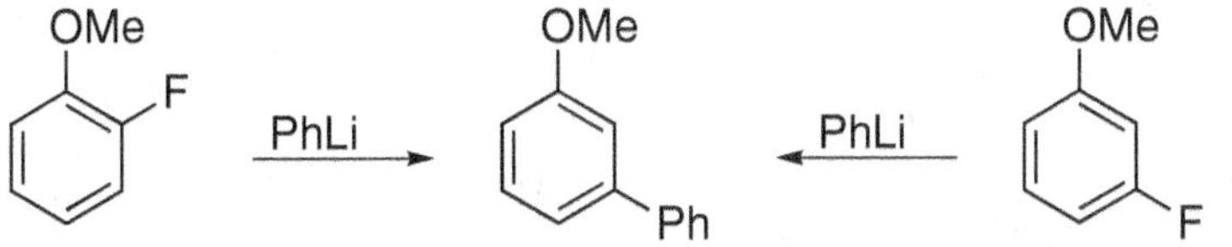

(13.61)

Benzyne is prepared by a wide variety of methods, several of which are summarized in Figure 13.7. The original method is the treatment of a halobenzene with strong base such as amide anion, but this suffers from the limitation that it cannot be used in the presence of base-sensitive functionality. The formation of an *o*-halophenyl anion during this reaction leads to the obvious extension of this method to the formation of benzyne by dehalogenation of *o*-dihalobenzenes. This is, indeed, observed when one attempts to prepare the Grignard reagent from *o*-bromofluorobenzene.[69] A similar method for the preparation of benzyne is the reaction between fluoride anion and *o*-trimethylsilylphenyl triflate.[70]

Benzyne may be generated from suitable precursors by either photolysis or thermolysis. Photolysis of *o*-diiodobenzene[71] and thermolysis of phthalic anhydrides[72] or benzothiadiazole dioxides[73] also provide methods for the formation of benzyne. The thermolysis or photolysis of zwitterionic precursors also provides an entry into benzyne chemistry. The internal salt of *o*-phenyliodoniumbenzoic acid[74] gives benzyne on thermolysis or photolysis, as does the zwitterion formed by diazotization of anthranilic acid with isoamyl nitrite.[75] Because the strongly electron-withdrawing diazonium group renders the *ortho* hydrogens especially acidic, the zwitterionic intermediate formed by decarboxylation of the diazonium carboxylate can also be generated by the reaction between an aryldiazonium ion and a weak base (e.g., potassium acetate).[76] Nitrene precursors to benzyne, which may also be viewed, formally, as having zwitterionic contributors to the resonance hybrid, may be formed (1) by oxidation of 1-aminobenzotriazole with lead tetraacetate[77] or (2) by reduction of 1-nitroso-benzotriazole with ethyl diphenylphosphonite.[78]

---

68. Huisgen, R.; Rist, H. *Naturwiss.* **1954**, *41*, 358.
69. Wittig, G.; Pohmer, L. *Chem. Ber.* **1956**, *89*, 1334.
70. (a) Himeshima, Y.; Sonoda, T.; Kobayashi, H. *Chem. Lett.* **1983**, 1211. (b) Liu, Z.; Larock, R.C. *Org. Lett.* **2003**, *5*, 4673.
71. Kampmeier, J.A.; Hoffmeister, E. *J. Am. Chem. Soc.* **1962**, *84*, 3787.
72. Fields, E.K.; Meyerson, S. *Chem. Commun.* **1965**, 474.
73. Wittig, G.; Hoffmann, R.W. *Chem. Ber.* **1962**, *95*, 2718.
74. Le Goff, E. *J. Am. Chem. Soc.* **1962**, *84*, 3786.
75. (a) Stiles, M.M.; Miller, R.G. *J. Am. Chem. Soc.* **1960**, *82*, 3802. (b) Friedman, L.; Longullo, F.M. *J. Am. Chem. Soc.* **1963**, *85*, 1549.
76. Rückhardt, C.; Tan, C.C. *Angew. Chem. Int. Ed. Engl.* **1970**, *9*, 522.
77. (a) Campbell, C.D.; Rees, C.W. *Proc. Chem. Soc.* **1964**, 296. (b) Campbell, C.D.; Rees, C.W. *J. Chem. Soc. C* **1969**, 742.
78. Cadogan, J.I.G.; Thomson, J.B. *Chem. Commun.* **1969**, 770.

Figure 13.7 Methods for the production of benzyne. Reagents: (a) room temp. (b) $NaNH_2$, PhLi, etc. (c) Mg, Et_2O. (d) CsF, MeCN. (e) RONO. (f) $Pb(OAc)_4$. (g) $(PhO)_2P(O)H$. (h) hν or Δ. (i) Δ. (j) Δ or hν.

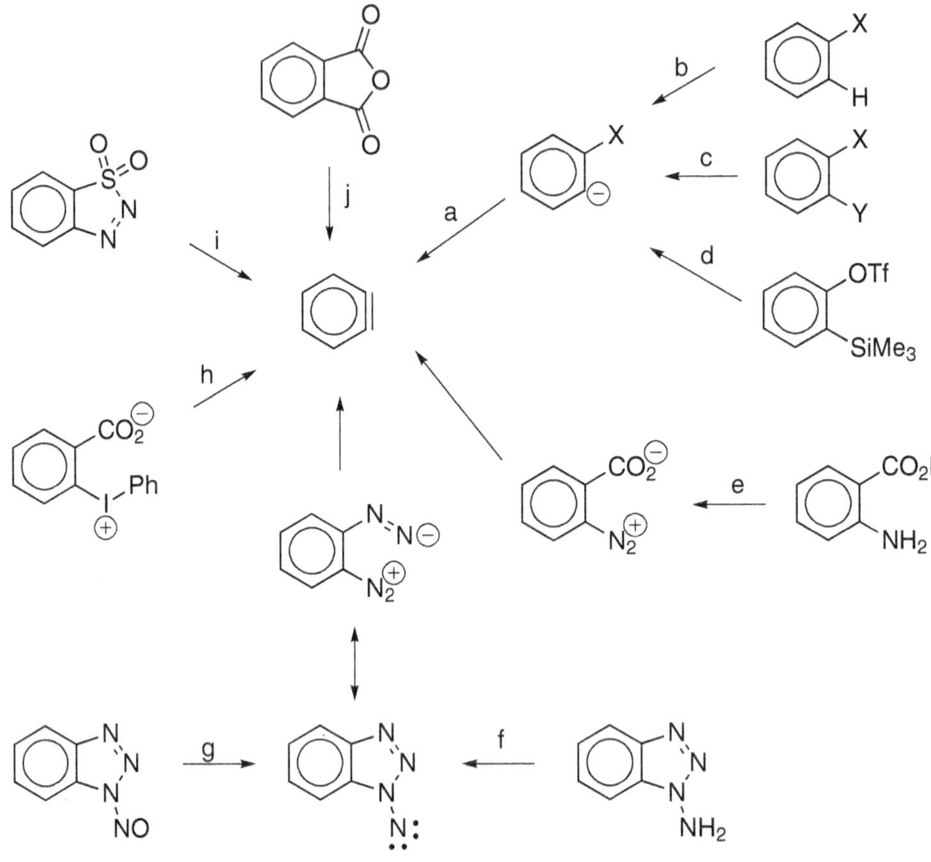

Structure and Bonding in Benzyne

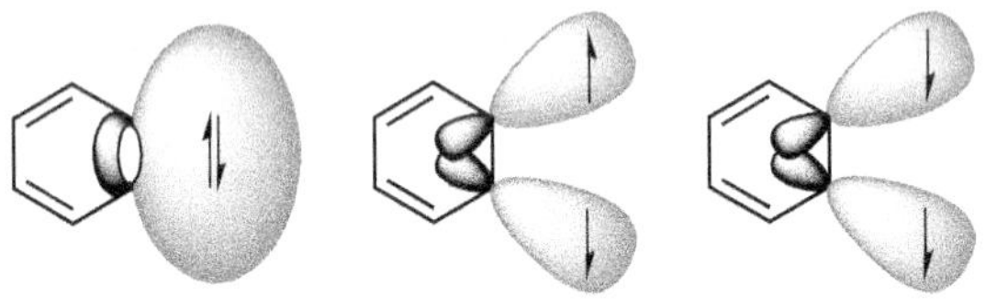

Singlet alkyne | Singlet diradical | Triplet diradical

Benzyne is characterized by an unusual bond formed by π-type overlap of sp^2 hybrid orbitals on adjacent carbons. There are two possible Kekulé structures for this species: one carrying a triple bond and the other carrying a cumulated triene moiety. The available spectra of benzyne seem to indicate (1) that the resonance contributor with the triple bond is more representative of the actual structure and (2) that benzyne is best viewed as a benzene molecule with an aromatic sextet, with distorted geometry due to shortening of one C—C bond.[79] The geometry of benzyne has been obtained from microwave measurements[80] and computationally.[81] The geometry shows that there is clearly a significant bonding interaction between the two benzyne carbons (the bond distances calculated at quite high levels of theory all have the distance between these two carbons

79. (a) Laing, J.W.; Berry, R.S. *J. Am. Chem. Soc.* **1976**, *98*, 6660. (b) Münzel, N.; Schweig, A. *Chem. Phys. Lett.* **1988**, *147*, 192.
80. Kukolich, S.G.; Tanjaroon, C.; McCarthy, M.C.; Thaddeus, P. *J. Chem. Phys.* **2003**, *119*, 4353.
81. de Visser, S.P.; Filatov, M.; Shaik, S. *Phys. Chem. Chem. Phys.* **2000**, *2*, 5046, and references therein.

close to 1.27Å, which lies between the triple bond length of 1.21 Å and the double bond length of 1.34 Å), but the geometry of the molecule also suggests that benzyne might also be described as a diradical rather than as a strained alkyne. This question has been resolved by measuring the singlet-triplet gap in benzyne; the value obtained is 37.5 kcal mol^{-1}.[82] This ensures that the ground state of benzyne—whether a strained alkyne or a diradical—is a singlet.

Reactivity of Benzyne

The reactive benzyne π bond is orthogonal to the aromatic π system, so there is no interaction between the two. Instead, the reactive π bond is best described as being part of the σ bonding system of the ring, so the σ bonding framework of the ring is the major determinant of regiochemistry in benzyne reactions. This reasoning leads to the conclusion that the most important factor determining the regiochemistry of benzyne reactions is the electronegativity of the substituent on the aromatic ring, whose effect is transmitted by induction through the σ framework.

Benzyne is an electrophile that reacts with a variety of nucleophiles to generate the phenyl anion. The outcomes of the reactions between a series of aromatic halides and amide anion,[83] each of which involves a benzyne intermediate, are summarized in Table 13.5. The choice of substituents in this study was predicated on the premise that the regiochemistry of the reaction could be affected at two stages: (1) the deprotonation step, controlled by the relative acidities of the hydrogens, and (2) the addition step, controlled by the relative energies of the isomeric aryl anions. Four substituents were used in the study: (1) a trifluoromethyl group, which is strongly electron-withdrawing by induction and weakly electron-withdrawing by resonance; (2) a methoxy group, which is electron-withdrawing by induction and strongly electron-releasing by resonance; (3) a fluoro group, which is electron-withdrawing by resonance, and relatively weakly electron-releasing by resonance; and (4) a methyl group, which is electron-releasing by both induction and resonance.

What is immediately apparent from Table 13.5 is that (1) *o*-bromoanisole and *m*-bromoanisole *both* give *m*-anisidine as the sole product of the reaction and that (2) (*o*-chlorophenyl)trifluoromethane and its *m*- isomer also *both* give *m*-trifluoromethylaniline as the sole product. This is consistent with the *o*- and *m*- halides both giving the same benzyne. Because there is only one proton situated *ortho* to the halogen, the *ortho* isomers can give only one benzyne but the *m*- isomer can give two possible benzynes. This means that the deprotonation of the *m*- isomer must be regiospecific at the position between the substituent and the halogen—the hydrogen that is rendered most acidic by the electron-withdrawing inductive effects of the substituent and the halogen. Because these benzynes each give only one product, they must undergo regiospecific addition of the nucleophile to give only one of the two possible aryl anions. Again, the regiochemistry is such that the nucleophile becomes bonded to the carbon further from the substituent, placing the negative charge on the carbon closer to the electron-withdrawing substituent.

In contrast to the halides carrying electronegative substituents, the *o*-halotoluenes do not give a single product but mixtures of both possible products, with the *m*- isomer of the product slightly predominating. Recalling that carbanions are not stabilized by adjacent alkyl groups, this regiochemical outcome could have been predicted. The *m*-halotoluenes give both possible benzynes and a mixture of all three possible products, which is consistent with the deductions made from the results with the simpler *o*-halotoluenes.

---

82. Wenthold, P.G.; Squires, R.R.; Lineberger, W.C. *J. Am. Chem. Soc.* **1998**, *120*, 5279.
83. Roberts, J.D.; Vaughan, C.W.; Carlsmith, L.A.; Semenow, D.A. *J. Am. Chem. Soc.* **1956**, *78*, 611.

Table 13.5 Regiochemistry of Formation and Addition Reactions of Benzyne

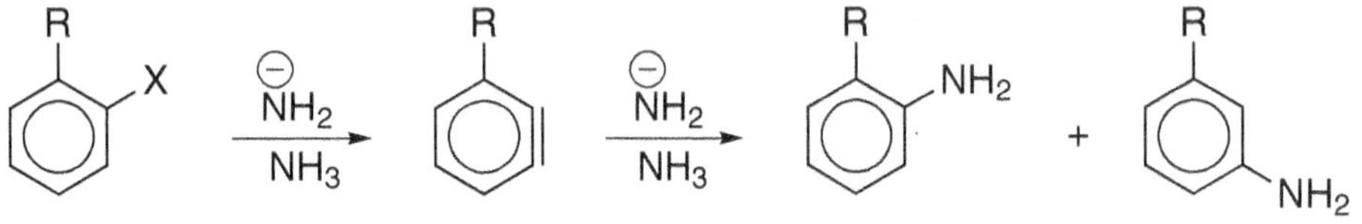

R = CF_3; X = Cl	0	100
R = CH_3; X = Cl	45 ± 4	55 ± 4
R = CH_3; X = Br	48.5 ± 2	51.5 ± 2
R = OCH_3; X = Br	0	100

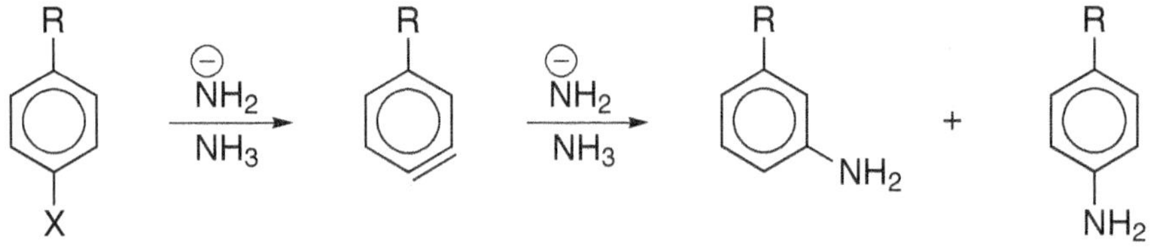

R = CF_3; X = Cl	50 ± 5	50 ± 5
R = CH_3; X = Cl	62 ± 4	38 ± 4
R = OCH_3; X = Br	49 ± 1	51 ± 1
R = F; X = Br	20 ± 1	80 ± 1

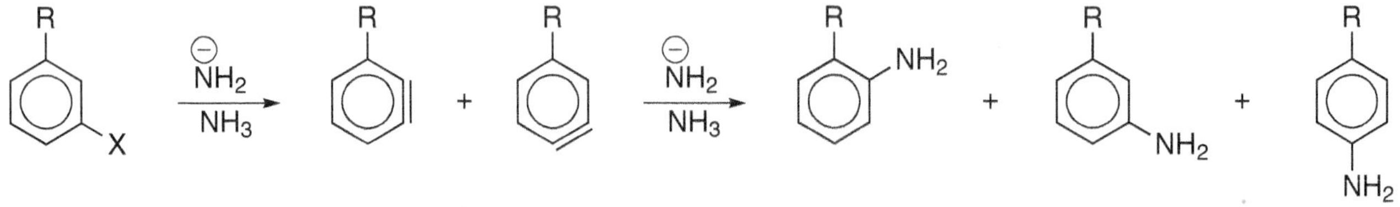

R = CF_3; X = Cl	0	100	0
R = CH_3; X = Cl	40 ± 4	52 ± 4	8 ± 4
R = CH_3; X = Br	22 ± 4	56 ± 4	22 ± 4
R = OCH_3; X = Br	0	100	0

Like the *ortho* isomers, the *p*-substituted halobenzenes can give only one regioisomeric benzyne. In this case, however, the substituent is located at least one bond further from the benzyne group, so its inductive effect is attenuated. Thus, Roberts and his coworkers found that only the most strongly electron-withdrawing substituent by induction (F; electronegativity = 4.0) gave predominantly the *p*-substituted aniline, whereas for *p*-choroanisole (O; electronegativity = 3.5), an essentially equal mixture of the two possible products is formed. This trend continues, with the alkyl group (C; electronegativity = 2.5) actually giving more of the *m*-toluidine.

As a highly strained alkyne, benzyne is a highly active dienophile in the Diels-Alder reaction. Thus, generation of benzyne in the presence of dienes such as furan (**13.62**)[84] and anthracene (**13.63**)[85] leads to the formation of Diels-Alder adducts. In the absence of nucleophiles or other reactive species, benzyne dimerizes to *o*-biphenylene (**13.64**).[86]

84. Wittig, G.; Polmer, L. *Angew. Chem.* **1955**, *67*, 348.
85. Friedman, L.; Longullo, F.M. *J. Org. Chem.* **1969**, *34*, 3089.
86. Longullo, F.M.; Seitz, A.H.; Friedman, L. *Org. Syn.* **1968**, *48*, 17.

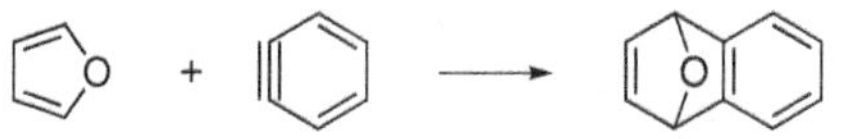

(13.62)

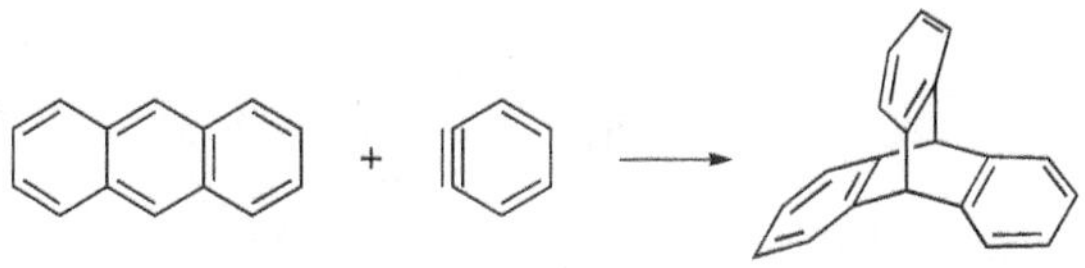

(13.63)

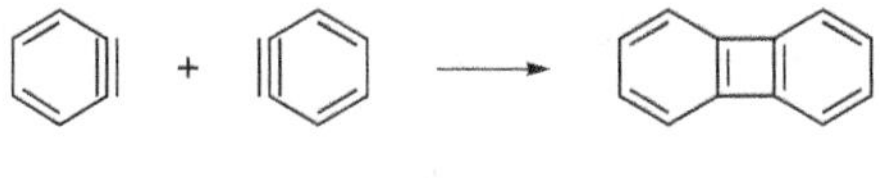

(13.64)

p-Benzyne and *m*-Benzyne

In addition to the benzyne intermediate just discussed, which is sometimes referred to as *o*-benzyne, there are two other dehydrobenzenes that have been implicated as intermediates in certain organic reactions. These are 1,4-didehydrobenzene, also known as ***p-benzyne***, and 1,3-didehydrobenzene, also known as ***m-benzyne***. The three didehydrobenzene isomers are shown in Figure 13.8, along with their highest energy occupied molecular orbitals (HOMOs) and lowest energy unoccupied molecular orbitals (LUMOs).[87] The HOMO of the first two isomers is the symmetric orbital, ϕ_S, relative to the σ_v plane of symmetry relating the dehydro carbon atoms, with the LUMO being the antisymmetric orbital, ϕ_A; in *p*-benzyne, this situation is reversed. This means that *p*-benzyne is best described as a singlet diradical in the ground state, rather than having an unusual transannular bond.

p-Benzyne has been much more widely studied than *m*-benzyne because of its relative ease of formation by the electrocyclic ring closure of *cis*-3-ene-1,5-diynes, called the ***Bergman cyclization***,[88] and its importance in the biological activity of antineoplastic agents such as the calicheamycins[89] and esperamicins.[90]

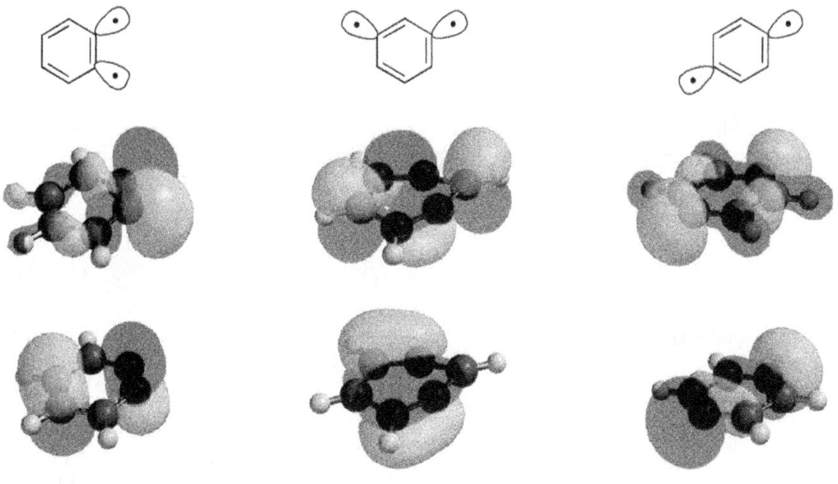

Figure 13.8 Frontier orbitals of isomeric benzynes. The lowest energy unoccupied molecular orbital is the upper orbital of each pair, and the highest energy occupied molecular orbital is the lower orbital of each pair.

87. Wierschke, S.G.; Nash, J.J.; Squires, R.R. *J. Am. Chem. Soc.* **1993**, *115*, 11958.

88. (a) Jones, R.R.; Bergman, R.G. *J. Am. Chem. Soc.* **1972**, *94*, 660. (b) Bergman, R.G. *Acc. Chem. Res.* **1973**, 6, 25. (c) Lockhart, T.P.; Comita, P.B.; Bergman, R.G. *J. Am. Chem. Soc.* **1981**, *103*, 4082. (d) Johnson, G.C.; Stofko, J.J.; Lockhart, T.P.; Brown, D.W.; Bergman, R.G. *J. Org. Chem.* **1979**, *44*, 4215.

89. (a) Lee, M.D.; Dunne, T.S.; Siegel, M.M.; Chang, C.C.; Morton, G.O.; Borders, D.B. *J. Am. Chem. Soc.* **1987**, *109*, 3464. (b) Lee, M.D.; Dunne, T.S.; Chang, C.C.; Ellestad, G.A.; Siegel, M.M.; Morton, G.O.; McGahren, W.J.; Borders, D.B. *J. Am. Chem. Soc.* **1987**, *109*, 3466. (c) Zein, N.; Sinha, A.M.; McGahren, W.; Ellestad, G.A. *Science (Washington, D.C.)* **1988**, *240*, 1198.

90. (a) Golik, J.; Clardy, J.; Dubay, G.; Groenewold, G.; Kawaguchi, H.; Konishi, M.; Krishnan, B.; Okhuma, H.; Saitoh, K.; Doyle, T.W. *J. Am. Chem. Soc.* **1987**, *109*, 3462. (b) Nicolaou, K.C.; Ogawa, Y.; Zuccarello, G.; Kataoka, H. *J. Am. Chem. Soc.* **1988**, *110*, 7247. (c) Nicolaou, K.C.; Smith, A.L. *Acc. Chem. Res.* **1992**, *25*, 497.

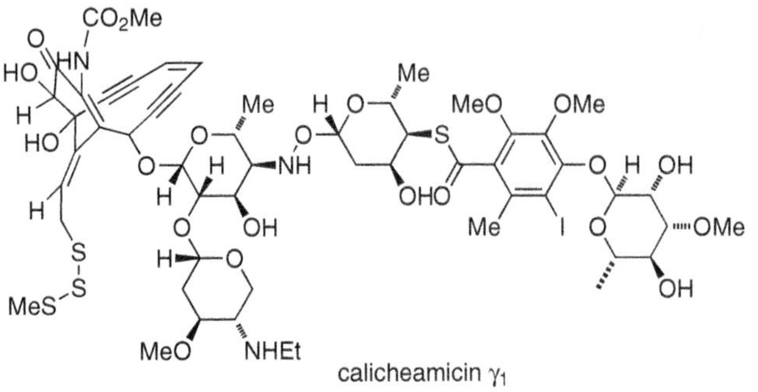

calicheamicin γ_1

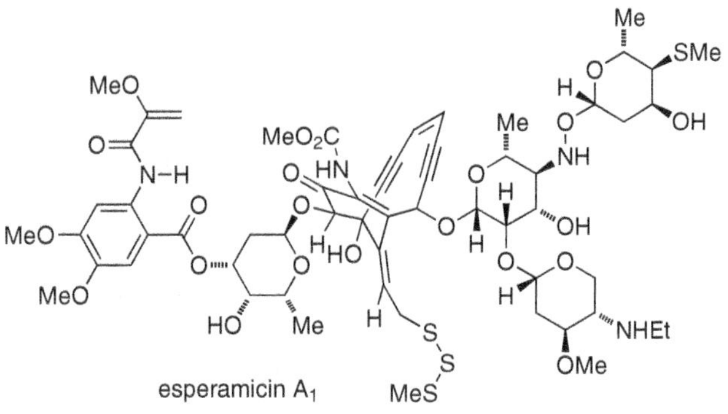

esperamicin A_1

In contrast to *o*-benzyne, there is ample evidence that *p*-benzyne is a singlet diradical rather than a bridged ion. Pyrolysis of *cis*-hex-3-ene-1,5-diyne under a variety of conditions gives only products of free radical atom abstraction, and no evidence of ionic intermediates.[91]

The mechanism of action of the calicheamicin and esperamicin antibiotics involves hydrogen atom abstraction from the backbone of the DNA, followed by chain scission of both chains. The essential features of the mechanism of DNA damage are shown in Figure 13.9. The key to this mechanism is the cleavage of the trisulfide by a nucleophile to give a sulfide anion that undergoes an intramolecular conjugate addition to the enone. This changes the hybridization of the β carbon of the enone from sp^2 to sp^3. This rehybridization removes a bridgehead π bond and, at the same time, brings the two ends of the enediyne system close enough together to allow the Bergman cyclization to proceed.

Figure 13.9 Involvement of the Bergman cyclization of enediynes in the action of esperamicin-type antineoplastic agents.

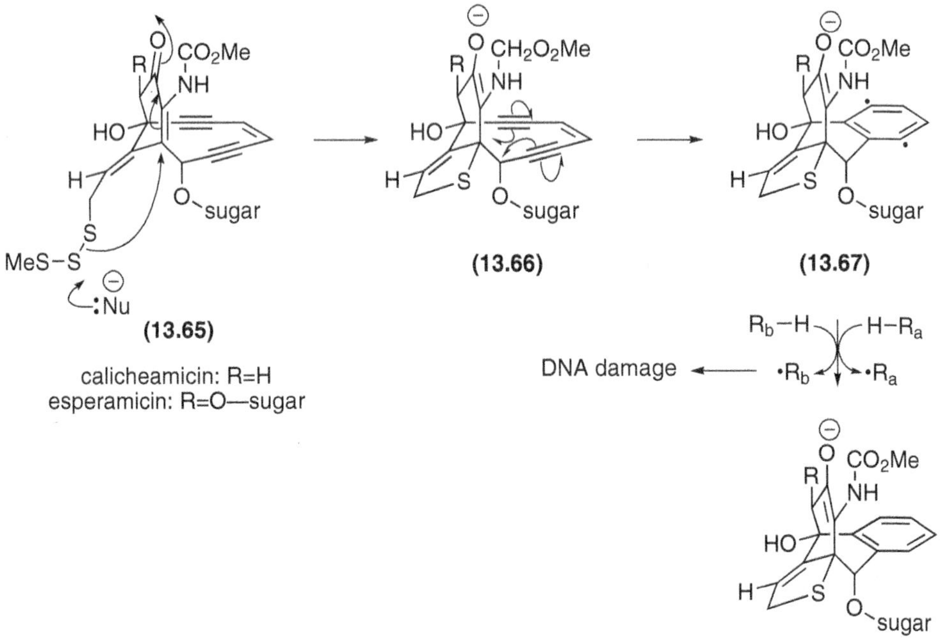

91. Jones, R.R.; Bergman, R.G. *J. Am. Chem. Soc.* **1972**, *94*, 660.

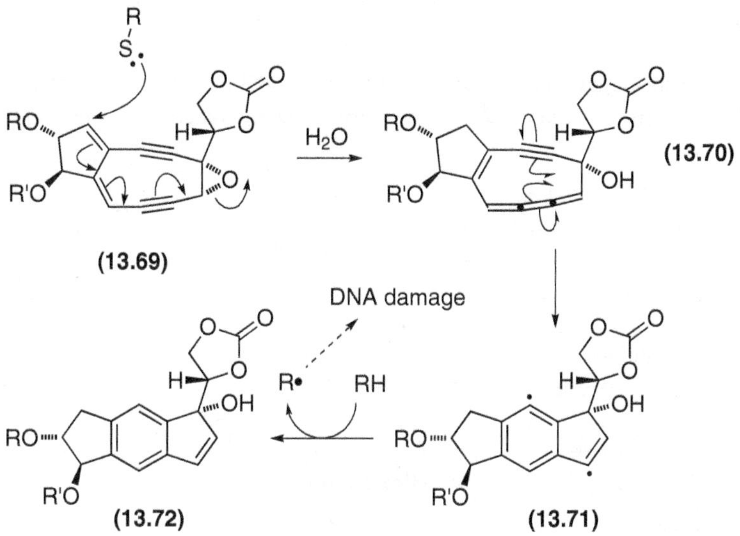

Figure 13.10 Involvement of the Myers-Saito cyclization of eneyne-allenes in the action of neocarzistatin chromophore antineoplastic agents.

A cycloaromatization reaction similar to the Bergman cyclization is the ***Myers-Saito cyclization*** of enyne-allenes[92] to give 3,α-didehydrotoluenes. The barrier to cyclization is low enough that cyclic enyne-allenes are virtually unknown (unlike cyclic enediynes), and they appear to cyclize to the diradical on formation.[93] Enyne-allenes are represented in nature by the neocarzistatin chromophore (**13.69**; this is the small prosthetic group bound to the apoprotein to give neocarzistatin itself) and its analogs.[94] The mechanism by which neocarzinostatin chromophore cleaves DNA involves the formation of the diradical from the enyne-cumulene formed by attack of a thiol on the double bond in the five-membered ring of **13.69** in an S_N2'-type ring opening of the epoxide to give the allenic alcohol **13.70**. Cyclization of this enyne-allene leads to the diradical **13.71**, which then abstracts hydrogen atoms from DNA, causing strand cleavage. This mechanism is summarized in Figure 13.10.

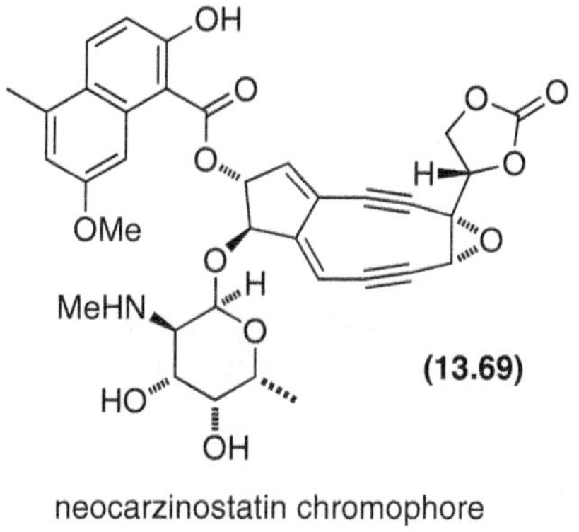

(13.69)

neocarzinostatin chromophore

92. (a) Myers, A.G.; Kuo, E.Y.; Finney, N.S. *J. Am. Chem. Soc.* **1989,** *111,* 8057. (b) Nagata, R.; Yamanaka, H.; Okazaki, E.; Saito, I. *Tetrahedron Lett.* **1989,** *30,* 4995. (c) Myers, A.G.; Dragovich, P.S.; Kuo, E.Y. *J. Am. Chem. Soc.* **1992,** *114,* 9369.

93. (a) Toshima, K.; Ohta, K.; Kano, T.; Nakamura, T.; Nakata, M.; Kinoshita, M.; Matsumura, S. *Bioorg. Med. Chem.* **1996,** *4,* 105. (b) Bekele, T.; Brunette, S.R.; Lipton, M.A. *J. Org. Chem.* **2003,** *68,* 8471.

94. (a) Ishida, N.; Miyazaki, K.; Kumagi, K.; Rikimaru, M. *J. Antibiot.* **1965,** *18,* 68. (b) Napier, M.A.; Holmquist, B.; Strydom, D.J.; Goldberg, I.H. *Biochem. Biophys. Res. Commun.* **1979,** *89,* 635. (c) Kappen, L.S.; Napier, M.A,; Goldberg, I.H. *Proc. Natl. Acad. Sei. U.S.A.* **1980,** *77,* 1970. (d) Povirk, L.**F.;** Goldberg, I.H. *Biochemistry* **1980,** *19,* 4773. (e) Napier, M.A.; Holmquiet, B.; Strydom, D.J.; Goldberg, I.H. *Biochemistry* **1981,** *20,* 5602. (f) Ohtauki, K.; Ishida, N. *J. Antibiot.* **1980,** *33,* 744. (f) Suzuki, H.; Mura, K.; Kumada, T.; Takeuchi, T.; Tanaka, N. *Biochem. Biophys. Res. Commun.* **1980,** *94,* 255.

Worked Problem

13-4 It has been suggested [*J. Org. Chem.* **1987**, *52*, 5685] that the reaction below involves a benzyne intermediate. Write a mechanism for the reaction consistent with this.

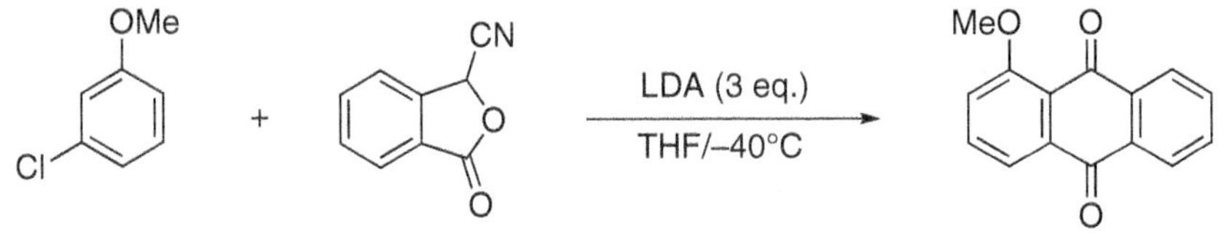

§Answers below.

Reaction Synopses

Methods for Generating Arynes
From Substituted Phenyl Anions

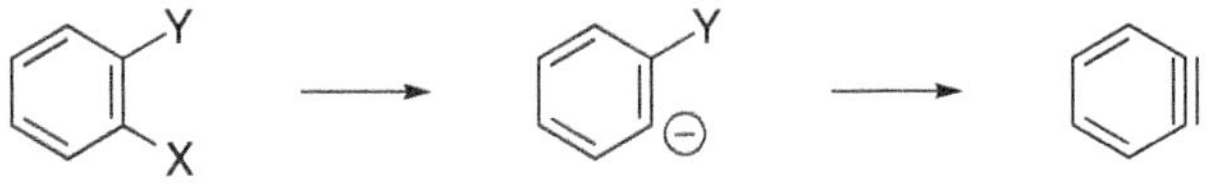

X=H, SiR_3 halogen; Y=halogen, OSO_2R, N_2^+, etc.

Reagents: X=H: $NaNH_2$, NH_3; LDA, THF; PhLi; etc.
X=SiR_3: CsF, MeCN
X=halogen: Mg, Et_2O; BuLi, THF; etc.

By Thermolysis or Photolysis

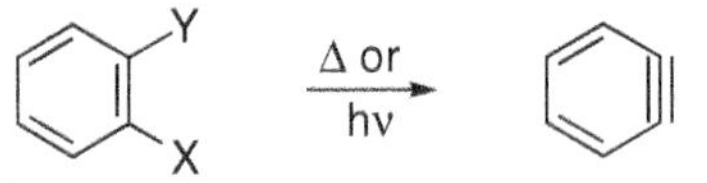

X, Y = CO-O-CO; N=N-SO_2; CO_2^- and PhI^+; CO_2^- and N_2^+

§ Answer to Worked Problem:
The LDA deprotonates both reactants—the *m*-chloroanisole between the substituents, to give 3-methoxybenzyne, and the nitrile at the α position to give an enolate anion that adds to the benzyne. The resulting phenyl anion now adds to the carbonyl group of the ester. The reaction is completed by elimination of cyanide anion.

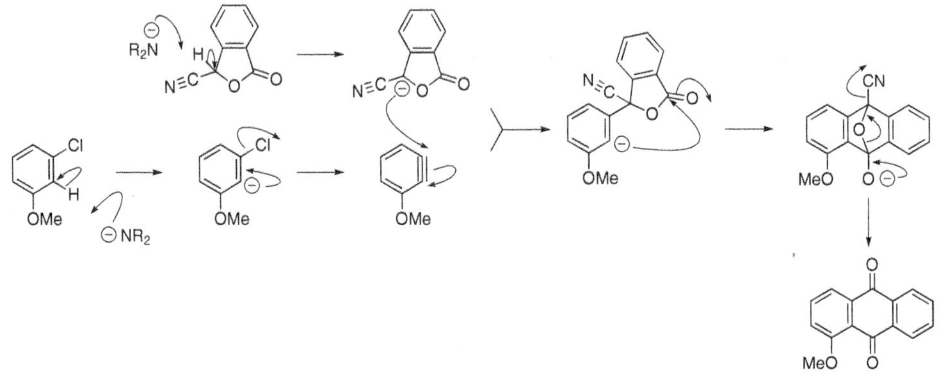

From Benzotriazole Derivatives

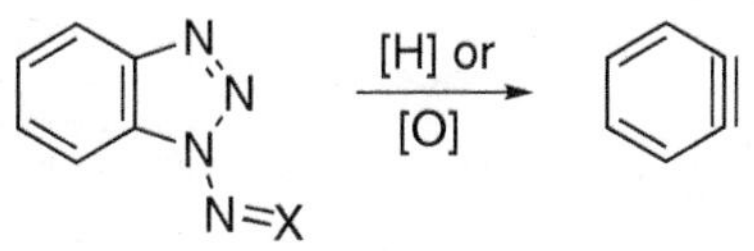

X=H_2, O

Reagents: X=O: $(MeO)_3P$;
X=H_2: $Pb(OAc)_4$

Bergman Cyclization of Enediynes

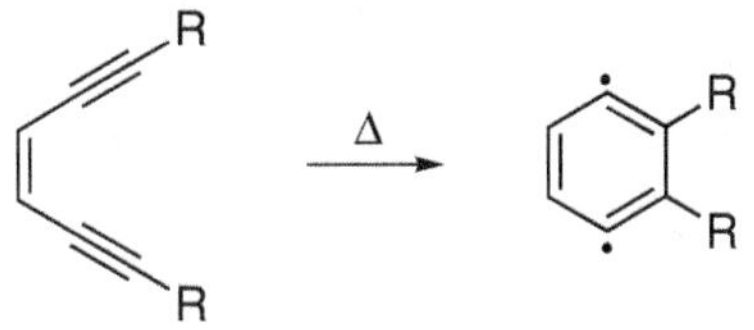

Reaction: proceeds spontaneously if the termini of the ene-diyne are close enough together

Myers-Saito Cyclization of Enyne-Allenes

Problems

13-12 The substitution of aromatic halides by strongly basic nucleophiles changes in the order Br > I > Cl >> F. The strengths of the carbon-halogen bonds increases in the order C—I < C—Br < C—Cl < C—F, whereas the electronegativity of the halogen increases in the order I < Br < Cl < F.

Using these data, rationalize the relative order of reactivity of halobenzenes in nucleophilic substitution by the benzyne mechanism. You might want to consider whether the rate-determining step of the reaction remains constant or changes.

13-13 The naphthalene derivative below reacts with lead tetraacetate and diethyl acetylenedicarboxylate to give the product shown. What is the most reasonable intermediate in this reaction and how is it formed?

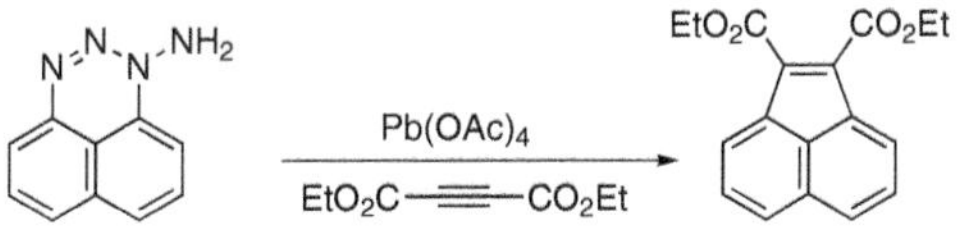

13-14 The reaction below is significantly affected by the nature of the substituent on the ten-membered ring. Provide a reasonable explanation for the effect of the substituent [*Org. Lett.* **2000**, *2*, 3761].

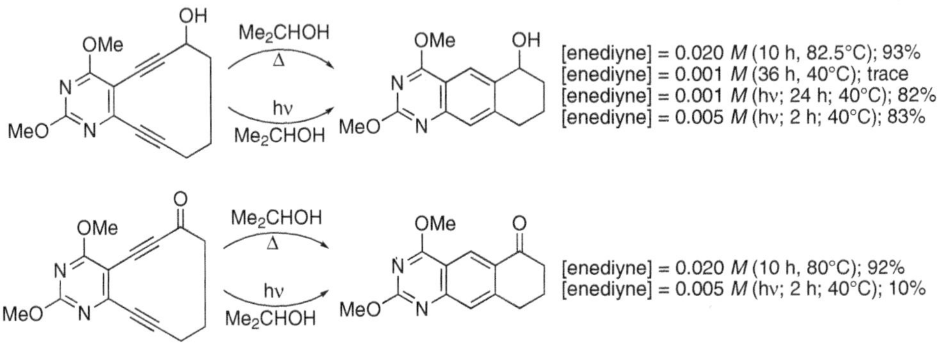

13-15 Suggest a reason why the Myers-Saito cyclization should be so much more facile than the Bergman cyclization.

Chapter Summary

In this chapter, we have discussed the chemistry of the remaining reactive intermediates of organic chemistry: free radicals, which possess at least one unpaired electron; carbenes, which are divalent carbon electrophiles with six valence electrons around carbon; nitrenes, which are monovalent nitrogen species with six valence electrons around nitrogen; and arynes, which are reactive intermediates possessing an aromatic ring lacking two of the C—H σ bonds. Free radicals are preferentially planar but are not restricted to planarity, which makes free radical reactions at bridgehead positions possible. Carbenes exist as both singlet and triplet forms; singlet carbenes possess characteristics of both a nucleophile and an electrophile, whereas triplet carbenes react as a typical diradical. Complexes of carbenes with metals, or organometallic precursors to carbenes, are termed carbenoids and react as singlet carbenes, including adding to alkenes to give cyclopropanes and reacting with alkenes via metallocyclobutanes in olefin metathesis. Like carbocations, carbenes do rearrange. The synthesis of stable carbenes (especially NHCs) has provided a wide variety of useful ligands for transition metal catalysis. Nitrenes are too reactive to have yet been made as an isolable species. Nitrenes and carbenes insert into C—H bonds with retention of configuration. Only *o*-benzyne (or simple "benzyne") and *p*-benzyne have been widely studied. Benzyne reacts as a singlet diradical and is an electrophilic species. *p*-Benzynes are formed in the Bergman cyclization of ene-diynes and are reactive diradicals capable of removing a hydrogen atom from, for example, DNA.

Key Terms

aryne
benzyne
Bergman cyclization
carbene
carbenoid
free radical
N-heterocyclic carbene (NHC)
Myers-Saito cyclization
nitrene

Additional Problems

13-16 Suggest a mechanism for the reaction below. The silicon reagent acts as a source of fluoride anion.

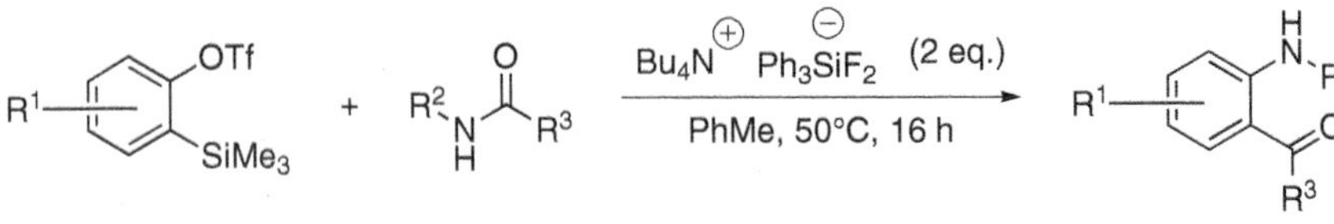

13-17 Photochemical reduction of benzophenone in isopropyl alcohol gives one molecule of benzpinacol for each photon absorbed. Suggest a mechanism consistent with this observation, given that the absorption of a photon excites the benzophenone molecule to a triplet diradical. What is the function of the isopropyl alcohol in this reaction? Suggest a reason why it is so effective in this role.

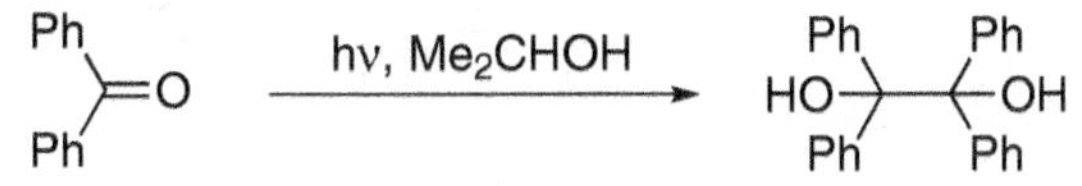

13-18 The reaction below was involved in Problem 11-17. What is the function of the rhodium (II) acetate in this reaction? [*Org. Lett.* **2006**, *8*, 2511.]

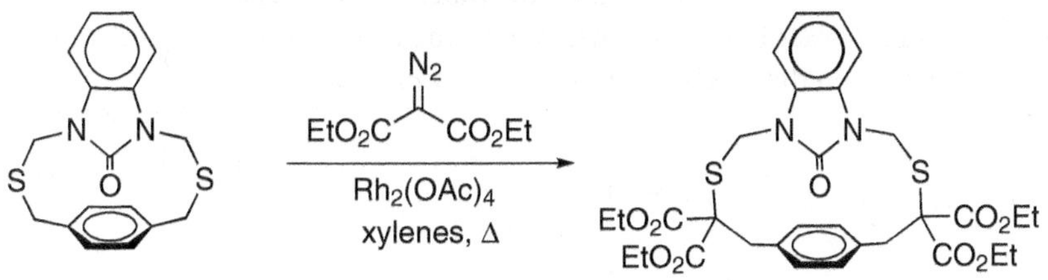

13-19 The reaction below was used as a key step in the synthesis of cytosporone B [*Chem. Lett.* **2010**, *39*, 508]. Suggest a mechanism for the reaction that accounts for the regiochemistry observed.

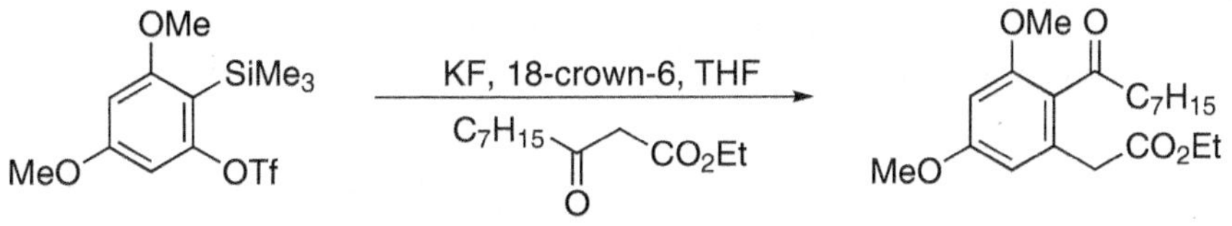

13-20 The reaction below is an example of a group transfer reaction. It occurs by a free radical mechanism. What are the propagation s.epsof the reaction? [*J. Am. Chem. Soc.* **1994**, *116*, 4279.]

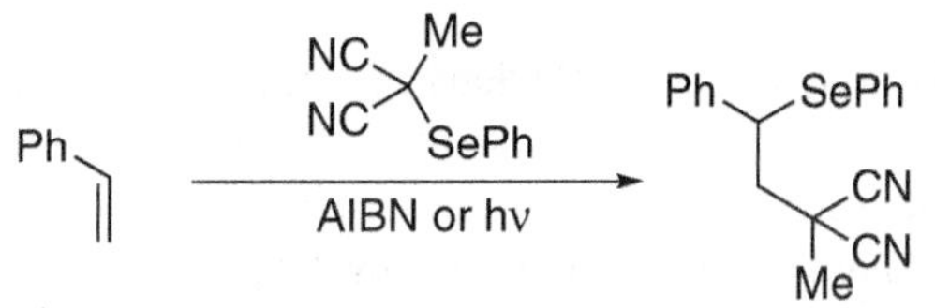

13-21 The reaction below is an interesting photochemical transformation of 1,4-benzoquinone derivatives carrying a side chain with a benzylic hydroxyl group. Suggest a mechanism that accounts for the formation of the products shown. [*J. Chem. Soc. (C)* **1971**, 2244.]

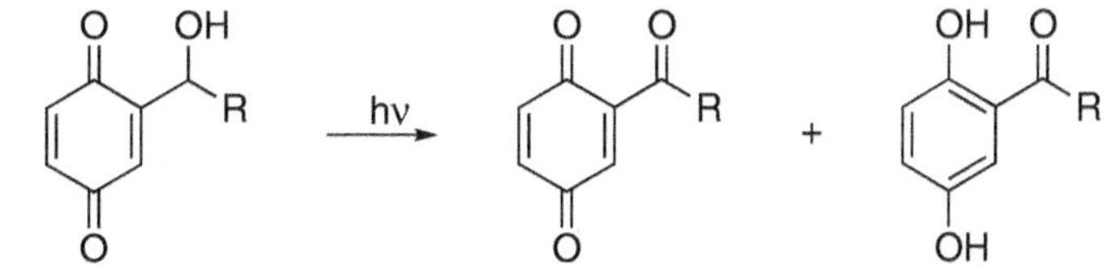

13-22 When 2-bromo-6-ethoxypyridine is treated with potassium amide in liquid ammonia, the two products shown below are obtained. What intermediate is implicated in their formation? [*Tetrahedron Lett.* **1969**, 1943.]

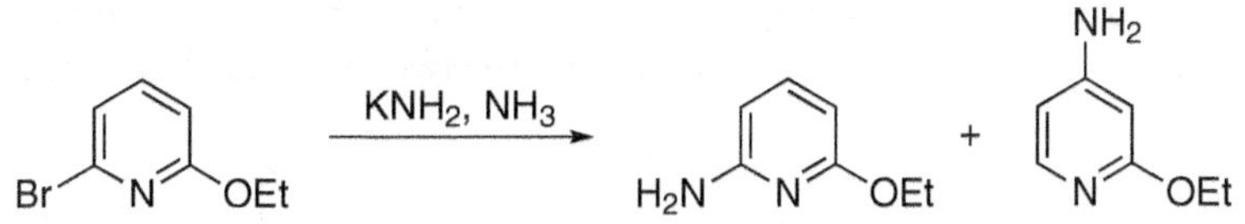

13-23 When the zwitterion below is heated in slightly moist tetrahydrofuran, the two products shown are formed. Provide a rationalization for this observation. [*Tetrahedron Lett.* **1970**, 407.]

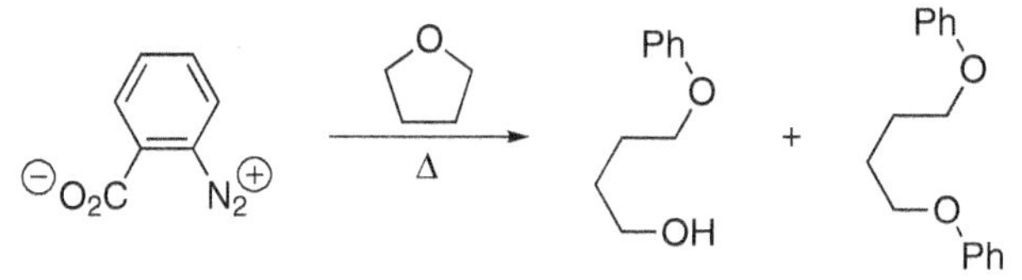

13-24 The dimerization of the stable carbene below to give the alkene shown only occurs in the presence of acid. The dimerization in the absence of acid is slow, and first order in the carbene. What conclusions about the mechanism of this reaction do these facts permit? [*Chem. Commun.* **1997**, 1513.]

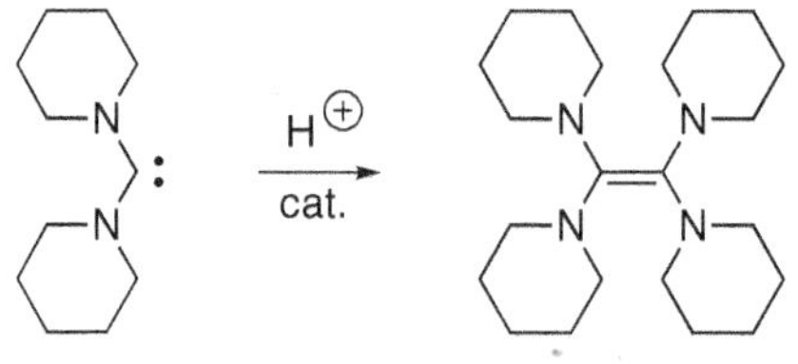

13-25 The treatment of 1,2-*bis*-trimethylsilylbenzene with $PhI(OAc)_2$ and trifluoromethanesulfonic acid gives the iodonium salt shown below [*J. Am. Chem. Soc.* **1999**, *121*, 11674]. When this salt is treated with tetrabutylammonium fluoride in the presence of 2-methylfuran, the product shown is obtained. Provide a mechanism for both reactions in the sequence.

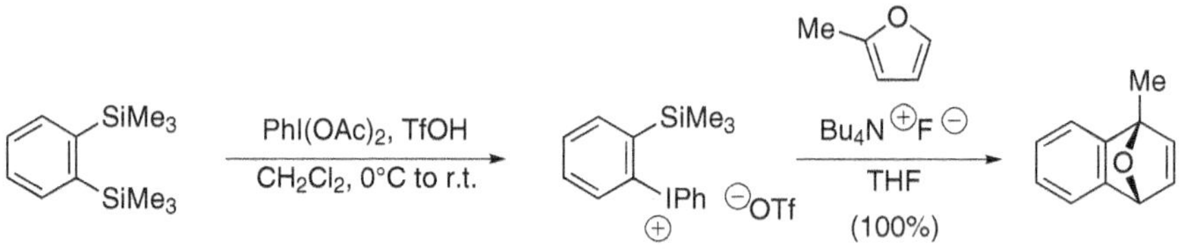

13-26 In a 1989 paper, Wiberg and Chaves [*J. Am. Chem. Soc.* **1989**, *111*, 8052] reported that the intramolecular addition of the cyclopropylidene (carbene) generated by the reaction between the dibromide below and methyllithium gave approximately 50% of the allene and two tricyclic products in the approximate ratios shown.

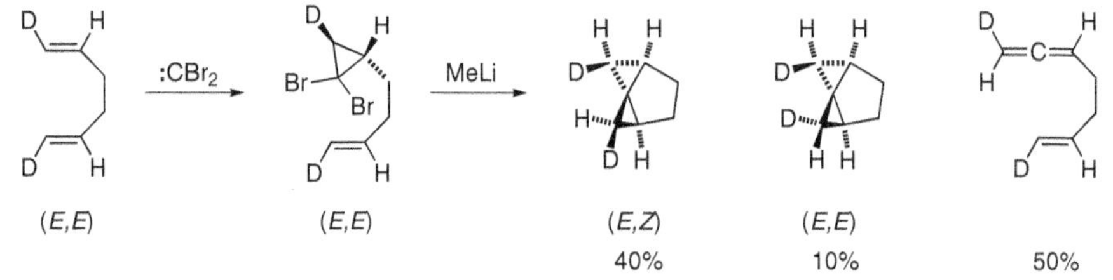

(a) How is the allene formed?

(b) Discuss the formation of the two tricyclic products with particular reference to the stereochemistry of the addition.

(c) The tricyclic products are likely to be quite strained, but they still constitute half of the total product. What conclusions do you draw from this?

13-27 Unlike carbocations, which readily undergo Wagner-Meerwein rearrangements, free radicals do not usually rearrange. The two reactions shown demonstrate that when free radicals do rearrange, the rearrangement may be reversible, and it may also involve migration over more than one carbon. Suggest a common mechanism for these two rearrangements that accounts for the formation of the products in both reactions.

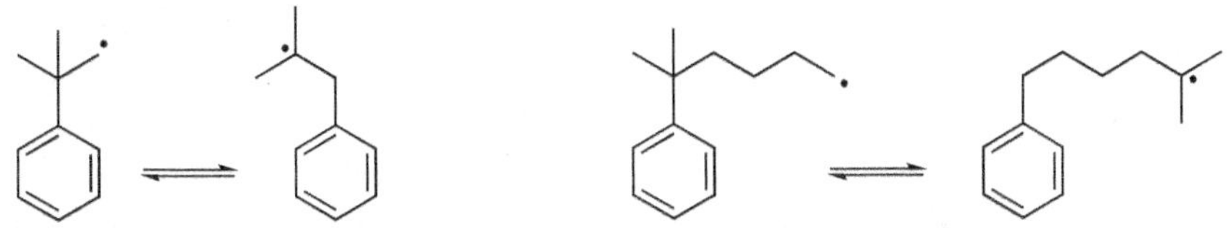

13-28 The insertion of dichlorocarbene into the Si—H bond of a series of aryldimethylsilanes has been studied, and fitting the observed relative rate constants to the Hammett equation gives a Hammett ρ value of −0.632 for the reaction. What inferences can you draw from this about the structure of the activated complex in this reaction? [*J. Am. Chem. Soc.* **1968**, *90*, 2944.]

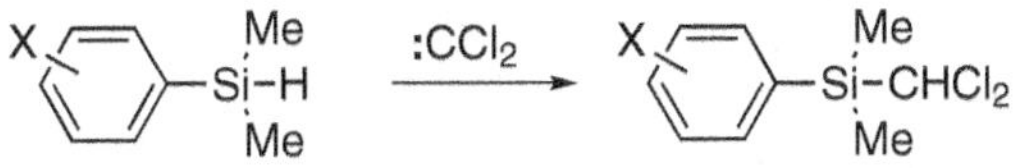

13-29 Provide a reasonable rationalization of the effect of solvent on the reactions shown. In each reaction, the major product is given, but both products are formed in both reactions. If the proportions of the two products in the top reaction change with the concentration of the base, what does this tell you about the mechanism of the reaction?

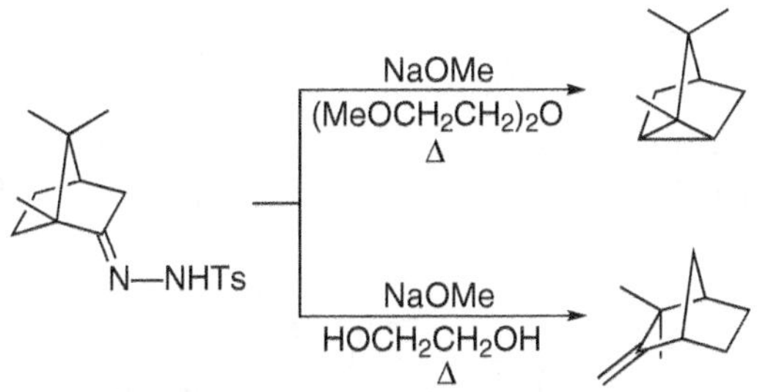

13-30 The Garratt-Braverman reaction, presented here, is analogous to the Bergman and Myers-Saito cyclizations, with *bis*-allenes as the reactants. The diradical formed can cyclize further, as shown in the example below. Suggest a mechanism for this reaction.

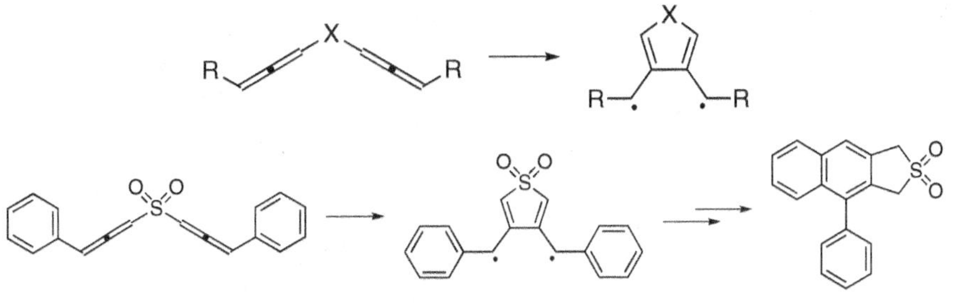

13-31 The reaction shown, which would be expected to give the carboxylic anhydride when the acid was treated with dicyclohexylcarbodiimide (a dehydrating agent) is common for a series of substituted 3-arylpropynoic acids, which give the substituted 1-phenylnaphthalene-2,3-dicarboxylic anhydrides [*Aust. J. Chem.* **1973**, *26*, 557]. Suggest a mechanism for this reaction.

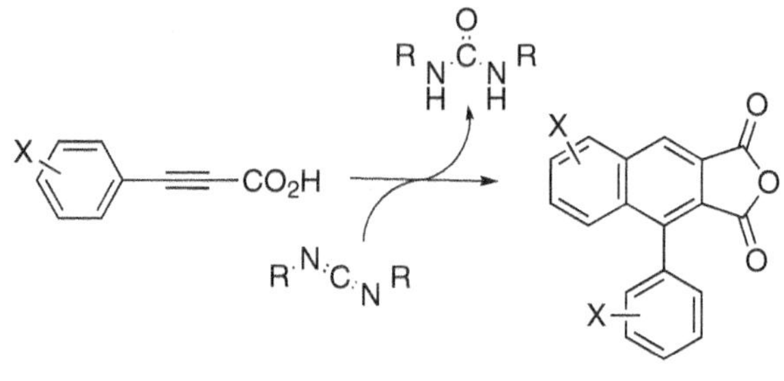

13-32 Nitroxyls are stable oxygen-centered free radicals that have found widespread use in biochemistry. One of the most widely used nitroxyls is 2,2,6,6-tetramethylpiperidyloxy (TEMPO), whose structure is shown. Why is this radical stable when alkoxy radicals are so reactive?

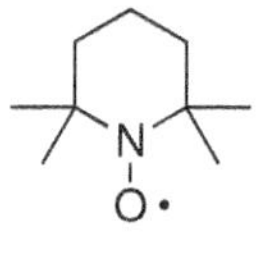

13-33 Suggest a reaction pathway whereby the reaction shown can take place. [*Tetrahedron Lett.* **1991**, *32*, 6735.]

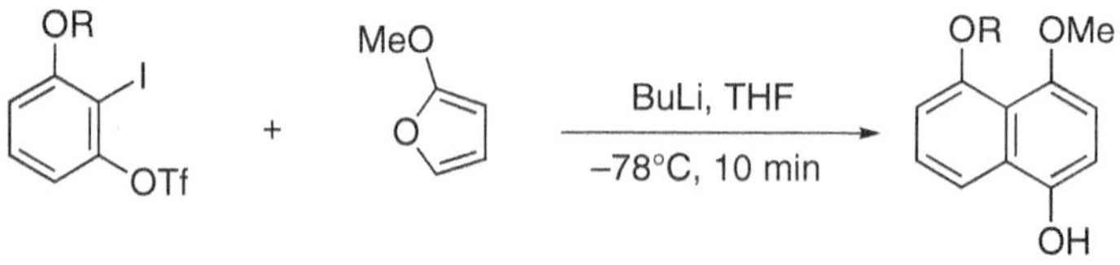

13-34 The reaction between an aryne and a urea gives the product of formal insertion into a C—N bond. Provide a mechanism that rationalizes the regiochemistry of this reaction. [*Angew. Chem. Int. Ed.* **2002**, *41*, 3247.]

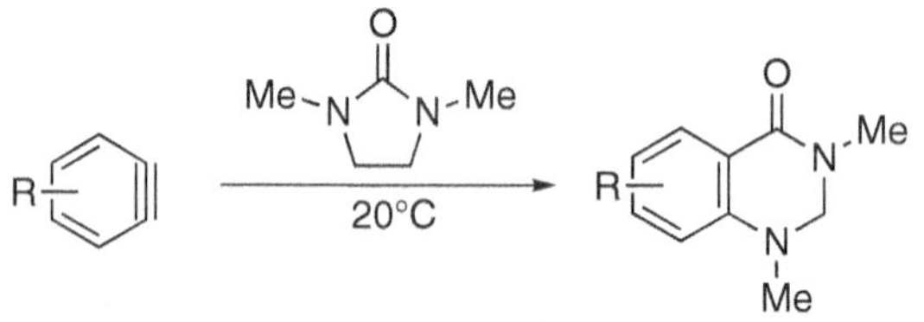

13-35 The oxidation reaction shown was used in a recent total synthesis of (–)-muramycin D2 and its epimer [*J. Org. Chem.* **2010**, *75*, 1366]. Suggest a mechanism by which the observed transformation may take place. Account for the stereochemistry of the reaction. [See also *J. Am. Chem. Soc.* **2007**, *129*, 14106.]

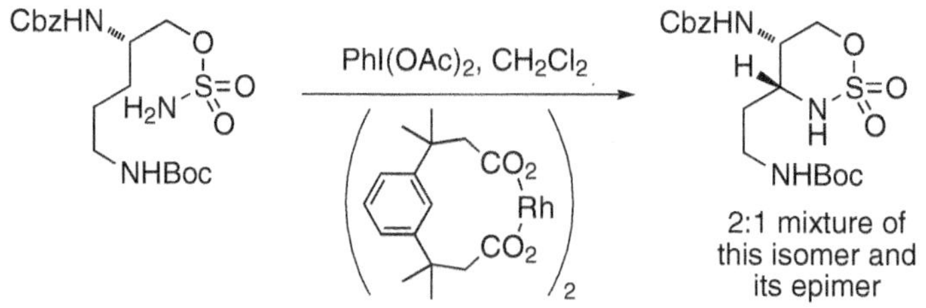

CHAPTER FOURTEEN

Organic Reactions IV

Applications of Free Radical Chemistry in Synthesis

14.1 Methods for Site-Specific Generation of Free Radicals[1]

In Chapter 13, we discussed the structure and reactions of free radicals. As implied in that chapter, much of the usefulness of free radical chemistry in synthesis is predicated on the ability to accomplish site-specific generation of radicals, so that we know where the free radical center is. There are several methods for doing this. Among these are fragmentation, atom abstraction, radical addition, and electron transfer, as well as the homolysis of nonradical precursors.

There are two major tasks in synthetic organic chemistry for which free radicals are well suited: the formation of carbon-carbon σ bonds and the activation of unactivated alkane hydrocarbons (the replacement of a carbon-hydrogen σ bond by a carbon-heteroatom σ bond, which requires the abstraction of a hydrogen atom from that position and its replacement by a heteroatom). The key to forming the specific radical necessary in each of these processes is often the initiator used in the reaction.

Initiators can be divided into two broad classes: those that abstract a hydrogen atom and those that abstract a heteroatom or add to a π bond. We can illustrate all these options with the bromination of cyclohexene, shown as the cycle diagram in Figure 14.1. The homolysis of azobisisobutyronitrile (AIBN) (**14.1**) leads to a free radical (**14.2**) that is not reactive enough to abstract a hydrogen atom from cyclohexene. However, this radical *is* reactive enough to generate a bromine atom, Br•, by abstracting a bromine atom from the bromine molecule itself. It can also add to cyclohexene to generate a new radical (**14.3**) that abstracts a bromine atom from the bromine molecule, thus generating the bromine atom indirectly. Thus, both these pathways lead to the generation of a bromine atom that then enters the chain reaction as a chain carrier. In contrast to the homolysis of AIBN, the homolysis of a peroxide leads to a radical (**14.4**) that is extremely reactive—certainly reactive enough to abstract a hydrogen atom from the alkene to give the allylic free radical, which then enters the cycle.

Simple Homolysis of Nonradical Precursors

The generation of free radicals by homolysis of weak bonds is one of the most general methods for forming radicals. The bonds most susceptible to homolysis are σ bonds between heteroatoms, especially X-X, N-N, N-O, O-O, O-X, and N^+-X bonds. Free radical halogenations, for example, are initiated by the homolysis of the halogen-halogen bond to

1. Reviews: (a) Schiesser, C.H.; Wild, L.M. *Tetrahedron* **1996**, *52*, 13265. (b) Beckwith, A.L.J. *Tetrahedron* **1981**, *37*, 3073. (c) Ramaiah, M. *Tetrahedron* **1987**, *43*, 3541. (d) Beckwith, A.L.J. *Chem. Soc. Rev.* **1993**, 143. (e) Schiesser, C.H. *Chem. Comm.* **2006**, 4055. (f) Curran, D.P. *Synthesis* **1988**, 417. (g) Curran, D.P. *Synthesis* **1988**, 489. (h) Walton, J.C. *Acc. Chem. Res.* **1998**, *31*, 99. (i) Porter, N.A.; Giese, B.; Curran, D.P. *Acc. Chem. Res.* **1991**, *24*, 296.

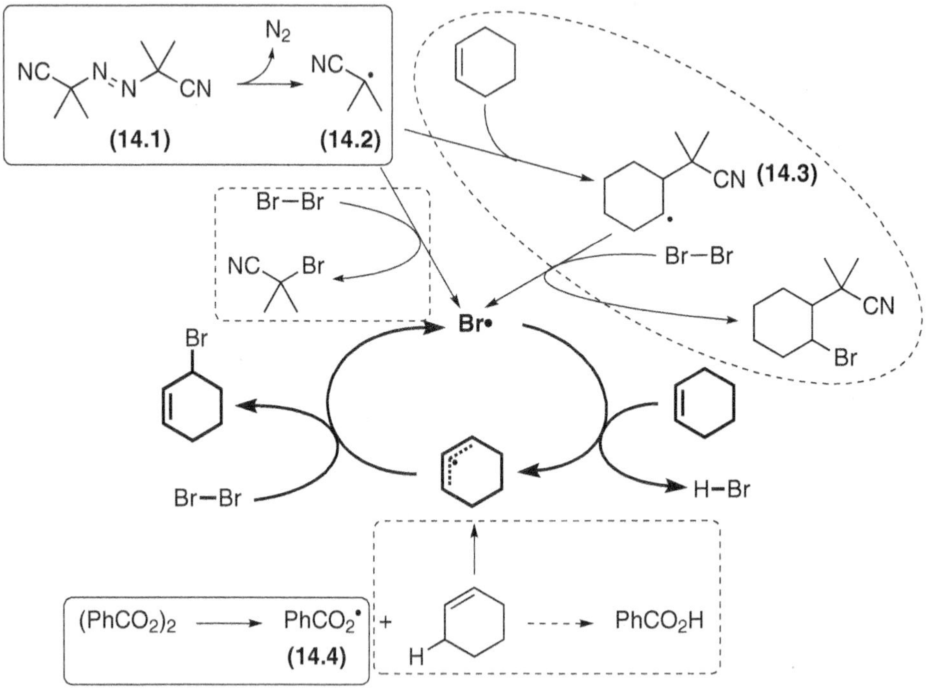

Figure 14.1 Processes for initiating free radical chain reactions. Note how initiation with azobisisobutyronitrile leads to the formation of bromine atoms, whereas initiation with peroxides generates an alkyl radical.

give halogen atoms. A number of typical radical precursors of this type are collected in Table 14.1. Many of these reactions are used to initiate other free radical reactions because they give highly reactive oxygen- or nitrogen-centered free radicals capable of abstracting hydrogen from carbon atoms to give synthetically useful alkyl radicals.

Let us now examine each of these major types of homolysis reactions in turn.

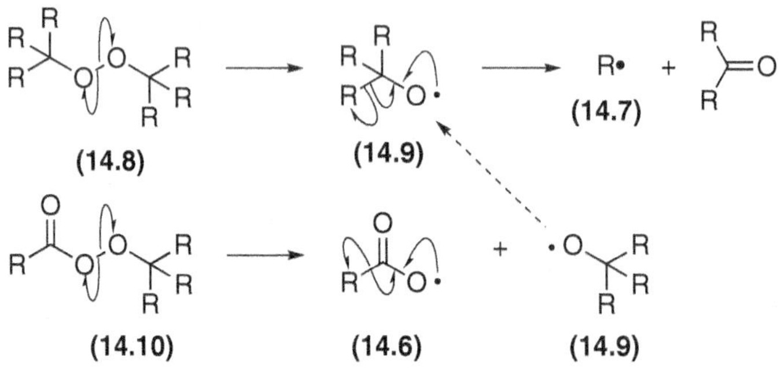

O—O Homolysis

The homolysis of a peroxide link, which can be accomplished under thermal or photochemical conditions, is a common method for the formation of highly reactive alkoxy or acyloxy radicals. The homolysis of the peroxide link occurs by more than one mechanism, depending on the solvent and the peroxide. Dialkyl peroxides (Example **14.8**) decompose by simple homolysis of the O—O bond.[2] The situation with diacyl peroxides (Example **14.5**), peroxy esters (Example **14.10**), and alkyl hydroperoxides is less straightforward, however. In solvents that do not react with free radicals, the cleavage is a simple O—O homolysis and obeys first-order kinetics.[3] However, in solvents capable of

2. (a) Raley, J.H.; Rust, F.F.; Vaughan, W.E. *J. Am.Chem. Soc.* **1948**, *70*, 88, 1336.
3. Hammond, G.S.; Soffer, L.M. *J. Am. Chem. Soc.* **1950**, *72*, 4711.

reacting with a free radical, the reaction occurs by an *induced decomposition* mechanism, where the solvent or other reagent participates in the homolysis mechanism by reacting with the acyloxy radicals formed in the simple homolysis.[4] Thus, the substrate for the free radical reaction initiated by a diacyl peroxide or a perester often accelerates the reaction by itself catalyzing the decomposition of the initiator to the reactive free radicals. A third source of free radicals by the cleavage of peroxides is the one-electron reduction of the peroxy linkage of an alkyl hydroperoxide by metal (usually Fe^{2+}) ions. In this reaction, the alkoxy radical is produced, with the hydroxy group appearing in the product as a hydroxide anion.[5]

Table 14.1 Nonradical Precursors of Free Radicals

Bond	General Type
X—X	Cl_2, Br_2
O—O	RCO_2—O_2CR, RO—OR, RCO_2—OR
O—N	R—O—N=O, R_2N—OCOR
O—X	RO—Cl, RCO—O—I
N—X	R_2NH^+—Cl, R_2NH^+—I
M—H	R—Hg—H, Bu_3Sn—H

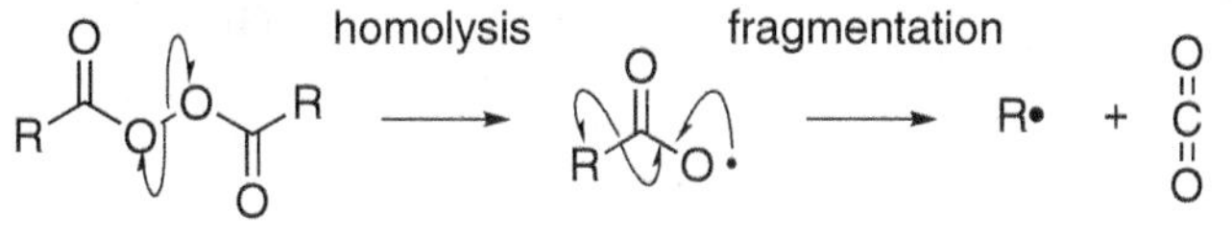

(14.5)

The formation of acyloxy radicals is especially useful as a means of initiating free radical reactions, not only because of this induced homolysis of the initiator, but also because these radicals undergo a facile loss of carbon dioxide (e.g., Example **14.6**) to give the alkyl radical (Example **14.7**),[6] with rate constants approaching 10^6 s^{-1} at −80°C.[7] This fragmentation of carboxy radicals to carbon dioxide and an alkyl radical is a manifestation of the high energy of oxygen-centered radicals and their propensity to fragment to carbon-centered radicals and carbonyl compounds in the absence of a source of hydrogen atoms.[8] In a manner similar to carboxy radicals, tertiary alkoxy radicals (Example **14.9**) decompose to ketones and alkyl radicals.

O—N Homolysis

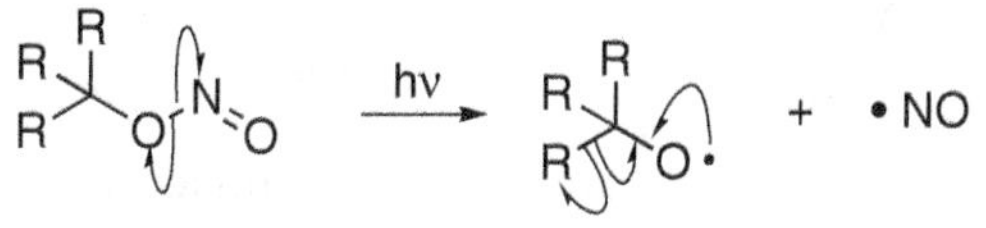

(14.11) **(14.6)**

The homolysis of an N—O bond as a source of free radicals is represented by the homolysis of nitrite esters (e.g., Example **14.11**)[9] and the homolysis of the N—O bond of esters of *N*-hydroxy-2-pyridinethione (Example **14.12**).[10] These are known as ***Barton esters***. Nitrite

4. Bartlett, P.D.; Nozaki, K. *J. Am. Chem. Soc.* **1947**, *69*, 2299.
5. Richardson, W.H. *J. Am. Chem. Soc.* **1965**, *87*, 247.
6. (a) Martin, J.C.; Taylor, J.W.; Drew, E.H. *J. Am. Chem. Soc.* **1967**, *89*, 129. (b) Greene, F.D.; Stein, H.P.; Chu, C.-C.; Vane, F.M. *J. Am. Chem. Soc.* **1964**, *86*, 2080. (c) Bartlett, P.D.; Hiatt, R.R. *J. Am. Chem. Soc.* **1958**, *80*, 1398.
7. Chateauneuf, J.; Lusztyk, J.; Ingold, K.U. *J. Am. Chem. Soc.* **1987**, *109*, 897.
8. (a) Gray, P.; Williams, A. *Chem. Rev.* **1959**, *59*, 239. (b) Greene, F.D.; Savitz, M.L.; Osterholtz, F.D.; Lau, H.H.; Smith, W.N.; Zanet, P.M. *J. Org. Chem.* **1963**, *28*, 55. (c) Walling, C.; Padwa, A. *J. Am. Chem. Soc.* **1963**, *85*, 1593. (d) Walling, C.; Padwa, A. *J. Am. Chem. Soc.* **1963**, *85*, 1597.
9. The best known application of this homolysis is the **Barton reaction**, used in the synthesis of aldosterone acetate: (a) Barton, D.H.R.; Beaton, J.M.; Geller, L.E.; Pechet, M.M. *J. Am. Chem. Soc.* **1960**, *82*, 2640. (b) Barton, D.H.R.; Beaton, J.M. *J. Am. Chem. Soc.* **1960**, *82*, 2641; **1961**, *83*, 4083. (c) Barton, D.H.R.; Basu, K.K.; Day, M.J.; Hesse, R.H.; Pechet, M.M.; Starratt, A.N. *J. Chem. Soc., Perkin Trans. 1* **1975**, 2243. (d) Hesse, R.H. *Adv. Free Radical Chem.* **1969**, *3*, 83. (e) Akhtar, M. *Adv. Photochem.* **1964**, *2*, 263.
10. (a) Barton, D.H.R.; Crich, D.; Motherwell, W.B. *J. Chem. Soc., Chem. Commun.* **1983**, 939; *Tetrahedron* **1985**, *41*, 3901. (b) Barton, D.H.R.; Lacher, B.; Zard, S.Z. *Tetrahedron Lett.* **1985**, *26*, 5939. For a review, see: (c) Barton, D.H.R. *Aldrichimica Acta* **1990**, *23*, 3.

esters undergo photochemical homolysis into an alkoxy radical (Example **14.6**) and the stable free radical, nitric oxide. In fact, much of the driving force for the reaction comes from the fact that one of the radicals formed, nitric oxide, *is* a stable free radical. The homolysis of Barton esters, which also results in a very stable free radical (the 2-pyridylmercapto radical; Example **14.13**), has become one of the most useful methods for the formation of carboxy radicals (e.g., Example **14.14**) and, therefore, of alkyl free radicals (e.g., Example **14.15**). The ready availability of these esters from acid chlorides has made this one of the methods of choice for the generation of alkyl radicals.

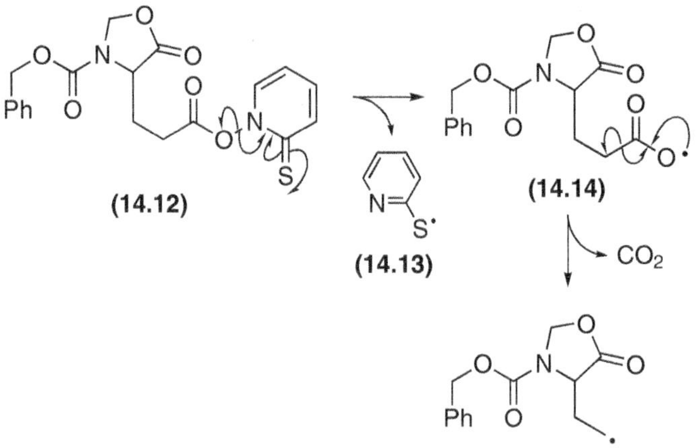

(14.12) **(14.13)** **(14.14)** **(14.15)**

O—Halogen and N—Halogen Homolysis

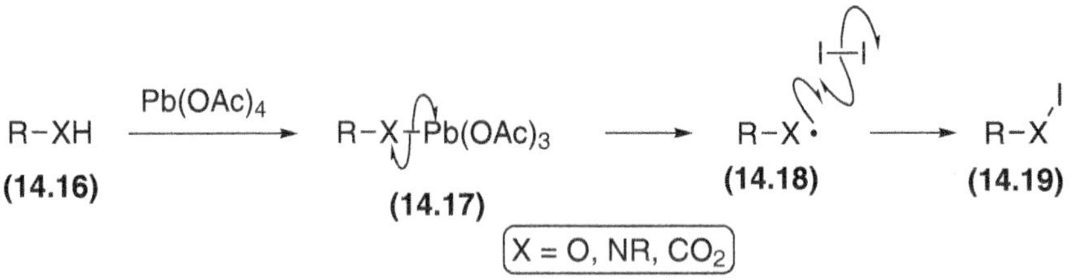

(14.16) **(14.17)** **(14.18)** **(14.19)**

The homolysis of alkyl and acyl hypohalites is another important method for the production of alkoxy and acyloxy free radicals, and the homolysis of *N*-haloammonium ions is an important method for the formation of the analogous highly reactive, nitrogen-centered aminium cation-radicals. The reactants for these homolyses are obtained a number of ways, almost always involving the oxidation of alcohol or amine precursors in the presence of the halogen. For example, the oxidation of an alcohol by lead tetraacetate in the presence of iodine (e.g., Example **14.16**; X=O) gives the corresponding hypoiodite (Example **14.19**),[11] which can then undergo further reaction. The function of the iodine in the reaction appears to be to intercept the initially formed radical (Example **14.18**) and thus suppress further oxidation of the radical by lead (III) or lead (IV).

Homolysis of Organometallic Compounds

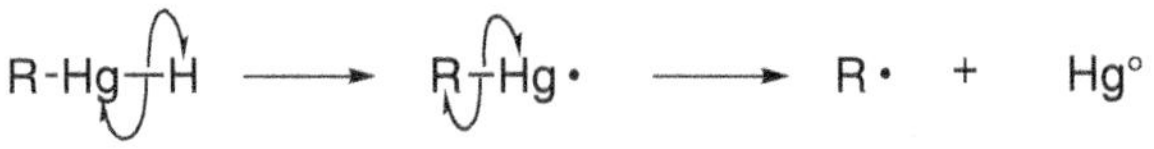

(14.20) **(14.21)** **(14.7)**

11. (a) Heusler, K.; Kavolda, J. *Angew. Chem. Int. Ed. Engl.* **1964**, 3, 525. (b) Akhtar, M.; Barton, D.H.R. *J. Am. Chem. Soc.* **1964**, *86*, 1528. (c) Kavolda, J. *Chem. Commun.* **1970**, 1002. (d) Heusler, K.; Wieland, P.; Meystre, C. *Org. Syn.* **1965**, *45*, 57. (e) Menkert, E.; Mylari, B.L. *J. Am. Chem. Soc.* **1967**, *89*, 174. (f) Shalon, Y.; Yanuka, Y.; Sarel, S. *Tetrahedron Lett.* **1969**, 957, 961.

Carbon-metal and metal-hydrogen bonds in organomercury[12] hydrides (e.g., Example **14.20**) and organocobalt compounds are susceptible to relatively facile homolysis, which makes these compounds useful precursors to free radicals. Organomercury hydrides, especially, are readily available by the sodium borohydride reduction of the corresponding halides and carboxylates, as occurs in the second step of the oxymercuration-demercuration sequence. In fact, the formation of radicals by this reduction is responsible for the loss of stereospecificity in the second step of that reaction.

Fragmentation of Radicals and Nonradical Precursors

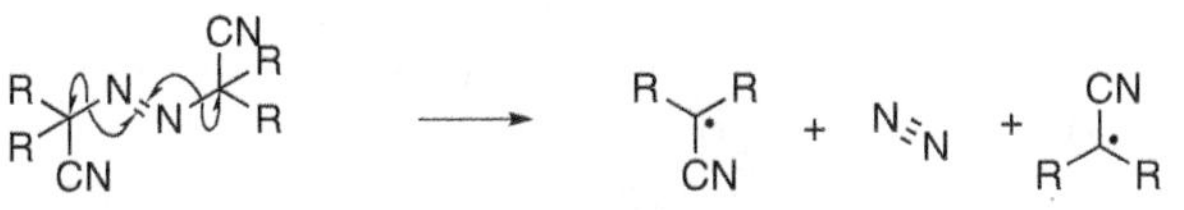

(14.22) **(14.23)** **(14.23)**

Another popular method for the generation of free radicals is the fragmentation of a nonradical precursor to give an especially stable product. We have already touched on the homolysis of heteroatom-heteroatom bonds. There are circumstances, however, when carbon-heteroatom bond cleavage can lead to the formation of free radicals. The most widely used of these reactions is the cleavage of azoalkanes (e.g., Example **14.22**) to give two alkyl free radicals (Example **14.23**) and a molecule of nitrogen; the driving force for the reaction comes from the formation of the especially stable N_2 molecule. There is evidence that the overall fragmentation occurs by a concerted pathway when the alkyl radicals formed are strongly stabilized and by a stepwise pathway when the free radicals formed are not.[13] Among the most important radical initiators that operate by this mechanism are the α,α'-azo-*bis*-alkanenitriles (e.g., Example **14.22**, which was used as an initiating reaction in Figure 14.1), which homolyze rapidly to strongly stabilized α-cyanoalkyl radicals.[14] A similar type of radical formation driven by the formation of an extremely stable by-product is the formation of alkyl radicals by the homolysis of diacyl peroxides. In this case, the formation of carbon dioxide provides the driving force.

Reduction Reactions to Generate Free Radicals

The transfer of an unpaired electron from an active metal to the σ* orbital of a carbon-heteroatom bond or to a π* orbital is known as ***single electron transfer***, or ***s.e.t.***. The product of s.e.t. is a free radical. However, the initial product of s.e.t. to a neutral molecule is a radical anion, and this species may or may not undergo further unimolecular reactions before reacting with itself or other species in the reaction mixture. When the receiving orbital is a carbon-heteroatom σ* orbital, the radical anion is usually unstable, and the s.e.t. is followed rapidly by loss of the heteroatom as an anion to yield the simple free radical. When the receiving orbital is a π* orbital, the radical anion is usually long-lived enough to react with itself or other reagents in the reaction mixture. Radical anions generated by s.e.t. are intermediates in a number of important reactions, including the Birch

12. (a) Giese, B.; Hueck, K. *Chem. Ber.* **1979**, *112*, 3759; *Tetrahedron Lett.* **1980**, *21*, 1829; *Chem. Ber.* **1981**, *114*, 1572. (b) Giese, B.; Horler, H.; Zwick, W. *Tetrahedron Lett.* **1982**, *23*, 931. (c) Giese, B. *Angew Chem. Int. Ed. Engl.* **1985**, *24*, 553. (d) Review: Baraluenga, J.; Yus, M. *Chem. Rev.* **1988**, *88*, 487.

13. (a) Keonig, T. In Kochi, J.K., Ed. *Free Radicals* (Wiley: New York, 1973), vol. 1, p.143. (b) Crawford, R.J.; Takagi, K. *J. Am. Chem. Soc.* **1972**, *94*, 7406. (c) Seltzer, S.; Dunne, F.T. *J. Am. Chem. Soc.* **1965**, *87*, 2628. (d) Prior, W.A.; Smith, K.J. *J. Am. Chem. Soc.* **1970**, *92*, 5403. (e) Engel, P.S.; Bishop, D.J. *J. Am. Chem. Soc.* **1975**, *97*, 6754. (f) Green, J.G.; Dubay, G.R.; Porter, N.A. *J. Am. Chem. Soc.* **1977**, *99*, 1264. (g) Porter, N.A.; Marnett, L.J.; Lochmüller, C.H.; Closs, G.L.; Shobataki, M. *J. Am. Chem. Soc.* **1972**, *94*, 3664. Neumann, R.C., Jr.; Lockyer, G.D., Jr. *J. Am. Chem. Soc.* **1983**, *105*, 3982.

14. (a) Overberger, C.G.; O'Shaughnessy, M.T.; Shalit, H. *J. Am. Chem. Soc.* **1949**, *71*, 2661. (b) Overberger, C.G.; Berenbaum, M.B. *J. Am. Chem. Soc.* **1951**, *73*, 2618. (c) Overberger, C.G.; Biltech, H.; Finestone, A.B.; Lilker, J.; Herbert, J. *J. Am. Chem. Soc.* **1953**, *75*, 2078. (d) Overberger, C.G.; Lebovitz, A. *J. Am. Chem. Soc.* **1954**, *76*, 2722. (e) Overberger, C.G.; Hale, W.F.; Berenbaum, M,.B.; Finestone, A.B. *J. Am. Chem. Soc.* **1954**, *76*, 6185.

reduction (via Example **14.24**),[15] the pinacol[16] and acyloin[17] reactions (via Example **14.26**), and the Wurtz coupling[18] of alkyl halides (via Example **14.27**).

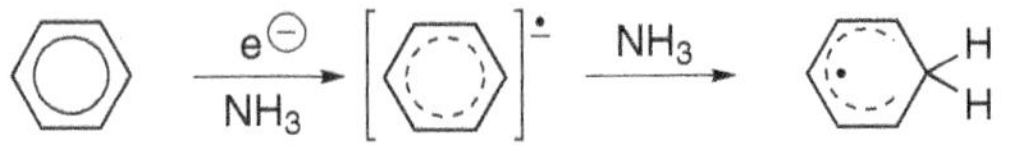

(14.24) **(14.25)**

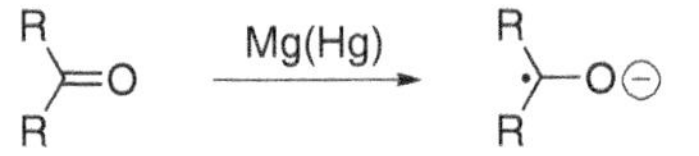

(14.26)

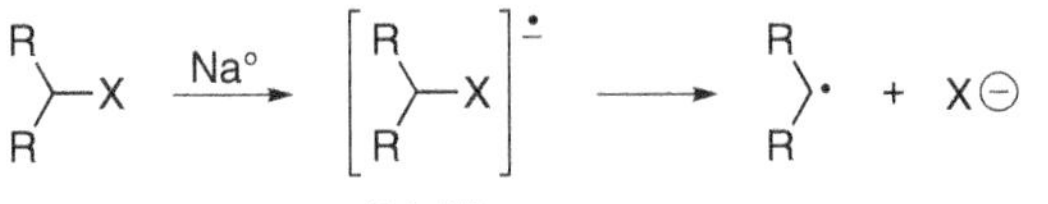

(14.27)

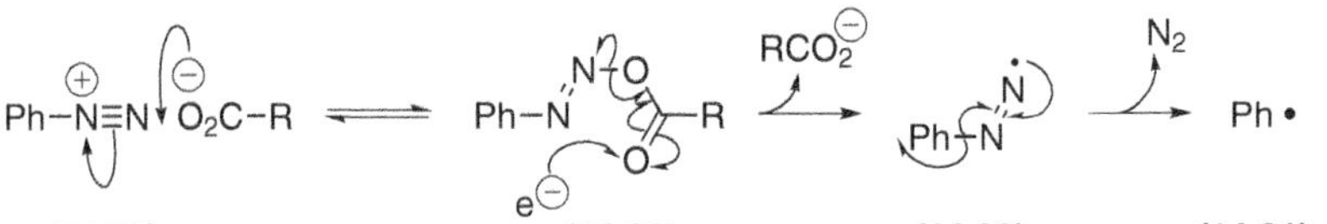

(14.28) **(14.29)** **(14.30)** **(14.31)**

The reduction of arenediazonium ions is an important method for the production of aryl free radicals. When an arenediazonium ion is treated with base or a carboxylate salt, a derivative of the diazoic acid (Example **14.29**) is formed in a reversible reaction. When the base is hydroxide anion, the reaction is known as the ***Gomberg-Bachmann reaction***.[19] Under the typical reaction conditions, the diazonium ion is reduced by s.e.t. (where the single electron comes from is often nebulous!) to start the radical chain reaction by forming the diazenyl radical (Example **14.30**), which then loses nitrogen to give the aryl radical (Example **14.31**). A related reaction occurs when *N*-nitrosoamides (Example **14.32**) are heated in a hydrocarbon solvent. In this reaction, the initial step involves the isomerization of the nitrosoamide to the ester by a formal 1,3-sigmatropic rearrangement; the ester then undergoes reversible ionization to the diazonium ion. The same reduction of the diazonium ion by s.e.t. gives the diazenyl radical, which loses nitrogen to give the aryl radical.[20] Because carbon dioxide is not isolated in this reaction, it is unlikely that the homolysis of the N—O bond occurs.

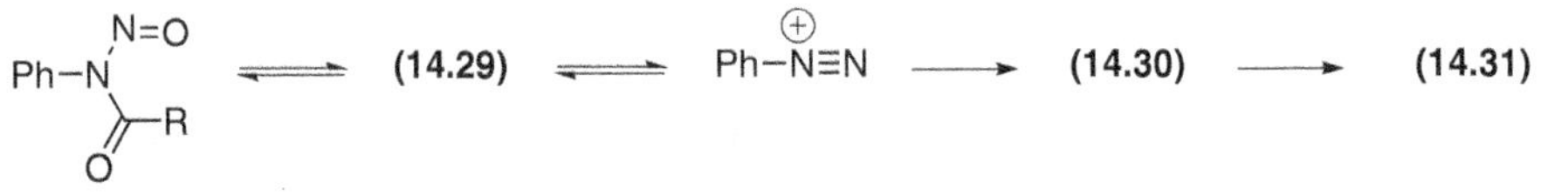

(14.32)

15. (a) Birch, A.J. *J. Chem. Soc.* **1944**, 430; **1945**, 809; **1946**, 593; **1947**, 1642. (b) Watt, G.W. *Chem. Rev.* **1950**, *46*, 317.

16. (a) Rausch, M.D.; McEwen, W.E.; Kleinberg, J. *Chem. Rev.* **1957**, *57*, 417. (b) Adams, R.; Adams, E.W. *Org. Syn., Coll. Vol. 1* **1944**, 459. (c) Vigevani, A.; Pasqualucci, R.; Gallo, G. *Tetrahedron* **1969**, *25*, 573. (d) Corey, E.J.; Danheiser, R.L.; Chandrasekaran, S.; Siret, P.; Keck, G.E.; Gras, J.-L. *J. Am. Chem. Soc.* **1978**, *100*, 8031. (e) Corey, E.J.; Danheiser, R.L.; Chandrasekaran, S.; Keck, G.E.; Gopsalom, B.; Larsen, S.D.; Siret, P.; Gras, J.-L. *J. Am. Chem. Soc.* **1978**, *100*, 8034.

17. (a) Bouveault, L.; Loquin, R. *Compt. rend.* **1905**, *140*, 1593. (b) McElvain, S.M. *Org. React.* **1948**, *4*, 256. (c) Finley, K.T. *Chem. Rev.* **1964**, *64*, 573. (d) Snell, J.M.; McElvain, S.M. *Org. Syn.* **1943**, *Coll. Vol. 2*, 114. (e) Ames, D.E.; Hall, G.; Warren, B.T. *J. Chem. Soc. C* **1968**, 2617. (f) Prelog, V. *J. Chem. Soc.* **1960**, 420. (g) Schaefer, J.P.; Bloomfield, J.J. *Org. React.* **1967**, *15*, 1.

18. (a) Wurtz, A. *Ann. Chim. Phys.* [3] **1855**, *44*, 275; *Ann. Chem. Pharm.* **1855**, *96*, 364. (b) Kosolapoff, G.M. *Org. React.* **1951**, *6*, 326.

19. (a) Gomberg, M.; Bachmann, W.E. *J. Am. Chem. Soc.* **1924**, *46*, 2339. (b) Hey, D.H.; Waters, W.A. *Chem. Rev.* **1937**, *21*, 169. (c) Elks, J.; Haworth, J.W.; Hey, D.H. *J. Chem. Soc.* **1940**, 1284. (d) Dermer, O.C.; Edmison, M.T. *Chem. Rev.* **1957**, *57*, 77. (e) Augood, D.R.; Williams, G.H. *Chem. Rev.* **1957**, *57*, 123. (f) Bachmann, W.E.; Hoffman, R.A. *Org. React.* **1944**, *2*, 224.

20. Cadogan, J.I.G. *Acc. Chem. Res.* **1971**, *4*, 186.

Oxidation Reactions to Generate Free Radicals

Conceptually, at least, one of the simplest methods for site-selective formation of alkyl free radicals by oxidation is the electrolytic oxidation of carboxylate anions (Example **14.33**), which forms the basis for the ***Kolbe electrolysis***.[21] Another major method for preparing free radicals in a site-selective manner is the reaction between an alcohol or amine and lead tetraacetate and iodine to generate an alkyl hypoiodite, whose homolysis was discussed earlier in this chapter.

$$R{-}CO_2^{\ominus} \xrightarrow{-e^{\ominus}} R{-}\dot{C}O_2 \longrightarrow R\bullet \qquad \textbf{(14.33)}$$

Manganese (III) Acetate Oxidation

Manganese (III) compounds are excellent one-electron oxidants for carbonyl compounds (especially β-dicarbonyl compounds). The mechanism of these reactions may be viewed as shown in Figure 14.2. In this mechanism, the initial complexation of the carbonyl oxygen (**14.34**) by manganese is followed by proton loss to give the Mn(III) enolate complex (**14.35**). The transfer of a single electron from the ligand to the metal results in loss of the metal and formation of the enolate radical (**14.36**). These radicals will react with manganese (III) acetate to give the corresponding acetates (**14.37**), as illustrated by Example **14.38**, or they will add to alkenes to give addition products.

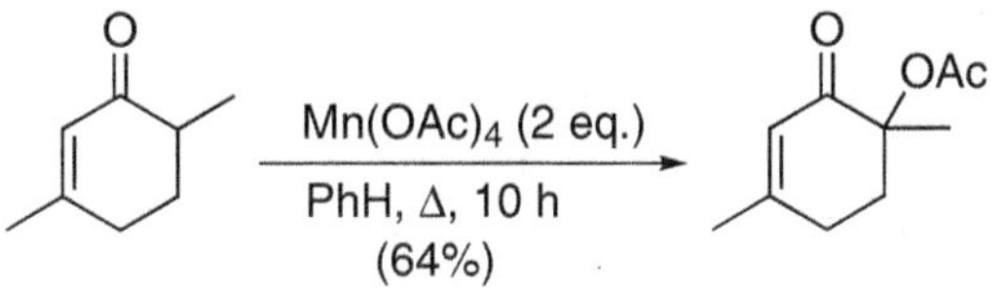

(14.38)[22]

The exact product depends to a degree on the reaction conditions (especially the reaction solvent). As the oxidation of **14.39** below shows, when a copper (II) co-oxidant is used, the initial radical (**14.40**) cyclizes to **14.41** and is then oxidized further to the carbocation. This cation is trapped by copper in the form of a copper (III) complex (Example **14.43**) that then either generates the alkene (**1** Example **4.45**) by deprotonation (favored by more basic solvents such as dimethyl sulfoxide) or a lactone (Example **14.44**) by intramolecular trapping of an ester carbonyl oxygen.

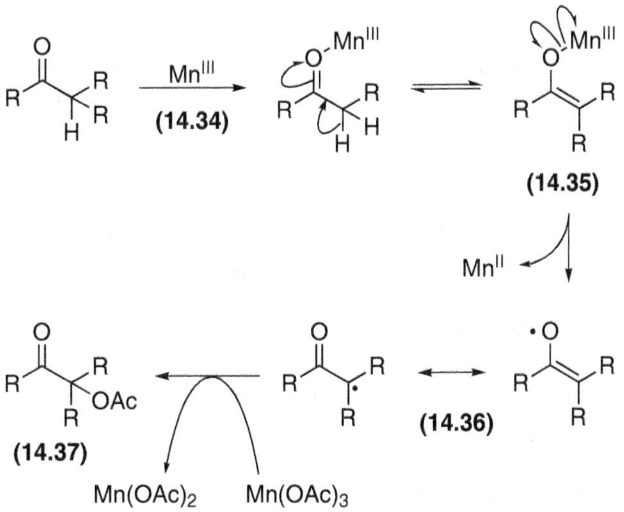

Figure 14.2 Mechanism of α-oxidation of ketones to radicals by Mn(III) acetate

21. (a) Kolbe, H. *Ann. Chem. Pharm.* **1849**, *69*, 257. (b) Crum Brown, A.; Walker, J. *Ann. Chem. Pharm.* **1891**, *261*, 107. (c) Glasstone, S.; Hickling, A. *Chem. Rev.* **1939**, *25*, 333. (d) Weedon, B.C.L. *Quart. Rev. (London)* **1952**, *6*, 380.
22. Tanyeli, C.; Iyigün, C. *Tetrahedron* **2003**, *59*, 7135.

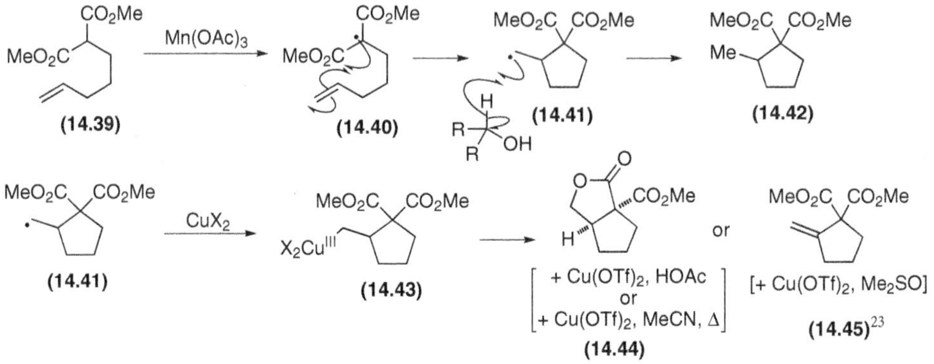

(14.39) (14.40) (14.41) (14.42)

(14.41) (14.43) (14.44) (14.45)[23]

Autoxidation of Trialkylboranes

Trialkylboranes (Example **14.46**) react with molecular oxygen to give peroxide-containing borane products (e.g., Example **14.47**), with the concomitant formation of free radicals.[24] In fact, the reaction between trialkylboranes and oxygen serves as an excellent method for initiating free radical reactions (especially radical additions to alkenes or radical cyclizations).

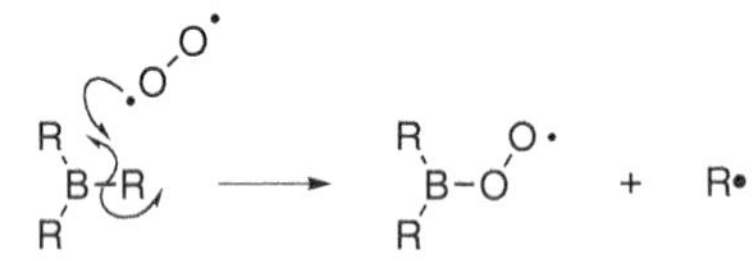

(14.46) (14.47)

Photochemical Generation of Radicals: Ketone Photolysis

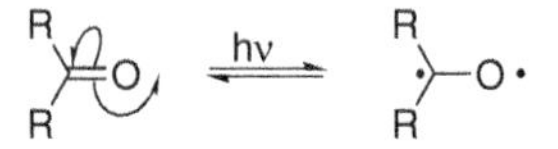

(14.48)

The irradiation of ketones leads to promotion of an electron from an occupied orbital to the unoccupied π* orbital of the carbonyl group (Example **14.48**). This process, which gives a ***diradical***, a species with two unpaired electrons, can occur in two ways: promotion of the electron from a lone pair on oxygen to the π* orbital (n-π*) and promotion of the electron from the C=O π orbital to the π* orbital (π-π*). The two processes are not equally facile. The π-π* transition is symmetry-allowed and occurs rapidly and easily. At the same time, the reverse process also occurs rapidly, so the diradical formed only survives for a short time (usually too short for any subsequent chemical reaction).

In contrast to the π-π* process, the n-π* transition is symmetry-forbidden, so the probability of the transition occurring is low, and the reaction is slow. At the same time, the reverse reaction, also, is symmetry-forbidden, so the diradical formed by this process is long-lived enough to participate in subsequent reactions.

The formation of useful diradicals in the π-π* process can be facilitated by the use of photosensitizers, which are compounds in which the singlet diradical initially formed (Example **14.48**) rapidly undergoes intersystem crossing (ISC; a symmetry-forbidden process that inverts of the spin of one of the electrons) to give the much longer-lived triplet diradical (Example **14.49**). Diradicals formed in this process necessarily have one of the unpaired electrons on oxygen. Because oxygen-centered radicals are very high energy,

---

23. Powell, L.H.; Docherty, P.H.; Hulcoop, D.G.; Kemmitt, P.D.; Burton, J.W. *Chem. Commun.* **2008**, 2559.
24. Review: Brown, H.C.; Midland, M.M. *Angew. Chem.Int. Ed.* **1972,** *11*, 692.

these diradicals are highly reactive. Typically, they tend to abstract hydrogen atoms from carbon (especially in intramolecular processes).

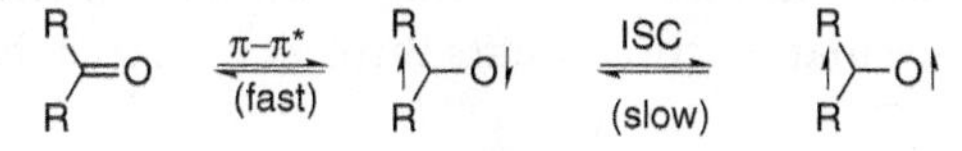

(14.48s) **(14.48t)**

Reaction Synopses

Homolysis

Peroxides

$$\text{RO–OR} \xrightarrow{\Delta \text{ or } h\nu} 2\ \text{RO}\bullet$$

Reagents: $(Me_3CO)_2$, $(PhCO_2)_2$, etc.

Aliphatic Azo Compounds

$$\text{RN=NR} \xrightarrow{\Delta \text{ or } h\nu} 2\ \text{R}\bullet + N_2$$

R= $Me_2C(CN)$, AIBN; c-$C_6H_{10}(CN)$, ACCN

Most widely used radical initiators

Nitrite Esters

$$\text{RO–N=O} \xrightarrow{\Delta \text{ or } h\nu} \text{RO}\bullet + \bullet\text{NO}$$

Barton Esters

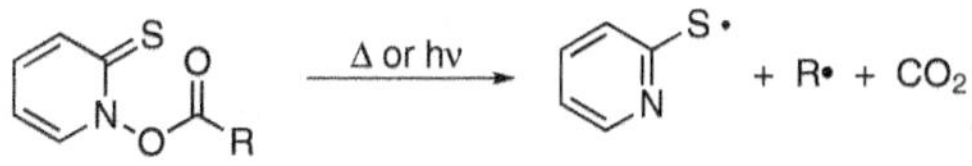

Organomercury Hydrides

$$\text{R-Hg–X} \xrightarrow[\text{NaOH}]{\text{NaBH}_4} \text{R}\bullet$$

X = Cl, Br, OCOR, etc.

Reagents: $NaBH_4$, NaOH, H_2O; etc.

Oxidation of Trialkylboranes

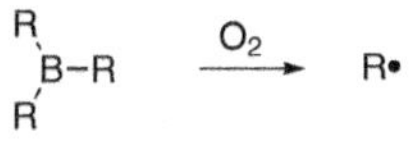

Oxidation with Mn(III)

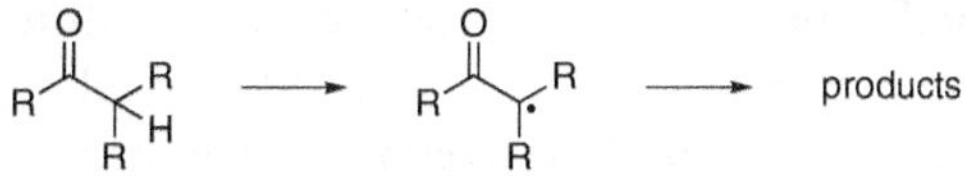

Reagents: $Mn(OAc)_3$, PhH, Δ; $Mn(acac)_3$, PhH, Δ; etc.

Atom Abstraction

Hydrogen

$$\text{R–H} \longrightarrow \text{R}\bullet$$

Reagents: ROOR, Δ or hν; etc.

(continues)

(**Reaction Synopses** continued)

Relatively nonspecific unless the C—H bond is activated or the reaction is intramolecular. Intramolecular reaction occurs through a six-membered cyclic transition state.

By Stannyl Radicals

$$\text{R–X} \longrightarrow \text{R}\bullet$$

X = Cl, Br, I, SR, OC(=S)–SR SeR, etc.

Reagents: Bu_3SnH, AIBN, Δ or hν; etc.

Single Electron Transfer

$$\text{R–X} \longrightarrow [\text{R–X}]^{\bullet -} \longrightarrow \bullet\text{R} + \text{X}^{-}$$

Reagents: Li (Na, K, Ca, etc.), NH_3; Li (Na, K, Mg, Ca, etc.), PhH; etc.

Photochemical Excitation of Carbonyl Compounds

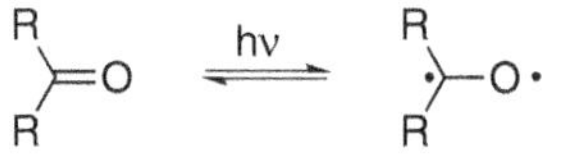

Diradical formed may be singlet or triplet. Symmetry-forbidden processes (n-π*, ISC) give the longest-lived, most synthetically useful diradicals.

14.2 **Reactions of Free Radicals**

With the ability to generate free radicals in a site-selective manner comes the ability to use the reactions of these useful intermediates in synthesis. The majority of free radicals react by fragmentation, addition, or rearrangement.

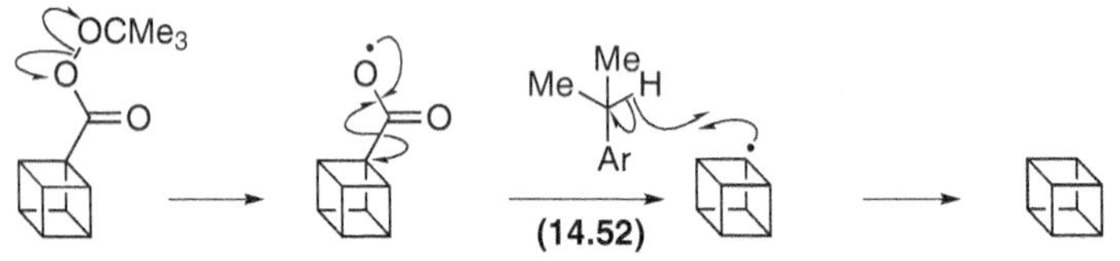

(14.49) **(14.50)** **(14.51)** **(14.53)**

Fragmentation

We have already touched on the generation of alkyl free radicals by the decarboxylation of carboxy radicals or by the fragmentation of *tert*-alkoxy radicals to give a ketone and the alkyl radical. Most carboxy free radicals are formed by initial homolysis of a relatively weak bond. Peresters, acyl peroxides, acyl hypohalites and Barton esters can all serve as precursors to acyl radicals, as can electrolysis of carboxylate anions (the Kolbe electrolysis). The homolysis of a *tert*-butyl perester (Example **14.49**) was a key step in Eaton's synthesis of cubane (Example **14.53**).[25]

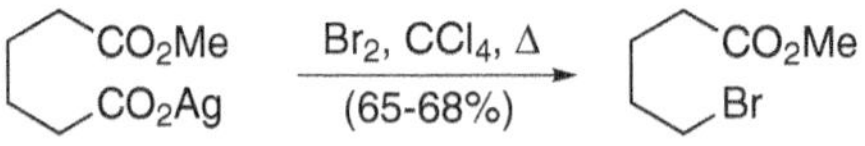

(14.54)[26]

25. (a) Eaton, P.E.; Cole, T.W., Jr. *J. Am. Chem. Soc.* **1964**, *86*, 962, 3157.
26. Allen, C.F.H.; Wilson, C.V. *Org. Syn., Coll. Vol. 3*, **1955**, 578.

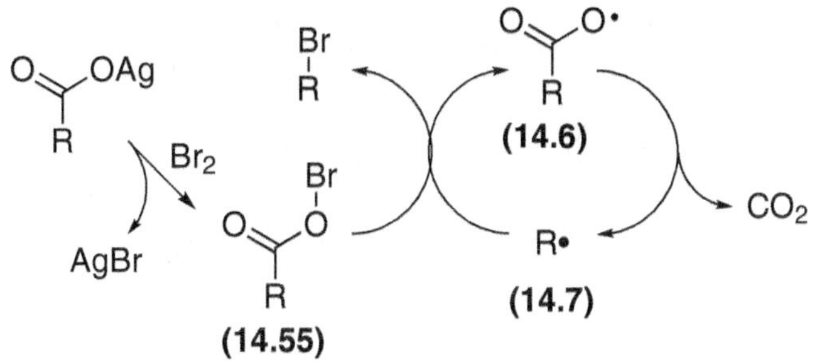

(14.55) (14.6) (14.7)

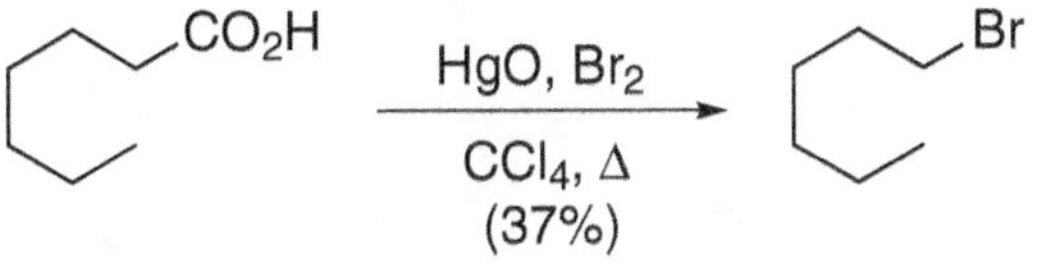

(14.56)[27]

The ***Borodin-Hunsdiecker reaction***[28] is a reaction that leads to the conversion of a silver[29] or mercury[30] salt of a carboxylic acid by elemental bromine into the alkyl bromide of the next-lower homolog (Example **14.54**). It is initiated by homolysis of the oxygen-halogen bond of the acyl hypobromite (Example **14.55**) to give the carboxy radical (Example **14.6**). In the first propagation step, the carboxy radical decarboxylates to the alkyl radical (Example **14.7**), which abstracts the halogen atom from a molecule of the acyl hypohalite in the second propagation step to give the alkyl halide and another carboxy radical. During the 1960s, the reaction was modified to replace the silver in the reaction by the less expensive (but more toxic) mercury. This modification (Examples **14.56** and **14.57**) involves adding bromine to the carboxylic acid and mercuric oxide heated in a halogenated solvent (although it is occasionally necessary to irradiate the reaction mixture).

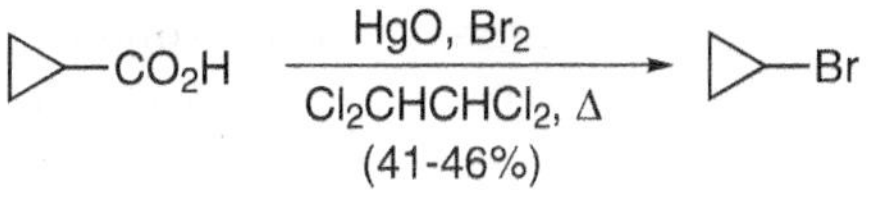

(14.57)[31]

Atom Abstraction (Atom Transfer)

An alternative method to thermolysis or photolysis of weak bonds for the generation of free radicals is atom abstraction—transfer of an atom from a molecule to a free radical. If the σ bond forming between the radical and the transferring atom is stronger than the original σ bond to the transferring atom, the atom transfer occurs rapidly.

---

27. Cristol, S.J.; Firth, W.C. *J. Org. Chem.* **1961**, *26*, 280.

28. Wilson, C.V. *Org. React.* **1957**, *9*, 332.

29. (a) Borodine, A. *Z, Chem.* **1861**, *4*, 5; **1869**, *12*, 342; *Ann. Pharm. Chem.* **1869**, *121*, 119. (b) Hunsdiecker, C.; Hunsdiecker, H. Vogt, E. U.S. Pat. No. 2,176,181 (1939). (c) Hunsdiecker, H.; Hunsdiecker, C. *Ber. Deut. chem. Ges.* **1942**, *75*, 291. (d) Wilson, C.V. *Org. React.* **1975**, *9*, 341.

Heinz Hunsdiecker (1904–1981) took his PhD in 1929 at Göttingen and worked in the laboratories of Dr. Vogt and Company, a chemical plant in Köln, where he was a chemist and partner. Hunsdiecker passed his career in industry, and he received several patents. With his wife, **Cläre Hunsdiecker**, née **Dieckmann** (1903–1995), he patented the reaction that now bears their joint names. This patent may hold the record for the time between public disclosure of a method and award of a patent for that method.

Aleksandr Porfir'evich Borodin (Borodine) (1833–1887) earned his MD from St. Petersburg University, and he then became Professor at the Medical-Surgical Academy in St. Petersburg after study in Germany and Italy. Borodin was a fine composer, although he always considered himself a chemist first. Much of his music was incomplete when he died, suddenly, of an aneurysm; it was completed by his friends. Borodin is the only chemist to have won a Tony award for best musical score in a Broadway show (*Kismet*, 1953). For more detail, see: Friedman, H.B. *J. Chem. Educ.* **1941**, *18*, 521.

30. (a) Ref. 27. (b) Davis, J.A.; Herynk, J.; Carroll, S.; Bunds, J.; Johnson, D. *J. Org. Chem.* **1965**, *30*, 415. (c) Jennings, P.W.; Ziebarth, T.D. *J. Org. Chem.* **1969**, *34*, 3216. (d) Meek, J.S.; Osuga, D.T. *Org. Syn.* **1963**, *43*, 9.

31. Meek, J.S.; Osuga, D.T. *Org. Syn., Coll. Vol. 5*, **1973**, 126.

Table 14.2 Bond Dissociation Energies of σ Bonds to Hydrogen

Bond	(kcal mol^{-1})	(kJ mol^{-1})	Bond	(kcal mol^{-1})	(kJ mol^{-1})
CH_3–H	104	435	HO–H	119	498
RCH_2–H	98	410	HOO–H	90	377
R_3C–H	92	385	RCH_2O–H	104	437
$C_6H_5CH_2$–H	85	356	RCOO–H	103–110	431–469
$H_2CCHCHR$–H	83	347	C_6H_5O–H	88	368
$HOCH_2$–H	94	393			
$MeOCH_2$–H	93	389	H_2N–H	102	427
$HOCMe_2$–H	91	381			
H_2C=CHCH(OH)–H	82	341	HS–H	95	397
			CH_3S–H	74	310
C_6H_5–H	110	461			
O=CR–H	87	364			

One can predict the probable outcome of atom-transfer reactions by determining the exothermicity of the reaction from bond dissociation energies. In the transfer of a bromine atom from molecular bromine to an alkyl radical, for instance, the bromine-bromine bond (46.3 kcal mol^{-1}, 194 kJ mol^{-1}) is broken and a carbon-bromine bond (70 kcal mol^{-1}, 293 kJ mol^{-1}) is formed as the reaction is written. Thus, as written, the reaction is exothermic by 23.7 kcal mol^{-1} (99 kJ mol^{-1}), and it is expected to be favorable.

The abstraction of a hydrogen atom is widely used for initiating free radical reactions. The abstraction of hydrogen from sp^3 carbon by an oxygen-centered radical or atom is usually an exothermic process, as can be calculated from the values in Table 14.2. However, unless one particular C—H bond is especially activated toward homolysis, hydrogen atom abstraction is a relatively nonselective method for generating of free radicals. Consequently, the use of hydrogen atom transfer reactions in laboratory synthesis is generally restricted to the generation of allylic free radicals. Abstraction from sp^2 or sp carbon is usually endothermic and seldom synthetically useful.

There is one exception to the generalization that the abstraction of a hydrogen atom from a nonallylic sp^3 carbon is nonselective; that is the intramolecular abstraction of hydrogen from carbon by oxygen- or nitrogen-centered radicals.[32] In these cases (e.g., Example **14.58**), the hydrogen that is abstracted can be predicted with a high degree of certainty because *intramolecular hydrogen atom transfer reactions occur preferentially through a six-membered cyclic transition state* unless there are unusual circumstances that prevent this or make other transition states more favorable.[33]

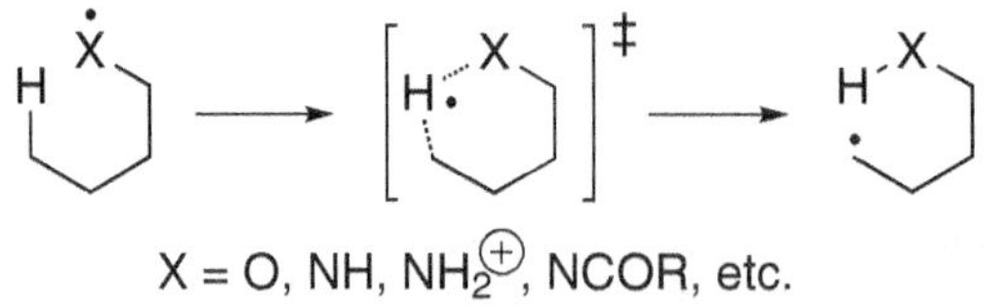

(14.58)

Intermolecular hydrogen atom abstraction is usually nonspecific. However, the abstraction of halogen atoms or sulfur- or selenium-based groups by tin-centered radicals, a

32. For a brief review of early work in this area, see: Acott, B.; Beckwith, A.L.J. *Aust. J. Chem.* **1964**, *17*, 1342.
33. (a) Hesse, R.H. *Adv. Free Radical Chem.* **1969**, 3, 83. (b) Barton, D.H.R. *Pure Appl. Chem.* **1968**, *16*, 1.

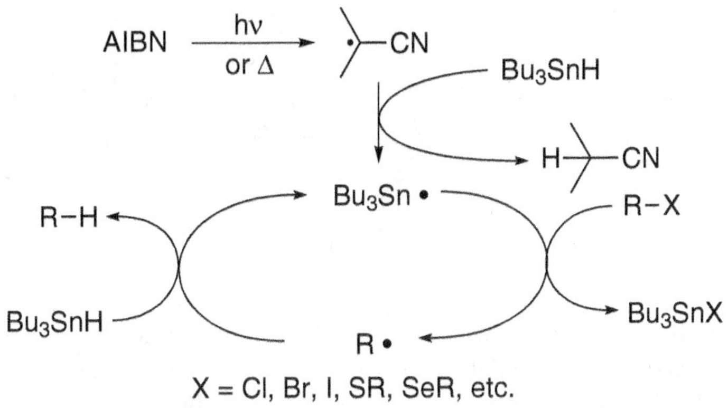

Figure 14.3
Stannane reduction

highly selective reaction, has the added advantage of giving a radical where the location of the free radical center is known. The trialkyltin radicals do not abstract hydrogen from organic compounds. The free radical reaction between an alkyl halide and tri-*n*-butyltin hydride (tri-*n*-butylstannane) is often the preferred method for the replacement of a halogen atom by hydrogen in organic compounds.[34] The reduction of an alkyl halide, sulfide, or selenide by tri-*n*-butylstannane may be represented by the cycle diagram in Figure 14.3; examples are shown below (Examples **14.59** and **14.60**).

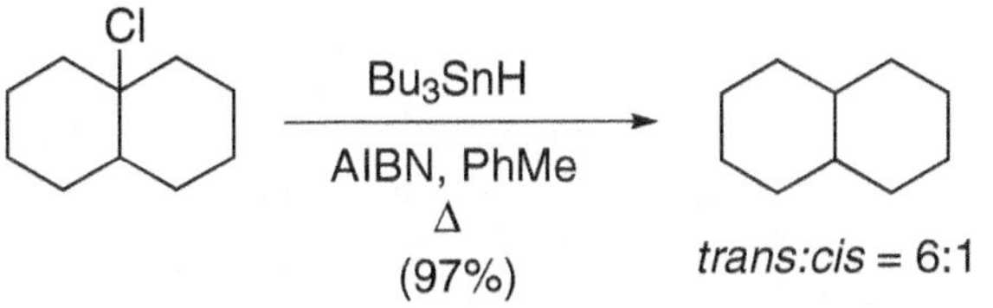

(14.59)[35]

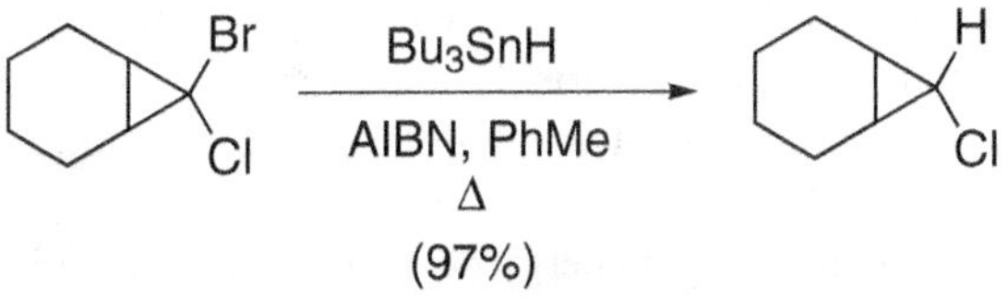

(14.60)[36]

Stannanes not only cleave C—X σ bonds, they also readily add to C=S π bonds. This reaction forms the basis for the reduction of xanthate esters or other thionoesters with tri-*n*-butylstannane, which has become an important method for the deoxygenation of alcohols[37] known as the ***Barton-McCombie reaction***. This use of stannanes to reduce alcohols is illustrated by Examples **14.61** and **14.62**.

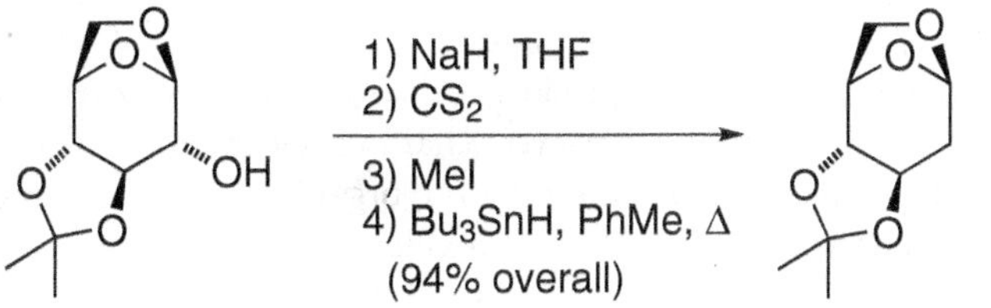

(14.61)

34. (a) Kuivila, H.G. *Acc. Chem. Res.* **1963**, *1*, 1968; *Synthesis* **1970**, 499.
35. Greene, F.D.; Lowry, N.N. *J. Org. Chem.* **1967**, 32, 882.
36. Seyferth, D.; Yamazaki, H.; Alleston, D.L. *J. Org. Chem.* **1963**, *28*, 703.
37. (a) Barton, D.H.R.; McCombie, S.W. *J. Chem. Soc. Perkin Trans. 1* **1975**, 1574. (b) Hartwig, W. *Tetrahedron* **1981**, *39*, 2609. (c) Robins, M.J.; Wilson, J.S. *J. Am.Chem. Soc.* **1981**, *103*, 933. (d) Robins, M.J.; Wilson, J.S.; Hansske, F. *J. Am.Chem. Soc.* **1983**, *105*, 4059. (e) RajanBabu, T.V. *J. Am.Chem. Soc.* **1987**, *109*, 609; *J. Org. Chem.* **1988**, *53*, 4522. (f) Prisbe, L.J.; Martin, J.C. *Synth. Commun.* **1985**, *15*, 401. (g) Barton, D.H.R.; Jaszberenyi, J.C. *Tetrahedron Lett.* **1989**, *30*, 2619.

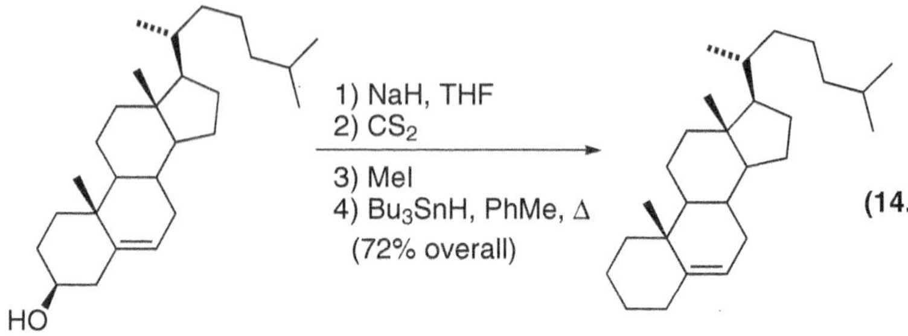

(14.62)

The reaction has been modified (Example **14.63**) for the deoxygenation of tertiary alcohols by means of a mixed oxalate ester with a Barton ester at one end of the molecule.[38] In this case, the tin reagent is not necessary for reduction, but a thiol is used instead.

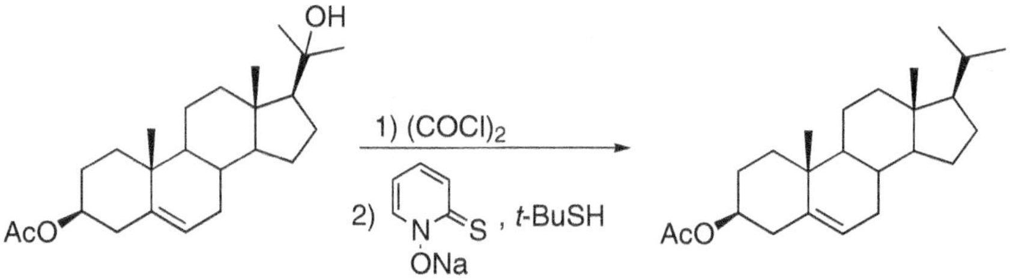

(14.63)

Addition

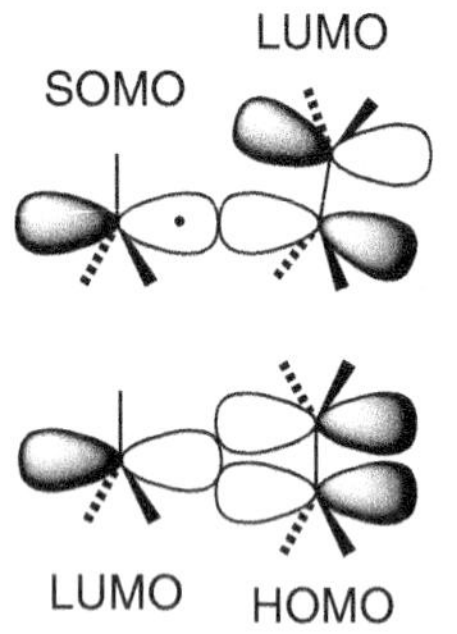

Figure 14.4 Frontier orbital overlap in addition of radicals (top) and cations (bottom) to alkenes. HOMO, highest energy occupied molecular orbital; LUMO, lowest energy unoccupied molecular orbital; SOMO, singly occupied molecular orbital.

In general, free radical additions to alkenes are favorable because the bond formed, which is a σ bond, is lower energy than the bond broken, which is a π bond. The addition of free radicals to alkenes is initiated by the overlap of the radical singly occupied molecular orbital (SOMO; 2p orbital) and the alkene lowest energy unoccupied molecular orbital (LUMO; the π* orbital). Because of the location of the nodal plane in the π* orbital perpendicular to the center of the σ bond, this frontier orbital overlap has one very important geometric consequence. The frontier overlap for the addition of a free radical to an alkene occurs at *one end* of the π bond and *not in the middle*, as occurs when carbocations add to alkenes, a reaction that involves the highest energy occupied molecular orbital (HOMO; π) of the alkene and the LUMO (2p) of the cation (Figure 14.4). This has important regiochemical consequences, especially in reactions such as radical cyclization.

A free radical addition, that was not recognized as such, puzzled organic chemists for decades until the solution was published by American chemist Morris Kharasch[39] in 1933.[40] The problem was a deceptively simple one: under certain circumstances, the addition of hydrogen bromide to alkenes proceeded according to Markovnikov's rule, whereas under other circumstances, it gave the anti-Markovnikov product. What was even more frustrating, workers often had difficulty reproducing their own results—even when using chemicals from the same bottles! In 1937, however, Kharasch succeeded in identifying the culprit for the lack of reproducibility in this reaction: organic peroxides formed by the reaction between the alkene and oxygen. These peroxides served as free radical initiators

---

38. Barton, D.H.R.; Crich, D. *J. Chem. Soc., Chem. Commun.* **1984**, 774.

39. **Morris Selig Kharasch (1895–1957)** was born in Ukraine, and migrated to the United States in 1908. He was educated at the University of Chicago (BS, 1917; PhD, 1919), where he was appointed to the faculty after graduation; he remained there for the rest of his life except for a hiatus at the University of Maryland from 1924–1928. Just prior to his death in 1957, the university created an Institute of Organic Chemistry, with Kharasch as its first director. For more biographical details, see: Westheimer, F.H. *Biogr. Mem. Natl. Acad. Sci.* **1960**, 123.

40. (a) Kharasch, M.S.; Mayo, F.R. *J. Am. Chem. Soc.* **1933**, *55*, 2468. (b) Kharasch, M.S.; McNab, M.C.; Mayo, F.R. *J. Am. Chem. Soc.* **1933**, *55*, 2531. (c) Kharasch, M.S.; Hinckley, J.A., Jr. *J. Am.Chem. Soc.* **1934**, *56*, 1642. (d) Kharasch, M.S.; Engelmann, H.; Mayo, F.R. *J. Org. Chem.* **1938**, *2*, 288. (e) Hey, D.H.; Waters, W.A. *Chem. Rev.* **1937**, *21*, 169.

in the reaction with hydrogen bromide, thus allowing the reaction to proceed by a radical rather than an ionic pathway (Example **14.64**).

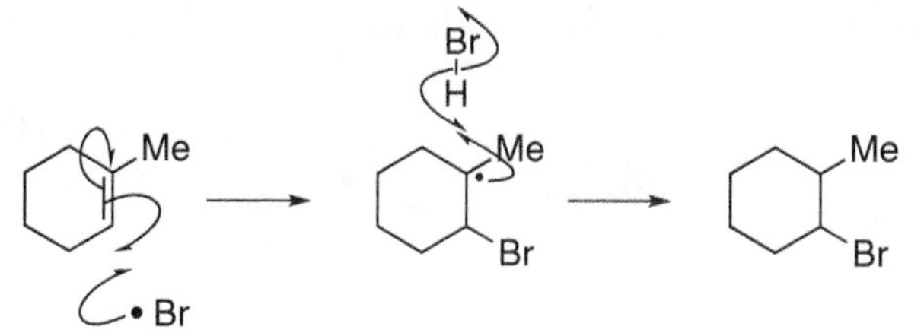

(14.64)

The addition of hydrogen halides to alkenes also illustrates the importance of enthalpy in determining the course of radical reactions. Of the four, only hydrogen bromide reacts by a free radical mechanism under any conditions. The radical addition of a hydrogen halide must occur by a two-step process involving (1) sequential addition of the halogen atom to the alkene π bond and (2) subsequent abstraction of the hydrogen atom from the hydrogen halide. In order for this reaction to occur at a reasonable rate, both propagation steps of the reaction must be exothermic. The values in Table 14.3 show that only for the reaction between hydrogen bromide and an alkene are both steps exothermic.

The free radical copolymerization of maleic anhydride and styrene leads to the formation of an alternating copolymer (Example **14.65**), where the styrene and maleic anhydride units alternate along the backbone. This result illustrates the fact that free radicals can exhibit either nucleophilic or electrophilic propensities. Simple alkyl radicals, for example, are generally nucleophilic in nature and will add more rapidly to electron-poor double bonds. In contrast, radicals carrying an electron-withdrawing (e.g., carbonyl or nitro) group directly bonded to the radical center tend to be electrophilic and add preferentially to electron-rich double bonds.

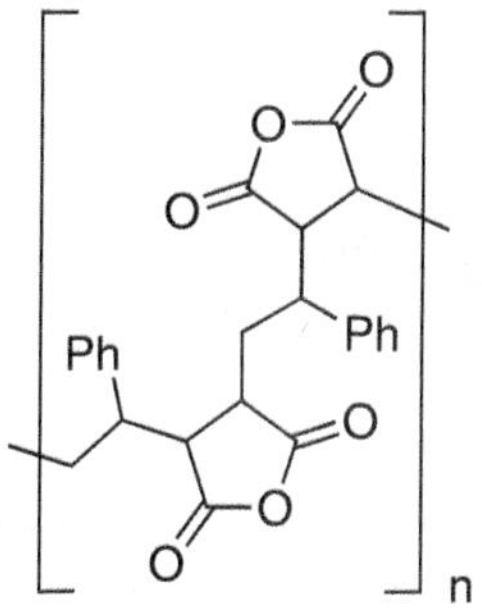

(14.65)

Table 14.3 Reaction Enthalpies for Free Radical Addition of HX to Alkenes*

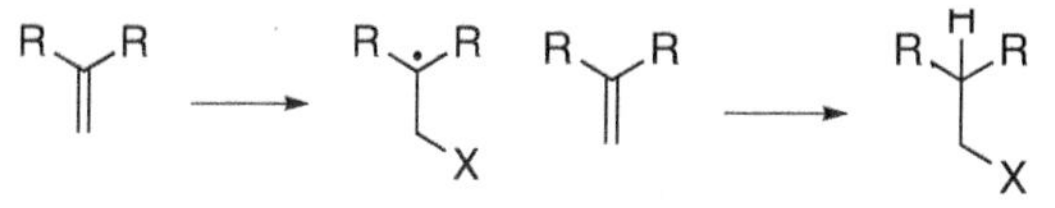

HX		
HF	−46	+36
HCl	−17	+4
HBr	−3	−11
HI	+12	−27

*Values adapted from Reference 41.

Problems

14-1 Draw a reaction cycle diagram (a "carousel" diagram) identifying the initiation step and the propagation steps of each of the following reactions.

(a) 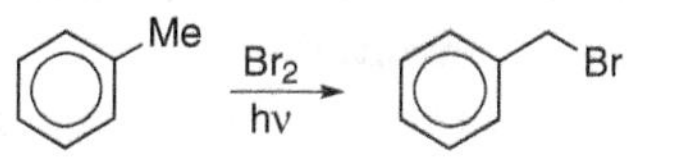

(b) 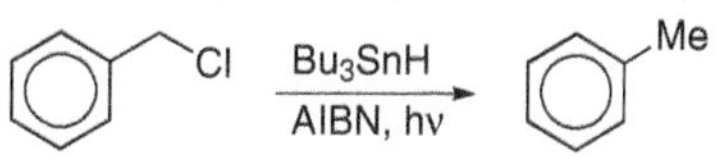

(c) 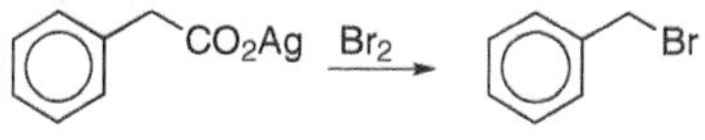

(d) 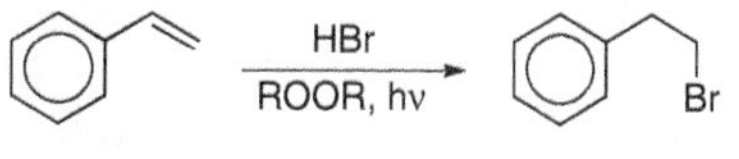

(e) 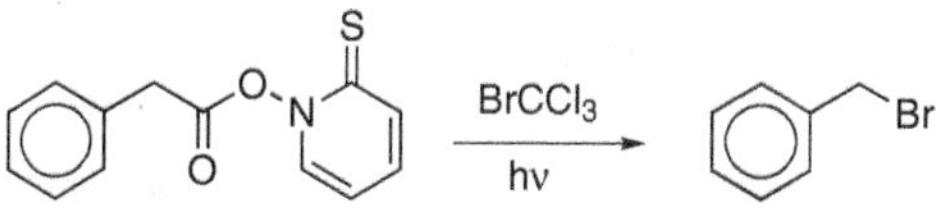

(continues)

41. Isenberg, N.; Grdinic, M. *J. Chem. Educ.* **1969**, *46*, 601.

(**Problems** continued)

14-2 What will be the final alkyl (carbon-centered) free radical formed when each of the following compounds is heated or photolyzed?

(a) 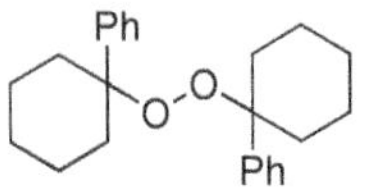

(b) 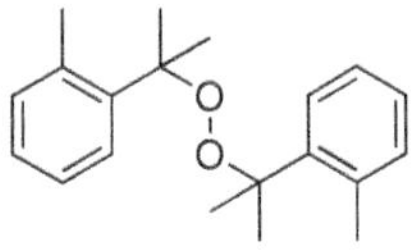

(c) 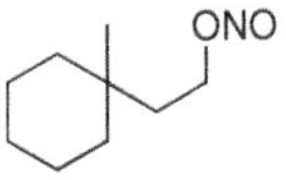

(d) 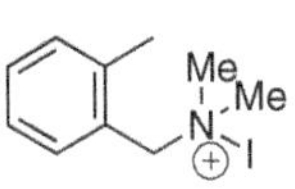

(e) 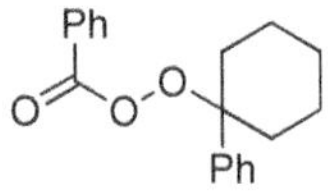

Reaction Synopses

Fragmentation
Alkoxy Radicals

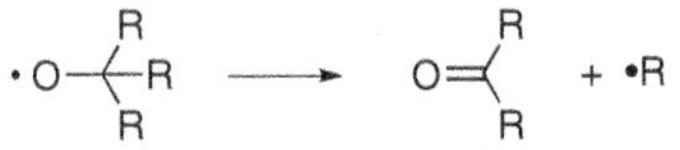

Reagents: $(t\text{-BuO})_2$, etc.

Carboxy Radicals

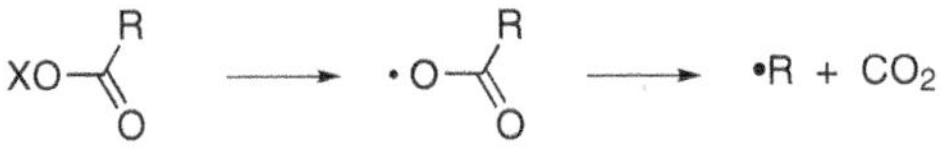

X = Br, OCOR, OR, etc.

Reagents: RCO_2OR'; RCO_2Ag, Br_2; $(RCO_2)_2$; etc.

Addition of Radicals to Alkenes

Regiochemistry: Markovnikov

14.3 Reactions of Free Radicals Useful for Synthesis[42]

Regardless of how they are generated, free radicals always tend to react by the most exothermic pathway available, so one can predict the probable outcome of a radical reaction based on the bond dissociation energies of the bonds involved. Synthetically important free radical addition reactions of alkenes to form new carbon-carbon bonds are relatively uncommon as yet, although the fact that many functional groups do not react with free radicals has provided an incentive for their development because there is often no need to protect sensitive functional groups during radical reactions.

---

42. Monographs: (a) Giese, B. *Radicals in Organic Synthesis: Formation of Carbon-Carbon Bonds* (Pergamon: Oxford, 1987). (b) Renaud, P.; Sibi, M. (Eds.) *Radicals in Organic Synthesis* (VCH-Wiley: Weinheim, 2001). (c) Zard, S.Z. *Radical Reactions in Organic Synthesis* (Oxford University Press: Oxford, 2003). (d) Togo, H. *Advanced Free Radical Reactions for Organic Synthesis* (Elsevier: Amsterdam, 2004).

Radical Decarboxylations: Substitution of a Carboxyl Group

The oxidative decarboxylation of carboxylic acids through carboxy radicals has become a popular method for the generation of free radicals under a wide variety of conditions. As we discussed in the previous section, there are three common precursors to carboxy radicals: acyl peroxides and peresters, acyl hypohalites, and Barton esters. In their reactions, all these precursors generate the carboxy free radical, which then rapidly decarboxylates to give the alkyl free radical.

Barton Modification of the Borodin-Hunsdiecker Reaction

One of the more useful modifications of the Borodin-Hunsdiecker reaction to be developed in recent years is the Barton modification, where the carboxy radical is obtained by homolysis of a Barton ester (Example **14.66**). This modification has far-reaching environmental implications. By removing the need for silver or mercury compounds, it obviates the necessity for disposal of these heavy metals in an environmentally sound manner. The typical procedure simply requires heating or photolyzing the Barton ester in the presence of a carbon tetrahalide (carbon tetrachloride to obtain chlorides, bromotrichloromethane to obtain bromides) or iodoform (to obtain an iodide). The alkyl radical (Example **14.67**) then abstracts the halogen with the weakest C—X bond to give the product halide (Example **14.68**).[43]

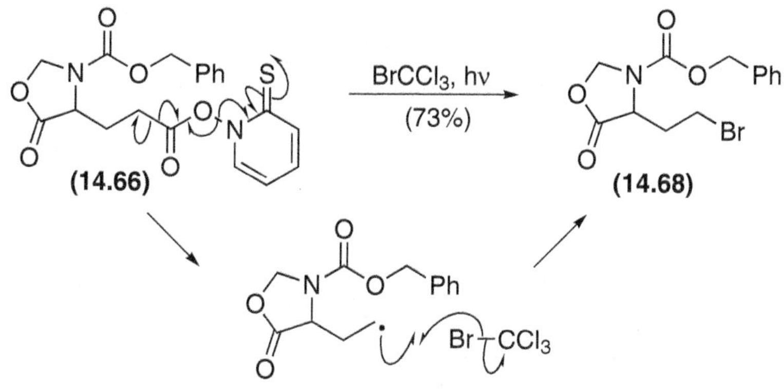

(14.66) **(14.68)** **(14.67)**

Another modification of the Borodin-Hunsdiecker reaction, also developed by Barton, yields alkyl iodides as the major product. This reaction involves irradiating a heated solution of the carboxylic acid, iodine and lead tetraacetate (Example **14.69**). The active intermediate in this reaction is the acyl hypoiodite. An alternative reagent, *tert*-butyl hypoiodite (formed from potassium *tert*-butoxide and iodine; used in Example **14.70**), was superior for the Borodin-Hunsdiecker decarboxylation of hindered and bridgehead carboxylic acids.

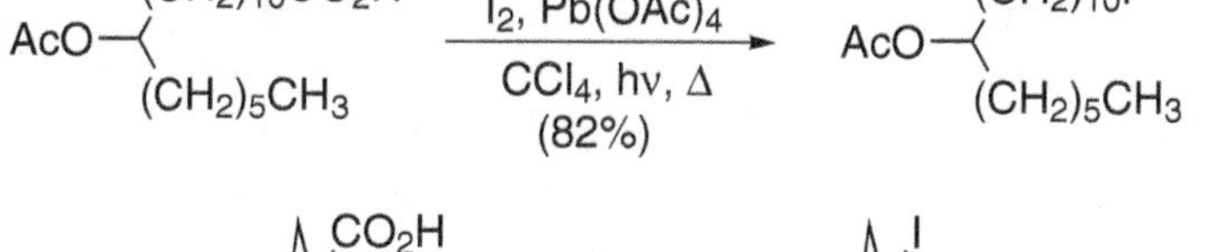

(14.69)[44]

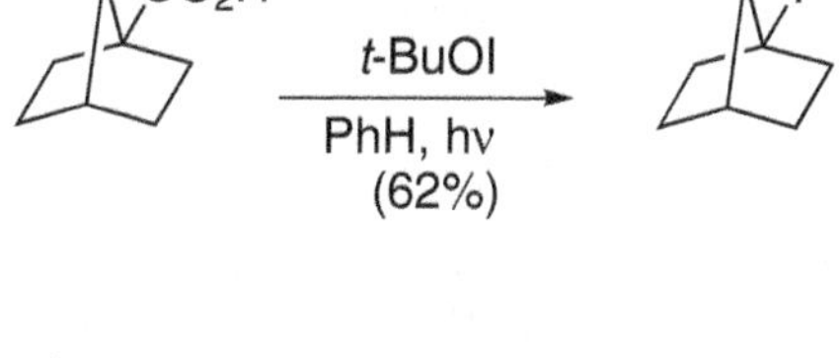

(14.70)[45]

---

43. (a) Barton, D.H.R.; Crich, D.; Motherwell, W.B. *Tetrahedron* **1985**, *41*, 3901. (b) Barton, D.H.R.; Lacher, B.; Zard, S.Z. *Tetrahedron Lett.* **1985**, *26*, 5939.

44. (a) Barton, D.H.R.; Faro, H.P.; Serebryakov, E.P.; Woolsey, N.F. *J. Chem. Soc.* **1965**, 2438. (b) Kochi, J.K. *J. Am. Chem. Soc.* **1965**, *87*, 2500; *J. Org. Chem.* **1965**, *30*, 3265. (c) Gunstone, F.D. *Adv. Org. Chem.* **1960**, *1*, 117.

45. Barton, D.H.R.; Faro, H.P.; Serebryakov, E.P.; Woolsey, N.F. *J. Chem. Soc.* **1964**, 2438.

The same combination of reagents (lead tetraacetate and iodine) can also be used to form other compounds with iodine-oxygen bonds, as well as compounds with iodine-nitrogen bonds. Such compounds are useful precursors to other oxygen-centered and nitrogen-centered free radicals.

Intramolecular Hydrogen Atom Transfer

Hydrogen atom transfer can serve as a method for the oxidation of alkanes, as we discussed in Chapter 11, and now let us discuss hydrogen atom transfer as a method for incorporating functional groups into unactivated (saturated hydrocarbon) locations in organic molecules. This involves the generation of a highly reactive radical in a position that allows the subsequent intramolecular transfer of a hydrogen atom to give a carbon-centered radical at the remote site. The typical reactive radicals in these reactions are electron-deficient and centered on nitrogen or oxygen, and the hydrogen atom transfer is driven by the formation of the stronger heteroatom-hydrogen bond. The atom transfer itself almost always occurs through a six-membered, cyclic transition state.

Hofmann-Löffler-Freytag Reaction and Related Reactions

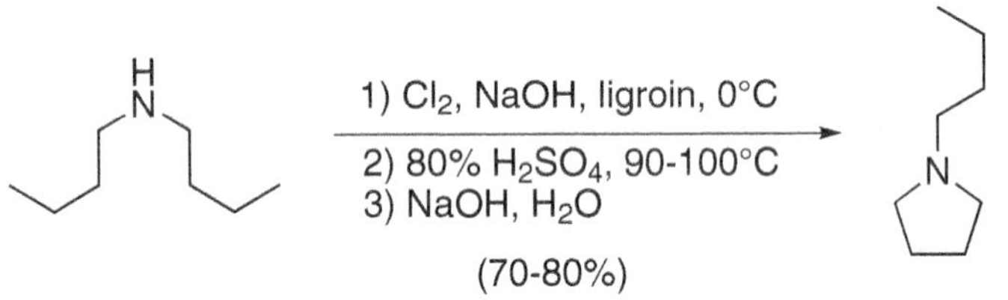

(14.71)[46]

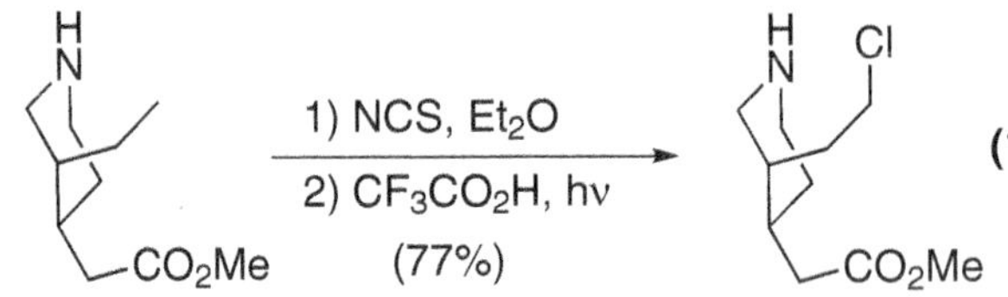

(14.72)[47]

In the Hofmann-Löffler-Freytag reaction,[48] the electron-deficient atom is the nitrogen of the aminium cation radical, formed by homolysis of the conjugate acid of an *N*-haloamine (e.g., Examples **14.71** and **14.72**). Because these amines are generally weak bases, the reaction requires strongly acidic conditions. A simple modification that allows the reaction to occur under much milder conditions involves the homolysis of the *N*-haloamine by transition metals (e.g., Fe^{II} or Cu^{I}). A similar reaction can be carried out by means of an amide radical, formed by homolysis of an *N*-haloamide, as illustrated in Figure 14.5. In this reaction, the *N*-haloamide (**14.73**) undergoes homolysis to give the amide radical (**14.75**) that then undergoes intramolecular hydrogen atom transfer to give the alkyl radical **14.76**. Trapping of this radical by transfer of the halogen atom from the starting haloamide completes the radical chain reaction, yielding the haloamide **14.74**. The advantage of this reaction is that it can be carried out under neutral conditions to give the alkyl halide. What is interesting about the haloamides that are formed in this reaction is that they give lactones (e.g., **14.78**), rather than lactams, on treatment with base.[49]

---

46. Coleman, G.H.; Nichols, G.; Martens, T.F. *Org. Syn., Coll. Vol. 3* **1955**, 159.

47. Uskokovic, M.R.; Henderson, T.; Reese, C.; Lee, H.L.; Grethe, G.; Gutzwiller, J. *J. Am. Chem. Soc.* **1978**, *100*, 571.

48. (a) Hofmann, A.W. *Ber. Dtsch. Chem. Ges.* **1883**, *16*, 558. (b) Löffler, K.; Freytag, C. *Ber. Dtsch. Chem. Ges.* **1909**, *42*, 3427. Reviews: (c) Wolff, M.E. *Chem. Rev.* **1963**, *63*, 55. (d) Neale, R.S. *Synthesis* **1971**, 1. (e) Minisci, F. *Synthesis* **1973**, 1. (f) Mackiewicz, P.; Furstoss, R. *Tetrahedron* **1978**, *34*, 3241. (g) Majetich, G. *Tetrahedron* **1995**, *51*, 7095.

49. (a) Barton, D.H.R.; Beckwith, A.L.J.; Goosen, A. *J. Chem.* Soc., **1965**, 181. (b) Petterson, R.C.; Wambsgans, A. *J. Am. Chem. Soc.* **1964**, *86*, 1648. (c) Beckwith, A.L.J.; Goodrich, J.E. *Aust. J. Chem.* **1965**, *18*, 747. (d) Neale, R.S. *Synthesis* **1971**, 1.

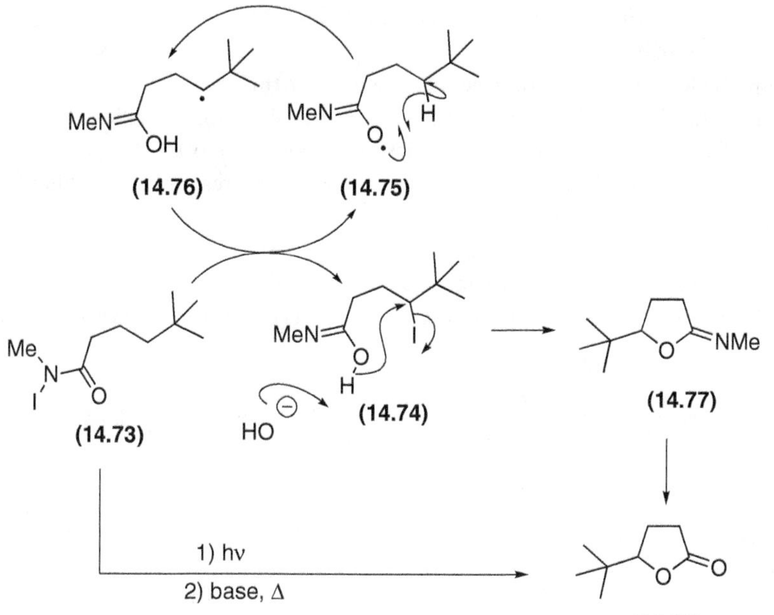

(14.76) (14.75) (14.73) (14.74) (14.77) (14.78)

Figure 14.5 Synthesis of δ-haloamides and δ-lactones from *N*-haloamides

Radical Coupling

Free radicals can be generated by oxidation or reduction under conditions where the radicals couple to form new products. We have already examined a number of such reactions: oxidative coupling of phenols, the pinacol reaction, and the McMurry reaction are three of the major reactions of this type.

Despite its longevity as a method for forming carbon-carbon bonds, the stereochemistry of the pinacol reaction was not subjected to systematic investigation until the 1980s, when McMurry carried out a series of studies of the titanium-promoted pinacol coupling of dialdehydes (Example **14.79**).[50] This work revealed that (1) the amount of *trans* (racemic) diol (Example **14.83**) increased with the size of the ring formed and (2) the formation of the six-membered ring led only to the *cis* (*meso*) diol. As the ring size becomes large enough to permit levels of flexibility similar to open-chain compounds, the ratio of the two stereoisomers gradually stabilizes near 30% *meso*: 70% *dl*. The inclusion of double bonds in the ring substantially increases the amount of the *dl* isomer. In intermolecular reactions, the stereochemistry is not so well controlled, and is often dependent on the exact metal used.[51]

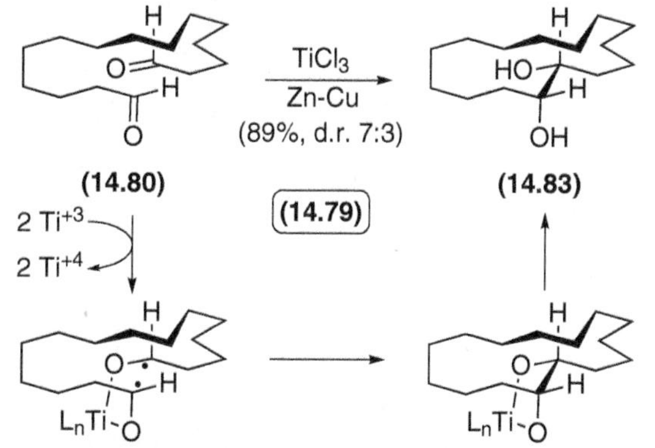

(14.80) (14.79) (14.83) (14.80) (14.83)

50. McMurry, J.E.; Rico, J.G. *Tetrahedron Lett.* **1989**, *30*, 1169.
51. (a) Mundy, B.P.; Srinivasa, R.; Kim, Y.; Dolph, T.; Warnet, R.J. *J. Org. Chem.* **1982**, *47*, 1657. (b) Mundy, B.P.; Bruss, D.R.; Kim, Y.; Larsen, R.D.; Warnet, R.J. *Tetrahedron Lett.* **1985**, *26*, 3927.

Samarium (II) Iodide: A Versatile Reagent for Radical Formation in Synthesis[52]

Samarium (II) iodide was introduced as a reagent for the one-electron reduction of carbonyl groups and halides by Kagan.[53] The reagent has since found widespread use for the formation of carbon-carbon bonds in reactions analogous to the Grignard and Reformatskii reactions as well as in pinacol-type reactions. The reagent is capable of adding a single electron to C—X σ bonds in alkyl halides, sulfonates, and sulfones—and to C—O π bonds. In its reactions with alkyl halides (Example **14.84**), an intermediate alkylsamarium diiodide is formed as an intermediate. This organometallic intermediate permits samarium (II) iodide to mediate Reformatskii- and Barbier-type additions to carbonyl acceptors.

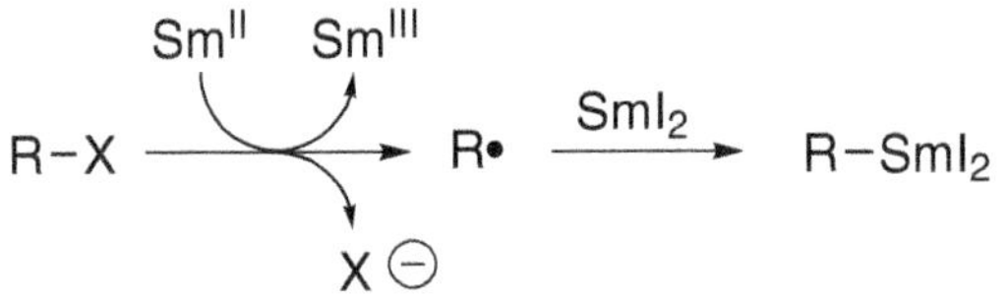

(14.84)

It is currently accepted (Figure 14.6) that the samarium (II)-mediated pinacol reaction takes place by an initial one-electron reduction of one of the carbonyl partners by samarium (II) to give a samarium (III) complex of the ketyl (**14.85**). This samarium (III) complex contains a fairly hard Lewis acid that can complex the carbonyl oxygen of the second reactant (**14.86**). Radical addition to the carbonyl π bond then gives a highly reactive oxygen-centered radical (**14.87**) that abstracts a hydrogen atom from the solvent to give the samarium (III) chelate; the cyclization and hydrogen abstraction steps may operate in concert with each other. This intramolecular coupling reaction has a further consequence. When the chelate is formed, it does so such that the two larger groups attached to the carbonyl carbons are *trans* to each other in the five-membered ring, leading to preferential formation of the *dl* pinacol (**14.88**) in an intermolecular reaction, and preferential formation of the *cis* diol in the intramolecular case.

Figure 14.6 Proposed mechanism of the SmI$_2$-promoted pinacol reaction that accounts for the observed stereochemistry of the reaction

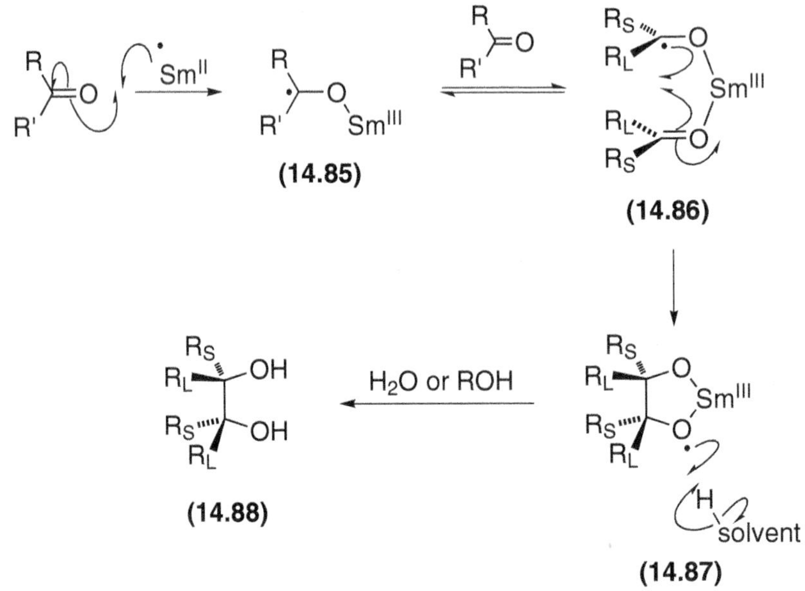

52. For reviews of Sm(II) in modern synthesis, see: (a) Molander, G.A. *Chem. Rev.* **1992**, *92*, 29; *Org. React.* **1994**, *436*, 211. (b) Molander, G.A.; Harris, C.R. *Chem. Rev.* **1996**, *96*, 307. (d) Molander, G.A. *Acc. Chem. Res.* **1998**, *31*, 603. (c) Edmonds, D.J.; Johnston, D.; Procter, D.J. *Chem. Rev.* **2004**, *104*, 3371.

53. (a) Namy, J.L.; Girard,P.; Kagan, H.B. *Nouv. J. Chim.* **1977**, *1*, 5. (b) Girard, P.; Namy, J. L.; Kagan, H. B. *J. Am. Chem. Soc.* **1980**, *102*, 2693.

Problem

14-3 The cyclization of γ-iodoamides with aqueous potassium hydroxide gives the γ-lactone rather than the γ-lactam. Write a reasonable mechanism that accounts for the formation of the product and give reasons why the lactam is not formed.

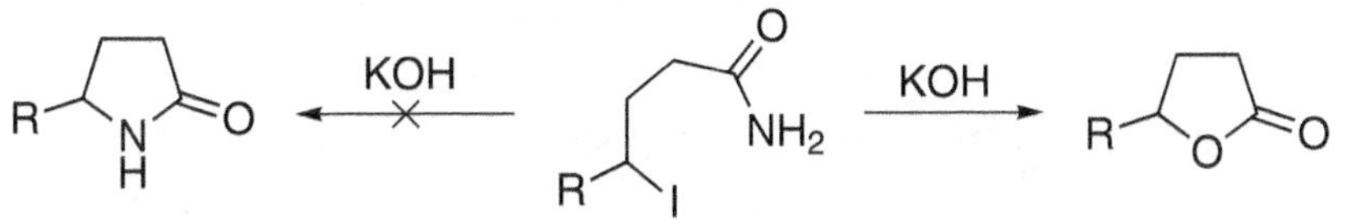

Phenol Coupling[54]

Phenols and phenoxide anions, especially, are susceptible to single electron oxidation by a wide range of reagents to give phenoxy radicals (or the related radical cations) that often undergo radical coupling. Aromatic amines undergo similar reactions to give cation radicals that then undergo analogous radical coupling reactions. The prototypical example of such reactions is the ***phenol coupling reaction***.

The phenoxy radical is resonance-stabilized, so the unpaired electron density is spread over several atoms of the radical, which allows the formation of a wide range of products, depending on the regiochemistry of the coupling reaction, as shown in Figure 14.7. The oxidation of a phenoxide anion gives a strongly stabilized free radical that adds to electron-rich π bonds, such as those in phenoxide anions, to form new carbon-carbon bonds. This reaction has been portrayed historically as the coupling of phenoxy radicals, but it actually involves the intramolecular addition of a radical to a phenoxide anion, followed by rapid oxidation of the adduct to give the observed product.[55] Among the most popular oxidizing agents are iron (III) compounds, lead (IV) compounds, vanadium (V) compounds, and manganese (III) compounds.

The examples in Figure 14.7 illustrate the variety of compounds that can be prepared by this reaction. It is no accident that several of the examples below are based on a 1-benzyl-1,2,3,4-tetrahydroisoquinoline ring system. This occurs because there are several important families of natural products (among them the morphine alkaloids) whose carbon skeletons are based on the phenol coupling products that can be obtained from this ring system. Indeed, there is good evidence that this reaction is the one that the opium poppy itself uses to construct the polycyclic ring systems of these compounds.[56]

Another important example of this coupling reaction is the coupling of 2-naphthol to give the dimer shown (Example **14.89**), often known as BINOL. This molecule is an important C_2-symmetric ligand for a wide range of asymmetric reactions and the starting

54. (a) Bacon, R.G.R.; Hill, H.A.O. *Quart. Rev.* **1965**, *19*, 95. (b) Scott, A.I. *Quart. Rev.* **1965**, *19*, 1. (c) Taimir, L.; Popisil, J. *Tetrahedron Lett.* **1971**, 2809. (d) Mahoney, L.R.; Weiner, S.A. *J. Am. Chem. Soc.* **1972**, *94*, 585. (e) Mahoney, L.R.; Weiner, S.A. *J. Am. Chem. Soc.* **1972**, *94*, 1412. (f) Anderson, R.A.; Dagleish, D.T.; Nonhebel, D.C.; Paulson, P.L. *J. Chem. Res. (S)* **1977**, 12. (g) Iqbal, J.; Bhatia, B.; Nayyar, N.K. *Chem. Rev.* **1994**, *94*, 519.

Monographs: (h) Battersby, A.R.; Taylor, W.I, Eds. *Oxidative Coupling of Phenols* (Dekker: New York, 1967). (i) Dhingra, O.P. In Trahanovsky, W.S., Ed. *Oxidations in Organic Chemistry* (Academic: New York, 1982), Part D, Ch. 4.

55. McDonald, P.D.; Hamilton, G.A. *J. Am. Chem. Soc.* **1973**, *95*, 7752.

56. (a) Kupchan, S.M.; Kim, K.-C.; Lynn, J.T. *J. Chem. Soc., Chem. Commun.*, **1976**, 86. (b) Gervay, G.E.; McCapra, F.; Money, T.; Sharma, G.M.; Scott, A.I. *Chem. Commun.*, **1966**, 142. (c) Barton, D.H.R.; James, R.; Kirby, G.W.; Turner, D.W.; Widdowson, D.A. *J. Chem. Soc. (C)*, **1968**, 1529. (d) Review: Herbert, R.B. *The Biosynthesis of Secondary Metabolites, 2nd ed.*; (Chapman & Hall: London, 1989); p. 146.

Figure 14.7 Regioisomeric products that can be formed by free radical coupling of phenols

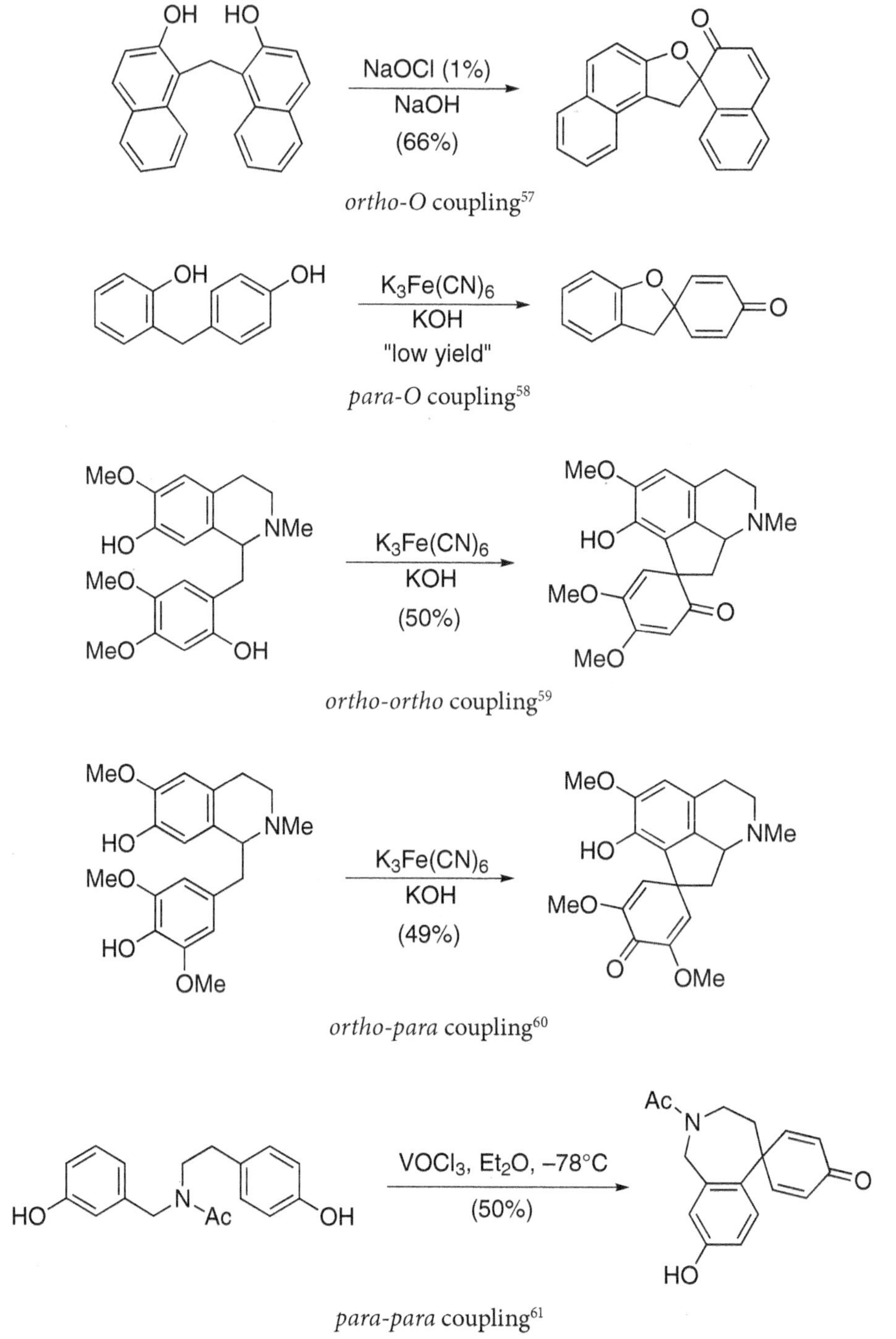

57. Shearing, E.A.; Smiles, S. *J. Chem. Soc.* **1937**, 1931.

58. (a) Barton, D.H.R.; Chow, Y.L.; Cox, A.; Kirby, G.W. *J. Chem. Soc.* **1965**, 3571. (b) Bacon, R.G.R.; Izatt, A.R. *J. Chem. Soc. (C)* **1970**, 2543. (c) Schulte-Frohlinde, D.; Erhardt, F. *Annalen* **1964**, *671*, 88.

59. (a) Jackson, A.H.; Martin, J.A. *J. Chem. Soc., Chem. Commun.* **1965**, 420. (b) Shamma, M.; Slusarchyk, W.A. *J. Chem. Soc., Chem. Commun.* **1965**, 528. (c) Jackson, A.H.; Martin, J.A. *J. Chem. Soc. C* **1966**, 2222.

60. Battersby, A.R.; Bradbury, R.B.; Herbert, R.B.; Monro, M.H.G.; Ramage, R. *J. Chem. Soc., Perkin Trans. 1* **1974**, 1394.

61. (a) Schwartz, M.A.; Holton, R.A.; Scott, S.W. *J. Am. Chem. Soc.* **1969**, *91*, 2800. (b) Tobinaga, S.; Kotani, E. *J. Am. Chem. Soc.* **1972**, *94*, 309. (c) Schwartz, M.A.; Rose, B.F.; Holton, R.A.; Scott, S.W.; Vishnuvajala, B. *J. Am. Chem. Soc.* **1977**, *99*, 2571.

material for the preparation of many important C_2-symmetric ligands for asymmetric catalytic hydrogenation and other asymmetric synthetic reactions.[62]

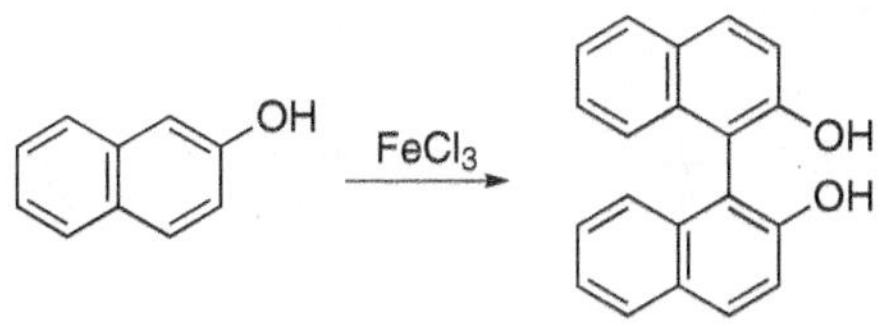

(14.89)

This general method of forming carbon-carbon bonds by coupling of free radicals obtained from easily oxidized anions has been extended to the oxidation of anions of phenols carrying aliphatic nitro groups (Example **14.90**)[63] or strongly enolized β-dicarbonyl groups (Example **14.91**)[64] at the end of a side chain. The range of oxidizing agents that have been applied to this reaction has gradually expanded, and it is now possible to oxidize arenes without a phenolic hydroxyl group. This is accomplished by carrying out the oxidation in strongly acidic media (often trifluoroacetic acid with co-solvents such as dichloromethane), using as the oxidizing agent thallium (III) trifluoroacetate,[65] iron (III) chloride and *m*-chloroperbenzoic acid,[66] vanadium (V) oxyhalides,[67] or lead tetraacetate.[68] A recent addition to this arsenal involves activation of manganese dioxide by trifluoroacetic acid (Example **14.92**).[69]

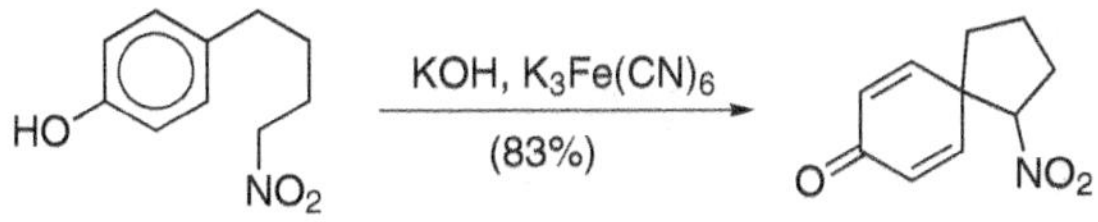

(14.90)

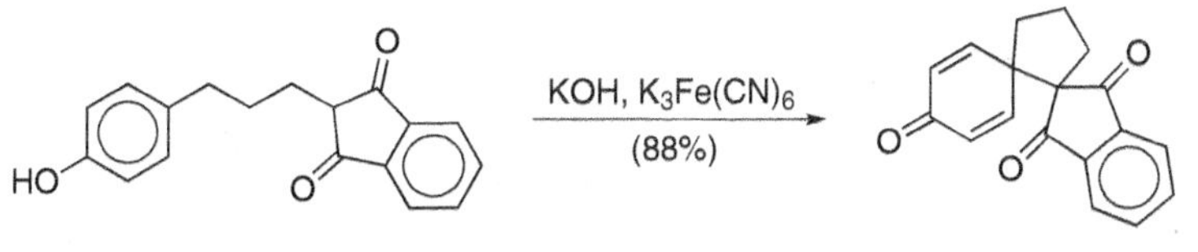

(14.91)

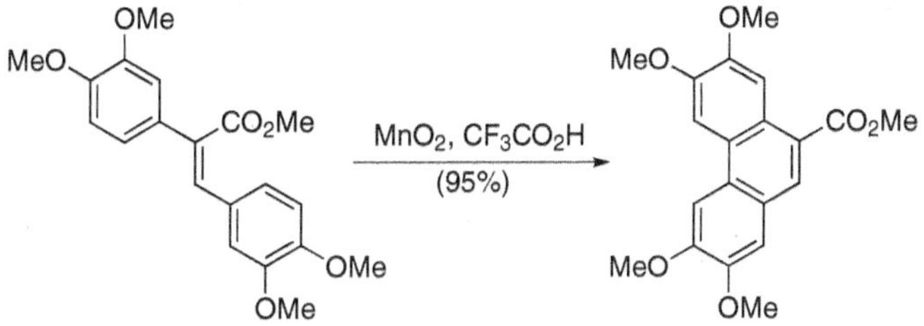

(14.92)

Problem

14-4 When *p*-cresol (4-methylphenol) is treated with potassium ferricyanide in aqueous base, a tricyclic ketone, Pummerer's ketone ($C_{14}H_{14}O_2$), is formed. This ketone contains one alkene π bond and an ether group, as well as a carbonyl group and an aromatic ring. What is its structure and how is it formed?

---

62. Review: Brunel, J.M. *Chem. Rev.* **2005**, *105*, 857.
63. Kende, A.S.; Koch, K. *Tetrahedron Lett.* **1986**, *27*, 6051.
64. Kende, A.S.; Koch, K.; Smith, C.A. *J. Am. Chem. Soc.* **1988**, *110*, 2210.
65. (a) Bringmann, G.; Walter, R.; Weirich, R. *Angew. Chem. Int. Ed. Engl.* **1990**, *29*, 977. (b) Taylor, E.C.; Andrade, J.G.; Rall, G.J.H.; McKillop, A. *J. Am. Chem. Soc.* **1980**, *102*, 6513.
66. Wang, K.L.; Lü, M.Y.; Yu, A.; Zhu, X.Q.; Wang, Q.M.; Huang, R.Q. *J. Org. Chem.* **2009**, *74*, 935.
67. (a) Liepa, A.J.; Summons, R.E. *J. Chem. Soc., Chem. Commun.* **1977**, 826. (b) Halton, B.; Maidment, A.I.; Officer, D.L.; Warner, J.M. *Aust. J. Chem.* **1984**, *37*, 2119. (c) Wang, K.; Wang, Q.; Huang, R. *J. Org. Chem.* **2007**, *72*, 8416.
68. Feldman, K.S.; Ensel, S.M. *J. Am. Chem. Soc.* **1994**, *116*, 3357.
69. Wang, K.; Hu, Y.; Li, Z.; Wu, M.; Liu, Z.; Su, B.; Yu, A.; Liu, Y.; Wang, Q. *Synthesis* **2010**, 1083.

Reaction Synopses

Borodin-Hunsdiecker Reaction

Reagents: Ag_2O, Br_2, Δ; etc. or
(1) $(COCl)_2$; (2) $Na^+ {}^-ONC_5H_4S$; (3) CCl_3Br or CCl_4, Δ; or
$Pb(OAc)_4$, I_2, hν; or
t-BuOI, hν

Hofmann-Löffler-Freytag and Related Reactions

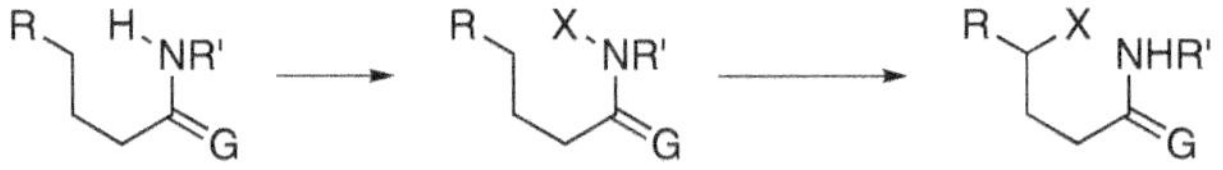

G=H_2, O; X=Cl, Br, I

Reagents:

$G{=}H_2$: (1) Cl_2, NaOH; 2) H_2SO_4, Δ; or
(1) Cl_2, NaOH; 2) CF_3CO_2H, hν; or
(1) NCS, Et_2O; 2) CF_3CO_2H, hν; etc.

G=O: (1) I_2, $Pb(OAc)_4$, CCl_4, Δ or hν; or
t-BuOI, PhH, hν

Pinacol and McMurry Reactions

Stereochemistry: *meso* isomer favored in cyclizations that give small (≤6 members) rings; *dl* isomer is favored from open-chain compounds and in cyclizations that give large rings.

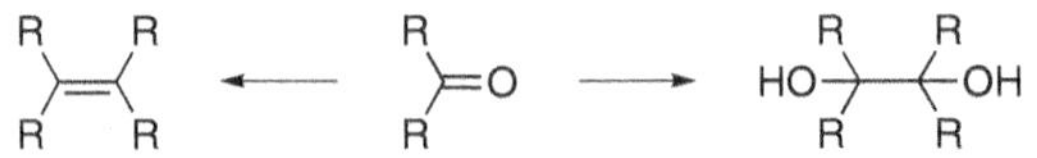

Reagents: SmI_2, THF
Mg(Hg), Et_2O; etc.
$TiCl_3$, Zn-Cu, THF; etc.

Phenol Coupling

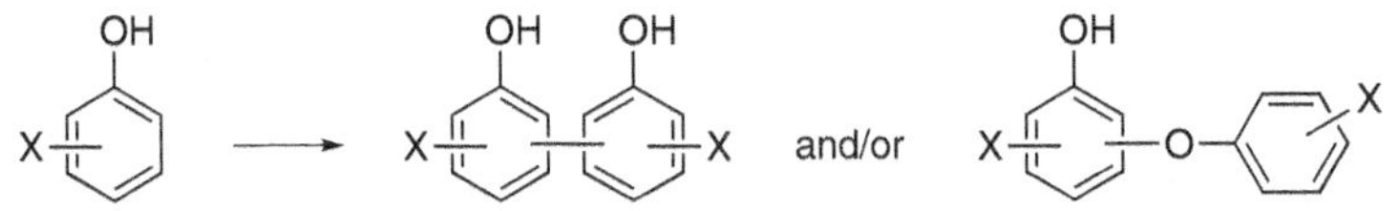

Reagents:

Fe (III): $K_3Fe(CN)_6$, KOH; $FeCl_3$, *m*-CPBA; $Fe(ClO_4)_3$; etc.

V (V): $VOCl_3$, Et_2O; VOF_3, CH_2Cl_2; etc.

Pb (IV): $Pb(OAc)_4$, AcOH; etc.

Mn (IV): MnO_2, CF_3CO_2H; MnO_2, MsOH; etc.

other: NaOCl, NaOH; etc.

14.4 Additions and Cyclizations of Free Radicals

Many of the radical reactions that are widely exploited in organic synthesis are addition reactions to carbon-carbon π bonds. For example, bromoform reacts with alkenes in the presence of a free radical initiator to give a 1,1,1-tribromoalkane with one carbon more than the original alkene, and carbon tetrachloride adds to alkenes under the same conditions to give a similar 1,1,1,2-tetrachloroalkane (e.g., Example **14.95**). Both reactions involve a trihalomethyl radical, which adds to an alkene to give an intermediate alkyl radical (Example **14.94**). This radical then abstracts a halogen or a hydrogen atom from the halogenated reactant. An example of this reaction is the radical addition of carbon tetrachloride to the alkene **14.93**; treatment of the product with two equivalents of base provides the cyclopropanecarboxylic acid part (Example **14.96**) of a synthetic pyrethroid insecticide.[70]

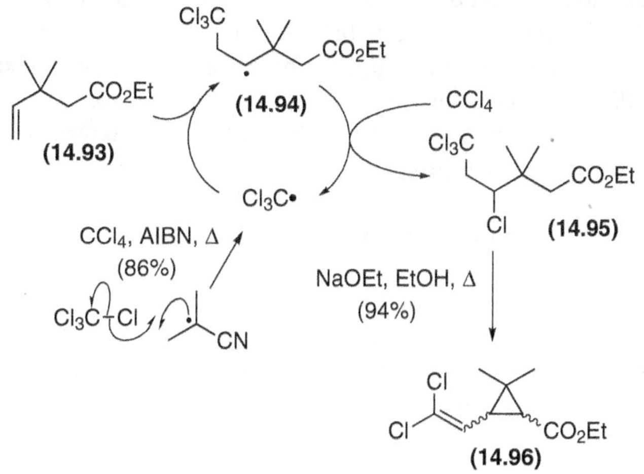

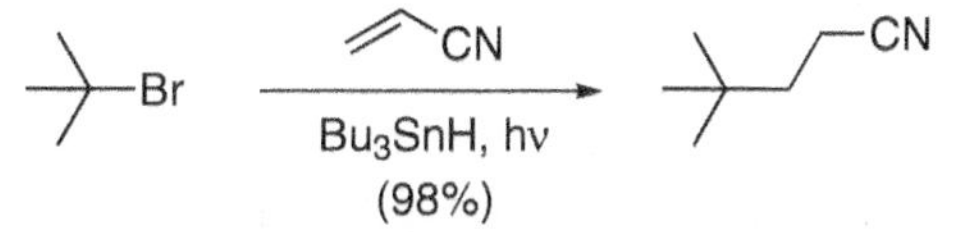

(14.97)

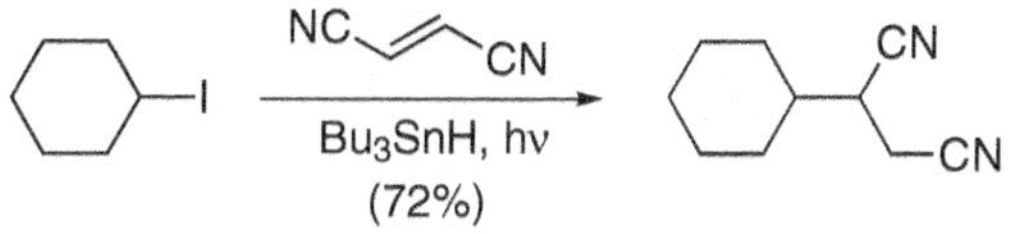

(14.98)

With the advent of newer methods for forming free radicals, it has become possible to form new carbon-carbon bonds by the addition of free radicals to a wide variety of alkenes. The most popular alkenes for use in this reaction tend to be conjugated with an electron-withdrawing group or an aromatic ring, because the free radicals used are almost always alkyl radicals (and, therefore, nucleophilic). Examples of this method for forming carbon-carbon bonds are provided by the reactions in Examples **14.97** to **14.99**.[71]

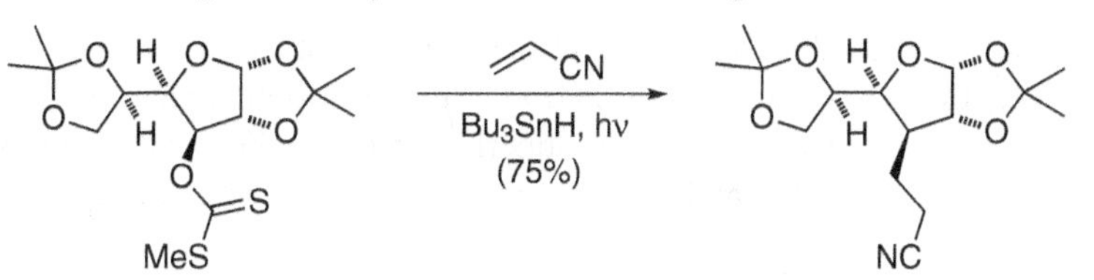

(14.99)

70. (a) Matsui, K.; Negishi, A.; Takahatake, Y.; Sugimoto, K.; Fujimoto, T.; Takashima, T.; Kondo, K. *Bull. Chem. Soc. Japan* **1986**, *59*, 221. (b) Kondo, K.; Matsui, K.; Negishi, A. In Elliott, M, Ed. *Synthetic Pyrethroids, ACS Symp. Ser.* **1977**, *42*, 128. (c) Kondo, K.; Matsui, K. **U.S. Patent No. 4,454,343** [June 12, 1984]. (d) Arlt, D.; Jautelat, M.; Lantzsch, R. *Angew. Chem. Int. Ed. Engl.* **1981**, *20*, 703.

71. Taken from Giese, B.; González-Gómez, J.A.; Witzel, T. *Angew. Chem. Int. Ed. Engl.* **1984**, *23*, 69.

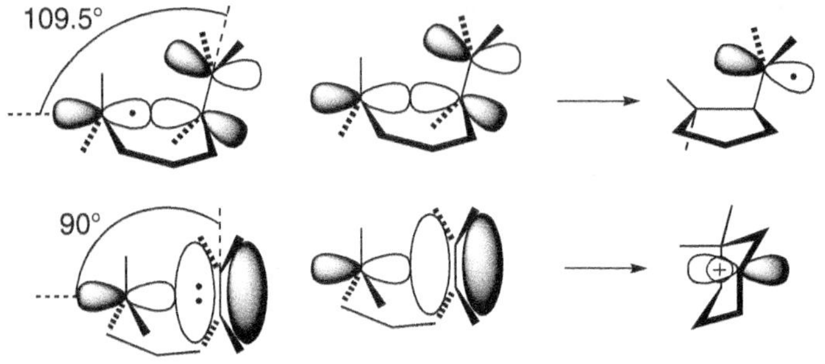

Figure 14.8 The cyclization of a free radical requires a different set of orbitals, and so leads to a different product from carbocation cyclization.

The intramolecular version of this reaction results in cyclization of unsaturated radicals and provides an especially useful method for forming cyclopentane derivatives. The outcome of this reaction is determined by the frontier orbital overlap. It is an example of a reaction where the products formed are determined by a stereoelectronic effect—the geometry of the participating orbitals controls which final product is formed, rather than the relative stabilities of the possible products.

In the addition of a carbocation to an alkene, the cation carbon approaches the center of the π bond from the orthogonal direction, whereas the addition of a radical to an alkene π bond occurs at one end of the bond and from a direction close to 109.5°.[72] This favors cyclopentane geometry by approximately 5 kcal mol^{-1}. Thus, in contrast to the cyclization of the 5-hexenyl cation, which gives the more stable cyclohexyl cation, the cyclization of the 5-hexenyl radical (which is essentially irreversible under normal reaction conditions unless the open-chain radical is very stable) gives the cyclopentylcarbinyl radical (Figure 14.8). When the radical reaction is carried out under reversible conditions the predominant products contain six-membered rings.[73]

Alkenes and alkynes can both serve as acceptor π bonds for the addition reaction, as illustrated by Examples **14.100** and **14.101**. The cyclization of free radicals with appropriate functionality can also be used to prepare heterocycles,[74] as also illustrated by **14.100**.

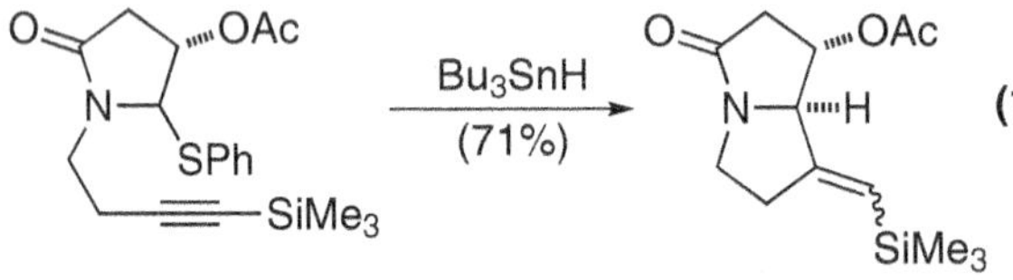

(14.100)[75]

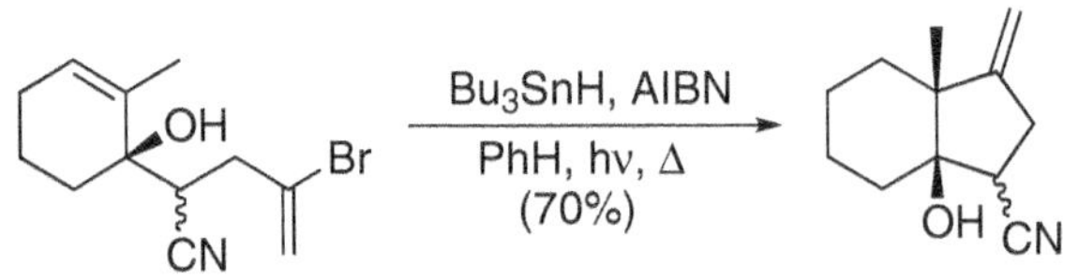

(14.101)[76]

The reaction has also served to form the basis of a tandem polycyclization reaction, as illustrated by the key reaction (Example **14.102**) from the synthesis of the natural product, capnellene.[77] An alternative method for the production of free radicals is shown by the cyclization of the iodoalkyne in Example **14.103**. In this reaction, the free radical reaction

72. Dewar, M.J.S.; Olivella, S. *J. Am. Chem. Soc.* **1978**, *100*, 5290.
73. Julia, M. *Acc. Chem. Res.* **1971**, *4*, 386; *Pure Appl. Chem.* **1974**, *40*, 553.
74. Review: Bowman, W.R.; Bridge, C.F.; Brookes, P. *J. Chem. Soc., Perkin Trans. 1*, **2000**, 1.
75. Choi, J.K.; Hart, D.J. *Tetrahedron* **1985**, *41*, 3959.
76. Stork, G.; Mook, R. *J. Am. Chem. Soc.* **1983**, *105*, 3720.
77. Curran, D.P.; Chen, M.H. *Tetrahedron Lett.* **1985**, *26*, 4991.

is initiated by the reaction between triethylborane and oxygen, with trimethylsilane as the hydrogen atom transfer agent.[78]

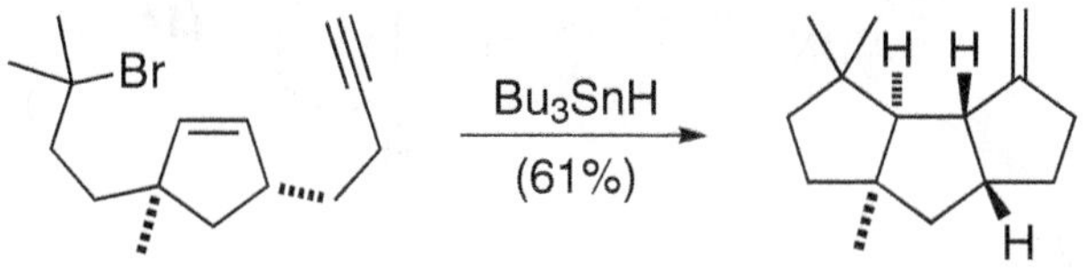

(14.102)

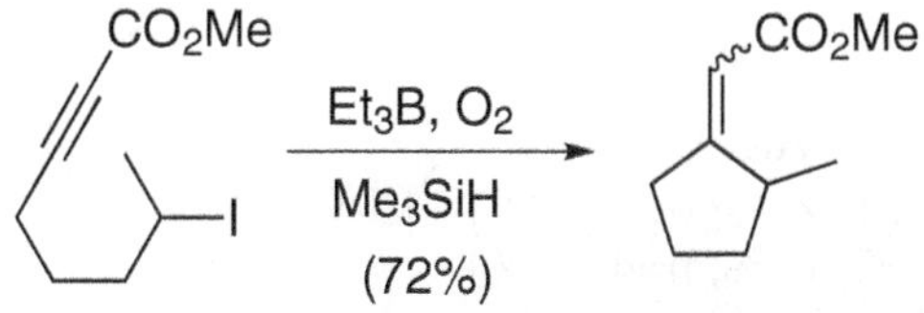

(14.103)

The cyclizations above all involve hydrogen atom transfer to the radical as the final step in the radical chain. However, the transfer of other atoms can also lead to the formation of useful products. For example, when a mixture of a halide and an alkene is heated or irradiated with a stannane reagent under conditions where the stannane is present in catalytic, rather than stoichiometric quantities, the product of the reaction is one in which there has been an effective insertion of the π bond into the carbon-halogen σ bond or addition of the carbon-halogen bond across the π bond. This is illustrated by Examples **14.104** and **14.105**, where the stannane serves as the initiating radical but where the chain is carried on by iodine atom transfer.

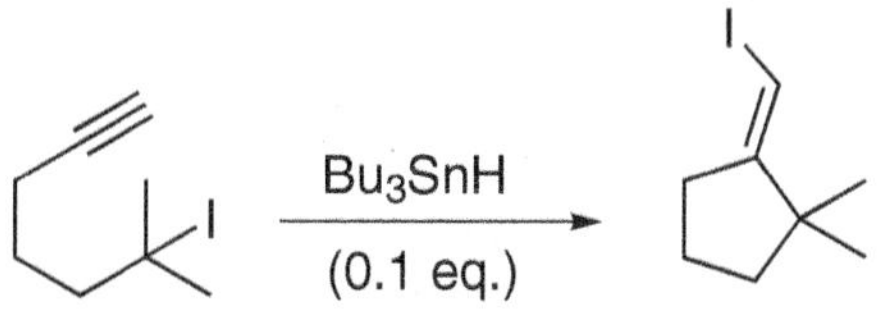

(14.104)[79]

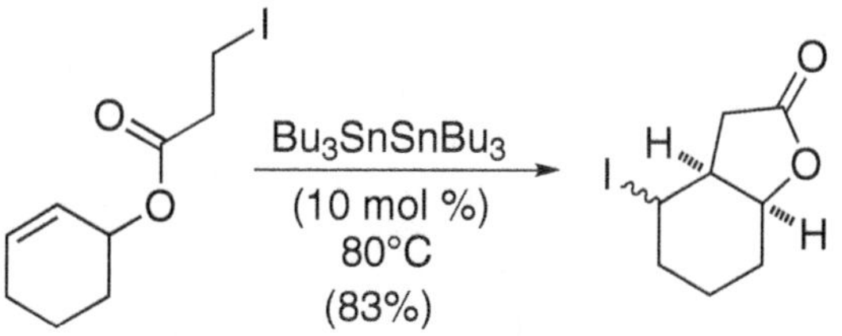

(14.105)[80]

When geometric constraints are removed and the stereoelectronic requirements for ring closure can be met by attack of the radical at either end of the alkene, the addition occurs preferentially to give the more stable of the two possible radicals. This has extended the value of radical cyclizations to the macrocyclization of ω-alkenyl radicals to large-ring cycloalkanes. For example, Pattenden has been able to construct large rings and polycyclic ring systems by radical cyclization pathways, as shown by Examples **14.106** to **14.108**.

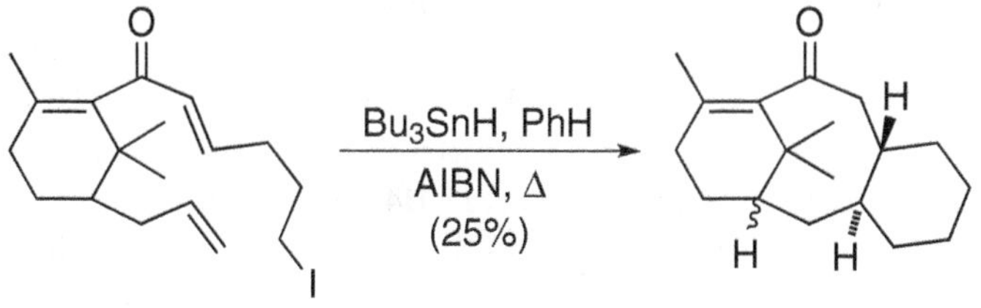

(14.106)[81]

---

78. (a) Lowinger, T.B.; Weiler, L. *J. Org. Chem.* **1992**, *57*, 6099. (b) Evans, P.A.; Roseman, J.D. *J. Org. Chem.* **1996**, *61*, 2252.

79. Curran, D.P.; Chen, M.-H.; Kim, D. *J. Am. Chem. Soc.* **1986**, *108*, 2489; **1989**, *111*, 6256.

80. Curran, D.P.; Tamine, J. *J. Org. Chem.* **1991**, *56*, 2746.

81. Hitchcock, S.A.; Pattenden, G. *Tetrahedron Lett.* **1992**, *33*, 4843.

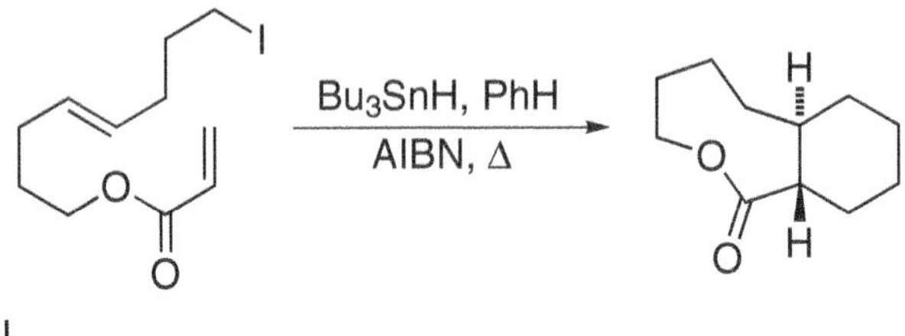

(14.107)[82]

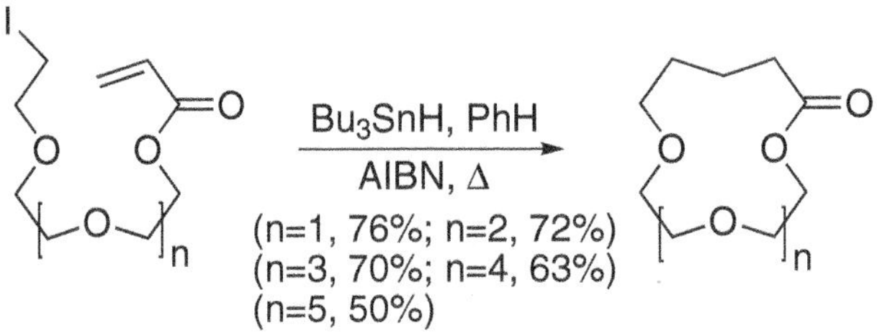

(14.108)[83]

Although tin hydrides have been the most widely used reagents to initiate radical cyclizations, samarium (II) iodide can also be used to promote free radical cyclization of unsaturated carbonyl compounds. One of the most useful characteristics of the SmI_2-promoted ketyl-alkene cyclization is its ability to form cyclooctanols in the presence of hexamethylphosphoramide (e.g., Example **14.109**).[84]

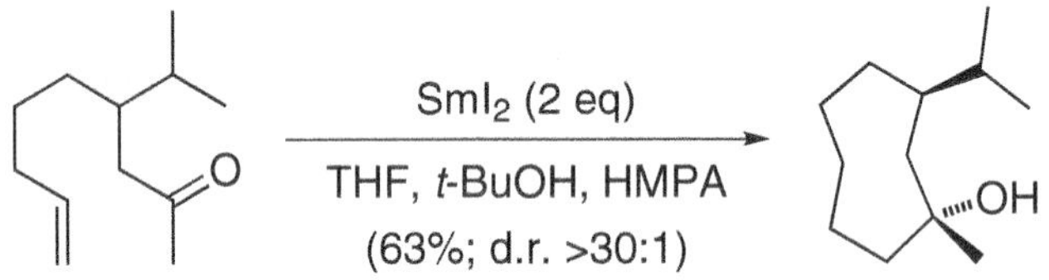

(14.109)

A similar cyclization to the less thermodynamically stable radical is observed with 3-butenyl (homoallyl) free radicals, which rapidly cyclize to the cyclopropylcarbinyl radical (Example **14.110**). Unlike the cyclization of the 5-hexenyl radical to the cyclopentylcarbinyl radical (Example **14.111**), however, the cyclization of the 3-butenyl radical to the cyclopropylcarbinyl radical is reversible and almost always favors the open-chain product. In fact, the ring opening of cyclopropylcarbinyl radicals is widely used as a test for the intervention of free radicals in a reaction mechanism and as a "clock reaction" to measure the rates of fast radical reactions.[85]

(14.110)

(14.111)

The ring opening of cyclopropylcarbinyl radicals allows the radical cyclization reaction to be extended to the synthesis, under kinetic conditions, of compounds containing six-membered rings. Two examples of this type of cyclization are shown as Examples **14.112**[86] and **14.113**. The alkyl free radical required to initiate the intramolecular cyclization reaction can be generated by single electron transfer to a carbonyl group. Two excellent examples of this are provided by the reactions **14.114** and **14.115**, where s.e.t. to the carbonyl group of the ketone gives a radical anion that adds to the alkyne triple bond,

82. Blake, A.J.; Hillingworth, G.J.; Pattenden, G. *Synlett* **1996**, 643.
83. Pattenden, G.; Wiedenau, P. *Tetrahedron Lett.* **1997**, *38*, 3647.
84. Molander, G.A.; McKie, J.A. *J. Org. Chem.* **1994**, *59*, 3186
85. (a) Newcomb, M.; Glenn, A.G. *J. Am. Chem. Soc.* **1989**, *111*, 275. (b) Bowry, V.W.; Lusztyk, J.; Ingold, K.U. *J. Am. Chem. Soc.* **1991**, *113*, 5687. (c) Newcomb, M. *Tetrahedron* **1993**, *49*, 1151. (d) Newcomb, M.; Toy, P.H. *Acc. Chem. Res.* **2000**, *33*, 449.
86. Kilburn, J.D.; Santagostino, M. *Tetrahedron Lett.* **1995**, *36*, 1365.

closing the ring. Nitrogen-centered radicals will also add to alkene π bonds to give cyclic amines, as illustrated by Examples **14.116** and **14.117**, below.

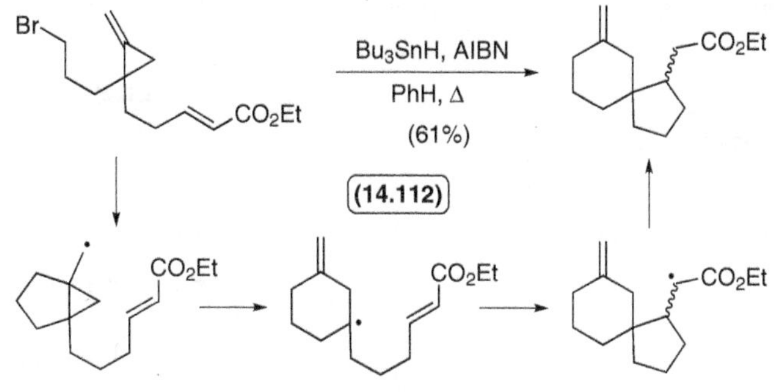

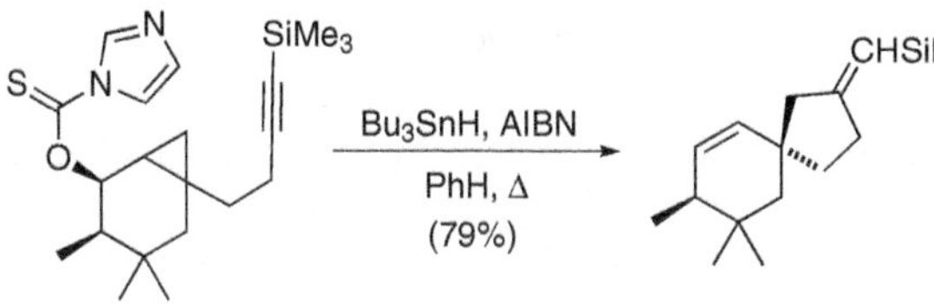

(14.113)[87]

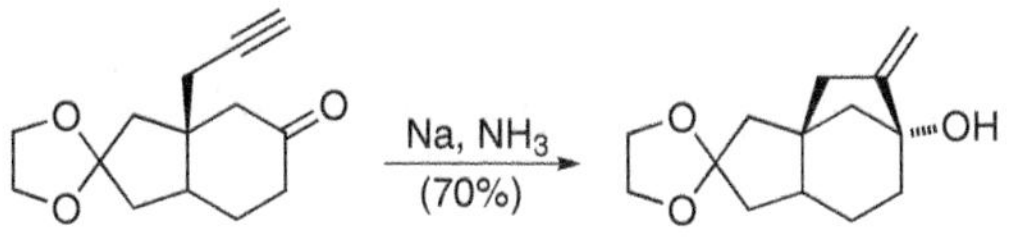

(14.114)[88]

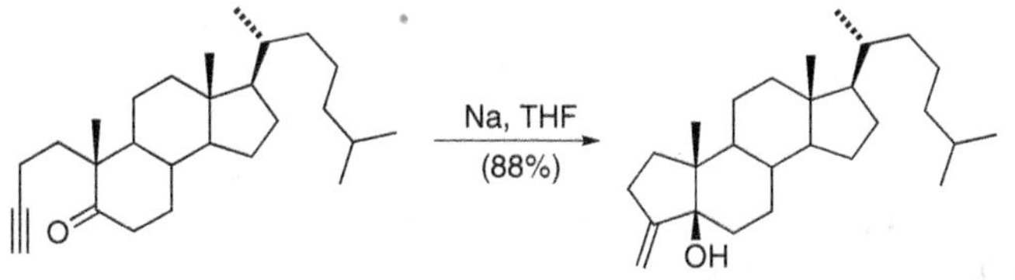

(14.115)[89]

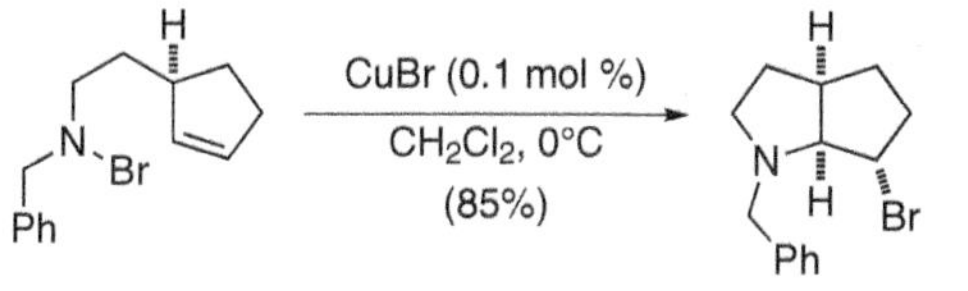

(14.116[90]

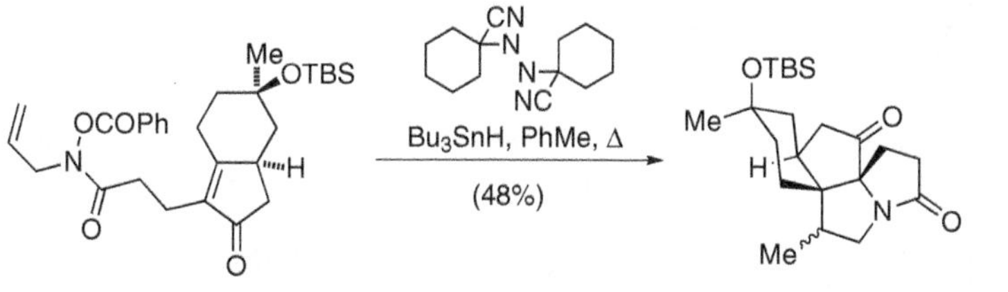

(14.117[91]

The formation of the product in Example **14.117** is especially instructive and illustrates just how well free radicals can enter into cascade reactions (Figure 14.9). Thus, the initial attack of the radical from the initiator or the tributylstannyl radical ($Bu_3Sn•$) is on the benzoyl oxygen and leads to formation of the amide radical **14.119**. This radical then cyclizes onto the cyclopentenone to give a cyclopentyl-carbinyl radical (**14.120**) that is set up to carry out a second cyclization to give the new cyclopentylcarbinyl radical **14.121**. This radical finally reacts with the stannane by hydrogen atom transfer to give the final product (**14.122**).

87. Pradhan, S.K.; Kadam, S.R.; Kolhe, J.N.; Radhakrishnan, T.V.; Sohani, S.V.; Thakker, V.B. *J. Org. Chem.* **1981**, *46*, 2622.

88. Motherwell, W.B.; Harling, D.J. *J. Chem. Soc., Chem. Commun.* **1988**, 1380.

89. Stork, G.; Boeckmann, R.K.; Taber, D.F.; Still, W.C.; Singh, J. *J. Am. Chem. Soc.* **1979**, *101*, 7107.

90. Corey, E.J.; Chen, C.-P.; Reichard, G.A. *Tetrahedron Lett.* **1989**, *30*, 5547.

91. Cassayre, J.; Gagosz, F.; Zard, S.Z. *Angew. Chem. Int. Ed.* **2002**, *41*, 1783.

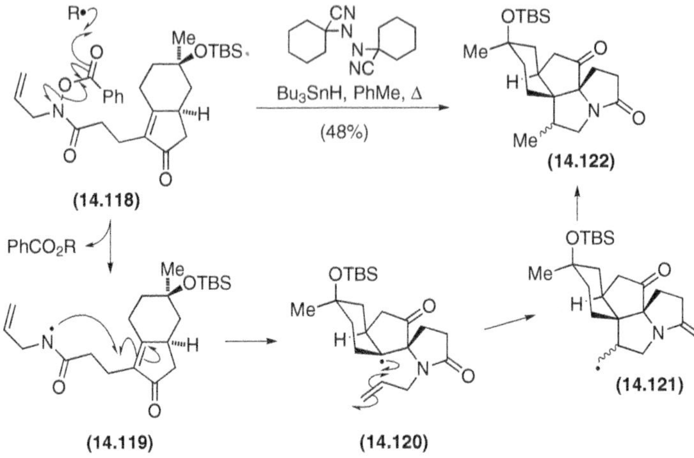

Figure 14.9 Cascade cyclization in Example **14.117**

Notice how the formation of the two rings must occur more rapidly than the hydrogen atom transfer reaction.

Worked Problem

14-1 (a) What is the structure of the major product of the following reaction? The product has only one aromatic ring.

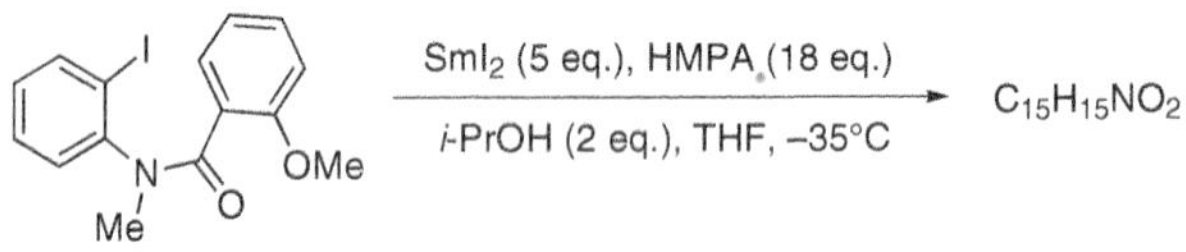

(b) Why does the final product in the reaction below still have a double bond? [*Org. Lett.* **2008**, *10*, 4049]

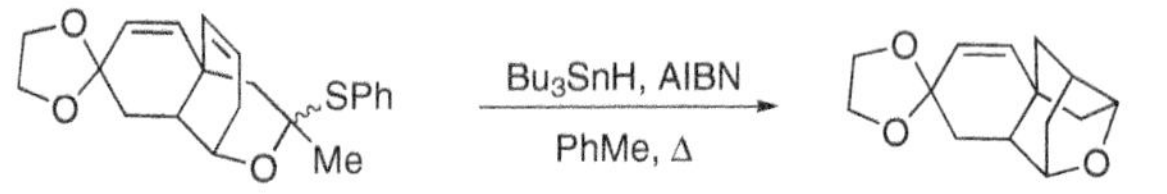

§Answers below.

§ Answers to Worked Problem:

(a) The samarium (II) iodide will reduce the carbon-iodine bond to give an aryl radical that will then cyclize to give the spirocyclic product [*J. Org. Chem.* **2008**, *73*, 7145]:

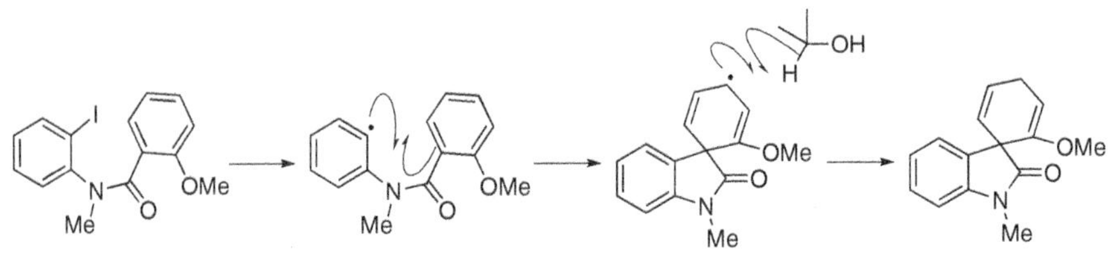

The carbon-iodine bond is reduced because this bond has the most accessible LUMO. What other reaction might occur if the bromide were used?

(b) The closure of the first ring gives a five-membered ether, and so is basically irreversible. The closure of the second ring, however, gives a cyclopropylcarbinyl radical that is thermodynamically less favorable than the butenyl radical.

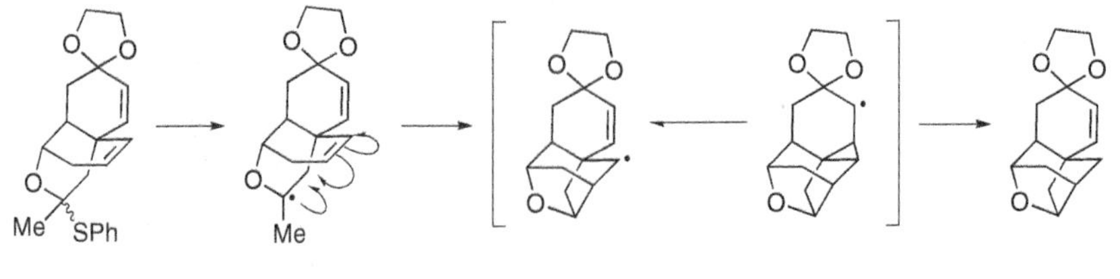

Problems

14-5 What is the major product formed when each of the of each of the following compounds is heated with tri-*n*-butylstannane and AIBN?

(a) 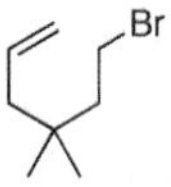 (b) 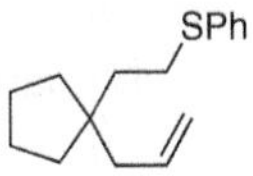 (c) 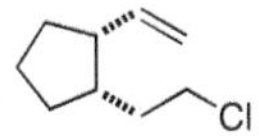 (d) 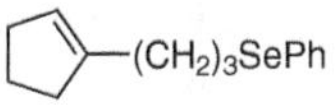

14-6 Draw the structure of the major organic product of each of the following reactions.

(a) 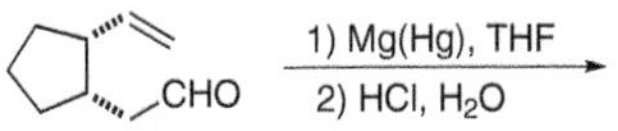

(b) 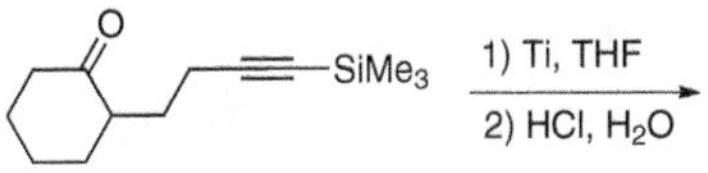

(c) 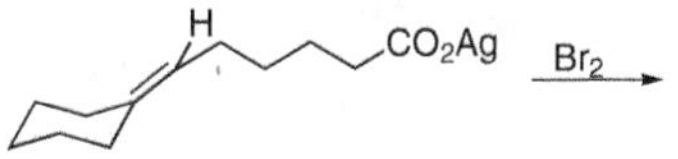

(d) 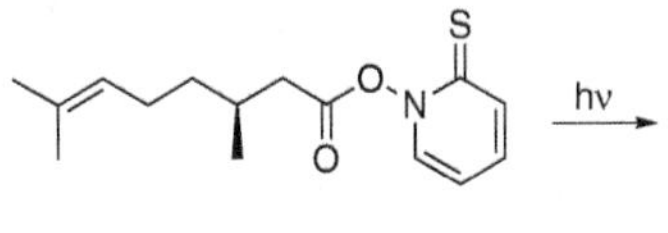

14-7 Write a reasonable mechanism for each of the following transformations:

(a) 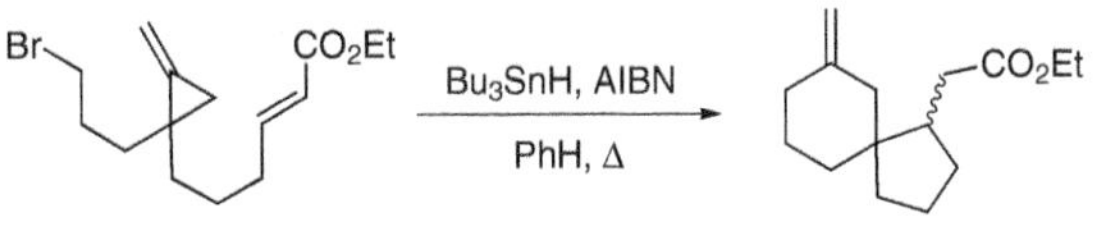

(b) 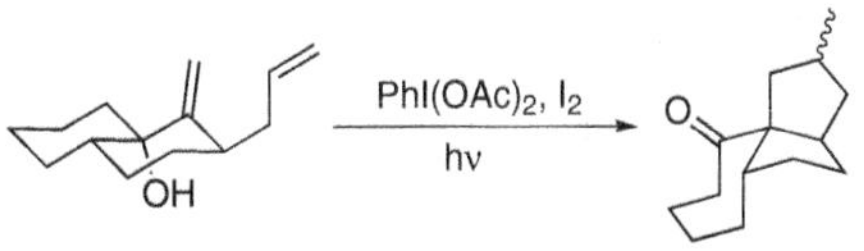

[This reagent converts an alcohol to an alkyl hypoiodite]

(c) 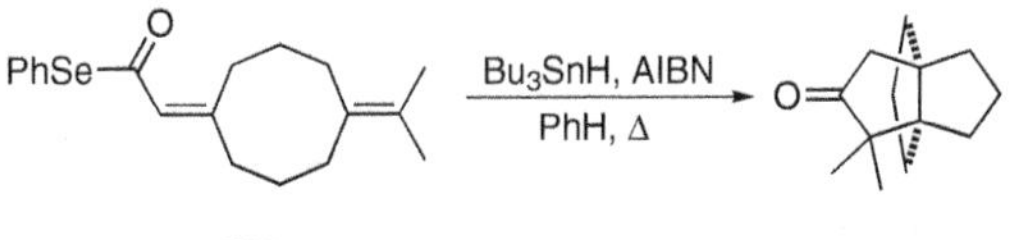

(d) 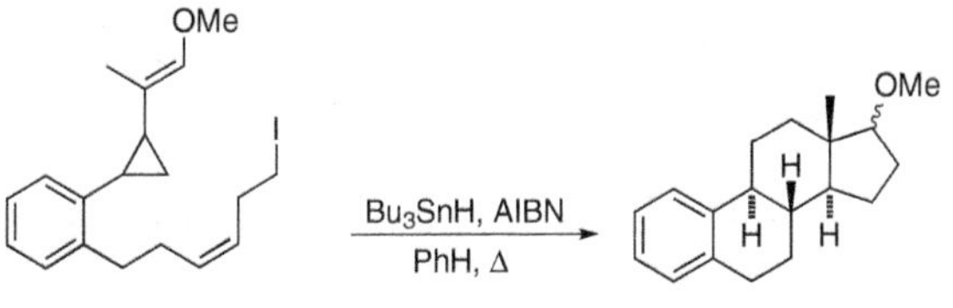

(e) 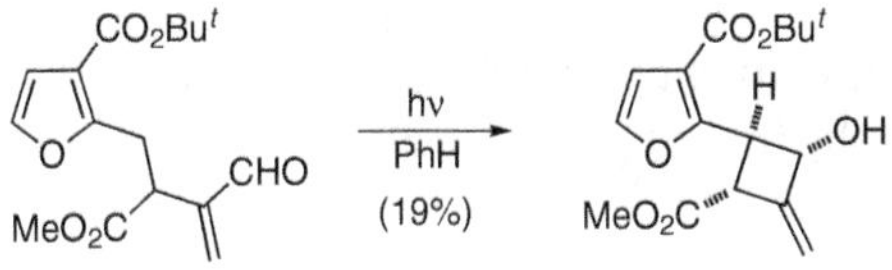

References: (a) *Tetrahedron Lett.* **1995**, *36*, 1365. (b) *Tetrahedron Lett.* **1993**, *34*, 127; *Tetrahedron Lett.* **1991**, *32*, 1591; *Aust. J. Chem.* **1995**, *48*, 381. (c) *Org. Biomol. Chem.* **2005**, *3*, 340. (d) *Tetrahedron Lett.* **2004**, *45*, 4027. (e) *Tetrahedron Lett.* **2006**, *47*, 3937.

Addition to Carbene- and Nitrene-like Compounds

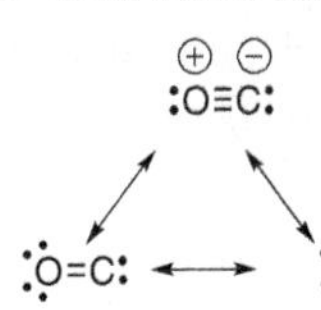

(14.123)

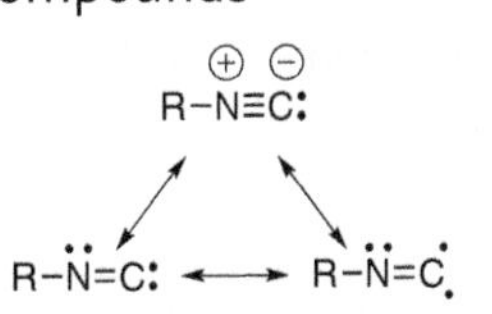

(14.124)

Carbon monoxide and its isoelectronic counterpart isocyanide (Example **14.124**) are molecules stabilized by resonance in which there is at least one carbenoid canonical form that can be represented as either a singlet or a triplet. Free radicals react with carbenes (it is easier to think of this reaction as a radical coupling reaction with a triplet carbene) to give new free radicals. One such reagent that reacts with free radicals in an atom transfer chain reaction is *tert*-butyl isocyanide, which reacts with radicals to give new free radicals containing one carbon atom more than the starting radical, as shown in cycle diagram **14.125**. Stork has used this addition as the basis for the formation of nitriles by free radical cyclization of bromoacetals (Examples **14.126** and **14.127**).[92] It is important in this reaction to use the distannane as the radical initiator, because hydrogen atom transfer from tri-*n*-butylstyannane is more rapid than addition to the isocyanide.

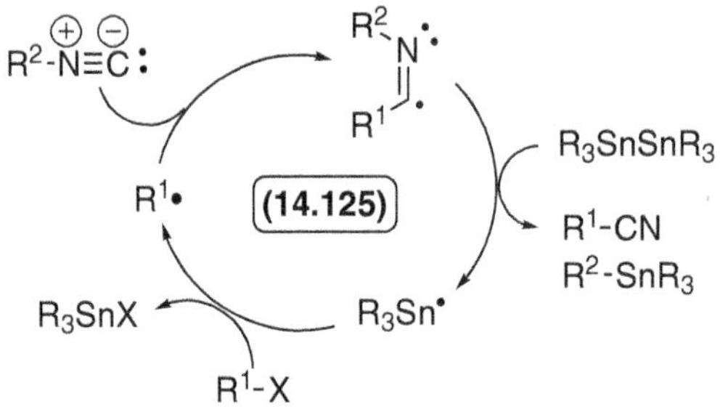

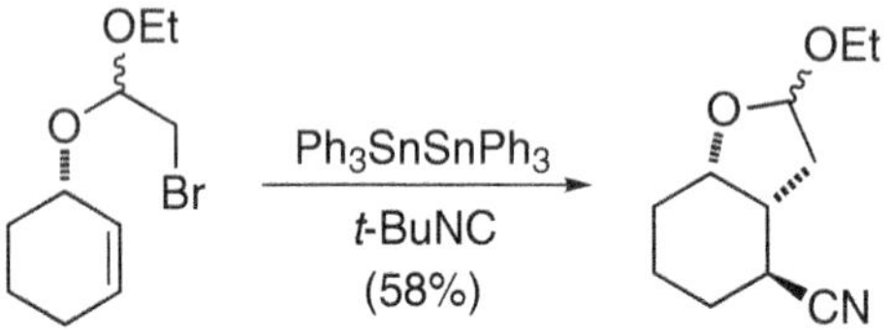

(14.126)

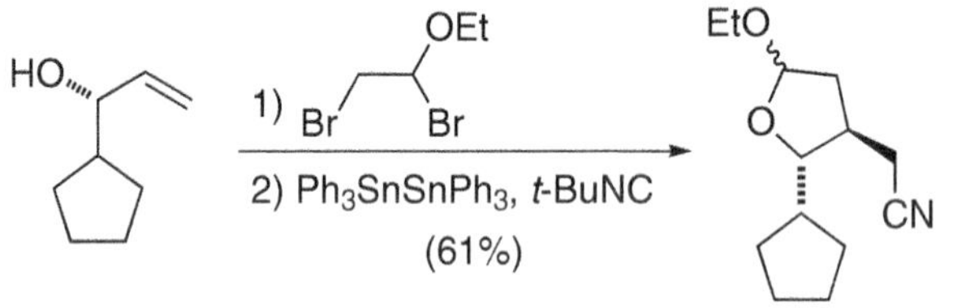

(14.127)

Carbonylation of free radicals with carbon monoxide is a reversible reaction (due to the high stability of carbon monoxide). However, when alkyl radicals are generated under carbon monoxide at high pressure, they react to give acyl radicals that can be trapped by a range of atom donors to give a variety of carbonyl compounds. For example, when the radical is a vinyl radical (formed by adding the tributyltin radical to the alkyne), free radical carbonylation can be followed by trapping with an amine to provide a versatile route to acrylamides (e.g., Example **14.128**).[93]

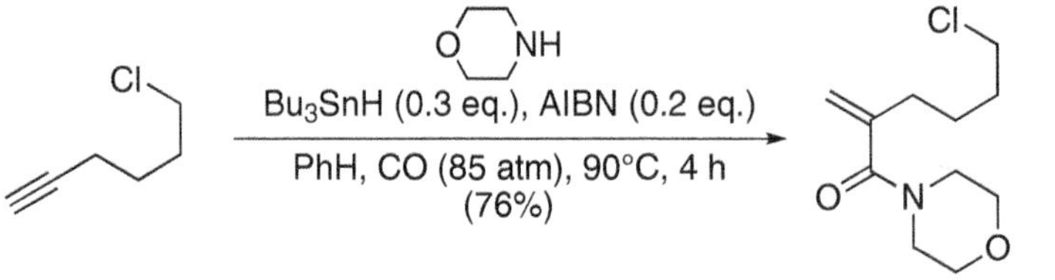

(14.128)

In similar fashion, radicals react with azides (e.g., Example **14.129**) to give what are formally products of a free radical reacting with a triplet nitrene. These nitrogen-centered radicals readily add to alkenes, so this provides an excellent method for the formation of

92. Stork, G.; Sher, P.M. *J. Am. Chem. Soc.* **1983**, *105*, 6765.
93. Uenoyama, Y.; Fukuyama, T.; Nobuta, O.; Matsubara, H.; Ryu, I. *Angew. Chem. Int. Ed.* **2005**, *44*, 1075.

complex amines, with the intramolecular addition providing a convenient entry into nitrogen heterocycles.[94]

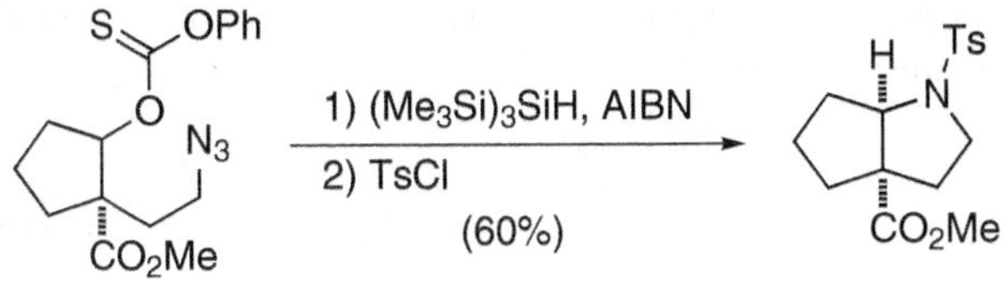

(14.129)

Problem

14-8 What are the expected products when the alcohols below are treated with the sequence of reagents given?

(a)

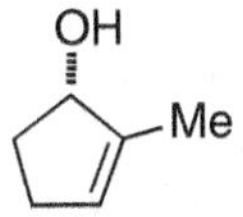

(b)

(1) $BrCH_2CH(OEt)Br$, base; (2) $Ph_3SnSnPh_3$, Me_3C-NC

Group Transfer Reactions and the Persistent Radical Effect

The ***persistent radical effect*** is a principle that explains the highly selective cross-coupling between a persistent and a transient radical when both species are formed at equal rates.[95] The Barton photolysis of nitrite esters (Example **14.130**) is an example of this type of reaction. In this reaction, two radicals are formed in the same step. One of these radicals, nitric oxide (•NO), is a stable, or ***persistent,*** free radical. The other radical formed in the Barton reaction, the alkoxy radical RO•, is a highly reactive, or ***transient,*** free radical.

(14.130)

It is a feature of persistent free radicals that they do not self-couple (or do so reversibly), so they must form cross-coupled products in a free radical reaction. This also means that the concentration of the persistent radical increases as the reaction progresses until a steady-state concentration is obtained. Although self-coupling of the transient radical does occur, the concentration of the transient radical is always low, so the dimerization (second order in the transient radical) is kinetically slow compared to the cross-coupling reaction (first order in the transient radical) except at the very beginning of the reaction. In the bimolecular coupling reaction, the cross-coupled product quickly becomes kinetically preferred.

The persistent radical effect can be exploited in synthesis because it guarantees that the cross-coupling product will dominate the product mixture. Typical persistent radicals are stable radicals such as nitric oxide (formed in the photolysis of nitrite esters or *N*-nitrosoamines) and 2,2,6,6-tetramethylpiperidyloxy (TEMPO) (shown in Example **14.133;** formed by homolysis of *N*-alkoxy-2,2,6,6-tetramethylpiperidines). The homolysis of these precursors provides reactive radicals that can be trapped by alkenes prior to recombination reactions to give the final product. This is illustrated by the thermolysis of the hydroxylamine derivative **14.131.** The initial homolysis gives the persistent radical

94. Kim, S.; Joe, G.H.; Do, J.Y. *J. Am. Chem. Soc.* **1994,** *116,* 5521,

95. Reviews: (a) Fischer, H. *Chem. Rev.* **2001,** *101,* 3581. (b) Studer, A. *Chem,. Eur. J.* **2001,** *7,* 1159; *Chem. Soc. Rev.* **2004,** *33,* 267.

TEMPO and a transient radical (Example **14.132**) that undergoes cyclization to the cyclopentylmethyl radical **14.134** and the cyclohexyl radical **14.135** before being trapped by TEMPO to give the final products.[96] In this particular reaction, a mixture of all four possible regio- and stereoisomers is formed.

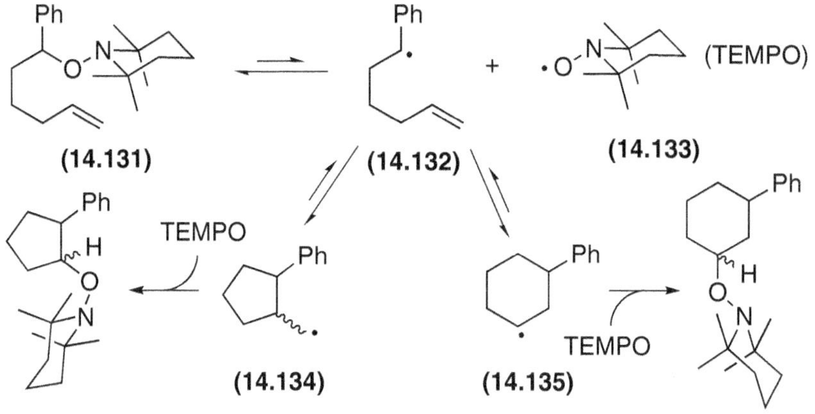

Reaction Synopses

Radical Additions

Intramolecular Radical Cyclization

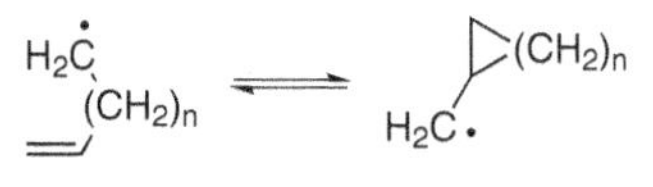

n = 1, 2, 3, 4

Closure to 3- to 6-membered rings gives smaller of two possible rings.

Intermolecular Additions to Alkenes

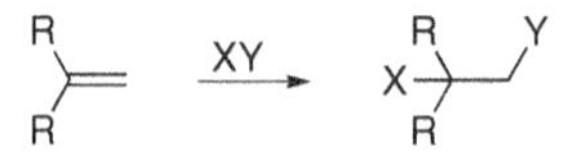

Reactants: CCl_4, ROOR; $BrCCl_3$, ROOR; HBr, ROOR; etc.

First step always gives more stable radical: Markovnikov addition for most reagents (anti-Markovnikov with HBr).

Additions to Carbene and Nitrene Equivalents

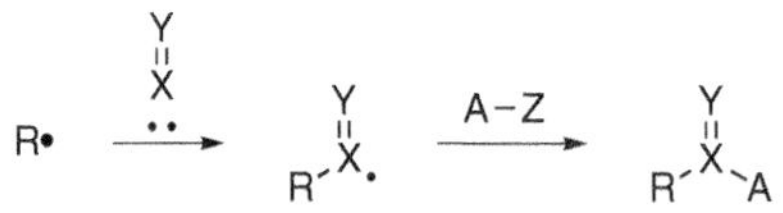

X=Y: :CO, R-NC:, R-N_3; etc.

Reactants: Bu_3SnH, CO, R_2NH; Bu_3SnH, RN_3; Bu_3SnH, RNC:; etc.

Group Transfer Reactions and the Persistent Radical Effect

$$\text{X–R} \rightleftharpoons \text{X}\bullet + \bullet\text{R} \rightleftharpoons \text{R}'\bullet \longrightarrow \text{X–R}'$$

R = alkyl, alkoxy, etc.; X=TEMPO, NO, I, 2-pyridylthio, etc.

---

96. Studer, A. *Angew. Chem. Int. Ed.* **2000**, 39, 1108.

14.5 Diradicals: Formation and Uses in Synthesis

In Section 14.1, we briefly discussed diradicals. In general, diradicals are not stable, but they occur as reactive intermediates that are formed and consumed during reactions. One important exception to this is molecular oxygen, which exists as a stable triplet diradical. A diradical is designated as either a ***singlet*** or a ***triplet***, depending on whether the two unpaired electron spins are parallel (triplet) or antiparallel (singlet). The triplet state of most diradicals is the low-energy state. Molecular oxygen exists as a stable triplet diradical, and linear carbenes also exist as triplets.

Diradicals react much like simple free radicals: they add to π bonds, oxygen-centered diradicals abstract hydrogen atoms from carbon atoms, and they combine with other radicals to give coupled products. If the diradical is a singlet, reactions that give nonradical products may be concerted. Triplet diradical reactions, however, cannot be concerted.

Ketone Photolysis

When a ketone is irradiated, the excitation of an electron from the nonbonding orbital of the oxygen to the π* orbital of the carbonyl group leads to a singlet diradical excited state with one of the unpaired electrons on oxygen. The fate of the diradical depends on the structure of the starting carbonyl compound, but generally two processes compete. In the first, there is cleavage of the bond between the carbonyl group and the α carbon to give an alkyl radical and an acyl radical that rapidly loses carbon monoxide to give a second alkyl radical. This process is known as the Norrish type I cleavage, after Nobel Prize-winning British chemist R. G. W. Norrish, who first described and rationalized it.[97]

The alternative fragmentation of this diradical is known as the Norrish type II fragmentation. In this pathway, as in the Barton reaction, the alkoxy radical abstracts a hydrogen atom from the δ carbon of the ketone to give a carbon-centered diradical. Coupling of the two radical centers gives a cyclobutanol (e.g., Example **14.136**),[98] or their fragmentation further gives an alkene and an enol. The Norrish type II reaction has been used to make some extremely strained systems, as illustrated by the formation of the bicyclo[1.1.1]pentane derivative in Example **14.137**.[99]

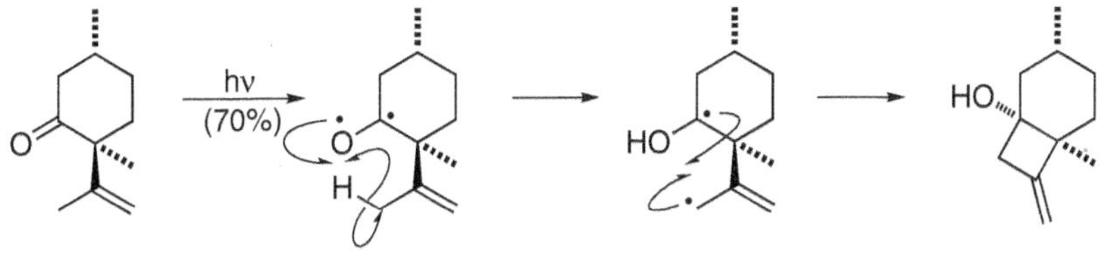

(14.136)

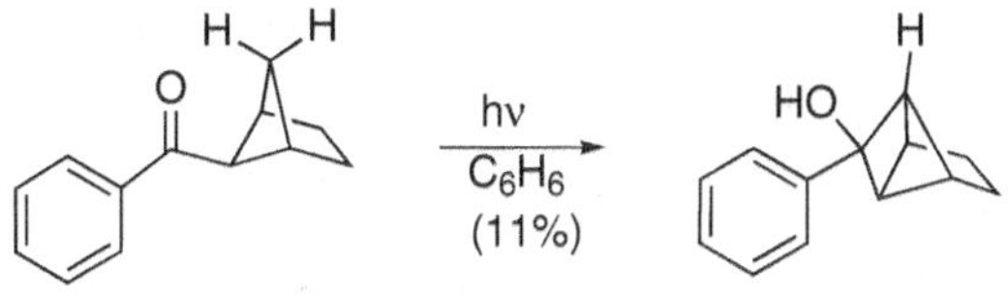

(14.137)

97. Bamford, C.H.; Norrish, R.G.W. *J. Chem. Soc.* **1935**, 1504.

Ronald George Wreyford Norrish (1897–1978) was educated at Cambridge (PhD, 1924), and remained there for his entire career. He was a pioneer in photochemistry and shared the 1967 Nobel Prize in Chemistry. For more detail, see: Dainton, F.; Thrush, B.A. *Biogr. Mem Fellows Roy. Soc.* **1981**, *27*, 379.

98. (a) Cookson, R.C.; Hudec, J.; Szabo, A.; Usher, G.E. *Tetrahedron* **1968**, *24*, 4353. (b) Cookson, R.C.; Rogers, N.R. *J. Chem. Soc., Chem. Commun.* **1972**, 809. (c) Cookson, R.C.; Rogers, N.R. *J. Chem. Soc., Perkin Trans 1* **1974**, 1037.

99. Padwa, A.; Eisenberg, W. *J. Am. Chem. Soc.* **1970**, *92*, 2590; **1972**, *94*, 5852, 5858.

Homolysis of Cyclic Azo Compounds

The homolysis of cyclic azo compounds provides the second major method for the formation of diradicals in organic chemistry. When these compounds are irradiated, they undergo loss of nitrogen to give a diradical. Depending on the conditions of the irradiation, the diradical may be a singlet or triplet. In the presence of a ***photosensitizer,*** irradiation almost always results in the formation of a triplet diradical. A photosensitizer is a molecule that absorbs light and transfers the energy absorbed to another molecule (usually the molecule that is actually to undergo the photochemical reaction); the most common photosensitizers are aromatic ketones such as benzophenone.

The five-membered azo compounds, which are readily available by the [3 + 2] cycloaddition of diazomethane to alkenes, decompose on irradiation to give cyclopropane derivatives (e.g., Example **14.138**) through intermediate diradicals.[100]

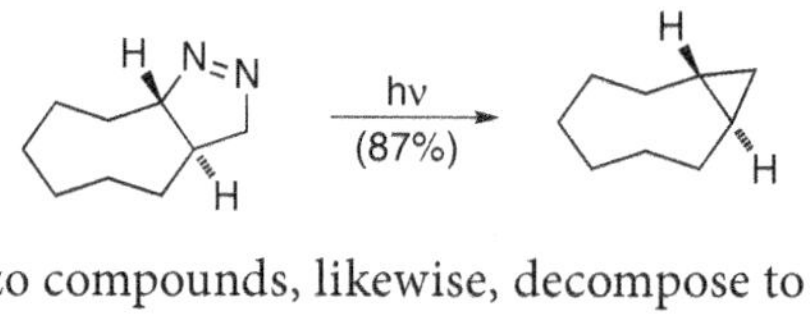

(14.138)

Six-membered azo compounds, likewise, decompose to diradicals that may cyclize to a cyclobutane derivative. However, these diradicals may also fragment to two alkene molecules. In simple six-membered azo compounds, irradiation favors fragmentation rather than cyclization to the cyclobutane (e.g., Example **14.139**).[101]

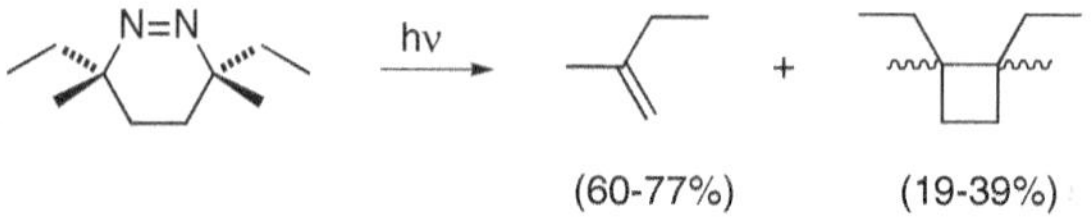

(14.139)

Regardless of the possible reaction outcomes of simple azo compounds, when the precursor to the diradical contains a suitably disposed alkene π bond, and especially when the diradical is stabilized by resonance to slow down competing fragmentation, the intramolecular trapping of the diradical can occur to give cyclized products. An excellent example of this is provided by the reaction in Example **14.140**, which was the key step in the synthesis of the natural product hirsutene.[102]

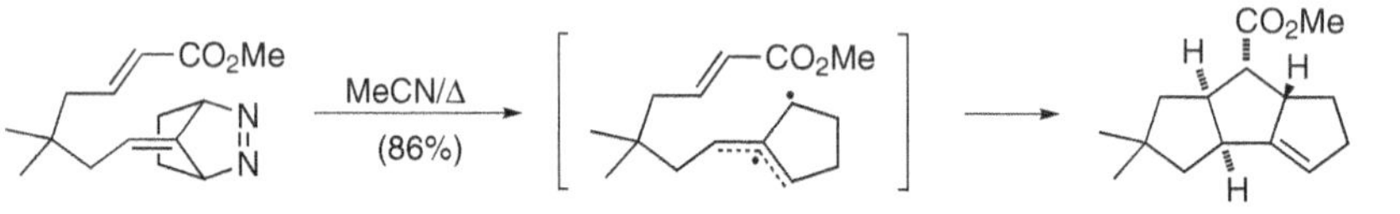

(14.140)

Bergman Cycloaromatization

In 1972, Bergman and Jones reported the conversion of *cis*-hex-3-en-1,5-diyne into the 1,4-phenylenediyl radical, or 1,4-dehydrobenzene, by heating the parent compound to approximately 200°C.[103] This reaction has since become known as the Bergman cycloaromatization reaction. Because temperatures of 200°C or more were required for the cyclization, it appeared that the reaction would remain simply an esoteric mechanistic study. However, this changed dramatically, when a series of potent anticancer natural products containing the enediyne moiety were discovered.[104] These compounds react with nucleophiles

100. Wiberg, K.B.; Lupton, E.C., Jr.; Wasserman, D.J.; de Meijere, A.; Kass, S.R. *J. Am. Chem. Soc.* **1984,** *106,* 1740.
101. Bartlett, P.D.; Porter, N.A. *J. Am. Chem. Soc.* **1968,** *90,* 5317.
102. Little, R.D.; Muller, G.W. *J. Am. Chem. Soc.* **1981,** *103,* 2744.
103. (a) Jones. R.R.; Bergman, R.G. *J. Am. Chem. Soc.* **1972,** *94,* 660. (b) Bergman, R.G. *Acc. Chem. Res.* **1973,** *6,* 25. (c) Lockhart, T.P.; Gomita, P.B.; Bergman, R.G. *J. Am.Chem. Soc.* **1981,** *103,* 4091.
104. Review: Nicolaou, K.C.; Dai, W.M. *Angew. Chem. Int. Ed. Engl.* **1991,** *30,* 1387.

to give the 1,4-phenylenediyl radical that then cleaves double-stranded DNA by abstracting a hydrogen atom from the sugar moieties of the DNA backbone.[105]

In Bergman's original study, temperatures of 200°C were required to drive the cyclization, but in the naturally occurring compounds, incorporation of the enediyne group into in a medium-sized ring results in a much more facile cyclization. As we saw in Chapter 13, and we again see in Example **14.141**, in the cyclization of calicheamicin γ_1, the nucleophilic cleavage of the trisulfide generates a thiolate anion that undergoes intramolecular Michael addition as the first step of the Bergman cyclization. The change in hybridization of the β-carbon of the enone brings the ends of the enediyne into close enough proximity to allow the ring closure to the diradical to occur at ambient temperatures.

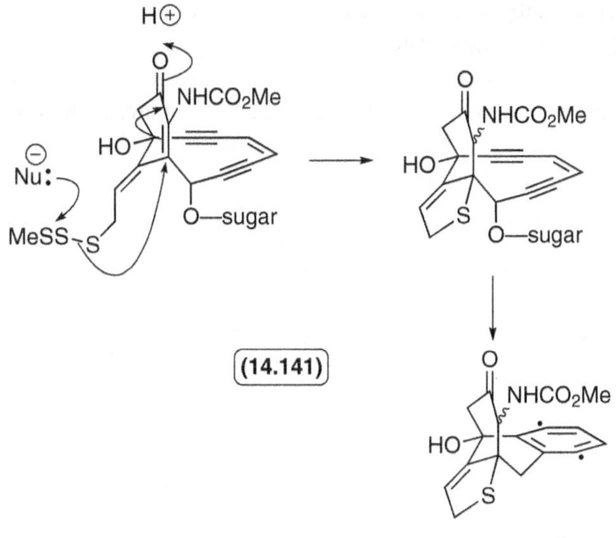

(14.141)

Reaction Synopses

Homolysis of Azo Compounds

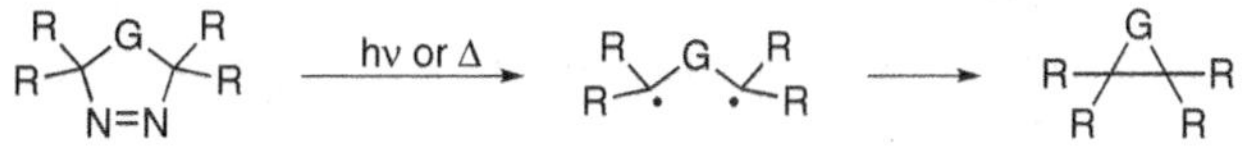

Bergman Cycloaromatization

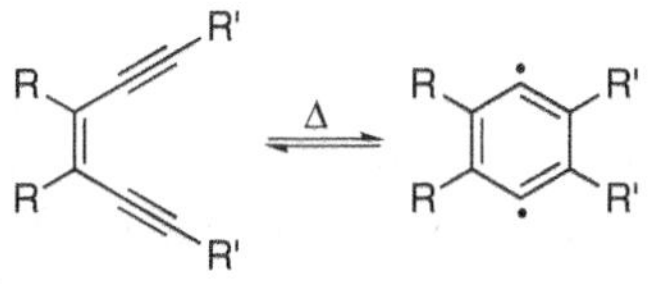

Chapter Summary

The site-specific generation of free radicals is an indispensable part of making these reactive intermediates useful in synthesis. This can be accomplished by means of addition reactions, site-specific hydrogen or halogen abstraction (as in stannane reductions of halides), or single-electron oxidation (e.g., oxidation of phenoxide anions to phenoxy radicals) or reduction (e.g., reduction of carbonyl compounds with metals to give ketyl radical

105. (a) Nicolaou, K.C.; Smith, A.L. *Acc. Chem. Res.* **1992**, *25*, 497. (b) Nicolaou, K.C.; Winssinger, N. *J. Chem. Educ.* **1998**, *75*, 1225.

anions). One especially useful method involves the homolysis and decarboxylation of Barton esters. Synthetically important reactions involving free radical intermediates include the Borodin-Hunsdiecker reaction, via an acyl hypobromite; the Barton-McCombie deoxygenation, via a xanthate ester that is reduced with a stannane reagent; and the Gomberg-Bachmann reaction, via a diazonium ion that is reduced to a diazenyl radical. Although radicals do not usually rearrange, they do cyclize. Radical cyclization under kinetic control is subject to a stereoelectronic requirement due to the SOMO-LUMO overlap required. This leads to the formation of cyclopentylmethyl radicals instead of the more stable cyclohexyl radical from the 5-hexenyl radical. Cyclopropylmethyl radicals undergo rapid ring opening to homoallyl radicals. Radicals react with carbene or nitrene equivalents to give acyl radicals or their equivalents that can be trapped to give synthetically useful products. Suitable unsaturated radicals undergo macrocyclization to give large- and medium-sized ring products. Diradicals may be generated by photolysis or thermolysis of aliphatic azo compounds and can be useful intermediates in synthesis.

Key Terms

Barton ester
Barton-McCombie deoxygenation
Barton reaction
Borodin-Hunsdiecker reaction
Gomberg-Bachmann reaction
Hofmann-Löffler-Freytag reaction
ketyl
Kolbe electrolysis
Norrish type I cleavage
Norrish type II cleavage
persistent radical effect
pinacol reaction

Additional Problems

14-9 The intermediate diradical (shown) in the cyclization reaction to give the triquinane in Example **14.140** can be represented by several canonical forms. Draw them. What isomeric product might be formed in this cyclization and why does the cyclization occur to give the product shown instead of the isomer?

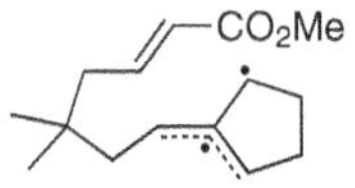

14-10 Draw mechanisms to show the formation of the cyclization and fragmentation products that will be obtained by irradiating the compound shown. Is the stereochemistry of the starting compound expected to affect the product distribution? How?

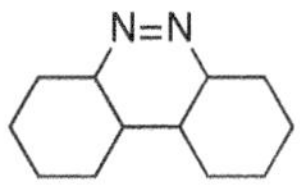

14-11 The reaction shown gives the amine indicated as the major product after treatment with base. Why is this product formed—and not the isomeric five-membered amine? (Build models here—it will help.)

14-12 Usnic acid is a compound that accounts for up to 2% of the dry weight of certain lichens. It was synthesized by dehydration of usnic acid hydrate, which was itself prepared by a phenolic coupling reaction. What phenol was used? Write a mechanism to show the conversion of this phenol to usnic acid hydrate.

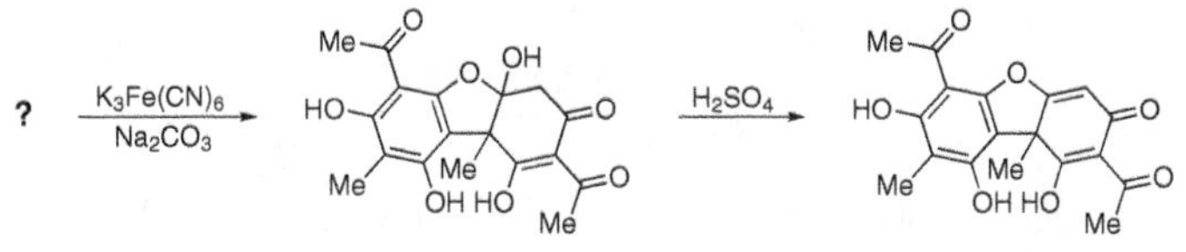

Usnic acid hydrate Usnic acid

14-13 Each of the reactions below has been used in the synthesis of a natural product or in model studies used to test critical steps in a proposed synthesis of a natural product. Draw the structure of the major organic product that will be formed in each reaction.

(a) 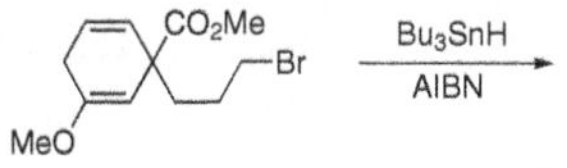 [*J. Am. Chem. Soc.* **1986**, *108*, 5893]

(b) 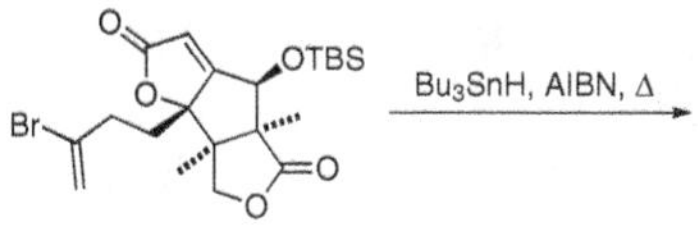 [*J. Am. Chem. Soc.* **2002**, *124*, 2080]

(c) 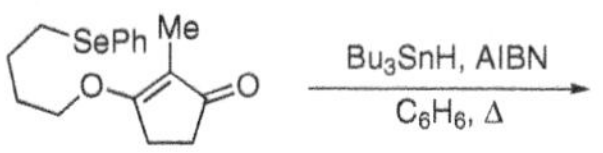 [*Tetrahedron* **1990**, *46*, 545]

(d) 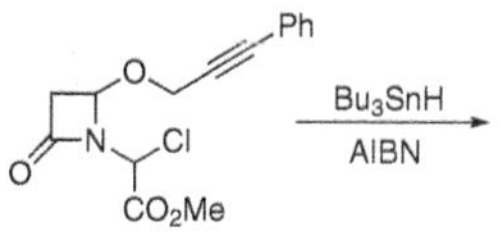 [*J. Org. Chem.* **1983**, *48*, 1841]

(e) 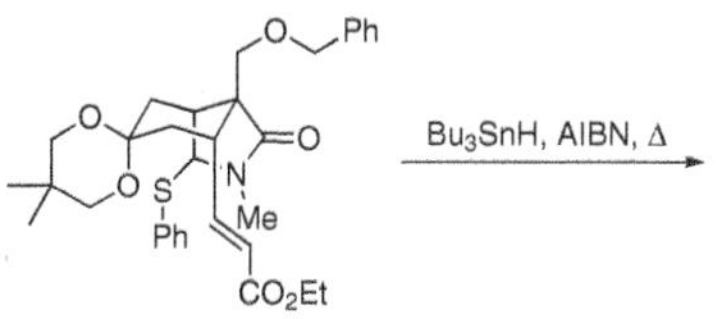 [*J. Am. Chem. Soc.* **1997**, *119*, 6226]

(f) 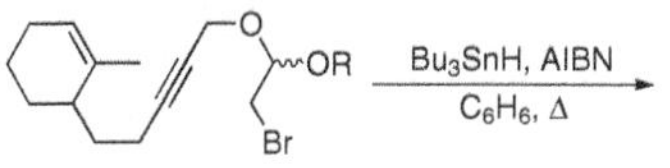 [*J. Am. Chem. Soc.* **1983**, *105*, 3720]

(g) 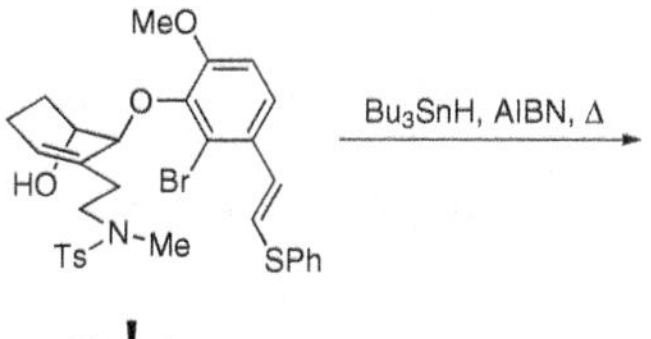 [*J. Org. Chem.* **2006**, *71*, 449]

(h) 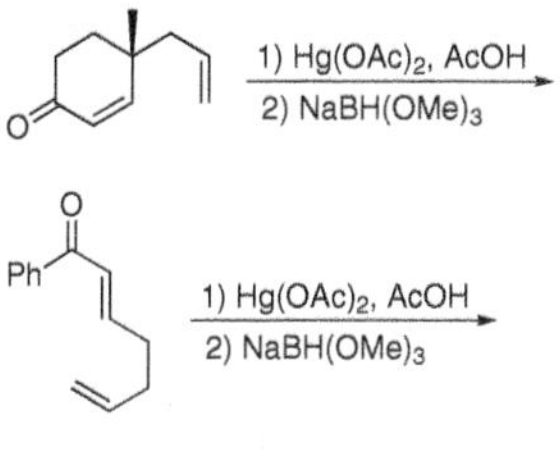 [*J. Org. Chem.* **1982**, *47*, 2231]

(i) 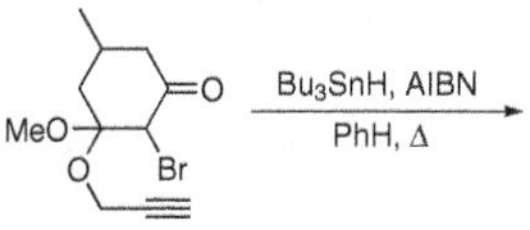 [*Tetrahedron Lett.* **1988**, *29*, 4995]

14-14 When the β-pinene is irradiated in carbon tetrachloride in the presence of benzoyl peroxide, the major organic product of the reaction is an addition compound that is monocyclic. What is the structure of this compound? Write a mechanism that accounts for its formation and explain why the expected addition compound is not isolated. If the pinene used is optically active, will the product be optically active also?

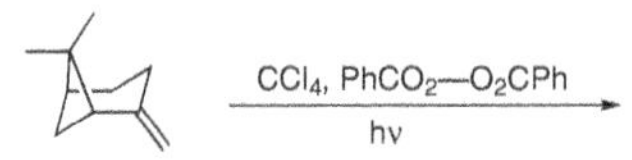

14-15 Draw the structure of the intermediate(s) formed when each of the compounds below is heated. What is the major product of each reaction (ignore stereochemistry)?

(a) 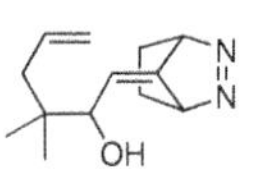

(b) 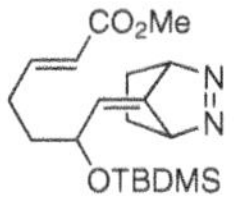

(c) 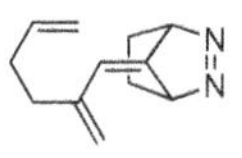

14-16 Gomberg's work, which culminated in the generation of triphenylmethyl, began as a project to prepare highly hindered hydrocarbons. He prepared tetraphenylmethane by a reaction in which a diazo compound was heated to high temperature, and it was to demonstrate that other similar hindered polyarylalkanes could exist that he attempted to prepare hexaphenylethane. What is the most likely structure of the azo compound that Gomberg prepared as the precursor to tetraphenylmethane and by what mechanism is it converted to tetraphenylmethane on heating?

14-17 What is the major organic product expected from each of the following reactions? Note that the reaction between Bu_3SnCl and $NaBH_3CN$ generates Bu_3SnH.

(a) 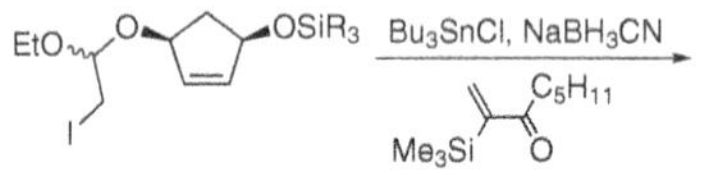

(b) 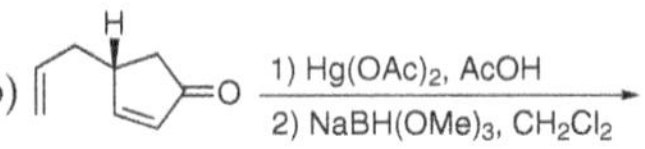

(c) 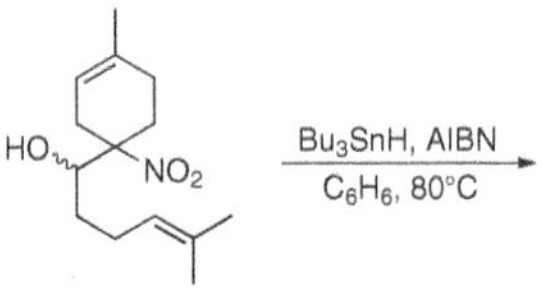

(d) 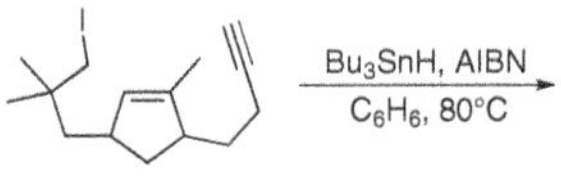

(e) 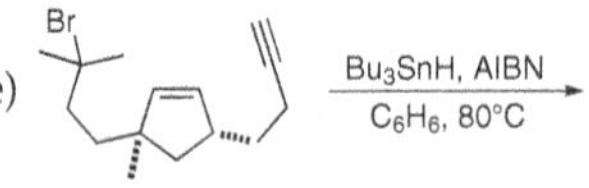

(f) 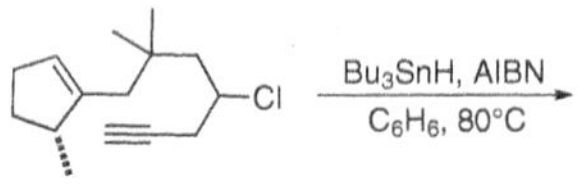

14-18 Suggest reasonable mechanisms for the following transformations.

(a) 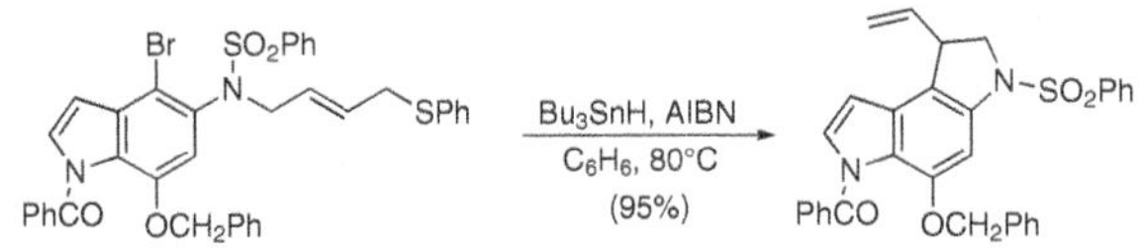

(b) 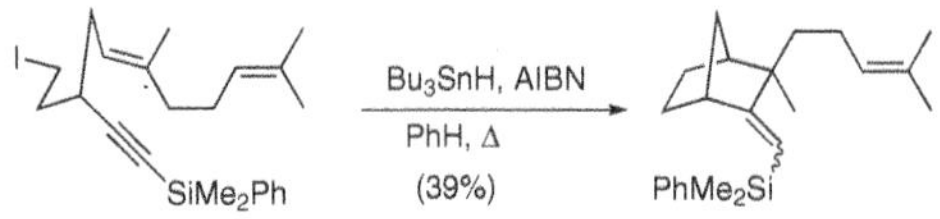

[*Synlett* **2005** 900]

(c) 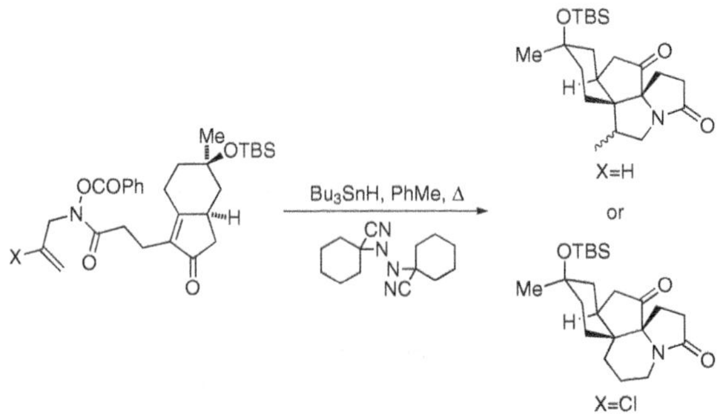

[*Angew. Chem. Int. Ed.* **2002**, *41*, 1783]

(d) 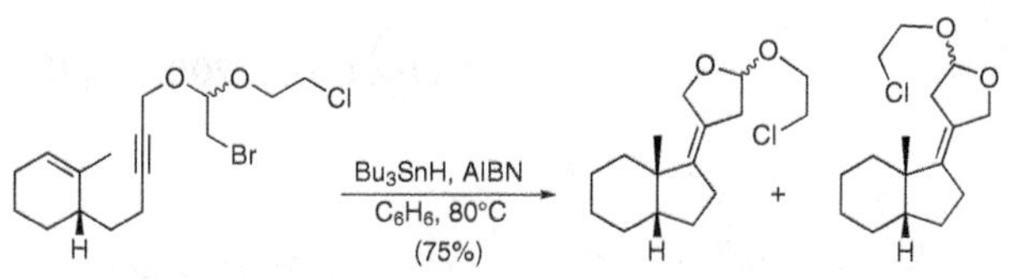

14-19 What is the major product of each of the following reactions?

(a) 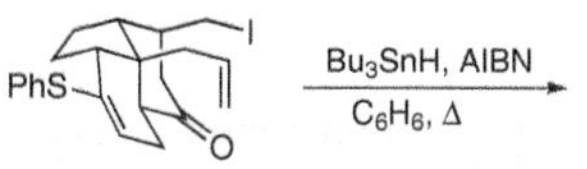

[*J. Am. Chem. Soc.* **1998**, *120*, 5128]

(b) 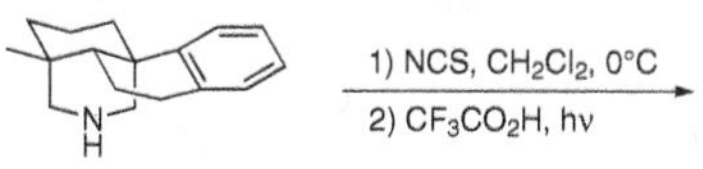

[*Chem. Pharm. Bull.* **1985**, *33*, 3187]

(c) 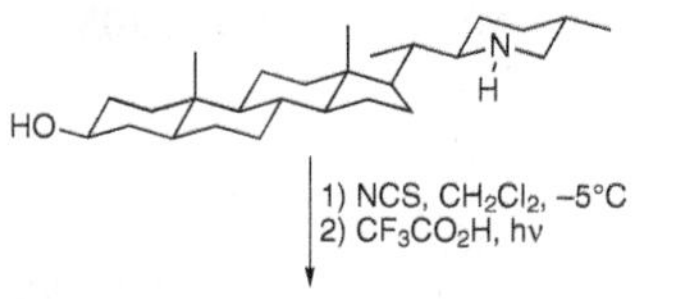

[*Tetrahedron* **1964**, *20*, 1719]

(d) 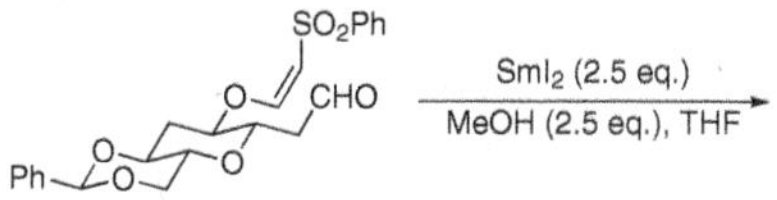

[*Tetrahedron Lett* **2007**, *48*, 43]

(e) 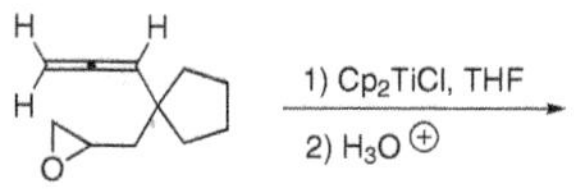

[*Tetrahedron Lett.* **2008**, *49*, 500]

(f) 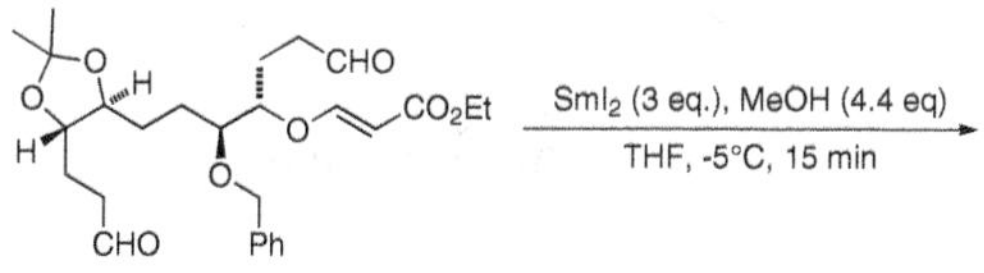

[*Angew. Chem. Int. Ed.* **2002**, *41*, 4751]

(g) 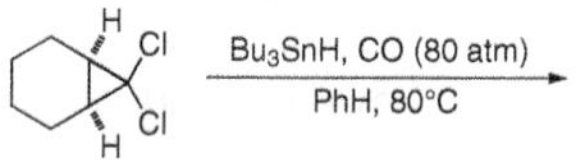

[*Org. Lett.* **2007**, *9*, 563]

What is the stereochemistry of the product?

(h) 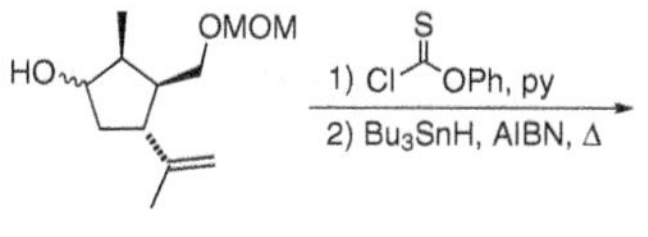

[*Org. Lett.* **2003**, *5*, 1321]

(i) 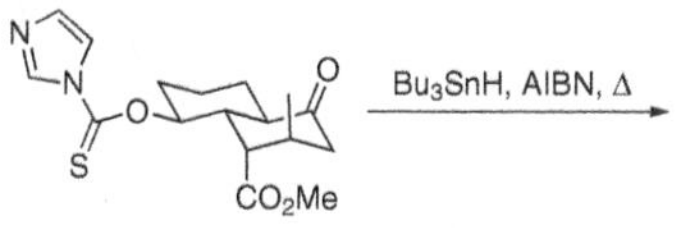

[*J. Org. Chem.* **2002**, *67*, 5969]

(j) 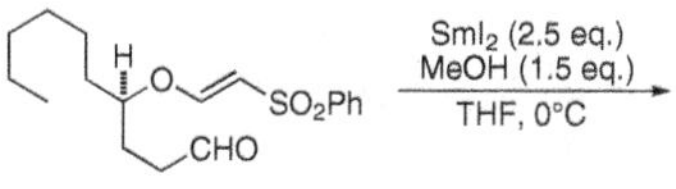

[*J. Org. Chem.* **2008**, *73*, 3040]

(k) 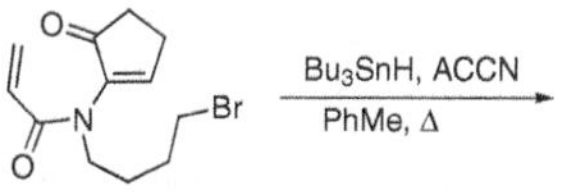

[*Org. Lett.* **2008**, *10*, 197]

(l) 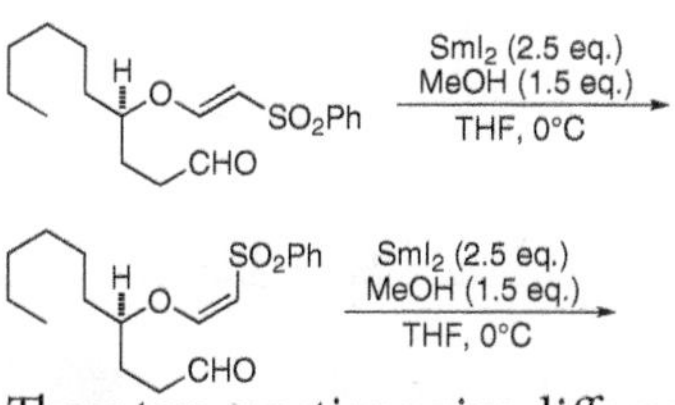

[*Tetrahedron Lett.* **2007**, *48*, 43]

These two reactions give different diastereoisomers

(m) 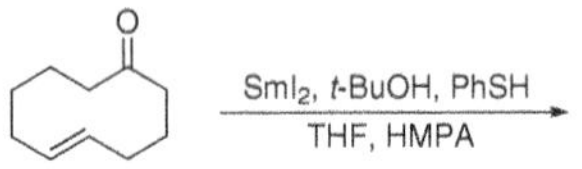

[*J. Org. Chem.* **2007**, *72*, 1755]

(n) 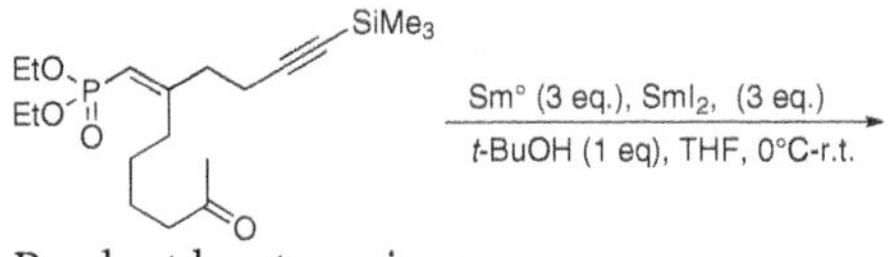

[*Org. Lett.* **2007**, *9*, 469]

Product has two rings.

(o) 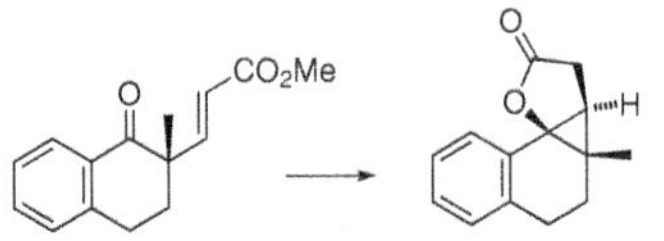

Supply the missing reagent
[*Tetrahedron Lett.* **2007**, *48*, 8234]

(p) 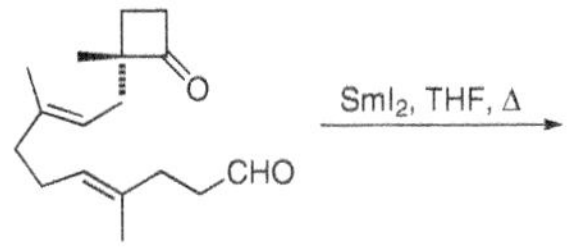

[*J. Am. Chem. Soc.* **2005**, *127*, 13813]

(q) 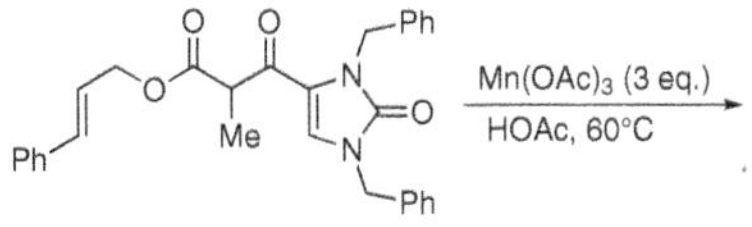

[*Angew. Chem. Int. Ed.* **2006**, *45*, 4345]

(r) 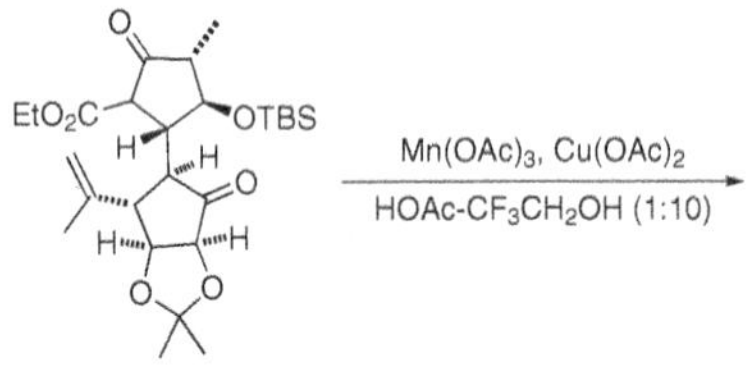

[*Org. Lett.* **2008**, *10*, 45]

(s) 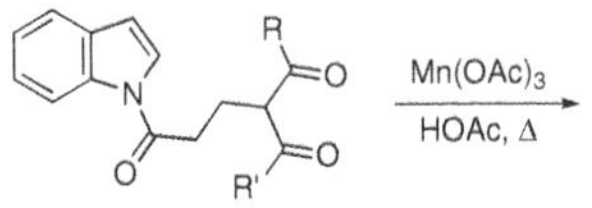

[*Org. Lett.* **2008**, *10*, 1437]

(t) 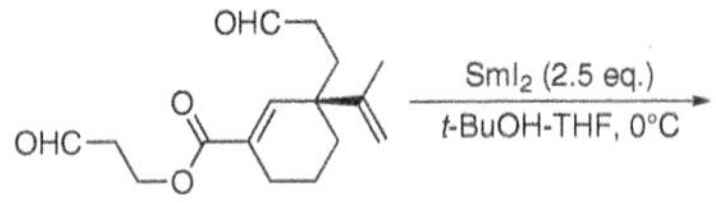

[*Tetrahedron Lett.* **2009**, *50*, 3224]

(u) 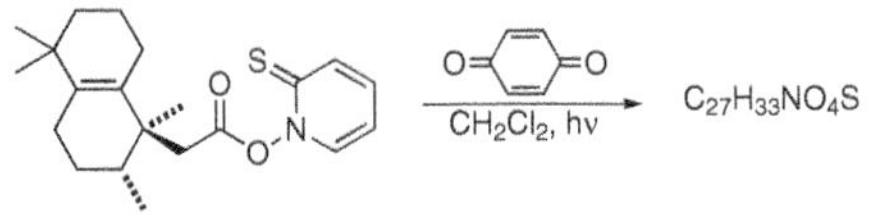

[*Tetrahedron* **2011**, *66*, 8280]

(v) 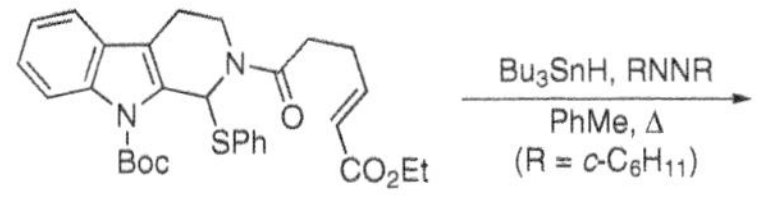

[*Tetrahedron Lett.* **2009**, *50*, 6342]

(w) 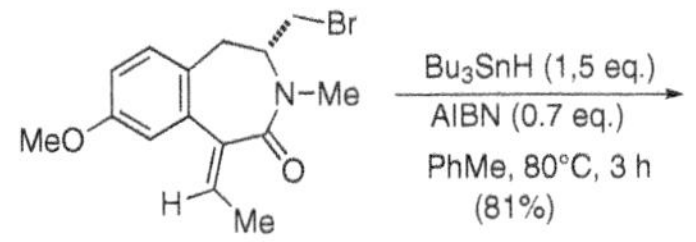

[*Eur. J. Org. Chem.* **2009**, 793]

(x) 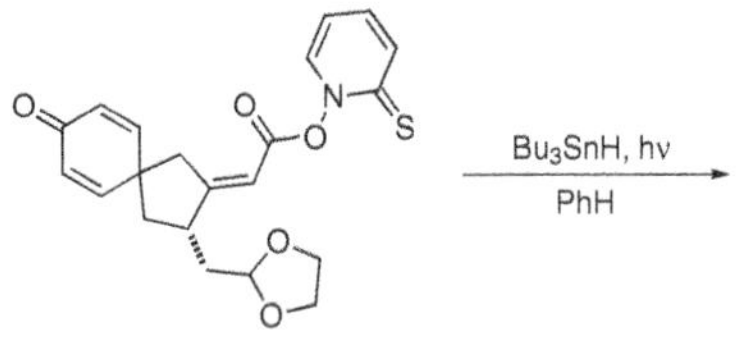

[*Angew. Chem. Int. Ed.* **2007**, *46*, 3942]

(y) 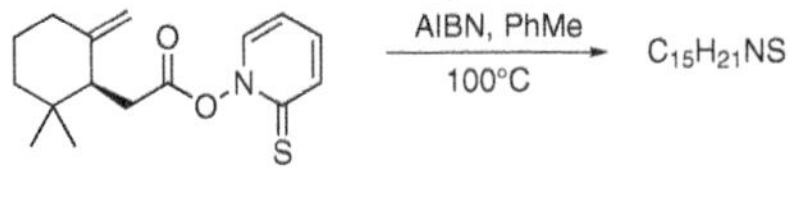

[*J. Am. Chem. Soc.* **2004**, *126*, 12565]

14-20 Suggest mechanisms for the transformations below.

(a) 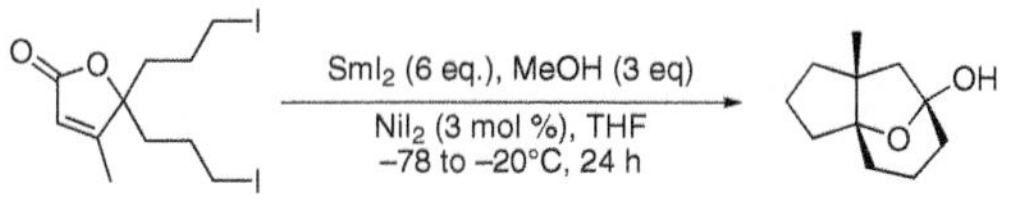

[*Tetrahedron Lett.* **2006**, *47*, 6225]

(b) 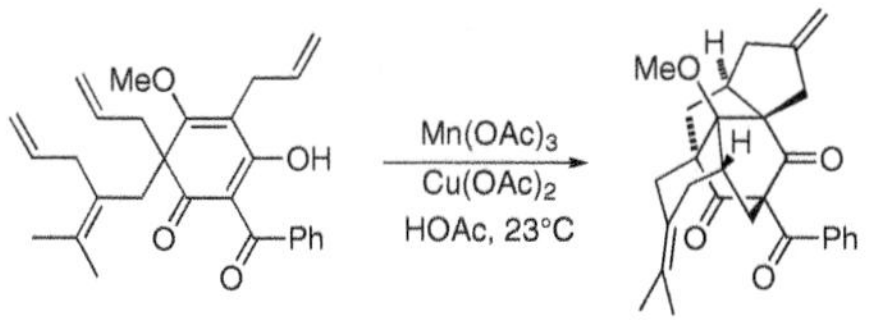

[*Org. Lett.* **2009**, *11*, 2285]

(c) 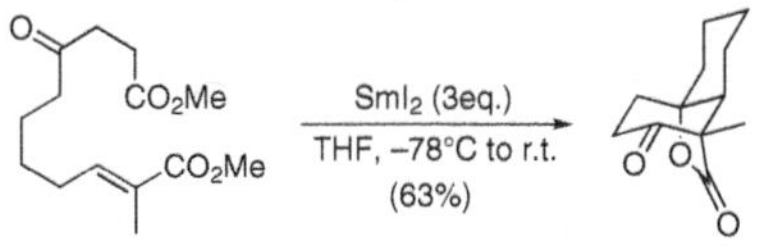

[*Tetrahedron Lett.* **2008**, *49*, 6393]

(d) 

[*Org. Lett.* **2006**, *8*, 831]

(e) 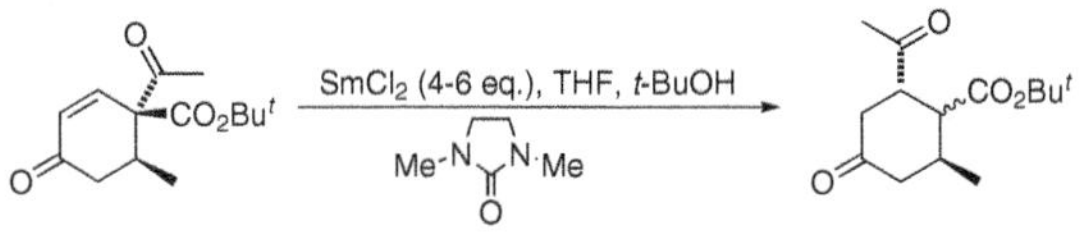

[*J. Org. Chem.* **2005**, *70*, 1497]

(f) 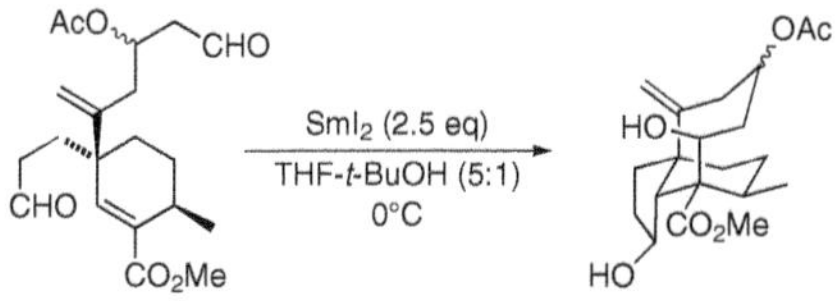

[*Angew. Chem. Int. Ed.* **2009**, *48*, 9315]

14-21 The reactants in the two reactions below differ by only a single chlorine atom. Write a mechanism for the reaction that accounts for the different outcomes using the two reactants. [*Angew. Chem. Int. Ed.* **2006**, *45*, 4345].

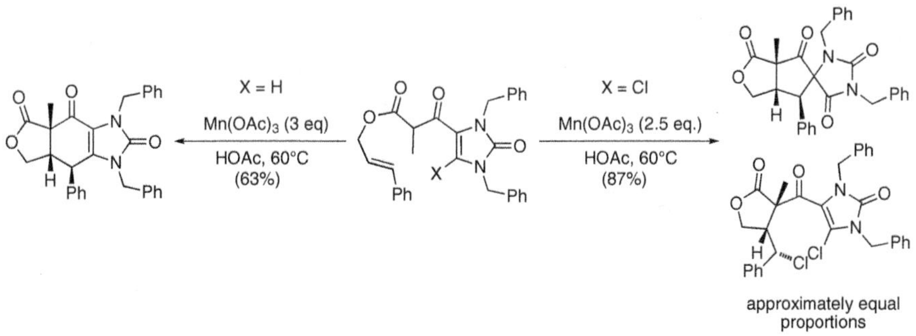

CHAPTER FIFTEEN

Organic Synthesis

Retrosynthetic Analysis, Protecting Groups, and the Strategy of Organic Synthesis

15.1 What Is Organic Synthesis?

Organic synthesis may well be the heart of organic chemistry, because everything we do as organic chemists ultimately requires the synthesis of organic compounds. Compounds are synthesized for a variety of reasons: to confirm structures assigned on the basis of spectroscopic studies or chemical degradation, to elucidate the stereochemistry of a compound whose stereochemistry cannot be established by other means, to modify an existing natural compound to make it more active, and to develop methods for the preparation of important medicinal compounds that may be in short supply from natural sources. Exploring the limits of structure and bonding has resulted in syntheses of compounds with multiple small rings, for example. The synthesis of cyclobutadiene (Example **15.1**)[1] was an important target for testing theories of aromaticity and the synthesis of prismane (Example **15.2**)[2] showed that this fully saturated isomer of benzene could, in fact, exist, which provided insights into the synthesis of highly strained molecules.

(15.1)

(15.2)

Organic synthesis is defined as *the process of constructing a complex molecule from less complex precursors*. In many ways, it resembles the process of building a fine piece of furniture. Just as one must learn to use the tools before one can build a piece of furniture, so one must learn to use the tools of organic synthesis—reactions—if one is to design the synthesis of a complex molecule. This means that the first task facing the synthetic chemist is to become thoroughly familiar with the tools of the discipline—the reactions of organic chemistry. That has been the focus of the last several chapters.

The design of a complex synthesis requires being able to combine reactions in such a way that the potential problems between the beginning of the synthesis and the formation of the final product are addressed. These problems will depend, in large part, on why the synthesis is being carried out (Figure 15.1). This can, in turn, be decided by answering several questions. How much of the product is needed? Must the product be a single compound, or will a mixture of isomers suffice? Does the product need to be a single diastereoisomer, or is a mixture of diastereoisomers adequate for the needs? Does the product need to be prepared as a single enantiomer, or can the racemate be used? Must certain reagents be avoided? Is the synthesis designed to demonstrate the wide synthetic potential of a new reaction? Every one of these questions will have a bearing on how the synthesis is designed and carried out. It is the goal of this chapter and, to some extent, the remainder of this book, to show how these questions are answered.

The questions above describe different levels of sophistication in synthesis, and the design of a synthesis depends very much on its ultimate goals. The least selective (and,

1. Review: Pettit, R. *Pure Appl. Chem.* **1968**, *17*, 253.
2. Katz T.J.; Acton N. *J. Am. Chem. Soc.* **1973**, *95*, 2738.

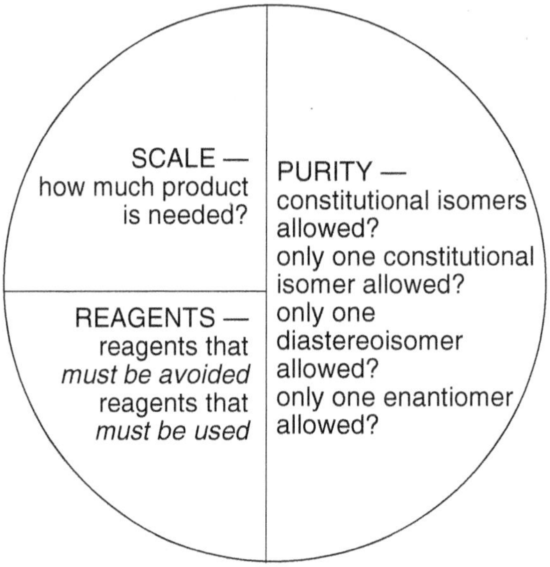

Figure 15.1 The three critical determinants of a synthesis

(15.3)

(15.4)

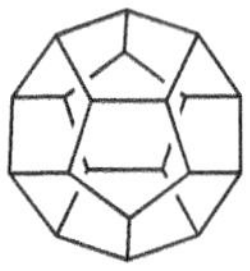
(15.5)

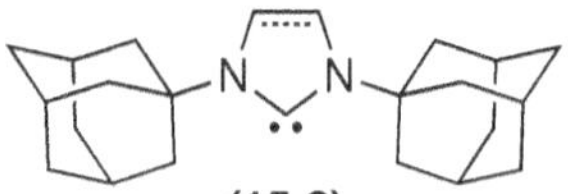
(15.6)

therefore, the least sophisticated) synthesis is one that can give a mixture of isomers without being unacceptable, and such a synthesis will differ from a synthesis where a single isomeric product is required. A synthesis designed to prepare a compound by the ton will be very different from a synthesis where milligrams of the final product will suffice.

The target compound for a synthesis almost always reveals the reasons for the synthesis, and can be as varied as the answers to the questions above. For example, chemists have long been fascinated by geometry, so the synthesis of the saturated hydrocarbon analogs of the Platonic solids (the tetrahedron, cube, octahedron, dodecahedron and icosahedron) has been the subject of numerous synthetic efforts. Of the Platonic solids, valence considerations limit the potential for a saturated hydrocarbon analog to tetrahedrane (Example **15.3**),[3] cubane (Example **15.4**),[4] and dodecahedrane (Example **15.5**).[5] Each of these compounds, in its own way, tests the limits of bonding capability. Arduengo's synthesis of the first stable *N*-heterocyclic carbenes (Example **15.6**)[6] provides another window into the extremes of bonding whose answers were supplied by synthesis.

Defining the ways that to make a particular target compound, and, more especially, identifying the *best* way to make a target compound, is the essence of organic synthesis. This is no trivial task, and some idea of the work involved can be gauged by noting that a 1994 textbook devoted to organic synthesis had more than 1500 pages, and the series *Comprehensive Organic Synthesis*, published in 1991, covered nine volumes averaging more than 1000 pages each. Obviously no single chapter, nor any sequence of chapters totaling just over 200 pages, can go into that kind of detail. This means that this chapter and those that follow it must serve just as an introduction to

---

3. Synthesis of tetra-*tert*-butyltetrahedrane: (a) Maier, G.; Pfriem, S.; Schaffer, U.; Matusch, R. *Angew. Chem. Int. Ed. Engl.* **1978**, *17*, 520. (b) Maier, G.; Pfriem, S. *Angew. Chem. Int. Ed. Engl.* **1978**, *17*, 519.

4. Eaton. P.E.; Cole. T. W.; Jr. *J. Am. Chem. Soc.* **1964**, *86*, 962, 3157.

5. (a) Ternansky, R.J.; Balogh, D.W.; Paquette, L.A. *J. Am. Chem. Soc.* **1982**, *104*, 4503. (b) Paquette, L.A.; Ternansky, R.J.; Balogh, D.W.; Kentgen, G. *J. Am. Chem. Soc.* **1983**, *105*, 5446. (c) Fessner, W.-D.; Murty, B.A.R.C.; Worth, J.; Hunkler, D.; Fritz, H.; Prinzbach, H.; Roth, W.D.; Schleyer, P.v.R.; McEwen, A.B.; Maier, W.F. *Angew. Chem., Int. Ed. Engl.* **1987**, *26*, 452.

6. (a) Arduengo, A.J., III; Harlow, R.L.; Kline, M. *J. Am. Chem. Soc.* **1991**, *113*, 361. (b) Regitz, M. *Angew. Chem. Int. Ed. Engl.* **1991**, *30*, 674.

organic synthesis and to the broad principles involved in designing a synthesis of an organic compound, rather than as an exhaustive treatise.

Elegance or, "What Constitutes a Good Organic Synthesis?"

The philosophical question of what constitutes an ideal synthesis is one that is being contemplated more frequently these days. The more reasonable—and pertinent—question may, in fact, be, "What constitutes a good synthesis?" As with all other largely philosophical questions of this sort, the answer becomes highly subjective. Every chemist has his or her own concept of just what constitutes a good synthesis (for this author, it was Woodward's synthesis of (±)-reserpine[7] that first defined elegance in synthesis). It will be for you to find your own.

Most synthetic chemists have a feel for what makes a good synthesis, although, like aromaticity, the concept can be hard to frame in words. There are, however, several common features of syntheses described as "elegant." For one, they usually exhibit a command of both stereochemistry and regiochemistry that results in a very short synthesis with very high levels of control. However, a quantitative measure of synthetic efficiency is still lacking.

The first to attempt to address the question of quantitating efficiency in an organic synthesis was Hendrickson who defined an "ideal synthesis" in 1975[8] as one that ". . . creates a complex molecule . . . in a sequence of only construction reactions involving no intermediary refunctionalizations, and leading directly to the target, not only its skeleton but also its correctly placed functionality." In 1975, this ideal was certainly far beyond contemporary abilities. In fact, many complex molecules that have since been synthesized were despaired of ever being prepared in the laboratory at that time. However, the ensuing decades have changed the landscape appreciably.

The last two decades of the 20th century provided a number of measures that could be used to quantitate synthetic efficiency. Among these was the concept of atom economy, first proposed by Trost.[9] In this measure of synthetic efficiency, the major goal is to minimize the number of atoms in the reactants that do not end up in the final product. What Trost pointed out in these articles was the change that transition metal-catalyzed reactions had wrought in organic synthesis. Prior to the advent of these catalytic reactions, few C—C bond-forming reactions were atom economical. The notable exceptions were pericyclic reactions such as the Diels-Alder reaction and the Claisen and Cope rearrangements. Strictly speaking, atom economical reactions should be additions or rearrangements, with any other species involved being catalytic. Other important C—C bond-forming reactions were decidedly not atom economical. For example the Claisen and Dieckmann condensations involve the loss of a molecule of alcohol, and the Wittig and Horner-Wadsworth-Emmons reactions lead to the loss of a phosphoryl-containing by-product. The concept of atom economy has achieved more prominence with the rise of ***"green" chemistry***,[10] which

---

7. (a) Woodward, R.B.; Bader, F.E.; Bickel, H.; Frey, A.J.; Kierstead, R.W. *J. Am. Chem. Soc.* **1956**, *78*, 2023, 2657; *Tetrahedron* **1958**, *2*, 1.

8. Hendrickson, J.B. *J. Am. Chem. Soc.* **1975**, *97*, 5784.

9. (a) Trost, B.M. *Science* **1991**, *254*, 1471. (b) Trost, B.M. *Angew. Chem. Int. Ed. Engl.* **1995**, *34*, 259. (c) Trost, B.M. *Acc. Chem. Res.*, **2002**, *35*, pp 695.

10. Reviews: (a) DeVito, S.C.; Garrett, R.L. *Green Chemistry for Pollution Prevention. ACS Symp. Ser.* (American Chemical Society, Washington, D.C.) **1996**, *640*. (b) Anastas, P.T.; Williamson, T.C. *Green Chemistry. Designing Chemistry for the Environment ACS Symp. Ser.* (American Chemical Society, Washington, D.C.) **1996**, *626*. (c) Anastas, P.T.; Kirchhoff, M.T. *Acc. Chem. Res.*, **2002**, *35*, 686. (d) Jessop, P.G.; Trakhtenberg, S.; Warner, J. In Flank, W.H.; Abraham, M.A.; Matthews, M.A., Eds. *Innovations in Industrial and Engineering Chemistry, ACS Symp. Ser.* (American Chemical Society: Washington, D.C.), **2009**, *1000*, ch. 12, 401.

is a deliberate attempt to design syntheses with a minimum of waste.[11] The search for atom economical methods of synthesis has led to the development of many transition metal-catalyzed C—C bond-forming reactions.

The concept of ***step economy*** was introduced by Wender in the last decade of the 20th century.[12] Step economy is an integral part of the development of organic synthesis as target molecules become more and more complex. Every step in a synthesis costs time, money, and effort, and every step of a synthesis has an economic and environmental impact. Consequently, reducing the number of steps in a synthesis is critically important. Too many steps, and a synthesis is no longer practical. Based on yield alone, if every step of a synthesis gives a 97% yield, a 10-step synthesis gives only a 76% overall yield, and a 20-step synthesis gives only a 58% yield. And these yields are unrealistic for every step in a synthesis. This is the one reason, of course, why convergent syntheses are so often viewed as being much more elegant than linear syntheses.

The relationship between target complexity and the number of steps required for its synthesis has changed over time. As we have developed new reactions capable of introducing greater complexity in a single step, the number of steps required or the synthesis has decreased, and the complexity of realistic target molecules has increased. This is illustrated graphically in the diagram on the left in Figure 15.2, which is an adaptation of the Wender diagram.[12d]

The diagram on the right in Figure 15.2 is a representation of how the length of total syntheses of the same target compound has changed over time, and it shows that there has been a general slow improvement in step economy until the discovery of important new reactions leads to a rapid increase in the value a brief period of time. The step economy then plateaus again until another synthetic breakthrough again rapidly improves the step economy of the synthesis. Of course, at some point, the number of steps will reach the absolute minimum number possible, at which point the synthesis will no longer be amenable to improvement in terms of step economy.

Molecular Complexity

Any discussion of step economy must also address the issue of molecular complexity. This has been the subject of a number of efforts over the years, beginning with Hendrickson's definition that we alluded to at the beginning of this section. In 1981, Bertz proposed one of the first general indexes of molecular complexity,[13] and he expanded on the definition in his subsequent works. Since Bertz' initial proposals, other measures of molecular complexity have been proposed.[14]

In its simplest form, the complexity of a molecule can be based on its line/polygon formula. In this procedure, one simply counts the number of valence angles possible in the formula, treating the component bonds of a multiple bond as separate entities. These are then summed to give the value of N, and the complexity index is calculated as $C_c = (2N \ln N)$.

---

11. "Green" chemistry has become so important that there are now journals devoted to it, including *Green Chemistry*, which is published by the Royal Society of Chemistry.

12. (a) Wender, P.A.; Bi, F.C.; Gamber, G.G.; Gosselin, F.; Hubbard, R.D.; Scanio, M.J.C.; Sun, R.; Williams, T.J.; Zhang, L. *Pure Appl. Chem.* **2002**, *74*, 25. (b) Wender, P.A.; Handy, S.T.; Wright, D.L. *Chem. Ind. (London)* **1997**, 765. (c) Wender, P.A.; Miller, B.L. In Hudlicky, T., Ed. *Organic Synthesis: Theory and Applications* (JAI: Greenwich, 1993); Vol. 2, pp 27–66. (d) Wender, P.A.; Croatt, M.P.;Witulski, B. *Tetrahedron* **2006**, *62*, 7505. (e) Wender, P.A.; Verma, V.A.; Paxton, T.J.; Pillow, T.H. *Acc. Chem. Res.* **2008**, *41*, 40. (cf Wender, P.A.; Miller, B.L. *Nature* **2009**, *460*, 197.

13. (a) Bertz, S.H. *J. Chem. Soc., Chem. Commun.* **1981**, 818. (b) Bertz, S.H. *J. Am. Chem. Soc.* **1981**, *103*, 3599. (c) Bertz, S.H.; Sommer, T.J. *Chem. Commun.* **1997**, 2409.

14. (a) Barone, R.; Chanon, M. *Tetrahedron* **2005**, *61*, 8916. (b) Chanon, M.; Barone, R.; Baralotto, C.; Julliard, M.; Hendrickson, J.B. *Synthesis* **1998**, 1559.

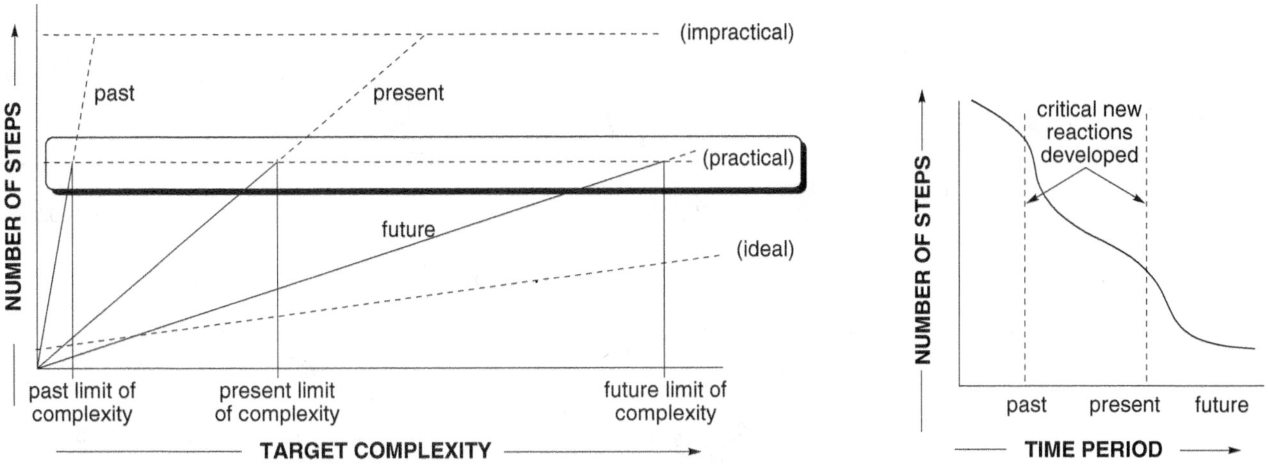

Figure 15.2 The relationship between target complexity and step economy over time

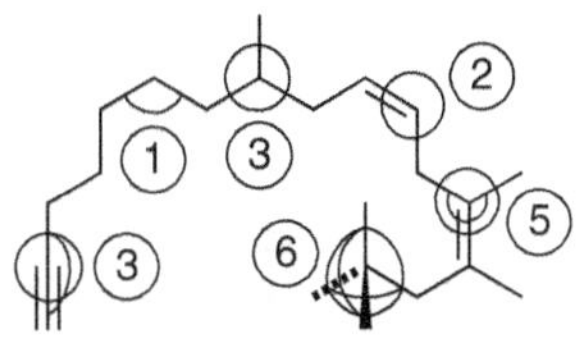

Figure 15.3 A simplified version of Bertz' index of molecular complexity

Each atom within an unbranched chain contributes one unit to the overall complexity index, a single branch point contributes three units, and a quaternary center contributes six. A fully substituted sp^2-hybridized carbon contributes five units, and a substituted sp-hybridized carbon atom contributes three (Figure 15.3).

Molecular symmetry reduces the complexity of the molecule. If a molecule possesses symmetry, the number of pairs of atoms related by symmetry (n_s) is counted, and the complexity index is reduced by the symmetry factor [$n_s \ln (n_s)$], to give $C_c = (2N \ln N - n_s \ln n_s)$. Heteroatoms also add to the complexity of a molecule, and they contribute an additional term that is calculated the same way as the carbon skeleton. If n_h is the total number of heteroatoms, the heteroatom coefficient, H, is given by $H = n_h + 1$; the coefficient for the number of symmetry-related pairs of heteroatoms is equal to h_s, defined analogously to n_s. Now, $C_h = (2H \ln H - h_s \ln h_s)$ and $C_{tot} = C_c + C_h$.

We can illustrate this form of the Bertz index by examining some simple molecules and reactions, shown in Figure 15.4. The first reaction is the Diels-Alder reaction between 1,4-naphthoquinone and cyclopentadiene. The complexity index calculated for the 1,4-naphthoquinone is derived from the total number of valence angles (34) and heteroatoms (2), but it is reduced by the number of pairs atoms (6) and heteroatoms (1) related by symmetry. This gives a complexity index for naphthoquinone of 235.6. The complexity indices for the cyclopentadiene and the cycloadduct may be calculated similarly and gives values of 35.6 and 351.9, respectively. These values allow us to determine that this Diels-Alder reaction increases the overall molecular complexity of the system by 80.7—a result very much in line with the qualitative view of the ability of this reaction to increase molecular complexity. Likewise, we find that aldol condensation of cyclohexanone results in an increase of (160.9 – 48.8 – 48.8) or 63.3 in molecular complexity. By comparison, the hydration of an alkene to an alcohol is a reaction that we do not usually view as adding significantly to molecular complexity of the system. In fact, the hydration of 1-methylcyclohexene leads to a small decrease (−10.9) in the total complexity index, and hydroboration of the same alkene leads to a slightly greater reduction in the complexity index (−14.5). The final example is an intramolecular photocycloaddition taken from Wender's synthesis of cedrene.[15] This reaction gives one of the largest increases in molecular complexity index (+54.3) for an intramolecular reaction.

15. Wender, P.A.; Howbert, J.J. *J. Am. Chem. Soc.* **1981**, *103*, 688.

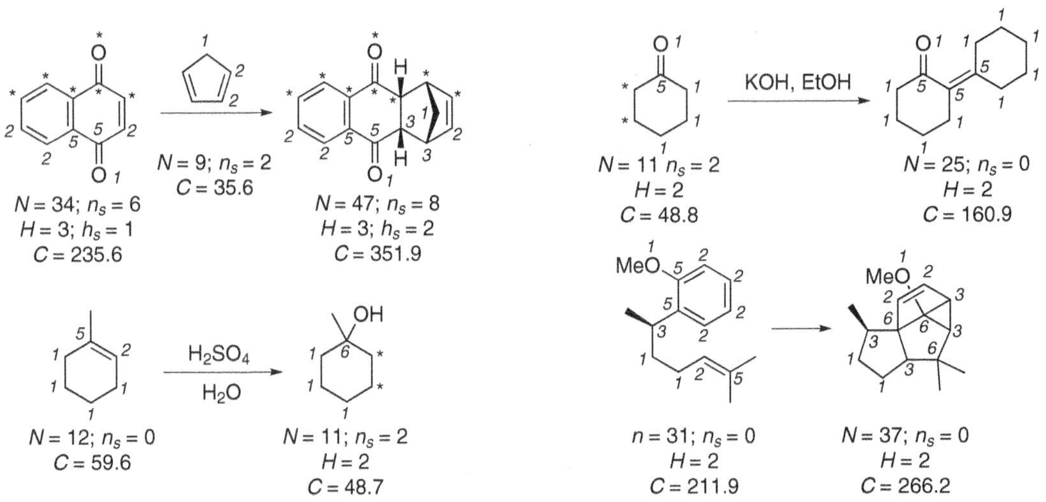

Figure 15.4 Complexity indices for some common reactions

Problem

15-1 What is the increase or decrease in the modified Bertz complexity index resulting from each of the following conversions?

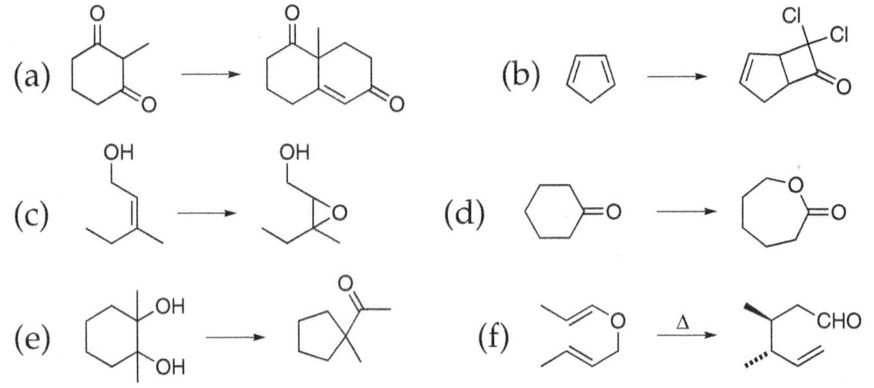

A third measure of synthetic efficiency may be found in the concept of redox economy.[16] Baran has extended this concept to define a measure of ideality based on the number and types of reactions in a synthesis[17]:

$$\%\ \text{ideality} = 100\left[\frac{(\#\ \text{of construction rxns}) + (\#\ \text{of strategic redox rxns})}{(\text{total}\ \#\ \text{of steps})}\right]$$

In this relationship, ***construction reactions*** are those that form C—C or C—heteroatom bonds in the skeleton of the target molecule, whereas ***strategic redox reactions*** are those that establish the correct functionality directly from the precursor. All other reactions are defined as ***concession steps***—steps that are necessary as part of the synthesis but do not directly lead to skeletal bonds or the correct final functionality in the final product. Reactions of this type are non-strategic redox reactions (e.g., conjugated ester to allyl

16. (a) Richter, J.M.; Ishihara, Y.; Masuda, T.; Whitefield, B.W.; Llamas, T.; Pohjakallio, A.; Baran, P.S. *J. Am. Chem. Soc.* **2008**, *130*, 17938. (b) Burns, N.Z.; Baran, P.S.; Hoffmann, R.W. *Angew. Chem., Int. Ed.* **2009**, *48*, 2854.
17. Gaich, T.; Baran, P.S. *J. Org. Chem.* **2010**, *75*, 4567.

alcohol to conjugated aldehyde), functional group interchanges (e.g., the conversion of alcohols to sulfonates prior to further reaction), and protecting group manipulations. Attaining ideality in a synthesis is not as simple as it might seem, and even the most elegant syntheses are often no better than 50% ideal under this definition.

So Which Metric Do You Use?

Good question. Even given the discussion above, there is still no consensus about how one evaluates two different syntheses of the same target compound. How they are ranked will depend on the perspective of the evaluator, and a combination of all three types of economy or ideality will almost certainly be taken into account. As you develop your own synthetic instincts, you will almost certainly also develop your own values for evaluating syntheses. Good luck![§]

15.2 **Organic Reactions: Tools of the Synthetic Chemist**

Let begin by examining the general form of an organic reaction once again. In introductory courses in organic chemistry, reactions are studied in terms of a starting compound reacting with a reagent to give a product: "A reacts with B to give C," as shown.

$$A + B \rightarrow C$$

For the synthetic organic chemist, however, the critical question is seldom "What is the product of the reaction between A and B?" but much more often "How can I convert compound A into compound B?" When one must plan a synthesis, one does not normally have the reactant and reagents, but one is usually given instead an end result—the target molecule—and must determine the best way to obtain it. The synthetic chemist is more often faced with the more general questions: "What reactions can be used to make C?" and "What is the *best* way to make C?"

$$? + ? \rightarrow C$$

This is not so uncommon a situation as you might think at first. In fact, being confronted by an end result and needing to piece together how it happened is so common in modern living that we have a term for it—we call it "troubleshooting." Designing a synthesis involves the same kinds of thought processes. We "troubleshoot" problems by working backward from the problem to its cause, and we design a synthesis by working backwards from the target compound to available precursors. Consequently, the study of organic reactions from the viewpoint of the synthetic chemist is usually more concerned with practical details than we have been so far. Small differences in reactivity or stereochemistry are relatively inconsequential when one is studying a single reaction; they can be critically important when that reaction is part of a sequence.

The three participants in the reaction above are members of a closely linked set, so this reaction can also be viewed in the light that "A and B always give C." This has other implications: it also means that "To get C from A, you must use B," and "It is A that reacts with B to give C," although we have not explicitly stated it this way yet. So any time one knows two of the three participating species in a reaction (reactant, reagent and product), one can

§ There is an unspoken caveat in our discussion of organic synthesis—this chapter and those that follow will contain the biases of the author, and I have chosen examples that I believe are both important and instructive. Not everyone will agree with these choices, so, like medical advice, it is probably worth your while to get a second opinion on some of these choices—talk to your instructors and colleagues, and read and appreciate syntheses for yourself.

deduce the third. Thus, one can easily categorize reactions according to one of the three protocols.

A + B → ?	[Answer: C]
? + B → C	[Answer: A]
A + ? → C	[Answer: B]

We can illustrate the preceding discussion with a relatively simple reaction that exhibits some of the more important characteristics that are useful to the chemist in the process of designing a synthesis—the reaction between 1-methylcyclohexene and aqueous bromine to give the racemic *trans*, Markovnikov bromohydrin (Example **15.7**).

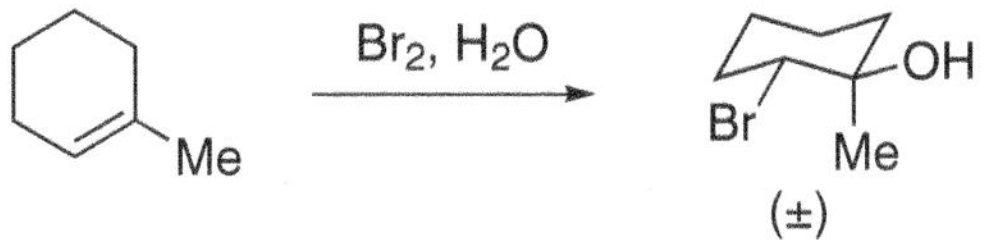

(15.7)

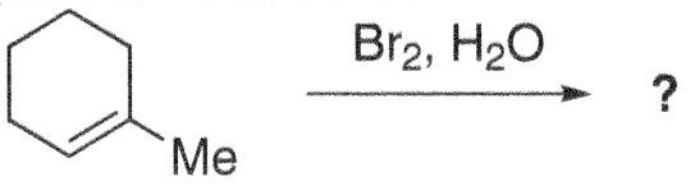

(15.8)

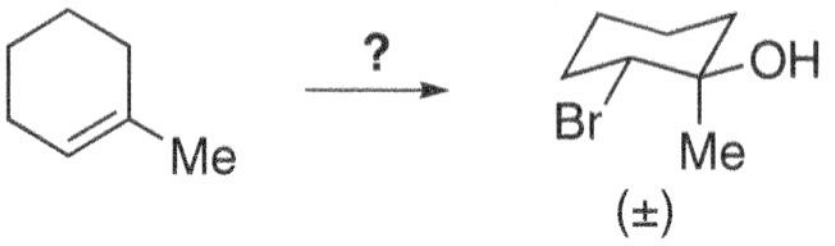

(15.9)

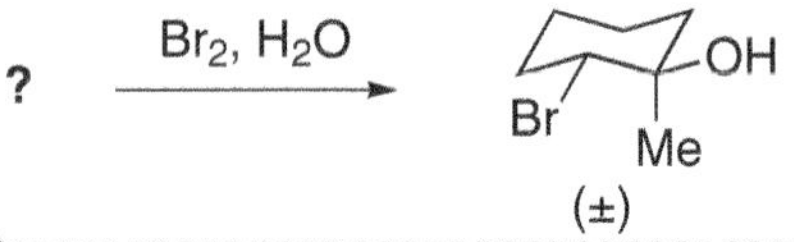

(15.10)

For this reaction, the three protocols above can now be represented as shown (Examples **15.8** to **15.10**).

Most of the time, one of these three ways of writing a reaction will suffice to pick the best way to make a molecule. Or, using the specific example above, we get the problem shown as Example **15.11**.

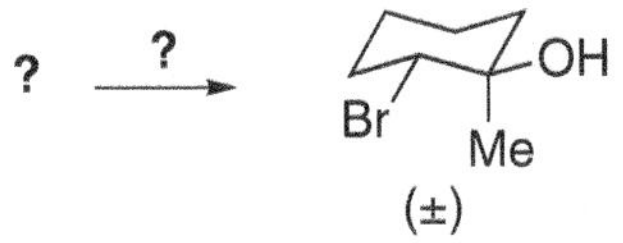

(15.11)

Historical Development of Organic Synthesis[18]

Following Wöhler's serendipitous synthesis of urea in 1828, chemists gradually turned to the process of deliberately building organic molecules. The first such successful designed synthesis was the total synthesis of acetic acid by Hermann Kolbe.[19] This synthesis went a long way toward ending the reign of vital force theory. From these humble beginnings, the edifice that is modern organic synthesis arose. The rise of selectivity in organic synthesis is shown in Figure 15.5.

---

18. For an entertaining review of the development of modern synthesis to the end of the twentieth century, see: Nicolaou, K.C.; Vourloumis, D.; Winssinger, N.; Baran, P.S. *Angew. Chem. Int. Ed.* **2000**, *39*, 44.
19. Kolbe, H. *Ann. Chem. Pharm.* **1845**, *54*, 145.

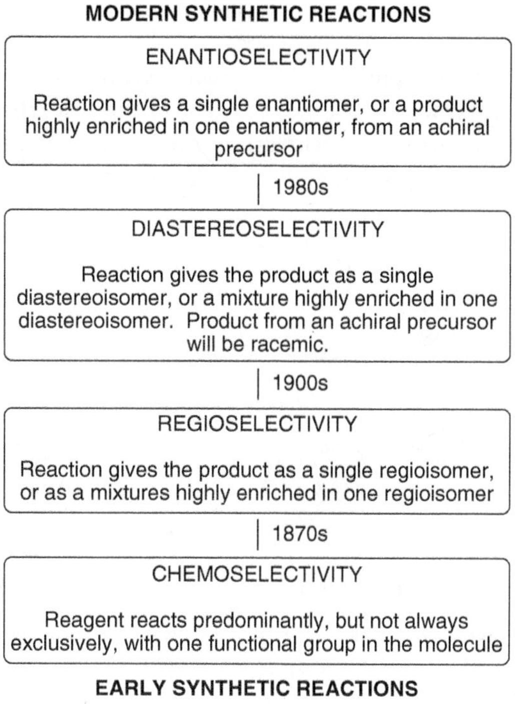

Figure 15.5 The rise of sophistication in organic synthesis

The first decades of organic synthesis were characterized by the development of a wide range of new organic reactions. During these decades, the questions addressed were in the area of ***chemoselectivity***—designing reactions that were selective for one functional group over another. This represents the lowest level of sophistication for selective transformations in organic compounds, and includes reactions such as the Tollens oxidation of aldehydes.[20]

As the number of selective reactions available to the organic chemist increased, the level of sophistication in their use also rose. In the 1870s, for example, the first investigations of reaction regiochemistry led to the formulation of Markovnikov's rule[21] for addition and the Zaitsev[22] and Hofmann[23] rules for elimination. With the formulation of these rules, chemists could begin to plan syntheses with reasonable levels of ***regioselectivity***.

By the turn of the 20th century, organic synthesis had become sufficiently sophisticated to allow the synthesis of moderately complex molecules as single diastereoisomers (e.g., the syntheses of camphor by Komppa[24] and glucose by Fischer[25] belong to this class). This period, when total syntheses of complex molecules were characterized by high levels of ***diastereoselectivity***, reached its zenith during the middle of the century, largely as a result of the work of Robert B. Woodward, who reported total syntheses of complex natural products such as strychnine[26] and reserpine.[27] In almost every case, however, the

20. Tollens, B. *Ber. Deut. chem. Ges.* **1882**, *15*, 1635.

21. (a) Markownikoff, W. *Ann. Chem. Pharm.* **1870**, *133*, 228. (b) Markovnikov, V. *Comptes rend.* **1875**, *82*, 668, 728, 776.

22. Saytzeff, A. *Ann. Chem. Pharm.* **1875**, *179*, 296.

23. Hofmann, A.W. *Ann. Chem. Pharm.* **1851**, *78*, 253; **1851**, *79*, 11.

24. Komppa, G. *Chem. Ber.* **1903**, *36*, 4332.

25. Fischer, E. *Ber. Deut. chem. Ges.* **1890**, *23*, 799.

26. (a) Woodward, R.B.; Cava, M.P.; Ollis, W.D.; Hunger, A.; Daeniker, H.U.; Schenker, K. *J. Am. Chem. Soc.* **1954**, *76*, 4749. (b) Woodward, R.B.; Cava, M.P.; Ollis, W.D.; Hunger, A.; Daeniker, H.U.; Schenker, K. *Tetrahedron* **1963**, *19*, 247.

27. Woodward, R.B.; Bader, F.E.; Bickel, H.; Frey, A.J.; Kierstead, R.W. *J. Am. Chem. Soc.* **1956**, *78*, 2023. (b) Woodward, R.B.; Bader, F.E.; Bickel, H.; Frey, A.J.; Kierstead, R.W. *J. Am. Chem. Soc.* **1956**, *78*, 2657. (c) Woodward, R.B.; Bader, F.E.; Bickel, H.; Frey, A.J.; Kierstead, R.W. *Tetrahedron* **1958**, *2*, 1.

synthesis reported the preparation of the racemate, and not the enantiomerically pure material.

The next advance in the sophistication of organic synthesis came in the last two decades of the 20th century, which saw the discovery of the first reliable methods for the synthesis from achiral or prochiral precursors of chiral molecules as single enantiomers[28] (or nearly so). With these discoveries, a new era of ***enantioselectivity*** in organic synthesis began.

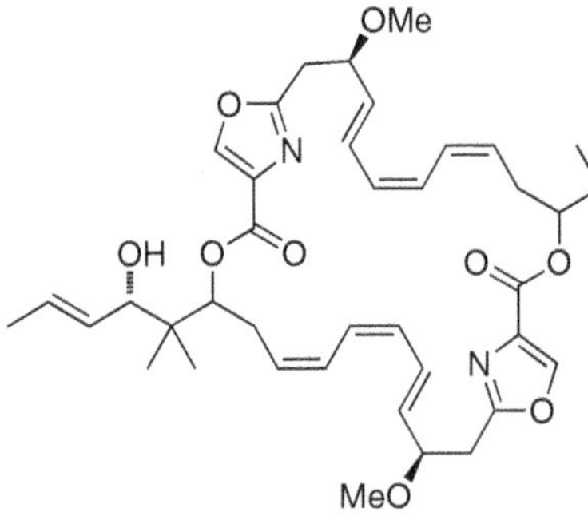

(15.12)

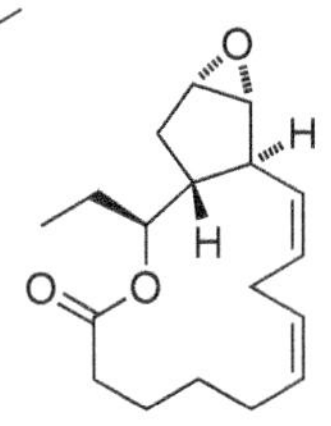

(15.13)

The turn of the 21st century has seen yet another advance in the toolkit available to the modern organic chemist—the development of catalysts for a wide range of carbon-carbon bond-forming reactions. It is worthwhile noting that the Nobel Prize in chemistry has been awarded three times since the turn of the century for reactions using catalysts. Just how powerful catalytic reactions can be in synthesis is illustrated by Wipf's synthesis of (–)-disorazole C_1 (Example **15.12**),[29] whose deceptively simple C_2-symmetric macrocycle was closed by means of a palladium-catalyzed Sonagashira coupling and whose conjugated (*E*,*Z*,*E*)-trienes must be introduced into the macrocycle very late in the synthesis, after much of the sensitive functionality is already in place—a serious challenge because one must find selective reactions that will not affect the sensitive groups in the molecule. An even more impressive illustration of the power of modern catalytic methods for synthesis is provided by Fürstner's synthesis of the ecklonialactones (Example **15.13**), a 13-step synthesis in which five catalytic reactions are used as key steps and in which *protecting groups are not required*.[30] In addition, it has also seen the rise of ***combinatorial chemistry*** or ***modular organic synthesis***, now a significant part of modern drug discovery.

Since the middle of the 20th century, the sheer number of organic reactions available to the modern synthetic organic chemist has been growing at an incredible rate, so some means of classifying reactions becomes essential if one is to be able to design a reasonable synthesis. There are many ways in which this can be accomplished, such as dividing reactions by the type of reagents involved or by the stereo- or regiochemical outcomes of the reactions involved. In Chapter 5, we categorized reactions based on the frontier orbitals involved. Table 15.1, categorizes reactions according to the number and type of bonds undergoing changes occurring during the reaction, and the types of changes occurring. Although this is a very simplistic method of characterization, it is one that I have personally found to be useful in organizing organic reactions. It is worthwhile noting that, as with any system to organize organic reactions, these categories are somewhat arbitrary and are not exclusive—a reaction may fall in more than one category, and many do. The important thing is to find a way for categorizing reactions that works for you.

---

28. (a) Katsuki, T.; Sharpless, K.B. *J. Am. Chem. Soc.* **1980**, *102*, 5974. (b) Knowles, W.S.; Sabacky, M.J.; Vineyard, B.D.; Weinkauff, D.J. *J. Am. Chem. Soc.* **1975**, *97*, 2567. (c) Knowles, W.S. *Acc. Chem. Res.* **1983**, *16*, 106. (d) Knowles, W.S. *J. Chem. Educ.* **1986**, *63*, 222.

29. Wipf, P.; Graham, T.H. *J. Am. Chem. Soc.* **2004**, *126*, 15346.

30. Hickmann, V.; Alcarazo, M.; Fürstner, A. *J. Am. Chem. Soc.* **2010**, *132*, 11042.

Table 15.1 Categorizing Organic Reactions

ONE-BOND (C—C) REACTIONS	
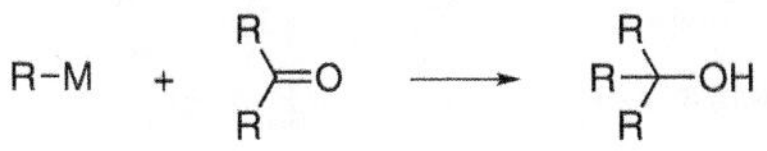	Grignard addition
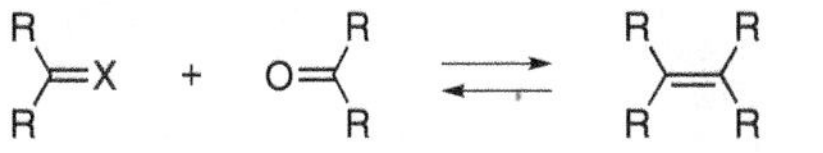	Wittig reaction (X=PR_3); McMurry reaction (X=O) ozonolysis; Lemieux-Johnson cleavage, etc.
	Friedel-Crafts acylation Gattermann-Koch formylation Vilsmeier-Haack formylation
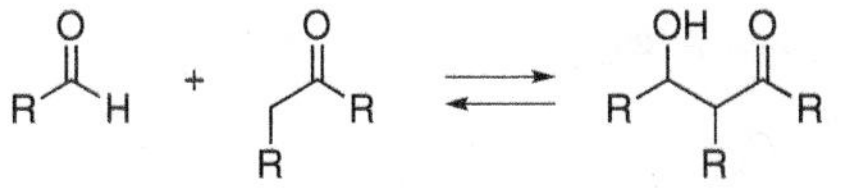	aldol addition; Bayliss-Hillman reaction; etc. retro-aldol fragmentation
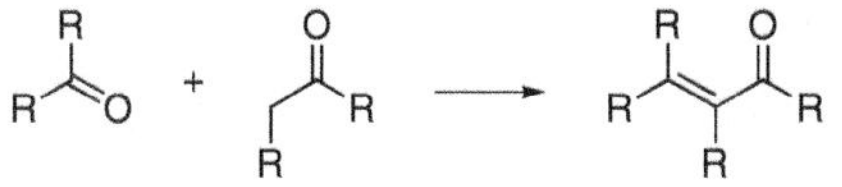	aldol condensation
	alkylation of enolates
	alkylation of alkynides
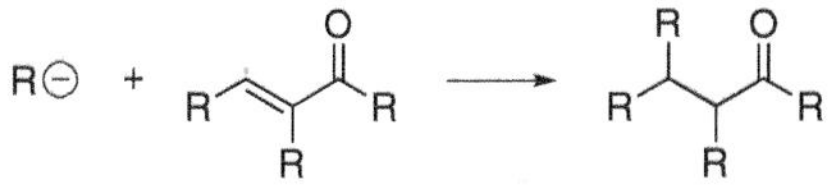	Michael addition
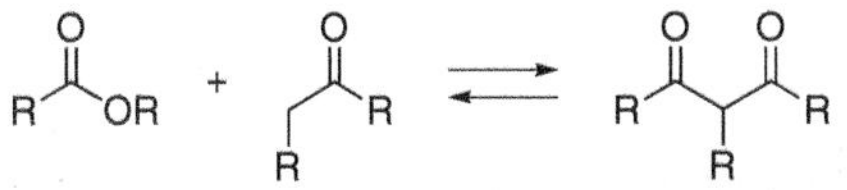	Claisen condensation retro-Claisen fragmentation
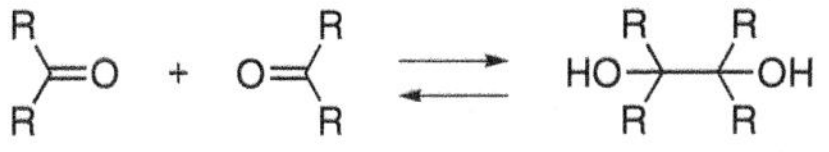	pinacol reaction glycol cleavage
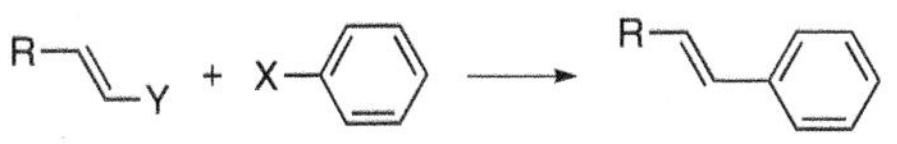	Heck coupling (X=halogen; Y=H) Suzuki coupling (X=halogen; Y=$B(OR_2)$)
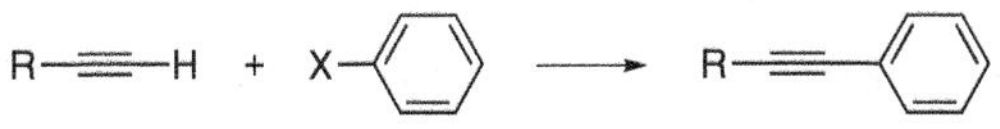	Sonogashira coupling

(continued)

Table 15.1 Categorizing Organic Reactions (continued)

ONE-BOND (C—X) REACTIONS	
	α-halogenation, sulfenylation, selenylation etc. α-hydrogenolysis
	allylic/benzylic halogenation, etc. allylic/benzylic hydrogenolysis, etc.
	nucleophilic substitution (S_N1, S_N2, etc.)
	aromatic nitration, halogenation, etc.
MULTI-BOND REACTIONS	
	acetal/thioacetal/ketal formation acetal/thioacetal/ketal hydrolysis
	electrophililc addition of HX base-promoted elimination (E1, E2, E1cb)
	electrophililc addition of XY, hydroxylation, etc. reductive elimination
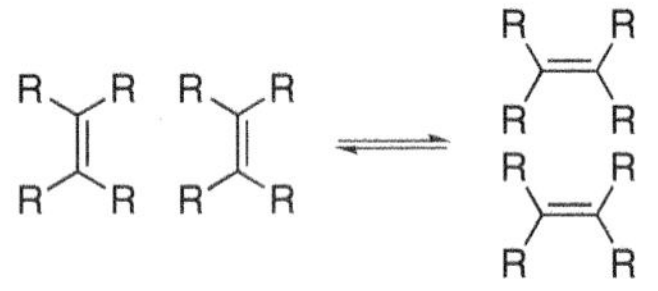	alkene metathesis
	Wittig/McMurry reactions ozonolysis, Lemieux-Johnson oxidation, etc.
	cyclopropanation (Simmons-Smith, etc.)
	epoxidation
	Darzens condensation; sulfur ylide addition, etc.
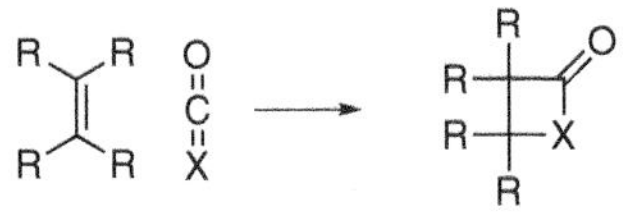	[2+2] cycloaddition
	[4+2] cycloaddition (Diels-Alder reaction)

(continued)

Table 15.1 Categorizing Organic Reactions (continued)

OXIDATION AND REDUCTION REACTIONS

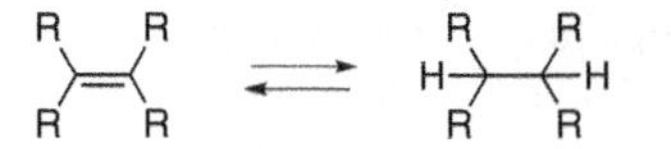	hydrogenation (→) dehydrogenation (←)
	epoxidation
	addition of halogens; hydroxylation, etc. (→) reductive elimination (←)
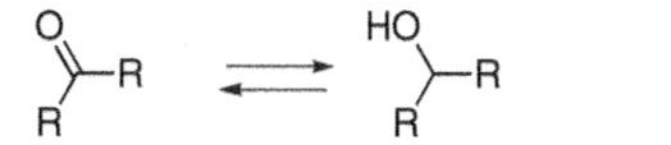	reduction of carbonyl compounds (→) oxidation of alcohols (←)
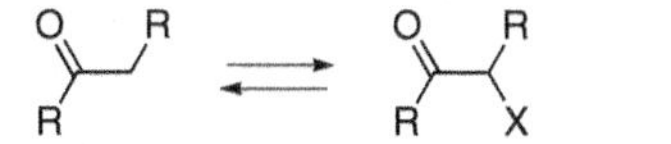	α-halogenation, sulfenylation, sulfonation, etc. (→) hydrogenolysis of α-halocarbonyl compounds, etc. (←)
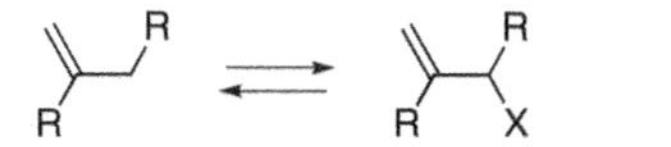	allylic halogenation, oxidation, etc. (→) allylic hydrogenolysis (←)
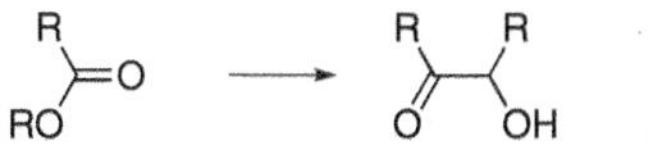	acyloin condensation
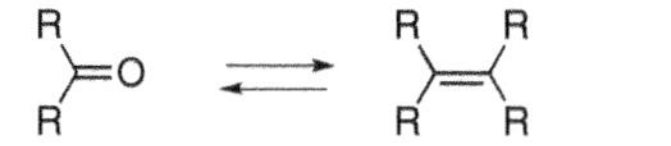	McMurry reaction (→) ozonolysis; Lemieux-Johnson oxidation (←)
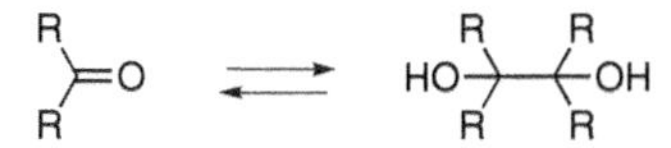	pinacol reduction (→) periodate oxidation (←)
	Wolff-Kishner, Clemmensen or related reductions
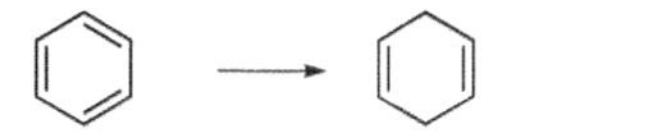	Birch reduction
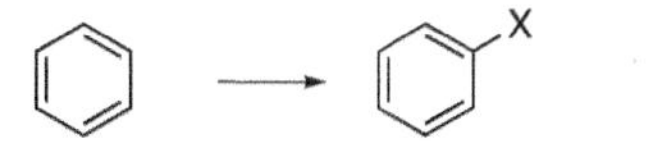	nitration, halogenation, etc. of aromatic rings

REARRANGEMENT REACTIONS

	Wagner-Meerwein rearrangement of cations
	pinacol rearrangement
	Favorskii rearrangement
	Hofmann, Curtius, Lossen rearrangements
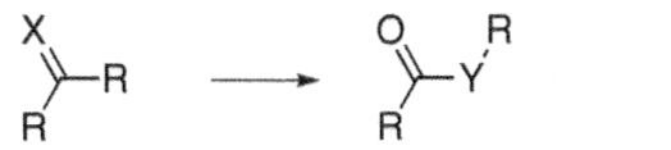	Baeyer-Villiger rearrangement (X=O; Y=O) Beckmann rearrangement (X=NOH; Y=NH) Schmidt reaction (X=O; Y=NH)
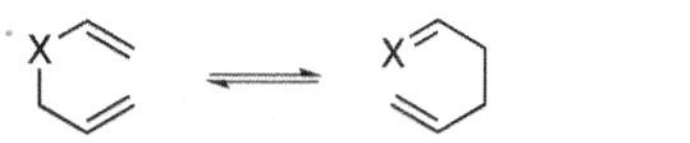	Cope (X=CR_2) and Claisen (X=O) rearrangements other sigmatropic rrearrangements
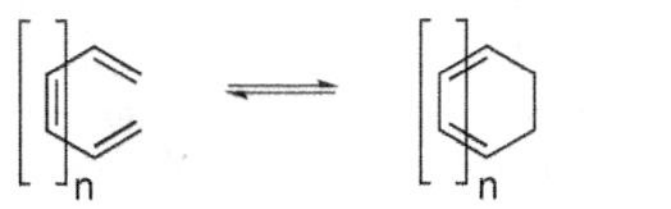	electrocyclization

Worked Problem

15-1 Categorize each of the reactions in the list below according to the following criteria: (1) chemoselectivity (what functional groups may be involved?); (2) regioselectivity (are regioisomers formed, and, if so, which is formed in major amount?); (3) stereoselectivity (what is the stereochemistry of the reaction, and which stereoisomer is formed in major amount?); and (4) interference (what functional groups interfere with the reaction, and are side reactions possible?)

(a) dichlorocarbene addition to an alkene

(b) S_N2 substitution

(c) addition of Br_2, H_2O to an alkene

§Answers below.

Problem

15-2 Categorize each of the reactions in the list below according to the following criteria: (1) chemoselectivity (what functional groups may be involved?); (2) regioselectivity (are regioisomers formed, and, if so, which is formed in major amount?); (3) stereoselectivity (what is the stereochemistry of the reaction, and which stereoisomer is formed in major amount?); and (4) interference (what functional groups interfere with the reaction; are side reactions possible?)

(a) Grignard addition to a carbonyl compound

(b) Diels-Alder cycloaddition

(c) E2 elimination from an alkane derivative

(d) Hydroboration-oxidation

(e) Alkylation of a ketone

(f) S_N1 substitution

(g) Acid-catalyzed dehydration

(h) Friedel-Crafts acylation of anisole

(i) Michael addition to an enone

(j) Wadsworth-Emmons reaction of aldehydes

§ Answers to Worked Problem:

(a) Dichlorocarbene addition to an alkene: (1) the reaction involves the π bond of an alkene; (2) because the reagent adds at both ends of the double bond simultaneously, there is no regiochemistry involved; (3) the reaction is stereospecific, and proceeds by *syn* (suprafacial) addition only; (4) the carbene will react with any strongly nucleophilic groups such as primary or secondary amines, and phenoxide anions.

(b) S_N2 substitution: (1) the reaction involves a nucleophile and a compound with a good leaving group bonded to a primary or secondary sp^3-hybridized carbon atom; (2) since no rearrangement is involved, regiochemistry is not a factor in this reaction; (3) the reaction is stereospecific and gives the product with inversion of configuration at carbon; (4) acidic groups often protonate the nucleophile and render it unreactive.

(c) Addition of Br_2, H_2O to an alkene: (1) The reaction is typical of reactions involving an electrophile with a lone pair on the electrophililc atom and an alkene π bond; (2) the reagent adds to the alkane with predominantly Markovnikov regiochemistry—the halogen ends up bonded to the atom less capable of bearing a formal positive charge; (3) the reaction is stereospecific, and gives overall *anti* (antarafacial) addition; the reaction may be diverted by other functional groups capable of reacting with the bromonium ion intermediate (e.g., electron-rich aromatics).

(k) Catalytic hydrogenation
(l) Simmons-Smith reaction
(m) Baeyer-Villiger oxidation
(n) Epoxide ring opening
(o) Allylic bromination with NBS
(p) Wittig reaction of aldehydes
(q) Aldol addition
(r) Olefin metathesis
(s) Suzuki coupling

Using Stereochemical Outcomes for Categorizing Organic Reactions

Another reaction characteristic that is extremely important, especially in modern syntheses, is the stereochemistry of the reaction, and all modern syntheses strive for some level of ***stereocontrol***. In stereospecific reactions, stereocontrol is complete. Only one stereoisomer of the product is formed, its stereochemistry is predictable based on the stereochemistry of the reactant, and its stereochemistry changes if we change the reactant stereochemistry. This makes stereospecific reactions important tools for controlling stereochemistry during a synthesis. In stereoselective reactions, the level of stereocontrol is not quite so high. Now more than one stereoisomer of the product is formed, and the stereochemistry of the product is not necessarily related to that of the reactant. Even so, many stereoselective reactions exhibit high levels of stereoselectivity and are useful tools for synthesis. Some of the more important stereocontrolled reactions available to the synthetic chemist are presented in Table 15.2.

Let us now end this section of the discussion with another piece of advice—also probably unsolicited. You should remember that organic synthesis is probably the most individualistic of the chemical sciences, so that you may find that an alternative method of classification that works better for you. If that is the case, do it! As you begin the process of designing your own syntheses, it is worthwhile expending some effort to develop your own method of classification of reactions for use in answering synthetic problems.

15.3 Retrosynthetic Analysis: An Introduction to the Vocabulary and Strategy of Organic Synthesis

Every organic structure contains within itself clues to how it might be assembled. The critical problem facing the organic chemist planning a synthesis is to choose the pathway that is most efficient. For the beginning student, this can be a daunting task—just choosing where to *start* can be frightening! So where *does* one start?

One of the most logical starting points is the one we have already adopted: categorizing organic reactions according to their effect on the reactant structure. By doing so, we become acquainted with the tools of organic synthesis and how to use them to control compound structure and stereochemistry.

Retrosynthetic Analysis

The process of analyzing the structural and functional features of a target molecule to see how it might be formed from simpler compounds is known as ***retrosynthetic analysis***. This process was first placed on a formal footing by Harvard chemist E. J. Corey, who

Table 15.2 Stereocontrolled Reactions

SUPRAFACIAL ADDITIONS	
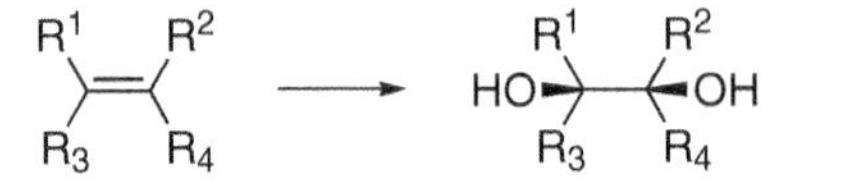	hydroxylation
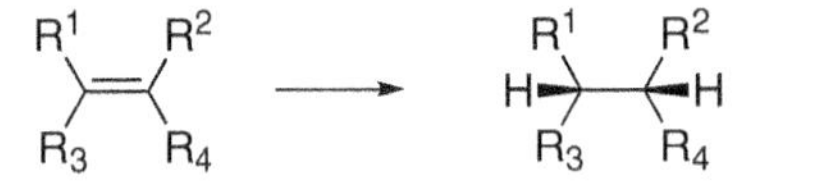	hydrogenation
	Lindlar hydrogenation
	epoxidation
	cyclopropanation
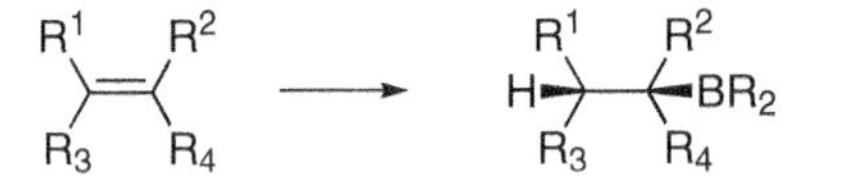	hydroboration
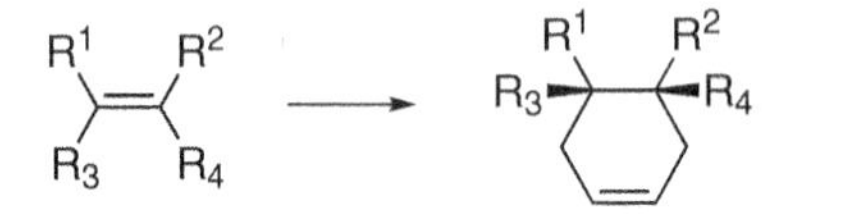	Diels-Alder cycloaddition
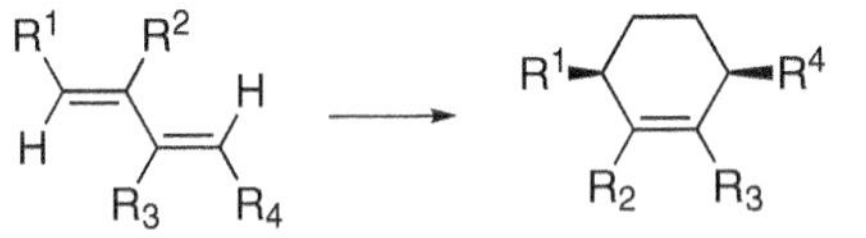	Diels-Alder cycloaddition
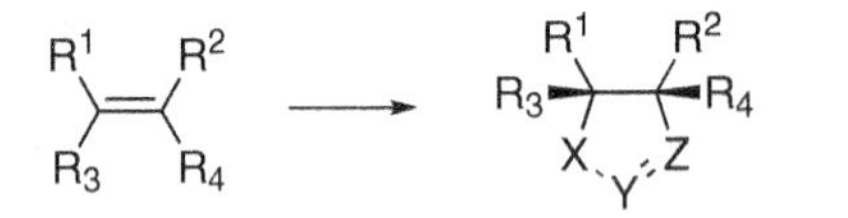	1,3-dipolar cycloaddition
	hydrometallation; M=Al, Zr, etc.
ANTARAFACIAL ADDITIONS	
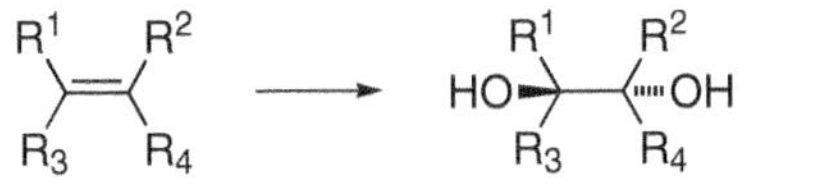	*anti*-hydroxylation
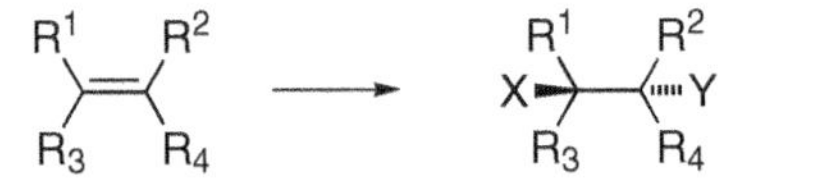	halogen addition, etc.
	Birch Reduction

(continued)

Table 15.2 Stereocontrolled Reactions (continued)

ELIMINATIONS

	E2 elimination (*anti*)
	reductive elimination *syn* or *anti*, depending on leaving group)
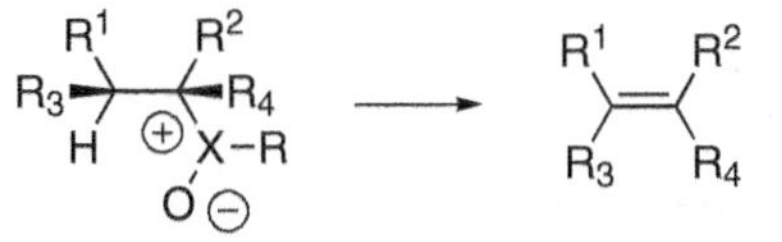	Cope elimination (X=N) (*syn* at low temperatures) selenoxide elimination (X=Se) (*syn*)
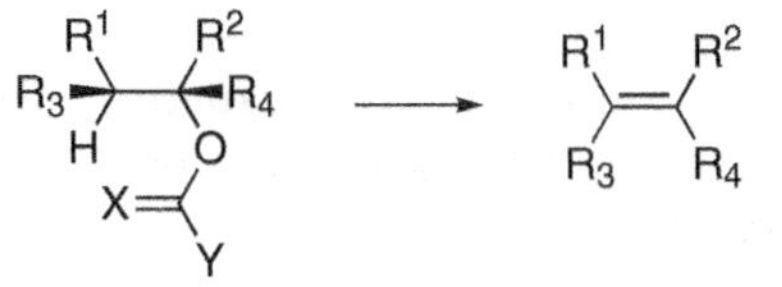	ester pyrolysis (X=O, Y=alkyl) xanthate pyrolysis (X=S, Y=SR) (*syn*)
	Diels-Alder cycloreversion (*syn*)

RETENTION OF CONFIGURATION

	electrophilic substitution of organometallics
	carbene (Y=CR_2) or nitrene (Y=NR) insertion
	hydrogenolysis over Ni
	Wagner-Meerwein rearrangement
	Baeyer-Villiger oxidation Beckmann rearrangement Schmidt rearrangement
	Hofmann rearrangement Curtius rearrangement Lossen rearrangement

(continued)

Table 15.2 Stereocontrolled Reactions (continued)

INVERSION OF CONFIGURATION

	S_N2 substitution; X=halogen, sulfonate Mitsunobu reaction (X=OH)
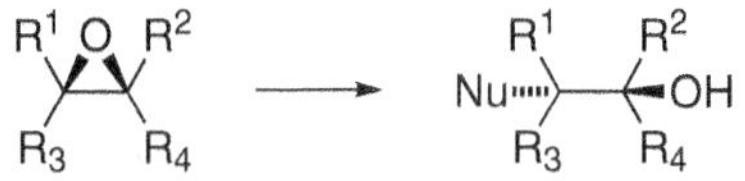	epoxide ring opening
	hydrogenolysis over Pd

developed much of the terminology of modern organic synthesis.[31] It is now routinely applied by organic chemists in designing syntheses of organic compounds.[32] In Corey's terminology, we define two directions: ***synthetic*** and ***antithetic*** (or ***retrosynthetic***).

> Processes that occur in the synthetic direction are called ***reactions***; they occur in the real world and describe the conversion of one *real* substance into another. One *can actually perform* a reaction in a laboratory and isolate the product or products.
>
> Processes that occur in the retrosynthetic direction are ***transforms***; they describe *possible* ways in which the target molecule might be disconnected to potential precursors, or ***synthons***. Transforms do not have an existence in the real world—they are conjectural intellectual constructs that simply describe possibilities. One *cannot perform* a transform in a laboratory.

Likewise, synthons are not necessarily stable molecules—they may be reactive intermediates such as a carbene, a carbocation or a carbanion or simply be structures that show reactivity patterns required. However, they do show the type of reagent that will be needed to effect the particular conversion in the synthetic direction. We distinguish between reactions and transforms by the type of arrow that we use to draw the process. The difference between synthesis and the way we have learned about organic reactivity is illustrated in Figure 15.6 by using a simple conjugate addition reaction as our example.

The verbal description of the *reaction* in Figure 15.6 would be "2-cyclohexenone reacts with lithium diethylcuprate in tetrahydrofuran, or THF, to give 3-ethylcyclohexanone." This process has an existence in the physical world. One can actually carry out this reaction and convert 2-cyclohexenone into 3-ethylcyclohexanone. The verbal description of the *transform*, on the other hand, makes it clear that one is describing only one possibility from a range of possibilities. "It is possible to disconnect the 3-ethylcyclohexanone molecule into an electrophilic

31. (a) Corey, E.J. *Pure Appl. Chem.* **1967**, *14*, 19. (b) Corey, E.J. *Chem. Soc. Rev.* **1988**, *17*, 111.

32. There are many textbooks and treatises dedicated to organic synthesis. Representative examples include: (a) Warren, S. *Organic Synthesis: The Disconnection Approach* (Wiley-Interscience: New York, 1982). (b) Corey, E.J.; Cheng, X.-M. *The Logic of Chemical Synthesis* (Wiley-Interscience: New York, 1989). (c) Ho, T.-L. *Tactics of Organic Synthesis* (Wiley-Interscience: New York, 1994). (d) Norman, R.O.C.; Coxon, J.M. *Principles of Organic Synthesis*, 3rd. ed. (Blackie Academic & Professional: Oxford, 1993). (e) Smith, M.B. *Organic Synthesis*, 2nd. ed (McGraw-Hill: Boston, 2002). (f) Boger, D.L. *Modern Organic Synthesis* (TSTI Press: La Jolla, 1999). (g) Fuhrhop, J.; Penzlin, G. *Organic Synthesis: Concepts, Methods Starting Materials* (Verlag Chemie: Weinheim, 1983). (h) Zweifel, G.S.; Nantz, M.H. *Modern Organic Synthesis: An Introduction* (W.H. Freeman & Co: New York, 2007). (i) Hudlicky, T.; Reed, J.W. *The Way of Synthesis* (Wiley-VCH: New York, 2007).

There are several multivlume treatises on aspects of synthesis, also: (a) Barton, D.H.R.; Ollis, W.D., Eds. *Comprehensive Organic Chemistry* (Pergamon Press: Oxford, 1979). (b) Trost, B.M.; Fleming, I., Eds. *Comprehensive Organic Synthesis* (Pergamon Press: Oxford, 1992). (c) *Science of Synthesis* (Thieme Chemistry: Stuttgart, 2000–2010). This multivolume treatise is an ongoing project with over 1700 contributing authors to date, and regular updates since then.

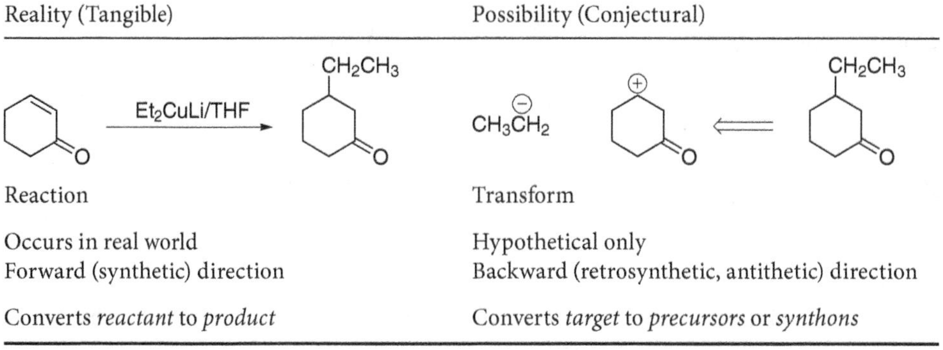

Figure 15.6 Comparing reactions and transforms

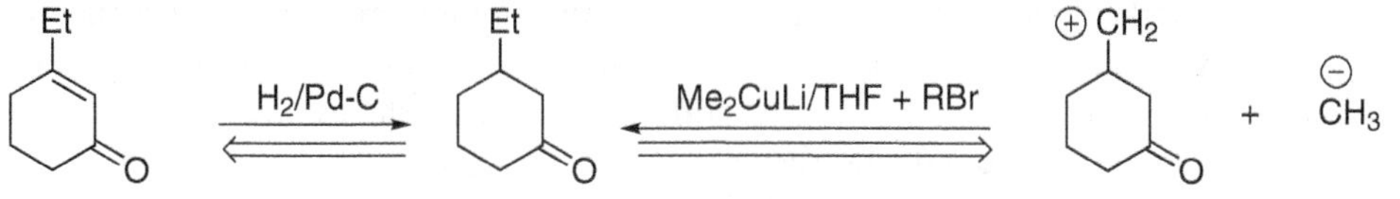

Figure 15.7 Alternative synthons for 3-ethylcyclohexanone

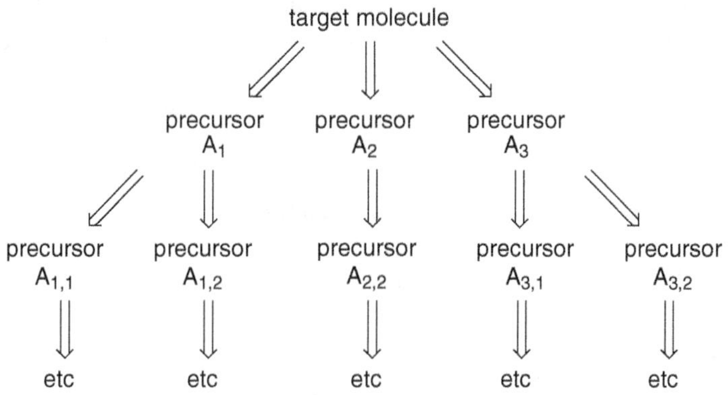

Figure 15.8 A retrosynthetic tree

cyclohexanone synthon with the electrophilic center β to the carbonyl group and a nucleophilic ethyl synthon." This retrosynthetic analysis is then translated into terms that one can use to select a reaction: the nucleophilic ethyl synthon could be an organometallic reagent such as lithium diethyl cuprate, and 2-cyclohexenone could serve as the electrophilic ketone synthon. Of course, there are other ways that one might envisage making this alkane. It might be prepared by catalytic hydrogenation of an alkene or by a nucleophilic substitution reaction, as shown in Figure 15.7, where the reactions corresponding to these transforms are also given.

Once you have become familiar with the principles of retrosynthetic analysis, the process of designing a synthesis simply becomes one of applying these principles to the target molecule and the successively generated precursors until one arrives at a point where the synthon generated corresponds to an available compound or one that has been prepared previously. This process quickly generates a ***retrosynthetic tree***, which is simply a collection of potential pathways radiating backward from the target compound to available starting materials (Figure 15.8).

In principle, one should be able to generate a retrosynthetic tree containing every possible synthetic pathway for the preparation of the target compound. Such an undertaking would, however, be extremely time-consuming, and few bother to complete the task once a viable route to the target compound is identified. In practice, many of the potential pathways are eliminated from consideration before they are completely described because the early steps do not result in any simplification in the problem. For example, if we examine the example on the right in Figure 15.7, we note that the disconnection used really does not simplify the problem in this synthesis at all—the electrophilic synthon still carries most of the structure and all of the functionality of the target compound. It is doubtful if it would

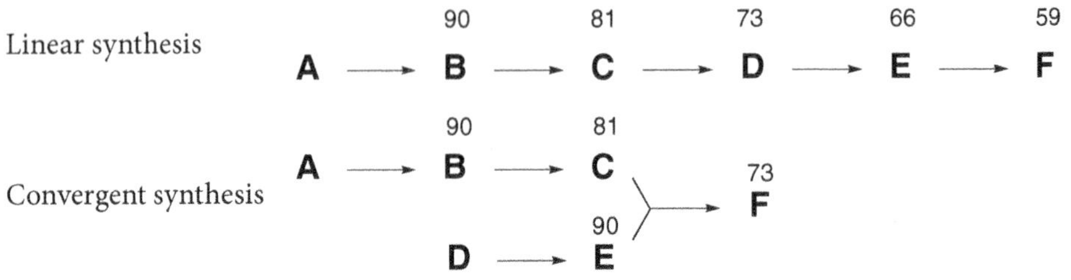

Figure 15.9 Comparing linear and convergent syntheses

actually be used in a real synthesis unless there was some specific reason for using it (e.g., to incorporate a labeled carbon atom from methyl iodide).

Syntheses may be ***linear*** or ***convergent***. In a linear synthesis, the target molecule is assembled by a single sequence of reactions from a single starting compound. In a convergent synthesis, the target molecule is assembled by two or more independent, parallel sequences of reactions to form major substructures of the target molecule that are assembled in the late stages of the synthesis. In general, convergent syntheses give higher overall yields of the target compound, and they require fewer steps. The comparisons of linear and convergent syntheses are illustrated schematically in Figure 15.9, where the numbers above the stages show the overall percent yield, assuming a 90% yield for each step.

Now let us study the process of retrosynthetic analysis in detail by examining Corey's original approach. Recall that this is the process of applying transforms to a target molecule to generate synthons that can then be used to generate a retrosynthetic tree. Every retrosynthetic transform corresponds to a reaction, and every reaction has its corresponding transform. The reactions that one commonly uses were summarized in Tables 15.1 and 15.2; they can be conveniently divided into six major types:

1. Reactions that remove a functional group. These reactions often occur near the end of a synthesis, when it is necessary to remove a functional group (e.g., a carbonyl group or double bond) that has been used in the construction of the molecule. These reactions, which are typified by catalytic hydrogenation or the Wolff-Kishner reduction, are represented in retrosynthetic analysis by ***functional group addition (FGA) transforms***—transforms that add a functional group to the target molecule without modifying its carbon skeleton.
2. Reactions that interchange functional groups. These reactions are among the most common and the simplest to use. Reactions such as hydroboration-oxidation, oxidation of alcohols to carbonyl compounds, and the formation and hydrolysis of acetals and ketals belong to this class. In retrosynthetic analysis, these reactions correspond to the ***functional group interchange (FGI) transforms***—transforms that alter the functional groups of the target structure without modifying its carbon skeleton.
3. Reactions that create a new functional group. These reactions often involve the replacement of benzylic or allylic hydrogens or electrophilic aromatic substitution by a heteroatom electrophile. Nitrene insertion reactions, which form a new C—N bond, also fall into this category. They are represented in retrosynthetic analysis by ***functional group removal (FGR) transforms***—transforms that remove a functional group from the target structure without modifying its carbon skeleton.
4. Carbon-carbon bond-forming reactions. These reactions are those that form the carbon-carbon bonds in a molecule, and they correspond to some of the most powerful synthetic methods available to the organic chemist. Reactions such as the Diels-Alder cycloaddition, Grignard and aldol additions, and Friedel-Crafts acylation are bond-forming reactions. In retrosynthetic analysis, they correspond to ***bond disconnection (DIS) transforms***—transforms that disconnect a bond in the target molecule. Bond disconnection almost always results in simplification of the target molecule. In Table 15.3, ring disconnection transforms are gathered in a separate group simply because the construction of rings is often an important step in a total synthesis.

Table 15.3 Major Transforms Used in Retrosynthetic Analysis

	Some Representative Examples
BOND DISCONNECTION (DIS)	
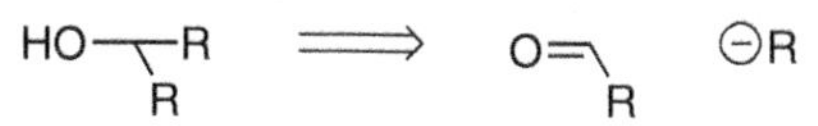	retro-Grignard addition retro-alkynide anion addition
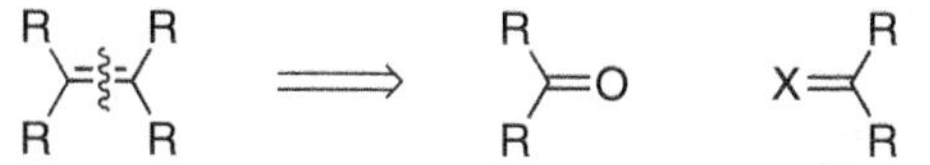	retro-Wittig transform retro-McMurry transform
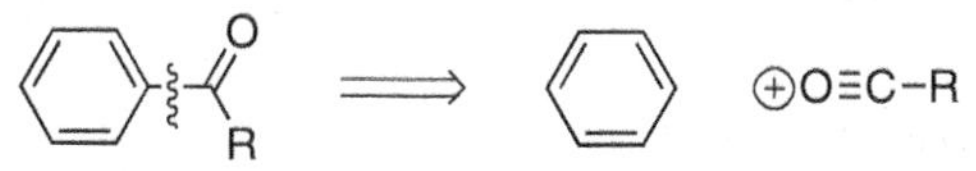	retro-acylation retro-formylation
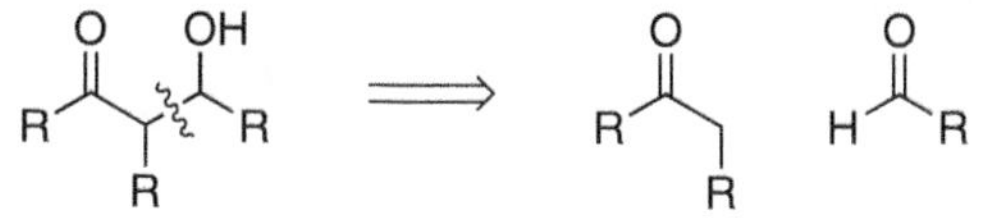	retro-aldol transform retro-Reformatskii transform
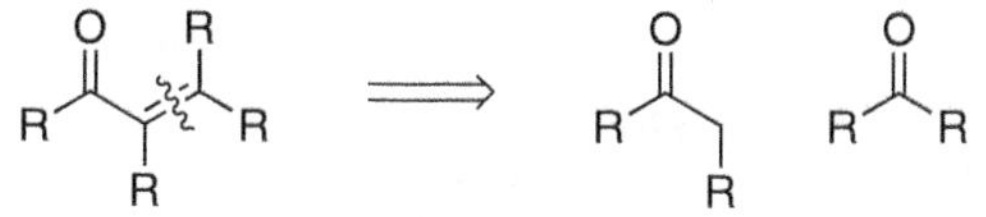	retro-aldol condensation transform
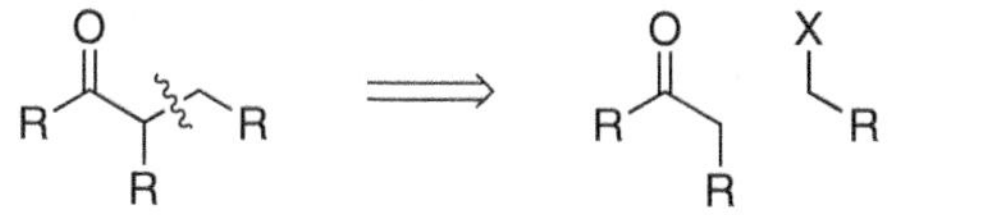	retro-alkylation of enolate
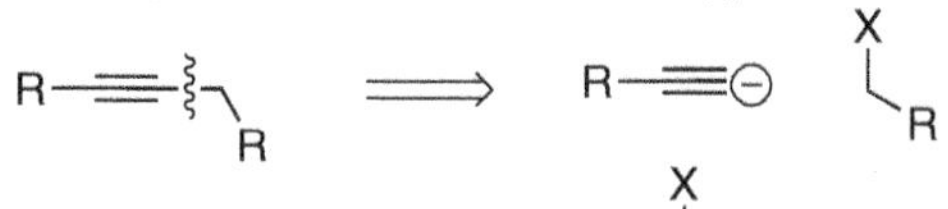	retro-alkylation of alkynide anion
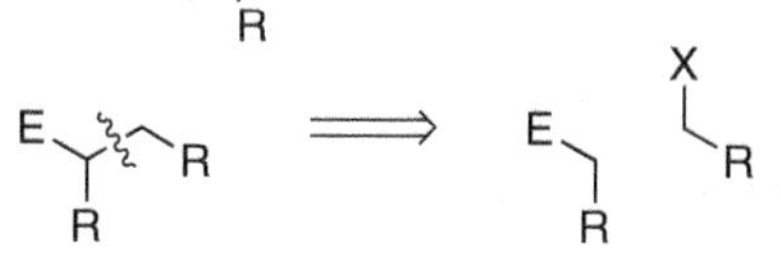	retro-alkylation of nitro compounds ($E=NO_2$), sulfoxides (E=SOR), dithianes ($E=(SR)_2$), etc.
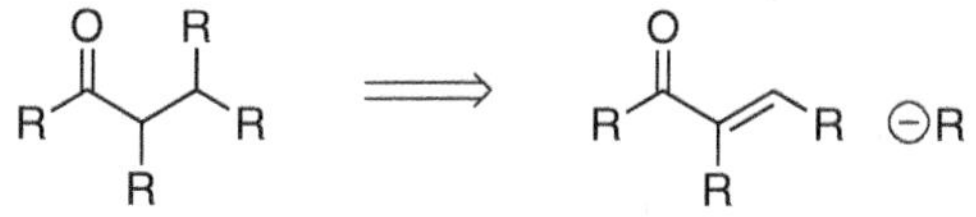	retro-Michael addition transform
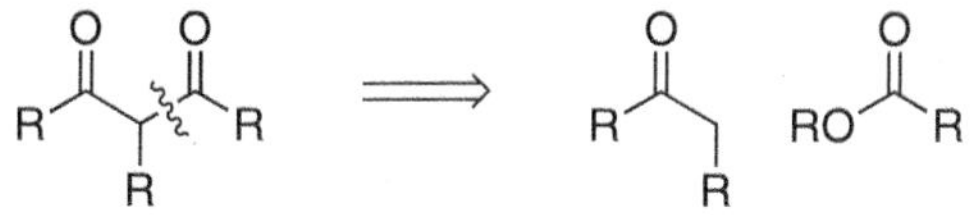	retro-Claisen condensation transform
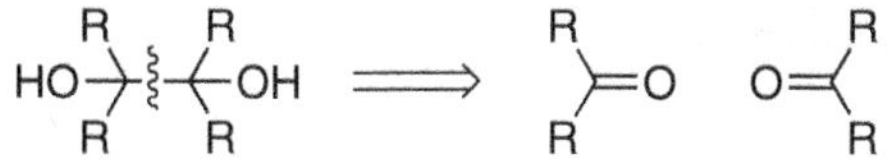	retro-pinacol transform
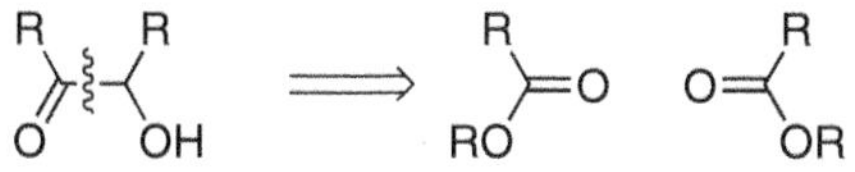	retro-acyloin condensation
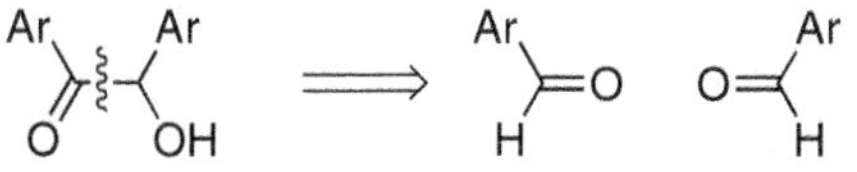	retro-benzoin transform
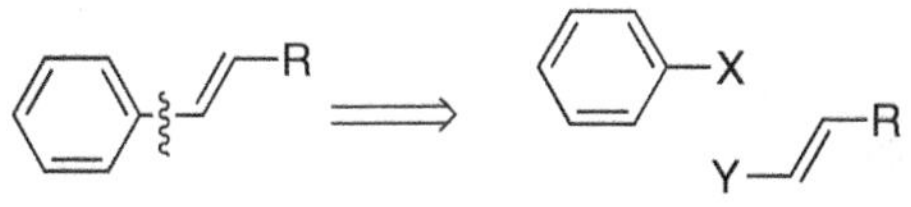	retro-Heck coupling retro-Suzuki coupling
	retro-Sonogashira coupling

(continued)

Table 15.3 Major Transforms Used in Retrosynthetic Analysis (continued)

Some Representative Examples

FUNCTIONAL GROUP INTERCHANGE (FGI)

retro-hydride reduction/reductive amination
retro-oxidation

retro-elimination
retro-electrophilic addition

retro-alkyne hydration

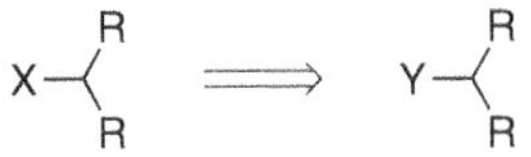

retro-S_N1, S_N2

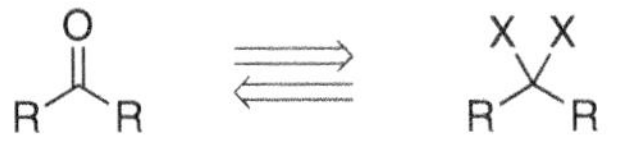

retro-ketal/thioketal hydrolysis
retro-ketal/thioketal formation

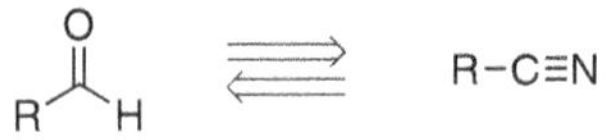

retro-nitrile reduction
retro-oxime dehydration

retro-Wharton reaction

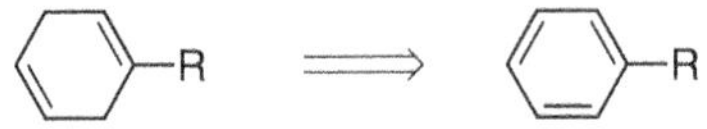

retro-Birch reduction

retro-epoxidation

retro-Lindlar hydrogenation

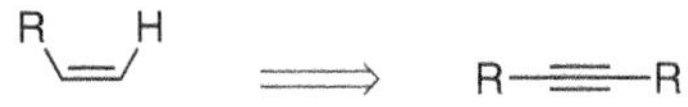

retro-Birch reduction

FUNCTIONAL GROUP REMOVAL (FGR)

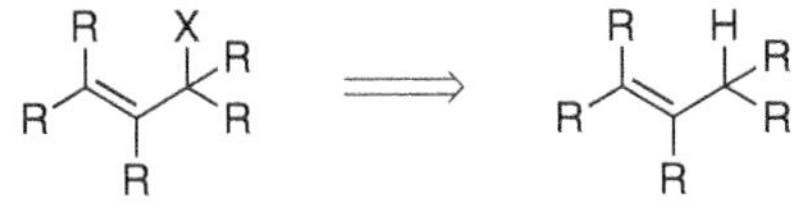

retro-allylic halogenation, oxidation, etc.

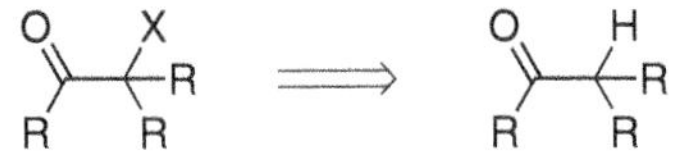

retro-α-halogenation, sulfenylation, selenylation, etc.

retro-nitration, halogenation, etc.

retro-Hofmann-Löffler-Freytag reaction;
retro-Barton reaction, etc.

retro-nitrene insertion

(continued)

Table 15.3 Major Transforms Used in Retrosynthetic Analysis (continued)

	Some Representative Examples
FUNCTIONAL GROUP ADDITION (FGA)	
R–H ⟹ R–M	retro-protonolysis
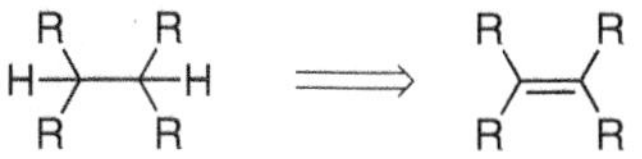	retro-hydrogenation
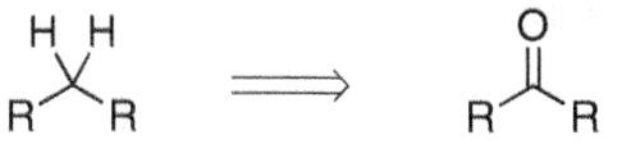	retro-deoxygenation
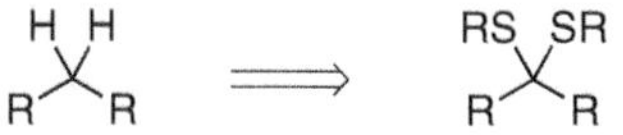	retro-desulfurization
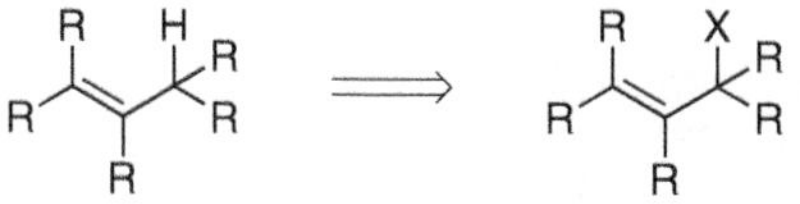	retro-hydrogenolysis
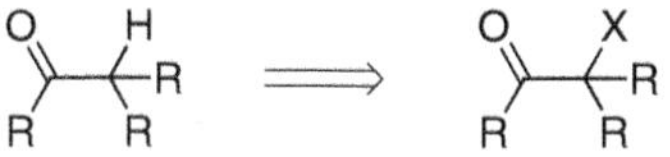	retro-α-reduction
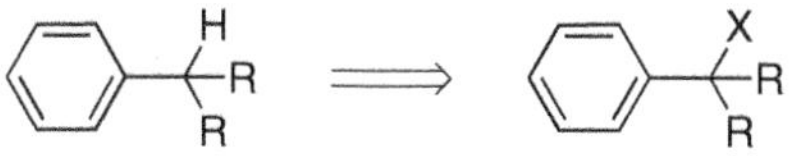	retro-hydrogenolysis
REARRANGEMENT (REARR)	
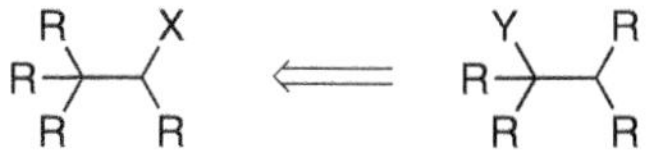	Wagner-Meerwein rearrangement of cations
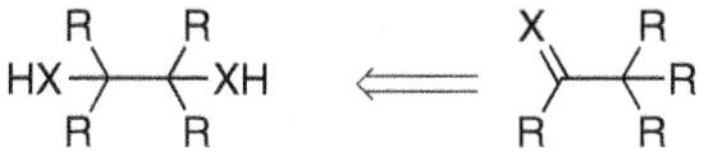	pinacol rearrangement
	Favorskii rearrangement
	Hofmann, Curtius, Lossen rearrangements
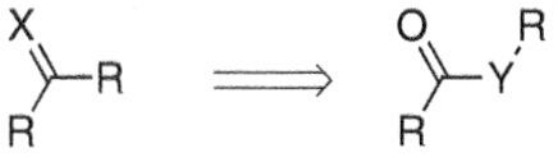	Baeyer-Villiger rearrangement (X=O; Y=O) Beckmann rearrangement (X=NOH; Y=NH) Schmidt reaction (X=O; Y=NH)
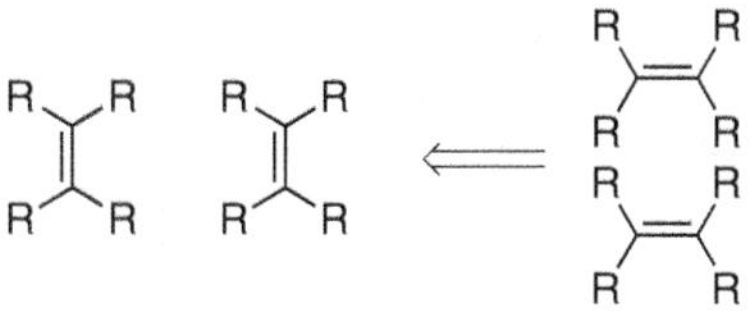	olefin metathesis
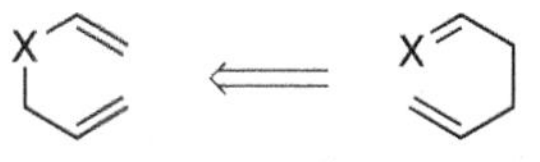	Cope (X=CR_2) and Claisen (X=O) rearrangements other sigmatropic rrearrangements
	electrocyclization

(continued)

Table 15.3 Major Transforms Used in Retrosynthetic Analysis (continued)

Some Representative Examples	
FUNCTIONAL GROUP REMOVAL (FGR)	
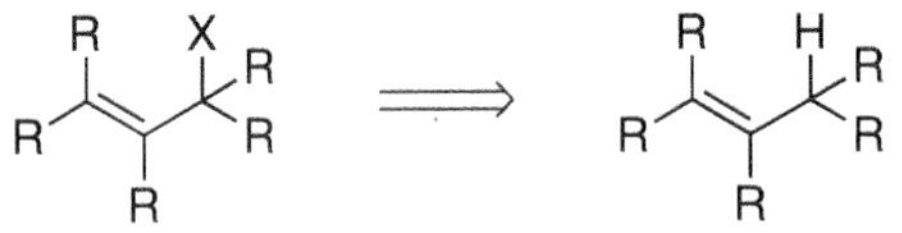	retro-allylic halogenation, oxidation, etc.
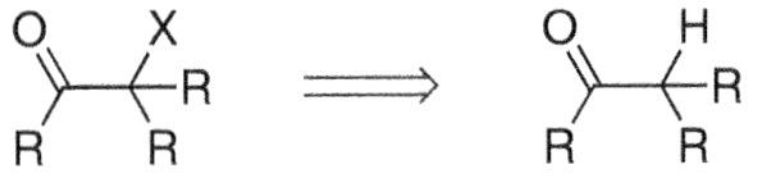	retro-α-halogenation, sulfenylation, selenylation, etc.
	retro-nitration, halogenation, etc.
	retro-Hofmann-Löffler-Freytag reaction; retro-Barton reaction, etc.
	retro-nitrene insertion
RING DISCONNECTION (DIS)	
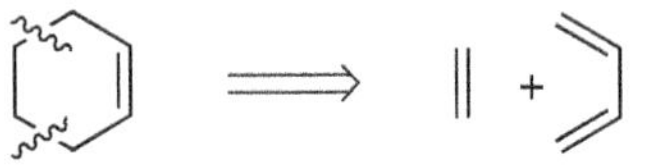	retro-Diels-Alder cycloaddition
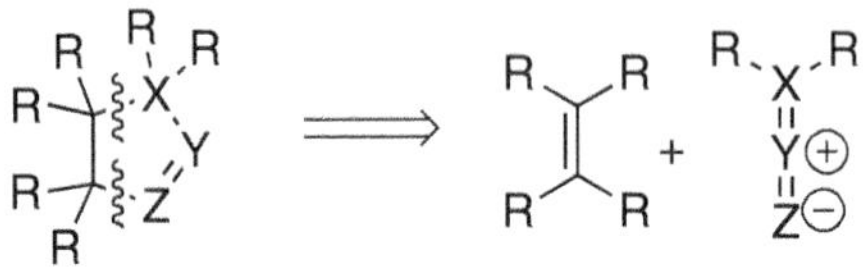	retro-[3+2] dipolar cycloaddition
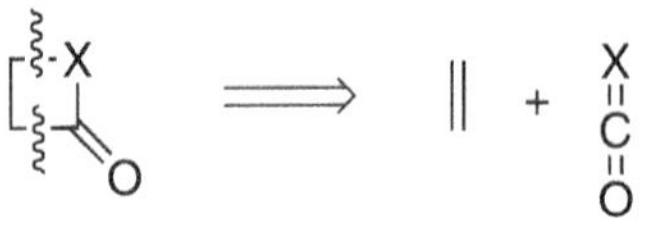	retro-[2+2] cycloaddition
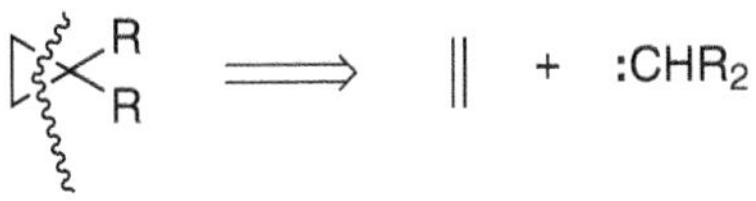	retro-carbene addition
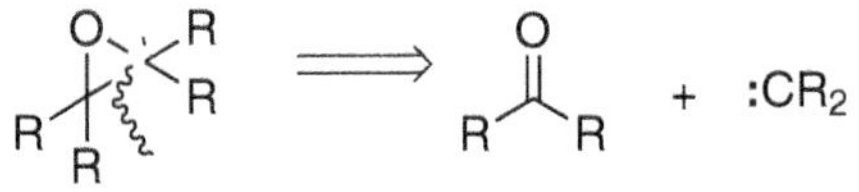	retro-Darzens condensation retro-sulfonium ylide addition

(continued)

Table 15.3 Major Transforms Used in Retrosynthetic Analysis (continued)

Some Representative Examples	
BOND RECONNECTION (REC)	
	retro-ozonolysis transform retro-Lemieux-Johnson transform
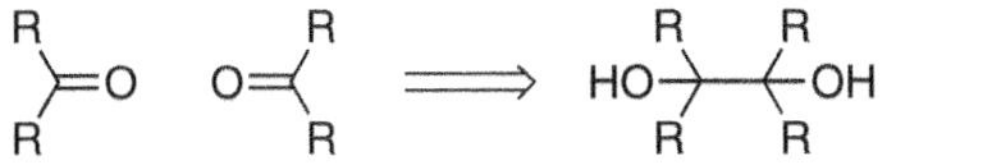	retro-glycol cleavage transform
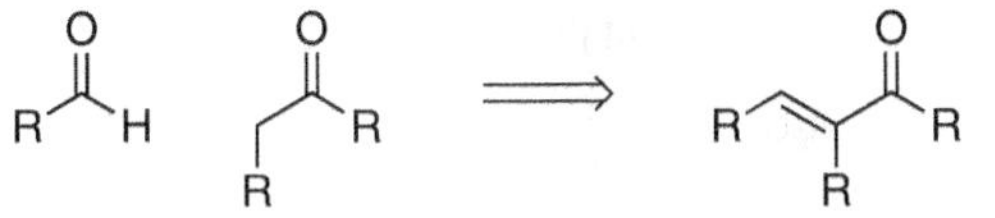	aldol condensation transform
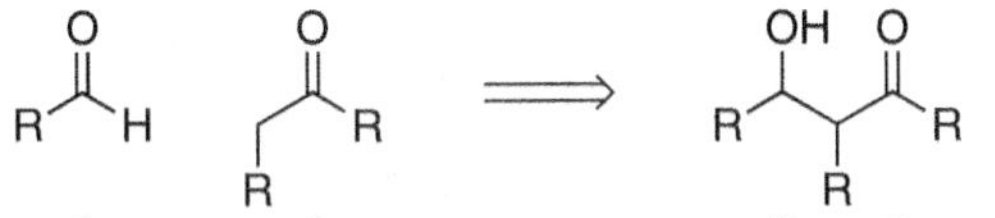	aldol addition transform
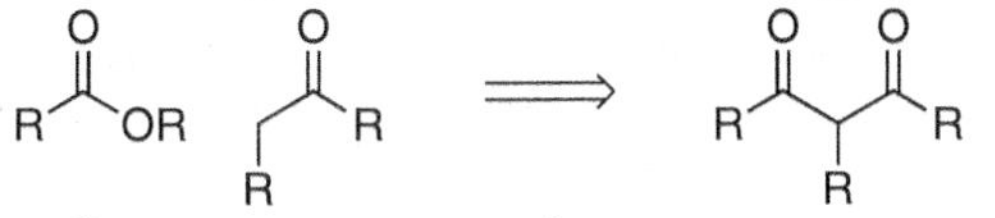	Claisen transform
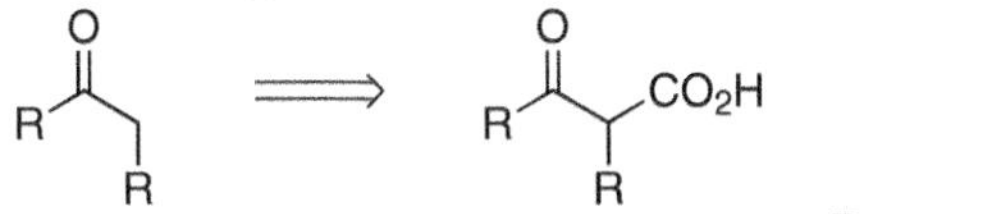	retro-decarboxylation transform
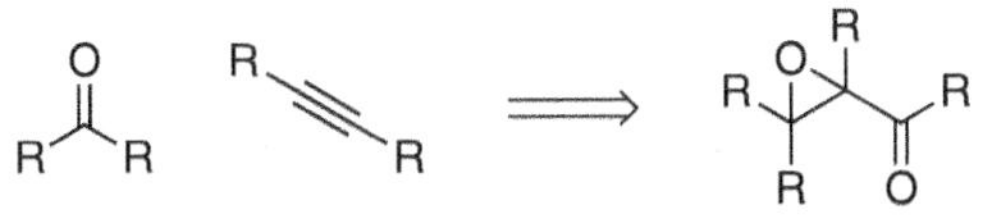	retro-Eschenmoser-Tanabe fragmentation

5. Carbon-carbon bond-cleaving reactions. These reactions cleave carbon-carbon bonds. Reactions such as ozonolysis or periodate cleavage of diols belong to this class of reactions, as does the decarboxylation of β-ketoacids and the reverse of the aldol addition. In retrosynthetic analysis, these reactions are represented by ***bond reconnection (REC) transforms***—transforms that reconnect a bond in the target molecule. Bond reconnection may result in a simplification or an increase in complexity of the target structure, depending on which bond is reconnected.
6. Rearrangement reactions. These reactions alter the carbon skeleton of a molecule; they include reactions such as the Wagner-Meerwein, Favorskii, Hofmann, and sigmatropic rearrangements. They are represented in retrosynthetic analysis by ***rearrangement (REARR) transforms.***

Most of the transforms in Table 15.3 correspond to reactions that are normally found in introductory organic chemistry textbooks. If we take the first two-bond disconnection (ring DIS) transforms as working examples, the first is the retro–Diels-Alder transform, which disconnects a cyclohexene into a diene and a dienophile. If this transform occurs as part of the retrosynthetic analysis, it indicates that a Diels-Alder reaction might be used to form the cyclohexene ring in the target molecule. The second bond disconnection transform in Table 15.3 disconnects the carbon-carbon bond to of an alcohol into a carbonyl synthon and a nucleophilic alkyl synthon. If this transform occurs as part of the retrosynthetic analysis, it indicates that one method for the formation of this bond of the target molecule might be the addition of an organometallic compound to an aldehyde or ketone (e.g., a Grignard reaction).

Table 15.4 Fuhrhop-Penzlin Designations of Synthons

Acceptor Synthons | Donor Synthons

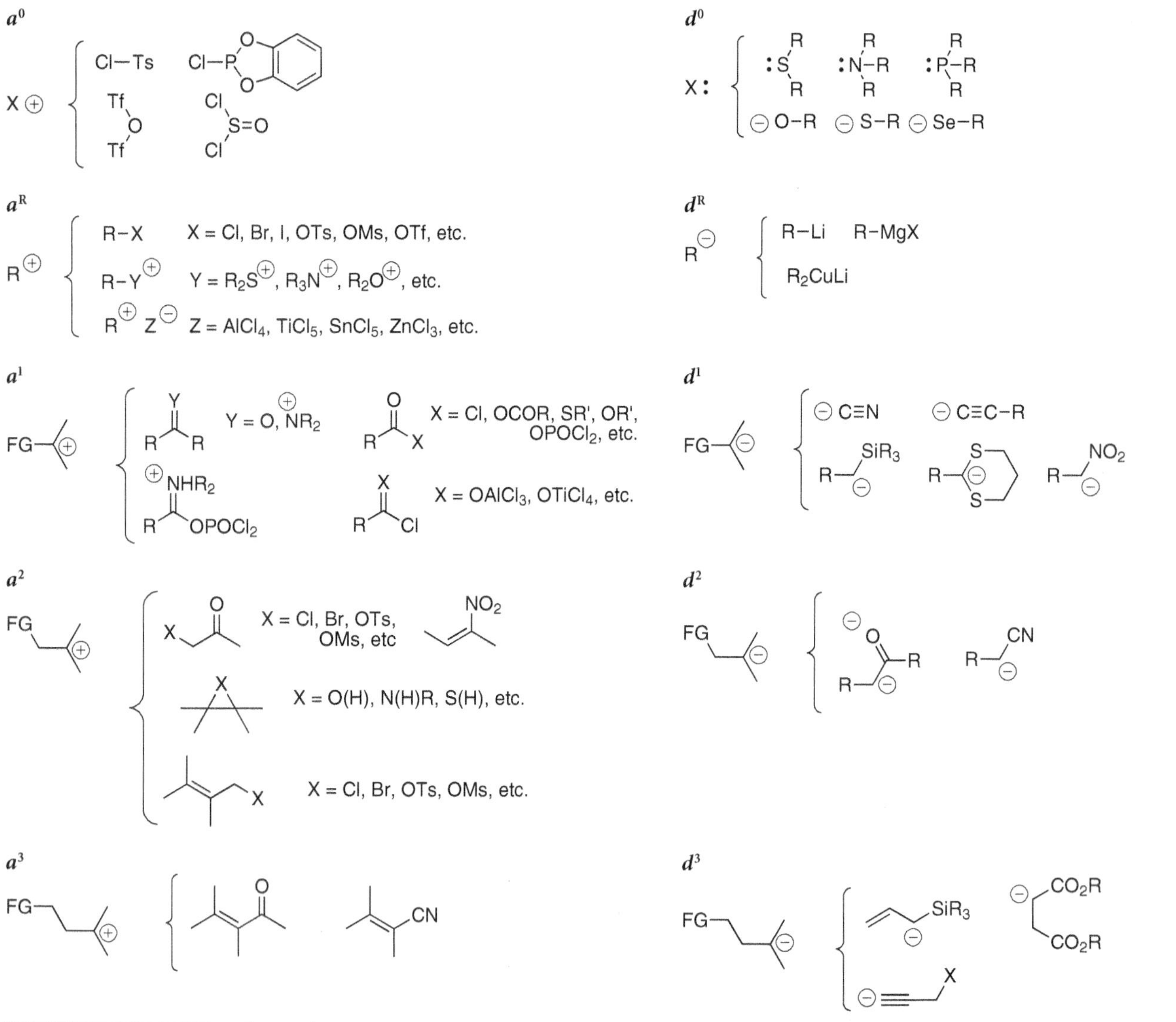

*Superscripts for synthons specify the relationship between the carbon atom that will react, and the carbon atom with the other functional group in the synthon

Other Retrosynthetic Strategies

One can also use synthons revealed by applying transforms to a target molecule as a basis for organizing reactions. This type of analysis forms the basis of the retrosynthetic terminology developed by Fuhrhop and Penzlin[33] (Table 15.4).

The real utility of this type of analysis comes when one realizes that the separation between the functional groups in the product is the sum of the two superscripts. A Michael addition, for example, is the reaction between of a d^2 synthon and an a^3 synthon to give a 1,5-dicarbonyl compound, as illustrated in Example **15.14**. The same type of transformation can also be accomplished by treating a γ-haloketone or ester (an a^4 synthon) with the conjugate base of nitromethane (a d^1 synthon), as illustrated in Example **15.15**.

---

33. Fuhrhop, J.; Penzlin, G. *Organic Synthesis. Concepts, Methods, Starting Materials*, 2nd ed. (VCH: Weinheim, 1994).

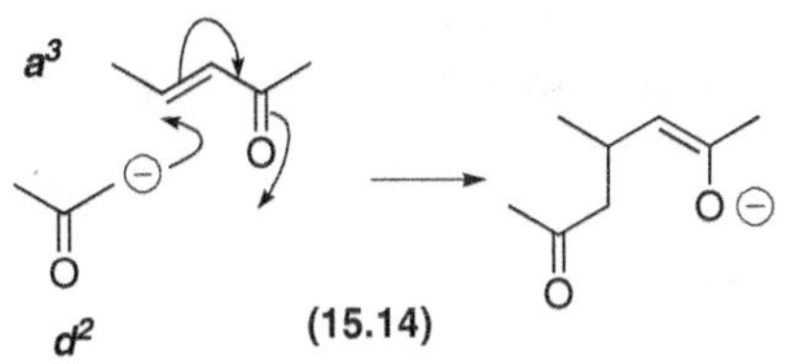

(15.14)

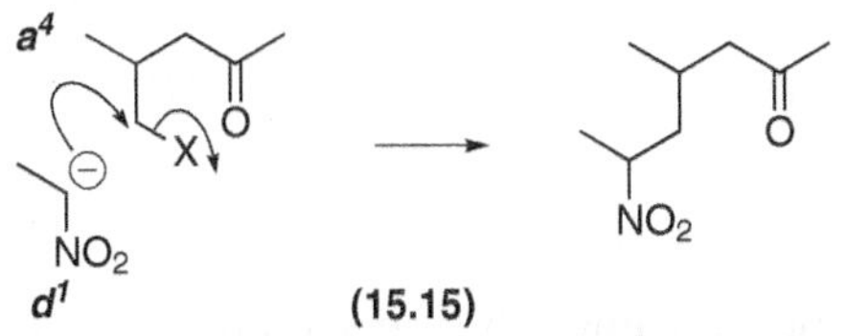

(15.15)

The two types of synthons are further subdivided into categories that denote the number of carbon atoms between the atom where the bond is formed and the functional group in the product, with that number shown as a superscript. Thus, an enolate anion in an alkylation reaction (Example **15.16**), for example, becomes a d^2 synthon—in the carbanion form of the ion, the anionic carbon is adjacent to (i.e., second to) the carbon atom of the carbonyl group. An allyl halide (Example **15.17**), on the other hand, is an a^2 synthon. Simple alkyl synthons that do not lead to a functional group in the final product of the reaction are designated as a^R or d^R synthons.

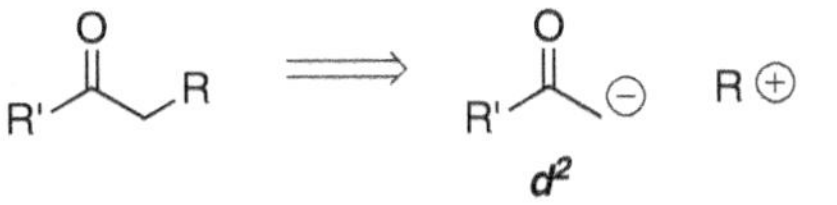

(15.16)

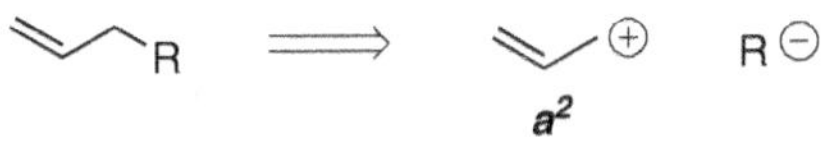

(15.17)

The Evans-Lapworth Alternating Polarity Disconnection Method

The ***alternating polarity disconnection method***[34] is extrapolated from the alternating polarity theory originally developed by British chemist Arthur Lapworth[35] in the 1920s for rationalizing bonding and reactivity in organic compounds.[36] It was transformed into a useful method for retrosynthetic analysis in the 1970s by American chemist David Evans. As an aside, Evans' initial seminar presentation,[37] which was rejected for publication when it was submitted to the journals, may be the most widely disseminated and extensively used unpublished paper of all time (the author of this book was given a copy of the manuscript in Australia while he was still a student!).

In this method, each atom of the molecule is assigned an ***affinity label***, either (+) or (−), with affinity labels alternating around the skeleton of the molecule as much as possible. In this chapter, we will use the labels (*e*) and (*n*), to avoid confusion that may arise by taking the (+) and (−) labels of the Lapworth-Evans system for formal electronic charges. Atoms given the affinity label (*e*) are part of an electrophilic synthon and they tend to be positions

34. (a) Evans, D.A. *Acc. Chem. Res.* **1974**, *7*, 147. (b) Ho, T.-L. *Polarity Control of Synthesis* (Wiley-Interscience: New York, 1991).

35. **Arthur Lapworth (1872–1941)** was educated in Birmingham and spent most of his career at Manchester. He became a pioneer of physical organic chemistry, especially in the elucidation of organic reaction mechanisms. For more detail, see: Robinson, R. *Obit. Notices Fell. Roy. Soc.* **1947**, 5, 554.

36. (a) Lapworth, A. *Mem. Manchester Phil. Soc.* **1920**, *64* [iii], 1. (b) Lapworth, A. *J. Chem. Soc.* **1922**, *121*, 416. (c) Lapworth, A.; Shoesmith, J.B. *J. Chem. Soc.* **1922**, *121*, 416. (d) Lapworth, A.; Robinson, R. *Nature* **1923**, *112*, 722. (e) Lapworth, A. *Chem. Ind.* **1924**, *43*, 1294; **1925**, *44*, 397. (f) Burkhardt, G.N.; Lapworth, A. *J. Chem. Soc.* **1925**, *127*, 1742. (g) Burkhardt, G.N.; Lapworth, A.; Walkden, J. *J. Chem. Soc.* **1925**, *127*, 2458.

37. Evans, D.A. "Consonant and Dissonant Relationships. An Organizational Model for Organic Synthesis" *UCLA Physical Organic Chemistry Seminar*, May 6, 1971.

in the molecule that are susceptible to attack by nucleophiles. Those given the affinity label (*n*) are part of a nucleophilic synthon and they tend to be positions in the molecule that are susceptible to attack by electrophiles. The affinity labels of functional groups are assigned first, and they are assigned so that they correspond to the polarity of the bonds of the functional group (see Examples **15.18** and **15.19**).

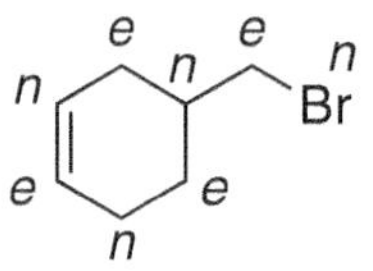

(15.18)

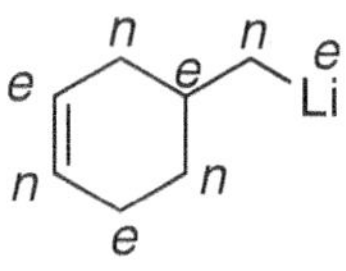

(15.19)

The real usefulness of the Lapworth-Evans affinity label method of retrosynthetic analysis becomes apparent when one looks at compounds having more than one functional group. In this situation, it allows one to deduce *reasonable* synthons that might be used to assemble a target compound. In a molecule with several functional groups, the relationship between *pairs* of functional groups can be particularly informative. It is in analyzing the relationships between functional groups in polyfunctional compounds that the real strength of the Evans method for analyzing the structures of complex organic compounds lies. This is done by defining two difunctional relationships: ***consonant difunctional relationships***, where the affinity labels alternate throughout the two functional groups and the space between them, and ***dissonant difunctional relationships***, where there is at least one site where two like affinity labels must be placed on adjacent atoms.

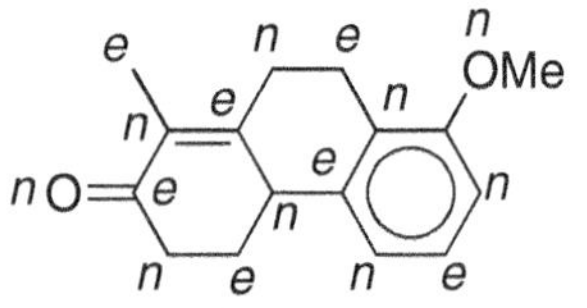

(15.20)

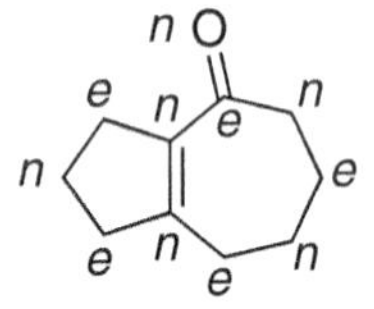

(15.21)

Figure 15.10 Consonant (upper) and dissonant (lower) molecules

The two molecules shown in Figure 15.10 are instructive. In molecule **15.20**, the affinity labels all alternate, and it is not necessary to have any pairs of atoms with like affinity labels adjacent; the molecule is consonant. Consonant difunctional compounds tend to be fairly easy to form because most of the common synthetic reactions in organic chemistry give consonant products (analyze the aldol addition, the Grignard addition and hydroboration-oxidation, for example).

In the molecule **15.21**, however, regardless of how one assigns the affinity labels, it is impossible to avoid at least one position in each ring where the same affinity label is on a pair of adjacent atoms; this molecule is dissonant. Inherently, there are fewer reactions that give rise to dissonant products than to consonant products, and most of those are redox reactions (e.g., epoxidation or hydroxylation, as well as the α-selenylation of a carbonyl compound).

It is often easiest to incorporate the dissonant element into the starting material. Where this cannot be done, strategies have been devised to accomplish the task of generating the dissonant element. One strategy involves the process of affinity reversal, now universally referred to by its German name of ***umpolung***.[38] In an ***umpoled synthon***, the normal affinity labels are reversed. The simplest umpoled synthon is cyanide ion, where the carbon carries the negative charge (and, therefore, the nucleophilic, or *n* affinity label), despite being less electronegative than the nitrogen. It is worth noting *that all umpoled synthons have a built-in dissonant relationship.*

---

38. Reviews: (a) Corey, E.J.; Seebach, D. *Angew. Chem. Int. Ed. Engl.* **1965**, *4*, 1075. (b) Gröbel, B.-T.; Seebach, D. *Synthesis* **1977**, 357. (c) Seebach, D. *Angew. Chem. Int. Ed. Engl.* **1979**, *18*, 239. (d) Hassner, A.; Lokanatha Rai, K.M. In Trost, B.M.; Fleming, I., Eds *Comprehensive Organic Synthesis* (Pergamon Press: Oxford, 1991), Vol. 1, p. 541. (e) Smith, A.B., III; Adams, C.M. *Acc. Chem. Res.* **2004**, *37*, 365.

Monograph: Hase, T.A. *Umpoled Synthons: A Survey of Sources and Uses in Synthesis* (John Wiley & Sons: New York, 1987).

15.4 Applying Retrosynthetic Analysis to a Real Example

It is all well and good to talk about retrosynthetic analysis in the semiabstract terms that we have used to this point, but let us now move the process forward by examining a real example. The hydrocarbon twistane (**15.22**) is an interesting molecule in that it is a C_2 hydrocarbon composed solely of cyclohexane rings in the skew conformation, which also has the potential to exist as enantiomers. In this case, the final product has no functional groups, so we have considerable freedom in how we approach the synthesis. Regardless of how we choose to proceed, it is clear that the first transform that we apply to the target molecule must be an FGA transform. There are actually relatively few FGA transforms that will lead to viable precursors, as shown in Figure 15.11, because the symmetry of the target molecule makes a number of positions equivalent. In the figure, three different additions of heteroatom-based functional groups (G) are shown (**15.23** to **15.25**), as is the inclusion of one unsaturation equivalent (**15.26**)—for a total of four potential synthons for this target molecule.

The inclusion of a functional group into the target now places more restrictions on how we can proceed with the retrosynthetic analysis. Let us now take synthon **15.23** further and see how this molecule, which possesses the complete twistane skeleton, can now be further transformed into an earlier generation synthon. There are two major approaches to this, depending on whether the group X is attached to the skeleton by a single bond (e.g., a cyano group, a sulfoxide, or a sulfide) or a double bond (e.g., a carbonyl group). In any case, it is now time to apply a ring disconnection transform to the target. Every such disconnection will lead to cleavage of one of the six-membered rings of the target. There are four simple possibilities for doing this, as illustrated in Figure 15.12. In disconnections a and b, the bond to the carbon carrying the singly bonded functional group is disconnected to reveal two different bifunctional synthons, both based on the bicyclo[2.2.2]octane ring system (**15.27** and **15.28**). Disconnection c results in a bicyclo[4.4.0]decane synthon (**15.29**), whereas disconnection d results in the bicyclo[3.3.1]nonane synthon (**15.30**). In all four of these new synthons, stereochemistry now plays an important role because the new groups (X and Y) must be arranged in a way that will allow the synthetic *reaction* to occur. The ring junction in synthon **15.29**, for example, must be *cis* for the twistane ring system to be formed successfully.

Some chemists can develop a rational retrosynthetic analysis to the starting compounds without having to check for reasonableness. The author of this book is not one of them. I

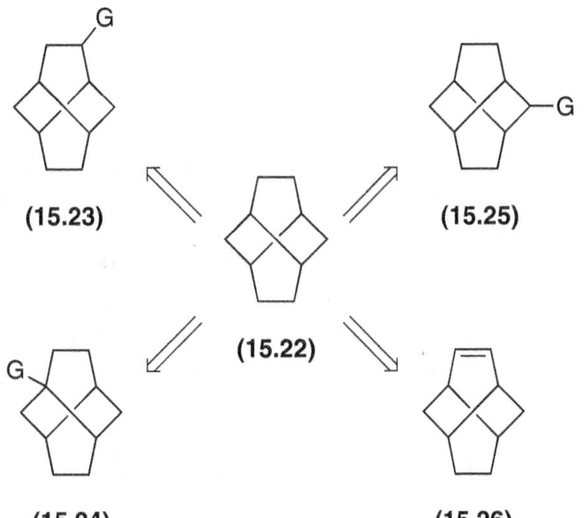

(15.23) (15.22) (15.25) (15.24) (15.26)

Figure 15.11 Functional group addition transforms applied to the twistane target molecule

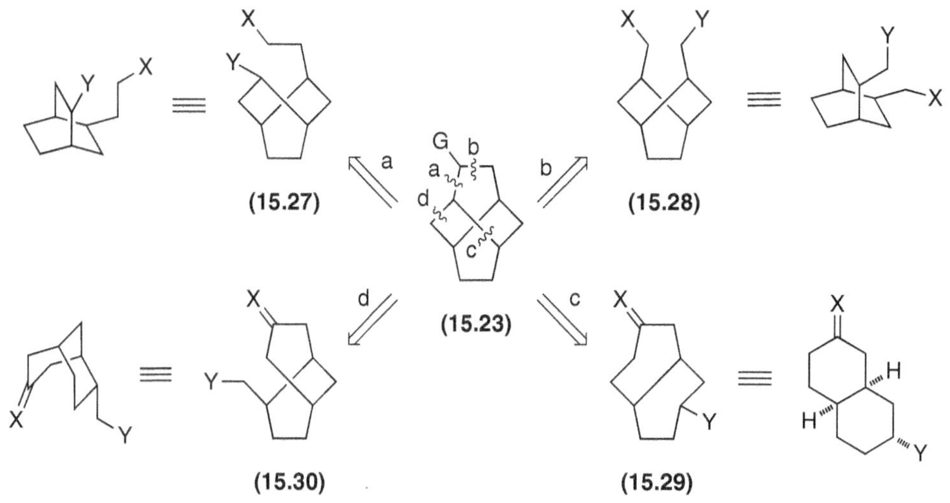

Figure 15.12 Possible bond disconnection transforms of the first-generation synthon 15.23

confess that I tend to become trapped by a certain level of tunnel vision when working in the retrosynthetic direction, so I find it wise to stop after major transforms to see if what I am proposing is *synthetically* reasonable. Let us do that here before we move forward in the retrosynthetic analysis, because it may affect our plans moving forward (or backward) with the retrosynthesis. At this point, one specifies the various groups in the synthon. For example, if we assume that group X in synthon **15.29** is oxygen, and group Y is a leaving group of some sort, we obtain an intermediate such as **15.31**. Looking at this target molecule, we see that there are now at least two problems that we will have to address at some point in the synthesis itself: (1) as a *cis*-decalin, this molecule is conformationally flexible, so we will have to fix it in one of the two conformers to get the correct leaving group stereochemistry; and (2) although it is fairly obvious that the twistane ring system will be most easily formed by means of a ketone alkylation, this means that we have to control which of the two possible enolate anions are formed or we will have to arrange for the formation of one of the two products to be strongly favored over the other. These problems do not generally reveal themselves during the retrosynthetic analysis, so I recommend that you follow the practice of checking regularly in the synthetic direction to see if there are any pitfalls or other problems that you may have missed. This often leads to a much better retrosynthetic analysis in the end.

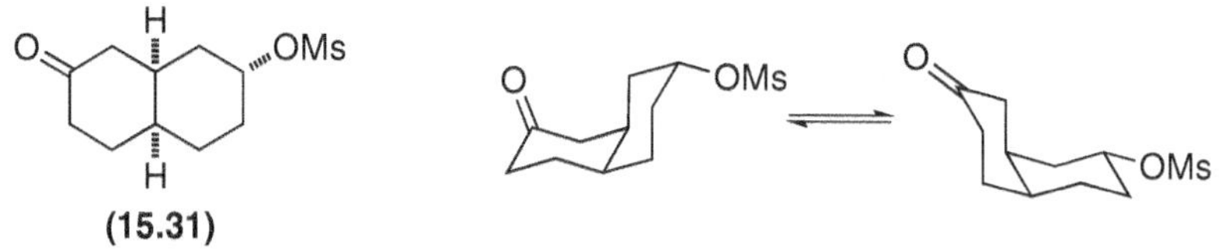

Let us now continue with the retrosynthetic analysis starting from synthon **15.29** (Figure 15.13). The one key feature of **15.29** is the *cis* fusion of the rings, and in the synthetic direction, this strongly suggests that catalytic hydrogenation of an alkene should be used. By applying an FGA transform to **15.29**, we obtain the new synthon **15.32**, and an FGI transform now gives the highly symmetrical synthon **15.33**. This type of dicarbonyl compound could, in principle, be obtained by reduction of the naphthalene **15.34**. In fact, this is exactly what Deslongchamps did in his synthesis of twistane,[39] obtaining

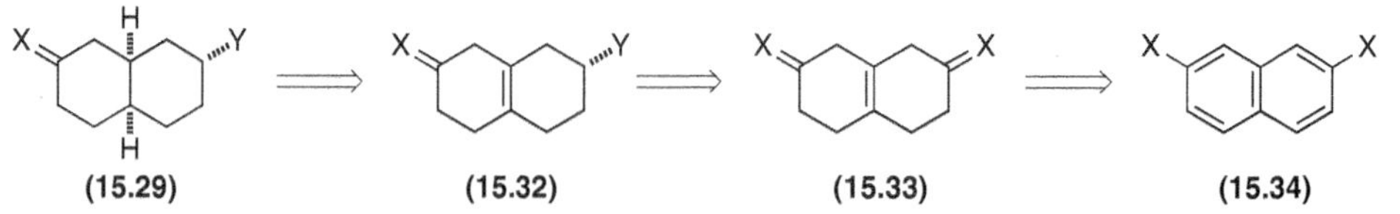

Figure 15.13 Retrosynthesis of synthon **15.29**

39. Gauthier, J.; Deslongchamps, P. *Can. J. Chem.* **1967**, *45*, 297.

the diketone **15.37** by catalytic hydrogenation of naphthalene-2,7-diol (**15.35**) and subsequent oxidation of the mixture or alcohols (**15.36**). These early stages of the synthesis followed earlier work by Anderson and Barlow[40] and gave the required diketone, albeit in relatively low yield.

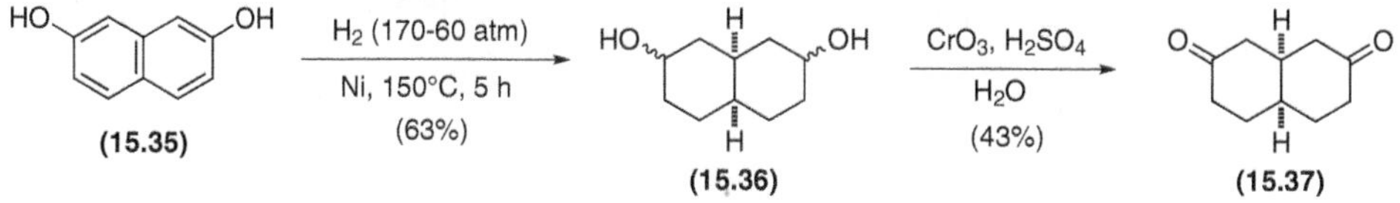

The problem now, of course was to prepare the required ketone, **15.31**, which involves differentiating the two carbonyl groups and reducing one to a particular diastereoisomer of the alcohol. Deslongchamps' approach to addressing this problem was to fix the decalin in one of the two possible conformations. He accomplished this by converting the diketone into the diethyl ketal, which exists as two conformers, **15.38** and **15.39**. As is evident from Figure 15.14, in one of the two conformers (**15.38**), there is an adverse steric interaction between one of the ethoxy groups and the axial α-hydrogen of the ketone. This means that the other conformer, which lacks this high-energy non-bonded interaction, is now strongly favored. What then remains is to reduce the carbonyl group to give the equatorial alcohol. Here again, however, we cannot use simple hydride reducing agents, because these will approach the carbonyl group from the less hindered *exo* face of the molecule, giving the axial alcohol. A Birch reduction of the ketone, however, gave the equatorial alcohol (**15.40**) as the sole product. The two-step conversion of the alcohol to the mesylate, and mild hydrolysis of the ketal with aqueous oxalic acid in a two-phase system gave the key ketone **15.31**.

The next step of the synthesis is interesting from a practical perspective. The ketone **15.31** has two α-methylene groups, each of which has an acidic axial hydrogen. In order for the twistane skeleton to be formed, the proton labeled "a" is the one that must be abstracted by the base. Fortunately, this proton is in a less hindered environment that the proton labeled "b," and it was found that this proton was removed much faster. Thus, heating the ketone **15.31** with sodium hydride in dioxane gave rise to a single product, the twistanone **15.41**. The remainder of the synthesis then consisted of the deoxygenation of this ketone, which was accomplished by the two-step reductive desulfurization shown in Figure 15.15.

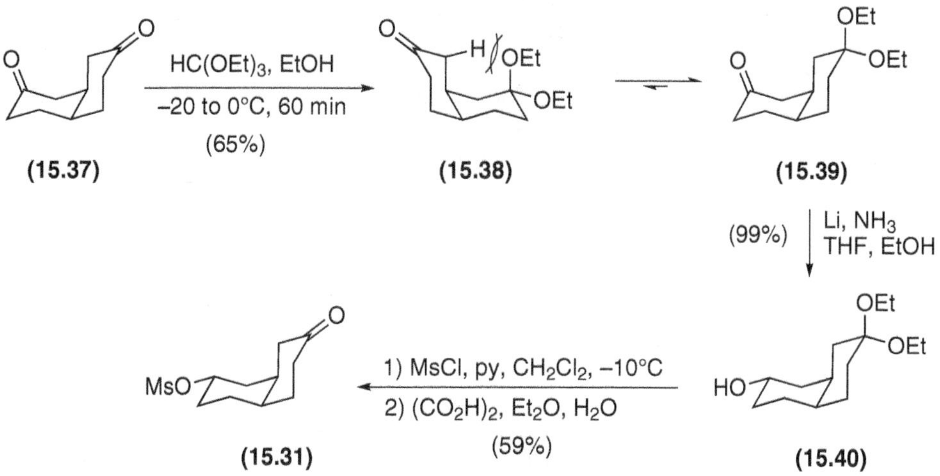

Figure 15.14 Synthesis of the key intermediate for cyclization to the twistane ring system

40. Anderson, A.G., Jr.; Barlow, D.O. *J. Am. Chem. Soc.* **1955**, *77*, 5165.

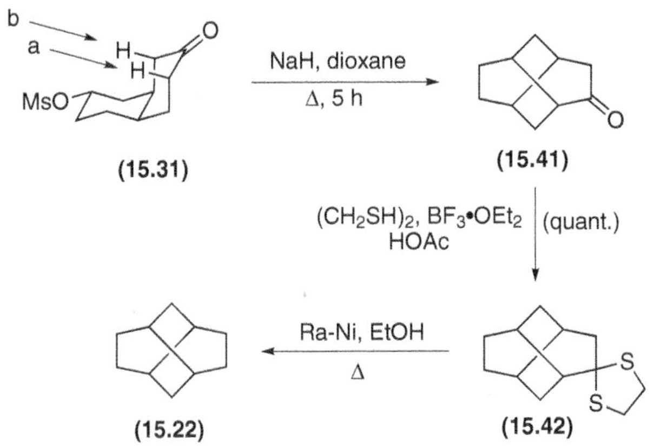

Figure 15.15 Completion of the synthesis of twistane

Worked Problem

15-2 The retrosynthetic analysis below represents the Martin analysis of tirandamycic acid, a degradation product of the antibiotic tirandamycin [*J. Org. Chem.* **1984**, *49*, 2512].

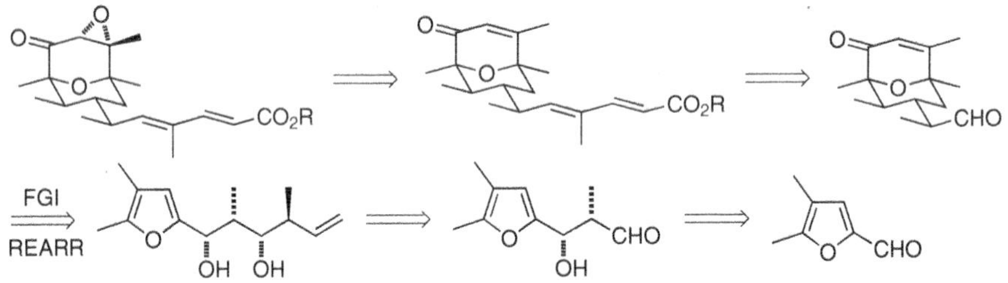

(a) To what class does each of the retrosynthetic transforms belong?

(b) Are all the intermediate compounds in the last bond disconnection specified explicitly? If not, which are missing?

§Answers on the next page.

Problems

15-3 Place each of the following retrosynthetic transforms into one of the classes defined in Table 15.3 and specify which member of the class the transform is. Give reasons for your choice where appropriate.

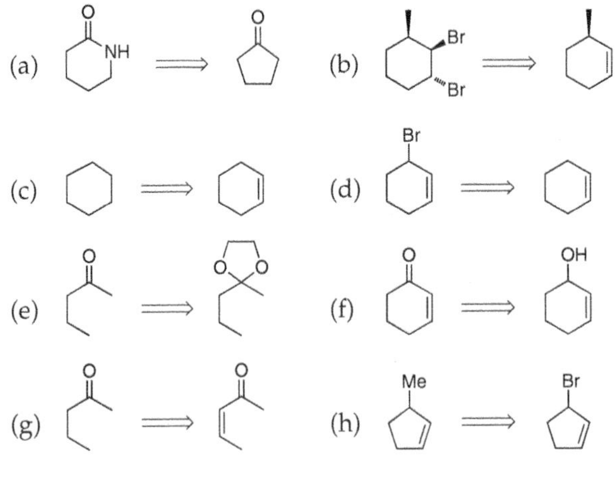

(i) 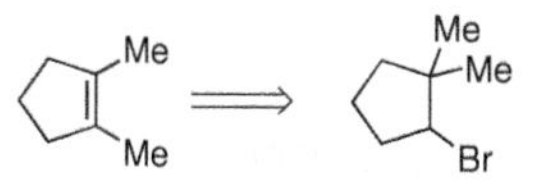 (j) 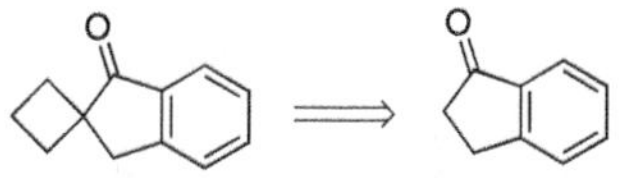

(k) 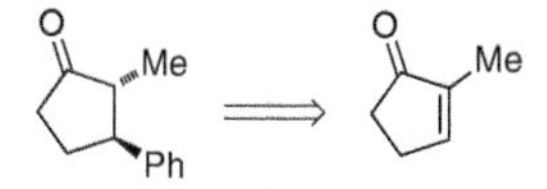 (l) 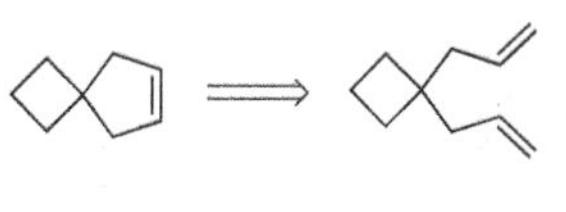

15-4 Place each of the transforms in the following retrosynthetic analysis into one of the classes defined in Table 15.3. The reaction corresponding to transform (c) occurs on heating with acid.

(a) Write a mechanism for the reaction corresponding to transform (c).

(b) Supply reagents that may be used to complete the synthesis of bicyclo[3.2.1] octane from anisole through the intermediates shown.

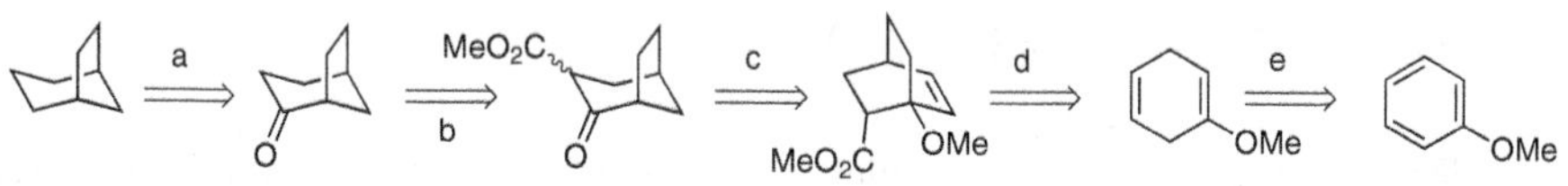

15-5 Design retrosynthetic analyses of the molecules below from the specified starting materials and any other needed reagent(s).

(a) 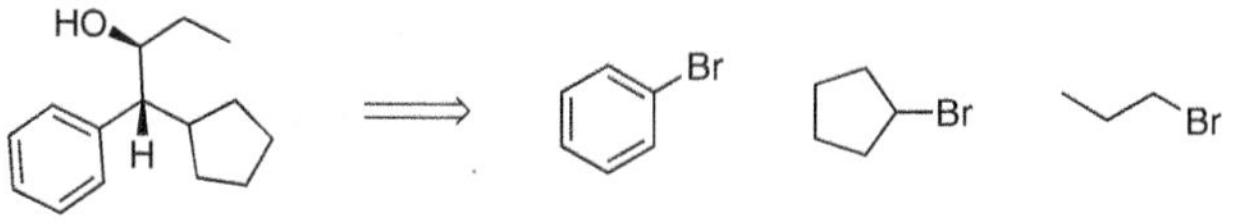

(continues)

....

§ Answers to Worked Problem:

(a) 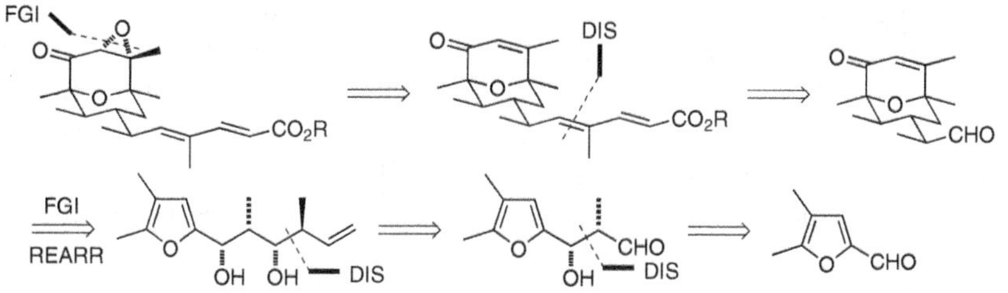

(b) Not all the intermediate compounds are shown. This is a retro-aldol addition transform that corresponds to an asymmetric aldol addition reaction. This reaction uses the boron enolate of an *N*-acyloxazolidinone to form the C—C σ bond, so the reaction product is also an *N*-acyloxazolidinone. To arrive at the aldehyde specified requires three more functional group interchange steps.

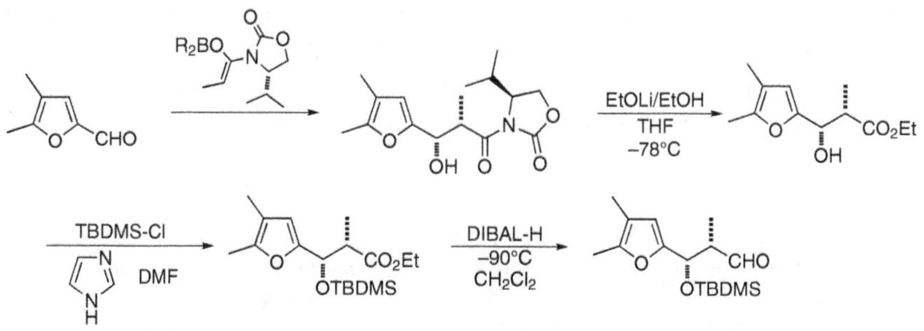

The silyl ether is required because the formation of the next bond will involve addition of an oranometallic compound, which might be compromised by the presence of the alcohol hydroxyl group.

(**Problems** continued)

(b) 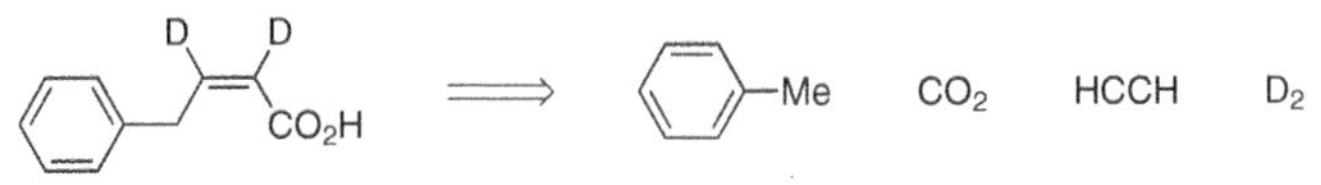

(c) 

15-6 Assign the affinity labels to each structure in Problems 15-3 and 15-4. Which of the reactions introduces a dissonant difunctional relationship into a consonant molecule? What type of reaction does this? Can you draw any generalizations about interconverting consonant and dissonant elements in a synthetic target?

15.5 **Selectivity Revisited**

Challenges in the synthesis of complex target compounds are associated with three types of selectivity.

1. ***Chemoselectivity***, which requires reactions that react preferentially with one of a set of different functional groups
2. ***Regioselectivity***, which requires reactions capable of selecting between two different forms or locations of the same functional group or those that will preferentially generate one of a set of possible regioisomers
3. ***Stereoselectivity***, which requires reactions that will react with the substrate to give preferentially one of the set of possible stereoisomers. In modern synthesis, this is often the need to control enantioselectivity.

Chemoselectivity

Chemoselectivity is usually the simplest of these problems to address; oxidation of aldehydes, for example, can be carried out in the presence of alcohols by means of reactions such as the Kraus-Pinnick oxidation.[41] Discriminating between identical functional groups in different environments may also be fairly easy; as early as 1878, Menshutkin had shown that substrate structure affects reaction rates.[42] This can be illustrated by the acetylation of the three alcohols in Example **15.43**. All three compounds carry the same functional group.

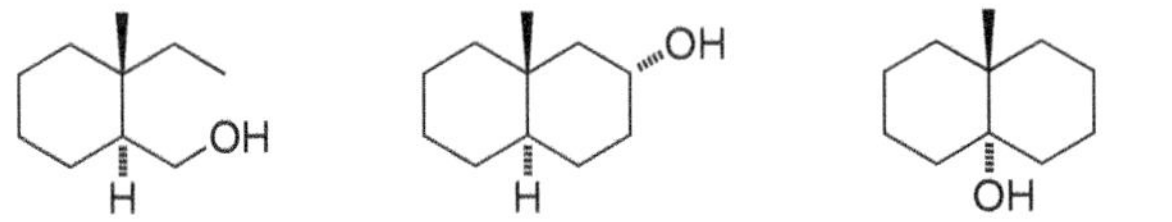

(15.43)

However, because primary alcohols are readily acetylated with acetic anhydride in pyridine and tertiary alcohols are not, the selective formation of a primary acetate is usually quite straightforward in the presence of a tertiary alcohol. In addition, the rate differences in the esterification of primary and secondary alcohols usually renders selecting

41. (a) Kraus, G.A.; Taschner, M.J. *J. Org. Chem.* **1980**, *45*, 1175. (b) Kraus, G.A.; Roth, B. *J. Org. Chem.* **1980**, *45*, 4825. (c) Bal, B.S.; Childers, W.E., Jr.; Pinnick, H.W. *Tetrahedron* **1981**, *37*, 2091.

42. Menschutkin. N. *Rec. Trav. Chim. Pays-Bas* **1883**, *2*, 117. (b) Menschutkin. N. *Ber. dtsch. chem. Ges.* **1878** *1*, 1507. (c) Menschutkin, N. *Ber. dtsch. chem. Ges.* **1878**, *11*, 2117. (d) Menschutkin, N. *Justus Liebigs Ann. Chem.* **1879**, *197*, 193. (e) Menschutkin, N. *J. Prakt. Chem.* **1882**, *26*, 103. (f) Menschutkin, N. *Justus Liebigs Ann. Chem.* **1879**, *195*, 334.

between these two types possible. Generally speaking, achieving chemoselectivity presents the least of the problems faced by the synthetic chemist.

Regioselectivity

Protecting Groups[43]

In contrast to chemoselectivity, regioselectivity is not always so simple to control. The easiest problems in regioselectivity occur when one is attempting to select between alternative positions within the same functional group, such as in the hydration of an alkene: oxymercuration-demercuration gives the Markovnikov alcohol as the major product, whereas hydration of the same alkene by hydroboration-oxidation gives the other regioisomer as the major product. Regioselectivity becomes much more problematic when one is attempting to select between functional groups that are identical and in similar environments. For example, ask yourself how the transformation in Example **15.44** could be accomplished.

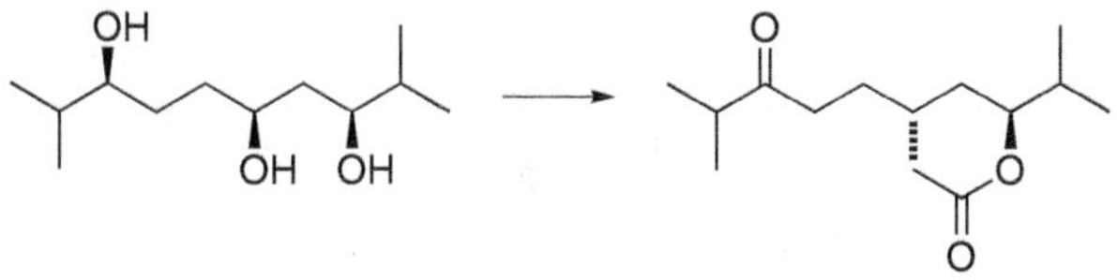

(15.44)

The problem here is that the three secondary hydroxyl groups end up as three different functional groups or structural features in the final product: one is oxidized to a ketone, one is replaced (with inversion of configuration) by an alkyl group, and one becomes part of a lactone (with retention of configuration). There is more than one way to accomplish these transformations selectively, but in order to do so, we need to use ***protecting groups***. A protecting group is chemically resistant derivative into which a functional group can be transformed in order to prevent it from reacting. Every protecting group must possess three fundamental characteristics to be useful.

1. It must be easily incorporated into the molecule without modifying other functional groups
2. It must not react with reagents that would otherwise react with the functional group that it modifies
3. It must be removable under conditions that will not affect the other functional groups of the molecule

Although the goal of total synthesis without the need for protecting groups remains, the simple fact is that organic synthesis has not yet matured to the point where protecting groups are unnecessary—indeed, the complexity of modern synthetic targets almost mandates their use except in very uncommon cases. Thus, the definitive work on protecting groups, Greene's *Protective Groups in Organic Synthesis*, expanded from 473 pages in the second edition (1991) to 1082 pages in the fourth (2007) as new protecting groups with greater selectivity were developed. Like them or not, protecting groups remain an indispensable part of (or necessary evil in) most modern syntheses.

One important concept underlying the use of protecting groups in modern organic synthesis is the concept of ***orthogonality***. The concept is illustrated in Figure 15.16.

43. Monographs: (a) Wuts, P.G.M.; Greene, T.W. *Greene' Protective Groups in Organic Synthesis*; 4th. ed. (John Wiley & Sons: New York, 2007). (b) Kocienski, P.J. *Protecting Groups*; 3rd. ed. (Georg Thieme Verlag: New York, 2005). The initial appearance of this book in 1994 was followed by a series of annual reviews: Jarowicki, K.; Kocienski, P. *Contemp. Org. Synth.* **1995**, *2*, 315; *Contemp. Org. Syn.* **1996**, *3*, 397; *Contemp. Org. Syn.* **1997**, *4*, 454; *J. Chem. Soc., Perkin Trans. 1* **1998**, 4005; *J. Chem. Soc., Perkin Trans. 1* **1999**, 1589; *J. Chem. Soc., Perkin Trans. 1* **2000**, 2495; *J. Chem. Soc., Perkin Trans. 1* **2001**, 2109.

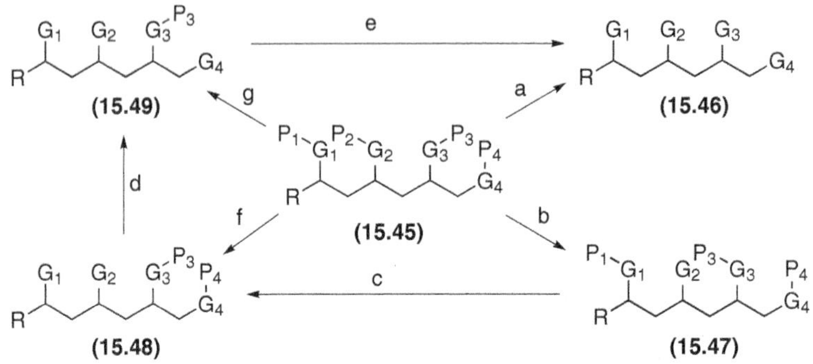

Figure 15.16 Orthogonality in protection of functional groups

A multifunctional molecule is said to have orthogonal protection of its functional groups if each can be deprotected in turn, without the other functional groups in the molecule also being deprotected. For example, compound **15.45** has four functional groups—G_1, G_2, G_3, and G_4—that are protected by protecting groups—P_1, P_2, P_3, and P_4. In pathway a, we see that every protecting group is lost in the deprotection step: this would be an example of non-orthogonal protection of the functional groups. The corollary to this is represented by the pathway b → c → d → e, where each of the protecting groups is removed in its own unique deprotection step. Here all the functional groups are protected orthogonally. Pathways f and g represent situations where more than one of the functional groups are protected using the same protecting group. In practical terms, it is extremely difficult to have more than four orthogonally protected functional groups in one molecule.

Directing Groups

Another strategy for the synthesis of complex target molecules involves the use of directing groups. Early on, it was recognized that substituents on an aromatic ring influence the orientation and rate of electrophilic aromatic substitution. In fact, this gave rise to the practice of deliberately incorporating groups such as amino or nitro groups in order to influence the orientation and rate of the subsequent electrophilic substitutions.

In later years of the 20th century, this same practice was incorporated into aliphatic chemistry as groups that could be used to influence the regiochemistry of reactions in alkenes and similar aliphatic systems were discovered. As in electrophilic aromatic substitution, the use of directing groups has been widespread in carbocation chemistry, where the directing group can ensure the formation of one regioisomeric alkene from a given carbocation.

In carbanion chemistry, the directing influences of substituent groups are often most evident in stabilization of carbanions. However, these groups can often so alter the chemistry of the anionic species that the normal reactivity (electrophilic or nucleophilic) of a particular location is reversed. This charge-type inversion is now known as ***umpolung***.[44] It has become an important part of the toolbox of the synthetic organic chemist because it allows one to use reagents that are surrogates for species such as carbonyl anions—in effect, to direct the addition of an *electrophile* to the *carbonyl carbon*.

Stereoselectivity

Starting in the middle of the 20th century, increasing attention has been paid to the control of stereochemistry during a synthesis—to *stereoselectivity*. The first generation of stereocontrolled syntheses were examples in which ***diastereoselectivity*** was the focus. These

44. Reviews: (a) Seebach, D. *Angew. Chem.* **1969**, *81*, 690. (b) Gröbel, B.-T.; Seebach, D. *Synthesis* **1977**, 357. (c) Seebach, D. *Angew. Chem. Int. Ed. Engl.* **1979**, *18*, 239.

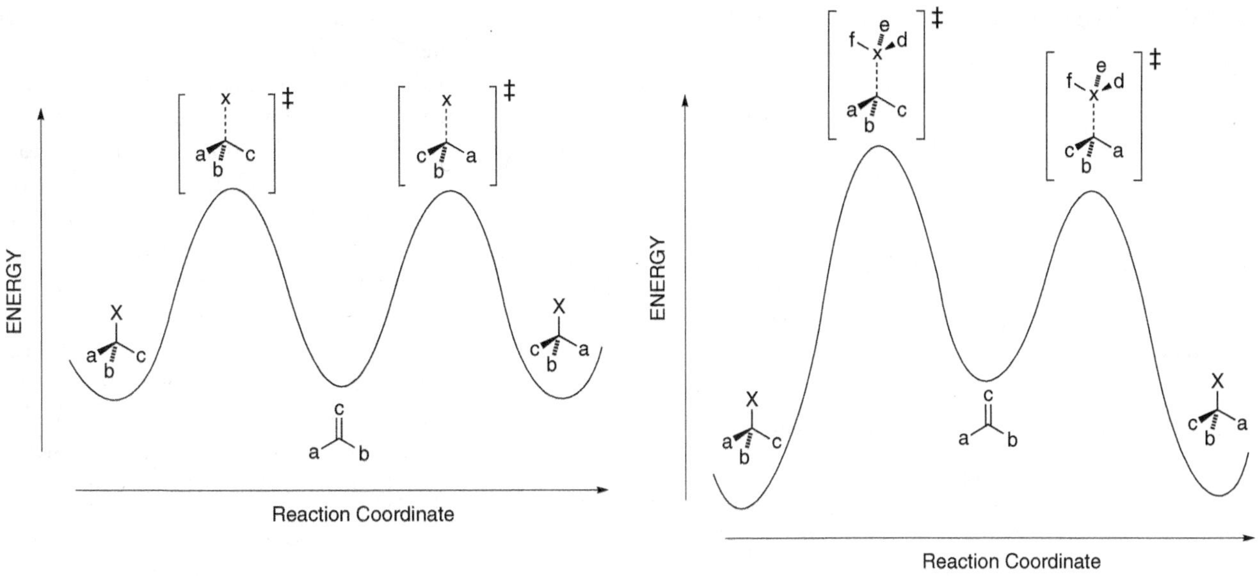

Figure 15.17 The reaction of a prochiral substrate with an achiral reagent (left) proceeds through enantiomeric transition states of equal energy and gives a racemic product. The same reaction in a chiral environment (e.g., with a chiral reagent or in the presence of a chiral catalyst) proceeds through diastereoisomeric transition states of different energy and thus gives a product enriched in one enantiomer.

syntheses invariably led to the synthesis of racemates, but in the best examples of this type, only one diastereoisomer of the product was obtained. The two necessary prerequisites to being able to complete syntheses at this level were:

1. Developing of a toolkit of stereoselective reactions to use
2. Developing a knowledge of the relationship between conformation and stereochemistry, especially in acyclic systems. Woodward's total synthesis of (±)-reserpine[45] remains one of the finest examples of this type of synthesis.

Beginning in the 1970s and gaining momentum after Sharpless' report of the discovery of the asymmetric epoxidation of allylic alcohols,[46] the focus changed to developing reactions that would control the absolute stereochemistry of the product—***enantioselectivity***. To achieve good levels of asymmetric induction, it is necessary to arrange for the two transition states leading to the two enantiomers of the product to be diastereoisomeric and of substantially different energy (Figure 15.17).

There are three ways in which this can be accomplished, and all involve the interaction of the prochiral substrate with a chiral entity. In the Sharpless asymmetric epoxidation, the reaction is catalyzed by a chiral catalyst assembled from the dialkyl tartrate and the titanium isopropoxide. An alternative approach involves the initial modification of the prochiral substrate by a ***chiral auxiliary***. This modified substrate, now being chiral, can react with a non-chiral reagent to give diastereoisomeric products. One of the most successful groups of chiral auxiliaries are the chiral oxazolidinones, first developed by Evans for use in what is now known as the ***Evans asymmetric aldol addition***.[47] It is worthwhile

45. (a) Woodward, R.B.; Bader, F.E.; Bickel, H.; Frey, A.J.; Kierstad, R.W. *J. Am. Chem. Soc.* **1956**, *78*, 2023, 2657; *Tetrahedron* **1958**, *2*, 1.
46. Katsuki, T.; Sharpless, K.B. *J. Am. Chem. Soc.* **1980**, *102*, 5974.
47. (a) Evans, D.A.; Takacs, J.M.; McGee, L.R.; Ennis, M.D.; Mathre, D.J.; Bartroli, J. *Pure Appl. Chem.* **1981**, *53*, 1109. (b) Evans, D.A.; Bartroli, J.; Shih, T.L. *J. Am. Chem. Soc.* **1981**, *103*, 2127.

Table 15.5 Protecting Groups for Alcohols

Class	Examples
Esters	-OAc, -OCOPh, -OCOBu^t, -OCOCH$_2$CCl$_3$ (Troc), -OCO$_2$Bu^t (Boc), -OCO$_2$CH$_2$Ph (Cbz), -OCO$_2$CH$_2$CH(*o*-C$_6$H$_4$)$_2$ (Fmoc)
Ethers	-OMe, -OCH$_2$Ph, -OCH$_2$C$_6$H$_4$OMe (PMB), -OCPh$_3$ (Tr)
Acetals and ketals	-OCH$_2$OMe (MOM), -OCH$_2$OCH$_2$CH$_2$OMe (MEM), -OCH(Me)OEt, -OCMe$_2$OMe (MOP), -O(C$_5$H$_9$O) (THP), -OCH$_2$OCH$_2$CH$_2$SiMe$_3$ (SEM)

Table 15.6 Conditions for Removing Protecting Groups from Alcohols

Class	Conditions
Esters	KOH, H$_2$O (base hydrolysis); HCl, H$_2$O (acid hydrolysis); LiAlH$_4$, Et$_2$O (reduction); H$_2$, Pd-C (Cbz esters)
Ethers	BBr$_3$, CH$_2$Cl$_2$; TiCl$_4$, CH$_2$Cl$_2$, -78°C; Ce^{4+}, EtOH (PMB ethers); H$_2$, Pd-C (benzyl ethers)
Acetals and ketals	H$_2$SO$_4$, H$_2$O, THF; Bu$_4$NF, THF (SEM)

noting that more than three decades after their initial use, the Evans auxiliaries are still in widespread use (even in reactions with chiral catalysts). The third motif for generating one enantiomer of a chiral product from a prochiral precursor involves the use of a chiral reagent. One early example of this approach was provided by Brown's asymmetric hydroboration of *cis* alkenes with diisopinocampheylborane.[48]

In the following sections, we will focus our attention of the role of protecting groups in organic synthesis. We will defer our discussion of directing groups and stereoselectivity to the next two chapters of this book.

15.6 Protection of Alcohols

The O–H group is acidic, with typical pK_a values varying from a low of 4 to 5 in carboxylic acids to a high of 18 to 19 in tertiary alcohols. As a consequence, this group often interferes in reactions involving nucleophiles because it is deprotonated by many nucleophilic reagents. This can interfere with the ability of the chemist to use particular reagents during a synthesis. Moreover, proton transfers from oxygen are among the fastest reactions known, so that it is almost impossible to avoid the reaction by attempting to exploit differences in reaction rates. Consequently, if a basic reagent is used, it is frequently necessary to add excess reagent to allow the initial reaction with the alcohol proton to proceed. Using a Grignard reagent under these circumstances, for example, requires an additional equivalent of the reagent to be used, so that the reaction actually proceeds via the magnesium alkoxide.

The protection of alcohols always involves the replacement of the acidic hydrogen by a less reactive surrogate. Prior to the last three decades of the 20th century, a variety of protecting groups in the form of esters and ethers were used to accomplish the selective protection of alcohols. A catalogue of widely used protecting groups for alcohols and phenols, along with their acronyms, is given in Table 15.5, and the common conditions for their removal are given in Table 15.6.

Esters

Their easy preparation and their (generally) simple removal by acid or base hydrolysis or ammonolysis has made esters popular as protecting groups for the hydroxyl groups of alcohols and phenols, especially in oxidation reactions. Selective formation of esters from primary alcohols has been accomplished by the use of hindered esters such as the pivalate ester (Example **15.50**). These highly hindered esters also address the susceptibility of simpler esters such as the acetate or benzoate toward acid or base hydrolysis. Mixed carbonate esters are sometimes used as protecting groups for alcohols (although carbonic acid derivatives are much more widely used for the protection of amines as carbamates). The three

48. Brown, H.C.; Ramachandran, V. *Pure Appl. Chem.* **1991**, *63*, 307.

most commonly used mixed carbonates are the alkyl *tert*-butyl carbonate (Boc), the alkyl benzyl carbonate (Cbz), and alkyl 9-fluorenylmethyl carbonate (Fmoc). The Fmoc group may be cleaved by base by an Elcb elimination pathway using triethylamine in pyridine (taking advantage of the aromatic character of the fluorenyl anion).

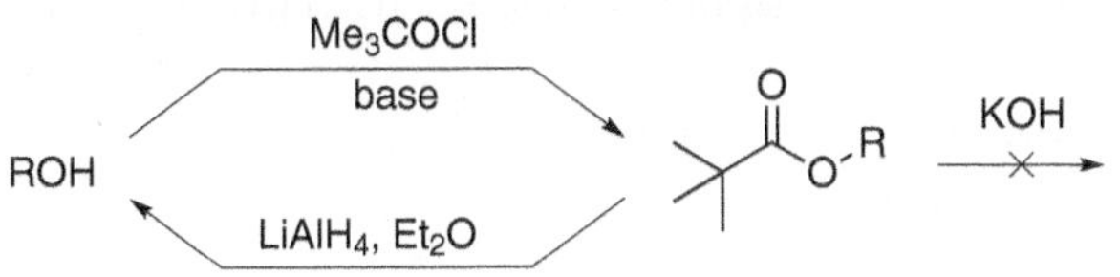

(15.50)

Representative conditions for the removal of ester-protecting groups from alcohols are gathered in Table 15.7. All esters can be cleaved by complex metal hydride reducing agents, but the use of these reagents is obviously restricted to the cleavage of esters in compounds without other reducible functional groups. Of the common esters in Table 15.6, the Boc ester is unique in being cleaved by anhydrous trifluoroacetic acid (TFA), and the Cbz ester is unique in being cleaved by catalytic hydrogenolysis. As referred to above, the hindered pivalate ester is resistant to acid or base hydrolysis, but it may be cleaved by reduction using complex metal hydrides such as lithium triethylborohydride. In general, benzoate esters react more slowly with nucleophiles than the acetate esters, so it is often possible to hydrolyze an acetate ester selectively in the presence of a benzoate ester.

Despite their ease of formation and cleavage, esters present one major drawback in their use as protecting groups: they exhibit a propensity to isomerize by internal migration of the ester group from one hydroxyl group to another under acid or base conditions. Such isomerizations compromise the integrity of the protection afforded by the ester group, which limits their usefulness.

Ethers

Generally, the least reactive derivative of an alcohol or phenol that one can make is a simple alkyl ether, which ought to make ethers ideal protecting groups for alcohols in a wide range of reactions. However, ethers are not easily cleaved under mild conditions, except for certain aryl methyl ethers. This makes the removal of the protecting group problematic in multifunctional molecules. The simplest method for cleaving simple alkyl ethers involves treating the ether with boron tribromide in dichloromethane, a reaction that works especially well for the cleavage of methyl ethers.

The difficulty inherent in using simple ethers to protect alcohols and phenols arises from the need to use strong Lewis acids for their cleavage, and this spurred the search for ethers that could be cleaved under milder conditions. One option is the benzyl ether, because the

Table 15.7 Representative Conditions for the Removal of Ester-Protecting Groups

	KOH, H_2O	RNH_2	H_2SO_4, H_2O	$LiEt_3BH$, THF	H_2, Pd-C	TFA
R–OAc	R–OH	R–OH	R–OH	R–OH	—	—
R–OCOPh	R–OH	R–OH	R–OH	R–OH	—	—
R–OCOBu^t	—	—	—	R–OH	—	—
R–OBoc	R–OH	R–OH	R–OH	R–OH	—	R–OH
R–OCbz	R–OH	R–OH	R–OH	R–OH	R–OH	—
R–OFmoc	R–OH	R–OH	—	R–OH	—	—

benzyl group can be removed by hydrogenolysis over palladium (Example **15.51**). In fact, because this group can be removed under very mild conditions, it has become a very popular protecting group for alcohols, amines, and amides. A wide range of benzyl ethers, benzyl esters, and benzylamines is now available commercially. The benzyl group has been especially widely used in the solid-state synthesis of proteins. The trityl (triphenylmethyl) ether can be used for the selective protection of primary alcohols in the presence of secondary alcohols; it is readily removed in acid or by hydrogenolysis.

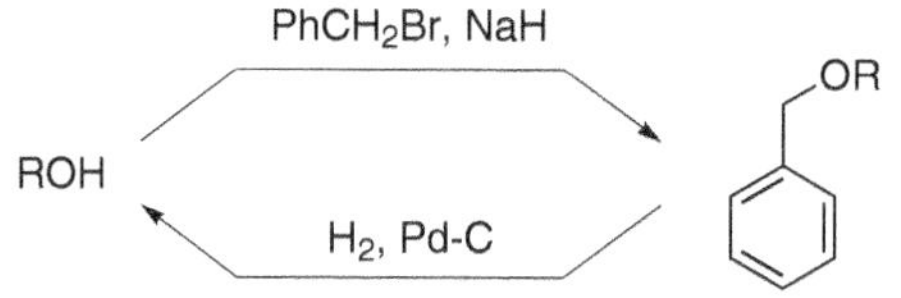

(15.51)

The protection of alcohols as ethers that can be cleaved by oxidation is a relatively modern development, with the *p*-methoxybenzyl (PMB) ether being the prototypical example of this type of protecting group. Cleavage of PMB ethers is accomplished fairly simply by oxidation with ceric ammonium nitrate (CAN) or 2,3-dichloro-5,6-dicyano-1,4-benzoquinone (DDQ) (Example **15.52**).

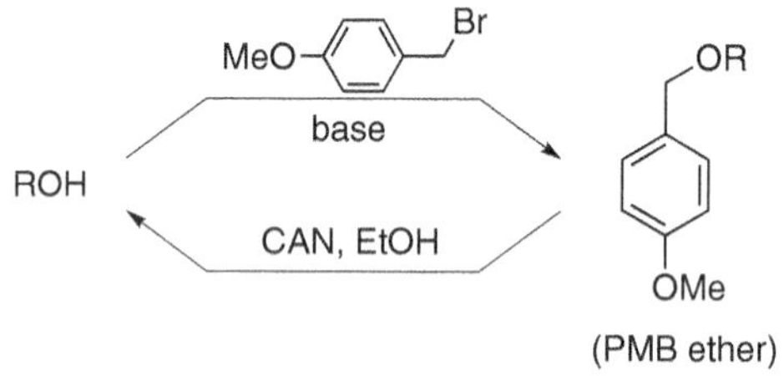

(15.52)

Acetals and Ketals

An alternative approach to solving the problem of cleaving the ether under mild conditions has been to use an acetal or ketal as the protecting group. Formaldehyde acetals can be readily prepared by treatment of the alcohol or phenol with a suitable chloromethyl ether and a base, and these compounds can then be cleaved by protic acids or Lewis acids under very mild conditions. Among the formaldehyde acetals used in synthesis, the methoxymethyl (MOM) ether (e.g., Example **15.53**) and the (2-methoxyethoxy)methyl (MEM) ether (e.g., Example **15.54**) are probably the most widely used.

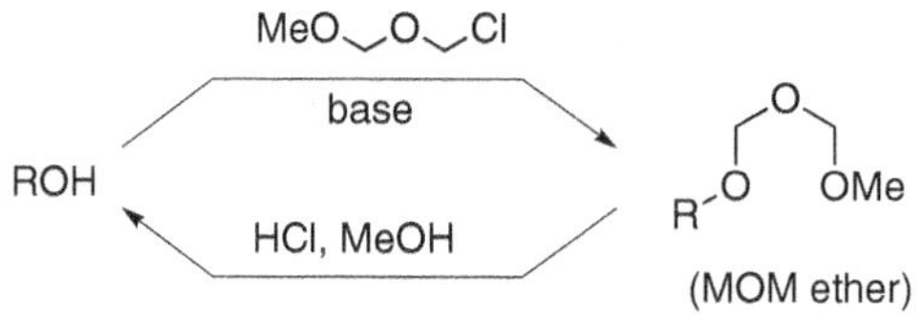

(15.53)

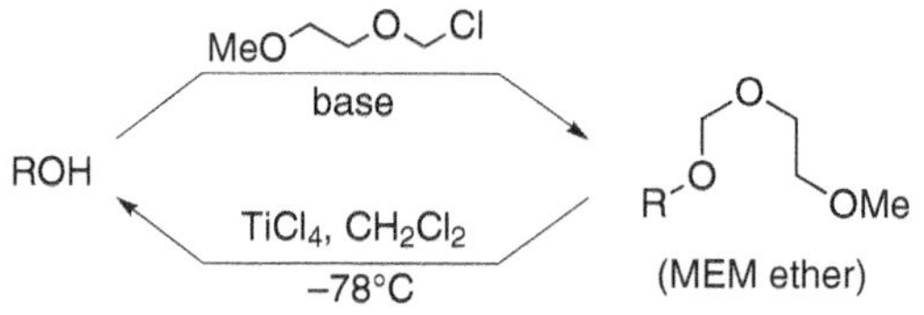

(15.54)

The cyclic acetals formed by the acid-catalyzed addition of the alcohol to dihydropyran have also been widely used to protect the alcohol. The only disadvantage that these tetrahydropyranyl (THP) ethers (Example **15.55**) have is that their formation involves the generation of a new chiral center in the molecule, and this often results in mixtures of diastereoisomers. Nevertheless, the ease of their formation and cleavage, combined with their robustness in the presence of strong nucleophiles and strong bases, has made them popular protecting groups. In similar fashion, the addition of the alcohol to ethyl vinyl

Table 15.8 Stability of protecting groups toward various reagents

Conditions→ Group↓	Aqueous acid	Aqueous base	TFA	Lewis acid	Strong base	Metal alkyls	Fluoride anion	Hydrogen, catalyst	Hydride reductant	Oxidation
—OR				a						
—OCH_2Ar				a				b		c
—OTr			d	a				b		
—OCOR	e	f			g	h			h	
—OCbz	e	f				h		b	h	
—OFmoc	e	f			i	h		b	h	
—OBoc	e	f	k	a		h			h	
—$OSiR_3$	e	f				l	m			
—OMOM	e				a					
—OMEM	e				a					
—OTHP	e									
—OSEM							n			

Where no indication is given, one may assume that the group is generally stable under the conditions specified.
a. Susceptible to strong Lewis acids (BBr_3, $TiCl_4$, etc.)
b. Susceptible to hydrogenolysis, especially over Pd
c. Electron-rich benzyl ethers can be cleaved by oxidation with reagents such as CAN and DDQ
d. Usually used with trifluoroacetic anhydride
e. Susceptible to acid-catalyzed hydrolysis or alcoholysis
f. Susceptible to base-promoted hydrolysis or alcoholysis
g. Only esters that have an α hydrogen; generally a problematic side reaction in reactions of ester enolates
h. Reduced by complex metal hydrides ($LiBH_4$, $LiAlH_4$, etc.); not generally susceptible to reduction by $NaBH_4$
i. Cleaved by Grignard reagents, alkyllithiums, etc.
j. Cleaved by bases, including tertiary amines; mechanism is E1cb
k. Isobutylene is formed in the E1 elimination of the *tert*-butyl group
l. Cleaved by alkyllithiums, but not by Grignard reagents
m. Ease of cleavage is slowed by increased steric bulk of the groups on silicon
n. Silyl fluoride and ethylene are formed in this fragmentation by the E1cb or E2 mechanism

ether gives a 1-ethoxyyethyl ether (a mixed acetaldehyde acetal) that can also be cleaved under mild conditions. Like the THP ether, the preparation of this ether generates a new chiral center in the molecule. A summary of protecting groups for monohydric alcohols and their susceptibility to reaction conditions is gathered in Table 15.8.

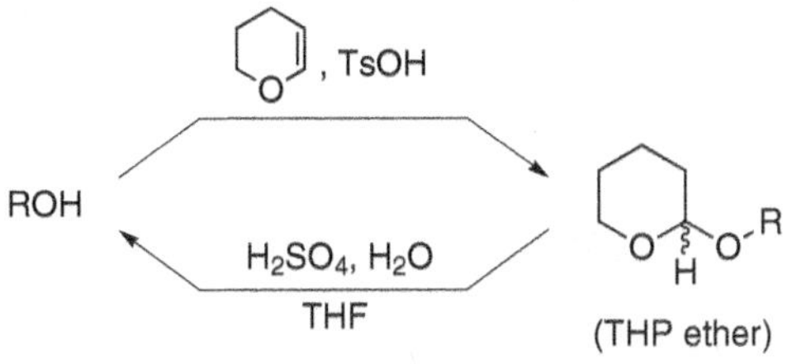

(15.55)

1,2- and 1,3-Diols

Ketals are much less widely used as protecting groups for monohydric alcohols, but the mixed ketals of acetone, formed by acid-catalyzed addition to methyl isopropenyl ether, known as MEP ethers, have been used. In contrast to the situation with monohydric alcohols, ketals are the protecting groups of choice for 1,2- and 1,3-diols. In fact, the acetonides of 1,2-diols are frequently commercially available. As shown in Example **15.56**, their synthesis by ketal exchange is usually straightforward.

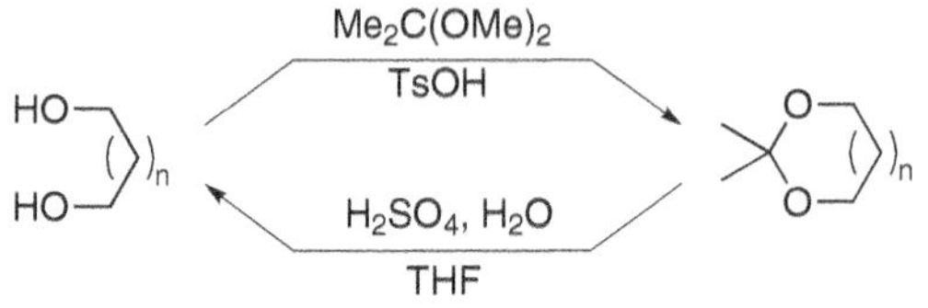

(15.56)

The protection of diol components of polyhydric alcohols is an important feature of carbohydrate chemistry, and the choice of the carbonyl component for the formation of the protected diol can be critical in determining exactly what protected derivative is formed. For example, the protection of D-glucose with acetone leads to the formation of the *bis*-acetonide where the six-membered pyranose ring of the glucose has contracted to the five-membered furanose form (Example **15.57**), whereas the same reaction with D-galactose, which also exists predominantly in the six-membered pyranose form, forms the *bis*-acetonide where the galactose ring remains in the six-membered pyranose form (Example **15.58**). The same protecting group thus leaves the 3-hydroxyl group free in D-glucose and the 6-hydroxyl group free in D-galactose. This example also serves as something of a cautionary tale. Often, exactly which isomeric product will be formed on incorporating a particular protecting group must be determined by experiment and cannot always be easily predicted. One of the general rules to come from the formation of acetonides from 1,2-diols in cyclic systems is the requirement that the two hydroxyl groups be *cis* to each other on the ring; the *trans* diols form a five-membered ring that is too strained.

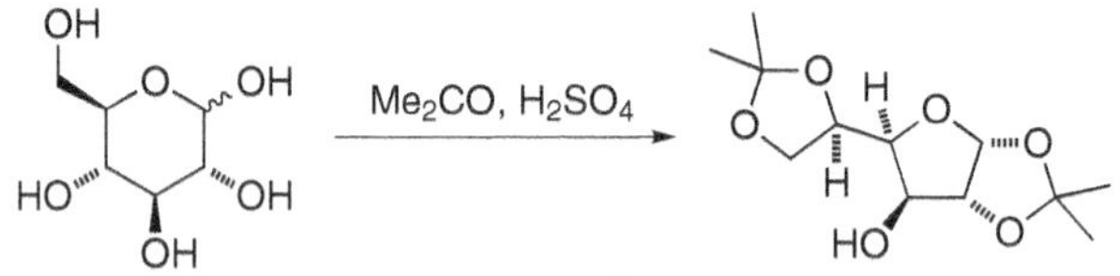

(15.57)

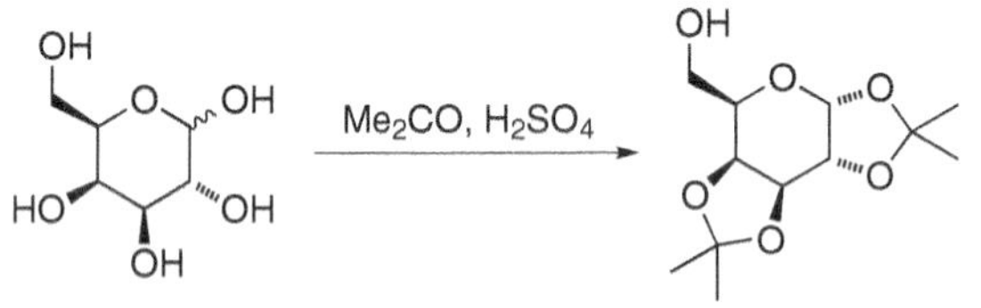

(15.58)

Wiesner's Synthesis of (±)-13-Desoxydelphonine: A Case Study Before the Widespread Use of Silyl Groups

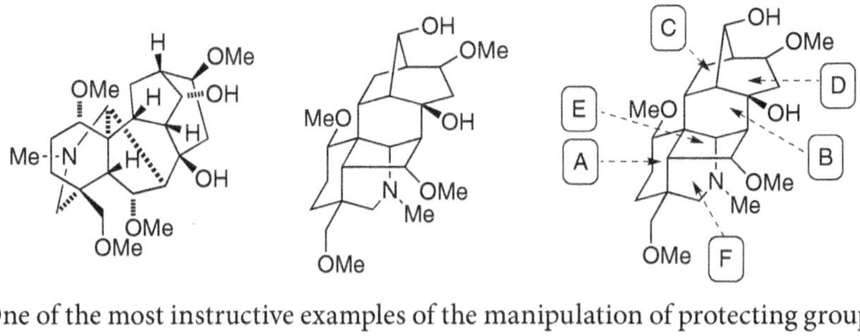

(15.59)

One of the most instructive examples of the manipulation of protecting groups in synthesis is provided by the "fourth generation" synthesis of the alkaloid (±)-13-desoxydelphonine (Example **15.59**), reported by Canadian chemist Karel Wiesner in 1979.[49] This

49. (a) Wiesner, K. *Pure Appl. Chem.* **1979**. *51*, 689. (b) Tsai, T.Y.R.; Nambiar, K.P.; Krikorian, D.; Botta, M.; Marini-Bettole, R.; Wiesner, K. *Can. J. Chem.* **1979**, *57*, 2124.

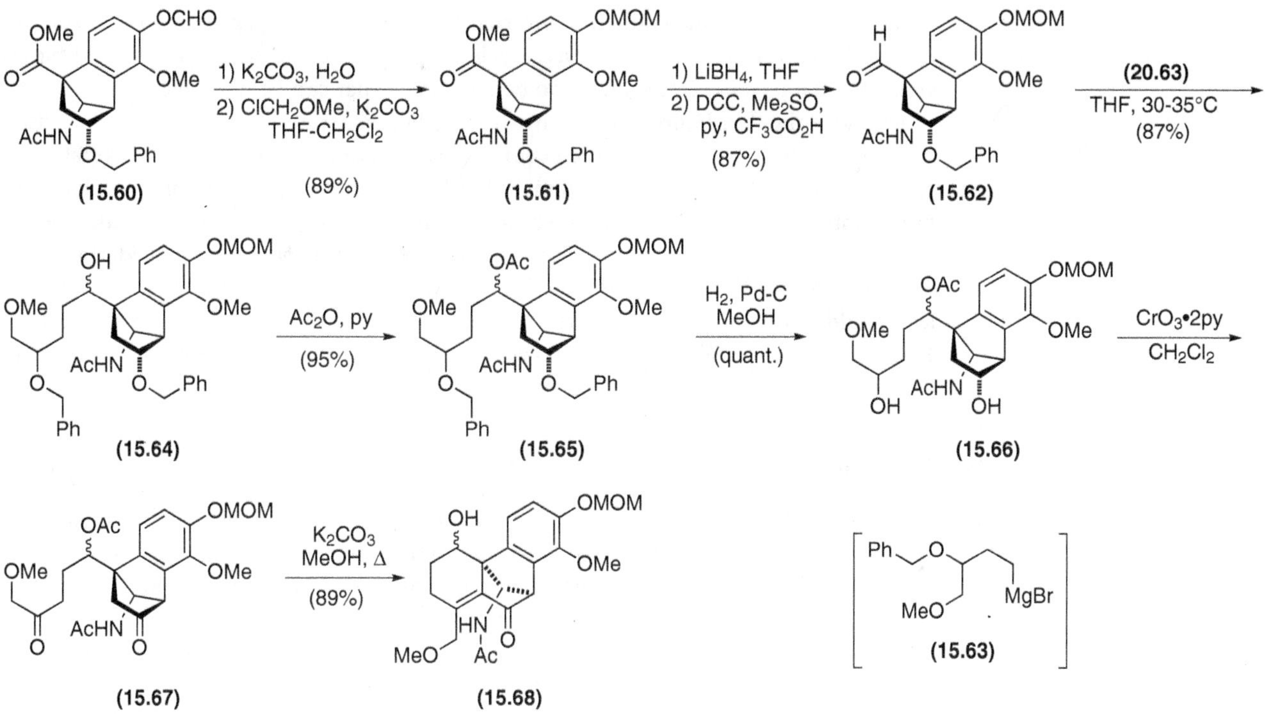

Figure 15.18 The use of protecting groups in Wiesner's synthesis of 13-desoxydelphonine

molecule contains a hexacyclic skeleton (rings A–F) adorned with 13 chiral centers (although only 4 are independent of the requirements imposed by the hexacyclic skeleton). The complete synthesis, which is well worth reading as an example of strategy in synthesis, required over 50 steps. Here, let us examine the use of protecting groups in a middle section of the sequence. Because this synthesis was carried out shortly before the rise of silyl ether protecting groups, it provides a "state of the art" view of the use of non-silyl protecting groups. The sequence in Figure 15.18 involves the protection of functional groups during the construction of the precursor containing the progenitors of the A, B, E, and F rings of the polycyclic framework of the target molecule.

In the first step of the sequence, the formate ester **15.60** (obtained by the Dakin oxidation[50] of the aromatic aldehyde) is selectively hydrolyzed with aqueous potassium carbonate. Neither the methyl ester nor the amide is hydrolyzed by this reagent, illustrating the greater reactivity of the formate ester toward hydrolysis. The phenol that is revealed is converted to the MOM ether **15.61**, which renders the phenol resistant to nucleophilic and basic reagents before the methyl ester is reduced with lithium borohydride to the corresponding primary alcohol. Lithium aluminum hydride cannot be used here because it would also reduce the amide functional group, whereas sodium borohydride does not reduce esters. The benzyl ether is resistant to the subsequent Moffatt oxidation[51] of the alcohol to give the aldehyde **15.62**.

The next reagent in the sequence (**15.63**) is an interesting one in its own right, because a Grignard reagent nucleophile was required that carried the MOM group of the final

50. (a) Dakin, H.D. *Am. Chem. J.* **1909**, *42*, 477. (b) Dakin, H.D. *Organic Syntheses.* **1923**, *3*, 28. For recent examples, see: (c) Bernini, R.; Coratti, A.; Provenzano, G.; Fabrizi, G.; Tofani, D. *Tetrahedron*, **2005**, *61*, 1821. (d) Varma, R.S.; Naicker, K.P. *Org. Lett.*, **1999**, *1*, 189. (e) Alamgir, M.; Mitchell, P.S.R.; Bowyer, P.K.; Kumar, N.; Black, D.St.C. *Tetrahedron* **2008**, *64*, 7136.

51. Reviews: (a) Moffatt, J.G. In Augustine, R.L.; Trecker, D.J., Eds. *Oxidation* (Dekker: New York, 1971), vol. 2, 1. (b) Tidwell, T.T. *Org. React.* **1990**, *39*, 297. (c) Lee, T.V. *Comp. Org. Syn.* **1991**, *7*, 291.

product and a group that could be converted to a ketone carbonyl group at a later stage in the synthesis. The benzyl ether served this latter purpose. When the Grignard addition is complete, the product (**15.64**) has two benzyl ethers that can be deprotected at the same time to give the diol **15.66**. The new hydroxyl group generated by the Grignard addition needed to be protected because the next major step in the sequence involved oxidation of **15.66** to give the diketone (**15.67**) required for the intramolecular aldol condensation. This protection was accomplished by conversion to the acetate ester, which is resistant both to the hydrogenation conditions needed to remove the benzyl groups, and to oxidation using Collins reagent.[52] Note how in Wiesner's strategy, the two carbons needed for the intramolecular aldol condensation to give the enone **15.68** are both protected as benzyl ethers in **15.65** so that they can be removed under conditions that do not compromise the integrity of any other protected hydroxyl group.

Silyl Ethers

Since the 1970s, silyl ethers have become by far the most common groups used for the protection of alcohols. In part, this is because of the ability to fine-tune the reactivity of the silyl ether by simply varying the size of the substituent groups on silicon. The range of commonly used silyl ethers for the protection of alcohols is given in Table 15.9, along with their relative reactivities in acidic and basic media. It is worthwhile noting that although larger alkyl groups on silicon do confer increased stability on the silyl ether toward acids and bases due to their steric bulk, the same steric bulk also makes the ethers more difficult to form. This has partly been overcome by means of the use of the more reactive silyl triflates, as shown in Table 15.9.[53] Similar data (in the form of reaction half-lives) is available for the acid- and base-promoted methanolysis of *n*-hexyl silyl ethers[54] and *n*-hexyl silyloxy ethers.[55]

Silyl ethers are formed by treating an alcohol with a silyl chloride or silylamine and a base. The rates of formation of the silyl ether and the ease of recovery of the hydroxyl group

Table 15.9 Relative Stabilities of Silyl Ether-Protecting Groups

Silyl Ether	Introduction	In Acid	In Base
R–OSiMe$_3$ (R–OTMS)	Me$_3$SiCl, imidazole, DMF	1	1
R–OSiEt$_3$ (R–OTES)	Et$_3$SiCl, imidazole, DMF	64	10–100
R–OSiMe$_2$Bu^t R–OTBS	Bu^{t}SiMe$_2$Cl, imidazole, DMF Bu^{t}SiMe$_2$OTf, 2,6-lutidine, CH$_2$Cl$_2$	2×10^4	2×10^4
R–OSi(Pr^i)$_3$ R–OTIPS	(*i*-Pr)$_3$SiCl, imidazole, DMF (*i*-Pr)$_3$SiOTf, 2,6-lutidine, CH$_2$Cl$_2$	7×10^5	1×10^5
R–OSi(Ph)$_2$Bu^t R–OTBDPS	*t*-Bu(Ph)$_2$SiCl, imidazole, DMF	2×10^4	5×10^6

52. (a) Collins, J.C.; Hess, W.W.; Frank, F.J. *Tetrahedron Lett.* **1968**, *9*, 3363. (b) Collins, J.C.; Hess, W.W. *Org. Synth.* **1988**, *Coll. Vol. 6*, 644. (c) Ratcliffe, R.; Rodehorst, R. *J. Org. Chem.* **1970**, *35*, 4000.

53. Wuts, P.G.M.; Greene, T.W. *Greene's Protective Groups in Organic Synthesis*, 4th ed. (Wiley-Interscience: New York, 2007).

54. Davies, J.S.; Higginbotham, L.C.L.; Tremeer, E.J.; Brown, C.; Threadgold, R.S. *J. Chem. Soc., Perkin Trans. 1* **1992**, 3043.

55. Gillard, J.W.; Fortin, R.; Motron, H.W.; Yiakim, C.; Quesnelle, C.A.; Daignault, S.; Guindon, Y. *J. Org. Chem.* **1988**, *53*, 2602.

from the silyl ether both depend on the alkyl groups on silicon. Trimethylsilyl ethers are formed readily from all alcohols by treating the alcohol with trimethylsilyl chloride and a base such as triethylamine, imidazole, or hexamethyldisilazane ($Me_3SiNHSiMe_3$), and they are cleanly hydrolyzed by dilute aqueous acid or base. They are the most labile of the silyl ethers.

Following their initial use for enols reported by Stork[56] and the use of imidazole in DMF as the base catalyst during their introduction,[57] the *tert*-butyldimethylsilyl (TBDMS) ethers have become extremely popular as protecting groups for alcohols due to their enhanced resistance to aqueous hydrolysis.[58] The TBDMS group has the added advantage that the steric hindrance around the silicon permits the selective silylation of primary alcohols with TBDMS chloride in the presence secondary alcohols[59] or the selective silylation of the less hindered of two hydroxyl groups. In fact, one of the most important advantages of silyl ethers is the way that they can be used for the selective protection and deprotection of the less hindered hydroxyl groups in polyhydric alcohols.

TBDMS ethers are resistant to hydrolysis by aqueous acids or bases, but the free alcohol can be recovered from the silyl ether by treatment with fluoride ion (the formation of the strong silicon-fluorine bond probably provides the driving force for this reaction). Like protection of hydroxyl groups as TBDMS ethers, deprotection of the least hindered hydroxyl group usually occurs most rapidly. For example, in compound **15.69** it is the TBDMS ether group flanked by only one branched chain (*a*) that is cleaved to give the monosilyl ether **15.70**; the silyl ether flanked by two branched chains (*b*) is not.[60]

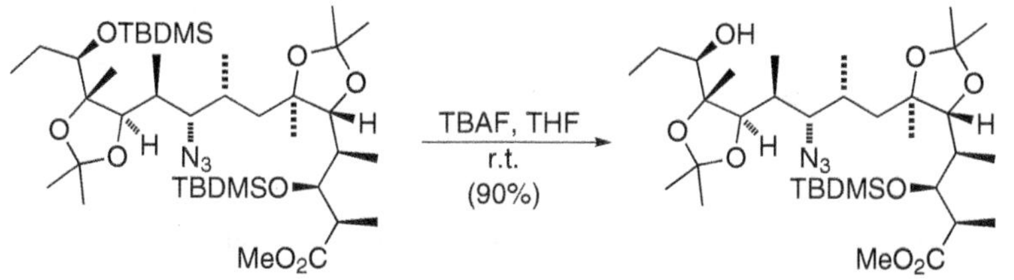

(15.69) **(15.70)**

An excellent model for the modern use of protecting groups for alcohols is provided by part of the synthetic sequence used in the early stages of Hashimoto's total synthesis of zaragozic acids A and C (Figure 15.19).[61] This sequence begins with a protected D-erythronic acid derivative. The three alcohol groups of the erythronic acid skeleton must eventually be converted into three different moieties, which means that they must be differentially protected from an early stage in the synthesis.

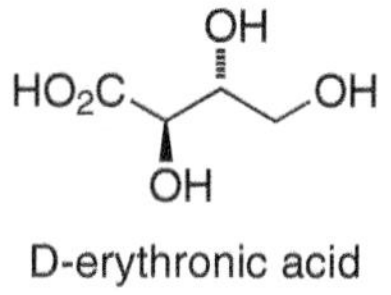

D-erythronic acid

The starting compound for the sequence is the ester **15.71**, in which the terminal diol is protected as the acetonide. Protection of the remaining hydroxyl group as the

56. Stork, G.; Hudrlik, P.F. *J. Am. Chem. Soc.* **1968**, *90*, 4462. See also: (b) Lalonde, M.; Chan, T.H. *Synthesis* **1985**, 817.

57. Corey, E.J.; Venkateswarlu, A. *J. Am. Chem. Soc.* **1972**, *94*, 6190.

58. For a discussion of the unusual stability of *tert*-butylmethylsilyl compounds, see: Sommer, L.H.; Tyler, L.J. *J. Am. Chem. Soc.* **1954**, *76*, 1030.

59. (a) Ogilvie, K.K.; Shifman, A.L.; Penney, C.L. *Can. J. Chem.* **1979**, *57*, 2230. (b) Kinzy, W.; Schmidt, R.R. *Liebigs Ann. Chem.* **1987**, 407.

60. Nakaba, T.; Fukui, M.; Oishi, T. *Tetrahedron Lett.* **1988**, *29*, 2219, 2223.

61. Hirata, Y.; Nakamura, S.; Watanabe, N.; Kataoka, O.; Kurosaki, T.; Anada, M.; Kitagaki, S.; Shiro, M.; Hashimoto, S. *Chem. Eur. J.* **2006**, *12*, 8898.

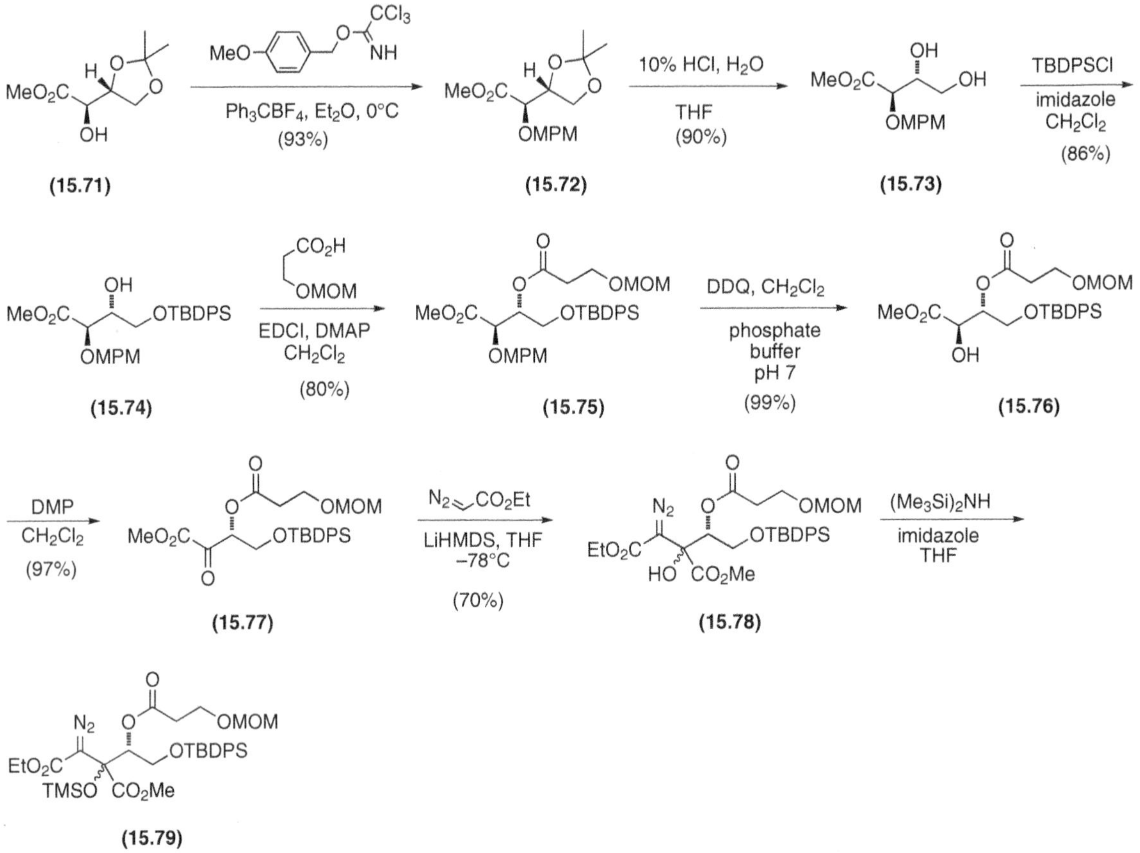

Figure 15.19 Sequence from Hashimoto's synthesis of the zaragozic acids

p-methoxybenzyl ether (or *p*-methoxyphenylmethyl, or MPM, ether) gave the ester (**15.72**). Unlike ester or silyl ether protecting groups, the MPM ether does not migrate under acid or base conditions. Hydrolysis of the acetonide gave the terminal diol (**15.73**), which was selectively monoprotected as the *tert*-butyldiphenylsilyl (TBDPS) ether. Like the TBDMS ether, the TBDPS ether of a primary alcohol can be formed selectively in the presence of a secondary hydroxyl group. This alcohol (**15.74**) now has only a single hydroxyl group left, so the chemical differentiation of the three hydroxyl groups is now accomplished, allowing their straightforward, selective transformation. Esterification of the free hydroxyl group of **15.74** with MOM-protected 3-hydroxypropanoic acid gave the ester **15.75**, which was next selectively deprotected with DDQ to give the α-hydroxyester **15.76**. Oxidation of the revealed alcohol with the Dess-Martin periodinane gave the α-ketoester **15.77**, which was then treated with the conjugate base of ethyl α-diazoacetate to give the addition product **15.78** as a mixture of diastereoisomers. Protection of the new hydroxyl group as the trimethylsilyl ether completed this part of the synthesis. Note how, in this densely functionalized molecule, all the alcohol hydroxyl groups are protected differently, and all three esters are also different.

Problems

15-7 What will be the major organic product formed when the alcohol below is treated with the reagents in the list below it? Unless otherwise specified, assume that the amount of the reagent is sufficient to react with all the hydroxyl groups that will react with it.

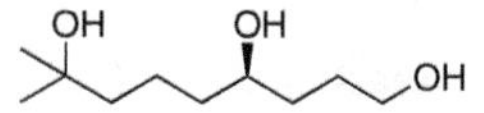

(a) Ac_2O, py

(b) TBDMSCl (1 eq), imidazole, DMF

(c) NaH, $PhCH_2Br$

(d) Ph_3CBr, py

(e) (1) TBDMSCl (2 eq), imidazole, DMF; (2) AcCl, $PhNMe_2$; (3) TBAF, THF

(f) (1) Ph_3CBr, py; (2) PMBCl, NaH; (3) DHP, TsOH; 4) DDQ

(g) Me_3COCl, py, DMAP

(h) (1) TBDMSCl (2 eq.), imidazole, DMF; (2) TBAF (1 eq), THF

(i) MOMCl, K_2CO_3, THF

(j) (1) TBDMSCl (1 eq), imidazole, DMF; (2) MEM-Cl, K_2CO_3, MeCN; (3) TBAF, THF

15-8 When acyclic 1,2,3-triols are protected as the acetone ketals (acetonides), the major product is almost always has a five-membered ring, and when benzaldehyde is used, the ring is usually six-membered. Suggest a reason why this should be the case.

15-9 Provide a reasonable rationalization for the observation that the acetonides of D-mannose and D-glucose are constitutional isomers, as shown, and not simply stereoisomers of each other.

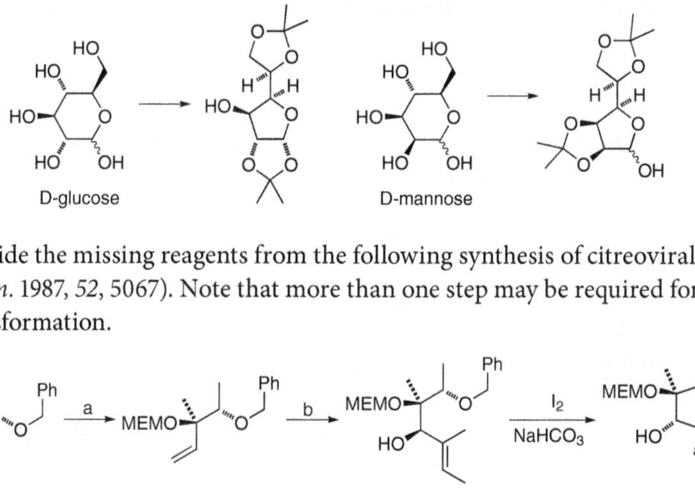

15-10 Provide the missing reagents from the following synthesis of citreoviral (*J. Org. Chem.* 1987, *52*, 5067). Note that more than one step may be required for each transformation.

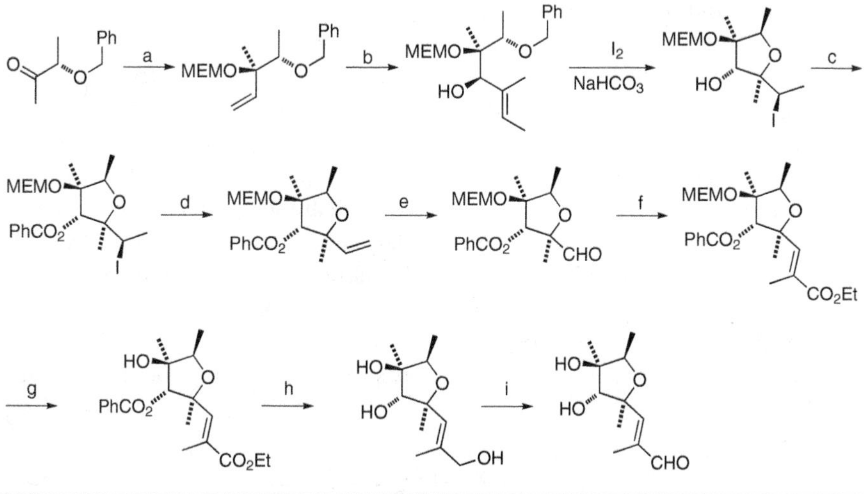

Reaction Synopses

Protecting Groups for Alcohols

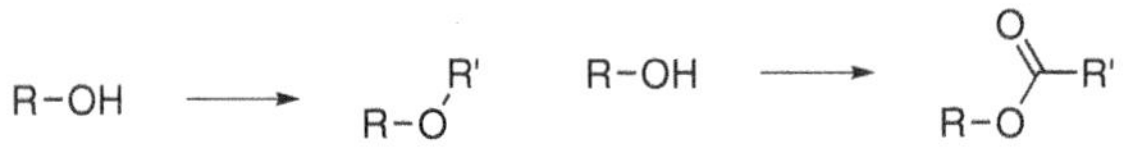

Alkyl Ethers

Formation: RX, base; R=Me, $PhCH_2$, p-$MeOC_6H_4CH_2$; etc.

Cleavage: BBr_3, CH_2Cl_2; H_2, Pd-C (allyl, benzyl); DDQ, CAN (PMB ethers); HCO_2H (trityl ethers)

Silyl Ethers

Formation: R_3SiX, $EtN(i\text{-}Pr)_2$; X=Cl, OTf

Cleavage: TBAF, THF; etc.

Acetals/Ketals

Formation: $ROCH_2$-Cl, base; R=Me, $Me(OCH_2CH_2)_2$, DHP, TsOH; etc.

Cleavage: H_3O^+; $TiCl_4$, CH_2Cl_2, etc.

Carboxylic Esters

Formation: Ac_2O, Py; Me_3COCl, py; PhCOCl, KOH, H_2O; etc.

Cleavage: K_2CO_3, MeOH; NaOH, H_2O, THF; NH_3, H_2O; etc.

Mixed Carbonate Esters

Formation: Boc_2O, Py; CbzCl, Py, etc.

Cleavage: TFA (Boc); H_2Pd-C (Cbz); etc.

15.7 Protection of Aldehydes, Ketones, and Carboxylic Acids

The carbonyl group is arguably the most versatile functional group in organic synthesis. A quick examination of the preceding 10 chapters of this book will reveal that more than half the reactions discussed involve carbonyl compounds as reactants, reagents, or products. Of course, this very reactivity presents a problem, and the protection of carbonyl groups has long been an important facet of multistep syntheses.

Protection of Aldehydes and Ketones

Acetals and Ketals

The simplest and most widely used protecting groups for aldehydes and ketones are the acetals and ketals. Although acetals and ketals derived from simple monohydric alcohols can be prepared and used, the corresponding compounds prepared from ethylene or propylene glycol derivatives are by far the most widely used.

The preparation of acetals is easier than the preparation of ketals, as should be expected given the difference in reactivity between the carbonyl precursors toward nucleophiles. Thus, the conversion of aldehydes to dialkyl acetals is usually straightforwardly accomplished by heating the aldehyde and the alcohol in the presence of an acid catalyst (e.g., Example **15.80**). When the alcohol is something other than methanol, provision must usually be made to remove the water formed in the equilibrium reaction either azeotropically or by means of a water scavenger. As the steric requirements of the alcohol and the carbonyl compound increase, the equilibrium for the formation of an acetal, and more so a ketal, becomes much less favorable.

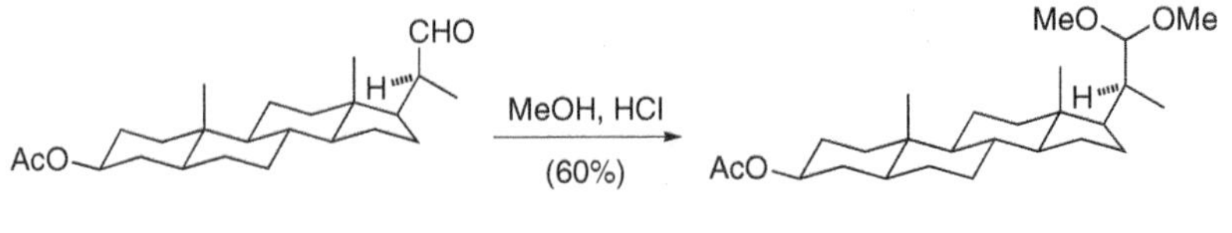

(15.80)

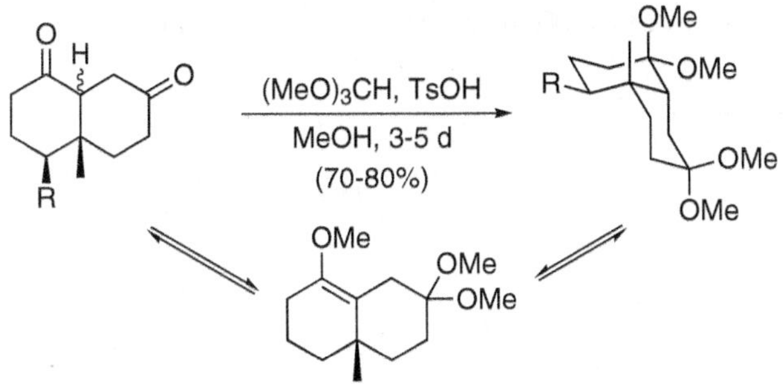

(15.81)

(15.82)

A more general method for preparing dialkyl acetals and ketals is to treat the carbonyl compound with the trialkyl orthoformate (Example **15.81**)[62] or with the alcohol and 2,2-dimethoxypropane[63] (Example **15.83**) in the presence of an acid catalyst, which may be a Lewis acid catalyst. In these reactions, the orthoformate or the acetone ketal functions as the water scavenger. These two reactions also illustrate another property of the ketalization reaction. The reaction in Example **15.81** is believed to proceed through the enol ether **15.82**, which accounts for the observation that both isomers of the diketone give the same *cis*-decalin derivative. Changing the alcohol in Example **15.83** to isopropyl alcohol leads to exclusive formation of the enol ether **15.84**.

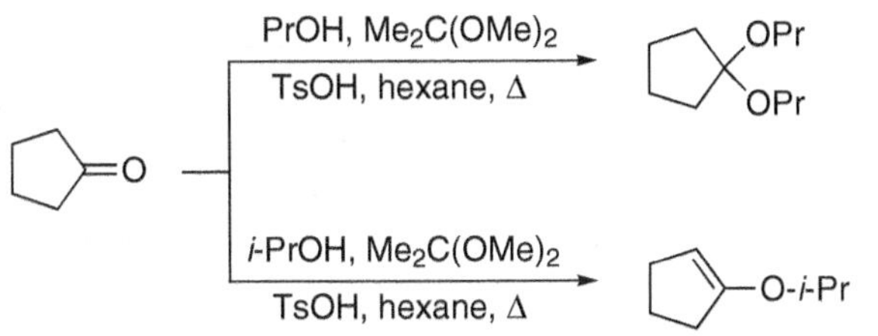

(15.83)

(15.84)

The replacement of open-chain alcohols by 1,2- or 1,3-diols has a dramatic effect on the position of the equilibrium for the formation of ketals from ketones. In part, this can be traced to the effects of entropy. In the conversion of a ketone to a dialkyl ketal (Example **15.85**), three molecules are converted to two, resulting in $\Delta S < 0$ for the reaction. This leads to an algebraic increase in the value of $\Delta G°$, and a reduction in the value of K_{eq} for the reaction. When a diol is used to form a cyclic ketal (Example **15.86**), the entropy change is much smaller and is due mainly to the loss of conformational freedom in the alkyl chain between the hydroxyl groups.

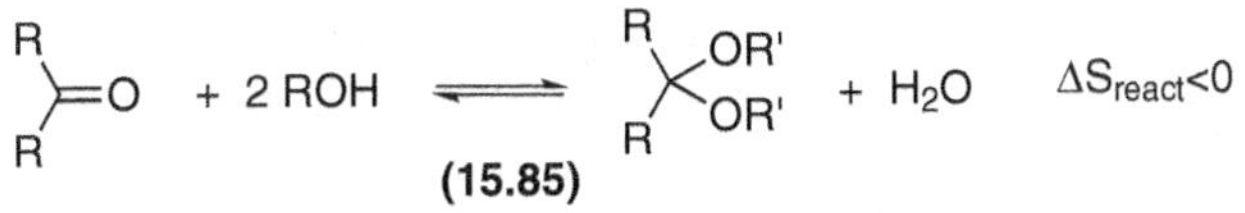

(15.85)

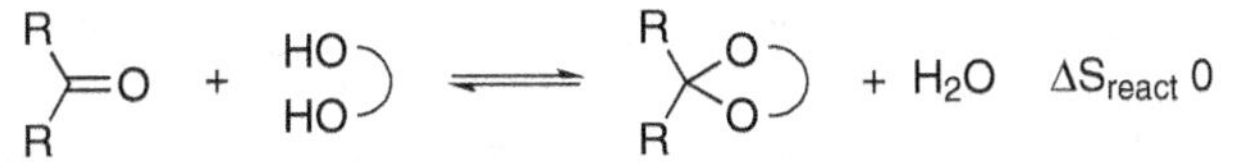

(15.86)

Problem

15-11 The formation of the *cis*-diketal in Example **15.89** may be rationalized in terms of both kinetic control and thermodynamic control. Which model gives the stronger rationalization and why?

---

62. Wijnberg, J.B.P.A.; Kesselmans, R.P.W.; de Groot, A. *Tetrahedron Lett.* **1986**, *27*, 2415
63. Rigby, H.L.; Lewis, D.E. *Synth. Commun.* **1993**, *23*, 993.

Table 15.10 Relatives rates for Hydrolysis of Cyclic Ketals by 0.003 *M* HCl in Dioxane-Water (70:30)

Ketal	Rate	Ketal	Rate	Ketal	Rate
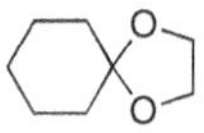	1	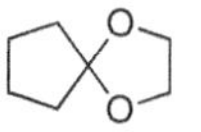	13.0	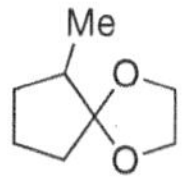	8.02
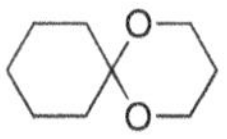	30.6	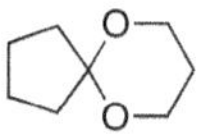	172	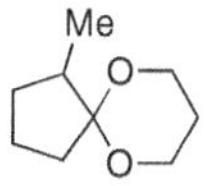	259
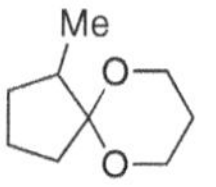	2.01	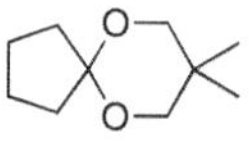	16.5	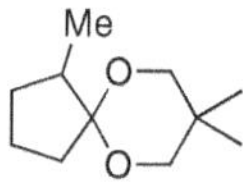	23.0
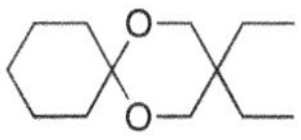	0.888	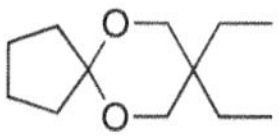	7.65	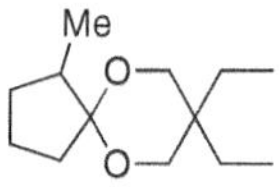	11.3
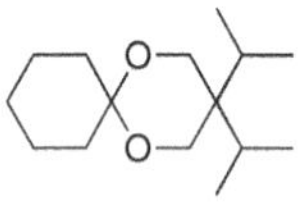	0.335	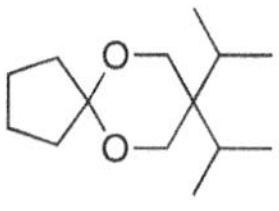	2.67	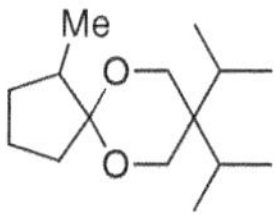	3.28

The rate of acid hydrolysis of ketals is strongly affected by the nature of the substituents on the ketal carbon[64] and in cyclic ketals by the size of the ring.[65] These effects are summarized in Table 15.10. From the data in the table, a number of generalizations can be drawn:

1. Cyclohexanone ketals are generally more resistant to hydrolysis than the corresponding cyclopentanone ketals.
2. 1,3-Dioxanes are more readily hydrolyzed than 1,3-dioxolanes.
3. Substitution at the α carbon of the ketone increases the rate of ketal hydrolysis.
4. 5,5-Dialkyl-1,3-dioxanes are more resistant to hydrolysis than simple 1,3-dioxanes, and the larger the substituents at the 5- position, the more stable the dioxane toward acid hydrolysis.

Table 15.11 Relative Rate Constants (L mol^{-1} s^{-1}) for the Acid-Catalyzed Hydrolysis of 1,3-Dioxolane and its Derivatives at 25°C

Dioxolane	*k*	Dioxolane	*k*	Dioxolane	*k*
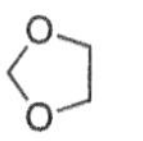	1	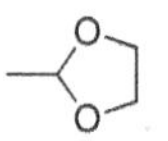	5130	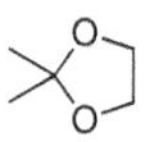	54,300
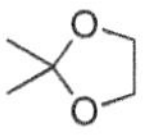	1.52	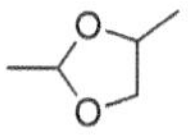	7540* 1910†	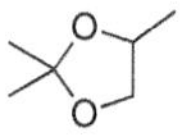	55,100
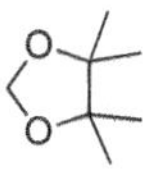	0.223	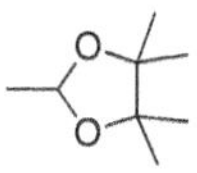	713		

**cis* isomer.
†*trans* isomer.

Not surprisingly, given how much more reactive aldehydes are toward nucleophiles than ketones, the acetals are generally more stable toward hydrolysis than the structurally similar ketals. Thus, the rate constants for the hydrolysis of acetals are significantly lower than the rate constants for the hydrolysis of structurally similar ketals, as is illustrated by the rate constants for the dioxolanes in Table 15.11.[66] These data show that (1) acetaldehyde acetals hydrolyze more than two orders of magnitude faster than the corresponding formaldehyde acetals and that (2) the acetone ketals hydrolyze another order of magnitude or more faster still. The same generalizations about the effects of ring size and substitution pattern in acetals can be made as were drawn for ketals: in general, 1,3-dioxolanes are more resistant to hydrolysis than are 1,3-dioxanes.

64. Satchell, D.P.N.; Satchell, R.S. *Chem. Soc. Rev.* **19990,** *19,* 55.
65. (a) Newman, M.S.; Harper, R.J., Jr. *J. Am. Chem. Soc.* **1958,** *80,* 6350. (b) Smith, S.W.; Newman, M.S. *J. Am. Chem. Soc.* **1968,** *90,* 1249, 1253.
66. Salomaa, P.; Kankaanperä, A. *Acta Chem. Scand.* **1961,** *15,* 871.

As the most widely used derivatives for the protection of ketones, the 1,3-dioxolanes have received a huge amount of attention in regard to both formation and hydrolysis. In compounds with two saturated carbonyl groups, the less hindered is preferentially protected with ethylene glycol and an acid catalyst, as illustrated by Example **15.87**.[67] When one of the two carbonyl groups is conjugated, the non-conjugated carbonyl group is selectively protected, as illustrated by Example **15.88**;[68] the formation of dioxolanes from α,β-unsaturated ketones is often accompanied by migration of the double bond into the β,γ-position, as shown in Example **15.89**.[69] The use of a weaker acid catalyst (e.g., oxalic acid) results in substantially less double bond migration during ketal formation.

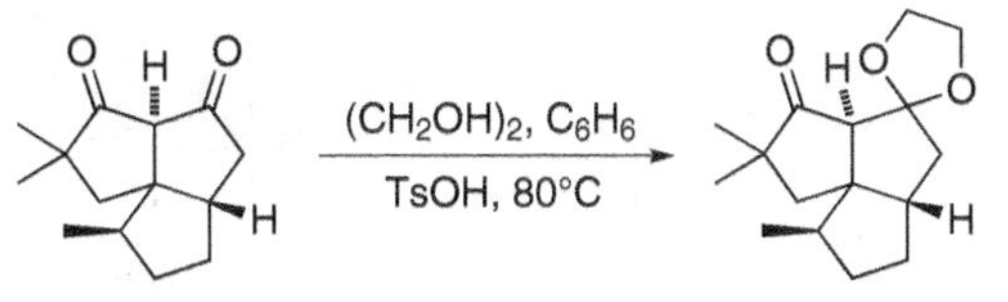

(15.87)

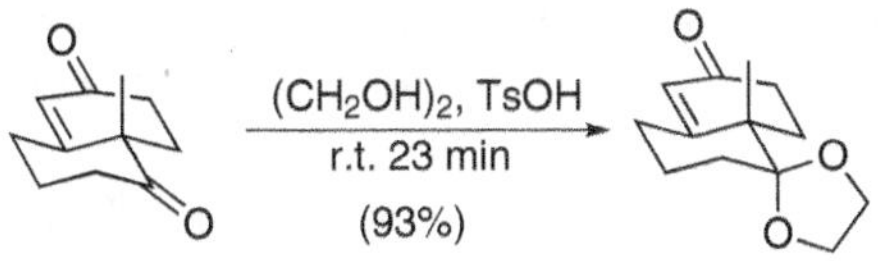

(15.88)

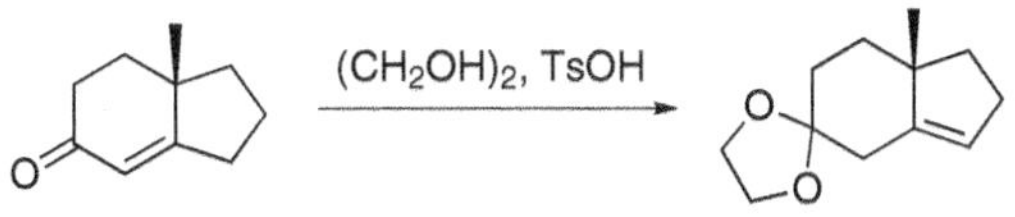

(15.89)

Dithioacetals and Dithioketals[70]

The dithioacetals and dithioketals are more robust toward aqueous acids than the corresponding acetals and ketals, which makes these protecting groups useful for the protection of aldehydes and ketones in these circumstances. The dithioketals are most commonly formed by the reaction between a dithiol and the carbonyl compound in the presence of a Lewis acid, as illustrated by reaction **15.90**, which is one step in Isobe's synthesis of the BCDE fragment of ciguatoxin 1B.[71] In a subsequent paper,[72] the BF_3 etherate was replaced by zinc triflate.

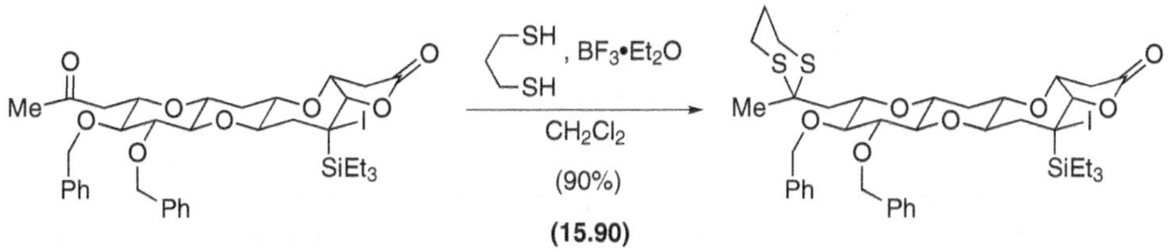

(15.90)

The traditional methods for the removal of dithioketal and dithioacetal protecting groups require the use of mercury and cadmium compounds, but these methods are now generally avoided on environmental grounds. Instead, strategies for removal of the functional group now focus on oxidation or alkylation to generate sulfonium or sulfoxide species that are more easily hydrolyzed by dilute acid or base. The limitations of many of these reagents were exposed by the attempted deprotection of the dithiane in Example **15.91**.[73] Although the desired product was obtained only in low yield and as part of complex mixtures using oxidative and alkylation methods, it could be obtained pure, albeit still in low yield, by means of mercuric perchlorate.

---

67. Crimmins, M.T.; DeLoach, J.A. *J. Am. Chem. Soc.* **1986**, *108*, 800.
68. Ciceri, P.; Demnitz, F.W.J. *Tetrahdedron Lett.* **1997**, *38*, 389.
69. de Leeuw, J.W.; de Waard, E.R.; Beetz, T.; Huisman, H.O. *Rec. Trav. Chim. Pays-Bas* **1973**, *92*, 1047.
70. For an extensive review of dithiane chemistry, see: Yus, M.; Nájera, C.; Foubelo, F. *Tetrahedron* **2003**, *59*, 6147.
71. Kira, K.; Isobe, M. *Tetrahedron Lett.* **2001**, *42*, 2821.
72. Kira, K.; Hamajima, A.; Isobe, M. *Tetrahedron* **2002**, *58*, 1875.
73. Ishihara, J.; Murai, A. *Synlett* **1996**, 363.

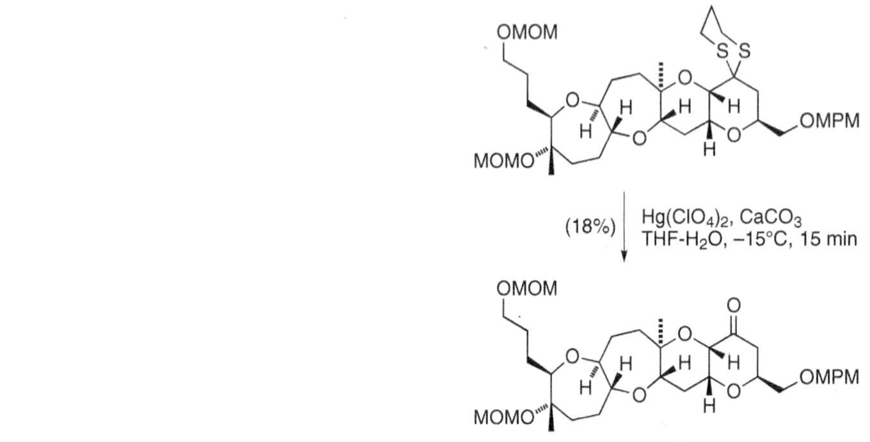

(15.91)

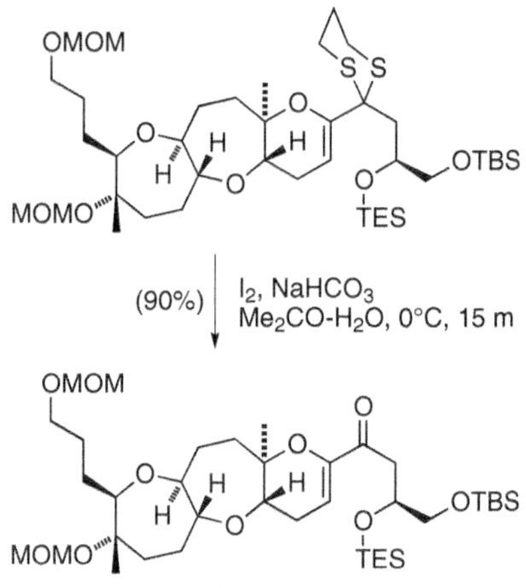

(15.92)

Based on their analysis of a range of failed reactions, the authors reasoned that the oxygen atom at position 4 was interfering with the deprotection. They were right. By carrying out the deprotection on the *seco* compound where this oxygen atom is part of an enol ether, essentially the same deprotection was accomplished in much higher yield with iodine under mildly basic conditions (Example **15.92**). This pair of reactions also demonstrates one of the frustrations in using protecting groups—often the correct protecting group must be determined in a rather more empirical fashion than the chemist would prefer.

Worked Problem

15-3 Design a reasonable sequence to accomplish the transformation in Example **15.44**.

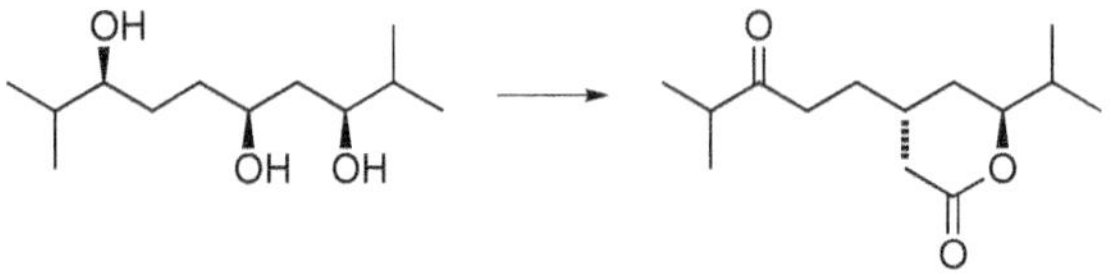

(15.44)

§Answer below.

§ Answer to Worked Problem:

The three hydroxyl groups in this molecule (2,9-dimethyldecane-2,4,8-triol), must be differentially protected during the sequence. The transformations involve the oxidation of the hydroxyl group at C-8, the displacement of the hydroxyl group at C-4 by an acetate enolate equivalent, with inversion of configuration, and the lactonization of the hydroxyl group at C-2. One possible approach to this problem is suggested below:

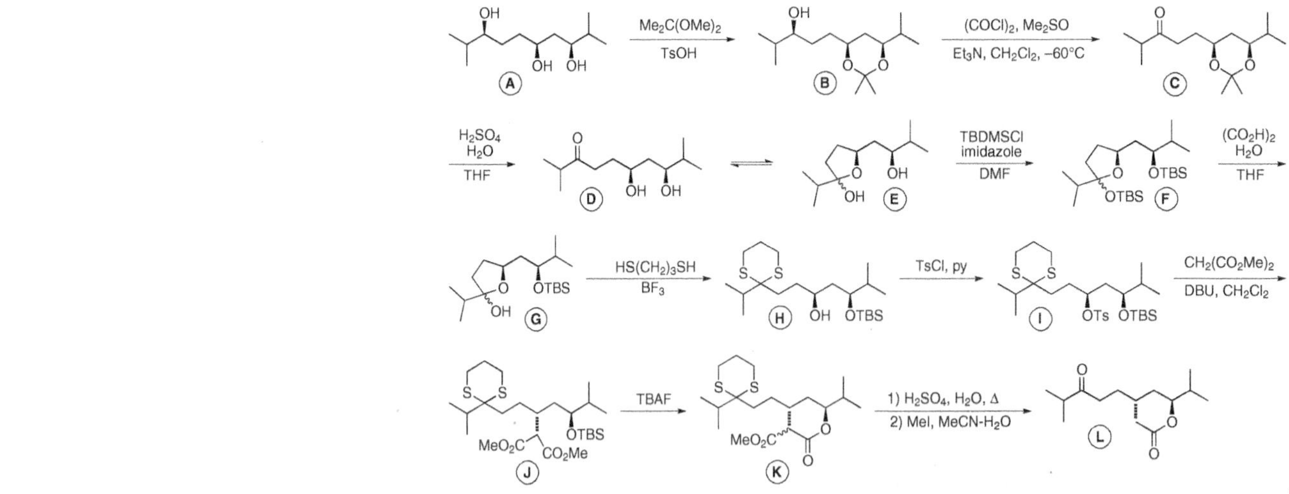

(continues)

Problems

15-12 Provide the missing reagents from the following steps in the synthesis of pleuromutilin (*Angew. Chem. Int. Ed.* **2009**, *48*, 9315). Note that more than one step may be required for each transformation.

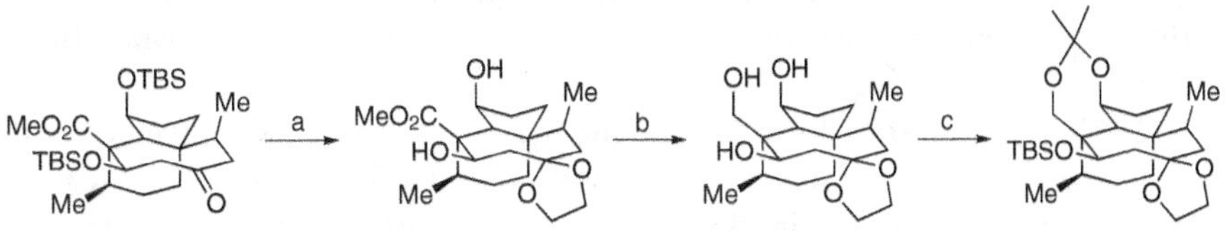

15-13 (a) The selective hydrolyses below, described in the same paper, have been reported to give the same outcome. Provide a reasonable rationalization for the fact that only one of the two acetonide groups is hydrolyzed in aqueous acetic acid. Your response should also provide a reason why each acetonide group is either susceptible to or resistant to hydrolysis under these conditions.

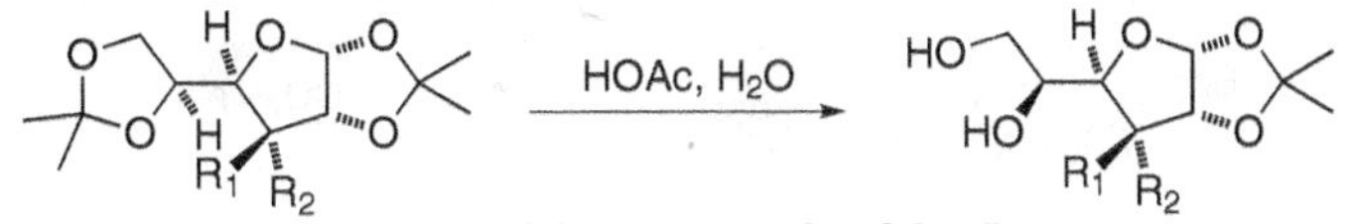

A: $R_1 = OCH_2Ph$; $R_2 = CH_2OCH_2Ph$
B: $R_1 = CH_2OCH_2Ph$; $R_2 = OCH_2Ph$

References: A: *J. Org. Chem.* **2008**, *73*, 1234; B: *Eur. J. Org. Chem.* **2008**, 5715.

(b) Provide reagents to carry out each of the following transformations, which were used in a synthesis of the marine toxin, tetrodotoxin (*J. Org. Chem.* **2008**, *73*, 1234).

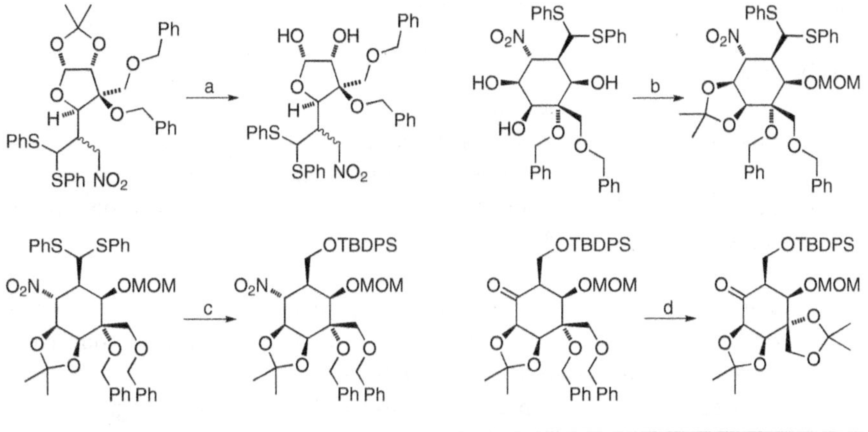

§ Answer to Worked Problem (Continued)

This process begins with protection of the 2,4-diol as the acetonide **B**, and this is followed by Swern oxidation of the C-8 hydroxyl group to give the ketone **C**. Hydrolysis of the acetonide should lead to formation of the cyclic hemiketal **E** as a mixture of anomers. Conversion of **E** to the *bis*-TBS ether **F** should be accomplished with TBS chloride and imidazole in DMF—anhydrous acidic conditions should be avoided to prevent the formation of the bicyclic ketal **M**, shown below. Mild acid hydrolysis of **F** should give a hemiketal, **G**, which could be converted to the dithioketal **H**; it is possible that the treatment of **F** itself with propanedithiol and BF_3 could lead directly to **H**. Conversion of **H** to the tosylate **I** would be followed by S_N2 displacement with the anion of dimethylmalonate to give the diester **J**. The deprotection of the TBS ether in **J** should be accompanied by intramolecular transesterification to give lactone **K**; ester hydrolysis and decarboxylation, followed by hydrolysis of the dithiane should complete the synthesis of the desired target **L**.

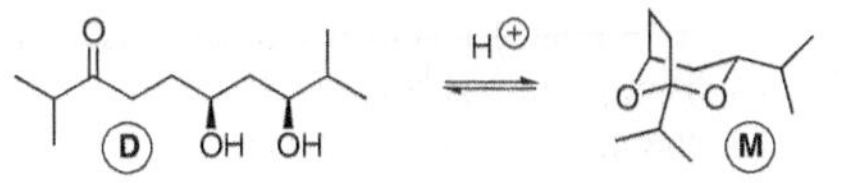

Carboxylic Acids

Esters

The acidic hydrogen of carboxylic acids is capable of interfering with a wide range of reactions involving nucleophiles, and so the protection of the acid as an ester is often carried out as a matter of course in syntheses. Although this does prevent the carboxylic acid acting as a proton donor, it generally does not protect against reactions where the carbonyl group is the electrophile.

The simplest esters to form are the methyl esters, which are conveniently prepared under very mild by treatment of the carboxylic acid with diazomethane (e.g., Example **15.93**). The fact that few hydroxyl groups interfere with this reaction (certain phenolic hydroxyl groups are converted to their methyl ethers) makes it especially suitable for use with sensitive substrates. A mild alternative to using diazomethane that can be applied to the synthesis of esters other than methyl involves the treatment of the acid with a carbodiimide, and the subsequent decomposition of the *N*-acylisourea with the alcohol. Although this reaction is not widely used in the synthesis of esters, it has become a mainstay of solid-phase peptide synthesis, where it is applied to the formation of amides.

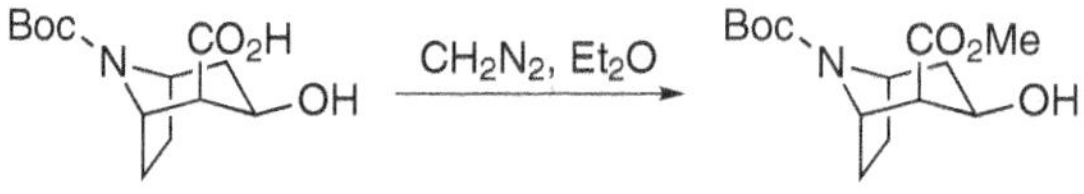

(15.93)

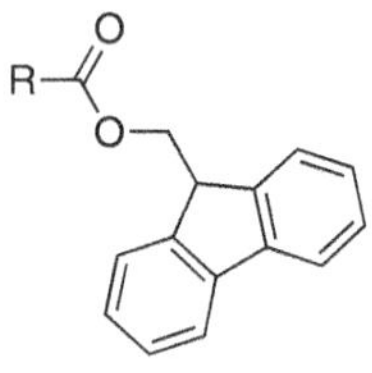

(15.94)

Much of the driving force for developing esters for the protection of acids has been provided by the need for deprotection under mild conditions: *tert*-butyl esters are deprotected under acidic conditions,[74] whereas 9-fluorenylmethyl (Fm) esters (Example **15.94**) are relatively stable to acid but cleaved by diethylamine or piperidine.[75] One interesting method for the cleavage of allyl esters is the Tsuji-Trost allylation of dimedone.[76] Also, *p*-methoxybenzyl esters can be deprotected by oxidation with CAN or DDQ or by hydrogenolysis over palladium, similar to the ethers.

1,3-Oxazolines

The 2-substituted-1,3-oxazoline ring system is a functional group in which the carboxylic acid group is present in a masked form. In both systems, the carbonyl and hydroxyl reactivity of the carboxylic acid is suppressed. Beginning around 1970, these compounds were developed as useful carboxylic acid surrogates in the formation of carbon-carbon bonds. Unlike esters, these compounds are resistant to the attack of Grignard reagents or lithium aluminum hydride (25°C, 2 hr), which makes them ideal reagents for the protection of the carboxyl functionality in the presence of strong nucleophiles. These derivatives can be formed directly from carboxylic acids[77] or acid chlorides[78] and can be converted back to the carboxylic acid under relatively mild conditions.

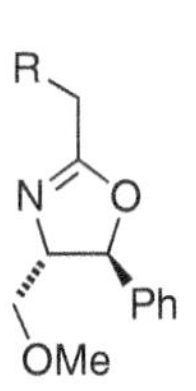

(15.95)

Relatively early in the development of 1,3-oxazolines as useful surrogates for carboxylic acids, the synthetic potential of chiral oxazolines in synthesis was realized. The 1,3-oxazolines derived from (1*S*,2*S*)-1-phenyl-2-aminopropane-1,3-diol (e.g., Example **15.95**) have been especially successful chiral carboxylic acid surrogates in this regard.[79]

---

74. (a) Anderson, G.W.; Callahan, F.M. *J. Am. Chem. Soc.* **1960**, *82*, 3359. (b) Chandrasekaran, S.; Kluge, A.F.; Edwards, J.A. *J. Org. Chem.* **1977**, *42*, 3972. (c) Bryan, D.B.; Hall, R.F.; Holden, K.G.; Huffman, W.F.; Gleason, J.G. *J. Am. Chem. Soc.* **1977**, *99*, 2353.

75. Kessler, H.; Siegmeier, R. *Tetrahedron Lett.* **1983**, *24*, 281.

76. Seki, M.; Yamanaka, T.; Kondo, K. *J. Org. Chem.* **2000**, *65*, 517.

77. Meyers, A.I.; Temple, D.L. *J. Am. Chem. Soc.* **1970**, *92*, 6644.

78. Schow, S.R.; Bloom, J.D.; Thompson, A.S.; Winzenberg, K.N.; Smith, A.B., III *J. Am. Chem. Soc.* **1986**, *108*, 2662.

79. Reviews: (a) Meyers, A.I.; Mihelich, E.D. *Angew. Chem. Int. Ed. Engl.* **1976**, *15*, 270. (b) Meyers, A.I. *Acc. Chem. Res.* **1978**, *11*, 375. (c) Meyers, A.I. In Eliel, E.L.; Otsuka, S., Eds. *Asymmetric Reactions and Processes in Chemistry. ACS Symp. Ser.* **1982**, *185*, 83. (d) Lutomski, K.A.; Meyers, A.I. In Morrison, J.D., Ed. *Asymmetric Synthesis* (Academic Press: New York, 1984), Vol. 3, 213. (e) Meyers, A.I. *J. Org. Chem.* **2005**, *70*, 6137.

Reaction Synopses

Protecting Groups for Aldehydes and Ketones

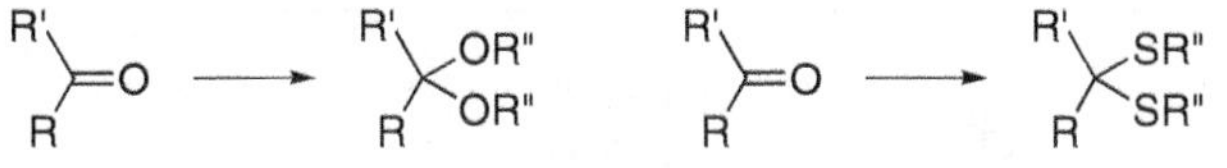

Acetals and Ketals

Formation: MeOH, TsOH, PhH, Δ; $HO(CH_2)_2OH$, TsOH, PhH, Δ; etc.

Cleavage: H_2SO_4, H_2O, THF; $(CO_2H)_2$, H_2O; etc.

Dithioacetals and Dithioketals

Formation: $HS(CH_2)_3SH$, BF_3; etc.

Cleavage: $HgCl_2$, $CaCO_3$; $HgCl_2$, $CdCO_3$; NBS, H_2O; MeI, MeCN-H_2O; etc.

Protecting Groups for Carboxylic Acids

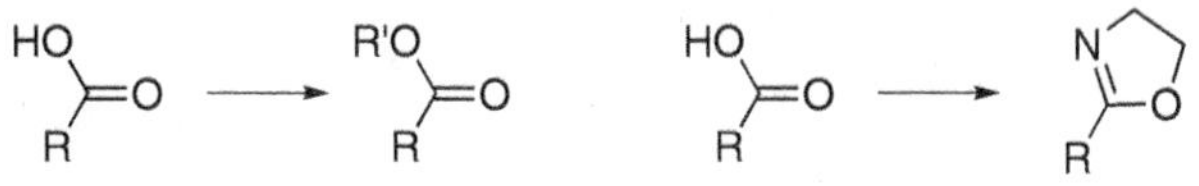

Esters

Formation: CH_2N_2, Et_2O (R'=Me); R'OH, DCC; $Me_2C{=}CH_2$, H_2SO_4 (R=Bu^t); etc.

Cleavage: K_2CO_3, MeOH; TFA (R'=Bu^t); Et_2NH (R'=Fm); dimedone, $Pd(OAc)_2$ (R'=allyl); $LiAlH_4$, Et_2O; H_2, Pd-C (R'=allyl, benzyl); etc.

1,3-Oxazolidines

Formation: (1) $(COCl)_2$, CH_2Cl_2; (2) $HOCH_2CMe_2NH_2$; etc.

Cleavage: HCl, H_2O; etc.

15.8 Protection of Amines

Much of the incentive to develop protecting groups for amines has come from the solid-phase synthesis of proteins and peptides. The synthesis of these compounds carries with it the special requirement that the reactions for incorporating and removing the protecting groups must not lead to racemization of the chiral α carbons, which places significant restrictions on the groups that can be used to protect these functional groups.

Amines are nucleophiles, and it is this reactivity toward a wide range of electrophiles that must usually be guarded against in a synthesis. The simplest way to reduce the nucleophilicity of the amine nitrogen is to convert it to an amide or an amide-like compound. By far the most widely used protecting groups of this type are the carbamates, all of which are obtained from the amine by treatment with an appropriate alkoxycarbonyl derivative (e.g., a chloroformate ester; see Example **15.96**). The variations available in the urethane protecting groups are illustrated by the examples in Table 15.12, where several common protecting groups for amines are gathered with representative conditions for their incorporation and cleavage. This list is by no means exhaustive, and you should consult a specialist monograph when choosing a protecting group for an amine.

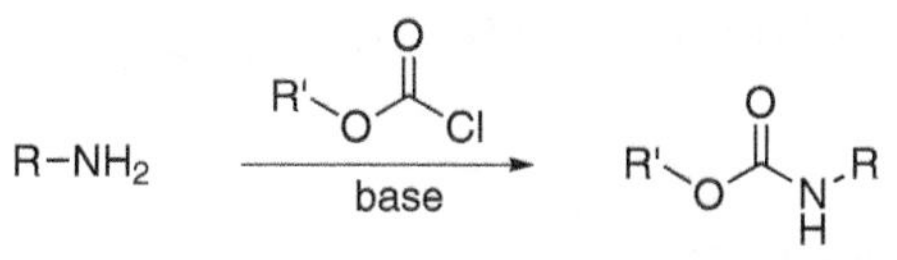

(15.96)

Unlike the carbamates, simple amides (usually acetamides) must be cleaved under relatively forcing conditions. The susceptibility of acetamides to hydrolysis increases

Table 15.12 Carbamate-Protecting Groups for Amines

Protected Amine	Acronym	Representative Conditions for Incorporation and Removal
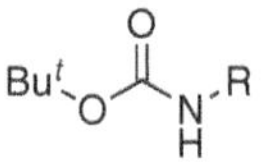	Boc	*On:* $(Boc)_2O$, NaOH, H_2O, 25°C *Off:* 3 *M* HCl, EtOAc; etc.
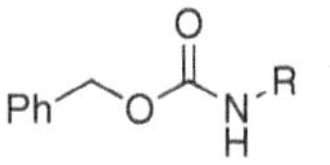	Cbz, or Z	*On:* Cbz-Cl, Na_2CO_3, H_2O *Off:* H_2, Pd-C
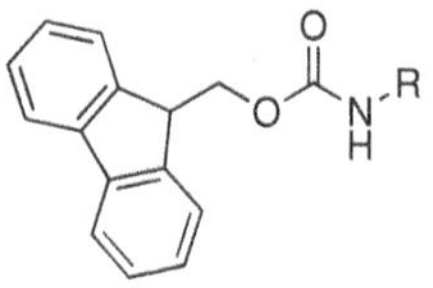	Fmoc	*On:* Fmoc-Cl, $NaHCO_3$, dioxane-H_2O *Off:* Piperidine, DMF; etc.
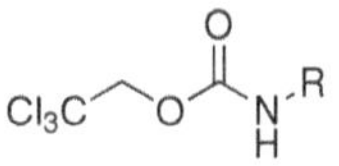	Troc	*On:* Troc-Cl, py *Off:* Zn, THF, H_2O (pH 4.2); etc.
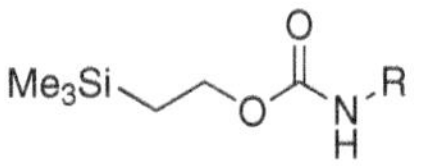	Teoc	*On:* Teoc-O-succinimide, $NaHCO_3$, dioxane-H_2O *Off:* TBAF, KF, MeCN
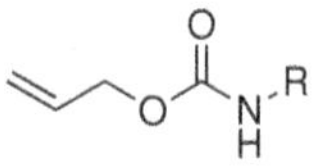	Aloc	*On:* $H_2C{=}CHCH_2OCOCl$, py *Off:* $(Ph_3P)_4Pd$, dimedone, THF

with substitution by halogens at the α carbon.[80] The rates of hydrolysis follow the order:

$$CH_3CONHR < ClCH_2CONHR < Cl_2CHCONHR < Cl_3CONHR < CF_3CONHR$$

Although acetamides require heating for several hours with hydrochloric acid, trifluoroacetamides can be hydrolyzed by sodium or potassium carbonate in aqueous methanol. (In fact, the *N*-trifluoroacetyl group can be hydrolyzed in the presence of a methyl ester[81]!) The hydrolysis of amides can also be facilitated by neighboring group participation by a hydroxyl group, as in 2-acyloxymethylbenzamides[82] (e.g., Example **15.97**). Primary amines can also be protected as the corresponding *N*-alkylphthalimides (e.g., Example **15.98**). These compounds, which are also intermediates in the Gabriel synthesis of primary amines,[83] can then be cleaved by hydrazinolysis in ethanol or base hydrolysis.

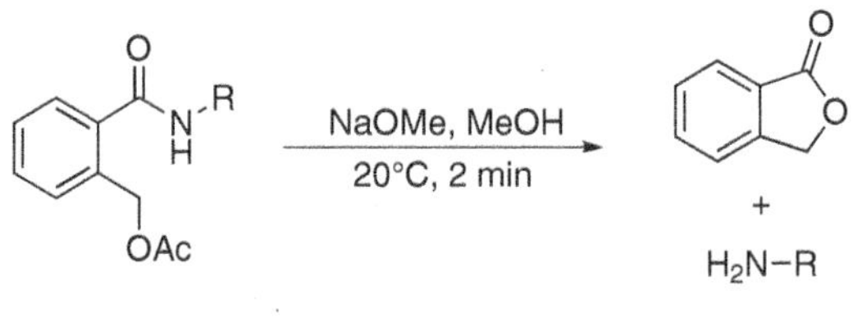

(15.97)

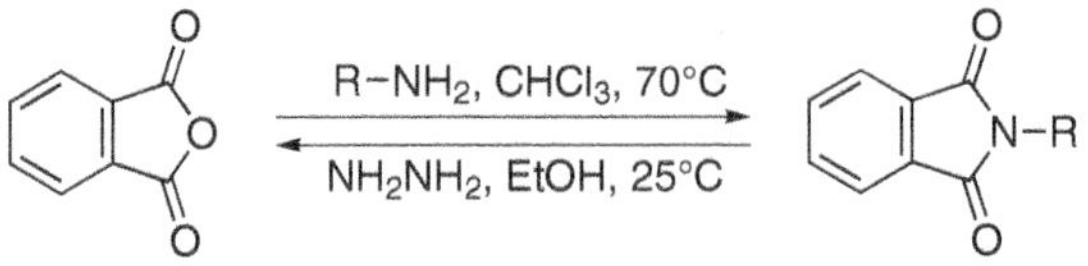

(15.98)

80. Goody, R.S.; Walker, R.T. *Tetrahedron Lett.* **1967**, 289.
81. Boger, D.L.; Yohannes, D. *J. Org. Chem.* **1989**, *54*, 2498.
82. Cain, B.F. *J. Org. Chem.* **1976**, *41*, 2029.
83. Review: Gibson, M.S.; Bradshaw, R.W. *Angew. Chem. Int. Ed. Engl.* **1968**, *7*, 919.

As an alternative to simple chemical hydrolysis or cleavage, the selective hydrolysis of a variety of simple amide can also be accomplished by means of acylases. This provides the opportunity to accomplish optical resolution of the amine at the same time, because the enzyme generally shows a marked preference for the cleavage of one of the two enantiomeric amides. For example, the hydrolysis of the ω-trifluoro-α-amino acid amides in Example **15.99** with hog kidney acylase results in hydrolysis of the *S* amide in 97% yield and 81% e.e., with the unhydrolyzed *R* amide being recovered in 92% yield and 78% e.e.[84]

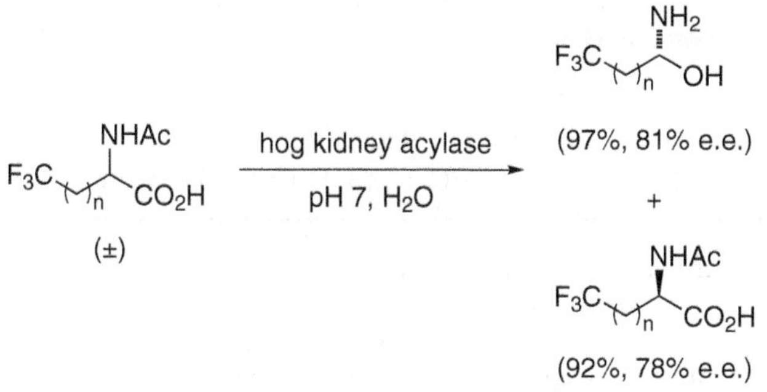

(15.99)

Problems

15-14 (a) There are two ways to deprotect the bicyclic alcohol below. What are the products of the two reactions specified below?

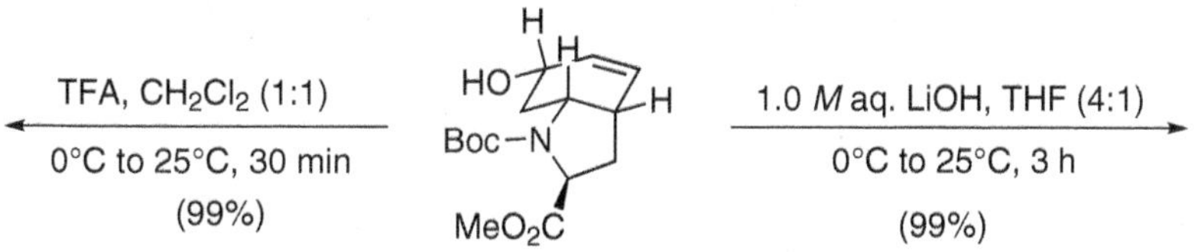

(b) What is the product of the deprotection reaction below?

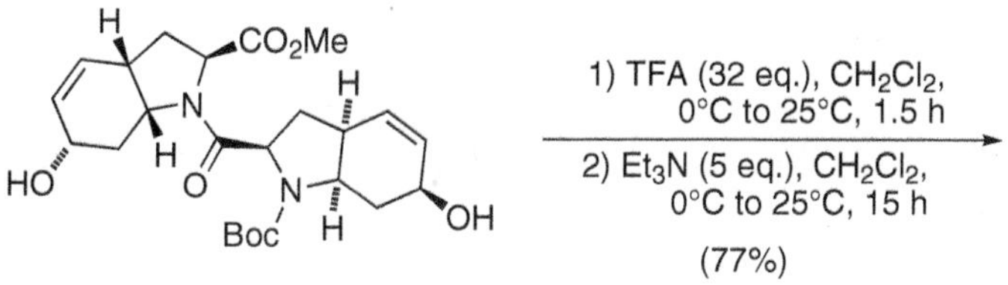

Reference: J. Am. Chem. Soc. **2012**, *134*, 17320.

15-15 Suggest a reason why phthalimides might be easier to hydrolyze with aqueous base than acetamides.

Reaction Synopses

Protecting Groups for Amines

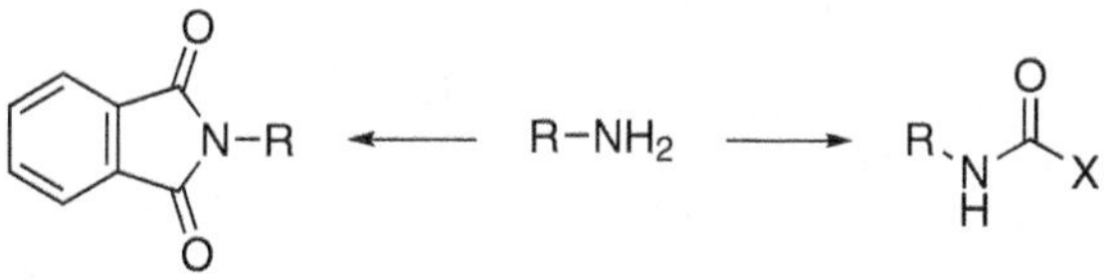

(continues)

84. Tsushima, T.; Kawada, K.; Ishihara, S.; Ucjida, N.; Shiratori, O.; Igaki, J.; Hirata, M. *Tetrahedron* **1988**, *44*, 5375.

(**Reaction Synopses** continued)

Amides	
Formation:	RCOCl, $(RCO)_2O$; etc.
Cleavage:	HCl, H_2O, Δ; KOH, H_2O-EtOH, Δ; etc. or acylases, buffer
Carbamates	
Formation:	ROCOCl, base (R=*t*-Bu, $PhCH_2$, etc.)
Cleavage:	HCl, H_2O; KOH, EtOH; etc. or TFA (R=*t*-Bu); H_2, Pd-C (R=$PhCH_2$); etc.
Phthalimides	
Formation:	phthalic anhydride, $CHCl_3$, Δ; etc.
Cleavage:	H_2NNH_2, EtOH, 25°C; KOH, H_2O, EtOH, Δ; etc.

Worked Problem

15-4 (a) Provide the missing reagents in the following steps from Sato's synthesis of tetrodotoxin (*J. Org. Chem.* **2008**, *73*, 1234).

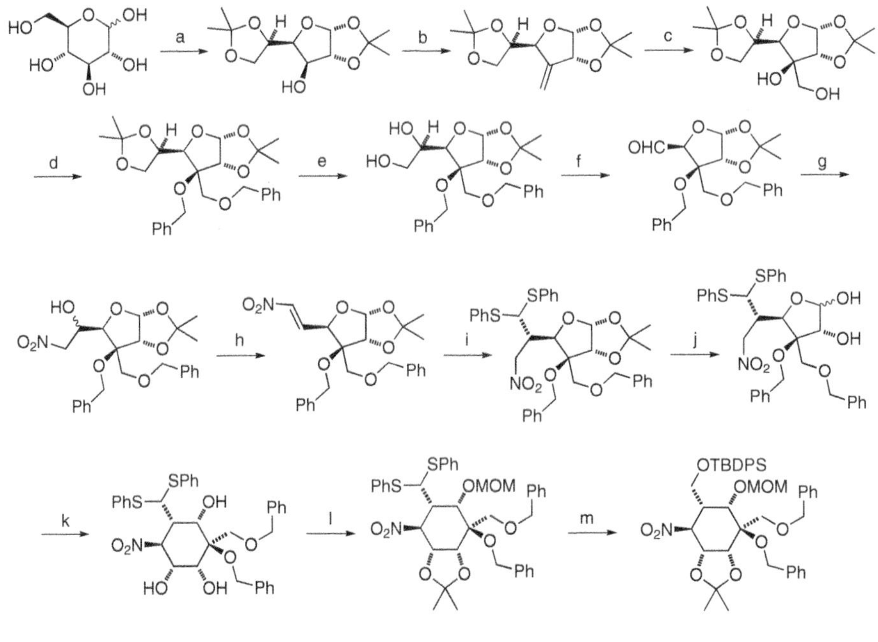

§Answers below.

§ Answers to Worked Problem:

(a) Me_2CO, H^+; the authors actually used I_2 as a Lewis acid.

(b) (i) $(CF_3CO)_2O$, DMSO, DBU; (ii) $Ph_3P{=}CH_2$. The first step is a version of the Moffatt oxidation, and is the best suited to oxidation of such a highly functionalized substrate. A Swern oxidation [$(COCl)_2$, Me_2SO, Et_3N, CH_2Cl_2, −60°C] would also work well here.

(c) (1) *m*-CPBA, $(CH_2Cl)_2$; (ii) NaOH, H_2O-THF. In this reaction, the epoxidation occurs from the less hindered *exo* face of the 1,3,5-trooxabicyclo[3.3.0]octane ring system, and the basic nucleophile attacks at the primary carbon to give the *exo* alcohol as the major product.

(d) NaH, $PhCH_2Br$, DMF (Williamson ether synthesis).

(e) HOAc, H_2O. This reagent hydrolyzes only the ketal derived from the 1,2-diol, and not the ketal whose hydrolysis would reveal the aldehyde.

(f) $NaIO_4$, MeOH-H_2O.

(g) $MeNO_2$, NaOMe, MeOH. This reaction gives Na^+ salt of nitromethane, which is a good carbon nucleophile that adds to the carbonyl group of the aldehyde. Since the next step of the sequence is dehydration, the stereochemistry of the new chiral center does not matter. This addition is known as the Henry reaction.

15.9 Protecting Groups for Hydrocarbons

The plethora of protecting groups for oxygenated functional groups and amines stands in stark contrast to the number of protecting groups available for protecting the π bonds of alkenes and alkynes. The strategy for protecting these functional groups is essentially the same as that used in protecting aldehydes and ketones as acetals and ketals: the reactive π bond is converted to a less reactive derivative by addition, which is then reversed at a later stage of the synthesis.

Alkenes

In contrast to the carbonyl π bond, the π bond of an alkene reacts with electrophiles and not with nucleophiles, so the protecting group needs to be resistant to the attack of an electrophile rather than the attack of a nucleophile. The most obvious derivative is the vicinal dibromide, formed by addition of bromine to the alkene. The alkene π bond can then be regenerated by reductive elimination, as shown in Example **15.100**.

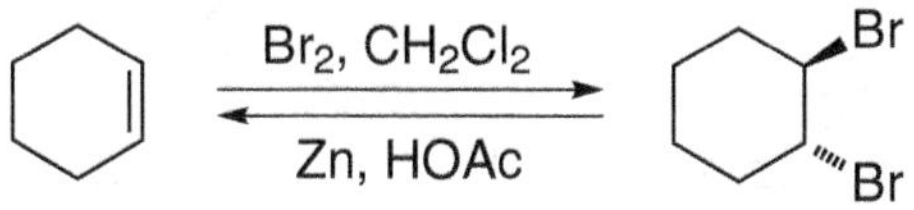

(15.100)

In 1984, Ganem proposed epoxides as protecting groups for alkenes, following his discovery that diazomalonate esters act as mild deoxygenating agents that preserve the stereochemistry of the original alkene in the presence of a rhodium catalyst (e.g., Example **15.101**). In a similar vein (e.g., Example **15.102**), a combination of epoxidation by alkaline hydrogen peroxide and subsequent deoxygenation by $Mo(CO)_6$ provides a convenient method for the protection of the alkene π bond of conjugated enones and esters.

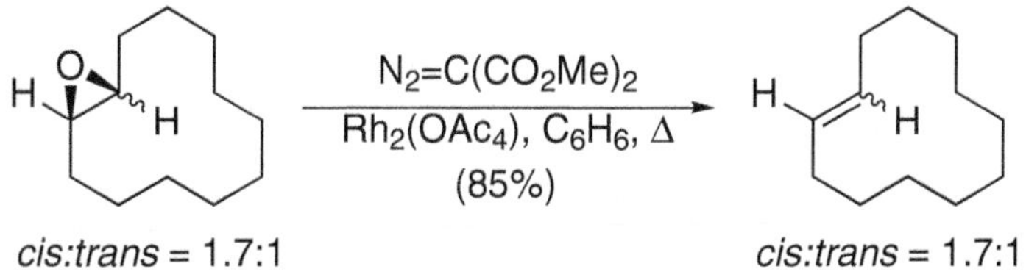

(15.101)[85]

85. Martin, M.G.; Ganem, B. *Tetrahedron Lett.* **1984**, *25*, 251.

§ Answer to Worked Problem (Continued)

(h) MsCl, Et_3N, CH_2Cl_2. The mesylate (any sulfonate would) is formed, and then undergoes E1cb elimination from the anion of the nitroalkane. Although the dehydration can be effected using stronger bases without making the sulfonate ester first, this two-stage approach is better suited to sensitive substrates.

(i) $(PhS)_2CH_2$, LDA, THF. The dithiane anion is formed first, and then it is added to the nitroolefin. The conjugate addition proceeds ≈10:1 in favor of the diastereoisomer shown. In the conformation drawn for the nitroolefin, this would correspond to addition to the less hindered face of the alkene.

(j) HOAc, H_2O, Δ. Under these more vigorous conditions than in (e), the aldehyde ketal is hydrolyzed.

(k) $NaHCO_3$, H_2O-MeOH. This is another Henry reaction, and previous work had shown that sodium bicarbonate was a sufficiently strong base to close the six-membered ring.

(l) (i) $Me_2C(OMe)_2$, TsOH, CH_2Cl_2; (ii) $CH_2(OMe)_2$, P_2O_5, CH_2Cl_2. There are two groups to be protected in this step: the *cis*-1,2-diol, and the secondary alcohol. They are protected in that order. The authors give no reason for the use of P_2O_5 as the acid catalyst in the second step to form the MOM ether, but it is worthwhile noting that P_2O_5 almost always contains some polyphosphoric acid, and that the hydroxyl group at this position is susceptible to elimination under E1cb conditions, which would preclude forming the MOM ether under the more usual basic conditions.

(m) (i) NBS, MeCN-H_2O; (ii) $NaBH_4$, MeOH. The aldehyde formed by hydrolysis of the dithioacetal is reported to be unstable, so both steps were carried out without isolating the aldehyde.

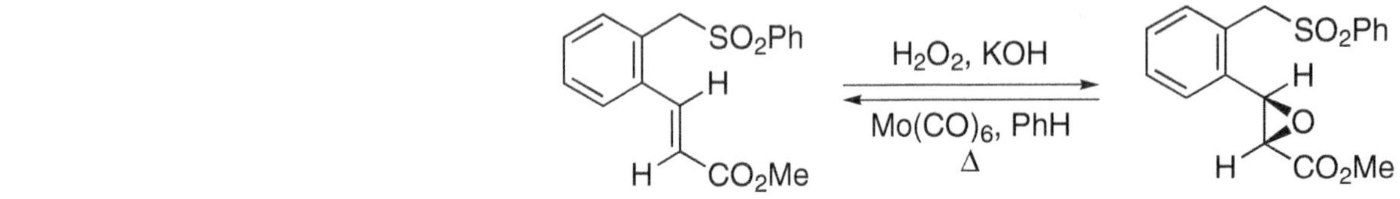

(15.102)[86]

Yet another method for protecting alkenes involves their conversion to 1,2-diols by oxidation with osmium tetroxide. These diols can be protected as acetonides, for example, and after the transformations have been completed, the diol can be revealed and reconverted to the alkene by means of the Corey-Winter reaction.[87]

A more recent method for the selective protection of the more reactive π bond of a non-conjugated diene, thus allowing selective hydrogenation of the less reactive π bond, has been developed by Goess.[88] In this method, the diene is treated with 9-borabicyclo[3.3.1]nonane (9-BBN), which reacts with the more reactive π bond to give the alkylborane and the less reactive π bond is then reduced. Treatment of the borane with 2-methyl-2-nitrosopropane then returns the alkene (e.g., Example **15.103**).

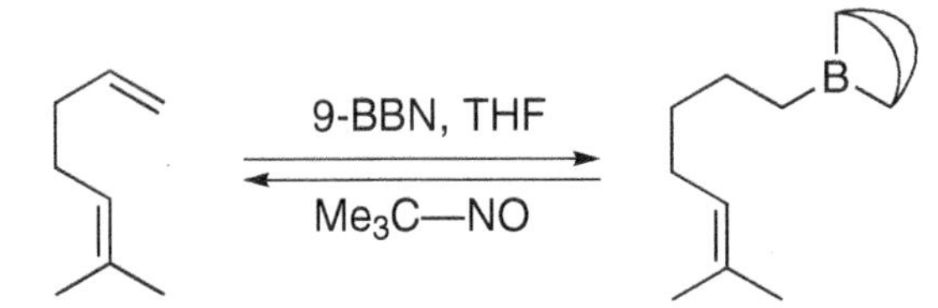

(15.103)

Another approach to the protection of alkene π bonds during synthesis has been to use a Diels-Alder adduct as a protecting group and then to reveal the alkene by flash vacuum pyrolysis. Although this method does work well, it suffers from the drawback that the Diels-Alder adduct is itself an alkene (unless the diene used was anthracene or a higher acene); thus, it is susceptible to the same reactions as the original alkene (e.g., Example **15.104**).

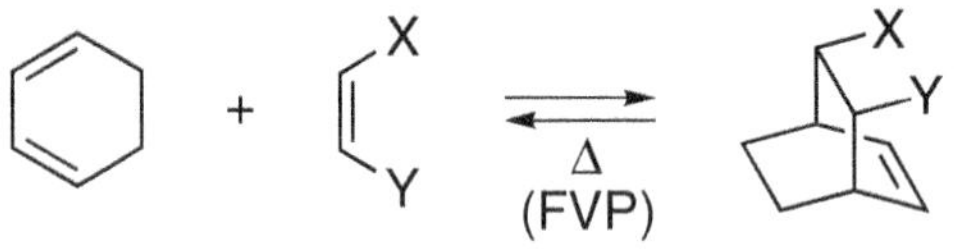

(15.104)

Alkynes

The protection of alkynes falls in two major areas: (1) protection of the terminal C—H bond, which is done using many of the same silane protecting groups as were used for alcohols and (2) protection of the C—C π bond. The protection of alkyne C—C π bonds relies, for the most part, on the formation of a transition metal complex of the alkyne from which the triple bond is later released by suitable means (as shown in Example **15.105**). The most common method for protecting the alkyne π bond involves the formation of a stable dicobalt hexacarbonyl complex of the alkyne,[89] from which the triple bond can be released by oxidation.[90] This approach has been especially successful in the synthesis of ethers from propargyl alcohols—the metal complex forms a very stable car-

86. Patra, A.; Bandyopadhyay, M.; Mal, D. *Tetrahedron Lett.* **2003**, *44*, 2355.

87. Corey, E.J.; Winter, R.A.E. *J. Am. Chem. Soc.* **1963**, *85*, 2677.

88. Graham, T.J.A.; Poole, T.H.; Reese, C.N.; Goess, B.C. *J. Org. Chem.* **2011**, *76*, 4132.

89. Greenfield, H.; Sternberg, H.W.; Friedel, R.A.; Wotiz, J.H.; Markby, R.; Wender, I. *J. Am. Chem. Soc.* **1956**, *78*, 120.

90. (a) Nicholas, K.M.; Pettit, R. *Tetrahedron Lett.* **1971**, 3475. (b) Seyferth, D.; Nestle, M.O.; Wehman, A.T. *J. Am. Chem. Soc.* **1975**, *97*, 7417.

bocation (a Nicholas cation[91]) that can be trapped by alcohol nucleophiles; intramolecular application of the Nicholas reaction leads to the formation of oxygen heterocycles. In a later report, a platinum complex of the alkyne was shown to be able to protect the alkyne π bond selectively in the presence of an alkene π bond, allowing the selective hydrogenation of the alkene.[92]

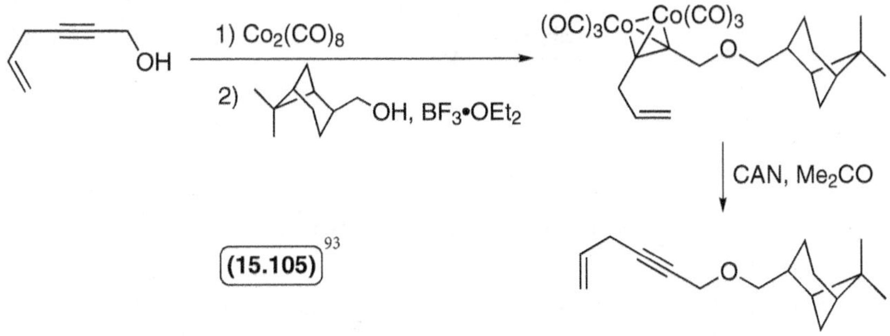

(15.105)[93]

Reaction Synopses

Protecting Groups for Alkenes

Halogen and Halogen-like Adducts

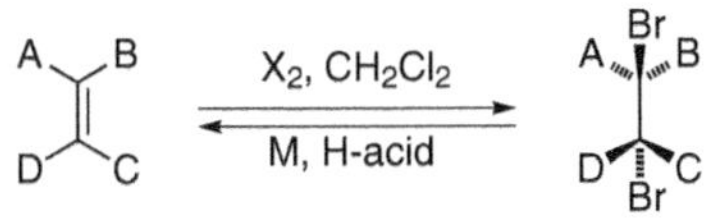

Protection: Br_2, CH_2Cl_2; etc.
Deprotection: Zn, HOAc; Mg, Et_2O; etc.

Epoxides

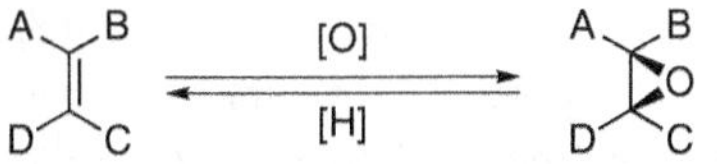

Protection: *m*-CPBA (simple alkenes); or H_2O_2, KOH (conjugated enones or esters)
Deprotection: $N_2{=}C(CO_2Me)_2$, $Rh_2(OAc)_4$, PhH, Δ; or $Mo(CO)_6$, PhH, Δ; etc.

Vicinal Diols and Acetonides

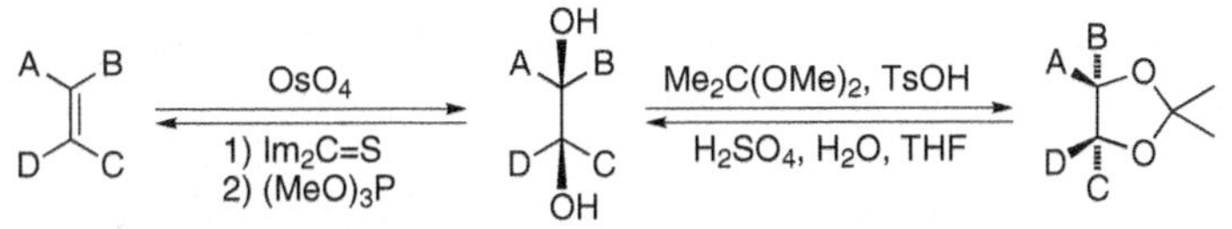

Protecting Groups for Alkynes

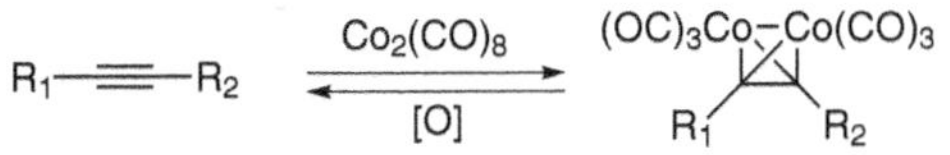

Protection: $Co_2(CO)_8$, CH_2Cl_2; etc.
Deprotection: $K_3Fe(CN)_6$; or CAN; etc.

---

91. Nicholas, K.M. *Acc. Chem. Res.* **1987**, *20*, 207.
92. Haskel, A.; Keinan, E. *Tetrahedron Lett.* **1999**, *40*, 7861.
93. Díaz, D.D.; Martín, V.S. *Tetrahedron Lett.* **2000**, *41*, 9993.

Problems

15-16 (a) Complete the following sequence of reactions by providing the structures of the missing species.

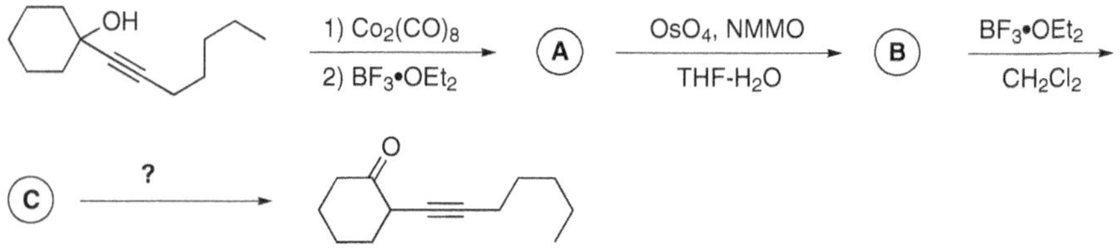

(b) How does one account for the regioselectivity of the reaction (B→C) to form the ketone?

Reference: Tetrahedron Lett. **1999**, *40*, 2815.

15-17 Complete the following reaction sequence by providing the structures of the missing compounds.

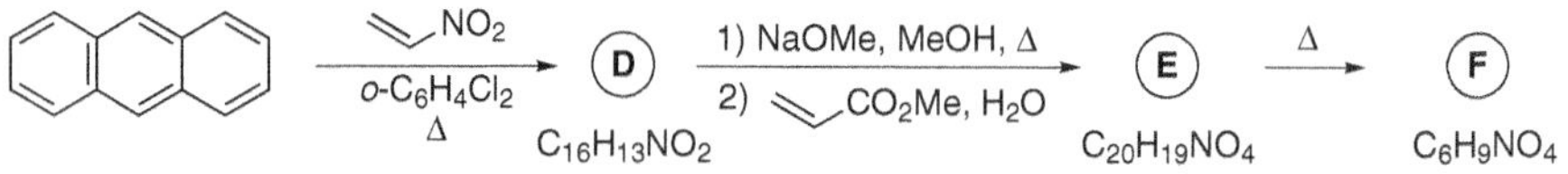

Reference: J. Org. Chem. **1955**, *20*, 650.

Chapter Summary

This chapter introduces organic synthesis along with the terminology of retrosynthetic analysis. A synthesis can be assessed in terms of the increasingly sophisticated levels of selectivity: chemoselectivity, regioselectivity, diastereoselectivity, and enantioselectivity. The tools of retrosynthetic analysis are transforms, which are intellectual constructs that convert a target compound to one or more potential precursors, known as synthons. The sequence of reactions in the synthesis may be linear, where the steps all follow in a single pathway from starting material to product, or convergent, where two separate paths are used to construct intermediates that are then coupled. Convergent syntheses often have higher overall yields.

Polyfunctional compounds may be subjected to the Evans-Lapworth analysis. This reveals consonant and dissonant difunctional relationships in the target compound. Although consonant relationships are easy to form, dissonant difunctional relationships can become a source of difficulty in a synthesis. One way to overcome this is to use reagents with charge-type inversion, or *umpolung*.

Modern synthesis of polyfunctional compounds almost always requires the use of protecting groups, and such species for alcohols (including 1,2- and 1,3-diols), amines, aldehydes, ketones, carboxylic acids, and hydrocarbons were introduced in this chapter.

Key Terms

acetal/ketal
chemoselectivity
consonant difunctional relationships
convergent synthesis
deconvolution
diastereoselectivity
dissonant difunctional relationships
dithioacetal/dithioketal
enantioselectivity
Evans-Lapworth analysis
linear synthesis
protecting group
regioselectivity
retrosynthetic analysis
silyl ethers
synthon
tetrahydropyranyl ethers
transform
umpolung

Additional Problems

15-18 (a) Carry out a Lapworth-Evans analysis of each of the following target molecules. Indicate any dissonant difunctional relationships that might be present in each target molecule and suggest key intermediates that might be used to introduce these dissonant relationships.

(b) Carry out a complete retrosynthetic analysis of each of the target molecules that would permit them to be prepared (in racemic form unless otherwise specified) from the indicated starting material. You may assume that any other reagents may be used unless limitations are imposed.

(c) Specify one sequence of reactions that would constitute a rational synthesis of each of the target compounds from the starting material given. Unless explicitly prohibited from doing so, you may use any other organic or inorganic compounds needed to accomplish the transformation requested. Where the relative configurations of chiral centers is not given, you may assume that a mixture of stereoisomers is permissible in the final product.

(a) 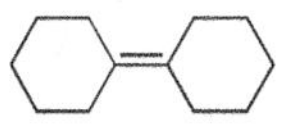 from cyclohexanol

(b) 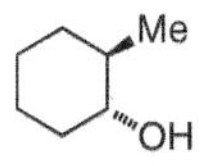

from cyclohexene

(c) 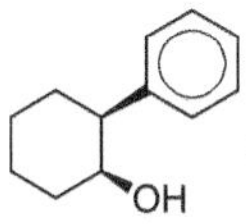

from 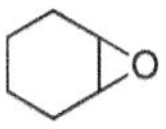

(d) 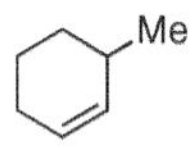

from cyclohexanol

(e) 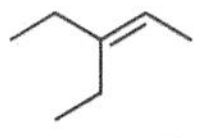 from ethanol

(f) 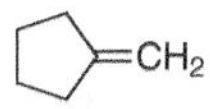

from cyclohexene

(g) 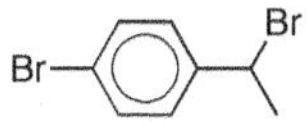

from benzene

(h) 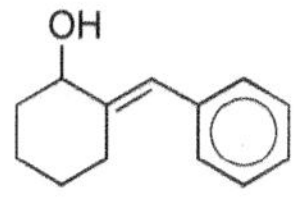

from benzyl alcohol

(i) 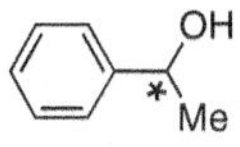

from K*CN (*C = ^{14}C)

(j) 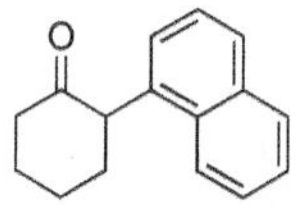

from cyclohexanone

(k) 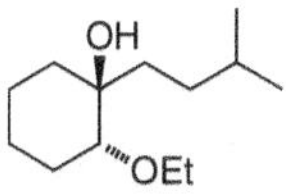

from cyclohexene

(l) 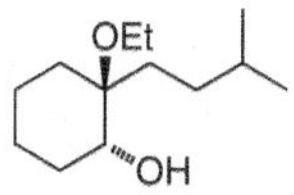

from cyclohexene

(m) 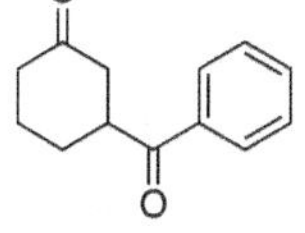

from 2-cyclohexenone

(n) 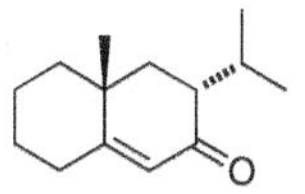

from 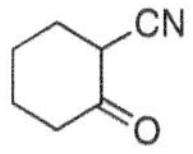

(o) 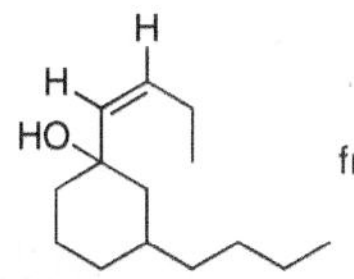

from 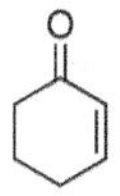

(p) 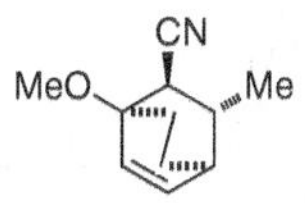

from anisole

(q) 1-dodecyne from compounds with four or less carbon atoms

(r) 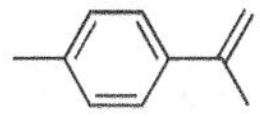 from benzene

(s) 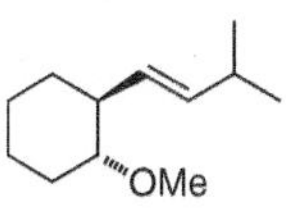

from cyclohexanol

(t) 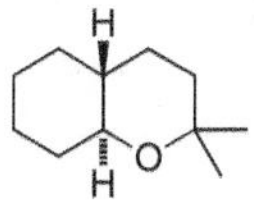

from cyclohexene

(u) 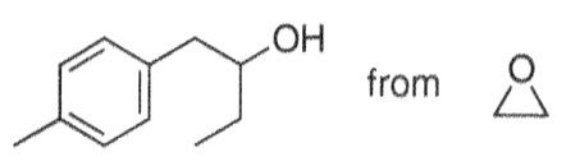

(v) 

(w) 

(x) 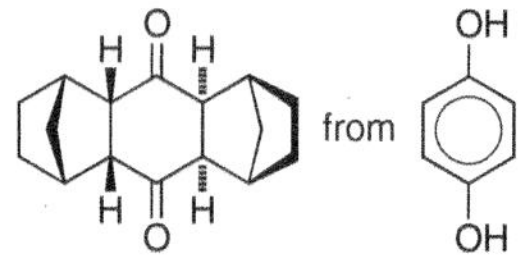

(y) 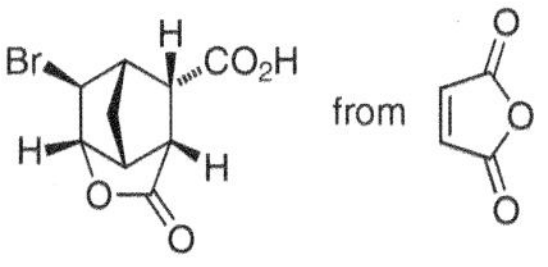

(z) 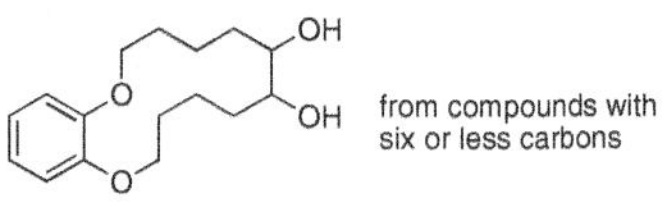

(aa) 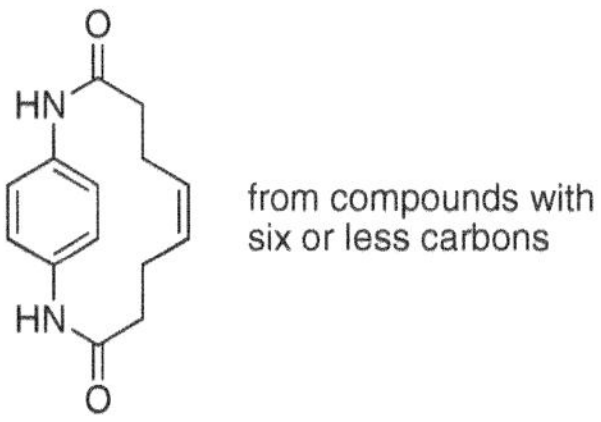

(bb) 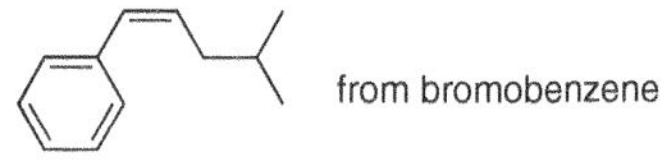

(cc) 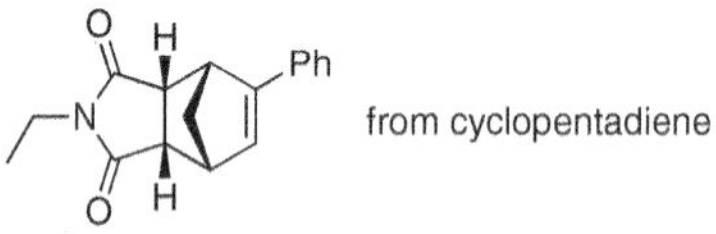

15-19 The compounds shown are all important industrial chemicals: bisphenol A is a key raw material in the epoxy resin industry, BHT and BHA are important preservatives, adipic acid is a key intermediate in the manufacture of nylon-6,6, and caprolactam is a key intermediate in the manufacture of nylon-6. Design a synthesis of each from the starting material indicated or from a phenol.

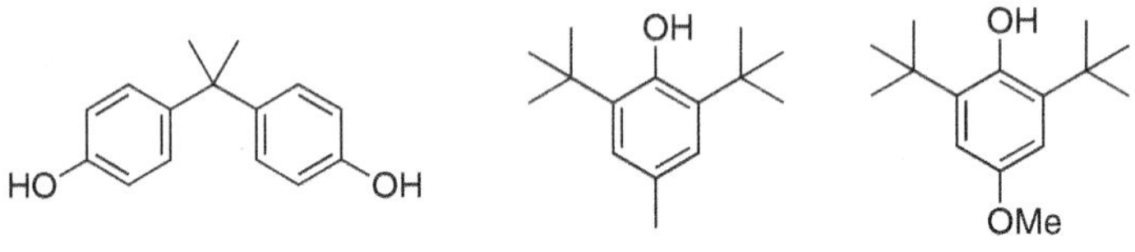

bisphenol A BHT BHA

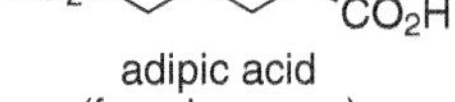

adipic acid
(from benzene)

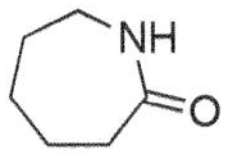

caprolactam
(from benzene)

15-20 What is the major organic product of each of the following reactions or reaction sequences?

(a) 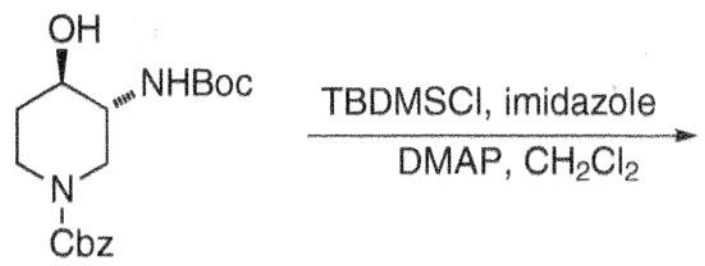

WO2010026122
(Novartis)

(b) 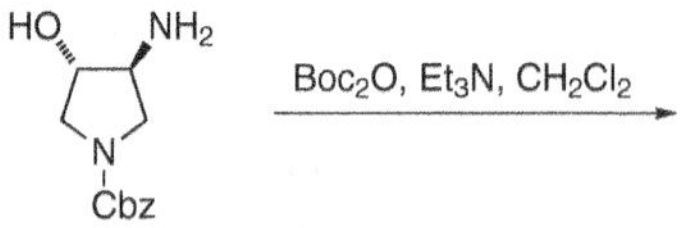

WO2010001169
(AstraZeneca)

(c)

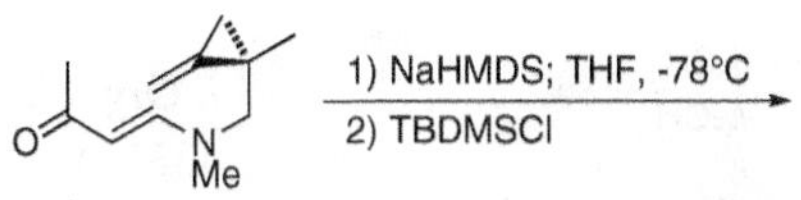

J. Am. Chem. Soc. **2011**, *133*, 7704.

(d)

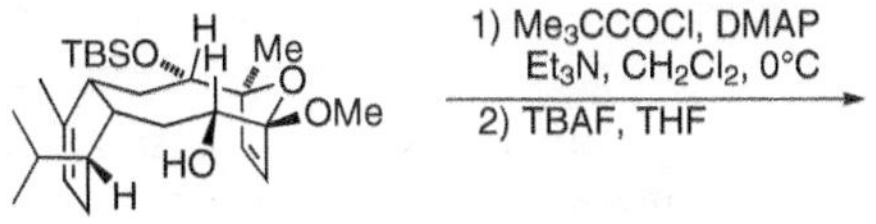

J. Am. Chem. Soc. **1999**, *121*, 6563.

(e)

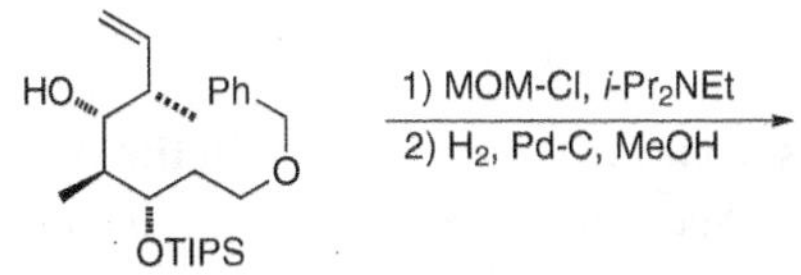

Org. Lett. **2004**, *6*, 3533.

(f)

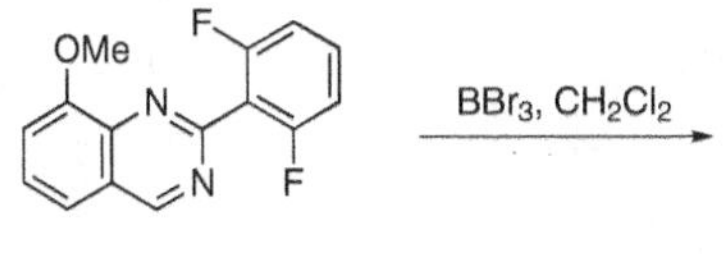

WO2010026122
(Novartis)

(g)

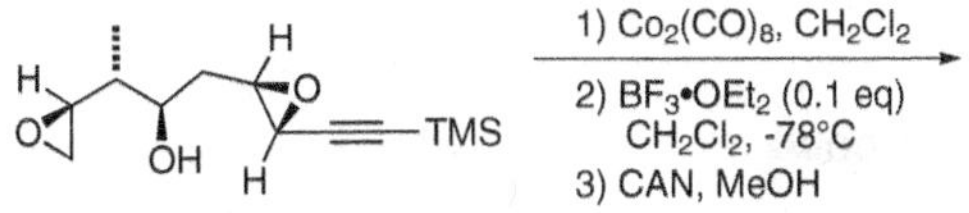

Tetrahedron **2011**, *67*, 9809.

(h)

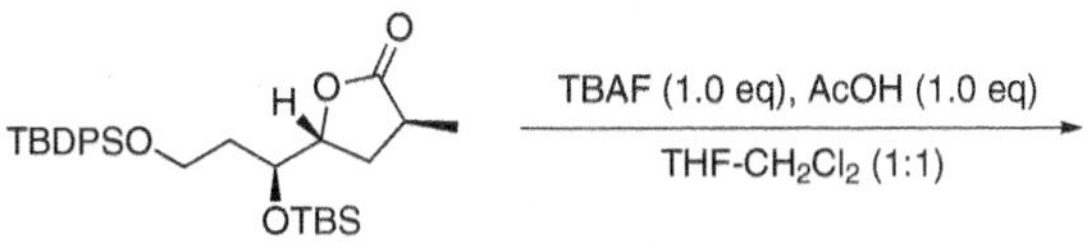

Org. Lett. **2008**, *10*, 1247.

(i)

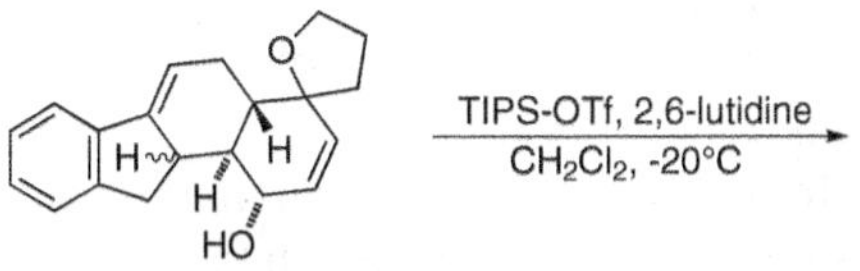

J. Am. Chem. Soc. **2008**, *130*, 2783.

(j)

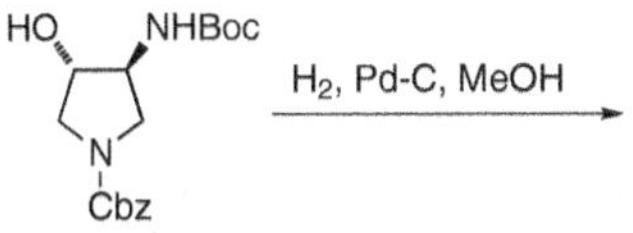

WO2010001169
(AstraZeneca)

(k)

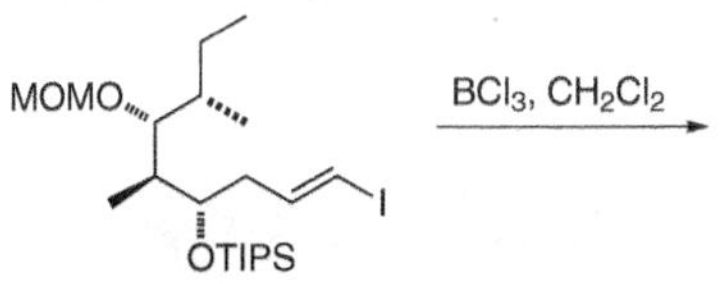

Org. Lett. **2004**, *6*, 3533.

(l)

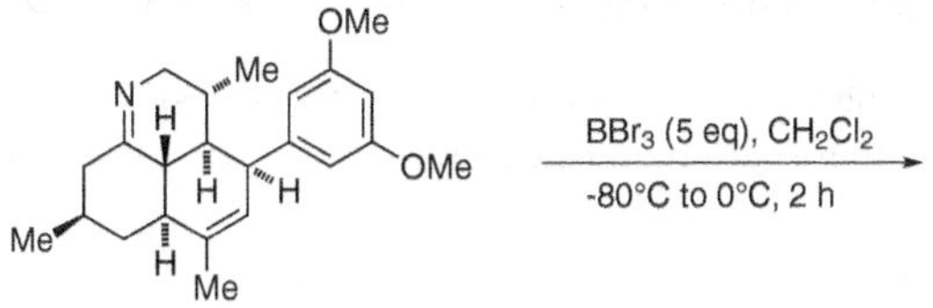

Org. Lett. **2007**, *9*, 1461.

(m)

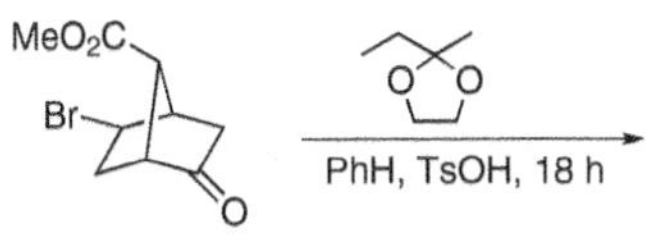

J. Org. Chem. **1975**, *40*, 150.

(n)

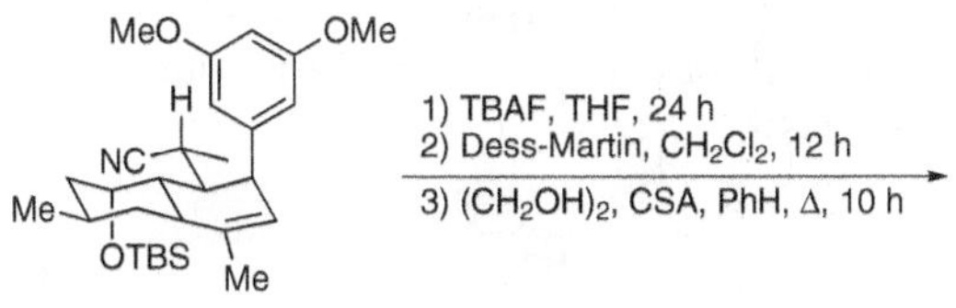

Org. Lett. **2007**, *9*, 1461.

(o) 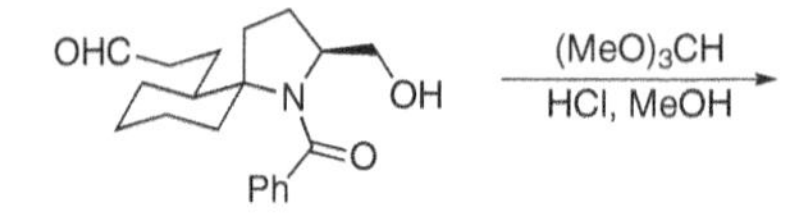

J. Org. Chem. **2002**, *67*, 4337.

(p) 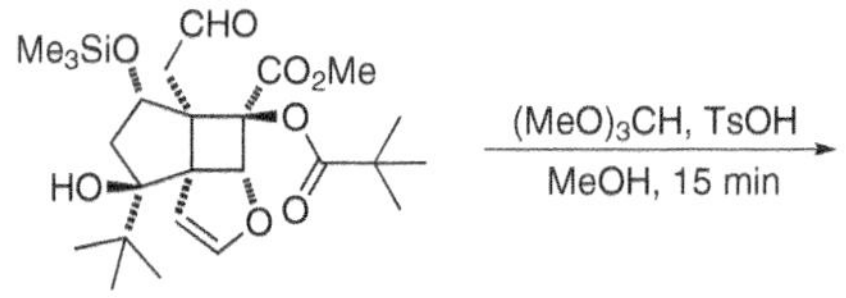

J. Am. Chem. Soc. **1993**, *115*, 3146.

15-21 (a) Provide the missing reagents from the following steps in the synthesis of hemibrevetoxin B (*J. Am. Chem. Soc.* **2003**, *125*, 7822). Note that more than one step may be required for each transformation.

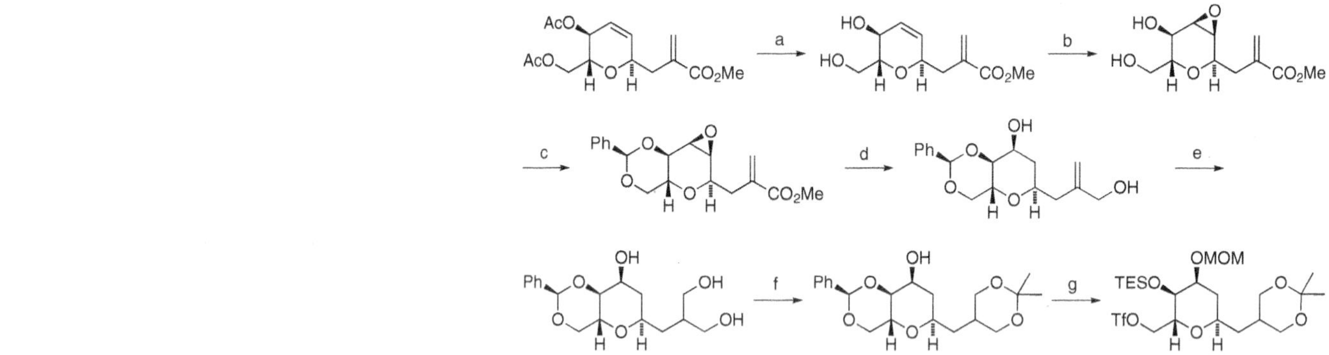

(b) Why was the hydrolysis of the diester in step **a** necessary before the formation of the epoxide in step **b**? (c) Why was the benzylidene acetal chosen in step **c** instead of the acetonide?

(d) Why were the operations in steps **d** and **e** carried out in the order shown?

(e) In step **g**, three hydroxyl groups are revealed, and all three are converted to different functional groups. Provide a reasonable rationale for the order in which the various operations are carried out.

15-22 (a) What are the intermediate compounds in the transformation shown (*Org. Lett.* **2003**, *5*, 4815)? (Hint: the mechanism of the first step should provide a strong clue here.)

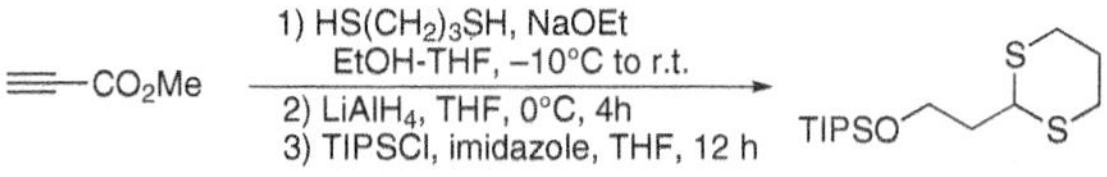

(b) Is there an alternative approach to this same compound?

15-23 Provide the missing reagents in the reaction sequence shown (*J. Am. Chem. Soc.* **2003**, *125*, 9554). Note that more than one step may be necessary to complete each transformation.

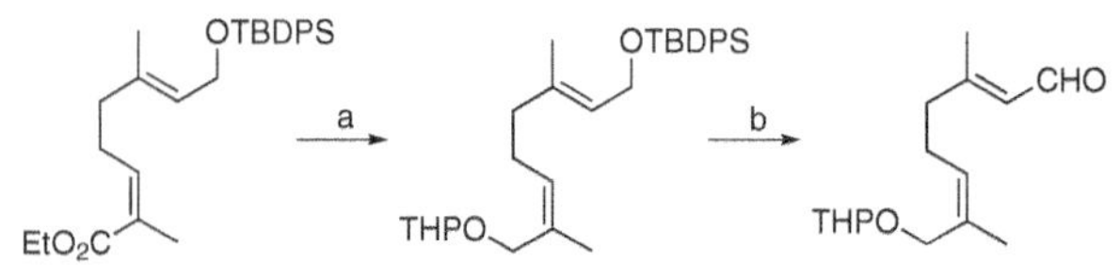

15-24 What will be the stereochemistry of the organic product of the reaction shown? How could predominantly one diastereoisomer of the product be obtained (*J. Org. Chem.* **2007**, *72*, 4098)?

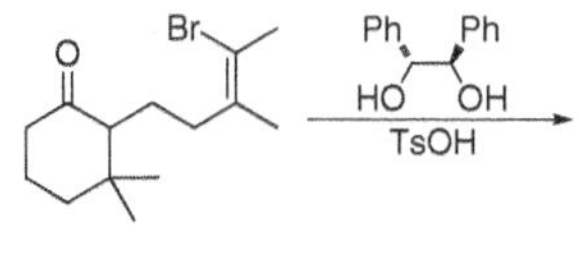

15-25 (a) Complete the following reaction sequence taken from Nicolaou's total synthesis of Taxol (*J. Am. Chem Soc.* **1995**, *117*, 634, 645).

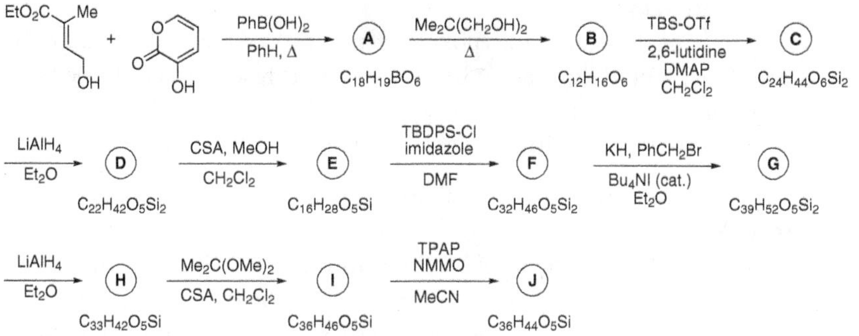

(b) What is the reason for using phenylboronic acid to give compound **A**?

(c) What is the role played by the neopentyl glycol in the transformation of **A** to **B**?

(d) Only one of the two allylic silyl protecting groups is removed in the conversion of **D** to **E**. Which one, and why?

(e) In the exhaustive reduction of **G** to **H**, one of the silyl protecting groups is lost. Which one, and why (suggest a mechanism)?

(f) There are two isomeric compounds consistent with the molecular formula for compound **I**. The isomer formed under kinetic control rearranges slowly and completely to the thermodynamic isomer. What is the structure of the kinetic isomer, and why is it formed more rapidly?

15-26 (a) The tetrodotoxin molecule is one of the most complex natural products whose synthesis has been completed. Complete the following reaction sequence, taken from Isobe's asymmetric total synthesis of tetrodotoxin (*J. Am. Chem. Soc.* **2003**, *125*, 8798).

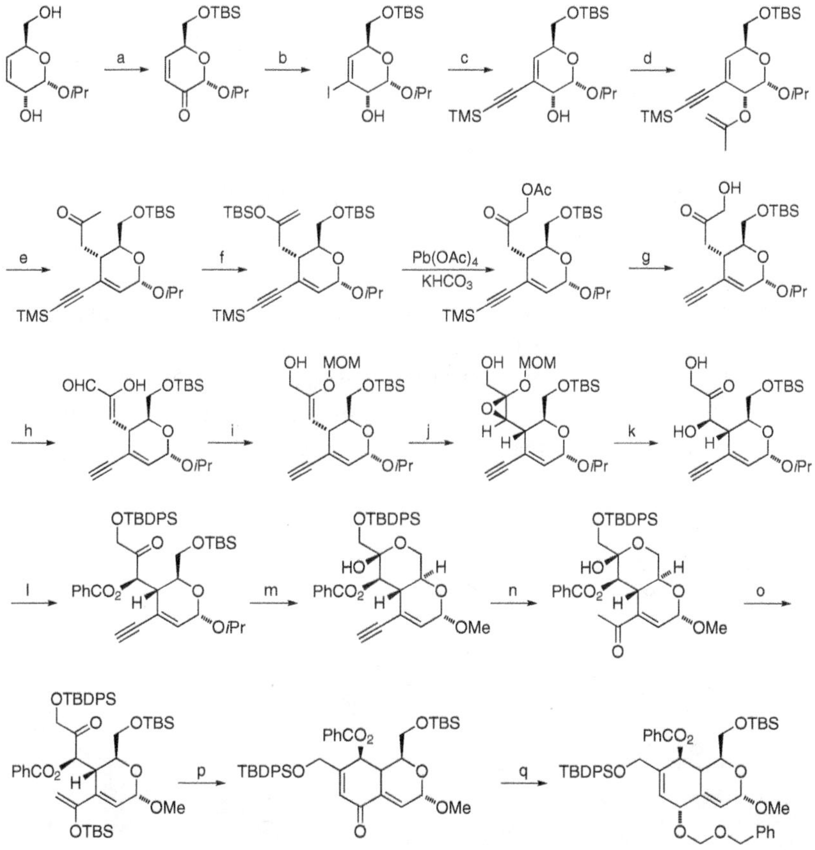

(b) Write a reasonable mechanism for step **m** of the sequence.

(c) Does the exact order in which the operations are carried out in step **i** matter? Why?

(d) Is the ring opening of the epoxide (step **k**) carried out under acidic or basic conditions? Would the product be the same if the reaction were carried out under the opposite (acidic instead of basic, or basic instead of acidic) conditions?

(e) step **o** actually required two separate reactions. Suggest a reason why.

15-27 Morphine is an intriguing natural product first synthesized in 1956 by Gates and Tschudi [*J. Am. Chem. Soc.* **1956**, *78*, 1380]. The following convergent formal synthesis is from the laboratory of David A. Evans in 1982 [*Tetrahedron Lett.* **1982**, *23*, 285]. Supply the missing reagents for each of the transformations shown. Note that more than one step may be necessary to complete each transformation.

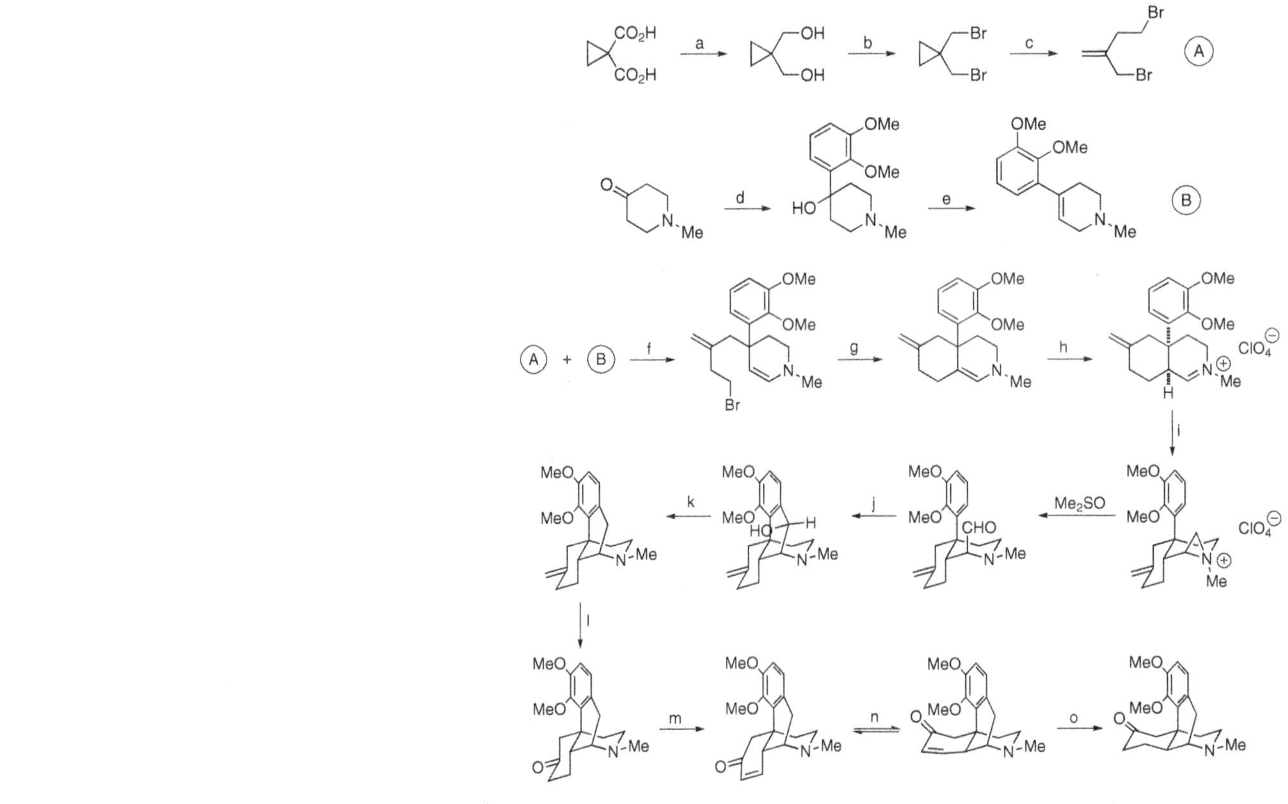

15-28 Eleven years after the Evans approach appeared, Overman published a more modern approach to the synthesis of optically active morphine derivatives [*J. Am. Chem. Soc.* **1993**, *115*, 11028], and a formal total synthesis of enantiomerically enriched morphine. Supply the missing reagents for each of the transformations shown. Note that more than one step may be necessary to complete each transformation.

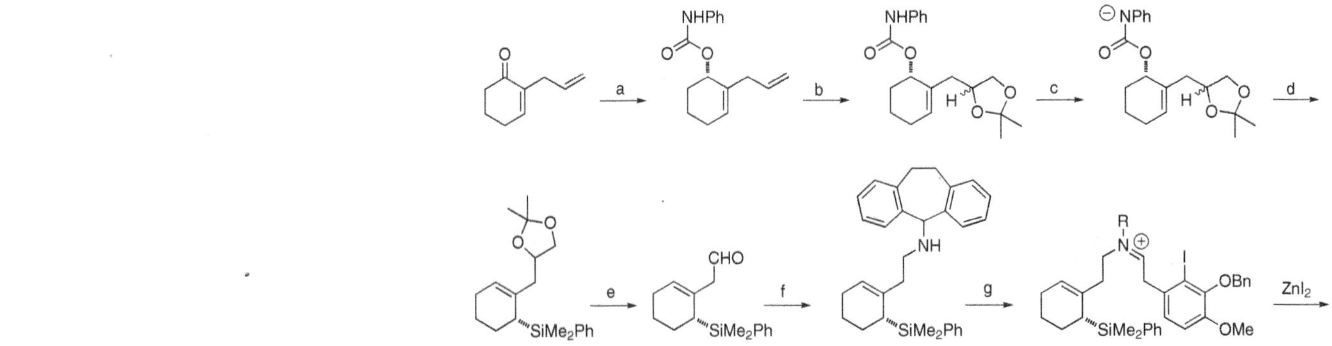

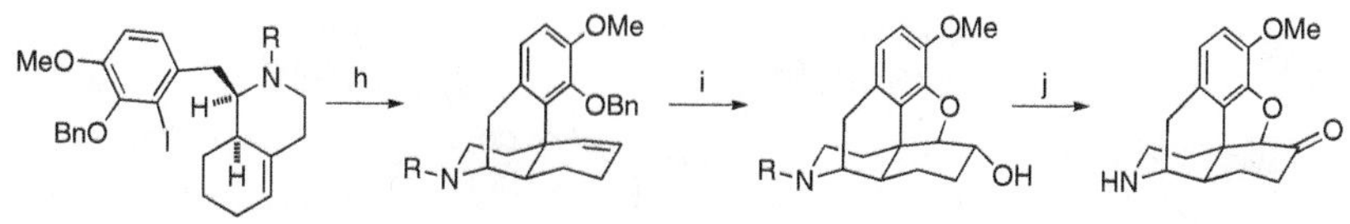

15-29 Cubane is the first of the "Platonic solids" to be prepared in hydrocarbon form [*J. Am. Chem. Soc.* **1964**, *86*, 3157]. Supply the missing reagents for each of the transformations shown. Note that more than one step may be necessary to complete each transformation.

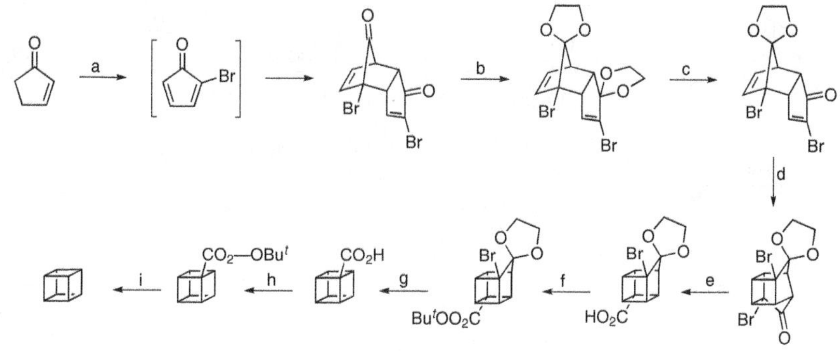

Why is the formation of the monoketal necessary for the synthesis to be successful? Is this step a concession step or not? Why or why not?

15-30 In 1979, Samir Chatterjee published two papers describing total syntheses. Both are now universally considered to be fabrications. The first was published in *Chem. Commun.* [**1979**, 620] as a total synthesis of the interesting sesquiterpene hydrocarbon, isocomene. The second was published in *Tetrahedron Lett.* [**1979**, 3249] as the synthesis of a synthon for the C_{19} diterpenoid alkaloids. There are several reactions described in both papers that contradict the known outcomes of such reactions in other systems, thus providing the hints that the syntheses are fraudulent. What are these reactions and what would be their expected outcome? What other evidence could be used to support the deduction that the syntheses as published are frauds (you should read the original papers to obtain this information)?

(a) *Chem. Commun.* **1979**, 620.

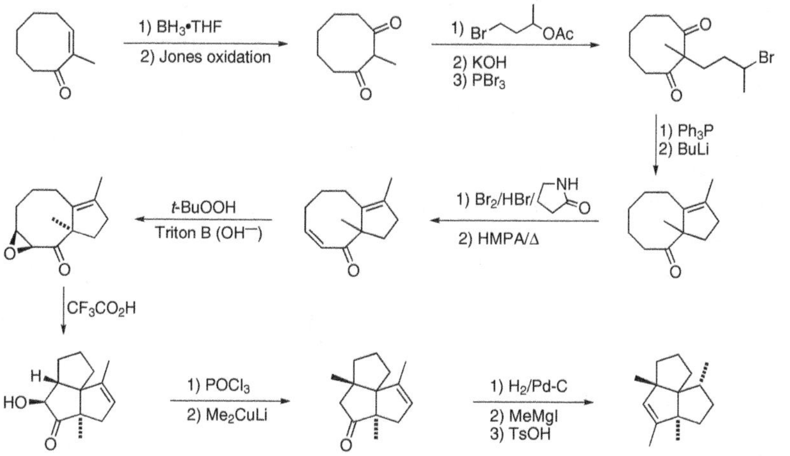

(b) *Tetrahedron Lett.* **1979**, *20*, 3249.

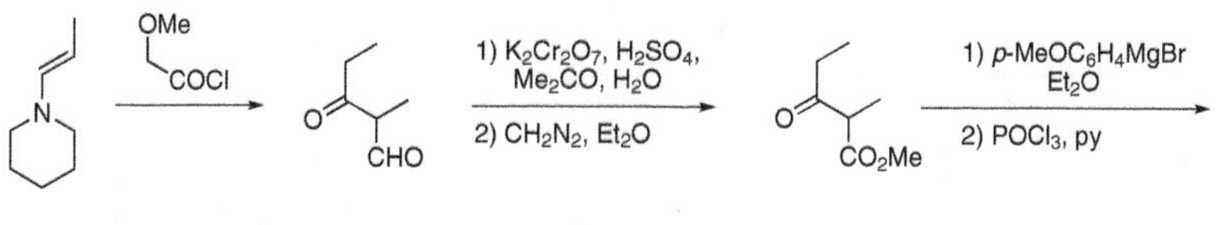

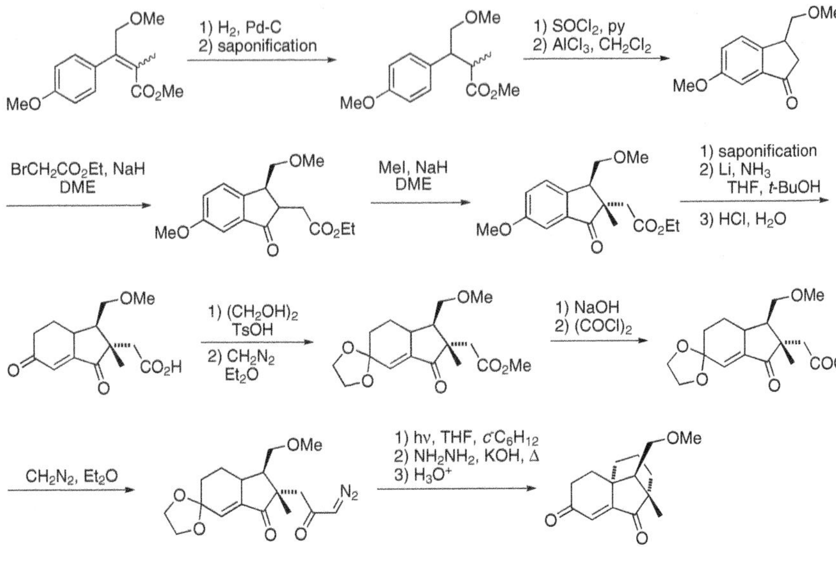
1) H_2, Pd-C
2) saponification
1) $SOCl_2$, py
2) $AlCl_3$, CH_2Cl_2
$BrCH_2CO_2Et$, NaH
DME
MeI, NaH
DME
1) saponification
2) Li, NH_3
THF, t-BuOH
3) HCl, H_2O
1) $(CH_2OH)_2$
TsOH
2) CH_2N_2
Et_2O
1) NaOH
2) $(COCl)_2$
CH_2N_2, Et_2O
1) hv, THF, c-C_6H_{12}
2) NH_2NH_2, KOH, Δ
3) H_3O^+
OMe
MeO
CO_2Me
CO_2Et
CO_2H
COCl
N_2

CHAPTER SIXTEEN

Organic Reactions V

Condensations and Cascade Reactions of Carbon Nucleophiles

16.1 Addition, Condensation, and Cascade Reactions: Definitions

Condensations differ from addition reactions in one very important aspect: during a condensation reaction, a small molecule (often water or an alcohol but occasionally a more complex species) is eliminated from the initial adduct. Condensation reactions occur by an addition-elimination sequence, so the product retains a π bond. Cascade reactions occur by an addition-intramolecular substitution sequence, so the product is net addition to the π bond (Example **16.1**); such reactions are also often called condensations. Either of the two steps may be rate determining, which can make it more difficult to control the stereochemistry, especially if the addition step is reversible. Questions of stereochemistry in condensation reactions are most often expressed in terms of *E/Z* or *cis/trans* instead of *R/S* or *syn/anti*.

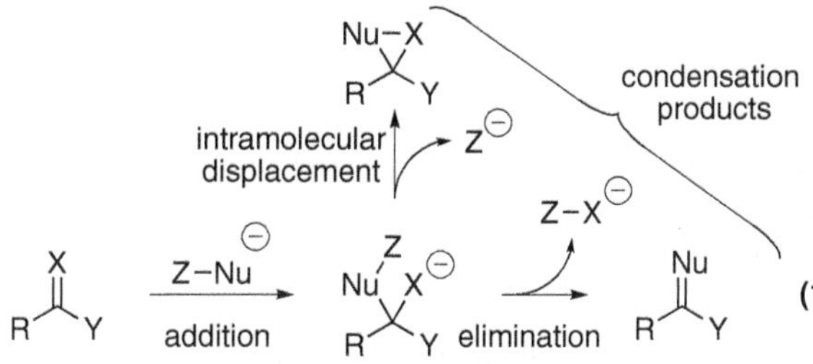

(16.1)

16.2 The Aldol Condensation and Related Reactions[1]

Although the aldol addition has become the more common form of this reaction in the past several decades, the aldol condensation, where the product is a conjugated carbonyl compound, still remains an important reaction. The aldol condensation is a two-stage reaction that involves initial aldol addition, followed by dehydration of the resultant β-hydroxycarbonyl compound under acidic or basic conditions to give the conjugated enone. Because modern methods for quantitative enolate generation make the aldol addition a much simpler reaction than before, the aldol condensation is now usually carried out in two separate steps: (1) aldol addition, followed by (2) acid- or base-catalyzed dehydration of the aldol. The acid-catalyzed dehydration is generally accepted to occur by the E1 mechanism, and the base-promoted elimination almost certainly involves an enolate anion and so occurs by the E1cb mechanism.

1. For a somewhat dated but still extremely useful review, see: House, H.O. *Modern Synthetic Reactions* (Benjamin/Cummings: Menlo Park, CA, 1972), ch. 10.

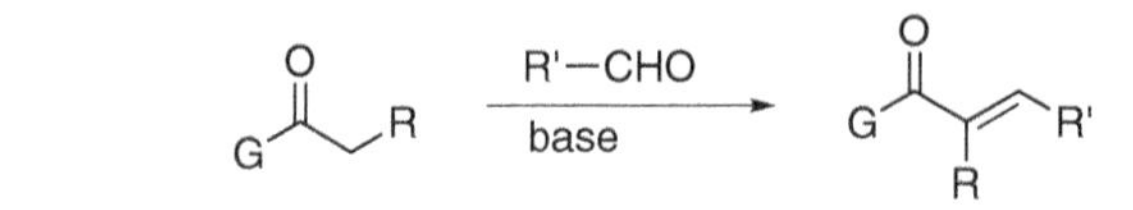

(16.2)

The Claisen-Schmidt Reaction

The reaction between a ketone enolate and an aldehyde is known as the Claisen-Schmidt condensation, after the two chemists (Ludwig Claisen and J. Gustav Schmidt) who reported it independently in 1881.[2] The product of the Claisen-Schmidt reaction is a conjugated enone. As with the aldol condensation, the Claisen-Schmidt reaction is most easily carried out today as a two-step process; it also works best when the aldehyde partner does not have an α-hydrogen—typically an aromatic aldehyde, as shown in Example **16.3**.[3]

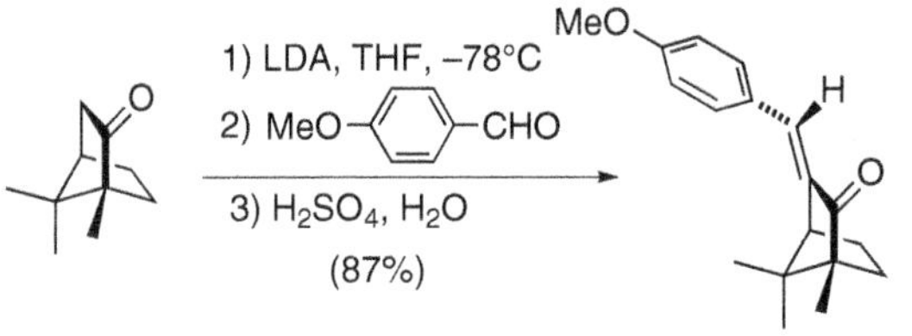

(16.3)

The Knoevenagel Condensation

When the source of the enolate anion is an active methylene compound—a compound where the CH_2 group is flanked by a pair of electron-withdrawing groups such as carbonyl groups, cyano groups, or nitro groups, so that it is acidic enough to be quantitatively converted to the enolate anion by alkoxide anion bases—the aldol condensation is frequently referred to as the ***Knoevenagel condensation***.[4] The reaction follows the same course as the aldol condensation (aldol addition followed by dehydration), but the presence of two groups capable of conjugating with the double bond renders the dehydration step much more facile. In 1902, Doebner[5] introduced a modification of the original Knoevenagel condensation in which an amine (usually piperidine or pyridine) and acetic acid replace more strongly basic catalysts.[6] The ***Doebner modification*** is now probably the most commonly used form of the reaction (as in Example **16.4**[7]). Alkylammonium carboxylate salts are known to be low-melting solids that can function as ionic liquids, which has led to the exploration of these salts as solvents for the Knoevenagel reaction. The results of some early work in this area are promising.

---

2. (a) Claisen, L.; Claparède. A. *Ber. dtsch. chem. Ges.* **1881**, *14*, 2460. (b) Schmidt, J.G. *Ber. dtsch. chem. Ges.* **1881**, *14*, 1459. (c) Claisen, L. *Ber. dtsch. chem. Ges.* **1887**, *20*, 655.

For biographical details of Claisen see Chapter 6. Biographical details about Schmidt are difficult to find; he was a friend of Traugott Sandmeyer at the ETH in Zurich.

3. Seo, B.-I.; Wall, L.K.; Lee, H.; Buttrum, J.W.; Lewis, D.E. *Synth. Commun.* **1993**, *23*, 15.

4. Knoevenagel. E. *Ber. dtsch. chem. Ges.* **1898**, *31*, 2596.

Heinrich Emil Albert Knoevenagel (1865–1921) was educated in Hannover (PhD, 1889). After working with Viktor Meyer at Göttingen and Heidelberg, he was appointed Extraordinary Professor at Heidelberg in 1896. He remained there until his death from peritonitis due to a burst appendix at age 56. For more biographical details, see *Angew. Chem.* **1922**, *35*, 29.

5. **Oscar Doebner (1850–1907)** was educated in Munich and Tübingen (Dr rer nat, 1873). After 18 months as Otto's assistant in Braunschweig, he moved to Berlin to work with Hofmann. He was made Professor at Halle in 1884. Doebner's accomplishments include the discovery of the two reactions that bear his name as well as the dye malachite green. For more information, see: Schotten, C. *Ber. dtsch. chem. Ges.* **1907**, *40*, 5131.

6. Doebner, O. *Ber. dtsch. chem. Ges.* **1902**, *35*, 1136.

7. Erdman, P.J.; Gosse, J.L.; Jacobson, J.A.; Lewis, D.E. *Synth. Commun.* **2004**, *34*, 1163.

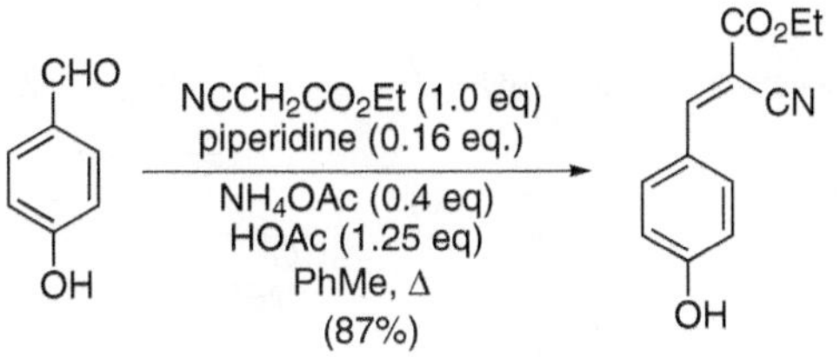

(16.4)

In a recent example of this reaction (Example **16.5**), the synthesis of either the *E* or *Z* isomers of nitroalkenes can be accomplished by appropriately choosing the solvent and temperature for the reaction.[8] In this example, the initial reaction is the nitro-aldol addition, or ***Henry reaction*** of nitromethane with aldehydes.[9] The use of molecular sieves is critical to the stereochemical outcome: in the absence of molecular sieves, the diastereoselectivity is poor.

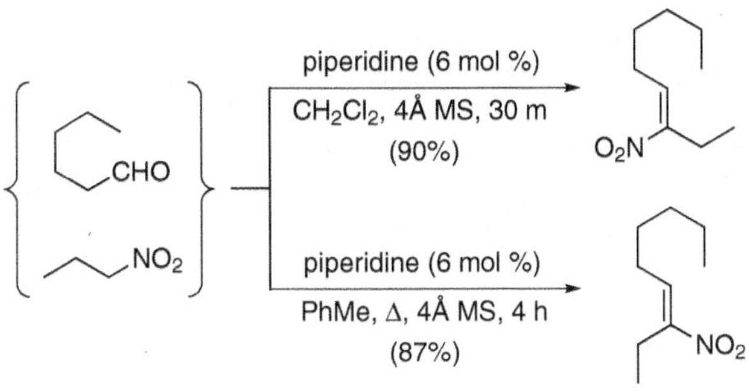

(16.5)

The Perkin Condensation

When an aromatic aldehyde is heated with an acid anhydride and the alkali metal salt of the acid, a cinnamic acid derivative is formed. This reaction was discovered by William Henry Perkin[10] and is now known by his name.[11] The reaction is thought to involve the formation of the enolate anion of the anhydride, which then reacts with the aldehyde to give a β-hydroxyacid that is acetylated and then undergoes elimination to give the cinnamic acid. However, this interpretation of the available evidence has recently been thrown into question.[12] More recent variants of the reaction use a tertiary amine as the base catalyst for the reaction.

Intramolecular Aldol Condensations *via* Enamines

Dibenzylamine trifluoroacetate was introduced as a reagent for the intramolecular aldol condensation by converting a dialdehyde to an enamine that would then undergo an intramolecular addition and subsequent dehydration. The reagent converts dialdehydes to conjugated cycloalkenecarboxaldehydes, as in Examples **16.6** and **16.7**.

---

8. Fioravanti, S.; Pellacani, L.; Tardella, P.A.; Vergari, M.C. *Org. Lett.* **2008**, *10*, 1449
9. Henry, L. *C. R. Hebd. Seances Acad. Sci.* **1895**, *120*, 1265.
10. **Sir William Henry Perkin (1838–1907)** was educated at the Royal College of Chemistry under Hofmann. Aged 17, he prepared "mauveine," the first commercially successful synthetic dye; before graduating, he left the College and set up a company to manufacture this dye; and at 36 he sold his company and became independently wealthy. He was elected FRS in 1866 and knighted in 1906. For more details, see: (a) Brightman, R. *Nature* **1956**, *177*, 805; (b) Holme, I. *Color. Technol.* **2006**, *122*, 235.
11. (a) Perkin, W.H. *J. Chem. Soc.* **1868**, *21*, 181; **1877**, *31*, 660. Reviews: (b) Johnson, J. R. *Org. React.* **1942**, *1*, 210. (c) Rosen, T. in Trost, B.M.; Fleming, I., Eds. *Comprehensive Organic Synthesis* (Pergamon: Oxford, 1991), vol. 2, **p.** 395.
12. Chandrasekhar, S.; Karri, P. *Tetrahedron Lett.* **2006**, *47*, 2249.

(16.6)[13]

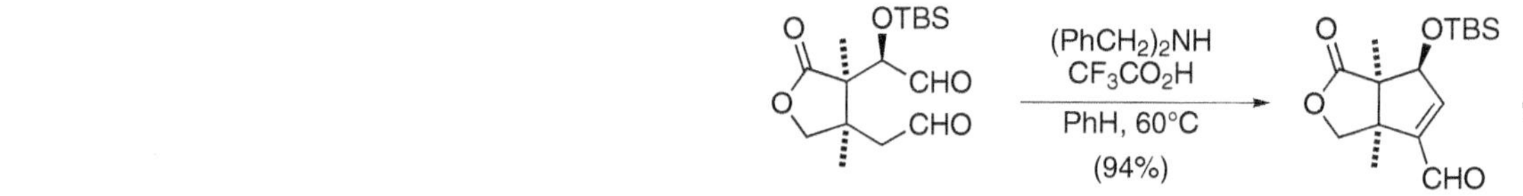

(16.7)[14]

Worked Problem

16-1 What will be the major organic product of each of the following reactions?

(a) 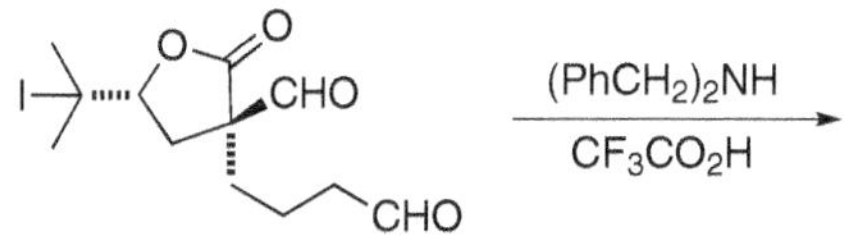

(b) 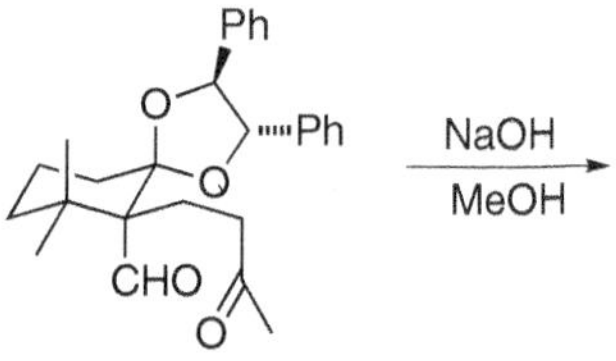

§Answers below

16.3 Condensations Where the Initial Adduct Is Intercepted

The initial adduct of an enolate anion and an aldehyde or ketone is an alkoxide that can be intercepted by an appropriately located electrophile such as a halide or an ester group (e.g., Example **16.8**). Both these types of reactions are known. Interception of the aldolate with an alkyl halide gives an epoxide in a reaction known as the ***Darzens glycidic ester condensation***, and interception by an ester group gives the ***Stobbe condensation***. The ***Corey-Chaykovsky reaction*** between a sulfonium ylide and an aldehyde to give an epoxide is closely related (mechanistically, at least) to the Darzens condensation.

13. Corey, E.J.; Danheiser, R.L.; Chandrasekharan, S.; Siret, P.; Keck, G.E.; Gras, J.-L. *J. Am. Chem. Soc.* **1978**, *100*, 8031.
14. Birman, V.B.; Danishefsky, S.J. *J. Am. Chem. Soc.* **2002**, *124*, 2080.

§ Answers to Worked Problem:

(a) 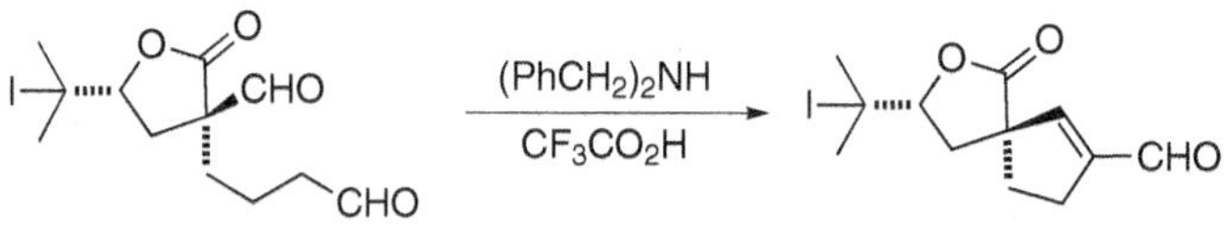

(b) 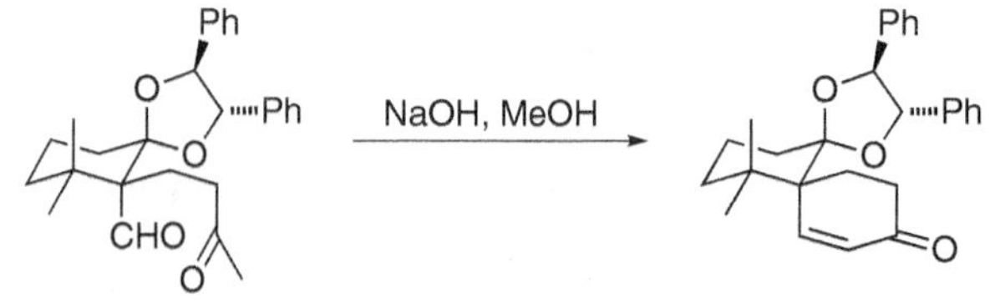

References: (a) *J. Org. Chem.* **2008**, *73*, 8049. (b) *J. Org. Chem.* **2007**, *72*, 4098.

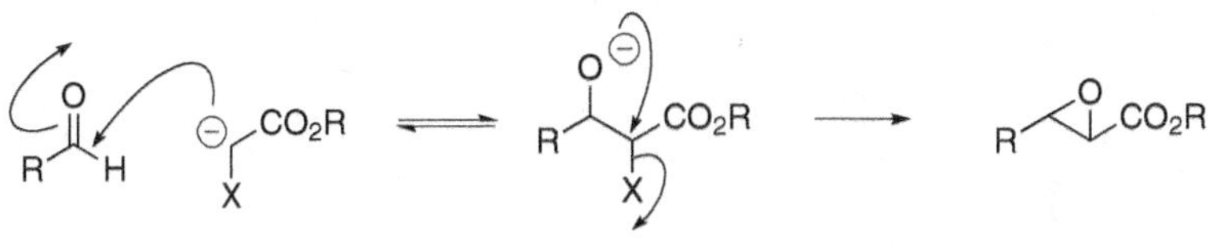

(16.8)

The Darzens Condensation[15]

The Darzens condensation is a reaction of α-haloesters and aldehydes in the presence of base to give an α,β-epoxyester (a glycidic ester), first reported by French chemist Auguste Georges Darzens[16] in 1904.[17] The reaction is known to proceed through a reversible initial addition of the enolate anion to the carbonyl group, followed by a rate-determining intramolecular displacement of the halide by the alkoxide anion[18] rather than the alternative mechanism, as was demonstrated by Zimmerman and Ahramjian.[19] The alternative mechanism involves formation of a carbene that then adds to the carbonyl group. In their study, they also found that the reaction in Example **16.9** gives predominantly the product with the two phenyl groups *cis*. The conformation of the intermediate alkoxide suggests that the major product is formed by addition of the *E(O)* enolate of the ester to the aldehyde through a six-membered Zimmerman-Traxler transition state.

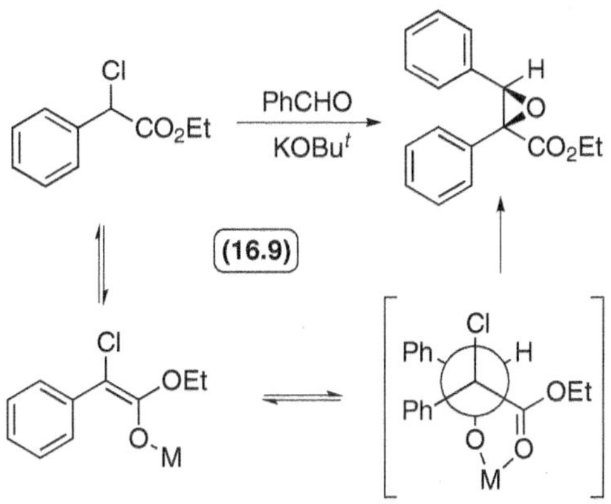

As with other reactions, efforts to control the stereochemistry of the Darzens condensation have become focused on absolute, as well as relative, control of the stereochemistry of the reaction. In this reaction, stereoselectivity must be managed at two levels: diastereoselectivity, which controls the *cis/trans* ratio, and enantioselectivity, which controls the enantiomer ratio. Particular success in the asymmetric Darzens condensation (e.g., Example **16.10**) has been achieved using chiral phase transfer catalysts[20] or chiral crown ethers.[21]

---

15. Reviews: (a) Newman, M.S.; Magerlein, B.J. *Org. React.* **1949**, *5*, 413. (b) Rosen, T. In Trost, B.M.; Fleming, I., Eds. *Comprehensive Organic Synthesis* (Pergamon: Oxford, 1991); vol 2, 409.

16. **Auguste Georges Darzens (1867–1954)** was born to French parents in Moscow, and educated in Paris at the École Polytechnique under Grimaux. He remained after graduation (1888; MD, 1890) to begin his career. Darzens eventually rose to Professor of Chemistry in 1913. Darzens and Grimaux were among the few at the École (a military school) who defended Alfred Dreyfus, and Grimaux lost his Professorship over it. Darzens remained as Professor at the École until his retirement in 1937. For more biographical details, see: Laszlo, P. *Bull. Hist. Chem.* **1994**, *15/16*, 59.

17. Darzens, G. *Compt. Rend.* **1904** *139*, 1214.

18. (a) Ballester, M.; Bartlett, P.D. *J. Am. Chem. Soc.* **1953**, *76*, 2042. (b) Ballester, M. *Chem. Rev.* **1955**, *55*, 283. (c) Ballester, M.; Perez-Blanco, D. *J. Org. Chem.* **1958**, *23*, 652.

19. Zimmerman, H.E.; Ahramjian, L. *J. Am. Chem. Soc.* **1960**, *82*, 5459.

20. (a) Arai, S.; Shioiri, T. *Tetrahedron Lett.* **1998**, *39*, 2145. (b) Arai, S.; Shirai, Y.; Ishida, T.; Shioiri, T. *Tetrahedron* **1999**, *55*, 6375. (c) Arai, S.; Shirai, Y.; Ishida, T.; Shioiri, T. *Chem. Commun* **1999**, 49. (d) Arai, S.; Tokumaru, K.; Aoyama, T. *Tetrahedron Lett.* **2004**, *45*, 1845. (e) Arai, S.; Ishida, T.; Shioiri, T. *Tetrahedron Lett.* **1998**, 39, 8299. (f) Arai, S.; Shioiri, T. *Tetrahedron* **2002**, *58*, 1407.

21. Bakó, P.; Szöllõzy, A.; Bombicz, P.; Tõke, L. *Synlett* **1997**, 291–292.

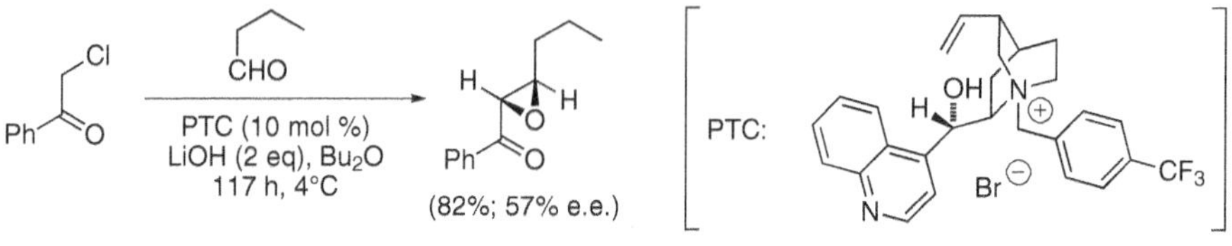

(16.10)

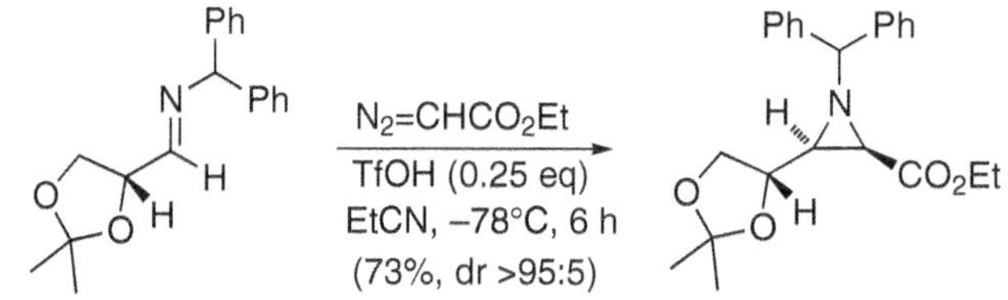

(16.11)

The aza-Darzens reaction, which gives aziridines as the final product, has been carried out using α-diazoketones under conditions of acid catalysis (Example **16.11**).[22] In this reaction, it is possible that the diazoketone becomes protonated to give a diazonium ion (Example **16.12**), which contains an excellent leaving group (N_2^+) that is displaced by the imine nitrogen to give iminium ion (Example **16.14**). This ion in turn is deprotonated to give an azomethine ylide (Example **16.13;** see Section 6.5: 1,3-dipolar cycloadditions, for more about these reagents). This ylide undergoes conrotatory ring closure. In an alternative pathway, the imine is protonated to give the iminium ion (Example **16.16**), which then reacts with the diazoalkane as a nucleophile to give the intermediate (Example **16.18**) carrying the leaving group.

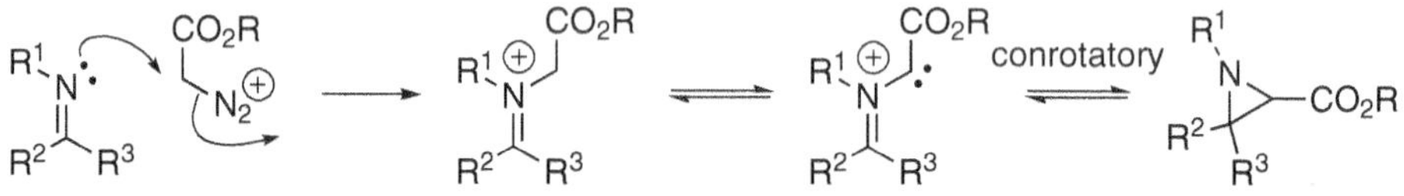

(16.12) (16.13) (16.14) (16.15)

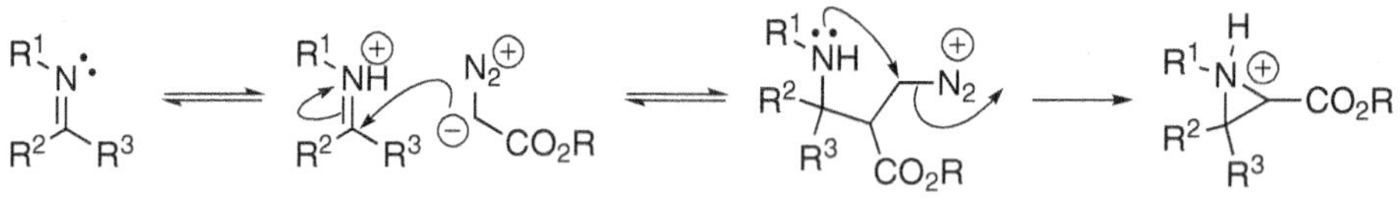

(16.16) (16.17) (16.18) (16.19)

The Corey-Chaykovsky Reaction[23]

A sulfur ylide is a species where the group stabilizing the carbanion center may also act as a leaving group, thus making it (in a formal sense, at least) analogous to the enolate anion of an α-halocarbonyl compound. Thus, when one replaces the α-haloester enolate of the Darzens condensation with a sulfur ylide, one also obtains an epoxide by the two-step process of addition to the aldehyde carbonyl group followed by intramolecular displacement of the dialkyl sulfide (Example **16.20**). The reaction can also be applied to the synthesis of aziridines from imines and to cyclopropane formation from α,β-unsaturated carbonyl compounds. The mechanism of all these reactions follows a common course: addition to the carbonyl group, followed by intramolecular displacement of the leaving group. Recently, an "instant sulfur ylide"—a dry mixture of the sulfonium or sulfoxonium salt and sodium hydride or potassium *tert*-butoxide—has been found to be stable for more than a year when dry, making it especially convenient for this reaction (e.g., Example **16.21**).[24]

---

22. Williams, A.L.; Johnston, J.N. *J. Am. Chem. Soc.* **2004,** *126,* 1612.
23. Corey, E.J.; Chaykovsky, M. *J. Am. Chem. Soc.* **1965,** *87,* 1353.
24. (a) Ciaccio, J.A.; Drahus, A.L.; Meis, R.M.; Tingle, C.T.; Smrtka, M.; Geneste, R. *Synth. Commun.* **2003,** *33,* 2135. (b) Ciaccio, J.A.; Aman, C.E. *Synth. Commun.* **2006,** *36,* 1333.

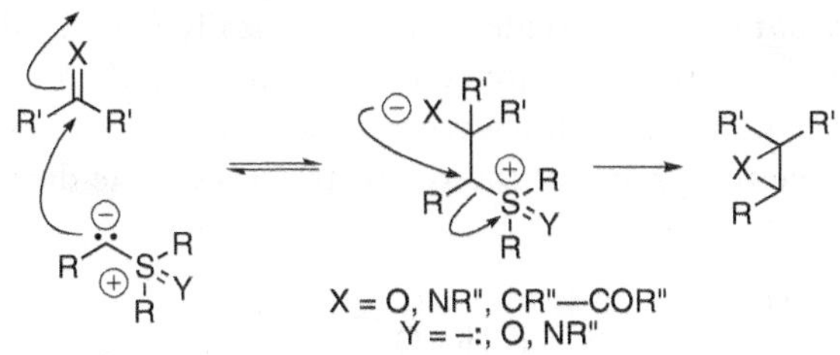

(16.20)

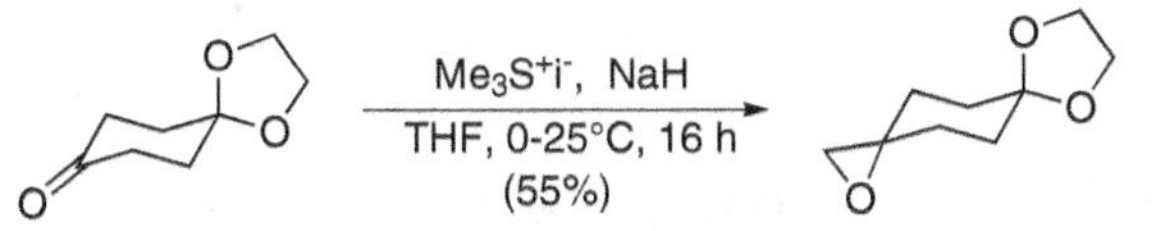

(16.21)

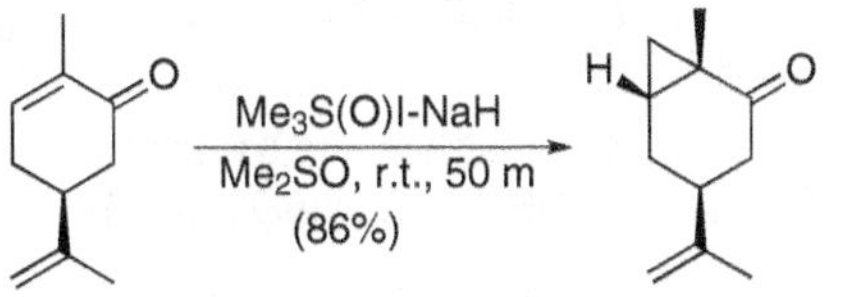

(16.22)

The course of the reaction with conjugated ketones, especially, is decided by the structure of the sulfur ylide. Sulfonium ylides, where the group Y on sulfur is a lone pair, are harder nucleophiles, and show a strong tendency to add to the carbonyl carbon of conjugated ketones, giving allylic epoxides. In contrast to this, sulfoxonium ylides (where the group Y on sulfur in Example **16.20** is oxygen) are softer nucleophiles, and tend to add to the β-carbon, giving cyclopropyl ketones as the major product (Example **16.22**).

The Stobbe Condensation

When the source of the enolate anion in an aldol condensation is a dialkyl succinate, the reaction is known as the ***Stobbe condensation***, first reported by German chemist Hans Stobbe[25] in 1893.[26] In the Stobbe condensation, as in the Darzens condensation, the intermediate alkoxide (**16.25**) is intercepted to form a lactone (**16.27**). The subsequent Elcb elimination of the carboxylate anion gives the half-ester of an alkylidenesuccinate, in which the ester group, rather than the carboxyl group, is conjugated with the double bond (Figure 16.1).

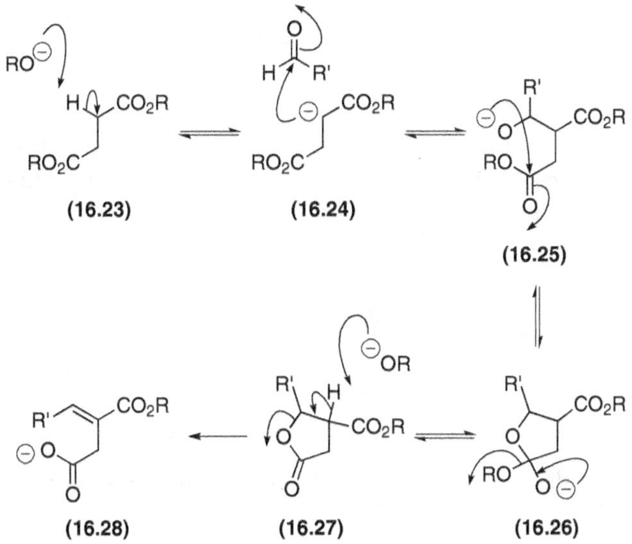

Figure 16.1 The course of the Stobbe condensation

25. **Hans Stobbe (1860–1938)** was educated at Heidelberg, Munich, Strasbourg, and then Leipzig, where he studied under Wislicenus. He remained at Leipzig, eventually succeeding Wislicenus in 1903. Stobbe was a pioneer in physical organic chemistry and in polymer chemistry. For more information, see his obituary (nekrolog): Weygand, C. *Kolloid-Z.* **1939**, *87*, 1.

26. Stobbe, H. *Ber. Dtsch. Chem. gs.* **1893**, *26*, 2312; *Justus Liebigs Ann. Chem.* **1894**, *282*, 280.

The stereochemistry of the Stobbe condensation is typically *E* with aldehydes, as in Example **16.29**,[27] but with ketones (e.g., Example **16.30**), a mixture of both stereoisomers (e.g., Examples **16.31** and **16.32**) is almost always obtained, as shown below.[28] Its ability to differentiate between the two groups of the symmetrical succinate diester makes the Stobbe condensation especially useful.

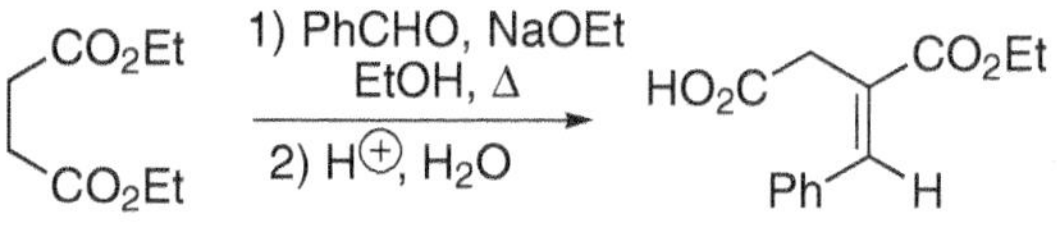

(16.29)

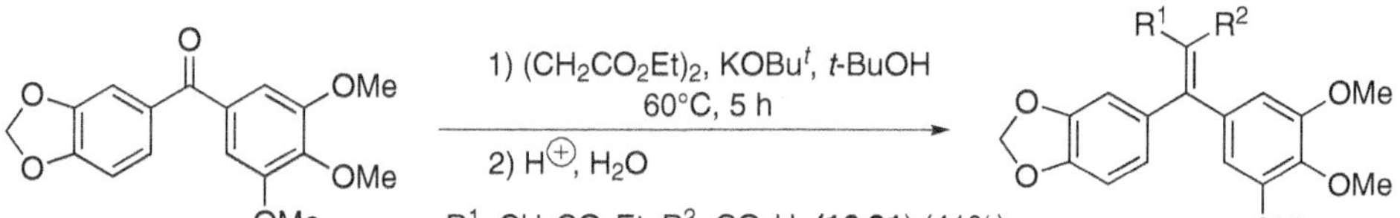

(16.30)

R^1=CH_2CO_2Et, R^2=CO_2H: (**16.31**) (41%)
R^1=CO_2H, R^2=CH_2CO_2Et: (**16.32**) 33%)

Reaction Synopses

Claisen-Schmidt Reaction

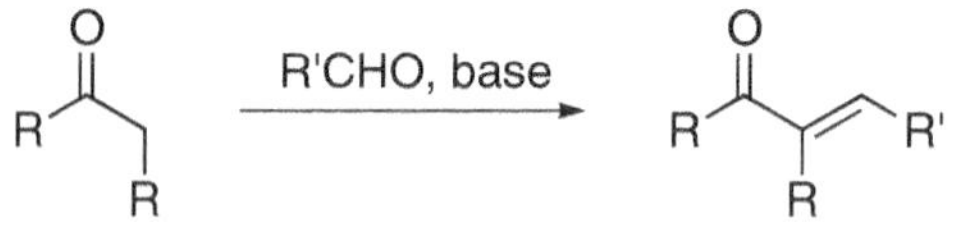

Reagents: NaOH, EtOH; KOH, EtOH; NaOEt, EtOH; etc.
Works best if the aldehyde is aromatic

Knoevenagel Condensation

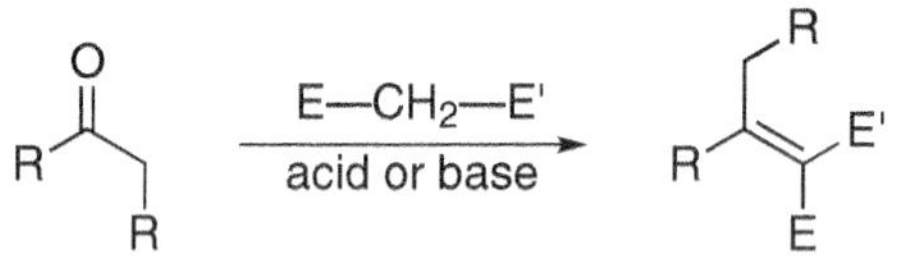

E, E' = COR', CO_2R', CN, NO_2, SO_2R', etc.

Reagents (all Doebner modification):

$CH_2(CO_2Et)_2$, NH_4OAc, piperidine, AcOH, Δ; N≡CCH_2CO_2Et, NH_4OAc, piperidine, AcOH, Δ; $MeCOCH_2CO_2Et$, NH_4OAc, piperidine, AcOH, Δ; etc.

Perkin Condensation

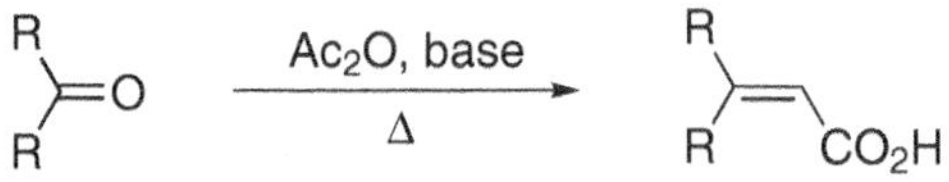

Reagents: Ac_2O, NaOAc, Δ; Ac_2O, Et_3N, Δ; etc.

Reaction proceeds more slowly with other anhydrides to give α-substituted cinnamic acids.

---

27. Plattner, J.J.; Marcotte, P.A.; Kleinert, H.D.; Stein, H.H.; Greer, J.; Bolis, G.; Fung, A.K.L.; Bopp, B.A.; Luly, J.R. *J. Med. Chem.* **1988**, *31*, 2277.

28. For a typical example, see: Smissman, E.E.; Portoghese, P.S.; Mode, R.A. *J. Org. Chem.* **1961**, *26*, 3628.

Darzens Condensation

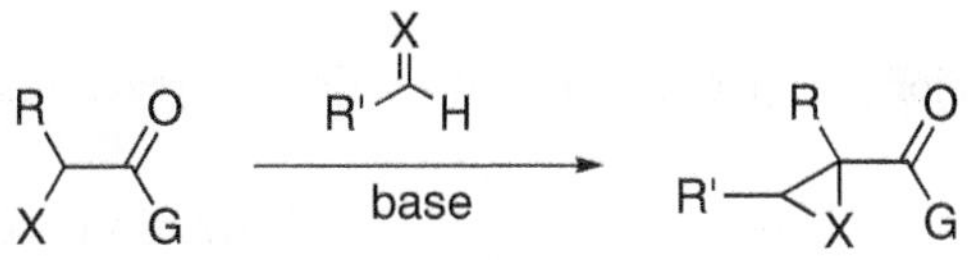

G=R, Ar, OR;

X=O, NR, C(R)—E (E=CO_2R, COR, CN, NO_2, etc,)

Reagents: RCH(Cl)CO_2Et, ArCHO, KOBu^t, THF; etc.

Corey-Chaykovsky Reaction

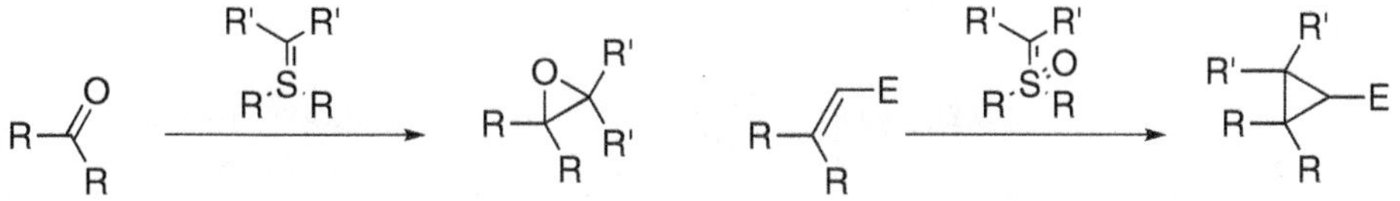

Reagents: $Me_3S^+I^-$, NaH, Me_2SO; $Me_3S^+I^-$, KOBu^t, THF; etc.

or $Me_3S(O)^+I^-$, NaH, Me_2SO; $Me_3S(O)^+I^-$, KOBu^t, THF; etc.

Stobbe Condensation

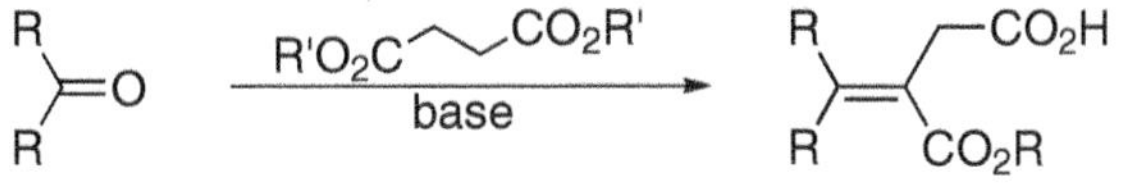

Reagents: $(CH_2CO_2Et)_2$, NaOEt, EtOH, Δ; etc.

Stereochemistry: with aldehydes, *E* isomer predominates; with ketones product is close to stereorandom

Worked Problem

16-2 What is the product of the following reaction?

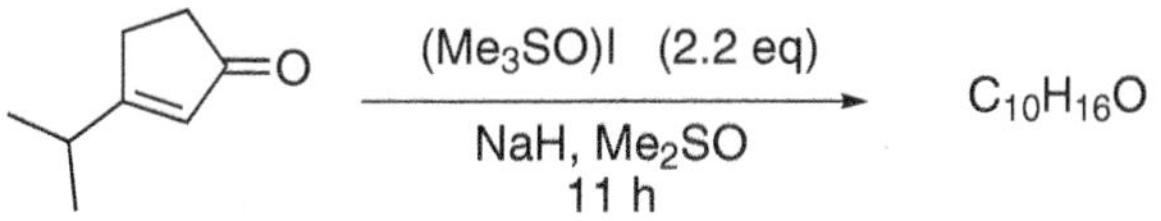

§Answer below

§ Answer to Worked Problem:

This Corey-Chaykovsky reaction almost certainly proceeds in two stages. In the first stage, the cyclopropyl ketone is formed as expected from a sulfoxonium ylide, and this cyclopropyl ketone then reacts more slowly with the second equivalent of the nucleophile to give the observed product. The bulky isopropyl group controls the stereochemistry of the second addition. It had been noted that when the reaction time was shorter, and the excess of the ylide less, a substantial amount of starting material was recovered. This presumably arises from the deprotonation of the ketone by the ylide before addition (a problem that often occurs in cyclopentanone derivatives).

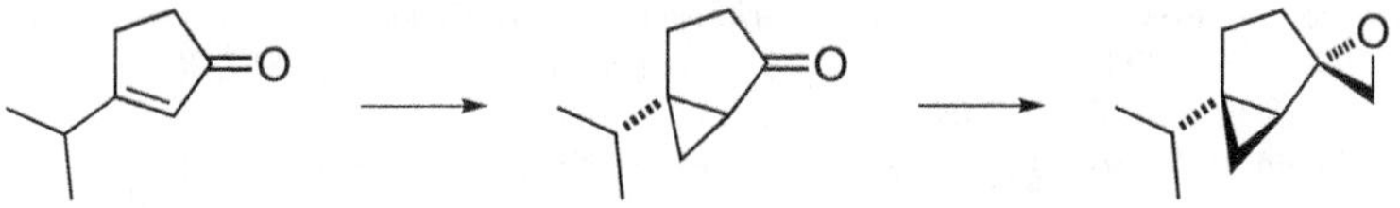

Reference: *Tetrahedron Lett.* **2001**, *42*, 5589.

Problem

16-1 What is the major product of each of the following reactions?

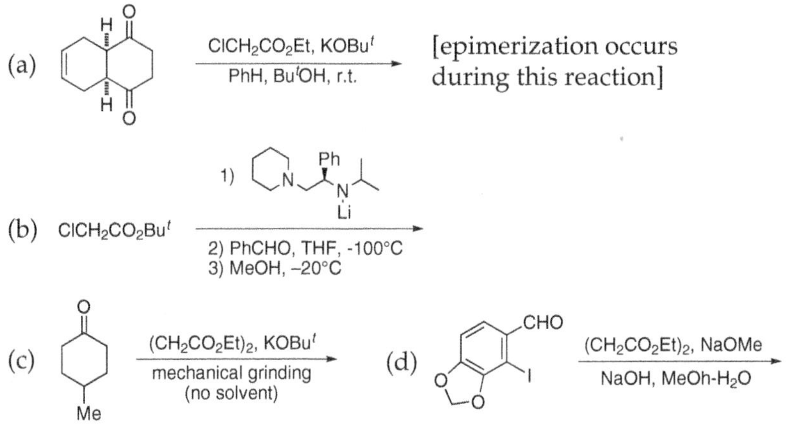

References: (a) *J. Am. Chem. Soc.* **1958**, *80*, 5006. (b) *Chem. Pharm. Bull.* **1995**, *43*, 1821. (c) *Green Chem.* **2000**, *2*, 303. (d) *Tetrahedron Lett.* **2001**, *42*, 437.

16.4 **The Claisen Condensation and Related Reactions**

The Claisen Condensation

The name of Ludwig Claisen has become attached to the prototypical base-promoted condensation of esters to give β-ketoesters.[29] However, the first observation of such a reaction—the condensation of ethyl acetate by sodium metal to give ethyl acetoacetate—was actually made by Frankland and Duppa[30] and further clarified by Geuther,[31] who first used an alkoxide anion as the base. In this reaction, the ester is converted to a small equilibrium concentration of the enolate anion (**16.34**) that then adds to the carbonyl group of a second molecule of ester to give an intermediate (**16.35**). This then loses the alkoxide (Figure 16.2) to give the β-ketoester (**16.36**), which is much more strongly acidic than the starting ester, and it is

Figure 16.2 The path of the Claisen condensation

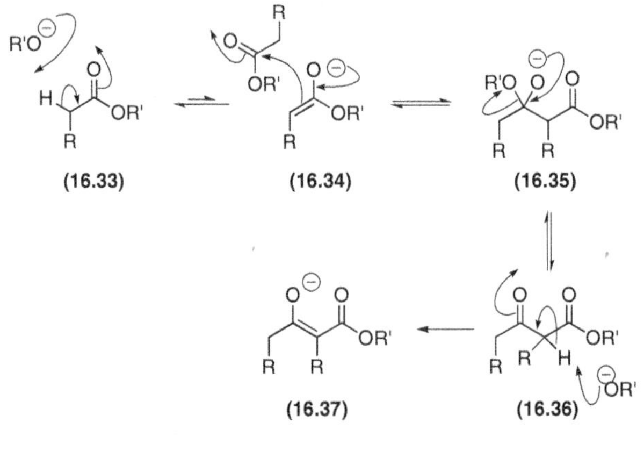

29. The first of a long series of papers by Claisen on the preparation and use of β-ketoesters in synthesis appeared in 1887: (a) Claisen, L.; Lowman, O. *Ber. dtsch. chem. Ges.* **1887**, *20*, 651. (b) Claisen, L. *Ber. dtsch. chem. Ges.* **1887**, *20*, 655. (c) Beyer, C.; Claisen, L. *Ber. dtsch. chem. Ges.* **1887**, *20*, 2178. (d) Claisen, L.; Stylos, N. *Ber. dtsch. chem. Ges.* **1887**, *20*, 2188. (e) Claisen, L.; Fischer, L. *Ber. dtsch. chem. Ges.* **1887**, *20*, 2191.

Biographical information about Claisen can be found in Chapter 6.

30. (a) Frankland, E.; Duppa, B.F. *J. Chem. Soc.* **1866**, *19*, 395. (b) Frankland, E.; Duppa, B.F. *Justus Liebigs Ann. Chem.* **1866**, *138*, 204, 328.

31. Geuther, A. *Z. Chem.* **1868**, *4*, 652.

quantitatively deprotonated by the alkoxide base in the final step of the reaction. Note that all previous the steps of the mechanism are reversible.

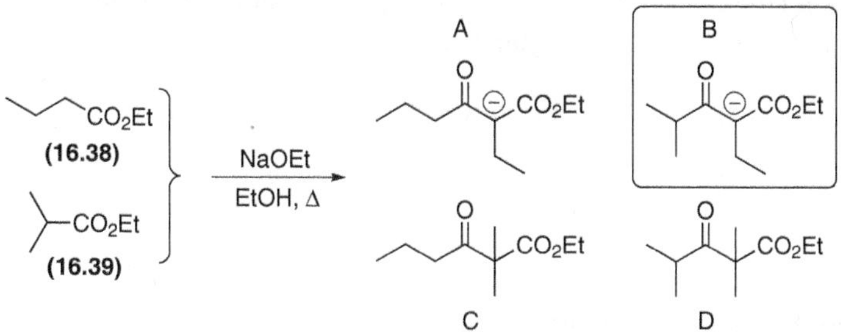

The Claisen condensation gives a mixture of products if both esters have α-methylene groups. However, if one can arrange the conditions such that only one of the possible products has a hydrogen atom flanked by two carbonyl groups, the reaction will ultimately proceed to give the conjugate base of that product alone. For example, if a 1:1 mixture of esters **16.38** and **16.39** is treated with base, there are four possible products: A, B, C, and D. However, neither of the esters C and D has a hydrogen atom flanked by two carbonyl groups, so both can revert to starting material in a retro-Claisen condensation. If the reaction mixture is heated long enough to go to completion, the anion B will be the major organic product of the reaction. This situation is even simpler when one of the two esters cannot form an enolate—for example, a carbonate ester (Example **16.40**), a benzoate ester (Example **16.41**), or a formate ester (Example **16.42**). Before the advent of methods to form simple enolates quantitatively using strong amide bases, the Claisen condensation between a carbonyl compound and ethyl formate, diethyl carbonate, or diethyl oxalate were used as a means to activate the α position of the carbonyl compound, because the activating group could be simply removed after the reaction by hydrolysis and decarboxylation, or a similar process.

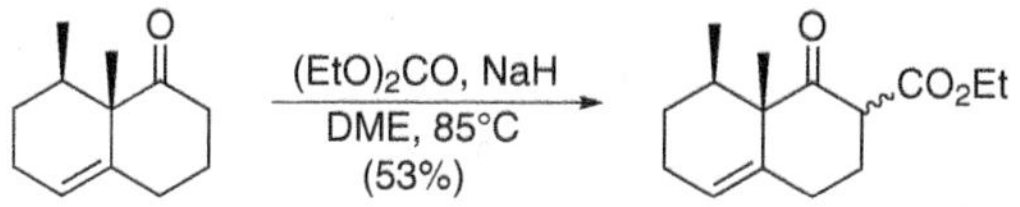

(16.40)[32]

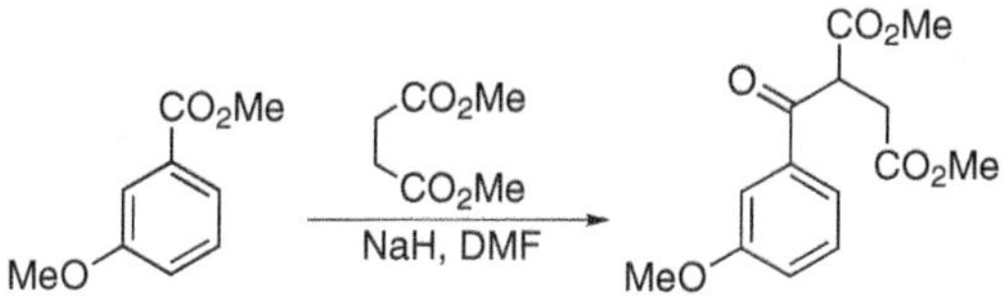

(16.41)[33]

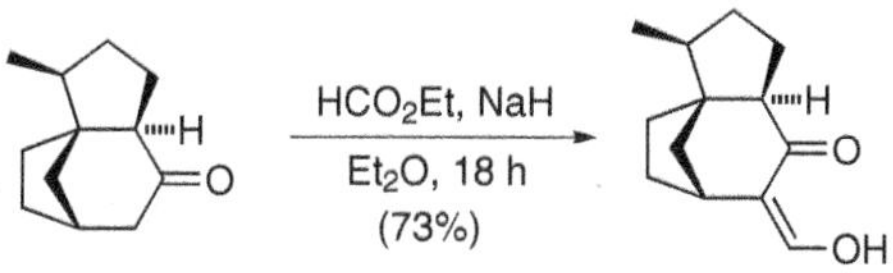

(16.42)[34]

The Dieckmann Condensation

The intramolecular version of the Claisen condensation is known as the ***Dieckmann condensation*** after its discoverer, German chemist Walter Dieckmann,[35] who disclosed it

32. Coates, R.M.; Shaw, J.E.J. *J. Am. Chem. Soc.* **1970**, *92*, 5657.
33. Conover, L.H.; Butler, K.; Johnston, J.D.; Korst, J.J.; Woodward, R.B. *J. Am. Chem. Soc.* **1962**, *84*, 3222.
34. Coates, R.M.; Sowerby, R.L. *J. Am. Chem. Soc.* **1972**, *94*, 5386.
35. **Walter Dieckmann (1869–1925)** took his PhD under Bamberger at Munich. He remained at Munich after graduation as Private Assistant to Adolf von Baeyer, before moving to industry, working for Badische Anilin- und Sodafabrik (BASF). He returned to academia in 1894 and completed his *habilitation*, which contained the reaction that now bears his name, in 1898. On completing his *habilitation*, he was appointed Extraordinary Professor of Chemistry at Munich.

in a series of papers between 1894 and 1901.[36] The Dieckmann condensation is a useful method for the production of cyclic ketoesters (e.g., Example **16.43**, where sodium metal is used as the base[37]). However, the reversibility of the reaction limits its use, in practical terms, to the formation of five- and six-membered rings.

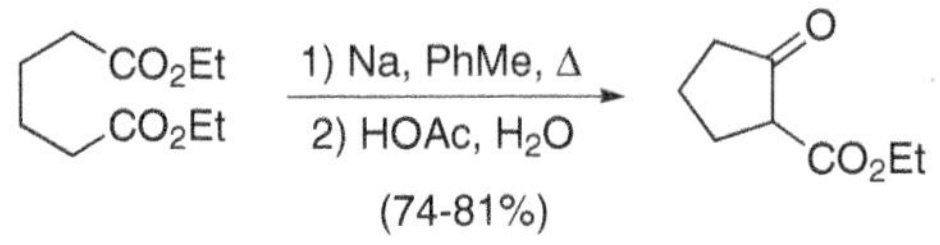

(16.43)

The Thorpe and Thorpe-Ziegler Reactions

A reaction analogous to the Claisen condensation based on nitriles instead of esters is known as the ***Thorpe reaction***.[38] In this case, the product is a β-amino-α,β-unsaturated nitrile. However, the reaction is hampered by the propensity of metalated nitriles to undergo rapid intermolecular proton transfer and for nitriles to undergo oligomerization during the reaction.[39] Therefore, the reaction is usually used in an intramolecular form, known as the ***Thorpe-Ziegler cyclization***,[40] which is the nitrile analog of the Dieckmann condensation. The one case where intermolecular condensation does not lead to oligomerization of the nitrile occurs when the electrophile used is a protected cyanohydrin (e.g., Example **16.44**); this allows the synthesis of densely functionalized products.[41] It is interesting to note that the Dieckmann condensation of the diester corresponding to the dinitrile in Example **16.45** fails to give more than 5% yield of product.

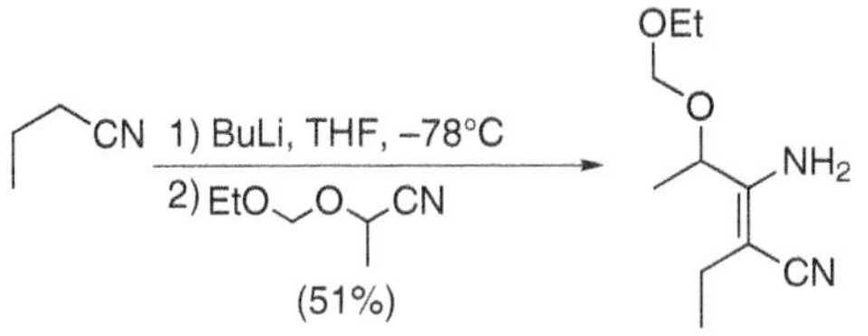

(16.44)[41]

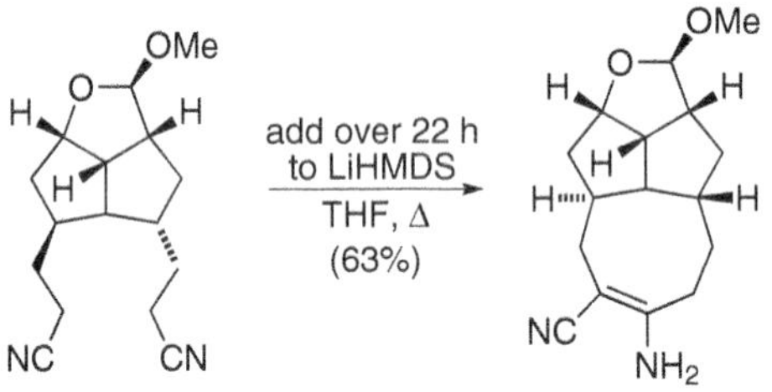

(16.45)[42]

Decarboxylation of β-Ketoacids

Carboxylic acids carrying a carbonyl group in the β position are susceptible to decarboxylation on heating. It is this facility that has made these compounds popular as a means to carry out monoalkylation of ketones by means of the sequence in Example **16.46**: (1) crossed Claisen condensation with diethyl carbonate or a similar compound, (2) alkylation of the

36. Dieckmann, W. *Ber. dtsch. chem. Ges.* **1894**, *27*, 102; **1900**, *33*, 595, 2670; *Justus Liebigs Ann. Chem.* **1901**, *317*, 51, 93.
37. Pinckney, P.S. *Org. Synth.* **1937**, *17*, 30; *Org. Synth.* **1943**, *Coll. Vol. 2*, 116.
38. (a) Baron, H.; Remfry, F.G.P.; Thorpe, J.F. *J. Chem. Soc.* **1904**, *85*, 1726. (b) Schaefer, J.P.; Bloomfield, J.J. *Org. Reactions* **1967**, *2*, 1.
39. Reynolds, G.A.; Humphlett, W.J.; Swamer, F.W.; Hauser, C.R. *J. Org. Chem* **1951**, *16*, 165.
40. For a review of this and other nitrile anion cyclizations, see: Fleming, F.F.; Shook, B.C. *Tetrahedron* **2002**, *58*, 1.
41. Kobayashi, K.; Hiyama, T. *Tetrahedron Lett.* **1983**, *24*, 3509.
42. Luyten, M.; Keese, R. *Tetrahedron* **1986**, *42*, 1687.

resultant β-ketoester, (3) acid hydrolysis of the product β-ketoester, and (4) decarboxylation. The hydrolysis step must generally be carried out under acidic, rather than basic, conditions. The reversibility of the Claisen condensation can lead to fragmentation of the β-ketoester instead of hydrolysis and decarboxylation.

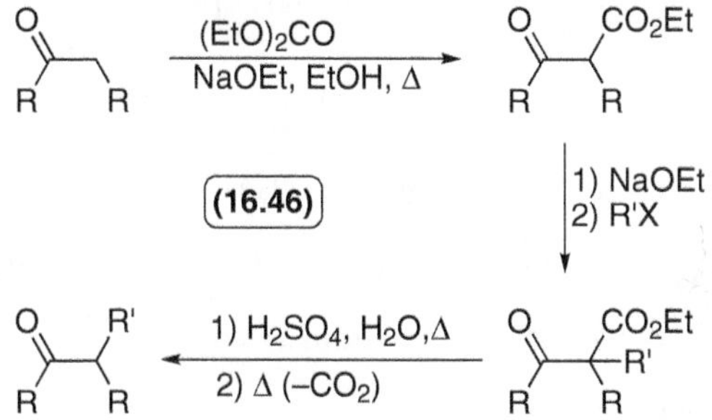

(16.46)

The decarboxylation is known to involve the formation of the enol, and the kinetics of the reaction suggest that the reaction is unimolecular, corresponding to a retro-ene reaction to eliminate carbon dioxide via an activated complex similar to Example **16.47.** There is evidence[43] that both the acid and carboxylate forms of the acid participate in the reaction concurrently.

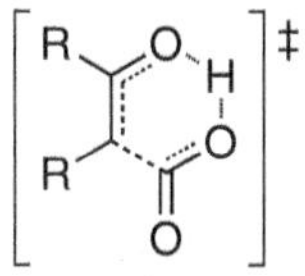

(16.47)

A useful variant of the decarboxylation reaction was introduced by Krapcho in 1982.[44] In this variant of the reaction, the ester is cleaved by halide in a dipolar aprotic solvent at elevated temperature, conditions under which the β-ketoacid decarboxylates as it is formed, as shown in Example **16.48.**[45]

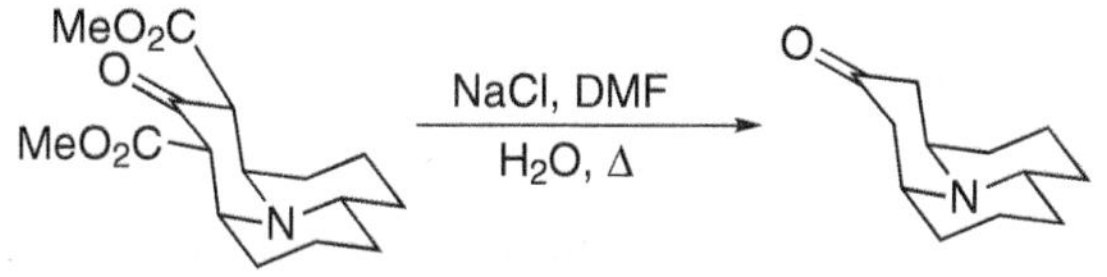

(16.48)

Reaction Synopses

Claisen Condensation

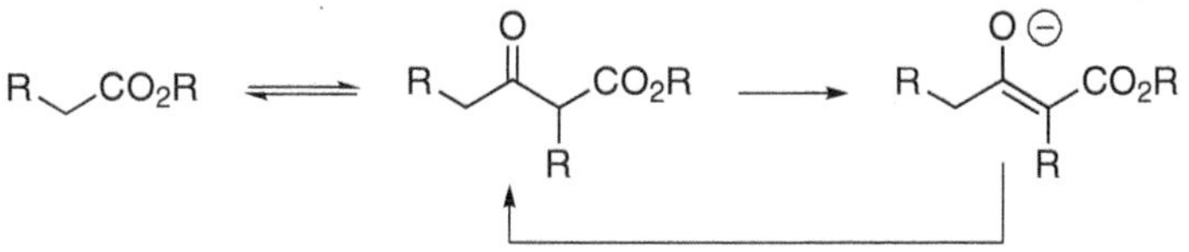

Reagents: (1) NaOEt, EtOH, Δ, then (2) HCl, H_2O; etc.

Equilibrium is favored by deprotonation of β-ketoester product.

Crossed Claisen condensation works best when only one product can deprotonate to shift equilibrium.

(continues)

43. Hall, G.A., Jr. *J. Am. Chem. Soc.* **1949**, *71*, 2691.
44. Krapcho, A.P. *Synthesis* **1982**, 805, 893.
45. Stevens, R.V.; Lee, A.W.N. *J. Am. Chem. Soc.* **1979**, *101*, 7032.

(**Reaction Synopses** continued)

Dieckmann Condensation

n=1, 2, 3, >11

Reagents: (1) NaOEt, EtOH, Δ, then (2) HCl, H_2O; etc.

Reaction works well to form 5- and 6-membered rings—and large rings.

Reaction does not work well to form medium-sized rings.

Thorpe-Ziegler Condensation

Reagents: (1) NaOEt, EtOH, Δ, then (2) HCl, H_2O; etc.

Reaction can be used to form medium-sized rings.

Decarboxylation

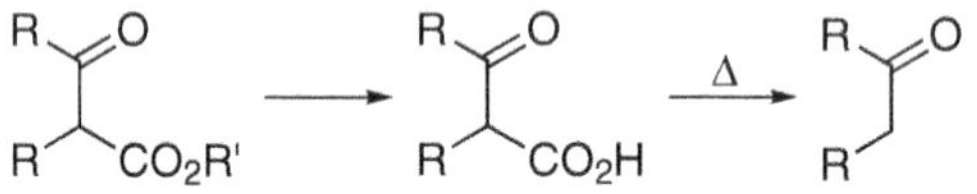

Reagents: (1) H_2SO_4, H_2O, Δ, then (2) Δ; etc. or LiI, DMF, Δ; LiBr, Me_2SO, H_2O, Δ; etc. (Krapcho)

Worked Problem

16-3 Give a reasonable mechanism for the transformation below.

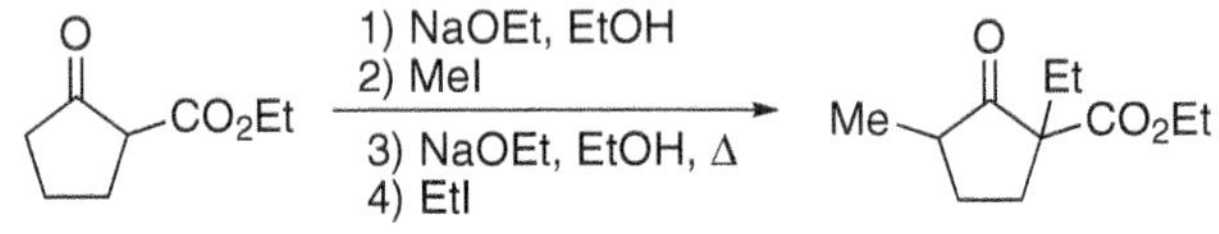

§Answer on the next page.

Problem

16-2 What is the major organic product of each of the following reactions?

(a) 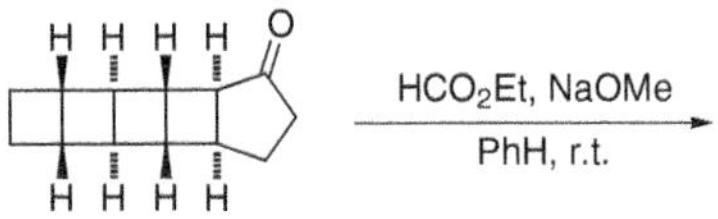

(b) 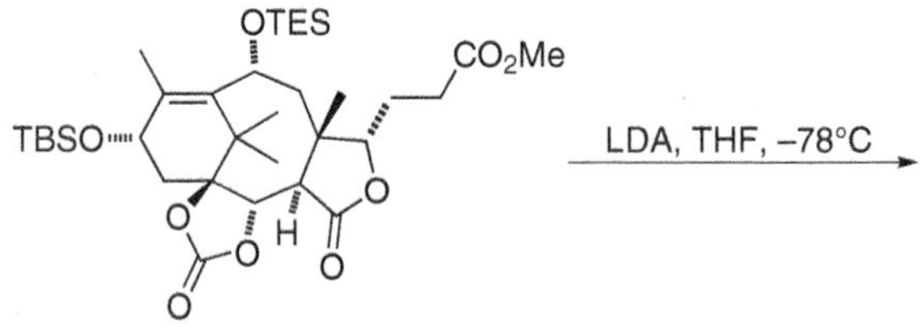

(c) 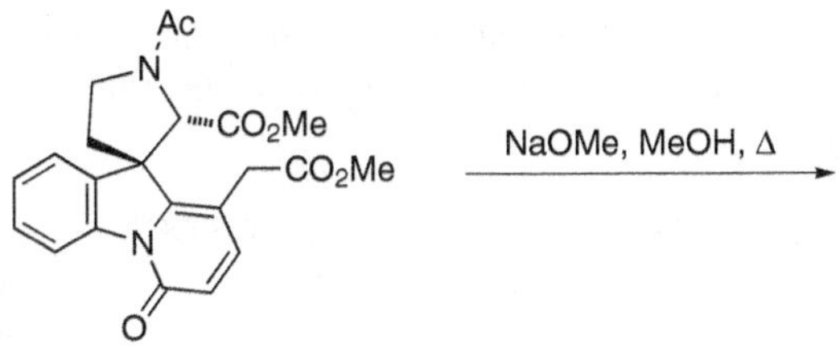

(d) 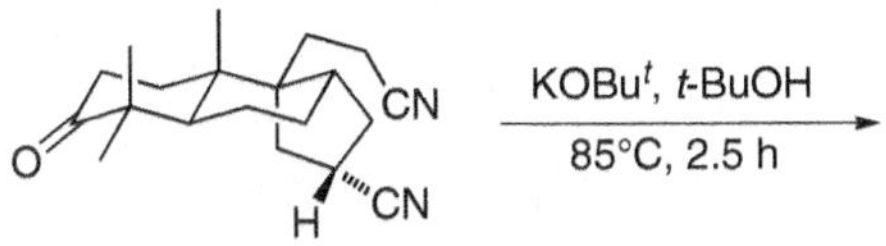

References: (a) *J. Am. Chem. Soc.* **2004**, *126*, 15664. (b) *J. Am. Chem. Soc.* **1994**, *116*, 1597, 1599. (c) *Tetrahedron* **1963**, *19*, 247. (d) *J. Am. Chem. Soc.* **2000**, *122*, 4526.

16.5 The Wittig, Horner-Wadsworth-Emmons, and Related Reactions

The regiochemical control of the location of the alkene π bond in reactions such as the aldol condensation and its analogs is a result of the fact that one product (the conjugated ketone or aldehyde) is stabilized by conjugation and the isomeric products are not. When forming a new alkene π bond in a molecule where there is not this possibility of conjugation, control of regiochemistry is often problematic. A good example is provided by the synthesis of methylene-cyclohexane by an elimination reaction. If one attempts to prepare this alkene by an E1 or E2 reaction (as in Example **16.49**), the major product formed is 1-methyl-cyclohexene (as predicted by Zaitsev's rule). Similarly, acid-catalyzed dehydration of cyclohexylmethanol also gives 1-methylcyclohexene as the major

....

§ Answer to Worked Problem

This reaction is deceptively simple. The first step must result in removal of the most acidic proton—from between the two carbonyl groups—to give anion **B** and *not* the removal of the less acidic α proton from the ketone to give the isomeric anion **B'** (even though this would appear to be the simplest way to get the alkylation pattern required). The alkylation of this anion will now put the methyl group between the carbonyl groups, in ketoester **C**. This means that somehow the ester and ketone carbonyl groups must be exchanged before the final product is obtained. This is, in fact, fairly straightforward when one remembers that the Dieckmann condensation—like the Claisen condensation—is *reversible*, and that *it is the deprotonation of the β-ketoester that renders the reaction effectively irreversible.* Consequently, the addition of ethoxide anion to the first alkylated product will lead to opening of the ring to give enolate **E**, which will undergo proton transfer (either intramolecularly or intermolecularly) to the enolate **F**, which closes to the β-ketoester **H**. Deprotonation of **H** gives the anion **I**, which then reacts with the ethyl iodide to give the observed product **J**.

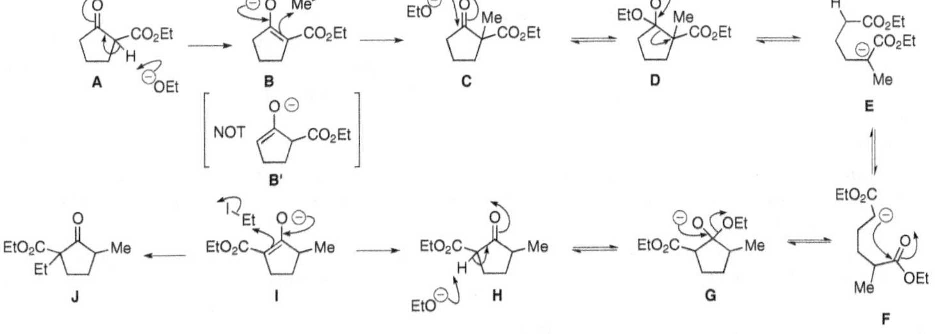

product. In fact, the only elimination reaction guaranteed to give the desired methylenecyclohexane is the base-promoted E2 elimination of a cyclohexylmethyl halide or its equivalent (Example **16.50**).

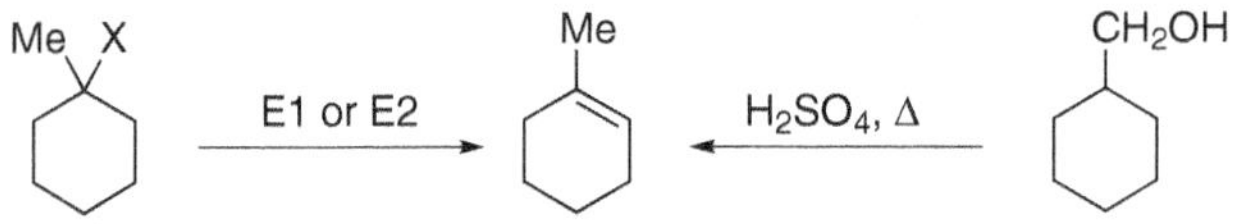

(16.49)

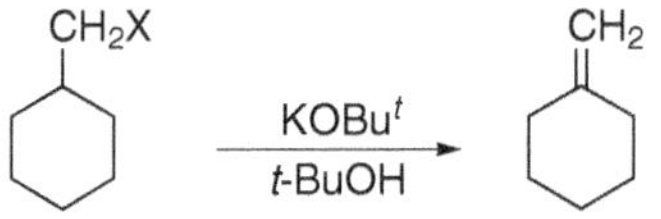

(16.50)

The solution to this problem lies in using a nucleophile that carries with it a leaving group that can be eliminated in an E2 reaction following the initial addition. As reported by Wittig[46] in the mid-1950s,[47] ***phosphonium ylides***, or ***phosphoranes***, are just such a class of nucleophiles. These reagents, which we discussed at some length in Section 11.3, are stabilized carbanions where the stabilization of the carbanion center may be due to d_π-p_π backbonding, or to delocalization of the lone pair on carbon into the low-lying σ* orbitals of the adjacent C—P bonds (Figure 16.3). The product of the reaction between a phosphonium ylide and an aldehyde or ketone is an alkene, where the new C—C double bond is formed between the carbonyl carbon and the ylide carbon of the phosphorane. The reaction has been reviewed.[48]

The first stage of the Wittig reaction involves the formation of a four-membered cyclic compound called an ***oxaphosphetane***. In the classical mechanism for the reaction, the formation of the oxaphosphetane is preceded by addition of the phosphorane to the carbonyl compound to give a ***betaine***.[49] The mechanism of oxaphosphetane formation is still open to interpretation today,[50] and both concerted and stepwise mechanisms have been proposed. Based on extensive nuclear magnetic resonance studies of the reaction, Vedejs has proposed that the oxaphosphetane is formed by an asynchronous, concerted [2 + 2] cycloaddition through a puckered transition state (**16.51**) that minimizes the steric hindrance between the two largest groups in the activated complex (Figure 16.4). In this model, the two larger groups are located to minimize the steric crowding of the phenyl groups on phosphorus, which places the group R of the aldehyde in a quasi-bowsprit position in the

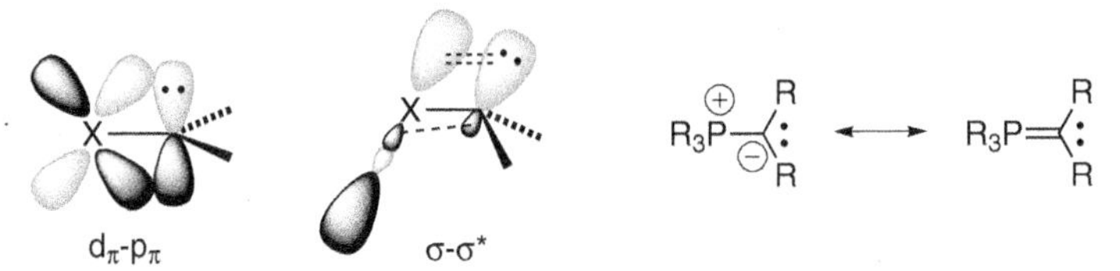

Figure 16.3 The models of the C—X bond in heteroatom-stabilized carbanions. The resonance description of the bonding is shown to the right of the orbital diagrams.

46. **Georg Wittig (1897–1987)** completed his graduate studies at Marburg (*habilitation* 1926) after returning from internment as a prisoner of war in England. He also began his independent career there before moving to to Braunschweig (1932), Freibourg (1937), Tübingen (1944), and Heidelberg (1956). He retired in 1967. Wittig shared the Nobel Prize in Chemistry for 1980 with H.C. Brown. More biographical information is available in his obituary (Tochtermann, W. *Liebigs Ann.* **1997**, I-XXI) and at the web site of the Nobel Foundation.

47. (a) Wittig, G.; Schöllkopf, U. *Chem. Ber.* **1954**, *87*, 1318. (b) Wittig, G.; Haag, W. *Chem. Ber.* **1955**, *88*, 1654.

48. (a) Maercker, A. *Org. React.* **1965**, *14*, 270. (b) Maryanoff, B.E.; Reitz, A.B. *Chem. Rev.* **1989**, *89*, 863. (c) Edmonds, M.; Abell, A. In Takeda, T., Ed. *Modern Carbonyl Olefination* (Wiley-VCH: Weinheim, 2004); 1.

49. Reviews: (a) Trippett, S. *Quart. Rev.* **1963**, *17*, 406. (b) Trippett, S. *Pure Appl. Chem.* **1964**, *9*, 255. (c) Wittig, G. *Pure Appl. Chem.* **1964**, *9*, 245.

50. For studies of the mechanism of the Wittig reaction, see: (a) Vedejs, E.; Marth, C.F.; Ruggeri, R. *J. Am. Chem. Soc.* **1988** *110*, 3940. (b) Vedejs, E.; Marth, C.F. *J. Am. Chem. Soc.* **1988**, *110*, 3948. (c) Vedejs, E.; Meier, G.P.; Snoble, K.A.J. *J. Am. Chem. Soc.* **1981**, *103*, 2823. (d) Vedejs, E.; Fleck, T.J. *J. Am. Chem. Soc.* **1989**, *111*, 5861. (e) Vedejs, E.; Marth, C.F. *J. Am. Chem. Soc.* **1990**, *112*, 3905.

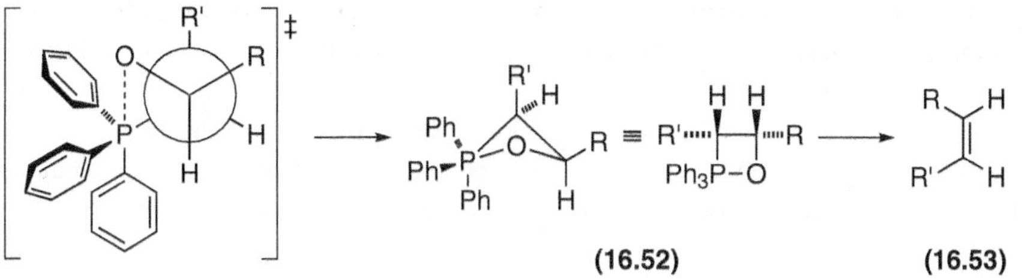

(16.51)

(16.52)

(16.53)

Figure 16.4 Rationalizing the stereochemistry of the concerted mechanism for the Wittig reaction

four-membered ring, and the group R' of the phosphorane in a quasi-flagpole position, *anti* to one phenyl group. The net result of this transition state model is that the *Z* alkene is predicted to be the major product of the reaction, which is what is observed. In one particularly interesting application of the Wittig reaction, White used the phosphorane first as a nucleophile to displace an allylic halide and then converted the resultant phosphonium salt to the phosphorane. He then carried out a Wittig olefination with that phosphorane to obtain the *Z* alkene during a synthesis of a number of related epothilones **(16.54)**.[51]

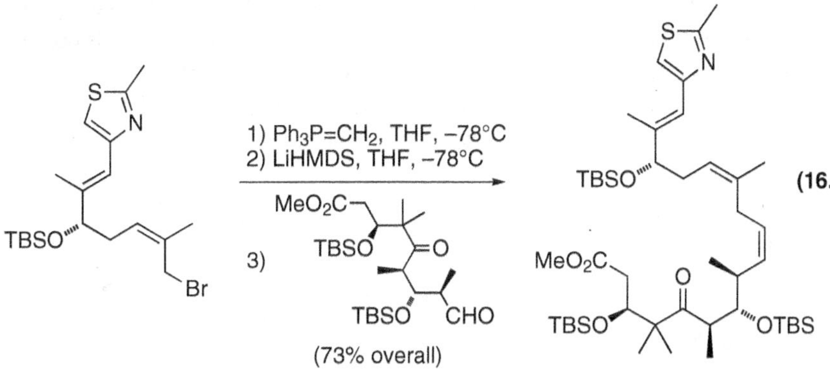

(16.54)

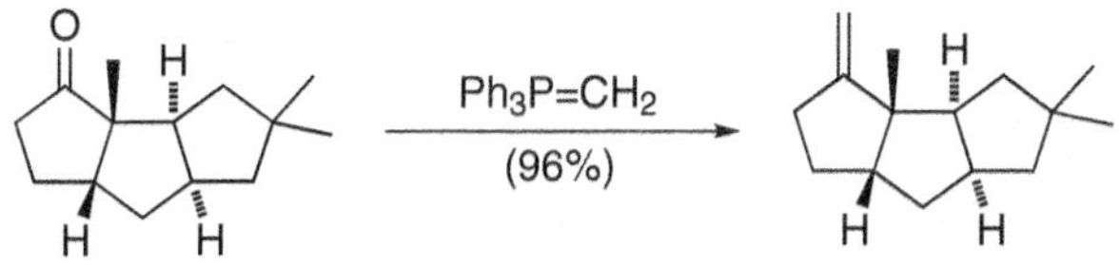

(16.55)[52]

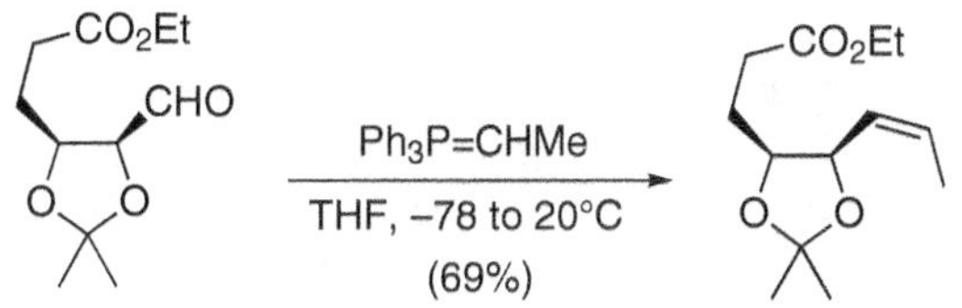

(16.56)[53]

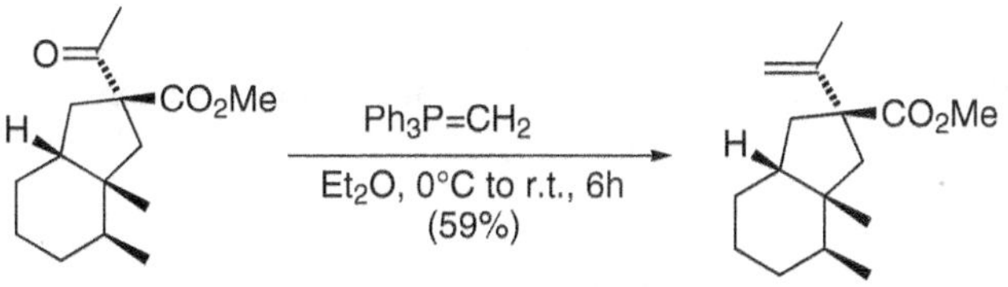

(16.57)[54]

The Wittig reaction has become the method of choice for forming alkenes. Its synthetic importance is due to its generality and to the fact that the alkene is formed with the double bond located in a predictable position—between the carbonyl carbon of the

51. White, J.D.; Carter, R.G.; Sundermann, K.F.; Wartmann, M. *J. Am. Chem. Soc.* **2001**, *123*, 5407.
52. Ramig, K.; Kuzemko, M.A.; McNamara, K.; Cohen, T. *J. Org. Chem.* **1992**, *57*, 1968.
53. Calo, F.; Richardson, J.; Barrett, A.G.M. *Org. Lett.* **2009**, *11*, 4910.
54. Srinivasa Reddy, D. *Org. Lett.* **2004**, *6*, 3345.

aldehyde or ketone and the carbanion carbon of the ylide. As illustrated by Examples **16.55** and **16.57**, it remains one of the best methods for the introduction of a terminal methylene group into a molecule.

When the aldehyde can form a cyclic hemiacetal, the compound will preferentially exist as the hemiacetal if the ring is five- or six-membered. Even so, when such aldehydes are treated with a phosphorane, the product formed is just what one would predict if the aldehyde existed as the open-chain hydroxyaldehyde form. Note, again, that the stereochemistry of the alkene double bond is *Z*, just as one would predict for the reaction of the open-chain hydroxyaldehyde with the phosphorane—as shown by a step in the synthesis of thromboxane B_2 (reaction **16.58**).[55]

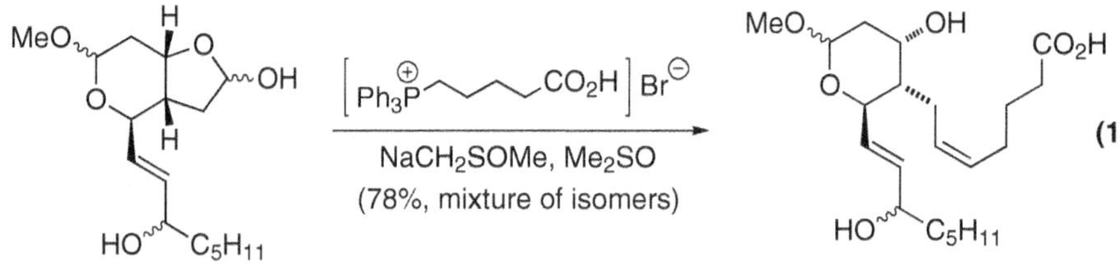

(16.58)

The ylides we have discussed to this point are stabilized only by the adjacent phosphonium group and are known as ***unstabilized ylides***. If the carbanion center is conjugated with an electron-withdrawing group such as a carbonyl group, a ***stabilized ylide*** is obtained. This relatively small change in the electronic structure of the ylide leads to profound changes in the outcome of the reaction: stabilized ylides strongly favor the formation of the *E* alkene, as illustrated by Example **16.59**.[56]

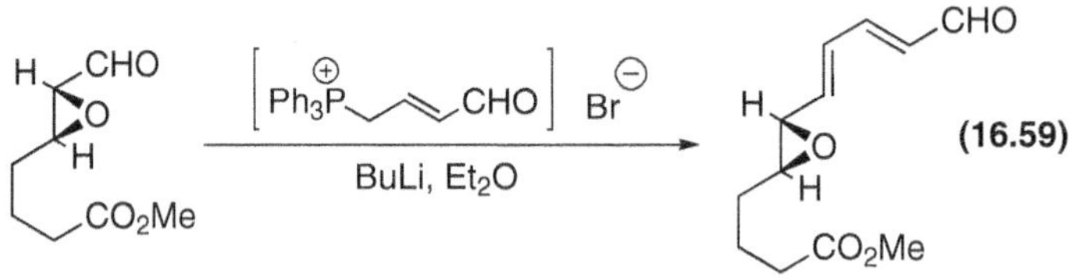

(16.59)

The Schlosser Modification

The propensity of unstabilized ylides to give the *Z* alkene can be reversed by carrying out the reaction in the presence of an excess of a soluble salt and one equivalent of an alkyllithium (preferably phenyllithium) at low temperature (where the fragmentation of the oxaphosphetane occurs only very slowly). This reaction was first described by Schlosser and Christmann.[57]

The rationalization of the Schlosser modification (Figure 16.5) is that adding the alkyllithium to the oxaphosphetane (**16.60**) leads to the formation of a β-oxidophosphorane (**16.61**) that on treatment with an electrophile (which may be a proton[58] or another electrophile[59]) gives the stereoisomeric oxaphosphetane (**16.62**), This then fragments normally to give the *E* alkene. The effect may be exaggerated by the addition of a soluble lithium salt, which is believed to open the oxaphosphetane to give a lithium betaine (**16.64**) that does not

55. Nelson, N.A.; Jackson, R.W. *Tetrahedron Lett.* **1976**, 3275.
56. Baker, S.R.; Clissold, D.W.; McKillop, A. *Tetrahedron Lett.* **1988**, *29*, 991.
57. (a) Schlosser, M.; Christmann, K.F. *Angew. Chem. Int. Ed. Engl.* **1966**, *5*, 126.
For reviews, see: (b) Gosney, I.; Rowley, A.G. In Cadogan, J.I.G, Ed. *Organophosphorus Reagents in Organic Synthesis* (Academic Press: New York, 1979) 17. (c) Maryanoff, B.E.; Reitz, A.B. *Chem. Rev.* **1989**, *89*, 863. (d) Vedejs, E.; Peterson, M.J. *Top. Stereochem.* **1994**, *21*, 1.
58. Wang, Q.; Deredas, D.; Huynh, C.; Schlosser, M. *Chem. Eur. J.* **2003**, *9*, 570.
59. (a) Schlosser, M.; Christmann, K.F. *Synthesis* **1969**, 38. (b) Corey, E.J.; Yamamoto, H. *J. Am. Chem. Soc.* **1970**, *92*, 226, corrigendum, 3523. (c) Hodgson, D.M.; Arif, T. *J. Am. Chem. Soc.* **2008**, *130*, 16500. (d) Hodgson, D.M.; Arif, T. *Org. Lett.* **2010**, *12*, 4204.

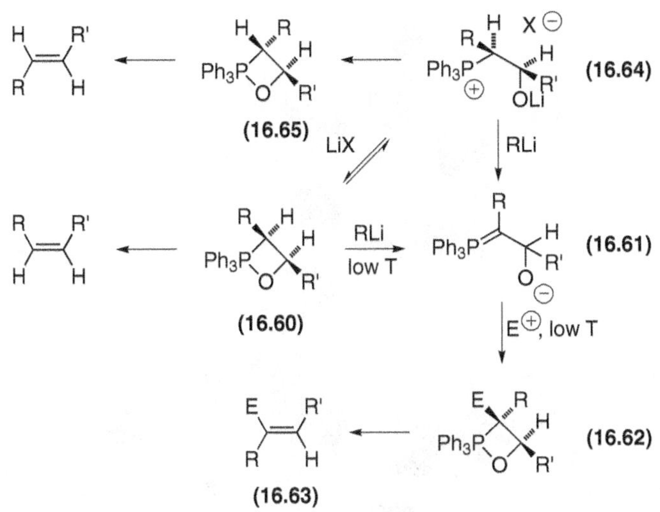

Figure 16.5 The possible outcomes of the Schlosser modification of the Wittig reaction

proceed to the alkene but can then be deprotonated to give the β-oxidophosphorane (**16.61**). Reprotonation of the β-oxidophosphorane, or trapping with an aldehyde[59b] and subsequent reprotonation, gives the oxaphosphetane (**16.65**). It then fragments to the alkene with the alkyl groups from the phosphorane and the aldehyde *trans* to each other. The same stereochemical outcome is obtained when the electrophile is a bromomethyl ester.[59d]

When the β-oxidophosphorane is trapped by a halogen donor (e.g., Example **16.66**) or by a carbon electrophile *larger than formaldehyde* (e.g., Example **16.67**), *E*-haloalkenes are obtained unless the original ylide was ethylidenetriphenylphosphorane. The origins of this size-dependent reversal of stereochemistry on changing the electrophile are not yet understood, but the reaction does provide an excellent source of *E*-haloalkenes by this reaction.[59c]

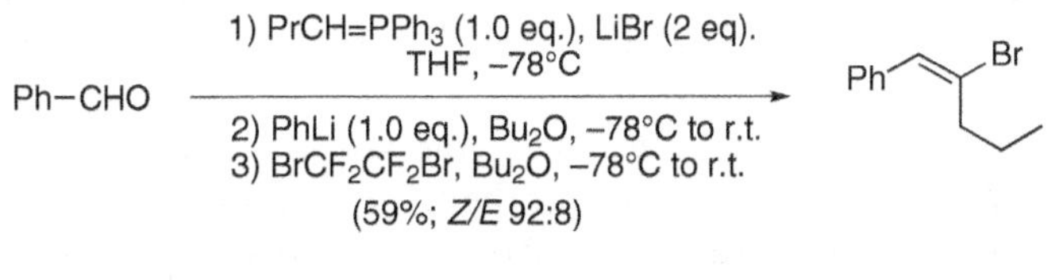

(16.66)

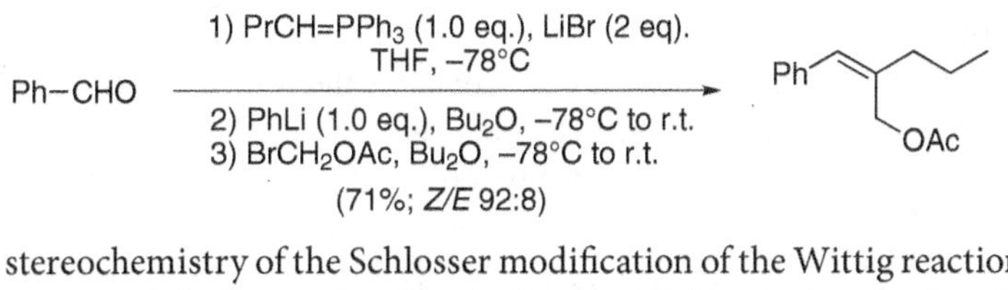

(16.67)

The stereochemistry of the Schlosser modification of the Wittig reaction is provided by the computational finding, using density functional theory, that the β-oxidophosphorane is not computationally stable at the B3LYP/6-311++G(2df,p) level, but minimizes to the oxaphosphetane anion. In that anion (Figure 16.6), the four-membered ring is close to planar, with the oxygen atom and one of the phenyl groups in the axial positions on the pentacoordinate phosphorus. The two equatorial phenyl groups are arranged approximately symmetrically about the plane of the four-membered ring. In this structure, the nucleophilic carbon is rigorously planar, so the adjacent alkyl group (derived from the aldehyde) directs the electrophile to the opposite face of the oxaphosphetane, which leads to the alkyl group from the aldehyde and the electrophile becoming *trans* to each other in the product alkene.[60]

60. Lewis, D.E. unpublished results.

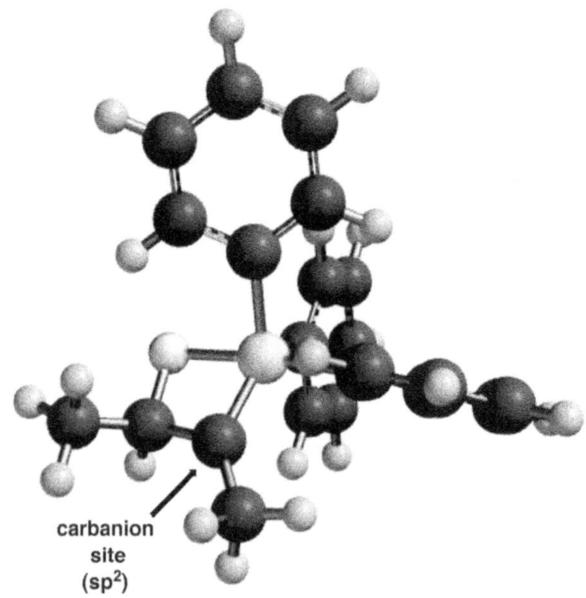

Figure 16.6 The computational model of an oxaphosphetane conjugate base

The Horner-Wadsworth-Emmons Reaction

One practical drawback of the Wittig reaction is the fact that the by-product, triphenylphosphine oxide, can be difficult to remove from the alkene. For this reason, considerable effort went into the development of alternative alkene-forming reactions in which the product was much easier to purify. One way to accomplish this was by means of an anionic phosphorus reagent that would give an anionic, water-soluble phosphorus-containing product. In addition, many ylides, especially stabilized ylides, are not as nucleophilic as one might expect, which also limits their usefulness. A carbanion stabilized by an adjacent neutral phosphorus-based group was expected to be a stronger nucleophile and this was, in fact, found to be the case.

The most widely used of this type of variant of the Wittig reaction is the ***Horner-Wadsworth-Emmons*** reaction, often referred to as the ***HWE reaction***,[61] in which the nucleophile is the anion of a phosphonate ester instead of a phosphorus ylide (Examples **16.68** and **16.69**). The original reaction was described by Horner,[62] but the scope of reaction was much expanded by Wadsworth and Emmons.[63] The reaction was originally known as the Wadsworth-Emmons reaction, but today it bears all three names. In contrast to the reactions of unstabilized phosphorus ylides, which tend to give the *Z* alkene, the HWE reaction usually gives the *E* isomer of the alkene as the predominant product (e. g., Example **16.70**).

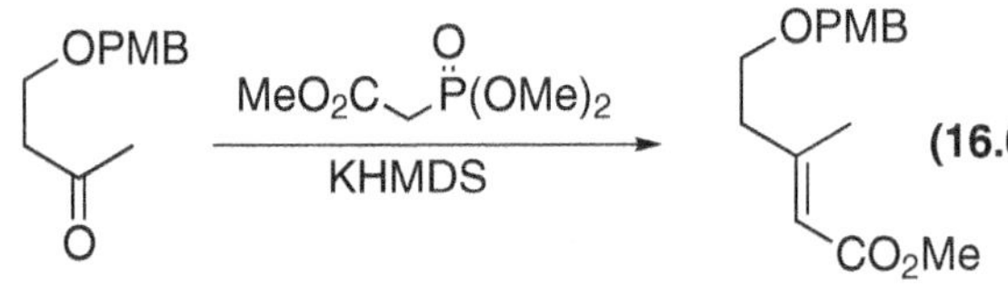

(16.68)[64]

61. Reviews: (a) Wadsworth, W. S., Jr. *Org. React.* **1977**, *25*, 73. (b) Boutagy, J.; Thomas, R. *Chem. Rev.* **1974**, *74*, 87. (c) Kelley, S.E. In. Trost, B.M.; Fleming, I., Eds. *Comprehensive Organic Synthesis* (Pergamon: Oxford, 1991); vol. 1, 729. (d) Maryanoff, B.E.; Reitz, A.B. *Chem. Rev.* **1989**, *89*, 863. (e) Walker, B.J. In Cadogan, J.I.G., Ed. *Organophosphorus Reagents in Organic Synthesis* (Academic Press: New York, 1979), 155. (f) Nicolaou, K.C.; Härter, M.W.; Gunzner, J.L.; Nadin, A. *Liebigs Ann./Recueil* **1997**, 1283.

Biographical sketches of Horner, Wadsworth, and Emmons can be found in Section 11.3 of this book.

62. (a) Horner, L.; Hoffmann, H.M.R.; Wippel, H.G. *Chem. Ber.* **1958**, *91*, 61. (b) Horner, L.; Hoffmann, H. M. R.; Wippel, H. G.; Klahre, G. *Chem. Ber.* **1959**, *92*, 2499.

63. Wadsworth, W.S., Jr.; Emmons, W.D. *J. Am. Chem. Soc.* **1961**, *83*, 1733; *Org. Synth.* **1973**, *Coll. Vol.* *5*, 547.

64. Menche, D.; Hassfels, J.; Li, J.; Mayer, K.; Rudolph, S. *J. Org. Chem.* **2009**, *74*, 7220.

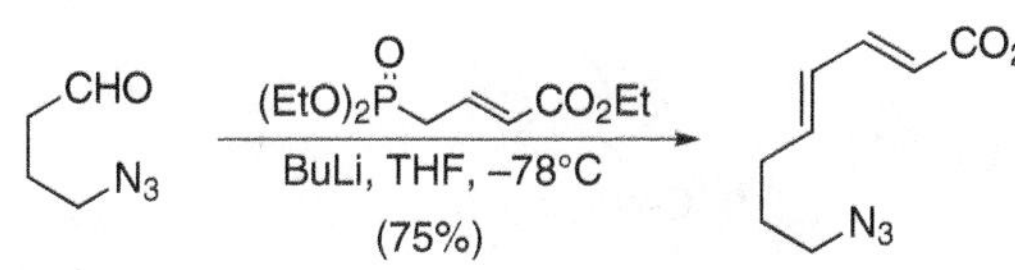

(16.69)[65]

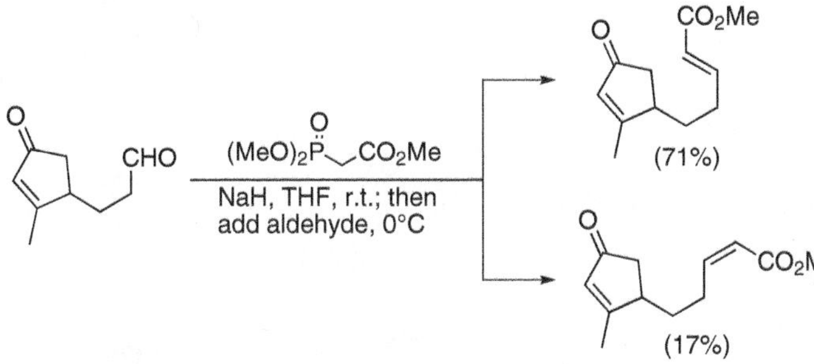

(16.70)[66]

Phosphorus-based Condensation Reactions to Give Alkynes

The reaction between an α-diazophosphonate and a carbonyl compound in the presence of a base gives an alkyne (e.g., Example **16.71**[67]). This reaction is generally known today as the ***Seyferth-Gilbert homologation***.[68] The strongly basic conditions of this reaction can be circumvented by using the ***Bestmann reagent***, which allows the reaction to be carried out in methanol solution, using potassium[69] or cesium[70] carbonate as the base (Example **16.72**[71]). Less reactive aldehydes require the use of cesium carbonate. The mechanism of the Bestmann variant of the reaction is shown in Figure 16.7. The addition of the Bestmann reagent (**16.73**) to the carbonyl compound gives a zwitterion (**16.74**) that closes to the oxaphosphetane (**16.75**). Fragmentation of the oxaphosphetane gives the vinyldiazonium ion **16.76**, which then undergoes a fragmentation by base, with concomitant loss of dinitrogen, to give the vinylalkylidene **16.77**. The rearrangement of this carbene then gives the alkyne product (Example **16.78**). The acetyl group, which is lost as methyl acetate, facilitates the addition to the aldehyde or ketone.

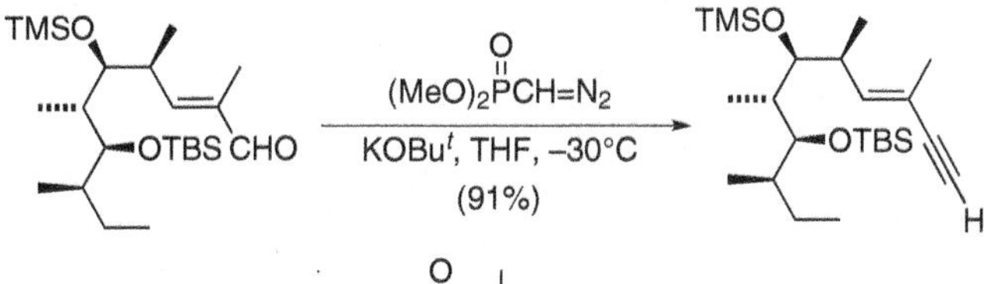

(16.71)

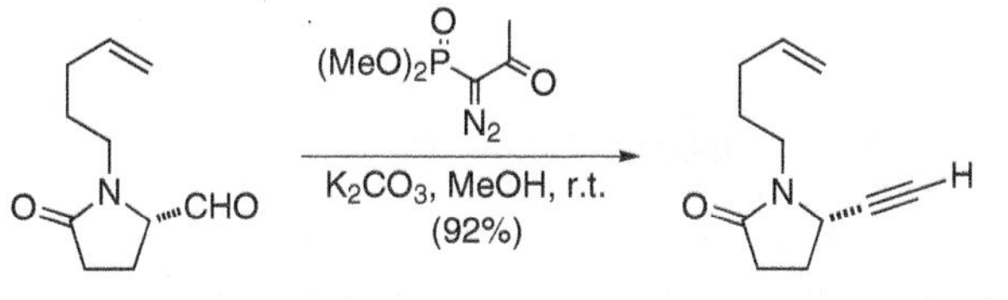

(16.72)

Another reaction sequence that can be used to convert an aldehyde into an alkyne is the ***Corey-Fuchs***[72] or ***Ramirez-Corey-Fuchs*** reaction.[73] In this reaction, the aldehyde is first

---

65. Hudlicky, T.; Frazier, J.O.; Seoane, G.; Tiedje, M.; Seoane, A.; Kwart, L.D.; Beal, C. *J. Am. Chem. Soc.* **1986**, *108*, 3755.

66. Snowden, R.L. *Tetarhedron* **1986**, *42*, 3277.

67. Marshall, J.A.; Bourbeau, M.P. *J. Org. Chem.* **2002**, *67*, 2751.

68. (a) Seyferth, D.; Marmor, R.S.; Hilbert, P. *J. Org. Chem.* **1971**, *36*, 1379. (b) Gilbert, J.C.; Weerasooriya, U. *J. Org. Chem.* **1982**, *47*, 1837. (c) Brown, D.G.; Velthuisen, E.J.; Commerford, J.R.; Brisbois, R.G.; Hoye, T.R. *J. Org. Chem.* **1996**, *61*, 2540.

69. (a) Müller, S.; Liepold, B.; Roth, G.; Bestmann, H.J. *Synlett* **1996**, 521. (b) Roth, G.; Liepold, B.; Müller, S.; Bestmann, H.J. *Synthesis* **2004**, 59.

70. Bondarenko, L.; Dix, I.; Hinrichs, H.; Hopf, H. *Synthesis* **2004**, 2751.

71. Torssell, S.; Wanngren, E.; Somfai, P. *J. Org. Chem.* **2007**, *72*, 4246.

72. (a) Corey, E.J.; Fuchs, P.L. *Tetrahedron Lett.* **1972**, *13*, 3769. (b) Mori, M.; Tonogaki, K.; Kinoshita, A. *Org. Synth.* **2005**, *81*, 1. (c) Marshall, J.A.; Yanik, M.M.; Adams, N.D.; Ellis, K.C.; Chobanian, H.R. *Org. Synth.* **2005**, *81*, 157.

73. Desai, N.B.; McKelvie, N.; Ramirez, F. *J. Am. Chem. Soc.* **1962**, *84*, 1745.

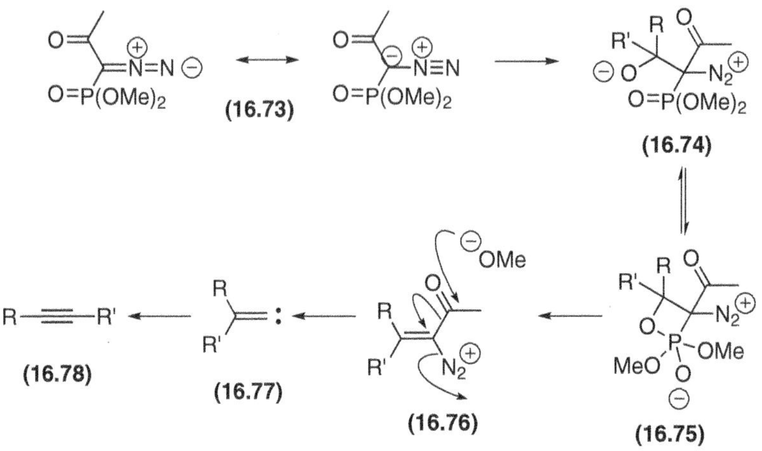

Figure 16.7 The reaction path of the Bestmann modification of the Seyferth-Gilbert homologation

converted into a 1,1-dibromoalkene, which is then treated with two equivalents of butyllithium to give the lithium alkynide. This may be protonated to give the terminal alkyne or trapped with another electrophile to give an internal alkyne, as shown in Example **16.79**,[74] where the original Seyferth-Gilbert reagent was ineffective.[75]

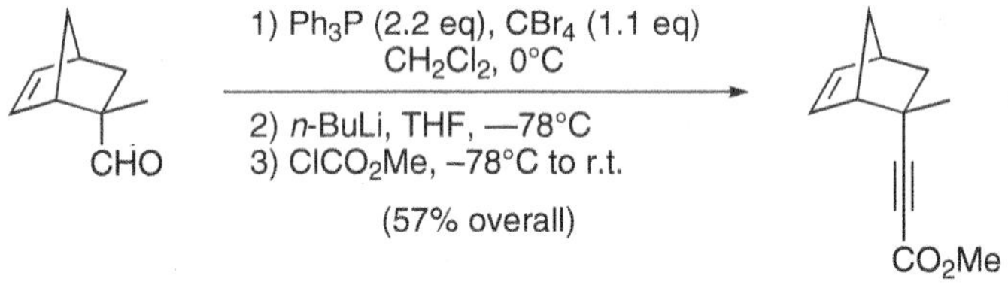

(16.79)

Reaction Synopses

Wittig Reaction

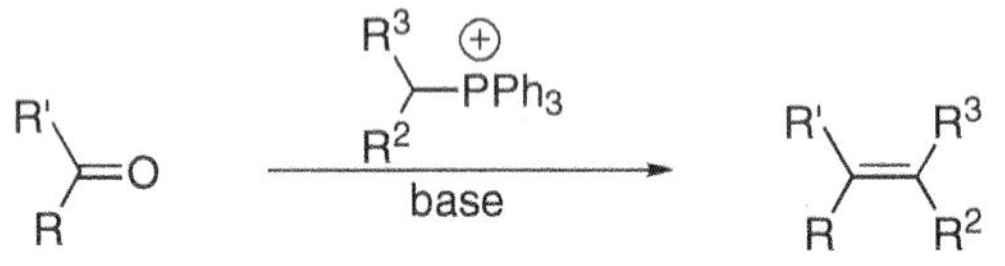

Base: NaH, Me_2SO; BuLi; NaH, THF; LDA, THF; etc.

Stereochemistry: R=alkyl, predominantly *Z*; R = COR, CO_2R, CHO, CN, Ph, etc.; predominantly *E*

Horner-Wadsworth-Emmons Reaction

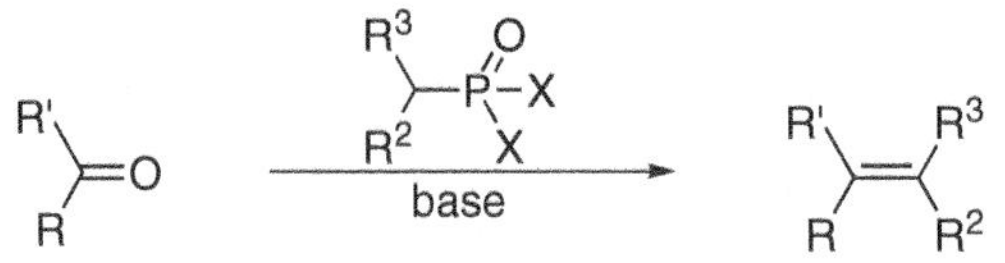

Base: NaH, Me_2SO; BuLi; NaH, THF; LDA, THF; etc.
X = alkyl, Ph (Horner); X = alkoxy (Wadsworth-Emmons)

Stereochemistry: predominantly *E*

---

74. Lewis, D.E.; Rigby, H.L. *Tetrahedron Lett.* **1985**, *26*, 3437.
75. Rigby, H.L. Ph.D. Dissertation, Baylor University, 1985.

Schlosser Modification of the Wittig Reaction

Reagents: (1) $Ph_3P{=}CHR'$, –78°C; (2) BuLi, –78°C; (3) HCl, Et_2O

Stereochemistry: predominantly (almost exclusively) *E*

Intermediate anion that can be intercepted by electrophiles ($H_2C{=}O$; $BrCH_2CO_2Et$; $BrCF_2CF_2Br$; etc.).

Seyferth-Gilbert Homologation

Reagents: $(MeO)_2P(O)CH{=}N_2$, $KOBu^t$, THF; or $(MeO)_2P(O)C({=}N_2)COMe$, K_2CO_3, MeOH

Corey-Fuchs Homologation

Reagents: (1) Ph_3P, CBr_4, CH_2Cl_2; (2) BuLi, THF, –78°C

Alkynide anion may be intercepted by a variety of electrophiles.

Worked Problem

16-4 What is the major organic product of each of the reactions involving isobutyraldehyde in the list below?

(a) (1) $MeCHCH_2PPh_3Br$, *n*-BuLi, THF, –30°C; (2) Me_2CHCHO, –30°C; (3) warm to 20°C

(b) (1) $MeCHCH_2PPh_3Br$, BuLi, LiBr (2 eq), Et_2O, –30°C; (2) Me_2CHCHO, –30°C; (3) warm to 20°C

(c) (1) $MeCHCH_2PPh_3Br$, BuLi, LiBr (2 eq), Et_2O, –30°C; (2) Me_2CHCHO, –78°C; (3) PhLi, Et_2O, –78°C; (4) HCl, Et_2O, –78°C to r.t.

(d) (1) $MeCHCH_2PPh_3Br$, BuLi, LiBr (2 eq), Et_2O, –30°C; (2) Me_2CHCHO, –78°C; (3) PhLi, Et_2O, –78°C; (4) CH_2O, –78°C; (5) HCl, Et_2O, –78°C to r.t.

(e) (1) $MeCHCH_2PPh_3Br$, BuLi, LiBr (2 eq), Et_2O, –30°C; (2) Me_2CHCHO, –78°C; (3) PhLi, Et_2O, –78°C; (4) $BrCF_2CF_2Br$, –78°C to r.t.

§Answers on the next page.

Worked Problem

16-5 Provide the missing reagents a to e in the diagram below. Where there is more than one alternative for a reagent to complete the transformation, give at least two alternative reagents to complete the transformation.

(continues)

(**Reaction Synopses** continued)

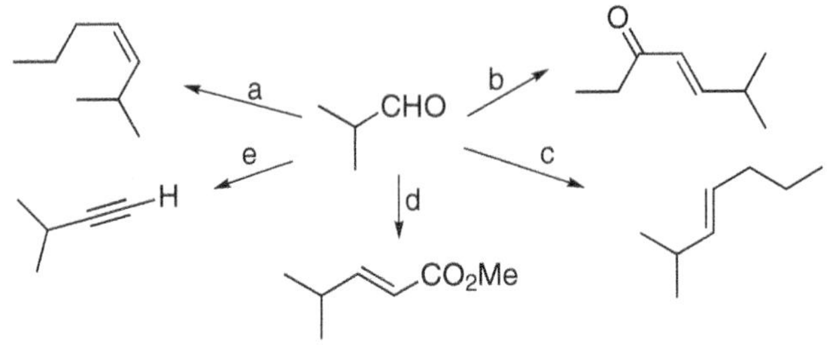

‡Answers below.

Problem

16-3 What is the major organic product expected from each of the reactions below?

(a)

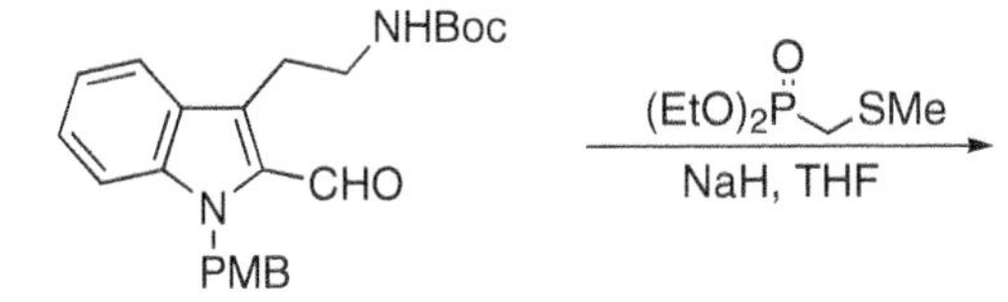

§ Answers to Worked Problem:

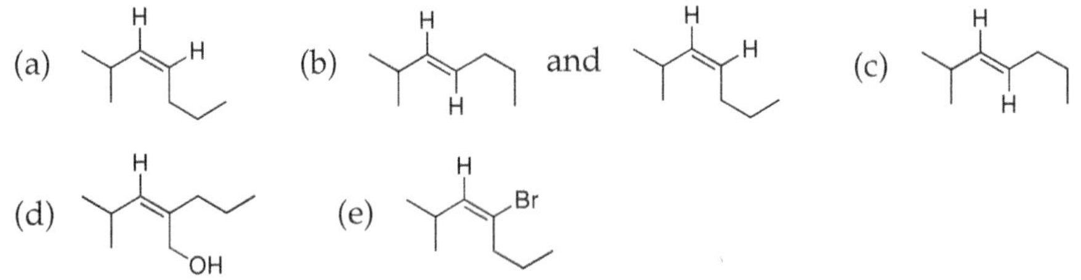

(a) This is a simple Wittig reaction between an aldehyde and an unstabilized ylide. The major product will be the *Z* isomer.

(b) The added lithium bromide will degrade the stereoselectivity of this reaction; the product will be a mixture of the *E* and *Z* isomers.

(c) This is the Schlosser modification of the Wittig reaction. It will give the *E* alkene as the major product.

(d) This is an intercepted Schlosser reaction, where the electrophile is formaldehyde. The major product will be the primary alcohol where the alkyl groups from the aldehyde and the ylide are *trans* to each other—the *Z* isomer in this case.

(e) This is an intercepted Schlosser reaction where the electrophile is a halogen. In these reactions, the product is the *E*-haloalkene.

‡ Answer to Worked Problem

(a) $Ph_3PCH_2CH_2CH_2CH_3Br$, BuLi, THF, –30°C. The *Z* isomer is required, so the Wittig reaction with an unstabilized ylide is the best choice.

(b) $EtCOCH{=}PPh_3$, CH_2Cl_2, r.t. or $EtCOCH_2P(O)(OEt)_2$, NaH, THF. The *E* isomer of the conjugated ketone can be formed by a Wittig reaction with a stabilized ylide or by the Horner-Wadsworth-Emmons reaction.

(c) 1) $Ph_3PCH_2CH_2CH_2CH_3Br$, BuLi, THF, –78°C; 2) BuLi, THF, –78°C; 3) HCl, Et_2O, –78 to –20°C. The *E* isomer is required, so the Schlosser modification of the Wittig reaction with an unstabilized ylide is the best choice.

(d) $MeOCOCH{=}PPh_3$, CH_2Cl_2, r.t. or $MeOCOCH_2P(O)(OEt)_2$, NaH, THF. The *E* isomer of the conjugated ester can be formed by a Wittig reaction with a stabilized ylide or by the Horner-Wadsworth-Emmons reaction.

(e) $(MeO)_2P(O)CH{=}N_2$, $KOBu^t$ or $(MeO)_2P(O)C(Ac){=}N_2$, K_2CO_3, MeOH, or (1) Ph_3P, CBr_4, CH_2Cl_2; (2) BuLi, THF, –78°C; (3) HCl, Et_2O, –78°C to r.t. The alkyne may be obtained using the Seyferth-Gilbert reagent, the Bestmann reagent, or by the Corey-Fuchs procedure.

(b)

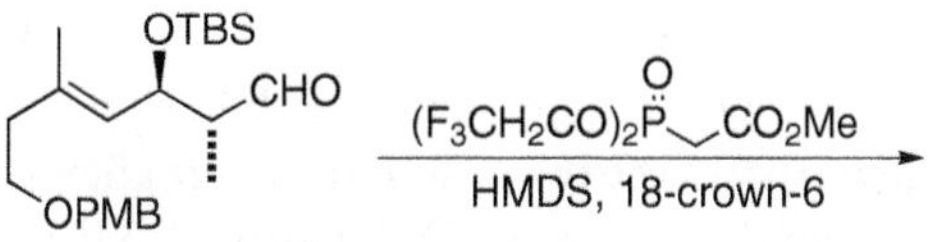

(c)

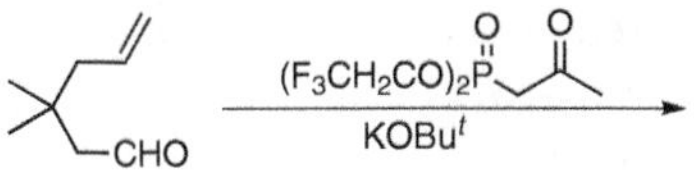

(d)

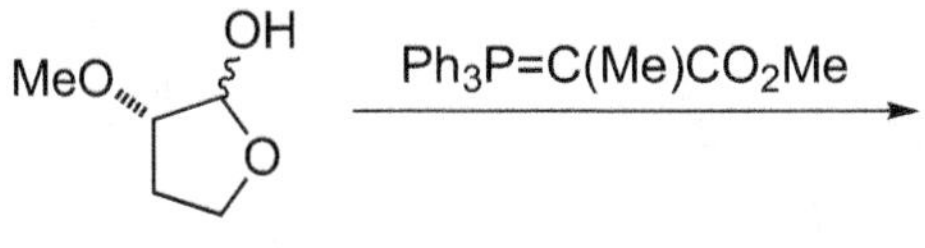

(e)

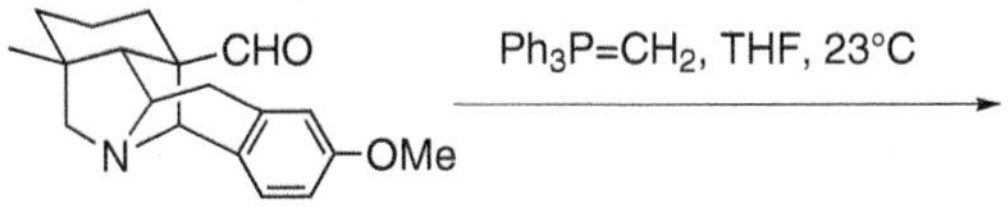

(f)

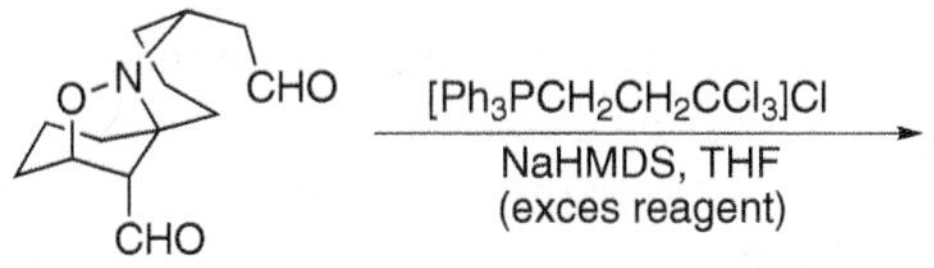

(g)

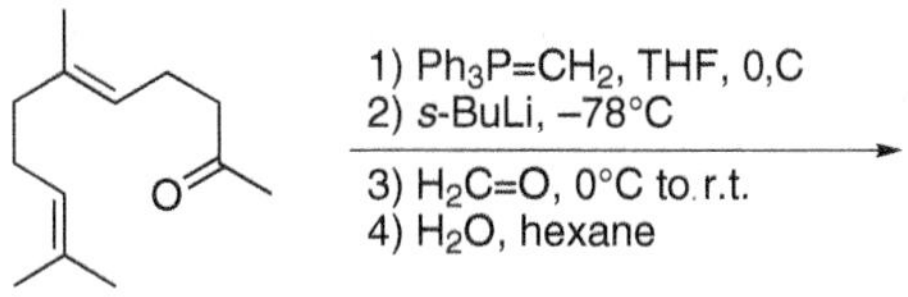

(h)

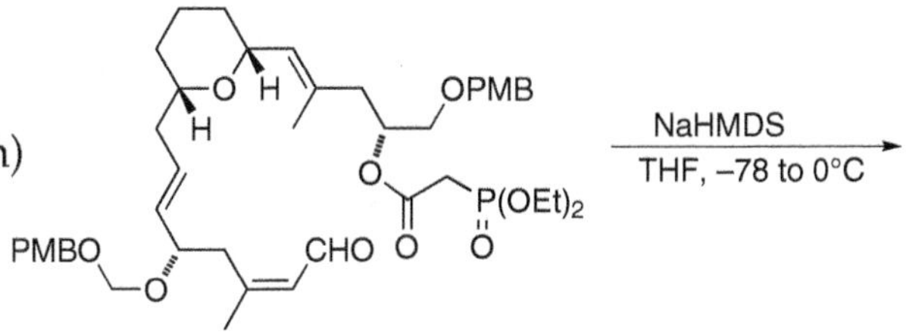

(i)

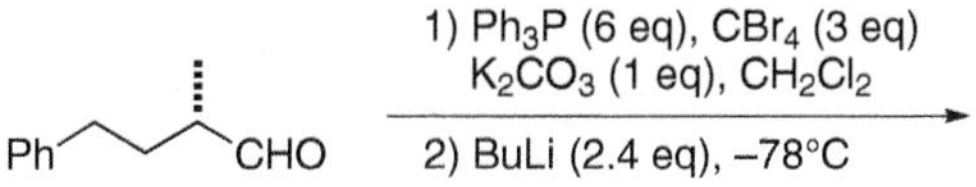

(j)

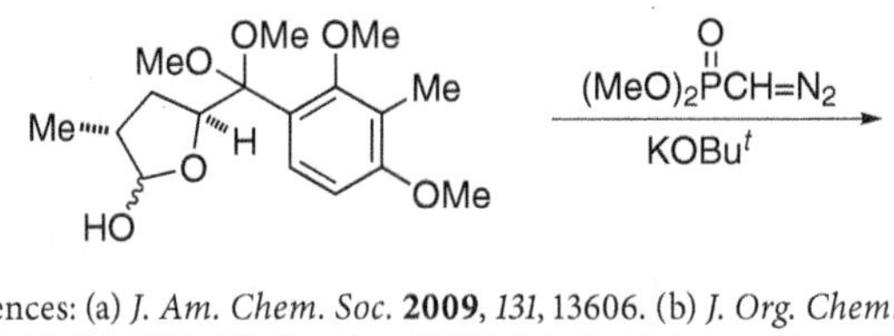

References: (a) *J. Am. Chem. Soc.* **2009**, *131*, 13606. (b) *J. Org. Chem.* **2009**, *74*, 7220. (c) *J. Am. Chem. Soc.* **2008**, *130*, 4421. (d) *Org. Lett.* **2007**, *9*, 4619. (e) *J. Am. Chem. Soc.* **2006**, *128*, 8734. (f) *J. Am. Chem. Soc.*, **2006**, *128*, 12656. (g) *Org. Lett.* **2005**, *7*, 4803. (h) *J. Am. Chem. Soc.* **2003**, *125*, 13022. (i) *Org. Lett.* **2004**, 6, 3245. (j) *Angew. Chem.* **2005**, *117*, 3551.

Silicon-based Olefination: The Peterson Reaction[76]

Much of the chemistry of silicon-stabilized carbanions was discussed in Chapter 11, so this discussion will be rather abbreviated, and you are referred to that chapter for more detail. One of the major uses of these anions is a two-step process involving sequential addition

---

76. (a) Peterson, D.J. *J. Org. Chem.* **1968**, *33*, 780.
Reviews: (b) Birkofer, L.; Stiehl, O. *Top. Curr. Chem.* **1980**, *88*, 58. (c) Ager, D. J. *Synthesis* **1984**, 384. (d) Ager, D. J. *Org. React.* **1990**, *38*, 1. (e) Chan, T.H. *Acc. Chem. Res.* **1977**, *10*, 442.
Biographical information about Peterson is provided in Section 11.3.

to the carbonyl group and elimination to give the alkene; the reaction is known as the ***Peterson olefination***. The initial adduct will fragment under either acidic or basic conditions, and the fact that the stereochemistry of the two reactions differs allows one to obtain a single stereoisomer of the alkene from the Peterson reaction provided that the two diastereoisomeric adducts can be separated. The Peterson olefination is less affected by steric hindrance than the Wittig reaction, so it offers a useful alternative to the Wittig reaction in reactions of hindered ketones (e.g., Example **16.80**[77]).

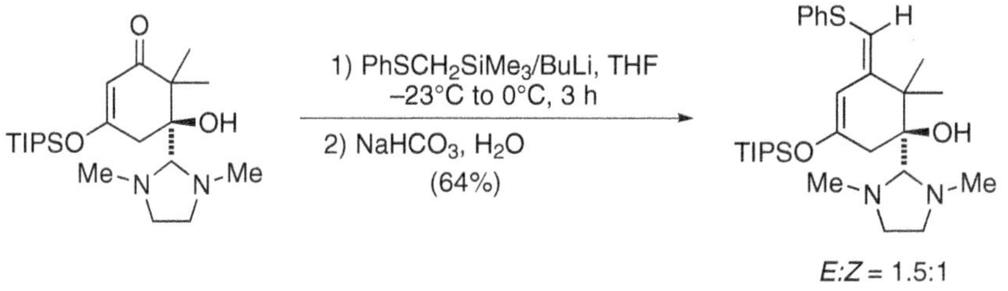

(16.80)

When the β-hydroxyalkylsilane is treated with strong base in the second step of the Peterson olefination, the elimination occurs by a concerted mechanism with *syn* stereochemistry. Conversely, when the adduct is treated with acid, there is a silicon-assisted E1 elimination that leads to overall *anti* elimination. The Peterson olefination has been used recently as a method for the introduction of enal functionality into a molecule by means of an intermediate β-hydroxy-α-silylimine, as shown in Example **16.81**.[78] Note how the trifluoroacetic acid serves both as the acidic reagent to generate the alkene π bond, as well as the acid catalyst for the hydrolysis of the imine.

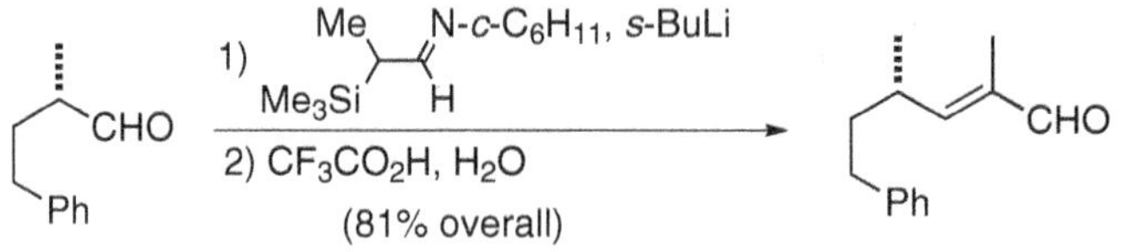

(16.81)

Sulfur-based Olefination

The Julia-Lythgoe reaction[79]

The Julia-Lythgoe olefination is a two-step sequence involving sequential nucleophilic addition of a sulfone anion to the carbonyl compound and reductive elimination of the β-hydroxysulfone. The reaction shows a strong propensity to give the alkene where the two largest groups are *trans* to each other. The mechanism of the reductive elimination is believed to occur by two sequential single-electron transfers, although the mechanism depends on the exact identity of the reducing agent.[80]

A more useful variant of this reaction, that avoids the need for the reducing agent, is the ***Julia-Kocieńsky olefination***,[81] whose mechanism is shown in Figure 16.8. In the reaction, the initial nucleophilic addition of the sulfone anion (**16.82**) to the carbonyl group is followed by intramolecular addition of the alkoxide anion (**16.83**) to the C—N π bond of

---

77. Kusama, H.; Hara, R.; Kawahara, S.; Nishimori, T.; Kashima, H.; Nakamura, N.; Morihira, K.; Kuwajima, K. *J. Am. Chem. Soc.* **2000**, *122*, 3811.

78. Zeng, X.; Zeng, F.; Negishi, E. *Org. Lett.* **2004**, *6*, 3245.

79. (a) Julia, M.; Arnould, D. *Bull.* Soc. *Chim. Fr.* **1973**, 743. (b) Julia, M.; Paris, J.-M. *Tetrahedron Lett.* **1973**, *49*, 4833. (c) Julia, M.; Launay, M.; Stacino, J.-P.; Verpeaux, J.-N. *Tetrahedron Lett.* **1982**, *23*, 2465. (d) Kocienski, P. J.; Lythgoe, B.; Ruston, S. *J. Chem. Soc., Perkin Trans. 1* **1978**, 829. (e) Kocienski , P.J. ; Lythgoe, B.; Waterhouse, I. *J. Chem. Soc., Perkin Trans. 1* **1980**, 1045.

Reviews: (f) Kocieński, P.J. *Phosphorus and Sulfur* **1985**, *24*, 97. (g) Kelley, S.E. In. Trost, B.M.; Fleming, I., Eds. *Comprehensive Organic Synthesis* (Pergamon: Oxford, 1991); vol. 1, 729.

80. Keck, G.E.; Savin, K.A.; Weglarz, M.A. *J. Org. Chem.* **1995**, *60*, 3194.

81. (a) Baudin, J.B.; Hareau, G.; Julia, S.A.; Ruel, O. *Tetrahedron Lett.* **1991**, *32*, 1175. (b) Blakemore, P.R.; Cole, W.J.; Kocieński, P.J.; Morley, A. *Synlett* **1998**, 26.

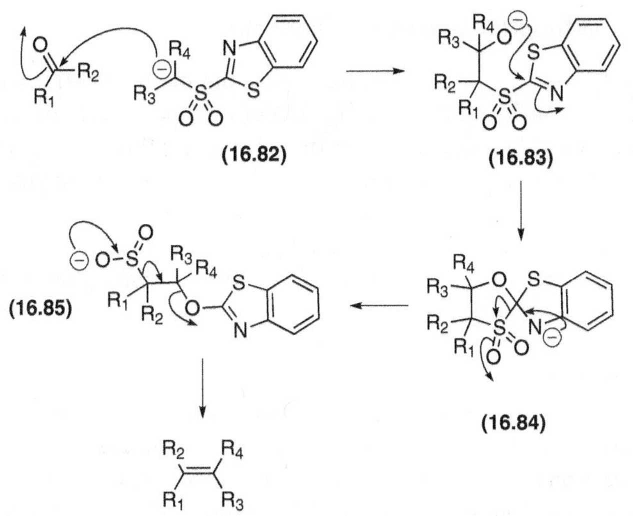

Figure 16.8 Mechanism of the Julia-Kocieński reaction

the heterocycle to give a cyclic orthocarbonic acid derivative (**16.84**) that then eliminates a sulfinic acid as the leaving group. Fragmentation of the β-alkoxysulfinate (**16.85**) then gives the product alkene. The conditions of this reaction allow it to be used with substrates sensitive to reduction, as illustrated by Example **16.86**.[82]

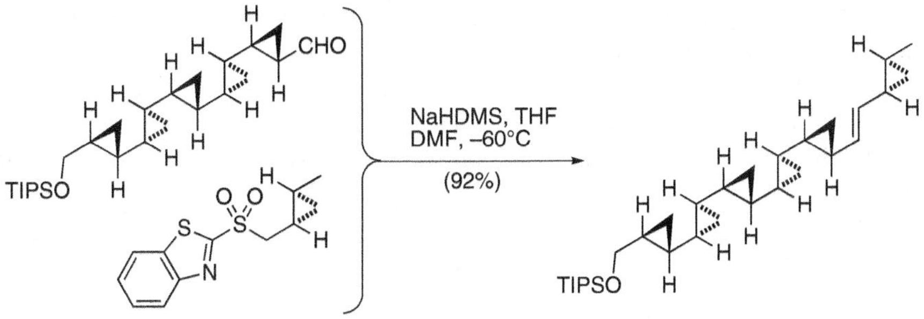

(16.86)

The Ramberg-Bäcklund Reaction[83]

In 1940, Swedish chemists Ludwig Ramberg and Birger Bäcklund reported that the treatment of α-halosulfones with strong base led to the formation of an alkene. This product arises from intramolecular displacement of the halide, followed by extrusion of sulfur dioxide from the resultant thiirane dioxide. The reaction usually yields a mixture of both *E* and *Z* isomers. In the modern procedure, the sulfone is treated with a halogenating agent (often CF_2Br_2) and strong base. The reaction can be applied to the synthesis of both small and large rings, as illustrated by Examples **16.87** and **16.88**.

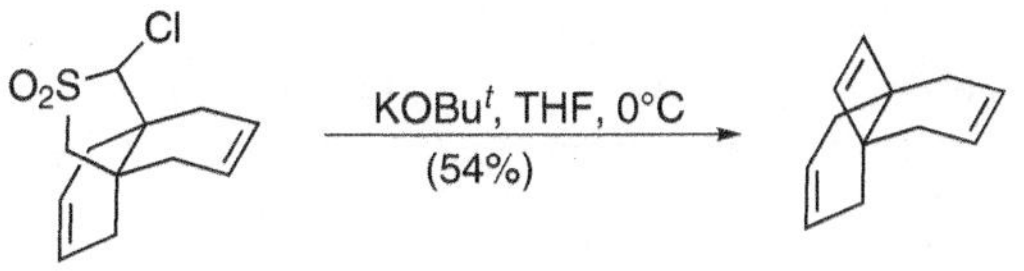

(16.87)[84]

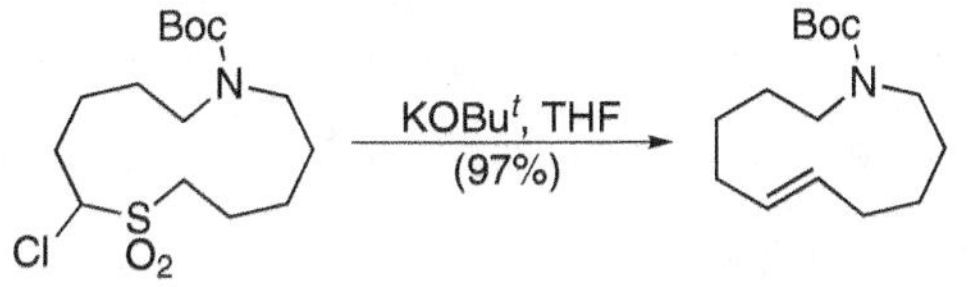

(16.88)[85]

82. Charette, A.B.; Lebel, H. *J. Am. Chem. Soc.* **1996**, *118*, 10327.
83. Review: Paquette, L. *Org. React.* **1977**, *25* 1.
84. Paquette, L.A.; Philips, J.C.; Wingard, R.E., Jr. *J. Am. Chem. Soc.* **1971**, *93*, 4516.
85. MaGee, D.I.; and Beck, E.J. *Can. J. Chem.* **2000**, *78*, 1060.

Transition Metal-based Reagents for Olefination

The strongly basic reaction conditions for generating methylene-triphenylphosphorane or phosphonate anions led to the investigation of methods involving transition metals, especially titanium, as an alternative to the Wittig reaction and its analogs. Two reagents, in particular, have come out of this work. One is the ***Tebbe reagent***,[86] which is converted by a modest Lewis base to a titanium carbenoid derivative that condenses with carbonyl groups to give methylene groups. The other is ***Lombardo's reagent***,[87] which is an aged zinc carbenoid-titanium tetrachloride reagent that performs much the same task.

The Tebbe Reaction

In 1978, Tebbe and coworkers reported a reaction in which a titanium complex reacted with a carbonyl compound to give an alkene in a regiospecific reaction formally analogous to the Wittig reaction (Figure 16.9). The initial complex (**16.89**) is prepared by the reaction between titanocene dichloride and trimethylaluminum in toluene,[88] followed by activation by treatment with a Lewis base (a tertiary amine or an ether solvent such as tetrahydrofuran) to liberate the active Schrock metallocarbene (**16.90**). An alternative method for the preparation of the reagent from titanocene dichloride and methyllithium or methylmagnesium halides gives the ***Petasis reagent***,[89] Cp_2TiMe_2, which loses methane on heating to give the same Schrock metallocarbene as the Tebbe reagent. This carbene complex then adds to the carbonyl group to give an oxatitanocyclobutane (**16.91**); this immediately eliminates the titanocene oxide to give the alkene (**16.92**).[90] Because the titanocene oxide product reacts with the active methylenation reagent to give the heterodimer **16.93**, two equivalents of these reagents are usually required.[91] Although the reaction is cleaner when the purified Tebbe reagent is used, the reaction has been carried out using the reagent generated in situ.[92]

Figure 16.9 Olefination with methylenetitanocene reagents

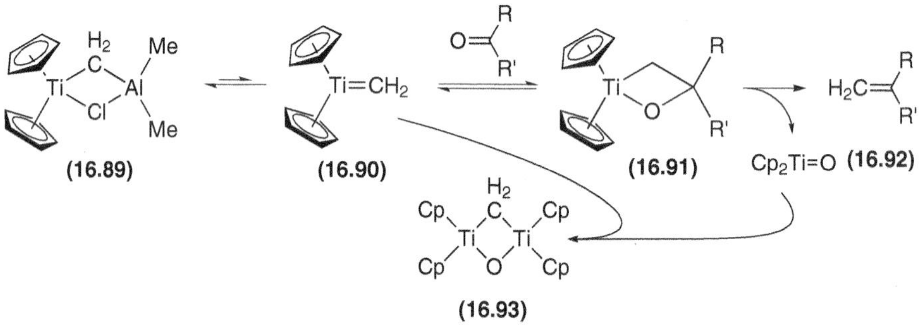

86. (a) Tebbe, F.N.; Parshall, G.W.; Reddy, G.S. *J. Am. Chem. Soc.* **1978**, *100*, 3611.
Reviews: (b) Hartley, R.C.; Li, J.; Main, C.A.; McKiernan, G.J. *Tetrahedron* **2007**, *63*, 4825. (c) Pine, S.H. *Org. React.* **1993**, *43*, 1. (d) Beadham, I.; Micklefield, J. *Curr. Org. Syn.* **2005**, *2*, 231.

87. (a) Lombardo, L. *Tetrahedron Lett.* **1982**, *23*, 4293. (b) Lombardo, L. *Org. Synth.* **1987**, *65*, 81.

88. (a) Herrmann, W.A. *Adv. Organomet. Chem.* **1982**, *20*, 195. (b) Strauss, D.A. "μ-Chlorobis(cyclopentadienyl)(dimethylaluminium)-μ-methylenetitanium." In *Encyclopedia of Reagents for Organic Synthesis.* (John Wiley: London, 2000).

89. (a) Petasis, N.A.; Bzowej, E.I. *J. Am. Chem. Soc.* **1990**, *112*, 6392. (b) Payack, J.F.; Hughes, D.L.; Cai, D.; Cottrell, I.F.; Verhoeven, T.R. *Org. Synth.* **2002**, *79*, 19.

90. (a) Meurer, E.C.; Santos, L.S.; Pilli, R.A.; Eberlin, M.N. *Org. Lett.* **2003**, *5*, 1391. (b) Meurer, E.C.; da Rocha, L.L.; Pilli, R.A.; Eberlin, M.N.; Santos, L.S. *Rapid Commun. Mass Spectrom.* **2006**, *20*, 2626.

91. (a) Hughes, D.L.; Payack, J.F.; Cai, D.; Verhoeven, T.R.; Reider, P.J. *Organometallics* **1996**, *15*, 663. (b) Payack, J.F.; Huffman, M.A.; Cai, D.W.; Hughes, D.L.; Collins, P.C.; Johnson, B.K.; Cottrell, I.F.; Tuma, L.D. *Org. Process Res. Dev.* **2004**, *8*, 256.

92. Cannizzo, L.F.; Grubbs, R.H. *J. Org. Chem.* **1985**, *50*, 2386.

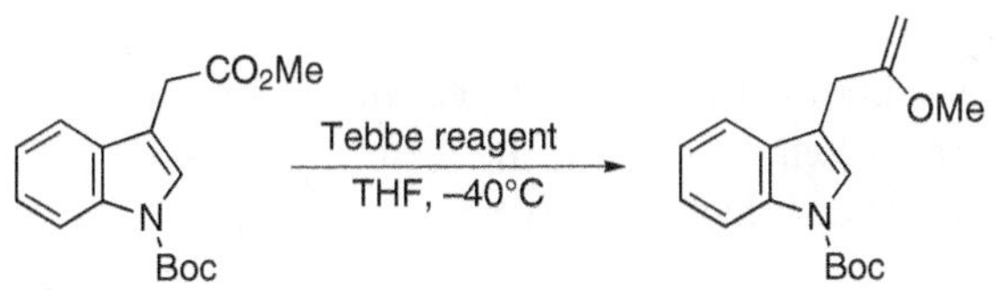

(16.94)

The strongly oxophilic nature of the titanium atom appears to provide the driving force for this reaction, which allows the methylenation of ketones, and aldehydes, but also (in contrast to the Wittig and HWE reactions) the methylenation of esters to enol ethers,[93] as shown in Example **16.94**,[94] and amides to enamines. Generally speaking, the reactivity of carbonyl compounds toward the Tebbe reagent is as expected for reactions with a carbon nucleophile:

$$RCH{=}O > R_2C{=}O > RCO_2R > RCONR_2$$

The mild conditions of the reaction mean that chiral aldehydes and ketones can be converted to the alkene without racemization. It also means that compounds with sensitive functionality can be converted to alkenes under conditions where the Wittig and similar reactions lead to undesired side reactions. Example **16.95**, which is taken from Evans' synthesis of (+)-azaspiracid-1,.[95] illustrates the use of these reagents with such a substrate.

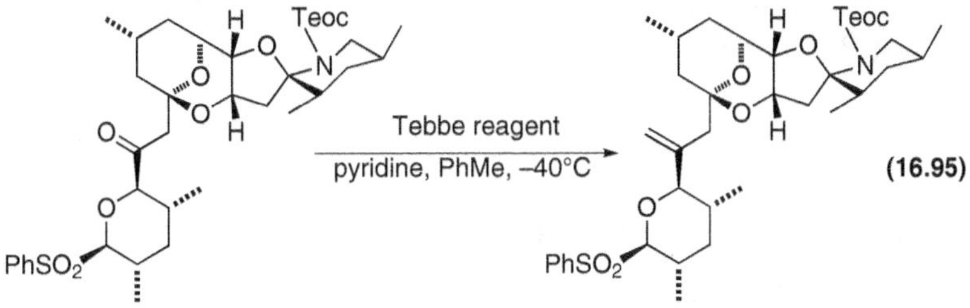

(16.95)

Lombardo's reagent reacts rapidly with ketones to give methylene compounds. Like the Tebbe reagent, it is a very mild reagent that tolerates sensitive functionality well, as shown by Example **16.96**.[96] With unhindered aldehydes, the reaction is often accompanied by pinacol coupling in a McMurry-type reaction, so the reaction is much more useful in reactions with ketones, or with hindered aldehydes, where the pinacol dimerization is slower.

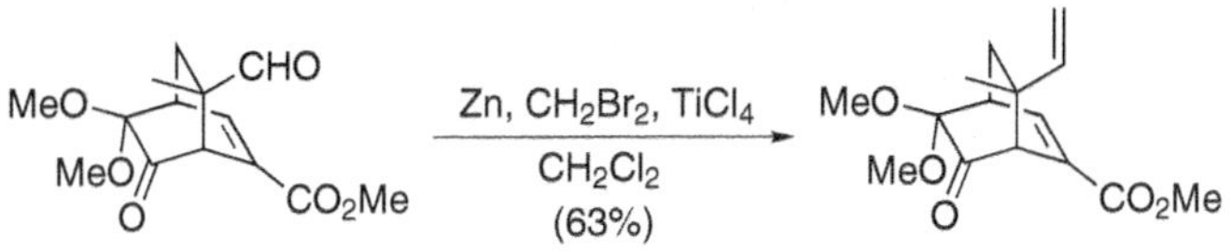

(16.96)

Reaction Synopses

Peterson Olefination

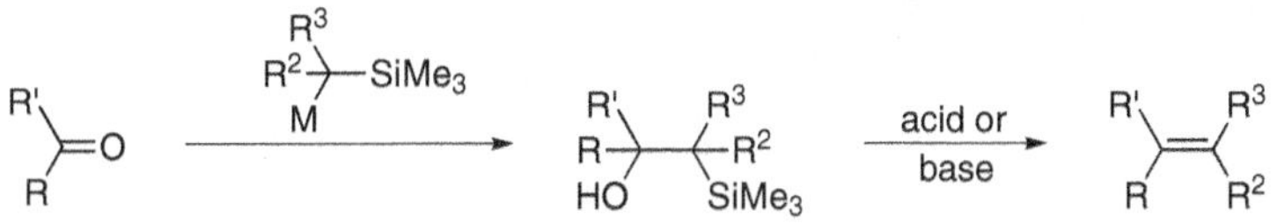

M = Li, Mg, etc.

(continues)

---

93. Pine, S.H.; Kim, G.; Lee, V. *Org. Synth.* **1993**, *Coll. Vol. 8*, 512.

94. Mann, E.; Moisan, L.; Hou, J.-L.; Rebek, J., Jr. *Tetrahedron Lett.* **2008**, *49*, 903.

95. (a) Evans, D.A.; Dunn, T.B.; Kværnø, L.; Beauchemin, A.; Raymer, B.; Olhava, E.J.; Mulder, J.A.; Juhl, M.; Kagechika, K.; Favor, D.A. *Angew. Chem. Int. Ed.* **2007**, *46*, 4698. (b) Evans, D.A.; Kværnø, L.; Dunn, T.B.; Beauchemin, A.; Raymer, B.; Mulder, J.A.; Olhava, E.J.; Juhl, M.; Kagechika, K.; Favor, D.A. *J. Am. Chem. Soc.* **2008**, *130*, 16295.

96. Hsu, D.-S.; Hsu, P.-Y.; Lee, Y.-C.; Liao, C.-C. *J. Org. Chem.* **2008**, *73*, 2554.

(**Reaction Synopses** continued)

Stereochemistry: elimination by base is *syn*; elimination by acid is *anti*. Alkene stereochemistry can be defined by choice of reagent to fragment the β-hydroxysilane.

Julia-Lythgoe Olefination

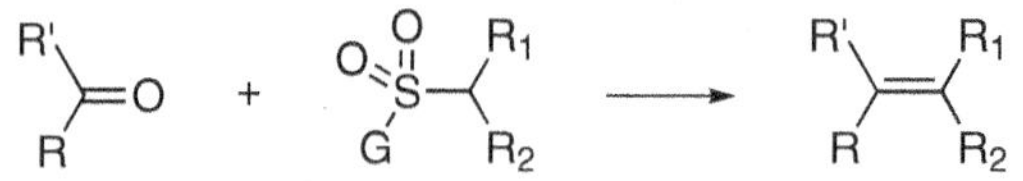

G = Ph or substituted phenyl

Reagents: (1) BuLi, THF; (2) add aldehyde; (3) Na(Hg), THF, ROH; etc.
or (1) BuLi, THF; (2) add aldehyde; (3) SmI_2, DMPU, THF; etc.

Stereochemistry: mainly *E*, with two largest groups *trans* to each other

Julia-Kocieńsky Olefination

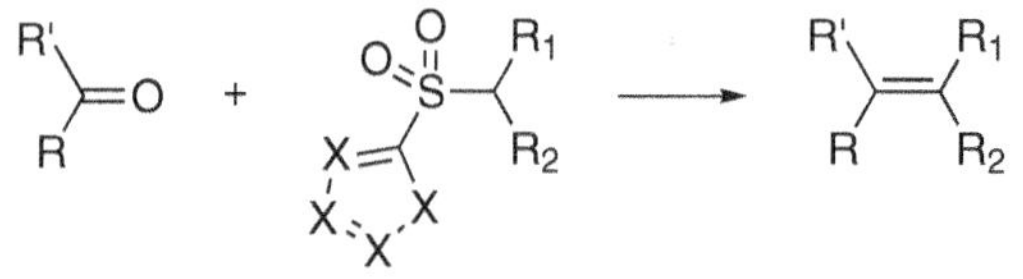

X = N, S (tetrazoles, thiazoles, benzothiazoles)

Reagents: strong, non-nucleophilic base (LDA, BuLi, etc.)

Reduction step: not needed with this reaction

Stereochemistry: as for the Julia-Lythgoe olefination

Ramberg-Bäcklund Reaction

Reagents:

from unsubstituted sulfone: KOBu^t, THF, CF_2Br_2; etc.
from α-halosulfone: KOBu^t, THF

Stereochemistry: mixture of *E* and *Z* isomers obtained

Tebbe, Petasis, and Lombardo Olefinations

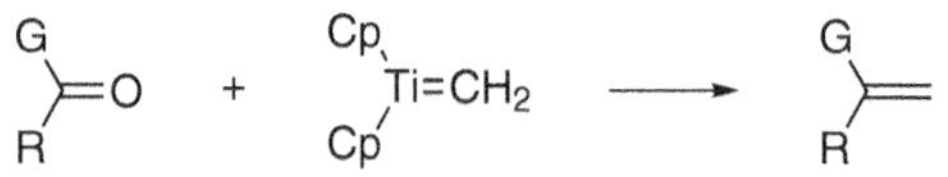

G = H, R', OR', NR'_2, etc.

Reagents:

Tebbe: Cp_2TiCl_2, Me_3Al, PhMe, THF; etc.

Petasis: Cp_2TiMe_2, PhMe, Δ

Lombardo: Zn, CH_2Br_2, $TiCl_4$, PhMe (age 3 days)

Worked Problem

16-6 In step above from the synthesis of azaspiracid-1 (Example **16.95**), what would be the possible problems if the Wittig reagent were used instead of the Tebbe reagent?

§Answer below.

Problem

16-4 What is the major organic product expected from each of the reactions below?

(a) 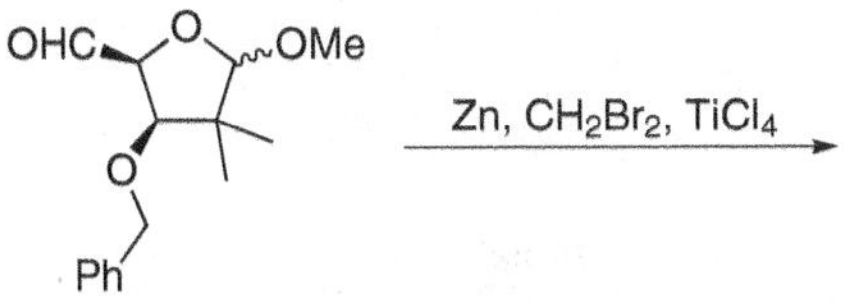

(b) 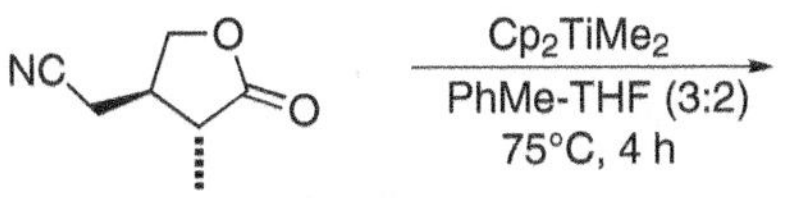

(c) 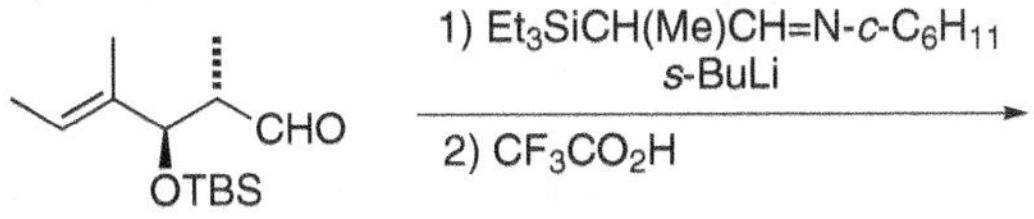

(d) 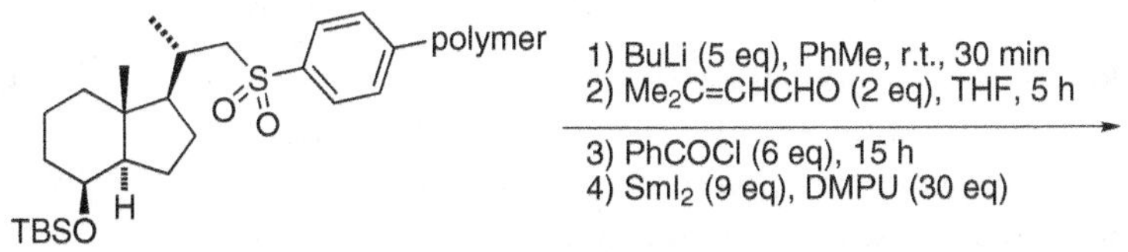

References: (a) *J. Org. Chem.* **2007**, *72*, 7125. (b) *Org. Lett.* **2008**, *10*, 1477. (c) *Org. Lett.* **2008**, *10*, 3223. (d) *Molecules* **2006**, *11*, 655.

16.6 Cascade Reactions Initiated by Carbon Nucleophiles

Cascade reactions initiated by carbon nucleophiles are important synthetic methods for the construction of complex molecules. Among the most important of these reactions are the ***Robinson annelation*** (or ***Robinson annulation***), the ***Morita-Baylis-Hillman reaction***, and the ***benzoin condensation***.

§ Answer to Worked Problem

This ketone has a pair of alkoxy groups at the β position, which would make the α hydrogens of the ketone significantly more acidic due to the inductive effect. Thus, the strongly basic ylide could, potentially, deprotonate the α carbon, resulting in the loss of the carbonyl group required for the methylenation. In addition, the neopentyl-like nature of the ketone carbonyl group makes it highly hindered, a situation in which the classical Wittig reaction frequently fails. In this case, the β-elimination to give an enone must also be considered, as must isomerization at the anomeric carbon of the sulfone.

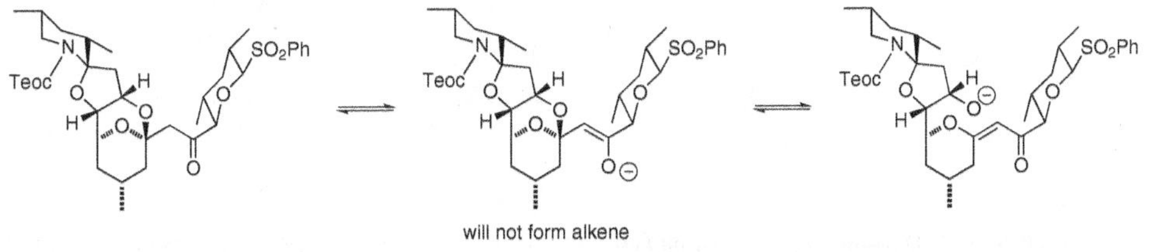

Michael-Initiated Ring Closure

By incorporating a suitable electrophile (e.g., an alkyl halide) and a Michael acceptor at appropriate locations in the same molecule (e.g., Example **16.97**), one is able to prepare cyclic products by a nucleophile-initiated domino reaction known as Michael-initiated ring closure (MIRC).[97] Essentially the same reaction occurs when a nucleophile carrying a leaving group (e.g., Example **16.98**) is used with a simple Michael acceptor—in the Corey-Chaykovsky reaction,[98] the leaving group (the sulfoxonium ion) and the nucleophile (the enolate) are generated in the Michael addition step.

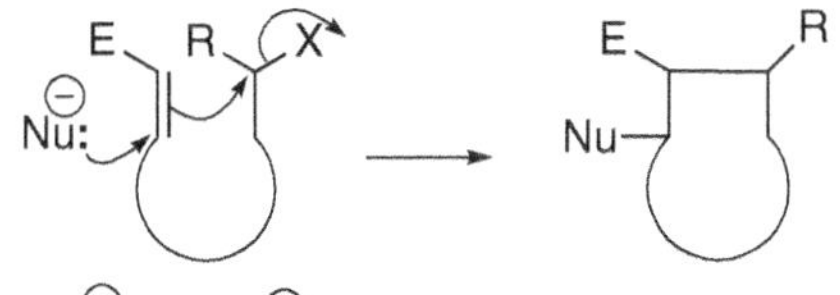

(16.97)

$\ominus$ $CN^{\ominus}$, $R_2Cu(CN)Li_2$, enolate, etc.
E: CN, COR, CO_2R; X: Cl, Br, I, OTs, epoxide, etc.

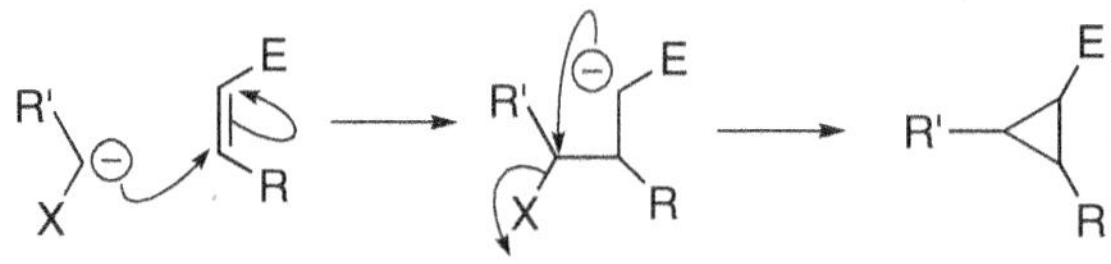

(16.98)

E: CN, COR, CO_2R, etc.
X: SMe_2, $S(O)Me_2$, Cl, Br, I, OTs, epoxide, etc.

The Robinson annelation is an MIRC in which a cyclohexenone ring is fused onto an existing ketone by treatment of the ketone with a base and an enone. The reaction proceeds by initial Michael addition of the enolate anion to the enone to give a new enolate anion (**16.99**) that undergoes a proton shift to give the isomeric enolate **16.100**. This enolate undergoes an intramolecular aldol addition to give the alkoxide anion **16.101**, which then gives the aldol **16.102**. Dehydration then gives the new enone (**16.103**; Figure 16.10). In the Robinson annelation, the enolate anion nucleophile may be replaced by an enamine, and this occasionally improves the yield by suppressing side reactions. Two sequential

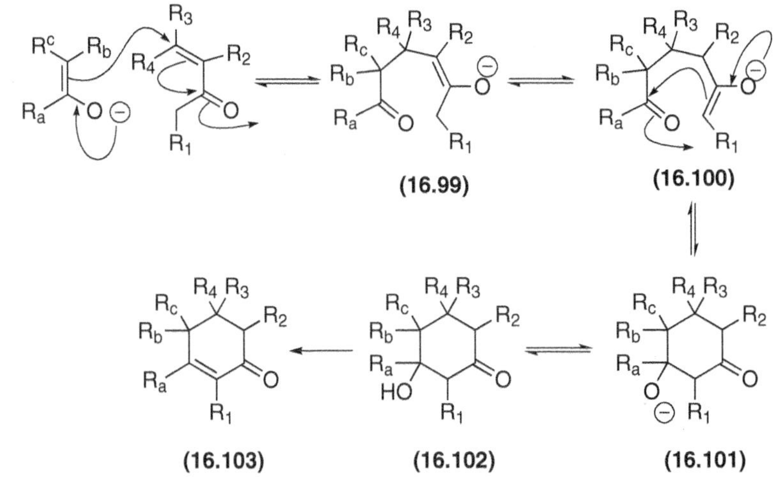

Figure 16.10 The mechanism of the Robinson annelation

97. (a) Little, R.D.; Dawson, J.R. *Tetrahedron Lett.* **1980**, *21*, 2609. (b) Amputch, M.A.; Matamoros, R.; Little, R.D. *Tetrahedron* **1994**, *50*, 5591. (c) Nakache, P.; Ghera, E.; Hassner, A.P. *Tetrahedron Lett.* **2000**, *41*, 5583.
98. Corey, E.J.; Chaykovsky, M. *J. Am. Chem. Soc.* **1965**, *87*, 1353.

Robinson annelations were key steps in building the tetracyclic framework (Example **16.106**) in Johnson's synthesis of testosterone.[99]

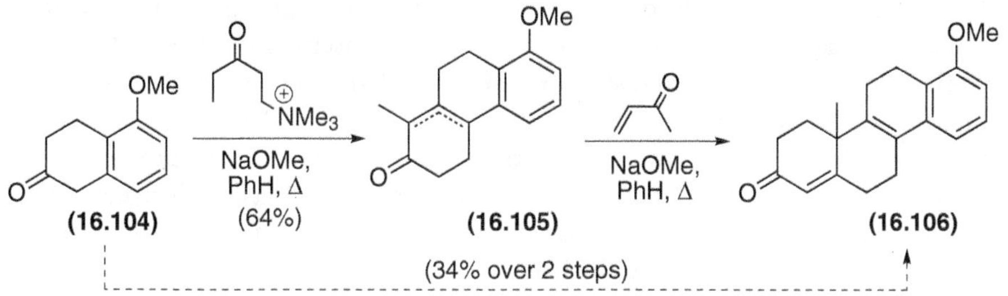

The Morita-Baylis-Hillman and Benzoin Condensation Reactions

The Morita-Baylis-Hillman Reaction[100]

One fascinating variant of enolate nucleophile chemistry is provided by the Morita-Baylis-Hillman reaction.[101] In this reaction, a Michael acceptor and an aldehyde react in the presence of a tertiary amine or phosphine catalyst to form an adduct that is densely functionalized. The mechanism of the reaction has been intensely studied, and recently McQuade has proposed the mechanism in Figure 16.11 based on the kinetics of the reaction.[102] The reaction is initiated by conjugate addition of the amine or phosphine nucleophile to the Michael acceptor to give a betaine (**16.107**), which then adds to the aldehyde to give a new zwitterion (**16.108**). The reaction of this zwitterion with another molecule of the aldehyde gives yet another new zwitterion (**16.109**), which now undergoes an intramolecular E2 elimination to give a hemiacetal (**16.110**); this then eliminates a molecule of aldehyde to give the product (**16.111**).

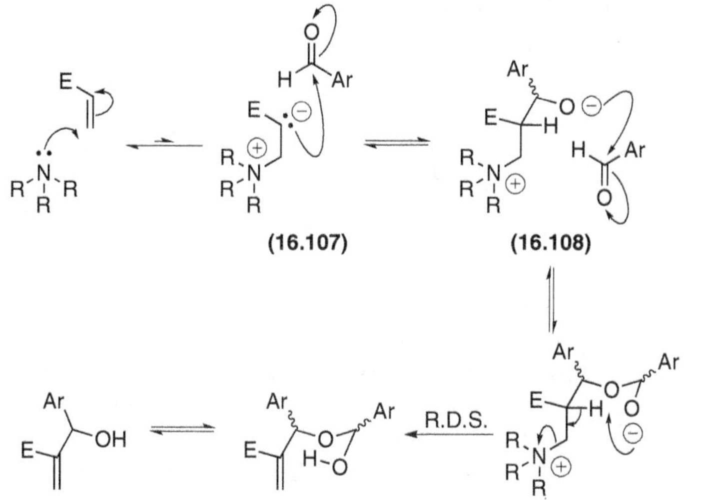

Figure 16.11 The mechanism of the Morita-Baylis-Hillman reaction

99. Johnson, W.S.; Szmuszkovicz, J.; Rogier, E.R.; Hadler, H.I.; Wynberg, H. *J. Am. Chem. Soc.* **1956**, *78*, 6285.

100. Reviews: (a) Basavaiah, D.; Rao, A.J.; Satyanarayana, T. *Chem. Rev.* **2003**, *103*, 811. (b) Masson, G.; Housseman, C.; Zhu, J. *Angew. Chem. Int. Ed.* **2007**, *46*, 4614. (c) Declerck, V.; Martinez, J.; Lamaty, F. *Chem. Rev.* **2009**, *109*, 1. (d) Basavaiah, D.; Reddy, B.S.; Badsara, S.S. *Chem. Rev.* **2010**, *110*, 5447. (e)

101. (a) Morita, K.; Suzuki, Z.; Hirose, H. *Bull. Chem. Soc. Japan* **1968**, *41*, 2815. (b) Baylis, A.B.; Hillman, M.E.D. **Offenlegungsschrift** 2155113, 1972; **U.S. Patent** 3,743,669, 1972; *Chem. Abstr.* **1972**, *77*, 34174q.

For reviews, see: (c) Basavaiah, D.; Rao, J.A.; Satyanarayana, T. *Chem. Rev.* **2003**, *103*, 811. (d) Ciganek, E. *Org. React.* **1997**, *51*, *201*. (e) Basavaiah, D.; Rao, P. D.; Hyma, R.S. *Tetrahedron* **1996**, *52*, 8001. (f) Drewes, S.E.; Roos, G.H.P. *Tetrahedron* **1988**, *44*, 4653.

102. (a) Price, K.E.; Broadwater, S.J.; Jung, H.M.; McQuade, D.T. *Org. Lett.* **2005**, *7*, 147. (b) Price, K.E.; Broadwater, S.J.; Walker, B.J.; McQuade, D.T. *J. Org. Chem.* **2005**, *70*, 3980.

Inherently, the reaction is slow, and obtaining high yields of products can take weeks. Nevertheless, its general lack of side reactions has made this reaction a popular method for synthesis—especially given the dense functionalization of the products formed. Efforts to accelerate the reaction by means of catalysts have led to observations that the reaction can be accelerated by protic additives, such as water or alcohols and by Lewis acids (often lanthanides).[103] Aggarwal has subsequently shown that a molecule of alcohol replaces the second molecule of aldehyde in the transition state of the reaction in protic solvents.[104] This change in the kinetic order of the reaction has been confirmed with glycol additives. The increase in the rate constant depends, unexpectedly, on the number of hydroxymethyl groups, rather than the number of hydroxyl groups, in the additive.[105] As with the aldol and Michael addition reactions, much of the modern focus on this reaction has now shifted to being able to carry out the reaction in an asymmetric manner. In reaction **16.112**, for example, the chiral catalyst derived from the quinine skeleton gives the *R* allylic alcohol in high enantiomeric excess.[106]

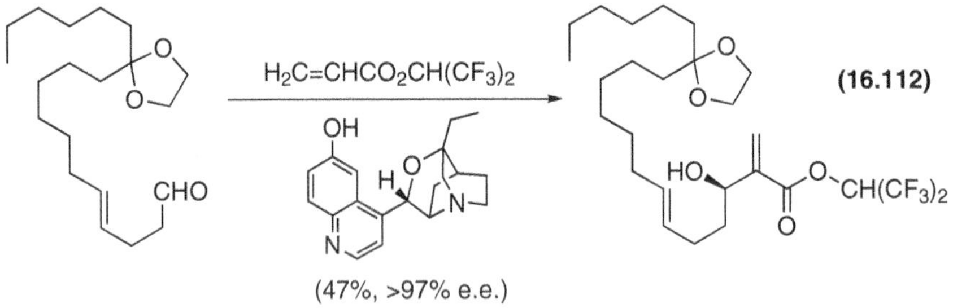

(16.112)

The Morita-Baylis-Hillman reaction has been expanded more recently by the use a relatively wide range of second acceptor molecules in place of the aldehyde, including imines and conjugated carbonyl compounds (in an intramolecular setting). One example of this is provided by a key cyclization step (Example **16.113**) in the synthesis of the spinosin A nucleus.[107]

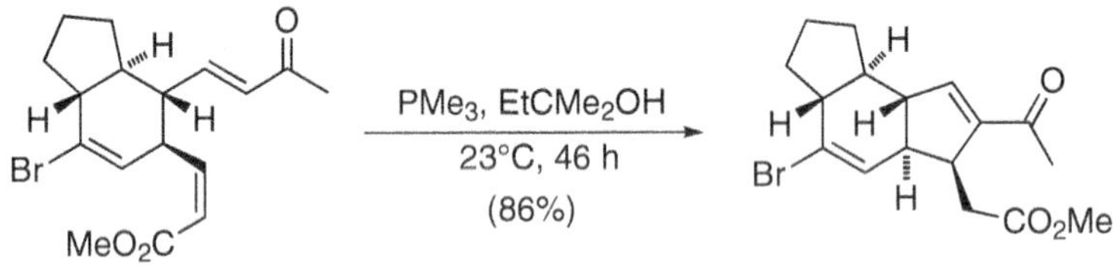

(16.113)

The Benzoin Condensation

The benzoin condensation is one of the oldest synthetic reactions known.[108] It is a dimerization reaction between two aldehydes to give an α-hydroxyketone (Example **16.114**). It is not formally a condensation reaction as we defined it in Section 16.1 because no small molecule is lost in the reaction. However, the reaction name is so long-standing that it is unlikely to be changed. In the original form of the reaction, the catalyst was cyanide ion, but in the modern variant, an *N*-heterocyclic carbene or thiazolium ylide acts as the catalyst. In the classic form of the reaction, the initial cyanohydrin anion (Example **16.115**)

---

103. (a) Aggarwal, V.K.; Mereu, A.; Tarver, G.J.; McCague, R. *J. Org. Chem.* **1998**, *63*, 7183. (b) Aggarwal, V.K.; Dean, D.K.; Mereu, A.; Williams, R. *J. Org. Chem.* **2002**, *67*, 510. (c) Ameer, F; Drewes, S.E.; Freese, S.; Kaye, P.T. *Synth. Commun.* **1988**, *18*, 495. (d) Drewes, S.E.; Freese, S.D.; Emslie, N.D.; Roos, G.H.P. *Synth. Commun.*, **1988**, *18*, 1565. (e) Basavaiah, D.; Sarma, P.K.S. *Synth. Commun.*, **1990**, *20*, 1611. (f) Auge, J.; Lubin, N.; Lubineau, A. *Tetrahedron Lett.*, **1994**, *35*, 7947. (g) Rafel, S.; Leahy, J.W. *J. Org. Chem.* **1997**, *62*, 1521.

104. (a) Robiette, R.; Aggarwal, V.K.; Harvey, J.N. *J. Am. Chem. Soc.* **2007**, *129*, 15513, and references therein. (b) Aggarwal, V.K.; Fulford, S.Y.; Lloyd-Jones, G.C. *Angew. Chem. Int. Ed.* **2005**, *44*, 1706–1708.

105. Mirzaei, J.; Lewis, D.E. unpublished results.

106. Iwabuchi, Y.; Furukawa, M.; Esumi T.; Hatakeyama, S. *Chem. Commun.* **2001**, 2030.

107. Mergott, D.J.; Frank. S.A.; Roush, W.R. *Org. Lett.* **2002**, *4*, 3157.

108. (a) Wöhler, F.; Liebig, J. *Ann. Chem. Pharm.* **1832**, *3*, 249. (b) Zinin, N. *Ann. Chem. Pharm.* **1839**, *31*, 329.

first isomerizes by a protonation-deprotonation sequence to give the isomeric nitrile α-anion (Example **16.116**). This nucleophile adds to a second molecule of aldehyde to give the alkoxide **16.117**, which also isomerizes to the cyanohydrin conjugate base (Example **16.118**). This then eliminates cyanide anion to regenerate the ketone carbonyl group (Example **16.114**). Example **16.119** shows the use of an *N*-heterocyclic carbene as the catalyst.[109]

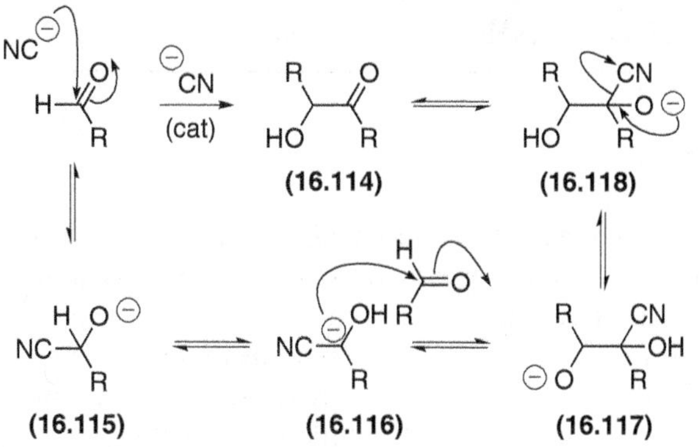

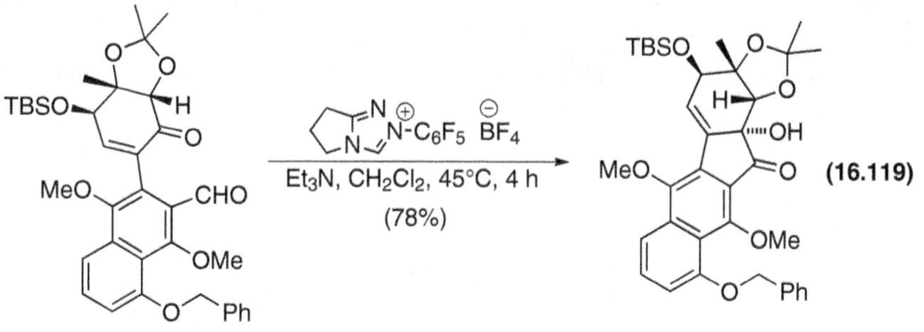

Reaction Synopses

Robinson Annulation (Robinson Annelation)

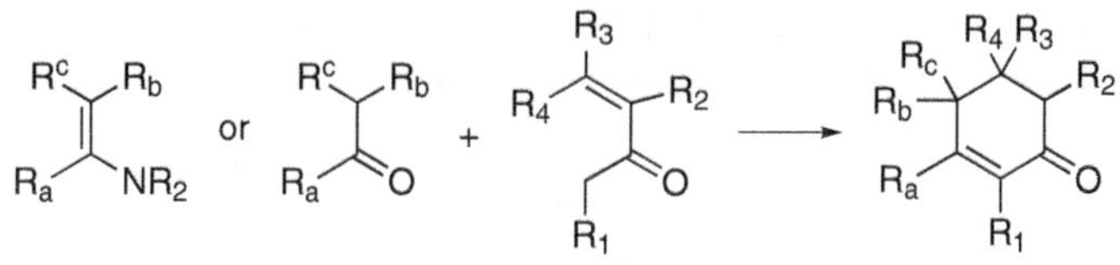

Reagents:

from ketone: KOBu^t, THF; NaOEt, EtOH; etc.
from enamine: PhH, Δ; etc.
Enone: may be replaced by Mannich base

Morita-Baylis-Hillman reaction

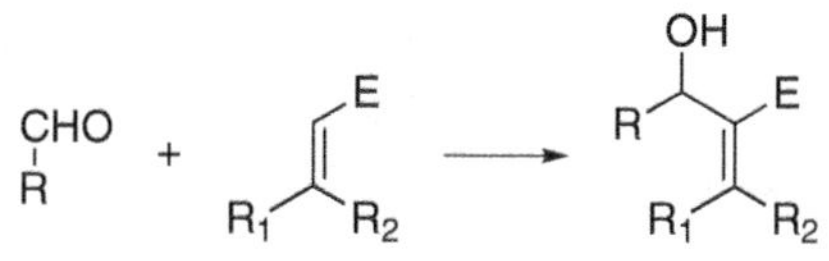

E: CO_2R, CN, COR, NO_2, etc.

Reagents: DABCO; DMAP; R_3P; etc.

Reaction: accelerated by protic solvents and by Lewis acids (e.g., Y^{+3})

(continues)

109. Nicolaou, K.C.; Li, H.; Nold, A.L.; Pappo, D.; Lenzen, A. *J. Am. Chem. Soc.* **2007**, *129*, 10356.

(**Reaction Synopses** continued)

Benzoin Condensation

Reagents: KCN, EtOH, Δ; NHC, CH_2Cl_2, Δ; etc.

Chapter Summary

Condensations proceed by a multistep mechanism that requires the elimination of a small molecule from the initial addition product. In the aldol condensation, the Henry reaction, and the Wittig and HWE reactions, a π bond is replaced by another π bond. The Stobbe condensation gives a conjugates half-ester of an alkylidenesuccinic acid, with the ester group conjugated to the alkene. With aldehydes, the stereochemistry of the double bond is *E*. The Claisen condensation is a net nucleophilic acyl substitution of an ester by an enolate anion to give a β-ketoester. The Claisen-Schmidt, Knoevenagel, and Doebner reactions are modifications of the original Claisen motif. The Dieckmann condensation is an intramolecular Claisen condensation to give a five- or six-membered β-ketoester. The related cyclization of dinitriles is the Thorpe-Ziegler cyclization; it is less subject to steric effects than the Dieckmann condensation and can be used to prepare medium-sized rings. Hydrolysis of β-ketoesters gives β-ketoacids that can be decarboxylated by heat; direct decarboxylation of the β-ketoesters can be accomplished using the Krapcho decarboxylation.

Alkenes can be formed using the Wittig, HWE, Peterson, Ramberg-Bäcklund, Julia, and Tebbe reactions, along with their derivative reactions. The Wittig reaction tends to produce the *Z* alkene when unstabilized ylides are used as the nucleophile and *E* alkenes when stabilized ylides are used. *E* alkenes are also the major product in the Schlosser modification of the Wittig reaction with unstabilized ylides. The HWE reaction gives the *E* isomer as the major product. The Peterson reaction is much less susceptible to steric hindrance than the Wittig reaction, and the stereochemistry of the alkene can be controlled by choosing the conditions under which the β-hydroxysilane is decomposed. The Julia-Lythgoe and Julia-Kocieńsky olefinations allow stereocontrolled synthesis of trisubstituted alkenes. The Ramberg-Bäcklund reaction leads to alkenes by extrusion of sulfur dioxide from intermediate thiirane dioxides. The Tebbe reaction is the prototypical synthesis of alkenes from carbonyl compounds using a metal-carbene complex. It permits the synthesis of enol ethers from esters and lactones.

The synthesis of alkynes by Wittig-type condensations can be accomplished by the Seyferth-Gilbert reaction or by the Corey-Fuchs procedures. The Seyferth-Gilbert reaction is facilitated by using the Bestmann reagent.

Cascade reactions are reactions where the product of the first reaction becomes a reactant in a second reaction to give the isolated product. Cascade reactions initiated by nucleophiles include the Robinson annelation (Robinson annulation), which is a Michael-initiated ring closure; the Morita-Baylis-Hillman reaction, which is a Michael-initiated aldol addition; and the benzoin condensation.

Key Terms

aldol condensation
benzoin condensation
Bestmann reagent
cascade reaction
Claisen condensation
Claisen-Schmidt reaction
condensation reaction
Corey-Chaykovsky reaction
Corey-Fuchs reaction
Darzens condensation
Decarboxylation
Dieckmann condensation

Doebner reaction
Henry reaction
Horner-Wadsworth-Emmons reaction
Julia-Kocieński olefination
Julia-Lythgoe olefination
Knoevenagel condensation
Krapcho decarboxylation
Lombardo reagent
MIRC
Morita-Baylis-Hillman reactionPerkin condensation
Petasis reagent
Peterson reaction
Ramberg-Bäcklund reaction
Robinson annelation (annulation)
Schlosser modification
Seyferth-Gilbert homologation
Stobbe condensation
Tebbe olefination
Tebbe reagent
Thorpe-Ziegler condensation
Wittig reaction

Additional Problems

16-5 What is the major organic product of each of the following reactions?

(a)
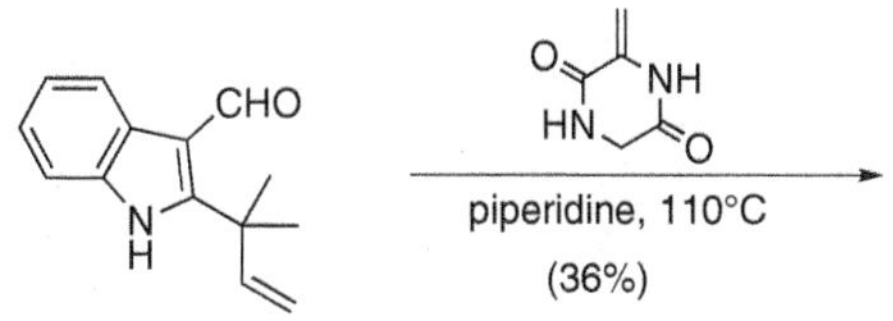

[*Angew. Chem. Int. Ed.* **2011**, *50*, 1402]

(b)
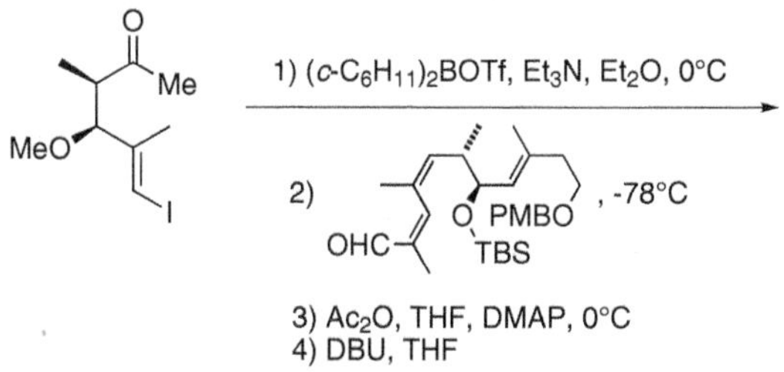

[*J. Org. Chem.* **2009**, *74*, 7220]

(c)
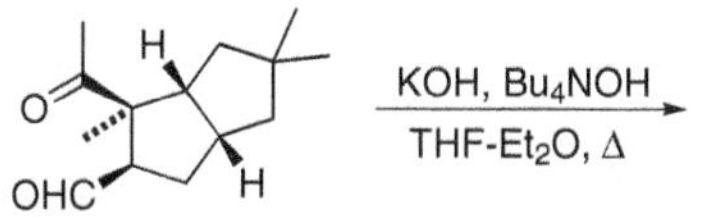

[*Angew. Chem. Int. Ed.* **2003**, *42*, 5855]

(d)
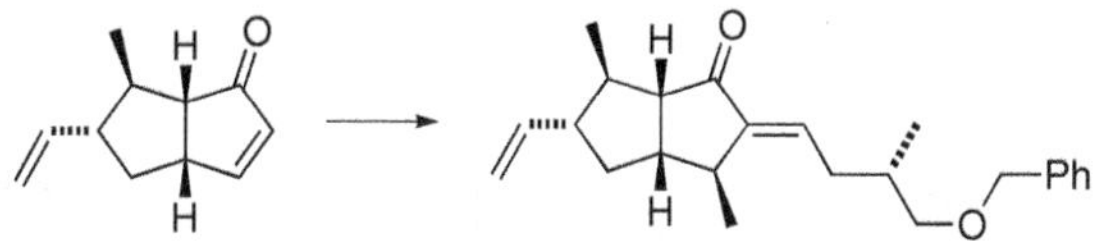

[*J. Am. Chem. Soc.* **2002**, *124*, 9974]

(e)
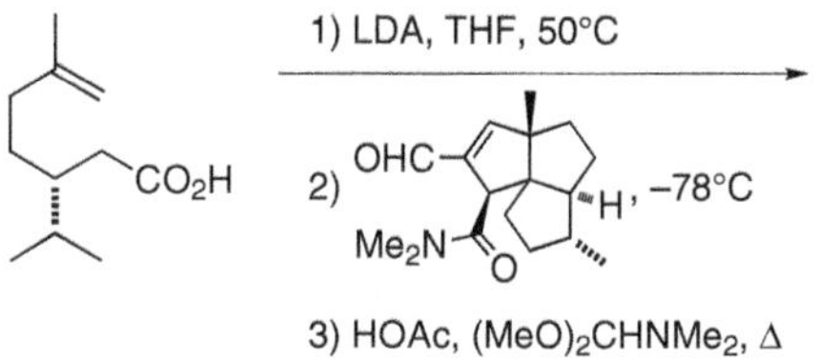

[*Tetrahedron Lett.* **1990**, *31*, 2517]

(f)
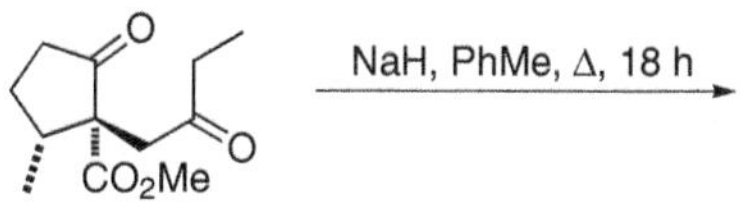

[*J. Am. Chem. Soc.* **1984**, *106*, 6690]

(g)
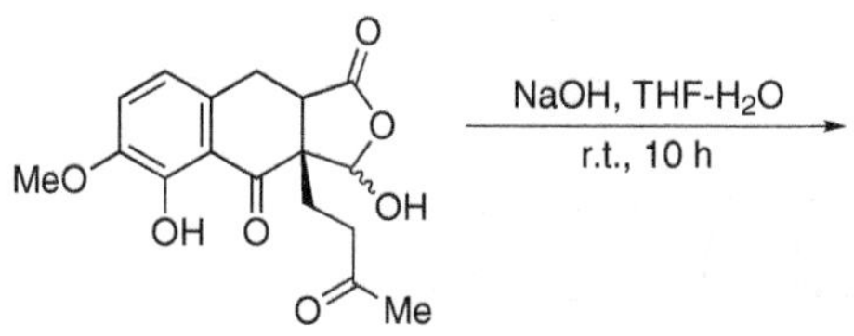

[*J. Org. Chem.* **1999**, *64*, 7871]

(h)
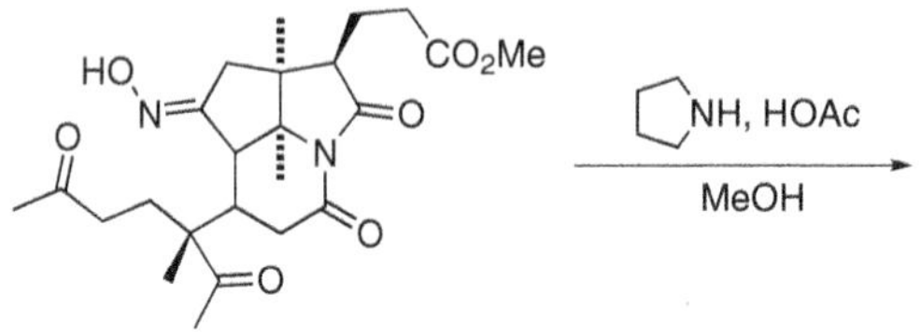

[*Pure Appl. Chem.* **1968**, *17*, 519]

(i)
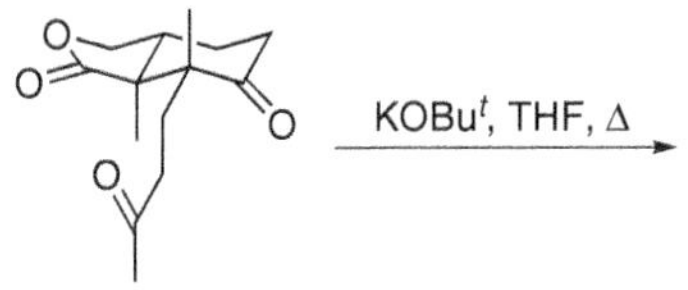

[*J. Org. Chem.* **1995**, *60*, 7387]

(j)
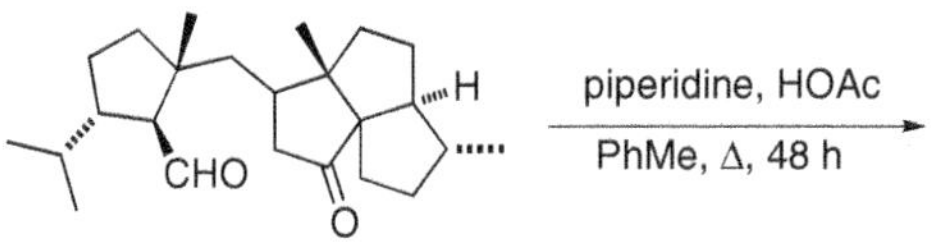

[*J. Org. Chem.* **1987**, *52*, 2960]

(k)
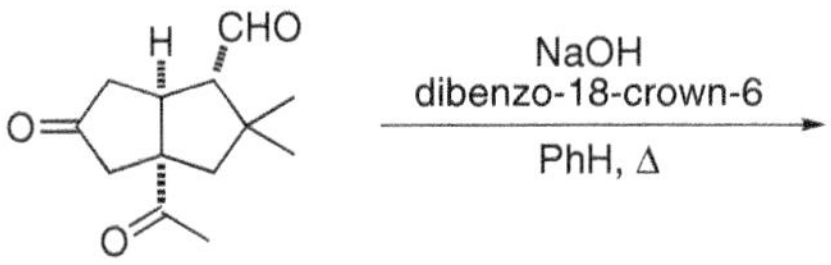

[*J. Am. Chem. Soc.* **1982**, *104*, 872]

(l)
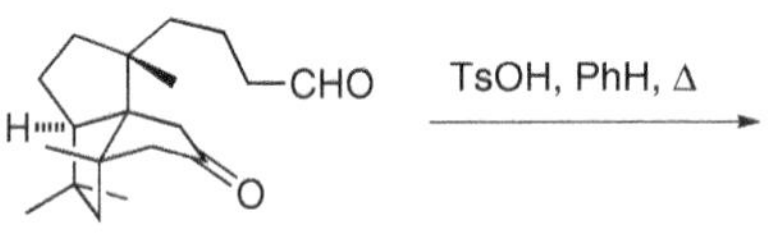

[*J. Am. Chem. Soc.* **1987**, *109*, 6199]

(m)
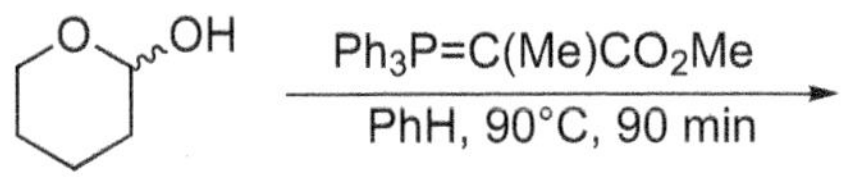

[*J. Org. Chem.* **2002**, *67*, 9443]

(n)
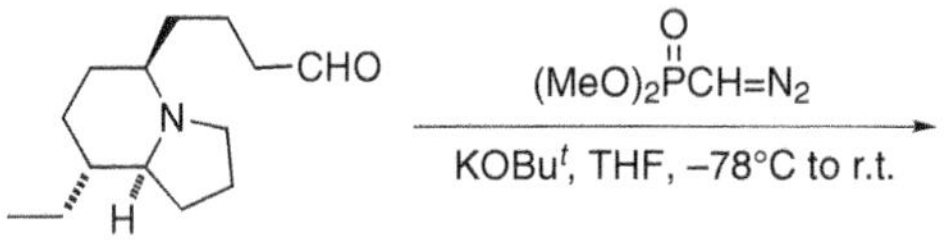

[*Beilstein J. Org. Chem.* **2007**, *3*, 29; doi:10.1186/1860-5397-3-29]

(o)
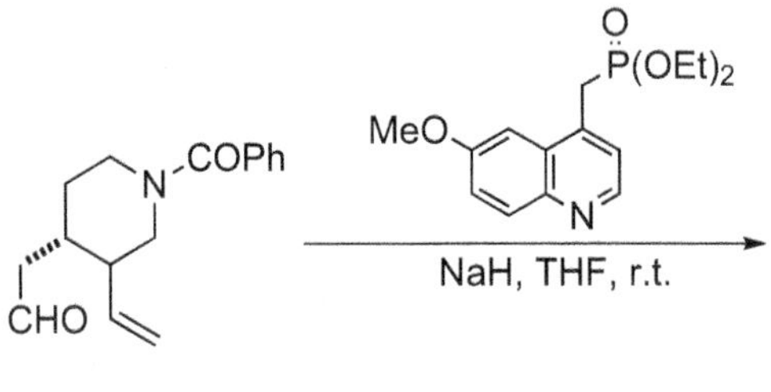

[*Tetrahedron Lett.* **2004**, *45*, 3783]

(p)
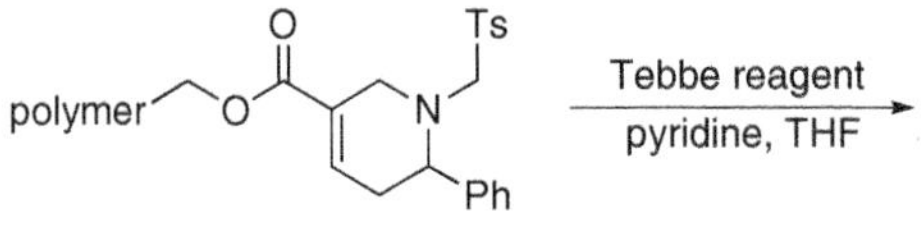

[*Proc. Natl. Acad. Sci., U.S.A.* **2011**, *108*, 6769]

(q)
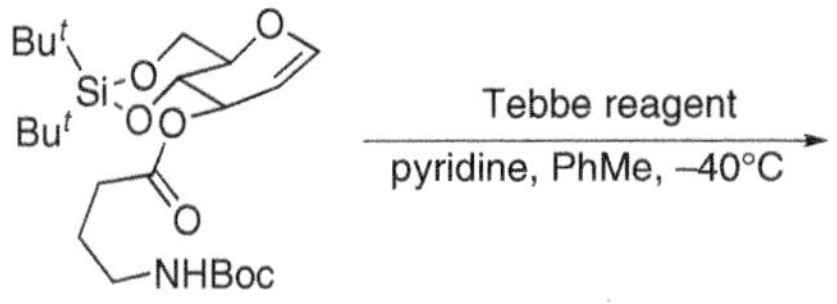

[*Tetrahedron Lett.* **2000**, *41*, 7589]

(r)
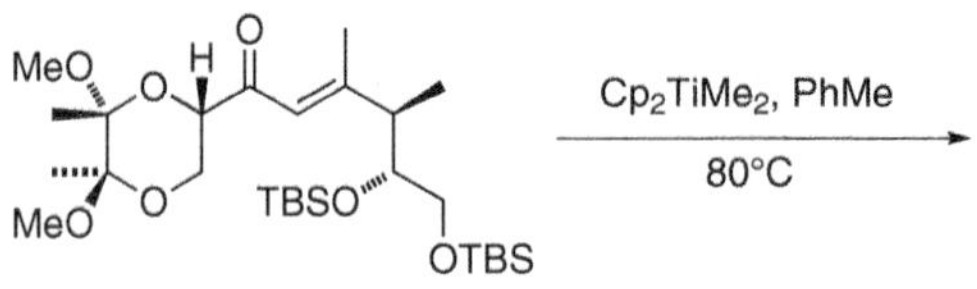

[*Org. Biomol. Chem.* **2009**, *7*, 4582]

(s) 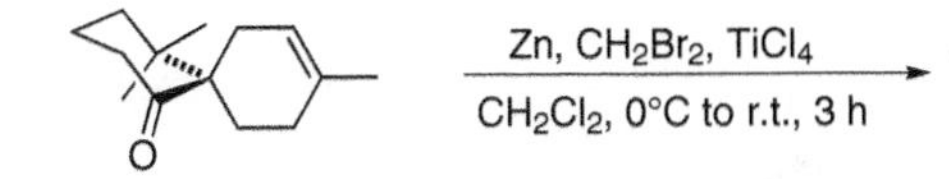

[*Tetrahedron Lett.* **2006**, *47*, 2103]

(t) 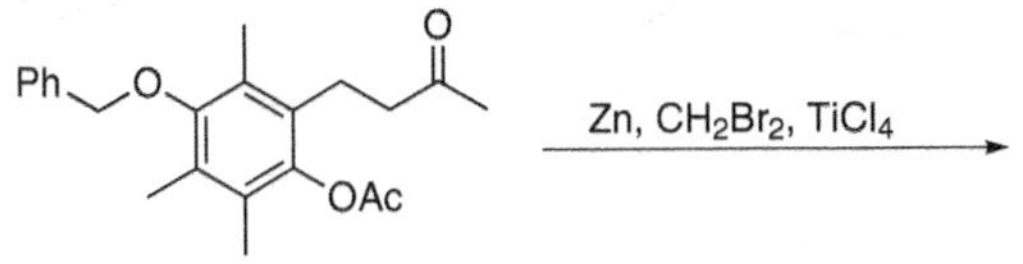

[*Pure Appl. Chem.* **2007**, *79*, 629]

(u) 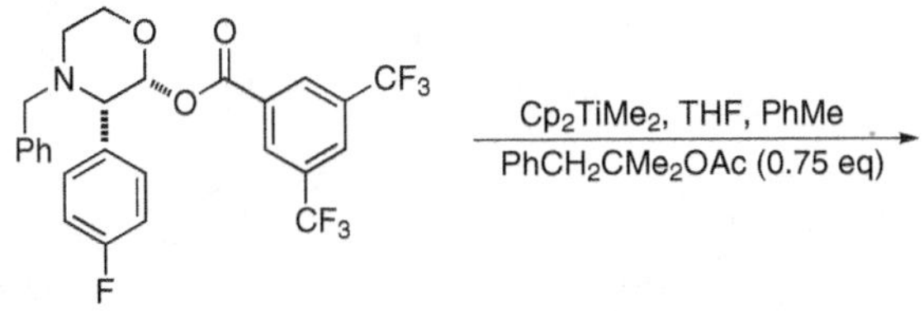

[*Org. Proc. Res. Dev.* **2004**, *8*, 256]

(v) 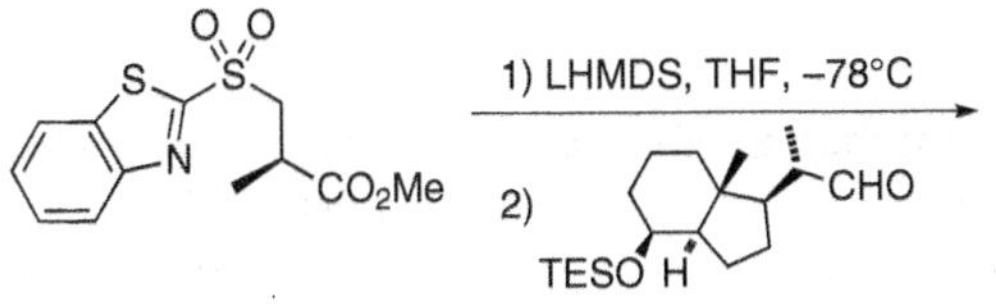

[*Tetrahedron Lett.* **2009**, *50*, 4874]

(w) 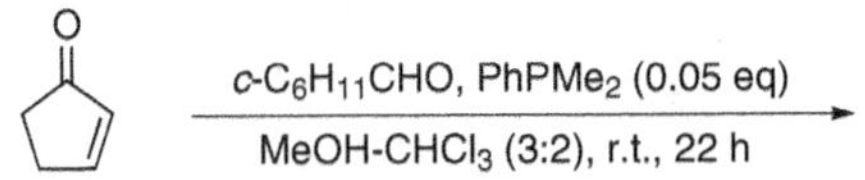

[*Synthesis* **2005**, 3035]

16-6 Suggest a mechanism for the step from MacMillan's synthesis of strychnine [*Nature* **2011**, *475*, 183] shown.

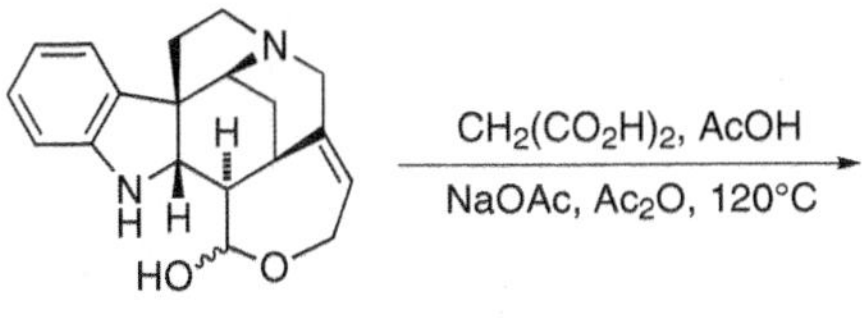

16-7 Suggest reasons for the different stereochemical outcome of the two reactions shown [*Tetrahedron Lett.* **2007**, *48*, 2961].

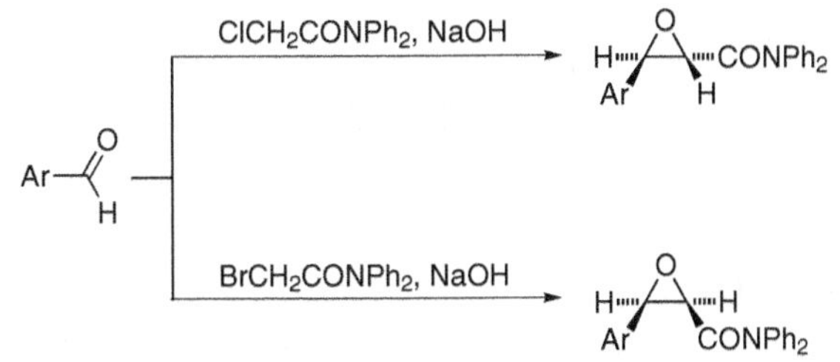

16-8 Complete the following transformations by supplying the missing reagent, sequence of reagents, or starting structure.

(a) 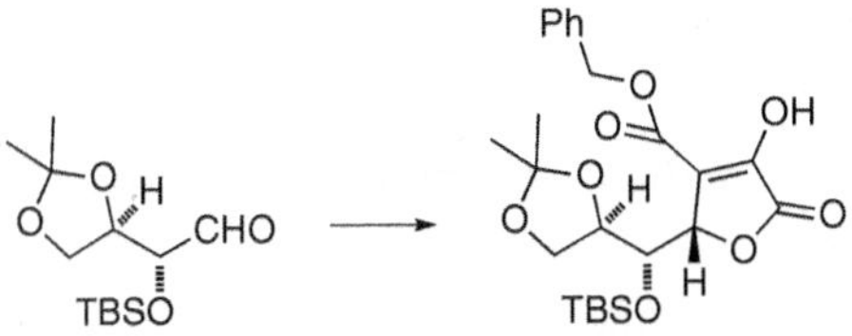

[*J. Am. Chem. Soc.* **2003**, *125*, 11510]

(b) 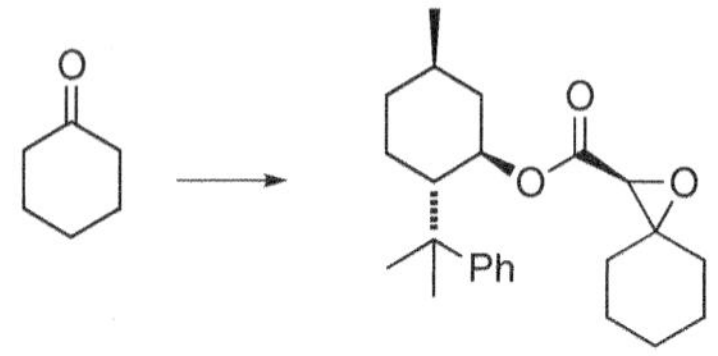

[*Chem. Commun.* **1996**, 2411]

(c) 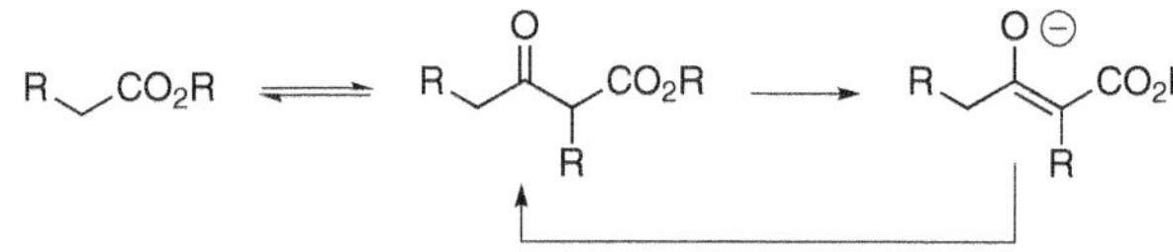

[*J. Chem. Soc., Perkin Trans. 1* **1998**, 689]

(d) ? 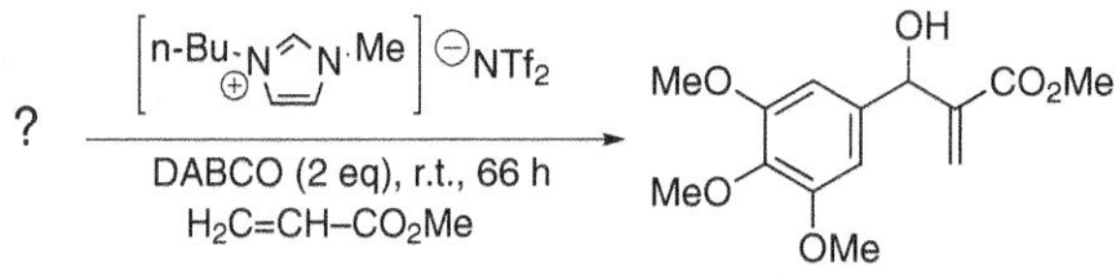

[*J. Org. Chem.* **2010**, *75*, 4183]

16-9 Suggest a reasonable mechanism for the tandem transformation shown (NHC is an *N*-heterocyclic carbene, and DBU is the non-nucleophilic amidine base, 1,8-diazabicyclo-[5.4.0]undec-7-ene.) [*Tetrahedron Lett.* **2012**, *53*, 453]

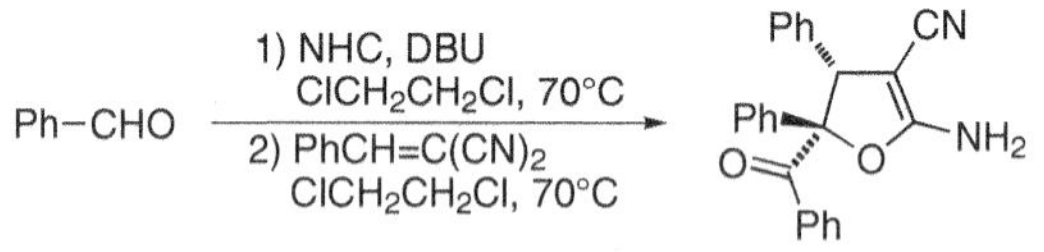

16-10 (a) Provide the missing reagents from the following steps from the total synthesis of morphine by Mulzer [*Angew. Chem. Int. Ed.* **1996**, *35*, 2830].

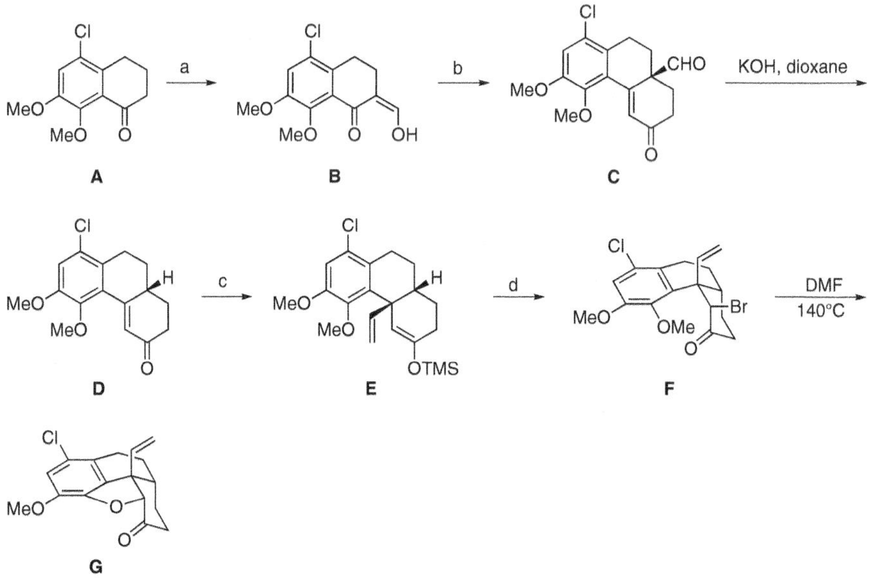

Suggest mechanisms for the conversions from compound **C** to compound **D**, and from compound **E** to compound **F**.

16-11 Provide the missing reagents from the steps in Wender's total synthesis of taxol, shown below [*J. Am. Chem. Soc.* **1997**, *119*, 2755, 2757].

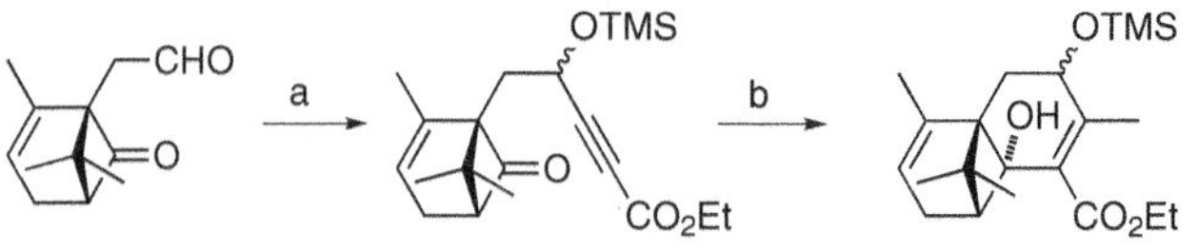

16-12 Rationalize the reasons for the different outcomes of the attempted Perkin condensations below [*Tetrahedron Lett.* **2003**, *44*, 7119].

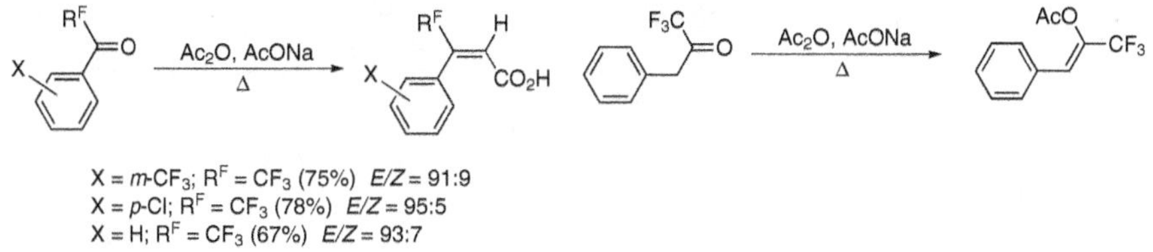

16-13 The alkaloid, yohimbine, has had a reputation among African natives as potent aphrodisiac since antiquity. The synthesis of (±)-yohimbine by van Tamelen [*J. Am. Chem. Soc.* **1969**, *91*. 7315] begins with the steps below. Provide the structures, including stereochemistry, of the intermediates **A-D**. Compound **C** exists as a mixture of two diastereoisomers. What are the structures of these two diastereoisomers, and why were they not separated during the synthesis?

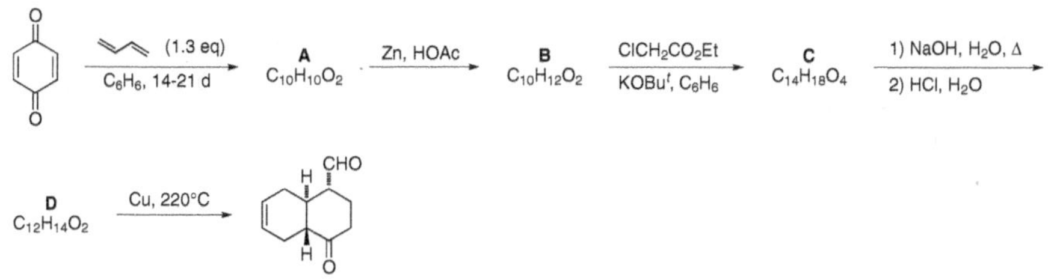

CHAPTER SEVENTEEN

Organic Reactions VI

Metal-Catalyzed Reactions for C—C Bond Formation

17.1 Catalysis and "Green" Chemistry

Monty Python, the British comedy team from the 1970s, often used a phrase for bridging segments of their television show that could easily serve as the subtitle for this chapter: ". . . And now for something completely different." In the previous chapters of this book, we have discussed the reactions of organic compounds and distilled much of this chemistry into a series of generalizations. The truth can now be told. Many of those generalizations are oversimplifications, and many should have appended to them the statement, "except if a transition metal complex is involved in the reaction." Two such generalizations for which this is especially true are, "In the absence of electron-withdrawing groups to activate the molecule toward nucleophilic displacement, the nucleophilic displacement of a vinyl halide or aryl halide will not occur except under extreme reaction conditions," and "Unless activated by an adjacent electron-withdrawing group, simple alkenes do not react with nucleophiles." In other words, this chapter is not so much an end as it is a beginning—a launching point from which one can begin to explore what modern organic chemistry has become and where it may be going.

The holy grail of organic synthesis is to prepare the target molecule with absolute control over regiochemistry and stereochemistry, and with perfect atom economy, a concept that was first defined by Trost during the early development of "green" chemistry.[1] In the ideal case, the target molecule is formed as a single, pure stereoisomer (or enantiomer, if it is chiral), and every atom used in the reactions of the synthetic sequence is incorporated into the target. Achieving this level of atom economy would, of course, restrict the synthetic chemist to using addition and rearrangement reactions only, because atoms are lost during substitution and elimination reactions. Obviously, this ideal is unattainable—even living cells, the most accomplished synthetic chemists of all, use substitution and elimination reactions in synthesizing complex molecules.

Nevertheless, the concept of atom economy has driven the move, begun during the last two decades of the 20th century, toward an increased use of catalysis as a means to reducing the environmental impact of organic synthesis. In 1999, the journal *Green Chemistry* was launched by the Royal Society of Chemistry to provide a place for the publication of articles addressing questions of environmentally friendly chemistry.

1. (a) Trost, B.M. *Science* **1991**, *254*, 1471. (b) Trost, B.M. *Angew. Chem. Int. Ed. Engl.* **1995**, *34*, 259. (c) Trost, B.M. *Acc. Chem. Res.* **2002**, *35*, 695.

The Twelve Principles of Green Chemistry[2] have been formalized as follows:

1. It is better to prevent waste than to treat or clean up waste after it is formed.
2. Synthetic methods should be designed to maximize the incorporation of all materials used in the process in the final product.
3. Wherever practical, synthetic methodologies should be designed to use and generate substances that possess little or no toxicity to human health and the environment.
4. Chemical products should be designed to preserve efficacy of function while reducing toxicity.
5. The use of auxiliary substances (e.g., solvents, separation agents) should be made unnecessary whenever possible and innocuous when used.
6. Energy requirements should be recognized for their environmental and economic impacts and should be minimized. Synthetic methods should be conducted at ambient temperature and pressure.
7. A raw material feedstock should be renewable rather than depleting whenever technically and economically practical.
8. Unnecessary derivatization (blocking group, protection-deprotection, temporary modification of physical/chemical processes) should be avoided whenever possible.
9. Catalytic reagents (as selective as possible) are superior to stoichiometric reagents.
10. Chemical products should be designed so that at the end of their function they do not persist in the environment and break down into innocuous degradation products.
11. Analytical methodologies need to be further developed to allow for real-time in-process monitoring and control prior to the formation of hazardous substances.
12. Substances and the form of a substance used in a chemical process should be chosen so as to minimize the potential for chemical accidents, including releases, explosions, and fires.

Not only do these principles explicitly (Principle 9) and implicitly (Principle 8) urge the use of catalytic chemistry, but they also promote the concept of atom economy (Principle 2).

Until the last third of the 20th century, catalytic reactions in organic chemistry had been largely restricted to catalytic hydrogenation. Then, in 1980, the whole field exploded as catalysts for asymmetric epoxidation[3] and hydrogenation[4] reactions were developed, work that resulted in the award of the 2001 Nobel Prize in Chemistry to Sharpless, Noyori, and Knowles. Since then, the Nobel Prize in Chemistry has been awarded twice more for catalysis applied to C—C bond-forming reactions.

The adoption of catalytic reactions for forming carbon-carbon bonds followed a slower path than that of the asymmetric redox reactions. Both the Sharpless asymmetric epoxidation and the asymmetric hydrogenation of α-amidoacrylates using a chiral 2,2'-*bis*-(diphenylphosphino)-1,1'-binaphthyl (BINAP) complex as catalyst appeared in 1980. Two decades later, both reactions were routine synthetic methods.

17.2 Bonding in Organometallic Complexes

One is unlikely to spark any controversy at all by asserting that few areas of modern organic chemistry have undergone the explosive growth that organometallic chemistry—especially catalysis—has. What is more, few areas of organic chemistry are showing such promise for new developments—and few have shown just how dangerous it is to make

2. Anastas, P.T,; Warner, J.C. *Green Chemistry Theory and Practice.* (Oxford University Press: New York, 1998).
3. Sharpless, K.B.; Katsuki, T. *J. Am. Chem. Soc.* **1980**.
4. (a) Knowles, W.S. *J. Am. Chem. Soc.* **1980**, (b) Noyori, R. *J. Am. Chem. Soc.* **1980**.

generalizations about the reactivity of organic compounds. Transition metal catalysis has become an important part of much modern industrial chemistry, as we shall see in the course of this chapter.

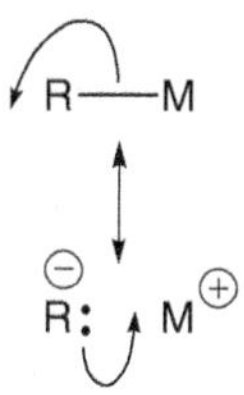

(17.1)

Organometallic compounds are all characterized by the presence of one or more metal-carbon bonds in the molecule. We have already discussed the chemistry of the Group IA and Group IIA metal alkyls such as the alkyllithiums and Grignard reagents, and we have also discussed, albeit briefly, the chemistry of the simple metal alkyls of the Group IB and Group IIB such as the dialkylzinc and organocuprate reagents. Although many of these organometallic compounds exist as aggregates, much of their chemistry can be rationalized in terms of a simple resonance picture (Example **17.1**) of a polar metal-carbon σ bond in which the carbon atom is the nucleophile.

These metal alkyls and metal "ate" complexes share several chemical properties. All these metal ions have a filled d subshell, so the metal d orbitals are not part of the carbon-metal bond, which is a more or less typical σ bond (although the alkali metal alkyls do form aggregates). They are all strong Lewis bases sensitive to protic acids, with which they react with to give the hydrocarbons. Most react with carbon electrophiles to form carbon-carbon bonds, as illustrated by the addition reactions between carbonyl compounds and Grignard (Example **17.2**), organolithium (Example **17.3**), organozinc (Example **17.4**) and organocopper compounds (Example **17.5**). Of course, as the electronegativity of the metal increases, the percent ionic character of the carbon-metal bond decreases and the nucleophilic strength and the base strength of the organometallic compound decreases.

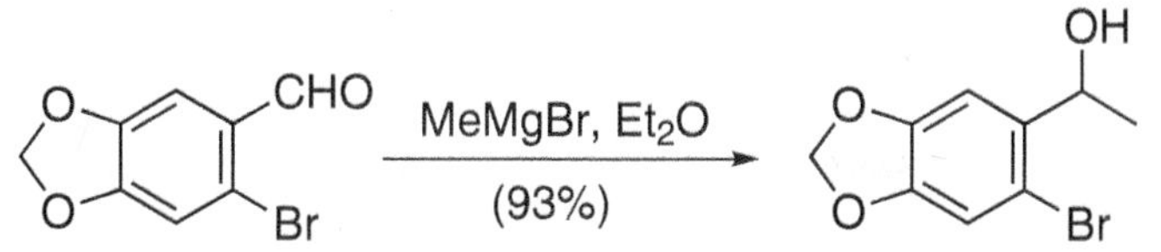

(17.2)[5]

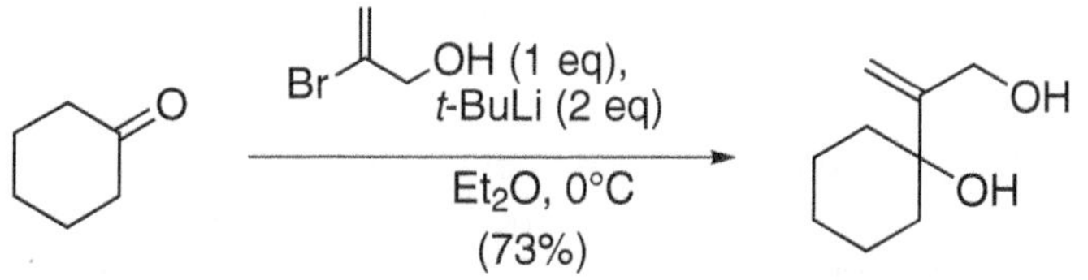

(17.3)[6]

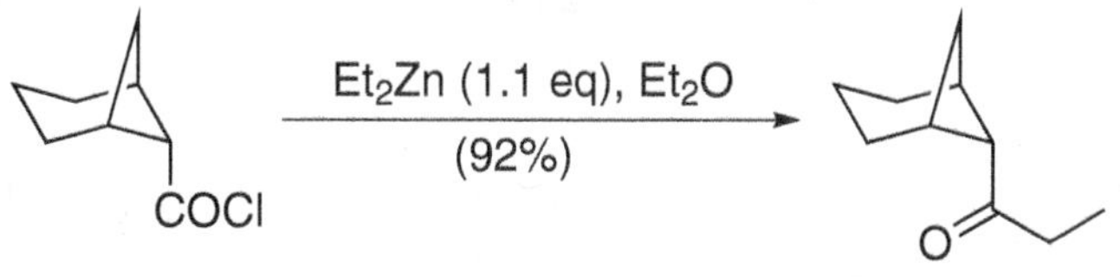

(17.4)[7]

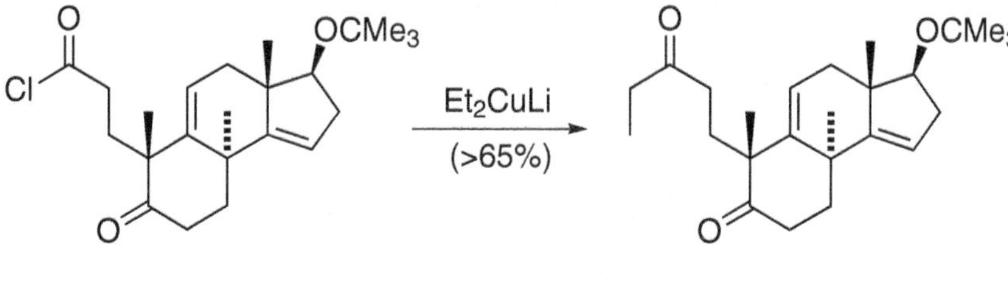

(17.5)[8]

5. Ziegler, F.E.; Schwartz, J.A. *J. Org. Chem.* **1978**, *43*, 985.
6. Corey, E.J.; Widiger, G.N. *J. Org. Chem.* **1975**, *40*, 2976.
7. Wiberg, K.B.; Hess, B.A., Jr. *J. Org. Chem.* **1966**, *31*, 2250.
8. Dauben, W.G.; Ahlgren, G.; Leitereg, T.J.; Schwarzel, W.C.; Yoshioko, M. *J. Am. Chem. Soc.* **1972**, *9*, 8593.

Worked Problem

17-1 What is the major organic product of the following reaction?

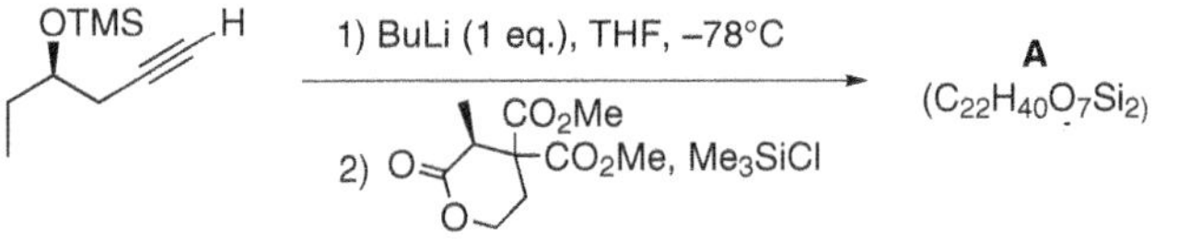

§Answer below.

Problem

17-1 What is the major organic product of each of the following reactions?

(a)

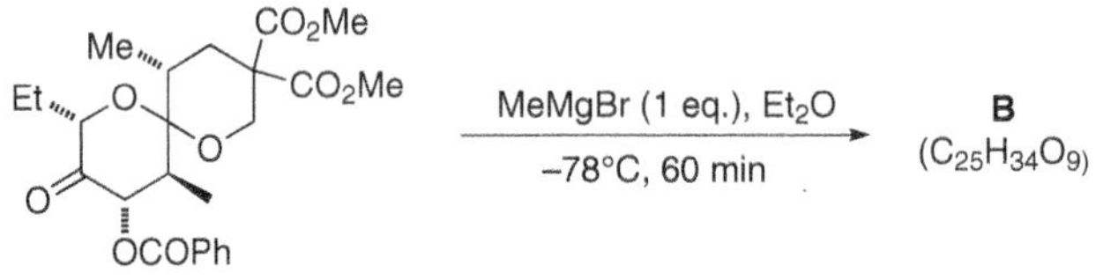

(b)

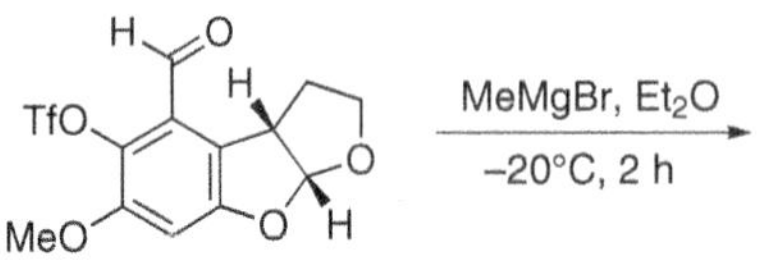

(c)

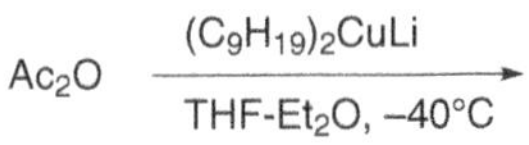

References: (a) *Can. J. Chem.* **1985**, *63*, 2810. (b) *J. Am. Chem. Soc.* **2005**, *127*, 11958. (c) *J. Am. Chem. Soc.* **1974**, *96*, 3654.

Transition Metal Organometallic Compounds

When the metal involved is a transition metal with vacant d orbitals, the chemistry of organometallic compounds frequently offers a stark contrast with the chemistry of those that have figured prominently in our discussions so far. The bonding in these compounds can be very different from that of the Group I and II organometallic compounds because the metal d orbitals are intimately involved in the formation of the metal-carbon bonds, which may be either σ bonds or π bonds or both.

These organometallic compounds are characterized by much more covalent character in the carbon-metal bond, so one generally finds that they are much weaker bases than the alkyllithiums or Grignard reagents. The alkyl groups bonded to the metal in these

§ Answer to Worked Problem:

In the initial step in this reaction, the alkyne is deprotonated to give the lithium alkynide, a powerful nucleophile. The other reactant has three carbonyl groups—all esters—capable of undergoing nucleophilic addition by the alkynide ion. When faced with this type of choice, it is worthwhile remembering that esters prefer to adopt a *syn* conformation, whereas the lactone is locked in the *anti* conformation, making its energy higher than in simple esters. Therefore, one would predict that the major product will be that formed by attack on the lactone carbonyl group [*Can. J. Chem.* **1985**, *63*, 2810].

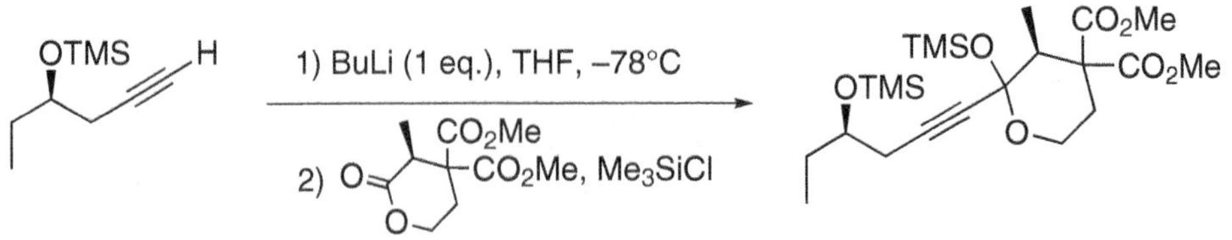

compounds also tend to be poorer nucleophiles than in the alkyllithiums and Grignard reagents, and, in fact, the *metal* atom may actually function as the nucleophile in these compounds.

The first synthetic organometallic complex of a transition metal was isolated by W.C. Zeise in Copenhagen around 1827.[9] The compound, $K[C_2H_4PtCl_3]{\bullet}H_2O$ (now known as Zeise's salt), was isolated from the reaction between potassium tetrachloroplatinate and ethanol in the presence of acid. Despite concerted efforts to elucidate the structure and bonding in this salt, more than a century would elapse before its true character was determined. During this same period, more organotransition metal complexes were isolated, but they were generally viewed as curiosities until after World War II, when the work of Reppe's group began to appear in the chemical literature.[10] Within 3 years of the publication of the first reports of the Reppe acetylene cyclooligomerizations to cyclooctatetraene, benzene, styrene, and other useful cyclic compounds (these papers began to appear in 1948), the field of organotransition metal chemistry had blossomed. In 1951, one of the most intriguing organometallic compounds known—ferrocene (Example **17.6**)—was discovered[11] (although its correct structure was not proposed until a year later[12]). The ensuing decades have seen the discovery of a multitude of organometallic compounds whose reactions have altered forever the way in which organic chemists view their discipline.

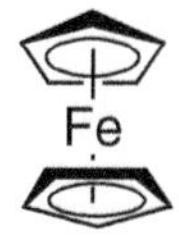

(17.6)

Of course, along with the mushrooming of the field comes the impossibility of covering it adequately in a single chapter. This chapter can only serve as an appetizer for the banquet that awaits those who are willing to consult one of the many textbooks[13] and comprehensive review series[14] devoted just to the subject of organotransition metal chemistry.

Electron Bookkeeping and the "18-Electron Rule"

Discussions of the structure and bonding in simple organic compounds were greatly simplified by the development of the concept of the valence-shell octet developed by G. N. Lewis. Based on this simple model, it became possible to rationalize the bonding in organic compounds and to correlate reactivity with structure. Of course, the theory has its limitations, and it has gradually been augmented (replaced) by molecular orbital theory, especially in compounds with extended π-bonding systems. Advances in computers have

---

9. Zeise, W.C. *Ann. Phys.* **1831**, *97*, 497.

William Christopher Zeise (1789–1847): for biographical information on Zeise and the history of the conformation of his proposed structure of this complex, see Hunt, L.B. *Platinum Metals Rev.* **1984**, *28* (2), 76.

10. (a) Reppe, W.; Schlichting, O.; Klager, K.; Toepel, T. *Justus Liebigs Ann. Chem.* **1948**, *560*, 1. (b) Reppe, W.; Schlichting, O.; Meister, H. *Justus Liebigs Ann. Chem.* **1948**, *560*, 93. (c) Reppe, W.; Schwekendiek, W.J. *Justus Liebigs Ann. Chem.* **1948**, *560*, 104.

Walter Julius Reppe (1892–1969) was a major figure in the German chemical industry (I.G. Farbenindustrie) during the Nazi period, and he continued at BASF after the war. His close association with the Nazi regime may have cost him the Nobel Prize in chemistry. Biographical information can be found at: (a) Oesper, R.A. *J. Chem. Educ.* **1950**, *27*, 648. See also: (b) Firmen- und Personennachrichten *Eur. J. Lipid Sci. Technol.* **1969**, *71*, 743; and (c) "Reppe, Julius Walter." *Complete Dictionary of Scientific Biography.* 2008. The article is available online: http://www.encyclopedia.com/doc/1G2-2830906041.html (accessed January 14, 2012).

11. (a) Kealy, T.J.; Pauson, P.L. *Nature* **1951**, *168*, 1039. (b) Miller, S.A.; Tebboth, J.A.; and Tremaine, J.F. *J. Chem. Soc.* **1952**, 632.

12. (a) Wilkinson, G.; Rosenblum, M.; Whiting, M.C.; Woodward, R.B. *J. Am. Chem. Soc.* **1952**, *74*, 2125. (b) Fischer, E.O.; Pfab, W. *Z. Naturforsch. B* **1952**, *7*, 377.

13. (a) Crabtree, R.H. *The Organometallic Chemistry of the Transition Metals*, 3rd ed. (Wiley-Interscience: New York, 2000). (b) Spessard, G.O.; Miessler, G.L. *Organometallic Chemistry*, 2nd ed. (Oxford University Press: New York, 2009). (c) Collman, J.P.; Hegedus, L.S.; Norton, J.R.; Fincke, R.G. *Principles and Applications of Organotransition Metal Chemistry* (University Science Books: Mill Valley, CA, 1987). (d) Hartwig, J. *Organotransition Metal Chemistry. From Bonding to Catalysis.* (University Science Books: Mill Valley, CA, 2010).

14. Crabtree, R.H.; Mingos, D.M.P., Eds. *Comprehensive Organometallic Chemistry, 3rd. ed.* (Elsevier: Amsterdam, 2007).

allowed the use of ab initio or density functional theory to largely supersede the Lewis structures, but, as pointed out in the early chapters of this book, the Lewis model, which is really little more than electron bookkeeping, still retains an attractive simplicity in terms of actually using it.

There is a similar electron bookkeeping model for the bonding in organotransition metal complexes, known as the "18-electron rule." The basic tenet of the 18-electron rule is the same as the Lewis octet rule: species are most stable once all the atoms involved have attained the outer electron configuration of a noble gas. The electron configuration of every transition metal corresponds to a set of core electrons with the configuration of the immediately preceding noble gas, with a valence shell composed of five d orbitals, one s orbital, and three p orbitals. Filling the valence shell to generate the electron configuration of the next noble gas requires a total of 18 electrons—hence the name. Organotransition metal complexes in which the metal atom has 18 valence electrons are described as ***coordinatively saturated***: like saturated organic molecules, their reactivity tends to be restricted to substitution reactions. Organotransition metal complexes in which the metal atom has less than 18 valence electrons are termed ***coordinatively unsaturated***. The metal atom in these complexes frequently participates in addition reactions that lead to the metal acquiring enough electrons to complete its valence shell.

To apply the 18-electron rule, one totals the number of valence electrons on the metal, adds the number of electrons contributed by the ligands, and then adjusts the total for the charge on the complex. The bonding between carbon and the metal in metal complexes may be either σ bonding or π bonding. Ligands bonded to the metal through σ bonds usually contribute either one or two electrons to the metal. The number of electrons donated by each π-bonded ligand is equal to the number of carbon atoms directly bonded to the metal, a number designated as the ***hapto number***, η. The hapto numbers of several common π-type ligands are shown in Figure 17.1.

Thus, in Zeise's salt, the two carbons of the ethylene are both simultaneously bound to the platinum: the alkene is an η^2 ligand. In ferrocene, each of the two cyclopentadienyl (frequently abbreviated as Cp) rings is simultaneously bound to the iron through all five carbon atoms, and the cyclopentadienyl groups are η^5 ligands. Although its use is widespread, the hapto nomenclature is by no means universal yet. Certain π-bonded ligands are known by trivial names that are much more widely used than the hapto names. One of the more common examples of the survival of a less systematic nomenclature for organometallic complexes concerns complexes containing η^3 ligands, which are widely referred to as π-allyl ligands.

To apply the 18-electron rule, one must know just how many electrons each ligand contributes to the metal in the complex. This can be determined fairly simply by noting that most ligands bonded to the metal through σ bonds donate one or two electrons and that π-bonded ligands donate a number of electrons equal to the hapto number of the ligand. The numbers of electrons donated by various ligands are given in Table 17.1, where the ligands are treated as neutral groups attached to a metal center, and Table 17.2, where the ligands are treated as ions bound to the metal center. Which convention is chosen will depend on the complex under study.

π-Allylpalladium complex **17.7** illustrates the use of this model for electron bookkeeping in metal complexes. This complex is neutral overall, and because the ligands are all

Figure 17.1 Hapto numbers for a series of common ligands

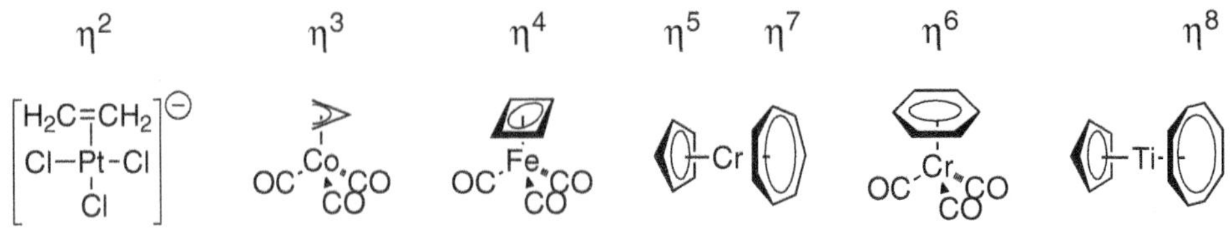

treated as monoanions, the metal atoms must each carry a +2 charge. In this version of the 18-electron rule, therefore, the configuration of the metal atom is d^8, and each metal atom carries 16 electrons. Both palladium atoms of the complex are coordinatively unsaturated, and both are in the +2 oxidation state—the complex is a palladium (II) complex.

Carbon-Metal Bonding in Organotransition Metal Complexes

The bonding between a transition metal and an organic ligand depends on both the metal and the nature of the organic ligand. When the ligand is a simple alkyl group, aryl group, or acyl group, the carbon-metal bond is a typical σ bond, which may be viewed as a pair of electrons shared between the carbon and the metal. In these complexes, the carbon-metal σ bond is formed by coaxial overlap of the metal orbital and the hybrid orbital on carbon to give σ and σ^* orbitals (Figure 17.2). In Figure 17.2, the process of bond formation is represented as being due over lap of the ligand highest energy occupied molecular orbital (HOMO) and the metal lowest energy unoccupied molecular orbital (LUMO). In terms of their reactivity, metal-carbon σ bonds behave as typical covalent bonds between carbon and an electropositive element.

The description of π bonding in organotransition metal complexes is less straightforward because the exact nature of the bonding depends on the exact type of the ligand. In the vast majority of cases, ligands that form π bonds with transition metals are themselves unsaturated—alkenes; arenes; and delocalized, conjugated radicals. The bonding between the metal and the ligand in these compounds is not simple two-center bonding, and the bonds cannot be described in terms of the simple Lewis electron pair model. Depending on the ligand, the metal may be simultaneously bonded to as few as two carbon atoms (η^2 ligands) or as many as seven or eight (η^7 or η^8 ligands). The X-ray crystal structure analysis of many organotransition metal complexes containing polyhapto alkene ligands has shown that the carbon atoms of the ligand are, where geometrically possible,

Table 17.1 Electrons Donated by Neutral Ligands for Electron Accounting by the 18-Electron Rule

Electrons	Typical Examples of Ligands
1	Alkyl, aryl, alkynyl, H, halogen
2	CO, phosphines, η^2 alkene, η^2 alkyne
3	η^3 allyl, NO
4	η^4 diene (e.g., 1,3-butadiene; cyclobutadiene; norbornadiene)
5	η^5 dienyl (e.g., cyclopentadienyl, Cp; cyclohexadienyl)
6	η^6 arene (e.g., benzene; mesitylene)
7	η^7 trienyl (e.g., cycloheptatrienyl)
8	η^8 tetraenyl (e.g., cyclooctatetraenyl)

Table 17.2 Electrons Donated by Ionic Ligands for Electron Accounting by the 18-Electron Rule

Electrons	Typical Examples of Ligands
2	Halide ion, alkoxide ion, carboxylate ion
4	η^3 allyl anion, η^4 diene
6	η^5 dienide anion, η^6 arene, η^7 trienyl cation
8	η^8 tetraenyl (e.g., cyclooctatetraenyl)

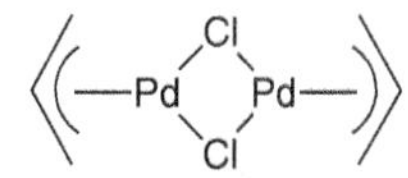

(17.7)

Pd^{2+} ($3d^8$)	8
C_3H_5(allyl anion 4 e^-)	4
Cl (anion, 2 @ 2e^-)	4
Total	16

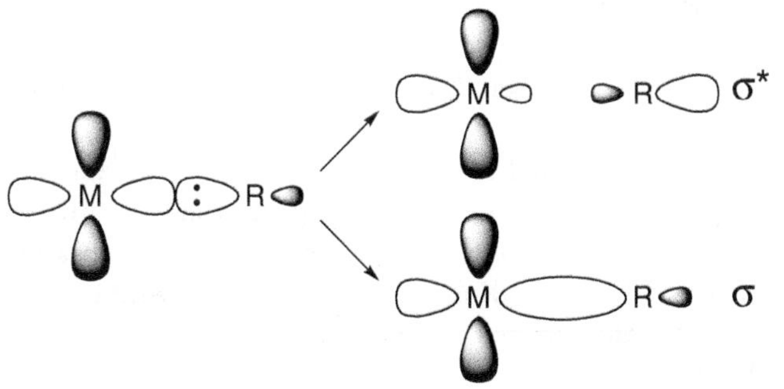

Figure 17.2 Orbital overlap between a metal atom lowest energy unoccupied molecular orbital and a ligand highest energy occupied molecular orbital to give the σ and σ^* orbitals of a metal-carbon bond

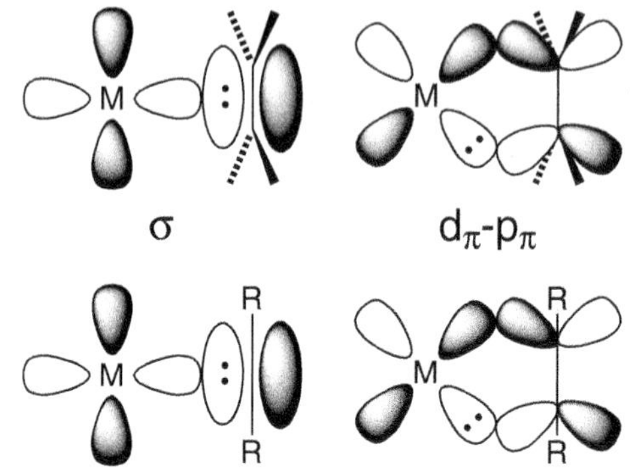

Figure 17.3 Metal-alkene (top) and metal-alkyne (bottom) σ bonding (left) and d_π-p_π backbonding (right)

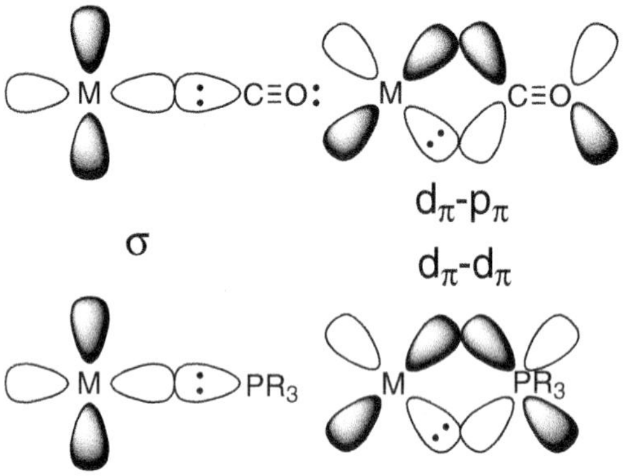

Figure 17.4 Bonding in carbonyl (top) and phosphine (bottom) complexes

equidistant from the metal. This means that all the carbons are equally involved in the bonding to the metal, which may be illustrated by the simplest example, a metal-alkene bond.

The metal–alkene bond (Figure 17.3) is rather more complex than the simple bonds that we have discussed to date because there are two components to the bonding process. Bonding in the "forward" direction involves the overlap of the ligand HOMO with an empty d orbital on the metal. Such bonds are almost invariably σ bonds. However, because most metal complexes of this type are formed by low-valent metals, the metal atom also has electrons available for donation. In addition, there is a component to the metal-carbon bonds that is due to overlap between the ligand LUMO and a filled d orbital ion the metal. The donation of electron density from the metal to the ligand is known as ***back-donation*** because the normal direction of electron donation in metal complexes is from the ligand to the metal; this type of bonding is called ***backbonding***, and it forms an important part of the bonding in many alkene complexes of transition metals. Backbonding is almost always π bonding, and because it involves the overlap of a metal d orbital and ligand p orbitals, it is often referred to as ***(d-p) π bonding*** or ***d_π-p_π bonding***. The bonding in alkyne complexes of transition metals is basically identical to the bonding in the alkene complexes.

Two important two-electron ligands for transition metals are carbon monoxide and trialkyl- or triarylphosphines. The compounds formed between a metal and carbon monoxide are known as ***carbonyl complexes***. In transition metal carbonyls, carbon monoxide acts as a two-electron donor, and it participates readily in backbonding through the C—O π* orbital (Figure 17.4). The bonding between a phosphine and a metal is almost identical to the bonding in a metal carbonyl, with the exception that the ligand orbital involved in the backbonding is an empty 3d orbital on phosphorus (so this bond is a d_π-d_π bond).

Worked Problem

17-2 Is the following metal complex coordinatively saturated or not? What is the oxidation state of the metal? Give your reasons.

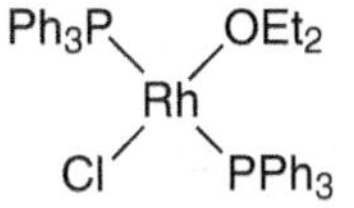

§ Answer below.

Problem

17-2 Which of the following complexes is(are) coordinatively unsaturated, and which is(are) coordinatively saturated? What is the oxidation state of the metal in each case?

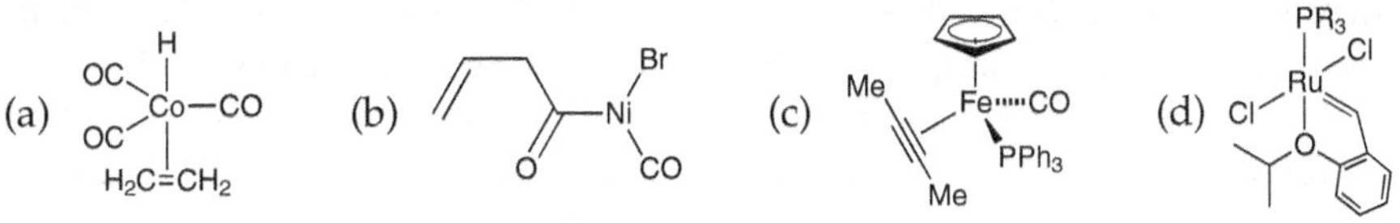

17.3 **Basic Organometallic Reactions I: Reactions at the Metal**

Ligand Substitution

Many of the reactions of organotransition metal complexes are due to the presence of accessible d orbitals on the transition metal. The simplest reactions of organometallic compounds are ligand substitution reactions where one ligand is replaced by another. These reactions may occur by either an associative mechanism (analogous to the S_N2 substitution) or a dissociative mechanism (analogous to the S_N1 substitution). Ligand substitution by a dissociative mechanism occurs through coordinatively unsaturated intermediates, and this reaction is therefore often a critical step in catalytic processes. A good example is provided by the initial stages of the ***hydroformylation reaction*** (Example **17.8**), an important method for the homologation of 1-alkenes into aldehydes by carbon monoxide using a cobalt catalyst.[15] Industrially, this reaction is known as the ***oxo process.***

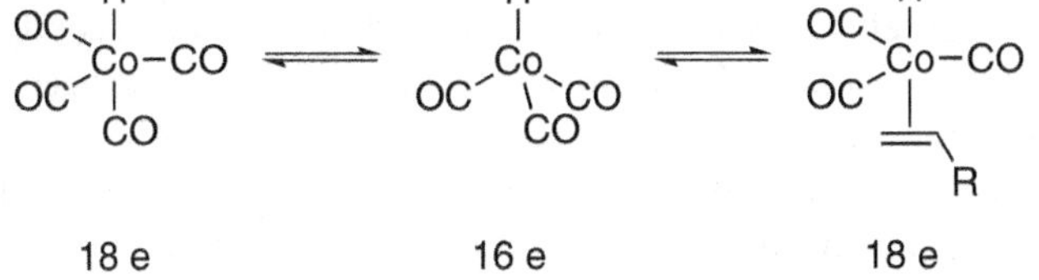

(17.8)

15. For a discussion of the mechanism of this reaction, see: (a) Heck, R.F.; Breslow, D.S. *J. Am. Chem. Soc.* **1961**, *83*, 4023. (b) van Leeuwen, P.W.N.M. *Homogeneous Catalysis: Understanding the Art* (Kluwer Academic: Dortrecht, Netherlands, 2004), ch. 7.

§ Answer to Worked Problem:
This complex has at least one ionic ligand (Cl^-), so we use the values in Table 17.2 to determine the total number of electrons. Because the overall complex is uncharged, this means that the rhodium must be in the form of Rh(I). It has 16 total electrons: Rh (8) + P (2 × 2) + Cl (2) + O(2), which makes it coordinatively unsaturated.

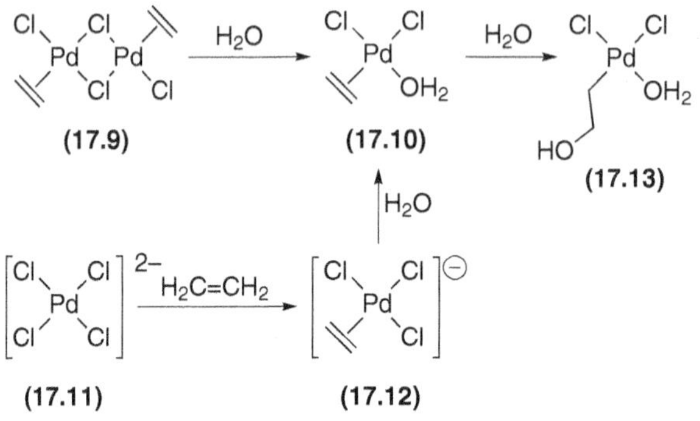

Figure 17.5 Metal substitution can dramatically affect ligand reactivity.

Ligand substitution can have a dramatic effect on the reactivity of a metal complex, and ligand substitution is an important method for modulating the reactivity of a metal complex. Thus, by appropriately choosing the ligands on the metal, one can stabilize reactive species (e.g., one can render air-sensitive complexes inert to all but the strongest oxidizing agents), or one can render ligands susceptible to attack by reagents toward which they are normally inert. These effects are well illustrated by the example in Figure 17.5. In the dimeric palladium complex (**17.9**), the alkene ligand is labile toward displacement but is not reactive toward nucleophilic attack. In the monomeric complexes (**17.11**, **17.12**), however, incorporating one water ligand and one alkene ligand gives a complex (**17.10**) where the alkene π bond is quite susceptible to attack by nucleophiles (e.g., water, as in **17.13**), as in the Wacker-Smidt[16] and related reactions.

Oxidative Addition and Reductive Elimination

Certain coordinatively unsaturated complexes react with neutral molecules to give coordinatively saturated products, a reaction accompanied by an increase in the formal oxidation number of the metal center. A lone pair on the metal is critical for this reaction to occur. This reaction is known as ***oxidative addition*** (Example **17.14**), and the reverse is known as ***reductive elimination*** (Example **17.15**). I have found it useful to write simple mechanisms, as shown, to rationalize electron movement in these reactions, but the involvement of the transition metal also means that these are almost certainly oversimplified.

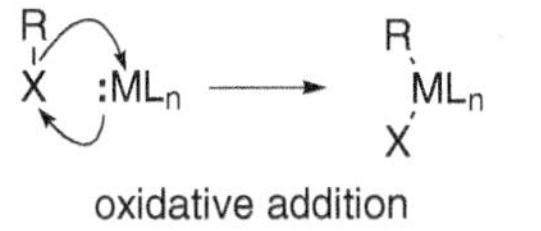

(17.14)

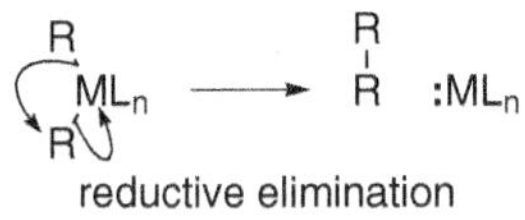

(17.15)

Examples of both types of reaction are provided by the catalytic cycle of hydrogenation using Wilkinson's catalyst (Figure 17.6).[17] In the first step of the catalytic cycle, the coordinatively unsaturated rhodium (I) complex (**17.16**) reacts with hydrogen to give the coordinatively saturated rhodium (III) hydride complex (**17.17**). In the last step of the catalytic cycle, the coordinatively unsaturated rhodium (I) complex is regenerated by reductive elimination of the alkane from the saturated alkyl complex (**17.18**).

---

16. (a) Smidt, J.;Hafner, W.; Jira, R.; Sedlmeier, J.; Sieber, R.; Rütlinger, R.; Kojer, H. *Angew. Chem.* **1959**, *71*, 176. (b) Smidt, J.; Hafner, W.; Jira, R.; Sieber, R.; Sedlmeier, J.; Sabel, S. *Angew. Chem. Int. Ed. Engl.* **1962**, *1*, 80.

17. (a) Tolman, C.A.; Meakin, P.Z.; Lindner, D.L.; Jesson, J.P. *J. Am. Chem. Soc.* **1974**, *96*, 2762. (b) Halpern, J.; Okamoto, T.; Zackiriev, A. *J. Mol. Calat.* **1976**, *2*, 65. (c) Collman, J.P.; Hegedus, L.S.; Norton, J.R.; Fincke, R.G. *Principles and Applications of Organotransition Metal Chemistry* (University Science Books: Mill Valley, CA, 1987), pp. 531–535. (d) Spessard, G.O.; Miessler, G.L. *Organometallic Chemistry*, 2nd ed. (Oxford University Press: New York, 2010), pp. 353–358.

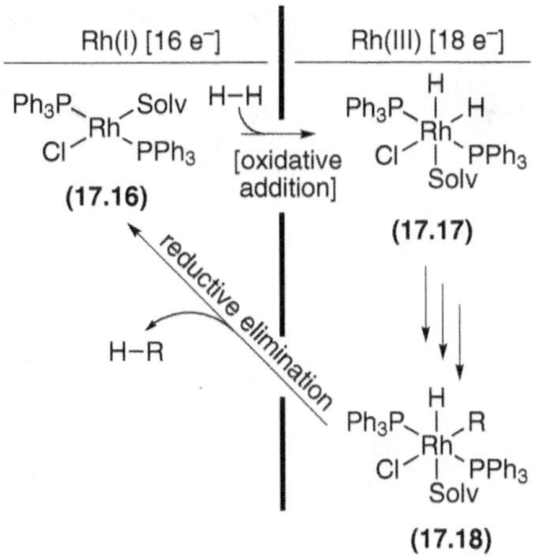

Figure 17.6 Oxidative addition and reductive elimination

Many simple molecules participate in oxidative addition reactions, and these may be divided into three broad classes: polar covalent molecules (e.g., alkyl or acyl halides; Example **17.19**), non-polar molecules (e.g., H_2 gas or C—H σ bonds; Example **17.20**), and unsaturated compounds (Example **17.21**). The overall result of the oxidative addition reaction is to form two new σ bonds to the metal, and to increase its oxidation number by two at the same time.

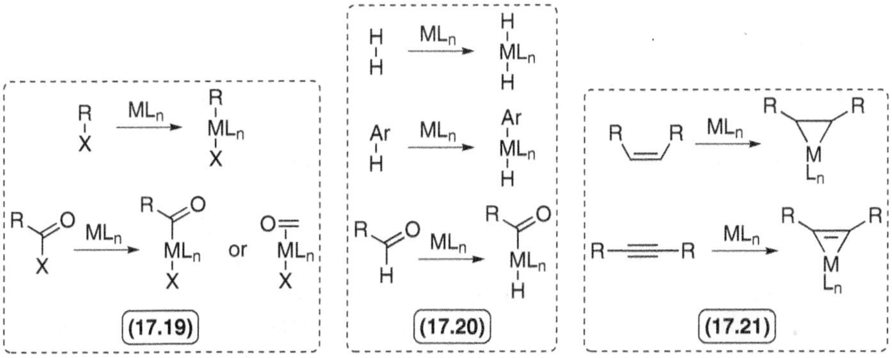

Insertion Reactions

Another reaction of organotransition metal complexes that is important in organic synthesis is known as ***migratory insertion*** (Example **17.22**). This reaction results in the insertion of one ligand into the metal-ligand bond of another. Two types of migratory insertion are commonly observed; in both cases, the migrating group formally adds to a π bond. In the one case, the bonding between the non-migrating ligand and the metal remains unaltered. In the second, the metal migrates to the other end of the π bond to give a net addition and conversion of a π complex to a σ complex. One of the most common examples of migratory insertion is the migratory insertion of carbon monoxide into a carbon-metal bond to give an acyl-metal bond that can be converted in turn into a variety of acyl compounds, including aldehydes, ketones, and carboxylic acids (as well as carboxylic acid derivatives such as amides and esters). The migratory insertion into carbon monoxide is assisted by external carbon monoxide, which maintains the coordination number of the metal atom. An excellent example of this reaction is provided by the carbonylation of allyl bromide by carbon monoxide in the presence of nickel tetracarbonyl (Figure 17.7). The migratory insertion step converts the tetracoordinate nickel carbonyl **17.24** into a

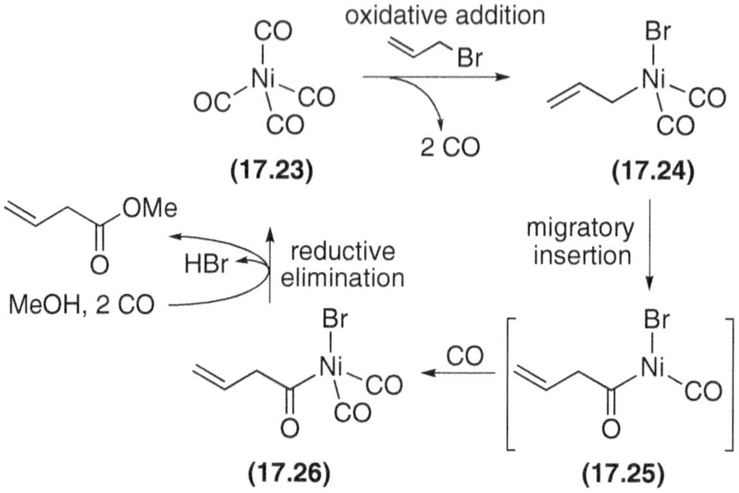

Figure 17.7 Carbonylation of allyl bromide by carbon monoxide in the presence of nickel tetracarbonyl

coordinatively unsaturated tricoordinate acylnickel complex (**17.25**), which then reacts further with carbon monoxide.

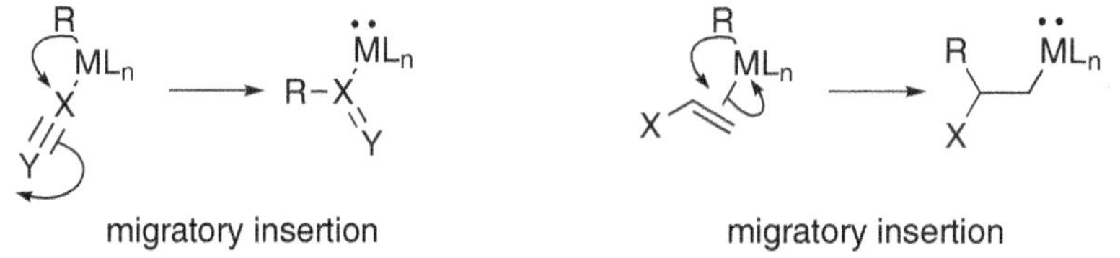

(**17.22**)

Migratory insertion is also an important step in the mechanism proposed[18] for the Ziegler-Natta polymerization reaction that proceeds through an η^2 titanium or vanadium alkene complex (Figure 17.8). In this mechanism, the polymer chain is extended by means of a migratory insertion reaction of the alkene into the metal-carbon σ bond of **17.28** to give **17.30**. Similar migratory insertion reactions of an alkene into a metal-hydrogen σ bond are believed to be critical steps in the mechanism of homogeneous catalytic hydrogenation.

The reverse of the migratory insertion reaction in Figure 17.7 (i.e., where a small molecule is *extruded from* a metal-carbon σ bond) is also important in organic synthesis, especially when that small molecule is carbon monoxide. This reaction is known as ***decarbonylation***. Decarbonylation of aldehydes by Wilkinson's catalyst is an important synthetic method, as well as being a significant (almost always undesirable) side reaction during hydrogenation of conjugated aldehydes with this catalyst. Decarbonylation occurs with retention of configuration (e.g., Example **17.31**).[19]

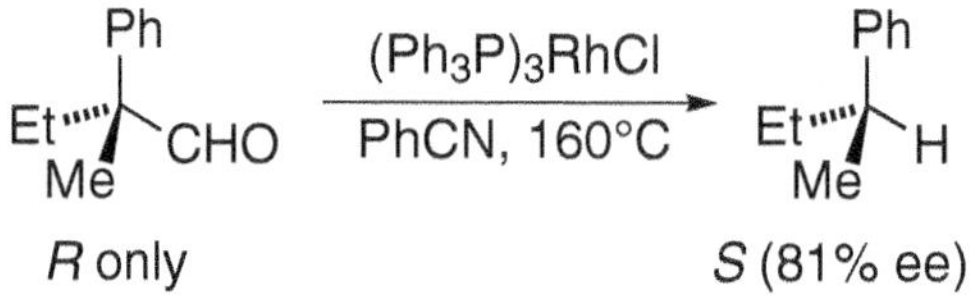

(**17.31**)

β-Hydride Elimination

The β-elimination of hydride from transition metal-alkyl σ-bonded complexes to give a metal hydride complex (or an η^2 alkene–metal hydride complex, as in Example **17.32**) and the unsaturated hydrocarbon also belong, in a formal sense, to this class of organometallic

18. (a) Cossee, P. *J. Catal.* **1964**, 3, 80. (b) Arlman, E.J.; Cossee, P. *J. Catal.* **1964**, 3, 99.
19. (a) Tsuji, J.; Ohno, K. *Synthesis* **1969**, 157. (b) Walborsky, H.M.; Allen, L.E. *J. Am. Chem. Soc.* **1971**, 93, 5465.

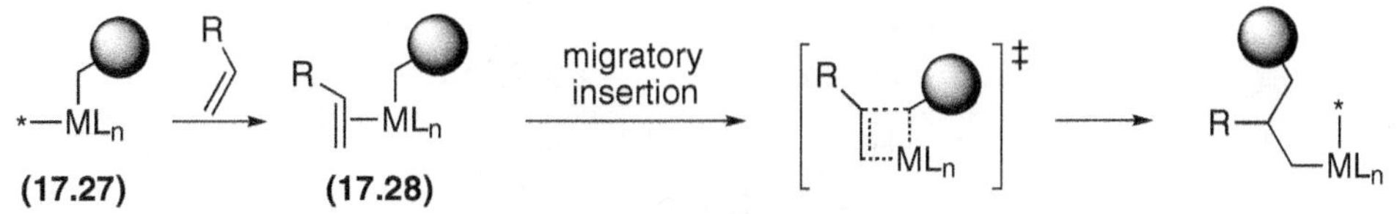

(17.27) (17.28) (17.29) (17.30)

Figure 17.8 Migratory insertion in the Ziegler-Natta polymerization. The asterisk is used here (and elsewhere) to denote a vacant coordination site on the metal atom.

reactions. The step of the Wacker process that converts Example **17.33** to the acetaldehyde enol η^2 complex of the palladium hydride (Example **17.34**) is a good example of this type of reaction.

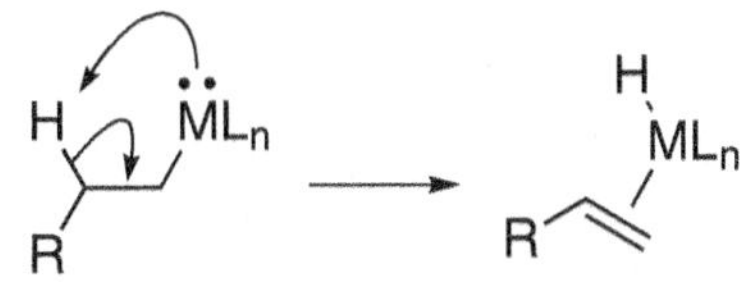

(17.32)

β-hydride elimination

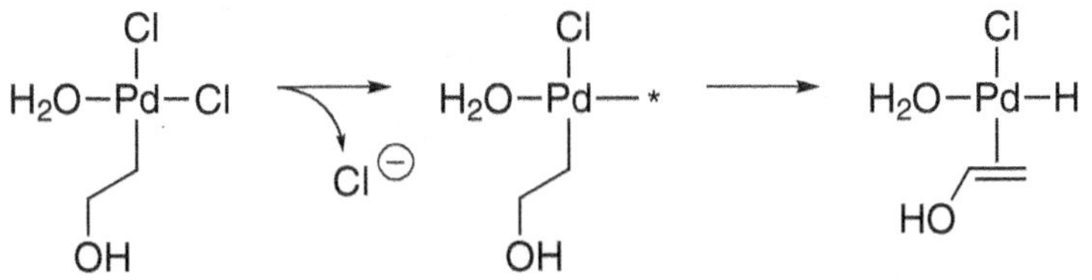

(17.13) (17.33) (17.34)

Transmetallation and Exchange with Metalloids

In Section 11.2, in the discussion of the metal alkyls of the alkali and alkaline earth metals, we considered the concept of metal exchange reactions and the formation of metal alkyls of the transition metals by metal-halogen exchange. This exchange is quite facile in these systems, but it also occurs quite easily when one or both of the metals involved is a transition metal. Thus, transmetallation is the key step of the ***Stille coupling***, a palladium-catalyzed reaction between a vinylstannane[20] (**17.38**, M=Sn) or alkylmercury compound (Example **17.38**, M=Hg)[21] and an alkenyl or aryl halide to give a conjugated diene or a styrene derivative. The reaction with carbon monoxide and methanol gives methyl esters.[22] Transmetallation can also be used with metal alkyls of magnesium,[23] zinc,[24] and aluminum.[25] We will consider these reactions in more detail at appropriate places later in this chapter. The exchange of palladium with boron in the boronic acids is the basis of the ***Suzuki coupling***,[26] and the exchange with silicon in tetraalkylsilanes is the basis of the ***Hiyama coupling***.[27]

---

20. Milstein, D.; Stille, J.K. *J. Am. Chem. Soc.* **1978**, *100*, 3636.
21. Reviews: Beletskaya, I.P. *J. Organomet. Chem.* **1983**, *250*, 551; *Pure Appl. Chem.* **2002**, *74*, 1327.
22. Stille, J.K.; Wong, P.K. *J. Org. Chem.* **1975**, *40*, 335.
23. Yamamoto, Y.; Kohara, T.; Yamamoto, A. *Chem. Lett.* **1976**, 1217.
24. (a) King, A.O.; Okukado, N.; Negishi, E.-i. *J. Chem. Soc. , Chem. Commun.* **1977**, 683. (b) Tamaru, Y.; Ochiai, H.; Yamada, Y.; Yoshida, Z. *Tetrahedron Lett.* **1983**, *24*, 3869.
25. Bumagin, N.A.; Ponomaryov, A.B.; Beletskaya, I.P. *Tetrahedron Lett.* **1985**, *39*, 4819.
26. Reviews: (a) Suzuki, A. *Pure Appl. Chem.* **1991**, *63*, 419; *J. Organomet. Chem.* **1999**, *576*, 147. (b) Miyaura, N.; Suzuki, A. *Chem. Rev.* **1995**, *95*, 2457. (c) Kotha, S.; Lahiri, K.; Kashinath, D. *Tetrahedron* **2002**, *48*, 9633.
27. Hatanaka, Y.; Hiyama, T. *J. Org. Chem.* **1988**, *563*, 918.

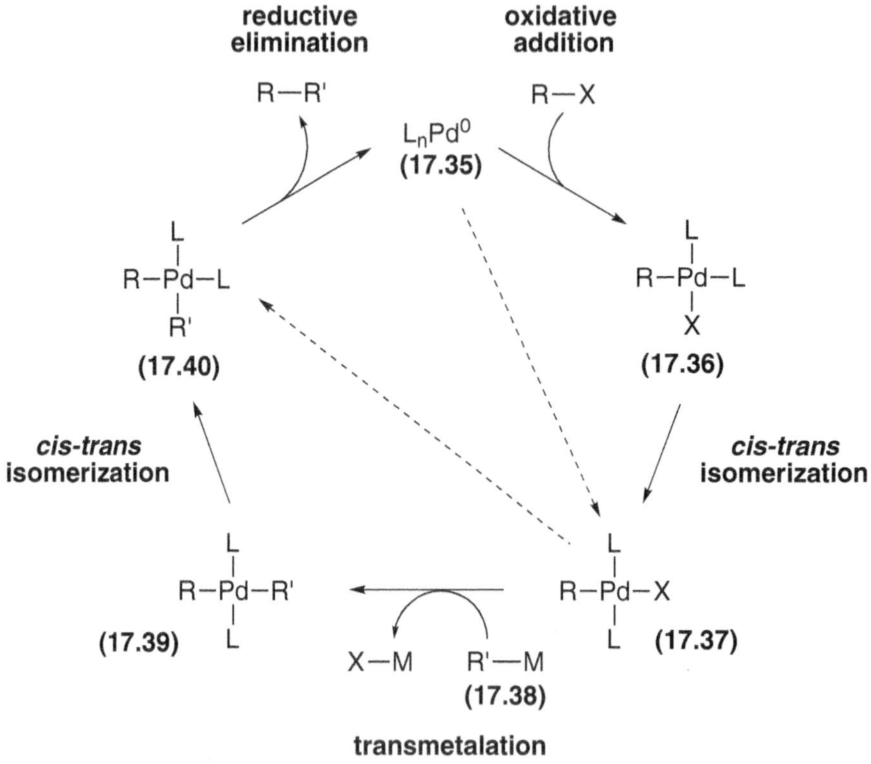

Figure 17.9 The catalytic cycle for metal-catalyzed coupling reactions

The foregoing discussion of reactions at the metal atom provides the backdrop for an introduction to the important synthetic reactions involving coupling of ligands on a metal. In most coupling reactions, it is a reductive elimination step involving two adjacent carbon ligands from the metal that leads to carbon-carbon bond formation. Metal-catalyzed coupling reactions all follow a general pattern where the final product arises in a reductive elimination reaction from a coordinatively saturated metal complex. The general catalytic cycle for coupling reactions is given in Figure 17.9. Beginning from the coordinatively unsaturated L_nPd (0) intermediate (**17.35**), the cycle consists sequentially of oxidative addition of the halide, isomerization to the *trans* isomer (if necessary), transmetallation, isomerization of the *trans* complex **17.39** to the *cis* isomer **17.40** (if necessary), and reductive elimination of the coupled product.

Reaction Synopses

Ligand Substitution

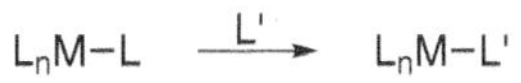

Reaction may proceed by a dissociative or an associative mechanism.

Oxidative Addition

Reaction converts coordinatively unsaturated complex to coordinatively saturated complex by effectively inserting metal into R—X σ bond.

Reductive Elimination

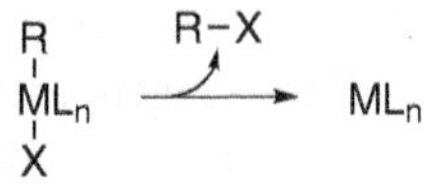

Reaction eliminates σ-bonded product from a coordinatively saturated complex to give a coordinatively unsaturated complex.

Migratory Insertion

Reaction results in coordinatively unsaturated complex in migration step; reaction is almost always assisted by additional ligand to satisfy need for coordinative saturation. It is most common with η^2 alkene or carbonyl ligands. With carbonyl ligands, the reaction may be reversible.

β-Hydride Elimination

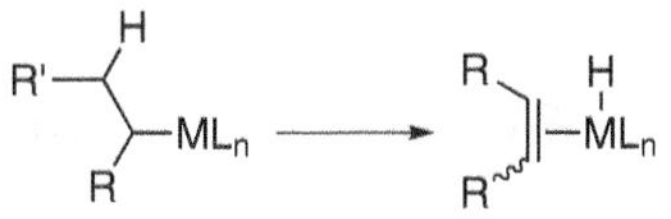

Reaction may lead to direct loss of alkene and formation of hydride complex, or to hydride complex with a bound to η^2 alkene ligand.

Transmetallation

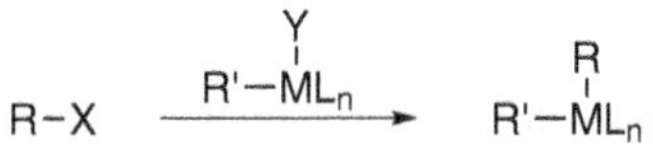

X: MgX, MgR, ZnX, ZnR, HgX, HgR, SnR_3, AlR_2, Cu, SiR_3, $B(OH)_2$ etc.

Problem

17-3 The reaction cycle diagram below is a common one for cross-coupling reactions catalyzed by transition metals (especially palladium). This particular example is for the Heck coupling reaction. Designate each of the steps of the cycle as belonging to one of the classes of reactions summarized in the Reaction Synopses immediately above. What is the oxidation state of the metal in each of the subsequent complexes, if complex A is Pd(0), and L is a neutral 2-electron ligand?

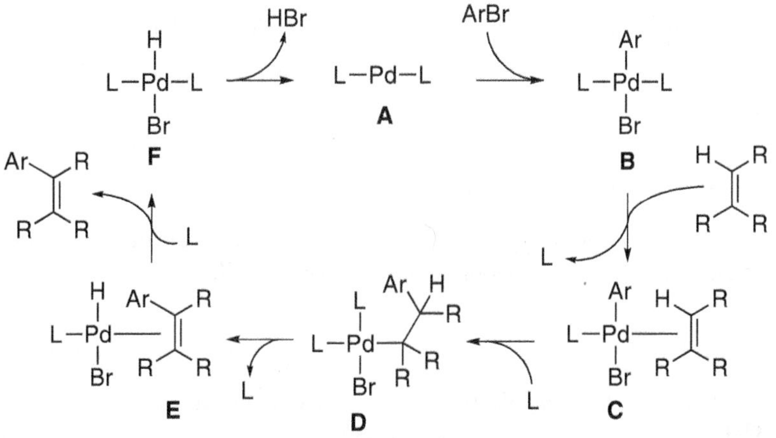

17.4 Basic Organometallic Reactions II: Reactions at the Ligand

In the reactions we have just discussed, the reagent attacks the metal atom directly. However, there is a rich chemistry of organic compounds coordinated to transition metal atoms where the attack of the reagent is at the coordinated ligand, instead of the metal. In the introductory section of this chapter, it was hinted that reactions in which transition metals play a role often disobey rules that have come to be viewed as axiomatic in introductory textbooks. In large part this is true: transition metal organometallic compounds exhibit many avenues of reactivity that are forbidden to simple organic compounds. This is nicely illustrated by the Wacker reaction. The first step of this reaction is a nucleophilic addition of water to the η^2 alkene ligand—and in introductory organic chemistry textbooks, the authors take great pains to make it clear that the activation energy of nucleophilic addition to the π bond of simple alkenes is prohibitively high. The major use of organotransition metal compounds in modern synthetic organic chemistry is, in fact, the formation of carbon-carbon bonds during the synthesis of complex organic compounds where the unusual reactivity of the ligands allows the construction of those bonds under extremely mild conditions (and often in violation of the "normal" reactivity).

Synthetic Reactions Involving Electrophilic Attack on Ligands

As might be expected, the nature of the metal atom (especially its oxidation state) profoundly affects the type of reactions that typically occur at the bound ligands. In general, metals in higher oxidation states render bound polyhapto ligands (η^2-alkene, η^3-allyl, η^4-dienyl, and so on) electrophilic and therefore susceptible to attack by nucleophilic species. In contrast, metals in lower oxidation states tend to render the same types of ligands (especially η^4-dienyl) more nucleophilic and therefore susceptible to attack by electrophiles. This is illustrated by the coordinatively saturated iron complexes **17.41**[28] and **17.42**.[29] In Example **17.41**, the iron atom is in the +2 oxidation state, and the ligand reacts with an enamine nucleophile. In Example **17.42**, the iron atom is in oxidation state zero, and the ligand reacts with an acylium ion electrophile. The treatment of the η^3-allyl product in Example **17.38** with a base—a general reaction of η^3-allyliron (II) complexes—leads to deprotonation and regeneration of an η^4 diene complex; the net effect is electrophilic acyl substitution of the diene.

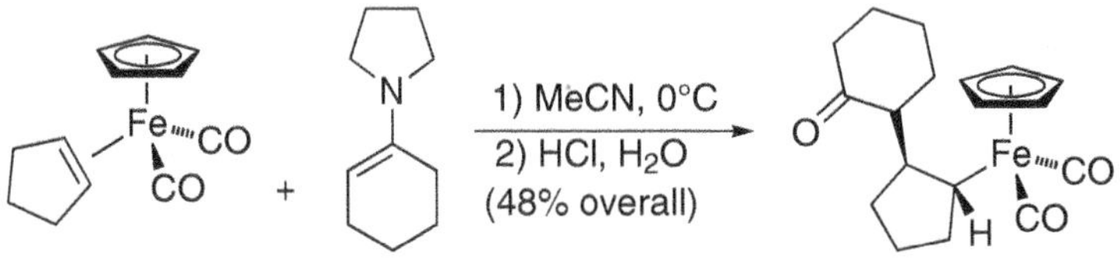

(17.41)

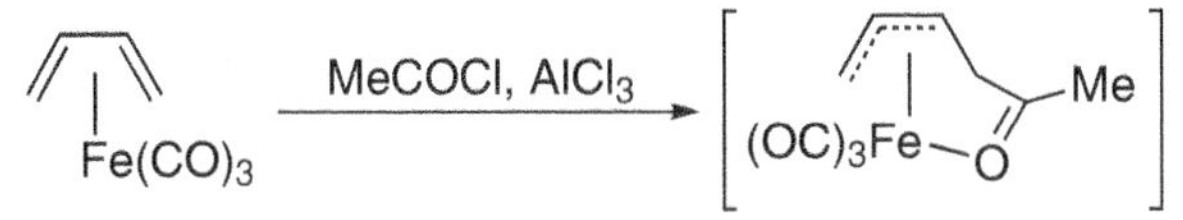

(17.42)

Synthetic Reactions Involving Nucleophilic Attack on Ligands

Alkene and alkyne complexes of iron (II), palladium (II), and platinum (II) are especially useful in reactions where a nucleophile attacks the hapto ligand. In the process, an η^2

28. (a) Rosenblum, M. *Acc. Chem. Res.* **1974**, *7*, 122. (b) Lennon, P.; Rosan, A.M.; and Rosenblum, M. *J. Am. Chem. Soc.* **1977**, *99*, 8426.

29. Greaves, E.O.; Knox, G.R.; Pauson, P.L.; Toma, S.; Sim, G.A.; Woodhouse, D.I. *J. Chem. Soc., Chem. Commun.* **1974**, 257.

ligand is converted to a σ ligand (Example **17.43**), and the hapto number of higher order hapto ligands is reduced by one (e.g., Examples **17.44** and **17.45**). The cationic (η^2-alkyne) iron (II) complex **17.46**, for example, reacts with a wide variety of nucleophiles, including malonic ester anions (e.g., Example **17.47**) and organocuprates (e.g., Example **17.48**) to give a σ-alkene complex by addition of the nucleophile *anti* to the iron.[30] The stereochemistry of addition in the corresponding complexes, in which the phosphine is replaced by a second carbonyl ligand, depends on the base strength of the nucleophile.[31]

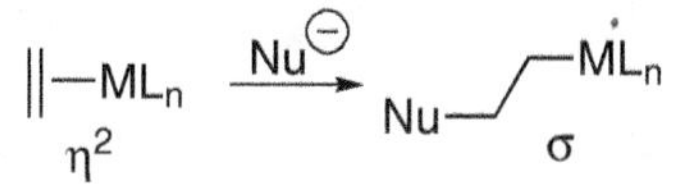

(17.43)

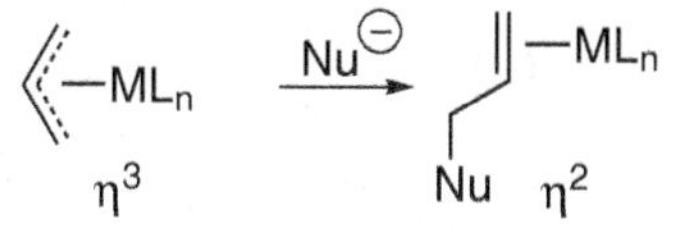

(17.44)

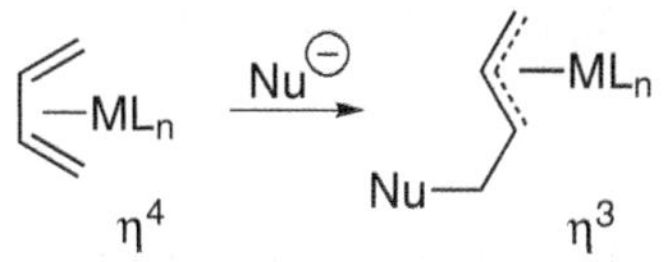

(17.45)

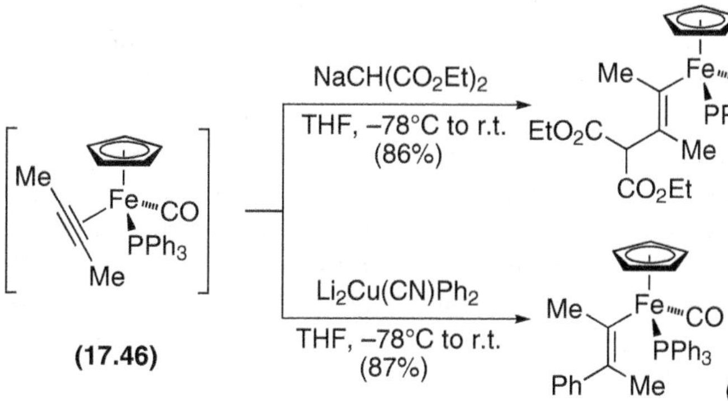

(17.46) **(17.47)** **(17.48)**

One of the most versatile applications of nucleophilic attack on the ligand occurs in π-allylpalladium (II) complexes, where the attack of a soft nucleophile occurs at the η^3 ligand from the face opposite the metal atom. This reaction, which can be represented by the catalytic cycle in Figure 17.10, quickly became an important synthetic reaction thanks to the pioneering contributions[32] of Jiro Tsuji[33] and Barry M. Trost,[34]

30. Reger, D.L.; Belmore, K.A.; Mintz, E.; McElligott, P.J. *Organometallics* **1984**, *3*, 134.

31. Akita, M.; Kakuta, S.; Sugimoto, S.; Terada, M.; Tanaka, M.; Moro-oka, Y. *Organometallics* **2001**, *20*, 2736.

32. (a) Tsuji, J.; Takahashi, H.; Morikawa, M. *Tetrahedron Lett.* **1965**, 4387. (b) Trost, B.M. *Tetrahedron* **1977**, *33*, 2615. (c) Trost, B.M. *Acc. Chem. Res.* **1980**, *13*, 385. (d) Tsuji, J. *Organic Synthesis with Palladium Compounds* (Springer-Verlag: New York, 1980). (e) Trost, B.M. In Bartmann, W.; Sharpless, K.B., Eds. *Stereochemistry of Organic and Bioorganic Transformations* (VCH: New York, 1986. (f) Trost, B.M. *Pure Appl. Chem.* **1981**, *53*, 2357. (g) Trost,B.M.; Verhoeven, T.R. In Wilkinson, G., Ed. *Comprehensive Organometallic Chemistry* (Pergamon Press: Oxford, 1982), Vol. 8, Ch. 57. (h) Tsuji, J. *Pure Appl. Chem.* **1982**, *54*, 197. (i) Tsuji, J. *Tetrahedron* **1986**, *42*, 4361. (j) Trost, B.M. *Angew. Chem. Int. Ed. Engl.* **1989**, *28*, 1173. (k) Godleski, S.A. In Trost, B.M.; Fleming, I., Semmelhack, M.F. Eds. *Comprehensive Organic Synthesis* (Pergamon Press: Oxford, 1991); Vol. 4, Ch. 3.3. (l) Tsuji, J. *Transition Metal Reagents and Catalysts. Innovations in Organic Synthesis* (John Wiley & Sons: Chichester, U.K., 2000). (m) Tsuji, J. *Palladium Reagents and Catalysts. New Perspectives for the 21st Century.* (Wiley: print-on-demand, ISBN 978-0-470-85032-9; 2004).

33. **Jiro Tsuji (1927–)** was educated at Kyoto (BS), Baylor (MS), and Columbia (PhD, 1960). After 14 years in industry, he joined the Tokyo Institute of Technology in 1974 as professor. In 1988, he moved to Okayama University of Science. Brief biographies are available: *Acc. Chem. Res.* **1973**, *6*, 8; **1987**, *20*, 140.

34. **Barry M. Trost (1941–)** was educated at the University of Pennsylvania (BA, 1962) and the Massachusetts Institute of Technology (PhD, 1965). He joined the University of Wisconsin, becoming Professor in 1969, and he moved to Stanford University in 1987. More details are available in *Acc. Chem. Res.* **2002**, *35*, 695.

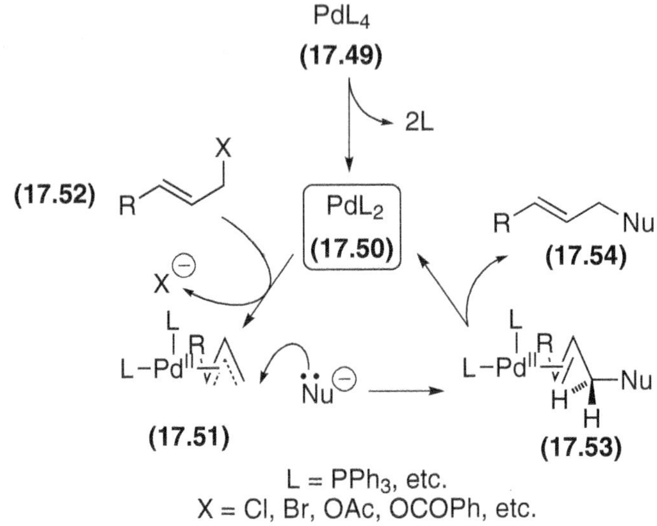

Figure 17.10 The catalytic cycle of the Tsuji-Trost reaction with soft nucleophiles

for whom the reaction is often named. Much of the work to adapt this reaction to asymmetric synthesis has been done by the Trost group.[35] The L_2Pd^0 complex (**17.50**) is formed from a coordinatively saturated palladium (0) complex, **17.49**—often $(Ph_3P)_4Pd^0$. The catalytic cycle is then initiated by an oxidative addition with the allylic halide or acetate (**17.52**).

The addition of alkyllithium reagents to carbon monoxide gives extremely unstable acyllithium reagents that are not useful reagents in organic synthesis. It is for this reason that many organic chemists have sought to generate "acyl anion equivalents" or ***umpoled*** reagents for use where an acyl anion would be useful.[36] In contrast to the acyllithium reagents, acyl complexes of transition metals are both stable enough to be useful reagents and also readily formed by the addition of alkyllithium reagents to coordinated carbon monoxide in metal carbonyls.

Unlike alkali metal acyl complexes, the transition metal acyl complexes have a resonance-stabilized structure, with both the acylmetal anion (Example **17.55**) and the metallocarbene (Example **17.56**) contributing to the resonance hybrid. The bond lengths of the C—O and C—M bonds in these complexes support this interpretation of the bonding in acylmetal complexes of the transition metals.

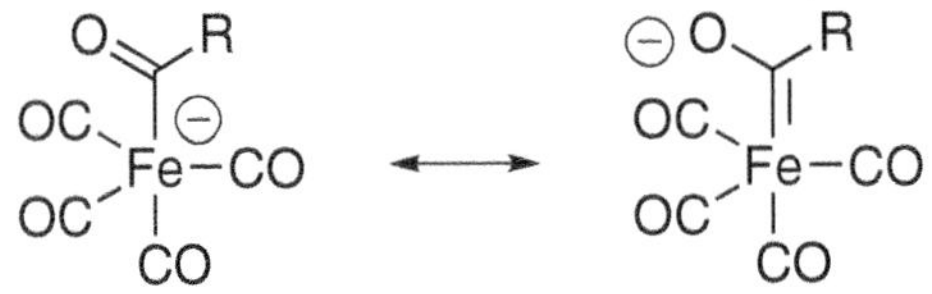

(17.55) **(17.56)**

Like an enolate anion, an acylmetal anion is also an ambident nucleophile. With protons and soft Lewis acids, the predominant reaction of these complexes is through the metal to give the product of oxidative addition. In the example in Figure 17.11, we see that

35. Reviews: (a) Trost, B.M.; Van Vranken, D.L. *Chem. Rev.* **1996**, *96*, 395. (b) Trost, B.M.; Crawley, M.L. *Chem. Rev.* **2003**, *103*, 2921.

36. For a useful compilation of early references to acyl anion equivalents, see: Hase, T.A.; Koskimies, J.K. *Aldrichimica Acta* **1981**, *14*, 73.

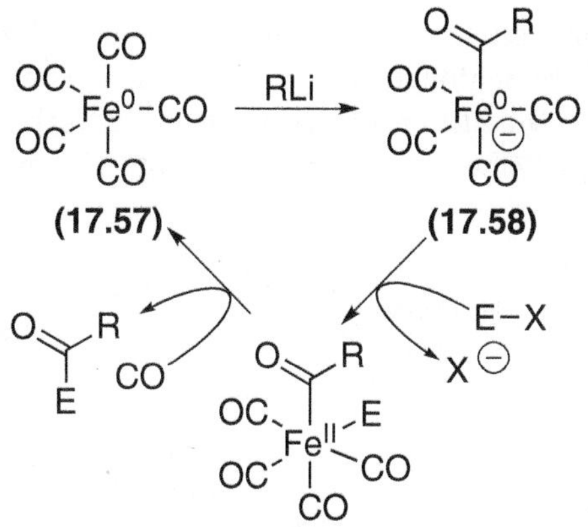

Figure 17.11 Acylmetal complexes react with soft electrophiles to give a metal-acyl product.

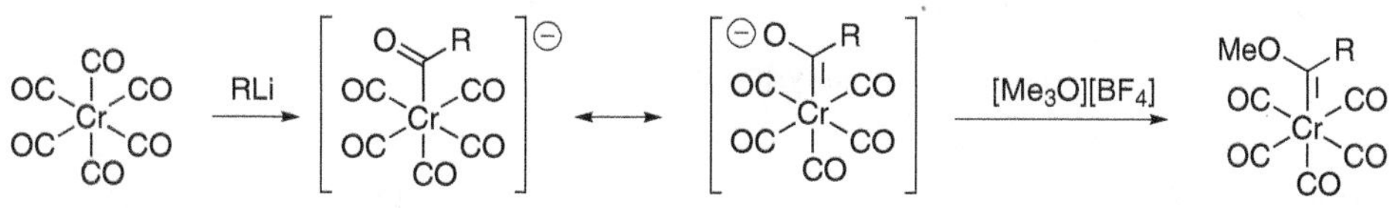

Figure 17.12 Formation of metal alkylidene complex from an acylmetal anion and a hard electrophile

these complexes (e.g., **17.54**) are formed quite readily by the addition of alkyllithium reagents to metal carbonyls (e.g., **17.57**). The resultant anionic acyl complexes react as metal-centered soft nucleophiles with a range of electrophiles to give a neutral complex (e.g., **17.59**) that gives the final carbonyl compound by reductive elimination. When a harder Lewis acid, such as a triflate or Meerwein's salt, is used to trap the anionic complex, the reaction occurs at the oxygen atom, the harder site, to give a metallocarbene, known as a ***Fischer complex***, after E. O. Fischer,[37] who first described them. One of the more synthetically useful Fischer complexes (**17.62**) is derived from chromium hexacarbonyl, as shown in Figure 17.12.

Synthetic Reactions Involving Electrophilic Attack on Ligands

Electrophilic substitution on η^n carbocyclic ligands has been known for decades. It is well illustrated by the acylation of the cyclobutadiene and cyclopentadienyl ligands in Examples **17.63**[38] and **17.64**.[39] Otherwise, the attack of electrophiles on ligands in transition metal complexes is less common, with the most frequent examples of the reaction being electrophilic substitution on the ligand, as in the earlier example of acylation of the η^4-butadieneiron tricarbonyl complex.[40] One interesting application of electrophilic addition to a metal complex is the formal [3 + 2] cycloaddition reaction between a σ-allyliron complex and diethyl methylenemalonate. The reaction proceeds by initial electrophilic addition of the very electron-deficient alkene (**17.65**) to

37. **Ernst Otto Fischer (1918–2007)** was a Nobel Prize-winning (1973) chemist. His education at the Technische Universität München (PhD, 1952) was delayed by "work service" under the Nazis and military service during World War II. He spent his career at the University of Munich (1957–1964) and then at the Technical University, where he retired in 1984. For more details, see the Nobel Foundation website.

38. Broadhead, G.B.; Osgerby, J.M.; Pauson, P.L. *J. Chem. Soc.* **1958**, 650.

39. Fitzpatrick, J.D.; Watts, L.; Emerson, G.F.; Pettit, R. *J. Am. Chem. Soc.* **1965**, *87*, 3524.

40. Greaves, E.O.; Knox, G.R.; Pauson, P.L.; Toma, S.; Sim, G.A.; Woodhouse, D.I. *J. Chem. Soc., Chem. Commun.* **1974**, 257.

the allyl group, with concomitant conversion of the σ- complex to an η^2-alkene complex, and formation of the malonate anion (**17.66**). This complex then undergoes a subsequent intramolecular addition of the nucleophile to the η^2-alkene complex to give a new σ complex (**17.67**) with a cyclopentane ring. The mechanism for this reaction is shown in Figure 17.13.

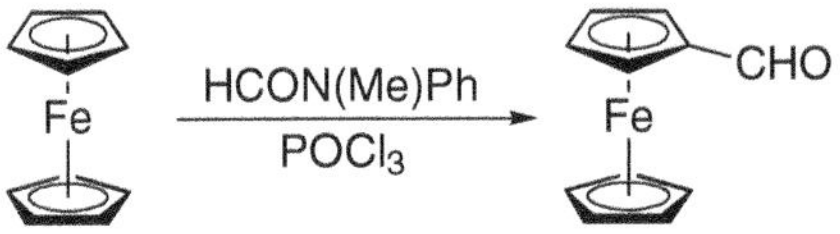

(17.63)

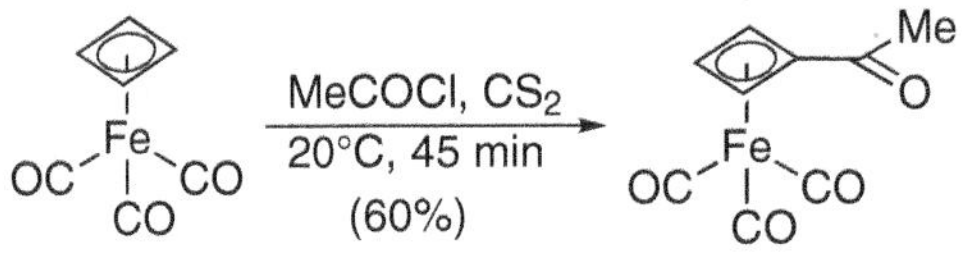

(17.64)

Oligomerization of Alkene and Alkyne η^2 Ligands

In 1940, German chemist Walter Reppe discovered that alkynes oligomerize to cyclic polyenes, often in high yields, in the presence of nickel catalysts. One of the earliest examples of this reaction—the nickel-catalyzed cyclopolymerization of acetylene to cyclic polyenes, especially cyclooctatetraene (Example **17.68**)—became an integral part of the German research efforts to use acetylene to circumvent the need for petroleum and rubber. The nickel-catalyzed reactions of acetylenes, now known collectively as ***Reppe chemistry***, were first reported after the end of World War II in three papers[41] occupied the first 116 pages of volume 560 of *Justus Liebigs Annalen der Chemie*! Although details of the reaction remain unclear, this reaction appears to be an example of a ***template reaction***, where all four alkynes are preassembled on the metal atom as a template (Example **17.69**) prior to the oligomerization. By using ligands on the nickel (e.g., triphenylphospine) that can compete with the acetylene for the metal, it is possible to use the reaction to prepare benzene derivatives instead of the cyclooctatetraene (Example **17.70**). In more recent forms of the reaction (e.g., starting from **17.71**, taken from the elegant synthesis of (±)-estrone by Funk and Vollhardt[42]), the nickel catalyst has been replaced by $CpCo(CO)_2$. Under these conditions, the reaction has generally been carried out with two of the alkyne groups within one of the two reactants.

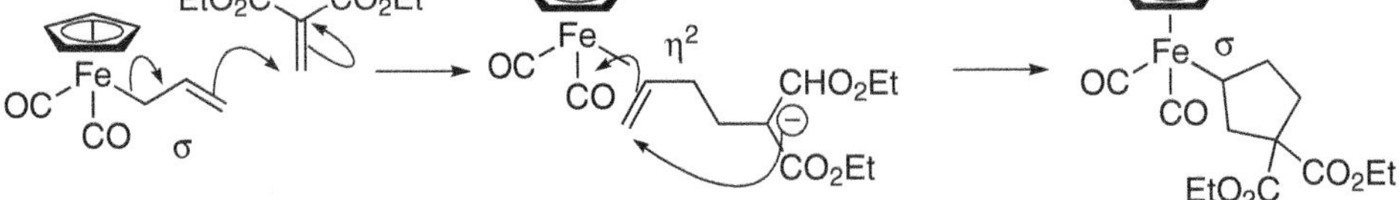

(17.65) **(17.66)** **(17.67)**

Figure 17.13 Formation of a cyclopentane by sequential electrophilic addition to a σ-allyl complex and intramolecular nucleophilic addition to an η^2-alkene complex

41. (a) Reppe, W.; Schlichting, O.; Klager, K.; Toepel, T. *Justus Liebigs Ann. Chem.* **1948**, *560*, 1. (b) Reppe, W.; Schlichting, O.; Meister, H. *Justus Liebigs Ann. Chem.* **1948**, *560*, 93. (c) Reppe, W.; Schwekendiek, W.J. *Justus Liebigs Ann. Chem.* **1948**, *560*, 104.
42. Funk, R.L.; Vollhardt, K.P.C. *J. Am. Chem. Soc.* **1979**, *101*, 215.

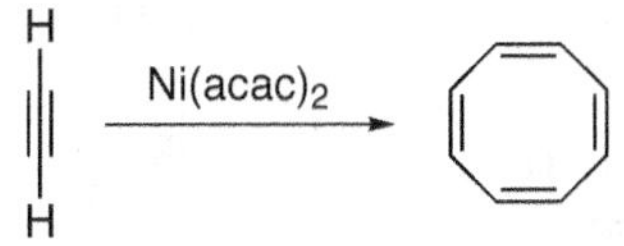

(17.68)

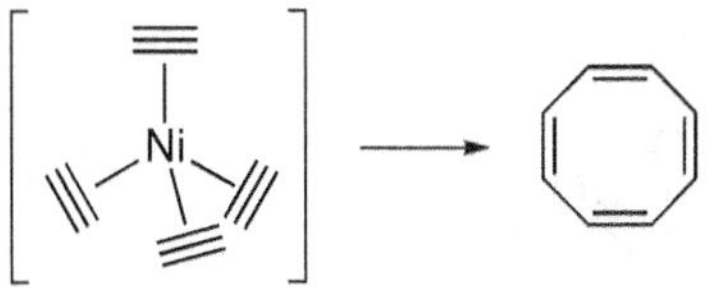

(17.69)

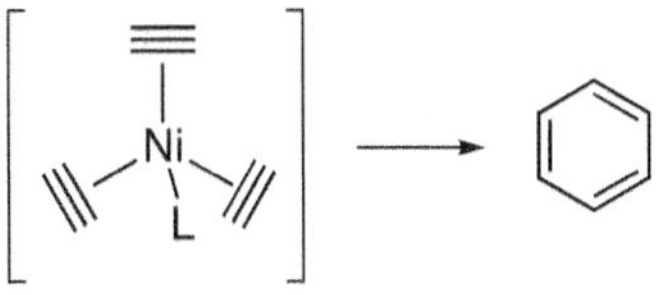

(17.70)

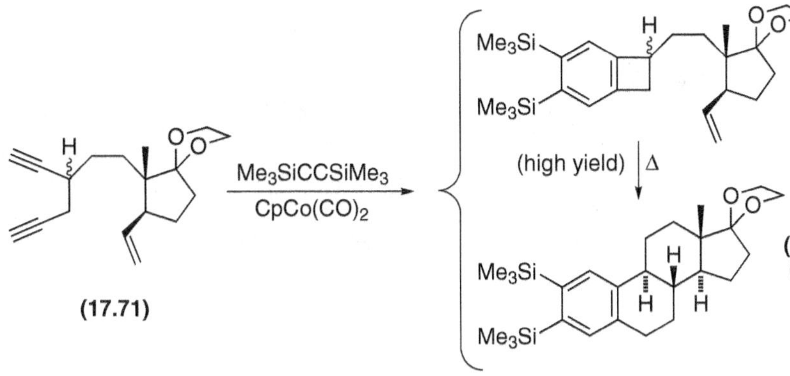

(17.71)

(17.72) (57%)

(17.73) (30%)

A related reaction is the ***Pauson-Khand*** reaction,[43] in which an alkene and an alkyne react with carbon monoxide under catalysis by a transition metal carbonyl such as $Co_2(CO)_8$ to give a cyclopentenone. The catalytic cycle of the reaction is given in Figure 17.14. It involves the complexation of the alkyne replacing the two μ-CO ligands of the $Co_2(CO)_8$ to give **17.75**, followed by displacement of one CO ligand by the alkene to give the η^2 complex **17.76**, which then undergoes migratory insertion of the η^2 ligand to give a cobaltocycle **17.77**. The mechanism as shown in Figure 17.14 then passes through migratory insertion into a carbonyl ligand with concomitant carbonylation to give the complex **17.78**, and a second carbonylation to give the cobaltocyclopropane **17.79**. This then reacts with a molecule of alkyne to release the enone (**17.80**) and regenerate the complex **17.75**. The formation of the cobalt carbonyl alkyne complex **17.75** changes the geometry about the triple bond and allows the reaction between remote groups on the chain that are not permitted while the triple bond is present.

Nucleophilic cyclizations promoted by Lewis acids on the alkyne complex of dinuclear cobalt carbonyls are known as the ***Nicholas reaction***.[44] An example of this is provided by the tandem Nicholas cyclization—Pauson-Khand cyclization of the propargyl ether **17.81** to give the tricyclic ether **17.84**.[45] Originally, the Pauson-Khand reaction's utility was limited by the need for stoichiometric amounts of the highly toxic, pyrophoric dicobalt octacarbonyl, the vigorous reaction conditions, and the low yields. Continued

---

43. (a) Khand, I.U.; Knox, G.R.; Pauson, P.L.; Watts, W.E. *J. Chem. Soc. D, Chem. Commun.* **1971**, 36a. (b) Khand, I.U.; Knox, G.R.; Pauson, P.L.; Watts, W.E.; Foreman, M.I. *J. Chem. Soc., Perkin Trans. 1* **1973**, 977. (c) Pauson, P.L.; Khand, I.U. *Ann. N.Y. Acad. Sci.* **1977**, *295*, 2. Reviews: (d) Schore, N.E. *Org. React.* **1991**, *40*, 1. (e) Gibson, S.E.; Stevenazzi, A. *Angew. Chem. Int. Ed.* **2003**, *42*, 1800. (f) Blanco-Urgoiti, J.; Añorbe, L.; Pérez-Serrano, L.; Domínguez, G.; Pérez-Castells, J. *Chem. Soc. Rev.* **2004**, *33*, 32. (g) Gibson, S.E.; Mainolfi, N. *Angew. Chem., Int. Ed.* **2005**, *44*, 3022.

44. For reviews, see: (a) Teobald, B.J. *Tetrahedron* **2002**, *58*, 4133. (b) Caffyn, A.J.M.; Nicholas, K.M. In *Comprehensive Organometallic Chemistry II*; Abel, E.W.; Stone, F.G.A.; Wilkinson, G. Eds. (Pergamon: Kidlington, 1995); Vol. 12, p 685.

45. Quintal, M.M.; Closser, K.D.; Shea, K.M. *Org. Lett.* **2004**, 6, 4949.

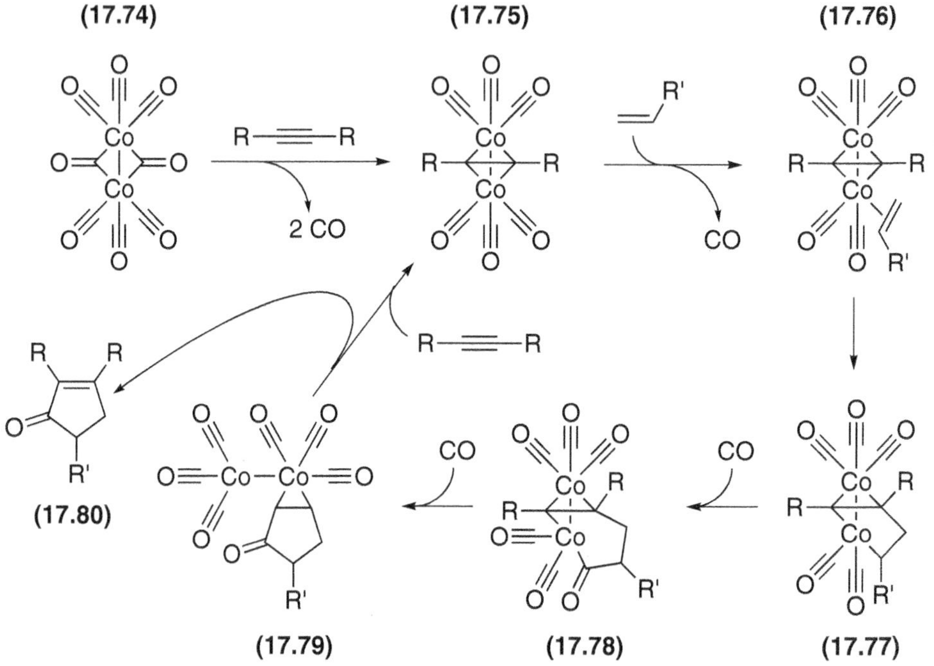

Figure 17.14 A catalytic cycle for the Pauson-Khand reaction

research, however, has resulted in a reaction that can be carried out under catalytic conditions, using rhodium, titanium, cobalt, iridium, and ruthenium carbonyl complexes as catalysts—under much milder reaction conditions. By appropriate use of chiral ligands (e.g., Example **17.87**), the reaction can be used to prepare chiral cyclopentenones such as Example **17.86**.[46]

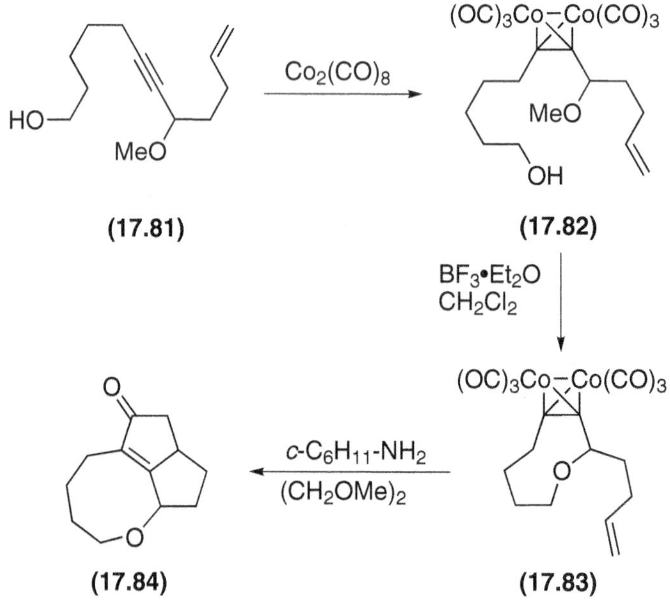

46. Kim, D.E.; Kim, I.S.; Ratovelomanana-Vidal, V.; Genêt, J.-P.; Jeong, N. *J. Org. Chem.* **2008**, *73*, 7985.

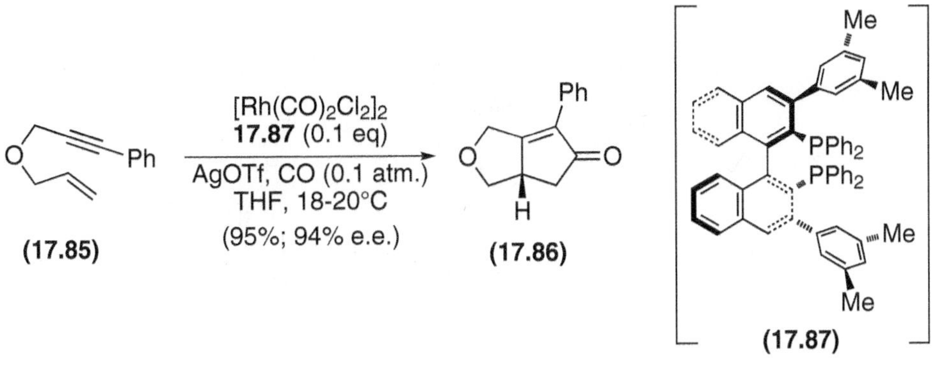

Metathesis and Metallocycle Formation

Olefin metathesis[47] has been known since the middle of the 20th century,[48] but it is only since the 1990s that its use has become widespread. The currently accepted mechanism of the reaction[49] was proposed in 1971 by Yves Chauvin,[50] who shared the 2005 Nobel Prize in Chemistry for the development of the reaction. The mechanism, shown in Figure 17.15, involves an initial [2 + 2] cycloaddition of a metallocarbene complex (**17.88**) to the first alkene to give a metallocyclobutane (**17.89**). This is followed by a retro-[2 + 2] cycloreversion to give a new metallocarbene complex (**17.90**) that then adds to the second alkene to give a second metallocyclobutane (**17.91**). This then expels the product alkene as a mixture of stereoisomers. The reaction proceeds most favorably when both participating alkenes are terminal, so that one of the alkene products is ethylene.

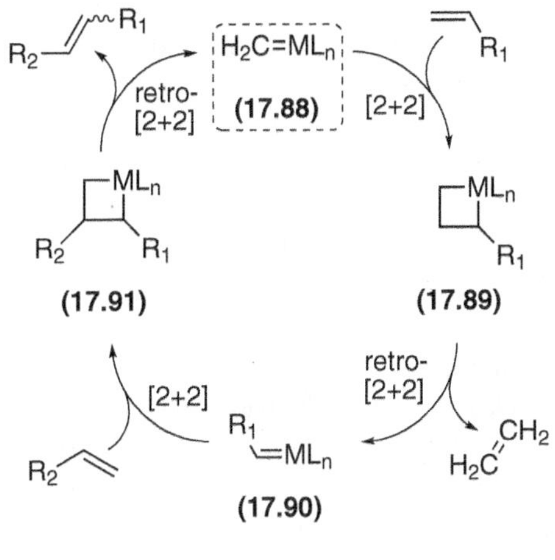

Figure 17.15 The catalytic cycle of the Chauvin mechanism for olefin metathesis

47. Reviews: (a) Hoveyda, A.H.; Zhugralin, A.R. *Nature* **2007**, *450*, 243. (b) Nicolaou, K C.; Bulger, P.G.; Sarlah, D. *Angew. Chem. Int. Ed.* **2005**, *44*, 4490. (c) Grubbs, R.H. *Tetrahedron* **2004**, *60*, 7117. (d) Chatterjee, A.K.; Choi, T.-L.; Sanders, D.P.; Grubbs, R.H. *J. Am. Chem. Soc.* **2003**, *125*, 11360. (e) Connon, S.J.; Blechert, S. *Angew. Chem. Int. Ed.* **2003**, 42, 1900. (f) Fürstner, A. *Angew. Chem. Int. Ed.* **2000**, 39, 3013. (g) Grubbs, R.H.; Chang, S. *Tetrahedron* **1998**, *54*, 4413. (h) Armstrong, S. K. *J. Chem. Soc., Perkin Trans. 1* **1998**, 371.

48. Anderson, A.W.; Merking, N.G. **U.S. Patent No. 2.721,189** [Oct. 18, 1955]

49. Hérisson, J.-L.; Chauvin, Y. *Makromol. Chem.* **1971**, *141*, 161.

50. **Yves Chauvin (1930–2015)**, who shared the Nobel Prize in Chemistry for 2005, spent his career with the Institut Français du Pétrole, from which he retired in 1995. His autobiography is available at the Nobel Foundation web site.

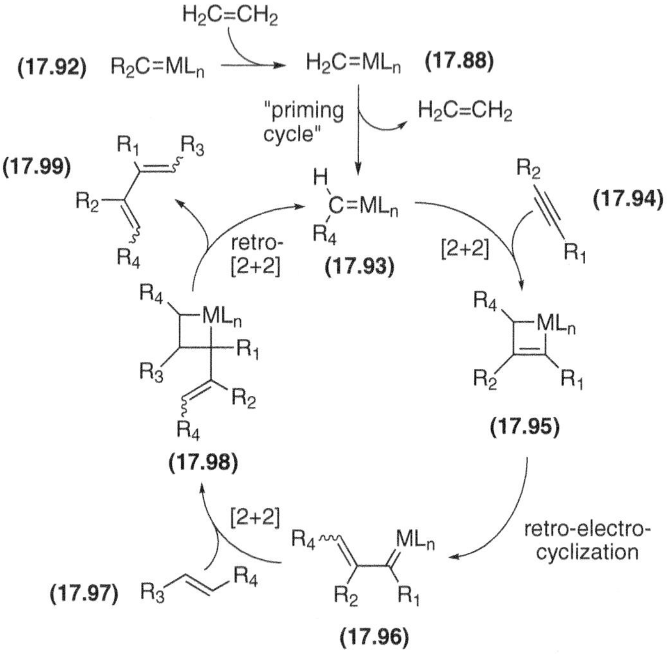

Figure 17.16 The catalytic cycle of the enyne metathesis reaction

A similar mechanism, shown in Figure 17.16, can be written for the related ***enyne metathesis*** reaction to give conjugated dienes. This reaction requires an initial priming cycle that generates the actual metallocarbene catalyst from the alkene partner in the reaction. By carrying out the reaction under an atmosphere of ethylene, the reaction is accelerated due to formation of the methylene complex[51] in a pre-equilibrium before entering the catalytic cycle.

The initial [2 + 2] cycloaddition gives a metallocyclobutene (**17.95**) that then undergoes a retro-electrocyclization to give a new metallocarbene (**17.96**). This then participates in the metathesis with the second alkene to give the diene (**17.99**). Note how, in contrast to olefin metathesis, this reaction leads only to reorganization of the atoms of the starting materials—all the atoms of the reactants end up in the product, and no molecule of a second alkene is produced.

The first work to show that a metal-alkylidene complex may be involved in the metathesis reaction was published by Casey,[52] and it was subsequently investigated by Katz[53] and then Schrock.[54] Catalysts such as **17.100**,[55] based on tungsten or molybdenum, are now generally referred to as ***Schrock catalysts***.

In parallel with Schrock's work, Grubbs had been investigating the use of Tebbe's reagent[56] (Example **17.101**) as a catalyst for olefin metathesis. Although it showed activity (and was important in elucidating the mechanism of the reaction), its reactivity limited its usefulness as a catalyst. Subsequently, Grubbs turned his attention to complexes of ruthenium, and this has provided a wide range of highly active catalysts.[57] The first truly active catalyst for general use developed by the Grubbs group is the alkylidene complex **17.102**, now known as the first-generation Grubbs catalyst (***Grubbs I***).[58] Subsequent research found that the

51. Mori, M.; Sakakibara, N.; Kinoshita, A. *J. Org. Chem.* **1998**, *63*, 6082.
52. Casey, C.P.; Burkhardt, T.J. *J. Am. Chem. Soc.* **1974**, *96*, 7808.
53. (a) Katz, T.J.; Lee, S.J.; Acton, N. *Tetrahedron Lett.* **1976**, 4247. (b) Katz, T.J.; Acton, N. *Tetrahedron Lett.* **1976**, 4251. (c) Katz, T.J.; Lee, S.J.; Shippey, M.A. *J. Mol. Catal.* **1980**, *8*, 219.
54. (a) Schrock, R.; Rocklage, S.; Wengrovius, J.; Rupprecht, G.; Fellmann, J. *J. Mol. Catal.* **1980**, *8*, 73. (b) Rocklage, S.M.; Fellmann, J.D.; Rupprecht, G.A.; Messerle, L.W.; Schrock, R.R. *J. Am. Chem. Soc.* **1981**, *103*, 1440.
55. Oskam, J.H.; Fox, H.H.; Yap, K.B.; McConville, D.H.; O'Dell, R.; Lichtenstein, B.J.; Schrock, R.R. *J. Organomet. Chem.* **1993**, *459*, 185.
56. Tebbe, F.N.; Parshall, G.W.; Reddy, G.S. *J. Am. Chem. Soc.* **1978**, *100*, 3611.
57. Review of catalyst development by the Grubbs group: Grubbs, R.H. *Tetrahedron* **2004**, *60*, 7117.
58. Schwab, P.; Grubbs, R.H.; Ziller, J.W. *J. Am. Chem. Soc.* **1996**, *118*, 100.

complex **17.103**, which carries an *N*-heterocyclic carbene ligand, is an even more active catalyst; this catalyst is now known as the second-generation Grubbs catalyst (***Grubbs II***).[59] A more stable (if less reactive) form of the Grubbs catalyst is the ***Hoveyda-Grubbs catalyst***,[60] in which the metal carries a chelating alkylidene ligand (Example **17.104**). Within a short time, a second-generation Hoveyda-Grubbs catalyst (Example **17.105**) had also been reported.[61]

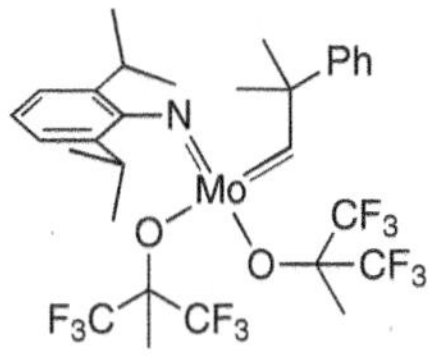

(17.100)

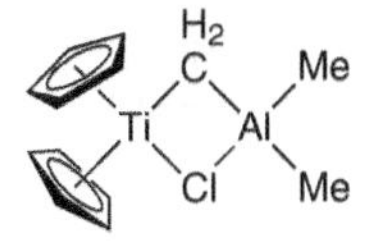

(17.101)

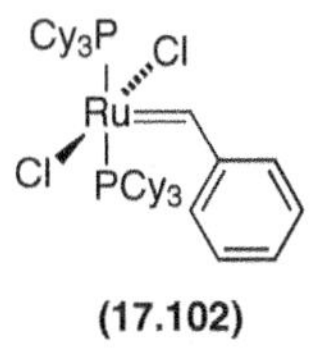

(17.102)

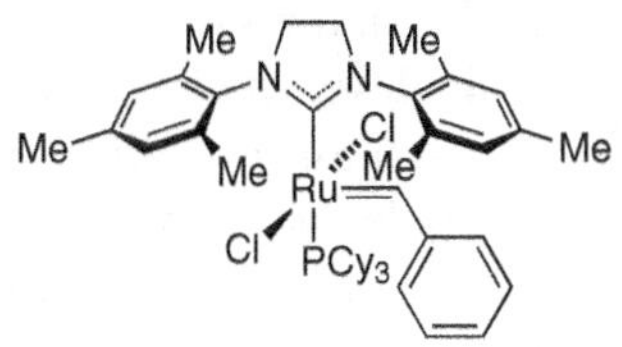

(17.103)

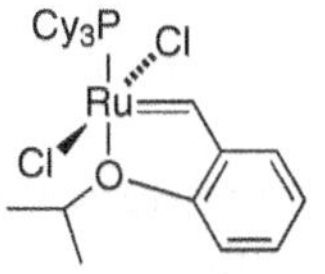

(17.104)

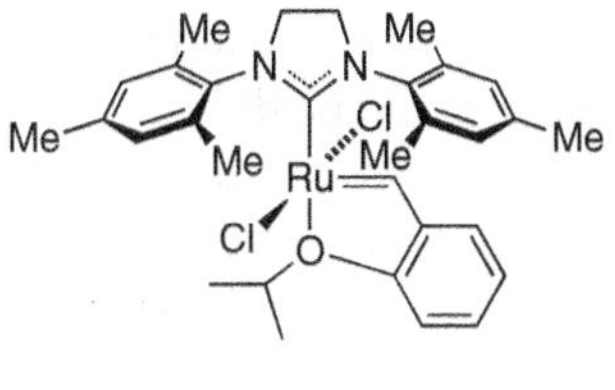

(17.105)

The original patent describing the reaction was for what is now known as ***ring-opening polymerization*** (***ROMP***) of norbornene. However, the major use of the reaction in modern synthesis is for the opposite task: closure of rings by elimination of a small alkene (usually ethylene). ***Ring- closing metathesis*** (***RCM***) has become a very powerful tool for the closure of medium to large rings, especially in active pharmaceuticals.[62] This is nicely illustrated by the closure of the 13-membered ring in Overman's synthesis of the marine alkaloid sarain-A, shown in Example **17.106**.[63] Although the regiochemistry of the enyne metathesis reaction is well defined, the stereochemistry of the intermolecular reaction is often poorly controlled.[64] For this reason, the intramolecular reaction, as illustrated in Example **17.107**,[65] is often preferred.

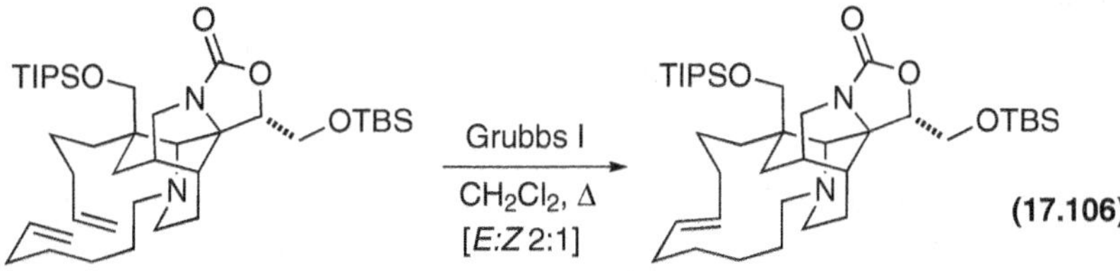

(17.106)

(82% after hydrogenation of double bond)

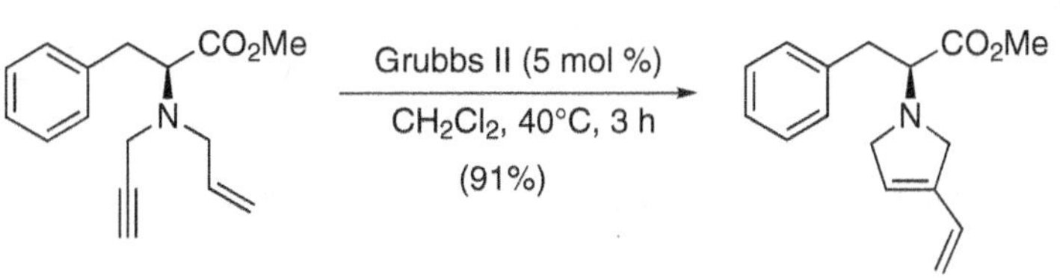

(17.107)

---

59. Scholl, M.; Trnka, T.M.; Morgan, J.P.; Grubbs, R.H. *Tetrahedron Lett.* **1999**, *40*, 2247.
60. Garber, S.B.; Kingsbury, J.S.; Gray, B.L.; Hoveyda, A.H. *J. Am. Chem. Soc.* **2000**, *122*, 8168.
61. Gessler, S.; Randl, S.; Blechert, S. *Tetrahedron Lett.* **2000**, *41*, 9973.
62. Reviews: (a) Phillips, A.J.; Abell, A.D. *Aldrichimica Acta* **1999**, 32, 75. (b) Vernall, A.J.; Abell, A.D. *Aldrichimica Acta* **2003**, *36*, 93.
63. Becker, M.H.; Chua, P.; Downham, R.; Douglas, C.J.; Garg, N.K.; Hiebert, S.; Jaroch, S.; Matsuoka, R.T.; Middleton, J.A.; Ng, F.W.; Overman, L.E. *J. Am. Chem. Soc.* **2007**, *129*, 11987.
64. For an extensive review, see: Diver, S.T.; Giessert, A.J. *Chem. Rev.* **2004**, *104*, 1317.
65. Yang, Q.; Alper, H.; Xiao, W.-J. *Org. Lett.* **2007**, *9*, 769.

Worked Problem

17-3 The following olefin metathesis reaction could give several possible products, but in practice, only one is isolated. What is that product and why is it preferred?

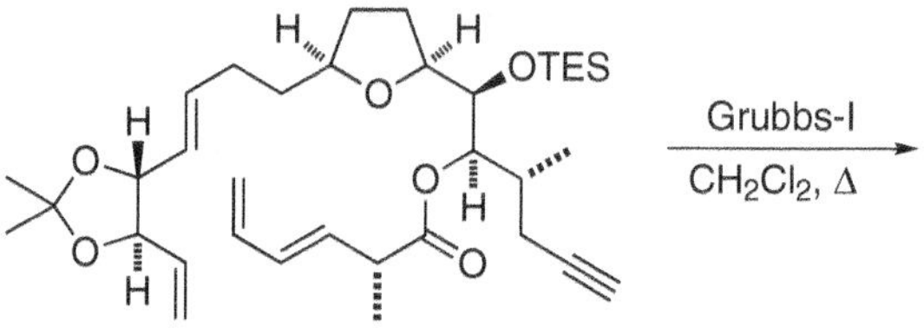

§Answer below.

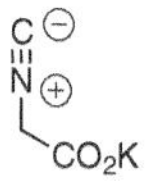

(17.108)

All metathesis reactions involving Grubbs or Hoveyda-Grubbs catalysts face the same problem: it can be difficult to remove the catalyst from the product, which often leads to further undesired metatheses during product isolation. Diver[66] has found a creative solution to this problem by using an isocyanide to stop the metathesis; by using an ionic isocyanide (Example **17.108**) in methanol, the metal could be removed quantitatively from the reaction mixture by simple chromatography.

Reaction Synopses

Nucleophilic Attack on Ligands

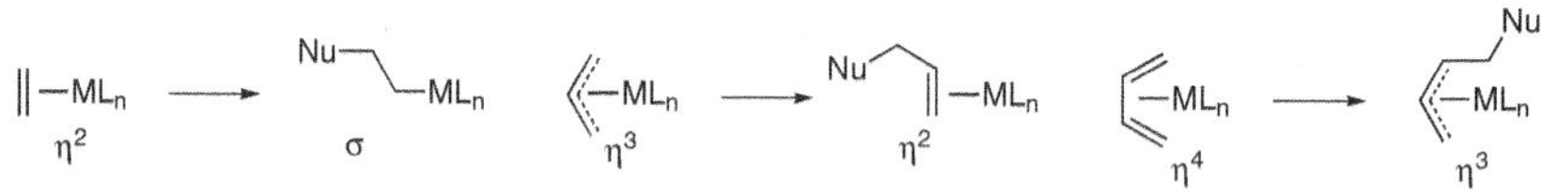

Reagents: RLi, RMgX, R_2Zn, etc.; enolate anions; oxyanions; etc.

This converts an η^n ligand to an η^{n-1} ligand (η^2 to σ) by attack of the nucleophile on the ligand.

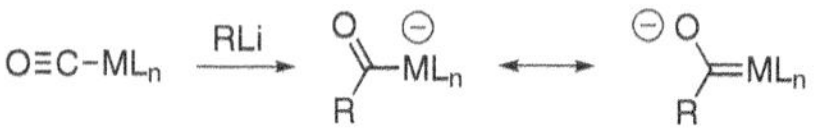

Attack on carbonyl ligands gives acylmetal complexes that are resonance-stabilized ambident nucleophiles.

Electrophilic Attack on Ligands

66. Galan, B.R.; Kalbarczyk, K.P.; Szczepankiewicz, S.; Keister, J.B.; Diver, S.T. *Org. Lett.* **2007**, *9*, 1203.

§ Answer to Worked Problem:
In this reaction, the preferred metathesis reaction occurs between the two monosubstituted alkene groups, which leads to the formation of ethylene as the other product. Since the ethylene is a volatile product that can escape, this route is also favored by Le Châtelier's Principle. [*J. Am. Chem. Soc.* **2006**, *128*, 15960].

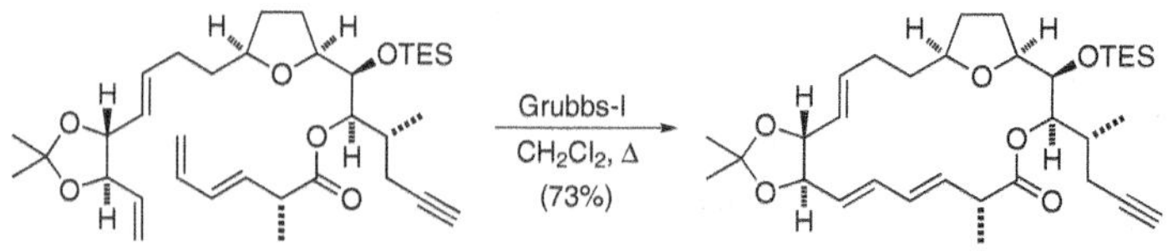

Reagents: RCOCl, $AlCl_3$; $HCONMe_2$, $POCl_3$; etc.

Reaction usually proceeds by conversion of the η^n complex to the cationic η^{n-1} complex, and elimination of the proton to give the new η^n complex with a substituted ligand.

Oligomerization

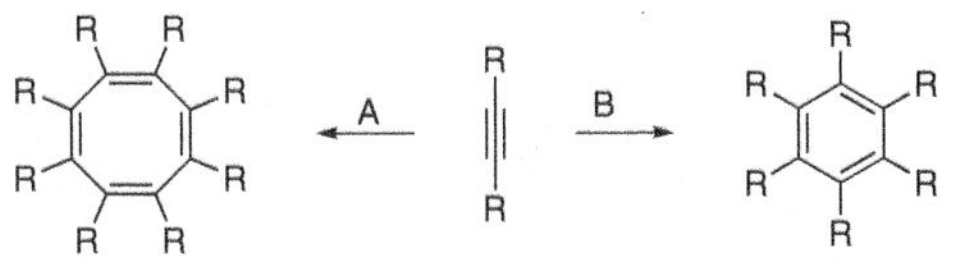

Reagents: A. Ni catalyst, frequently $Ni(CO)_4$
B. Ni catalyst with a strongly binding ligand (e.g., Ph_3P)

Reaction of alkynes with diynes in the presence of $CpCo(CO)_2$ gives benzenes.

Metathesis and Metallocyclobutane Formation

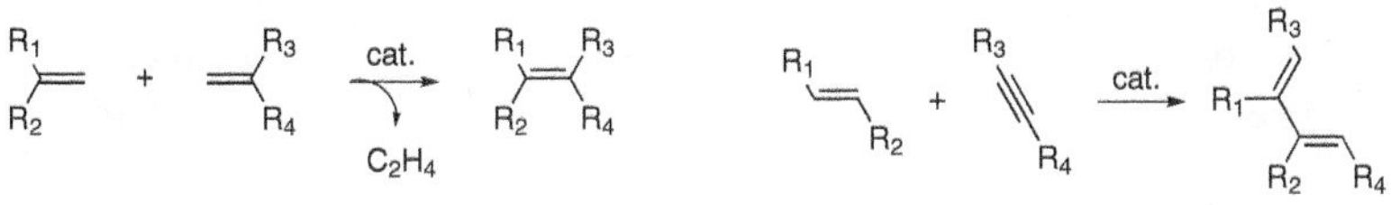

Reagents: catalyst is a ruthenium (Grubbs-type), or tungsten or molybdenum (Schrock-type) carbene complex carrying hindered ligands

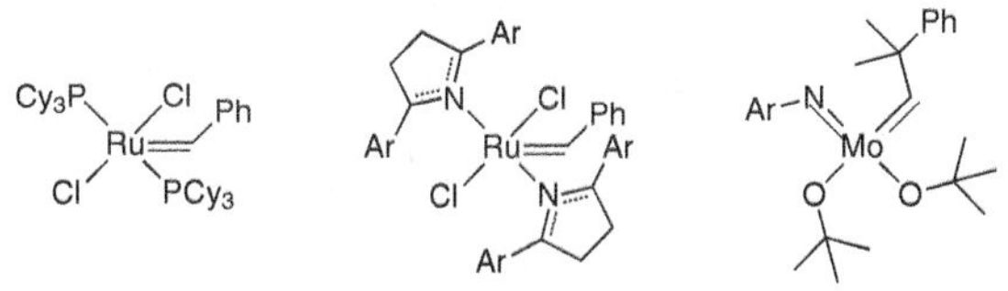

Grubbs I Grubbs II Schrock

Problem

17-4 What is the major product formed in each of the following reactions?

(a)

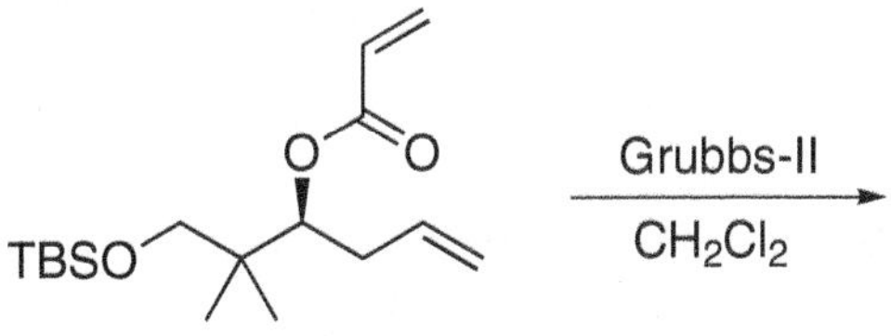

(b)

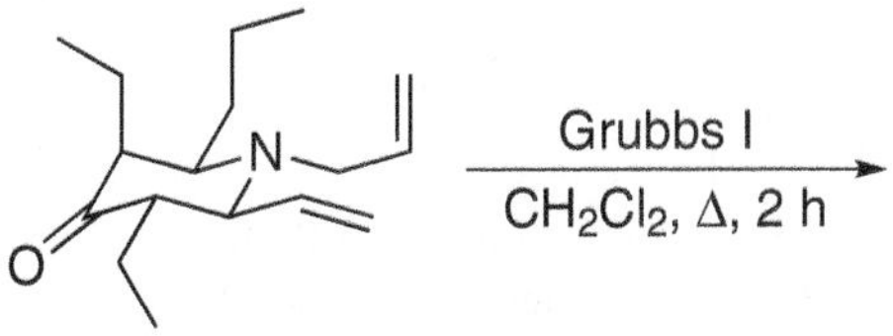

(c)

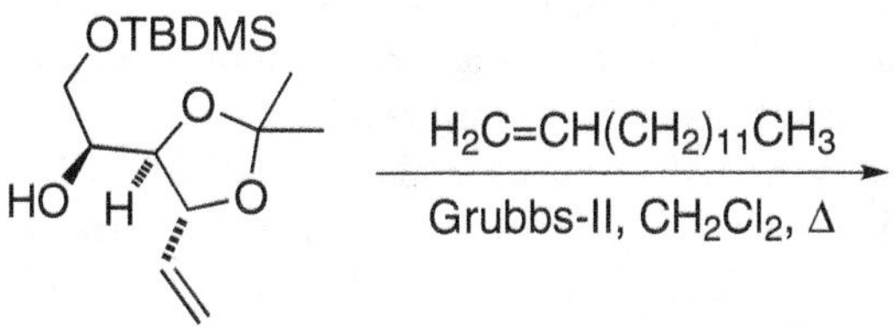

(continues)

(**Problem** continued)

(d)

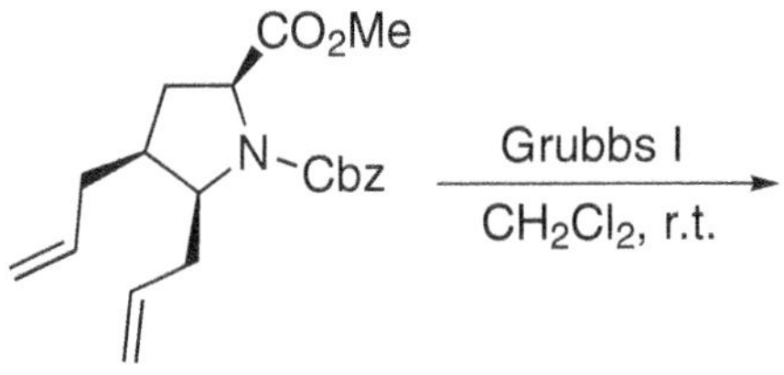

References: (a) *Tetrahedron* **2011**, *67*, 9376. (b) *J. Am. Chem. Soc.* **2005**, *127*, 8398. (c) *Tetrahedron* **2011**, *67*, 9283. (d) *J. Am. Chem. Soc.* **2002**, *124*, 13342.

17.5 Hydrometallation: Making the Organometallic Reagents by Addition

The recurring motif in cross-coupling reactions is one of metathesis between an organic derivative of a metal or metalloid and a hydrocarbon or halide equivalent (Example **17.109**). One key step in most of these reactions involves the transfer of an alkyl group from the metalloid or a metal to the transition metal at some point during the catalytic cycle. These ***transmetallation (metal exchange) reactions*** proceed, in most cases, with retention of configuration of the migrating organic ligand (as in Example **17.110**). There are two major ways for preparing the organometallic reagents required as reactants in the cross-coupling reactions: (1) transmetallation between less reactive metal halides and more reactive alkali metal organometallic reagents and (2) ***hydrometallation*** and ***carbometallation*** of unsaturated hydrocarbons with metal hydrides and metal alkyls (e.g., Example **17.111**).

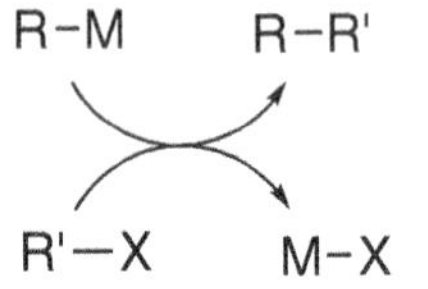

(17.109)

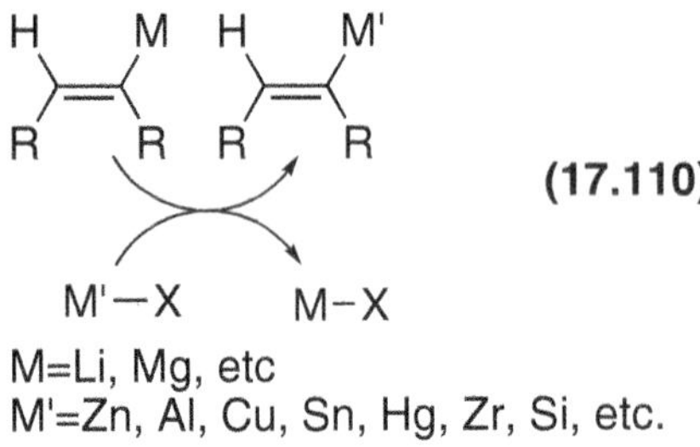

(17.110)

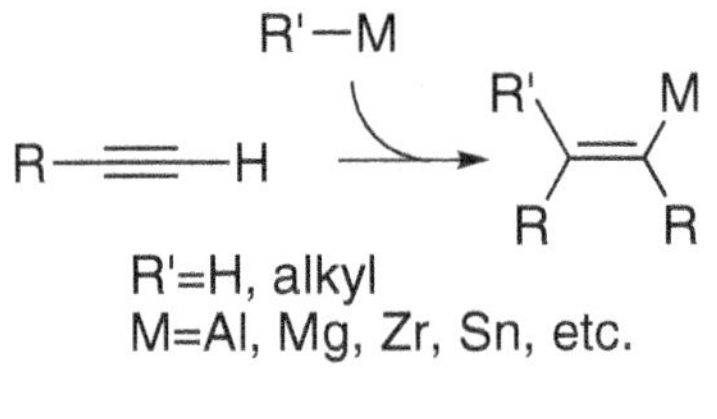

(17.111)

The transmetallation of alkali metal alkyls by transition metal halides was discussed in detail in Section 11.2. It involves the formation of a metal alkyl of a more active metal, followed by the exchange with a halide salt of a less active metal. The method can be used to prepare not only transition metal alkyls such as cuprates and zincates but also alkyl derivatives of non-metals and metalloids (e.g., the formation of phosphines and arsines from phosphorus or arsenic trihalides and Grignard reagents).

Hydrometallation involves the addition of a metal (or metalloid) hydride across the π bond of an alkene or an alkyne in a process that is formally the reverse of the β-hydride elimination. The reactive σ bond to hydrogen is polarized such that the hydrogen is $\delta-$ and the heteroatom is δ+. The closely related carbometallation reaction involves the addition of a similarly polarized metal-carbon σ bond across the π bond of an alkene or alkyne. It is somewhat ironic that the prototypical addition reaction of this type does not involve the addition of a *metal*-hydrogen σ bond across a C—C π bond at all. It involves, instead, the addition of a B—H σ bond across the C—C π bond: ***hydroboration***. The hydroboration reaction, however, exhibits many of the fundamental characteristics of hydrometallation: (1) the addition gives *syn* addition of B-H across the π bond, (2) the regiochemistry of addition is generally anti-Markovnikov (in the formal sense, based on the location of the hydrogen atom), and (3) the resultant borane can be used in a wide variety of synthetic reactions. It is instructive to note how organoboranes have undergone a resurgence of importance in organic synthesis with the development of metal-catalyzed cross-coupling reactions such as the ***Suzuki-Miyaura*** and ***Sonogashira*** reactions.

A wide range of metal and metalloid hydrides will add to C—C π bonds. The most important among these are the hydrides of silicon, tin, zirconium, and aluminum (Example **17.112**). All these hydrides will react with alkenes to give saturated organometallic compounds and with alkynes to give vinyl derivatives with well-defined stereochemistry. In the absence of a directing (usually chelating) group, the addition of the M—H bond is usually *syn*, and the regiochemistry of the addition is anti-Markovnikov. This is in keeping with the metal atom being the electrophilic end of the M—H bond.

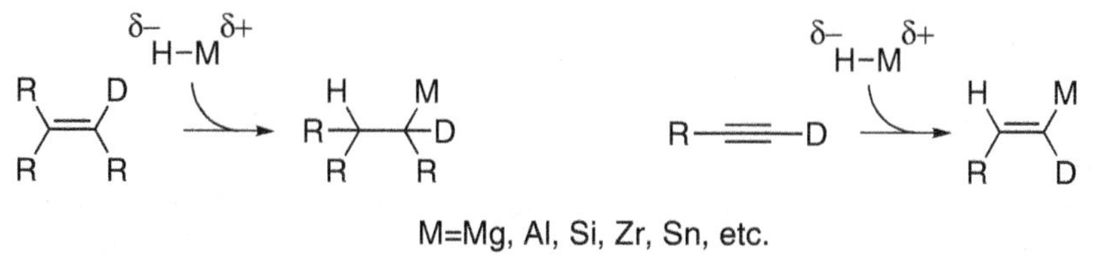

(17.112)

Hydroboration

The discovery that the hydroboration of alkenes is catalyzed by ethers and other modest Lewis bases[67] was followed by decades of research into the synthesis and reactions of boranes. Early on, much of the importance of hydroboration in organic synthesis was due to the facile conversion of the resultant boranes to alcohols, amines, and halides. These reactions all occur with anti-Markovnikov regiochemistry and are stereospecific, giving the *syn* adduct (amines and alcohols), or the *anti* adduct (halides).

Hydroboration-oxidation is the longest used variant of the reaction. It is an extremely useful method for the formation of alcohols due to its regiochemical control and its stereospecificity (all additions of borane are *syn*). Historically, the oxidation of the intermediate trialkylborane was effected by alkaline hydrogen peroxide, but organic oxidants such as *tert*-butyl hydroperoxide and base, and triethylamine *N*-oxide (TEAO) or *N*-methylmorpholine *N*-oxide (NMMO) have become popular; the amine *N*-oxides avoid the necessity for strongly basic reagents.

The regiochemical preferences of the BH_3 reagent can be enhanced by incorporating one or more sterically demanding groups on the boron. There are three borane derivatives that have been particularly widely used in organic synthesis (**17.113**).

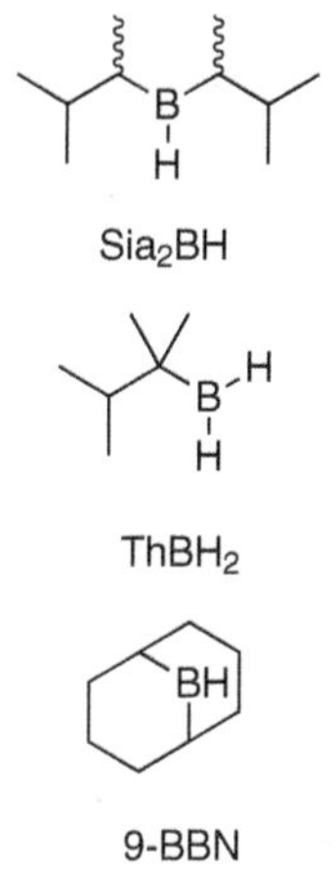

(17.113)

1. ***Disiamylborane (Sia_2BH)***, which is formed in situ by the addition of two equivalents of 2-methyl-2-butene to BH_3•THF prior to the hydroboration of the alkene of interest

67. Review: Brown, H.C. *Tetrahedron* **1961**, *12*, 117.

2. ***Thexylborane ($ThBH_2$)***, formed in situ by the addition of one equivalent of 2,3-dimethyl-2-butene to BH_3•THF prior to the hydroboration of the alkene of interest
3. ***9-Borabicyclo[3.3.1]nonane [9-BBN]***, which is a stable, commercially available solid prepared from 1,5-cyclooctadiene by hydroboration with BH_3•THF and subsequent thermal equilibration of the product mixture

The high levels of both regioselectivity and diastereoselectivity attainable with these hindered borane reagents led to the study of asymmetric hydroboration. These early efforts identified the ***diisopinocampheylborane (Ipc_2BH)*** system, a chiral analog of disiamylborane available as either enantiomer from α-pinene, as a useful reagent for the enantioselective synthesis of secondary alcohols from *cis* alkenes.[68] In these reactions, e.e's are often greater than 99%. The major enantiomer formed using diisopinocampheylborane may be predicted on the basis of the simple model for the hydroboration of *cis*-2-butene given in Figure 17.17. It is important to recognize that this model is a simple mnemonic (much like the Cram's rule model)—useful for prediction but not necessarily representative of the actual transition state structure in the hydroboration reaction. In this model, the conformation of the borane is written so that the large and small groups (R_L, R_S) are arranged with the large group on one isopinocampheyl group approximately eclipsing the hydrogen on the other. The alkene now approaches the boron to give approximate sp^3 hybridization at boron, with the methyl groups oriented away from R_L, which brackets the two hydrogens. This leads to the borane from 1*S*-α-pinene giving the *S* alcohol from a symmetrical *Z* alkene. This model also predicts that lower levels of enantioselectivity should be expected for the *E*-alkene due to increased interference between the methyl group of the alkene and R_L of the borane.

In the ensuing decades, other chiral hydroborating agents have been developed, most noteworthy among them being Masamune's chiral borolane,[69] although the lengthy synthesis required to generate the reagent has limited its use. In 2008, Soderquist reported the synthesis (Figure 17.18) of a new class of chiral cyclic boranes[70] (10-alkyl-9-borabicyclo[3.3.2] decanes, **17.118**) and their use in hydroboration of 1,1-disubstituted alkenes with high levels of enantioselectivity. These alkenes had previously given poor results in asymmetric hydroboration reactions. The results in Table 17.3 are typical of those obtained with chiral borane reagents. Soderquist reported that the in situ generation of the active hydroborating agent from the borohydride was the most convenient method for carrying out hydroborations with these reagents.

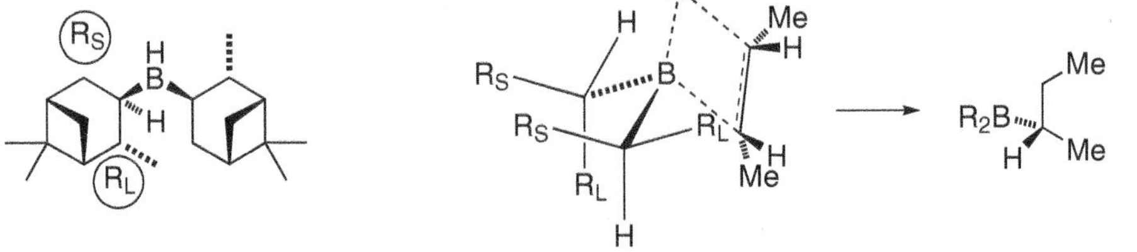

Figure 17.17 A model for predicting the stereochemistry of addition of diisopinocampheylborane to *cis*-2-butene (a representative *Z* alkene)

68. Reviews: (a) Srebnik, M.; Ramachandran, P.V. *Aldrichimica Acta* **1987**, *20*, 9. (b) Brown, H.C.; Ramachandran, P.V. *Pure Appl. Chem.* **1991**, *63*, 307.

69. Masamune, S.; Kim, B.M.; Petersen, J.S.; Sato, T.; Veenstra, S.J.; Imai, T. *J. Am. Chem. Soc.* **1985**, *107*, 45–49.

70. (a) Gonzalez, A.Z.; Romàn, J.G.; Gonzalez, E.; Martinez, J.; Medina, J.R.; Matos, K.; Soderquist, J.A. *J. Am. Chem. Soc.* **2008**, *130*, 9218. (b) Burgos, C.H.; Canales, E.; Matos, K.; Soderquist, J.A. *J. Am. Chem. Soc.* **2008**, *130*, 8044. (c) Canales, E.; Ganeshwar, P.; Soderquist, J.A. *J. Am. Chem. Soc.* **2005**, *127*, 11572.

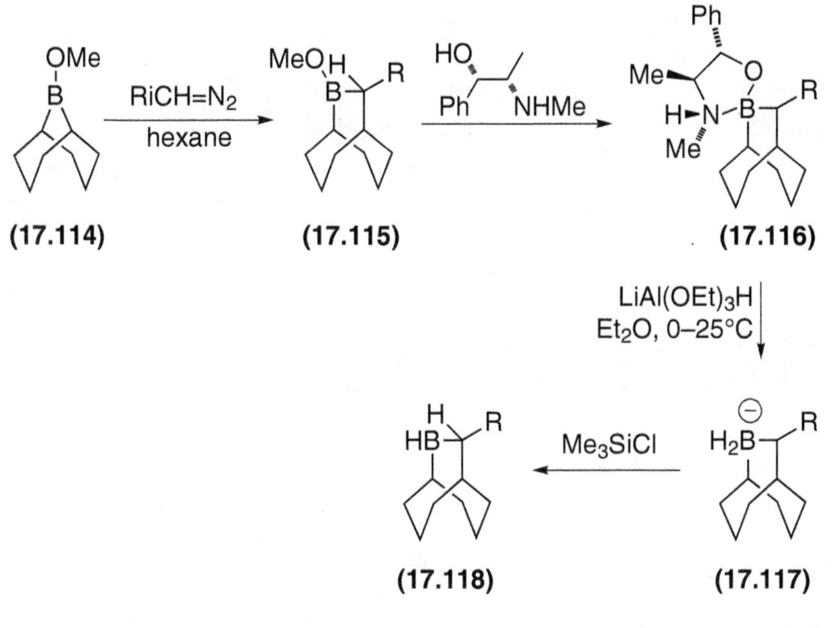

Figure 17.18 The synthesis of the Soderquist boranes

Table 17.3 Hydroboration with Chiral Boranes: Effects of Alkene and Borane Structure on Enantioselectivity

Representative Alkene	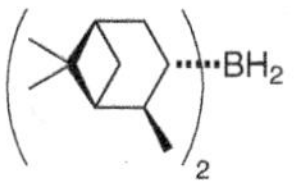 Ipc_2BH	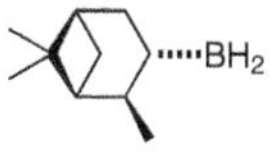 $IpcBH_2$	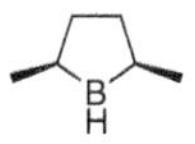 DMB	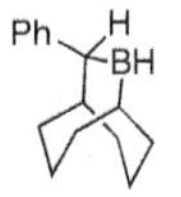	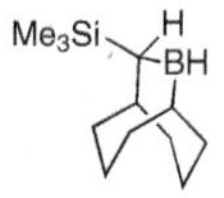
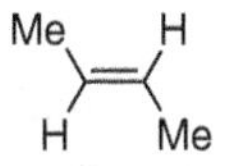	14	73	99.5	96	95
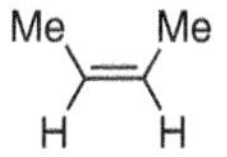	99.1	24	97.6	32	84
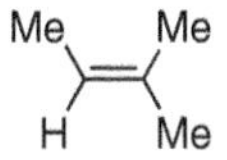	15	53	97.6	74	—
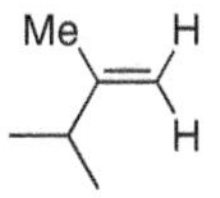	32	—	1.5	38	52
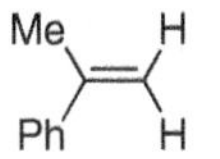	—	5	—	78	66

Values are e.e.

The stereochemistry of the product of hydroboration with the Soderquist reagents can be predicted relatively straightforwardly using the model in Figure 17.19. In this model, the alkene approaches the boron from the direction opposite the substituent at the 10- position, with the large group oriented away from the bicycle (**17.119**). In this model, the 1*R*

Figure 17.19 Model for predicting the stereochemistry of addition of the Soderquist borane to a 1,1-disubstituted alkene

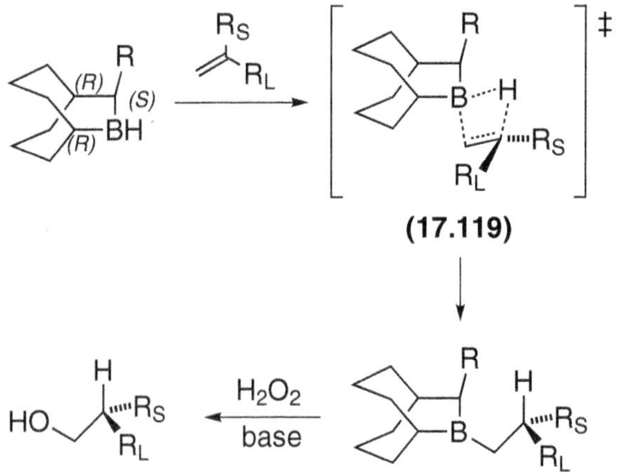

(17.119)

(17.121)

(17.120)

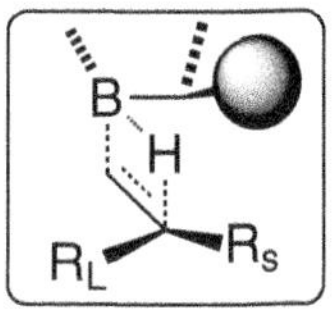

enantiomer of the borane is predicted to give the *R* enantiomer of 2-methyl-1-butanol (**17.121**, R_L=Et, R_S=Me) from 2-methyl-1-butene. Thomas and Aggarwal subsequently reported high levels of asymmetric induction in the hydroboration of 2-alkyl-1-alkenes with the Soderquist boranes and the Suzuki-Miyaura coupling of the intermediate boranes.[71] With the advent of these new chiral hydroborating agents, the enantioselective synthesis of chiral alcohols by hydration of alkenes has now been raised to levels similar to those obtainable with asymmetric epoxidation and dihydroxylation reactions. The origins of the enantioselectivity have been explored computationally for both diisopinocampheylborane[72] and the Soderquist reagent.[73]

A recent application of boranes in synthesis has been their use as a protecting group for alkenes to allow the selective semihydrogenation of dienes.[74] In this approach, the less hindered double bond of the diene reacts with one equivalent of a hindered borane such as 9-BBN, and the resultant unsaturated borane is subjected to catalytic hydrogenation. The oxidation of the saturated product with a tertiary nitroso compound returns the original alkene. As Examples **17.122** and **17.123** show, the reaction can be applied to either conjugated or non-conjugated dienes.

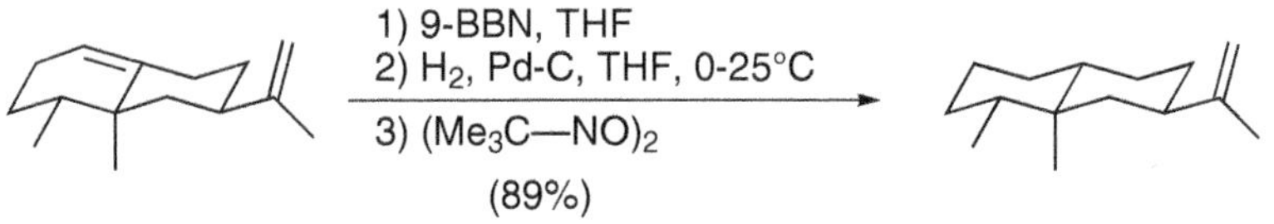

(17.122)

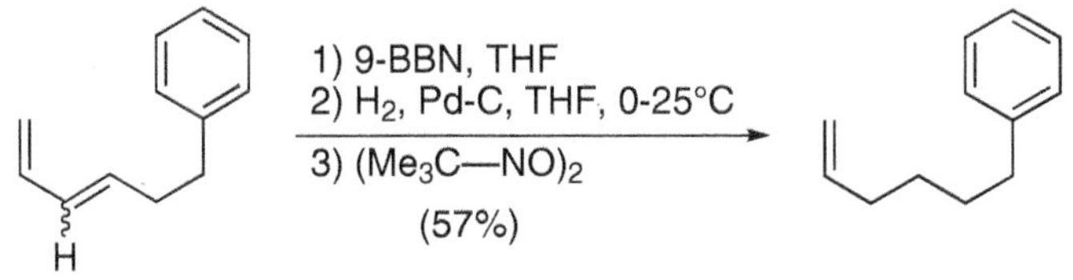

(17.123)

Migration of Groups from Boron

Despite the fact that boranes are generally (but not always!) stable with respect to rearrangements, the migration of groups from boron is a key feature of most organoborane chemistry and is essential to the formation of useful products from boranes. The

71. Thomas, S.P.; Aggarwal, V.K. *Angew. Chem. Int. Ed.* **2009**, *48*, 1896.
72. Houk, K.N.; Rondan, N.G.; Wu, Y.D.; Metz, J.T.; Paddon-Row, M.N. *Tetrahedron* **1984**, *40*, 2257.
73. Ess, D.H.; Kister, J.; Chen, M.; Roush, W.R. *J. Org. Chem.* **2009**, *74*, 8626.
74. Graham, T.J.A.; Poole, T.H.; Reese, C.N.; Goess, B.C. *J. Org. Chem.* **2011**, *76*, 4132.

driving force for these migrations almost invariably comes from the elimination of the formal negative charge on boron in an "ate" complex by migration of an alkyl group from boron to an acceptor electrophile. These reactions can occur in either an intramolecular or an intermolecular sense. In the absence of other factors, intramolecular reactions (Example **17.124**) generally proceed with retention of configuration at the migrating carbon, whereas intermolecular reactions usually proceed with inversion at the migrating carbon (Example **17.125**).

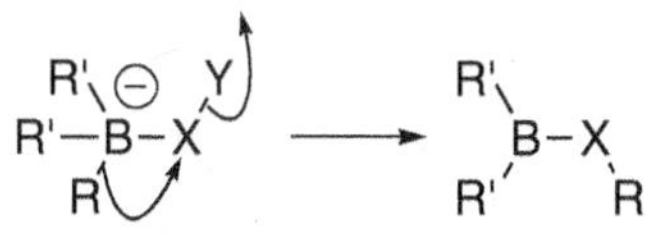

retention of configuration at R

(17.124)

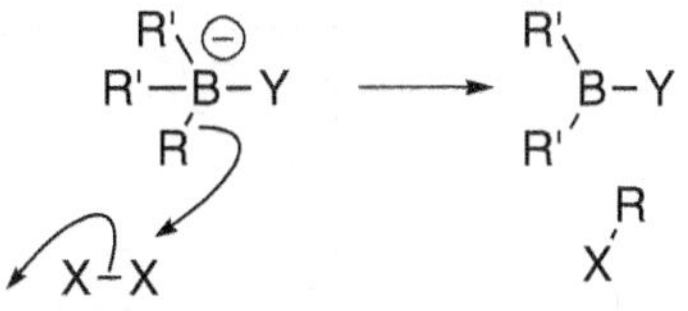

inversion of configuration at R

(17.125)

Trialkylboranes react slowly with halogens, but the reaction is accelerated by the addition of an alkoxide base, which leads to an intermediate "ate" complex (Example **17.129**) that then reacts rapidly with the halogen.[75] The reaction in this case proceeds with net inversion of configuration, as illustrated by the conversion of an *exo*-2-norbornylborane (Example **17.128**) to *endo*-2-bromo-norbornane (Example **17.127**).[76] As shown in this example, the stereochemistry is consistent with the halogen attacking the borate from the back side (i.e., with inversion of configuration at the migrating carbon).

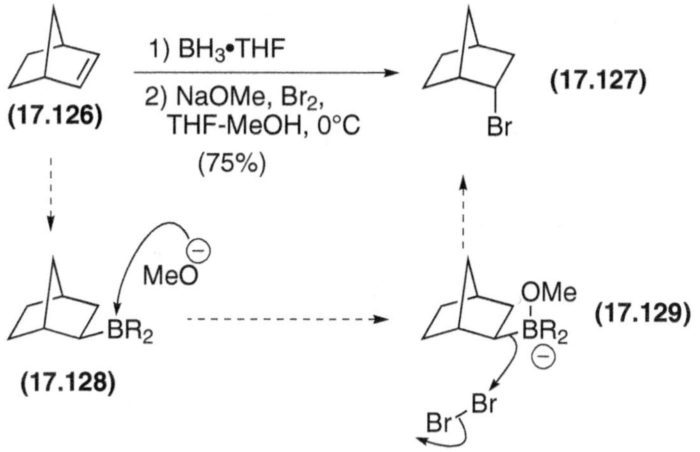

When one of the groups on boron is unsaturated, the reaction of a trialkylborane with iodine in the presence of base takes a dramatically different course. On treatment with iodine and base, the borane undergoes a rearrangement with concomitant formation of a carbon-carbon bond, rather than a carbon-iodine bond to give *cis* alkenes from *trans* vinyldialkylboranes and conjugated dienes from divinylalkylboranes.[77] In this reaction, Zweifel and his coworkers found that replacing the iodine by cyanogen bromide resulted in a reversal of the stereochemistry of the double bond in the product.

---

75. (a) Brown, H.C.; Rathke, M.W.; Rogic, M.M. *J. Am. Chem. Soc* **1968**, *90*, 5038. (b) Brown, H.C.; Lane, C.F. *J. Am. Chem. Soc.* **1970**, *92*, 6660.

76. Brown, H.C.; Lane, C.F. *J. Chem. Soc., Chem. Commun.* **1971**, 521.

77. (a) Zweifel, G.; Arzoumanian, H.; Whitney, C.C. *J. Am. Chem. Soc.* **1967**, *89*, 3652. (b) Zweifel, G.; Polston, N.L.; Whitney, C.C. *J. Am. Chem. Soc.* **1968**, *90*, 6243. (c) Zweifel, G.; Fisher, R.P.; Snow, J.T.; Whitney, C.C. *J. Am. Chem. Soc.* **1971**, *93*, 6309. (d) Zweifel, G.; Fisher, R.P.; Snow, J.T.; Whitney, C.C. *J. Am. Chem. Soc.* **1972**, *94*, 6560. (e) Evans, D.A.; Crawford, T.C.; Thomas, R.C.; Walker, J.A. *J. Org. Chem.* **1976**, *41*, 3947.

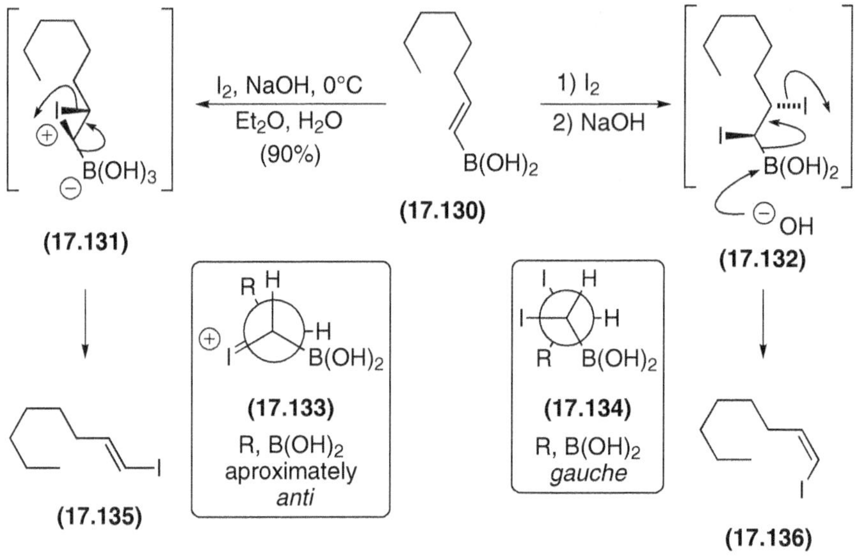

Figure 17.20 Variation of stereochemistry in iodination of vinylboronic acids

The situation with terminal vinylboronic acids (Figure 17.20) is more interesting because, depending on the exact order of the addition of the base and the electrophile (almost always iodine), one may obtain either geometric isomer of the vinyl iodide.

Thus, when the base is added to *E*-1-octeneboronic acid (**17.130**) first, followed by iodine, the product is the *E* iodide (**17.135**).[78] When the iodine is added first, followed by the base, *Z*-1-iodo-1-octene (**17.136**) is obtained.[79] The stereochemistry of these reactions is rationalized in terms of the timing of the cleavage of the B—C bond. When the iodine and base are added simultaneously, the cleavage occurs at the iodonium ion stage (**17.133**), leading to apparent retention of configuration at carbon, whereas when excess halogen is added first, it is proposed that the B—C bond is cleaved in a concerted *anti* elimination from the diiodide (**17.134**). A variant of this reaction that gives 2-iodo-1-alkenes as the major product involves the initial iodoboration of a terminal alkyne with 9-iodo-9-BBN, followed by protonolysis of the B—C bond,[80] as illustrated by a key step in Overman's synthesis of alcyonin, shown as Example **17.137**. [81]

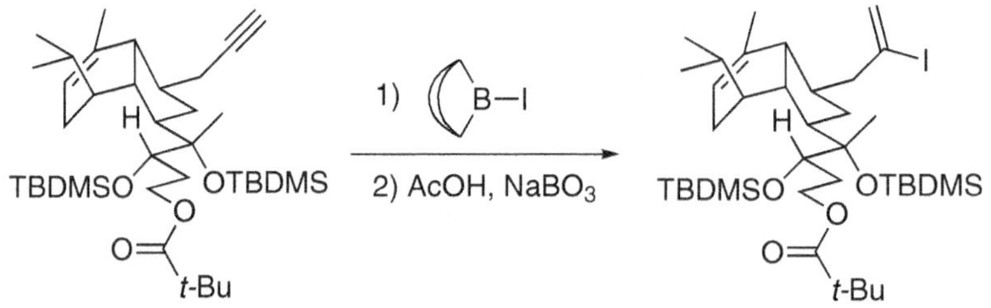

(**17.137**)

Hydroalumination[82]

Aluminum is located directly below boron in the periodic table, so one might reasonably expect that compounds of aluminum could exhibit some of the chemical properties of the

78. Brown, H.C.; Hamaoka, T.; Ravindran, N. *J. Am. Chem. Soc.* **1973**, *95*, 5786.
79. Brown, H.C. *Pure Appl. Chem.* **1976**, *47*, 49.
80. Hara, S.; Dojo, H.; Takinami, S.; Suziki, A. *Tetrahedron Lett.* **1983**, *24*, 731.
81. Corminboeuf, O.; Overman, L.E.; Pennington, L.D. *Org. Lett.* **2003**, *5*, 1543.
82. Reviews and monographs: (a) Dzhmenilev, U.M.; Ibragimov, A.G. In Andersson, P.G.; Munslow, I.J. *Modern Reduction Methods* (Wiley-VCH: Weinheim, 2008), ch. 18, p. 447.

analogous boron compounds. Lithium aluminum hydride does add to unsaturated hydrocarbons, but these addition reactions typically require high temperatures and pressures, and ether solvents. The neutral aluminum hydrides (e.g., diisobutylaluminium hydride [DIBAL-H]), on the other hand, are electrophilic and add much more easily to terminal alkynes, with *syn* stereochemistry to give the *E*-vinylalanes.[83] The regioselectivity with internal alkynes is generally poor unless there is a bias in the form of the electronic characteristics of the two groups bonded to the triple bond.[84]

The stereochemistry of the hydroalumination reaction depends markedly on the formal charge on aluminum. Neutral alanes lead to *syn* addition of the Al—H bond, while anionic aluminates lead to *anti* addition of the Al—H bond. The stereochemistry of the addition of neutral alanes to terminal alkynes is most easily interpreted in terms of a concerted addition through a four-membered cyclic transition state, as in Example **17.138**. The stereochemistry with aluminates, on the other hand, requires a two-step mechanism, with hydride addition as the first step to give the anion **17.139**, which is in equilibrium with its *cis* isomer (Example **17.140**), and trapping of the vinyl anion by the neutral alane as the second. Because the intermediate vinyl anion exists preferentially as the *trans* isomer, the overall addition becomes *trans* (Example **17.141**).

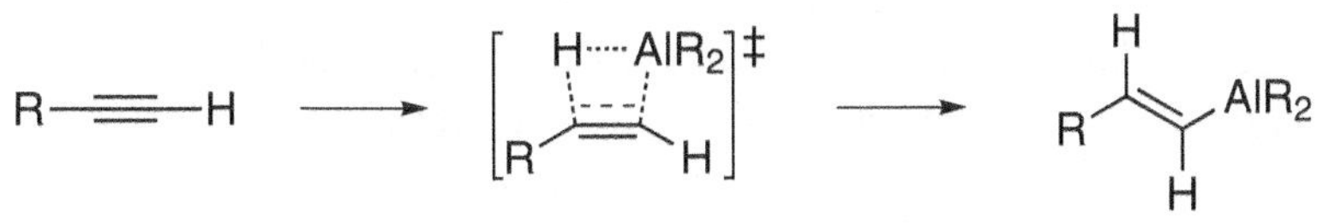

(17.138)

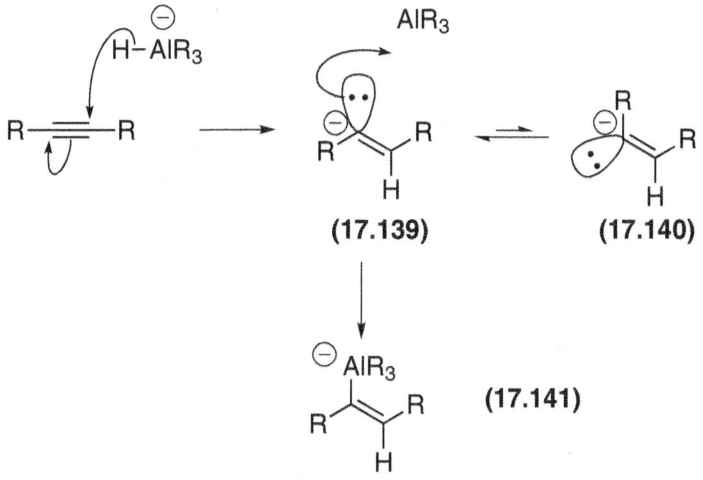

(17.139) **(17.140)**

(17.141)

The one drawback of hydroalumination of terminal alkynes comes in the form of competing deprotonation to generate the corresponding alkynylaluminum species. This can be suppressed, in part, by using the trimethylsilyl group as a surrogate for the acetylenic hydrogen (e.g., as in Example **17.142**). Interestingly, by changing the solvent from heptane (a non-coordinating solvent) to *N*-methylpyrrolidine (a good Lewis base), the stereochemistry of the hydroalumination can be reversed. In heptane, the major product is from *syn* addition (Example **17.144**), whereas in mixed heptane–*N*-methylpyrrolidine, the isomer from *anti* addition (Example **17.143**) predominates. With diethyl ether as the solvent (a weaker Lewis base than *N*-methylpyrrolidine), a mixture of stereoisomers is

83. (a) Wilke, G.; Müller, H. *Chem. Ber.* **1956**, *89*, 444. (b) Eisch, J.J.; Kaska, W.C. *J. Am. Chem. Soc.* **1963**, *85*, 2165. (c) Bundens, J.W.; Francl, M.M. *Organometallics* **1992**, *12*, 1608.

84. Eisch, J.J.; Gopal, H.; Rhee, S.-G. *J. Org. Chem.* **1975**, *40*, 2064.

formed.[85] The unexpected stereochemistry of the addition in heptane was ascribed[86] to catalysis of the *cis-trans* isomerization by the alane itself. Researchers found that this isomerization also required the trialkylsilyl group in order to occur: the simple vinylalanes did not isomerize under the same conditions. Obviously, these results mandate caution in making predictions about the stereochemistry of this reaction in systems other than simple hydrocarbons.

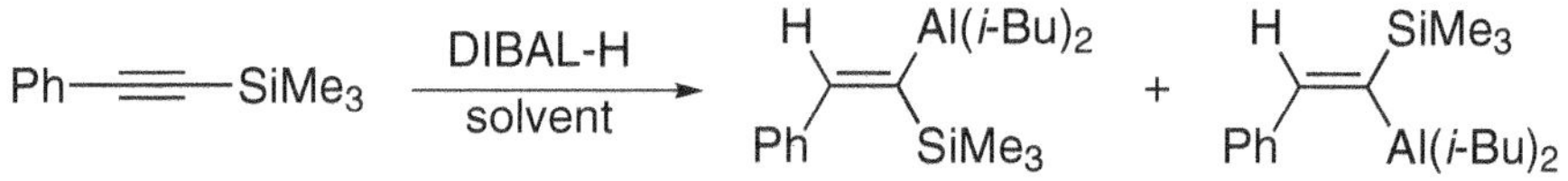

(17.142) **(17.143)** **(17.144)**

solvent = heptane	4	:	96
solvent = heptane, *N*-Me-pyrrolidine)	96	:	4
solvent = Et_2O	65	:	35

Hydrozirconation[87]

Organozirconium compounds have emerged over the past four decades as extremely useful reagents in the functionalization of alkenes and alkynes and as key intermediates in transmetallation reactions to give the organometallic partner in a wide range of cross-coupling reactions. One of the most important zirconium complexes is the Schwartz reagent[88] (Example **17.145**), which reacts with alkenes and alkynes to give products of *syn* addition. Unlike many other hydrometallations, hydrozirconation generally proceeds under very mild conditions. The relatively facile transmetallation thus makes a wide range of organometallic complexes with defined stereochemistry readily available. As an example, although hydroalumination of alkenes generally requires elevated temperatures and pressures, the addition of small amounts of zirconium chloride or of Schwartz reagent allows the reaction to be carried out at room temperature.

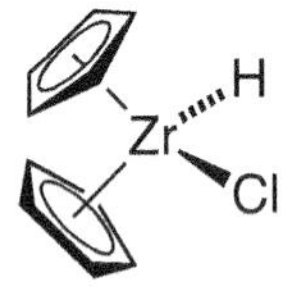

(17.145)

One of the interesting features of the hydrozirconation of alkenes is the facile rearrangement of the initially formed alkylzirconocene chloride (regardless of the alkene used initially) by Zr–H elimination and subsequent addition to give a less hindered alkylzirconocene chloride. By simply warming a tetrahydrofuran, or THF, solution of the alkene and Schwartz reagent to 50°C, an internal alkylzirconocene chloride rearranges quantitatively to the terminal alkylzirconocene chloride. A similar rearrangement occurs in the addition of Schwartz reagent to internal alkynes (e.g., Example **17.146**) in THF, provided that the Schwartz reagent is present in excess. In the presence of one equivalent of the Schwartz reagent, the kinetic product mixture of Examples **17.147** and **17.148** is obtained.[89] With excess Schwartz reagent at 50°C in THF, the addition is *syn*, with zirconium bonded to the carbon bearing the smaller substituent (**17.148**).

---

85. Eisch, J.J.; Foxton, M W. *J. Org. Chem.* **1971**, *36*, 3520.

86. Eisch, J.J.; Rhee, S.-G. *J. Am. Chem. Soc.* **1975**, *97*, 4673.

87. Reviews: (a) Wipf, P.; Jahn, H. *Tetrahedron* **1996**, *52*, 12853. (b) Wipf, P.; Takahashi, H.; Zhuang, N. *Pure Appl. Chem.* **1998**, *70*, 1077. (c) Wipf, P. *Top. Organomet. Chem.* **2005**, *8*, 1.

88. (a) Hart, D.W.; Schwartz, J. *J. Am. Chem. Soc.* **1974**, *96*, 8115. (b) Schwartz, J.; Labinger, J.A. *Angew. Chem. Int. Ed. Engl.* **1976**, *15*, 333. (c) Hart, D.W.; Blackburn, T.F.; Schwartz, J. *J. Am. Chem. Soc.* **1975**, *97*, 679. (d) Buchwald, S.L.; LaMaire, S.J.; Nielsen, R.B.; Watson, B.T.; Kin, S.M. *Org. Synth.* **1998**, *Coll. Vol. 9*, 162.

89. Panek, J.S.; Hu, T. *J. Org. Chem.* **1997**, *62*, 4912.

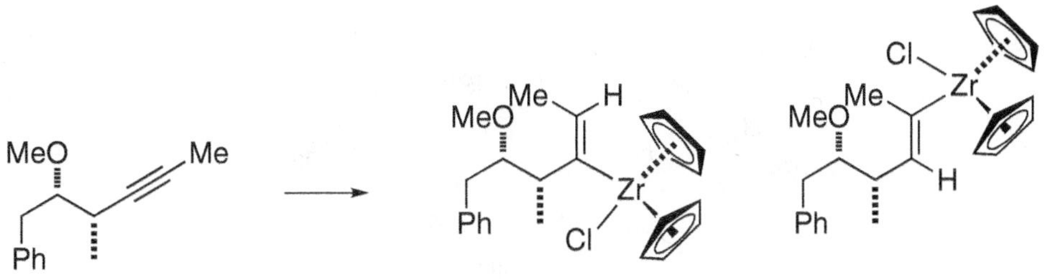

	(17.146)	(17.147)	(17.148)
	$Cp_2Zr(H)Cl$ (1.2 eq), PhMe, 45°C, 2 h	37	63
	$Cp_2Zr(H)Cl$ (1.4 eq), THF, 50°C, 1 h	0	100

Hydrostannylation and Hydrosilation

Silicon frequently serves as a hydrogen surrogate in organic synthesis, and this reactivity extends to the addition of Si—H bonds across an alkene or alkyne π bond. Thus, although hydrosilation is a slow reaction in the absence of a metal catalyst, in the presence of a palladium catalyst, the reaction proceeds at a useful rate to give alkyl or vinylsilanes, depending on the unsaturated hydrocarbon partner.

The stereochemistry of the addition in the metal-catalyzed hydrosilation is dependent on the metal catalyst chosen. With platinum catalysts,[90] the overall addition is *syn* and is thus analogous to catalytic hydrogenation. A catalytic cycle proposed by Chalk and Harrod in 1965[91] involved sequential addition of the silane to the coordinatively unsaturated metal to give the hydride **17.150**, coordination of the alkene (alkyne) to give **17.152**, migratory insertion into the M—H bond to give **17.153**, and reductive elimination of the final alkane (alkene) product (**17.154**). Later work by Roy and Taylor,[92] using 1,5-cyclooctadiene as a non-reactive ligand toward the silane, led to the proposal of the alternative catalytic cycle shown in Figure 17.21. The key change from the original catalytic cycle is that the active form of the catalyst (**17.150**), formed from the pre-catalyst by multiple hydrosilations, already carries multiple silyl groups. The only irreversible step in this catalytic cycle is the reductive elimination step leading to formation of the product.

The reaction catalyzed by other metals (Ru, Rh, Ir) leads to overall *anti* addition, and a slight modification of the original Chalk and Harrod mechanism has been proposed to account for this observation. In this mechanism (Figure 17.22), the initial *syn* adduct (**17.157**) undergoes isomerization (presumably driven by relief of steric crowding between the alkyl substituent and the metal) prior to the reductive elimination step. The regiochemistry can be controlled by the electronic nature of the substituents on the π bond. In the absence of strongly orienting groups, the tendency is for the silicon to become bonded to the less hindered carbon, which usually corresponds to anti-Markovnikov addition. Although there are studies on the mechanism of this reaction and on the factors affecting its stereochemistry, the overall stereochemical control is still somewhat situation-specific.

In similar vein, addition of Sn—H bonds across an alkene or alkyne π bond can also be catalyzed by palladium. Hydrostannylation is the addition of the Sn—H bond across an alkene or alkyne π bond. The reaction can be carried out under free radical conditions

90. (a) Speier, J.L.; Webster, J.A.; Barnes, G.H. *J. Am. Chem. Soc.* **1957**, *79*, 974. (b) Speier, J.L. In Stone, F.G.A.; West, R., Eds. *Advances in Organometallic Chemistry* (Academic Press: New York, 1979) Vol. 17, p. 407.

91. Chalk, A.J.; Harrod, J.F. *J. Am. Chem. Soc.* **1965**, *87*, 16.

92. Roy, A.K.; Taylor, R.B. *J. Am. Chem. Soc.* **2002**, *124*, 9510.

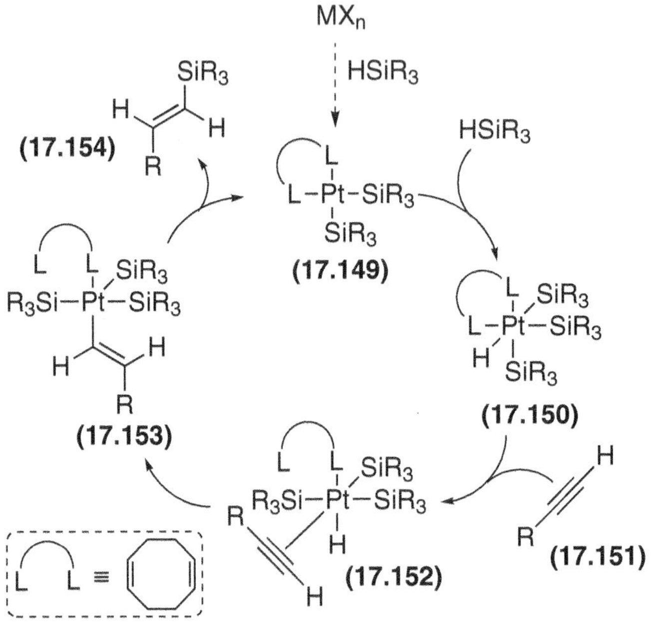

Figure 17.21 Roy and Taylor mechanism for hydrosilation of alkynes

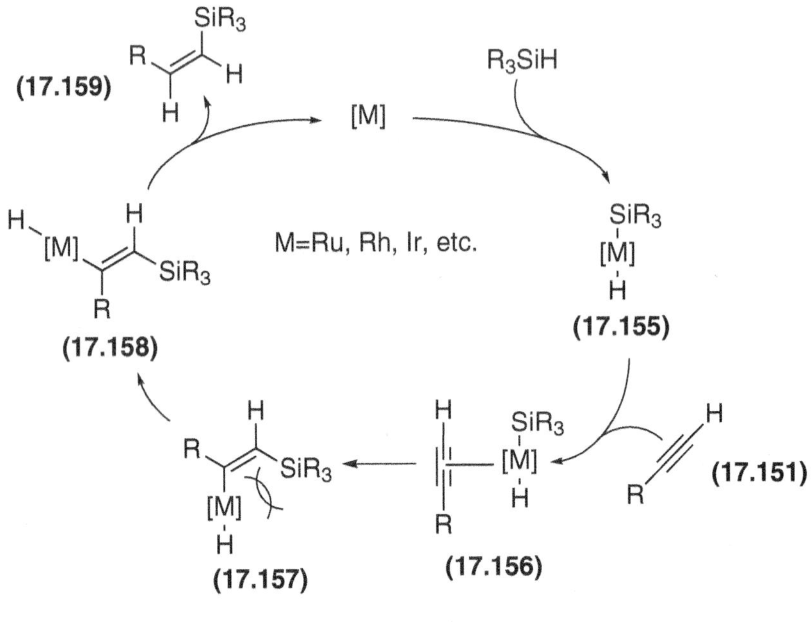

Figure 17.22 A modified Chalk and Harrod mechanism that accounts for the formation of the *Z*-vinylsilane from an alkyne

or under metal (usually palladium) catalysis.[93] Under free radical conditions, the regiochemistry of the addition is controlled by the relative stability of the two possible radicals, but stereochemical control is often problematic—in direct contrast to the metal-catalyzed reaction, which is reliably *syn*. One reaction that can be used to form C—Sn bonds is stannylmetallation-protonolysis (Example **17.160**). The second metal may be aluminum, zinc, magnesium, or boron,[94] but copper (I) is by far the most popular choice. In

93. Review: Smith, N.D.; Mancuso, J.; Lautens, M. *Chem. Rev.* **2000**, *100*, 3257.

94. (a) Hibino, J.; Matsubara, S.; Morizawa, Y.; Oshima, K. *Tetrahedron Lett.* **1984**, *25*, 2151. (b) Matsubara, S.; Hibino, J.; Morizawa, Y.; Oshima, K. *J. Organomet. Chem.* **1985**, *285*, 163. (c) Aksela, R.; Oehlschlager, A.C. *Tetrahedron* **1991**, *47*, 1163. (d) Nonaka, T.; Okuda, Y.; Matsubara, S.; Oshima, K.; Utimoto, K.; Nozaki, H. *J. Org. Chem.* **1986**, *51*, 4716.

stannylcuprations, the regiochemistry of the addition can be controlled by the choice of the stannylcopper reagent (as in Example **17.161**).[95] Note how using the higher order cuprate leads to higher levels of the 1-stannyl-1,3-diene from terminal alkynes.

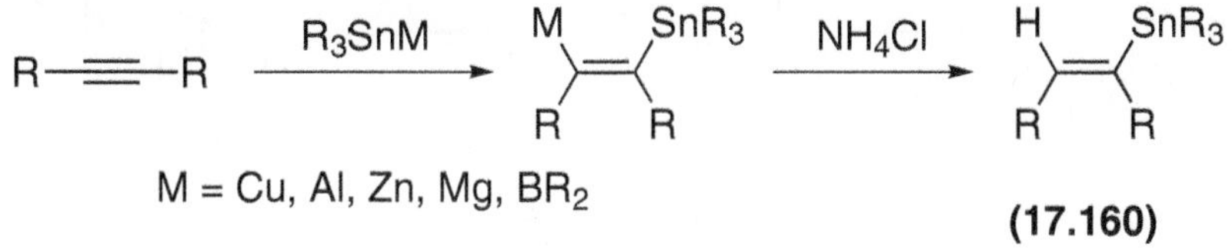

(17.160)

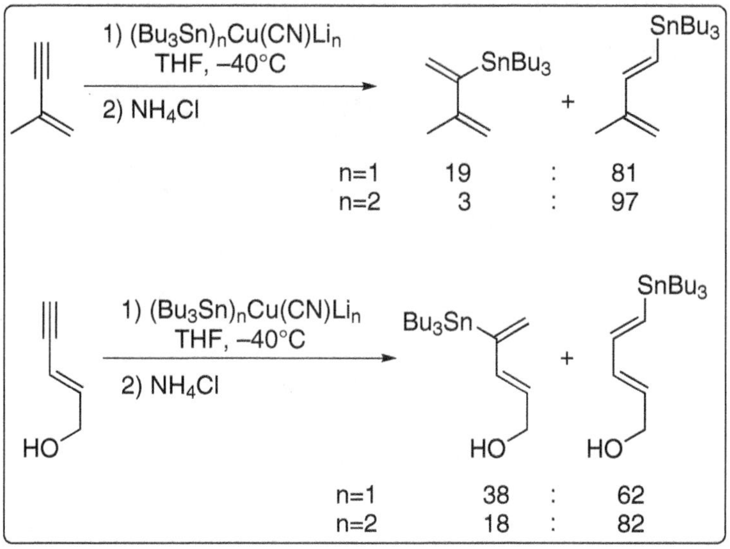

(17.161)

Carbometallation[96]

The addition of a carbon-metal σ bond across a carbon-carbon π bond is most widely used in the preparation of vinylmetal complexes with defined stereochemistry. Many carbometallation reactions involve copper, either in the form of a stoichiometric organocopper reagent or as a catalyst.[97] In the carbocupration of alkynes, the stereochemistry is almost universally *syn*, and the regiochemistry tends to place the metal at the terminal carbon, although the regioselectivity in simple alkynes is not always good. In contrast, the inclusion of a group capable of complexing the copper (e.g., a propargyl ether) leads to high levels of regioselectivity, as shown in Example **17.162**.[98]

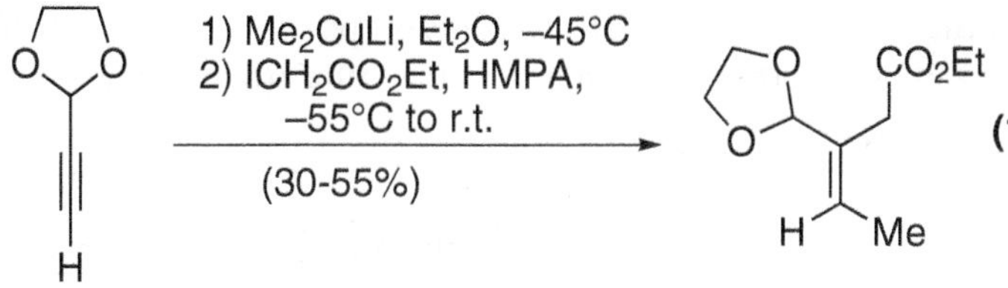

(17.162)

The combination of an aluminum alkyl and titanium or zirconium chloride provides an active catalyst for the Ziegler-Natta polymerization of alkenes. A similar catalyst can be generated from an aluminum halide and a zirconocene. In the initial reaction, there is

95. (a) Barbero, A.; Cuadrado, P.; Fleming, I.; Gonzalez, A.M.; Pulido, F.J. *J. Chem. Soc., Perkin I* **1992**, 351. (b) Aksela, R.; Oehlschlager, A.C. *Tetrahedron* **1991**, *47*, 1163. (c) Betzer, J.- F.; Delaloge, F.; Muller, B.; Pancrazi, A.; Prunet, J. *J. Org. Chem.* **1997**, *62*, 7768.

96. Review: Basheer, A.; Marek, I. *Beilstein J. Org. Chem.* **2010**, *6*, 77.

97. Review: Normant, J.F.; Alexakis, A. *Synthesis* **1981**, 841.

98. Magnuson, S.R.; Sepp-Lorenzino, L.; Rosen, N.; Danishefsky, S.J. *J. Am. Chem. Soc.* **1998**, *120*, 1615.

Figure 17.23 The Cossee mechanism for the Ziegler-Natta polymerization of an alkene

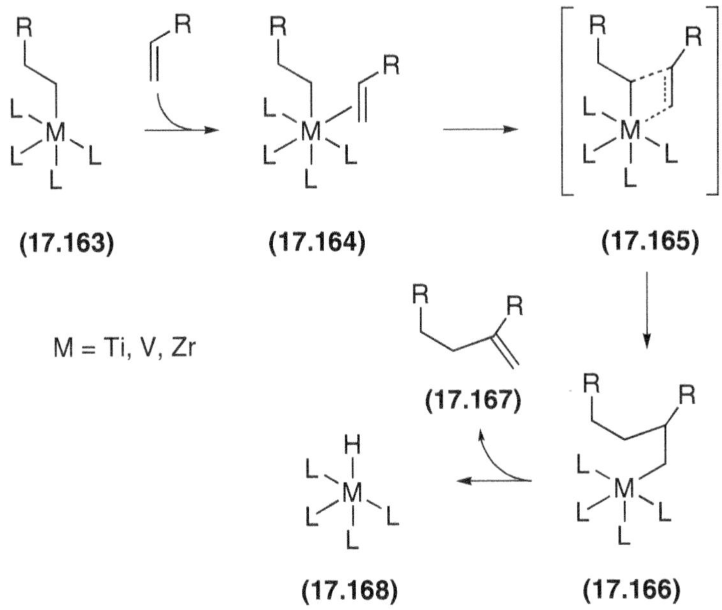

an alkyl transfer from aluminum to the Group IVB metal, and this alkylzirconium or alkyltitanium species reacts with the alkene to give the polymeric product. As illustrated in Figure 17.23, the currently accepted mechanism of the Ziegler-Natta polymerization, first proposed by Cossee,[99] involves the initial formation of a coordinatively unsaturated metal alkyl (e.g., **17.163**) that then actively catalyzes the addition polymerization. Complexation of the alkene to the vacant site on the metal gives the species **17.164**, which then undergoes migratory 1,2-insertion through a four-membered transition state (**17.165**) to generate a new coordinatively unsaturated complex (**17.166**). By repetition of these steps, one can obtain a high polymer bound to the metal. During polymerization, the reaction may be terminated by β-hydride elimination to give an alkene (**17.167**) and a metal hydride complex (**17.168**) capable of adding to a new alkene to begin a new polymer chain.

In 1978,[100] Negishi reported a zirconium-catalyzed single-stage carboalumination of alkynes to give a vinylalane by *syn* addition of the C—Al bond across the alkyne. For many years, the mechanism of the reaction was assumed to proceed by a carbozirconation, followed by transmetallation. However, later work by Negishi demonstrated that—with trialkylaluminum reagents bearing alkyl groups larger than methyl, at least—the mechanism is rather more complicated[101] and probably involves direct carboalumination of the alkyne, catalyzed by a zirconium species. One such species is the metallocycle **17.169**, in which there is both an electrophilic zirconium and a nucleophilic aluminum.

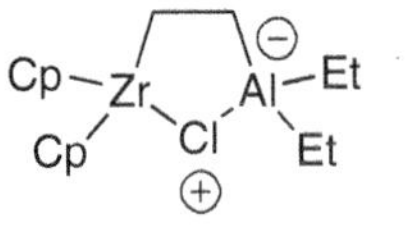

(17.169)

Why Are These Hydrometallation Reactions Important?

Much of the usefulness of hydrometallation with hydrides of zirconium, aluminum, and tin, especially, derives from the stereospecific replacement of the C—metal bond by a

99. (a) Cossee, P. *J.Catal.* **1964**, *3*, 80. (b) Arlman, E.J.; Cossee, P. *J. Catal.* **1964**, *3*, 99. An alternative mechanism proposed by Green and Rooney [(c) Ivin, K.J.; Rooney, C.D.; Stewart, C.D.; Green, M.L.H.; Mahtab, J.R. *J. Chem. Soc., Chem. Commum.* **1978**, 604. (d) Green, M.L.H. *Pure Appl. Chem.* **1978**, *100*, 2079] was later ruled out in an ingenious isotope effect/stereochemistry experiment by Grubbs [(e) Clawson, L.; Soto, J.; Buchwald, S.L.; Steigerwald, M.L.; Grubbs, R.H. *J. Am. Chem. Soc.* **1985**, *107*, 3377].

100. (a) Van Horn, D.E.; Negishi, E. *J. Am. Chem. Soc.* **1978**, *100*, 2252. (b) Negishi, E.; Van Horn, D.E.; Yoshida, T. *J. Am. Chem. Soc.* **1985**, *107*, 6639. (c) Negishi, E. *Pure Appl. Chem.* **1981**, *53*, 2333.

101. Negishi, E.-i.; Kondakov, D.Y.; Choueiry, D.; Kasai, K.; Takahashi, T. *J. Am. Chem. Soc.* **1996**, *118*, 9577.

C—X bond by simple treatment with the halogen. The reaction proceeds with retention of configuration with vinylmetal compounds, which allows the preparation of vinyl halides of known stereochemistry for the formation of C—C bonds by palladium-catalyzed cross-coupling reactions.

Problem

17-5 What is the major organic product of each of the following reactions?

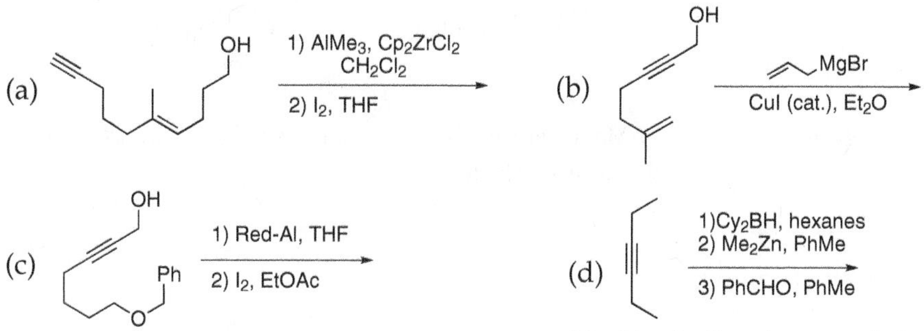

[step 2 is a B—Zn exchange reaction]

References: (a) *Org. Lett.* **2005**, *6*, 4297. (b) *J. Org. Chem.* **1998**, *63*, 7945. (c) *J. Am. Chem. Soc.* **2010**, *132*, 8219. (d) *J. Org. Chem.* **2007**, *72*, 5935.

Reaction Synopses

Hydroboration

Reagents: for alkenes: BH_3•THF; Sia_2BH, THF; $ThBH_2$, THF; 9-BBN, THF;
for alkynes: [o-$C_6H_4O_2$]BH, THF; $(Me_2CO)_2BH$, THF; etc.

Regiochemistry: Markovnikov; regiochemical control is improved with bulky boranes.

Stereochemistry: strictly *syn*

Reaction may be carried out asymmetrically using $IpcBH_2$, Ipc_2BH, and Soderquist boranes:

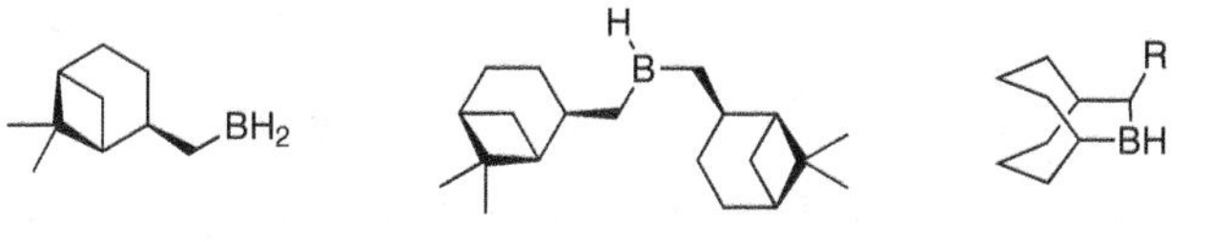

Hydroalumination

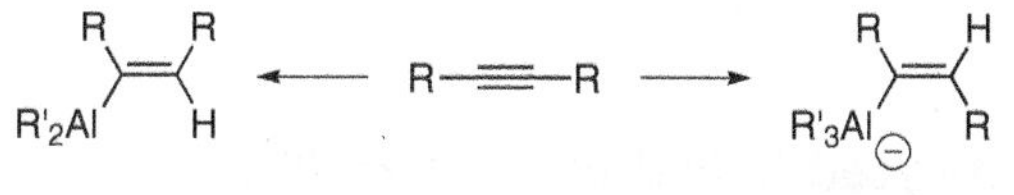

Reagents: *syn* addition: DIBAL-H, PhMe; DIBAL-H, Et_2O; DIBAL-H, $MeN(CH_2)_4$, etc.
anti addition: $LiAlH_4$, Et_2O; $LiEt_3AlH$, Et_2O; etc.

(continues)

(**Reaction Synopses** continued)

Regiochemistry: Markovnikov with electrophilic aluminum reagents, but can be altered by complexation of Al with a strong Lewis base prior to the hydrometallation to give a more nucleophilic aluminum reagent

Hydrozirconation

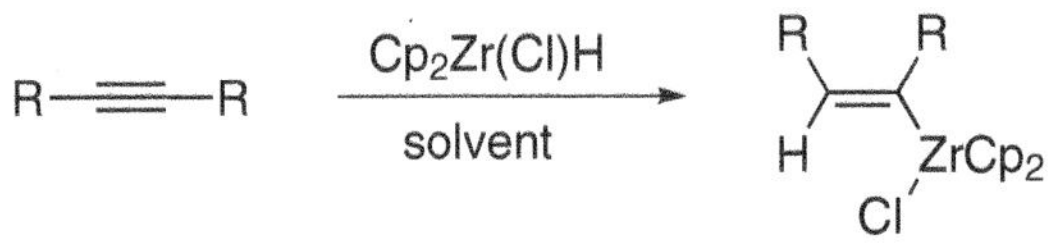

Solvent: PhMe (mixture of regioisomers); THF (Zr bonds to less hindered carbon); etc.

Stereochemistry: *syn*

Hydrostannylation

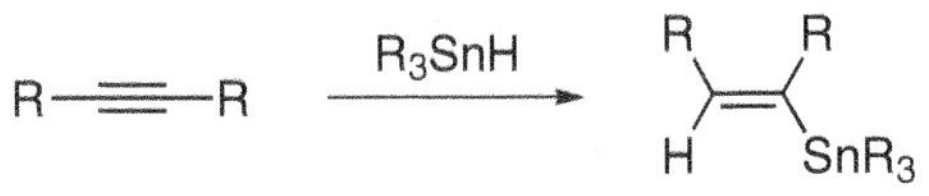

Reagents: R_3SnH, AIBN, hν; etc. (free radical mechanism); R_3SnH, Pd (not widely used)

Copper: most widely used metal

Hydrosilation

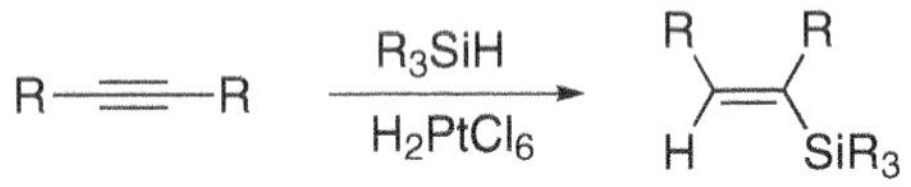

Stereochemistry: *syn* when Pt catalysts are used

Carbometallation

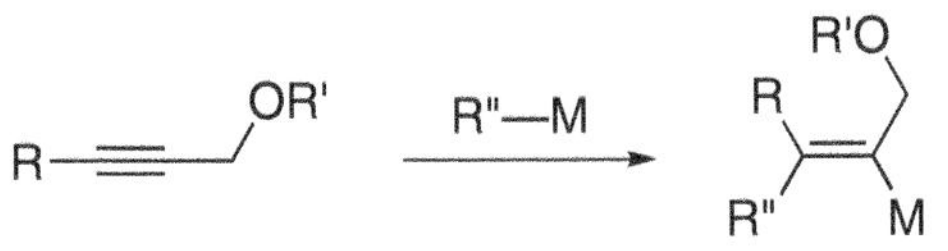

Reagents: R''_2CuLi, THF; $R''_2Cu(CN)Li_2$, THF (carbocupration); R''_3Al, $ZrCl_4$; R_3Al, $Cp_2Zr(Cl)H$, (carboalumination); etc.

17.6 Palladium-Catalyzed Substitution: Tsuji-Trost and Buchwald-Hartwig Reactions

Despite the fact that no textbook chapter can keep up with developments as rapid as those taking place in modern catalytic organic synthesis, such a chapter is not without value. It can still serve as a jumping-off point for your own reading about organic synthetic reactions catalyzed by transition metal complexes. Let us, therefore, begin by looking at palladium-catalyzed analogs of simple nucleophilic displacement reactions.

One of the oldest coupling reactions of this type is the ***Tsuji-Trost reaction***.[102] The reaction between an allylic compound and a nucleophile catalyzed by a palladium complex, discovered by Jiro Tsuji and developed by Barry M. Trost, relies on the oxidative addition of an allylic ester, halide, or similar species to a palladium (0) precursor to generate an electrophilic η^3-allylpalladium (II) species. The reactivity of common leaving groups follows the general order:

$$\text{Cl, Br} > \text{OCO}_2\text{R} > \text{OCOR}$$

Allylic carbonate esters have achieved popularity as precursors to the electrophile because the oxidative addition of these esters to the palladium leads to an anion that loses carbon dioxide to make the addition irreversible. The usual regiochemistry of the reaction gives the product with the nucleophile bound to the allyl group at the less substituted end.

The stereochemistry of the reaction depends on the nature of the nucleophile. Hard nucleophiles (e.g., R_2Zn) attack the η^3 allyl complex at the metal atom first, giving a complex (**17.176**) that lead to inversion of configuration, whereas soft nucleophiles (e.g., enolates) attack the ligand first, giving a complex (**17.174**) that leads to retention of configuration. The two divergent stereochemical outcomes can be accommodated by the catalytic cycles in Figure 17.24.

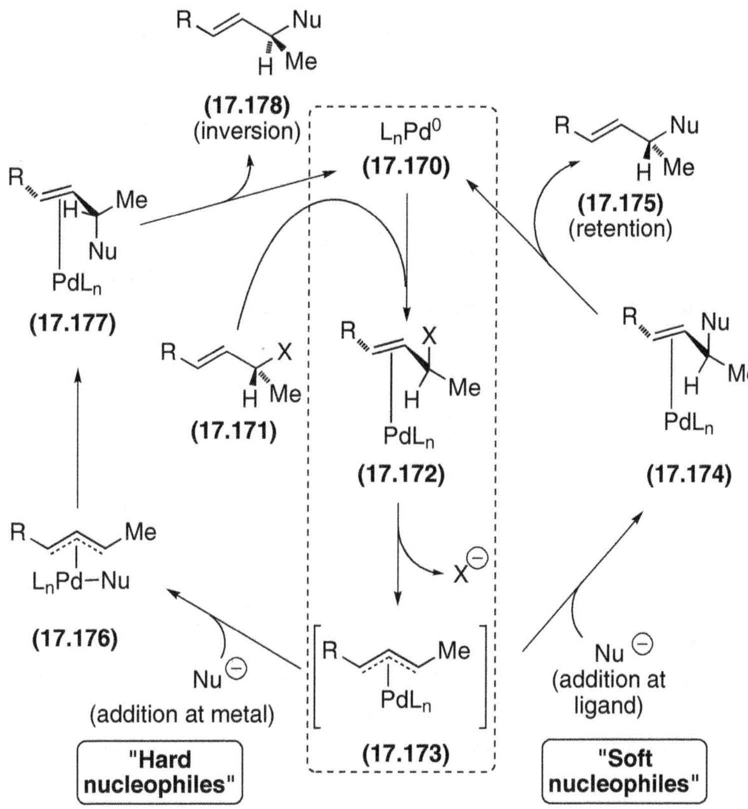

Figure 17.24 Catalytic cycles of the Tsuji-Trost allylation reaction with hard and soft nucleophiles, showing reversal of stereochemistry

102. (a) Tsuji, J.; Takahashi, H.; Morikawa, M. *Tetrahedron Lett.* **1965**, 4387. (b) Trost, B.M. *Tetrahedron* **1977**, *33*, 2615. (c) Trost, B.M. *Acc. Chem. Res.* **1980**, *13*, 385. (d) Trost, B.M. *Pure Appl. Chem.* **1981**, *53*, 2357. (e) Tsuji, J. *Pure Appl. Chem.* **1982**, *54*, 197. (f) Tsuji, J. *Tetrahedron* **1986**, *42*, 4361. (g) Trost, B.M. *Angew. Chem. Int. Ed. Engl.* **1989**, *28*, 1173. (h) Tsuji, J. *Palladium Reagents and Catalysts. New Perspectives for the 21st Century.* (Wiley: print-on-demand, ISBN 978-0-470-85032-9; 2004).

The allylpalladium (II) species are susceptible to attack by a wide range of anionic nucleophiles, as illustrated by Examples **17.179**,[103] **17.180**,[104] and **17.181**.[105] The Tsuji-Trost allylation has been the subject of several modifications, some to allow its use with organocatalysis, where the enolate anion becomes replaced by an enamine generated in situ, and also by a creative combination with the Morita-Baylis-Hillman reaction to form cyclized enones (Example **17.180**). Using allylic epoxides in the reaction is illustrated by Example **17.181**, which is an early coupling step in Fukuyama's synthesis of strychnine.

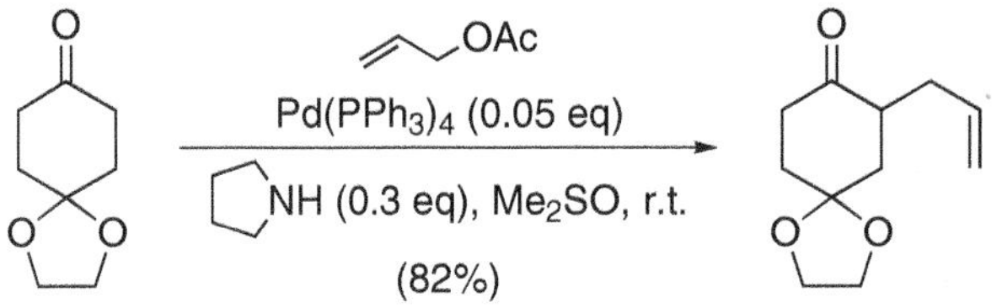

(17.179)

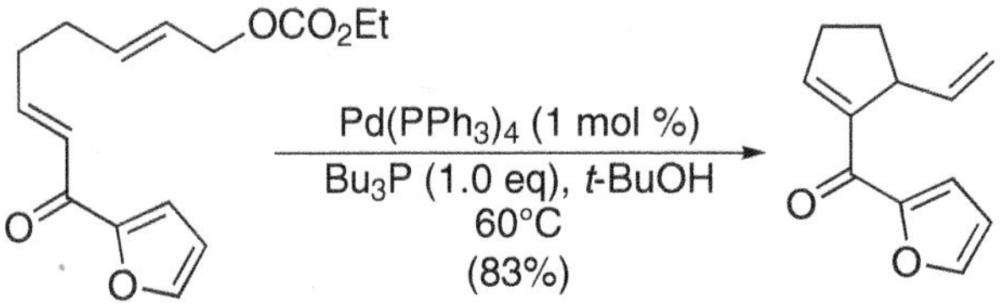

(17.180)

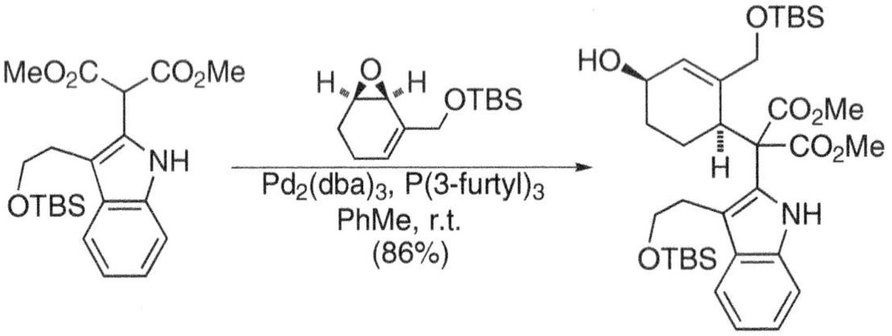

(17.181)

More recent work[106] has led to the development of useful chiral ligands for use in asymmetric allylation, especially of ethers.[107] A typical example of a chiral diphosphine ligand suitable for asymmetric allylation is given as Example **17.182**. With C_2-symmetric diphosphine ligands such as **17.182**, Trost has established a mnemonic to predict the stereochemistry of the alkylation of *meso*-diesters.[108]

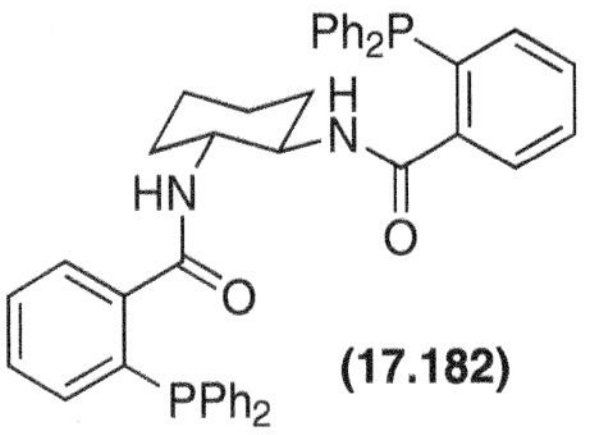

(17.182)

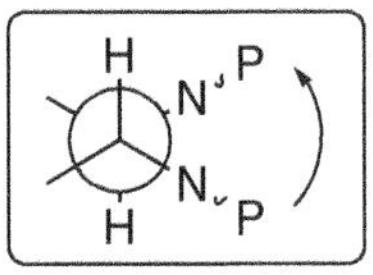

In this mnemonic, shown in Figure 17.25, the diphosphine is viewed along the axis between the two phosphines, and the direction of rotation is determined to be clockwise or counterclockwise. (In this example, it is counterclockwise.) The initial η^2 complexation

103. Ibrahem, I.; Córdova, A. *Angew. Chem. Int. Ed.*, **2006**, *45*, 1952.
104. Jellerichs, B.G.; Kong, J.-R.; Krische, M.J. *J. Am. Chem. Soc.* **2003**, *125*, 7758.
105. Kaburagi, Y.; Tokuyama, H.; Fukuyama, T. *J. Am. Chem. Soc.* **2004**, *126*, 10246.
106. Review: Trost, B.M.; Crawley, M.L. *Chem. Rev.* **2003**, *103*, 2921.
107. Trost, B.M.; Toste, F.D. *J. Am. Chem. Soc.* **2003**, *125*, 3090.
108. (a) Trost, B.M.; Van Vranken, D.L. *Chem. Rev.* **1996**, *96*, 395. (b) Trost, B.M.; Lee, C. In Ojima, I, Ed. *Catalytic Asymmetric Synthesis*, 2nd Ed. (Wiley-VCH: New York, 2000), pp. 593–649.

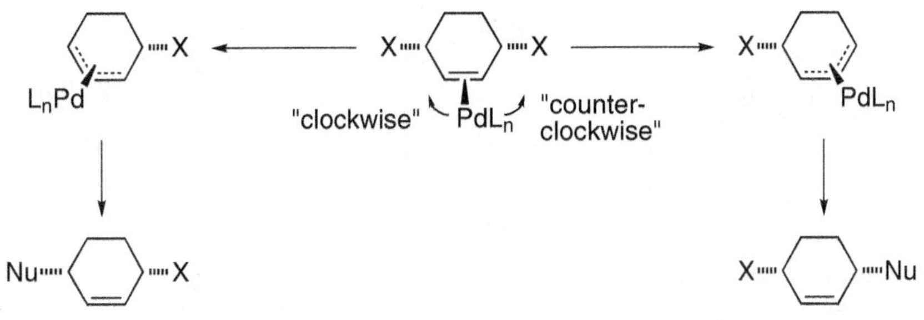

Figure 17.25 The mnemonic for predicting stereochemistry in the asymmetric allylation of *meso* substrates using C_2-symmetric "Trost" diphosphine ligands

of the alkene to the metal occurs on the face opposite to the leaving groups. The alkene is oriented so that the metal is above the plane of the π bond. The initial complexation centers the palladium above the alkene, and conversion of the η^2-alkene complex to the η^3-allyl complex requires movement of the palladium to the left (clockwise) or the right (counterclockwise) to center the metal atom above the ligand.

The clockwise ligand leads to clockwise movement of the palladium. The nucleophile then attacks the bottom face of the ligand to give the observed product (e.g., Example **17.183**). There is one caveat in this mnemonic: when the nitrogen and carbonyl groups of the amides are reversed (i.e., the diamide of a diamine is replaced by the isomeric diamide of a dicarboxylic acid; Example **17.184**), the stereochemistry of the reaction is also reversed, despite the phosphines having the same relative disposition (here, counterclockwise). There is, as yet, no general model that one can use to predict the stereochemistry in this reaction.

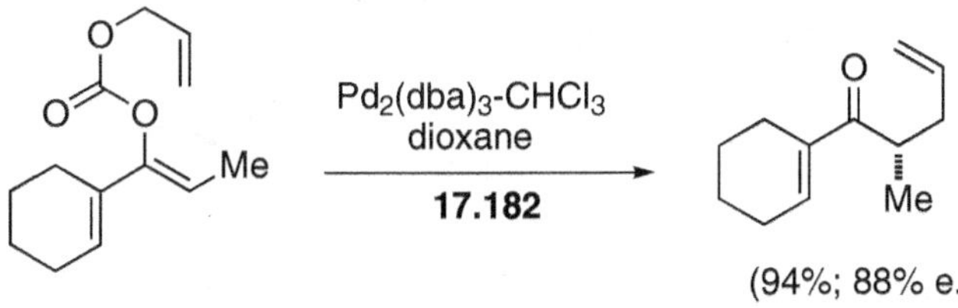

(17.183)

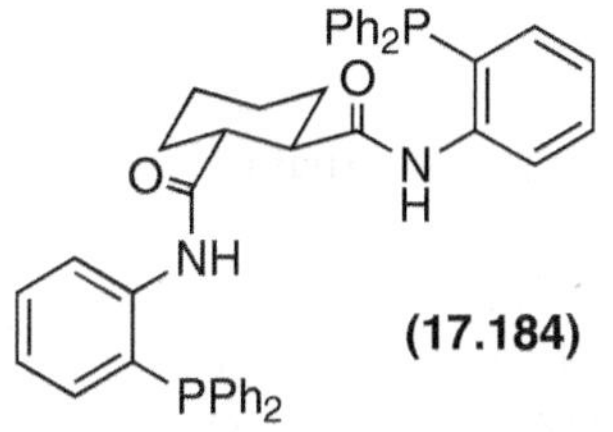

(17.184)

The Tsuji-Trost reaction has also been useful in the asymmetric construction of quaternary centers. One useful example of this process is given by the allylation reaction in Example **17.185**.[109] In this reaction, a ketone is first converted to a mixed enol allyl carbonate, which is then treated with a palladium catalyst in the presence of a chiral phosphine. A range of phosphine ligands has been used, with the PHOS ligands, especially *t*-Bu-PHOS, being the most effective. The reaction also applies to intermolecular reactions using enol trimethylsilyl ethers and diallyl carbonate.[110] Interestingly, these workers also found that the best ligands for this reaction were not chiral diphosphines but rather the chiral PHOS ligands, with the enantioselectivity rising as the steric bulk of the group on the PHOS ligand increased. The scope of the reaction has been further extended since the first report, with its application to acyclic enol allyl carbonates.[111,112]

109. McFadden, R. M.; Stoltz, B. M. *J. Am. Chem. Soc.* **2006**, *128*, 7738.
110. Behenna, D.C.; Stoltz, B.M. *J. Am. Chem. Soc.* **2004**, *126*, 15044.
111. Trost, B.M.; Xu, J. *J. Am. Chem. Soc.* **2005**, *127*, 17180.
112. Mohr, J.T.; Stoltz, B.M. *Chem. Asian J.* **2007**, *2*, 1476.

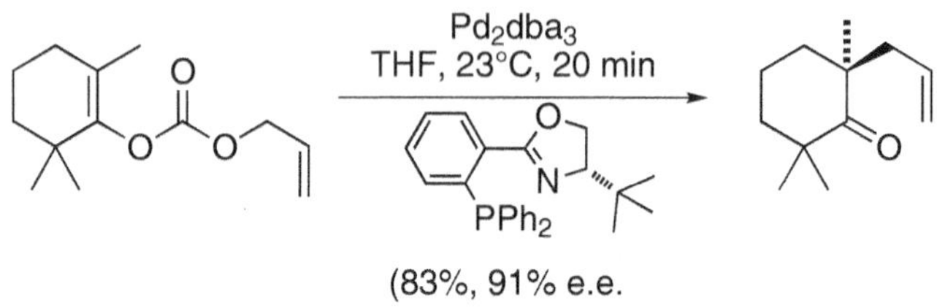

(17.185)

In contrast to the palladium-catalyzed allylation, which gives the less substituted regioisomer, Hartwig has reported[113] that the use of an iridium-based catalyst leads to the formation of the more substituted allyl regioisomer with hard nucleophiles, as illustrated in Example **17.186**. The active catalyst (Example **17.187**) is formed by treating $[Ir(cod)Cl]_2$ with a chiral phosphoramidite ligand and an amine.

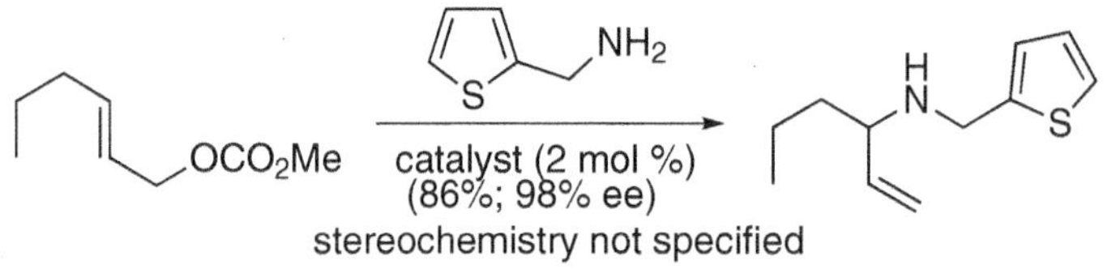

(17.186)

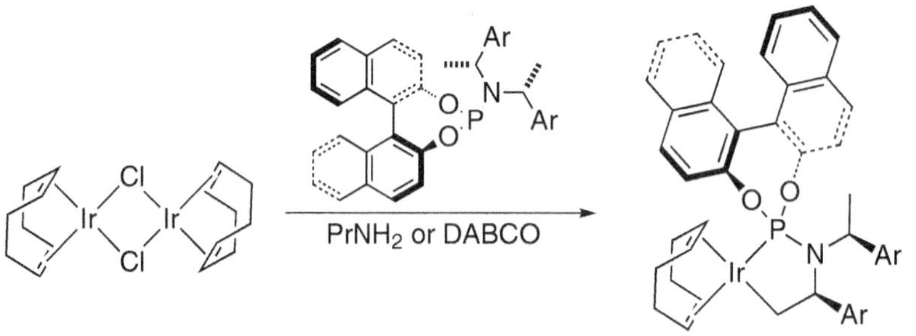

(17.187)

The Buchwald-Hartwig and Related Cross-Coupling Reactions

The amination of aryl halides catalyzed by palladium complexes was described early in the 1980s but was not really developed until the mid-1990s, when Buchwald and Hartwig independently began the work that defined the scope of the reaction. The reaction is now known by both chemists' names.[114] The catalytic cycle of the Buchwald-Hartwig coupling (Figure 17.26) follows a fairly standard scheme for palladium-catalyzed coupling at sp^2 carbon: (1) oxidative addition of the aryl halide to a coordinatively unsaturated Pd^0 complex (**17.188**) to give a saturated complex (**17.189**), followed by (2) replacement of the halide ligand by the other coupling partner to give the complex **17.191** (this may require more than one step), and (3) a final reductive elimination of the product (**17.192**). In its most widely used form, the nucleophile is a primary amine, although simple ammonia has proved to be problematic, requiring the use of surrogates for ammonia in the reaction due to the very strong complexation of the palladium by ammonia itself.

Like most palladium-catalyzed cross-coupling reactions, the Buchwald-Hartwig coupling is tolerant of a wide range of functional groups—esters, nitriles, tertiary

113. (a) Shu, C.; Hartwig, J.F. *Angew. Chem. Int. Ed.* **2004**, *43*, 4794. (b) Shu, C.; Leitner, A.; Hartwig, J.F. *Angew. Chem. Int. Ed.* **2004**, *43*, 4797. (c) Leitner, A.; Shu, C.; Hartwig, J.F. *Org. Lett.* **2005**, *7*, 1093.

114. Reviews: (a) Hartwig, J.F. *Synlett* **1997**, *4*, 329. (b) Hartwig, J.F. *Angew. Chem. Int. Ed.* **1998**, *37*, 2046. (c) Wolfe, J.P.; Wagaw, S.; Marcoux, J.F.; Buchwald, S.L. *Acc. Chem. Res.* **1998**, *31*, 805. (d) Hartwig, J.F. *Acc. Chem. Res.* **1998**, *31*, 852. (e) Hartwig, J.F. *Pure Appl. Chem.* **1999**, *71*, 1416. (f) Muci, A.R.; Buchwald, S.L. *Top. Curr. Chem.* **2002**, *219*, 131. (g) Hartwig, J.F. *Acc. Chem. Res.* **2008**, *41*, 1534. (h) Surry, D.S.; Buchwald, S.L. *Angew. Chem. Int. Ed.* **2008**, *47*, 6338. (i) Surry, D.S.; Buchwald, S.L. *Chem. Sci.* **2011**, *2*, 27.

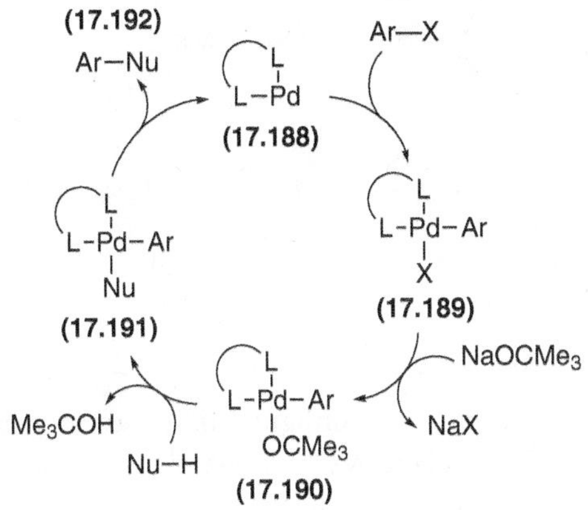

Figure 17.26 The catalytic cycle of the Buchwald-Hartwig coupling reaction

amines, and ethers—which makes this reaction an excellent method for the formation of highly functionalized aniline derivatives. This is illustrated by the formation of the carbazole ring system of (±)-murrayazoline (Example **17.193**).[115] The same synthesis contains the intramolecular formation of an ether using the same Buchwald-Hartwig strategy (Example **17.194**).

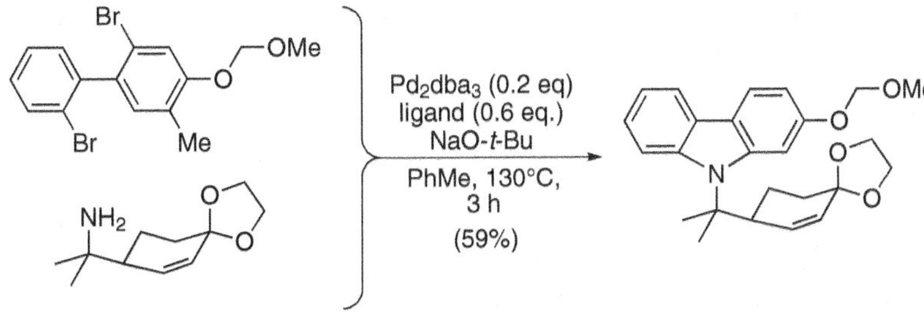

(17.193)

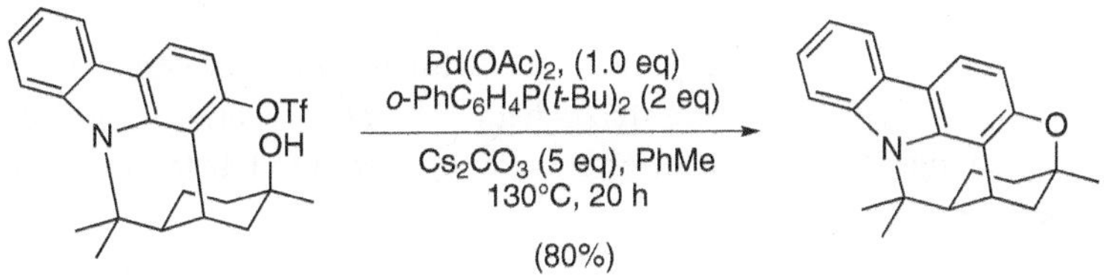

(17.194)

Considerable work has been expended in the search for suitable ligands for use in cross-coupling reactions. The results of many investigations are summarized in structure **17.195**, where the effects of each of the structural features of the ligand on its performance in palladium-catalyzed aminations are summarized.[116]

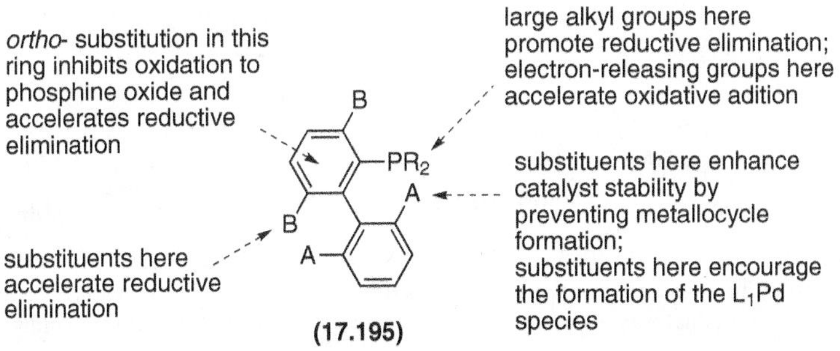

(17.195)

115. Ueno, A.; Kitawaki, T.; Chida, N. *Org. Lett.* **2008**, *10*, 1999.
116. Surry, D.S.; Buchwald, S.L. *Chem. Sci.* **2011**, *2*, 27.

Reaction Synopses

Tsuji-Trost Reaction

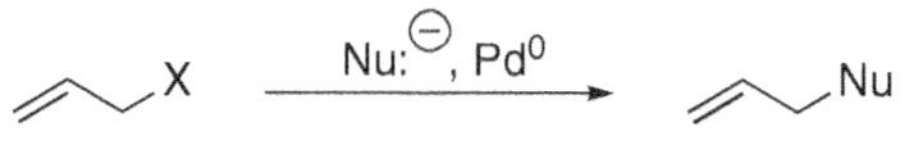

X: Cl, Br, OAc, OCOPh, OCO_2Et, etc.

Pd^0: $Pd(PPh_3)_4$, etc.

Nu^-: enolate anions, metal alkyls, etc.

Stereochemistry: net retention of configuration at carbon with soft nucleophiles (e.g., enolates); net inversion with hard nucleophiles (e.g., dialkylzinc reagents)

Chiral ligands: allow enantioselective reactions of *meso* diesters to give chiral products

Buchwald-Hartwig Coupling

$$\text{Ar–X} \xrightarrow[\text{Pd, ligand, base}]{\text{H–Nu}} \text{Ar–Nu}$$

X: Br, OTf, I, etc. (Cl and OTs require more active catalysts)

Pd: Pd_2dba_3, $Pd(OAc)_2$, etc.

Ligand: dppf, *o*-$ArC_6H_4PR_2$, etc.

Nu: Ar'NR, Ar'O, etc. (with stronger bases, enolate anions can be used)

Worked Problem

17-4 What is the major organic product of the following reaction? Provide a rationalization for the regiochemistry and stereochemistry of the reaction. How would the reaction change if the malonate ester were replaced by a harder nucleophile (e.g., a phenoxide anion)?

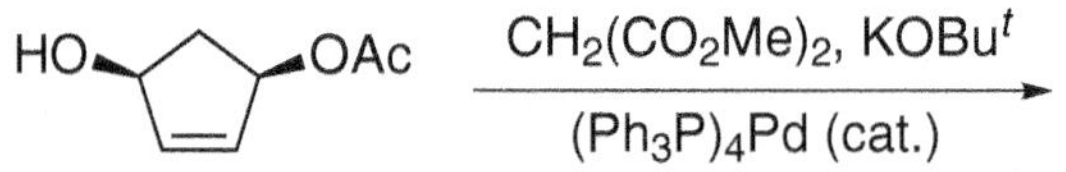

§Answer below.

§ Answer to Worked Problem:

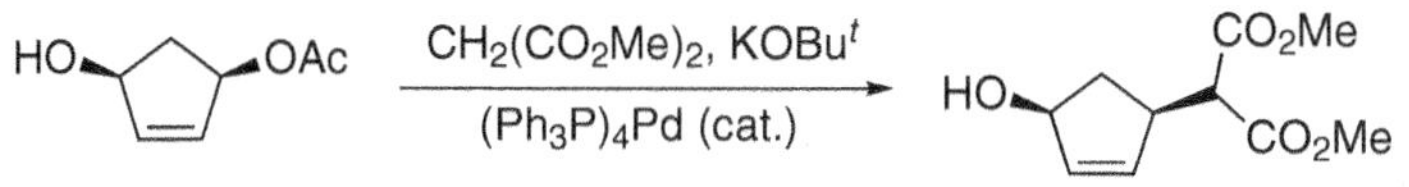

This reaction proceeds by means of the η^3-allyl complex, formed by displacement of the acetate with inversion of configuration. The nucleophile in this reaction is a soft carbon nucleophile, so it attacks the face opposite the metal, leading to a product whose stereochemistry corresponds to overall retention of configuration.

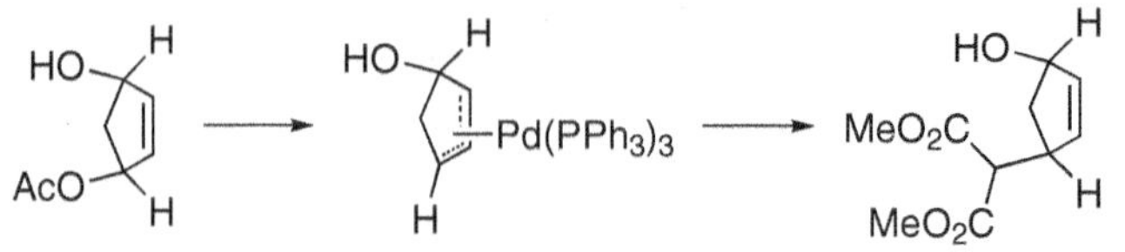

Problems

17-6 What is the major organic product of each of the following reactions?

(a)

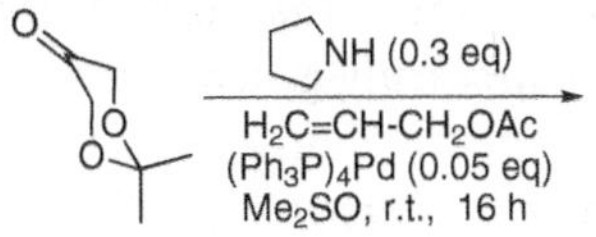

(b)

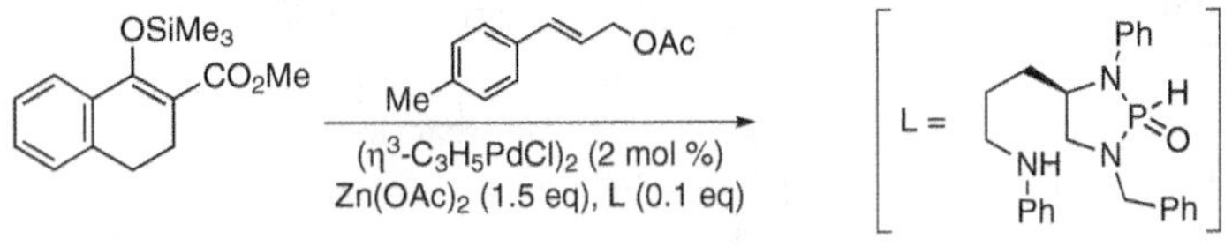

(c)

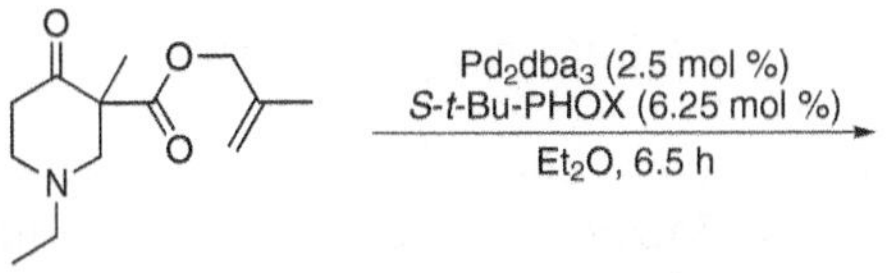

References: (a) *J. Am. Chem. Soc.* **2006**, *128*, 4232. (b) *J. Am. Chem. Soc.*, **2004**, *126*, 3690. (c) *Angew. Chem. Int. Ed.*, **2005**, *44*, 6924.

17-7 Suggest a reasonable model that will permit the rationalization of the stereochemistry of the following reaction.

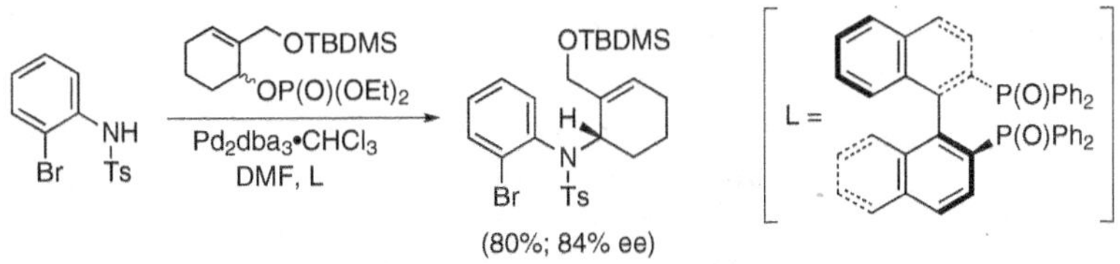

References: Angew. Chem. Int. Ed. **2002**, *41*, 1934; *J. Am. Chem. Soc.* **2003**, *125*, 9801.

17.7 Cross-Coupling Reactions to Form C—C Bonds

Now that we have discussed the general features of reactions involving organotransition metal complexes, it is time to turn our attention to the ways in which these reactions have transformed modern organic synthesis. Again, one should be aware of the fact that this is an extremely rapidly growing area of organic chemistry, so the material here will almost certainly be somewhat dated by the time that you read it. As a guide to the importance of this field, you should consult Table 17.4. This table lists the number of articles appearing in 2011 in journals published by national chemical societies in which the specified transition metal-catalyzed reaction was mentioned.

Table 17.4 Number of Metal-Catalyzed C—C Bond-Forming Reactions Cited in 2011

Reaction	Number of Mentions
Suzuki-Miyaura cross-coupling	10,721
Heck cross-coupling	3,157
Sonogashira cross-coupling	2,302
Stille cross-coupling	2,176
Negishi cross-coupling	1,461
Hiyama-Denmark cross-coupling	647
Tsuji-Trost	124
Olefin metathesis	7,397

Replacement of the malonate ester by a phenoxide anion, which is a harder oxygen nucleophile, would lead to the product of overall inversion of configuration.

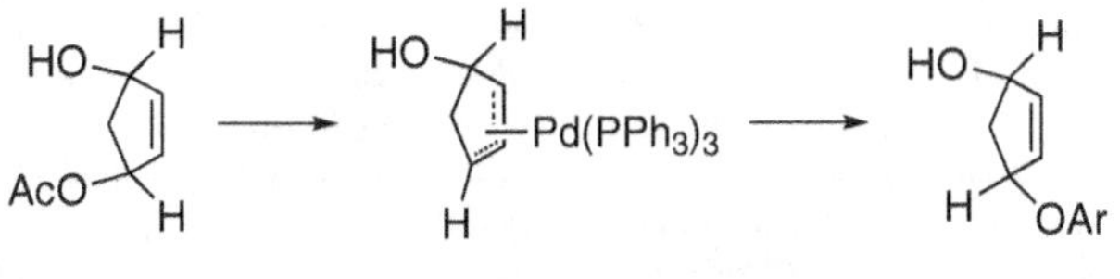

Reference: Tetrahedron Lett. **2004**, *45*, 3783.

Table 17.5 Transition Metal–Catalyzed Cross-Coupling Reactions

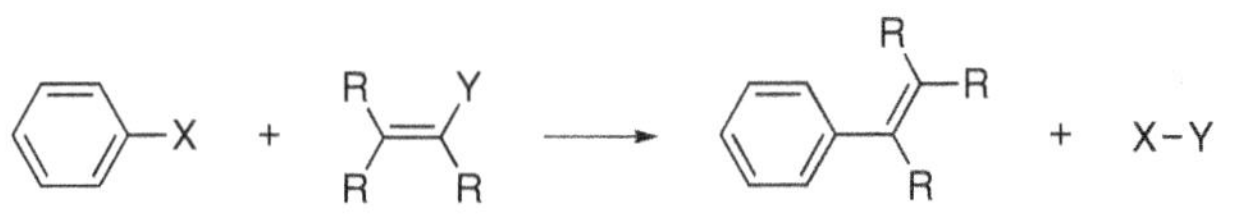

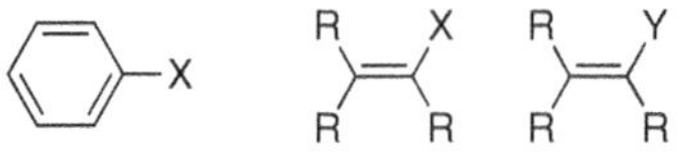

		Catalyst	Reaction Name
X=I, OTf, Br [Cl]	Y=H	Pd	Heck
X=HgX, SnR_3, PbR_3,	Y=H	Pd	Heck
X=I, OTf, Br [Cl]	$Y=SnR_3$	Pd	Stille
X=I, OTf, Br [Cl]	Y=ZnX, AlX_2, ZrL_n	Pd, Ni	Negishi
X=I, OTf, Br [Cl]	$Y=SiR_3$, F^-	Pd	Hiyama
X=I, OTf, Br [Cl]	$Y=Si(OR)_3$, RO^-	Pd	Hiyama-Denmark
X=I, OTf, Br [Cl]	Y=MgX	Ni, Pd	Corriu-Kumada
X=I, OTf, Br [Cl]	$Y=B(OR)_2$	Pd	Suzuki-Miyaura
X=I, OTf, Br [Cl]	RC≡CH	Pd-Cu	Sonogashira

The rapid rise of transition metal–catalyzed synthetic reactions can be traced to the early work of Japanese chemist Tsutomu Mizoroki[117] and Nobel laureates Richard Heck[118] and Ei-ichi Negishi.[119] The work of these chemists identified one of the first truly general cross-coupling reactions for forming a new carbon-carbon bond. Since the initial reports by these workers were published, the field of cross-coupling reactions has literally exploded, with several permutations of substrate and nucleophile now being well-known name reactions. A collection of the most widely used reactions is gathered in Table 17.5.

The Heck Coupling Reaction

The reaction now known as the ***Heck reaction***[120] was first reported by Mizoroki[121] a year before Heck's first report,[122] but Mizoroki's untimely death meant that developing the reaction rested squarely with Heck. The Heck reaction is the prototype general reaction involving the formal nucleophilic substitution of a halogen or triflate bonded to an aromatic ring or a double bond, or a formal metathesis of an aryl halide and an alkene to generate the hydrogen halide and the vinylarene.

The reaction is catalyzed by palladium and occurs through a catalytic cycle shown in Figure 17.27. The reaction is usually initiated by reduction of palladium (II) acetate to a palladium (0) complex (**17.196**) that is the functional catalyst. Oxidative addition of the halide to this complex gives a palladium complex (**17.197**) that then undergoes ligand exchange to give the η^2 olefin complex (**17.198**). Migratory insertion of the aryl group gives a σ alkyl complex (**17.199**) that undergoes β-hydride elimination to give a new η^2 olefin complex (**17.200**). Ligand substitution and reductive elimination of the hydrogen halide completes the catalytic cycle. The reaction is facilitated by the addition of base to neutralize the hydrogen halide.

For most Heck reactions, the reactivity of the substrate bearing the leaving group decreases in the order I > OTf > Br >> Cl. The reaction is also susceptible to steric hindrance: alkenes react in the approximate order:

$$H_2C{=}CH_2 > H_2C{=}CHR > R_2C{=}CH_2 \approx RCH{=}CHR > R_2C{=}CHR_2$$

117. **Tsutomu Mizoroki (1933–1980)** was a young Associate Professor at Tokyo Institute of Technology when he discovered the palladium-catalyzed reaction of iodoarenes with alkenes. He died at age 47 of pancreatic cancer, but his efforts laid the groundwork for the Nobel Prizes in chemistry three decades later.

118. **Richard Fred Heck (1931–)** was educated at the University of California, Los Angeles (UCLA) (BS, 1952; PhD, 1954 with Winstein), and then, after postdoctoral study at ETH, Zurich, and at UCLA, he joined Hercules Corporation. From 1971–1989, was at Delaware. For more biographical details see he Nobel Foundation web site.

119. **Ei-ichi Negishi (1935–)** was born in China, and educated at the University of Tokyo (BE, 1955) and the University of Pennsylvania (PhD, 1963). His postdoctoral study with H.C. Brown, at Purdue, led to a permanent appointment to the Purdue faculty, where he has spent his career. For more details, see Negishi, E.-i. *Angew. Chem. Int. Ed.* **2011**, *50*, 6738.

120. Reviews: (a) Heck, R.F. *Acc. Chem. Res.* **1979**, *12*, 146; *Org. React.* **1982**, *27*, 345. (b) de Meijere, A.; Meyer, F.E. *Angew. Chem. Int. Ed. Engl.* **1994**, *33*, 2379. (c) Cabri, W.; Candiani, I. *Acc. Chem. Res.* **1995**, *28*, 2. (d) Shibasaki, M.; Boden, C.D.J.; Kojima, A. *Tetrahedron* **1997**, *53*, 7371. (e) Beletskaya, I.P.; Cheprakov, A.V. *Chem. Rev.* **2000**, *100*, 3009. (e) Whitcombe, N.J.; Hii, K.K.; Gibson, S.E. *Tetrahedron* **2001**, *57*, 7431. (g) Link, J.T. *Org. React.* **2002**, *60*, 157. (h) Dounay, A.B.; Overman, L.E. *Chem. Rev.* **2003**, *103*, 2945. (i) Gurry, D.J.; Kiely, D. *Curr. Org. Chem.* **2004**, *8*, 781. (j) Heck, R.F. *Synlett.* **2006**, 2855. (k) Knowles, J.P.; Whiting, A. *Org. Biomol. Chem.* **2007**, *5*, 31.

121. Mizoroki, T.; Mori, K.; Ozaki, A. *Bull. Chem. Soc. Jpn.* **1971**, *44*, 581.

122. Heck, R.F.; Nolley, J.P., Jr. *J. Org. Chem.* **1972**, *37*, 2320.

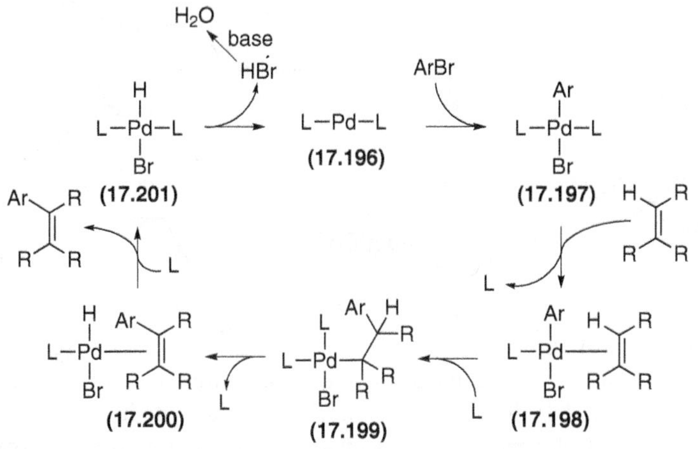

Figure 17.27 The catalytic cycle of the Heck reaction

The reaction is therefore most often used with monosubstituted alkenes. One of the major features of the intermolecular Heck reaction is its high *trans* selectivity. The intramolecular form of the Heck reaction has been widely used to make cyclic compounds, as shown by Examples **17.202**[123] and **17.203**.[124]

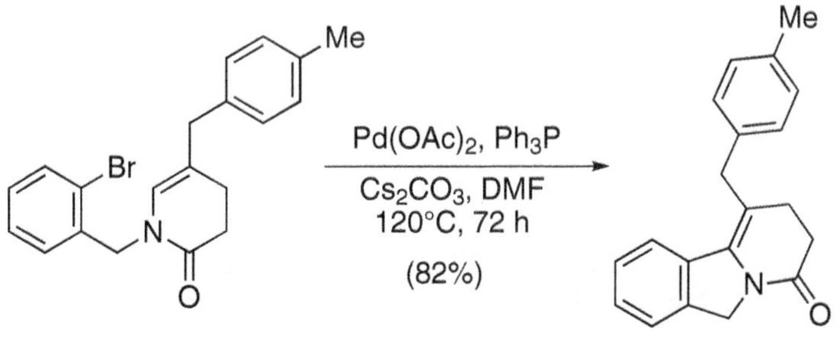

(17.202)

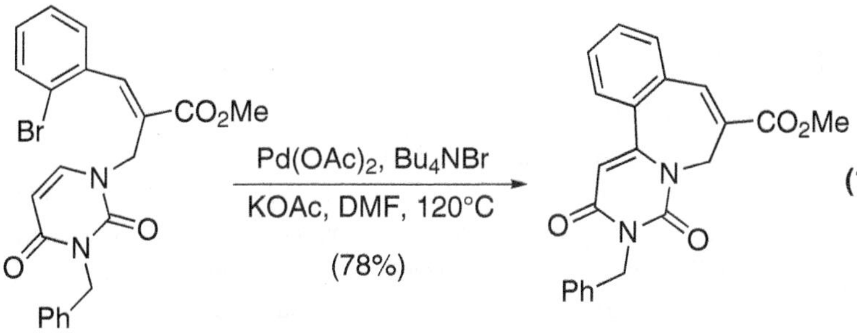

(17.203)

As with all other reactions since the end of the 20th century, the Heck reaction has also been the subject of intense research to develop asymmetric versions of the reaction.[125] Because the reaction does not form a new chiral center at either end of the new C—C bond, the adaptation of this reaction to asymmetric synthesis relies on the desymmetrization of compounds with C_s symmetry, such as the diene in Example **17.204**, where reaction at the two double bonds of the cyclohexadiene ring gives opposite enantiomers of the product. When this same diene was treated with a palladium catalyst in the presence of chiral BINAP, the product obtained was significantly enantio-enriched (Example **17.205**).[126]

123. Satyanarayana, D.; Maier, M.E. *Tetrahedron* **2012**, *68*, 1745.

124. Lee, H.S.; Kim, K.H.; Kim, S.H.; Kim, J.N. *Tetrahedron Lett.* **2012**, *53*, 497.

125. Reviews: (a) Shibasaki, M.; Boden, C.D.J.; Kojima, A. *Tetrahedron* **1997**, *53*, 7371. (b) Dounay, A.B.; Overman, L.E. *Chem. Rev.* **2003**, *103*, 2945. (c) Shibasaki, M.; Vogl, E.M.; Ohshima, T. *Adv. Synth. Catal.* **2004**, *346*, 1533. (d) McCartney, D.;Guiry, P.J. *Chem. Soc. Rev.* **2011**, *40*, 5122.

126. Sato, Y.; Sodeoka, M.; Shibasaki, M. *J. Org. Chem.* **1989**, *54*, 4738.

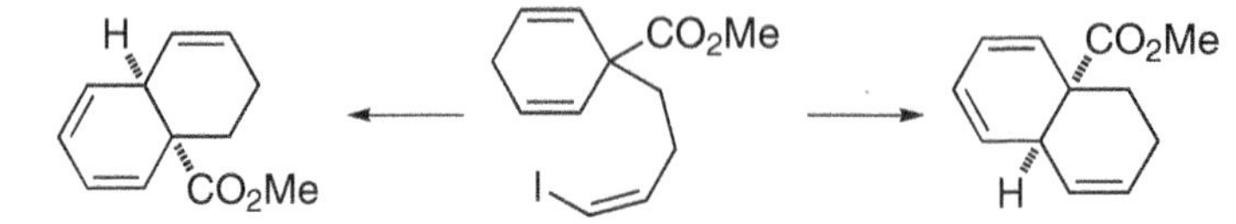

(17.204)

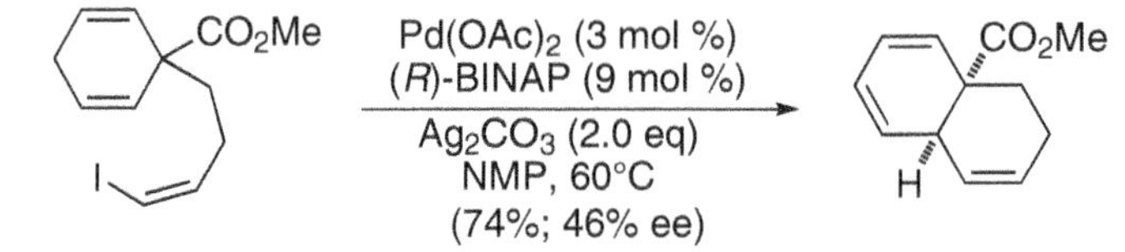

(17.205)

The application of the asymmetric Heck cyclization to cascade reactions was first demonstrated in 1989 by Overman (Example **17.206**).[127] More recently, the Heck coupling has been applied to aldehydes (Example **17.207**). The aldehyde hydrogen is the one lost during this formal Heck-type reaction, which thus provides a palladium-catalyzed approach to the acylation of arenes.[128] The Heck coupling of vinyl sulfides, sulfoxides, and sulfones with aryl bromides shows that the isolated yield of the reaction is increased as the alkene becomes more electrophilic.[129]

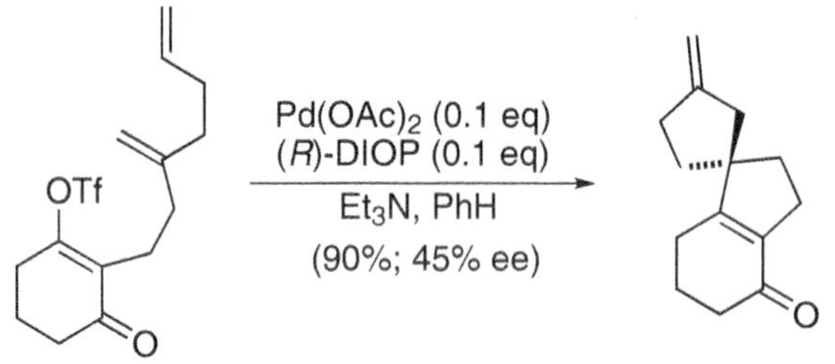

(17.206)

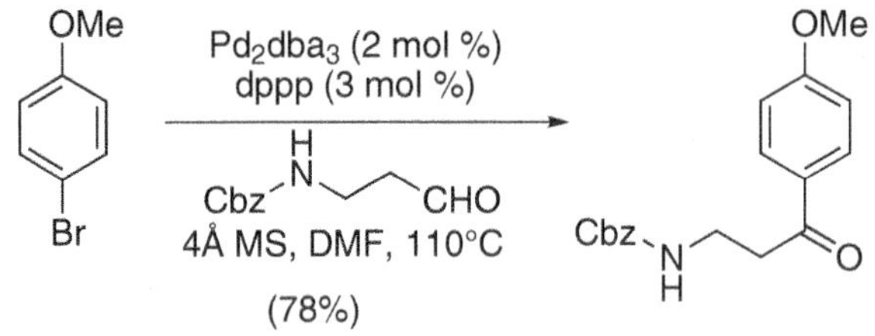

(17.207)

In a series of papers, Jeffrey introduced a ligand-free version of the Heck reaction that involves using added quaternary ammonium salts to accelerate the reaction and control its regioselectivity.[130] By combining a quaternary salt with added potassium formate, one can directly obtain the product of coupling and transfer hydrogenation, as illustrated in Example **17.208**.[131] A similar one-pot Heck coupling and hydrogenation cascade has also been described.[132]

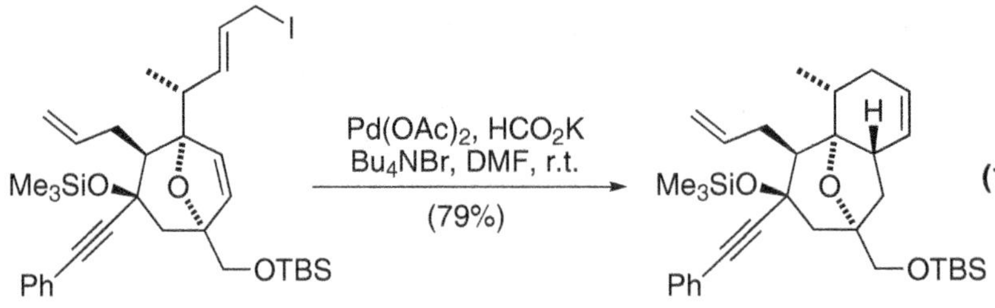

(17.208)

---

127. Carpenter, N.E.; Kucera, D.J.; Overman, L.E. *J. Org. Chem.* **1989**, *54*, 5846.
128. Ruan, J.; Saidi, O.; Iggo, J.A.; Xiao, J. *J. Am. Chem. Soc.* **2008**, *130*, 10510.
129. Battace, A.; Zair, T.; Doucet, H.; Santelli, M. *Synthesis* **2006**, 3495.
130. (a) Jeffery, T. *Tetrahedron Lett.* **1994**, *35*, 3051. (b) Jeffery, T.; Galland, J.C. *Tetrahedron Lett.* **1994**, *35*, 4103. (c) Jeffery T. *Tetrahedron*, **1996**, *52*, 10113. (d) Jeffery, T.; David, M. *Tetrahedron Lett.* **1998**, *39*, 5751.
131. Lee, K.; Cha, J.K. *J. Am. Chem. Soc.* **2001**, *123*, 5590.
132. Geoghegan, K.; Kelleher, S.; Evans, P. *J. Org. Chem.* **2011**, *76*, 2187, and references therein.

The Stille, Negishi, Hiyama, and Kumada Coupling Reactions

In Heck's earliest work,[133] he reported the stoichiometric reaction using organomercury, organotin, and organolead compounds where the halide (almost always an iodide or bromide) is now used in the modern variant of reaction. The by-product of the reaction was a metal hydride derivative that either decomposed to the free metal or to a reduced organometallic species (Example **17.209**). Within a short space of time, however, this work was superseded by the catalytic reaction now used, which couples a halide with the alkene.

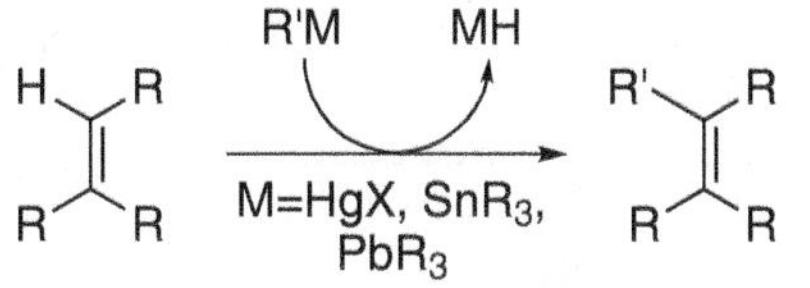

(17.209)

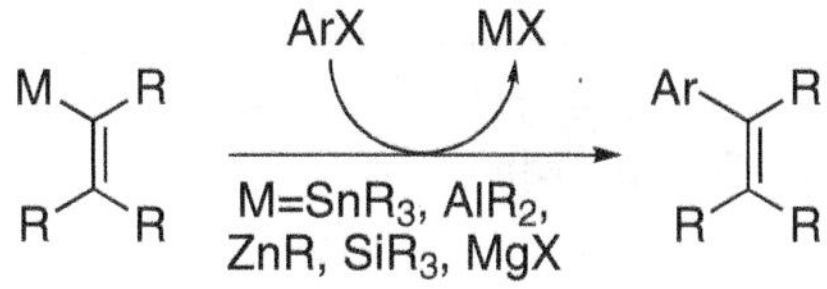

(17.210)

Subsequent versions of these palladium-catalyzed coupling reactions have focused, instead, on replacing the alkene by a suitable alkylmetal compound capable of metal-metal exchange (as shown in Example **17.210**). Thus, replacement of the alkene in the Heck reaction by a stannane gives the ***Stille reaction***,[134] by aluminum or zinc alkyls gives the ***Negishi coupling***,[135] and by silicon gives the ***Hiyama coupling***.[136] When a nickel catalyst is used with a Grignard reagent as the organometallic participant, the reaction is known as the ***Kumada coupling***,[137] or the ***Corriu-Kumada coupling***.

The Stille Coupling Reaction

The reaction between a stannane and a halide or triflate to give coupled products was first reported by American chemist John K. Stille,[138] whose untimely death may well have deprived him of a Nobel Prize. Due to the wide range of substrates that can be used and its tolerance to a wide variety of functional groups, the Stille reaction quickly became one of the most widely used catalytic cross-coupling reactions. (In fact, it was used in half the cross-coupling reactions reported in 1992.) Some idea of the wide range of substrates that can be used with this reaction can be obtained from the examples gathered in Table 17.6.

Table 17.6 Substrates for the Stille Coupling Reaction

Electrophile	Electrophile	Stannane	Stannane
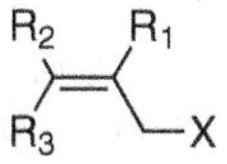	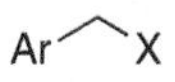	$R_1-\equiv-SnR_3$	
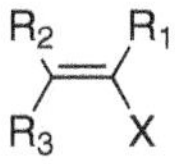	Ar−X	$Ar-SnR_3$	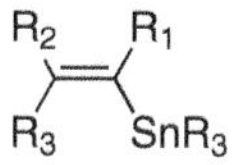
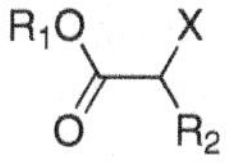	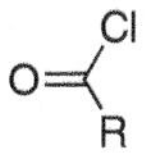	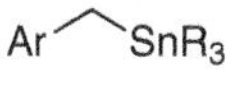	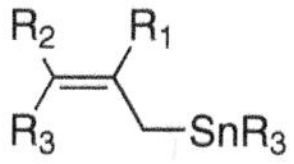
		$R-SnMe_3$	$H-SnR_3$

X, Br, I, OTf, Cl; Ar, aromatic residue.

The key steps of the catalytic cycle (Figure 17.28) are basically the same as in all other cross-coupling reactions: (1) oxidative addition of the halide to the coordinatively

133. Heck, R.F. *J. Am. Chem. Soc.* **1968**, *90*, 5518.

134. Reviews: (a) Stille, J.K. *Angew. Chem. Int. Ed. Engl.* **1985**, *25*, 508. (b) Mitchell, T.N. *Synthesis* **1992**, 803. (c) Farina, V.; Krishnamurthy, V.; Scott, W.J. *Org. React.* **1998**, *50*, 1. (d) Espinet, P.; Echavarren, A.M. *Angew. Chem. Int. Ed.* **2004**, *43*, 4704.

135. King, A.O.; Okukado, N.; Negishi, E.-i. *J. Chem. Soc., Chem. Commun.* **1977**, 683.

136. Hatanaka, Y.; Hiyama, T. *J. Org. Chem.* **1988**, *53*, 918.

137. (a) Corriu, R.J.P.; Masse, J.P. *J. Chem. Soc., Chem. Commun.* **1972**, 144a. (b) Tamao, K.; Sumitani, K.; Kumada, M. *J. Am. Chem. Soc.* **1972**, *94*, 4374.

138. **John Kenneth Stille (1930–1989)** was educated at the Universities of Arizona (BA, 1952; MA, 1953) and Illinois (PhD, 1957). He followed with faculty positions at Iowa and Colorado State University. Stille, who was one of the victims of the United Airlines crash in Sioux City, Iowa, pioneered palladium-catalyzed stannane coupling chemistry. More information is available at: *Organometallics* **1990**, *9*, 3007; *Macromolecules* **1990**, *23*, 2417.

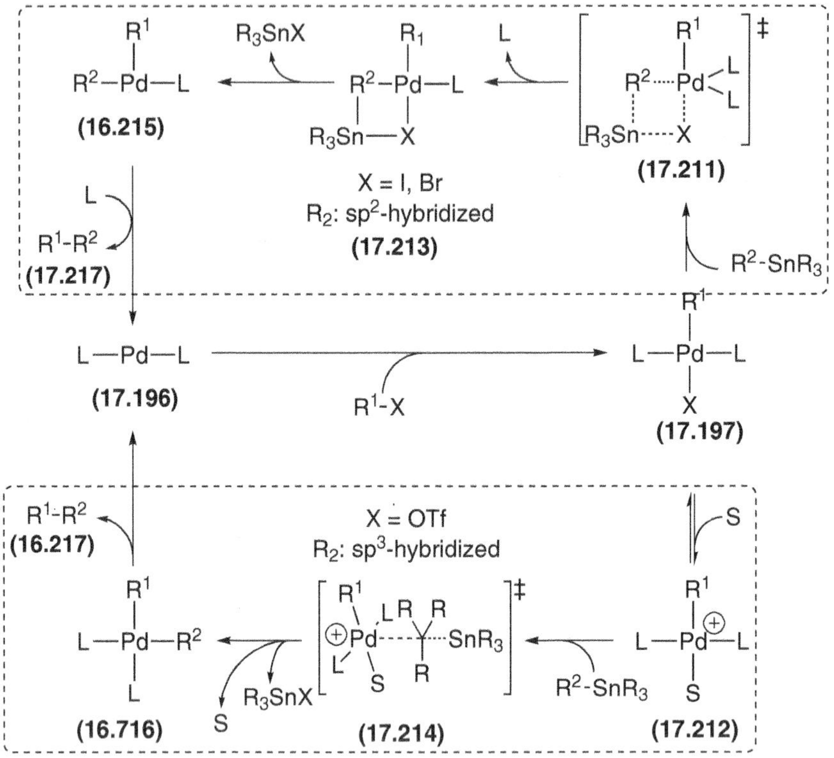

Figure 17.28 The catalytic cycle of the Stille coupling. S represents a molecule of solvent

unsaturated palladium (0) precursor (**17.196**) gives a coordinatively saturated complex (**17.197**), (2) transmetallation gives the palladocycle **17.213**, (3) elimination of the tin, and finally (4) reductive elimination of the coupled product (**17.217**) regenerates **17.196**. Alternatively, solvent displaces the halide in **17.197**, followed by transmetallation and elimination of the tin via an activated complex similar to **17.214**. Reductive elimination of the product then regenerates **17.193**. In the transmetallation step, it has been found that the rate of ligand transfer of groups from tin to palladium is approximately:

$$RC{\equiv}C > RCH{=}CH > Ar > allyl \approx benzyl > RCOCH_2 > R$$

There is, however, one major difference between the Stille coupling and other metal-catalyzed cross-coupling reactions. The mechanism of the reaction appears to depend on both the identity of the leaving group and also on the hybridization of the electrophilic carbon. The results of computational studies suggest that alkyl halides react through a cyclic transition state (**17.211**), whereas triflates react preferentially through an open transition state (**17.214**). In both cases, the transmetallation step is proposed to be associative.[139]

The relative order of reactivity of the tin ligands has been exploited to ensure the transfer of the desired ligand from tin by using methyl or butyl groups as the non-transferring ligands. The low reactivity of alkylstannanes, compared to vinyl-, aryl- and alkynylstananes, can be overcome by using strong donor solvents such as hexamethylphosphoric triamide (HMPT), dimethylformamide (DMF), and dioxane, which also suppress

139. (a) Nova, A.; Ujaque, G.; Maseras, F.; Lledós, A.; Espinet, P. *J. Am. Chem. Soc.* **2006**, *128*, 14571. (b) Pérez-Temprano, M.H.; Nova, A.; Casares, J.A.; Espinet, P. *J. Am. Chem. Soc.* **2008**, *130*, 10518. (c) Álvarez, R.; Pérez, M.; Faza, O.N.; de Lera, A.R. *Organometallics* **2008**, *27*, 3378. (d) Álvarez, R.; Faza, O.N.; de Lera, A.R.; Cárdenas, D.J. *Adv. Synth. Catal.* **2007**, *349*, 887.

homocoupling of the stannane. Selectivity and reactivity in the Stille reaction can be improved by the use of Cu (I)[140] or Mn (II) salts in stoichiometric quantities (e.g., Example **17.218**[141]). The use of Cu (I) salts also allows the use of palladium on carbon as the catalyst.[142] To accelerate the reaction, ligands are needed that stabilize both the nucleophilic palladium (0) complex and the electrophilic palladium (II) complex. These two effects require opposite donicity in the ligand. Therefore, ligands of intermediate donicity, such as tri-(2-furyl)phosphine and triphenylarsine, have been used with success in promoting Stille reactions that failed under the standard conditions.[143] Adding lithium chloride to the reaction mixture is often found to increase the yield of the reaction (e.g., Example **17.219**[144]).

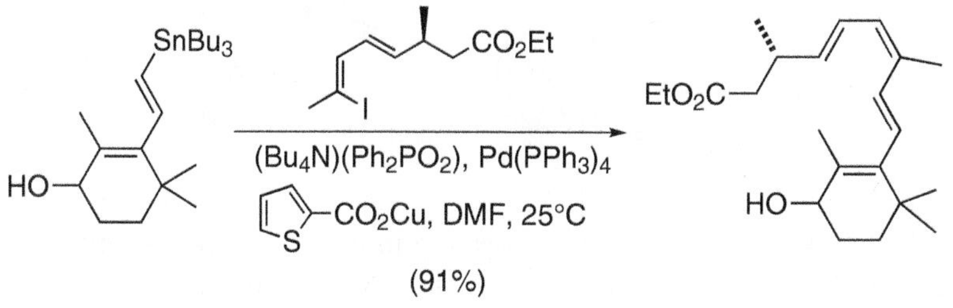

(17.218)

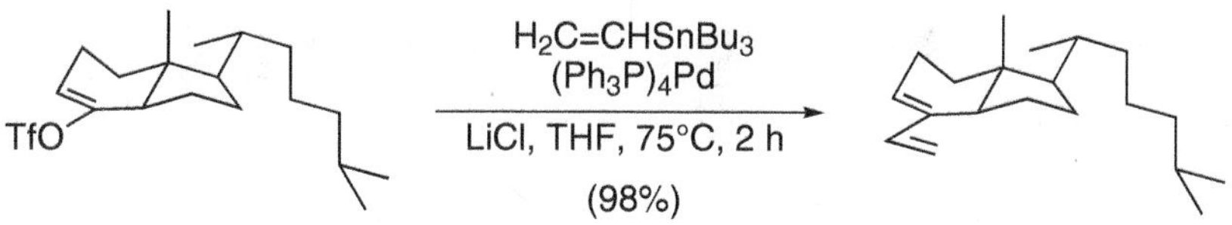

(17.219)

The Negishi Cross-Coupling Reaction[145]

The corecipient of the Nobel Prize in 2010 was Ei-ichi Negishi, who developed a nickel- or palladium-catalyzed cross-coupling reaction based on organozinc, organoaluminum, and organozirconium reagents. The reaction rose to prominence due to its position as the first catalytic cross-coupling reaction capable of forming unsymmetrical biaryls.

The catalytic cycle of the Negishi reaction is shown in Figure 17.29. In the reaction, both dialkylzinc reagents and alkylzinc halides can be used. In one study, the alkylzinc halide was found to give the *cis*-dialkylpalladium (II) complex (**17.216**) directly after the transmetallation, and the dialkylzinc reagent gave the *trans*-dialkylpalladium (II) complex (**17.220**) first, and this then underwent a slow *trans*→*cis* isomerization. These observations strongly suggest that the Negishi coupling is less susceptible to side reactions when the alkylzinc halide is used instead of the dialkylzinc reagent.[146] Perhaps one of the most spectacular applications of the Negishi coupling was its use in the synthesis of hexa(ferrocenyl)benzene (Example **17.221**).[147] The orientations of the six ferrocenyl groups in this molecule alternate above and below the plane of the benzene ring, giving the molecule an

140. (a) Liebeskind, L.S.; Fengl, R.W. *J. Org. Chem.* **1990**, *55*, 5359. (b) Farina, V.; Kapadia, S.; Krishnan, B.; Wang, C.; Liebeskind, L.S. *J. Org. Chem.* **1994**, *59*, 5905. (c) Piers, E.; Romero, M.A. *J. Am. Chem Soc.* **1996**, *118*, 1215. (d) Takeda, T.; Matsunaga, K.; Kabawasa, Y.; Fujiwara, T. *Chem. Lett.* **1995**, 771. (e) Allred, G.D.; Liebeskind, L.S. *J. Am. Chem. Soc.* **1996**, *118*, 2748.

141. Domínguez, M.; Álvarez, S.; Álvarez, R.; de Lera, Á.R. *Tetrahedron* **2012**, *68*, 1756.

142. Liebeskind, L. S.; Peña-Cabrera, E. *Org. Synth., Coll. Vol. 10* **2009**, 9; *Org. Synth.* **2000**, *77*, 135.

143. (a) Farina, V.; Krishnan, B. *J. Am. Chem. Soc.* **1991**, *113*, 9585. (b) Farina, V.; Baker, S.R.; Benigni, D.A.; Hauck, S.I.; Sapino, C., Jr. *J. Org. Chem.* **1990**, *55*, 5833.

144. Mayer, C.D.; Allmendinger, L.; Bracher, F. *Tetrahedron* **2012**, *68*, 1810.

145. Review: Negishi, E.-i.; Hi, Q.; Huang, Z.; Qian, M.; Wang, G. *Aldrichimica Acta* **2005**, *38*, 71.

146. Casares, J.A.; Espinet, P.; Fuentes, B.; Salas, G. *J. Am. Chem.Soc.* **2007**, *129*, 3508.

147. Yu, Y.; Bond, A.D.; Leonard, P.W.; Lorenz, U.J.; Timofeeva, T.V.; Vollhardt, K.P.C.; Whitener G.D.; Yakovenko, A.A. *Chem. Commun.* **2006**, 2572.

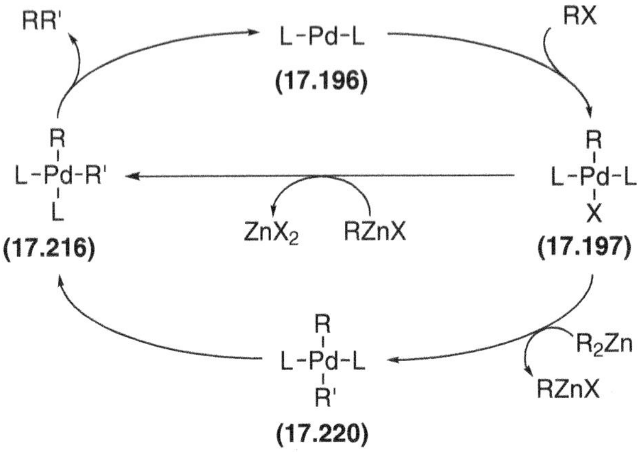

Figure 17.29 The catalytic cycle of the Negishi cross-coupling

S_6 improper axis. Despite the extreme steric crowding in this molecule, its synthesis by means of the Negishi coupling was still accomplished in a yield of 4%.

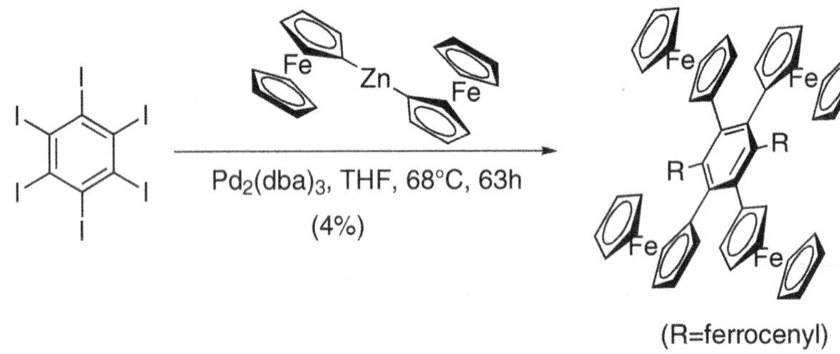

(17.221)

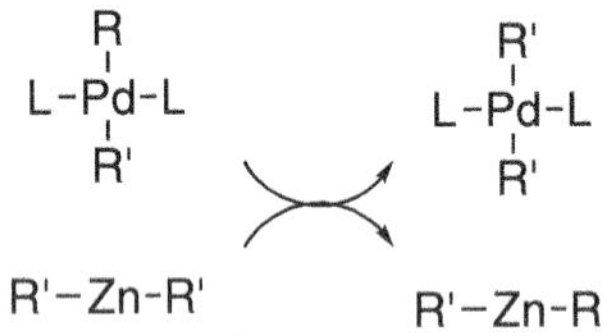

(17.222)

One of the major side reactions in the Negishi cross-coupling reaction is the formation of homocoupled products. These products arise when the reductive elimination step is slow, from a second transmetallation prior to the reductive elimination of the product (Example **17.222**). The fact that the reductive elimination step is rapid when alkylzinc halides are used makes these compounds the more attractive organometallic partner in the Negishi coupling.

In addition to the organometallic partner, the phosphine ligand for palladium also exerts a strong influence over the course of the Negishi coupling. By far the most work has been done with phosphines, with the Buchwald[148] hindered biphenyldicyclohexylphosphines carrying potential chelating groups at the *ortho* positions of the biphenyl (e.g., Example **17.223**) being especially useful for difficult couplings.

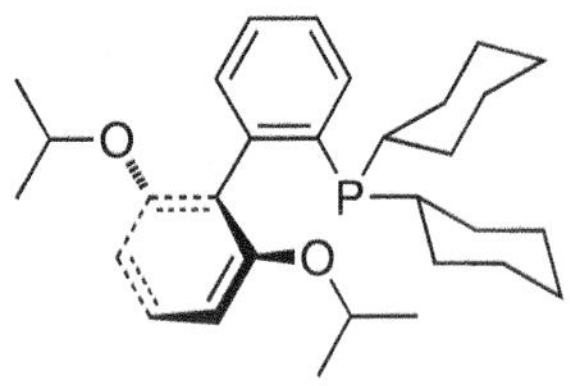

(17.223)

148. (a) Milne, J.E.; Buchwald, S.L. *J. Am. Chem. Soc.* **2004**, *126*, 13028. (b) Barder, T.E.; Walker, S.D.; Martinelli, J.R.; Buchwald, S.L. *J. Am. Chem. Soc.* **2005**, *127*, 4685.

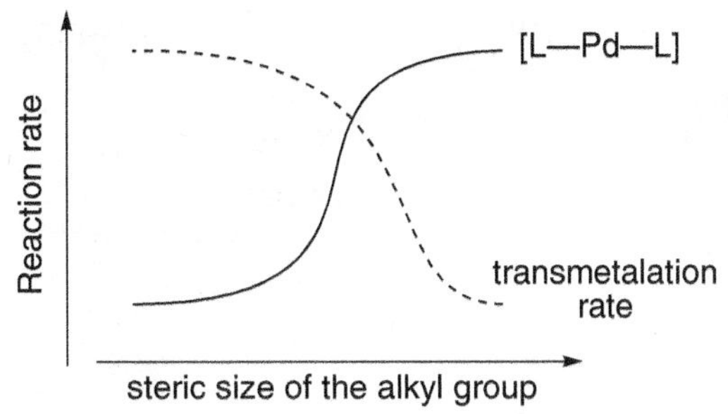

Figure 17.30 Effects of group size in the catalyst on the two key stages of the Negishi coupling in reaction 17.224

Buchwald's group has demonstrated the utility of these phosphine ligands in a wide range of cross-coupling reactions. One of the observations from their study has been that the steric bulk of the phosphine for the Negishi coupling must meet two criteria for maximum efficiency. One, it must be sterically congested enough to permit a significant quantity of the L—Pd—L species (the reactive form of palladium) to be present in the reaction mixture. Two, it must not be so sterically bulky that the transmetallation step becomes prohibitive. This is illustrated by the effects of substituent size in the phosphine ligands in the sterically demanding cross-coupling in Example **17.224.** Here the reaction using ligands carrying the smaller methoxy and ethoxy groups is slower than that with the isopropoxy groups, and the ligand carrying the sterically more demanding triisopropylsilyl (TIPS) groups is now hindered enough to significantly retard the transmetallation (Figure 17.30).

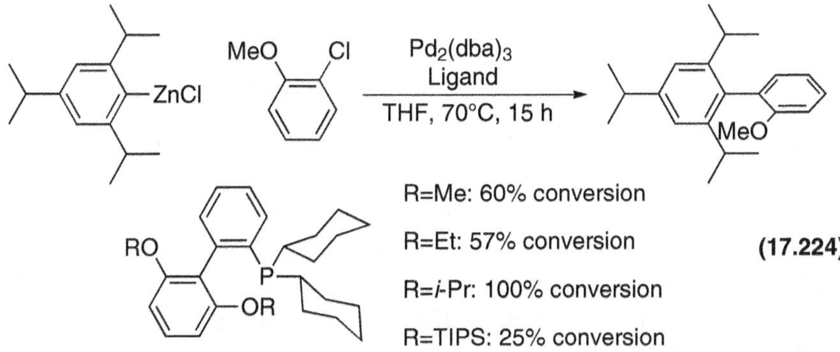

(17.224)

The Hiyama and Related Cross-Couplings

The high toxicity of stannane reagents is not generally shared by their silicon analogs. Consequently, that someone would explore the palladium-catalyzed cross-coupling of silanes with alkenes (a silicon version of the Stille coupling) was inevitable. However, silanes proved to be very unreactive under the standard Stille-type conditions, and it was not until 1988 that the reaction became synthetically useful. This, followed Hatanaka and Hiyama's discovery[149] that the silane required activation by fluoride anion (or a fluoride anion surrogate such as $Me_3SiF_2^-$ ion, as in Example **17.225**) for the reaction to proceed.

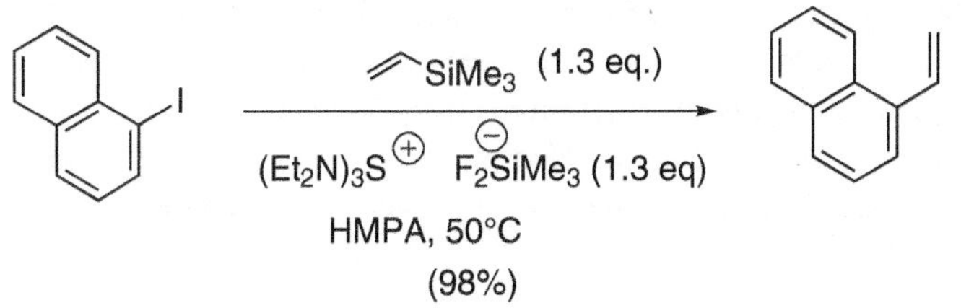

(17.225)

149. Hatanaka, Y.; Hiyama, T. *J. Org. Chem.* **1988**, *53*, 918.

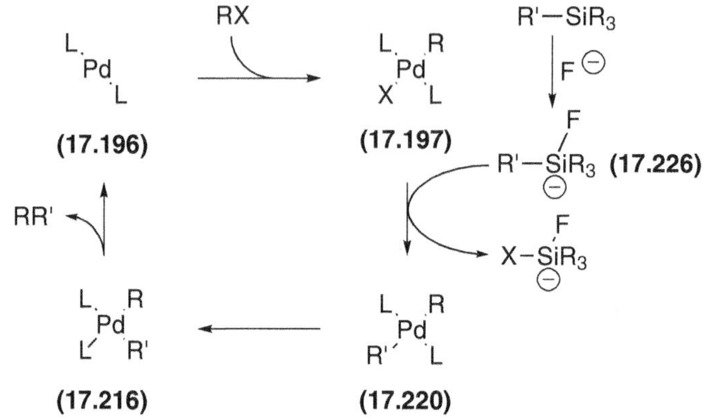

Figure 17.31 The catalytic cycle of the Hiyama cross-coupling

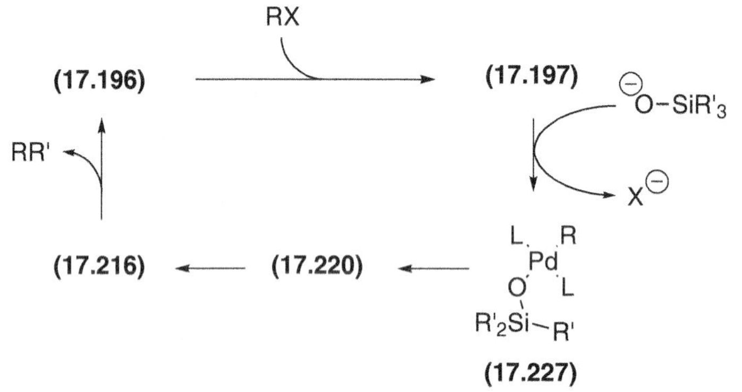

Figure 17.32 The catalytic cycle of the Hiyama-Denmark reaction

The catalytic cycle of the Hiyama reaction is shown in Figure 17.31. The fluoride anion is critical to the success of the reaction because it increases the electron density at carbon in the Si—C bond, so it resembles the Sn—C bond more closely. This is accomplished by forming a pentacoordinate silicon anion (**17.226**) as the first step of silane activation. The results of a study of the mechanism of the reaction by Denmark[150] led to the development of the ***Hiyama-Denmark reaction***,[151] in which silyloxide anions are used as the silane component of the reaction and in which fluoride anion is no longer needed as an activator. Mechanistic studies have shown that a pentavalent silicon species is not involved in the Hiyama-Denmark reaction but that the silyloxide palladium complex (**17.227**) undergoes unimolecular elimination of the silicone, with concomitant migration of the alkyl group to generate the dialkylpalladium intermediate (**17.216**). The catalytic cycle of the Hiyama-Denmark reaction is shown in Figure 17.32.

The Kumada Coupling

In 1900, the Zaitsev-Wagner synthesis of alcohols from aldehydes or formate esters and organozinc reagents[152] (especially alkylzinc iodides) was superseded by the Grignard synthesis, based on the corresponding organomagnesium reagents. Thus, it was only a matter of time before an organomagnesium analog of the Negishi cross-coupling, which is based on organozinc reagents, would be examined. The resulting reaction, discovered independently by

150. Denmark, S.E.; Wehrli, D.; Choi, J.Y. *Org. Lett.* **2000**, *2*, 2491.
151. Review: Denmark, S.E.; Regens, C.S. *Acc. Chem. Res.* **2008**, *41*, 1486.
152. Lewis, D.E. *Bull. Hist. Chem.* **2002**, *27*, 37.

Corriu and Kumada[153] is now generally known as the *Kumada cross-coupling*. Unlike the Negishi cross-coupling, the reaction is not widely used, although it does have applications in the synthesis of some important drugs.

The Suzuki-Miyaura and Related Coupling Reactions

Despite its flexibility, the Stille coupling suffers from the disadvantage that toxic organotin compounds are critical to the reaction. Recently, this problem has been addressed by developing methods where the tin is used in catalytic amounts.[154] The development of polymeric stannane reagents[155] also promises to make the reaction easier to use without generating large quantities of difficult-to-handle, toxic organotin residues. Among the most successful replacements for the tin compounds are alkylboron compounds, especially alkylboronic acid derivatives, in a reaction now generally referred to as the ***Suzuki-Miyaura coupling***.[156]

Suzuki-Miyaura Coupling

Based on the data in Table 17.4, the Suzuki-Miyaura coupling is the most widely used metal-catalyzed C—C bond forming reaction in use at the present time. This reaction follows a similar catalytic cycle to other coupling reactions. However, the presence of the oxyanion base results in two different catalytic cycles (Figure 17.33) depending on whether it attacks boron, to generate the borate complex **17.228** (cycle **A**), or the metal, to generate the complex **17.229** (cycle **B**). The reactivity of the halide participant, R_aX, in the reaction follows the general order:

$$RI > ROTf > RBr >> RCl$$

The oxidative addition of a vinyl halide to the metal proceeds with retention of configuration. With allylic and benzylic halides, it occurs with inversion (by a formal displacement of the halogen from carbon by an S_N2-like process).[157] The oxidative addition initially gives a *cis* metal complex that then rapidly isomerizes to the *trans* (**17.196**).[158] The reductive elimination step also proceeds with retention of configuration.[159] By far the most common

---

153. (a) Corriu, R.J.P.; Masse, J.P. *J. Chem. Soc., Chem. Commun.* **1972**, 144a. (b) Tamao, K.; Sumitani, K.; Kumada, M. *J. Am. Chem. Soc.* **1972**, *94*, 4374.

154. Gallagher, W.P.; Malezcka, R.E., Jr. *J. Org. Chem.* **2005**, *70*, 841.

155. (a) Chrétien, J.-M.; Mallinger, A.; Zammattio, F.; Le Grognec, E.; Paris, M.; Montavon, G.; Quintard, J.-P. *Tetrahedron Lett.* **2007**, *48*, 1781. (b) Kerric, G.; Le Grognec, E.; Zammattio, F.; Paris, M.; Quintard, J.-P. *J. Organomet. Chem.* **2010**, *695*, 103. (c) Albeniz, A.C.; Carrera, N. *Eur. J. Inorg. Chem.* **2011**, 2347. (d) Carrera, N.; Salinas-Castillo, A.; Albeniz, A.C.; Espinet, P.; Mallavia, R. *J. Organomet. Chem.* **2011**, *696*, 33126.

156. Reviews: (a) Miyaura, N.; Suzuki, A. *Chem. Rev.* **1995**, *95*, 2457. (b) Chemler, S.R.; Trauner, D.; Danishefsky, S.J. *Angew. Chem., Int. Ed.* **2001**, *40*, 4544. (c) Suzuki, A.; Brown, H.C. *Organic Syntheses Via Boranes* (Aldrich Chemical Co., Inc.: Milwaukee, 2002); Vol. 3. (e) Kotha, S.; Lahiri, K.; Kashinath, D. *Tetrahedron* **2002**, *58*, 9633. (d) Bellina, F.; Carpita, A.; Rossi, R. *Synthesis* **2004**, 2419. (f) Miyaura, N., Ed. *Cross-Coupling Reactions. A Practical Guide.* (Springer-Verlag: Heidelberg, 2002). (g) Miyaura, N. In Diederich, F.; de Meijere, A., Eds. *Metal-Catalyzed Cross-Coupling Reactions* (Wiley-VCH: New York, 2004), ch. 2. (h) Hall, D.G. In Hall, D.G., Ed. *Boronic Acids-Preparation, Applications in Organic Synthesis and Medicine* (Wiley-VCH; Weinheim: 2005), pp. 1–99. (i) Bai, L.; Wang, J.-X. *Curr. Org. Chem.* **2005**, *9*, 535.

Norio Miyaura (1946–) was educated at Hokkaido University under Suzuki and is currently Professor at the same university. For more details, see: *Chem. Rev.* **1995**, *95*, 2457.

Akiro Suzuki (1930–) was educated at Hokkaido University and served there from 1961–1994. After retiring, he joined the faculty of Okayama University of Science. He shared the 2010 Nobel Prize in Chemistry. More detail may be found at the Nobel Foundation web site.

157. Stille, J.K.; Lau, K.S.Y. *Acc. Chem. Res.* **1977**, *10*, 434.

158. Casado, A.L.; Espinet, P. *Organometallics* **1998**, *17*, 954.

159. (a) Ridgway, B.H.; Woerpel, K.A. *J. Org. Chem.* **1998**, *63*, 458. (b) Matos, K.; Soderquist, J.A. *J. Org. Chem.* **1998**, *63*, 461.

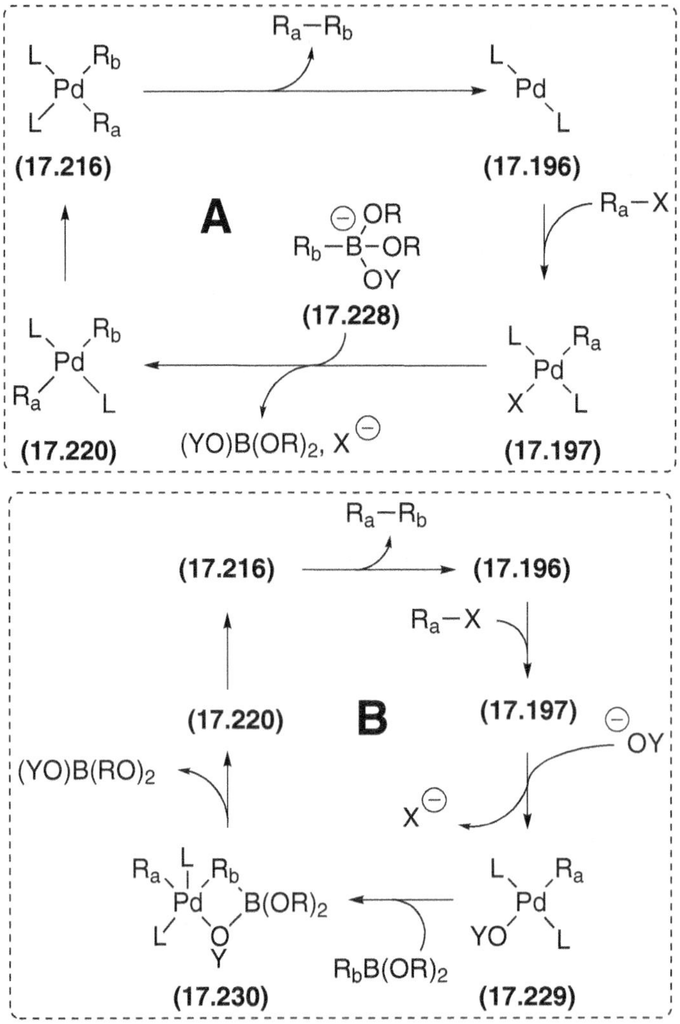

Figure 17.33 The catalytic cycles for the Suzuki-Miyaura coupling. OY is an oxyanion base.

catalysts are based on palladium, which has been shown in a computational study[160] to be the most reactive of the noble metals with respect to the reductive elimination step of the Suzuki-Miyaura reaction. In this study, the order of reactivity was found to be:

$$\text{Pd} > \text{Pt} > \text{Rh} > \text{Ir, Ru, Os}$$

By far, the major applications of the Suzuki-Miyaura cross-coupling are in the synthesis of biaryls and styrenes from aryl halides and aryl- or vinylboronic acids as illustrated by Example **17.231**.[161] One of the major strengths of the reaction is the fact that in the synthesis of styrenes, for example, the stereochemistry of the vinylborane is retained during the migration step, allowing the stereochemistry of the product to be predicted. The same holds true in the synthesis of simple dienes by means of the Suzuki coupling between vinyl halides and vinylboronic acid derivatives. For example, the synthesis of caparratriene (Example **17.232**) by means of a Wittig reaction provided the *E,E* isomer only as the major component of a 2:1 mixture. When the Suzuki-Miyaura coupling was used with the *E* bromide and the

160. Ananikov, V.P.; Musaev, D.G.; Morokuma, K. *J. Am. Chem. Soc.* **2002**, *124*, 2839; *Organometallics* **2005**, *24*, 715.

161. Al-Lakkis-Wehbe, M.; Roux, L.; Charrier, C.; Alavi, S.; Le Nouën, D.; Defoin, A.; Tarnus, C.; Albrecht, S. *Tetrahedron* **2012**, *62*, 6447.

E vinylboronic acid, however, only the natural *E,E* isomer was produced.[162] An interesting example of the intramolecular application of the Suzuki coupling is provided by Example **17.233**, from the formal total synthesis of strychnine by Bodwell and Li.[163] In this case, the required boron derivative was formed by hydroboration of an alkene immediately prior to the addition of the catalyst.

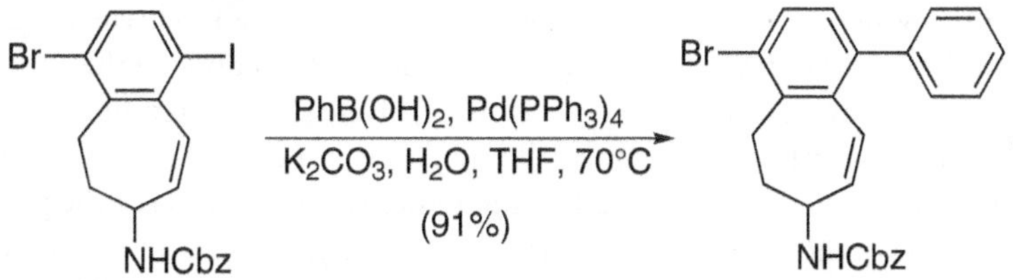

(17.231)

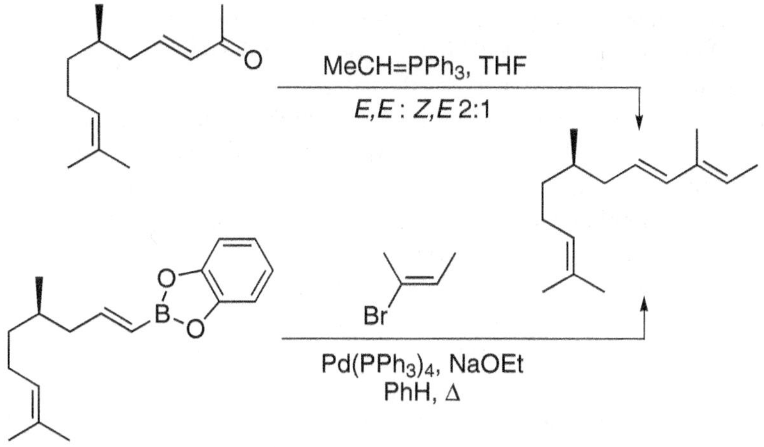

(17.232)

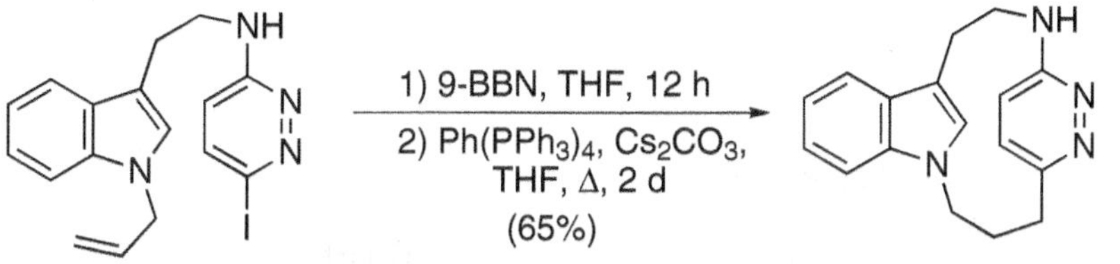

(17.233)

The boronic acids in the Suzuki coupling may be successfully replaced by alkyltrifluoroborates, which offer advantages in terms of reagent stability compared to other boronic acid derivatives. This can permit a ligand-less version of the Suzuki coupling to be used.[164] Example **17.234** gives a recent illustration of the application of these boron reagents in the Suzuki coupling.[165]

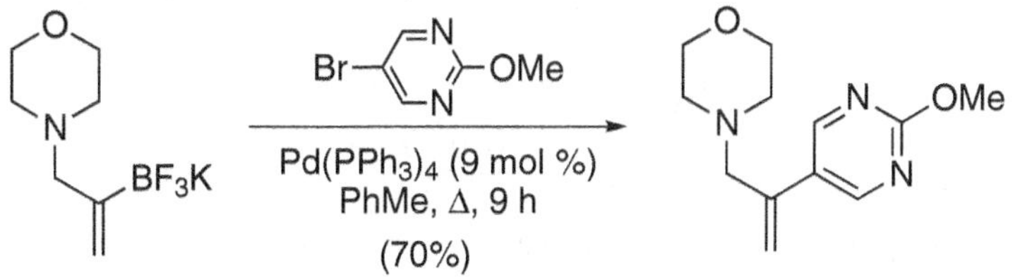

(17.234)

---

162. Vyvyan, J.R.; Peterson, E.A.; Stephan, M.L. *Tetrahedron Lett.* **1999**, *40*, 4947.
163. Bodwell, G.J.; Li, J. *Angew. Chem. Int. Ed.* **2002**, *41*, 3261.
164. (a) Molander, G.A.; Rodriguez Rivero, M. *Org. Lett.* **2002**, *4*, 107. (b) Molander, G.A.; Bernardi, C.R. *J. Org. Chem.* **2002**, *67*, 8424. (c) Molander, G. A.; Ito, T. *Org. Lett.* **2001**, *3*, 393. (d) Molander, G.A.; Biolatto, B. *Org. Lett.* **2002**, *4*, 1867.
165. Math, S.K.; Mokri, H.H.; LaMunyon, J.B.; Cefalo, D.R.; Testa, C.R. *Tetrahedron Lett* **2012**, *53*, 2487.

Sonogashira Coupling

When the nucleophilic partner in the coupling reaction is a copper (I) alkynide, formed either stoichiometrically or catalytically, the reaction is known as the ***Sonogashira coupling.***[166] When both metals are used in catalytic amounts, the catalytic cycle for the reaction involves two intermeshed cycles, one for the palladium, and one for the copper. The catalytic cycle of the Sonogashira cross-coupling is shown in Figure 17.34. The palladium cycle begins with the familiar pattern of oxidative addition of the halide to the coordinatively unsaturated palladium (**17.196**) to give the coordinatively saturated complex (**17.197**), ligand exchange with copper to give the *trans* alkynylpalladium complex **17.235**, isomerization of the complex to the *cis* isomer (**17.236**), and reductive elimination of the product. The copper cycle involves the formation of an initial η^2 alkynyl complex of the copper (I) halide (**17.237**). This reacts with the base to give the corresponding copper (I) alkynide (**17.238**), which then undergoes ligand exchange with the palladium to regenerate the copper (I) halide. More recently, a nickel-catalyzed version of the reaction[167] has been developed, reducing the expense associated with the palladium.

Like the Suzuki reaction, the Sonogashira reaction is highly tolerant of a wide range of functional groups, as shown by the synthesis of the phenylalanine-biotin conjugate **17.239**, where the sulfide, ester and urethane groups are all present in the starting materials. The order of reactivity of the halide participant in the Sonogashira cross-coupling reaction follows the order in the Suzuki coupling, with the added observation that aryl halides and triflates are less reactive than similar vinyl halides and triflates.

The copper in the Sonogashira coupling does occasionally lead to one undesired side reaction: the Glaser homocoupling of the alkyne.[168] For this reason, an alternative, copper-free variant of the reaction has been developed,[169] in which a very hindered phosphine ligand, such as tri-*tert*-butylphosphine, is used in place of the copper; the resulting palladium complex reacts directly with the alkyne to give the alkynylpalladium complex.

Figure 17.34 The catalytic cycle of the Sonogashira cross-coupling reaction

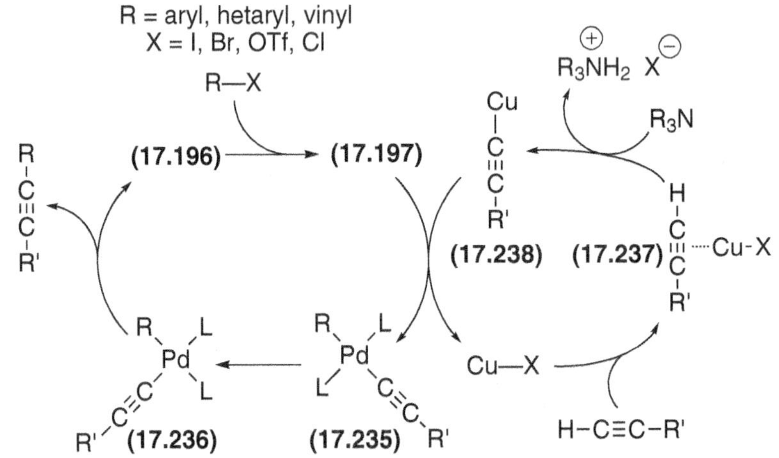

166. (a) Sonogashira, K.; Tohda, Y.; Hagihara, N. *Tetrahedron Lett.* **1975**, *16*, 4467. (b) Sonogashira, K. *J. Organomet. Chem.* **2002**, *263*, 46. (c) Chinchilla, R.; Nájera, C. *Chem. Rev.* **2007**, *107*, 874. (d) Chinchilla, R.; Nájera, C. *Chem. Soc. Rev.* **2011**, *40*, 5084.

167. Vechorkin, O.; Barmaz, D.; Proust, V.; Hu, X. *J. Am. Chem. Soc.* **2009**, *131*, 12078.

168. (a) Glaser, C. *Justus Liebigs Annm. Chem.* **1870**, *154*, 137; *Ber. dtsch. chem. Ges.* **1869**, *2*, 422. (b) Johansson Seechurn, C.C.C.; Kitching, M.O.; Colacot, T.J.; Snieckus, V. *Angew. Chem. Int. Ed.* **2010**, *51*, 5062.

169. (a) Böhm, V.P.W.; Herman, W.A. *Eur. J. Org. Chem.* **2000**, 3679. (b) Méry, D.; Heuzé, K.; Astruc, D. *Chem. Commun.* **2003**, 1934.

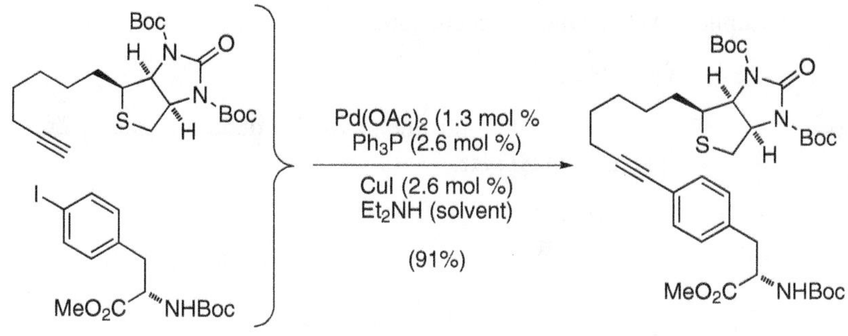

(17.239)

Reaction Synopses

Heck Reaction

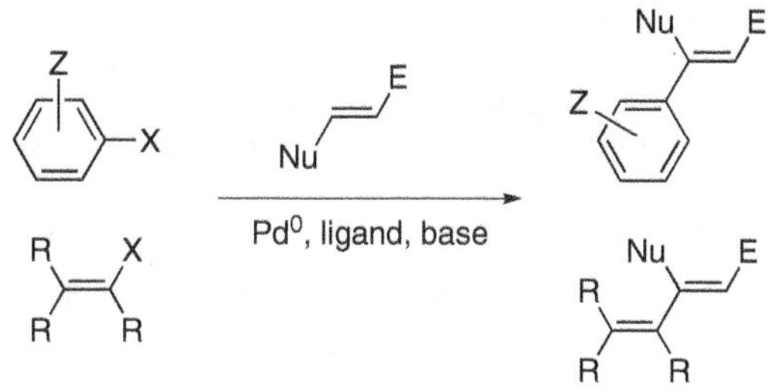

E = COR, CO_2R, SOR, SO_2R, CN, Ar, etc.
Nu = OR, OCOR, NR_2, N(R)COR, etc.

Reagents: $Pd(PPh_3)_4$, Cs_2CO_3, DMF, 100°C; $Pd(OAc)_2$, dppp, NEt_3, Me_2SO, 115°C; etc.

Regiochemistry: reaction occurs β to electron-withdrawing substituents, and α to electron-releasing substituents; reaction is much more favorable with electron-deficient alkenes.

Stereochemistry: reaction exhibits strong preference for *trans* product.

Stille Reaction

G = aromatic or hindered alkyl
R = aryl, vinyl, alkyl, acyl
R" = vinyl, aryl, alkyl, alkynyl

Reagents: $PdCl_2$, $(t\text{-Bu})_3P$, CuI, CsF, DMF, Δ; $Pd(OAc)_2$, DABCO, KF or Bu_4NF, dioxane, Δ; etc.

Migratory preference in stannane: vinyl, aryl > alkyl; this permits the use of reagents such as $ArSnBu_3$, $RCH{=}CHSnBu_3$.

Negishi Coupling

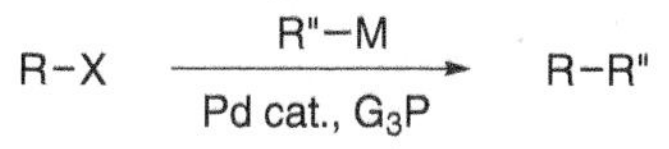

R: usually aryl

R"—M: R''_2Zn, R"ZnX, R''_3Al, $R''ZrCp_2L$, etc; organozinc reagents are the most widely used.

G_3P: Ph_3P, 1,1'-(diphenylphosphino)ferrocene [dppf]

o-$Cy_2PC_6H_4$—2,6-$(MeO)_2C_6H_4$; etc.

Reagents: RZnBr, (*t*-$Bu_3P)_2Pd$, dioxane; R_2Zn, Pd_2dba_3, dioxane; R_2Zn, (dppf)$PdCl_2$, THF; etc.

Hiyama and Hiyama-Denmark Couplings

R'_3Si: $SiMe_3$, $Si(OR')_3$, $Me_{3-n}SiF_n$, etc.

Pd cat.: $PdCl_2$, $Pd(OAc)_2$, Pd_2dba_3, $[C_3H_5PdCl]_2$, etc.

Reagents: Hiyama coupling

R"$Si(OMe)_3$, $PdBr_2$, $(t\text{-}Bu)_2PMe$, Bu_4N^+ F^- (TBAF), THF; R"$Si(OMe)_3$, $Pd(OAc)_2$, DABCO, TBAF, dioxane, Δ.; etc.

Reagents: Hiyama-Denmark coupling

R"$Si(OK)Me_2$, $[(t\text{-}Bu)_3P]_2Pd$, PhMe, Δ; R"$Si(OG)Me_2$, Me_3SiOK, Pd_2dba_3, dioxane; etc.

Kumada Coupling

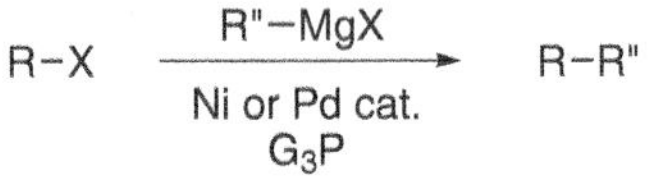

Ni cat: (dppb)$NiCl_2$; etc. Pd cat.: $(Ph_3P)_4Pd$; etc.

Reagents: R"MgBr, $Ni(acac)_2$, *o*-[MeCH(OH)]$C_6H_4PPh_2$; etc.

Suzuki-Miyaura coupling

Molander modification

X: Br, I, OTf, etc.; R':H, alkyl; R": alkyl, aryl

Reagents: $(Ph_3P)_4Pd$, Na_2CO_3, R—$B(OCMe_2)_2$, C_6H_6-EtOH, Δ; $Pd(OAc)_2$, *o*-(2,6-MeO)$_2C_6H_3$—$C_6H_4PCy_2$, K_3PO_4, THF; etc.

Sonogashira Coupling

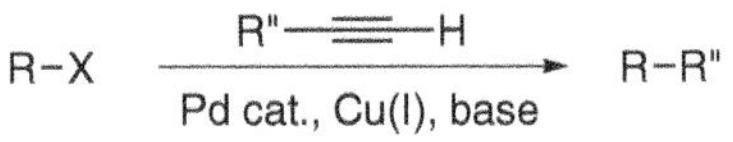

X: Cl, Br, I, OTf.

Reagents: $(Ph_3P)_4Pd$, CuI, *n*-$PrNH_2$, PhH; $(Ph_3P)_2PdCl_2$, CuI, Et_3N (solvent); etc.

Chapter Summary

This chapter has concerned catalytic reactions for the formation of new carbon-carbon bonds, a field that has grown dramatically as the 12 Principles of Green Chemistry move into the mainstream of organic synthesis. Most of these reactions involve the use of organometallic complexes as the catalytic species; reactions in these complexes, which can be coordinatively saturated (18 electrons formally on the metal atom) or coordinatively unsaturated (less than 18 electrons formally on the metal atom); bonding to ligands may be σ or π—especially (d-p)π bonding. Bonding between the metal and ligands in π complexes is frequently described using the hapto number, η. Reactions of these species can involve reactions at either the metal or a ligand. Reactions at the metal include ligand exchange; migratory insertion, which is a rearrangement occurring at the metal atom; oxidative addition, where the coordination number and the oxidation number of the metal both increase by two; and reductive elimination, where the reverse occurs. These reactions figure prominently in the catalytic cycles of a number of important synthetic reactions.

Addition of neutral metal hydrides across C—C π bonds is often facile and occurs with *syn* stereochemistry. The addition of organometallic reagents across alkynes (carbometallation) also occurs with predominant *syn* stereochemistry. Transition metal carbene complexes react with both alkenes and alkynes to generate new metal carbene complexes, with concomitant elimination of the alkene.

Important bond-forming reactions discussed in this chapter include the Tsuji-Trost reaction; olefin metathesis; the Reppe oligomerization reactions of alkynes; Heck reaction and its congeners (the Stille, Negishi, Kumada, Hiyama and Hiyama-Denmark couplings); and the Suzuki-Miyaura reaction and its congener, the Sonogashira coupling.

Key Terms

Buchwald-Hartwig reaction
carbometallation
carbonylation
coordinatively saturated
coordinatively unsaturated
decarbonylation
ene-yne metathesis
Fischer complexes
"green" chemistry
hapto number
Heck reaction
β-hydride elimination
Hiyama coupling
Hiyama-Denmark coupling
hydroformylation
hydrometallation
Kumada coupling
ligand
migratory insertion
Negishi coupling
olefin metathesis
oxidative addition
reductive elimination
Reppe chemistry
Sonogashira coupling
Stille coupling
Suzuki (Suzuki-Miyaura) coupling
transmetallation
Tsuji-Trost reaction

Additional Problems

17-8 What is the major organic product of each of the following reactions?

(a)

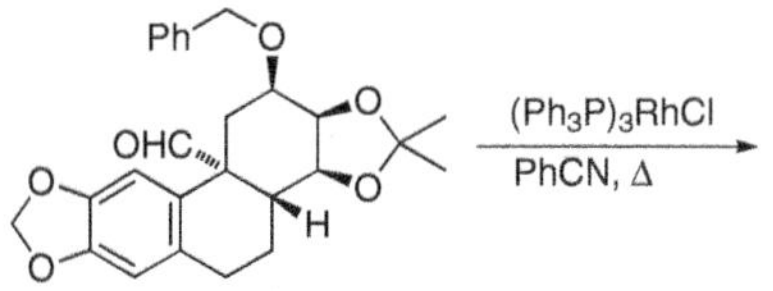

[*Tetrahedron Lett.* **2006**, *47*, 3905]

(b)

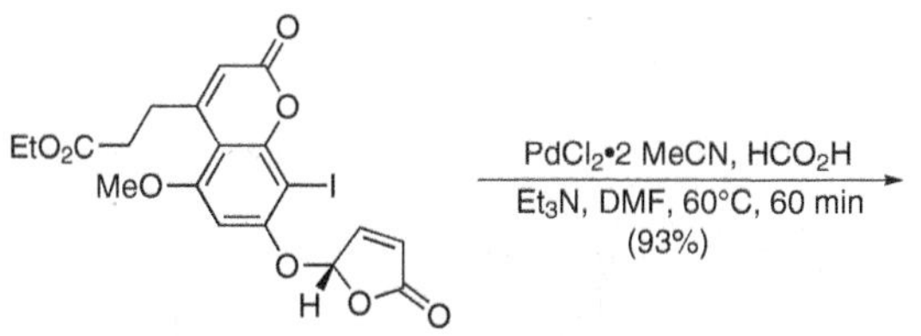

[*J. Am. Chem. Soc.* **2003**, *125*, 3090]

(c)

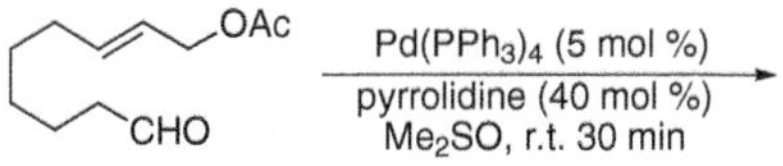

[*Org. Lett.* **2007**, *9*, 5063]

(d)

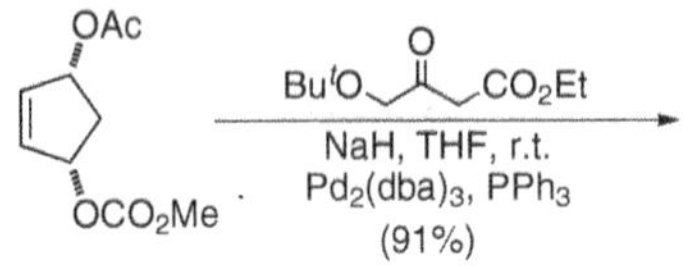

[*J. Am. Chem. Soc.* **1993**, *115*, 9293]

(e)

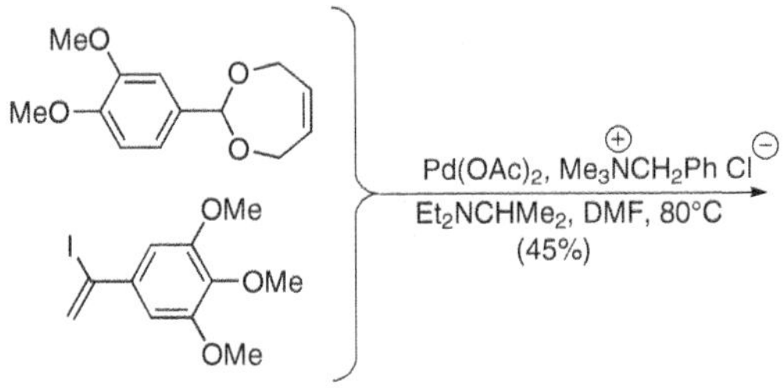

[*Synlett* **2008**, *126*]

(f)

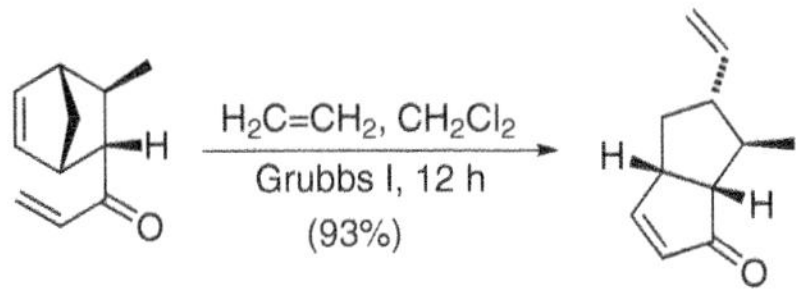

[*J. Am. Chem. Soc.* **2002**, *124*, 9974]

(g)

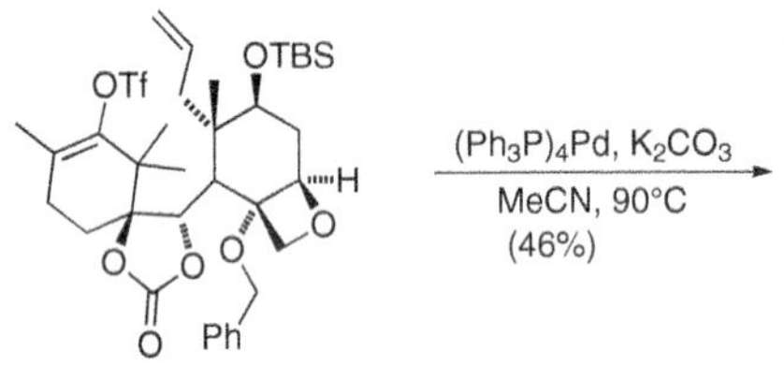

[*J. Am. Chem. Soc.* **1996**, *118*, 2843]

(h)

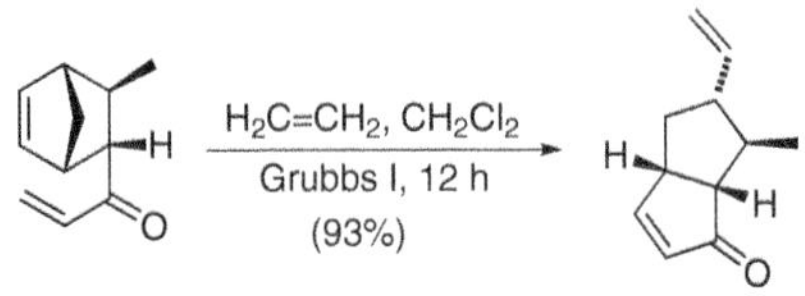

[*J. Am. Chem. Soc.* **2002**, *124*, 9974]

(i)

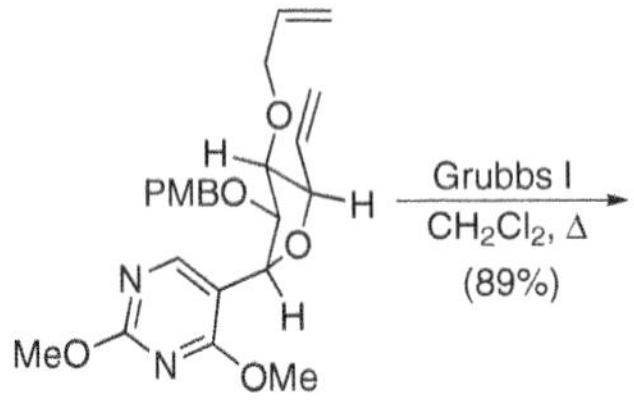

[*Org. Lett.* **2003**, *5*, 4277]

(j)

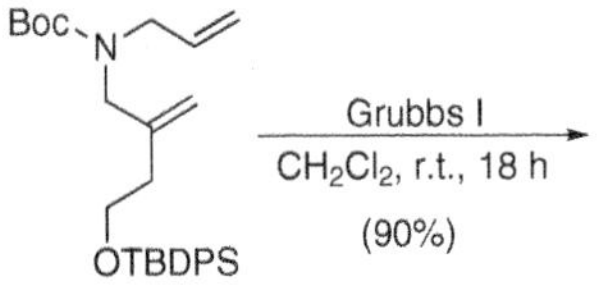

[*J. Am. Chem. Soc.* **2002**, *124*, 13342]

(k)

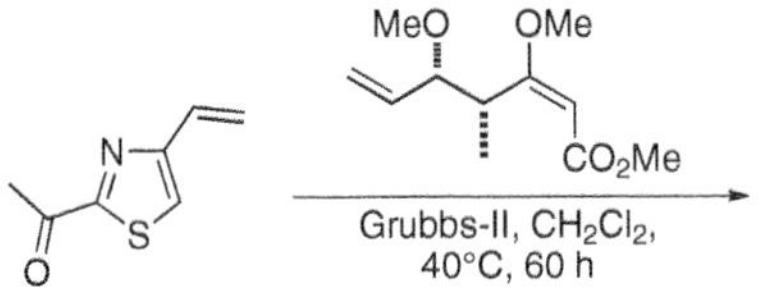

[*Org. Lett.* **2007**, *9*, 3425]

(l)

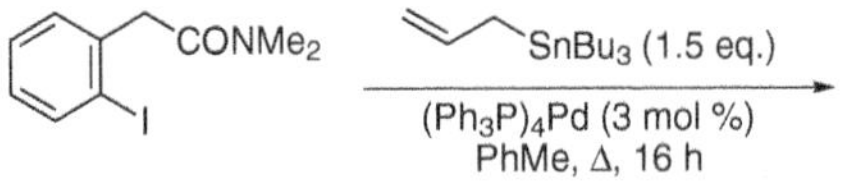

[*Tetrahedron Lett.* **2012**, *53*, 4514]

(m) 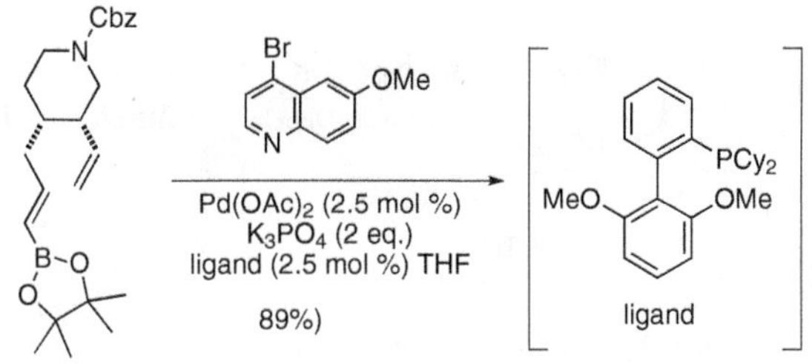

[*J. Am. Chem. Soc.* **2004**, *126*, 706]

(n) 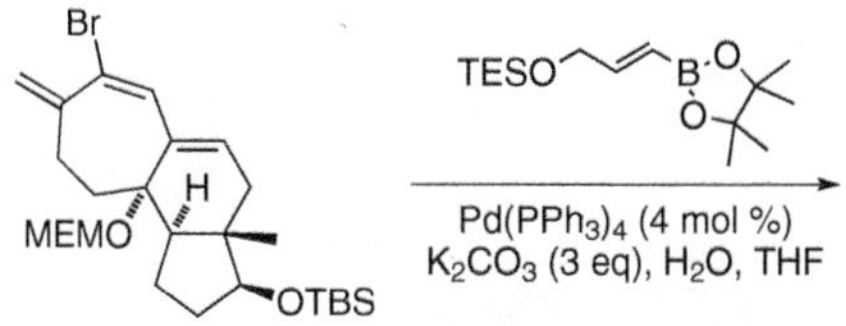

[*J. Am. Chem. Soc.* **2008**, *130*, 16864]

(o) 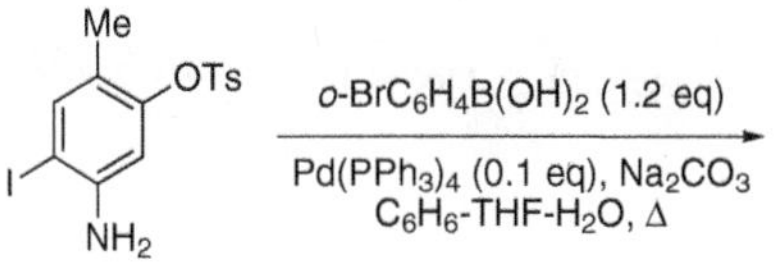

[*Org. Lett.* **2008**, *10*, 1999]

(p) 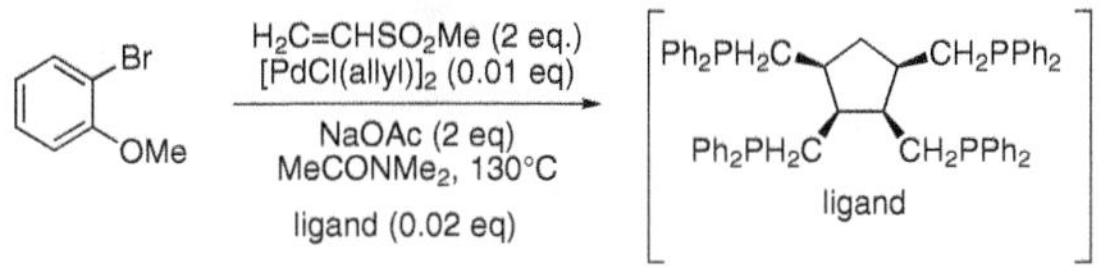

[*Synthesis*, **2006**, 3495]

(q) 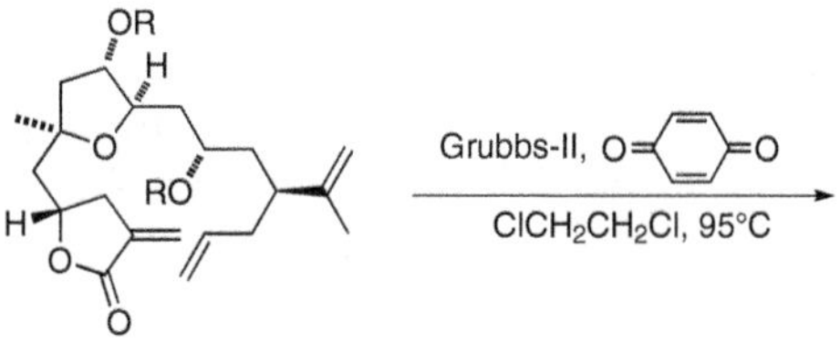

[*Tetrahedron* **2011**, *67*, 10179]

(r) 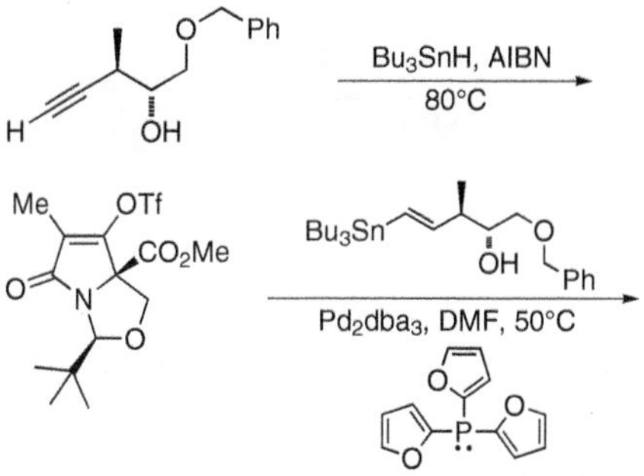

[*Tetrahedron* **2011**, *67*, 10026]

(s) 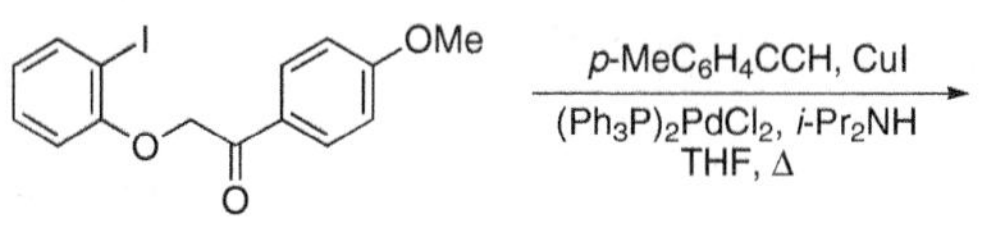

[*Tetrahedron* **2011**, *67*, 9291]

(t) 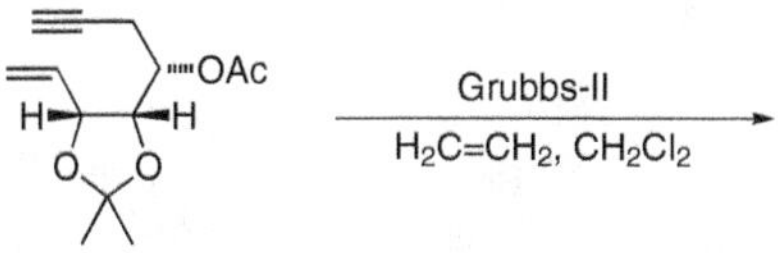

[*J. Org. Chem.* **2002**, *67*, 4441]

(u) 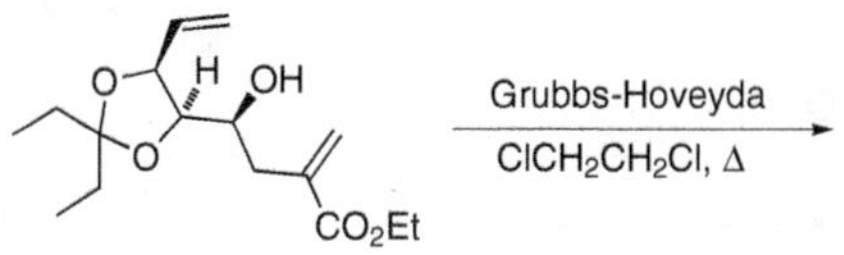

[*Org. Lett.* **2010**, *12*, 60]

(v) 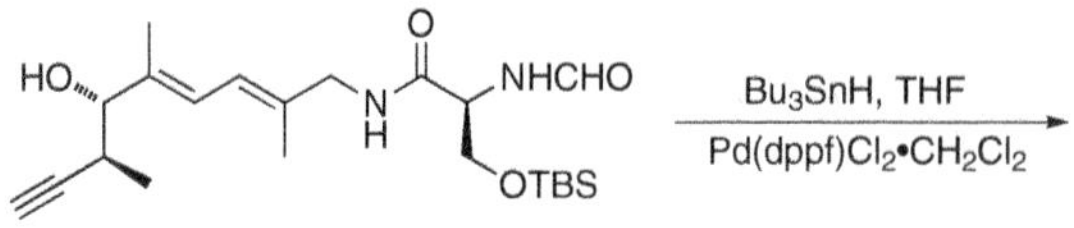

[*Org. Lett.* **2007**, *9*, 4619]

(w) 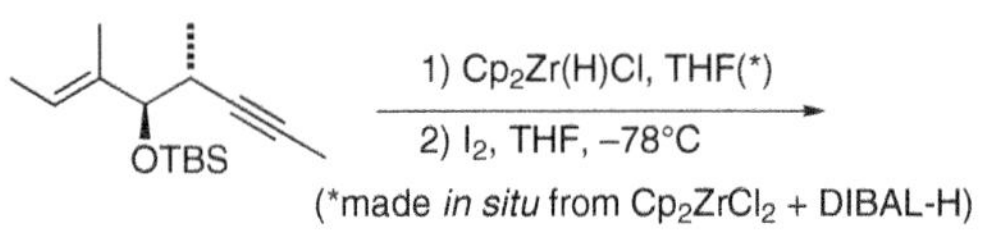

[*J. Am. Chem. Soc.* **2007**, *129*, 3826]

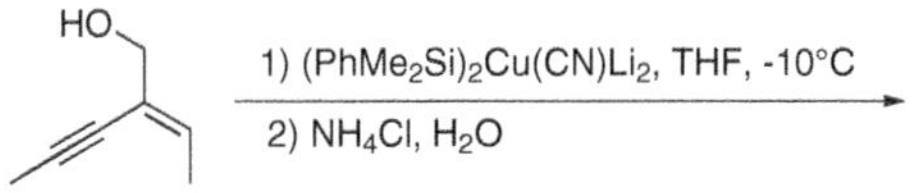

(x) 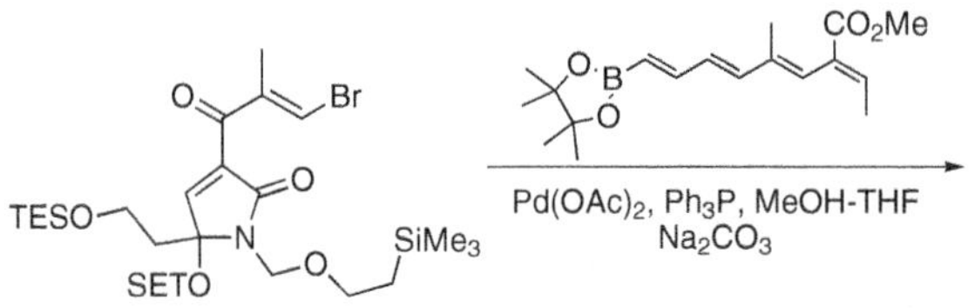

[*J. Am. Chem. Soc.* **2005**, *127*, 16038]

(y) 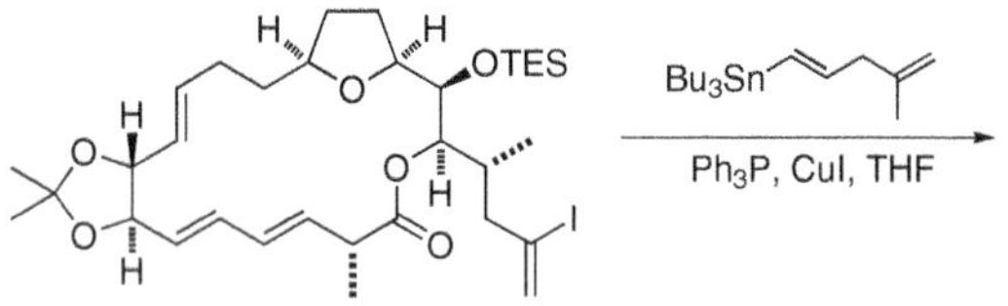

[*J. Am. Chem. Soc.* **2006**, *128*, 15960]

(z) 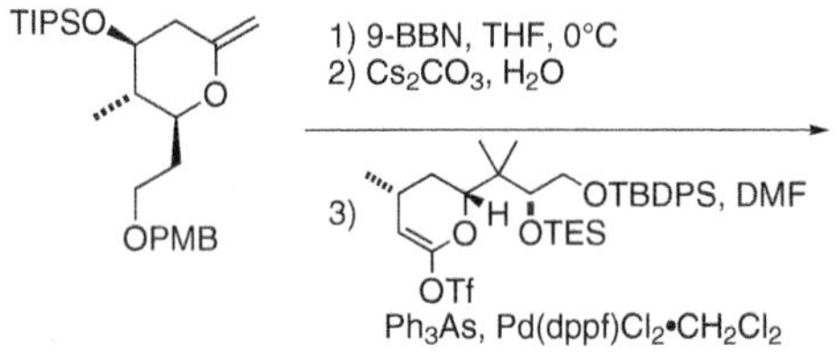

[*Org. Lett.* **2012**, *14*, 3186]

(aa) 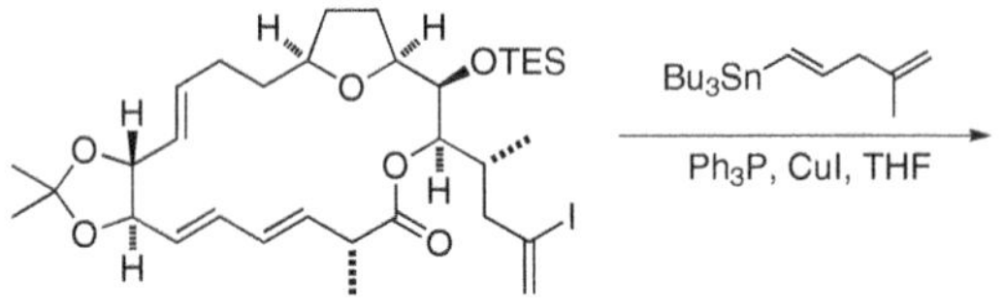

[*J. Am. Chem. Soc.* **2006**, *128*, 15960]

(bb) 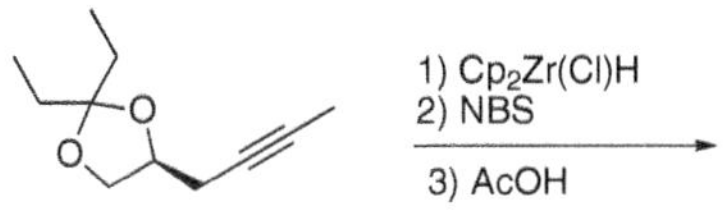

[*J. Am. Chem. Soc.* **1997**, *119*, 10935]

(cc) 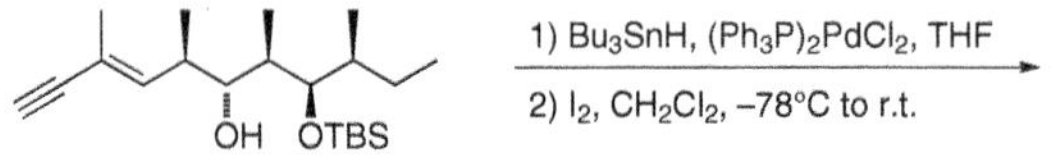

[*J. Org. Chem.* **2002**, *67*, 2751]

(dd) 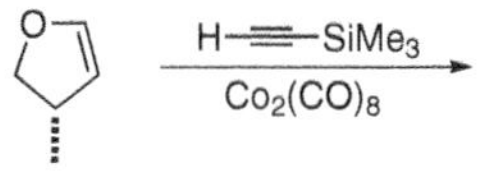

[*J. Am. Chem. Soc.* **2003**, *125*, 11514]

(ee) 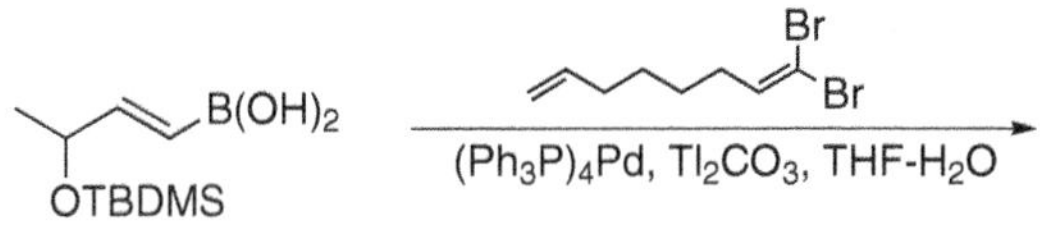

[*J. Am. Chem. Soc.* **2006**, *128*, 8126]

(ff) 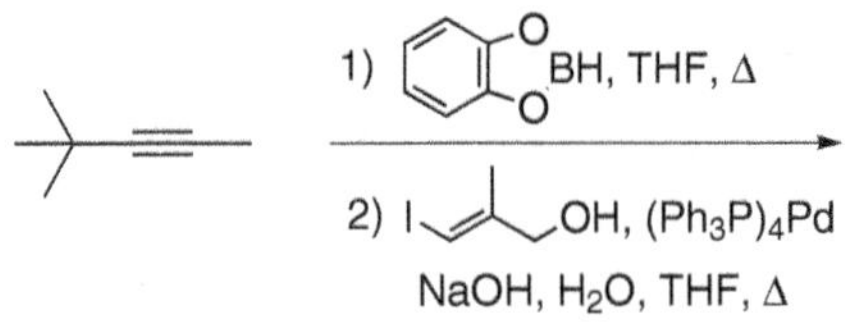

[*J. Org. Chem.* **1998**, *63*, 8638]

(gg) 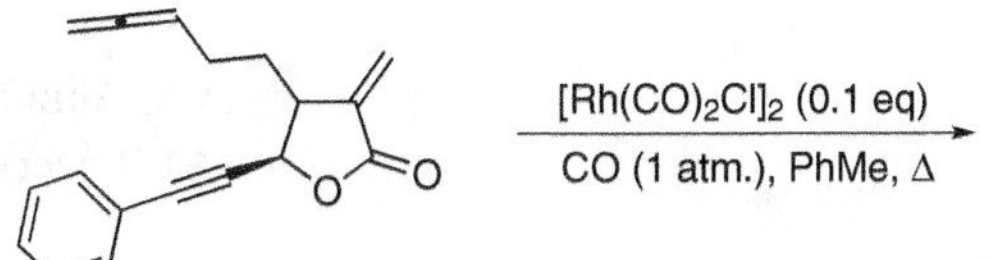

[*Org. Lett.* **2011**, *13*, 6304]

[Which double bond of the allene reacts?]

(hh) 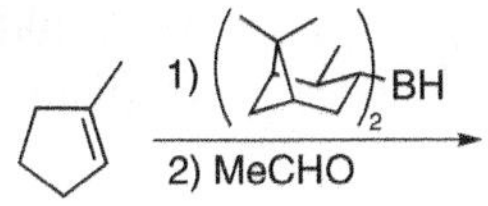

[*Pure Appl. Chem.* **1991**, *63*, 307]

(ii) 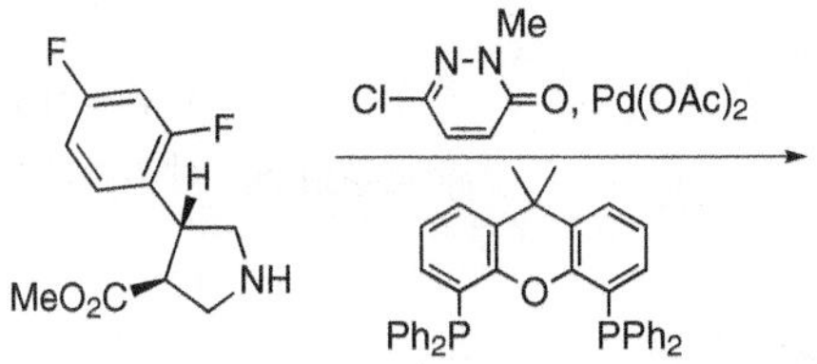

[**WO 2010/015972 A1**] (Pfizer)

17-9 The following reactions from applications for international patents by pharmaceutical companies in 2010 illustrate the impact of transition metal- (especially palladium-) catalyzed reactions on modern synthesis. Give the structure of the major organic product of each reaction.

(a) 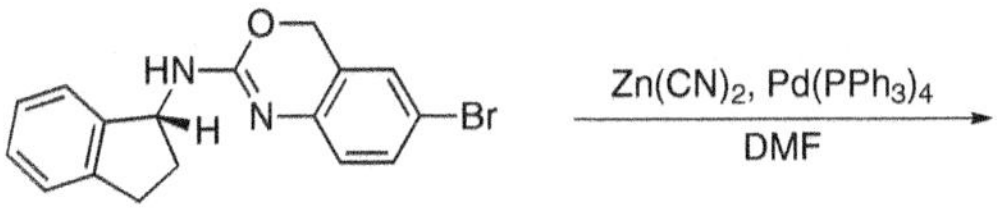

WO 2010/026110 A2 (Hofmann-La Roche)

(b) 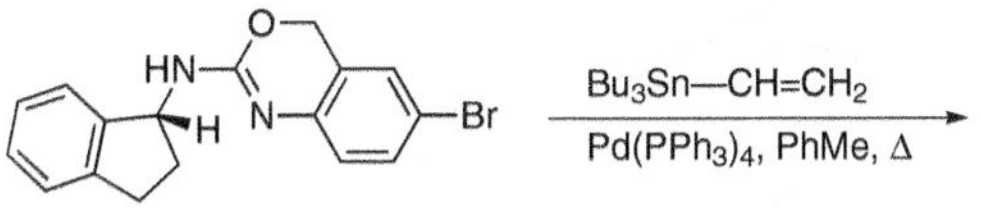

WO 2010/026110 A2 (Hofmann-La Roche)

(c) 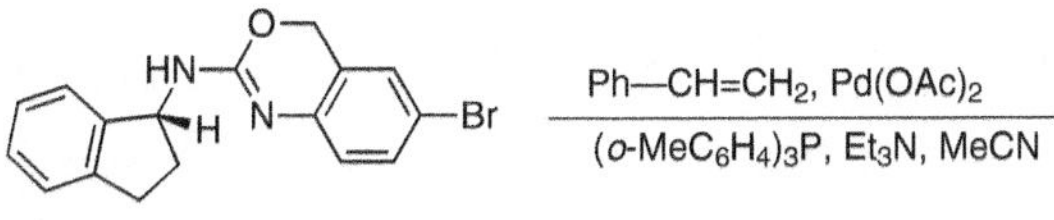

WO 2010/026110 A2 (Hofmann-La Roche)

(d) 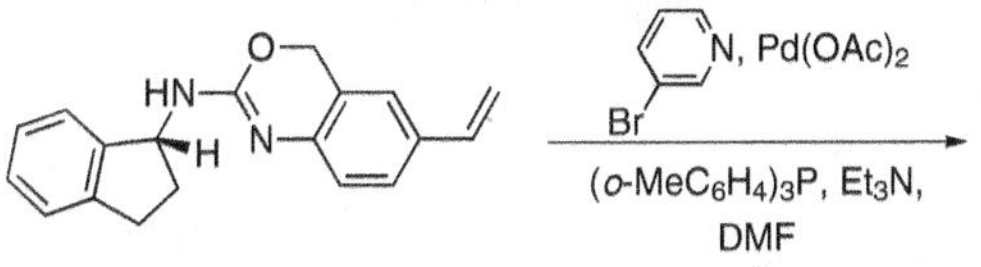

WO 2010/026110 A2 (Hofmann-La Roche)

(e) 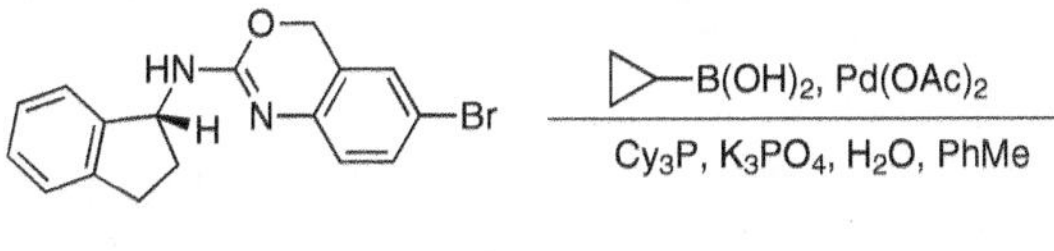

WO 2010/026110 A2 (Hofmann-La Roche)

(f) 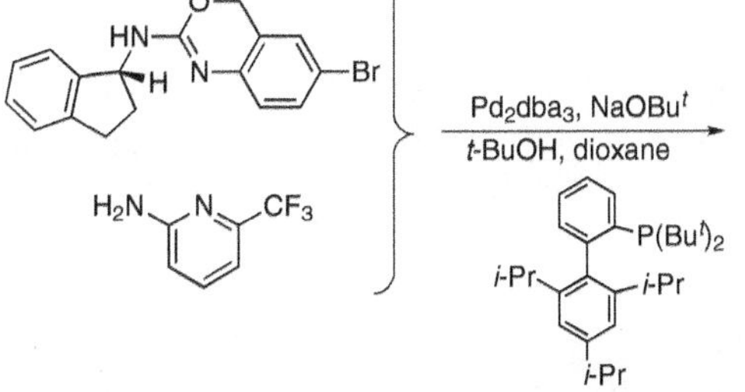

WO 2010/026110 A2 (Hofmann-La Roche)

(g) 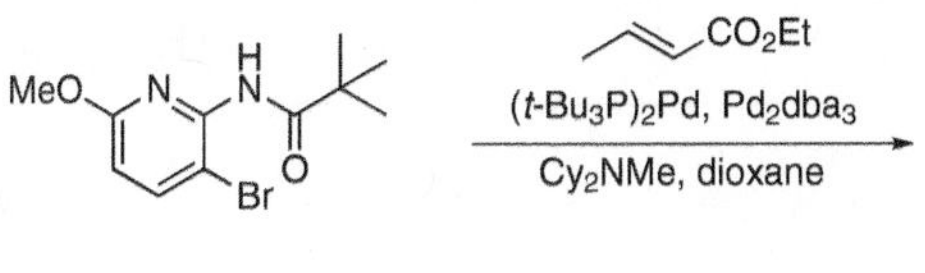

WO 2010/043714 A1 (Glaxo)

(h) 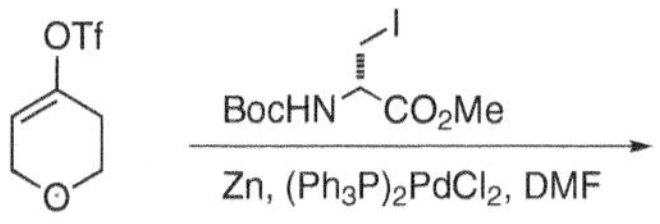

WO 2010/013161 A1 (Pfizer)

(i) 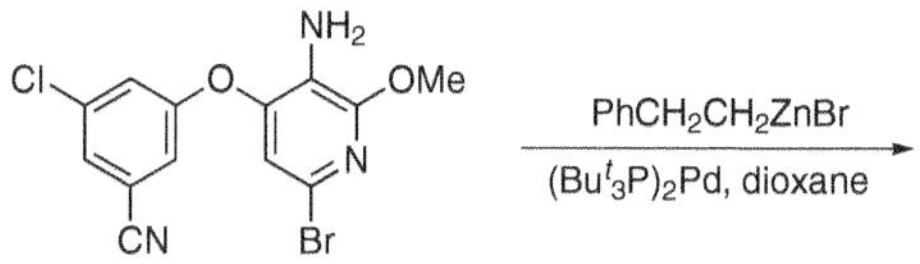

WO 2010/026075 A1 (Hofmann-La Roche)

(j) 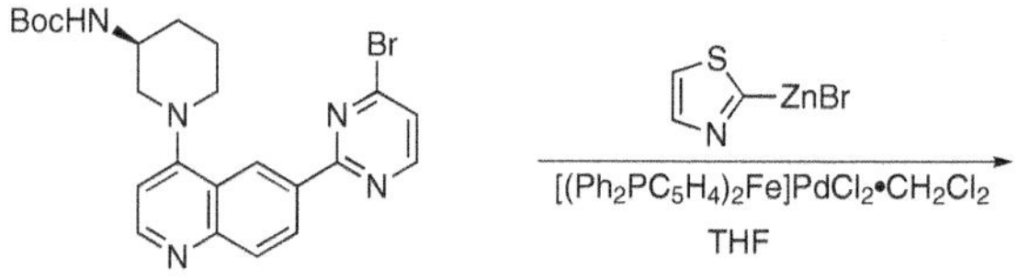

WO 2010/026121 A1 (Novartis)

(k) 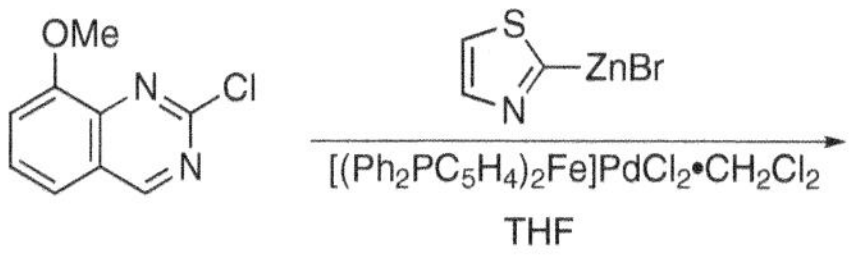

WO 2010/026122 A1 (Novartis)

(l) 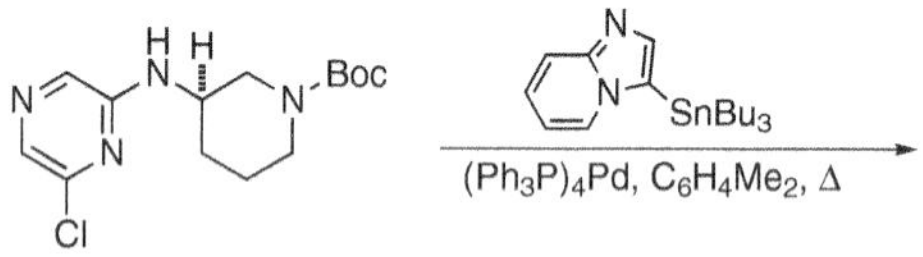

WO 2010/016005 A1 (Pfizer)

(m) 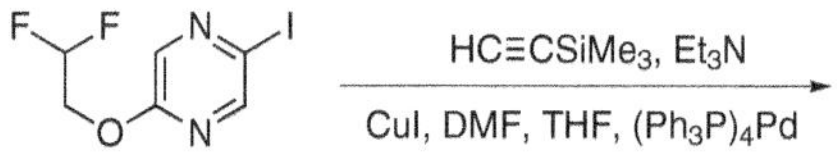

WO 2010/016005 A1 (Pfizer)

(n) 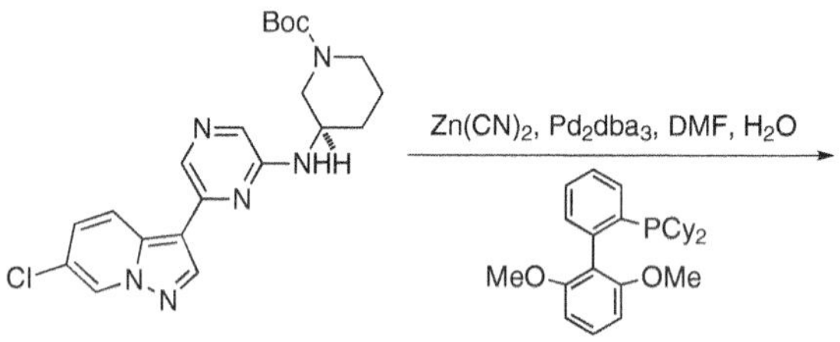

WO 2010/016005 A1 (Pfizer)

(o) 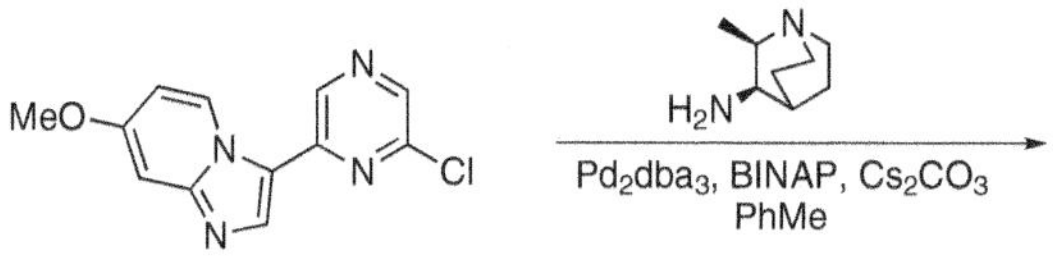

WO 2010/016005 A1 (Pfizer)

(p) 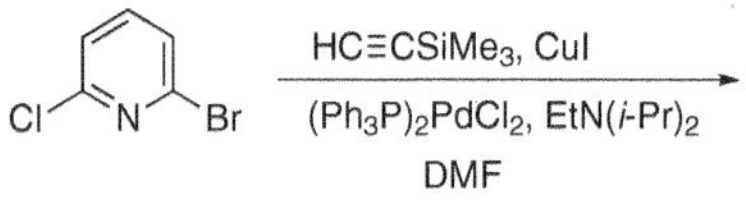

WO 2010/016005 A1 (Pfizer)

(q) 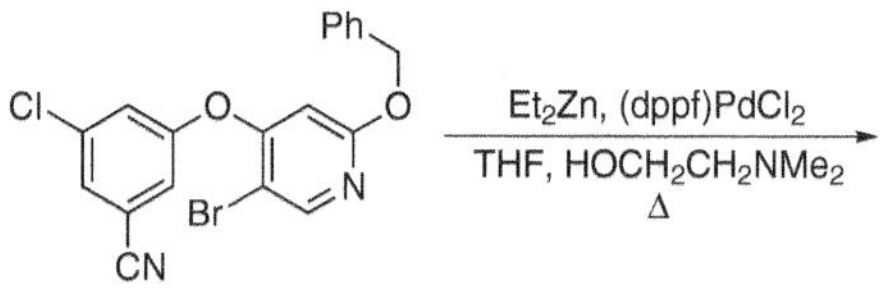

WO 2010/026075 A1 (Hofmann-La Roche)

(r) 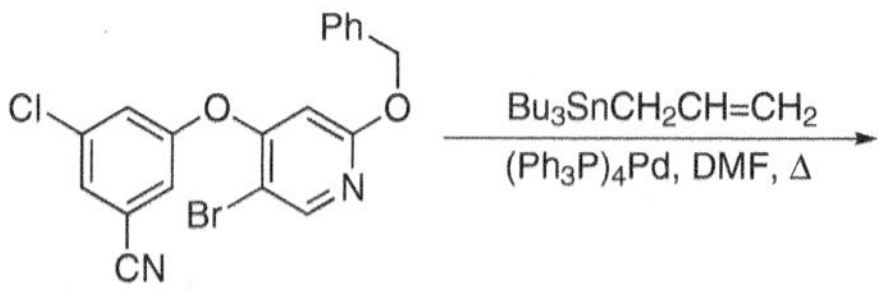

WO 2010/026075 A1 (Hofmann-La Roche)

(s) 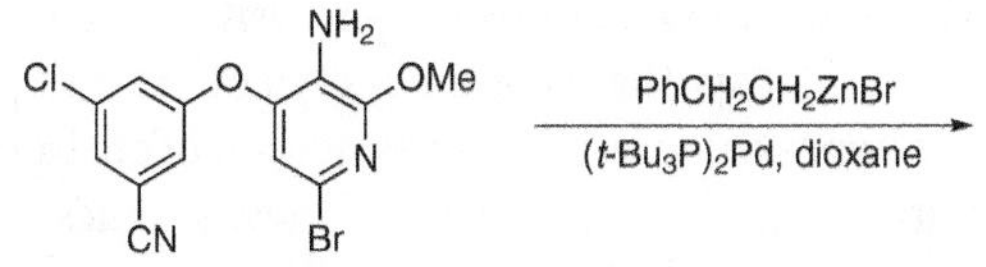

WO 2010/026075 A1 (Hofmann-La Roche)

(t) 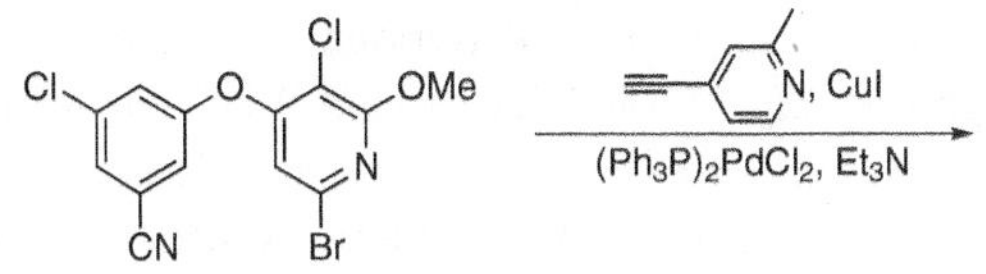

WO 2010/026075 A1 (Hofmann-La Roche)

(u) 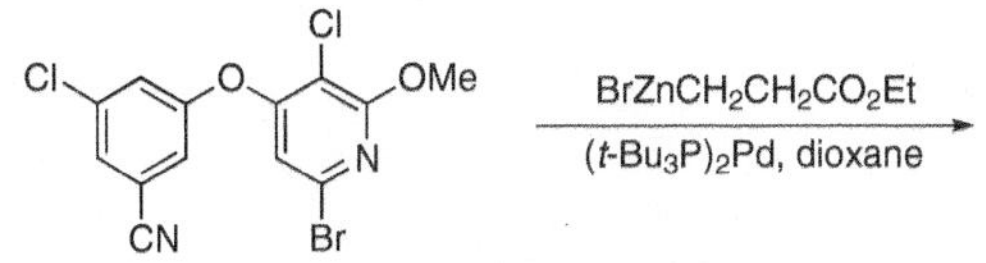

WO 2010/026075 A1 (Hofmann-La Roche)

(v) 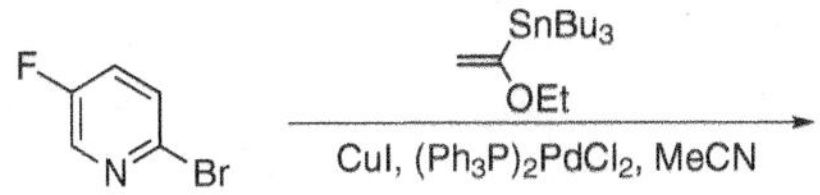

WO 2010/020810 A1 (AstraZeneca)

17-10 Suggest a mechanism for the reaction shown. What is the role of the pyrrolidine? [*Angew. Chem. Int. Ed.*, **2006**, *45*, 1952].

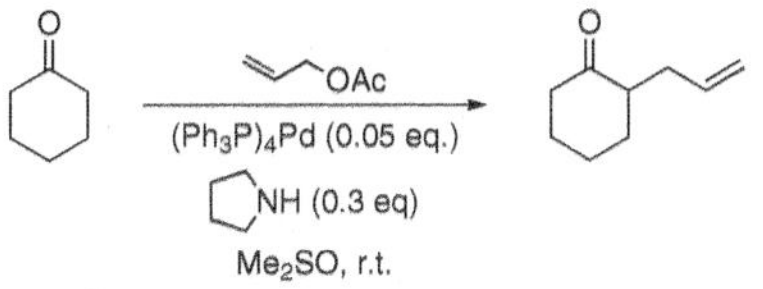

17-11 What does the stereochemistry in the reaction below [*J. Am. Chem. Soc.* **2004**, *126*, 10246] suggest about the mechanism of the reaction?

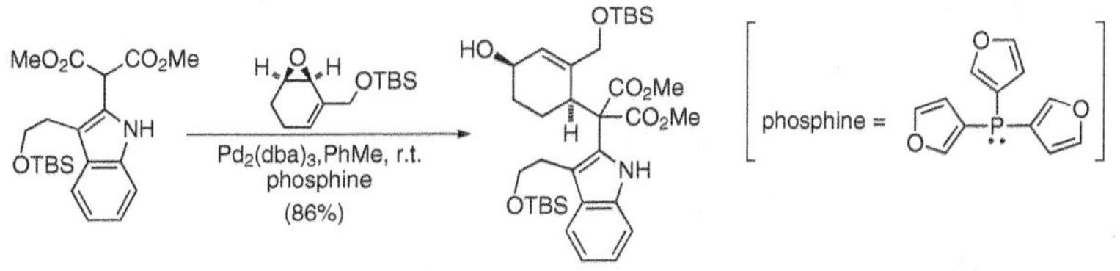

17-12 Suggest reagents that can be used for the conversion in each of the reactions below. Where possible, carry out the transformation in a single step.

(a) 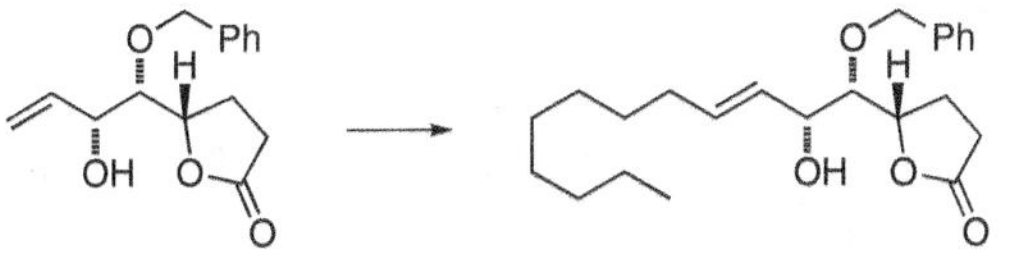

[*Tetrahedron* **2011**, *67*, 9358]

(b) 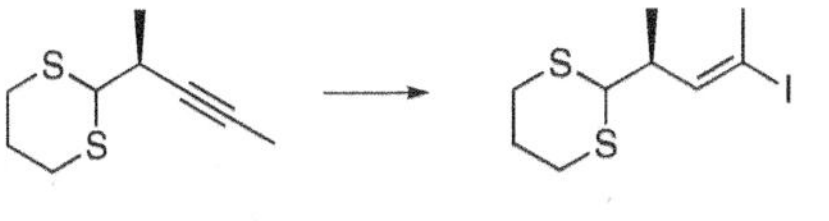

[*J. Am. Chem. Soc* **2003**, *125*, 350]

(c) 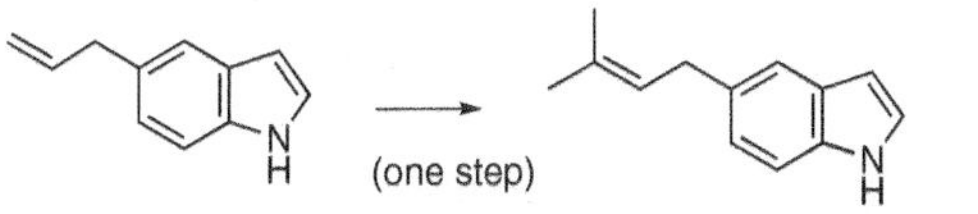

[*Angew. Chem. Int. Ed.* **2011**, *50*, 1402]

(d) 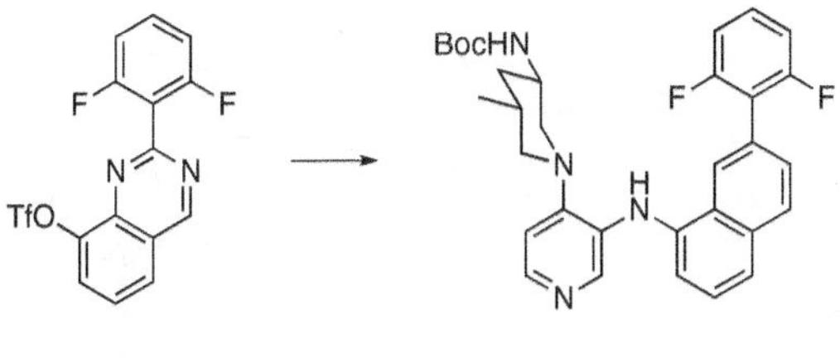

[WO 2010/026122 A1 (Novartis)]

17-13 The power of modern cross-coupling reactions in synthesis is nicely demonstrated by two stereospecific total syntheses of the acetylenic sesquiterpene, freelingyne. The two syntheses appeared within months of each other in 1997 and show remarkable similarities. The stereospecificity of these syntheses may be contrasted with the lack of stereospecificity in the previous two syntheses, which also appeared in close succession [Knight, D.W.; Pattenden, G. *J. Chem. Soc., Chem. Commun.* **1974**, 188. Ingham, C.F.; Massy-Westropp, R.A.; Reynolds, G.D. *Aust. J. Chem.* **1974**, *27*, 1477]. When you have completed each synthesis by supplying the missing reagents or reaction products, compare the two syntheses. How are they similar and how do they differ?

(a) Katsumura's synthesis

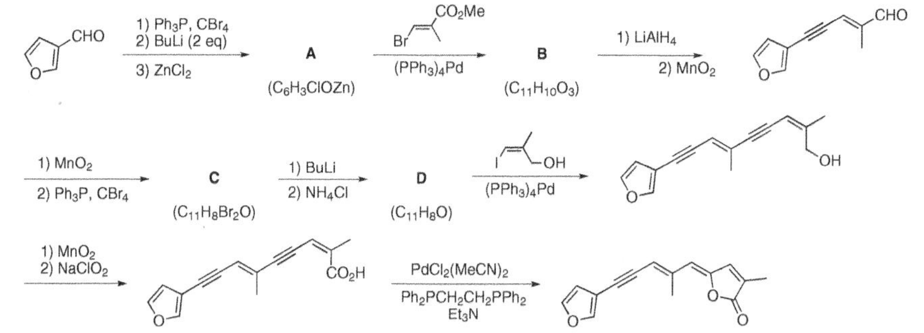

Reference: Tetrahedron Lett. **1997**, *38*, 5311.

(b) Negishi's synthesis

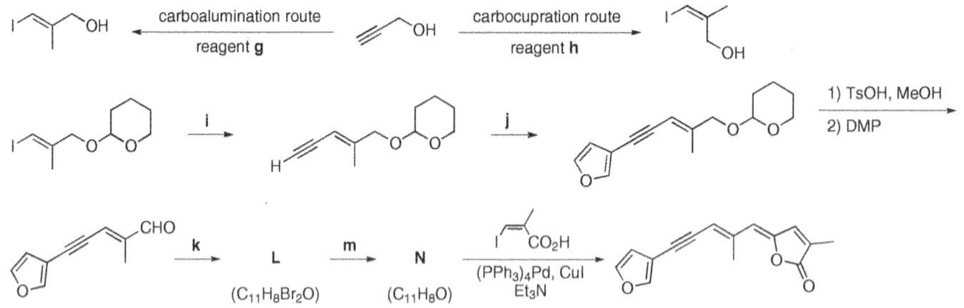

Reference: *J. Org. Chem.* **1997**, *62*, 8591.

17-14 There are three steps in Somfai's short synthesis of the alkaloid, stemoamide. Supply the missing reagents and products in this three-reaction sequence in the middle of the synthesis.

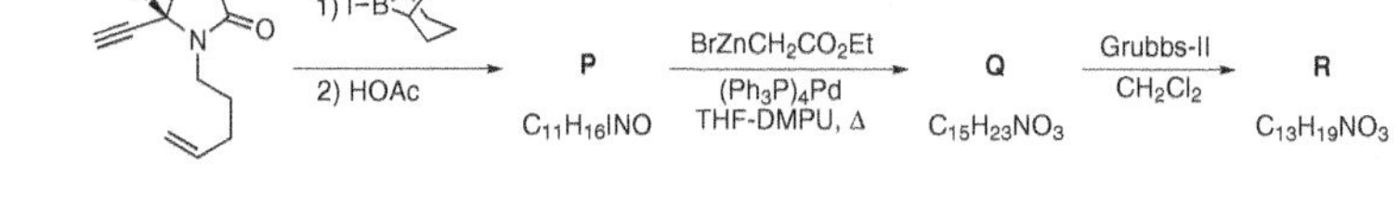

Reference: J. Org. Chem. **2007**, *72*, 4246.

CHAPTER EIGHTEEN

Redox Reactions I

Oxidation

18.1 Introduction

Oxidation and reduction reactions can comprise more than 25% the reactions used in the synthesis of a complex molecule, so they are important processes in organic chemistry. This also means that a large number of reagents have been developed for carrying out these reactions. Oxidation and reduction may be defined a number of ways. The first is the definition that is taught in introductory courses: *oxidation is a net loss of electrons, and reduction is a net gain in the number of electrons.* An alternative, older definition still remains useful: *oxidation is a net gain in the number of oxygen (or other electronegative) atoms or a net loss in the number of hydrogen (or other electropositive) atoms.* Reduction is the reverse of oxidation: *a net gain in the number of hydrogen (or other electropositive) atoms or a net loss in the number of oxygen (or other electronegative) atoms.*

Based on these definitions, let us categorize a familiar reaction, the conversion of a primary alcohol to a carboxylic acid:

$$R\text{—}CH_2OH \rightarrow R\text{—}CO_2H$$

In this transformation, there is a net loss of two hydrogen atoms and a net gain of an oxygen atom. The reaction is oxidation. We can also write an oxidation half-equation for this transformation:

$$R\text{—}CH_2OH + H_2O \rightarrow R\text{—}CO_2H + 4H^+ + 4e^-$$

Likewise, the conversion of an alkyl halide to a Grignard reagent is a representative reduction: there is a net loss of one electronegative atom (the halogen) and a net gain of one electropositive element (the magnesium). Note how it is much easier to see the change in the number of electrons when one uses the minor contributor to the resonance hybrid.

$$R\text{—}X + Mg \rightarrow R\text{—}Mg\text{—}X \text{ (or } R^- \, Mg^{2+} \, X^-\text{)}$$

Again, we can write oxidation and reduction half-equations for this reaction:

$$R\text{—}X + 2e^- \rightarrow R^- + X^- \qquad \text{[reduction half-equation]}$$

$$Mg \rightarrow Mg^{2+} + 2e^- \qquad \text{[oxidation half-equation]}$$

Oxidation Numbers: Use and Limitations

One convenient method for monitoring oxidation and reduction reactions is by the use of oxidation numbers. Oxidation numbers are useful when applied to the inorganic reactants in oxidation and reduction reactions. We readily refer to potassium permanganate as manganese (VII) reagents and various reagents based on chromic acid as chromium (VI) reagents.

The oxidation numbers of carbon in a series of organic compounds with a range of commonly found functional groups are presented in Table 18.1. X, Y, and Z refer to electronegative elements (e.g., O, N, halogens, S); the hydrogen atoms in the structures may be replaced by atoms of a similarly electropositive element (e.g., metals, Si, B), without affecting the oxidation number of carbon. Oxidation formally corresponds to movement up or to the right and reduction to movement down or to the left. The oxidation numbers are calculated using the rule that hydrogen is +1 in organic compounds, halogens are −1, oxygen is −2, and that alkyl groups contribute 0.

One consequence of this way of computing oxidation numbers, is that the substitution of hydrogen (+1) by alkyl (0) leads to an algebraic increase in the oxidation number of carbon of +1—formally an oxidation. This shows up in Table 18.1. The formal oxidation

Table 18.1 The Oxidation Number of Carbon in Representative Organic Compounds

Formal Oxidation Number of Carbon

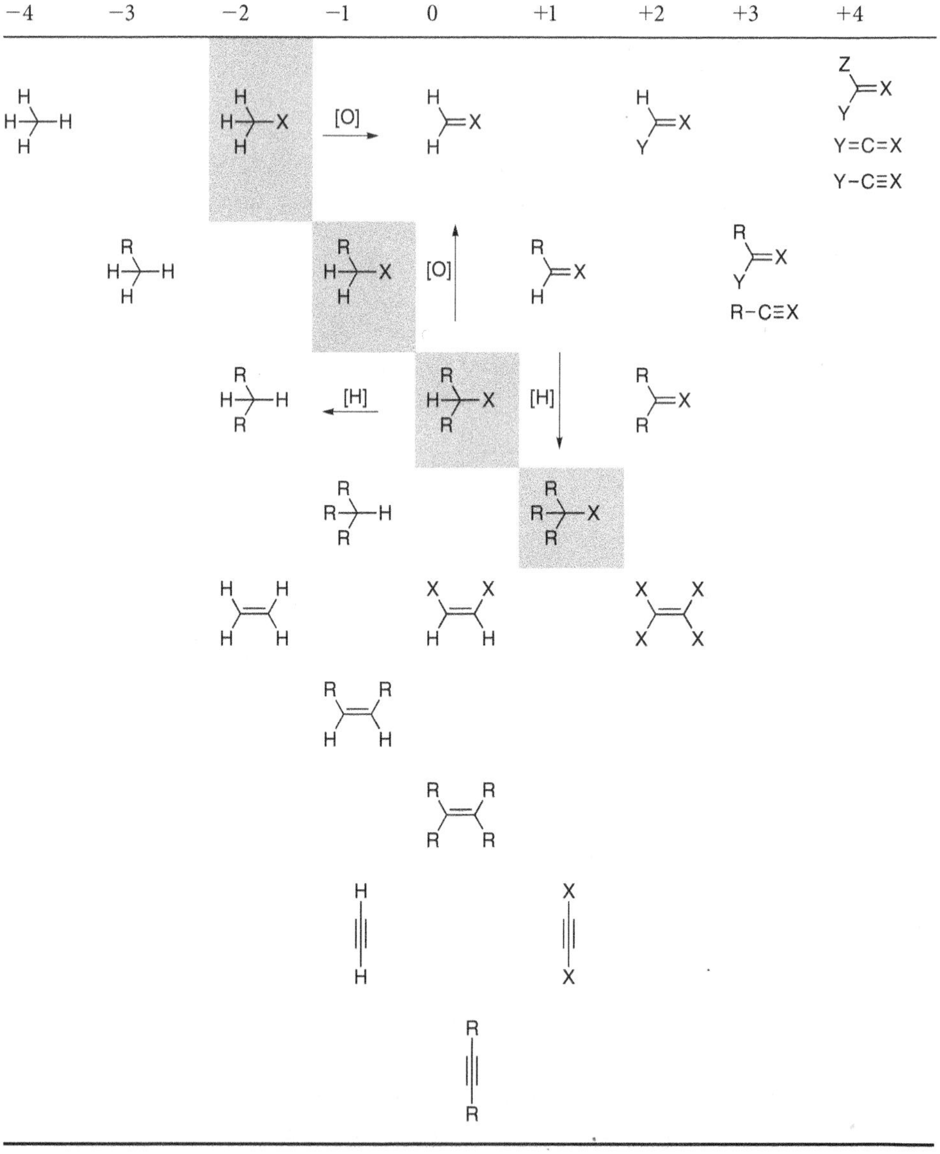

number for a particular functional group is not constant but depends on the other groups attached to the carbon. This is illustrated by the oxidation number of the carbon bearing the functional group in the series of compounds in the accented boxes (e.g., alcohols or alkyl halides): moving through this series from methanol to tertiary alcohols increases the formal oxidation number of the carbinol carbon by +1 each step. So formally this is an oxidation, although this structural change is seldom viewed in this way. Instead, this series gives the potential for oxidation: carbon has a maximum oxidation number of +4, so the difference between this and the value of the oxidation number in the compound under scrutiny tells us something about its potential for oxidation. Methanol can be oxidized all the way to carbon dioxide, primary alcohols to carboxylic acids, and secondary alcohols to ketones. Each of these oxidations leads to a change of +2 in the oxidation number of carbon. Tertiary alcohols resist oxidation.

Oxidizable and Reducible Groups in Organic Compounds

Carbon-based oxidizable groups in organic chemistry fall into two major classes: σ bonds to hydrogen and C—C π bonds. Oxidation of these groups is achieved by (1) replacement of two σ bonds to hydrogen by a π bond (Example **18.1**), (2) replacement of a σ bond to hydrogen by a σ bond to an electronegative element (Example **18.2**), and (3) replacement of a carbon-carbon π bond by two σ bonds (or π bonds) to electronegative elements (Examples **18.3** and **18.4**). Examples of the more common oxidation reactions in organic chemistry are presented in Table 18.2.

(18.1) **(18.2)**

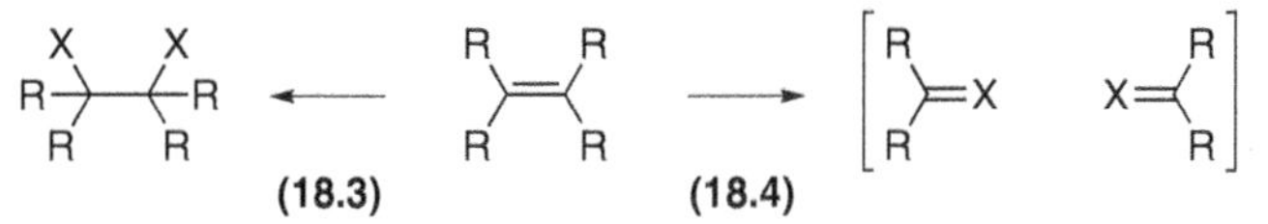

(18.3) **(18.4)**

Reducible groups in organic chemistry fall into two major classes: π bonds and σ bonds to an electronegative atom. Reduction of these is achieved by (1) replacement of a π bond by two σ bonds to hydrogen (Example **18.5**), (2) replacement of a σ bond to an electronegative element a σ bond to hydrogen or an electropositive element (Example **18.6**), and (3) replacement of σ (or π) bonds to a pair of electronegative elements by a new C—C π bond (Examples **18.7** and **18.8**). Examples of the more common types of reduction reactions in organic chemistry are presented in Table 18.3.

(18.5) **(18.6)**

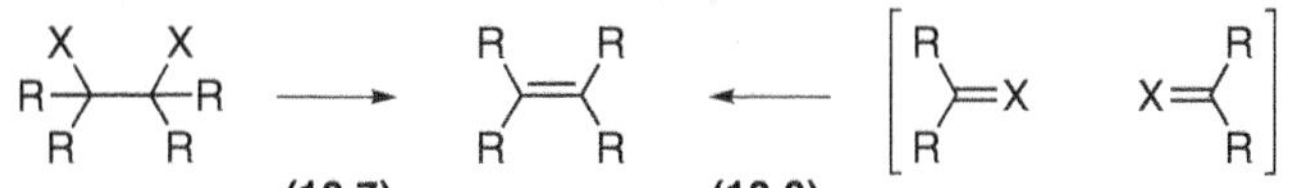

(18.7) **(18.8)**

Problems

18-1 Using molecular formulas, verify that each of the examples in Tables 18.2 and 18.3 are actually oxidations and reductions.

Table 18.2 Some Common Oxidation Reactions

Reaction Type	Example	
Alcohol oxidation	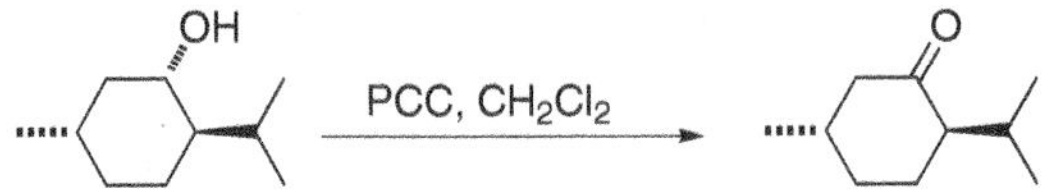	(18.9)
Dehydrogenation	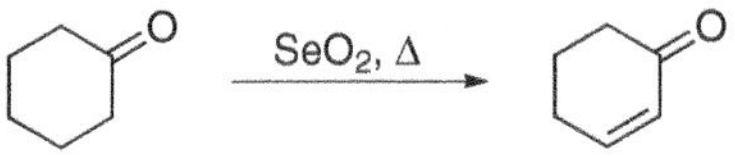	(18.10)
	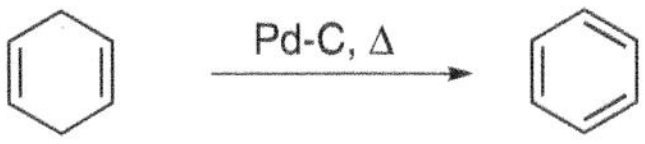	(18.11)
Oxidative cleavage of glycols	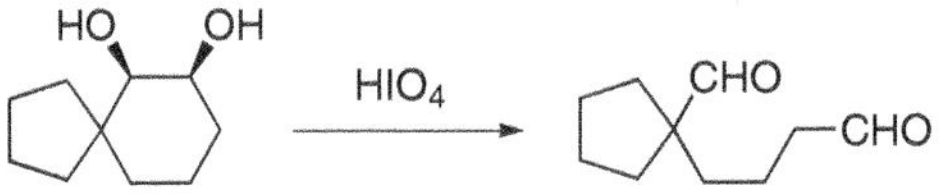	(18.12)
α-Selenylation	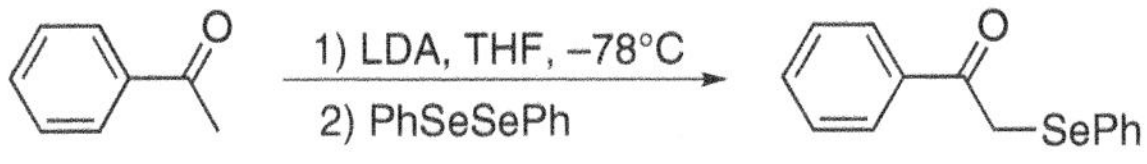	(18.13)
Hydroxylation	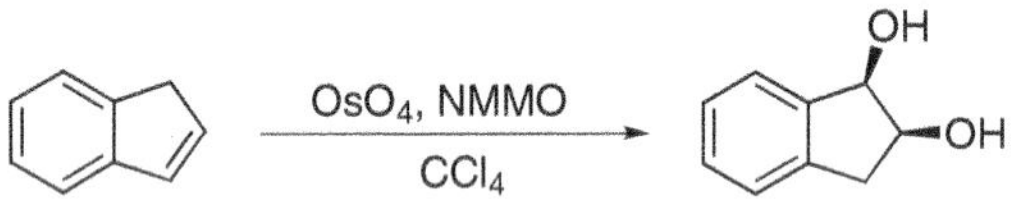	(18.14)
Epoxidation	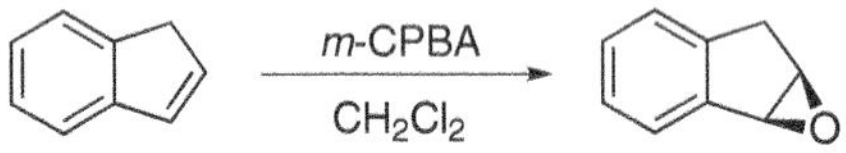	(18.15)
Ozonolysis	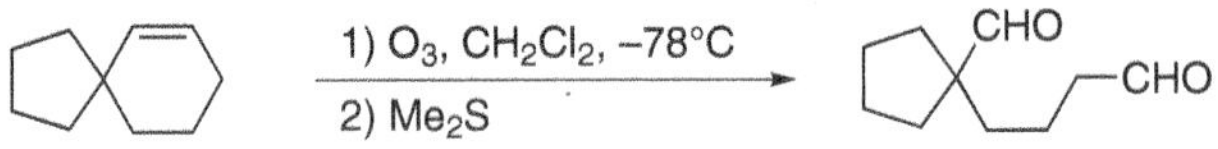	(18.16)

Table 18.3 Some Common Reduction Reactions

Reaction Type	Example	
Catalytic hydrogenation	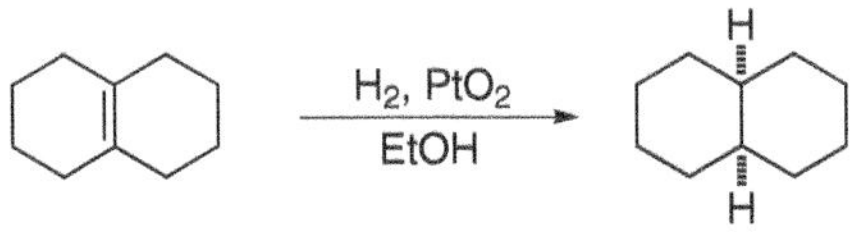	(18.17)
Metal hydride reduction	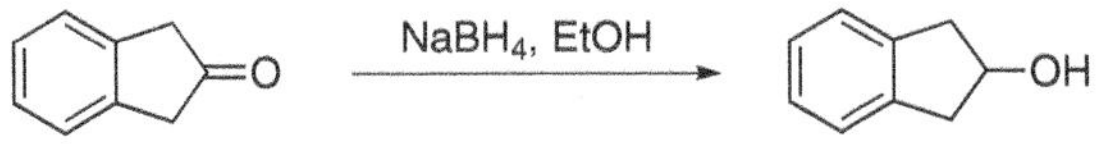	(18.18)
Reductive elimination	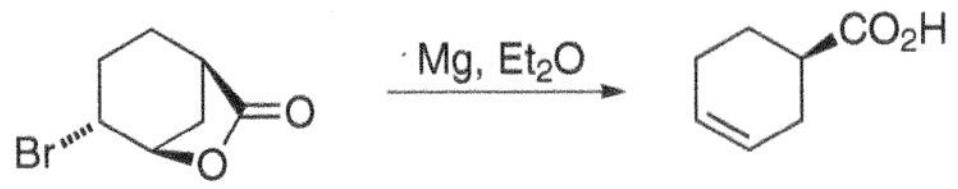	(18.19)
Dissolving metal (Birch) reduction	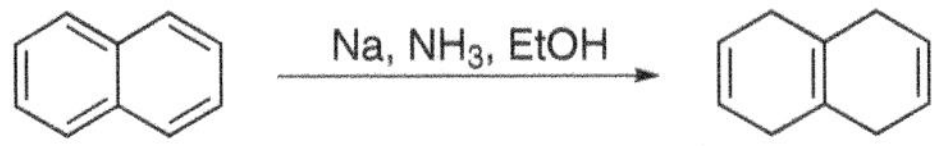	(18.20)
Deoxygenation	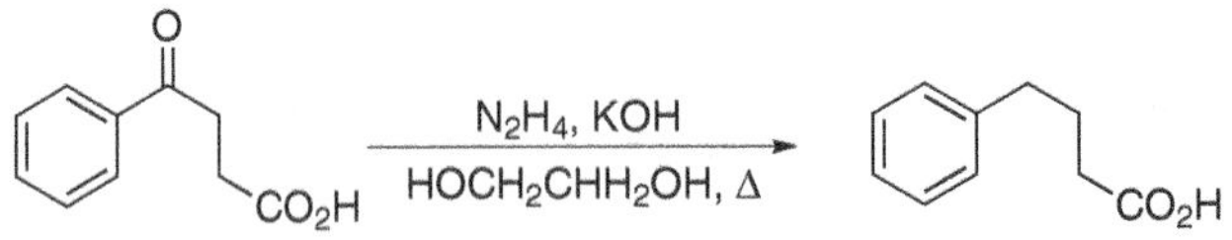	(18.21)
Reductive substitution	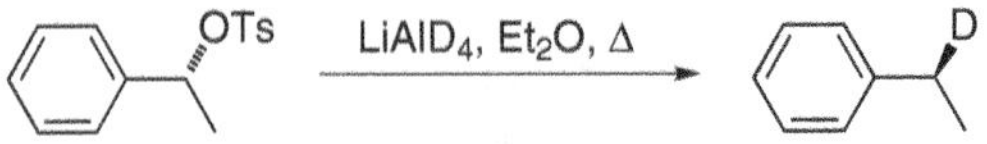	(18.22)

18-2 Write balanced oxidation or reduction half-equations for each of the following reactions. You may use H^+ or H_2O to provide hydrogen or oxygen atoms as needed.

(a) $MnO_4^- \rightarrow MnO_2$

(b) $Cr_2O_7^{2-} \rightarrow Cr^{3+}$

(c) $RCH_2OH \rightarrow RCHO$

(d) $RCONH_2 \rightarrow RCH_2NH_2$

(e) $R_2C{=}CR_2 \rightarrow 2\ R_2C{=}O$

(f) $R_2C{=}CR_2 \rightarrow R_2C(OH){-}C(OH)R_2$

(g) $R_2C{=}CR_2 \rightarrow R_2CHCHR_2$

(h) $BrCH_2CH_2OR \rightarrow CH_2{=}CH_2 + Br^-$

(i) $RCOR' \rightarrow RCO_2R'$

(j) $ClCrO_3^- \rightarrow Cr^{3+}$

18-3 Indicate the formal oxidation number of each carbon atom indicated in each of the following compounds.

(a) 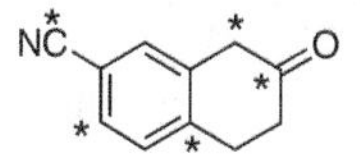

(b) 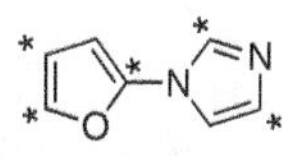

(c) 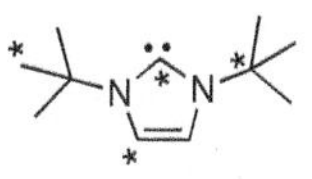

(d) 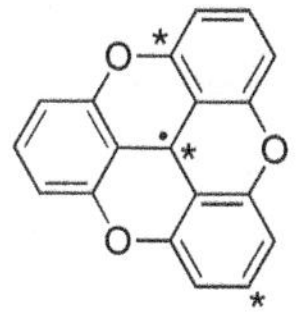

(e) 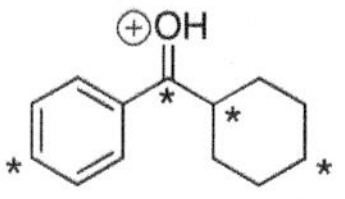

(f) 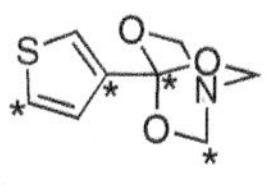

18-4 Classify each of the following reactions as oxidation, reduction, or neither. It may be necessary to consider atoms within the reacting species individually to obtain a satisfactory answer.

(a) 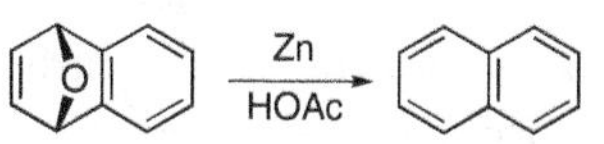

(b) 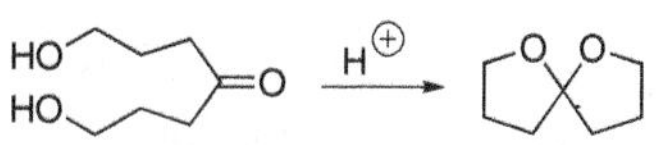

(c) 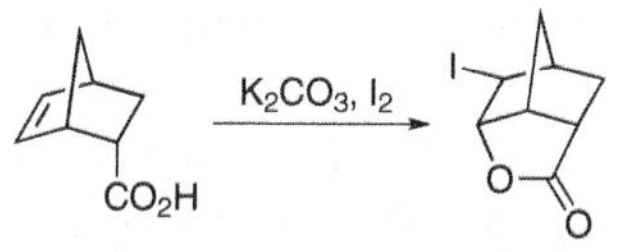

(d) 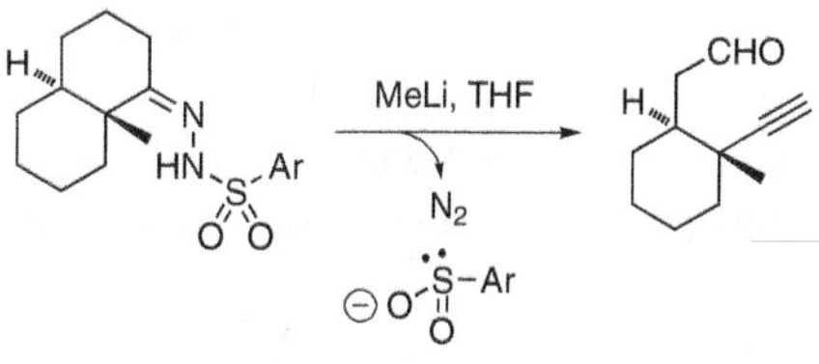

18.2 Overview of Oxidation[1]

There are a number of types of oxidations that it is important to be able to carry out selectively in organic chemistry. These are shown schematically in Figure 18.1. In the diagram, there are five different classes of oxidations shown.

1. Books on oxidation reactions: (a) Wiberg, K.B. *Oxidation in Organic Chemistry. Part A* (Academic Press: New York, 1965). (b) Trahanovsky, W.S. *Oxidation in Organic Chemistry. Part B* (Academic Press: New York, 1973); *Part C* (Academic Press: New York, 1978); *Part D* (Academic Press: New York, 1982). (c) Hudlicky, M. *Oxidations in Organic Chemistry* (American Chemical Society: Washington, D.C., 1990). (d) Haines, A.H. *Methods for the Oxidation of Organic Compounds*, Vol. 1 (Academic Press: New York, 1985); Vol. 2 (Academic Press: New York, 1988). (e) Chinn, L.J. *Selection of Oxidants in Organic Synthesis* (Marcel Dekker: NewYork, 1971). (f) Augustine, R.L.; Trecker, D.J. *Oxidation*, Vol. 1 (Marcel Dekker: New York, 1969); Vol. 2 (Marcel Dekker: New York, 1971). (g) Mijs, W.J.; de Jonge, C.R.J.I. *Organic Synthesis, by Oxidation with Metal Compounds* (Plenum: NewYork, 1986). (h) Cainelli, G.; Cardillo, G. *Chromium Oxidations in Organic Chemistry* (Springer Verlag: New York, 1984). (i) Arndt, D. *Manganese Compounds as Oxidizing Agents in Organic Chemistry* (Open Court Publishing Company: La Salle, IL, 1981). (j) Lee, D.G. *The Oxidation of Organic*

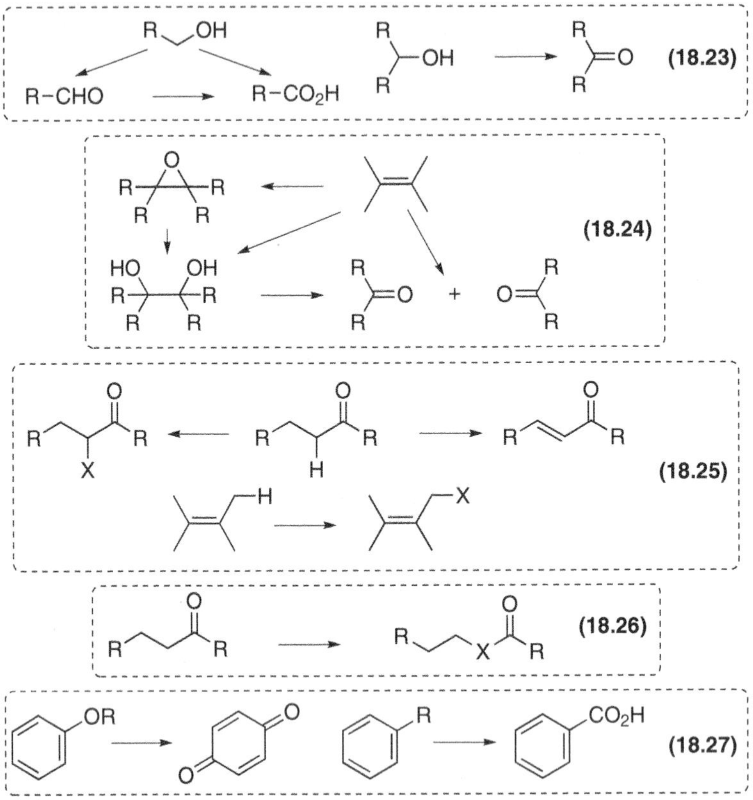

Figure 18.1 Major classes of oxidation reactions in organic chemistry

The first class of oxidation reactions (**18.23**) is the oxidation of alcohols to aldehydes, ketones, and carboxylic acids, as well as the oxidation of aldehydes to carboxylic acids. These reactions are oxidations that lead to the incorporation of a new π bond into the molecule without the cleavage of any groups other than bonds to hydrogen. The second class of oxidation reactions (**18.24**) is the oxidation of alkenes that lead to the cleavage of carbon-carbon π bonds and/or carbon-carbon σ bonds. The third class of oxidation reactions (**18.25**) includes oxidations of carbon-hydrogen bonds adjacent to carbon-carbon or carbon-oxygen π bonds. These types of reactions always lead to products with a new functional group. The fourth class of oxidation reactions (**18.26**) includes reactions in which a heteroatom is inserted into a carbon-carbon σ bond. The final class of oxidations (**18.27**) involves the oxidation of aromatic compounds, either in the ring, or in the side chain.

18.3 Oxidation of Alcohols to Aldehydes and Ketones with Stoichiometric Metal-Based Reagents[2]

The oxidation of alcohols is perhaps one of the most studied of organic reactions due to the importance of the oxidation products in organic synthesis. Depending on the reagent chosen, primary alcohols may be oxidized either to aldehydes or to carboxylic acids, and secondary alcohols are oxidized to ketones. Tertiary alcohols are generally resistant to oxidation (indeed, *tert*-butyl alcohol is often chosen as a solvent for oxidations).

The oxidation of secondary alcohols to ketones is generally fairly easy to accomplish, because the further oxidation of the ketone usually requires a powerful oxidizing agent

Compounds by Permanganate Ion and Hexavalent Chromium (Open Court Publishing Company: La Salle, IL, 1980).

2. Monograph: Tojo, G.; Fernández, M. *Oxidation of Alcohols to Aldehydes and Ketones* (Springer: New York, 2006).

under rather forcing conditions. The same cannot be said, however, for the oxidation of primary alcohols, especially if one is attempting to stop the oxidation at the level of the aldehyde. Oxidation of primary alcohols to carboxylic acids is generally easier than oxidation to the aldehyde level. For these reasons, much of the effort behind the development of selective oxidation of alcohols has been driven by the need for oxidizing agents and reactions that will halt at the aldehyde stage.

Most reagents that can be used to oxidize a primary alcohol to an aldehyde or carboxylic acid can be used to oxidize a secondary alcohol to a ketone. For this reason, the discussion that follows will be focused on methods for the selective oxidation of primary alcohols to aldehydes.

Metal-Based Reagents

Chromium (VI) Reagents

Aldehydes can be prepared by the oxidation of primary alcohols with compounds of a high-valent metal such as chromium (VI) or manganese (VII). The mechanism currently accepted for the chromium (VI) oxidation of alcohols was first proposed by American chemist Frank Westheimer in 1949.[3] It is given in Figure 18.2. In the first step of the mechanism, the alcohol reacts rapidly and reversibly with the chromium (VI) compound to form an intermediate chromate ester. This ester is unstable and undergoes an E2-type elimination reaction to give the carbonyl compound and a chromium (IV) by-product (**A**). The identity of the base in the elimination step is not known, but there is a large hydrogen isotope effect for the reaction,[4] indicating that the C—H bond is broken in the rate determining step. The elimination may also occur by an E*i* mechanism, where the anionic oxygen attached to chromium functions as the base.

The Cr (IV) is an active oxidant that is believed to react directly with the alcohol in a free radical reaction (**B**) to give the alkyl radical and a Cr (III) species. Evidence for the direct oxidation of the alcohol by Cr (IV) was obtained by observing the polymerization of acrylamide and concomitant reduction in the yield of acetone in the oxidation of

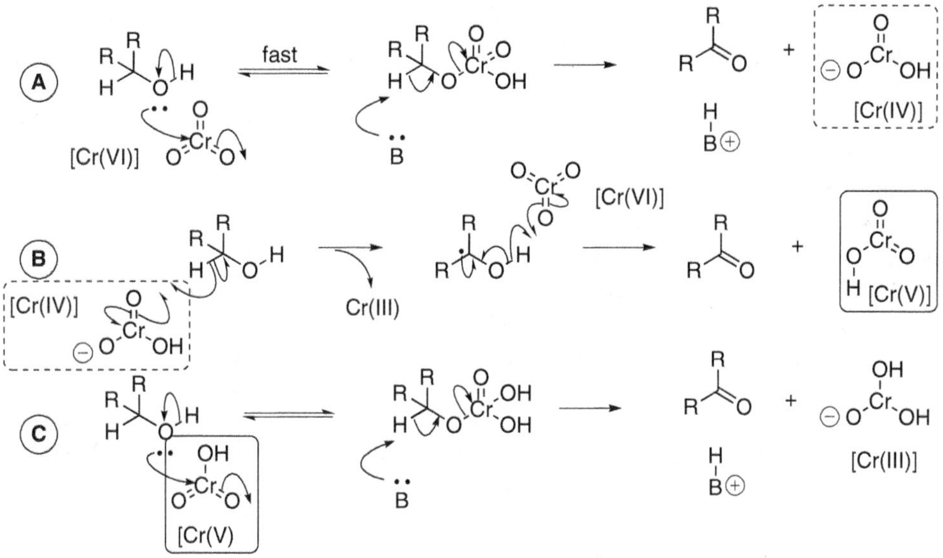

Figure 18.2 The Westheimer mechanism for oxidation of alcohols by chromium (VI)

3. (a) Westheimer, F.H. *Chem. Rev.* **1949**, *45*, 419. (b) Watanabe, W.; Westheimer, F.H. *J. Chem. Phys.* **1949**, *17*, 61. (c) Holloway, F.; Cohen, M.; Westheimer, F.H. *J. Am. Chem. Soc.* **1951**, *73*, 65. (d) Cohen, M.; Westheimer, F.H. *J. Am. Chem. Soc.* **1952**, *74*, 4387.

4. Westheimer, F.H.; Nicolaides, N. *J. Am. Chem. Soc.* **1949**, *71*, 25.

isopropyl alcohol by Cr (IV).[5] The organic radical then reacts with another Cr (VI) species to give the carbonyl compound and a Cr (V) species, which is believed to react with another molecule of the alcohol by a mechanism like that in the first step (**C**) to give the carbonyl compound and a Cr (III) species. Interestingly, the reaction between Cr (IV) and Cr (VI) to give Cr (V) appears to be much slower than the oxidation of the alcohol by Cr (IV).

Chromium (VI) compounds are among the most widely used oxidants for oxidizing alcohols (e.g., Examples **18.28** to **18.30**), and an idea of the variety of chromium (VI) reagents now available may be gauged from the entries in Table 18.4, which summarize some of the more commonly used Cr (VI) reagents.

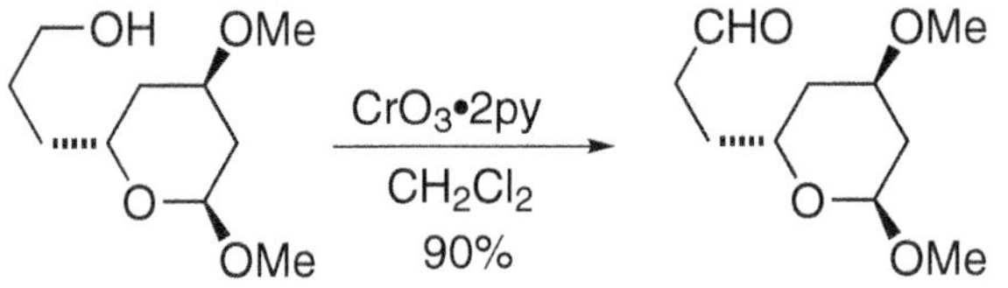

(18.28)

Collins oxidation[6]

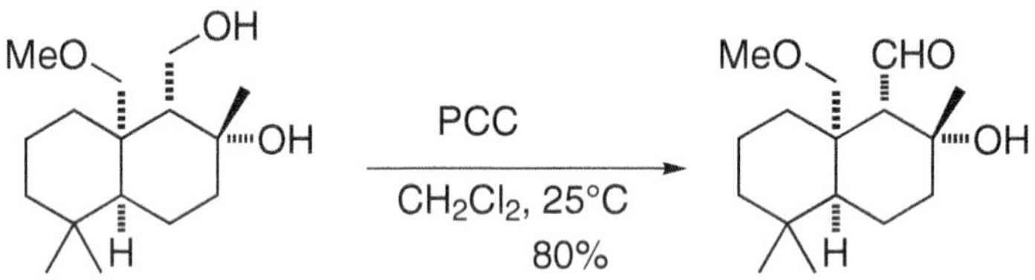

(18.29)

Pyridinium chlorochromate[7]

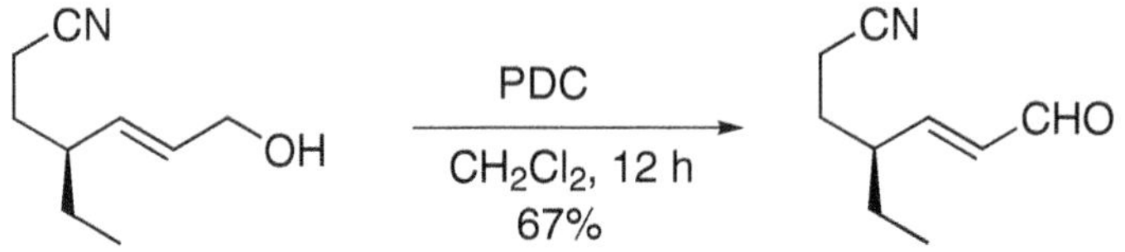

(18.30)

Pyridinium dichromate[8]

In the absence of water, chromium (VI) compounds can be used to oxidize primary alcohols to aldehydes. Oxidation of primary alcohols by aqueous chromic acid almost always results in oxidation to the carboxylic acid, presumably through the aldehyde hydrate. To suppress this further oxidation of the aldehyde, reagents such as the Sarett and Collins (Example **18.28**) reagents and the various pyridinium chlorochromate and dichromate salts (all of which are used under anhydrous conditions so that hydrate formation is prevented) were developed.

The two most widely used Cr (VI) oxidizing agents today are pyridinium chlorochromate (PCC) and pyridinium dichromate (PDC), both of which were developed for the oxidation of primary alcohols under mild conditions. Of the two reagents, PCC is the more acidic oxidant, so PDC is often used for more sensitive applications. The acidic character of pyridinium chlorochromate is revealed in reactions where a cation might be formed during the reaction. A good example of this is provided by the oxidation of

5. (a) Rahman, M.; Roček, J. *J. Am. Chem. Soc.* **1971**, *93*, 5455. (b) Rahman, M.; Roček, J. *J. Am. Chem. Soc.* **1971**, *93*, 5462.

6. Grieco, P.A.; Lis, R.; Zelle, R.E.; Finn, J. *J. Am. Chem. Soc.* **1986**, *108*, 5908.

7. (a) Kato, M.; Heima, K.; Matsumura, Y.; Yoshikoshi, A. *J. Am. Chem. Soc.* **1981**, *103*, 2434. (b) Kato, M.; Matsumura, Y.; Heima, K.; Fukamiya, N.; Kabuto, C.; Yoshikoshi, A. *Tetrahedron* **1987**, *43*, 717.

8. (a) Kato, M.; Heima, K.; Matsumura, Y.; Yoshikoshi, A. *J. Am. Chem. Soc.* **1981**, *103*, 2434. (b) Kato, M.; Matsumura, Y.; Heima, K.; Fukamiya, N.; Kabuto, C.; Yoshikoshi, A. *Tetrahedron* **1987**, *43*, 717.

(*S*)-(−)-citronellol to (*S*)-(−)-isopulegone by PCC.[18] In the course of the reaction, the aldehyde formed initially (Example **18.32**) undergoes acid-catalyzed cyclization (probably an ene reaction, as shown below) to give the cyclized alcohol (Example **18.33**) that then undergoes a second oxidation to give the ketone (Example **18.34**) as the final product.

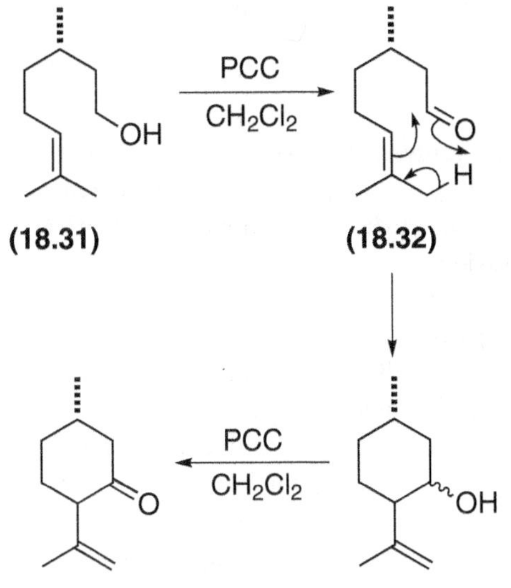

(18.31) **(18.32)** **(18.34)** **(18.33)**

Pyridinium dichromate, on the other hand, is a much less acidic (and, therefore, less aggressive) reagent. For example, it also oxidizes citronellol, but unlike the oxidation with pyridinium chlorochromate, the oxidation stops at the aldehyde stage without the subsequent acid-catalyzed cyclization reaction.[19] It can be used to oxidize primary alcohols to aldehydes in the presence of acid-sensitive functional groups such as enol ethers (e.g., Example **18.35**).[20] Pyridinium dichromate also shows enhanced selectivity for oxidation of allylic alcohols compared to saturated alcohols, as illustrated by the oxidation of the glycal in Example **18.36.** Here the secondary allylic hydroxyl group is oxidized in preference to secondary hydroxyl group, again without affecting the acid-sensitive enol ether.[21]

Table 18.4 Oxidizing Agents Based on Chromium (VI)

Reagent	Compound Name	Acronym or Reaction Name	Ref.
$K_2Cr_2O_7$, H_2SO_4, H_2O	Chromic acid	—	
$K_2Cr_2O_7$, H_2SO_4, H_2O, Et_2O	Chromic acid	***Brown oxidation***	9
CrO_3, H_2SO_4, H_2O, acetone	Chromic acid	***Jones oxidation***	10
CrO_3, pyridine [C_5H_5N, py]	Chromic anhydride	***Sarett oxidation***	11
CrO_3•2py	Chromic anhydride	***Collins oxidation***	12
$pyHCrO_3Cl$	Pyridinium chlorochromate	PCC	13
$pyHCrO_3F$	Pyridinium fluorochromate	PFC	14
$(pyH)_2Cr_2O_7$	Pyridinium dichromate	PDC	15
$(C_5H_4N)_2HCrO_3Cl$ [$bpyHCrO_3Cl$]	Bipyridyl chlorochromate	BPCC	16
$Me_3SiOCrO_2Cl$	Trimethylsilyl chlorochromate	TMSCC	17

9. Brown, H.C.; Garg, C.P.; Liu, K.-T. *J. Org. Chem.* **1971**, *36*, 387.
10. (a) Bowden, K.; Heilbron, I.M.; Jones, E.R.H.; Weedon, B.C.L. *J. Chem. Soc.* **1946**, 39. (b) Meinwald, J.; Cranfall, J.; Hymans, W.E. *Org. Synth.* **1965**, *45*, 77.
11. Poos, G.I.; Arth, G.E.; Beyler, R.E.; Sarett, L.H. *J. Am. Chem. Soc.* **1953**, *75*, 422.
12. (a) Collins, J.C.; Hess, W.W.; Frank, F.J. *Tetrahedron Lett.* **1968**, 3363. (b) Ratcliffe, R.; Rodehorst, R. *J. Org. Chem.* **1970**, *35*, 4000. (c) Collins, J.C.; Hess, W.W. *Org. Synth.* **1972**, *52*, 5. (d) Ratcliffe, R. *Org. Synth.* **1975**, *55*, 84.
13. (a) Corey, E.J.; Suggs, J.W. *Tetrahedron Lett.* **1975**, 2647. (b) Piancatelli, G.; Scettri, A.; D'Auria, M. *Synthesis* **1982**, 245. (c) Brown, H.C.; Rao, C.G.; Kulkarni, S.U. *J. Org. Chem.* **1979**, *44*, 2809.
14. (a) Bhattacharjee, M.N.; Chaudhuri, M.K.; Dasgupta, H.S.; Roy, N.; Khathing, D.T. *Synthesis*, **1982**, 588. (b) Lewis, D.E.; Rigby, H.L. *Tetrahedron Lett.* **1985**, *26*, 3437.
15. (a) Coates, W.M.; Corrigan, J.R. *Chem. Ind.* **1969**, 1594. (b) Corey, E.J.; Schmidt, G. *Tetrahedron Lett.* **1979**, 399.
16. Guziec, F.S., Jr.; Luzzio, F.A. *Synthesis* **1980**, 691.
17. Aizpurua, M.; Palomo, C. *Tetrahedron Lett.* **1983**, *24*, 4637.
18. Corey, E.J.; Ensley, H.E.; Suggs, J.W. *J. Org. Chem.* **1976**, *41*, 380.
19. Corey, E.J.; Schmidt, G. *Tetrahedron Lett.* **1979**, 399.
20. Corey, E.J.; Schmidt, G. *Tetrahedron Lett.* **1979**, 399
21. Czernecki, S.; Vijayakumaran, K.; Ville, G. *J. Org. Chem.* **1986**, *51*, 5472.

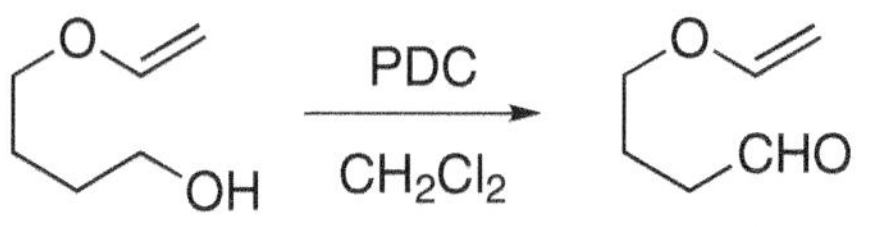

(18.35)

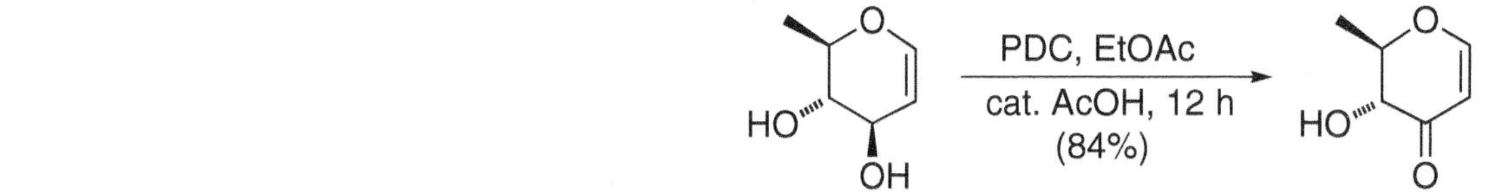

(18.36)

Manganese-Based Reagents

Two manganese-based reagents are commonly used for the oxidation of alcohols: potassium permanganate, which contains Mn(VII), and manganese dioxide, which contains Mn(IV). More recently, barium manganate, $BaMnO_4$, which contains Mn(VI), has been used as an alternative to manganese dioxide.

Potassium permanganate is one of the most powerful of the common oxidizing agents, and oxidations with this reagent are often difficult to control. The oxidation of alcohols by potassium permanganate occurs reasonably readily, although it is known that other functional groups, such as alkene π bonds and aldehydes, are actually oxidized more rapidly then the alcohol. That potassium permanganate can oxidize alkene π bonds faster than a primary hydroxyl group is illustrated by the oxidation of methyl cinnamate in 90% ethanol (Example **18.37**), where the final product is isolated in 67% yield, despite the potential for competition for the oxidizing agent from the solvent.[22] Aldehydes are usually oxidized much more rapidly by permanganate than alcohols, so potassium permanganate oxidizes primary alcohols to carboxylic acids (Example **18.38**).[23]

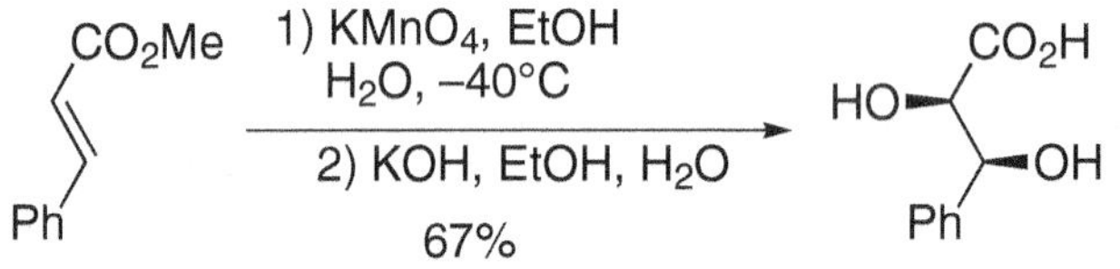

(18.37)

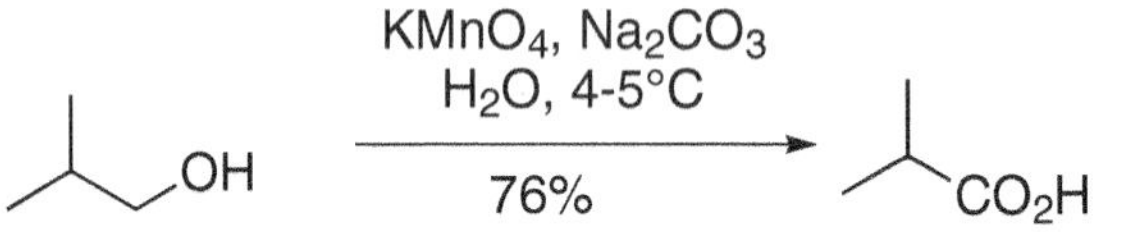

(18.38)

Bonds at allylic positions and α to a carbonyl group tend to be unusually reactive, and this extends to allylic C—H bonds. Allyl alcohols can often be oxidized under very mild conditions. Activated manganese dioxide, for example, selectively oxidizes allyl and benzyl alcohols to the conjugated aldehydes and ketones (e.g., Examples **18.39** to **18.41**). Because this reagent does not rapidly oxidize saturated alcohols, it can be used to oxidize allyl or benzyl hydroxyl groups in the presence of saturated primary and secondary hydroxyl groups.[24] However, this reagent is not always straightforward to use for two

---

22. Riiber, C.N. *Ber. dtsch. chem. Ges.* **1915**, *48*, 823.

23. It has been reported that this can be reversed by using potassium permanganate and copper (II) sulfate pentahydrate on alumina in the absence of a solvent [Shaabani, A.; Lee, D.G. *Tetrahedron Lett.* **2001**, *42*, 5833]. The alumina has the effect of reversing the normal order of ease of oxidation, making the alcohol more easily oxidized. The reaction in the absence of alumina leads to a mixture of aldehyde and carboxylic acid.

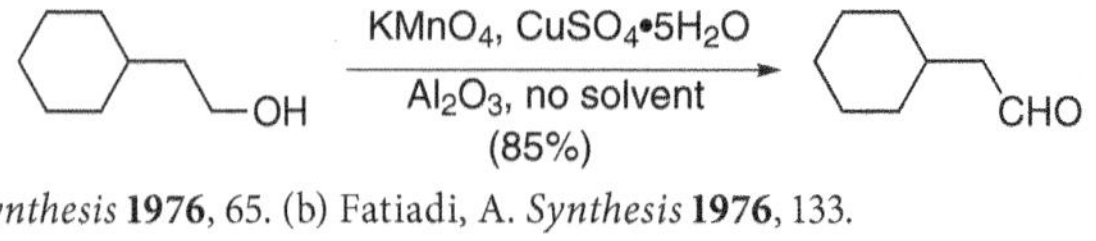

24. (a) Fatiadi, A. *Synthesis* **1976**, 65. (b) Fatiadi, A. *Synthesis* **1976**, 133.

reasons: (1) its activity depends on its precise mode of preparation and drying[25] and (2) large excesses of the reagent are sometimes needed.

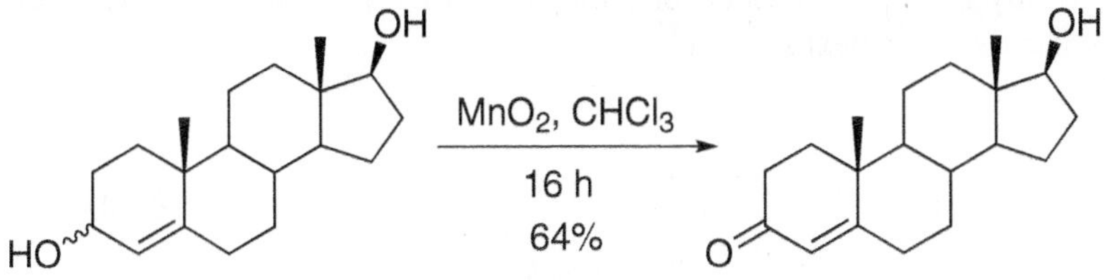

(18.39)[26]

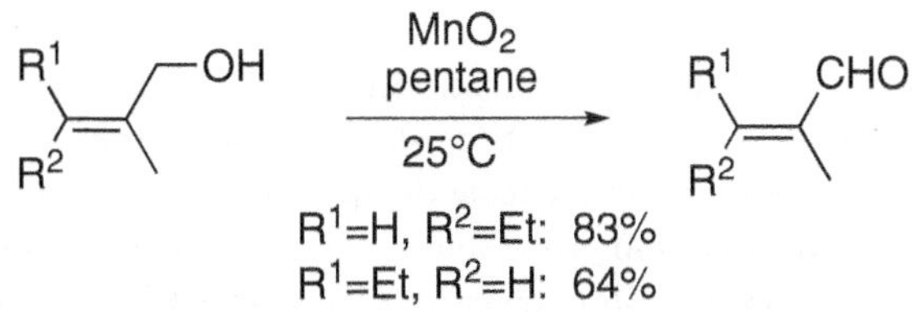

(18.40)[27]

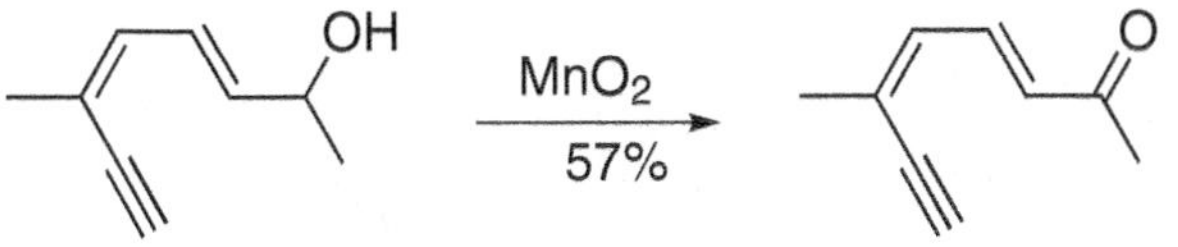

(18.41)[28]

A reagent with similar selectivity is barium manganate,[29] an insoluble compound formed by the reduction of permanganate ion with iodide ion in the presence of barium chloride. This reagent selectively oxidizes allyl and benzyl hydroxyl groups (e.g., Example **18.42**) but is more reproducibly made and used than manganese dioxide. Allylic 1,4-diols can be oxidized to lactones (e.g., Example **18.43**) by large amounts of the reagent.[30]

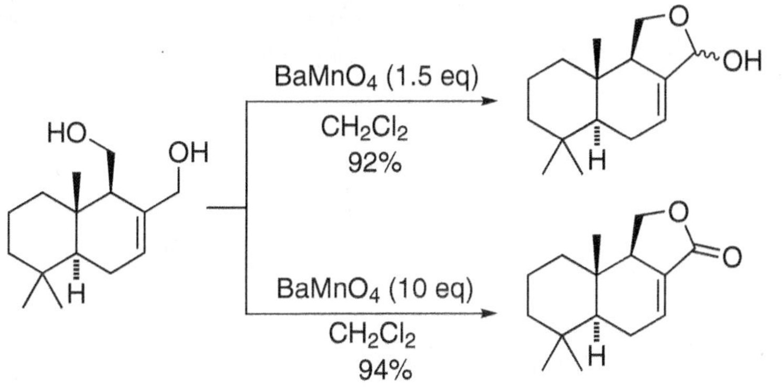

(18.42)

(18.43)

Other Transition Metal-Based Oxidants

Tetrapropylammonium perruthenate[31] (TPAP, $Pr_4N^+RuO_4^-$, Ley's reagent]) rapidly and selectively oxidizes primary alcohols to aldehydes, as illustrated by Example **18.44**.[32] It can

25. (a) Attenburrow, J.; Cameron, A.F.B.; Chapman, J.H.; Evans, R.M.; Hems, B.A.; Jansen, A.B.A.; Walker, T. *J. Chem. Soc.* **1952**, 1094. (b) Goldman, I.M. *J. Org. Chem.* **1969**, *34*, 1979.

26. Johnson, W.S.; Vredenburgh, W.A.; Pike, J.E. *J. Am. Chem. Soc.* **1960**, *82*, 3409.

27. Chan, K.C.; Jewel, R.A.; Nutting, W.H.; Rapoport, H. *J. Org. Chem.* **1968**, *33*, 3382.

28. Boeckmann, R.K., Jr.; Enholm, E.J.; Demko, D.M.; Charette, A.B. *J. Org. Chem.* **1986**, *51*, 4743

29. Firouzabadi, H.; Ghaderi, E. *Tetrahedron Lett.* **1978**, 839.

30. (a) Garigipati, R.S.; Freyer, A.J.; Whittle, R.R.; Weinreb, S.M. *J. Am. Chem. Soc.* **1984**, *106*, 7861. (b) Hollinshead, D.M.; Howell, S.C.; Ley, S.V.; Mahon, M.; Ratcliffe, N.M.; Worthington, P.A. *J. Chem. Soc., Perkin Trans. 1* **1983**, 1579.

31. (a) Griffith, W.P.; Ley, S.V.; Whitcombe, G.P.; White, A.D. *J. Chem. Soc., Chem. Commun.* **1987**, 1625. (b) Ley, S.V.; Norman, J.; Griffith, W.P.; Marsden, S.P. *Synthesis* **1994**, 639. (c) Markó, I.E.; Giles, P.R.; Tsukazaki, M.; Chellé-Regnaut, I.; Urch, C.J.; Brown, S.M. *J. Am. Chem. Soc.* **1997**, *119*, 12661. (d) Hinzen, B.; Lenz, R.; Ley, S.V. *Synthesis* **1998**, 977.

32. Yang, Z.; Attygale, A.B.; Meinwald, J. *Synthesis* **2000**, 1936.

be used catalytically with a co-oxidant such as *N*-methylmorpholine *N*-oxide (NMMO) so that the amount of the relatively expensive ruthenium reagent needed is minimized. TPAP oxidation frequently gives higher yields of product than the popular Swern oxidation,[33] which we will discuss in the next section.

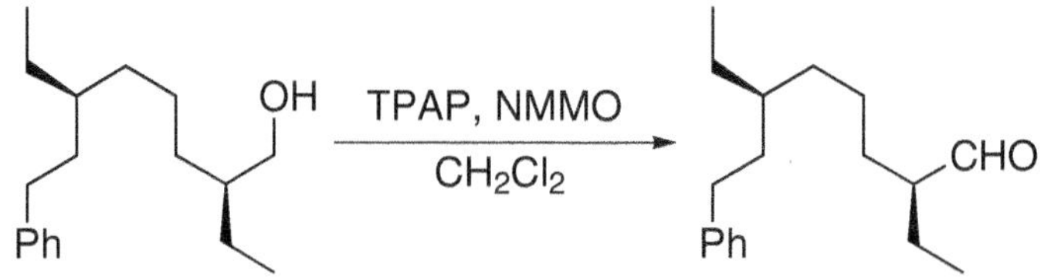

(18.44)

In a series of papers beginning in 1968, Fétizon introduced silver carbonate on celite (the ***Fétizon reagent***) as a mild oxidizing agent for alcohols and lactols.[34] The reagent is expensive but is particularly useful for small-scale oxidations due to the ease of workup (simple filtration of the reagent suffices to give a solution of the product that can be evaporated). The final step in Grieco's synthesis of quassin (Example **18.45**)[35] illustrates the use of this reagent in sensitive systems.

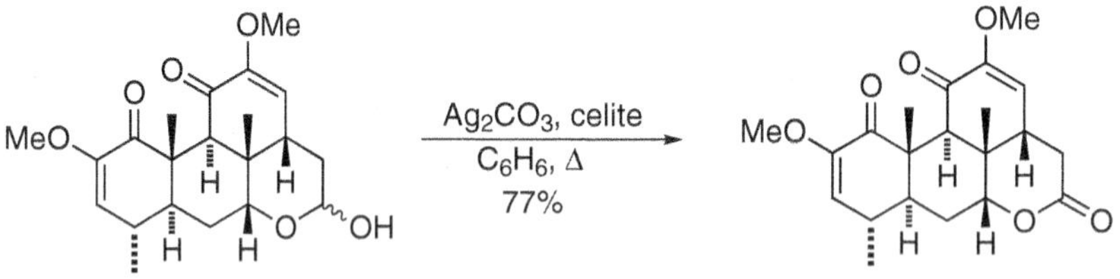

(18.45)

Vanadium (V) compounds have long been used in industry as oxidation catalysts. (Their use in the epoxidation of alkenes will be discussed at length in the next chapter.) Recently, however, catalytic methods that use vanadium (V) catalysts and molecular oxygen as the stoichiometric oxidant have been developed for the oxidation of alcohols. Vanadium pentoxide in refluxing toluene with potassium carbonate as a buffer exhibits a preference for the oxidation of secondary alcohols or benzylic alcohols, as was shown by a competition between cyclohexanol and 1-heptanol (Example **18.46**).[36]

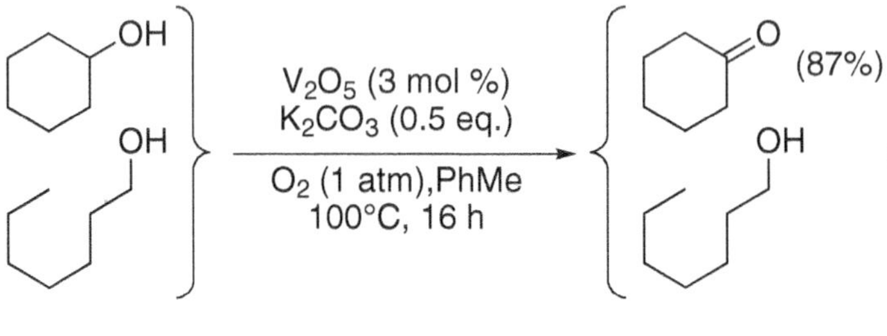

(18.46

Oxidation of Tertiary Allylic Alcohols to Ketones: Transposition of the Carbonyl Group

The oxidation of saturated tertiary alcohols by chromic acid requires forcing conditions and is preceded by dehydration, so that the alkene is the compound actually oxidized. β,γ-Unsaturated tertiary alcohols carrying a hydrogen at the γ carbon, however, are

33. Griffith, W.P.; Ley, S.V. *Aldrichimica Acta* **1990**, *23*, 13.

34. (a) Fétizon, M.; Golfier, M. *Compt. rend. (C)* **1968**, *267*, 900. (b) Fétizon, M.; Golfier, M.; Louis, J.-M. *Chem. Commun.* **1969**, 1102. (c) Fétizon, M.; Golfier, M.; Louis, J.-M. *Chem. Commun.* **1969**, 1118. (d) Kakis, F.J.; Fétizon, M.; Douchkine, N.D.; Golfier, M.; Mourgues, P.; Prange, T. *J. Org. Chem.* **1974**, *39*, 523. (e) Fétizon, M.; Moreau, N. *Compt. rend. (C)* **1972**, *275*, 621. (f) Bastard, J.; Fétizon, M.; Gramain, J.C. *Tetrahedron* **1973**, *29*, 2867.

35. Grieco, P.A.; Ferriño, S.; Vidari, G. *J. Am. Chem. Soc.* **1980**, *102*, 7586.

36. Velusamy, S.; Punniyamurthy, T. *Org. Lett.*, **2004**, *6*, 217.

smoothly oxidized by less acidic chromium (VI) reagents (which do not catalyze the dehydration of the alcohol) on prolonged exposure (e.g., Examples **18.47** and **18.48**).[37]

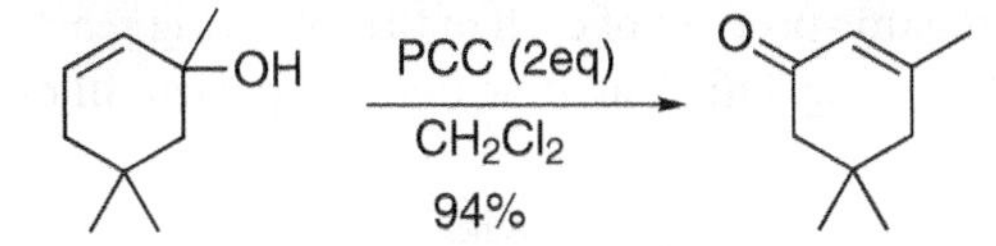

(18.47)

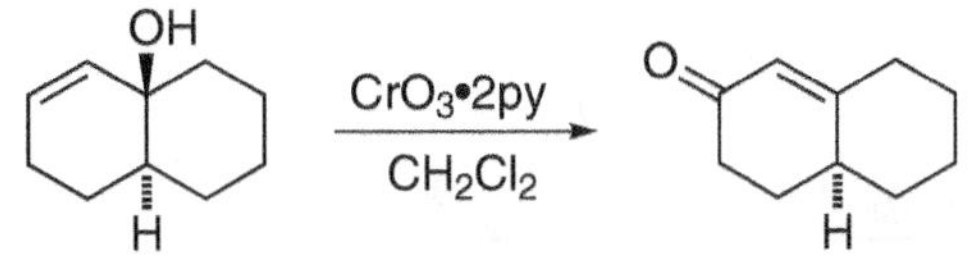

(18.48)

These oxidations occur with allylic rearrangement: during the reaction, the oxygen atom in chromate ester **18.50** becomes bonded to the other end of the allylic system to give the molecule in Example **18.51**, which then completes the oxidation to give the rearranged ketone **18.52**. The reaction is occasionally known as the ***Babler oxidation***. The rearrangement step probably occurs at the chromate ester stage of the reaction; its mechanism has not yet been unambiguously determined, but it most likely involves the carbocation formed by heterolysis of the carbon-oxygen bond. It may also be a [3,3] sigmatropic rearrangement of the chromate ester.

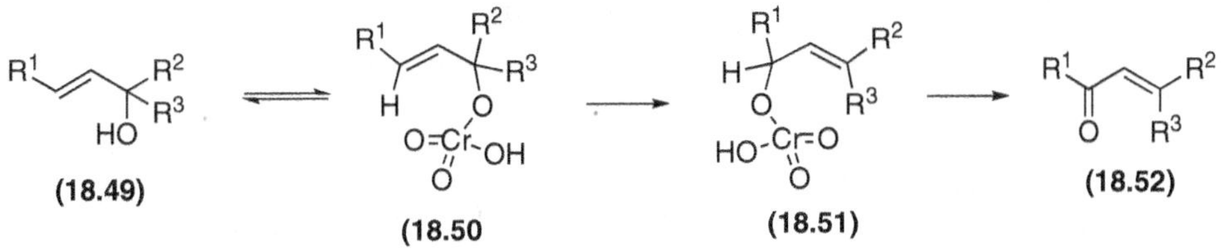

Worked Problem

18-1 What is the major organic product obtained from the oxidation of the alcohol below with each of the reagents that follow? Where appropriate, give your reasons.

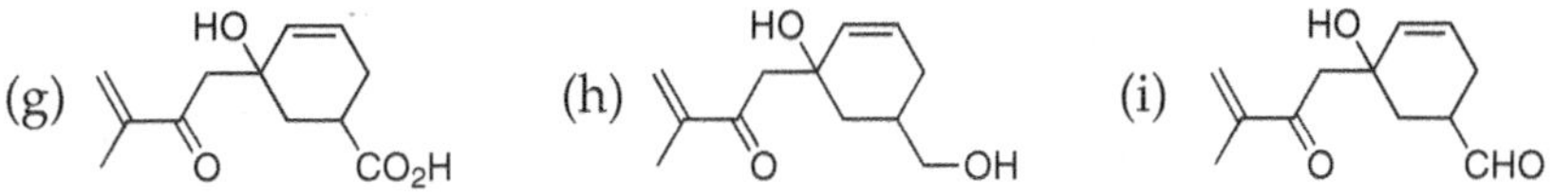

(a) PCC, CH_2Cl_2, 25°C, 30 min

(b) MnO_2, pentane, 25°C

(c) $BaMnO_4$, CH_2Cl_2

(d) TPAP, NMMO, CH_2Cl_2, 25°C

(e) PCC, CH_2Cl_2, 25°C, 24 h

(f) Ag_2CO_3, celite, PhH, Δ

(g) CrO_3, H_2SO_4, H_2O, acetone

(h) MnO_2, NaCN, MeOH

(i) PDC (1 eq), CH_2Cl_2, 30 min

§Answers on the next page.

37. (a) Babler, J.H.; Coghlan, M.J. *Synth. Commun.* **1976**, *6*, 469. (b) Dauben, W.G.; Michno, D.M. *J. Org. Chem.* **1977**, *42*, 682. (c) Sundaraman, P.; Herz, W. *J. Org. Chem.* **1977**, *42*, 813.

Problem

18-5 What is the major organic product of each of the following reactions? Where the quantity of reagent is not specified, assume that it is present in excess.

(a) 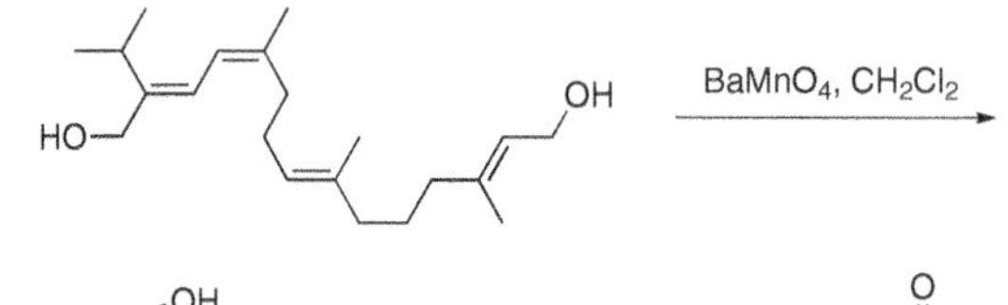

(b) 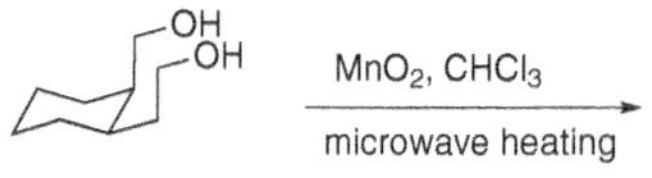

(c) 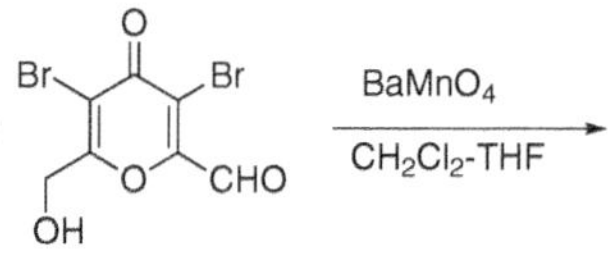

(d) 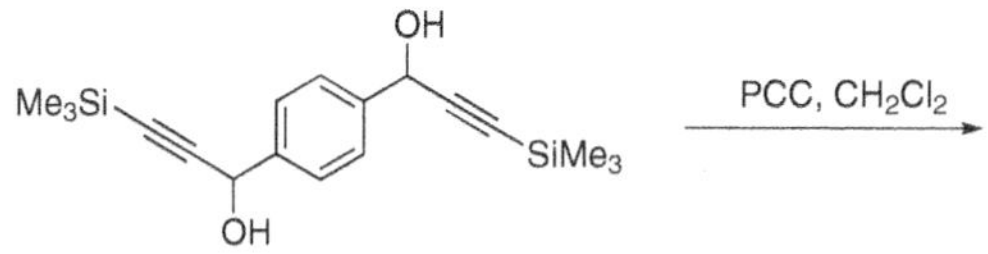

(e) 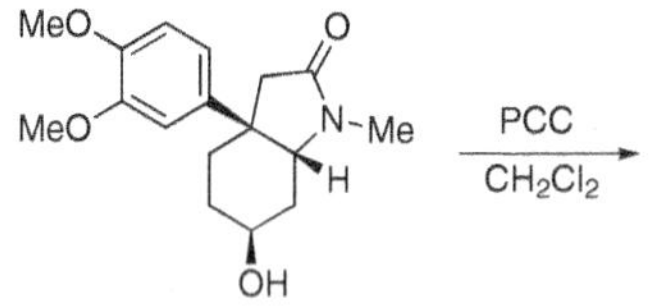

(f) 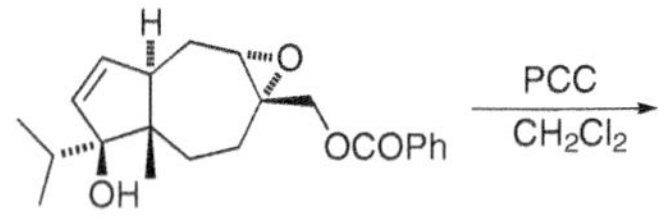

(g) 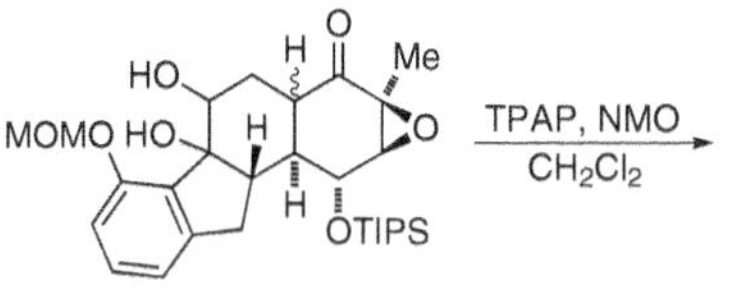

(h) 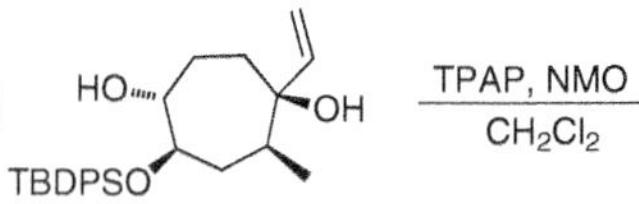

References for problems: (a) *Tetrahedron Lett.* **1989**, *30*, 1173. (b) *Tetrahedron Lett.* **2009**, *50*, 6823. (c) *Can. J. Chem.* **2007**, *85*, 352. (d) *J. Org. Chem.* **2003**, *68*, 1339. (e) *J. Org. Chem.* **1982**, *47*, 1302. (f) *J. Org. Chem.* **2010**, *75*, 8337. (g) *J. Am. Chem. Soc.* **2008**, *130*, 2783. (h) *Org. Lett.* **2008**, *10*, 1247.

....

§ Answers to Worked Problem:

(a) 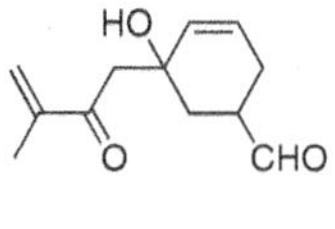

(b) 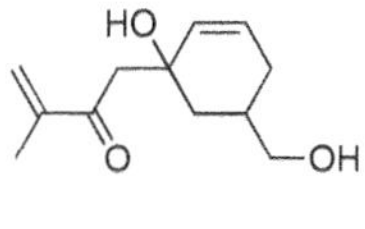

(c) 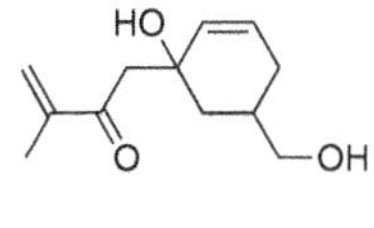

(d) 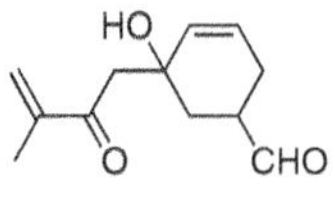

(e) 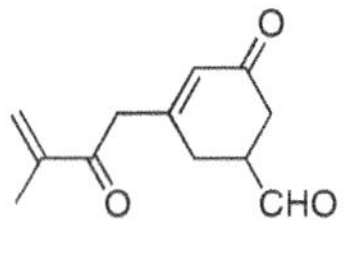

(f) 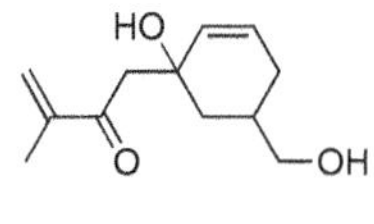

(g) 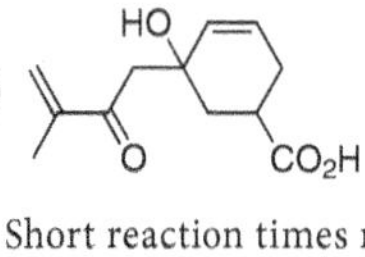

(h) 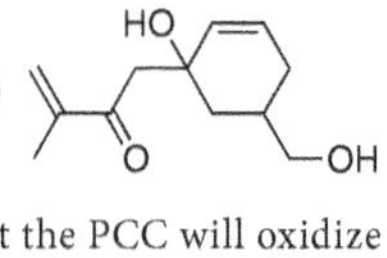

(i) 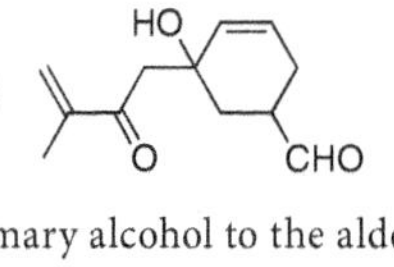

Reasons: (a) Short reaction times mean that the PCC will oxidize the primary alcohol to the aldehyde and the secondary alcohol to the ketone, but no further transformation will occur. (b) and (c) Only the secondary allylic alcohol will oxidize. (d) TPAP will oxidize the primary alcohol to the aldehyde and the secondary alcohol to the ketone. (e) In this reaction, the PCC will oxidize the primary and secondary alcohol groups to the aldehyde and ketone, respectively, but the prolonged reaction time will also permit the Babler oxidation of the tertiary allylic alcohol. (f) and (h) The reagent is selective for oxidation of allylic alcohols. (g) Jones reagent is aqueous and acidic; it will lead to the carboxylic acid from the primary alcohol and will oxidize the secondary alcohol to the ketone. (i) PDC is a less aggressive reagent than PCC, so it will oxidize the primary and secondary alcohols, but not the tertiary allylic alcohol.

18.4 Oxidations of Alcohols Using Non-Metal-Based Reagents

Although they are convenient to use, metal-based oxidants are sometimes much too powerful to use in the synthesis of sensitive molecules. Not only can they oxidize alcohols to aldehydes and ketones, but they can also oxidize sulfur-based functional groups (e.g., thioacetals and thioketals), and they can cause oxidative cleavage of carbon-carbon bonds when there are two oxidizable functional groups sufficiently close to each other (as in vicinal diols, for example, which are cleaved by Cr (VI) reagents). As an alternative to high-valent metal oxidants, several oxidation methods based on non-metal oxides (particularly oxides of sulfur, iodine and nitrogen) as the oxidizing agent have been developed. More recently, catalytic methods with molecular oxygen as the terminal oxidant have come under fairly intense study.[38]

Oxidations Based on Sulfoxides[39]

The earliest of the oxidation methods based on non-metal oxides is the oxidation of alcohols with dimethyl sulfoxide, for which a variety of reagents are now available (Table 18.5).

Each of these reactions follows the same basic protocol: a solution of the alcohol in dimethyl sulfoxide is first treated with a dehydrating agent, and then with a base. The active oxidizing agent in all these oxidations is a trivalent sulfonium ion carrying a sulfur-heteroatom bond (or a sulfur-carbon bond in a sulfonium amidine). This ion is formed by the attack of the dimethyl sulfoxide on the dehydrating agent. This sulfonium ion reacts with the alcohol by an addition-elimination mechanism or in what may be direct S_N2 displacement at sulfur to give a sulfoxonium ion derived from the alcohol (e.g., Example **18.53**). E2 elimination of dimethyl sulfide as the leaving group by the base gives the carbonyl compound.[43]

Table 18.5 Oxidizing Agents Based on Dimethyl Sulfoxide

Reagent	Name of Reaction
Ac_2O, Me_2SO, 25°C	
$C_6H_{11}N{=}C{=}NC_6H_{11}$, CF_3CO_2H, C_6H_6, Me_2SO, py, 25°C	
$C_6H_{11}N{=}C{=}NC_6H_{11}$, H_3PO_4, Me_2SO, 25°C	***Moffatt***(or ***Moffatt-Pfitzner***)[40]
py•SO_3, Et_3N, Me_2SO, 25°C	***Parikh-Doering***[41]
$(COCl)_2$, Et_3N, Me_2SO, CH_2Cl_2, −78°C	***Swern***[42]

$$Me_2SO + (COCl)_2 \rightarrow \mathbf{Me_2S^+}{-}Cl + CO + CO_2 + Cl^-$$

$$Me_2SO + RN{=}C{=}NR + H^+ \rightarrow \mathbf{Me_2S^+}{-}[C(NHR){=}NR]$$

$$Me_2SO + Ac_2O \rightarrow \mathbf{Me_2S^+}{-}OAc + OAc^-$$

$$Me_2SO + SO_3 \rightarrow \mathbf{Me_2S^+}{-}OSO_3^-$$

38. (a) Lee, A.F.; Gee, J.J.; Theyers, H.J. *Green Chem.* **2000**, *2*, 279. (b) Mattal, T.; Baiker, A. *Chem. Rev.* **2004**, *104*, 3037. (c) Gangwal, V.R.; van der Schaaf, J.; Kuster, B.F.M.; Schouten, J.C. *J. Catalysis* **2005**, *232*, 432.

39. Reviews: (a) Mancuso, A.J.; Swern, D. *Synthesis* **1981**, 165. (b) Tidwell, T.T. *Org. React.* **1990**, *39*, 297. (c) Tidwell, T.T. *Synthesis* **1990**, 857. (d) Moffatt, J.G. In Augustine, R.L.; Trecker, D.J., Eds. *Oxidation* (Marcel Dekker: New York, 1971), p. 1.

40. (a) Pfitzner, K.E.; Moffatt, J.G. *J. Am. Chem. Soc.* **1965**, *87*, 5661. (b) Albright, J.D.; Goldman, L. *J. Am. Chem. Soc.* **1967**, *89*, 2416.

41. Parikh, J.R.; Doering, W.v.E. *J. Am. Chem. Soc.* **1963**, 89, 5505.

42. (a) Mancuso, A.J.; Huang, S.-L.; Swern, D. *J. Org. Chem.* **1978**, *43*, 2480. (b) Omura, K.; Swern, D. *Tetrahedron*, **1978**, *34*, 1651.

Daniel A. Swern (1916–1982) was educated at the City College of New York, Columbia University, and the University of Maryland. In 1937, he joined the East Regional Research Center of the US Department of Agriculture in Philadelphia, serving as a research chemist here until 1963, when he was appointed to the faculty of Temple University. Swern was an eminent lipid chemist, and published more than 280 books and articles; among other awards, he received the 1968 AOCS Award in Lipid Chemistry.

43. This mechanistic pathway was first proposed by: Albright, J.D.; Goldman, L. *J. Am. Chem. Soc.* **1967**, *89*, 2146.

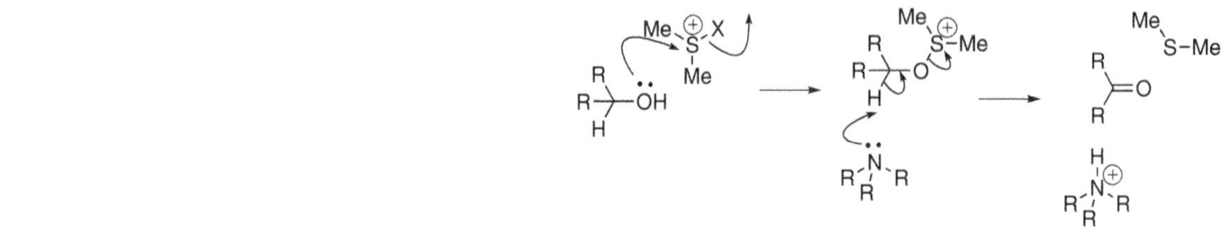

(18.53)

The Swern oxidation is by far the most widely used of these reactions. It is admirably suited to the synthesis of aldehydes containing both sensitive functionality and labile stereochemistry, as is illustrated by Examples **18.54** and **18.55** of Swern oxidations used in Evans' total synthesis of rutamycin B.[44] In this synthesis, the retention of labile stereocenters adjacent to aldehyde carbonyl groups was essential, and the Swern oxidation was the reaction of choice for forming the required aldehydes. Ireland[45] has used the Swern oxidation as a method for the manipulation of highly reactive carbonyl compounds by trapping the crude product of the Swern oxidation in situ with nucleophiles (Examples **18.56** and **18.57**).

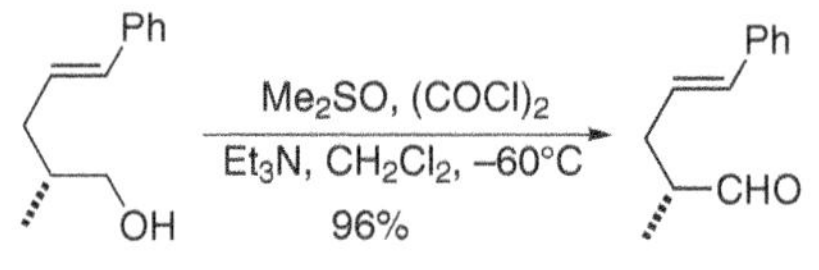

(18.54)

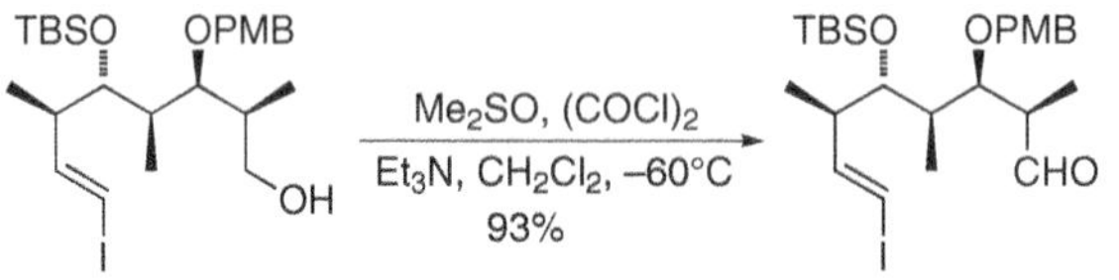

(18.55)

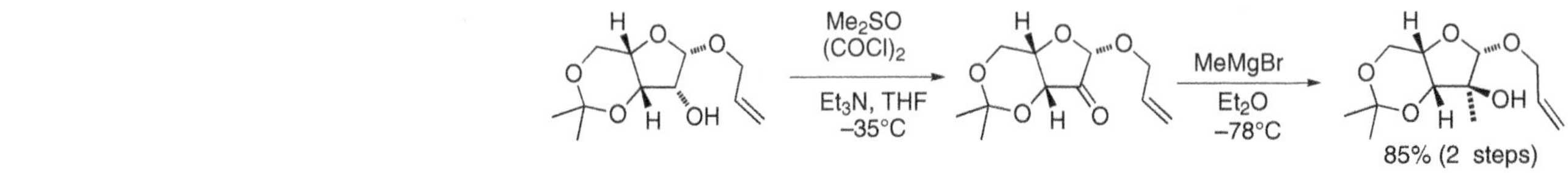

(18.56)

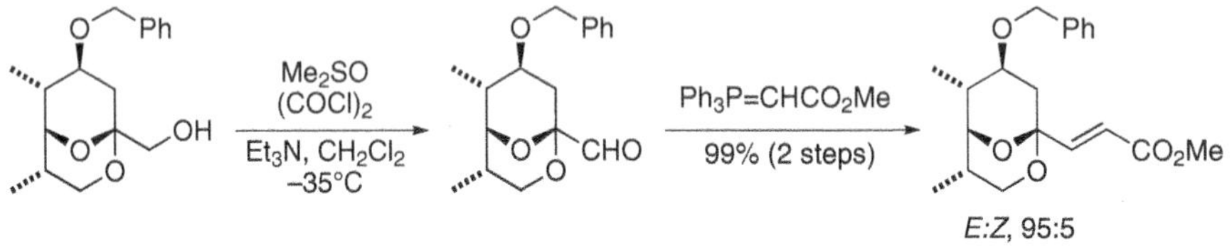

(18.57)

The involvement of the chlorodimethylsulfonium ion in the Swern oxidation led a number of workers to experiment with other sulfonium ions carrying a leaving group at sulfur. The most widely used of these new methods for oxidation is the ***Corey-Kim oxidation***,[46] in which a mixture of dimethyl sulfide and *N*-chlorosuccinimide, which generates the actual oxidant, *N*-(dimethylsulfonium)succinimide. This reaction is an excellent method for the selective oxidation of primary alcohols to aldehydes, but the need to use dimethyl sulfide has inhibited its widespread use. More recently, efforts have been focused on replacing dimethyl sulfide with more benign, less volatile sulfides (e.g., decyl methyl sulfide, as in Example **18.58**).[47] A fluorous analog of the reaction has also been developed.[48]

44. Evans, D.A.; Ng, H.P.; Rieger, D.L. *J. Am. Chem. Soc.* **1993**, *115*, 11446.

45. Ireland, R.E.; Norbeck, D.W. *J. Org. Chem.* **1985**, *50*, 2198.

46. (a) Corey, E.J.; Kim, C.U. *J. Am. Chem. Soc.* **1972**, *94*, 7586. (b) Corey, E.J.; Kim, C.U. *Tetrahedron Lett.* **1974**, *15*, 287. (c) Katayama, S.; Fukuda, K.; Watanabe, T.; Yamauchi, M. *Synthesis* **1988**, 178. (d) Review: Tidwell, T.T. *Synthesis* **1990**, 857. (e) Pulkkinen, J.T.; Vepsäläinen, J.J. *J. Org. Chem.* **1996**, *61*, 8604.

47. (a) Ohsugia, S.-I.; Nishidea, K.; Oonob, K.; Okuyamab, K.; Fudesakaa, M.; Kodamaa, S.; Node, M. *Tetrahedron* **2003**, *59*, 8393. (b) Crich, D.; Neelamkavil, S. *Tetrahedron* **2002**, *58*, 3865.

48. Crich, D.; Neelamkavil, S. *Tetrahedron* **2002**, *58*, 3865.

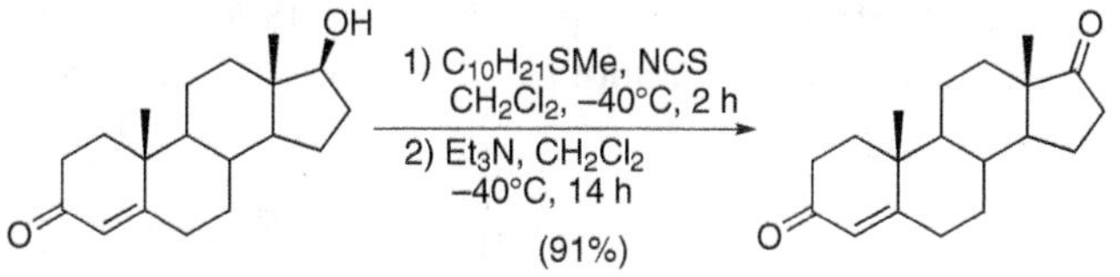

(18.58)

The Parikh-Doering oxidation can be carried out at room temperature, which makes it especially convenient. This has led to a resurgence of its popularity in recent years, as illustrated by Examples **18.59**[49] and **18.60**.[50]

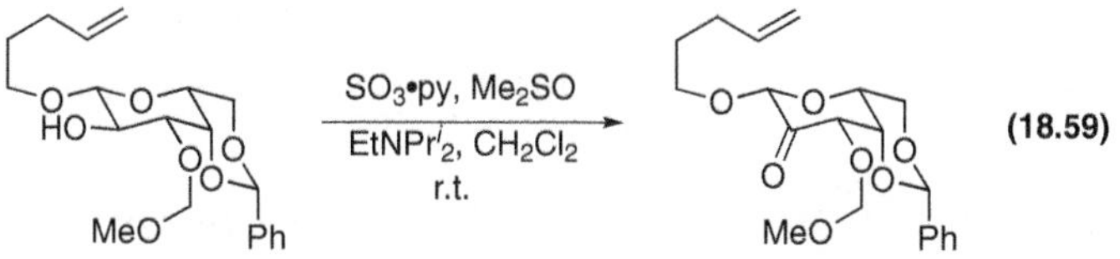

(18.59)

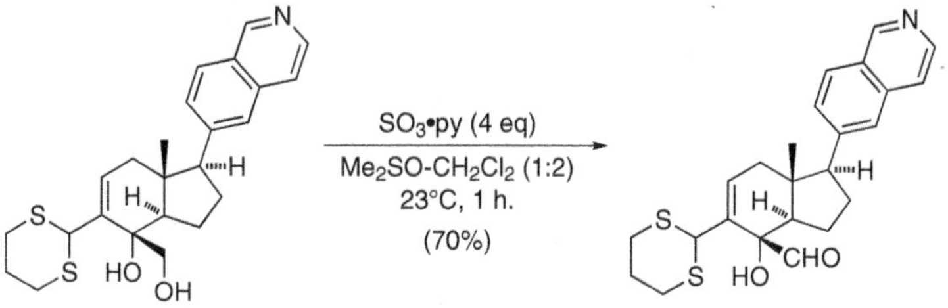

(18.60)

Oxidations with the Dess-Martin Periodinane[51,52]

Among the more useful reagents for the oxidation of alcohols under mild conditions are the ***Dess-Martin periodinane*** (Example **18.61;** DMP),[53] a hypervalent iodine species readily prepared from *o*-iodobenzoic acid,[54] and the related *o*-iodoxybenzoic acid (Example **18.62**; IBX). Of the two, DMP is easier to use because of its greater solubility in organic solvents.

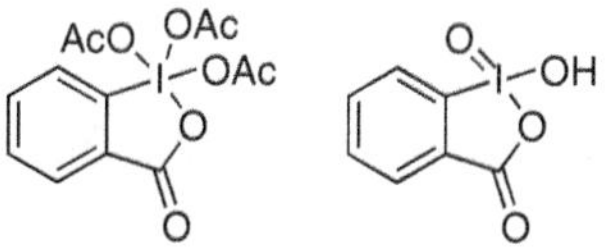

DMP **(18.61)**   IBX **(18.62)**

Treatment of an alcohol with DMP in dichloromethane gives the carbonyl compound (Example **18.64**) via an oxygen-linked periodinane (Example **18.63**). The reaction rate is markedly accelerated by the presence of small amounts of water (e.g., Example **18.66**).[55] The reaction is particularly suited to the oxidation of compounds bearing a wide range of

---

49. Veitch, G.E.; Beckmann, E.; Burke, B.J.; Boyer, A.; Maslen, S.L.; Ley, S.V. *Angew. Chem. Int. Ed.* **2007**, *46*, 7629.

50. Nicolaou, K.C.; Peng, X.-S.; Sun, Y.-P.; Polet, D.; Zou, B.; Lim, C.S.; Chen, D.Y.-K. *J. Am. Chem. Soc.* **2009**, *131*, 10587.

51. (a) Dess, D.B.; Martin, J.C. *J. Am. Chem. Soc.* **1991**, *113*, 7277. (b) Boeckmann, R.K., Jr.; Shao, P.; Mullins, K. *Org. Synth.* **2000**, *77* 141.

52. Reviews: (a) Stang, P.J.; Zhdankin, V.V. *Chem. Rev.* **1996**, *96*, 1123. (b) Kitamura, T.; Fujiwara, Y. *Org. Prep. Proc. Int.* **1997**, *29*, 409.

53. (a) Dess, D.B.; Martin, J.C. *J. Org. Chem.* **1983**, *48*, 4155.

54. The Dess-Martin periodinane has been reported to be shock-sensitive under some conditions, and to explode above 200°C under confined conditions: Plumb, J.B.; Harper, D.J. *Chem. Eng. News* **1990** *68* (No. 28, July 16), p. 3.

For an improved synthesis of this reagent, see: (c) Ireland, R.E.; Liu, L.J. *J. Org. Chem.* **1993**, *58*, 2899. (d) Frigerio, M.; Santagostino, M.; Sputore, S. *J. Org. Chem.* **1999**, *64*, 4537.

James Cullen Martin (1928–1999) was educated at Vanderbilt University (MS, 1952) and Harvard University (PhD, 1956). He served on the faculty at Illinois (1956–1985) and Vanderbilt (1985–1992). For more detail, see: Akiba, K.-Y. *Phosphorus, Sulfur, Silicon Relat. Elem.* **2006**, *181*, 1201.

55. Meyer, S.D.; Schreiber, S. *J. Org. Chem.* **1994**, *59*, 7549.

sensitive functionalities, as illustrated by Examples **18.67** and **18.68**.[56] Ionic liquids have also been shown to be excellent solvents for both DMP and IBX oxidations of a wide variety of alcohols.[57]

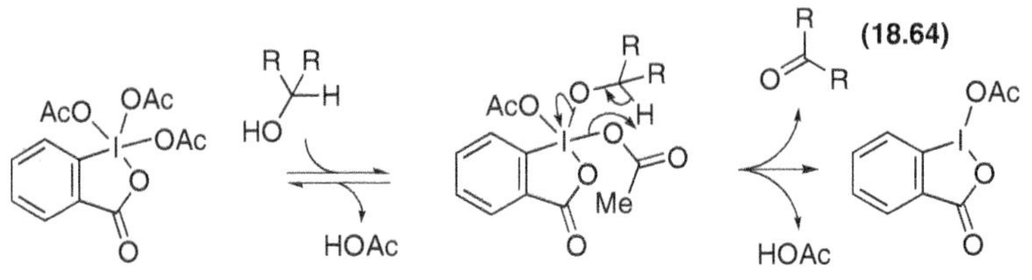

(18.61) (18.63) (18.64) (18.65)

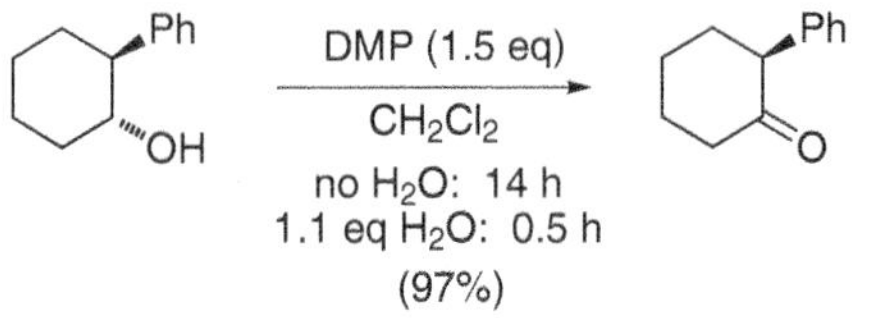

(18.66)

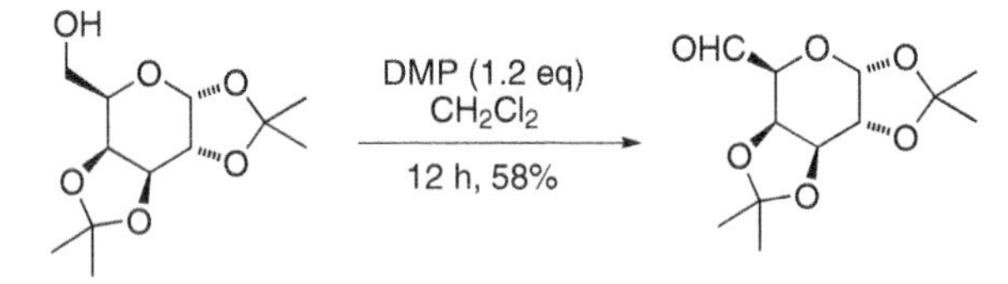

(18.67)

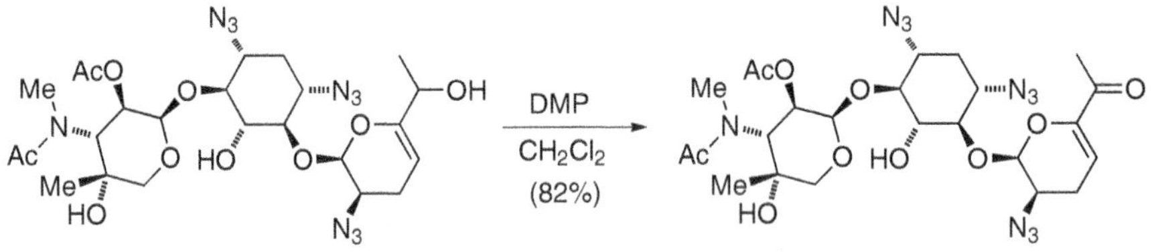

(18.68)

Oxidation by Hydride Transfer to an Acceptor π Bond

Aluminum-Based Reagents

Aluminum alkoxides react with carbonyl compounds to give products involving hydrogen transfer from carbon, with the concomitant generation of carbonyl compounds. The first observation of this type of reactivity in aluminum compounds was by Russian chemist V. E. Tishchenko.[58] He published a series of eight papers (four in Russian, four in German) in 1906, where he described the conversion of aromatic aldehydes (Example **18.69**) to benzyl benzoates (Example **18.71**) catalyzed by aluminum ethoxide.[59] The reaction is now known as the ***Tishchenko reaction.*** It is formally a disproportionation of the aldehyde with the first molecule of aldehyde transferring a hydrogen atom to the second and probably involves a hemiacetal derivative of the type shown as Example **18.70**.

---

56. Hanessian, S.; Szychowski, J.; Pablo, J. *Org. lett.* **2009**, *11*, 429.

57. Yadav, J.S.; Reddy, B.V.S.; Basak, A.K.; Venkat Narsaiah, A. *Tetrahedron* **2004**, *60*, 2131, and references therein.

58. **Vyacheslav Yevgenievich Tishchenko (1861–1941)** graduated from St. Petersburg in 1883 and was appointed lecturer at the university in 1891. Much of his research was in applied organic chemistry. For more detail, see: Lewis, D.E. *Early Russian Organic Chemists and Their Legacy* (Springer: Heidelberg, 2012), p. 114.

59. (a) Tishchenko, V.E. *Zh. Russ. Fiz.-Khim. Obshch.* **1906**, *38*, 355, 482, 540, 547. (b) Tishchenko, V. *Chem. Zentr.* **1906**, *II*, 1309, 1552, 1555,1556.

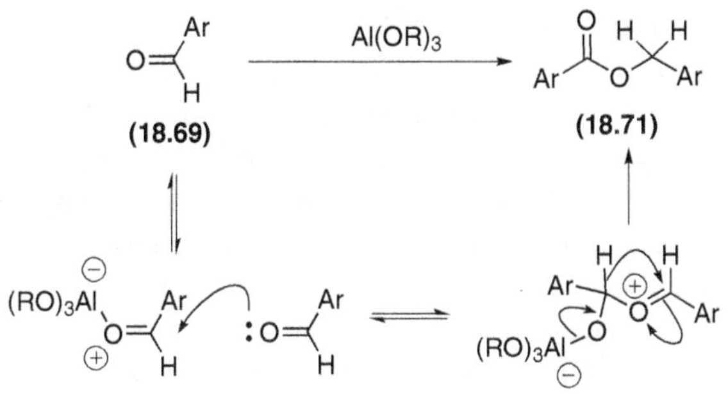

(18.69) (18.71) (18.70)

In 1937, Rupert Viktor Oppenauer[60] published a method for the oxidation of alcohols by heating their aluminum salts with acetone.[61] The Oppenauer oxidation occurs by means of hydride transfer from the aluminum alkoxide (Example **18.72**) to the carbonyl compound through a cyclic, six-membered transition state (Example **18.73**). The reaction is reversible, and the position of the equilibrium is determined by the reduction potentials of the two carbonyl compounds.[62] Carbonyl acceptors (i.e., the oxidizing participant in the reaction) are most effective if the groups bonded to the carbonyl carbon are electron-withdrawing. The reaction tolerates a wide range of functional groups. In oxidations of homoallylic alcohols such as cholesterol, the migration of the alkene double bond into the position of conjugation is usually observed.[63] This isomerization is likely catalyzed by the alkoxide anion bases present in the reaction mixture.

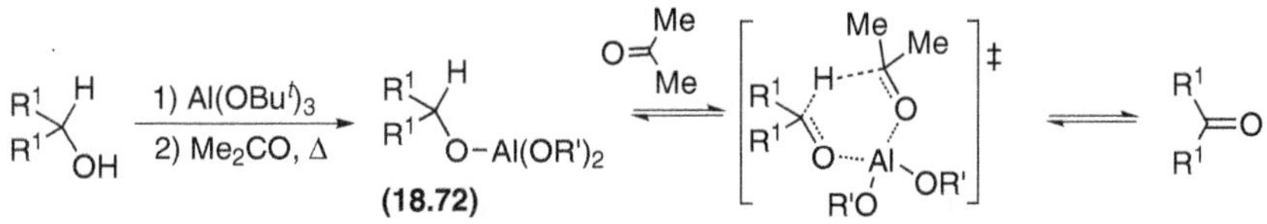

(18.72) (18.73)

In 2002, Maruoka reported a modified aluminum catalyst (Example **18.74**) for the reaction.[64] In 2006, Nguyen and his coworkers reported that using trimethylaluminum as the source of the aluminum alkoxide and nitrated benzaldehydes as the hydride acceptor[65] gives generally high (> 95%) yields when two to three equivalents of the hydride acceptor are used in toluene at room temperature. The same authors also noted that the reaction displays a significant preference for the oxidation of a secondary alcohol over a primary alcohol. Likewise, 1,1,1-trifluoroacetone has been reported as a superior hydride acceptor.[66]

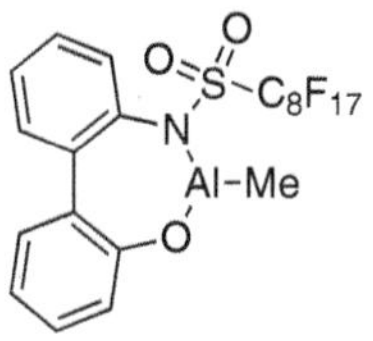

(18.74)

60. **Rupert Viktor Oppenauer (1910–1969)** was educated in Zürich at the ETH under two Nobel laureates: Leopold Ružička (Chemistry, 1939), and Tadeus Reichstein (Physiology or Medicine, 1950). Oppenauer entered industry, where he spent his entire career. Beginning in the late 1930's Oppenauer's work appears primarily in the patent literature, with assignments to Ciba and Roche Pharmaceutical companies, with Oppenauer's address listed as Amsterdam. His final patent, which issued after his death, was a German patent assigned to Merck. During the last five years of his life, he moved from Innsbruck to Buenos Aires.

61. Oppenauer, R.V. *Rec. Trav. Chim. Pays-Bas* **1937**, *56*, 137.

62. (a) Graves, C.R.; Campbell, E.J.; Nguyen, S.T. *Tetrahedron:Asymmetry* **2005**, *16*, 3460. (b) de Graauw, C.F.; Peters, J.A.; van Bekkum, H.; Huskens, J. *Synthesis* **1994**, 1007.

63. Eastham, J.F.; Teranishi, R. *Org. Syn., Coll. Vol. 4*, **1963**, 192; *Org. Syn.* **1955**, *39*, 39.

64. Ooi, T.; Otsuka, H.; Miura, T.; Ichikawa, H.; Maruoka, K. *Org. Lett.* **2002**, *4*, 2669.

65. Graves, C.R.; Zheng, B.-S.; Nguyen, S.T. *J. Am. Chem. Soc.* **2006**, *128*, 12596.

66. Mello, R.; Martinez-Ferrer, J.; Asensio, G.; González-Núñez, M.E. *J. Org. Chem.* **2007**, *72*, 9376.

Ruthenium-Catalyzed Reactions

Ruthenium has become an important element in a wide range of catalytic reactions in organic chemistry.[67] Among other applications of this element in synthesis, low-valent ruthenium complexes (especially ***Shvo's catalyst;***[68] Example **18.75**) react with alcohols by oxidative addition of the O—H bond, followed by β-elimination of hydrogen to give carbonyl compounds. Although the oxidation of secondary alcohols to ketones is without complication, the ruthenium-catalyzed formation of aldehydes is usually followed by Tishchenko-type disproportionation of the aldehydes to esters (e.g., Example **18.76**).[69] A variety of hydrogen acceptor molecules have been used in this reaction, including acetone, benzylideneacetone, and diphenylacetylene (used in Example **18.77**).

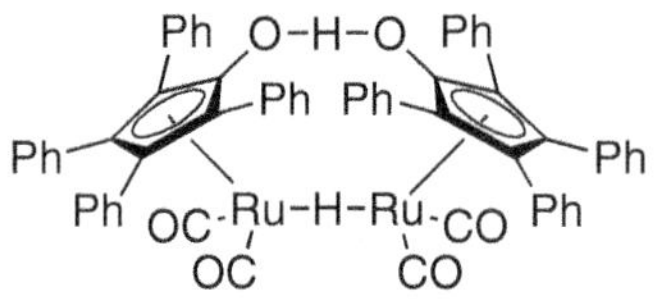

(18.75)

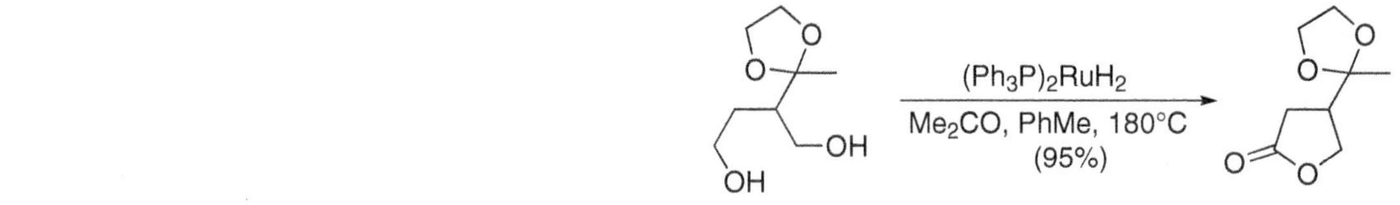

(18.76)

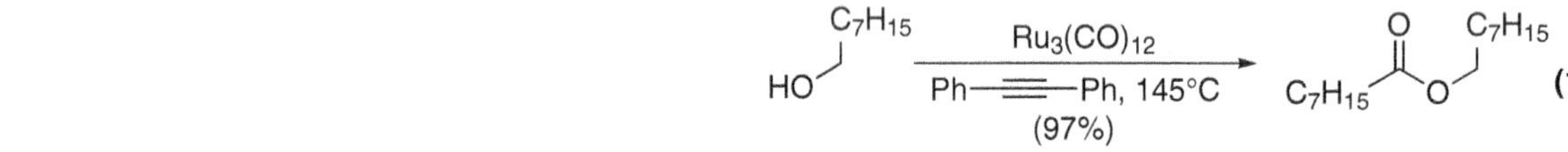

(18.77)

As pointed out above, the Oppenauer oxidation of homoallylic alcohols such as cholesterol is often accompanied by migration of the double bond into conjugation. This also occurs using the ruthenium-catalyzed hydrogen transfer reaction, as illustrated by the oxidation of the steroid in Example **18.78**.[70] In this case, the potassium carbonate used is basic enough to catalyze the isomerization of the double bond.

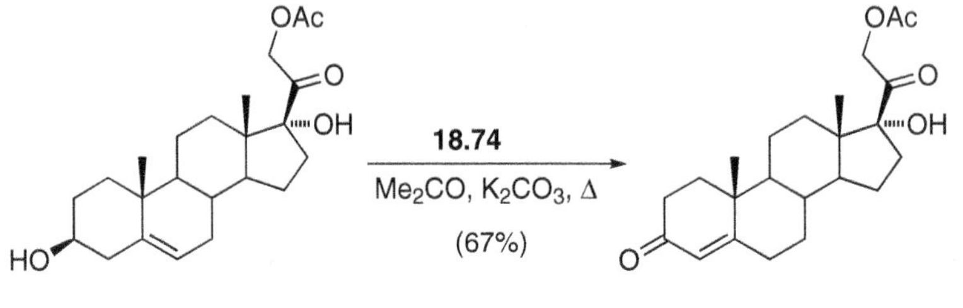

(18.78)

Oxidations Catalyzed by Stable Nitroxyl Free Radicals

Relatively recently, *N*-oxonitronium ions, prepared in situ by oxidation of nitroxyl radicals with a secondary oxidant, have found application in the oxidation of alcohols (e.g., Examples **18.79** and **18.80**).[71] In these reactions, the nitroxyl radical is an organocatalyst.

---

67. Naota, T.; Takaya, H.; Murahashi, S.-I. *Chem. Rev.* **1998**, *98*, 2599.
68. Shvo, Y.; Czarkie, D.; Rahamim, Y. *J. Am. Chem. Soc.* **1986**, *108*, 7400.
69. (a) Blum, Y.; Reshef, D.; Shvo, Y. *Tetrahedron Lett.* **1981**, *22*, 1541. (b) Blum, Y.; Shvo, Y. *J. Organomet. Chem.* **1984**, *263*, 93. (c) Blum, Y.; Shvo, Y. *Isr. J. Chem.* **1984**, *24*, 144. (d) Menashe, N.; Shvo, Y. *Organometallics* **1991**, *10*, 3885.
70. Almeida, M.L.S.; Kočovský, P.; Bäckvall, J.-E. *J. Org. Chem.* **1996**, *61*, 6587.
71. Review: Vogler, T.; Studer, A. *Synthesis* **2008**, *13*, 1979.

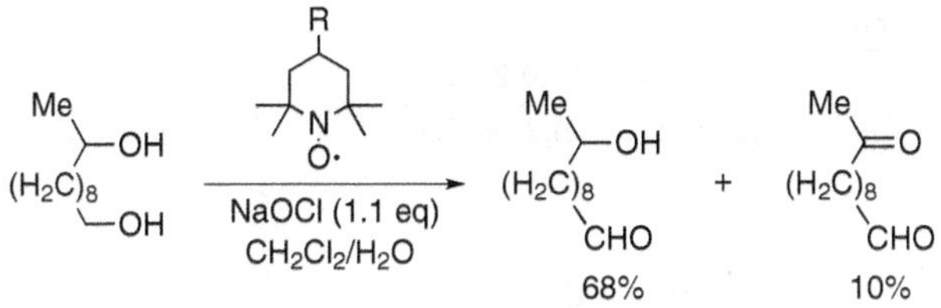

(18.79)

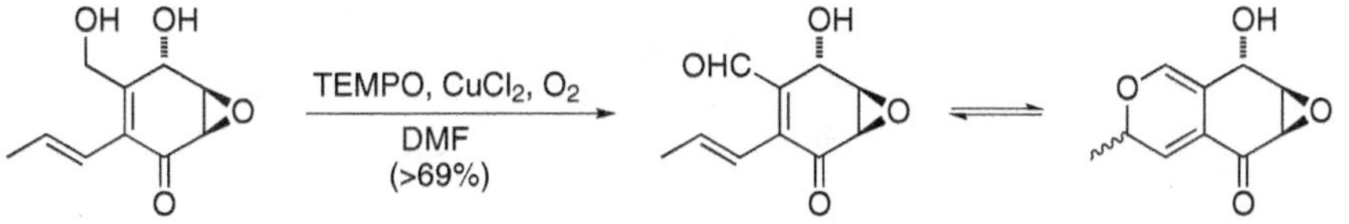

(18.80)

One useful feature of this oxidation is that it permits the oxidation of a primary hydroxyl group in preference to a secondary hydroxyl group,[72] as illustrated by the oxidation of the unprotected glycoside derivatives shown in Example **18.81**.[73] The identity of the co-oxidant is not critical, with molecular oxygen serving as the oxidant when 2,2,6,6-tetramethylpiperi dinyloxy (TEMPO) is used in dimethylformamide (DMF) solution with a catalytic amount of copper (I) chloride.[74] Because of its ease of use and the high yields that it can give, oxidation with TEMPO as an organocatalyst using inexpensive terminal oxidants such as sodium hypochlorite, trichloroisocyanuric acid, or oxygen is becoming an important industrial method for the selective oxidation of alcohols.[75] In these applications, the more expensive TEMPO can often be replaced by the much less expensive 4-hydroxy-TEMPO and 4-acetamido-TEMPO. The TEMPO-catalyzed oxidation of 1,2-diols to give α-diketones (e.g., Example **18.82**)[76] provides a mild method for the synthesis of these important precursors to medicinally important heterocycles. Fluorinated TEMPO derivatives with multiple triazole ring systems have recently been proposed as improved catalysts with catalytic sodium bromide and using sodium hypochlorite as the stoichiometric oxidant (e.g., Example **18.83**),[77] conditions first used by Agnelli.[72] In the same paper, the use of a mixed manganese (II)-cobalt (II) catalyst with the same fluorinated TEMPO and molecular oxygen as the oxidant has been found to be efficient for the oxidation of primary alcohols to aldehydes, using conditions developed by Minisci (e.g., Example **18.84**).[78]

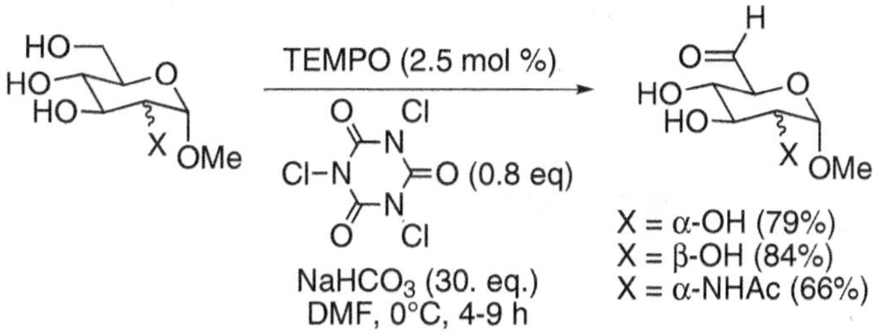

(18.81)

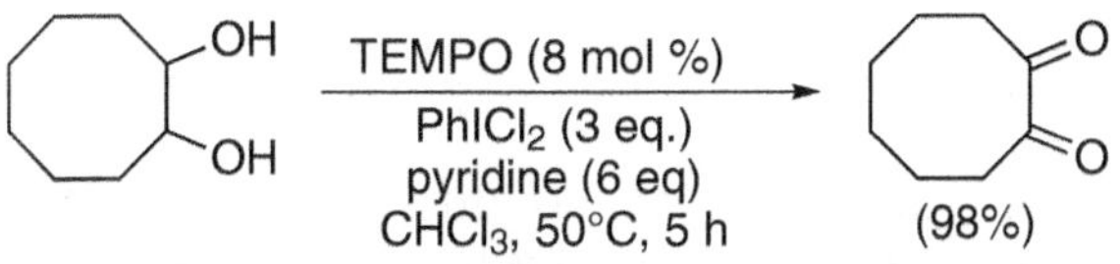

(18.82)

72. (a) Anelli, P. L.; Biffi, C.; Montanari, F.; Quici, S. *J. Org. Chem.* **1987**, *52*, 2559. (b) Anelli, P.L.; Banfi, S.; Montanari, F.; Quici, S. *J. Org. Chem.* **1989**, *54*, 2970.

73. Angelion, M.; Hermansson, M.; Dong, H.; Ramström, O. *Eur. J. Org. Chem.* **2006**, 4323.

74. (a) Semmelhack, M.F.; Schmid, C.R.; Cortes, D.A.; Chou, C.S. *J. Am. Chem. Soc.* **1984**, *106*, 3374. (b) Li, C.; Bardhan, S.; Pace, E.A.; Liang, M.-C.; Gilmore, T.D.; Porco, J.A., Jr. *Org. Lett.* **2002**, *4*, 3267.

75. Review: Ciriminna, R.; Pagliaro, M. *Org. Proc. Res. Dev.* **2010**, *14*, 245.

76. Zhao, X.-F.; Zhang, C. *Synthesis* **2007**, 551.

77. Gheorghe, A.; Chinnusamy, T.; Cuevas-Yañez, E.; Hilgers, P.; Reiser, O. *Org. Lett.*, **2008**, *10*, 4171.

78. (a) Cecchetto, A.; Fontana, F.; Minisci, F.; Recupero, F. *Tetrahedron Lett.* **2001**, *42*, 6651. (b) Minisci, F.; Recupero, F.; Rodino, M.; Sala, M.; Schneider, A. *Org. Proc. Res. Dev.* **2003**, *7*, 794.

(18.83)

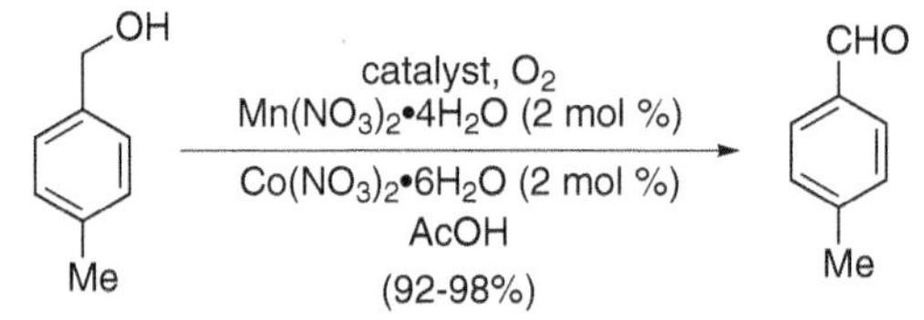

(18.84)

Oxidations with Molecular Oxygen

The direct oxidation of alcohols using oxygen as the terminal oxidant has received considerable impetus from the movement toward "green" chemistry. The process of oxidizing alcohols with oxygen is not trivial, however, and much of the work in this area is focused on the development of catalysts for the reaction.

The first reaction in which a palladium catalyst was used for the oxidation of alcohols by molecular oxygen was the 1996 report by Kaneda that palladium cluster complexes such as $Pd_4Phen_2(CO)(OAc)_4$ catalyzed the dehydrogenation of allylic and benzylic alcohols with molecular oxygen as the stoichiometric oxidant (Example **18.85**).[79] Nevertheless, the Kaneda oxidation did not result in what may be termed a breakthrough in the use of molecular oxygen for the oxidation of alcohols. That breakthrough came with the reports by Uemura's group that oxidation of alcohols by palladium (II) acetate was catalyzed by pyridine derivatives (e.g., Example **18.86**).[80] This stimulated a rapid exploration of asymmetric oxidation using molecular oxygen as the terminal oxidant and chiral amines as promoters. This reaction will be covered in much more detail when we discuss asymmetric oxidation reactions.

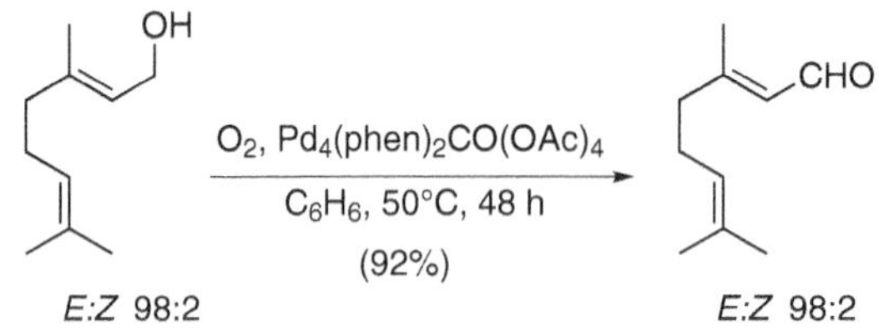

(18.85)

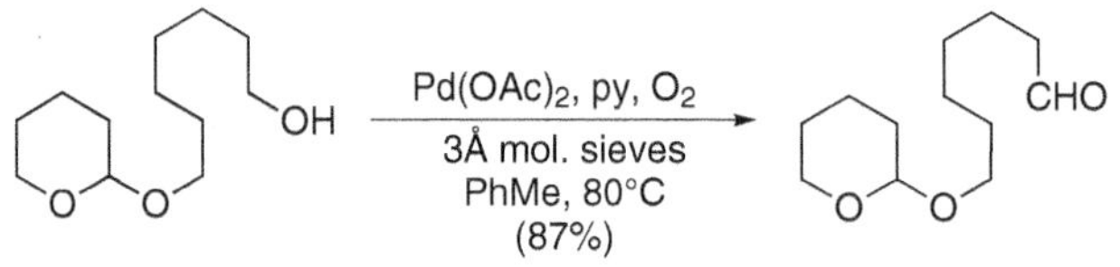

(18.86)

Drill Exercises

What is the major organic product obtained from the oxidation of the alcohol below with each of the following reagents? Where appropriate, give your reasons.

(1) PCC, CH_2Cl_2, 25°C, 30 min

(2) MnO_2, pentane, 25°C

(3) CrO_3, H_2SO_4, H_2O, acetone, 25°C

(4) $BaMnO_4$, CH_2Cl_2

79. Kaneda, K.; Fujii, M.; Morioka, K. *J. Org. Chem.* **1996**, *61*, 4502.

80. (a) Nishimura, T.; Onoue, T.; Ohe, K.; Uemura, S. *Tetrahedron Lett.* **1998**, *39*, 6017. (b) Nishimura, T.; Onoue, T.; Ohe, K.; Uemura, S. *J. Org. Chem.* **1999**, *64*, 6750.

(5) TPAP, NMMO, CH_2Cl_2, 25°C

(6) $(COCl)_2$, Et_3N, CH_2Cl_2, −78°C, 25°C

(7) TEMPO, CH_2Cl_2, NaOCl, H_2O

(8) Me_2S, NCS, Et_3N, CH_2Cl_2, −25°C

Substrates:

(a) 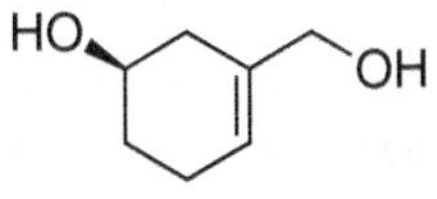

(b) 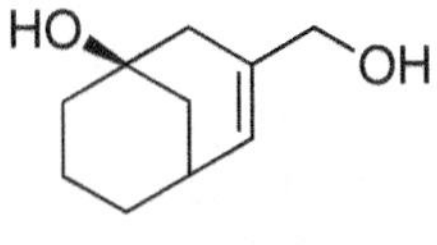

(c) 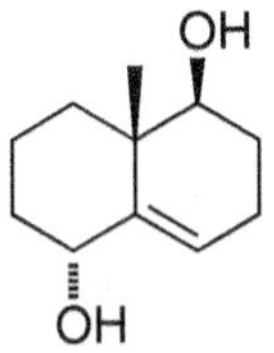

(d) 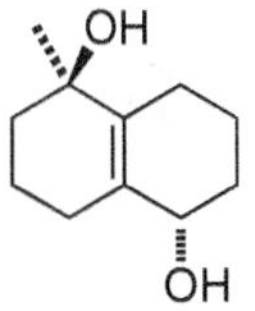

(e) 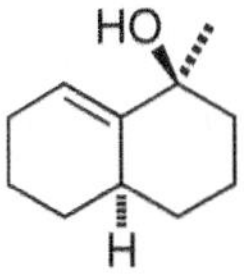

(f) 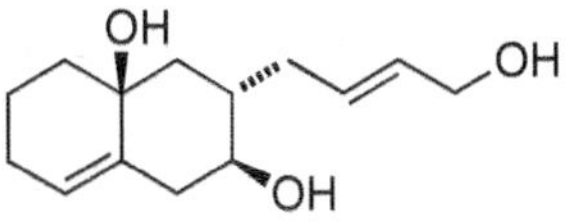

Problem

18-6 Which reagent(s) from the list in the Drill Exercises above can be used to effect the changes below, and which cannot? Give reasons.

(a) 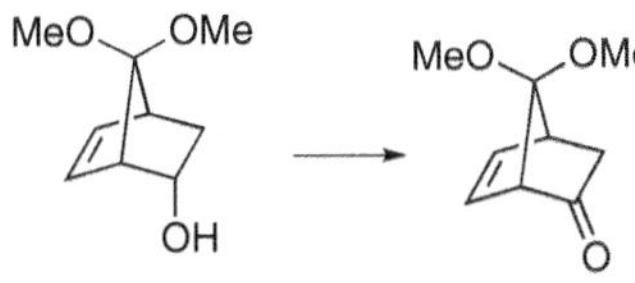

(b) 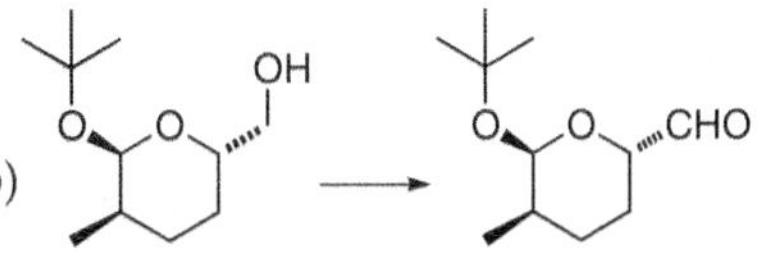

(c) 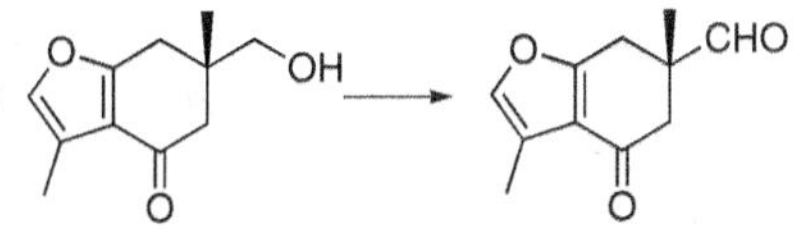

(d) 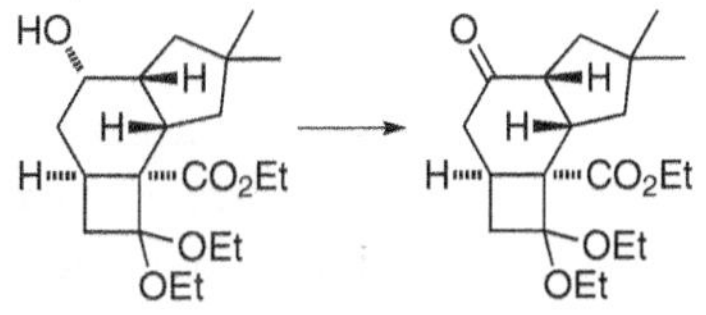

(e) 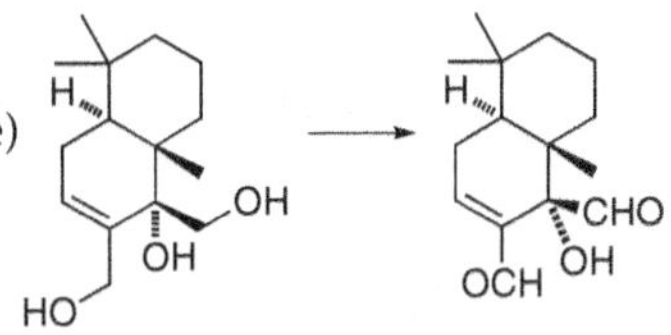

(f) 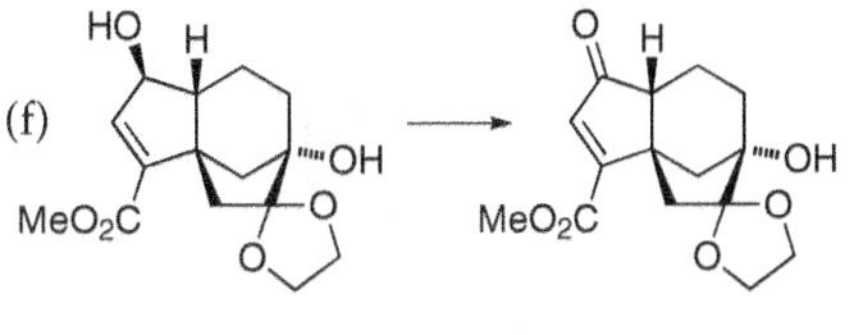

Reaction Synopses

Oxidation of Alcohols to Aldehydes and Ketones

Reagents:

Cr (VI):	PCC, CH_2Cl_2; PDC, CH_2Cl_2; CrO_3•2py, CH_2Cl_2; etc.
Mn (VI):	$BaMnO_4$, CH_2Cl_2
Mn (IV):	MnO_2, C_6H_6, Δ (allylic and benzylic alcohols only)
Mn (VII):	$KMnO_4$, $CuSO_4$•$5H_2O$, Al_2O_3, r.t. (solvent-free)

(continues)

(**Reaction Synopses** continued)

Ru (VII):	TPAP, NMMO, CH_2Cl_2, r.t.
Ag (I):	Ag_2CO_3, celite, C_6H_6, Δ
Al (III):	Al(O-*i*-Pr)$_3$, Me_2CO, Δ (Oppenauer);
Ru (II), R_2CO:	$(Ph_3P)_2RuH_2$, PhMe, Me_2CO, 180°C; Shvo's catalyst, K_2CO_3, Me_2CO, Δ; etc.
Me_2SX^+:	$(COCl)_2$, Me_2SO, CH_2Cl_2, Et_3N, −60°C (Swern); DCC, Me_2SO, CH_2Cl_2 (Pfitzner-Moffatt); Me_2S, NCS, Et_3N, CH_2Cl_2, −25°C (Corey-Kim); etc.
Dess-Martin:	DMP, CH_2Cl_2(, H_2O); selective for 2° ROH over 1°
TEMPO:	TEMPO, CH_2Cl_2, NaOCl, H_2O; TEMPO, CuCl, O_2, DMF; etc.
O_2:	O_2, $Pd(OAc)_2$, pyridine, PhMe, 80°C

Oxidation of Primary Alcohols to Esters

Reagents:

(R=aryl):	$Al(OR_3)$ (Tishchenko);
(R=akyl):	$Ru_3(CO)_{12}$, PhC≡CPh, Δ; Shvo's catalyst, PhCH=CHCOMe, Δ; etc.

18.5 Oxidations to Carboxylic Acids

The reactivity of chromium and manganese agents can be significantly modified by the solvent used. Thus, although pyridinium dichromate in non-polar solvents oxidizes primary alcohols to aldehydes, in DMF solution it can be used to oxidize a primary alcohol to a carboxylic acid (Example **18.87**).[81] Similarly, although potassium permanganate often reacts only sluggishly with primary alcohols, when solubilized in benzene as a quaternary ammonium salt[82] or by means of a crown ether[83] ("purple benzene"), it is an excellent reagent for oxidizing primary alcohols to carboxylic acids. However, the reagent also rapidly cleaves alkene π bonds (Example **18.88**)[84] and oxidizes aldehydes.

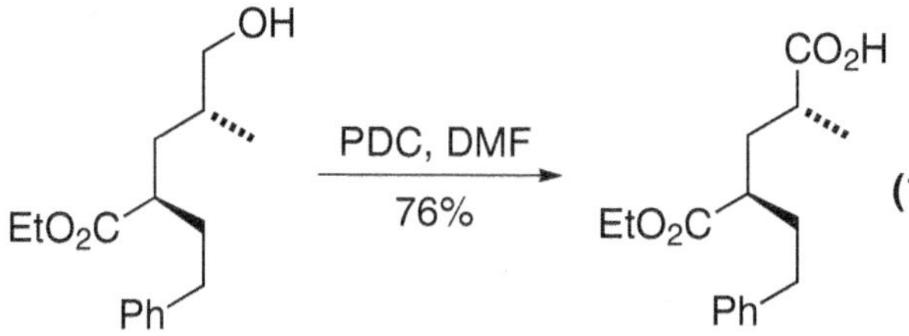

(18.87)

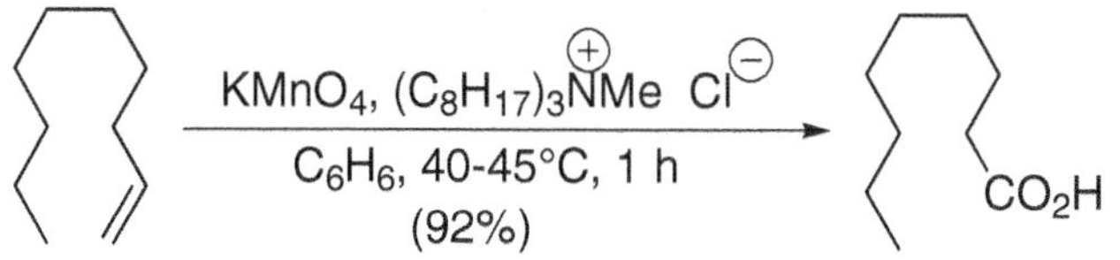

(18.88)

---

81. Gruefeld, N.; Stanton, J.L.; Yuan, A.M.; Ebetino, F.H.; Browne, L.J.; Gude, C.; Huebner, C.F. *J. Med. Chem.* **1983**, *26*, 1277.

82. Herriot, A.W.; Picker, D. *Tetrahedron Lett.* **1974**, 1511.

83. Sam, D.J.; Simmons, H.E. *J. Am. Chem. Soc.* **1972**, *94*, 4024.

84. Starks, C.M. *J. Am. Chem. Soc.* **1971**, *93*, 195.

Of the many chromium (VI) reagents, the Jones reagent (e.g., Example **18.89**) is the least selective. On occasion, however, its very reactivity and lack of selectivity allows multiple steps to be accomplished in one pot. For example, the strongly acidic conditions of the Jones oxidation permit the simultaneous deprotection of silyl ethers and oxidation of the revealed alcohols.[85] In the oxidation in Example **18.90**,[86] the Jones reagent oxidized both silyl ethers directly to the corresponding carboxylic acids. Interestingly, the adsorption of Jones reagent on silica gel allows it to be used to oxidize primary benzyl alcohols to aldehydes.[87]

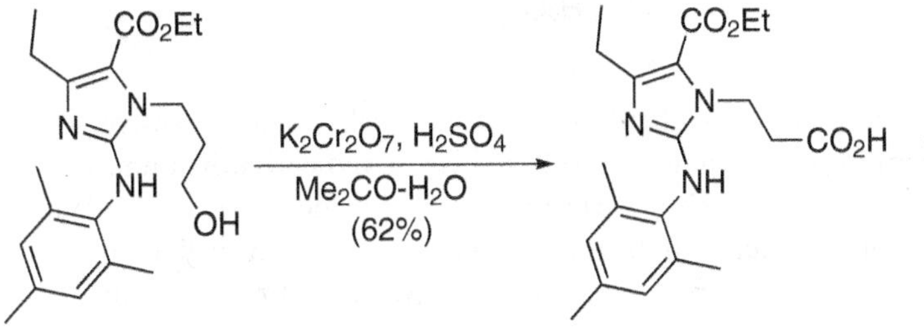

(18.89)[88]

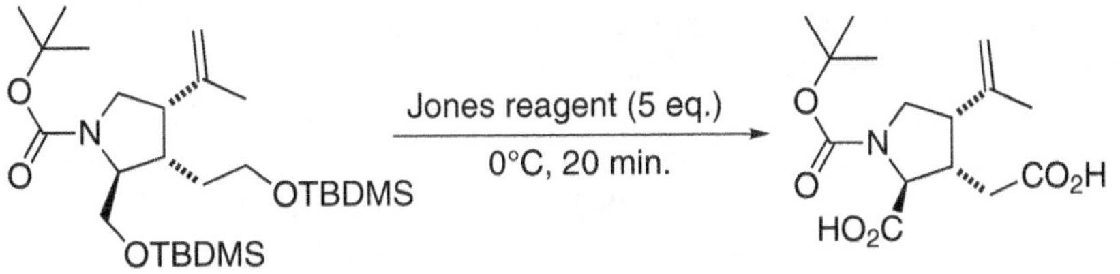

(18.90)

The oxidation of primary alcohols and aldehydes to carboxylic acids can also be accomplished using chromium (VI) reagents (CrO_3 and PCC have both been used) as catalysts and periodic acid as the stoichiometric oxidant (Example **18.91**).[89] In these reactions, the Cr (VI) reagent is responsible for the oxidation, with the periodic acid functioning to recycle the Cr (VI) catalyst.

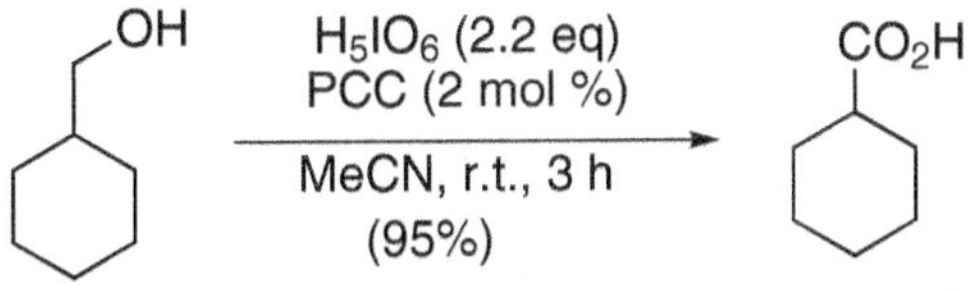

(18.91)

There is a reagent for oxidizing alcohols to carboxylic acids that is a stronger oxidizing agent than any reagent we have discussed to this point: ruthenium tetroxide.[90] This extremely powerful oxidant also attacks carbon-carbon π bonds (including aromatic rings, as we shall discuss later). In addition, it attacks ethers and amines, but esters and amides tend to be resistant (but not completely inert) to ruthenium tetroxide oxidation, and, oddly enough, even epoxides are oxidized only slowly by this reagent in comparison to alcohols (Example **18.92**).[91] It has been used for decades as a method for the oxidation of carbohydrate derivatives, such as the glucose derivative in Example **18.93**.[92]

---

85. (a) Evans, P.A.; Roseman, J.D.; Garber, L.T. *Synth. Commun.* **1996**, *26*, 4685. (b) Evans, P.A.; Murthy, V.S.; Roseman, J.D.; Rheingold, A.L. *Angew. Chem., Int. Ed.* **1999**, *38*, 3175. (c) Liu, H.-J.; Han, I.-S. *Synth. Commun.* **1985**, *15*, 759.
86. Farwick, A.; Helmchen, G. *Org. Lett.* **2010**, *12*, 1108.
87. Ali, M.H.; Wiggins, C.J. *Synth. Commun.* **2001**, *31*, 1389, 3383.
88. Vrudhula, V.M.; Dasgupta, B.; Pin, S.S.; Burris, K.D.; Balanda, L.A.; Fung, L.K.; Fiedler, T.; Browman, K.E.; Taber, M.T.; Zhang, J.; Macor, J.E.; Dubowchik, G.M. *Bioorg. Med. Chem. Lett.* **2010**, *20*, 1905.
89. (a) Zhao, M.; Li, J.; Song, Z.; Desmond, R.; Tschaen, D.M.; Grabowski, E.J.J.; Reider, P.J. *Tetrahedron Lett.*, **1998**, *39*, 5323. (b) Hunson, M. *Synthesis* **2005**, 2487.
90. Review: Lee, D.G.; van den Engh, M. In Trahanovsky, W.S., Ed. *Oxidation in Organic Chemistry, Part B* (Academic Press: New York, 1973), p. 177.
91. Carlsen, P.H.J.; Katsuki, T.; Martin, V.S.; Sharpless, K.B. *J. Org. Chem.* **1981**, *46*, 3936.
92. (a) Beynon, P.J.; Collins, P.M.; Overend, W.G. *Proc. Chem. Soc.* **1964**, 342. (b) Parikh, V.M.; Jones, J.K.N. *Can. J. Chem.* **1965**, *43*, 3452.

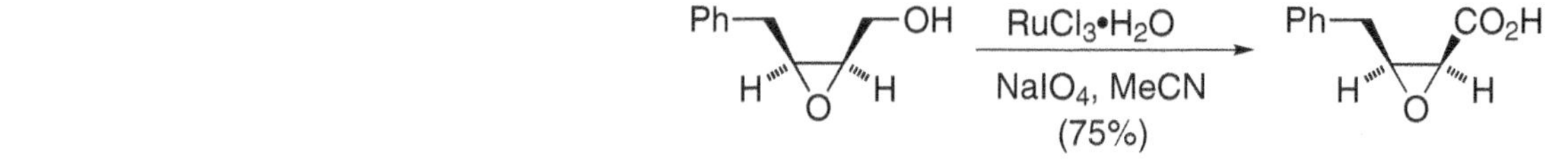

(18.92)

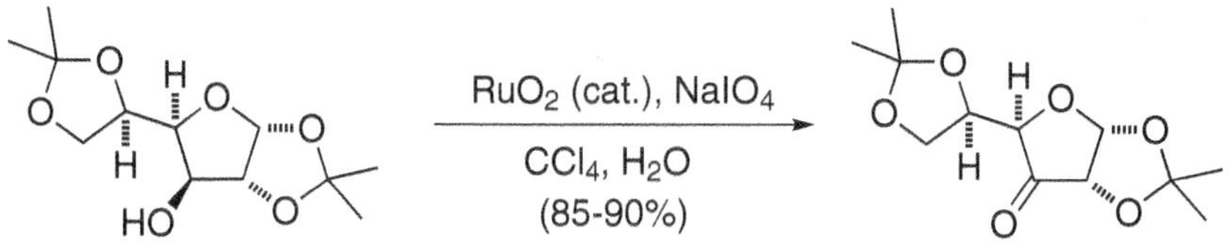

(18.93)

As with osmium tetroxide, ruthenium tetroxide is volatile and expensive, so considerable effort has gone into developing methods that are catalytic in ruthenium. These almost always start with ruthenium trichloride or ruthenium dioxide. The reagent is black, which allows the course of the reaction to be followed visually; when the yellow color of the ruthenium tetroxide persists, the reaction is complete. What is also interesting with the ruthenium reagent is that its reactivity can be quite muted under catalytic conditions; this raises the possibility (as yet unaddressed) that the +8 oxidation state of ruthenium may not be the active oxidant under these conditions.

In 1882, Bernard Tollens[93] described a reaction between aldehydes and ammoniacal silver (I) hydroxide that led to reduction of the metal, and oxidation of the aldehyde to the corresponding carboxylic acid (e.g., Example **18.94**).[94] This became the basis of the ***Tollens test*** for aldehydes. The reaction is very mild and almost specific for the aldehyde functional group, but it suffers from low yields in many cases. In addition, the high cost of the reagent, which is used stoichiometrically rather than catalytically, makes it prohibitively expensive except on the small scale.

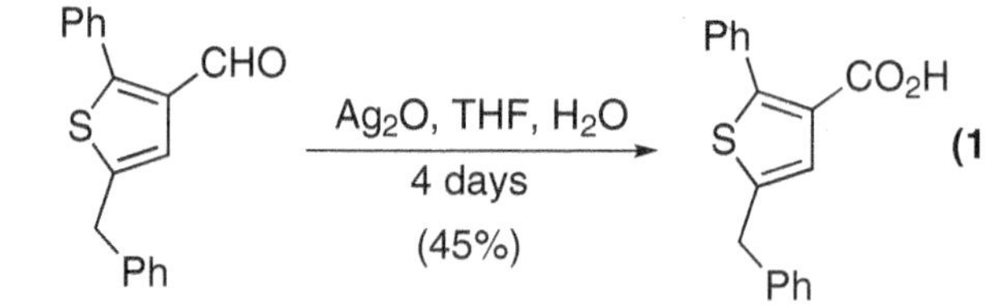

(18.94)[95]

In 1973,[96] Lindgren and Nilsson introduced sodium chlorite as a much more economical and predictable reagent for oxidizing aldehydes to carboxylic acids. The method was adapted by Kraus and Taschner,[97] who used 2-methyl-2-butene as a chlorine scavenger. The Kraus procedure was thoroughly tested by Pinnick,[95c] and the reaction is now often described as a ***Pinnick-type oxidation***. This improvement in the chlorite oxidation of aldehydes has now made the two-stage oxidation of primary alcohols to carboxylic acids (as in Examples **18.95** to **18.97**) the method of choice for carrying out this transformation. An alternative method for the oxidation of benzylic aldehydes to methyl esters involves using vanadium pentoxide and hydrogen peroxide in methanol containing perchloric acid.[98] The oxidation of primary alcohols to acids by a composite catalyst of vanadium pentoxide, copper (II) 2-ethylhexanoate, and 1,4-diazabicyclo[2.2.2]octane

---

93. **Bernhard Christian Gottfried Tollens (1841–1918)** studied under Wöhler and Fittig in Göttingen (PhD, 1864). He traveled to Heidelberg, Paris (1868), and Coimbra, Portugal (1869) before returning to Göttingen in 1870, where he was later appointed as Professor and Director of the agricultural chemistry laboratory. For more detail, see: Wallach, O. *Ber. dtsch. chem. Ges.* **1918**, *51*, 1539.

94. Tollens, B. *Ber. dtsch. chem. Ges.* **1882**, *15*, 1635.

95. LaLonde, R.T.; Florence, R.A.; Horenstein, B.A.; Fritz, R.C.; Silveira, L.; Clardy, J.; Krishnan, B.S. *J. Org. Chem.* **1985**, *50*, 85.

96. Lindgren, B.O.; Nilsson, T. *Acta Chem. Scand.* **1973**, *27*, 888.

97. (a) Kraus, G.A.; Taschner, M.J. *J. Org. Chem.* **1980**, *45*, 1175. (b) Kraus, G.A.; Roth, B. *J. Org. Chem.* **1980**, *45*, 4825. (c) Bal, B.S.; Childers, W.E., Jr.; Pinnick, H.W. *Tetrahedron* **1981**, *37*, 2091.

98. Gopinath, R.; Patel, B. *Org. Lett.* **2000**, *2*, 577.

(DABCO) in an ionic liquid, as well as using molecular oxygen as the stoichiometric reagent, has recently been reported.[99]

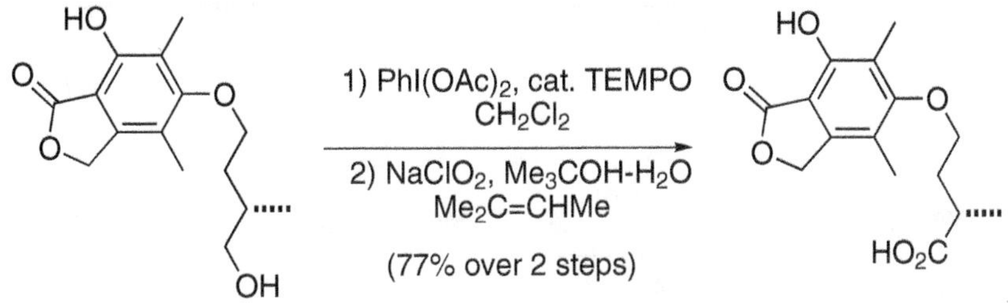

(18.95)[100]

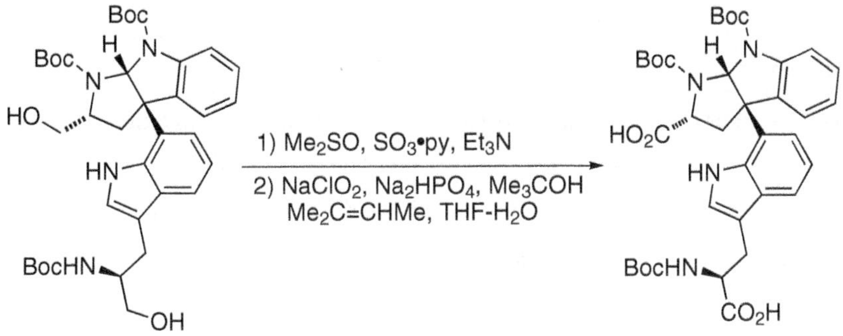

(18.96)[101]

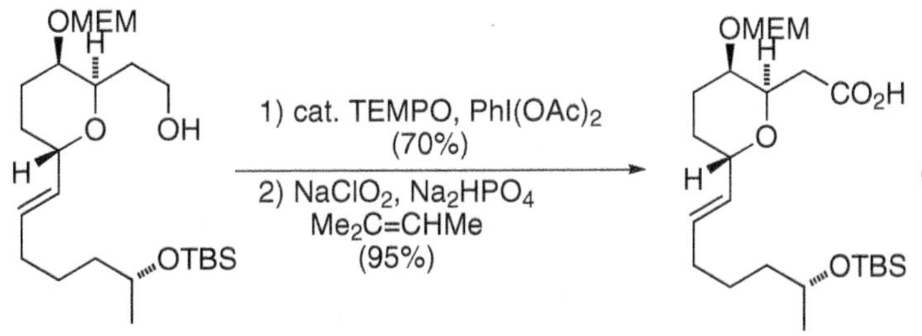

(18.97)[102]

Manganese dioxide oxidation of primary allylic alcohols leads to the formation of α,β-unsaturated aldehydes. The addition of a nucleophile to the aldehyde leads to a new allylic alcohol that can be further oxidized by manganese dioxide. This has led to a number of useful extensions to this oxidation, due in part to the relative lack of reactivity by manganese dioxide toward a wide range of functional groups. The use of manganese dioxide in tandem oxidation reactions has been reviewed.[103]

The first of these adaptations of the manganese dioxide oxidation of allylic alcohols is provided by the oxidation of primary allyl alcohols to methyl esters by manganese dioxide and sodium cyanide in methanol (Example **18.98**).[104] In this reaction, the primary alcohol is first oxidized to the aldehyde, which then reacts with cyanide to give a cyanohydrin, which is further oxidized to the acyl cyanide. This compound, which is a less reactive analog of an acyl chloride or acid anhydride, then reacts with methanol to regenerate the cyanide. A similar adaptation has been developed by replacing the toxic cyanide anion by an *N*-heterocyclic carbene (NHC),[105] which adds to the aldehyde to give a 2-(1-hydroxyalkyl)imidazolium ion that is oxidized to the 2-acylimidazolium ion, which can then react with the alcohol to release the NHC, giving the ester (e.g., Example **18.99**).

99. Jiang, N.; Ragauskas, A.J. *J. Org. Chem.* **2007**, *72*, 7030.
100. Katoh, N.; Nakahata, E.; Kuwahara, S. *Tetrahedron* **2008**, *64*, 9073.
101. Govek, S.P.; Overman, L.E. *Tetrahedron* **2007**, *63*, 8499.
102. Mueller Hendrix, A.J.; Jennings, M.P. *Tetrahedron Lett.* **2010**, *51*, 4260.
103. Taylor, R.J.K.; Reid, M.; Foot, J.; Raw, S.A. *Acc. Chem. Res.* **2005**, *38*, 851.
104. Corey, E.J.; Gilman, N.W.; Ganem, B.E. *J. Am. Chem. Soc.* **1968**, *90*, 5616.
105. (a) Maki, B.E.; Chan, A.; Phillips, E.M.; Scheidt, K.A. *Org. Lett.* **2007**, *9*, 371. (b) Maki, B.E.; Chan, A.; Phillips, E.M.; Scheidt, K.A. *Tetrahedron* **2009**, *65*, 3102.

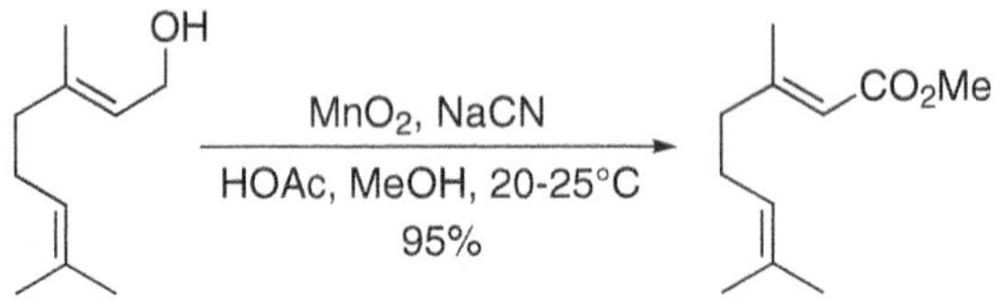

(18.98)

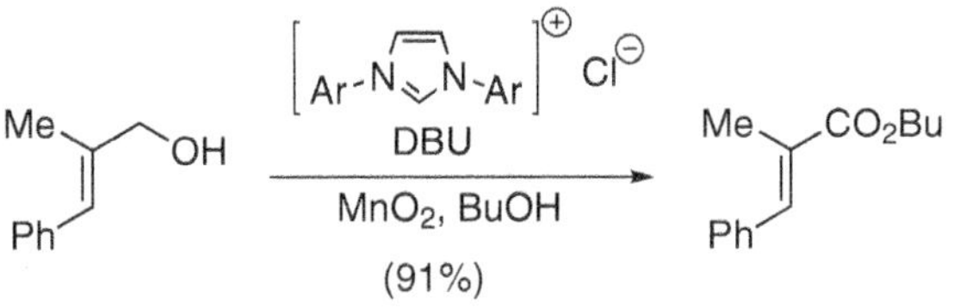

(18.99)

Worked Problem

18-2 What reagent or reagents could be used to carry out each of the oxidations specified?

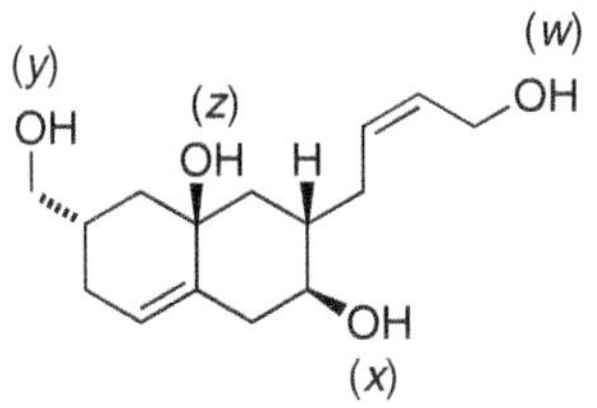

(a) Oxidize hydroxyl group *w* to the aldehyde without any other oxidation.

(b) Oxidize the primary hydroxyl groups (*w* and *y*) to aldehydes and the secondary hydroxyl group (*x*) to a ketone.

(c) Oxidize the primary hydroxyl groups (*w* and *y*) to aldehydes without any other oxidation.

(d) Oxidize the hydroxyl group *w* to a methyl ester group without affecting any other functional group.

(e) Oxidize the primary hydroxyl groups (*w* and *y*) to aldehydes, the secondary hydroxyl group (*x*) to a ketone, and the tertiary hydroxyl group to a ketone with allylic rearrangement of the double bond.

(f) Oxidize the primary hydroxyl groups (*w* and *y*) to carboxylic acids and the secondary hydroxyl group (*x*) to a ketone.

(g) Oxidize the secondary hydroxyl group (*x*) to a ketone without any other change.

(h) Oxidize the primary hydroxyl groups (*w* and *y*) to aldehydes and the secondary hydroxyl group (*x*) to a ketone, with rearrangement of the double bonds into positions conjugated with the carbonyl groups.

§Answers below.

§ Answer to Worked Problem:

A: MnO_2, CH_2Cl_2; or $BaMnO_4$, CH_2Cl_2; or O_2, $Pd_4Phen_2(CO)(OAc)_4$, C_6H_6, 50°C; etc.

B: PCC, CH_2Cl_2; or PDC, CH_2Cl_2; or $(COCl)_2$, Me_2SO, Et_3N, CH_2Cl_2, −60°C; or R_2S, NCS, CH_2Cl_2, NEt_3; or DMP, CH_2Cl_2(, H_2O); etc.

C: TEMPO, NaOCl, CH_2Cl_2, H_2O; or O_2, TEMPO, CuCl, DMF; etc.

D: MnO_2, NaCN, MeOH; or MnO_2, NHC, MeOH; etc.

E: This is a Babler-type oxidation—PCC, CH_2Cl_2, 24 h.

F: Jones' reagent—CrO_3, H_2SO_4, H_2O, Me_2CO; or 1) cat. TEMPO, $PhI(OAc)_2$, CH_2Cl_2; then $NaClO_2$, NaH_2PO_4, $Me_2C{=}CHMe$, H_2O; or cat. PCC, H_5IO_6, MeCN.

G: O_2, V_2O_5 (3 mol %), K_2CO_3 (0.5 eq), PhMe, 100°C.

H: Oppenauer oxidation—$Al(OCMe_3)_3$, Me_2CO, Δ; or Me_3Al, Me_2CO, Δ.

Drill Exercises

What will be the major organic product formed when each of the compounds in the list below is treated with each of the reagents specified?

(a) 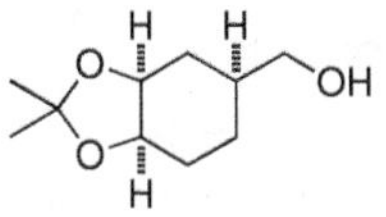 (b) 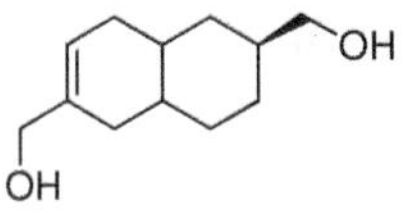 (c) 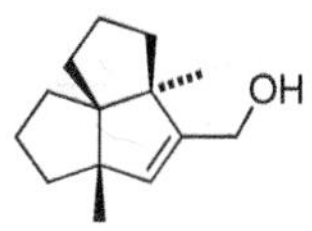

(d) 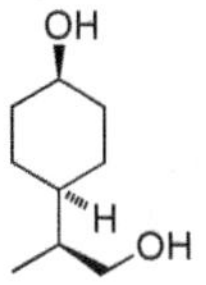 (e) 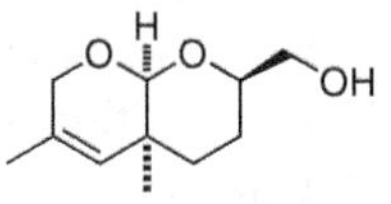 (f) 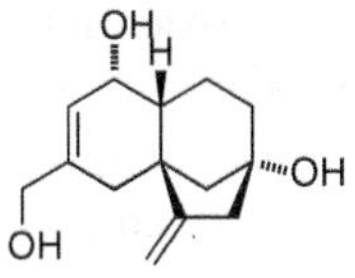

Reagents:
A: MnO_2, NHC, EtOH
B: PDC, DMF, r.t.
C: $KMnO_4$, H_2O
D: CrO_3, H_2SO_4, H_2O, acetone
E: (1) $PhI(OAc)_2$, CH_2Cl_2, TEMPO; (2) $NaClO_2$, $NaHCO_3$, THF, H_2O
F: (1) DMP, CH_2Cl_2; (2) Ag_2O, NH_3, H_2O

Reaction Synopses

Oxidation of Alcohols to Carboxylic Acids

$$R{-}CH_2OH \xrightarrow{[O]} R{-}CO_2H$$

Reagents:

Cr (VI):	$K_2Cr_2O_7$, H_2SO_4, H_2O; $K_2Cr_2O_7$, H_2SO_4, acetone, H_2O; CrO_3, H_2SO_4, H_2O, acetone, 0°C (Jones reagent); PDC, DMF; PCC, H_5IO_6MeCN; etc.
Mn (VII):	$KMnO_4$, H_2O; $KMnO_4$, H_2O, Bu_4NCl, C_6H_6; $KMnO_4$, 18-C-6. PhH; etc
Ru (VIII):	RuO_4; $RuCl_3{\bullet}H_2O$, $NaIO_4$, MeCN; RuO_2, $NaIO_4$, CCl_4, H_2O; etc.

Oxidation of Aldehydes to Acids or Esters

$$R{-}CHO \xrightarrow{[O]} R{-}CO_2H$$

Reagents:	$Ag(NH_3)_2OH$, H_2O; Ag_2O, THF, H_2O; etc.
or	$NaClO_2$, NaH_2PO_4, $Me_2C{=}CHMe$, THF, H_2O
or	V_2O_5, H_2O_2, $HClO_4$, MeOH; etc.

18.6 Oxidation of Ethers and Amines

The C—H σ bond α to the oxygen of an ether is amenable to oxidation under free radical conditions (e.g., Example **18.100**) because the radical formed is strongly stabilized by resonance, as we discussed in Chapter 13. In fact, this very susceptibility of primary or secondary alkyl ethers to free radical oxidation by atmospheric oxygen[106] makes care in

106. Steere, N.V. In Patai, S., Ed. *The Chemstry of the Ether Linkage* (Interscience: London, 1967), ch. 16.

storing these solvents especially important, because the products of these atmospheric oxidations are highly explosive and shock-sensitive peroxides that must be removed before the ether can be used safely as a solvent.[107]

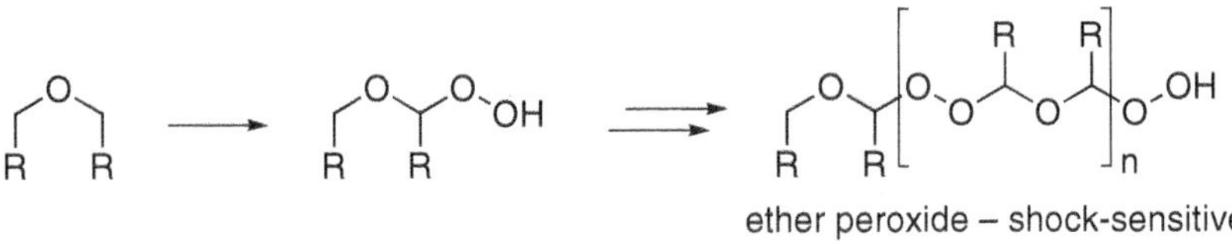

(18.100)

The oxidation of ethers can be accomplished in a controlled manner, however, by means of ruthenium tetroxide, which oxidizes simple ethers to esters and cyclic ethers to lactones (e.g., Example **18.101**).[108] Under conditions of a two-phase reaction with water and excess oxidant, γ-lactones may be further oxidized to γ-ketoacids.[108d]

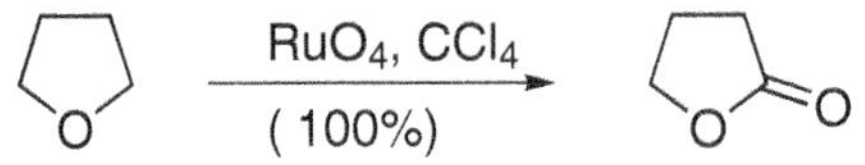

(18.101)

Amination α to an ether by nitrene insertion has been used in the formation of highly functionalized stereochemically complex amine derivatives (Example **18.102**) that can be elaborated to a wide range of target molecules.[109]

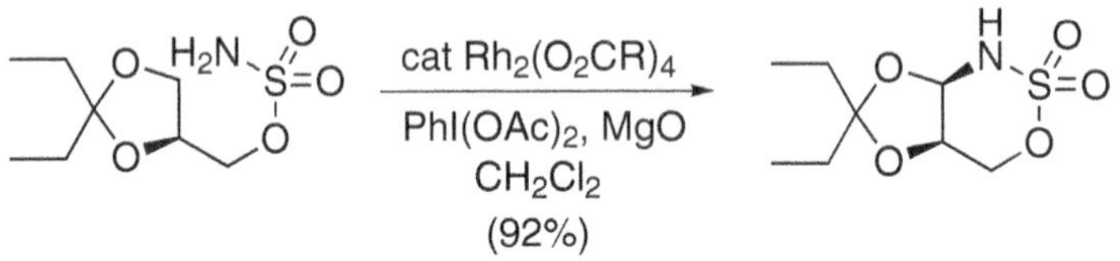

(18.102)

There are two locations where oxidation of an amine can occur: at carbon or at nitrogen. The oxidation of amines at carbon ultimately leads to the conversion of the amine to an amide or an imide, and the reaction is generally (but not exclusively) restricted to methylene groups attached to nitrogen. Again, one reagent that is highly effective for this oxidation is ruthenium tetroxide, which oxidizes open-chain amines to amides and oxidizes cyclic amines to give cyclic imides (e.g., Example **18.103**)[110] or lactams.[111]

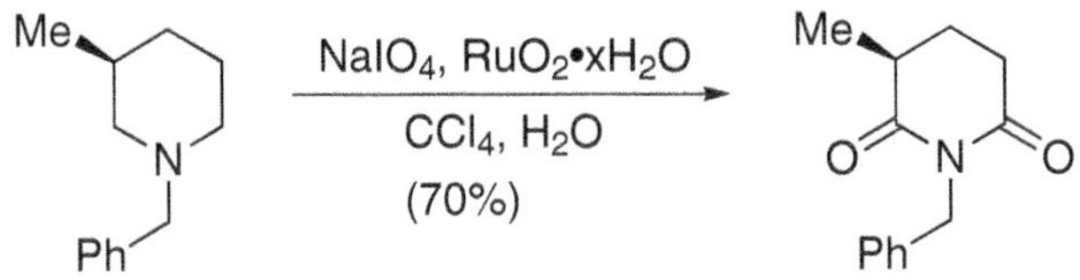

(18.103)

The oxidation of tertiary amines with mercuric acetate (e.g., Example **18.104**)[112] provides a convenient entry into both enamines and iminium ions. The oxidation has a stereoelectronic requirement that makes the oxidation of amines where the β-hydrogen and

107. Vogel, A.I. *Practical Organic Chemistry*, 3rd. ed. (Longmans: London, 1956), p. 163.

108. (a) Berkowitz, L.M.; Rylander, P.N. *J. Am. Chem. Soc.* **1958**, *80*, 6682. (b) Wolff, M.E.; Kerwin, J.F.; Owings, F.F.; Lewis, B.B.; Blank, B. *J. Org. Chem.* **1963**, *28*, 2729. (c) Gopal, H.; Adams, T.; Moriarty, R.M. *Tetrahedron* **1972**, *28*, 4259. (d) Smith, A.B., III; Scarborough, R.M., Jr. *Synth. Commun.* **1980**, *10*, 205.

109. Fleming, J.J.; Fiori, K.W.; Du Bois, J. *J. Am. Chem. Soc.* **2003**, *125*, 94305.

110. Bettoni, G.; Franchini, C.; Morlacchi, F.; Tangari, N.; Tortorella, V. *J. Org. Chem.* **1976**, *41*, 2780.

111. Lin, G.; Midha, K.K.; Hawes, E.M. *J. Het. Chem.* **1991**, *28*, 215.

112. (a) Leonard, N.J.; Hay, A.S.; Fulmer, R.W.; Gash, V.W. *J. Am. Chem. Soc.* **1955**, *77*, 439. (b) Leonard, N.J.; Sauers, R.R. *J. Am. Chem. Soc.* **1957**, *79*, 6210. (c) Leonard, N.J.; Middleton, W.J.; Thomas, P.D.; Choughury, D. *J. Org. Chem.* **1956**, *21*, 344.

the nitrogen lone pair can adopt an *anti* arrangement much more facile than the oxidation of amines, where this cannot happen. For example, 11-methyl-11-azabicyclo[5.3.1]decane (Example **18.105**) cannot adopt this arrangement except at the methyl group, and the oxidation occurs exclusively at the methyl group[113].

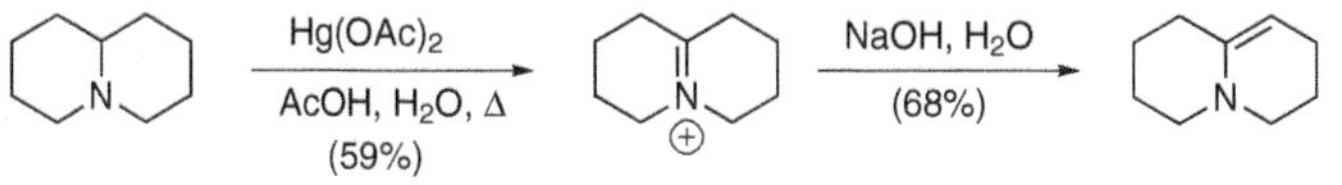

(18.104)

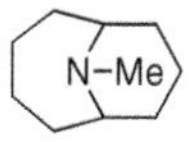

(18.105)

Tertiary amines react with peracids to form amine *N*-oxides. These compounds (e.g., *N*-methylmorpholine-*N*-oxide; NMO or NMMO) are often used as co-oxidants in reactions such as hydroboration-oxidation, instead of hydrogen peroxide anion. In 1926, the French chemists Max and Michel Polonovski reported that tertiary amine *N*-oxides reacted with acetic anhydride or acetyl chloride to give the acetamide of the secondary amine; the reaction is now known as the ***Polonovski reaction***.[114] The reaction is considered to proceed through the mechanism shown in Figure 18.3. The procedure was modified by French chemist, Pierre Potier, who replaced the acetic anhydride by trifluoroacetic anhydride, which allowed the iminium ion intermediate to be isolated due to the lower nucleophilicity of the carboxylate anion. This modification of the reaction is occasionally referred to as the ***Polonovski-Potier reaction***.[115] Aqueous workup of the Polonovski-Potier reaction gives the ketone as the product, as shown below; adding perchloric acid to the reaction mixture permits the iminium perchlorate to be isolated. Figure 18.4 compares the two variants of the reaction. The Polonovski reaction gives the acetamide (**18.112**) as the major product, whereas the major product in the Polonovski-Potier reaction is the ketone **18.113**.

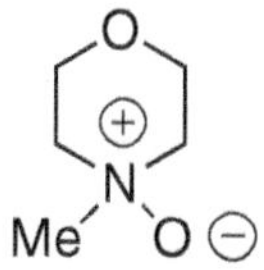

NMMO

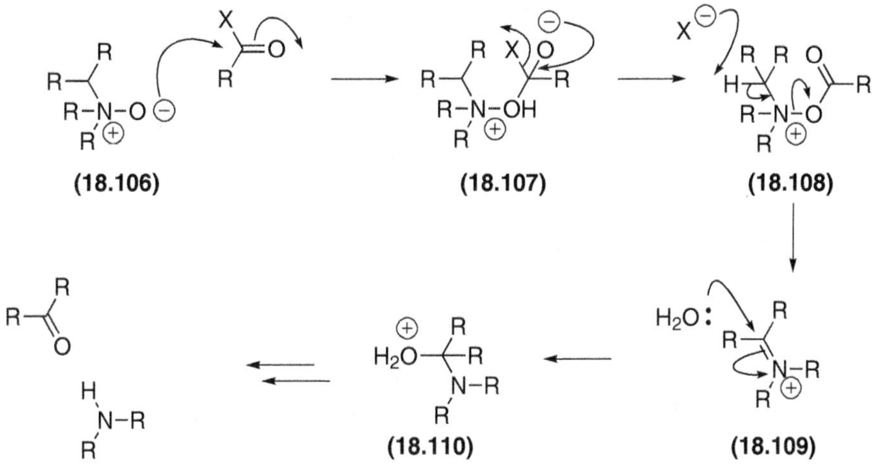

Figure 18.3 The mechanism of the Polonovski reaction

113. Leonard, N.J.; Morrow, D.F. *J. Am. Chem. Soc.* **1958**, *80*, 371.

114. (a) Polonovski, M.; Polonovski, M. *Bull. Soc. Chim. France* **1926**, 39, 147; **1927**, 41, 1190. (b) Huisgen, R.; Bayerlein, F.; Heydkamp, W. *Chem. Ber.* **1959**, *92*, 3223. (c) Oae, S.; Kitao, T.; Kitaoka, Y. *J. Am. Chem. Soc.* **1962**, *84*, 3366. (d) Metlesics, W.; Silverman, G.; Sternbach, L.H. *J. Org. Chem.* **1964**, *29*, 1621.

Reviews: (e) Katritzky, A.R.; Lagowski, J.N. *Chemistry of Heterocyclic N-Oxides* (Academic Press: New York, 1971) p 279, 362. (f) Grierson, D. *Org. React.* **1990**, *39*, 85. (g) Grierson, D.S.; Husson, H.-P. In Trost, B.M.; Fleming, I., Eds. *Comprehensive Organic Synthesis.* (Pergamon: Oxford, 1991) vol 6, 909–924.

115. (a) Cave, A.; Kan-Fan, C.; Potier, P.; Le Men, J. *Tetrahedron* **1967, 23,** 4681. (b) Tamminen, T.; Jokela, R.; Tirkkonen, B.; Lounasmaa, M. *Tetrahedron* **1989**, *45*. 2683. (c) Sundberg, R.J.; Gadamasetti, K.G.; Hunt, P. J. *Tetrahedron* **1992**, *48*, 277.

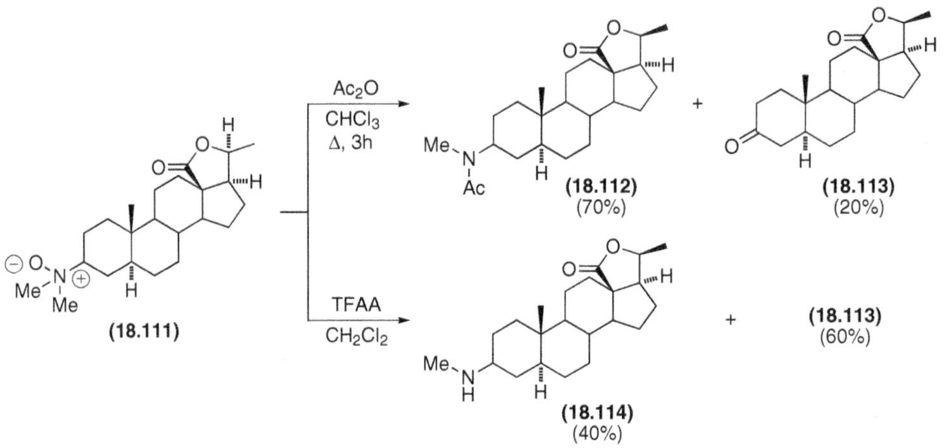

Figure 18.4 Comparison of the Polonovski and Potier-Polonovski reactions

Reaction Synopses

Oxidation of Ethers and Amines to Esters and Amides

Reagents: RuO_4; $RuCl_3{\bullet}H_2O$, $NaIO_4$, MeCN; RuO_2, $NaIO_4$, CCl_4, H_2O; etc.
or $Hg(OAc)_2$, HOAc, H_2O, Δ (3° amines)

Polonovski Reaction

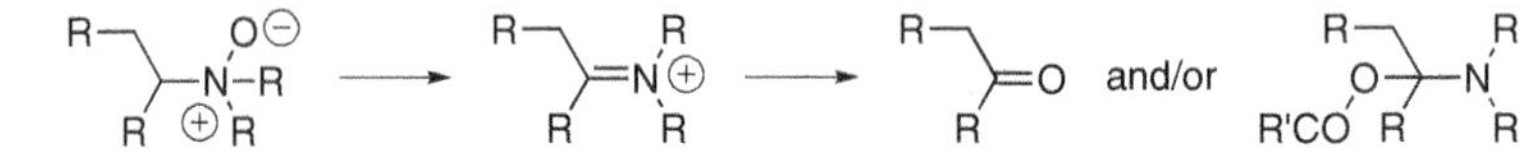

Reagents: Ac_2O, $CHCl_3$, Δ; AcCl, $CHCl_3$; etc (Polonovski reaction); $(CF_3CO)_2O$, CH_2Cl_2 (Polonovski-Potier); etc.

18.7 Oxidative Rearrangements[116]

The final type of oxidation in Figure 18.1 leads to the cleavage of a σ bond between a carbonyl carbon and the α carbon, and its replacement by two carbon-heteroatom σ bonds, which is equivalent to the insertion of a heteroatom between two carbon atoms. The process results in an increase in the number of bonds to electronegative atoms at each end of the C—C σ bond and so corresponds formally to an oxidation. Many named reactions fall within this overall category, and some of the more important ones are summarized in Table 18.6.

Most of these reactions proceed through a variant of the mechanisms shown in Figure 18.5. The most important feature of these mechanisms is that the rearrangement is intramolecular, with synchronous migration and loss of the leaving group through transition states such as **18.115** for aldehydes and ketones and **18.116** for acid derivatives. Because the migrating group never becomes completely detached from the rest of the

116. Review: Smith, P.A.S. In de Mayo, P, Ed. *Molecular Rearrangements, Part 1* (Interscience Publishers: New York, 1963), p. 457.

Table 18.6 Oxidative Rearrangements

Substrate	Reagent	Product	Reaction Name
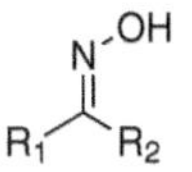	H_2SO_4; H_3PO_3; etc.	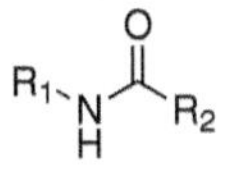	Beckmann rearrangement[117]
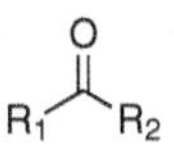	HN_3; etc.	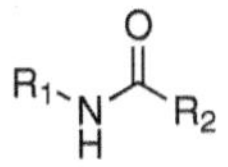	Schmidt reaction[118]
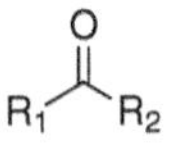	RCO_3H; etc.	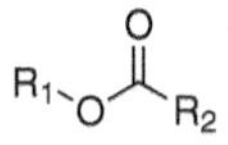	Baeyer-Villiger oxidation[119]
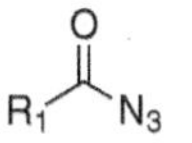	Δ or hν	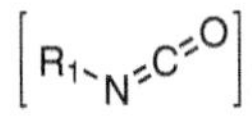	Curtius rearrangement[120]
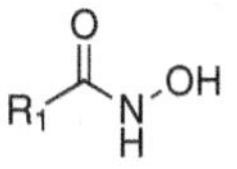	H_2SO_4; TsCl; etc.	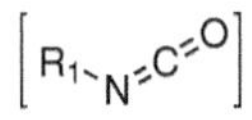	Lossen rearrangement[121]
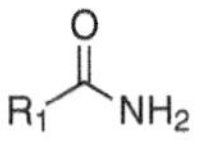	Br_2, KOH	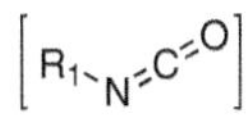	Hofmann rearrangement[122]

molecule or ion, almost all these reactions occur with complete (or almost complete) retention of configuration at the migrating center.[123]

---

117. (a) Beckmann, E. *Ber. dtsch. chem. Ges.* **1886**, *19*, 988. (b) Blatt, A.H. *Chem. Rev.* **1933**, *12*, 215. (c) Jones, B. *Chem. Rev.* **1944**, *35*, 335. (d) Popp, F.D.; McEwen, W.E. *Chem. Rev.* **1958**, *58*, 370. (e) Heldt, W.Z.; Donaruma, L.G., *Org. React.* **1960**, *11*, 1. (b) Gawley, R.E. *Org. React.* **1988**, *35*, 1. (c) McCarty, C.G. In Patai, S., Ed. *Chemistry of the Carbon-Nitrogen Double Bond* (Wiley: New York, 1970), p. 408.

118. (a) Schmidt, K.F. Z. *angew. Chem.* **1923**, *36*, 511. (b) Schmidt, K.F. *Chem. Ber.* **1924**, *57*, **704**. (c) Wolff, H. *Org. React.* **1946**, *3*, 307. (d) Benson, F.R. *Chem. Rev.* **1947**, *41*, 48. (e) Popp, F.D.; McEwen, W.E. *Chem. Rev.* **1958**, *58*, 370.

119. (a) Baeyer, A.; Villiger, V. *Ber Deut. chem. Ges.* **1899**, *32*, 3625. (b) Baeyer, A.; Villiger, V. *Ber Deut. chem. Ges.* **1900**, *33*, 858. (c) Hassall, C.H. *Org. React.* **1957**, *9*, 73. (d) Krow, G.R. *Org. React.* **1993**, *43*, 251. (e) Rentz, M.; Meunier, B. *Eur. J. Org. Chem.* **1999**, 737.

120. (a) Curtius, T. *Ber. dtsch. chem. Ges.* **1890**, *23*, 3023. (b) Curtius, T. *J. Prakt. Chem.* [2] **1894**, *50*, 275. (b) Smith, P.A.S. *Org. React.* **1946**, *3*, 337. (c) Saunders, J.H.; Slocombe, R.J. *Chem. Rev.* **1948**, *43*, 205.

121. (a) Lossen, W. *Ann. Chem. Pharm.* **1872**, *161*, 347. (b) Lossen, W. *Ann. Chem. Pharm.* **1875**, *175*, 271, 343. (c) Yale, H.l. *Chem. Rev.* **1943**, *33*, 209. (d) Popp, F.D.; McEwen, W.E. *Chem. Rev.* **1958**, *58*, 370.

122. (a) Hofmann, A.W. *Ber. dtsch. chem. Ges.* **1881**, *14*, 2725. (b) Hofmann, A.W. *Ber. dtsch. chem. Ges.* **1882**, *15*, 407, 762. (c) Hofmann, A.W. *Ber. dtsch. chem. Ges.* **1884**, *17*, 1406. (d) Hofmann, A.W. *Ber. dtsch. chem. Ges.* **1885**, *18*, 2734. (e) Hofmann, A.W. *Ber. dtsch. chem. Ges.* **1882**, *15*, 752. (f) Franklin, E.C. *Chem. Rev.* **1934**, *14*, 219. (g) Wallis, E.S.; Lane, J.F. *Org. React.* **1946**, *3*, 267.

123. For stereochemical studies of these reactions, see:

Baeyer-Villiger reaction: (a) Turner, R.B. *J. Am. Chem. Soc.* **1950**, *72*, 879. (b) Gallagher, T.F.; Kristschevsky, T.H. *J. Am. Chem. Soc.* **1950**, *72*, 882. (c) Mislow, K.; Brenner, J. *J. Am. Chem. Soc.* **1953**, *75*, 2319.

Beckmann rearrangement: (a) Kenyon, J.; Young, D.P. *J. Chem. Soc.* **1941**, 263. (b) Campbell, A.; Kenyon, *J. J. Chem. Soc.* **1946**, 25. (c) For a contradictory result, see Hill, R.K.; Chortyk, O.T. *J. Am. Chem. Soc.* **1962**, *84*, 1064.

Curtius rearrangement: (a) Jones, L.W.; Wallis, E.S. *J. Am. Chem. Soc.* **1926**, *48*, 169. (b) Schrecker, A.W. *J. Org. Chem.* **1957**, *22*, 33

Hofmann rearrangement: (a) Wallis, E.S.; Nagel, S.C. *J. Am. Chem. Soc.* **1931**, *53*, 2787. (b) Wallis, E.S.; Moyer, W.W. *J. Am. Chem. Soc.* **1933**, *55*, 2598. (c) Kenyon, J.; Phillips, H.; Pittman, V.P. *J. Chem. Soc.* **1939**, 1072. (d) Arcus, C.L.; Kenyon, J. *J. Chem. Soc.* **1939**, 916. (e) Archer, S. *J. Am. Chem. Soc.* **1940**, *62*, 1872.

Schmidt rearrangement: von Braun, J.; Friehmelt, E. *Chem. Ber.* **1933**, *66*, 684.

Lossen rearrangement: Wallis, E.S.; Dripps, R.D. *J. Am. Chem. Soc.* **1933**, *55*, 1701.

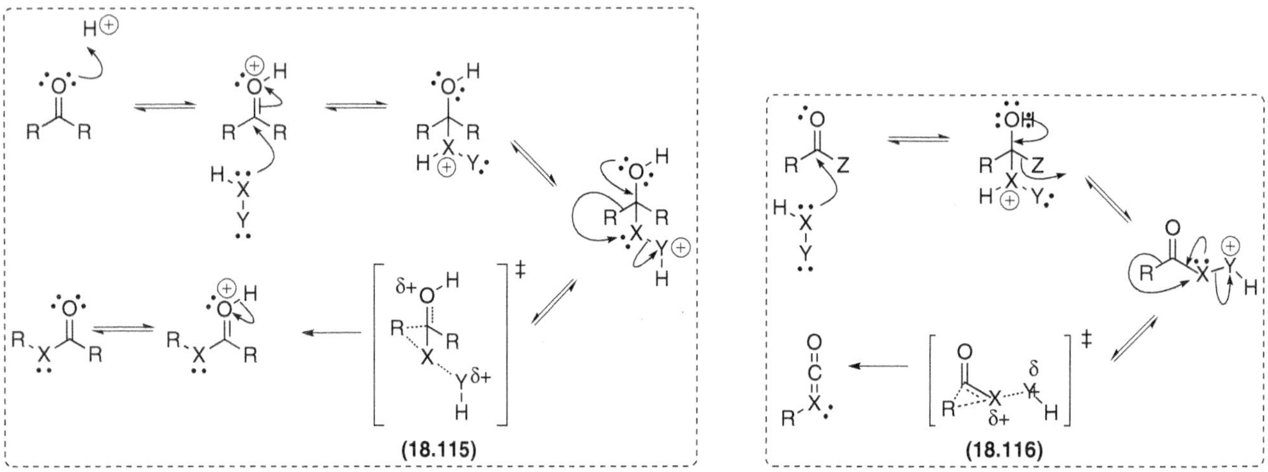

Figure 18.5 General mechanisms of oxidative rearrangements

Insertion of Oxygen into a C—C σ Bond: The Baeyer-Villiger Oxidation

The Baeyer[124]-Villiger[125] oxidation is the reaction between an aldehyde or ketone with a peroxidic reagent to give an ester by the formal insertion of an oxygen atom into the σ bond between the carbonyl carbon and the α carbon. The Baeyer-Villiger reaction has been especially useful for the synthesis of lactones from cyclic ketones (e.g., the synthesis of δ-lactones from cyclopentanones[126]).

The mechanism of the reaction, shown in Figure 18.6, involves two separate stages: (1) the addition of the peroxidic nucleophile to the carbonyl group (**18.117**) and (2) the rearrangement, with cleavage of the peroxide bond (**18.118**). Either of these stages may be rate determining, although the rearrangement step is slower in most cases.[127] It is important to note that there is a stereoelectronic requirement in the Baeyer-Villiger oxidation: the alkyl group *anti* to the departing carboxylate is the one that migrates. As illustrated by

Figure 18.6 The mechanism of the Baeyer-Villiger oxidation

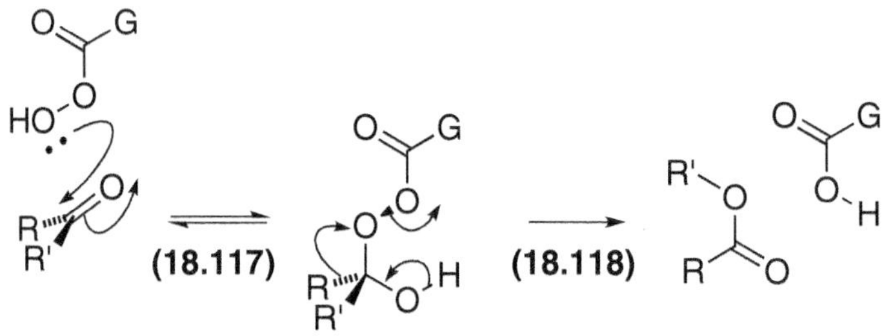

124. **Johann Friedrich Wilhelm Adolf von Baeyer (1835–1917)** studied under Bunsen and Kekulé (PhD, Berlin, 1858). His academic career took him to Berlin, Strasbourg, and Munich. His work in organic chemistry ranged from organic synthesis to physical organic chemistry. He was awarded the Nobel Prize in Chemistry in 1905. For more information, see: de Meijere, A. *Angew. Chem. Int. Ed.* **2005**, *44*, 7836, and the Nobel Foundation web site.

125. **Victor Villiger (1868–1934)** obtained his PhD in 1893 under Baeyer. Until 1904 he was Baeyer's personal assistant, then 1904 he left to work for Badische Anilin- und Sodafabrik, now BASF, and ultimately became a leader of the laboratories of the German chemical giant, I.G. Farben. For more detail, see: Hofmann, K.A. *Ber. dtsch. chem. Ges.* **1934**, *67*, A111.

126. Sager, W.F.; Duckworth, A. *J. Am. Chem. Soc.* **1955**, *77*, 188.

127. Rearrangement rate determining: (a) Hawthorne, M. F.; Emmons, W. D. *J. Am. Chem. Soc.* **1958**, *80*, 6398. (b) Okuno, Y. *Chem. Eur. J.* **1997**, *3*, 212. Addition rate determining: Singleton, D. A.; Szymanski, M. J. *J. Am. Chem. Soc.* **1999**, *121*, 9455.

the oxidation of optically active 2-*exo*-acetylnorbornane, the reaction occurs with retention of configuration (Example **18.119**).[128]

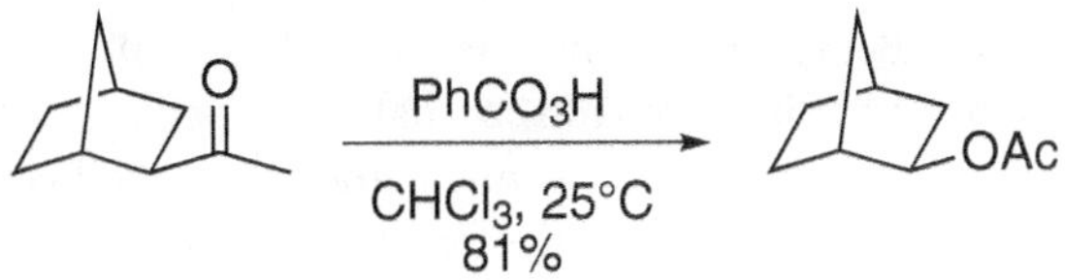

(18.119)

As one might expect from the fact that addition to the carbonyl group is involved and that addition to a carbonyl group is catalyzed by acids, the Baeyer-Villiger oxidation is facilitated by acids. By far the most common reagents for Baeyer-Villiger oxidation of ketones are organic peracids, and the ability of these acids, R—CO_2OH, to promote the reaction decreases as the electron deficiency of the α-carbon (i.e., the acid strength) decreases:[129]

$$CF_3 > c\text{-}CH{=}CHCO_2H > o\text{-}C_6H_4CO_2H > p\text{-}C_6H_4NO_2 > H \approx m\text{-}C_6H_4Cl > C_6H_5 > Me >> Bu$$

However, the move toward "green" chemistry and "green" oxidants has led to a resurgence of the examination of hydrogen peroxide as an oxidant for this reaction.[130] The first successes using hydrogen peroxide for the Baeyer-Villiger oxidation of cyclic ketones were with strained cycloalkanones such as cyclobutanones.[131] Arsonic acids have also been found to catalyze the reaction with unstrained cycloalkanones,[132] and certain platinum complexes also catalyze the reaction.[133] More recently, the use of fluorinated alcohols with[134] or without[135] Brønsted acids in the Baeyer-Villiger reaction of cycloalkanones has allowed the use of hydrogen peroxide as the terminal oxidant. The use of Oxone® in ionic liquids as solvents[136] has also proved to be a useful "green" modification of the reaction. The use of heterogeneous catalysts in the Baeyer-Villiger reaction has been reviewed.[137]

When a ketone is unsymmetrical and has the conformational flexibility to obtain more than one conformation suitable for rearrangement, the question of which group actually migrates becomes important. The ***migratory aptitudes***, or propensity of alkyl groups to migrate in the Baeyer-Villiger oxidation, is:[138]

$$3°R > \text{cyclohexyl} \approx 2°R \geq PhCH_2 \approx Ph \approx \text{vinyl} > 1°R > \text{cyclopropyl} > Me$$

In general, the migratory aptitude of a saturated alkyl group, R parallels the ease with which the corresponding alkyl bromide R-Br undergoes S_N1 substitution reactions. Thus, the methyl group almost never migrates when a methyl ketone is treated with a peracid, and secondary alkyl groups usually migrate in preference to primary alkyl groups. However, this general rule of thumb applies only to saturated alkyl groups. In general, unsaturated groups (vinyl groups and aryl groups) migrate approximately as readily as a secondary alkyl group and more readily than hydrogen, which, in turn, migrates more readily than a primary alkyl group.

---

128. (a) Berson, J. A.; Suzuki, S. *J. Am. Chem. Soc.* **1969**, 81, 4088. (b) Gallagher, T.F.; Kritchevsky, T.H. J. *Am. Chem. Soc.* **1960**, *72*, 882. (c) Turner, R.B. *J. Am. Chem. Soc.* **1960**, *72*, 878. (d) Mislow, K.; Brenner, J. *J. Am. Chem. Soc.* 1963, *75*, 2318.

129. Krow, G.C. In *Comprehensive Organic Synthesis*; Trost, B. M., Fleming, I., Eds.; (Pergamon: Oxford, 1991), vol. 7, pp 671–688.

130. Review: ten Brink, G.-J.; Arends, I.W.C.E.; Sheldon, R.A. *Chem. Rev.* **2004**, *104*, 4105.

131. Bogdanowicz, M.J.; Ambelang, T.; Trost, B.M. *Tetrahedron Lett.* **1973**, 14, 923.

132. Jacobson, S.E.; Mares, F.; Zambri, P.M. *J. Am. Chem. Soc.* **1979**, 101, 6938.

133. Strukul, G. *Angew. Chem., Int. Ed. Engl.* **1998**, 37, 1198.

134. Berkessel, A.; Andreae, M.R.M. *Tetrahedron Lett.* **2001**, *42*, 2293.

135. Neimann, K.; Neumann, R. *Org. Lett.* **2000**, 2, 2861.

136. Chrobok, A. *Tetrahedron* **2010**, *66*, 6216.

137. Jiménez-Sanchidrián, C.; Ruiz, J.R. *Tetrahedron* **2008**, *64*, 2011.

138. (a) Doering, W.v.E.; Speers, L. *J. Am. Chem. Soc.* **1950**, *72*, 5515. (b) Friess, S.L.; Soloway, A.H. *J. Am. Chem. Soc.* **1951**, *73*, 3968. (c) Sauers, R.R.; Ubersax, R.W. *J. Org. Chem.* **1965**, *30*, 3939.

Insertion of Nitrogen into a C—C σ Bond of an Aldehyde or Ketone: The Beckmann and Schmidt Rearrangements

The accepted mechanism of the Beckmann[139] rearrangement (Figure 18.7) proceeds from a protonated oxime (**18.120**) to a nitrilium ion (**18.122**) that is trapped by water to give the amide (**18.125**). Oximes can exist as geometric isomers, and when the Beckmann rearrangement is carried out under conditions where the two geometric isomers are not interconverted, the group *anti* to the oxime hydroxyl group migrates (**18.120**). By contrast, when the Beckmann reaction is carried out under strongly ionizing conditions, the two isomers of the oxime are rapidly interconverted,[140] and a mixture of the possible products is formed. In certain cases, the isomerization occurs more rapidly than the rearrangement, making the product independent of the reactant stereochemistry.[141] Aliphatic ketoximes isomerize so rapidly that practically all the mechanistic (and stereochemical) studies on the Beckmann rearrangement have been carried out on aromatic systems. The solvent also has a role to play in the Beckmann rearrangement, as confirmed by computational studies.[142] Isomerization of the oxime is less pronounced in low dielectric solvents with Lewis acids and is facilitated by proton acids in protic solvents.[143] In an ironic twist,[144] the Beckmann rearrangement of cyclohexanone oxime to caprolactam (Example **18.126**, an important intermediate in the synthesis of Nylon®) is catalyzed by the ionic liquid formed between caprolactam itself and tetrafluoroboric acid! More recently, the soft Lewis acid, mercuric chloride, has been found to promote the Beckmann rearrangement in aprotic solvents (e.g., Example **18.127**).[145]

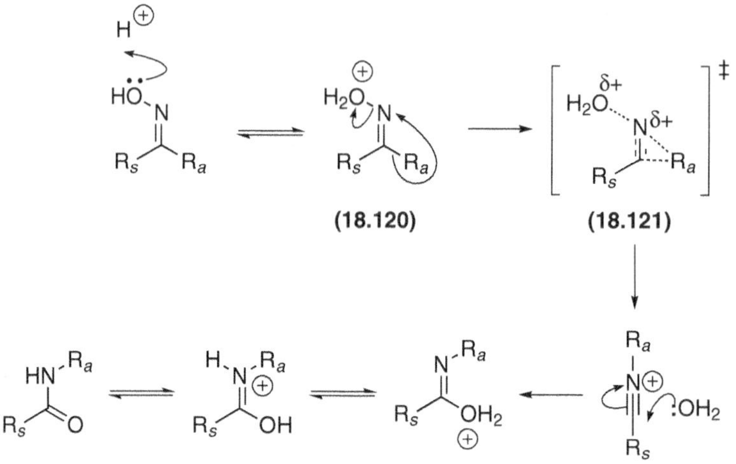

Figure 18.7 Mechanism of the Beckmann rearrangement

139. **Ernst Otto Beckmann (1853–1923)** was educated at Leipzig, (PhD; 1878). A journeyman in many ways, his career took him from Strasbourg (1878) to Brunswick (1879), to Leipzig (1884), to Giessen (1891), and to Erlangen (1892). He then returned to Leipzig (1897) and ended his career in Berlin (1902). His research covered areas of organic synthesis and natural products chemistry. For more biographical information, see: Oesper, R.E. *J. Chem. Educ* **1944**, *21*, 470.

140. It has long been known that the isomerization of oximes is promoted by acids: Hantzsch, A.; Lucas, A. *Ber. dtsch. chem. Ges.* **1895**, *28*, 744.

141. Smith, P.A.S.; Antoniades, E.P. *Tetrahedron* **1960**, *9*, 210.

142. (a) Nguyen, M.T.; Raspoet, G.; Vanquickenborne, L.G. *J. Am. Chem. Soc.* **1997**, *119*, 2552. (b) Yamabe, S.; Tsuchida, N.; Yamazaki, S. *J. Org. Chem.* **2005**, *70*, 10638.

143. Brown, R.F.; van Gulick, N.M.; Schmidt, G.H. *J. Am. Chem. Soc.* **1955**, *77*, 1094.

144. Guo, S.; Du, Z.; Zhang, S.; Li, D.; Li, Z.; Deng, Y. *Green Chem.* **2006**, *8*, 296.

145. Ramalingan, C.; Park, Y.-T. *J. Org. Chem.* **2007**, *72*, 4536.

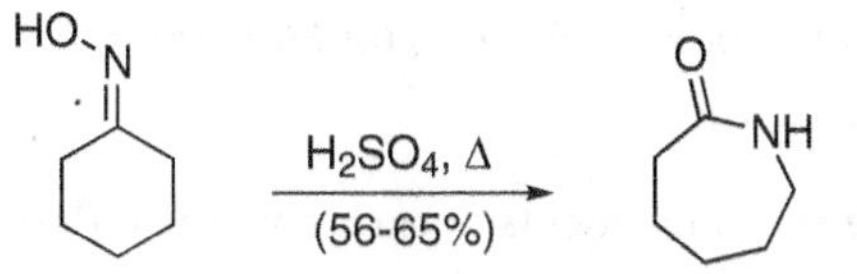

(18.126)[146]

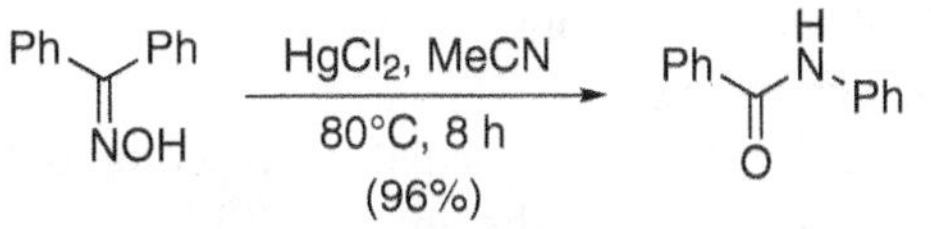

(18.127)[145]

In mechanistic terms (Figure 18.8), the Schmidt[147] reaction of azides is formally the nitrogen analog of the Baeyer-Villiger oxidation, with molecular nitrogen functioning as the leaving group. In the Schmidt reaction, the carbonyl compound is treated with hydrazoic acid and a strong acid catalyst to give an adduct (**18.128**) that then loses nitrogen to give the amide. The reaction occurs by one of two mechanisms: (1) by concerted rearrangement of **18.128** with concomitant loss of nitrogen and formation of the conjugate acid of the amide (**18.124**) and (2) by dehydration of **18.128**, with rearrangement of the cationic intermediate (**18.129**) to a nitrilium ion (**18.122**). The exact mechanism followed depends on the carbonyl compound, but—regardless of mechanism—at no time does the migrating group dissociate completely from the rest of the molecule: it rearranges with retention of configuration.

The possibility of replacing the hydrazoic acid with an alkyl azide has been explored off and on for many years with little or no success. Recently, however, an intramolecular analog of the Schmidt reaction has been developed. In this reaction, an ω-azidoketone (Example **18.130**) is treated with a strong acid or Lewis acid, and an intramolecular rearrangement through the cyclic *N*-diazonium ion **18.132** ensues. Loss of the Lewis acid then provides the final amide (Example **18.133**). The rearrangement proceeds with retention of configuration at the migrating carbon.[148]

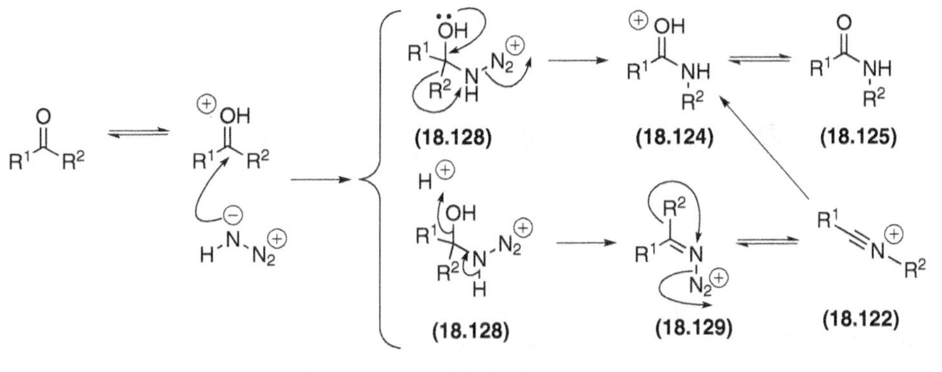

Figure 18.8 Mechanisms of the Schmidt reaction

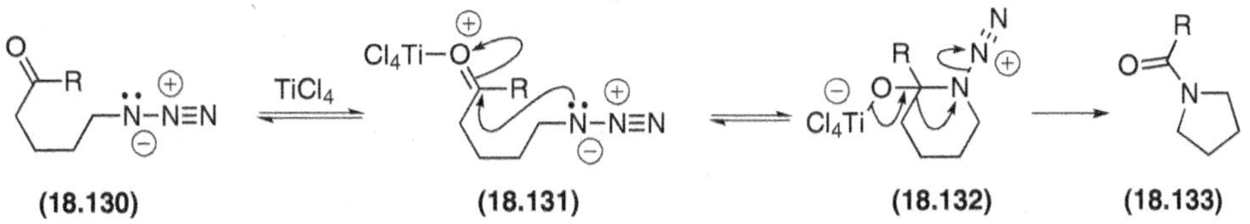

146. Marvel, C.S.; Eck, J.C. *Org. Synth., Coll. Vol. 2* **1943**, 371.

147. **Karl Friedrich Schmidt (1887–1971)** was educated at Heidelberg (PhD; 1912, under Curtius). From 1912 to 1921, he was Assistant to Theodor Curtius, except for World War I. His career then took him to the Swedish University of Åbo in Turku, Finland (1921) and then Heidelberg (1925). From 1926 to 1936 Schmidt was a director of Knoll AG, and in 1936 he opened his own research laboratory (but remained associated with Knoll). Schmidt's work consisted of adapting earlier reactions to industrial use and production.

148. Milligan, G.L.; Mossman, C.J.; Aube, J. *J. Am. Chem. Soc.* **1995**, *117*, 10449.

Oxidative Rearrangements in Carboxylic Acids: The Hofmann, Curtius, and Lossen Rearrangements

The most widely used oxidative rearrangements of carboxylic acid derivatives all give rise to amine derivatives by migration of carbon to nitrogen to form an intermediate isocyanate. As we saw in Figure 18.6, all three reactions share a common mechanism, with only the identity of the leaving group changing.

In the Hofmann rearrangement,[149] the participating reactant is the anion of an *N*-haloamide. This is formed by treatment of the amide with halogen and base, which first gives an *N*-haloamide (Example **18.134**) that is then rapidly deprotonated by the base. The conjugate base then rearranges with loss of the halide anion to give the isocyanate (Example **18.135**). The isocyanate then reacts further to give a carbamate anion (Example **18.137**), whose spontaneous decarboxylation yields the product amine. Since its discovery, the Hofmann rearrangement has been the subject of continuous study, especially with respect to the oxidant used. Thus, over the second half of the 20th century, a wide range of oxidizing reagents was studied, including lead tetraacetate,[150] benzyltrimethylammonium perbromide and sodium hydroxide,[151] and *N*-bromosuccinimide in the presence of mercury or silver salts in DMF.[152] In the last two decades or so of the century, the use of iodine (III) reagents was extended to the Hofmann rearrangement, permitting this reaction to be carried out under mildly acidic conditions.[153] Iodine (III) reagents permit the reaction to be carried out without using highly toxic reagents or heavy metals, making the reaction much "greener."

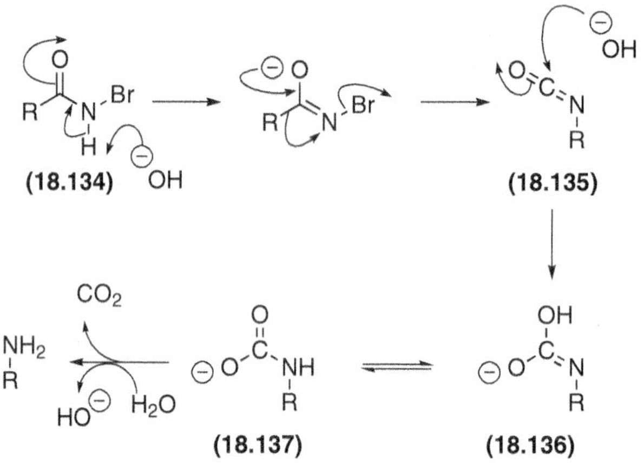

149. Review: Wallis, E.S.; Lane, J.F. *Org. React.* **1946**, *3*, 267.

August Wilhelm von Hofmann (1818–1892) was educated at Giessen under Liebig. His career took him from Bonn (1841) to the Royal College of Chemistry in London (1845). He returned to Bonn in 1864 and in 1865 he moved to Berlin, where he remained until his death in 1892. He was ennobled (made *von* Hofmann) in 1888. For more biographical details, see: Beer, J.J. *J. Chem. Educ.* **1960**, *37*, 248; Meinel, C. *Angew. Chem. Int. Ed. Engl.* **1992**, *31*, 1265; Travis, A.S. *Endeavour* **1992**, *16*, 59.

150. (a) Baumgartan, H.E.; Smith, H.L.; Staklis, A. *J. Org. Chem.* **1975**, *40*, 3554. (b) Acott, B.; Beckwith, A.L.J.; Hassanali, A. *Aust.J.Chem* **1968**, *21*, 197. (c) Acott, B.; Beckwith, A.L.J.; Hassanali, A. *Aust. J. Chem.* **1968**, *21*, 185. (d) Acott, B.; Beckwith, A.L.J. *J. Chem. Soc., Chem. Commun.* **1965**, 161.

151. Kajigaeshi, S.; Asano, K.; Fujasaki, S.; Kakinami, T.; Okamoto, T. *Chem. Lett.* **1989**, 463.

152. Jew, S.A.; Park, H.G.; Park, H.-J.; Park, M.A. ; Cho, Y.-S. *Tetrahedron Lett.* **1990** ,*31*, 1559.

153. (a) Radhakrishna, A.S.; Parham, M. E.; Riggs, R.M.; Loudon, G.M. *J. Org. Chem.* **1979**, *44*, 1746. (b) Loudon, G.M.; Radhakrishna, A.S.; Almond, M.R.; Blodgett, J.K.; Boutin, R.H. *J. Org. Chem.* **1984**, *49*, 4272. (c) Loudon, G.M.; Boutin, R.H. *J. Org. Chem.* **1984**, *49*, 4277. (d) Almond, M.R.; Stimmel, J.B.; Thompson, A.E.; Loudon, G.M. *Org. Synth.* **1988**, 67, 132. (e) Tamura, M.; Jacyno, J.; Stammer, C.H. *Tetrahedron Lett.* **1986**, *27*, 5435. (f) Radhakrishna, A.S.; Rao, C.D.; Varma, R.K.; Sing, B.B.; Bhatnagar, S.P. *Synthesis* **1983**, 538. (g) Wasielewki, C.; Topolski, M.; Dembkowski, M. *J. Prakt. Chem.* **1989**, *331*, 507. (h) Lazbin, I. M.; Koser, G.F. *J. Org. Chem.* **1986**, 51, 2669. (k) Vasudervan, A.; Koser, G.F. *J. Org. Chem.* **1988**, *53*, 5158. (l) Moriarty, R.M.; Khoshrowshahi, J.S.; Awaathi, A.K.; Penmaata, R. *Synth. Commun.* **1988**,*18*, 1179. (m) Moriarty, R.M.; Chany, C.J., II; Vaid, R.K.; Prakash, O.; Tuladhar, S.M. *J. Org. Chem.* **1993**, *58*, 2478.

In the Curtius[154] rearrangement, an acyl azide (Example **18.138**) is the participating reactant. The azide is formed by the reaction of an acid chloride or mixed anhydride with azide ion or by the reaction of an acyl hydrazide with nitrous acid. On heating, or on irradiation with ultraviolet light, it loses molecular nitrogen to give the isocyanate. One major advantage of the Curtius rearrangement is that it permits the isocyanate (Example **18.135**) to be generated under anhydrous conditions, where it may be isolated.

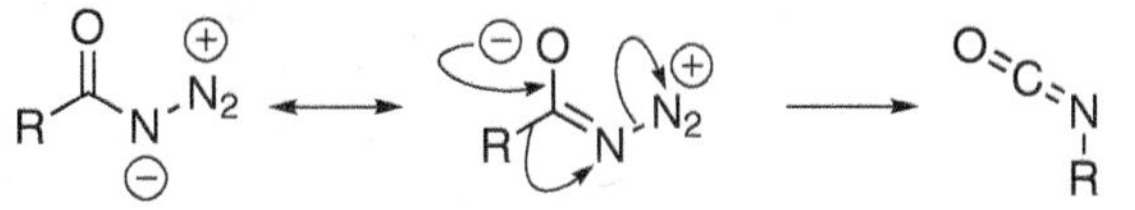

(18.138) **(18.135)**

In the Lossen[155] rearrangement, the conjugate base of an *O*-acylhydroxamic acid or an *O*-sulfonyl-hydroxamic acid (Example **18.139**) is the participating reactant. To obtain this reactant, the carboxylic acid is first converted to a hydroxamic acid, which is then acylated or sulfonated on oxygen to give the *O*-acyl or *O*-sulfonylhydroxamate. The treatment of this hydroxamate derivative with base gives the conjugate base that then loses either a carboxylate or sulfonate anion to give the isocyanate. The Lossen rearrangement can also be carried out directly on the hydroxamic acid itself by heating with strong acid. The general move toward "green" chemistry has also had a major impact on the Lossen rearrangement. Although this reaction had not been developed as rapidly as the Curtius, Schmidt, and Hofmann rearrangements, carbonyl-diimidazole has been found to be an effective reagent for the Lossen rearrangement of hydroxamic acids (as in Example **18.140**),[156] as has cyanuric chloride.[157] In an intriguing twist on this reaction, it has recently been reported that certain arylhydroxamic acids actually rearrange with potassium carbonate in dimethyl sulfoxide.[158] In this reaction, the hydroxamic acid also furnishes the leaving group.

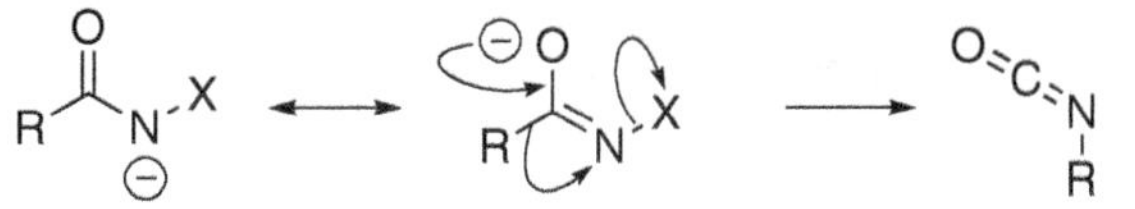

(18.139) **(18.135)**

X = OCOR, OSO_2R

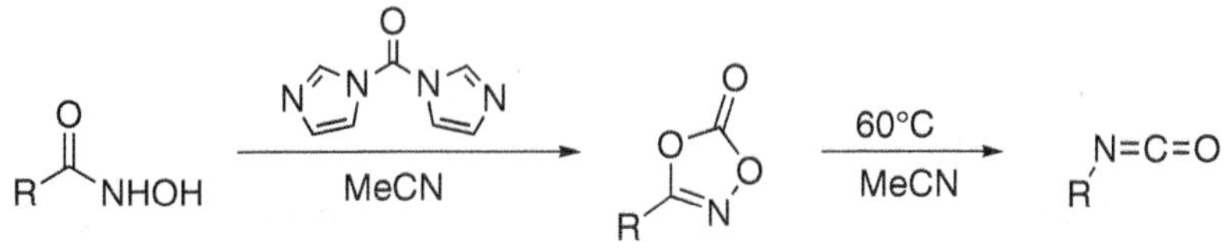

(18.140)

Like the analogous reactions in aldehydes and ketones, these reactions proceed with retention of configuration at the migrating carbon. Their utility in synthesis is illustrated

154. **Julius Wilhelm Theodor Curtius (1857–1928)** was educated at Heidelberg and Leipzig (PhD; 1882). Curtius immediately moved to Munich to study under Baeyer. His career then took him to Erlangen (1886), Kiel (1889), Bonn (1897), and Heidelberg (1898), where he remained for the rest of his career. For more biographical information, see: Freudenberg, K. *Chem. Ber.* **1963**, *96*, I.

155. **Wilhelm Lossen (1838–1906)** studied under Albert Niemann, the first chemist to isolate cocaine, at Göttingen. His career took him to Karlsruhe (1862), Halle (1863), Heidelberg (1866), and the Albertus University in Königsburg, Prussia (1877), where he taught until 1904. Lossen's work included the discovery of hydroxylamine and deducing the first correct structure for cocaine. For additional detail, see: Priesner, C. *Neue Deutsche Biographie*, vol. 15, p. 202.

156. Dubé, P.; Fine, N.F.; Vitelino, M.; Couturier, M.; Abbousafy, C.L.; Pichette, S.; Jorgensen, M.L.; Hardinck, M. *Org. Lett.* **2009**, *11*, 5622.

157. Hamon, F.; Prié, G.; Lecornué, F.; Papot, S. *Tetrahedron Lett.* **2009**, *50*, 6800.

158. Hoshino, Y.; Okuno, M.; Kawamura, E.; Honda, K.; Inoue, S. *Chem. Commun.* **2009**, 2281.

by the historical examples below, where the absolute configurations of (+)-hydratropic acid (Example **18.141**) and (−)-α-phenylethylamine (Example **18.144**) were correlated.[159]

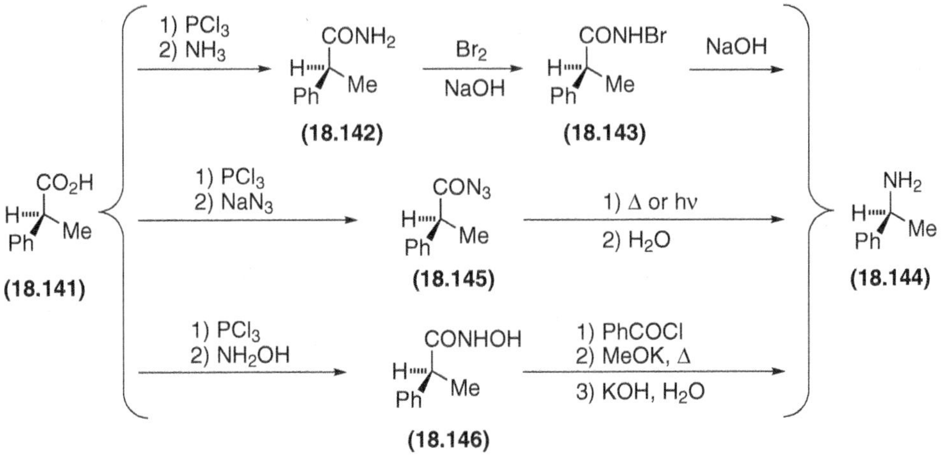

Problem

18-7 Draw the structure of the major product of each of the following reactions. Give reasons for your answer.

(a) 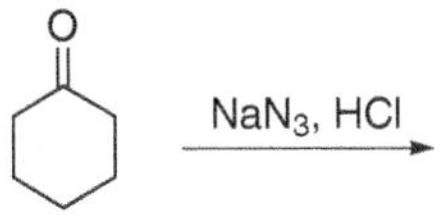

(b) 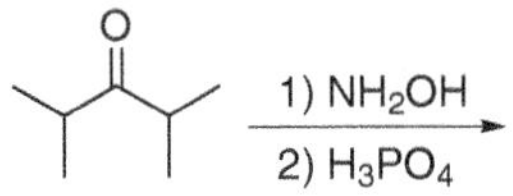

(c) 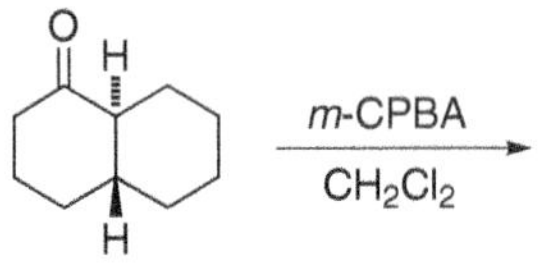

(d) 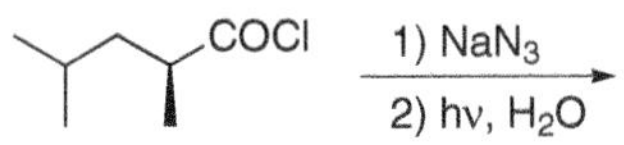

(e) 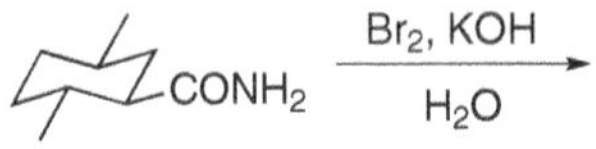

(f) 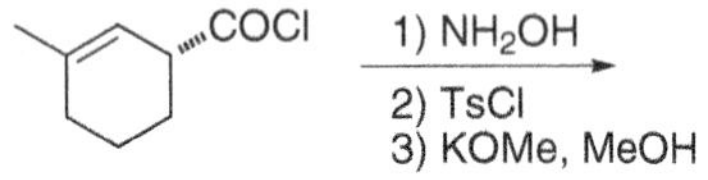

(g) 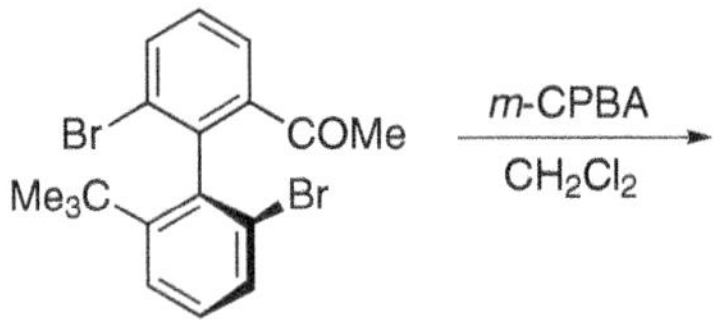

(h) 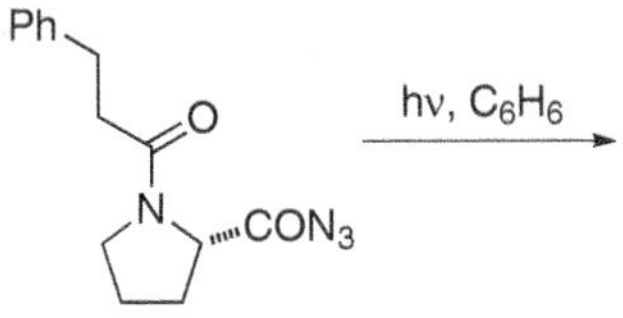

Reaction Synopses

Baeyer-Villiger Oxidation

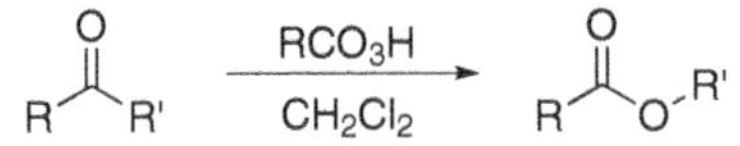

Reagents:	HCO_3H; $MeCO_3H$; *m*-CPBA; etc.
or	H_2O_2, R_FOH, $PhAsO_3H$; H_2O_2, R_FOH, TsOH; etc.
or	H_2O_2, zeolites; etc.

159. This correlation has been accomplished by all five methods (Hofmann, Curtius, Lossen, Schmidt, and Beckmann rearrangements): (a) Arcus, C.L.; Kenyon, J. *J. Chem. Soc.* **1939**, 916. (b) Kenyon, J.; Young, D.P. *J. Chem. Soc.* **1941**, 263. (c) Campbell, A.; Kenyon, J. *J. Chem. Soc.* **1946**, 25.

Beckmann Rearrangement

Reagents: H_2SO_4, H_3PO_4 or $POCl_3$, PCl_5, P_2O_5, $SOCl_2$, $(COCl)_2$, cyanuric chloride, Ac_2O, RN=C=NR, RN=C=O; etc.

Schmidt Reaction

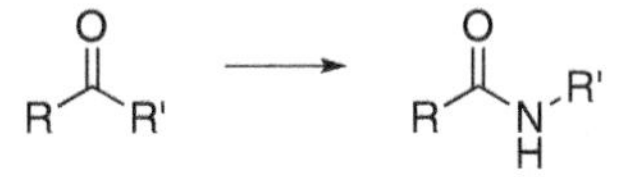

Reagents: HN_3; NaN_3, HCl; Me_3SiN_3, ZnI_2; etc.

Hofmann Rearrangement

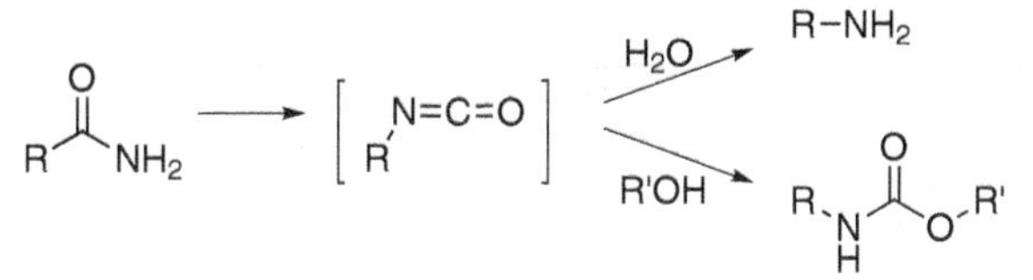

Reagents: Br_2, KOH, H_2O; NaOCl, H_2O; etc. or Br_2, ROH, RONa

Curtius Rearrangement

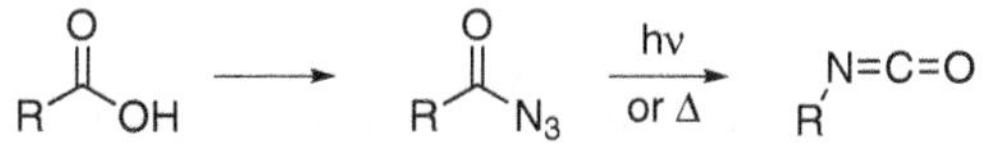

Reagents: (1) PCl_3; (2) NaN_3; (3) Δ

or (1) $ClCO_2Et$, Et_3N, THF; (2) NaN_3, H_2O; (3) hν, PhH

or NaN_3, H_2SO_4, H_2O, Δ

or (1) PCl_3; (2) H_2NNH_2; (3) HONO

Lossen Rearrangement

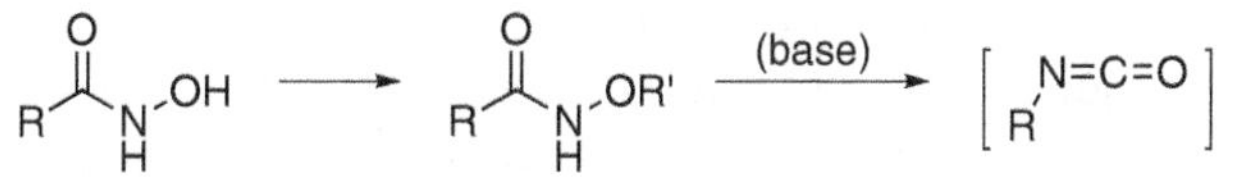

Reagents: H_2SO_4; H_3PO_4; etc.

or (1) PhCOCl; (2) KOMe, MeOH, Δ; Ac_2O, NaOAc, Δ;

or (1) TsCl, py; (2) $KOCMe_3$; etc.

or K_2CO_3, Me_2SO, Δ; etc.

18.8 Oxidation of Unsaturated Hydrocarbons

Up to now, we have discussed the oxidation of organic compounds at functional groups with a heteroatom: alcohols, ethers, aldehydes, ketones and amines. Now, we will change focus and concentrate on the oxidation of hydrocarbons, instead. Hydrocarbon oxidations generally fall into two major classes (Table 18.7). The first of these is oxidative substitution, which may involve replacement of the C—H σ bond by a σ bond to a more electronegative element, or replacement of a C—C σ bond by bonds to an electronegative

Table 18.7 Forms of Oxidation Reactions

Oxidative Substitution — Dehydrogenation

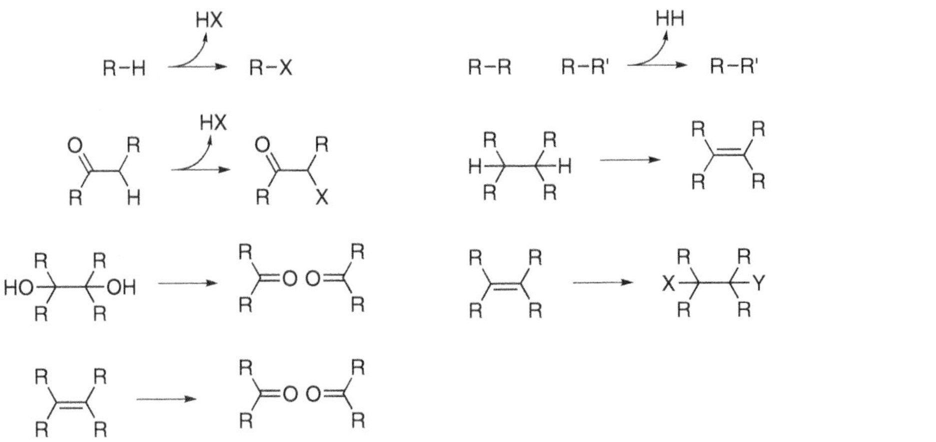

element. The second is dehydrogenation, where a pair of hydrogen atoms is removed from adjacent or non-adjacent carbon atoms to give a new carbon-carbon bond.

The presence of a π bond in a molecule renders the adjacent groups unusually reactive because the intermediates generated by cleavage of the σ bond adjacent to the π bond are resonance-stabilized. Allyl and benzyl halides are much more susceptible to displacement by both S_N1 and S_N2 mechanisms. Allyl and benzyl alcohols are oxidized by reagents that do not oxidize similar saturated alcohols, and these reactions may proceed through allyl and benzyl free radicals. Allylic hydrogens can be replaced by other groups under relatively mild conditions. The hydrogen atoms α to a carbonyl, cyano, or similar group are unusually acidic because the conjugate base formed (an enolate-type anion) is stabilized by resonance.

Oxidation of Alkenes by Addition

Epoxidation

The most important reaction that oxidizes an alkene π bond may well be epoxidation if one is to judge by the frequency with which it is used as a critical step in a synthesis. The most common method for forming epoxides is by oxidation of alkenes with organic peroxyacids (or peracids),[160] a reaction occasionally known as the ***Prilezhaev reaction*** after the Russian chemist who first reported it.[161] One important peroxyacid is *m*-chloroperbenzoic acid (*m*-CPBA), whose thermal stability and commercial availability made it a reagent of choice for the preparation of epoxides. Recent difficulties in obtaining this reagent due to regulations for its transportation have led to its frequent replacement by magnesium

160. Reviews: (a) Plesnicar, B. In Trahanovsky, W.S., Ed. *Oxidation in Organic Chemistry. Part C.* (Academic Press: New York, 1978), 211. (b) Hudlicky, M. *Oxidations in Organic Chemistry. Am. Chem. Soc. Monogr. 186* (American Chemical Society: Washington, D.C., 1990).

161. Prileschajew, N. *Ber. dtsch. chem. Ges.* **1909**, *42*, 4811.

Nikolai Aleksandrovich Prilezhaev (1872–1944) was destined for the priesthood by his father, but studied in chemistry under Wagner at Warsaw (M Chem, 1912). His career took him from Warsaw (1912); to Kiev Polytechnic Institute (1915), where he was Rector after the Russian Revolution; and to Minsk (1924), where he helped to organize the chemistry faculties at Belarus State University and Belarus Polytechnic Institute. He was elected a Corresponding Member of the USSR Academy of Sciences in 1933 and of the Belarus Academy of Sciences in 1940. For more biographical detail, see: Akhrem, A.A.; Prilezhaeva, E.N.; Meshcheriakov, A.P. *Zh. Obshch. Khim.* **1951**, *21* (11).

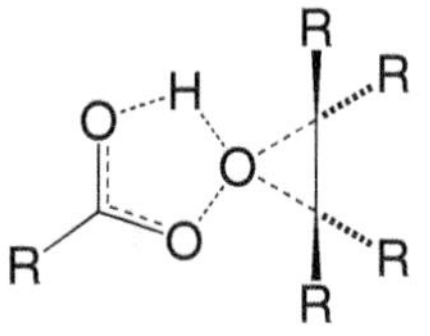

Figure 18.9 Transition state for concerted transfer of oxygen in epoxidation

monoperphthalate (MMPP),[162] which is much less shock sensitive and whose transportation is less heavily regulated.

The reaction in non-hydroxylic solvents is concerted and is generally considered to involve the transfer of the hydroxyl oxygen to the alkene from the intramolecularly hydrogen-bonded form of the peroxyacid (Figure 18.9).[163] It gives the epoxide where the oxygen atom is added to the less hindered face of the π bond in a suprafacial manner, thus preserving the stereochemistry of the original alkene in the epoxide product. In the reaction **18.147**, for example, the epoxidation occurs on the less hindered *exo* face of the molecule. In dienes and polyenes, the oxygen is added to the most substituted, most electron-rich π bond first.

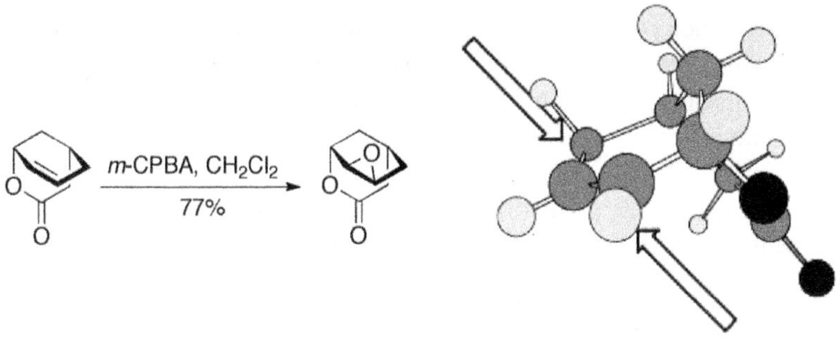

(18.147)

Epoxidation of alkenes by peracids is facilitated by electron-withdrawing groups on the peracid[164] and by electron-releasing groups on the alkene.[165] Epoxidation by peracids is also facilitated by non-polar solvents and retarded by solvents that can form hydrogen bonds to the peracid.[166] One of the limitations of epoxidation reactions with organic peracids is the potential for further reactions of the epoxide catalyzed by the carboxylic acid by-product. For this reason, sodium bicarbonate is often used as a buffering agent in epoxidations.

The epoxidation of alkenes can be carried out using organic peroxides (e.g., *tert*-butyl hydroperoxide, *t*-BuOOH) and transition metal compounds as catalysts. *tert*-Alkyl hydroperoxides have three properties that make them attractive oxidants: they are stable, they are commercially available, and they do not form acidic reduction products. Thus, the treatment of an alkene with molybdenum hexacarbonyl and *tert*-butyl or amyl hydroperoxide (Example **18.148**) gives the epoxide; if the alkene contains two π bonds (e.g., *S*-limonene), the more substituted double bond is oxidized first. Predicting the stereochemistry of epoxidation is not always straightforward, but it is usually safe to predict that oxygen will be delivered to the less hindered face of the π bond, as illustrated by the epoxidation of β-pinene (Example **18.149**). Here, access to the top face of the double bond is hindered by the methyl group on the bridge.

---

162. Review: Heaney, H.H. *Aldrichimica Acta* **1993**, *26*, *35*.

163. Bartlett, P.D. *Rec. Chem. Progr.* **1957**, *18*, 111.

164. (a) Vilkas, M. *Bull. Soc. Chim. France* **1959**, 1401. (b) Ogata, Y.; Tabushi, I. *J. Am. Chem. Soc.* **1961**, *83*, 3440. (c) Ogata, Y.; Tabushi, I. *J. Am. Chem. Soc.* **1961**, *83*, 3444.

165. (a) Lee, J.B.; Uff, B.C. *Quart. Rev. Chem. Soc.* **1967**, *21*, 431. (b) Medvedev, S.; Blokh, O. *J. Phys. Chem. (Moscow)* **1933**, *4*, 721. (c) Boeksen, J.; Stuurman, J. *Rec. Trav. Chim. Pays-Bas* **1937**, *56*, 1034. (d) Boeksen, J.; Nonegraaff, C.J.A. *Rec. Trav. Chim. Pays-Bas* **1942**, *61*, 69. (e) Swern, D. *J. Am. Chem. Soc.* **1947**, *69*, 1692. (f) Frostick, F.C., Jr.; Phillips, B.; Starcher, P.S. *J. Am. Chem. Soc.* **1959**, *81*, 3350.

166. (a) Renolen, P.; Ugelstad, J. *J. Chim. Phys.* **1960**, *57*, 1976. (b) Itier, J.I.; Tournaire, M.; Casadevall, A. *Compt. rend. Hebd. Seances Acad. Sci.*, *Ser. C* **1970**, *271*, 878. (c) Kavčič, R.; Plesničar, B. *J. Org. Chem.* **1970**, *35*, 2033.

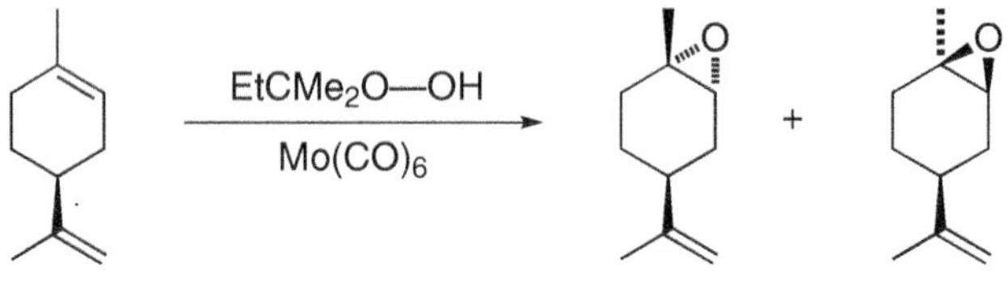

(18.148)

21% 50%

S-Limonene

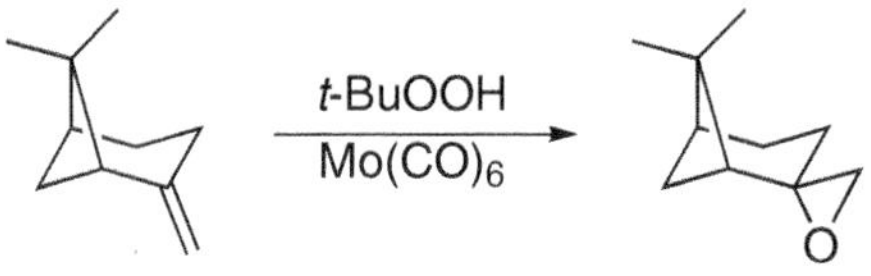

(18.149)

β-Pinene

Epoxidation of Allyl Alcohols

The functional group of an allylic alcohol affects both the regiochemistry and the stereochemistry of the epoxidation reaction due to its ability to form hydrogen bonds to the reagent. The hydroxyl group affects regiochemistry by directing the epoxidizing agent to the double bond nearest to itself whether the oxidizing agent is a peracid or a metal-based reagent.

The effect of the allylic hydroxyl group on the stereochemistry of epoxidation is nicely illustrated by cyclic allylic alcohols. Peracids oxidize cyclic allylic alcohols to the epoxide where the epoxide oxygen is *cis* to the hydroxyl group. When the hydroxyl group is esterified, the stereochemistry of the epoxidation is determined by steric factors and the reagent approaches the less hindered face of the molecule. For instance, 2-cyclohexenol reacts with perbenzoic acid to give predominantly (90:10) the epoxide where the epoxide oxygen is *cis* to the hydroxyl group (Example **18.150**).[167] Its acetate ester, on the other hand, gives a 57:43 mixture of products in which the major isomer is the one with the epoxide oxygen *trans* to the ester group (Example **18.151**).

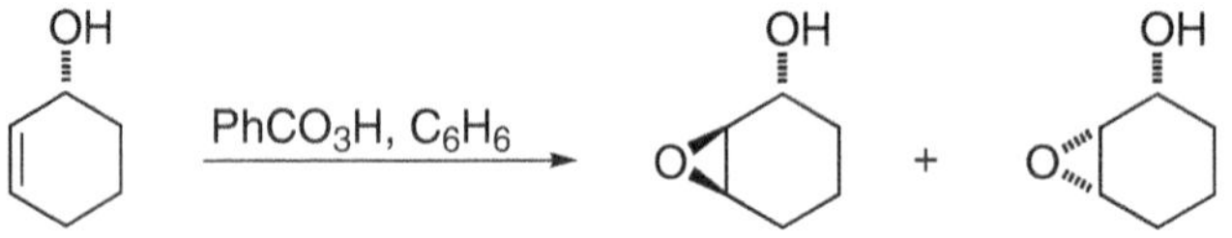

(18.150)

(8%) (78%)

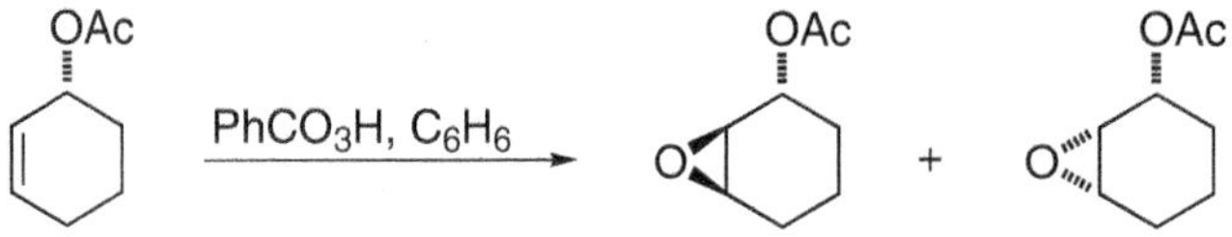

(18.151)

(38%) (30%)

These observations have been rationalized on the basis of a transition state (Example **18.152**) in which the peracid is hydrogen bonded to the alcohol hydroxyl group.[168] According to this model, the peracid forms a hydrogen bond with one of the peroxy oxygens of the peracid, resulting in the peracid being directed to the same face of the double bond.

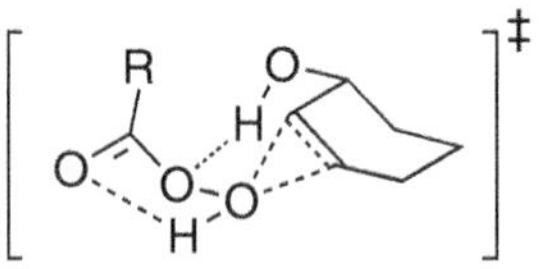

(18.152)

167. Chamberlain, P.; Roberts, N.L.; Whitham, G.H. *J. Chem. Soc. B* **1970**, 1374.

168. (a) Henbest, H.B. *Proc. Chem. Soc., London* **1963**, 159. (b) Cerefice, S.A.; Fields, E.K. *J. Org. Chem.* **1976**, *41*, 355.

Transition metal catalyzed epoxidation is especially useful when the double bond is part of an allyl alcohol. In these reactions, a vanadium complex, VO(acac)$_2$, has proved to be particularly useful, normally giving the desired product in yields of 80% or more. By using this reagent, it is possible to epoxidize an allyl alcohol to the exclusion of other double bonds (e.g., Example **18.153**).[169]

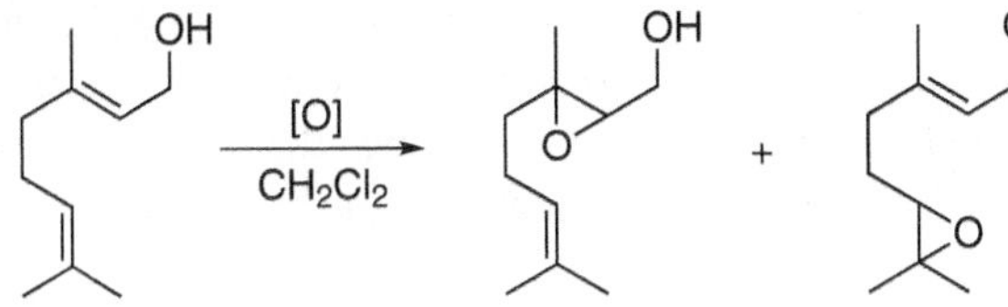

(18.153)

m-CPBA	33%	67%
t-BuOOH, VO(acac)$_2$	99%	1%

Epoxidation of Conjugated Carbonyl Compounds

The alkene π bond conjugated to a carbonyl group is electron deficient, so epoxidation with electrophilic oxidizing agents is often slow. However, epoxidation of α,β-unsaturated carbonyl compounds can be readily effected using a nucleophilic oxidizing agent. By far the most widely used such reagents are the conjugate bases of hydrogen peroxide and alkyl hydroperoxides (e.g., Examples **18.154** and **18.155**).[170] These reactions proceed by a two-step mechanism[171] involving Michael addition of the peroxide nucleophile, followed by intramolecular displacement of the leaving group from oxygen. As a consequence, they are not stereospecific like the reactions discussed earlier but are stereoselective, as illustrated by epoxidation of the geometric isomers of 3-methyl-3-penten-2-one (Examples **18.156** and **18.157**), which give the same major product (Example **18.158**).[174]

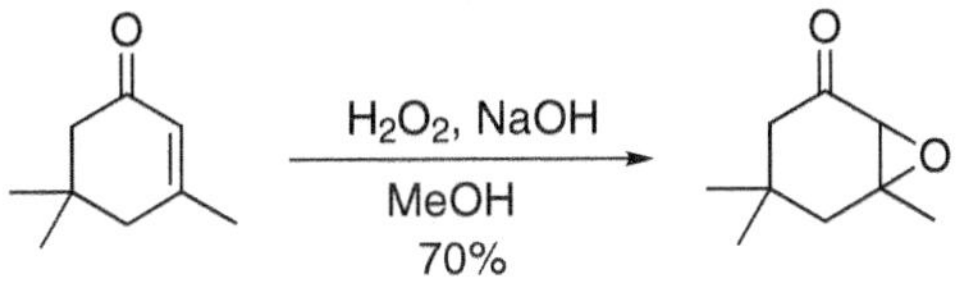

(18.154)[172]

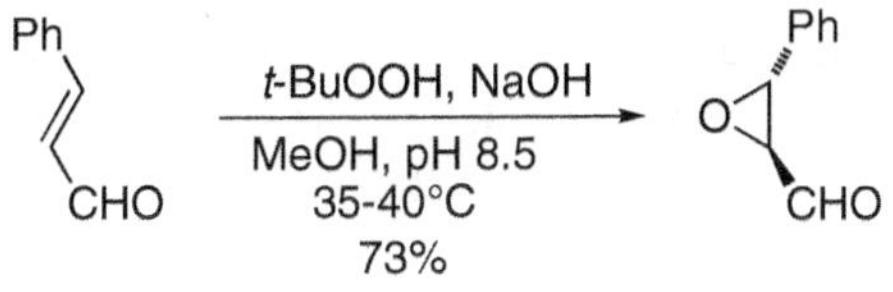

(18.155)[173]

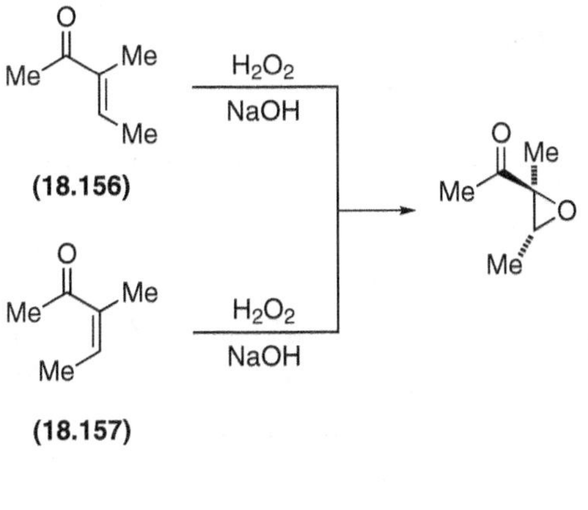

(18.158)[174]

169. Sharpless, K.B.; Verhoeven, T.R. *Aldrichimica Acta* **1979**, *12*, 63.
170. (a) Hassall, C.H. *Org. React.* **1957**, *9*, 73. (b) Yang, N.C.; Finnegan, R.A. *J. Am. Chem. Soc.* **1958**, *80*, 5845.
171. Bunton, C.A.; Minkoff, G.O. *J. Chem. Soc.* **1949**, 665.
172. Wasson, R.L.; House, H.O. *Org. Syn. Coll. Vol. 4.* **1963**, 552.
173. Payne, G.B. *J. Org. Chem.* **1960**, *25*, 275.
174. House, H.O.; Ro, R.S. *J. Am. Chem. Soc.* **1958**, *80*, 2428.

A reagent that has become gradually more popular for epoxidizing alkenes, including conjugated enones, is dimethydioxirane.[175] Dioxiranes in general are readily prepared from ketones and a persulfate salt. The most popular oxidant is Oxone®, $2KHSO_5{\bullet}KHSO_4{\bullet}K_2SO_4$, which oxidizes a wide range of ketones to the corresponding dioxiranes. In the presence of an alkene, the dioxirane effectively transfers an oxygen atom to give the epoxide in a high-yield, stereospecific reaction (e.g., Examples **18.160** and **18.161**). The reaction is stereospecific, involving *syn*, or suprafacial, addition of oxygen to the alkene π bond,[176] although there is evidence that the mechanism may involve homolysis of the three-membered ring instead of an oxygen atom insertion mechanism.[177]

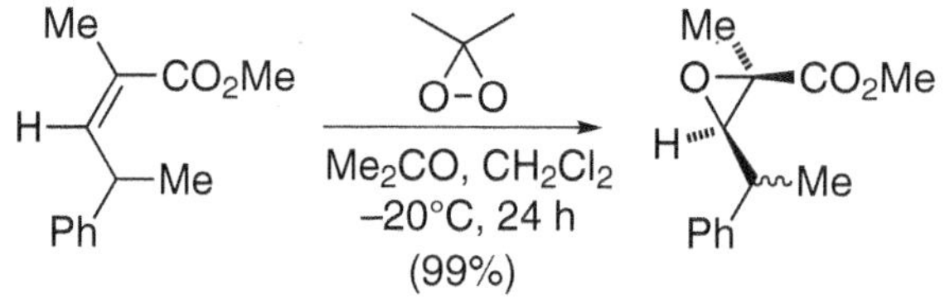

(18.159)[178]

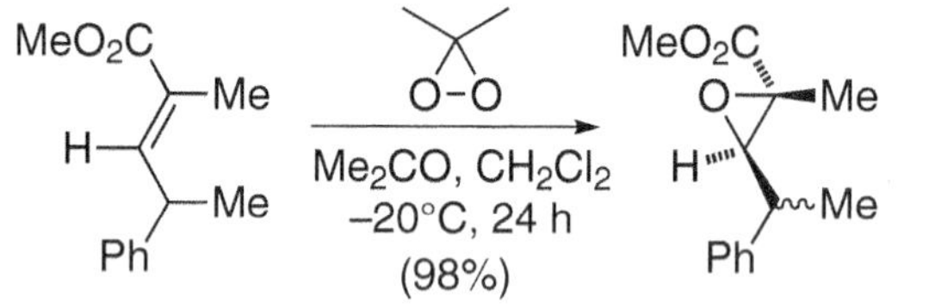

(18.160)[173]

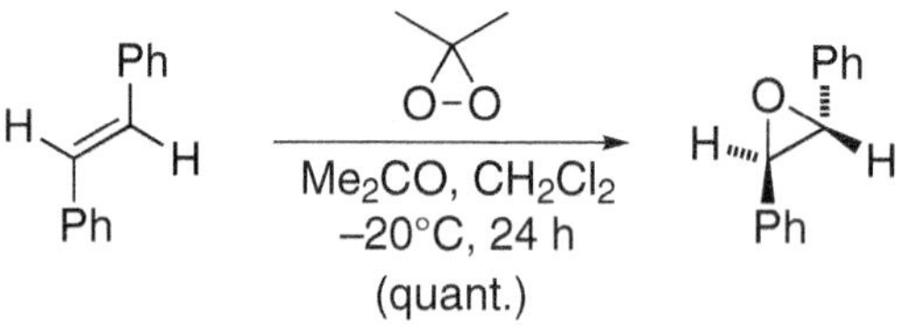

(18.161)[179]

Hydroxylation

The synthesis of 1,2-diols from alkenes (formally, the addition of hydrogen peroxide across the π bond) is known as hydroxylation. The reaction can be carried out with either *anti*, (antarafacial, **18.162**) or *syn* (suprafacial, **18.163**) stereochemistry, depending on the reagent. See Table 18.8.

Table 18.8 Stereochemistry of Hydroxylation

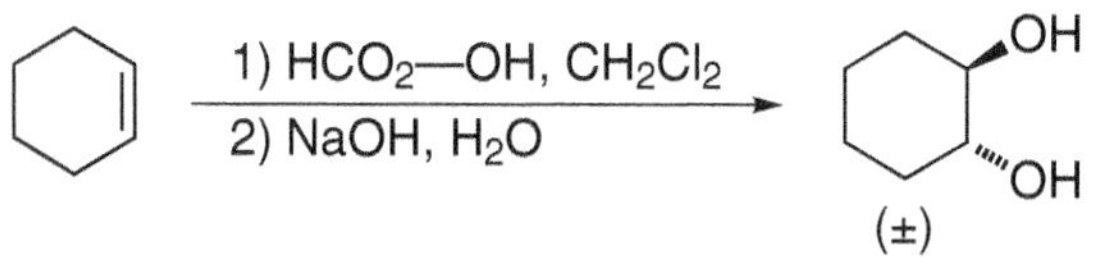	**(18.162)**	*anti* (antarafacial) Hydroxylation
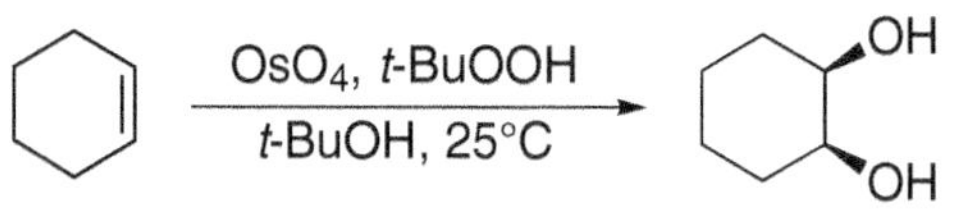	**(18.163)**	*syn* (suprafacial) Hydroxylation

175. Reviews: (a) Murray, R. W. *Chem. Rev.* **1989**, *89*, 1187. (b) Adam, W.; Curci, R.; Edwards, J.O. *Acc. Chem. Res.* **1989**, *22*, 205. (c) Adam, W.; Hadjiarapoglou, L.P. *Top. Curr. Chem.* **1993**, *164*, 45.
176. Baumstark, A.L.; McCloskey, C.J. *Tetrahedron Lett.* **1987**, *28*, 3311.
177. Bravo, A.; Fontana, F.; Fronza, G.; Minisci, F.; Zhao, L. *J. Org. Chem.* **1998**, *63*, 254.
178. Adam, W.; Hadjiarapoglou, L.; Nestler, B. *Tetrahedron Lett.* **1990**, *31*, **331.**
179. Murray, R.M.; Singh, M. *Org. Syn.* **1996**, *74*, 91.

Note that in six-membered rings, *anti* hydroxylation occurs preferentially to give the product of diaxial addition (Example **18.165**). However, the diequatorial isomer (Example **18.166**) can still be observed as a minor pathway due to the greater conformational flexibility of the cyclohexene oxide, which allows *anti* ring opening through the other half-chair conformation.

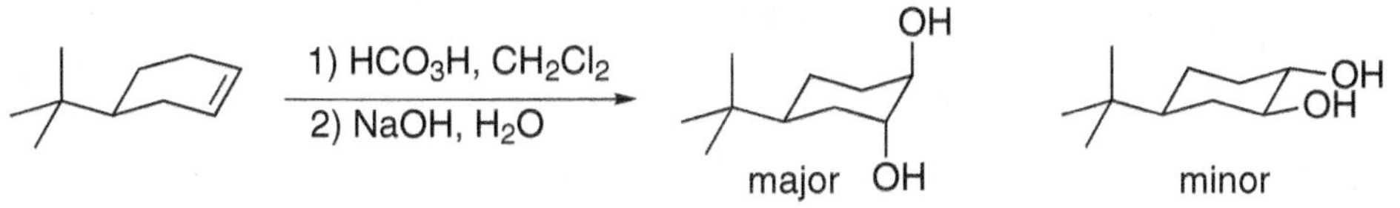

(18.164) **(18.165)** **(18.166)**

syn Hydroxylation

Several reagents accomplish the suprafacial addition of the elements of hydrogen peroxide to an alkene. High-valent compounds of manganese and osmium are the most common.

The controlled oxidation of alkenes with alkaline potassium permanganate was first developed as a useful reaction in a series of papers by the Russian chemist Wagner[180] (of Wagner-Meerwein rearrangement fame), following preliminary reports of the oxidation of oleic acid (Example **18.167**).[181] Wagner also suggested[182] that a cyclic manganate ester was probably involved in the reaction. Thus, the reaction of an alkene with alkaline permanganate at low temperature under carefully controlled conditions gives the dihydroxy compound (diol) in which the two hydroxy groups have both been added to the less hindered face of the double bond. The yields in potassium permanganate hydroxylations can often be improved by using a phase-transfer catalyst (e.g., the yield of cyclooctanediol from cyclooctene is increased from 7% to 50% by the addition of benzyltriethylammonium chloride and the use of a two-phase reaction system[183]). It is important that the reaction be carried out with *dilute, alkaline* permanganate and that it be carried out at *low temperature*[184] because the diol is itself susceptible to oxidation by potassium permanganate under neutral or acidic conditions. Over-oxidation is a frequent side reaction with this reagent.

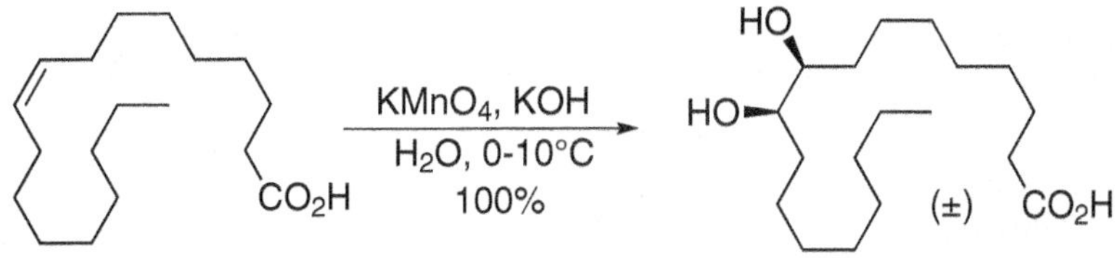

(18.167)[184]

One of the most useful alternatives to permanganate oxidation for the preparation of *syn* diols is the reaction between an alkene and osmium tetroxide, first reported in 1852.[185] Osmium tetroxide adds to the less hindered face of the π bond to form an intermediate osmate (VI) ester[186] that is subsequently cleaved with a reducing agent such as sodium bisulfite or hydrogen sulfide.

---

180. Wagner, G. *Ber. dtsch. chem. Ges.* **1888**, *21*, 1230, 3343, 3347, 3356. Review: Fatiadi, A.J. *Synthesis* **1987**, 85.
181. (a) Gröger, M. *Ber. dtsch. chem. Ges.* **1885**, *18*, 1268. (b) Saytzeff, A. *J. Prakt. Chem.* **1885**, *31*, 541. (c) Saytzeff, A. *J. Prakt. Chem.* **1886**, *33*, 300. (d) Zaitsev, A.M. *Zh. Russ. Fiz.-Khim. Obshch.* **1885**, *17*, 417.
182. (a) Vagner, E.E. *Zh. Russ. Fiz.-Khim. Obshch.* **1895**, *27*, 219. (b) Boesken, J. *Rec. Trav. chim.* **1922**, *41*, 199. (c) Boesken, J. *Rec. Trav. chim.* **1928**, *47*, 683.
183. Weber, W.P.; Shepherd, J.P. *Tetrahedron Lett.* **1972**, 4907.
184. (a) Robinson, G.M.; Robinson, R. *J. Chem. Soc., Trans.* **1925**, *127*, 175. (b) Lapworth, A.; Mottram, E.N. *J. Chem. Soc., Trans.* **1925**, *127*, 1628.
185. (a) Butlerov, A. *Bull. Phys. Math. Acad. St. Petersburg* **1852**, 177. (b) Butlerow, A. *Ann. Chem. Pharm.* **1853**, *84*, 278.
186. (a) DelMonte, A.J.; Haller, J.; Houk, K.N.; Sharpless, K.B.; Singleton, D.A.; Strassner, T.; Thomas, A.A. *J. Am. Chem. Soc.* **1997**, *119*, 9907. (b) Pidun, U.; Boehne, C.; Frenking, G. *Angew Chem. Int. Ed. Engl.* **1997**, *35*, 2817.

Figure 18.10 Cyclic ester intermediates in hydroxylation

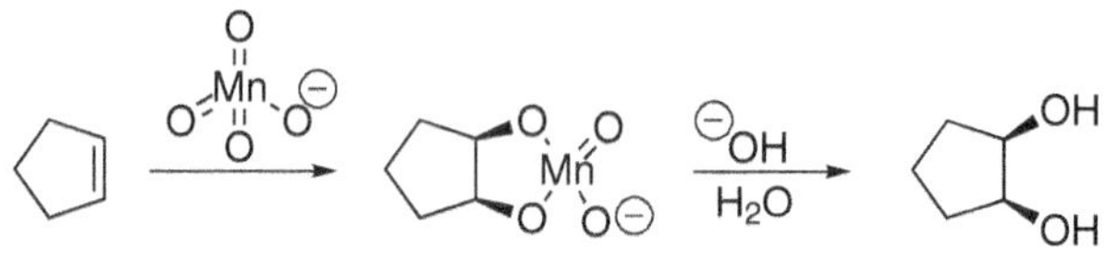

(18.168)

manganate ester

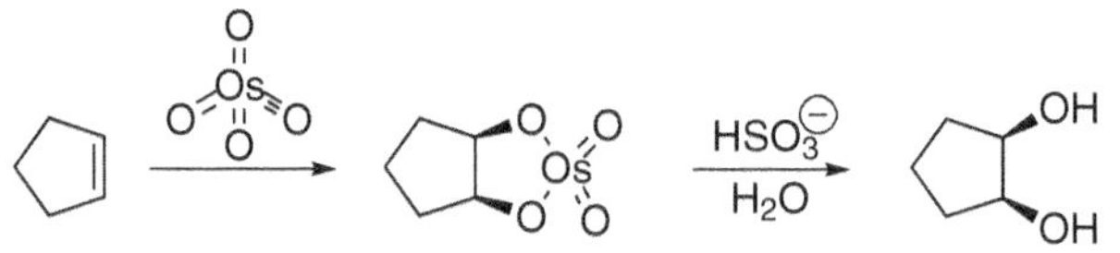

(18.169)

osmate ester

Potassium permanganate and osmium tetroxide oxidations are both thought to occur by suprafacial addition of the oxidant to the double bond (either concerted or stepwise) to give a cyclic manganate [Mn(V), **18.168**] or osmate [Os(VI), **18.169**] ester (Figure 18.10). In the permanganate oxidation, the function of the aqueous base is to free the diol from the manganate ester by hydrolysis before the ester reacts further to give products of over-oxidation. In the osmium tetroxide oxidation, the product can be released from the osmate ester by reduction. Unlike the cyclic manganate ester, the osmate ester does not react further if not cleaved immediately.

Osmium tetroxide reacts with alkenes to the exclusion of virtually all other functional groups, and it gives only *syn* addition.[187] There are two disadvantages to using osmium tetroxide as a stoichiometric reagent on all but the smallest scales or with the most valuable alkenes. One, it is extremely costly—on a gram-for-gram basis, osmium (VIII) oxide and gold (III) oxide were almost the same price between 2010 and 2013, for example. Two, it combines high volatility with reputed high toxicity,[188] so it may constitute a serious personal exposure hazard. Thus, much effort has gone into developing *catalytic* reactions that release the diol by oxidizing the osmate ester back to osmium tetroxide rather than reducing it. Most reactions of this type use peroxides (e.g., Example **18.170**),[189] amine-*N*-oxides,[190] or aqueous sodium[191] or barium chlorate[192] as the co-oxidant. The problems inherent in the volatility of the active oxidant have been obviated, in part, by the use of osmium tetroxide encapsulated in a poly(phenoxyethoxy-methylstyrene-co−styrene) carrier, using aqueous acetone as solvent and $K_3Fe(CN)_6$ as the co-oxidant.[193]

---

187. Schröder, M. *Chem. Rev.* **1980**, *80*, 187.

188. However, the chronic toxicity of osmium compounds may not be as serious as commonly portrayed: (a) McLaughlin, A.I.G.; Milton, R.; Perry, M.K.A. *Br. J. Ind. Med.* **1946**, *3*, 183; *Chem. Abstr.* **1946**, *40*, 5841. (b) Hinckley, C.C.; Bemiller, J.N.; Strack, L.E.; Russell, L.D. In Lippard, S.J., Ed. *Platinum, Gold and Other Metal Chemotherapeutic Agents. ACS Symp. Ser.* **1983**, *209*, 421.

189. Hydrogen peroxide: (a) Milas, N.A.; Sussman, S. *J. Am. Chem. Soc.* **1936**, *58*, 1302. (b) Milas, N.A.; Sussman, S. *J. Am. Chem. Soc.* **1937**, *59*, 2345.

tert-Butyl hydroperoxide: (c) Sharpless, K.B.; Akashi, K. *J. Am. Chem. Soc.* **1976**, *98*, 1986. (d) Akashi, K.; Palermo, R.E.; Sharpless, K.B. *J. Org. Chem.* **1978**, *43*, 2063.

190. *N*-Methylmorpholine-*N*-oxide (NMMO): VanRheenen, V.; Kelly, R.C.; Cha, D.Y. *Tetrahedron Lett.* **1976**, 1973. Trimethylamine-*N*-oxide (TMO): Ray, R.; Matteson, D.S. *Tetrahedron Lett.* **1980**, 449.

191. (a) Hofmann, K.A.; Ehrhart, O.; Schneider, O. *Ber. dtsch. chem. Ges.* **1913**, *46*, 1657. (b) Milas, N.A.; Terry, E.M. *J. Am. Chem. Soc.* **1925**, *47*, 1415. (c) Braun, G. *J. Am. Chem. Soc.* **1929**, *51*, 228. (d) Braun, G. *J. Am. Chem. Soc.* **1930**, *52*, 3176, 3188. (e) Criegee, R. *Justus Liebigs Ann. Chem.* **1936**, *522*, 75

192. (a) Plaha, L.; Weichert, J.; Zvacek, J.; Smolik, S.; Kakac, B. *Coll. Czech. Chem. Commun.* **1960**, *25*, 237. (b) Kende, A.S.; Bentley, T.V.; Mader, R.A.; Ridge, D. *J. Am. Chem. Soc.* **1974**, *96*, 4332. (c) Danishesfky, S.; Schuda, P.F.; Kitahara, T.; Etheredge, S.J. *J. Am. Chem. Soc.* **1977**, *99*, 6066.

193. Ishida, T.; Akiyama, R.; Kobayashi, S. *Adv. Synth. Catal.* **2003**, *345*, 576 and references therein.

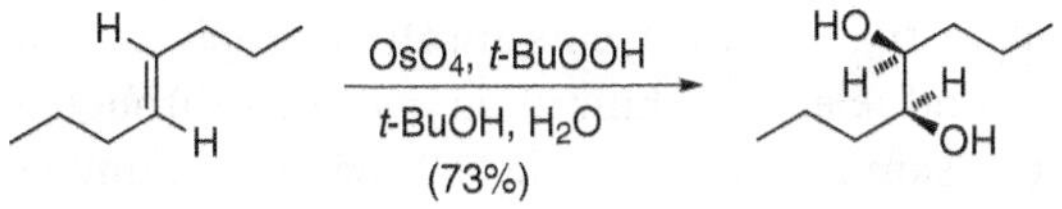

(18.170)

Table 18.9 Oxidizing Electrophiles That Add to Alkenes to Give *anti* Adducts

Electrophile (E—Nu)	Typical Reagents	Ref.
X—X'	Br_2, CH_2Cl_2; BrCl, CH_2Cl_2	
X—OH (X—OR)	Br_2, H_2O; NBS, Me_2SO, H_2O; Br_2, MeOH;	196
Br—OH, I—OH	$NaBrO_3$ (or $NaIO_4$), $NaHSO_3$, H_2O, MeCN;	197
X—OCOR	Br_2, HOAc; Br_2, NaOAc, MeCN	
RS—X	PhSCl, CH_2Cl_2; PhSCl, NaOAc, THF	198
RSe—X	PhSeCl, CH_2Cl_2; PhSeCl, NaOAc, THF	199
$I—N_3$	IN_3, CH_2Cl_2	200
$Br—N_3$	HN_3, Br_2	201
I—NCO	AgNCO, I_2	202
X—SCN	I_2, $(NCS)_2$; Cl_2, $Pb(SCN)_2$	203

anti *Hydroxylation*

The hydroxylation of an alkene with antarafacial (*anti*) stereochemistry cannot be accomplished in a single step. It almost always involves the ring opening of an epoxide, a reaction that occurs with inversion of configuration at the reacting carbon.[194] This leads to overall antarafacial stereochemistry of addition, although the reaction does not always lead to a single stereoisomer in the product, as shown by the ring opening of epoxides **18.171** and **18.174**.[195] The epoxide exists in a half-chair conformation, so the products both arise from antarafacial addition; the di-equatorial diol is formed in the twist boat conformation through a higher energy activated complex than that leading to the diaxial product. In the twist-boat, the two hydroxyl groups are *anti* to each other, and the ring then reverts to the more stable chair conformation with the two hydroxyl groups equatorial.

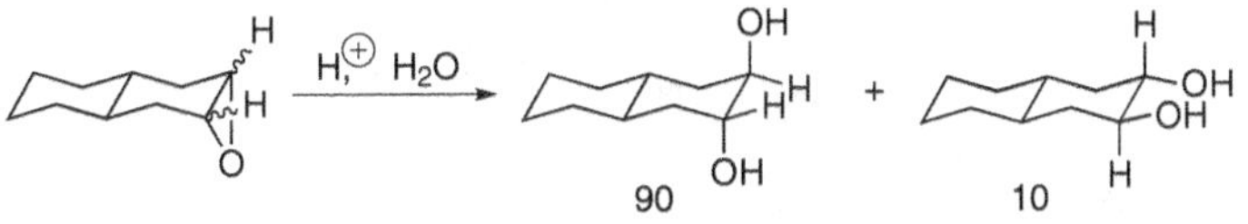

(18.171) (18.172) : (18.173)

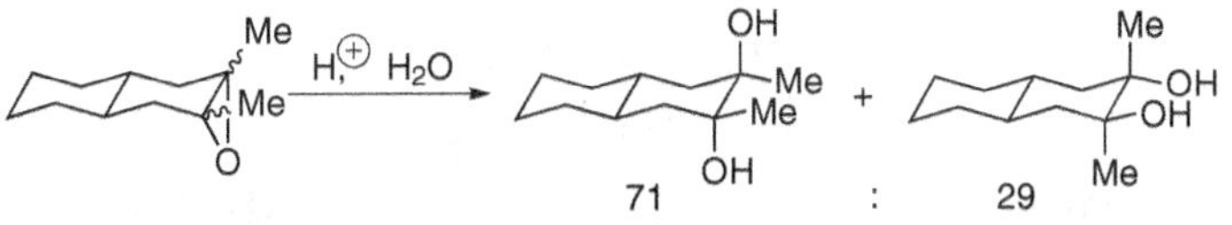

(18.174) (18.175) : (18.176)

Other Oxidative *anti* Additions to Alkenes

A wide array of reagents adds to alkenes with *anti* stereochemistry. The prototypical reagents for this are the halogens and hypohalic acids. However, there are many reagents

194. (a) Böeseken, J. *Rec. trav. chim.* **1928**, *47*, 683. (b) Criegee, R.; Stanger, H. *Ber. dtsch. chem. Ges.* **1936**, *69B*, 2753.
Review: (c) Parker, R.E.; Isaacs, N.S. *Chem. Rev.* **1959**, *59*, 737.

195. Henbest, H.B.; Smith, M.; Thomas, A. *J. Chem. Soc.* **1958**, 3293.

196. (a) Guss, C.O.; Rosenthal, R. *J. Am. Chem. Soc.* **1955**, *77*, 2549. (b) Dalton, D.R.; Dutta, V.P.; Jones, D.C. *J. Am. Chem. Soc.* **1968**, *90*, 5498. (c) Sisti, A.J.; Meyers, M. *J. Org. Chem.* **1973**, *38*, 4431. (d) Langman, A.W.; Dalton, D.R. *Org. Syn.* **1979**, *59*, 16.

197. Masuda, H.; Takase, K.; Nishio, M.; Hasegawa, A.; Nishiyama, Y.; Ishii, Y. *J. Org. Chem.* **1994**, *59*, 5550.

198. (a) Tuladhar, S.M.; Fallis, A.G. *Tetrahedron Lett.* **1987**, *28*, 523. (b) Muehlstaedt, M.; Schubert, C.; Kleinpeter, E. *J. Prakt. Chem.* **1985**, *327*, 270. (c) Muehlstaedt, M.; Widera, R.; Olk, B. *J. Prakt. Chem.* **1982**, *324*, 362.

199. (a) Bennett, F.; Knight, D.W. *Tetrahedron Lett.* **1988**, *29*, 4625. (b) Danishefsky, S.; DeNinno, S.; Lartey, P. *J. Am. Chem. Soc.* **1987**, *109*, 2082. (c) Toshimitsu, A.; Terao, K.; Uemura, S. *J. Org. Chem.* **1986**, *51*, 1724. (d) Toshimitsu, A.; Terao, K.; Uemura, S. *J. Org. Chem.* **1987**, *52*, 2018.

200. Fowler, F.W.; Hassner, A.; Levy, L.A. *J. Am. Chem. Soc.* **1967**, *89*, 540.

201. Hassner, A.; Boerwinkle, F.P.; Levy, A.B. *J. Am. Chem. Soc.* **1970**, *92*, 4879.

202. (a) Hassner, A.; Lorber, M.; Heathcock, C. *J. Org. Chem.* **1967**, *32*, 540. (b) Hassner, A.; Hoblitt, R.P.; Heathcock, C.; Kropp, J.E.; Lorber, M. *J. Am. Chem. Soc.* **1970**, *92*, 1326

203. (a) Maxwell, R.J.; Silbert, L.S. *Tetrahedron Lett.* **1978**, 4991. (b) Guy, R.G.; Pearson, I. *J. Chem. Soc., Perkin Trans. 1* **1973**, 281. (c) Guy, R.G.; Pearson, I. *J. Chem. Soc., Perkin Trans. 2* **1973**, 1359.

that will give products with *anti* stereochemistry. These reagents are all electrophiles that share a general structure with a lone pair on the electrophilic atom and react through a three-membered onium ion intermediate (Example **18.177**). Table 18.9 contains a number of reagents that are variations on this theme.

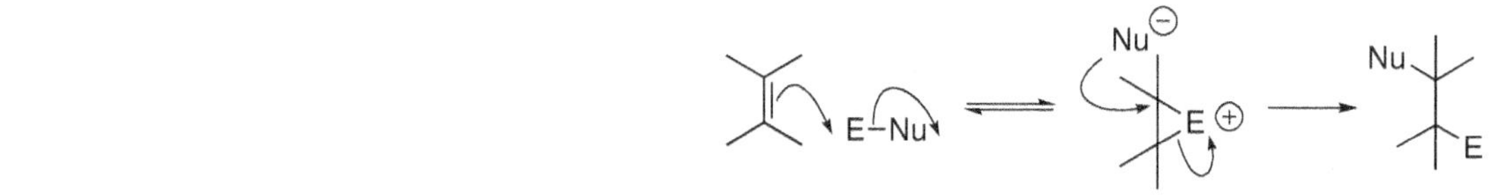

(18.177)

Among the most valuable of these reactions are cyclizations such as the halolactonization (especially iodolactonization of unsaturated carboxylic acids, esters, and amides, which have been widely used for the stereocontrolled formation of chiral centers in a molecule (e.g., Examples **18.178** and **18.179**).[204] Ring size in lactones is governed by stereoelectronic factors: five-membered lactones are favored kinetically, and six-membered lactones are favored thermodynamically (Example **18.180**).[205]

(18.178)[206]

(18.179)[207]

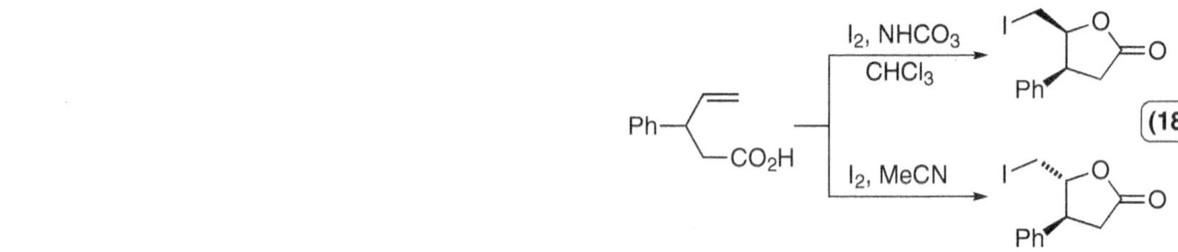

(18.180)

The change from kinetic to thermodynamic control in the addition reaction can also manifest itself in changed stereochemistry for the reaction, as shown in Example **18.180**.[208] Reagents other than halogens, such as sulfenyl and selenenyl halides, can be used to initiate the cyclization, as is illustrated by Examples **18.181** to **18.184**.

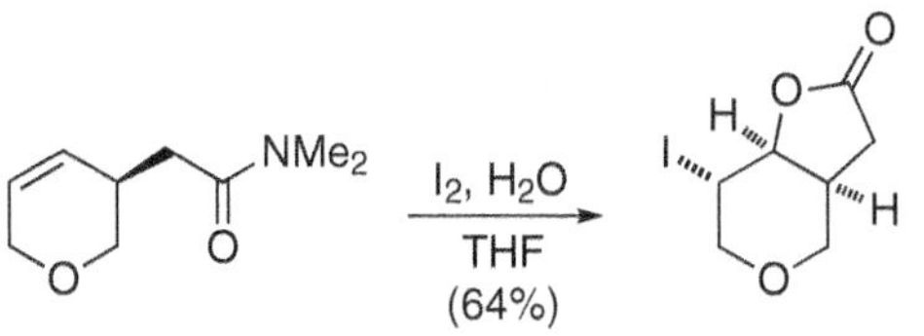

(18.181)[209]

---

204. Review: Bartlett, P.A. In Morrison, J.D., Ed. *Asymmetric Synthesis* (Academic Press: New York, 1984), vol. 3, p. 411.

205. (a) Bartlett, P.A.; Myerson, J. *J. Am. Chem. Soc.* **1978**, *100*, 3950. (b) Gonzàlez, F.B.; Bartlett, P.A. *Org. Syn.* **1985**, *64*, 175.

206. Neukom, C.; Richardson, D.P.; Myerson, J.H.; Bartlett, P.A. *J. Am. Chem. Soc.* **1986**, *108*, 5559.

207. Bartlett, P.A.; Meadows, J.D.; Brown, E.G.; Morimoto, A.; Jernstedt, K.K. *J. Org. Chem.* **1982**, *47*, 4013.

208. Gonzàlez, F.B.; Bartlett, P.A. *Org. Syn.* **1985**, *64*, 175; **1990**, *Coll. Vol. 7*, 164.

209. Fleet, G.W.J.; Gough, M.J. *Tetrahedron Lett.* **1982**, *23*, 4509.

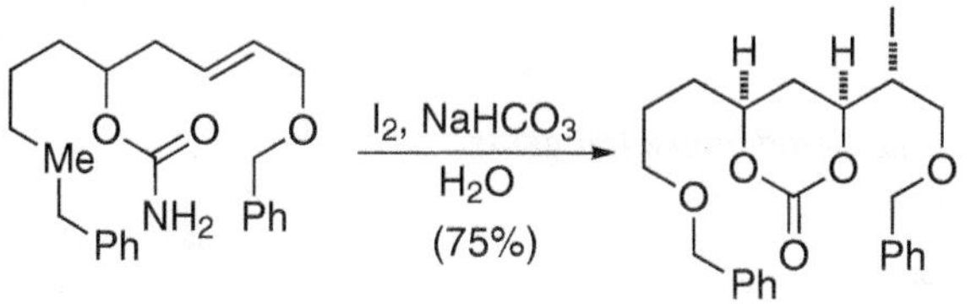

(18.182)[210]

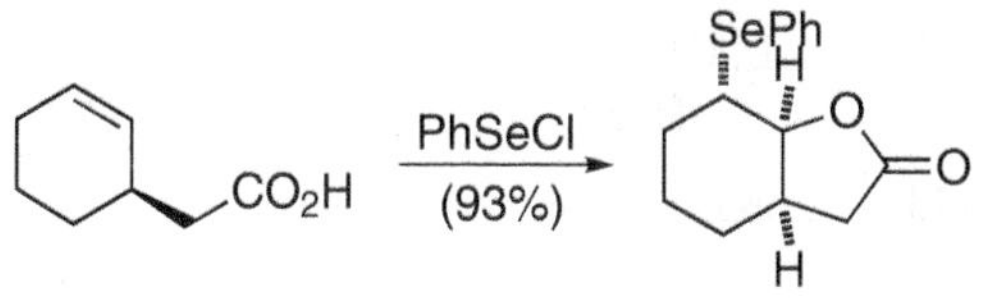

(18.183)[211]

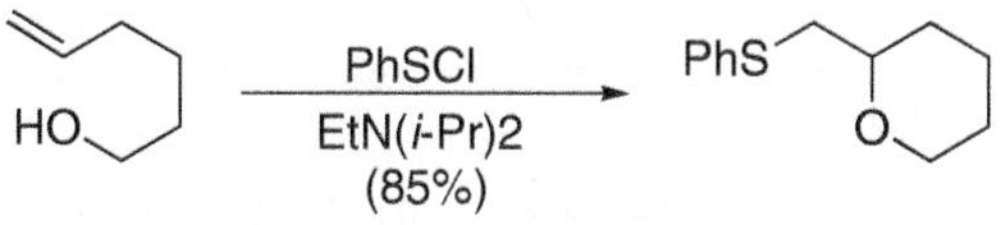

(18.184)[212]

The use of molecular bromine for addition reactions is complicated by its volatility, its reactivity, and its toxicity. In recent years, molecular bromine has been replaced by *N*-bromosuccinimide (NBS) with species capable of opening the three-membered bromonium ion such as amides,[213] sulfonamides,[214] and lithium bromide.[215]

Addition of Metal-Based Reagents

The addition of metals as part of the converting alkenes to other products is well established. For example, in the first step of the oxymercuration-demercuration sequence for Markovnikov hydration of an alkene π bond (Example **18.185**), the mercury acts as an electrophile to give a cyclic mercurinium ion that is then attacked by water, giving overall *anti* addition. The replacement of mercury by hydrogen gives a product where neither overall oxidation nor overall reduction has taken place. However, its replacement by another electronegative element permits the net oxidation of the molecule.[216] This same reactivity motif, slightly modified, is also exhibited by palladium (II), which adds to alkenes to give β-hydroxyalkylpalladium compounds by overall *anti* addition through a π complex of the alkene with the palladium. Unlike the organomercury compounds, however, the palladium group is subject to β-elimination, with concomitant reduction to palladium (0). This has the net effect of replacing a carbon-hydrogen bond with a carbon-heteroatom bond (if the nucleophile in the addition is a heteroatom nucleophile. This reaction provides the basis of the ***Wacker reaction***, which we will discuss later in this chapter.

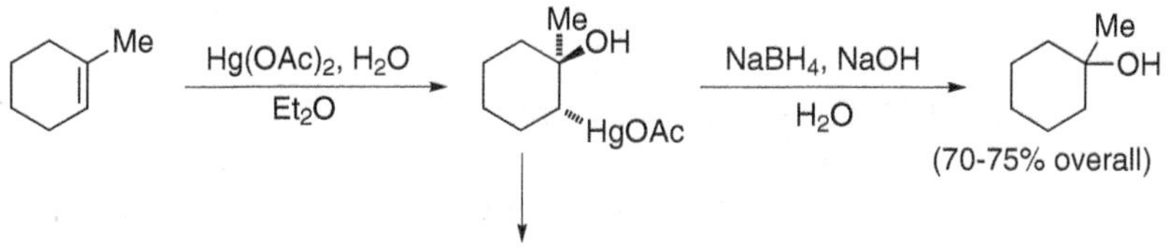

(18.185)[217]

---

210. Hirama, M.; Uei, M. *Tetrahedron Lett.* **1982**, *23*, 5307.

211. Nicolaou, K.C.; Seitz, S.P.; Sipio, W.J.; Blount, J.F. *J. Am. Chem. Soc.* **1979**, *101*, 3884.

212. Tuladhar, S.M.; Fallis, A.G. *Tetrahedron Lett.* **1987**, *28*, 523.

213. Wang, Z.; Zhang, Y.; Fu, H.; Jiang, Y.; Zhao, Y. *Synlett* **2008**, 2667.

214. (a) Thakur, V.V.; Talluri, S.K.; Sudalai, A. *Org. Lett.* **2003**, *5*, 861. (b) Huang, X.; Fu, W.-J. *Synthesis* **2006**, 1016.

215. Shao, L.-X.; Shi, M. *Synlett* **2006**, 1269.

216. For a review of the uses of organomercury compounds in synthesis, see: Larock, R.C. *Angew. Chem. Int. Ed. Engl.* **1978**, *17*, 27.

217. Jerkumica, J.M.; Taylor, T.G. *Org. Synth.* **1988**, *Coll. Vol. 6*, 766; **1973**, *53*, 94.

Reaction Synopses

Addition of Halogens and Halogen-like Reagents

Reagents: Br_2, CH_2Cl_2; Cl_2, CH_2Cl_2; IN_3; Br_2, H_2O; etc.
or (Y=CO_2H): Br_2, $NaHCO_3$, H_2O; I_2, $NaHCO_3$, H_2O; I_2, MeCN; etc.

Stereochemistry: *anti* addition predominates

Oxymetallation

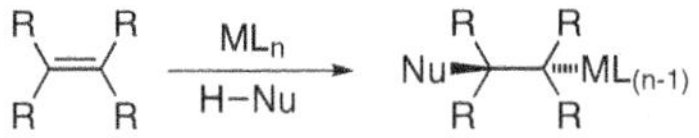

Reagents: $Hg(OAc)_2$, H_2O, THF; $Hg(OCOR)_2$, ROH; $Hg(OCOR)_2$, RNHCOR'; $Pd(OAc)_2$, H_2O, MeCN; etc.

Replacement of metal by heteroatom: leads to net oxidation

Epoxidation of Alkenes (Prilezhaev Reaction)

Reagents: *m*-CPBA, CH_2Cl_2; HCO_3H, CH_2Cl_2; CF_3CO_3H, CH_2Cl_2; MMPP, CH_2Cl_2; 1) Br_2, H_2O, 2) K_2CO_3, acetone; etc.

Stereochemistry: overall suprafacial addition to the π bond

Epoxidation of Conjugated Carbonyl Compounds

E=COR, CO_2R, CN, etc.

Reagents: H_2O_2, KOH; Me_3COOH, KOH; Me_2CO_2, Me_2CO; Me_2CO_2, Me_2CO; Me_2CO, Oxone®; etc.

Hydroxylation of Alkenes

Stereochemistry: may be either *syn* (suprafacial) or *anti* (antarafacial)

Reagents: *syn*: $KMnO_4$, KOH, H_2O; $KMnO_4$, R_4N^+, H_2O, CH_2Cl_2; (1) OsO_4, (2) KOH;
or $NaHSO_3$, H_2O; OsO_4, $NaClO_3$, H_2O, CH_2Cl_2; etc.
anti: (1) HCO_3H, CH_2Cl_2, (2) KOH, H_2O; etc.

Worked Problem

18-3 Give the structure of the starting material in the reaction below. Be sure to include stereochemistry.

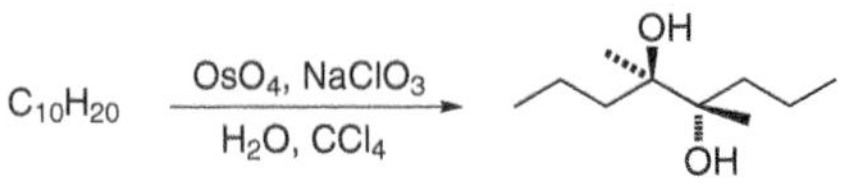

§Answer on the next page.

Drill Exercises

Draw the structure of the major product expected when carvone is treated with one equivalent of each of the following reagents:

A. Br_2 (1 eq.), CH_2Cl_2

B. OsO_4, NMMO, CH_2Cl_2.

C. (1) *m*-CPBA, CH_2Cl_2; (2) NaOAc, DMF

D. H_2O_2, Me_3COH, KOH

E. Me_3COOH, $Mo(CO)_6$

F. IN_3, CH_2Cl_2, 0°C

G. PhSeBr, NaOAc

H. Me_3COOH, $VO(acac)_2$

I. $PhCO_3H$, PhH

Problems

18-8 Which reagent or sequence of reagents should be used to accomplish the following transformations?

(a) 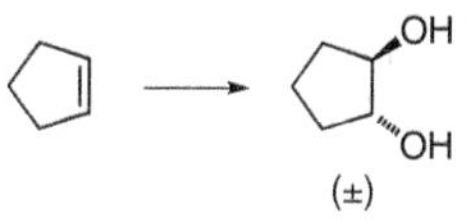

(b) 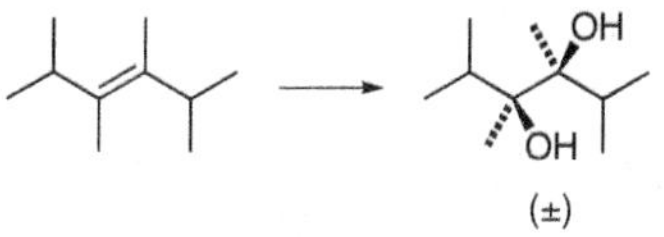

(c) 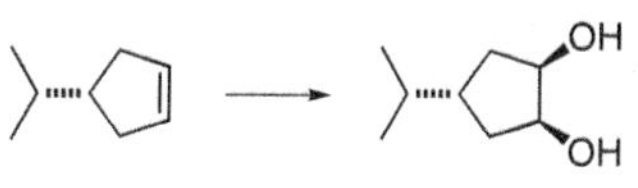

(d) 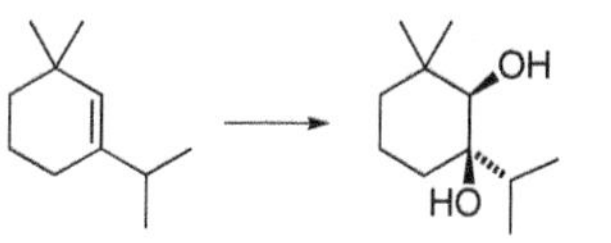

(e) 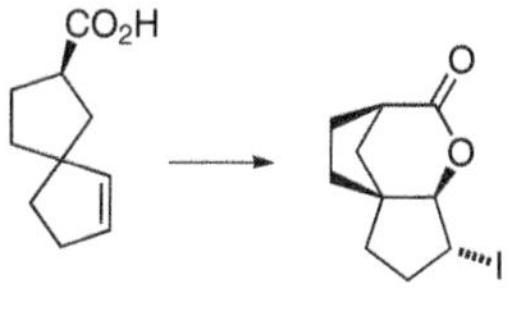

(f) 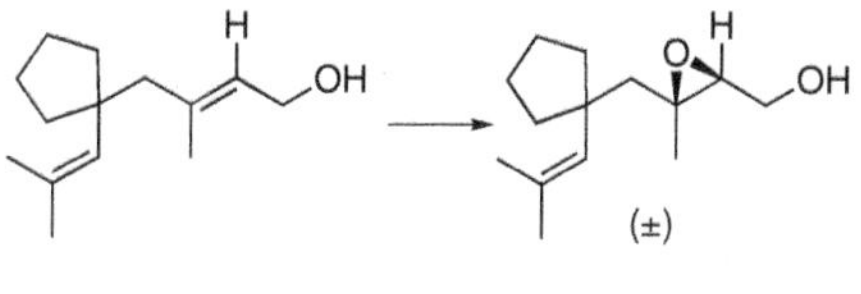

(continues)

§ Answer to Worked Problem:
It is important to consider the stereochemistry in this reaction before answering. The product here is a *meso* diol, drawn with the two hydroxy groups *anti* to each other. It is very tempting to simply write the *E*-alkene as the starting material but this is wrong, since that alkene would require a reagent that adds the two hydroxyl groups *anti*. Osmium tetroxide, however, results in *syn* addition of the two hydroxy groups to the alkene. This means that the stereochemistry of the alkene must not be *E*, but *Z*, and that the product has simply been drawn in a conformation different from that in which it is initially produced; one may change the stereochemistry of the product by changing the stereochemistry of the starting alkene or by changing the stereochemistry of the reaction:

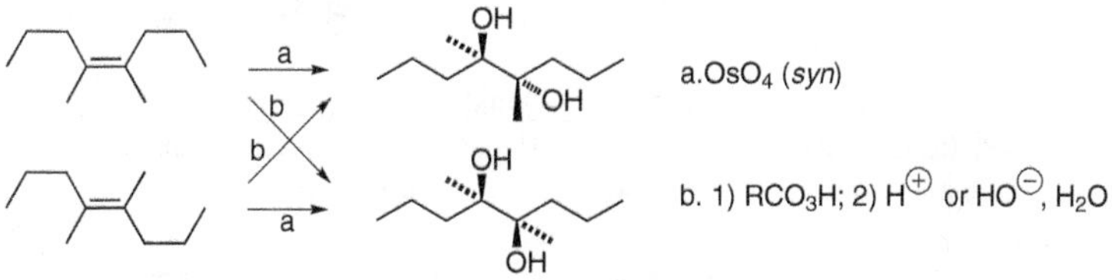

(**Problems** continued)

(g) 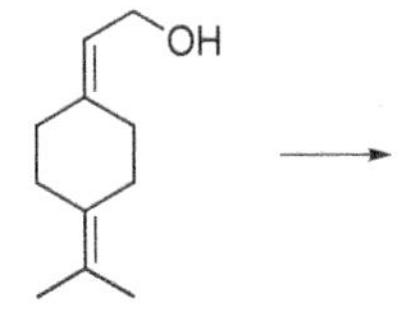

(h) 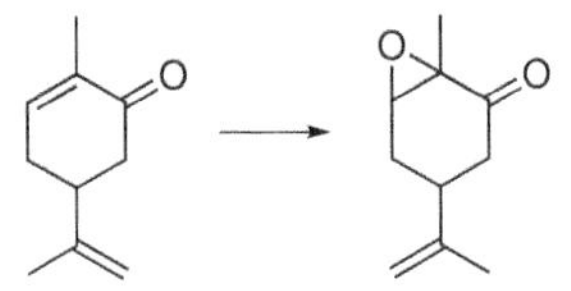

18-9 An alternative *syn*-hydroxylation of alkenes, known as the Woodward hydroxylation, consists of treating the alkene with iodine and silver (I) or copper (II) acetate in wet acetic acid. The key feature of this reaction is that the hydroxyl groups end up on the *more hindered face* of the π bond (in contrast to osmium tetroxide, which adds to the less hindered face). When the reaction is carried out under anhydrous conditions, the diesters of the *anti* diol are formed. Suggest a mechanism for this reaction that accounts for the formation of products under both anhydrous and hydrated conditions.

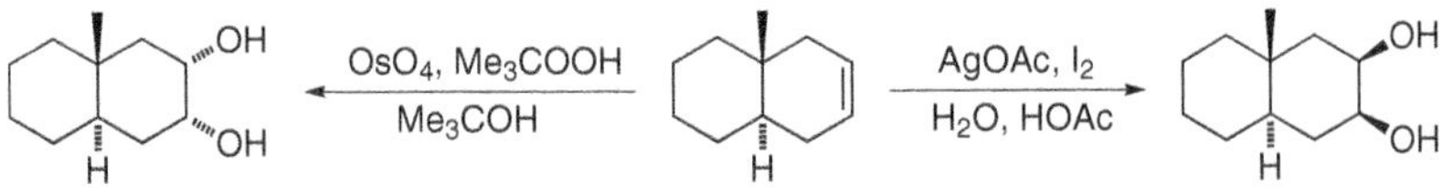

18-10 Draw a mechanism that accounts for the differences in the two reactions below.

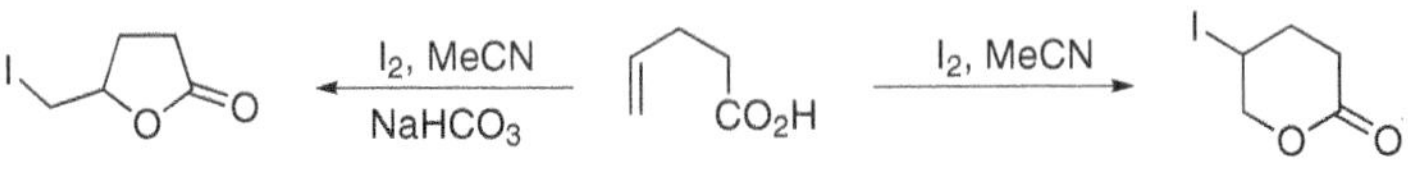

18-11 The following reactions were used in the total synthesis of several natural products. Give the structure of the major organic product formed in each reaction.

(a) 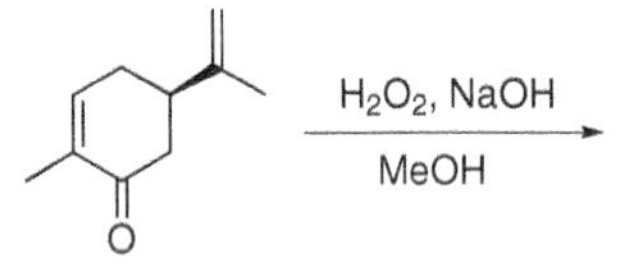

(b) 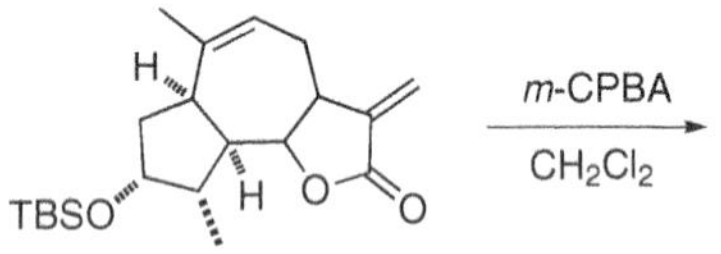

(c) 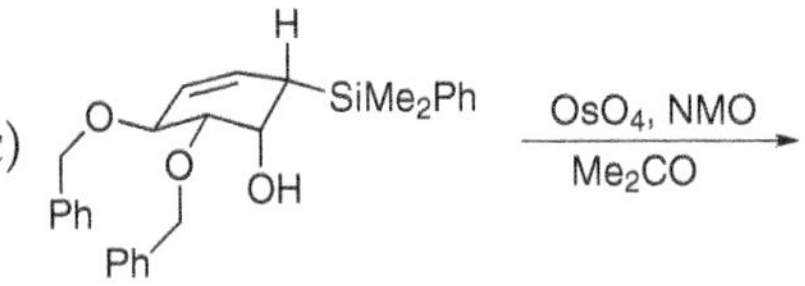

(d) 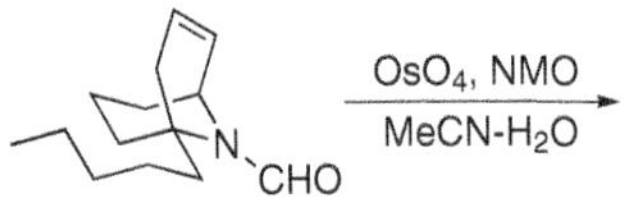

(e) 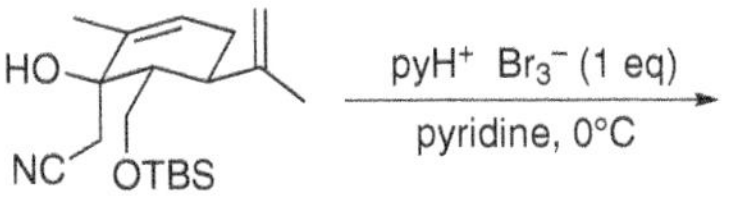

(f) 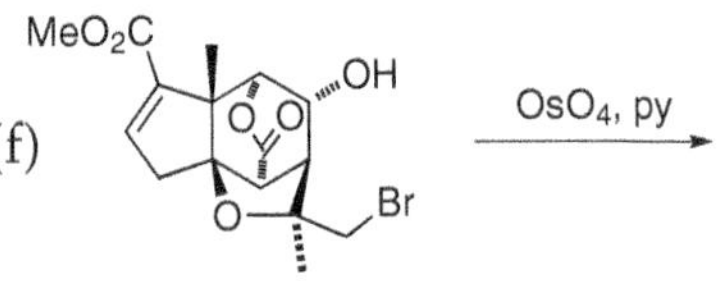

(g) 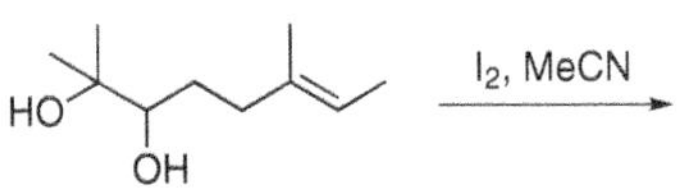

(h) 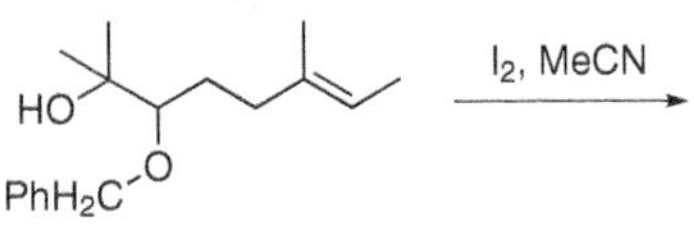

(i) 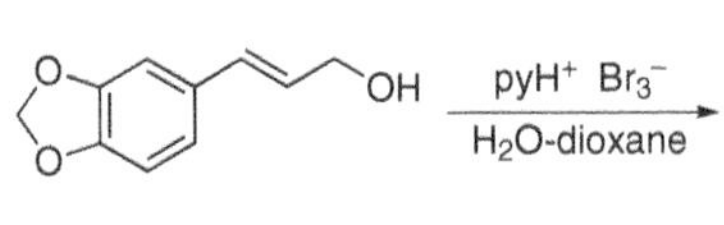

(j) 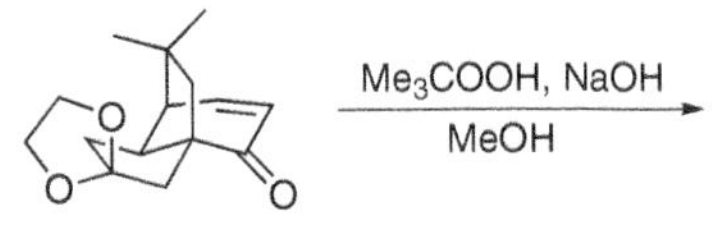

References: (a), (b) [*J. Am. Chem. Soc.* **2010**, *132*, 1488]. (c) [*Org. Lett.* **2003**, *5*, 1697]. (d) [*Org. Lett.* **2002**, *4*, 2469]. (e), (f) [*J. Am. Chem. Soc.* **1996**, *118*, 233]. (g), (h) [*J. Am. Chem. Soc.* **1981**, *103*, 3963]; which reaction gives which isomer, and why? (i) [*Tetrahedron Lett.* **1989**, *30*, 907]. (j) [*J. Am. Chem. Soc.* **1982**, *104*, 872].

18.9 Oxidative Cleavage of Carbon-Carbon Bonds

Reactions where carbon-carbon σ bonds are cleaved under mild conditions are relatively rare, with two notable exceptions: the cleavage of vicinal diols, and the oxidative cleavage of alkenes.

The oxidative cleavage of 1,2-diols is often a side reaction during oxidations with chromium (VI) reagents, although the reaction is seldom useful in a preparative sense. The one place where oxidative cleavage with a chromic acid oxidant is actually useful is as part of the Barbier-Wieland degradation[218] of carboxylic acids (Example **18.186**). In this two-reaction sequence to convert a carboxylic acid to its next lowest homolog, the ester is treated with phenylmagnesium bromide to give the diphenylcarbinol, which is then oxidized with chromic acid to the carboxylic acid with one less carbon than the starting compound. Given the extremely vigorous conditions of the oxidation, this is obviously limited to very robust substrates, and relatively difficult to control.

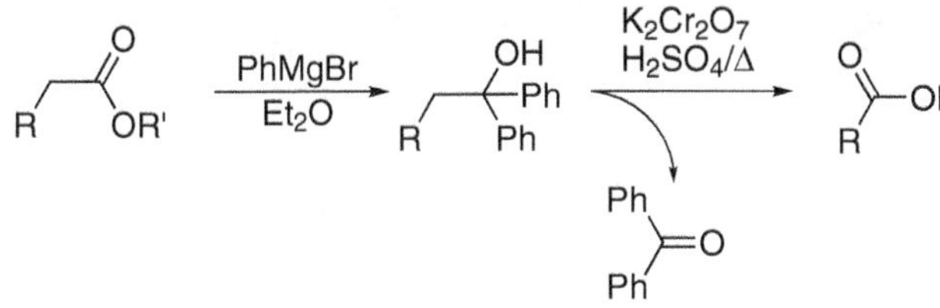

(18.186)

Oxidative Cleavage of Vicinal Diols

In contrast to their monohydric counterparts, the vicinal diols (the 1,2-diols) are susceptible to oxidative cleavage by reagents such as sodium periodate (e.g., Example **18.187**)[219] and lead tetraacetate (e.g., Example **18.188**).[220] The periodate cleavage of vicinal diols has been quite extensively studied, and there is good evidence to support a mechanism involving a five-membered cyclic periodate ester. The lead tetraacetate cleavage of vicinal diols is sometimes called the ***Criegee oxidation***.

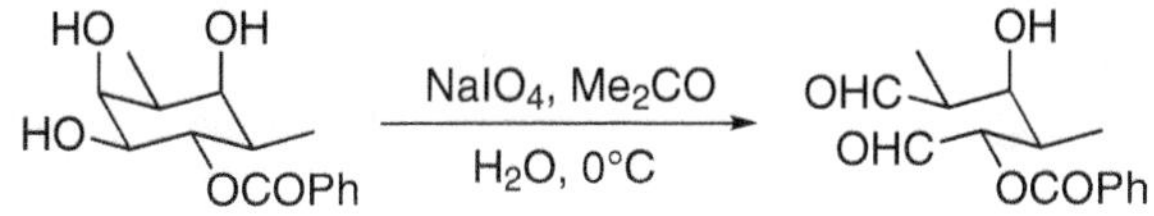

(18.187)[221]

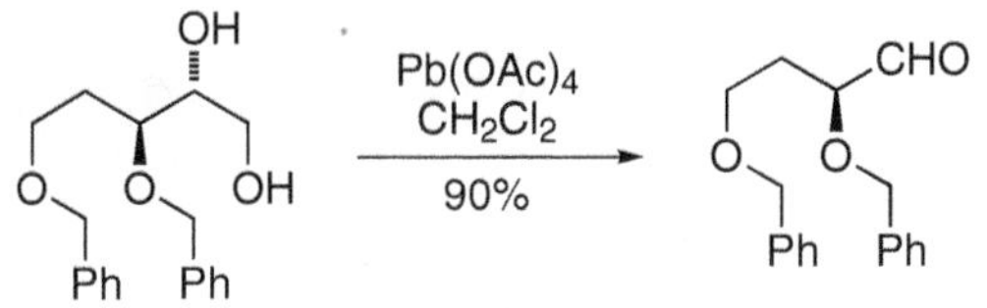

(18.188)[222]

Evidence supporting a mechanism involving a cyclic intermediate comes from kinetic studies of periodate cleavage of related diols. Although *cis*-1,2-cyclohexanediol (which can readily from a cyclic periodate ester; Example **18.189**) undergoes rapid cleavage with periodate, the *trans* isomer (which does not readily from a cyclic periodate ester; Example **18.190**) cleaves

218. (a) Wieland, H. *Chem. Ber.* **1912**, *45*, 484. (b) Barbier, P.; Locquin, R. *Compt. Rend.* **1913**, *156*, 1443. (c) Wieland, H.; Schlichting, O.; Jacobi, R. *Z. Physiol. Chem.* **1926**, *161*, 80. (d) Shoppee, C.W. *Ann. Rep. Chem. Soc., (London)* **1947**, *44*, 184. (e) Baker, W.; Curtis, R.F.; McOmie, J.F.W.; Oliol, L.W.; Rogers, V. *J. Chem. Soc.* **1958**, 1007.

219. (a) Malaprade, L. *Bull. Soc. chim. France* **1928** *[iv]*, *43*, 683. (b) Malaprade. L. *Bull. Soc. chim. France* **1934** *[v]*, *1*, 833. Reviews: (c) Jackson, E.L. *Org. React.* **1942**, *2*, 341. (d) Bunton, C.A. In Wiberg, K.B., Ed. *Oxidation in Organic Chemistry, Part B* (Academic Press: NewYork, 1965), p. 367. (e) Perlin, A.S. In Augustine, R.L., Ed. *Oxidation* (Marcel Dekker: New York, 1969), vol. 1, p. 189.

220. Reviews: (a) Rubottom, G.M. In Trahanovsky, W.S., Ed. *Oxidation in Organic Chemistry, Part D.* (Academic Press: New York, 1982), p. 1. (b) Aylward, J.B. *Quart. Rev. Chem. Soc.* **1971**, *25*, 407.

221. Chida, N.; Yoshinaga, M.; Tobe, T.; Ogawa, S. *Chem. Commun.* **1997**, 1043.

222. Mulzer, J.; Scharp, M. *Synthesis* **1993**, 615.

30 times more slowly.[223] Periodate cleavage has been especially widely applied in the carbohydrate field, where the water solubility of the organic substrate is an advantage, and it has been used in aqueous acetone for systems that do not have appreciable water solubility. Like the periodate cleavage, the lead tetraacetate cleavage of diols appears to require a cyclic intermediate in most cases, and there is substantial evidence that the formation of the cyclic intermediate is rate determining. However, there is also evidence that the cleavage may occur by an acyclic mechanism where a cyclic intermediate cannot form, such as glycols **18.191** and **18.192**, both of which are cleaved by lead tetraacetate and neither of which is cleaved by periodic acid.[224] For this reason, the Criegee oxidation is sometimes preferred over the periodate method.

(18.189)

(18.190)

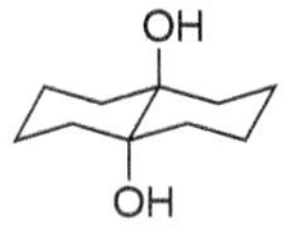

(18.191)

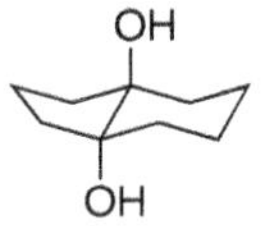

(18.192)

Ozonolysis[225] and Its Alternatives

Ozonolysis is one of the few reactions of alkenes that leads to the cleavage of both bonds of the double bond—the π bond *and* the σ bond—without any rearrangements, so that one can use it to locate double bonds within a molecule. This was one of its most important applications historically. More recently, however, ozonolysis has been used as a means to permit alkenes to be used as masked aldehydes.

In almost all of its reactions, ozone reacts as an electrophile. The currently accepted mechanism for ozonolysis (Figure 18.11) was first proposed by Criegee.[226] In the Criegee

Figure 18.11 The Criegee mechanism of ozonolysis

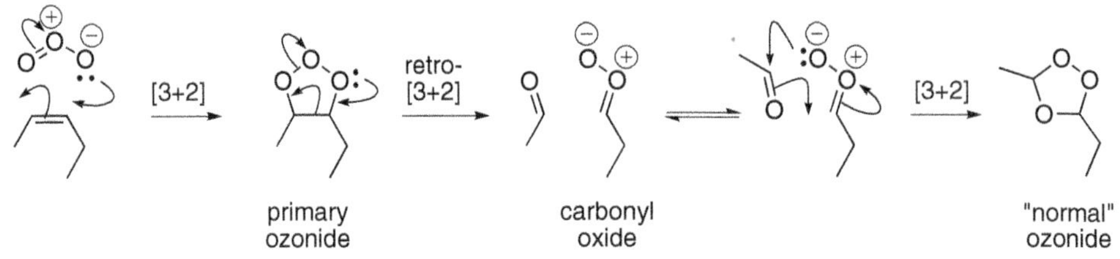

223. (a) Price, C.C.; Kroll, T.J. *J. Am. Chem. Soc.* **1942**, *60*, 2727. For other papers on the effects of stereochemistry on the reaction rate, see also; (b) Criegee, R.; Büchner, E.; Walther, W. *Ber. dtsch. chem. Ges.* **1940**, *73*, 571. (c) Bulgrin, V.C. *J. Phys. Chem.* **1957**, *61*, 702. (d) Bulgrin, V.C.; Dahlgren, G. *J. Am. Chem. Soc.* **1958**, *80*, 3883. (e) Bunton, C.A.; Carr, M.D. *J. Chem. Soc.* **1963**, 770.

224. Criegee, R.; Büchner, E.; Walther, W. *Ber. dtsch. chem. Ges.* **1940**, *73*, 571.

225. (a) Bailey, P.S. *Ozonization in Organic Chemistry*, (Academic Press: New York, 1981), vol. 1. (b) Belew, J.S. In Augustine, R.L. Ed. *Oxidation*, (Marcel Dekker: New York, 1969), vol. 1, p. 262.

226. (a) Criegee, R. *Rec. Chem. Progr*, **1957**, *18*, 111. (b) Criegee, R. *Angew. Chem. Int. Ed. Engl.* **1975**, *14*, 745. (c) Kuczkowski, R.L. *Acc. Chem. Res.* **1983**, *16*, 42. (d) Pryor, W.A.; Giamalva, D.; Church, D.F. *J. Am. Chem. Soc.* **1985**, *107*, 2793.

Rudolf Criegee (1902–1975) was educated at Würzburg (PhD, 1926). His independent work, which he started at Marburg and continued after his move to Karlsruhe in 1930, concerned the oxidation of organic compounds. For much more detail, see: Maier, G. *Chem. Ber.* **1977**, *110*, XXVII, Huisgen, R. (Gilde, H.G., transl.) *J. Chem. Educ.* **1979**, *56*, 369.

mechanism, ozone, a ***1,3-dipolarophile***, first adds to the alkene in a 1,3-dipolar cycloaddition (also termed a [3 + 2]-cycloaddition) to give the primary ozonide, a 1,2,3-trioxolane. This highly unstable intermediate undergoes a retro-[3 + 2]-cycloaddition, or [3 + 2]-cycloreversion, to generate a carbonyl compound and a new 1,3-dipolarophile known as a carbonyl oxide. The final step of the reaction is a second 1,3-dipolar cycloaddition to give the 1,2,4-trioxolane, or "normal" ozonide. Note how all three reactions in this sequence are pericyclic reactions.

Ozonides are stable compounds, but most decompose explosively when heated or subjected to shock. Although there are some known ozonide natural products, one seldom isolates the ozonide itself, but oxidizes or reduces it prior to isolation of the products. Depending on the isolation procedure, one may use ozonolysis of alkenes to prepare carboxylic acids, aldehydes, or alcohols. By far the most common isolation procedure involves the addition of dimethyl sulfide, which reduces the ozonide to a pair of carbonyl compounds (Example **18.194**). Reduction with sodium borohydride gives a pair of alcohols instead (Example **18.193**). Alternatively, the ozonide may be oxidized with hydrogen peroxide or a peracid to give the carboxylic acid (Example **18.195**). Oxidative workup also leads to oxidation of aldehyde products to the carboxylic acid. Despite its reactivity, many common functional groups are resistant to oxidation by ozone. Among these are ethers, ketones, nitriles, amides, esters, and carboxylic acids.

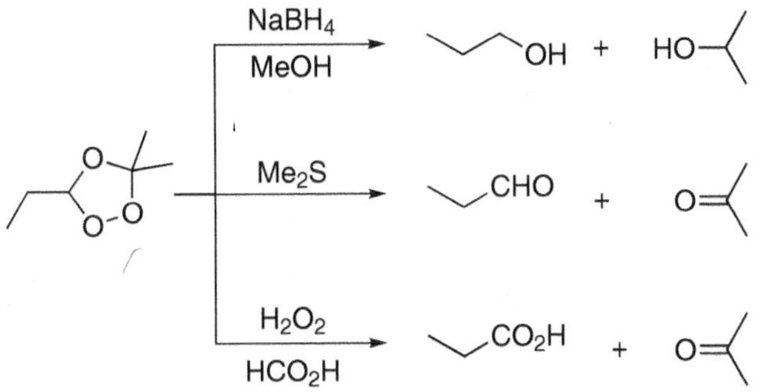

(18.193)

(18.194)

(18.195)

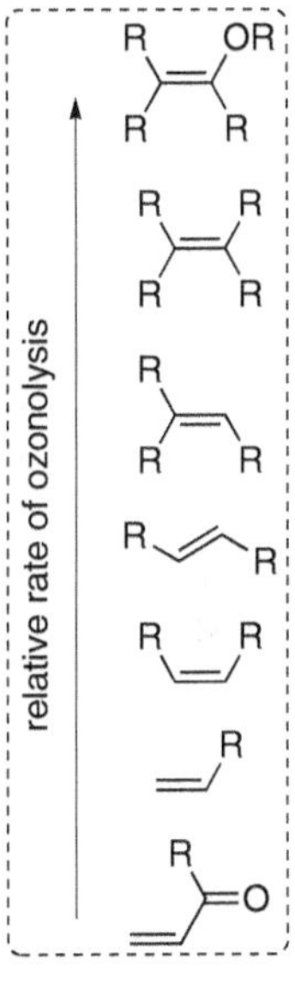

Electron-rich alkenes react most rapidly, and the more substituted usually react more rapidly. *Trans*-disubstituted alkenes tend to react more rapidly than their *cis*-disubstituted isomers. The approximate order of reactivity is shown for a series of alkenes. As shown by Examples **18.196** and **18.197**, it is possible to achieve selective cleavage of one π bond in a diene by limiting the amount of ozone.

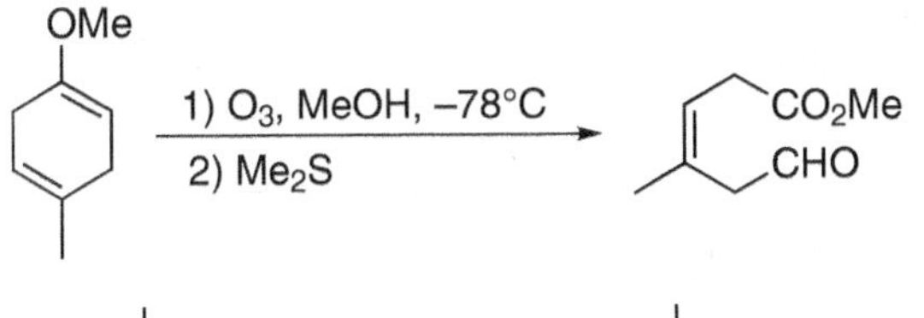

(18.196)[227]

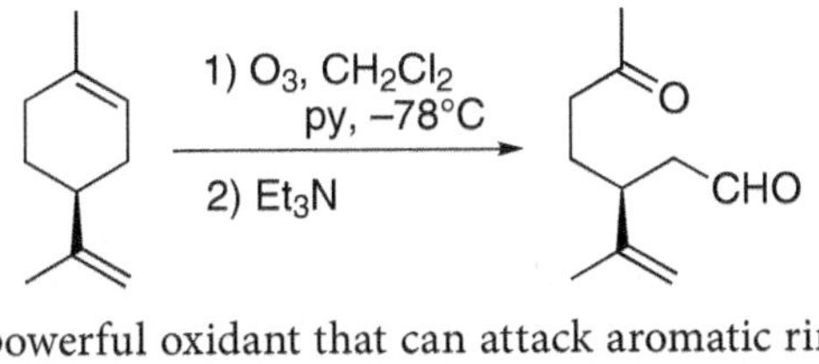

(18.197)[228]

Ozone is a powerful oxidant that can attack aromatic rings, giving the α-dicarbonyl products that would be expected from the Kekulé structures following the reduction of

227. Corey, E.J.; Katzenellenbogen, J.A.; Gilman, N.W.; Roman, S.A.; Erickson, B.W. *J. Am. Chem. Soc.* **1968**, *90*, 5618.

228. (a) Knöll, W.; Tamm, C. *Helv. Chim. Acta* **1975**, *58*, 1162. (b) Heath, R.R.; Doolittle, R.E.; Sonnet, P.E.; Tumlinson, J.H. *J. Org. Chem.* **1979**, *45*, 2910. (c) Griesbaum, K.; Hilß, M.; Bosch, J. *Tetrahedron* **1996**, *52*, 14813.

the ozonide. An effective alternative reaction for the cleavage of alkenes that does not affect aromatic rings is the ***Lemieux-Johnson oxidation***.[229]

In this reaction, the alkene is treated with a catalytic amount of osmium tetroxide in the presence of sodium metaperiodate. The periodate functions in much the same way as the co-oxidants in the catalytic osmium tetroxide hydroxylations discussed earlier—it cleaves the osmate ester and regenerates the osmium tetroxide. However, rather than allowing the diol to be isolated, as do such co-oxidants as sodium chlorate, the sodium periodate cleaves the diol to give aldehydes and ketones. Thus, the Lemieux-Johnson oxidation is the equivalent of ozonolysis followed by reduction of the ozonide with dimethyl sulfide. In certain cases (e.g., Example **18.198**[230]), the Lemieux-Johnson oxidation is superior to selective ozonolysis.

Using periodate as the oxidant to cleave the osmate ester (e.g., Example **18.198**) gives aldehydes when one end of the alkene is monosubstituted (e.g., cyclohexene gives hexanedial as the oxidation product under Lemieux-Johnson conditions). The alkene can be converted to the carboxylic acids by replacing the periodate oxidant with Oxone®; *trans*-stilbene, for example, is oxidized by this reagent to benzoic acid.[231] The reagent is selective for electron-rich alkenes, so nootkatone can be oxidized to the diketone shown in Example **18.199**.

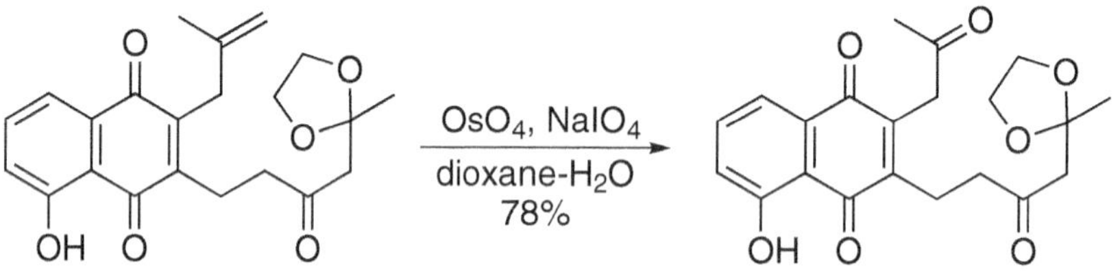

(18.198)

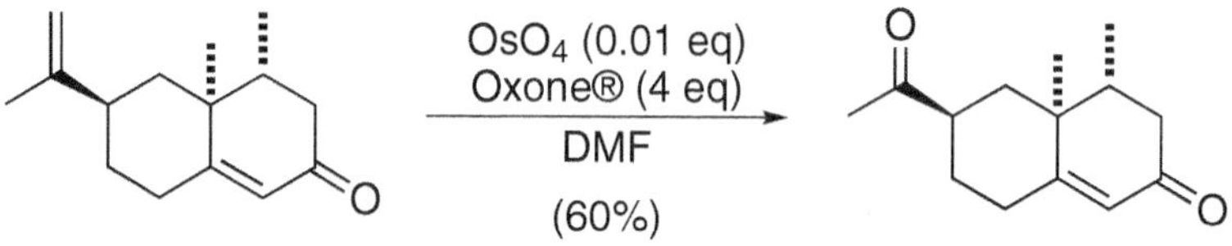

(18.199)

Earlier, we discussed ruthenium tetroxide,[232] a much more powerful oxidant, and its use in oxidizing alcohols to acids. This reagent is reactive enough to cleave both bonds (σ and π) of an alkene to give carboxylic acids and ketones instead of aldehydes and ketones as products. In contrast to sodium periodate, ruthenium tetroxide also rapidly oxidizes α-dicarbonyl compounds with cleavage of the carbon-carbon bond to give carboxylic acid derivatives.

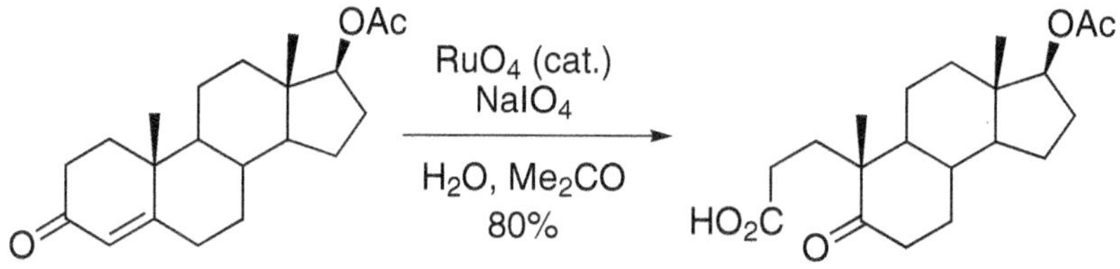

(18.200)

229. Pappo, R.; Allen, D.S., Jr.; Lemieux, R.U.; Johnson, W.S. *J. Org. Chem.* **1956**, *21*, 478.

William Sumner Johnson (1913–1995) was educated at Amherst College (AB, 1936) and Harvard (AM, 1938; PhD, 1940). His academic career started at Amherst (1936) and continued at Wisconsin (1940) and Stanford (1960). For more biographical detail, see: Stork, G. *Biogr. Mem.* **2001**, *80*, 1.

Raymond Urgel Lemieux (1920–2000) took his PhD in 1946 (McGill). After a year at Ohio State, he returned to Canada. His career took him from Saskatchewan (1947) to the National Research Council (1949) to the University of Ottawa (Dean, 1954), to Alberta, where he retired in 1985. In 1967 he was made a Fellow of the Royal Society. For more biographical information, see: *Biogr. Mem. Fell. Roy. Soc. London* **2002**, *48*, 251.

230. (a) Krohn, K.; Böker, N.; Flörke, U.; Freund, C. *J. Org. Chem.* **1997**, *62*, 2350. (b) Krohn, K. *Eur. J. Org. Chem.* **2002**, 1351.

231. Travis, B.R.; Narayan, R.S.; Borhan, B. *J. Am. Chem. Soc.*, **2002**, *124*, 3824.

232. Review: Lee, D.G.; van den Engh, M. In Trahanovsky, W.S., Ed. *Oxidation in Organic Chemistry, Part B* (Academic Press: New York, 1973), p. 177.

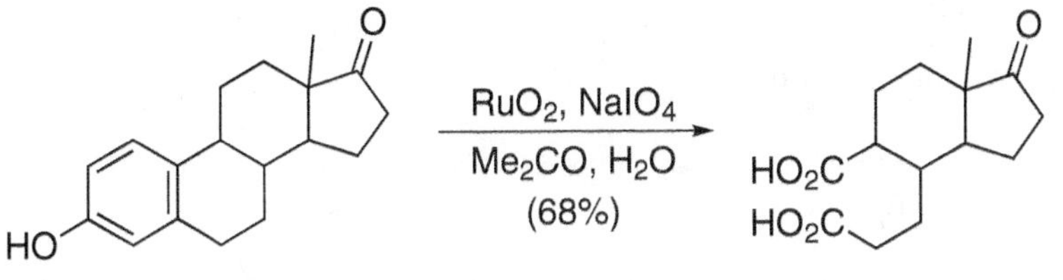

(18.201)

Example **18.200**, the oxidation of testosterone acetate,[233] is testament to the power of this reagent: the double bond in this compound is unaffected under Lemieux-Johnson conditions. This reagent can also attack aromatic rings, which are oxidized to carboxylic acids, as illustrated by the oxidation of estrone (Example **18.201**).[234] Ethers are oxidized by this reagent to esters, and amines are oxidized to amides. Its reactivity limits its use to solvents such as carbon tetrachloride, water, and aqueous acetone. The substitution in the aromatic ring *does* affect the outcome of the oxidation, as shown by the series of reactions in Example **18.202**. When the *para* substituent is strongly electron withdrawing (e.g., NO_2, CO_2Me, or 2-pyridyl), the starting ester is recovered unchanged.[235]

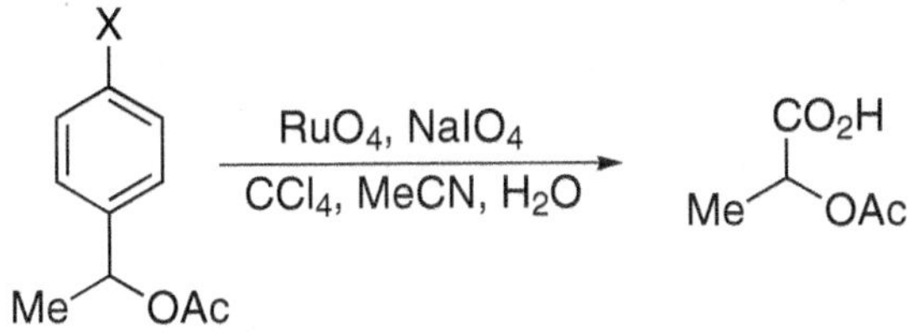

(18.202)

Reaction Synopses

Oxidative Cleavage of 1,2-Diols

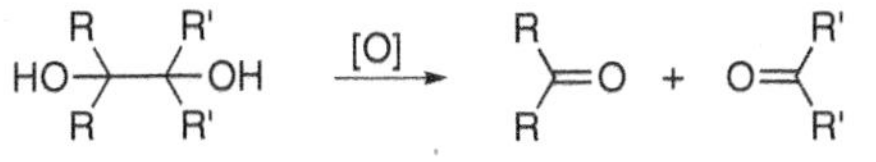

Reagents: HIO_4; $Pb(OAc)_4$, CH_3CO_2H; CrO_3; etc.
In cyclic systems, *cis*-diols react faster than their *trans* isomers.

Ozonolysis

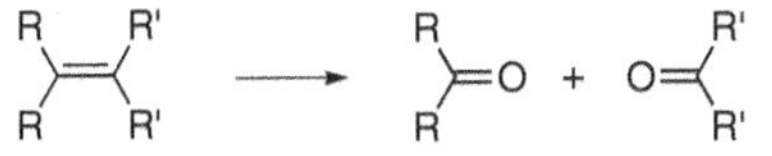

Reagents: (1) O_3, CH_2Cl_2, −78°C; (2) Zn, H_2O;
or (1) O_3, CH_2Cl_2, −78°C; (2) Me_2S; etc.
or (1) O_3, CH_2Cl_2, −78°C; (2) CH_3CO_2OH; etc.

Lemieux-Johnson Oxidation Reactions

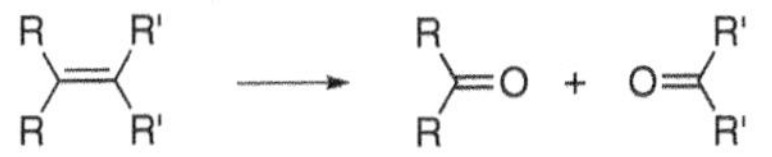

Reagents: OsO_4, $NaIO_4$, CCl_4, H_2O; etc. to give aldehydes;
or OsO_4, Oxone®, DMF; etc. to give acids

(continues)

---

233. Patiak, D.M.; Bhat, H.B.; Caspi, E. *J. Org. Chem.* **1969**, *34*, 112.
234. Piatak, D.M.; Herbst, G.; Wicha, J.; Caspi, E. *J. Org. Chem.* **1969**, *34*, 116.
235. Kasai, M.; Ziffer, H. *J. Org. Chem.* **1983**, *48*, 2346.

(**Reaction Synopses** continued)

Ruthenium Tetroxide Oxidation

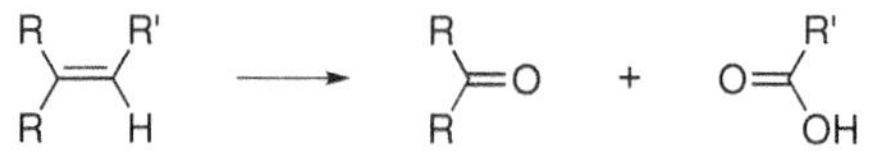

Reagents:	RuO_4, CCl_4; RuO_4, Me_2CO, H_2O;
or	(RuO_2 or $RuCl_3$), $NaIO_4$, CCl_4, H_2O

Worked Problem

18-4 What are the major products of the following reactions?

(a) 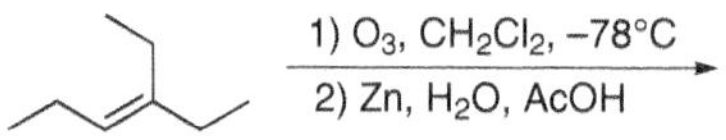

(b) 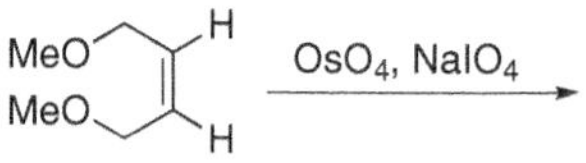

(c) 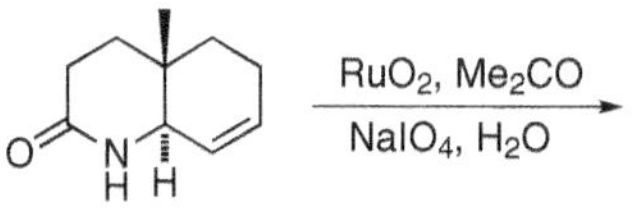

(d) 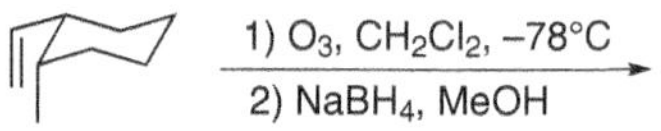

(e) 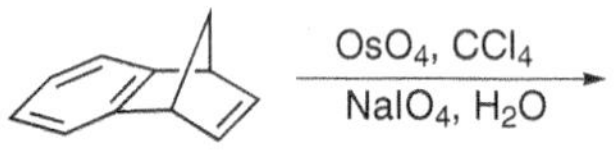

(f) 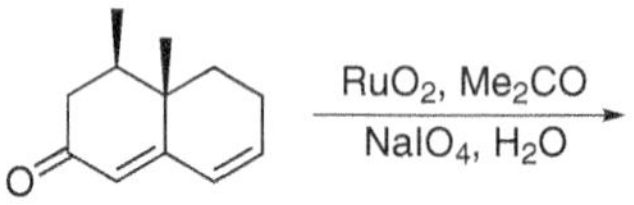

§Answers below.

Problems

18-12 The oxidative cleavage of *meso*-2,3-butanediol by lead tetraacetate and periodic acid occurs much more slowly than the oxidative cleavage of the (±)-isomer under the same reaction conditions. Suggest a reason why this should be so (it will help to use models here).

18-13 What reagent or reagents should be used to complete each of the following transformations? Where more than one reagent can be used to complete the

§ Answers to Worked Problem:

(a) 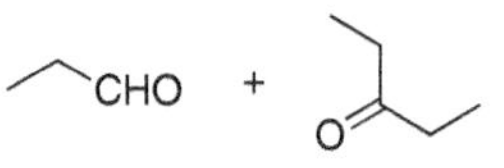

(b) 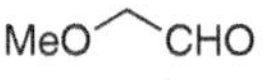

(c) 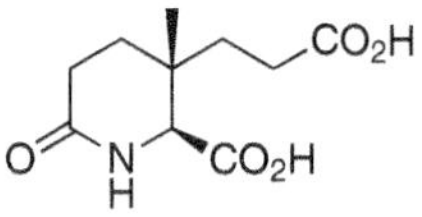

(d) 

(e) 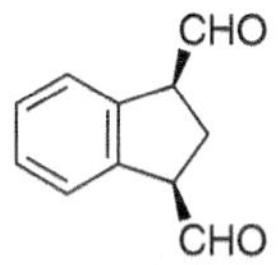

(f) 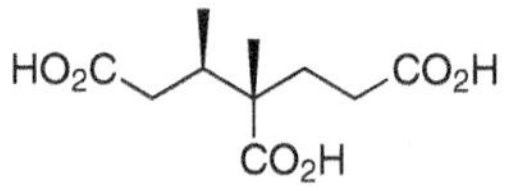

(a) The reduction of the ozonide by zinc in acetic acid gives the aldehyde and/or ketone. (b) This is a straightforward Lemieux-Johnson cleavage of the alkene π bond. (c) The ruthenium tetroxide generated by oxidation of the ruthenium dioxide will cleave the double bonds to generate carboxylic acids. (d) Reduction of the ozonide with sodium borohydride gives the primary alcohol rather than the aldehyde. (e) This Lemieux-Johnson oxidation is straightforward. (f) The α-dicarbonyl compounds formed by the initial oxidation of the alkene π bonds are further oxidized with C—C cleavage to carboxylic acids.

transformation, give all options and suggest which reagent may be best (with reasons).

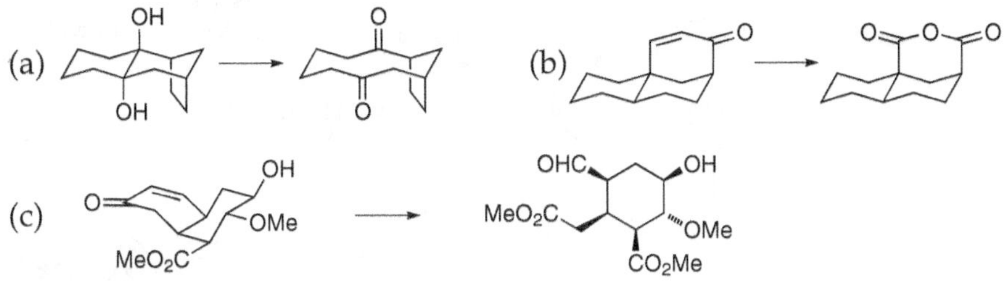

18.10 Oxidative Substitution of Carbon-Hydrogen σ Bonds

Oxidative substitution of a carbon-hydrogen bond is one of the "holy grails" of organic synthesis, especially when the position is unactivated. Although progress has been made in that arena, by far the greatest number of oxidative substitution reactions currently known involve unsaturated substrates, so that is where we will start.

The Wacker Oxidation and Related Reactions

The oxidation of an alkene to an enol derivative by replacement of the vinyl hydrogen with oxygen is known as the ***Wacker oxidation***. In the classical Wacker oxidation, the enol or enol acetate is formed by oxopalladation of the alkene, followed by β-hydride elimination. The reaction was first used for the industrial oxidation of ethylene to acetaldehyde[236] but is generally applicable to the conversion of terminal alkenes to methyl ketones (e.g., Example **18.203**) by way of the corresponding enol. The most common reagent for carrying out the Wacker oxidation is a mixture of palladium (II), with copper (I) or copper (II) as a catalyst for reoxidation of the palladium to the active +2 oxidation state, with molecular oxygen as the stoichiometric oxidant.[237] The mechanism has been studied extensively, and the catalytic cycle currently accepted for the Wacker process is shown in Figure 18.12. Although it had originally been proposed that the hydroxylation of the double bond was accomplished by intramolecular migration of OH from palladium to carbon,[238] it is now generally accepted that the hydroxypalladation occurs by addition of water to the η^2 alkene complex (**18.204**) with *trans* stereochemistry.[239]

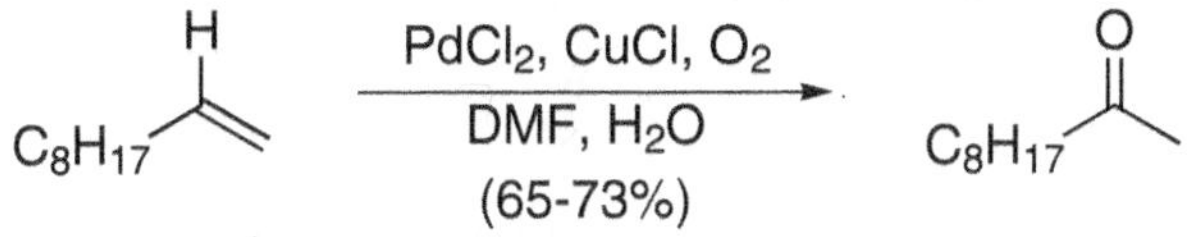

(18.203)

Oxidation Adjacent to π Bonds

Allylic Halogenation

Historically the most common allylic oxidation reaction of alkenes has been halogenation. This can be carried out using the halogen itself, but it is more often achieved by using a halogen surrogate such as *N*-bromosuccinimide. As we have already seen (in the

236. (a) Smidt, J.; Hafner, W.; Jira, R.; Sedlmeier, J.; Sieber, R.; Ruttinger, R.; Kojer, H. *Angew. Chem.* **1959**, *71*, 176. (b) Smidt, J.; Hafner, W.; Jira, R.; Sieber, R.; Sedlmeier, J.; Sabel, A. *Angew. Chem. Int. Ed. Engl.* **1962**, *1*, 80.
237. Tsuji, J.; Nagashima, H.; Nemoto, H. *Org. Synth., Coll. Vol. VII* **1990**, 137.
238. Henry, P.M. *Acc. Chem. Res.* **1973**, *6*, 16.
239. Bäckvall, J.-E.; Åckermark, B.; Ljunggren, S.O. *J. Chem. Soc., Chem. Commun.* **1977**, 264.

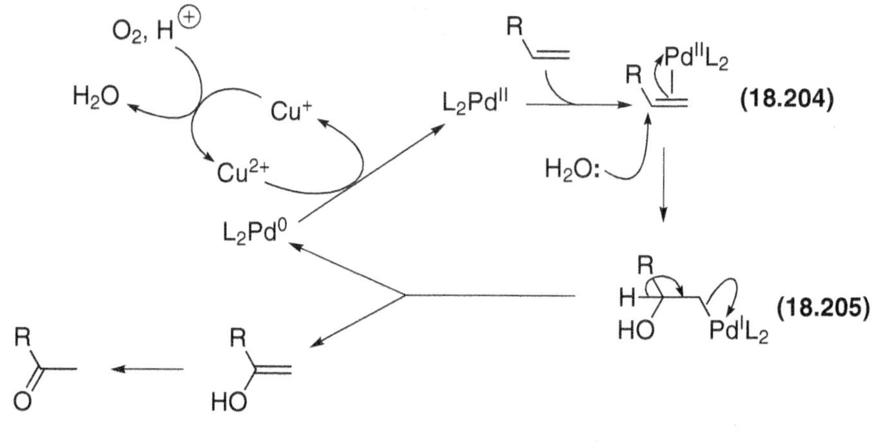

Figure 18.12 The catalytic cycle of the Wacker oxidation of 1-alkenes

Corey-Kim oxidation of alcohols), *N*-halosuccinimides are just such a convenient replacement for the halogen itself.[240] The favored reagent for allylic bromination is NBS,[241] which is a crystalline solid that reacts with alkenes in a non-polar solvent in the presence of a free radical initiator to give the allylic bromide (e.g., Example **18.206**). The advantages of avoiding the toxicity and corrosiveness of the free halogen are improved by the fact that the by-product of the NBS oxidation is succinimide, which is neutral, rather than hydrogen bromide. Thus, side reactions caused by excess acid are also avoided.

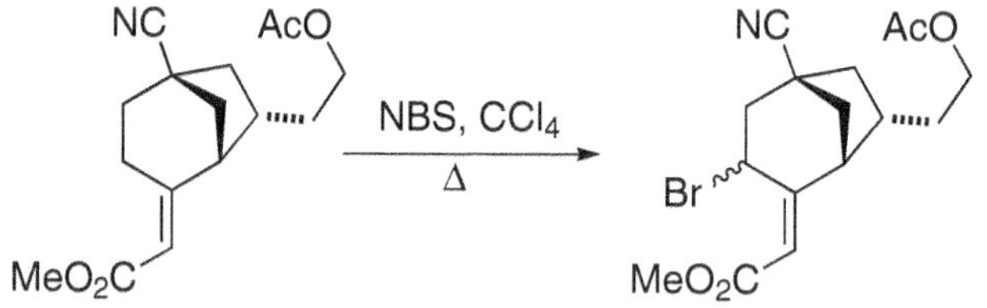

(18.206)[242]

The brominating agent under these conditions is actually still elemental bromine,[243] which is formed by the rapid ionic reaction between the NBS and hydrogen bromide (Example **18.207**; the HBr probably comes from hydrolysis of the NBS by a trace of water). The reaction proceeds only as long as there is a trace of hydrogen bromide present. More recently, lithium bromide has been used as a reagent to convert NBS to molecular bromine.

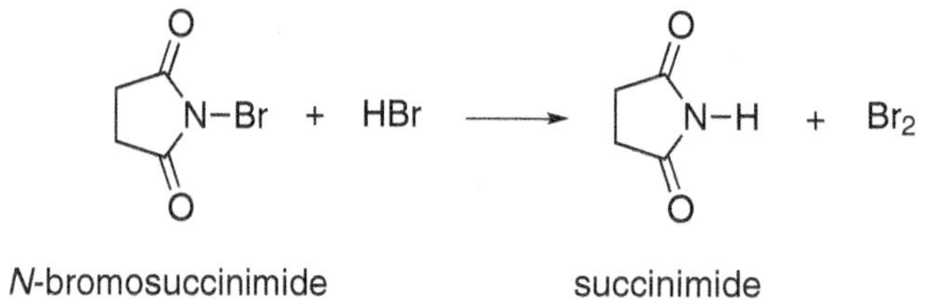

(18.207)

Allylic Oxidation of Alkenes to Give Alcohols, Esters, and Enones

Allylic bromination has been by far the most commonly used method for the allylic oxidation of alkenes. However, the very reactivity of the allyl bromides produced often results in side reactions, leading to unsatisfactory yields of the desired product. For example, most NBS oxidations give yields below 80%, and many give yields below 40%. It is partly

240. Corey, E.J.; Kim, C.U. *J. Am. Chem. Soc.* **1972**, *94*, 7586.

241. Djerassi, C. *Chem. Rev.* **1948**,*43*, 271.

242. Trost, B.M.; Shuey, C.D.; DiNinno, F., Jr.; McElvain, S.S. *J. Am. Chem. Soc.* **1979**, *101*, 1284.

243. (a) Adam, J.; Gosselain, P.A.; Goldfinger, P. *Nature* **1953**, *171*, 704. (b) Adam, J.; Gosselain, P.A.; Goldfinger, P. *Bull. soc. chim. Belges* **1956**, *65*, 533. (c) Skell, P.S.; Tuleen, D.L.; Readio, P.D. *J. Am. Chem. Soc.* **1963**, *85*, 2850. (d) Walling, C.; Rieger, A.L.; Tanner, D.D. *J. Am. Chem. Soc.* **1963**, *85*, 3129. (f) Russell, G.A.; Desmond, K.M. *J. Am. Chem. Soc.* **1963**, *85*, 3139. (g) Pearson, R.E.; Martin, J.C. *J. Am. Chem. Soc.* **1963**, *85*, 3142.

for this reason that there has been considerable effort expended to develop alternative methods for the oxidation of alkenes at the allyl position. Oxidation at allylic positions can be arranged to give a wide range of substitution products by taking advantage of π-allylpalladium (II) complexes (η^3-allylpalladium complexes). In the presence of an external base, such as pyridine, the η^2 palladium (II) complex initially formed in the Wacker process can be deprotonated to give the η^3-allylpalladium (II) complex. This complex can then be intercepted by a wide range of nucleophiles, including oxygen and nitrogen nucleophiles, (***oxidative heterocyclization***[244]), as well as electron-rich π bonds such as those in the heterocyclic ring of furans and indoles (see Examples **18.208** to **18.210**). The attack of the nucleophile gives a new η^3-allylpalladium (II) complex that provides the product where the allylic hydrogen has been substituted by a nucleophile, in what is a formal oxidation process with oxygen as the stoichiometric oxidant.

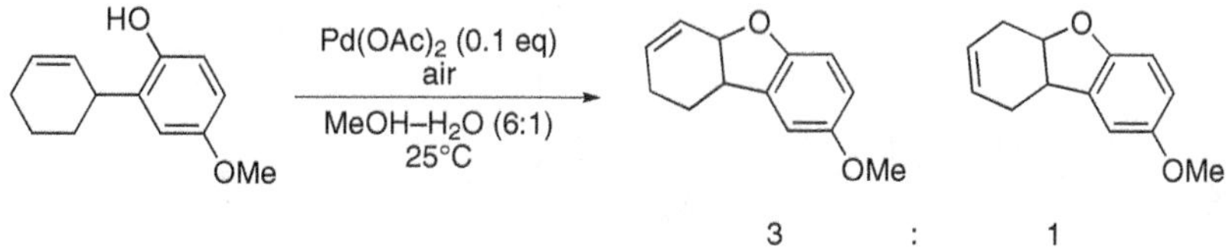

(18.208)[245]

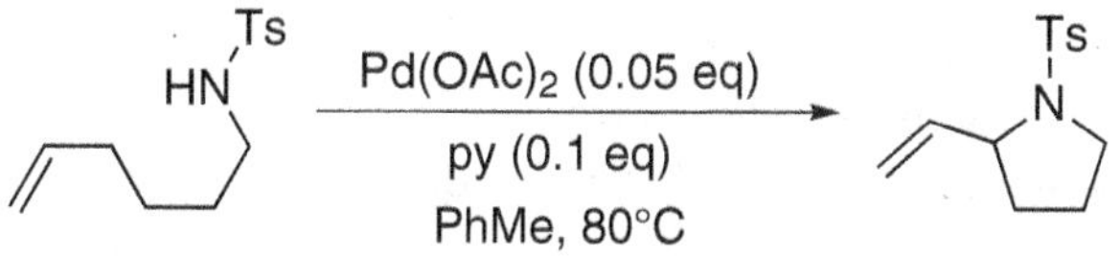

(18.209)[246]

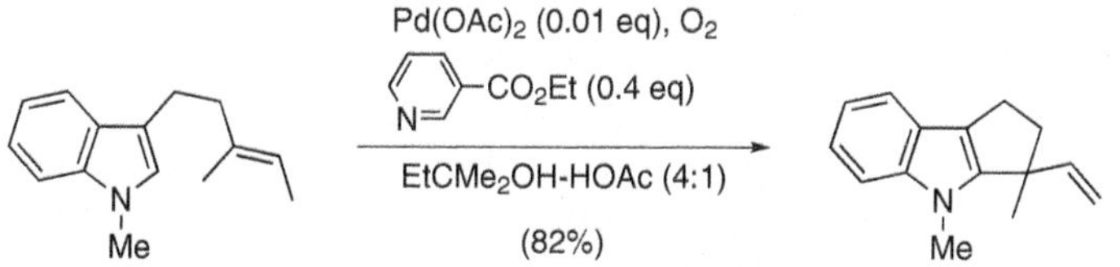

(18.210)[247]

Another successful method for the oxidation of allylic C—H bonds involves treating the alkene with a *tert*-butyl perester and a cuprous salt; the product of this reaction is the corresponding allyl ester (e.g., Example **18.211**).[248] The reaction is believed to involve a copper (III) species such as the one in Example **18.212**, formed by the addition of an allyl radical to copper (II) benzoate.[249] This then undergoes an intramolecular rearrangement to give the observed product. This has the consequence that where two regioisomeric esters are possible, the product with the less substituted double bond tends to predominate (but not always; see Example **18.213**).[250] By the use of a chiral ligand for copper, this reaction can be used to prepare chiral esters from alcohols.[251]

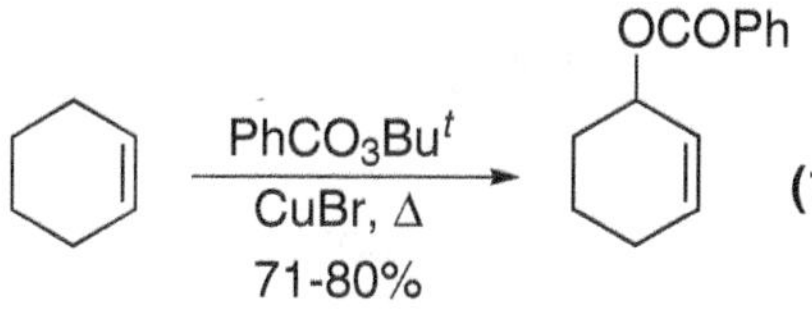

(18.211)

244. Reviews: (a) Zeni, G.; Larock, R.C. *Chem. Rev.* **2006**, *106*, 4644. (b) Stoltz, B.M. *Chem. Lett.* **2004**, 33, 362. (c) Stahl, S. S. *Angew.Chem., Int. Ed.* **2004**, *43*, 3400. (d) Sigman, M.S.; Schultz, M.J. *Org. Biomol. Chem.* **2004**, *2*, 2551.

245. Hosokawa, T.; Miyagi, S.; Murahashi, S.-I.; Sonoda, A. *J. Org. Chem.* **1978**, *43*, 2752.

246. Fix, S.R.; Brice, J.L.; Stahl, S.S. *Angew. Chem. Int. Ed.* **2002**, *41*, 164.

247. Ferreira, E.M.; Zhang, H.; Stoltz, B.M. *Tetrahedron* **2008**, *6*, 5987.

248. Pedersen, K.; Jakobson, P.; Lawesson, S.-O. *Org. Syn., Coll. Vol. 5* **1973**, 70.

249. Kochi, J.K.; Bemis, A. *Tetrahedron* **1968**, *24*, 5099.

250. Rawlinson, D.J.; Sosnovsky, G. *Synthesis* **1972**, 1.

251. (a) Andrus, M.B.; Chen, X. *Tetrahedron* **1997**, *53*, 16229. (b) Sekar, G.; DattaGupta, A.; Singh, V.K. *J.Org. Chem.* **1998**, *63*, 2961. (c) Malkov, A.V.; Bella, M.; Langer, V.; Kocovsky, P. *Org. Lett.* **2000**, *2*, 3047.

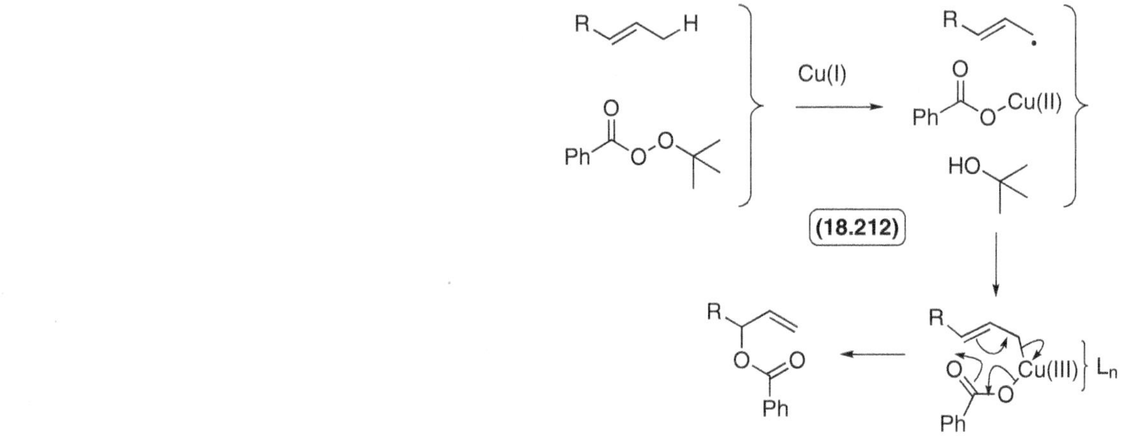

(18.212)

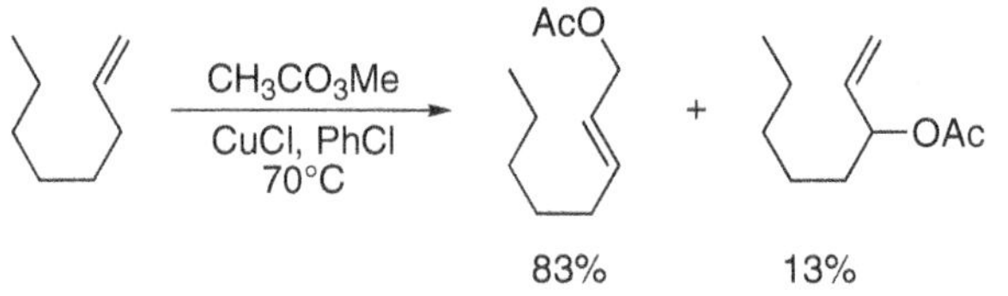

(18.213)

A variation of this approach involves the oxidation of alkenes with palladium (II) acetate in acetic acid and dimethyl sulfoxide (DMSO), with benzoquinone as the terminal oxidant. In the traditional Wacker process, DMSO is absent, and the reaction gives ketones or enol acetates. The addition of DMSO as cosolvent, however, or the use of a *bis*-sulfoxide ligand for the metal, leads to the formation of allylic acetates. The ratio of allylic isomers may be controlled by adjusting the reaction conditions, as shown in Example **18.214**.[252] Sanford has reported regioselective palladium-catalyzed oxidation at the benzylic hydrogen of heterocyclic aromatics by using the directing influence of a nitrogen atom in the heterocycle (e.g., Example **18.215**).[253]

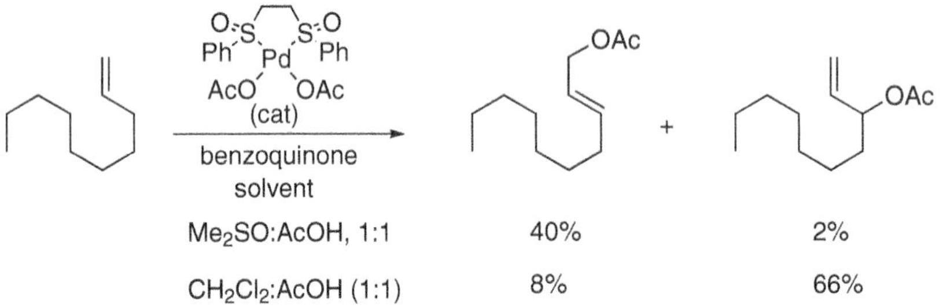

(18.214)

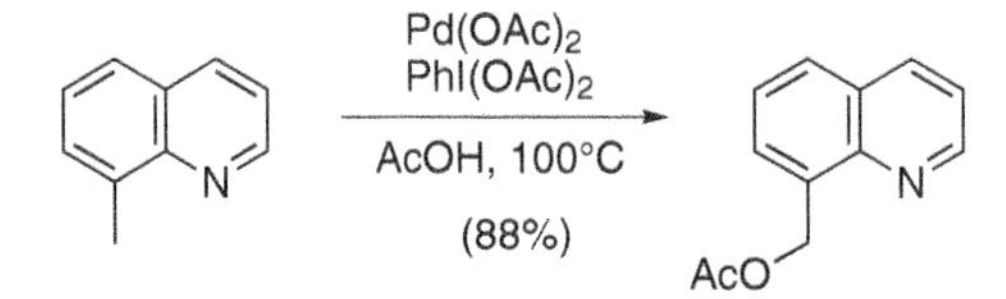

(18.215)

Manganese (III) acetate has proved to be a particularly mild oxidant for the synthesis of conjugated enones from alkenes, as shown in the oxidation of α-pinene to verbenone (Example **18.216**).[254] The same oxidation can be accomplished by oxidation with molecular oxygen in the presence of a cobalt (II) catalyst.[255] An interesting dichotomy of reactivity has been observed, depending on the exact nature of the cobalt (II) catalyst.[256] The same

252. Chen, M.S.; White, M.C. *J. Am. Chem. Soc.* **2004**, *126*, 1346.
253. Dick, A.R.; Hull, K.L.; Sanford, M.S. *J. Am. Chem. Soc.* **2004**, *126*, 2300.
254. Shing, T.K.L.; Yeung, Y.-Y.; Su, P.L. *Org. Lett.* **2006**, *8*, 3149. (b)
255. Lajunen, M.; Koskinen, A.M.P. *Tetrahedron Lett.* **1994**, *35*, 4461.
256. Reddy, M.M.; Punniyamurthy, I.; Iqbal, J. *Tetrahedron Lett.* **1995**, *36*, 159.

reagent can be used for the regioselective oxidation of conjugated enones to give α'-hydroxyketones.[257] The oxidation of aromatic compounds by this reagent generally proceeds by a free radical mechanism.[258]

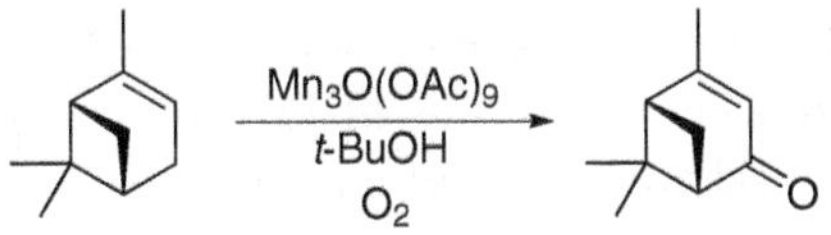

(18.216)

Similar oxidations at allylic positions are accomplished by using *tert*-butyl hydroperoxide and a metal catalyst. Doyle and his coworkers have reported[259] that, in the presence of a rhodium (II) catalyst, this oxidant oxidizes the allylic position of alkenes by a free radical mechanism to give conjugated carbonyl compounds. This has been especially useful in the synthesis of conjugated dicarbonyl compounds (e.g., Example **18.217**).[260] The regiochemistry of oxidation of Δ^5-steroid derivatives with the catalytic reagents above almost always gives the C-7 ketone. Crich has shown that this regiochemistry can be reversed when a fluorous seleninic acid and iodoxybenzene are used to effect the oxidation (Example **18.218**).[261]

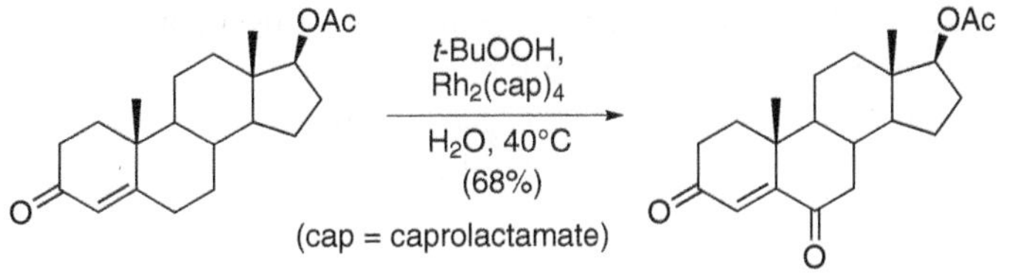

(18.217)

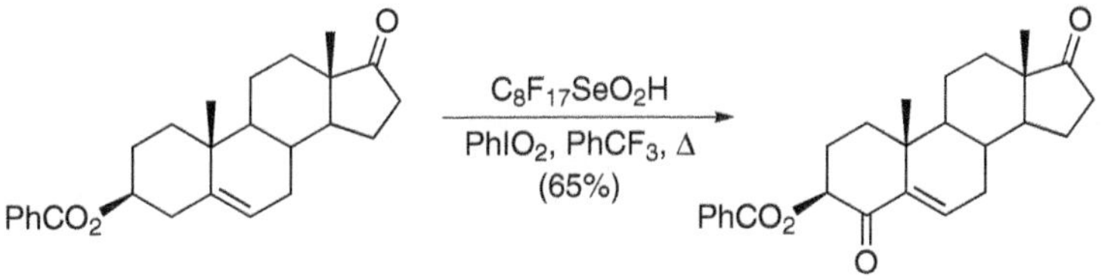

(18.218)

Selenium Dioxide Oxidations and Photosensitized Singlet Oxygen Oxidations

Oxidation at allylic and benzylic positions can also be accomplished using selenium dioxide (Example **18.219**) or by the photosensitized ene reaction with singlet oxygen (Example **18.220**). In the selenium dioxide reaction, the products are allylic alcohols or α,β-unsaturated carbonyl compounds, where the allylic hydrogen is replaced without rearrangement of the double bond. The reaction with singlet oxygen gives the rearranged hydroperoxide, which is usually reduced in situ. The mechanism of the reaction remains elusive.[262]

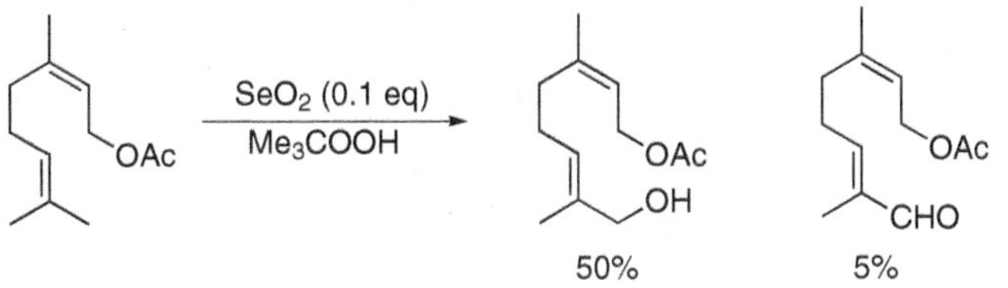

(18.219)

257. Tanyeli, C.; Iyigün, Ç. *Tetrahedron* **2003**, *59*, 7135.

258. Bryant, J.R.; Taves, J.E.; Maer, J.M. *Inorg. Chem.* **2002**, *41*, 2769.

259. (a) Choi, H.; Doyle, M.P. *Org. Lett.* **2007**, *9*, 5349. (b) McLaughlin, E.C.; Choi, H.; Wang, K.; Chiou, G.; Doyle, M.P. *J. Org. Chem.* **2009**, *74*, 730.

260. Salvador has reported a similar oxidation using bismuth (III) as the catalyst in the presence of the clay, montmorillonite K10, which facilitates the isolation of the product: Salvador, J.A.R.; Silvestre, S.M. *Tetrahedron Lett.* **2005**, *46*, 2581

261. Crich, D.; Zou, Y. *Org. Lett.* **2004**, *6*, 775.

262. (a) Frimer, A,H. *Chem. Rev.* **1979**, *79*, 359. (b) Singleton, D.A.; Hang, C.; Szymanski, M.J.; Meyer, M.P.; Leach, A.G.; Kuwata, K.T.; Chen, J.S.; Greer, A.; Foote, C.S.; Houk, K.N. *J. Am. Chem. Soc.* **2003**, *125*, 1319.

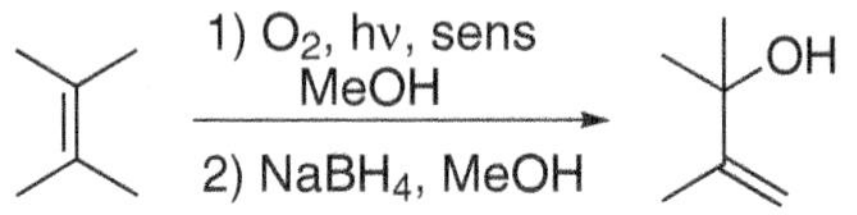

(18.220)

The mechanism of the selenium dioxide oxidation (Figure 18.13) involves an initial ene reaction of the alkene with selenium dioxide (**18.221**), followed by [2,3]-sigmatropic rearrangement (**18.222**) of the seleninic acid and hydrolysis of the selenium-oxygen bond of the product (**18.223**) to give the alcohol.[263] This reaction may be made catalytic in selenium dioxide by using *tert*-butyl hydroperoxide as a co-oxidant.[264] The oxidation of 2-methyl-2-alkenes exhibits a useful selectivity for the formation of the *E* product.[265]

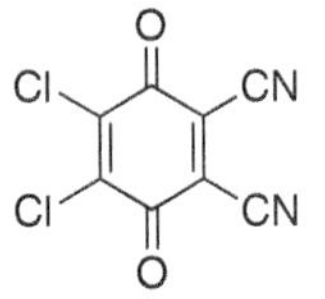

DDQ

Dehydrogenation of Alkenes and Dienes

The dehydrogenation of alkenes and dienes (especially 1,3- or 1,4-cyclohexadienes) can be readily accomplished by oxidation with 2,3-dichloro-5,6-dicyano-1,4-benzoquinone (DDQ). This quinone is a powerful oxidant that can lead to insertion of a double bond in conjugation with an aromatic ring.

The substrate must, however, have a benzylic or allylic hydrogen, or a carbonyl group with an available α-hydrogen. An example of this reactivity is provided by the synthesis of the chromene (Example **18.224**).[266] The same reagent can be used to introduce a double bond in conjugation with a carbonyl group[267] and the oxidative deprotection of benzyl[268] and electron-rich silyl[269] ethers.

(18.224)

Figure 18.13 Mechanism of selenium dioxide oxidation of alkenes

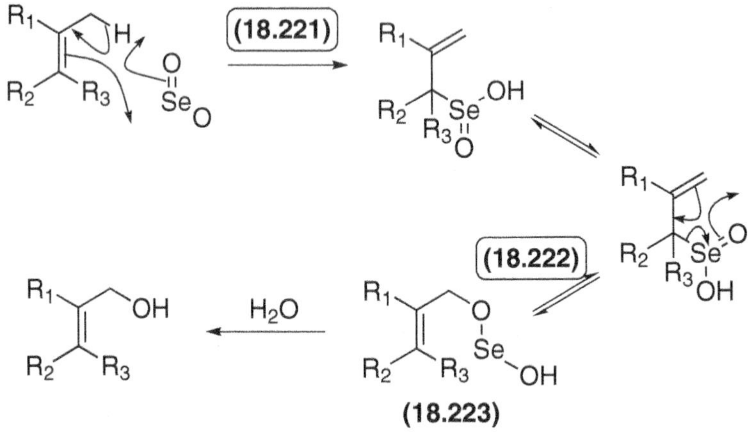

263. Sharpless, K.B.; Lauer, R.F. *J. Am. Chem. Soc.* **1972**, *94*, 7154.
264. Umbreit, M.A.; Sharpless, K.B. *J. Am. Chem. Soc.* **1977**, *99*, 5526.
265. (a) Bhalerao, U.T.; Rapoport, H. *J. Am. Chem. Soc.* **1971**, *93*, 4835. (b) Büchi, G.; Wüerst, H.W. *Helv. Chim. Acta* **1967**, *50*, 2440.
266. Ahluwalia, V.K. ; Jolly, R.S. *Synlett*, **1982**, 74.
267. Walker, D.; Hiebert, J.D. *Chem. Rev.*, **1966**, *66*, 153.
268. Wang, W.; Li, T.; Attardo, G. *J. Org. Chem.*, **1997**, *62* , 6598.
269. Paterson, I.; Cowden, C.J.; Rahn, V.S.; Woodrow, M.D. *Synlett*, **1998**, 915.

Worked Problem

18-5 What is (are) the major organic product(s) formed when indane is treated with each of the reagents in the list? Assume that one equivalent of the reagent is used.

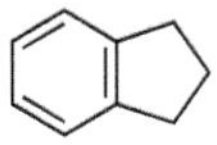

Reagents:

(a) Br_2, hν

(b) NBS, ROOR, CCl_4, Δ or hν

(c) $PhCO_3CMe_3$, CuBr, Δ

(d) SeO_2, Me_3COOH

(e) DDQ, CH_2Cl_2

§Answers below.

18-6 What is the major organic product when indene is treated with each of the reagents in the list below?

Reagents:

(a) $PdCl_2$, CuCl, O_2, DMF, H_2O

(b) $Mn(OAc)_3$, Me_3COH, O_2

(c) $(PhSOCH_2)_2Pd(OAc)_2$, benzoquinone, AcOH:Me_2SO (1:1)

¶Answers below.

Drill Exercises

What is (are) the major organic product(s) formed when each of the alkenes in the first list below is treated with each of the reagents in the second list? Assume that one mole equivalent of the oxidant is consumed unless otherwise noted.

Alkenes:

(a) 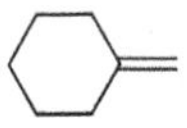 (b) 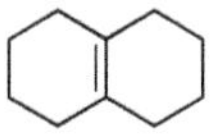 (c) 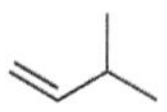 (d) 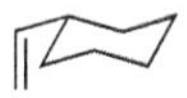

(continues)

§ Answers to Worked Problem 8-5:

(a) 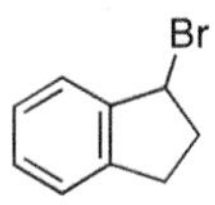

(b) 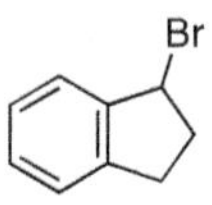

(c) 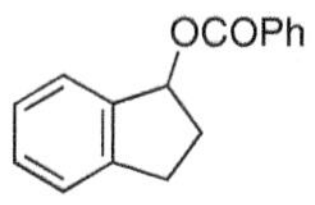

(d) 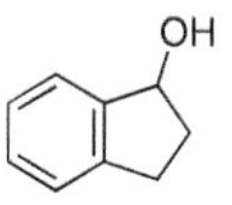

(e) 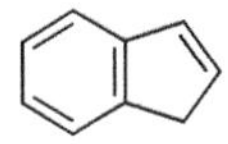

¶ Answers to Worked Problem 8-6:

(a) 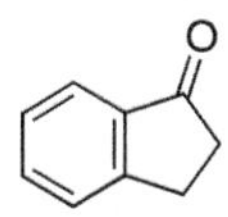

(b) 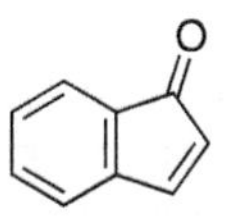

(c) 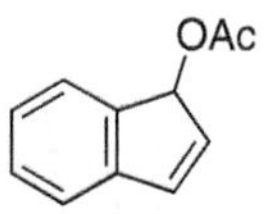

(Drill Exercises continued)

(e) 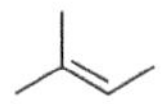 (f) 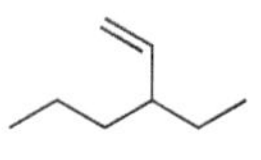 (g) 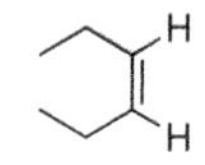

(h) 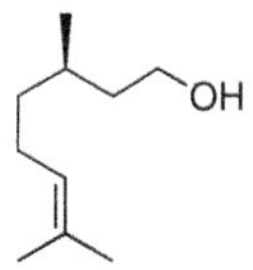

Reagents:

(a) Br_2, hν

(b) NBS, ROOR, CCl_4, Δ or hν

(c) $PhCO_3CMe_3$, CuBr, Δ

(d) SeO_2, Me_3COOH

(e) $PdCl_2$, CuCl, O_2, DMF, H_2O

(f) $Mn(OAc)_3$, Me_3COH, O_2

(g) $PhSOCH_2)_2Pd(OAc)_2$, benzoquinone, AcOH:CH_2Cl_2 (1:1)

Oxidation at the α Carbon of Carbonyl Compounds and Nitriles

The α carbon of carbonyl compounds shares much of the reactivity of the allyl position of alkenes, and this extends to the oxidation at this position. Replacement of one of these hydrogens by a heteroatom is often the first step in introducing a π bond in conjugation with the carbonyl group. The most useful oxidative substitutions at the α carbon involve replacement of hydrogen by halogen, sulfur, or selenium. All three of these reactions occur through the enol or enolate anion of the carbonyl compound, as illustrated in Example **18.225**.

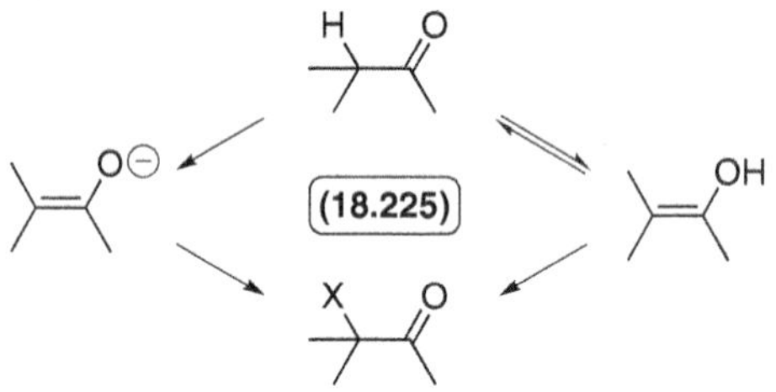

X = Cl, Br, I, SR, SeR

Halogenation is usually carried out in acid solution, through the enol as the reactive species (formation of the enol is usually the rate-limiting step). Treatment of a ketone with the halogen, or with the halogen in acetic acid, gives controlled monohalogenation to give the α-haloketone.[270] The halogen may also be generated in situ from the *N*-halosuccinimide or by oxidation of the halide anion with a secondary oxidant (such as bromate; Example **18.226**[271]).

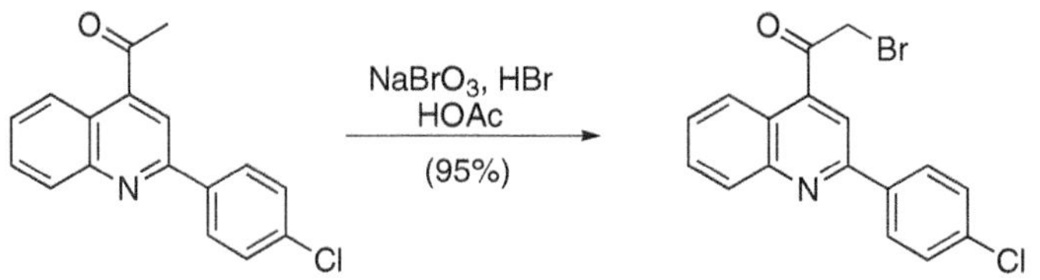

(18.226)

270. (a) Levene, P.A. *Org. Syn., Coll. Vol. 3* **1943**, 88. (b) Cowper, R.M.; Davidson, L.H. *Org. Syn., Coll. Vol. 3* **1943**, 480. (c) Newman, M.S.; Farbman, M.B.; Hipsher, H. *Org. Syn.* **1945**, *22*, 22.

271. Winstein, S.; Jacobs, T.L.; Linden, G.B.; Seymour, D.; Levy, E.F.; Day, B.F.; Robson, J.H.; Henderson, R.B.; Florsheim, W.H. *J. Am. Chem. Soc.* **1946**, *68*, 1831.

The α-bromination of carboxylic acids carried out using bromine in the presence of a catalytic amount of red phosphorus is known as the ***Hell-Volhard-Zelinskii reaction***.[272] The reaction actually proceeds through an acid bromide or anhydride formed by the reaction of the carboxylic acid with phosphorus tribromide formed in situ by the reaction between bromine and phosphorus (Example **18.227**). The acid derivative plays an important role for two reasons. One, it is much more readily enolized than the carboxylic acid or the anhydride. Two, the bromination itself occurs by addition of bromine to the enol double bond.

(18.227)

The halogenation of carbonyl compounds under basic conditions proceeds through the enolate anion, which leads to polyhalogenation, because the initial monohalogenated product compound is more acidic than the starting carbonyl compound. In methyl ketones, this results in the formation of α,α,α-trihalomethyl ketones that react with the base to give the carboxylic acid with one less carbon and the haloform.[273] Controlled hydroxylation of an enolate anion, on the other hand, can be accomplished using the MoO_5•py•HMPA complex (MoOPH, HMPA = hexamethylphosphoramide; Example **18.228**).[274]

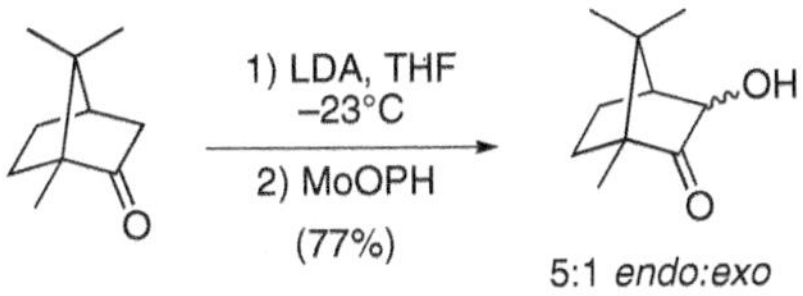

(18.228)

A more controlled oxidation at the α carbon of carbonyl compounds is provided by selenylation and sulfenylation. The treatment of a carbonyl compound with two equivalents of a strong base (typically a hindered lithium amide) and phenylselenyl chloride gives the conjugate base of the α-phenylseleno derivative. The reagents used to introduce the selenium into the molecule depend on the nature of the carbonyl group itself. With esters, diselenides (RSeSeR) can be used (e.g., Example **18.229**), but with ketones, whose enolate anion is less basic, selenyl halides (RSeX) must be used to avoid an unfavorable equilibrium (e.g., Example **18.230**).[275]

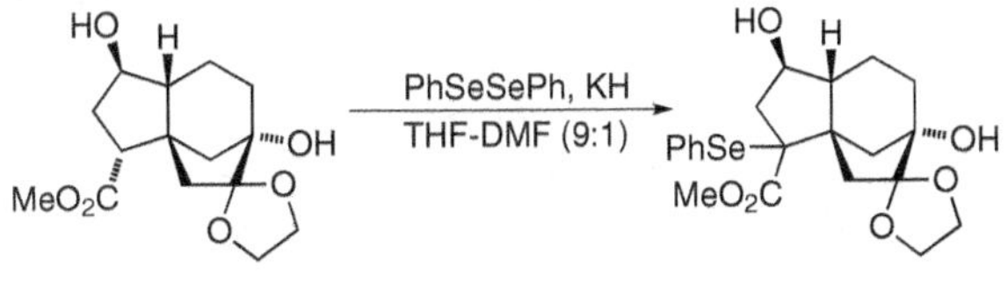

(18.229)[276]

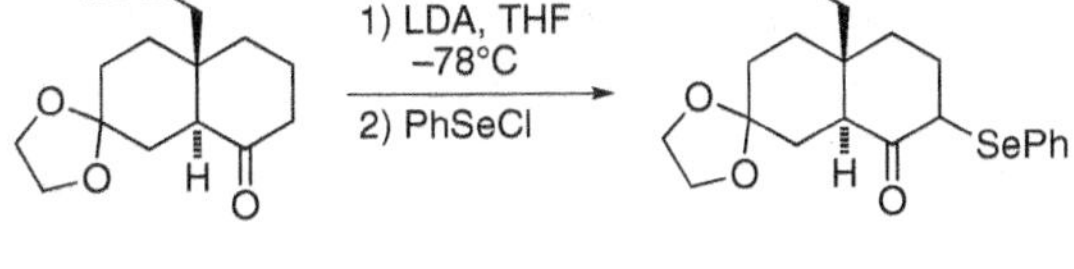

(18.230)[277]

---

272. (a) Hell, C. *Ber. dtsch. chem. Ges.* **1881**, *14*, 891. (b) Volhard, J. *Ann. Chem. Pharm.* **1887**, *242*, 141. (c) Zelinsky, N. *Ber. dtsch. chem. Ges.* **1887**, *20*, 2026. (d) Watson, H.B. *Chem. Rev.* **1930**, *7*, 180.

273. (a) Lieben, A. *Ann. Chem. Pharm. (Suppl.)* **1870**, *7*, 218. (b) Fuson, R.C.; Bull, B.A. *Chem. Rev.* **1934**, *15*, 275.

274. (a) Vedejs, E.; Engler, D.A.; Telschow, J.E. *J. Org. Chem.* **1978**, *43*, 188. (b) Vedejs, E.; Telschow, J.E. *J. Org. Chem.* **1976**, *41*, 740. (c) Vedejs, E.; Larsen, S. *Org. Syn.* **1990**, *Coll. Vol. 7*, 277.

275. Reich, H.J.; Renga, J.M.; Reich, I.L. *J. Am. Chem. Soc.* **1975**, *97*, 5434.

276. Lombardo, L.; Mander, L.N.; Turner, J.V. *J. Am. Chem. Soc.* **1980**, *102*, 6626.

277. (a) Grieco, P.A.; Nishizawa, M.; Burke, S.D.; Marinovic, N. *J. Am. Chem. Soc.* **1976**, *98*, 1612. (b) Grieco, P.A.; Nishizawa, M.; Oguri, T.; Burke, S.D.; Marinovic, N. *J. Am. Chem. Soc.* **1977**, *99*, 5773.

The analogous reaction with a sulfur electrophile (Example **18.231**) is sulfenylation.[278] The sulfides produced can participate in several reactions similar to the selenides, although the reactions usually require more vigorous conditions than the analogous selenium reactions. For this reason, the selenium compounds are usually preferred unless cost is a factor.

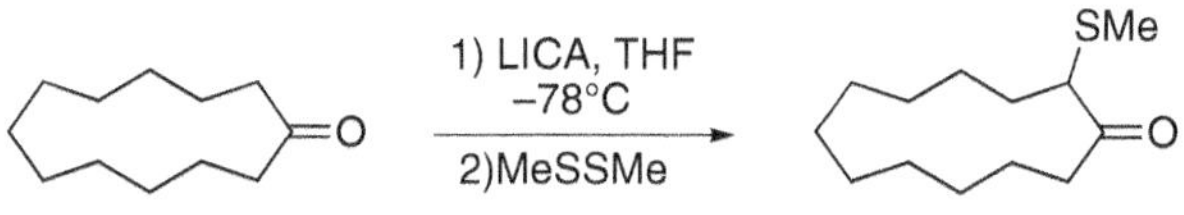

(18.231)[278 (a)]

One situation where sulfenylation is particularly useful is the *bis*-sulfenylation reaction (Example **18.232**), where the reaction is used to introduce a carbonyl group α to an existing carbonyl or cyano group by making a dithioketal. Liberation of the new carbonyl group from the dithioketal is effected by hydrolysis accompanied by alkylation, oxidation, or mercury desulfurization.[279] The reaction typically involves activating the α carbon by preparing the hydroxymethylene compound, which then reacts with propylene dithiotosylate with concomitant loss of the formyl group to give the α-acyl-1,3-dithiane.

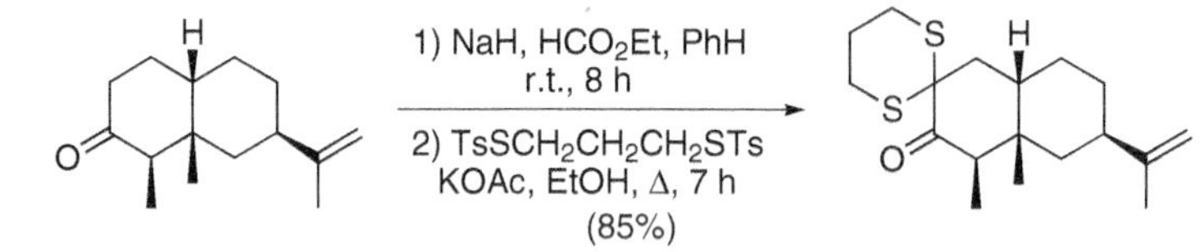

(18.232)

Dehydrogenation of Saturated Carbonyl Compounds

Although the most obvious method for introducing a double bond into a position of conjugation with a carbonyl group is the β-elimination from an α-halocarbonyl compound, itself prepared by halogenation as discussed above, the reaction is seldom easy to carry out cleanly. The dehydrohalogenation of α-haloaldehydes and α-haloketones, especially, has the potential for competing Favorskii rearrangement.[280] A wide variety of bases have been used to carry out this reaction, including magnesium oxide in DMF,[281] calcium carbonate in dimethylacetamide,[282] lithium bromide and lithium carbonate in DMSO[283] or DMF,[284] lithium chloride in DMF,[285] HMPA,[286] and collidine (2,4,6-trimethylpyridine).[287] Nevertheless, alternative methods for the dehydrogenation of carbonyl compounds have long been the subject of research because of the unsatisfactory nature of the dehydrohalogenation reaction and its potential for side reactions.

Until recently, the same dehydrogenation was almost invariably carried out by means of the thermal *syn* elimination from sulfoxides and selenoxides obtained by oxidation of the corresponding sulfides or selenides with peroxide-based oxidants (e.g., Example **18.233**),[288]

278. (a) Trost, B.M.; Salzmann, T.N. *J. Am. Chem. Soc.* **1973**, *95*, 6840. (b) Seebach, D.; Teschner, M. *Tetrahedron Lett.* **1973**, 5113. (c) Zoretic, P.A.; Soja, P. *J. Org. Chem.* **1976**, *41*, 3587. (d) Gassman, P.G.; Balchunis, R.J. *J. Org. Chem.* **1977**, *42*, 3236.

Reviews: (e) Trost, B.M. *Pure Appl. Chem.* **1975**, *43*, 563. (f) Caine, D. In Augustine, R.L. *Carbon-Carbon Bond Formation* (Marcel Dekker: New York, 1979), vol. 1, p. 278.

279. (a) Woodward, R.B.; Patchett, A.A.; Barton, D.H.R.; Ives, D.A.J.; Kelly, R.B. *J. Chem. Soc.* **1957** , 1131. (b) Woodward, R.B.; Pachter, I.J.; Scheinbaum, M.L. *J. Org. Chem.* **1971** , *36*, 1137. (c) Marshall, J.A.; Roebke, H. *J. Org. Chem.* **1969**, *34*, 4188.

280. (a) Favorskii, A. *Zh. Russ. Fiz.-Khim. Obshch.* **1905**, *37*, 643. (b) Wallach, O. *Justus Liebigs Annn. Chem.* **1918**, *414*, 296. (c) Kende, A.S. *Org. React.* **1960**, *11*, 261.

281. Miyano, M.; Dorn, C.R. *J. Org. Chem.* **1972**, *37*, 268.

282. Green, G.F.H.; Long, A.G. *J. Chem. Soc.* **1961**, 2532.

283. Grimshaw, J.; Grimshaw, J.T.; Juneja, H.R. *J. Chem. Soc., Perkin Trans. 1* **1972**, 50.

284. Payne, T.G.; Jefferies, P.R. *Tetrahedron* **1973**, *29*, 2575.

285. Taylor, K.G.; Hobbs, W.E.; Clark, M.S.; Chaney, J. *J. Org. Chem.* **1972**, *37*, 2436.

286. Hanna, R. *Tetrahedron Lett.* **1968**, 2105.

287. Shoppee, C.W.; Roy, S.K.; Goodrich, B.S. *J. Chem. Soc.* **1961**, 1583.

288. Hove, T.H.; Kurth, M.J. *J. Org. Chem.* **1978**, *43*, 3693.

or in steroidal systems especially, by dehydrogenation with DDQ. However, the last decade has seen a strong move toward catalytic reactions using more benign terminal oxidants. This has extended to the dehydrogenation of saturated carbonyl compounds to their conjugated analogs.

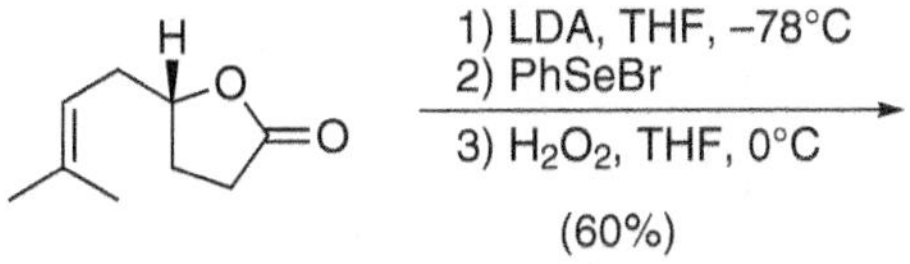

(18.233)

In 1978, Saegusa reported a method, now known as the ***Saegusa oxidation***, for dehydrogenating a ketone to give the conjugated enone by treating the trimethylsilyl enol ether with palladium (II) chloride in the presence of a hydrogen atom acceptor—typically a benzoquinone derivative.[289] On the small scale, it has been found to be more convenient to carry the reaction out using palladium (II) acetate as the stoichiometric oxidant. Its mild conditions have made the reaction a popular method for introducing unsaturation into a ketone in complex systems, as illustrated by Examples **18.234** to **18.237**.

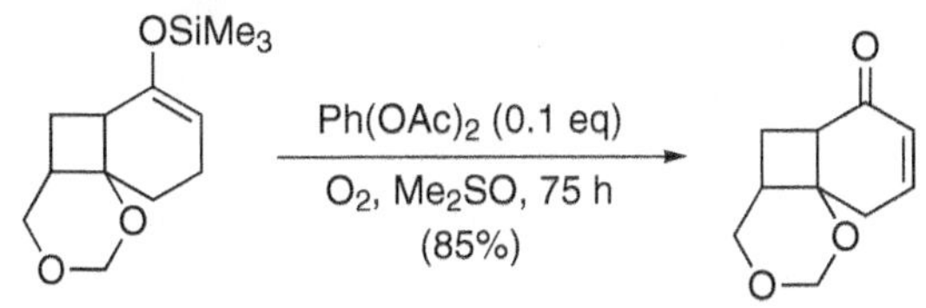

(18.234)[294]

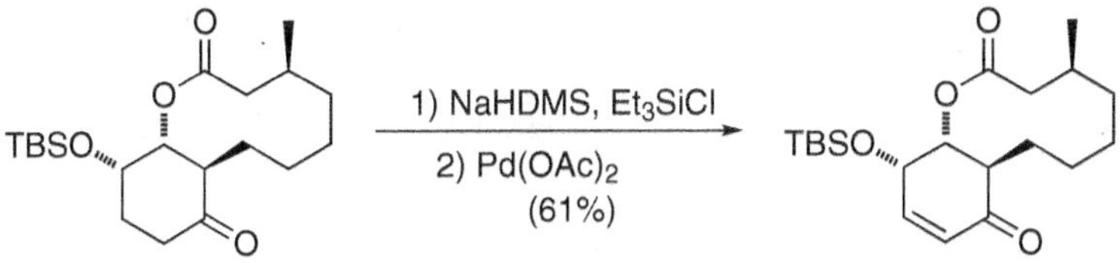

(18.235)[290]

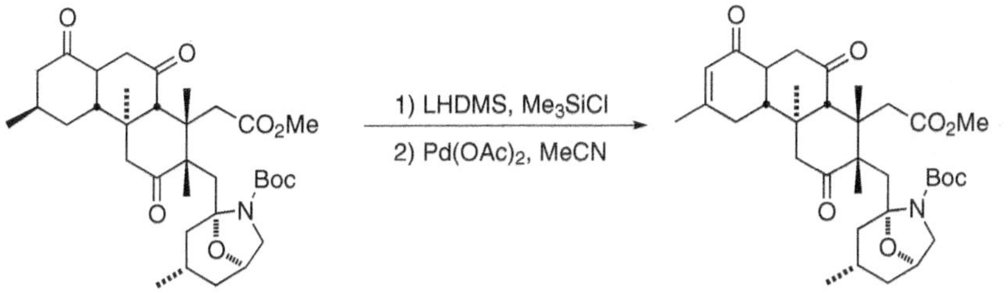

(18.236)[291]

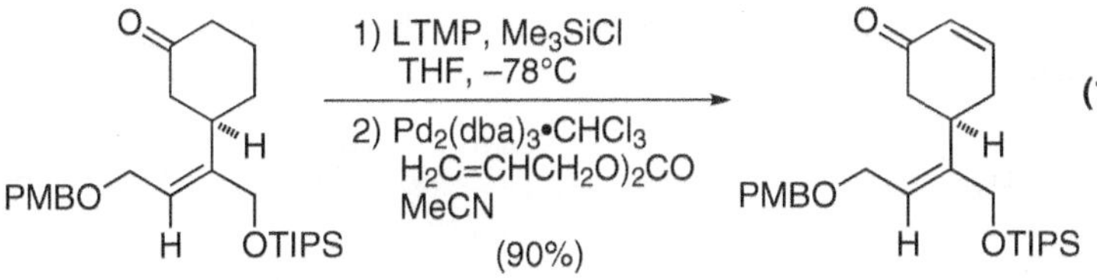

(18.237)[292]

The reaction pathway (Figure 18.14) is similar to that of the Wacker oxidation of alkenes. It involves electrophilic addition of the palladium to the electron-rich enol ether π bond to give the oxonium ion **18.238**, which is then desilylated to give the ketone **18.239**. β-Hydride elimination from this σ complex leads to the π complex **18.240**, from which the unsaturated ketone and acetic acid are liberated, to produce a palladium (0) species.[293]

---

289. (a) Ito, Y.; Hirao, T.; Saegusa, T. *J. Org. Chem.* **1978,** *43,* 1011. (b) Review: Muzart, J. *Eur. J. Org. Chem.* **2010,** 3779.
290. Wilson, E.M.; Trauner, D. *Org. Lett.* **2007,** *9,* 1327.
291. (a) Miyashita, M.; Sasaki, M.; Hattori, I.; Sakai, M.; Tanino, K. *Science* **2004,** *305,* 495. (b) Miyashita, M. *Pure Appl. Chem.* **2007,** *79,* 651.
292. Oshima, T.; Xu, Y.; Takita, R.; Shimizu, S.; Zhong, D.; Shibasaki, M. *J. Am. Chem. Soc.* **2002,** *124,* 14546.
293. Bäckvall, J.E.; Akermark, B.; Ljunggren, S.O. *J. Am. Chem. Soc.* **1979,** *101,* 2411.

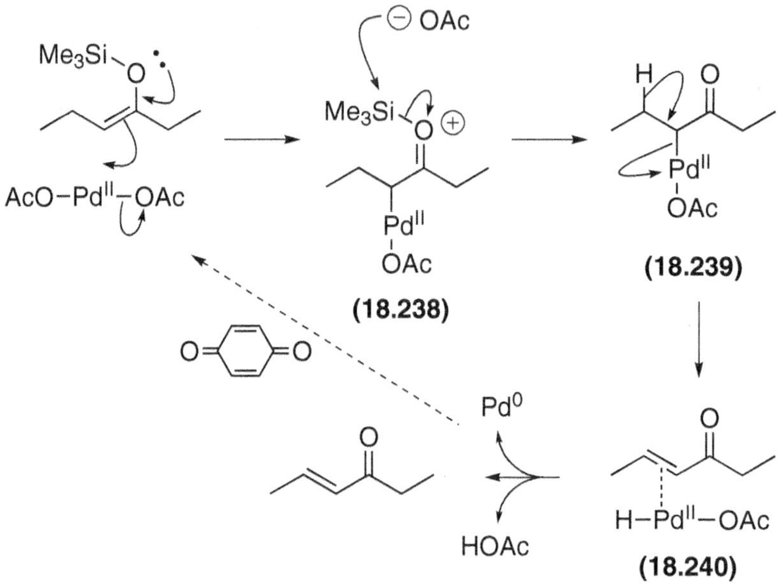

Figure 18.14 Reaction pathway proposed for the Saegusa oxidation

Larock has reported that the palladium (0) product may be reoxidized to palladium (II) using molecular oxygen if dimethylsulfoxide is used as the reaction solvent without added acetate anion.[294]

Work by Tsuji has allowed the use of *tris*(dibenzylideneacetone)dipalladium (0), as in Example **18.241**, abbreviated $Pd_2(dba)_3$, as a catalyst for the reaction, with diallyl carbonate as the oxidant. The reaction proceeds in acetonitrile solvent to give the conjugated product. In other solvents, α-allylation is the major pathway.[295] The procedure has been extended to the use of enol acetates and alkyl allyl carbonates.[296]

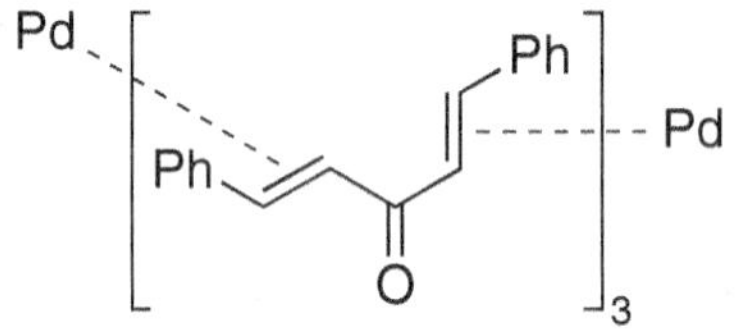

(18.241)

Drill Exercises

Suggest two or more methods for carrying out each of the following transformations.

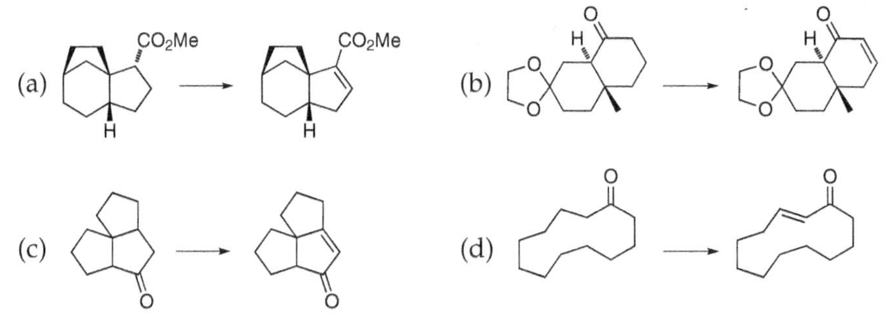

294. Larock, R.C.; Hightower, T.R.; Kraus, G.A.; Hahn, P.; Zheng, D. *Tetrahedron Lett.* **1995**, *36*, 2423.

295. (a) Tsuji, J.; Minami, I.; Shimizu, I. *Tetrahedron Lett.* **1983**, *24*, 5635. (b) Minami, I.; Takahashi, K.; Shimizu, I.; Kimura, T.; Tsuji, J. *Tetrahedron* **1986**, *42*, 2971.

296. Tsuji, J.; Minami, I.; Shimizu, I. *Tetrahedron Lett.* **1983**, *24*, 5639.

(e) 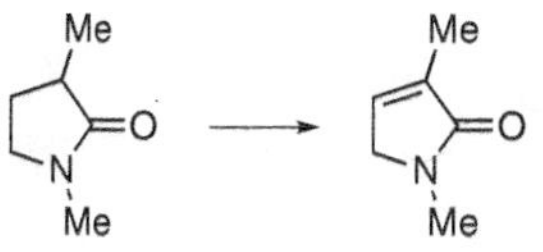

(f) 

(g) 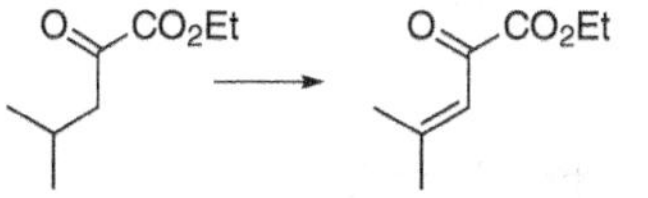

(h) 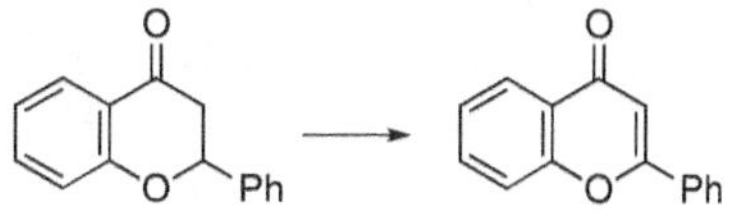

(i) 

Reaction Synopses

Allylic Halogenation

Reagents: Br_2, $h\nu$, Cl_2, $h\nu$; etc.
or NBS, CCl_4, Δ; NCS, CCl_4, Δ; etc.

Allylic Oxidation

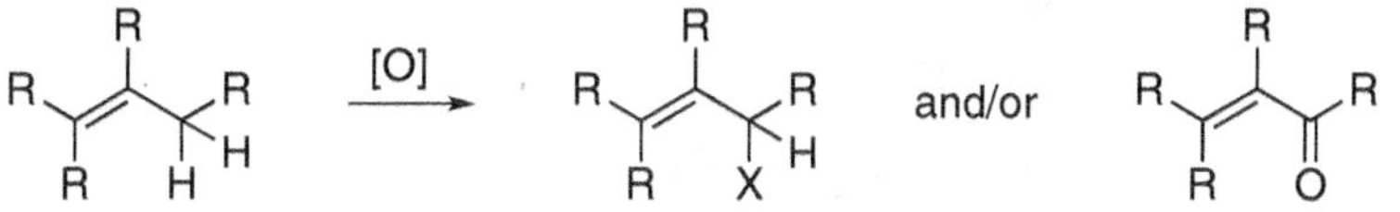

Reagents: RCO_2OBu^t, Cu^+, Δ (R'=OCOR; with allylic rearrangement)
SeO_2 (R' = H; without allylic rearrangement); SeO_2, Me_3COOH; SeO_2, NaOCl; etc.
R_FSeO_2H, $PhIO_2$, $PhCF_3$, Δ; Bu^tOOH, $BiCl_3$, MeCN;
(similar reactions with Cu, Co, Mo, V, etc. as metal);
O_2, $h\nu$, rose bengal (R'=OOH; with allylic rearrangement)
DDQ, CH_2Cl_2;
$Mn_3O(OAc)_9$, Bu^tOH, O_2;
$(PhSOCH_2)_2Pd(OAC)_2$, benzoquinone, solvent
(solvent = AcOH, Me_2SO 1:1 allylic rearrangement)
(solvent = AcOH, Ch_2Cl_2 1:1 no allylic rearrangement);
O_2, $Co(py)_2Br_2$, MeCN; or O_2, CoL_2, Me_2CHCHO, MeCN

α-Halogenation

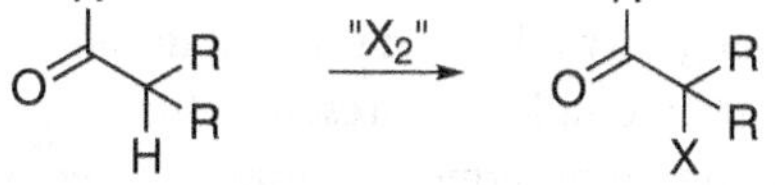

Reagents: Br_2, HOAc; Cl_2, HOAc; etc. (halogenation)

(continues)

(**Reaction Synopses** continued)

α-Selenylation and Sulfenylation

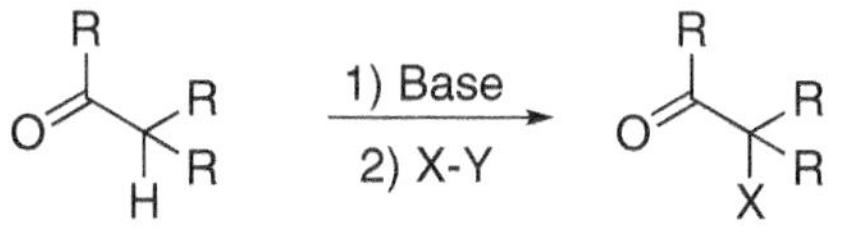

Base: LDA, THF, −78°C; NaH, THF; KH, THF; etc.
X—Y: PhSeCl; PhSeBr; PhSeSePh; PhSCl; PhSBr; RSSR

α-Hydroxylation

Reagents: (1) LDA, THF, −23°C, (2) MoOPH

Dehydrogenation of Carbonyl Compounds

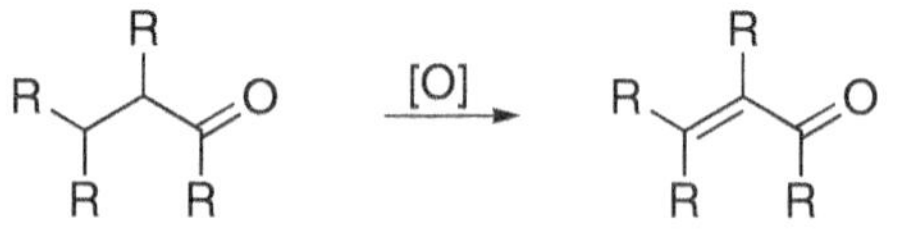

Reagents: (1) LDA, THF, −78°C, (2) RSeBr, (3) H_2O_2
or (1) LDA, THF, −78°C, (2) RSSR, (3) H_2O_2, (4) Δ
or (1) Br_2, HOAc, (2) base, Δ
or DDQ, CH_2Cl_2; etc.

Dehydrosilylation of Enol Ethers (Saegusa Oxidation)

Reagents: $Pd(OAc)_2$, MeCN, *p*-benzoquinone; $Pd(OAc)_2$, Me_2SO, O_2
or $Pd(OAc)_2$, MeCN
or $Pd_2(dba)_3$•$CHCl_3$, MeCN, $(H_2C{=}CH\text{-}CH_2O)_2CO$

18.11 Oxidation of Alkanes

And now we return to the "holy grail." One of the functions of cytochrome P450 in living cells is the oxidation of hydrocarbons by direct insertion of an oxygen atom from molecular oxygen into a C—H σ bond, while the flavin enzymes carry out dehydrogenation of saturated alkyl chains. Finding catalytic methods to mimic these activities has been a subject of intense research in the last decade or so.[297] Significant research into catalytic dehydrogenation has also been carried out in the first decade of the 21st century.[298]

The direct functionalization of alkanes presents a unique problem for organic chemists because these hydrocarbons are unique in having neither an accessible HOMO nor an accessible LUMO—they are neither nucleophilic nor electrophilic. All other organic

297. Reviews: (a) Ishihara, Y.; Baran, P.S. *Synlett* **2010**, 1733. (b) Newhouse, T.; Baran, P.S. *Angew. Chem. Int. Ed.* **2011**, *50*, 3362.

298. (a) Crabtree, R.H.; Mihelcic, J.M.; Quirk, J.M. *J. Am. Chem. Soc.* **1979**, *101*, 7738.

compounds have at least one accessible frontier orbital that renders them reactive toward a complementary species.

The lack of an accessible HOMO or LUMO makes the C—H σ bond in alkanes one of the least reactive bonds in a molecule. The effect is to make many (most) of the reactions involving its cleavage and replacement inherently endothermic. Oxidations of C—H σ bonds in alkanes must therefore be coupled with a strongly exothermic reduction reaction, or they must be carried out under conditions where energy is provided, often in the form of photons.

It is for these reasons that the number of reactions that allow selective oxidation of alkanes is actually quite small and that the earliest direct oxidation methods for alkanes were based on free radicals. Following the formation of the initiating free radical (often effected by photochemical cleavage of a suitable precursor), a free radical chain reaction will proceed to completion, provided that the net reaction (the sum of the propagation steps) is exothermic. The radical oxidation of alkanes by halogens was the first of these reactions to be studied in depth. It was an essential part of the early work that led to the reactivity-selectivity principle.[299] To date, the most successful functionalizations of unactivated C—H bonds have involved intramolecular transfer of a hydrogen atom, and we will focus on these reactions here.

The transfer of a hydrogen atom from a remote site to a reactive center in the same molecule has been known for more than a century. Intramolecular hydrogen atom transfer[300] to reactive free radicals provided the first reliable methods for the regioselective oxidation of alkane hydrocarbons. One of the first such reactions was the Norrish type II photochemical reaction of ketones (Figure 18.15)[301] In this reaction, the excited state of the carbonyl group (**18.243**) abstracts a hydrogen from the γ carbon to generate the diradical (**18.244**) that then collapses by radical recombination to give cyclobutanols (**18.245**) or by fragmentation to generate an enol and an alkene (**18.246**).

It was established fairly early on that the most facile hydrogen atom transfer would occur through a six-membered transition state.[302] Work by Houk has shown that the transfer occurs through a transition state where the angle O—H—C is large (> 150°).[303] As we saw in Chapter 14, this was put to use in spectacular fashion by Barton, whose synthesis of aldosterone[304] relied on a key intramolecular hydrogen atom transfer from an unactivated methyl group to an alkoxy radical (Example **18.248**) generated by photolysis of a nitrite ester (Example **18.247**).[305] Trapping of the resultant alkyl radical (Example **18.249**) by nitric oxide gave a nitroso compound that tautomerized to the oxime (Example **18.250**). When Barton carried out this reaction, the product that crystallized in his reaction flask had a market value of more than a million dollars. Of course, the price immediately plummeted.

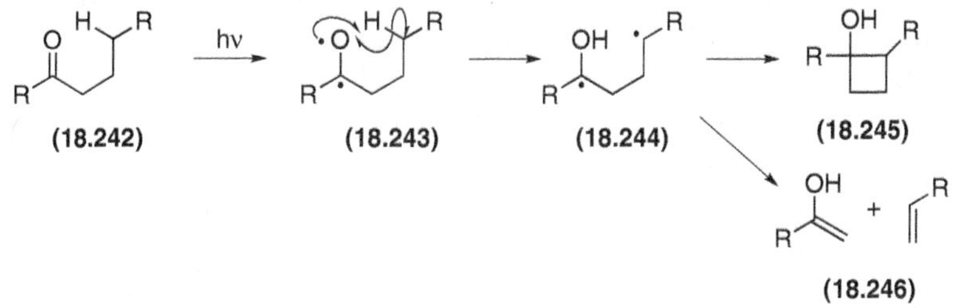

Figure 18.15 The Norrish type II photochemical reaction of a ketone

299. Review: Pross, A. *Adv. Phys. Org. Chem.* **1977**, *14*, 69.

300. Review: Feray, L.; Kuznetsov, N.; Renaud, P. In Renaud, P.; Sibi, M.P., Eds. *Radicals in Organic Synthesis* (Wiley-VCH: Weinheim, 2001), Part II, ch. 3.6, p.246.

301. (a) Norrish, R.G.W. *Trans. Faraday Soc.* **1937**, *33*, 1521. (b) Yang, N.C.; Yang, D.-H. *J. Am. Chem. Soc.* **1958**, *80*, 2913. (c) Wagner, P.J. *Acc. Chem. Res.* **1971**, *4*, 168. (d) Wagner, P.; Park, B-S. In Padwa, A., Ed. *Organic Photochemistry* (Marcel Dekker: New York, 1991) Volume 11; Chapter 4.

302. Barton, D.H.R.; Hesse, R.H.; Pechet, M.M.; Smith, L.C. *J. Chem. Soc., Perkin Trans. 1* **1979**, 1159.

303. (a) Dorigo, A.E.; Houk, K.N. *J. Am. Chem. Soc.* **1987**, *109*, 2195. (b) Dorigo, A.E.; McCarrick, M.A.; Loncharich, R.J.; Houk, K.N. *J. Am. Chem. Soc.* **1990**, *112*, 7508. (c) Ihmels, H.; Scheffer, J.H. *Tetrahedron* **1999**, *55*, 885.

304. Barton, D.H.R.; Beaton, J.M. *J. Am. Chem. Soc.* **1961**, *83*, 4083.

305. Barton, D.H.R.; Beaton, J.M.; Geller, L.E.; Pechet, M.M. *J. Am. Chem. Soc.* **1961**, *83*, 4076.

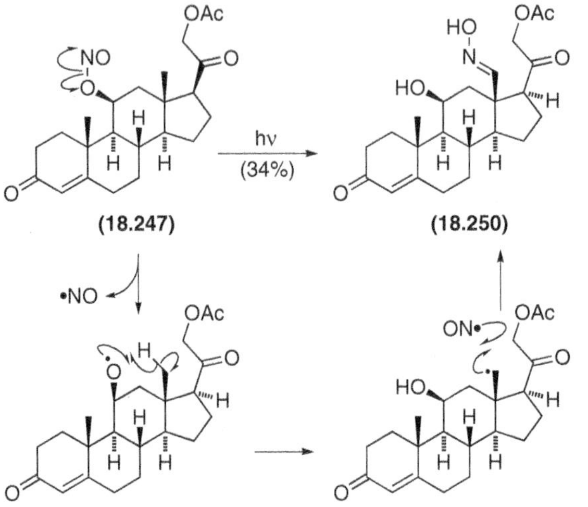

(18.247) (18.250) (18.248) (18.249)

An alternative source of oxygen-centered free radicals that can abstract the remote hydrogen is an alkyl hypoiodite, generated by oxidation of the alcohol with lead tetraacetate and iodine or, more recently, with diacetoxyiodobenzene and iodine (Example **18.251**).[306] These compounds undergo facile homolysis on irradiation to give the corresponding alkoxy radicals that abstract the hydrogen atom from the γ position. The final product of the hypoiodite reaction is usually a tetrahydrofuran derivative.

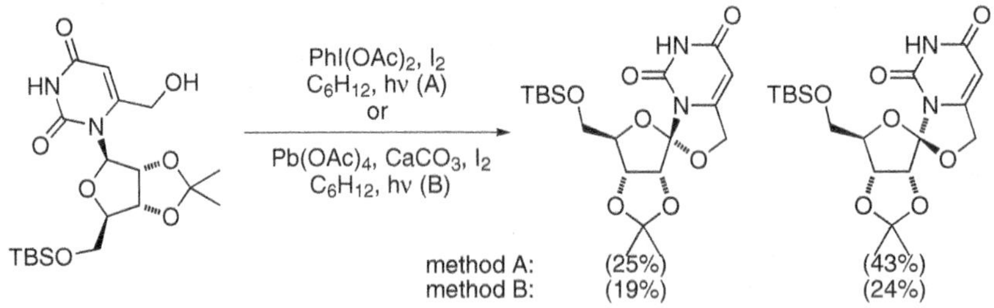

(18.251)[307]

Another reaction that involves the intramolecular transfer of a hydrogen atom from an unactivated alkyl group to a reactive free radical is the ***Hofmann-Löffler-Freytag reaction***.[308] The reaction involves the homolysis of an *N*-haloammonium ion, with hydrogen atom transfer to the nitrogen, followed by trapping of the radical by the halogen. Treatment of the product with base provides a convenient entry into pyrrolidine rings. The one limitation of the Hofmann-Löffler-Freytag reaction is that it requires strongly acidic conditions; for example, the conversion of L-leucine to a 1:1 mixture of *cis*- and *trans*-4-methyl-L-proline involves a Hofmann-Löffler-Freytag reaction carried out in 80% sulfuric acid (Example **18.252**).[309]

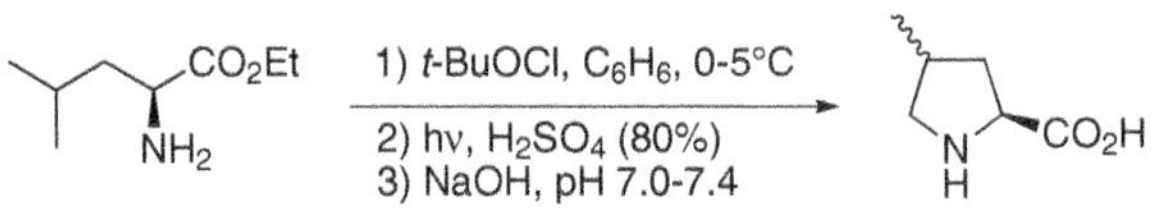

(18.252)

306. (a) Concepción, J.I.; Francisco, C.G.; Hernández, R.; Salazar, J.A.; Suárez, E. *Tetrahedron Lett.* **1984**, *25*, 1953. (b) Baker, R.; Brimble, M.A. *J. Chem. Soc., Chem. Commun.* **1985**, 78. (c) Martin, A.; Salazar, J. A.; Suárez, E. *J. Org. Chem.* **1996**, *61*, 3999. (d) Dorta, R.L.; Martin, A.; Salazar, J. A.; Suárez, E.; Prangé, T. *Tetrahedron Lett.* **1996**, *37*, 6021.

307. Kittaka, A.; Kato, H.; Tanaka, H.; Nonaka, Y.; Amano, M.; Nakamura, K.T.; Miyasaka, T. *Tetrahedron* **1999**, *55*, 5319.

308. (a) Hofmann, A.W. *Ber. dtsch. chem. Ges.* **1883**, *16*, 558. (b) Löffler, K.; Freytag, C. *Ber. dtsch. chem. Ges.* **1909**, *42*, 3427.

Reviews: (c) Wolff, M.E. *Chem. Rev.* **1963**, *63*, 55. (d) Neale, R.S. *Synthesis* **1971**, 1. (e) Minisci, F. *Synthesis* **1973**, 1. (f) Mackiewicz, P.; Furstoss, R. *Tetrahedron* **1978**, *34*, 3241. (g) Majetich, G. *Tetrahedron* **1995**, *51*, 7095.

309. Titouani, S.L.; Lavergne, J.-P.; Viallefont, P.; Jacquier R. *Tetrahedron* **1980**, *36*, 2961

The reaction can also be carried out under neutral conditions by the reaction of an amide with iodine and lead tetraacetate. This reaction gives the *N*-iodoamide, which undergoes homolysis on irradiation to give the reactive radical that then abstracts the hydrogen from the remote site. The use of phosphoramidates (e.g., Example **18.253**) facilitates this version of the reaction.

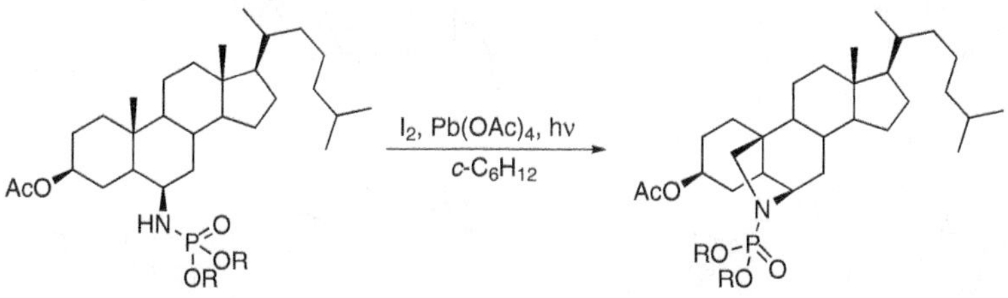

(18.253)[310]

Converting an amine to the *N*-bromo-trifluoroacetamide (Example **18.254**) also provides a much more reactive intermediate on photolysis; this has permitted the synthesis of selectively brominated L-isoleucine derivatives in good yields by the Hofmann-Löffler-Freytag route. Nitrogen-centered free radicals in which nitrogen is bonded to a third-row or higher element such as tin tend to be much more reactive in hydrogen atom transfer reactions. These radicals can be formed by reduction of azides with stannanes (e.g., Example **18.255**).

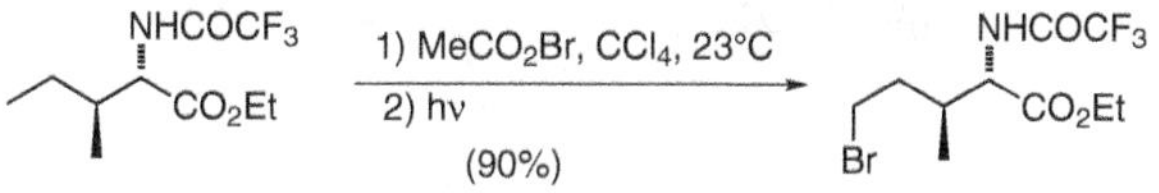

(18.254)[311]

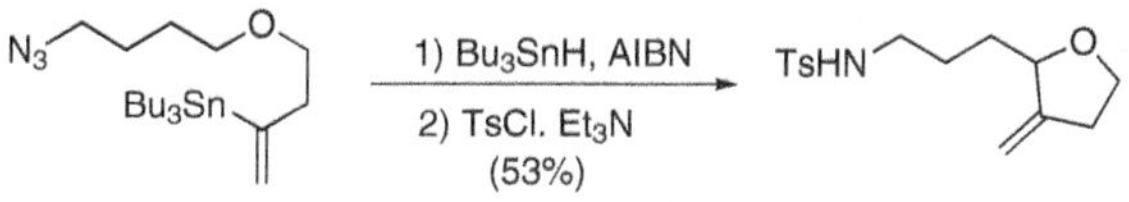

(18.255)[312]

Problems

18-14 The Barton synthesis of aldosterone gave the compound below as a second major product from the nitrite ester irradiation. Write a mechanism to account for the formation of this product.

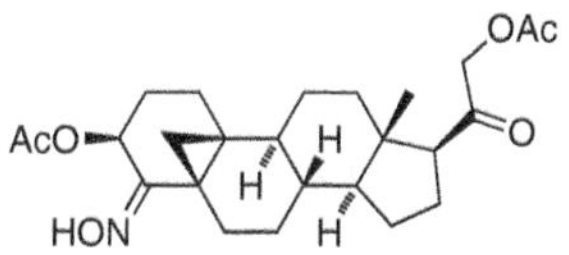

18-15 The two oximes below are the major organic products of irradiation of the nitrite ester shown. Provide a reasonable mechanism that accounts for the formation of both products. [See *Tetrahedron* 1995, *51*, 7095]

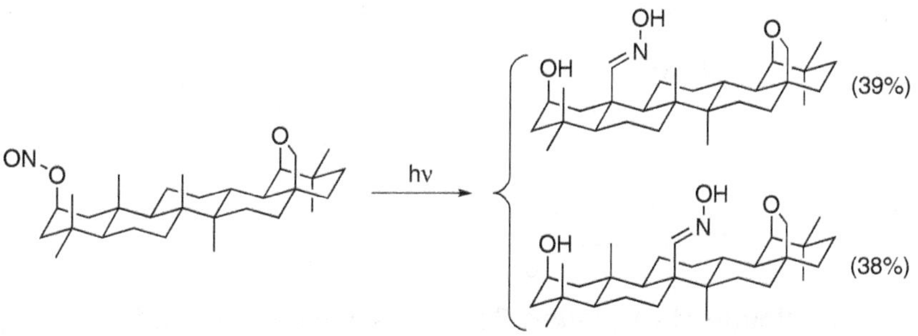

(continues)

310. Betancor, C.; Conceptión, J.I.; Hernandez, R.; Salazar, J.A.; Suárez. E. *J. Org. Chem.* **1983,** *48*, 4430.
311. Rajender Reddy, L.; Subba Reddy, B.V.; Corey, E.J. *Org. Lett.* **2006,** *8*, 2819.
312. Kim, S.; Yeon, K.M.; Yoon, K.S. *Tetrahedron Lett.* **1997,** *38*, 3919.

(**Problems** continued)

18-16 Supply the structure of the expected organic product in each of the following reactions:

(a) 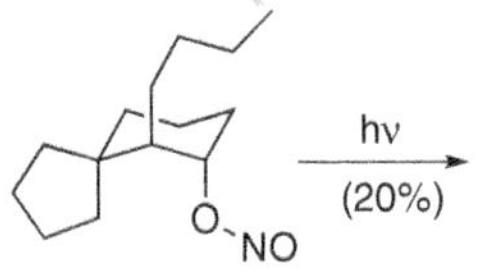

(b) 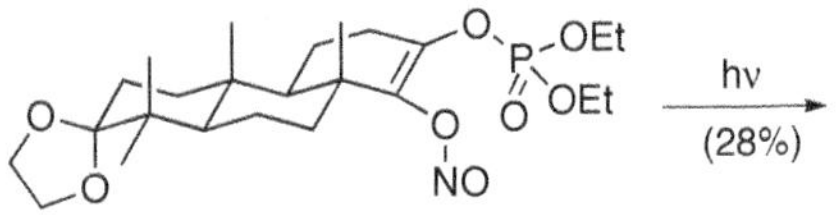

(c) 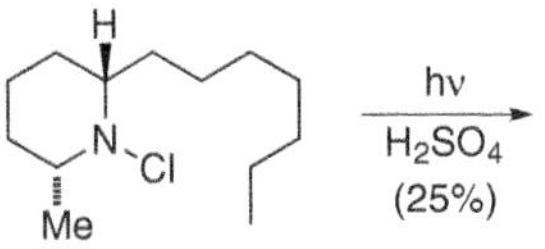

(d) 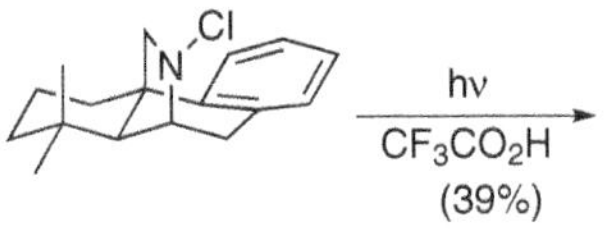

References: (a) *J. Am. Chem. Soc.* **1975**, *97*, 430. (b) *Tetrahedron Lett.* **1989**, *30*, 3023. (c) *J. Heterocycl. Chem.* **1975**, *12*, 289. (d) *Chem. Pharm. Bull.* **1985**, *33*, 3187.

Reaction Synopses

Intramolecular Hydrogen Atom Transfer

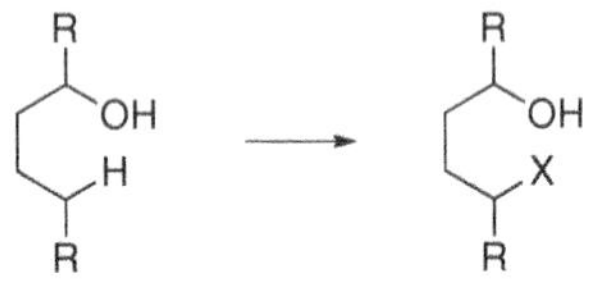

Reagents:	(1) NOCl, pyridine; (2) hν (Barton reaction);
or	(1) I_2, $Pb(OAc)_4$; (2) hν (hypoiodite reaction);
or	I_2, $PhI(OAc)_2$, cyclohexane, hν; etc.

Products may react with base to give tetrahydrofuran derivatives.

Norrish Type II

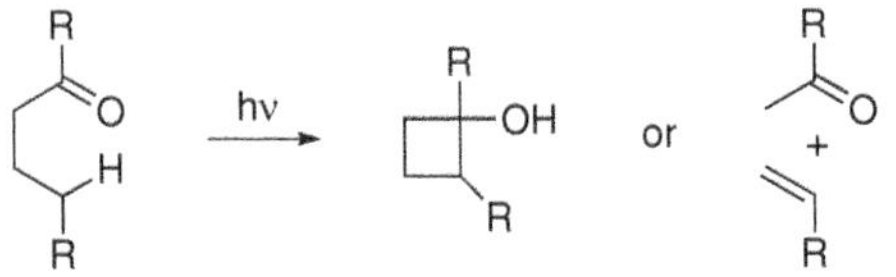

Hofmann-Löffler-Freytag Reaction

Reagents:	(1) *t*-BuOCl; (2) hν (X=H_2, Y=alkyl, Z=halogen);
or	(1) $MeCO_2Br$; (2) hν (X=O, Y=H, Z=halogen);
or	$Pb(OAc)_4$, I_2, cyclohexane, hν (X=O, Y=H, Z=halogen)

Products may react with base to give pyrrolidine derivatives.

18.12 A Catalogue of Oxidation Reactions: Oxidation at a Glance

In this chapter, we have been concerned with oxidation reactions, and it is now time to put all these reactions together in a way that will allow us to see the available choices. This is done in Table 18.10, where oxidation methods and the functional groups that are incompatible with them are listed.

Table 18.10 Catalogue of Oxidation Reactions

Transformation	Reagents	Incompatible Functional Groups	Other Limitations or Features
	PCC, PDC; CrO_3, py; CrO_3, py, CH_2Cl_2; etc.	Thiol, sulfide, 1,2-diols can be a problem; 1,3-diols can be a problem	
	Swern reagent, Moffatt-Pfitzner reagent, Corey-Kim reagent, etc.	Primary and secondary amines can interfere by reacting with the dehydrating agent	
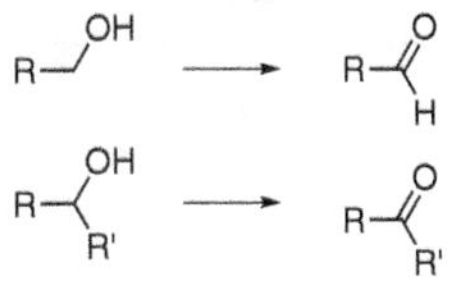	Al(O-*t*-Bu)$_3$, Me_2CO, Δ		Reaction is reversible
	TPP (Ley reagent)		Often gives higher yields than Swern oxidations
	DMP, IBX, etc.		Can be used in ionic liquids; tolerates most functional groups
	TEMPO, NaOCl; etc.		
	Ag_2CO_3, celite		Expensive—usually reserved for small-scale reactions
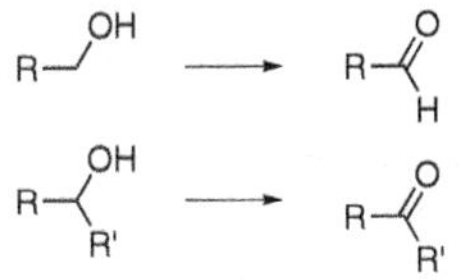	$Pd(OAc)_2$, py, O_2		Reaction is catalytic in Pd; py is essential
	Jones reagent		1,2- and 1,3-Diols; aldehydes
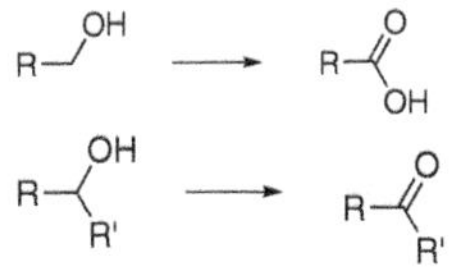	$KMnO_4$; Bu_4NMnO_4, PhH; etc.		Alkenes; aldehydes; "purple benzene" is an extremely powerful oxidant
	CrO_3 (cat), H_5IO_6		Same as Jones reagent
	RuO_4; $RuCl_3$, $NaIO_4$; etc.		Alkenes; ethers and amines
	PDC, DMF		
	$NaClO_2$, *t*-BuOH, H_2O, $Me_2C{=}CHMe$		Two-stage oxidation of primary alcohols is preferred way to make a carboxylic acid
	Ag_2O, NH_3, H_2O		Highly selective, but expensive

(continued)

(**Table 18.10** continued)

Transformation	Reagents	Incompatible Functional Groups	Other Limitations or Features
	MnO_2		Reagent can be difficult to make with reproducible activity
	$BaMnO_4$		Reagent is easier to make and needs less than MnO_2
	PDC, EtOAc		Reagent can be used to oxidize 1,2-diols where one OH is allylic
	PCC, CH_2Cl_2, ≥ 24 h	Primary and secondary alcohols	
	Cr (VI) reagents	Usually an unwanted side reaction in alcohol oxidations	
	HIO_4	Does not cleave diols that cannot form a cyclic periodate ester	
	$Pb(OAc)_4$	Cleaves all 1,2-diols	
	$KMnO_4$, KOH, H_2O	Aldehydes, sulfides	
	OsO_4 (catalytic or stoichiometric)		Can be carried out enantioselectively using AD-mix reagents
	Br_2, ROH; I_2, TOH; IN_3; RSX, RSeX; etc.		Reaction conditions may change product
	m-CPBA, MgMPP, etc.;	Sulfides, selenides	
	Mn(salen), NaOCl		Works best with *cis* alkenes; chiral complexes give chiral epoxides
	Oxone®, Me_2CO; Oxone®, fructose derivative		Works best with trisubstituted alkenes; epoxneide is chiral with fructose derivative
	OsO_4, $NaIO_4$; etc.		
	(1) O_3 then (2) Me_2S; etc.	Aromatic rings are cleaved	
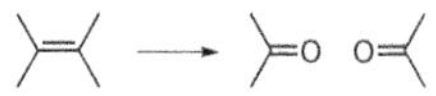	RuO_4; $RuCl_3$, HIO_4; etc.	Alcohols, aldehydes	Aromatic rings are cleaved
	Bu_4NMnO_4, C_6H_6; etc.	Alcohols, aldehydes	

(**Table 18.10** continued)

Transformation	Reagents	Incompatible Functional Groups	Other Limitations or Features
	NBS, CCl_4; NBS, hν; etc.	Thiols, sulfides	Electron-rich aromatics undergo substitution
	RCO_2OR', CuI; etc.		
	SeO_2, t-BuOOH; etc.		
	O_2, hν, sensitizer		
	$Pd(OAc)_2$, py, MeOH-H_2O; etc.		
	t-BuOOH, RhII or BiIII, etc.	Sulfides	
	$Mn(OAc)_3$, O_2	Sulfides	
	CF_3SeO_2H, $PhIO_2$; etc.	Sulfides	
	DDQ, CH_2Cl_2		X must be C=O or Ar
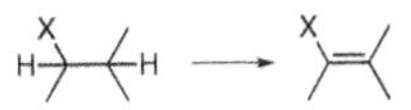	(1) LDA, (2) PhSeCl, (3) *m*-CPBA (X=CO)		X must be C=O or similar anion-stabilizing group
	(1) Me_3SiCl, i-$PrNEt_2$; (2) $Pd(OAc)_2$, benzoquinone		
	t-BuOOH, M		M=$VO(acac)_2$, $Mo(CO)_6$; etc. reaction is enantioselective when Ti(O-*i*-Pr)$_4$ and a dialkyl tartate are used as the catalyst
	(1) LDA, then (2) Y_2 or YBr		X=O; Y=RS, RSe; etc.; *cannot be used to halogenate carbonyl compounds*
	X_2, HOAc		X=Cl, Br
	(1) LDA, then (2) oxaziridine		Y=OH; product is chiral if a chiral oxaziridine is used
	*m*-CPBA (X=O); (1) NH_2OH, then (2) H_2SO_4 or TsCl, py; HN_3	Alkenes	Reactions proceed with retention of configuration at migrating center
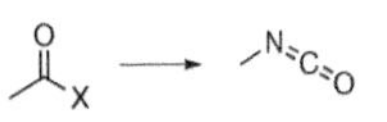	KOH, Br_2 (X=NH_2) (1) NH_2OH, (2) TsCl, (3) base (X=Cl, OR) (1) $(COCl)_2$, (2) NaN_3, (3) Δ or hν (X=OH)		Reactions proceed with retention of configuration at migrating center

Chapter Summary

This chapter has introduced oxidation in organic chemistry, beginning with the concepts of oxidation and reduction and the use and limitations of oxidation numbers in organic chemistry. The oxidations of alcohols with metal-based reagents (e.g., PCC, $KMnO_4$, MnO_2, Ag_2O, TPAP.) has been discussed, and alcohol oxidation by non−metal-based oxidants (e.g., DMP, Me_2SO, TEMPO) has been introduced. Dehydrogenation of carbonyl compounds has been considered, and rearrangements involving insertion of an electronegative atom (e.g., N, O) into C—C bonds have been introduced. The oxidation of hydrocarbons—hydroxylation, epoxidation, allylic oxidation, and oxidative cleavage of alkenes—has been introduced. Free radical intramolecular hydrogen atom transfers as methods for the oxidation of an unactivated C—H bond have also been introduced.

Key Terms

allylic oxidation
Babler oxidation
Baeyer-Villiger rearrangement
Barton reaction
Beckmann rearrangement
Collins reagent
Corey-Kim oxidation
Criegee oxidation
Curtius rearrangement
dehydrogenation
Dess-Martin periodinane (DMP)
Fétizon reagent
halolactonization
Hell-Volhard-Zelinskii reaction
Hofmann-Löffler-Freytag reaction
Hofmann rearrangement
hydroxylation
Jones reagent
Lemieux-Johnson cleavage
Lossen rearrangement
Meerwein-Ponndorf-Verley reduction
Moffatt-Pfitzner oxidation
Norrish type II reaction
Oppenauer oxidation
oxidation
oxymetallation
ozonolysis
Prilezhaev reaction
reduction
Sarett reagent
Saegusa oxidation
Schmidt rearrangement
Swern oxidation
Tishchenko reaction
Tollens reagent
TPAP
Uemura oxidation
Wacker oxidation

Additional Problems

18-17 What is the major product of each of the following reactions?

(a) 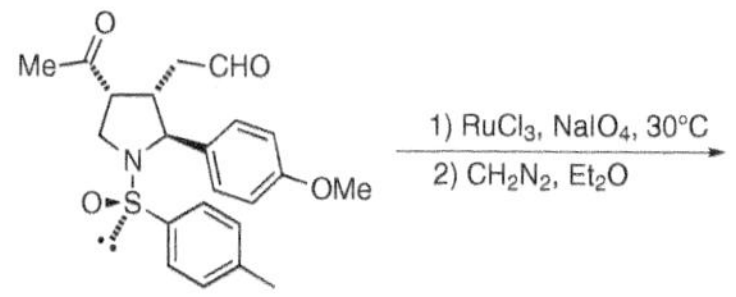

[*Org. Lett.* **2005**, *7*, 3045]

(b) 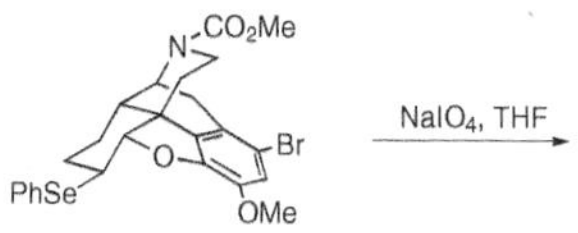

[*J. Am. Chem. Soc.* **2009**, *131*, 16045]

(c) 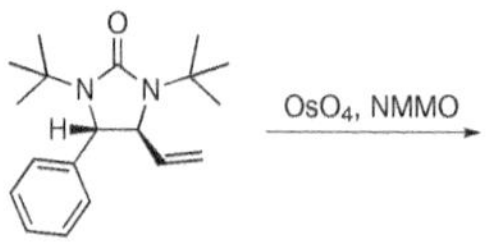

[*J. Org. Chem.* **2009**, *74*, 7577]

(d) 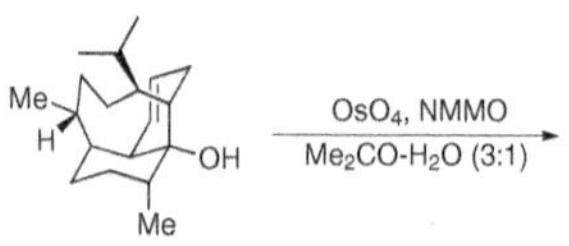

[*J. Am. Chem. Soc.* **2009**, *131*, 17066]

(e) 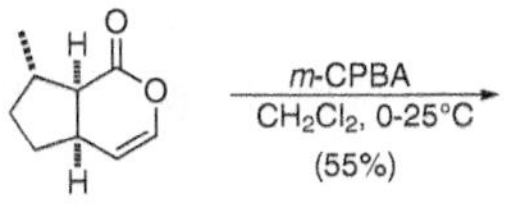

[*Angew. Chem. Int. Ed.* **2009**, *48*, 9105]

(f) 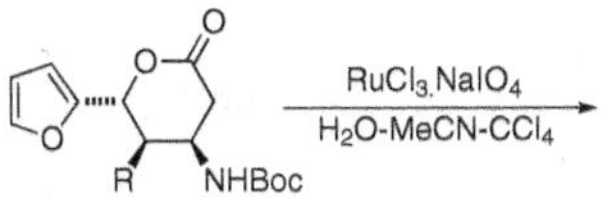

[*Org. Lett.* **2007**, *9*, 1231]

(g) 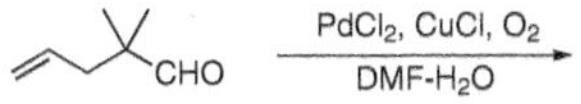

[*Org. Synth.* **1988**, *67*, 121]

(h) 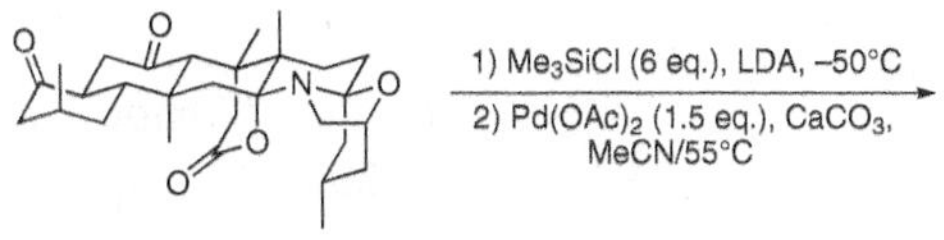

[*Angew. Chem. Int. Ed.* **2009**, *48*, 8905]

(i) 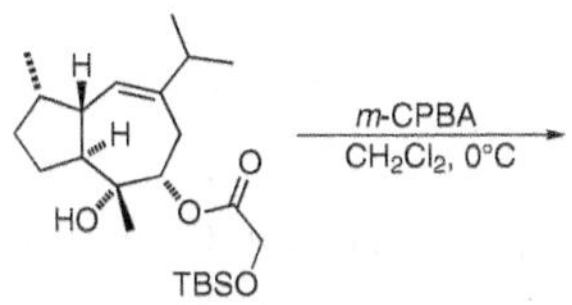

[*Angew. Chem. Int. Ed.* **2009**, *48*, 9105]

(j) 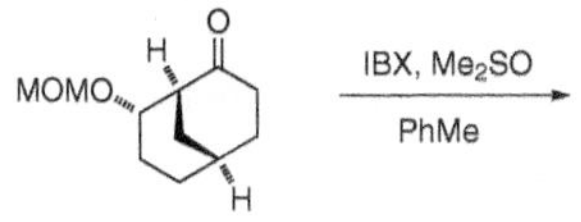

[*Org. Lett.* **2005**, *7*, 4181]

(k) 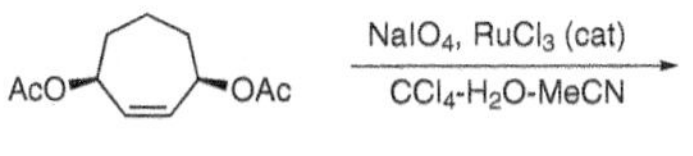

[*Bioorg. Med. Chem. Lett.* **2007**, *17*, 5894]

(l) 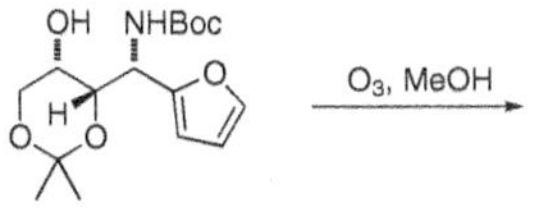

[*Synthesis* **2006**, 2155]

(m) 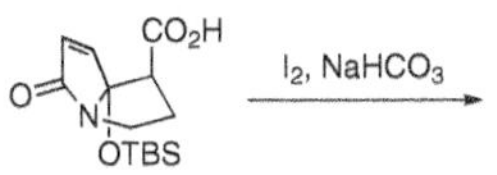

[*Angew. Chem. Int. Ed.* **2007**, *46*, 2883]

(n) 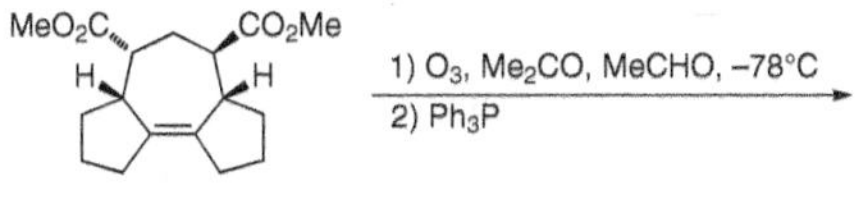

[*Chem. Comm.* **2004**, 2404]
[*Org. Biomol. Chem.* **2005**, *3*, 1557]

(o) 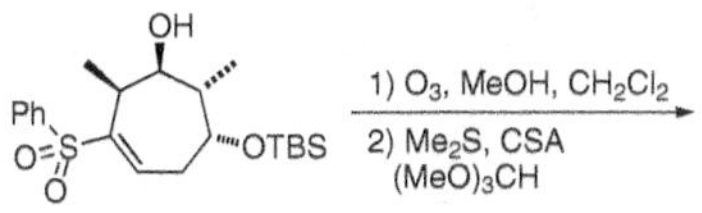

[*Org. Lett.* **2006**, *8*, 2905]
[Hint: $RCOSO_2R$ + MeOH → RCO_2Me]

(p) 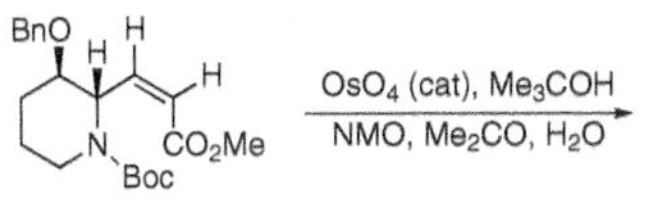

[*J. Org. Chem.* **2005**, *70*, 2325]

(q) 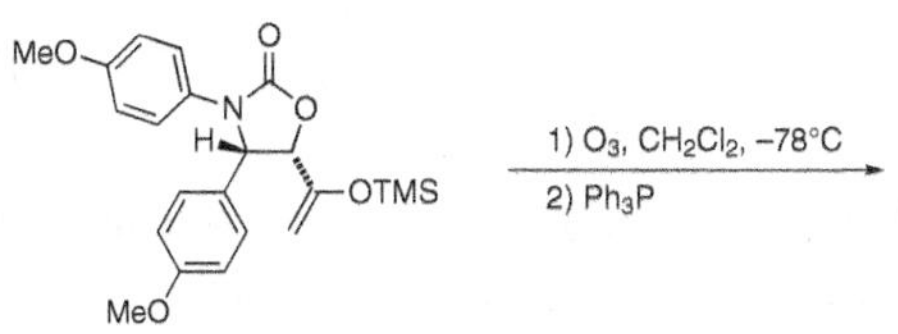

[*Tetrahedron* **2006**, *62*, 5756]

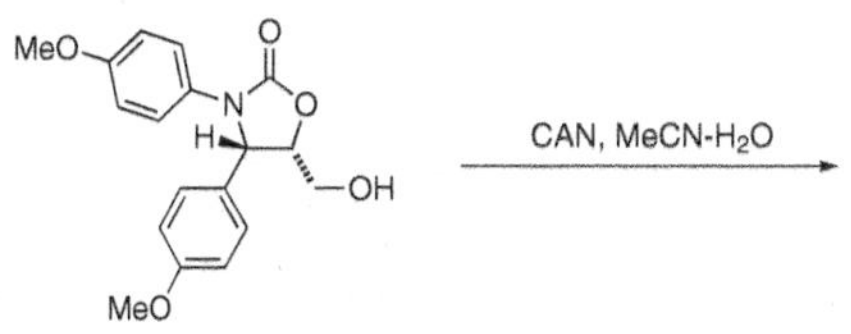

(r) 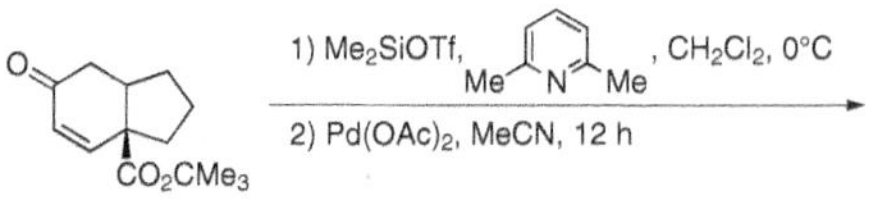

[*Org. Lett.* **2007**, *9*, 2677]

(s) 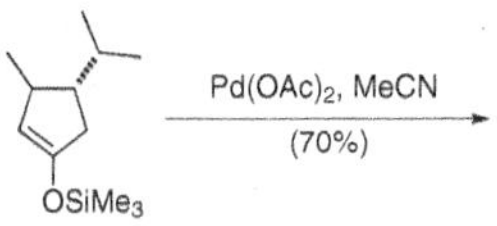

[*Org. Lett.* **2005**, *7*, 3425]

(t) 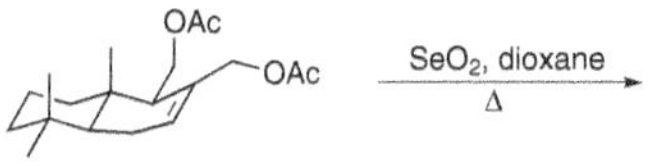

[*J. Chem. Soc., Chem. Comm.* **1981**, 507]

(u) 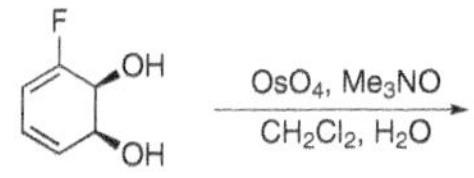

[*Org. Biomol. Chem.* **2007**, *5*, 2267]

(v) 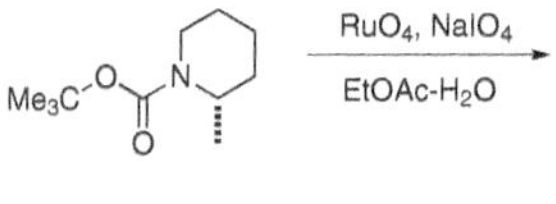

[*Org. Lett.* **2005**, *7*, 585]

(w) 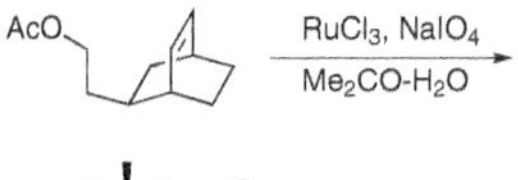

[*J. Am. Chem. Soc.* **1983**, *105*, 5688]

(x) 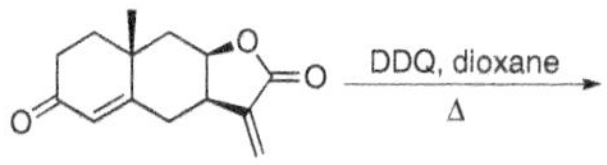

[*Tetrahedron Lett.* **1975**, 743]

(y) 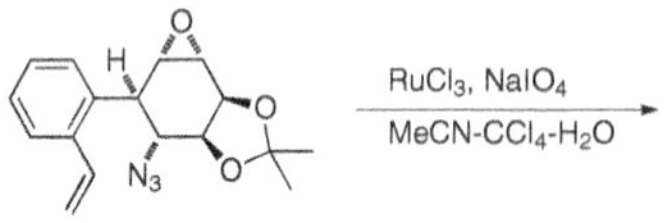

[*Org. Lett.* **2000**, *2*, 3683]

(z) 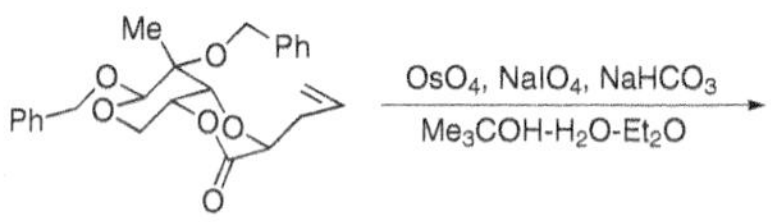

[*J. Am. Chem. Soc.* **2003**, *125*, 7822]

(aa) 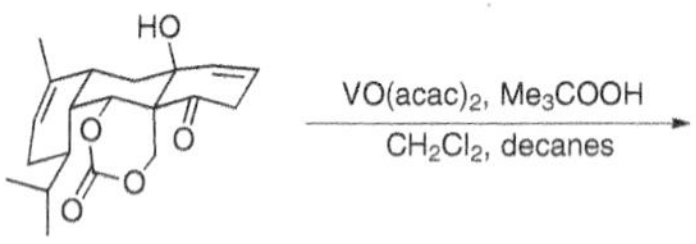

[*Org. Lett.* **2003**, *5*, 1805]

(bb) 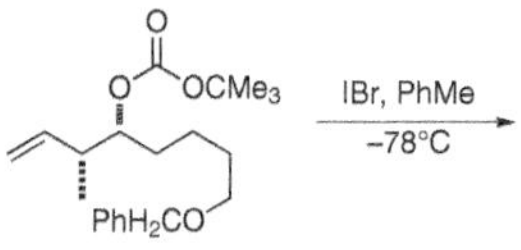

[*Tetrahedron Lett.* **1997**, *38*, 8667]

(cc) 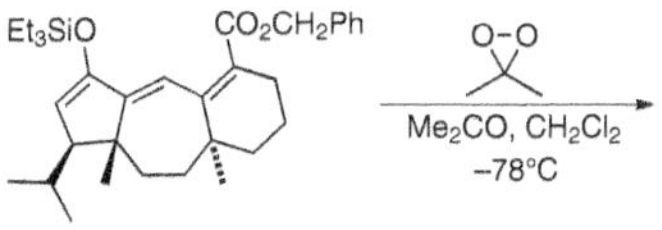

[*J. Am. Chem. Soc.* **2006**, *128*, 13095]

(dd) 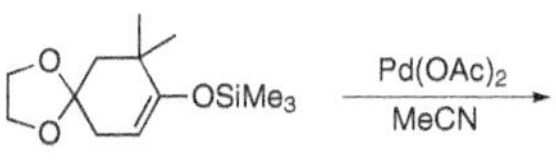

[*J. Am. Chem. Soc.* **2005**, *127*, 5342]

18-18 In Corey's classic synthesis of prostaglandin $F_{2\alpha}$, the oxidation of the ketoalkene with *m*-chloroperbenzoic acid occurs to give the lactone shown instead of the epoxide. Rationalize both the chemoselectivity (i.e., functional group selectivity) of this reaction, and its regioselectivity [*J. Am. Chem. Soc.* **1969**, *91*, 5675].

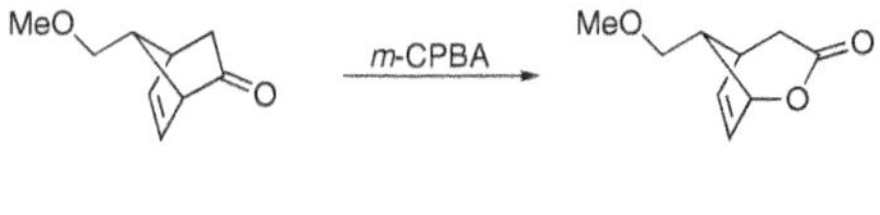

18-19 The allylic oxidation of Δ^5-steroids by most allylic oxidants results in oxidation at C-7, whereas the fluorinated seleninic acid discussed above leads to oxidation at C-4. Suggest a plausible reason why this should occur. (Hint: compare this selenium-based reagent with SeO_2.)

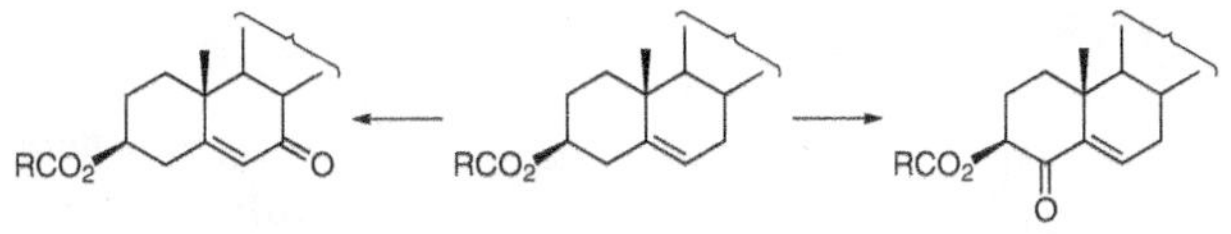

18-20 The steps shown below occur in a synthesis of pinitol [*Tetrahedron* **1989**, *45*, 3463]. The molecule has six chiral centers, all introduced in a stereocontrolled manner from the starting *meso* diol.

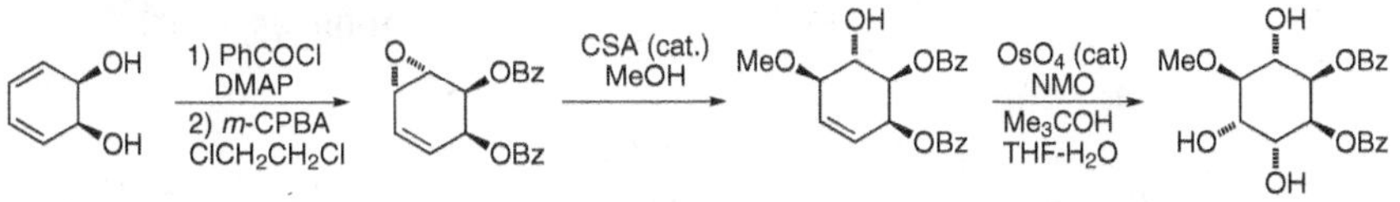

(a) What would be the product of the epoxidation reaction if the diol were not converted to the dibenzoate ester first?

(b) The acid-catalyzed ring opening of the epoxide proceeds with the regiochemistry and stereochemistry shown. Rationalize both.

(c) Suggest a reason why the diol formed in the osmium tetroxide hydroxylation has the stereochemistry shown.

18-21 Suggest intermediates in the following transformation [*Org. Lett.* **2000**, *2*, 3217].

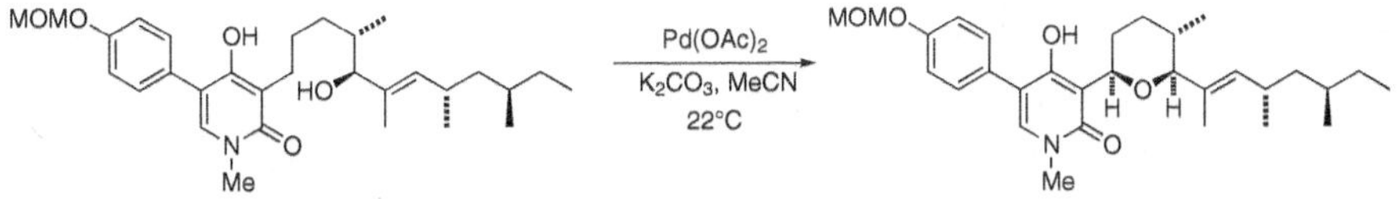

18-22 Supply the missing reagent from each of the following transformations.

(a) 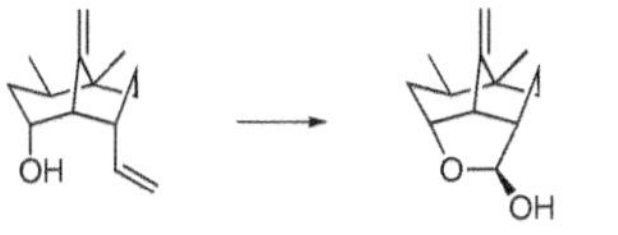

[*Angew. Chem. Int. Ed.* **2008**, *47*, 131]

(b) 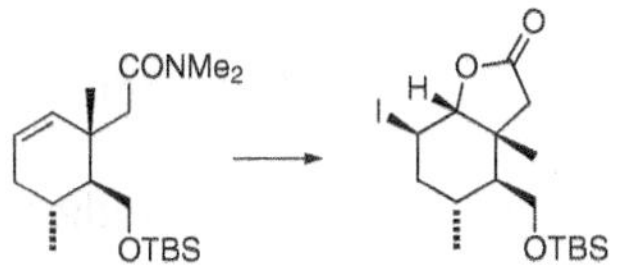

[*Tetrahedron* **2010**, *66*, 6340]

(c) 

[*J. Am. Chem. Soc.* **2008**, *130*, 13765]

(d) 

[*Angew. Chem. Int. Ed.* **2008**, *47*, 131]

(e) 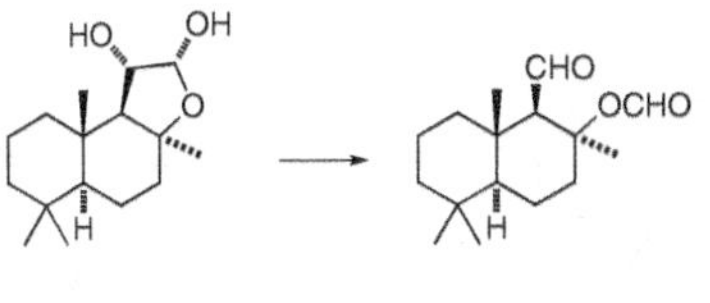

[*Tetrahedron* **2010**, *66*, 6321]

(f) 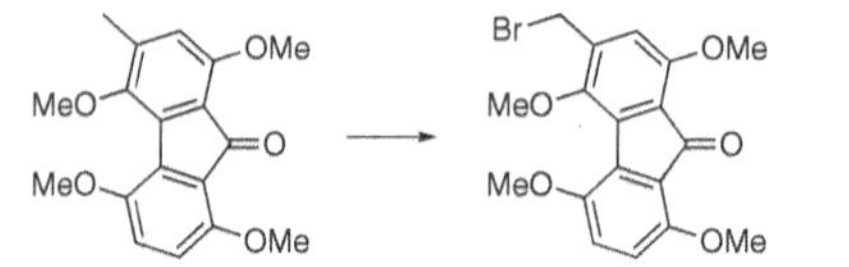

[*Tetrahedron Lett.* **2006**, *47*, 2535]

(g) 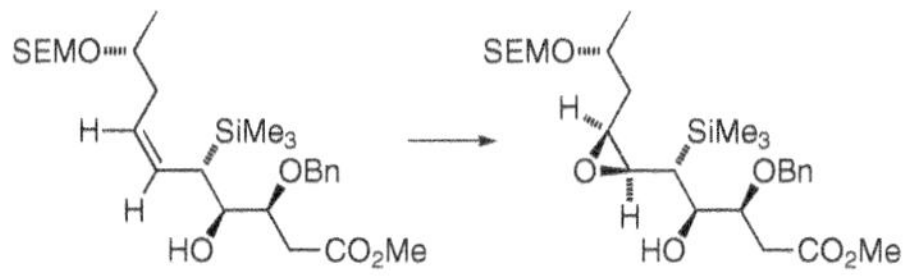

[*Angew. Chem. Int. Ed.* **2010**, *49*, 4264]

(h) 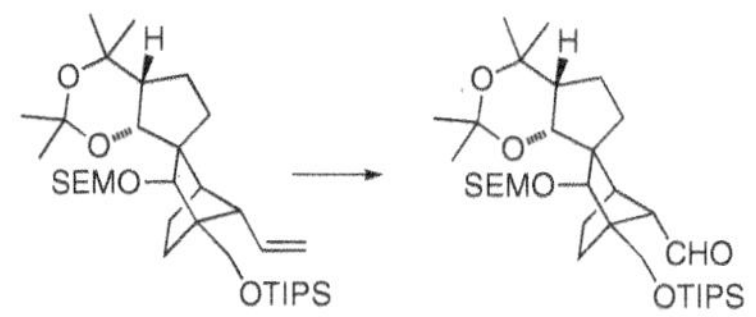

[*Angew. Chem. Int. Ed.* **2008**, *45*, 8605]

(i) 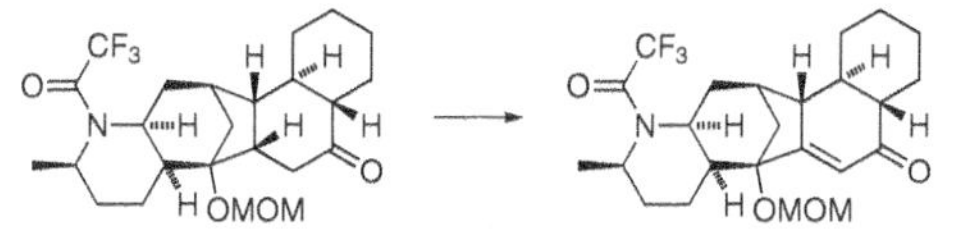

[*J. Am. Chem. Soc.* **2003**, *125*, 2400]

(j) 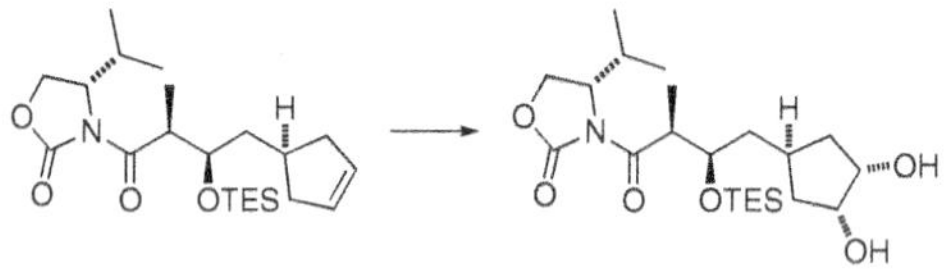

[*Org. Lett.* **2004**, *6*, 1445]

(k) 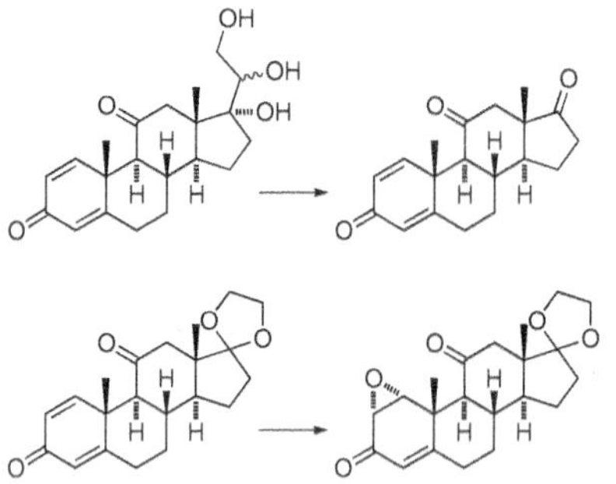

[*J. Am. Chem. Soc.* **2011**, *133*, 8014]

(l) 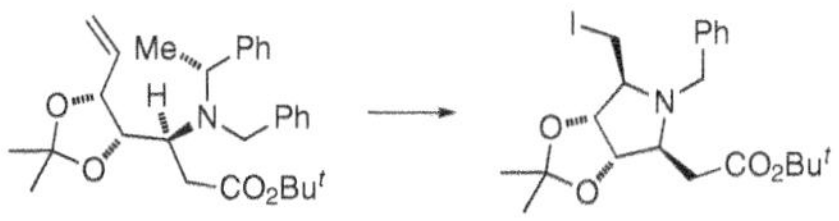

[*Synlett* **2004**, 901; *Tetrahedron: Asymmetry* **2009**, *20*, 758]

18-23 The taxane skeleton has been a popular target for synthesis due to the highly useful anticancer properties of Taxol®, especially against solid tumor, and the limited supplies of the natural material. Each of the following oxidations has been carried out on the taxane ring system. Suggest the reagent for each transformation. Where more than one option is available, suggest which would be best and tell why.

(a) 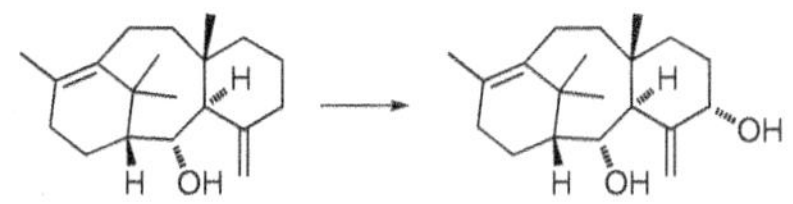

[*J. Org. Chem.* **2000**, *65*, 7865]

(b) 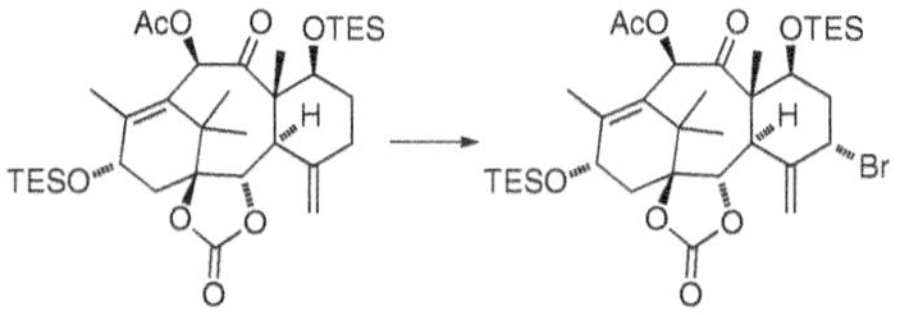

[*Chem. Eur. J.* **1999**, *5*, 121]

(c) 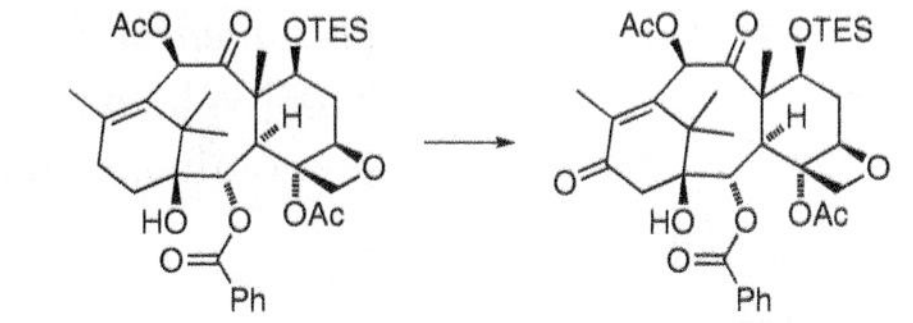

[*Nature* **1994**, *367*, 630]

(d) 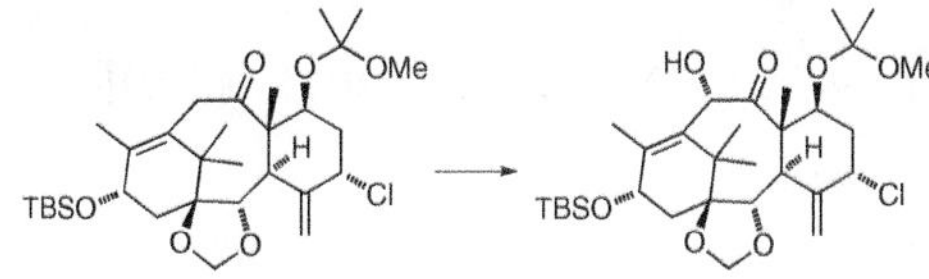

[*J. Am. Chem. Soc.* **2000**, *122*, 3811]

(e) 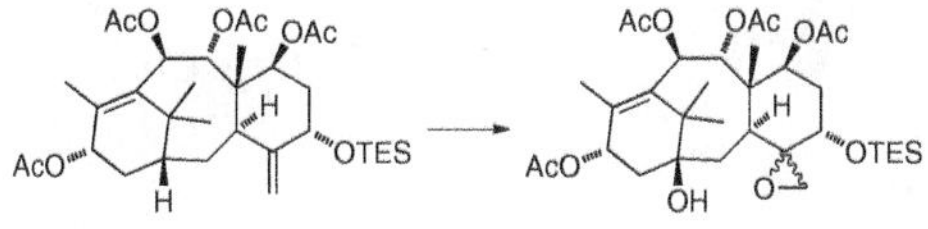

[*Tetrahedron Lett.* **2000**, *41*, 3907]

18-24 When certain secondary allylic alcohols are treated with an excess of the Swern reagent, the major product isolated is the corresponding halogenated enone. It is believed that the halogenation reaction involves the conjugated enone as the reactant. Suggest a mechanism for this transformation. [*J. Org. Chem.* **2007**, *72*, 7054].

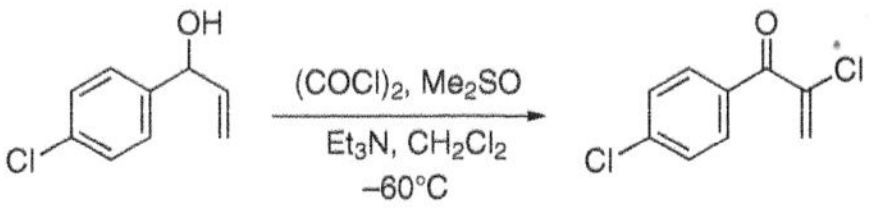

18-25 What is the major product of each of the following reactions?

(a) 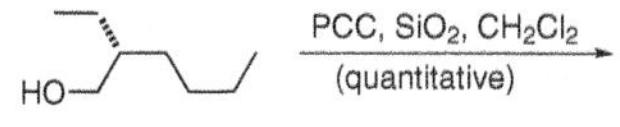

[*J. Org. Chem.* **2010**, *75*, 2333]

(b) 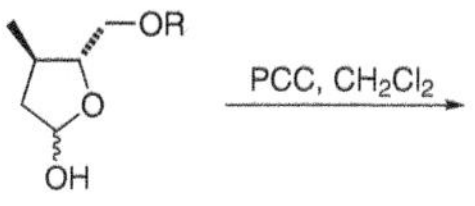

[*Angew. Chem. Int. Ed.* **2009**, *48*, 9549]

(c) 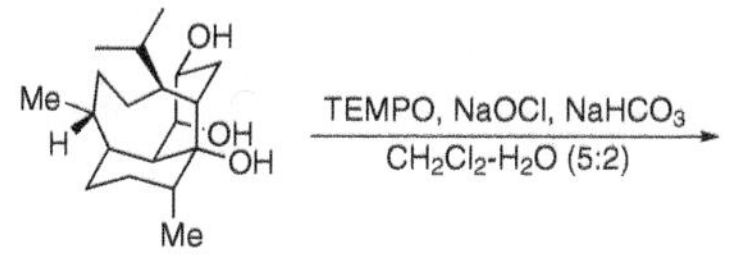

[*J. Am. Chem Soc.* **2009**, *131*, 17066]

(d) 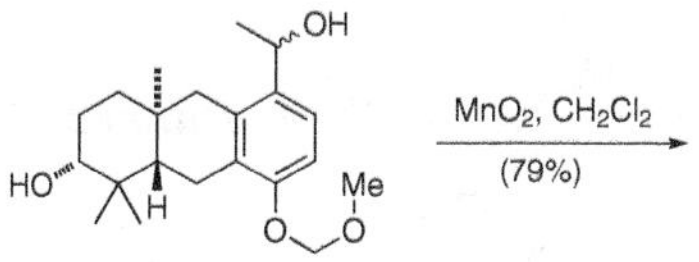

[*J. Am. Chem Soc.* **2009**, *131*, 14630]

(e) 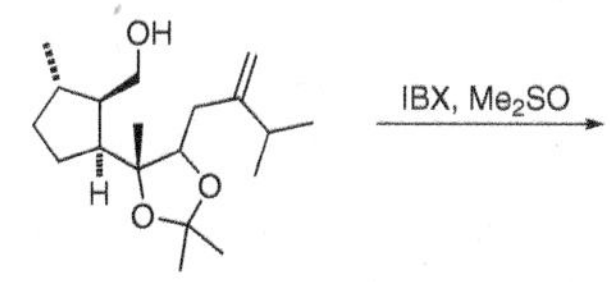

[*Angew. Chem. Int. Ed.* **2009**, *48*, 9105]

(f) 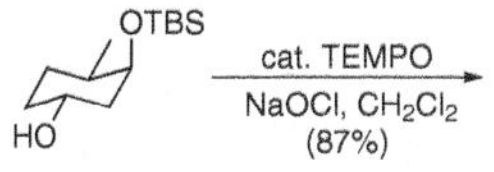

[*Tetrahedron Lett.* **1998**, *39*, 8405]

(g) 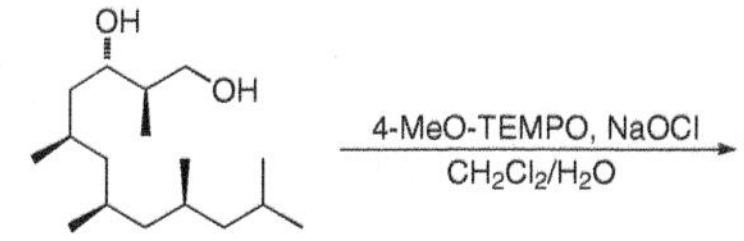

[*Org. Lett.* **2010**, *11*, 4767]

(h) 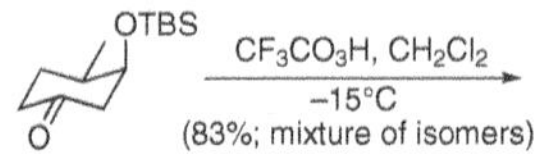

[*Tetrahedron Lett.* **1998**, *39*, 8405] Which isomer predominates, and why?

(i) 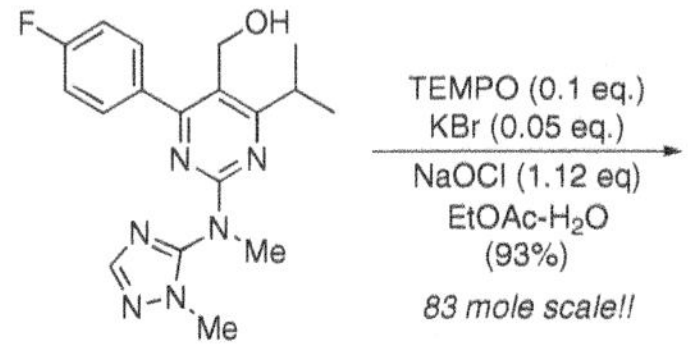

[*Org. Process Res. Dev.* **2010**, *14*, 441]

(j) 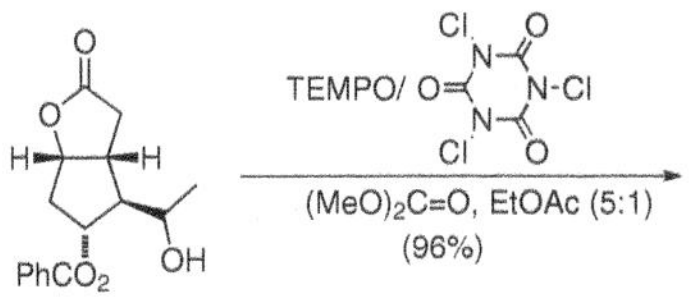

[*J. Org. Chem.* **2009**, *74*, 8298]

(k) 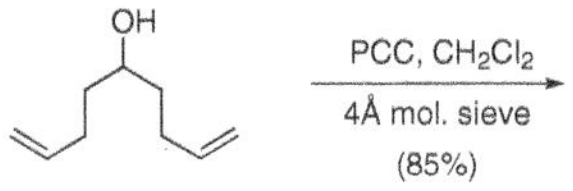

[*Tetrahedron* **2010**, *66*, 6445]

(l) 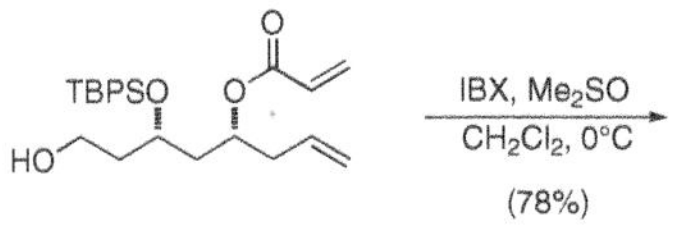

[*Synthesis* **2010**, 1171]

(m) 

[*Synthesis* **2010**, 1217]

(n) 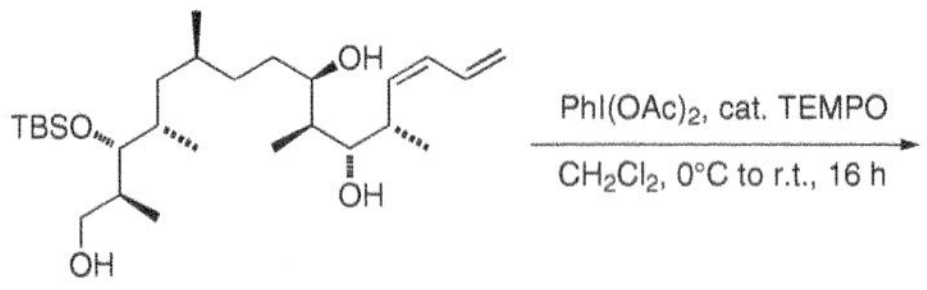

[*Tetrahedron* **2010**, *66*, 6534]

(o) 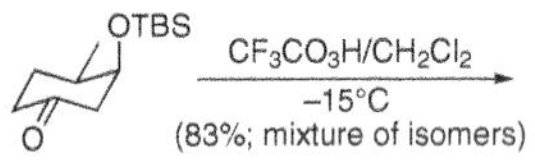

[*Synthesis* **2010**, 2533]

(p) 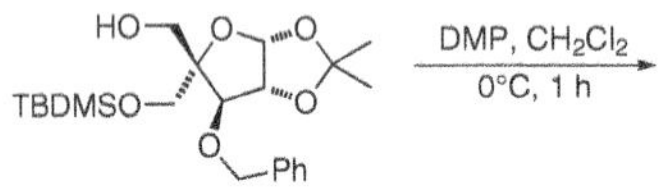

[*Synthesis* **2010**, 2643]

(q) 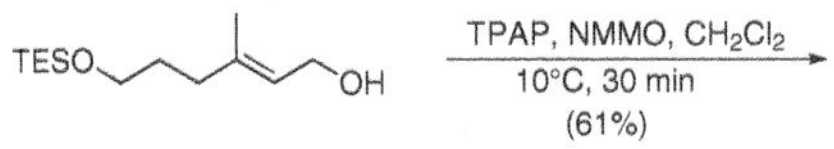

[*Tetrahedron* **2010**, *66*, 6462]

(r) 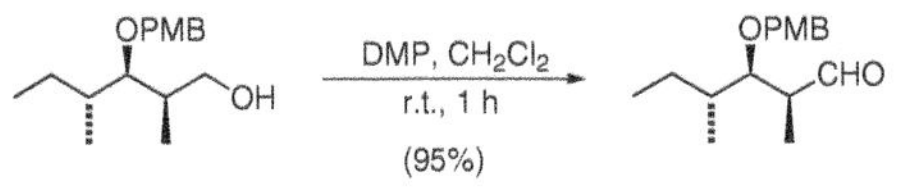

18-26 The reaction below proceeds with the stereochemistry shown, with ee's between 66 and 89%. Write a reasonable mechanism for the reaction. What is the function of the $SnCl_4$?

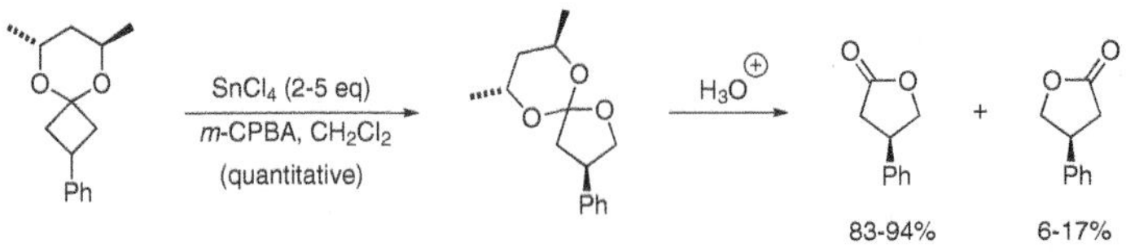

[*Tetrahedron Lett.* **1997**, *38*, 6019]

18-27 The Beckmann rearrangement of cyclohexanone oxime under "green" conditions involves the use of the ionic liquid formed between caprolactam and tetrafluoroboric acid. In the paper [*Green Chem.* **2006**, *8*, 296], the structure of the ionic liquid is written as shown. Is this a likely structure for the ionic liquid? Why or why not?

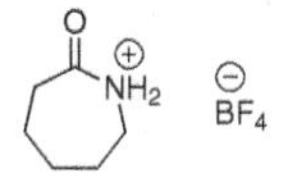

18-28 The following synthesis of fluostatins [*J. Am. Chem. Soc.* **2008**, *130*, 2783] illustrates the application of a wide range of oxidation reactions. Supply the missing reagents a-f in the synthesis below.

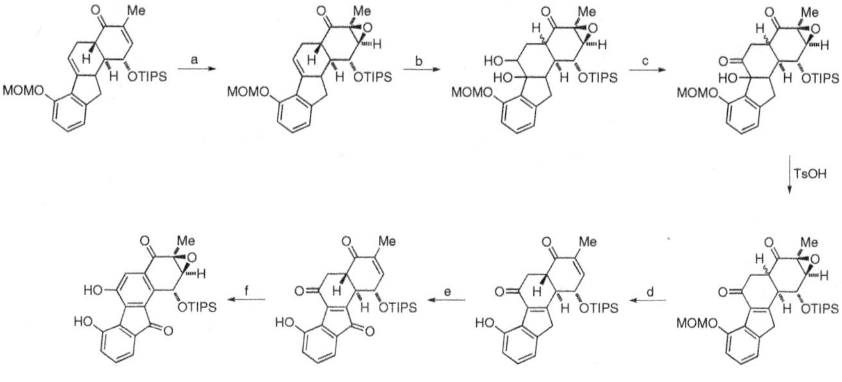

18-29 Each of the following oxidations occurs in the synthesis of the C(1)-C(25) southern hemisphere of the spirastrellolides A and B, reported by Smith *et al.* [*Tetrahedron* **2010**, *66*, 6597]. Provide reasonable alternative reagents to accomplish each of the transformations shown. Where more than one reagent may be used to effect the transformation, provide all reasonable alternatives. Suggest why the authors may have chosen the reagent they did.

(a)

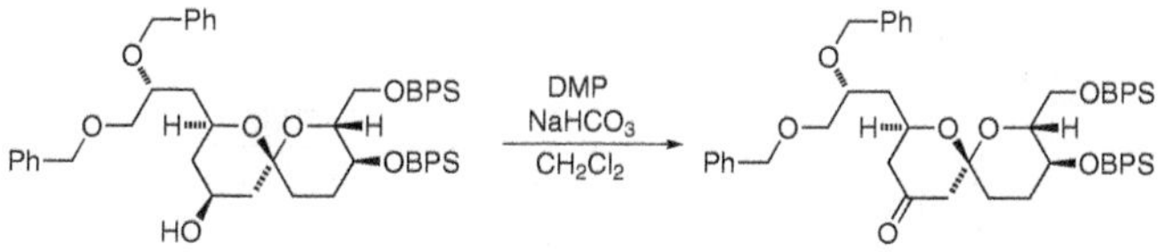

(b)

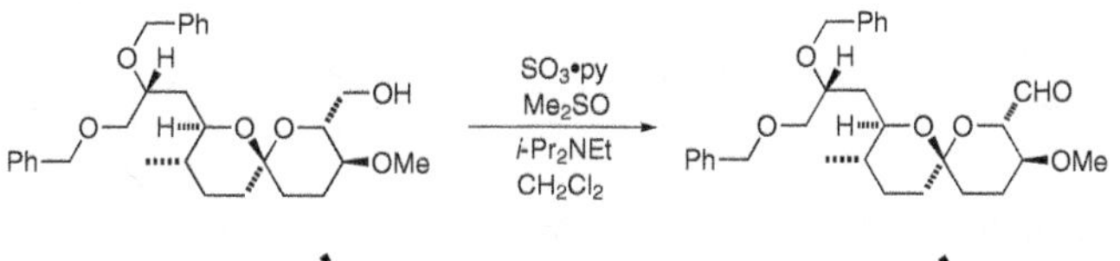

(c)

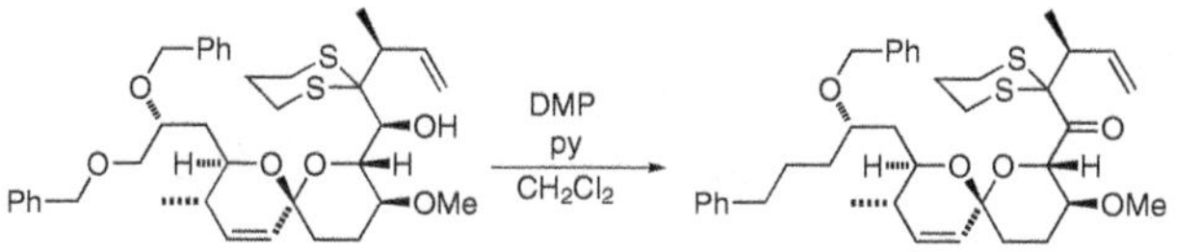

(d)

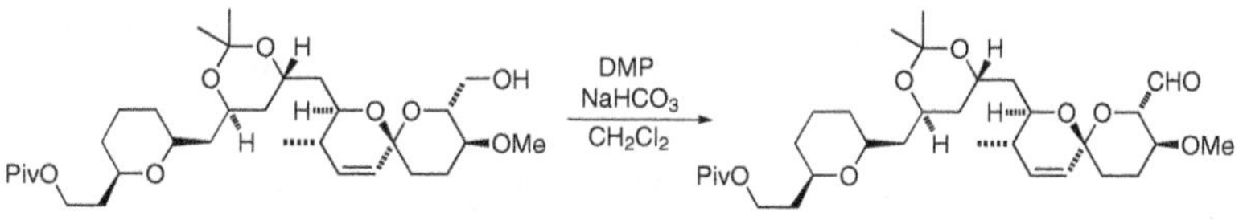

18-30 The Baeyer-Villiger oxidation of the two isomeric ketones shown proceeds with differing regiochemistry, depending only on the stereochemistry of the ketone.

Suggest a reason why this should happen. What can be learned from this experiment? [*Angew. Chem., Int. Ed.* **2000**, *39*, 2852].

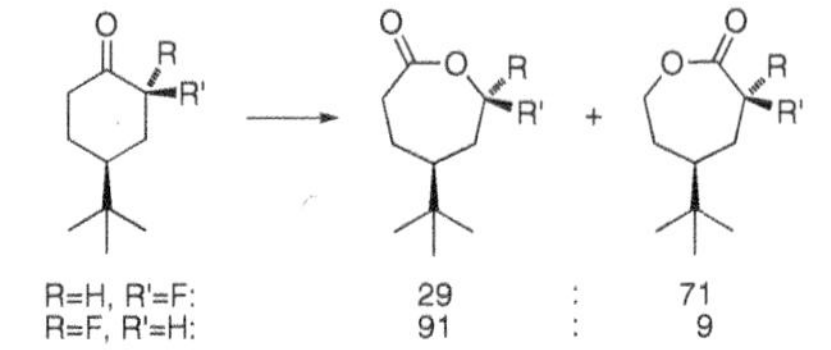

18-31 The oxidations below were used by Cooksey, *et al.* in their synthesis of (−)-ebelactones A and B [*Tetrahedron* **2010**, *66*, 6462]. Provide reagents to accomplish each of the transformations shown. Provide reasonable alternatives to the reagent chosen, and suggest why the authors may have chosen the reagent they did.

(a) 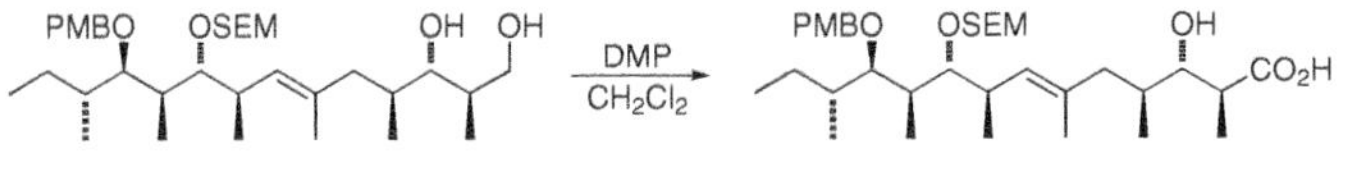

(b) 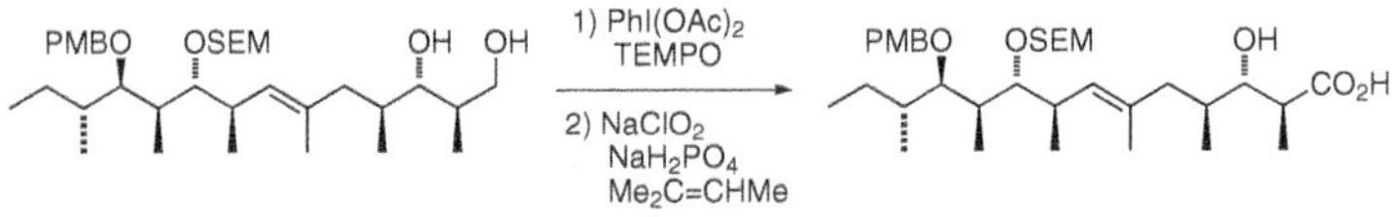

18-32 The 2009 Tetrahedron Prize for Creativity in Organic Chemistry was awarded to Steven V. Ley of Cambridge University. In his award article [*Tetrahedron* **2010** *66*, 6270], the following alcohol oxidation reactions were highlighted. Give the product of each.

(a) 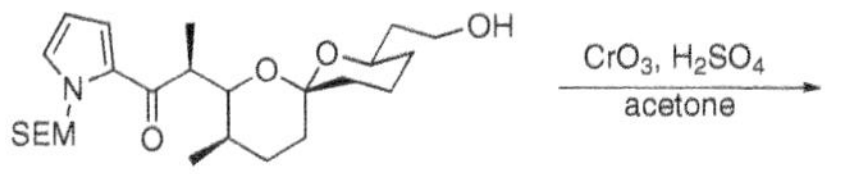

(b) 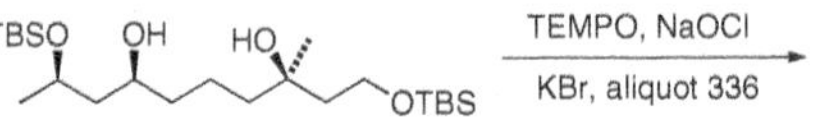

(c) 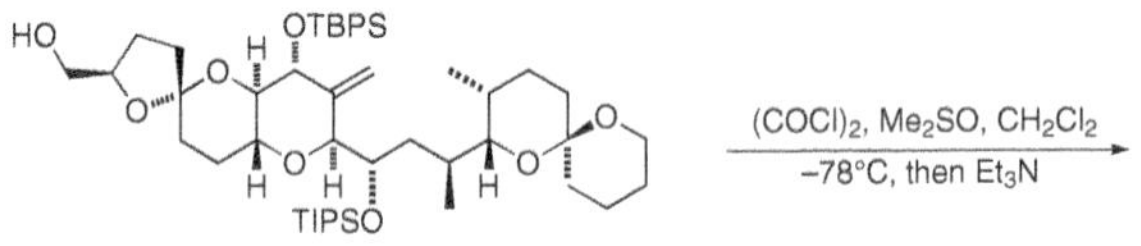

(d) 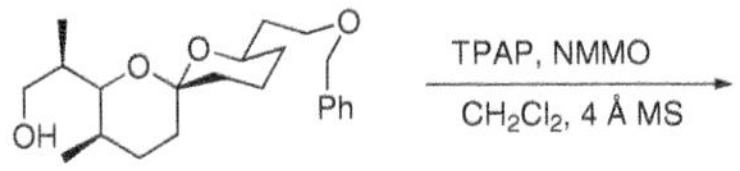

18-33 What is the major organic product of each of the following reactions?

(a) 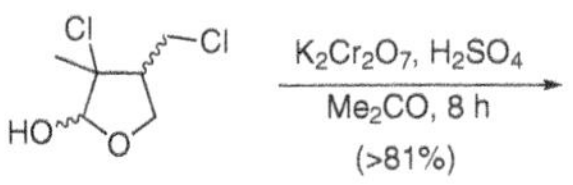 [*Tetrahedron* **2009**, *65*, 10323]

(b) 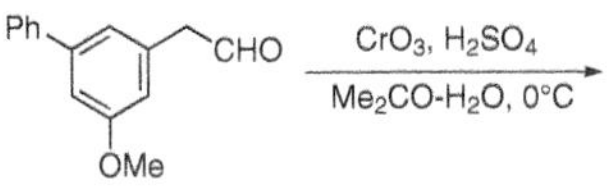 [*J. Med. Chem.* **2008**, *51*, 6839]

(c) 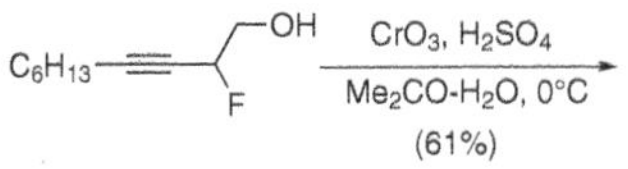 [*J. Org. Chem.* **2006**, *71*, 3518]

(d) 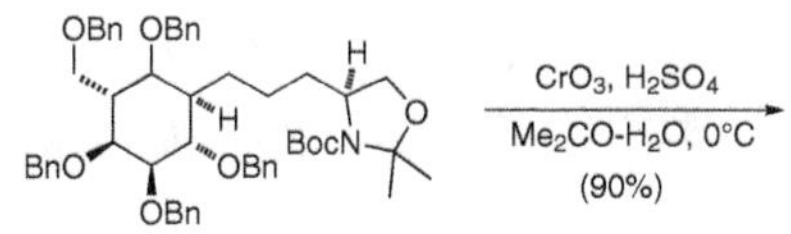

[*J. Org. Chem.* **2005**, *70*, 5508]

(e) 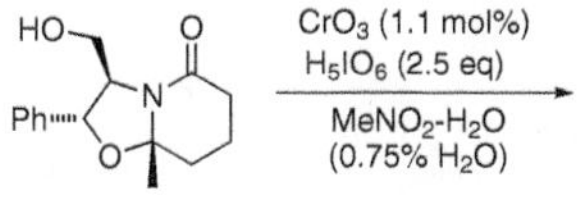

[*Tetrahedron Lett.* **1998**, *39*, 5323]

(f) 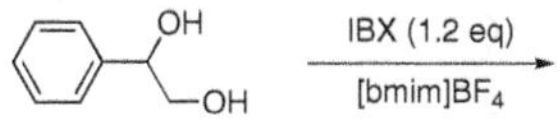

[*Tetrahedron* **2004**, *60*, 2131]

(g) 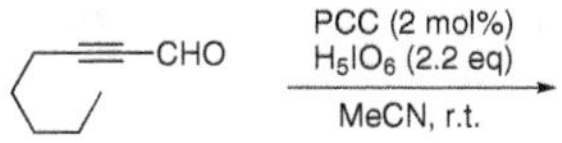

[*Synthesis* **2005**, 2487]

(h) 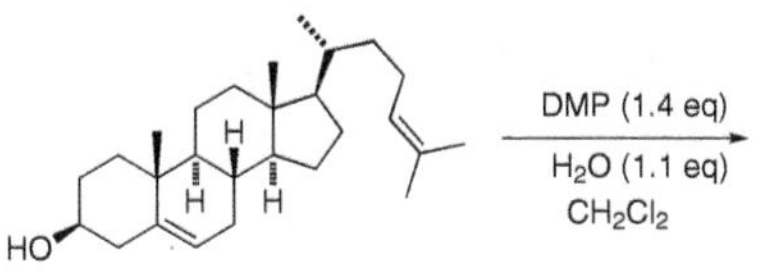

[*J. Org. Chem.* **1994**, *59*, 754]

(i) 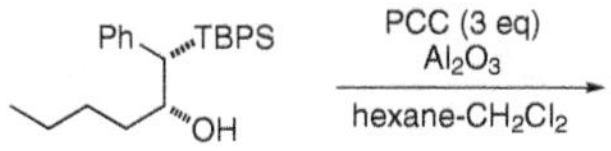

[*Synthesis* **2000**, 1223]

(j) 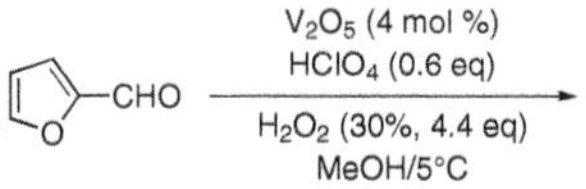

[*Org. Lett.* **2000**, *2*, 577]

(k) 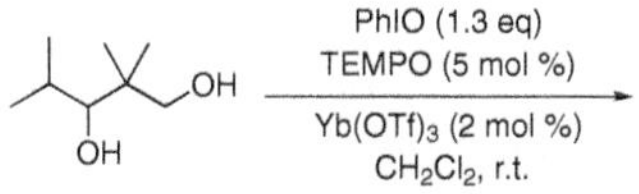

[*Synlett* **2006**, 2055]

(l) 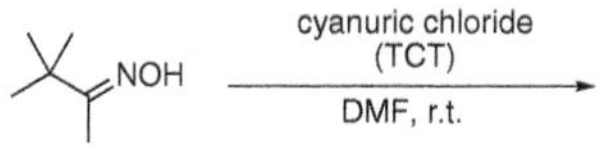

[*J. Org. Chem.* **2002**, *67*, 6272]

(m) 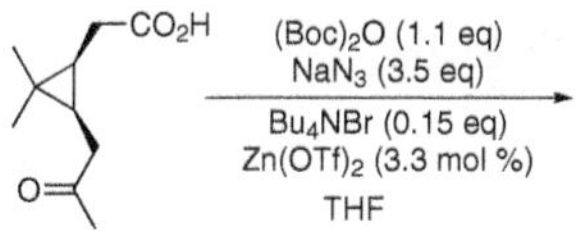

[*Org. Lett.* **2005**, *7*, 4107]

(n) 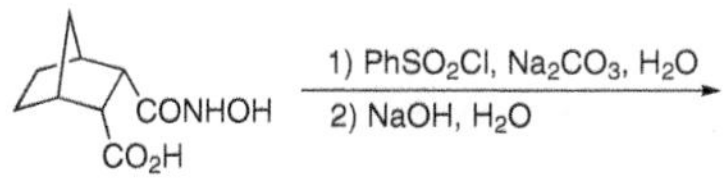

[*J. Org. Chem.* **1959**, *24*, 1293]

(o) 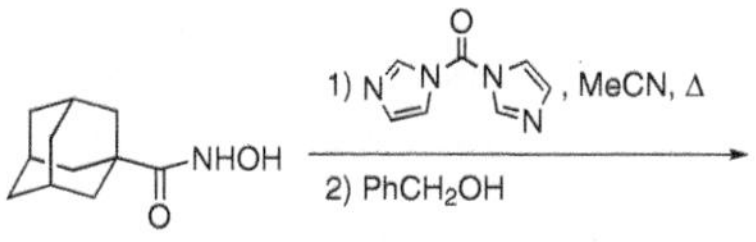

[*Org. Lett.* **2009**, *11*, 5622]

(p) 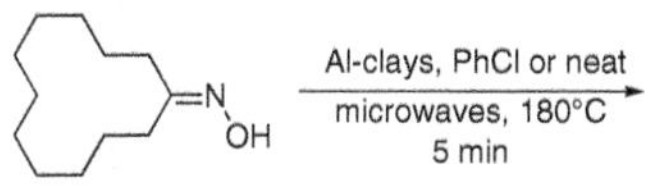

[*Green Chem.* **2007**, *9*, 1109]
Note: the Al-based reagent is acidic

(q) 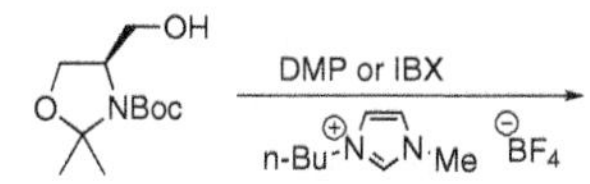

[*Tetrahedron* **2004**, *60*, 2131]

(r) 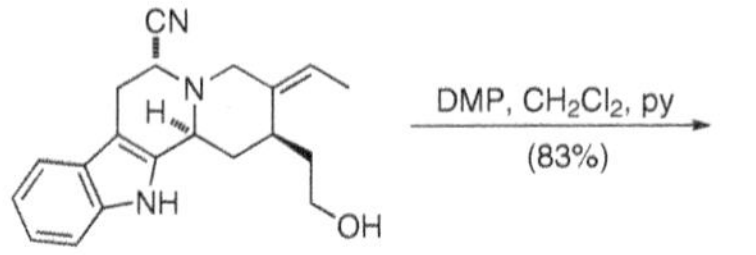

[*Tetrahedron* **2006**, *62*, 8655]

(s) 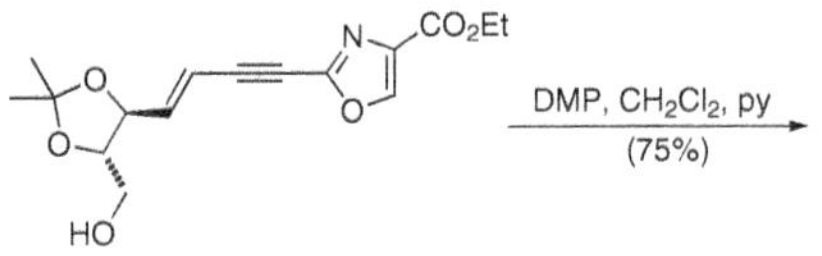

[*Tetrahedron* **2003**, *59*, 6967]

(t) 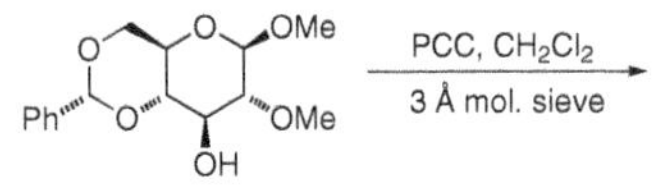

[*Tetrahedron* **2009**, *65*, 9378]

(u) 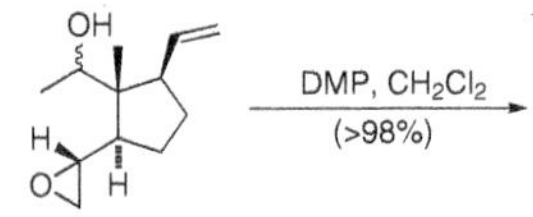

[*Tetrahedron* **2009**, *65*, 7001]

(v) 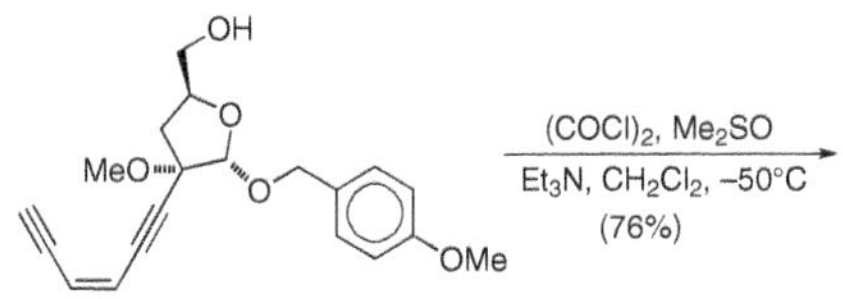

[*Tetrahedron* **1994**, *50*, 1435]

(w) 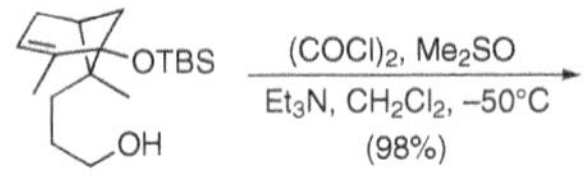

[*Tetrahedron* **2006**, *62*, 8933]

(x) 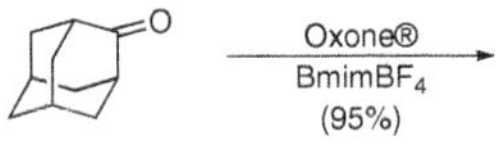

[*Tetrahedron* **2010**, *66*, 6212]

18-34 Rationalize the formation of the product shown in the oxidation of the allylic alcohol by manganese dioxide [*Tetrahedron Lett* **2009**, *50*, 6597].

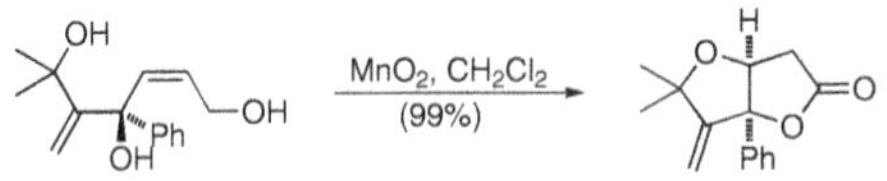

18-35 The benzyl group is frequently used as a protecting group for alcohols, amines, and amides. It has been reported [*Synthesis* **2007**, 3129] that the *N*-benzyl group can be removed from a carboxamide by treatment with NBS. Suggest a mechanism for this.

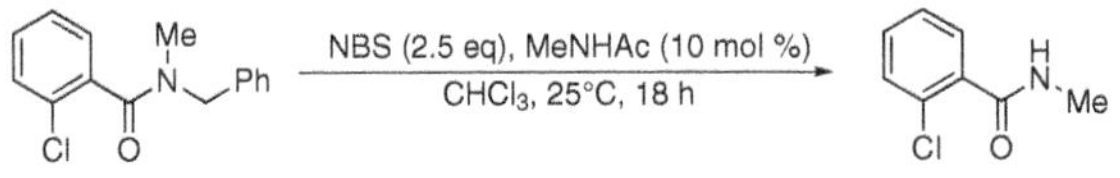

18-36 What is the major organic product of each of the following reactions?

(a) 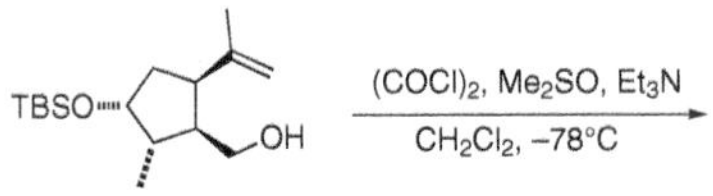

[*J. Am. Chem. Soc.* **2010**, *132*, 1488]

(b) 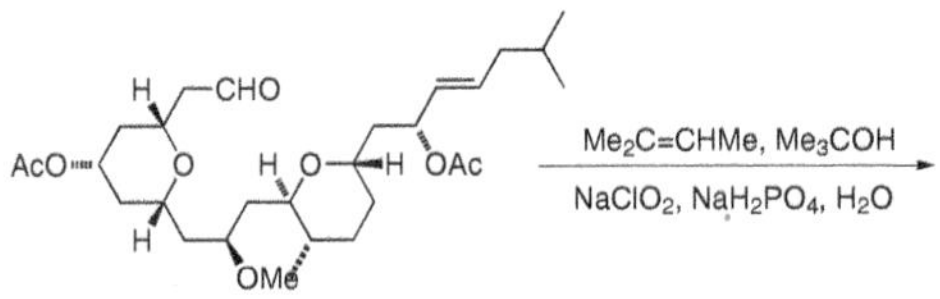

[*Org. Lett.* **2003**, *5*, 5035]

(c) 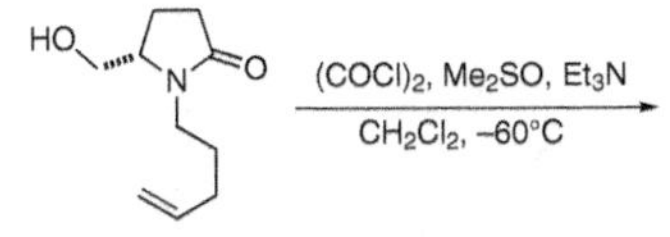

[*J. Org. Chem.* **2007**, *72*, 4246]

(d) 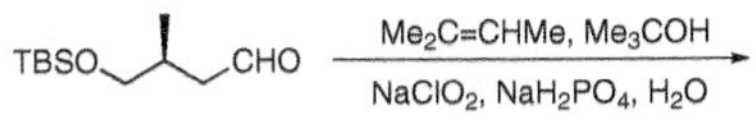

[*Org. Lett.* **2005**, *7*, 3371]

(e) 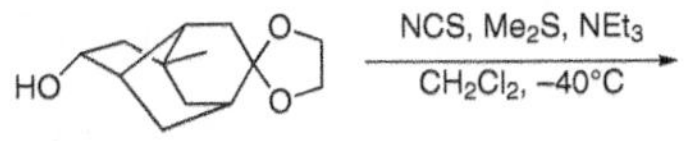

[*J. Am. Chem. Soc.* **1979**, *101*, 1609]

(f) 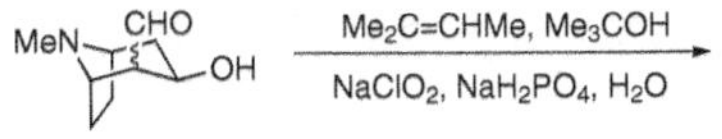

[*Org. Lett.* **2004**, *6*, 3305]

(g) 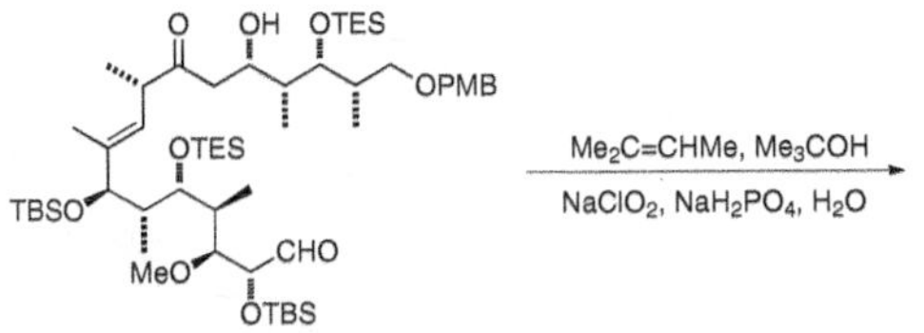

[*Synthesis* **2005**, 1183]

(h) 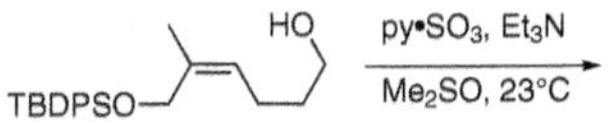

[*Org. Lett.* **2007**, *9*, 4619]

18-37 Each of the following transformations may be accomplished by at least two reagents. Suggest reagents to accomplish the transformation, and rank them in order of probable effectiveness. One lead reference to a reagent that could be used to effect this transformation is given.

(a) 
[*Tetrahedron Lett.* **1996**, *37*, 4191]

(b) 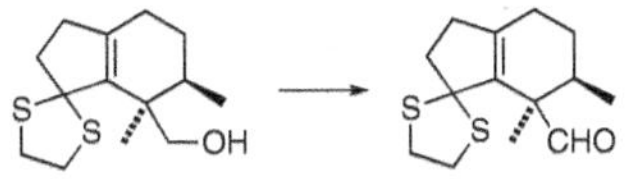
[*J. Org. Chem.* **1988**, *53*, 1441]

(c) 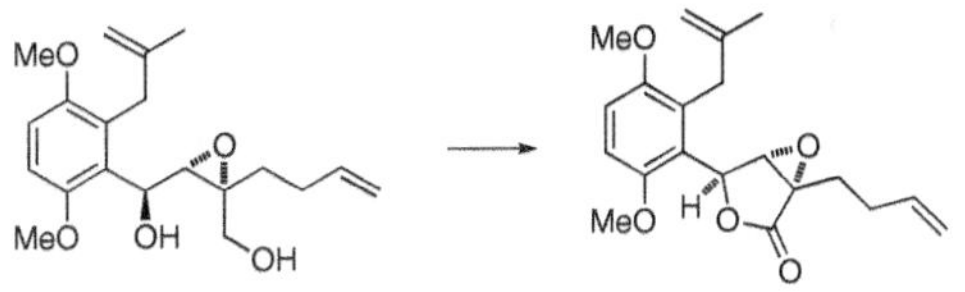
[*J. Am. Chem. Soc.* **2006**, *128*, 14042]

(d) 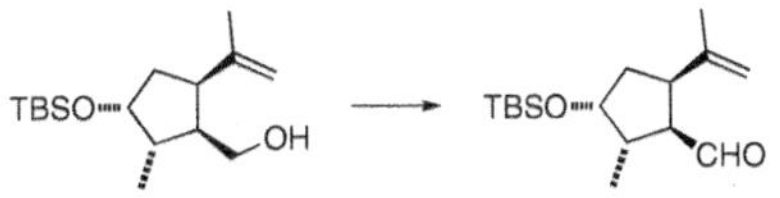
[*J. Am. Chem. Soc.* **2010**, *132*, 1488]

(e) 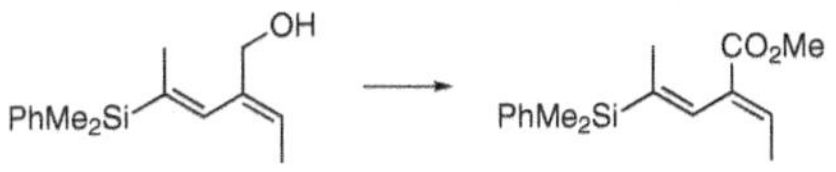
[*J. Am. Chem. Soc.* **2006**, *127*, 16038]

(f) 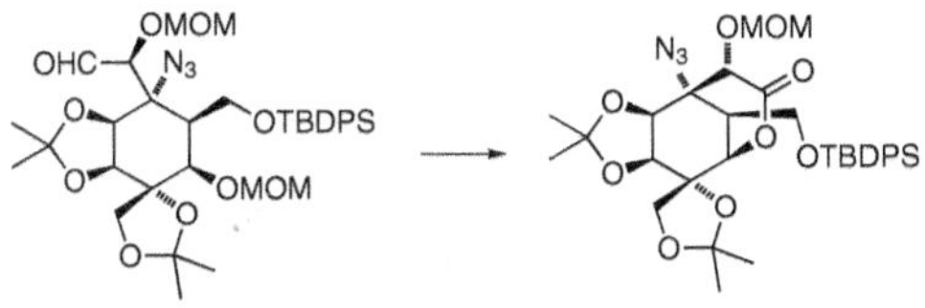
[*J. Org. Chem.* **2008**, *73*, 1234]

CHAPTER NINETEEN

Redox Reactions II

Reduction with Molecular Hydrogen or Its Equivalent

19.1 Overview of Reduction

In Figure 19.1, four common, fundamental types of reduction used in organic chemistry are represented schematically. The reactions in Example **19.1** are reductions of carbon-carbon π bonds in alkenes and alkynes. These reactions lead to products containing at least two more carbon-hydrogen σ bonds and at least one less carbon-carbon π bond. The reactions in Example **19.2** are the reduction of carbonyl compounds to alcohols. These reactions are reductions of polar π bond in the molecule without the cleavage of any groups. In Example **19.3** are the reductive elimination reactions: reductions that lead to the cleavage of polar σ bonds and their replacement by a non-polar π bond. The fourth type of reductions, shown in Example **19.4**, involves reductions of carbon-heteroatom σ bonds adjacent to carbon-carbon or carbon-oxygen π bonds.

A closer examination of the reactions in Figure 19.1 reveals that they can be subdivided into three major classes: (1) the reduction of a non-polar π bond and its replacement by two σ bonds to hydrogen (**19.1**), (2) the reduction of a polar π bond and its replacement by two σ bonds to hydrogen (**19.2**), and (3) the reduction of one or more polar σ bonds and their replacement by σ bonds to hydrogen or a non-polar π bond (**19.3, 19.4**). What reagent is best used is determined by which of these classes of reduction is required.

Since 1980, the chemistry of reductions has focused intensely on the development and use of methods for carrying out the reaction in an enantioselective manner. However, it is important that we first discuss these basic reactions without considering stereochemical control, because they provide the basis for understanding much of the asymmetric redox chemistry that will then be discussed in Chapter 21.

19.2 Catalytic Hydrogenation

The reduction of non-polar π bonds is most frequently carried out by ***catalytic hydrogenation***,[1] a reaction that was first studied in depth by the French chemists Paul Sabatier[2]

1. Monographs and reviews: (a) Augustine, R.L. *Catalytic Hydrogenation* (Marcel Dekker: New York, 1965). (b) Rylander, P.N. *Catalytic Hydrogenation over Platinum Metals* (Academic Press: NewYork, 1967). (c) Freifelder, M. *Practical Catalytic Hydrogenation* (Wiley-Interscience: New York, 1971). (d) House, H.O. *Modern Synthetic Reactions*, 2nd ed (Benjamin: New York, 1972), ch. 1. (e) Thomas, C.L. *Catalytic Processes and Proven Catalysts* (Academic Press: New York, 1970), p. 125. (f) Rideal, E.K. *Concepts in Catalysis* (Academic Press: London, 1968), p.109. (g) Burwell, R.L. *Chem. Rev.* **1957**, *57*, 895. (h) Brieger, G.; Nestrick, T.J. *Chem. Rev.* **1974**, *74*, 567. (h) Rylander, P.N. *Aldrichimica Acta* **1997**, *12*, 53. (i) Siegel, S. In Trost, B.M.; Fleming, I., Eds. *Comprehensive Organic Synthesis* (Pergamon: Oxford, 1991), vol. 8, p. 417.

2. **Paul Sabatier (1854–1941)** was educated at the Collège de France (Dr ès Sc, 1880). In 1882, he joined the Université de Toulouse, where he remained for his entire career. He received the Nobel Prize in 1913 for his work in hydrogenation. For more biographical details, see: the web site of the Nobel Foundation and Rideal, E.K. *Biogr. Mem. Fell. Roy. Soc.* **1942**, *4*, 63.

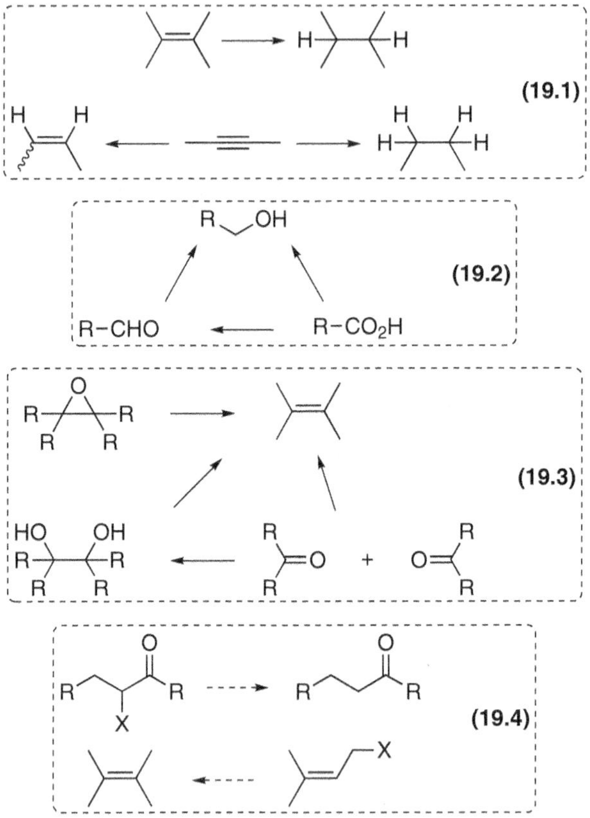

Figure 19.1 Major classes of reduction reactions in organic chemistry

and Jean-Baptiste Senderens,[3] who demonstrated its generality.[4] The addition of hydrogen to a carbon-carbon π bond (e.g., Examples **19.5** to **19.7**) is an exothermic process, but the activation energy for the uncatalyzed process is very high. So high, in fact, that the reaction occurs only extremely slowly—more on the geological time scale than the laboratory time scale. In the presence of a catalyst, however, the addition of hydrogen occurs quite readily, and the reaction has therefore become a major method of reduction. Most functional groups are susceptible to reduction by hydrogen in the presence of a catalyst, although their relative reactivity varies widely.

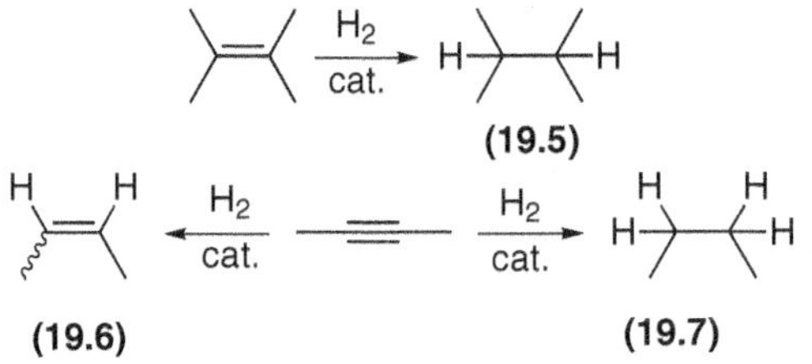

3. **Jean-Baptiste Senderens (1856–1937)**, chemist and priest, took his Dr ès Sc in 1892 at Toulouse. From 1883–1913 he was director of the «École superieure des Sciences» and Professor of Chemistry at the «Institut catholique» at Toulouse, as well as a collaborator with Sabatier (who later refused to acknowledge Senderens' contribution to the Nobel-winning work). In 1913 he began work with the Poulenc chemical company, and in 1922 Poulenc built in a personal laboratory in his home village. In 1905 he won the Prix Jecker, and in 1923 he was made a chevalier of the Légion d'Honneur. For more biographical information, see: Couderc, F. *Jean-Baptiste Senderens* (Institut catholique: Toulouse, 2009).

4. Sabatier, P.; Senderens, J.B. *Compt. rend.* **1899**, *128*, 1173; **1900**, *130*, 1761; **1901**, *132*, 210, 566, 1254; **1902**, *134*, 1127; **1902**, *135*, 87.

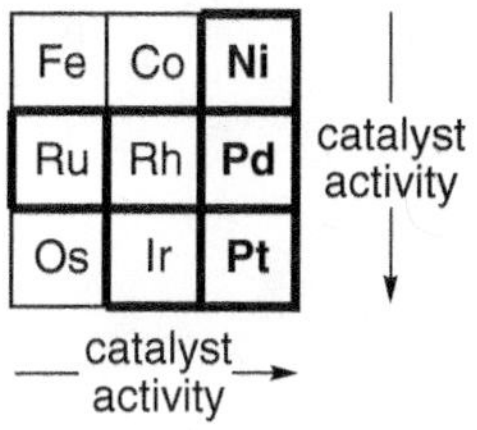

Practically all known active hydrogenation catalysts are based on transition metals. The most common catalysts used are based on the Group VIIIB metals: nickel, palladium, and platinum are most widely used However, other noble metals (especially rhodium, ruthenium, and iridium) can be used to effect catalytic hydrogenations under specific circumstances. The catalysts can be either the finely divided metal or metal oxide, in which case the reaction is known as ***heterogeneous catalytic hydrogenation***, or soluble metal complexes, in which case the reaction is known as ***homogeneous catalytic hydrogenation***. Heterogeneous hydrogenation is much more widely used, but more is known about the course of the hydrogenation under homogeneous catalysis.

Factors Affecting Reactivity in Catalytic Hydrogenation

The ease of hydrogenation of a π bond tends to increase with increasing electron density of the π bond and decrease with increasing polarity of the π bond. Thus, the non-polar C–C π bonds of alkynes hydrogenate rapidly, followed by the C–C π bonds of alkenes, the C–N π bonds of nitriles and imines, and the C–O π bonds of carbonyl compounds. The alkene π bonds of conjugated enones can be reduced selectively over more electron-rich alkenes by using rhodium as the catalyst.[5] The most difficult π bonds to hydrogenate are those of aromatic rings. The hydrogenation of alkynes can be arrested at the alkene stage by using an appropriate "poisoned" palladium catalyst. The hydrogenation of C–N and C–O π bonds leads to amines or alcohols, regardless of the starting compound. Among carbonyl compounds, the aldehydes are the most readily hydrogenated, and carboxylate anions do not hydrogenate under any circumstances at all. This order of reactivity permits the chemist a degree of control over the reaction. For example, hydrogenation of α,β-unsaturated ketones over palladium on carbon leads to the saturated ketone, especially in the presence of a small amount of diphenyl sulfide as a catalyst poison (e.g., Example **19.8**).[6]

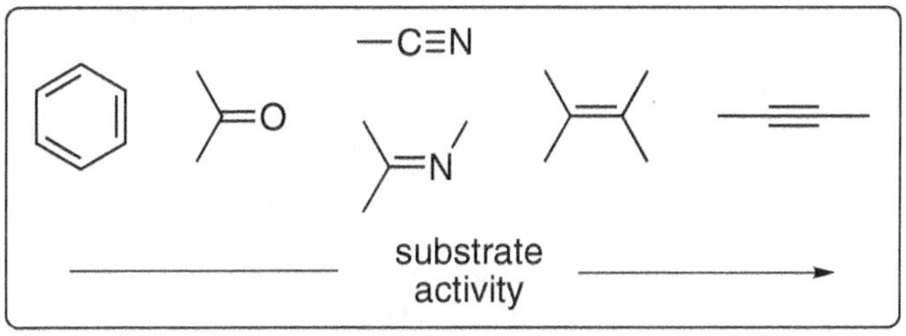

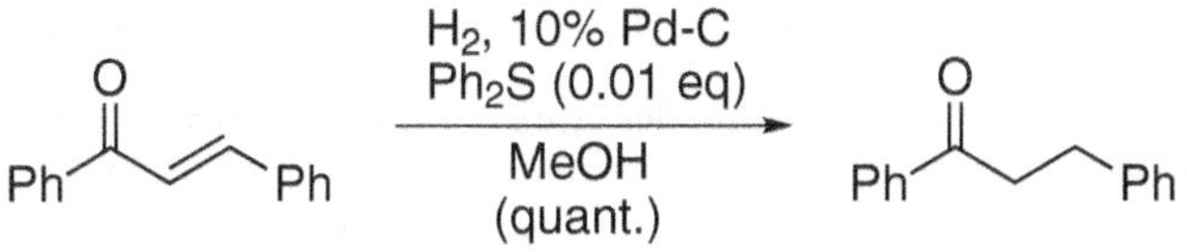

(19.8)

The activity of the catalyst is an important determining factor in the rate of a hydrogenation reaction. Platinum catalysts are the most reactive, and they are usually prepared as the finely divided metal by in situ reduction of a platinum compound, most commonly platinum dioxide, PtO_2, known as Adams catalyst[7] after American chemist Roger

5. Roy, S.K.; Wheeler, D.M.S. *J. Chem. Soc.* **1963**, 2155.
6. Mori, A.; Miyakawa, Y.; Ohashi, E.; Haga, T.; Maegawa, T.; Sajiki, H. *Org. Lett.* **2006**, *8*, 3279.
7. (a) Adams, R.; Voorhees, V. *Org. Syn., Coll. Vol. 1* **1932**, 61. (b) Adams, R.; Voorhees, V.; Shriner, R.L. *Org. Syn., Coll. Vol. 2* **1941**, 463.

Adams,[8] and chloroplatinic acid, H_2PtCl_4. Rather less reactive, and therefore more selective, are the palladium catalysts, prepared by precipitating the finely divided metal on an inert support. Palladium on charcoal and palladium on alumina are particularly popular catalysts for hydrogenating alkenes under mild conditions (1 to 3 atmospheres pressure and temperatures below 100°C). Nickel is usually used as Raney nickel,[9] discovered by Murray Raney[10] and developed by Homer Adkins[11] and his students.[12] Raney nickel is prepared by dissolving the aluminum from an aluminum-nickel alloy (Raney alloy) in sodium hydroxide; when prepared in this way, it has a large amount of chemisorbed hydrogen.

Nickel is the least active of the common metal catalysts, and it must be used under pressure and often at high temperatures. It does, however, have the advantage of being particularly resistant to deactivation by catalyst "poisons." The most common catalyst "poisons" are sulfur compounds, amines, and lead compounds. Platinum and palladium catalysts are especially vulnerable to poisoning; poisoning by amines can be reduced by carrying out the reduction in acidic solution. Occasionally, catalyst poisoning can be useful. Poisoning of Raney nickel by sulfur compounds has been exploited in the reductive desulfurization of dithioketals. Catalytic hydrogenation of alkynes over a palladium catalyst poisoned with quinoline and lead (Lindlar's catalyst, named for British-Swiss chemist, Herbert Lindlar[13]) gives *Z* alkenes.

Industrially, copper chromite[14] is used for the hydrogenation of fatty esters. The five most common hydrogenation catalysts and their uses in hydrogenation reactions are summarized in Table 19.1.

The outcome of a heterogeneous catalytic hydrogenation can be highly dependent on the catalyst chosen, the method of catalyst preparation, the solvent, the catalyst loading, the reaction temperature, and the pressure. Indeed, most reviews of the reaction emphasize the difficulty of predicting the performance of a particular catalyst in a specific application and the importance of trial-and-error in optimizing the reaction conditions.

---

8. **Roger Adams (1889–1971)** was educated at Harvard (PhD, 1912), and then worked with Diels and Willstätter in Germany. He joined the University of Illinois in 1916 and spent his entire career there. For more biographical information, see: Tarbell, D.S.; Tarbell, A.T. *Biogr. Mem.* (National Academy of Sciences: Washington, D.C., 1982), p. 1.

9. Raney, M. **U.S. Patent No. 1,628,190** [May, 1927].

10. **Murray Raney (1885–1966)** was a mechanical engineer. He worked as a teacher (and heating and lighting supervisor) at Eastern Kentucky State Normal College until 1910. His research career began in 1915, when he joined the Lookout Oil & Refining Company; here he began work on nickel alloys that would later yield "Raney catalysts." In 1925, he took a sales manager position with Gilman Paint and Varnish Co., rising to president of the company. In 1950, he left Gilman to found the Raney Catalyst Company and devote full-time effort to the production of his catalysts. The company was acquired by W.R. Grace & Company in 1963.

11. **Homer Burton Adkins (1892–1949)** received his education at Denison University (BS, 1915) and the Ohio State University (MS, 1916; PhD, 1918). After temporary appointments during 1918–1919 in the War Department, Du Pont, and Ohio State, he joined the University of Wisconsin, where he spent the rest of his life. For more biographical information, see: Daniels, F. *Biogr. Mem., Natl. Acad. Sci.* **1952**, 291.

12. Adkins, H. *Reactions of Hydrogen with Organic Compounds over Copper Chromium Oxide and Nickel* (University of Wisconsin Press: Madison, WI, 1937).

Among later papers by Adkins and his students, the following are illustrative: (b) Durland, J.R.; Adkins, H. *J. Am. Chem. Soc.* **1932**, *60*, 1501. (c) Wojcik, B.; Adkins, H. *J. Am. Chem. Soc.* **1934**, *56*, 2424. (d) Adkins, H.; Krsek, G. *J. Am. Chem. Soc.* **1948**, *70*, 412. (e) Pavlic, A.A.; Adkins, H. *J. Am. Chem. Soc.* **1946**, *68*, 1471. (f) Adkins, H.; Billica, H.R. *J. Am. Chem. Soc.* **1948**, *70*, 695. (g) Billica, H.R.; Adkins, H. *Org. Syn., Coll. Vol. 3* **1955**, 176.

13. **Herbert Wilson Lindlar (1909–2009)** was a British-Swiss chemist (PhD, Bern 1939). He spent his entire career at Hofmann-La Roche, in Basel. He retired in 1974, and then worked as British Vice-Consul in Zurich. He died three months after his 100th birthday.

14. Connor, R.; Folkers, K.; Adkins, H. *J. Am. Chem. Soc.* **1931**, *53*, 2012; **1932**, *54*, 1138.

Table 19.1 Heterogeneous Catalysts for Hydrogenation

Catalyst	Reaction	Comments
Pd		Can cause π bond migration during reaction. To prevent this, use Ni or Ru.

		To avoid this reaction, Rh catalysts should be used.

		This may be halted at alkene stage by limited catalyst and hydrogen.
		This uses a Lindlar catalyst, or Pd/polyethyleneimine.
		Pd catalyst is poisoned with S and/or quinoline.

Pt		
		Reduction of aldehydes usually requires Fe^{2+} as an accelerator.

Ni		1 atmosphere; ambient temperature
		100–200 atmosphere; 100–200°C
		50–100°C; high pressure

(continued)

(**Table 19.1** continued)

Catalyst	Reaction	Comments
		1–4 atmosphere; 50–100°C
	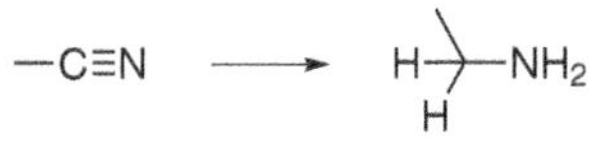	50–100°C; high pressure; NH_3 added to minimize yield of secondary amine
		1 atmosphere; ambient temperature
Rh		5–10 atmospheres; 50–100°C; does not usually lead to loss of benzyl substituents
		50–100°C; high pressure; NH_3 added to minimize yield of secondary amine

Ru		Only heterogeneous catalyst for this transformation; 300–600 atmospheres; H_2 required in H_2O

Despite the obvious complexity of the reaction, certain general facts about the reaction mechanism have emerged. It is generally accepted that the first event of the reaction is adsorption of the reactants onto the metal surface. This occurs by a process known as ***chemisorption***, which is adsorption accompanied by covalent bond formation. The chemisorption of the alkene on the catalyst surface is susceptible to steric hindrance, and evidence for this comes from the hydrogenation of alkenes in which the two faces of the double bond are not equivalent. In these reactions, the hydrogen is selectively added to the less hindered face of the π bond.

The stereochemistry of reduction of such compounds results in the major product of hydrogenation invariably being the alkane formed by *syn* addition of the hydrogen atoms to the less hindered surface of the π bond. This is shown in Examples **19.9** and **19.10** and Figure 19.2.

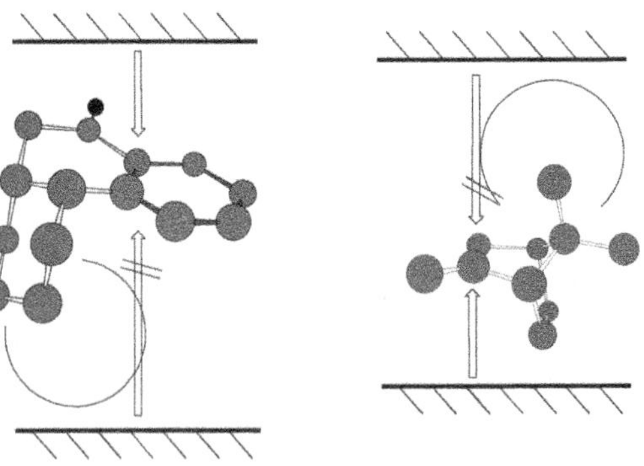

Figure 19.2 The initial step of catalytic hydrogenation is the chemisorption of the alkene on the catalyst surface. If one face of the double bond is blocked from close approach to the metal surface so efficient orbital overlap between the π bond and the metal atoms cannot occur, hydrogenation will occur from the other side.

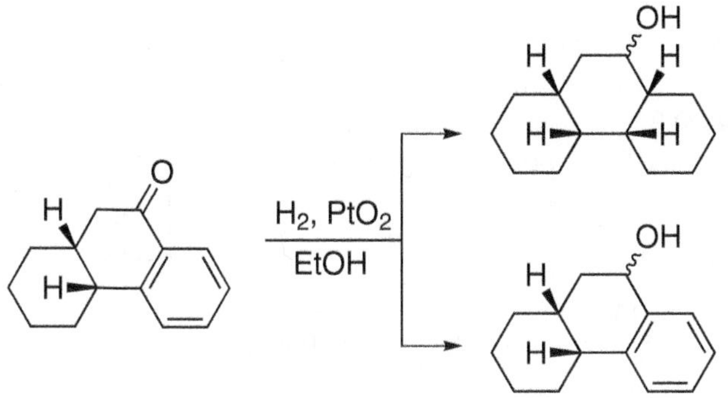

(19.9)[15]

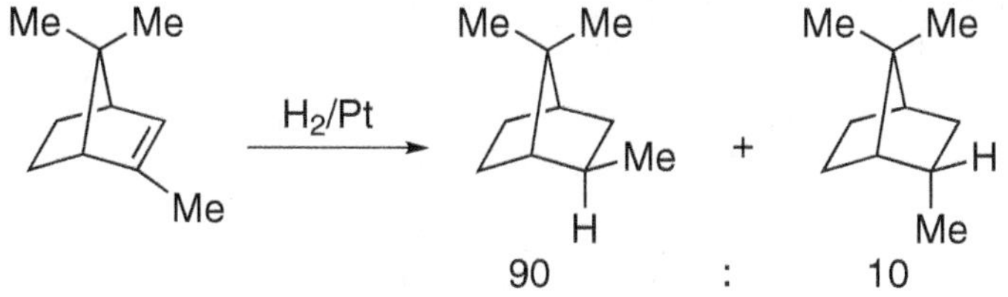

(19.10)[16]

Even today, a century after the development of the reaction, neither the exact bonding involved in the chemisorption of the reactants to the catalyst surface nor the mechanistic details of heterogeneous catalytic hydrogenation are well understood. In large part, this is because of the heterogeneity of the reacting system. Thus, many questions critical to understanding what actually occurs remain unanswered.

Nevertheless, some general conclusions about the mechanism have been drawn. It is generally accepted that the process involves the sequence of steps in Figure 19.3, which are a slight modification of the original proposal by Horiuti and Polanyi[17]:

1. Chemisorption of hydrogen on the metal surface to give metal-hydrogen bonds (**19.11**)
2. Chemisorption of the alkene on the surface to give metal-carbon bonds by *syn* addition to the π bond (this also anchors the alkene to the metal so that the two faces of the alkene are differentiated) (**19.12**)
3. Stereospecific replacement of the metal in one metal-carbon bond by hydrogen to give a carbon-hydrogen bond with retention of configuration at carbon (**19.13**)
4. Stereospecific replacement of the metal in the other metal-carbon bond by hydrogen with retention of configuration at carbon (**19.14**)

One of the important features of this model is that all steps of this mechanism are reversible. This provides a means for isomerizing the π bond during hydrogenation and a method for the net *anti* addition of hydrogen by desorption (**19.17**) and resorption of the isomerized alkene (**19.18**) on the other face of the π bond (Figure 19.4). Alternatively, it allows for shifting of the π bond during hydrogenation, leading to other unexpected hydrogenation products.[18]

It is important to remember that, despite almost a century of study, the actual nature of the bonding of the hydrogen and the alkene to the metal remains unknown.[19] The

15. Linstead, R.P.; Doering, W.v.E.; Davis, S.B.; Levine, P.; Whetstone, R.R. *J. Am. Chem. Soc.* **1942**, *64*, 1985, 1991, 2003, 2007, 2009, 2014, 2022.

16. (a) Bell, H.M. PhD Thesis, Purdue University, 1964. Cited in: (b) Brown, H.C.; Kawakami, J.H. *J. Am. Chem. Soc.* **1970**, *92*, 201. (c) Brown, H.C.; Kawakami, J.H.; Liu, K.-T. *J. Am. Chem. Soc.* **1973**, *95*, 2009.

17. Horiuti, I.; Polanyi, M. *Trans. Faraday Soc.* **1934**, *30*, 1164.

18. Siegel, S.; Smith, G.V. *J. Am. Chem. Soc.* **1960**, *81*, 6082, 6087.

19. (a) Burwell, R.L., Jr.; Schrage, K. *J. Am. Chem. Soc.* **1965**, *87*, 5234. (b) McKee, D.W. *J. Am. Chem. Soc.* **1962**, *84*, 1109. (c) Siegel, S. *Adv. Catal.* **1966**, *16*, 123. (d) Ledoux, M.J. *Nouv. J. Chim.* **1978**, *2*, 9. (e) Clarke, J.K.A.; Rooney, J.J. *Adv. Catal.* **1976**, *25*, 125. (f) Bautista, F.M.; Campelo, J.M.; Garcia, A.; Guardeño, R.; Luna, D.; Marinas, J.M. *J. Chem. Soc., Perkin Trans. 2* **1989**, 493.

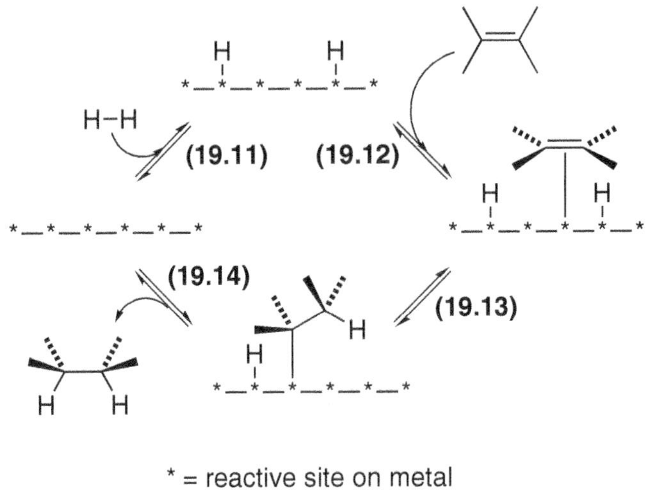

Figure 19.3 Accepted mechanism of heterogeneous catalytic hydrogenation

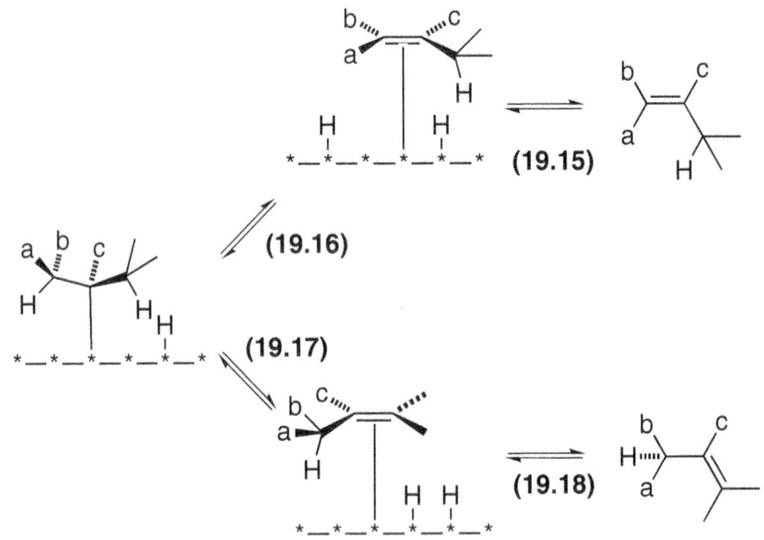

Figure 19.4 Mechanism whereby isomerization of the π bond regiochemistry or stereochemistry may occur

Horiuti-Polanyi model in Figure 19.3 does, however, predict that the overall stereochemistry of the addition to the chemisorbed face of the π bond should be suprafacial. Experimental observations confirm this. Hydrogenation of the isomeric 2,3-diphenyl-2-butenes over palladium in acetic acid gives exclusively the (±) isomer from the *E* alkene (Example **19.19**) and almost exclusively the *meso* isomer from the *Z* alkene (Example **19.20**).[20]

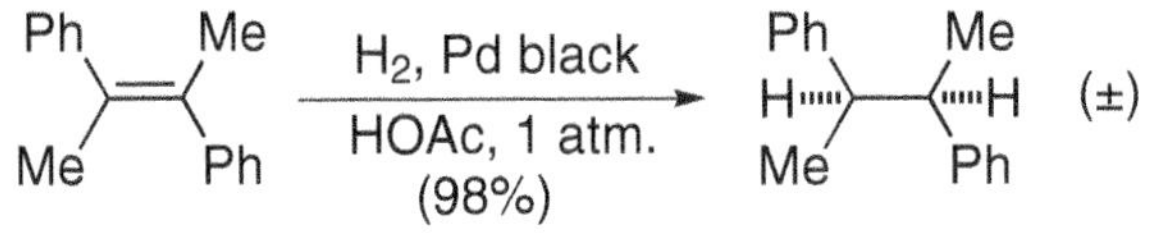

(19.19)

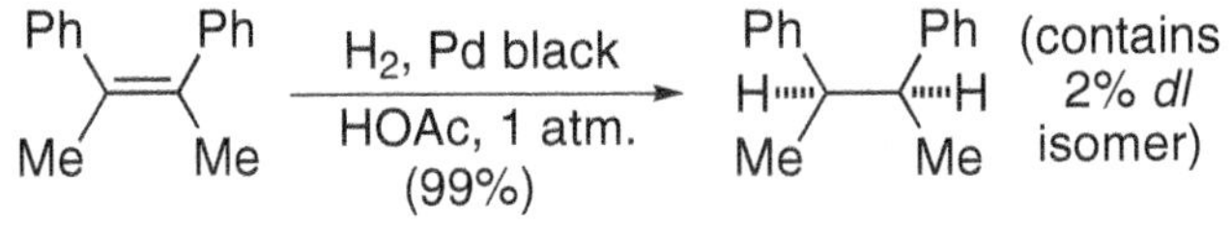

(19.20)

However, there is significant work that has identified several different types of sites on the metal surface: (g) Augustine, R.L.; Yaghmaie, F.; Van Peppen, J.F. *J. Org. Chem.* **1984**, *49*, 1865. (h) Brower, W.E.; Matyjaszczyk, M.S.; Pettit, T.L.; Smith, G.V. *Nature (London)*, **1983**, *301*, 497. (i) Somorjai, G. *Acc. Chem.Res.* **1976**, *9*, 248. (j) Smith, G.V.; Zahraa, O.; Molnar, A.; Khan, M.M.; Richter, B.; Brower, W.E. *J. Catal.* **1983**, *83*, 238.

20. von Wessely, F.; Welleba, H. *Chem.Ber* **1941**, *74*, 777.

Table 19.2 Relative Ease of Hydrogenation of Functional Groups*

Approximate Rank	Functional Group	Reduction Product	Approximate Rank	Functional Group	Reduction Product
1	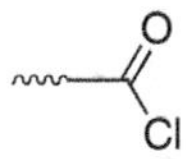	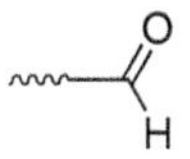	7	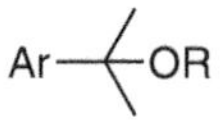	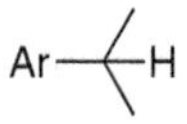
2	$-NO_2$	$-NH_2$	8	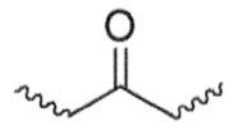	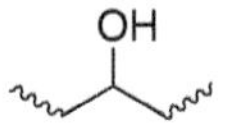
3†		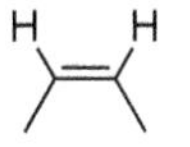	9	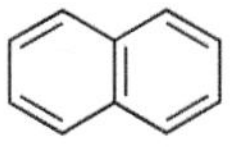	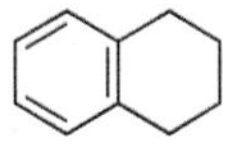
4	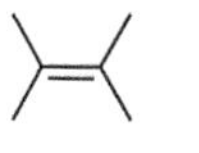	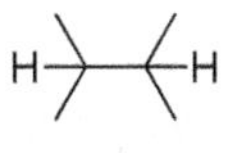	10	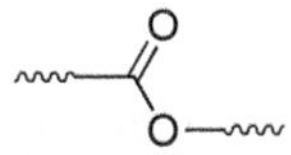	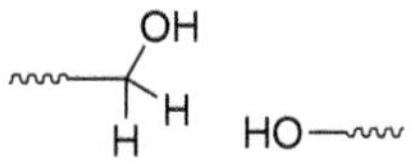
5‡	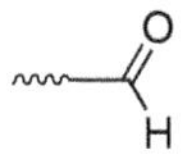	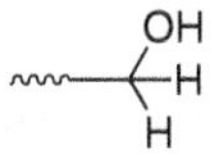	11	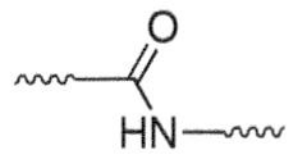	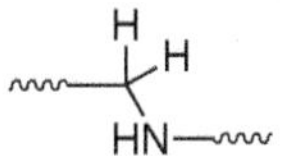
6	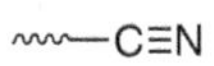	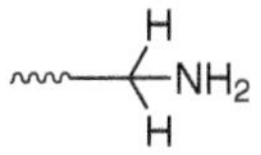	12		

*The order is catalyst-dependent. Functional groups not listed (e.g., carboxylate anions) are considered resistant to hydrogenation.
†The high level of reactivity of alkynes is due to the much stronger chemisorption of the alkyne on the catalyst surface.
‡Fe^{2+} or Fe^{3+} is required as an accelerator with Pt. Pd often leads to decarbonylation competing with hydrogenation.

The ease of reduction of functional groups depends in large part on the catalyst used. Thus, palladium catalysts are excellent for use in hydrogenation of non-polar π bonds and in hydrogenolysis of benzyl substituents but are seldom the catalyst of choice for the hydrogenation of carbonyl groups. To avoid hydrogenolysis, rhodium catalysts are preferred. In fact, rhodium allows the hydrogenation of aromatic rings to cyclohexanes without hydrogenolysis of the benzyl substituent. Table 19.2 contains a composite picture of the general order of reactivity of functional groups toward hydrogen in the presence of palladium, platinum, or nickel. Of course, this is only an approximate order, and as the preceding discussion implies, it may be significantly modified by the actual choice of catalyst.

Homogeneous Catalytic Hydrogenation of Alkenes

Few advances have had as profound an impact on the development of a synthetic method as the introduction of soluble transition metal catalysts (which allow the reaction to be carried out under homogeneous, rather than heterogeneous conditions) have had on hydrogenation. Homogeneous catalytic hydrogenation is now a major method for the reduction of alkenes.[21] The prototypical catalyst for homogeneous catalytic hydrogenation

21. (a) James, B.R. *Homogeneous Hydrogenation* (Wiley: New York, 1973). (b) Strohmeier, W. *Fortschr. Chem. Forsch.* **1972**, *25*, 71. (c) Harmon, R.E.; Gupta, S.K.; Brown, D.J.; *Chem. Rev.* **1973**, *73*, 21. (d) Birch, A.J.; Williamson, D.H. *Org. React.* **1976**, *24*, 1. (e) Collman, J.P.; Hegedus, L.S.; Norton, J.R.; Finke, R.G. *Principles and Applications of Organotransition Metal Chemistry* (University Science Books: Mill Valley, CA, 1987), p. 523. (f) Takaya, H.; Noyori, R. In Trost, B.M.; Fleming, I., Eds. *Comprehensive Organic Synthesis* (Pergamon: Oxford, 1991), vol. 8, p. 443.

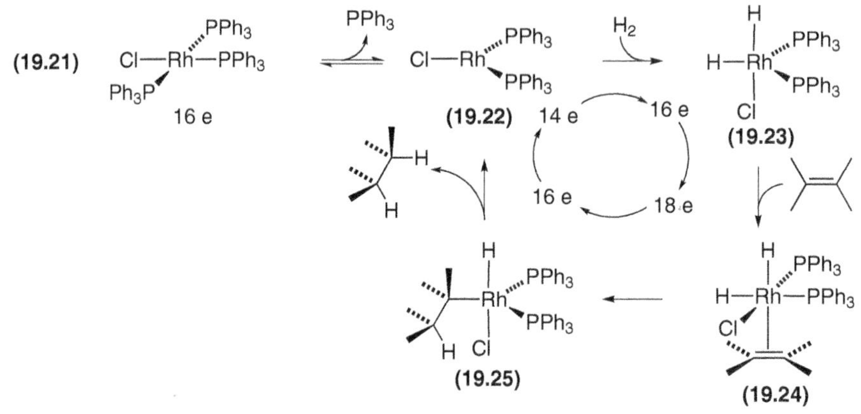

Figure 19.5 The proposed catalytic cycle for homogeneous hydrogenation over Wilkinson's catalyst

was the rhodium (I) catalyst, $[(C_6H_5)_3P]_3RhCl$, called ***Wilkinson's catalyst***[22] after its developer, Sir Geoffrey Wilkinson. Subsequently, it was found that ruthenium (II) catalysts also worked very well.[23]

Wilkinson's catalyst is a 16-electron rhodium complex (**19.21**) that loses one triphenylphosphine ligand under the reaction conditions to give a 14-electron complex, **19.22**, which is the active form of the catalyst during the reaction. The accepted mechanism for the reaction[24,25] is shown in Figure 19.5. It is worth noting that as the reaction proceeds, the geometry around the central rhodium atom becomes more sterically crowded (**19.22** [14 electrons] is trigonal planar, **19.23** and **19.25** [16 electrons] are trigonal bipyramidal, and **19.24** [18 electrons] is octahedral). This means that steric hindrance in the alkene should have a noticeable effect on the rate of the reaction. It does, and homogeneous catalysis is an excellent method for reducing the less hindered of two double bonds. Like heterogeneous catalytic hydrogenation, hydrogenation of alkenes over Wilkinson's catalyst involves suprafacial addition of the two hydrogen atoms to the less hindered face of the alkene.

Because the intermediates in a homogeneous catalytic hydrogenation are discrete molecules or ions, rather than regions of a surface, the geometry of the intermediates in the reaction can be much more easily understood and controlled. It is often found, for example, that both the regioselectivity and the stereoselectivity of hydrogenation of alkenes is much higher when a homogeneous catalyst is used in the reaction.

---

22. (a) Jardine, F.H.; Osborn, J.A.; Wilkinson, G.; Young, J.F. *Chem. Ind. (London)* **1965**, 560. (b) Osborn, J.A.; Jardine, F.H.; Young, J.F.; Wilkinson, G. *J. Chem. Soc., A* **1966**, 1711. (c) Bennett, M.A.; Longstaff, P.A. *Chem. Ind. (London)* **1965**, 846. (d) Osborn, J.A.; Wilkinson, G. *Inorg. Synth.* **1967**, *10*, 67. (e) Biellmann, J.F. *Bull. Soc. Chim. France* **1968**, 3055. (f) van Bekkum, H.; van Rantwijk, F.; van de Putte, T. *Tetrahedron Lett.* **1969**, 1.

Sir Geoffrey Wilkinson (1921–1996) took his PhD from Imperial College in 1946 after service in World War II. After postdoctoral work at Berkeley and the Massachusetts Institute of Technology, he joined Harvard (1951) and Imperial College (1955). Wilkinson shared the 1973 Nobel Prize in Chemistry. For more biographical detail, see: Green, M.L.H.; Griffith, W.P. *Biogr. Mem. Fell. Roy. Soc.* **2000**, *46*, 593, and the web site of the Nobel Foundation.

23. (a) Hallman, P.S.; McGarvey, B.R.; Wilkinson, G. *J. Chem. Soc., A*, **1968**, 3143. (b) Jardine, F.H.; McQuillin, F.J. *Tetrahedron Lett.* **1968**, 5189.

24. Reviews: (a) Crabtree, R.H. *Organometallic Chemistry of the Transition Metals* (Wiley: New York, 1988), p. 190. (b) Jardine, F.H. In Hartley, F.R. *The Chemistry of the Metal-Carbon Bond* (Wiley: New York, 1987), vol. 4, p. 1049.

25. (a) Collman, J.P. *Acc. Chem. Res.* **1968**, *1*, 136. (b) Vaska, L. *Acc. Chem. Res.* **1968**, *1*, 335. (c) Halpern, J. *Quart. Rev.* **1956**, *10*, 463. (d) Halpern, J. *J. Phys. Chem.* **1959**, *63*, 398. (e) Halpern, J. *Ann. Rev. Phys. Chem.* **1965**, *16*, 103. (e) Tolman, C.A.; Meakin, P.Z.; Lindner, D.L.; Jesson, J.P. *J. Am. Chem. Soc.* **1976**, *96*, 2762.

In 1976, Crabtree published the first in a series of papers describing iridium catalysts for the homogeneous catalytic hydrogenation of unactivated alkenes.[26] Crabtree's catalyst, (cod)Ir(PCy_3)py•PF_6, has proved to be an extremely efficient catalyst for the hydrogenation of tetrasubstituted alkenes, compounds which are otherwise difficult to hydrogenate. The catalyst has also been highly useful in directed hydrogenations (e.g., Example **19.26**,[27] where an allylic hydroxyl group directs the reaction to the same face of the adjacent π bond, similar to the way in which an allylic hydroxyl group directs a peracid to the adjacent π bond). This has been taken advantage of in the synthesis of taxanes (e.g., Example **19.27**).[28]

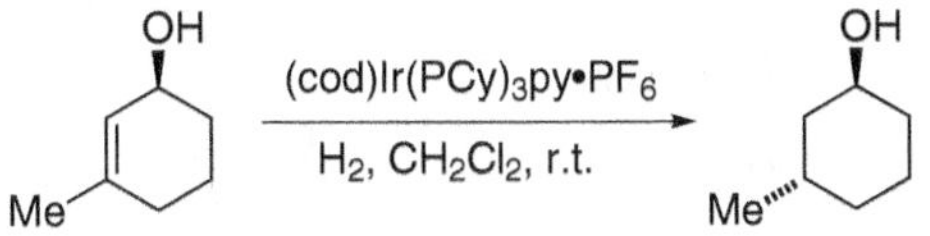

(19.26)[27]

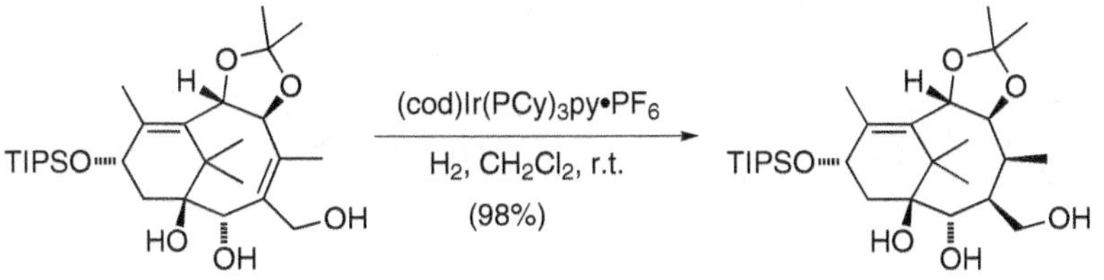

(19.27)[28]

Reaction Synopses

Catalytic Hydrogenation

[H]:	H_2; NH_4OCHO, HCO_2H; cyclohexene; R_2CHOH; etc.
Catalyst:	Ni; PtO_2; Pd; Os; Ru; etc; heterogeneous catalysts may be powders or supported metals (on C, $BaSO_4$, $CaCO_3$, etc.) $(Ph_3P)_3RhCl$; etc.
Stereochemistry:	*syn* (suprafacial)

Worked Problem

19-1 What will the major organic product of the following reaction be? [*J. Am. Chem. Soc.* **1980**, *102*, 889]

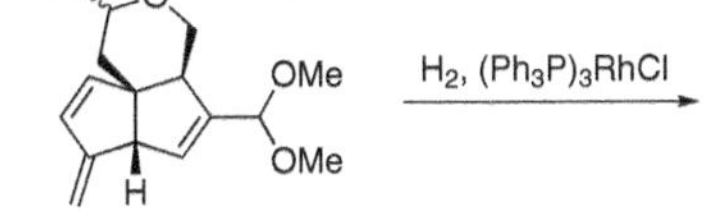

§Answer on next page.

---

26. (a) Crabtree, R.H.; Felkin, H.; Morris, G.E.; King, T.J.; Richards, J.A. *J. Organometal. Chem.* **1976**, *113*, C7. (b) Crabtree, R.H.; Morris, G.E. *J. Organometal. Chem.* **1977**, *135*, 395. (c) Crabtree, R.H.; Felkin, H.; Morris, G.E. *J. Organometal. Chem.* **1977**, *141*, 205. (d) Crabtree, R.H. *Acc. Chem. Res.* **1979**, *12*, 331. (d) Crabtree, R.H.; Davis, M.W. *J. Org. Chem.* **1986**, *51*, 2655.

27. Evans, D.A.; Morrissey, M.M. *Tetrahedron Lett.* **1984**, *25*, 4637.

28. Wender, P.A.; Badham, N.F.; Conway, S.P.; Floreancig, P.E.; Glass, T.E.; Gränicher, C.; Houze, J.B.; Jänichen, J.; Lee, D.; Marquess, D.G.; McGrane, P.L.; Meng, W.; Mucciaro, T.P.; Mühlebach, M.; Natchus, M.G.; Paulsen, H.; Rawlins, D.B.; Satkofsky, J.; Shuker, A.J.; Sutton, J.C.; Taylor, R.E.; Tomooka, K. *J. Am. Chem. Soc.* **1997**, *119*, 2755.

Problem

19-1 What is the major organic product of each of the following reactions?

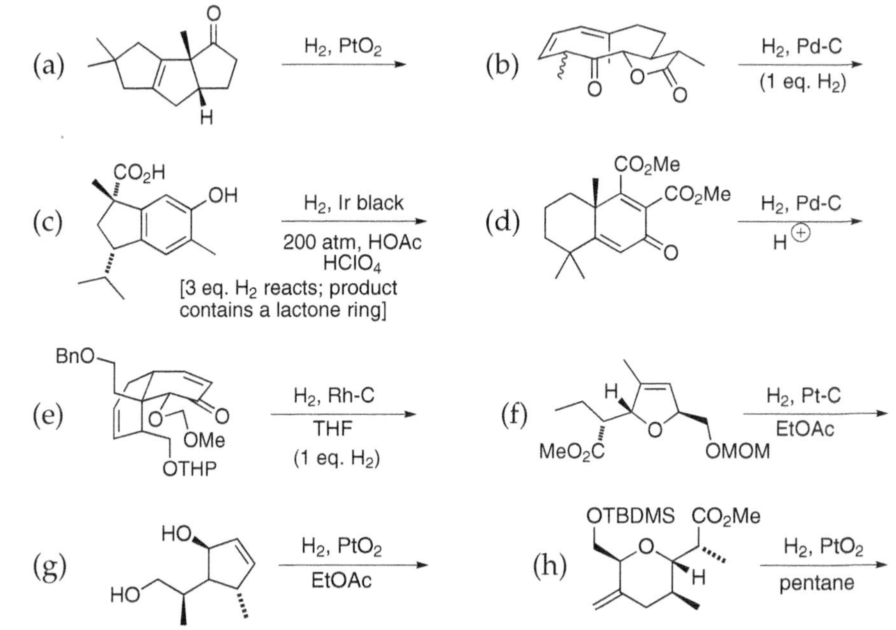

References for problems: (a) *J. Am. Chem. Soc.* **1980**, *102*, 6351. (b) *Tetrahedron Lett.* **1976**, 2041. (c) *J. Am. Chem. Soc.* **1979**, *101*, 1608. (d) *J. Am. Chem. Soc.* **1979**, *101*, 4398. (e) *J. Am. Chem. Soc.* **1978**, *100*, 8031. (f) *J. Am. Chem. Soc.* **1983**, *105*, 1988. (g) *J. Am. Chem. Soc.* **1979**, *101*, 4749. (h) *J. Org. Chem.* **1981**, *46*, 479.

Hydrogenation of Alkynes

One of the most important catalytic hydrogenations of functional groups other than alkenes is the hydrogenation of alkynes to alkenes over a palladium catalyst poisoned with lead and quinoline ("Lindlar" catalysts).[29] This reaction owes its importance to the fact that the alkene produced is almost entirely the *Z* isomer. This becomes even more important when one considers that the asymmetric epoxidation by either the Jacobsen-Katsuki[30]

29. (a) Lindlar, H. *Helv. Chim. Acta* **1952**, *35*, 446. (b) Lindlar, H.; Dubuis, R. *Org. Syn.* **1966**, *46*, 89.
30. Review: Matsumoto, K.; Katsuki, T. In Ojima, I., Ed. *Catalytic Asymmetric Synthesis*, 3rd ed. (John Wiley & Sons: New York, 2010), ch. 11, p. 839.

§ Answer to Worked Problem:
The homogeneous rhodium catalyst will reduce the least hindered double bond first, from the most accessible face. The perspective representation of this molecule, based on an examination of molecular models, reveals that the exocyclic methylene group is the least hindered and that the *exo* face of the bicyclo[3.3.0]octane ring system will be the less hindered in terms of being able to bind the metal. Thus, we predict that the major product will be as shown below:

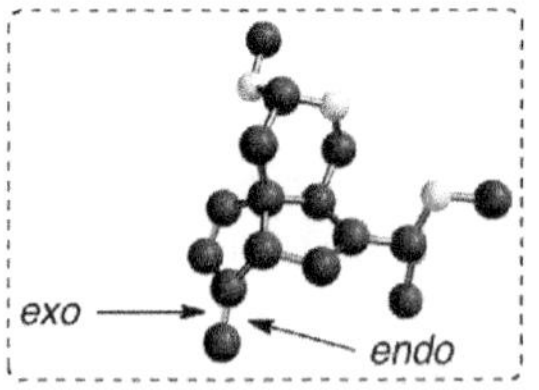

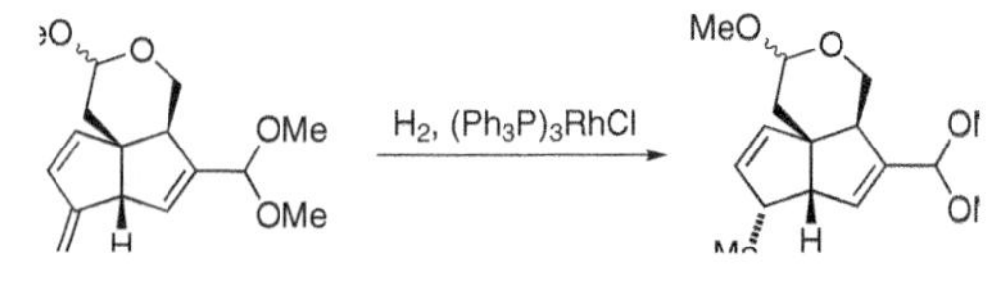

or Shi[31] catalytic methods, which we will discuss in the next chapter, works best with *cis* alkenes.

The classical Lindlar catalyst is made by adding a lead salt (the acetate, sulfate, oxide, and carbonate have all been used) to palladium supported on calcium carbonate or barium sulfate. Palladium precipitated on a support in the presence of quinoline has been reported to be superior to the original Lindlar catalyst in terms of preparation, stereoselectivity, and reproducibility (e.g., Example **19.28**). Pre-prepared Lindlar catalysts are now commercially available (e.g., Example **19.29**). In cases where the Lindlar catalyst is ineffective, P2-nickel, prepared by reduction of nickel (II) acetate by sodium borohydride in ethanol under an atmosphere of hydrogen, in the presence of ethylenediamine as a catalyst poison has provided a useful alternative.[32] Recently, a "green" alternative to the lead-based Lindlar catalyst for the same hydrogenation of alkynes to *Z*-alkenes has been reported.[33] In this procedure, the catalyst is a palladium-polyethyleneimine (PEI) complex (Example **19.30**).

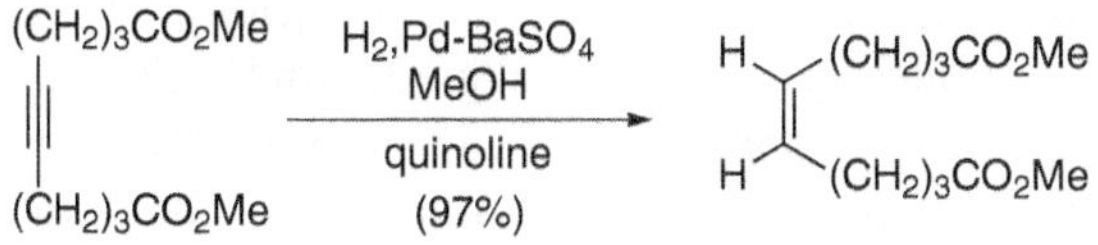

(19.28)[34]

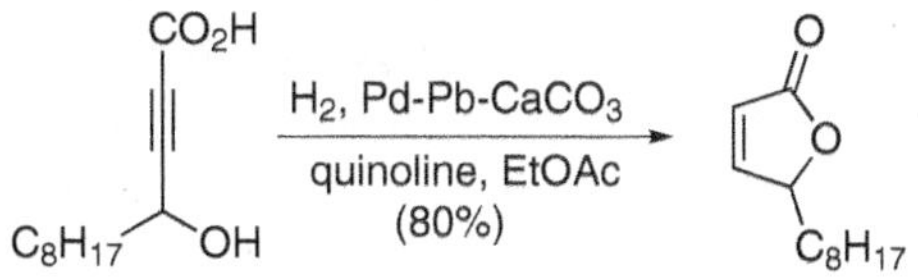

(19.29)[35]

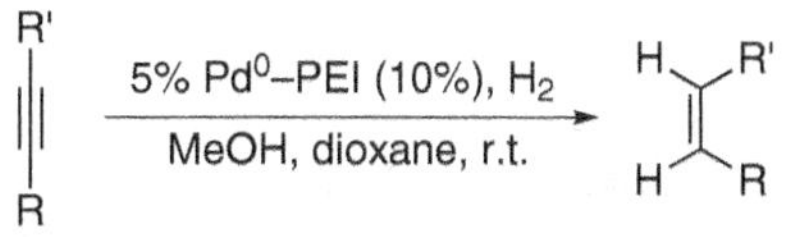

(19.30)

Problem

19-2 Give the major organic product of each of the reactions below.

(a) 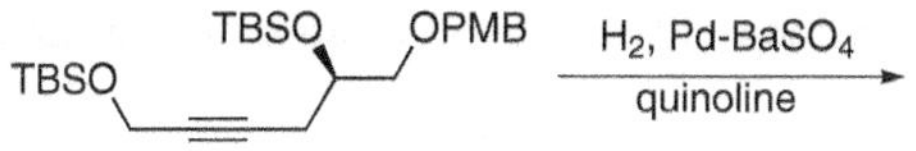

(b) 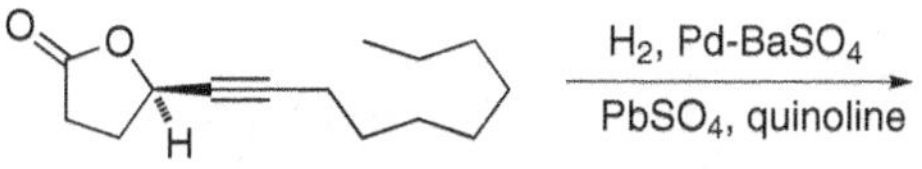

(c) 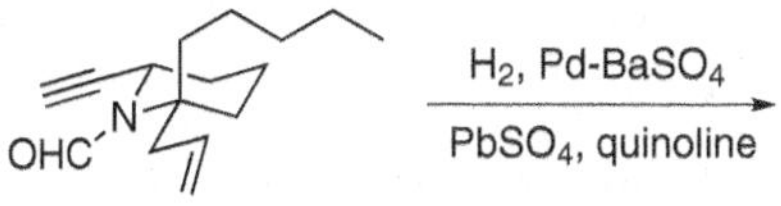

(d) 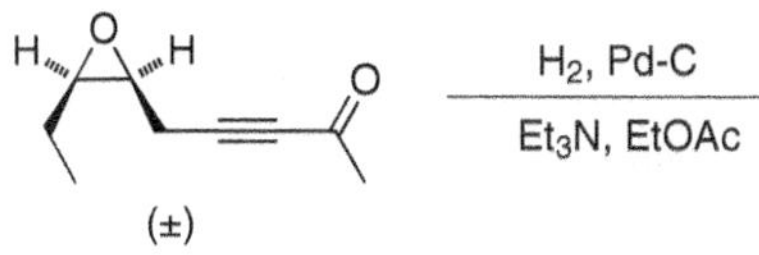

(continues)

---

31. Review: Shi, Y. *Acc. Chem. Res.* **2004**, *37*, 488.
32. Ganame, D.; Quach, T.; Poole, C.; Rizzacasa, M.A. *Tetrahedron Lett.* **2007**, *48*, 5481.
33. Sajiki, H.; Mori, S.; Ohkubo, T.; Ikawa, T.; Kume, A.; Maegawa, T.; Monguchi, Y. *Chem. Eur. J.* **2010**, *14*, 5109.
34. Cram, D.J.; Allinger, N.L. *J. Am. Chem. Soc.* **1956**, *78*, 2518.
35. Jakubowski, A.A.; Guziec, F.S., Jr.; Sugiura, M.; Tam, C.C.; Tishler, M.; Omura, S. *J. Org. Chem.* **1982**, *47*, 1221.

(**Problem** continued)

(e) 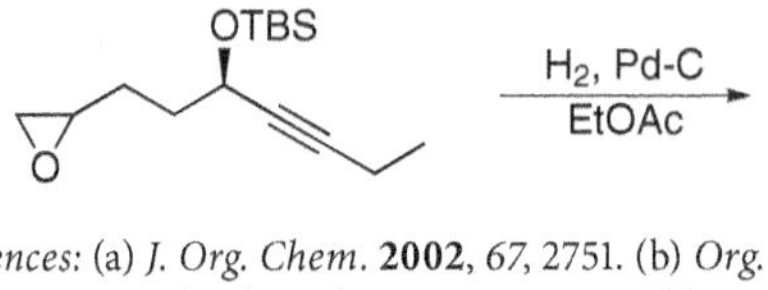

References: (a) *J. Org. Chem.* **2002**, *67*, 2751. (b) *Org. Lett.* **2002**, *4*, 119. (c) *Org. Lett.* **2002**, *4*, 2469. (d) *Biosci. Biotechnol Biochem.* **2011**, *75*, 976. (e) *Org. Lett.* **2004**, *6*, 1493.

Hydrogenation of Carbonyl Compounds

In 1969, McAlees and McCrindle established that the ease of hydrogenation of carboxylic acid derivatives followed the order: anhydrides > esters > carboxylic acids > amides.[36] The reduction of amides by hydrogen is especially difficult. It almost always requires high temperatures and high pressures,[37] and it seldom gives good yields of the desired products. Therefore, the reduction of amides is still carried out predominantly by means of hydride-reducing agents ($LiAlH_4$ or B_2H_6), despite the replacement of such reagents being one of the priorities for "greening" of pharmaceutical manufacture.[38] Some of the problems associated with direct hydrogenation of amides have been overcome by using silanes as hydrogen donors in metal-catalyzed reductions.[39] In general, however, the catalytic hydrogenation of carbonyl groups is usually more conveniently carried out using complex metal hydride reducing agents. Thus, the hydrogenation of carbonyl compounds is not often a useful synthetic method.

The one exception to this generalization is the ***Rosenmund reduction*** of acid chlorides to aldehydes over a modified supported palladium catalyst (e.g., Example **19.31**).[40] In the presence of a number of modifiers containing sulfur (thiourea, tetramethylthiourea, and sulfur-quinoline), acid chloride groups can be reduced selectively in the presence of alkene π bonds.[41] A more modern version of this reaction (Example **19.32**) involves the hydrogenation of esters of hydroxytriazine. The reagent used in this reaction reacts with the carboxylic acid by an addition-elimination mechanism (can you write a mechanism for this reaction?) to give the nucleophilic aryl substitution product. Aromatic acids in this modification of the reaction tend to give over-reduction to the alcohol, but the reaction can be used to reduce Boc-protected α-amino acids to the corresponding aldehydes without epimerization.[42]

36. McAlees, A.J.; McCrindle, R. *J. Chem. Soc. C* **1969**, 2425.

37. (a) Wojcik, B.; Adkins, H. *J. Am. Chem. Soc.* **1934**, *56*, 247. (b) Paden, J. H.; Adkins, H. *J. Am. Chem. Soc.* **1936**, *58*, 2487. (c) Sauer, J. C.; Adkins, H. *.I. Am. Chem. Soc.* **1938**, *60*, 402. (d) D'Ianni, J. D.; Adkins, H. *J. Am. Chem. Soc.* **1939**, *61*, 1675. (e) Broadbent, H.S.; Bartley, W J. *J. Org. Chem.* **1963**, *28*, 2345. (f) Guyer, A.; Bieler, A.; Gerliczy, G. *Helv. Chim. Acta* **1955**, *38*, 1649. (g) Galinovsky, F.; Stem, E. *Chem. Ber.* **1943**, *76*, 1034.

Even modern bimetallic catalysts that give good yields of the desired amine products require temperatures above 150°C and pressures of 100 atm: Hirosawa, C.; Wakasa, N.; Fuchikami, T. *Tetrahedron Lett.* **1996**, *37*, 6749.

38. Constable, D.J.C.; Dunn, P.J.; Hayler, J.D.; Humphrey, G.R.; Leazer, J.L., Jr.; Linderman, R.J.; Lorenz, K.; Manley, J.; Pearlman, B.A.; Wells, A.; Zaks, A.; Zhang, T.Y. *Green Chem.* **2007**, *9*, 411.

39. Igarashi, M.; Fuchikami , T. *Tetrahedron Lett.* **2001**, *42*, 1945.

40. (a) Rosenmund, K.W. *Ber. dtsch. chem. Ges.* **1918**, *51*, 585. (b) Rosenmund, K.W.; Zetzsche, F. *Chem. Ber.* **1921**, *54*, 425. (c) Mosettig, E.; Mozingo, R. *Org. React.* **1948**, *4*, 362.

Karl Wilhelm Rosenmund (1884–1965) took his PhD under Otto Diels at Berlin in 1906. He was then appointed as Assistant in the Pharmacy Institute at Berlin. In 1925, he joined the University of Kiel as Professor of Chemistry and Director of the Pharmacy Institute. Following his retirement, he returned to Berlin as an unofficial Extraordinary Professor of Chemistry.

41. (a) Weygand, C.; Meusel, W. *Chem. Ber.* **1943**, *76*, 503. (b) Affrossman, S.; Thomson, S.J. *J. Chem. Soc.* **1962**, 2024.

42. Falorni, M.; Giacomelli, G.; Porcheddu, A.; Taddei, M. *J. Org. Chem.* **1999**, *64*, 8962.

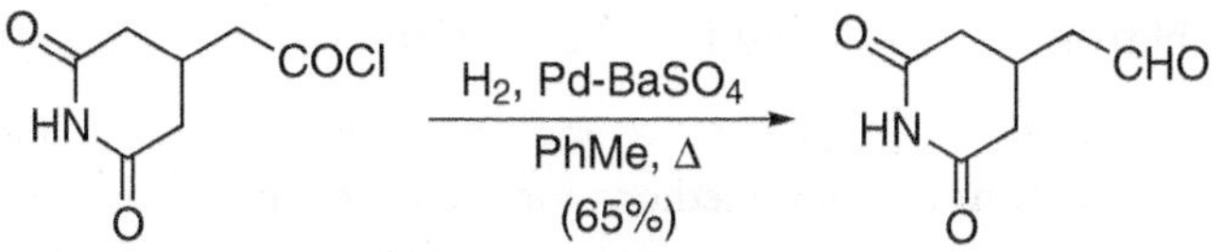

(19.31)

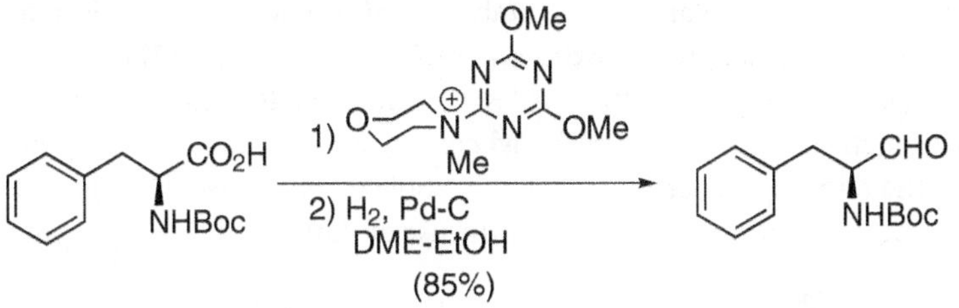

(19.32)

The major problem with the Rosenmund reduction is that the results tend to be highly variable, with the catalyst and the catalyst "poison" both having major effects on the reaction outcome. One rationalization for this irreproducibility was the fact that hydrogen chloride is produced in the reaction. In the 1970s, Burgstahler[43] and Van Bekkum[44] reported that tertiary amines could be used as modifiers in the Rosenmund reduction (e.g., Example **19.33**), thus eliminating the need for the quinoline-sulfur "poison." Van Bekkum found, also, that this would permit hydrogenation of acid chlorides in the presence of aromatic nitro groups (e.g., Example **19.34**). More recently, the deactivation of palladium on carbon by prolonged heating in xylene at 140°C under an atmosphere of hydrogen has been shown to be sufficient to produce a good catalyst for the Rosenmund reduction (e.g., Example **19.35**).[45] The one caveat with using the deactivated palladium on carbon catalyst is that the pressure of hydrogen must be kept high enough to prevent decarbonylation of the aldehyde product.

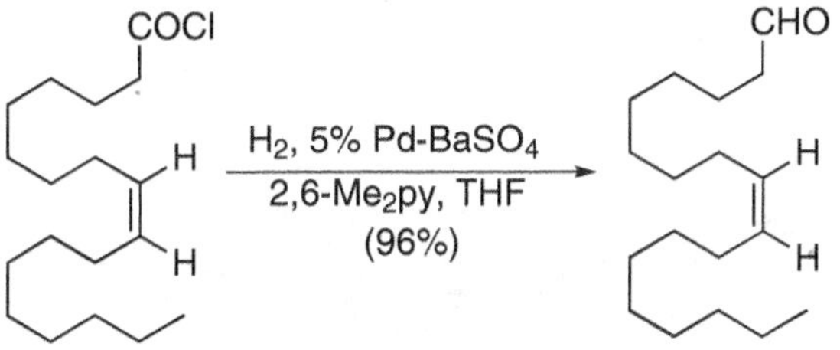

(19.33)[66]

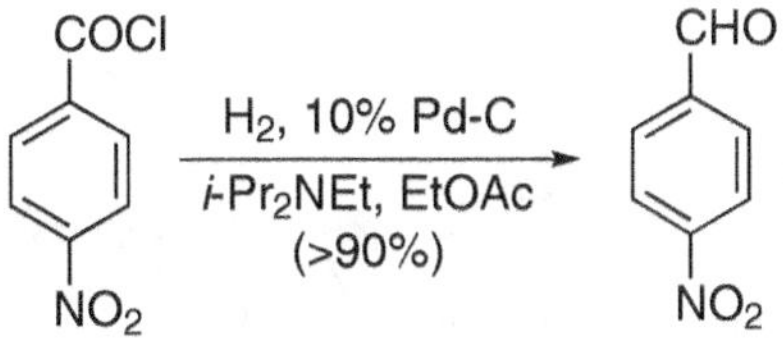

(19.34)[67]

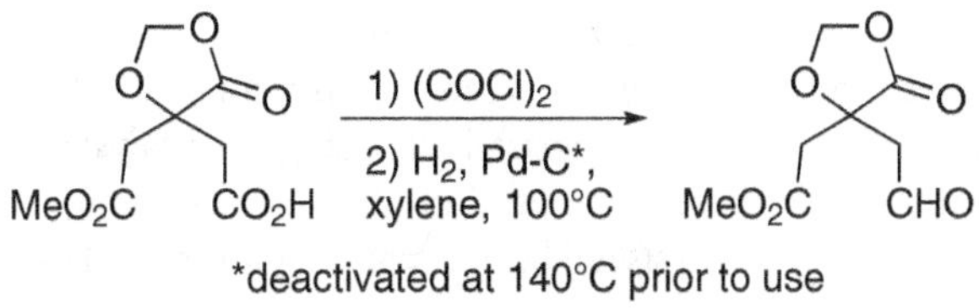

(19.35)[68]

43. Burgstahler, A.W.; Weigel, L.O.; Schaefer, C.G. *Synthesis* **1976**, 767.
44. Peters, J.A.; Van Bekkum, H. *Rec. Trav. Chim. Pays-Bas* **1971**, *90*, 1323.
45. Ancliff, R.A.; Russell, A.T.; Sanderson, A.J. *Chem. Commun.* **2006**, 3243.

Hydrogenation of Nitrogen-Containing Functional Groups

Amines are almost universally biologically active and occupy by far a dominant position among pharmaceuticals. This means that methods for the formation of amines assume a high level of economic importance, and techniques for doing so have thus been the subject of intensive research. The synthesis of amines by catalytic hydrogenation can be effected starting from five different functional groups: oximes (Examples **19.36** and **19.37**), nitro groups (Example **19.38**), azides (Example **19.39**), nitriles (Example **19.40**), and imines (usually formed in situ in reductive aminations of carbonyl compounds). In all five cases, poisoning of the catalyst by the amine product is a potential problem that is usually suppressed by carrying out the reduction in acidic solvents or by the use of Raney nickel as the catalyst.

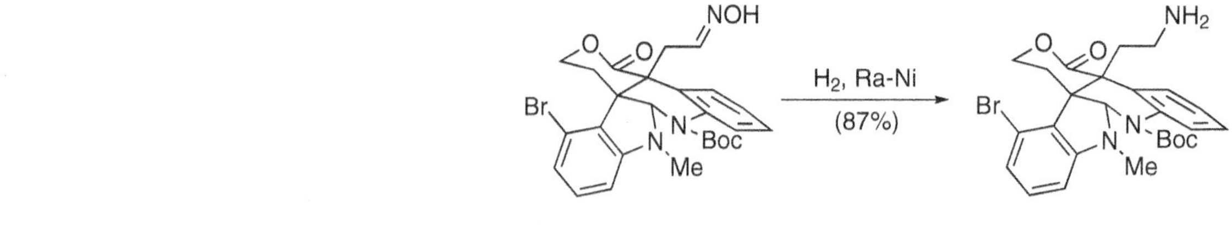

(19.36)[46]

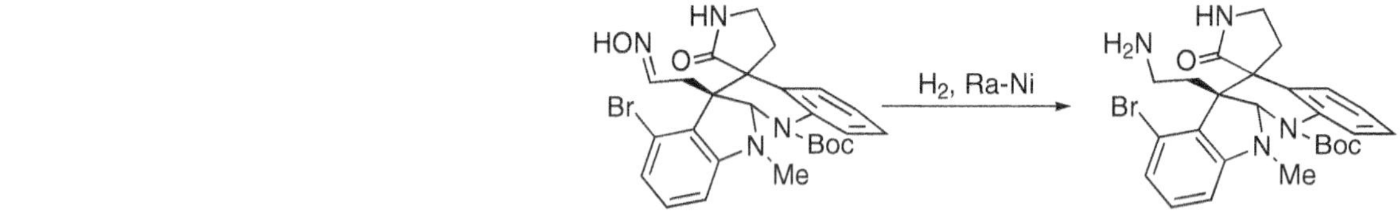

(19.37)

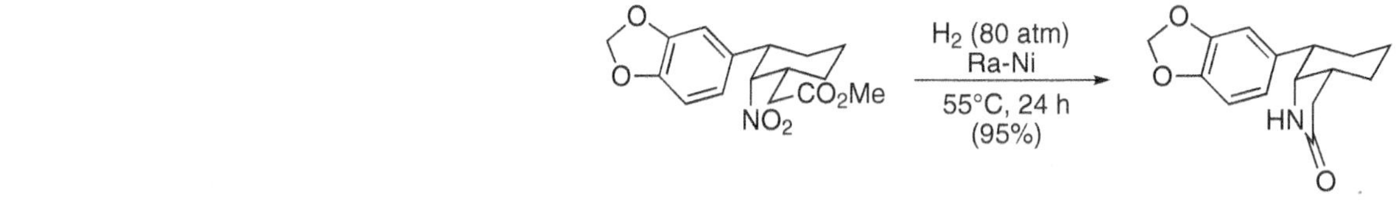

(19.38)[47]

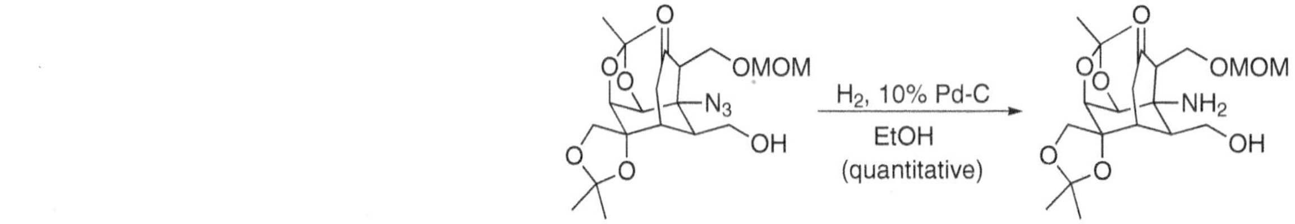

(19.39)[48]

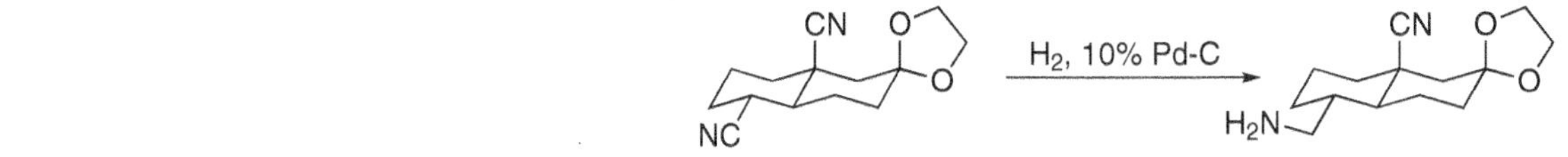

(19.40)[49]

The hydrogenation of nitriles is complicated by the propensity of the primary amine to react with the nitrile to give products that are then reduced to secondary or tertiary amines. This type of activity can be modulated by appropriately choosing the catalyst. Ru and Ni catalysts tend to favor the formation of primary amines, Cu and Rh catalysts tend to favor secondary amine formation, and Pd and Pt catalysts tend to favor tertiary amines.[50] Hydrogenation of

46. Yang, J.; Wu, H.; Shen, L.; Qin, Y. *J. Am. Chem. Soc.* **2007**, *129*, 13794.
47. Dong, L.; Xu, Y.-J.; Cun, L.-F.; Cui, X.; Mi, A.-Q.; Jiang, Y.-Z.; Gong, L.-Z. *Org. Lett.* **2005**, *7*, 4285.
48. Sato, K.; Akai, S.; Shoji, H.; Sugita, N.; Yoshida, S.; Nagai, Y.; Suzuki, K.; Nakamura, Y.; Kajihara, Y.; Funabashi, M.; Yoshimura, J. *J. Org. Chem.* **2008**, *73*, 1234.
49. Minato, H.; Nagasaki, T. *Chem. Comm.* **1965**, 377.
50. (a) Volf, J.; Pašek, J. In Červený, L., Ed. *Catalytic Hydrogenation*; (Elsevier: Amsterdam, 1986) p. 105. (b) Barrault, J.; Pouilloux, Y. *Catal. Today* **1997**, *37*, 137. (c) Huang, Y.; Sachtler, W.M.H. *Appl. Catal. A* **1999**, *182*, 365.

acetonitrile over heterogeneous Mo/P catalysts has shown that the Mo/P ratio can dramatically affect the outcome of the reduction: high Mo/P surface ratios favor ethylamine, and low Mo/P surface ratios favor secondary and tertiary amines.[51] Homogeneous catalysis of the hydrogenation of nitriles usually leads to secondary amines as the major products.[52] This can be changed by carrying out the hydrogenation in CO_2-expanded solvents.[53]

Worked Problem

19-2 What will the major organic product of the following reaction be? [*Tetrahedron Lett.* **2009**, *50*, 3654]

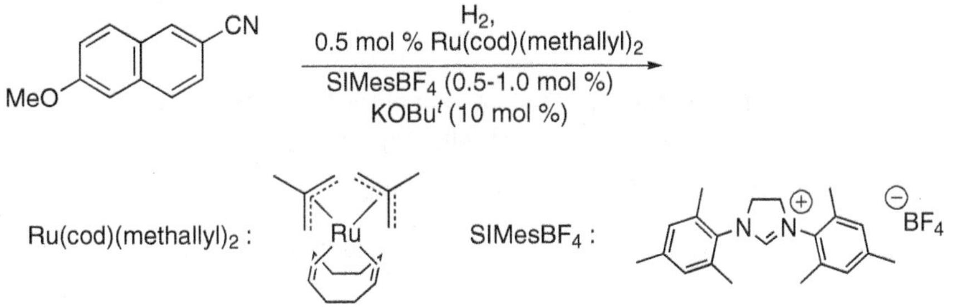

§Answer below.

Reaction Synopses

Catalytic Hydrogenation of Alkynes

(a) To saturated systems

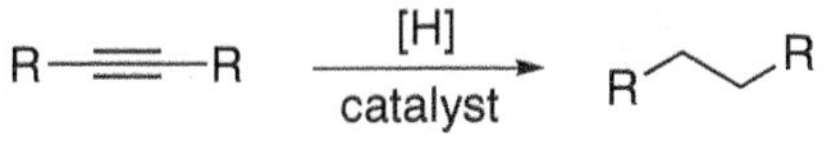

Reagents: see hydrogenation of alkenes

(b) To Z-alkenes

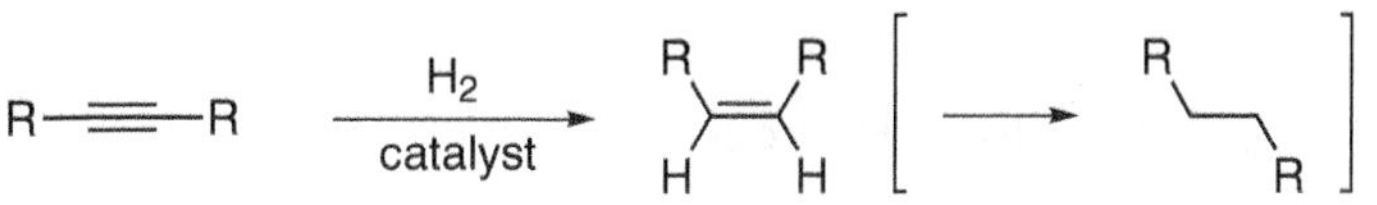

Catalyst: Pd-$CaCO_3$, $PbCO_3$; Pd-$CaCO_3$, $PbSO_4$; Pd-Al_2O_3, $PbSO_4$; etc. H_2, Pd-$BaSO_4$, $PbSO_4$, quinoline (Lindlar Pd); H_2, Pd-$CaCO_3$, $PbSO_4$, quinoline (Lindlar Pd); Pd-PEI, MeOH, dioxane: a "green" alternative

(continues)

51. Yang, P.; Jiang, Z.; Ying, P.; Li, C. *J. Catal.* **2008**, *253*, 66.
52. De Bellefon, C.; Fouilloux, P. *Catal. Rev.-Sci. Eng.* **1994**, *36*, 459.
53. Xie, X.; Liotta, C.L.; Eckert, C.A. *Ind. Eng. Chem. Res.* **2004**, *43*, 7907.

§ Answer to Worked Problem:
The homogeneous catalyst is based on ruthenium, so we predict that the reaction will lead to formation of the primary amine. Thus, we predict that the major product will be as shown below:

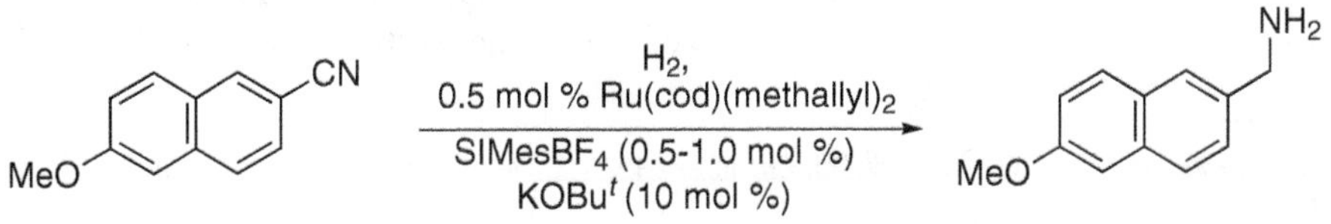

(**Reaction Synopses** continued)

Rosenmund Reduction

Reagent: X=Cl, H_2, Pd-$BaSO_4$, S, quinoline; H_2, Pd-C, R_3N; H_2, Pd-C*; etc.
X=triazine ester, H_2, Pd-$BaSO_4$; etc.

Problem

19-3 Give the major organic product of each of the following reactions or reaction sequences. Specify stereochemistry where appropriate.

(a)

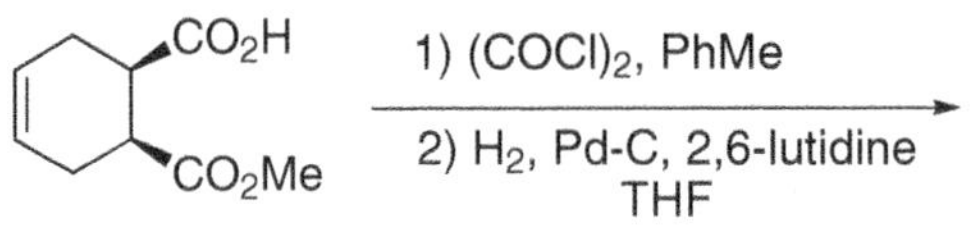

(b)

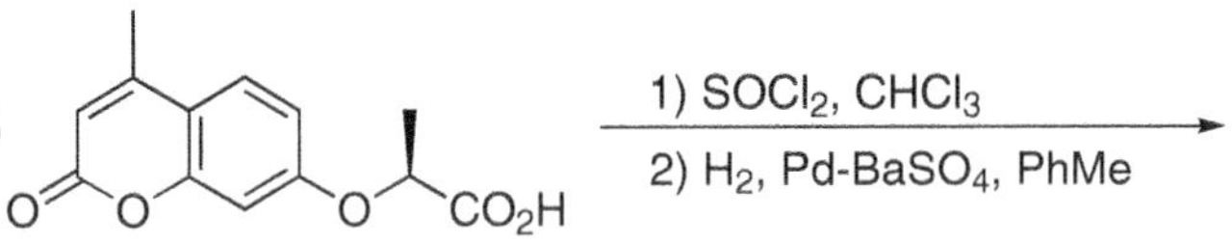

(c)

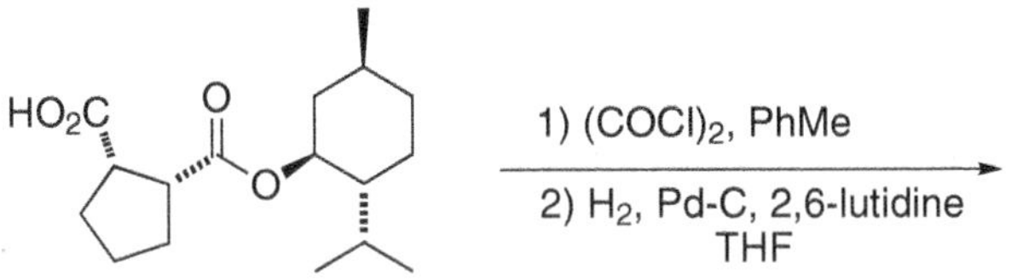

References: (a) *Tetrahedron Lett.* **1993**, *34*, 2593. (b) *Tetrahedron Lett.* **2006**, *47*, 2405. (c) *J. Am. Chem. Soc.* **2003**, *125*, 1458.

Transfer Hydrogenation, Ionic Hydrogenation, and Organocatalysis

Hydrogenation catalyzed by transition metals is a cornerstone of organic synthesis. However, there are problems, such as the cost of the catalysts involved, their occasional toxicity and air sensitivity, and the dangers inherent in handling hydrogen at high pressures. These have prompted efforts to develop methods for the reduction of carbon-carbon and carbon-heteroatom π bonds that do not require a transition metal catalyst and/or molecular hydrogen. Reactions that do not involve a transition metal catalyst are now often described using the term ***organocatalytic***, meaning that the catalysis of the reaction is provided by a non-metal-containing molecule or ion. Most organocatalytic reactions involve proton transfer as one of the steps.

Transfer Hydrogenation[54]

The first technique that we will discuss as a replacement for traditional catalytic hydrogenation is ***transfer hydrogenation***. In this technique, a molecule such as an ammonium salt of formic acid, cyclohexene, 1,4-cyclohexadiene, or isopropyl alcohol serves as the source of hydrogen for the reduction. It is itself oxidized during the reaction (formic acid derivatives

---

54. Reviews: (a) Brieger, G.; Nestrick, T.J. *Chem. Rev.* **1974**, *74*, 567. (b) Ram, S.; Ehrenkaufer, R.E. *Synthesis* **1988**, 91.

to carbon dioxide, isopropyl alcohol to acetone, and cyclohexene and cyclohexadiene to benzene). The elimination of molecular hydrogen from the system has made this a popular reaction in process chemistry, where reactions are often carried out on the ton scale. The reaction has been widely used for the reduction of nitro compounds, which are reduced with ammonium formate over a number of supported metals (Examples **19.41** to **19.44**).

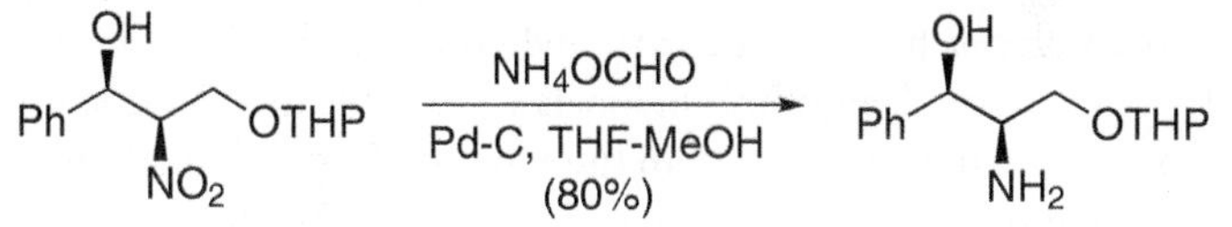

(19.41)[55]

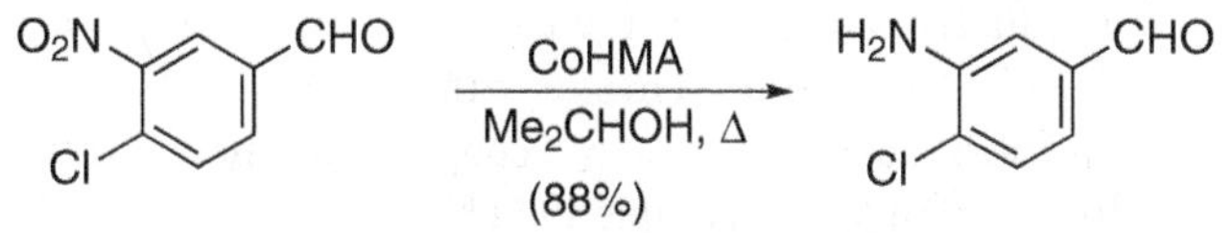

(19.42)[56]

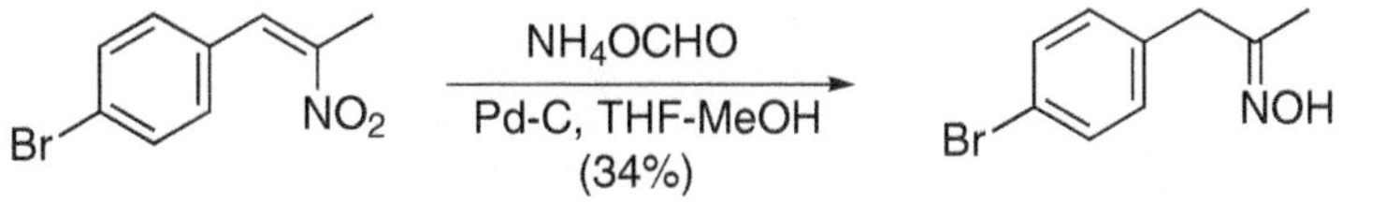

(19.43)[57]

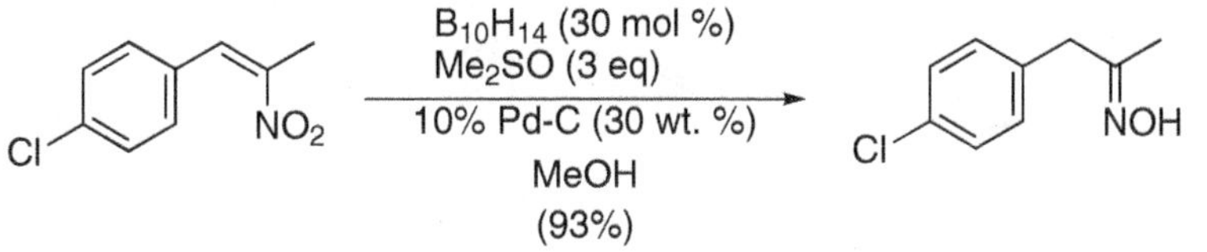

(19.44)[58]

The reduction occurs with retention of configuration,[59] which suggests that the substrate is not released from the catalyst until the reduction is complete. Conjugated nitroalkenes are reduced using the same method to oximes (presumably formed from the corresponding saturated nitroso compounds).[60] The effects of the support for certain metals can be quite striking. Using mesoporous aluminosilicate molecular sieves, for example, leads to selectivity that allows the chemoselective transfer hydrogenation of nitro groups, aldehydes, and halides.[61] The activity of the ruthenium catalyst in Example **19.45** shows a fairly strong dependence on pH, with an optimum pH of 4.[62]

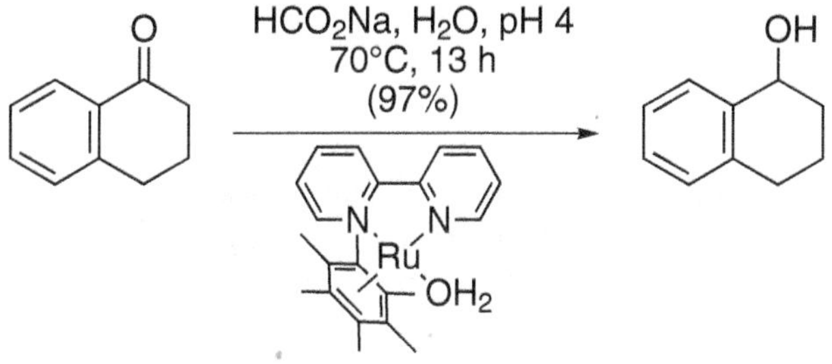

(19.45)

55. (a) Ram, S.; Ehrenkaufer, R.E, *Tetrahedron Lett.* **1984**, *25*, 3415. (b) Ram, S.; Ehrenkaufer, R.E. *Synthesis* **1986**, 133. (c) Barrett, A.G.M.; Spilling, C.D. *Tetrahedron Lett.* **1988**, *29*, 5733.

56. Mohapatra, S.K.; Sonavane, S.U.; Jayaram, R.V.; Selvam, P. *Tetrahedron Lett.* **2002**, *43*, 8527; *Tetrahedron Lett.* **2003**, *44*, 1107.

57. Kabalka, G.W.; Goudgaon, N.M. *Synth. Commun.* **1988**, *18*, 693.

58. Lee, S.H.; Park , Y.J.; Yoon, C.M. *Org. Biomol. Chem.* **2003**, *1*, 1099.

59. Barrett, A.G.M.; Spilling, C.D. *Tetrahedron Lett.* **1988**, *29*, 5733.

60. (a) Varma, R.S.; Varma, M.; Kabalka, G.W. *Synth. Commun.* **1986**, *16*, 91. (b) Varma, R.S.; Varma, M.; Kabalka, G.W. *Synth. Commun.* **1985**, *15*, 1325.

61. (a) Mohapatra, S.K.; Sonavane, S.U.; Jayaram, R.V.; Selvam, P. *Tetrahedron Lett.* **2002**, *43*, 8527; *Tetrahedron Lett.* **2003**, *44*, 1107 (Co). (b) Selvam, P.; Mohapatra, S.K.; Sonavane, S.U.; Jayaram, R.V. *Tetrahedron Lett.* **2004**, *45*, 2003 (Ni). (c) Selvam, P.; Sonavane, S.U.; Mohapatra, S.K.; Jayaram, R.V. *Tetrahedron Lett.* **2004**, *45*, 3071 (Pd).

62. Ogo, S.; Abura, T.; Watanabe, Y. *Organometallics* **2002**, *21*, 2964.

Reductive Amination of Aldehydes and Ketones

The synthesis of amines from aldehydes and ketones by treatment with a nitrogen nucleophile and a reducing agent is known as ***reductive amination***.[63] The condensation between the aldehyde or ketone and ammonia or a primary amine gives an imine (Example **19.47**) as the major product, and the condensation with a secondary amine gives an enamine (Example **19.48**). Like the polar π bond of the carbonyl compounds, the polar C=N π bond of an imine or its conjugate acid, an iminium ion, is susceptible to reduction by complex metal hydrides. However, being less polar than the π bond of the carbonyl group, it is also susceptible to hydrogenation. Both reduction pathways lead to the same final product. In modern variants of the reductive amination, the carbonyl compound is either converted to the iminium ion, which is then reduced in situ with a hydride-reducing agent (we will discuss this reaction in more detail later in this chapter), or the neutral imine or enamine is reduced by hydrogenation (e.g., Example **19.50**[64]).

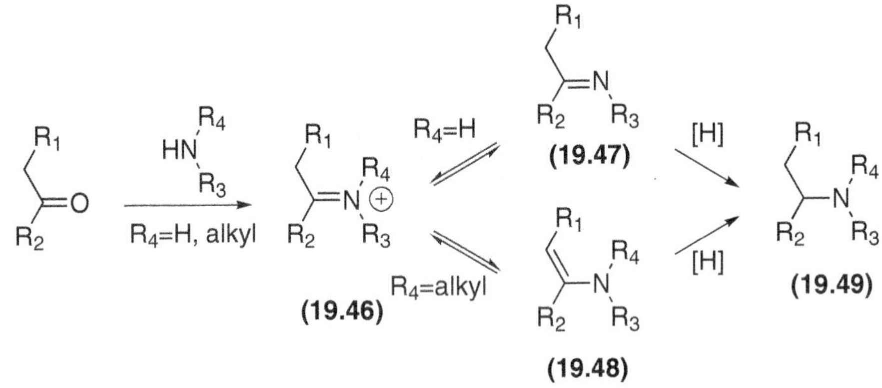

(19.46) **(19.47)** **(19.48)** **(19.49)**

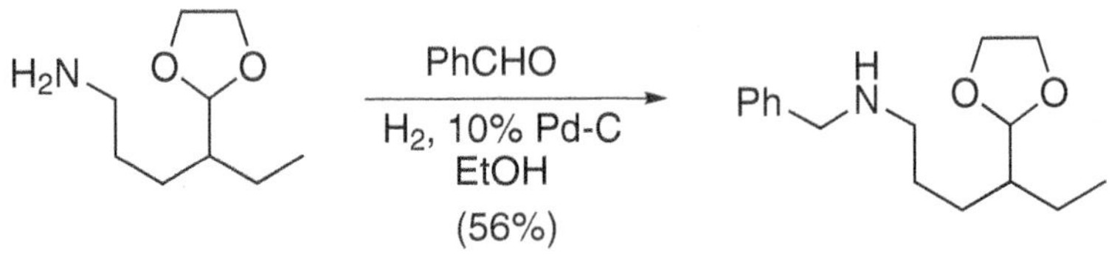

(19.50)

Ionic Hydrogenation[65]

Ionic hydrogenation is defined broadly as the reduction of a π bond by means of a complementary pair of reagents comprising a proton donor and a hydride ion donor (Example **19.51**). The two parts of the reagent may be two separate molecules, or they may be part of the same molecule. In general, the key task in an ionic hydrogenation is to identify the hydride donor; the proton donor is almost always a strong Brønsted acid.

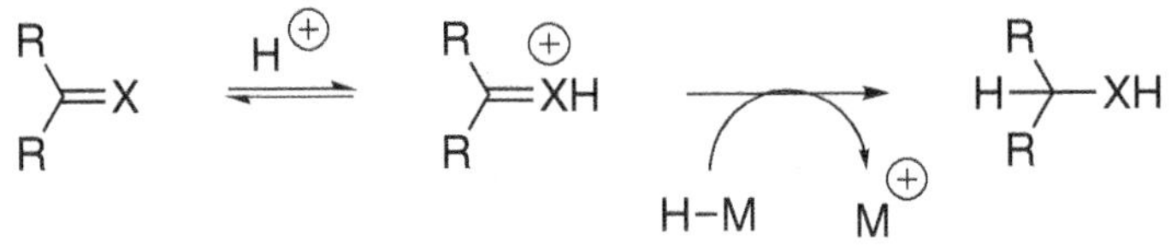

(19.51)

X = CR_2, NR, O   M = Si, transition metals

---

63. Reviews: (a) Kobayashi, S.; Ishitani, H. *Chem. Rev.* **1999**, *99*, 1069. (b) Shimizu, H.; Nagasaki, I.; Matsumura, K.; Sayo, N.; Saito, T, *Acc. Chem. Res.* **2007**, *40*, 1385. (c) Tararov, V.I.; Kadyrov, R.; Riermeier, T.H.; Dingerdissen, U.; Boerner, A. *Org. Prep. Proced. Int.* **2004**, *36*, 99. (b) Tararov, V.I.; Boerner, A. *Synlett* **2005**, 203.

64. Ziegler, F.E.; Kloek, J.A.; Zoretic, P.A. *J. Am. Chem. Soc.* **1969**, *91*, 2342.

65. Review: Kursanov, D.N.; Parnes, Z.N.; Loim, N.M. *Synthesis* **1974**, 633.

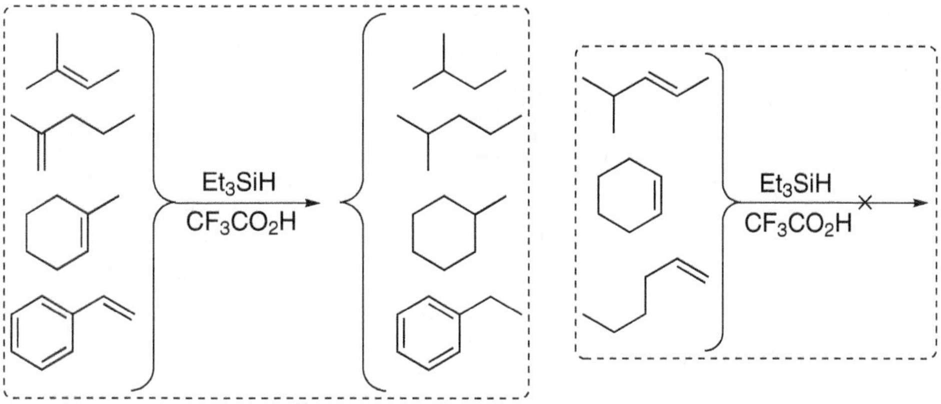

Figure 19.6 Ionic hydrogenation of representative alkenes shows that only those alkenes that can give relatively unreactive carbocations will give hydrogenation products.

Silanes were the first hydride donors used in ionic hydrogenations, and this early work has since been expanded to involve hydride complexes of a wide range of transition metals. The common feature of these reactions is that there is an initial protonation of the π bond followed by hydride transfer from the metal hydride. The treatment of an alkene with a silane in trifluoroacetic acid leads to reduction of the alkene if the double bond can be protonated to give a relatively stable carbocation. Alkenes that give tertiary or benzyl cations on protonation are reduced with triethylsilane and trifluoroacetic acid, but alkenes that give secondary carbocations are not (Figure 19.6)[66] The reaction can be used to reduce carbonyl groups and the alkene π bond of conjugated carbonyl compounds, but it is also a useful method for the reduction of (1) hemiacetals to ethers and (2) aminals and α-aminoethers to amines (e.g., Example **19.52**).[67]

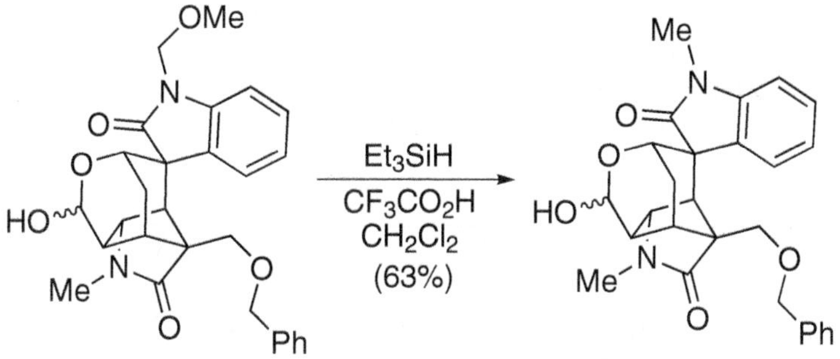

(19.52)

More recently, ionic hydrogenations have been carried out using Hantzsch esters as the hydride donor, in a reaction that is strongly reminiscent of the biological reduction of carbonyl compounds by NADH (e.g., Example **19.53**). Activation of the π bond, either by means of a strong Brønsted acid or by conjugation with a strongly electron-withdrawing group, is necessary for this reaction to proceed. When a Brønsted acid is used, it is important that the conjugate base of the acid be a poor nucleophile, so that it does not compete successfully with the Hantzsch ester for the cation. One of the most useful approaches to the reduction of enones and enals has been to convert them into iminium ions by means of a catalytic amount of a secondary amine; the carbonyl group is regenerated during the reaction.

66. Kursanov, D.N.; Parnes, Z.N.; Bassova, G.I.; Loim, N.M.; Zdanovich, D.I. *Tetrahedron* **1967**, *23*, 2235.

67. Ataraschi, S.; Choi, J-K.; Ha, D-C.; Hart, D.J.; Kuzmich, D.; Lee, C-S.; Ramesh, S.; Wu, S.C. *J. Am. Chem. Soc.* **1997**, *119*, 6226.

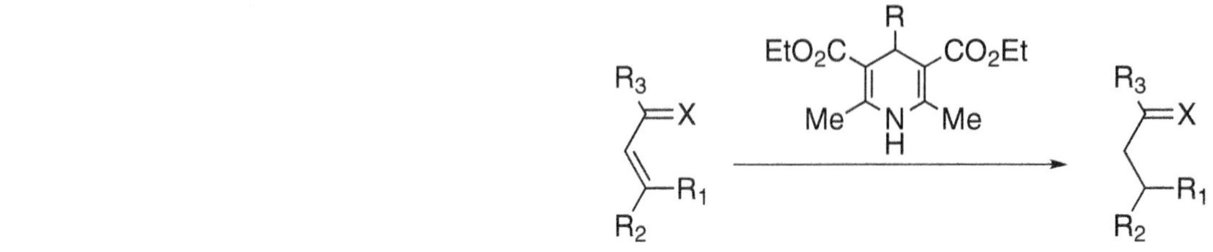

(19.53)

Reaction Synopses

Transfer Hydrogenation

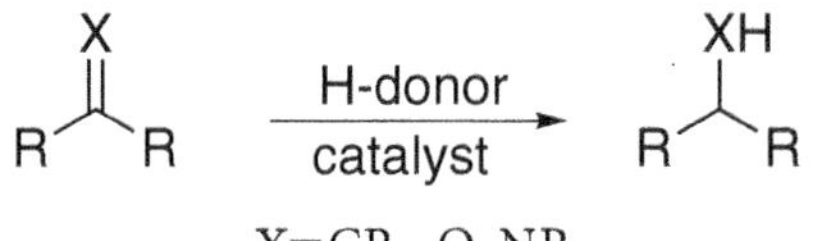

X=CR_2, O, NR

Reagents:

H-donor: cyclohexene, cyclohexadiene, phellandrene, etc.; *i*-PrOH, etc.; HCO_2H, HCO_2NH_4, etc.

Catalyst: heterogeneous: Pd-C preferred; cozeolites; etc.

homogeneous: Ru, Rh, Pd complexes; etc.

Ionic Hydrogenation

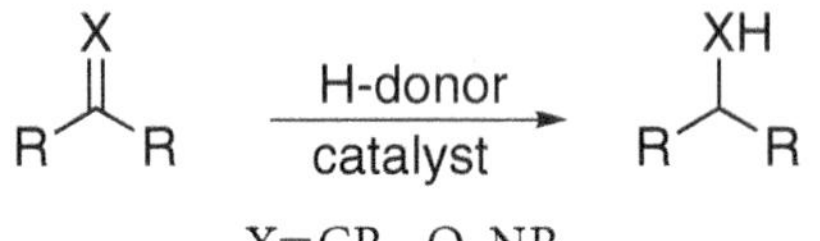

X=CR_2, O, NR

Reagents:

H-donor: R_3SiH; Hantzsch ester; etc.

Catalyst: CF_3CO_2H; $BF_3{\bullet}Et_2O$; $(RO)_2PO_2H$; etc.

Problems

19-4 Give the major organic product of each of the following reactions or reaction sequences. Specify stereochemistry where appropriate.

(a) 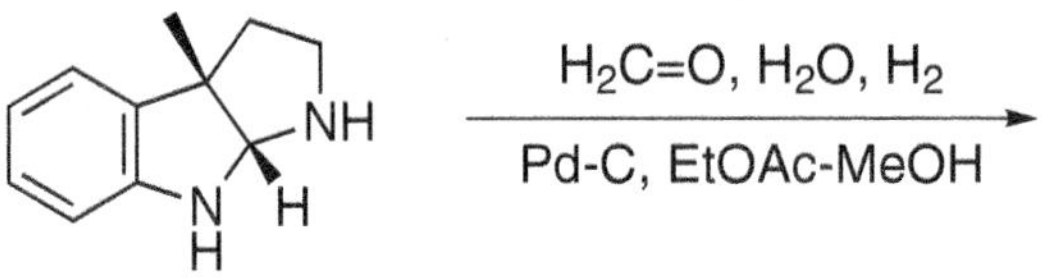

(b) 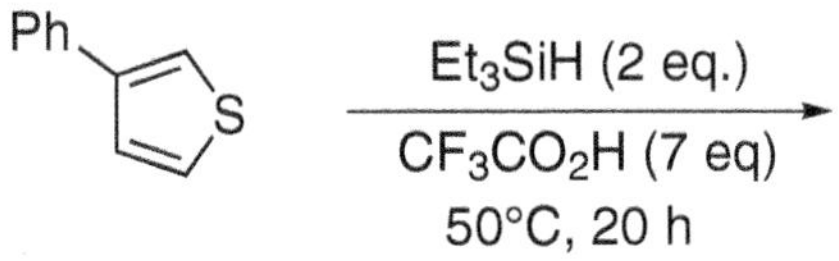

(c) 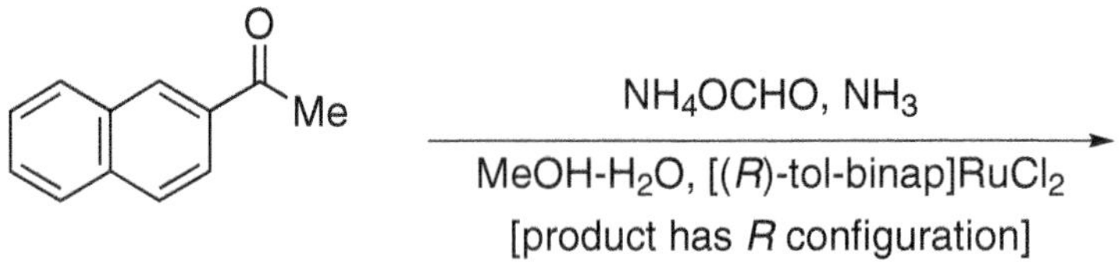

References: (a) *J. Am. Chem. Soc.* **2010**, *132*, 8282. (b) *Tetrahedron* **1975**, *31*, 311. (c) *Angew. Chem. Int. Ed.* **2003**, *42*, 5472.

19.3 Hydrogenolysis

Carbon-heteroatom bonds located adjacent to a π bond tend to be much more reactive than their saturated counterparts. This extraordinary reactivity applies to most allyl and benzyl functional groups, as well as to α-substituted carbonyl compounds. All these compounds are susceptible to reactions in which the allylic carbon-heteroatom σ bond is replaced by a carbon-hydrogen σ bond (Examples **19.54** to **19.56**). This reaction, which can be effected either with hydrogen and a catalyst or a dissolving metal (which will be discussed in more detail in the next chapter) is called ***hydrogenolysis*** ("cleavage by hydrogen"). However, the name is often restricted to the reactions with hydrogen using a palladium or nickel catalyst.[68]

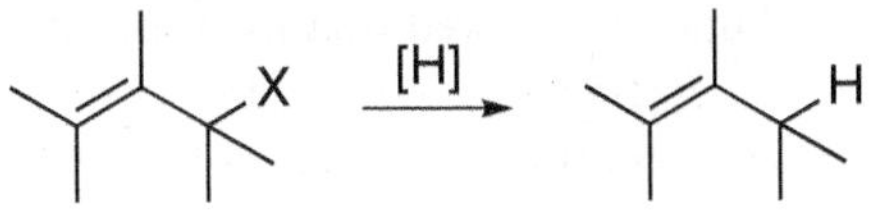

(19.54)

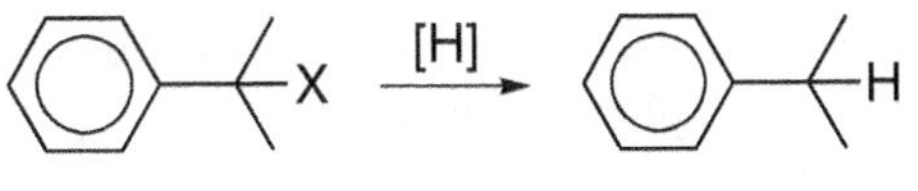

(19.55)

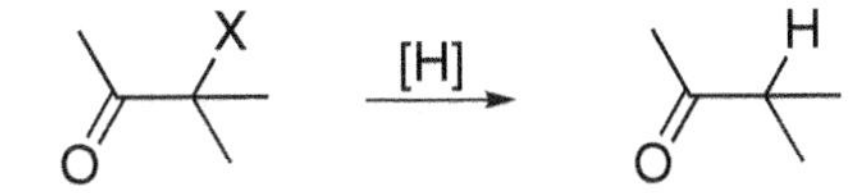

(19.56)

X = Cl, Br, I, OH, OR, NR_2, OCOR, NHCOR, OTs, $OP(O)(OR)_2$, etc.

[H] = H_2, Pd-C; H_2, Ni; etc.

The preferred catalyst for hydrogenolysis of benzyl groups is palladium on carbon because competing ring hydrogenation occurs only very slowly with this catalyst. When hydrogenolysis of benzyl and allyl substituents is to be minimized, rhodium and ruthenium catalysts are preferred.[69] Although the hydrogenolysis of benzyl alcohols, ethers, and amines is synthetically useful, the corresponding hydrogenolysis of allyl compounds is often less satisfactory due to competing hydrogenation of the double bond.

In general, the rate of hydrogenolysis of a benzyl carbon-heteroatom bond increases with the ability the heteroatom to carry an additional unit of negative charge.[70] For example, the ease of hydrogenolysis of benzyl carbon-oxygen bonds parallels the approximate base strength of the leaving groups. The relative order for hydrogenolysis of benzyl carbon-oxygen bonds in a series of benzyl compounds over palladium is:

$$\text{HO-CH}_2\text{Ph} < \text{RO-CH}_2\text{Ph} < \text{ArO-CH}_2\text{Ph} < \text{R}\overset{\oplus}{\underset{|}{\text{O}}}\text{-CH}_2\text{Ph} < \text{H}_2\overset{\oplus}{\text{O}}\text{-CH}_2\text{Ph} < \text{AcO-CH}_2\text{Ph} < \text{F}_3\text{CCO}_2\text{-CH}_2\text{Ph}$$
$$\text{(H on the RO}^{\oplus}\text{ oxygen)}$$

In similar fashion, the relative rates of hydrogenolysis of other carbon-heteroatom bonds follow the order: $NR_3^+ > OR > NR_2$. This is illustrated by the hydrogenolysis of the

68. (a) Hartung, W.H.; Simonoff, R. *Org. React.* **1953**, *7*, 263. (b) Entwhistle, I.D.; Wood, W.W. In Trost, B.M.; Fleming, I., Eds. *Comprehensive Organic Synthesis* (Pergamon: Oxford, 1991), vol. 8, p. 955.

69. (a) Burgstahler, A.W.; Bithos, Z.J. *Org. Syn.* **1962**, *42*, 62. (b) Stocker, J.H. *J. Org. Chem.* **1962**, *27*, 2288. (c) Barkdoll, A.E.; England, D.C.; Gray, H.W.; Kirk, W., Jr.; Whitman, G.M. *J. Am. Chem. Soc.* **1953**, *75*, 1156. (d) Ham, G.E.; Coker, W.P. *J. Org. Chem.* **1964**, *29*, 194. (e) Meyers, A.I.; Beverung, W.; Garcia-Muñoz, G. *J. Org. Chem.* **1964**, *29*, 3427. (f) Kaye, I.A.; Matthews, R.S. *J. Org. Chem.* **1963**, *28*, 325.

70. (a) Khan, A.M.; McQuillin, F.J.; Jardine, I. *J. Chem. Soc.* **1967**, 136. (b) Kieboom, A.P.G.; deKreuk, J.J.; van Bekkum, H. *J. Catal.* **1971**, *20*, 58.

ammonium (Example **19.57**). In this compound, the quaternary salt is selectively debenzylated in the presence of the benzyl ether.[71]

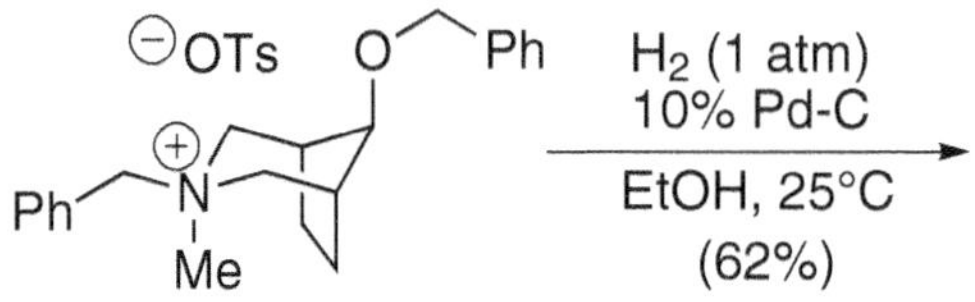

(19.57)

The hydrogenolysis of benzyl and allyl groups proceeds with either retention or inversion of configuration, depending on the catalyst used. Early work with atrolactic acid (e.g., Example **19.58**)[72] and 2-phenyl-2-butanol[73] showed that hydrogenolysis over nickel proceeds largely with retention of configuration; the typical ratios of retention to inversion in these studies vary from 95:5 to 82:18. In contrast to hydrogenolysis over nickel, hydrogenolysis over palladium occurs with inversion of configuration,[74,75] as illustrated by the hydrogenolysis of the 9-oxabicyclo[3.3.1]nonane derivative in Example **19.59**. This is not always the case, however. The results of a series of experiments using derivatives of the conformationally locked 4-*tert*-butyl-1-phenylcyclohexyl system[75a] are shown in Table 19.3.

Table 19.3 Stereochemical Outcome of Hydrogenolysis Experiments

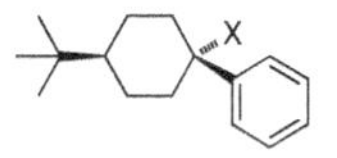	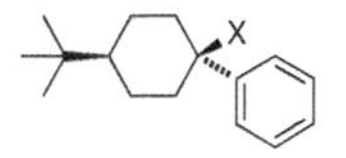	Catalyst	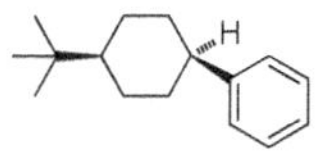	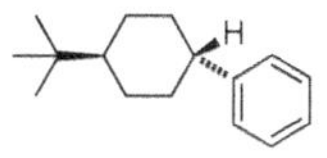
X=OH		Pt	76	24
	X=OH	Pt	74	26
X=OH		10% Pd-C	80	20
	X=OH	10% Pd-C	90	10
X=OH		Ra-Ni	100	0
	X=OH	Ra-Ni	0	100
X=OAc		Pt	0	100
	X=OAc	Pt	100	0
X=OAc		10% Pd-C	0	100
	X=OAc	10% Pd-C	100	0
X=OAc		Ra-Ni	0	100
	X=OAc	Ra-Ni	100	0
X=OMe		Ra-Ni	100	0
	X=OMe	Ra-Ni	0	100

71. House, H.O.; Wickham, P.P.; Müller, H.C. *J. Am. Chem. Soc.* **1962**, *84*, 3139.
72. Bonner, W.A.; Zderic, J.A.; Casaletto, G.A. *J. Am. Chem. Soc.* **1952**, *74*, 5086.
73. Cram, D.J.; Allinger, N.L. *J. Am. Chem. Soc.* **1954**, *76*, 4516.
74. (a) Cope, A.C.; McKervey, M.A.; Weinshenker, N.M. *J. Am. Chem. Soc.* **1967**, *89*, 2932. (b) Cope, A.C.; McKervey, M.A.; Weinshenker, N.M.; Kinnel, R.B. *J. Org. Chem.* **1970**, *35*, 2918. (c) Ojima, I.; Shimizu, N. *J. Am. Chem. Soc.* **1986**, *108*, 3100.
75. (a) Bonner, W.A.; Zderic, J.A. *J. Am. Chem. Soc.* **1956**, *78*, 3218. (b) Garbisch, E.W., Jr. *J. Org. Chem.* **1962**, *27*, 3363. (c) Garbisch, E.W., Jr. *Chem. Commun.* **1967**, 806. (d) Garbisch, E.W., Jr.; Schreader, L.; Frankel, J.J. *J. Am. Chem. Soc.* **1967**, *89*, 4233. (e) Zderic, J.A.; Rivera, M.E.C.; Limon, D.C. *J. Am. Chem. Soc.* **1960**, *82*, 6373.

These experiments reveal that the course of the hydrogenolysis depends on both the catalyst and the benzyl substituent.

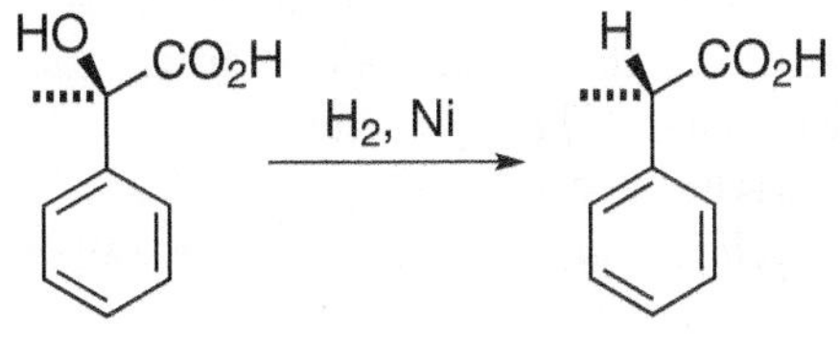

(19.58)

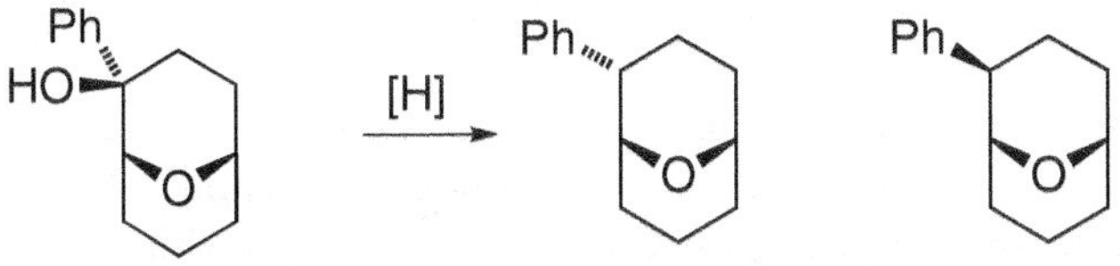

(19.59)

H_2, Ra-Ni, EtOH	100		0
H_2, Pd-C, $HClO_4$, EtOH	11	:	89

In general, reactions over Raney nickel are stereospecific. The alcohols and ethers react with retention of configuration, and the acetates react with inversion of configuration. Over palladium and platinum, the acetates react with inversion of configuration, but the alcohols react to give the same mixture of stereoisomers regardless of the stereochemistry of the starting material. The reduction of 4-*tert*-butyl-1-phenylcyclohexene gives 40% to 50% of *cis*-1-*tert*-butyl-4-phenylcyclohexane, regardless of the catalyst.

Hydrogenolysis of benzyl ethers and benzyl esters proceeds readily, so the benzyl group is a useful protecting group. An example of how this reaction can be applied in synthesis is provided by the synthesis of the pyrrolizidine derivative shown in Example **19.60.** The reductive amination of the ketone occurs once the free amine is revealed by the hydrogenolysis of the benzyl urethane.

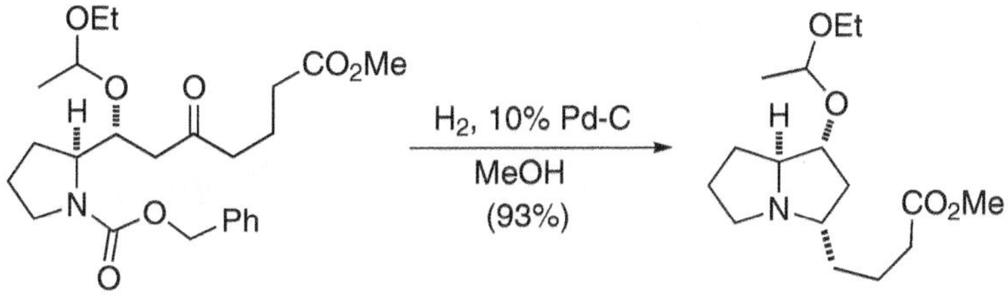

(19.60)[76]

Hydrogenolysis often competes with reduction of alkenes and alkynes, and it is often an undesired side reaction under those circumstances. For this reason, considerable effort has been expended in developing methods for the selective reduction of alkenes and alkynes in the presence of benzylic ethers, esters, and halides. One recent approach, mentioned earlier in this chapter uses diphenyl sulfide as a catalyst "poison" to prevent hydrogenolysis (Example **19.61**).[77]

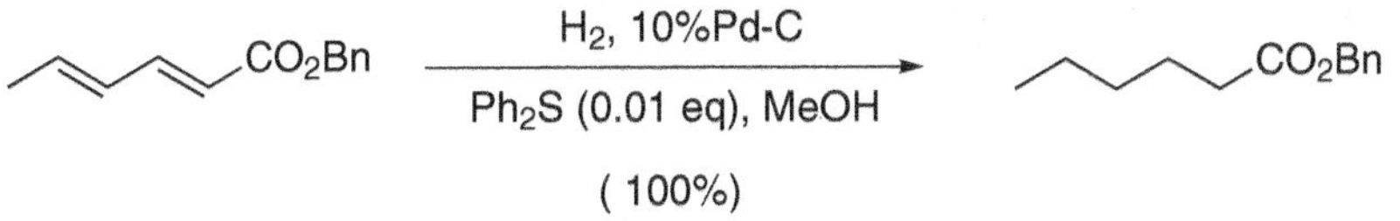

(19.61)

Recall that the C–C σ bonds in cyclopropane differ from the C–C σ bonds in other cycloalkanes by having a substantial part of the electron density in the bond outside the internuclear region (i.e., outside the perimeter of the ring). These σ bonds therefore exhibit some of the reactivity of C–C π bonds. They will, for example, react with halogens and other

76. Snider, B. B.; Gao, X. *Org. Lett.* **2005**, *7*, 4419.
77. Mori, A.; Miyakawa, Y.; Ohashi, E.; Haga, T.; Maegawa, T.; Sajiki H. *Org. Lett.*, **2006**, *8*, 3279.

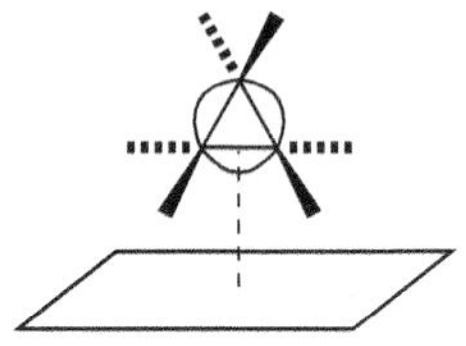

Figure 19.7 Chemisorption of a cyclopropane onto a metal surface

electrophiles to give 1,3-disubstituted propane derivatives. This reactivity extends to reacting with hydrogen in the presence of a catalyst, but because it is a σ bond rather than a π bond that is broken, these reactions are formally hydrogenolysis reactions.

As with hydrogenation of alkenes, the course of the hydrogenation of cyclopropanes is controlled by which of the three sides of the ring can best complex (chemisorb) to the metal surface (Figure 19.7). This is decided, in turn, by steric and electronic considerations. Sterically congested sides of a cyclopropane cannot chemisorb well onto the metal surface, so they often do not hydrogenate as effectively. In contrast, an atom that can form a complex with metal atoms (e.g., oxygen or nitrogen) can often facilitate the hydrogenolysis of the adjacent ring bond.

Which bond of the three-membered ring is cleaved is determined by a number of factors, including steric hindrance and the conjugating or non-conjugating nature of the ring substituents.[78] The mechanism of the hydrogenolysis is not known with certainty. However, in studies of the hydrogenolysis of 1,1-difluorocyclopropanes (e.g., Example **19.62**) over palladium, Bessard and Schlosser postulated[79] a metallocyclobutane intermediate (e.g., Example **19.63**) in the reaction to account for dehalogenation that occurred at the same time. A similar palladium insertion into the ring bond was proposed by Jackson and his coworkers in the late 1970s, based on their studies of the stereochemistry of the hydrogenolysis of propellenes containing three-membered rings.[80] These same workers found that the hydrogenolysis over platinum and palladium is not stereoselective (Example **19.65**).

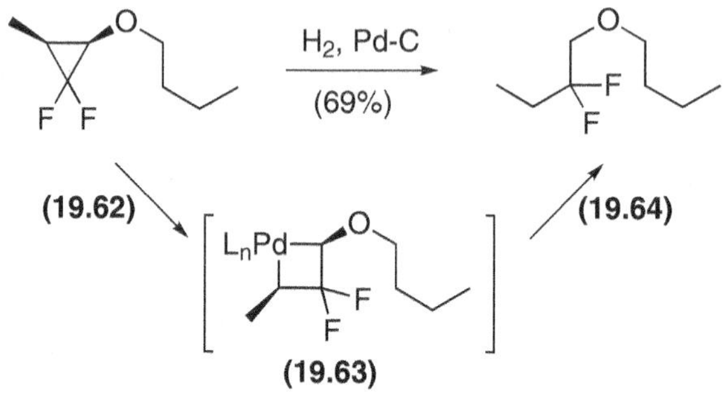

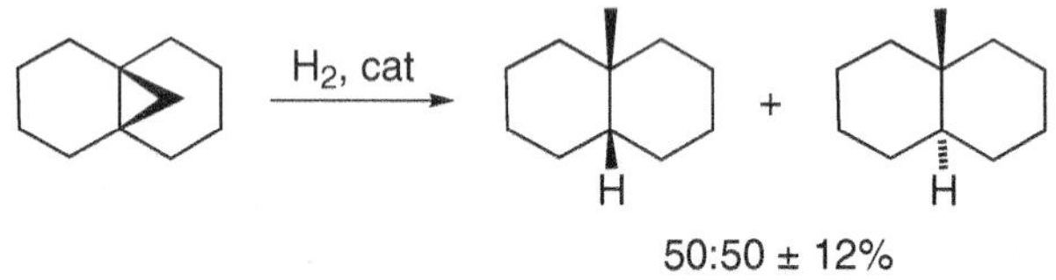

(19.65)

The inclusion of a conjugating group (e.g., aryl, carbonyl) on the ring activates the adjacent ring bonds toward hydrogenolysis, Thus, one finds that hydrogenolysis of cyclopropyl ketone and cyclopropanecarboxylic acid derivatives leads to the cleavage of the α,β-bond of the carbonyl compound,[81] as illustrated by the hydrogenolysis in Example **19.66**. Like several other hydrogenation methods, the hydrogenolysis of cyclopropanes has been shown to be amenable to asymmetric reduction.[82]

78. Roth, J.A. *J. Catal.* **1971**, *26*, 97.
79. Bessard, Y.; Schlosser, M. *Tetrahedron* **1991**, *47*, 1231.
80. (a) Akhtar, M.N.; Jackson, W.R.; Rooney, J.J. *J. Chem. Soc., Perkin Trans. 2*, **1976**, *12*, 141. (b) Akhtar, M.N.; Rooney, J.J.; Summan, N.G.; Hügel, H.M.; Jackson, W.R.; Nicholson, D.M.; Rash, D. *Aust. J. Chem.* **1977**, *30*, 1509. (c) Gooding, K.R.; W Jackson, W.R.; Pincombe, C.F.; Rash, D. *Aust. J. Chem.* **1977**, *30*, 1517.
81. Schultz, A.L. *J. Org. Chem.* **1971**, *36*, 383.
82. Barluenga, J.; Suárez-Sobrino, A.L.; Tomás, M.; García-Granda, S.; Santiago-García, R. *J. Am. Chem. Soc.* **2001**, *123*, 10494.

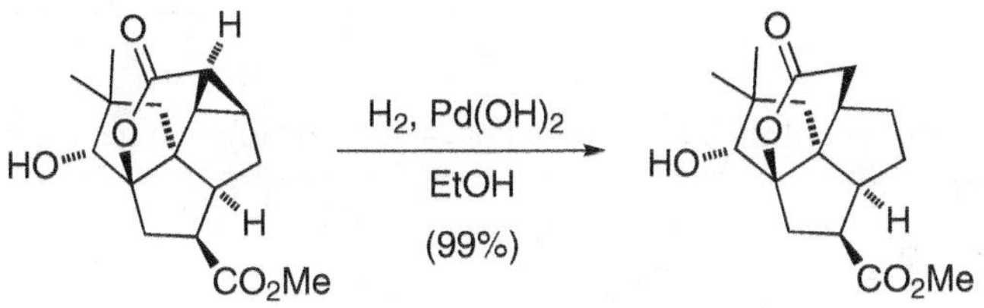

(19.66)[83]

Reductive Desulfurization of Dithioketals and Dithioacetals

Both sulfur atoms of a dithioketal or dithioacetal can be replaced by hydrogen when an appropriate reducing agent is used, and this sequence of dithioacetal formation and desulfurization is often used as a simple two-step alternative to the Wolff-Kishner reduction (discussed in Chapter 20) and its variants. The most widely used methods for desulfurization of the dithioketals are hydrogenation over Raney nickel (Examples **19.67** and **19.68**), and Birch reduction (discussed in Chapter 20). The most effective catalyst is W-2 Raney nickel, although nickel boride catalysts are also highly effective for effecting desulfurization.

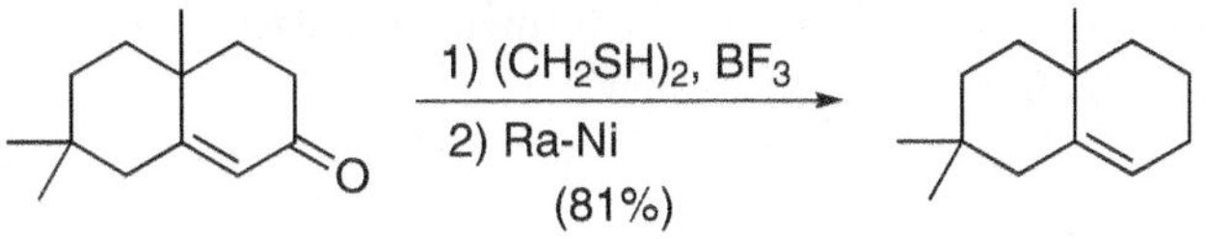

(19.67)[84]

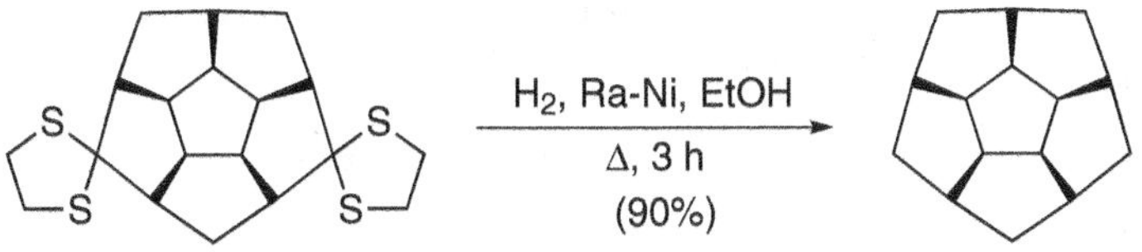

(19.68)[85]

Worked Problem

19-3 What is the major organic product of the following reaction? [*J. Am. Chem. Soc.* **2002**, *124*, 14546]

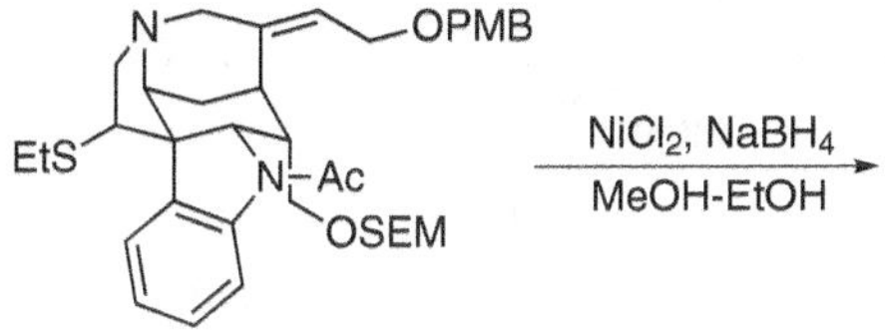

§ Answer below.

83. Rigby, J.H.; Laxmisha, M.S.; Hudson, A.R.; Heap, C.H.; Heeg, M.J. *J. Org. Chem.* **2004**, *69*, 6751.
84. Sondheimer, F.; Wolfe, S. *Can. J. Chem.* **1959**, *37*, 1870.
85. Eaton, P.E.; Mueller, R.H.; Carlson, G.R.; Cullison, D.A.; Cooper, G.F.; Chou, T.C.; Krebs, E.-P. *J. Am. Chem. Soc.* **1977**, *99*, 2751.

§ Answer to Worked Problem:

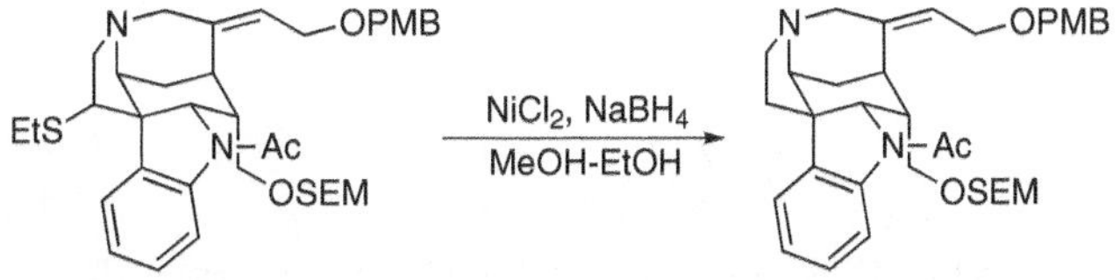

The sodium borohydride reduces the nickel chloride to an especially active form of the metal that then catalyzes the reaction of sodium borohydride with solvent to generate to hydrogen. The substrate is then added to the metal, and hydrogenolysis of the C–S bonds occurs to generate the product. Note how this reaction shows that *all* C–S bonds (not just those in dithioacetals and dithioketals) are susceptible to hydrogenolysis.

Problems

19-5 What is the structure of the final product of each of the following reactions or reaction sequences?

(a) 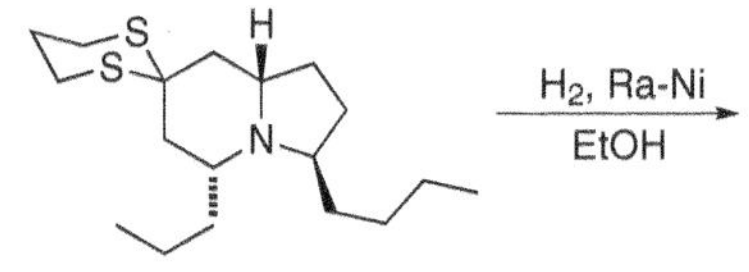

(b) 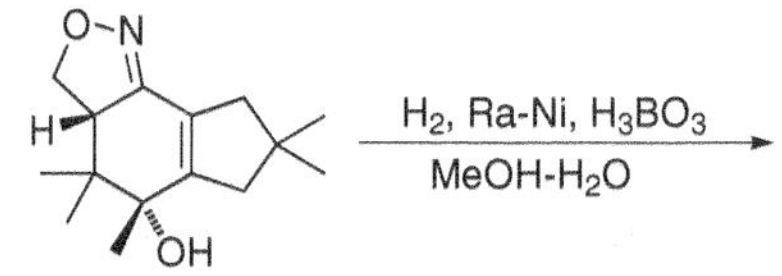

References: (a) *Org. Lett.* **2004**, *6*, 1493. (b) *Org. Lett.* **2001**, *3*, 2611.

19-6 Give the major organic product that will be formed when each of the compounds in the list below is hydrogenated over Raney nickel.

(a) 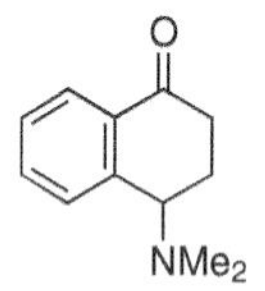

(b) 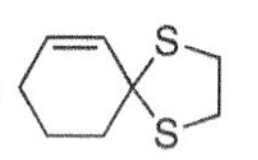

(c) 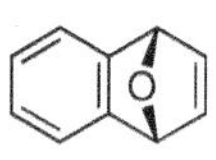

(d) 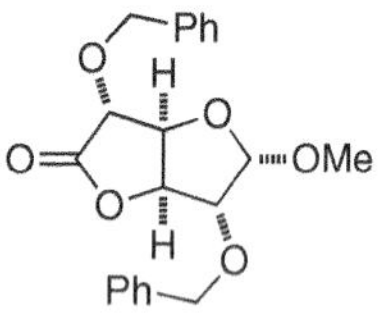

(e) 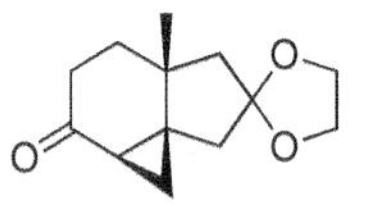

(f) 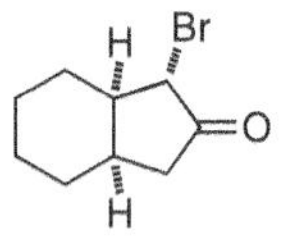

19-7 What is the major product of each of the following reactions?

(a) 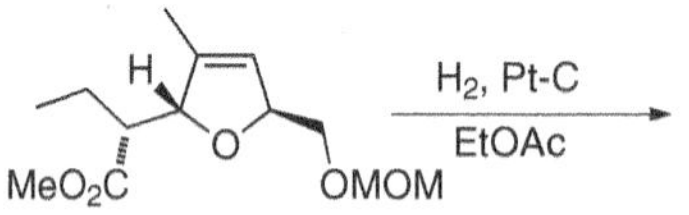

(b) 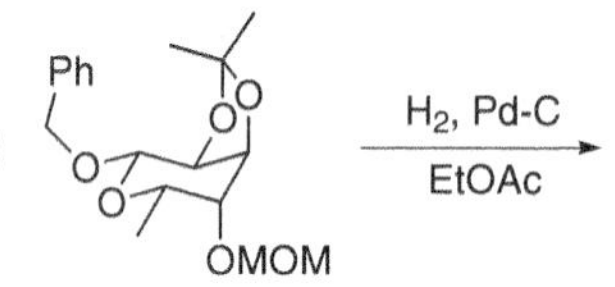

(c) 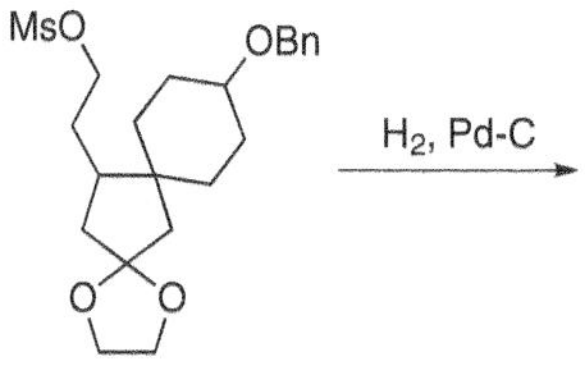

References: (a), (b) *J. Am. Chem. Soc.* **1983**, *105*, 1988. (c) *J. Am. Chem. Soc.* **1979**, *101*, 1038.

Reaction Synopses

Hydrogenolysis

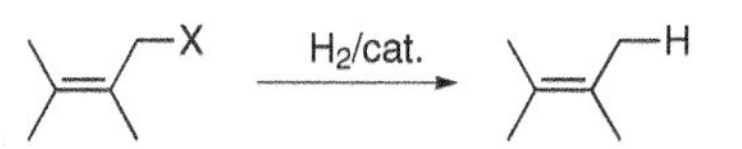

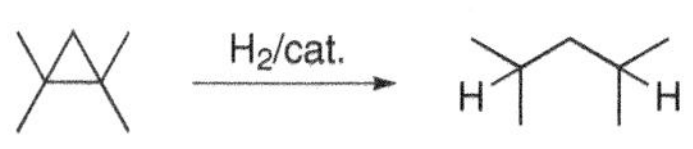

X: NR_3^+, CF_3CO_2 > OH_2^+ > OAc > OR > OH ≈ NR_2

Catalyst: Pd preferred; Ni, Pt may be used.

To suppress hydrogenolysis: use Rh, Ru.

Regiochemistry of hydrogenolysis of cyclopropane controlled by substituent groups on ring

Stereochemistry: inversion with Pd, retention Ni; Pt varies

Chapter Summary

This chapter has introduced reduction in organic chemistry—by means of with heterogeneous and homogeneous catalytic hydrogenation, which are most useful for the reduction of C–C π bonds by the suprafacial addition of H–H across the π bond. Hydrogenation of polar bonds includes the Rosenmund reduction, transfer and ionic hydrogenation of carbonyl compounds, and reductive amination. Hydrogenolysis of allyl and benzyl substituents can be carried out with either retention (Ni) or inversion (Pd) of configuration. Hydrogenolysis of cyclopropanes occurs on the most electron-rich ring bond, or the sterically most accessible bond.

Key Terms

catalytic hydrogenation
Hantzsch esters
heterogeneous catalytic hydrogenation
homogeneous catalytic hydrogenation
hydrogenolysis
ionic hydrogenation
Lindlar hydrogenation
reductive amination
reductive desulfurization
Rosenmund reduction
transfer hydrogenation
Wilkinson's catalyst

Additional Problems

Comprehensive Reduction Review Problem Set. What is the major organic product formed when each of the reactants in the first list is treated with each of the reducing agents in the second list? If a reagent can react more than once with the particular reducing agent, show the product of each step in the reduction when an excess of the reducing agent is used (e.g., when ethyl acetate reacts with lithium aluminum hydride, the first product is acetaldehyde, and this is further reduced to ethanol).

Note that not all the reactants will react with every reducing agent.

Compounds:

(a) 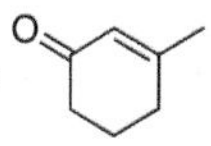
(b) 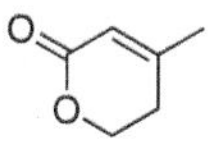
(c) 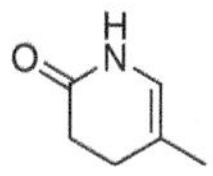

(d) 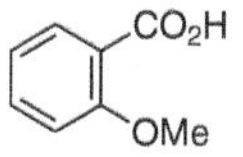

(e) 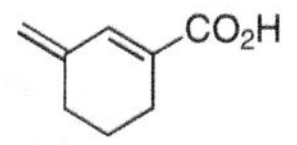

(f) 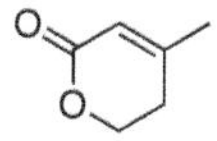
(g) 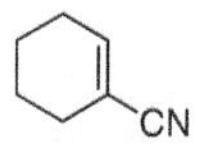

(h) 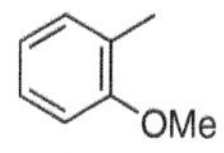

(i) 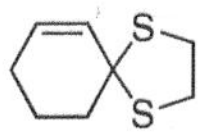
(j) 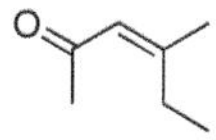
(k) 
(l) 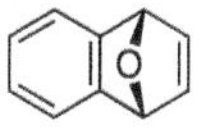

(m) 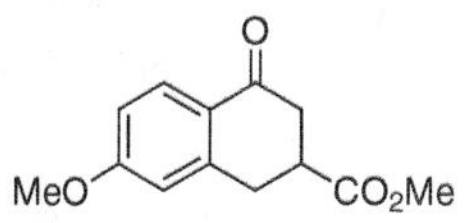

(n) 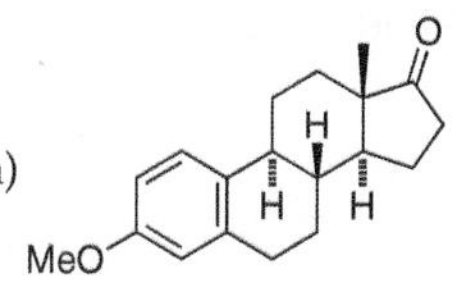

(o) 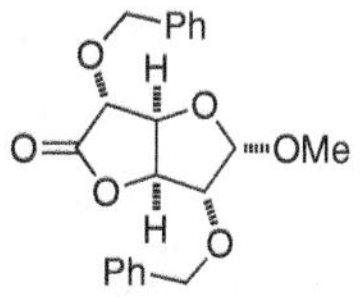

(p) 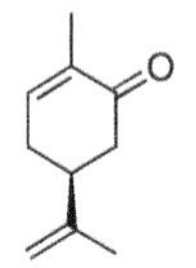 (q) 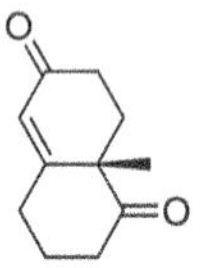 (r) 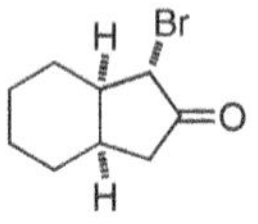

(s) 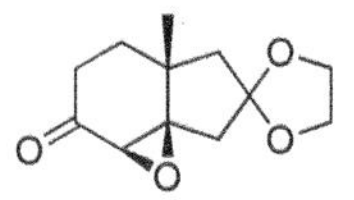

(t) 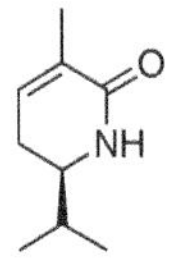

(u) 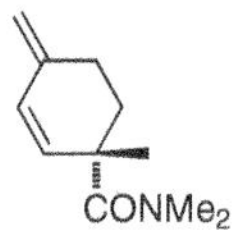

(v) 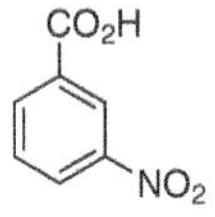

Reagents:

(i) H_2, PtO_2. (ii) H_2, Pd-$BaSO_4$, $PbSO_4$, quinoline.

(iii) H_2, Pd-C. (iv) $BuSiH_3$, CF_3CO_2H, CH_2Cl_2.

(v) H_2, Rh-C. (vi) NH_4OCHO, Pd-C, MeOH, THF.

19-8 What is wrong with the proposed reaction below?

(a) 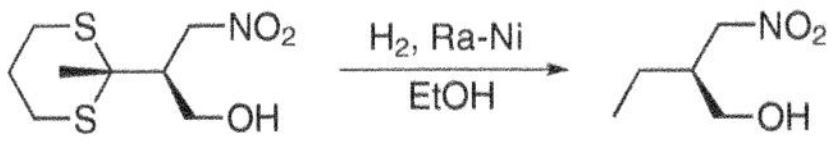

(b) 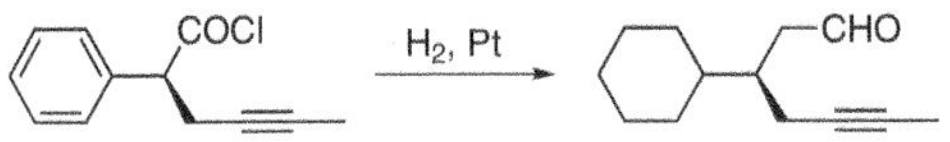

(c) 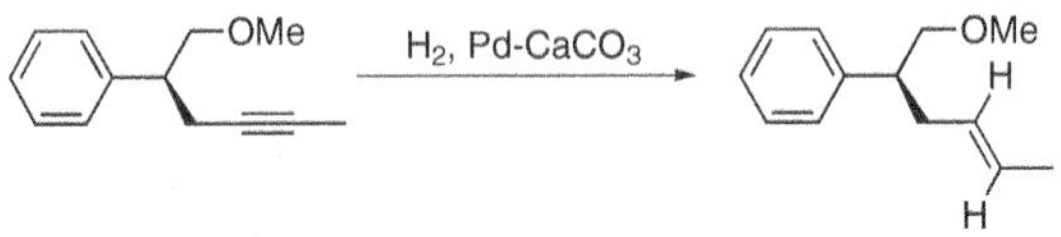

(d) 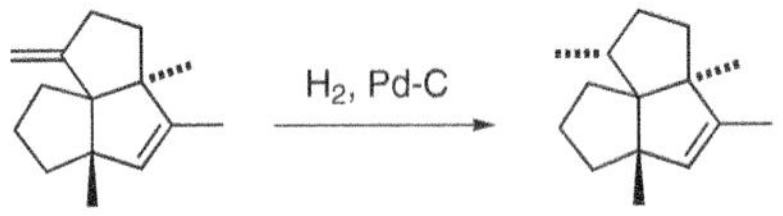

[molecular models will be useful here]

19-9 What is the major organic product expected from each of the following reactions? Where appropriate, specify stereochemistry.

(a) 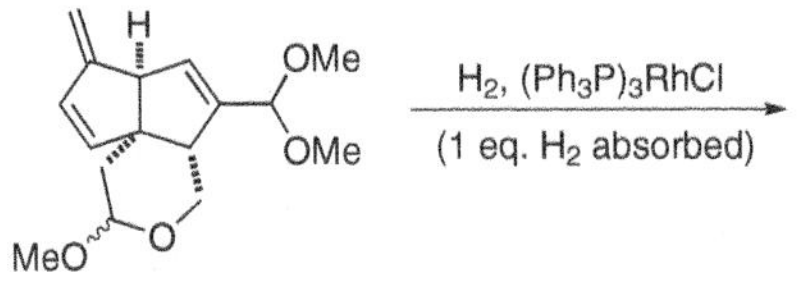

[*J. Am. Chem. Soc.* **1980**, 102, 889; *Bull. Soc. Chim. Fr.* 1980, 327]

(b) 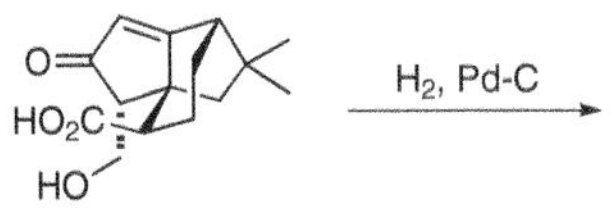

[*J. Am. Chem. Soc.* **1980**, 102, 4262; 1981, 103, 4136]

(c) 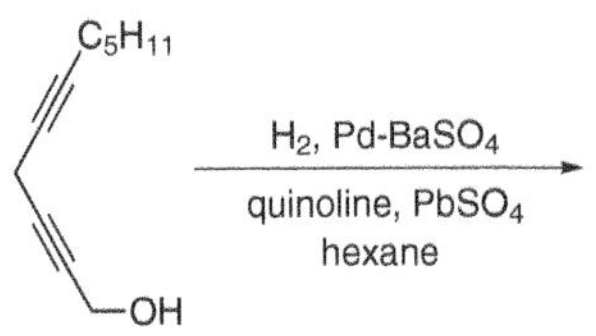

[*Tetrahedron Lett.* **1980**, 1129]

(d) 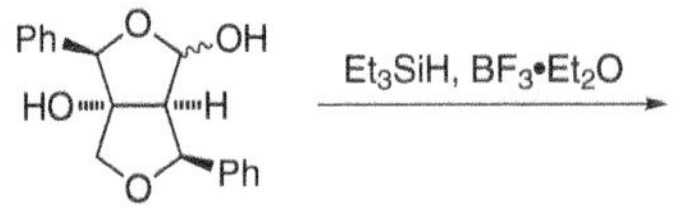

[*Tetrahedron Lett.* **2011**, 52, 4720]

19-10 What is the major organic product expected from each of the following reactions? Note that stereochemistry may be important in many of these reactions.

(a) 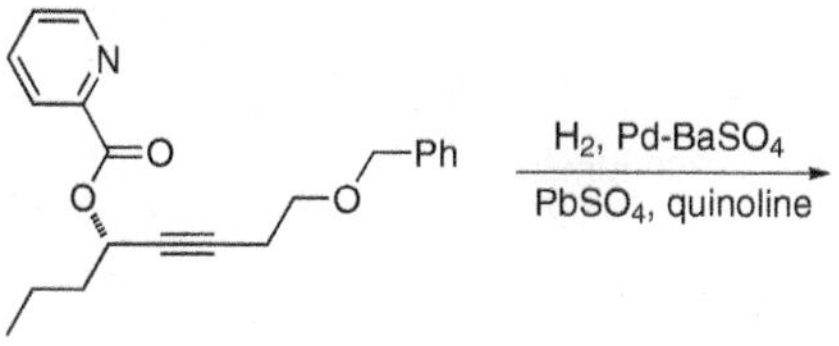

[*Tetrahedron Lett.* **2010**, *51*, 6018]

(b) 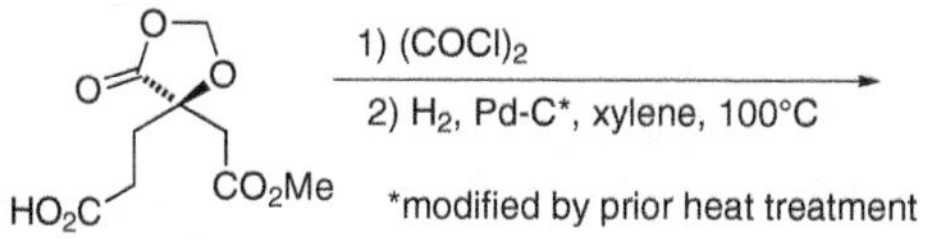

[*Chem. Commun.* **2006**, 3243]

(c) 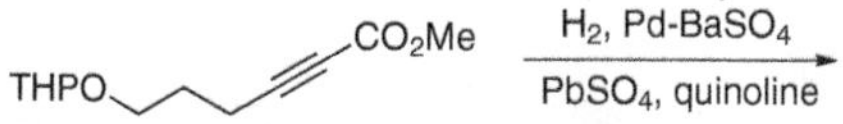

[*Synlett* **2008**, 1500]

(d) 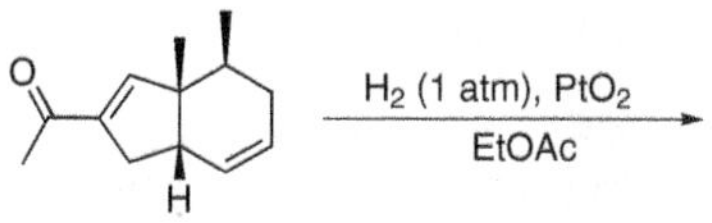

[*Org. Lett.* **2004**, *6*, 3345]

(e) 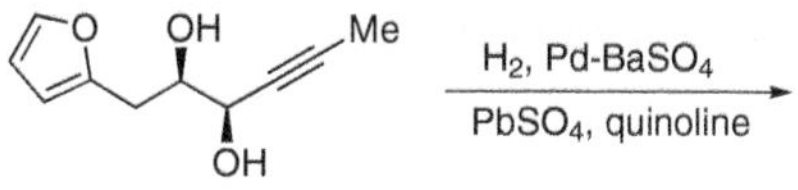

[*Org. Lett.* **2009**, *11*, 4556]

(f) 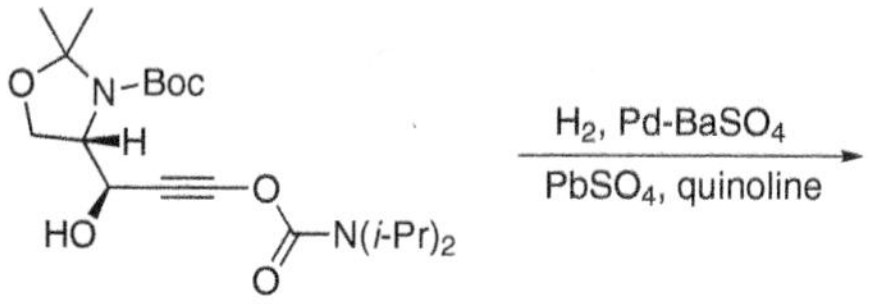

[*Tetrahedron Lett.* **2009**, *50*, 1236]

(g) 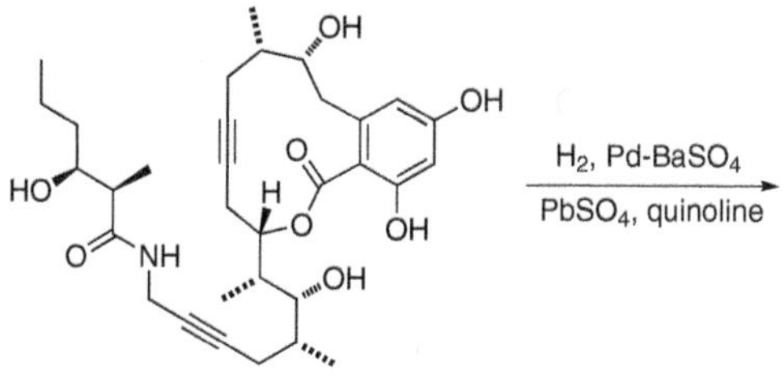

[*ARKIVOC* **2008**, *xiv*, 314]

(h) 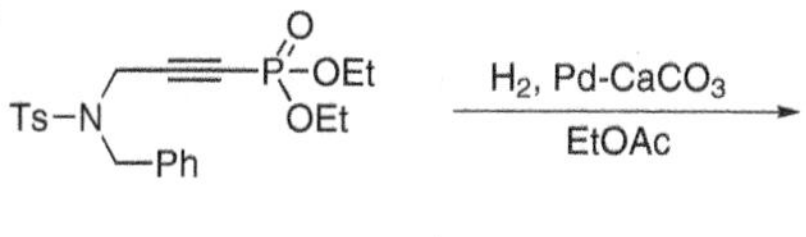

[*Tetrahedron Lett.* **2010**, *51*, 60]

(i) 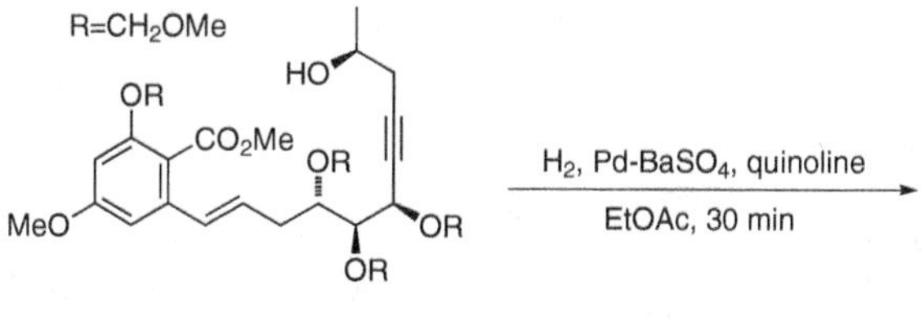

[*C. R. Chimie* **2008**, *11*, 1318]

(j) 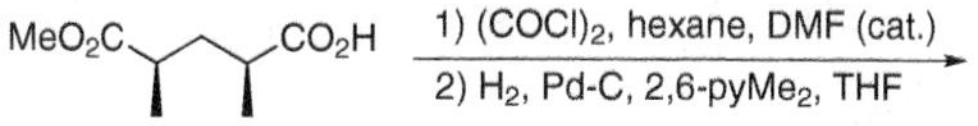

[*Tetrahedron Lett.* **1984**, *25*, 5881]

(k) 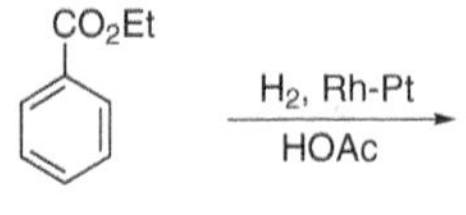

[*Bull. Chem. Soc. Japan* **1960**, *33*, 566]

(l) 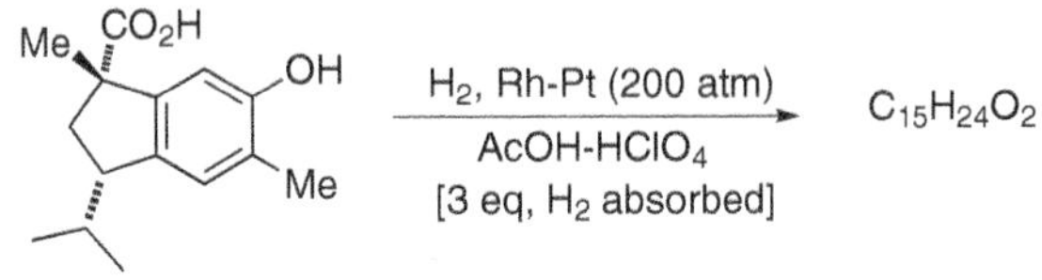

[*J. Org. Chem.* **2004**, *69*, 5219]

(m) 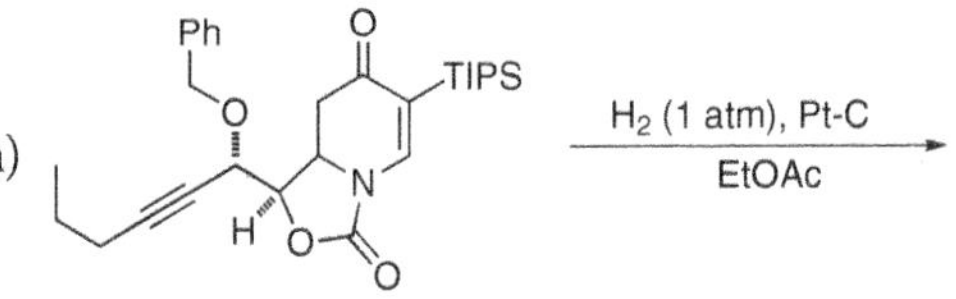

[*J. Am. Chem. Soc.* **2009**, *131*, 5635]

(n) 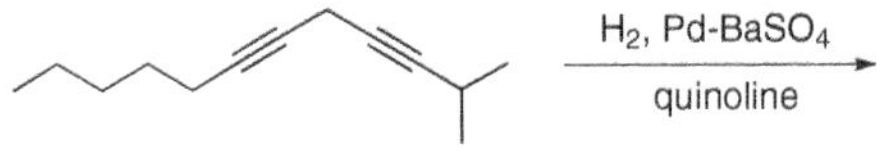

[*Tetrahedron* **2011**, *67*, 6659]

(o) 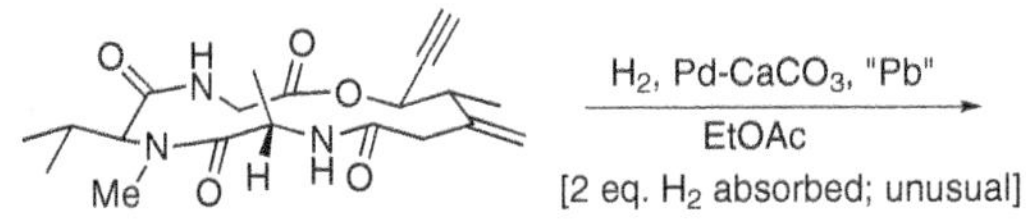

[*Org. Lett.* **2001**, *3*, 3927]

(p) 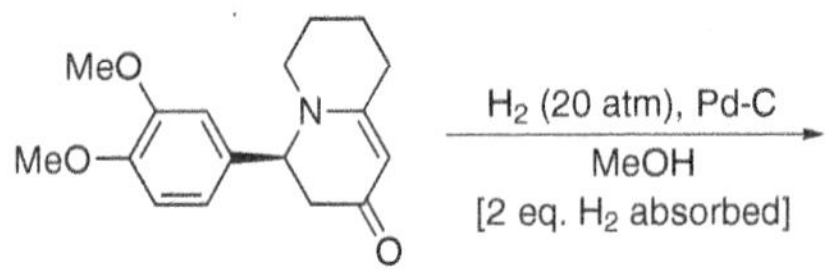

[*Org. Lett.* **2001**, *3*, 2555]

(q) 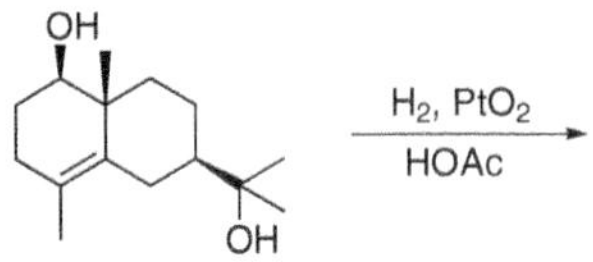

[*Tetrahedron Lett.* **2011**, *52*, 4619]

(r) 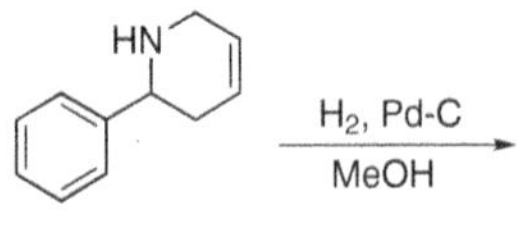

[*Tetrahedron* **2011**, *67*, 6547]

(s) 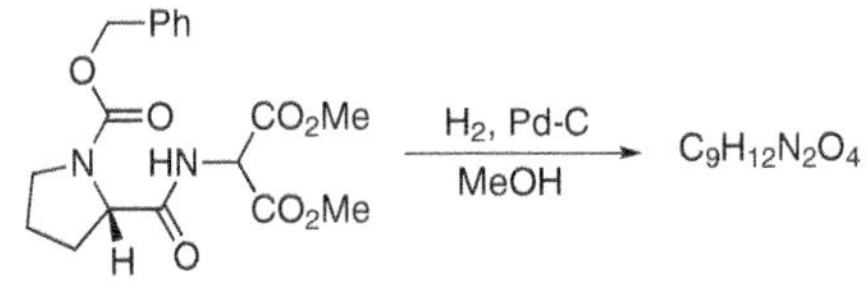

[*Tetrahedron Lett.* **2011**, *52*, 4570]

(t) 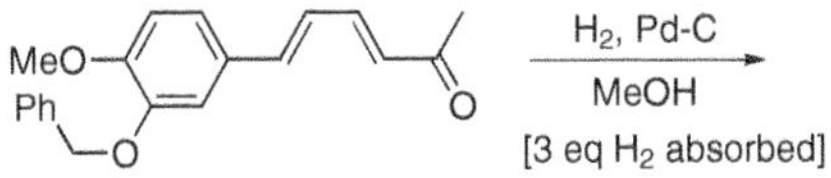

[*Tetrahedron* **2011**, *67*, 6697]

(u) 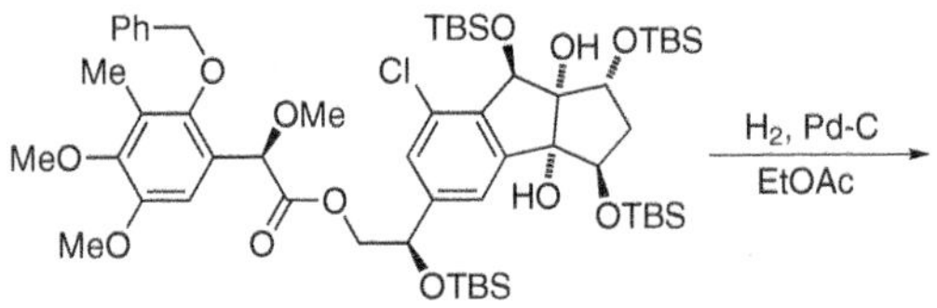

[*Tetrahedron* **2011**, *67*, 6206]

(v) 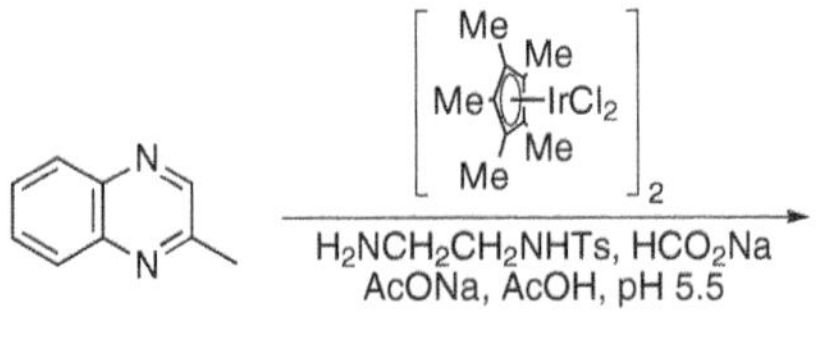

[*J. Org. Chem.* **2010**, *75*, 933]

(w) 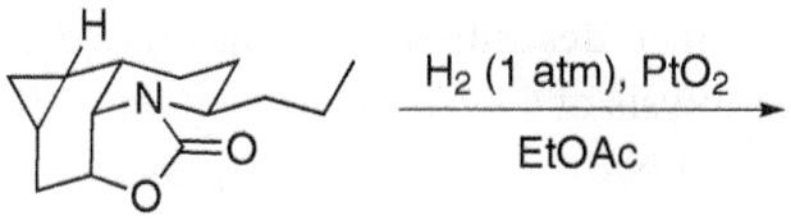

[*J. Org. Chem.* **2010**, *75*, 933]

(x) 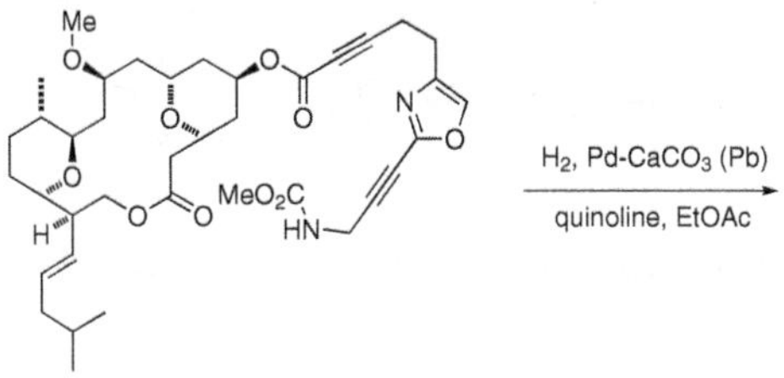

[*Angew. Chem. Int. Ed.* **2003**, *42*, 343]

(y) 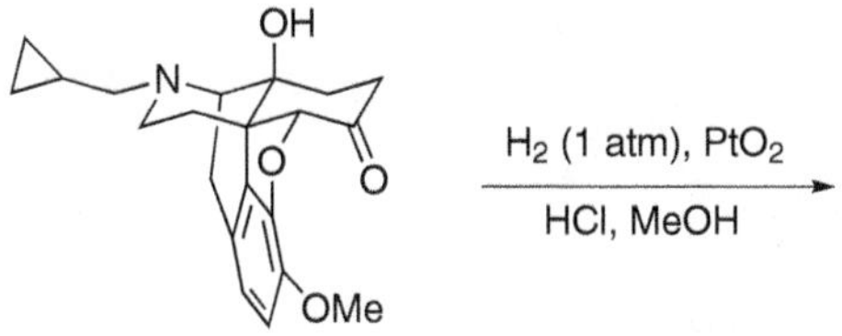

[*Tetrahedron* **2009**, *65*, 10623]

(z) 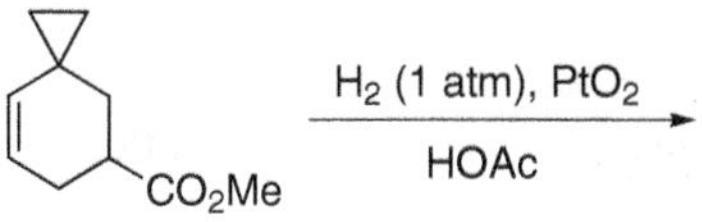

[*J. Org. Chem.* **1983**, *48*, 1135]

(aa) 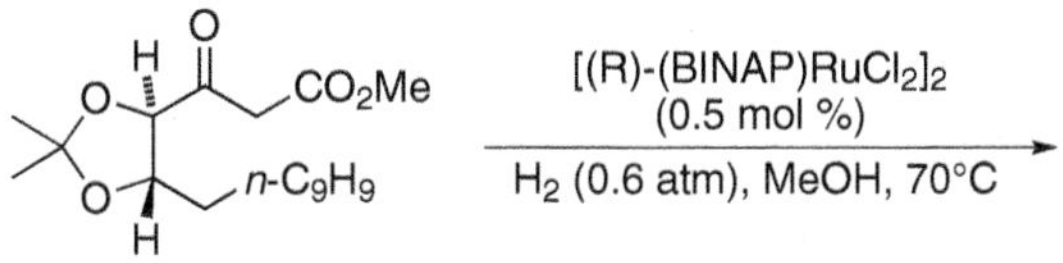

[*Tetrahedron* **2011**, *67*, 6281]

(bb) 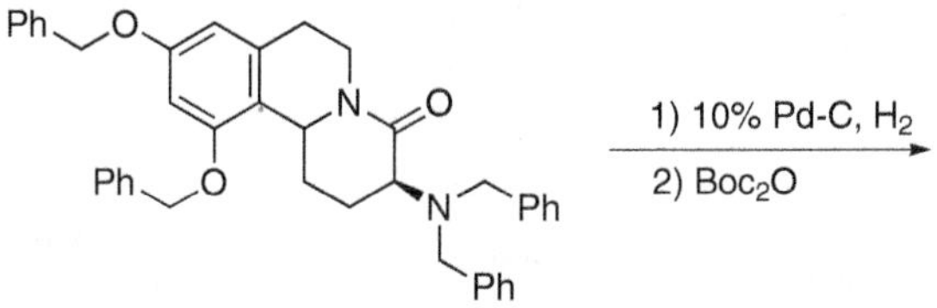

[*Tetrahedron* **2011**, *67*, 6281]

(cc) 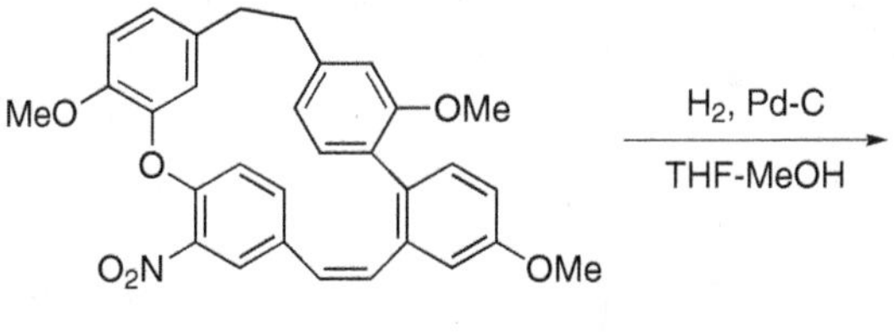

[*Tetrahedron* **2011**, *67*, 6073]

(dd) 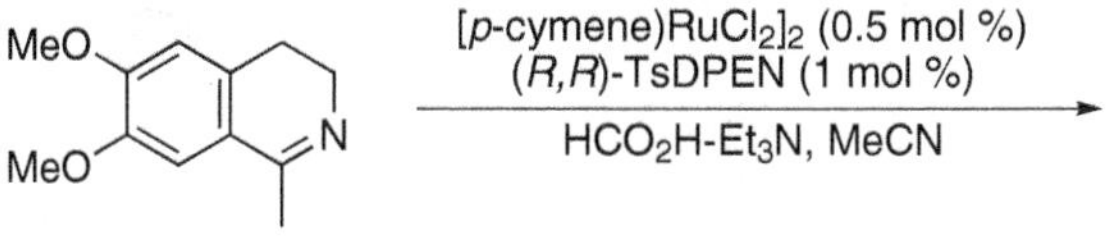

[*Org. Lett.* **2003**, *5*, 4227]

(ee) 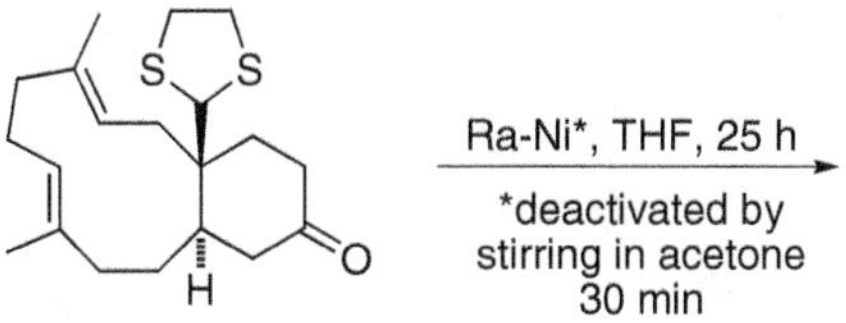

[*J. Am. Chem. Soc.* **2006**, *128*, 740]

(ff) 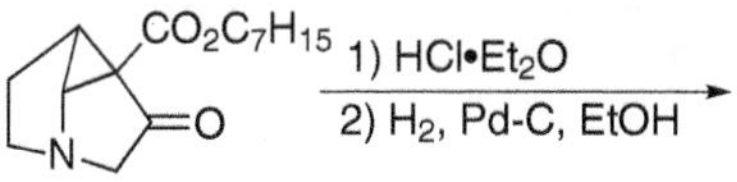

[*Org. Lett.* **2003**, *5*, 3811]

19-11 The interesting reaction shown has been described as being carried out under typical conditions for the hydrogenolysis of a cyclopropane [*Org. Lett.* **2008**, *10*, 3381]. However, there is no net gain of hydrogen during the reaction. Suggest a mechanism.

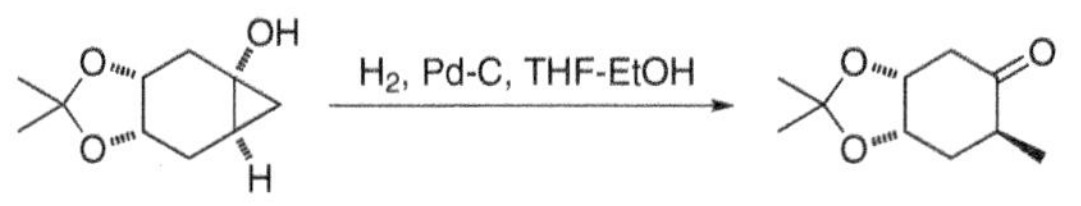

19-12 Suggest reasonable reagents that could be used to carry out the following transformations:

(a)

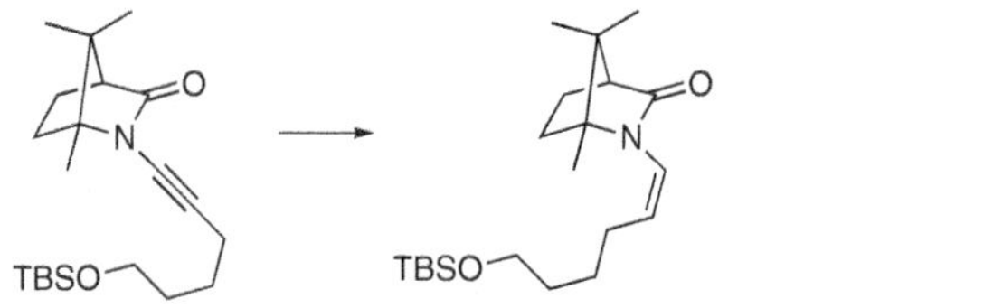

[*J. Org. Chem.* **2006**, *71*, 4170]

(b)

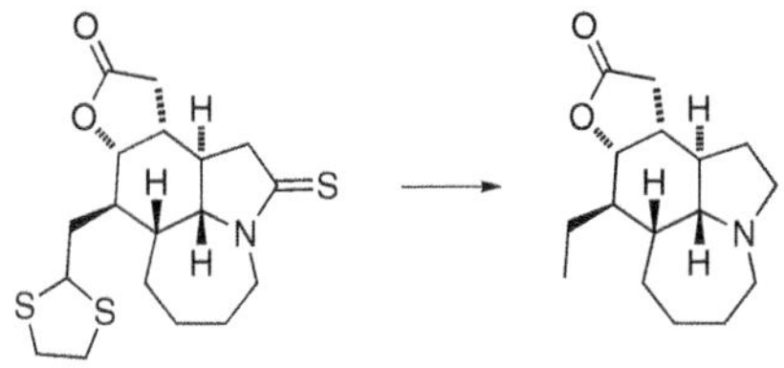

[*J. Org. Chem.* **2005**, *70*, 5197]

(c)

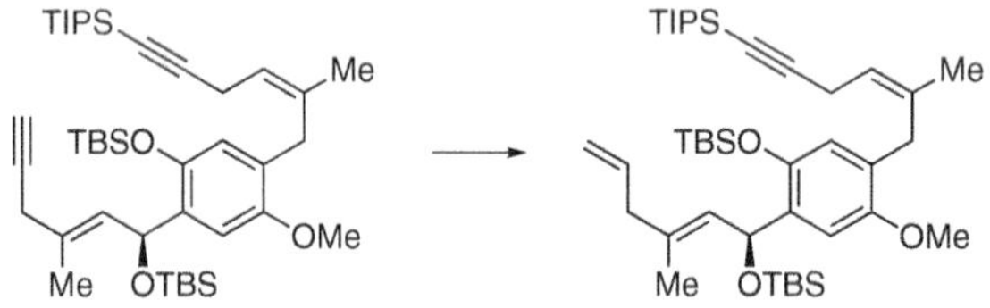

[*J. Am. Chem. Soc.* **2002**, *124*, 773]

(d)

[*Tetrahedron* **2010**, *66*, 9340]

(e)

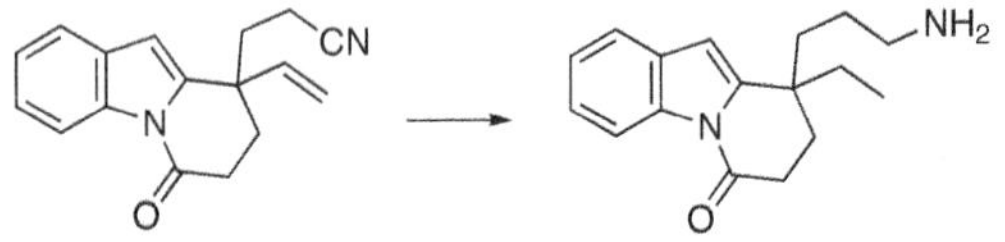

[*Org. Lett.* **2008**, *10*, 1437]

(f)

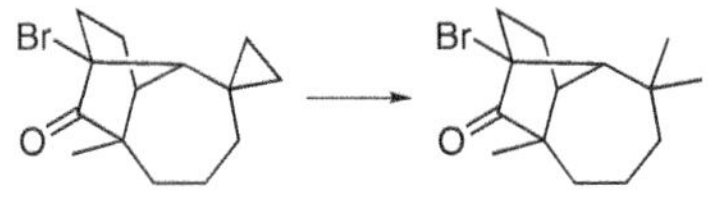

[*J. Nat. Prod* **2003**, *66*, 520]

(g)

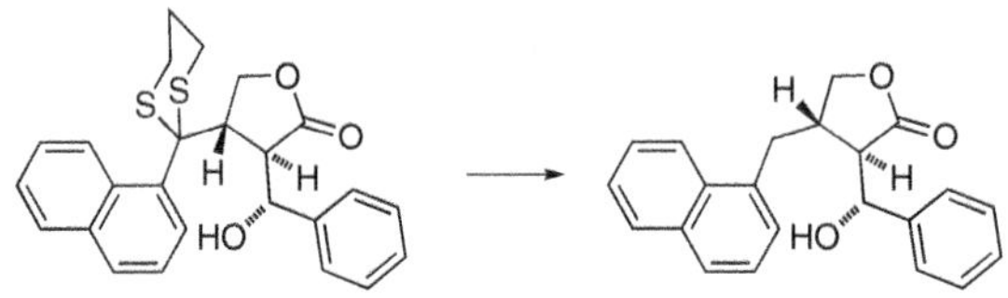

[*J. Org. Chem.* **2003**, *68*, 854]

(h)

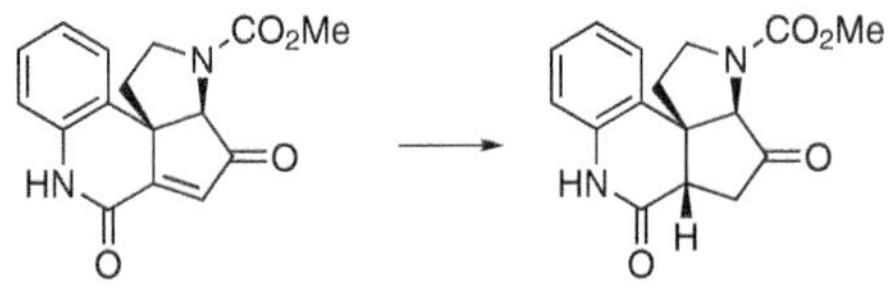

[*Org. Lett.* **2011**, *13*, 1778]

19-13 (a) Rationalize the outcome in Problem 19-12 (c). Only one equivalent of hydrogen is absorbed, so why does it add where it does?

(b) Why is the heterocyclic double bond in Problem 19-12 (e) resistant to hydrogenation?

19-14 Suggest a mechanism for the transformation shown.

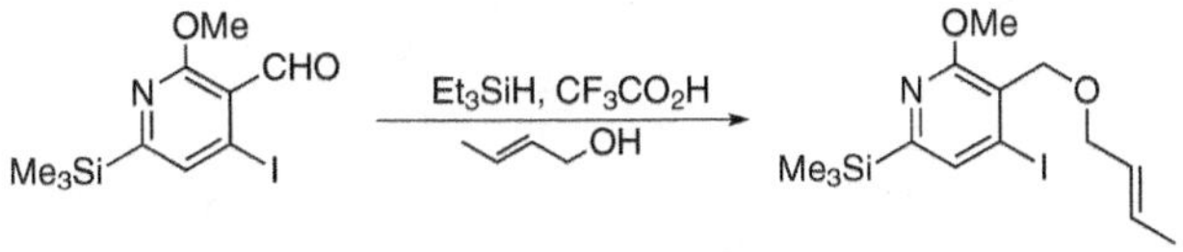

[*Angew. Chem. Int. Ed. Engl.* **1996**, *34*, 2683]

19-15 The hydrogenation of alkene π bonds in the presence of C–S σ bonds is not a simple process due to the easy hydrogenolysis of the C–S bond and the tendency of sulfur compounds to act as catalyst "poisons." How might the reaction at right be carried out without extensive loss of sulfur?

[*Tetrahedron* **1975**, *31*, 311]

CHAPTER TWENTY

Redox III

Reduction with Complex Metal Hydrides and Active Metals

20.1 Reductions with Metal Hydrides

As we saw in the previous chapter, catalytic hydrogenation is often poorly suited to the reduction of polar π bonds (despite the inroads of transfer hydrogenation in this area). For the reduction of carbonyl groups, imines, and nitriles, the use of complex metal hydrides, especially hydrides based on Group IIIA metals, is the method of choice. This reaction has been the subject of intense study over several decades, and therefore there is a huge literature covering the topic.[1] Metal hydride reduction complements catalytic hydrogenation, which tends to proceed most rapidly with electron-rich, non-polar π bonds. Alkenes, whose π bonds are non-polar, are readily reduced by hydrogen in the presence of a palladium catalyst. In general, ketones and, to a lesser extent, aldehydes, whose π bonds are very polar, are not. Conversely, carbonyl compounds are typically reduced very rapidly by complex metal hydrides, whereas alkenes are usually inert. The CN π bonds of nitriles and imines, being of intermediate polarity, tend to be readily reducible by both methods.

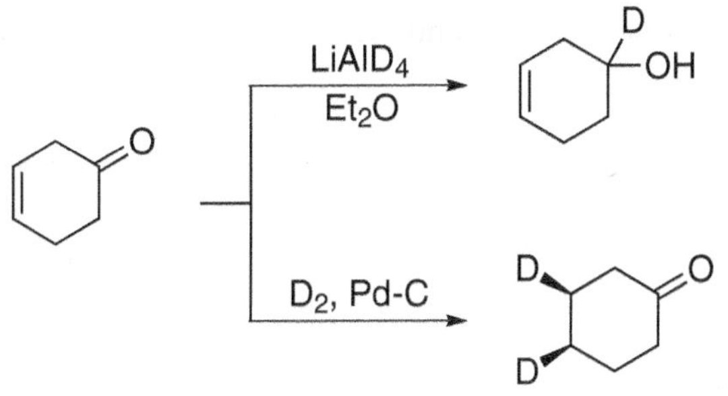

(20.1)

Nucleophilic Complex Metal Hydrides

In a formal sense, hydride anion, H^-, is a nucleophile, but in a practical sense it seldom (if ever) functions in this way because it reacts so rapidly as a base. The basicity of the hydride anion can be reduced by complexing it with a suitable Lewis acid to give a less basic nucleophile. Two of the most widely used reducing agents, sodium borohydride ($NaBH_4$) and lithium aluminum hydride ($LiAlH_4$), both contain complex anions of this type. The AlH_4^- anion, also known as tetrahydridoaluminate or aluminum hydride anion, is the complex

1. Monographs and reviews: (a) Gaylord, N.G. *Reduction with Complex Metal Hydrides* (Interscience: New York, 1956). (b) Málek, J. *Org. React.* **1985**, *34*, 1. (c) Málek, J. *Org. React.* **1988**, *36*, 249. (c) Brown, H.C.; Krishnamurthy, S. *Tetrahedron* **1979**, *35*, 567. (d) Brown, H.C. *Boranes in Organic Chemistry* (Cornell University Press: Ithaca, NY, 1972). (e) Hajos, A. *Complex Hydrides and Related Reducing Agents in Organic Synthesis* (Elsevier: Amsterdam, 1979). (f) Pelter, A.; Smith, K.; Brown, H.C. *Borane Reagents* (Academic Press: London, 1988). (g) Hudlicky, M. *Reductions in Organic Chemistry, 2nd. ed., ACS Monograph 188* (American Chemical Society: Washington, DC, 1996). (h) Brown, H.C.; Ramachandran, P.V. In Abdel-Magid, A.F., Ed. *Reductions in Organic Synthesis, ACS Symp. Ser. 641* (American Chemical Society: Washington, DC, 1996).

of hydride anion with aluminum hydride (Example **20.2**), and the BH_4^- anion, also known as the tetrahydridoborate or borohydride anion, is its complex with borane (Example **20.3**). Both anions react as nucleophilic donors of hydride anion with the polar π bonds (carbonyl groups and cyano groups). Generally speaking, neither anion reacts with an isolated alkene π bond.

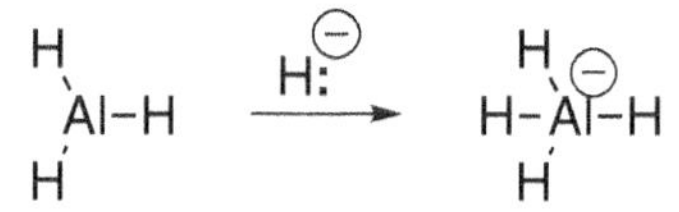

(20.2)

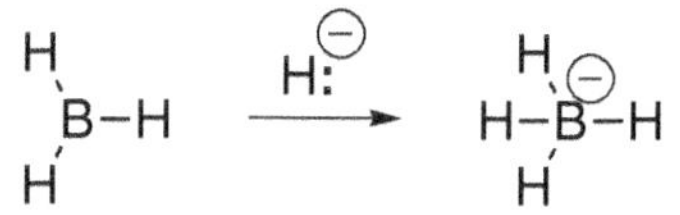

(20.3)

The steric and nucleophilic properties of the complex metal hydrides are affected by several factors: solvent, counterion, and the substitution of hydrogen atoms with alkoxy or cyano groups. The reactivity of several common hydride reducing agents is summarized in Table 20.1. The data in Table 20.1 show that (1) aluminum-based reagents are usually stronger reducing agents than borohydrides and that (2) the more effective the cation in forming a complex with the carbonyl oxygen, the stronger the reducing agent.

Aluminum-Based Reagents

Lithium aluminum hydride is the strongest and least selective reducing agent for routine reductions of carbonyl compounds and nitriles. Because it reacts explosively with water, it must be used in anhydrous ether solvents. Its power as a reducing agent has made it a popular reducing agent for reductions where chemoselectivity is not an issue. Its reactions are characterized by the reduction of the substrate to the lowest oxidation level. Carboxylic acid derivatives are reduced to primary alcohols, amides are reduced to primary amines by effective deoxygenation of the amide carbonyl group, and aldehydes and ketones are reduced to alcohols. Even carboxylic acids and carboxylate anions are reduced to primary alcohols by this reagent.

Sodium borohydride, on the other hand, reacts only very slowly with water or alcohols, so it can be easily used in these solvents. It reacts with only aldehydes, ketones, and the more reactive derivatives of carboxylic acids—it will not react with nitriles. Lithium borohydride exhibits reactivity intermediate between these two extremes. Solvent plays an important role in borohydride reductions, in particular. Reactions of borohydrides tend to be more stereoselective in alcohol solvents.

The substitution of hydrogen by alkoxy reduces the reactivity of the aluminum hydrides.[2] With primary and secondary alkoxyaluminum hydrides, there is a rapid disproportionation to $LiAlH_4$ and $LiAl(OR)_4$, a reaction that is suppressed in lithium tri-*tert*-alkoxyaluminum hydrides, presumably due to the difficulty of forming the highly hindered $Al(OR)_4^-$ species with four *tert*-alkoxy groups.[3] The modification of lithium aluminum hydride with *tert*-butyl alcohol to give lithium tri-*tert*-butoxyaluminum hydride leads to a substantial reduction in reactivity and reducing power.[4] This reagent will reduce the more reactive acid derivatives, such as acid chlorides and anhydrides, but it reduces esters only slowly. It can be used can be used, for example, to reduce an acid chloride in the presence of other reducible groups, as shown in Example **20.4**.[5] Sodium *bis*-(2-methoxyethoxy)aluminum hydride, available under the trade names Red-Al® and Vitride®, is a nucleophilic aluminum hydride that is soluble in organic solvents and less

2. (a) Brown, H.C.; McFarlin, R.F. *J. Am. Chem. Soc.* **1958**, *80*, 5372. (b) Shoaf, C.J. PhD Dissertation, Purdue University, 1957; *Diss. Abstr.* **1957**, *17*, 1904. (c) Hesse, G.; Schrödel, R. *Ann. Chem.* **1957**, *607*, 24.

3. Haubenstock, H.; Eliel, E.L. *J. Am. Chem. Soc.* **1962**, *84*, 2363.

4. Reviews: (a) Málek, J. *Org. React.* **1985**, *34*, 1. (b) Málek, J. *Org. React.* **1988**, *36*, 249.

5. Roush, W.R. *J. Am. Chem. Soc.* **1978**, *100*, 3599.

Table 20.1 Common Hydride Reducing Agents and Products of Their Reactions

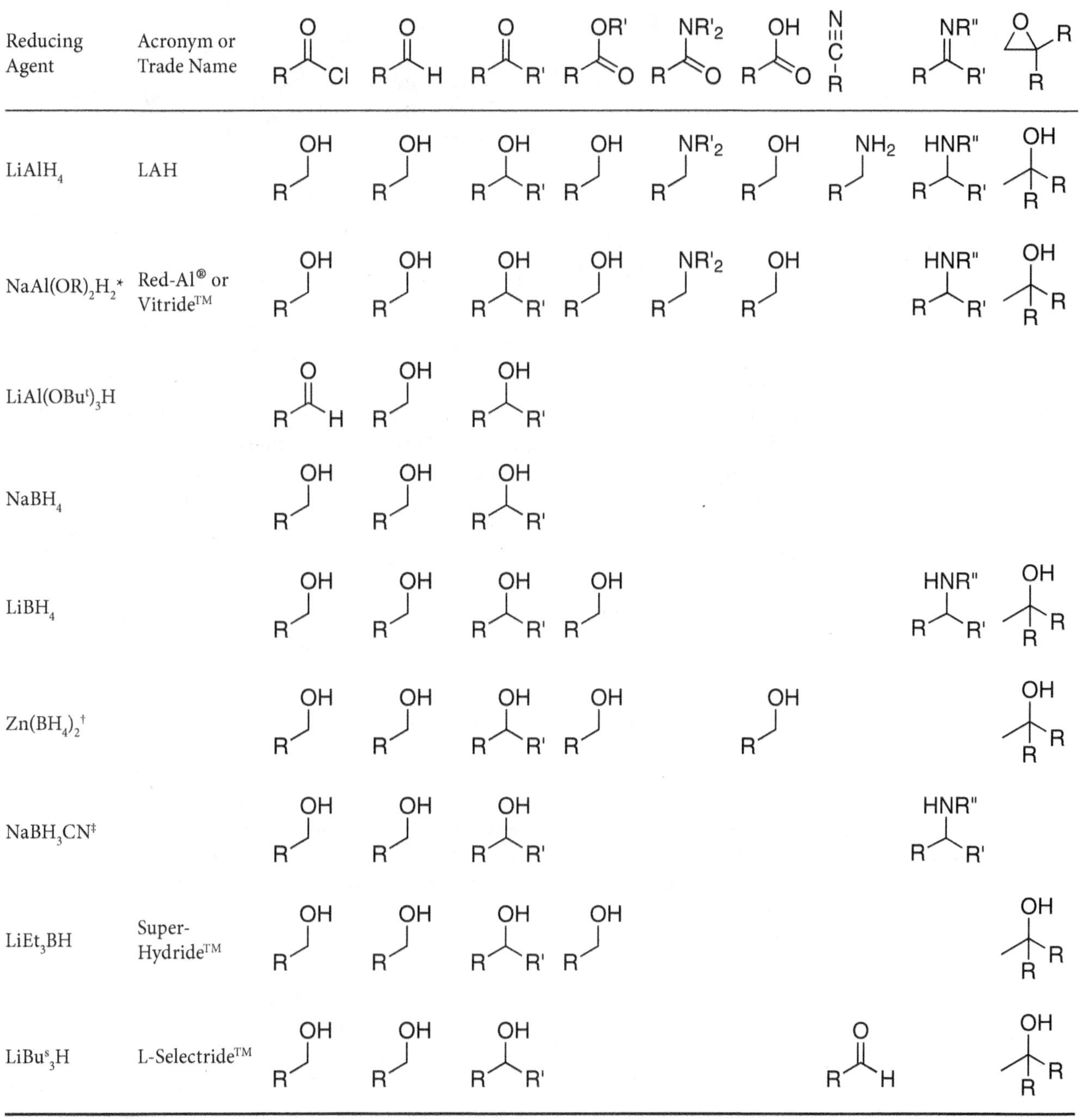

*R=CH_2CH_2OMe.
†Not commercially available.
‡Acidic conditions usually required for reduction.

prone to disproportionation than other alkoxyaluminum hydride reagents. Its reactivity is slightly less than that of lithium aluminum hydride, although it still reduces esters under mild conditions, which allows its use for the reduction of esters in the presence of amides, for example. It has also been used for the regioselective reduction of chiral epoxy alcohols (e.g., Example **20.5**).[6] The source of the regioselectivity may be the ability of the

6. Gao, Y.; Sharpless, K.B. *J. Org. Chem.* **1988**, *53*, 4081.

alcohol hydroxyl group to displace one of the alkoxy groups on the reducing agent, thus rendering the reaction intramolecular instead of intermolecular.

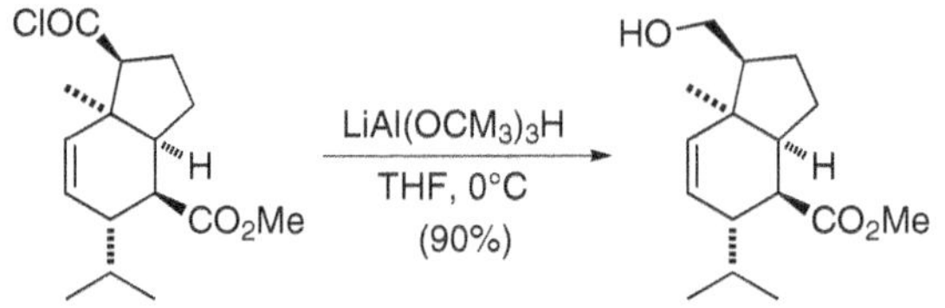

(20.4)

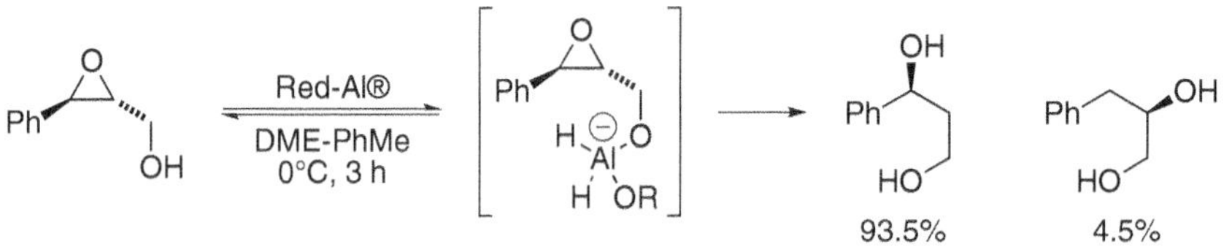

(20.5)

Borohydrides

In contrast to lithium aluminum hydride, sodium borohydride is one of the least reactive and most selective reducing agents available for the reduction of carbonyl compounds. It reacts rapidly with aldehydes in aqueous or ethanol solution and with acid chlorides in aprotic solvents to give primary alcohols. It also reacts with saturated ketones (but not aryl ketones) to give secondary alcohols. It does not reduce acids, amides, or esters. The mechanism of reduction by borohydride anion appears to require prior complexation of the carbonyl oxygen by the metal cation or a hydrogen bond to a protic solvent (e.g., Example **20.6**). When the metal cation is completely complexed by a cryptand, no reduction of the carbonyl compound occurred until a protic solvent such as water was added.[7]

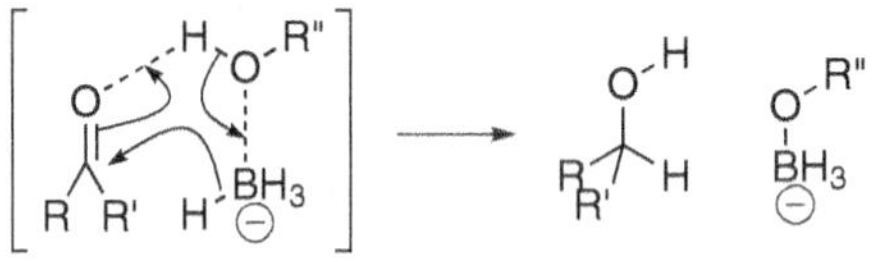

(20.6)

One useful property of sodium borohydride as a reducing agent is its ability to distinguish between saturated ketones and α,β-unsaturated ketones or aryl ketones. In the reduction of the ketones in Examples **20.7** and **20.8**, for example, the saturated ketone carbonyl group is reduced much faster than the conjugated carbonyl group in the same molecule, even though the saturated carbonyl group in Example **20.7** is in a more hindered position.

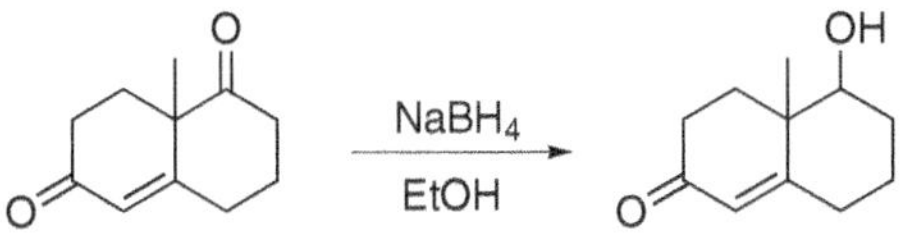

(20.7)[8]

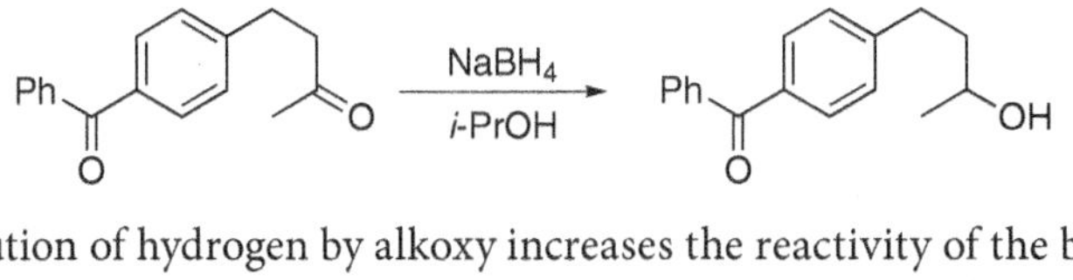

(20.8)[9]

Substitution of hydrogen by alkoxy increases the reactivity of the borohydrides.[10] The array of complex metal hydride reducing agents now commercially available includes the

7. (a) Pierre, J.-L.; Handel, H. *Tetrahedron Lett.* **1974**, 2317. (b) Handel, H.; Pierre, J.-L, *Tetrahedron Lett.* **1976**, 741. (c) Handel, H.; Pierre, J.-L, *Tetrahedron Lett.* **1976**, 2029. (d) Costes, E.; Bénard, C.; Lattes, A. *Tetrahedron Lett.* **1976**, 1185.

8. Cocker, J.D.; Halsall, T.G. *J. Chem. Soc.* **1957**, 3441.

9. Lansbury, P.T.; Peterson, J.D. *J. Am. Chem. Soc.* **1962**, *84*, 1756, 1963; **1963**, *85*, 2236.

10. (a) Brown, H.C.; Mead, B.J.; Subba Rao, B.C. *J. Am. Chem. Soc.* **1955**, *77*, 6209. (b) Brown, H.C.; Mead, B.J.; Shoaf, C.J. *J. Am. Chem. Soc.* **1956**, *78*, 3616. (c) Garrett, E.R.; Lyttle, D.A. *J. Am. Chem. Soc.* **1953**, *75*, 6051.

trialkylborohydrides,[11] which are derived from the metal hydride and a trialkylborane. The presence of the three alkyl groups on boron increases the reactivity of these reagents beyond that of lithium borohydride itself, making these reagents air-sensitive compounds that must be handled in ether solvents. The three most popular variants (lithium tri-*sec*-butylborohydride, or L-Selectride®, potassium tri-*sec*-butylborohydride, or K-Selectride®, and lithium triethylborohydride, or Super-Hydride™) are available commercially as solutions in tetrahydrofuran (THF). Their steric bulk has made them very popular reagents for the stereoselective reduction of carbonyl groups.

Lithium triethylborohydride is a particularly effective nucleophilic reagent for reduction at sp^3-hybridized carbon, especially the reduction of primary alkyl halides and tosylates to methyl groups, even in relatively hindered locations.[12] The reduction of 1-adamantylmethyl tosylate (Example **20.9**)[13] provides an interesting comparison of the reactivity of lithium triethylborohydride and lithium aluminum hydride toward sp^3-hybridized carbon atoms. The borohydride reagent gives 1-methyladamantane in 91% isolated yield, whereas lithium aluminum hydride gives only 25% of the methyladamantane.[14] Final oxidative workup is required due to the formation of triethylborane as a by-product of the reaction.

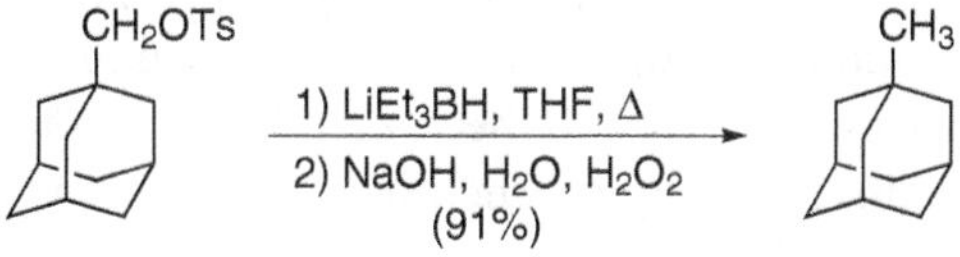

(20.9)

Lithium triethylborohydride and the Selectride reagents can also be used to reduce conjugated carbonyl compounds by 1,4-addition of the nucleophile provided that the β position of the conjugated system carries a hydrogen atom. Lithium triethylborohydride in *tert*-butyl alcohol-THF can be used effectively to carry out the reduction of α,β-unsaturated esters to the saturated esters without competing condensation of the intermediate enolate anion to give the Claisen condensation product.[15]

Sodium borohydride and the Selectride™ reagents provide an interesting comparison in terms of the stereochemistry of the reduction. With sodium borohydride in alcohol solvents, the approach of the nucleophile to the carbonyl carbon in cyclohexanones is from the axial direction, leading to the equatorial alcohol as the major product. With the Selectride reagent, however, the large alkyl groups on boron lead to much more severe 1,3-diaxial interactions in the activated complex, leading to mainly equatorial attack and to the axial alcohol as the major product. This is nicely illustrated by the reactions in Figure 20.1.[16] Both carbonyl groups in this molecule are hindered, but the carbonyl group in the bicyclo[3.2.1]-octane ring system is the more hindered. Reduction of the less hindered ketone with sodium borohydride gives the expected equatorial alcohol as the major product of the reaction; with L-Selectride™, on the other hand, a 1:1 mixture of the axial and equatorial diastereoisomers is obtained. An examination of models shows that the nucleophile can approach the bicyclo[3.2.1]octanone only from the *exo* face, so reduction of this functional group gives the axial alcohol.

The reactivity of sodium borohydride can be modified by the use of transition metal cations. In the presence of 0.2 mole equivalents of cobalt (II) chloride in methanol solution, for example, sodium borohydride becomes an excellent reagent for the reduction of

11. Reviews: (a) Lane, C.F. *Aldrichimica Acta* **1974**, *7*, 32. (b) Krishnamurthy, S. *Aldrichimica Acta* **1974**, *7*, 55.

12. (a) Brown, H.C.; Krishnamurthy, S. *J. Am. Chem. Soc.* **1973**, *95*, 1669. (b) Brown, H.C.; Kim, S.C.; Krishnamurthy, S. *J. Org. Chem.* **1980**, *45*, 1. (c) Krishnamurthy, S.; Schubert, R.M.; Brown, H.C. *J. Am. Chem. Soc.* **1973**, *95*, 8486.

13. Krishnamurthy, S. *J. Organomet. Chem.* **1978**, *156*, 171.

14. Stetter, H.; Schwartz, M.; Hirschhorn, A. *Chem. Ber.* **1959**, *92*, 1629.

15. (a) Ganem, B.; Fortunato, J.M. *J. Org. Chem.* **1975**, *40*, 2846. (b) Ganem, B. *J. Org. Chem.* **1975**, *40*, 146. (c) Fortunato, J.M.; Ganem, B. *J. Org. Chem.* **1976**, *41*, 2194.

16. Zou, Y.; Chen, C.-H.; Taylor, C.D.; Foxman, B.M.; Snider, B.B. *Org. Lett.* **2007**, *9*, 1825.

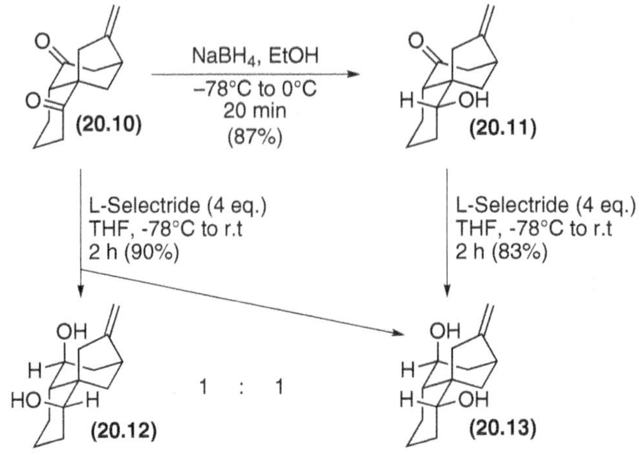

Figure 20.1 Reduction of an intermediate in the total synthesis of (±)-platensimycin

nitriles to primary amines.[17] This reaction is especially applicable to the reduction of cyanohydrins to β-aminoalcohols (e.g., Example **20.14**). Because acrylonitrile is reduced to allylamine under these conditions, it is unlikely that the reaction involves hydrogenation over a cobalt catalyst. Other metals that have been used to modify the reactivity of the borohydride anion are aluminum,[18] copper,[19] nickel,[20] tin,[21] titanium,[22] and zinc.[23] Zinc borohydride gives a mild reagent that is still more reactive than sodium borohydride and soluble in solvents such as THF.[24]

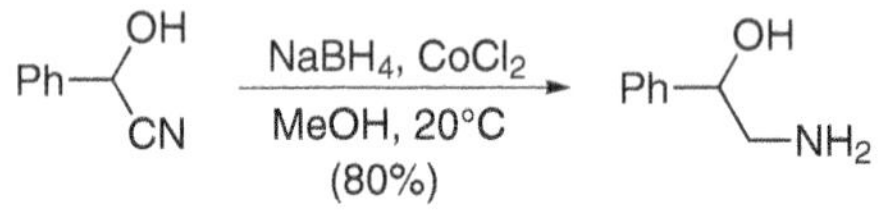

(20.14)

Tetraalkylammonium triacetoxyborohydrides (also known as Gribble reagents)[25] are a particularly useful addition to the arsenal of reducing agents, especially for the chemoselective reduction of aldehydes and the stereoselective reduction of α- and β-hydroxyketones to *anti* 1,2- and 1,3-diols (e.g., Example **20.19**). The reduction presumably occurs through a five- or six-membered complex of the substrate and the boron hydride so the hydrogen is delivered to the face of the carbonyl group *syn* to the hydroxyl group (Figure 20.2). Similar chelation of the reducing agent by α-alkoxyketones has been invoked to rationalize the tendency of open-chain compounds to give high levels of *anti* 1,2-diols (e.g., Example **20.21**) with zinc borohydride and Red-Al®, with

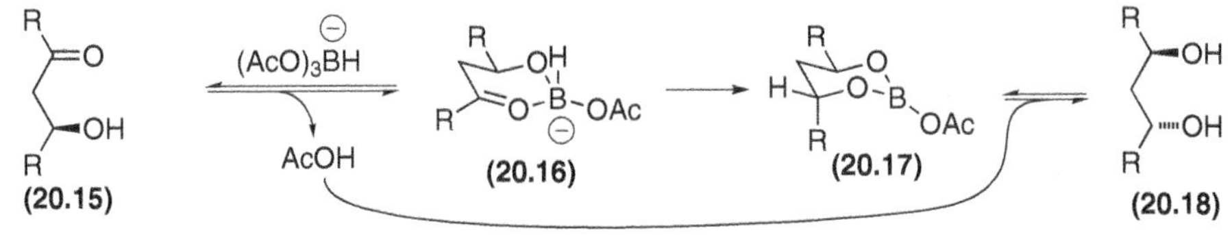

Figure 20.2 Putative mechanism for the reduction of β-hydroxyketones by triacetoxyborohydrides

17. (a) Satoh, T.; Suzuki, S.; Suzuki, Y.; Miyaji, Y.; Imai, Z. *Tetrahedron Lett* **1969**, 4555. (b) Satoh, T.; Suzuki, S.; Kikuchi, T.; Okada, T. *Chem. Ind. (London)*, **1970**, 1626.

18. Brown, H.C.; Subba Rao, B.C. *J. Am. Chem. Soc.* **1956**, *78*, 2582.

19. (a) Fleet, G.W.J.; Fuller, C.J.; Harding, P.J.C. *Tetrahedron Lett.* **1978**, 1437. (b) Sorrell, T.N.; Pearlman, T.S. *J. Org. Chem.* **1980**, *45*, 3449.

20. Lin, S.T.; Roth, J.A. *J. Org. Chem.* **1979**, *44*, 309.

21. Kano, S.; Yuasa, Y.; Shibuya, S. *Chem. Commun.* **1979**, 796.

22. Tanaka, Y.; Hibino, S. *Chem. Commun.* **1980**, 415.

23. Review: Narasimhan, S.; Balakumar, R. *Aldrichimica Acta* **1998**, *31*, 19.

24. Review of zinc borohydride reactivity: Narasimhan, S.; Balakumar, R. *Aldrichimica Acta* **1998**, *31*, 19.

25. Nutaitis, C. F.; Gribble, G. W. *Tetrahedron Lett.* **1983**, *24*, 4287.

the opposite selectivity (e.g., Example **20.22**) when K-Selectride™ (which does not chelate) is used.

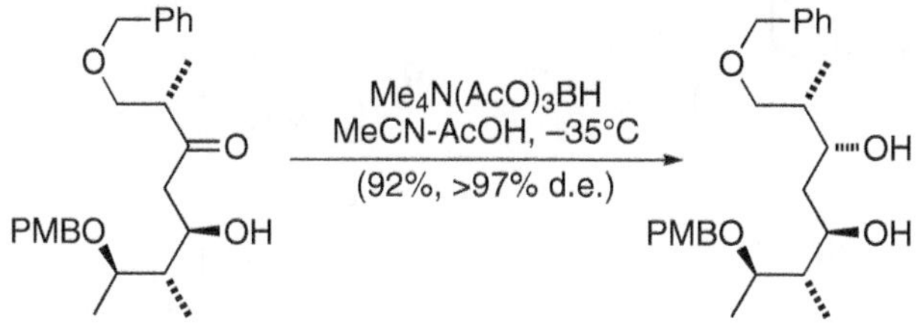

(20.19)[26]

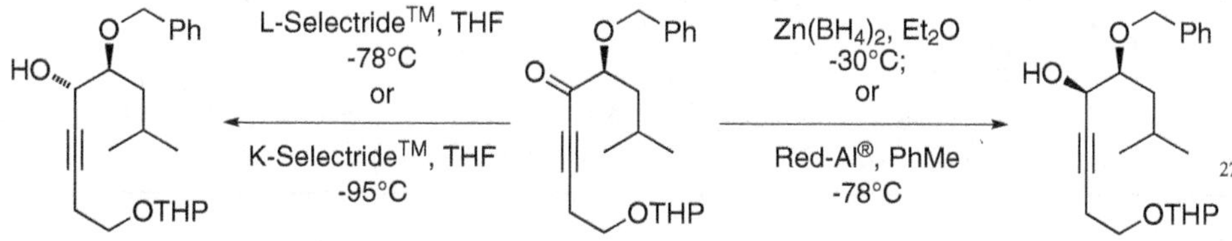

[27]

(20.22) **(20.20)** **(20.21)**

Worked Problem

20-1 What will be the major organic product of the following reactions?

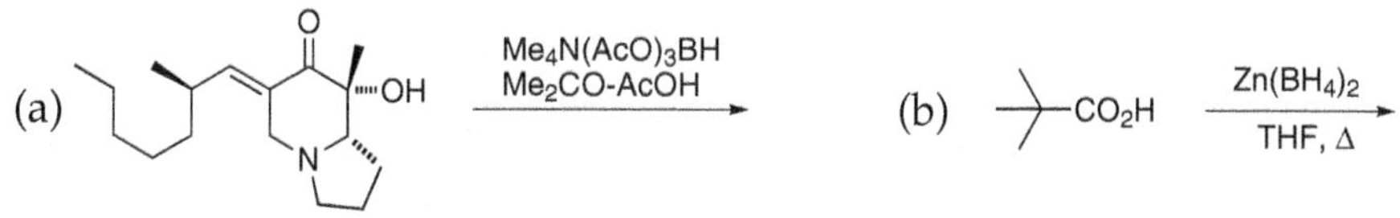

§Answers below.

Problem

20-1 What will be the major organic product of each of the following reactions?

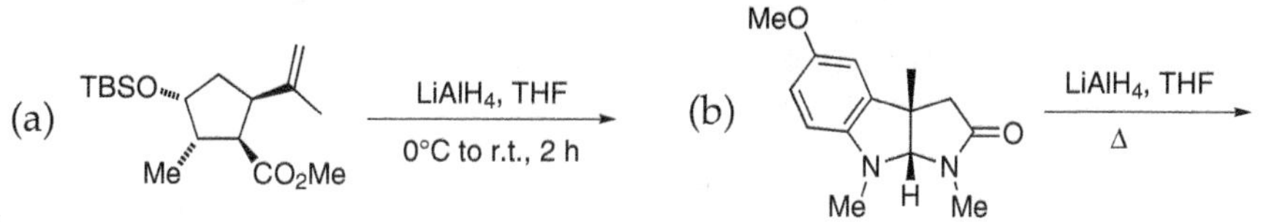

(continues)

26. Paterson, I.; Lombart, H. -G.; Allerton, C. *Org. Lett.* **1999**, *1*, 19.
27. Takahashi, T.; Miyazawa, M.; Tsuji, J. *Tetrahedron Lett.* **1985**, *26*, 5139.

§ Answers to Worked Problem:
(a) The reduction proceeds through a five-membered cyclic transition state to give the *trans*-cyclohexanediol derivative. [*Org. Lett.* **2001**, *3*, 469]

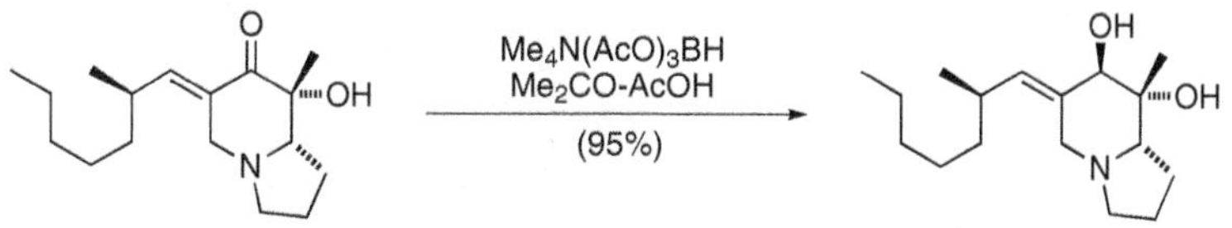

(b) Unlike other borohydride reagents, zinc borohydride will reduce carboxylic acids in THF under reflux. The product is the primary alcohol. [*J. Org. Chem.* **1995**, *60*, 5314]

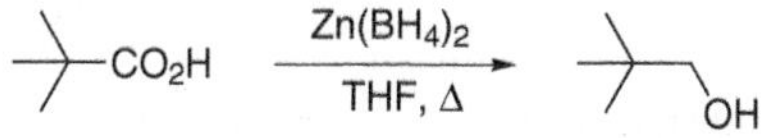

(**Problem** continued)

(c) 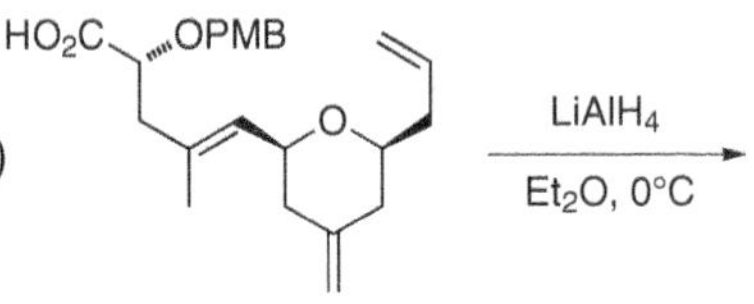

(d) 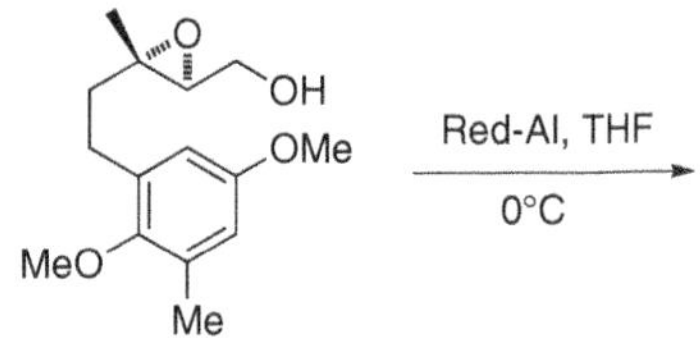

(e) 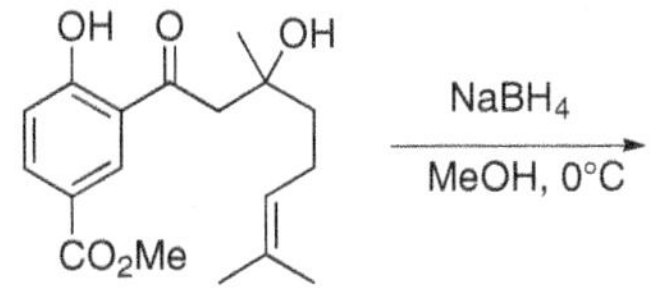

(f) 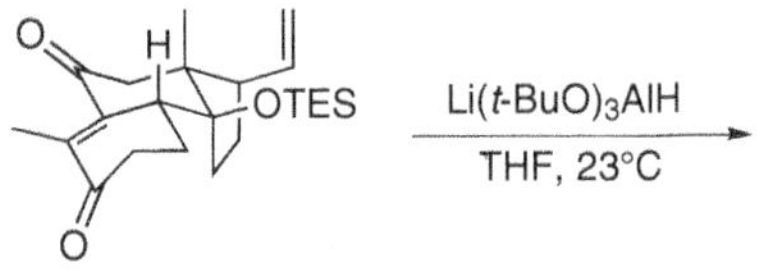

References for Problems: (a) *J. Am. Chem. Soc.* **2010**, *132*, 1488. (b) *Org. Lett.* **2007**, *9*, 1219. (c) *Org. Lett.* **2006**, *8*, 1117. (d) *Org. Let.* **2005**, *7*, 4297. (e) *Tetrahedron* **2011**, *67*, 6300. (f) *Tetrahedron Lett.* **2011**, *52*, 4512.

Regiochemistry of Reduction of α,β-Unsaturated Carbonyl Systems

The α,β-unsaturated carbonyl group allows two possible regiochemical modes of reaction: 1,2-addition, where the product is an allylic alcohol, and 1,4-addition, where the product is the saturated carbonyl compound. The results of systematic studies of the reduction of conjugated systems by sodium borohydride and lithium aluminum hydride (Figure 20.3) show that lithium aluminum hydride favors 1,2-reduction except in fairly hindered systems, whereas sodium borohydride often gives substantial amounts of the 1,4-reduction product. The work of Anh and coworkers[28] correlates the tendency of the reducing agent to promote 1,4-reduction of conjugated ketones with the softness of the hydride reagent: the softer the reagent, the higher the proportion of 1,4-reduction. Thus, sodium borohydride gives more 1,4-reduction than lithium borohydride (sodium is a softer Lewis acid than lithium), whereas lithium trimethoxyaluminum hydride gives more 1,2-reduction than lithium aluminum hydride (the alkoxy groups render the aluminum hydride anion a harder base). In a systematic study of carbonyl compounds with sodium

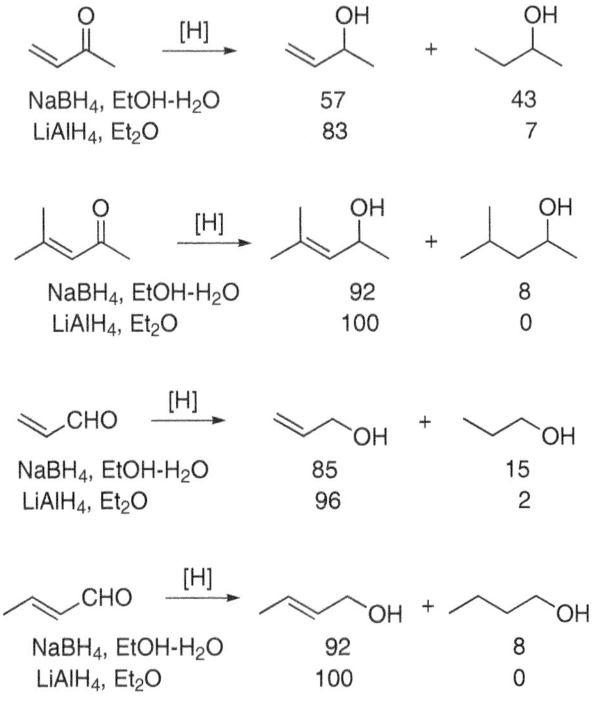

Figure 20.3 Regiochemistry of reduction of representative conjugated aldehydes and ketones

28. Bottin, J.; Eisenstein, O.; Minot, C.; Anh, N.T. *Tetrahedron Lett.* **1972**, 3015.

borohydride, Johnson and Rickborn found that α,β-unsaturated ketones are much more prone to 1,4-reduction with sodium borohydride.[29] The effects of other additives (e.g., pyridine) on the regiochemistry and stereochemistry of reduction of a number of 3-alkylidenecamphor derivatives by sodium borohydride have been quantified.[30]

Neither lithium aluminum hydride nor sodium borohydride is universally reliable for carrying out 1,2-reduction, so modifications of these reagents with metal ions have been studied in an effort to obtain a reagent with exclusive, or almost exclusive, 1,2-reduction regiochemistry. This has been accomplished by the use of Ce^{3+} ions as additives with sodium borohydride. In the presence of cerium (III) chloride, sodium borohydride becomes an effective reagent for the reduction of α,β-unsaturated aldehydes and ketones (e.g., Example **20.23**) to the corresponding allylic alcohols (e.g., Examples **20.24** and **20.25**). In the presence of 0.25 mole equivalents of Ce^{3+} ions per mole of sodium borohydride, a reagent is obtained that cleanly reduces the carbonyl group of the conjugated system.[31]

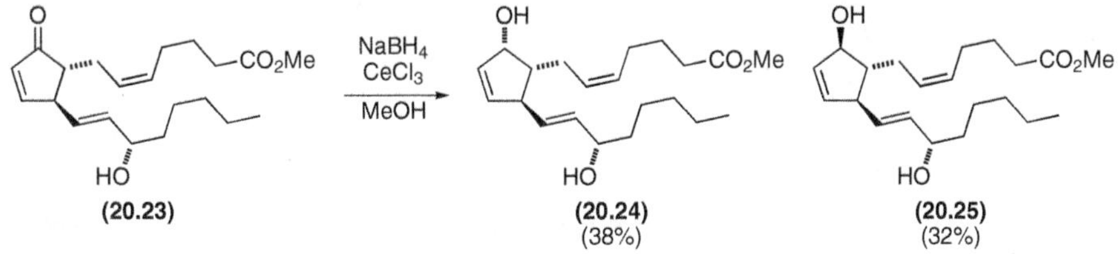

Reductive Amination

In the previous chapter, we discussed reductive amination by means of catalytic hydrogenation. However, this reaction is more widely carried out using complex metal hydrides. Among these, sodium cyanoborohydride, $NaBH_3CN$,[32] is probably most widely used—despite its high toxicity—because it reduces iminium ions rapidly in alcohol or mildly acidic solvents (e.g., Example **20.26**) but reacts only sluggishly with carbonyl compounds. Interestingly, this reagent will also reduce α,β-unsaturated iminium ions derived from conjugated aldehydes to the saturated systems (e.g., Example **20.27**). More recently, sodium triacetoxyborohydride[33] has gained popularity as a nontoxic replacement for this toxic reagent, which can leave residual cyanide in the product.

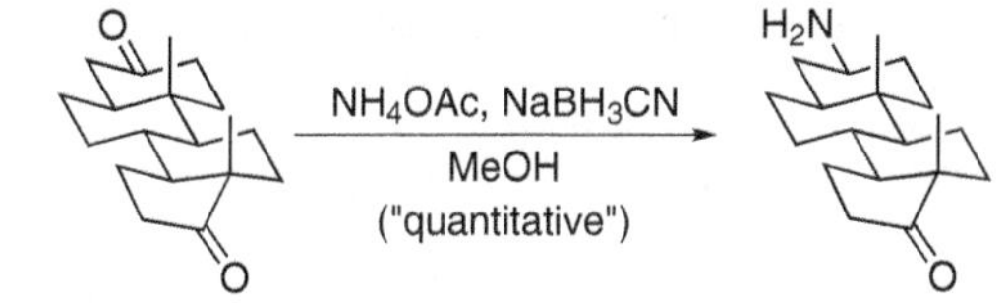

(20.26)[34]

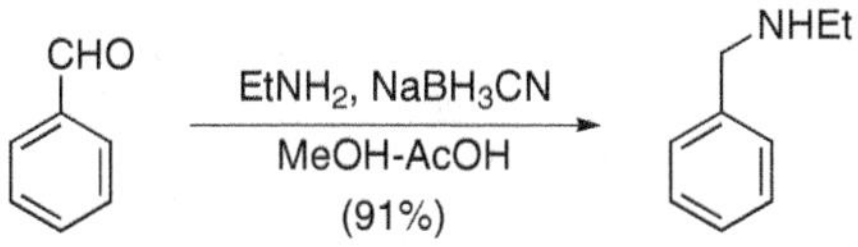

(20.27)[35]

29. Johnson, M.R.; Rickborn, B. *J. Org. Chem.* **1970**, *35*, 1041.

30. Richer, J.-C.; Rossi, A. *Can. J. Chem.*, **1975**, *50*, 1376.

31. (a) Gemal, A.L.; Luche, J.-L. *J. Am. Chem. Soc.* **1981**, *103*, 5454. (b) Luche, J.-L. *J. Am. Chem. Soc.* **1978**, *100*, 2226. (c) Luche, J.-L.; Rodriguez-Hahn, L.; Crabbé, P. *Chem. Commun.* **1978**, 601. (d) Gemal, A.L.; Luche, J.-L. *J. Org. Chem.* **1979**, *44*, 4187.

32. (a) Borch, R.F.; Bernstein, M.D.; Durst, H.D. *J. Am. Chem. Soc.* **1971**, *93*, 2897. (b) Mattson, R.J.; Pham, K.M.; Leuck, D.J.; Cowan, K.A. *J. Org. Chem.* **1990**, *55*, 2552. (c) Barney, C.L.; Huber, E.W.; McCarthy, J.R. *Tetrahedron Lett.* **1990**, *31*, 5547. (d) Hutchins, R.O.; Natale, N.R. *Org. Prep. Proc. Int.* **1979**, *11*, 201. (e) Lane, C.F. *Synthesis*, **1975**, 135. (f) Lane, C.F. *Aldrichimica Acta* **1975**, *8*, 3.

33. (a) Abdel-Magid, A.F.; Maryanoff, C.A.; Carson, K.G. *Tetrahedron Lett.* **1990**, *31*, 5545. (b) Abdel-Magid, A.F.; Carson, K.G.; Harris, B.D.; Maryanoff, C.A.; Shah, R.D. *J. Org. Chem.* **1996**, *61*, 3849.

34. Boutique, M.-H.; Jacquesy, R. *Bull. Soc. Chim. France* **1973**, 750.

35. Borch, R.F.; Bernstein, M.D.; Durst, H.D. *J. Am. Chem. Soc.* **1971**, *93*, 2897.

Before the advent of complex metal hydrides, the reducing agent of choice was formic acid, and under these conditions the reductive amination occurs by a mechanism similar to that of the crossed Cannizzaro reaction.[36] Depending on the exact conditions used, this reaction has several names. If ammonium formate is the reagent, the product is the primary amine and the reaction is called the ***Leuckart reaction.***[37] Extension of the Leuckart reaction to using primary or secondary amines with the formic acid is known as the ***Wallach reaction.***[38] In its most widely used form today, the carbonyl compound in the Leuckart reaction is formaldehyde, and the product is usually the tertiary amine where all the amine hydrogens have been replaced by methyl groups; this reaction is known as the ***Eschweiler-Clarke reaction*** (e.g., Example **20.28**).[39] These reactions will be discussed in greater detail later in this chapter.

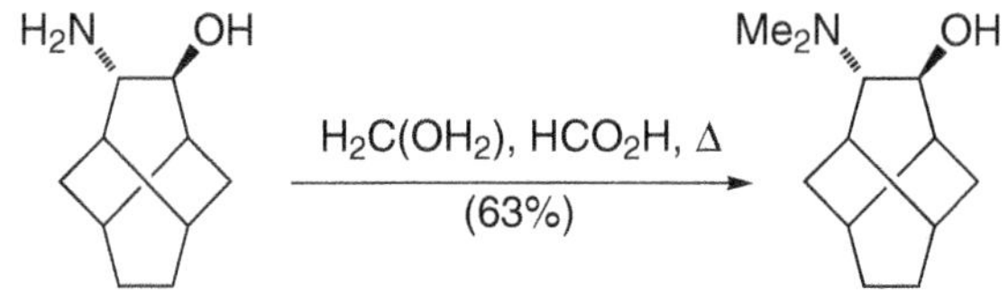

(20.28)[40]

Electrophilic Hydride Reducing Agents

Borohydrides and hydroaluminates are nucleophilic reducing agents whose reactivity is promoted by Lewis acids. However, reduction of carbonyl compounds and nitriles can also be effected by electrophilic reducing agents. Two of the most widely used electrophilic reducing agents are diisobutylaluminum hydride (DIBAL-H)[41] and borane (BH_3) in the form of its complex with a suitable Lewis base (e.g., THF or dimethyl sulfide). The key difference between electrophilic and nucleophilic hydride reducing agents, mechanistically, is the ability of the metal atom in the hydride to form a Lewis acid–Lewis base complex (**20.29**) with the carbonyl oxygen atom of the carbonyl group (Figure 20.4). This

---

36. For a discussion of the reductive amination with ammonium formate, see: (a) Lukasiewicz, A. *Tetrahedron* **1963**, *19*, 1789. (b) Awachie, P.I.; Agwada, V.C. *Tetrahedron* **1990**, *46*, 1899.

37. (a) Leuckart, R. *Ber. Dtsch. chem. Ges.* **1885**, *18*, 2341. (b) Moore, M.L. *Org. React.* **1949**, *5*, 301.

Carl Louis Rudolf Alexander Leuckart (1854–1889) was educated at Leipzig (PhD, 1879) and stayed until 1880 as assistant to Kolbe before moving to Munich as an assistant to Baeyer. In 1883 he moved to Göttingen and was promoted to professor in 1889. He died at Göttingen of an apparent stroke a scant month after his thirty-fifth birthday. For more detail, see: *Ber. dtsch. chem. Ges.* **1889**, *22*, (Referate, Patente, Nekrologe), 885.

38. Wallach, O. *Ann. Chem. Pharm.* **1892**, *272*, 100.

Otto Wallach (1847–1931) was educated at Göttingen (PhD, 1869). His career took him to Bonn (1870), where he became Professor of Pharmacy in 1876, and then back to Göttingen (1889). While at Bonn, Wallach became interested in essential oils and formulated the isoprene rule for terpenes. He received the Nobel Prize in Chemistry in 1910. For more biographical detail, see; (a) Ruzicka, L. *J. Chem. Soc.* **1932**, 1582; (b) Christmann, M. *Angew. Chem. Int. Ed.* **2010**, *49*, 9580. Wallach's autobiography is available on the Nobel Foundation web site.

39. (a) Eschweiler, V. *Ber. Dtsch. chem. Ges.* **1905**, *38*, 880. (b) Clarke, H.T.; Gillespie, H.B.; Weisshaus, S.Z. *J. Am. Chem. Soc.* **1933**, *55*, 4571. For mechanistic and competitive reactions, see: (c) Baltzly, R. *J. Am. Chem. Soc.* **1953**, *75*, 6083. (d) Cope, A.C.; Burrows, W.D. *J. Org. Chem.* **1965**, *30*, 2163.

Wilhelm Eschweiler (1860–1936) began his studies at the Technische Hochschule in Hannover in 1887, as Assistant in the chemistry laboratory. This work led his PhD from Rostock in 1889 under Kraut. In 1892 he was promoted to privat-dozent and rose through the ranks to become Professor of Analytical Chemistry at the Technische Hochschule.

Hans Thacher Clarke (1887–1972) was educated at University College, London (BSc, 1908; DSc, 1914). His career began at Eastman Kodak in Rochester, New York (1914), and he became Professor of Biological Chemistry at the College of Physicians and Surgeons at Columbia University in 1928. For more detail, see: Vickery, H.B. *Biogr. Mem.* **1975**, 1.

40. Tichy, M.; Vasickova, S. *Coll. Czech. Chem. Commun.* **1974**, *39*, 555.

41. For an overview of DIBAL-H reductions of functional groups, see: (a) Yoon, N.M.; Gyoung, Y.S. *J. Org. Chem.* **1985**, *50*, 2443. (b) Winterfeldt, E. *Synthesis* **1975**, 617.

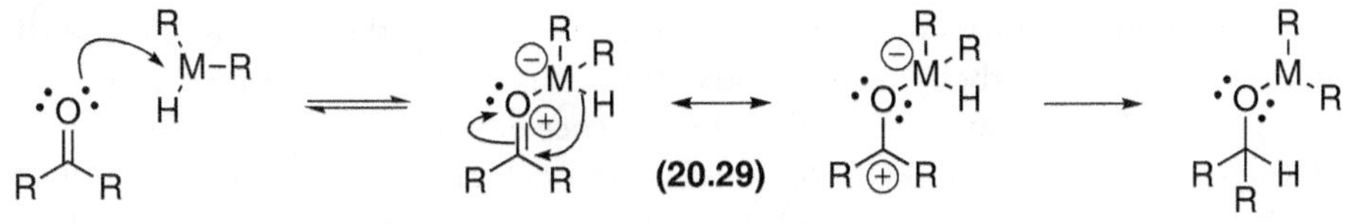
(20.29)

Figure 20.4 Reduction of carbonyl compounds with electrophilic reducing agents

process both renders the metal-hydrogen bond more strongly nucleophilic and the carbonyl carbon more electrophilic.

DIBAL-H has achieved considerable popularity for the reduction of conjugated carbonyl compounds to allylic alcohols (e.g., Examples **20.30** and **20.31**) and nitriles to imines (and hence to aldehydes; e.g., Example **20.32**). With two equivalents of the reducing agent, the reduction of esters gives the primary alcohol (e.g., Examples **20.30** and **20.31**; example **20.31** also shows that the reduction of the carbonyl group is faster than the hydroalumination of the alkyne). With one equivalent of the reducing agent at low temperature (usually −70°C or lower), the hemiacetal formed in the initial addition of hydrogen to the ester carbonyl group is stable, and the aldehyde is the major product isolated. This reduction to an aldehyde has been especially useful for the reduction of lactones to cyclic hemiacetals (e.g., Example **20.33**).

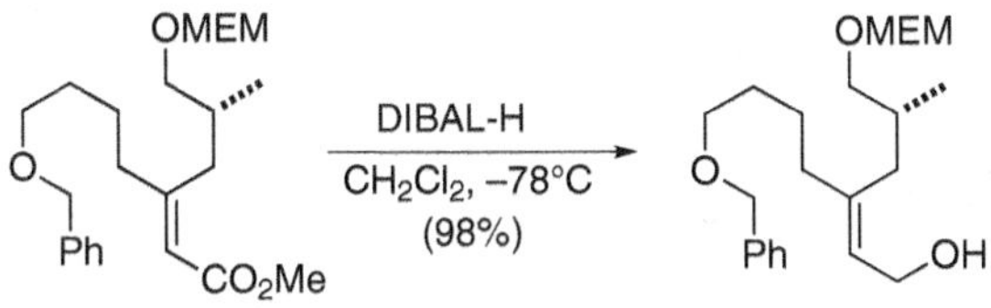

(20.30)[42]

(20.31)[43]

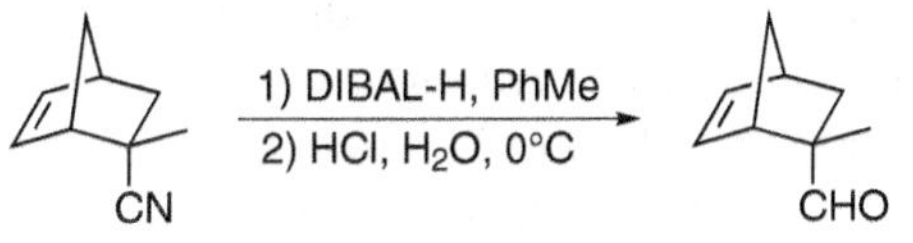

(20.32)[44]

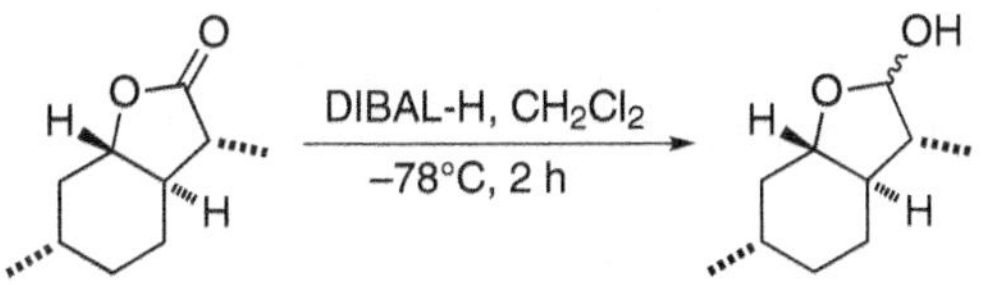

(20.33)[45]

Diborane is another electrophilic reducing agent that is highly selective for carboxylic acids (e.g., Example **20.34**)[46] and amides[47]—two functional groups that usually resist reduction. The borane-THF and borane–dimethyl sulfide complexes, both of

42. Williams, D.R.; Brown, D.L.; Benbow, J.W. *J. Am. Chem. Soc.* **1989**, *111*, 1923.
43. Meyers, A.G.; Harrington, P.M.; Kuo, E.Y. *J. Am. Chem. Soc.* **1991**, *113*, 694.
44. Lewis, D.E.; Rigby, H.L. *Tetrahedron Lett.* **1985**, *26*, *3437*.
45. Corey, E.J.; Carpino, P. *J. Am. Chem. Soc.* **1989**, *111*, 5472.
46. (a) Brown, H.C.; Subba Rao, B.C. *J. Am. Chem. Soc.* **1960**, *82*, 681. (b) Brown, H.C.; Korytnik, W. *J. Am. Chem. Soc.* **1960**, *82*, 3866. (c) Brown, H.C.; Heim, P.; Yoon, N.M. *J. Am. Chem. Soc.* **1970**, *92*, 1637. (d) Yoon, N.M.; Pak, C.S.; Brown, H.C.; Krishnamurthy, S.; Stocky, T.P. *J. Org. Chem.* **1973**, *38*, 2786.
47. Brown, H.C.; Heim, P. *J. Org. Chem.* **1973**, *38*, 912.

which function as sources of borane, BH_3, react with the carboxylic acid much faster than with all other carbonyl compounds. Thus, this reagent can be used to reduce a carboxylic acid to a primary alcohol in the presence of most other carbonyl groups, including esters (e.g., Example **20.35**), amides, and ketones. Amides are reduced faster than ketones by borane. The one problem encountered in borane reduction of amides is the formation of very stable borane-amine complexes with the product of the reduction. Fortunately, it has been found that these complexes can be decomposed using supported palladium or Raney nickel in methanol. An interesting side effect of this cleavage reaction is the simultaneous hydrogenolysis of benzyl groups in the molecule, as shown by Example **20.36.**[48]

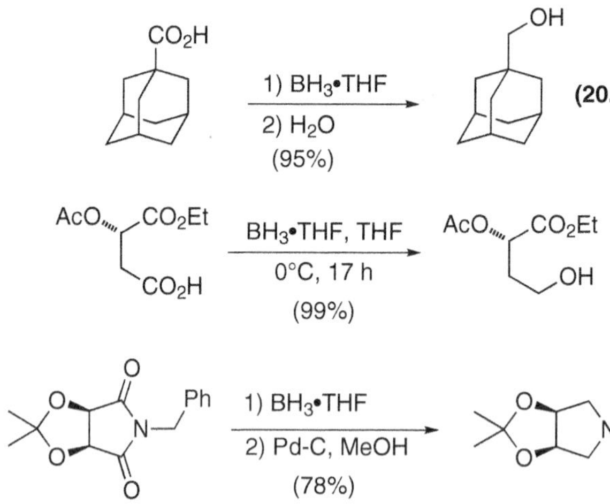

(20.34)

(20.35)

(20.36)

Hydroalumination of Alkynes

In contrast to the π bonds of unactivated alkenes, the triple bond of alkynes can be reduced by aluminum hydride reagents, which add to the triple bond to give organoaluminum intermediates that may be protonated or trapped with electrophiles to give alkenes with defined stereochemistry. When the aluminum hydride is a neutral aluminum species such as DIBAL-H, the reaction is known as ***hydroalumination*** (by analogy with hydroboration), and the addition occurs with the anti-Markovnikov regiochemistry and *syn* stereochemistry (e.g., Example **20.37**). When the aluminum species is negatively charged, however, the stereochemistry of the hydroalumination reaction is reversed (e.g., Example **20.38**). Thus, with lithium aluminum hydride and lithium alkylaluminum hydrides the overall addition of hydrogen becomes *anti*.[49]

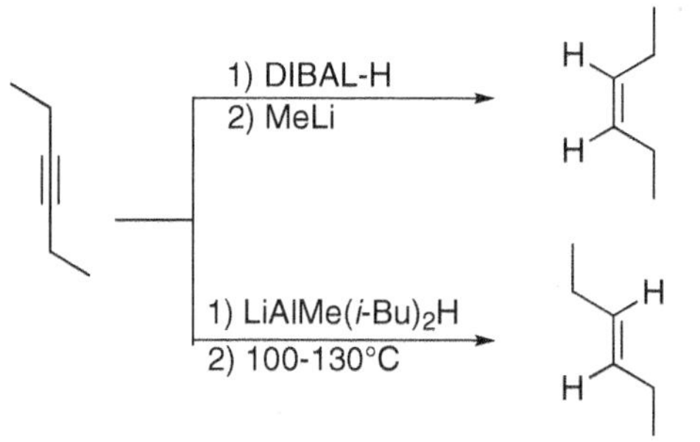

(20.37)

(20.38)

48. Couturier, M.; Tucker, J.L.; Andresen, B.M.; Dubé, P.; Negri, J.T. *Org. Lett.*, **2001**, *3*, 465.

49. (a) Zwiefel, G.; Steele, R.B. *Tetrahedron Lett.* **1966**, 6021. (b) Zwiefel, G.; Steele, R.B. *J. Am. Chem. Soc.* **1967**, *89*, 2754. (c) Zwiefel, G.; Steele, R.B. *J. Am. Chem. Soc.* **1967**, *89*, 5085. (d) Magoon, E.F.; Slaugh, L.H. *Tetrahedron* **1967**, *23*, 4509. (e) Molloy, B.B.; Hauser, K.L. *J. Chem. Soc., Chem. Commun.* **1968**, 1017.

The hydroalumination of alkynes with anionic aluminum hydride species has been used as a method for the formation of stereochemically pure trisubstituted alkenes from propargyl alcohols. In this case, the aluminum adds to the end of the triple bond furthest from the carbinol carbon. As expected for an anionic aluminum nucleophile, the stereochemistry of the addition is *anti*. A typical example of this useful reaction is the synthesis of the iodoallyl alcohol in Example **20.39**; the product of the reaction is the *Z* isomer.[50]

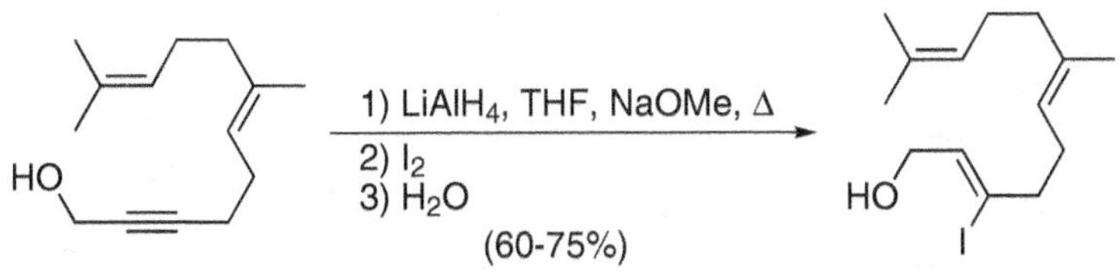

(20.39)

Worked Problem

20-2 What is the major product expected from the following reaction? What major advantage does the reagent used possess?

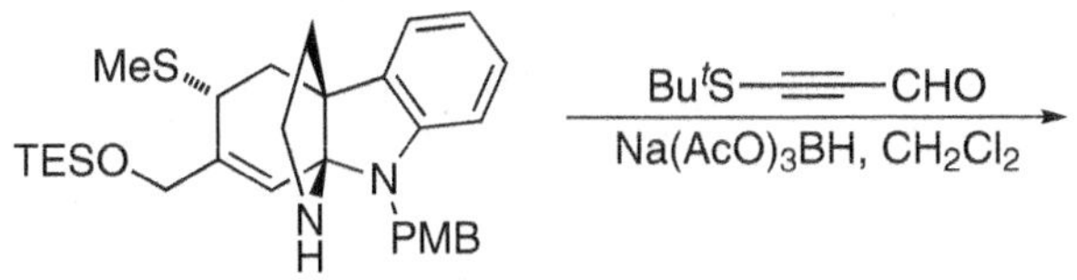

§Answer below.

Problems

20-2 Suggest a reasonable mechanism for the following transformation:

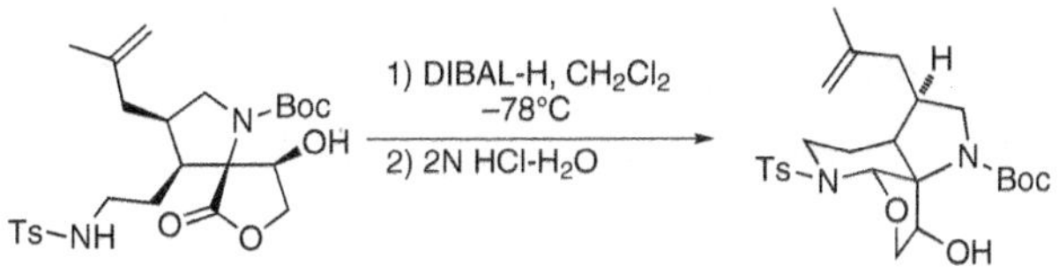

[*J. Am. Chem. Soc.* **2007**, *129*, 11987]

(continues)

50. Corey, E.J.; Katzenellenbogen, J.A.; Posner, G.H. *J. Am. Chem. Soc.* **1967**, *89*, 4245.

§Answer to Worked Problem:
This is a reductive amination of the aldehyde by the secondary amine group of the starting material. The reducing agent used will reduce only the imine, which makes it ideally suited to this reaction, where there are so many sensitive functional groups involved [*J. Am. Chem. Soc.* **2009**, *131*, 13606].

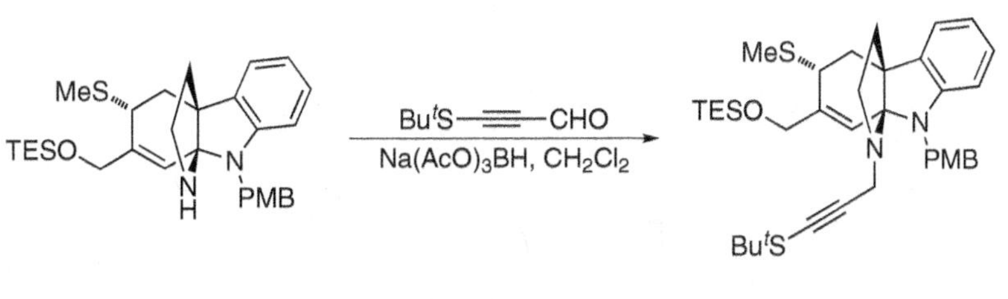

(**Problem** continued)

20-3 What is the structure of major organic product of each of the following reactions?

(a) 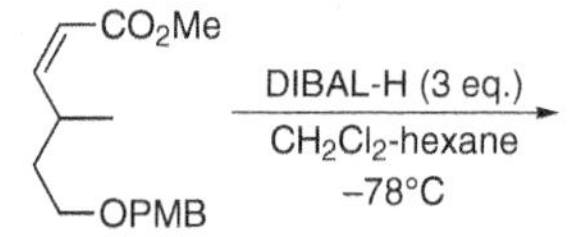

(b) 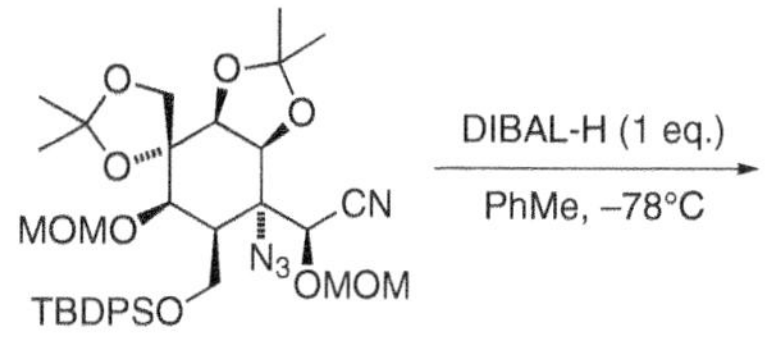

(c) 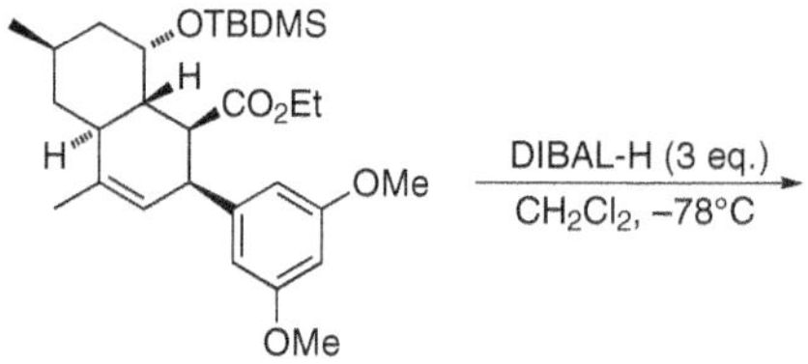

(d) 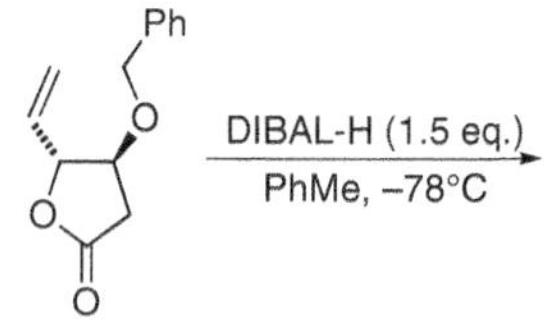

(e) 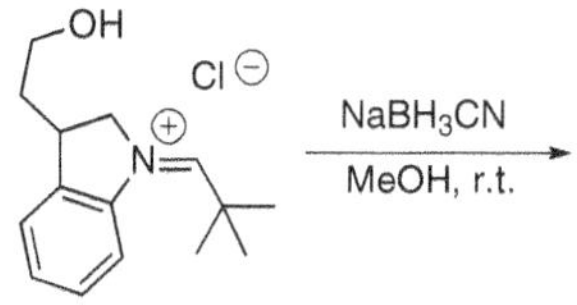

(f) 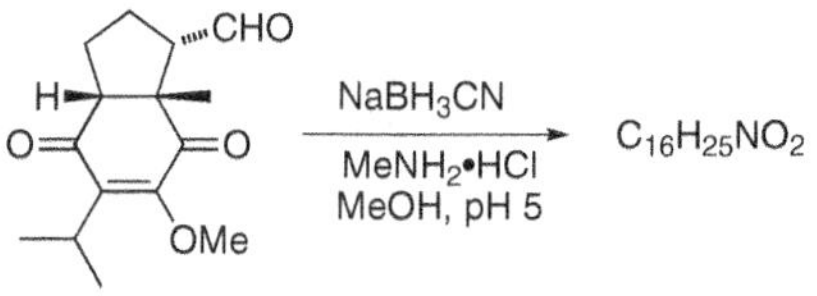

References: (a) *J. Org. Chem.* **2009**, *74*, 7220. (b) *J. Org. Chem.* **2008**, *73*, 1234. (c) *Org. Lett.* **2007**, *9*, 1461. (d) *Tetrahedron* **2011**, *67*, 4036. (e) *Tetrahedron* **2011**, *61*, 6015. (f) *J. Am. Chem. Soc.* **1974**, *96*, 4332.

Reaction Synopses

Hydride Reduction of Aldehydes and Ketones

Reagents:	$LiAlH_4$, Et_2O (or THF); Red-Al®; etc.
or	$NaBH_4$, CH_3OH; LiR_3BH, THF; etc.
or	DIBAL-H, hexane
or	BH_3•THF; BH_3, oxazaborolidine, THF; etc.

Hydride Reduction of Conjugated Carbonyl Compounds
To Allylic Alcohols

Reagents:	$LiAlH_4$, Et_2O, low temp., short time;
or	DIBAL-H, hexane; etc.
or	$NaBH_4$, $CeCl_3$, CH_3OH; etc.

To Saturated Alcohols

Reagents: $NaBH_4$, EtOH, extended time

Hydride Reductions of Acyl Compounds and Nitriles

To Alcohols

X = Cl, OCOR, OR, OH, O^-

Reagents: $LiAlH_4$, Red-Al®; BH_3•THF (X=OH);
or DIBAL-H (X=Cl, OCOR, OR)

To Amines

Reagents: $LiAlH_4$; Red-Al®; etc.
or BH_3•THF

To Aldehydes

Reagents: DIBAL-H, PhMe, 0°C, etc.

Reductive Amination

Reagents: R''_2NH, HOAc, MeOH, $NaBH_3CN$
or NH_3, HCO_2H, Δ (Leuckart)
or R''_2NH, HCO_2H, Δ (Wallach if R, R' = alkyl; Eschweiler-Clarke if R, R' = H)

20.2 Reduction by Hydrogen Transfer from Carbon

The complex metal hydrides based on aluminum or boron are the most popular reagents for reducing carbonyl groups. However, carbonyl compounds can also be reduced by a process where the hydrogen atom transferred to the carbonyl group is originally bound to carbon rather than boron or aluminum. Reduction by chiral boranes, in particular, has become a popular method for asymmetric reduction of ketones.

The common feature of all these reductions is that they involve a six-membered cyclic transition state (Example **20.40**)[51] in which the hydrogen atom β to the electropositive element in a metal alkyl or a metal alkoxide is transferred to the carbonyl carbon,[52] formally as a hydride ion. The mechanism may involve single electron transfer and free radical intermediates.[53] When the electropositive element is bonded to carbon in the reducing agent, the most common reducing agents used are Grignard reagents, in which case that element is magnesium, and trialkylboranes, in which case that element is boron. When an alkoxide is the reducing agent, the electropositive elements most commonly used are aluminum and magnesium.

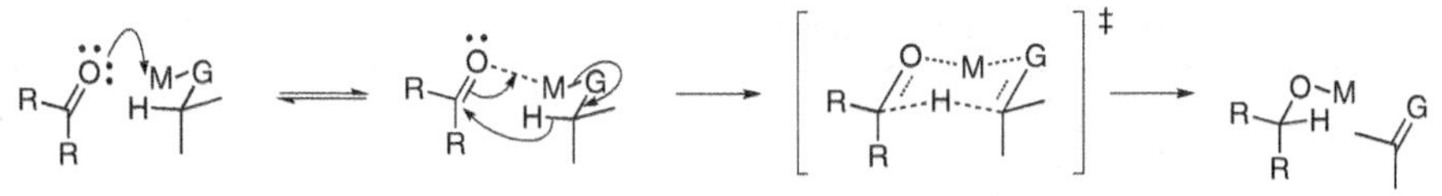

(20.40)

M = B, Al, Mg, Zn
G = O, CR_2

During the reaction, the organometallic reducing agents (e.g., G = CR_2, M = Mg, Zn) and boranes (G = CR_2, M = BR_2) eliminate an alkene as the hydrogen is transferred to the

51. Woodward, R.B.; Wendler, N.L.; Brutschy, F.K. *J. Am. Chem. Soc.* **1945**, *67*, 1425.

52. For example, aluminum alkoxides: (a) Doering, W.v.E.; Ashner, T.C. *J. Am. Chem. Soc.* **1953**, *75*, 393. (b) Williams, E.D.; Krieger, K.A.; Day, A.R. *J. Am. Chem. Soc.* **1953,** *75*, 2404. (c) Lutz, R.E.; Gillespie, J.S., Jr. *J. Am. Chem. Soc.* **1950**, *72*, 344.

53. For example, Grignard reagents: Ashby, E.C.; Goel, A.B. *J. Am.Chem. Soc.* **1981**, *103*, 4983.

carbonyl carbon. In reductions with metal alkoxides (e.g., G = O, M = Al, Mg, Zn,), the reagent is converted into a carbonyl compound, which makes these reactions reversible.

Reduction by Grignard Reagents (G = CR_2, M = Mg)[54]

Normally, a Grignard reagent adds to a carbonyl compound to give the product of nucleophilic addition to the carbonyl group. However, if the Grignard reagent is bulky (e.g., isopropylmagnesium halides or *tert*-butyl-magnesium halides) or if the carbonyl compound is sterically hindered, the major product of the reaction is often due to reduction of the ketone by transfer of a β hydrogen from the Grignard reagent to the carbonyl carbon. The mechanism of the reaction has been the subject of numerous studies,[55] which reveal that this reaction may be suppressed by using hydrocarbon solvents for the addition of Grignard reagents to carbonyl compounds. The reaction is most often observed when the carbonyl compound is a ketone, but this is not an absolute restriction. Two typical examples are shown as **20.41** and **20.42**.

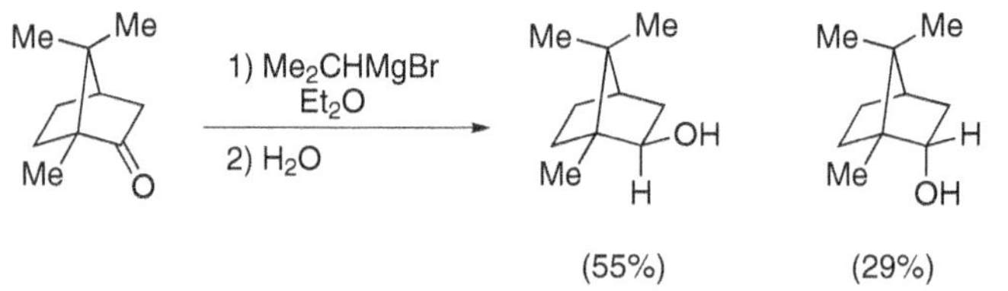

(20.41)[56]

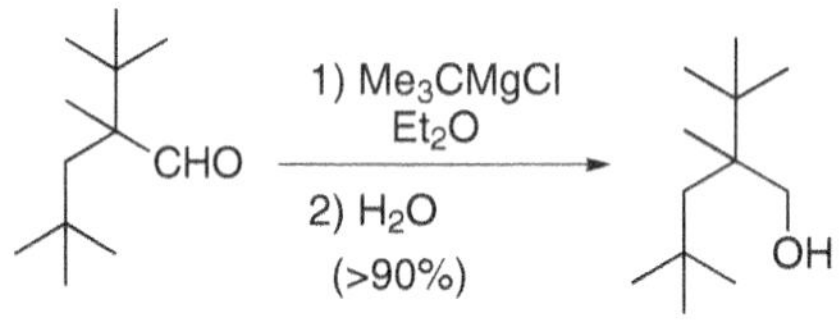

(20.42)[57]

The Meerwein-Ponndorf-Verley Reduction (G = O, M = Al)[58]

In the mid-1920s, German chemists Hans Meerwein[59] and Wolfgang Ponndorf,[60] along with French chemist Albert Verley,[61] all independently reported that aldehydes and

---

54. (a) Whitmore, F.C.; George, R.S. *J. Am. Chem. Soc.* **1942**, *64*, 1239. (b) Conant, J.B.; Blatt, A.H. *J. Am. Chem. Soc.* **1929**, *51*, 1227. (c) Conant, J.B.; Webb, C.N.; Mendum, W.C. *J. Am. Chem. Soc.* **1929**, *51*, 1246. (d) Blatt, A.H.; Stone, J.F., Jr. *J. Am. Chem. Soc.* **1932**, *54*, 1495. (d) Kharasch, M.S.; Weinhouse, S. *J. Org. Chem.* **1936**, *1*, 209. (e) Greenwood, F.L.; Whitmore, F.C.; Crooks, H.M. *J. Am. Chem. Soc.* **1938**, *60*, 2028. (f) Copwan, D.O.; Mosher, H.S. *J. Org. Chem.* **1962**, *27*, 1. (g) Lajis, N.Hj.; Khan, M.N.; Hassan, H.A. *Tetrahedron* **1993**, *49*, 3405.

55. (a) Singer, M.S.; Salinger, R.M.; Mosher, H.S. *J. Org. Chem.* **1967**, *32*, 3821. (b) Denise, B.; Fauvarque, J.-F.; Ducom, J. *Tetrahedron Lett.* **1970**, 335. (c) Cabaret, D.; Welvart, Z. *J. Organomet. Chem.* **1974**, *80*, 199. (d) Morrison, J.D.; Tomaszewski, J.E.; Mosher, H.S.; Dale, J.; Miller, D.; Elsenbaumer, R.L. *J. Am. Chem. Soc.* **1977**, *99*, 3167.

56. (a) Whitmore, F.C.; Whitaker, J.S.; Mosher, W.A.; Breivik, O.N.; Wheeler, W.R.; Miner, C.S., Jr.; Sutherland, L.H.; Wagner, R.B.; Clapper, T.W.; Lewis, C.E.; Lux, A.R.; Popkin, A.H. *J. Am. Chem. Soc.* **1941**, *63*, 643. (b) Hamelin, A. *Bull. Soc. Chim. France* **1961**, 926.

57. Malkonen, P.J. *Suomen. Kem.* **1965**, *38B*, 89 [*Chem. Abstr.* **1965**, *63*, 8411b].

58. Wilds, A.I. *Org. React.* **1944**, *2*, 174.

59. Meerwein, H.; Schmidt, R. *Liebigs Ann. Chem.* **1925**, *444*, 221.

60. Ponndorf, W. *Angew. Chem.* **1926**, *39*, 138.

Wolfgang Ponndorf (1894–1948) obtained his PhD from the University of Berlin in 1918. He was working at Anton Deppe & Sons when he reported his reduction. He later took his Dr Med (1931) from the Medical University Clinic of Hamburg-Eppendorf Hospital.

61. (a) Verley, A. *Bull. Soc. Chim. France [4]* **1925**, *37*, 537. (b) Verley, A. *Bull. Soc. Chim. France [4]* **1925**, *37*, 871.

Albert Verley (*ca.* 1896–1959?) spent his entire career in industry, founding his own company and establishing himself as a perfumer in Paris. Albert Verley, Inc., was incorporated in 1932 in Chicago, and became Albert Verley and Co. during the later 1930s (possibly when Verley's son, Guy, became active in the running of the company). He appears to have returned to France after 1944. In his youth, Verley was the only composition student of French composer, Erik Satie. His career in music was cut short by a laboratory accident that permanently injured his right hand.

ketones could be reduced to alcohols by aluminum alkoxides. The reaction is now known by their names: the ***Meerwein-Ponndorf-Verley (MPV) reduction.*** Unlike reductions with complex metal hydrides, Grignard reagents or boranes, this reaction is reversible, so that the final product is the more stable stereoisomer of the alcohol. It is worth noting that if the MPV reduction is stopped after a short time—before the equilibrium mixture of products is formed—it gives a higher proportion of the *endo* isomer. In similar fashion, the reduction of the ketone gives only the β alcohol (Example **20.44;** derived by attack of the reagent from the less hindered *exo* direction) with lithium aluminum hydride but only the α alcohol (Example **20.43**) under MPV conditions.[62] The reversibility of the reaction has also been exploited for oxidation, under the name of the ***Oppenauer oxidation***.[63]

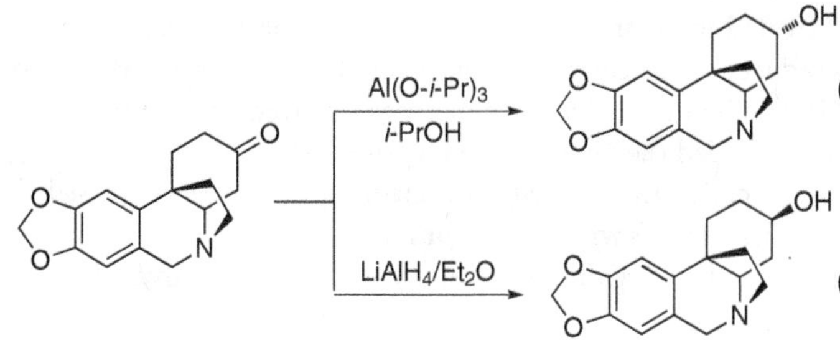

(20.43)

(20.44)

The reversibility of the MPV reduction affects the stereochemistry of the reduction in hindered systems, as is illustrated by the reductions of camphor with lithium aluminum hydride in THF and aluminum isopropoxide in isopropyl alcohol. Lithium aluminum hydride, which reacts irreversibly from the less hindered face of the carbonyl group, leads to the more hindered (less stable) *exo* isomer as the major product of the reaction. In contrast, MPV reduction, being reversible, eventually gives the less hindered (more stable) *endo* isomer. When the energy difference between the two isomers of the product is small, the difference between the use of lithium aluminum hydride and the MPV reduction are also less pronounced, as shown by the reduction of 3-methylcyclohexanone. A milder, modern variant of the MPV reduction that avoids the strongly basic reaction conditions of the original reaction involves the use of samarium (II) iodide (e.g., Example **20.45**).[64]

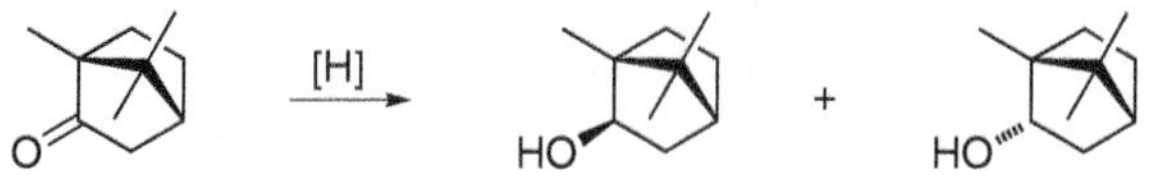

[H] = $LiAlH_4$	89	:	11
[H] = Al(O-*i*-Pr)$_3$, *i*-PrOH	10	:	90

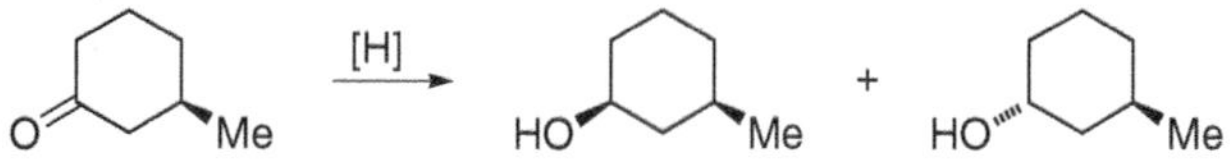

[H] = $LiAlH_4$	76	:	24
[H] = Al(O-*i*-Pr)$_3$, *i*-PrOH	70	:	30

---

62. (a) Irie, H.; Uyeo, S.; Yoshitake, A. *J. Chem. Soc., Chem. Commun.* **1966**, 635. (b) Irie, H.; Uyeo, S.; Yoshitake, A. *J. Chem. Soc., C* **1968**, 1802.

63. (a) Oppenauer, R.V. *Rec. Trav. Chim.* **1937**, *1937*, 137. (b) Djerassi, C. *Org. React.* **1951**, *6*, 207. (c) Opppenauer, R.V. *Org. Syn. Coll. Vol. 3*, **1943**, 207.

Rupert Viktor Oppenauer (1910–1969) was born in Austria, and educated at the ETH, in Zürich (PhD, 1934, under Nobel laureates Leopold Ruzicka and Tadeusz Reichstein). Following several academic appointments in Europe, he worked for a time at Hofmann–La Roche and then moved to Buenos Aires, where he worked for the Ministry of Public Health.

64. (a) Collins, J.; Namy, J.-L.; Hagan, H.F. *Nouv. J. Chim.* **1986**, *10*, 229. (b) Evans, D.A.; Kaldor, S.W.; Jones, T.K.; Clardy, J.; Stout, T.J. *J. Am. Chem. Soc.* **1990**, *112*, 7001.

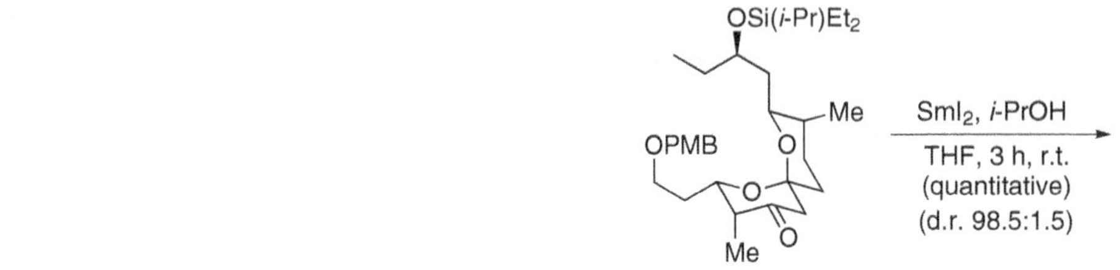

(20.45)

The Cannizzaro and Tishchenko Reactions (G = O, M = Na or Al)

In 1853, the Italian chemist Stanislao Cannizzaro noted that when aldehydes that lack an α hydrogen were treated with strong base, a disproportionation reaction occurred. Half the aldehyde was reduced to the corresponding alcohol, and the other half was oxidized to the carboxylic acid.[65] When the aldehyde molecules involved are identical (e.g., Example **20.46**), the reaction is called the ***Cannizzaro reaction***. When they are not (e.g., Example **20.47**), the reaction is called the ***crossed Cannizzaro reaction***. The crossed Cannizzaro reaction with formaldehyde was widely used for reducing aldehydes, although it has been largely superseded in modern organic synthesis by complex metal hydrides.

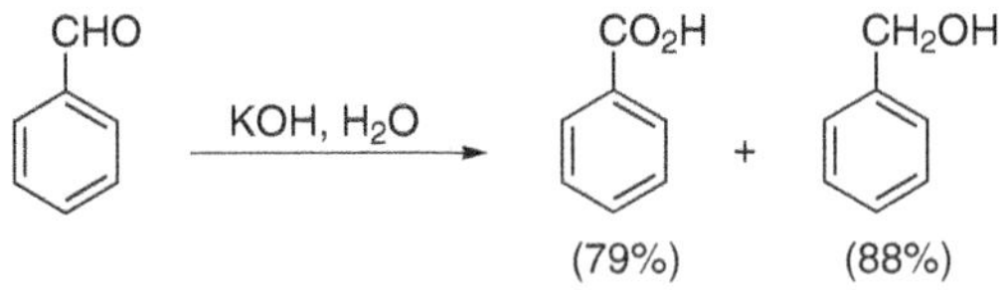

(20.46)

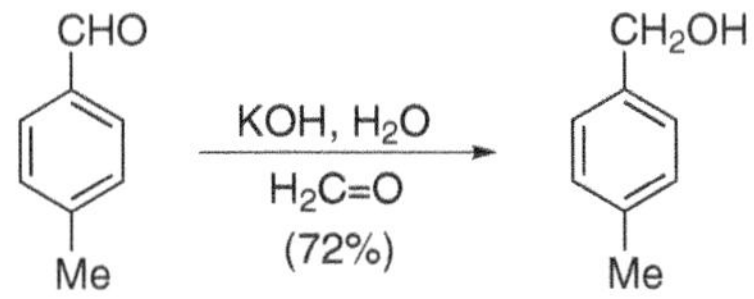

(20.47)

The crossed Cannizzaro reaction is still used in the industrial preparation of certain alcohols. The industrial synthesis of pentaerythritol (Example **20.48**), an important raw material for plastics and explosives, from acetaldehyde and formaldehyde provides an excellent example. The reaction involves a sequential combination of three aldol additions and a final crossed Cannizzaro reaction.

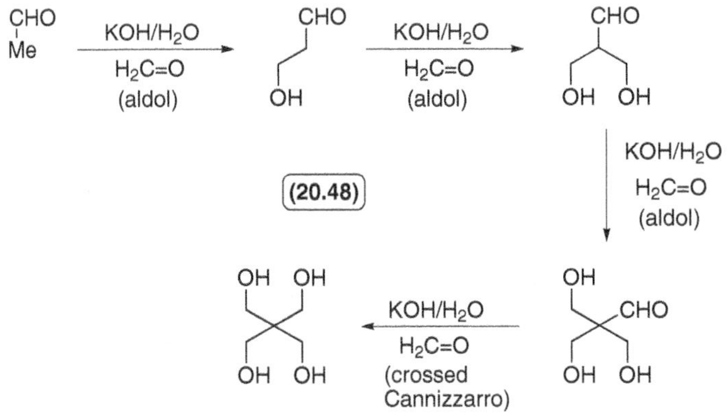

65. (a) Cannizzaro, S. *Ann. Chem. Pharm.* **1863**, *88*, 129. (b) List, K.; Limpricht, H. *Ann. Chem. Pharm.* **1864**, *90*, 180. **Stanislao Cannizzaro (1826–1910)** began as a medical student at Palermo, moving to Pisa in 1845. After the collapse of the Sicilian revolution in 1849, he was condemned to death but escaped to France and entered Chevreul's laboratory. His subsequent career took him from Alessandria (1851) to Genoa (1855), Palermo (1861), and finally Rome (1871). He ultimately became vice president of the Italian Senate. For a more complete biography, see: Tilden, W.A. *J. Chem. Soc., Trans.* **1912**, *101*, 1677.

In a series of papers in 1906, the Russian chemist V. E. Tishchenko described a similar redox reaction of aldehydes catalyzed by aluminum isopropoxide.[66] In this case, the product of the reaction is an ester derived by condensation of two molecules of the aldehyde with each other (e.g., Example **20.49**). As in the Cannizzaro reaction, one molecule of aldehyde is reduced to the alcohol level and the other is oxidized to the carboxylic acid level. A similar reaction, using the sodium alkoxide rather than the aluminum alkoxide, had been noted by Claisen in 1887.[67] A modern variant, known as the ***Evans-Tishchenko reaction***, involves the cross-reaction of an aldehyde and a β-hydroxyketone catalyzed by samarium (II) iodide to give the monoester of a 1,3-*anti*-diol.[68] Although the ester group usually ends up at the site of the ketone carbonyl (e.g., Example **20.50**), this is not always the case (e.g., Example **20.51**), so one must be prepared to determine its precise location experimentally.

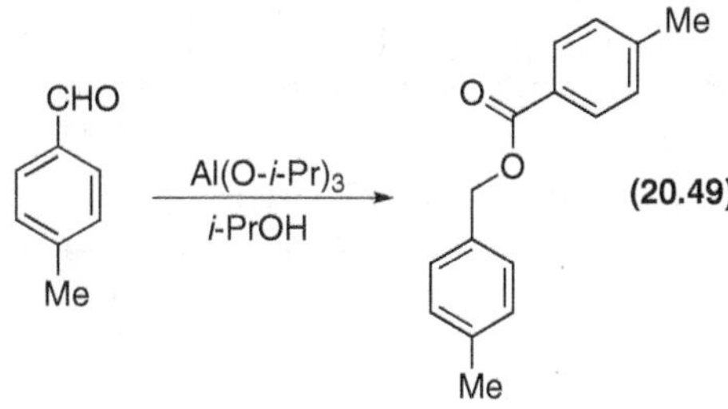

(20.49)

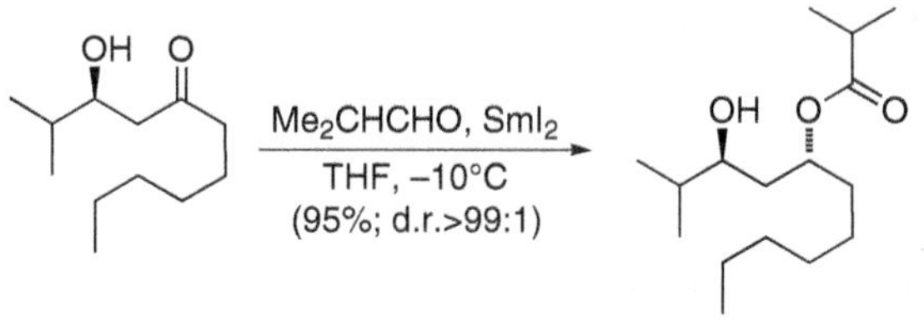

(20.50)

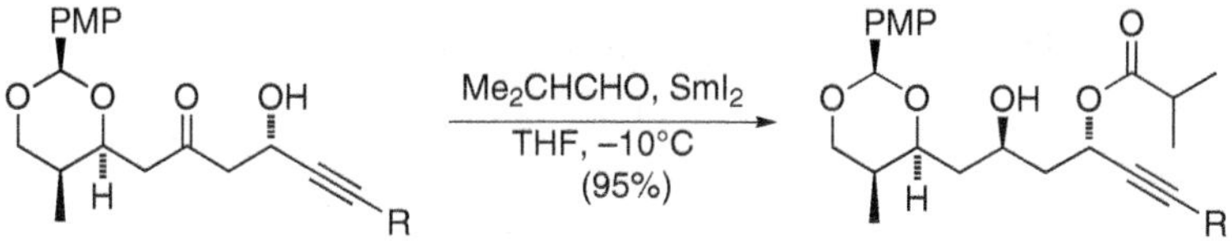

(20.51)

Worked Problem

20-3 What is the major product expected from the following reaction?

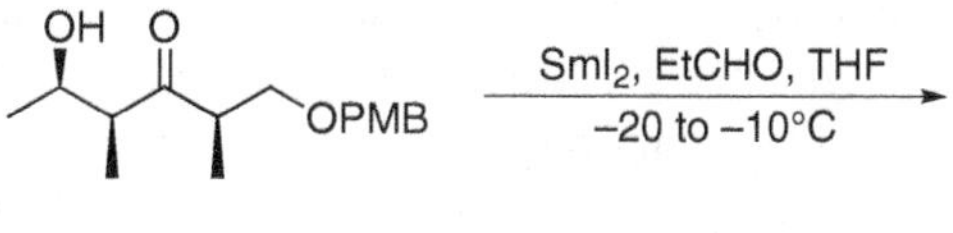

§Answer on next page.

66. (a) Tishchenko, V. *Zh. Russ. Fiz.-Khim. Obshch.* **1906**, *38*, 355. (b) Tishchenko, V. *Zh. Russ. Fiz.-Khim. Obshch.* **1906**, *38*, 482. (c) Tishchenko, V. *Zh. Russ. Fiz.-Khim. Obshch.* **1906**, *38*, 540. (d) Tishchenko, V. *Zh. Russ. Fiz.-Khim. Obshch.* **1906**, *38*, 547. (e) Tishchenko, V. *Chem. Zentr. II* **1906**, 1309. (f) Tishchenko, V. *Chem. Zentr. II* **1906**, 1552. (g) Tishchenko, V. *Chem. Zentr. II* **1906**, 1555. (h) Tishchenko, V. *Chem. Zentr. II* **1906**, 1556.

Vyacheslav Yevgenievich Tishchenko (1861–1941) graduated from St. Petersburg in 1883 and was appointed to the faculty in 1891. In 1901 he was appointed Professor of Chemistry at the Women's Medical Institute concurrent with his position at the University. He remained in Russia after the revolution, becoming Director of the Chemical Institute of Leningrad University. For more detail, see: Lewis, D.E. *Early Russian Organic Chemists and Their Legacy* (Springer: Heidelberg, 2012), p. 114.

67. Claisen, L. *Ber. Dtsch. chem. Ges.* **1887**, *20*, 646.

68. Evans, D.A.; Hoveyda, A.H. *J. Am. Chem. Soc.* **1990**, *112*, 6447.

Problems

20-4 What is the structure of the product of the following reaction (only a catalytic amount of DIBAL-H instead of a stoichiometric quantity is used)? How is the product formed?

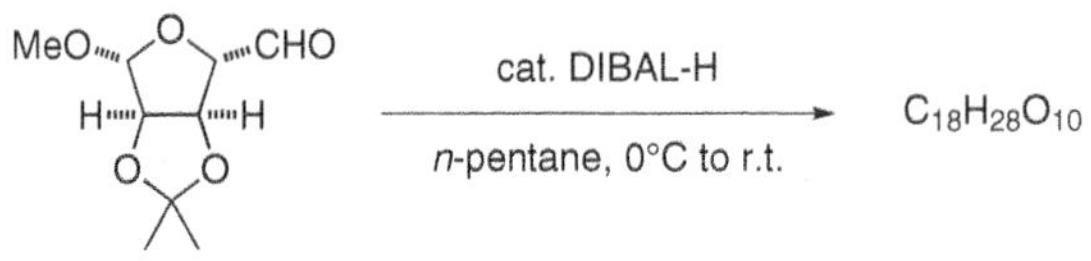

[*Tetrahedron* **2007**, *63*, 11325]

20-5 What is the major organic product of each of the following reactions?

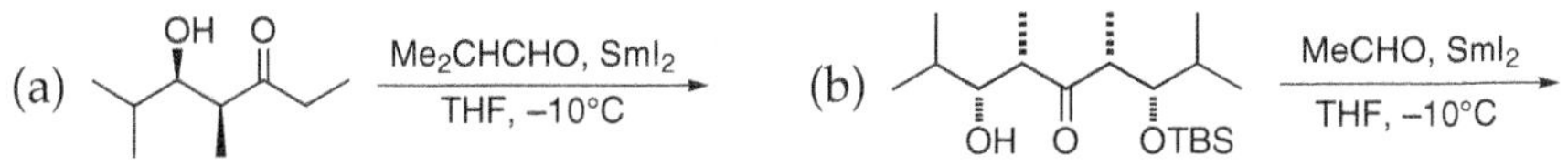

Reference for problems: J. Am. Chem. Soc. **1990**, *112*, 6447.

Reaction Synopses

Reduction of Aldehydes and Ketones by Hydrogen Transfer

Reagents: $R_2CH\text{-}CR_2\text{-}MgX$; $R_2CH\text{-}CR_2\text{-}ZnX$; etc.
or $R_2CH\text{-}CR_2\text{-}BR_2$;
or $(R_2CH\text{-}O)_3Al$, R_2CHOH; SmI_2, *i*-PrOH, THF; etc.
or $H_2C{=}O$, KOH (Cannizzaro)

Reduction with organometallic reagents: occurs when the systems are hindered

20.3 Stereochemistry and Stereoselectivity in Hydride Reductions

The addition of hydride to a ketone carbonyl group has the potential for generating a new stereogenic center, so the reduction of ketones has been the subject of extensive stereochemical studies, and there is a voluminous literature on the stereochemistry of reduction by hydride reagents. Despite this intensive activity, the origins of the stereochemistry of complex hydride reductions remain in some doubt, although the stereochemical preferences themselves are now fairly clear. Thus, it is apparent that sodium borohydride (as well as other complex metal hydrides) exhibits an inherent preference

....

§ Answer to Worked Problem:
This is an Evans-Tishchenko reaction, and it gives the monopropionate of the *anti*-1,3-diol. In this particular reaction, the ester group actually ends up at the position of the original alcohol (i.e., the lefthand product is the one isolated).
[*Org. Lett.* **2013**, *15*, 3118]

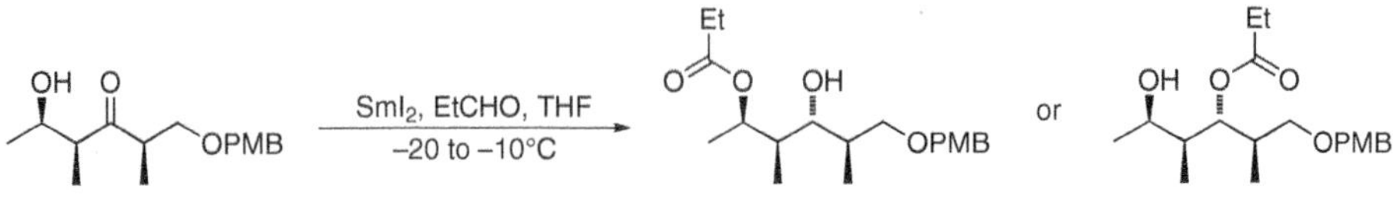

for axial attack on the carbonyl group of a cyclohexanone, a generalization first proposed by Barton.[69]

The first serious attempts to rationalize why there is this stereochemical bias in the reduction of cyclohexanone derivatives led to two models of reduction: the "product development control" model and the "steric approach control" model.[70] Both these models have been strongly criticized.[71] Explanations based on parameters including torsional strain,[72] orbital symmetry control,[73] dissymmetry in the π orbital of the carbonyl group,[74] and charge-transfer stabilization of the transition state[75] have been advanced to rationalize the stereochemistry obtained. The controversy over the stereochemistry of nucleophilic addition to the carbonyl group of cyclohexanone derivatives continues.[76]

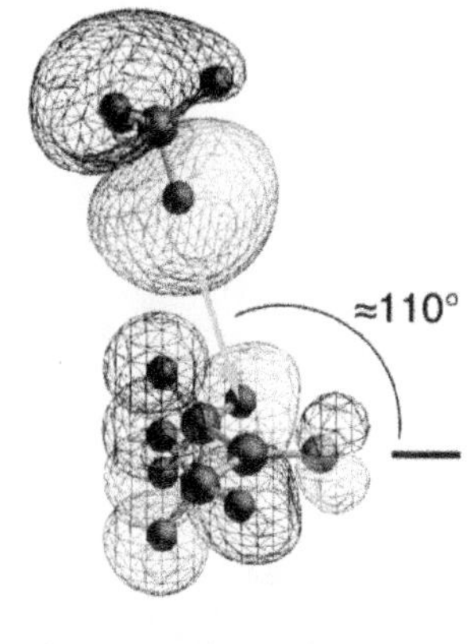

Figure 20.5 Bürgi-Dunitz trajectory for the addition of borohydride anion to acetone

Hydride addition to the carbonyl group may be viewed as a typical nucleophilic addition to the π bond. Bürgi and Dunitz studied nucleophilic addition to carbonyl compounds and observed that there was a favored trajectory of approach of the nucleophile to the carbonyl group.[77] In this preferred approach trajectory, the nucleophile makes an O—C—Nu angle of approximately 110° with the C=O axis perpendicular to the plane of the carbonyl group. The Bürgi-Dunitz trajectory (Figure 20.5) has become one of the foundations on which stereochemical predictions in the reduction of carbonyl groups is based.

The reactions of acyclic ketones are generally adequately predicted by applying the Cram[78] and Felkin-Anh[79] models that were discussed at length in Chapter 2. In the case of complex metal hydrides, both the reducing agent and the substrate affect the stereochemical outcome of the reaction. This is illustrated by the results in Table 20.2, where the reduction of a series of structurally related α-substituted acyclic ketones is summarized.[80]

The data in Table 20.2 show that, as the reducing agent becomes more sterically demanding, the proportion of the Cram isomer increases rapidly. As the steric bulk of the α′ substituent increases, there is also an increase in the proportion of the Cram isomer in the product. However, it is much more modest in scope (each methyl branch at the α′ position increases the percent of the major isomer by approximately 10%). As with other nucleophilic additions to chiral ketones and aldehydes, a substituent at the α position capable of coordinating the cation by chelation leads to a general reversal of the observed stereochemical bias, and the anti-Cram product is usually observed as the major product.

69. Barton, D.H.R. *J. Chem. Soc.* **1953**, 1927.

70. (a) Dauben, W.G.; Fonken, G.J.; Noyce, D.S. *J. Am. Chem. Soc.* **1956**, *78*, 2579. (b) Dauben, W.G.; Blanz, E.J., Jr.; Jiu, J.; Micheli, R.J. *J. Am. Chem. Soc.* **1956**, *78*, 3752. (c) Dauben, W.G.; Bozak, R.E. *J. Org. Chem.* **1959**, *24*, 1596. (d) Dauben, W.G.; Bozak, R.E.; Ellis, R.; Willey, F. *Rev. Chim. Acad. Roumaine* **1962**, *7*, 803 [*cf Chem. Abstr.* **1964**, *61*, 4424].

71. Wigfield, D.C.; Phelps, D.J. *Can. J. Chem.* **1972**, *50*, 388.

72. Chérest, M.; Felkin, H. *Tetrahedron Lett.* **1971**, 383.

73. (a) Klein, J. *Tetrahedron Lett.* **1973**, 4307. (b) Klein, J. *Tetrahedron* **1974**, *30*, 3349.

74. (a) Liotta, C.L. *Tetrahedron Lett.* **1975**, 519. (b) Anh, N.T.; Eisenstein, O.; Lefour, J.-M.; Trân Huu Dâu, M.-E. *J. Am. Chem. Soc.* **1973**, *95*, 6146. (c) Ashby, E.C.; Noding, S.A. *J. Am. Chem. Soc.* **1976**, *98*, 2010.

75. Cieplak, A.S. *J. Am. Chem. Soc.* **1981**, *103*, 4540.

76. (a) Gung, B.W. *Tetrahedron* **1996**, *52*, 5263. (b) Tomoda, S.; Senju, T. *Tetrahedron* **1999**, *55*, 3871. (c) Yadav, V.K.; Jeyaraj, D.A.; Balamurungan, R. *Tetrahedron* **2000**, *56*, 7581.

77. (a) Bürgi, H.B.; Dunitz, J.D.; Schefter, E. *J. Am. Chem. Soc.* **1973**, *95*, 5065. (b) Bürgi, H.B.; Dunitz, J.D.; Lehn, J.M.; Wipff, G. *Tetrahedron* **1974**, *30*, 1563. (c) Bürgi, H.B.; Dunitz, J.D. *Acc. Chem. Res.* **1983**, *16*, 153.

78. (a) Cram, D.J.; Abd Elhafez, F.A. *J. Am. Chem. Soc.* **1952**, *74*, 5828. (b) Cram, D.J.; Knight, J.D. *J. Am. Chem. Soc.* **1952**, *74*, 5835. (c) Curtin, D.Y.; Harris, E.E.; Meislich, E.K. *J. Am. Chem. Soc.* **1952**, *74*, 2901. (d) Leitereg, T.J.; Cram, D.J. *J. Am. Chem. Soc.* **1968**, *90*, 4019.

79. (a) Chérest, M.; Felkin, H.; Prudent, N. *Tetrahedron Lett.* **1968**, 2199. (b) Anh, N.T.; Eisenstein, O. *Nouv. J. Chem.* **1977**, *1*, 61.

80. Yamamoto, Y.; Matsuoka, K.; Nemoto, H. *J. Am. Chem. Soc.* **1988**, *110*, 4475.

Table 20.2 Reduction of α-Substituted ketones with Complex Hydrides

Reducing Agent	Ketone	Major (Cram)	Minor (Anti-Cram)	Major: Minor	Yield (%)
$LiAlH_4$, Et_2O, 0°C	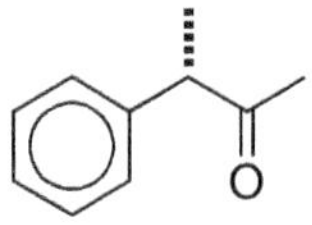	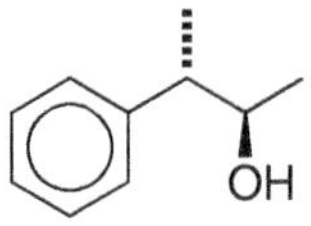	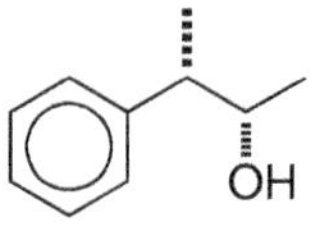	74:26	97
L-Selectride	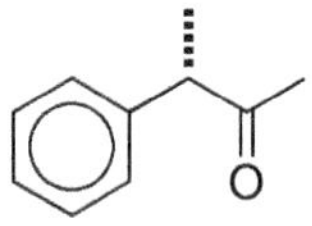	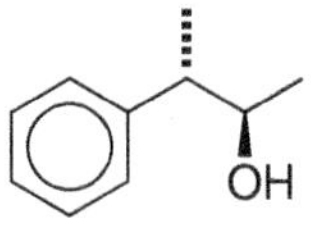	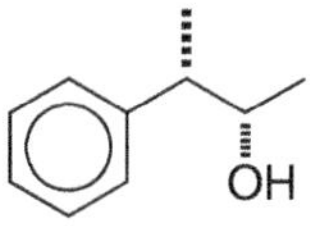	> 99:1	80
$LiAlH_4$, Et_2O, 0°C	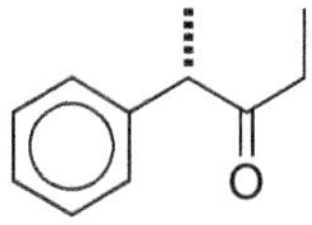	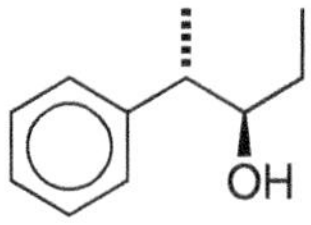	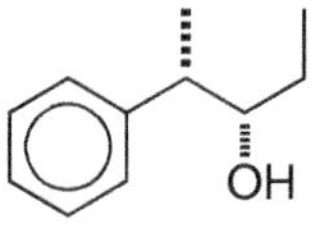	77:23	74
L-Selectride	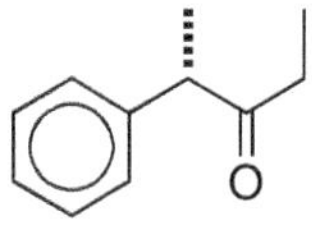	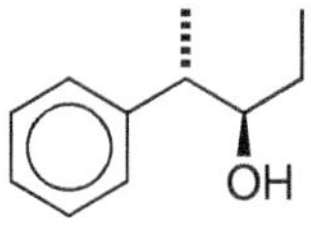	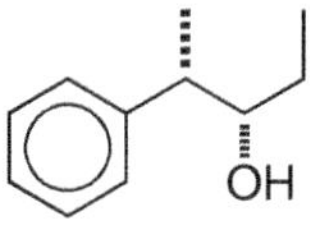	> 99:1	80
$LiAlH_4$, Et_2O, 0°C	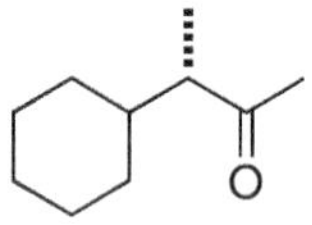	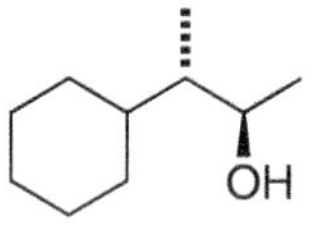	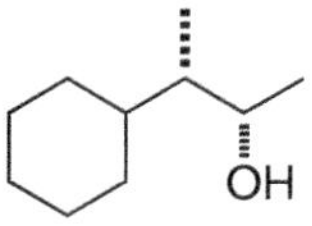	58:42	92
$LiAlH_4$, Et_2O, 0°C	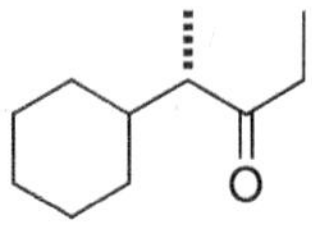	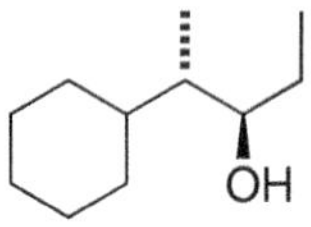	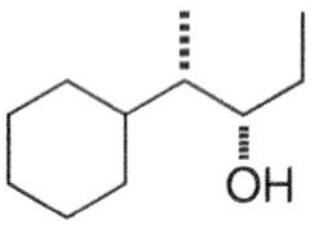	68:32	99
$LiAlH_4$, Et_2O, 0°C	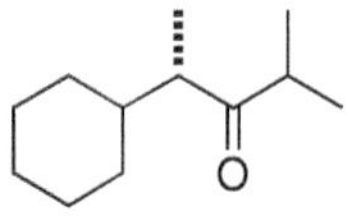	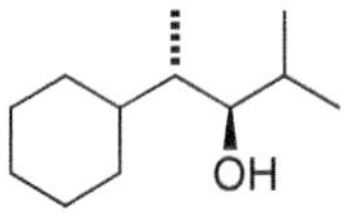	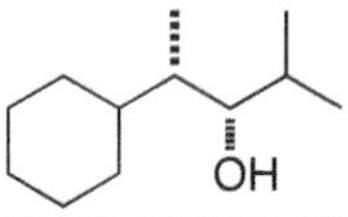	81:19	99

20.4 **Reduction Using Metals**[81]

The reduction of carbonyl compounds with active metals is a reaction of considerable long standing. Sodium,[82] sodium amalgam,[83] zinc,[84] and magnesium[85] have been used to reduce ketones and carboxylic acid derivatives for more than a century. In the ensuing time, several low-valent metal species (e.g., Li°, Na°, K°, Zn°, Mg°, Ca°, Ti°, Cr^{+2}, Sm^{+2}) have been used in organic synthesis. The common feature of dissolving metal reductions is that they proceed through free radical intermediates. Depending on the exact reaction conditions,

81. Monographs: (a) House, H.O. *Modern Synthetic Reactions*, 2nd ed. (Benjamin Cummings: Menlo Park, 1972), ch. 3.

82. (a) Bouvealt, L.; Blanc, G. *Compt. rend.* **1903**, *136*, 1676. (b) Bouveault, L.; Blanc, G. *Bull. Soc. Chim. France* **1904**, *31*, 666. (c) Adkins, H.; Gillespie, R.H. *Org. Syn., Coll. Vol. 3* **1955**, 671. (d) Bouveault, L.; Loquin, R. *Compt. rend.* **1905**, 140, 1593. (e) Finley, K.T. *Chem. Rev.* **1964**, *64*, 573. (f) Rühlmann, K. *Synthesis* **1971**, 236.

83. (a) Saytzeff, A. *Ann. Chem. Pharm.* **1873**, *171*, 258. (b) Saytzeff, A. *Z. Chem., N.F.* **1869**, *5*, 551. (c) Saytzeff, A. *Z. Chem., N.F.* **1870**, *6*, 105. (d) Saytzeff, A. *J. prakt. Chem.* [2] **1871**, *3*, 76. (e) Saytzeff, A. *J. prakt. Chem.* [2] **1871**, *3*, 427.

84. (a) Clemmensen, E. *Ber. Dtsch. chem. Ges.* **1913**, *46*, 1837. (b) Clemmensen, E. *Ber. Dtsch. chem. Ges.* **1914**, *47*, 51. (c) Clemmensen, E. *Ber. Dtsch. chem. Ges.* **1914**, *47*, 681. (d) Martin, E.L. *Org. React.* **1942**, *1*, 155. (e) Vedejs, E. *Org. React.* **1975**, *22*, 401.

85. Fittig, R. *Ann. Chem. Pharm.* **1859**, *110*, 17.

these radicals may yield either the alcohol or other reduction product, or a dimeric product formed by combination of two radicals.

The initial step of these reductions (Example **20.52**) requires the transfer of a single electron from the metal atom to the lowest energy unoccupied molecular orbital (π*) of the carbonyl compound, resulting in rupture of the π bond and formation of a radical anion. Note that the singly occupied 2p orbital on carbon is coplanar with the orbital carrying the non-bonding lone pair on oxygen. Thus, the possibility for electron delocalization (similar to hyperconjugation in alkanes) exists.

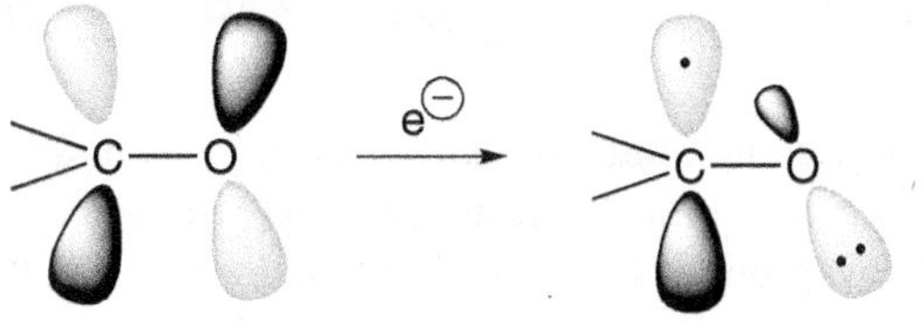

(20.52)

If the metal is a very strong reducing agent (typically a Group IA metal or a Group IIA metal below magnesium), the radical anion may accept another electron to give a dianion that is protonated on carbon by a protic solvent (including such poor proton donors as liquid ammonia). The acid added during product isolation protonates the oxygen to give the alcohol product (Example **20.53**). In the absence of a proton source, the radical anion may dimerize to give the dianion of a diol; protonation during isolation gives the diol (Example **20.54**).

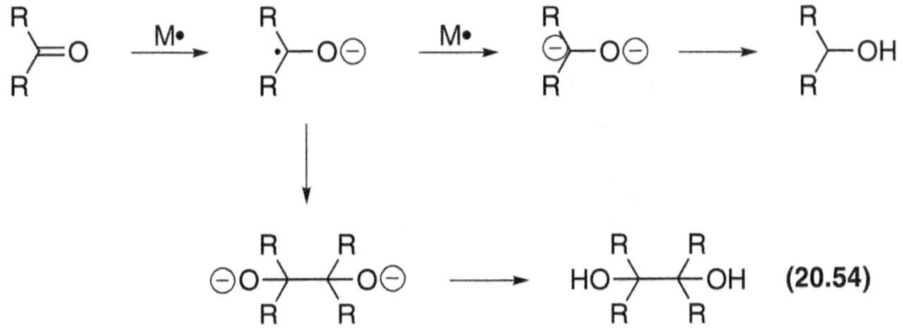

(20.53)

(20.54)

Dissolving Metal Reduction of Carbonyl Compounds

The oldest useful reduction of carbonyl compounds with metals is the reduction of esters by sodium metal in refluxing ethanol (e.g., Example **20.55**), first published by French chemist Louis Bouveault[86] and his students in a series of papers beginning in 1903.[87] It is now known as the ***Bouveault-Blanc reduction***. This permits the selective reduction of the ester group of diacid half-esters (e.g., Example **20.56**) and the reduction of conjugated ketones to the saturated alcohol (e.g., Example **20.57**).

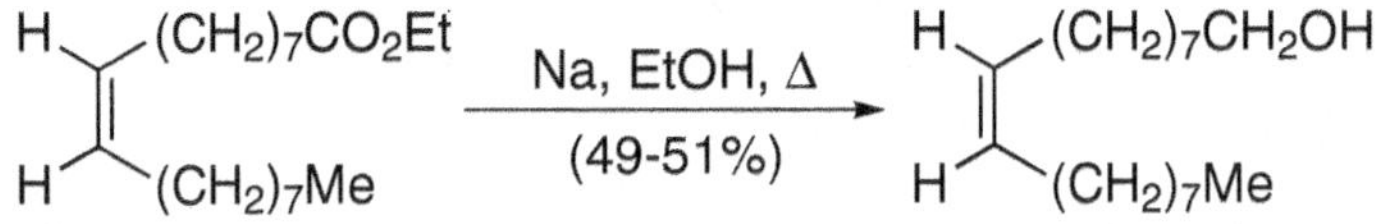

(20.55)[88]

86. **Louis Bouveault (1864–1909)** was born in Nevers, and educated at the Sorbonne (Dr ès-sc phys, 1890). On graduating, Bouveault became Professor of Organic Chemistry at the Institut Chimique de Nancy, and then Maître de Conférences at the University of Lyons. In 1894 Bouveault was appointed to the University of Nancy and in 1902 to the Sorbonne. In 1904 he was appointed as Maître de Conférences. He died prematurely in 1909 in the midst of a flourishing career. For more details, see; Nye, M.J. *Science in the Provinces* (University of California Press: Berkeley, 1986), p. 166 ff.

87. (a) Bouvealt, L.; Blanc, G. *Compt. rend.* **1903**, *136*, 1676. (b) Bouveault, L.; Blanc, G. *Bull. Soc. Chim. France* **1904**, *31*, 666. (c) Bouveault, L.; Loquin, R. *Compt. rend.* **1905**, 140, 1593.

88. Adkins, H.; Gillespie, R.H. *Org. Syn., Coll. Vol. 3* **1955**, 671.

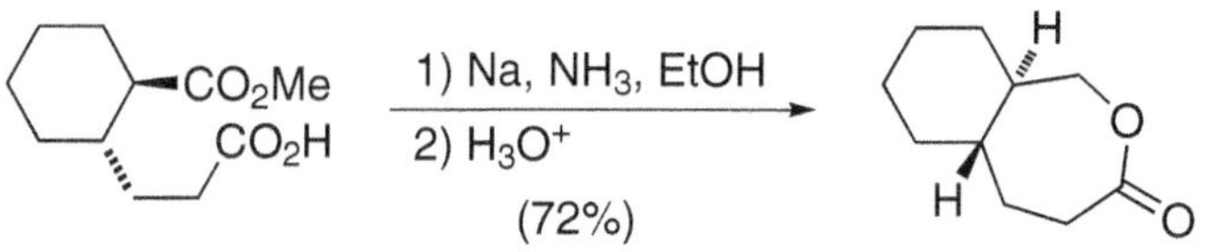

(20.56)[89]

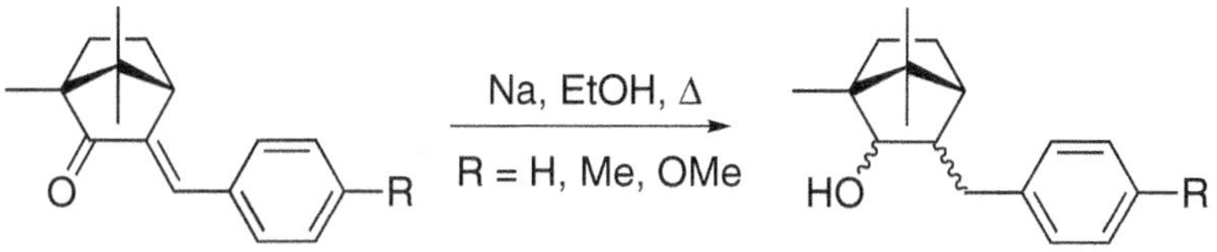

(20.57)[90]

The reduction of carbonyl compounds by solutions of metals in ammonia gives the corresponding primary or secondary alcohol if the carbonyl group is not conjugated with a double bond. If the carbonyl group is conjugated with a double bond, the products are derived by initial reduction of the carbon-carbon double bond. Reduction of α-substituted ketones proceeds predominantly with anti-Cram selectivity, although the level of diastereoselectivity is usually lower than the Cram selectivity with lithium aluminum hydride.[91] Simple alkenes are not reduced by dissolving metals in alcohol or ammonia solvent. However, reduction of α,β-unsaturated carbonyl compounds with lithium in ammonia in the absence of a stronger proton source gives a single regioisomer of the lithium enolate, which can then be used in an alkylation reaction or an aldol addition reaction (e.g., Example **20.58**). In the presence of a proton source, the saturated alcohol is the usual product (e.g., Example **20.59**). When the enone is a cyclohexenone, the two hydrogens are introduced from the axial direction, so the stereochemistry of the reaction can be predicted (e.g., Example **20.60**).

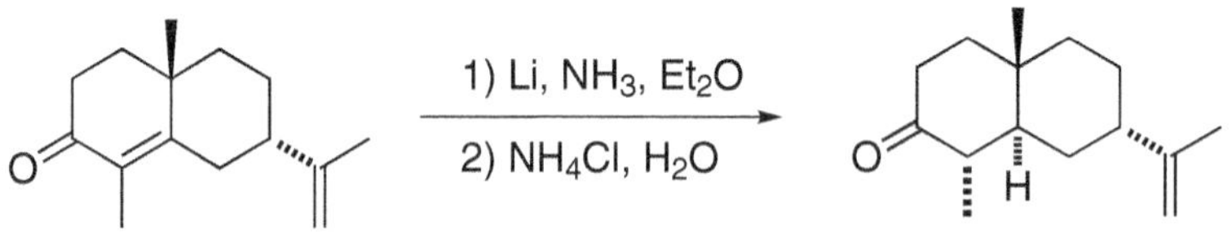

(20.58)[92]

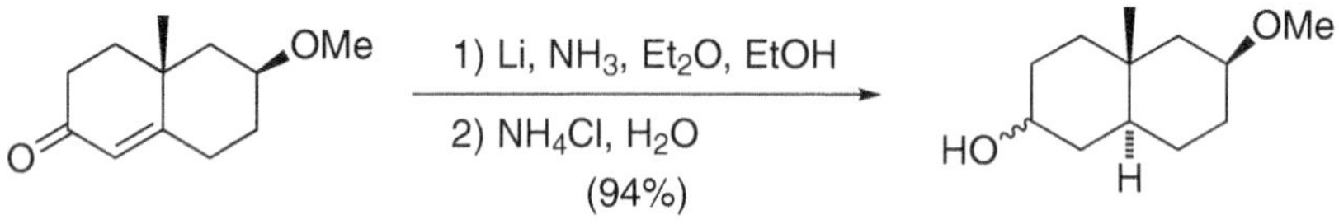

(20.59)[93]

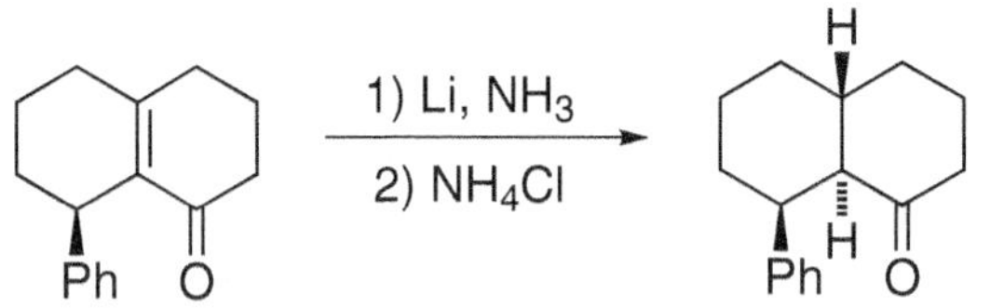

(20.60)[94]

The Pinacol Reaction and Related Reactions

The reaction between a carbonyl compound and an active metal in a protic solvent usually results in reduction to the primary alcohol if the carbonyl compound is an aldehyde or to

---

89. Paquette, L.A.; Nelson, N.A. *J. Org. Chem.* **1962**, *27*, 2272.

90. (a) Seo, B.-I.; Lewis, D.E.; Wall, L.K.; Lee, H.; Buttrum, J.W. *Synth. Commun.* **1993**, *23*, 15. (b) Seo, B.-I.; Suh, I.-H.; Jensen, W.P.; Lewis, D.E.; Wall, L.K.; Jacobson, W.P. *Tetrahedron: Asymmetry*, **1992**, *3*, 367.

91. Yamamoto, Y.; Matsuoka, K.; Nemeto, H. *J. Am. Chem. Soc.* **1988**, *110*, 4475.

92. (a) Halsall, T.G.; Theobald, D.W.; Walshaw, K.B. *J. Chem. Soc.* **1964**, 1029. (b) Chen, Y.; Xiong, Z.; Zhou, G.; Yang, J.; Li, Y. *Chem. Lett.* **1997**, 1289.

93. Stork, G.; Darling, S.D. *J. Am. Chem. Soc.* **1964**, *86*, 1761.

94. House, H.O.; Thompson, H.W. *J. Org. Chem.* **1963**, *28*, 360.

the secondary alcohol if the carbonyl compound is a ketone. However, if the reaction is carried out in an aprotic solvent such as diethyl ether, another reaction can occur. In this reaction, two molecules of the carbonyl compound become bonded together at the carbonyl carbons, as discovered by Rudolf Fittig (Examples **20.61** to **20.63**).[95] The product is a vicinal diol called a ***pinacol*** (Greek πιναξ, *pinax*, a table, referring to the tabular form of the crystals obtained from acetone).

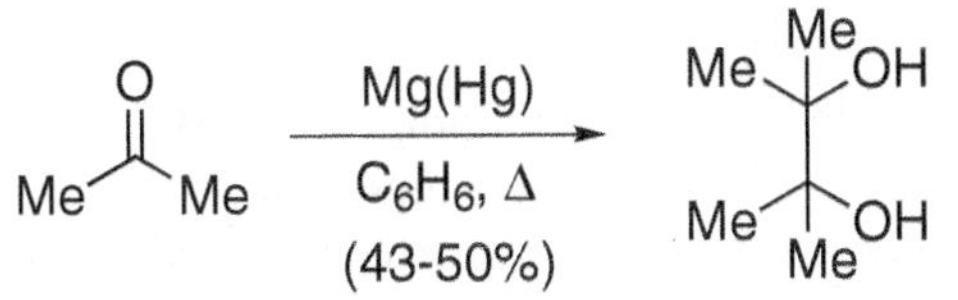

(20.61)[96]

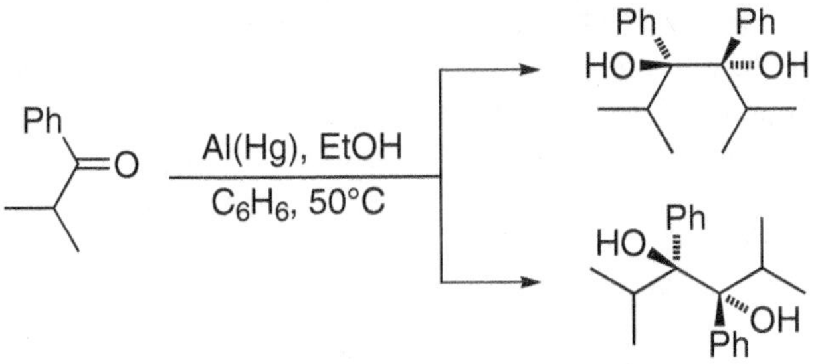

(20.62)[97]
(18%)

(20.63)
(27%)

Historically, the most popular metal for carrying out the pinacol reaction has probably been magnesium. The modern variants of the pinacol reaction are most frequently based on metals other than magnesium, and titanium-based reagents are especially popular for carrying out the reaction. The key step (Example **20.64**) in the first total synthesis of the plant hormone gibberellic acid (gibberellin A_3), for example, involved a pinacol reaction using titanium metal.[98] Where the double bond does not occupy a position prohibited by Bredt's rule, titanium metal can also reduce carbonyl compounds to alkenes, a reaction known as the ***McMurry reaction***.[99] As illustrated in the reactions of cyclohexanone, titanium-based reactions (Example **20.65**) often give much higher yields than the corresponding magnesium-based pinacol reactions (Example **20.66**). For this reason, the titanium-based reagents have gained considerable popularity.

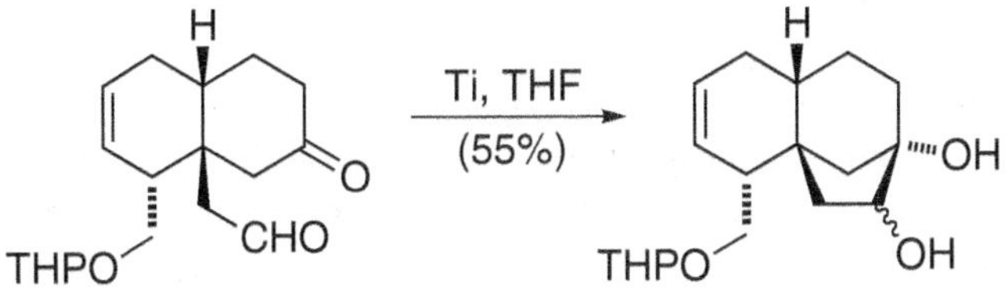

(20.64)

95. **Wilhelm Rudolf Fittig (1835–1910)** was educated at Göttingen (PhD, 1858); he was appointed to the faculty in 1860. He subsequently moved to Tübingen (1870) and Strasbourg (1876). Fittig was awarded the Davy medal by the Royal Society in 1906. He was also a teacher of some note, publishing two widely-used textbooks. For more information, see: Witt, O.N. *J. Chem. Soc., Trans.* **1911**, *99*, 1646.

96. Adams, R.; Adams, E.W. *Org. Syn., Coll. Vol. 1* **1944**, 459.

97. Vigevani, A.; Pasqualucci, R.; Gallo, G. *Tetrahedron* **1969**, *25*, 573.

98. Corey, E.J.; Danheiser, R.L.; Chandrasekaran, S.; Siret, P.; Keck, G.E.; Gras, J.-L. *J. Am. Chem. Soc.* **1978**, *100*, 8031. (b) Corey, E.J.; Danheiser, R.L.; Chandrasekaran, S.; Keck, G.E.; Gopsalom, B.; Larsen, S.D.; Siret, P.; Gras, J.-L. *J. Am. Chem. Soc.* **1978**, *100*, 8034.

99. (a) McMurry, J.E. *Acc. Chem. Res.* **1974**, *7*, 281. (b) McMurry, J.E.; Rico, J.G. *Tetrahedron Lett.* **1989**, *30*, 1169.

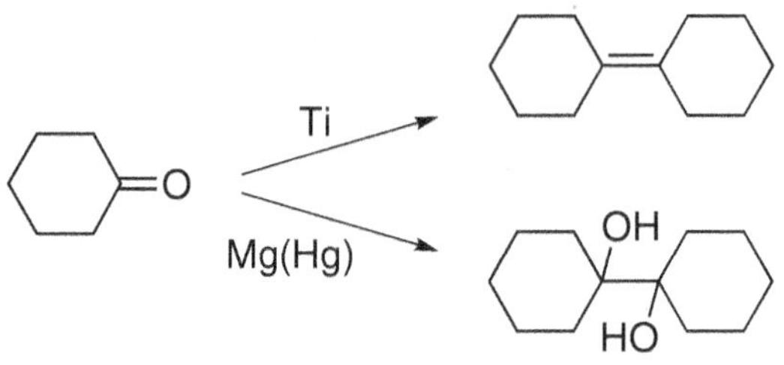

(20.65) (85%)

(20.66) (30%)

One particularly useful application of the titanium-based reductive coupling of carbonyl compounds is the reduction of dicarbonyl compounds with titanium (III) chloride and a zinc-copper couple. This reaction provides an excellent method for the formation of compounds with medium-sized rings that are extremely difficult to obtain by other cyclization methods (e.g., Example **20.67**).[100] Similar reactions of aldehydes with low-valent titanium (prepared from titanium tetrachloride and manganese metal) give *anti*-1,2-diols in a crossed pinacol reaction (e.g., Example **20.68**).[101]

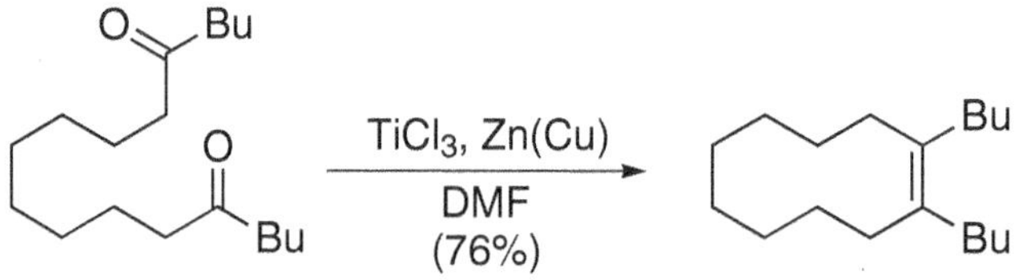

(20.67)

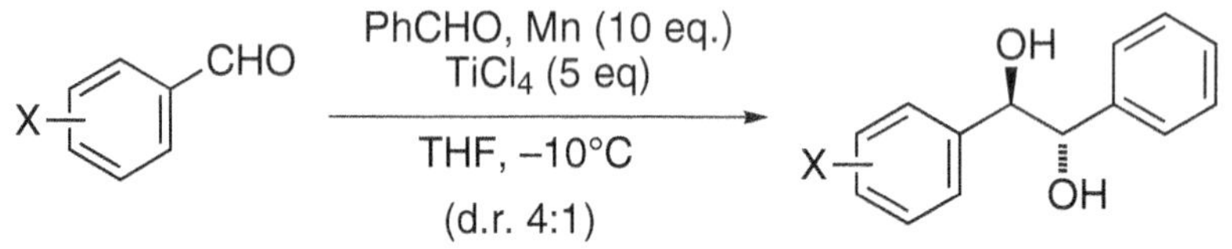

(20.68)

Reductive Elimination: The E1cb Elimination

If an organometallic reagent is to be formed either by direct combination or by metal-halogen exchange with another metal alkyl, it is also important that there be no group at the β carbon that might conceivably act as a leaving group—even a poor leaving group. In such compounds, the reaction with the metal leads not to the formation of the expected organometallic compound but rather to elimination of the halogen and the leaving group, a reaction known as ***reductive elimination***. The reaction is also promoted by low-valent metal ions (e.g., Sm^{2+}).

Several groups can function as the leaving group at the β carbon, including halogens, oxygen-containing functional groups, and even some nitrogen functional groups; they are summarized in Table 20.3. When the leaving group is a halogen, reductive elimination proceeds readily with *anti* stereochemistry by a mechanism that is basically the same as the E2 mechanism. However, when the leaving group is relatively poor, other factors come into play. As the leaving group becomes more difficult to remove, the reaction mechanism changes to E1cb. As a result, the stereochemistry becomes less rigidly defined. This is illustrated in Example **20.69**, where the product ratio is independent of the stereochemistry of the starting bromoether. Reductive elimination of β-haloethers often occurs during the attempt to prepare organometallic reagents (e.g., Example **20.70**). Reductive elimination does not always require a metal. For example, the formation of

100. McMurry, J.E.; Fleming, M.P.; Kees, K.L.; Krepski, L.R. *J. Org. Chem.* **1978**, *43*, 3255.
101. Duan, X.-F.; Feng, J.-X.; Zi, G.-F.; Zhang, Z.-B. *Synthesis*, **2009**, 277.

Table 20.3 Reactant and Reagent Combinations for Reductive Elimination

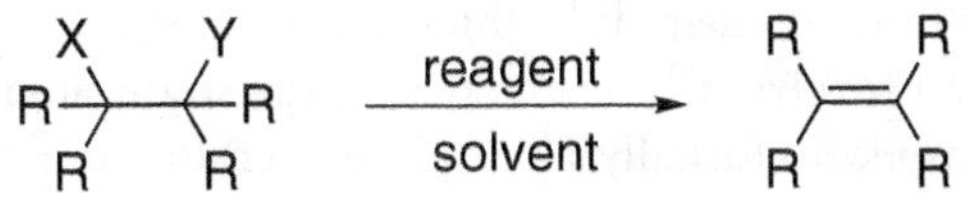

X	Y	Reagent/Solvent
Halogen	Halogen	Li, NH_3; Na, NH_3
	OR	Na, ROH; Mg, ROH
	OH (acid conditions)	Mg, Et_2O; Li, Et_2O
	NHCOR	Zn, AcOH, Δ; *n*-BuLi, THF
OH	OH	(1) MeLi; (2) K_2WCl_6, THF, Δ
OH	OH	(1) $Im_2C{=}S$; (2) $P(OR)_3$, Δ
OH	OH	TsOH, NaI
OH	OH	Ti, THF
OTs	OTs	NaI, DMF

1,2-diodoalkanes is always followed by attack of iodide ion to give iodine and the alkene (Example **20.71**).

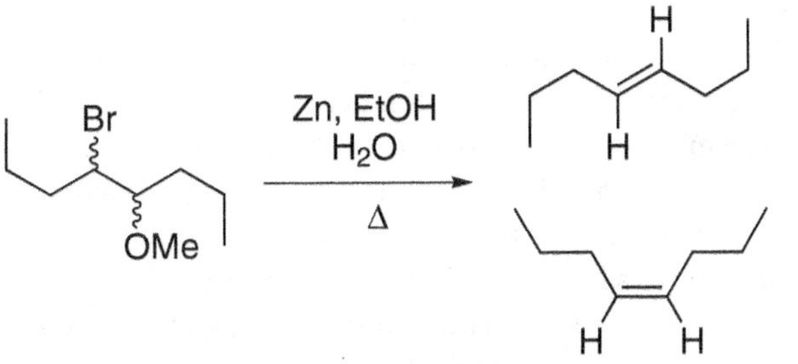

11 .. 9 **(20.69)**[102]

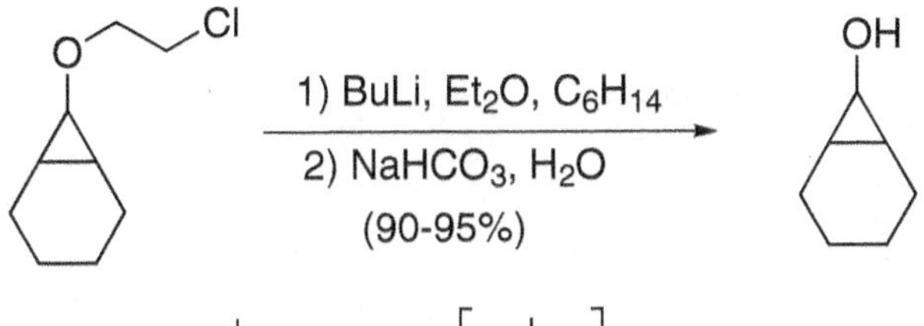

(20.70)[103]

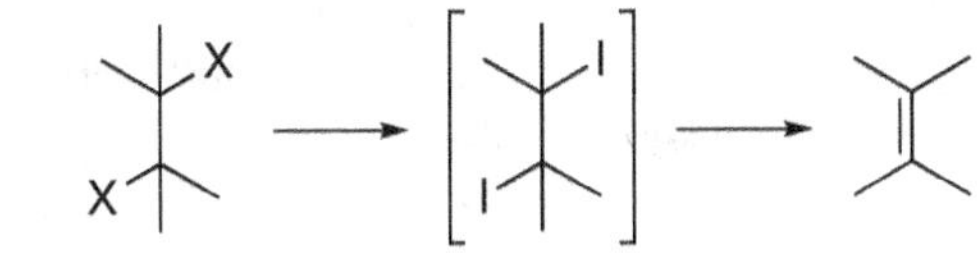

(20.71)

X = Cl, Br, OTs.
OH_2^+, etc.

102. (a) House, H.O.; Ro, R.S. *J. Am. Chem. Soc.* **1958**, *80*, 182. (b) Cristol, S.J.; Rademacher, L.E. *J. Am. Chem. Soc.* **1959**, *81*, 1600.

103. Schöllkopf, U.; Paust, J.; Patsch, M.R. *Org. Syn.* **1969**, *49*, 86.

An important method for the stereospecific formation of alkenes is the reductive elimination of vicinal diols through their cyclic thionocarbonates, known as the ***Corey-Winter reaction***.[104] The reaction involves the abstraction of the sulfur atom from the thionocarbonate by trivalent phosphorus (usually a phosphite ester) to generate a cyclic carbene that rearranges to the alkene with loss of carbon dioxide. The overall reaction is stereospecific and suprafacial (*syn*), as illustrated by the formation of *trans*-cyclooctene from *trans*-cyclooctane-1,2-diol (Example **20.72**).[105]

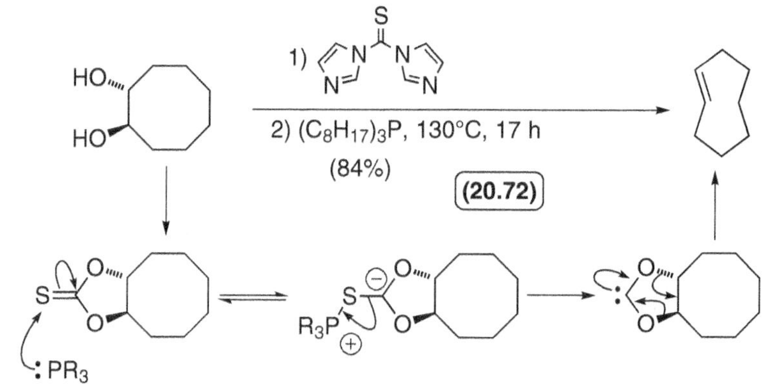

(20.72)

Wurtz Coupling

The reaction between an alkyl halide and sodium metal is almost always complicated by a coupling reaction called the ***Wurtz coupling*** (after the French chemist Charles Adolphe Wurtz,[106] who discovered it). The net result of the Wurtz coupling is that two of the alkyl groups of the alkyl halide become bonded to each other to form the alkane. Thus, if one attempts to prepare butylsodium by the direct reaction between sodium and 1-bromobutane, one obtains octane as the major organic product, not butylsodium. The intramolecular version of the Wurtz coupling is a reductive elimination of a non-vicinal dihalide, and it has been used to form cycloalkanes by the reaction between a dihalide and sodium metal or a sodium-potassium alloy in a hydrocarbon solvent. This reaction is an extremely important method for the preparation of small-ring compounds (e.g., Example **20.73**) and compounds with extreme ring strain (e.g., Example **20.74**).

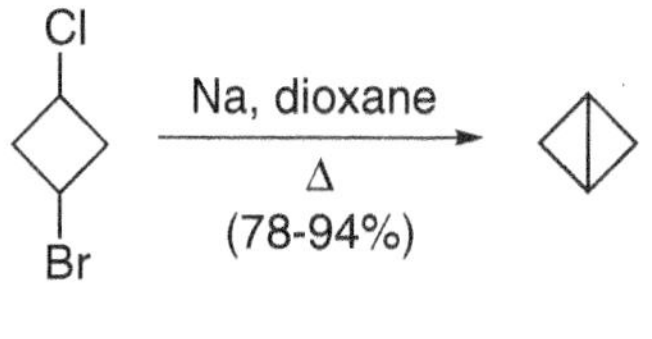

(20.73)[107]

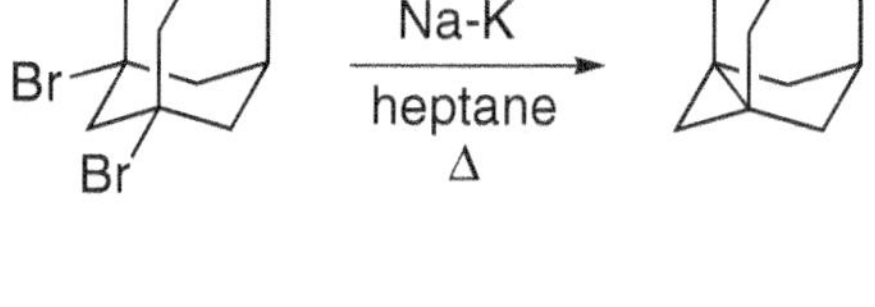

(20.74)[108]

---

104. (a) Corey, E.J.; Winter, R.A.E. *J. Am. Chem. Soc.* **1963**, *85*, 2677. (b) Block, E. *Org. React.* **1984**, *30*, 457. (c) Sonnet, P.E. *Tetrahedron* **1980**, *36*, 557.

105. Corey, E.J.; Shulman, J.I. *Tetrahedron Lett.* **1968**, 3655

106. **Charles Adolphe Wurtz (1817–1884)** graduated from Strasbourg (MD, 1843). After study with Liebig at Giessen and with Dumas in Paris, he became Dumas' assistant in 1845 and his successor in 1853. In 1879, he became the first professor of organic chemistry at the Sorbonne. For much more detail, see: (a) Rocke, A.J. *Nationalizing Science: Adolphe Wurtz and the Battle for French Chemistry* (MIT Press: Cambridge, 2001); (b) Williamson, A.W. *Proc. Roy. Soc.* **1885**, *38*, xxiii.

107. Lampman, G.M.; Aumiller, J.C. *Org. Syn., Coll. Vol. 6* **1988**, 133.

108. Pincock, R.E.; Torupka, E.J. *J. Am. Chem.Soc.* **1969**, *91*, 4593.

Birch Reduction

Carbon-carbon π bonds are usually inert to reduction by metals. However, in several papers starting in 1944, Australian chemist Arthur Birch[109] described the reduction of aromatic rings by a solution of an alkali metal in liquid ammonia in the presence of a proton donor such as ethanol or *tert*-butyl alcohol.[110] This reaction, which is carried out with lithium (occasionally sodium or potassium) in liquid ammonia in the presence or absence of a proton source or an organic cosolvent is now known as the ***Birch reduction***.[111] More than 60 years after its development, the Birch reduction remains one of the most useful methods for the reduction of aromatic rings. The key to the Birch reduction is the fact that alkali metals dissolve in liquid ammonia to give deep blue, electrically conductive solutions containing the metal ion and solvated electrons.

In the absence of the proton donor, the reduction of aromatic rings is quite slow because the addition of the second electron to the cyclohexadienyl radical anion generates a dianion—an extremely powerful base. Protonation of the radical anion (**20.75**), however, leads to the cyclohexadienyl radical (**20.76**), which readily accepts a second electron to give the cyclohexadienyl anion (**20.77**). Depending on the exact reaction conditions, this anion may be protonated to give the 1,4-cyclohexadiene or it may give the 1,3-cyclohexadiene, which is then reduced to the cyclohexene. Under kinetic conditions (in the presence of a proton donor strong enough to irreversibly protonate the cyclohexadienyl anion), the thermodynamically less stable nonconjugated diene is produced (Figure 20.6).

Ring substituents also play an important role in determining the regiochemistry and rate of a Birch reduction. Electron-releasing groups result in products that carry the substituent on one of the residual π bonds (e.g., Example **20.78**); alkyl groups tend to retard

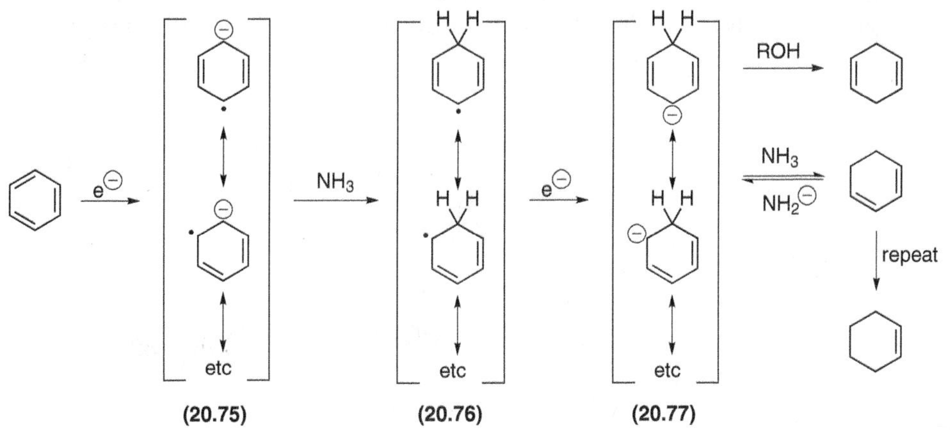

(20.75) **(20.76)** **(20.77)**

Figure 20.6 The mechanism of the Birch reduction of aromatic rings in the presence and absence of a proton source

109. **Arthur John Birch (1915–1995)** was educated at Sydney (MSc, 1937) and Oxford (D Phil, 1941). He stayed in Britain until 1952, working on the synthesis of steroid hormones. His career then took him to Sydney (1952), Manchester (1955), and the Australian National University (1967). For more biographical detail, see: (a) Birch, A.J. *To See the Obvious.* (Oxford University Press: New York, 1995); (b) Rickards, R.W.; Cornforth, J. *Biogr. Mem. Fell. Roy. Soc.* **2007**, *53*, 21.

110. (a) Birch, A.J. *J. Chem. Soc.* **1944**, 430. (b) Birch, A.J. *J. Chem. Soc.* **1945**, 809. (c) Birch, A.J. *J. Chem. Soc.* **1946**, 593. (d) Birch, A.J. *J. Chem. Soc.* **1947**, 102. (e) Birch, A.J. *J. Chem. Soc.* **1947**, 1642. (f) Birch, A.J. *J. Chem. Soc.* **1949**, 2531.

111. Reviews: (a) Birch, A.J. *Quart. Rev. (London)* **1950**, *4*, 69. (b) Birch, A.J.; Smith, H. *Quart. Rev. (London)* **1958**, *12*, 17. (c) Watt, G.W. *Chem. Rev.* **1950**, *46*, 317. (d) Caine, D. *Org. React.* **1976**, *23*, 1. (e) Hook, J.M.; Mander, L.N. *Nat.Prod. Rep.* **1986**, 35. (f) Rabideau, P.W. *Tetrahedron* **1989**, *45*, 1579. (g) Rabideau, P.W.; Marcinow, Z. *Org. React.* **1992**, *42*, 1. (h) Subba Rao, G.S.R. *Pure Appl. Chem.* **2003**, *75*, 1443. (i) For a review of the mechanism, see: Zimmerman, H.E. *Acc. Chem. Res.* **2012**, *45*, 164.

the rate of the reduction, whereas alkoxy and amino groups tend to accelerate the reduction slightly.[112] Electron-withdrawing substituents on the ring change the regiochemistry of the reduction, with the major product being that in which the substituent is not on a double bond but on one of the two sp^3-hybridized carbons (Example **20.79**). Phenols are rapidly deprotonated under Birch reduction conditions to give phenoxide anions that resist reduction with lithium. Yet, the substituted ring of β-naphthols is rapidly reduced to give the corresponding tetralone (Example **20.80**).

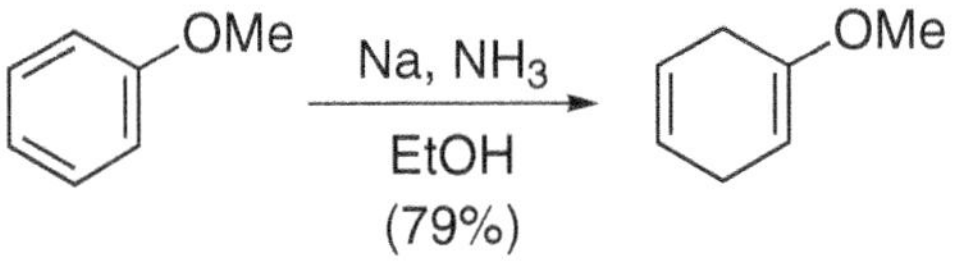

(20.78)

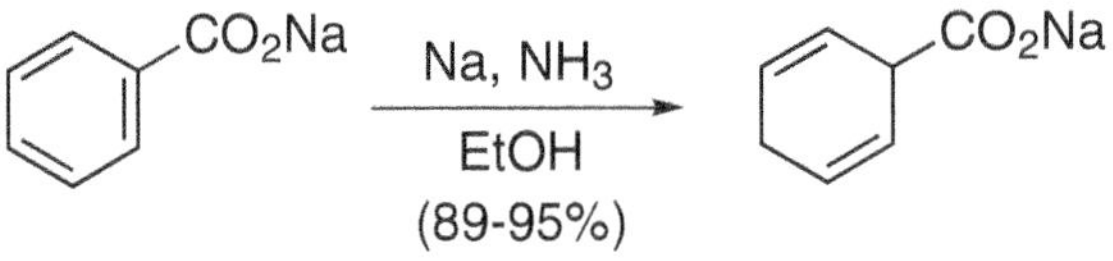

(20.79)

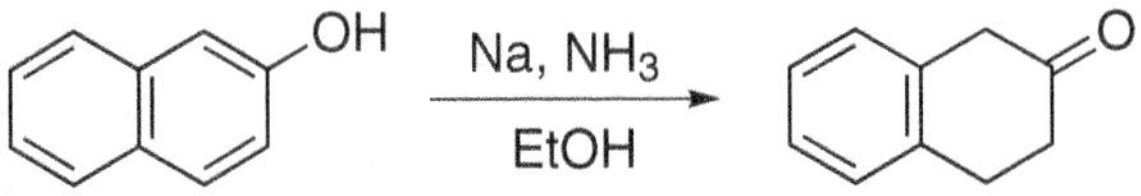

(20.80)

The Birch reduction of aromatic ethers is an especially important reaction synthetically because the products of the reduction are enol ethers. These compounds may be hydrolyzed to cyclohexenone derivatives, thus making aromatic ethers a useful "masked" enone for synthesis where the reactivity of the carbonyl group or the alkene double bond would be undesirable. Hydrolysis of the enol ether with a weak acid such as aqueous oxalic acid gives the non-conjugated enone (Example **20.81**). Hydrolysis of the enol ether with a stronger acid usually results in migration of the double bond into the conjugated position (Example **20.82;** the more common situation).

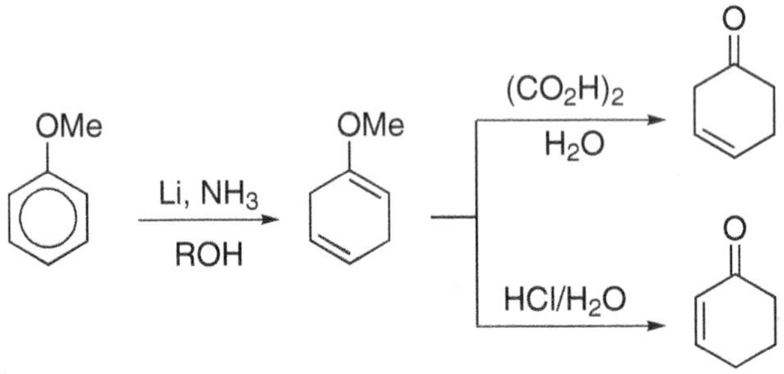

(20.81)

(20.82)

In the absence of a proton source, the reduction of benzoate esters with lithium in liquid ammonia gives an intermediate cross-conjugated ester enolate anion. The similar reduction of a benzoate anion gives the analogous cross-conjugated dianion of the carboxylic acid. Like all enolates, these anions can be trapped with an alkyl halide to give the alkylated product (e.g., Example **20.83**).[113]

---

112. Krapcho, A.P.; Bothner-By, A.A. *J. Am. Chem. Soc.* **1959**, *81*, 3658.

113. For examples, see: (a) Overman, L.E.; Ricca, D.J.; Tran, V.D. *J. Am. Chem. Soc.* **1997**, *119*, 12031. (b) Hook, J.M.; Mander, L.N.; Woolias, M. *Tetrahedron Lett.* **1982**, *23*, 1095. (c) Taber, D.F. *J. Org. Chem.* **1976**, *41*, 2649. (d) Schultz, A.G.; Pettus, L. *J. Org. Chem.* **1997**, *62*, 6855.

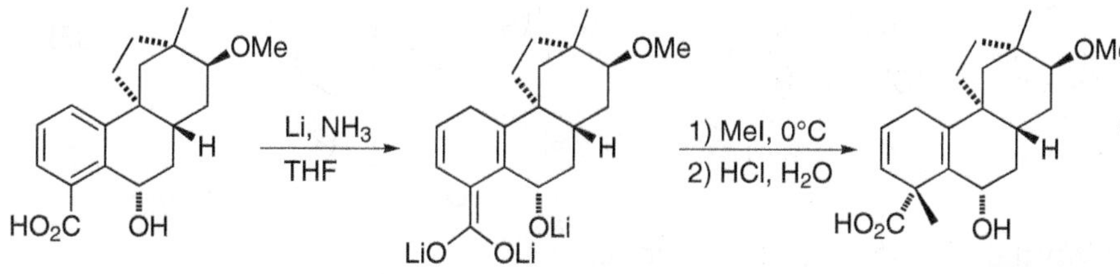

(20.83)

One of the most spectacular applications of the Birch reduction in synthesis is from a steroid synthesis by W. S. Johnson. This chemist used the Birch reduction to introduce no less than five new chiral centers into a molecule in a single step with complete control over the relative stereochemistry (Example **20.84**).[114]

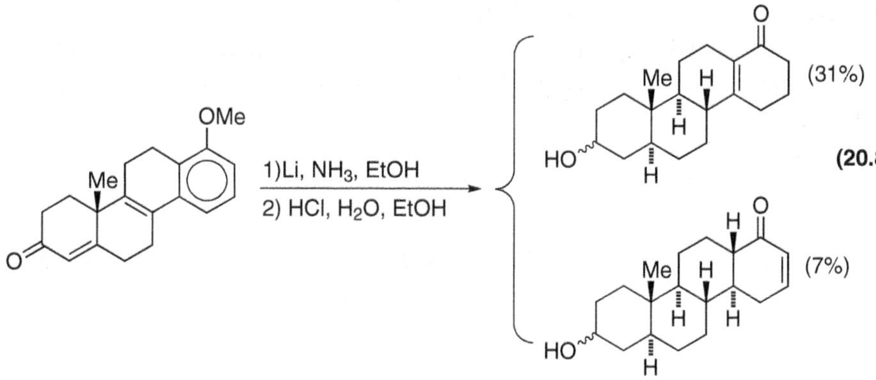

(20.84)

The Birch reduction of conjugated dienes gives simple alkenes, whereas the Birch reduction of nitriles and imines gives amines. Reduction with sodium or lithium in liquid ammonia may be used for hydrogenolysis of benzyl esters, ethers, and amines as well as for hydrogenolysis of alkyl halides to hydrocarbons. Internal alkynes are reduced to *E* alkenes using the Birch reduction. Here, virtually none of the *Z* isomer is produced, as illustrated by the reduction of the C_2-symmetric alkyne shown in Example **20.85**. Beginning in 1952, a useful modification of the original Birch procedure was developed by Benkeser, who initially replaced the liquid ammonia with a low-boiling amine[115] and then replaced the alkali metal with calcium (Example **20.86**).[116] This latter version of the ***Benkeser reduction*** has become very popular due to its much improved performance over the classical Birch reduction in large-scale applications.

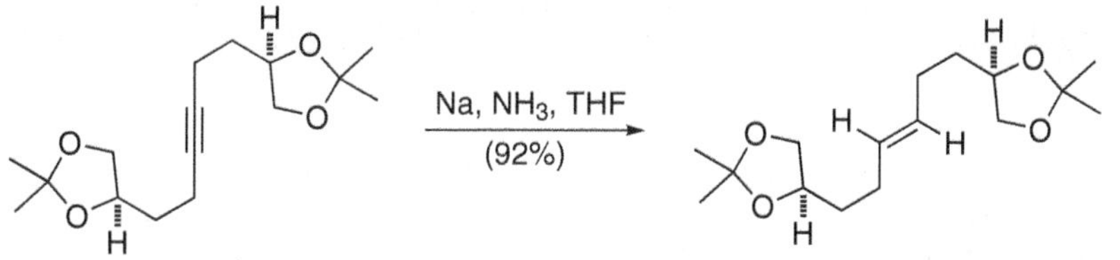

(20.85)

114. (a) Johnson, W.S.; Bannister, B.; Pappo, R. *J. Am. Chem. Soc.* **1956**, *78*, 6331. Earlier papers in this series address the stereochemistry of the reduction of the enone and styrene moieties: (b) Johnson, W.S. *J. Am. Chem. Soc.* **1956**, *78*, 6278. (c) Johnson, W.S.; Szmuszkovicz, J.; Rogier, E.R.; Hadler, H.I.; Wynberg, H. *J. Am. Chem. Soc.* **1956**, *78*, 6285. (d) Johnson, W.S.; Ackerman, J.; Eastham, J.F.; DeWalt, H.A., Jr. *J. Am. Chem. Soc.* **1956**, *78*, 6302. (e)

115. (a) Benkeser, R.A.; Robinson, R.E.; Landesman, H. *J. Am. Chem. Soc.* **1952**, *74*, 5699. (b) Benkeser, R.A.; Robinson, R.E.; Sauve, D.M.; Thomas, O.H. *J. Am. Chem. Soc.* **1955**, *77*, 3230. (c) Benkeser, R.A.; Agnihotri, R.K.; Burrous, M.L. *Tetrahedron Lett.* **1960** (16), 1.

116. Benkeser, R.A.; Belmonte, F.G.; Kang, J. *J. Org. Chem.* **1983**, *48*, 2796.

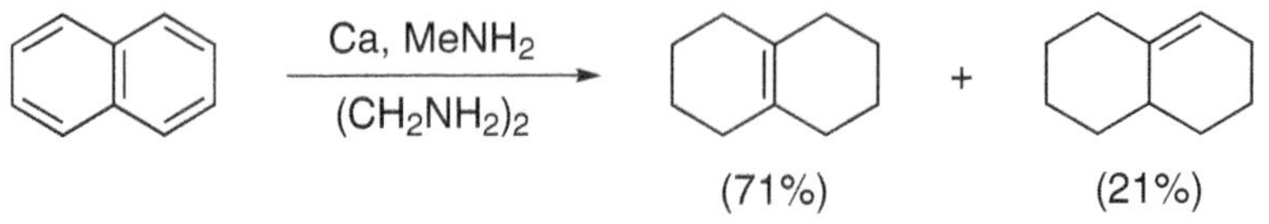

(20.86)

A Final Word on Dissolving Metal Reductions

The alkali metals always pose a threat of fire or explosion, so for process chemistry, especially, a less pyrophoric alternative is always being sought. In recent years, the use of silica-supported alkali metals has begun to supersede the use of alkali metals in many reactions because the need for cryogenic conditions and solvents such as liquid ammonia is eliminated by these reagents. The momentum toward using these reagents as the first choice in reactions such as the Bouveault-Blanc and Birch reductions, as well as in reductive dehalogenation and desulfurization is increasing. More of these supported alkali metals, which are mobile black powders, are becoming commercially available.[117]

Worked Problem

20-4 What are the intermediate compounds in the following sequence? Why is step **B** → **C** necessary?

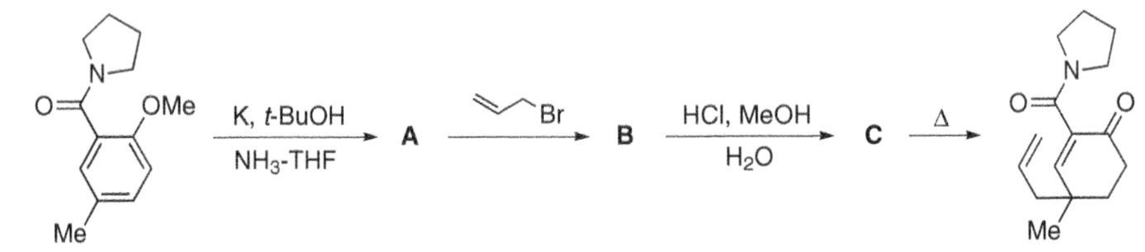

§Answer below.

117. (a) Shatnawi, M.; Paglia, G.; Dye, J.L.; Cram, K. C.; Lefenfeld, M.; Billinge, S.J.L. *J. Am. Chem. Soc.* **2007**, *129*, 1386. (b) Dye, J.L.; Cram, K.D.; Urbin, S.A.; Redko, M.Y.; Jackson, J.E.; Lefenfeld, M. *J. Am. Chem. Soc.* **2005**, *127*, 9338. (c) Nandi, P.; Redko, M.Y.; Petersen, K.; Dye, J.L.; Lefenfeld, M.; Vogt, P.F.; Jackson, J.E. *Org. Lett.* **2008**, *10*, 5441. (d) Bodnar, B.S.; Vogt, P.F. *J. Org. Chem.* **2009**, *74*, 2598. (e) Nandi, P.; Dye, J.L.; Jackson, J.E. *J. Org. Chem.* **2009**, *74*, 5490. (f) Costanzo, M.J.; Patel, M.N.; Petersen, K.A.; Vogt, P.F. *Tetrahedron Lett.* **2009**, *50*, 5463.

§ Answer to Worked Problem:

The initial step is the Birch reduction to give the potassium salt of the amide enolate, **A**, which is then alkylated to give the product **B**. The next step is hydrolysis of the enol ether, which is essential to prevent the possibility of regioisomeric rearrangement products. The [3,3]-sigmatropic rearrangement of the 1,5-diene then completes the sequence. [*Tetrahedron Lett.* **2004**, *45*, 8183]

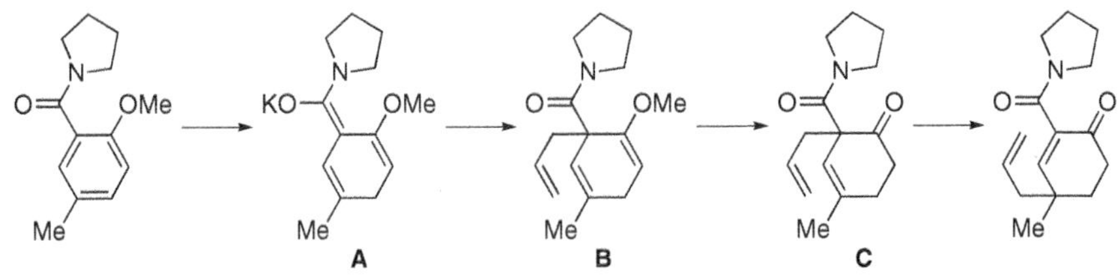

Problem

20-6 What is the major product of each of the following reactions?

(a)

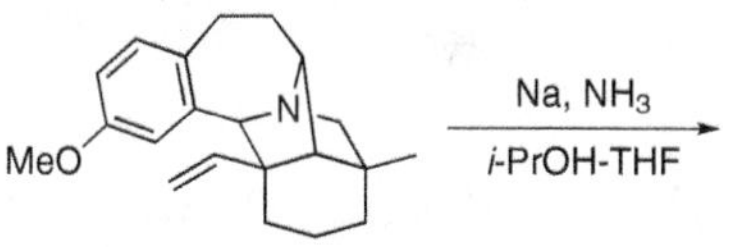

(b)

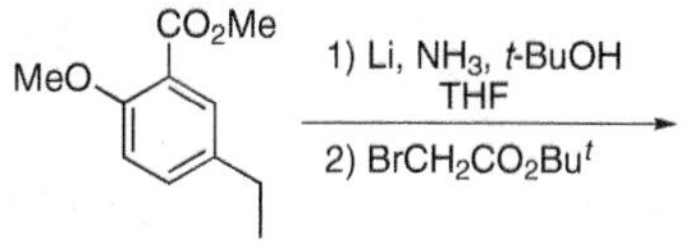

(c)

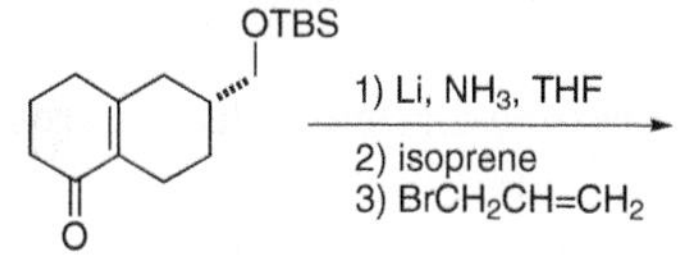

References: (a) *J. Am. Chem. Soc.* **2006**, *128*, 8734. (b) *Org. Lett.* **2006**, *8*, 831. (c) *J. Org. Chem.* **2009**, *74*, 6623.

Reaction Synopses

Reductive Elimination

X, Y: Cl, Br, I, OSO_2R, NR_3^+, etc.

Reagents: Li, NH_3; Mg, Et_2O; Zn, EtOH, Δ; Zn, AcOH, Δ;
or BuLi, Et_2O; etc.
or NaI, DMF, Δ; etc.

Mechanism: E1cb or E2

Stereochemistry: *anti* (E2); mainly *anti* (E1cb)

Birch and Benkeser Reductions

M: Li, Na, K, etc.; ROH: EtOH, Me_3COH, etc.

Reagents: Li, NH_3, *t*-BuOH, THF; etc. (Birch reduction)
or Ca, $MeNH_2$, $(CH_2NH_2)_2$; etc. (Benkeser)

Reduction of Conjugated Carbonyl Compounds

(a) To Saturated Ketones

Reagents: H_2, Pd-C; etc.
or Li, NH_3 (no proton source); etc.

(b) To Saturated Alcohols

Reagents: Li, NH_3, ROH; etc.
or $NaBH_4$, MeOH; $LiAlH_4$, Et_2O, extended time

(continues)

(**Reaction Synopses** continued)

Pinacol Reaction

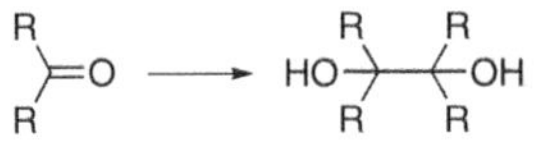

Reagents:	Mg(Hg), Et_2O; Zn(Hg), Et_2O; Al(Hg), Et_2O; etc.
or	Ti, THF, $TiCl_3$, K, THF; etc. (provided any alkene would violate Bredt's rule)
or	Mn, $TiCl_4$, THF; etc. (allows the diastereoselective crossed pinacol reaction of aromatic aldehydes)

McMurry Reaction

Reagents:	Ti, THF, $TiCl_3$, K, THF; etc.

20.5 **Reduction of Carbonyl Compounds to Hydrocarbons**

Deoxygenation is carried out using three major reactions. The first is the ***Wolff-Kishner reduction***, in which hydrazone is treated with a strong base at elevated temperature; the second is the ***Clemmensen reduction***, in which the carbonyl compound is reduced with zinc amalgam and concentrated hydrochloric acid; and the third is ***reductive desulfurization***, in which a dithioketal is reduced under catalytic hydrogenation or Birch reduction conditions.

The Wolff-Kishner Reduction and Related Reactions

The Wolff-Kishner reduction[118] was originally carried out by heating the hydrazone of the carbonyl compound with potassium hydroxide or by heating the carbonyl compound with potassium hydroxide in hydrazine. However, in 1946 Chinese chemist Huang Ming-long reported a modified version of the reaction,[119] now known by his name, which involved a high-boiling protic solvent (he used ethylene glycol, but diethylene glycol is now more widely used). The overall yields are higher than the original Wolff-Kishner procedure (e.g., Examples **20.87** and **20.88**). The Huang Minlon procedure (the name of the reaction is based on the name used in the original paper, not on more modern transliterations of the Chinese) has itself been modified by Barton, who

118. Todd, D. *Org. React.* **1948**, *4*, 378.

Ludwig Wolff (1859–1919) was educated at Würzburg, Munich, and Strasbourg (PhD, 1882 under Fittig). Wolff became Fittig's assistant on graduating, and he was appointed privatdozent at Strasbourg in 1885. In 1891, he was appointed extraordinary professor of analytical chemistry at the University of Jena, where he spent the rest of his working life. His major works involved the study of hydrazines and diazo compounds. A plaque in his honor was unveiled at Jena in 2012.

Nikolai Matveyevich Kizhner (1867–1935) was educated in Moscow (Dr Khim, 1900) and was immediately appointed inaugural Professor of Organic Chemistry at the Tomsk Technological Institute, in Siberia. His decade at Tomsk was tumultuous. He resigned under duress in 1912 and left Tomsk in 1914. He taught at Shanyavskii People's University in Moscow until 1918, when he became director of the Aniline Trust Research Institute, in Moscow. For more detail, see: Lewis, D.E. *Angew. Chem. Int. Ed.* **2013**, *52*, 11704.

119. (a) Huang-Minlon *J. Am. Chem. Soc.* **1946**, *68*, 2487. (b) Huang-Minlon *J. Am. Chem. Soc.* **1949**, *71*, 3301. The fortuitous discovery of this reaction is related in: Ma, S.; Craig, G.W. *Helv. Chim. Acta* **2013**, *96*, 1822.

Huang Ming-long (1898–1979) was educated at Zhejiang Medical College (graduated 1918) and at the University of Berlin (PhD, 1924). He worked in the West during 1935–1940 (Schering AG in Berlin) and 1946–1952 (Harvard and Merck). From 1925–1934 he served as professor of chemistry at Zhejiang Medical College and from 1940–1946 was at the Academia Sinica in Kunming. In 1952 he returned to the PLA Academy of Medical Sciences, and in 1956, he moved to the Academia Sinica in Shanghai. For more detail, see: (a) Han, G.; Jin, S.; Wu, Y. *Progr. Chem.* **2012**, *24*, 1229 (in Chinese); (b) Ma, S.; Craig, G.W. *Helv. Chim. Acta* **2013**, *96*, 1822.

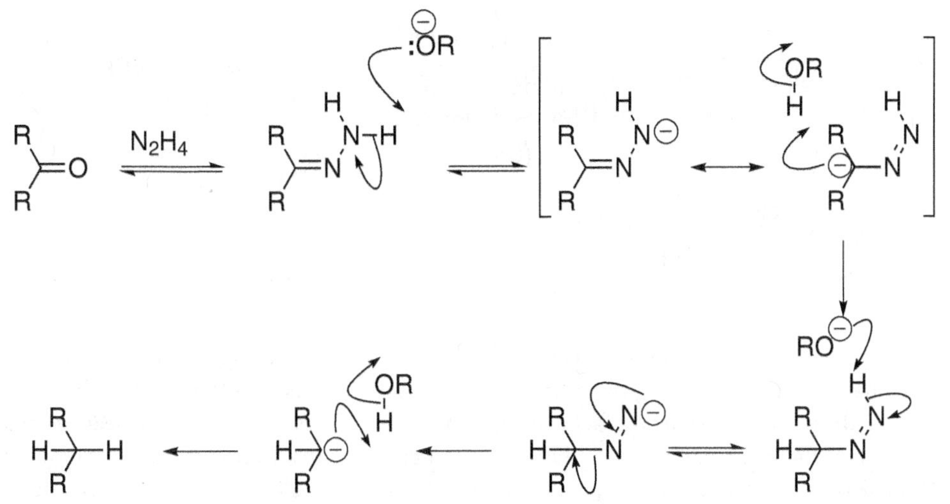

Figure 20.7 The generally accepted mechanism of the Wolff-Kishner reduction of aldehydes and ketones

used anhydrous hydrazine and sodium metal in diethylene glycol for the reduction of highly hindered ketones.[120]

The generally accepted mechanism of the Wolff-Kishner reaction is given in Figure 20.7.

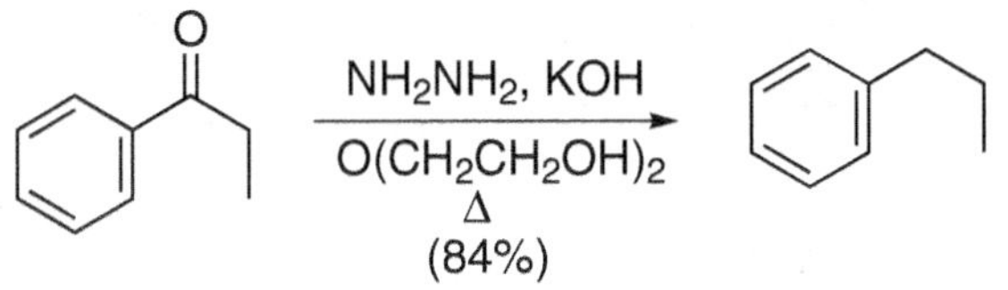

(20.87)[122]

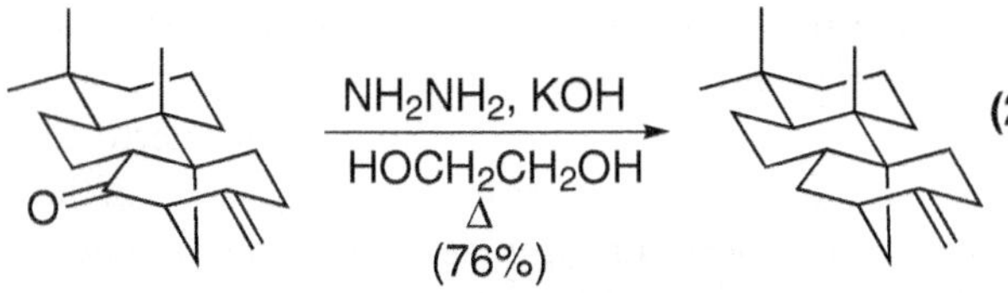

(20.88)[121]

The strongly basic conditions of the Wolff-Kishner reduction can make it unsuitable for use in the presence of sensitive functional groups. In such cases, a modification of this reaction based on the tosylhydrazone (e.g., Examples **20.89** and **20.90**) is often the method of choice for deoxygenation.[122] In this variation of the Wolff-Kishner reaction, the tosylhydrazone is formed first, and this is then subjected to reduction by a nucleophilic hydride nucleophile or a borane[123] (the sulfur is actually reduced from the +6 to the +4 oxidation state during this reaction). The overall effect is to allow the Wolff-Kishner reduction to be carried out under much milder reaction conditions. When this reaction is used to reduce α,β-unsaturated ketones, the product is an alkene. The double bond is located between the original carbonyl carbon and the α carbon, even if this results in loss of conjugation of the double bond with an aromatic ring.[124]

120. (a) Barton, D.H.R.; Ives, D.A.J.; Thomas, B.R. *J. Chem. Soc.* **1954**, 903. (b) Barton, D.H.R.; Ives, D.A.J.; Thomas, B.R. *J. Chem. Soc.* **1955**, 2056.

121. White, J.D.; Somers, T.C. *J. Am. Chem. Soc.* **1994**, *116*, 9912.

122. (a) Hutchins, R.O.; Maryanoff, B.E.; Milewski, C.A. *J. Am. Chem. Soc.* **1971**, *93*, 1793. (b) Hutchins, R.O.; Milewski, C.A.; Maryanoff, B.E. *J. Am. Chem. Soc.* **1973**, *95*, 3662.

123. Greco, M.N.; Maryanoff, B.E. *Tetrahedron Lett.* **1992**, *33*, 5009.

124. (a) Kabalka, G.W.; Yang, D.T.C.; Baker, J.D., Jr. *J. Org. Chem.* **1976**, *41*, 574. (b) Taylor, E.J.; Djerassi, C. *J. Am. Chem. Soc.* **1976**, *98*, 2275. (c) Hutchins, R.O.; Natale, N.R. *J. Org. Chem.* **1978**, *43*, 2299. (d) Greene, A.E. *Tetrahedron Lett.* **1979**, 63.

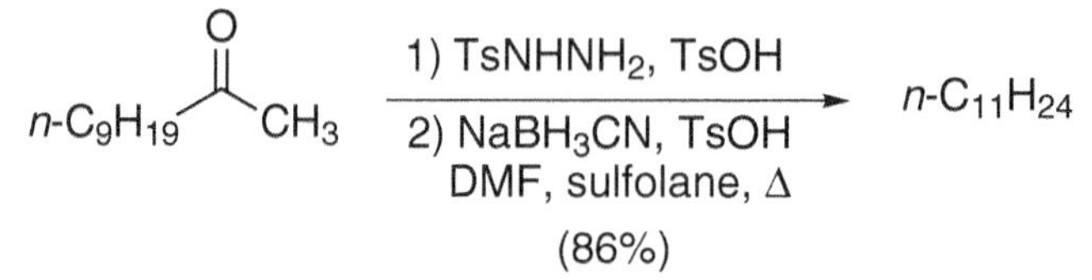

(20.89)[125]

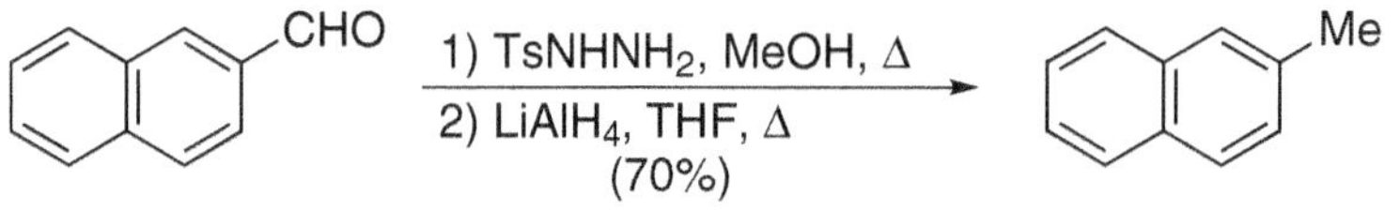

(20.90)[123(b)]

A more recent approach to effecting the Wolff-Kishner reduction under mild conditions has involved exploiting two facts. First, silyl compounds are often more stable than their hydrogen counterparts (a fact routinely exploited in synthesis). Second, cleavage of a σ bond to silicon by oxyanions or fluoride anion gives anions. In this modification of the Wolff-Kishner reduction, a *tert*-butyldimethylsilyl hydrazone is formed and, in the same vessel, is then treated with potassium *tert*-butoxide at temperatures more than 70°C lower than most Wolff-Kishner reductions (e.g., Example **20.91**).[126] In this system, aromatic ketones and aldehydes are reduced at 100°C, whereas aliphatic ketones are reduced at room temperature.

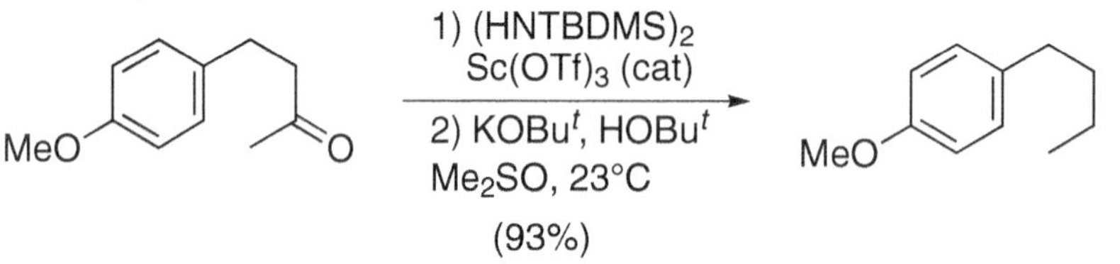

(20.91)

Wolff-Kishner reductions of carbonyl compounds carrying a heteroatom substituent at the α carbon lead to the formation of alkenes via alkenyl anion intermediates (Example **20.92**). This results from a simple modification of the reaction mechanism. The initial anion formed by deprotonation of the hydrazone leads to elimination of the leaving group to give the azoalkane, which then loses nitrogen through either anionic (shown) or free radical intermediates to give the alkene. When the leaving group at the α position is an epoxide oxygen, the reaction leads to ring opening, with the formation of an allylic alcohol. This variation of the Wolff-Kishner reduction is known as the ***Wharton reaction*** (e.g., Example **20.93**).[127]

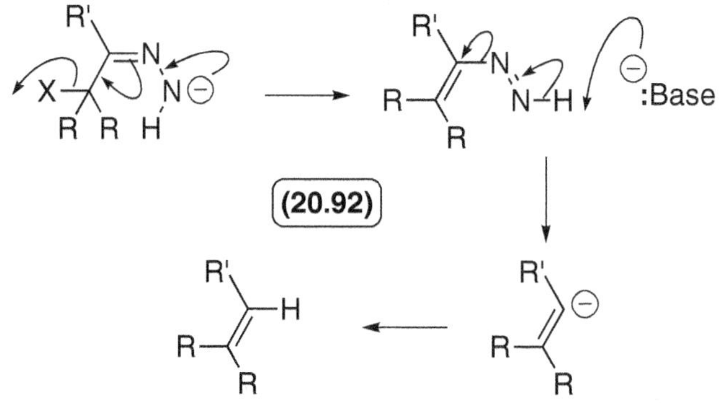

(20.92)

125. Caglioti, L.; Magi, M. *Tetrahedron*, **1963**, *19*, 1127.
126. Furrow, M.E.; Myers, A.G. *J. Am. Chem. Soc.* **2004**, *126*, 5436.
127. Wharton, P.S.; Bohlen, D.H. *J. Org. Chem.* **1961**, *26*, 3615.

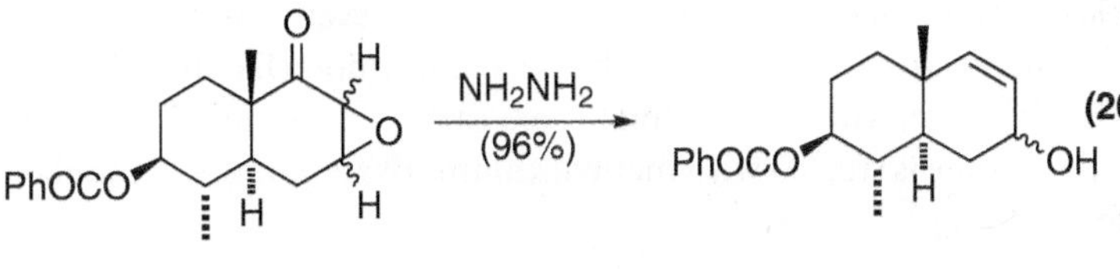

(20.93)[128]

A similar reaction is observed when the hydrazone carries a substituent on nitrogen capable of elimination as a leaving group. One such class of compounds is the sulfonylhydrazones. Treating a sulfonylhydrazone with a base gives an alkene with rearrangement of one of the groups bonded to the carbonyl carbon. When the base used is an alkoxide, the reaction is known as the ***Bamford-Stevens reaction***,[129] and the more substituted alkene is formed as the major product. When an alkylithium is used as the base, the reaction is known as the ***Shapiro reaction***,[130] and the less substituted alkene is obtained as the major product. It appears clear that the mechanism of the Shapiro reaction (Example **20.94**) proceeds through a vinyllithium formed by loss of nitrogen from an azovinyl anion.[131] In fact, this reaction is a useful method for the preparation of vinyllithium reagents.[132] The mechanism of the Bamford-Stevens reaction is less clearcut,[133] and it may involve carbene (shown as Example **20.95**) or carbocation intermediates.

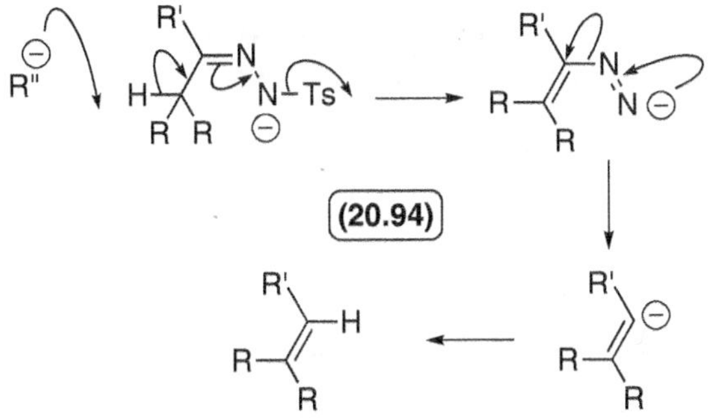

(20.94)

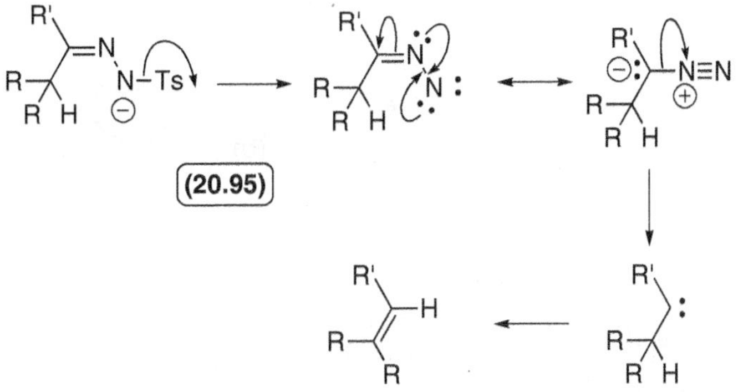

(20.95)

128. Ziegler, F.E.; Hwang, K.-J.; Kadow, J.F.; Klein, S.I.; Pati, U.K.; Wang, T.F. *J. Org. Chem.* **1986**, *51*, 4573.

129. Bamford, W.R.; Stevens, T.S. *J. Chem. Soc.* **1952**, 4735.

130. (a) Shapiro, R.H. *Tetrahedron Lett.* **1968**, 345. (b) Shapiro, R.H.; Heath, M.J. *J. Am. Chem. Soc.* **1967**, *89*, 5734. (c) Shapiro, R.H. *Org. React.* **1976**, *23*, 405. (d) Adlington, R.M.; Barrett, A.G.M. *Acc. Chem. Res.* **1983**, *16*, 55. (e) Chamberlin, A.R.; Bloom, S.H. *Org. React.* **1990**, *39*, 1.

131. (a) Casanova, J.; Waegell, B. *Bull. Soc. Chim. France* **1975**, 922. (b) Lipton, M.F.; Shapiro, R.H. *J. Org. Chem.* **1978**, *43*, 1409.

132. See, for example, (a) Traas, P.C.; Boehlens, H.; Takken, H.J. *Tetrahedron Lett.* **1976**, 2287. (b) Stemke, J.E.; Chamberlin, A.R.; Bond, F.T. *Tetrahedron Lett.* **1976**, 2947.

133. (a) Powell, J.W.; Whiting, M.C. *Tetrahedron* **1959**, *7*, 305. (b) Powell, J.W.; Whiting, M.C. *Tetrahedron* **1961**, *12*, 168. (c) De Puy, C.H.; Froemsdorf, D.H. *J. Am. Chem. Soc.* **1960**, *82*, 634. (d) Bayless, J.H.; Friedman, L.; Cook, F.B.; Shechter, H. *J. Am. Chem. Soc.* **1968**, *90*, 531. (e) Nickon, A.; Werstiuk, N.H. *J. Am. Chem. Soc.* **1972**, *94*, 7081.

The possible involvement of carbenes in the Bamford-Stevens reaction is illustrated by the reaction of camphor tosylhydrazone with hydroxide anion (Example **20.96**). One of the products is a tricyclane, which could come from carbene insertion. In contrast to this, when the same hydrazone is treated with methyllithium (Example **20.97**), the alkene is the only product formed.

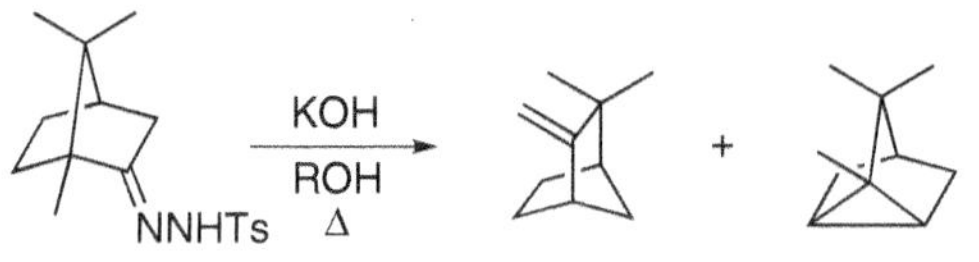

(20.96)[134]

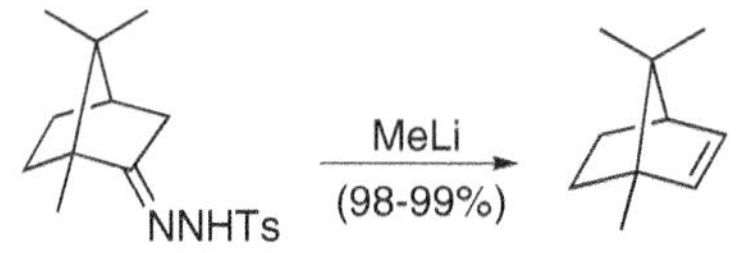

(20.97)[135]

The reaction between a sulfonylhydrazide and an α,β-epoxyketone represents the hybrid of both these types of reduction. In the reaction (e.g., Example **20.98**), α-substituted carbonyl compounds and functionalized hydrazines are both used. In this situation, the reaction leads to fragmentation to give an alkyne and a ketone. The reaction is generally known as the ***Eschenmoser-Tanabe fragmentation.***[136] It has been especially successfully applied to the synthesis of medium-ring alkynes.

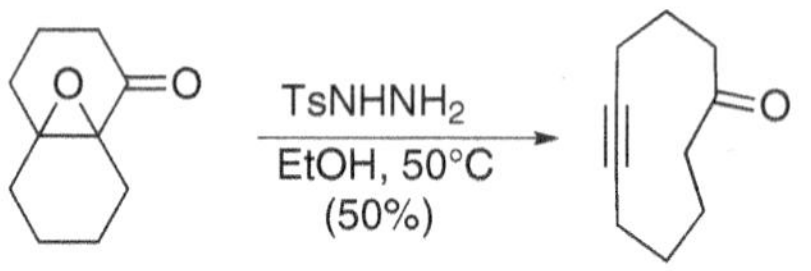

(20.98)[137]

Problem

20-7 What is the major organic product of each of the following reactions?

(a)

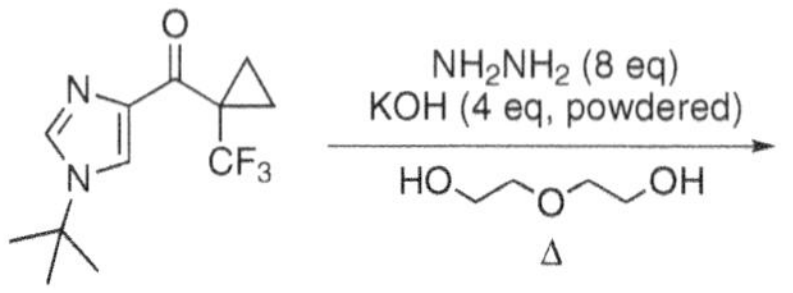

(b)

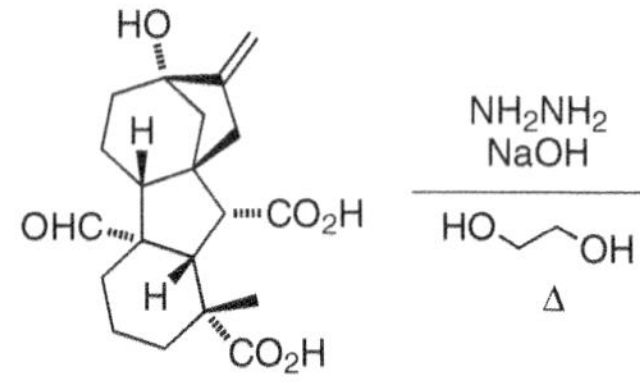

(c)

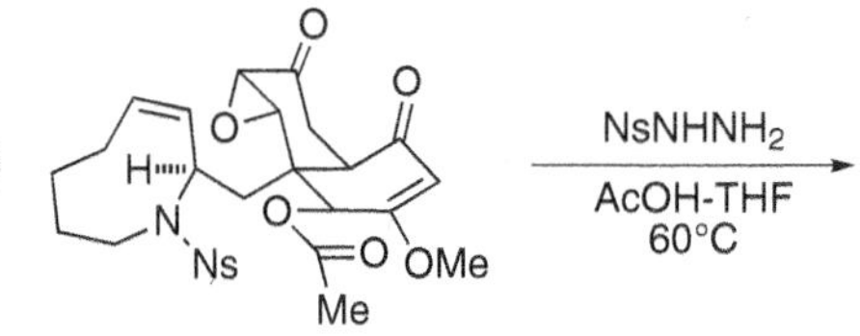

References: (a) *Org. Proc. Res. Dev.* **2009**, *13*, 576. (b) *Aust. J. Chem.* **1996**, *49*, 249. (c) *Org. Lett.* **2008**, *10*, 3251.

---

134. Clark, P.; Whiting, M.C.; Papenmeier, G.; Reusch, W. *J. Org. Chem.* **1962**, *27*, 3356.
135. Shapiro, R.H.; Duncan, J.H. *Org. Syn.* **1971**, *51*, 66.
136. Eschenmoser, A.; Felix, D.; Ohloff, G. *Helv. Chim. Acta* **1957**, *50*, 708.
137. Tanabe, M.; Crowe, D.F.; Dehn, R.L. *Tetrahedron Lett.* **1967**, 3943.

The Clemmensen Reduction

The final method of deoxygenation of aldehydes and ketones that we will discuss is the Clemmensen reduction,[138] which involves heating the aldehyde or ketone with zinc amalgam and concentrated hydrochloric acid (e.g., Example **20.99**).[139] The strongly acidic conditions of this reaction are tolerated by relatively few functional groups, and they tend to promote rearrangement reactions.

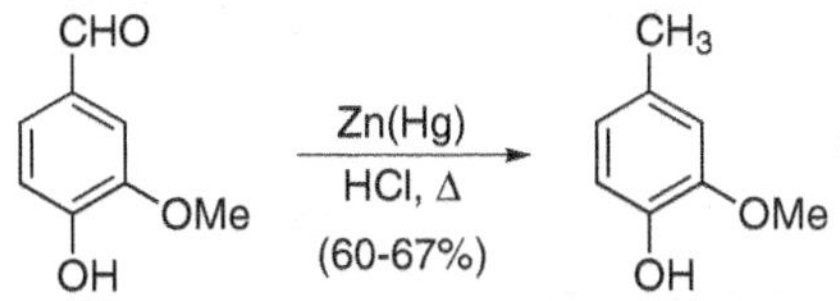

(20.99)

A more modern version of the reaction (Example **20.100**) involves the use of hydrogen chloride and zinc dust in an organic solvent.[140]

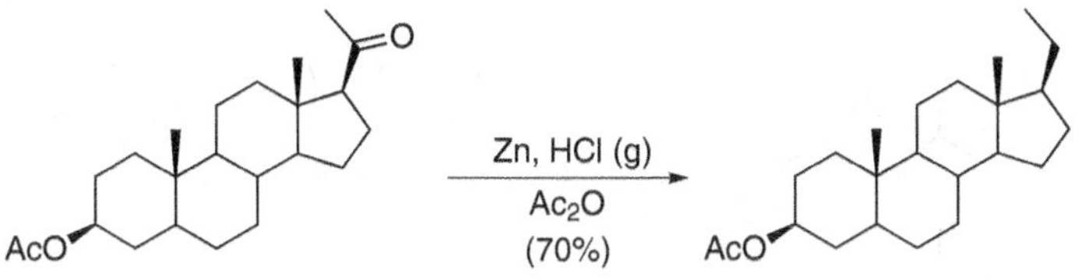

(20.100)

Reductive Desulfurization of Dithioketals and Dithioacetals

In the previous chapter, we saw that both sulfur atoms of a dithioketal or dithioacetal can be replaced by hydrogen when the dithioketal is reduced with hydrogen-saturated Raney nickel. The same reduction can also be accomplished using the Birch reduction, as shown in Example **20.101**.[141] This combination of dithioacetal formation and desulfurization is often used as a simple two-step alternative to the Wolff-Kishner and Clemmensen reductions and their variants.

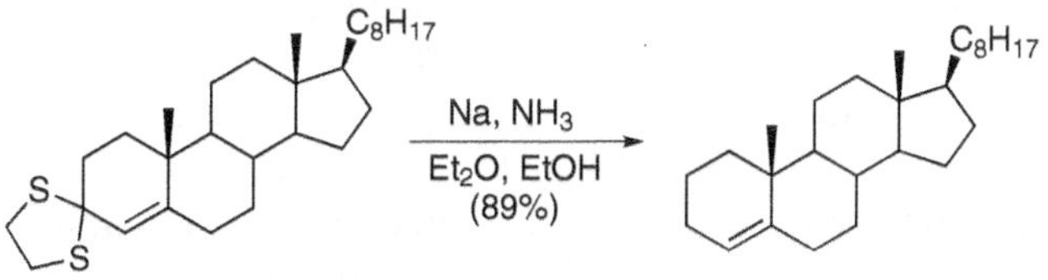

(20.101)

Silicon-Based Deoxygenations

A potentially very useful deoxygenation reaction has been reported by Nimmagadda and McRae.[142] In this reaction, oxygenated organic compounds are treated with excess butylsilane

---

138. (a) Martin, E.L. *Org. React.* **1942**, *1*, 155. (b) Vedejs, E. *Org. React.* **1975**, *22*, 401. (c) Buchanan, J.G.St.C.; Woodgate, P.D. *Quart. Rev. Chem. Soc.* **1969**, *23*, 522.

Erik Christian Clemmensen (1876–1941) was born in Odensee, Denmark and educated at the Royal Polytechnic Institute. He left school in 1891, at age 15 and in 1900, he emigrated to the United States, after his graduation. In 1903, he joined Parke, Davis & Company in Detroit, where he discovered (but never patented!) the reaction named for him; it did, however, earn him a PhD from the University of Copenhagen in 1913. Clemmensen's career was spent entirely in industry. In 1914, he founded the Commonwealth Chemical Corporation, which was acquired by Monsanto in 1929. In 1935, he left Monsanto and founded the Clemmensen Chemical Corporation. Up to his death at age 65, he was carrying out research at the Clemmensen Chemical Company.

139. Schwartz, R.; Hering, H. *Org. Syn., Coll. Vol. 4* **1963**, 203.

140. Yamamura, S.; Hirata, Y. *J. Chem. Soc. C* **1968** 2887.

141. Ireland, R.E.; Wrigley, T.I.; Young, W.G. *J. Am. Chem. Soc.* **1958**, *80*, 4604.

142. Nimmagadda, R.D.; McRae, C. *Tetrahedron Lett.* **2006**, *47*, 5755.

in the presence of 5 to 10 mol % *tris*-(pentafluorophenyl)borane in dichloromethane. The reaction showed remarkable selectivity for the reductive cleavage of carbon-oxygen bonds, as shown in Examples **20.102** to **20.104**. It is noteworthy that even the carboxyl group is reduced to a methyl group by this reagent.

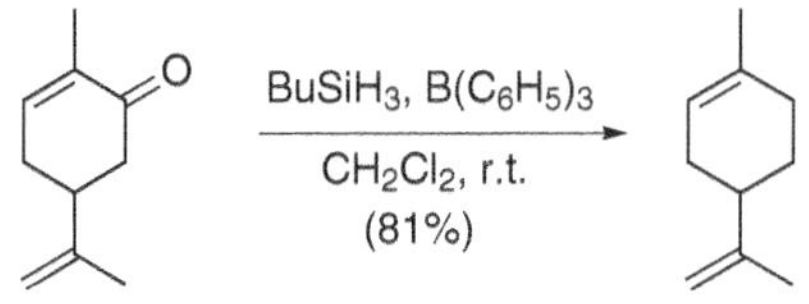

(20.102)

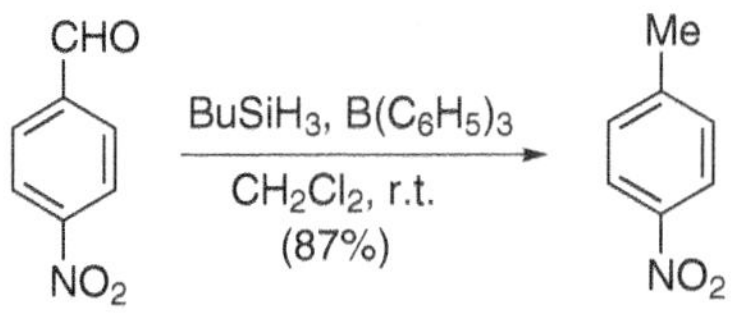

(20.103)

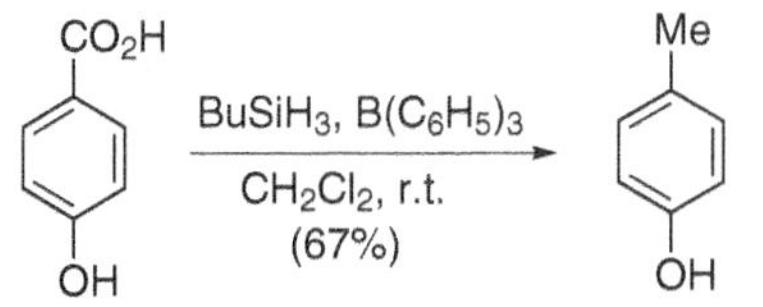

(20.104)

Reduction of Halides and Alcohols to Alkanes

The cleavage of carbon-heteroatom σ bonds is not restricted to dithioketals and sulfides. Alkyl halides can be reduced chemoselectively by means of tributylstannane; the reaction occurs by a free radical mechanism through tributylstannyl radicals (Figure 20.8). The reaction is driven by the fact that the tin-halogen bond is much stronger than the tin-hydrogen bond. A similar situation occurs with sulfur (C=S) and selenium compounds, which also react with stannanes to give alkyl free radicals.

Stannane reductions are especially useful for deoxygenation of alcohols through thiocarbonyl intermediates, a reaction known as the ***Barton-McCombie reaction*** or the ***Barton deoxygenation***.[143] In these reductions (Examples **20.105** and **20.106**), the tributylstannyl radical adds to the C—S π bond, and the alkyl free radical is generated by elimination of a carbonyl compound. Example **20.105** is particularly interesting. The attempted Wolff-Kishner deoxygenation of the corresponding ketone failed under Huang-Minlon conditions, and the preparation of the dithioketal for reductive desulfurization also failed.

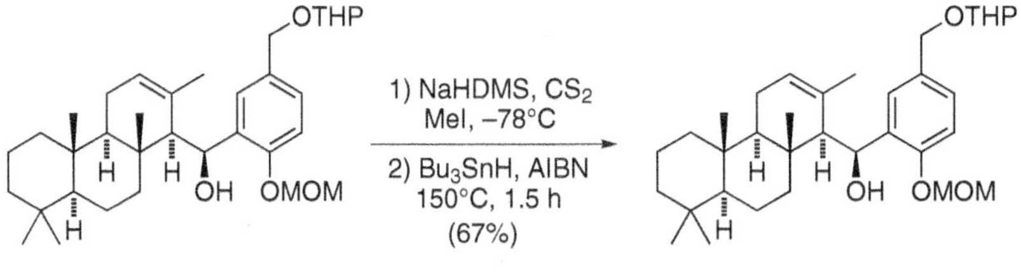

(20.105)[144]

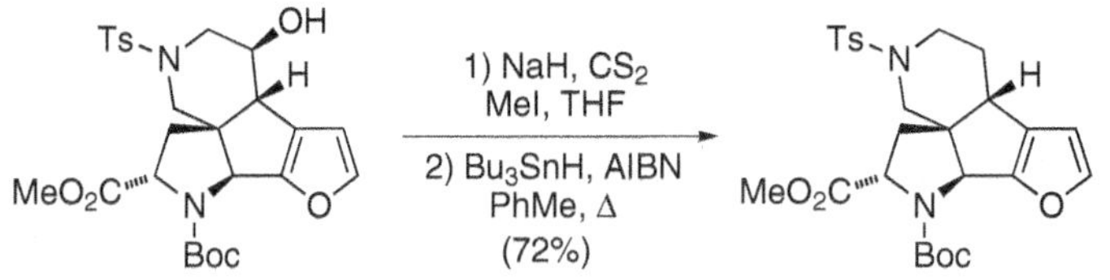

(20.106)[145]

---

143. Barton, D.H.R.; McCombie, S.W. *J. Chem. Soc., Perkin Trans. 1* **1975**, 1974.
144. Basabe, P.; Martín, M.; Bodero, O.; Blanco, A.; Marcos, I.S.; Díez, D.; Urones, J.G. *Tetrahedron* **2010**, *66*, 6008.
145. Deng, H.; Yang, X.; Tong, Z.; Li, Z.; Zhai, H. *Org. Lett.* **2008**, *10*, 1791.

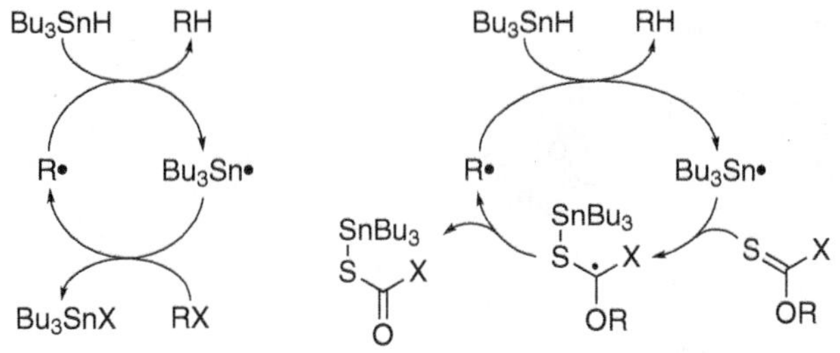

Figure 20.8 The free radical chain reaction for reduction of halides (left) or thioesters (right) with tributylstannane

Reaction Synopses

Deoxygenation to Alkanes

By Wolff-Kishner Reduction Variants

Reagents: N_2H_4, KOH, $HOCH_2CH_2OH$, Δ (Huang Minlon)
N_2H_4, Na, $O(CH_2CH_2OH)_2$ (Barton);
or (1) $TsNHNH_2$, MeOH, Δ, (2) $NaBH_3CN$, MeOH, Δ; etc.

By Clemmensen Reductions

Reagents: Zn(Hg), HCl, H_2O, Δ; Zn, HCl (g), Et_2O; Zn, HCl (g), Ac_2O; etc.

By Reductive Sulfurization

Reagents: (1) $HSCH_2CH_2SH$, BF_3, (2) H_2, Ra-Ni; etc.
or (1) $HSCH_2CH_2SH$, BF_3, (2) Na, NH_3; etc.

By Silane Reduction

Reagents: $BuSiH_3$ (2–10 eq), $B(C_6F_5)_3$ (5–10 mol %), CH_2Cl_2

Deoxygenation to Alkenes

Reagents: (1) $TsNHNH_2$, MeOH, Δ, (2) MeLi, THF (Shapiro)
or (1) $TsNHNH_2$, MeOH, Δ, (2) NaOR, THF (Bamford-Stevens)

Regiochemistry: Shapiro reaction gives less substituted alkene; Bamford-Stevens reaction gives more substituted alkene.

Wharton Reaction

Reagents: (1) $TsNHNH_2$, MeOH, Δ, (2) base

(continues)

(**Reaction Synopses** continued)

Eschenmoser-Tanabe Fragmentation

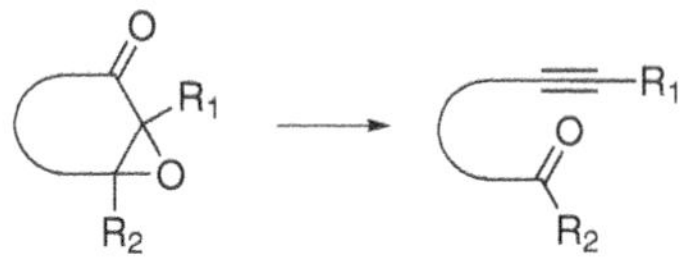

Reagents:	TsNHNH$_2$, ROH, Δ
Restrictions:	β position may not carry a hydrogen.

Barton-McCombie Reaction; Barton Deoxygenation

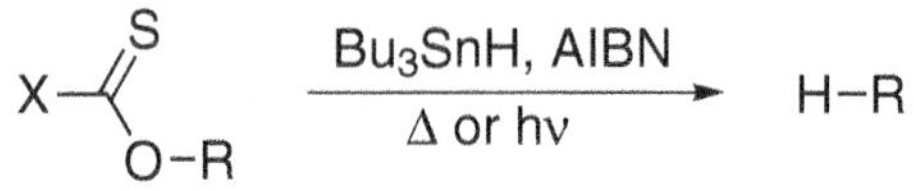

X: Ar, R, SR, imidazole, ArO, etc.

20.6 A Catalog of Reduction Reactions: Reduction at a Glance

Now it is time to put the reduction reactions we have studied together in a way that will allow us to see the available choices. This is done in Table 20.4.

Table 20.4 Catalog of Reduction Reactions

Transformation	Reagents	Incompatible Functional Groups	Other Limitations or Features
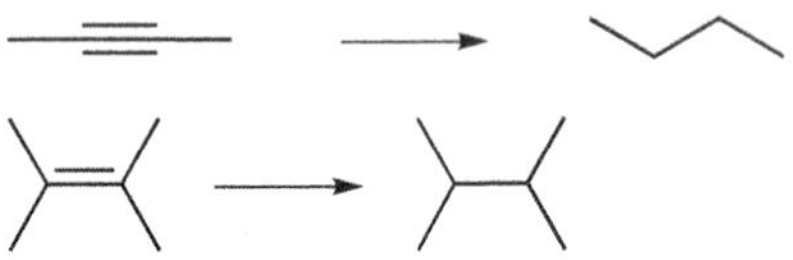	H$_2$, PtO$_2$; H$_2$, Pd-C; H$_2$, Ra-Ni; H$_2$, Ru	Alkynes, C=N bonds	Addition of H–H is *syn*. Pt and Pd catalysts can cause π bond migration
	H$_2$, (Ph$_3$P)$_3$RhCl, and similar Rh, Ru, Ir compounds		Chiral diphosphine ligands on catalyst allow enantioselective reduction
	R$_3$SiH, CF$_3$CO$_2$H;		Alkene must be able to give a stable carbocation (3°, allyl, benzyl)
	HCO$_2$H, Pd-C; 1,4-cyclohexadiene, Pd-C; NH$_4$OCHO, Pd-C; etc.		Transfer hydrogenation; tolerates a wide range of functional groups
	H$_2$, Pd-BaSO$_4$, Pb, quinoline		Gives *Z* alkene
	Li, NH$_3$, ROH		Gives *E* alkene
	Hantzsch ester, Brønsted acid		
	Li, NH$_3$, THF; Na, EtOH; etc.	Saturated aldehydes, esters, ketones, nitriles; conjugated dienes; benzyl derivatives	
	H$_2$, catalyst	Alkenes, alkynes, benzyl derivatives	

Transformation	Reagents	Incompatible Functional Groups	Other Limitations or Features
	DIBAL-H	Other carbonyl compounds, nitriles	
	$NaBH_4$, $CeCl_3$		Can also be used to reduce nitriles to primary amines
	H_2, Pd-C; H_2, Ni; etc.		Stereochemistry depends on metal catalyst
	Li, NH_3, ROH	Carbonyl compounds; conjugated systems	
	$LiAlH_4$; Red-Al®; etc.		Reduces esters to primary alcohols
	$NaBH_4$		Does not reduce esters; reduces conjugated ketones very slowly
	Al(O-*i*-Pr)$_3$, *i*-PrOH, Δ		Reaction is reversible; thermodynamically most stable alcohol predominates
	Na, EtOH; Li, NH_3, ROH; etc.	Conjugated carbonyl compounds	
	Et_3SiH, CF_3CO_2H		If conjugated with alkene, alkene π bond reacts first
	HCO_2Na, Ru-TsDPEN	Alkenes in conjugated systems	Allows enantioselective reduction
	DIBAL-H etc.		Reduces esters to hemiacetals at low temperature
	NH_2NH_2, KOH, $HOCH_2CH_2OH$, Δ	Groups susceptible to strong bases	
			$TsNHNH_2$, $NaBH_3CN$, HOAc-MeOH
	Zn(Hg), HCl, Δ	Groups susceptible to strong acids	
	$BuSiH_3$, $B(C_6F_5)_3$		
	(1) $HSCH_2CH_2SH$, BF_3; (2) H_2, Ni or Li, NH_3		
	(1) $TsNHNH_2$, (2) MeLi; or (1) $TsNHNH_2$; (2) $KOBu^t$		MeLi gives vinyl anion; carbenes may be involved when $KOBu^t$ is used as base
	H_2, Pd-C		X=halogen
	Bu_3SnH, AIBN, hν		X=halogen, OC(S)SMe
	L-Selectride®; Super-Hydride™; etc.		X=halogen, OSO_2R; reaction occurs with inversion

(continued)

Table 20.4 (continued)

Transformation	Reagents	Incompatible Functional Groups	Other Limitations or Features
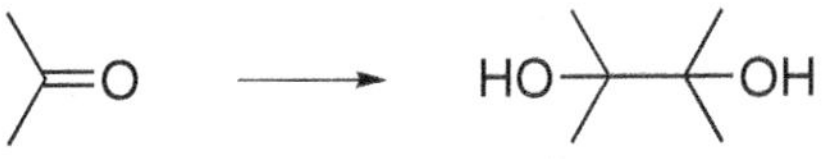	Ti, THF; $TiCl_3$, K, THF; etc.		
	Mg(Hg), Et_2O; Zn(Hg), Et_2O; Al(Hg), Et_2O; etc.		
	Mn, $TiCl_4$, THF		
	Na, THF; Na-K, THF	Carbonyl compounds	
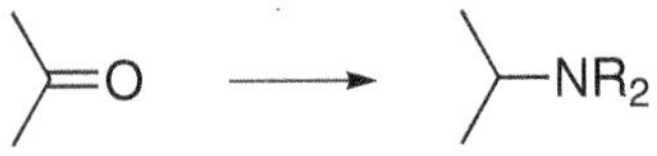	R_2NH, $NaBH_3CN$, HOAc; R_2NH, $Na(OAc)_3BH$		Reduction works best under acidic conditions
	(1) NH_2OH; H_2, Ra-Ni		
	R_2NH, HCO_2H, Pd-C; etc.		Ammonium formate may also be used
	HCHO, HCO_2NH_4, Δ; etc.		Gives product where R=Me
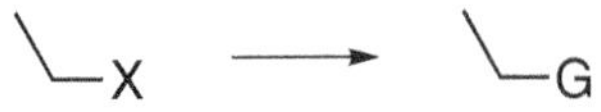	$LiAlH_4$, Et_2O, Δ; etc.		X=CN → G=CH_2NH_2; X=$CONR_2$ → G=CH_2NR_2;
	H_2, Pd-C; etc.		X=N_3 → G=NH_2; X=NO_2 → G=NH_2; etc.
	BH_3•THF; etc.		X=CN → G=CH_2NH_2
	$NaBH_4$, $CeCl_3$		X=CN → G=CH_2NH_2
	H_2Pd-C; etc.		
	Li, NH_3, *t*-BuOH; etc.	Carbonyl compounds, conjugated compounds, alkynes	
	(1) $TsNHNH_2$, (2) base		
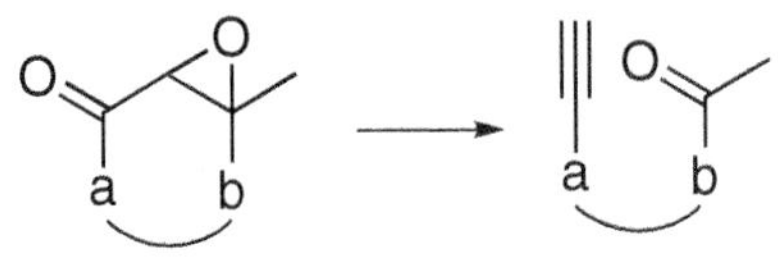	$TsNHNH_2$, EtOH, Δ; etc.		β Position of epoxyenone must be fully substituted
	$Al(OR)_3$; SmI_2; Shvo's Ru catalyst, ROH; etc.		
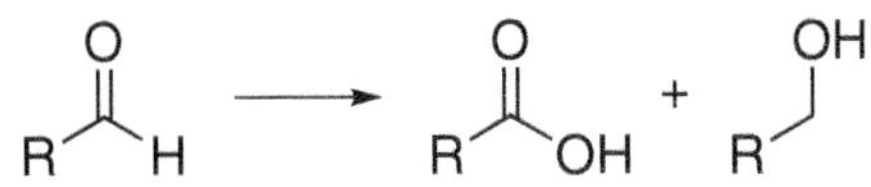	NaOH; H_2CO, NaOH; etc.		

Transformation	Reagents	Incompatible Functional Groups	Other Limitations or Features
	(X=Cl): $LiAl(OBu^t)_3$; H_2, Pd-$BaSO_4$, quinoline-S; DIBAL-H, −78°C (ester);		
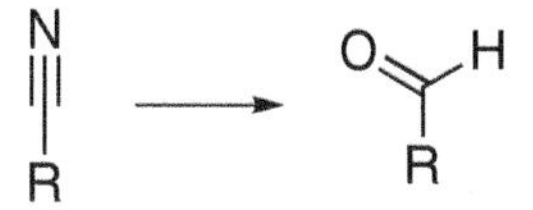	DIBAL-H, 0°C	Aldehydes, ketones, acid chlorides	
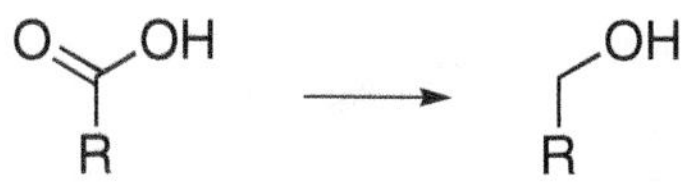	BH_3•THF	Alkenes, amides, nitriles	Reaction with carboxylic acids is fastest
	$LiAlH_4$, Et_2O, Δ; etc.	Any other carbonyl compound or nitrile	

Chapter Summary

This chapter has continued the discussion of reduction in organic chemistry, focusing on methods other than hydrogenation. Complex metal hydrides provide a complementary method for reduction and are most useful for reducing polar π bonds. The nucleophilic complex hydrides vary in strength from lithium aluminum hydride, which reduces most every polar π bond, to sodium cyanoborohydride, which requires acidic conditions and is best adapted to reducing iminium ions. DIBAL-H is a typical electrophilic hydride and is similar to lithium aluminum hydride in activity. Addition of the hydride generally conforms to Cram's rule. Reductions using elemental metals include reductive elimination, the Birch reduction, the Wurtz coupling, and the pinacol and McMurry reactions. Deoxygenation of carbonyl compounds can be effected by the Clemmensen and Wolff-Kishner reductions, as well as their modifications—the Bamford-Stevens, Shapiro, and Wharton reactions, as well as the Eschenmoser-Tanabe fragmentation. Deoxygenation can also be accomplished by the two-step sequence of thioketal formation and reductive desulfurization. Deoxygenation of alcohols can be effected by the Barton-McCombie deoxygenation.

Key Terms

Bamford-Stevens reaction
Barton-McCombie reaction
Benkeser reductionBirch reduction
Bouveault-Blanc reduction
Cannizzarro reaction
Clemmensen reduction
electrophilic hydrides
Eschenmoser-Tanabe fragmentation
Huang Minlon modification
hydrogen transfer reduction
McMurry reaction
Meerwein-Ponndorf-Verley reduction
nucleophilic hydrides
pinacol reaction
reductive elimination
Shapiro reaction
Tishchenko reaction
Wharton reaction
Wolff-Kishner reduction
Wurtz coupling

Additional Problems

Comprehensive Reduction Review Problem Set 1. What is the major organic product formed when each of the reactants in the first list is treated with each of the reducing agents in the second list? If a reagent can react more than once with the particular reducing agent, show the product of each step in the reduction when an excess of the reducing agent is used (e.g., when ethyl acetate reacts with lithium aluminum hydride, the first product is acetaldehyde, and this is further reduced to ethanol). Note that not all the reactants will react with every reducing agent.

Compounds:

(a) 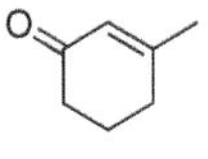 (b) 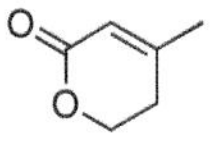 (c) 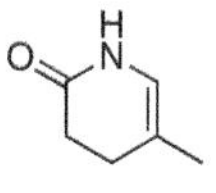 (d) 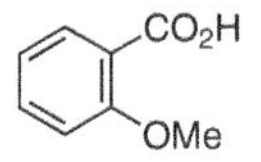

(e) 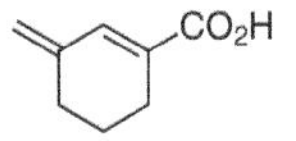

(f) 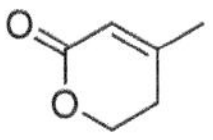 (g) 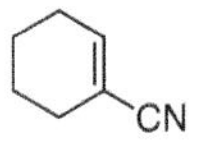

(h) 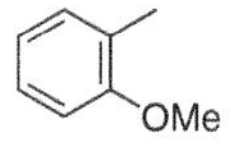

(i) 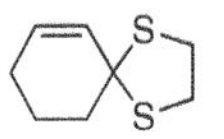 (j) 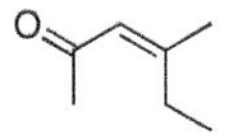 (k)  (l) 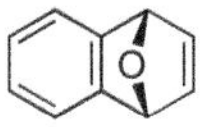

(m) 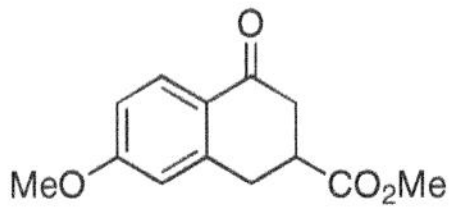

(n) 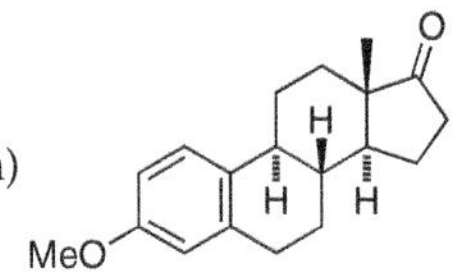

(o) 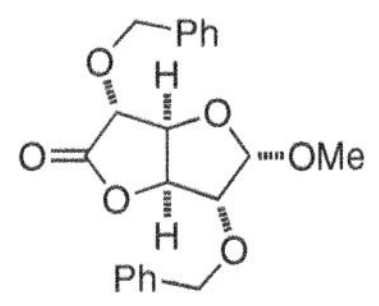

(p) 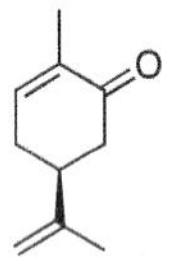 (q) 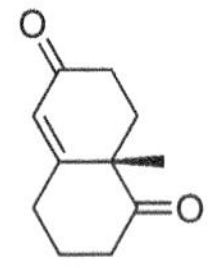 (r) 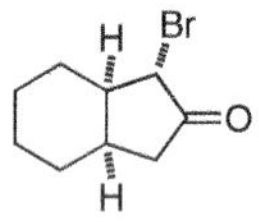

(s) 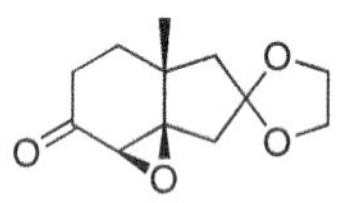

(t) 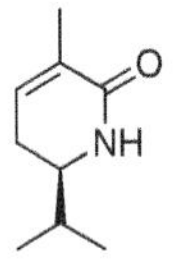

(u) 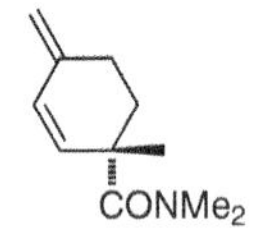

(v) 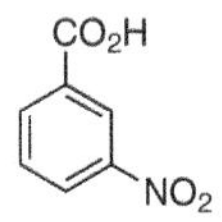

Reagents:

(i) $LiAlH_4$, Et_2O
(ii) Li, NH_3
(iii) $NaBH_4$, EtOH
(iv) (1) DIBAL-H, PhMe; (2) HCl, H_2O
(v) Zn(Hg), HCl (g), AcOH
(vi) Na, EtOH, Δ
(vii) N_2H_4, KOH, $O(CH_2CH_2OH)$, Δ
(viii) (1) $TsNHNH_2$, MeOH; (2) $NaBH_3CN$, MeOH
(ix) Zn(Cu), $TiCl_3$, THF
(x) (*S*)-Alpine-borane®, THF
(xi) $BuSiH_3$, $B(C_6F_5)_3$, CH_2Cl_2

Comprehensive Reduction Review Problem Set 2. What reagent or sequence of reagents should be used to convert the ketoester below into each of the products shown? Note that more than one step may be required in some cases; in addition, there may be more than one reagent, or set of reagents, that can be used to make each of the products.

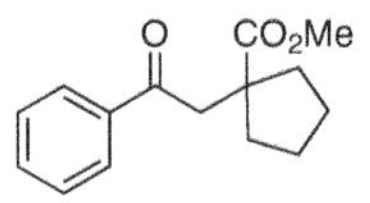

(a) 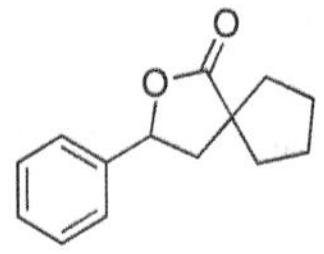 (b) 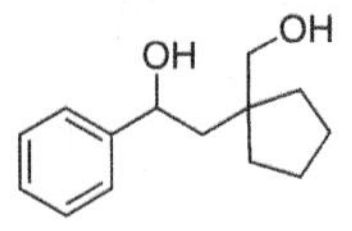

(c) 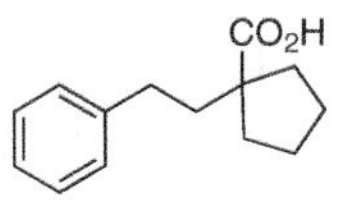

(d) 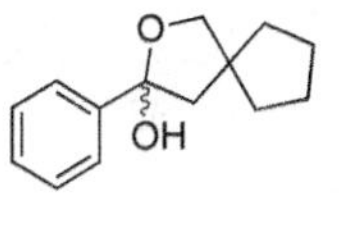
(e) 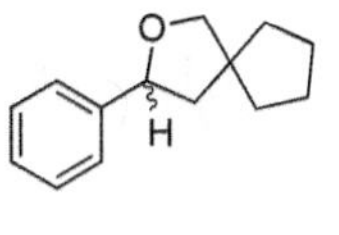
(f) 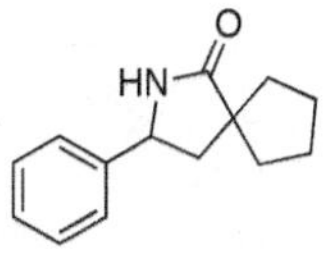

(g) 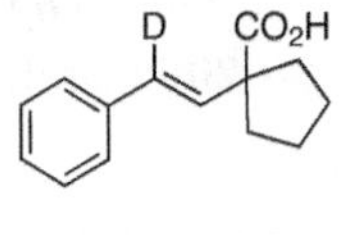
(h) 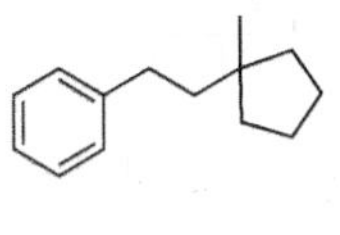
(i) 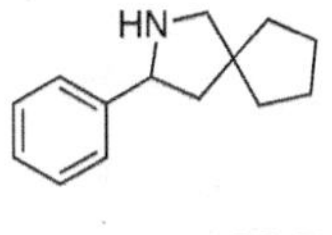

(j) 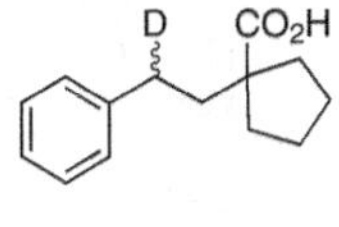
(k) 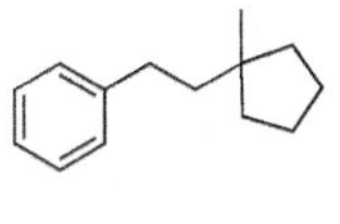
(l) 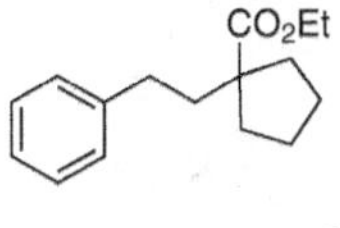

(m) 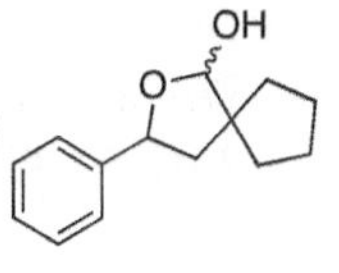
(n) 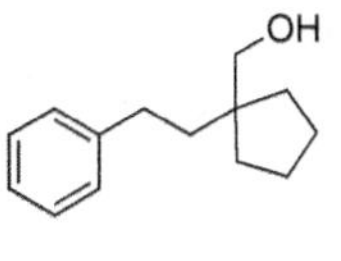
(o) 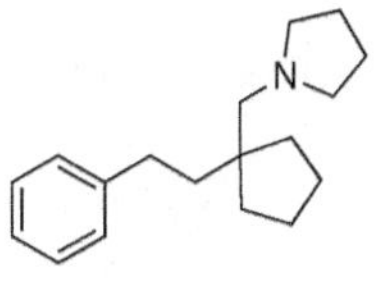

20-8 What is wrong with each of the proposed reactions below?

(a) 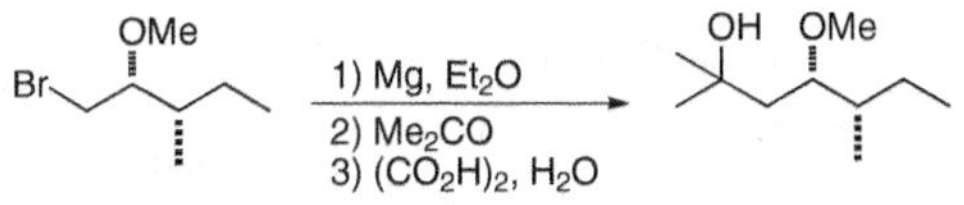

(b) 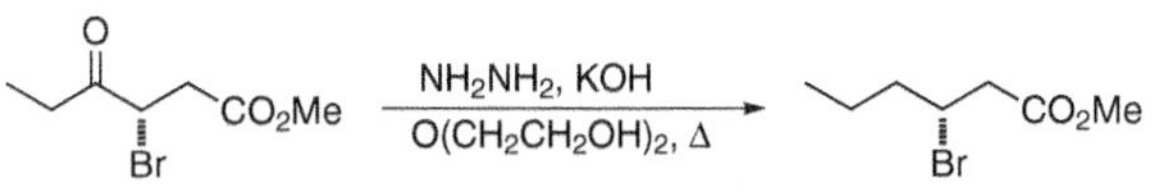

20-9 What is the major organic product expected from each of the following reactions. Where appropriate, specify stereochemistry.

(a) 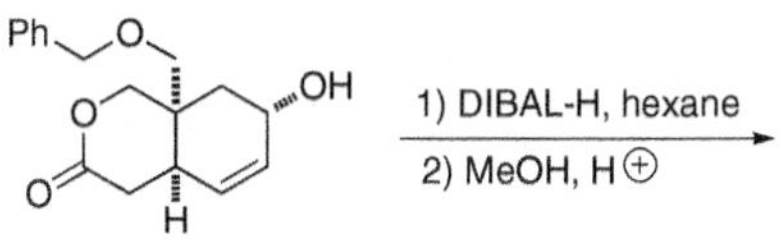

[*Tetrahedron* **1979**, *35*, 2389]

(b) 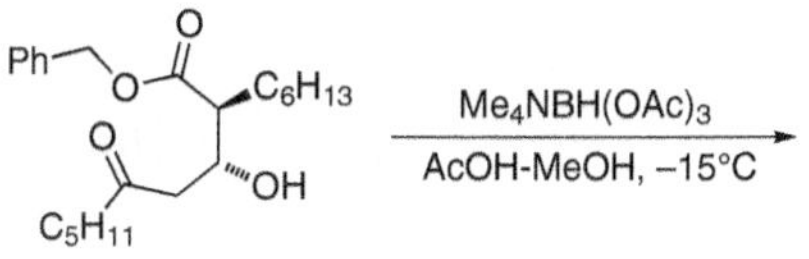

[*J. Org. Chem* **2006**, *71*, 5748]

(c) 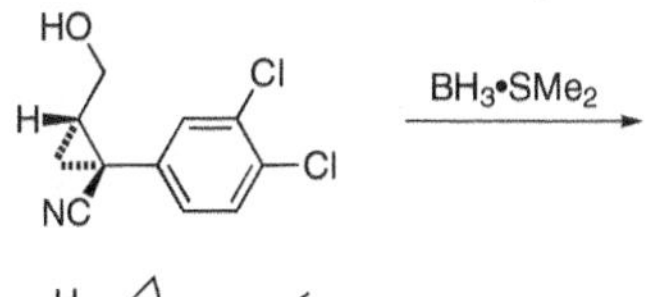

[*Org. Lett.* **2006**, *8*, 3885]

(d) 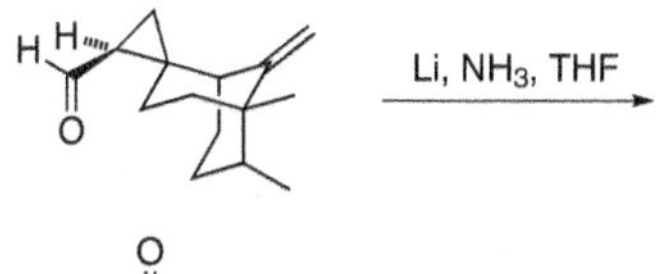

[*J. Org. Chem.* **1995**, *60*, 5746]

(e) 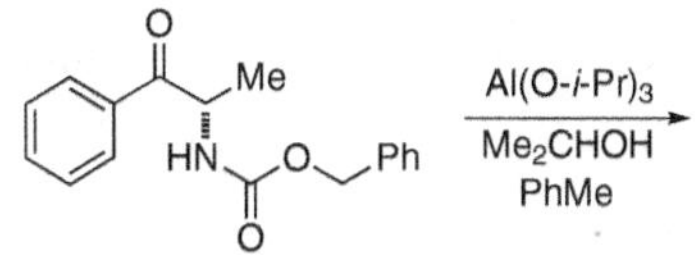

[*J. Org. Chem.* **2006**, *71*, 840]

(f) 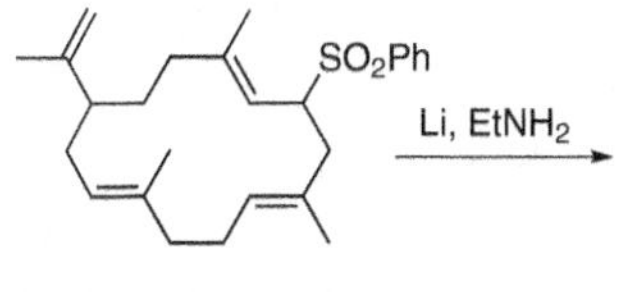

[*Chem. Commun.* **1979**, 359]

(g) 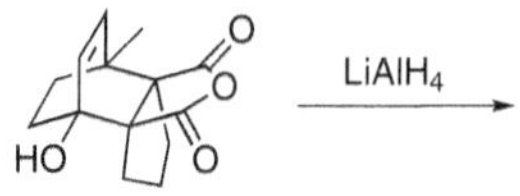

[*Can. J. Chem.* **1979**, *57*, 3343]

(h) 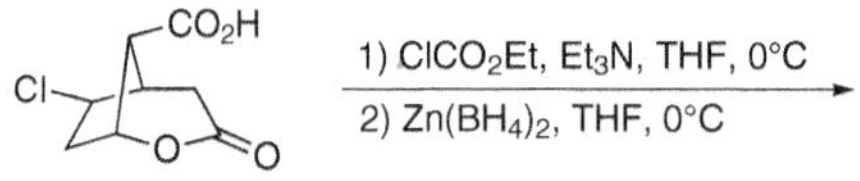

[*Eur. J. Org. Chem.* **2008**, 2551]

(i) 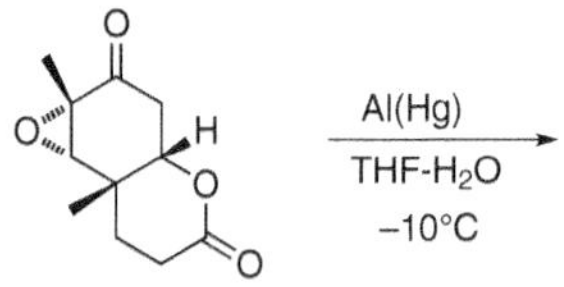

[*J. Am. Chem. Soc.* **1978**, *100*, 4618]

(j) 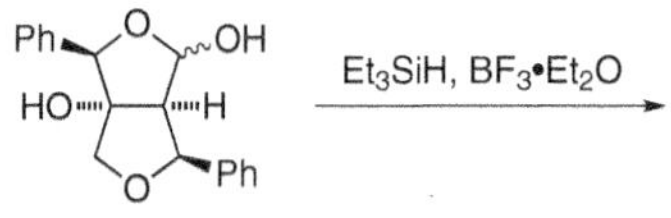

[*Tetrahedron Lett.* **2011**, *52*, 4720]

20-10 What is the major organic product of each of the following reactions?

(a) 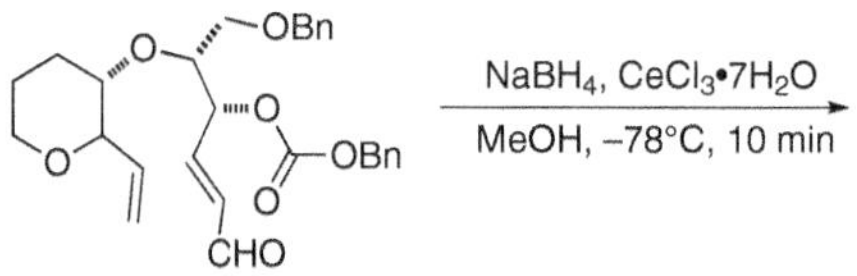

[*Tetrahedron Lett.* **2009**, *50*, 1236]

(b) 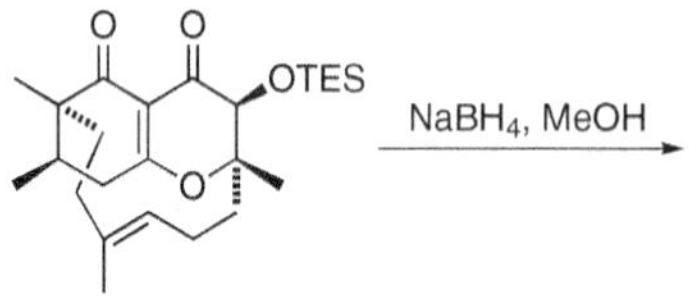

[*Org. Lett.* **2009**, *11*, 1591]

(c) 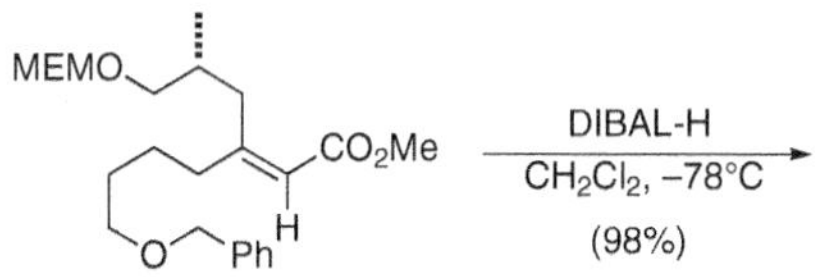

[*J. Am. Chem. Soc.* **1989**, *111*, 1923]

(d) 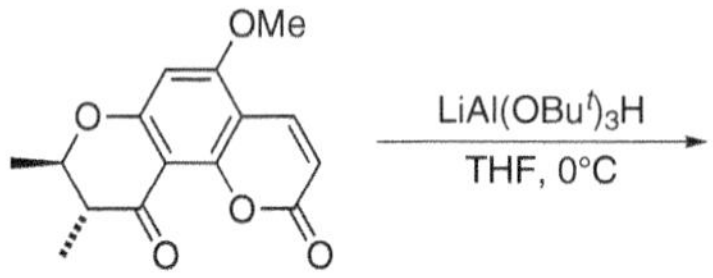

[*Tetrahedron* **1997**, *53*, 14915]

(e) 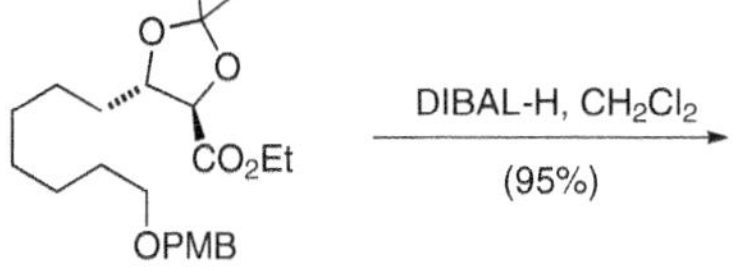

[*Tetrahedron* **2007**, *63*, 7624]

(f) 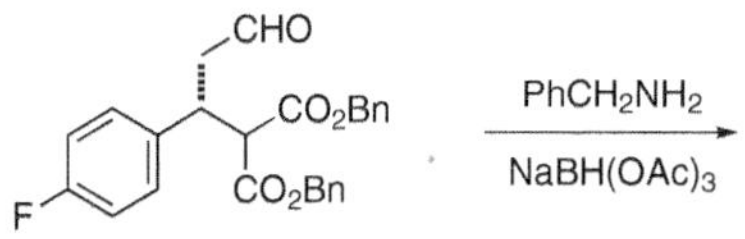

[*Angew. Chem. Int. Ed. Engl.* **2006**, *45*, 4305]

(g) 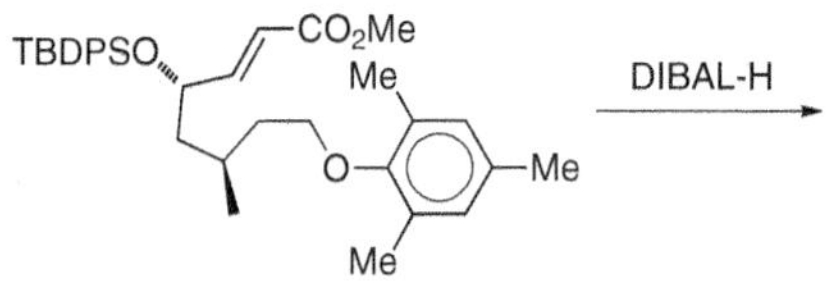

[*J. Am. Chem. Soc.* **2005**, *127*, 3696]

(h) 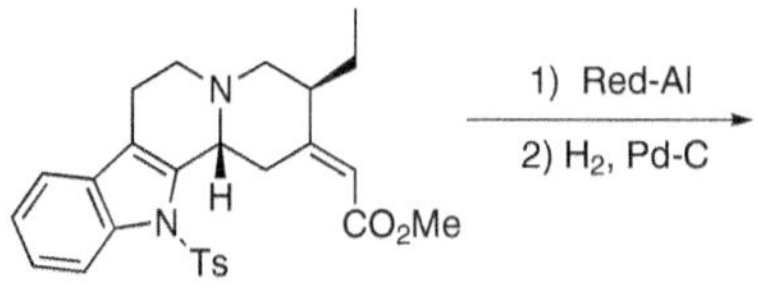

[*Org. Lett.* **2006**, *8*, 1533]

(i) 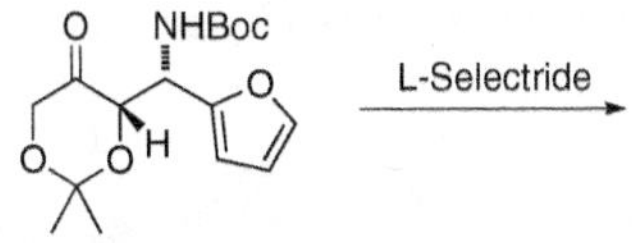

[*Synthesis* **2006**, 2155]

(j) 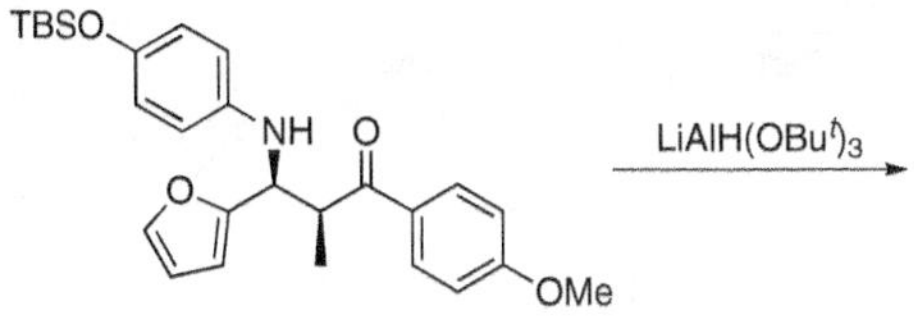

[*Tetrahedron* **2005**, *61*, 11393]

(k) 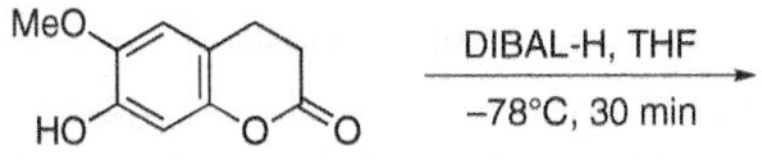

[*Tetrahedron* **2002**, *58*, 2163]

(l) 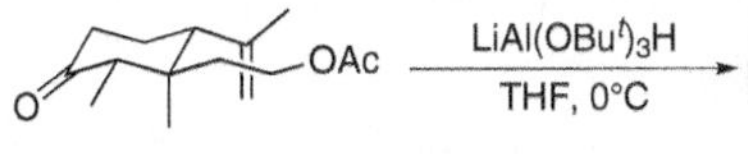

[*Tetrahedron* **2001**, *57*, 8243]

(m) 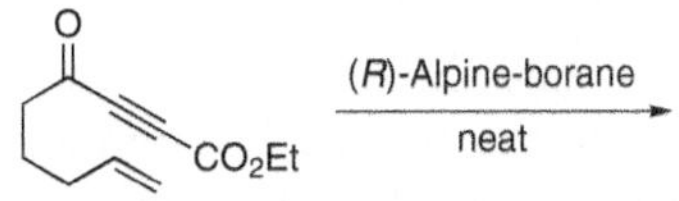

[*Tetrahedron* **2002**, *58*, 2523]

(n) 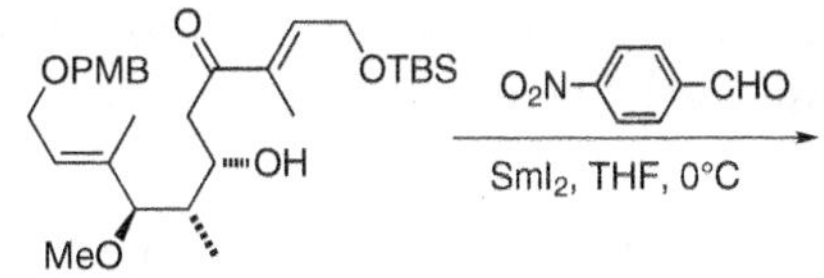

[*Org. Lett.* **2004**, *6*, 1445]

(o) 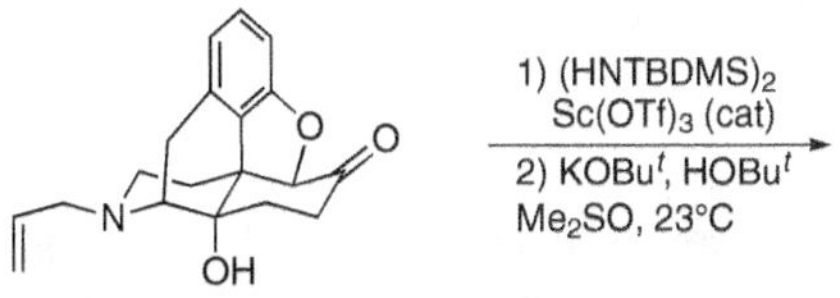

[*J. Am. Chem. Soc.* **2004**, *126*, 5436]

(p) 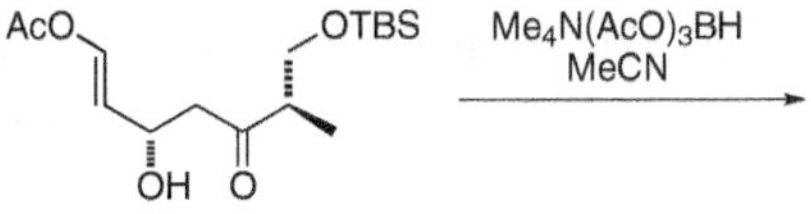

[*J. Am. Chem. Soc.* **2004**, *126*, 2194]

(q) 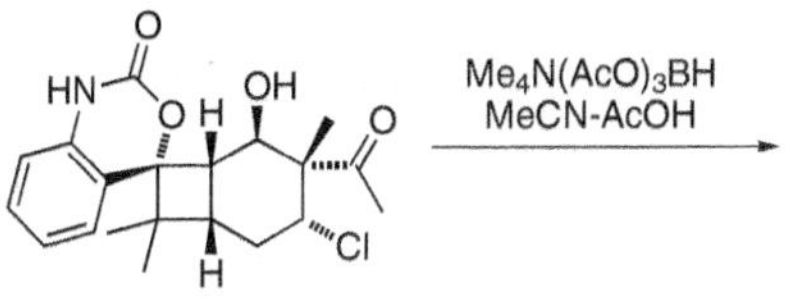

[*J. Am. Chem. Soc.* **2006**, *126*, 1448]

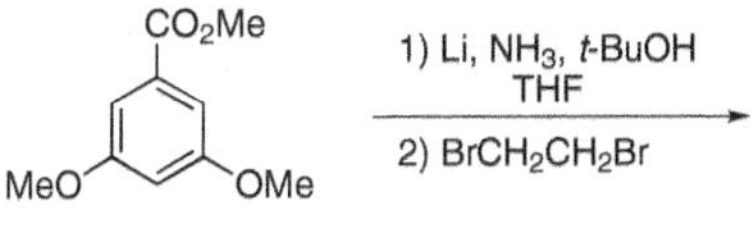

(r) 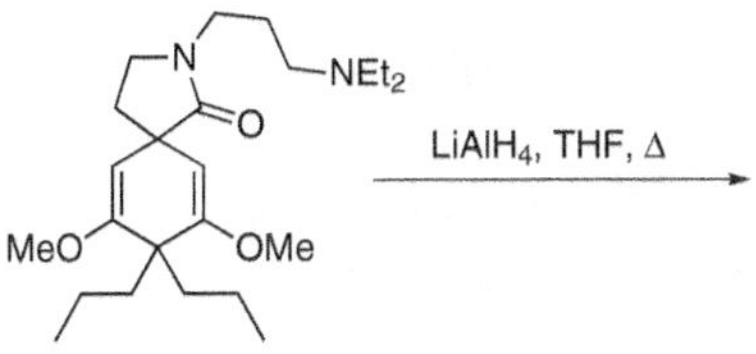

[*Tetrahedron Lett.* **2006**, *47*, 6733]

(s) 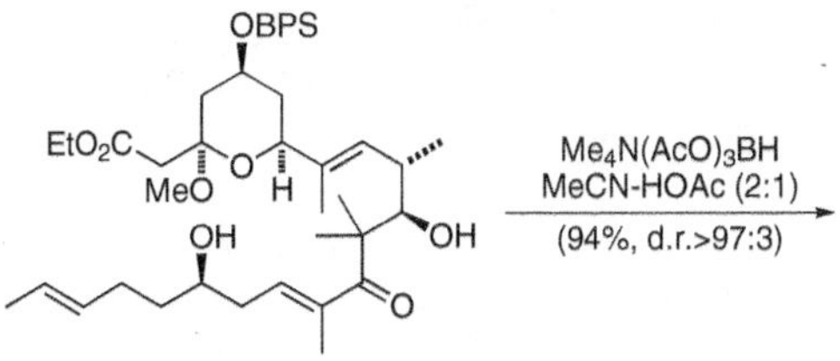

[*J. Am. Chem. Soc.* **1997**, *119*, 19035]

20-11 Consider the two sequential reduction reactions below [*J. Am. Chem. Soc.* **2007**, *129*, 11987].

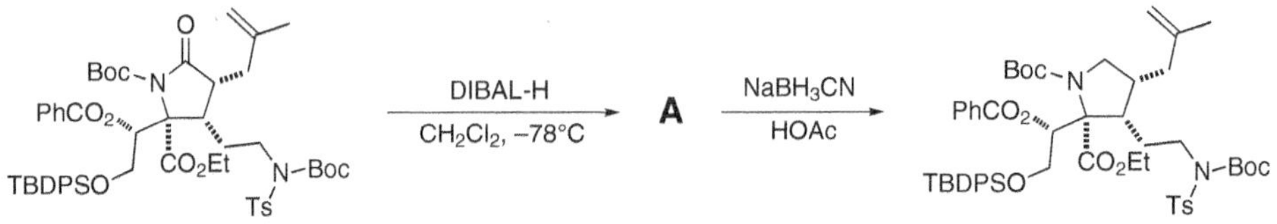

(a) What is the structure of compound **A**?

(b) There are five carbonyl groups in the starting material. Even if only one equivalent of the reducing agent is used, why does the reagent in the first step attack only one of the carbonyl groups, and why does it attack that particular one?

(c) Why is the second stage of the reduction necessary, and why was that particular reagent chosen?

20-12 (a) Suggest a mechanism for the Eschenmoser-Tanabe fragmentation using the example shown to illustrate your proposed mechanism. Are there any special requirements for the reaction (especially stereoelectronic requirements) that your proposed mechanism highlights?

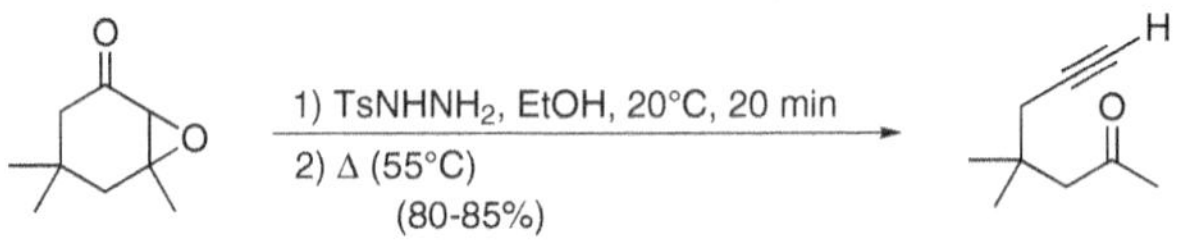

(b) When a semicarbazone is heated with potassium hydroxide or potassium *tert*-butoxide in DMSO, an alkane is formed, where the imine group of the semicarbazone has been replaced by a methylene group. Write a plausible mechanism for this reaction. Of what named reaction is it a modification? Suggest practical reasons for why it might be a useful synthetic modification of this reaction.

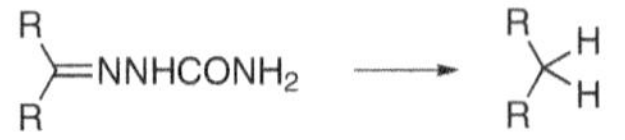

20-13 Suggest a mechanism for the transformation shown.

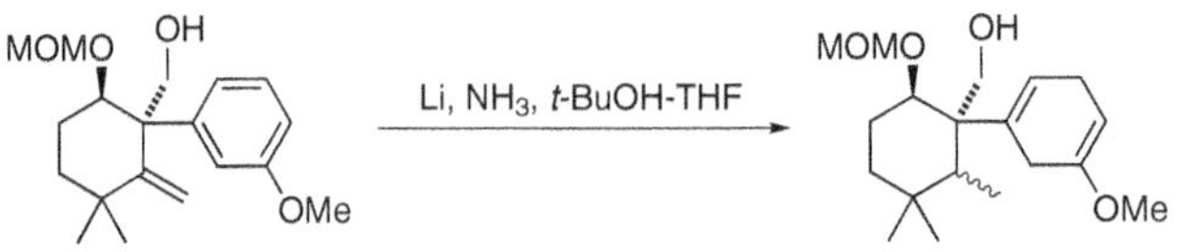

[*Tetrahedron Lett.* **2009**, *50*, 6586]

20-14 The interesting reaction shown has been described as being carried out under typical conditions for the hydrogenolysis of a cyclopropane [*Org. Lett.* **2008**, *10*, 3381]. However, there is no net gain of hydrogen during the reaction. Suggest a mechanism.

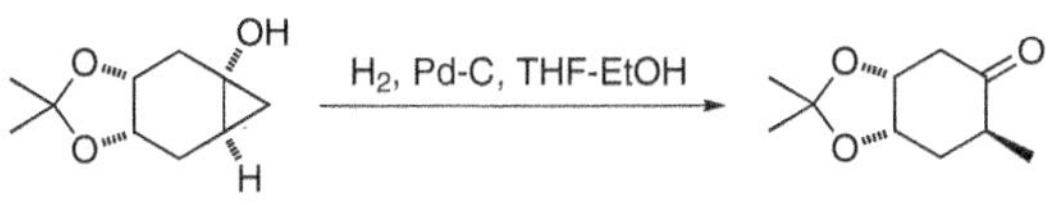

20-15 Suggest reasonable reagents that could be used to carry out the following transformations:

(a) 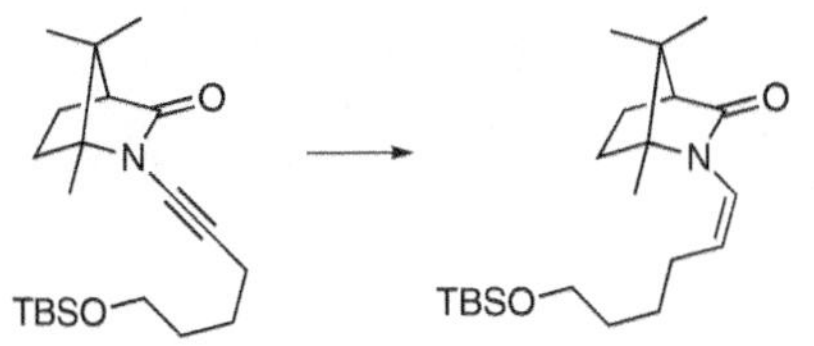 [*J. Org. Chem.* **2006**, *71*, 4170]

(b) 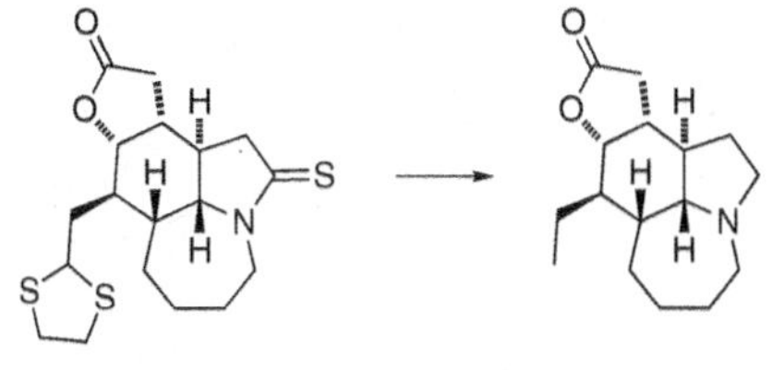 [*J. Org. Chem.* **2005**, *70*, 5197]

(c) 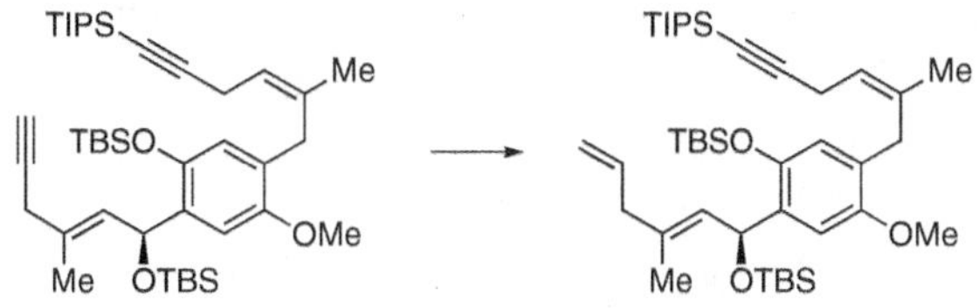 [*J. Am. Chem. Soc.* **2002**, *124*, 773]

(d) 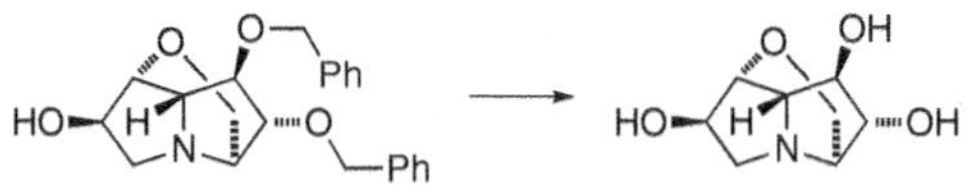 [*Tetrahedron* **2010**, *66*, 9340]

(e) 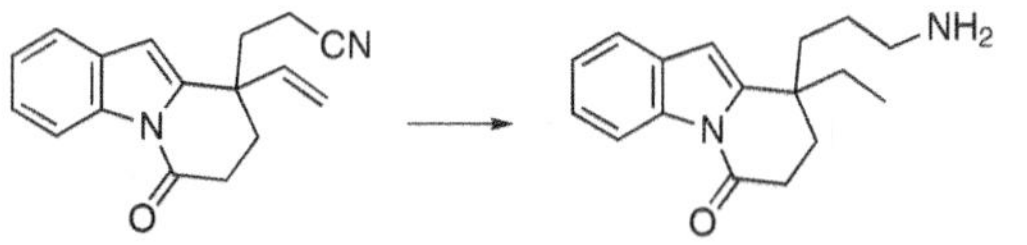 [*Org. Lett.* **2008**, *10*, 1437]

(f) 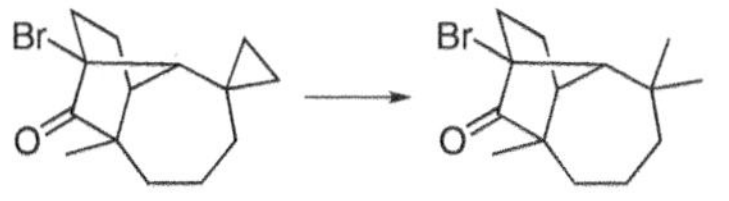 [*J. Nat. Prod* **2003**, *66*, 520]

(g) 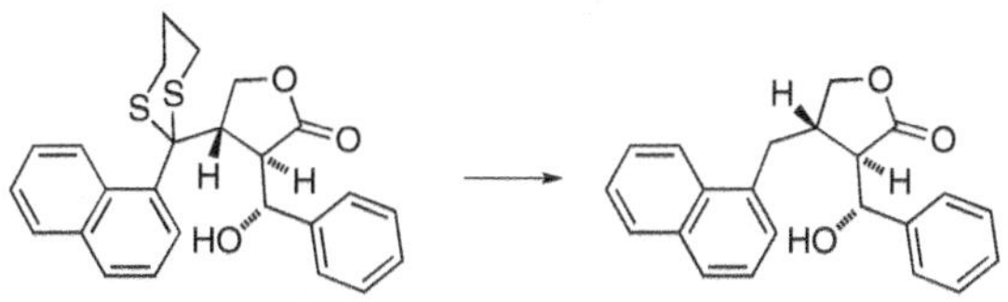 [*J. Org. Chem.* **2003**, *68*, 854]

(h) 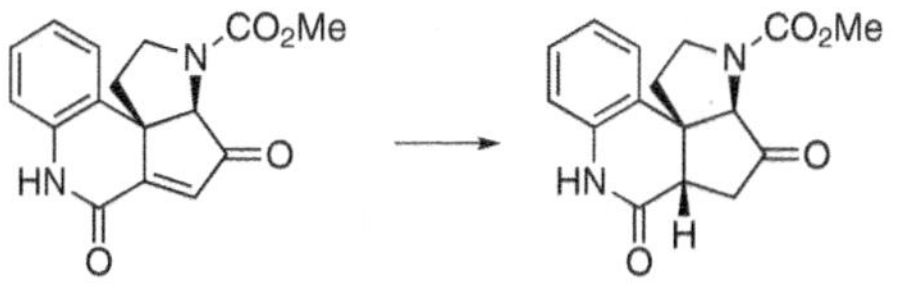 [*Org. Lett.* **2011**, *13*, 1778]

20-16 What reagent would you use if you also wished to reduce the alcohol hydroxyl group in the starting compound in Problem 20-15 (g) in the same step as reduction of the dithiane group?

20-17 Suggest a mechanism for the reaction shown. How is this reaction related to other reactions discussed in this chapter?

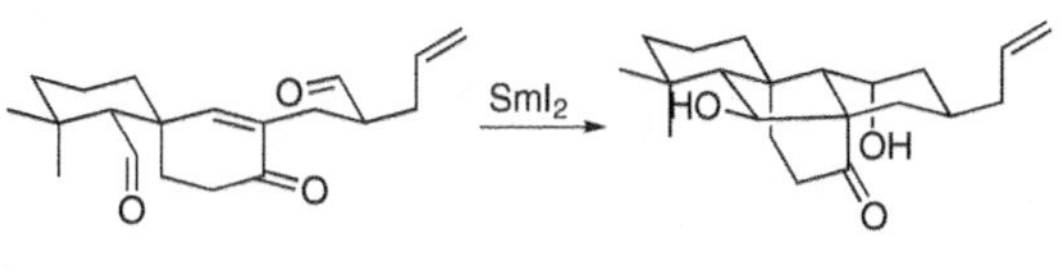

CHAPTER TWENTY-ONE

Asymmetric Oxidation and Reduction

21.1 Asymmetric Redox Reactions: An Overview

In the preceding three chapters, we have explored oxidation and reduction without considering stereochemistry other than in terms of diastereoselectivity of the reaction. This was the norm until 1980, when Sharpless and Katsuki[1] reported their dramatic findings for the asymmetric epoxidation of alkenes that eventually led to Sharpless sharing the Nobel Prize in Chemistry.

Since that first discovery of a method that could be used to exercise useful levels of predictable enantioselectivity in oxidation, considerable effort has been expended to develop methods for asymmetric oxidation and reduction reactions—to the point that *using enantioselective methods has become the default option for carrying out many organic oxidations and reductions.* Furthermore, Sharpless' successful use of transition metal catalysts to accomplish the asymmetric epoxidation of allylic alcohols led to an explosion of work in the area of catalyst design and generation for asymmetric synthesis. Today, *many of the major steps in a total synthesis are amenable to being carried out enantioselectively by using chiral catalysts.*

The enantioselective generation of a chiral product by a redox reaction can be accomplished by only two general strategies. In the first option (Examples **21.1** to **21.4**), the enantioselective reaction generates one or more *new* chiral centers enriched in one of the two enantiomeric forms by addition to a π bond. This is the process by which chirality is incorporated into the product in, for example, asymmetric epoxidation and dihydroxylation of alkenes, asymmetric hydride reduction of ketones, and asymmetric hydrogenation of alkenes. Alternatively (Examples **21.5** to **21.7**), the reaction may reveal *existing, latent* chirality in an achiral precursor by eliminating a mirror plane of symmetry in the molecule, a process usually called ***desymmetrization***. In this case, only one of the two equivalent functional groups related by the mirror plane reacts, thus lifting the mirror plane relationship between existing chiral enters (as occurs in a *meso* compound, for example). Desymmetrization is most frequently accomplished by substitution at an existing prochiral sp^3 center. Nevertheless, methods for the desymmetrization of *meso* compounds by redox reactions are gradually gaining in popularity. These two strategies are summarized in Figure 21.1. In the first part of this chapter, we will focus our attention on asymmetric oxidation reactions.

---

1. **Karl Barry Sharpless (1940–)** was educated at Dartmouth College (BA, 1963) and Stanford (PhD, 1968). After postdoctoral study at Stanford and Harvard, his career took him to the Massachusetts Institute of Technology (MIT; 1970) and Stanford (1977) before returning to MIT (1980) and finally going to to the Scripps Institute (1986). Sharpless shared the 2001 Nobel Prize for his work in asymmetric epoxidation. For more detail, see his autobiography at the Nobel Foundation web site.

Tsutomu Katsuki (1946–) was educated at Kyushu University (DSc, 1976), and spent 2 years as a postdoctoral fellow with Sharpless at Stanford and MIT. During this time, he discovered the asymmetric epoxidation of allylic alcohols using titanium tartrate catalysts. Since 1988, Katsuki has been Professor at Kyushu University. In 2005, he was the first Japanese chemist to win the Ryoji Noyori Prize. He continues to develop catalytic asymmetric reactions based on transition metals.

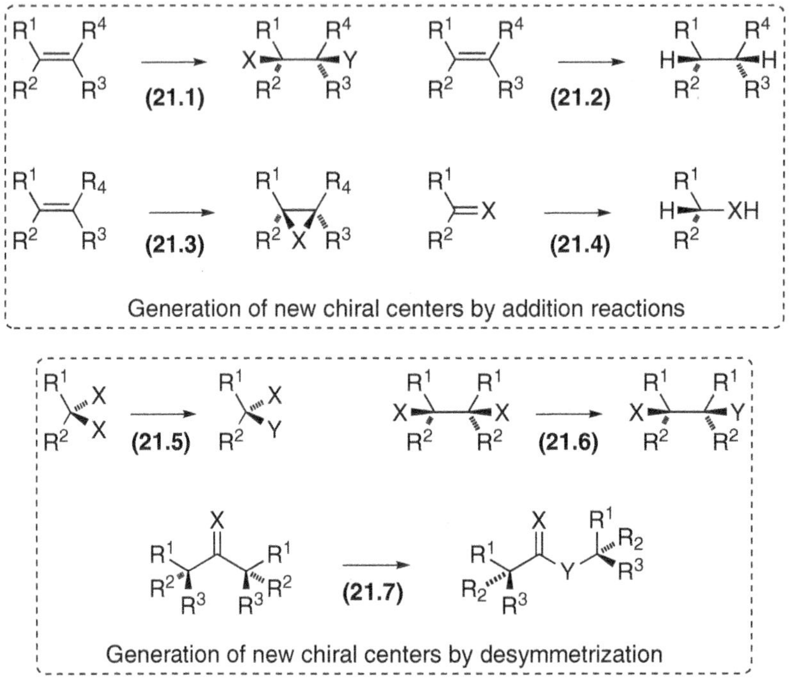

Figure 21.1 Strategies for enantioselective redox

21.2 **Sharpless Asymmetric Epoxidation of Allylic Alcohols**

Perhaps the most important reaction of allylic alcohols to have emerged in the past 50 years is the (now venerable) ***Sharpless asymmetric epoxidation***. In this reaction, an allylic alcohol is oxidized with a hydroperoxide and a titanium alkoxide in the presence of a chiral dialkyl tartrate ester (which is also a chiral diol; e.g., Example **21.8**). This reaction was first reported in 1980, in a series of papers reporting that *E* allylic alcohols could be converted to *almost entirely one enantiomer* of the corresponding epoxide using *tert*-butyl hydroperoxide as the oxidizing agent and a catalyst formed from titanium tetraisopropoxide and a dialkyl tartrate ester.[2,3] The efficiency of the Sharpless epoxidation (in terms of chemical yield and reaction rate) is comparable to the vanadium-catalyzed oxidation of geraniol, but the Sharpless procedure gives an enantiomer ratio of 39:1.

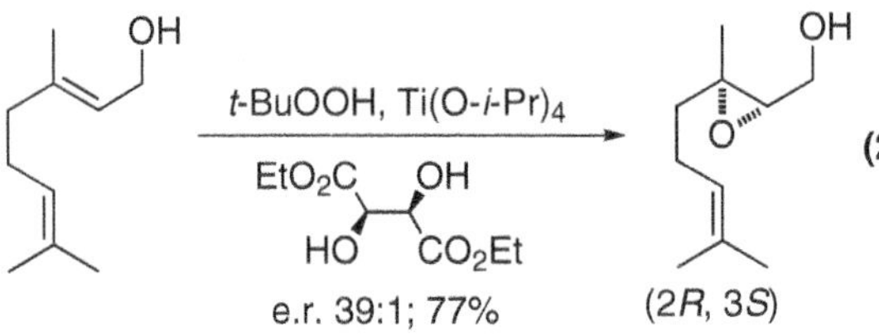

(21.8)

2. (a) Katsuki, T.; Sharpless, K.B. *J. Am. Chem. Soc.* **1980**, *102*, 5974. (b) Hill, J.G.; Sharpless, K.B. *Org. Synth.* **1984**, *63*, 66.

Reviews: (c) Pfenniger, A. *Synthesis* **1986**, 89. (d) Johnson, R.A.; Sharpless, K.B. In Ojima, E, Ed. *Catalytic Asymmetric Synthesis*, 2nd. ed. (VCH Publishers, NewYork: 2003), 231. (e) Katsuki, T.; Martin, V.S. *Org. React.* **1996**, *48*, 1. (f) Rossiter, B.E. In Morrison, J.D., Ed. *Asymmetric Synthesis* (Academic Press: New York, 1985), vol. 5, 193. (g) Finn, M.G.; Sharpless, K.B. In Morrison, J.D., Ed. *Asymmetric Synthesis* (Academic Press: New York, 1985), vol. 5, 247.

3. The real importance of this reaction was that, for the first time in the history of organic synthesis, organic chemists had a reaction of general scope that gave useful functionality, and it gave the chemist control over the absolute configuration of the new chiral centers by allowing the reliable prediction of the absolute stereochemistry of the product.

What was even more revolutionary about these papers was that they also reported that the absolute configurations of the two new chiral centers could be *predicted* from the configuration of the alkene and the absolute configuration of the tartrate ester (Figure 21.2). The stereochemistry of the epoxide can be predicted on the basis of the simple model in Figure 21.2. First the allylic alcohol is drawn as shown, with the chain extended and the carbon bearing the hydroxyl group at the bottom right-hand corner with the OH group eclipsing the π bond. Based on this orientation, the oxygen is delivered to the top face of the π bond when the D-(−)-tartrate ester (from 2*S*,3*S* tartaric acid, the unnatural enantiomer) is used as the chiral component of the catalyst, and to the bottom face when the L-(+)-tartrate ester (from 2*R*,3*R* tartaric acid, the natural enantiomer) is used.

Effects of Alcohol Structure

The substituents on the allyl alcohol base structure can exercise a profound effect on the yield, rate, and enantioselectivity of the Sharpless asymmetric oxidation. Some typical results are gathered in Table 21.1, where the yields from closely-related reactions are gathered. These reactions have been carried out under both stoichiometric conditions (1 mol eq Ti) and catalytic conditions (5–10 mol % Ti). What is immediately apparent from Table 21.1 is that the *Z*-2-alken-1-ols show a much larger variation in the levels of enantioselectivity than any of the other alcohols. In addition, it has been found that branching at the γ and δ positions significantly lowers the level of asymmetric induction, although moving the point of branching to the homoallylic (ε) position has a much smaller effect.

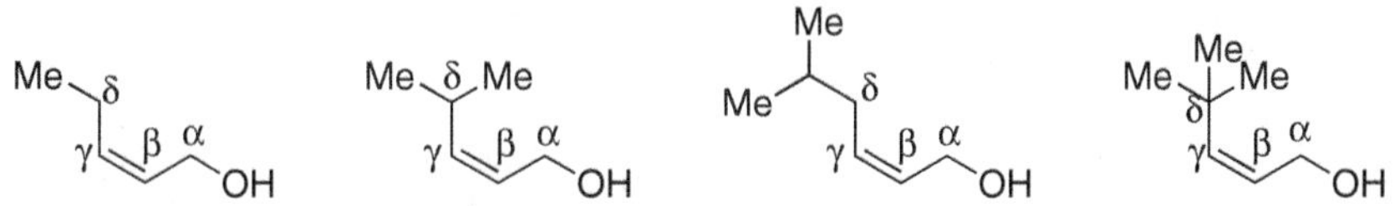

80 66 95 25

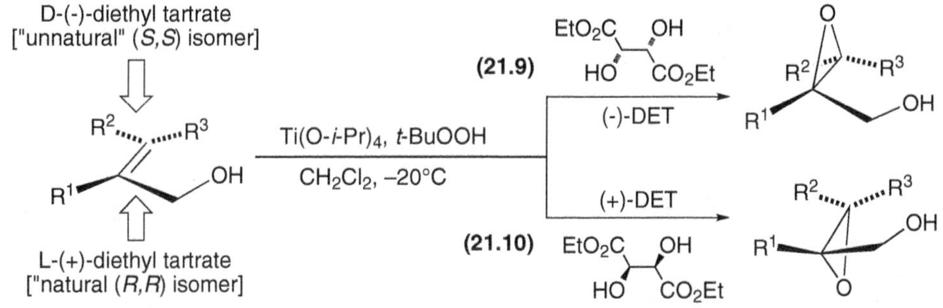

Figure 21.2 Prediction of epoxide stereochemistry in the Sharpless asymmetric epoxidation

Table 21.1 Levels of Asymmetric Induction in Sharpless Epoxidation of Allylic Alcohols

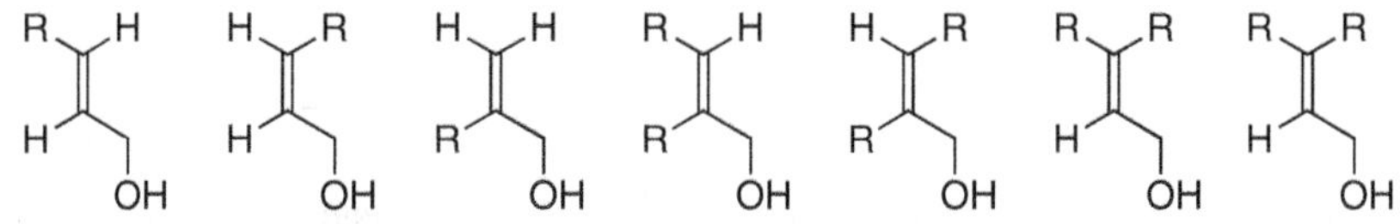

Alcohol Structure							
% Yield, (R=Me)	70	68	47	77	n.a.	25	n.a.
Range of % e.e.	92–98	25–95	85–95+	90–98+	78–91	86–95+	53–95

Catalyst Structure

The rate law for the reaction is complex. Pseudo–first-order kinetic studies show that the reaction exhibits (1) first-order dependence on the concentrations of the titanium tartrate complex, the hydroperoxide, and the allylic alcohol and (2) an inverse second-order dependence on the concentration of the original ligand alcohol (e.g., *i*-PrOH).

$$\text{Rate} = k\,\frac{[(i\text{-PrO})_2\text{Ti(tartrate)}][\text{ROOH}][\text{R}_2\text{C}=\text{CRCH}_2\text{OH}]}{[i\text{-PrOH}]^2}$$

Sharpless interpreted the results of the kinetic studies and the effects of alcohol structure on the course of the reaction in terms of the reaction proceeding through a dimeric species, shown in the dynamic equilibrium in Figure 21.3.[4] In this reaction system, the whole system is in rapid equilibrium with one tartrate ester ligand bonded to each titanium atom. Sharpless proposed that the reactive complex is formed by replacement of both isopropoxy ligands on the same titanium atom by the allylic alcohol and the hydroperoxide.

The reaction of *E*-2-alken-1-ols proceeds readily and tolerates a wide range of functional groups, as illustrated by Examples **21.13** to **21.16**. Water is deleterious to the reaction, and its removal has the added advantage of allowing the titanium complex to be used in catalytic amounts. Both activated 4Å molecular sieves and calcium hydride with silica gel[5] have been used as dehydrating agents for the reaction.

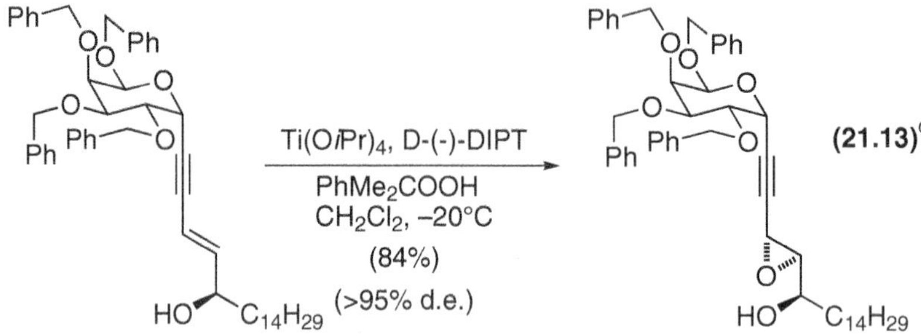

(21.13)[6]

Figure 21.3 Proposed active species in the Sharpless asymmetric epoxidation

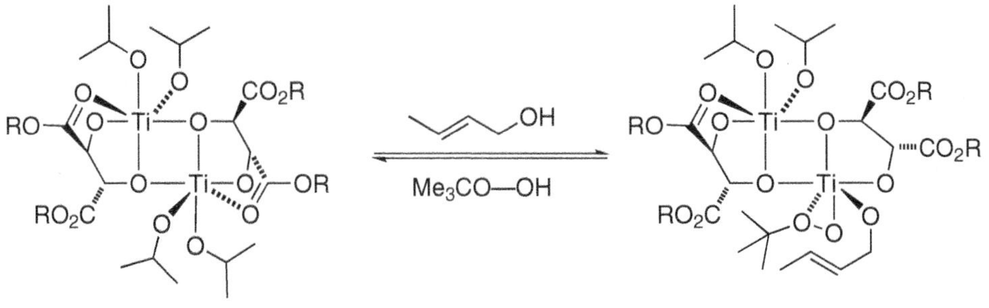

(21.11) **(21.12)**

4. (a) Woodard, S.S.; Finn, M.G.; Sharpless, K.B. *J. Am. Chem. Soc.* **1991**, *113*, 106. (b) Finn, M.G.; Sharpless, K.B. *J. Am. Chem. Soc.* **1991**, *113*, 113. See also: (c) Potvin, P.G.; Bianchet, S. *J. Org. Chem.* **1992**, *57*, 6629.
5. (a) Wang, Z.-M.; Zhou, W.-S.; Lin, G.-Q. *Tetrahedron Lett.* **1985**, *26*, 6221. (b) Wang, Z.-M.; Zhou, W.-S. *Synth. Commun.* **1989**, *19*, 2627.
6. Liu, Z.; Byun, H.-S.; Bittman, R. *Org. Lett.* **2010**, *12*, 2974.

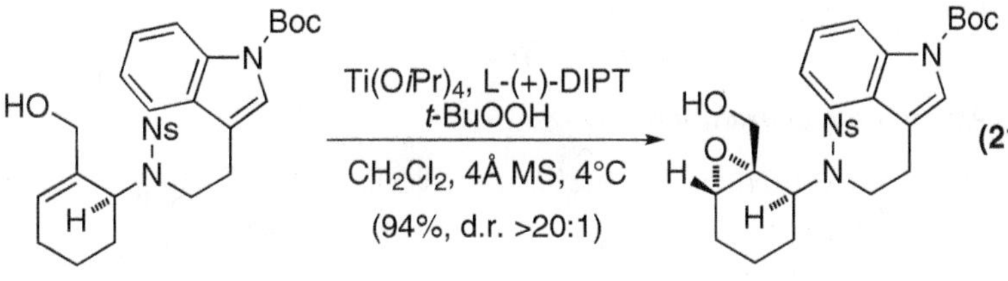

(21.14)[7]

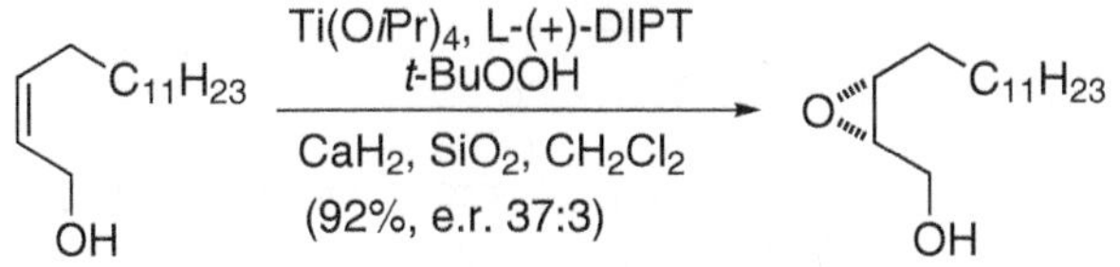

(21.15)[8]

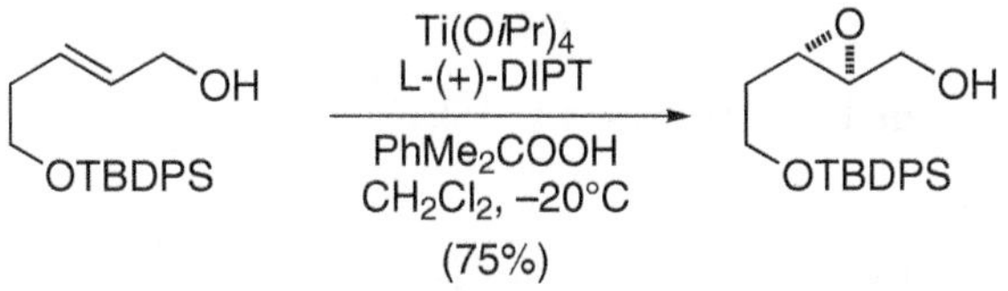

(21.16)[9]

Kinetic Resolution of Racemic Allylic Alcohols

Sharpless also found that the reaction between a racemic allylic secondary alcohol is often rapid for only one of the two enantiomers. As in the epoxidation of primary allylic alcohols, the preferred direction of attack of the oxidant is determined by the configuration of the tartrate ligand. This leads to the situation shown in Figure 21.4, where the reactive conformation of the alcohol means that, in one of the two enantiomers of the chiral secondary alcohol, the oxidant must approach the π bond along an axis that leads to strong steric interactions between the reagent and the substituent, whereas in the other, this does not occur. Thus, one can predict which of the two enantiomers of the racemate will react rapidly to give the epoxide. In most cases, this leads to large differences in the rates of epoxidation of the two enantiomers of the allylic alcohol (varying from 16:1 in the least favorable cases to 700:1 in the most favorable cases[10]). Again, the ability to predict the absolute configuration of the major product is an important feature of the reaction.

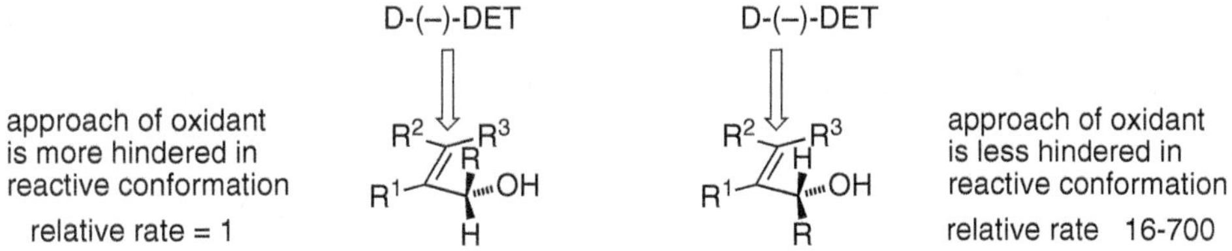

Figure 21.4 Prediction of stereochemistry epoxidation of chiral allylic alcohols based on rate differential

7. Nemoto, T.; Yamamoto, E.; Franzén, R.; Fukuyama, T.; Wu, R.; Fukamachi, T.; Kobayashi, H.; Hamada, Y. *Org. Lett.* **2010**, *12*, 872.
8. Du, Y.; Zheng, J.-F.; Wang, Z.-G.; Jiang, L.-J.; Ruan, Y.-P.; Huang, P.-Q. *J. Org. Chem.* **2010**, *75*, 4619.
9. Krishna, P.R.; Jagannadha Rao, T. *Tetrahedron Lett.* **2010**, *51*, 4017.
10. (a) Johnson, R.A.; Sharpless, K.B. In Trost, B.M.; Fleming, I., Eds. *Comprehensive Organic Synthesis* (Pergamon: Oxford, 1991), vol. 7, p. 389. (b) Johnson, R.A.; Sharpless, K.B. In Ojima, I., Ed. *Catalytic Asymmetric Synthesis*, (VCH Publishers: New York, 1993), p. 103; *2nd ed.* (Wiley-VCH: New York, 2000), p. 231. (c) Finn, M.G.; Sharpless, K.B. In Morrison, J.D., Ed. *Asymmetric Synthesis* (Academic Press: New York, 1985), vol. 5, 247.

Reaction Synopses

Sharpless Asymmetric Epoxidation

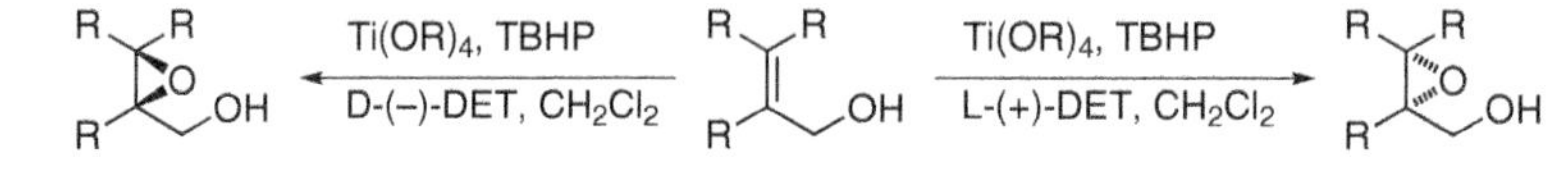

Reagents:

Titanium alkoxide: $Ti(OCHMe_2)_4$, $Ti(OCMe_3)_4$

Tartrate esters: diethyl tartrate (DET), diisopropyl tartrate (DIPT)

Oxidant: *tert*-butyl hydroperoxide (Me_3COOH, TBHP); cumyl hydroperoxide ($PhCMe_2OOH$); etc.

Adjuvants: 4Å molecular sieves (4Å MS); CaH_2, SiO_2; etc.

Kinetic Resolution of Allylic Alcohols

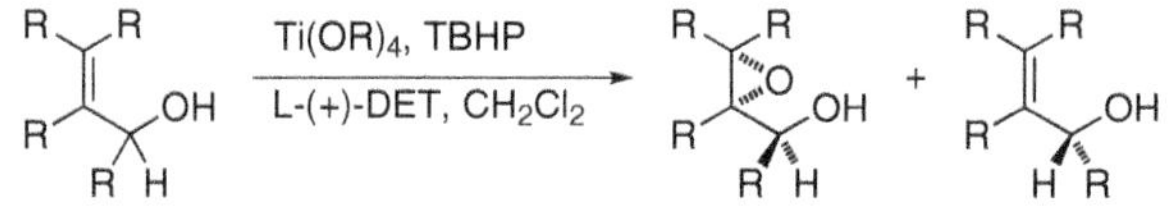

Reagents: as for asymmetric epoxidation

Worked Problem

21-1 What is the major product of this reaction?

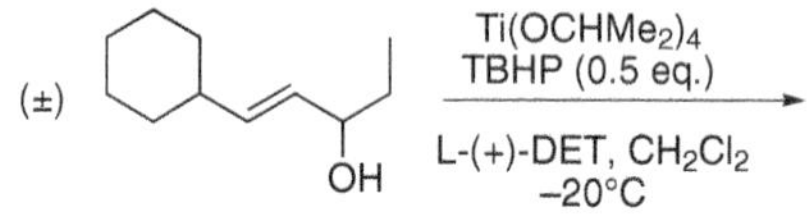

§Answer below.

Problems

21-1 How does the model in Figure 21.3 account for the much lower enantioselectivity of *Z*-2-alken-1-ols in the Sharpless asymmetric epoxidation? (Build models!)

§ Answer to Worked Problem:

We begin by orienting the alcohol in the standard orientation, with the CH_2OH group in the lower right hand. In this orientation, the reagent based on (+)-DET delivers the oxygen atom to the bottom face of the double bond. Because we are using only 0.5 equivalents of the oxidant, the major organic product will be determined by the relative rates of epoxidation. When the alcohols are drawn with the hydroxyl group eclipsing the double bond, the alcohol where the alkyl group is on the face opposite to that attacked by the oxidant will react more rapidly, while the other enantiomer will react more slowly. We predict that the epoxide shown will be the major organic product and that the remaining enantiomer of the alcohol will be recovered largely unchanged.

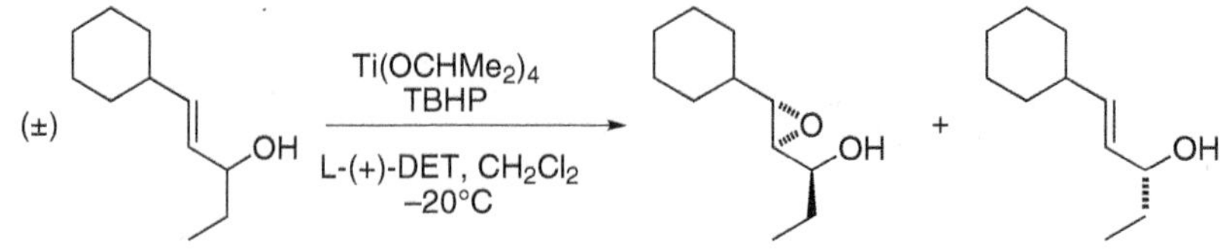

21-2 Draw the structure of the major organic product expected from each of the following reactions.

(a) 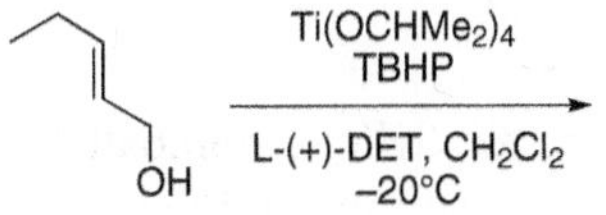

(b) 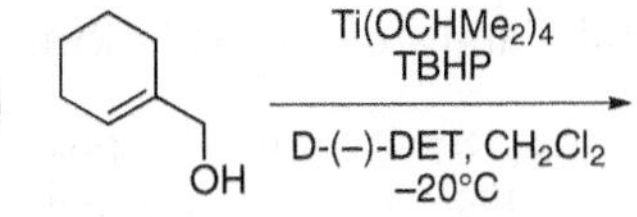

(c) 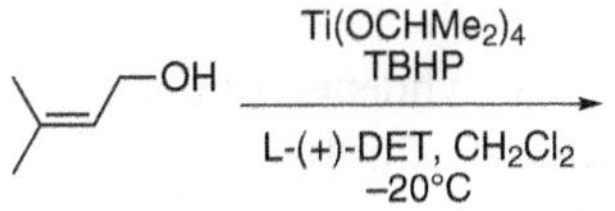

(d) 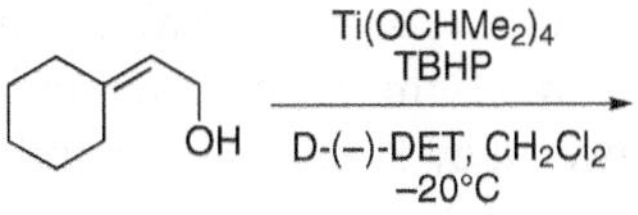

(e) 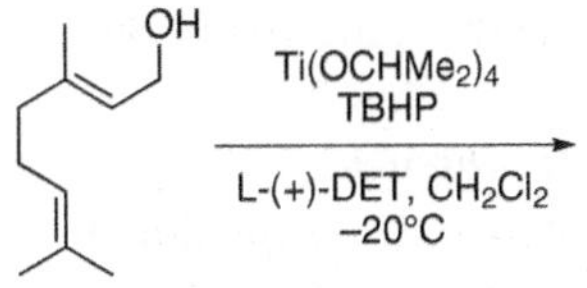

(f) 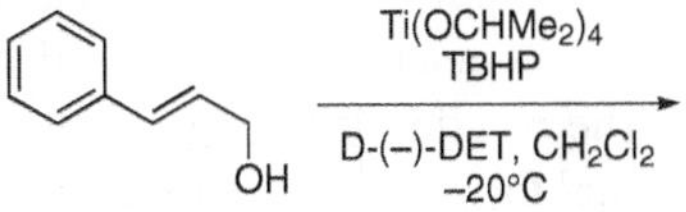

(g) 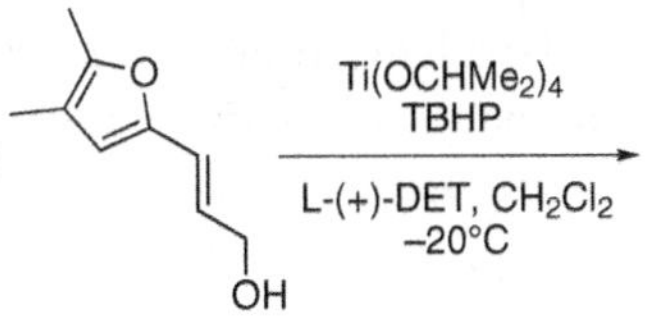

(h) 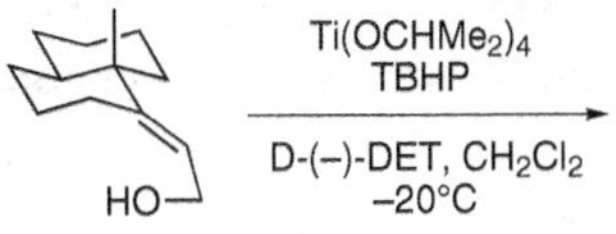

(i) 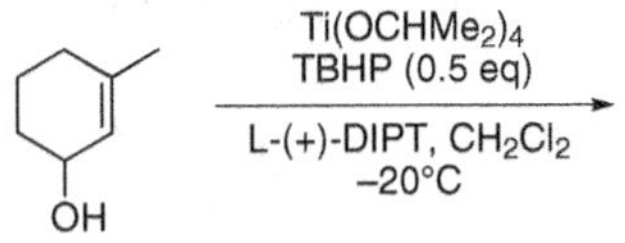

(±)

(j) 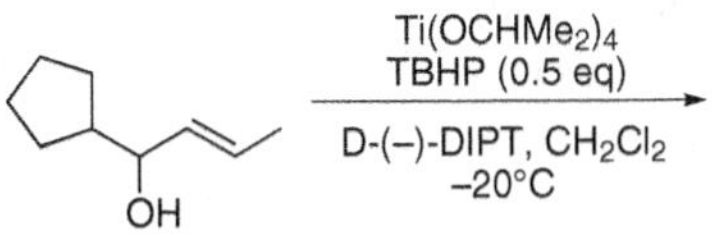

(±)

(k) 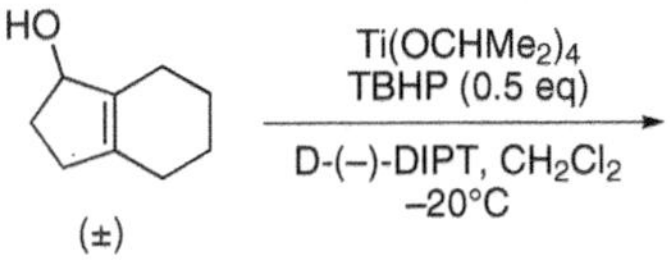

(±)

(l) 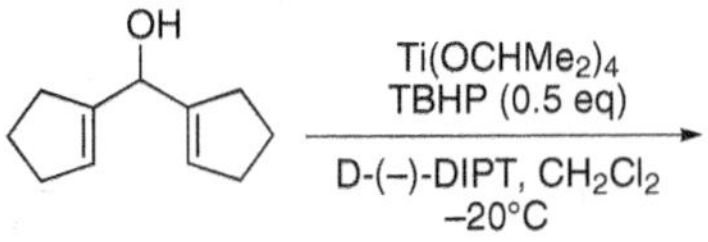

21-3 Suggest reagents that can be used to effect the following transformations:

(a) 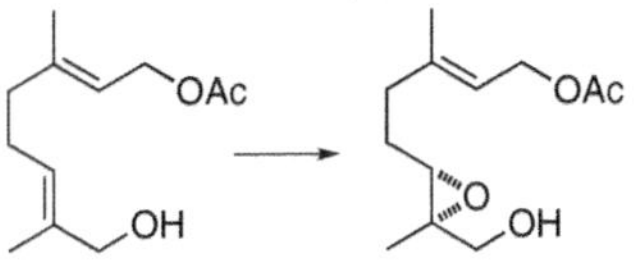

(b) 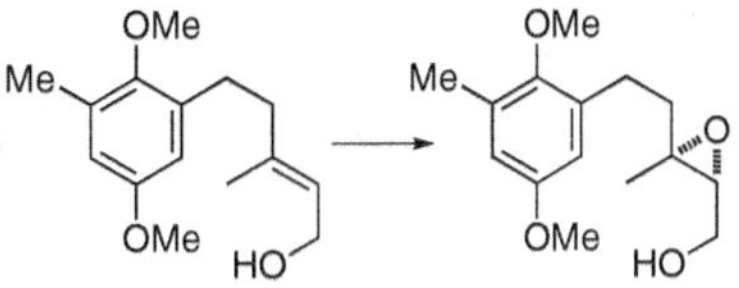

(c) 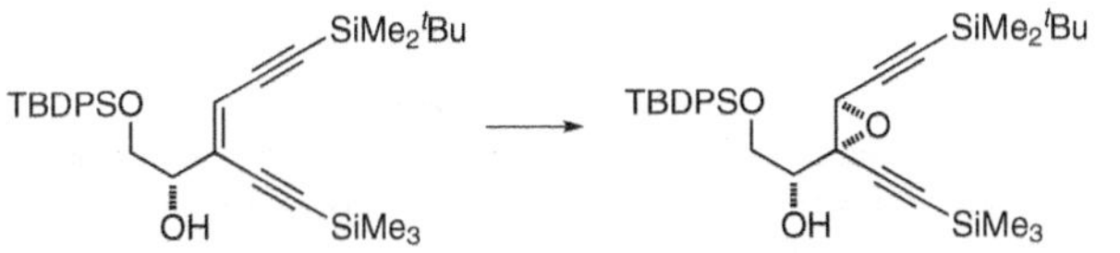

(d) 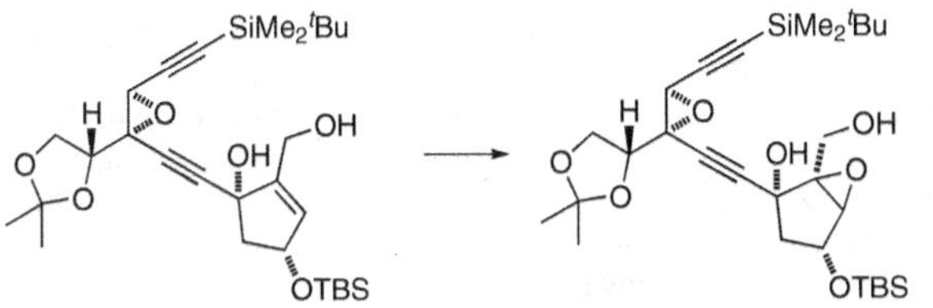

(e) 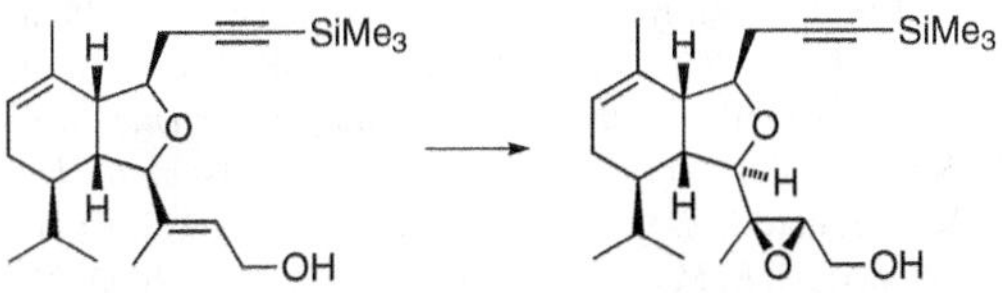

References: (a) *J. Am. Chem. Soc.* **1980**, *102*, 5974. (b) *Org. Lett.* **2005**, *6*, 4297. (c), (d) *J. Am. Chem. Soc.* **2002**, *124*, 5380. (e) *J. Am. Chem. Soc.* **1995**, *117*, 10391.

21.3 Asymmetric Epoxidation of Unfunctionalized Alkenes

Metal-Catalyzed Epoxidations

The Sharpless asymmetric epoxidation is restricted to allylic or homoallylic alcohols, so very quickly, considerable effort was put into finding methods for the asymmetric epoxidation of simple alkene hydrocarbons. Success came in the form of the asymmetric modification of the chromium and manganese *bis*-salicylideneimine ("salen") catalysts discovered by Kochi.[11] The enantioselective epoxidation of unfunctionalized alkenes was first accomplished by using chiral catalysts developed independently by Eric Jacobsen at Harvard[12] and Katsuki at Kyushu. Both sets of catalysts are based on manganese (III) complexes of a C_2-symmetric chiral salen ligand.

The reaction is now generally known as the ***Jacobsen-Katsuki epoxidation***.[13] In the original form of this reaction, alkenes were converted to chiral epoxides using iodosobenzene as the stoichiometric oxidant. In the asymmetric form, both iodosobenzene and sodium hypochlorite have been used as the terminal oxidant. Amine *N*-oxides, especially pyridine *N*-oxides, have been found to enhance the enantioselectivity of the reaction.

Jacobsen Catalysts

In the case of the Jacobsen catalysts (Example **21.17**), the chirality is imparted by the groups labeled R_1. The presence of these groups in a chiral arrangement imparts a twist to the five-membered chelate ring. This results in a characteristic "stepped" conformation of the complex as a whole, as was illustrated by the X-ray structures of a series of related oxo(salen)Cr (V) complexes.[14]

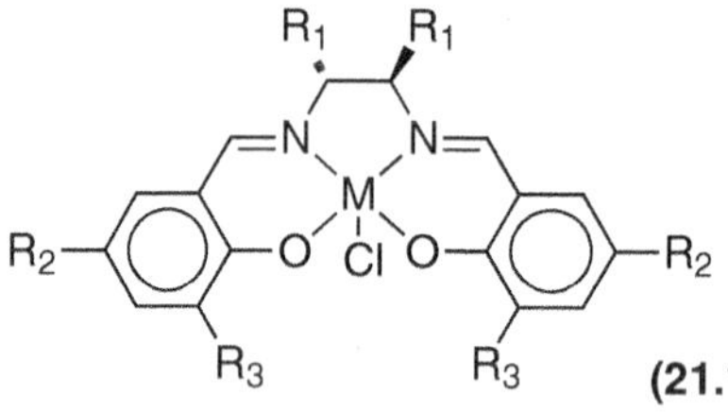

(21.17)

M = Mn or Cr
R_1 = Ph or —$(CH_2)_4$—
R_2 = R, Ar, OR
R_3 = *t*-Bu, $CHAr_2$

---

11. Srinivasan, K.; Michaud, P.; Kochi, J. K. *J. Am. Chem. Soc.* **1986**, *108*, 2309.

12. **Eric N. Jacobsen (1960–)** was educated at New York University (BS, 1982) and Berkeley (PhD, 1986). He took an NIH postdoctoral fellowship with Sharpless at the Massachusetts Institute of Technology, during which he developed catalysts fro the asymmetric dihydroxylation. He joined the faculty of the University of Illinois at Urbana-Champaign in 1988 and moved to Harvard as Professor in 1993. Jacobsen is best known for his work in developing catalysts for enantioselective organic reactions.

13. (a) Zhang, W.; Loebach, J.L.; Wilson, S.R.; Jacobsen, E.N. *J. Am. Chem. Soc.* **1990**, *112*, 2801. (b) Jacobson, E.N.; Zhang, W.; Güler, M.L. *J. Am. Chem. Soc.* **1991**, *113*, 6703. (c) Jacobsen, E.N.; Zhang, W.; Muci, A.R.; Ecker, J.R.; Deng, L. *J. Am. Chem. Soc.* **1991**, *113*, 7063. (d) Zhang, W.; Jacobsen, E.N. *J. Org. Chem.* **1991**, *56*, 2296. (e) Irie, R.; Noda, K.; Ito, Y.; Matsumoto, N.; Katsuki, T. *Tetrahedron Asymm.* **1991**, *2*, 481. (f) Hatayama, A.; Hosoya, N.; Irie, R.; Ito, Y.; Katsuki, T. *Synlett.* **1992**, 407.

Reviews: (g) Jacobsen, E.N. In Ojima, I.M, Ed. *Catalytic Asymmetric Synthesis*, (VCH Publishers: New York, 1993), p. 159. (h) Katsuki, T. *Coord. Chem. Rev.* **1995**, *140*, 189. (i) Jacobsen, E.N. In Wilkinson, G.W.; Stone, F.G.A.; E.W. Abel, E.W.; Hegedus, L.S., Eds., *Comprehensive Organometallic Chemistry II*, Vol. 12 (Pergamon: New York, 1995), Vol. 12, ch. 11.1. (j) Katsuki, T. In Ojima, I.M, Ed. *Catalytic Asymmetric Synthesis, 2nd ed.* (Wiley-VCH: New York, 2000), p. 287. (k) Katsuki, T. *Synlett* **2003**, 281. (l) Larrow, J. F.; Jacobsen, E. N. *Top. Organomet. Chem.* **2004**, *6*, 123. (m) McGarrigle, E. M.; Gilheany, D. G. *Chem. Rev.* **2005**, *105*, 1563.

14. Srinivasan, K.; Kochi, J.K. *Inorg. Chem.* **1985**, *24*, 4671.

The factors affecting the mechanism and enantioselectivity of the reaction may be summarized as follows:

1. *Z*-Disubstituted or trisubstituted cyclic alkenes conjugated with a π system capable of stabilizing a positive charge are the best substrates for the reaction.
2. *E*-Disubstituted alkenes give lower levels of enantioselectivity than the *Z* isomers.[15]
3. Epoxides derived from *trans* alkenes are sometimes observed in the epoxidation of acyclic *Z* alkenes.
4. Formation of *trans* epoxides may be enhanced by appropriate choice of the stoichiometric oxidant and additives such as quaternary ammonium ions.
5. Substitution of the salen ligand by electron-donating groups at R_2 enhance enantioselectivity, whereas electron-withdrawing groups at this position degrade enantioselectivity.

The mechanism of the reaction is complex and continues to be the subject of controversy,[16] although the consensus favors the involvement of a Mn (V) oxo species.[17,18] How this species attacks the alkene remains unknown. It may react through both manganese and oxygen to give a manganoxetane, possibly by means of a [2 + 2] cycloaddition,[19] or through oxygen alone to give an intermediate oxiranyl species.[20] The reaction through the oxygen atom may be synchronous, forming an oxirane with the manganese complexed to the oxiranyl oxygen, or it may occur by a stepwise process involving free radical intermediates.[21] The most recent proposal invokes the possibility of cationic intermediates and a mechanism with a key step reminiscent of a 1,3-dipolar cycloaddition has been proposed.[22] Earlier computational results suggested that the salen ligand, specifically the oxygen atom of the salen ligand, may be more involved in the reaction than previously believed.[23] The mechanistic options are summarized in Figure 21.5.

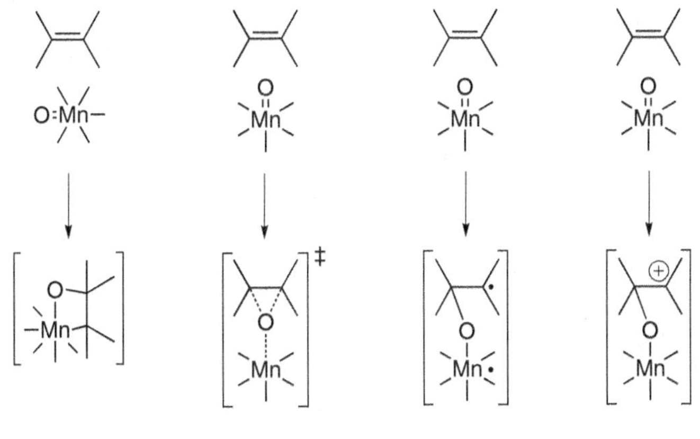

Figure 21.5 Possible mechanisms for delivery of oxygen to the alkene

15. (a) Zhang, W.; Jacobsen, E.N. *J. Org. Chem.* **1991**, *56*, 2296. (b) Irie, R.; Noda, K.; Ito, Y.; Katsuki, T. *Tetrahedron Lett.* **1991**, *32*, 1055.
16. Linker, T. *Angew. Chem. Int. Ed. Engl.* **1997**, *36*, 2060.
17. (a) Palucki, M.; Finney, N.S.; Pospisil, P.J.; Güler, M.L.; Ishida, T.; Jacobson, E.N. *J. Am. Chem. Soc.* **1998**, *120*, 948. (b) Campbell, K.A.; Lashley, M.R.; Wyatt, J.K.; Nantz, M.H.; Britt, R.D. *J. Am. Chem. Soc.* **2001**, *123*, 5710. (c) Schmitt, H.; Lomoth, R.; Magnuson, A.; Park, J.; Fryxelius, J.; Kritikos, M.; Mårtensson, M.; Hammarström, L.; Sun, L.; Åkermark, B. *Chem. Eur. J.* **2002**, *8*, 3757.
18. Feichinger, D.; Plattner, D.A. *Angew. Chem. Int. Ed. Engl.* **1997**, *36*, 1718.
19. (a) Linde, C.; Arnold, M.; Norrby, P.-O.; Åkermark, B. *Angew. Chem. Int. Ed. Engl.* **1997**, *36*, 1723. (b) El-Bahraoui, J.; Wiest, O.; Feichtinger, D.; Plattner , D.A. *Angew. Chem. Int. Ed.* **2001**, *40*, 2073.
20. Finney, N.S.; Pospisil, P.J.; Chang, S.; Palucki, M.; Konsler, R.G.; Hansen, K.B.; Jacobsen, E.N. *Angew. Chem.* **1997**, *109*, 1798; *Angew. Chem. Int. Ed. Engl.* **1997**, *36*, 1720.
21. Cavallo, L.; Jacobsen, H. *Angew. Chem. Int. Ed.* **2000**, *39*, 589.
22. Kürti, L.; Blewett, M.M.; Corey, E.J. *Org. Lett.* **2009**, *11*, 4592.
23. Abashkin, Y.G.; Burt, S.K. *Org. Lett.* **2004**, *6*, 59.

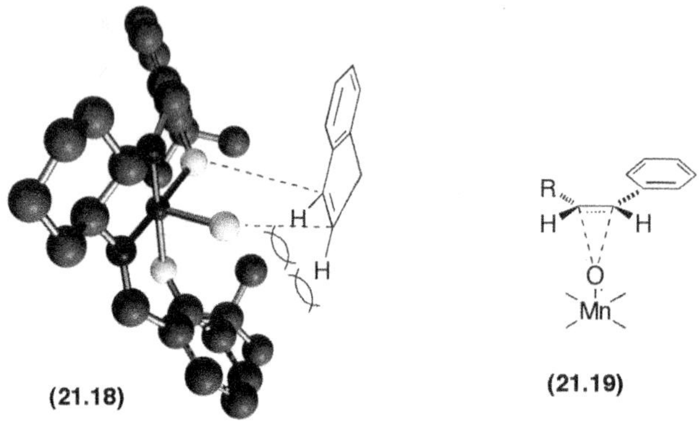

(21.18) (21.19)

Figure 21.6 Predicting the stereochemistry of olefin epoxidation using Mn (*S*,*S*)-salen complexes

The salen complex adopts a "stepped" structure, which results in one of the aromatic rings occupying a position where it interacts unfavorably with all but the smallest substituents (**21.18**). The sense of the stepping is determined by the chirality in the diamine part of the complex. A simplified version of this model is shown in Figure 21.6. To predict the sense of epoxidation, the alkene is oriented as shown in **21.19**, with the aromatic ring in the upper right quadrant and the alkyl group in the upper left quadrant. In this orientation, the manganese atom and the oxo oxygen of the catalyst are below the plane of the alkene with the (*S*,*S*) catalyst and above the plane with the (*R*,*R*) catalyst. Based on this model, *S*,*S*-salen complexes should oxidize indene to (1*R*,2*S*)-1,2-epoxyindane (**21.18**), and *cis*-1-phenylpropene should give the (1*R*,2*S*) enantiomer of the epoxide (Example **21.20**). In addition, 1-methylcyclohexene should also give the (1*R*, 2*S*) enantiomer (Example **21.21**) as the major product when the (*R*,*R*) complex is used.

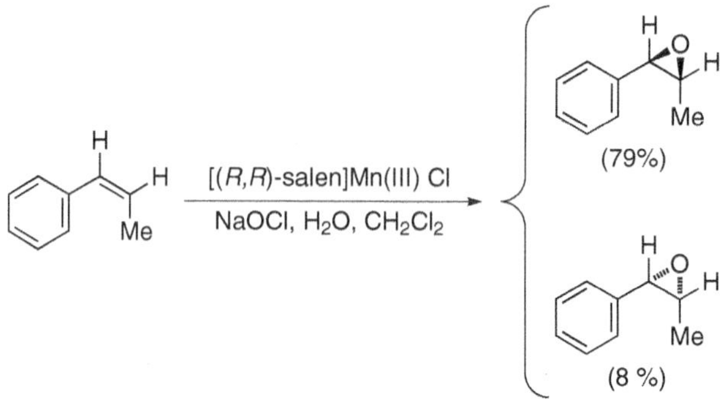

(21.20)

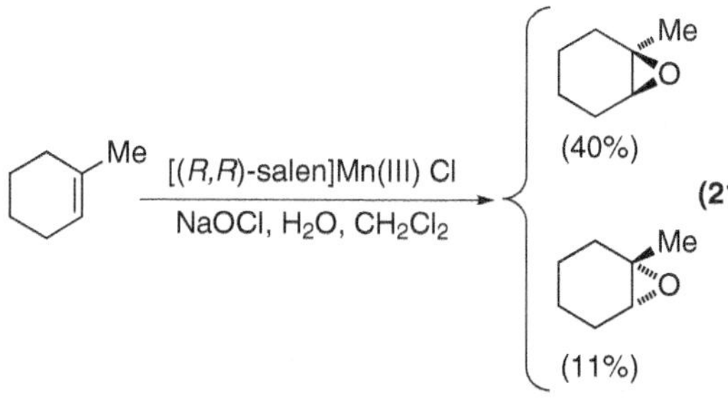

(21.21)

The model in Figure 21.6 also suggests a reason for the poor levels of enantioselectivity in *trans*-1,2-disubstituted alkenes and tetrasubstituted alkenes, because these alkenes, when oriented as described, have significant steric interactions with R_1 or R_3 of the complex. In more recent applications using the first-generation Jacobsen catalyst, hydrogen peroxide in the presence of ammonium acetate has been found to be a good replacement for sodium hypochlorite, as shown in the oxidation of the diene in Example **21.22.**[24]

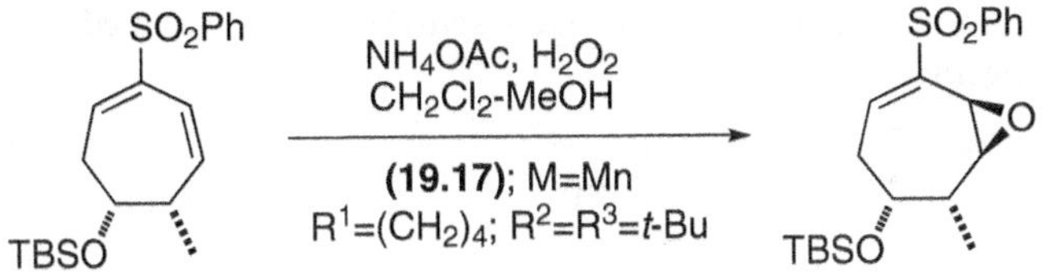

(21.22)

Katsuki Catalysts

In the Katsuki catalysts, the chirality is incorporated into the groups labeled R_3 in the generic Jacobsen catalyst seen in Example **21.17.** In the original Katsuki catalyst (Katsuki-I), these groups were substituted chiral 1-phenylpropyl groups, and the salen backbone contained a chiral 1,2-diaminocyclohexane moiety. As work to elucidate the mechanism of the reaction has evolved, new, second-generation salen complexes have been developed. In the most successful of these complexes (Katsuki-II; e.g., Example **21.23**), the chiral directing influence is exercised by a binaphthyl unit rather than by the ethylenediamine moiety. These complexes, developed by Katsuki,[25] represent an extension of the original Katsuki complexes. Complex **21.23** exhibits markedly improved enantioselectivity over the first-generation salen complexes in both homogeneous and two-phase reactions, with enantiomeric excess (e.e.) values typically above 80%—in Example **21.24**, for example, the e.e. is 96% (enantiomeric ratio [e.r.] 49:1). Katsuki, in particular, continues to experiment with salen-type ligands and a variety of metals, with the result that a large range of potential new catalysts for asymmetric epoxidation of simple alkenes is emerging.[26]

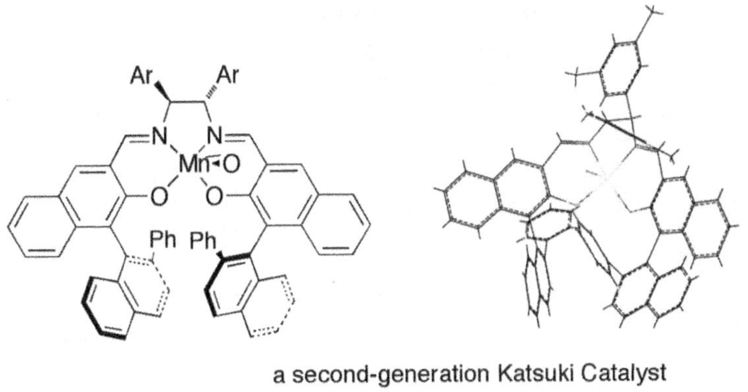

(21.23)

a second-generation Katsuki Catalyst

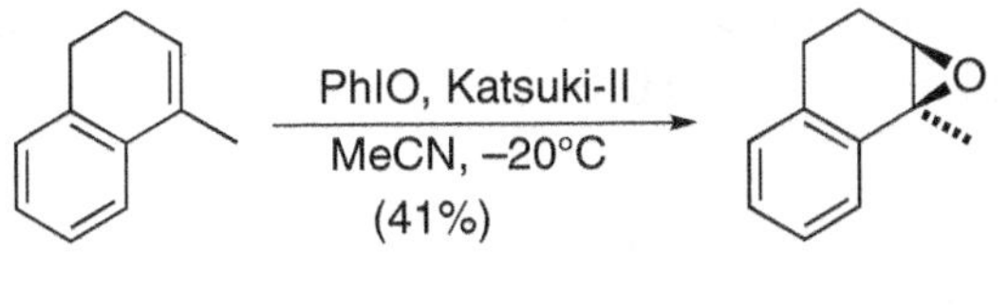

(21.24)

---

24. El-Awa, A.; Fuchs, P. *Org. Lett.* **2006**, *8*, 2905.
25. (a) Hosoya, N.; Hatayama, A.; Irie, R.; Sasaki, H.; Katsuki, T. *Tetrahedron* **1994**, *50*, 4311. (b) Sasaki, H.; Irie, R.; Katsuki, T. *Synlett* **1994**, 356. (c) Sasaki, H.; Irie, R.; Hamada, T.; Suzuki, K.; Katsuki, T. *Tetrahedron* **1994**, *50*, 11827. (d) Fukuda, T.; Irie, R.; Katsuki, T. *Synlett.* **1995**, 197.
26. For a useful review, see: Matsumoto, K.; Katsuki, T. In Ojima, I., Ed. *Catalytic Asymmetric Synthesis*, 3rd ed. (John Wiley & Sons: New York, 2010), ch. 11, p. 839.

Worked Problem

21-2 Predict the stereochemistry the major organic product when each of the alkenes below is oxidized using the Mn-salen (Jacobsen-I) catalyst indicated, with the co-oxidant specified.

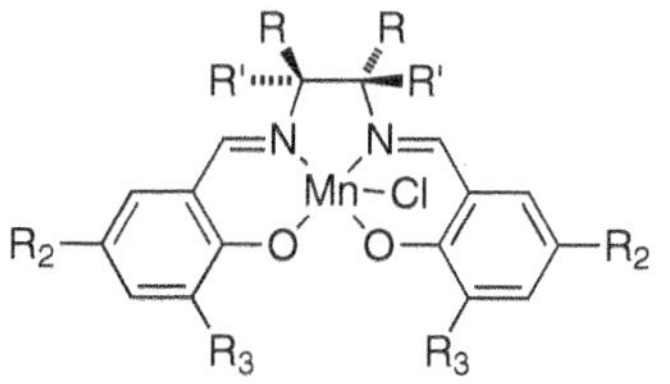

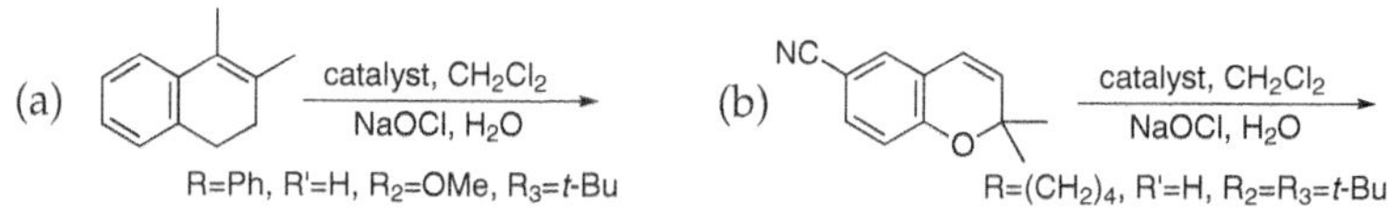

§Answers below.

Reaction Synopses

Jacobsen-Katsuki Epoxidation of Unactivated Alkenes

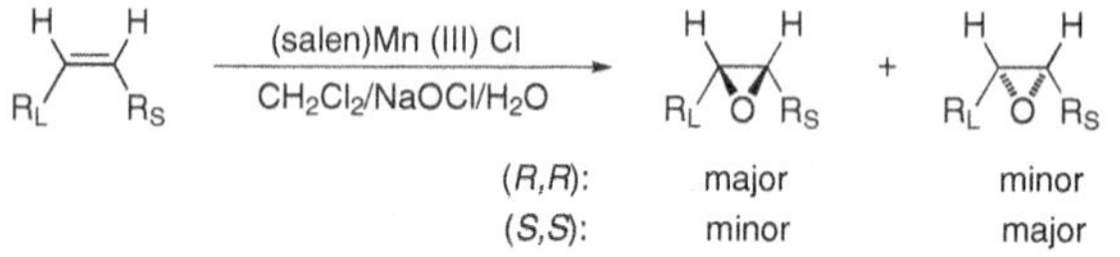

Enantioselectivity is higher for *cis*-disubstituted alkenes.
Enantioselectivity is enhanced by donor ligands (e.g., NMMO, pyridine-*N*-oxide, etc.).
Reactivity often enhanced by ammonium ion additives.

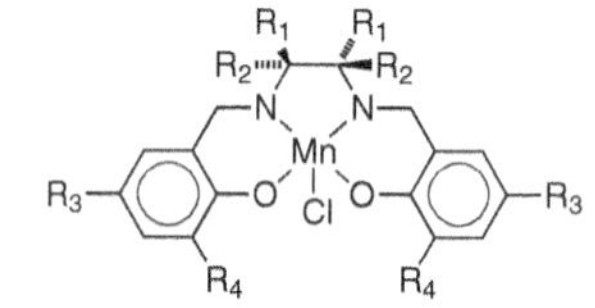

Jacobsen Catalyst	Katsuki Catalyst
R_1=Ar, $(CH_2)_4$ OR H	R_1=Ar, $(CH_2)_4$ OR H
R_2=H OR Ar, $(CH_2)_4$	R_2=H OR Ar, $(CH_2)_4$
R_3=R, OR	R_3=R, OR
R_4=t-Bu	R_4=CH(Ar)Ar')

Problem

21-4 Suggest the major organic product when each of the alkenes below is oxidized using the Jacobsen Mn-salen catalyst from Worked Problem 21-2, as indicated, with the co-oxidant specified. Note that the *trans* olefin epoxide may be the major product in some cases.

§ Answers to Worked Problem:

Both these stereochemical outcomes can be predicted using the model in Figure 21.6. The enantioselectivity in Example (b) is high; because the double bond is tetrasubstituted, the enantioselectivity in Example (a) is predicted to be lower. It is.

References: (a) *Tetrahedron Lett.* **1996**, *36*, 5123. (b) *J. Am. Chem. Soc.* **1991**, *113*, 7063.

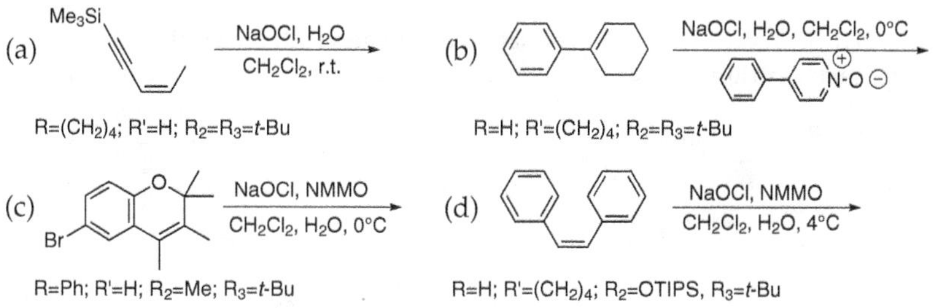

References: (a) *Tetrahedron Lett.* **1991**, *32*, 6533. (b) *J. Org. Chem.* **1994**, *59*, 4378. (c) *Tetrahedron Lett.* **1995**, *36*, 5123. (d) *J. Am. Chem. Soc.* **1994**, *116*, 6937.

21.4 **Oxidation with Chiral Three-membered Heterocycles**

Chiral Oxaziridines

Another strategy for asymmetric oxidation relies on the reactivity of three-membered rings containing two heteroatoms: the *N*-sulfonyloxaziridines and the dioxiranes. The oxaziridine ring system is rendered reactive by the presence of both a strained three-membered ring and a relatively weak N—O bond in the ring.[27] Despite these advantages, however, these reagents tend to be weakly electrophilic oxidizing agents, so their reactions with all but the most electron-rich of alkenes are fairly sluggish. The tendency of oxaziridines to react as oxygen transfer agents (as opposed to nitrogen transfer agents) is improved by incorporating a bulky substituent on nitrogen.

The electrophilicity of the oxaziridine ring system in oxygen transfer reactions can be markedly improved by incorporating electron-withdrawing groups into the molecule. Fluorination, for example, renders these compounds much more reactive as epoxidizing agents—comparable to the much more reactive dioxiranes.[27c] The use of *N*-sulfonyloxaziridines as mild oxidizing agents was pioneered by American chemist Franklin Davis. It is interesting to note that although sulfonyloxaziridines such as the ***Davis oxaziridine***[28] (Example **21.25**) can be used to oxidize alkenes, the ***Davis chiral sultam***[29] (Example **21.26**), which carries the sulfonyloxaziridine moiety, does not oxidize simple alkenes. However, it does serve as an excellent reagent for the α-hydroxidation of enolate anions.[27b]

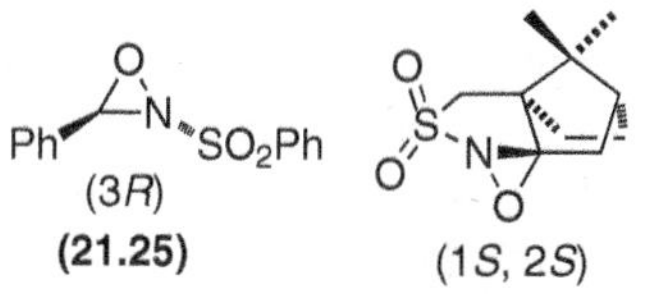

(3*R*)
(21.25)

(1*S*, 2*S*)
(21.26)

The Davis oxaziridine permits the α-hydroxylation of ketone enolates with high levels of asymmetric induction (e.e. > 95%, e.r. > 39:1 in favorable cases), although the level of

27. Reviews: (a) Davis, F.A.; Sheppard, A.C. *Tetrahedron* **1989**, *45*, 5703. (b) Davis, F.A.; Chen, B.-C. *Chem. Rev.* **1992**, *92*, 919. (c) Petrov, V.A.; Resnati, G. *Chem. Rev.* **1996**, *96*, 1809. (d) Adam, W.; Saha-Möller, C.R.; Ganeshpure, P.A. *Chem. Rev.* **2001**, *101*, 3499.

28. (a) Davis, F.A.; Stringer, O.D. *J. Org. Chem.* **1982**, *47*, 1774. (b) Davis, F.A.; Lamendola, J., Jr.; Nadir, U.; Kluger, E.W.; Sedergran, T.C.; Panunto, T.W.; Billmers, R.; Jenkins, R., Jr.; Turchi, I.J.; Watson, W.H.; Chen, J.S.; Kimura, M. *J. Am. Chem. Soc.* **1980**, *102*,2000. (c) Vishwakarma, L.C.; Stringer, O.D.; Davis, F.A. *Org. Syn.* **1987**, *66*, 203.

29. (a) Davis, F.A.; Towson, J.C.; Weismiller, M.C.; Lal, S.; Carroll, P.J. *J. Am. Chem. Soc.* **1988**, *110*, 8477. (b) Towson, J.C.; Weismiller, M.C.; Lal, G.S.; Sheppard, A.C.; Davis, F.A. *Org. Syn.* **1990**, *69*, 158. (c) Cermak, S.C.; Wiemer, D.F. *J. Chem. Educ.* **1999**, 76, 1715.

Table 21.2 Stereochemical Outcome of Oxidation of Enolates with the Davis Camphorsulfonyloxaziridine*

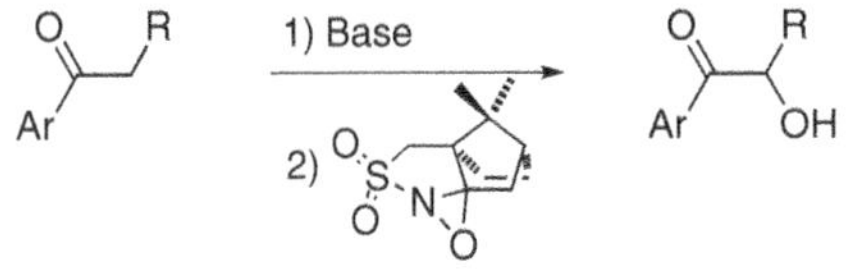

Ketone	Base	Product	e.e.	Ketone	Base	Product	e.e.
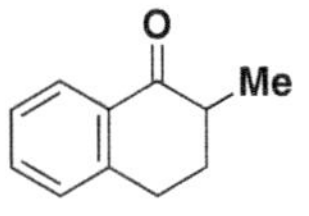	NaHMDS	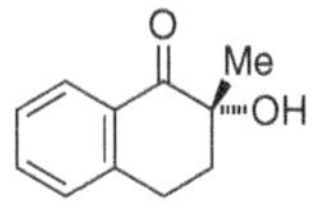	95	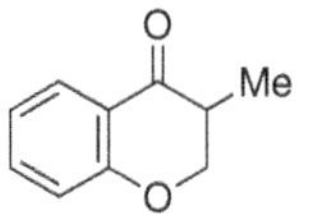	LDA	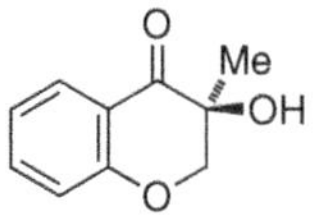	30
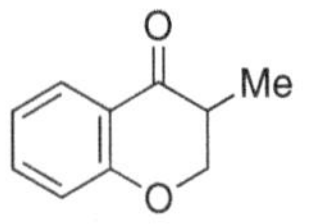	NaHMDS	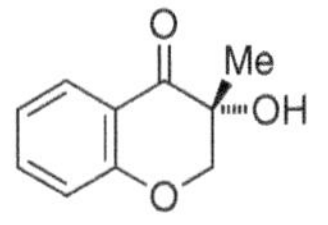	71	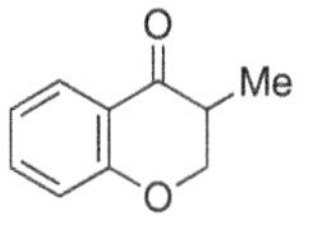	LDA	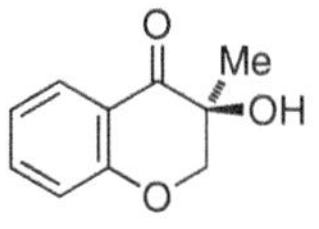	64
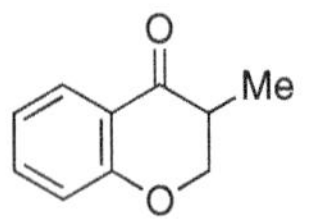	NaHMDS	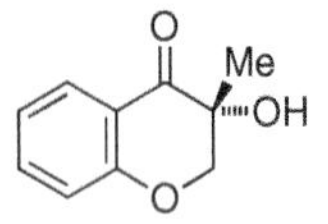	96	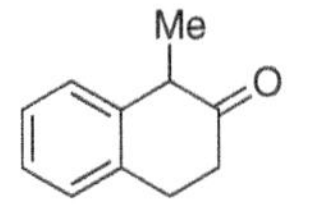	NaHMDS	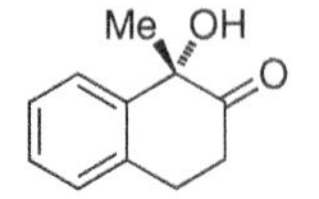	76
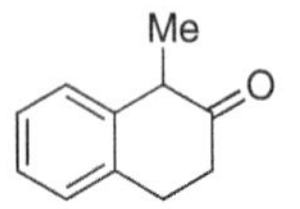	NaHMDS HMPA	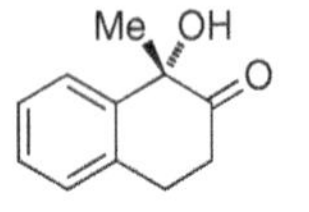	65	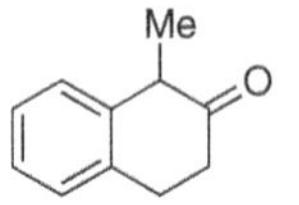	LDA	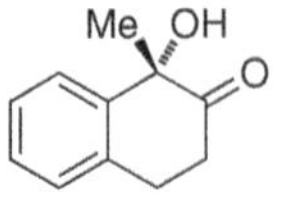	10
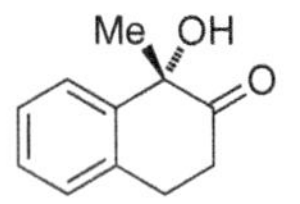	NaHMDS	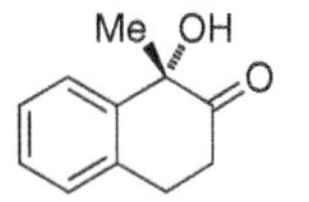	16	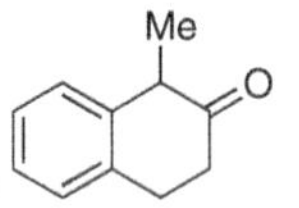	LDA	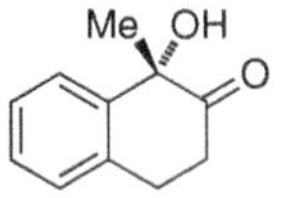	54

*Data taken from Davis, F.A.; Weismiller, M.C. *J. Org. Chem.* **1990**, 55, 3715.

asymmetric induction is often only modest. To obtain high levels of asymmetric induction, it has been found that sodium enolates, generated using sodium hexamethyldisilazide (NaHMDS), appear to be superior to lithium, potassium, and zinc enolates. However, the generalization is far from perfect, as shown by the reactions in Table 21.2.

With open-chain enolates, the stereochemistry of the oxidation with chiral oxaziridines depends on the stereochemistry of the enolate. Deprotonation of *tert*-butyl phenylacetate in tetrahydrofuran (THF) gives the *E* enolate preferentially. The two possible approaches of this enolate to the oxidant are shown in Figure 21.7. When the *re* face of the prochiral center approaches the oxygen atom of the (–)-(2*S*,8a*R*)-oxaziridine, there is significant steric interaction between the phenyl group and the *endo* hydrogens at position 6 of the camphor ring. When the *si* face of the prochiral center approaches the oxygen, this

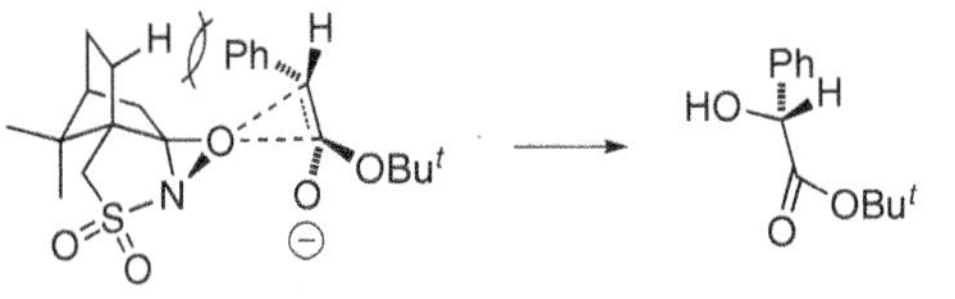

(21.27)

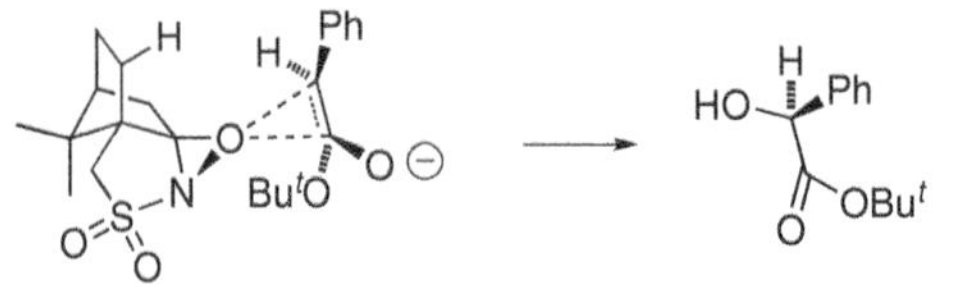

(21.28)

Figure 21.7 Preferred approach of the enolate anion to the chiral oxaziridine avoids steric congestion between the alkene substituent and the camphor framework.

steric interaction is absent. The major product of this oxidation is the (+)-*S*-hydroxyester.[30] When hexamethylphosphoramide (HMPA) is added to the reaction solvent, the stereochemistry of the enolate generation is reversed,[31] and the absolute configuration of the hydroxyester product is also reversed.

As can be seen from Examples **21.29** to **21.32**, the Davis camphorsulfonylaziridine (CSO) has been used in the oxidation of both prochiral and chiral enolates. As shown by the first example, there is a caveat with the use of the Davis CSO for the oxidation of ester enolates because this reaction does not always give high enantioselectivities. When chirality is already present in the enolate, however, the diastereoselectivity of the oxidation can be quite high. As shown in one of the examples below, the Davis CSO is an excellent reagent for effecting hydroxylation, even when no new chiral center is formed. Enolates from deprotonation of saturated compounds or conjugate addition to α,β-unsaturated carbonyl compounds are both amenable to oxidation.

Another important application of chiral oxaziridines in organic synthesis is in the preparation of enantiomerically enriched sulfoxides. As we discussed in Chapter 2, chiral sulfoxides show significant configurational stability, and chiral oxaziridines have proved to be efficient reagents for the chiral oxidation of sulfides, as shown in the example below.[32] Occasionally, the use of the chiral oxaziridine reagent allows the stereochemistry of oxidation by an achiral oxidizing agent (e.g., *m*-chloroperoxybenzoic acid) to be reversed.[33]

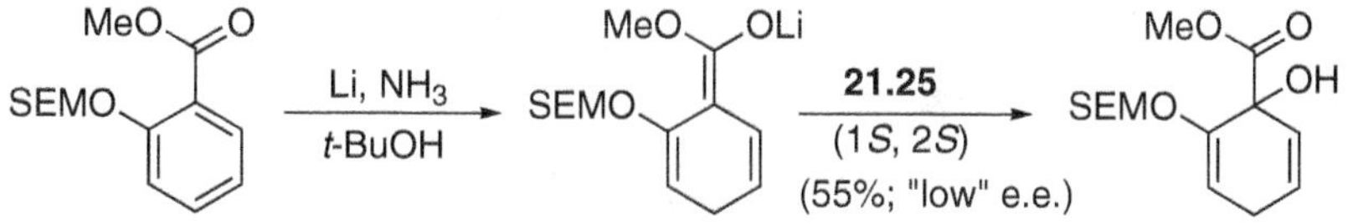

(21.29)[34]

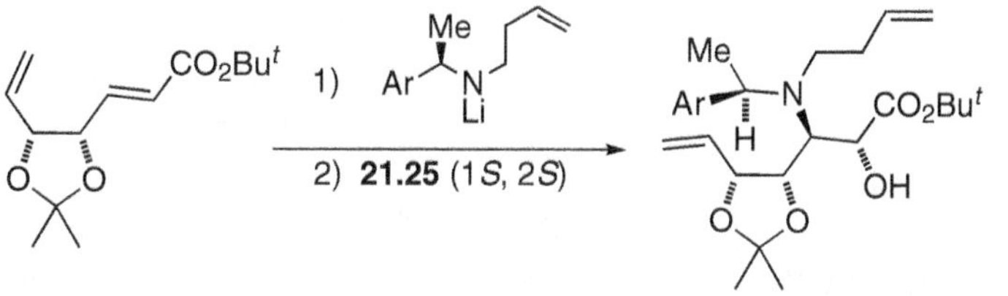

(21.30)[35]

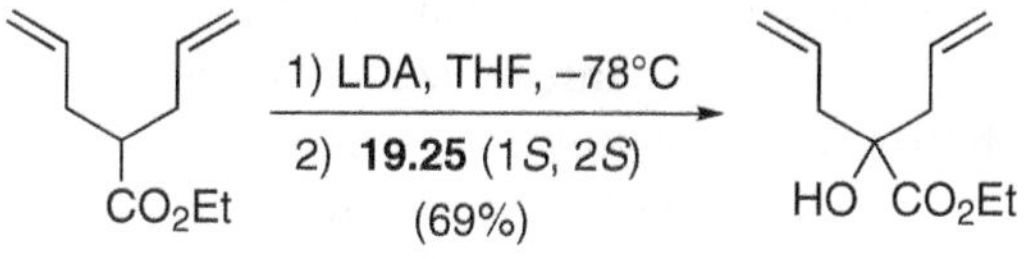

(21.31)[36]

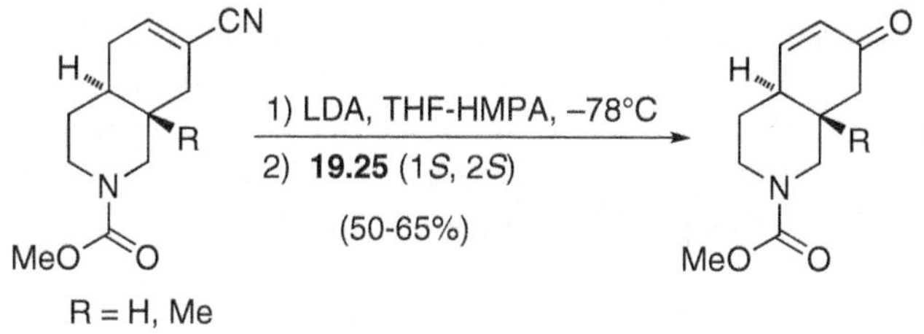

(21.32)[37]

30. Davis, F.A.; Haque, M.S.; Ulatowski, T.G.; Towson, J.C. *J. Org. Chem.* **1986**, *51*, 2402.
31. Ireland, R.E.; Muller, R.H.; Willard, A.K. *J. Am. Chem. Soc.* **1976**, *98*, **2868.**
32. Renard, M.; Ghosez, L.A. *Tetrahedron* **2001**, *57*, 2597.
33. McKittrick, B.A.; Ma, K.; Huie, K.; Yumibe, N.; Davis, H. Jr.; Clader, J.W.; Czarniecki, M.; McPhail, A.T. *J. Med. Chem.* **1998**, *41*, 752.
34. Richardson, A.M.; Chen, C.-H.; Snider, B.B. *J. Org. Chem.* **2007**, *72*, 8099.
35. Brock, E.A.; Davies, S.G.; Lee, J.A.; Roberts, P.M.; Thomson, J.E. *Org. Lett.* **2011**, *13*, 1594.
36. Van Ornum, S.G.; Cook, J.M. *Tetrahedron Lett.* **1996**, *37*, 7185.
37. MaGee, D.I.; Lee, M.L. *Tetrahedron Lett.* **2001**, *42*, 7177.

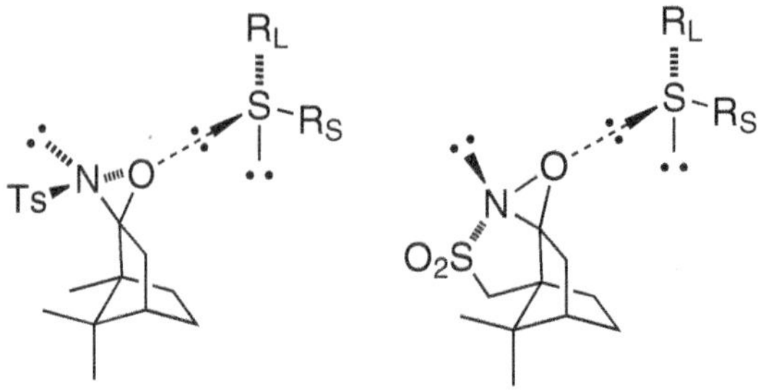

(21.34) (21.35)

Figure 21.8 A model for predicting the stereochemistry of the major epoxide by oxidation of sulfides with chiral oxaziridines

In the reaction in Example **21.33**, for example, one can predict the stereochemistry of the major sulfoxide by aligning the oxaziridine oxygen, the sulfur, and one lone pair on sulfur in a linear arrangement with the largest group eclipsing the lone pair on nitrogen, and orienting the other lone pair on sulfur into the *endo* cavity of the bornane ring system. In this orientation, the absolute configuration of the major sulfoxide formed in the reaction is predicted to be that shown. Similar logic suggests that the Davis camphorsultam should give the same chirality of the major sulfoxide, as shown on the right in Figure 21.8. In this case, the large group should orient itself away from position C-3 of the bornane ring system.

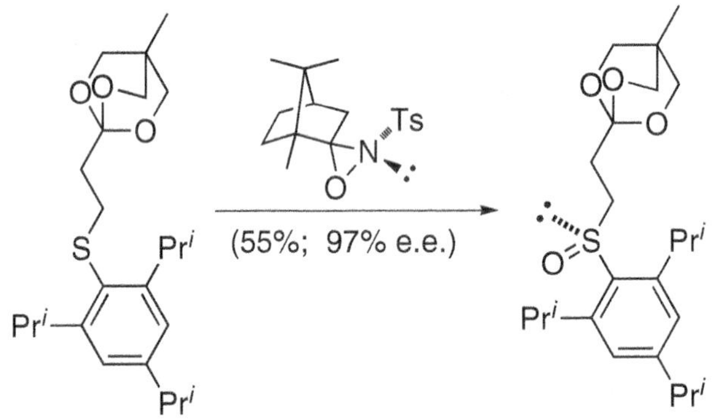

(21.33)

We have already seen that the reactivity of oxaziridines as oxygen atom transfer agents can be enhanced by substituting the ring with electron-withdrawing substituents. In 1976, another way to increase the reactivity of the three-membered ring was disclosed: the oxaziridinium ions, formed by oxidation of iminium ions with Oxone® ($2KHSO_5{\cdot}KHSO_4{\cdot}K_2SO_4$), are extremely reactive reagents for the transfer of oxygen to an alkene.[38] The predominant drawback with using these reagents is the need for Oxone® as the terminal oxidant, because this reagent must be used in aqueous acetonitrile, which freezes at –8°C. By using chiral iminium ions, epoxidations of alkenes with e.e. values from 60% to 97% can be realized.[39] Recently, the procedure has been modified to use tetraphenylphosphonium monopersulfate (Ph_4P^+ HSO_5^-) as the co-oxidant, which has allowed the elimination of both water and base.[40] Although this non-aqueous reaction system gives improved

38. (a) Picot, A.; Millet, P.; Lusinchi, X. *Tetrahedron Lett.* **1976**, *17*, 1573. (b) Hanquet, G.; Lusinchi, X.; Milliet, P. *Tetrahedron Lett.* **1987**, *28*, 6061. (c) Hanquet, G.; Lusinchi, X.; Milliet, P. *Tetrahedron Lett.* **1988**, *29*, 3941.

39. (a) Page, P.C.B.; Rassias, G.A.; Bethell, D.; Schilling, M.B. *J. Chem. Soc., Perkin Trans. 1* **2000**, 3325. (b) Page, P.C.B.; Rassias, G.A.; Barros, D.; Ardakani, A.; Buckley, B.; Bethell, D.; Smith, T.A.D.; Slawin, A.M.Z. *J. Org. Chem.* **2001**, *66*, 6926. (c) Page, P.C.B.; Rassias, G.A.; Barros, D.; Ardakani, A.; Bethell, D.; Merrifield, E. *Synlett* **2002**, 580. (d) Page, P.C.B.; Buckley, B.R.; Rassias, G.A.; Blacker, A.J. *Eur. J. Org. Chem.* **2006**, 803. (e) Page, P.C.B.; Buckley, B.R.; Heaney, H.; Blacker, A.J. *Org. Lett.* **2005**, *7*, 375. (f) Page, P.C.B.; Buckley, B.R.; Blacker, A.J. *Org. Lett.* **2004**, *6*, 1543.

40. (a) Page, P.C.B.; Barros, D.; Buckley, B.R.; Ardakani, A.; Marples, B.A. *J. Org. Chem.* **2004**, *69*, 3595. (b) Page, P.C.B.; Buckley, B.R.; Barros, D.; Blacker, A.J.; Heaney, H.; Marples, B.A. *Tetrahedron* **2006**, *62*, 6607.

yields relative to the aqueous acetonitrile reaction, for most alkenes it still gives only modest e.e. values.

Chiral Dioxiranes

The dioxirane ring system is formed by the reaction between a ketone and persulfate anion (usually in the form of Oxone®). This highly strained ring system is an excellent oxidizing agent in its own right, and it can be used for the epoxidation of alkenes. Early work on the asymmetric epoxidation of alkenes with dioxiranes such as Example **21.37**, formed from C_2-symmetric ketones such as Example **21.36**, gave useful results with *trans*-cinnamate esters (especially those carrying an electron-releasing group at the *p*- position).[41] However, the results were less satisfactory with the *cis*-isomers or with β-alkylcinnamates.[42]

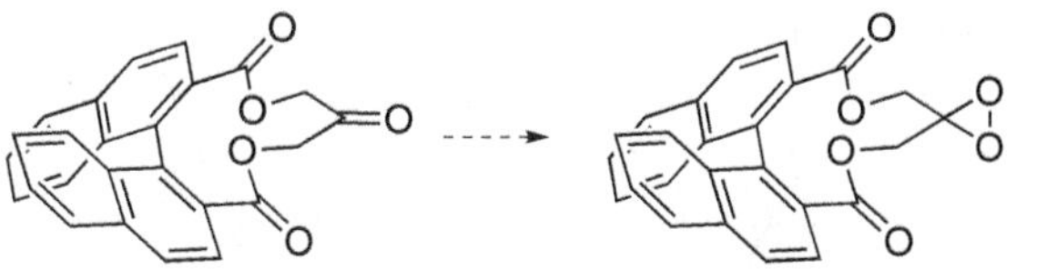

(21.36) **(21.37)**

The reaction has become much more useful since Shi and his coworkers discovered that by using fructose derivatives (e.g., Epoxone; Example **21.38**, which is produced industrially on the kilogram scale[43]) as the chiral catalyst and Oxone® as the terminal oxidant,[44] alkenes can be epoxidized with high levels of asymmetric induction. In alkenes with more than one double bond, the most electron-rich, least hindered double bond is oxidized preferentially, and alkenes are oxidized in preference to alkynes. What makes this chiral catalyst especially useful is the fact that its enantiomeric form can be prepared from L-sorbose, which can be converted to L-fructose.[45] The reaction is now generally known as the ***Shi epoxidation***.[46] To suppress competing Baeyer-Villiger oxidation of the catalyst, the reaction is usually carried out at relatively high pH (typically 10.5, using potassium carbonate as the base in acetonitrile).

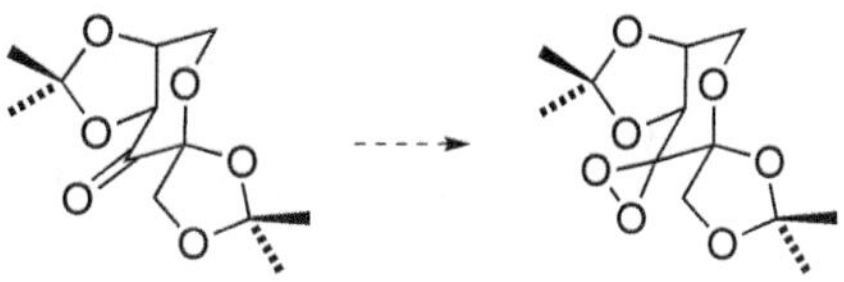

(21.38) **(21.39)**

The epoxidation by Epoxone proceeds through two competing transition states. In one, the oxygen is delivered to the alkene through an orthogonal spirocyclic activated complex.[47] In the other, the oxygen is delivered through a planar spirocyclic activated complex. In both activated complexes, the alkane approaches the dioxirane on the *exo* face of the bicyclic ketal. This places one of the groups on the alkene in close proximity to the axial oxygen of the spirocyclic ketal ring, and this alkyl group (labeled R_1 in

41. Seki, M.; Furutani, T.; Imashiro, I.; Kuroda, T.; Yamanaka, T.; Harada, N.; Arakawa, H.; Kusama, M.; Hashiyama, T. *Tetrahedron Lett.* **2001**, *42*, 8201.

42. Imashiro, I.; Seki, M. *J. Org. Chem.* **2004**, *69*, 4216.

43. Ager, D.J.; Anderson, K.; Oblinger, E.; Shi, Y.; VanderRoest, J. *Org. Proc. Res. Dev.* **2007**, *11*, 44.

44. (a) Wang, Z.-X.; Tu, Y.; Frohn, M.; Zhang, J.-R.; Shi, Y. *J. Am. Chem. Soc.* **1997**, *119*, 11224. (b) Shi, Y. *Synthesis* **2000** 1979. (c) Tian, H.; She, X.; Shu, L.; Yu, H.; Shi, Y. *J. Am. Chem. Soc.* **2001**, *122*, 11551. (d) Tian, H.; She, X.; Xu J.; Shi, Y. *Org. Lett.* **2001**, *3*, 1929.

45. (a) Wang, Z.-X.; Tu, Y.; Frohn, F.; Zhang, J.-R.; Shi, Y. *J. Am. Chem. Soc.* **1997**, *119*, 11224. (b) Chen, C.-C.; Whistler, R. L. *Carbohydr. Res.* **1988**, *175*, 265.

46. Review: Wong, O.A.; Shi, Y. *Chem. Rev.* **2008**, *108*, 3958.

47. Baumstark, A.L.; McCloskey, C.J. *Tetrahedron Lett.* **1987**, *28*, 3311.

Figure 21.9 Competing activated complexes for the Shi epoxidation of alkenes

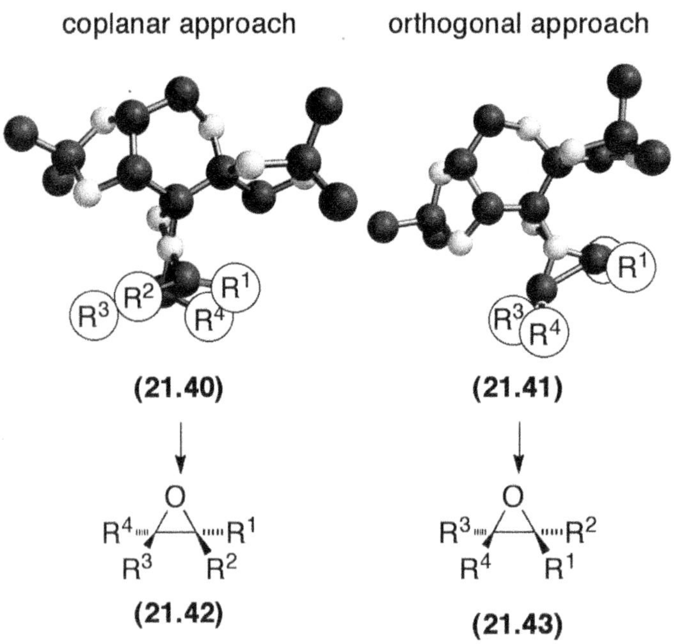

Table 21.3 Effects of Alkene Structure on the Enantioselectivity of the Shi Oxidation of Trisubstituted Alkenes

R_1	R_2	R_3	% e.e. (e.r.)
Ph	H	Ph	> 95 (> 39:1)
Ph	H	Me	88 (47:3)
Ph	—	$(CH_2)_4$-	98 (99:1)
Ph	H	CH=CHPh	97 (197:3)
$PhCO_2$	—	$(CH_2)_4$-	93 (193:7)
Ph	H	CH_2Cl	93 (193:7)
Ph	H	$(CH_2O)_2CH$	93 (193:7)
$C_{10}H_{21}$	Et	Et	89 (189:11)
Me_3C	—	$(CH_2)_4$-	26 (63:37)
n-Bu	—	$(CH_2)_4$-	79 (179:21)
Me	—	$(CH_2)_4$-	81 (181:19)

Figure 21.9) will be the smallest substituent on the alkene. This smallest substituent is almost always hydrogen: tetrasubstituted alkenes do not react well in the Shi epoxidation. As is shown in Figure 21.9, the two isomeric activated complexes lead to enantiomeric stereochemical results, with the planar activated complex (**21.40**) and the orthogonal spirocyclic transition state (**21.41**).[48] The enantioselectivities observed in trisubstituted alkenes—that higher levels of enantioselectivity are obtained when R_1 is small and when R_3 is large (Table 21.3)—are consistent with the orthogonal activated complex being preferred. However, the overall enantioselectivity of the reaction still comes down to the competition between these two different modes of reaction.

More recently, it has been found that Oxone® can be replaced with hydrogen peroxide in acetonitrile. It is thought that the actual oxidant under these conditions is a peroxyimidic acid, which produces a stoichiometric amount of acetamide as the by-product. The proposed catalytic cycle for this oxidation is shown in Figure 21.10.

The original Shi catalyst does not perform particularly well with terminal alkenes or with *cis*-disubstituted alkenes. For these substrates, alternative oxidation catalysts containing *N*-substituted oxazolidinone functionality, as shown in Example **21.47**,[49] have been developed. Using this catalyst, it has been possible to oxidize styrene with an e.e. in excess of 80%, with the *S* epoxide predominating. In both cases, the chemical yield is good (≥ 90%), but the oxazolidinone catalyst delivers superior enantioselectivity.

48. Tu, Y.; Wang, Z.-X.; Shi, Y. *J. Am. Chem. Soc.* **1996**, *118*, 9806.

49. (a) Tian, H.; She, X.; Shu, L.; Yu, H.; Shi, Y. *J. Am. Chem. Soc.* **2000**, *122*, 11551. (b) Tian, H.; She, X.; Xu, J.; Shi, Y. *Org. Lett.* **2001**, *3*, 1929. (c) Tian, H.; She, X.; Yu, H.; Shu, L.; Shi, Y. *J. Org. Chem.* **2002**, *67*, 2435. (d) Shu, L.; Shen, Y.-M.; Burke, C.; Goeddel, D.; Shi, Y. *J. Org. Chem.* **2003**, *68*, 4963.

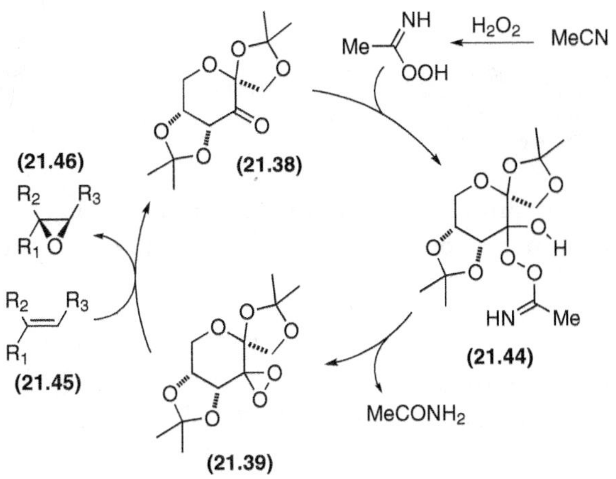

Figure 21.10 Catalytic cycle for the Shi oxidation using hydrogen peroxide-acetonitrile as the stoichiometric oxidant

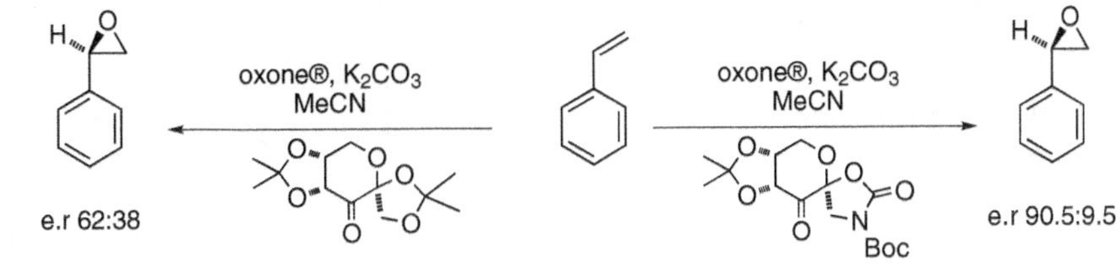

(21.47)

The original Shi catalyst tends to undergo Baeyer-Villiger oxidation (Example **21.48**) faster than it reacts with electron-deficient alkenes, making it unsuitable for epoxidizing conjugated esters, for example. This has been addressed by replacing one of the dioxolane rings with a pair of acetate groups (Example **21.49**). This modified catalyst permits the oxidation of ethyl cinnamate in good chemical yield, and with high e.r. values (98:2).

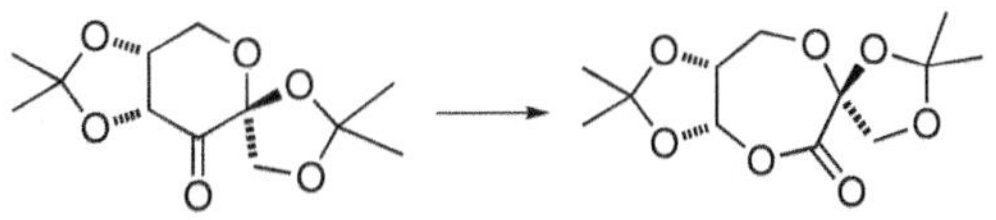

(21.48)

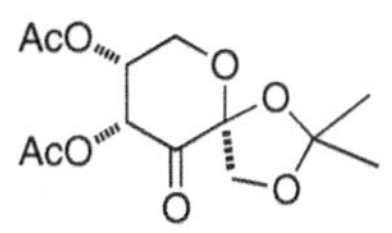

(21.49)

Asymmetric Epoxidation of Conjugated Carbonyl Compounds Using Chiral Phase Transfer Catalysts

The oxidation of enones and similar conjugated carbonyl compounds by alkaline hydrogen peroxide is well established. Yet moving this chemistry to reliable asymmetric applications has been less straightforward, and the oxidation with second-generation Shi-type catalysts still tends to give the best results in terms of enantioselectivity. However, it has been discovered that quaternary ammonium cations formed from the *Cinchona* alkaloids can be very effective catalysts for the asymmetric epoxidation of conjugated carbonyl compounds by hydrogen peroxide (e.g., Example **21.50**).[50]

---

50. (a) Lygo, B.; Wainwright, P.G. *Tetrahedron* **1999**, *55*, 6289. (b) Corey, E.J.; Zhang, F.-Y. *Org. Lett.* **1999**, *1*, 1287. (c) Arai, S.; Tsuge, H.; Oku, M.; Miura, M.; Shioiri, T. *Tetrahedron* **2002**, *58*, 1623. (d) Ooi, T.; Ohara, D.; Tamura, M.; Maruoka, K. *J. Am. Chem. Soc.* **2004**, *126*, 6844.

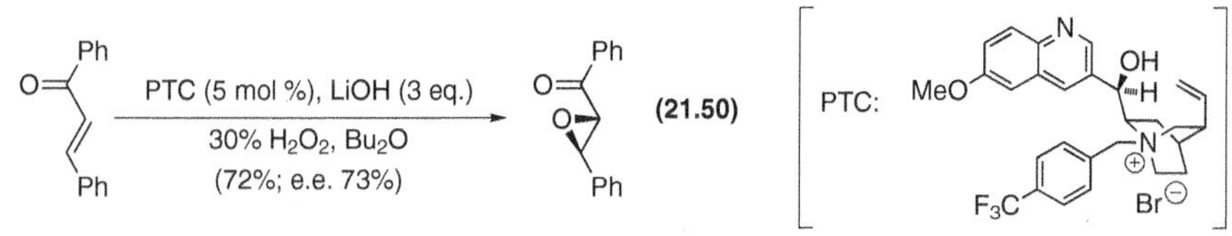

(21.50)

Dimeric quinine or quinidine derivatives, such as in Example **21.51**, are even more effective, especially in the presence of a small quantity of surfactant such as Triton-X100.[51]

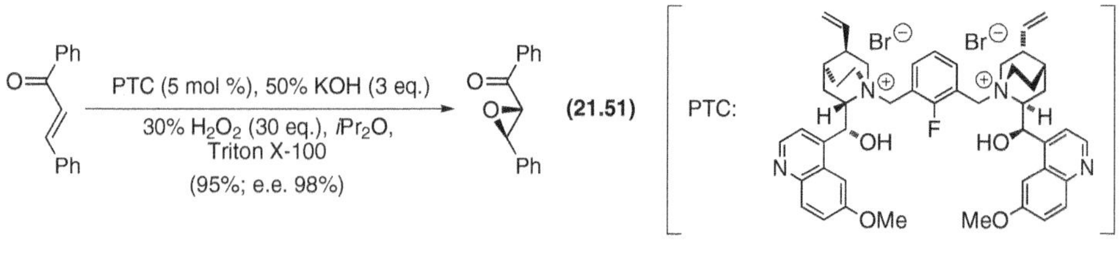

(21.51)

Worked Problem

21-3 Suggest reagents to effect the following transformations with the stereochemical outcomes shown.

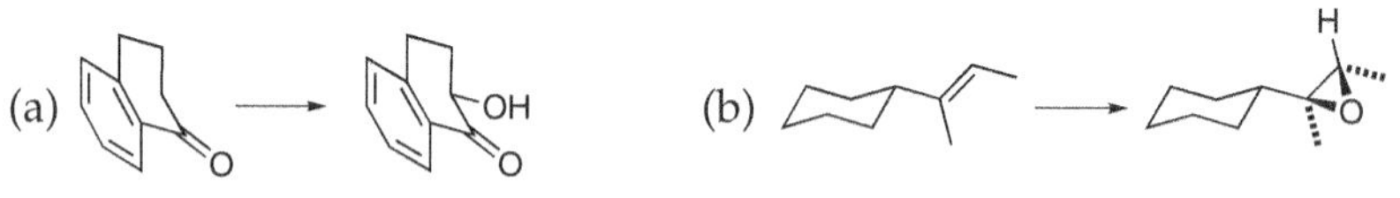

§Answers below.

51. Jew, S.-s.; Lee, J.-H.; Jeong, B.-S.; Yoo, M.-S.; Kim, M.-J.; Lee, Y.-J.; Lee, J.; Choi, S.-h.; Kyungjae Lee, K.; Lah, M.S.; and Park, H.-g. *Angew. Chem. Int. Ed.* **2005**, *44*, 1383.

§ Answers to Worked Problem:

(a) This reaction involves the formation of the enolate and then oxidation of the enolate to give the *R*-2-hydroxyketone. This will be accomplished by using a Davis oxaziridine to oxidize the sodium enolate, which will be formed, in turn, using sodium hexamethyldisilazide. Using the model in Figure 21.7 as a guide, we predict that the lowest-energy arrangement of the oxidant and the enolate should be as shown below. This model shows that the camphorsulfonyloxaziridine that we should use is the one based on (1*R*)-camphor.

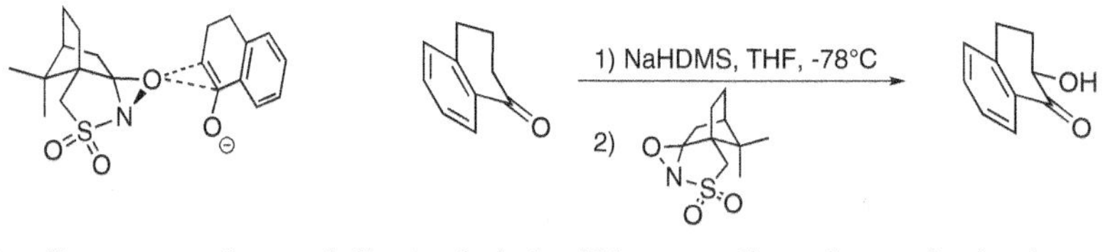

(b) This alkene is a trisubstituted alkene, which should be an excellent substrate for the Shi epoxidation. Based on the absolute stereochemistry of the product epoxide, with R_1 as cyclohexyl, and R_2 and R_3 both methyl, we should use the catalyst based on D-fructose.

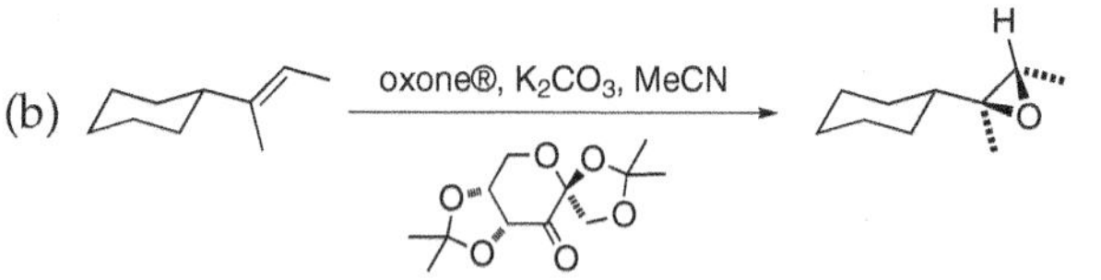

Reaction Synopses

Oxidation with Chiral Oxaziridines

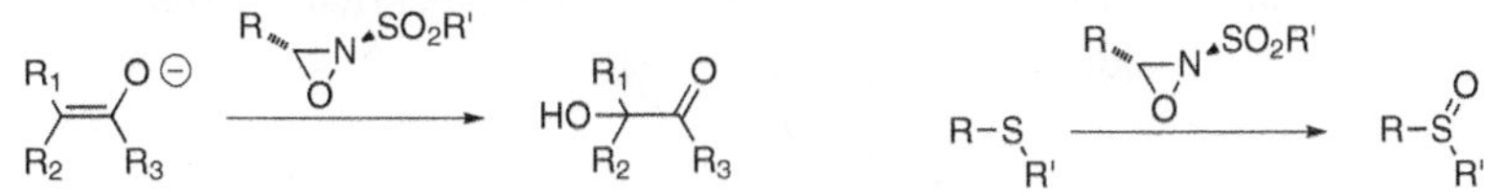

Reagents: 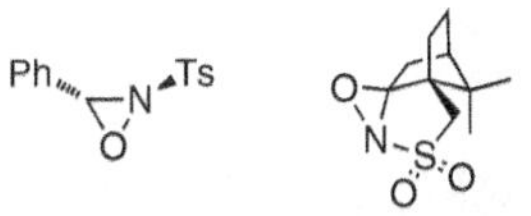 and enantiomers

or RR'C=NR", Oxone®, $NaHCO_3$, $MeCN\text{-}H_2O$; RR'C=NR", $(Ph_4P)(HSO_5)$, MeCN

Solvent: usually THF at low (−90 to −50°C) temperatures due to generation of enolate and oxidation being carried out in a single operation

Enantioselectivity varies from high with ketone enolates to marginal with ester enolates. Reactivity is greater with oxaziridinium ions, but enantioselectivity may be modest.

Oxidation with Dioxiranes (Shi Oxidation)

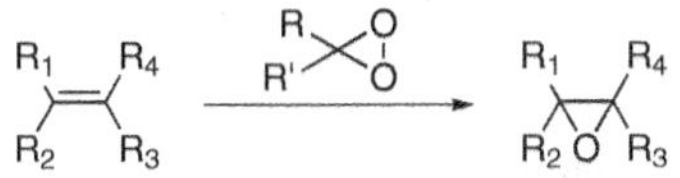

Reagents: $R_1R_2C{=}O$, oxone®, K_2CO_3, MeCN; $R_1R_2C{=}O$, H_2O_2, K_2CO_3, MeCN etc.

Enantioselective Oxidation of Conjugated Carbonyl Compounds

Reagents: H_2O_2, chiral phase transfer catalyst, R_2O, KOH; etc.

Table 21.4 Choosing an Asymmetric Epoxidation Method

Alkene Structure	Method
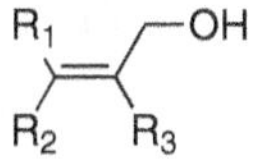	Sharpless epoxidation
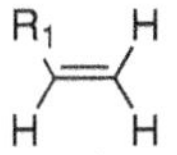	Shi epoxidation
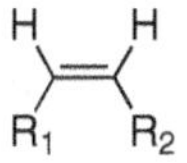	Jacobsen-Katsuki epoxidation
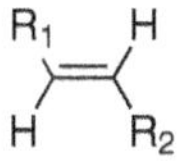	Shi epoxidation
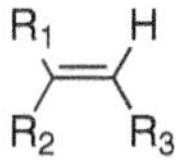	Shi epoxidation
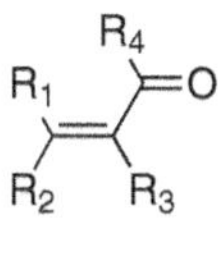	Epoxidation with dimeric *Cinchona* alkaloid phase transfer catalyst; Shi (modified catalyst) epoxidation
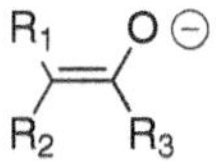	Davis oxaziridine epoxidation

A Final Word on Epoxidation: What's Missing, and Which Method?

If you examine the contents of this section carefully, you will see that there is one method of epoxidation that does not appear in the discussion: epoxidation by chiral organic peracids. The reason for this omission is simple. As of the time of this writing, methods using chiral peracids are generally not giving levels of enantioselectivity competitive with the other methods discussed. Even so, the range of chiral epoxidation methods available has grown in the past two decades into an impressive array of synthetic methods. So, it is worth asking the question, "Which asymmetric epoxidation methods should be used for which type of substrate?" Some ideas can be gleaned from Table 21.4.

Problems

21-5 Suggest the major organic product when each of the alkenes below is oxidized using the catalyst and terminal oxidant specified. Note that the *trans* olefin epoxide may be the major product in some cases.

(a)
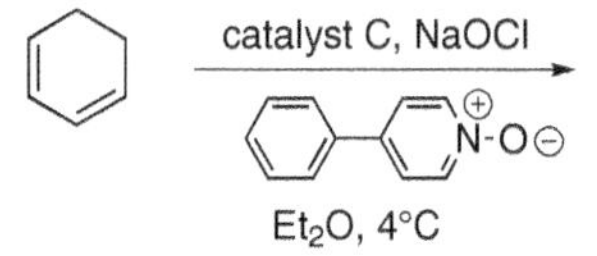

(b)
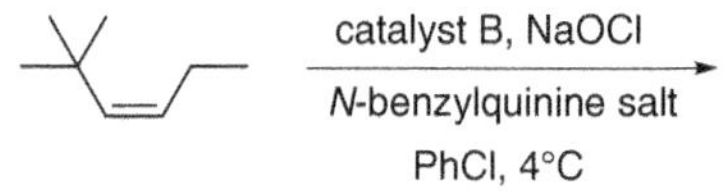

(c)
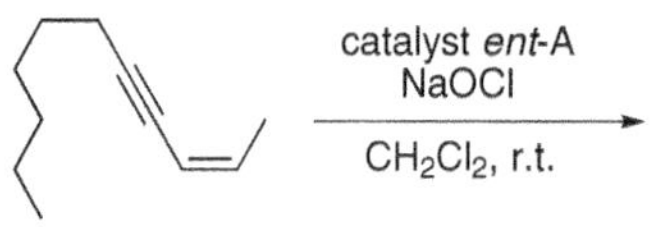

(d)
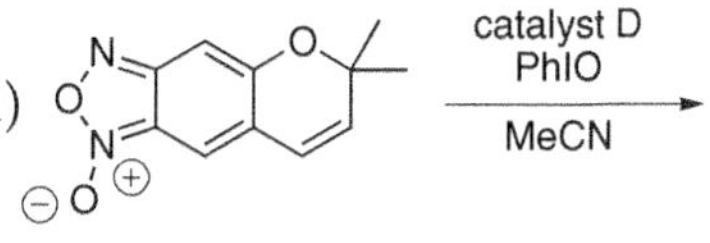

Catalysts:

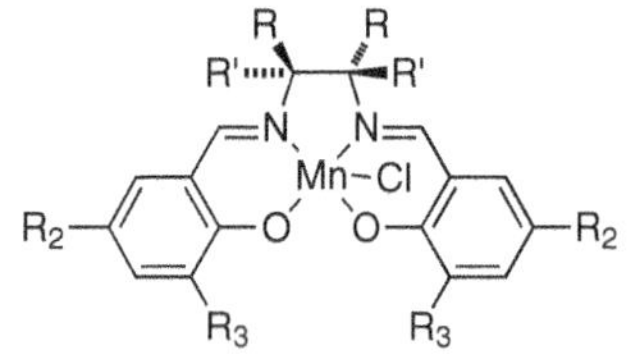

A: R = H; R' = $(CH_2)_4$; $R_2 = R_3 = t$-Bu
B: R = Ph; R' = H; R_2 = OTIPS; $R_3 = t$-Bu
C: R = H; R' = $(CH_2)_4$; R_2 = OTIPS; $R_3 = t$-Bu
D: R = $(CH_2)_4$; R' = H; R_2 = H (Me adjacent to R_3);
R_3 = *R*-EtCHPh

References: (a) *Tetrahedron Lett.* **1994**, *35*, 669. (b) *J. Am. Chem. Soc.* **1994**, *116*, 6937. (c) *Tetrahedron Lett.* **1991**, *32*, 6533. (d) *Coord. Chem. Rev.* **1995**, *140*, 189.

21-6 What is the expected major organic product of the following oxidations using the second-generation Katsuki catalyst **21.23**?

(a)
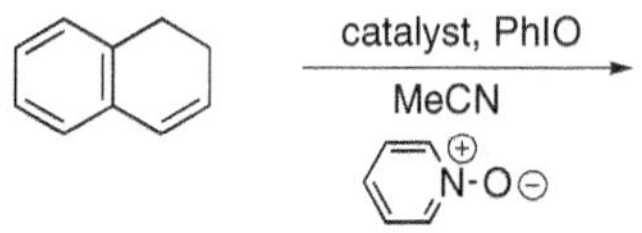

(b)
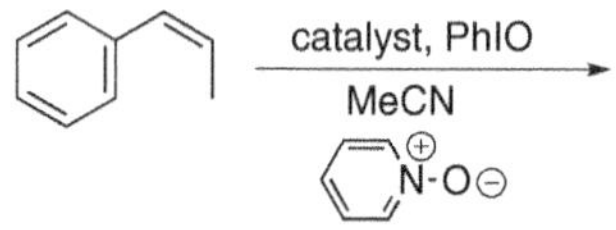

(c)
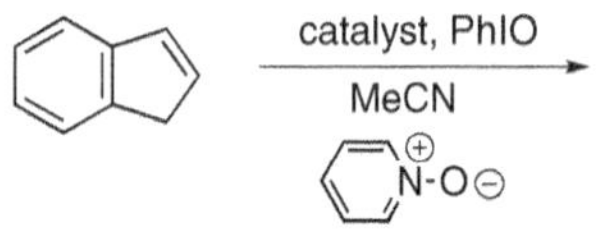

(d)
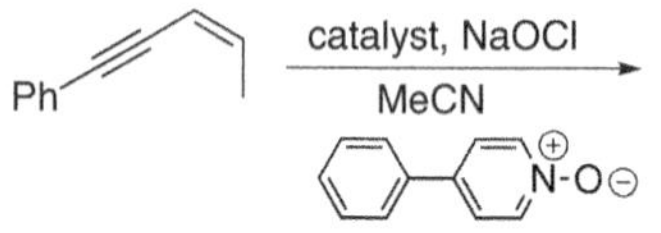

References: (a), (b) *Tetrahedron* **1994**, *50*, 4311. (c), (d) *Tetrahedron* **1994**, *50*, 11827.

21-7 What is the major product expected from each of the following reactions?

(a)
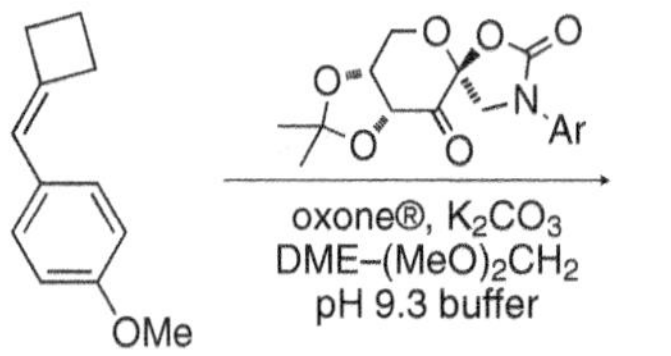

(b)
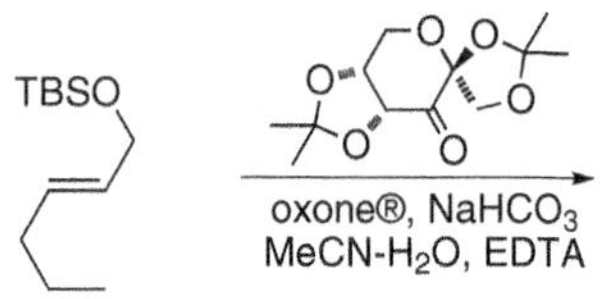

(c)
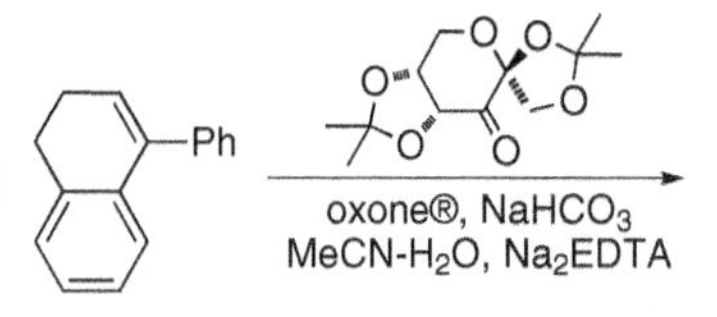

(d)
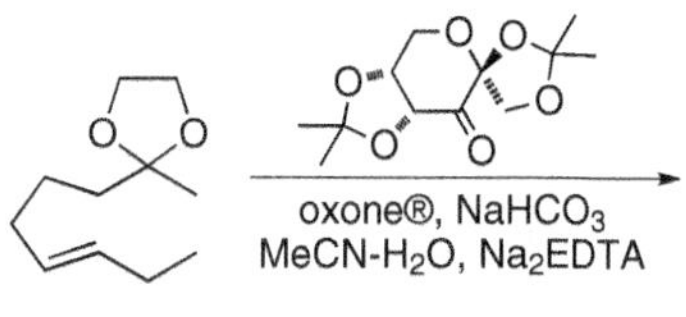

References: (a) *Angew. Chem. Int. Ed.* **2006**, *45*, 1429. (b), (c) *J. Am. Chem. Soc.* **1996**, *118*, 9806. (d) *J. Am. Chem. Soc.* **1997**, *119*, 11224.

21.5 Asymmetric Dihydroxylation

It has been known for many decades that pyridine catalyzes the addition of osmium tetroxide to alkenes[52] and that the osmium tetroxide-pyridine complex reacts selectively with electron-rich alkenes. This observation ultimately led to the development of the asymmetric hydroxylation reaction, in which the addition is catalyzed by C_2-symmetric dihydroquinine or dihydroquinidine dimers.[53] The most effective catalysts are the dimeric species where the two cinchona alkaloid moieties are separated by a phthalazine spacer, abbreviated as $[DHQ]_2PHAL$ for the dihydroquinine-based catalyst and $[DHQD]_2PHAL$ for the dihydroquinidine-based catalyst. The two catalysts give stereochemically complementary results.

These catalysts are commercially available in pre-mixed reagent mixtures with $K_2OsO_2(OH)_4$ (the osmium oxidant), $K_3Fe(CN)_6$ (the secondary oxidant to regenerate the active osmium oxidation state), and K_2CO_3 (to hydrolyze the osmium intermediate). These mixtures are known by the trade names *AD-mix-α* (based on $[DHQ]_2PHAL$) and *AD-mix-β* (based on $[DHQD]_2PHAL$), respectively. Provided that the largest group attached to the double bond can be reasonably easily identified, one may use the simple mnemonic in Figure 21.11 to predict the stereochemistry of hydroxylation. Thus, when the large alkyl group is oriented to the front and on the left-hand side of the double bond, AD-mix-α hydroxylates the lower, or α, face of the π bond, and AD-mix-β hydroxylates the upper, or β, face of the π bond (see Examples **21.51** to **21.54**).

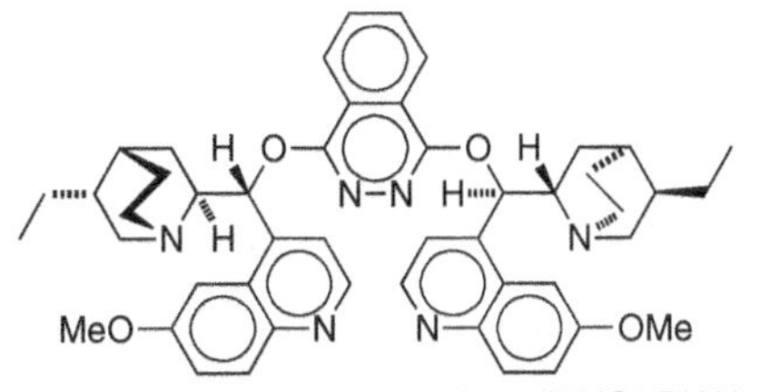

dihydroquinine-phthalazine dimer, $[DHQ]_2PHAL$ ("AD-mix-α")

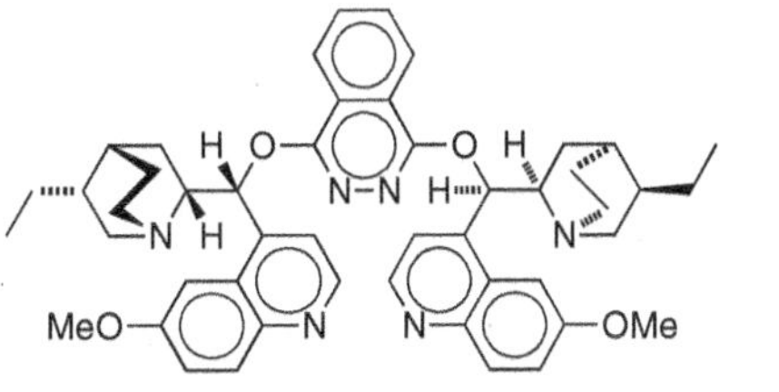

dehydroquinidine-phthalazine dimer, $[DHQD]_2PHAL$ ("AD-mix-β")

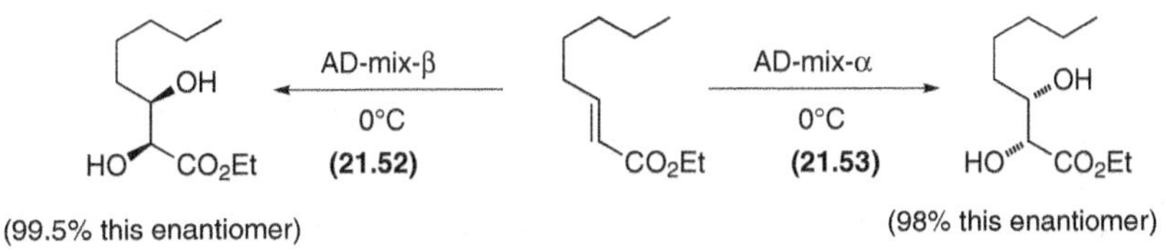

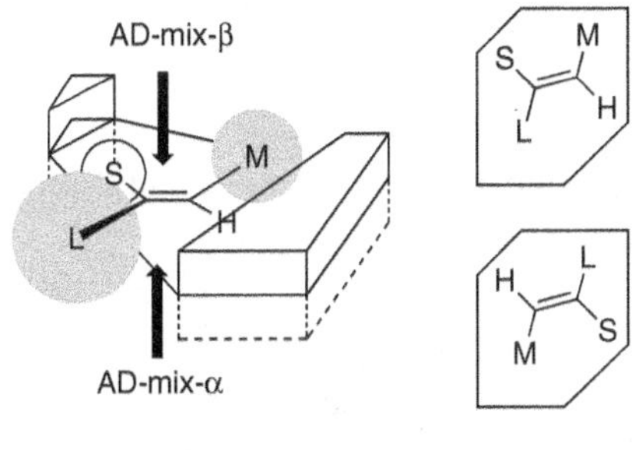

Figure 21.11 Mnemonic for predicting absolute stereochemistry in hydroxylation of alkenes bearing large (L), medium-sized (M), and small (S) groups attached to the double bond

52. Criegee, R.; Marchand, B.; Wannowius, H. *Justus Liebigs Ann. Chem.* **1942**, *550*, 99.

53. (a) Kolb, H.C.; VanNieuwenhze, M.S.; Sharpless, K.B. *Chem. Rev.* **1994**, *94*, 2483. (b) Crispino, G.A.; Jeong, K.-S.; Kolb, H.C.; Wang, D.; Xu, D.; Sharpless, K.B. *J. Org. Chem.* **1993**, *58*, 3785. (c) Crispino, G.A.; Makita, A.; Wang, Z.-M.; Sharpless, K.B. *Tetrahedron Lett.* **1994**, *35*, 543. (d) Sharpless, K.B.; Amberg, W.; Bennani, Y.L.; Crispino, G.A.; Hartung, J.; Jeong, K.S.; Kwong, H.-L.; Morikawa, K.; Wang, Z.-M.; Xu, D.; Zhang, X.-L. *J. Org. Chem.* **1992**, *57*, 2768. (e) Amberg, W.; Bennani, Y.L.; Chada, R.K.; Crispino, G.A.; Davis, W.D.; Hartung, J.; Jeong, K.-S.; Ogino, Y.; Shibata, T.; Sharpless, K.B. *J. Org. Chem.* **1993**, *58*, 844. (f) Becker, H.; King, S.B.; Taniguchi, M.; Vanhessche, K.P.M.; Sharpless, K.B. *J. Org. Chem.* **1995**, *60*, 3940.

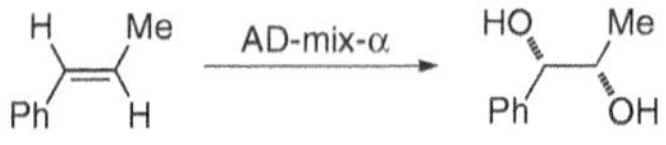

(21.54)

(98.5% this enantiomer)

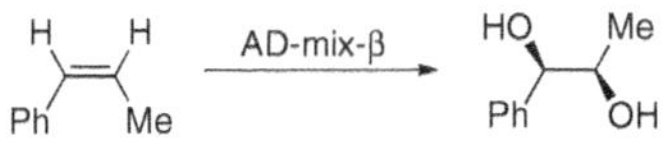

(21.55)

(64.5% this enantiomer)

Alkene structure plays an important role in determining the stereochemistry of the hydroxylation. The C_2 stereochemistry of the ligand imposes a preferred geometry for the alkene, so that alkenes with an approximate C_2 axis of symmetry tend to give higher levels of chiral induction. The mnemonic model shows that there are two reasonable orientations of the alkene in the complex and that both of these give the same absolute configuration in the dihydroxylation reaction. More importantly, perhaps, this mnemonic predicts that in alkenes where the two larger groups are on the same side of the alkene, the enantioselectivity should be lower. This is reflected in the values shown at left for some typical alkenes, where R_S is the smallest and R_L is the largest alkyl group.[54]

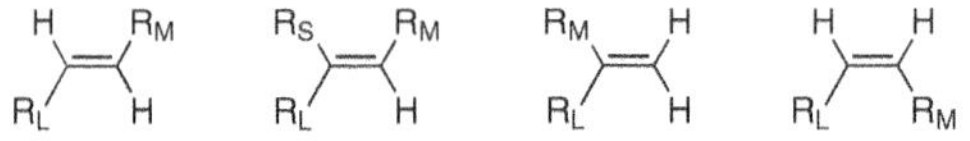

99.5:0.5	90:10	85:15	65:35

Reaction Synopses

Asymmetric Dihydroxylation of Alkenes

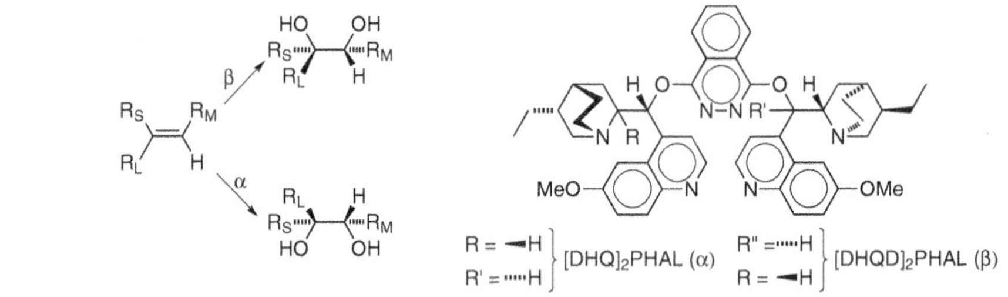

Reagents: L*, $K_2OsO_4(OH)_2$, $K_3Fe(CN)_6$; L*, K_2CO_3
L* = $(DHQ)_2$PHAL (AD-mix-α) or $(DHQD)_2$PHAL (AD-mix-β)

Problem

21-8 What is the major product expected from each of the following reactions?

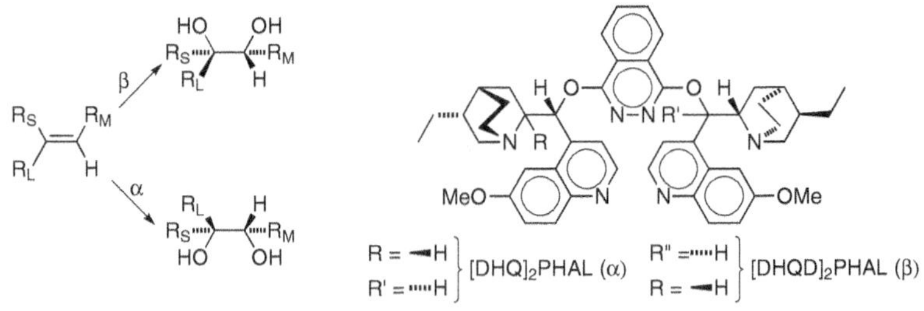

54. These values are a synopsis of results compiled in: Johnson, R.A.; Sharpless, K.B. In Ojima, I, Ed. *Catalytic Asymmetric Synthesis*, 2nd. ed (Wiley-VCH: New York, 2000), p. 375 (especially Tables 6D.2–6D.6)

(b) 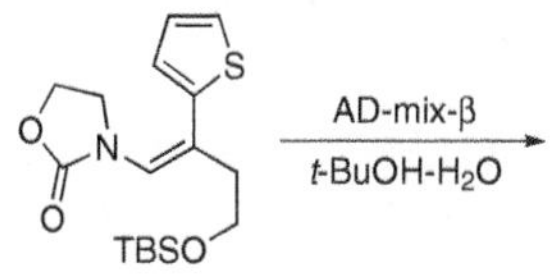

(c) 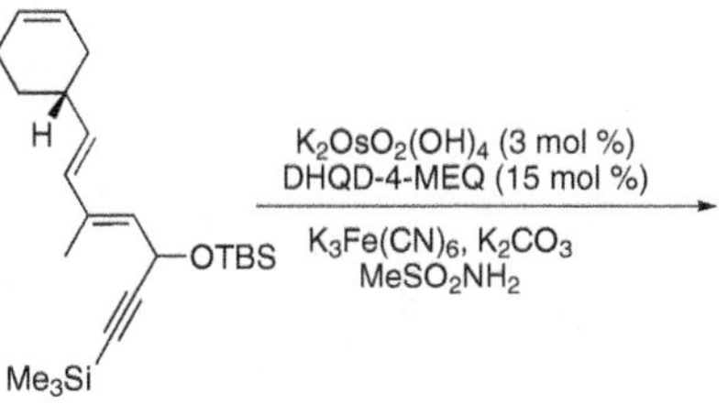

(d) 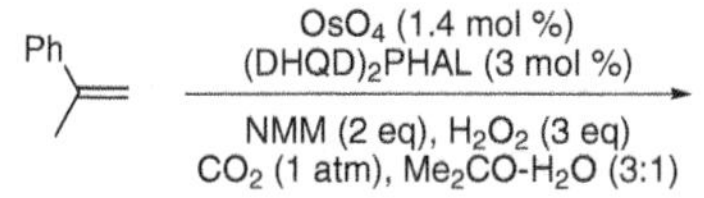

(e) 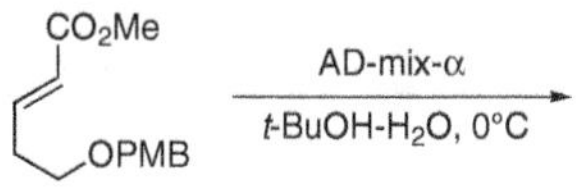

(f) 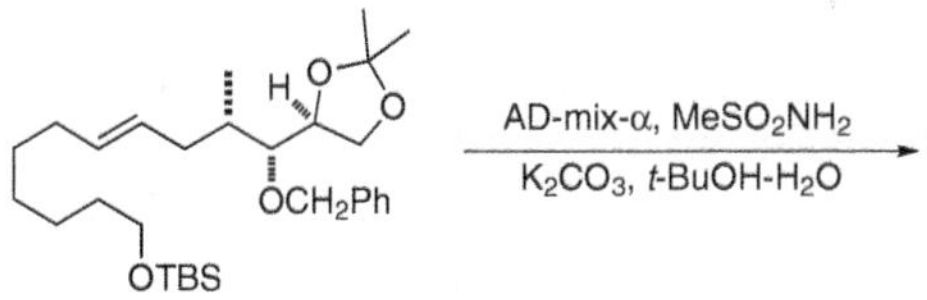

References: (a) *Angew. Chem. Int. Ed.* **2005**, *44*, 580. (b) *Angew. Chem. Int. Ed.* **2010**, *49*, 8733. (c) *Org. Lett.* **2009**, *11*, 5498 (see also: *J. Org. Chem.* **1991**, *56*, 4585). (d) *Tetrahedron Lett.* **2008**, *49*, 1071. (e) *Chem. Pharm. Bull.* **2007**, *55*, 1284. (f) *Org. Lett.* **2007**, *9*, 4757.

21.6 **Kinetic Resolution of Alcohols**

Because many synthetic reactions require a carbonyl compound as one participant, the oxidation of alcohols has long been a pivotal process in organic chemistry. The ability to incorporate chirality in this process, however, does not lie in the generation of a new chiral center. It lies in the selective destruction of an existing chiral center (as in kinetic resolution of racemates, or desymmetrization of *meso* compounds) to reveal underlying chirality.

Kinetic Resolution of Racemates

Kinetic resolution works only when the reaction is not carried out to completion, so that the unchanged reactant gradually increases in enantiomeric excess (Figure 21.12). The enantiomeric excess in the residual reactant depends on the relative rates of the two competing reactions and the degree of conversion of reactant to product. The larger the ratio of rates for the two competing reactions, the more efficient the kinetic resolution and the closer to 50% conversion the reaction can be stopped (i.e., the higher the yield of enantiomerically enriched compound). The lower this ratio, the closer to completion the reaction must be taken to achieve good levels of enantiomeric excess, as illustrated by Figure 21.12. To take typical values from Figure 21.12, for a reaction at 50% completion, the e.e. of the residual reactant is 24% if the rate ratio is 2:1, 52% if 5:1, 67% if 10:1, 79% if 20:1, and 89% if 50:1.[55] Generally speaking, rate ratios above 10:1 are necessary to achieve good levels of enantioselectivity (e.e. > 90% at 55%–60% conversion) in the product and reasonable recovery of the enantiomerically enriched starting compound.

55. For a full mathematical treatment of the relationship between the enantiomeric excess (e.e.), and the extent of conversion (C), which gives the stereoselectivity factor *s*, defined as $s = (k_{fast}/k_{slow}) = \ln[(1 - C)(1 - e.e,)]/\ln[(1 - C)(1 + e.e.)]$, see: Kagan, H.B.; Fiaud, J.C. *Top. Stereochem.* **1988**, *18*, 249.

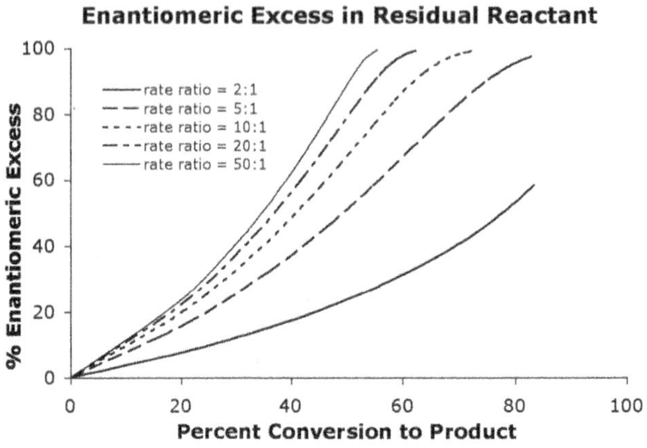

Figure 21.12 Effect of rate ratio on the enantiomeric excess of the residual reactant in a kinetic resolution

Hydrolytic Kinetic Resolution: Epoxides

Although it is not formally an asymmetric oxidation reaction, one useful advance in the synthesis of optically active diols and epoxides has been the development of hydrolytic kinetic resolution of epoxides by Jacobsen and his coworkers.[56] In this reaction, a cobalt (III) salen complex (Figure 21.13; M=Co) catalyzes the S_N2 ring opening of the epoxide with water as the limiting reagent to give a highly enantiomerically enriched diol, leaving unchanged a highly enantiomerically enriched epoxide. The preferred binding of the epoxide to the catalyst can be predicted using the mnemonic given below the complex in Figure 21.13. The active form of the complex is usually prepared in situ by air oxidation of the (salen)cobalt (II) precursor in the presence of acetic acid or *p*-toluenesulfonic acid. The use of this technique is illustrated by the hydrolytic kinetic resolution of methyl glycidate with the (*R,R*) catalyst. This catalyst preferentially promotes the hydrolysis of the *R* epoxide to yield methyl (*S*)-glycidate (> 99% e.e. at 70% conversion; Example **21.56**). In addition to water, other nucleophiles have been used to open epoxides in the presence of a chiral cobalt (III) salen complex—as in Example **21.57**, where the epoxide ring is opened by trimethylsilyl azide in the presence of a chromium (III) salen catalyst to give the chiral azide.

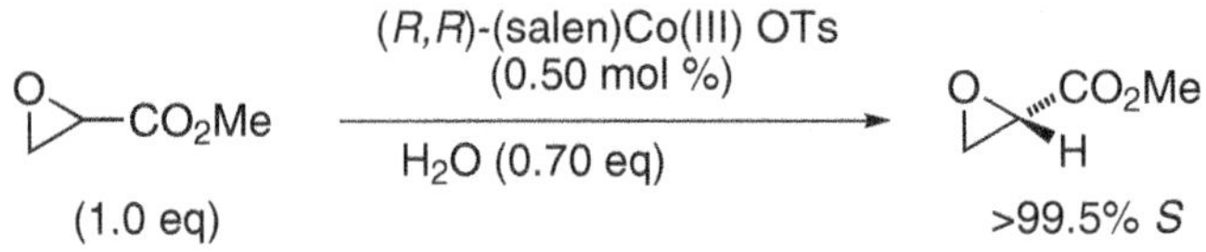

(21.56)[57]

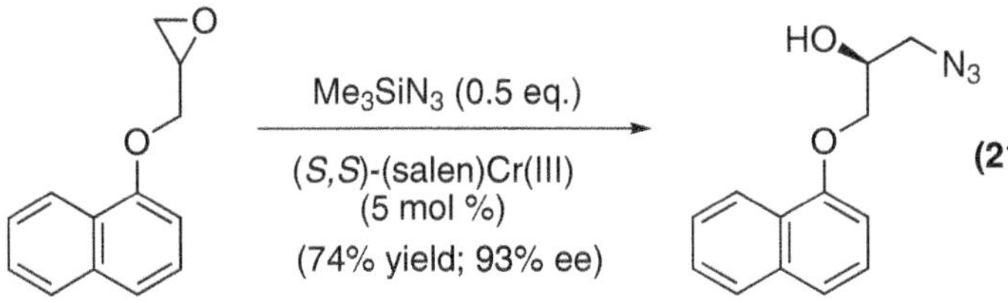

(21.57)[58]

56. (a) Tokunaga, M.; Larrow, J.F.; Kakiuchi, F.; Jacobsen, E.N. *Science* **1997**, *277*, 936. (b) Furrow, M.E.; Schaus, S.E.; Jacobsen, E.N. *J. Org. Chem.* **1998**, *63*, 6776. (c) Schaus, S.E.; Brandes, B.D.; Jacobsen, E.N. *J. Am. Chem. Soc.* **2002**, *124*, 1307. (d) Nielsen, L.P.C.; Stevenson, C.P.; Blackmond, D.G.; Jacobsen, E.N. *J. Am. Chem. Soc.* **2004**, *126*, 1360. (e) Schaus, S.E.; Brånalt, J.E.; Jacobsen, E.N. *J. Org. Chem.* **1998**, *63*, 4876. (f) Liu, P.; Panek, D.S. *J. Am. Chem. Soc.* **2000**, *122*, 1235. (g) Chavez, D.E.; Jacobsen, E.N. *Angew. Chem.Int. Ed. Engl.* **2001**, *40*, 3667. (h) Holland, J.M.; Lewis, M.; Nelson, A. *Angew. Chem. Int. Ed. Engl.* **2000**, *40*, 4082. (i) Bode, J.W.; Carreira, J.M. *J. Org. Chem.* **2001**, *66*, 6410. (j) Fürstner, A.; Thiel, O.R.; Ackermann, L. *Org. Lett.* **2001**, *3*, 449.

57. Stevenson, C.P.; Nielsen, L.P.C.; Jacobsen, E.N. *Org. Syn.* **2006**, *83*, 162.

58. Larrow, J.F.; Schaus, S.E.; Jacobsen, E.N. *J. Am. Chem. Soc.* **1996**, *118*, 7420.

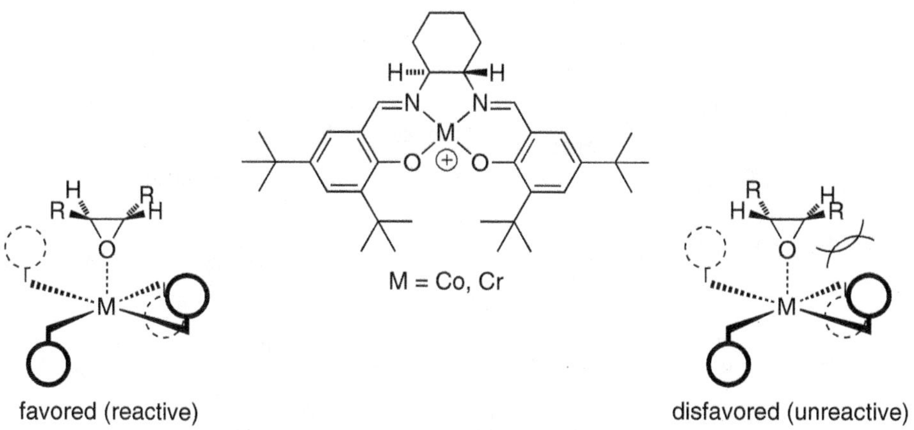

Figure 21.13 (*S*,*S*)-(salen) catalysts and a mnemonic model for the binding of the epoxide to the catalyst

Example **21.58** shows just how useful this can be: the mixed isomers (± and *meso*) of the diepoxide react with water to give three chiral products, each with known configuration. The epoxide rings in the products all have the *R* configuration at the chiral center, indicating that the *S* epoxide is hydrolyzed preferentially. In the diol moieties of the alcohol products, the secondary hydroxyl group has the *S* configuration, consistent with S_N2 ring opening through an S_N1-like activated complex.

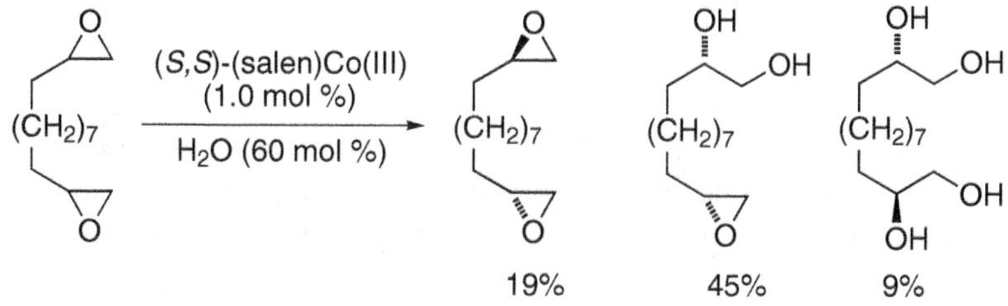

(21.58)[59]

Oxidative Kinetic Resolution: Alcohols

The use of environmentally more benign oxidants has been coupled with the effort to effect efficient asymmetric oxidations. One of the most heavily studied methods for the oxidation of secondary alcohols with molecular oxygen involves the use of a palladium (II) catalyst and a base, especially in the form of the Uemura oxidation.[60]

This method has been adapted successfully[61] to the kinetic resolution of racemic alcohols. By the oxidation of secondary benzylic alcohols with molecular oxygen using the chiral diamine, sparteine, as the chiral adjuvant.[62] The generally accepted catalytic cycle for this oxidation is shown in Figure 21.14. The catalytic cycle involves the displacement of one of the ligands on palladium to give the complex **21.61**, followed by β-hydride elimination to give the ketone (**21.63**) and the hydridopalladium (II) complex (**21.64**) that eliminates HX to give a palladium (0) complex. There are two pathways for the regeneration of the palladium (II)

59. Chow, S.; Kitching, W. *Tetrahedron: Asymmetry* **2002,** *13,* 779.

60. (a) Nishimura, T.; Onoue, T.; Ohe, K.; Uemura, S. *Tetrahedron Lett.* **1998,** *39,* 6011. (b) Peterson, K.C.; Larock, R.C. *J. Org. Chem.* **1998,** *63,* 3185.

61. (a) Jensen, D.R.; Pugsley, J.S.; Sigman, M.S. *J. Am. Chem. Soc.* **2001,** *123,* 7475. (b) Ferreira, E.M.; Stoltz, B.M. *J. Am. Chem. Soc.* **2001,** *123,* 7725.

62. Reviews: (a) Muzart, J. *Tetrahedron* **2003,** *59,* 5789. (b) Stahl, S.S. *Angew. Chem. Int. Ed.* **2004,** *43,* 3400. (c) Ikariya, T.; Murata, K.; Noyori, R. *Org. Biomol. Chem.* **2006,** *4,* 393. (d) Sigman, M.S.; Jensen, D.R. *Acc. Chem. Res.* **2006,** *39,* 221.

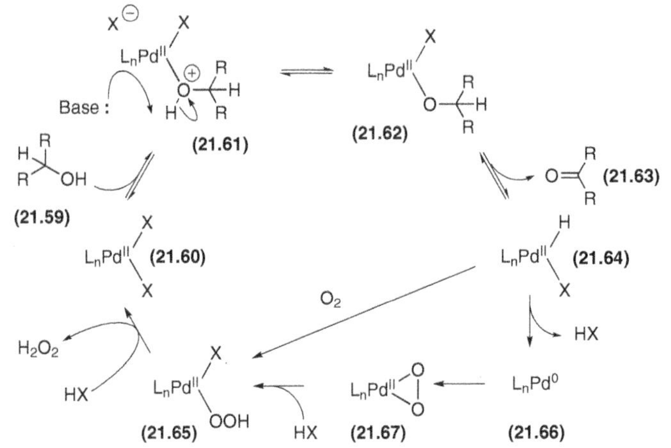

Figure 21.14 Putative catalytic cycle for the oxidation of a secondary alcohol by molecular oxygen, catalyzed by a palladium (II) complex

form of the catalyst by molecular oxygen: (1) by direct insertion of the O_2 molecule into the Pd–H bond to give **21.65** directly or (2) through a peroxometallocycle (e.g., **21.67**).

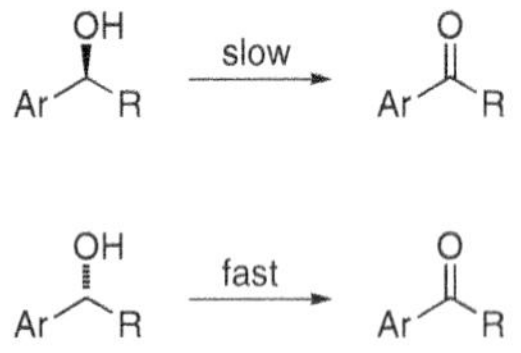

(21.68)

(21.69)

When (–)-sparteine is used as the adjuvant, the *R* enantiomer of the benzyl alcohol is oxidized (Example **21.69**) more rapidly under the reaction conditions (Example **21.68**). The result is that the residual alcohol becomes enriched in the *S* enantiomer, as shown by Example **21.70**. Sparteine acts as more than just a chiral ligand for the palladium in this reaction. It has been observed that the pre-formed complex from (–)-sparteine and palladium (II) chloride does not oxidize 1-phenylethanol in the absence of free sparteine.[63] The same approach to selective oxidation of *meso* diols leads to enantiomeric enrichment in the monohydroxy product, as shown by the oxidation of *meso*-1,3-diphenylpropane-1,3-diol (Example **21.71**).[64] Using chloroform as solvent and adding cesium carbonate to the reaction mixture leads to an increase in the selectivity of the reaction, especially for allylic alcohols (e.g., Example **21.72**), which do not show the same levels of selectivity as benzylic alcohols. Allowing the use of air as the oxidant for benzylic alcohols, rather than pure oxygen, helps as well.[65]

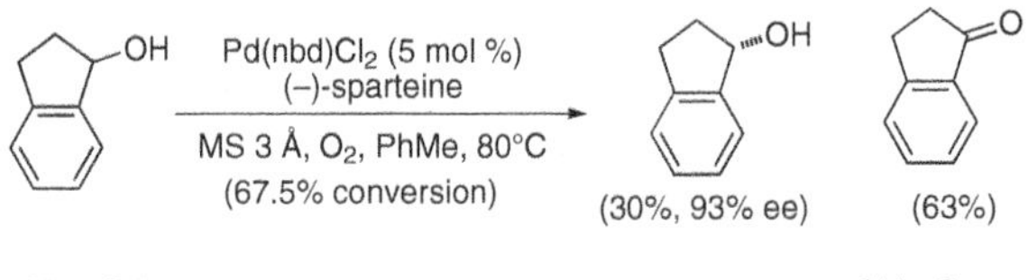

(21.70)[66]

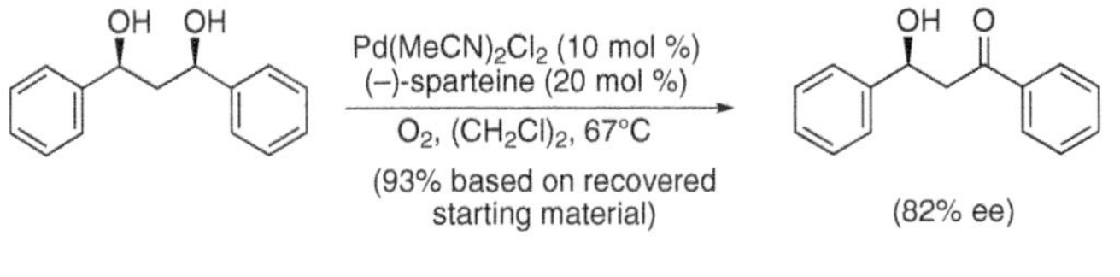

(21.71)[67]

63. Mueller, J.A.; Jensen, D.R.; Sigman, M.S. *J. Am. Chem. Soc.* **2002**, *124*, 8202.
64. Jensen, D.R.; Pugsley, J.S.; Sigman, M.S. *J. Am. Chem. Soc.* **2001**, *123*, 7475.
65. Bagdanoff, J.T.; Stoltz, B.M. *Angew. Chem. Int. Ed.* **2004**, *43*, 353.
66. Ferreira, E.M.; Stoltz, B.M. *J. Am. Chem. Soc.* **2001**, *123*, 7725.
67. Jensen, D.R.; Pugsley, J.S.; Sigman, M.S. *J. Am. Chem. Soc.* **2001**, *123*, 7475.

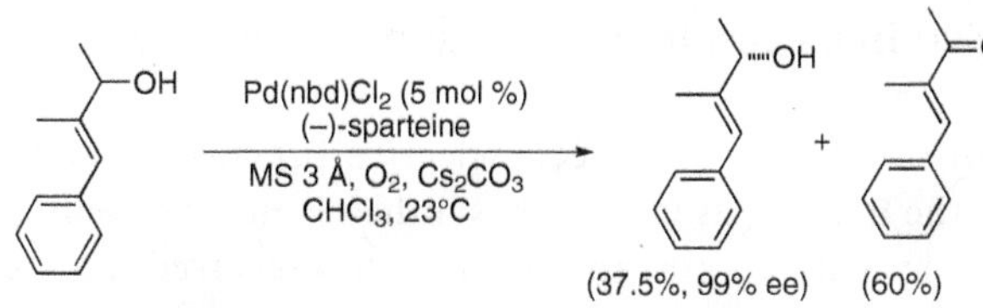

(37.5%, 99% ee) (60%)

(21.72)[68]

A similar reaction involves the use of high-valent vanadium as the catalytic oxidant, with oxygen as the stoichiometric oxidant (Example **21.73**). In this reaction, vanadyl tri-isopropoxide is treated with a chiral salen ligand derived from *tert*-leucine to generate the chiral reagent. The reagent leads to high e.e. (90%–99%) in the unreacted starting compound at relatively low (52%–62%) levels of conversion.[69]

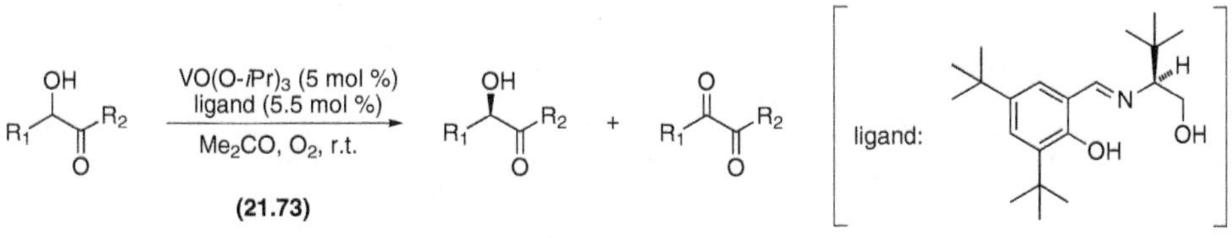

(21.73)

Reaction Synopses

Hydrolytic Kinetic Resolution of Epoxides

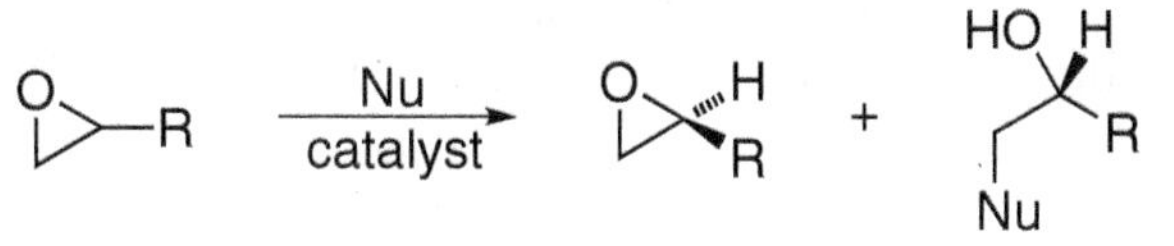

Reagents:
 Catalyst: chiral (salen)Co(III) complex (0.5 mol %)
 Nu: H—OH (70 mol %); Me_3SiN_3; etc.
 No solvent used with liquid epoxides

Enantioselective Oxidation of Secondary Alcohols (Pd)

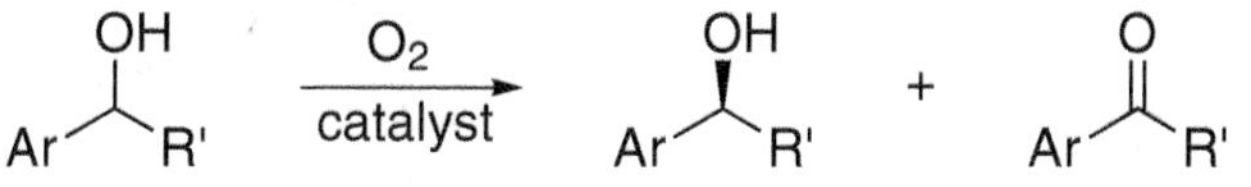

Reagents:
 Catalyst: $Pd(MeCN)_2Cl_2$; $Pd(nbd)Cl_2$; etc.
 Adjuvants: (–)-sparteine, other chiral diamines perform poorly; $CsCO_3$
 Solvent: PhMe, $(CH_2Cl_2)_2$, $CHCl_3$

Enantioselective Oxidation of Secondary Alcohols (V)

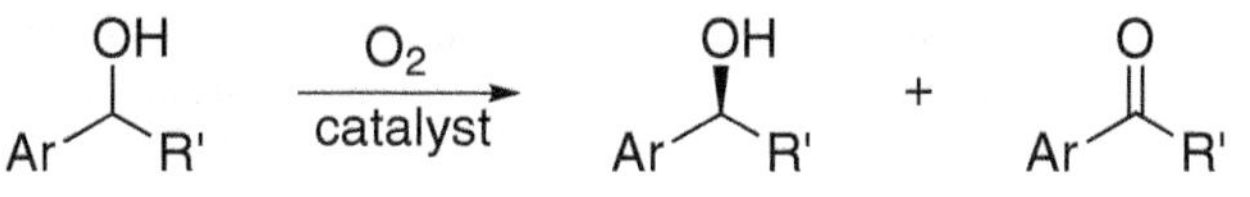

Reagents:
 Catalyst: $VO(O\text{-}i\text{-}Pr)_3$;
 Adjuvants: chiral salen ligand based on *tert*-leucine
 Solvent: MeCN, Me_2CO; etc.

---

68. Bagdanoff, J.T.; Stoltz, B.M. *Angew. Chem. Int. Ed.* **2004**, *43*, 353.
69. Radosevich, A.T.; Musich, C.; Toste, F.D. *J. Am. Chem. Soc.* **2005**, *127*, 1090.

21.7 Enantioselective Insertion into C—C and C—H Bonds

As specified earlier in this book, the direct, selective functionalization of C—H σ bonds remains one of the holy grails of organic synthesis, and the enantioselective replacement of hydrogen by both carbon and nitrogen has received intense study since the turn of the 21st century.[70] The status of nitrogen insertion into C—H σ bonds received a significant boost when Che[71] and Du Bois[72] found that sulfamate esters would react in the presence of a rhodium (II) catalyst and phenyliodine diacetate to give products of insertion with complete retention of configuration at carbon (e.g., Example **21.74**). This reaction has been elegantly applied by Du Bois in the synthesis of complex natural products such as tetrodotoxin[73] and saxitoxin.[74]

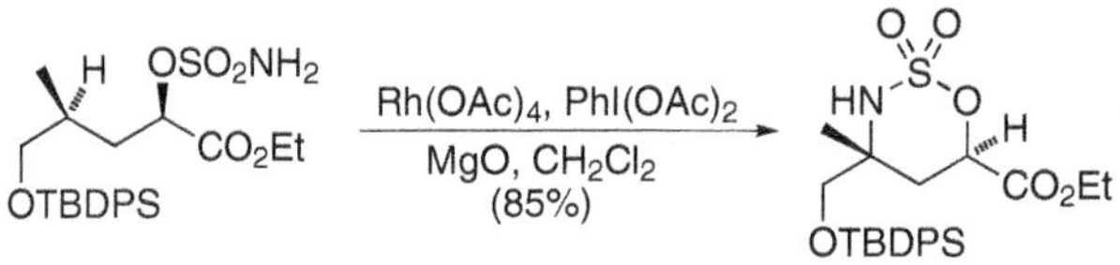

(21.74)

The reaction has been modified into enantioselective form by using the chiral complex $Rh_2(TCPTAD)_4$ as the catalyst, with *p*-nitrobenzenesulfonamide as the nitrogen donor and phenyliodine diacetate as the oxidant to generate the nitrene (Example **21.75**). With this reaction system, Davies has obtained enantiomerically enriched amines in e.e. values from 43% to above 90%.[75] Although the same insertion reaction can be catalyzed by chiral copper complexes, the levels of enantioselectivity are still below useful levels.[76] Another rhodium complex, $Rh_2(S\text{-nap})_4$, has also shown promise as a catalyst for the intramolecular asymmetric amination of alkanes (Example **21.76**).[77]

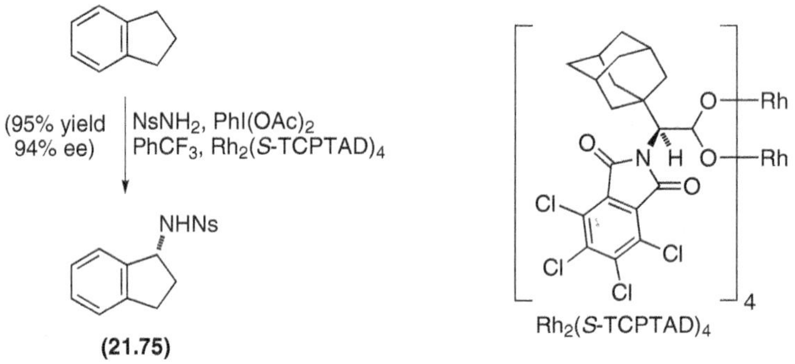

(21.75)

70. Reviews: (a) Müller, P.; Fruit, C. *Chem. Rev.* **2003**, *103*, 2905. (b) Davies, H.M.L.; Long, M.S. *Angew. Chem. Int. Ed.* **2005**, *44*, 3518. (c) Dick, A.R.; Sanford, M.S. *Tetrahedron* **2006**, *62*, 2439. (d) Davies, H.M.L.; Manning, J.R. *Nature* **2008**, *451*, 417. (e) Díaz-Requejo, M.M.; Pérez, P.J. *Chem. Rev.* **2008**, *108*, 3379. (f) Zalatan, D.N.; Du Bois, J. *Top. Curr. Chem.* **2010**, *292*, 347. (g) Davies, H.M.L.; Morton, D. *Chem. Soc. Rev.* **2011**, *40*, 1857.

71. (a) Liang, J.-L.; Yuan, S.-X.; Chan, P.W.H.; Che, C.-M. *Org. Lett.* **2002**, *4*, 4507. (b) Yu, X.-Q.; Huang, J.-S.; Zhou, X.-g.; Che, C.-M. *Org. Lett.* **2000**, *2*, 2233. (c) Au, S.-M.; Zhang, S.-B.; Fung, W.-H.; Yu, X.-Q.; Che, C.-M.; Cheung, K.-K. *Chem. Commun.* **1998**, 2677. (d) Au, S.-M.; Huang, J.-S.; Che, C.-M.; Yu, W.-Y. *J. Org. Chem.* **2000**, *65*, 7858.

72. (a) Espino, C.G.; Wehn, P.M.; Chow, J.; Du Bois, J. *J. Am. Chem. Soc.* **2001**, *123*, 6935. (b) Espino, C.G.; Du Bois, J. *Angew. Chem., Int. Ed.* **2001**, *40*, 598. (c) Fiori, K.W.; Fleming, J.J.; Du Bois, J. *Angew. Chem., Int. Ed.* **2004**, *43*, 4349. (d) Espino, C.G.; Du Bois, J. *Angew. Chem., Int. Ed.* **2001**, *40*, 598. (e) Kim, M.; Mulcahy, J.V.; Espino, C.G.; Du Bois, J. *Org Lett.* **2006**, *8*, 1073. (f) Espino, C.G.; Fiori, K.W.; Kim, M.; Du Bois, J. *J. Am. Chem. Soc.* **2004**, *126*, 15378. (f) Wehn, P.M.; Lee, J.; Du Bois, J. *Org. Lett.* **2003**, *5*, 4823.

73. Hinmann, A.; Du Bois, J.H. *J. Am. Chem. Soc.* **2003**, *125*, 11510.

74. Fleming, J.J.; Du Bois, J.H. *J. Am. Chem. Soc.* **2006**, *128*, 3926.

75. Reddy, R.P.; Davies, H.L.M. *Org. Lett.* **2006**, *8*, 5013.

76. Barman, D.N.; Nicholas, K.M. *Tetrahedron Lett.* **2010**, *51*, 1815.

77. Zalatan, D.N.; Du Bois, J. *J. Am. Chem. Soc.* **2008**, *130*, 9220.

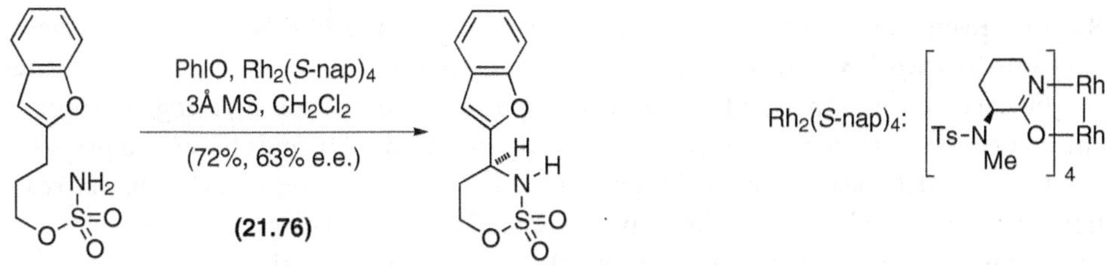

(21.76)

The one oxidation where a heteroatom is inserted into a C—C bond that has been the subject of significant research to develop an enantioselective form of the reaction is the Baeyer-Villiger reaction. Although much progress has been made toward an asymmetric Baeyer-Villiger reaction, good enantioselectivity has only been obtained to date with cyclobutanones (e.g., Example **21.77**), and even then, the enantioselectivity still falls far short of the levels attainable with enzymes. Thus, this area of organic synthesis must still be considered a "work in progress."

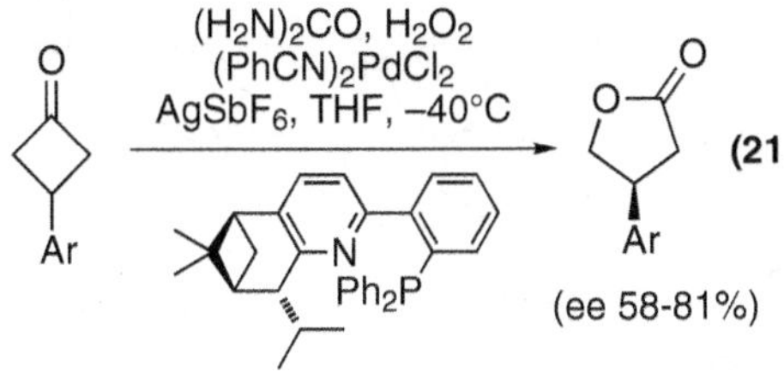

(21.77)[78]

Reaction Synopses

Asymmetric Nitrene Insertion into C—H Bonds

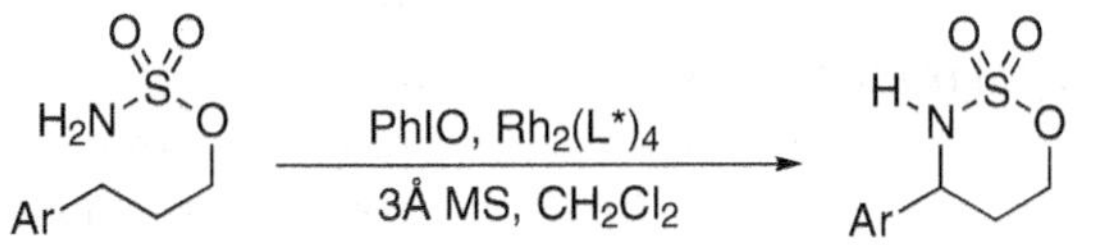

Configuration of product: L* = *S*-TCPTAD, *R*; L* = *S*-nap, *S*

21.8 Asymmetric Hydrogenation

The stereochemical control available with homogeneous catalytic hydrogenation had a major impact in organic synthesis by making possible the technique of ***asymmetric hydrogenation.***[79] This area was pioneered by French chemist Henri

78. Ito, K.; Ishiia, A.; Kuroda, T.; Katsuki, T. *Synlett* **2003**, 643.

79. Reviews and monographs: (a) Noyori, R.; Kitamura, M. In Scheffold, R., Ed. *Modern Synthetic Methods* (Springer: Berlin, 1989), vol. 5, p. 115. (b) Takaya, H.; Ohta, T.; Noyori, R. In Ojima, I., Ed. *Catalytic Asymmetric Synthesis* (VCH: New York, 1993), ch. 1., p. 1. (c) Noyori, R. *Asymmetric Catalysis in Organic Synthesis* (Wiley: New York, 1994), ch. 2. (d) Genêt, J.-P. In Abdel-Magid, A.F. (Ed.) *Reductions in Organic Synthesis: Recent Advances and Practical Applications, ACS Symp. Ser. 641* (American Chemical Society: Washington, D.C., 1996), ch. 2. (e) Albrecht, J.; Nagel, U. *Angew. Chem. Int. Ed. Engl.* **1996**, *35*, 407. (f) Halpern, J.A. In Morrison, J.D., Ed. *Asymmetric Synthesis* (Academic Press: New York, 1985), vol. 5, p. 41. (g) Koenig, K.H. In Morrison, J.D., Ed. *Asymmetric Synthesis* (Academic Press: New York, 1985), vol. 5, p. 71. (h) Seyden-Penne, J. *Chiral Auxiliaries and Ligands in Asymmetric Synthesis* (Wiley-Interscience: New York, 1995), ch. 7.1, p. 367. (i) Ohkuma, T.; Kitamura, M.; Noyori, R. In Ojima, I., Ed. *Catalytic Asymmetric Synthesis, 2nd Ed.* (Wiley-VCH: New York, 2000), ch. 1, p. 1. (j) Knowles, W.S. *Acc. Chem. Res.* **1983**, *16*, 106. (k) Noyori, R.; Takaya, H. *Acc. Chem. Res.* **1990**, *23*, 345. (l) Blystone, S.L. *Chem. Rev.* **1989**, *89*, 1663. (m) Knowles, W.S. *Angew. Chem. Int. Ed. Engl.* **2002**, *41*, 1998. (n) Noyori, R. *Angew. Chem. Int. Ed. Engl.* **2002**, *41*, 2008. (o) Shang, G.; Li, W.; Zhang, X, In Ojima, I., Ed. *Catalytic Asymmetric Synthesis, 3rd ed.* (John Wiles & Sons: New York, 2010), ch. 7, p. 343.

Kagan.[80] Ryoji Noyori[81] and William S. Knowles[82] shared the 2001 Nobel Prize in Chemistry for their work on this technique. Today, the asymmetric hydrogenation of alkenes can be carried out with very high levels of asymmetric induction by using appropriate chiral ligands on the appropriate metal. Literally, hundreds of ligands have been prepared and tested, which means that there is an abundance of information available on this reaction, as is evidenced by the reviews in Reference 79. Four major groups of bidentate ligands have been the major focus of research: (1) the C_2-symmetric chiral diphosphines with a flexible backbone, where the chirality is centered in backbone (DIOP) or on phosphorus (DIPAMP); (2) the C_2-symmetric 1,1'-binaphthyl-based diphosphines (BINAP); (3) the C_2-symmetric *bis*-phospholane ligands, where the chirality is centered in the chiral 2,5-dialkylphospholane rings (DuPHOS); and (4) ligands where the two chelating atoms are different (PHOX).

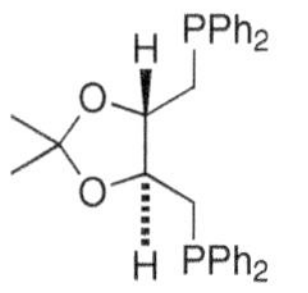

(21.78)
(*S,S*)-DIOP

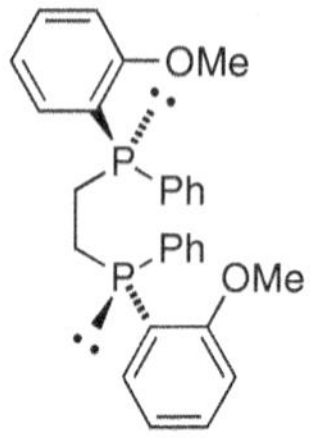

(21.79)
(*R,R*)-DIPAMP

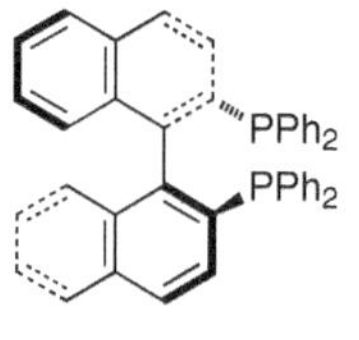

(21.80)
(*S*)-BINAP

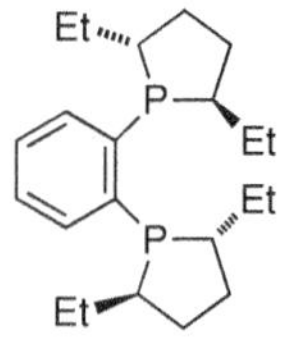

(21.81)
(*R,R*)-Et-DuPHOS

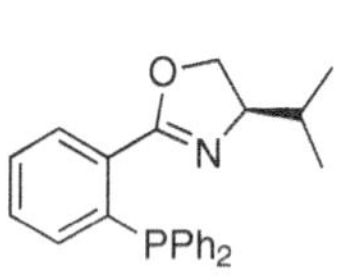

(21.82)
(*R*)-PHOX

Catalysts based on rhodium and ruthenium have been most widely applied. The synthesis of derivatives of *N*-acetylphenylalanine and other *N*-acetyl-α-amino acids by hydrogenation of the corresponding *Z* acrylic acid derivatives has long been a benchmark reaction for testing the efficacy of new chiral ligands. An initial idea of the scope of this reaction may be gauged from the list of catalysts in Table 21.5.[83]

The origins of the chiral induction during asymmetric hydrogenation of α-amidocinnamic acids over rhodium chelated by chiral diphosphine ligands may be accounted for by the asymmetric arrangement of the aryl groups attached to phosphorus in the complex (Example **21.83**). In the case of C_2-symmetric diphosphines, one aromatic ring (shaded) at each phosphorus occupies a quasiaxial position. It has been shown that the enantioselectivity of the reaction increases as the "axial" group becomes larger, and more nearly in the classical axial orientation.[84]

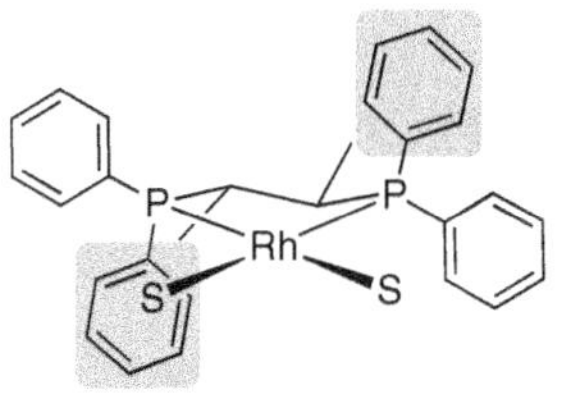

(21.83)

80. **Henri Boris Kagan (1930–)** was educated at the Collège de France in (PhD, 1960). After postdoctoral study with A. Horeau in Paris and with T. Mabry at Texas, he spent his career at the Université de Paris-Sud, in Orsay, France, where he spent almost 40 years. Kagan was a pioneer in the study of chiral induction, absolute reaction stereochemistry, and asymmetric catalysis. He received the Wolf Prize in 2001. For more details, see: *Chem. Rev.*, **1992**, *92*, 1007; *Tetrahedron* **2001**, *57*, vii; *Tetrahedron:Asymmetry* **2010**, *21*, 1017.

81. **Ryoji Noyori (1938–)**: see biographical note in Chapter 2.

82. **William Standish Knowles (1917–2012)**: see biographical note in Chapter 2.

83. For extensive tabulations of data on ligands and reactions, see refs. 77 (i) and 77 (l).

84. Nagel, U.; Rieger, B. *Organometallics* **1989**, *8*, 1534.

Table 21.5 Asymmetric Hydrogenation of *Z*-2-(*N*-Acetylamino)cinnamic Acid with Representative Rh(I) Complexes

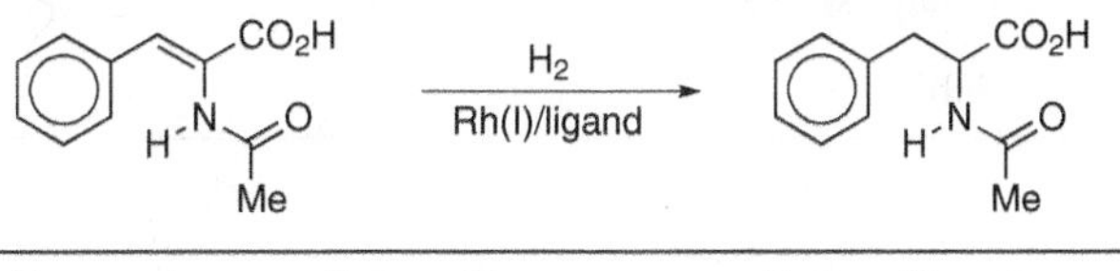

Catalyst	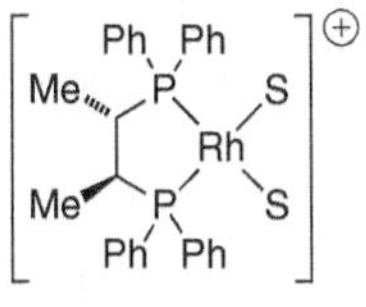 (S,S)-CHIRAPHOS	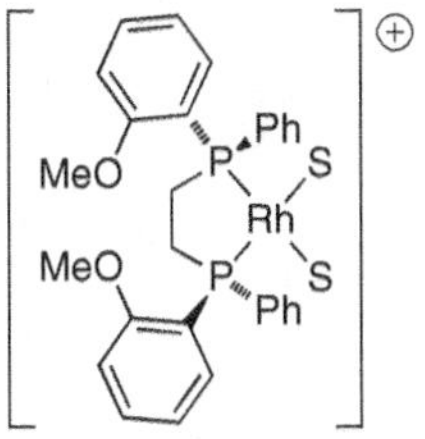 (R,R)-diPAMP	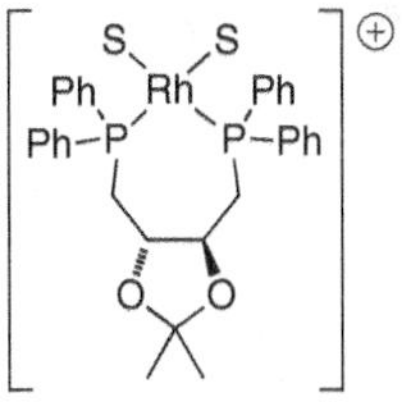 (R,R)-DIOP	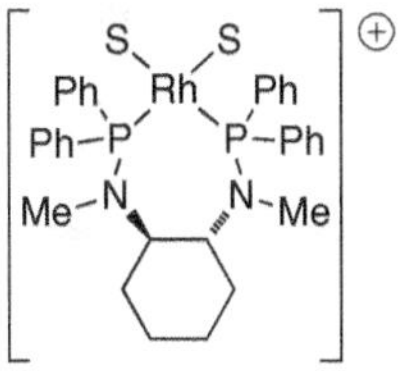 (R,R)	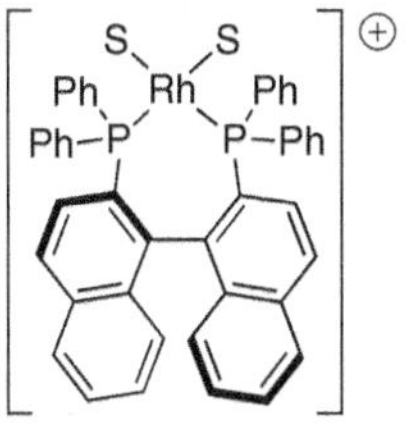 (S)-BINAP
Product	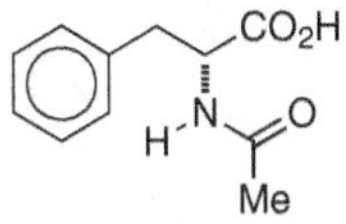	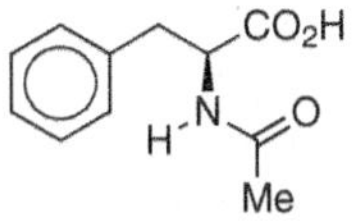	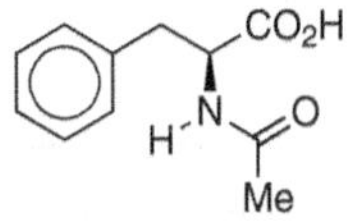	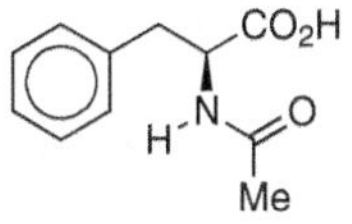	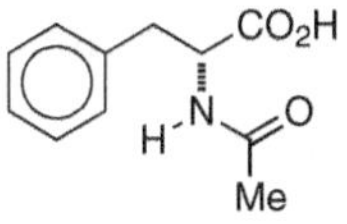
e.r.	99.5:0.5	98:2	91:9	96:4	92:8
Ref.	85	86	87	88	89

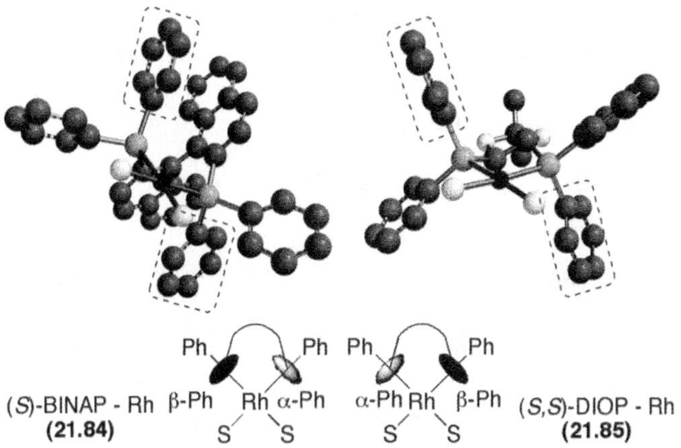

Figure 21.15 The (*S*)-BINAP and (*S*,*S*)-DIOP ligands have "axial" phenyl groups (in boxes) bound to phosphorus (shown by the shaded ellipses; the darker one projects above the plane of the paper) in enantiomeric alignment when bound to rhodium (I). Thus, we expect opposite chirality in the products of hydrogenation.

Because the stereochemistry of the hydrogenation is determined, at least in part, by the arrangement of phenyl groups bonded to phosphorus, analyzing the conformation (pseudoaxial or pseudoequatorial) of the phenyl groups bound to phosphorus presents a convenient way to determine the relative stereochemistry of a number of related hydrogenations. When we do this, we find that the phenyl rings are arranged as in the simple model in Figure 21.15, which predicts that the (*S*)-BINAP–Rh and (*S*,*S*)-DIOP–Rh catalysts should give opposite enantioselectivities with the same prochiral substrates. It is

85. Fryzuk, M.D.; Bosnich, B. *J. Am. Chem. Soc.* **1977**, *99*, 6262.
86. Vineyard, B.D.; Knowles, W.S.; Sabacky, M.J.; Bachman, G.L.; Weinkauff, D.J. *J. Am. Chem. Soc.* **1977**, *99*, 5946.
87. Sinou, D. *Tetrahedron Lett.* **1981**, *22*, 2987.
88. Onuma, K.; Ito, T.; Nakamura, A. *Bull. Chem. Soc.* Japan **1980**, *53*, 2016.
89. Miyashita, A.; Yasuda, A.; Takaya, H.; Toriumi, K.; Ito, T.; Souchi, T.; Noyori, R. *J. Am. Chem. Soc.* **1980**, *102*, 7932.

Table 21.6 Stereochemical Outcome of Hydrogenations with Chiral Rhodium (I)-BINAP and -DIOP Catalysts*

Substrate	Rh–(S)-BINAP	Rh–(*R*,*R*)-DIOP
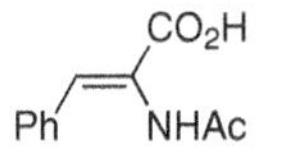	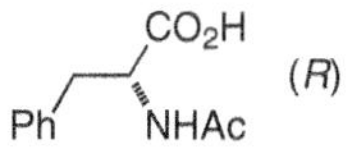 100% e.e.	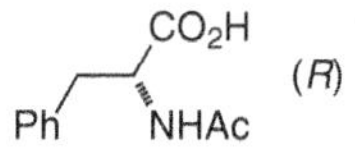 85% e.e.
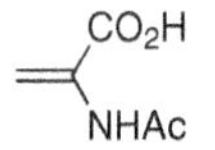	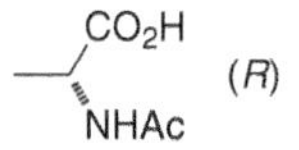 98% e.e.	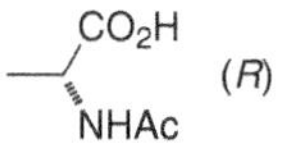 73% e.e.
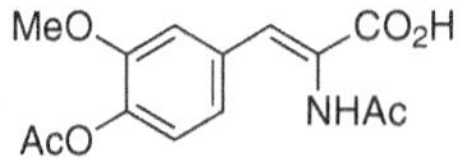	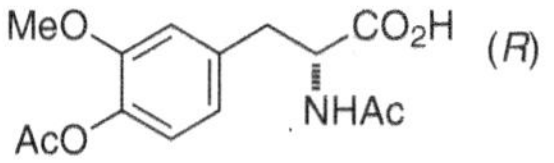 79% e.e.	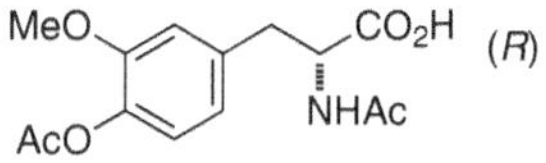 84% e.e.

*Data taken from Takaya, H.; Ohta, T.; Noyori, R. In Ojima, I, Ed. *Catalytic Asymmetric Synthesis* (VCH: Weinheim, 1993), ch. 1.

encouraging to see that this has been observed with the (*Z*)-α-amidoacrylic acids (Table 21.6). This generalization is useful, because it means that one can extrapolate with some confidence from one catalyst system to another. Taken with the observation that the ruthenium (II) (Example **21.86**) and rhodium (I) catalysts (Example **21.87**) formed with the same enantiomer of the ligand give opposite enantiomers of the product, this begins to permit the organization of the huge amount of information about asymmetric hydrogenation now available. However, this model is limited in its applicability, and *it is still not a replacement for consulting the primary literature.*

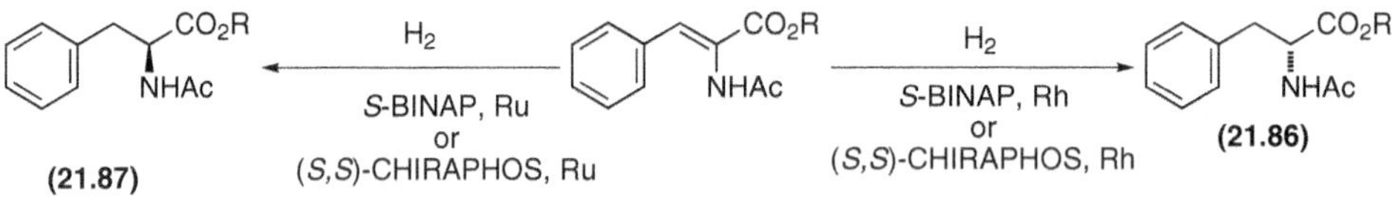

(21.87) **(21.86)**

The initial complexation of an α-amidoacrylic acid to the rhodium atom occurs through the carbonyl oxygen of the enamide to give two diastereomeric complexes. In the favored diastereoisomer, there is π stacking between the axial phenyl groups and the carboxyl and phenyl groups of the substrate, and in the disfavored diastereoisomer, this interaction is absent (Figure 21.16). Work by Halpern[90] has provided evidence that the stereochemical outcome of the hydrogenation is counterintuitive. The more favored of the two diastereomeric complexes is much less reactive toward addition of hydrogen, with the result that the major stereoisomer of the product arises from the minor diastereoisomer of the initial complex.[91] This may also permit rationalization of the intriguing observation that, under

90. **Jack Halpern (1925–)** was born in Poland, but migrated to Canada while a child. He was educated at McGill University (PhD, 1949) and spent a year as a postdoctoral fellow at the University of Manchester before joining the University of British Columbia in 1950. In 1962, he moved to the University of Chicago as Louis Block Distinguished Service Professor of Chemistry, a position he held until his retirement. In 1974, Halpern was elected to the National Academy of Sciences, and was elected a Fellow of the Royal Society. Halpern's elucidation of the origins of enantioselectivity in asymmetric hydrogenation was critical to our understanding of how this important reaction works.

91. (a) Chan, A.S.C.; Halpern, J. *J. Am. Chem. Soc.* **1980**, *102*, 838. (b) Chua, P.S.; Roberts, N.K.; Bosnich, B.; Okrasinski, S.J.; Halpern, J. *J. Chem. Soc., Chem. Commun.* **1981**, 1278. (c) Halpern, J. In Augustine, R.L., Ed. *Catalysis of Organic Reactions* (Marcel Dekker: NewYork, 1985), p. 3. (d) Halpern, J. In Morrison, J.D., Ed. *Asymmetric Synthesis* (Academic Press: New York, 1985), vol. 5, ch. 2, p. 41. (e) Landis, C.R.; Halpern, J. *J. Am. Chem. Soc.* **1987**, *109*, 1746.

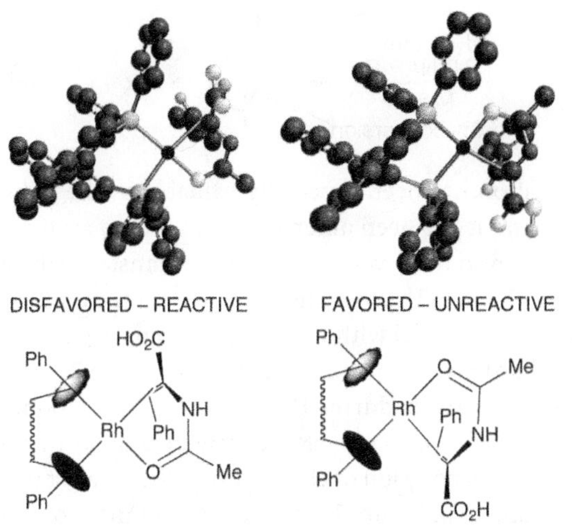

Figure 21.16 The two possible complexes between a rhodium (I) catalyst and an amidoacrylic acid. In the favored isomer, the complex does not react readily with hydrogen.

certain conditions, the same chiral diphosphine gives products of opposite absolute configuration when the two different metals are used.[92] This difference has been rationalized in terms of the different mechanisms of the two reactions. In the rhodium-catalyzed addition, a rhodium dihydride species is involved.[93] In contrast, in the ruthenium-catalyzed addition, a monohydride intermediate is involved.[94] However, a simpler rationalization may be that the hydrogenation with the ruthenium complex fits the intuitive model, so the major stereoisomer of the product actually comes from the major diastereoisomer of the complex.

The effects of alkene geometry on asymmetric hydrogenation are particularly evident in the hydrogenation of β-alkyl-α-acetamidoacrylates. These compounds are usually prepared as mixtures of *E* and *Z* isomers,[95] and it has generally been found that although hydrogenation of the *Z* isomer is fast and highly enantioselective, the *E* isomer often reacts slowly and with less selectivity.[96] With rhodium (I) catalysts, the *E* and *Z* substrates usually give enantiomeric products, indicating that the face selectivity of hydrogen addition is determined by the double bond geometry.[97] DuPHOS complexes are remarkably enantioselective, but both geometric isomers of the alkene give the same enantiomer of the product.[98]

Asymmetric hydrogenation also provides a method for the kinetic resolution of racemic allylic alcohols. Thus, when the racemic allylic alcohol in Example **21.88** is hydrogenated using a ruthenium *S*-BINAP catalyst, the unaltered alcohol at 68% conversion is the *S* enantiomer in 98% e.e.[99]

---

92. (a) Ikariya, T.; Ishii, Y.; Kawano, H.; Arai, T.; Saburi, M.; Yoshikawa, S.; Akutagawa, S. *J. Chem. Soc., Chem. Commun.* **1985**, 922. (b) Kawano, H.; Ikariya, T.; Ishii, Y.; Saburi, M.; Yoshikawa, S.; Uchida, Y.; Kumobayashi, H. *J. Chem. Soc., Perkin Trans. 1* **1989**, 1571. (c) James, B.R.; Pacheco, A.; Rettig, S.J.; Thorburn, I.S.; Ball, R.G.; Ibers, J.A. *J. Mol. Catal.* **1987**, *41*, 147.

93. Brown, J.M. *Chem. Soc. Rev.* **1993**, 25.

94. (a) Kitamura, M.; Yoshimura, M.; Tsukamoto, M.; Noyori, R. *Enantiomer* **1996**, *1*, 281. (b) Wiles, J.A.; Bergens, S.H.; Young, V.G. *J. Am. Chem. Soc.* **1997**, *119*, 2940. (c) Wiles, J.A.; Bergens, S.H. *Organometallics* **1998**, *17*, 2228.

95. Burk, M.J.; Bienewald, F. In Beller, M.; Bolm, C., Eds. *Transition Metals for Organic Synthesis* (Wiley-VCH: Weinheim, 1998), ch. 2, p. 13.

96. (a) Scott, J.W.; Kieth, D.D.; Nix, D.; Parrish, D.R.; Remington, S.; Townsend, G.P.; Valentine, D.; Yang, R. *J. Org. Chem.* **1981**, *46*, 5086. (b) Vineyard, B.D.; Knowles, W.S.; Sabacky, M.J.; Bachman, G.L.; Weinkauff, D.J. *J. Am. Chem. Soc.* **1977**, *99*, 5946.

97. Miyashita, A.; Takaya, H.; Souchi, T.; Noyori, R. *Tetrahedron* **1984**, *40*, 1245.

98. (a) Burk, M.J./ *J. Am. Chem. Soc.* **1991**, *113*, 8518. (b) Burk, M.J.; Feaster, J.E.; Nugent, W.A.; Harlow, R.L. *J. Am. Chem. Soc.* **1993**, *115*, 10125. (c) Burk, M.J. *Acc. Chem. Res.* **2000**, *33*, 363.

99. Kitamura, M.; Kasahara, I.; Manabe, K.; Noyori, R.; Takaya, H. *J. Org. Chem.* **1988**, *53*, 708.

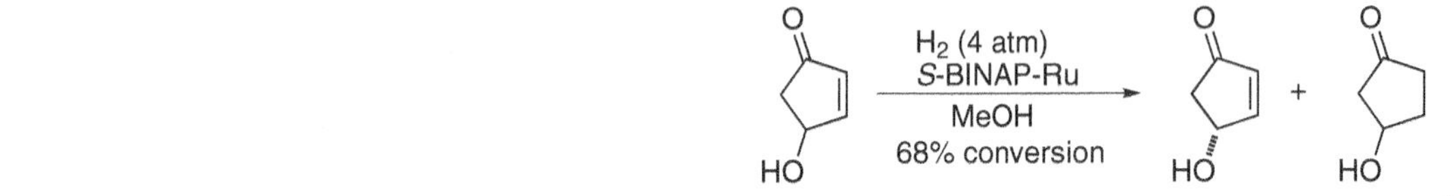

(21.88)

Transfer hydrogenation has been especially valuable in the reduction of imines and carbonyl compounds, and it has been amenable to use in an asymmetric form.[100] Ruthenium complexes have proved to be very useful in the transfer hydrogenation of carbonyl compounds, as well as alkenes. The most useful ruthenium catalysts are those where the metal carries: (1) a bidentate ligand with hard donor atoms (e.g., 2,2'-bipyridyl); (2) an η^6-arene ligand; and (3) a hard monodentate ligand (e.g., water, in **21.89**), The hard monodentate ligand is replaced by hydrogen during the course of the reaction. The bidentate ligand allows for chirality in the form of a C_2 axis or pseudoaxis to be incorporated into the complex and allows the incorporation of chirality into the catalyst (e.g., **21.90**). Two representative examples are shown in Figure 21.17. As the size of the group labeled "R" increases, the activity of the catalyst decreases.[101]

One of the most important chiral ligands in this series of catalysts is the *N*-tosyl-1, 2-diamino-1,2-diphenylethane (TsDPEN) ligand.[102] The stereochemistry of the hydrogenation using these chiral ruthenium catalysts can be predicted on the model **(21.91)** in Figure 21.17. This model reflects that the binding of the substrate to the catalyst is through a hydrogen bond between oxygen and the hydrogen of the amine group of the ligand and by face-edge π-stacking of the aryl rings. This H—π interaction is the critical factor determining the absolute configuration of the product.[103] Hydrogen transfer occurs through a six-membered ring. The mechanism has been probed using fluorinated substrates, and the

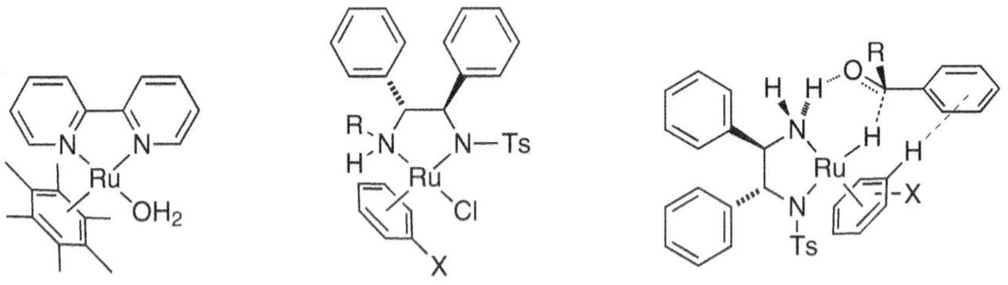

(21.89) **(21.90)** **(21.91)**

Figure 21.17 η^6-Areneruthenium (II) catalysts for transfer hydrogenation of ketones and imines (top) and model for predicting stereochemistry of the hydrogenation of ketones (bottom)

---

100. Reviews: (a) Ikariya, T.; Blacker, A.J. *Acc. Chem. Res.* **2007**, *40*, 1300. (b) Joseph, S.M.; Samec, J.S.; Bäckvall, J.-E.; Andersson, P.G.; Brandt, P. *Chem. Soc. Rev.* **2006**, *35*, 237. (c) Gladiali, S.; Alberico, E. *Chem. Soc. Rev.* **2006**, *35*, 226. (d) Blacker, A.J. In de Vries, J.G.; Elsevier, C.J., Eds. *The Handbook of Homogeneous Hydrogenation* (Wiley- VCH: Weinheim, 2007); pp 1215. (e) Ikariya, T.; Murata, K.; Noyori, R. *Org. Biomol. Chem.* **2006**, *4*, 393. (e) Clapham, S.E.; Hadzovic, A.; Morris, R.H. *Coord. Chem. Rev.* **2004**, *248*, 2201. (f) Blaser, H.-U.; Malan, C.; Pugin, B.; Spindler, F.; Steiner, H.; Studer, M. *Adv. Synth. Catal.* **2003**, *345*, 103. (g) Everaere, K.; Mortreux, A.; Carpentier, J.-F. *Adv. Synth. Catal.*, **2003**, *345*, 67. (h) Saluzzo, C.; Lemaire, M. *Adv. Synth. Catal.*, **2002**, *344*, 915. (i) Fan, Q.H; Li, Y.M.; Chan, A.S.C. *Chem. Rev.*, **2002**, *102*, 3385. (f) Palmer, M.J.; Wills, M. *Tetrahedron: Asymmetry*, **1999**, *10*, 2045. (j) Noyori, R.; Hashiguchi, S. *Acc. Chem. Res.*, **1997**, *30*, 97.

Recent examples: (h) Wu, X.; Li, X.; Zanotti-Gerosa, A.; Pettman, A.; Liu, J.; Mills, A.J.; Xiao, J. *Chem. Eur. J.* **2008**, *14*, 2209. (i) Zhou, S.; Fleischer, S.; Junge, K.; Das, S.; Addis, D.; Beller, M. *Angew. Chem. Int. Ed.* **2010**, *49*, 8121.

101. Martins, J.E.D.; Clarkson, G.J.; Wills, M. *Org. Lett.* **2009**, *11*, 847.

102. Ohkuma, T.; Utsumi, N.; Tsutsumi, K.; Murata, K.; Sandoval, C.; Noyori, R. *J. Am. Chem. Soc.* **2006**, *128*, 8724.

103. (a) Yamakawa, M.; Yamada, I.; Noyori, R. *Angew. Chem., Int. Ed.* **2001**, *40*, 2818. (b) Noyori, R.; Yamakawa, M.; Hashiguchi, S. *J. Org. Chem.* **2001**, *66*, 7931. (c) Yamakawa, M.; Ito, H.; Noyori, R. *J. Am. Chem. Soc.* **2000**, *122*, 1466. (d) Alonso, D.A.; Brandt, P.; Nordin, S.J.M.; Andersson, P.G. *J. Am. Chem. Soc.* **1999**, *121*, 9580. (e) Brandt, P.; Roth, P.; Andersson, P.G. *J. Org. Chem.* **2004**, *69*, 4885.

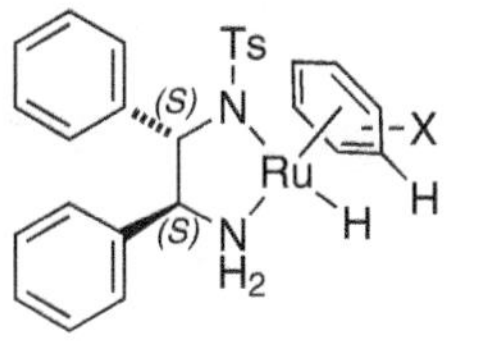

(21.92)

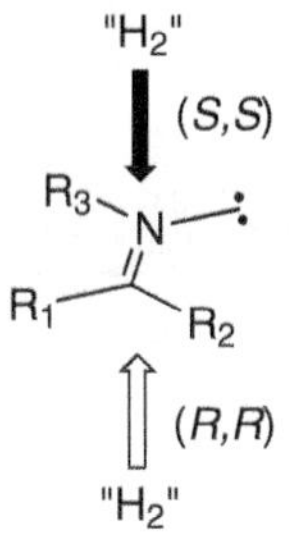

(21.93)

Figure 21.18 Model for predicting outcome of asymmetric transfer hydrogenation of imines by TsDPEN complexes

results show that the electronic nature of the carbonyl carbon is important in obtaining high levels of enantioselectivity as well as the π-π interactions.[104] Catalyst **21.89** shows a fairly strong dependence on pH, with an optimum pH of 4.[105]

As noted above, transfer hydrogenation has become especially popular for this reaction, and has been found to be especially useful in the synthesis of chiral amines by carrying out the reduction of the imine with chiral catalysts.[106] In particular, the chiral η^6-arene-ruthenium-(TsDPEN) catalysts in Figure 21.17 can also be used for transfer hydrogenation of imines. In this case, the absolute configuration of the amine product (Figure 21.18) is fixed by chirality of the TsDPEN backbone.[107] Interestingly, the NH_2 group on the ligand and the toluenesulfonyl group are both essential for high catalyst activity in this application. Replacing the NH_2 group with NMe_2 dramatically lowers catalyst efficiency, and replacing the toluenesulfonyl group with a trifluoromethylsulfonyl or an acyl group also lowers catalyst efficiency.[108]

Although isopropyl alcohol has long been the solvent of choice for transfer hydrogenations, using aqueous sodium formate (e.g., Example **21.94**) or the formic acid–triethylamine azeotrope (Example **21.95**) as both solvent and hydrogen source has been advocated and illustrated using the representative chiral catalyst above.[109] The authors make the point that the participation of the 2-propanol in alcohol-ketone equilibria can compromise enantioselectivity. This was demonstrated by Noyori, who observed that the e.e. of the chiral 1-phenylethanol formed by reduction of acetophenone with an η^6-mesityleneruthenium (TsDPEN) catalyst in isopropyl alcohol decreased from more than 99% initially to less than 97% at 75% conversion.[110]

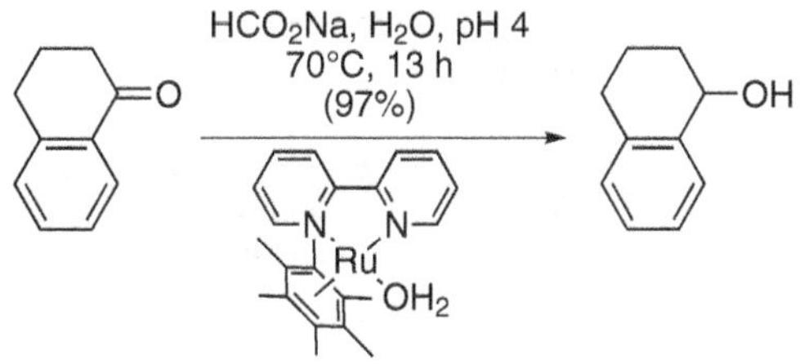

(21.94)

104. Slungård, S.V.; Krakeli, T.-A.; Thvedt, T.H.K.; Fuglseth, E.; Sundby, E.; Hoff, B.H. *Tetrahedron* **2011**, 67, 5642.

105. Ogo, S.; Abura, T.; Watanabe, Y. *Organometallics* **2002**, *21*, 2964.

106. Review: Okuma, T.; Kitamura, M.; Noyori, R. In Ojima, I. *Catalytic Asymmetric Synthesis*, 2nd ed. (Wiley-VCH: New York, 2000), ch. 1 (esp. pp. 83–91).

107. (a) Hashiguchi, S.; Fujii, A.; Takehara, J.; Ikariya, T.; Noyori, R. *J. Am. Chem. Soc.* **1995**, *117*, 7562. (b) Fujii, A.; Hashiguchi, S.; Uematsu, N.; Ikariya, T.; Noyori, R. *J. Am. Chem. Soc.* **1996**, *118*, 2521. (c) Haack, K.-J.; Hashiguchi, S.; Fujii, A.; Ikariya, T.; Noyori, R. *Angew. Chem., Int. Ed. Engl.* **1997**, *36*, 285. (d) Matsumura, K.; Hashiguchi, S.; Ikariya, T.; Noyori, R. *J. Am. Chem. Soc.* **1997**, *119*, 8738. (e) Noyori, R.; Hashiguchi, S. *Acc. Chem. Res.* **1997**, *30*, 97. (f) Watanabe, M.; Murata, K.; Ikariya, T. *J. Org. Chem.* **2002**, *67*, 1712. (g) Ma, Y.; Liu, H.; Chen, L.; Cui, X.; Zhu, J.; Deng, J. *Org. Lett.* **2003**, *5*, 2103.

108. Uematsu, N.; Fujii, A.; Hashiguchi, S.; Ikariya, T.; Noyori, R. *J. Am. Chem. Soc.* **1996**, *118*, 4916.

109. Wu, X.; Li, X.; Hems, W.; King, F.; Xiao, J. *Org. Biomol. Chem.* **2004**, *2*, 1818.

110. Hashiguchi, S.; Fujii, A.; Takehara, J.; Ikariya, T.; Noyori, R. *J. Am. Chem. Soc.* **1995**, *117*, 7562.

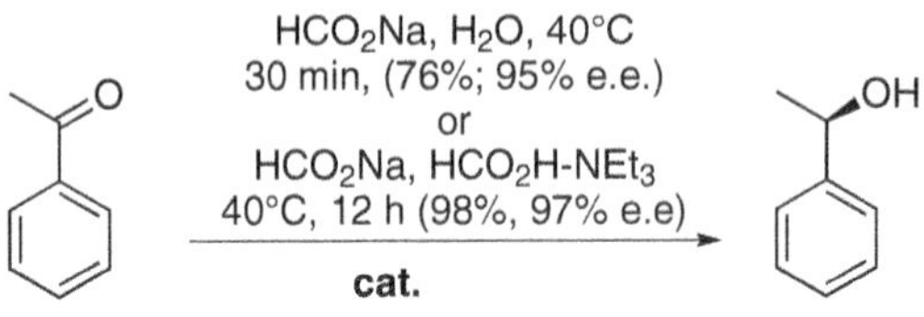

(21.95)

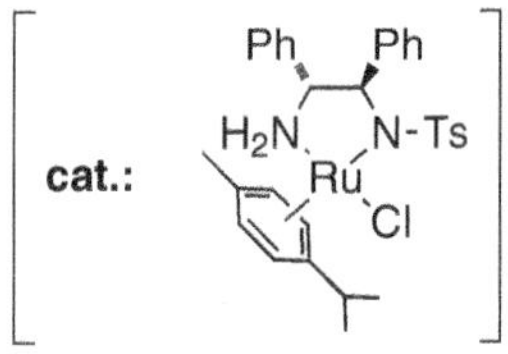

A Final Word . . .

From the start, the potential (and real) financial benefits of asymmetric hydrogenation have driven the effort to develop new and better ligands for use with transition metal catalysts. This means that by the time you read this chapter, it will already be out of date, so you are encouraged to consult any one of a number of specialist books and reports on the subject for the latest ligands to use.[111]

Problem

21-9 In addition to the results in Table 21.6, an analysis of the stereochemical outcome of enantioselective hydrogenations suggests that the hydrogenation of alkenes using Rh-DuPHOS complexes gives the opposite stereochemical outcome to the Rh-BINAP complex. Predict the major product of each of the following reactions when the reagent is changed as indicated.

(a)

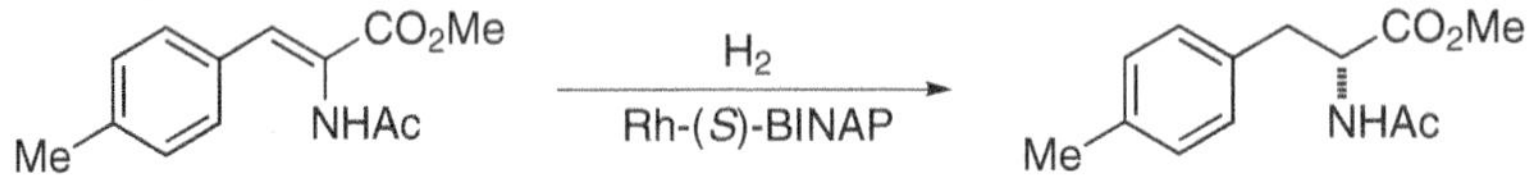

Changes: (i) change Rh to Ru; (ii) Change (*S*)-BINAP to (*S,S*)-DIOP; (iii) change Rh to Ru, and change (*S*)-BINAP to (*S,S*)-DIOP

(b)

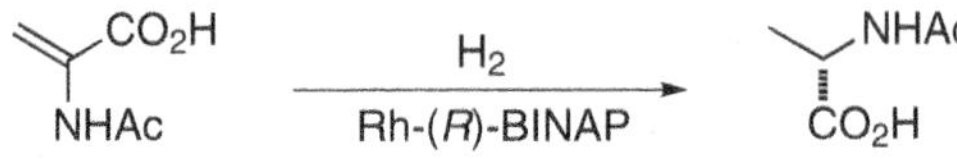

Changes: (i) change Rh to Ru; (ii) Change (*R*)-BINAP to (*S,S*)-DIOP; (iii) change Rh to Ru, and change (*R*)-BINAP to (*R,R*)-DIOP

(c)

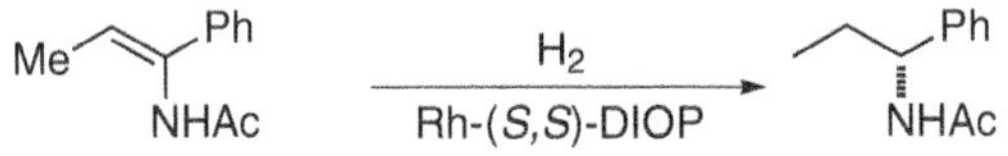

Changes: (i) change Rh to Ru; (ii) Change(*S,S*)-DIOP to (*R*)-BINAP; (iii) change Rh to Ru, and change (*R,R*)-DIOP to (*R*)-BINAP; (iv) change (*S,S*)-DIOP to (*S,S*)-Et_2-DuPHOS

21.9 Asymmetric Reductions with Metal Hydrides and Metal Alkyls

Oxazaborolidine-Catalyzed Asymmetric Reduction by Borane

In 1985, Itsuno and his coworkers reported that chiral β-aminoalcohols could be used to reduce aryl alkyl ketones to chiral alcohols in with excellent enantioselectivity.[112] The

---

111. One especially useful source for information on catalysts and ligands are the technical notes on the web site of Strem Chemicals, Inc.: http://www.strem.com

112. Itsuno, S.; Nakano, M.; Miyazaki, K.; Matsuda, M.; Ito, K.; Hirao, A.; Nakahama, S. *J. Chem. Soc., Perkin Trans. 1* **1985**, 2039.

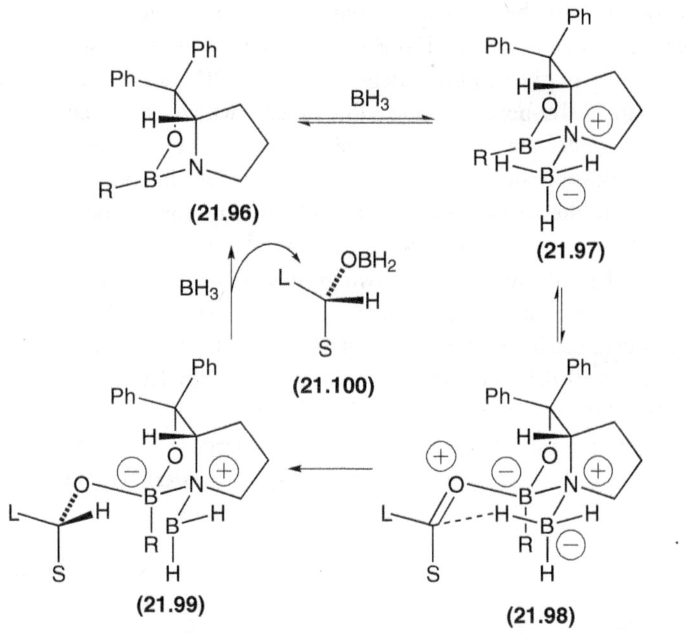

Figure 21.19 A model for the reduction of ketones by borane with a chiral oxazaborolidine catalyst

active reagent was identified by Corey and his coworkers as a chiral oxazaborolidine (**21.96**). This molecule contains both nucleophilic (nitrogen) and electrophilic (boron) sites in the same molecule (Figure 21.19). This renders the molecule capable of simultaneously coordinating the boron of the reducing borane, forming an "ate" complex (**21.97**), and the oxygen of the carbonyl group, thus providing activation of the carbonyl carbon toward nucleophilic addition. These species are useful catalysts for the asymmetric reduction of carbonyl compounds by boranes (e.g., **21.101**).[113]

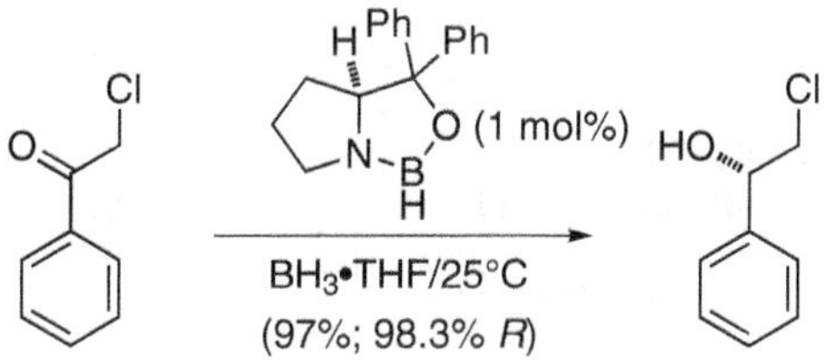

(21.101)

The model in Figure 21.19 (**21.98**, in which the oxazaborolidine is complexed to a molecule of the borane and a molecule of the ketone) allows the stereochemistry of the product of the reduction to be predicted. The ketone orients itself preferentially as shown, with the larger group in the equatorial position.

Reduction with Trialkylboranes ($G{=}CR_2$, $M{=}B$)[114]

Trialkylboranes, especially *B*-alkyl-9-BBN derivatives, react with aldehydes and ketones to give alcohols by transfer of a β hydrogen atom of the borane and elimination of the

113. (a) Corey, E.J.; Bakshi, R.K.; Shibata, S. *J. Am. Chem. Soc.* **1987**, *109*, 5551. (b) Corey, E.J.; Shibata, S.; Bakshi, R.K. *J. Org. Chem.* **1988**, *53*, 2861. (c) Corey, E.J.; Bakshi, R.K.; Shibata, S.; Chen, C.-P.; Singh, V.K. *J. Am. Chem. Soc.* **1987**, *109*, 7925. (d) Corey, E.J. *Pure Appl. Chem.* **1990**, *62*, 1209. (e) Corey, E.J.; Cheng, X.M.; Cimprich, K.A.; Sarshar, S. *Tetrahedron Lett.* **1991**, *32*, 6835. (f) Corey, E.J.; Kigoshi, M. *Tetrahedron Lett.* **1991**, *32*, 5025. (g) Corey, E.J.; Rao, K.S. *Tetrahedron Lett.* **1991**, *32*, 4623. (h) Jones, D.K.; Liotta, D.C.; Shinkai, I.; Mathre, D.J. *J. Org. Chem.* **1993**, *58*, 799.

Monograph: Lohrag, B.B.; Bhusan, V. *Oxazaborolidines* (Wiley-VCH: New York, 1998).

114. Reviews: (a) Srebnik, M.; Ramachandran, P.V. *Aldrichimica Acta* **1987**, *20*, 9. (b) Dhar, R.K. *Aldrichimica Acta* **1994**, *27*, 43. (c) Cho, B.T. *Aldrichimica Acta* **2002**, *35*, 3.

alkene. This reaction has been especially useful in the asymmetric reduction of aldehydes and ketones, and there are several examples known where the use of an optically pure borane reducing agent gives a chiral alcohol in close to 100% enantiomeric excess.

9-Isopinocampheyl-9-borabicyclo-[3.3.1]nonane,[115] which is available commercially as both enantiomers under the trade name Alpine borane®, is one of the most effective borane reducing agents, possibly because of the relief of steric strain on release of the alkene by-product. It is highly effective for the reduction of carbonyl compounds where there is a substantial size difference between the two groups bonded to the carbonyl group. However, it tends to be much less effective when the two groups are similar in size. The reagent has been especially useful for reducing acetylenic ketones (e.g., Example **21.102**). Important applications of this reagent are its use in preparing chiral primary alcohols by reduction of deuterated aldehydes (e.g., Example **21.103**) or by reduction of aldehydes with deuterated Alpine borane®. The stereochemistry of the reaction may be predicted by means of the empirical model (Example **21.104**) for reductions of an aldehyde or ketone with a large (L) and small (S: hydrogen or an alkyne) group attached to the carbonyl group, by the borane derived from (1*R*)-α-pinene.

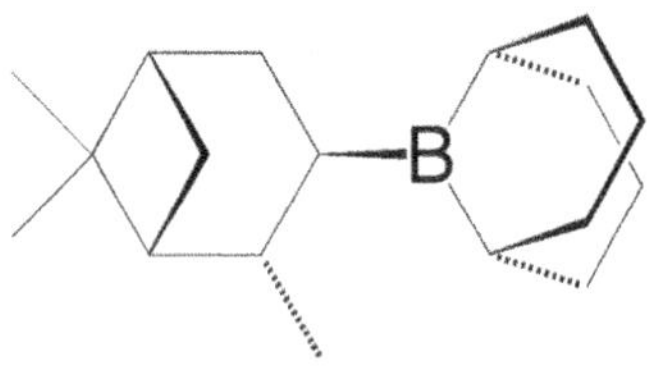

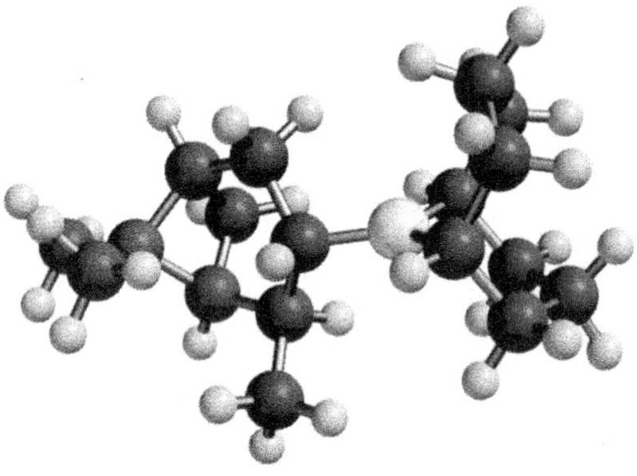

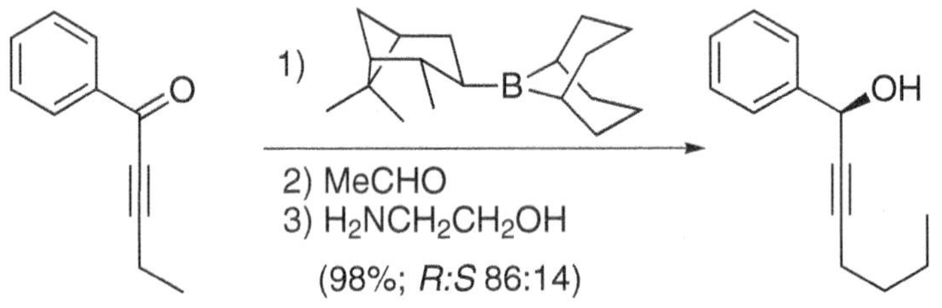

(21.102)

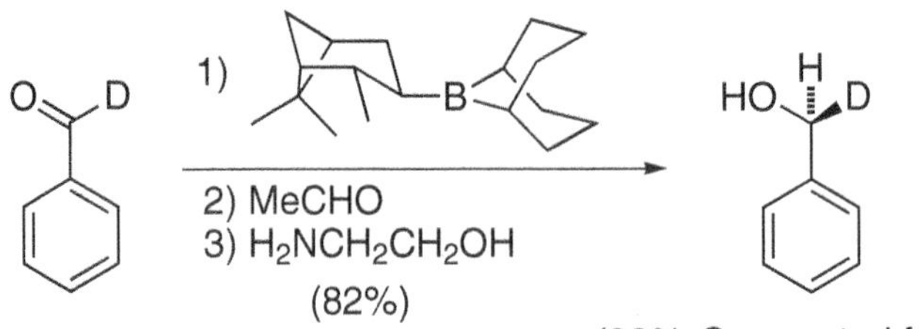

(21.103)

(98% *S* corrected for optical purity of pinene)

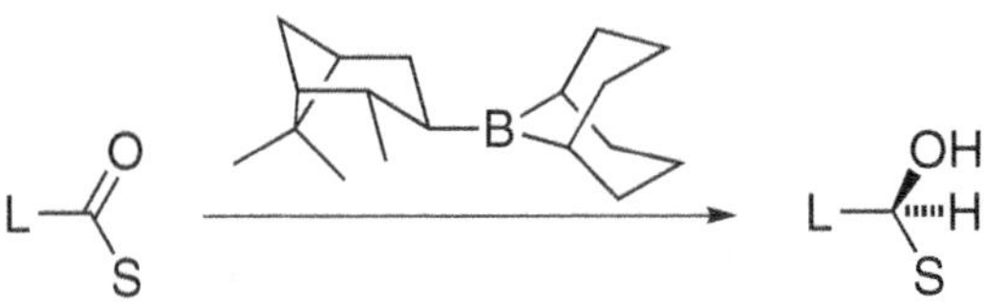

(21.104)

115. (a) Midland, M.M.; Greer, S.; Tramontano, A.; Zderic, S.A. *J. Am. Chem. Soc.* **1979**, *101*, 2352. (b) Midland, M.M.; McDowell, D.C.; Hatch, R.L.; Tramontano, A. *J. Am. Chem. Soc.* **1980**, *102*, 867. (c) Midland, M.M.; Graham, R.S. *Org. Syn.* **1985**, *63*, 57. (d) Midland, M.M. *Chem. Rev.* **1989**, *89*, 1553.

Problems

21-10 What is the structure of major organic product of each of the following reactions?

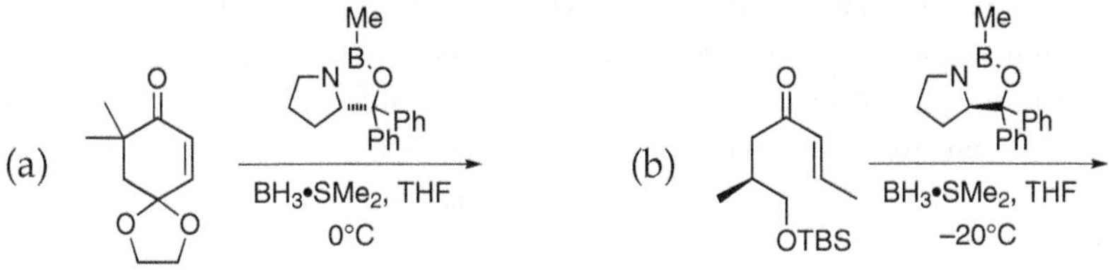

References: (a) *J. Am. Chem. Soc.* **2005**, *127*, 5342. (b) *Org. Lett.* **2005**, *7*, 3371.

21-11 What is the major organic product of each of the following reactions?

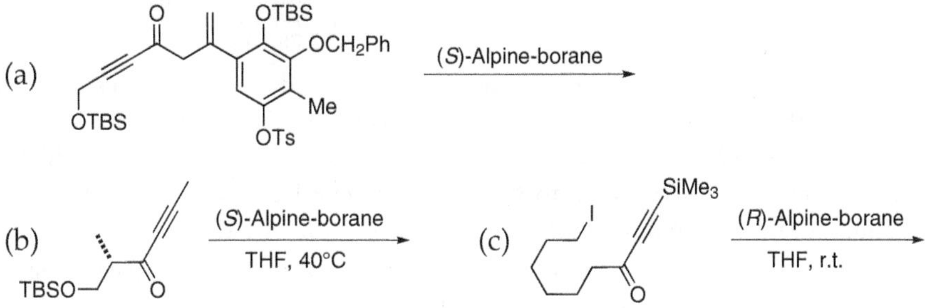

References for problems: (a) *Tetrahedron Lett.* **2008**, *49*, 5906. (b) *J. Am. Chem. Soc.* **2007**, *129*, 8960. (c) *Tetrahedron Lett.* **2001**, *42*, 1941.

21.10 Enzymes: Biocatalysis of Redox Reactions

Practically every reaction that one might want to use in a synthesis can be carried out enzymatically. Six classes of enzymes are broadly used in synthesis.

1. Oxidoreductases, which obviously catalyze redox reactions
2. Transferases, which transfer groups from one molecule to another; these enzymes are usually too specific to be of general use synthetically
3. Hydrolases, which hydrolyze a wide range of functional groups
4. Lyases, which catalyze additions to π bonds and the eliminations that generate π bonds
5. Isomerases, whose function is to isomerize the substrate (e.g., double-bond migration or racemization). Of all the enzyme-catalyzed reactions, these reactions are the most easily carried out without an enzyme.
6. Ligases, or synthetases, which catalyze the formation of C—X and C—C bonds

Enzymes can be used as isolated enzymes, as is done with many of the alcohol dehydrogenases, for example. Alternatively, the microorganism from which they are isolated can be used without explicit isolation of the enzyme, as in fermentation processes. Of particular interest in this chapter is the use of enzymes in oxidation reactions.[116]

Desymmetrization of prochiral or *meso* precursors can be readily accomplished by using an appropriate alcohol dehydrogenase. Thus, the oxidation of 3-methylpentane-1,5-diol catalyzed by horse liver alcohol dehydrogenase (HLADH) occurs to give the *S* alcohol preferentially. Cyclization of this alcohol to the hemiacetal gives another substrate of the same

116. Reviews: (a) Brown, S.M. In Trost, B.M.; Fleming, I.; Ley, S.V., Eds. *Comprehensive Organic Synthesis* (Pergamon: Oxford, 1991), vol. 7, ch. 1.4, p. 53. (b) Martin, G.D.A. *Curr. Org. Chem.* **2010**, *14*, 1.

enzyme, so the product isolated is the lactone.[117] The examples in **21.105** illustrate another important aspect of many enzyme-catalyzed reactions: many enzymes require cofactors. In this case, the cofactor nicotinamide adenine dinucleotide (NAD^+) is essential, but this cofactor is also very expensive. The authors therefore used the much less expensive cofactor flavin mononucleotide to permit recycling of the much more expensive NAD^+. The results also illustrate a limitation on enzyme-catalyzed reactions—the specificity of the enzyme usually means that structural changes are not without their costs (note the difference in e.e. when the methyl group is changed to phenyl).

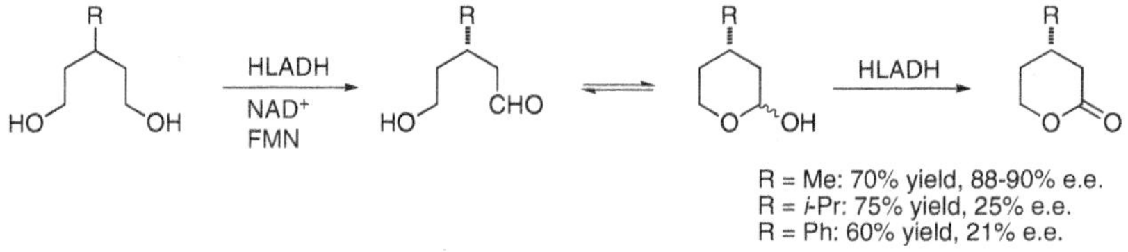

(21.105)

The oxidation of aromatic rings to non-aromatic products presents a significant challenge (with the notable exception of oxidation to quinones) because of the loss of aromaticity. This very oxidation, however, is routinely carried out by living cells, which oxidize aromatic rings to arene oxides or to 3,5-cyclohexadiene-1,2-diols. An especially useful application of microbial oxidation is the oxidation of aromatic rings to substituted *cis*-3,5-cyclohexadien-1,2-diols (Example **21.106**).[118] The reaction is stereospecific, giving one enantiomer of the *cis*-diol and thus providing extremely valuable intermediates for organic synthesis of single enantiomers. Such synthesis is illustrated by the stereospecific oxidation of toluene to (1*S*,2*R*)-3-methylcyclohexa-3,5-dien-1,2-diol by *Pseudomonas putida* 39/D, a mutant that lacks the enzyme responsible for dehydrating the diol to *o*-cresol (Example **21.107**).[119]

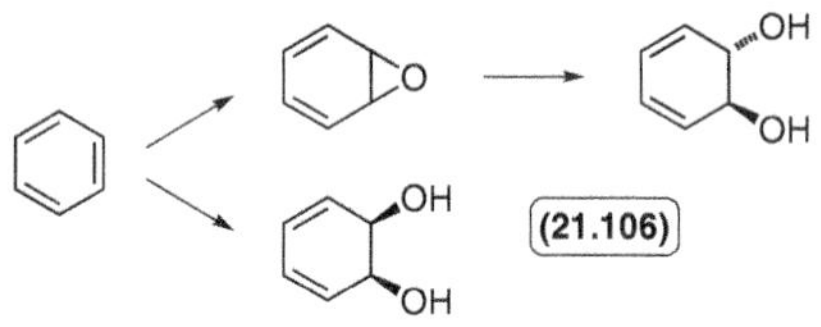

(21.106)

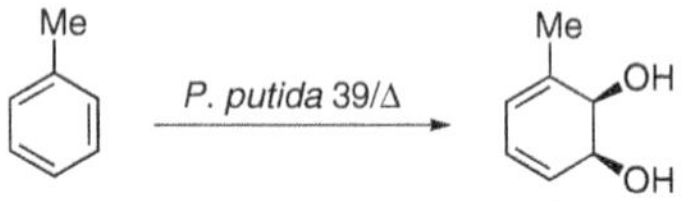

(21.107)

The Baeyer-Villiger reaction is another reaction where our efforts in the laboratory have not yet matched those of biological systems, where the reaction is catalyzed by flavin-dependent Baeyer-Villiger monooxygenases.[120] Nevertheless, considerable progress has been made in using recombinant techniques to design organisms that overexpress monooxygenases such as cyclohexanone monooxygenases (CHMOs) and cyclopentanone monooxygenases that are capable of accepting a wide range of substrates. These enzymes

117. Irwin, A.J.; Jones, J.B. *J. Am. Chem. Soc.* **1977**, *99*, 556.

118. Reviews: (a) Hudlicky, T. *Chem. Rev.* **1996**, *96*, 3. (b) Hudlicky, T.; Entwistle, D.A.; Pitzer, K.K.; Thorpe, A.J. *Chem. Rev.* **1996**, 96, 1195. (c) Hudlicky, T. In Anastas, P.T., Williamson, T.C., Eds. *Green Chemistry: Designing Chemistry for the Environment.* (American Chemical Society: Washington, DC, 1996) *ACS Symp. Ser.* **1996**, *626*, 180. (d) Hudlicky, T. In Anastas, P. T., Williamson, T. C., Eds. *Green Chemistry: Frontiers in Benign Chemical Syntheses and Processes* (Oxford University Press: Oxford, UK, 1998); Ch. 10, p 166. (e) Hudlicky, T.; Gonzalez, D.; Gibson, D.T. *Aldrichimica Acta* **1999**, *32*, 35. (f) Johnson, R.A. *Org. React.* **2004**, *63*, 117. (g) Hudlicky, T.; Reed, J.W. *Synlett.* **2009**, 685.

119. (a) Gibson, D.T.; Hensley, M.; Yoshioka, H.; Mabry, T.J. *Biochemistry* **1970**, *9*, 1626. (b) Ziffer,H.; Jerina, D.M.; Gibson, D.T.; Kobal, V.M. *J. Am. Chem. Soc.* **1973**, *95*, 4048.

120. (a) Walsh, C.T.; Chen, Y.-C.J. *Angew. Chem.* **1988**, *100*, 342. (b) Roberts, S.M.; Wan, P.W.H. *J. Mol. Catal. B: Enzym.* **1998**, *4*, 111. (c) Mihovilovic, M.D.; Müller, B.; Stanetty, P. *Eur. J. Org. Chem.* **2002**, 3711. (d) Kamerbeek, N.M.; Janssen, D.B.; van Berkel, W.J.H.; Fraaije, M.W. *Adv. Synth. Catal.* **2003**, *345*, 667.

Table 21.7 Enantioselectivity in the Baeyer-Villiger Oxidation of 4-Substituted Cyclohexanones by CHMO in Recombinant *Escherichia coli*

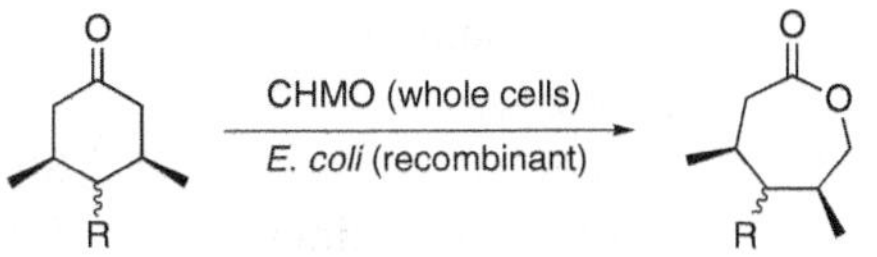

Substituent	Yield (%)	e.e. (%)	Substituent	Yield (%)	e.e. (%)
cis-OH	77	99	*trans*-OH	80	96
cis-Cl	40	> 99	*Trans*-Cl	54	92
$=CH_2$	63	99	$(CH_2)_2$	65	> 99
H	58	91			

allow the Baeyer-Villiger oxidation of cyclic ketones to lactones, and these oxidation products are often very useful in synthesis. Examples of the usefulness of this reaction are given in Table 21.7.

Insertion of oxygen into unactivated C—H σ bonds is a well-known microbial oxidation reaction. For example, practically every position on the steroid nucleus is amenable to microbial oxidation, and the importance of steroid pharmaceuticals has led to extensive research into steroid oxidation by microbes.[121] The enantioselectivity of oxidation of prochiral compounds is not always good, however, which means that one must often screen many microorganisms to find the one that is optimal for the application.

Chapter Summary

In this chapter, the modern face of redox chemistry has been discussed, with emphasis on the control of absolute stereochemistry in oxidation and reduction reactions. Asymmetric addition reactions include the Sharpless asymmetric epoxidation of allylic alcohols and alkene dihydroxylation reactions; epoxidation of simple alkenes by use of the Jacobsen and Katsuki salen-type catalysts or by means of chiral dioxiranes (the Shi oxidation); and asymmetric hydrogenation with molecular hydrogen (alkenes) or by transfer hydrogenation (ketones and imines). Enolate anions can be enantioselectively oxidized by means of chiral oxaziridines. The effects of the structure on substrate reactivity and the level of asymmetric induction have been discussed. The potential role of microbial oxidations in synthesis has been introduced.

The formation of chiral compounds by desymmetrization of *meso* compounds by methods such as kinetic hydrolysis of epoxides or by enantioselective oxidation of *meso* diols provide a route to enantiomerically enriched products. The kinetic resolution of racemic alcohols by enantioselective oxidation also serves the same purpose.

121. Reviews and monographs: (a) Charney, W.; Herzog, H.L. *Microbial Transformations of Steroids* (Academic Press: New York, 1967). (b) Rehm, H.A.; Reed, G., Eds. *Biotechnology* (Verlag Chemie: Weinheim, 1984), vol. 6a. (c) Rose, A.H., Ed. *Economic Microbiology* (Academic Press: New York, 1985), vol. 6. (d) Fonken, G.S.; Johnson, R.A. In Belew, J.S., Ed. *Oxidation in Organic Chemistry* (Marcel Dekker: New York, 1972), vol. 2. (e) Johnson, R.A. In Trahanovsky, W.S., Ed. *Oxidatioj in Organic Chemistry* (Academic Press: New York, 1978), vol. 5, p. 131. (f) Sariaslani, F.S.; Dalton, H. *Crit. Rev. Biotechnol.* **1989**, *9*, 171. (g) Holland, H.L. *Acc. Chem. Res.* **1984**, *17*, 398.

Key Terms

Alpine borane®
BINAP
Corey oxazaborolidine
Davis oxaziridines
desymmetrization
DIOP
DiPAMP
Jacobsen catalyst
Jacobsen-Katsuki epoxidation
Katsuki catalystkinetic resolution
Sharpless asymmetric epoxidation
Sharpless asymmetric dihydroxylation
Shi oxidation
Uemura oxidation

Additional Problems

21-12 What is the major organic product of each of the following reactions?

(a)
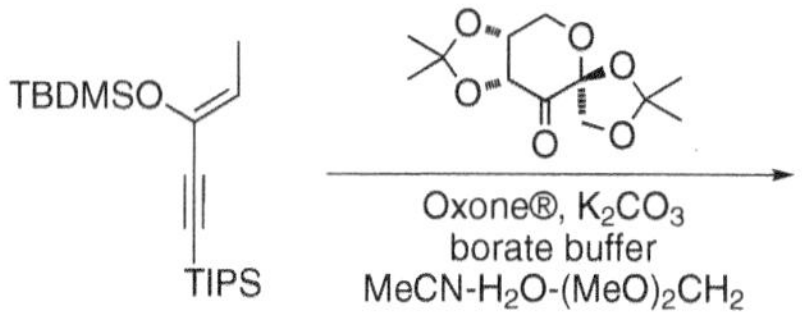

[*J. Am. Chem. Soc.* **2009**, *131*, 5763]

(b)
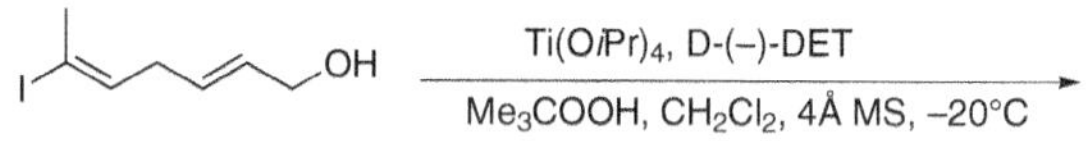

[*J. Am. Chem. Soc.* **2011**, *133*, 214]

(c)
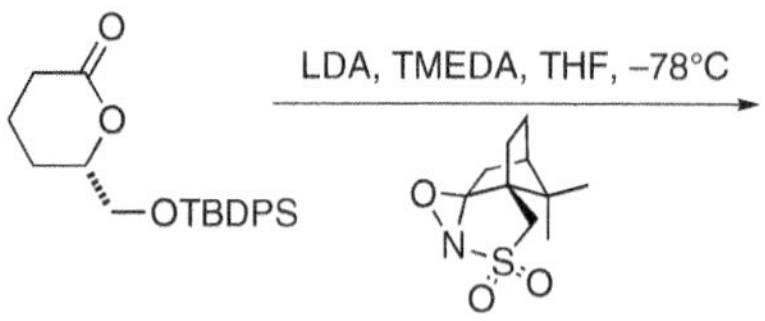

[*J. Org. Chem.* **2001**, *66*, 925]

(d)
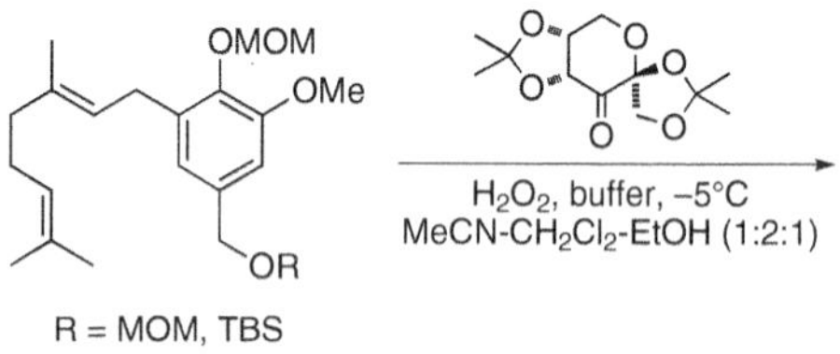

[*Tetrahedron Lett.* **2008**, *49*, 516]

(e)
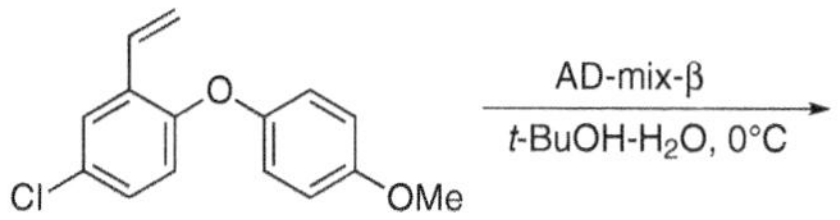

[*Tetrahedron* **2010**, *66*, 6826]

(f)
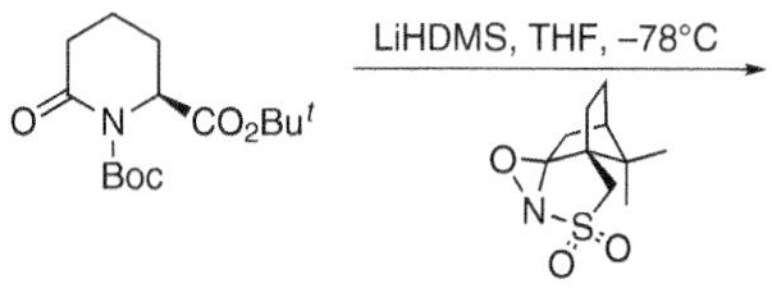

[*J. Org. Chem.* **2002**, *67*, 8440]

(g)
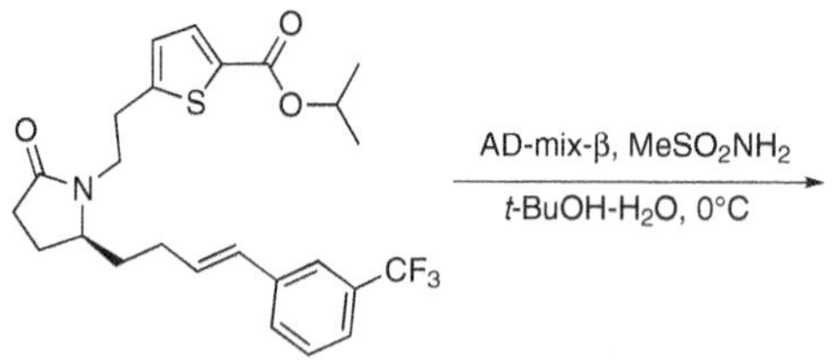

[*Tetrahedron Lett.* **2010**, *51*, 1451]

(h)
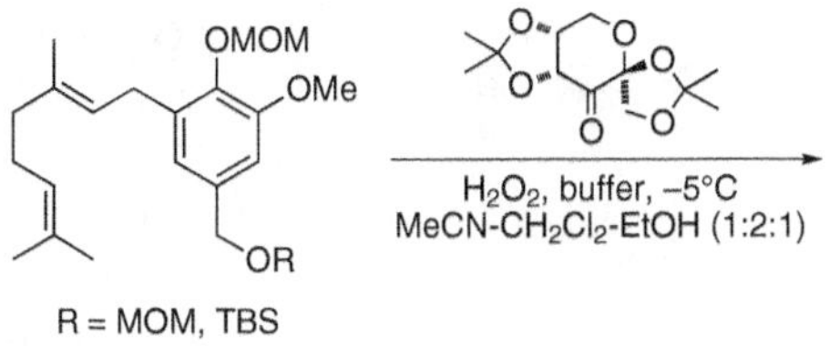

[*Org. Lett.* **2003**, *5*, 4823]

(i)
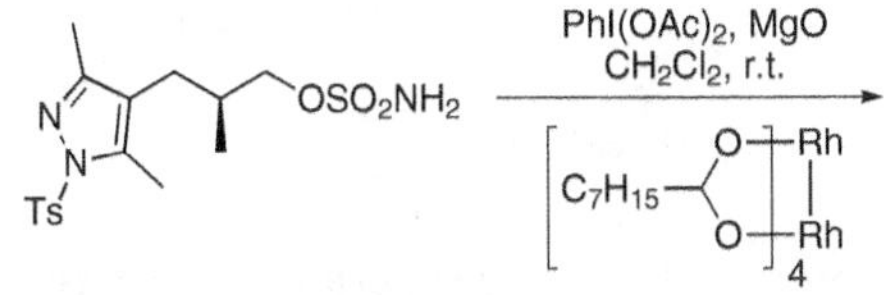

[*Synlett* **2009**, 417]

(j)
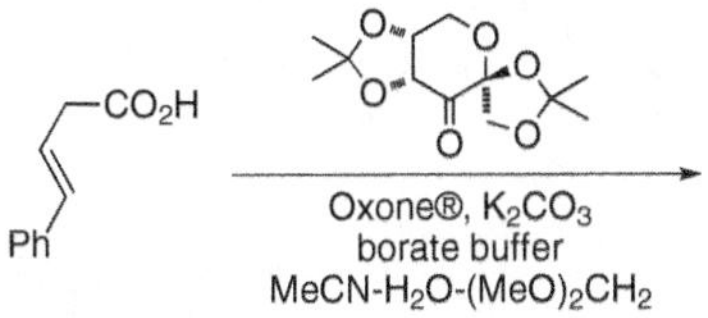

[*J. Org. Chem.* **2010**, *75*, 2333]

(k)
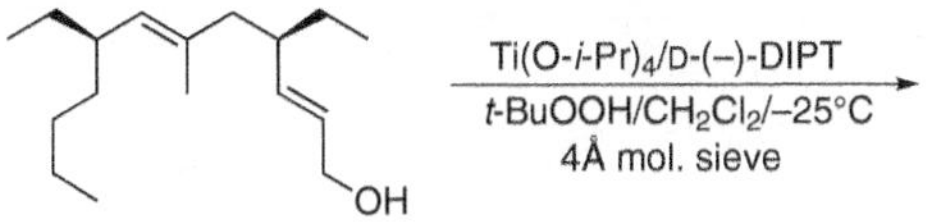

[*J. Am. Chem. Soc.* **1993**, *115*, 10400; **1996**, *118*, 6806]

(l)
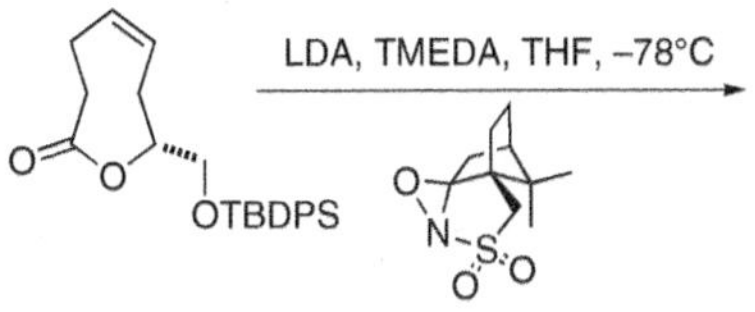

[*J. Am. Chem. Soc.* **2008**, *130*, 9220]

(m)
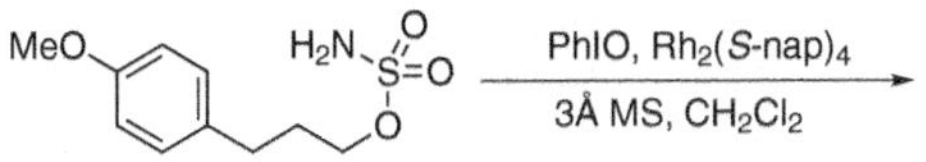

[*Tetrahedron Lett.* **2010**, *51*, 4058]

(n)
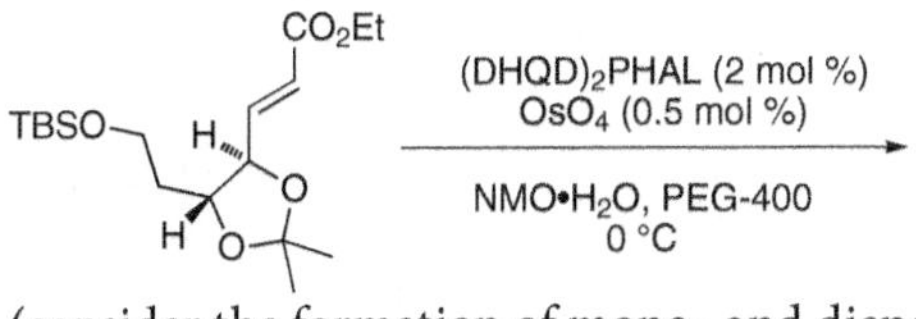

[*J. Org. Chem.* **2009**, *74*, 8407]

(consider the formation of mono- and diepoxides)

(o)
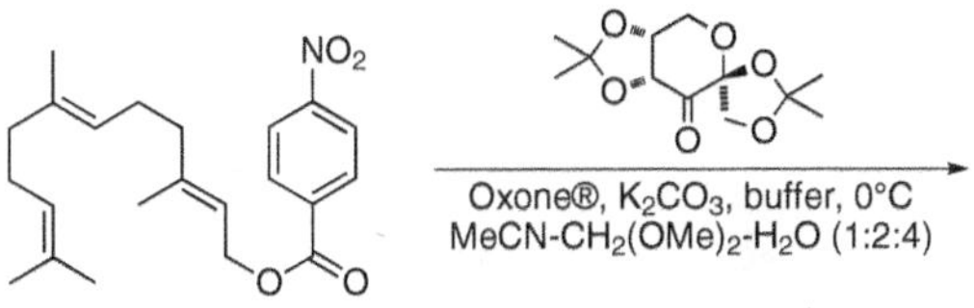

[*Tetrahedron Lett.* **2008**, *49*, 3899]

(p)
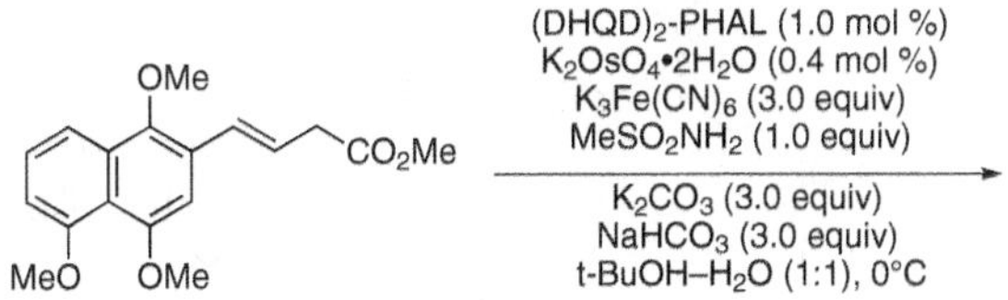

[*Org. Lett.* **2008**, 10, 5123]

(q)
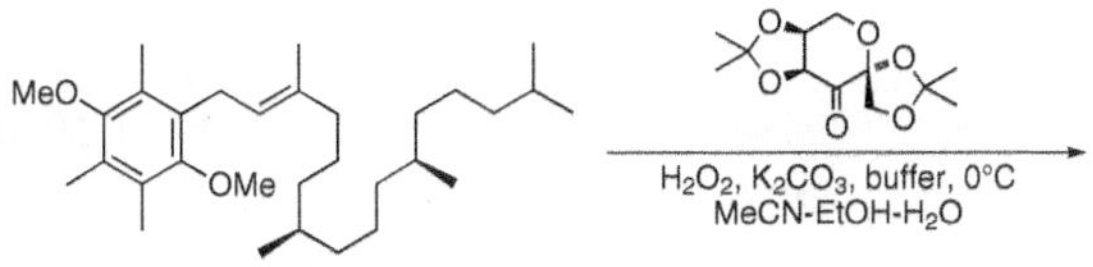

[*Synthesis* **2005**, 1183]

(r) 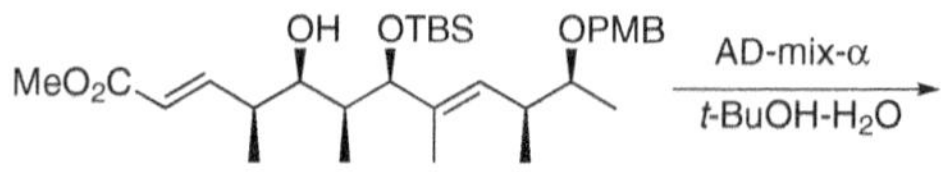

[*Org. Lett.* **2002**, *4*, 4507]

(s) 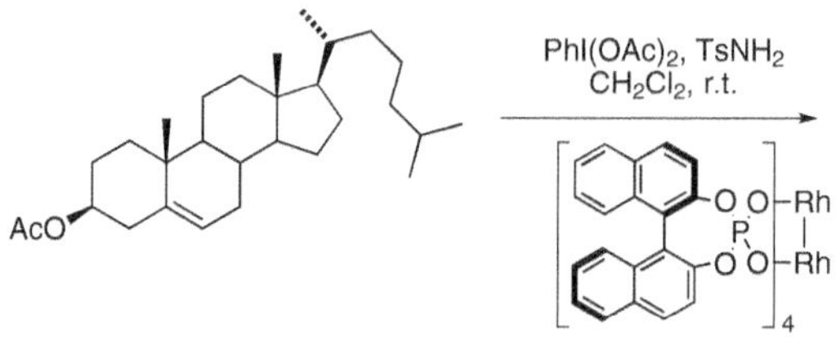

[*Tetrahedron Lett.* **2008**, *49*, 5736]

21-13 The reaction below has been proposed as being superior to other asymmetric amination reactions. Suggest reasons for why this should be so, and comment on the factors that might affect the level of asymmetric induction in the reaction. [*Angew. Chem. Int. Ed.* **2006**, *45*, 4641]

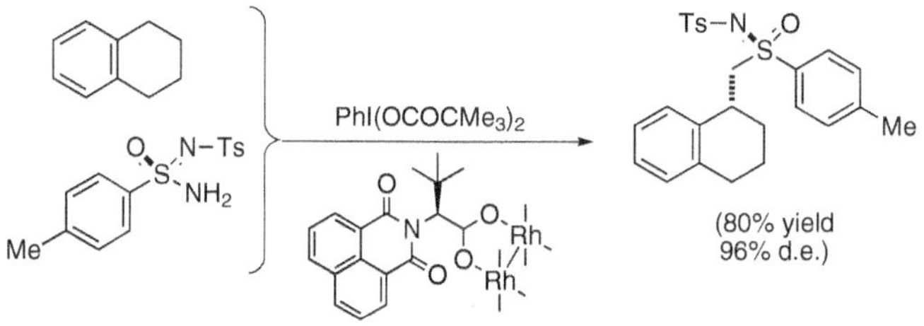

21-14 What reagent (or what choice of reagents) should be used to effect the following transformations? Where the authors chose one particular reagent where a choice of different reagents is possible, suggest reasons why the particular reagent was chosen.

(a) 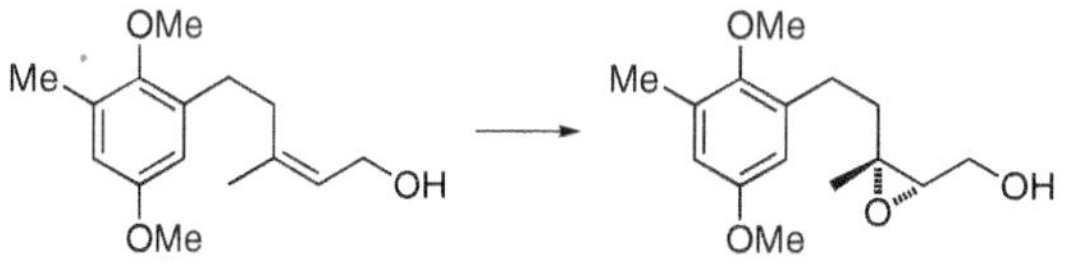

[*Org Lett.* **2005**, *6*, 4297]

(b) 

[*J. Org. Chem.* **1999**, *64*, 6005]

(c) 

[*J. Org. Chem.* **2008**, *73*, 9539]

(d) 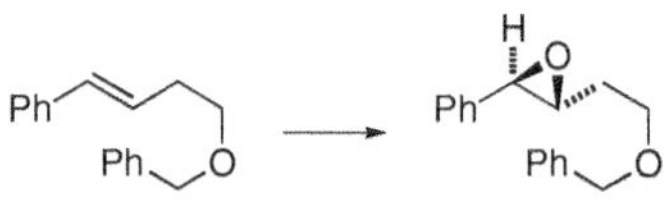

[*Synlett* **2008**, 2856]

(e) 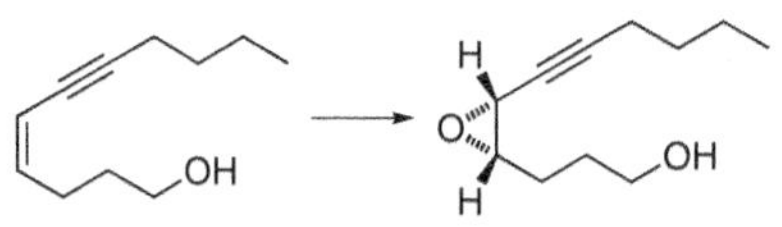

[*J. Org. Chem.* **2007**, *72*, 4093]

(f) 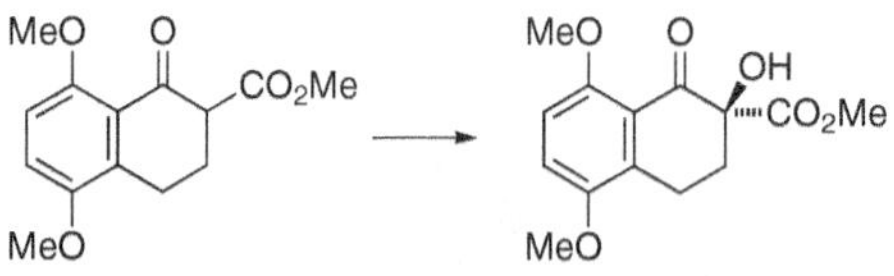

[*J. Org. Chem.* **1994**, *59*, 1184]

(g) 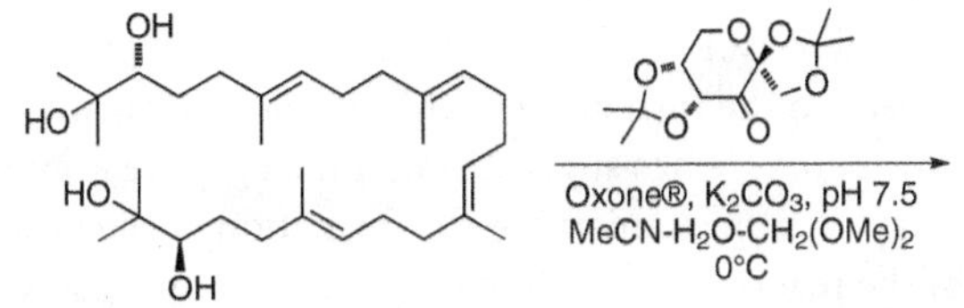

[*J. Am. Chem. Soc.* **2000**, *122*, 9328]

21-15 The mannose-derived ketones shown have been found to be useful as an organocatalyst in asymmetric Shi-type epoxidations. The results of asymmetric epoxidations of 1,1,2-triphenylethene are given in the table below. What does this tell about the transition state for the epoxidation? Is there anything that appears to be surprising or counterintuitive in the stereochemical outcome of this reaction? [*Tetrahedron* **2011**, *67*, 7057]

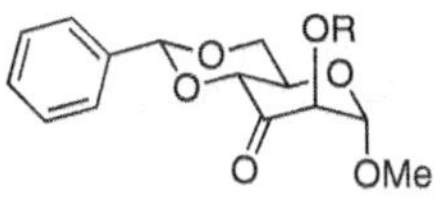

R:	Me	Et	i-Pr	i-Bu	$PhCH_2$
e.e.	90%	78%	71%	73%	64%

21-16 The α-hydroxylation of the lactam below under a variety of conditions by a variety of oxidants gives very different diastereoisomer ratios. [*Tetrahedron* **1995**, *51*, 11831]

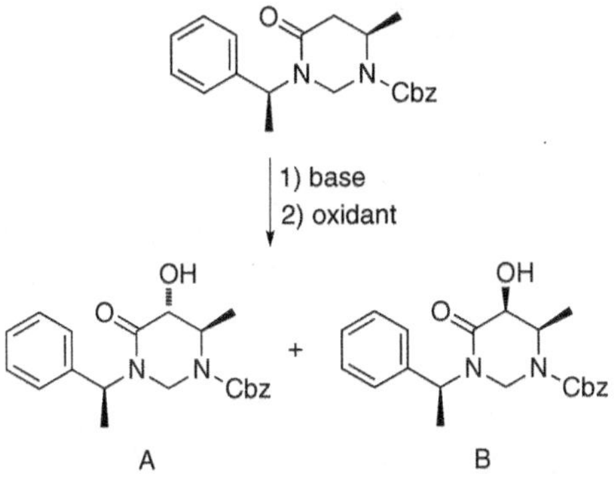

(a) Give a rationale for the observed stereochemistry of these reactions.

(b) Estimate the diastereoisomer ratio expected when the base in the last entry is changed to sodium hexamethyldisilazide (NaHMDS)

Base	Oxaziridine	Ratio A:B (Yield, %)	Base	Oxaziridine	Ratio A:B (Yield, %)
LiHMDS	MoOPH	80:20 (61%)	NaHMDS	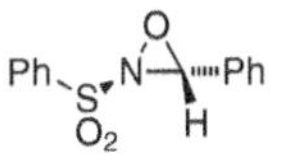 (±)	25:75 (77%)
NaHMDS	MoOPH	76:24 (66%)	LiHMDS	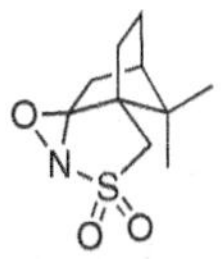	66:34 (78%)
LiHMDS	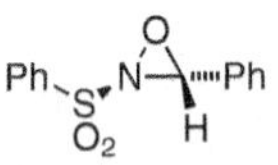 (±)	47:53 (41%)	LiHMDS	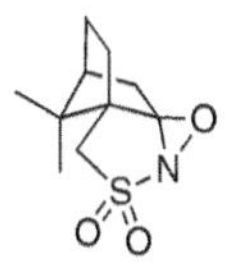	78:22 (66%)

21-17 The treatment of enolized α-cyclopentanediones with the Sharpless asymmetric epoxidation reagent leads to the formation of chiral γ-lactonic acids. The use of labeled *tert*-butyl hydroperoxide has suggested that the epoxide is an intermediate in the reaction. Suggest a mechanism for the formation of the observed product. At what point in the reaction is the stereochemistry of the observed product fixed? [*Tetrahedron* **2011**, *67*, 5942].

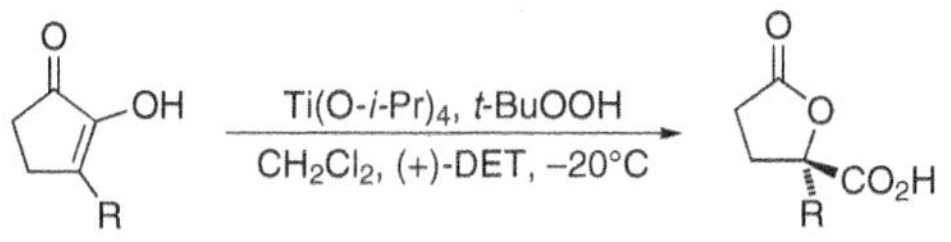

21-18 Consider the two closely related reactions shown below.

(a) What is the most likely reason for the wide disparity in the chemical yields of the two reactions?

(b) What is the most likely reason for the wide disparity in the enantioselectivities of the two reactions?

(c) What would you predict for the reaction where the 3-methyl-3-phenylpropanoate ester is used instead of the 2-methyl-3-phenylpropanoate ester?

[*Tetrahedron Lett.* **1986**, *27*, 3999]

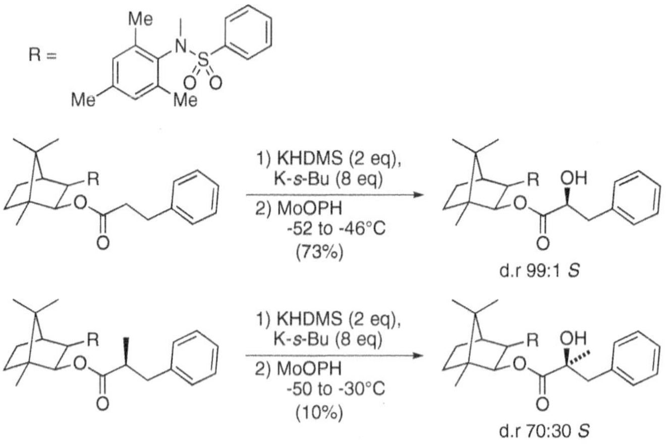

21-19 The two reactions below are believed to proceed by a common mechanism. In the same paper, it is reported that a chelating diphosphine ligand leads to a much slower reaction and loss of regioselectivity. Propose a mechanism that accounts for the observed stereochemistry and the effect of replacing the ligand by a chelating diphosphine.

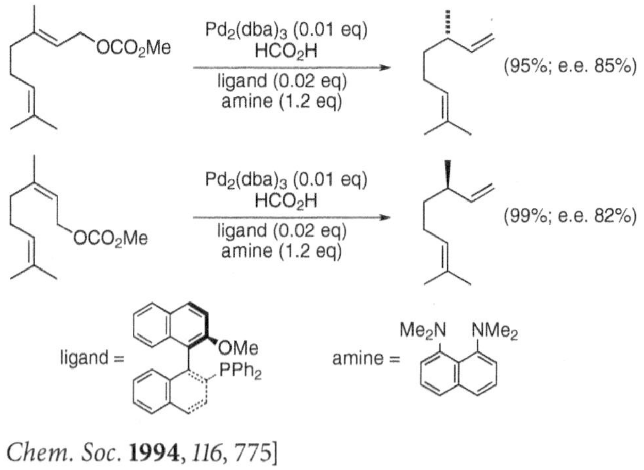

[*J. Am. Chem. Soc.* **1994**, *116*, 775]

21-20 What is the major organic product of each of the following reactions?

(a) 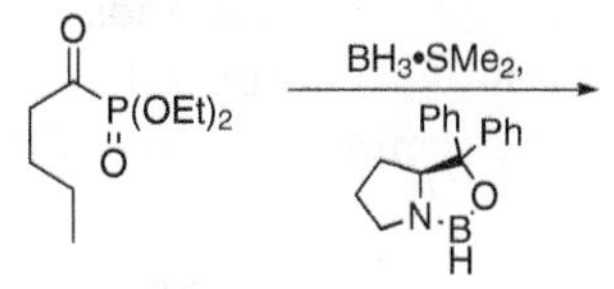

[*Tetrahedron:Asymmetry* **1994**, 5, 1965]

(b) 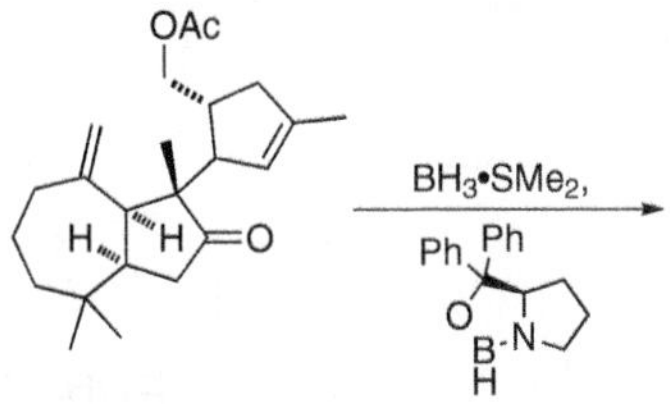

[*J. Am. Chem. Soc.* **2001**, *123*, 4851]

(c) 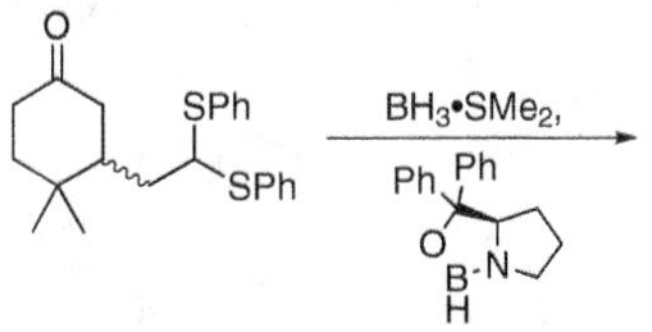

[*J. Am. Chem. Soc.* **2001**, *123*, 4851]

(d) 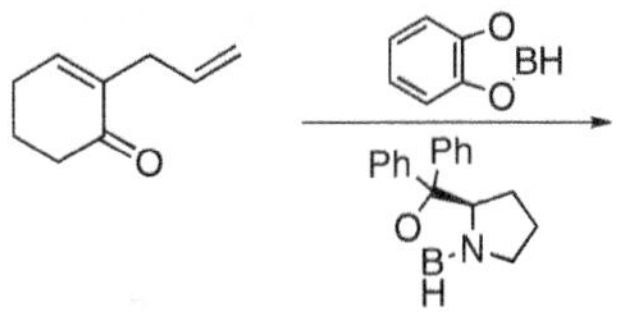

[*J. Am. Chem. Soc.* **1993**, *115*, 11028]

(e) 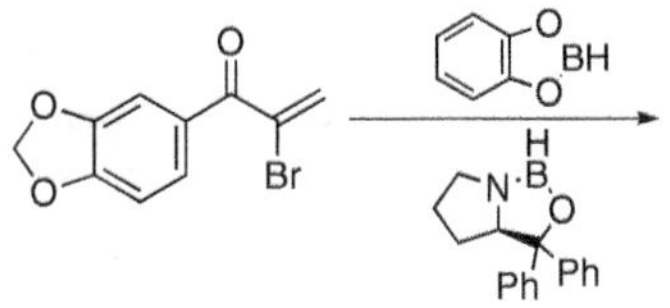

[*Org. Lett.* **1999**, *1*, 1871]

(f) 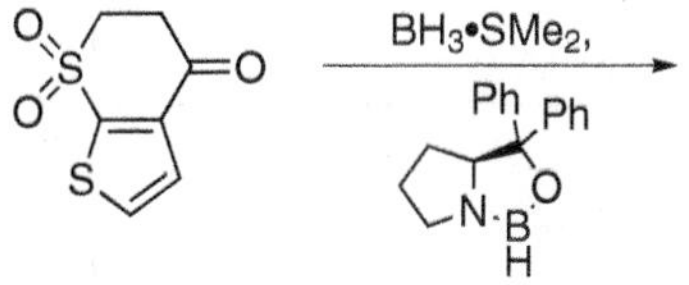

[*J. Org. Chem.* **1991**, *56*, 763]

(g) 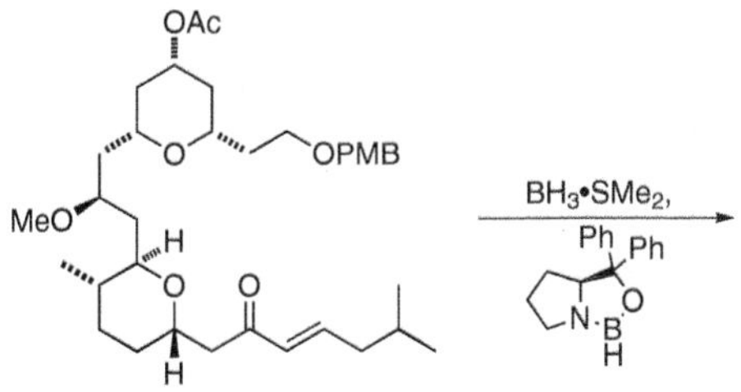

[*Org. Lett.* 2003, 5, 5035]

21-21 Based on the stereochemical outcomes for the asymmetric hydrogenations of the ester shown using rhodium complexes of the diphosphine ligands shown, can one design a simple mnemonic model to predict the outcome of each hydrogenation? If so, what is it? Are C_2-symmetric and non-C_2-symmetric ligands different, or can they be fit to the same model?

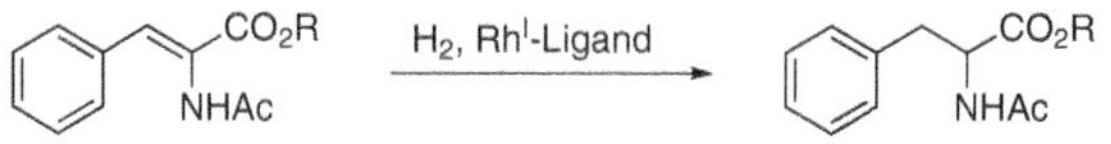

Ligand	Product	Ligand	Product	Ligand	Product
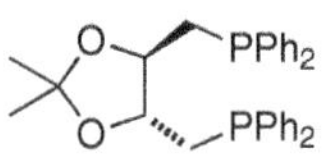 (*R,R*)-DIOP	*R*	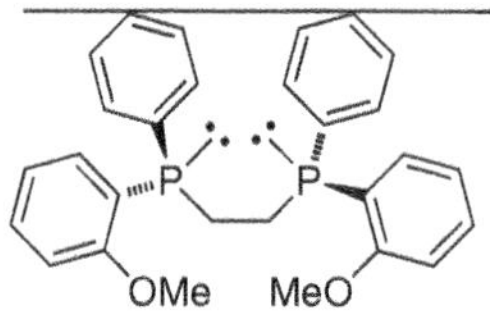 (*R,R*)-DiPAMP	*S*	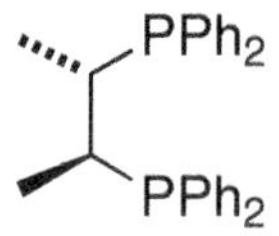 (*S,S*)-CHIRAPHOS	*R*
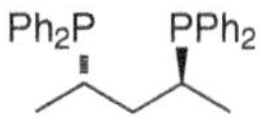 (*S,S*)-BDPP	***R***	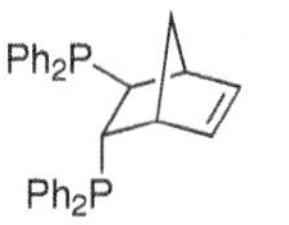 (*S,S*)-NORPHOS	*S*	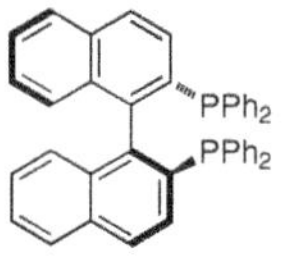 *S*-BINAP	*R*
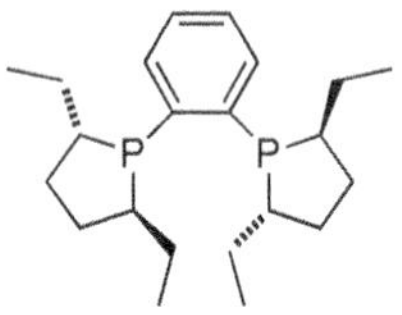 (*S,S*)-Et-DuPHOS	*S*	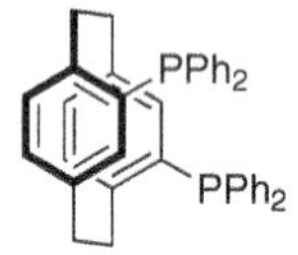 (*R*)-[2.2]PhanePhos	*R*	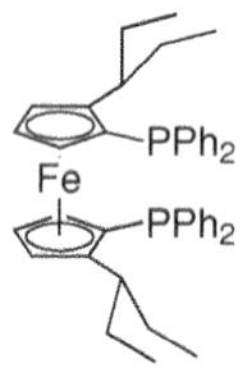 (*S,S*)-FerroPHOS	*R*
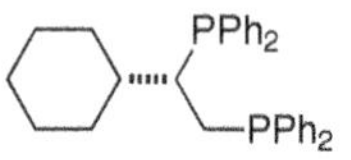 (*R*)-CycPHOS	*S*				

CHAPTER TWENTY-TWO

The Organic Compounds of Silicon, Phosphorus, Sulfur, Selenium, and Tin

22.1 Introduction

In the last half of Chapter 15, we focused our attention on controlling reaction chemoselectivity and regioselectivity by modifying the reactivity of functional groups by means of protecting groups. By their judicious use, one is able to reverse the relative reactivity of pairs of functional groups toward reagents—that is, one can force an ester to react with a Grignard reagent in preference to a ketone by first forming the ketal.

Using protecting groups represents a negating strategy for controlling reactivity—the undesired reactivity is blocked prior to the reaction, which means that the functional group must be regenerated later in the sequence. There is, however, also a proactive strategy for controlling reaction regiochemistry, which involves the use of ***directing groups***. These groups are functional groups whose presence in a substrate tends to stabilize a reactive intermediate in such a way that it controls the regioselectivity of a subsequent reaction.

The effects of substituents on the aromatic ring on the regiochemistry of electrophilic aromatic substitution are well known. Substituents with lone pairs on the atom directly attached to the ring tend to be *o,p*- directors and tend to activate the ring toward electrophiles, whereas the reverse is true for substituents bonded to the aromatic ring through an atom that is π-bonded to an electronegative second-row element (e.g., nitro, cyano, carbonyl, and sulfonyl groups). In the late 1970s, researchers found that certain substituents that ought to react with alkyllithium reagents (e.g., dialkylamide groups, Example **22.1**; and heterocyclic substituents such as the oxazolidinyl group, Example **22.2**) do not do so but instead direct metallation of arenes by alkyllithiums to the *ortho* position of the arene. It was possible to trap the resultant *o*-lithioarenes by a variety of electrophiles. In this chapter, however, we will not discuss these reactions at length but will focus instead on the organic chemistry of the heavier main group elements. Reactions **22.1** and **22.2** provide lead references for you to pursue the *o*-metallation chemistry further.

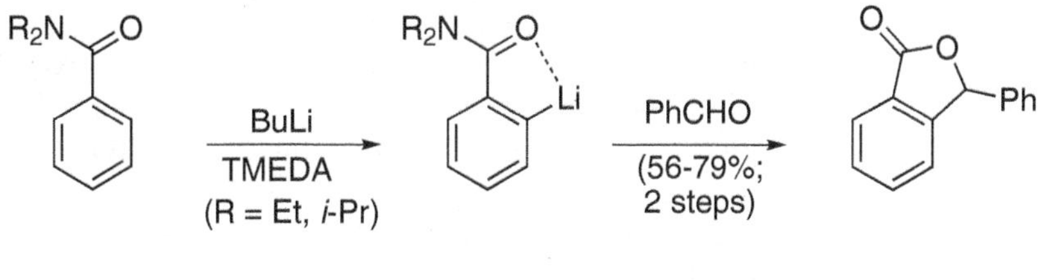

(22.1)[1]

1. (a) Beak, P.; Brown, R.A. *J. Org. Chem.* **1977**, *42*, 1823; **1982**, *47*, 34. (b) Beak, P.; Tse, A.; Hawkins, J.; Chen, C.-W.; Mills, S. *Tetrahedron* **1983**, *39*, 1983.

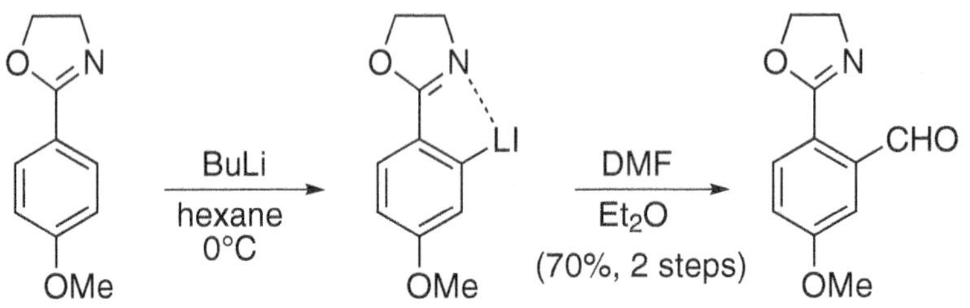

(22.2)[2]

In focusing on the organic chemistry of the heavier main group elements, we will also shift our focus from using protecting groups to more proactive methods for controlling organic reactivity—how to reverse the typical electrophilic or nucleophilic character of a functional group, how to use surrogates to stabilize unstable species, and how to direct the regiochemistry of a reaction. Let us begin with a brief introduction to the organic compounds of silicon, phosphorus, sulfur, selenium, and tin—the heavier main group elements most widely used in organic synthesis.

One place to begin our discussion of the organic chemistry of the heavier main group elements is by reexamining the bonding between carbon and these elements in terms of electronegativity. Recall that electronegativity is the ability of an atom *in a molecule* to attract electrons to itself; in an isolated atom, this becomes the electron affinity of the atom. Numerous electronegativity scales have been proposed since Pauling's first scale,[3] including the Allred-Rochow scale,[4] proposed 29 years later, and a later modification of Pauling's original scale by Allred 3 years later still.[5] The modified Pauling values developed by Allred are now the most commonly used. On the modified Pauling electronegativity scale (Table 22.1 contains the electronegativities of the main group elements), carbon has an electronegativity of 2.55 and hydrogen has an electronegativity of 2.20. These values confirm that the C—H bond is effectively non-polar but still very slightly polarized toward carbon.

This area of organic synthesis is vast, so here we will restrict our discussion to silicon, phosphorus, sulfur, selenium, and tin. Even with this restriction, what follows is, of necessity, only a brief introductory survey of the uses of these elements in modern organic chemistry. Fortunately, there are numerous excellent books covering the organic chemistry of each of these elements in depth, and it is worthwhile consulting them after you have completed this introductory material.

Table 22.1 Electronegativities of Main Group Elements

IIIA	IVA	VA	VIA	VIIA
B 2.04	C 2.55	N 3.04	O 3.44	F 3.98
Al 1.61	Si 1.90	P 2.19	S 2.58	Cl 3.16
Ga 1.81	Ge 2.01	As 2.18	Se 2.55	Br 2.96
In 1.78	Sn 1.96	Sb 2.05	Te 2.10	I 2.66
Tl 2.04	Pb 2.33	Bi 2.02	Po 2.00	At —

Classification, Structure, and Bonding

Carbon, nitrogen, oxygen, and boron are second-row elements. This very fact restricts the types of reactions in which these atoms can participate due to their inability to expand their outer shell octet of electrons. Except for compounds where aggregation is an important part of structure stabilization (e.g., the lithium alkyls), there are no stable compounds of carbon, nitrogen, or oxygen where the center atom is surrounded by more than four groups. In contrast to this, the heavier elements of the p block of the periodic table all have vacant d orbitals that may accommodate additional electrons beyond the octet required to fill the s and p subshells of any energy level.

2. Gschwend, H.W.; Hamdan, A. *J. Org. Chem.* **1975**, *40*, 2008.
3. Pauling, L. *J. Am. Chem. Soc.* **1932**, *54*, 3570.
4. Allred, A.L.; Rochow, E.G. *J. Inorg. Nucl. Chem.* **1958**, *5*, 264.
5. Allred, A.L. *J. Inorg. Nucl. Chem.* **1961**, *17*, 215.

These elements can and do form stable compounds where the center atom is surrounded by more than four groups.

π Bonding to Heavier Main Group Elements

The effect on bond strength of having to overlap orbitals from different electron shells to form a bond is even more pronounced in π bond formation than in σ bond formation. Unlike the π bonds between carbon and the second-row elements, π bonds between carbon and third-row elements are weak. Stable compounds with carbon-silicon π bonds are unknown, despite much research in an effort to prepare one (the silicones were given that name because it was believed that they possessed a silicon-oxygen π bond like the ketones). Most π bonds to phosphorus involve the phosphorus 3d orbitals, not the 3p orbitals, and there has been considerable debate about the exact nature of the carbon-phosphorus bond (formally a π bond) in the phosphorus ylides. Compounds with carbon-selenium π bonds are known, but they are difficult to make and extremely rare, and compounds with simple thiocarbonyl groups (e.g., thioaldehydes and thioketones) usually form cyclic trimers. In fact, thiocarbonyl derivatives of carboxylic acids are the only compounds with carbon-heteroatom π bonds that are widely found in organic chemistry. This means that unlike the reactions of alkyl halides, where elimination often competes with substitution, elimination does not compete with substitution if the elimination reaction would result in the formation of a carbon-heteroatom π bond. The formation of carbon-heteroatom π bonds is rare enough for us to omit it from our discussions in this chapter.

Group IVA

Silicon and tin probably have the simplest organic chemistry of any of the elements we will discuss in this chapter. Organic compounds of silicon are called ***silanes*** and organic compounds of tin are called ***stannanes***. In both the silanes and the stannanes, the heteroatom is sp^3 hybridized. Thus, the molecule is tetrahedral, and the only bonds to the heteroatom are σ bonds. The stability of π bonds involving silicon and tin is sufficiently low that we can also ignore the chemistry of π-bonded organosilicon or organotin species—we can restrict our discussion in this chapter to the chemistry of the σ-bonded compounds.

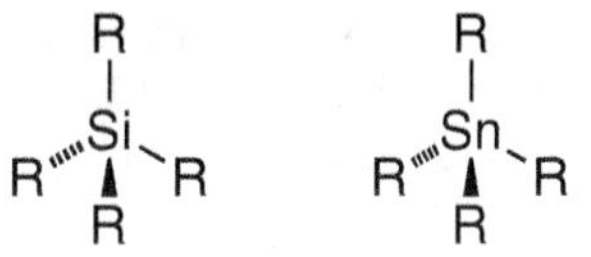

silane stannane

Their positions in the periodic table usually result in silicon being classed as a metalloid and tin as a metal, and the silanes and stannanes exhibit many of the properties of organometallic compounds in their chemistry. Silicon and tin are of comparable electronegativity, and both elements are less electronegative than hydrogen and considerably less electronegative than carbon (the modified Pauling electronegativities of silicon, tin, hydrogen, and carbon are 1.90, 1.96, 2.20, and 2.55, respectively). Carbon-silicon and carbon-tin bonds are polarized such that the carbon carries the partial negative charge. The peak electron density of the C—Si bond is closer to carbon than to the heteroatom. This has two major effects. First, it makes the C—Si σ bond an excellent electron donor by hyperconjugation. The presence of a silicon atom in a carbocation β to the electron-deficient carbon atom stabilizes the carbocation (Figure 22.1). Second, because the carbon atom is the negative end of the dipole and the heteroatom is the positive end, nucleophiles tend to attack silanes at the heteroatom and not at carbon. This is further facilitated by the fact that the heteroatom is much larger than carbon, making addition of the nucleophile to silicon a less sterically demanding process.

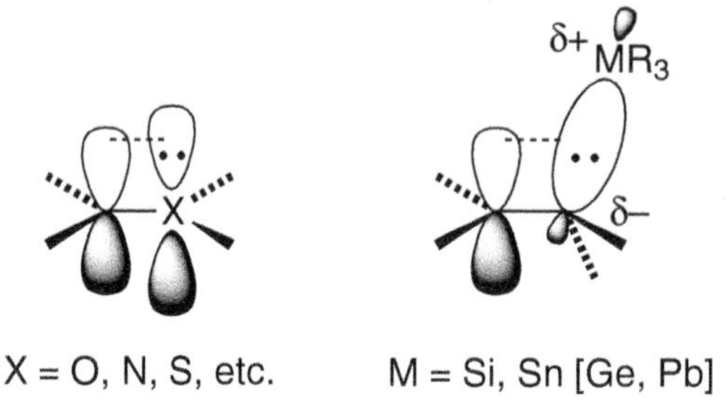

Figure 22.1 Stabilization of a carbocation by an atom bearing lone pair (left) and a methyl group carrying a Group IVA substituent (right)

Table 22.2 Bond Energies to Group IVA Elements (kcal mol^{-1})

Element	C—X	Si—X	Sn—X
H	88–98	90	60
C	82–88	88–94	46
N	80	100	—
O	80	100–136	95
F	116	139–158	—
Cl	80	109–117	82
Br	64	90–102	65
I	51	71–82	45

Much of the chemistry of organosilicon compounds is driven by the extreme thermodynamic stability of the Si—O and Si—F bonds, as we shall see later in this chapter. Table 22.2 contains values for average bond energies of a number of σ bonds to the Group IVA elements.[6] The values for the corresponding bonds to tin are approximately 10 to 30 kcal mol^{-1} lower than those of the bonds to silicon. The most important value in the table is the Si—F dissociation energy, which is the largest value in the table. This large bond energy is consistent with the formation of a Si—F bond being very favorable. It rationalizes the fact that fluoride ion is capable of displacing a number of anions (e.g., enolate anions and alkynide anions) from trimethylsilyl derivatives.

The data in Table 22.2 illustrate several important relationships about bond strengths of some of the more important σ bonds involving the Group IVA elements. It is worth noting that C—H and C—C σ bonds are similar in strength to Si—H and Si—C σ bonds, whereas σ bonds between silicon and the other elements in Table 22.2 are stronger than the corresponding σ bonds between carbon and the same electronegative atoms. The silicon-fluorine σ bond is one of the strongest σ bonds known. A similar pattern emerges for the strengths of σ bonds to tin: most Sn—X bonds are weaker than the C—X bonds, although Sn—O and Sn—Cl bonds are stronger than C—O and C—Cl bonds. Unlike silicon, tin forms a weaker bond to iodine than does carbon.

Group VA

Phosphorus, in the form of a staggering array of phosphine ligands, has become an indispensable partner in modern organic synthesis. Phosphorus forms tricoordinate organic compounds that are completely analogous to the corresponding nitrogen compounds. The simplest hydride of phosphorus, PH_3, is known as ***phosphine***, and each of the P—H bonds can be replaced to give ***alkylphosphines***, ***dialkylphosphines***, or ***trialkylphosphines***; the phosphorus analogs of ammonium ions are ***tetraalkylphosphonium ions***. Like the amines, the phosphines are pyramidal, with typical bond angles around phosphorus of 93° to 100°, increasing to 109.5° when the alkyl groups are extremely bulky. However, the barrier to inversion of pyramidal phosphorus atom in PH_3 (17 kcal mol^{-1}) is much larger than the corresponding barrier to inversion of pyramidal nitrogen in NH_3 (5.9 kcal mol^{-1}. The trend continues with AsH_3 (32.1 kcal mol^{-1}).[7] Measurement of racemization rates of the unimo-

6. Data from (a) Cotton, F.A.; Wilkinson, G. *Advanced Inorganic Chemistry*, 5th ed. (Wiley-Interscience: New York, 1988), p. 266. (b) Walsh, R. *Acc. Chem. Res.* **1981**, *14*, 246. (c) Walsh, R. http://www.gelest.com/goods/pdf/Library/10BondDiss.pdf

7. Costain, C.C.; Sutherland, G.B.B.M. *J. Phys. Chem.* **1952**, *56*, 321.

lecular inversion of chiral phosphines gives activation energies between 29 and 36 kcal mol^{-1},[8] and nuclear magnetic resonance measurements give inversion barriers of 35.7 kcal mol^{-1} in six-membered phosphinanes.[9]

The P—C bonds in trialkylphosphines can be replaced by P—X bonds, where X is a heteroatom. The derivatives of tricoordinate phosphorus are presented in Figure 22.2. Derivatives of this type are derivatives of (hypothetical) tribasic phosphorous acid, $P(OH)_3$. The phosphorus trihalides have long been used to convert alcohols to alkyl halides; the intermediate proposed in these reactions is the *trialkyl phosphite*. The reaction of a phosphorus trihalide with a secondary amine gives a ***hexaalkylphosphorous triamide***.

As noted above, tribasic phosphorous acid does not exist, except in the form of its derivatives. Instead, the phosphorus atom in phosphorous acid, H_3PO_3, is not tricoordinate. It is tetracoordinate, with a P—H bond and a ***phosphoryl group***, P=O. Compounds with the phosphoryl group are abundant: the phosphorus atom in ***phosphine oxides***, ***phosphonate esters***, and ***phosphate esters*** is of this type. Replacement of the alkyl groups of the trialkylphosphine oxide with dialkylamino groups gives the ***hexaalkylphosphoric triamides***. We have seen one such compound in the form of the aprotic dipolar solvent hexamethylphosphoramide (HMPA; R=Me). Structures of a range of representative phosphoryl compounds are presented in Figure 22.3.

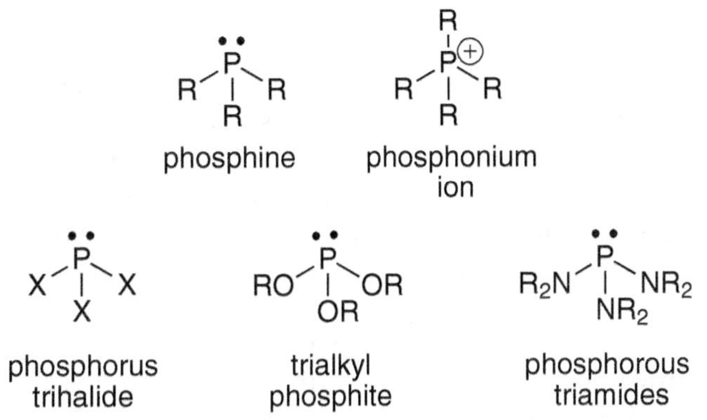

Figure 22.2 Tricoordinate organophosphorus compounds

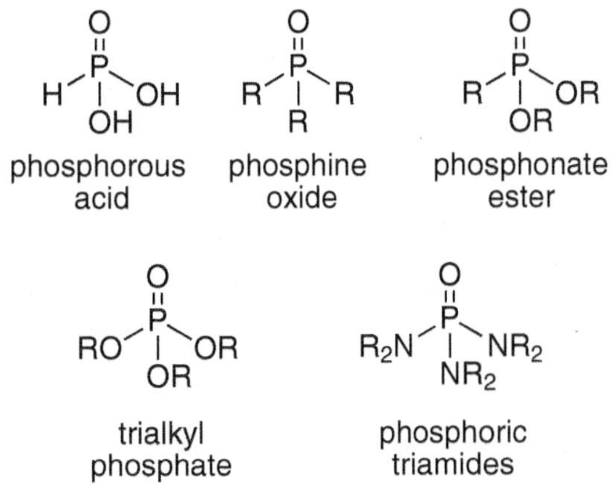

Figure 22.3 Phosphoryl compounds

---

8. Baechler, R.D.; Mislow, K. *J. Am. Chem. Soc.* **1970**, *92*, 3090.
9. Macdonell, G.D.; Berlin, K.D.; Baker, J.R.; Ealick, S.E.; Van der Helm, D.; Marsi, K.L. *J. Am. Chem. Soc.* **1978**, *100*, 4535.

Much of the chemistry of organophosphorus compounds is driven by the extreme thermodynamic stability of the phosphoryl group. Bond energies for the phosphoryl group vary from 125 kcal mol^{-1} in $Br_3P{=}O$ to 151 kcal mol^{-1} in $(EtO)_3P{=}O$, with values near 130 to 134 kcal mol^{-1} for most compounds.[10] Thus, the formation of a phosphoryl group from a trivalent phosphorus compound, or its formation by replacement of the formal P=C bond in ylides, provides considerable driving force for the reaction. This almost certainly provides much of the driving force for the Wittig, Horner-Wadsworth-Emmons, and Mitsunobu reactions, for example.

Although arsenic compounds are not as widely used in organic chemistry as their phosphorus counterparts, you should note that there is an arsenic analog of most of the phosphorus compounds in Figures 22.2 and 22.3 (e.g., arsines are structurally analogous to phosphines), but, unlike phosphorous acid, which has a phosphoryl group, arsenous acid exists as the tricoordinate, tribasic acid, $As(OH)_3$.

Group VIA

The two elements directly below oxygen in the periodic table, sulfur and selenium, have become invaluable tools in organic chemistry. Although many of the sulfur and selenium compounds were discovered during the 19th century, much of their synthetic utility was not developed until the 1960s and 1970s.

Stable organosulfur compounds are known where the heteroatom is in more oxidation states than any other element, and all are used in modern organic chemistry. The simplest organic sulfur compounds are organic derivatives of hydrogen sulfide—the ***thiols*** (or *mercaptans*, from the Latin phrase describing their ability to form crystalline mercury compounds by reacting with mercuric oxide, *mercurium captans*) and ***thioethers*** (or *sulfides*). Mild oxidation of the thiols gives the ***dialkyl disulfides***, and replacement of the hydrogen of a thiol by a halogen atom gives ***sulfenyl halides***. The ***trialkylsulfonium ions***, which are precursors to sulfonium ylides, are formed by alkylation of thioethers with strong electrophiles. The functional groups based on bicoordinate sulfur are presented in Figure 22.4.

Oxidation of divalent sulfur leads to the formation of a thionyl group, S=O, and it is possible to oxidize the sulfur to a level where the sulfur is part of one or two thionyl moieties. Oxidation of the sulfur atom of a thioether gives first a ***dialkyl sulfoxide***, with one thionyl group, then a ***dialkyl sulfone***, with two. Oxidation of the sulfur of a thiol ultimately gives a ***sulfonic acid*** and its derivatives (e.g., a *sulfonyl chloride*). Alkylation of a sulfoxide gives the ***sulfoxonium ion***, which is a precursor to the sulfoxonium ylides. As with the trivalent phosphorus compounds, the barrier to inversion in tricoordinate sulfur compounds such as the sulfoxides is large enough to permit the isolation of enantiomerically pure

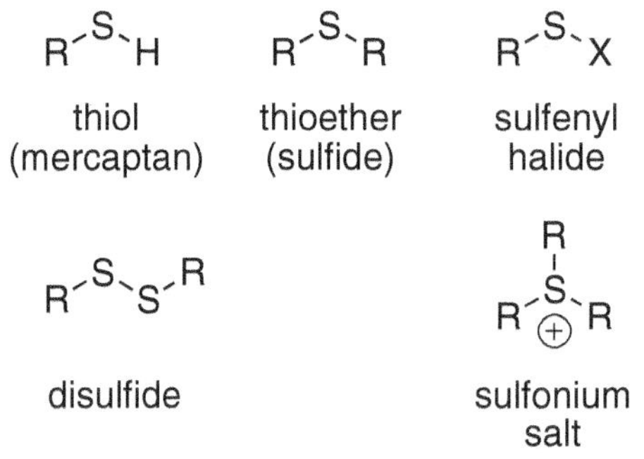

Figure 22.4 Functional groups based on bicoordinate sulfur

10. (a) Emsley, J.; Hall, D. *The Chemistry of Phosphorus* (Halsted Press: New York, 1976); p. 42. (b) Goldwhite, H. *Introduction to Phosphorus Chemistry* (Cambridge Univresity Press: Cambridge, 1981); p. 30.

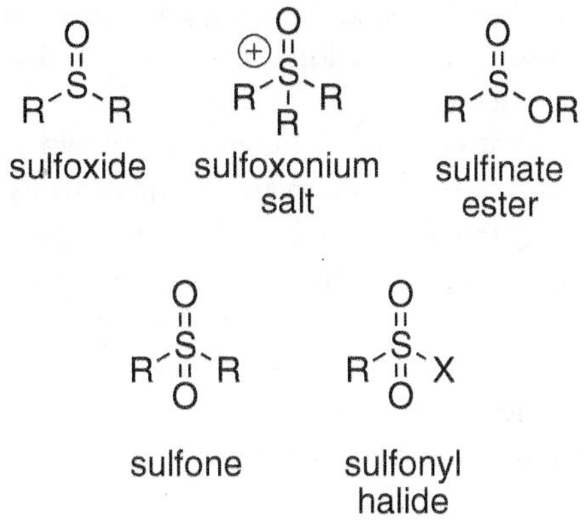

Figure 22.5 Functional groups containing the thionyl moiety

sulfoxides where the chiral center is the sulfur.[11] Functional groups containing the thionyl moiety are gathered in Figure 22.5.

Table 22.3 Approximate Average Bond Energies of σ Bonds to Group VIA Elements (Bond Energy (kcal mol^{-1}) [kJ kcal mol^{-1}])

Element	H	C	O	S	Se
O	111 [464]	80 [336]	33 [138]	—	—
S	88 [368]	65 [272]	—	63 [264]	
Se	67 [230]	58 [243]	—	—	44 [184]

Practically every structure in Figures 22.4 and 22.5 can be replicated by replacing sulfur by selenium to give compounds that are completely analogous to the corresponding sulfur compounds. One obtains selenols by analogy with thiols, diselenides by analogy with disulfides, selenenic acids by analogy with sulfenic acids, and so on. Chemically speaking, the major difference between the organoselenium compounds and their organosulfur counterparts is usually one of reaction rate, not reaction outcome. Like the organosulfur compounds, organoselenium compounds—especially the selenides and selenoxides—have become routinely used in modern organic synthesis.

The modified Pauling electronegativities of sulfur and selenium (2.58 and 2.56, respectively) are practically identical to that of carbon (2.55), so carbon-sulfur and carbon-selenium σ bonds are effectively non-polar and much less efficient than carbon-silicon σ bonds at stabilizing a positive charge at the β position by hyperconjugation. Sulfur and selenium do stabilize these carbocations, but the stabilization is achieved by bond formation to give a three-membered heteronium ion. These cyclic cations are important intermediates in the addition reactions of sulfur and selenium electrophiles.

The bond strengths of important σ bonds involving the Group VIA elements are compared in Table 22.3. These bond energies show much the same trend as those for the Group IVA elements: the strength of the σ bonds to carbon decrease as one descends the periodic table.

22.2 Silanes and Stannanes as Directing Groups: Carbocation Chemistry

Silicon as a Surrogate for Hydrogen Bound to Carbon

The chemistry of organic silicon compounds is rather fortunately covered by a series of reviews by British chemist Ian Fleming.[12] Despite their age, they are well worth reading. One of the most common reactions of alkylsilanes is the replacement of the carbon-silicon bond with a carbon-electrophile bond. This reaction is especially common in the reactions

11. Rayner, D.R.; Gordon, A.J.; Mislow, K. *J. Am. Chem. Soc.* **1968**, *90*, 4854.

12. (a) Fleming, I.; Barbero, A.; Walter, D. *Chem. Rev.* **1997**, *97*, 2063. (b) Fleming, I.; Dunoguès, J.; Smithers, R. *Org. React.* **1989**, *37*, 57; *Chem. Soc. Rev.* **1981**, *10*, 83; *Comprehensive Org. Synth.* **1991**, *2*, 563. (c) Fleming, I. *Comprehensive Org. Chem.* **1980**, *4*, 542.

Table 22.4 σ^+ Constants for Silicon- and Tin-containing Substituents

Group	σ^+ constant
$—CH_3$	-0.30
$—CH_2SiPh_3$	-0.4
$—CH_2SiMe_3$	-0.62
$—OCH_3$	-0.78
$—CH_2SnMe_3$	-0.92
$—CH_2SnPh_3$	-1.0

between electrophiles and allylsilanes, arylsilanes, and vinylsilanes. In all three of these compounds, the regiochemistry of the addition of electrophiles to the alkene π bond is directed by the Group IVA element so that the new carbocation center is formed in the position β to the Group IVA atom. This phenomenon, known as the ***β effect***, occurs because the carbocation centers at this position are stabilized by hyperconjugation with adjacent carbon-silicon σ bonds (recall Figure 22.1). The partial positive charge on the silicon atom, its larger size, and its ability to expand its valence shell all lead to nucleophiles' tending to attack silanes at silicon rather than at carbon or hydrogen. This results in electrophilic substitution of the silicon often being the dominant reaction pathway. This may be contrasted with the reactions of simple alkenes with electrophiles, where electrophilic addition is the dominant reaction pathway (cyclohexene reacts with bromine to give *E*-1,2-dibromocyclohexane, for example).

The effectiveness of silicon- and tin-containing substituents at stabilizing a carbocation at the β position can be gauged from the data in Table 22.4, where the σ^+ substituent constants (see Section 8.5 for review) for a group of electron-releasing substituents are presented. As is evident, all the silylalkyl and stannylalkyl substituents are significantly more effective at stabilizing a positive charge than a methyl group and are actually comparable to a methoxy group.

Much of the chemistry of allylsilanes was discussed in Chapter 10 (Section 10.4), so the discussion here will be rather abbreviated. Unsaturated alkylsilanes and arylsilanes react with electrophiles by initial addition of the electrophile to the alkene to generate a silicon-stabilized carbocation (Figure 22.6). Attack of the nucleophile on the silicon-stabilized cation occurs at silicon rather than at carbon to generate the product where the silyl group has been substituted— directly by the electrophile in the case of the vinyl- and arylsilanes and with allylic rearrangement in the case of the allylsilanes. This result with allylsilanes contrasts with the results of nucleophilic substitution of allylic halides, where the S_N2 and S_N2' mechanisms often compete. This leads to mixtures of products with allylic rearrangement and without it. The electrophile in reactions with allyl- and vinylsilanes can be a proton, a halogen or halogen-like molecule, or a carbon electrophile.

The reaction between an allylsilane and an electrophile is also highly regioselective. In the synthesis of artemisia ketone by means of Friedel-Crafts acylation of the allylsilane

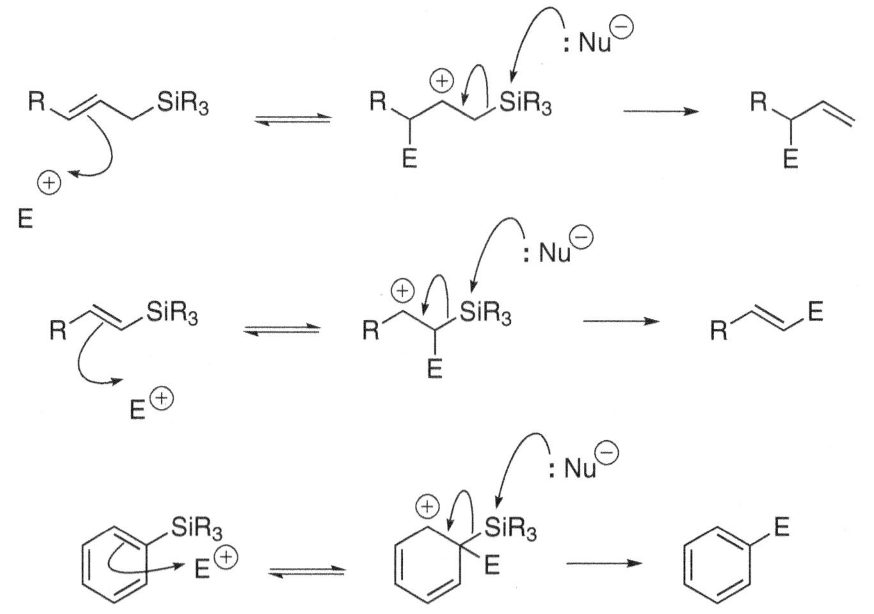

Figure 22.6 Reactions of allylsilanes (top), vinylsilanes (middle), and arylsilanes (bottom) with electrophiles.

shown in Example **22.3**,[13] the electrophile attacks the silane to give the β-silyl cation in preference to the tertiary carbocation. This occurs despite the fact that this requires attack at the more sterically hindered end of the double bond. The subsequent regiospecific allylic rearrangement leads to simultaneous generation of a new quaternary center and an unsubstituted vinyl group. In fact, the reactions of similar allylic silanes can be used to generate quaternary centers in highly hindered environments, as illustrated by Example **22.4**,[14] where a *tert*-butyl group is incorporated into the molecule to give a product with adjacent quaternary centers as well as the vinyl group.

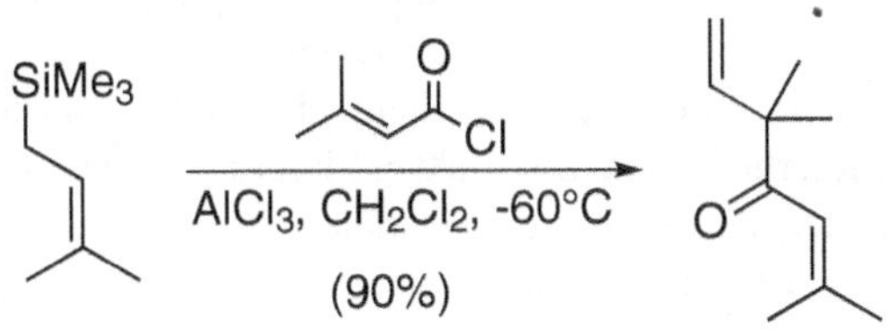

(22.3)

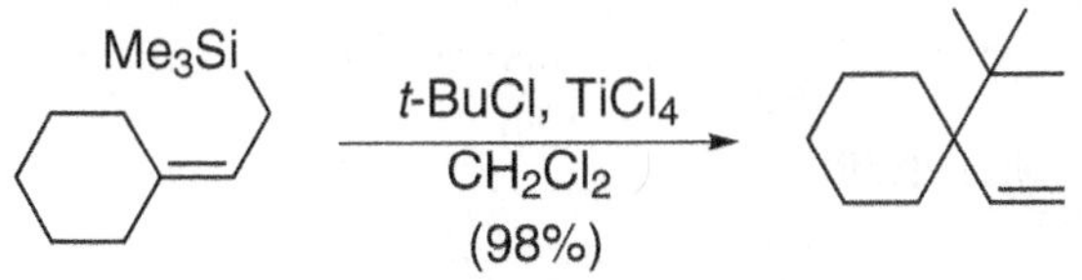

(22.4)

The electrophilic substitution of allylsilanes has been shown to occur by way of an *anti*-S_E2' mechanism.[15] The stereochemistry is most readily viewed in terms of a preferred conformation, where (1) the alkene π* orbital is eclipsed by the C—Si σ orbital to maximize the hyperconjugative stabilization and (2) the smaller group bonded to the silane carbon, eclipses the double bond. For both steric and electronic reasons, the electrophile attacks the face opposite the silicon, as shown in Figure 22.7. In this example, which uses a chiral allylsilane, we see that there is also efficient transfer of chirality to the new chiral center during the reaction.

The corresponding reaction of vinylsilanes occurs with retention of configuration, by the mechanism shown in Figure 22.8. In this case, the attack of the electrophile on the alkene π orbital from a direction perpendicular to the σ framework leads to an initial cation that becomes stabilized by the silicon only after rotation around the σ bond. As is evident from the diagram, there is a preferred direction of rotation (requiring rotation through only 60° rather than 120°, in keeping with the principle of least motion[16]). This brings the electrophile into a position fairly closely approximating the original position of the silicon in the silane. Desilylation leads to the product in which the configuration of the double bond is retained.

In addition to simple alkyl cations, there are many carbon electrophiles that will react with silanes and stannanes. All show the same level of regiochemical control by the silicon. These include cations generated (Example **22.5**) by the reaction of organic oxygen compounds (e.g., aldehydes, ketones, acetals, ketals, epoxides, esters, acid chlorides) with

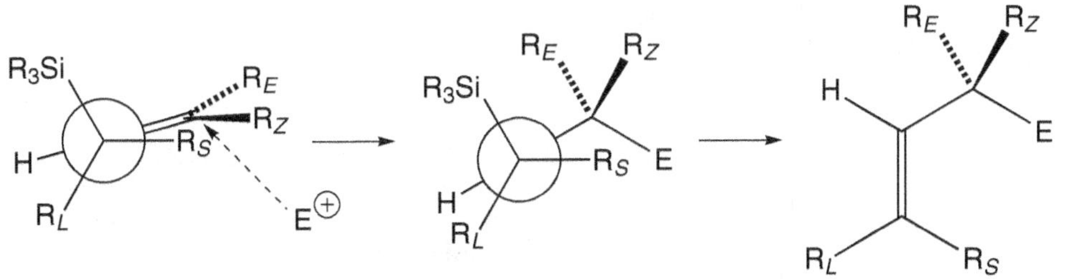

Figure 22.7 The stereochemistry of the reaction between an allylsilane and an electrophile. R_S is the smaller substituent; R_L is the larger substituent.

13. Pillot, J.-P.; Dunogues, J.; Calas, R. *Tetrahedron Lett.* **1976**, 1871.
14. Fleming I.; Paterson, I. *Synthesis* **1979**, 446.
15. Koenig, K.E.; Weber, W.P. *J. Am. Chem. Soc.* **1973**, *95*, 3416.
16. Hine, J. *Adv. Phys. Org. Chem.* **1977**, *15*, 1.

Figure 22.8 The stereochemistry of the reaction of a vinylsilane with an electrophile

Lewis acids. The reaction of these cations with allylsilanes is known as the ***Hosomi-Sakurai reaction***.[17] The electrophilic complexes of enones with Lewis acids react with allylsilanes regioselectively through the β carbon of the enone, thus providing a non-basic alternative to the Michael addition of cuprate reagents. In the Hosomi-Sakurai reactions of allylsilanes with aldehydes, the alcohol obtained exhibits a strong preference for the 1,2-*syn* stereochemistry when the allylsilane is the *E* isomer. The *syn/anti* stereoselectivity of the *Z* allylsilanes is often lower in this reaction.[18]

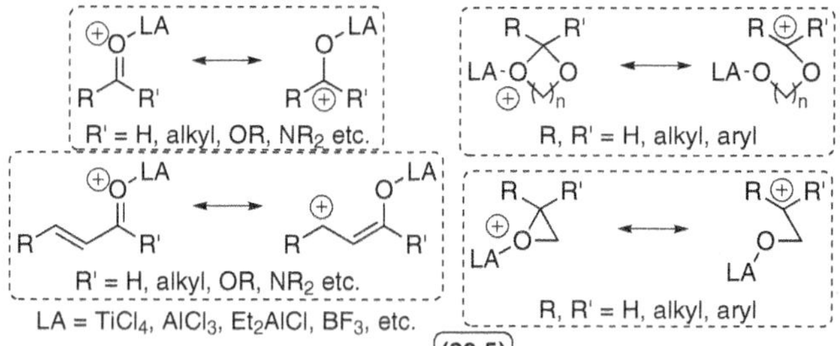

(22.5)

Worked Problem

22-1 Give a reasonable pathway for the following transformation. How does your mechanism rationalize the observed stereochemistry? [*Tetrahedron Lett.* **2010**, *51*, 3542]

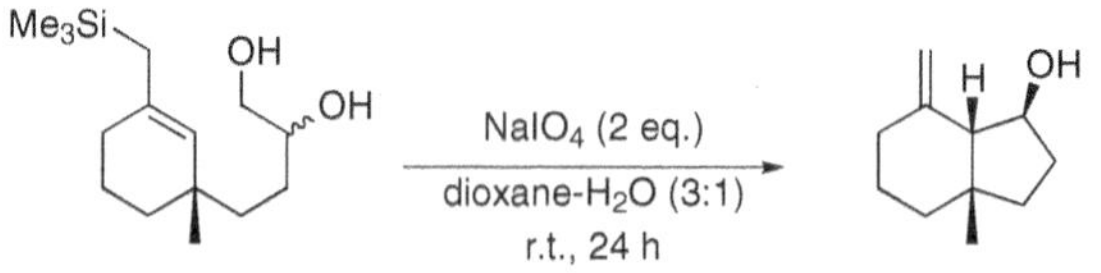

§Answer below.

17. (a) Hosomi, A.; Sakurai, H. *Tetrahedron Lett.* **1976**, 1295; *J. Am. Chem. Soc.* **1977**, *99*, 1673. (b) Hosomi, A.; Endo, M.; Sakurai, H. *Chem. Lett.* **1976**, *5*, 941.
Reviews: (c) Hosomi, A. *Acc. Chem. Res.* **1988**, *21*, 200. (d) Fleming, I.; Dunogues, J.; Smithers, R. *Org. React.* **1989**, *37*, 57. (e) Fleming, I. *Comp. Org. Synth.* **1991**, *2*, 563.
18. Green, J.R.; Majewski, M.; Alo, B.I.; Snieckus, V. *Tetrahedron Lett.* **1986**, *27*, 535.

§ Answer to Worked Problem:
The first step in the mechanism is the cleavage of the diol to the aldehyde. The Hosomi-Sakurai addition of the carbonyl group to the allylsilane can be catalyzed by traces of acid or a Lewis acid in the mixture. The addition should take place through the less hindered *exo* orientation shown, and the resultant silicon-stabilized carbocation will then be attacked by water to give the observed final product.

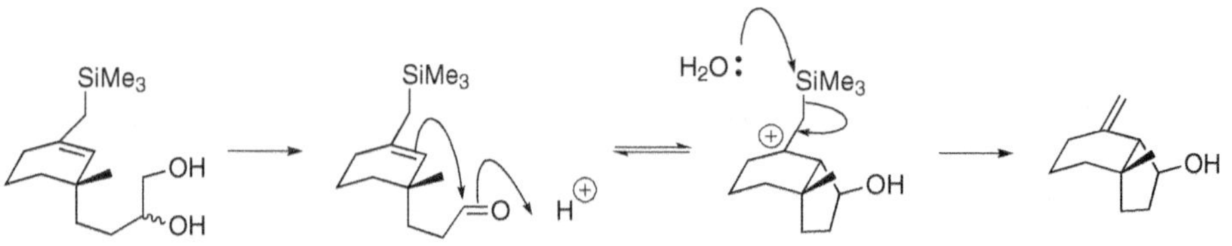

Problem

22-1 What is the structure of the major organic product of each of the following reactions? Where appropriate specify the stereochemistry of the reaction.

(a)
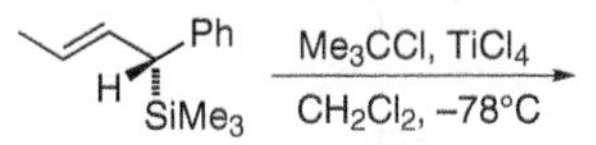

(b)
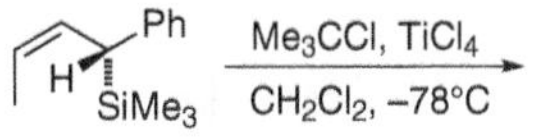

(c)
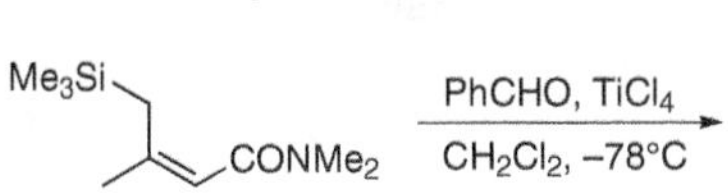

(d)
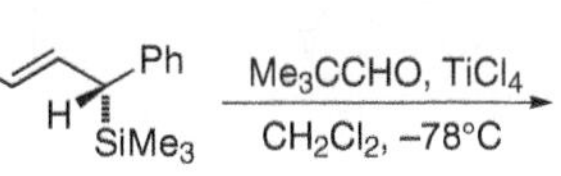

(e)
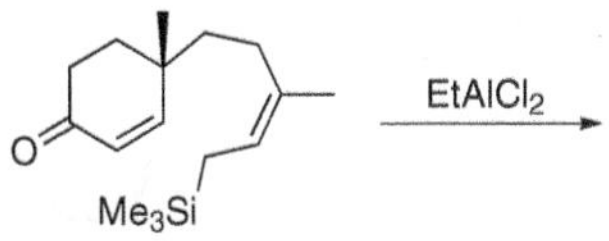

(f)
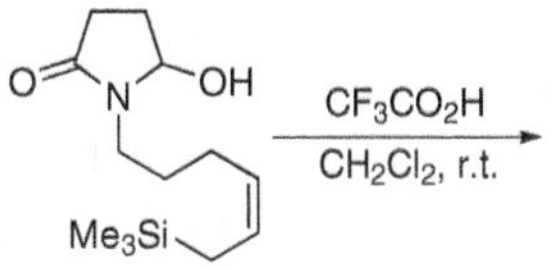

References: (a) (b) *J. Am. Chem. Soc.* **1982**, *104*, 4962. (c) *Tetrahedron Lett.* **1986**, *27*, 535. (d) *J. Am. Chem. Soc.* **1982**, *104*, 4963. (e) *Tetrahedron Lett.* **1985**, *26*, 2755. (f) *J. Org. Chem.* **1985**, *50*, 4014.

Silicon as a Surrogate for Hydrogen on Oxygen: Silyl Ethers

If one is seeking a surrogate for an element, one would be looking first at elements with similar electronegativities—surrogates for hydrogen, for example, should have values between 2.2 and 1.9. Elements that have been successfully used as surrogates for the hydrogen can be seen in the many variants of the Heck cross-coupling reaction: boron (2.04, used in the Suzuki and related coupling reactions); silicon (1.90, used in the Hiyama-Denmark cross-coupling); and tin (1.96, used in the Stille coupling). Two elements not in Table 22.1 that have also served as hydrogen surrogates include the palladium that is essential to these catalyzed reactions (2.20) and the copper essential to the Sonogashira coupling (1.90). In yet another example—the oxymercuration reaction—the mercury that is a surrogate for electrophilic hydrogen also has an electronegativity (2.00) close to that of hydrogen.

Since the late 1960s, silicon has evolved into a routine part of modern organic synthesis. Much of the rise in the popularity of silicon compounds as reagents and intermediates in modern chemistry can be traced to the ability of silicon to replace other electropositive elements—most notably hydrogen and the alkali metals—in organic compounds. In these compounds, the silicon-heteroatom bond is usually much less labile than the corresponding hydrogen-heteroatom bond, which makes the silyl derivative an excellent protected version of the reactive functional group. Moreover, the relatively facile cleavage of the silicon-heteroatom bond under mild conditions (Si—O bonds are susceptible to acid or base hydrolysis by water, for example) also makes the silyl group an excellent surrogate for the reactive hydrogen. As we saw in Chapter 15, silyl ethers are widely used as protecting groups for both alcohols and enols.

The substitution of hydrogen by silicon often dramatically increases the chemical stability of compounds, as illustrated by the examples in Figure 22.9. Trimethylsilyl azide, for example, is quite thermally stable, whereas the corresponding compound containing hydrogen, hydrazoic acid, is dangerously explosive. *It should be remembered, however, that the replacement of hydrogen by silicon seldom changes the biological activity of a compound because the silyl group is usually lost in moist tissues—trimethylsilyl azide is still a virulent poison.*[19] The trialkylsilyl enol ethers provide another good example of example of the effects of replacing hydrogen by silicon. Although the enols of simple aldehydes and ketones

19. Birkofer, L.; Wegner, P. *Org. Synth.* **1970**, *50*, 107; **1988**, *Coll. Vol. 6*, 1030.

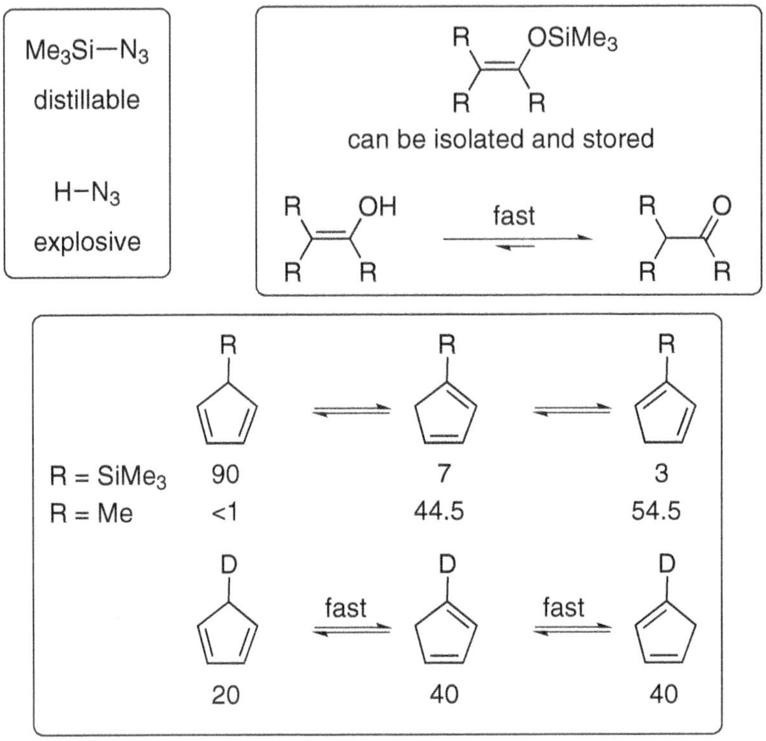

Figure 22.9 Effects of replacing hydrogen by trimethylsilyl on reactivity and stability of compounds

cannot be isolated because they rapidly tautomerize to the keto forms, enol trimethylsilyl ethers of simple ketones can be easily prepared, isolated, and purified. As a surrogate for hydrogen, the trimethylsilyl group suppresses the [1,5]-sigmatropic rearrangement of hydrogen in cyclopentadiene and alters the position of the equilibrium to give the mixture shown, where the silicon is preferentially bound to the sp^3 carbon.[20] By comparison, the methylcyclopentadienes undergo rapid sigmatropic shift of hydrogen to give a mixture of 1- and 2-methyl-1,3-cyclopentadienes with almost no 5-methyl-1,3-cyclopentadiene.[21] The migration of hydrogen in the silylated compound is much slower than in the parent diene.[22] In similar fashion, 5-deuterio-1,3-cyclopentadiene rapidly isomerizes to a statistical mixture of all three isotopomers.[23]

Silyl Enol Ethers

The silicon-oxygen bond is a very strong bond, and its formation is usually exothermic. However, despite its thermodynamic stability, the silicon-oxygen bond is also fairly easy to break (it is kinetically labile), a property that it shares with the O—H bond. Much useful chemistry is the result of this combination of stability and reactivity. Silyl ethers do not transfer silicon to nucleophilic species nearly as readily as the hydroxyl group transfers a proton, so the silyl ethers are much more resistant to nucleophilic attack. This property makes silyl ethers excellent protecting groups for alcohols, as discussed in Chapter 15. The hydroxyl group can be recovered from the silyl ether by mild acid hydrolysis or treatment with fluoride ion. The same chemistry can also be applied to the protection of enols and phenols as silyl ethers.

---

20. Ashe, A.J., III *J. Am. Chem. Soc.* **1970**, *92*, 1233.
21. McLean, S.; Haynes, P. *Tetrahedron* **1965**, *21*, 2313.
22. Taha, A.N.; Moreno, P.O.; LeMaster, C.B.; LeMaster, C.L.; True, N.S. *J. Mol. Struct.* **2000**, *553*, 37.
23. McLean, S.; Haynes, P. *Tetrahedron* **1965**, *21*, 2329.

Silyl enol ethers are formed by the reaction between an enolate anion, which is an ambident nucleophile, with a hard electrophile such as a silyl halide or triflate. Under these conditions, the enolate anion acts as an oxygen nucleophile. The enolate anion itself can be prepared by any of the methods shown in Example **22.6**: (1) kinetic or thermodynamic deprotonation of a carbonyl compound; (2) Birch reduction of a conjugated enone in the absence of a proton donor; or (3) conjugate addition of an organometallic nucleophile (e.g., a cuprate) to a conjugated enone.

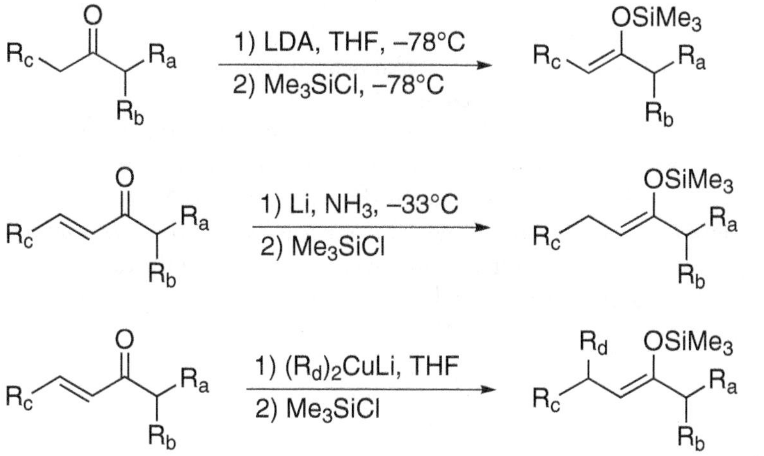

(22.6)

Silylation can be an important way to moderate extreme reactivity. Enolate anions, for example, are strong bases—one cannot alkylate the lithium enolates of carbonyl compounds with tertiary halides because the strongly basic enolate anion invariably causes E2 elimination from the halide. One can, however, alkylate a carbonyl compound with a tertiary halide by converting the lithium enolate into the enol trimethylsilyl ether first, then treating this nucleophilic alkene with the tertiary halide and a Lewis acid catalyst,[24] usually titanium tetrachloride (Example **22.7**). This reaction also works well with allylic and benzylic halides. This shows that enol silyl ethers are excellent surrogates for the free enol, so that reactions that might involve the strongly basic enolate anion can be run under non-basic conditions (e.g., the Mukaiyama aldol addition[25]).

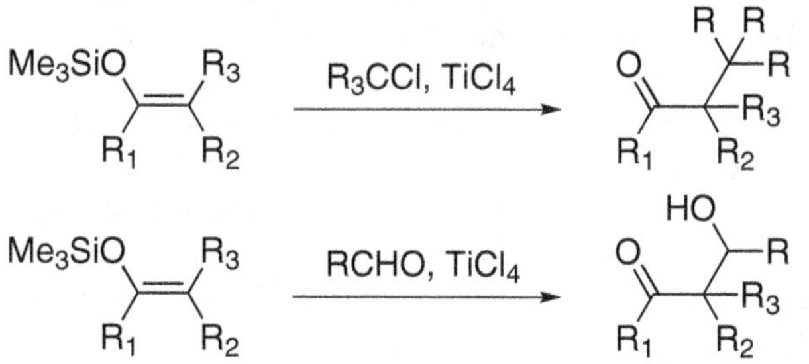

(22.7)

It is well established that Diels-Alder reactions work best with an electron-rich diene and an electron-deficient dienophile, and therefore one might expect that the enol of a conjugated aldehyde or ketone would make an excellent diene for this purpose. Unfortunately, the very instability of the enol toward tautomerization renders the enol unsuited to this purpose: its concentration is too low, and its lifetime is too short to be useful. This problem can be circumvented, however, by first converting the enol to the enol trimethylsilyl ether and then using this electron-rich diene in the Diels-Alder reaction (Example **22.8**).

---

24. Examples: (a) Reetz, M.T.; Chatziiosifidis, I.; Hübner, F.; Heimbach, H. *Org. Synth.* **1984**, *6*, 95. (b) Kraus, G.A.; Kirihara, M. *Tetrahedron Lett.* **1992**, *33*, 7727.

25. Mukaiyama, T.; Narasaka, K.; Banno, K. *Chem. Lett.* **1973**, 1011. (b) Mukaiyama, T.; Banno, K.; Narasaka, K. *J. Am. Chem. Soc.* **1974**, *96*, 7503. (c) Mukaiyama, T.; Narasaka, K. *Org. Synth.* **1986**, *62*, 6.

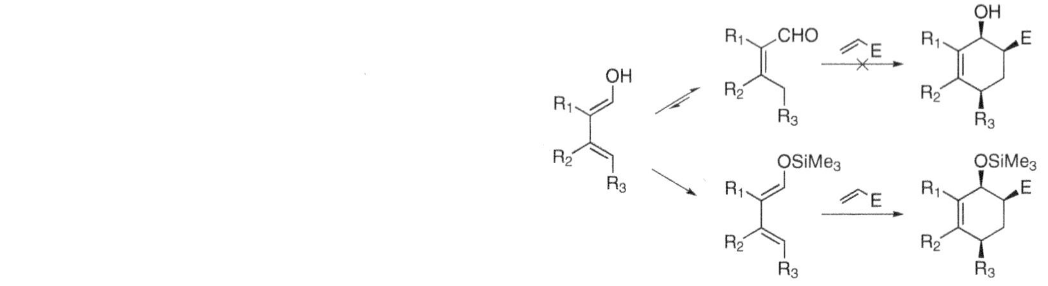

(22.8)

One of the most reactive Diels-Alder dienes known is 1-methoxy-3-trimethylsilyloxy-1,3-butadiene (Example **22.9**), known as ***Danishefsky's diene.***[26] Danishefsky's diene and its analogs (e.g., the *tert*-butyldimethylsilyl [TBS] ether in place of the trimethylsilyl [TMS] ether) react with most common dienophiles to give cyclohexenes that are easily hydrolyzed by dilute aqueous acid to cyclohexenones as shown in Example **22.10**. Occasionally, the elimination does not occur when the enol TMS ether is hydrolyzed with dilute acid. Under these circumstances, the elimination-desilylation can be catalyzed by TMS triflate.[27] The Diels-Alder reactions can be catalyzed by rare earth triflates to give the enone as the final isolated product (e.g., Example **22.11**[28]).

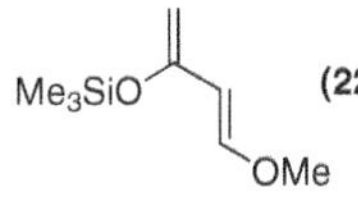

(22.9)

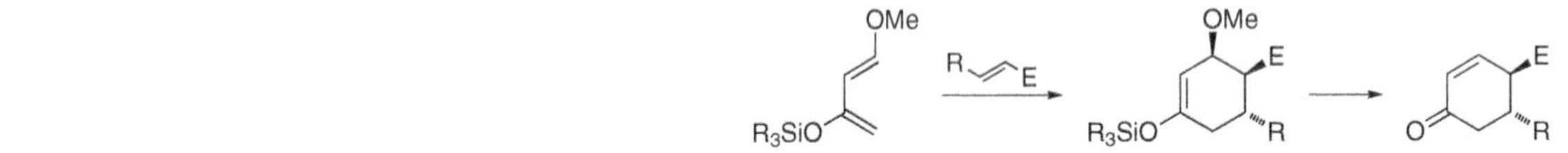

(22.10)

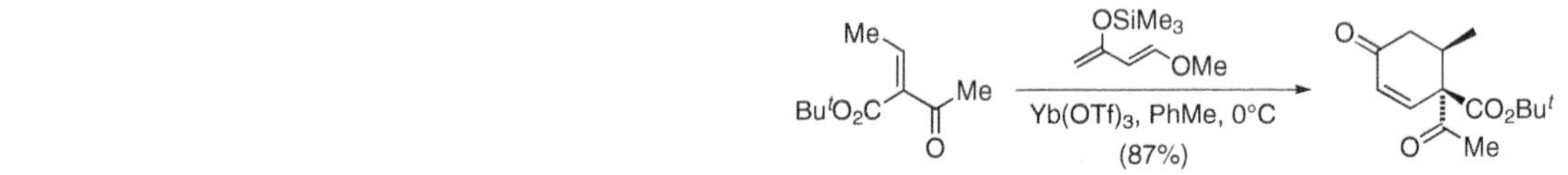

(22.11)

Danishefsky dienes also react with aldehydes in a hetero-Diels-Alder reaction[29] to give dihydropyran derivatives. These products have become popular intermediates in the synthesis of carbohydrates. Danishefsky dienes also react with imines to give dihydropyridine derivatives that are useful pharmaceutical intermediates. An example of these hetero-Diels-Alder reactions under the influence of a chiral catalyst is given in Example **22.12**.[30] There is evidence that the hetero-Diels-Alder reactions with aldehydes catalyzed by strong Lewis acids may actually proceed through initial Mukaiyama aldol addition, followed by ring closure in an acid-catalyzed Michael addition and elimination process, as shown in Example **22.13**.[29e-f, 30]

---

26. (a) Danishefsky, S.; Kitahara, T. *J. Am. Chem. Soc.* **1974**, *96*, 7807. (b) Danishefsky, S. *Acc. Chem. Res.* **1981**, *14*, 400. (c) Bertinato, P.; Sorensen, E.J.; Meng, D.; Danishefsky, S.J. *J. Org. Chem.* **1996**, *61*, 8000.

Samuel J. Danishefsky (1936–) was educated at Yeshiva University (BS, 1956) and Harvard (PhD, 1962). After postdoctoral study at Columbia, he held appointments at Pittsburgh and Yale. In 1994 he returned to Columbia, where he now holds a joint appointment with the Memorial Sloan-Kettering Cancer Center. For more information, see the web site of the Benjamin Franklin Award.

27. Vorndam, P.E. *J. Org. Chem.* **1990**, *55*, 3693.

28. Inokuchi, T.; Okano, M.; Miyamoto, T. *J. Org. Chem.* **2001**, *66*, 8059.

29. (a) Bednarski, M.; Maring, C.; Danishefsky, S. *Tetrahedron Lett.* **1983**, *24*, 3451. (b) Maruoka, K.; Itoh, T.; Shirasaka, T.; Yamamoto, H. *J. Am. Chem. Soc.* **1988**, *110*, 310. (c) Gao, Q.; Maruyama, T.; Mouri, M.; Yamamoto, H. *J. Org .Chem.* **1992**, *57*, 1951. (d) Giao, Q.; Ishihara, K.; Maruyama, T.; Mouri, M.; Yamamoto, H. *Tetrahedron* **1994**, *50*, 979. (e) Corey,E.J.; Cywin, C.L.; Roper, T. D. *Tetrahedron Lett.* **1992**, *33*, 6907. (f) Keck, G.E.; Li, X.-Y.; Krishnamurthy, D. *J. Org. Chem.* **1995**, *60*, 5998.

30. Cox, J.M.; Rainier, J.D. *Org. Lett.* **2001**, *3*, 2919.

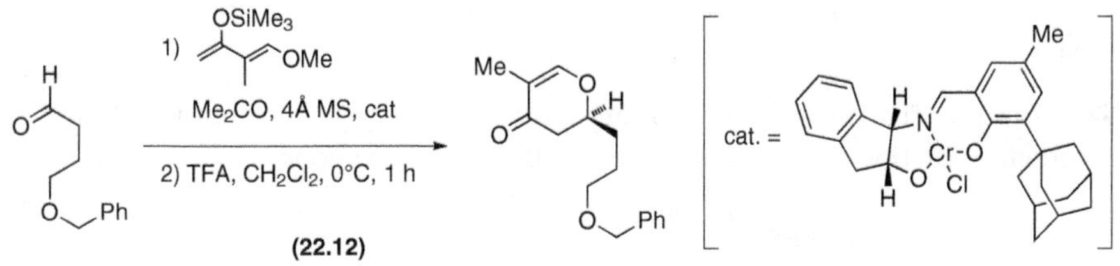

(22.12)

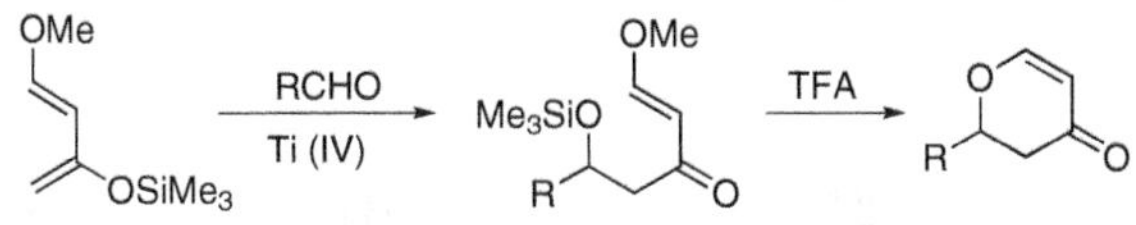

(22.13)

Worked Problem

22-2 Suggest a reasonable mechanism for the transformation in Example **22.13**. How could one distinguish between the two-step process illustrated and a true hetero–Diels-Alder reaction?

§Answer below.

---

§ Answer to Worked Problem:

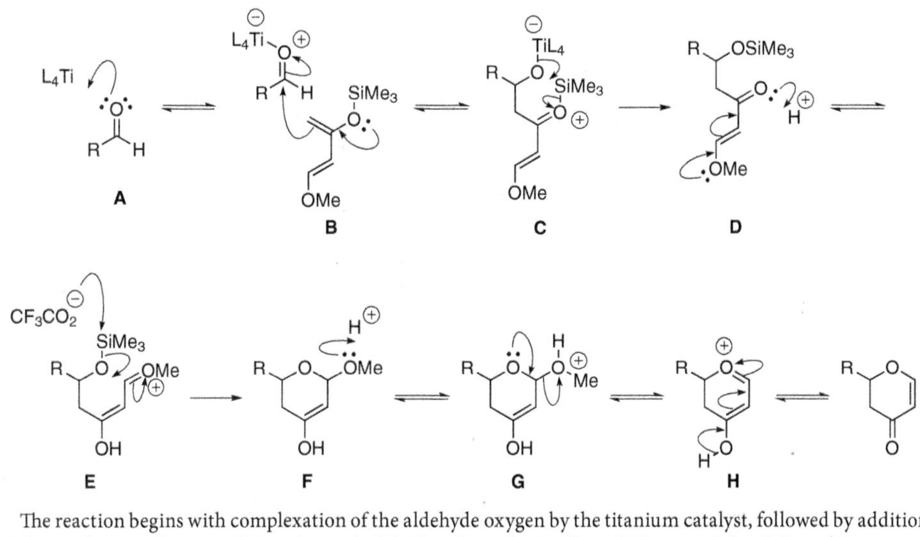

The reaction begins with complexation of the aldehyde oxygen by the titanium catalyst, followed by addition of the resultant carbocation (**B**) to the enol silyl ether (here the β effect of silicon and the ability of the oxygen lone pair to stabilize the product cation reinforce each other) to give an oxonium ion (**C**) that then loses the titanium with concomitant transfer of silicon to give the silyloxyketone **D**. Protonation of **D** occurs at the carbonyl oxygen to give the resonance-stabilized cation **E** (can you draw the other contributing structures to the resonance hybrid?) that is attacked by any nucleophile capable of attacking the silicon to give the hemiacetal **F**. The formation of the final product is simply the E1 elimination of methanol from **F** via cations **G** and **H**.

To distinguish this reaction, which occurs stepwise, from a true hetero–Diels-Alder reaction, it is necessary to show that the reaction does not occur in a single step, and the easiest way to do this is to carry out the reaction without subsequent hydrolysis. Under those conditions, any isolable intermediate rules out the concerted reaction. In addition, one might use a diene such as K, where there is a stereochemical probe of mechanism. The concerted addition to the diene should give only the 1,4-cis diastereoisomers shown as L. A stepwise reaction should not give these diastereoisomers preferentially.

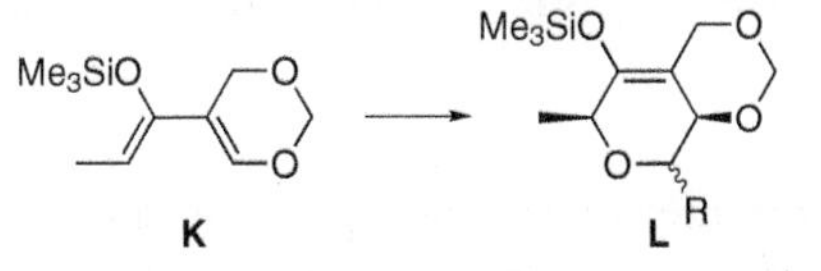

Problem

22-2 Draw the structure of the major organic product expected from each of the following reaction sequences that start with 2-cyclohexenone.

(a) 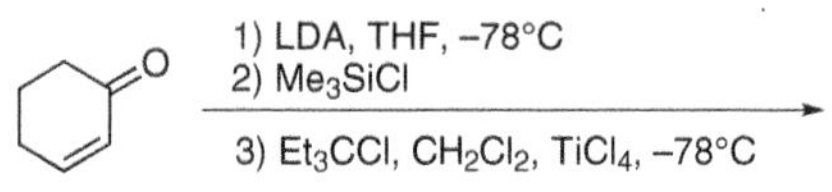

(b) 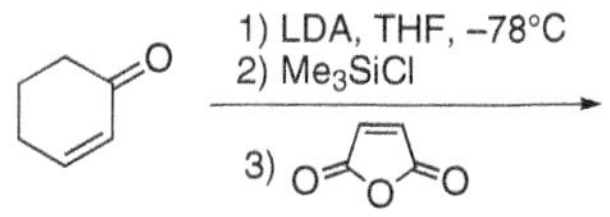

(c) 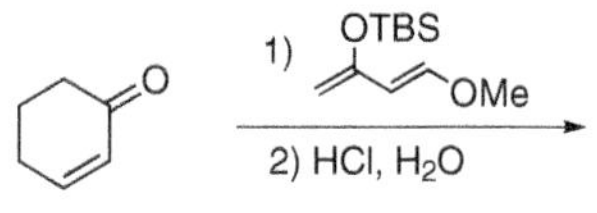

(d) 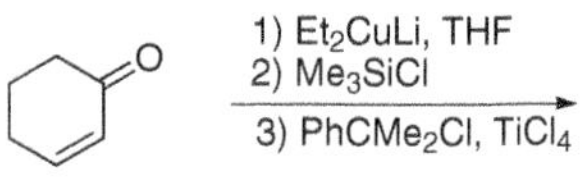

(e) 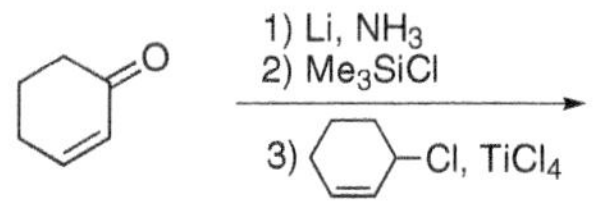

(f) 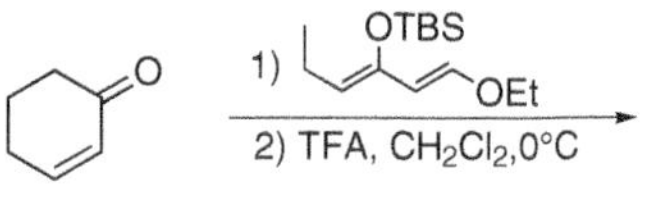

Reaction Synopses

Electrophilic Substitution of Vinylsilanes

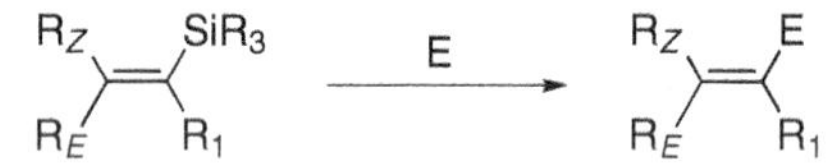

Reagents:	H_3O^+; etc.;
or	Br_2, CH_2Cl_2; Br_2, H_2O; etc.
or	RCl, $AlCl_3$; RCOCl, $AlCl_3$; RCOCl, $TiCl_4$; etc.;
or	RCHO, $SnCl_4$; R_2CO, $TiCl_4$; $RCH(OR')_2$, $TiCl_4$; etc.
or	RCH=CHCOR, Et_2AlCl; etc.
or	epoxide, $SnCl_4$; etc.

Electrophilic Substitution of Allylsilanes and Allylstannanes (Hosomi-Sakurai Reaction)

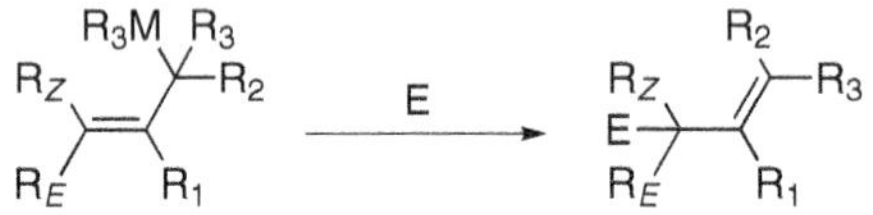

M = Si or Sn

Reaction occurs with allylic rearrangement; chirality transfer occurs with chiral allylsilanes and allylstannanes.

Reagents:	H_3O^+; etc.;
or	Br_2, CH_2Cl_2; Br_2, H_2O; etc.
or	RCl, $AlCl_3$; RCOCl, $AlCl_3$; RCOCl, $TiCl_4$; etc.;
or	RCHO, $SnCl_4$; R_2CO, $TiCl_4$; $RCH(OR')_2$, $TiCl_4$; etc.
or	RCH=CHCOR, Et_2AlCl; etc. or epoxide, $SnCl_4$; etc.

22.3 Heteroatom-Centered Nucleophiles: Silyllitihiums, Stannyllithiums, Sulfides, and Selenides

These anions may be prepared from the corresponding hydrides by reaction with bases: thiols and selenols are relatively strong acids. Thiols have pK_as in the range 10 to 12, so they are readily deprotonated by hydroxide ion or alkoxide ions. The pK_as of selenols are typically 1 to 2 units lower than those of the corresponding thiols. The bonds between hydrogen and phosphorus, silicon, and tin are less polar than those between sulfur or selenium and hydrogen, so the silanes, stannanes, and phosphines are much weaker acids than the thiols and selenols (the pK_a of PH_3 is approximately 29). Extremely strong bases are needed to form the phosphorus-, silicon-, or tin-centered anions from the heteroatom hydride, but the ready availability of alkyllithium reagents and lithium diisopropylamide makes these anions accessible by deprotonation of their hydrides.

The simplest method for preparing trialkylsilyl- or trialkylstannyllithiums is the reaction between the silyl or stannyl halide and lithium metal. The reaction between hexaalkyldistannanes, dialkyl disulfides, and dialkyl diselenides and lithium metal or lithium alkyls also provides a convenient method for the production of these heteroatom-centered nucleophiles.

Conjugate Additions

Tin-,[31] sulfur-, and selenium-centered anions are all strong nucleophiles whose 1,4-addition to conjugated ketones and esters occurs readily. The stannyl anion also reacts with conjugated SAMP/RAMP hydrazones to give products of conjugate addition that can be cleaved to give chiral β-trialkylstannylketones.[32] Silicon-centered anions, on the other hand, tend to add to conjugated carbonyl compounds with 1,2- regiochemistry to give α-silylalcohols. For the trialkylstannyl anion, conjugate addition to conjugated carbonyl compounds to give the enolate of the corresponding β-trialkylstannyl compound is probably their most important reaction. The cuprate reagents derived from the stannyl anions[33] are especially useful nucleophiles (Example **22.14**[34]). The enolate anion produced by the conjugate addition may be trapped by most electrophiles to give the product.

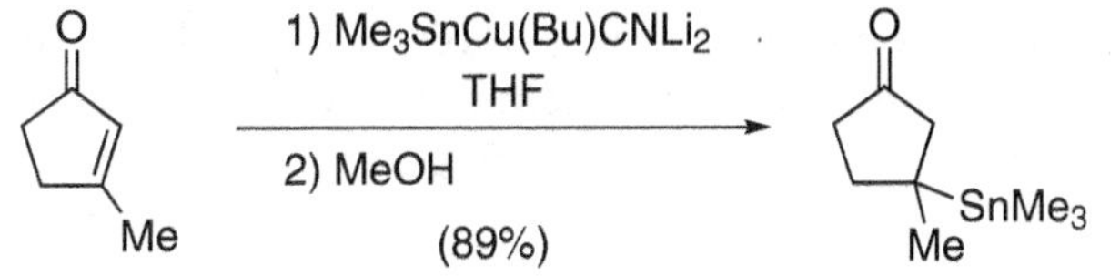

(22.14)

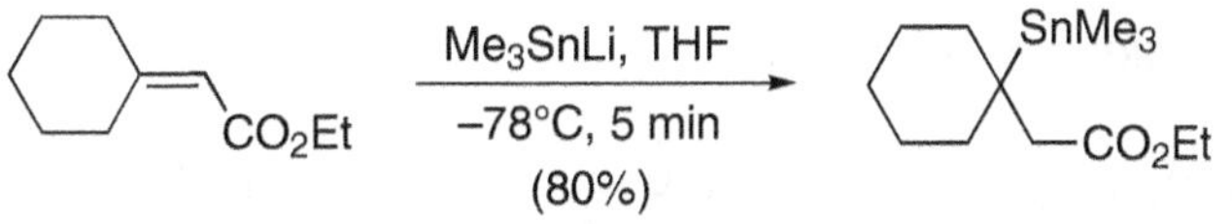

(22.15)

The conjugate addition of trialkylstannyl anions to conjugated ketones (e.g., Example **22.15**) is one of the most successful such reactions. Unlike many other conjugate addition reactions, it is relatively insensitive to steric hindrance at the β carbon of the enone, so that the conjugate addition of trialkylstannyllithiums to β,β-disubstituted carbonyl compounds occurs in high yields. Although trialkylsilyl anions also react with enones to give the products of conjugate addition, the reaction with the silicon nucleophile often fails due to competing 1,2-addition becoming the dominant reaction pathway.

31. Review: Sato, T. *Synthesis* **1990**, 259. (a) Posner, G.H.; Webb, K.S.; Asirvatham, E.; Jew, S.; Del'Innocenti, A. *J. Am. Chem. Soc.* **1988**, *110*, 4754. (b) Still, W.C. *J. Am. Chem. Soc.* **1977**, *99*, 4836. (c) Still, W.C.; Mitra, A. *Tetrahedron Lett.* **1978**, 2659. (d) Fleming, I.; Barbero, A.; Walter, D. *Chem. Rev.* **1997**, *97*, 2063. (e) Takano, S.; Higashi, Y.; Kamikubo, T.; Moriya, M.; Ogasawara, K. *J. Chem. Soc., Chem. Commun.* **1993**, 788.

32. Enders, D.; Helder, K.-J.; Raabe, G. *Angew. Chem. Int. Ed. Engl.* **1993**, *32*, 598.

33. Review: Dieter, R.K. In Krause, N, Ed. *Modern Organocopper Chemistry* (Wiley-VCH: Weinheim, 2002), ch. 3, p. 79.

34. Lipshutz, B.H.; Sharma, S.; Reuter, D.C. *Tetrahedron Lett.* **1990**, *31*, 7253.

Like the O—H group of alcohols, the S—H group of thiols and the Se—H group of selenols are acidic. These compounds are readily converted to their conjugate bases by bases such as sodium hydroxide or alkoxide ions. The anions formed are powerful nucleophiles that react readily in S_N2 reactions, nucleophilic additions, or conjugate addition reactions. Their extremely high nucleophilicity makes mercaptide anions the reagents of choice for making dialkyl sulfides. It has also allowed them to be used for carrying out the S_N2 displacement of carboxylate anions from reactive esters, permitting the hydrolysis of these esters by a mechanism other than the traditional addition-elimination mechanism.[35] Lithium *n*-propylmercaptide in HMPA has been widely used for this transformation (as in Example **22.16**[36]). Thiolate anions in HMPA also cleave aryl methyl ethers (as in Example **22.17**[37]) and also demethylate methyltrialkylammonium salts by an S_N2 mechanism. Alkylselenide anions, likewise, are powerful nucleophiles that react well with alkyl halides by the S_N2 mechanism to give dialkyl selenides. This reaction is almost always the method of choice for making these compounds.

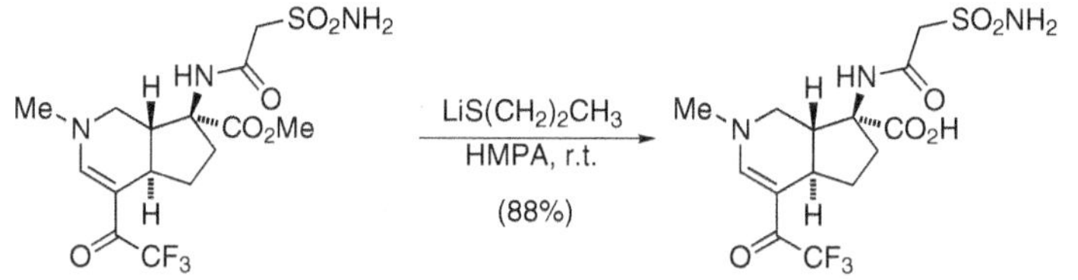

(22.16)

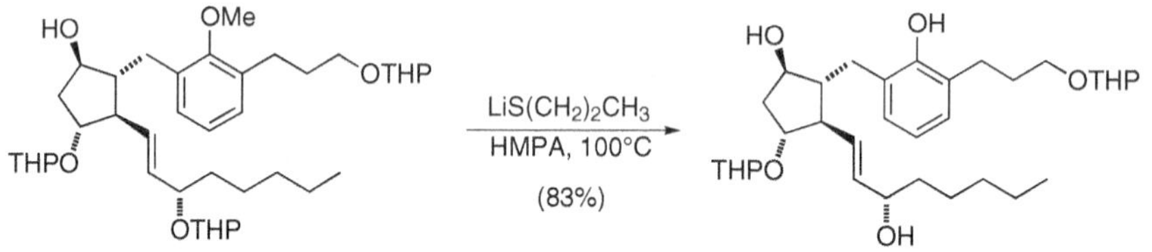

(22.17)

The conjugate addition of thiols and selenols to conjugated carbonyl compounds is catalyzed by relatively weak bases such as piperidine. This occurs because the S—H and Se-H bonds are sufficiently acidic for these bases to generate a significant nucleophile concentration.

22.4 Reactions of Organophosphorus, Organosulfur, and Organoselenium Compounds

The widespread use of phosphorus compounds in organic chemistry is a long-standing one. However, their use in forming carbon-carbon π bonds dates only from the development of the Wittig and Horner-Wadsworth-Emmons reactions. Mixtures of phosphorus and a halogen, which rapidly give the phosphorus trihalide, have been used since the 19th century for converting alcohols to alkyl bromides and iodides, and the reactions of trialkyl phosphites have been studied since the end of the 19th century.

The formation of a phosphoryl group from a trivalent phosphorus precursor is a highly exothermic process, and many of the uses of phosphorus compounds in modern organic synthesis owe their success to this fact. Two modern examples of this that we have already studied are the Wittig and Horner-Wadsworth-Emmons reactions. In both reactions,

35. Early papers: (a) Elsinger, F.; Schreiber, J.; Eschenmoser, A. *Helv. Chim. Acta* **1960**, *43*, 113. (b) Bartlett, P.A.; Johnson, W.S. *Tetrahedron Lett.* **1970**, 4459. (c) Sheehan, J.C.; Daves, G.D. *J. Org. Chem.* **1964**, *29*, 2006. (d) Node, M.; Nishida, K.; Sari, M.; Fujita, E. *Tetrahedron Lett.* **1978**, 5211. (e) Kelly, T.R.; Dali, H.M.; Tsang, W.-G. *Tetrahedron Lett.* **1977**, 3859.

36. Kende, A.S.; Liu, K.; Jos Brands, K.M. *J. Am. Chem. Soc.* **1995**, *117*, 10597.

37. Aristoff, P.A.; Harrison, A.W.; Huber, A.M. *Tetrahedron Lett.* **1984**, *25*, 3955.

which we have discussed previously, the formation of a phosphoryl group during the course of the reaction provides much of the driving force for the reaction. Another important reaction driven by the formation of the P=O bond is the Mitsunobu reaction, which is one of the most widely used S_N2 reactions in modern organic synthesis.

S_N2 Displacement of Alcohols

By far the most common stereochemical adjustment required in organic syntheses is the inversion of configuration of a stereocenter, a reaction that is obviously carried out by S_N2 substitution of a suitable derivative. This process is carried out on alcohols to a far larger extent than any other functional group, probably because there are more methods for the synthesis of alcohols and alcohols are by far the most widely available starting materials.

Until 1967, the inversion of configuration at a carbinol carbon required a two-stage process. In the first stage, the hydroxyl group (a poor leaving group) was converted into a sulfonate ester (a good leaving group approximately on a par with a bromide). This reaction, which does not affect the configuration at carbon, was then followed by displacement with a suitable nucleophile (e.g., acetate anion in dimethylformamide), and subsequent liberation of the hydroxyl group by hydrolysis. A carboxylate anion was often used as the nucleophile to reduce the potential for competing E2 elimination that is possible if the more strongly basic hydroxide anion is used as the nucleophile.

Since its first report by Japanese chemist Oyo Mitsunobu[38] in 1967, the Mitsunobu reaction,[39] which is a single-stage S_N2 displacement of alcohols driven by oxidation of a phosphine to the phosphine oxide, has become the method of choice for the substitution of the alcohol hydroxyl group with inversion of configuration. The paper containing the report of the reaction has become one of the most cited articles in the history of organic chemistry. (A *Scifinder Scholar* search of the literature using the general term "Mitsunobu reaction" for the period 1992–2012 returned 2,434 citations containing the exact term and 3,066 citations containing the concept, and a *Web of Science* search of the same term over the same time period returned 1,981 citations.)

The mechanism of the Mitsunobu reaction is not especially simple. However, the results of several studies during the 1980s[40] have led to one generally accepted mechanism (shown in Figure 22.10). In this mechanism, the reaction is initiated by the addition of triphenylphosphine to the azodicarboxylate ester to give the betaine **22.18**. This betaine deprotonates the nucleophile to give a phosphonium ion (**22.19**) carrying a reasonably good leaving group. The position of this equilibrium (marked **A**) depends on the pK_a value of the nucleophile and the polarity of the solvent; the reaction works best with nucleophiles whose pK_a values are below approximately 16. This makes it an excellent method for the production of esters from carboxylic acids (p$K_a \approx$ 4–6), azides (pK_a of HN_3 = 4.6), *N*-alkylimides from phthalimide (pK_a = 8.3), *N*-alkylsulfonamides from sulfonamides (pK_as ≈ 10.5), and ethers from phenols (pK_as ≈ 8–11). Displacement of the leaving group from phosphorus by the alcohol gives an alkoxyphosphonium ion (**22.20**), which then reacts with the nucleophile conjugate base by an S_N2 mechanism to give the final product (**22.21**) with inverted configuration.

---

38. **Oyo Mitsunobu (1934–2003)** studied under Mukaiyama at Tokyo Institute of Technology, where he became a faculty member. He subsequently moved to Aoyama Gakuin University.

39. (a) Mitsunobu, O.; Yamada, M. *Bull. Chem. Soc. Japan* **1967**, *40*, 2380. (b) Mitsunobu, O. *Synthesis* **1981**, 1. Reviews: (c) Castro, B.R. *Org. React.* **1983**, *29*, 1. (d) Hughes, D.L. *Org. React.* **1992**, *42*, 335. (e) Hughes, D.L. *Org. Prep. Proc. Int.* **1996**, *28*, 127. (f) But, T.Y.S.; Toy, P.H. *Chem. Asian J.* **2007**, *2*, 1340. (g) Swamy, K.C.; Kumar, N.N.; Balaraman, E.; Kumar, K.V. *Chem. Rev.* **2009**, *109*, 2551.

40. (a) Grochowski, E.; Hilton, B.D.; Kupper, R.J.; Michejda, C.J. *J. Am. Chem. Soc.* **1982**, *104*, 6876. (b) Camp, D.; Jenkins, I.D. *J. Org. Chem.* **1989**, *54*, 3045, 3049. (c) Hughes, D.L.; Reamer, R.A.; Bergan, J.J.; Grabowski, E.J.J. *J. Am. Chem. Soc.* **1989**, *110*, 6487.

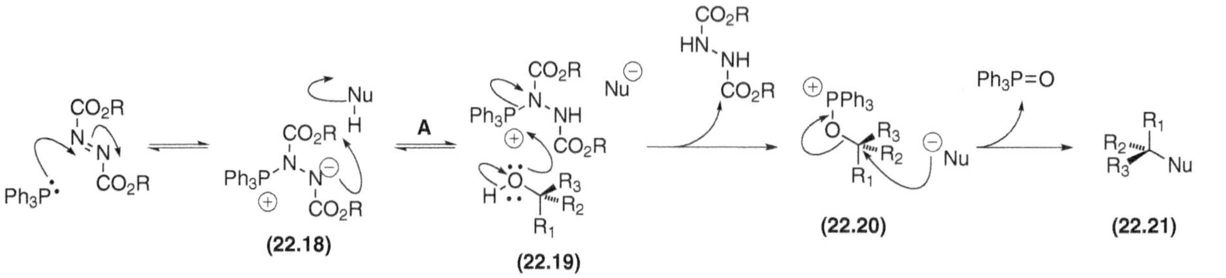

Figure 22.10 The mechanism of the Mitsunobu reaction. The position of the equilibrium marked as A depends on the pK_a value of H—Nu

Figure 22.11 The mechanism of the one-step S_N2 conversion of alcohols to selenides

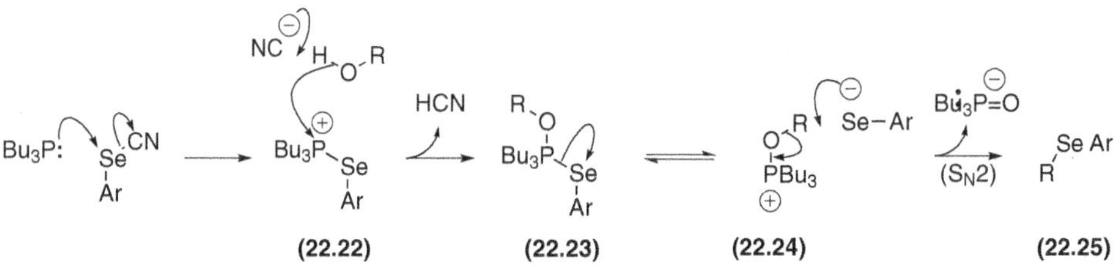

Formation of Selenides with Inversion

In a reaction similar to the Mitsunobu reaction, it has been found that the treatment of alcohols with substituted phenyl selenocyanates (ArSeCN) and tributylphosphine leads to the formation of phenylselenides in which the stereochemistry at the carbinol carbon has been inverted.[41] Like the Mitsunobu reaction, the driving force for the reaction comes from the formation of a phosphoryl group from the phosphine. The mechanism of the reaction is summarized in Figure 22.11. The key step of the reaction is the initial displacement of cyanide anion from the selenocyanate by the phosphine to give the phosphonium ion **22.22**, which then reacts with the alcohol to give the phosphorane **22.23** and hydrogen cyanide. Ionization of the P—Se bond then gives an intermediate alkoxyphosphonium salt (**22.24**) that reacts with the phenylselenide anion to give the S_N2 product (**22.25**).

The Corey-Winter Elimination

The abstraction of an oxygen or sulfur atom from an organic molecule by phosphite esters is an important step of several organic reactions. In the Corey-Winter reaction (Figure 22.12), for example, a cyclic thiocarbonate ester (**22.26**, X=S) generated by the reaction between a vicinal diol and thiophosgene or an equivalent of thiophosgene is heated with a trialkyl phosphite.[42] The phosphorus abstracts the sulfur from the ester to give the trialkyl thiophosphate and a carbene (**22.29**).[43] The carbene decomposes to give carbon dioxide and an alkene (**22.27**) that has the same stereochemistry as the cyclic thiocarbonate ester. As alluded to in Chapter 15, the Corey-Winter reaction may provide a means for the deprotection of an alkene that has been protected by *syn*-hydroxylation.

---

41. Grieco, P.A.; Gilman, S.; Nishizawa, M. *J. Org. Chem.* **1976**, *41*, 1485.
42. (a) Corey, E.J.; Winter, R.A.E. *J. Am. Chem. Soc.* **1963**, *85*, 2677. (b) Corey, E.J.; Hopkiss, P.B. *Tetrahedron Lett.* **1982**, *23*, 1979. (c) Block, E. *Org. React.* **1984**, *30*, 457.
43. Horton, D.; Tyndall, C.G., Jr. *J. Org. Chem.* **1970**, *35*, 3558.

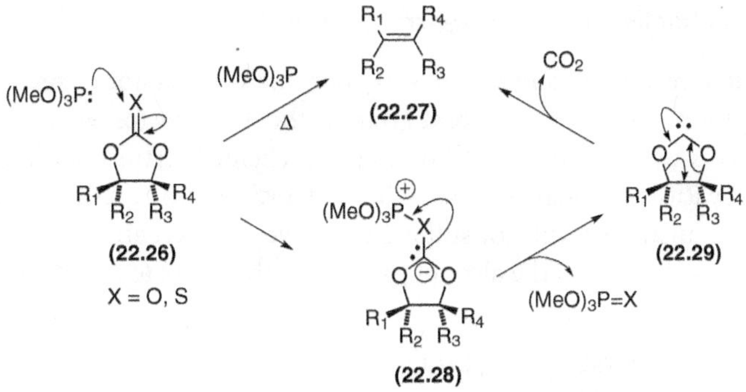

Figure 22.12 The Corey-Winter elimination

Worked Problem

22-3 What is the major organic product of the following reaction if one equivalent of benzoic acid is used?

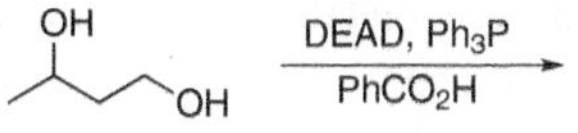

§Answer below.

Problems

22-3 Based on the discussion above, what is the expected major monoalkylation product of the reaction below?

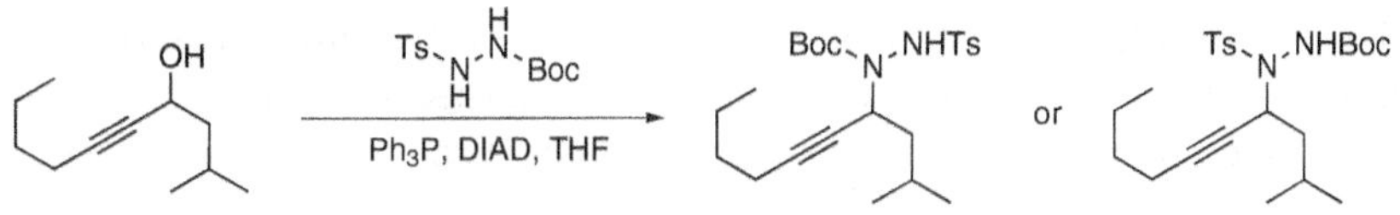

Given that only one monoalkylation product is formed, why is this that product?

References: (a) *Chem. Soc. Rev.* **2001**, *30*, 205. (b) *Tetrahedron Lett.* **2012**, *53*, 7006.

22-4 What is the major organic product expected from each of the following reactions?

(a)

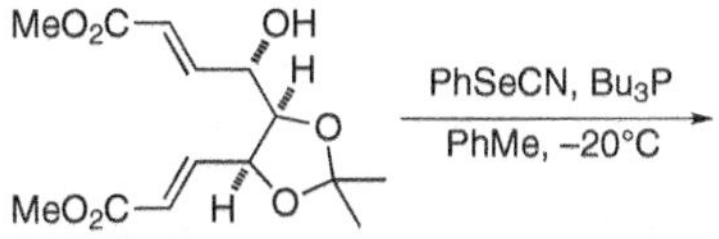

(b)

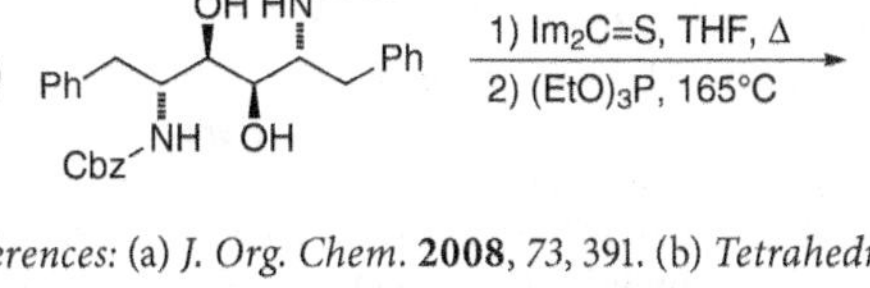

References: (a) *J. Org. Chem.* **2008**, *73*, 391. (b) *Tetrahedron Lett.* **2009**, *50*, 552.

§ Answer to Worked Problem:

This Mitsunobu reaction proceeds normally, but because there are two hydroxyl groups present in the alcohol, there are two ways that the oxygen could attack at phosphorus. The favored product is the one where the less hindered hydroxyl group reacts.

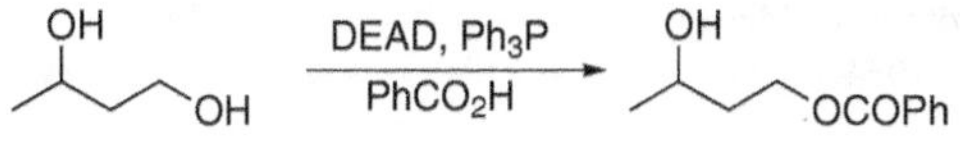

The Arbuzov (Michaelis-Arbuzov) Rearrangement[44]

One of the earliest reactions converting a tricoordinate phosphorus precursor into a tetracoordinate product with a phosphoryl group is the ***Arbuzov*** (or ***Michaelis-Arbuzov***) ***rearrangement***. This was first reported[45] by German chemist Arthur Michaelis[46] but explored in depth[47] by Russian chemist Aleksandr Arbuzov.[48] In this reaction, a trialkyl phosphite is heated with an alkyl halide (or sulfonate) to give a dialkyl alkylphosphonate and a halide or sulfonate derived from the alcohol part of the phosphite (e.g., Example **22.30**[49]).

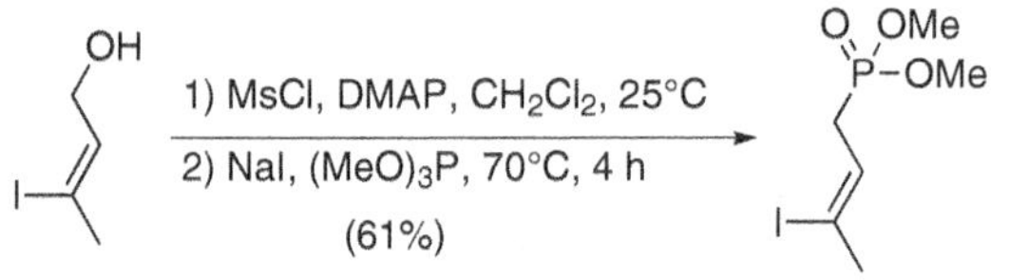

(22.30)

The reaction is believed to occur in two steps (Figure 22.13). The first step is an S_N2 reaction where the phosphite ester (**22.31**) displaces the halogen from the halide. The trialkoxyphosphonium salt formed (**22.32**) is structurally very similar to the alkoxyphosphonium salts formed during the Mitsunobu reaction. Like the Mitsunobu reaction intermediates, the phosphorus-containing group in the molecule functions as an excellent leaving group in an S_N2 reaction, and it is easily displaced by the halide ion generated in the first step of the reaction. The final product of the reaction is the alkyl halide derived from the original trialkyl phosphite and a dialkyl alkylphosphonate (**22.33**) where the phosphonic acid moiety is derived from the original alkyl halide. With the development of the Horner-Wadsworth-Emmons reaction as a method for preparing *E* alkenes, the Arbuzov rearrangement has assumed a much more important place in modern organic chemistry.

The Perkow Rearrangement

In 1954, German chemist Werner Perkow[50] described a reaction between chloral and trimethyl phosphite that led not to the expected Michaelis-Arbuzov rearrangement product but instead, to a vinyl phosphate.[51] This initially unexpected finding has been further generalized to reactions of α-halocarbonyl compounds (Example **22.34**) where S_N2 attack

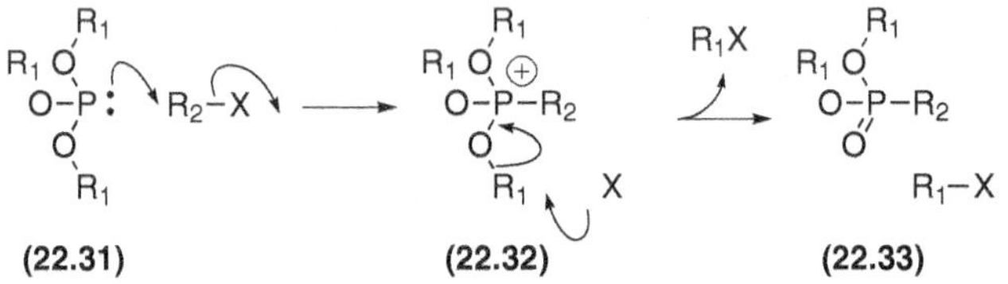

(22.31) **(22.32)** **(22.33)**

Figure 22.13 The mechanism of the Michaelis-Arbuzov rearrangement

44. Reviews: (a) Arbusow, B.A. *Pure Appl. Chem.* **1964**, *9*, 307. (b) Bhattacharya, A. K.; Thyagarajan, G. *Chem. Rev.* **1981**, *81*, 415.

45. Michaelis, A.; Kaehne, R. *Ber. dtsch. chem. Ges.* **1898**, *31*, 1048.

46. **Karl Arnold August Michaelis (1847–1916)** studied at Göttingen and Jena, and he served at the Technische Hochschulen in Karlsruhe and Aachen before moving to Rostock in 1890. Here he spent his career. For more biographical detail, see: Wichelhaus, H. *Ber. dtsch. chem. Ges.* **1916**, *49*, 468.

47. (a) Arbuzov, A.E. *Zh. Russ. Fiz.-Khim. O-va.* **1906**, *38*, 161, 293, 687. (b) Arbusow, A. *Chem. Zentr II.* **1906**, 1639.

48. **Aleksandr Yerminingel'dovich Arbuzov (1877–1968)** was educated at Kazan (M Chem, 1905). As a student under Zaitsev he began a study of phosphorous acid and its esters. He joined the Novo-Aleksandriya Institute in 1901, and in 1910 he returned to Kazan as Zaitsev's successor. The Organic & Physical Chemistry Institute of the Russian Academy of Science at Kazan is named after him. For more biographical information, see: Lewis, D.E. *Early Russian Organic Chemists and Their Legacy* (Springer: Heidelberg, 2012), p. 89.

49. Domínguez, M.; Álvarez, S.A.; Álvarez, R.; de Lera, A.R. *Tetrahedron* **2012**, *68*, 1756.

50. **Werner Perkow (1915–1994)** was an industrial chemist employed by the Norddeutsche Affinerie (now Aurubis), in Hamburg, the largest copper refinery in Europe, at the time he reported his discovery. Much of his career was spent working with pesticides.

51. Perkow, W. *Chem. Ber.* **1954**, *87*, 755.

of phosphorus on the carbon carrying the halogen is hindered; these reactions give the vinyl phosphate (Example **22.35**) as the major product.[52] The reaction is proposed to occur[53] by means of initial attack of the phosphite ester on the carbonyl group, followed by rearrangement to give the vinyl phosphate ester (**22.37**), as shown in Figure 22.14. The addition gives a zwitterion (**22.38**) that can undergo direct rearrangement of phosphorus from carbon to oxygen, with concomitant elimination of the halide anion to give the phosphonium ion **22.40**, or it gives ring closure to the oxaphosphirane **22.39**, which then fragments to give **22.40**. This cation then reacts with the halide anion to generate the phosphoryl group and an alkyl halide. Support for this mechanism comes from two important observations. First is that the reaction is faster with α-chlorocarbonyl compounds than with the corresponding α-bromo compounds. Second is that increasing substitution at the α carbon decreases the amount of phosphonate ester formed and increases the amount of vinyl phosphate. The stereochemistry of the reaction is consistent with the initial attack of the phosphite on the carbonyl group occurring through the conformation where the carbonyl and the halogen are eclipsed, as in Figure 22.14.

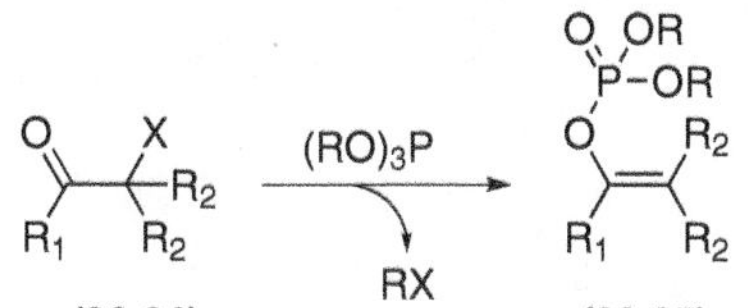

(22.34) **(22.35)**

Problem

22-5 What is the major organic product of each of the following reactions?

(a) 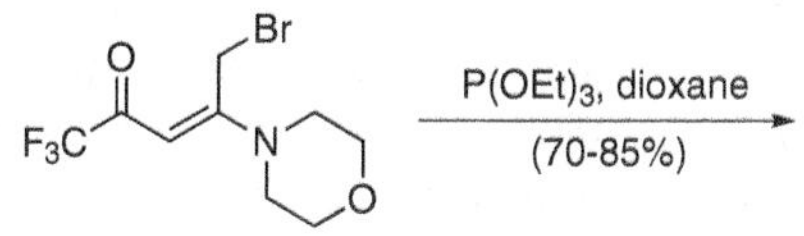

(b) 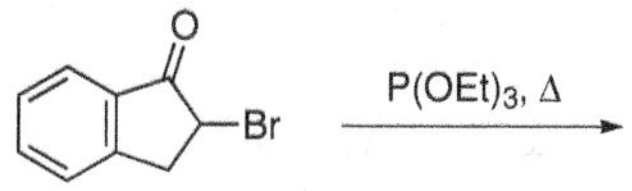

(c) 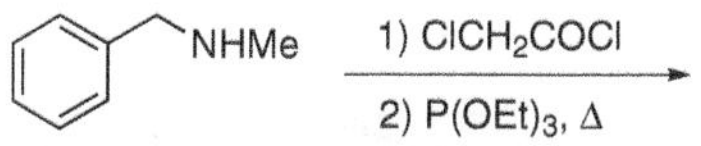

References: (a) *Tetrahedron Lett.* **2010**, *51*, 4623. (b) *Tetrahedron Lett.* **2009**, *50*, 7314. (c) *Tetrahedron* **2008**, *64*, 10123.

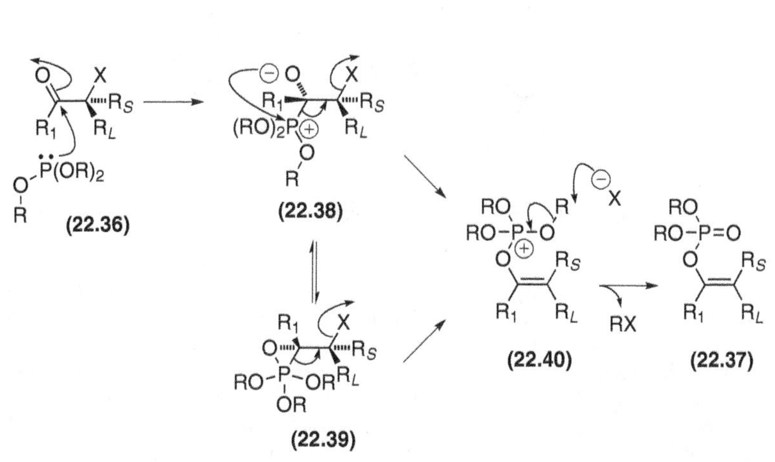

Figure 22.14 The mechanism of the Perkow reaction. R_L is the sterically more demanding substituent, and R_S is the sterically less demanding substituent at the α carbon.

52. Borowitz, I.J.; Anschel, M.; Firstenberg, S. *J. Org. Chem.* **1967**, *32*, 1723.
53. Borowitz, I.J.; Firstenberg, S.; Borowitz, G.B.; Schuessler, D. *J. Am. Chem. Soc.* **1972**, *94*, 1623.

Sigmatropic Rearrangements of Allylic Sulfoxides, Sulfonium Ylides, and Selenoxides: The Mislow-Evans Rearrangement

Allylic sulfoxides and selenoxides (Example **22.41**) undergo a reversible suprafacial [2,3]-sigmatropic rearrangement to an allyl sulfenate or allyl selenenate ester (Example **22.42**).[54] This reaction is named after American chemists Kurt Mislow[55] and David A. Evans, who discovered and developed it. Mislow's early work showed that the racemization of optically active allylic sulfoxides (chiral sulfur) occurred with a much lower (≈ 20 kcal mol^{-1}) activation energy than in the corresponding saturated sulfoxides.[56] In the same study, the sulfenate esters were identified as the other compound involved in the racemization reaction, and the intramolecular nature of the reaction was established. The thermodynamics and kinetics of the rearrangement in the prenyl systems (R=Me)[57] show that the selenenate ester is approximately 11 kcal mol^{-1} lower energy than the selenoxide, whereas the energy gap in the corresponding sulfur system favors the sulfoxide by approximately 1.5 kcal mol^{-1}.

(22.41) (22.42)

X = SR, SeR Y = O, CR_2

Selenenate esters are susceptible to hydrolysis to the corresponding allylic alcohol, but sulfenate esters are not so readily hydrolyzed. Fortunately, although the sulfoxide is not attacked by thiophilic agents, sulfenate esters react readily with a thiophilic reagent such as a trialkyl phosphite to give the allylic alcohol (see Example **22.43**[58]). Vinyl sulfoxides isomerize to the allylic sulfoxides when heated with base, and this allows the introduction of oxygen into an allylic position, as shown by the conversion of Examples **22.44** through **22.45** to **22.46**.[59] The reactions of sulfoxides and selenoxides have been used to convert allylic alcohols and halides to their allylic regioisomers. This sulfoxide-sulfenate equilibrium has been used to good effect to invert the stereochemistry of the secondary alcohol **22.47** in a synthesis of (±)-14-deoxyisoamijiol (Example **22.51**).[60]

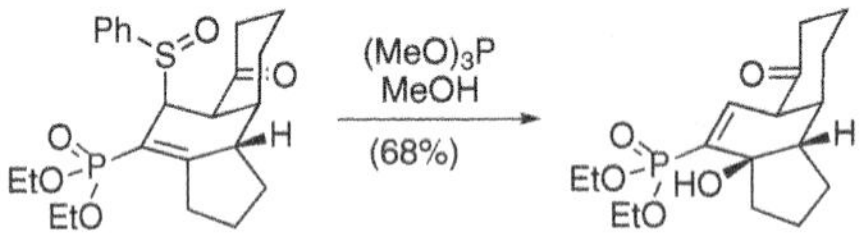

(22.43)

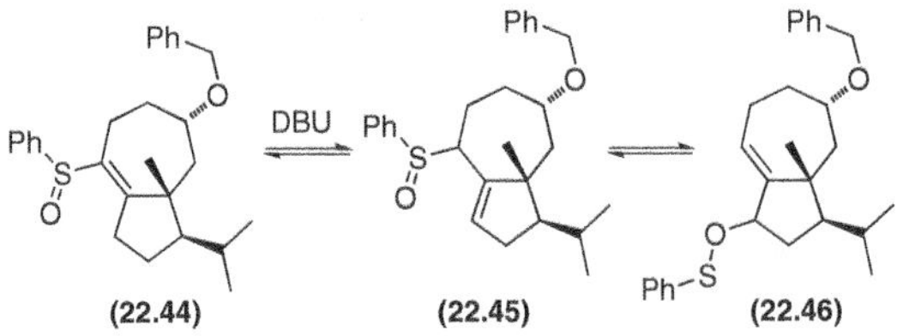

(22.44) (22.45) (22.46)

54. (a) Evans, D.A.; Andrews, G.C.; Sims, C.L. *J. Am. Chem. Soc.* **1971**, *93*, 4956. (b) Evans, D.A.; Andrews, G.C. *Acc. Chem. Res.* **1974**, *7*, 147.
Review: Parker, A.J.; Kharasch, N. *Chem. Rev.* **1959**, *59*, 583.
55. **Kurt Mislow (1923–)** was born in Germany, and educated at Tulane (BS, 1944) and the California Institute of Technology (PhD, 1947). His career was spent at New York University (1947–1964) and Princeton (1964–1988). He is best known for his contributions to stereochemistry. For more detail, see: (a) Siegel, J.S. *Chirality*, **1998**, *10*, 3.
56. (a) Rayner, D.R.; Miller, E.G.; Bickart, P.; Gordon, A.J.; Mislow, K. *J. Am. Chem. Soc.* **1966**, *88*, 3138. (b) Miller, E.G.; Rayner, D.R.; Mislow, K. *J. Am. Chem. Soc.* **1966**, *88*, 3139. (c) Mislow, K. *Rec. Progr. Chem.* **1966**, *28*, 217. (d) Bickart, P.; Carson, F.W.; Jacobus, J.; Miller, E.G.; Mislow, K. *J. Am. Chem.* Soc. **1968**, *90*, 4869. (e) Tang, R.; Mislow, K. *J. Am. Chem. Soc.* **1970**, *92*, 2100.
57. Reich, H.J.; Yelm, K.E.; Wollowitz, S. *J. Am. Chem. Soc.* **1983**, *105*, 2503.
58. Koprowski, M.; Krawczyk, E.; Skowrońska, A.; McPartlin, M.; Choi, N.; Radojevic, S. *Tetrahedron* **2001**, *57*, 1105.
59. Magnus, P.; Ollivier, C. *Tetrahedron Lett.* **2002**, *43*, 9605.
60. Majetich, G.; Song, J.S.; Ringold, C.; Nemeth, G.A. *Tetrahedron Lett.* **1990**, *31*, 2239.

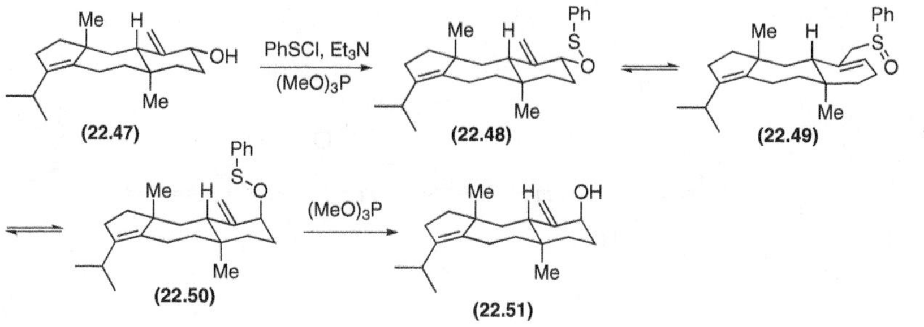

Because configurationally stable sulfoxides can be prepared and used, this reaction has also been applied to the introduction of chirality into a molecule by transfer of chirality from sulfur to carbon. Mislow's early work[61] had shown that the rearrangement is suprafacial with respect to both reacting species, a finding confirmed by subsequent workers.[62] The reaction is now believed to proceed through the envelope conformation of a five-membered cyclic transition state. The *exo* and *endo* diastereoisomers of these transition states give rise to opposite chirality in the product (Figure 22.15). Reich's study[63] of the stereochemistry of rearrangements in chiral sulfoxides showed that difference in the activation energies of the *exo* and *endo* transition states for the sulfoxide rearrangement is small, with the consequence that the levels of asymmetric induction are generally rather low. In a study of allylic selenoxides, Davis and his coworkers[64] found that chiral selenoxides, which can be obtained by oxidation of the selenide with a chiral oxaziridine, rearrange through an *exo* transition state to give optically active allylic alcohols.

The Pummerer Rearrangement[65]

The reaction between a sulfoxide and a strong electrophile (e.g., an anhydride or acid chloride) at low temperatures gives an intermediate *S*-acylsulfoxonium ion (**22.52**), which we have seen already as a key intermediate in the Swern and Moffatt-Pfitzner oxidations. When this ion is heated, however, a formal rearrangement occurs with reduction at sulfur and oxidation at the adjacent carbon. This reaction is known as the Pummerer rearrangement,

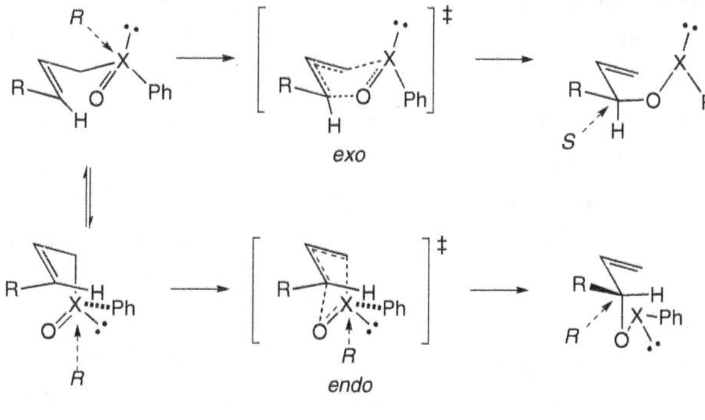

Figure 22.15 *Exo* and *endo* transition states for the concerted [2,3]-sigmatropic rearrangements of chiral sulfoxides (X=S) and selenoxides (X=Se)

61. Bickart, P.; Carson, F.W.; Jacobus, J.; Miller, E.G.; Mislow, K. *J. Am. Chem. Soc.* **1968**, *90*, 4869.
62. (a) Rautenstrauch, V. *J. Chem. Soc., Chem. Commun.* **1970**, 526. (b) Goldmann, S.; Hoffmann, R.W.; Maak, N., Geueke, K-J. *Chem. Ber.* **1980**, *117*, 831.
63. Reich, H.J.; Yelm, K.E.; Wollowitz, S. *J. Am. Chem. Soc.* **1983**, *105*, 2503.
64. Davis, F.A.; Stringer, O.D.; McCauley, J.P., Jr. *Tetrahedron* **1985**, *41*, 4747.
65. Reviews: (a) De Lucchi, O.; Miotti, U.; Modena, G. *Org. React.* **1991**, *40*, 157. (b) Padwa, A.; Gunn, D. E., Jr.; Osterhout, M. H. *Synthesis* **1997**, 1353. (c) Padwa, A.; Bur, S. K.; Danca, M. D.; Ginn, J. D.; Lynch, S. M. *Synlett* **2002**, 851. (d) Bur, S.K.; Padwa, A. *Chem. Rev.* **2004**, *104*, 2401. (e) Feldman, K.S. *Tetrahedron* **2006**, *62*, 5003. (f) Akai, S.; Akita, Y. *Top. Curr. Chem.* **2007**, *274*, 35. (f) Smith, L.H.S.; Coote, S.C.; Sneddon, H.F.; Procter, D.J. *Angew. Chem. Int. Ed.* **2010**, *49*, 5832.

Figure 22.16 The currently accepted mechanism of the Pummerer rearrangement

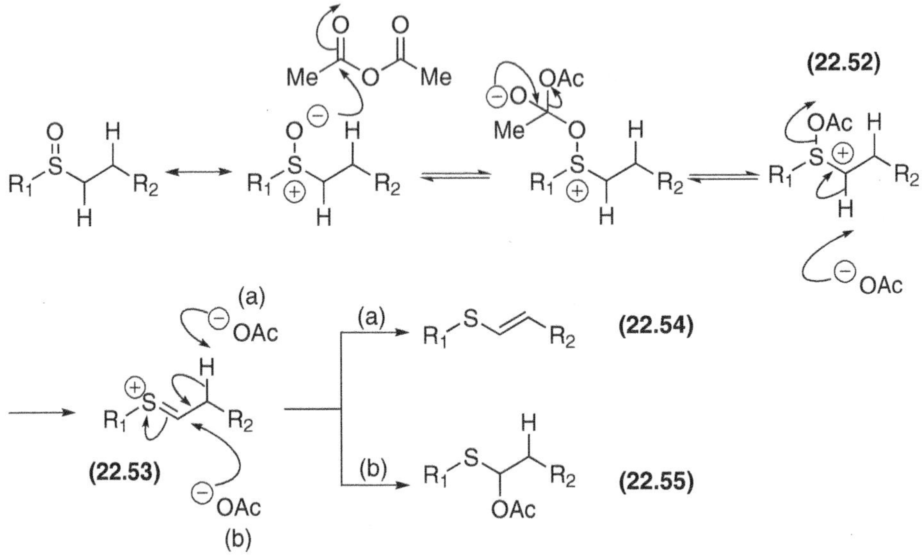

and it was first reported in 1909[66] by German chemist Rudolph Pummerer.[67] The currently accepted mechanism of the reaction is shown in Figure 22.16 with acetic anhydride as the electrophile. After the initial formation of the *S*-acylsulfonium ion, an E2 elimination of acetic acid takes place to give the thionium ion **22.53**. As is shown in Figure 22.16, this thionium ion may react with the nucleophile in two ways: it may lose a proton (path a) to give a vinyl sulfide (**22.54**) or it may undergo nucleophilic addition to the carbon-sulfur π bond (path b) to give the acyl derivative of a hemithioketal (**22.55**). In both products, the carbon atom is at the oxidation level of a carbonyl group.

This conversion of a sulfoxide into a monothioacetal or monothioketal, or a thioenol ether—a masked aldehyde or ketone—has become important in synthesis, because it makes the sulfoxide group itself an equivalent of a carbonyl group. Even more interesting is the observation that the α carbon of the sulfoxide, which corresponds to the carbonyl carbon after the Pummerer rearrangement, can be deprotonated to give the sulfoxide anion. This makes the sulfoxide anion the synthetic equivalent of an acyl anion. This charge-type inversion is now known as ***umpolung***.[68] It has become an important part of the toolbox of the synthetic organic chemist because it allows one to use reagents that are surrogates for species such as carbonyl anions—in effect, to direct the addition of an ***electrophile*** to the ***carbonyl carbon***.

In addition to the "normal" Pummerer rearrangement pathway, a secondary pathway, known as the ***Pummerer fragmentation***, is available to the intermediate *S*-acylsulfonium ion if there is a group at the α position of the sulfoxide (group **G** in compound **22.56**, Figure 22.17) that could give a very stable carbocation. In this mechanistic pathway, the stable carbocation is lost to give the thionium ion (**22.57**). The limitations on this mechanistic pathway have been explored by Lacour and coworkers,[69] who find that when group

66. Pummerer, R. *Ber. dtsch. chem. Ges.* **1909**, *42*, 2282; **1910**, *43*, 1401.

67. **Rudolf Pummerer (1882–1973)** was educated at Munich (PhD, 1905). After study under Baeyer, then at the Badische Anilin und Soda Fabrik (BASF), he returned to the University of Munich. After moving to Griefswald (1923), he moved to Erlangen (1925), where he spent the rest of his career. For a more detailed biography, see Oesper, R.E. *J. Chem. Educ.* **1951**, *28*, 243.

68. Reviews: (a) Seebach, D. *Angew. Chem.* **1969**, *81*, 690. (b) Gröbel, B.-T.; Seebach, D. *Synthesis* **1977**, 357. (c) Seebach, D. *Angew. Chem. Int. Ed. Engl.* **1979**, *18*, 239.

69. Laleu, B.; Santarem, M.; Lacour, J. *Chem. Commun.* **2006**, 2786. This paper contains several useful leading references.

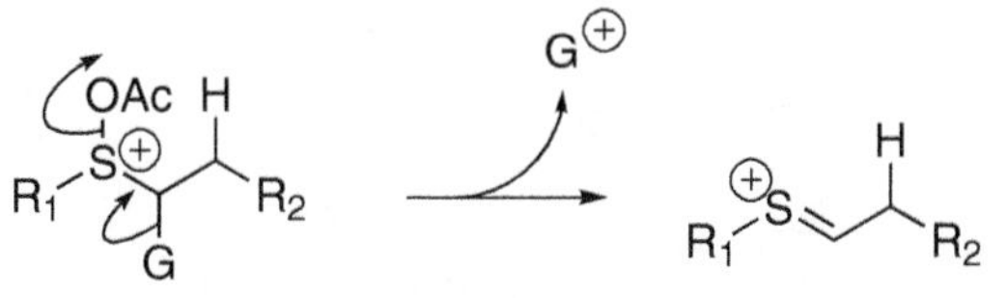

(22.56) **(22.57)**

Figure 22.17 The key step of the Pummerer fragmentation

G can give a very stable cation (e.g., **22.58**), the fragmentation becomes the dominant pathway.

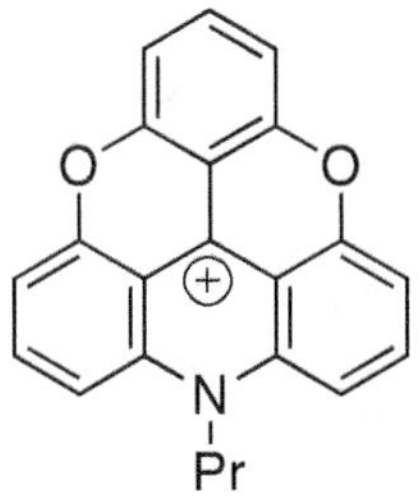

(22.58)

The thionium ion intermediate in the Pummerer reactions is a modest electrophile that can react with electron-rich olefins to give alkylated sulfides. In particular, the intermediate thionium ion generated during the Pummerer rearrangement of β-ketosulfoxides has been trapped by electron-rich aromatic rings to give cyclized products.

Sulfur Dioxide Extrusion: Ramberg-Bäcklund Reaction[70]

Most of the reactions of sulfones are basically the same as the corresponding reactions of sulfoxides. Sulfones can be alkylated, and the C—S bonds of sulfones can be reduced to C—H bonds by active metals. However, the sulfones can also undergo reactions where both C—S bonds are cleaved simultaneously with the loss of sulfur dioxide, a reaction generally known as extrusion.

One important application of sulfur dioxide extrusion from a sulfone is the Ramberg-Bäcklund[71] reaction, which was first reported in 1940.[72] It gives an alkene by extrusion of sulfur dioxide from a three-membered cyclic sulfone (Figure 22.18). When α-halosulfone with at least one hydrogen on the α carbon (**22.59**) is treated with strong base, it is deprotonated. This is followed immediately by intramolecular displacement of the leaving group to give the transient three-membered thiirane dioxide (**22.60**), which loses sulfur dioxide in a concerted, suprafacial extrusion reaction to give the alkene (**22.61**). In Example **22.62**, the formation of the α-halosulfone is carried out at the same time that the Ramberg-Bäcklund reaction is. The carbanion formed initially displaces the CCl_3^- anion from the carbon tetrachloride to give the α-chlorosulfone, which then reacts with a second mole of base.

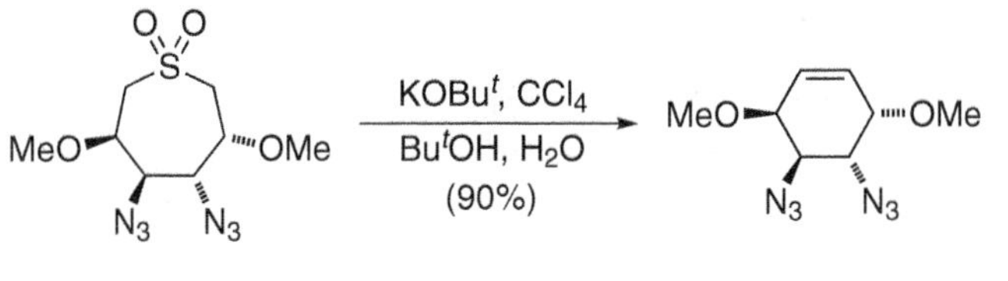

(22.62)

70. Review: Taylor, R.J.K.; Casey, G. *Org. React.* **2003**, *62*, 357.

71. **Ludwig Ramberg (1874–1940)** was educated at Lund (PhD, 1902), where he joined the faculty after graduating. In 1918, he moved to Uppsala, where he spent the remainder of his career. For more biographical detail, see: *Ber. dtsch. chem. Ges.* **1941**, *74*, A106.

Birger Bäcklund (1908–1997) was Ramberg's student at the time of the discovery of the reaction. No further biographical information is available.

72. Ramberg, L.; Bäcklund, B. *Arkiv. Kemi Mineral Geol.* **1940**, *13A*, 50.

Figure 22.18 The mechanism of the Ramberg-Bäcklund reaction

(22.59)

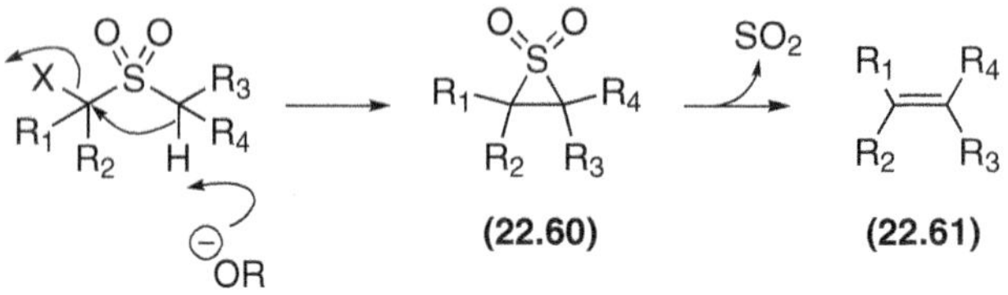

(22.60) (22.61)

Reaction Synopses

Mitsunobu Reaction

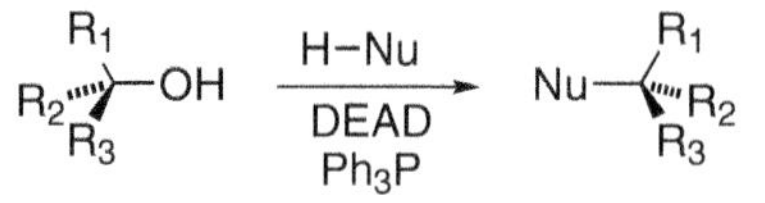

H—Nu: RCO_2H; phthalimide; RSO_2NH_2; ArOH; etc. DEAD may be replaced by the methyl (DMAD) or isopropyl (DIAD) analogs

Stereochemistry: inversion of configuration

Grieco Selenylation Reaction

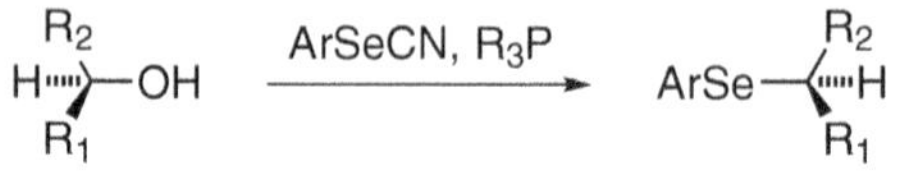

Reagents: PhSeCN, Bu_3P, PhMe; *o*-$O_2NC_6H_4$SeCN, Bu_3P, PhMe; etc.

Stereochemistry: inversion of configuration at carbon

Corey-Winter Elimination

Reagents: (1) $Cl_2C{=}X$, py; (2) $(MeO)_3P$, Δ (X=O, S);
or (1) $Im_2C{=}X$, CH_2Cl_2; (2) $(MeO)_3P$, Δ (X=O, S)

Stereochemistry: *syn* elimination through a cyclic carbene

Arbuzov (Michaelis-Arbuzov) Rearrangement

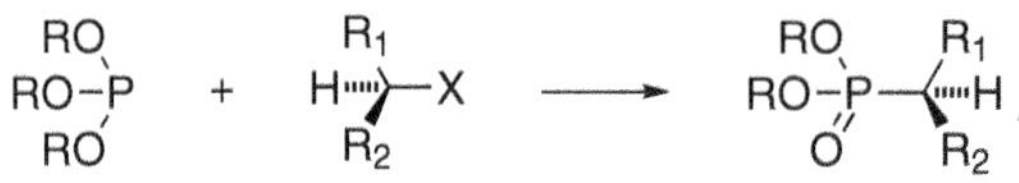

Stereochemistry: inversion of configuration in alkyl halide

Perkow Rearrangement

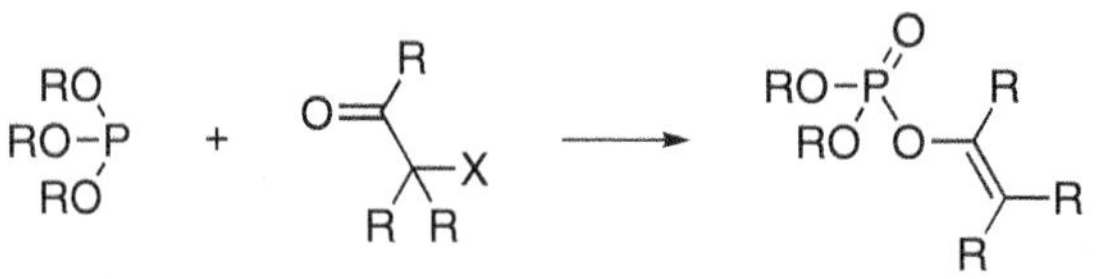

Reaction occurs when the S_N2 displacement of the halogen required by the Michaelis-Arbuzov rearrangement is inhibited or prevented by steric hindrance.

The Pummerer Rearrangement

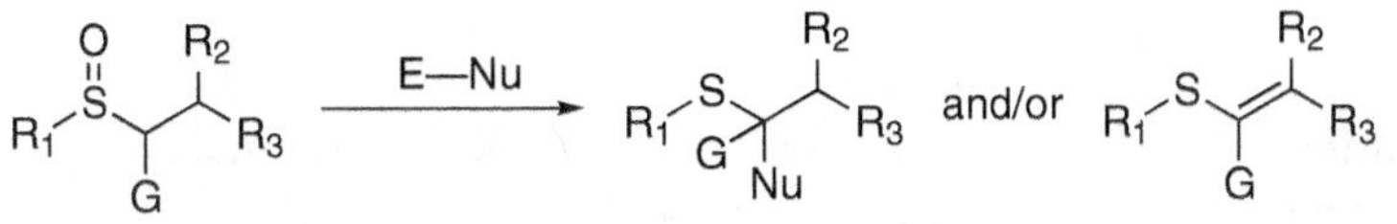

Reagents: Ac_2O, Δ; Ph-N=C=O; H^+; etc.

If group G can leave as a highly stable cation, the Pummerer fragmentation may dominate the reaction.

The Ramberg-Bäcklund Reaction

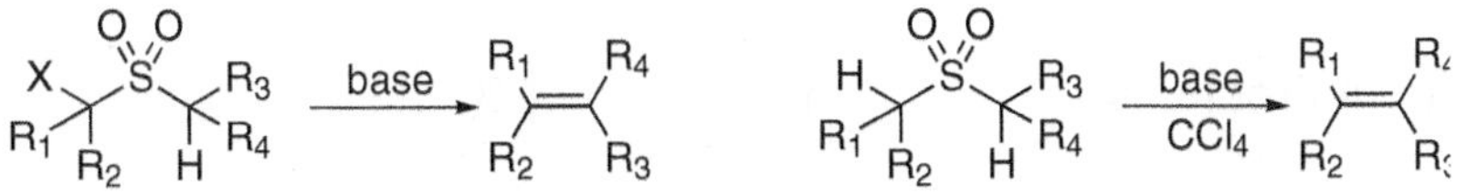

Reagents: $KOBu^t$, THF; KOH, Me_2SO; etc. or $KOBu^t$, t-BuOH-H_2O, CCl_4; etc.

Rearrangements of Allylic Sulfoxides and Selenoxides; Mislow-Evans rearrangement

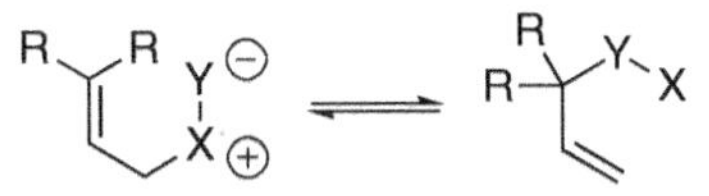

X = SR, SeR; Y=O

Reaction may be extended to sulfonium ylides (Y=CR_2), as well as sulfilimines (Y=NR).

Chapter Summary

In this chapter, the various organic derivatives of the main group non-metals (Si, P, S, Se) and metals (Sn) have been introduced. These compounds are all characterized by having a bond between carbon and an element carrying energetically accessible empty d orbitals, which allows them to stabilize adjacent anions. Silicon and tin, in addition, are efficient at stabilizing a carbocation at the β position by hyperconjugation; this finds use in the Hosomi-Sakurai reaction. Silicon often serves as a surrogate for hydrogen in organic chemistry, thus permitting the reactions of enol equivalents where the actual enols are not stable enough to use (e.g., the Danishefsky diene used in Diels-Alder reactions, enol silyl ethers in the Mikaiyama reactions). Allylic sulfoxides and selenides undergo a facile, suprafacial [2,3] sigmatropic rearrangement to give sulfenate or selenenate esters, thus providing a method for the stereocontrolled formation of allylic alcohols by means of the Mislow-Evans rearrangement. Sulfoxides rearrange in the presence of electrophilic reagents to α-acyloxysulfides, which can be converted to carbonyl compounds. Sulfones can be used to couple two smaller fragments to give an alkene by the Ramberg-Bäcklund reaction. Unsaturation may be introduced into a molecule by sulfoxide or selenoxide elimination. Trialkyl phosphites rearrange in the presence of an alkyl halide to give phosphonate esters in the Arbuzov (Michaelis-Arbuzov) rearrangement. Phosphonium species generated from alcohols are excellent substrates for S_N2 reactions, as in the Mitsunobu reaction and the Grieco selenylation.

Key Terms

mercaptan
Mislow-Evans rearrangement
Mitsunobu reaction
Mukaiyama reactions
Perkow reaction
phosphine
phosphine oxide
phosphonic acid derivatives
phosphorous acid
Pummerer rearrangement
Ramberg-Bäcklund reaction
selenide
selenoxide
silane
stannane
sulfide
sulfone
sulfoxide
thioether
thiol
trialkyl phosphite

Additional Problems

22-6 What is the major organic product of each of the following reactions?

(a) 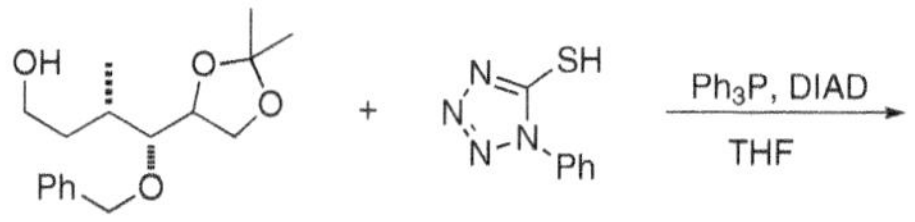

Org. Lett. **2007**, *9* 4757.

(b) 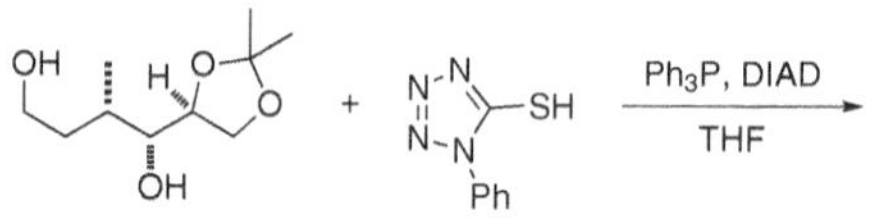

Org. Lett. **2006**, *8*, 1117.

(c) 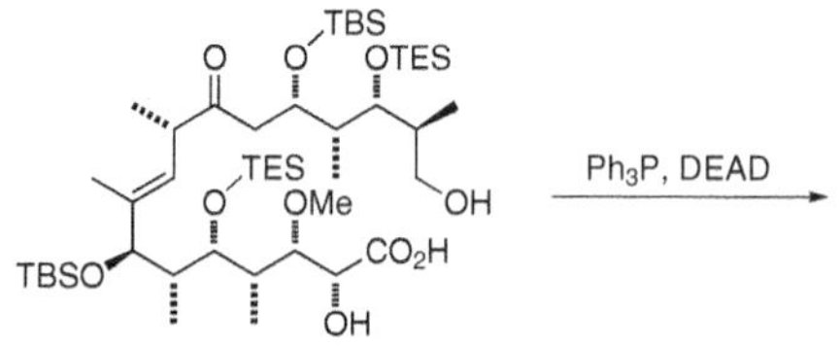

Synthesis **2005**, 1183.

(d) 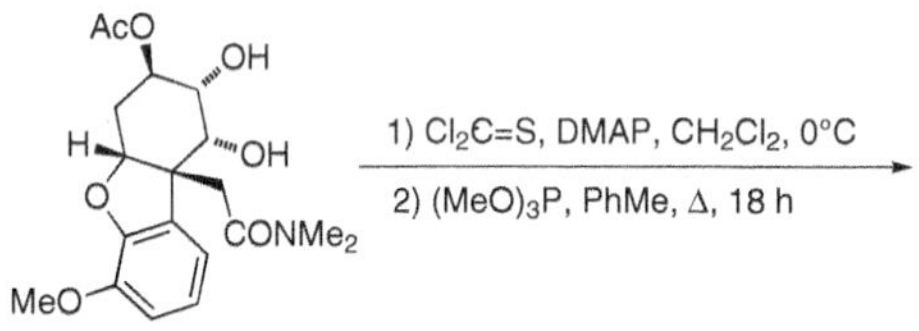

Aust. J. Chem. **2010**, *63*, 1437.

(e) 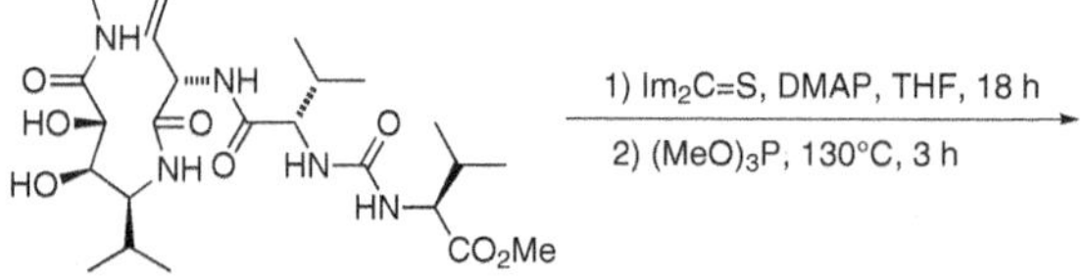

Proc. Natl. Acad. Sci. U.S.A. **2009**, *106*, 6507.

(f) 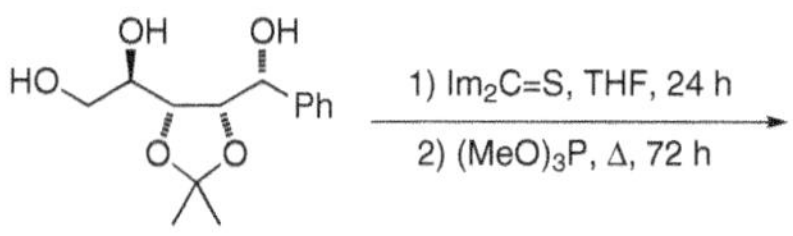

Tetrahedron **2010**, *66*, 2363.

(g) 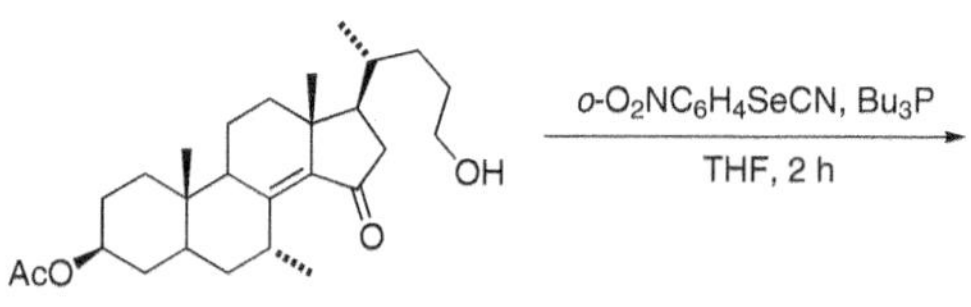

J. Lipid Res. **1995**, *36*, 767.

(h) 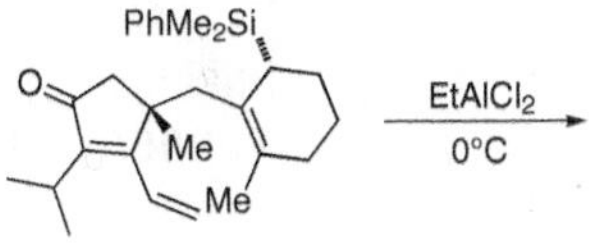

Tetrahedron Lett. **1990**, *31*, 2239.

(i) 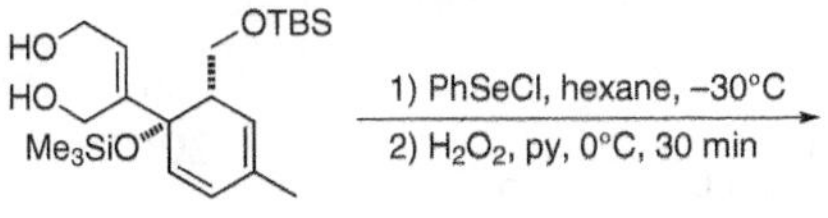

[step 1 results in an electrophilic addition to give a product with a molecular formula of $C_{27}H_{45}ClO_4SeSi_2$]

Tetrahedron Lett. **1987**, *28*, 3647.

(j) 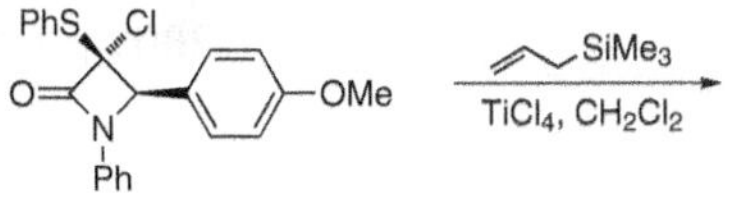

Tetrahedron Lett. **2003**, *44*, 895.

(k) 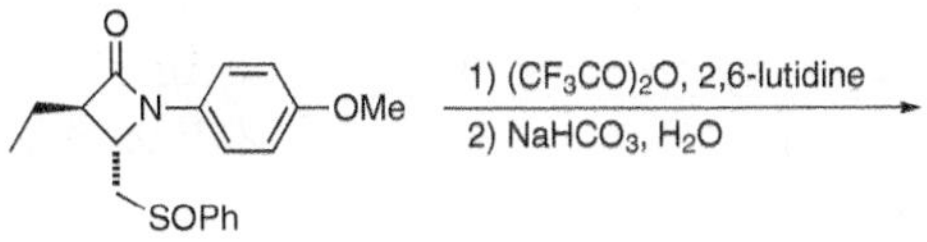

Tetrahedron **2012**, *68*, 10696.

(l) 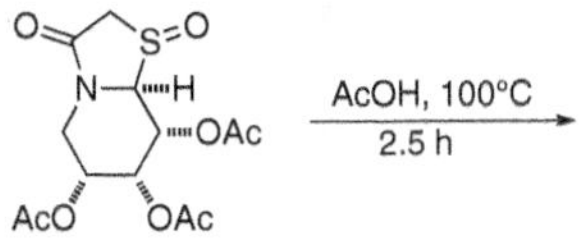

Bioorg. Med. Chem. Lett. **2012**, *22*, 7041.

(m) 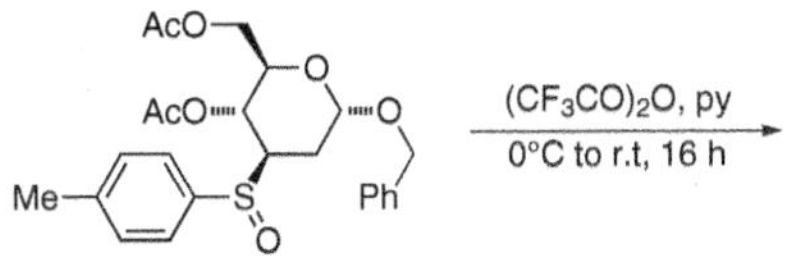

Tetrahedron **2012**, *68*, 8746.

(n) 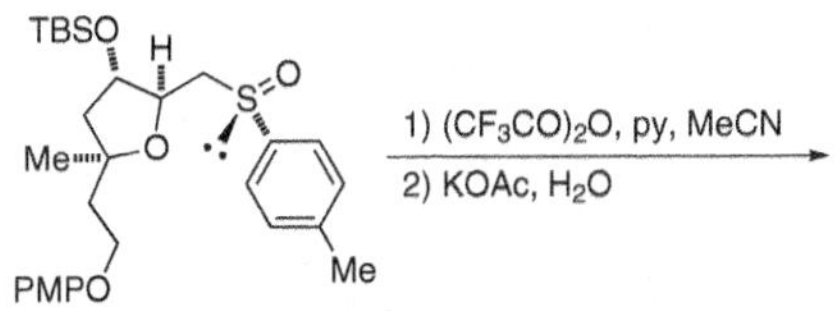

Tetrahedron **2011**, *67*, 10179.

(o) 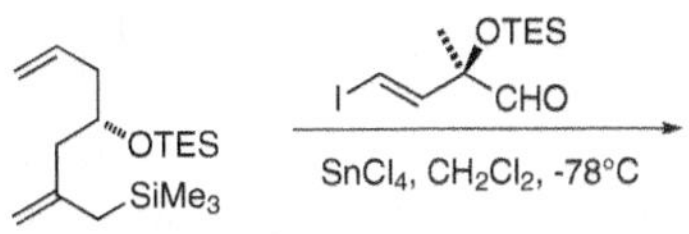

Tetrahedron **2011**, *67*, 7485.

(p) 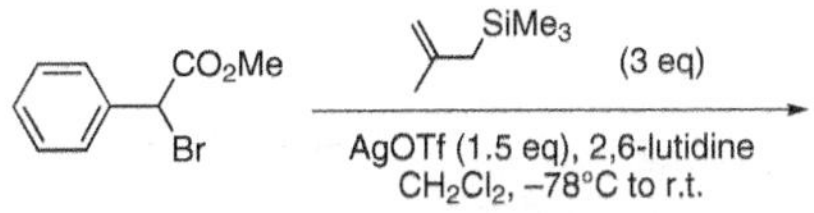

Tetrahedron **2011**, *67*, 7586.

(q) 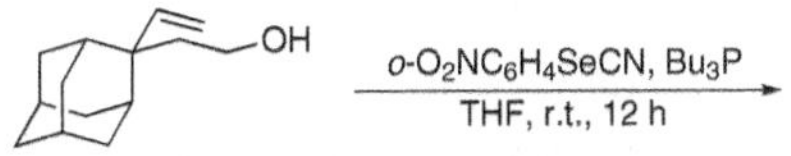

Tetrahedron **1998**, *54*, 11899.

(r) 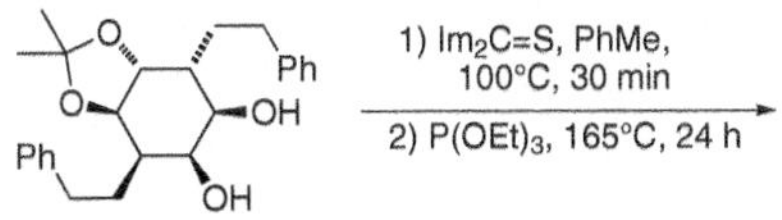

J. Org. Chem. **1997**, *62*, 1916.

(s) 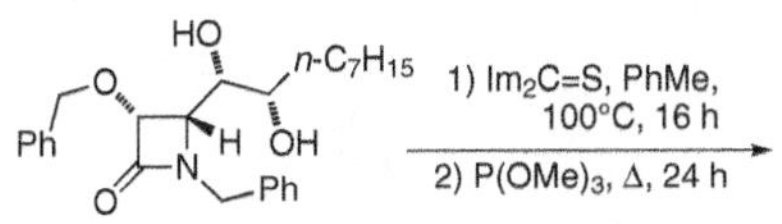

J. Org. Chem. **2001**, *66*, 4180.

(t)
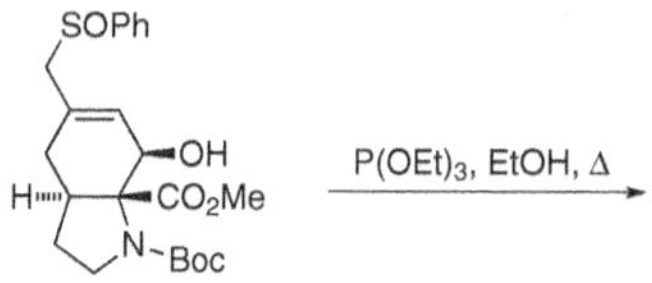

Org. Lett. **2006**, *8*, 4971.

(u)
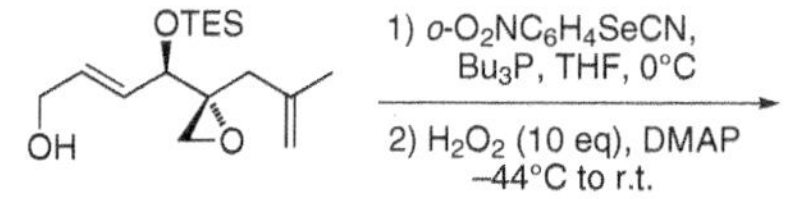

J. Am. Chem. Soc. **2007**, *129*, 2648.

(v)
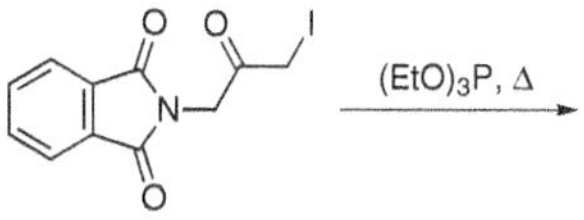

Tetrahedron Lett. **2009**, *50*, 3050.

(w)
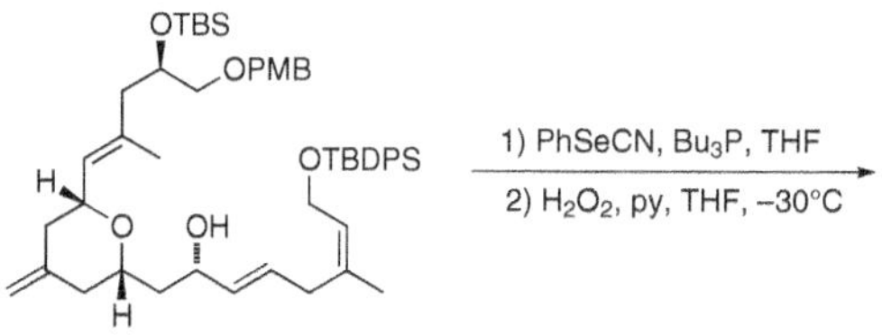

Angew. Chem. Int. Ed. **2005**, *44*, 3485.

(x)
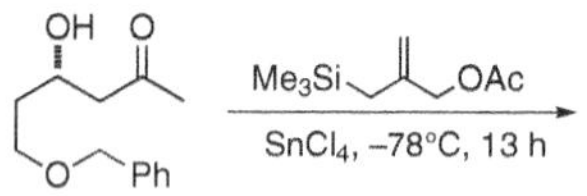

Chem. Eur. J. **2010**, *16*, 14074.

(y)
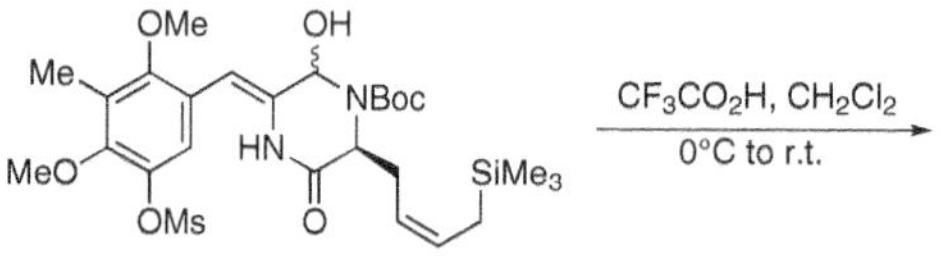

Chem. Eur. J. **2012**, *18*, 11192.

22-7 Suggest a reasonable mechanism for the transformation shown (*Chem. Eur. J.* **2003**, *9*, 262).

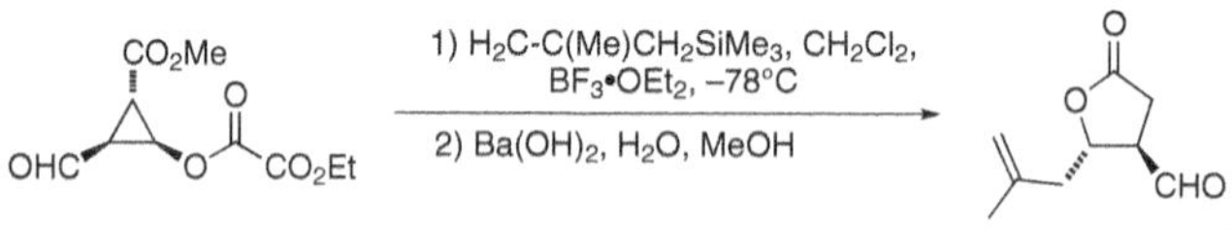

22-8 Complete the following reactions by supplying the missing reagent or reagents. Note that more than one step may be required to complete some of these transformations.

(a)
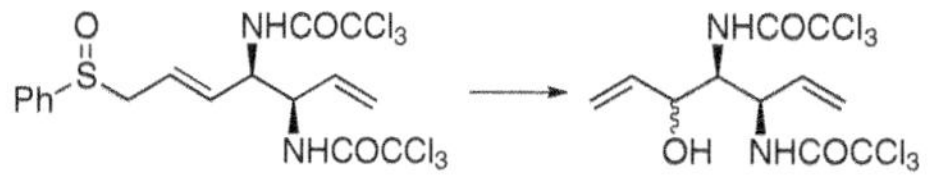

Org. Lett. *2009, 11, 2687.*

(b)
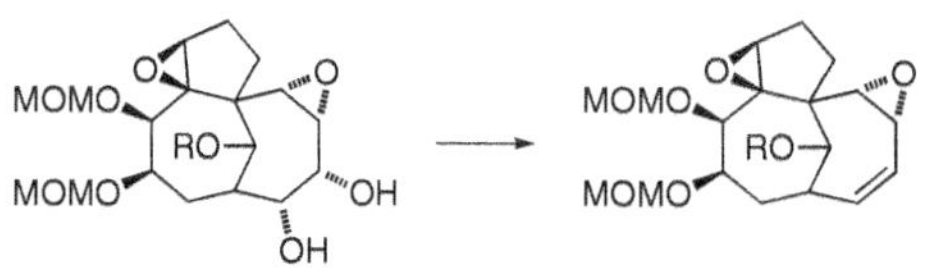

Org. Lett. 2002, 4, 799.

(c)
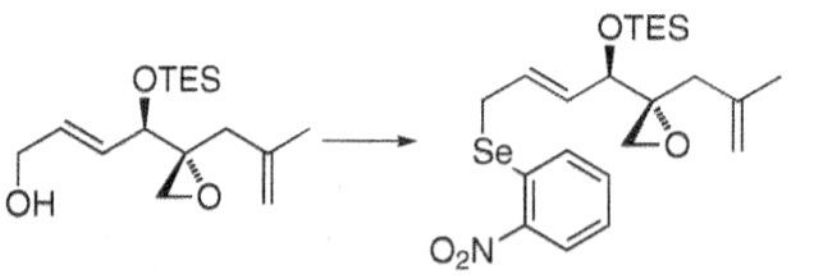

J. Am. Chem. Soc. **2007**, *129*, 2648.

(d)

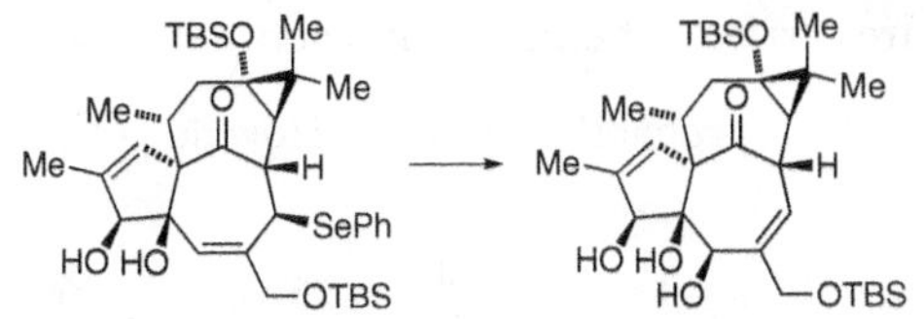

Angew. Chem. Int. Ed. **2012**, *51*, 4972.

(e)

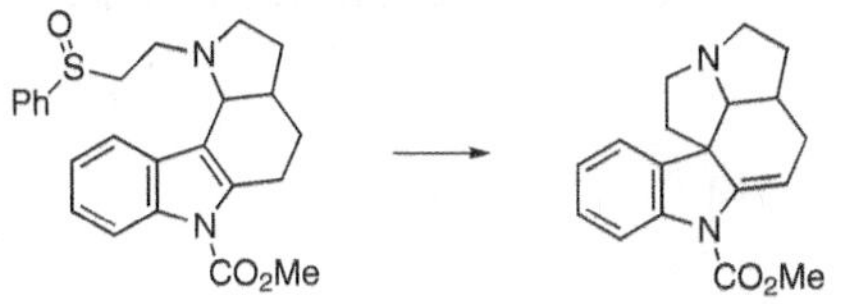

J. Org. Chem. **1996**, *61*, 7106.

22-9 Suggest a reasonable mechanism for the reaction below (*Tetrahedron Lett.* **2011**, *52*, 6260). How does your mechanism account for the stereochemistry of the reaction?

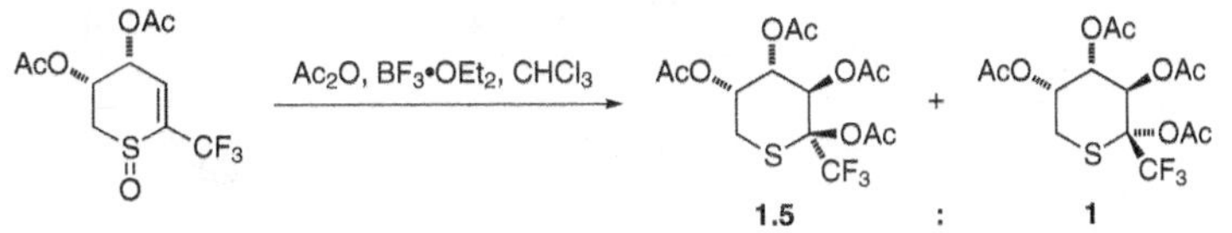

22-10 Suggest a reasonable mechanism for the transformation below (*Org. Lett.* **2005**, *7*, 1629).

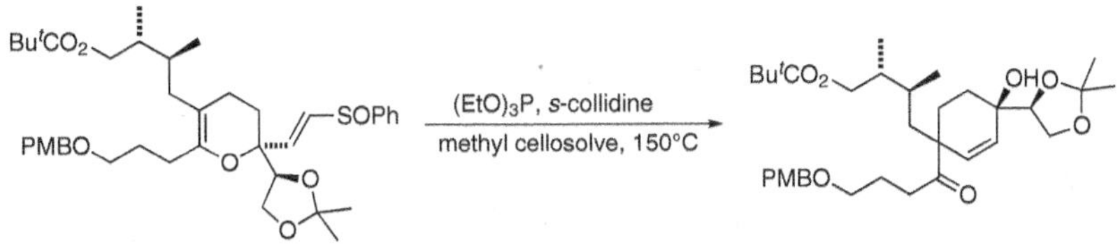

22-11 Complete the following reaction sequence from the synthesis of otteliones A and B (*Org. Lett.* **2003**, *5*, 3903) by supplying the missing reagents.

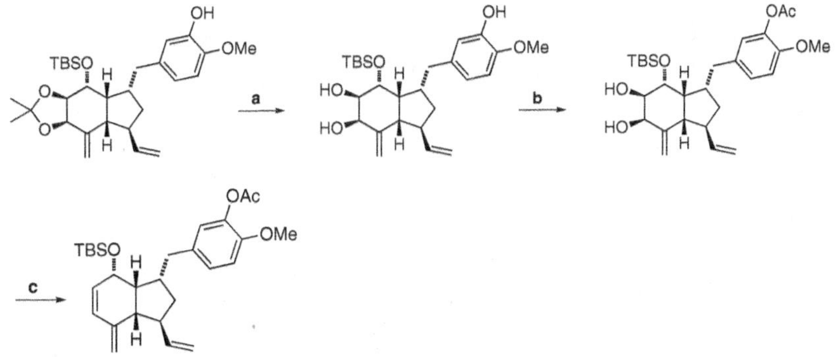

How can you ensure that in step **b**, only the hydroxyl group shown is esterified?

22-12 What are the structures of the missing compounds in this sequence from Fukuyama's synthesis of (±)-morphine (*Org. Lett.* **2006**, *8*, 5311)?

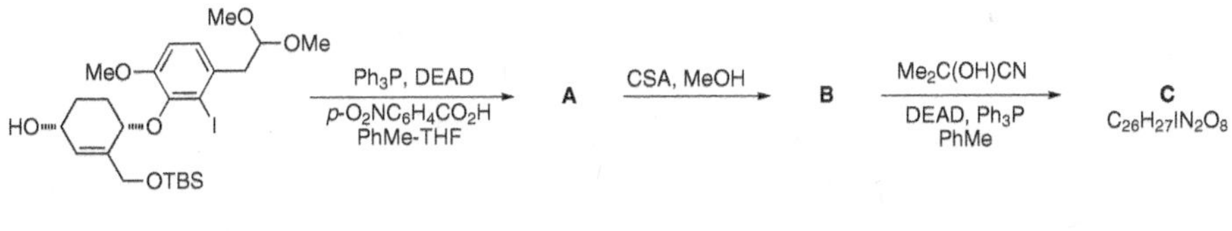

In a practical sense, what is the function of the acetone cyanohydrin [$Me_2C(OH)CN$] in this sequence?

22-13 The two reactions below are carried out under identical conditions.

(a) Which of the two reactions is probably concerted? How do you know that is the other not concerted?

(b) In one of these reactions, silica gel on a thin-layer chromatography plate was a sufficiently strong Lewis acid to promote the reaction. Which one, and why?

[*Tetrahedron Lett.* **1978**, 3513]

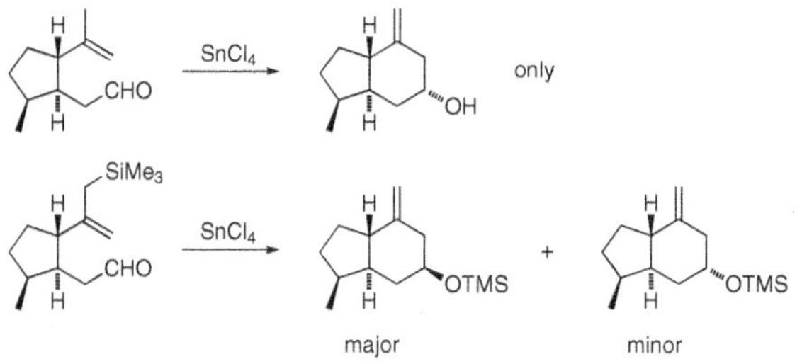

CHAPTER TWENTY-THREE

Modern Asymmetric Synthesis

23.1 Controlling Absolute Stereochemistry

As has been pointed out numerous times in this book, the control of stereochemistry in the absolute sense has become a hallmark—even requirement—of modern organic synthesis. From relatively modest beginnings,[1] the field of asymmetric synthesis has grown to provide the synthetic chemist with a diverse toolbox of reactions that give high levels of enantioselectivity. This is attested to by the explosion of reference works dedicated to the subject of asymmetric synthesis, which includes several books and book series.[2]

The generation of a new chiral center in a molecule will always lead to the racemic product unless that new chiral center is generated in an environment that is already chiral. Thus, if one is to generate a single enantiomer of a chiral compound from an achiral precursor, it is imperative that one arrange for the transition state of the reaction that generates the new chiral center to be not only chiral, *but a diastereoisomer of the transition state leading to the other enantiomer of the product.* There are three general ways in which this can be accomplished, and these were addressed at some length in Chapter 2 (Section 2.5) of this book. These three strategies for generating a chiral environment for the reaction are (1) the chiral auxiliary strategy, (2) the chiral reagent strategy, and (3) the chiral catalyst strategy. Each has its strengths and weaknesses, as will be discussed in the following sections of this chapter.

Chiral Auxiliaries

The first of these methods is the ***chiral auxiliary*** approach, in which the substrate is first covalently modified by a suitable chiral adjuvant that is removed after the reaction is complete (Figure 23.1, which is a reproduction of Figure 2.15). This approach is one of the longest established methods for the asymmetric synthesis of organic compounds. With an appropriate auxiliary, one can routinely obtain enantiomeric excess (e.e.) values above

---

1. Early books devoted to the subject include: (a) Eliel, E.; Otsuka, S., Eds. *Asymmetric Reactions and Processes in Chemistry.* (American Chemical Society: Washington, D.C.). *ACS Symp. Ser.* **1981**, *185.* (b) Morrison, J.D., Ed. *Asymmetric Synthesis* (Academic: New York); this series went through five volumes between 1983 and 1985. (c) Hanessian, S. *Total Synthesis of Natural Products: The 'Chiron' Approach* (Pergamon: Oxford, 1983).

2. Some books devoted to the subject from the 1990s: (a) Ojima, I., Ed. *Catalytic Asymmetric Synthesis* (Wiley-VCH: Weinheim). This work has been through three editions between 2000 and 2010. (b) Christmann, M.; Bräse, S., Eds. *Asymmetric Synthesis—The Essentials* (Wiley-VCH: Weinheim, 2007). (c) Enders, D.; Jaeger, K.-E., Eds. *Asymmetric Synthesis with Chemical and Biological Methods.* (Wiley-VCH: Weinjeim, 2007). (d) Jacobsen, E.N.; Pfaltz, A.; Yamamoto, H., Eds. *Comprehensive Asymmetric Catalysis* (Springer: Berlin, 1999), vols. 1-3. (e) Gawley, R.E.; Aubé, J. *Principles of Asymmetric Synthesis* (Pergamon: Oxford, 1996). (f) Procter, G. *Asymmetric Synthesis* (Oxford Press: Oxford, 1996). (g) Seyden-Penne, J. *Chiral Auxiliaries and Ligands in Asymmetric Synthesis* (John Wiley & Sons: New York, 1995). (h) Noyori, R. *Asymmetric Catalysis in Organic Synthesis* (John Wiley & Sons: New York, 1994). (h) Aitken, A.; Kilenyi, S.N., Eds. *Asymmetric Synthesis* (Blackie Academic & Professional: London, 1992.

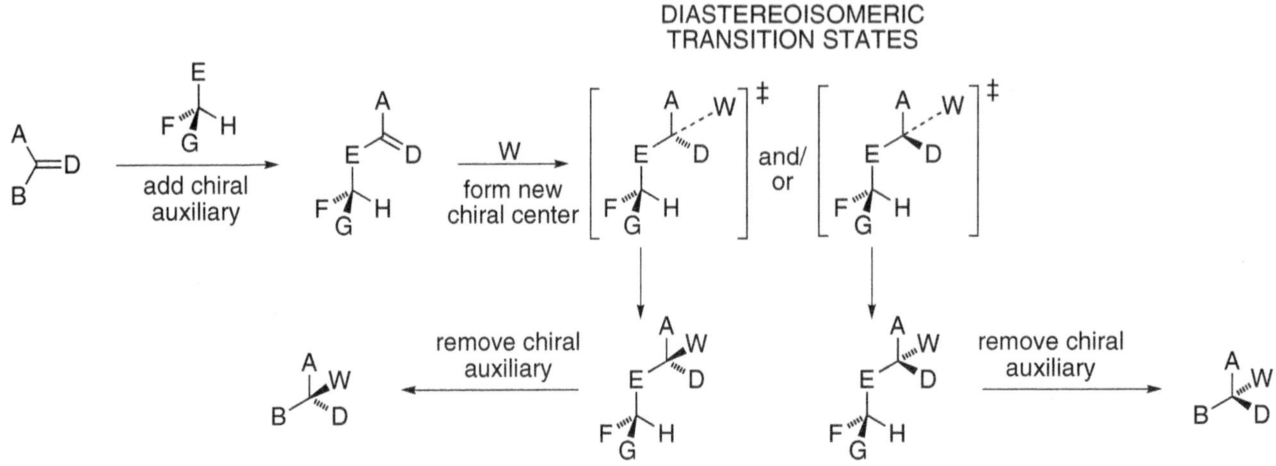

Figure 23.1 Formation of a chiral product enriched in one enantiomer by means of a chiral auxiliary

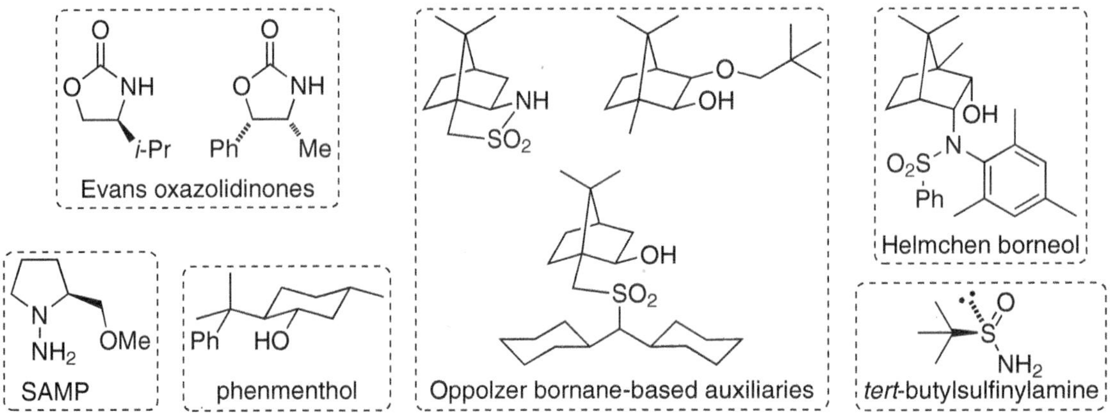

Figure 23.2 Representative successful chiral auxiliaries

90% to 95%. The criteria that should be characteristic of a useful chiral auxiliary were described by Eliel in 1974.[3] They are:

1. The chiral auxiliary must be readily separated from the chiral product without racemization of the new chiral center.
2. The chiral auxiliary should preferably be recoverable in good yield and without racemization after being used.
3. The chiral auxiliary must be available in optically pure form.

Typical examples (Figure 23.2) are the oxazolidinones,[4] which are derived from α-amino acids and norephedrine; a series of chiral auxiliaries including Oppolzer sultam, the Oppolzer isoborneol, and the Helmchen borneols, which are all derived from camphor;[5] and phenmenthol, which is derived from pulegone.[6] More recently,

3. Eliel, E.L. *Tetrahedron* **1974**, *30*, 1503.
4. Review: Evans, D.A. *Adrichimica Acta* **1982**, *15*, 23.
5. Review: Oppolzer, W. *Tetrahedron* **1987**, *43*, 1969.
6. Corey, E.J.; Ensley, H.E. *J. Am. Chem. Soc.* **1975**, *97*, 6908.

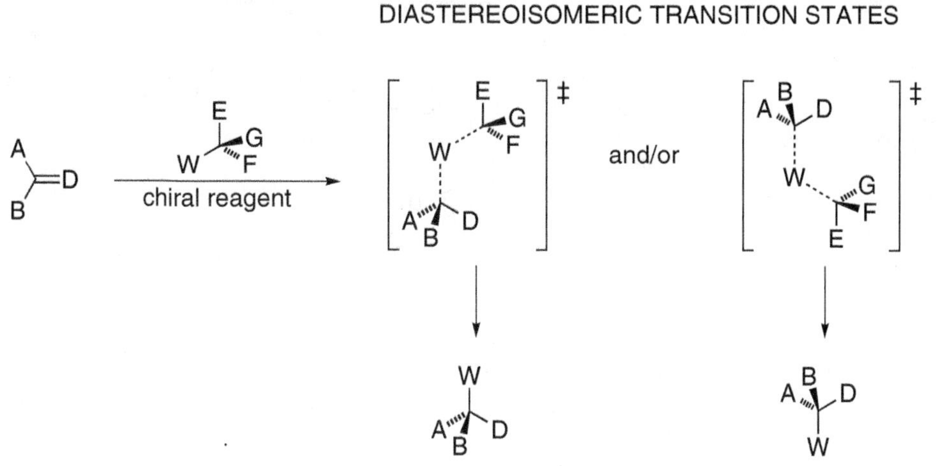

Figure 23.3 Formation of a chiral product enriched in one enantiomer by means of a chiral reagent

tert-butylsulfinylamines[7] have found extensive use as chiral auxiliaries for the asymmetric synthesis of amines and for asymmetric conjugate additions.

Chiral Reagents

The second way in which a non-racemic chiral product can be generated from a prochiral substrate involves the use of a chiral reagent to carry out the transformation (Figure 23.3, which is a reproduction of Figure 2.19). Unlike the chiral auxiliary approach, in this methodology, there is no need to carry out a post-reaction transformation to recover the chiral product. However, forming suitable chiral reagents has been a difficult task, and there are still relatively few reagents that can be used to give high e.e. values of the chiral product. To date, borane reagents (usually from α-pinene) and the Davis oxaziridine have been the most successful.

Chiral Catalysis

Asymmetric synthesis underwent a paradigm shift in the early 1980s, with the reports of the Sharpless asymmetric epoxidation[8] and the asymmetric hydrogenation of α-aminoacrylate ester derivatives.[9] These reactions relied on the third approach to generating a non-racemic chiral product from a reaction of an achiral precursor: the use of a chiral catalyst (Figure 23.4, which is a reproduction of Figure 2.20).

Like the chiral reagent approach, this methodology for generating non-racemic chiral compounds does not require any post-reaction modifications to the product, but it has the added advantage that neither of the participants in the reaction (the substrate and the reagent) need be chiral. Asymmetric catalytic reactions were the focus of much of Chapter 21, and you should refer to that chapter for a more in-depth introduction to the field.

The growth of catalytic asymmetric reactions over the past two decades has been explosive. Here are some examples:

- Ojima's *Catalytic Asymmetric Organic Synthesis* grew from a 476-page first edition in 1993 to an 864-page second edition in 2000 to a 998-page third edition in 2010—a growth rate of effectively one 30-page review article per year

7. The explosive growth in the use of this auxiliary is attested by the size of the 2010 review—140 pages! Robak, M.T.; Herbage, M.A.; Ellman, J.A. *Chem. Rev.* **2010**, *110*, 3600.

8. Katsuki, T.; Sharpless, K.B. *J. Am. Chem. Soc.* **1980**, *102*, 5974.

9. Reviews: (a) Noyori, R. *Adv. Synth. Catal.* **2003**, *345*, 15. (b) Knowles, W.S. *Angew. Chem. Int. Ed.* **2002**, *41*, 1998.

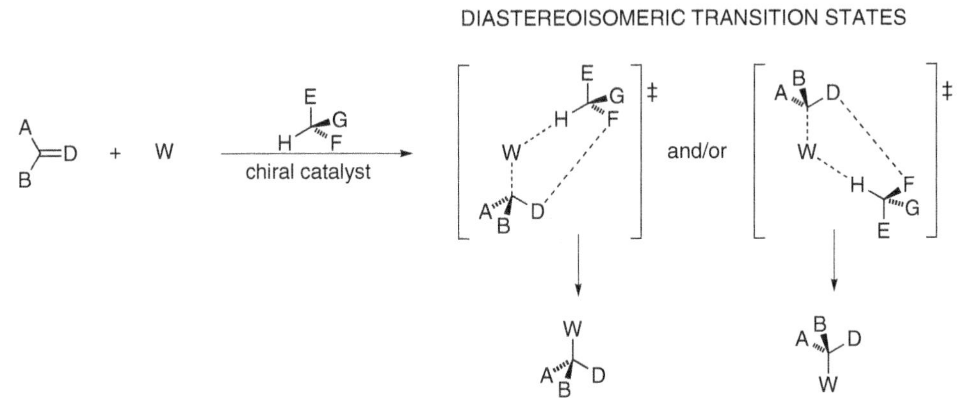

Figure 23.4 Formation of a chiral product enriched in one enantiomer by means of a chiral catalyst

- A search of the American Chemical Society web site using the term "asymmetric catalysis" returned 671 research articles from 2010 to 2013
- A search of the Royal Society of Chemistry web site returned 596 research articles and books for the same time period.
- In 2001, the *Journal für Praktische Chemie* changed its focus and became *Advanced Synthesis & Catalysis.*

To obtain an up-to-date perspective on this rapidly changing field, you should consult one of the many specialist books, monographs or reviews in the area.

Chiral catalysts fall into two major classifications: transition metal catalysts, where the chirality is derived from the ligands, and organocatalysts, which contain no metal. The ultimate organocatalysts are, of course, enzymes, whose control over the configurations of the products is absolute for the natural substrates for the enzyme and still very high for unnatural substrates. Recall how acylases can be used to liberate amines from amide protecting groups and, at the same time, allow the resolution of the amine.

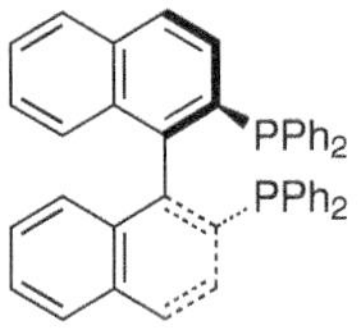

BINAP

Unlike most chiral auxiliaries and chiral reagents, many chiral transition metal complexes are based on the chiral 2,2'-disubstituted-1,1'-binaphthyl (e.g., BINAP[10]) unit or a closely related atropisomeric species. This same carbon skeleton has also provided a scaffold for the preparation of organocatalysts, such as chiral Brønsted acids.[11]

Simple Diastereoselection and Double Diastereoselection

When a new chiral center is formed in a reaction involving a compound already containing a chiral center (e.g., a prochiral substrate conjugated with a chiral auxiliary), the new chiral center leads to the formation of a mixture of diastereoisomers. The relative amounts of the two diastereoisomers in the product (the ***diastereoisomer ratio** [**d.r.**]* or the ***diastereoselectivity*** of the reaction) depends on the difference in activation energies of the two competing reactions (provided, of course, that the reactions are under kinetic control). When only one of the reacting species is chiral, the diastereoselectivity of the reaction is described as being ***simple diastereoselection***, and the only new interaction is between the existing chiral center and the new center being generated. This is the situation illustrated in Figure 23.5. The products of the reaction, after removal of the chiral auxiliary, are enantiomers, but they are not formed in equal amounts.

10. Noyori, R.; Takaya, H. *Acc. Chem. Res.* **1990**, *23*, 345.

11. (a) Rueping, M.; Sugiono, E.; Azap, C.; Theissmann, T.; Bolte, M. *Org. Lett.* **2005**, *7*, 3781. (b) Rueping, M.; Azap, C.; Sugiono, E.; Theissmann, T. *Synlett* **2005**, 2367.

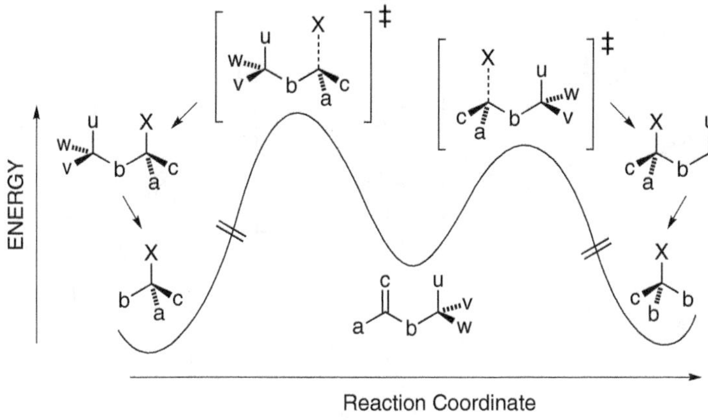

Figure 23.5 Simple diastereoselection in a reaction between an achiral reactant and a prochiral substrate modified by a chiral auxiliary

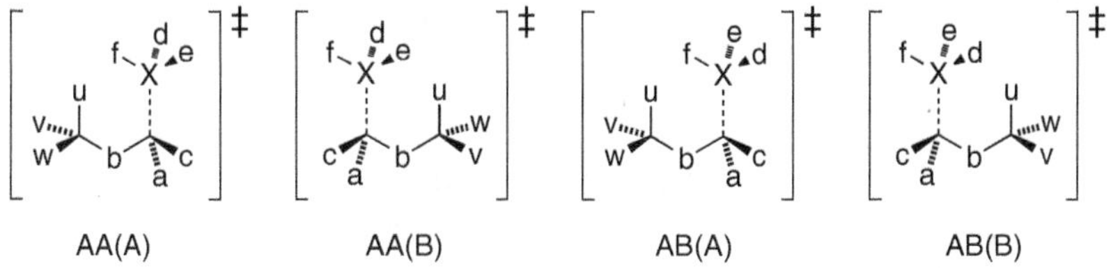

Figure 23.6 The four diastereomeric transition states for the reaction between two chiral molecules to generate a new chiral center

When two chiral species react to give a product containing a new chiral center, four possible diastereoisomeric products can be generated (Figure 23.6). These products and the transition states leading to them are all (in principle, at least) of different energy, which means that they will not be generated in equal amounts. Usually—but not always—one of the four diastereoisomeric transition states is significantly lower in energy than the others, as shown in Figure 23.7, with the result that that particular diastereoisomer of the product will be strongly favored. In this situation, the chiral substrate and the chiral reagent are said to be ***matched***. In the converse situation, there is little difference in the energies of the diastereoisomeric transition states, with the result that much more nearly equal amounts of the possible products are formed. In this situation, the chiral substrate and the chiral reagent are said to be ***mismatched***. These possible situations are summarized in Figure 23.7, where the transition state labeled AB(B) clearly has the lowest energy and should therefore lead to the major product.

The Cram and Felkin-Anh Rules

Let us begin by reviewing the additions of nucleophiles to chiral aldehydes. These additions lead to the formation of a pair of diastereoisomers, whose stereochemical outcome was first rationalized on the basis of a conformational model proposed by Donald Cram and modified by Felkin and Anh. Applying the Cram[12] or the Felkin-Anh rules[13] to a chiral aldehyde, one obtains the situation shown in Figure 23.8 (illustrated using Cram's rule). Thus, when phenylmagnesium bromide is added to (*R*)-2-phenylpropanal, the major product of the reaction is (1*R*,2*R*)-1,2-diphenyl-1-propanol. That is, there is an intrinsic stereochemical bias where the *R* configuration is favored for the new chiral center in additions of carbon nucleophiles to this aldehyde.

12. (a) Cram, D.J.; Abd Elhafez, F.A. *J. Am. Chem. Soc.* **1952**, *74*, 5828. (b) Cram, D.J.; Knight, J.D. *J. Am. Chem. Soc.* **1952**, *74*, 5835. (c) Curtin, D.Y.; Harris, E.E.; Meislich, E.K. *J. Am. Chem. Soc.* **1952**, *74*, 2901. (d) Leitereg, T.J.; Cram, D.J. *J. Am. Chem. Soc.* **1968**, *90*, 4019.

13. (a) Chérest, M.; Felkin, H.; Prudent, N. *Tetrahedron Lett.* **1968**, 2199. (b) Anh, N.T.; Eisenstein, O. *Nouv. J. Chem.* **1977**, *1*, 61.

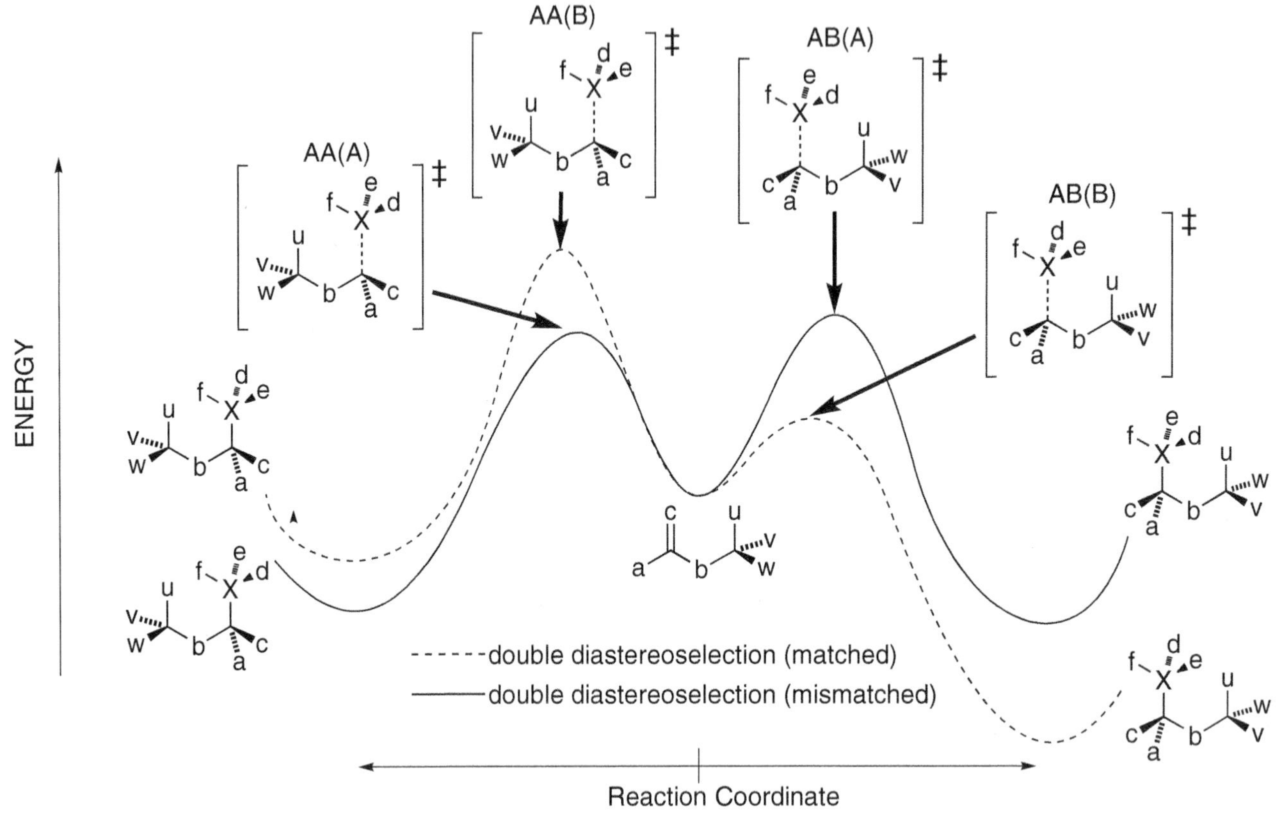

Figure 23.7 The reaction energy profiles for the reaction between two chiral compounds to generate a new chiral center gives rise to four possible transition states, all of different energy. Based on this diagram, one would expect the product formed via AB(B) to dominate the product mixture, with a small amount of the product formed via AA(A) and negligible amounts of the other two diastereoisomers.

Figure 23.8 Cram's rule may be used to predict the stereochemistry of addition of nucleophiles to the carbonyl group of chiral aldehydes and ketones.

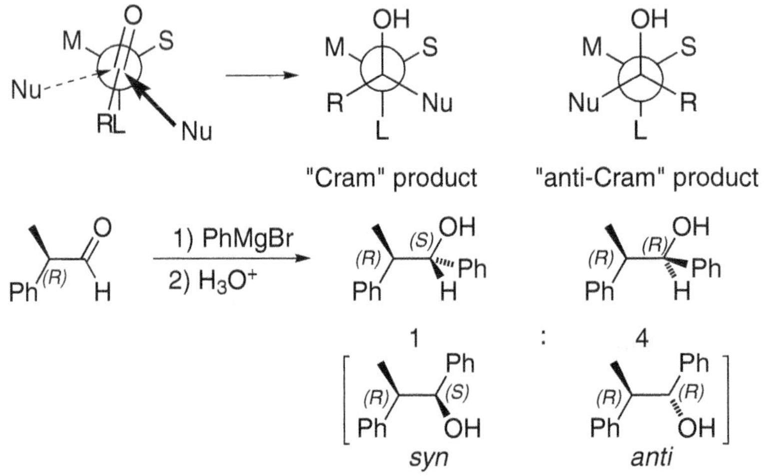

The Aldol Addition

The aldol addition reaction provides another excellent example with which to discuss modern asymmetric synthesis. The stereochemistry of the kinetic aldol addition product in reactions where two new chiral centers are formed simultaneously depends on the stereochemistry of the enolate and on the metal or non-metal of the enolate. When a very hindered ketone is used, the enolate generated is almost invariably the *Z(O)* enolate, and there is a very strong bias toward the *syn* aldol, as shown in Example **23.1**.[14] When the aldehyde carries a group at the α carbon capable of complexing with the lithium ion, the chelation effect takes over, and the stereochemical preference is reversed: now the *anti* aldol becomes the dominant product. The highest levels of diastereoselectivity in this reaction occur when the transition state is tight (i.e., when the metal-oxygen bond is short). For this reason, boron enolates have become popular reagents for carrying out stereoselective aldol addition reactions.

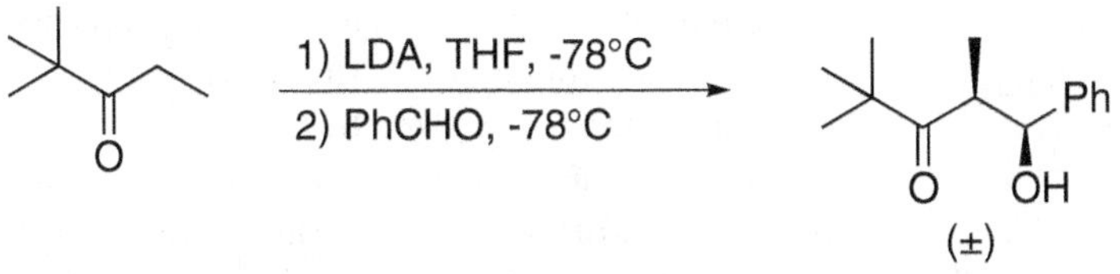

(23.1)

The Cram/Felkin-Anh analysis also applies to reactions such as the aldol addition. The addition of the kinetic lithium enolate of an achiral methyl ketone to (*R*)-2-phenylpropanal leads to the formation of an aldol where the preferred configuration of the new chiral center is also *R* (Figure 23.9).

The concept of matched and mismatched chiral partners in a reaction can be illustrated especially well by examining the Evans' asymmetric aldol addition reaction of chiral boron enolates to chiral aldehydes. In 1981, Evans disclosed[15] chiral oxazolidinones as practical chiral auxiliaries for carrying out enantioselective aldol additions to aldehydes. In Example **23.2**, the dramatic levels of diastereoselectivity provided by the Evans chiral boron enolates (which give the *Z(O)* enolate) is illustrated by the reaction between the *Z(O)* boron enolate of (4*S*)-*N*-propanoyl-1,3-oxazolidin-2-one and isobutyraldehyde is shown. What is initially apparent is that the same stereochemical bias toward the *syn* aldol is present in these reactions: the *syn* aldols now make up all the detectable products.

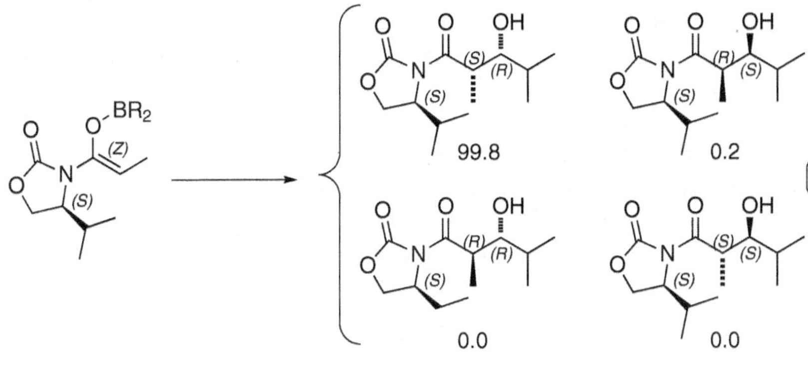

(23.2)

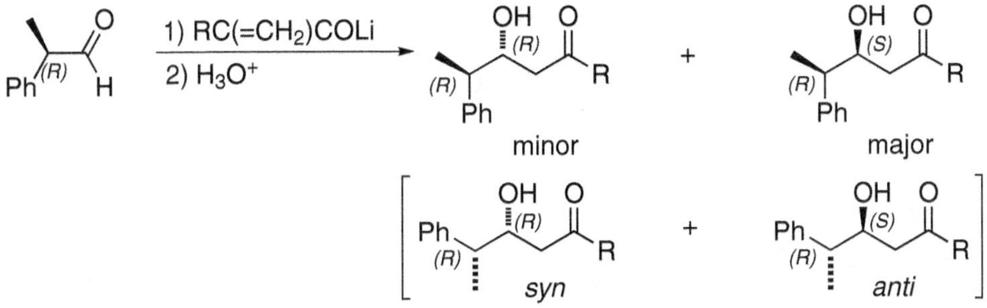

Figure 23.9 The kinetic aldol addition of the enolate of a chiral aldehyde to the enolate of an achiral methyl ketone gives the Cram (*anti*) isomer of the product as the major product.

14. Dubois, J.-E.; Dubois, M. *Tetrahedron Lett.* **1967**, 4215; *Chem. Commun.* **1968**, 1567; *Bull. Cos. Chim. France* **1969**, 3120, 3553.
15. Evans, D.A.; Bartroli, J.; Shih, T.L. *J. Am. Chem. Soc.*, **1981**, *103*, 2127.

Table 23.1 Effects of Reactant Stereochemistry on the Stereochemistry of the Aldol Addition

Enolate Type	Aldehyde Type	Product Stereochemistry
Z(O), achiral	Achiral aldehyde	*syn*
Acetyl enolate [$H_2C{=}C(OM)R$]	Chiral aldehyde {RCHR'CHO}	Cram, *anti*
Z(O), chiral	Achiral aldehyde	*syn*; chirality fixed by chirality of enolate

The example shows that there is an extremely high inherent stereochemical bias in this reaction, with the absolute configuration of the new chiral centers being defined by the configuration of the chiral auxiliary. The preferred aldol adducts are both *syn* aldols, with the (2*S*,3*R*)-2,4-dimethyl-3-hydroxybutanoyl product being preferred by an almost 500:1 ratio over the (2*R*,3*S*) isomer. The other two diastereoisomers are even less favored. The effects of the stereochemistry of the participating reagents on the stereochemical outcome of the aldol addition reaction can be summarized as given in Table 23.1.

Now let us ask the question: what would happen in the aldol addition between the Evans chiral boron enolate in Example **23.20** and 2-phenylpropanal? The analysis of the problem with respect to the *S* aldehyde may be found in Figure 23.10. As we have seen, there is an inherent bias in additions to this aldehyde where the new chiral center (the carbinol carbon) has the same configuration as the existing chiral center in the aldehyde, which should lead to the preferential formation of the Cram isomers **B** and **D**. There is also a stereochemical bias imposed by the enolate, which should lead to the *syn* diastereoisomers **A** and **B** being preferred. However, there is a third stereochemical influence that we have not yet taken into account. The chiral auxiliary also exerts an influence that would lead to the product where the configuration at the α carbon of the aldol matches the configuration of the chiral center at position 4 of the oxazolidinone. This would lead to isomers **A** and **D** being favored. We see from this rather simplistic analysis that in the reaction of the *S* aldehyde with the *S* enolate (the upper example), all three biases disfavor product **C**, which we would therefore expect to be formed in least amount. Which product is the major one will depend on the relative importance of the three influences.

We can perform a similar analysis for the reaction between the same enolate and the *R* enantiomer of the aldehyde (Figure 23.11). Again, the stereochemical bias due to the aldehyde obeys Cram's rule, favoring products **W** and **Y.** The stereochemical bias due to the chiral auxiliary prefers the *S* configuration at the α carbon of the aldol, favoring products **W** and **Z,** and the inherent stereochemical bias of the aldol addition product itself would favor the *syn* aldol, favoring products **W** and **X**. Here, the situation is also fairly clearcut, and we would reasonably expect that compound **W**, which is the favored product according to all three criteria, should be the major product.

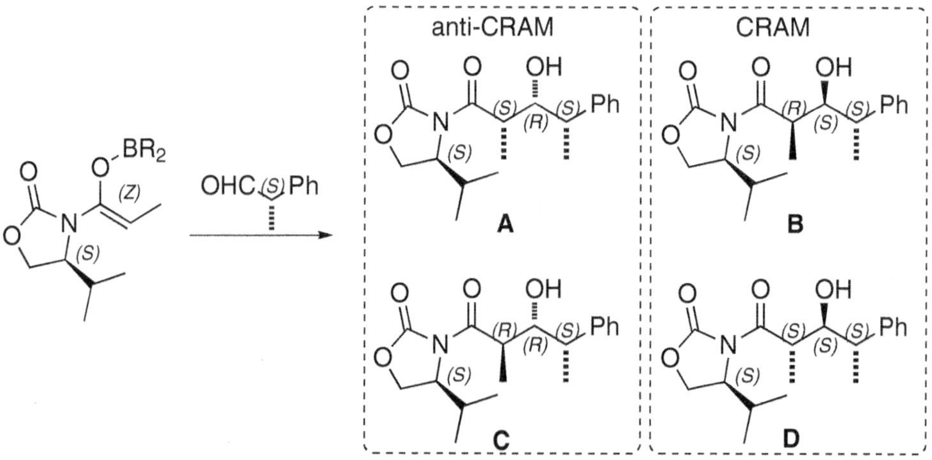

Figure 23.10 The possible diastereoisomeric products of the addition of the *S*-Evans chiral boron enolate to *S*-2-phenylpropanal

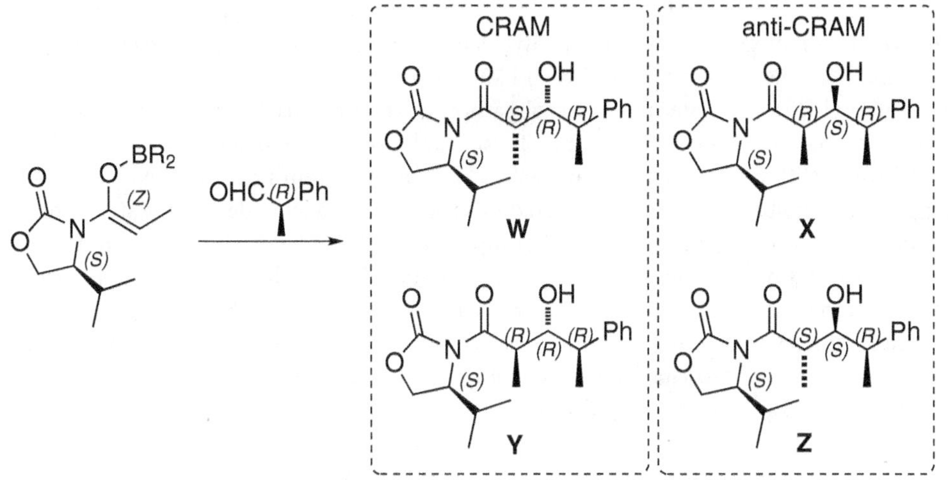

Figure 23.11 The possible diastereoisomeric products of the addition of the *S*-Evans chiral boron enolate to *R*-2-phenylpropanal

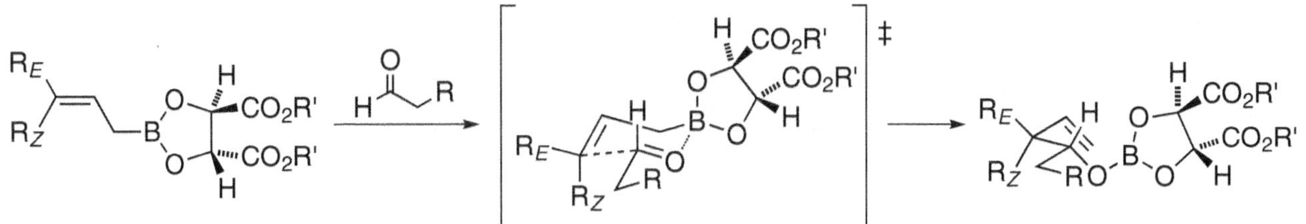

(23.3) **(23.4)** **(23.5)**

Figure 23.12 Chiral crotylboration

In the foregoing analysis, the absolute configurations of the two new chiral centers in the major products of the reactions are predicted to depend on which enantiomer of the aldehyde is used. However, this is, as indicated, a simplistic view, because there are two different species exerting a stereochemical influence on the reaction: (1) the aldehyde, which imposes the Cram's rule selectivity on the reaction, and (2) the inherent *syn* bias of the aldol addition, with the further modification of absolute configuration imposed by the chiral auxiliary. In other words, the above analysis is only fully correct if these stereochemical influences operate to approximately equal extents. This is certainly not the case in this reaction, where the stereochemical bias of the aldol addition to give the *syn* addition product with the configuration of the new α carbon matching the configuration of the chiral center in the boron enolate overwhelms all other stereochemical influences. This leads to the product **A**, rather than the products **B** or **D** being the major product with the *S* aldehyde. There are relatively few reactions where the configuration of the major product is as easily predicted as the Evans' asymmetric aldol addition. Thus, *it is important when attempting to predict the stereochemical outcome of a reaction to know the relative importance of the various stereochemical influences on the course of that reaction.*

Crotylboration of Aldehydes

Chiral crotylboration of a chiral aldehyde provides a very nice example of matched versus mismatched stereochemistry in addition reactions. When the chiral *E*-crotylboronate ester based on a tartrate ester reacts with an achiral aldehyde, the reaction is known as the ***Roush crotylboration***.[16] It proceeds through a six-membered cyclic transition state (**23.4**), as shown in Figure 23.12.

16. Roush, W.R.; Ando, K.; Powers, D.B.; Palkowitz, A.D.; Halterman, R.L. *J. Am. Chem. Soc.* **1990**, *112*, 6339.

The addition reaction is highly stereoselective, with d.r. values typically above 49:1; the *E*-crotylboronate gives the *anti* homoallyl alcohol, and the *Z*-crotylboronate gives the *syn* alcohol. As shown in Examples **23.6** and **23.7**, where the chiral boronate is derived from diisopropyl (*R*,*R*)-tartrate, the carbinol carbon of the product has the *R* configuration, with e.e. values between 83% and 86%. These results are typical of a wide range of achiral aldehydes. With chiral aldehydes and achiral boronates, Cram's rule selectivity generally applies in crotylboration; the level of enantioselectivity is independent of the stereochemistry of the double bond of the crotyl group. This is illustrated by the conversion of the aldehydes **23.8** into the homoallyl alcohols **23.9** and **23.10** by the reaction with pinacolyl crotylboronate (product ratios are given; where the sum is not 100%, minor products account for the rest).[17] The only real difference between the results obtained with the geometric isomers of the boronate ester arise from the fact that the *Z* crotylboronate could be obtained only 90% stereochemically pure.

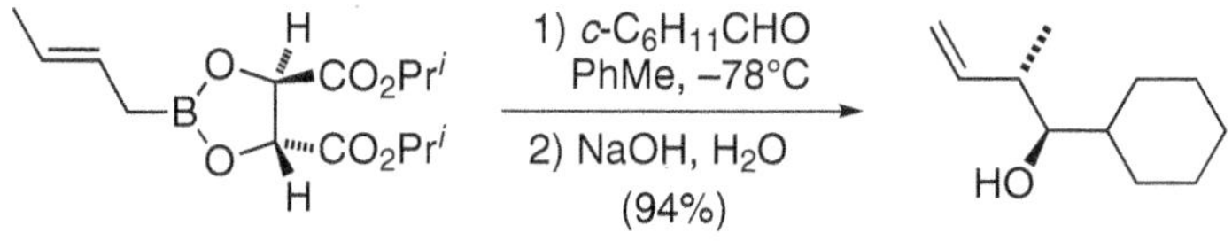

(23.6)

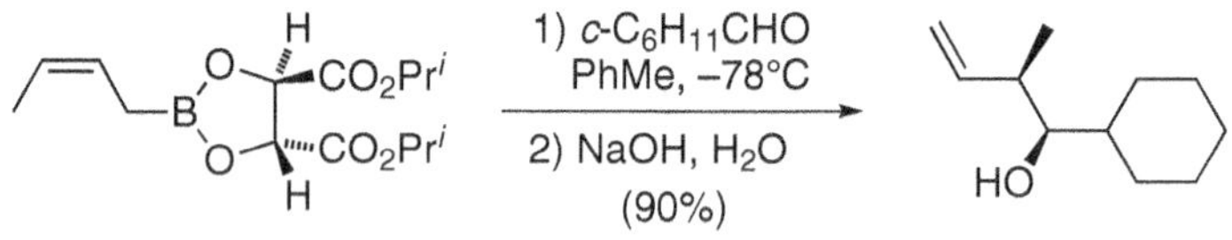

(23.7)

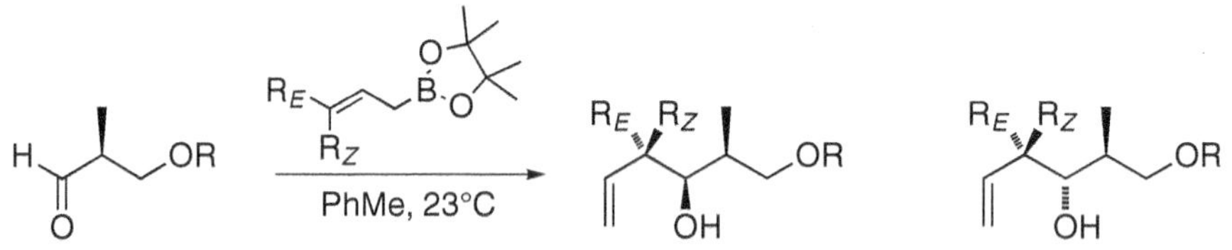

(23.8) **(23.9)** **(23.10)**

	23.9		23.10
R = TBDMS; R_E = Me; R_Z = H	61	:	39
R = TBDPS; R_E = Me; R_Z = H	62	:	38
R = CH_2Ph; R_E = Me; R_Z = H	68	:	32
R = TBDMS; R_E = H; R_Z = Me	55	:	29
R = TBDPS; R_E = H; R_Z = Me	61	:	32
R = CH_2Ph; R_E = H; R_Z = Me	61	:	32

The stereochemical bias in the asymmetric Roush crotylboration of the achiral aldehydes was for the carbinol to have the opposite absolute configuration to the tartrate group: the (*R*,*R*)-tartrate giving the *S* alcohol. The question of whether the stereochemical influence of the chiral crotylboronate could overcome the inherent Cram's rule selectivity of the aldehyde was answered by treating the chiral aldehyde **23.11** with the (*R*,*R*)- and (*S*,*S*)- enantiomers of the *E*-crotylboronate. In the case of the *R*,*R*-boronate and the 2*S*-aldehyde, the diastereoselectivity was high, with a d.r. of more than 49:1, indicating that this constitutes a matched pair of reactants—the selectivity of the aldehyde is reinforced by the selectivity of the boronate for the Cram (*R*) carbinol. The same aldehyde and the *S*,*S*-boronate, however, constitute a mismatched pair, because the Cram selectivity of the aldehyde opposes the enantioselectivity of the boronate, with the result that the d.r. drops to 5.25:1. Even in the mismatched pair, however, the crotylboronate is still the major determining factor of the absolute configuration of the major stereoisomer of the product.

17. Roush, W.R.; Palkowitz, A.D.; Ando, K. *J. Am. Chem. Soc.* **1990**, *112*, 6348.

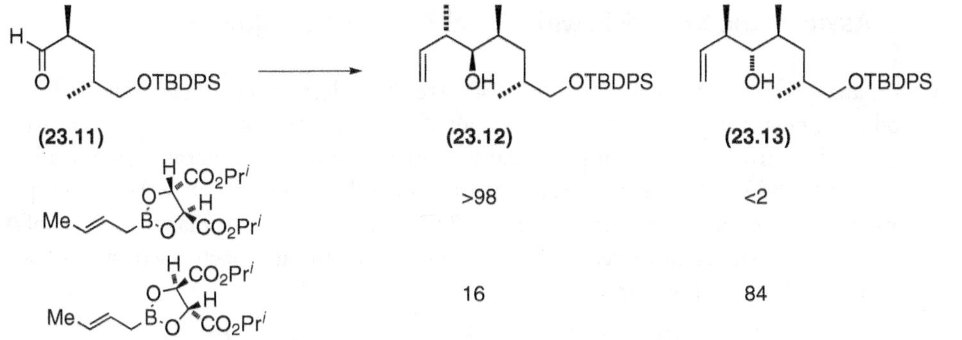

(23.11)	(23.12)	(23.13)
Roush (R,R)-tartrate crotylboronate	>98	<2
Roush (S,S)-tartrate crotylboronate	16	84

As an alternative to using chiral boronate esters for crotylboration, one can also use dialkylcrotylboranes. When diisopinocampheylcrotylboranes (Examples **23.14** and **23.15**) are used in the reaction, it is known as the ***Brown crotylboration***.[18] Again, the stereochemistry of the borane reagent is the dominant factor affecting the stereochemistry of the reaction. One advantage of the Brown crotylboration over the Roush procedure is the observation that in mismatched reactions, the boranes sometimes give superior results to the corresponding Roush boronates (e.g., in Example **23.16**, the d.r. from the reaction of the matched pair is 49:1 but from the mismatched pair it is still 19:1).

(23.14) **(23.15)**

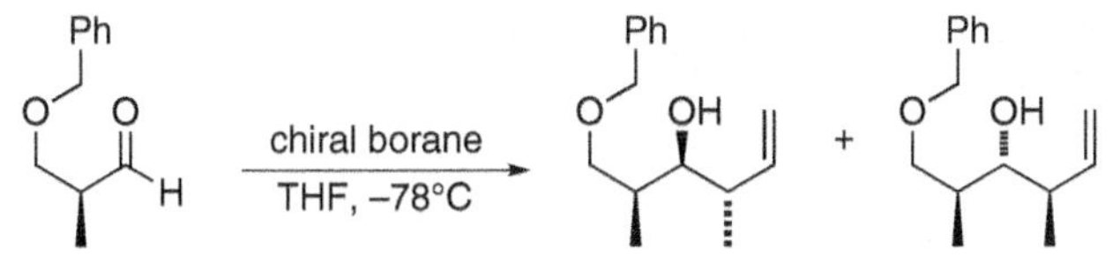

(23.16)

borane: **22.14**	98 :	2
borane: **22.15**	5 :	95

Sample Problem

23-1 One of the two reactions below gave disappointing results in the stereoselective synthesis of the product.

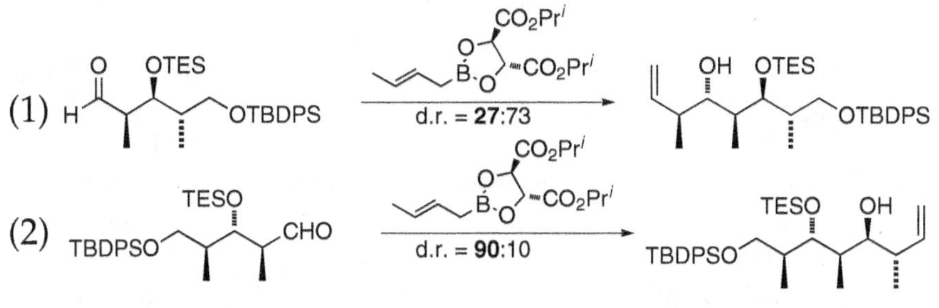

Account for the different outcomes of the two reactions.

References: J. Am. Chem. Soc. **1990**, *112*, 6348. *J. Org. Chem.* **2013**, *78*, 3.

§Answer on the next page.

---

18. (a) Brown, H.C.; Bhat, K.S. *J. Am. Chem. Soc.* **1986**, *108*, 293, 5919. (b) Brown, H.C.; Bhat, K.S.; Randad, R.S. *J. Org. Chem.* **1987**, *52*, 3701; **1989**, *54*, 1570.

23.2 Asymmetric Synthesis with Chiral Auxiliaries: Additions

Chiral auxiliary-based asymmetric reactions are the oldest of the efficient methods for introducing new chirality into a reaction product. There are two ways in which the chiral center can be introduced into the prochiral substrate: (1) it may be generated during addition to a π bond or (2) it may be generated by substitution of one of two enantiotopic groups in the prochiral substrate (Example **23.17**). To date, the primary function of a chiral auxiliary in an addition reaction has been to hinder the approach of a reagent to one face of the π bond, as shown in Example **23.18**.

(23.17)

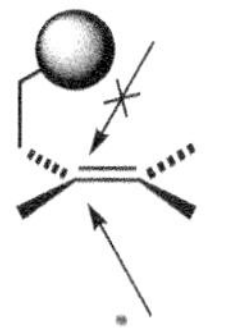

(23.18)

Because of the efficiency of the chiral auxiliary approach (developed over long experience), many chiral auxiliaries give high levels of asymmetric induction.[19] Given the recent emphasis on methods based on a chiral reagent or a chiral catalyst for asymmetric synthesis, it may seem surprising at first that chiral auxiliary techniques are still widely used in asymmetric synthesis. Part of the reason for this is practical—often, it is so easy to separate

---

19. Monographs and reviews: (a) Seyden-Penne, J. *Chiral Auxiliaries and Ligands in Asymmetric Synthesis* (Wiley-Interscience: New York, 1996). (b) Evans, D.A.; Helmchen, G.; Rüping, M. In Christmann, M.; Bräse, S. *Asymmetric Synthesis—The Essentials* (Wiley-VCH: Weinheim, 2007), p. 3.

....

§ Sample Problem:

Both products specified in the problem follow the rule that the absolute configuration of the new carbinol chiral center (here, *S*, is the same as the chiral centers in the chiral boronate, and that the stereochemistry of the new chiral centers is *anti*. In other words, these are the products that would be formed in the event that the absolute configuration of the product is fixed by the absolute configuration of the borane.

(1)

(2)

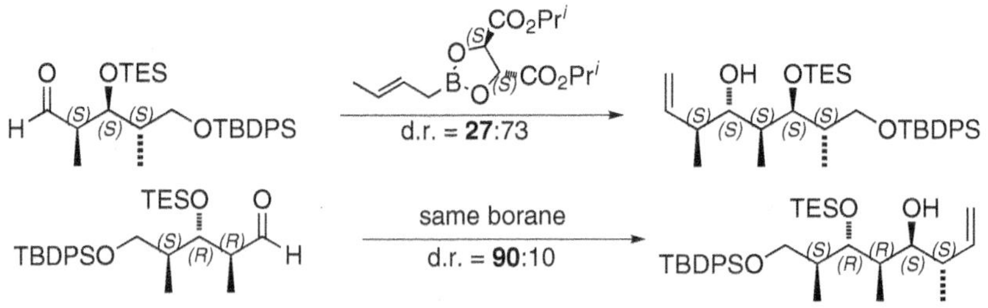

When we look at the addition reaction in terms of Cram's rule, we obtain the corresponding results in Equations 3 and 4.

(3)

(4)

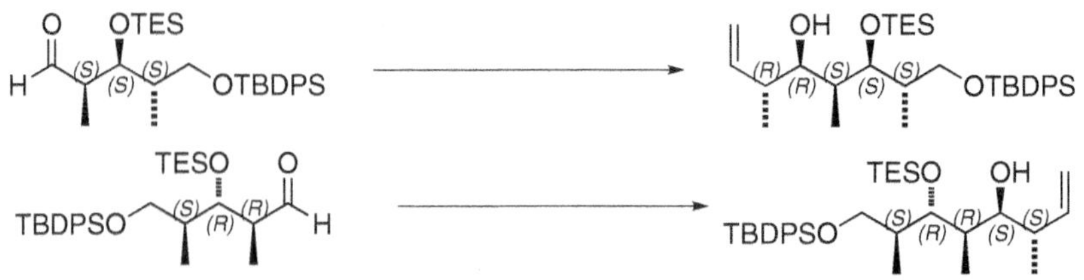

In this case, we see that the product in Equation 4 is identical in all respects to the product in Equation 2, which means that this pair of reagents constitutes a matched pair. On the other hand, the products in Equations 1 and 3 are diastereoisomers, not identical. This can only be interpreted as this pair of reactants being mismatched, with the stereochemical preference of the aldehyde being the dominant stereochemical influence.

the diastereoisomeric products of the chiral auxiliary reaction that it allows both enantiomers to be obtained pure after removal of the auxiliary. Moreover, the d.r. values can often be determined directly by ^{1}H nuclear magnetic resonance (NMR) before removal of the auxiliary—without the need for chiral derivatization prior to the analysis.

The Asymmetric Aldol Addition

Evans Chiral Boron Enolates

One addition reaction for which the chiral auxiliary approach remains the state of the art—despite the development of chiral organocatalysts for the reaction—is the aldol addition reaction. In 1981,[20] Evans' first report of the extremely high levels of asymmetric induction and *syn* selectivity in the aldol additions of the *Z(O)* boron enolates of chiral *N*-acyl-1,3-oxazolidin-2-ones (recall Example **23.2**) provided a dramatic incentive for the development of new auxiliaries of this general type, with the result that today there is a whole class of similar heterocyclic auxiliaries available to the chemist. Table 23.2 contains a variety of five-membered heterocycles developed over two decades following the disclosure of the Evans oxazolidines, including cyclic carbamate,[21] urea,[22] and thiocarbamate[23] auxiliaries for use in this and a wide variety of other reactions.

The original Evans aldol addition provided the first truly reliable method for the asymmetric formation of new carbon-carbon bonds. In the ensuing decades, a semi-systematic

Table 23.2 *N*-Acyloxazolidinone-Style Chiral Auxiliaries Used in Asymmetric Synthesis

Oxazolidinones	Developer(s)	Analogs	Developer(s)
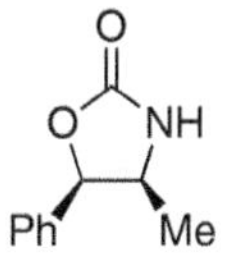 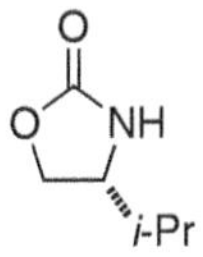	Evans	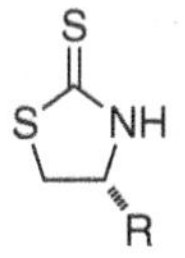	Nagao, Fujita (Crimmins)
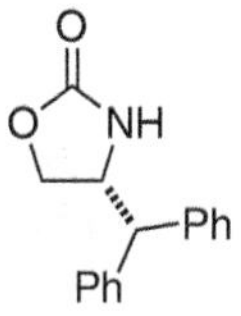	Sibi	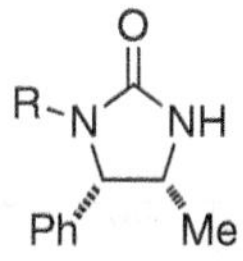	Helmchen
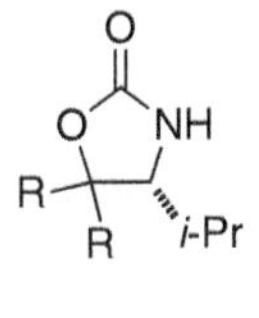	Seebach	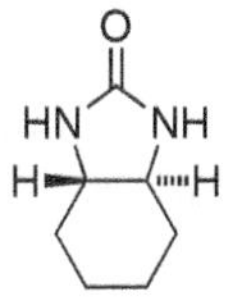	S. G. Davies
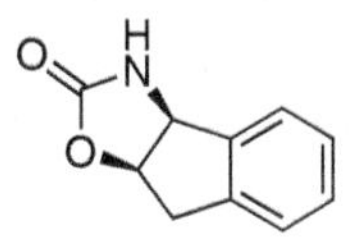	Askin, I. W. Davies		

20. Evans, D.A.; Bartroli, J.; Shih, T.L. *J. Am. Chem. Soc.*, **1981**, *103*, 2127.

21. (a) Sibi, M.P.; Deshpande, P.K.; Ji, J. *Tetrahedron Lett.* **1995**, *36*, 8965. (b) Hintermann, T.; Seebach, D. *Helv. Chem. Acta* **1998**, *81*, 2093. (c) Askin, D.; Eng, K.K.; Rossen, K.; Purick, R.M.; Well, K.M.; Volante, R.P.; Reider, P.J. *Tetrahedron Lett.* **1994**, *35*, 673. (d) Davies, I.W.; Senanayake, C.H.; Castonguay, L.; Larsen, R.D.; Verhoeven, T.R.; Reider, P.J. *Tetrahedron Lett.* **1995**, *36*, 7619. (e) Ghosh, A.K.; Onishi, M. J. *Am. Chem. Soc.* **1996**, *118*, 2527.

22. (a) Roder, H.; Helmchen, G.; Peters, E.-M.; Peters, K.; von Schnering, H. *Angew. Chem., Int. Ed. Engl.* **1984**, *23*, 898. (b) Davies, S.G.; Mortlock, A.A. *Tetrahedron Lett.* **1991**, *36*, 4787.

23. (a) Nagao, Y.; Yamada, S.; Kumagai, T.; Ochiai, M.; Fujita, E. *Chem. Commun.* **1985**, 1418. (b) Crimmins, M.T.; Emmitte, K.A. *J. Am. Chem. Soc.* **1997**, *119*, 7883.

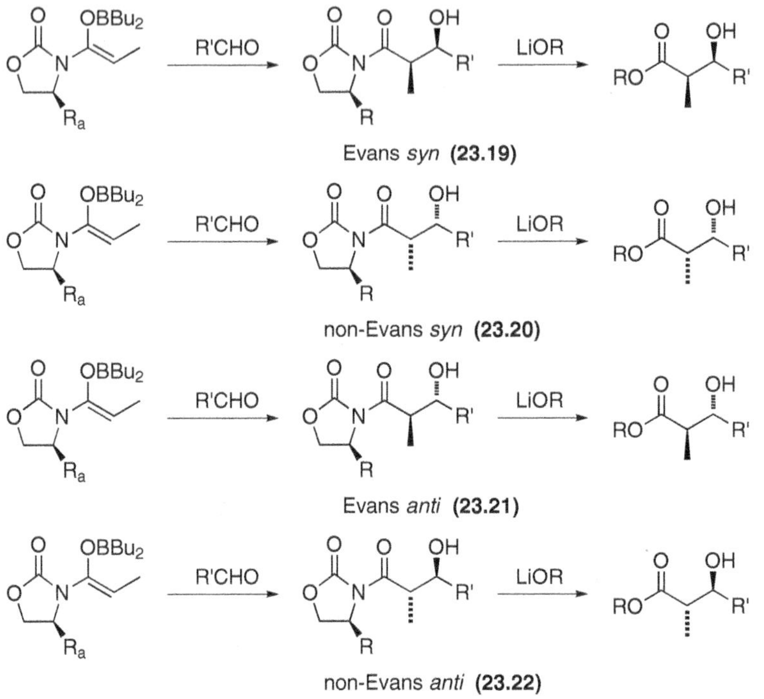

Figure 23.13 The possible stereochemical outcomes of asymmetric aldol addition reactions. In the Evans stereochemistry, the alkyl group *cis* to boron ends up *syn* to the directing group in the conformation drawn.

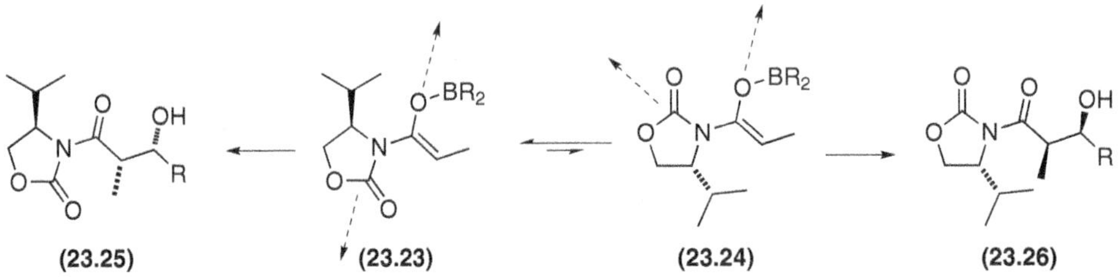

Figure 23.14 Rationalizing the stereochemistry of the Evans aldol addition of boron enolates

nomenclature for the aldol products obtained from Evans *N*-acyloxazolidinones has been developed (Figure 23.13). The aldol products are all named based on the relationship between the configuration at the α carbon of the aldol and the chiral center at position 4 of the heterocycle. If these two chiral centers have the same configuration (**23.19**, **23.21**), the aldol is designated an "Evans aldol," whereas if the configurations differ (**23.20**, **23.22**), the product is a "non-Evans aldol." The other important stereochemical relationship is the relative configuration of the α and β carbons of the aldol, and this is usually designated as *syn* (**23.19**, **23.20**) or *anti* (**23.21**, **23.22**).[24] Thus, there are four possible aldol addition products, and much research has gone into developing methods for the formation of the non-Evans aldols and the *anti* aldols. As is evident from Figure 23.14, after removal of the chiral auxiliary, the Evans and non-Evans aldols are enantiomeric, and this has spurred considerable research into the preparation of non-Evans aldols.

The key to the control over the face selectivity of the aldol addition is controlling the location of the steric blocking group in the transition state for the addition. The model used to rationalize the stereochemistry of the aldol additions with the boron enolates is given in Figure 23.14. Because boron cannot be more than tetracoordinate, the complexation of the aldehyde oxygen by the boron in the transition state for the reaction means that the oxygen atom in the auxiliary cannot be simultaneously coordinated to boron with the enolate oxygen—chelation is not possible. This, in turn, means that there is a conformational flexibility available to the transition state that is not available in a chelated

24. Masamune, S.; Kaiho, T.; Garvey, D.S. *J. Am. Chem. Soc.* **1982**, *104*, 5521.

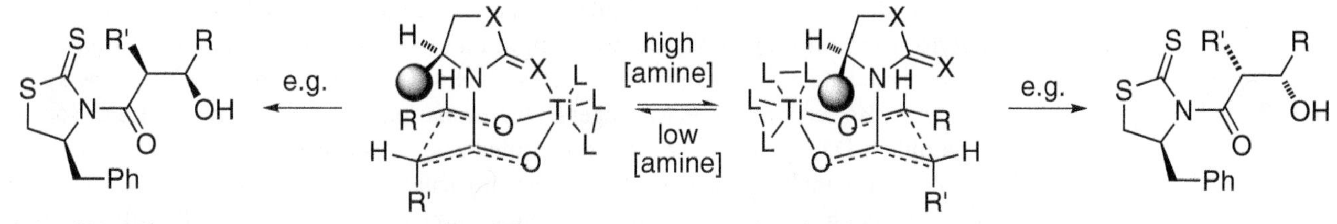

(23.30) non-Evans *syn* aldol; (23.28); X = O, S; (23.27); (23.29) Evans *syn* aldol

Figure 23.15 Stereochemical reversal by diamines in the aldol additions of titanium enolates

transition state. Under these conditions, the boron enolate can adopt two conformations, **23.23** and **23.24.** The observed result of the Evans aldol addition implies that the equilibrium between the two conformations favors the conformer where the C—O dipoles are arranged antiparallel (**23.23**) and the dipole moment is minimized, in preference to the conformer where the dipoles are nearer to parallel (**23.24**) and the dipole moment is maximized. This conformation makes the *si* face of the enolate β carbon more accessible to the aldehyde and thus leads to the Evans *syn* aldol product.

If one considers the stereochemistry of the aldol addition further, one comes to the conclusion that the simplest way to control the face selectivity of the enolate is by fixing the conformation of the enolate by chelation of the carbonyl group of the chiral auxiliary and the enolate oxygen. Under those conditions, the product of the addition of the *Z(O)* enolate is the non-Evans aldol (Figure 23.15).

The role of chelation in determining the face selectivity of these reactions has been elegantly demonstrated by Crimmins,[25] who showed that the *Z(O)* titanium enolate of a chiral *N*-acyl-1,3-oxazolidin-2-one gave opposite stereochemical results depending on the mole ratio of titanium to a diamine, usually (–)-sparteine. In the presence of two equivalents of the diamine, the Evans *syn* aldol (**23.29**) predominated, whereas in the presence of one equivalent of the diamine, the non-Evans *syn* aldol (**23.30**) was obtained. In the Crimmins model, the larger diamine-to-metal ratio allowed all the metal coordination sites to be filled (**23.27**) without the need for coordination by the acyl oxygen, with the result that no chelation of the auxiliary to the metal occurs. This situation then parallels the situation with the chiral boron enolates of the original Evans asymmetric aldol addition. In contrast, the lower diamine-to-metal ratio results in the heteroatom of the auxiliary being necessary to complete the coordination sphere of the metal (**23.28**), and the chelated model for the aldol addition then pertains.

Chiral Sulfinylamines

The success of the Evans chiral oxazolidinones as auxiliaries for the aldol addition has recently been eclipsed by the use of chiral sulfinylamines in this reaction. The sulfur-based auxiliary, *tert*-butylsulfinylamine,[26] possesses a chiral sulfur atom and is one of the most (if not the most) widely used chiral auxiliaries today. It permits the formation of β-hydroxycarbonyl compounds from methyl ketones or acetaldehyde with high levels of diastereoselectivity—something the Evans oxazolidinones cannot do. Importantly, the sulfur in this compound and its derivatives is quite configurationally stable, so it can be used as a chiral auxiliary in a wide range of applications. One of these is the asymmetric addition of the corresponding metal azaenolates to aldehydes to give chiral β-hydroxyaldimines that can be elaborated to a wide range of other products.

25. Crimmins, M.T.; King, B.W.; Tabet, E.A.; Chaudhary, K. *J. Org. Chem.* **2001**, *66*, 894.

26. Reviews: (a) Ellman, J.A.; Owens, T.D.; Tang, T.P. *Acc. Chem. Res.* **2002**, *35*, 984. (b) Ellman, J.A. *Pure Appl. Chem.* **2003**, *75*, 39. (c) Ferreira, F.; Botuha, C.; Chemla, F.; Pérez-Luna, A. *Chem. Soc. Rev.* **2009**, *38*, 1162. (d) Roback, M.T.; Herbage, M.A.; Ellman, J.A. *Chem. Rev.* **2010**, *110*, 3600.

The aldimine or ketimine required (Example **23.31**) is formed from the chiral *tert*-butylsulfinylamine and the carbonyl compound under Lewis acid catalysis by anhydrous copper sulfate (aldehydes) or titanium tetraethoxide (ketones).[27] Note how both Lewis acid catalysts also serve as water scavengers. Deprotonation of the imine by lithium diisopropylamide (LDA) occurs on the less substituted side. This is usually followed by exchange with magnesium bromide to give a magnesium azaenolate that then adds to the carbonyl group through a chair-type transition state (Example **23.32**) to give the β-hydroxyimine (Example **23.33**). Yields in this reaction are typically above 50% (usually above 75%), and d.r. values range from 73:27 to 98:2, depending on the exact substitution pattern of the ketone and aldehyde partners. The β-hydroxyimine products can be converted to a wide range of products by relatively simple chemistry, as shown in Examples **23.34** to **23.36**. Note how the stereochemistry of the secondary amine formed by reduction can be controlled by the appropriate choice of reducing agent.

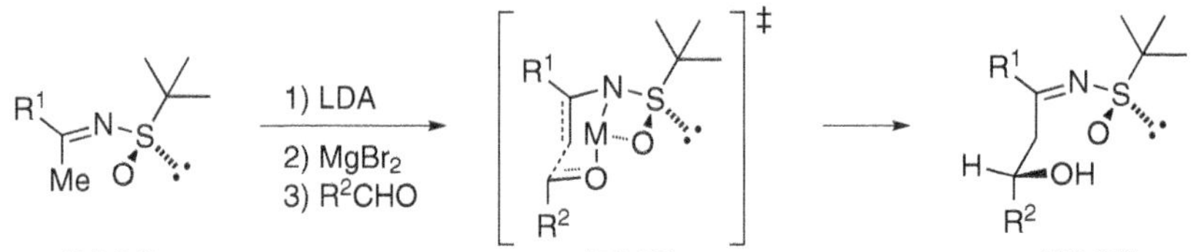

(23.31) **(23.32)** **(23.33)**

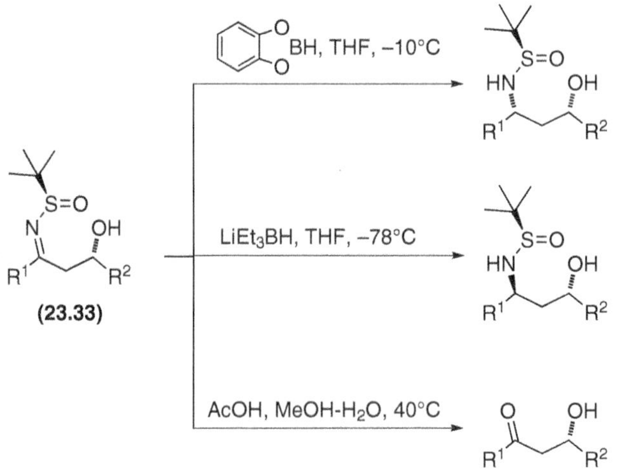

(23.34)

(23.35)

(23.36)

Problems

23-1 The ansa chain of the macrolide antibiotic rifamycin S, shown below, consists of a polypropionate-derived backbone. One retrosynthetic analysis of this ansa chain leads to the unsaturated aldehyde synthon shown. This synthon has been assembled by Roush using a sequence of asymmetric crotyl- and allylborations.

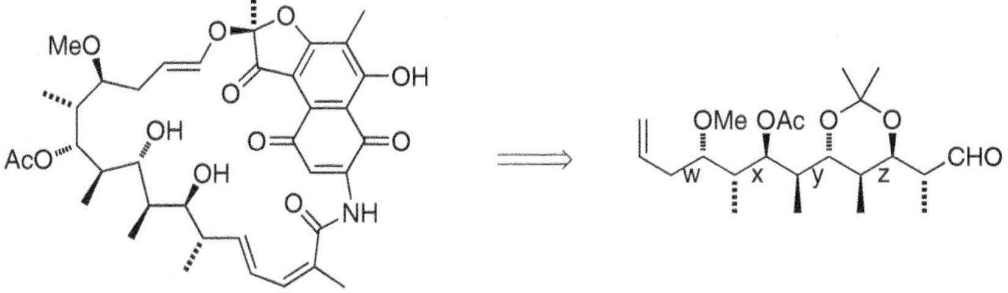

27. (a) Liu, G.; Cogan, D.A.; Ellman. J.A. *J. Am. Chem. Soc.* **1997**, *119*, 9913. (b) Cogan, D.A.; Ellman, J.A. *J. Am. Chem. Soc.* **1999**, *121*, 268.

There are four key bonds (w, x, y, z) in this synthon that were formed by asymmetric allyl- or crotylboration reactions in the Roush synthesis. A major part of designing the successful synthesis involved finding a suitable sequence of these reactions. There are 24 possible sequences; analyze each of them as matched/mismatched reactions. Which sequences does your analysis suggest have the best potential for success?

Reference: J. Am. Chem. Soc. **1990**, *112*, 6348.

23-2 (1) The key step in the synthesis of each of the following compounds involves an asymmetric aldol addition. Draw two different aldol additions that will give a product with the absolute configuration shown. You may assume that the stereochemistry of the product is not altered by removal of the chiral auxiliary.

(a) 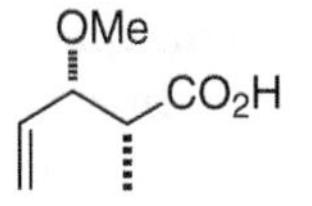

(b) 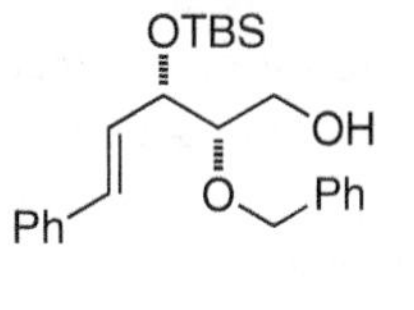

(c) 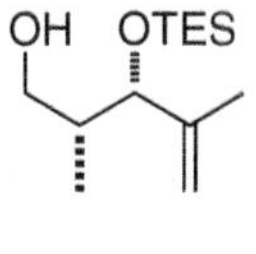

(d) 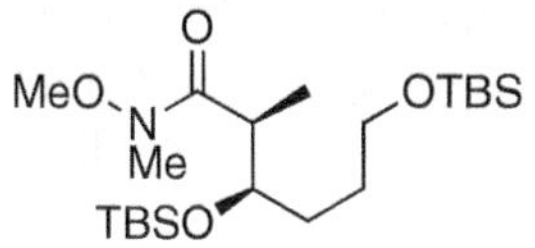

(2) Could any of these products be obtained by means of an asymmetric allyl- or crotylboration reaction? Which reagent would you use? Would additional steps be necessary?

References: (a) *Org. Lett.* **2007**, *9*, 3425. (b) *J. Am. Chem. Soc.* **1994**, *116*, 12111. (c) *J. Org. Chem.* **1997**, *62*, 454. (d) *Tetrahedron Lett.* **1993**, *34*, 8403.

23-3 Discuss the formation of this alcohol by Roush asymmetric crotylboration of the aldehyde shown. What is the d.r. reported for the reaction? Is this value what one would predict from the reagents used? Why or why not?

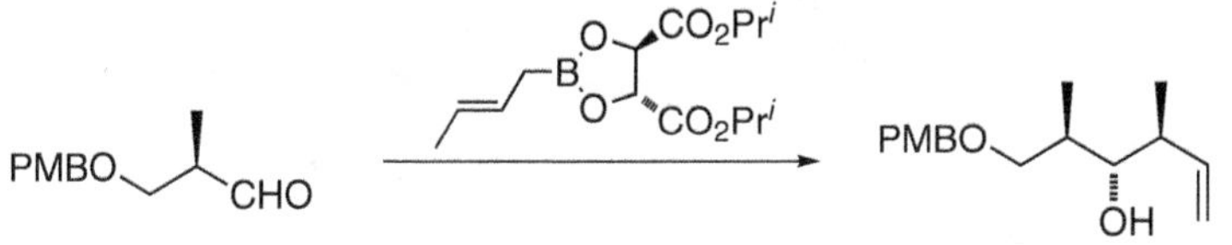

Reference: Angew. Chem. Int. Ed. **1999**, *38*, 1652.

23-4 Suggest a mechanism to account for each step in the following transformation.

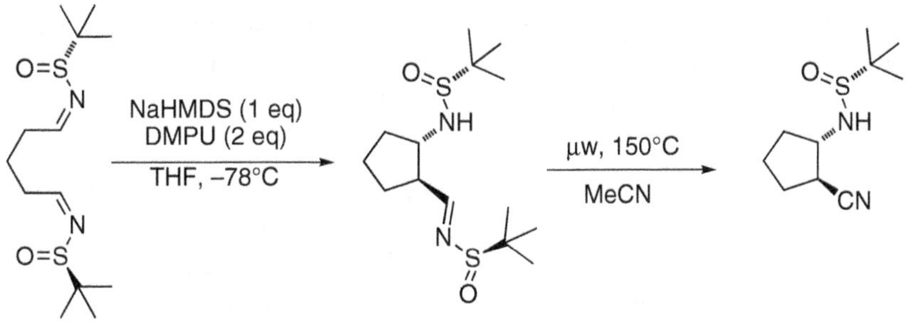

Reference: Org. Lett. **2004**, *6*, 3621.

The Asymmetric Diels-Alder Reaction[28]

Like the aldol addition, the Diels-Alder reaction has been successfully adapted to asymmetric synthesis by means of chiral auxiliaries. Although auxiliaries for both the diene and the dienophile are known, by far the greatest number of successful auxiliaries for asymmetric Diels-Alder reactions modify the dienophile. This is where we will focus our attention in this section.

8-Phenylmenthol and Similar Auxiliaries

The Diels-Alder reaction is accelerated by electron-withdrawing groups on the dienophile, so acrylate esters of chiral alcohols were among the first successful chiral auxiliaries for this reaction. In 1975, Corey and Ensley introduced 8-phenylmenthol (Example **23.37**) as a chiral auxiliary for asymmetric Diels-Alder reactions.[29] The Lewis acid complexes of the acrylate esters of this auxiliary give reproducibly high levels of asymmetric induction. This has been rationalized in terms of a π stacking interaction in the complexes ester that leads to one face of the acrylate π bond being shielded from attack. In Example **23.38**, the π stacking of the phenyl group and the alkene in the *s-trans* conformation of the acrylate ester of (1*S*, 2*R*, 5*S*)-8-phenylmenthol is illustrated; in this conformation, the *re* face of the double bond is the one accessible to the diene. This face selectivity was demonstrated by Corey and Ensley in their initial publication, which showed the synthesis of a chiral intermediate (Example **23.39**) for the asymmetric synthesis of prostaglandins. The phenmenethyl ester is highly *endo* selective and gave the *endo* ester as a 98.5:1.5 mixture of diastereoisomers.

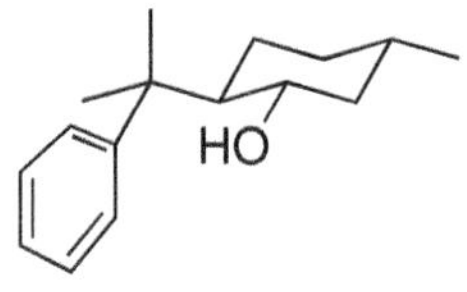

(23.37)

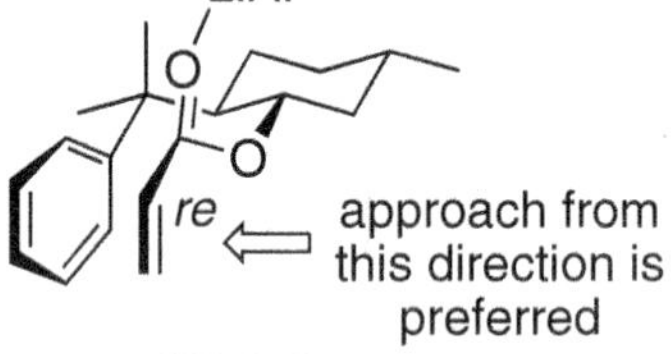

(23.38)

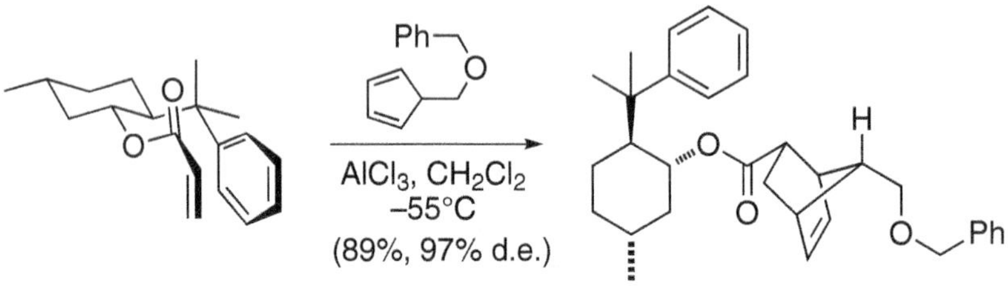

(23.39)

This auxiliary is generally prepared by the procedure that Corey and Ensley reported, but the synthesis does lead to approximately 7% of the product containing the axial methyl group, to which Whitesell gave the semi-systematic name, epi-*ent*-phenmenthol.[30] Because this product has the opposite chirality at positions 1 and 2 of the cyclohexane ring with

---

28. Reviews: (a) Oppolzer, W. In Trost, B.M.; Fleming, I, Eds. *Comprehensive Organic Synthesis* (Pergamon: Oxford, 1991); Vol. 5, 315. (b) Kagan, H.B.; Riant, O. *Chem. Rev.* **1992**, 92, 1007. (c) Helmchen, G.; Karge, R.; Weetman, J. In Scheffold, R., Ed. *Modern Synthetic Methods* (Springer-Verlag: Berlin, Heidelberg, 1986); Vol. 4, 262. (d) Reilly, M.; Oh, T. *Org. Prep. Proc. Int.* **1994**, 26, 131. (e) Maruoka, K.; Yamamoto, H. In Ojima, I, Ed. *Catalytic Asymmetric Synthesis* (Wiley-VCH: New York, 1993), 413. (f) Evans, D.A.; Johnson, J.S. In Jacobsen, E.N.; Pfaltz, A.; Yamamoto, H., Eds. *Comprehensive Asymmetric Catalysis* (Springer: New York, 1999); Vol III, 1177. (g) Mikami, K.; Aikawa, K. In Ojima, I, Ed. *Catalytic Asymmetric Synthesis*, 3rd ed. (John Wiley & Sons: Hoboken, N.J., 2010), 683. (h) Oppolzer, W. *Angew. Chem. Int. Ed.* **1984**, *23*, 876.

29. (a) Corey, E.J.; Ensley, H.E. *J. Am. Chem. Soc.* **1975**, *97*, 6908. (b) Ensley, H.E.; Parnell, C.A.; Corey, E.J. *J. Org. Chem.* **1978**, *43*, 1610.

30. Whitesell, J.K.; Liu, C.-L.; Buchanan, C.M.; Chen, H.-H.; Minton, M.A. *J. Org. Chem.* **1986**, 51, *551*.

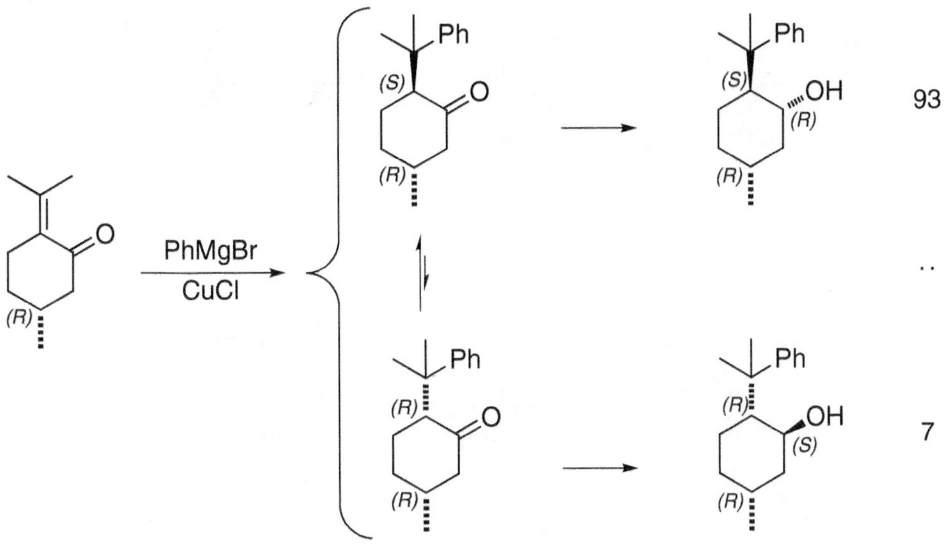

Figure 23.16 Preparation of the 8-phenylmenthol auxiliary from pulegone

respect to 8-phenylmenthol (see Figure 23.16), it leads to the opposite chirality in the product. Therefore, it must be removed chromatographically prior to use of the auxiliary.

The difficulties associated with 8-phenylmenthol as a chiral auxiliary—particularly the lack of ready availability of both enantiomers and the problems associated with the concomitant formation of epi-*ent*-phenmenthol—quickly prompted the development of alternatives. Among the more successful of these were the *trans*-2-phenylcyclohexanols, developed by Whitesell.[31] These alcohols are easily obtained optically pure by enzymatic hydrolysis of the chloroacetate esters,[32] either by Sharpless asymmetric dihydroxylation of 1-phenylcyclohexene and subsequent hydrogenolysis over nickel[33] or by hydroboration of 1-phenylcyclohexene with a chiral borane.[34] These auxiliaries have proved to be excellent replacements for 8-phenylmenthol in a variety of reactions, especially in the ene reactions of glyoxylate esters. The ene reactions of these glyoxylates with alkenes give chiral α-hydroxyketones with high e.e. values. One interesting facet of these ene reactions (Figure 23.17) is that the absolute configuration of the new chiral center in the ene product from the phenylmenthyl ester is the opposite of that obtained from the 2-phenylcyclohexyl ester with the same configuration.

Evans 1,3-Oxazolidin-2-one Amides

The *N*-acryloyl-1,3-oxazolidin-2-one derived from phenylalanine contains an aromatic ring in the same relationship to the acryloyl group as the phenyl ring in 8-phenylmenthyl acrylate. Thus, it was logical to expect that the oxazolidinone might function as a chiral auxiliary in the Diels-Alder reaction in much the same as phenylmenthol. By chelating the two oxygen atoms, the acyl group is fixed in the *s-cis* conformation because the *s-trans* conformation would lead to untenable steric interactions with the substituent at the 4-position of the oxazolidinone ring. In this conformation, the phenyl ring shields one face of the π bond (π stacking).[35] In Example **23.40**, the *R* acrylate is shown after chelation

31. (a) Whitesell, J.K.; Chen, H.-H.; Lawrence, R.M. *J. Org. Chem.* **19885**, *50*, 4664. (b) Whitesell, J.K. *Chem. Rev.* **1992**, *92*, 953.

32. Schwartz, A.; Madan, P.; Whitesell, J.K.; Lawrence, R.M. *Org. Syn.* **1993**, *Coll. Vol. 8*, 516; *Org. Syn.* **1990**, *69*, 1.

33. King, S.B.; Sharpless, K.B. *Tetrahedron Lett.* **1994**, *35*, 5611.

34. (a) Brown, H.C.; Prasad, J.V.N.V.; Gupta, A.K.; Bakshi, R. K. *J. Org. Chem.* **1987**, *52*, 310. (b) Brown, H.C.; Jadhav, P.K.; Mandal, A.K. *J. Org. Chem.* **1982**, *47*, 5074.

35. Evans, D.A.; Chapman, K.T.; Hung, D.T.; Kawaguchi, A.T. *Angew. Chem. Int. Ed. Engl.* **1987**, *26*, 1184.

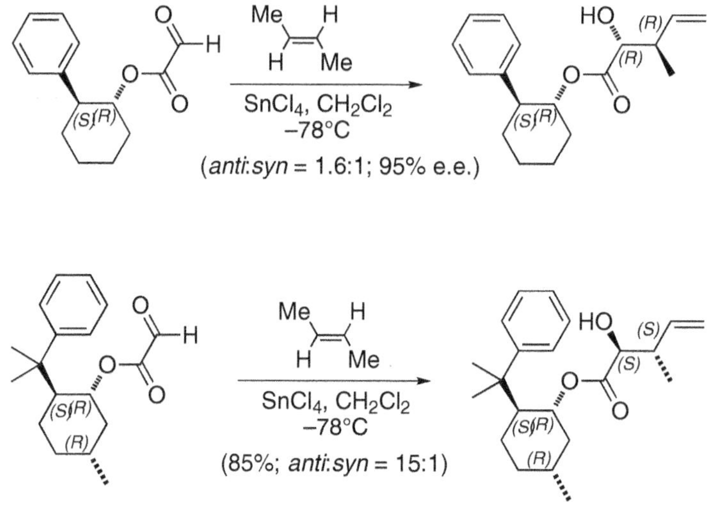

Figure 23.17 Asymmetric ene reactions of glyoxylate esters

by diethylaluminum chloride. The products of this complexation reaction are the cationic diethylaluminum chelate and the diethyldichoroaluminate anion. This cation is an extremely electrophilic dienophile that reacts readily at –100°C with cyclopentadiene. It is the *re* face of the acrylate π bond that is accessible to the diene in the chelated cation.[36] The Evans oxazolidinones form dienophiles with *E*-β-substituted acrylates that react well in Diels-Alder reactions, with both high levels of enantioselectivity and high levels of *endo* selectivity. In Example **23.41**, the *N*-crotonyl (*S*)-oxazolidinone reacts with cyclopentadiene to give the *endo* adduct as the major product of a 56:1 *endo:exo* mixture; the *endo* isomer is itself a 97:3 mixture of diastereoisomers, but purification of the crude product by recrystallization gives the *endo* isomer shown as a single isomer. The same auxiliary has been used to promote enantioselective intramolecular Diels-Alder reactions, as shown in Example **23.42**.[37]

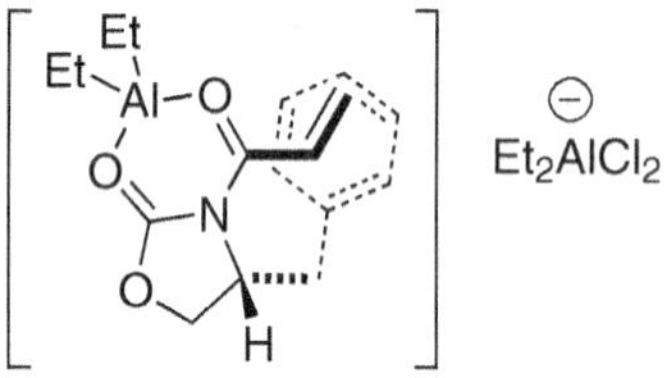

(23.40)

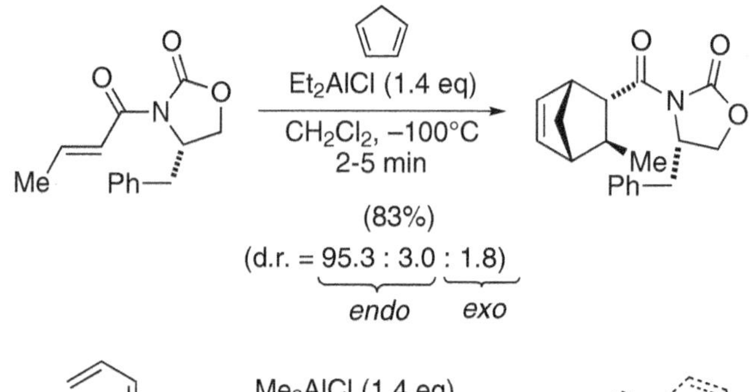

(23.41)

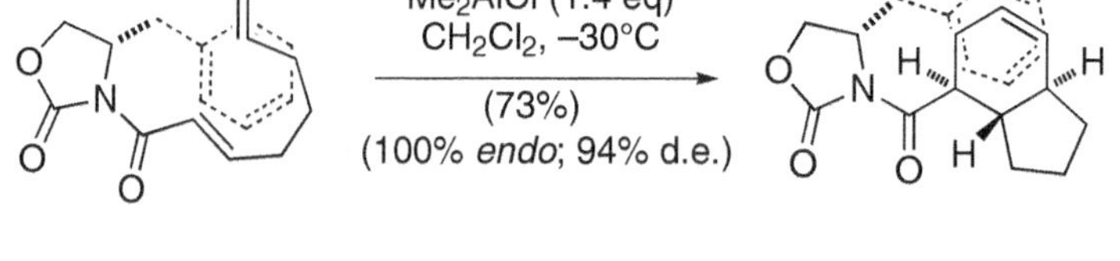

(23.42)

36. Evans, D.A.; Chapman, K.T.; Bisaha, J. *J. Am. Chem. Soc.* **1984**, *106*, 4261; **1988**, *110*, 1238.
37. Evans, D.A.; Chapman, K.T.; Bisaha, J. *Tetrahedron Lett.* **1984**, *25*, 4071.

Oppolzer Bornane-Based Auxiliaries

The research group of Oppolzer[38] in Geneva was particularly active in developing chiral auxiliaries based on the camphor skeleton.[39] The first camphor-based chiral auxiliary that they reported was the mononeopentyl ether of *exo-cis*-bornane-2,3-diol (Example **23.43**). The use of these auxiliaries in the Lewis acid–catalyzed Diels-Alder reactions have been especially successful, as illustrated in Example **23.44**.

(23.43)

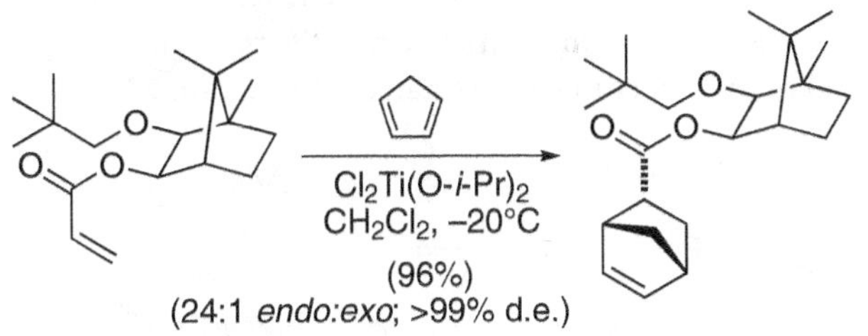

(23.44)

In 1984, the enantiomeric chiral auxiliaries based on camphorsulfonic acid,[40] now known as the Oppolzer camphorsultam (Example **23.45**) were described. These compounds can be prepared easily from the camphorsulfonyl chloride.[41] Because the sulfonyl chloride is commercially available as both enantiomers, the preparation and use of this auxiliary is particularly facile. The enoyl derivatives of these auxiliaries react in Diels-Alder reactions catalyzed by dialkylaluminum chlorides at very low temperatures, in a manner similar to the enoyl derivatives of the benzyl Evans oxazolidinone discussed above. The *endo:exo* ratios are very similar to those obtained using the Evans' auxiliary, and the diastereoselectivity between *endo* isomers is also very high. Moreover, the products are frequently highly crystalline, which means that purification to stereochemical homogeneity can be effected by recrystallization. It is instructive to compare the intramolecular Diels-Alder reaction of the *N*-enoyl (1*R*)-camphorsultam in Example **23.46**[42] with that given in Example **23.42.** Both reactions give exclusive *endo* addition and 93% to 94% d.e.

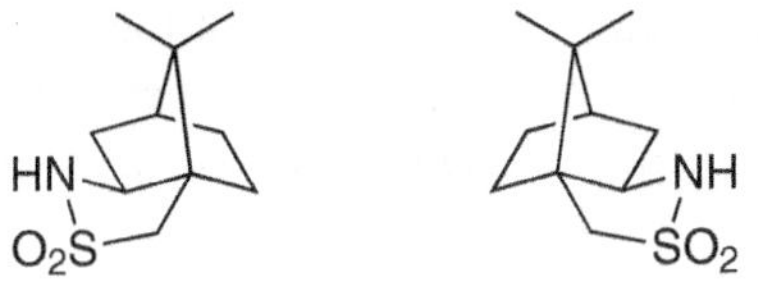

(23.45)

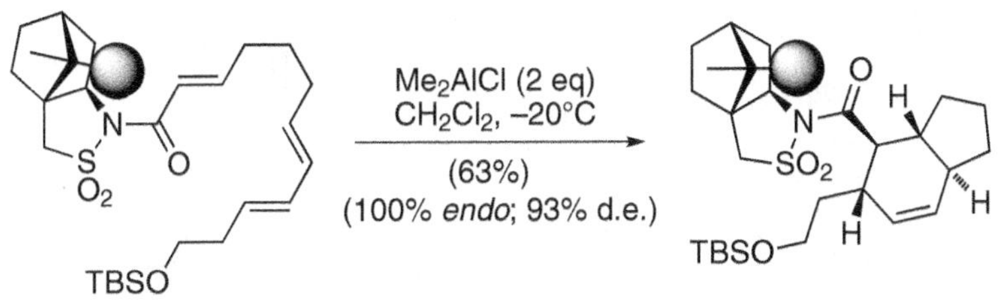

(23.46)

38. **Wolfgang Oppolzer (1937–1996)** was born and educated in Vienna before taking his PhD in 1964 at the ETH in Zürich in 1964. After postdoctoral study under Corey and Woodward, he joined Sandoz, and then the University of Geneva in 1974. For more detail, see the University of Geneva web site and Kündig, E.P. *Tetrahedron* **1996,** *52* (35), xiii.

39. Review: Oppolzer, W. *Tetrahedron* **1987,** *43,* 1969.

40. (a) Oppolzer, W.; Chapuis, C.; Bernardelli, G. *Helv. Chim. Acta* **1984,** *67,* 1397. (b) Vandewalle, M.; Van der Eycken, J.; Oppolzer, W.; Vullioud, C. *Tetrahedron* **1986,** *42,* 4035.

41. (a) Towson, J.C.; Weismiller, M.C.; Sankar Lal, G.; Sheppard, A.C.; Kumar, A.; Davis, F.A. *Org. Syn.* **1993,** *Coll. Vol. 8,* 104; *Org. Syn.* **1990,** *60,* 158. (b) Weismiller, M.C.; Towson, J.C.; Davis, F.A. *Org. Syn.* **1993,** *Coll. Vol. 8,* 110; *Org. Syn.* **1990,** *69,* 154.

42. Oppolzer, W.; Dupuis, D.; Poli, G.; Raynham, T.D.; Bernardinelli, G. *Tetrahedron Lett.* **1988,** *29,* 5885.

Asymmetric Additions to Imines

One of the most important applications of chiral sulfinylimines is in the synthesis of chiral amines by addition of organometallic nucleophiles. The sulfinylamine auxiliary is approximately as reactive toward nucleophiles as a Boc group, which means that it can serve not only as a chiral auxiliary but also as a protecting group for the product amine. The high levels of diastereoselectivity in the addition reactions come from the pronounced conformational preference of the sulfinylimine group itself. Computations[43] reveal that there is a strong conformational preference for the S=O bond and the N=C bond to be *syn*-coplanar (Example **23.47**). In this conformation, there is the possibility of significant π delocalization of the lone pair on nitrogen into the S—O σ* orbital. A second low-energy conformation (Example **23.48**) places the two lone pairs in an *antiperiplanar* arrangement; this is consistently some 6 kcal mol^{-1} higher in energy than conformation **23.47**.

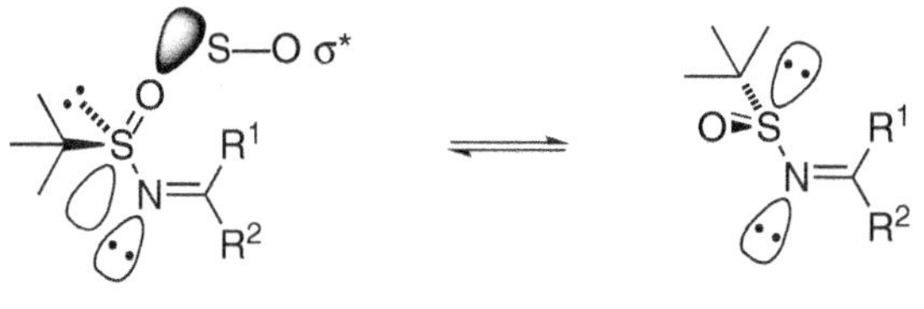

(23.47) **(23.48)**

The addition of the Grignard reagent occurs through a cyclic activated complex similar to that shown in Example **23.49**. The formation of this type of chelated activated complex suggests that the enantioselectivity of this reaction in solvents capable of coordinating the magnesium atom should be lower than in hydrocarbon or chlorocarbon solvents. It is reassuring to find that this is the case.[44] Interestingly, the stereochemistry of the addition is reversed when alkyllithium reagents are used as the nucleophiles in the presence of Lewis acid additives and in coordinating solvents. It is proposed that in this case, the reaction proceeds through an open transition state, making the addition intermolecular rather than intramolecular. The complexation of the sulfinyl oxygen by the Lewis acid should further stabilize the conformation shown as Example **23.51**, with the result that the metal alkyl now attacks from the face opposite the *tert*-butyl group, giving the stereochemistry shown in Example **23.52**. The effects of changing the conditions of the addition are illustrated by Examples **23.53** and **23.54**. In all cases, the free amine is liberated from the sulfinylamine as the hydrochloride salt by treatment with methanolic hydrogen chloride.

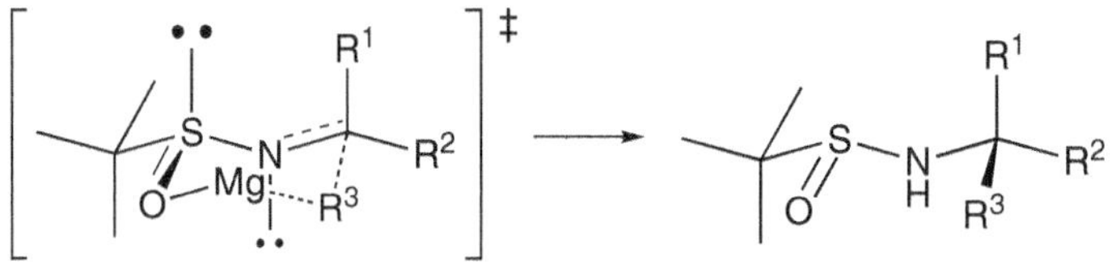

(23.49) **(23.50)**

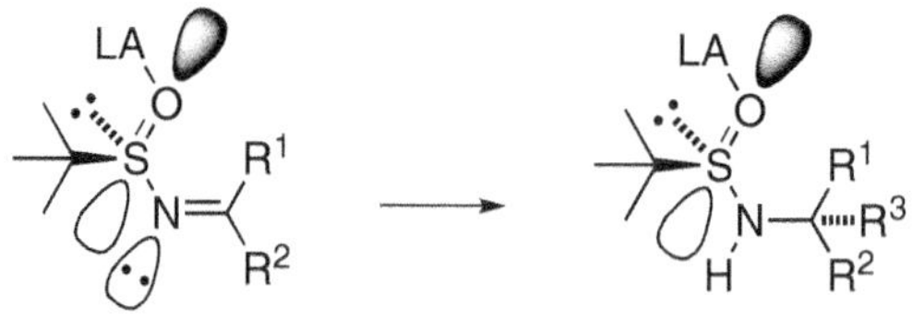

(23.51) **(23.52)**

43. (a) Bharatam, P.V.; Uppal, P.; Kaur, A.; Kaur, D. *J. Chem. Soc., Perkin Trans. 2*, **2000**, 43. (b) Bharatam, P.V.; Kaur, A.; Kaur, D. *J. Phys. Org. Chem.* **2002**, *15*, 197.
44. Cogan, D.A.; Liu, G.; Ellman, J. *Tetrahedron* **1999**, *55*, 8883.

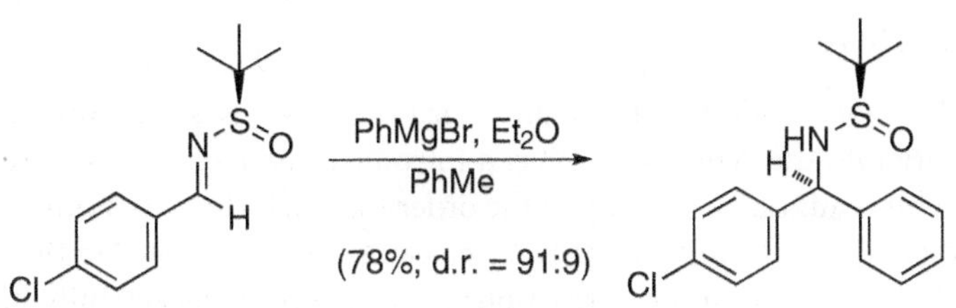

(23.53)[45]

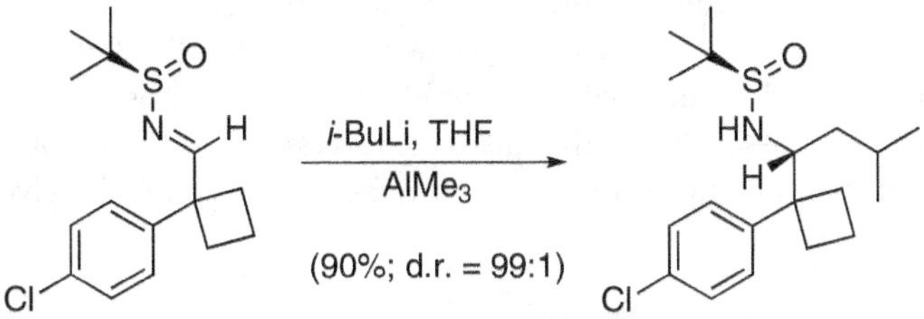

(23.54)[46]

The key to using the *tert*-butylsulfinylamines as chiral auxiliaries is, of course, the ability to prepare them optically pure form. This has been accomplished in several ways, including synthesis using asymmetric oxidizing agents and by oxidation of chiral sulfides.[47] The products of these methods are sulfinate esters or thioesters, which are converted to the sulfinylamine by treatment with lithium amide or lithium hexamethyldisilazide (LHMDS). The substitution proceeds with inversion of configuration at sulfur. The other major problem with using these auxiliaries—recovering and reusing the auxiliary—is a major problem due to the rapid racemization of the sulfinyl chloride. This has now been overcome by taking advantage of the fact that the sulfinyl chloride can be converted to the sulfinate esters under mild conditions. The treatment of the chloride with an achiral alcohol in the presence of proton sponge and a chiral tertiary amine such as quinidine[47] gives the chiral sulfinate ester in good yields. This can then be converted to the sulfinylamine of the opposite configuration by simple treatment with an amide salt (Example **23.55**). An alternative approach[48] requires the formation of diastereoisomeric esters with a chiral alcohol (Example **23.56**). These esters are generally amenable to separation by chromatography.

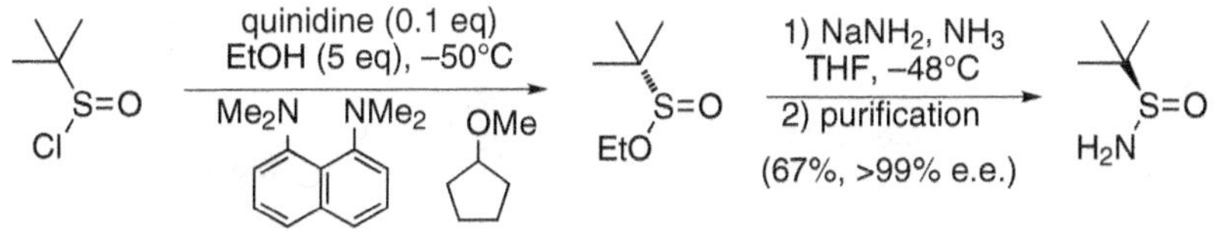

(23.55)[48]

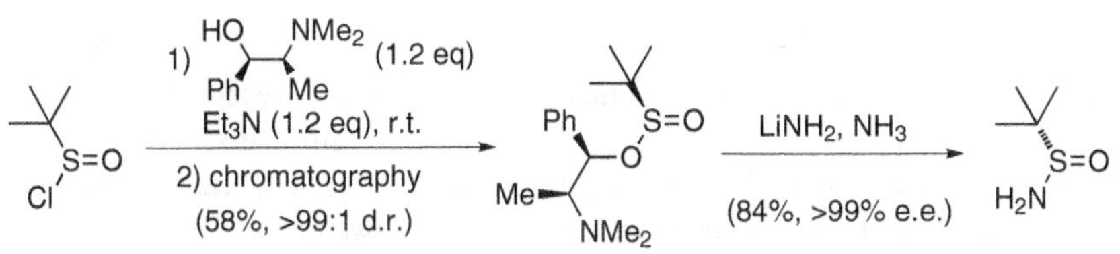

(23.56)[49]

The similar easy separation of diastereoisomers by chromatography was used by Ellman in the early synthesis of the optically active sulfinylamines.[50]

45. Pflum, D.A.; Krishnamurthy, D.; Han, Z.; Wald, S.A.; Senanayake, C.H. *Tetrahedron Lett.* **2002**, *43*, 923.
46. Han, Z.; Krishnamurthy, D.; Pflum, D.; Grover, P.; Wald, S.A.; Senanayake, C.H. *Org. Lett.* **2002**, *4*, 4025.
47. (a) Cogan, D.A.; Liu, G.; Kim, K.; Backes, B.J.; Ellman, J.A. *J. Am. Chem. Soc.* **1998**, *120*, 8011. (b) Weix, D.J.; Ellman, J.A.; Wang, X.; Curran, D.P. *Org. Synth.* **2005**, *82*, 157. (c) Colonna, S.; Gaggero, N.; Carrea, G.; Pasta, P.; Alphand, V.; Furstoss, R. *Chirality* **2001**, *13*, 40. (d) Han, Z.; Krishnamurthy, D.; Grover, P.; Fang, Q.K.; Senanayake, C.H. *J. Am. Chem. Soc.* **2002**, *124*, 7880.
48. Wakayama, M.; Ellman, J. A. *J. Org. Chem.* **2009**, *74*, 2646.
49. Aggarwal, V.K.; Barbero, N.; McGarrigle, E.M.; Mickle, G.; Navas, R.; Suarez, J.R.; Unthank, M.G.; Yar, M. *Tetrahedron Lett.* **2009**, *50*, 3482.
50. Cogan, D.A.; Liu, G.; Kim, K.; Backes, B.J.; Ellman, J.A. *J. Am. Chem. Soc.* **1998**, *120*, 8011.

Asymmetric Conjugate Additions

The real value of the conjugate addition in asymmetric synthesis arises from the way that it introduces flexibility into the synthesis. One can introduce complementary stereocenters using the same chiral auxiliary by altering the order in which the two β substituents are introduced. The same chiral substrates that have been successfully used in the asymmetric Diels-Alder cycloaddition and ene reactions have also been successfully used in enantioselective conjugate additions (Examples **23.57** to **23.60**). An encyclopedic review of the work to 1992 was published by Rossiter and Swingle.[51] Example **23.60** is especially interesting. Here the 4-benzyl auxiliary gives the product with the opposite configuration and with much reduced (62%) d.e. Recall that the same reversal of configuration occurred in the ene reactions of phenmenthyl glyoxylate and *trans*-2-phenylcyclohexyl glyoxylate.[52]

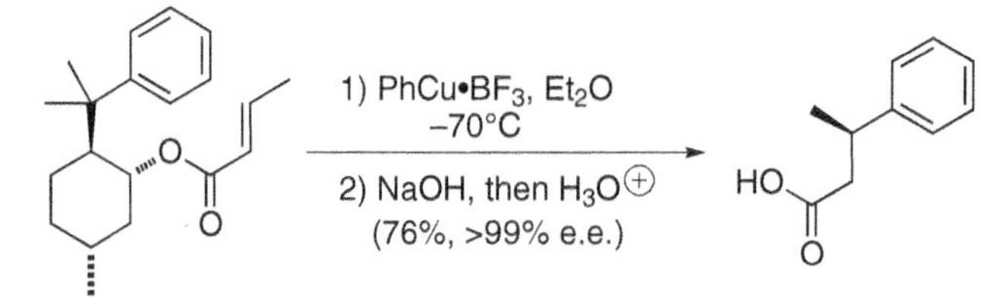

(23.57)[53]

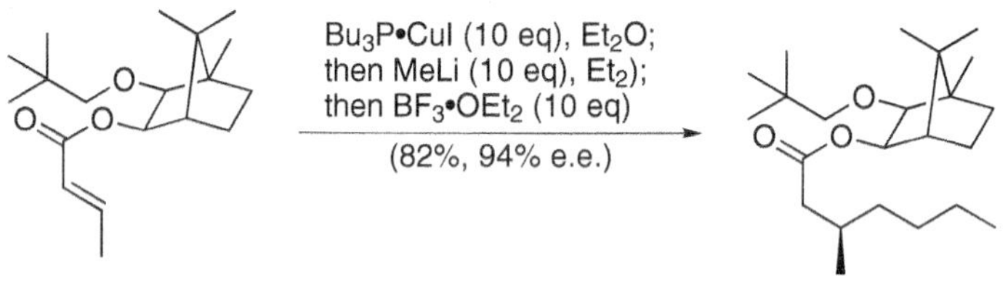

(23.58)[54]

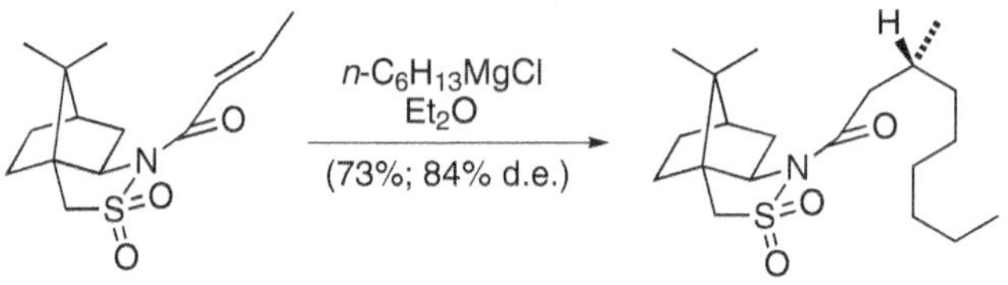

(23.59)[55]

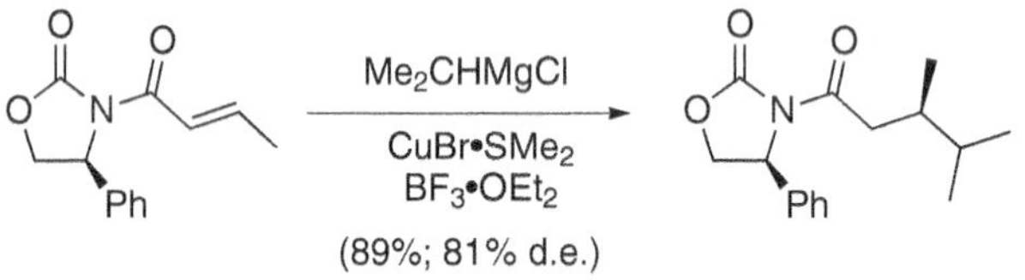

(23.60)[56]

The enantioselective conjugate addition of metal alkyls to Michael receptors can be accomplished by means using a chiral α-(*p*-tolyl)sulfinyl group as a chiral auxiliary. The zinc bromide chelates of these enones undergo conjugate additions with Grignard reagents and organocuprates to give chiral β-substituted ketones after reductive removal of the auxiliary.[57] The reaction is especially useful for the production of chiral β-alkylcyclopentanones, which are obtained in significantly higher e.e. values than the reaction of acyclic systems. This reaction has been used in the synthesis of intermediates for the synthesis of the D ring of steroids, as shown in Example **23.61**. In this area, also, the use of chiral sulfinyl imines of

---

51. Review: Rossiter, B.E.; Swingle, N.M. *Chem. Rev.* **1992**, *92*, 771.
52. Whitesell, J.K.; Bhattacharya, A.; Buchanan, C.M.; Chen, H.H.; Deyo, D.; James, D.; Liu, C.-L.; Minton, M.A. *Tetrahedron* **1986**, *42*, 2993.
53. Oppolzer, W.; Löher, H.J. *Helv. Chim. Acta* **1981**, *64*, 2808.
54. Oppolzer, W.; Moretti, R.; Godel, T.; Meunier, A.; Löher, H. *Tetrahderon Lett.* **1983**, *24*, 4971.
55. Oppolzer, W.; Poli, G.; Kingma, A.J.; Starkemann, C.; Bernardinelli, G. *Helv. Chim. Acta* **1987**, *70*, 2201.
56.Williams, D.R.; Kissel, W.S.; Li, J.J. *Tetrahedron Lett.* **1998**, *39*, 8593.
57. (a) Posner, G.H.; Mallamo, J.P.; Miura, K. *J. Am. Chem. Soc.* **1981**, *103*, 2886. (b) Posner, G.H.; Mallamo, J.P.; Miura, K.; Hulce, M. *Pure Appl. Chem.* **1981**, *53*, 2307.

conjugated ketones (e.g., Example **23.62**) has proved to be a useful method for obtaining chiral ketones and aldehydes with high levels of diastereoselectivity.

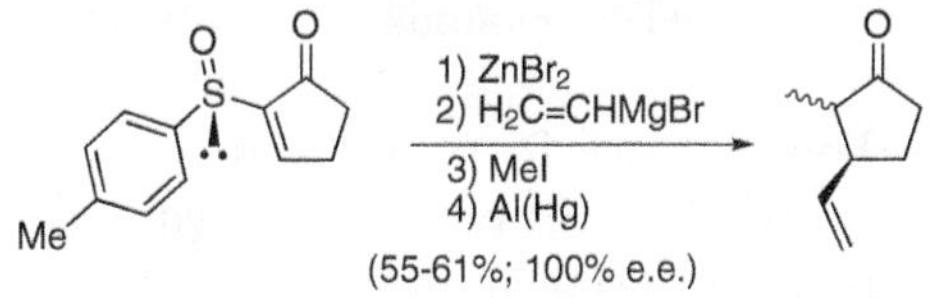

(23.61)[58]

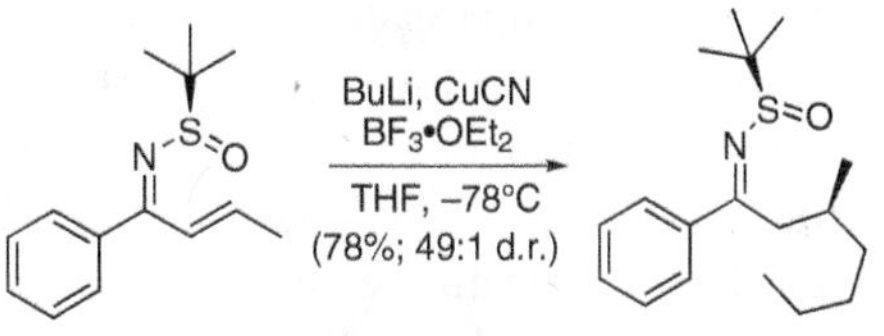

(23.62)[59]

As with many other reactions, advances in enantioselective conjugate additions are now being made by designing chiral ligands for the metal atoms in the achiral organometallic reagent, making the reaction catalytic.

Problem

23-5 What is the stereochemistry of each of the following reactions? What is the major product of the reaction if the phenyl group in the chiral auxiliary is changed toi a benzyl group [*Tetrahedron Lett.* 1998, *39*, 8593]?

(a)
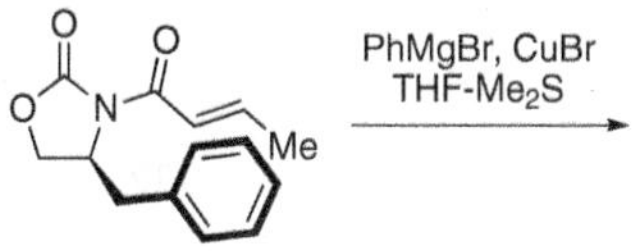

(b)
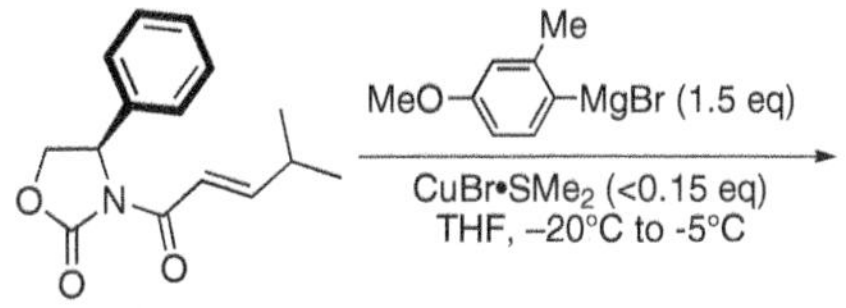

(c)
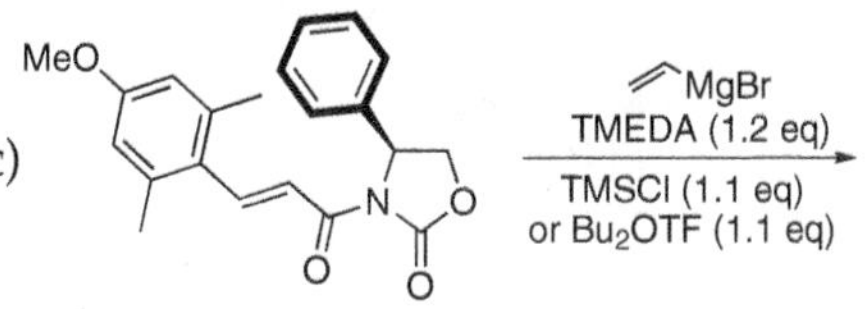

References: (a) *J. Org. Chem.* **1993**, *58*, 766. (b) *Tetrahedron Lett.* **1996**, *37*, 7917. (c) *Tetrahedron Lett.* **1997**, *38*, 7317.

23.3 Asymmetric Substitution at Carbon: Alkylation of Enolates and Related Nucleophiles

The replacement of hydrogen at a prochiral sp^3 center by a fourth different group will lead to enantiomeric products, and this reaction is therefore a candidate for use in asymmetric synthesis. The most common versions of this reaction are alkylation at the α carbon of a prochiral carbonyl compound and asymmetric oxidation at a similar position (Example **23.63**). Of the two, alkylation is by far the more established, and we will focus our attention there in this section.

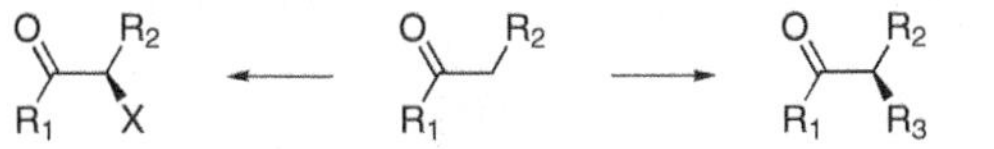

(23.63)

As with the asymmetric aldol addition, asymmetric alkylation of enolates is most frequently mediated by chiral auxiliaries. Generally speaking, the enolate equivalents used

---

58. Posner, G.H.; Hulce, M.; Mallamo, J.P.; Drexler, S.A.; Clardy, J. *J. Org. Chem.* **1981**, *46*, 5244,
59. McMahon, J.P.; Ellman, J.A. *Org. Lett.* **2005**, *7*, 5393.

in asymmetric alkylation are much stronger nucleophiles than the imide derivatives used in the asymmetric aldol addition and Diels-Alder reactions because alkyl halides are generally weaker electrophiles than carbonyl compounds (especially when the latter are complexed by a Lewis acid).

The replacement of a carbonyl oxygen by nitrogen in the *N,N*-dialkylhydrazone leads to an increase of approximately 10 in the pK_a of the α hydrogens.[60] This makes the azaenolate a much stronger base than an enolate and a more potent carbon nucleophile. Simple alkylimines have long been used as surrogates for ketones and aldehydes in synthesis, where their relative lack of reactivity toward nucleophiles often makes them easier to handle. The preferred deprotonation of the imine by Grignard reagents occurs *syn* to the *N*-alkyl group,[61] which permits a level of regiocontrol. The deprotonation of the *N*-cyclohexylimine of 2-methylcyclohexanone by ethylmagnesium bromide and trapping of the magnesium azaenolate by methyl iodide and subsequent hydrolysis gave 2,6-dimethylcyclohexanone. Deprotonation of the dimethylhydrazone of the same ketone by *sec*-butyllithium and trapping of the lithium azaenolate by methyl iodide gave 2,2-dimethylcyclohexanone after hydrolysis (Example **23.64**).[62] This has also allowed the generation of chiral imine derivatives that can be selectively alkylated to give chiral alkylation products after removal of the auxiliary.

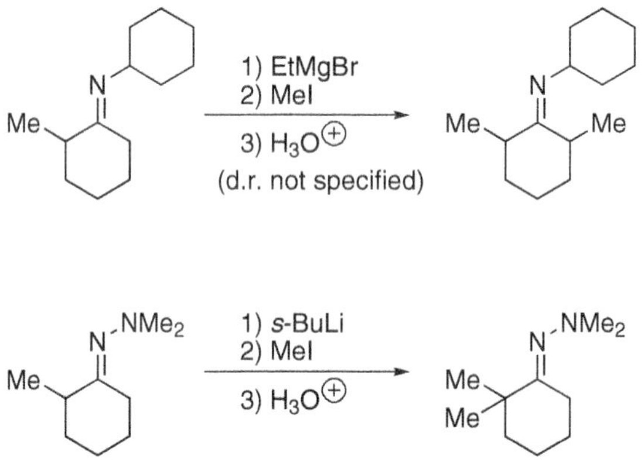

(23.64)

SAMP/RAMP Hydrazones

The first truly successful chiral auxiliaries for the alkylation of ketones and more especially, aldehydes, were the (*S*)-1-amino-2-methoxymethylpyrrolidine (SAMP) and (*R*)-1-amino-2-methoxymethylpyrrolidine (RAMP) hydrazines (Example **23.65**),[63] developed by Enders.[64] The deprotonation of the hydrazones derived from these hydrazines requires a lithium amide base. As with other asymmetric C—C bond-forming reactions, the control of the product stereochemistry requires control of the reagent stereochemistry. In the case of the SAMP/RAMP hydrazones, the stereochemical elements that must be controlled are the stereochemistry of the azaenolate double bond and the conformation of the azaenolate C—N bond. It has been found that deprotonation of the SAMP/RAMP hydrazones of

---

60. See: Lazny, R.; Nodzewska, A. *Chem. Rev.* **2010**, *110*, 1386.

61. (a) Stork, G.; Dowd, S.R. *J. Am. Chem. Soc.* **1963**, *85*, 2178. (b) Wittig, G.; Frommeld, H.D.; Suchanek, P. *Angew. Chem.* **1963**, *75*, 978.

62. Hosomi, A.; Araki, Y.; Sakurai, H. *J. Am. Chem. Soc.* **1982**, *104*, 2081.

63. Reviews: (a) Enders, D. In Morrison, J.D., Ed. *Asymmetric Synthesis* (Academic: New York, 1984), *Vol. 3*, 275. (b) Enders, D.; Wortmann, L.; Peters, R. *Acc. Chem. Res.* **2000**, *33*, 157. (c) Job, A.; Janeck, C.F.; Bettray, W.; Peters, R.; Enders, D. *Tetrahedron* **2002**, *58*, 2253.

64. **Dieter Enders (1946–)** was educated at Giessen (PhD, 1974 under Seebach), Harvard under Corey, and obtained his *habilitation* at Giessen in 1979. He became Professor at Bonn (1980) and Aachen (1985). He is editor-in-Chief of *Synthesis*. More biographical information is available at http://www.oc.rwth-aachen.de/akenders/enders_e.html.

acyclic ketones leads to the *E* double bond and the eclipsed conformation of the C—N bond, with nitrogen and carbon eclipsed. In this conformation, stabilized by chelation of the lithium cation, the quasi-1,3-diaxial interaction between the methylene group of the pyrrolidine ring and the approaching alkyl halide hinders the approach to one face of enolate (Example **23.66**). The removal of the chiral auxiliary from the product can be carried out by ozonolysis[65] or by alkylation of the imine nitrogen and hydrolysis of the resulting iminium salt.[66] The use of this method is illustrated in Examples **23.67** and **23.68**.

(23.65)

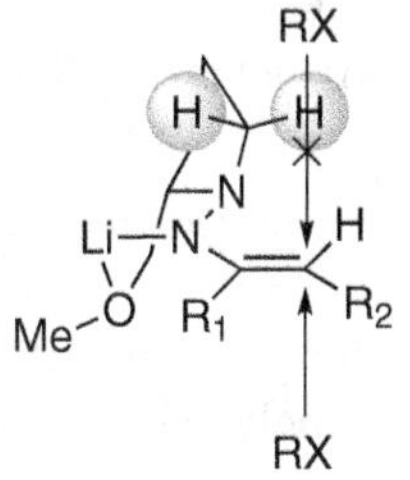

(23.66)

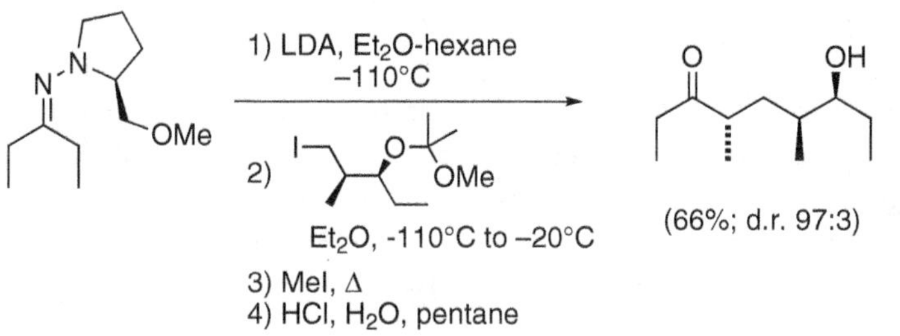

(23.67)[67]

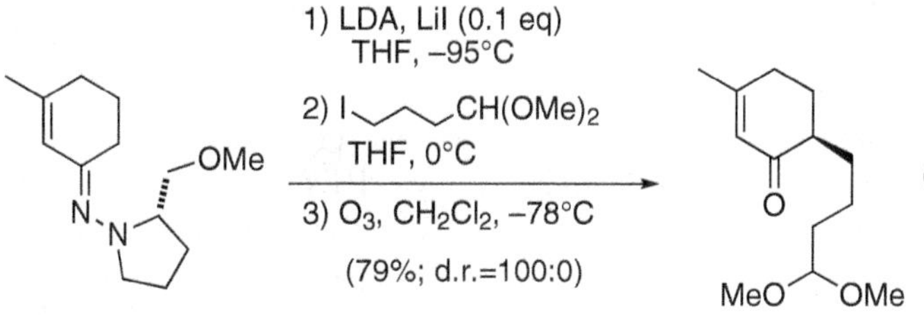

(23.68)[68]

A related alkylation of ketones, this time through chiral imines, was reported by Meyers[69] in 1981.[70] Like SAMP/RAMP hydrazones, the imines in this alkylation reaction carry a group capable of chelating the lithium atom. The most successful imines were those derived from the amine **23.69**, which is derived from phenylalanine. Following condensation of the amine with a cyclic ketone, the imine was deprotonated using LDA to give the chelated lithioenamine **23.70**. The stereochemistry of the reaction may be rationalized in terms of the model in Figure 23.18, which is based on the *S* auxiliary. In this model, the chelate ring is perpendicular to the plane of the C=C bond of the lithiated imine, and the benzyl group hinders the approach of the alkyl halide to one face the π bond. This leads to alkylation on the face adjacent to lithium. Other lithiated imines also undergo alkylation on the face of the π bond adjacent to lithium,[71] possibly because complexation of the incipient halide anion by lithium, such as that shown in **23.70**, lowers the

65. Enders, D.; Eichenauer, H. *Chem. Ber.* **1979**, *112*, 2933.

66. Avaro, M.; Levisalles, J.; Rudler, H. *J. Chem. Soc., Chem. Comm.* **1969**, 445.

67. Mori, K.; Nomi, H.; Chuman, T.; Kohno, M.; Kato, K.; Noguchi, M. *Tetrahedron* **1982**, *38*, 3705.

68. Schwaebe, M.; Little, R.D. *J. Org. Chem.* **1996**, *61*, 3240.

69. **Albert I. Meyers (1932–2007)** was educated at New York University (PhD, 1957), and Harvard before joining Louisiana State University (1959), Wayne State University (1970), and Colorado State University (1972). For a more complete biographical sketch, see: Tomioka, K. *Heterocycles* **2002**, *58*, 5.

70. Meyers, A.I.; Williams, D.R.; Erickson, G.W.; White, S.; Druelinger, M. *J. Am. Chem. Soc.* **1981**, *103*, 3081.

71. (a) Hoobler, M.A.; Bergbreiter, D.E.; Newcombe, M. *J. Am. Chem. Soc.* **1978**, *100*, 8182. (b) Meyers, A.I.; Snyder, E.S.; Ackerman, J.J.H. *J. Am. Chem. Soc.* **1978**, *100*, 8186.

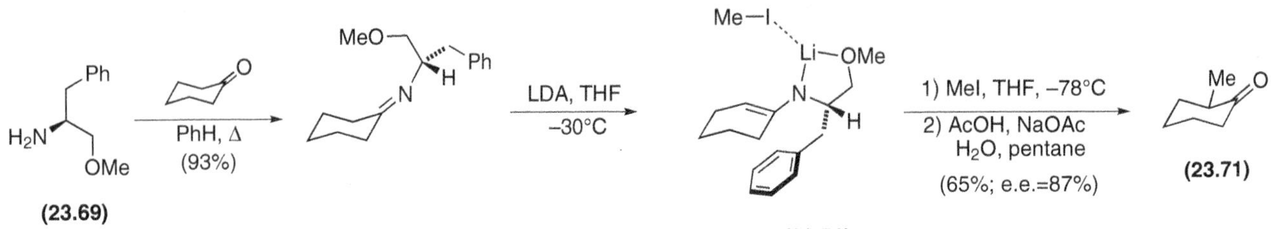

(23.69) (23.70) (23.71)

Figure 23.18 Alkylation of chiral imines

activation energy for this reaction pathway. Alkylation and hydrolysis of the product gives the chiral 2-alkylcycloalkanones in acceptable to good chemical yields (e.e. > 79%). The exceptions to this generalization were the imines of cyclooctanone and 1-indanone, which gave disappointingly poor (e.e. < 20%) asymmetric induction.

Chiral Oxazolines[72]

In Chapter 15, we discussed the use of oxazolines as protecting groups for carboxylic acids and esters. In a logical extension of these observations, it was inevitable that chiral oxazolines would emerge as chiral surrogates for carboxylic acids and esters, especially because the 2-alkyloxazolines can be deprotonated readily by a variety of strong lithium bases (*t*-BuLi, *n*-BuLi, and LDA have all been popular). The resulting lithiated oxazoline is a powerful azenolate nucleophile.

The structural similarities with simple chiral imines suggested that oxazolines formed from a chiral aminoalcohol should possess good potential for asymmetric induction in alkylation reactions. This is exactly what was observed. Sequential alkylation of the chiral oxazoline with an alkyl iodide and acid-catalyzed hydrolysis of the oxazoline gave the chiral carboxylic acid in good chemical and optical yield. Example **23.72** illustrates the reaction sequence by means of the synthesis of *S*-(+)-2-methylhexanoic acid from the chiral 2-ethyloxazoline.[73] Bromides and chlorides generally gave lower chemical and optical yields (both) than the iodides. Given that a chiral α-alkylcarboxylic acid may be prepared two ways by alkylation (see Example **23.73**), the ability to do so in an asymmetric sense also allows the two enantiomers of the final acid to be prepared from the same enantiomer of the chiral auxiliary.

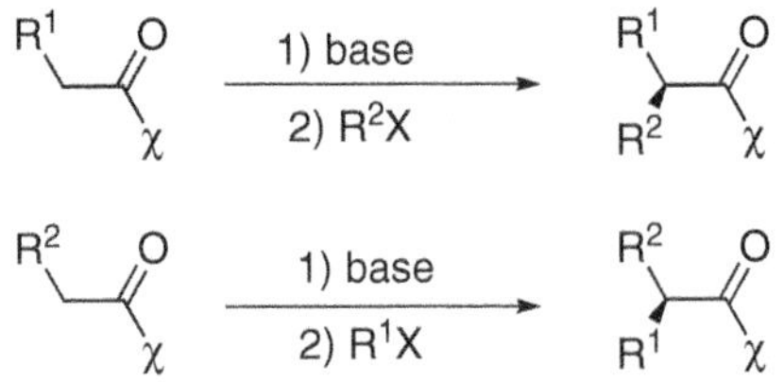

(23.73)

The one drawback of the chiral oxazolines compared to the chiral imines we discussed earlier is the fact that the deprotonation of the oxazoline may give either the *E* or the *Z* azaenolate (see Example **23.74**). Because the face-selectivity of the alkylation reaction is determined by the chelate, with the alkyl iodide preferentially attacking the face proximal to the lithium, this means that the reaction can, in principle, lead to both diastereoisomers of the alkylated oxazoline and thus to both enantiomers of the final carboxylic acid. It is

72. Reviews: (a) Meyers, A.I. *Acc. Chem. Res.* **1978**, *11*, 375. (b) Meyers, A.I. In Eliel, E.L.; Otsuke, S., Eds. *Asymmetric Reactions and Processes in Chemistry.* (American Chemical Society: Washington, D.C., 1982); *ACS Symp. Ser. 185* 83. (c) Lutomski, K.; Meyers, A.I. In Morrison, J.D., Ed. *Asymmetric Synthesis* (Academic: New York, 1984), *Vol. 3*, 213. (d) Meyers, A.I. *J. Heterocyclic Chem.* **1998**, *35*, 991. (e) Meyers, A.I. *J. Org. Chem.* **2005**, *70*, 6137.

73. Meyers, A.I.; Knaus, G.; Kamata, K.; Ford, M.E. *J. Am. Chem. Soc.* **1976**, *98*, 567.

clear that there is a reasonably high level of stereoselectivity in both the deprotonation and the alkylation steps. In Example **23.74**, the d.r. in the deprotonation with LDA is 95:5 and the enantiomer ratio of the final carboxylic acid is 89:11.

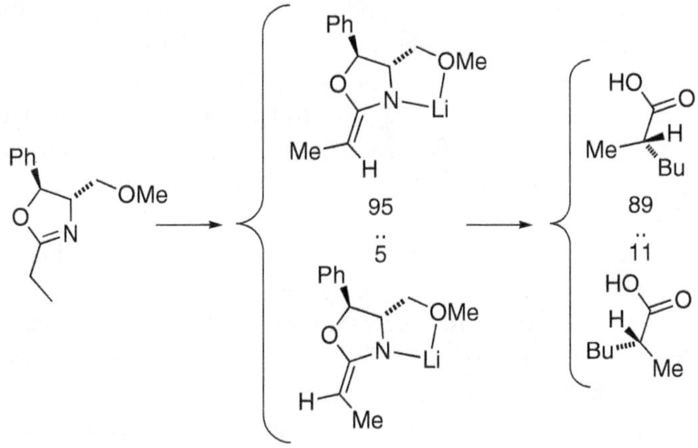

(23.74)

Chiral 2-alkenyloxazolines readily undergo conjugate addition of organolithium to give, after acid-catalyzed hydrolysis, chiral β-substituted carboxylic acids (e.g., Example **23.75**) in good chemical yields and excellent optical yields.[74] One of the most interesting forms of this reaction has involved the addition of alkyllithium reagents to chiral oxazolines based on 1-naphthoic acid. When the intermediate azaenolate is trapped by an alkyl halide, substituted dihydronaphthalene derivatives with a quaternary center are generated (e.g., Example **23.76**).[75]

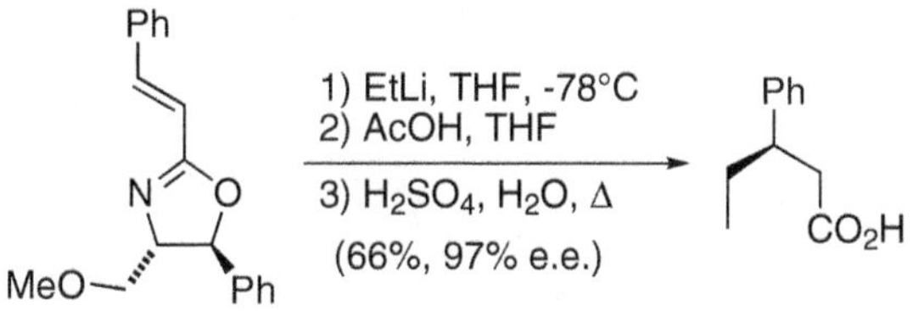

(23.75)

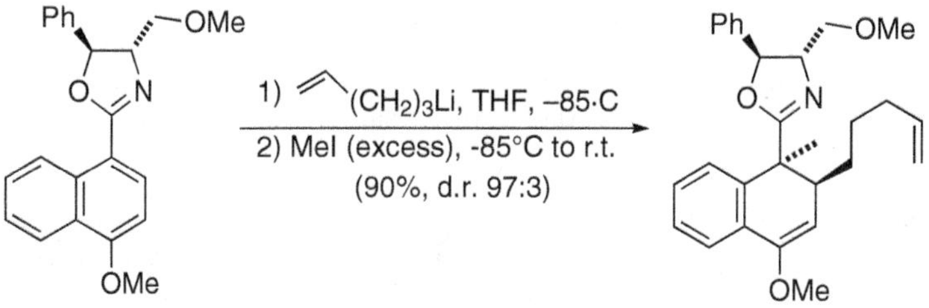

(23.76)

23.4 **Catalytic Asymmetric Synthesis**

The field of asymmetric catalysis has undergone a genuine explosion since 1980, as attested by the volume of literature on the subject, and even the extent of our discussion that has been devoted to it. Other than to point out where our discussion has already taken place, we will not cover this material again here.[76]

---

74. (a) Meyers, A.I.; Whitten, C.E. *J. Am. Chem. Soc.* **1975**, *97*, 6266. (b) Meyers, A.I.; Smith, R.K.; Whitten, C.E. *J. Org. Chem.* **1979**, *44*, 2250.

75. Meyers, A.I.; Roth, G.P.; Hoyer, D.; Barner, B.A.; Laucher, D. *J. Am. Chem. Soc.* **1988**, *110*, 4611.

76. A brief selection of the books on this subject: (a) Ojima, I., Ed. *Catalytic Asymmetric Synthesis* (Wiley-VCH: Weinheim). This work has been through three editions between 2000 and 2010. (b) Jacobsen, E.N.; Pfaltz, A.; Yamamoto, H., Eds. *Comprehensive Asymmetric Catalysis* (Springer: Berlin, 1999), vols. 1-3. (c) Noyori, R. *Asymmetric Catalysis in Organic Synthesis* (John Wiley & Sons: New York, 1994).

The important asymmetric redox reactions, which were discussed at length in Chapter 21, include the Sharpless asymmetric epoxidation of allylic alcohols;[77] the Jacobsen-Katsuki epoxidation,[78] which permitted the reliable asymmetric epoxidation of alkenes; the Sharpless asymmetric dihydroxylation;[79] and the Shi asymmetric epoxidation.[80] In this author's opinion, the Sharpless asymmetric epoxidation was the truly revolutionary development in modern organic chemistry. By permitting organic chemists to synthesize just one enantiomer of a chiral product from a prochiral precursor and to be able to predict its absolute configuration with confidence, this reaction moved epoxides from the fringes of organic synthesis to the center. It also paved the way for many of the asymmetric catalytic reactions that have followed.

Likewise, asymmetric reduction, which was also discussed in detail in Chapters 19, 20 and 21, was driven initially by asymmetric hydrogenation that gave high e.e. values of products with predictable configurations.[81] Success in these reductions was eventually followed by asymmetric hydride reductions, including oxazaborolidine-catalyzed reductions with borane[82] and reductions with chiral boranes.[83]

Following the emergence of chiral catalysts for oxidation and reduction reactions, it is logical to deduce that catalysts for use in reactions that form carbon-carbon bonds would be the next major target in asymmetric synthesis. Until very recently, however, chiral auxiliary approaches have been more successful in Diels-Alder and conjugate addition reactions. Nevertheless, catalysts for Diels-Alder reactions[84] and conjugate addition reactions

---

77. Katsuki, T.; Sharpless, K.B. *J. Am. Chem. Soc.* **1980**, *102*, 5974.

Reviews: (b) Katsuki, T.; Martin, V.S. *Org. React.* **1996**, *48*, 1. (c) Johnson, R.A.; Sharpless, K.B. In Ojima, I., Ed. *Catalytic Asymmetric Synthesis, 1st ed.*, (VCH: New York, 1993), 103; *2nd ed.* (VCH: New York, 2000), 231. (d) Katsuki, T. In Jacobsen, E.N.; Pfaltz, A.; Yamamoto, H. Eds. *Comprehensive Asymmetric Catalysis*, 1st ed. (Springer: New York, 1999), Vol. 2 621. (e) Johnson, R.A.; Sharpless, K.B. In Trost, B.M.; Fleming, I, Eds. *Comprehensive Organic Synthesis* (Pergamon Press: New York, 1991), Vol. 7, 389. (f) Pfenninger, A. *Synthesis* **1986**, 89. (g) Jacobsen, E.N. In Abel, E.W.; Stone, F.G.A.; Wilkinson, G., Eds. *Comprehensive Organometallic Chemistry II, 1st ed.* (Pergamon: New York, 1995), Vol. 12, 1097.

78. (a) Zhang, W.; Loebach, J.L.; Wilson, S.R.; Jacobsen, E.N. *J. Am. Chem. Soc.* **1990**, *112*, 2801. (b) Jacobsen, E.N.; Zhang, W.; Muci, A.R.; Ecker, J.R.; Deng, L. *J. Am. Chem. Soc.* **1991**, *113*, 7063. (c) Irie, R.; Noda, K.; Ito, Y.; Matsumoto, N.; Katsuki, T. *Tetrahedron: Asymmetry* **1991**, *2*, 481. (d) Linker, T. *Angew. Chem. Int. Ed. Engl.* **1997**, *36*, 19.

79. (a) Jacobsen, E.N.; Marko, I.; Mungall, W.S.; Schroeder, G.; Sharpless, K.B. *J. Am. Chem. Soc.* **1988** *110*, 1968. Review: (b) Kolb, H.C.; VanNieuwenhze, M.S.; Sharpless, K.B. *Chem. Rev.* **1994**, *94*, 2483.

80. (a) Wang, Z.-X.; Tu, Y.; Frohn, M.; Zhang, J.-R.; Shi, Y. *J. Am. Chem. Soc.* **1997**, *119*, 11224. (b) Frohn, M.; Shi, Y. *Synthesis* **2000**, 1979. (c) Tian, H.; She, X.; Shu, L.; Yu, H.; Shi, Y. *J. Am. Chem. Soc.* **2000**, *122*, 11551. (d) Tian, H.; She, X.; Xu, J.; Shi, Y. *Org. Lett.* **2001**, 3, *1929*.

81. Reviews and monographs: (a) Noyori, R.; Kitamura, M. In Scheffold, R., Ed. *Modern Synthetic Methods* (Springer: Berlin, 1989), vol. 5, p. 115. (b) Takaya, H.; Ohta, T.; Noyori, R. In Ojima, I., Ed. *Catalytic Asymmetric Synthesis* (VCH: New York, 1993), ch. 1., p. 1. (c) Noyori, R. *Asymmetric Catalysis in Organic Synthesis* (Wiley: New York, 1994), ch. 2. (d) Genêt, J.-P. In Abdel-Magid, A.F. (Ed.) *Reductions in Organic Synthesis: Recent Advances and Practical Applications, ACS Symp. Ser. 641* (American Chemical Society: Washington, D.C., 1996), ch. 2. (e) Albrecht, J.; Nagel, U. *Angew. Chem. Int. Ed. Engl.* **1996**, *35*, 407. (h) Halpern, J.A. In Morrison, J.D., Ed. *Asymmetric Synthesis* (Academic Press: New York, 1985), vol. 5, p. 41. (i) Koenig, K.H. In Morrison, J.D., Ed. *Asymmetric Synthesis* (Academic Press: New York, 1985), vol. 5, p. 71. (k) Seyden-Penne, J. *Chiral Auxiliaries and Ligands in Asymmetric Synthesis* (Wiley-Interscience: New York, 1995), ch. 7.1, p. 367. (l) Ohkuma, T.; Kitamura, M.; Noyori, R. In Ojima, I., Ed. *Catalytic Asymmetric Synthesis, 2nd Ed.* (Wiley-VCH: New York, 2000), ch. 1, p. 1. (m) Knowles, W.S. *Acc. Chem. Res.* **1983**, *16*, 106. (n) Noyori, R.; Takaya, H. *Acc. Chem. Res.* **1990**, *23*, 345. (o) Blystone, S.L. *Chem. Rev.* **1989**, *89*, 1663. (p) Knowles, W.S. *Angew. Chem. Int. Ed. Engl.* **2002**, *41*, 1998. (q) Noyori, R. *Angew. Chem. Int. Ed. Engl.* **2002**, *41*, 2008. (r) Shang, G.; Li, W.; Zhang, X, In Ojima, I., Ed. *Catalytic Asymmetric Synthesis, 3rd ed.* (John Wiles & Sons: New York, 2010), ch. 7, p. 343.

82. a) Corey, E.J.; Bakshi, R.K.; Shibata, S. *J. Am. Chem. Soc.* **1987**, *109*, 5551. (b) Corey, E.J.; Shibata, S.; Bakshi, R.K. *J. Org. Chem.* **1988**, *53*, 2861. (c) Corey, E.J.; Bakshi, R.K.; Shibata, S.; Chen, C.-P.; Singh, V.K. *J. Am. Chem. Soc.* **1987**, *109*, 7925. (d) Corey, E.J. *Pure Appl. Chem.* **1990**, *62*, 1209. (e) Corey, E.J.; Cheng, X.M.; Cimprich, K.A.; Sarshar, S. *Tetrahedron Lett.* **1991**, *32*, 6835. (f) Corey, E.J.; Kigoshi, M. *Tetrahedron Lett.* **1991**, *32*, 5025. (g) Corey, E.J.; Rao, K.S. *Tetrahedron Lett.* **1991**, *32*, 4623. (h) Jones, D.K.; Liotta, D.C.; Shinkai, I.; Mathre, D.J. *J. Org. Chem.* **1993**, *58*, 799.

Monograph: Lohrag, B.B.; Bhusan, V. *Oxazaborolidines* (Wiley-VCH: New York, 1998).

83. Reviews: (a) Srebnik, M.; Ramachandran, P.V. *Aldrichimica Acta* **1987**, *20*, 9. (b) Dhar, R.K. *Aldrichimica Acta* **1994**, *27*, 43. (c) Cho, B.T. *Aldrichimica Acta* **2002**, *35*, 3.

84. (a) Corey, E.J.; Lo, T.-P. *J. Am. Chem. Soc.*, **1991**, *113*, 8966. (b) Corey, E.J.; Ishihara, K. *Tetrahedron Lett.* **1992**, *33*, 6807. (c) Corey, E.J. *J. Am. Chem. Soc.* **1991**, *113*, 728. (d) Evans, D.A.; Miller, S.J.; Lectka, T.; von Matt, P. *J. Am. Chem. Soc.* **1999**, *121*, 7559.

(e.g., the use of phosphoramidate ligands in conjugate additions to enones[85]) have been developed. It is rather interesting that the chiral oxazolines developed by Meyers discussed in the previous section. have found new prominence in the form of chiral *bis*-oxazoline (BOX) ligands for asymmetric synthesis.

Ligands for Transition Metal Catalysts

Chiral catalysts are almost always formed from an achiral precursor and a chiral ligand. For example, the chiral catalysts **23.77** and **23.78** are formed from the reaction between the chiral ligand and copper (II) triflate (Example **23.77**[86]) or a titanium halide or alkoxide (Example **23.78**[87]).

The one common structural feature of many successful transition metal catalysts for asymmetric synthesis is the presence of C_2-symmetric (or quasisymmetric) ligands on the metal so that the entire catalyst has approximate C_2 symmetry. The activity of the catalyst can usually be modulated by altering the donor atoms of the ligand from hard donors (e.g., oxygen or nitrogen) to soft donors (e.g., sulfur and phosphorus). For example, copper (II) and titanium (IV) catalysts are generated from ligands with hard donor atoms illustrated by the BOX complex **23.77**; the 1,1'-bi-2-naphthol (BINOL) complexes **23.78**; and the quasi-C_2-symmetric Carreira complex **23.79**, which was used in the Mukaiyama aldol addition reaction.[88]

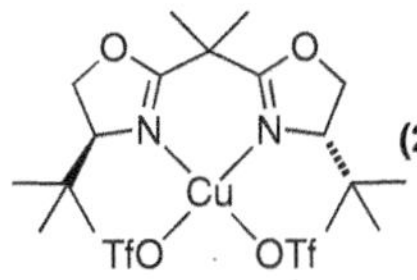

(23.77)

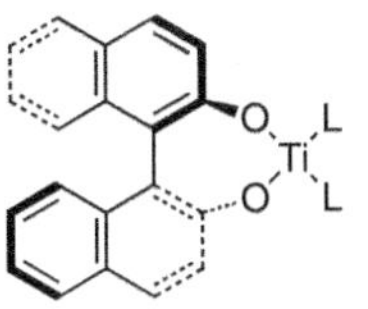

(23.78)

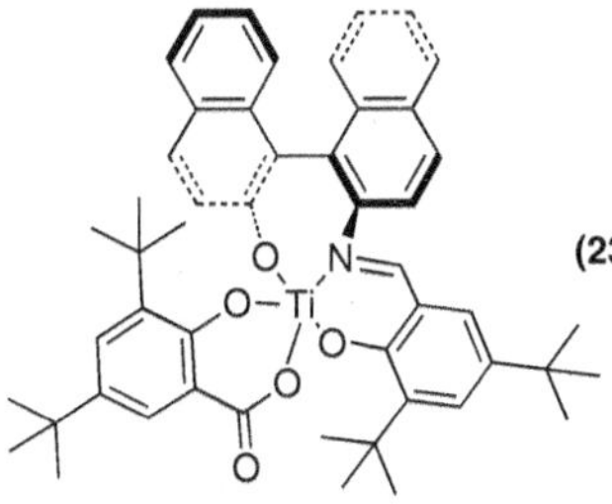

(23.79)

Noble metal catalysts[89] for asymmetric hydrogenation, on the other hand, are generally formed using soft, phosphine ligands. As we discussed in Chapter 19, there are four major groups of C_2-symmetric bidentate ligands that have been used for generating these catalysts. The first three are (1) the C_2-symmetric chiral diphosphines with a flexible backbone, where the chirality is centered in backbone (DIOP, **23.80**) or on phosphorus (DIPAMP, **23.81**); (2) the C_2-symmetric 1,1'-binaphthyl-based diphosphines (BINAP, **23.82**); and (3) the C_2-symmetric *bis*-phospholane ligands, where the chirality is centered in the chiral 2,5-dialkyl-phospholane rings (DuPHOS, **23.83**). These examples are given in Figure 23.19. The fourth class (e.g., PHOX, **23.84**) differs from the others in having one hard and one soft ligating atom, as well as in lacking C_2 symmetry.

Chiral Activation and Deactivation[90]

Many ligands are available in optically active form, and this makes the formation of the chiral catalytic complex a reasonably trivial undertaking. The question to be answered

85. (a) Feringa, B.L. *Acc. Chem. Res.* **2000**, *33*, 346. (b) Harutyunyan, S.R.; den Hartog, T.; Geurts K.; Minnaard, A.J.; Feringa, B.L. *Chem. Rev.* **2008**, *108*, 2824.

86. Evans, D.A.; Miller, S.J.; Lecta, T. *J. Am. Chem. Soc.* **1993**, *115*, 6460.

87. (a) Mikami, K.; Matsukawa, M. *J. Am. Chem. Soc.* **1994**, *116*, 4077. (b) Mikami, K.; Matsukawa, S. *J. Am. Chem. Soc.* **1993**, *115*, 7039. (c) Mikami, K. *Pure Appl. Chem.* **1996**, *68*, 639. (d) Keck, G.E.; Krishnamurthy, D. *J. Am. Chem. Soc.* **1995**, *117*, 2363.

88. Carreira, E.M.; Singer, R.A.; Lee, W. *J. Am. Chem. Soc.* **1994**, *116*, 8837; **1995**, *117*, 3649 (b) Singer, R.A.; Carreira, E.M. *J. Am. Chem. Soc.* **1995**, *117*, 12360.

89. For extensive lists of chiral phosphine ligands for hydrogenation catalysts, see: (a) Koenig, K.H. In Morrison, J.D., Ed. *Asymmetric Synthesis* (Academic Press: New York, 1985), vol. 5, p. 71. (b) Ohkuma, T.; Kitamura, M.; Noyori, R. In Ojima, I., Ed. *Catalytic Asymmetric Synthesis, 2nd Ed.* (Wiley-VCH: New York, 2000), ch. 1, p. 1.

90. Review: Mikami, K.; Terada, M.; Korenaga, T.; Matsumoto, Y.; Ueki, M.; Angelaud, R. *Angew. Chem. Int. Ed.* **2000**, *39*, 3532.

Figure 23.19 C_2-symmetric diphosphine ligands used in making asymmetric hydrogenation catalysts (top row). The C_2 axis common to these ligands is indicated by the dashed arrow.

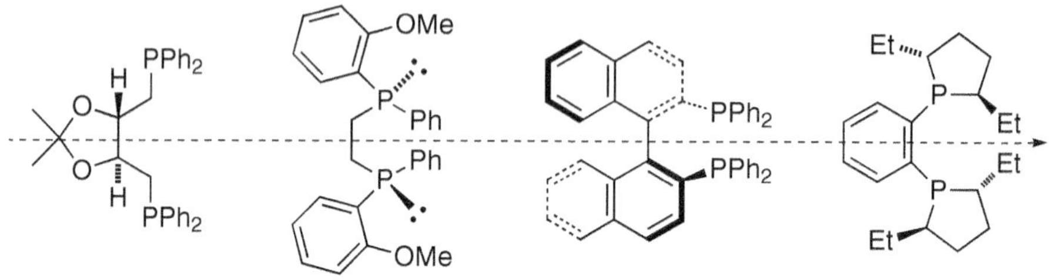

(23.80) (*S,S*)-DIOP **(23.81)** (*R,R*)-DIPAMP **(23.82)** (*S*)-BINAP **(23.83)** (*R,R*)-Et-DuPHOS

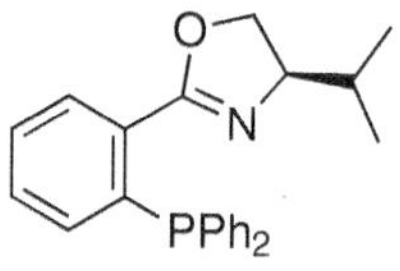

(23.84) (*R*)-PHOX

Figure 23.20 Nonlinear effects in asymmetric reactions catalyzed by chiral adjuvants that are not enantiomerically pure

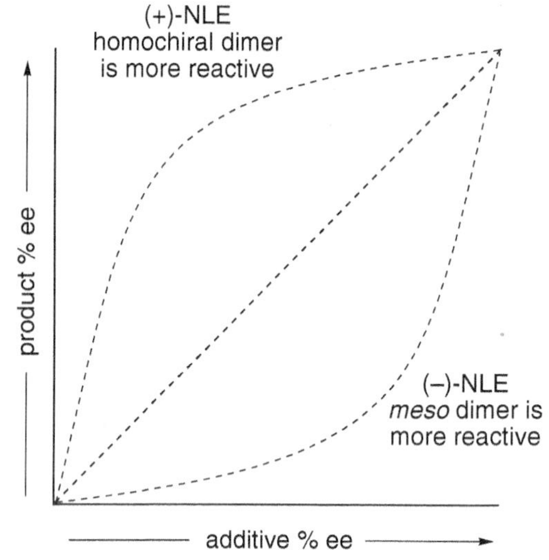

was "How can one carry out an asymmetric synthesis using a racemic catalyst?" This is an important question, and the possibility that this might be accomplished at all was first suggested by the studies of Oguni[91] and Noyori[92] in the catalyzed addition of dialkylzinc reagents to aldehydes (Example **23.85**). The 3-dimethylminoborneol catalyst was available in just 15% e.e., which would suggest that the e.e. of the final product should also be low. It was not, and this provided an excellent example of what has become known as ***asymmetric amplification***.[93] These non-linear effects of chirality can be discussed in terms of the diagram in Figure 23.20. Thus, if there is no nonlinear effect, the optical purity of the chiral

91. Oguni, N.; Matsuda, Y.; Kaneko, T. *J. Am. Chem. Soc.* **1988**, *110*, 7877.

92. (a) Kitamura, M.; Okada, S.; Suga, S.; Noyori, R. *J. Am. Chem. Soc.* **1989**, *111*, 4028. (b) Kitamura, M.; Suga, S.; Niwa, M.; Noyori, R. *J. Am. Chem. Soc.* **1995**, *117*, 4832. (c) Kitamura, M.; Yamakawa, M.; Oka, H.; Suga, S.; Noyori, R. *Chem. Eur. J.* **1996**, *2*, 1173. (d) Kitamura, M.; Suga, S.; Oka, H.; Noyori, R. *J. Am. Chem. Soc.* **1998**, *120*, 9800. (e) Kitamura, M.; Oka, H.; Noyori, R. *Tetrahedron* **1999**, *55*, 3605.

93. For reviews, see: (a) Girard, C.; Kagan, H.B. *Angew. Chem., Int. Ed. Engl.* **1998**, *37*, 2922. (b) Kagan, H.B.; Girard, C.; Guillaneux, D.; Rainford, D.; Samuel, O.; Zhang, S.Y.; Zhao, S.H. *Acta Chem. Scand.* **1996**, *50*, 345. (c) Avalos, M.; Babiano, R.; Cintas, P.; Jiménez, J. L.; Palacios, J.C. *Tetrahedron: Asymmetry* **1997**, *8*, 2997. (d) Reggelin, M. *Nachr. Chem., Technol. Lab.* **1997**, *45*, 392, 622. (e) Bolm, C. In Stephenson, G.R., Ed. *Advanced Asymmetric Catalysis* (Blackie Academic and Professional: New York, 1996) pp 9–26. (f) Noyori, R.; Kitamura, M. *Angew. Chem., Int. Ed. Engl.* **1991**, *30*, 49.

product will be equal to the optical purity of the chiral adjuvant. This is the normal outcome when the reagent generated from the chiral adjuvant is monomeric.

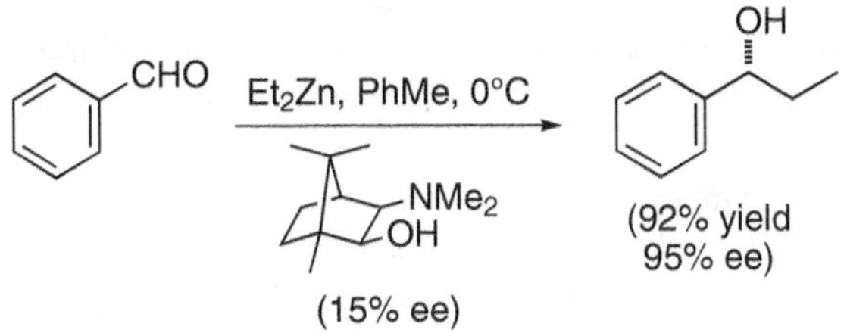

(23.85)

If the reagent generated is dimeric, however, the possibility exists for the formation of both homochiral and heterochiral (*meso*) reagents. If the homochiral reagent is the more reactive, then the e.e. of the product will be higher than the e.e. of the chiral adjuvant—a positive non-linear effect, or (+)-NLE. On the other hand, if the *meso* form of the reagent is the more reactive, the e.e. of the product will be less than that of the chiral adjuvant—a negative non-linear effect, or (–)-NLE.

As would be predicted by Figure 23.20, the origin of the chiral amplification in the Noyori system was traced to the fact that the heterochiral (*meso*) dimeric reagent (Example **23.86**) was not actually an active nucleophile toward the aldehyde—only the C_2-symmetric, homochiral dimeric reagent (Example **23.87**) was active. Several examples of (+)-NLE reactions have since been observed. In addition, as the size of the homochiral oligomer increases, the NLE should also increase. In an interesting observation, when formed from the metal and ligand of relatively low e.e., the complex **23.88** appears to generate the homochiral form selectively. Moreover, only homochiral forms appear to be an active catalyst for the hetero-Diels-Alder reaction between Danishefsky's diene and benzaldehyde to give the chiral dihydropyran **23.89**.[94] In this case, ligand with just 20% e.e. gave a product with the limiting value of 90% e.e.—the same level attained with optically pure catalyst in this system.

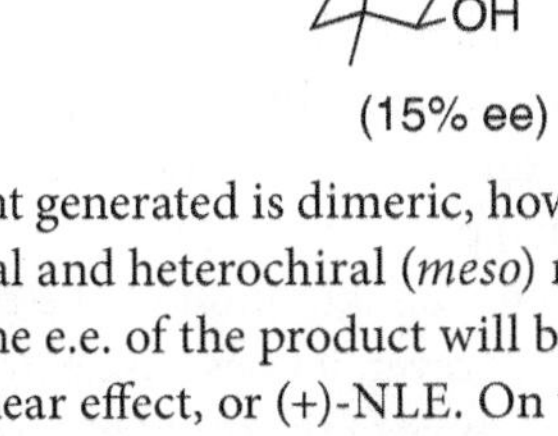

(23.86)
meso—inactive

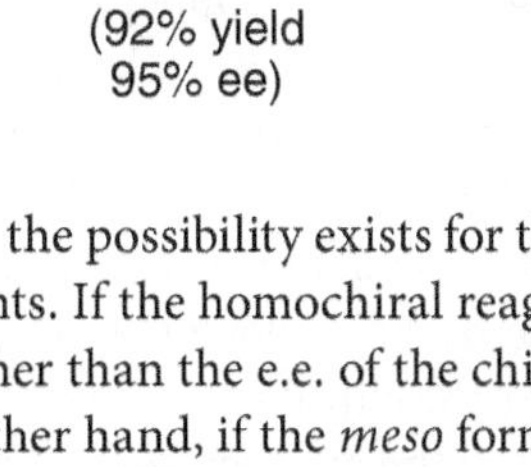

(23.87)
homochiral—active

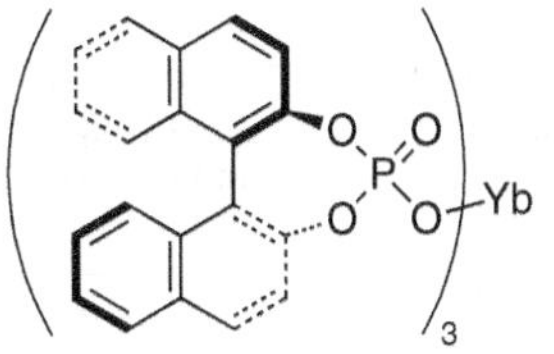

(23.88)

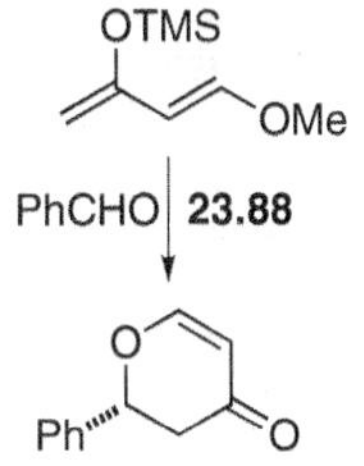

(23.89)

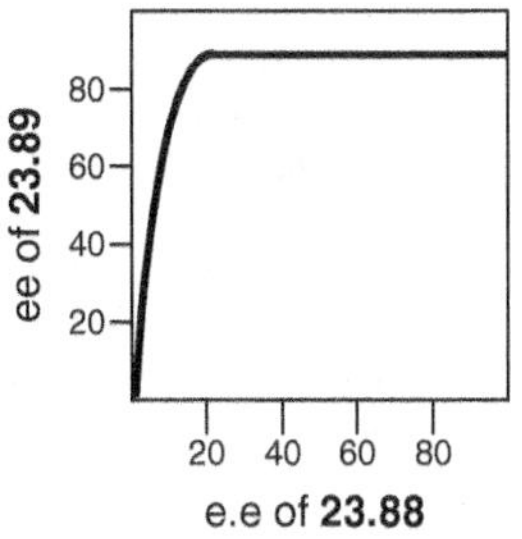

Asymmetric Cycloadditions

The power of the Diels-Alder cycloaddition as a synthetic method—it permits the formation of up to four new chiral centers in a single step, as well as generating a six-membered ring—has provided strong motivation for the search for catalysts to enable this reaction to be carried out enantioselectively. It is known that complexation of the dienophile by a

94. Furuno, H.; Hanamoto, T.; Sugimoto, Y.; Inanaga, J. *Org. Lett.* **2000**, *2*, 49.

Lewis acid accelerates the reaction, so research has focused for some time on finding a suitable chiral Lewis acid catalyst for the reaction. The search has yielded compounds capable of catalyzing the reaction with quite high levels of asymmetric induction. By the end of the 1990s, chiral catalysts for this reaction based on titanium,[95] aluminum,[96] boron,[97] lanthanide metals,[98] ruthenium,[99] manganese,[100] iron,[101] copper,[102] magnesium,[103] and zinc[102e] had been developed.

Thus, Evans has used a copper (II) BOX complex **23.77** with success in Diels-Alder cycloadditions (Example **23.90**) with cyclopentadiene.[104] What is interesting in this reaction is the fact that the replacement of the achiral oxazolidinone group with a chiral version (the Evans chiral auxiliaries), as shown in Example **23.91**, does not change the absolute configuration of the major enantiomer of the final product, regardless of the absolute configuration of the oxazolidinone (i.e., the absolute configuration of the product is determined mainly by the chirality of the catalyst). However, it does have a dramatic effect on the yield and e.e. of the product—an excellent example of the result of carrying out a reaction with mismatched reactants (in this case, the catalyst and the chiral acrylamide). A literature search conducted in late 2013 showed that this Evans paper had been cited 277 times since its publication.

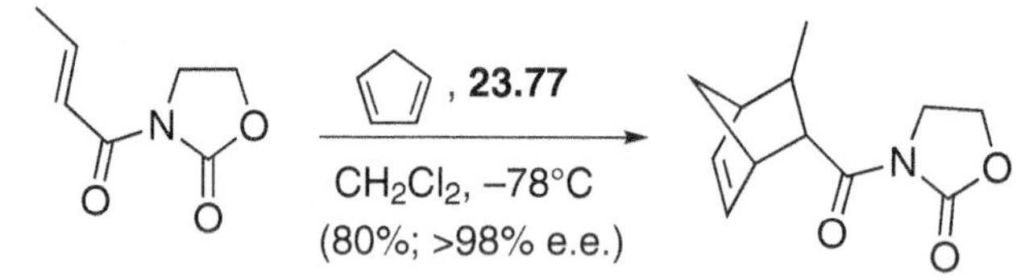

(23.90)

95. (a) Narasaka, K.; Inoue, M.; Okada, N. *Chem. Lett.* **1986**, 1109. (b) Narasaka, K.; Inoue, M.; Yamada, T. *Chem. Lett.* **1986**, 1967. (c) Narasaka, K.; Iwasawa, N.; Inoue, M.; Yamada, T.; Nakashima, M.; Sugimori, J. *J. Am. Chem. Soc.* **1989**, *111*, 5340. (d) Corey, E.J.; Matsumura, Y. *Tetrahedron Lett.* **1991**, 32, 6289. (e) Haase, C.; Sarko, C.R.; DiMare, M. *J. Org. Chem.* **1995**, *60*, 1777. (f) Seebach, D.; Dahinden, R.; Marti, R.E.; Bek, A.K.; Plattner, D.A.; Kuehnle, N.M. *J. Org. Chem.* **1995**, *60*, 1788. (g) Gothelf, K.V.; Hazell, R.G.; Jorgensen, K.A. *J. Am. Chem. Soc.* **1995**, *117*, 4435. (h) Gothelf, K.V.; Jorgensen, K.A. *J. Org. Chem.* **1995**, *60*, 6847. (i) Seebach, D.; Marti, R.E.; Hintermann, T. *Helv. Chim. Acta* **1996**, *79*, 1710.

96. (a) Rebiere, F.; Riant, O.; Kagan, H.B. *Tetrahedron:Asymmetry* **1990**, *1*, 199. (b) Corey, E.J.; Imai, N.; Pikul, S. *Tetrahedron Lett.* **1991**, *32*, 7517. (c) Corey, E.J.; Sarshar, S.; Bordner, J. *J. Am. Chem. Soc.* **1992**, *114*, 7938. (d) Corey, E.J.; Sarshar, S.; Lee, D.H. *J. Am. Chem. Soc.* **1994**, *116*, 12089. (e) Maruoka, K.; Oishi, M.; Shiohara, K.; Yamamoto, H. *Tetrahedron* **1994**, *50*, 8983.

97. (a) Furuta, K.; Shimizu, S.; Miwa, Y.; Yamamoto, H. *J. Org. Chem.* **1989**, *54*, 1481. (b) Corey, E.J.; Loh, T.-P. *J. Am. Chem. Soc.* **1991**, *113*, 8966. (c) Corey, E.J.; Loh, T.-P.; Roper, T.D.; Azimioara, M.D.; Noe, M.C. *J. Am. Chem. Soc.* **1992**, *114*, 8290. (d) Ishihara, K.; Yamamoto, H. *J. Am. Chem. Soc.* **1994**, *116*, 1561. (e) Loh, T.-P.; Wang, R.-B.; Sim, K.-Y. *Tetrahedron Lett.* **1996**, *37*, 2989. (f) Hayashi, Y.; Rohde, J.J.; Corey, E.J. *J. Am. Chem. Soc.* **1996**, *118*, 5502.

98. (a) Kobayashi, S.; Hachiya, I.; Araki, M.; Ishitani, H. *Tetrahedron Lett.* **1993**, *34*, 3755. (b) Kobayashi, S.; Hachiya, I.; Ishitani, H.; Araki, M. *Tetrahedron Lett.* **1993**, *34*, 4535. (c) Kobayashi, S.; Araki, M.; Hachiya, I. *J. Org. Chem.* **1994**, *59*, 3758. (d) Kobayashi, S.; Ishitani, H.; Hachiya, I.; Araki, M. *Tetrahedron* **1994**, *50*, 11623. (e) Kobayashi, S.; Ishitani, H. *J. Am. Chem. Soc.* **1994**, *116*, 4083. (f) Kobayashi, S.; Ishitani, H.; Araki, M.; Hachiya, I. *Tetrahedron Lett.* **1994**, *35*, 6325.

99. Carmona, D.; Cativiela, C.; Garcia-Correas, R.; Lahoz, F.J.; Lamata, M.P.; Lopez, J.A.; Lopez-Ram de Viu, M.P.; Oro, L.A.; San Josè, E.; Viguri, F. *J. Chem. Soc., Chem. Commun.* **1996**, 1247.

100. Yamashita,Y.; Katsuki, T. *Synlett* **1995**, 829.

101. (a) Corey, E.J.; Imai, N.; Zhang, H.-Y. *J. Am. Chem. Soc.* **1991**, *113*, 728. (b) Khiar, N.; Fernandez, I.; Alcudia, F. *Tetrahedron Lett.* **1993**, *34*, 123.

102. (a) Evans, D.A.; Miller, S.J.; Lectka, T. *J. Am. Chem. Soc.* **1993**, *115*, 6460. (b) Evans, D.A.; Lectka, T.; Miller, S.J. *Tetrahedron Lett.* **1993**, *34*, 7027. (c) Evans, D.A.; Murry, J.A.; von Matt, P.; Norcross, R.D.; Miller, S.J. *Angew. Chem. Int., Ed. Engl.* **1995**, *34*, 798. (d) Davies, I.W.; Senanayake, C.H.; Larsen, R.D.; Verhoeven, T.R.; Reider, P.J. *Tetrahedron Lett.* **1996**, *37*, 1725. (e) Evans, D.A.; Kozlowski, M.C.; Tedrow, J.S. *Tetrahedron Lett.* **1996**, *37*, 7481. (f) Davies, I.W.; Gerena, L.; Castonguay, L.; Senanayake, C.H.; Larsen, R.D.; Verhoeven, T.R.; Reider, P.J. *J. Chem. Soc., Chem. Commun.* **1996**, 1753.

103. (a) Corey, E.J.; Ishihara, K. *Tetrahedron Lett.*, **1992**, *33*, 6807. (b) Desimoni, G.; Faita, G.; Righetti, P.P. *Tetrahedron Lett.*, **1996**, *37*, 3027. (c) Ordonez, M.; Guerrero-de la Rosa, V.; Labastida, V.; Llera, J.M. *Tetrahedron:Asymmetry*, **1996**, 7, 2675.

104. Evans, D.A.; Miller, S.J.; Lecta, T. *J. Am. Chem. Soc.* **1993**, *115*, 6460.

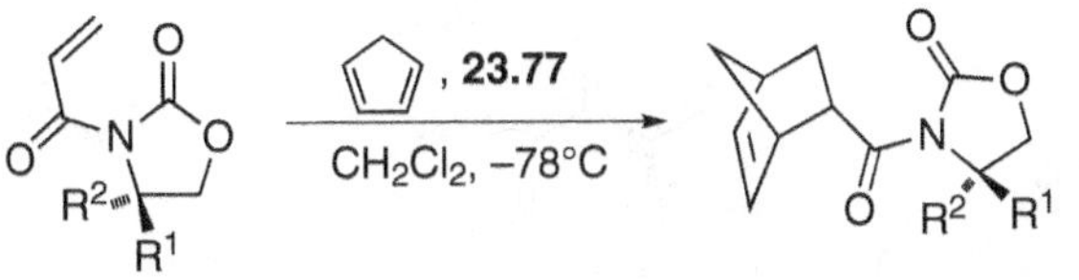

(23.91)

R^1= H, R^2 = CH_2Ph: 20% conversion, 36% e.e.
R^1= CH_2Ph, R^2 = H: 100% conversion, >99% e.e.

Asymmetric Aldol Reactions

The Evans aldol addition was one of the first C—C bond-forming reactions to give high levels of asymmetric induction. It should be no surprise, therefore, that it has been followed by efforts to develop a catalytic version of the reaction.[105] The same copper (II) BOX complexes developed by Evans for the asymmetric catalysis of the Diels-Alder reaction have also been used for asymmetric catalysis of the Mukaiyama aldol addition.[106]

The direct metal-catalyzed aldol addition—the reaction where no pre-derivatization of either participating compound is necessary—has been a rather more elusive target. However, reactions that mimic the enzyme aldolase, which catalyzes the condensation of an α-hydroxyketone with an aldehyde to give the α,β-dihydroxyketone, have allowed some success. Zinc-based catalysts have been developed that mimic the reactivity of the metalloenzyme (class II) aldolases in vitro. Shibasaki and his coworkers[107] have undertaken a systematic study of a series of complexes of dimeric BINOL ligands (e.g., Example **23.92**) with diethylzinc and found them to be effective catalysts for the aldolase-type aldol addition reaction of α-hydroxy-*o*-methoxyacetophenone with a wide range of aldehydes. The preformed catalyst is oligomeric, with two ligands and three zinc atoms in the complex (Example **23.93**), and the *o*-methoxy substituent in **23.94** is essential for high enantioselectivities. The reaction was found to favor the formation of the *syn* product (Example **23.95**) and to be strongly enantioselective, with e.e. values in the *syn* diol typically above 90% and e.e. values of the *anti* diol (Example **23.96**) above 80%.

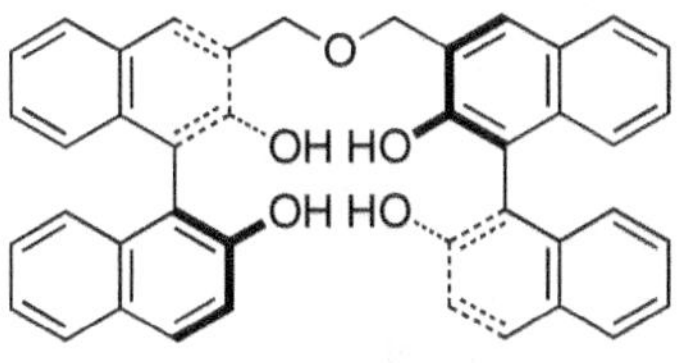

(23.92)

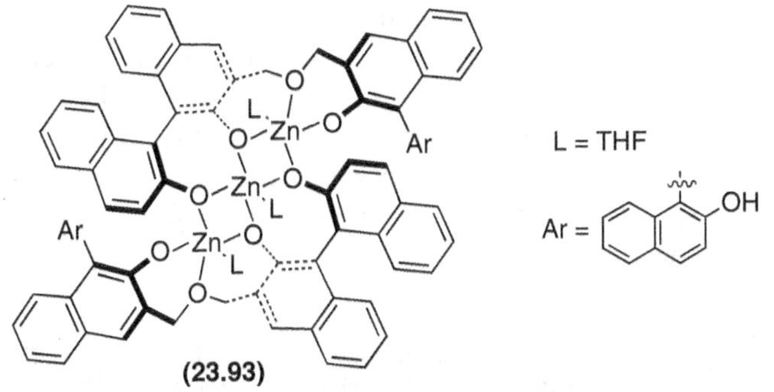

(23.93)

105. Review: Trost, B.M.; Brindle, C.S. *Chem. Soc. Rev.* **2010**, *39*, 1600.

106. (a) Evans, D.A.; Murry, J.A.; Kozlowski, M.C. *J. Am. Chem. Soc.* **1996**, *118*, 5814. (b) Evans, D.A.; Kozlowski, M.C.; Tedrow, J.S. *Tetrahedron Lett.* **1997**, *42*, 7481. (c) Evans, D.A.; Kozlowski, M.C.; Burgey, C.S.; MacMillan, D.W.C. *J. Am. Chem. Soc.* **1997**, *119*, 7893. (d) Evans, D.A.; MacMillan, D.W.C.; Campos, K.R. *J. Am. Chem. Soc.* **1997**, *119*, 10859. Reviews: (e) Pfalz, A. *Acc., Chem. Res.* **1993**, *26*, 339. (f) Ghosh, A.K.; Mathivanan, P.; Cappiello, J. *Tetrahedron:Asymmetry.* **1998**, *9*, 1. (g) Jorgensen, K. A.; Johannsen, M.; Yao, S.; Audrain, H.; Thorhauge, J. *Acc. Chem. Res.* **1999**, *32*, 605. (g) Johnson, J.S.; Evans, D.A. *Acc. Chem. Res.* **2000**, *33*, 325.

107. Kumagai, N.; Matsunaga, S.; Kinoshita, T.; Harada, S.; Okada, S.; Sakamoto, S.; Yamaguchi, K.; Shibasaki, M. *J. Am. Chem. Soc.* **2003**, *125*, 2169. Review: (b) Shibasaki, M.; Matsunaga, S. *J. Organomet. Chem.* **2006**, *691*, 2089.

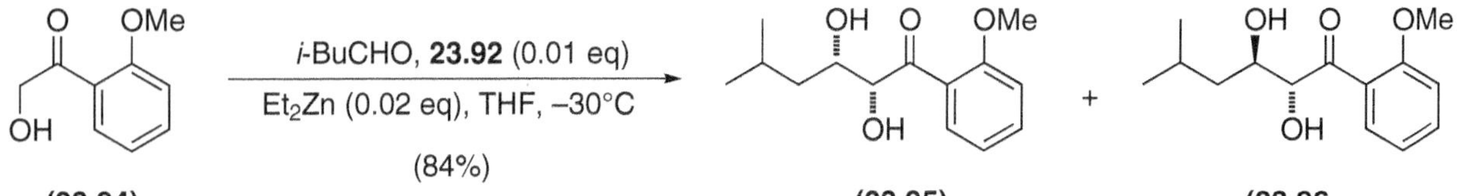

(23.94) **(23.95)** **(23.96**

	(23.95)		(23.96)
dr:	84	:	16
ee:	93%		87%

The same catalyst was also found to catalyze the addition of the same hydroxyacetophenone to imines to give chiral α-hydroxy-β-aminophenones.[108]

A similar phenolic ligand (Example **23.97**) developed by Trost has been successfully used in the zinc-catalyzed direct aldol addition reaction of aryl methyl ketones, with e.e. values varying from 53% with butyraldehyde to above 97% with branched aldehydes.[109] The use of this ligand in asymmetric synthesis is illustrated in Example **23.98**.

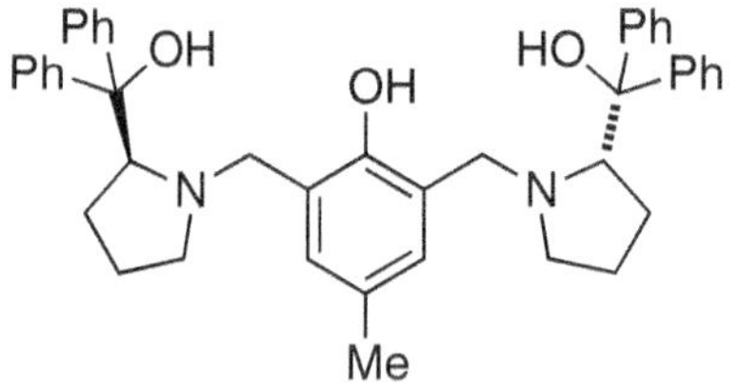

(23.97)

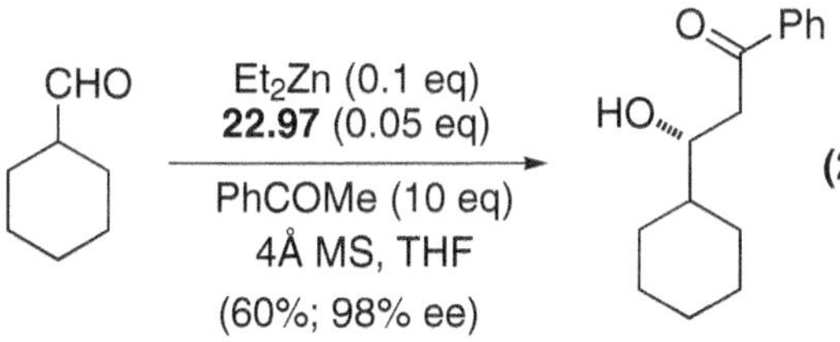

(22.98)

Organocatalysis

Until the end of the 20th century, the field of ***organocatalysis*** (catalysis of a reaction by a small-molecule, metal-free species) remained largely unexplored, but that situation changed around the beginning of the 21st century.[110] Since then, the development of organocatalysis has been one of the more explosive developments in the area of asymmetric catalysis[111]—one review[112] states that a SciFinder search of the literature to 2009 yielded 3,100 articles in which organocatalysis is a key word. Of course, it is impossible to cover so broad a field in the space available here, so the examples cited should be used as starting points for your own further searches.

---

108. (a) Matsunaga, S.; Kumagai, N.; Harada, S.; Shibasaki, M. *J. Am. Chem. Soc.* **2003**, *125*, 4712. (b) Matsunaga, S.; Yoshida, T.; Morimoto, H.; Kumagai, N.; Shibasaki, M. *J. Am. Chem. Soc.* **2004**, *126*, 8777.

109. Trost, B.M.; Ito, H. *J. Am. Chem. Soc.* **2000**, *122*, 12003.

110. Reviews: (a) Watson, A.J.B.; MacMillan, D.W.C. In Ojima, I, Ed., *Catalytic Asymmetric Synthesis* (John Wiley & Sons: New York, 2010), Ch. 2A, p. 39. (b) Yang, Y.; Deng, L. In Ojima, I, Ed., *Catalytic Asymmetric Synthesis* (John Wiley & Sons: New York, 2010), Ch. 2B, p. 59. (c) Shirakawa, S.; Maruoka, K. In Ojima, I, Ed., *Catalytic Asymmetric Synthesis* (John Wiley & Sons: New York, 2010), Ch. 2C, p. 95. (d) Pellisier, H. *Tetrahedron* **2007**, *63*, 9267. (e) Mukherjee, S.; Yang, J.W.; Hoffmann, S.; List, B. *Chem. Rev.* **2007**, *107*, 5471. (f) Dalko, P.I.; Moisan, L. *Angew. Chem. Int. Ed.* **2004**, *43*, 5138.

111. Reviews: (a) Gröner, H.; Vogl, E.M.; Shibasaki, M. *Chem. Eur. J.* **1998**, *4*, 1i37. (b) Notz, W.; Tanaka, F.; Barbas, C.F., III *Acc. Chem. Res.* **2004**, *37*, 580.

112. Zhong, C.; Shi, X. *Chem. Eur. J.* **2010**, 2999.

It has been known since the 1970s that proline catalyzes the asymmetric aldol condensation[113] and that the reaction likely proceeds via an enamine intermediate.[114] The reaction of acetone with a variety of aromatic aldehydes (**23.99**) gives acceptable chemical yields and modest e.e. values. The corresponding reaction with isobutyraldehyde gives high chemical (97%) and optical (96% e.e.) yields. In the same study, it was shown that both functional groups of the proline—the acid and the base—are important in the catalysis.[115]

Basic Types of Organocatalysis

The largest number of organocatalytic reactions are based on activation of a carbonyl precursor as an electrophile in one of three ways: (1) by conversion to an iminium ion (**23.100**); (2) by hydrogen bonding of the carbonyl oxygen to a Brønsted acid (**23.101**—this is effectively the mode of activation by transition metal catalysts [**23.102**]—and becomes complete transfer of the proton to the carbonyl oxygen in the limiting case); or (3) by enhancement of the nucleophilicity of the α carbon by conversion to an enamine or a similar species (**23.103**). These options are illustrated in Figure 23.21. In every case, the design of the chiral catalyst requires that the activated carbonyl group be constrained to a single conformation in which access to one face of the complex is blocked.

Two similar transition state models for the proline-catalyzed aldol addition (**23.104**) have been proposed. In the earlier model (Example **23.105**), proposed by List,[116] there is a bifurcated hydrogen bond to both the amine nitrogen and the aldehyde oxygen. Yet in a later model (Example **23.106**), based on that proposed for the intramolecular reaction by Houk,[117] this hydrogen bond is absent. What both these models illustrate, however, is the finding that *both nucleophilic (the secondary amine) and electrophilic (the carboxylic acid) groups are essential for good organocatalyst reactivity.*

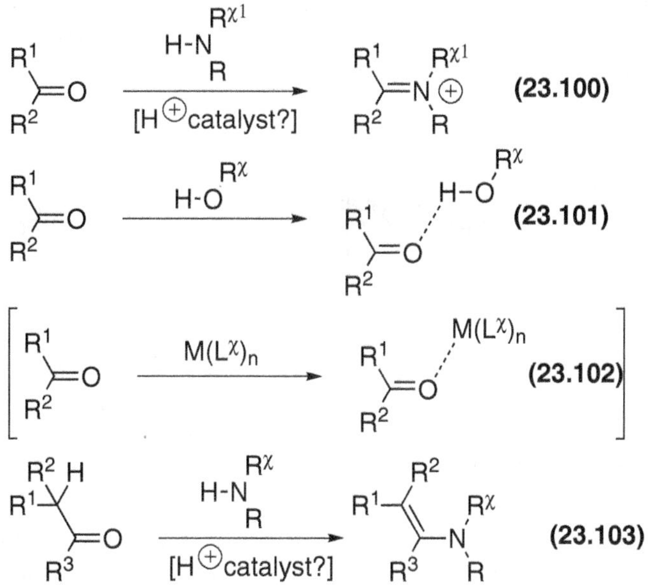

Figure 23.21 Basic activation of a carbonyl group by chiral catalysis

113. (a) Hajos, Z.G.; Parrish, D.R. *J. Org. Chem.* **1974**, *39*, 1615. (b) Eder, U.; Sauer, G.; Wiechert, R. *Angew. Chem., Int. Ed. Engl.* **1971**, *10*, 496. (c) Agami, C.; Platzer, N.; Sevestre, H. *Bull. Soc. Chim. Fr.* **1987**, *2*, 358.

114. (a) Agami, C.; Puchot, C.; Sevestre, H. *Tetrahedron Lett.* **1986**, *27*, 1501. (b) Agami, C. *Bull. Soc. Chim. Fr.* **1988**, *3*, 499.

115. List, B.; Lerner, R.A.; Barbas, C.F., III *J. Am. Chem. Soc.* **2000**, *122*, 2395.

116. List, B.; Lerner, R.A.; Barbas, C.F., III *J. Am. Chem. Soc.* **2000**, *122*, 2395.

117. (a) Bahmanyar, S.; Houk, K.N. *J. Am. Chem. Soc.* **2001**, *123*, 9922. (b) Bahmanyar, S.; Houk, K.N. *J. Am. Chem. Soc.* **2001**, *123*, 11273, 12911. (c) Allemann, C.; Gordillo, R.; Clemente, F.R.; Cheong, P.H.-Y.; Houk, K.N. *Acc. Chem. Res.* **2004**, *37*, 558. (d) Clemente, F.R.; Houk, K.N. *J. Am. Chem. Soc.* **2005**, *127*, 11294.

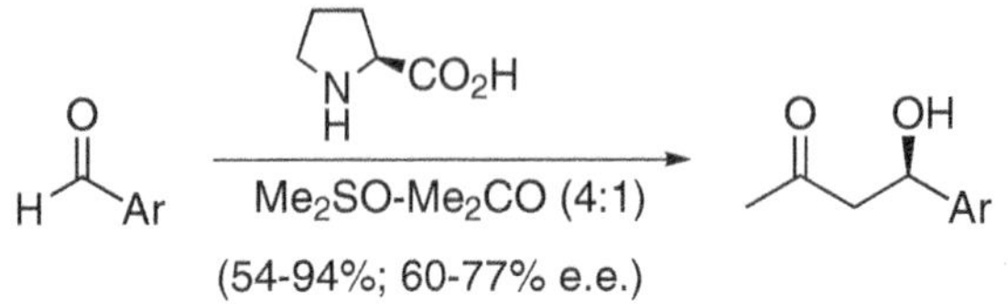

(23.104)

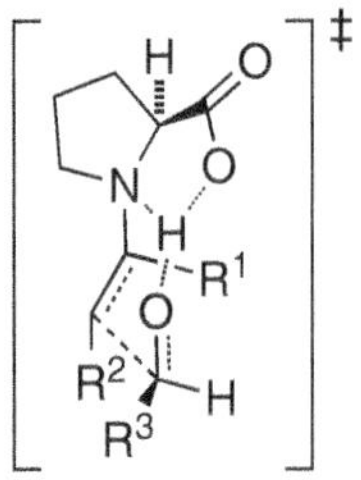

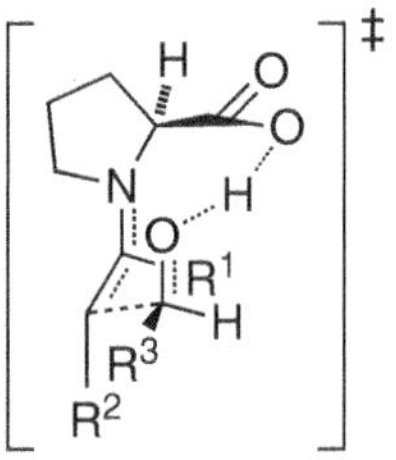

(23.105) (23.106)

This particular feature has become an important part of the design of catalysts for the aldol addition, in particular. One basic design of an organocatalyst for asymmetric aldol-type addition reactions is that shown in Example **23.107**. In this design, the secondary amine is drawn as being in a ring, but any secondary amine that exists in a stereochemically rigid relationship with respect to the electrophilic hydrogen is permissible. The group E—H may be any electrophilic group: an alcohol, a phenol, a carboxylic acid, or a carboxamide; a phosphoric, phosphonic, or phosphinic acid derivative; or a sulfonamide. Most of the organocatalysts based on this model are based on a pyrrolidine (five-membered) ring. Two such organocatalysts developed by Barbas (Example **23.108**, where the electrophilic group is a carboxylic acid, and Example **23.109**, where the electrophilic group is an ammonium ion)[115,118] show this pattern. Organocatalyst **23.109** has been shown to be a good bifunctional catalyst for the aldol addition reactions of α-branched aldehydes with aromatic aldehydes.[119] Organocatalysts **23.110** and **23.111** are diastereoisomers. They are both derived from the same enantiomer of natural hydroxyproline but differ in the absolute configuration of the α carbon of the acid, which means that opposite enantioselectivity can be expected—and is observed[120]—when they are used.

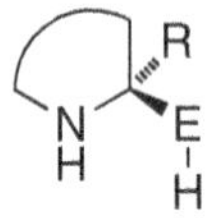

ring size: usually 5
E = O, ArO, CO_2, CONR, $(RO)_2PO_2$, $RO(R)PO_2$, R_2PO_2, CR_2NCOR, CR_2NSO_2R

(23.107)

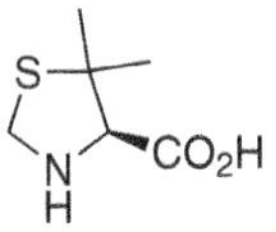

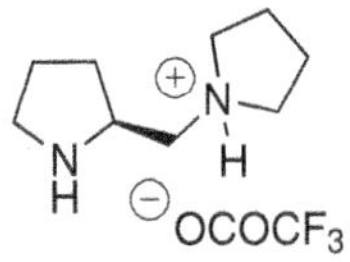

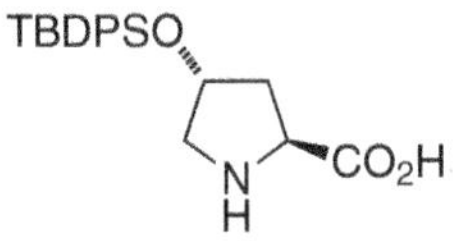

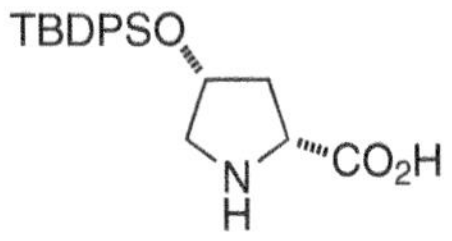

(23.108) (23.109) (23.110) (23.111)

Although proline and proline analogs work well as organocatalysts in organic solvents, they tend to be much less effective in aqueous environments. In aqueous solvents, however, proline amides and sulfonamides work quite well as organocatalysts for asymmetric aldol additions. The range of groups that have been used may be gauged from the structures in Figure 23.22. It is interesting to note that although many of these organocatalysts are amides of C_2-symmetric diamines, only the binaphthyl derivative **23.116** has a formal C_2 axis of symmetry. Catalysts **23.115**, **23.117**, and **23.118** all differ in the acyl groups

118. Sankthivel, K.; Notz, W.; Bui, T.; Barbas, C.F., III *J. Am. Chem. Soc.* **2001**, *123*, 5260.
119. Mase, N.; Tanaka, F.; Barbas, C.F., III *Org. Lett.* **2003**, *5*, 4369; *Angew. Chem. Int. Ed.* **2004**, *43*, 2420.
120. Itagaki, N.; Kimura, M.; Sugahara, T.; Iwabuchi, Y. *Org. Lett.* **2005**, *7*, 4185.

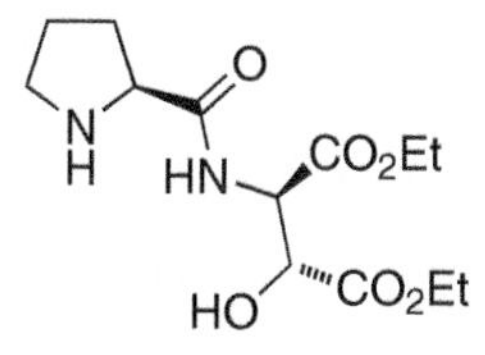

(23.112)

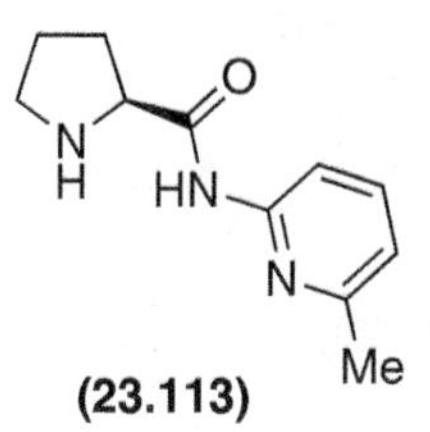

(23.113)

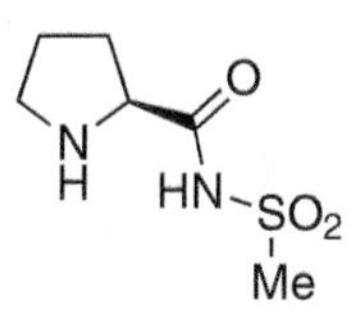

(23.114)

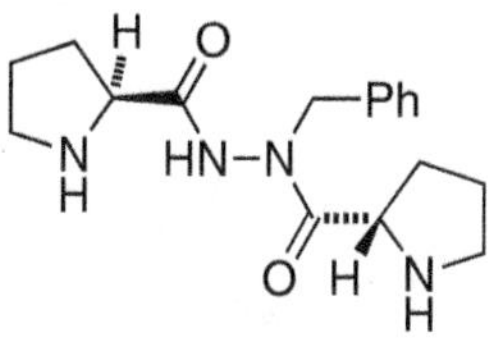

(23.115)

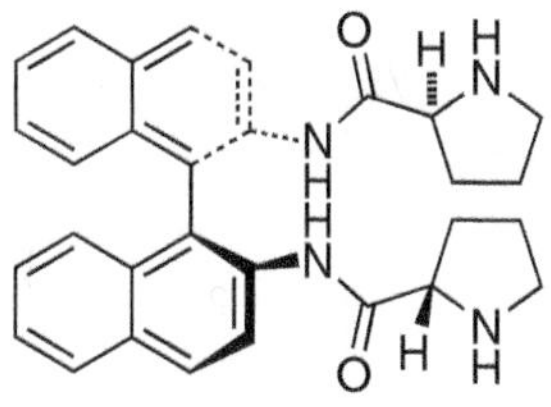

(23.116)

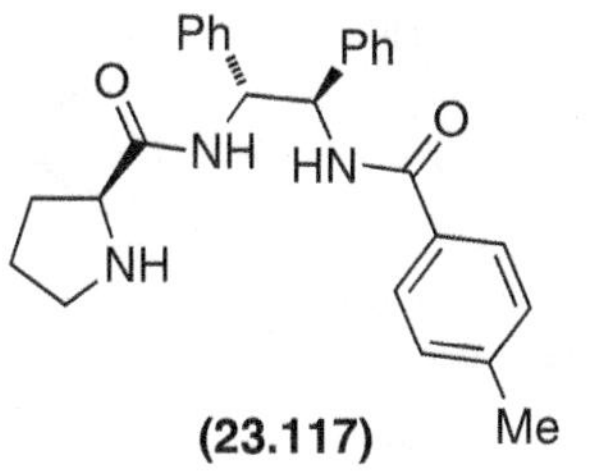

(23.117)

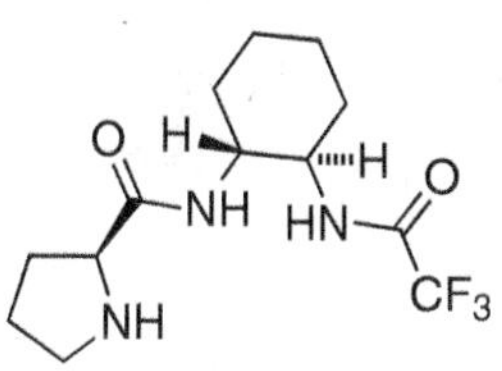

(23.118)

Figure 23.22 Proline amide organocatalysts for aldol additions in aqueous solution

of the diamide, which removes the overall symmetry of the molecule, although the local symmetry of the amide functional groups is largely preserved. However, the ability to replace the proline at one end of the molecule by another acyl group allows one to "tune" the catalyst activity—the aldol additions catalyzed by **23.117** in tetrahydrofuran (THF)-acetic acid proceed with both high yield and exceptionally high e.e. (≥ 99%). The role of the amide groups in ordering the reacting species is shown by the transition state model **23.119**, which shows the reaction between acetone and pyruvic acid to give a chiral tertiary alcohol. Note how the chirality of the proline subunit controls the absolute configuration of the new chiral center.

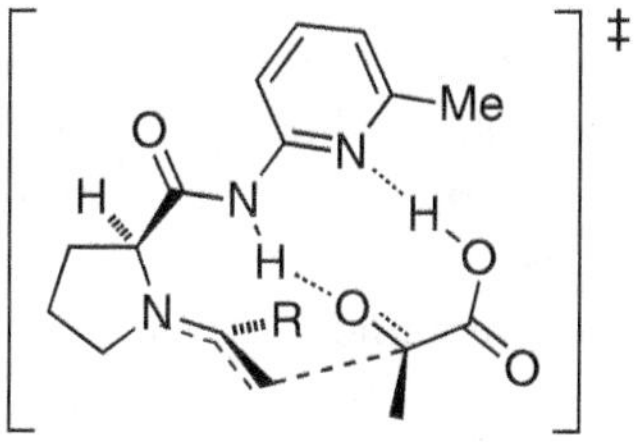

(23.119)

The ability to activate carbonyl groups as iminium ions, in a reaction reminiscent of similar activation in biological systems, has been one of the most enduring features of successful organocatalysts. Among the first such compounds specifically designed for use as catalysts was the MacMillan imidazolidinone **23.120**. In addition to the initial report, which concerned the Diels-Alder cycloaddition of aldehydes,[121] these organocatalysts have shown their versatility by being applied to the Diels-Alder reaction with ketones,[122] 1,3-dipolar cycloaddition reactions,[123] formal Friedel-Crafts substitutions in electron-rich aromatic compounds,[124] and in Mukaiyama-Michael reactions of silyloxyfurans (e.g., Example **23.121**).[125] The success of these organocatalysts is due to the steric constraints placed on the conformation of the iminium ion (Example **23.122**) by the adjacent *gem*-dimethyl moiety, which must adopt a conformation where the benzyl group selectively shields one face of the conjugated system (π-stacking).

121. Ahrendt, K.T.; Borths, C.J.; MacMillan, D.W.C. *J. Am. Chem. Soc.* **2000**, *122*, 4243.

122. Northrup, A.B.; MacMillan, D.W.C. *J. Am. Chem. Soc.* **2002**, *124*, 2458.

123. Jen, W.S.; Wiener, J.J.M.; MacMillan, D.W.C. *J. Am. Chem. Soc.* **2000**, *122*, 9874.

124. (a) Paras, N.A.; MacMillan, D.W.C. *J. Am. Chem. Soc.* **2001**, *123*, 4370; **2002**, *124*, 7894. (b) Austin, J.F.; MacMillan, D.W.C. *J. Am. Chem. Soc.* **2002**, *124*, 1172.

125. Brown, S.P.; Goodwin, N.C.; MacMillan, D.W.C. *J. Am. Chem. Soc.* **2003**, *125*, 1192.

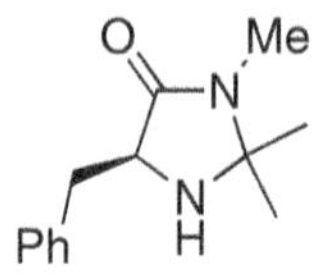

(23.120)

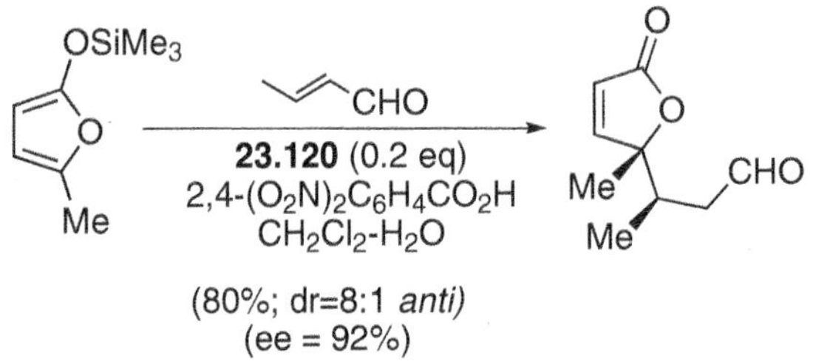

(23.121)

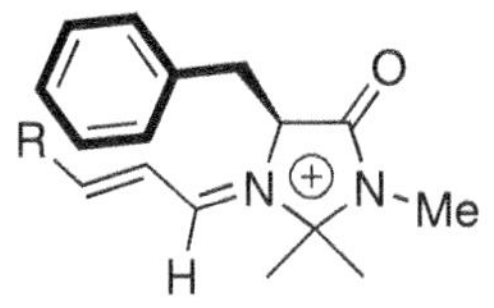

(23.122)

23.5 Representative Asymmetric Syntheses

The advice given in Chapter 15, when we began our study of organic synthesis, remains true here. The only real way to develop a feel for the process of organic synthesis and to begin to develop your own style is to study examples of successful syntheses. A number of enjoyable compendia[126] have been published over the years, and these can be consulted to see how practitioners of the art have approached the problem of total synthesis of complex molecules over time. There are several targets that stand out as being particularly attractive to organic chemists as target compounds. They are:

1. Strychnine,[127] for which there are now no fewer than ten total syntheses following Woodward's first synthesis of (±)-strychnine
2. Reserpine,[128] for which there are also ten total syntheses discussed in a recent review
3. Cortisone, for which the effects of chiral cationic catalysts on the stereochemistry of several classical syntheses have been recently explored by Corey[129]

The examples chosen for this chapter have been chosen less for their status as iconic syntheses and more for what the author perceives as their instructive value. The iconic achievements in synthesis have been well covered in the various compendia of syntheses, and you should consult these books frequently as you study the art of organic synthesis.

---

126. (a) Nicolaou, K.C.; Sorensen, E.J. *Classics in Total Synthesis* (Wiley-VCH: Weinheim, 1996). (b) Nicolaou, K.C.; Snyder, S.A. *Classcs in Total Synthesis II.* (Wiley-VCH: Weinheim, 2003). (c) Nicolaou, K.C.; Chen, J.S. *Classics in Total Synthesis III.* (Wiley-VCH: Weinheim, 2011). (d) Corey, E.J.; Chen, X.-M. *The Logic of Chemical Synthesis.* (e) The Elsevier series, *Strategy and Tactics in Total Synthesis* now contains eight volumes under the editorship of Thomas Lindberg, and then Michael Harmata. (f) Hudlický, T.; Reed, J.W. *The Way of Synthesis* (Wiley-VCH: Weinheim, 2007).

127. For a review discussing the various total syntheses of strychnine from a historical standpoint, see: Cannon, J.S.; Overman, L.E. *Angew. Chem. Int. Ed.* **2012**, *51*, 4288.

128. For an informative review discussing the various total syntheses of reserpine from a historical standpoint, see: Chen; F.-E.; Huang, J. *Chem. Rev.* **2005**, *105*, 4671.

129. Hu, Q.-Y.;Zhou, G.; Corey, E.J. *J. Am. Chem. Soc.* **2004**, *126*, 13708.

The Enantioselective Total Synthesis of Salicylihalamides A and B

Let us first examine the 2001 synthesis of the cytotoxic macrolides salicylihalamides A and B (Example **23.123**, with the enamide double bond being either *E* or *Z*).[130] An examination of these molecules reveals the two key structural features that must be addressed by the enantioselective synthesis: (1) there is a 12-membered unsaturated lactone ring containing a *trans* double bond at C-3 and three chiral centers, at C-12, C-13, and C-15; and (2) there is a (2*Z*,4*Z*)-2,4-heptadienoyl acyl enamide that can be either *E* or *Z* at the enamine double bond (C-17). There is a 1,2-*anti* relationship (which appears to be *syn* because of the conformation of the ring) between the substituents on the two adjacent chiral centers at C-12 and C-13 in the ring.

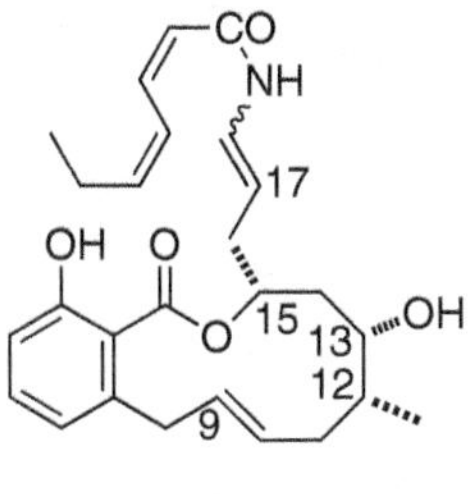

(23.123)

The retrosynthetic analysis of these target molecules is shown in Figure 23.23 and the synthesis in Figure 23.24. Only the key steps of the retrosynthetic analysis are shown in Figure 23.23, as has become the norm over the years since the process was first formalized by Corey. There are four especially critical transforms that have been highlighted. The first of these a retrometathesis bond disconnection (DIS) transform that disconnects the macrocycle (**23.125**) into a pair of terminal olefins (**23.126**). This step is key and reflects the dominant position of the ring-closing metathesis (RCM) in the synthesis of macrocycles. The second is a functional group interchange (FGI) transform corresponding to a retro-Mitsunobu esterification of a chiral alcohol (**23.127**) to give the ester with the opposite configuration (**23.126**). This reaction fixes the stereochemistry of the lactone oxygen, and this determines which enantiomer of the starting material will be required. The third is another DIS transform: a retrocrotylboration transform of the 12,13-*anti* homoallylic alcohol (**23.128**). Using a chiral borane makes its possible to control the absolute configuration

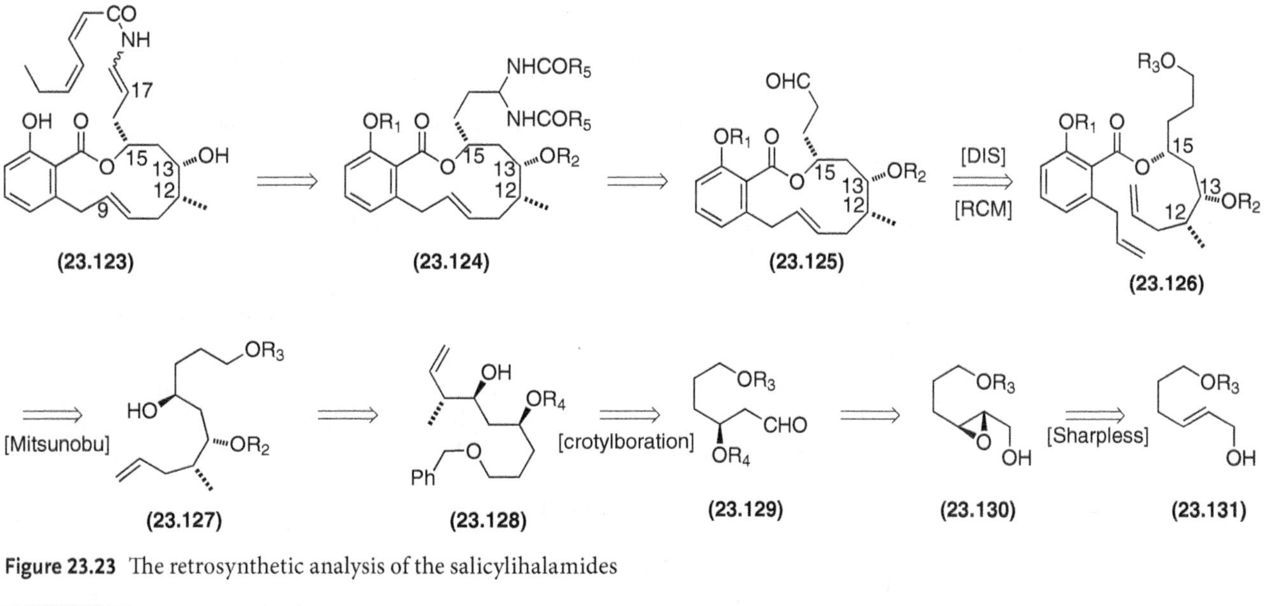

Figure 23.23 The retrosynthetic analysis of the salicylihalamides

130. Labrecque, D.; Charron, S.; Rej, R.; Blais, C.; Lamothe, S. *Tetrahedron Lett.* **2001**, *42*, 2645.

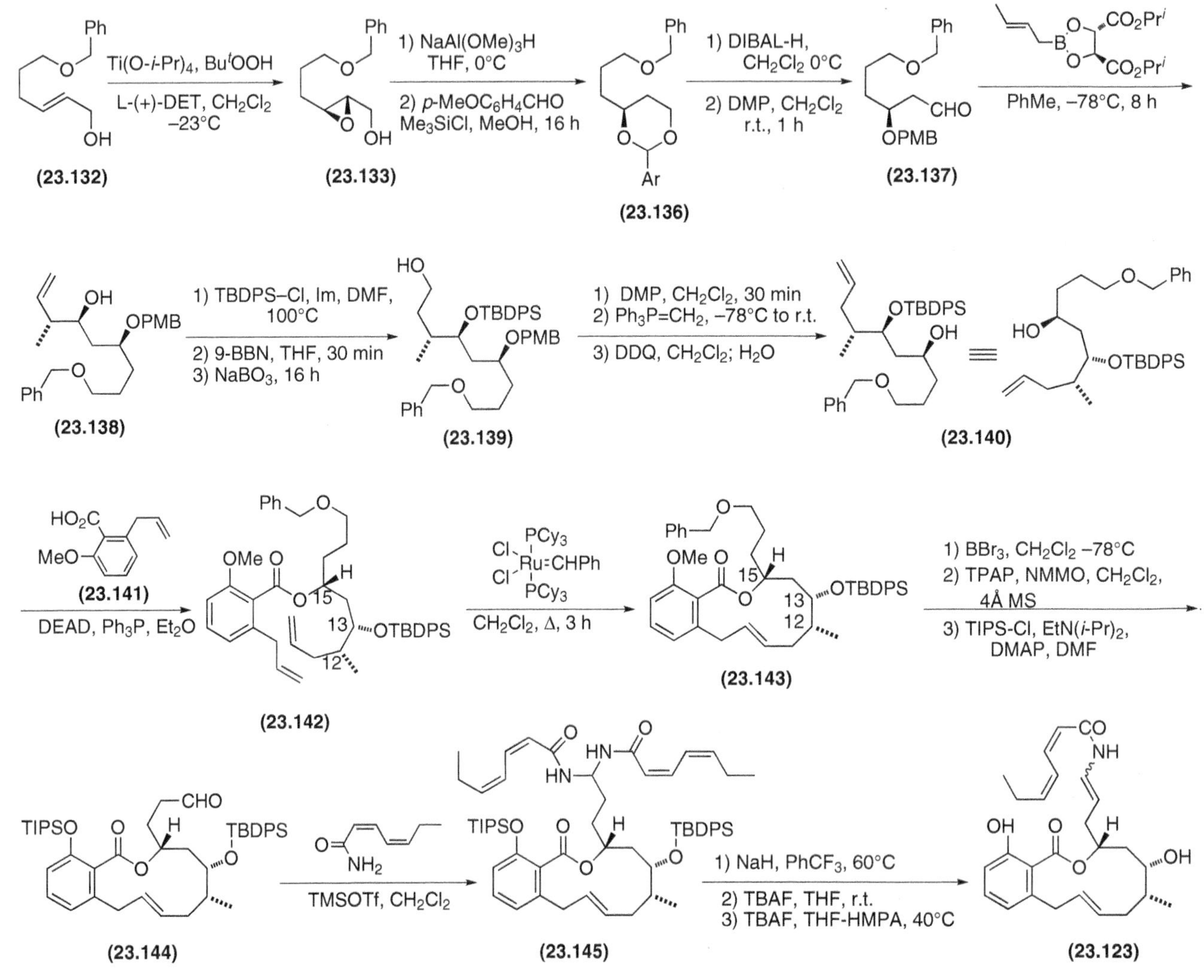

Figure 23.24 The synthesis of salicylihalamides A and B

of the *anti* alcohol formed. The final critical transform is another FGI transform: a retro-Sharpless asymmetric epoxidation. The Sharpless asymmetric epoxidation of allylic alcohols has become a mainstay of asymmetric synthesis, and in this molecule it establishes the chirality of the lactone oxygen.

The synthesis in Figure 23.24 begins, as have many enantioselective syntheses since 1981, with the Sharpless asymmetric epoxidation of a *trans* allyl alcohol. In the presence of L-(+)-diisopropyl tartrate as the chiral ligand, the epoxide **23.133** is obtained as the major product of the reaction. In this step, the absolute configuration of the chiral center at C-15 is fixed. The reduction of the epoxide was carried out by means of a directed hydride reduction: the aluminum reagent exchanges an alkoxide ligand with the epoxyalcohol to give the complex aluminate **23.134.** This places the hydride in a position (5-*exo-tet*) to reduce the proximal bond of the epoxide to give the 1,3-diol derivative **23.135.** Protection of the resultant 1,3-diol with p-methoxybenzaldehyde leads to the acetal **23.136.**

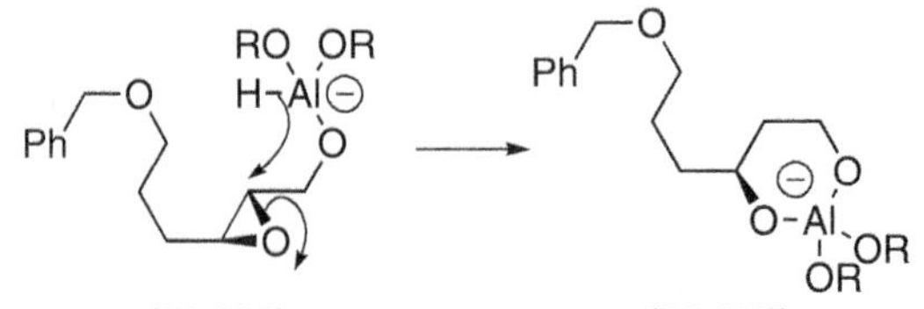

(23.134) (23.135)

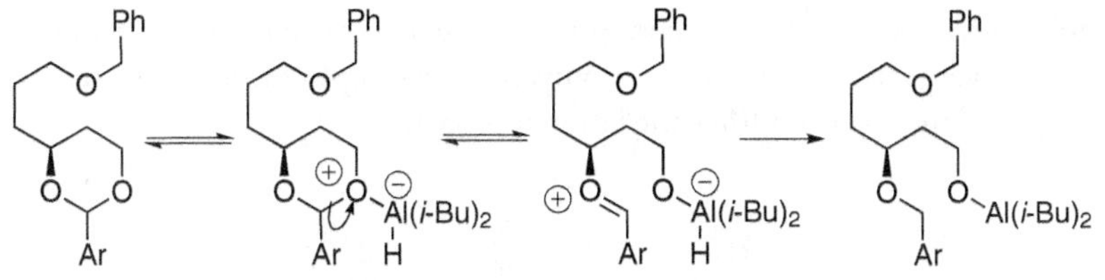

(23.136) (23.146) (23.147) (23.148)

Figure 23.25 Diisobutylaluminium hydride (DIBAL-H) reduction of the acetal **23.68**

Reductive cleavage of **23.136** with the electrophilic reducing agent, diisobutylaluminum hydride (DIBAL-H), results in cleavage of one of the two benzylic C—O bonds (Figure 23.25). The observed product (**23.148**) corresponds to that obtained when the aluminum binds to the less hindered of the two acetal oxygens to give **23.146.** Its formation is almost certainly aided by the additional stabilization of the oxonium ion **23.147** by the *p*-methoxy substituent on the aromatic ring.

The primary alcohol obtained from the reduction is then oxidized with the Dess-Martin periodinane to give the aldehyde (**23.137**) that is then subjected to the Roush crotylboration to give the 12,13-*anti* homoallylic alcohol **23.138**. The absolute configurations of the two new chiral centers (C-12 and C-13) are fixed by the chirality of the borane reagent, as we discussed earlier in this chapter. The homologation of the alkene was accomplished by sequential protection of the alcohol as the *tert*-butyldiphenylsilyl (TBDPS) ether and hydroboration-oxidation of the terminal alkene to give the primary alcohol **23.139.** Oxidation with the Dess-Martin periodinane and Wittig methylenation then gave the homologous terminal alkene. Oxidative deprotection of the *p*-methoxybenzyl (PMB) ether then provided the secondary alcohol **23.140.**

This secondary alcohol has the wrong stereochemistry at C-15 for its eventual conversion to the salicylihalamides, but this was rectified during the coupling reaction to incorporate the aromatic ester of the target by using the acid **23.141** in a Mitsunobu esterification, because the necessary inversion of configuration was accomplished at the same time. The resultant ester **23.142** was then subjected to RCM using the Grubbs II catalyst to give the precursor **23.143** containing the complete macrocyclic ring system.

The final product of this synthesis carries a highly acid-sensitive *N*-acyl enamine group, so at this point in the synthesis, it was necessary to replace the aromatic methoxy group by one that did not require such strongly acidic conditions for its cleavage. Cleavage of both ether groups with boron tribromide was followed by oxidation of the primary alcohol to the aldehyde with tetrapropylammonium perruthenate (TPAP). Protection of the phenol as the triisopropylsilyl ether then gave the aldehyde **23.144**. The reaction of this aldehyde with (2*Z*,4*Z*)-heptadienamide catalyzed by trimethylsilyl triflate as an oxophilic Lewis acid led to the *gem*-diamide **23.145**. The *gem*-diamide was then converted into the *N*-acyl enamine by elimination refluxing with sodium hydride in trifluorotoluene, a solvent that has been recommended as a replacement for dichloromethane.[131] In this particular synthesis, the desilylation had to be effected in two stages to obtain a good yield of the target compound.

The Enantioslective Synthesis of Aigialomycin D

In the previous synthesis, the absolute stereochemistry in the final product was established by a Sharpless asymmetric epoxidation early in the synthesis and by a Roush asymmetric crotylboration later in the sequence. The other approach to fixing absolute stereochemistry is what Hanessian called the ***chiron approach***.[132] In this approach, at least some of the chiral elements of the final product are derived from naturally occurring

131. Ogawa, A.; Curran, D.P. *J. Org. Chem.* **1997**, *62*, 450.
132. Hanessian, S. *Total Synthesis of Natural Products: The 'Chiron' Approach* (Pergamon: Oxford, 1983).

chiral molecules of known absolute configuration, such as carbohydrates and amino acids. The highly convergent synthesis of aigialomycin D (Example **23.149**) reported by Harvey and coworkers[133] nicely exemplifies the latter approach.

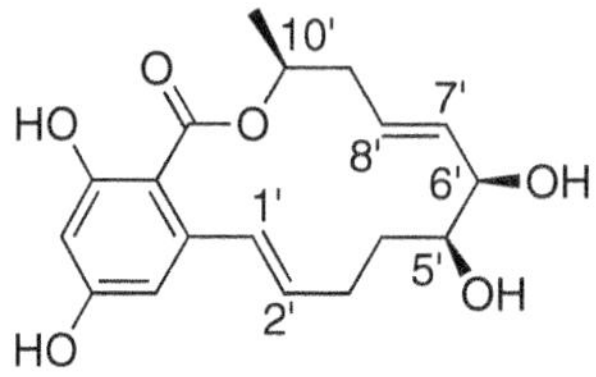

(23.149)

The retrosynthetic analysis of the target molecule is shown in Figure 23.26. In this synthesis, unlike the synthesis of the salicylhalamides just discussed, the chirality at C-5', C-6' and C-10' is fixed by using chiral starting compounds of known absolute configuration—the chiral centers at C-5' and C-6' are derived from D-ribose, and the alcohol **23.155** is commercially available. The fact that so many chiral compounds are now commercially available as single enantiomers is yet another measure of the impact of asymmetric synthesis on modern synthetic chemistry.

There are two double bonds in the macrocyclic lactone, and, as has become almost routine in modern synthesis, one of the two is generated by means of an RCM. However, as specified in Figure 23.26, the triene **23.151** cannot be used as a substrate in the macrolactonization by RCM (why not?). Consequently, in this synthesis, it is the C-7'—C8' double bond that is formed by RCM. The C1'—C2' double bond could have been introduced any number of ways, but the authors chose to use a Ramberg-Bäcklund reaction for this purpose. It is represented in the retrosynthetic analysis as a bond reconnection transform (**23.149** ⇒ **23.150**), which would conform to the known intermediacy of the episulfone in the reaction. The ring-opened synthon **23.152** can be readily disconnected into the ribose-derived synthon **23.153**; the orsellinic acid derivative **23.154**; and, as stated previously, the commercially available alcohol **23.155**.

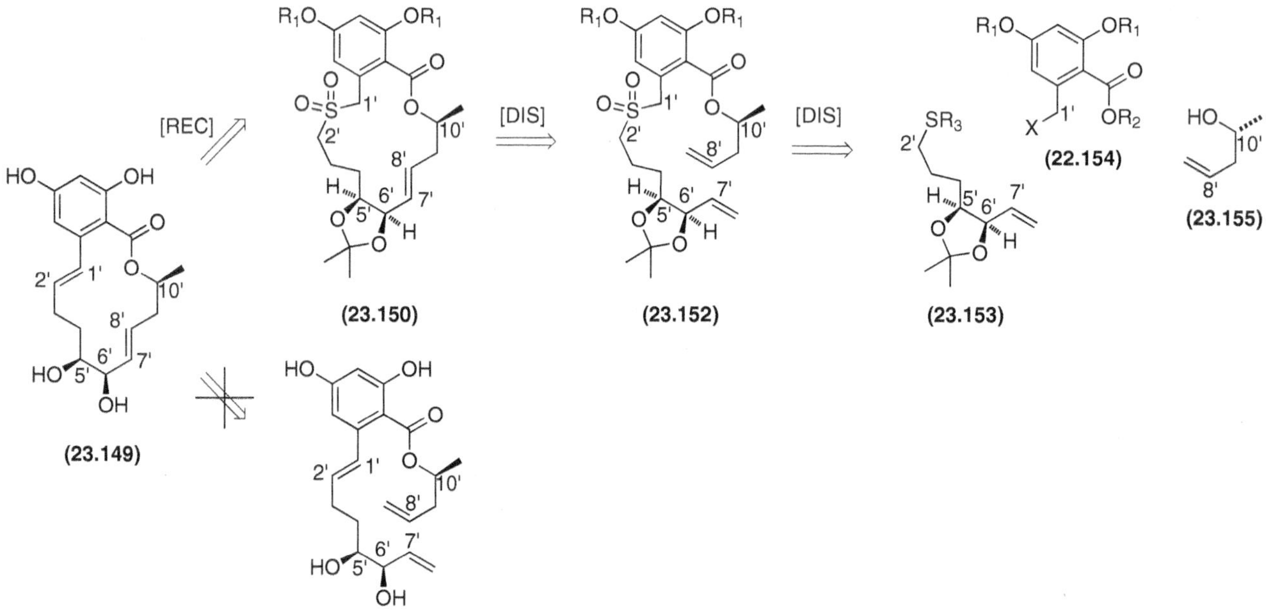

Figure 23.26 Retrosynthetic analysis of the Harvey synthesis of aigialomycin

133. Baird, L.J.; Trimmer, M.S.M.; Teesdale-Spittle, P.H.; Harvey, J.E. *J. Org. Chem.* **2009**, *74*, 2271.

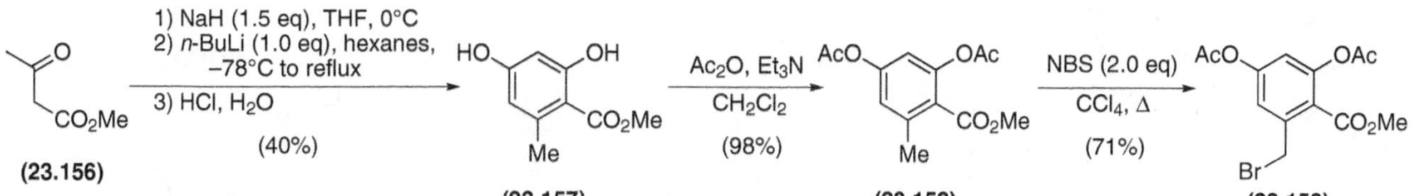

(23.156) (23.157) (23.158) (23.159)

Figure 23.27 Synthesis of the orsellinic acid synthon

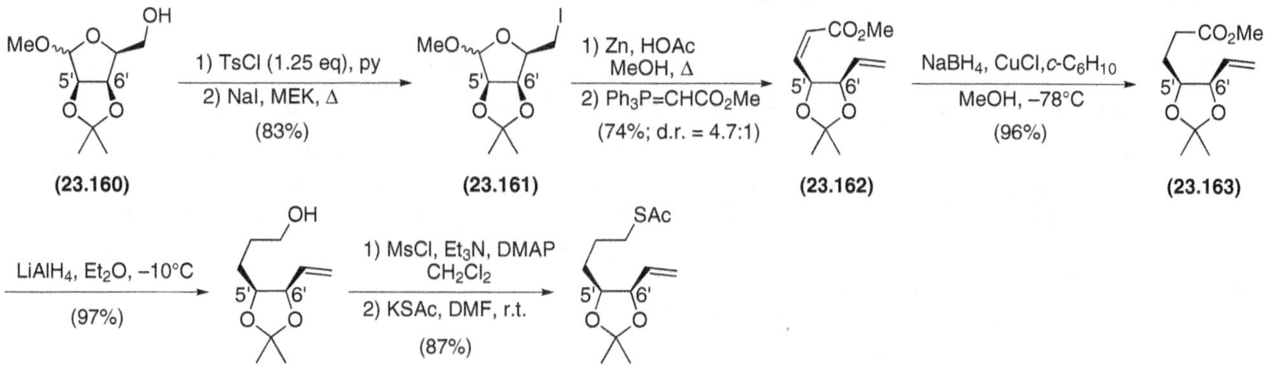

(23.160) (23.161) (23.162) (23.163) (23.164) (23.165)

Figure 23.28 Synthesis of the ribose-based thiol synthon

Problem

23-6 Why is the triene 23.151 an unsuitable substrate for the RCM for the formation of the macrocyclic lactone ring of aigialomycin?

The two starting materials that are not commercially available are the mercaptan **23.153** (R_3=H) and the orsellinic acid derivative **23.154**. The synthesis of the orsellinic acid synthon was carried out as shown in Figure 23.27. The base-promoted self-condensation of methyl acetoacetate (**23.156**) gave methyl orsellinate (**23.157**) in relatively low yield. However, the fact that this reaction required only a single step and had very inexpensive starting materials makes it a useful in this synthesis. Protection as the diacetate **23.158** and allylic bromination with *N*-bromosuccinimide gave the benzylic bromide **23.159**.

The ribose-derived synthon was prepared (Figure 23.28) from D-ribose by initial protection as the acetonide of methyl riboside **23.160**. This protected ribose was then converted to the iodide **23.161** by tosylation and iodide displacement in refluxing methyl ethyl ketone. Reductive elimination of the resulting iodide with zinc and acetic acid in methanol gave a vinyl aldehyde that was immediately converted to a 4.7:1 *Z*:*E* mixture of the conjugated ester **23.162** by Wittig reaction with a stabilized ylide.

The next step of the synthesis required the reduction of the electrophilic enoate double bond without concomitant reduction of the vinyl double bond or hydrogenolysis of the ether at C-5'. This required a nucleophilic hydride reducing agent, with sodium borohydride being the most logical choice. The conditions for the reduction in this case required considerable experimentation. However, finally the combination of sodium borohydride and copper (I) chloride in the presence of excess cyclohexene to prevent over-reduction proved to be the reagent of choice; without the cyclohexene, there was significant over-reduction. With this reagent, the mixed enoate esters were reduced to the ester **23.163**. Lithium aluminum hydride reduction of the saturated ester gave the primary alcohol **23.164**. This was then converted into the thioester **23.165** by sequential treatment with methanesulfonyl chloride and potassium thioacetate.

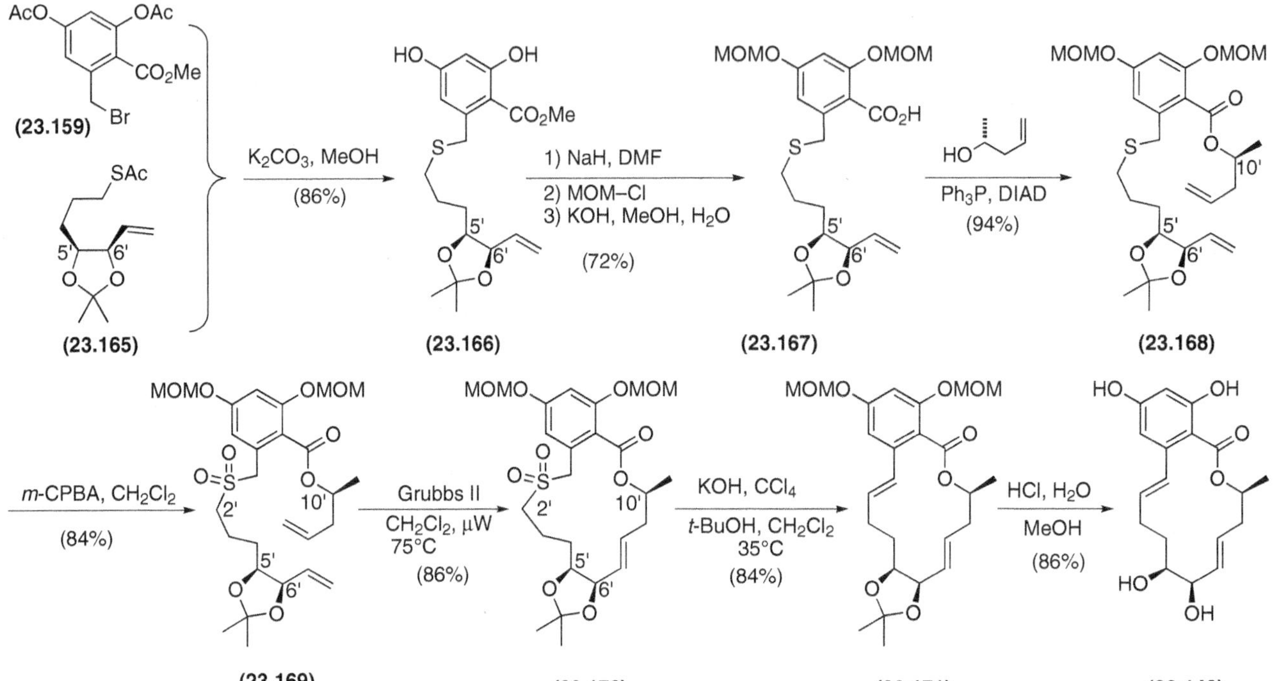

Figure 23.29 Completion of the synthesis of aigialomycin

The coupling of the three synthons and the completion of the synthesis is shown in Figure 23.29. The choice of the thioacetate **23.165** as the thiol synthon became necessary because the obvious synthon—the thiol itself—was unstable to storage. The coupling of the halide **23.159** and the thioester **23.165** was accomplished by displacing the bromide with the thiolate anion revealed in situ by base hydrolysis of the thioester. The phenolic hydroxyl groups of the resultant sulfide (**23.166**) were protected as the methoxymethyl (MOM) ethers (**23.167**) prior to base hydrolysis of the methyl ester. When the hydrolysis was attempted without protecting the phenolic groups, decarboxylation was the major reaction.

The Mitsunobu coupling of *R*-4-penten-2-ol with the acid **23.167** gave the ester **23.168**, with the expected inversion of configuration at what becomes C-10' of the target compound. Oxidation of the sulfide **23.168** to the sulfone **23.169** achieved two major goals: (1) it provided the sulfone needed for the later Ramberg-Bäcklund reaction and (2) it removed the thioether as a potential ligand (poison) for the ruthenium in the metathesis catalyst. The RCM under microwave irradiation provided the macrocyclic lactone **23.170**, which was then subjected to a one-pot Ramberg-Bäcklund procedure to give the protected aigialomycin derivative **23.171**. Acid hydrolysis of the protecting groups completed the synthesis.

Problem

23-7 Suggest a reasonable mechanism for the decarboxylation of the resultant carboxylic acid during attempted base hydrolysis of the ester **23.166**.

The Enantioselective Synthesis of Chinesiolide B

Chinensiolide B (Example **23.172**) is a sesquiterpene lactone that has also been synthesized by a strategy where chirality is incorporated by using an optically pure natural product of known absolute configuration as the starting compound. In this synthesis,[134] the

134. Elford, T.G.; Hall, D.G. *J. Am. Chem. Soc.* **2010**, *132*, 1488.

starting compound is *R*-carvone. The target has a tricyclic structure, with six chiral centers and a reactive α-methylene lactone. The retrosynthetic analysis of this target is presented in Figure 23.30.

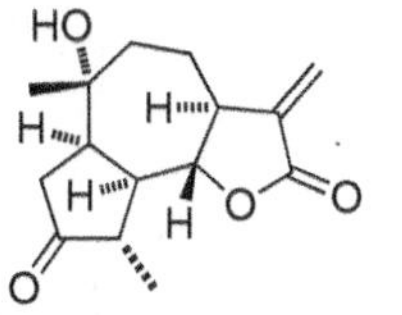

(23.172)

In Figure 23.30, the final steps of the synthesis are envisaged as involving functional group manipulations. The FGI transform on the final target yields a protected epoxy alcohol synthon (**23.173**), which is, in turn, converted by another FGI transform into the alkene synthon **23.174**. At this stage, two key bond disconnection transforms come into play. The first DIS transform, corresponding to a retro-RCM, converts synthon **23.174** into the triene synthon **23.175**, which is itself disconnected into a hydroxyester synthon **23.176** by a retro-transesterification. This alcohol has the requisite functionality to be disconnected in another key DIS transform into substituted allylborane (**23.177**) and cyclopentane-carboxaldehyde (**23.178**) synthons. Applying a retro-Favorskii rearrangement to this aldehyde synthon converts it into a cyclohexanone synthon (**23.179**) that can be prepared from *R*-carvone (**23.180**).

The synthesis corresponding to this retrosynthetic analysis is presented in Figure 23.31. Thus, carvone is converted to the chlorohydrin **23.179** by a previously established sequence of enone epoxidation and acid-catalyzed epoxide ring opening with chloride ion. After protection of the alcohol as the *tert*-butyldimethylsilyl (TBS) ether **23.181**, the Favorskii rearrangement of this α-chloroketone with methoxide anion gave the methyl ester **23.182**, which was reduced to the primary alcohol **23.183** that was then subjected to a Swern oxidation to give the key aldehyde **23.184**.

At this point in the synthesis, the stage was set for the first key bond-forming reaction. An analysis of the stereochemistry of **23.185** shows that the relationship between the hydroxyl group and the long alkyl group in the side chain is *anti*, which lends itself to incorporation by a suitable crotylboration. However, instead of working to ensure the absolute configurations of the two new chiral centers by using a Roush or a Brown crotylboration reaction, the authors chose to use an achiral boronate to add the side chain. In this case, despite the potential for a large number of stereoisomers being formed, the desired isomer was obtained from the *Z* boronate as a 19:1 mixture of *anti* diastereoisomers. The *E* boronate was unreactive under the conditions used (although it comprised

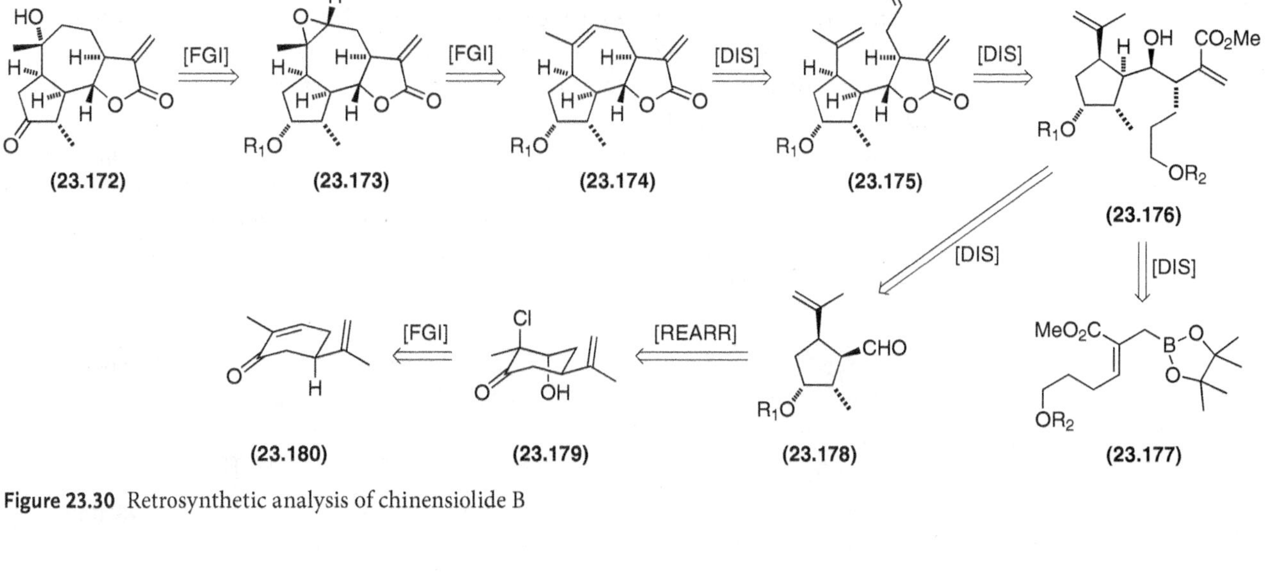

Figure 23.30 Retrosynthetic analysis of chinensiolide B

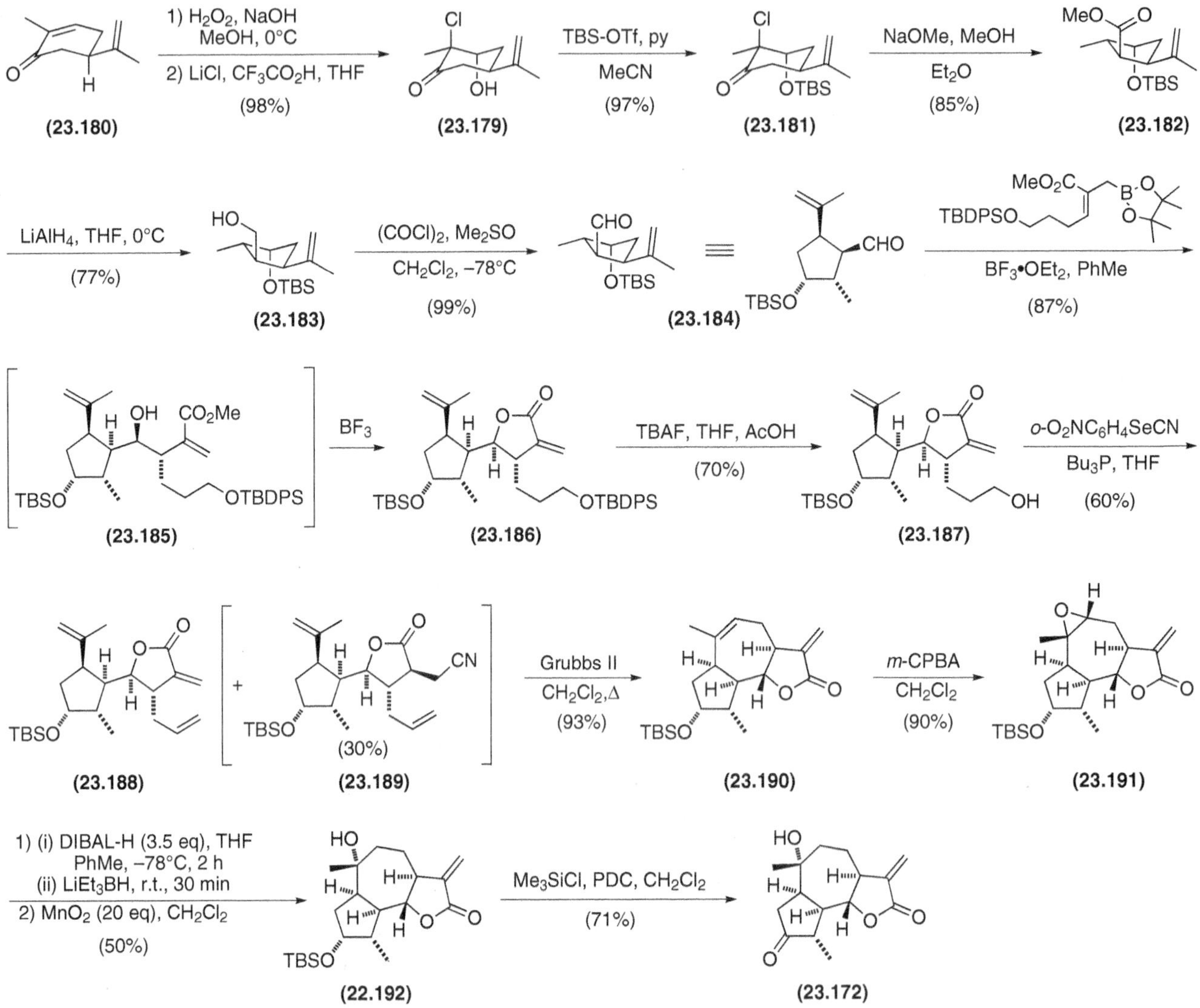

Figure 23.31 The synthesis of chinensiolide B

approximately 22% of the boronate used). This hydroxyester was not isolated because the BF_3 used to catalyze the crotylboration also catalyzes the lactonization to give the α-methylene lactone **23.186**. The silyl ether was then deprotected with fluoride ion to give the primary alcohol **23.187**, which was, in turn, subjected to Grieco elimination to give the key triene **23.188**. This was accompanied by a by-product (Example **23.189**) formed by conjugate addition of the hydrogen cyanide, formed in the elimination reaction, to the α-methylene lactone.

The next step in the synthesis involved the closure of the seven-membered ring by RCM. When we examine the triene **23.188**, there are at least two potential ways for this reaction to occur. The more reactive alkene π bond from the α-methylene lactone could react to form a six-membered ring (Example **23.193**), or the terminal π bond of the side chain may react to form the seven-membered ring (Example **23.190**). Only compound **23.190** is formed.

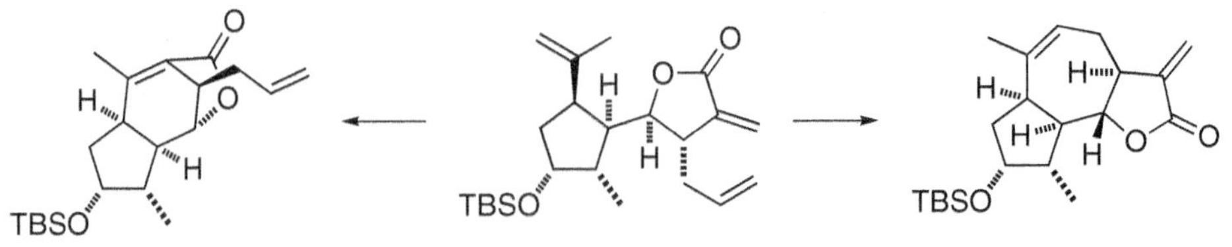

(23.193) (23.188) (23.190)

Problems

23-8 (a) Why is the sequence of epoxidation and acid-catalyzed epoxide ring opening required to ensure the formation of the desired chlorohydrin **23.179**?

(b) What product is expected to dominate if the hydroxyl group of **23.179** is not protected first as the TBS ether?

23-9 It could have been predicted that compound **23.193** would be unlikely to be formed in the ring closing metathesis reaction. Why?

With compound **23.190** in hand, the carbon skeleton of the target compound is in place, and it required only adjustment of the functionality to complete the synthesis. The tertiary alcohol was incorporated by reduction of an epoxide with lithium triethylborohydride, and the epoxide **23.191** was obtained straightforwardly by oxidation of the alkene with *m*-chloroperoxybenzoic acid. However, when this epoxide was treated with lithium triethylborohydride, the α-methylene lactone was also reduced to the saturated ester, meaning that protection of the lactone group would be required before the reduction of the epoxide. This was accomplished by pre-reducing the lactone to the diol with DIBAL-H and then reducing the epoxide ring of the aluminum alkoxide complex with the borohydride reagent. Reoxidation of the resultant triol was accomplished with activated manganese dioxide to give the required α-methylene lactone **23.189** in acceptable yield. The final oxidation of the TBS ether was accomplished in one pot by means of pyridinium dichromate and trimethylsilyl chloride under anhydrous conditions.

The Enantioselective Synthesis of Plakortethers F and G

Finding symmetry or pseudosymmetry in a molecule can dramatically simplify the synthesis, especially if the desymmetrization step can be accomplished well. The plakortethers are just such a group of target molecules. These ethers constitute a family of polyketide natural products whose synthesis was reported in 2009.[135] The molecules are interesting in that retrosynthetic analysis of the molecules (e.g., plakortethers F and G; Example **23.194**) reveals a latent C_2 symmetry in the molecule, as is manifest in synthons **23.195** and **23.196**. A simplified retrosynthetic analysis of the plakortethers A and B (Example **23.194**) is given in Figure 23.32. The disconnection of the cyclic acetal and a retro-aldol disconnection on the other side of the molecule reveal a C_2-symmetric dialdehyde (**23.195**) that can be subjected to a dehydration transform to give another C_2-symmetric dialdehyde (**23.196**). This can, in turn, be disconnected to the dihalide **23.197**. Note how the retro-desymmetrization step occurs relatively early in the retrosynthesis.

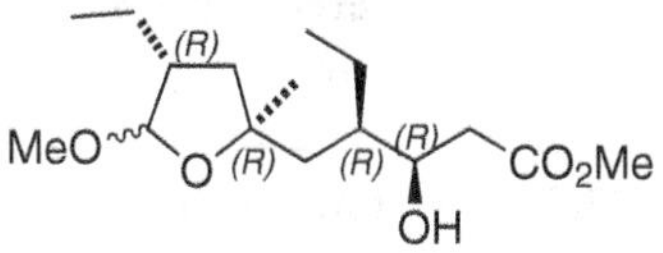

(23.194)

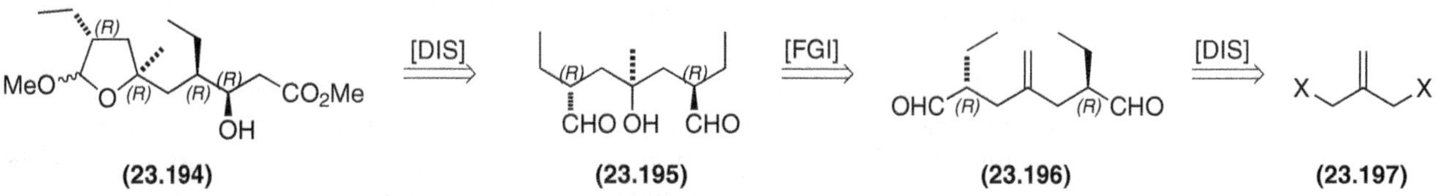

Figure 23.32 Simplified retrosynthetic analysis of plakortethers F and G

135. John, J.P.; Jost, J.; Novikov, A.V. *J. Org. Chem.* **2009**, *74*, 6083.

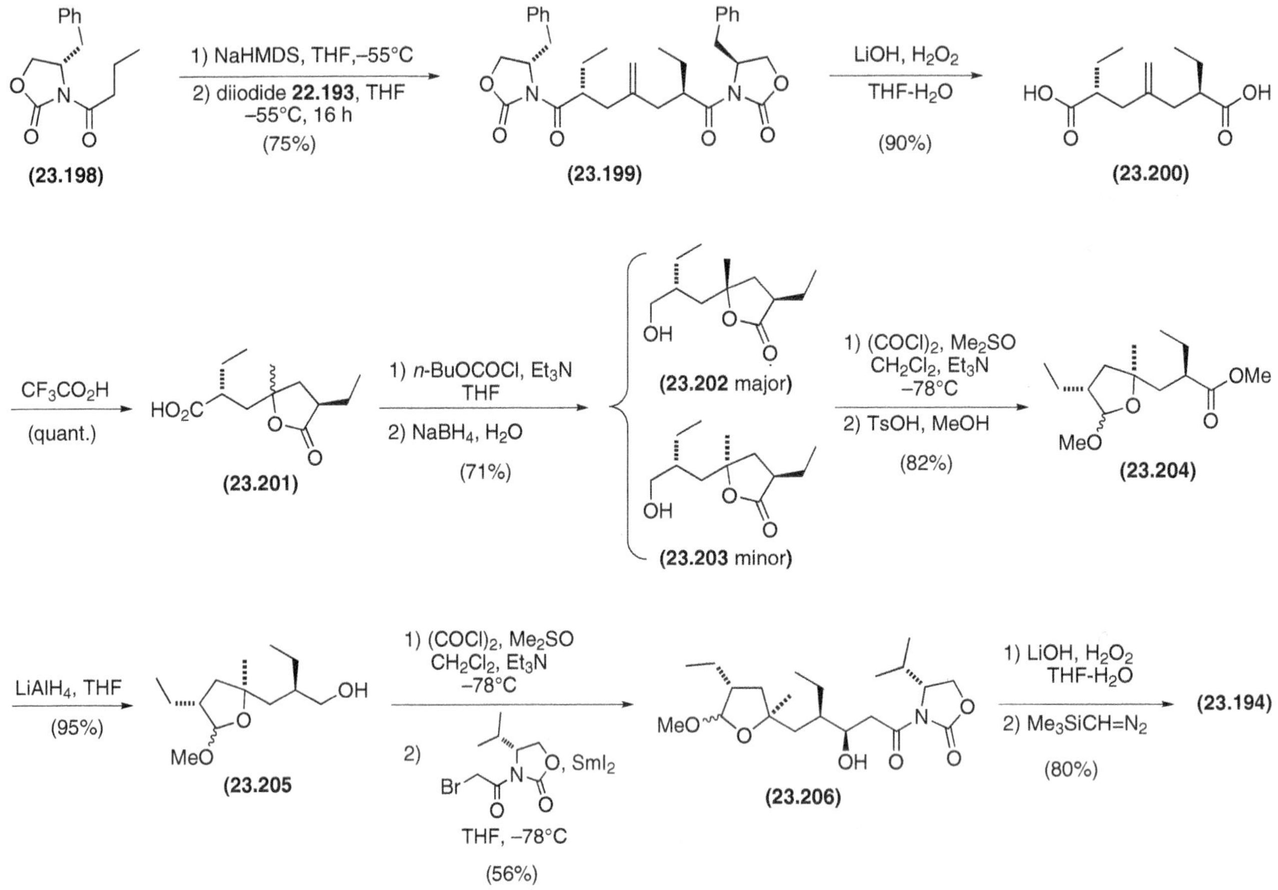

Figure 23.33 The total synthesis of plakortethers F and G

The synthesis of the plakortethers is shown in Figure 23.33. The chirality in this synthesis is incorporated early by alkylation of the sodium enolate of the *N*-butyryl (*S*)-Evans oxazolidinone (**23.194**) derived from phenylalanine. When the reaction temperature was kept low throughout the reaction, a good yield of the *R*,*R*-dialkyl-*bis*-oxazolidinone **23.199** was obtained. Base hydrolysis of the *bis*-oxazolidinone gave the C_2-symmetric unsaturated diacid **23.200**, which was then desymmetrized by acid-catalyzed lactonization with trifluoroacetic acid to lactone **23.201**. This desymmetrization gave a mixture of separable diastereoisomers. The mixture of *cis* and *trans* carboxylic acids was reduced to the mixture of primary alcohols by conversion to the mixed anhydride with *n*-butyl chloroformate and reduction with sodium borohydride. At this stage, the major (*cis*, **23.202**) and minor (*trans*, **23.203**) diastereoisomers were conveniently separated by chromatography. Swern oxidation of the major isomer gave an aldehyde that was treated with methanol in the presence of *p*-toluenesulfonic acid. The product of the reaction was a mixture of the diastereoisomers of the cyclic acetal **23.204**. The ester group of **23.204** was reduced with lithium aluminum hydride to the primary alcohol **23.205**.

Although the β-hydroxyester **23.204** might have been incorporated by means of an aldol addition reaction, the authors chose a samarium (II) iodide-mediated Reformatskii reaction of an *N*-bromoacetyl Evans oxazolidinone to complete the construction of the carbon skeleton of the target (**23.206**). The hydrolysis of the oxazolidinone and esterification of the acid with trimethylsilyldiazomethane completed the synthesis.

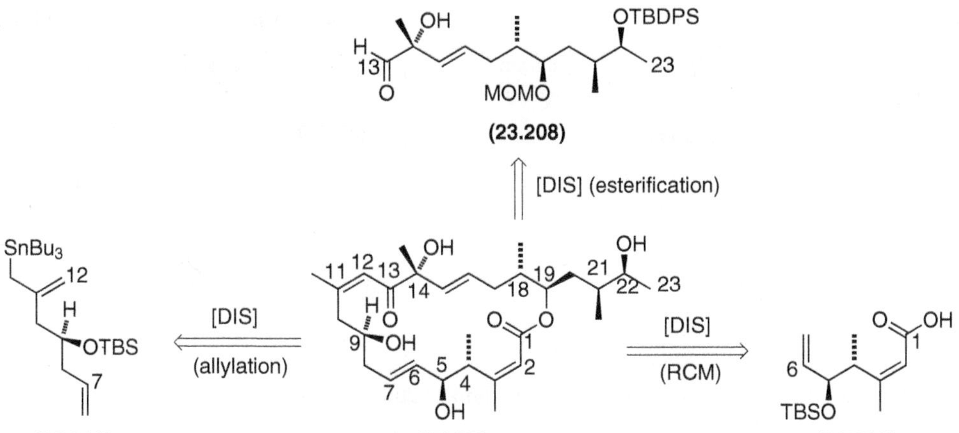

Figure 23.34 Major disconnections of the iriomoteoilide-1b molecule

The Asymmetric Synthesis of a Proposed Structure of Iriomoteolide-1b

Iriomoteolide-1b (proposed structure, **23.203**) is a macrocyclic lactone. The molecule contains a 20-membered ring with six chiral centers (C-4, C-5, C-9, C-14, C-18, and C-19) in the ring; a side chain at C-19 carrying two chiral centers (C-21 and C-22); and four alkenes, including two *Z* trisubstituted alkenes at C-2 and C-11. These stereochemical features constitute a major synthetic challenge. In addition, the structure of the natural product (which was determined by ^{1}H NMR spectroscopy) is still in doubt, which means that the synthesis of candidate compounds must be carried out to establish the correct structure of the compound.[136]

The retrosynthetic analysis of the target molecule reveals three major disconnections, shown in Figure 23.34, that lead to a convergent approach to the target. The retro-esterification and retro-allylation transforms yield the synthon **23.208,** the retro-esterification and retro-RCM transforms yield the synthon **23.209,** and the retro-allylation and retro-RCM transforms yield the synthon **23.210.** Each of these synthons has its own complexities, including chiral centers and dense functionalization. We will discuss the retrosynthetic analysis of each in turn, at the time we discuss the synthesis of the synthon.

The C13–C23 Synthon 23.208

Stereochemically, this is the most complex of the synthons, with five chiral centers and an *E* double bond. In addition, it possesses four oxygen-containing functional groups. The retrosynthetic analysis of the synthon is given in Figure 23.35. In this particular case, the desired starting compounds were the chiral ester **23.215** and the chiral glyceraldehyde derivative **23.212**, both of which had been prepared previously.[137]

The first disconnection is a retro-olefination, which disconnects **23.208** into the tetrazolyl sulfone **23.211** and the aldehyde **23.212**. Subjecting the sulfone to a series of FGI transforms leads to the synthon **23.213**, which is, in turn, simplified by application of a retro-crotylboration transform to give the aldehyde synthon **23.214**. This synthon could be envisaged as being prepared from the available chiral ester **23.215**.

The synthesis of **23.208** is summarized in Figure 23.36. Starting from the ester **23.215**, the unsaturated aldehyde **23.216** was prepared in two steps by DIBAL-H reduction of the

136. Ye, Z.; Gao, T.; Zhao, G. *Tetrahedron* **2011**, *67*, 5979.
137. Ye, Z.; Deng, L.; Qian, S.; Zhao, G. *Synlett* **2009**, 2469.

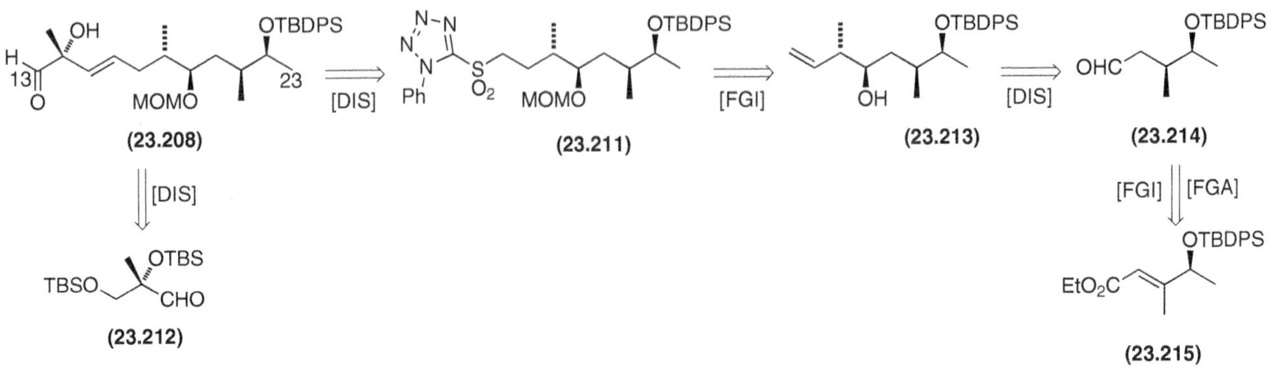

Figure 23.35 The retrosynthetic analysis of synthon **23.208**

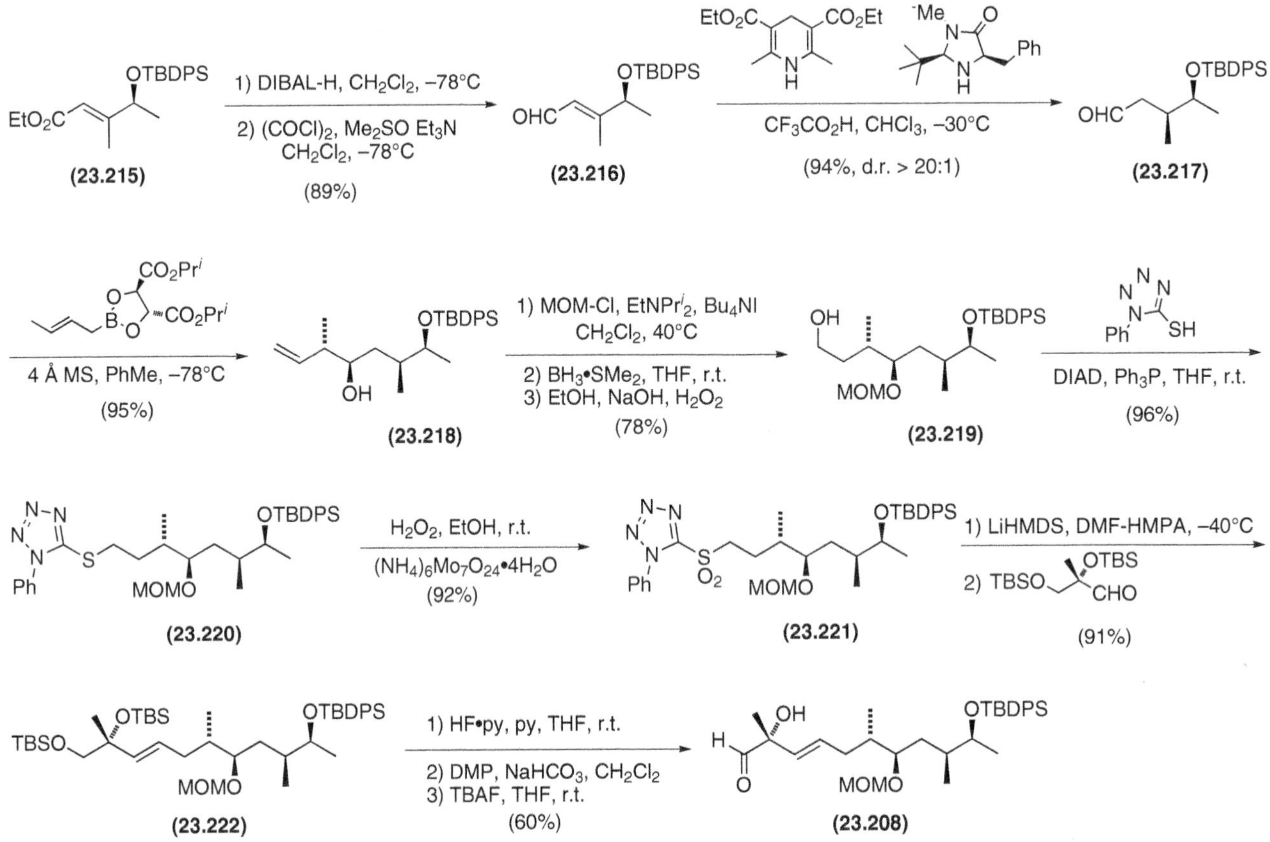

Figure 23.36 The synthesis of synthon **23.139**

ester to the allylic alcohol and subsequent Swern oxidation. Asymmetric transfer hydrogenation of **23.216** using a chiral imidazolidinone catalyst and a Hantzsch ester as the hydrogen source gave the saturated aldehyde **23.217** in good yield and with good stereocontrol. Roush crotylboration of the aldehyde led to the desired *anti* homoallylic alcohol **23.218**. This reaction did give a mixture of diastereoisomers, but the diastereoselectivity was still good (d.r. > 10). The homoallylic alcohol was converted into the tetrazolyl sulfide **23.220** by sequential hydroboration-oxidation to the primary alcohol **23.219** and Mitsunobu coupling with the thiol, and the sulfide was then oxidized with hydrogen peroxide in the presence of an ammonium molybdate catalyst to complete the synthesis of the sulfone synthon **23.221**. Condensation of the anion of this sulfone with the aldehyde **23.212** gave

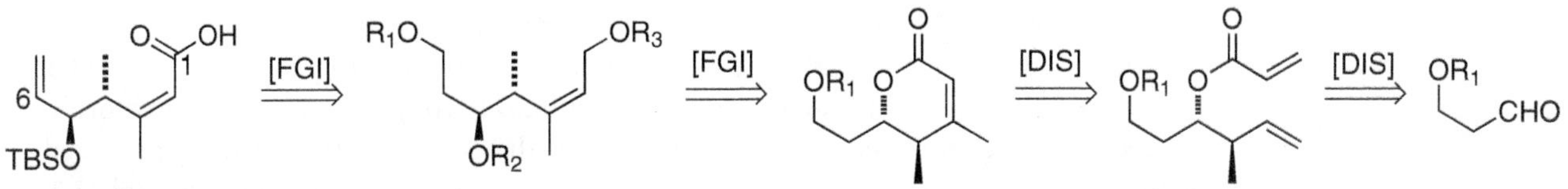

(23.209) (23.223) (23.224) (23.225) (23.226)

Figure 23.37 Retrosynthetic analysis of synthon **23.209**

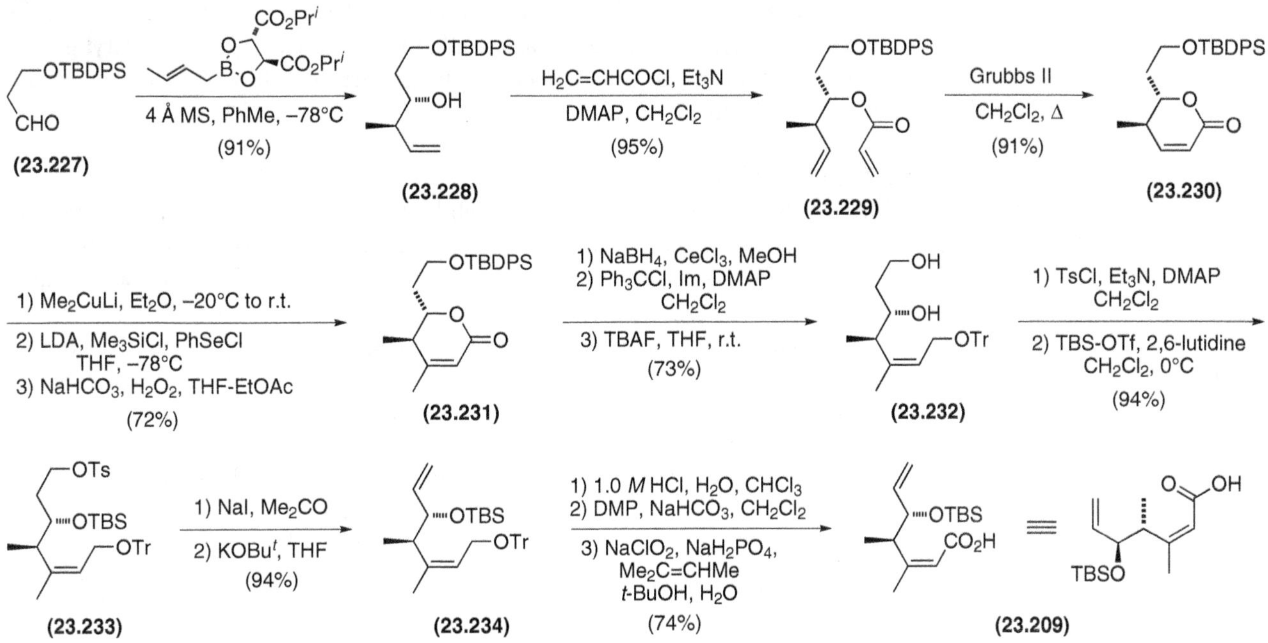

Figure 23.38 The synthesis of the C1-C6 synthon **23.209**

the alkene **23.222**. Selective deprotection of the primary TBS ether was accomplished using pyridinium fluoride in pyridine. The resultant primary alcohol was oxidized with the Dess-Martin periodinane. The tertiary TBS ether was then deprotected to give the synthon **23.208** in 30% overall yield from the starting ester and aldehyde.

The C1–C6 Synthon 23.209

This section of the macrocyclic ring can be traced back to the propanal derivative **23.226** by the pathway given in Figure 23.37. The authors chose to establish the *Z* configuration of the C2—C3 double bond by means of an unsaturated lactone intermediate. A series of FGI transforms converts the target synthon into the conjugated lactone synthon **23.224** by way of synthon **23.223**. Applying a bond DIS (retro-RCM) transform to **23.224**, one obtains the acrylate synthon **23.225**, itself disconnected by means of a retro-crotylboration transform into the aldehyde synthon **23.226**.

The synthesis of **23.29** is shown in Figure 23.38. The sequence begins with the TBDPS ether of 3-hydroxypropanal (**23.227**), which is subjected to Roush crotylboration with the boronate ester derived from diisopropyl (*S*,*S*)-tartrate. This reaction gives the *anti* homoallylic alcohol (**23.228**) with the *S* configuration at the carbinol carbon, thus establishing the absolute configuration at C5 of the iriomoteolide-1b. This alcohol was then esterified with acryloyl chloride to give the key acrylate **23.229**, which serves as the substrate for the RCM. The RCM of the homoallyl acrylate **23.229** using the second-generation Grubbs catalyst gave the key unsaturated lactone **23.230**. The incorporation of the methyl group at C3 was accomplished by a standard three-step sequence of conjugate addition, selenylation, and selenide oxidation and elimination to give the lactone **23.231**. Reduction of the unsaturated lactone with sodium borohydride and cerium chloride in methanol gave the allylic alcohol. The primary hydroxyl group was selectively protected as the trityl ether, and the TBDPS ether was then cleaved with fluoride to give the diol **23.232**. To incorporate the terminal double bond at C6, the two hydroxyl groups needed to differentiated. The primary alcohol was selectively sulfonated. The secondary alcohol was then protected as the TBS ether, yielding the sulfonate ester **23.233**. The elimination of the sulfonate was not carried out directly, but the sulfonate was first converted into the iodide by a Finkelstein reaction, which was then eliminated to the alkene **23.234** with potassium *tert*-butoxide in THF. The synthesis was completed by acid cleavage of the trityl group, sequential oxidation of the primary alcohol to the acid by Dess-Martin periodinane oxidation to the aldehyde, and Kraus-Pinnick oxidation to the acid **23.209**.

*The C7–C12 Synthon **23.210***

The third synthon from which the iriomoteolide-1b target is assembled is the allylstannane **23.210**. The retrosynthetic analysis of the stannane is summarized in Figure 23.39. The initial transform in the synthesis involves a functional group interconversion to give the bromide synthon **23.235**. The authors chose a creative application of cyclopropanol chemistry to obtain this bromide synthon. Thus, recognizing that cyclopropanols can be ring-opened to give allyl products, the authors proposed that the allylic bromide be obtained from the cyclopropyl derivative **23.236**. This, in turn, was ultimately to be obtained from the known ester **23.238** *via* the nitrile **23.237**.

The sequence used to prepare **23.210** is summarized in Figure 23.40. Initially, the hydroxyl group of the known chiral ester **23.239** was protected as the TBS ether. The protected ester was then converted to the cyclopropanol **23.240** by the method of Kulinkovich and de Meijere.[138] The cyclopropanol was next converted to its tosylate ester, and the TBS

Figure 23.39
The retrosynthetic analysis of the C7-C12 allylstannane synthon 23.210

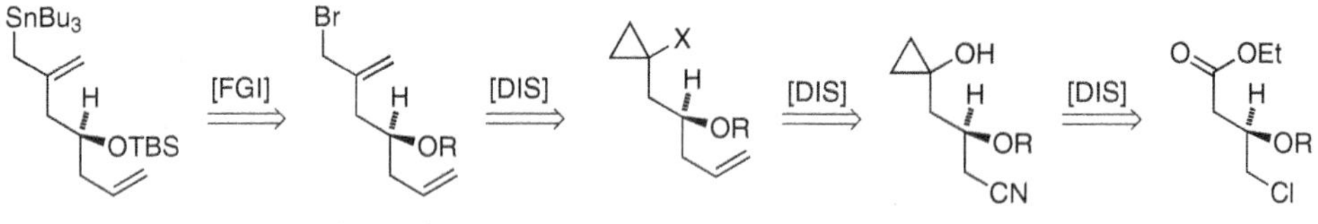

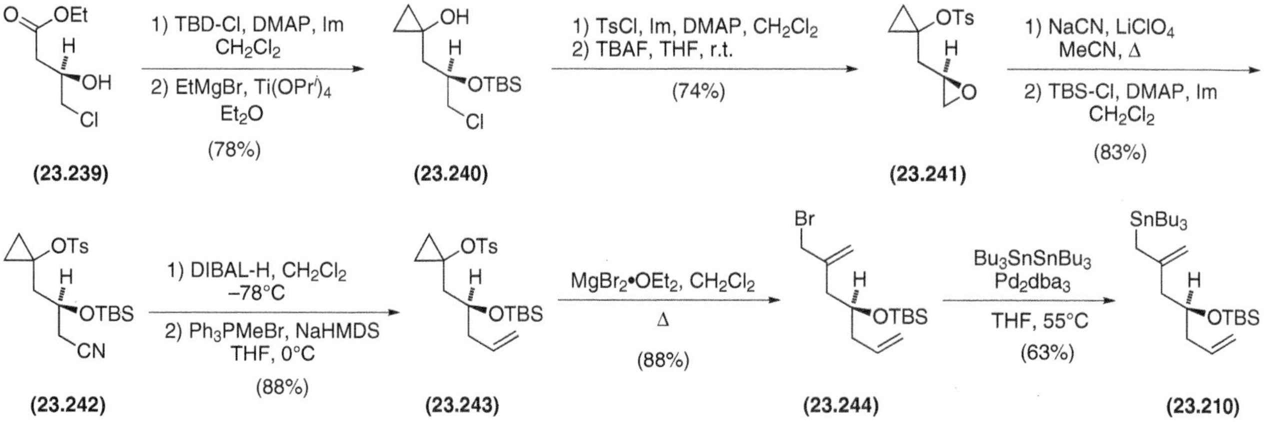

Figure 23.40 Synthesis of the C7-C12 allylstannane synthon **23.206**

138. (a) Kulinkovich, O.G.; de Meijere, A. *Chem. Rev.* **2000**, *100*, 2789. (b) Kulinkovich, O.G. *Chem. Rev.* **2003**, *103*, 2597.

ether was deprotected with fluoride. The product of this deprotection was almost certainly the epoxide **23.241**, which was ring-opened with cyanide anion to give the β-hydroxynitrile. The ring closure of the epoxide and the ring opening with cyanide anion both proceed with retention of configuration at the carbinol carbon. Protection of the hydroxyl group completes the conversion of epoxide **23.241** to nitrile **23.242**. The elaboration of the monosubstituted alkene was completed by the two-step process of DIBAL-H reduction of the nitrile to the aldehyde and subsequent Wittig methylenation to **23.243**. The final stages of the synthesis were the solvolytic ring-opening of the cyclopropane to give an allyl cation that was trapped by bromide anion, yielding the allylic bromide **23.244** and subsequent palladium-catalyzed stannylation of the bromide.

Final Assembly of the Iriomoteolide Target Structure

With all three chiral synthons in hand, the final stage was set for assembly of the macrolactone. This was accomplished by the sequence in Figure 23.41. The first step in the assembly process was the yttrium-catalyzed allylstannylation of the aldehyde **23.208** with the stannane **23.210**. This gave the homoallylic alcohol **23.245** as a mixture of diastereoisomers. The rather unusual conditions, involving both yttrium triflate and *p*-nitrobenzoic acid as catalysts for the reaction, avoided the decomposition of the aldehyde **23.208**, which occurred in the presence of each of these catalysts alone.

Oxidation of the secondary alcohol to the ketone by iodoxybenzoic acid was followed by selective cleavage of the MOM ether using the *B*-bromocatecholborane to give the diol **23.246**. The resultant non-conjugated ketone was esterified with the acid **23.209** using the Yamaguchi procedure,[139] which gave the desired ester but with concomitant migration of the double bond into conjugation with the ketone. The product, **23.247**, was obtained as a 5:2 mixture of geometric isomers, in which the desired *Z* isomer was the major component. Removal of the two silyl protecting groups and RCM gave the target compound, **23.207**.

This synthesis was undertaken to test whether the structure proposed for iriomoteolide-1b was correct. After the synthesis had been completed, the spectroscopic properties of the synthetic and the natural material were compared. The comparison showed unequivocally that the two compounds were not identical, thus demonstrating that the structure assigned to the natural product is incorrect (and leaving yet another target for total synthesis).

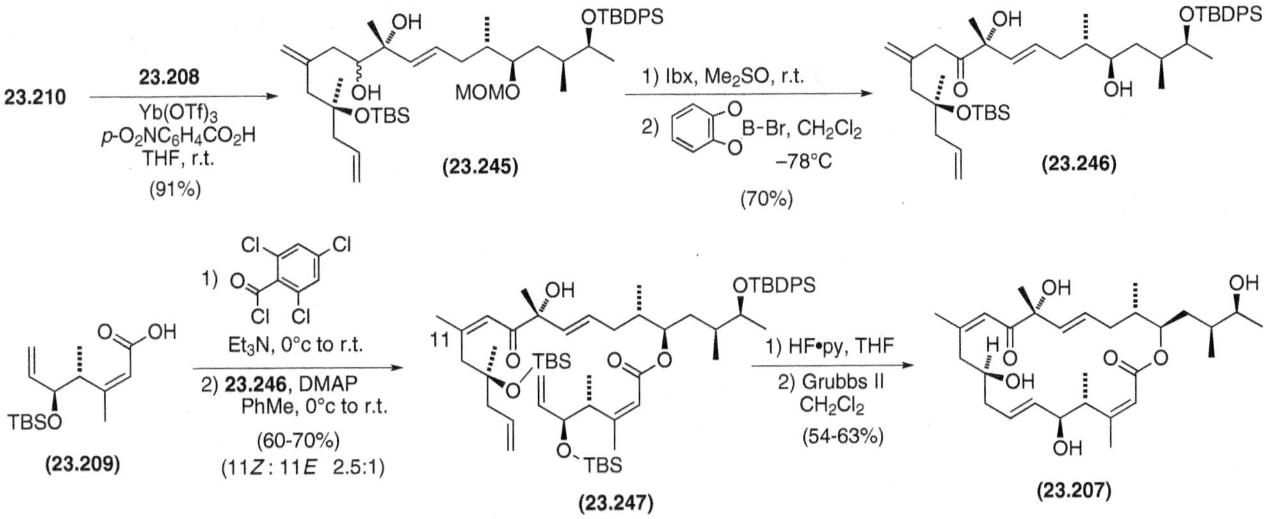

Figure 23.41 Completion of the synthesis of the structure proposed for iriomoteolide-1b

139. (a) Inanaga, J.; Hirata, K.; Saeki, H.; Katsuki, T.; Yamaguchi, M. *Bull. Chem. Soc. Jpn.* **1979**, *52*, 1989. (b) Kawanami, Y.; Dainobu, Y.; Inanaga, J.; Katsuki, T.; Yamaguchi, M. *Bull. Chem. Soc. Jpn.* **1981**, *54*, 943.

Figure 23.42 Retrosynthetic analysis of manzacidin-C

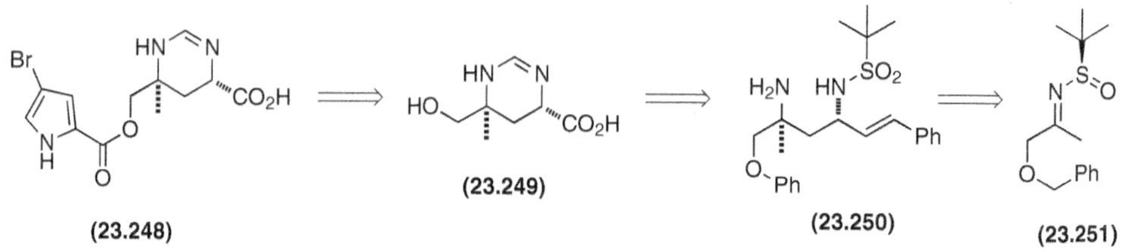

An Enantioselective Synthesis of Manzacidin-C[140]

The brominated natural product manzacidin C (Example **23.248**) provides an illustration of the synthetic utility of chiral *tert*-butylsulfinylimine anions in the asymmetric synthesis of 1,3-diamino compounds. The retrosynthetic analysis of the target compound is shown in Figure 23.42.

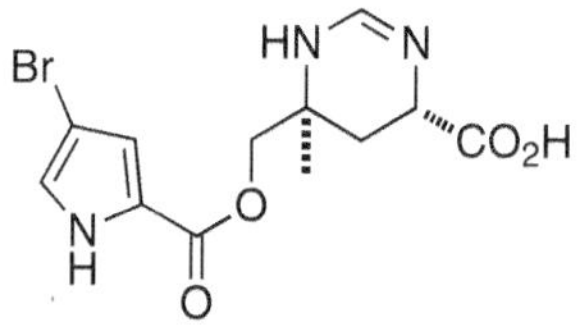

(23.248)

The obvious first disconnection yields the alcohol **23.249**, which, in turn, can be simplified to the diamine derivative **23.250** by a ring DIS transform. The diamine can, in turn, be constructed from the chiral *tert*-butylsulfinylimine **23.251**. The stereochemistry of the final product indicates that the *R* enantiomer of the sulfinylamine auxiliary should be used.

The synthesis of the target compound is summarized in Figure 23.43. Thus, the chiral imine **23.251** was deprotonated with LHMDS to give the chiral nucleophile that then added to the imine group of the *tert*-butylsulfonamide **23.252** to give the allylic sulfonamide **23.253**.

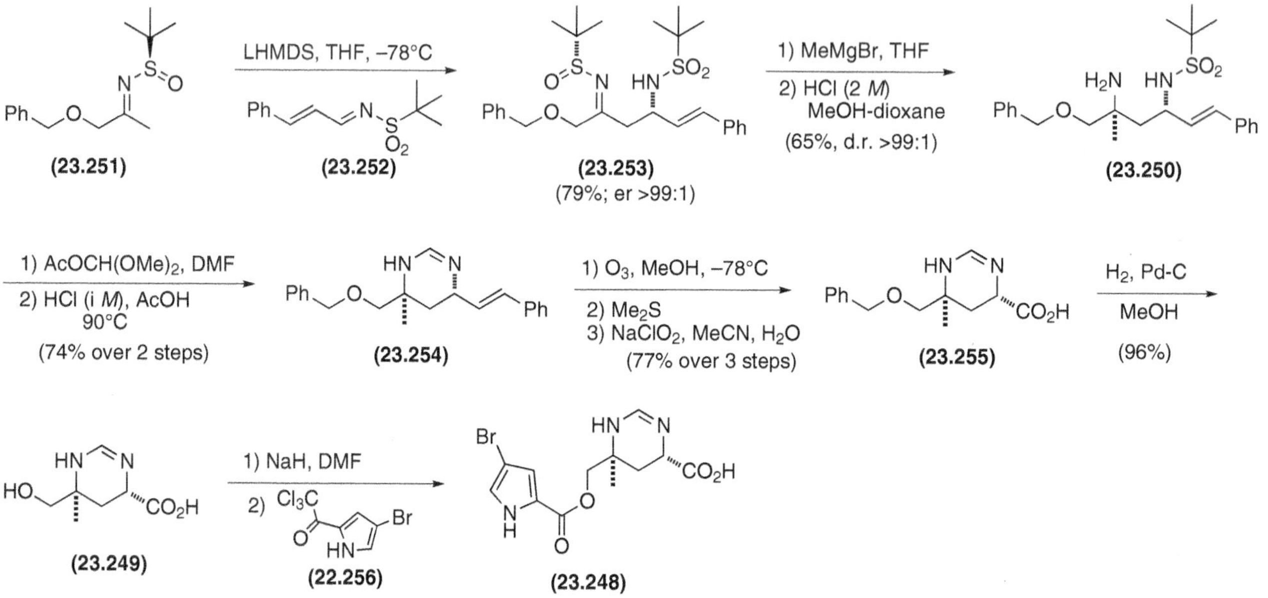

Figure 23.43 Synthesis of manzacidin-C by means of chiral control by a sulfinylamine chiral auxiliary

140. Lanter, J.C.; Chen, H.; Zhang, X.; Sui, Z. *Org. Lett.* **2005**, *7*, 5905.

In this case, the propensity for the sulfonylimine anion to self-condense had to be overcome, and this was accomplished by using the more electrophilic sulfonamide as the electrophile. The addition of methylmagnesium bromide to the imine double bond, followed by hydrolysis of the sulfinylimine gave the sulfonamide **23.250**.

The next step involved the reaction of the amine group of **23.250** with the orthoformate derivative, acetoxydimethoxymethane, in dimethylformamide. After hydrolysis of the initial product with hydrochloric acid in acetic acid, the amidine **23.254** was obtained. It is worthwhile noting that both chiral centers of this compound were fixed by the face selectivity of the chiral sulfinyl group. Ozonolysis of the alkene double bond and Kraus-Pinnick oxidation of the resultant aldehyde gave the carboxylic acid **23.255**. Deprotection of the benzyl ether was accomplished by hydrogenolysis, and the hydroxyacid **23.249** was then esterified with 2-trichoroacetyl-4-bromothiophene (**23.256**) in an unusual application of the haloform reaction.

Problem

23-10 Suggest a reasonable mechanism for the esterification of the alcohol **23.249**.

A Final Chapter: The Enantioselective Synthesis of Ecklonialactones

At the beginning of Chapter 15, we discussed protecting groups in organic synthesis and spoke about the ultimate goal of synthesis without the need for them. Although this goal is still largely unrealized, it is not unrealistic as illustrated by the asymmetric synthesis of ecklonialactones A (**23.257-A**) and B (**23.257-B**) by Fürstner's group.[141] The ecklonialactones are macrocyclic lactones with a 13-membered lactone ring, with either one or two *Z* double bonds, *trans*-fused to a cyclopentane that carries an epoxide. There are five contiguous chiral centers in the molecule. The molecule was assembled in two halves, with the macrocycle being closed by an alkyne metathesis reaction.

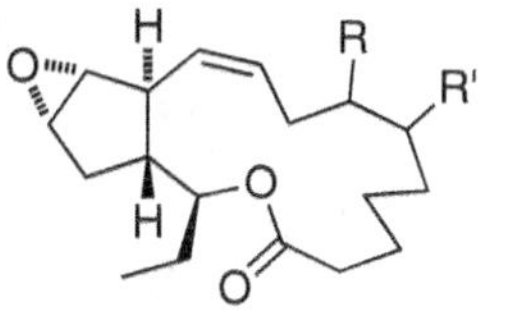

(**23.257-A**: R, R' = π bond)
(**23.257-B**: R = R' = H)

The retrosynthetic analysis of the ecklonialactones is shown in Figure 23.44. The *Z* double bond is transformed to an alkyne (**23.258**) by FGI, and then the structure is simplified by a series of bond disconnections. Retro–ring closing acetylene metathesis (RCAM) gives the diyne synthon **23.259**, which is disconnected by retro-esterification to give the key cyclopentene synthon **23.260**. This synthon carries three of the five contiguous chiral centers. It is further simplified by DIS transforms through **23.261** to the open-chain diene synthon **23.262** by another retro-RCAM transform. This, in turn, can be derived from the butenolide **23.263**.

The synthesis of the first synthon, the alkynyl alcohol **23.260**, is shown in Figure 23.45. The synthesis begins with an enantioselective rhodium-catalyzed conjugate addition of a styrylboronate (**23.264**) to butenolide (**23.263**). The addition proceeds with a carvone-derived triene[142] as the ligand to give a 52% chemical yield of the lactone **23.265** with 80%

141. Hickmann, V.; Alcarazo, M.; Fürstner, A. *J. Am. Chem. Soc.* **2010**, *132*, 11042.
142. Defieber, C.; Paquin, J.-F.; Serna, S.; Carreira, E.M. *Org. Lett.* **2004**, *6*, 3873.

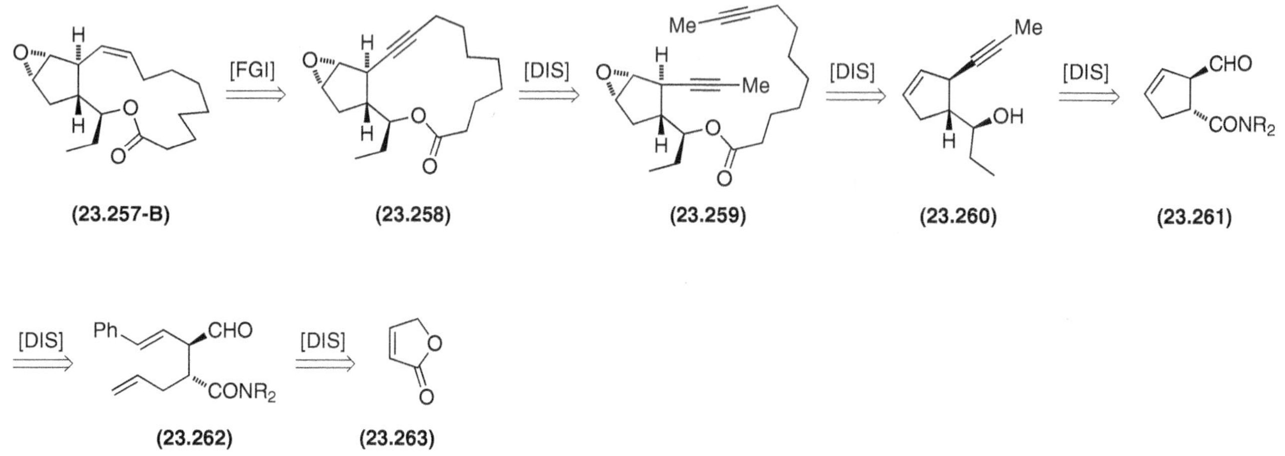

Figure 23.44 Retrosynthetic analysis of ecklonialactone B

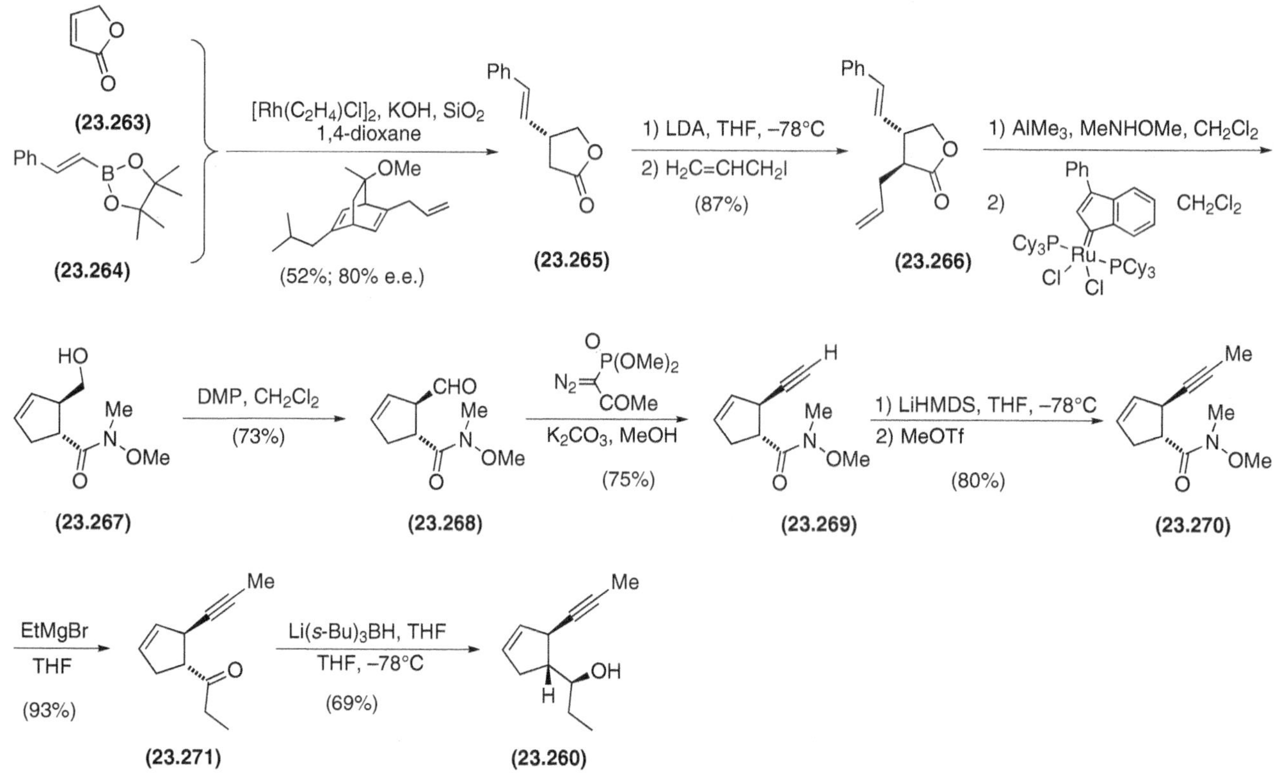

Figure 23.45 Synthesis of the alkynyl alcohol synthon **23.256**, carrying three contiguous chiral centers

e.e. The e.e. could be raised to 90% by recrystallization at the expense of 20% of the isolated yield. Alkylation of the lactone gave the expected *trans*-disubstituted lactone **23.266**, which was converted into the Weinreb amide under Lewis acid catalysis by trimethylaluminum. RCM gave the cyclopentene **23.267**. Oxidation of the primary alcohol gave the aldehyde **23.268.** Due to its instability, this aldehyde was immediately converted to the terminal alkyne **23.269** by means of the Ohira-Bestmann alkynylation.[143] Capping of

143. (a) Ohira, S. *Synth. Commun.* **1989**, *19*, 561. (b) Müller, S.; Liepold, B.; Roth, G.J.; Bestmann, H.J. *Synlett* **1996**, 521.

the terminal alkyne to give the amide **23.270** was straightforward. The ethyl carbinol side chain was elaborated by the two-step sequence of first adding ethylmagnesium bromide to give the ethyl ketone **23.271**, which was then reduced with L-Selectride to give the desired stereoisomer of the alcohol (**23.260**).

Problem

23-11 Based on the observed product of the reduction of 23.271 with L-Selectride, what is the most likely transition state for the reduction? Does this transition state conform to the Cram or Felkin-Anh rules?

The remainder of the synthesis of ecklonialactone B is summarized in Figure 23.46. Epoxidation of the cyclopentene **23.260** with *tert*-butyl hydroperoxide and vanadyl acetylacetonate gave the α epoxide (**23.272**) as the only stereoisomer isolated. This epoxide was then esterified with 9-undecynoic acid using a carbodiimide as the dehydrating agent to give the ester **23.273**, and the diyne was then subjected to RCAM to give the cyclic alkyne **23.258**. Lindlar hydrogenation completed the synthesis of **23.257-B**.

Problem

23-12 The synthesis of ecklonialactone B described above begins by the epoxidation of the alkene, followed by the esterification of the alcohol and the RCAM. These reactions can also be carried out in a different order, as shown below.

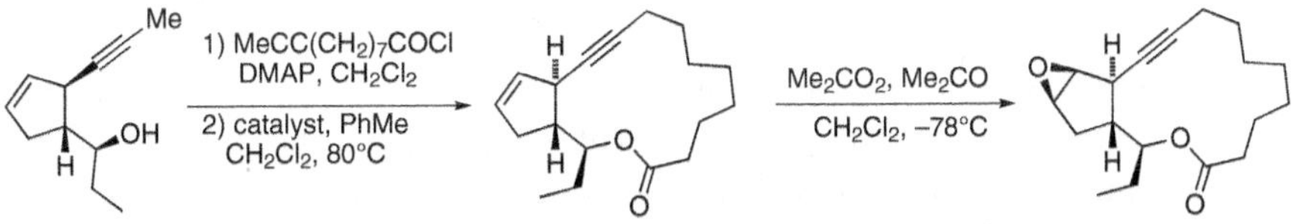

The steps are (1) esterification with 9-undecynoyl chloride, (2) RCAM, and (3) epoxidation with dimethyldioxirane. As is shown, the major product of the epoxidation is the β-epoxide, not the α-epoxide. Suggest why the epoxide stereochemistry changes between the two reaction sequences.

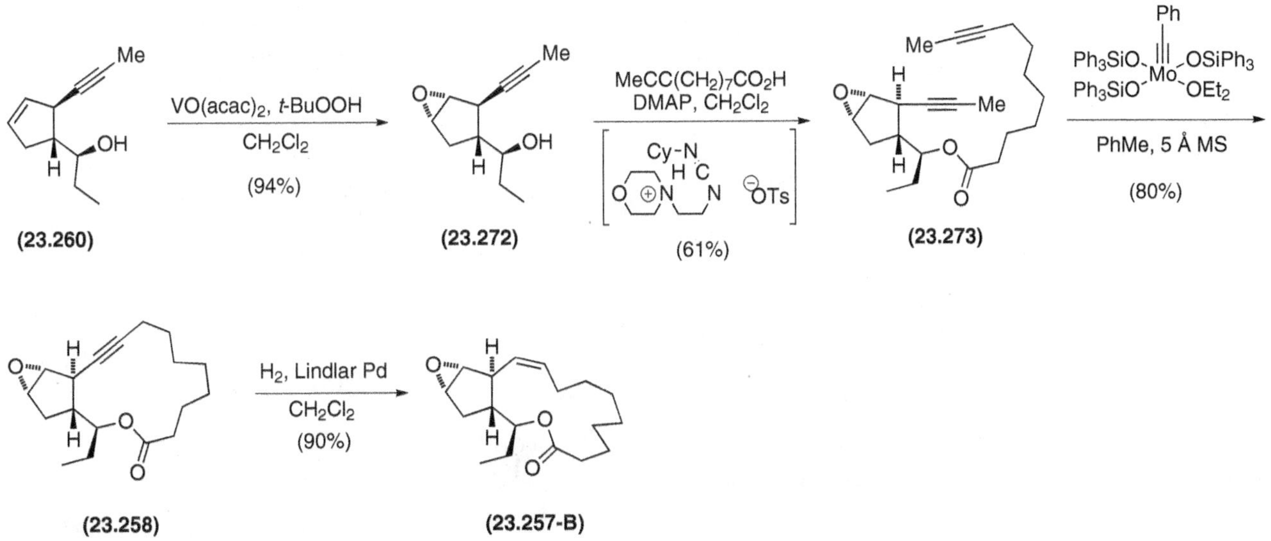

Figure 23.46 Completing the synthesis of ecklonialactone B

Chapter Summary

The construction, from achiral precursors, of chiral molecules enriched in a single enantiomer is known as asymmetric synthesis. The formation of an enantiomerically enriched product depends on having the two enantiomers formed via diastereoisomeric transition states. This can be accomplished by (1) modifying the precursor with a chiral auxiliary prior to the reaction, (2) using a chiral reagent to carry out the reaction, or (3) using a chiral catalyst to promote the reaction. Chiral auxiliaries include the Evans oxazolidinones, the Oppolzer camphorsultam, phenmenthol, SAMP/RAMP hydrazines, and *tert*-butylsulfinylamines. Stereoselectivity in reactions generating a new chiral center is simple diastereoselection when only one of the reactants is chiral and double diastereoselection when both reactants are chiral. Systems with double diastereoselection may be either matched, which usually leads to very high e.e. values, or mismatched, which usually leads to significantly lower e.e. values.

Key Terms

asymmetric synthesis
chiral auxiliary
double diastereoselection
Evans oxazolidinones
matched reactant stereochemistry
mismatched reactant stereochemistry
non-linear effect (NLE)
SAMP/RAMP hydrazones
simple diastereoselection

Additional Problems

23-13 Suggest methods for accomplishing the following conversions. Unless indicated otherwise, the product has the absolute configuration shown, and it is obtained as a single enantiomer or with high e.e.

(a)
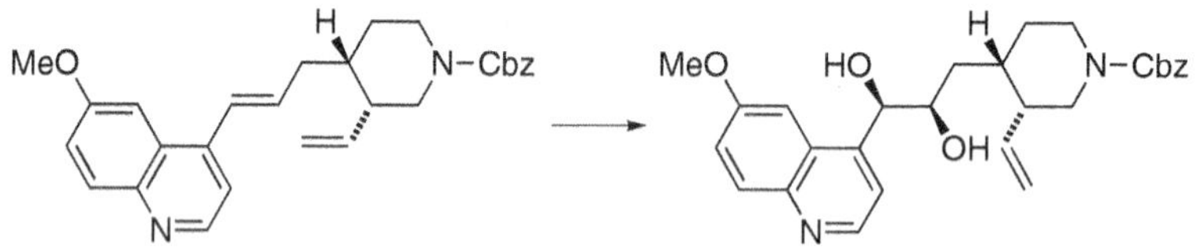

(b)

(c)
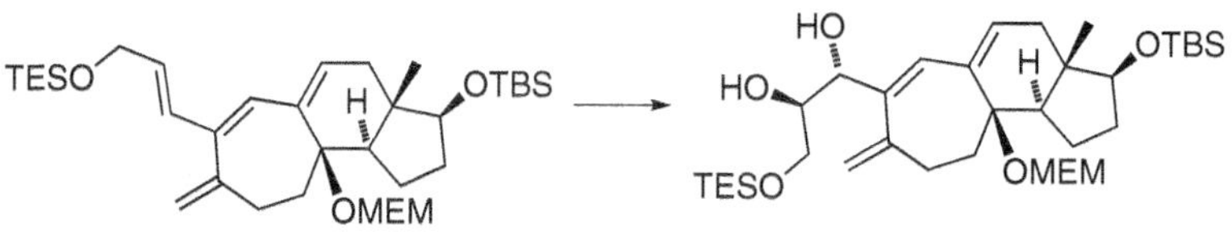

(d)

(e)

(f)
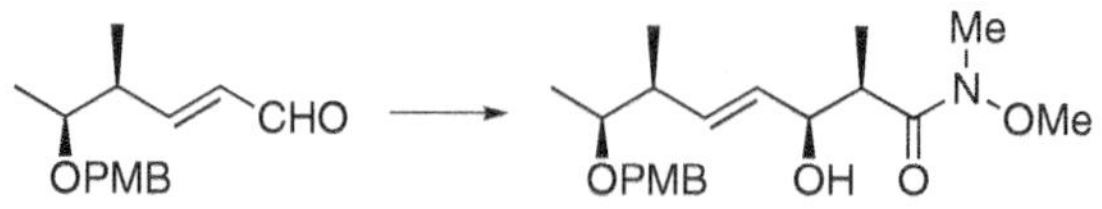

(g) 

(h) 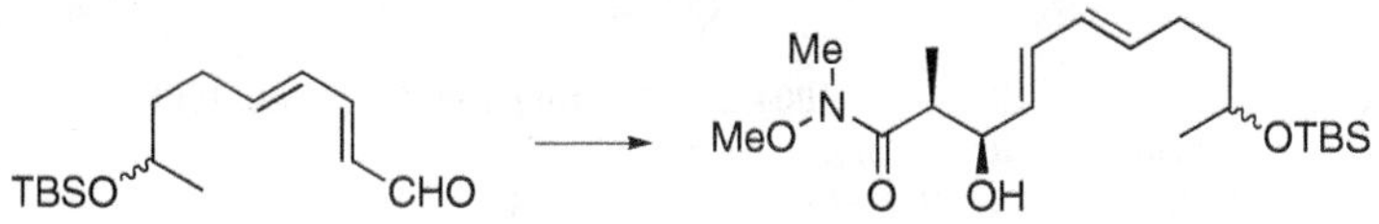

(i) 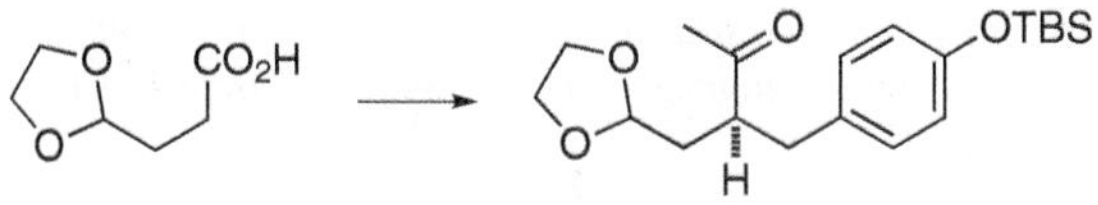

(j) 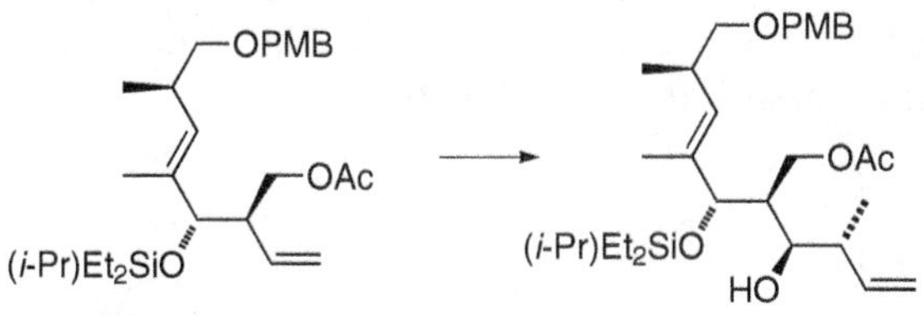

(k) 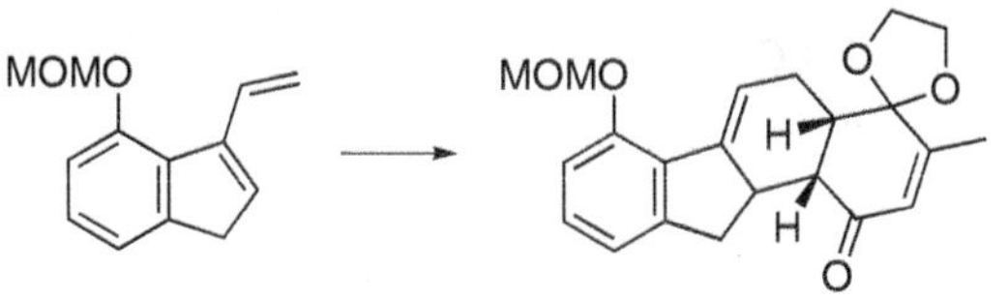

(l) 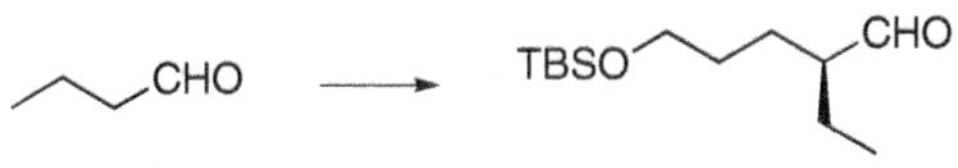

(m) 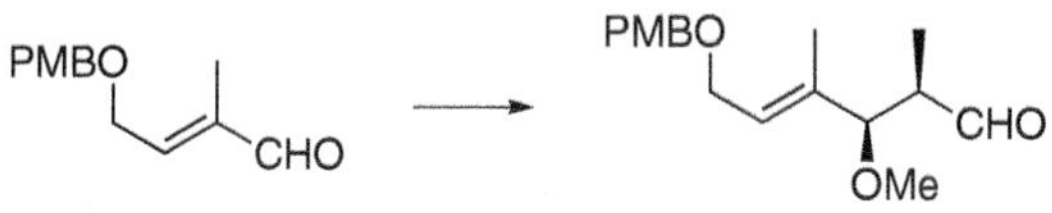

(n) 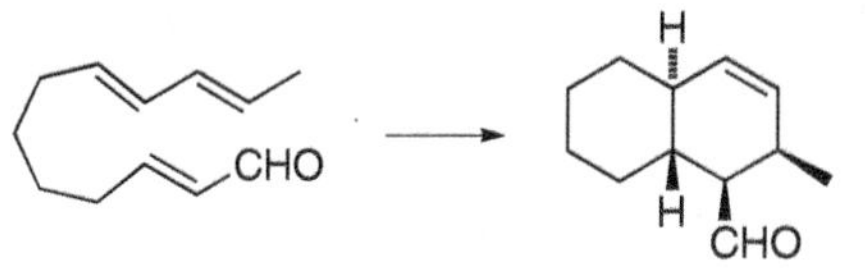

(o) 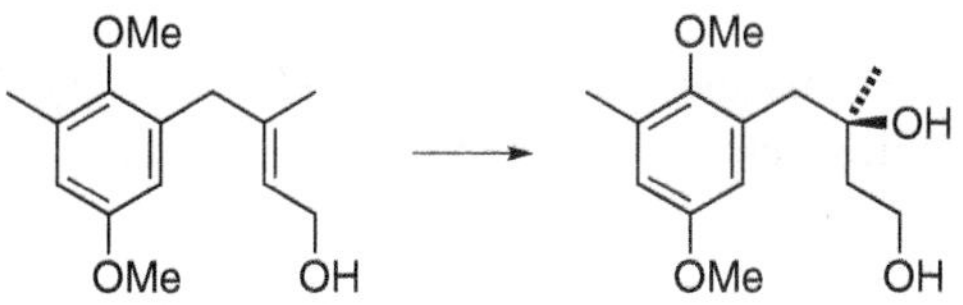

(p) 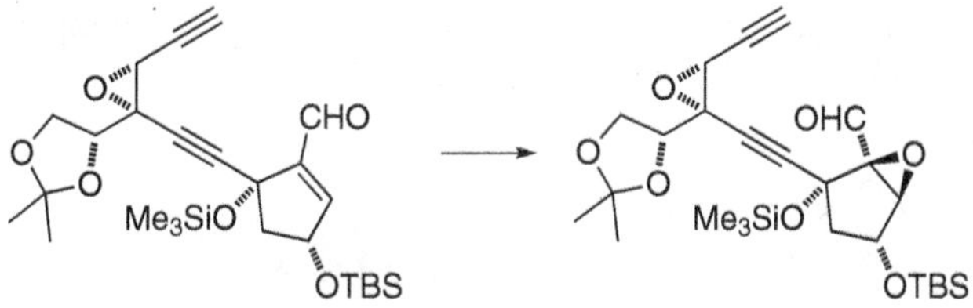 (this diastereoisomer)

(q) 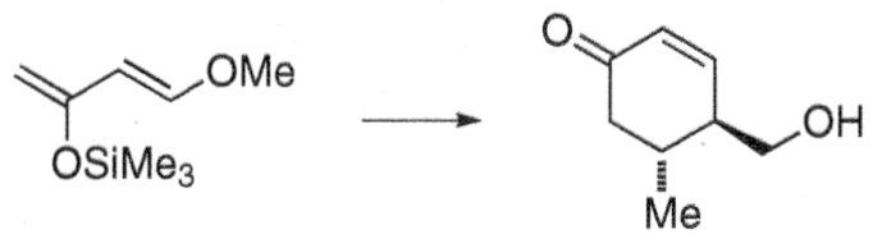

(r) 

(s) 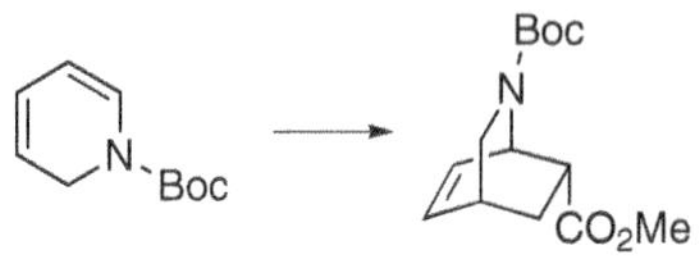

References: (a) *J. Am. Chem. Soc.* **2004**, *126*, 706. (b) *J. Am. Chem. Soc.* **2000**, *122*, 3811. (c) *J. Am. Chem. Soc.* **2008**, *130*, 16864. (d) *J. Am. Chem. Soc.* **201-**, *132*, 2540. (e, f) *Synthesis* **2005**, 1183. (g) *Org. Lett.* **2001**, *3*, 3021. (h) *Org. Lett.* **2004**, *6*, 3889. (i) *J. Am. Chem. Soc.* **2009**, *131*, 16905. (j) *J. Am. Chem. Soc.* **2003**, *125*, 350. (k) *J. Am. Chem. Soc.* **2008**, *130*, 2783. (l) *J. Am. Chem. Soc.* **1981**, *103*, 6967. (m) *Org. Lett.* **2004**, *6*, 1445. (n) *J. Am. Chem. Soc.* **2005**, *127*, 11616. (o) *Org. Lett.* **2005**, *7*, 4297. (p) *J. Am. Chem. Soc.* **2002**, *124*, 5380. (q) *Tetrahedron Lett.* **2003**, *44*, 3915. (r) *Tetrahedron Lett.* **1994**, *35*, 3509. (s) *Tetrahedron Lett.* **1992**, *33*, 3181.

23-14 Complete the following synthesis of (−)-decarestrictine D by supplying the missing reagents. Note that some of the transformations shown may require more than one step. [*Tetrahedron Lett.* **1998**, *39*, 4421.]

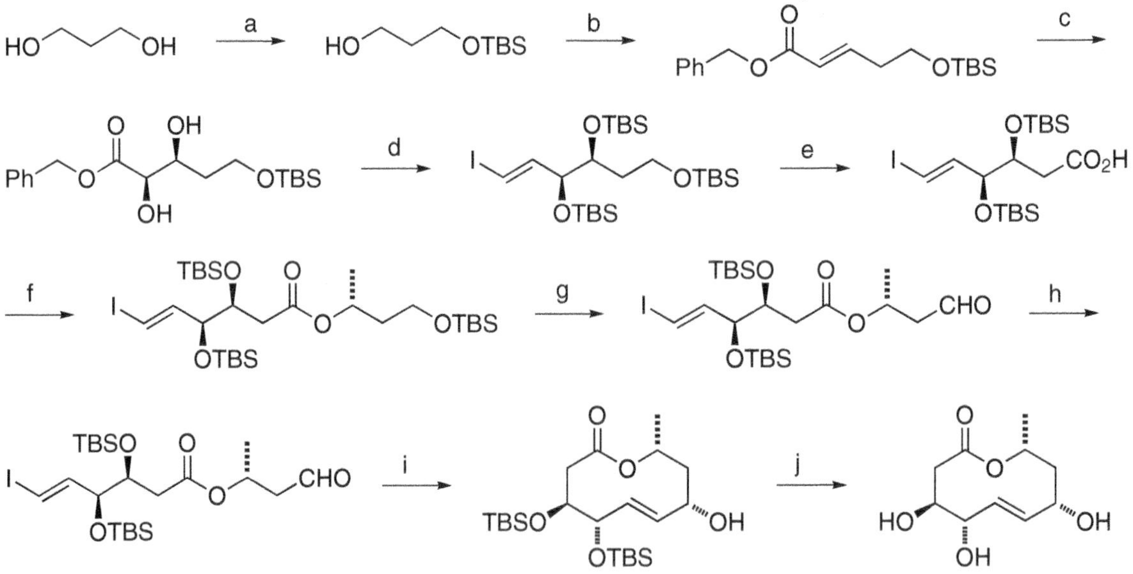

The cyclization reaction on the aldehyde using reagent **h** gives a single diastereoisomer of the product. How can this result be rationalized?

23-15 Answer the following questions about the total synthesis of (−)-sessilifoliamide, whose four contiguous chiral centers present a significant challenge. [*Org. Lett.* **2011**, *10*, 2634]

(a) The broad-brush retrosynthetic analysis of the target molecule is provided below. Provide a more detailed retrosynthetic analysis, paying particular attention to the problems of stereochemistry.

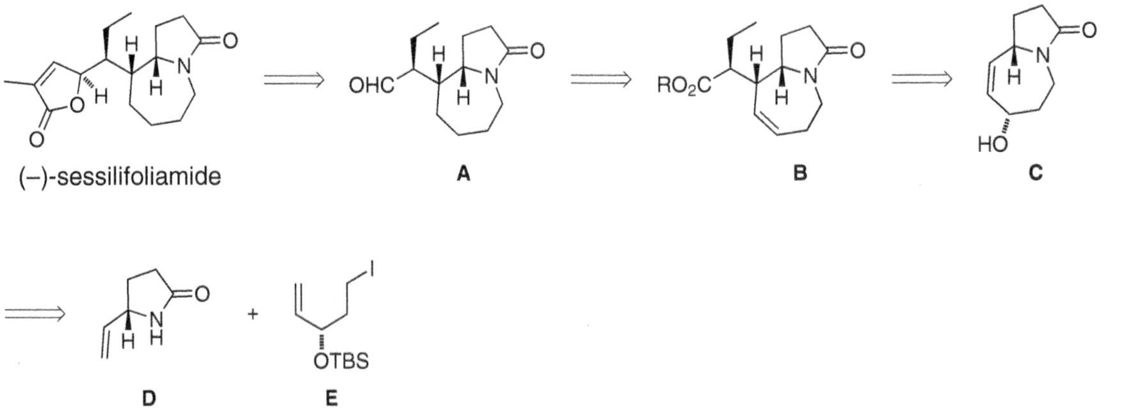

(b) Does your analysis of the authors' retrosynthetic analysis suggest any particularly critical steps in the synthesis? How is the chirality in the final product introduced?

(c) Complete the synthesis by providing the missing structures in the scheme below.

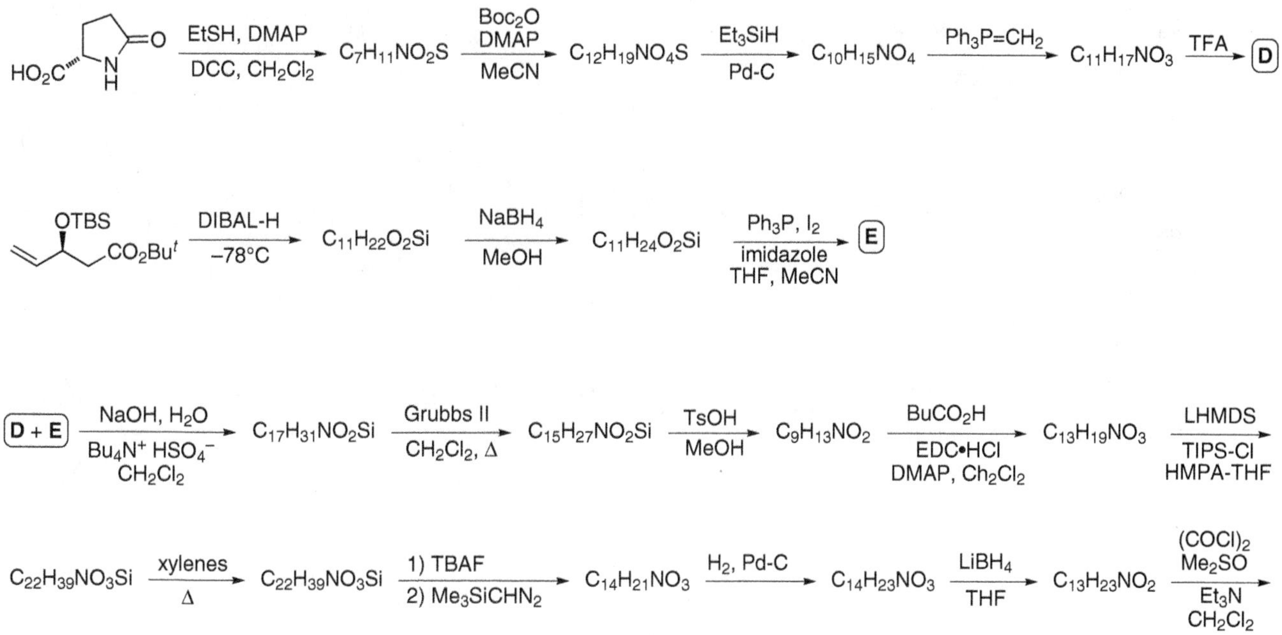

23-16 In 1993, Overman published an approach to the synthesis of optically active morphine derivatives [*J. Am. Chem. Soc.* **1993**, *115*, 11028] and a formal total synthesis of enantiomerically enriched morphine. Supply the missing reagents in the sequence below.

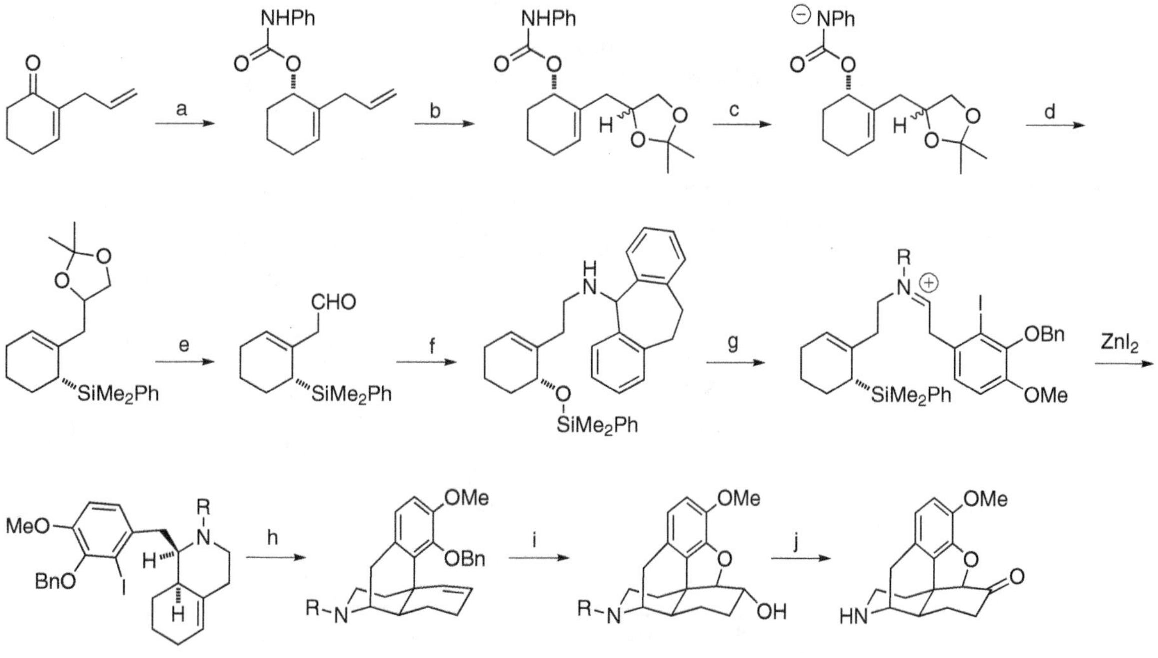

23-17 (a) Provide the missing reagents and structures in the synthesis of the left side (C1-C10) of hennoxazole A. [*J. Am. Chem. Soc.* **1995**, *117*, 558]

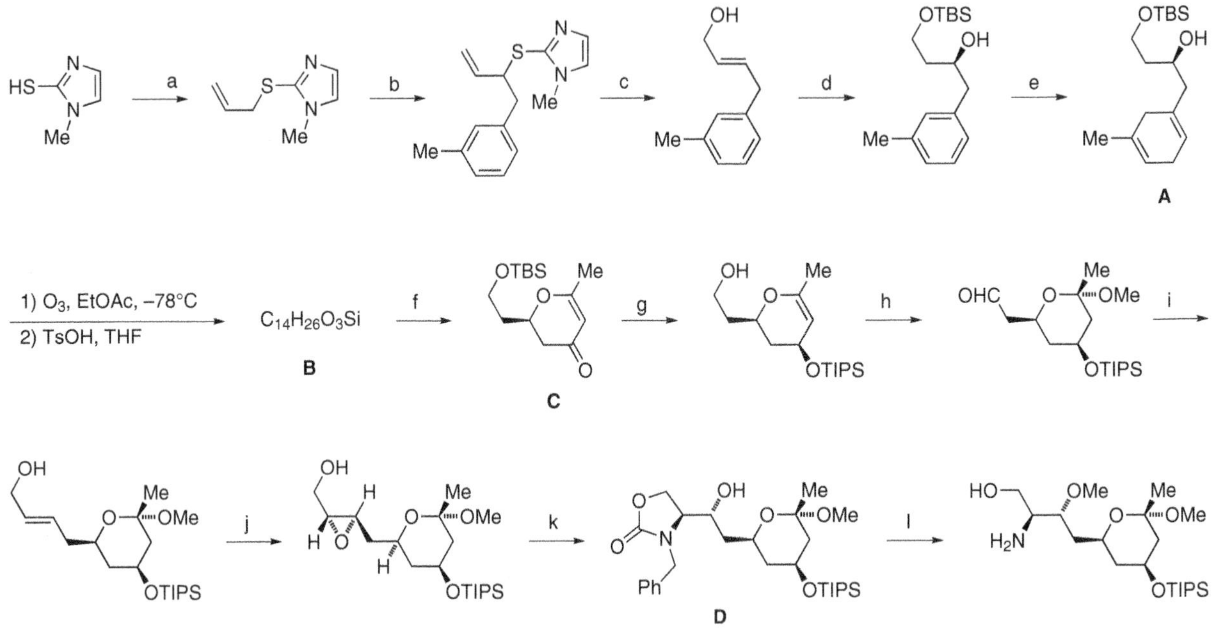

(b) Write a reasonable mechanism for the conversion of compound **B** into compound **C**.

(c) Compound **D** is accompanied by a minor isomer. What is the structure of this isomer?

(d) Write a mechanism for the formation of compound **D** that accounts for the stereochemistry of the product. Does this mechanism also account for the formation of the minor isomer?

23-18 Oxylipins are natural products that have wide-ranging biological effects, including stimulation of the immune system. The synthesis of oxylipin-3, (6*R*, 9*R*, 10*R*, 7*E*)-6, 9, 10-trihydroxyoctadec-7-enoic acid, is given below.

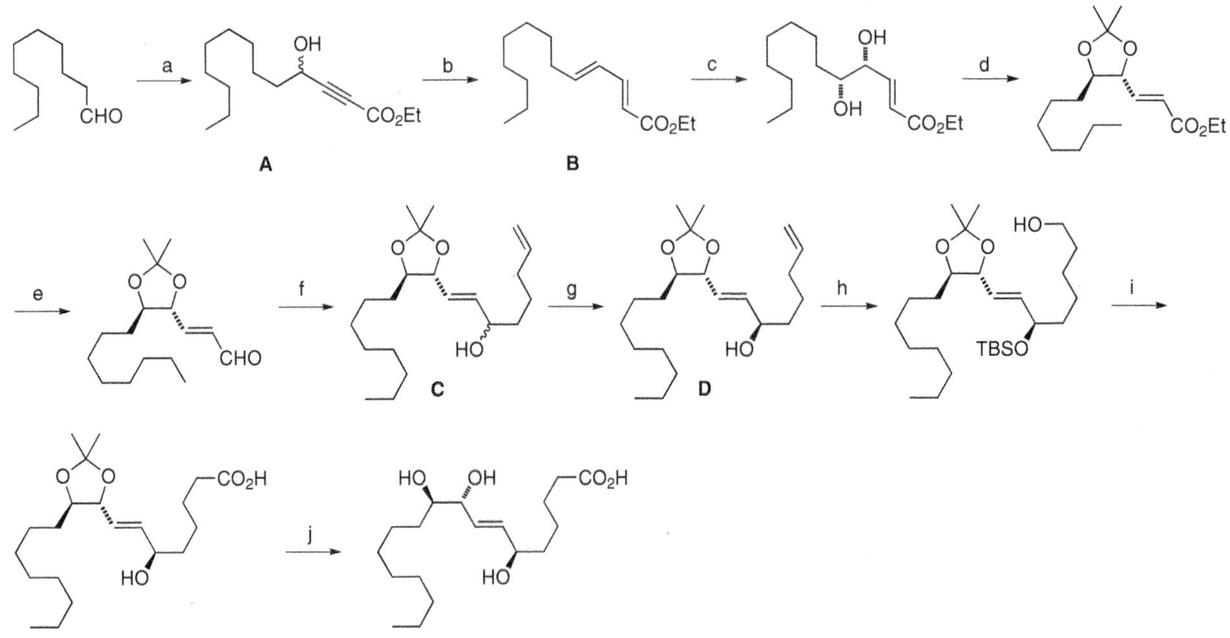

(a) Supply the missing reagents. [*Tetrahedron Lett.* **2012**, *53*, 3955]

(b) What transformations are required to complete the transformation of **A** to **B**? What characteristics would be necessary in a reagent to allow this transformation to be completed in a single step?

(c) Suggest a reason for preparing the mixture of alcohols (**C**) and then converting it to the *R* diastereoisomer (**D**). (The reaction used by the authors proceeds with d.r. 95%).

23-19 Solomonamides are peptide-based marine natural products isolated from a sponge. The solomonamide A molecule can be disconnected as shown: the DIS transforms consist of disconnection of the peptide bonds to give a serine-based synthon, a synthon based on valylglycine, and an unusual, highly functionalized segment, enclosed in the box (*Tetrahedron Lett.* **2013**, *54*, 2128).

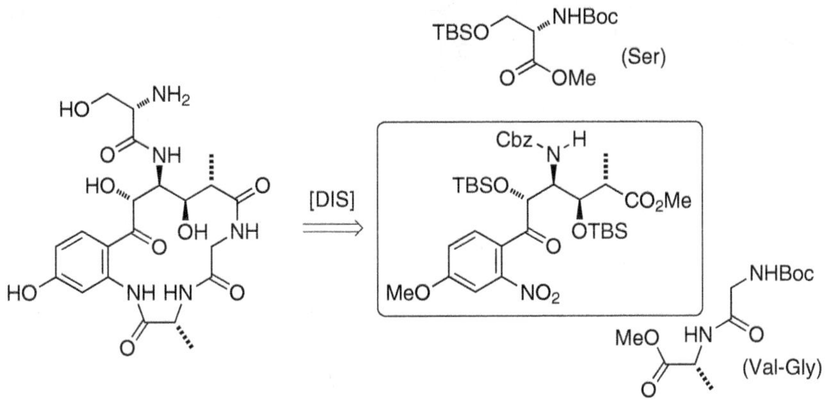

(a) Develop a retrosynthetic analysis of the highly functionalized segment of the solomonamide A, so that the synthesis begins with the Roche ester (methyl *R*-3-hydroxy-2-methylpropanoate), which is commercially available as both enantiomers

(b) Complete the synthesis of this fragment as reported by the authors by providing the structures of the missing compounds.

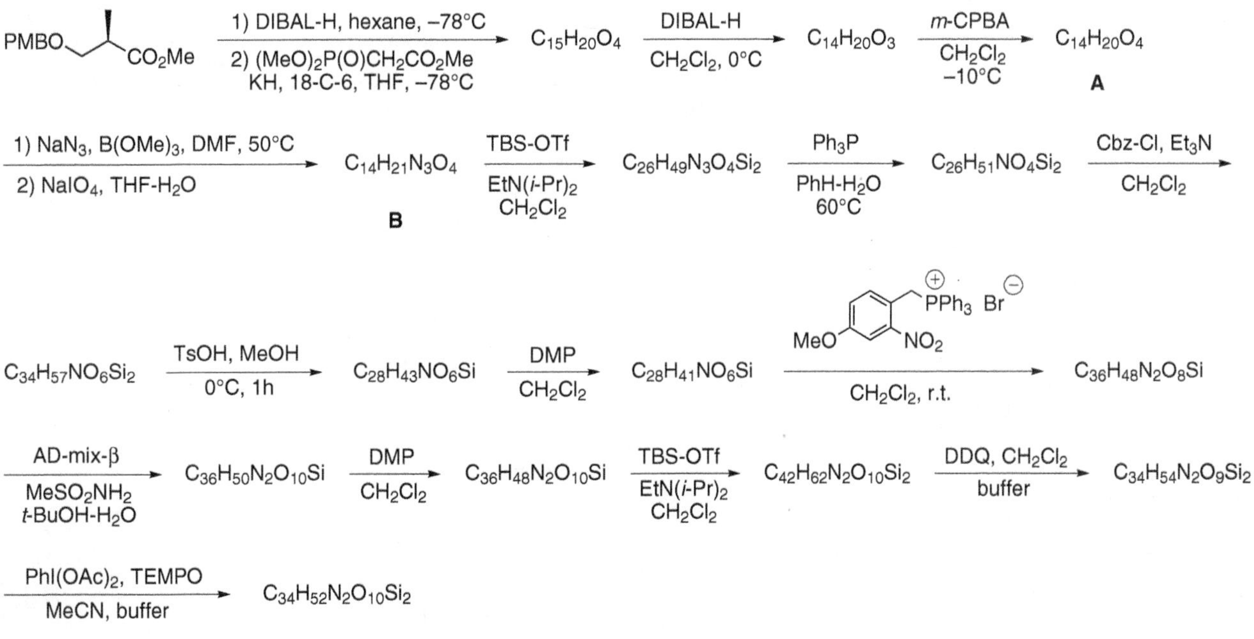

(c) What is the role of the trimethyl borate in the conversion of compound **A** to compound **B**?

23-20 The decahydrofluorene skeleton shown forms the basis for a number of antimicrobial natural products isolated from a variety of sources. The broad-brush retrosynthetic analysis of this molecule is given below. [*Tetrahedron Lett.* **2011**, *52*, 6242]

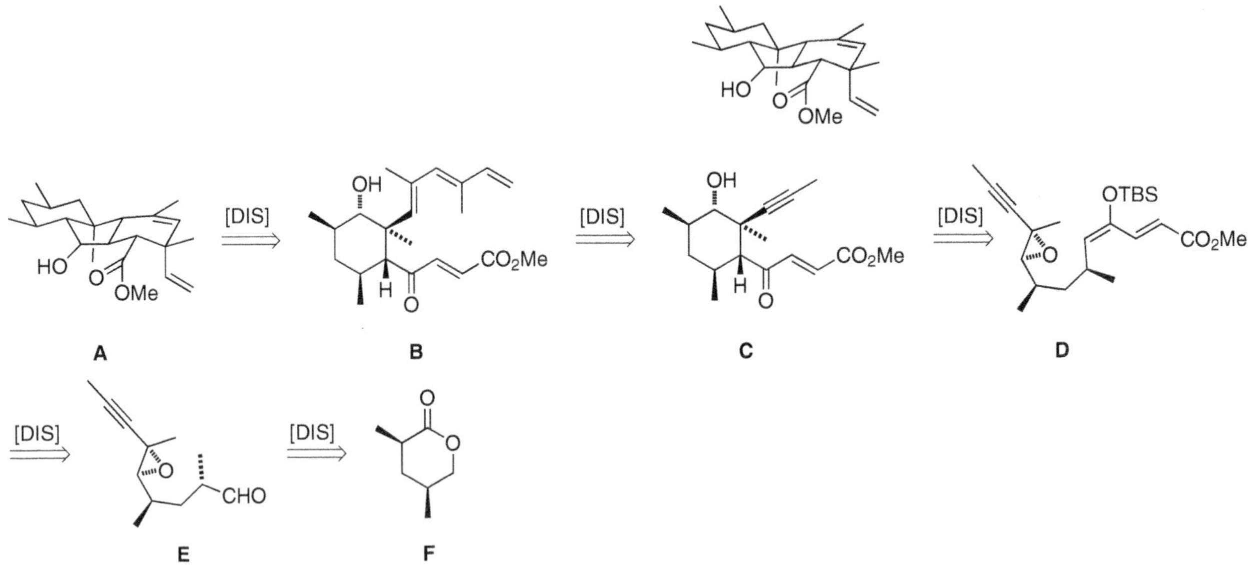

(a) Provide a detailed retrosynthetic analysis for the transformation of intermediate **B** to intermediate **C**.

(b) Complete the authors' synthesis by providing structures for each of the missing compounds. Molecular formulas are provided for guidance.

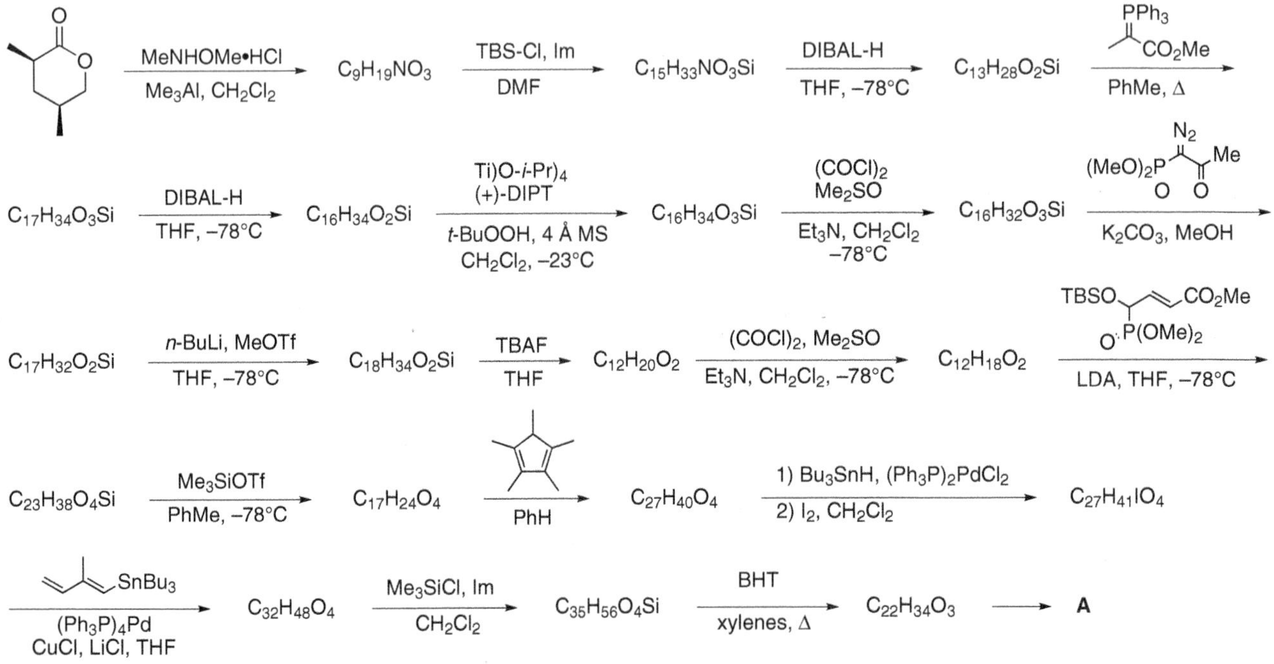

23-21 Isooncinotine is a representative of a series of macrocyclic spermidine alkaloids that contains one chiral center and a 22-membered lactam ring. The synthesis of the alkaloid by Fürstner provides another example from this laboratory of the use of catalytic reactions. The key step in Fürstner's synthesis [*Proc. Nat. Acad. Sci. U.S.A.* **2004,** *101,* 11960] involved the catalytic hydrogenation of a pyridinium ion, with concomitant release of the chiral auxiliary. Suggest a mechanism for this reaction that provides a rationalization for the stereochemistry observed.

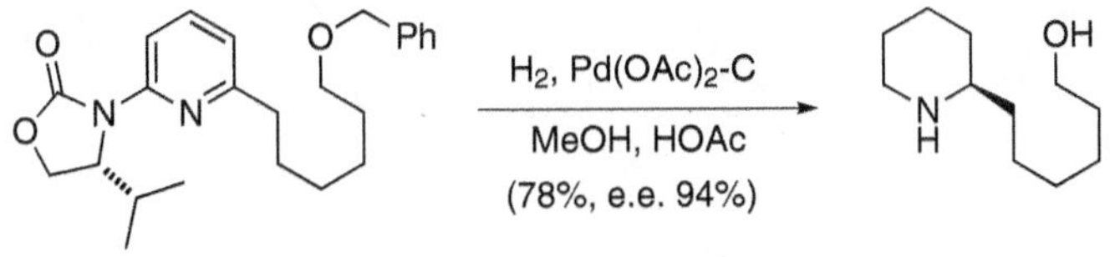

APPENDIX 1

Named Reactions, Named Reagents, and Named Rules

A

acetylene zipper: rearrangement of an internal alkyne to a terminal alkyne with a strong amide anion base (Ch. 11)

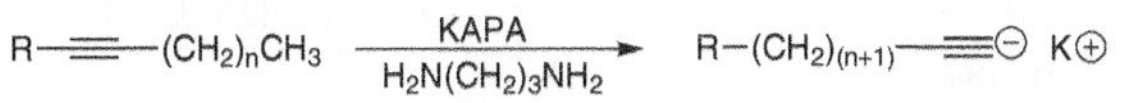

Adams catalyst: a highly active hydrogenation catalyst used as PtO_2 (Ch. 19)

aldol addition: the addition of an enolate anion to an aldehyde or ketone to give a β-hydroxycarbonyl compound as the product (Ch. 12)

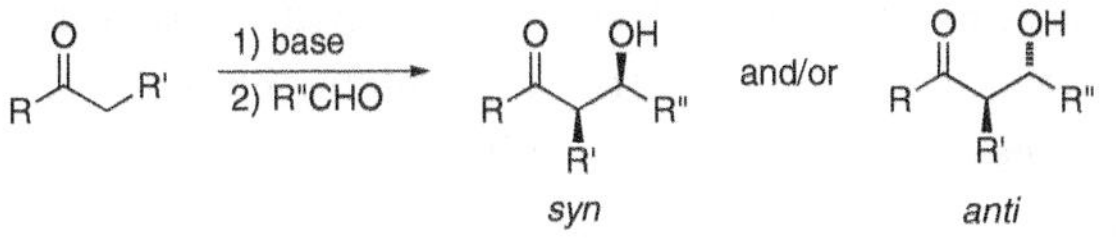

aldol condensation: the addition of an enolate anion to a carbonyl compound, followed by dehydration to give a conjugated product (Ch. 16)

allylboron allylation: an addition reaction between an aldehyde or ketone and an allylborane to give homoallylic alcohols. The reaction may be carried out asymmetrically using *B*-allyldiisopinocamphenylboranes as reagents. (Ch. 10)

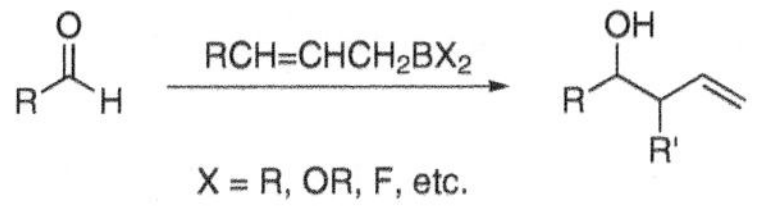

Alpine borane®: a chiral *B*-alkyl-9-borabicyclo(3.3.1)nonane derivative derived from α-pinene (Ch. 21)

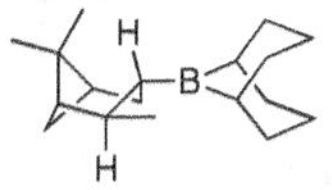

anchimeric assistance (neighboring group participation): participation by a neighboring group that stabilizes a carbocation or incipient carbocation during a reaction; results in increased rates of ionization reactions and retention of configuration at the carbon carrying the leaving group (Ch. 9)

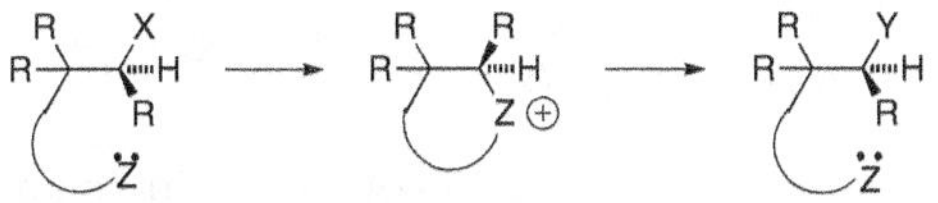

Arbuzov (Michaelis-Arbuzov) rearrangement: the conversion of a trialkyl phosphite to a dialkyl alkylphosphonate by treatment with an alkyl halide. (Ch. 22)

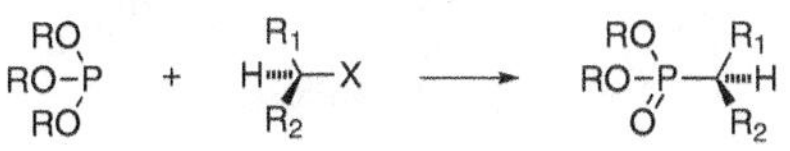

Arrhenius equation: the relationship that relates the rate constant for a reaction to the Kelvin temperature: $k = Ae^{-Ea/RT}$ (Ch. 8)

aza-Prins reaction: a Prins reaction where the oxonium ion is replaced by an iminium ion; works best in an intramolecular mode (Ch. 10)

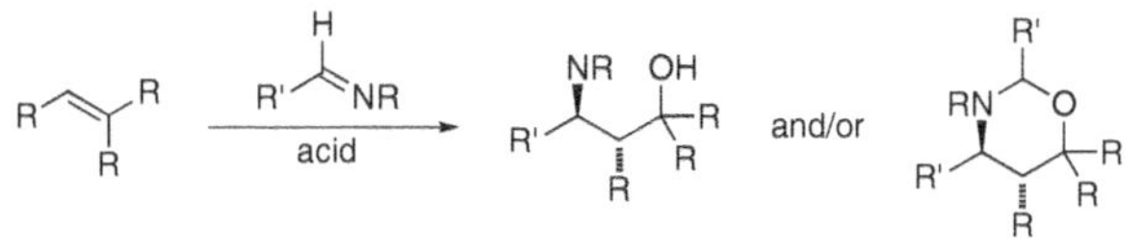

B

Babler oxidation: oxidation of a tertiary allylic alcohol to an α,β-unsaturated ketone by prolonged treatment with pyridinium chlorochromate (PCC); involves an allylic rearrangement (Ch. 18)

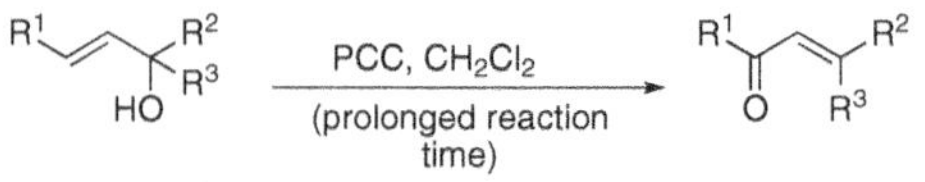

Baeyer-Villiger rearrangement: rearrangement of the α-hydroxyperester formed by addition of the peracid to a ketone carbonyl group to give an ester by net insertion of oxygen into the adjacent C—C bond (Ch. 18)

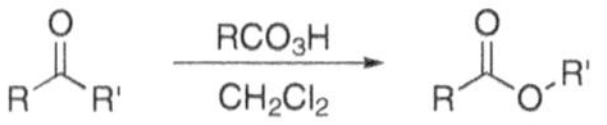

Baker-Nathan effect: the observation that alkyl substituents on a benzyl halide facilitate displacement in an order opposite to their inductive effects; correlates with the ability of the alkyl group to stabilize a carbocation by hyperconjugation (Ch. 9)

$$CH_3— > CH_3CH_2— > (CH_3)_2CH— > (CH_3)_3C—$$

Baldwin's rules: a series of empirical rules for nucleophilic ring closure that can be used to predict the regiochemistry and the ease of nucleophilic ring closures (Ch. 5)

Bamford-Stevens reaction: heating a tosylhydrazone with an alkoxide base to give alkenes and cyclopropanes; probably occurs through carbene intermediates. See Shapiro reaction. (Ch. 20)

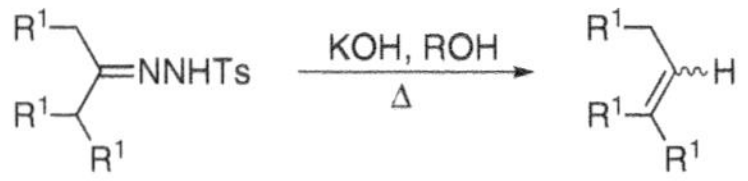

Barton ester: an ester of *N*-hydroxy-2-pyridinethione; used as a precursor to alkyl radicals (Ch. 14)

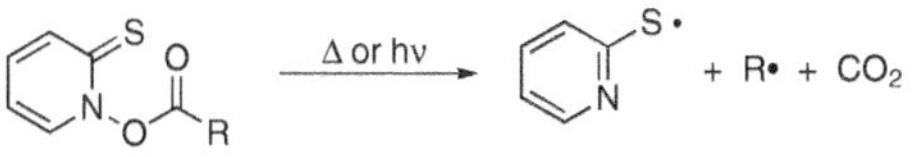

Barton-McCombie deoxygenation: the reduction of an alcohol to an alkane by means of stannane reduction of the xanthate or similar thionoester (Ch. 14)

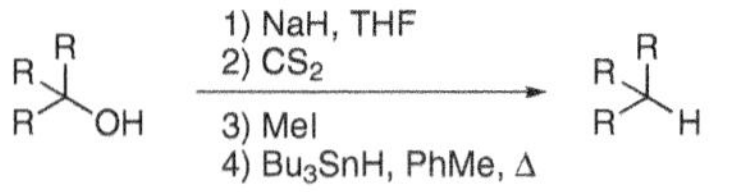

Barton reaction: remote functionalization of an unactivated position by homolysis of a nitrite ester and intramolecular hydrogen atom transfer from carbon to oxygen through a six-membered transition state. The product of the reaction is a δ-nitrosoalcohol that tautomerizes to an oxime. (Ch. 17, 18)

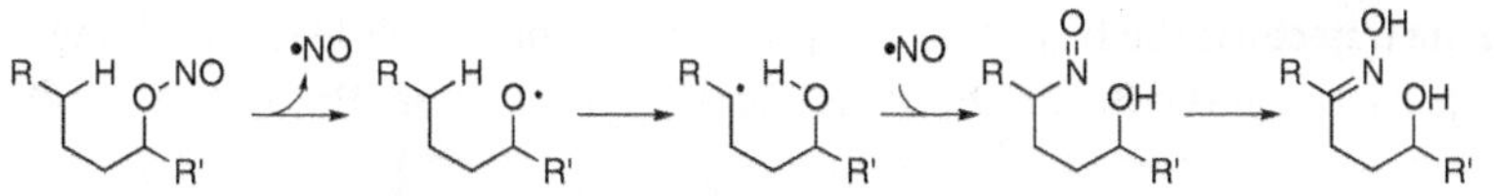

Beckmann rearrangement: rearrangement of an oxime under acidic conditions to give an amide by net insertion of nitrogen into the adjacent C—C bond (Ch. 18)

Benkeser reduction: variant of the Birch reduction in which the substrate is reduced with calcium metal in a low-boiling amine solvent (Ch. 20)

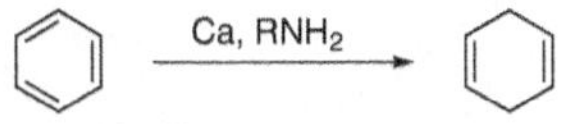

benzoin condensation: the self-condensation of two molecules of an aldehyde to give an α-hydroxyketone; catalyzed by cyanide ion or by *N*-heterocyclic carbenes (Ch. 16)

Bergman cyclization: the cyclization of a conjugated ene-diyne to a *p*-benzyne (Ch. 14)

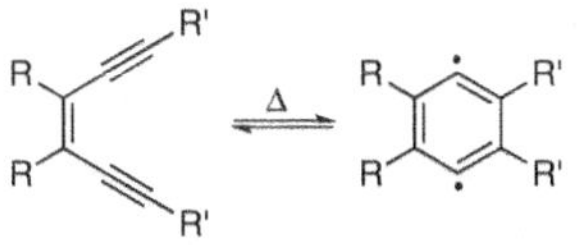

Bestmann reagent: diethyl 1-diazo-2-oxopropylphosphonate (Ch. 16)

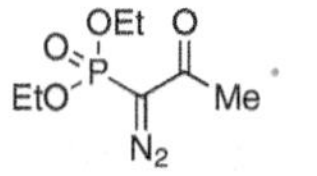

Birch reduction: reduction with an alkali metal (usually Li) in liquid ammonia with or without an organic cosolvent or proton source (Ch. 20)

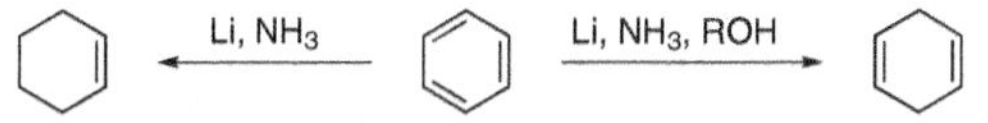

Borodin-Hunsdiecker reaction: the formation of the alkyl bromide of the next lower homolog by treating a silver or mercury carboxylate with elemental bromine. The reaction now has a variation using the Barton esters in a halogenated solvent. (Ch. 14)

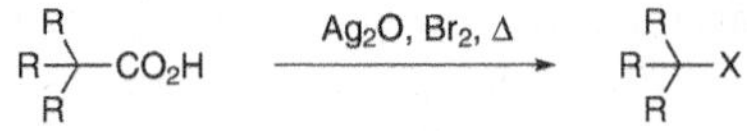

Bouveault-Blanc reduction: reduction with sodium metal in boiling ethanol. The reaction reduces esters to primary alcohols and enones to saturated secondary alcohols. (Ch. 20)

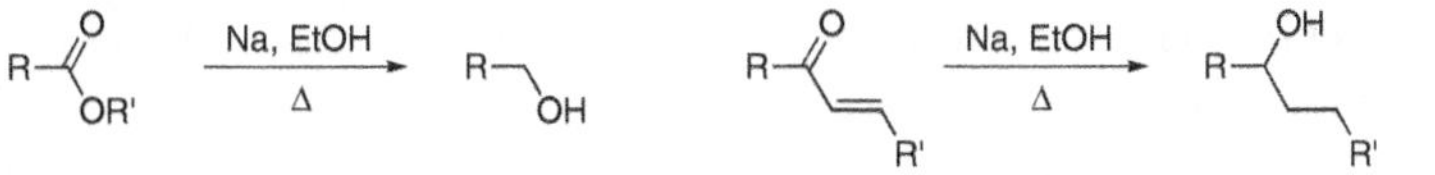

Brønsted catalysis law: a relationship describing the effects of acid or base on the rate constant of a reaction: $\log k = \alpha \log K_a + \log C = \log C - pK_a$. The value of a determines the nature of the acid catalysis—$\alpha = 0$, no catalysis; $\alpha = 1$, specific acid catalysis; $0 < \alpha < 1$, general acid catalysis (Ch. 8)

Buchwald-Hartwig reaction: palladium-catalyzed displacement of an allylic ester or other allylic group by an amine or similar nucleophile (Ch. 17)

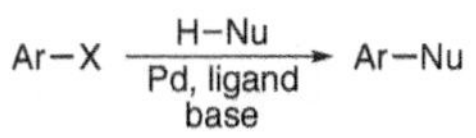

Bürgi-Dunitz trajectory: the trajectory of approach of a nucleophile to a carbonyl group. The approach is at the Bürgi-Dunitz angle of approximately 110° to the C—O bond (Ch. 5)

C

Cahn-Ingold-Prelog systems: systems for specifying stereochemistry (*R/S*, *E/Z*, *M/P*) based on sequence rules (Ch. 2)

Cannizzaro reaction: internal redox reaction of aromatic aldehydes promoted by base. Crossed Cannizzaro reactions with formaldehyde and base allow reduction of the aldehyde. (Ch. 20)

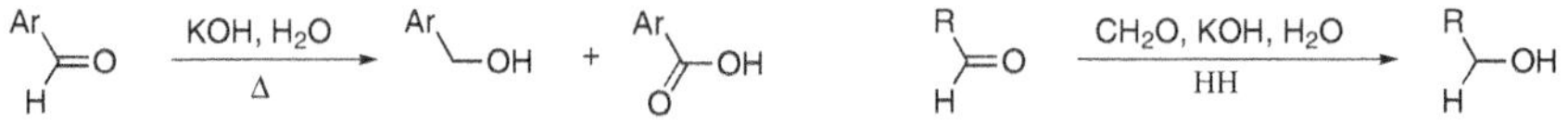

Carroll variant of the Claisen rearrangement: a variant of the Claisen rearrangement in which the reactive 1,5-diene is formed by enolization of a β-ketoester. The reaction is rendered irreversible by the loss of carbon dioxide. (Ch. 6)

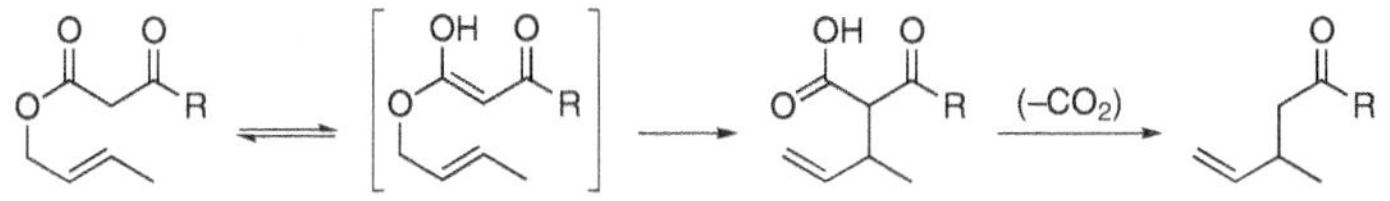

Claisen condensation: the base-promoted condensation of two esters to give a β-ketoester. The two esters may be the same or different. (Ch. 16)

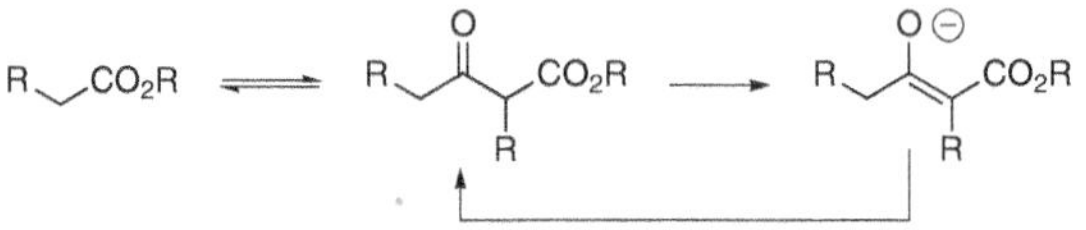

Claisen rearrangement: the (3,3)-sigmatropic rearrangement of an allyl vinyl ether to a γ,δ-unsaturated carbonyl compound (Ch. 6)

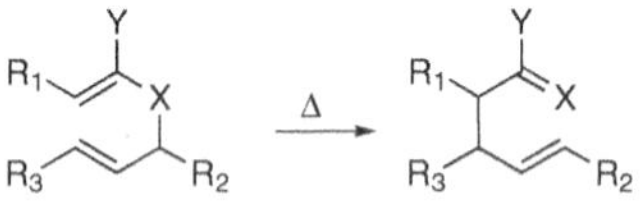

Carroll variant: a variant of the Claisen rearrangement in which the reactive 1,5-diene is formed by enolization of a β-ketoester. The reaction is rendered irreversible by the loss of carbon dioxide

Eschenmoser variant: a variant of the Claisen rearrangement in which the reactive 1,5-diene is formed from an allylic alcohol and an orthoamide

Ireland ester enolate variant: a variant of the Claisen rearrangement in which the reactive 1,5-diene is the lithium or silyl enolate of an allylic ester

Johnson orthoester variant: a variant of the Claisen rearrangement in which the reactive 1,5-diene is formed from an allylic alcohol and an orthoester

Claisen-Schmidt reaction: condensation of a ketone enolate with an ester to give a β-diketone (Ch. 16)

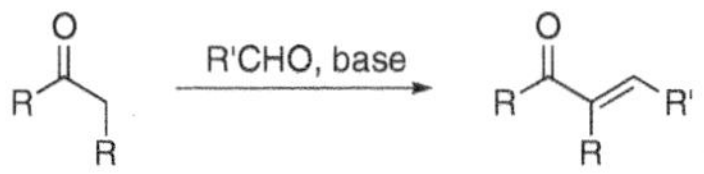

Clar sextet: the classical electron sextet in a six-membered ring. Maximum aromatic resonance stabilization is associated with the maximum number of Clar rings. (Ch. 7)

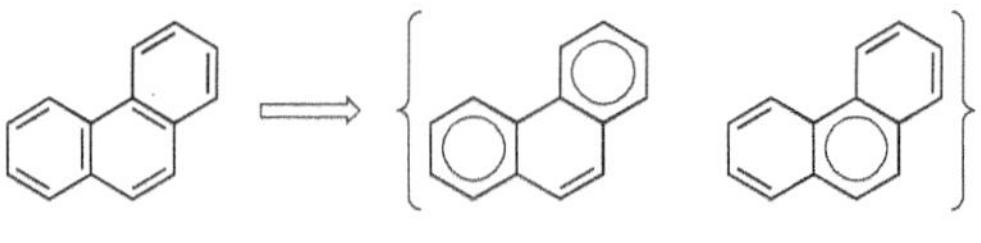

two Clar sextets one Clar sextet

Clemmensen reduction: deoxygenation of an aldehyde or ketone by heating with zinc amalgam in concentrated hydrochloric acid. (Ch. 20)

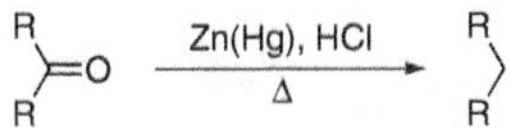

click chemistry: metal-catalyzed 1,3-dipolar cycloadditions to give five-membered heterocyclic products (Ch. 6)

Collins reagent: CrO_3•2py, CH_2Cl_2; oxidizes primary alcohols to aldehydes (Ch. 18)

Cope rearrangement: the (3,3)-sigmatropic rearrangement of a 1,5-diene. The reaction may come to an equilibrium. (Ch. 6)

Corey-Bakshi-Shibata (CBS) oxazaborolidine: a chiral catalyst for reductions of ketones with borane. The catalyst contains both nucleophilic (N) and electrophilic (B) sites. (Ch. 21)

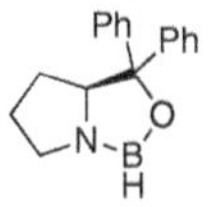

Corey-Chaykovsky reaction: the formation of a three-membered ring product using a sulfur ylide. Epoxides ($R_2S{=}CR_2$ + $R_2C{=}O$) and cyclopropanes ($R_2S(O){=}CR_2$ + RCH=CHCOR) may be formed. (Ch. 11; Ch. 16)

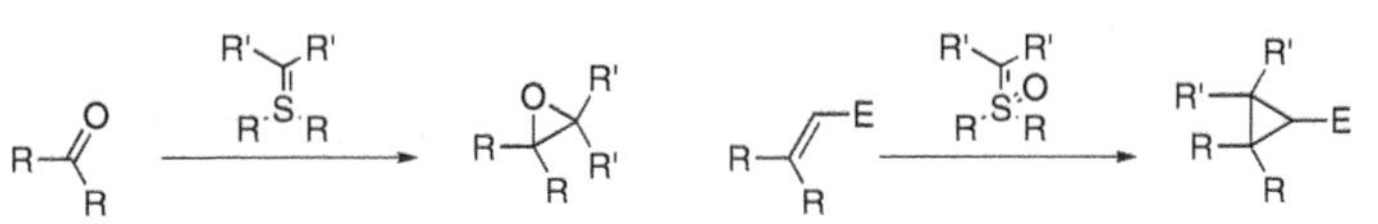

Corey-Fuchs reaction: synthesis of an alkyne from an aldehyde by sequential Wittig dibromoalkenylation and dehalogenation with butyllithium (Ch. 16)

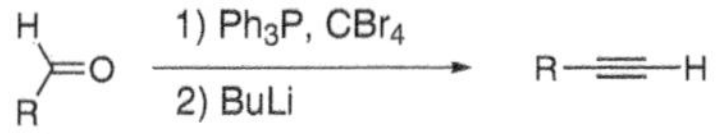

Corey-Kim oxidation: oxidation of primary alcohols to aldehydes using an alkyl methyl sulfide and *N*-chlorosuccinimide in dichloromethane at low temperature in the presence of triethylamine (Ch. 18)

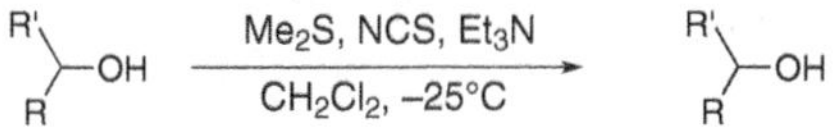

Corey-Winter elimination: the suprafacial elimination of a cyclic carbonate or thiocarbonate to give an alkene when treated with a trialkylphosphine or trialkyl phosphite (Ch. 22)

Cram's rule: an empirical rule for predicting the absolute configuration of the major diastereoisomer formed by addition of nucleophiles to chiral aldehydes. See Felkin-Anh rule. (Ch. 2)

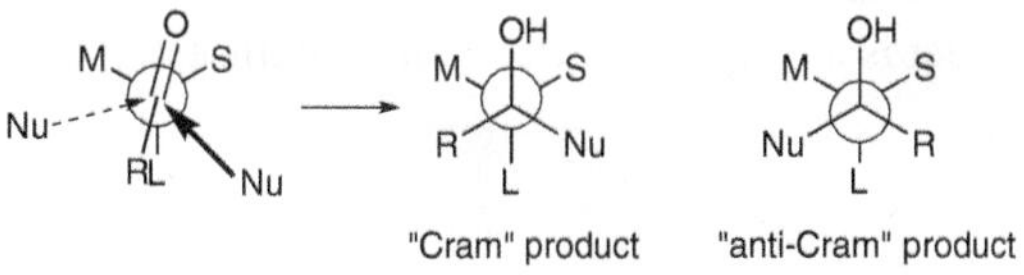

Criegee oxidation: oxidative cleavage of a 1,2-diol to a pair of carbonyl compounds with lead tetraacetate (Ch. 18)

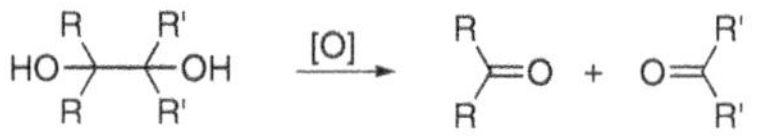

Curtius rearrangement: rearrangement of an acyl azide to give an isocyanate by insertion of nitrogen into the adjacent C—C bond (Ch. 18)

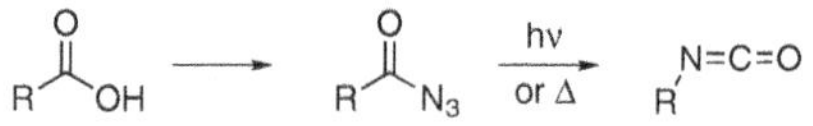

[2 + 1] cycloaddition (carbene cycloaddition): a concerted suprafacial cycloaddition of a carbene to an alkene to give a cyclopropane. See Simmons-Smith reaction. (Ch. 6)

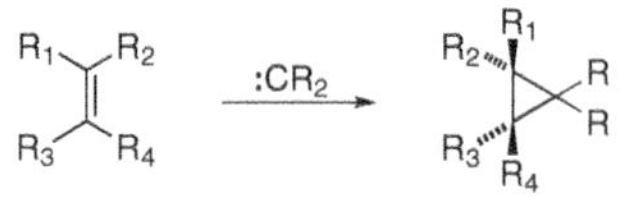

[2 + 2] cycloaddition (ketene addition): a cycloaddition of a ketene to an alkene to give a cyclobutanone; may not be concerted (Ch. 6)

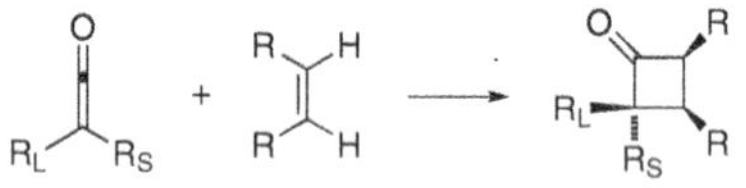

[2 + 3] cycloaddition (1,3-dipolar cycloaddition): concerted suprafacial addition of a resonance-stabilized zwitterion to an alkene or alkyne to give a five-membered heterocyclic product (Ch. 6)

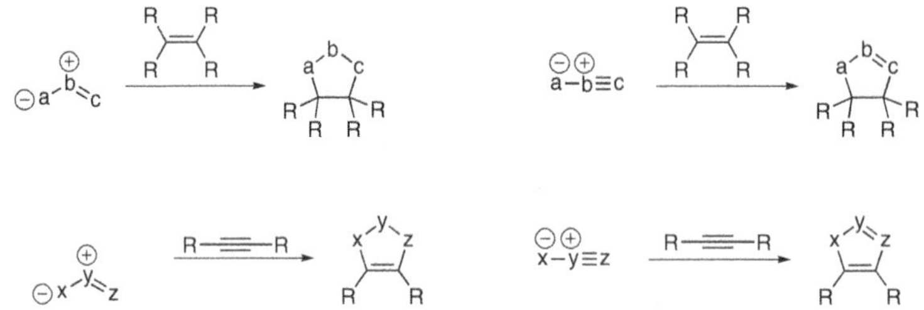

[2 + 4] cycloaddition (Diels-Alder reaction): a concerted cycloaddition reaction between a diene and an alkene (dienophile). The reaction is suprafacial with respect to both reactants. (Ch. 6)

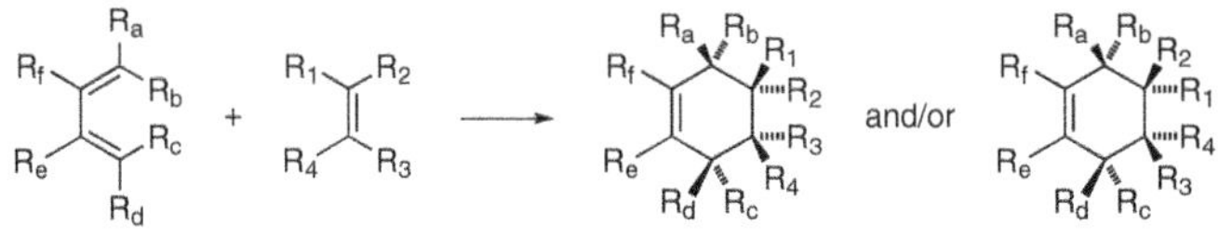

D

Danishefsky diene: a highly reactive diene in the Diels-Alder reaction; 1-methoxy-3-trimethylsilyloxy-1,3-butadiene (Ch. 22)

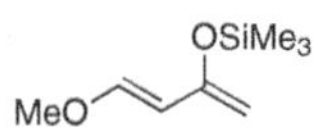

Darzens condensation: the condensation of the enolate of an α-haloester with an aldehyde or ketone. The product is an α,β-epoxyester, or glycidic ester. (Ch. 16)

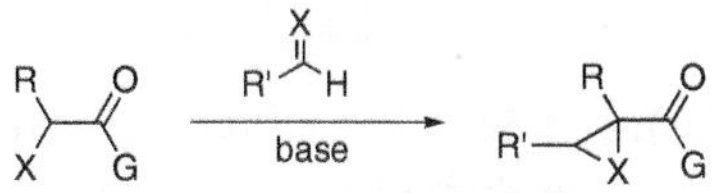

Davis oxaziridines: chiral oxaziridines that can be used for chiral oxidation of ketone enolate anions (Ch. 21)

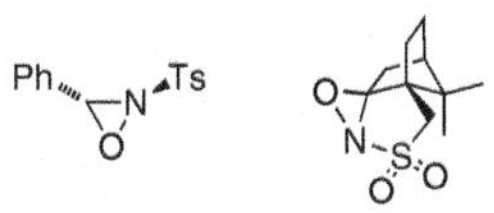

Dess-Martin periodinane (DMP): an iodine (V) oxidant prepared from *o*-iodobenzoic acid (Ch. 18)

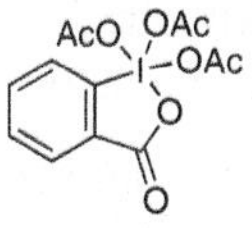

Dewar resonance energy (DRE): a value for the resonance energy of a ring calculated using the equation derived by M .J. S. Dewar (Ch. 7)

Dieckmann condensation: an intramolecular Claisen condensation (Ch. 16)

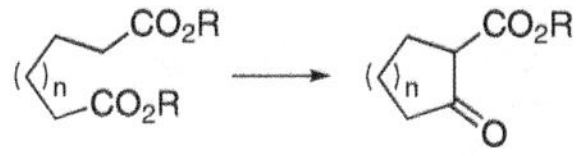

n = 1, 2, 3, and > 11

Diels-Alder reaction: a [4 + 2] cycloaddition reaction between a diene and an alkene (dienophile). The reaction is suprafacial with respect to both reactants. (Ch. 6)

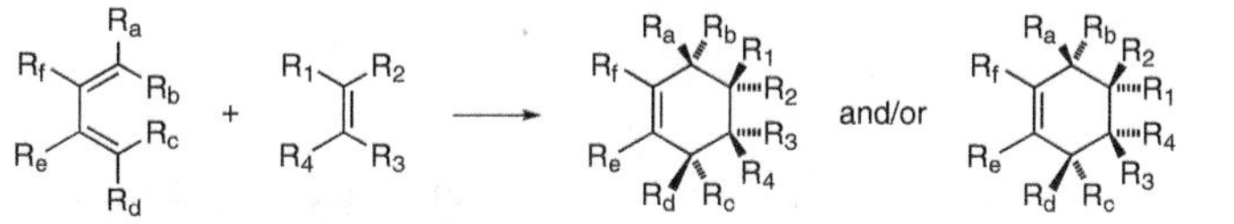

1,3-dipolar cycloadditions: concerted, suprafacial addition of a resonance-stabilized zwitterion to an alkene or alkyne to give a five-membered heterocyclic product (Ch. 6)

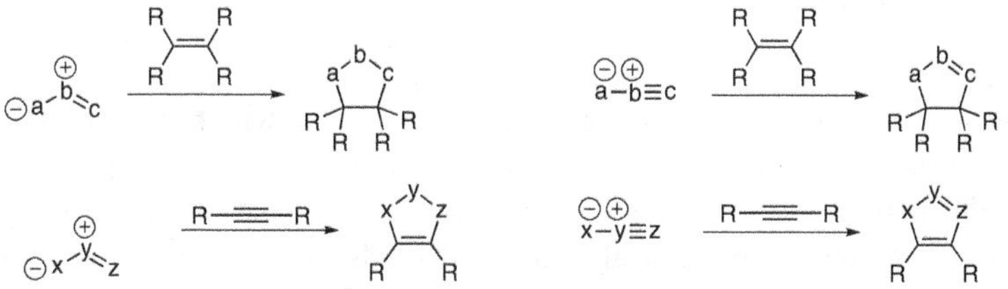

Doebner reaction: a Knoevenagel reaction catalyzed by a secondary amine and a carboxylic acid (Ch. 16)

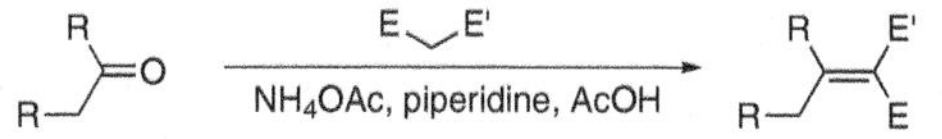

Duff reaction: conversion of an arene to the corresponding aldehyde by treatment with hexamethylenetetramine and acetic acid, followed by hydrolysis of the resulting iminium salt (Ch. 10)

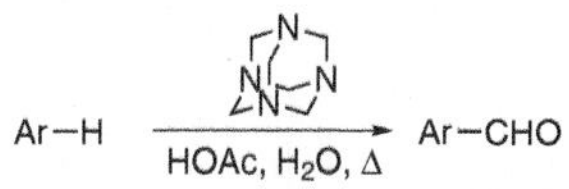

E

ene reaction (Alder ene reaction): a pericyclic reaction between two alkenes (the "ene" and the "enophile") that leads to migration of an allylic hydrogen and formation of a new C—C σ bond (Ch. 6)

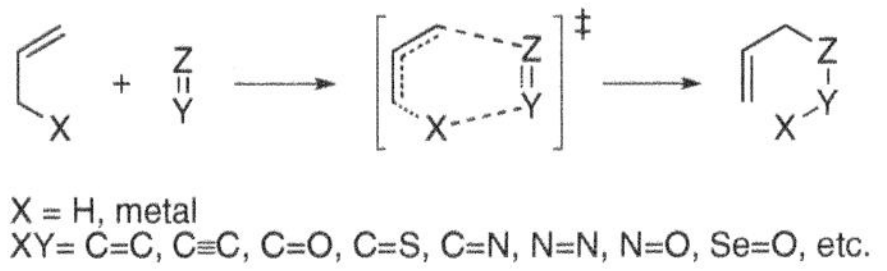

ene-yne metathesis: the reaction between an alkene and an alkyne to give a conjugated diene; catalyzed by metal-carbene complexes

Eschenmoser-Tanabe fragmentation: fragmentation of the tosylhydrazone of an α,β-epoxyketone by heat. See Wharton reaction. (Ch. 20)

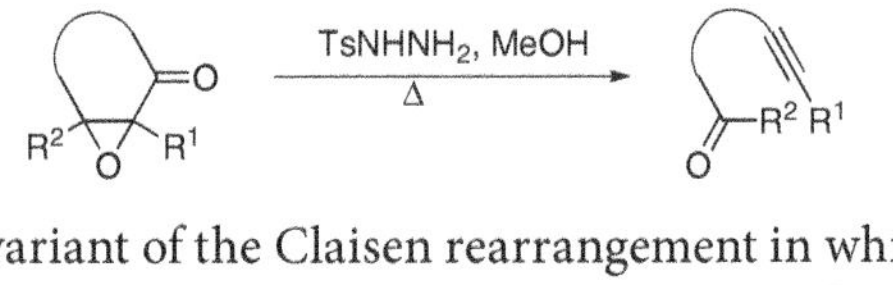

Eschenmoser variant: a variant of the Claisen rearrangement in which the reactive 1,5-diene is formed from an allylic alcohol and an orthoamide (Ch. 6)

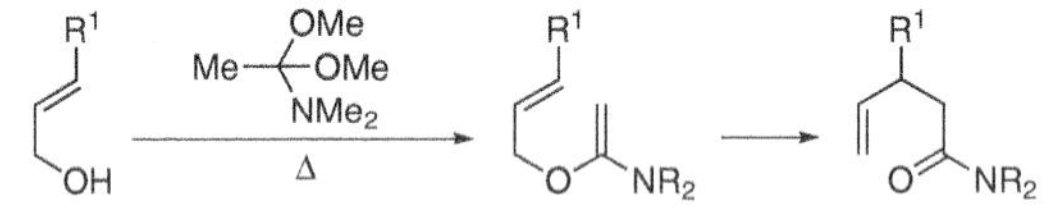

Eschweiler-Clark methylation: conversion of a primary or secondary amine to a tertiary amine (the *N,N*-dimethylamine, or the *N*-methylamine) with formaldehyde and formic acid (Ch. 20)

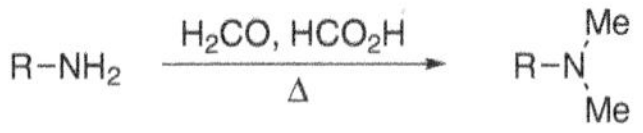

Evans asymmetric aldol addition: the aldol addition between an aldehyde and the boron enolate derived from a chiral *N*-acyloxazolidinone. This reaction gives exceptional control over stereochemistry of the aldol product. (Ch. 12)

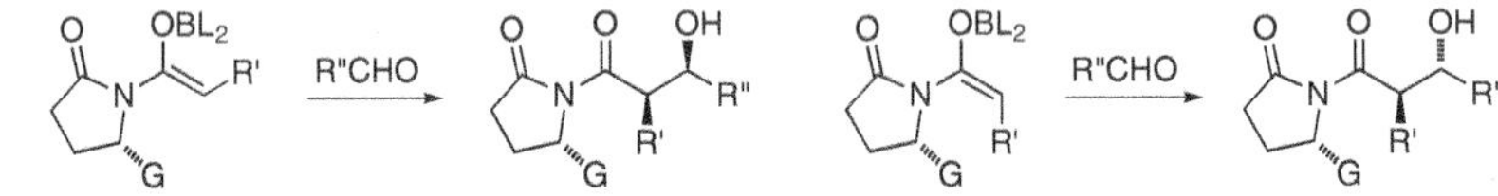

Evans-Lapworth analysis: a form of retrosynthetic analysis in which the inherent charge affinity labels are used to identify potential synthons (Ch. 15)

Evans oxazolidinones: cyclic urethanes formed from chiral β-aminoalcohols, derived either from amino acids or natural β-aminoalcohols (e.g., norephedrine). They are useful chiral adjuvants in a wide range of reactions. (Ch. 12)

F

Felkin-Anh rule: an empirical rule for predicting the absolute configuration of the major diastereoisomer formed by addition of nucleophiles to chiral aldehydes. See Cram's rule. (Ch. 2)

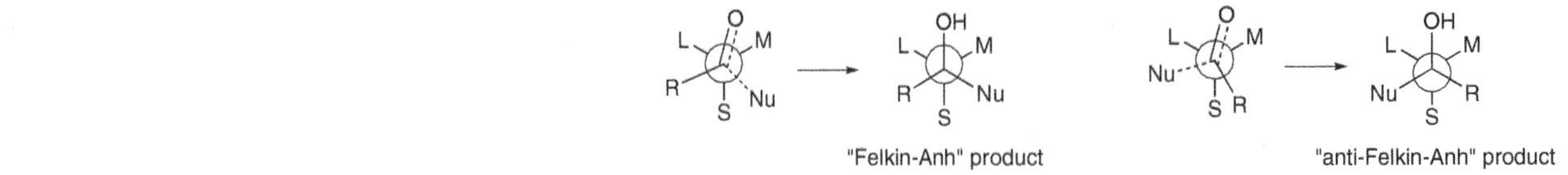

Fétizon reagent: Ag_2CO_3, celite; oxidizes primary alcohols and lactols (Ch. 18)

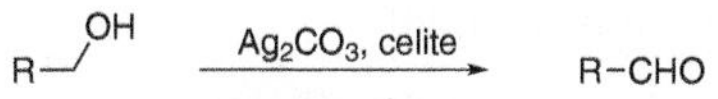

Fischer complexes: chromium-carbene complexes. A common example is shown below. (Ch. 17)

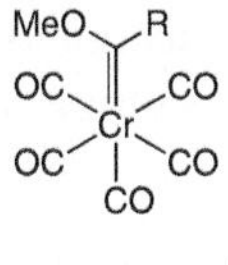

Friedel-Crafts acylation: the reaction between an acid chloride or anhydride (including mixed anhydrides) and an arene in the presence of Lewis acid catalyst to give an aryl ketone. The reaction does not work with electron-deficient arenes. (Ch. 10)

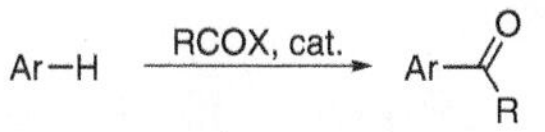

Friedel-Crafts alkylation: the reaction between an alkyl halide and an arene in the presence of a Lewis acid catalyst to give an alkylarene. It does not work with electron-deficient arenes, and is often difficult to control due to polyalkylation. Using chiral catalysts has allowed the reaction to be carried out asymmetrically. (Ch. 10)

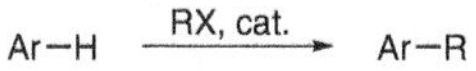

Frost-Musulin circle: a mnemonic for predicting orbital degeneracy and relative orbital energies in Hückel aromatic species (Ch. 7)

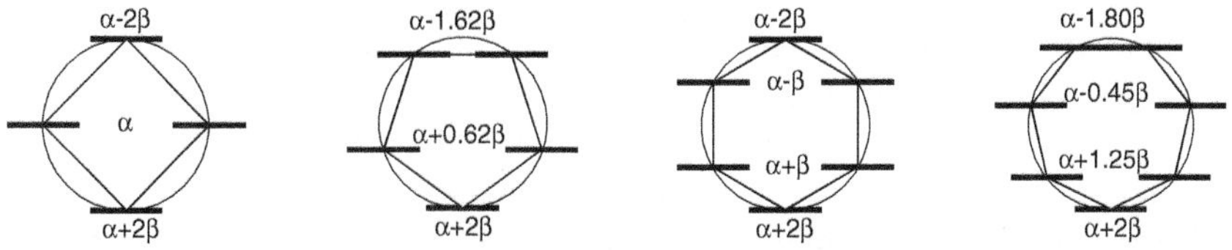

G

Gattermann formylation: conversion of an arene to the corresponding aldehyde by treatment with a metal cyanide, hydrogen chloride, and a Lewis acid catalyst; the metal cyanide often also serves as the Lewis acid. The reaction depends on the H_0 value of the solution. It does not work with electron-deficient arenes. (Ch. 10)

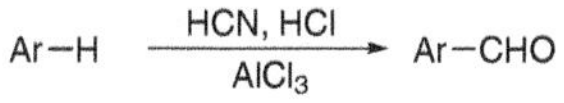

Gattermann-Koch formylation: conversion of an arene to the corresponding aldehyde by treatment with carbon monoxide and hydrogen chloride in the presence of a copper (I) salt and a strong Lewis acid. It does not work with electron-deficient arenes. (Ch. 10)

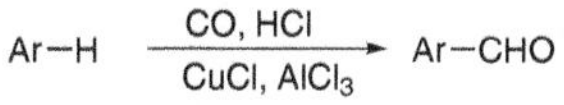

Gattermann-Olah formylation: conversion of an arene to the corresponding aldehyde by treatment with formyl fluoride and boron trifluoride. It does not work with electron-deficient arenes. (Ch. 10)

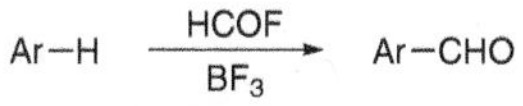

Gilman reagent: a lithium dialkylcuprate formed by treatment of two equivalents of an alkyllithium with one equivalent of a cuprous halide (Ch. 11)

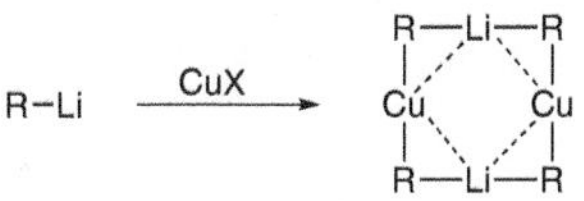

Gomberg-Bachmann reaction: the formation of aryl free radicals by one-electron reduction of aryldiazonium ions (Ch. 14)

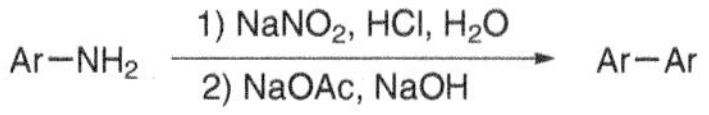

Grieco selenylation: the conversion of an alcohol to a selenide by treatment of the alcohol with a selenocyanate and a trialkylphosphine. The reaction proceeds with inversion of configuration. (Ch. 22)

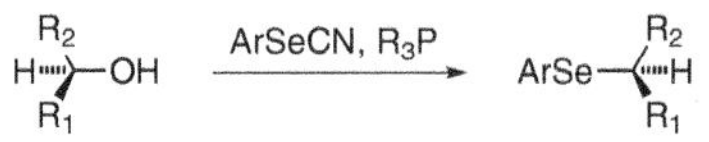

Grubbs catalysts: ruthenium carbene catalysts for metathesis reactions (Ch. 17)

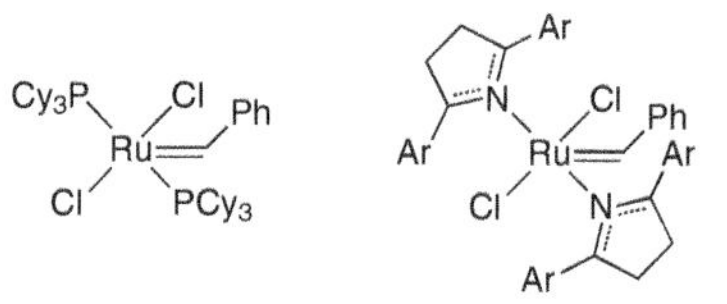

Grubbs-I Grubbs-II

Grunwald-Winstein relationship: a relationship that defines the effects of solvent on the rate constant of a reaction; log $(k/k_0) = mY$ (Ch. 8)

H

Hammett constants: substituent constants (Ch. 8)

σ: based on the ionization of substituted benzoic acids

σ^+: based on the rates of solvolysis of substituted cumyl chlorides

σ^-: based on the ionization of substituted phenols

σ_I: the contribution to σ from inductive effects, ionization of 4-substituted bicyclo(2.2.2)octane-1-carboxylic acids

σ_R: the contribution to σ from resonance effects; $\sigma_R = \sigma - \sigma_I$

Hammett equation: a linear free energy relationship: log $(K/K_o) = \sigma\rho$; log $(k/k_o) = \sigma\rho$. (Ch. 8)

Hammond postulate: a postulate that predicts that in a series of related reactions, the more exothermic (exergonic) a reaction, the lower its activation energy, the faster the reaction, and the more closely the structure of the activated complex resembles the structure of the reactants. The converse is true for endothermic (endergonic) reactions, where the activated complex most closely resembles the structure of the products. (Ch. 1)

Hantzsch esters: dialkyl 4-aryl-2,6-dimethyl-1,4-dihydropyridine-3,5-dicarboxylates; used as hydride donors in ionic hydrogenation (Ch. 19)

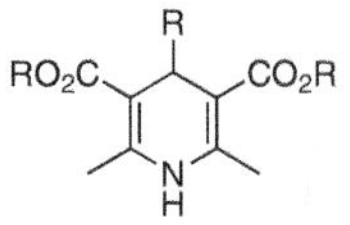

Heck reaction: palladium-catalyzed coupling of a halide with a hydrocarbon in the presence of a Lewis base to react with the hydrogen halide formed (Ch. 17)

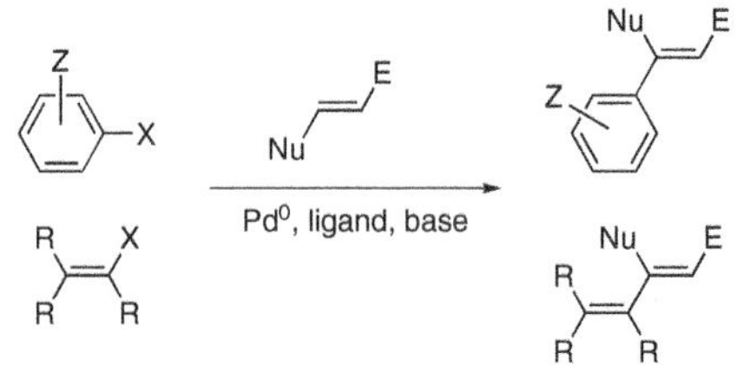

Hell-Volhard-Zelinskii reaction: the replacement of an α-hydrogen of a carboxylic acid by bromine and red phosphorus. The reaction proceeds through the enol of the acid bromide. (Ch. 18)

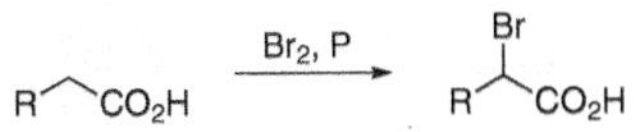

Henry reaction: an analog of the aldol condensation reaction where the enolate is derived from a nitroalkane (Ch. 16)

Hiyama coupling: palladium-catalyzed coupling of a silane and a halide in the presence of stoichiometric fluoride ion (Ch. 17)

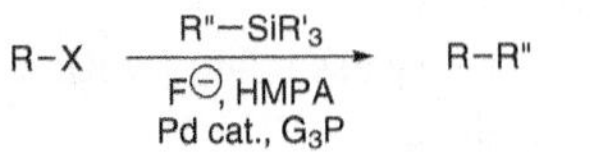

Hiyama-Denmark coupling: palladium-catalyzed coupling of a silanol and a halide in the presence of base (Ch. 17)

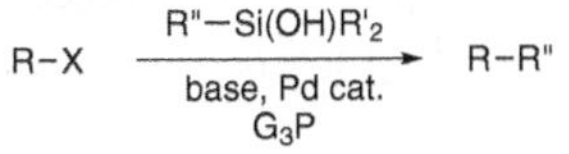

Hofmann elimination: the E2 elimination of quaternary ammonium ions, sulfonium ions, or alkyl fluorides (which proceed by an Elcb-like E2 mechanism). The least substituted alkene will generally predominate, subject to the requirement of *anti* elimination. (Ch. 8)

Hofmann-Löffler-Freytag reaction: remote functionalization of an unactivated position by homolysis of an *N*-haloammonium ion and intramolecular hydrogen atom transfer from carbon to nitrogen through a six-membered transition state. The product of the reaction is a δ-haloamine. (Ch. 14, 18)

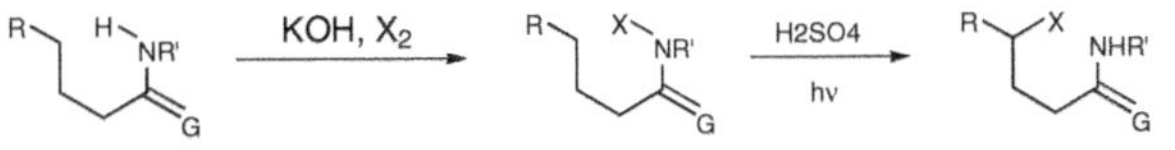

Hofmann rearrangement: rearrangement of an *N*-haloamide anion to give an isocyanate by insertion of nitrogen into the adjacent C—C bond; attack of the base on the isocyanate gives carboxamates. (Ch. 18)

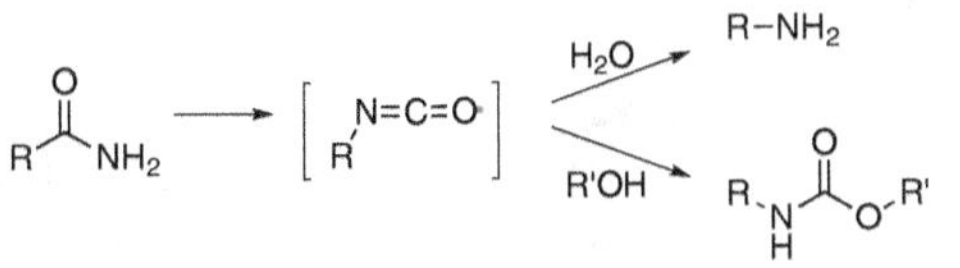

Horner-Wadsworth-Emmons (HWE) reaction: an analog of the Wittig reaction in which the phosphonium ylide is replaced by the conjugate base of a phosphonate ester; gives predominantly *E* alkenes (Ch. 11)

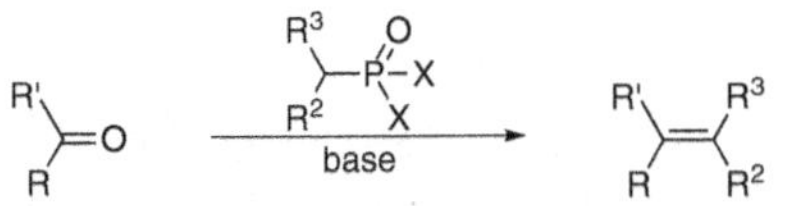

Hosomi-Sakurai reaction: formation of a new C—C bond by treating an allylsilane or allylstannane with a carbon electrophile. The reaction occurs with allylic rearrangement. (Ch. 22)

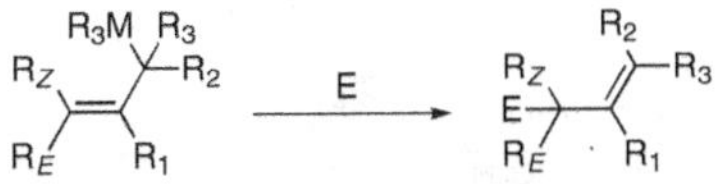

Houben-Hoesch acylation: conversion of an arene to the aryl ketone by treatment with a nitrile and a Lewis acid. The reaction probably involves the doubly protonated *sym*-triazene as an intermediate. (Ch. 10)

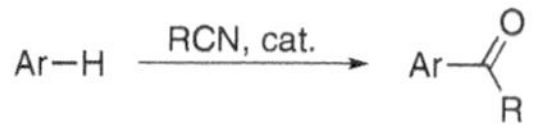

Huang Minlon modification: Wolff-Kishner reduction using KOH and hydrazine in a high-boiling glycol solvent (Ch. 20)

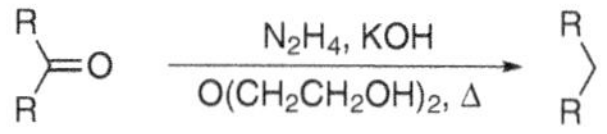

Hückel theory: a simplified form of molecular orbital theory that can be used to treat extended π-bonded systems (Ch. 4)

Hückel antiaromaticity: added reactivity associated with a cyclic, delocalized, conjugated π electron system with (4*n*) π electrons (Ch. 7)

Hückel aromaticity: added stability associated with a cyclic, delocalized, conjugated π electron system with (4*n* + 2) π electrons (Ch. 7)

Hund's rule: states that the lowest energy atom or molecule is the one with the greatest number of unpaired electrons in degenerate orbitals consistent with the Aufbau principle (Ch. 4)

I

Ireland ester enolate variant: a variant of the Claisen rearrangement in which the reactive 1,5-diene is the lithium or silyl enolate of an allylic ester (Ch. 6)

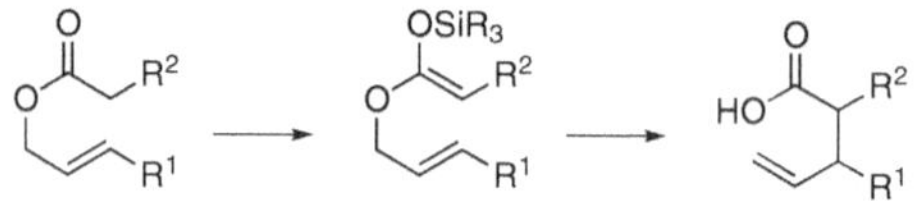

J

Jacobsen catalyst: a chiral metal salen complex with chirality in the salen backbone useful for asymmetric epoxidation of simple alkenes. The highest enantiomeric excesses are obtained with *Z* alkenes. (Ch. 21)

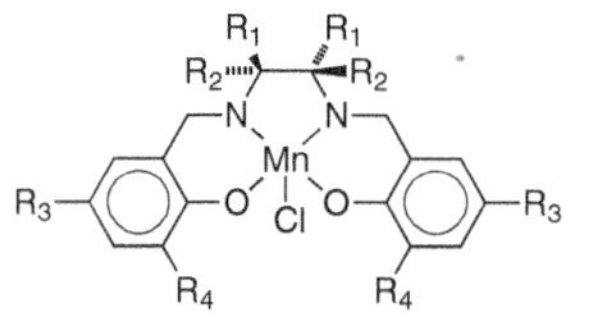

Jacobsen Catalyst
R_1=Ar, $(CH_2)_4$ OR H
R_2=H OR Ar, $(CH_2)_4$
R_3=R, OR
R_4=*t*-Bu

Jacobsen-Katsuki epoxidation: epoxidation of alkenes by hypochlorite and chiral metal salen complexes (Ch. 21)

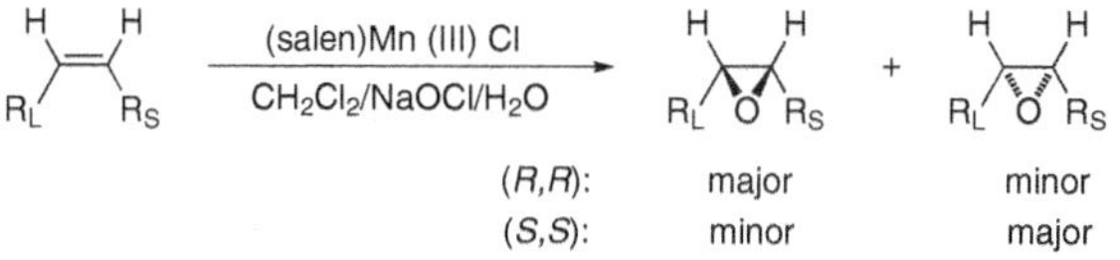

Johnson orthoester variant: a variant of the Claisen rearrangement in which the reactive 1,5-diene is formed from an allylic alcohol and an orthoester (Ch. 6)

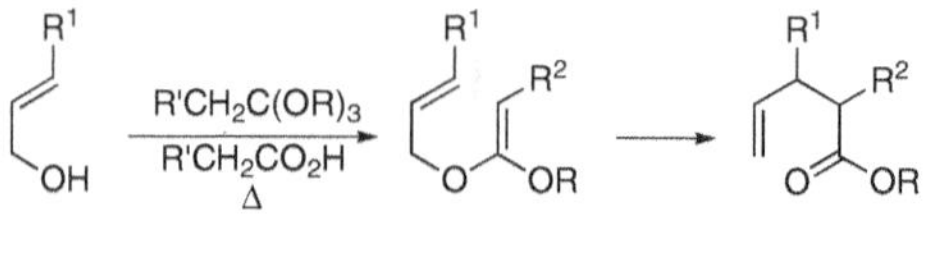

Jones reagent: $K_2Cr_2O_7$, H_2SO_4, H_2O, acetone; oxidizes primary alcohols to acids (Ch. 18)

Julia-Kocieńsky olefination: a variant of the Julia-Lythgoe olefination in which the reductive elimination step is not necessary (Ch. 16)

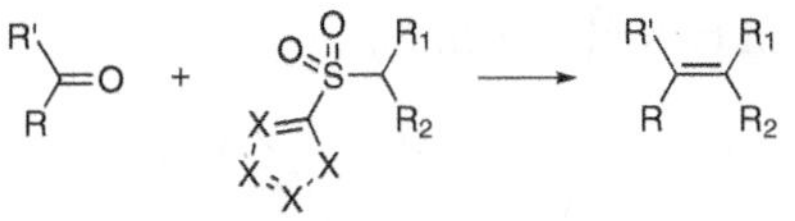

Julia-Lythgoe olefination: condensation of aldehydes and ketones to give alkenes by treatment of a carbonyl compound with a sulfone anion and reductive elimination of the product (Ch. 16)

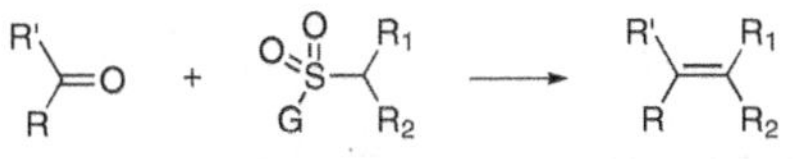

K

Katsuki catalyst: a chiral metal salen complex with chirality in the terminal groups of the salen ligand useful for asymmetric epoxidation of simple alkenes. The highest enantiomeric excesses are obtained with *Z* alkenes. (Ch. 21)

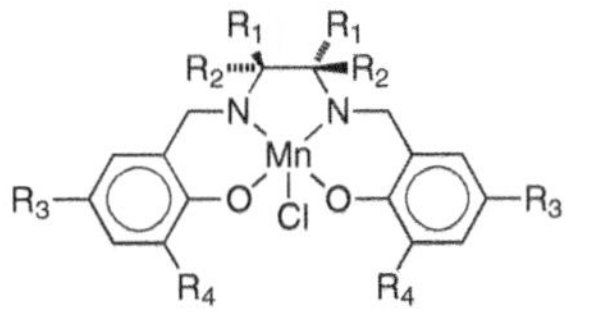

Katsuki Catalyst
R_1=Ar, $(CH_2)_4$ OR H
R_2=H OR Ar, $(CH_2)_4$
R_3=R, OR
R_4=CH(Ar)Ar')

Kharasch reaction: anti-Markovnikov addition of HBr to alkenes in the presence of peroxides (Ch. 14)

Knoevenagel condensation: condensation of a β-dicarbonyl or similar activated methylene compound with a ketone to give a conjugated product (Ch. 16)

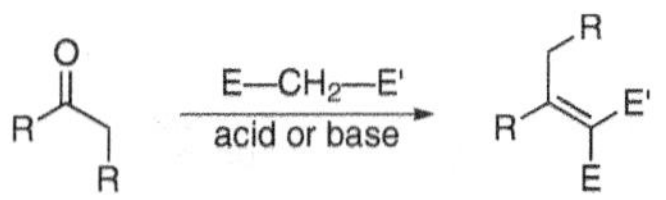

Kolbe electrolysis: electrolysis of a carboxylate anion to give a carboxy radical that decarboxylates to the alkyl radical (Ch. 14)

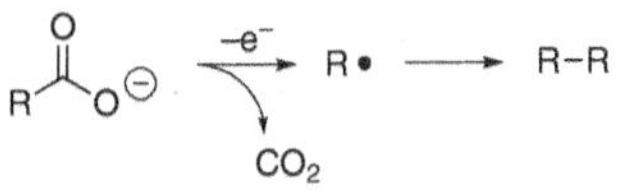

Krapcho decarboxylation: decarboxylation of a β-ketoester by heating with a lithium halide in dimethylformamide (DMF) (Ch. 16)

Kraus-Pinnick oxidation: oxidation of an aldehyde to a carboxylic acid using sodium chlorite as the terminal oxidant and 2-methyl-2-butene as a chlorine scavenger (Ch. 18)

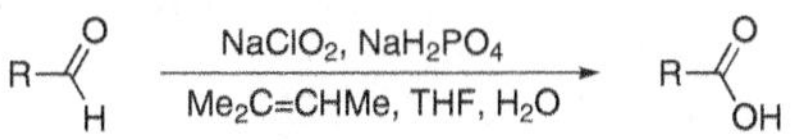

Kumada coupling: coupling of a Grignard reagent with an alkyl halide in the presence of a nickel or palladium catalyst (Ch. 17)

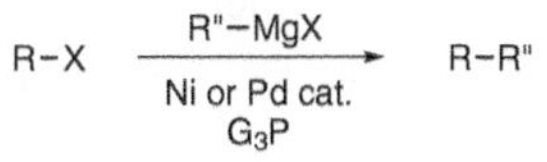

L

Lemieux-Johnson cleavage: oxidative cleavage of an alkene by osmium tetroxide (catalytic) and periodate. The periodate oxidizes the intermediate osmate ester to the carbonyl compounds and osmium tetroxide (Ch. 18)

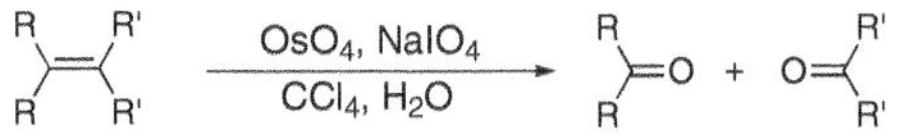

Leuckart reaction: reductive amination of an aldehyde or ketone to give a primary amine, using ammonium formate as the reagent (Ch. 20)

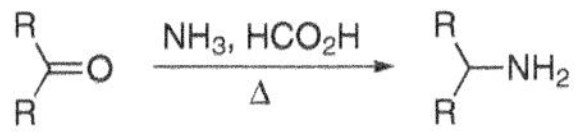

LICKOR: Schlosser's reagent; a strong base formed by mixing potassium *tert*-butoxide and butyllithium in equal proportions (Ch, 11, 12)

Lindlar hydrogenation: hydrogenation of alkynes over a palladium catalyst poisoned with lead and quinoline (Ch. 19)

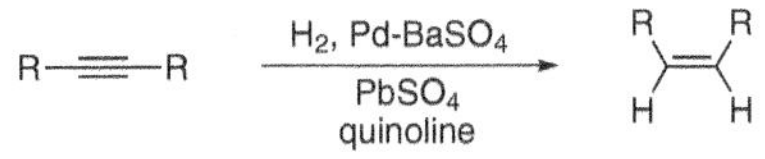

Lipshutz cuprate: a highly nucleophilic cuprate formed by treatment of two equivalents of an alkyllithium with one equivalent of cuprous cyanide (Ch. 11)

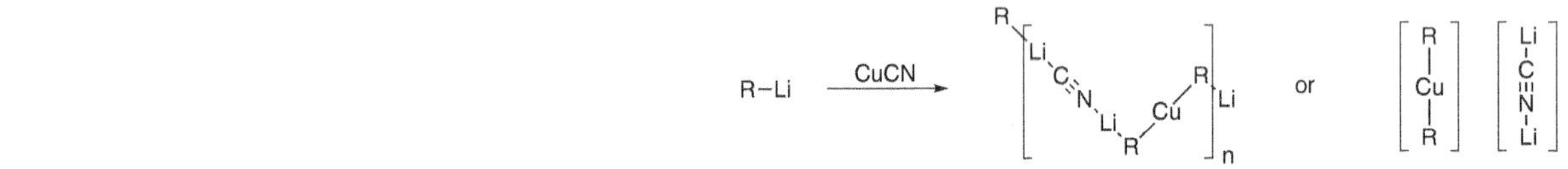

Lombardo reagent: a variant of the Tebbe reagent formed from titanocene dichloride, zinc, methylene bromide, and titanium tetrachloride. The reagent must be aged before using. (Ch. 16)

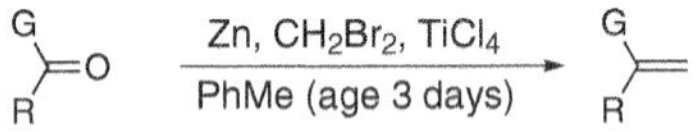

Lossen rearrangement: rearrangement of an *N*-acylhydroxamate anion to give an isocyanate by insertion of nitrogen into the adjacent C—C bond (Ch. 18)

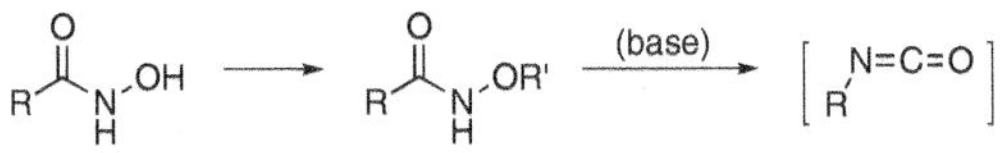

Lowry-Brønsted acid: a proton donor (Ch. 1)

Lowry-Brønsted base: a proton acceptor (Ch. 1)

M

Mannich reaction: the reaction between an enol and an iminium ion to give a β-dialkylamino ketone. The enol and the iminium ion may both be generated in situ. The reaction may also be carried out asymmetrically. (Ch. 10)

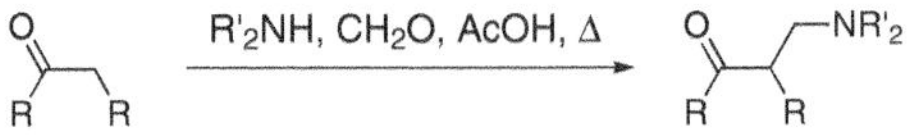

Markovnikov's (Markownikoff's) rule: states that the major product of the addition of an unsymmetrical electrophile to an unsymmetrical alkene will be the product in which the attacking electrophile adds to the alkene π bond to give the more stable of the two possible carbocations (Ch. 8)

McMurry reaction: reductive dimerization of an aldehyde or ketone to an alkene by means of titanium metal or titanium (III) salts. With ketones, a further deoxygenation occurs to give the alkene. (Ch. 14, 20)

Meerwein-Ponndorf-Verley reduction: equilibrium reduction of a ketone by aluminum isopropoxide in isopropyl alcohol; functional reverse of the Oppenauer oxidation (Ch. 20)

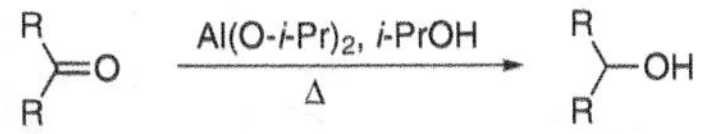

Metathesis: coupling of alkenes (usually terminal alkenes) with elimination of a second alkene (usually ethylene), or coupling of an alkene and an alkyne to give a conjugated diene. Metathesis of alkynes is less common. (Ch. 17)

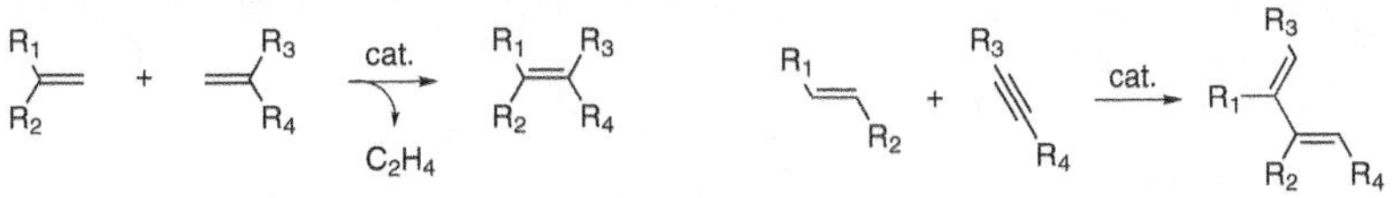

Michael addition: originally, 1,4-addition of an enolate anion nucleophile or its equivalent to a conjugated carbonyl compound. The definition now usually includes all nucleophiles, including heteroatom nucleophiles and organometallic (especially organocopper) reagents. (Ch. 12)

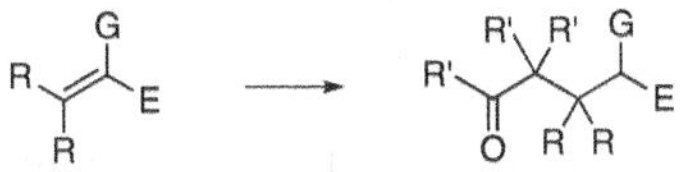

Mislow-Evans rearrangement: the suprafacial (2,3) sigmatropic rearrangement of an allylic sulfoxide or selenoxide to the corresponding sulfenate or selenenate ester (Ch. 22)

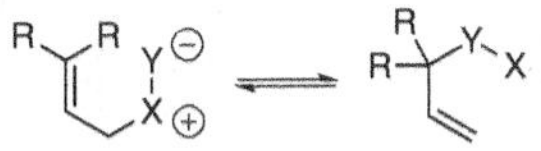

Mitsunobu reaction: the S_N2 reaction with an external nucleophile of an alkoxyphosphonium salt generated by treating an alcohol with diethyl diazodicarboxylate and triphenylphosphine (Ch. 22)

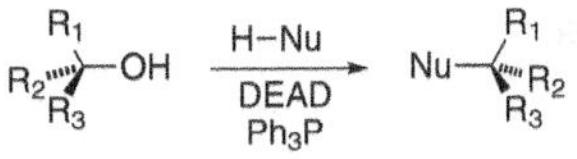

Möbius antiaromaticity: added reactivity associated with a cyclic, delocalized, conjugated π electron system with $(4n + 2)$ π electrons (Ch. 7)

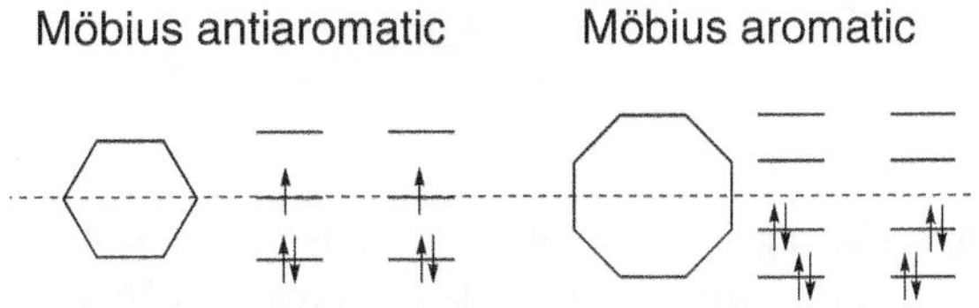

Möbius aromaticity: added stability associated with a cyclic, delocalized, conjugated π electron system with $(4n)$ π electrons (Ch. 7)

Moffatt-Pfitzner oxidation: oxidation of primary alcohols to aldehydes using dimethyl sulfoxide and acetic anhydride (Ch. 18)

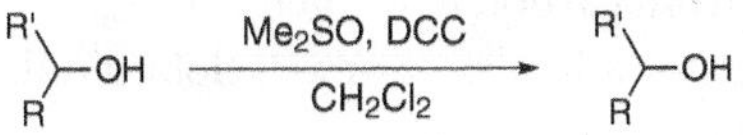

More O'Ferrall-Jencks diagram: a simplified potential surface diagram that omits the energy contour lines. The reaction coordinate is shown as a dashed line. (Ch. 8)

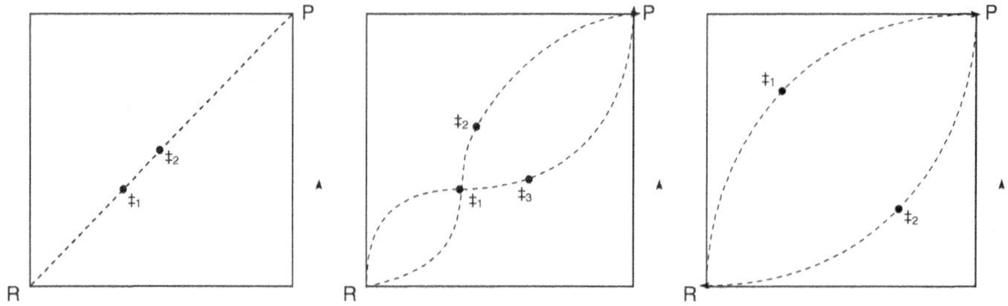

Morita-Baylis-Hillman reaction: the reaction between an acrylic acid derivative and an aldehyde in the presence of a tertiary amine or phosphine to give an α-hydroxyalkylacrylic acid derivative (Ch. 16)

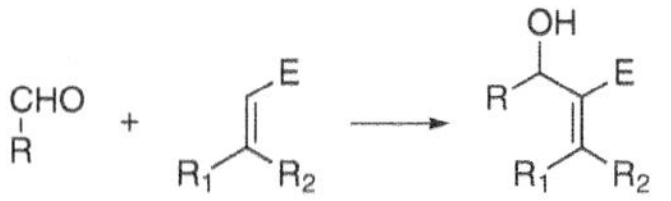

Mukaiyama aldol addition: formation of a β-hydroxycarbonyl compound by treatment of an enol trimethylsilyl ether with a carbonyl compound in the presence of a strong Lewis acid (usually titanium tetrachloride) (Ch. 10)

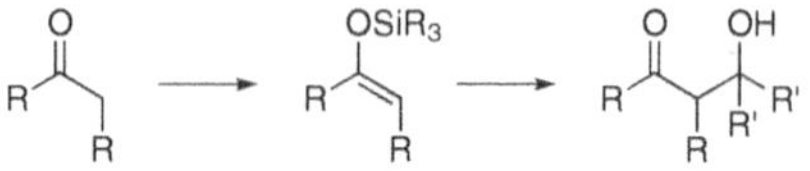

Mukaiyama alkylation: formation of an α-alkylated ketone by treatment of an enol trimethylsilyl ether with an alkyl halide in the presence of a strong Lewis acid (usually titanium tetrachloride) (Ch. 10)

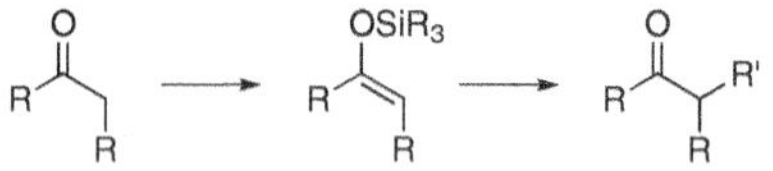

Myers-Saito cyclization: cycloaromatization of an enyne-allene to give a 1,3-diradical (Ch. 13)

N

Negishi coupling: palladium-catalyzed coupling of an alkyl halide with an organozinc (or organoaluminum or zirconocene) reagent (Ch. 17)

$$\text{R-X} \xrightarrow[\text{Pd cat., G}_3\text{P}]{\text{R''-M}} \text{R-R''}$$

Norrish type I cleavage: cleavage of the photochemically excited state of a ketone into an alkyl radical and an acyl radical that then decarbonylates. (Ch. 14)

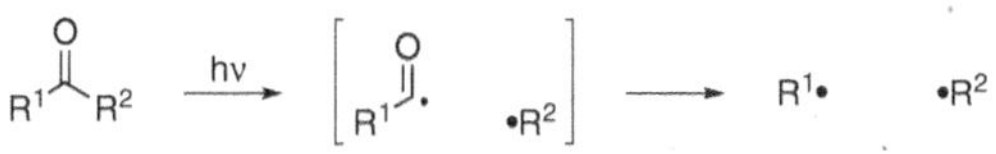

Norrish type II reaction: reaction of the excited state of a ketone involving intramolecular hydrogen atom transfer from carbon to oxygen through a six-membered transition state. The product is a ketone and alkene or a cyclobutanol. (Ch. 14)

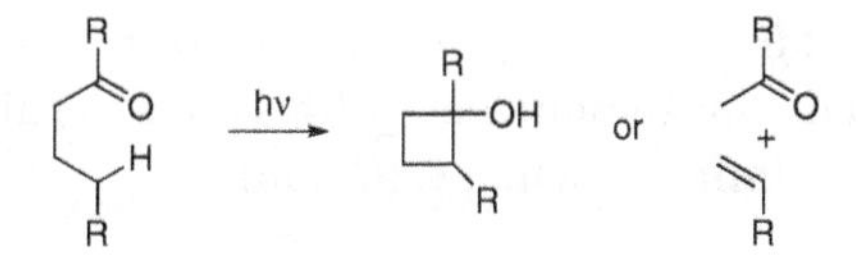

O

olefin metathesis: the interchange of terminal groups of two alkenes catalyzed by a metal-carbene complex (Ch. 17)

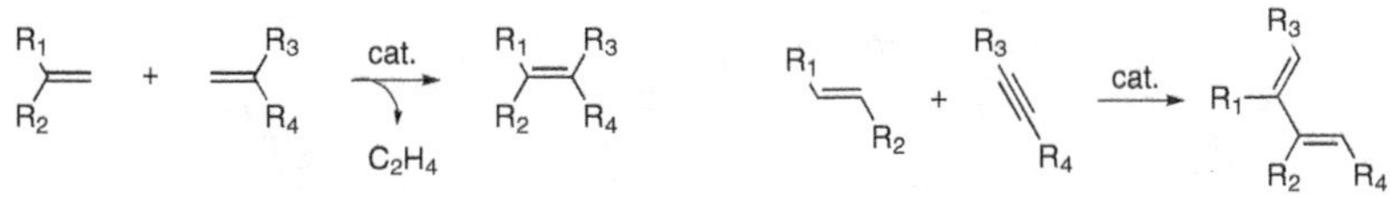

Oppenauer oxidation: equilibrium oxidation of an alcohol by aluminum *tert*-butoxide and acetone. See Meerwein-Ponndorf-Verley reduction. (Ch. 18)

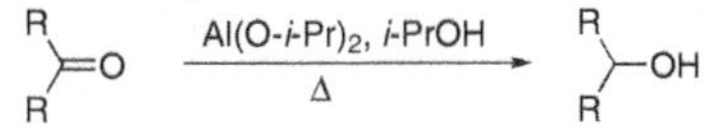

oxyanion Cope rearrangement: the (3,3)-sigmatropic rearrangement of a 1,5-diene, accelerated by the presence of an alkoxide group at the 3- position (Ch. 6)

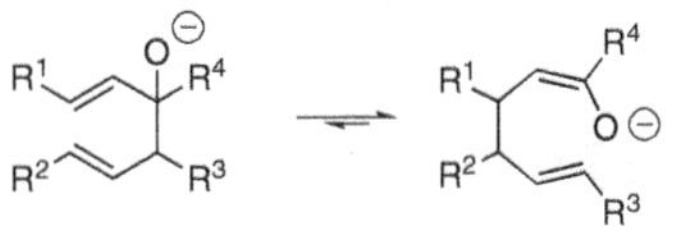

oxymetallation: the addition of a metal and a hydroxyl group to an alkene. The stereochemistry is usually *anti* (antarafacial). (Ch. 18)

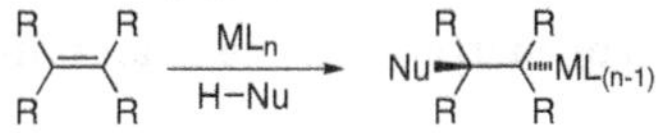

ozonolysis: oxidative cleavage of an alkene with ozone. The products obtained may be altered by the way that the intermediate ozonide is worked up. (Ch. 18)

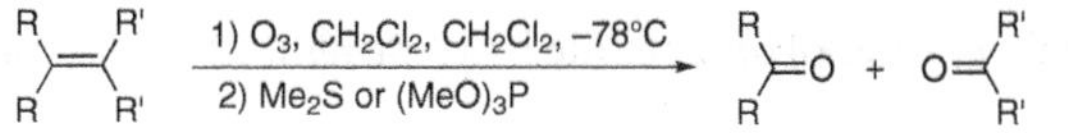

P

Parikh-Doering oxidation: oxidation of primary alcohols to aldehydes using dimethyl sulfoxide, the pyridine-sulfur trioxide complex, and triethylamine (Ch. 18)

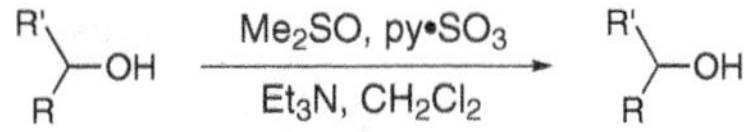

Perkin condensation: the reaction between acetic anhydride and potassium acetate with an aldehyde to give a cinnamic acid (Ch. 16)

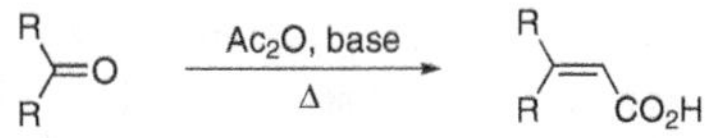

Perkow reaction: the conversion of an α-haloketone to an enol phosphate by treatment with a trialkyl phosphite (Ch. 22)

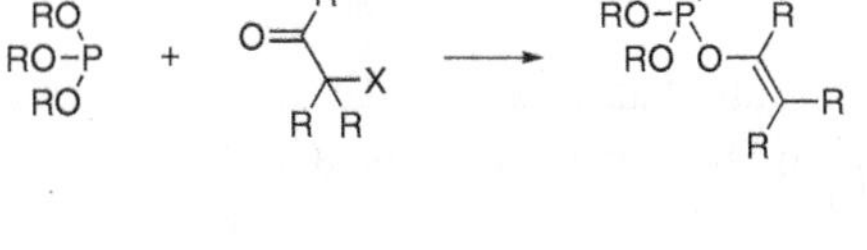

Petasis reagent: a variant of the Tebbe reagent prepared from titanocene dichloride and methyllithium or a methylmagnesium halide (this gives the dimethyltitanocene). It is activated by heating to eliminate methane. (Ch. 16)

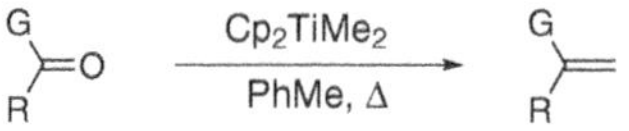

Peterson reaction: formation of an alkene from an aldehyde or ketone by reaction with a trialkylsilylalkyl anion and either acid or base. The stereochemistry of the final alkene is controlled by the conditions used in the second step of the reaction. (Ch. 11, 16)

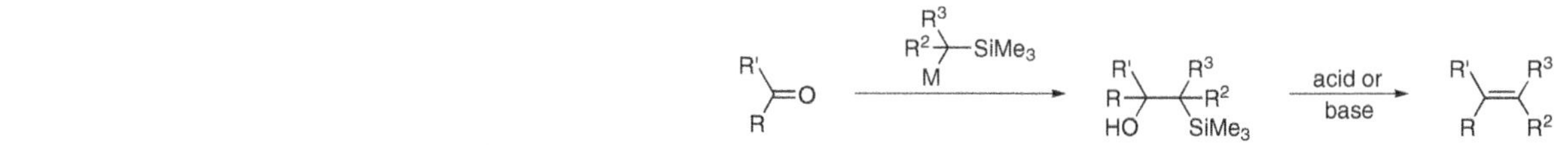

pinacol reaction: reductive dimerization of an aldehyde or ketone by metals (Mg, Al) in the absence of a proton source to give a 1,2-diol (Ch. 14)

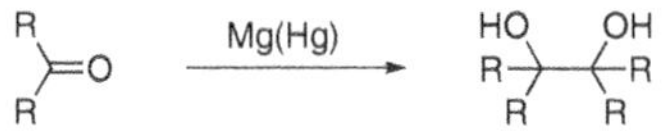

pinacol-terminated Prins reaction: an acid-catalyzed reaction between a vicinal diol with the grouping, C=C—C(OH)—C(OH), and an aldehyde or ketone, and a strong acid to give a tetrahydrofuryl ketone as the product. The reaction applies equally well to allylic 1,3-dioxlanes. (Ch. 10)

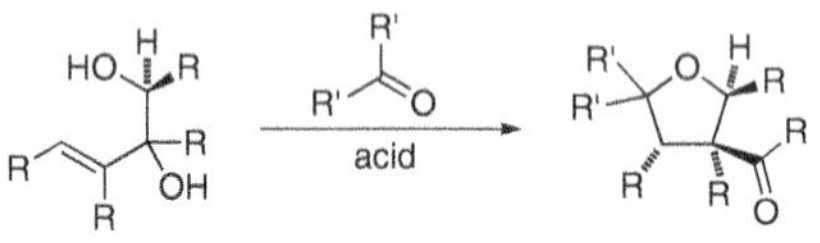

Pinnick oxidation: see Kraus-Pinnick oxidation (Ch. 18)

Polonovski reaction: conversion of a tertiary amine to an iminium ion *via* the amine *N*-oxide (Ch. 18)

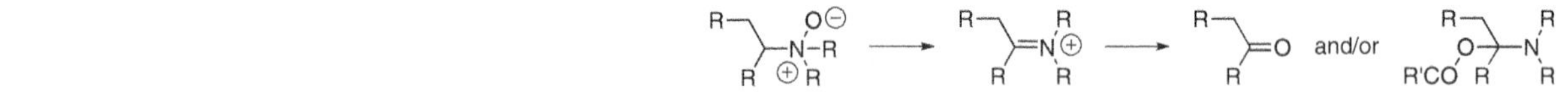

Prilezhaev reaction: formation of an epoxide by treatment of an alkene with a peracid (Ch. 18)

Prins reaction: the acid-catalyzed reaction between an alkene and an aldehyde catalyzed by strong acid. The product is usually a 1,3-dioxane derivative. The stereochemistry of the reaction is *anti*. (Ch. 10)

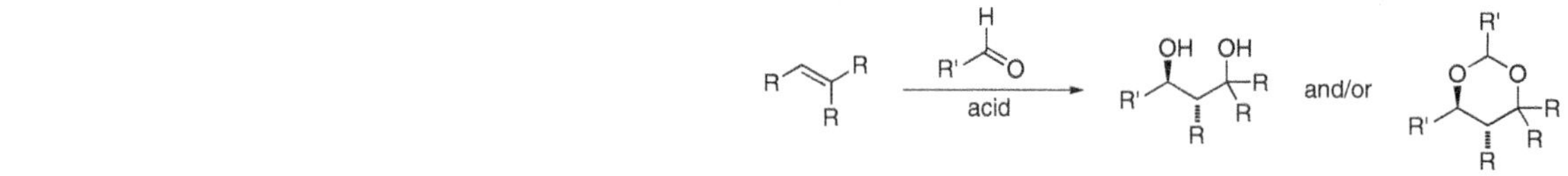

Pummerer rearrangement: the rearrangement of a sulfoxide under the influence of a strong dehydrating agent to give an α-substituted sulfide that is formally the addition product of a thiol to an aldehyde or ketone (Ch. 22)

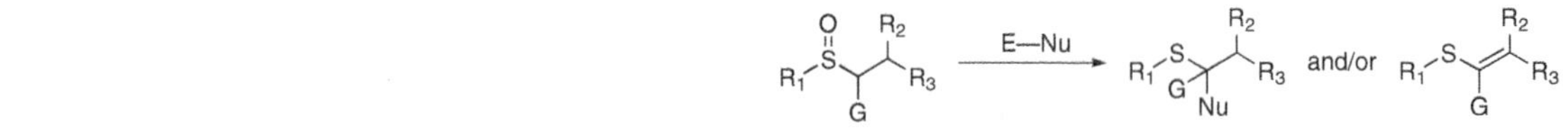

pyrolytic elimination: formation of an alkene by thermal decomposition of the precursor. These reactions are often retro-ene reactions. (Ch. 6)

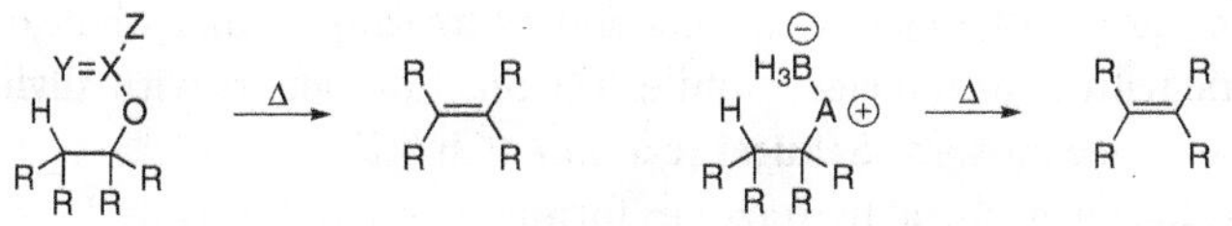

R

Ramberg-Bäcklund reaction: the formation of an alkene by coupling the two alkyl groups of a sulfone by treatment with base and a halogen. The reaction proceeds by extrusion of SO_2 from a thiirane dioxide. (Ch. 16)

Raney nickel: a finely divided nickel catalyst obtained by leaching the aluminum from a Ni-Al alloy with sodium hydroxide (Ch. 19)

Reformatskii reaction: the addition of the organozinc nucleophile obtained from the reaction between zinc metal and an α-ester to give a β-hydroxyester; especially useful in the synthesis of highly hindered products (Ch. 12)

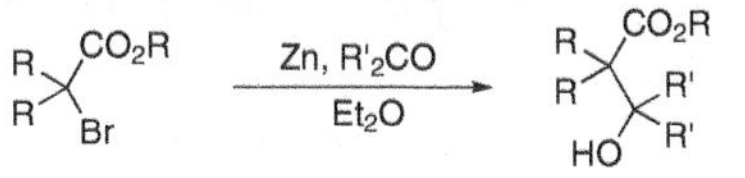

Reich-Gross formylation: conversion of an arene to the corresponding aldehyde by treatment with methyl dichloromethyl ether and a Lewis acid, followed by hydrolysis of the α-haloether. It does not work with electron-deficient arenes. (Ch. 10)

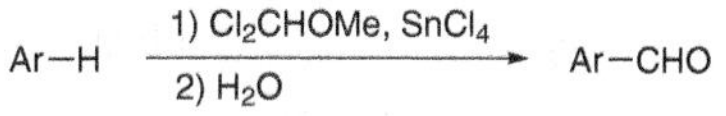

Reppe chemistry: the templated oligomerization of alkynes to give cyclic polyene products (Ch. 17)

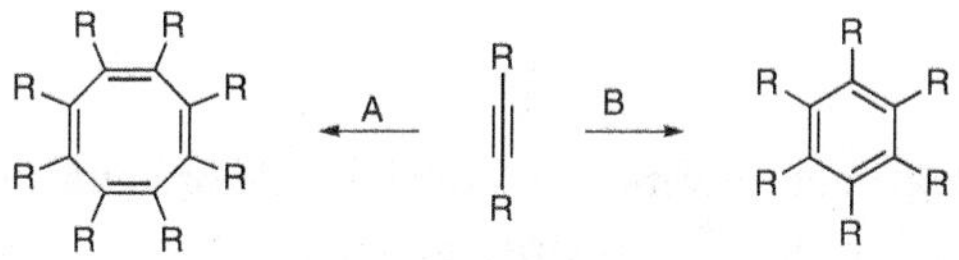

Robinson annelation (annulation): a reaction that adds a fused cyclohexenone to a ketone by the Michael-initiated ring closure (MIRC) reaction with an enone (Ch. 16)

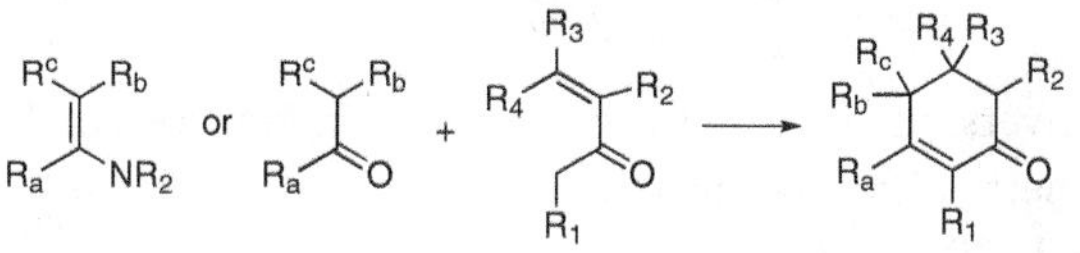

Rosenmund reduction: hydrogenation of an acid chloride to an aldehyde over a palladium catalyst poisoned with sulfur or an amine (Ch. 19)

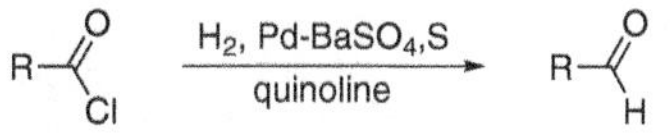

S

Saegusa oxidation: a dehydrogenation of a ketone by oxidation of its enol trialkylsilyl ether with palladium (II) acetate and a terminal co-oxidant (e.g., benzoquinone) (Ch. 18)

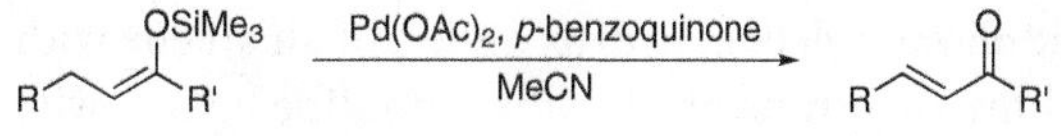

Sakurai reaction: formation of a new C—C bond by treating an allylsilane or allylstannane with a carbon electrophile. The reaction occurs with allylic rearrangement. See Hosomi-Sakurai reaction. (Ch. 22)

SAMP/RAMP hydrazones: chiral hydrazones formed from ketones and *S*- or *R*-1-amino-2-methoxymethylpyrrolidine. The aza-enolate anions of these hydrazones are useful nucleophiles for asymmetric alkylation of ketones. (Ch. 12)

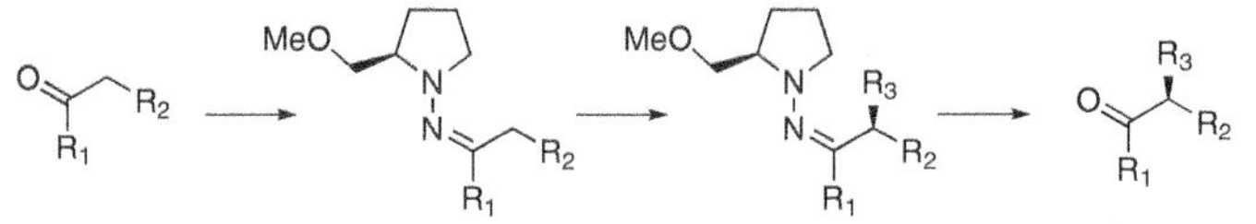

Sarett reagent: CrO_3, py; oxidizes primary alcohols to aldehydes (Ch. 18)

Schlosser modification of the Wittig reaction: modification of the Wittig reaction at low temperature by adding lithium salts and butyllithium; gives predominantly *E* alkenes (Ch. 16)

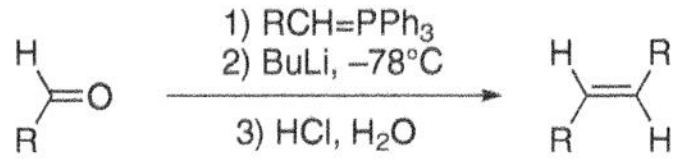

Schlosser's reagent: a strong base formed by mixing potassium *tert*-butoxide with butyllithium in equal proportions; probably corresponds to butylpotassium; also known as LICKOR (Ch. 12)

Schmidt rearrangement: treatment of an aldehyde or ketone with hydrazoic acid to give an isocyanate by insertion of nitrogen into the adjacent C—C bond (Ch. 18)

Schrock catalyst: a molybdenum-based metathesis catalyst (Ch. 17)

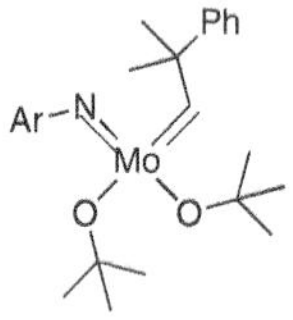

Seyferth-Gilbert homologation: conversion of an aldehyde or ketone to an alkyne with one more carbon atom using an α-diazophosphonate and base. The reaction is believed to involve a carbene intermediate. (Ch. 16)

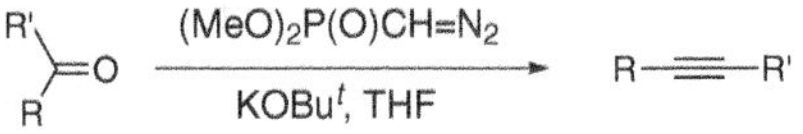

Shapiro reaction: conversion of a tosylhydrazone to an alkene or vinyllithium by treatment with methyllithium. The less substituted alkene usually predominates. (Ch. 20)

Sharpless asymmetric epoxidation: epoxidation of an allyl alcohol with *tert*-butyl hydroperoxide and titanium isopropoxide and a chiral tartrate ligand. The highest enantiomeric excesses are obtained with *E* alcohols. (Ch. 21)

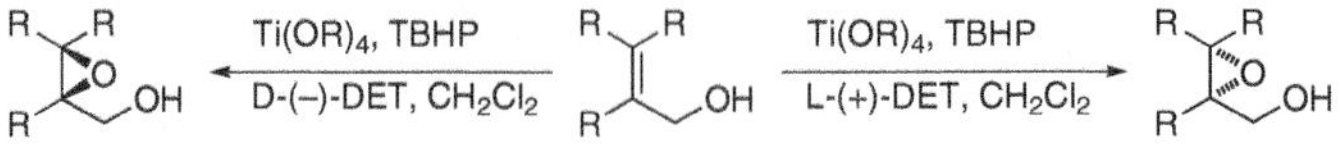

Sharpless asymmetric dihydroxylation: hydroxylation of an alkene with potassium osmate and a chiral *bis*-dihydroquinine or -dihydroquinidine ligand, which determines the configuration of the diol produced. The highest enantiomeric excesses are obtained with *E* alkenes. (Ch. 21)

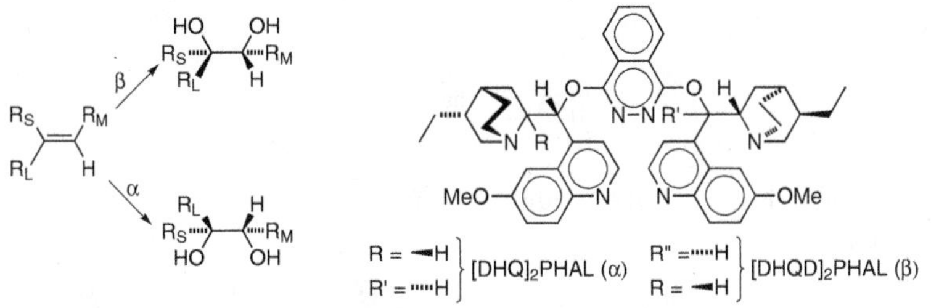

Shi oxidation: epoxidation of an alkene with a chiral dioxirane derived from a carbohydrate; best current method for oxidizing terminal alkenes (Ch. 21)

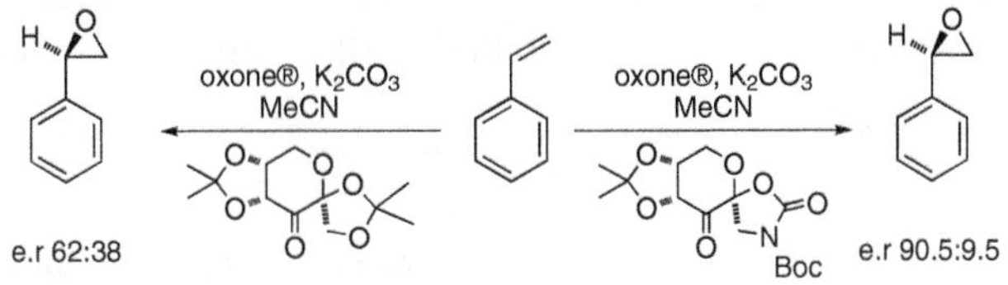

Shvo's catalyst: a low-valent ruthenium hydride catalyst for the oxidation of secondary alcohols to ketones, using acetone as the oxidant. The oxidation of primary alcohols to aldehydes is usually complicated by an accompanying Tishchenko reaction. (Ch. 18)

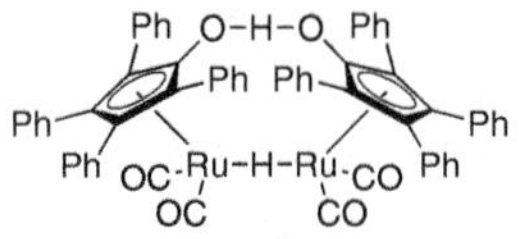

sigmatropic rearrangement: a pericyclic reaction in which an atom to a remote site in the molecule occurs during the reaction with concomitant migration of π bonds. See also Claisen rearrangement and Cope rearrangement. (Ch. 6)

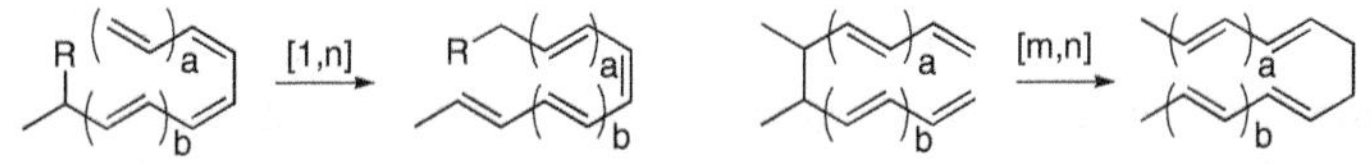

Simmons-Smith reaction: a [2 + 1] cycloaddition reaction where the carbene is generated by the reaction between methylene iodide and zinc metal or a dialkylzinc (Ch. 6)

Soderquist boranes: bulky boranes that may be used as chiral replacements for 9-9-borabicyclo(3.3.1)nonane (Ch. 17)

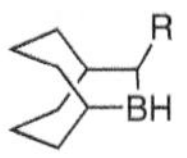

Sommelet-Hauser rearrangement: rearrangement of a benzyltrialkylammonium ion to an *o*-alkylbenzylamine by strong base (usually alkyllithium). (Ch. 11)

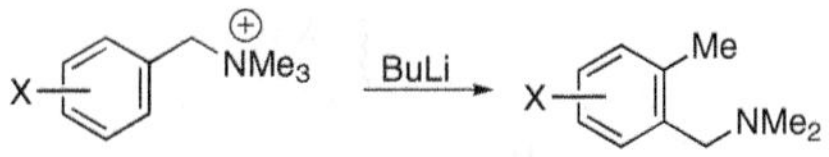

Sonogashira coupling: palladium-catalyzed coupling of a terminal alkyne with a halide in the presence of copper (I) and a base (Ch. 17)

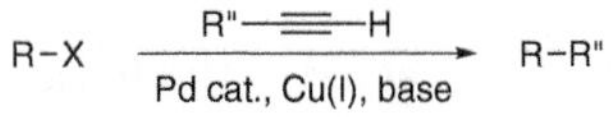

Stevens rearrangement: base-promoted [1,2]- or [2,3]-sigmatropic rearrangement of a quaternary ammonium salt to give a tertiary amine (Ch. 11)

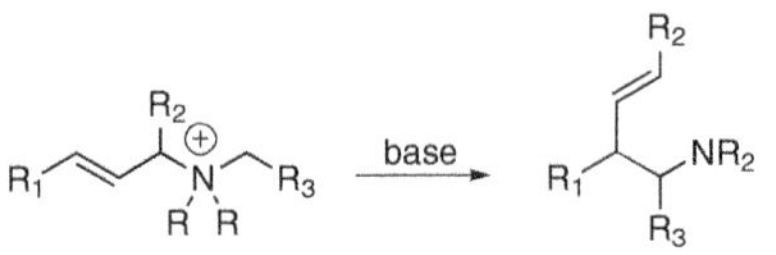

Stille coupling: palladium-catalyzed coupling of an alkylstannane with an alkyl halide (Ch. 17)

Stobbe condensation: the base-promoted condensation between a succinate diester and an aldehyde or ketone to give the half-ester of an alkylidenesuccinic acid (Ch. 16)

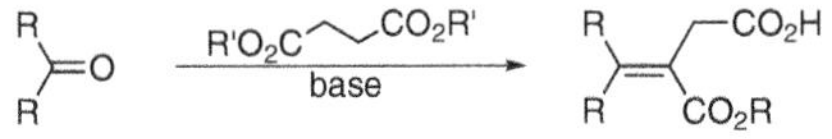

Suzuki (Suzuki-Miyaura) coupling: palladium-catalyzed coupling of a halide with an alkylboronic acid or alkylboronic ester in the presence of a base. In the *Molander modification*, the boronic acid is replaced by a potassium alkyltrifluoroborate salt. (Ch. 17)

Swern oxidation: oxidation of primary alcohols to aldehydes using dimethyl sulfoxide and oxalyl chloride in dichloromethane in the presence of triethylamine (Ch. 18)

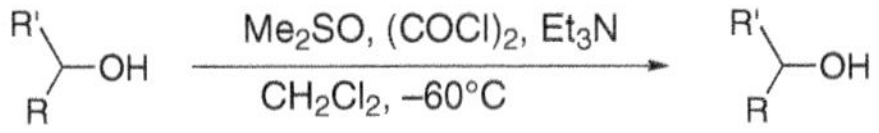

T

Tebbe olefination: conversion of a carbonyl compound to an alkene with a titanium carbene complex. It may be used to form enol ethers from esters. (Ch. 16)

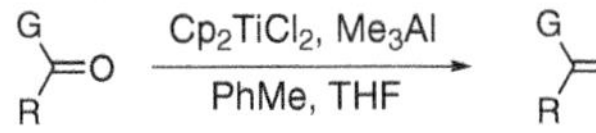

Tebbe reagent: a metal carbene reagent prepared from titanocene dichloride and trimethylaluminum, activated by a Lewis base (Ch. 16)

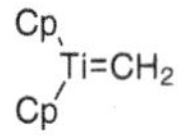

Thorpe-Ziegler condensation: intramolecular base-promoted condensation of a dinitrile to give a β-aminonitrile. The reaction can be used to form medium-sized rings. (Ch. 16)

Tishchenko reaction: an aluminum-catalyzed redox disproportionation reaction of aldehydes to esters of primary alcohols. It is most commonly used to convert aromatic aldehydes to benzyl benzoates. (Ch. 18)

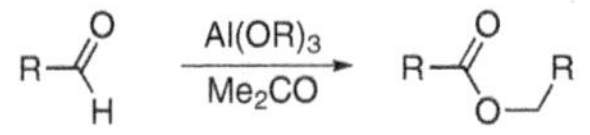

Tollens reagent: Ag_2O, NH_3, H_2O; oxidizes aldehydes to carboxylic acids (Ch. 18)

$$\text{R–CHO} \xrightarrow{[O]} \text{R–CO}_2\text{H}$$

TPAP: tetrapropylammonium perruthenate (Ley reagent) (Ch. 18)

Tsuji-Trost reaction: palladium-catalyzed displacement of an allylic ester by a carbanion nucleophile (Ch. 17)

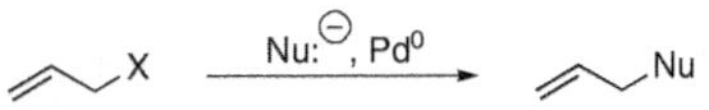

U

Uemura oxidation: oxidation of a primary alcohol to an aldehyde using palladium (II) acetate and molecular oxygen in the presence of pyridine; also oxidation of a secondary alcohol by oxygen in the presence of a base and a palladium catalyst. The reaction is enantioselective in the presence of sparteine. (Ch. 21)

$$\text{R–CH}_2\text{OH} \xrightarrow[\text{3Å MS, PhMe, 80°C}]{\text{Pd(OAc)}_2\text{, py, O}_2} \text{R–CHOH}$$

umpolung: charge affinity inversion, which leads to reversal of normal electrophilic/nucleophilic behavior (e.g., a carbonyl anion synthon) (Ch. 15)

V

Vilsmeier-Haack formylation: conversion of an arene to the corresponding aldehyde by treatment with dimethylformamide and phosphorus oxychloride (or oxalyl chloride); does not work with electron-deficient arenes. (Ch. 10)

$$\text{Ar–H} \xrightarrow[\text{DMF}]{\text{POCl}_3} \text{Ar–CHO}$$

W

Wacker oxidation: oxidation of an alkene to a ketone; terminal alkenes give methyl ketones. The reaction passes through the enol. (Ch. 18)

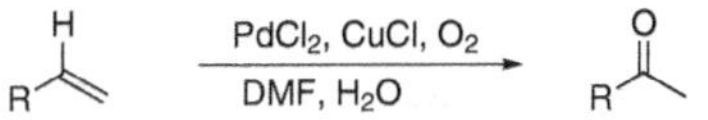

Wagner-Meerwein rearrangement: a suprafacial 1,2-rearrangement of hydrogen or alkyl in a carbocation; the reaction is rapid only if it leads to a more stable carbocation. Migration of alkyl occurs with retention of configuration at the migrating carbon. (Ch. 9)

Wallach reaction: reductive amination of an aldehyde or ketone with a secondary amine and formic acid (Ch. 20)

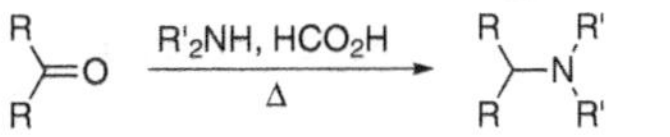

Wharton reaction: modification of the Wolff-Kishner reduction in which the tosylhydrazone of an α,β-epoxyketone is converted to an allylic alcohol. See Eschenmoser-Tanabe fragmentation. (Ch. 20)

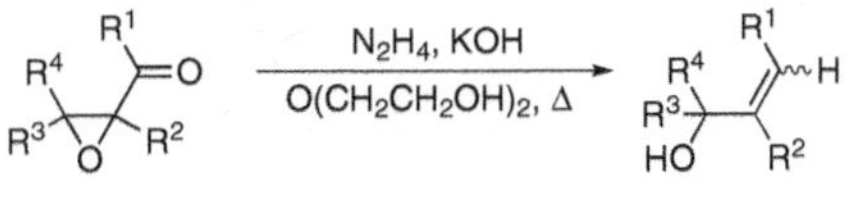

Wilkinson's catalyst: $ClRh(PPh_3)_3$. (Ch. 19)

Wittig reaction: condensation of a phosphorus ylide (phosphorane) with an aldehyde or ketone to give an alkene. Aldehydes give mainly *Z* alkenes with unstabilized ylides and *E* alkenes with stabilized ylides. (Ch. 11, 16)

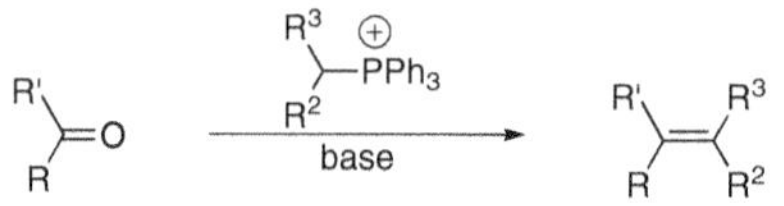

Schlosser modification: modification of the Wittig reaction by adding lithium salts and butyllithium; gives predominantly *E* alkene()

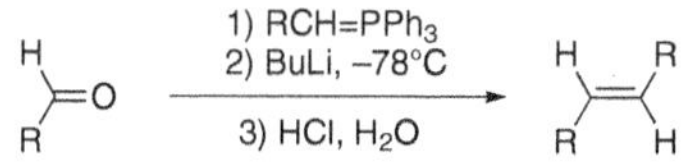

Wittig rearrangement: base-promoted [1,2]- or [2,3]-sigmatropic rearrangement of an ether to give an alkoxide anion (Ch. 11)

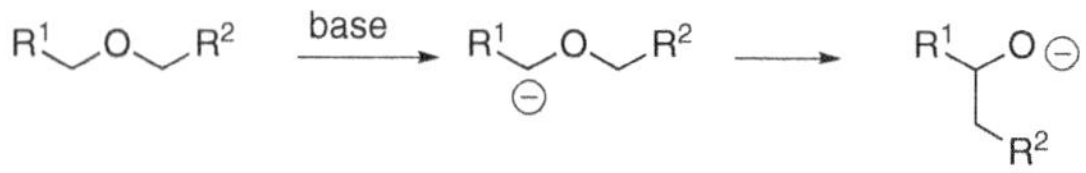

[1,2]-Wittig rearrangement

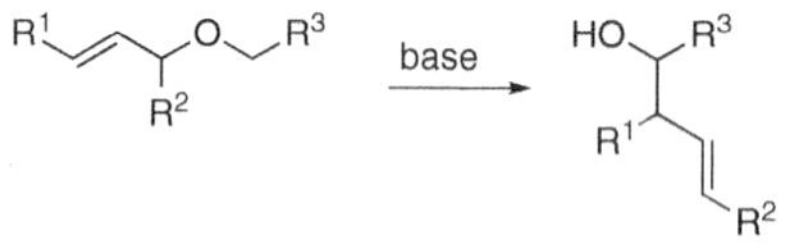

[2,3]-Wittig rearrangement

Wolff-Kishner reduction: deoxygenation of an aldehyde or ketone by heating the corresponding hydrazone with base (Ch. 20)

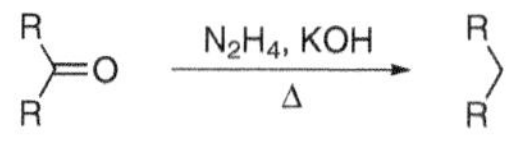

Wurtz coupling: reductive coupling of two alkyl halides by an active metal (Na, Na-K alloy, etc.). (Ch. 20)

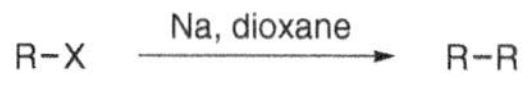

Z

Zaitsev's (Saytzeff's) Rule: states that in the E2 eliminations of alkyl halide or sulfonates by a sterically nondemanding base, the major product will be the most substituted alkene that can be formed by *anti* elimination. (Ch. 8)

Zimmerman circle: a mnemonic for predicting orbital degeneracy and the relative energies in Möbius aromatic systems (Ch. 7)

Zimmerman-Traxler model: a chair-like six-membered model of the transition state for the aldol addition reaction that can be used to predict the stereochemistry of the major product of the addition (Ch. 12)

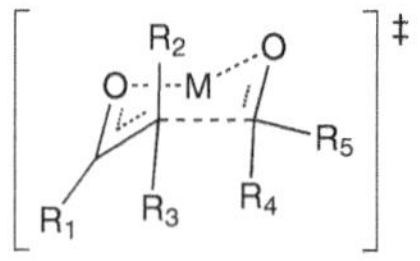

APPENDIX 2

Glossary of Key Terms

A

absolute configuration: description of a chiral molecule that specifies the three-dimensional arrangement of all atoms in the molecule. The absolute configuration may be expressed in the Cahn-Ingold-Prelog systems (*R/S*) or by the DL conventions for certain biological molecules.
absolute hardness: half the energy gap between the HOMO and LUMO
acetal/ketal: a *gem*-diether used to protect aldehydes and ketones
acid: any species that will generate hydronium ion in water (Arrhenius) or act as a proton donor (Lowry-Brønsted) in a chemical reaction
activated complex: formally, the structure of the high-energy species between two minima on the potential energy surface; often used as a full synonym of a transition state. An activated complex is not a stable state for a molecule.
activation energy: the energy required to form the activated complex for the rate-limiting step from the reactants
alkynide anion: conjugate base of an alkyne. Alkali metal alkynides are ionic, whereas transition metal alkynides are covalent.
allylic oxidation: oxidative substitution of an allylic C—H bond by a C—X bond. The reaction may involve free radicals (e.g., *N*-bromosuccinimide) or allyl-metal complexes.
anchimeric assistance: participation by a neighboring group that stabilizes a carbocation or incipient carbocation during a reaction; results in increased rates of ionization reactions and retention of configuration in solvolysis reactions
anomeric effect: the propensity of the heteroatom bonded to the acetal carbon in a pyran derivative to occupy an axial position in much higher proportion than in the corresponding cyclohexane derivative
anomers: epimers where the difference in configuration occurs at an acetal or ketal carbon
antarafacial: reaction stereochemistry where the new bonds are formed on opposite faces of the participating π system (e.g., *anti* addition)
***anti*, *gauche*:** staggered conformations about a σ bond in hydrocarbons and other compounds with at least four non-hydrogen atoms in the chain. The *anti* conformation is lower in energy.
antiaromatic: possessing chemical and physical properties that indicate unusual reactivity relative to a non-aromatic unsaturated compound
aprotic dipolar solvent: a polar solvent lacking O—H or N—H bonds
aromatic: possessing chemical and physical properties that indicate unusual stability relative to a non-aromatic unsaturated compound
aromatic resonance stabilization energy: the added stabilization of an aromatic compound over that calculated for a hypothetical structure without electron delocalization
Arrhenius equation: the relationship that relates the rate constant for a reaction to the Kelvin temperature: $k = Ae^{-E_a/RT}$
aryne: an aromatic derivative of an aromatic hydrocarbon (usually benzene) lacking two C—H σ bonds
asymmetric induction: the formation of unequal amounts of the enantiomeric products from a prochiral precursor

asymmetric synthesis: generally refers to enantioselective synthesis of one enantiomer of a chiral target compound
ate complexes: transition metal alkyls where the metal bears a formal negative charge
atomic orbital: a solution to the wave equation for an atom; the region around the nucleus where the probability of finding an electron is greater than zero
***s orbitals*:** spherically symmetrical orbitals for which $l = 0$; they have spherical nodal surfaces. There is one s orbital in each electron shell.
***p orbitals*:** cylindrically symmetrical orbitals for which $l = 1, 0,$ or -1; they have one planar node at the nucleus, and all other nodes are spherical. There are three degenerate p orbitals in each p subshell.
***d orbitals*:** degenerate orbitals for which $l = 2, 1, 0, -1$ or -2; they have two nodes at the nucleus. There are five degenerate d orbitals in each d subshell.
atropisomers: enantiomers that arise due to restricted rotation about a σ bond
axial, equatorial: locations of groups on a cyclohexane in the chair conformation
axis of chirality: an axis about which the arrangement of groups leads to two non-superimposable structures (e.g., the C=C=C axis of 2,3-pentadiene)
Aufbau principle: the principle that lower energy orbitals are filled before higher energy orbitals. This principle applies to both molecular and atomic orbitals.

B

Baker-Nathan effect: the observation that alkyl substituents on a benzyl halide facilitate displacement in an order opposite to their inductive effects; correlates with the ability of the alkyl group to stabilize a carbocation by hyperconjugation
base: any species that will generate hydroxide ion in water (Arrhenius) or act as a proton acceptor (Lowry-Brønsted) in a chemical reaction
base system: in perturbation theory, the hydrocarbon whose π bonding system most closely matches that of the species under investigation
benzyne: any of three species, C_6H_4, based on the benzene ring. All three isomers (*o*-benzyne, *m*-benzyne, and *p*-benzyne) are known.
BINAP: a C_2-symmetric chiral *bis*-phosphine ligand based on the chiral 2,2'-*bis*-diarylphosphino-1,1'-binaphthyl ring system
boat, chair: extreme conformations of the cyclohexane ring; the chair is the low-energy conformation
Brønsted catalysis law: a relationship describing the effects of acid or base on the rate constant of a reaction: $\log k = \alpha \log K_a + \log C = \log C - \alpha\, pK_a$. The value of α determines the nature of the acid catalysis—$\alpha = 0$, no catalysis; $\alpha = 1$, specific acid catalysis; $0 < \alpha < 1$, general acid catalysis.

C

canonical forms: Lewis structures that can be written following the rules for writing Lewis structures but that do not alone adequately describe the structure of the species because at least two such structures can be written for that species
carbene: a neutral, electrophilic species with a divalent carbon carrying six valence electrons. The species may be a singlet, which has zwitterion character, or a triplet (diradical). Carbenes react with alkenes to give cyclopropanes.
carbenium ion: a term suggested by Olah for carbocations of the type R_3C^+; not widely used
carbenoid: a complex of a carbene with a metal or an organometallic precursor to a singlet carbene
carbocation: an electron-deficient species with a trivalent carbon carrying a formal positive charge, R_3C^+

carbometallation: addition of a metal-carbon bond across a C—C π bond; usually occurs with *syn* stereochemistry and works best with copper as a catalyst
carbonium ion: a carbocation with a pentacoordinate carbon such as CH_5^+. In older literature, this is used for simple carbocations, R_3C^+.
carbonylation: the formal insertion of carbon monoxide into a C—H bond to give an aldehyde
cascade reaction: a reaction sequence carried out in a single synthetic operation where the product of one reaction in the sequence is the starting material for the next
catalysis: a process whereby the rate of a reaction is increased by means of an alternative pathway from reactant(s) to product(s) that has a lower activation energy
catalytic hydrogenation: addition of molecular hydrogen across a π bond in the presence of a catalyst
center of chirality: a non-planar atom in a molecule carrying at least four different groups
center of inversion: a point in a molecule that is the center of a line joining every pair of identical atoms in the molecule
center of pseudochirality: a center in a *meso* compound (e.g., C-3 of pentane-2,3,4-triol) where a change in the configuration leads to a different *meso* compound
chair, boat: extreme conformations of the cyclohexane ring; the chair is the low-energy conformation
chemoselectivity: selectivity based on differences in reactivity between different functional groups (e.g., the Tollens oxidation of aldehydes).
chiral auxiliary: a chiral molecule that can be used to derivatize a prochiral substance prior to generating a new chiral center in the prochiral substance
chirality: the property of possessing a non-superimposable mirror image
circular dichroism: the difference in the absorptivity of left and right circularly polarized light as a function of wavelength
Clar sextet: the classical electron sextet in a six-membered ring. Maximum aromatic resonance stabilization is associated with the maximum number of Clar rings.
classical carbocation: a carbocation that is not stabilized by σ-bridging or similar remote electron delocalization
click chemistry: metal-catalyzed 1,3-dipolar cycloadditions to give five-membered heterocyclic products
complete structural formula: the structural formula of a molecule where every bond is explicitly shown
concerted: reactions where all the bonding changes take place in a single step
condensation reaction: a multistep reaction that involves an initial addition to a carbonyl group, followed by loss of a small molecule
condensed structural formula: a structural formula of a molecule where the bonds between groups are shown but groups are shown as formulas (e.g., CH_3 for methyl, C_6H_5 for phenyl)
conformation: three-dimensional structures of a compound that can be interconverted by rotation around σ bonds
conformational preference: the energy difference between two different conformations of the same molecule
connectivity formula: a complete structural formula with hydrogen atoms bonded to carbon omitted
conrotatory: in electrocyclization, formation of the ring by rotation of the termini of the π system in the same sense (clockwise or counterclockwise).
consonant difunctional relationships: relationships between pairs of functional groups in a target where the charge affinity labels alternate between the functional groups.
convergent synthesis: a synthesis carried out such that the target molecule is assembled from two or more advanced intermediates that are synthesized separately

coordinate covalent bond: in Lewis theory, a covalent bond where both electrons formally come from only one of the two atoms
coordinatively saturated: a metal complex in which the metal formally carries 18 electrons
coordinatively unsaturated: a metal complex in which the metal formally carries less than 18 electrons
covalent bond: in Lewis theory, a covalent bond is a pair of electrons shared between two or more nuclei
crossover experiment: a reaction where two closely related reactant molecules are mixed prior to carrying out the reaction. The observation of crossover products is evidence that the reaction occurs by a dissociative mechanism.
cycloaddition: a pericyclic reaction that result in formation of a new ring; the reaction between two π-bonded systems leads to a ring with two fewer π bonds than the reactants

D

dative bond: a coordinate covalent bond
dealkylation: loss of an alkyl group from a benzenonium ion to regenerate the dealkylated arene
decarbonylation: the extrusion of carbon monoxide from an aldehyde to give a hydrocarbon
decarboxylation: loss of a carboxylic acid group as carbon dioxide; most widely used with β-ketoacids or β-diacids
dehydrogenation: oxidation by removal of two hydrogen atoms from the precursor; may be effected by the Saegusa oxidation or using 2,3-dichloro-5,6-dicyano-1,4-benzoquinone (DDQ).
desymmetrization: a process whereby latent chirality in a molecule is revealed by modification of one of a pair of functional groups in a *meso* compound
Dewar benzene: hexagonal structure with two double bonds and one transannular "long" bond; bicyclo[2.2.0]hexa-4,5-diene
Dewar resonance energy (DRE): a value for the resonance energy of a ring calculated using the equation derived by M. J. S. Dewar
dialkyl phosphite: a tetracoordinate compound with two alkoxy groups and a hydrogen atom bonded to a central P=O group
diastereoisomers: stereoisomers that are not related as object and non-superimposable mirror image
diatropic ring current: the ring current induced in the π electrons of an aromatic ring; leads to deshielding around the perimeter of the ring and shielding in the center of the ring
dielectric constant: a constant that is related inversely to the force of attraction between ions in a medium
dihydroxylation (hydroxylation): addition of HO—OH across a C—C π bond
dimsyl anion: the conjugate base of dimethyl sulfoxide; a popular strong base for deprotonating sulfonium and phosphonium salts
DIOP: a C_2-symmetric chiral *bis*-phosphine ligand based on a chiral acetonide
DiPAMP: a C_2-symmetric chiral *bis*-phosphine ligand with chirality at phosphorus
1,3-dipole: a resonance-stabilized species that contains both electrophilic and nucleophilic sites
disrotatory: in electrocyclization, formation of the ring by rotation of the termini of the π system in the opposite sense (clockwise or counterclockwise)
dipole moment: a quantitative measure of the asymmetry in the distribution of partial charges in a polar covalent molecule

dithiane: a six-membered cyclic compound with two sulfur atoms in the ring. 1,3-Dithianes are deprotonated between the sulfur atoms to give a useful nucleophile.
diastereoselectivity: selectivity that leads to the formation of one diastereoisomer of the product preferentially (e.g., *anti* addition of bromine to an alkene)
dissonant difunctional relationships: relationships between pairs of functional groups in a target where at least one pair of like charge affinity labels are on adjacent atoms. All odd-membered rings are dissonant.
dithioacetal/dithioketal: *gem*-dithioethers used as protecting groups for aldehydes and ketones. Dithioacetals and dithioketals are resistant to acid.
domino reactions: sequential reactions where each step in the sequence generates the starting reagents for the next step; also known as tandem reactions.
double diastereoselection: the preferential formation of one diastereoisomer of a product in which one or more chiral centers are added to the chiral starting compound

E

eclipsed, staggered: extreme conformations about a σ bond in an open-chain compound; the eclipsed conformation is the high-energy conformation
electrocyclization: cyclization of a conjugated polyene to give a cyclic product with one less π bond in the ring.
electrophilic hydrides: neutral hydride reagents based on boron or aluminum (e.g., diisobutylaluminum hydride [DIBAL-H]).
element effect: the change in rate constant of a reaction caused by changing the identity of the element at a location undergoing a bonding change (e.g., changing a chloride to an iodide in a nucleophilic substitution)
enamines: enolate surrogates formed by heating an aldehyde or ketone with a secondary amine in the presence of an acid catalyst and removing the water formed; an alkene with an amine substituent that reacts as a soft carbon nucleophile similar to an enolate anion
enantiomeric excess: the amount of one enantiomer present in a mixture in excess of the racemate; used to specify the efficiency of asymmetric induction
enantiomers: stereoisomers that are related as object and non-superimposable mirror image
enantioselectivity: selectivity that leads to the formation of predominantly one enantiomer of a chiral product from an achiral precursor
ene reaction: a pericyclic reaction between two alkenes (the "ene" and the "enophile") that leads to migration of an allylic hydrogen and formation of a new C—C σ bond
ene-yne metathesis: the reaction between an alkene and an alkyne to give a conjugated diene; catalyzed by metal-carbene complexes
epimers: diastereoisomers with two or more chiral centers that differ in the stereochemistry of only one chiral center
equilibrium isotope effect: the ratio of equilibrium constants for a reversible reaction where the reactants have different isotopic compositions

F

first-order kinetics: describes a reaction where the rate law shows a linear dependence of the rate on the concentration of only one of the reactants
formal charge: an electronic charge assigned to an atom in a molecule or ion that arises from the difference between the number of valence electrons under the formal control of the atom in the molecule, and that same atom when neutral in the gas phase
formylation: substitution of hydrogen by a formyl (–CHO) group.

free radical: a species with at least one unpaired valence electron.
frontier orbitals: the highest energy occupied and lowest energy unoccupied molecular orbitals
HOMO: the highest-energy occupied molecular orbital.
LUMO: the lowest energy unoccupied molecular orbital.
SOMO: in free radicals, a singly occupied molecular orbital.
Frost-Musulin circle: a mnemonic for predicting orbital degeneracy and relative orbital energies in Hückel aromatic species

G

general acid catalysis: catalysis of a reaction by acids other than hydrogen ion
general base catalysis: catalysis of a reaction by bases other than hydroxide ion
geometric isomers: diastereoisomers differing in the arrangement of groups about a double bond, or a ring; designated by *cis/trans* or *E/Z* descriptors
"green" chemistry: chemistry carried out in such a way that the environmental impact of processes is minimized: processes conform to the 12 Principles of Green Chemistry
Grunwald-Winstein relationship: a relationship that defines the effects of solvent on the rate constant of a reaction; $\log (k/k_0) = mY$.

H

halolactonization: intramolecular interception of the three-membered halonium ion formed in the reaction between an alkene and a halogen, by a carboxylate anion. The stereochemistry is almost always *anti* (antarafacial).
Hammett constants: substituent constants
σ: based on the ionization of substituted benzoic acids.
σ^+: based on the rates of solvolysis of substituted cumyl chlorides.
σ^-: based on the ionization of substituted phenols.
σ_I: the contribution to σ from inductive effects, ionization of 4-substituted bicyclo[2.2.2]octane-1-carboxylic acids.
σ_R: the contribution to σ from resonance effects.; $\sigma_R = \sigma - \sigma_I$.
Hammett equation: a linear free energy relationship: $\log (K/K_o) = \sigma\rho$; $\log (k/k_o) = \sigma\rho$
hapto number: the number of sites on a ligand directly bonded to the metal; designated by the letter η, with the number superscripted. This corresponds to the number of electrons formally contributed to the metal by the neutral ligand.
***N*-heterocyclic carbene**: a stable carbene that can be isolated and used as a nucleophilic ligand for transition metals
heterogeneous catalytic hydrogenation: hydrogenation using a metal powder (supported or unsupported) as the catalyst
heterolysis: cleavage of a covalent bond so that both electrons end up on just one of the atoms. This results in the formation of two ions of opposite charge when starting from a neutral molecule.
heteronium ion: a cation in which the formal positive charge is localized on a heteroatom rather than carbon; usually the major canonical form of carbocations with a heteroatom adjacent to the electron-deficient carbon
HOMA: the Harmonic Oscillator Model of Aromaticity. The index is a measure of bond distance equalization. It has a value close to 1.0 if the ring is aromatic and close to 0 if the ring is non-aromatic.
HOMO: the highest energy occupied molecular orbital of a molecule
homoaromatic: a species containing a cyclic delocalized array of π electrons with at least one intervening sp^3-hybridized atom

homogeneous catalytic hydrogenation: hydrogenation using a soluble metal complex as the catalyst

homoisodesmic reaction: an *in silico* reaction that allows the resonance energy of a system to be calculated (e.g., two butadiene molecules giving one molecule of ethylene and one of benzene)

homolysis: cleavage of a covalent bond so that one electron from the bond ends up on each atom. This type of cleavage does not alter the formal charges on either of the atoms of the bond.

Hückel antiaromaticity: added reactivity associated with a cyclic, delocalized, conjugated π electron system with $(4n)$ π electrons

Hückel aromaticity: added stability associated with a cyclic, delocalized, conjugated π electron system with $(4n + 2)$ π electrons

hybrid atomic orbitals: degenerate atomic orbitals formed by mathematically combining the wave functions of the s orbital and one or more p orbitals of an atom. Hybrid atomic orbitals have a large front lobe and a small back lobe, and so possess directionality (i.e., they are vectors).

- ***sp hybrid orbitals***: two degenerate orbitals formed by hybridization of the s orbital with one p orbital; sp hybrid orbitals have linear geometry. The sp hybrid orbitals are orthogonal to the remaining unhybridized p orbitals.
- ***sp^2 hybrid orbitals***: three degenerate orbitals formed by hybridization of the s orbital with a pair of p orbitals; sp^2 hybrid orbitals have trigonal planar geometry. The remaining unhybridized p orbital is orthogonal to the plane defined by the sp^2 hybrid orbitals.
- ***sp^3 hybrid orbitals***: four degenerate orbitals formed by hybridization of the s orbital with all three p orbitals of a p subshell; sp^3 hybrid orbitals have tetrahedral geometry.

β-hydride elimination: a rearrangement of a coordinatively unsaturated complex where a σ alkyl ligand is converted to an η^2 alkene ligand and a hydrido ligand

hydroformylation: the formal addition of formaldehyde (H—CHO) across an alkene double bond by the cobalt-catalyzed reaction with carbon monoxide and hydrogen

hydrogenolysis: cleavage of a C—X bonds by hydrogen

hydrogen transfer reduction: reduction by transfer of a hydrogen from one carbon atom to another (e.g., reduction by hindered Grignard reagents), usually through a six-membered transition state

hydrometallation: the addition of a metal-hydrogen bond across a C—C π bond, including hydroalumination (H—Al), hydrostannylation (H—Sn), hydrosilation (H—Si), hydrozirconation (H—Zr). Hydroboration may also be viewed as a hydrometallation. It usually proceeds with *syn* stereochemistry.

hydroxylation (dihydroxylation): addition of HO—OH across a C—C π bond

I

improper axis of rotation: an axis of rotation and a mirror plane perpendicular to it such that rotation about the axis followed by reflection in the mirror returns the original representation; not compatible with chirality

ionic bond: electrostatic attraction between cations and anions in a crystal lattice

ionic hydrogenation: hydrogenation by addition of protons and hydride ions across the π bond

ionic solvent: a low-melting ionic compound (e.g., [bmim][BF_4]) that can be used as a non-volatile, highly polar solvent for organic reactions

isotopomers: compounds related by isotopic substitution (e.g., D for H) at one or more positions in the molecule

K

Kekulé structure: the hexagonal structure of benzene with alternating double bonds in the ring.

ketene: a cumulene in which the group C=C=O is present. Ketenes are electrophilic.

ketene addition: a [2 + 2] cycloaddition to give a cyclobutanone

ketyl: an α-hydroxyalkyl free radical

kinetic isotope effect: the ratio of rate constants for a reaction where the reactants have different isotopic compositions

kinetic order: the sum of the exponents of the concentration terms in a rate law

kinetic resolution: a process for the resolution of racemates by the selective reaction of one enantiomer with a chiral reagent. It may be applied to epoxide hydrolysis or to alcohol oxidation.

kinetics: the study of reaction rates and the factors that affect them

Körner proof: proof of the hexagonal structure of benzene by determining the number of different substitution products formed from substituted benzenes

L

Ladenburg benzene: the benzene structure devoid of double bonds; prismane

LCAO: linear combination of atomic orbitals; a mathematical process whereby atomic orbitals are transformed into hybrid atomic orbitals (atomic orbitals of the same atom) or molecular orbitals (atomic orbitals of different atoms)

ligand: a group bonded to a metal atom

line formula: a structural formula where C—H bonds are omitted and where the symbol for carbon is omitted so that alkyl chains become zigzag lines and rings become polygons

linear free energy relationships: linear relationships that correlate the rate constants or equilibrium constants for closely related reactions.

linear synthesis: a synthesis carried out where all the steps follow in a single sequence

London dispersion force: general attractive forces between molecules that are non-covalent in nature.

LUMO: the lowest energy unoccupied molecular orbital of a molecule

M

magnetic exaltation: an increase in the molar magnetic susceptibility that can be attributed to electron delocalization

major contributor: the canonical form that contributes most to the structure of the resonance hybrid. It correlates most closely with the structure and stability of the molecule.

matched reactant stereochemistry: the situation that pertains when chiral reactants react through diastereoisomeric transition states of significantly different energy so that the yield of one stereoisomer of the product far exceeds the other

mechanism: a detailed, step-by-step description of the electron reorganization that converts reactants to products in a chemical reaction

mechanistic spectrum: the concept that the position of the transition state of a reaction on the potential surface for the reaction may be altered without fundamentally altering the nature of the reaction mechanism (e.g., E2 reactions may be E1-like, E1cb-like, or in between)

mercaptan: a thiol

***meso* isomers:** achiral molecules containing chiral centers. These compounds have an internal mirror plane or inversion center of symmetry.

migratory insertion: a rearrangement at a metal atom where one ligand migrates to one terminus of an η^2 ligand to give a new σ ligand
minor contributor: the canonical form that contributes less (least) to the structure of the resonance hybrid. It correlates most closely with the reactivity of the molecule.
mirror plane of symmetry: a plane in which one half of a molecule generated from the other half by reflection in the mirror plane. In planar molecules, the plane of the molecule is also a mirror plane of symmetry. This is not compatible with chirality.
mismatched reactant stereochemistry: the situation that pertains when chiral reactants react through diastereoisomeric transition states of comparable energy so that the major diastereoisomer is not always very dominant in the product mixture
Möbius antiaromaticity: added reactivity associated with a cyclic, delocalized, conjugated π electron system with $(4n + 2)$ π electrons
Möbius aromaticity: added stability associated with a cyclic, delocalized, conjugated π electron system with $(4n)$ π electrons
molecular formula: the simplest type of formula, showing the total number of atoms of each type in a molecule without specifying any distinction between them
molecular orbital: a solution to the wave equation for a molecule; the region around and between nuclei in a molecule where the probability of finding an electron is greater than zero; formed by overlap of atomic orbitals
σ orbitals: molecular orbitals formed by coaxial overlap of atomic orbitals on different atoms
π orbitals: molecular orbitals formed by coplanar (parallel) overlap of p atomic orbitals on different atoms
molecularity: a property of a reaction that describes the number of particles that participate in the rate determining step to form the activated complex for the reaction. This is not the same as the kinetic order of a reaction.
More O'Ferrall-Jencks diagram: a simplified potential surface diagram that omits the energy contour lines

N

N: a solvent nucleophilicity parameter.
N-heterocyclic carbene: a stable carbene that can be isolated and used as a nucleophilic ligand for transition metals
neighboring group participation: see anchimeric assistance
NHC: an *N*-heterocyclic carbene
NICS: nucleus-independent chemical shift
nitrene: a neutral species with a monovalent nitrogen carrying six valence electrons; the species may be a singlet, which has zwitterion character, or a triplet (diradical) species.
NLE: non-linear effect
node: a region in an orbital where the electron density is zero. Nodes may be planar or curved.
non-classical cation: a carbocation in which the stabilization of the cation is accomplished by delocalization of electrons from a remote (non-adjacent) σ (σ-bridging) or π (π-bridging) bond.
non-linear effect: in asymmetric catalysis, formation of the chiral product with an enantiomeric excess greater than that of the chiral catalyst (positive NLE) or less than that of the chiral catalyst (negative NLE)
nucleophilic hydrides: hydride reducing agents based on tetracoordinate aluminum or boron anions.

nucleus-independent chemical shift: a computational value for the chemical shift of a proton in the center of a ring, or a set distance (usually 1Å) above the center of the ring. The value is negative for aromatic rings and positive for antiaromatic rings.

O

olefin metathesis: the interchange of terminal groups of two alkenes catalyzed by a metal-carbene complex

optical activity: the property of rotating the plane of plane polarized light; requires a chiral arrangement of atoms in the substance

optical rotatory dispersion: the optical rotation of a chiral substance with respect to wavelength

orbital correlation diagram: a diagram that maps orbitals of the reactant(s) onto corresponding orbitals of the product(s)

oxidation: removal of electrons, addition of one or more electronegative atoms, or loss of one or more hydrogen atoms from the substrate

oxidative addition: the addition of a molecule (usually a halide or hydride) across a coordinatively unsaturated metal complex such that the coordination number and the oxidation number of the metal both increase by two

oxymetallation: the addition of a metal and a hydroxyl group to an alkene. The stereochemistry is usually *anti* (antarafacial).

ozonolysis: oxidative cleavage of an alkene with ozone; the products obtained may be altered by the way that the intermediate ozonide is worked up.

P

paratropic ring current: the ring current induced in the π electrons of an antiaromatic ring; leads to shielding around the perimeter of the ring and deshielding in the center of the ring

Pauli Exclusion principle: the limitation that no two electrons in an atom may simultaneously possess identical values for all four quantum numbers; has the effect of limiting occupancy of an orbital to two electrons or less. This limitation on occupancy applies to molecular orbitals as well as atomic orbitals.

pericyclic reactions: concerted reactions that proceed through a fully conjugated cyclic array of continuously bonded atoms. The cyclic transition state requires electron reorganization at every position in the activated complex.

persistent radical effect: the observation that when two radicals—one very stable (persistent) and one very reactive (transient)—are formed at the same rate in a reaction, the product formed is almost entirely derived by cross-coupling of the persistent and transient radicals

perturbation theory: a theory that allows a qualitative estimate of the coefficients in the HOMO and LUMO of a conjugated system to be obtained by combining the π molecular orbitals of the base system with those of a perturbing system

perturbing influence: in perturbation theory, the cation or anion whose π bonding system can be used to modify the π orbitals of the base system to give the orbitals of the species under investigation

phosphine: an organophosphorus compound where the phosphorus is tricoordinate

phosphine oxide: an organophosphorus compound with a tetracoordinate phosphorus and three P—C bonds and one P=O group

phosphonic acid derivatives: tetracoordinate organophosphorus compounds with a central P=O group bonded to an alkyl group and two alkoxy or hydroxy groups

phosphorous acid: a tetracoordinate compound with two hydroxyl groups and a hydrogen atom bonded to a central P=O group

plane of chirality: a plane in a molecule lacking chiral centers or axes about which groups are disposed dissymmetrically, thus conferring chirality on the molecule (e.g., *trans*-cyclooctene)
polygon formula: a line formula
potential energy surface: a map of a reaction where the energy of the species involved is plotted against structural (bond lengths, bond orders) parameters. The potential energy surface is usually represented in the form of a contour map.
pre-equilibrium: an equilibrium step in a multistep reaction that occurs prior to the rate determining step
primary kinetic isotope effect: a kinetic isotope effect in a reaction where the labeled atom is part of a bond being broken (or formed)
prochiral center: a planar atom in a molecule bonded to three different groups, so that the addition of a fourth group to give a non-planar product generates a new chiral center
proper axis of rotation: an axis about which rotation of the molecule leads to an identical representation of the molecule; compatible with chirality
protecting group: a removable group that can be used to temporarily mask the reactivity of a functional group toward a reagent
protocenters of stereochemistry: centers or groups bonded to two different groups. Two protocenters in a molecule may lead to stereoisomers, depending on the orientation and location of the two protocenters relative to each other
pseudo–first-order kinetics: reaction kinetics of a second-order reaction measured under conditions of a large excess of one of the participating reactants
pyrolytic elimination: formation of an alkene by thermal decomposition of the precursor. These reactions are often retro-ene reactions.

Q

quantum numbers: numbers that occur in the solutions of the wave function for an atom
principal quantum number, n: the quantum number that specifies the electron shell. It has integer values from 1 to n
azimuthal quantum number, l: the quantum number that specifies the electron subshell. It has integer values from 0 to $(n - 1)$.
magnetic quantum number, m_l: the quantum number that specifies the orbital within a subshell. It has integer values from $-l$ to $+l$.
spin quantum number, m_s: the quantum number that specifies the spin on the electron; it has a value of ±½.

R

racemate: an equal mixture of two enantiomers; optically inactive
rate law: an algebraic expression that relates the rate constant of the reaction to the concentrations of the reactants
reaction coordinate diagram: a diagram that shows the change in energy of a single reactant molecule as it undergoes the electron reorganization to give the product
reaction energy profile: reaction coordinate diagram
rate constant: the term in a rate law that is not a concentration term
rate law: the mathematical relationship between the rate of a reaction and the concentrations of the participating species.
rate-limiting step (rate-determining step): the slowest step in a multistep reaction; defines the maximum possible rate for the reaction
reaction coordinate diagram: a plot of the energy of a molecule as it undergoes the bonding changes from reactant to product through activated complex(es) and reactive intermediates in stepwise reactions; a reaction energy profile

reaction energy profile: a plot of the energy of a molecule as it undergoes the bonding changes from reactant to product through activated complex(es) and reactive intermediates in stepwise reactions; a reaction coordinate diagram

reaction kinetics: the study of reaction rates and the factors (e.g., concentration, temperature, solvent) that affect them

reactive intermediate: a stable, high-energy species between reactants and products in a reaction; flanked on both sides by transition states

reactivity-selectivity principle: the principle that reactivity and selectivity are complementary

reduction: addition of electrons, removal of one or more electronegative atoms, or addition of one or more hydrogen atoms to the substrate

reductive amination: conversion of an aldehyde or ketone to the corresponding amine by reduction of the imine or enamine

reductive desulfurization: conversion of a sulfide or thioketal to a hydrocarbon by hydrogenation or dissolving metal reduction

reductive elimination: (1) the elimination of a molecule (usually a hydrocarbon) from a metal complex such that the coordination number and the oxidation number of the metal both decrease by two; (2) formation of an alkene by reducing a 1,2-dihalide or similar substrate with a metal

regioselectivity: selectivity based on the propensity of a reaction to give mainly one regioisomer of the product (e.g., Markovnikov addition)

REPB (resonance energy per π bond): the value obtained by dividing the DRE by the number of π bonds

REPE (resonance energy per π electron): the value obtained by dividing the DRE by the number of π electrons

resonance: a theory whereby the structure of a species that cannot be adequately represented by a single Lewis structure is described as the hybrid of two or more such structures

resonance contributors: canonical forms

resonance hybrid: the structure of a molecule that can be deduced by taking the weighted average of the canonical forms of the molecule

retro-ene reaction: a pericyclic reaction that leads to two new π bonds by migration of a hydrogen and cleavage of a C—C σ bond

retrosynthetic analysis: the process of generating a rational pathway for the synthesis of a target compound by successively applying transforms to the target structure and the generated synthons until a starting material is identified

S

SAMP/RAMP hydrazones: hydrazones formed from ketones with *S*- or *R*-1-amino-2-methoxymethylpyrrolidine. Deprotonation of these hydrazones allows asymmetric alkylation of aldehydes and ketones.

selenide: the selenium analog of an ether

selenoxide: an organoselenium compound containing one Se=O group

sigmatropic rearrangement: a pericyclic reaction in which an atom migrates to a remote site in the molecule with concomitant migration of π bonds.

silane: an organosilicon compound with at least one Si—C bond

silyl enol ether: a derivative of a carbonyl compound formed by treating the carbonyl compound sequentially with a strong base (e.g., lithium diisopropylamide) and trimethylsilyl chloride; serves as a non-basic surrogate for an enolate anion

silyl ethers: protecting groups for alcohols. The ease of cleavage can be modulated by choosing the appropriate size and electronic characteristics for the groups on silicon.

simple diastereoselection: the preferential formation of one diastereoisomer of a product in which a new chiral center is added to the chiral starting compound
solvolysis: S_N1 substitution or E1 elimination of a compound where the solvent is the nucleophile or base
SOMO: in free radicals, a singly occupied molecular orbital
stannane: an organotin compound with at least one Sn—C bond
stereoelectronic effect: an effect where the stereochemical or regiochemical outcome of a reaction is dictated by the orientation of the participating orbitals and where some orbital orientations cannot give rise to products
sulfide: a thioether
sulfinylamines: chiral amides of sulfinic acids, $RS(=O)NR'_2$. the chiral sulfur is configurationally stable in sulfinylamines.
sulfinylimines: chiral imine derivatives formed by condensation of a chiral sulfinylamine, $RS(=O)NH_2$, with an aldehyde or ketone. These compounds are useful electrophiles that react with Grignard reagents to give amines. The aza-enolates from these species are good nucleophiles for S_N2 substitution and additions to aldehydes and ketones.
sulfone: an organosulfur compound containing the SO_2 group
sulfoxide: an organosulfur compound containing one S=O group
suprafacial: reaction stereochemistry where the new bonds are formed on the same face of the participating π system (e.g., *syn* addition)
superacid: a highly reactive acid capable of protonating alkanes (e.g., FSO_3H-SbF_5).
symmetry-allowed: reactions where the symmetry of the participating orbitals is preserved throughout the reaction, leading to a bonding interaction where the new bond is formed
symmetry-forbidden: reactions where the symmetry of the participating orbitals is not preserved throughout the reaction, leading to an antibonding interaction where the new bond should be formed
synthon: a potential precursor to a target compound generated by application of a transform to the target structure. Synthons may or may not be explicit compounds.

T

tandem and domino reactions: sequential reactions where each step in the sequence generates the starting reagents for the next step
tetrahydropyranyl ethers: protecting groups for alcohols that can be cleaved by aqueous acid
topism: the stereochemical characteristic of chemically identical groups in a molecule that predicts the outcome of replacing one of the groups. Groups may be *homotopic* (e.g., the hydrogens of a methyl group), where the change leads to the same stereoisomer in the product; *enantiotopic* (e.g., the methylene hydrogens of methane), where the change leads to enantiomeric products; or *diastereotopic* (e.g., the methylene hydrogens of 2-chlorobutane), where the change leads to diastereoisomers.
transfer hydrogenation: hydrogenation in which a hydrogen atom transfer agent (e.g., HCO_2H) is used in place of molecular hydrogen
transform: an intellectual construct that converts a target compound into one or more potential precursors, or synthons
transmetallation: a metathesis reaction between a metal alkyl, and a metal alkyl or a metal salt of a less electropositive metal
thioether: the sulfur analog of an ether
thiol: the sulfur analog of an alcohol
trialkyl phosphite: a tricoordinate compound with three alkoxy groups bonded to a central phosphorus

Y

ylide: a carbanion nucleophile where the charged carbon is bonded to a heteroatom carrying a formal positive charge. Bonding in ylides indicates that there is considerable C=X double-bond character.

Z

Zimmerman circle: a mnemonic for predicting orbital degeneracy and the relative energies in Möbius aromatic systems

APPENDIX 3

Selected Physical Constants

Table A3.1 Hammett Substituent Constants for Some Common Substituent Groups

Substituent	σ_m	σ_p	σ^{+}*	σ^{-}*	σ_I	σ_R	F	R
NH_2	−0.16	−0.66	−1.3		0.12	−0.82	0.38	−2.52
Me	−0.07	−0.17	−0.31		−0.04	−0.11	0.01	−0.41
Ph	0.06	−0.01	−0.17		0.10	−0.11	0.25	−0.37
OH	0.12	−0.37	−0.92		0.25		0.46	−1.89
OMe	0.12	−0.27	−0.78	−0.2	0.27	−0.61	0.54	−1.68
F	0.34	0.06	−0.07	0.02	0.50	−0.45	0.74	−0.60
I	0.35	0.18	0.13		0.39	−0.16	0.65	−0.12
CHO	0.35	0.42		1.04	0.25			
CO_2H	0.37	0.45	0.42				0.44	0.66
Cl	0.37	0.23	0.11		0.46	−0.23	0.72	−0.24
COMe	0.38	0.50		0.87	0.28	0.16	0.50	0.90
Br	0.39	0.23	0.15		0.44	−0.19	0.72	
CO_2R	0.37	0.45	0.48	0.68	0.30	0.14	0.47	0.67
CF_3	0.43	0.54			0.45	0.08	0.64	0.76
CN	0.56	0.66	0.66	0.90	0.56	0.13	0.90	0.71
NO_2	0.71	0.78	0.79	1.24	0.65	0.15	1.00	1.00

*These values are for *para* substituents only.

Table A3.2 Typical pK_a Ranges for Hydrogen Atoms in Representative Organic Molecules

Functional Group	Acidic Bond	Acid	Examples	Conjugate Base	Examples	pK_a Range
Alcohol	$O\text{–}\underline{H}$	$RO\text{–}\underline{H}$	$(CH_3)_3C\text{–}O\underline{H}$, $\underline{H}_2O$	RO^-	$(CH_3)_3C\text{–}O^-$, OH^-	15–19
Carboxylic acid	$COO\text{–}\underline{H}$	$RCOO\text{–}\underline{H}$	$HCO\text{–}O\underline{H}$, $CH_3CO\text{–}O\underline{H}$	$RCO\text{–}O^-$	$HCO\text{–}O^-$, $CH_3CO\text{–}O^-$	4–6
Amine	$N\text{–}\underline{H}$	$R_2N\text{–}\underline{H}$	$(CH_3)_2CH\text{–}N\underline{H}_2$, $N\underline{H}_3$	R_2N^-	NH_2^-, $(CH_3)_2CH\text{–}NH^-$	33–38
Amide	$CON\text{–}\underline{H}$	$RCONR\text{–}\underline{H}$	$CH_3CO\text{–}N\underline{H}_2$	$RCONR^-$	$CH_3CO\text{–}NH^-$	15–17
Alkane	$C\text{–}\underline{H}$	$R\text{–}\underline{H}$	$C\underline{H}_4$, $(C\underline{H}_2)_6$	R^-	CH_3^-, $[(CH_2)_5CH]^-$	45–60
Alkyne	$C{\equiv}C\text{–}\underline{H}$	$RC{\equiv}C\text{–}\underline{H}$	$CH_3\text{–}C{\equiv}C\text{–}\underline{H}$	$RC{\equiv}C^-$	$CH_3\text{–}C{\equiv}C^-$	25–30
Aldehyde, ketone	$COC\text{–}\underline{H}$	$RCOCR_2\text{–}\underline{H}$	$CH_3COC\underline{H}_3$	$RCOCR_2^-$	$CH_3COCH_2^-$	25–30

Table A3.3 Dielectric Constants and Dipole Moments of Common Organic Solvents

Solvent	D	μ
Acetic acid	6.17	1.68
Acetone	20.6	2.69
Acetonitrile	35.9	3.53
Benzene	2.27	0
Chloroform	4.80	1.15
Dichloromethane	8.9	1.14
Diethyl ether	4.2	1.15
Dimethylformamide (DMF)	36.7	3.24
Dimethyl sulfoxide (Me_2SO)	46.5	4.06
Ethanol	24.5	1.66
Ethyl acetate	6.02	1.82
Hexane	1.89	0.085
Hexamethylphosphoric triamide (HMPA)	29.3	4.31
Methanol	32.7	2.87
Pyridine	12.9	2.37
Tetrahydrofuran	7.58	1.75
Water	78.4	1.8

Table A3.4 Solvent Ionizing Power and Nucleophilicity Parameters

Solvent	Y	Y_{OTs}	N_{OTs}
F_3CCO_2H	1.84	4.57	-5.56
H_2O	3.49	4.1	-0.44
$(F_3C)_2CHOH-H_2O$ (97:3)	2.46	3.61	-4.27
F_3CCH_2OH	1.04	1.80	-3.0
$F_3CCH_2OH-H_2O$ (85:15)		1.92	-2.01
$F_3CCH_2OH-H_2O$ (50:50)		2.14	-0.93
HCO_2H	2.05	3.04	-2.35
$MeCO_2H$	-1.64	-0.61	-2.35
EtOH	-2.03	-1.75	0.00
$EtOH-H_2O$ (80:20)	0.00	0.00	0.00
$EtOH-H_2O$ (50:50)	1.66	1.29	-0.09
$EtOH-H_2O$ (20:80)	3.05	3.32	-0.34
MeOH		-0.92	-0.04
$MeOH-H_2O$ (80:20)		0.47	-0.05

(continues)

Table A3.4 Solvent Ionizing Power and Nucleophilicity Parameters (*Continued*)

Solvent	Y	Y_{OTs}	N_{OTs}
MeOH–H_2O (50:50)		2.00	−0.19
MeOH–H_2O (20:80)		3.39	−0.35
Me_2CO–H_2O (80:20)		−0.94	−0.42
Me_2CO–H_2O (50:50)	1.40	1.26	−0.39
Me_2CO–H_2O (20:80)	2.91	3.05	−0.38
Dioxane–H_2O (90:10)	-2.03	-2.41	−0.51
Dioxane–H_2O (80:20)		-1.30	−0.29
Dioxane–H_2O (20:80)	2.88		

Table A3.5 Selectivity of Radicals Toward C–H bonds

Radical (X•)	CH_3–H	RCH_2–H	R_2CH–H	R_3C–H	D (X–H)
F•	0.5	1	1.2	1.4	135
Cl•	0.004	1	4.3	7.0	103
Br•	0.002	1	780	1700	87
I•		1	18		71
HO•	0.04	1	7	45	118
H•		1	4.8	40	105
CF_3•	0.06	1	8.3	90	105
CH_3•	0.09	1	10	80	104

INDEX